PHYSIOLOGY

PHYSIOLOGY

Edited by

ROBERT M. BERNE, M.D., D.Sc. (Hon.)

Chairman and Charles Slaughter Professor of Physiology,
Department of Physiology,
University of Virginia School of Medicine,
Charlottesville, Virginia

MATTHEW N. LEVY, M.D.

Chief of Investigative Medicine, Mount Sinai Medical Center;
Professor of Physiology and Biophysics and of Biomedical Engineering,
Case Western Reserve University,
Cleveland, Ohio

Second Edition

with 1264 *illustrations*

The C. V. Mosby Company

ST. LOUIS • WASHINGTON, D.C. • TORONTO 1988

MOSBY

A TRADITION OF PUBLISHING EXCELLENCE

Editor: Stephanie Bircher
Developmental editor: Elaine Steinborn
Assistant editor: Anne Gunter
Manuscript editors: Jennifer Collins, Kathy Lumpkin
Design: Gail Morey Hudson
Production: Jennifer Collins, Celeste Clingan, Jeanne Gulledge

SECOND EDITION

Printed in the United States of America

The C.V. Mosby Company
11830 Westline Industrial Drive, St. Louis, Missouri 63146

Library of Congress Cataloging in Publication Data

Physiology.

Bibliography: p.
Includes index.
1. Human physiology. I. Berne, Robert M., 1918-
II. Levy, Matthew N., 1922- [DNLM: 1. Physiology.
QT 4 P5783]
QP34.5.P496 1988 612 87-14114
ISBN 0-8016-1697-2

C/VH/VH 9 8 7 6 5 4 3 2 1 02/B/268

Contributors

MURRAY D. ALTOSE, M.D.

Associate Professor of Medicine, Case Western Reserve University; Associate Director, Department of Medicine, and Chief, Pulmonary Division, Cleveland Metropolitan General Hospital, Cleveland, Ohio

ROBERT M. BERNE, M.D., D.Sc. (Hon.)

Chairman and Charles Slaughter Professor of Physiology, Department of Physiology, University of Virginia School of Medicine, Charlottesville, Virginia

NEIL S. CHERNIACK, M.D.

Professor of Medicine, Case Western Reserve University, and Chief, Pulmonary Division, University and Cleveland VA Hospitals, Cleveland, Ohio

DAVID H. COHEN, Ph.D.

Vice President for Research and Dean of the Graduate School, Professor of Neurobiology and Physiology, Northwestern University, Evanston, Illinois

BRIAN R. DULING, Ph.D.

Professor, Department of Physiology, University of Virginia School of Medicine, Charlottesville, Virginia

SAUL M. GENUTH, M.D.

Professor of Medicine, Case Western Reserve University School of Medicine, Mount Sinai Medical Center, Cleveland, Ohio

STEVEN G. KELSEN, M.D.

Associate Professor of Medicine, Case Western Reserve University, Cleveland, Ohio

HOWARD C. KUTCHAI, Ph.D.

Professor of Physiology, University of Virginia School of Medicine, Charlottesville, Virginia

MATTHEW N. LEVY, M.D.

Chief of Investigative Medicine, Mount Sinai Medical Center; Professor of Physiology and Biophysics and of Biomedical Engineering, Case Western Reserve University, Cleveland, Ohio

RICHARD A. MURPHY, Ph.D.

Professor, Department of Physiology, University of Virginia School of Medicine, Charlottesville, Virginia

OSCAR D. RATNOFF, M.D.

Professor of Medicine, Case Western Reserve University; Career Investigator of the American Heart Association; Co-Director, Division of Hematology/Oncology, Department of Medicine, University Hospitals of Cleveland, Cleveland, Ohio

S. MURRAY SHERMAN, Ph.D.

Professor, Department of Neurobiology and Behavior, State University of New York at Stony Brook, Stony Brook, New York

Preface

The second edition of this text, like the first, is designed to emphasize broad concepts and to minimize the compilation of isolated facts. Each chapter in this edition has been altered significantly to make the text as accurate and current as possible. Another goal was to further clarify and simplify complex information. To help attain this goal, hundreds of new and revised illustrations have been added to visually assist the reader in grasping difficult physiological information. Finally, in keeping with our emphasis on broad principles, homologies of important mechanisms are grouped wherever possible. We hope these alterations will further enhance the text's value as a teaching text.

At the beginning of the book itself, as well as at the beginning of several of the sections, many of the important physicochemical principles of physiology are covered in considerable detail. When these principles could be represented profitably by equations, the bases of the equations and the major underlying assumptions have been discussed. This approach provides students with a satisfactory understanding of these basic principles, so that a mastery of certain topics will involve a minimum of pure memorization. This coverage of physicochemical principles is most evident in the cellular, cardiovascular, and respiratory sections.

Throughout the section on muscle physiology, the three types of muscle are not described in sequence but are consistently considered together. It is emphasized that the basic mechanisms of contraction of skeletal, cardiac, and smooth muscles are very similar and that differences lie mainly in the relative importance of certain components of the underlying process.

In the renal physiology section, homologies influence the presentation of material. The mechanisms whereby the kidneys handle a few important solutes are described in detail. The specific details of the transport of the myriad individual substances that pass through the kidneys are intentionally ignored.

The section on the nervous system reveals a functional neuroanatomical approach while still capturing the spirit of contemporary cellular neurophysiology. A major portion of the section is devoted to sensory and motor systems because of their relevance to clinical problems. The theoretical framework common to all sensory systems is constructed so as to facilitate the learning of the various components.

The section on hematology covers blood composition and provides a complete and up-to-date picture of blood coagulation and its aberrations.

To provide a clearer understanding of cardiovascular physiology, the entire system is initially dissected into its major components, and the functions of these components are examined in detail. The system is then reconstructed and considered as a whole to indicate how the various parts interact in physiological and pathophysiological states.

In the endocrine section, homologies again influence the presentation of material. Discussions of the male and female gonads are included in the same chapter to highlight the similarities between the Sertoli cell functions in spermatogenesis and the granulosa cell functions in oogenesis.

Again, the framework of this textbook comprises firmly established facts and principles. Isolated phenomena generally are ignored unless they are considered to be highly significant, and few experimental methods are described unless they are essential for the comprehension of a specific topic. We have not documented most of the assertions made throughout the book but we have provided references at the end of each chapter. These references have been selected because they provide a current and comprehensive review of the topic, a clear and detailed description of important mechanisms, or a complete and up-to-date bibliography of the subject.

Although theoretical controversies exist in virtually all areas of physiology, such controversies are not described unless they provide a deeper understanding of the subject. Each author has described what he believes to be the most likely mechanism responsible for the phenomenon under consideration. While we recognize that future advances may prove many of our conjectures wrong, we have decided to make this compromise to achieve brevity, clarity, and simplicity.

We wish to express our appreciation to our colleagues and students who provided constructive criticism during the revision of this book. We also want to give special thanks to Frances S. Langley, who skillfully drew the illustrations for the entire volume.

Robert M. Berne
Matthew N. Levy

Contents

Introduction

As the reader embarks on the study of human physiology, it is instructive to consider briefly what physiology is and how it relates to other disciplines.

For centuries physiology and anatomy were the only recognized basic biomedical sciences, having originated in the western world with the ancient Greeks. Comparatively recently, physiology and anatomy, together with biology, chemistry, physics, psychology, and other sciences, have given rise to other biomedical disciplines, whose areas of inquiry overlap to considerable degrees. Among the newer biomedical sciences are biochemistry, genetics, pharmacology, biophysics, biomedical engineering, molecular biology, cell biology, and neuroscience. Because the overlap among the interests of the different biomedical sciences is so extensive, it is often difficult, and sometimes not even useful, to decide where one discipline begins and the other ends.

Physiology may be distinguished from the other basic biomedical sciences by its concern with the function of the intact organism and its emphasis on the processes that control and regulate important properties of living systems. In the healthy human many variables are actively maintained within relatively narrow physiological limits. The list of controlled variables is long; it includes body temperature, blood pressure, the ionic composition of blood plasma, blood glucose levels, the oxygen and carbon dioxide content of blood, and a host of other properties. The tendency to maintain the relative constancy of certain variables, even in the face of significant environmental changes, is known as *homeostasis*. A central goal of physiological research is the elucidation of the mechanisms responsible for homeostasis.

In studying a homeostatic mechanism, physiologists attempt to characterize the components of the control system. What is the *sensor* that detects the difference between a physiological variable and its *set point?* How does the sensor operate? What is the *integrating center* that receives information from the sensor via an *afferent pathway* and communicates with *effectors* by *efferent pathways?* The afferent and efferent pathways frequently involve nerves or hormones. What are the effectors that function to bring the level of the controlled variable closer to its set point? The *steady-state* value of a controlled variable typically results from a dynamic balance between certain effectors that function to increase the value of the variable and other effectors that act to decrease it.

We are in the midst of an explosion of knowledge of biological systems. This information expansion is characterized by a progressively greater understanding of the behavior of individual cells and of the molecules that make up those cells. The precise nucleotide sequences of many genes have been determined. The physiological regulation of the oxygen affinity of the blood is understood in terms of precise knowledge of conformational changes in the hemoglobin molecule. The structure of certain ion channels and the nature of the processes by which the channels open and close are being unraveled. The mechanisms that control important variables in individual cells, such as the level of free Ca^{++} in the cytosol, are being elucidated. Many physiologists now study biological processes at the level of single cells or even single molecules in order to explain their *in vivo* functions. Scientists in related disciplines also concentrate their inquiries on the behavior of cells and biological molecules. What distinguishes physiology as a scientific discipline is its emphasis on homeostatic mechanisms, at all levels of organization, and its concern with synthesizing knowledge of the functions of tissues, cells, and molecules to achieve a more complete understanding of the behavior of the intact organism.

This textbook describes what is now known about the function of the major organ systems of mammals and the mechanisms that control and regulate their behavior. It may be emphasized that physiological knowledge is being continually broadened, and especially deepened. "The fabric of knowledge is being continuously woven, and altered by the introduction of new threads, the texture and device even are mobile, yet it is the object of a comprehensive work to attempt to capture the fleeting pattern. That the design is of growing complexity, and less and less easy of discernment, may be matter for regret though not for surprise.''* In our attempt to ''capture the fleeting pattern'' we have tried to integrate the descriptions of individual organ systems and homeostatic mechanisms in order to provide a broad appreciation for the physiological function of the whole organism. The authors hope that this text will stimulate its readers to actively pursue the study of physiology.

*Evans, C.L.: Preface. In Evans, C.L., and Hartridge, H., editors, Starling's principles of human physiology, ed. 7, Philadelphia, 1936, Lea & Febiger.

SECTION

I

CELLULAR PHYSIOLOGY

Howard C. Kutchai

CHAPTER

1

Cellular Membranes and Transmembrane Transport of Solutes and Water

Cellular Membranes

Each cell is surrounded by a plasma membrane that separates it from the extracellular milieu. The plasma membrane serves as a permeability barrier that allows the cell to maintain a cytoplasmic composition far different from the composition of the extracellular fluid. The plasma membrane contains enzymes, receptors, and antigens that play central roles in the interaction of the cell with other cells and with hormones and other regulatory agents in the extracellular fluid.

The membranes that enclose the various organelles divide the cell into discrete compartments and allow the localization of particular biochemical processes in specific organelles. Many vital cellular processes take place in or on the membranes of the organelles. Striking examples are the processes of electron transport and oxidative phosphorylation, which occur on, within, and across the mitochondrial inner membrane.

Most biological membranes have certain features in common. However, in keeping with the diversity of membrane functions, membrane composition and structure differ from one cell to another and among the membranes of a single cell.

Membrane Structure

Proteins and phospholipids are the most abundant constituents of cellular membranes. A phospholipid molecule has a polar head group and two very nonpolar, hydrophobic fatty acyl chains. In an aqueous environment it is most energetically stable for phospholipids to form structures that allow the fatty acyl chains to be kept from contact with water. One such structure is the *lipid bilayer* (Fig. 1-1). Many phospholipids, when dispersed in water, spontaneously form lipid bilayer structures. Most of the phospholipid molecules in biological membranes have a lipid bilayer structure.

The proteins of biological membranes are associated with the membrane phospholipids in two major ways: (1) by charge interactions between the polar head groups of the phospholipids and acidic or basic amino acid residues of the protein, and (2) by hydrophobic interactions of the phospholipid acyl chains with hydrophobic amino acid residues of the proteins.

Fig. 1-2 depicts the "fluid mosaic" model of membrane structure. This model is consistent with many of the properties of biological membranes. Note the bilayer structure of most of the membrane phospholipids. The membrane proteins can be divided into two major classes: (1) *integral or intrinsic* membrane proteins that are embedded in the phospholipid bilayer, and (2) *peripheral or extrinsic* membrane proteins that are associated with the surface of the phospholipid bilayer. The peripheral membrane proteins interact with membrane lipids predominantly by charge interactions with integral membrane proteins. Thus sometimes they may be removed from the membrane by altering the ionic composition of the medium. Integral membrane proteins have important hydrophobic interactions with the interior of the membrane. These hydrophobic interactions can be disrupted only by detergents that solubilize the integral proteins by forming their own hydrophobic interactions with nonpolar amino acid side chains.

Cellular membranes are fluid structures in which many of the constituent molecules are free to diffuse in the plane of the membrane. Most lipids and proteins are

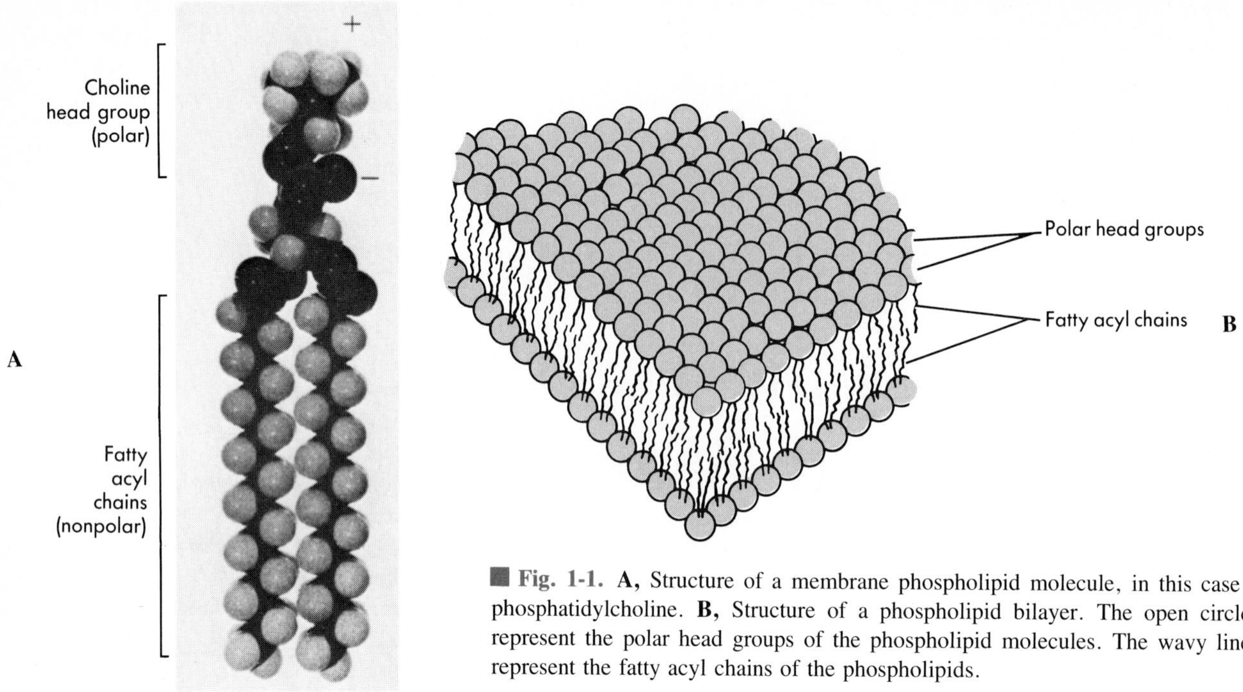

Fig. 1-1. A, Structure of a membrane phospholipid molecule, in this case a phosphatidylcholine. **B,** Structure of a phospholipid bilayer. The open circles represent the polar head groups of the phospholipid molecules. The wavy lines represent the fatty acyl chains of the phospholipids.

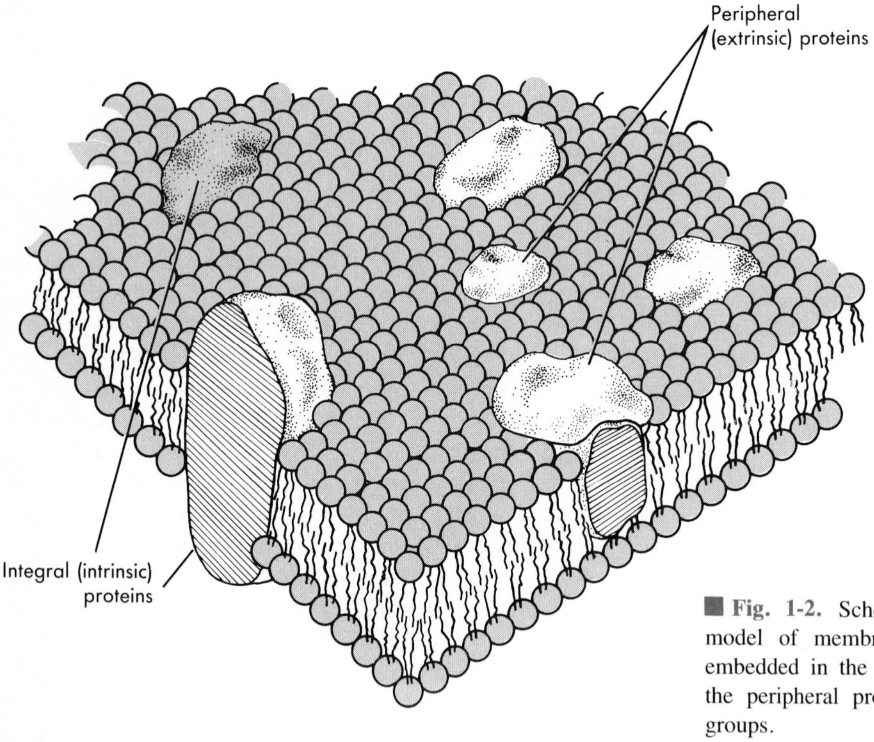

Fig. 1-2. Schematic representation of the fluid mosaic model of membrane structure. The integral proteins are embedded in the lipid bilayer matrix of the membrane and the peripheral proteins are associated with the polar head groups.

free to move in the bilayer plane, but they "flip-flop" from one phospholipid monolayer to the other at much slower rates. This is less likely to occur when a large hydrophilic moiety must be dragged through the nonpolar interior of the lipid bilayer.

In some cases membrane components clearly are *not* free to diffuse in the plane of the membrane. Examples of this motional constraint are the sequestration of acetylcholine receptors (integral membrane proteins) at the motor endplate of skeletal muscle and the presence of different membrane proteins in the apical and basolateral plasma membranes of epithelial cells. At present little is known about the ways in which membrane constituents are restrained from lateral diffusion. The cytoskeleton may help to anchor certain membrane proteins.

Membrane Composition

Lipid Composition

Major phospholipids. In animal cell membranes the most abundant phospholipids are often the choline-containing phospholipids: the lecithins (phosphatidylcholines) and the sphingomyelins. Next in abundance are usually the amino phospholipids: phosphatidylserine and phosphatidylethanolamine. Other important phospholipids that are present in smaller amounts are phosphatidylglycerol, phosphatidylinositol, and cardiolipin.

Cholesterol. Cholesterol is a major constituent of animal cell plasma membranes. The steroid nucleus of cholesterol lies parallel to the fatty acyl chains of membrane phospholipids. Thus cholesterol alters the molecular packing of the membrane phospholipids. In natural membranes cholesterol diminishes the lateral mobility of the lipids and proteins of the membrane.

Glycolipids. Glycolipids are present in small quantities, but they have important functions. Glycolipids are found mostly in plasma membranes, where their carbohydrate moieties protrude from the external surface of the membrane. The blood group antigens and certain other antigens are the carbohydrate side chains of specific glycolipids or glycoproteins.

Asymmetry of lipid distribution in the bilayer. In many membranes the lipid components are not distributed uniformly across the bilayer. As just mentioned, the glycolipids of the plasma membrane are located exclusively in the outer monolayer and thus show absolute asymmetry. Asymmetry of phospholipids occurs but is not absolute. In the red blood cell membrane, for example, the outer monolayer contains most of the choline-containing phospholipids, whereas the inner monolayer is enriched in the amino phospholipids.

Membrane Proteins

The protein composition of membranes may be simple or complex. The highly specialized membranes of the sarcoplasmic reticulum of skeletal muscle and the disks of the rod outer segment of the retina contain only a few different proteins. Plasma membranes perform many functions and may have more than 100 different protein constituents. Membrane proteins include enzymes (such as adenylate cyclase), transport proteins (such as the Na, K-ATPase), hormone receptors, receptors for neurotransmitters, and antigens.

Glycoproteins. Some membrane proteins are glycoproteins with covalently bound carbohydrate side chains. As with glycolipids, the carbohydrate chains of glycoproteins are located exclusively on the external surfaces of plasma membranes. Cell surface carbohydrate has important functions. The negative surface charge of cells is almost entirely the result of the negatively charged sialic acid of glycolipids and glycoproteins. Receptors for viruses may involve surface carbohydrate. Certain surface antigenic determinants reside in carbohydrate moieties on the cell surface. Surface carbohydrate participates in cellular aggregation phenomena and other forms of cell-cell interactions.

Asymmetry of membrane proteins. The absolute asymmetry of glycolipids and glycoproteins was mentioned earlier. The Na, K-ATPase of the plasma membrane and the Ca^{++} pump protein (Ca^{++}-ATPase) of the sarcoplasmic reticulum membrane are other examples of the asymmetrical functions of membrane proteins. In both cases ATP is split on the cytoplasmic face of the membrane, and some of the energy liberated is used to pump ions in specific directions across the membrane. In the case of the Na, K-ATPase, K^+ is pumped into the cell, and Na^+ is pumped out, whereas the Ca^{++}-ATPase actively pumps Ca^{++} into the sarcoplasmic reticulum. Integral membrane proteins are inserted into the membrane lipid bilayer during protein synthesis. The configuration of the protein in the membrane is the result of this specific insertion and the secondary and tertiary structures the protein assumes in the membrane.

Membranes as Permeability Barriers

Biological membranes serve as permeability barriers. Most of the molecules present in living systems have high solubility in water and low solubility in nonpolar solvents. Such molecules have low solubility in the nonpolar environment in the interior of the lipid bilayer of biological membranes. As a consequence, biological membranes pose a formidable permeability barrier to most water-soluble molecules. The plasma membrane is a permeability barrier between the cytoplasm and the extracellular fluid. This permeability barrier allows the maintenance of cytoplasmic concentrations of many substances that differ greatly from their concentrations in the extracellular fluid. The localization of various cellular processes in certain organelles depends on the barrier properties of cellular membranes. For example, the inner mitochondrial membrane is impermeable to the enzymes and substrates of the tricarboxylic acid cycle, allowing the localization of the tricarboxylic cycle in the mitochondrial matrix. The spatial organization of chemical and physical processes in the cell depends on the barrier functions of cellular membranes.

The passage of important molecules across membranes at controlled rates plays a central role in the life of the cell. Examples are the uptake of nutrient molecules, the discharge of waste products, and the release of secreted molecules.

In some cases molecules move from one side of a

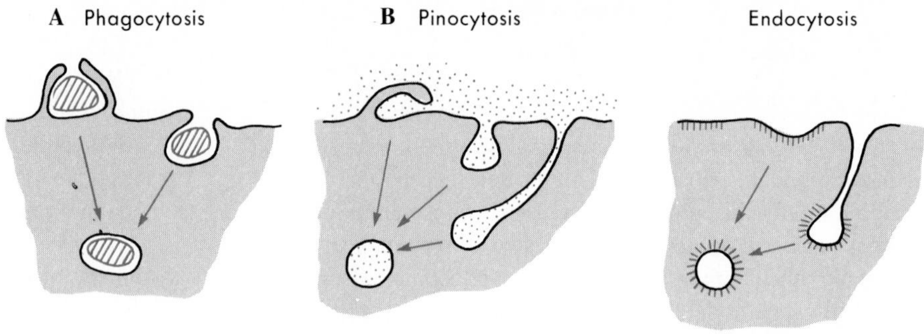

Fig. 1-3. Schematic depiction of endocytotic processes. **A,** Phagocytosis of a solid particle. **B,** Pinocytosis of extracellular fluid. **C,** Receptor-mediated endocytosis by coated pits. (Redrawn from Silverstein, S.C., et al.: Ann. Rev. Biochem. **46:**669, 1977. Reproduced, with permission, from the Annual Review of Biochemistry, © 1977 by Annual Reviews Inc.)

membrane to another without actually moving through the membrane itself. *Endocytosis* and *exocytosis* are examples of processes that transfer molecules across, but not through, biological membranes. In other cases molecules cross a particular membrane by actually moving *through* the membrane by passing through or between the molecules that make up the membrane.

Transport Across, But Not Through, Membranes

Endocytosis. Endocytosis allows material to enter the cell without passing through the membrane (Fig. 1-3). The uptake of particulate material is termed *phagocytosis* (Fig. 1-3, *A*). The uptake of soluble molecules is called *pinocytosis* (Fig. 1-3, *B*). Sometimes special regions of the plasma membrane, whose cytoplasmic surface is covered with bristles made primarily of a protein called *clathrin,* are involved in endocytosis. These bristle-covered regions are called *coated pits,* and their endocytosis gives rise to *coated vesicles* (Fig. 1-3, *C*). The coated pits appear to be involved primarily in *receptor-mediated endocytosis*. Specific proteins to be taken up are recognized and bound by specific membrane receptor proteins in the coated pits. The binding often leads to aggregation of receptor-ligand complexes, and the aggregation triggers endocytosis in ways that are not yet understood. Endocytosis is an active process that requires metabolic energy. Endocytosis also can occur in regions of the plasma membrane that do not contain coated pits.

Exocytosis. Molecules can be ejected from cells by exocytosis, a process that resembles endocytosis in reverse. The release of neurotransmitters, which is considered in more detail in Chapter 4, takes place by exocytosis. Exocytosis is responsible for the release of secretory proteins by many cells; the release of pancreatic zymogens from the acinar cells of the pancreas is a well-studied example. In such cases the proteins to be secreted are stored in secretory vesicles in the cytoplasm. A stimulus to secrete causes the secretory vesicles to fuse with the plasma membrane and to release the vesicle contents by exocytosis.

Fusion of membrane vesicles. The contents of one type of organelle can be transferred to another organelle by fusion of the membranes of the organelles. In some cells secretory products are transferred from the endoplasmic reticulum to the Golgi apparatus by fusion of endoplasmic reticulum vesicles with membranous sacs of the Golgi apparatus. Fusion of phagocytic vesicles with lysosomes allows intracellular digestion of phagocytosed material to proceed.

Transport of Molecules Through Biological Membranes

The traffic of molecules through biological membranes is vital for most cellular processes. Some molecules move through biological membranes simply by diffusing among the molecules that make up the membrane, whereas the passage of other molecules involves the mediation of *specific transport proteins* in the membrane.

Oxygen, for example, is a small molecule with fair solubility in nonpolar solvents. It crosses biological membranes by diffusing among membrane lipid molecules. Glucose, on the other hand, is a much larger molecule with low solubility in the membrane lipids. Glucose enters cells via a specific glucose transport protein in the plasma membrane.

Diffusion

Diffusion is the process whereby atoms or molecules intermingle because of their random thermal (Brownian) motion. Imagine a container divided into two compartments by a removable partition. A much larger number of molecules of a compound is placed on side A than on side B, and then the partition is removed. Every molecule is in random thermal motion. It is equally probable that a molecule which begins on side A will move to side B in a given time as it is that a molecule beginning on side B will end up on side A. Since there are many more molecules present on side A, the total *number* of molecules moving from side A to side B will be greater than the number moving from side B to side A. In this way the number of molecules on side A will decrease, while the number of molecules on side B will increase. This process of *net diffusion* of

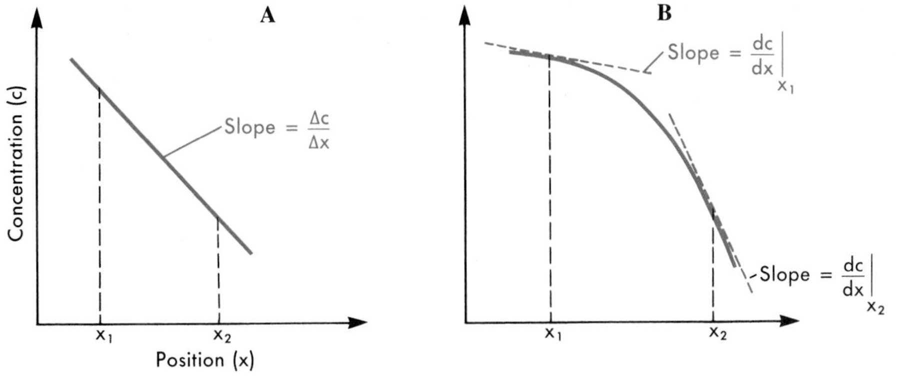

Fig. 1-4. **A,** Linear concentration profile. **B,** Nonlinear concentration profile. The concentration gradient at any point is the slope of the concentration profile at that point.

molecules will continue until the number of molecules on side A equals that on side B. Thereafter the rate of diffusion of molecules from A to B will equal that from B to A, so no further net movement will occur. A state of dynamic *equilibrium* exists when concentrations on side A and side B are equal.

Diffusion leads to a state in which the concentration of the diffusing species is constant in space and time. Thus diffusion across cellular membranes tends to equalize the concentrations on the two sides of the membrane. The diffusion rate across a particular planar surface is proportional to the area of the plane and to the difference in concentration of the diffusing substance on the two sides of the plane. *Fick's first law of diffusion* states that

$$J = -DA\frac{dc}{dx} \tag{1}$$

where

J = net rate of diffusion in moles or grams per unit time
A = area of the plane
dc/dx = concentration gradient across the plane
D = constant of proportionality called *diffusion coefficient*

The nature of the concentration gradient dc/dx in Fick's first law requires further explanation. The concentration profile of the diffusing substance may take many forms. In Fig. 1-4 two possible forms of the concentration profile are shown, and the locations of planes across which the rates of diffusion are to be determined are represented by x_1 and x_2. In Fig. 1-4, *A*, the concentration profile is a straight line. In this case the value of the concentration gradient is simply the slope Δc/Δx of the line, and the rate of flux is the same at both planes, that is:

$$J = -DA\frac{dc}{dx} = -DA\frac{\Delta c}{\Delta x} \tag{2}$$

In Fig. 1-4, *B*, however, the concentration profile is nonlinear. At each plane, x_1 and x_2, the value of dc/dx is equal to the slope of the tangent to the curve at that point. Since the slopes of those tangents are different, the rates of diffusion across planes placed at x_1 and x_2 will be different at this time.

Note also that the equation for Fick's first law contains a minus sign. The minus sign indicates the *direction* of diffusion. In Fig. 1-4 the slope of the concentration profile is *negative*, but the direction of diffusion is in the *positive* x direction. The need for the minus sign in the equation arises because molecules flow down a concentration gradient, that is, from higher to lower concentration.

The diffusion coefficient. The diffusion coefficient, D, has units of square centimeters per second. Solving equation 1 for D and expressing all the variables on the right-hand side of the resulting equation in cgs units, we obtain

$$D = \frac{J}{A\frac{dc}{dx}} = \frac{\text{moles/second}}{\text{cm}^2\,\frac{\text{moles/cm}^3}{\text{cm}}} = \text{cm}^2\text{/second} \tag{3}$$

D can be thought of as proportional to the speed with which the diffusing molecule can move in the surrounding medium. D is smaller the larger the molecule and the more viscous the medium.

For spherical solute molecules that are much larger than the surrounding solvent molecules Albert Einstein obtained the following equation:

$$D = kT/(6\pi r\eta) \tag{4}$$

where

k = Boltzmann's constant
T = absolute temperature (kT is proportional to the average kinetic energy of a solute molecule)
r = molecular radius
η = viscosity of the medium

The equation is called the *Stokes-Einstein relation,* and the molecular radius defined by this equation is known as the Stokes-Einstein radius.

For large molecules equation 4 predicts that D will be inversely proportional to the radius of the diffusing molecule. Since the molecular weight (MW) is approximately proportional to r^3, D should be inversely proportional to $(MW)^{1/3}$, so that a molecule which is ⅛ the mass of another molecule will have a diffusion coefficient only twice as large as the other molecule. For smaller solutes, with a molecular weight less than about 300, D is inversely proportional to $(MW)^{1/2}$ rather than $(MW)^{1/3}$.

Diffusion is a rapid process when the distance over which it must take place is small. This can be appreciated from another relation derived by Einstein. He considered the random movements of molecules that are originally located at $x = 0$. Since a given molecule is equally likely to diffuse in the $+x$ or $-x$ direction, the *average* displacement of all the molecules that begin at $x = 0$ will be zero. The average displacement squared, $\overline{(\Delta x)^2}$, which is a positive quantity, is represented by

$$\overline{(\Delta x)^2} = 2\,Dt \qquad (5)$$

where t is the time elapsed since the molecules started diffusing. The *Einstein relation* (equation 5) tells us how far the average molecule will diffuse in time (t), and it is useful as a rough estimate of the time scale of a particular diffusion process.

Einstein's relation shows us that the time required for a diffusion process increases with the square of the distance over which diffusion occurs. Thus a tenfold increase in the diffusion distance means that the diffusion process will require about 100 times longer to reach a given degree of completion. Table 1-1 shows the results of calculations using Einstein's relation for a typical, small, water-soluble solute. It can be seen that diffusion is extremely rapid on a microscopic scale of distance. For macroscopic distances diffusion is rather slow. A cell that is 100 μm away from the nearest capillary can receive nutrients from the blood by diffusion with a time lag of only 5 seconds or so. This is sufficiently fast to satisfy the metabolic demands of many cells. However, a nerve axon that is 1 cm long cannot rely on diffusion for the intracellular transport of vital metabolites, since the 14 hours it would take for diffusion over the 1 cm distance is too long on the time scale of cel-

■ **Table 1-1.** The time required for diffusion to occur over various diffusion distances*

Diffusion distance (μm)	*Time required for diffusion*
1	0.5 msec
10	50 msec
100	5 seconds
1000 (1 mm)	8.3 minutes
10,000 (1 cm)	14 hours

*The time required for the "average" molecule (with diffusion coefficient taken to be 1×10^{-5} cm²/second) to diffuse the required distance was computed from the Einstein relation.

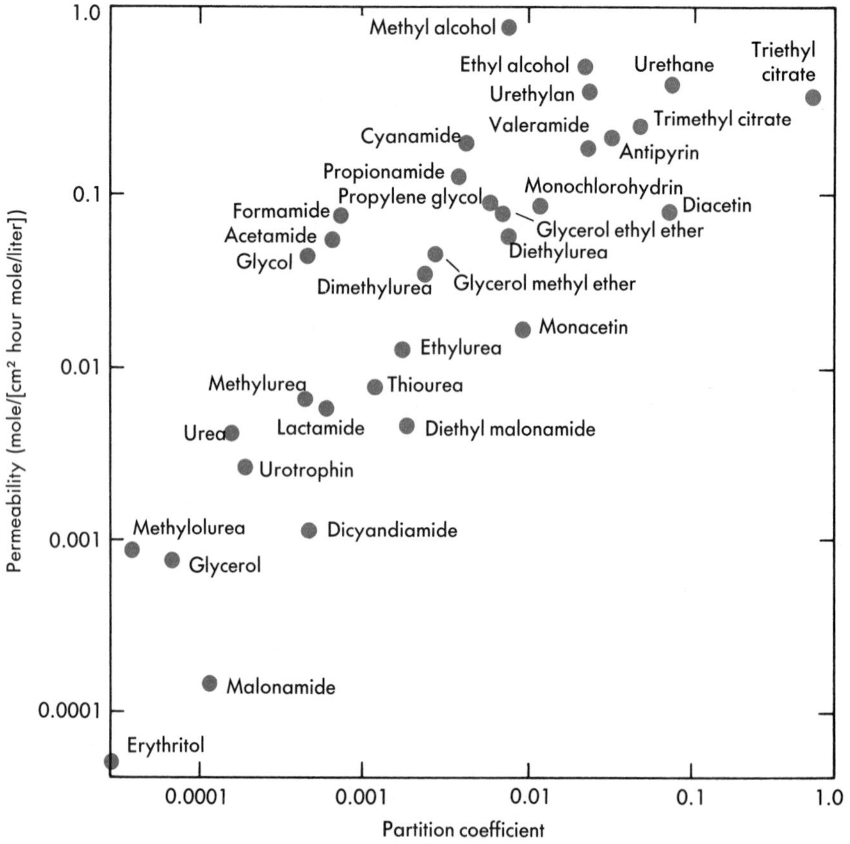

■ **Fig. 1-5.** The permeability of the plasma membrane of the alga *Chara ceratophylla* to various nonelectrolytes as a function of the lipid solubility of the solutes. Lipid solubility is represented on the abscissa by the olive oil/water partition coefficient. (Redrawn from Christensen, H.N.: Biological transport, ed. 2, Menlo Park, Calif., 1975, W.A. Benjamin. Data from Collander, R.: Trans. Faraday Soc. **33:** 985, 1937.)

lular metabolism. Some nerve fibers are as long as 1 m; it is no wonder then that intracellular axonal transport systems are involved in transporting important molecules along nerve fibers. Because of the slowness of diffusion over macroscopic distances, it is not surprising that even rather small multicellular organisms have evolved circulatory systems to bring the individual cells of the organism within reasonable diffusion range of nutrients.

Diffusive Permeability of Cellular Membranes

Permeability of lipid-soluble molecules. The plasma membrane serves as a diffusion barrier enabling the cell to maintain cytoplasmic concentrations of many substances that differ markedly from their extracellular concentrations. As early as the turn of the century, the relative impermeability of the plasma membrane to most water-soluble substances was attributed to its "lipoid nature."

The hypothesis that the plasma membrane has a lipoid character is supported by experiments showing that compounds that are soluble in nonpolar solvents (such as ether or olive oil) enter cells more readily than do water-soluble substances of similar molecular weight. Fig. 1-5 shows the relationship between membrane permeability and lipid solvent solubility for a number of different solutes. The ratio of the solubility of the solute in olive oil to its solubility in water is used as a measure of solubility in nonpolar solvents. This ratio is called the olive oil/water partition coefficient. The permeability of the plasma membrane to a particular substance increases with the "lipid solubility" of the substance. For compounds with the same olive oil/water partition coefficient, permeability decreases with increasing molecular weight. As described previously, the fluid mosaic model of membrane structure envisions the plasma membrane as a lipid bilayer with proteins embedded in it. The data of Fig. 1-5 support the idea that the lipid bilayer is the principal barrier to substances that permeate the membrane by simple diffusion.

The positive correlation between lipid solvent solubility and membrane permeability suggests that lipid-soluble molecules can dissolve in the plasma membrane and diffuse across it. Consider a substance that dissolves in the lipid bilayer and then diffuses across the plama membrane to the other side (Fig. 1-6). If the substance equilibrates at the edges of the lipid bilayer, its concentration at the outer face of the bilayer, $C_m(o)$, will be βC_o: the partition coefficient (β) times the concentration in the extracellular medium. Its concentration at the inner face of the membrane, $C_m(i)$, will be βC_i: β times the concentration in the cytoplasm. The concentration difference within the membrane itself is not $C_o - C_i$, but $C_m(o) - C_m(i)$. Since $C_m(o) - C_m(i) = \beta [C_o - C_i]$, the concentration gradient relevant to diffusion across the membrane is

$$\frac{\Delta C_m}{\Delta x} = \beta \frac{C_o - C_i}{\Delta x} \tag{6}$$

The more lipid soluble the substance, the larger is β, and the larger is the effective concentration gradient within the membrane that causes it to diffuse. According to Fick's first law the flux of the substance across the membrane is

$$J = -DA \frac{\Delta C_m}{\Delta x} = -DA\beta \frac{C_o - C_i}{\Delta x} = -\frac{D\beta}{\Delta x} A (C_o - C_i) \tag{7}$$

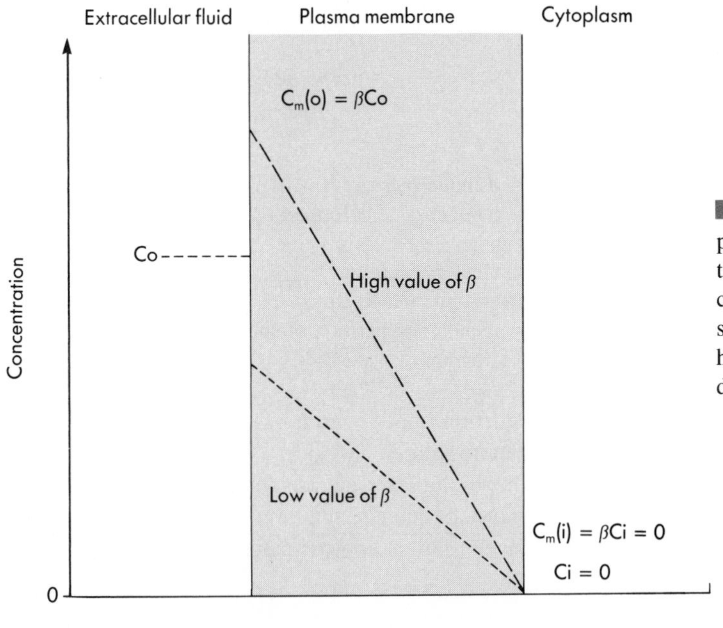

Fig. 1-6. Influence of the membrane lipid/water partition coefficient (β) of a solute on the concentration profile of the solute within the membrane. The cytoplasmic concentrations of the solutes are assumed to be zero for the sake of illustration. The higher the value of β, the steeper the intramembrane diffusion gradient.

where D is the diffusion coefficient within the membrane. $D\beta/\Delta x$ is a constant for a particular substance and a particular membrane, and it is called the *permeability coefficient* (k_p). In terms of the permeability coefficient:

$$J = -k_pA(C_o - C_i) \quad (8)$$

Since $k_p = \frac{D\beta}{\Delta x}$, β is dimensionless, the units of D are expressed in square centimeters per second, the units of k_p are centimeters per second.

Permeability of water-soluble molecules. Very small, uncharged, water-soluble molecules pass through cell membranes much more rapidly than predicted by their lipid solubility. For example, water permeates the cell membrane much more readily than do larger molecules with similar olive oil/water partition coefficients. The reason for the unusually high permeability of water is controversial. Some investigators think that cell membranes contain small pores that allow water and small water-soluble molecules to pass. Others think that very small water-soluble molecules can pass between adjacent phospholipid molecules without actually *dissolving* in the region occupied by the fatty acid side chains. Some experiments suggest that membrane proteins are responsible for the high-membrane permeability of water.

As the size of uncharged, water-soluble molecules increases, their membrane permeability decreases. Most plasma molecules are essentially impermeable to water-soluble molecules whose molecular weights are greater than about 200.

Because of their charge, ions are relatively insoluble in lipid solvent and thus have low permeability to most ions. Ionic diffusion across membranes is thought to occur through protein ''channels'' that span the membrane. Some channels are highly specific with respect to the ions allowed to pass, whereas others allow all ions below a certain size to pass. Some ion channels are controlled by the voltage difference across the membrane, whereas others are regulated by neurotransmitters or certain other molecules.

Certain water-soluble molecules such as sugars and amino acids, which are essential for cellular survival, do not cross plasma membranes at appreciable rates by simple diffusion. Plasma membranes have specific proteins that allow the transfer of vital metabolites into or out of the cell. The characteristics of membrane protein-mediated transport are discussed later.

Osmosis

Definitions. *Osmosis* is the flow of water across a semipermeable membrane from a compartment in which the solute concentration is lower to one in which the solute concentration is greater. A *semipermeable membrane* is defined as a membrane permeable to water but impermeable to solutes. Osmosis takes place because the presence of solute results in a decrease of the *chemical potential* of water. Water tends to flow from where its chemical potential is higher to where its chemical potential is lower. Other effects caused by the decrease of the chemical potential of water (because of the presence of solute) include reduced vapor pressure, lower freezing point, and higher boiling point of the solution as compared with pure water. Because these properties, and osmotic pressure as well, depend on the *concentration* of the solute present rather than on its chemical properties, they are called *colligative properties*.

Osmotic pressure. In Fig. 1-7 a semipermeable membrane separates a solution from pure water. Water flow from side B to side A by osmosis occurs because the presence of solute on side A reduces the chemical potential of water in the solution. Pushing on the piston will increase the chemical potential of the water in the solution of side A and slow down the net rate of osmosis. If the force on the piston is increased gradually, a pressure is reached eventually at which net water flow stops. Application of still more pressure will cause water to flow in the opposite direction, from side A to side B. The pressure on side A that is just sufficient to keep pure water from entering is called the *osmotic pressure* of the solution on side A.

The osmotic pressure of a solution depends on the number of particles in solution. Thus the degree of ionization of the solute must be taken into account. A 1 M solution of glucose, a 0.5 M solution of NaCl, and a 0.333 M solution of $CaCl_2$ theoretically should have the same osmotic pressure. (Actually their osmotic pressures will differ because of the deviations of real solutions from ideal solution theory.) Important equations that pertain to osmotic pressure and the other colligative properties were derived by van't Hoff. One form of *van't Hoff's Law* for calculation of osmotic pressure is

$$\pi = iRTm \quad (9)$$

where

π = osmotic pressure
i = number of ions formed by dissociation of a solute molecule
R = ideal gas constant
T = absolute temperature
m = molal concentration of solute (moles of solute per kilogram of water)

This equation applies more exactly as the solution becomes more dilute.

In the biological sciences molar concentrations are used more frequently than molal concentrations, and van't Hoff's law is approximated by

$$\pi = iRTc \quad (10)$$

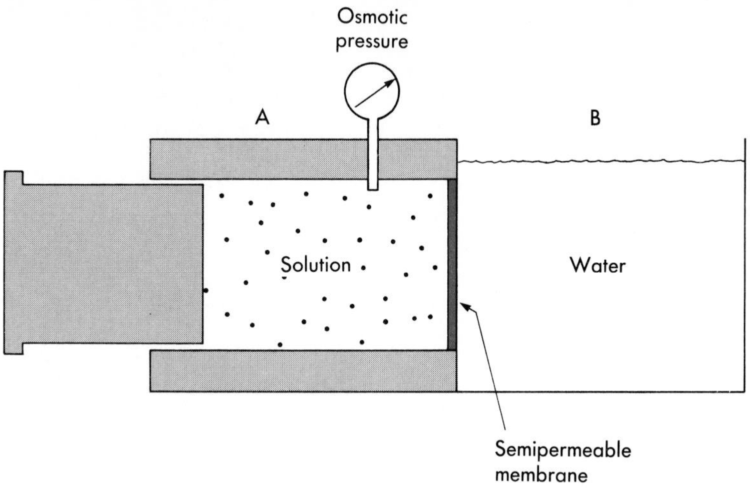

Fig. 1-7. Schematic representation of the definition of osmotic pressure. When the hydrostatic pressure applied to the solution in chamber *A* is equal to the osmotic pressure of that solution, there will be no net water flow across the membrane.

where c is the molar concentration of solute (moles of solute per liter of solution). The discrepancy between theory and reality is a bit larger for this form of van't Hoff's law than for equation 9.

Equations 9 and 10 do not predict precisely the osmotic pressures of real solutions. At the concentrations of many substances in cytoplasm and extracellular fluids, the deviations from ideality may be substantial. For example, sodium is the principal cation of the extracellular fluids, and chloride is the main anion. Na^+ is present at about 150 mEq/L and Cl^- at about 120 mEq/L. NaCl solutions in this concentration range differ considerably in their osmotic pressures from the predictions of van't Hoff's law.

One way of correcting for the deviations of real solutions from the predictions of van't Hoff's law is to use a correction factor called the *osmotic coefficient* (ϕ). Including the osmotic coefficient, equation 10 becomes

$$\pi = \phi iRTc \qquad (11)$$

The osmotic coefficient may be greater or less than one. It is less than one for electrolytes of physiological importance, and for all solutes it approaches one as the solution becomes more and more dilute. The term ϕic can be regarded as the osmotically effective concentration, and ϕic often is referred to as the *osmolar concentration,* with units in osmoles per liter. Values of the osmotic coefficient depend on the concentration of the solute and on its chemical properties. Table 1-2 lists osmotic coefficients for several solutes. These values apply fairly well at the concentrations of these solutes in the extracellular fluids of mammals. The value of ϕ may be highly concentration dependent. More precise values of ϕ can be obtained from tables in handbooks that list values of ϕ for different substances as functions of concentration. Solutions of proteins deviate greatly from van't Hoff's law, and different proteins may deviate to different extents. The deviations from ideality are frequently more concentration dependent for proteins than for smaller solutes.

Table 1-2. Osmotic coefficients (ϕ) of certain solutes of physiological interest

Substance	*i*	*Molecular weight*	ϕ
NaCl	2	58.5	0.93
KCl	2	74.6	0.92
HCl	2	36.6	0.95
NH_4Cl	2	53.5	0.92
$NaHCO_3$	2	84.0	0.96
$NaNO_3$	2	85.0	0.90
KSCN	2	97.2	0.91
KH_2PO_4	2	136.0	0.87
$CaCl_2$	3	111.0	0.86
$MgCl_2$	3	95.2	0.89
Na_2SO_4	3	142.0	0.74
K_2SO_4	3	174.0	0.74
$MgSO_4$	2	120.0	0.58
Glucose	1	180.0	1.01
Sucrose	1	342.0	1.02
Maltose	1	342.0	1.01
Lactose	1	342.0	1.01

Reproduced with permission from Lifson, N., and Visscher, M.B.: Osmosis in living systems. In Glasser, O., editor: Medical physics, vol. 1. Copyright © 1944 by Year Book Medical Publishers, Inc., Chicago.

Sample calculations

1. What is the osmotic pressure (at 0° C) of a 154 mM NaCl solution?

$$\pi = \phi iRTc \qquad (12)$$

Taking ϕ for NaCl from Table 1-2, we obtain

$$\pi = 0.93 \times 2 \times 22.4 \text{ L-atm/mol} \times 0.154 \text{ mol/L} = 6.42 \text{ atm} \qquad (13)$$

2. What is the osmolarity of this solution?

Osmolarity = ϕic = 0.93 × 2 × 0.154 = 0.286 osmole/L = 286 mOsm

Measurement of osmotic pressure. The osmotic pressure of a solution can be obtained by determining the pressure required to prevent water from entering the solution across a semipermeable membrane (Fig. 1-7).

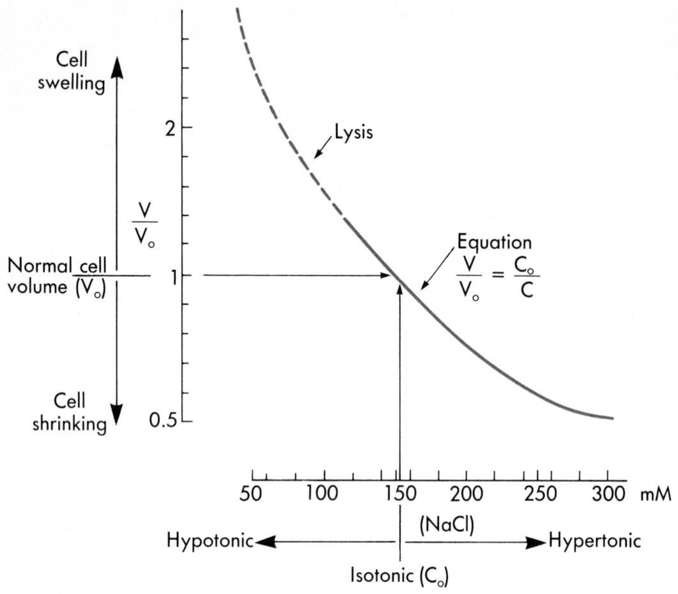

Fig. 1-8. The osmotic behavior of human red blood cells in NaCl solutions. At 154mM NaCl (isotonic) the red cell has its normal volume. It shrinks in more concentrated (hypertonic) solutions and swells in more dilute (hypotonic) solutions.

However, this method is time consuming and technically difficult. Consequently the osmotic pressure more often is estimated from another colligative property, such as depression of the freezing point. The relation that describes the depression of the freezing point of water by a solute is

$$\Delta T_f = 1.86\ \phi ic \quad (14)$$

where ΔT_f is the freezing point depression in degrees centigrade. Thus the effective osmotic concentration (in osmoles per liter) is

$$\phi ic = \frac{\Delta T_f}{1.86} \quad (15)$$

When the freezing point depression of a multicomponent solution is determined, the effective osmolar concentration (in osmoles per liter) of the solution as a whole can be obtained.

If the total osmotic pressures of two solutions (as measured by freezing point depression or by the osmotic pressure developed across a true semipermeable membrane) are equal, the solutions are said to be *isosmotic* (or iso-osmotic). If solution A has greater osmotic pressure than solution B, A is said to be *hyperosmotic* with respect to B. If solution A has less total osmotic pressure than solution B, A is said to be *hypoosmotic* to B.

Osmotic Swelling and Shrinking of Cells

The plasma membranes of most of the cells of the body are relatively impermeable to many of the solutes of the interstitial fluid but are highly permeable to water. Therefore when the osmotic pressure of the interstitial fluid is increased, water leaves the cells by osmosis, the cells shrink, and cellular solutes become more concentrated until the effective osmotic pressure of the cytoplasm is again equal to that of the interstitial fluid. Conversely, if the osmotic pressure of the extracellular fluid is decreased, water enters the cells, and the cells will swell until the intracellular and extracellular osmotic pressures are equal.

Red blood cells often are used to illustrate the osmotic properties of cells because they are readily obtained and are easily studied. Within a certain range of external solute concentrations the red cell behaves as an osmometer, since its volume is inversely related to the solute concentration in the extracellular medium. In Fig. 1-8 the red cell volume, as a fraction of its normal volume in plasma, is shown as a function of the concentration of NaCl solution in which the red cells are suspended. At a NaCl concentration of 154mM (308mM particles) the volume of the cells is the same as their volume in plasma; this concentration of NaCl is said to be *isotonic* to the red cell. A concentration of NaCl greater than 154mM is called *hypertonic* (cells shrink), and a solution less concentrated than 154mM is termed *hypotonic* (cells swell). When red cells have swollen to about 1.4 times their original volume, some cells *lyse* (burst). At this volume the properties of the red cell membrane abruptly change, and hemoglobin leaks out of the cell. The membrane becomes transiently permeable to other large molecules at this point as well.

The intracellular substances that produce an osmotic pressure which just balances that of the extracellular fluid include hemoglobin, K^+, organic phosphates (such as ATP and 2,3-diphosphoglycerate), and glycolytic intermediates. Regardless of the chemical nature of its contents, the red cell behaves as though it were filled with a solution of *impermeant* molecules with an osmotically effective concentration of 286 milliosmolar (154mM NaCl = 286 milliosmolar).

Osmotic effects of permeant solutes. Permeating solutes eventually equilibrate across the plasma membrane. For this reason permeating solutes exert only a transient effect on cell volume.

Consider a red blood cell placed in a large volume of 0.154M NaCl, containing 0.050M glycerol. Initially, because of the extracellular NaCl and glycerol, the osmotic pressure of the extracellular fluid will exceed that of the cell interior, and the cell will shrink. With time, however, glycerol will equilibrate across the plasma membrane of the red cell, and the cell will swell back toward its original volume. The steady-state volume of the cell will be determined only by the *impermeants* in the extracellular fluid. In this case the impermeants (NaCl) have a total concentration that is isotonic, so that the final volume of the cell will be equal to the normal red cell volume. Because the red cell ultimately returns to its normal volume, the solution (0.050 M glycerol in 0.154 M NaCl) is *isotonic*. Since the red cell initially shrinks when put in this solution, the solution is *hyperosmotic* with respect to the normal red cell. The transient changes in cell volume depend on equilibration of glycerol across the membrane. Had we used urea (a more rapidly permeating substance), the cell would have reached steady-state volume sooner.

The following rules help predict the volume changes a cell will undergo suspended in solutions of permeant and impermeant solutes:

1. The steady-state volume of the cell is determined only by the concentration of impermeant particles in the extracellular fluid.
2. Permeant particles cause only transient changes in cell volume.
3. The time course of the transient changes is more rapid the greater the permeability of the permeant molecule.

The magnitudes of osmotic flows caused by permeating solutes. In the preceding example it was explained that permeants, such as glycerol, exert only a transient osmotic effect. It is sometimes important to determine the magnitude of the osmotic effect exerted by a particular permeant.

When a difference of *hydrostatic* pressure (ΔP) causes water flow across a membrane, the rate of water flow ($\dot{V}_w$) is

$$\dot{V}_w = L\Delta P \quad (16)$$

where L is a constant of proportionality called the *hydraulic conductivity*.

Osmotic flow of water across a membrane is directly proportional to the osmotic pressure difference ($\Delta\pi$) of the solutions on the two sides of the membrane; thus

$$\dot{V}_w = L\Delta\pi \quad (17)$$

Equation 17 holds only for osmosis caused by impermeants; permeants cause less osmotic flow. The greater the permeability of a solute, the less the osmotic flow it causes. Table 1-3 shows the osmotic water flows induced across a porous membrane by solutes of different molecular size. The solutions have identical freezing points, so the total osmotic pressures are the same. Note that the larger the solute molecule, and thus the more impermeable it is to the membrane, the greater the osmotic water flow it causes.

Equation 17 can be rewritten to take solute permeability into account by including σ, the *reflection coefficient*.

$$\dot{V}_w = \sigma L\Delta\pi \quad (18)$$

σ is a dimensionless number that ranges from 1 for completely impermeant solutes down to 0 for extremely permeant solutes. σ is a property of a particular solute and a particular membrane and represents the osmotic flow induced by the solute as a fraction of the theoretical maximum osmotic flow (Table 1-3).

Table 1-3. Osmotic water flow across a porous dialysis membrane caused by various solutes*

Gradient producing the water flow	*Net volume flow (μl/minute)**	*Solute radius (Å)*	*Reflection coefficient (σ)*
D_2O	0.06	1.9	0.0024
Urea	0.6	2.7	0.024
Glucose	5.1	4.4	0.205
Sucrose	9.2	5.3	0.368
Raffinose	11	6.1	0.440
Inulin	19	12	0.760
Bovine serum albumin	25.5	37	1.02
Hydrostatic pressure	25		

Data from Durbin, R.P.: J. Gen. Physiol. **44**:315, 1960. Reproduced from The Journal of General Physiology by copyright permission of The Rockefeller University Press.

*Flow is expressed as microliters per minute caused by a 1M concentration difference of solute across the membrane. The flows are compared with the flow caused by a theoretically equivalent hydrostatic pressure.

Protein-Mediated Membrane Transport

Certain substances enter or leave cells by way of specific carriers or channels that are intrinsic proteins of the plasma membrane. Transport via such protein carriers or channels is called *protein-mediated transport* or simply *mediated transport*. Specific ions or molecules may cross the membranes of mitochondria, endoplasmic reticulum, and other organelles by mediated transport. Mediated transport systems include *active transport* and *facilitated transport* processes. As discussed later, active transport and facilitated transport have a number of properties in common. The principal distinction between these two processes is that active transport is capable of "pumping" a substance against a gradient of concentration (or electrochemical potential), whereas facilitated transport tends to equilibrate the substance across the membrane.

Properties of mediated transport. The basic properties of mediated transport are the following:

1. Transport is more rapid than that of other molecules of similar molecular weight and lipid solubility that cross the membrane by simple diffusion.
2. The transport rate shows saturation kinetics: as the concentration of the transported compound is increased, the rate of transport at first increases, but eventually a concentration is reached after which the transport rate increases no further. At this point the transport system is said to be saturated with the transported compound.
3. The mediating protein has *chemical specificity:* only molecules with the requisite chemical structure are transported. The specificity of most transport systems is not absolute, and, in general, it is broader than the specificity of most enzymes. Glucose is transported into red blood cells by facilitated transport. Glucose is the preferred substrate, but mannose, galactose, xylose, L-arabinose, and certain other sugars also are transported by the same membrane protein. D-Arabinose, however, is a poor substrate for the glucose system, and sorbitol and mannitol do not enter red cells at all. Mediated processes have *stereospecificity* as well. In the red cell sugar transport system, D-glucose is transported well, but L-glucose is barely transported at all.
4. Structurally related molecules may compete for transport. Typically the presence of one transport substrate will decrease the transport rate of a second substrate by competing for the transport protein. The competition is analogous to *competitive inhibition* of an enzyme.
5. Transport may be inhibited by compounds that are not structurally related to transport substrates. An inhibitor may bind to the transport protein in a way that decreases its affinity for the normal transport substrate. The compound phloretin does not resemble a sugar molecule, yet it strongly inhibits red cell sugar transport. Active transport systems, which require some link to metabolism, may be inhibited by metabolic inhibitors. The rate of Na^+ transport out of cells by the Na^+, K^+-ATPase is decreased by substances that interfere with ATP generation.

Facilitated Transport

Facilitated transport, sometimes called facilitated diffusion, occurs via a transport protein that is not linked to metabolic energy. Facilitated transport has the properties discussed previously, except that facilitated transport is not generally depressed by metabolic inhibitors. Because they have no link to energy metabolism, facilitated transport processes cannot move uncharged substances against concentration gradients or ions against electrochemical potential gradients. Facilitated transport systems act to equalize concentrations (or electrochemical potentials for ions) in the cytoplasm and extracellular fluid of the substances they transport.

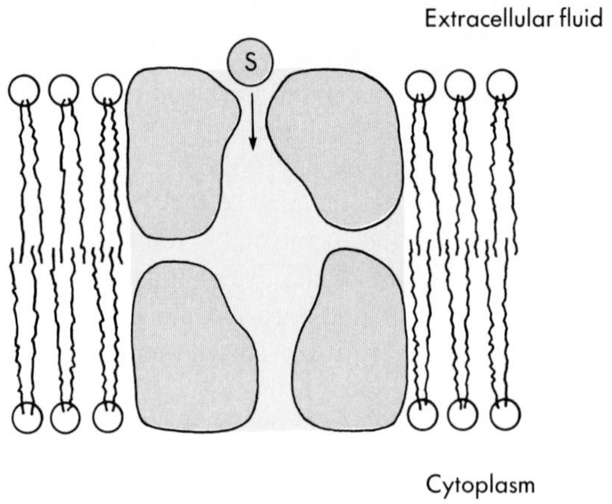

■ **Fig. 1-9.** Hypothetical model of a transport protein. This is a model of the mechanism of monosaccharide transport by the sugar transport protein of the human red blood cell. The protein is postulated to be a tetramer. Binding of sugar is proposed to cause conformational changes that allow sugar molecules to enter and leave the central cavity of the transport protein. (Redrawn from Jones, M.N., and Nickson, J.K.: Biochim. Biophys. Acta **650:**1, 1981.)

Monosaccharides enter muscle cells by a facilitated transport process. Glucose, galactose, arabinose, and 3-*O*-methylglucose compete for the same carrier. The rate of transport shows saturation kinetics. The nonphysiological stereoisomer L-glucose enters the cells very slowly, and nontransported sugars such as mannitol or sorbose enter muscle cells very slowly if at all. Phloretin inhibits sugar uptake. The sugar transport system in muscle cells is stimulated by insulin. In the absence of insulin, glucose transport is rate limiting for glucose use, so that insulin is a major physiological regulator of muscle glucose metabolism.

The molecular details of protein-mediated transport processes are not well understood. Current evidence suggests that most transport proteins span the membrane and are multimeric. Fig. 1-9 depicts a hypothetical model that has been proposed for the monosaccharide transport protein of the membrane of the human red blood cell. The protein is postulated to be a tetramer. Conformational changes of the protein, induced by monosaccharide binding, may allow a sugar molecule to enter and leave the central cavity.

Active Transport

Active transport processes have most of the properties of facilitated transport and, in addition, can concentrate their substrates against electrochemical potential gradients. This requires energy, so active transport processes must be linked to energy metabolism in some way. Active transport systems may use ATP directly, or they may be linked more indirectly to metabolism. Because of their dependence on metabolism, active transport processes may be inhibited by any substance that interferes with energy metabolism.

One possible general mechanism for active transport. Mediated transport is analogous in many respects to an enzyme-catalyzed chemical reaction. When the transport process is in the steady state across the plasma membrane, the unidirectional flux from cytoplasm to extracellular fluid is exactly equal to the unidirectional flux from extracellular fluid to cytoplasm. Since influx equals efflux, there is no net flux across the plasma membrane, and the cellular concentration is constant in the steady state.

For a facilitated transport process across the plasma membrane, influx and efflux are often symmetrical processes. In that case, in the terms of enzyme kinetics, influx and efflux have the same K_m and V_{max}. (V_{max} is the maximal rate of transport, and K_m is the concentration of the transported substance that gives a transport rate of $V_{max}/2$.) Imagine that the K_m of a glucose transport system is 1 mM. When the glucose concentration, in cytoplasm and extracellular fluid are 1 mM, influx and efflux will be equal (each being equal to $V_{max}/2$). This example illustrates that the steady state of a facilitated transport system is characterized by equilibration of substrate across the membrane.

With an active transport system the influx and efflux processes are not symmetrical. Consider an imaginary transport system for glycine that has a K_m for influx of 0.5 mM, a K_m for efflux of 5 mM, and the same V_{max} for both influx and efflux. When the intracellular glycine concentration is 5 mM and the extracellular glycine concentration is 0.5 mM, influx of glycine will equal glycine efflux (both being equal to $V_{max}/2$), and the steady state will be obtained. Note that the system is in steady state with a glycine concentration 10 times higher inside the cell than outside, so this is an active transport system.

In creating the tenfold concentration difference of glycine the transport system does work. The ultimate source of the energy to do transport work is metabolic energy. How is metabolic energy harnessed to do transport work? One way might involve the cyclical phosphorylation and dephosphorylation of the transport protein. In the glycine transport system just described imagine that phosphorylation of the protein at the inner aspect of the membrane (by a protein kinase-using ATP) converts the protein to the form in which K_m at the inner surface of the plasma membrane is 5 mM. If dephosphorylation occurs, the protein "reverts" to the form in which K_m at the outer face of the plasma membrane is 0.5 mM. In this way the K_m for efflux is made greater than that for the influx, allowing a tenfold concentration difference to be created. Since the imaginary glycine active transport system described uses ATP directly, this is termed a *primary active transport system,* as would be any transport system that is rather directly linked to any high-energy metabolic intermediate.

Another way in which active transport is powered involves a less direct link to metabolic energy. The transmembrane concentration difference of a second molecule that itself is transported actively may be used to allosterically modify the affinity of the transport protein for its substrate. Sodium is actively extruded from most cells by the Na, K pump (Na^+, K^+-ATPase) by primary active transport, so that the extracellular concentration of Na^+ is often about tenfold higher than its intracellular concentration. In the imaginary glycine transport system described previously, imagine that at the high Na^+ concentrations present in the extracellular fluid Na^+ binds to the transport carrier at the external face of the membrane and that Na^+ binding converts the protein to the form in which K_m at the outer face of the membrane is 0.5 mM. At the inner surface of the membrane the Na^+ concentration is much lower, so much less Na^+ will bind to the transport protein, and it thus will be predominantly in the form in which K_m at the inner membrane surface is 5 mM. In this way the transmembrane gradient of Na^+ (which is created by primary active transport) is harnessed to bring about a transmembrane concentration difference of glycine (in this hypothetical example). In this case glycine transport is said to occur by *secondary active transport.*

Examples of active transport processes

Transport powered by the phosphorylation of a transport protein. In the cytoplasm of most animal cells the concentration of Na^+ is much less and the concentration of K^+ much more than the extracellular concentrations of these ions. This situation is brought about by the action of a Na^+,K^+ pump in the plasma membrane that pumps Na^+ out of the cell and K^+ into the cell. The Na^+,K^+ pump activity is the result of an integral membrane protein called the Na^+,K^+-ATPase. The Na^+,K^+-ATPase consists of a "catalytic" α-subunit of about 100,000 daltons and a glycoprotein β-subunit of about 50,000 daltons. When operating near its capacity for ion transport, the Na^+,K^+-ATPase transports three sodium ions out of the cell and transports two potassium ions into the cell for each ATP hydrolyzed. The cyclical phosphorylation and dephosphorylation of the protein causes it to alternate between two conformations, E1 and E2. In E1 the ion-binding sites of the protein have high affinity for Na^+ and face the cytoplasm. In E2 the ion-binding sites favor the binding of K^+ and face the extracellular fluid.

A simplified view of the ion transporting cycle of the Na^+,K^+-ATPase is as follows (Fig. 1-10). The E1 conformer has high affinities for intracellular Na^+, Mg^{++},

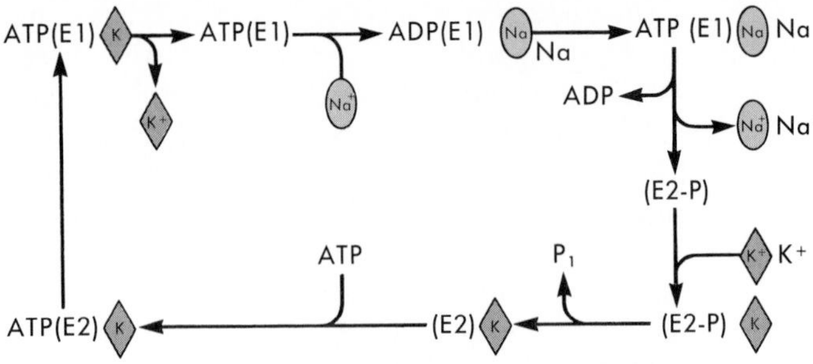

■ **Fig. 1-10.** A simplified view of the reaction cycle of the Na^+,K^+-ATPase. Ligands shown on the inside of the cycle bind or are released on the inside of the plasma membrane; ligands shown on the outside of the cycle bind or are released on the outside of the plasma membrane. (Adapted from Post, R.L.: A perspective on sodium and potassium ion transport adenosine triphosphatase. In Mukohata, Y., and Packer, L., editors: Cation fluxes across biomembranes, New York, 1979, Academic Press, Inc.)

and ATP. When these three ligands are bound, the protein is phosphorylated at an aspartate residue near the ATP binding site, and the Na^+ binding site becomes "occluded", that is, inaccessible from either side of the membrane. The ATPase then is thought to undergo a conformational change that makes the ion binding sites accessible to the extracellular fluid and reduces their affinity for Na^+. Na^+ then dissociates on the extracellular side of the membrane, and the protein undergoes a conformational change to E2-P. E2-P has a high affinity for K^+ at the extracellular side of the membrane and proceeds to bind two K^+ ions. The binding of K^+ is thought to promote the hydrolysis of the acylphosphate bond, and the K^+ binding sites may move to an occluded position. Then the binding of ATP to the E2•K_2 promotes the conformational change back to the El state, and the dissociation of K^+ occurs at the cytoplasmic face of the membrane. In this way, the Na^+,K^+-ATPase is thought to alternate between the El and the E2 conformations, a process termed *molecular peristalsis*.

Because the Na^+,K^+-ATPase uses the energy in the terminal phosphate bond of ATP to power its transport cycle, it is said to be a *primary active transport* system. A transport process powered by some other high-energy metabolic intermediate or linked directly to a primary metabolic reaction would also be classified as primary active transport.

Transport powered by the gradient of another species: secondary active transport. The last section emphasized that energy is required to create a concentration gradient of a transported substance. Once created, a concentration gradient (or the electrochemical potential gradient of an ion) represents a store of chemical potential energy that can be used to perform work. When the substance flows down its gradient, it releases energy that can be harnessed to do work. For example, in the mitochondrion, the electrochemical potential gradient of H^+ ion created across the inner mitochondrial membrane by electron transport is used to perform the work of ATP synthesis. In many cell types the electrochemical potential (see p. 22) gradient of Na^+, created by the Na^+,K^+-ATPase, is used to actively transport certain other solutes into the cell. Many cells take up the neutral, hydrophilic amino acids by a membrane transport protein that links the inward transport of Na^+ down its electrochemical potential gradient to the inward transport of amino acids against their gradients of electrochemical potential (Fig. 1-11). The energy for the transport of the amino acid is not provided directly by ATP or some other high-energy metabolite, but indirectly, from the gradient of another species that is itself actively transported. Hence the amino acid is said to be transported by *secondary active transport*. In the secondary active transport of amino acids, both the rate of amino acid transport and the extent to which the amino acid is accumulated (Fig. 1-12) are dependent on the electrochemical potential gradient of Na^+.

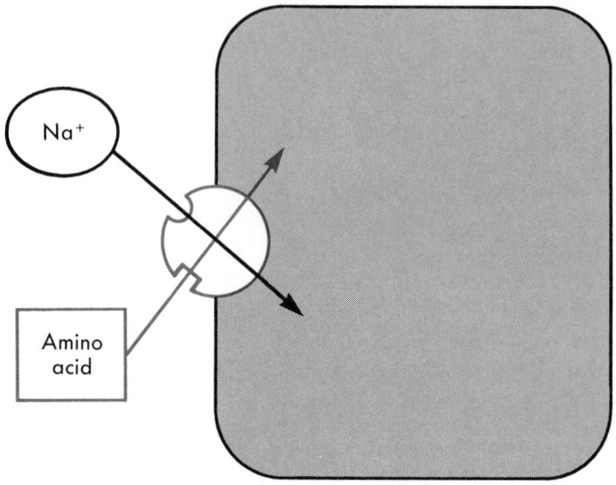

■ **Fig. 1-11.** Many cells take up certain neutral amino acids by secondary active transport. The transport protein binds both Na^+ and the amino acid. Na^+ is transported down its electrochemical potential gradient, and the protein uses the energy released by Na^+ flux to transport the amino acid against a concentration gradient.

■ *Some Other Important Membrane Transport Processes*

Calcium transport. Under most circumstances the concentration of Ca^{++} in the cytosol of cells is maintained at rather low levels, below 10^{-7} M, whereas the concentration of Ca^{++} in extracellular fluids is of the order of a few mM. The plasma membranes of most

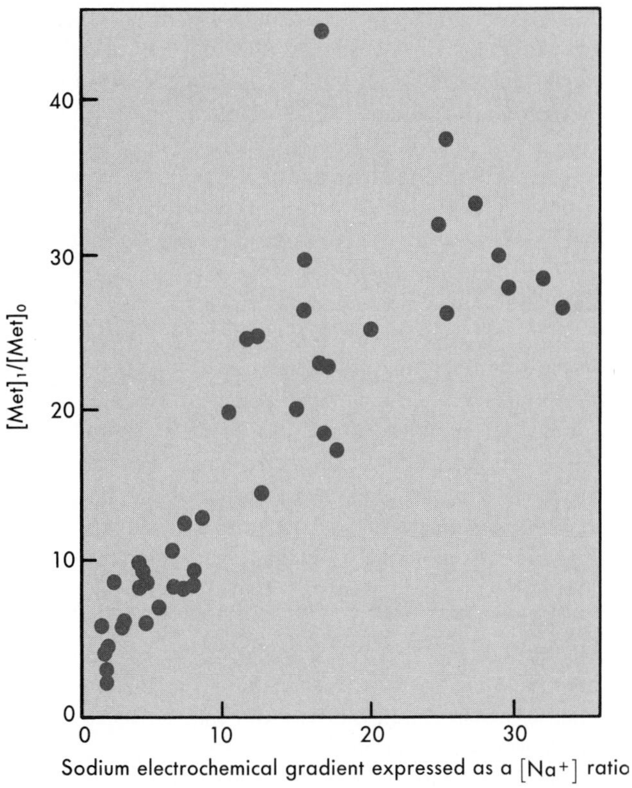

Fig. 1-12. The extent to which methionine can be accumulated by mouse ascites-tumor cells depends on the gradient of the electrochemical potential of Na^+ across the plasma membrane. The ratio of steady-state intracellular to extracellular concentrations of methionine is plotted against the steady-state ratio of the electrochemical gradient of the Na^+ ion across the membrane. (Modified and reprinted by permission from Philo, R., and Eddy, A.A.: Biochem. J. **174:**811, copyright © 1978 The Biochemical Society, London.)

cells contain a Ca^{++}-ATPase that helps to establish the large gradient of Ca^{++} across the plasma membrane. The plasma membrane Ca^{++}-ATPase is different from the Ca^{++}-ATPase that is responsible for sequestering Ca^{++} in the sarcoplasmic reticulum of muscle (see Chapter 22). However, the plasma membrane Ca^{++}-ATPase shares a number of important properties with the Ca^{++}-ATPase of the sarcoplasmic reticulum (and with the Na^+,K^+-ATPase of plasma membranes). Both proteins carry out the primary active transport of Ca^{++} across a membrane, and they use the energy of the terminal phosphate bond of ATP to accomplish this task. The transport cycle of both proteins is driven by the cyclical phosphorylation and dephosphorylation of an acidic amino acid residue of the transport protein. Both proteins alternate between two distinct classes of conformational states: one in which a high affinity binding site for Ca^{++} faces the cytosol, and the other in which the binding site is of lower affinity and faces the other side of the membrane.

A plasma membrane Ca^{++}-ATPase occurs in the red blood cells, cardiac muscle, smooth muscle, neurons, liver cells, kidney tubule cells, intestinal epithelial cells, and several other cell types. The Ca^{++}-ATPase of the plasma membrane of the human red blood cell is the best characterized plasma membrane Ca^{++}-ATPase, and it serves as a model of the Ca^{++}-ATPases of the plasma membranes of most other cells. The plasma membrane Ca^{++}-ATPase is an integral membrane protein of 130,000 to 140,000 molecular weight. The number of Ca^{++} ions pumped for each ATP hydrolyzed is still controversial; estimates range from 1 to 2 Ca^{++}/ATP. In keeping with the vital role of Ca^{++} as a controller of cellular processes, the activity of the Ca^{++}-ATPase is highly regulated. Intracellular *calmodulin,* the ubiquitous calcium-dependent regulatory protein, in the presence of micromolar levels of Ca^{++} increases the affinity of the ATPase for Ca^{++} at the inner surface of the plasma membrane and increases its rate of turnover. The activity of the Ca^{++}-ATPase is also regulated by the phosphatidylinositol cycle: phosphatidylinositol-4,5-bisphosphate enhances the activity of the Ca^{++}-ATPase.

Many excitable cells, those of the heart for example, have an additional mechanism for controlling the level of intracellular Ca^{++}. A sodium/calcium exchange protein in the plasma membrane uses the energy in the Na^+ gradient to extrude Ca^{++} from the cell. In heart cells it appears that rapid, transient changes in intracellular Ca^{++} are mediated by the sodium/calcium exchange protein, whereas the resting level of intracellular Ca^{++} is set primarily by the Ca^{++}-ATPase (see Chapter 29).

In most cells the rate at which Ca^{++} leaks into the cell down its electrochemical potential gradient is rather slow, so that the energy cost of maintaining a low intracellular level of Ca^{++} is not high. This is in contrast to the cost of pumping Na^+ and K^+; running the Na^+, K^+-ATPase is a major item in the energy budget of many cells.

Sugar transport. Glucose is a primary fuel for most of the cells of the body, but glucose diffuses across plasma membranes rather slowly. The plasma membranes of many cell types contain a protein that mediates the facilitated transport of glucose and related monosaccharides across the membrane. Red blood cells, hepatocytes, adipocytes, and muscle cells (skeletal, cardiac, and smooth) all possess facilitated diffusion systems for uptake of glucose. The uptake of glucose in these cell types does not depend on the electrochemical potential difference of Na^+ across the plasma membrane nor in any direct way on cellular metabolism. In adipocytes and muscle cells transport of glucose across the plasma membrane is increased in the presence of physiological levels of insulin. Insulin may increase glucose transport by causing more glucose transport proteins to be inserted into the plasma membrane. The source of the newly inserted protein is a preformed pool of transporters that is present in the membranes of microsomes within the cell.

Amino acid transport. Most of the cells in the body synthesize proteins and therefore require amino acids. A number of different amino acid transport systems are present in plasma membranes. Among the amino acid transport systems that may be present in the plasma membrane of a given cell type are three distinct systems for neutral amino acids, a system for basic amino acids, and a system for acidic amino acids. Amino acid transport systems overlap significantly in specificities, and the distribution of the different systems varies from one cell type to another.

The three transport systems for neutral amino acids have been named the A system (prefers alanine and other small, hydrophilic amino acids), the L system (prefers leucine and the more hydrophobic amino acids), and the ASC system (prefers alanine, serine, and cysteine). Systems A and ASC cotransport Na^+ along with a neutral amino acid and use the energy in the electrochemical potential gradient of Na^+ to actively transport the neutral amino acids into the cell against significant concentration gradients. The A and ASC systems are thus secondary active transport systems that use the gradient of Na^+ created by the Na^+, K^+-ATPase (a primary active transport system). The L amino acid transport system, on the other hand, does not depend on the Na^+ gradient or on metabolic energy. The L system is a facilitated transport system and functions to equilibrate the concentrations of the transported amino acids across the plasma membrane. The L system prefers the bulkier and more hydrophobic neutral amino acids, namely leucine, isoleucine, valine, phenylalanine, and methionine. Other transport systems for neutral amino acids may exist in particular cell types, but these other transport systems are less well characterized, and probably less widely distributed, than the three transport systems just described.

Transport systems for basic and acidic amino acids have been identified in several cell types, but such transport systems are not as well understood as those for neutral amino acids. A facilitated (independent of Na^+) transport system for basic amino acids (lysine, arginine, and ornithine) has been described in certain cell types. An active (Na^+ dependent) transport system for acidic amino acids (glutamate and aspartate) exists in a number of cell types.

Transport of nucleosides and nucleic acid bases. Nucleotides enter most cells very slowly. Nucleosides and nucleic acid bases enter mammalian cells by facilitated transport. After entering the cell, nucleosides are phosphorylated, and bases are phosphoribosylated. The rates of transport are greatly in excess of the rates of the intracellular reactions. A single facilitated transport system, with broad specificity, may be responsible for the transport of nucleosides. The transport system has higher affinities for purine nucleosides than for pyrimi-

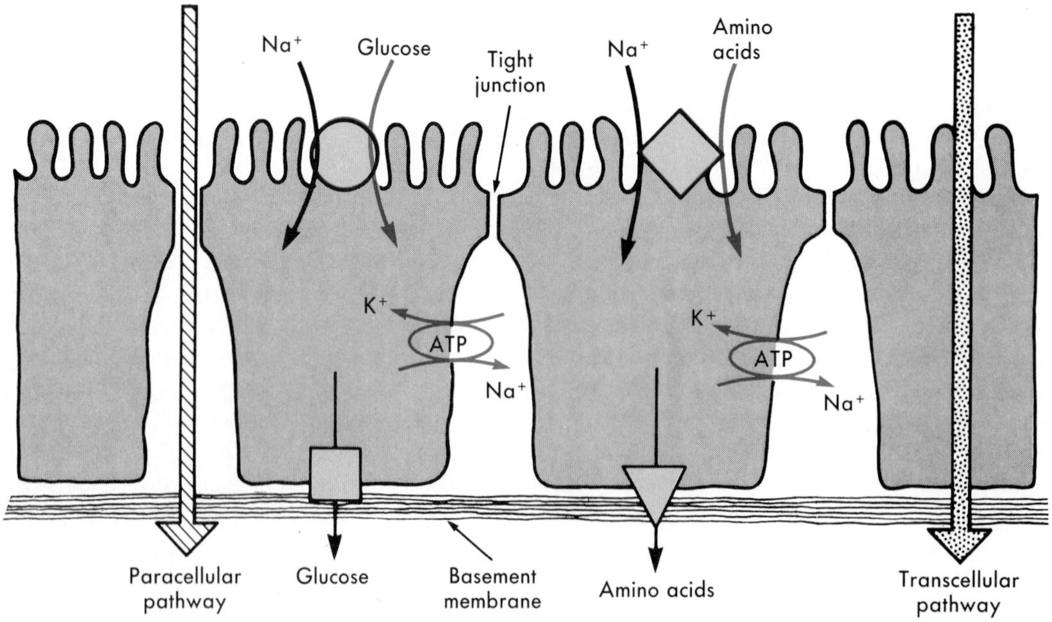

Fig. 1-13. Epithelial transport processes that occur in the small intestine and renal tubules. Certain epithelia are polarized such that the transport processes that take place on one side of the cell differ from those on the other side. Active fluxes are shown in color.

dine nucleosides; cytosine nucleosides have especially low affinity for the transport system.

The transport of the bases of nucleic acids is less well characterized. The number of distinct transport systems for nucleobases is a matter of controversy. Purines are transported much more rapidly than pyrimidines. Cytosine is the most poorly transported base; its transport may not be mediated by a membrane protein at all.

The uptake of purines and purine nucleosides by cells is vital for those cells that are incapable of de novo synthesis of purines. Membrane transport may be viewed as the first step in the salvage pathways for purine nucleotides. As mentioned, the rates at which extracellular purines and purine nucleotides are incorporated by cultured cells are not limited by the rate of membrane transport but rather by the rates of the intracellular phosphorylation or phosphoribosylation reactions.

Transport across epithelia. Frequently epithelial cells are polarized with respect to their transport properties; that is, the transport properties of the plasma membrane facing one side of the epithelial cell layer are different from those of the membrane facing the other side.

The epithelial cells of the small intestine and the proximal tubule of the kidney are good examples of this polarity. The complement of membrane transport proteins in the brush border that faces the lumen of the small bowel or the renal tubule differs from the transport protein composition of the basolateral plasma membrane of the cell. The tight junctions that join the epithelial cells side to side prevent mixing of the transport proteins of the luminal and basolateral plasma membranes. The brush border plasma membranes of these epithelia contain very few Na^+, K^+-ATPase molecules, which reside primarily in the basolateral plasma membrane. Glucose (and galactose) and neutral amino acids enter these epithelial cells at the brush border by Na^+ gradient-driven secondary active transport systems but leave the cells at the basolateral membrane via facilitated transport systems (Fig. 1-13). The tight junctions that join the cells are leaky to water and small water-soluble molecules and ions. There are thus two types of pathways for transport across the epithelia—*transcellular,* through the cells, and *paracellular,* in between the cells (Fig. 1-13). The nature and significance of transport across the epithelia of the intestine and the renal tubule are discussed in Chapters 44 and 46, respectively.

Bibliography

Carafoli, E., and Scarpa, A., editors: Transport ATPases. In Annals of the New York Academy of Sciences, vol. 402, New York, 1982, New York Academy of Sciences.

Christensen, H.N.: Biological transport, ed. 2, Menlo Park, Calif., 1975, W.A. Benjamin, Inc.

Davson, H.: A textbook of general physiology, ed. 4, Baltimore, 1970, Williams & Wilkins.

Finean, J.B., Coleman, R., and Michell, R.H.: Membranes and their cellular functions, ed. 2, New York, 1978, John Wiley & Sons, Inc.

Finean, J.B., and Michell, R.H., editors: Membrane structure, New York, 1981, Elsevier/North-Holland Biomedical Press.

Fox, C.F., and Keith, A., editors: Membrane molecular biology, Stamford, Conn., 1972, Sinauer Associates, Inc.

Hatefi, Y., and Djavadi-Ohaniance, L., editors: The structural basis of membrane function, New York, 1976, Academic Press, Inc.

Hoffman, J.F., and Forbush, B., editors: Structure, mechanism, and function of the Na/K pump. In Current topics in membranes and transport, vol. 19, New York, 1983, Academic Press, Inc.

Kotyk, A., and Janacek, K.: Cell membrane transport: principles and techniques, ed. 2, New York, 1975, Plenum Publishing Corp.

Martonosi, A.N., editor: Membranes and transport, vol. 2, New York, 1982, Plenum Press.

Quinn, P.J.: The molecular biology of cell membranes, Baltimore, 1976, University Park Press.

Solomon, A.K., and Karnovsky, M., editors: Molecular specialization and symmetry in membrane function, Cambridge, Mass., 1978, Harvard University Press.

CHAPTER

2

Ionic Equilibria and Resting Membrane Potentials

Most animal cells have an electrical potential difference (voltage) across their plasma membranes. The cytoplasm is usually electrically negative relative to the extracellular fluid. Since the electrical potential difference across the plasma membrane is present even in resting cells, it sometimes is referred to as the *resting membrane potential*. The resting membrane potential plays a central role in the excitability of nerve and muscle cells and in certain other cellular responses.

The major purpose of this chapter is to discuss the ways that electrochemical potential gradients of certain ions across the plasma membrane generate the resting membrane potential. The first part of the chapter deals with some fundamental definitions and concepts that describe the flow of ions across membranes.

Ionic Equilibria

Electrochemical Potentials of Ions

A membrane separates aqueous solutions in two chambers (A and B). Na^+ is at a higher concentration on side A than on side B. If there is no electrical potential difference between side A and side B, Na^+ will tend to diffuse from side A to side B, just as if it were an uncharged molecule. If, however, side A is electrically negative with respect to side B, the situation is more complex. The tendency for Na^+ to diffuse from side A to side B because of the concentration difference remains, but now Na^+ also tends to move in the opposite direction (from B to A) because of the electrical potential difference across the membrane. The direction of net Na^+ movement depends on whether the effect of the concentration difference or the effect of the electrical potential difference is larger. By comparing the two tendencies—concentration and electrical—one can predict the direction of net Na^+ movement.

The quantity that allows us to compare the relative contributions of ionic concentration and electrical potential is called the *electrochemical potential* (μ) of an ion. The electrochemical potential is defined as

$$\mu = \mu^\circ + RT\ln C + zFE \qquad (1)$$

where

- μ° = electrochemical potential of the ion at some reference state (say 1M concentration at 0° C with zero electrical potential)
- R = gas constant
- T = absolute temperature
- lnC = natural log of concentration
- z = charge number of the ion (+ 2 for Ca^{++}, − 1 for Cl^-, etc.)
- F = Faraday's number
- E = electrical potential

The units of μ and of each term in equation 1 are energy per mole. The electrochemical potential represents the *chemical potential energy* possessed by a mole of ions resulting from their concentration and the electrical potential.

The net flow of an ion will be from where its electrochemical potential is higher to where its electrochemical potential is lower. Consider a membrane separating two chambers (A and B), each of which contains ion x in solution. The tendency of the ion to move from A to B is proportional to $\mu(x)$ on side A, and the tendency of the ion to move from B to A is proportional to $\mu(x)$ on side B. The *net* tendency for x to flow from A to B is $\mu_A(x) - \mu_B(x)$. This difference is the *electrochemical potential difference* of x across the membrane ($\Delta\mu$).

$$\Delta\mu(x) = \mu_A(x) - \mu_B(x) \qquad (2)$$

Substituting into equation 2 the values of $\mu_A(x)$ and $\mu_B(x)$ derived from equation 1, we obtain

$$\mu_A(x) = \mu^\circ(x) + RT\ln[x]_A + zFE_A$$
$$\mu_B(x) = \mu^\circ(x) + RT\ln[x]_B + zFE_B$$

so that

$$\Delta\mu(x) = \mu_A(x) - \mu_B(x) = RT\ln\frac{[x]_A}{[x]_B} + zF(E_A - E_B) \quad (3)$$

The first term ($RT\ln [x]_A/[x]_B$) on the right-hand side of equation 3 is the tendency for the ion x to move from A to B because of the concentration difference, and the second term [$zF(E_A - E_B)$] is the tendency for the ion to move from A to B because of the electrical potential difference. The first term represents the chemical potential difference between a mole of x ions on side A and a mole of x ions on side B as a result of the concentration difference. The second term represents the chemical potential difference between a mole of x ions on side A and a mole of x ions on side B caused by the electrical potential difference between A and B. Thus $\Delta\mu$ describes the difference in chemical potential between a mole of x ions on side A and a mole of x ions on side B resulting from *both* concentration and electrical potential difference.

The x ions will tend to move from higher to lower electrochemical potential. $\Delta\mu$ is defined as the electrochemical potential of the ion on side A minus that on side B. If $\Delta\mu$ is positive, the ions will tend to move from A to B; if $\Delta\mu$ is zero, there is no net tendency for the ions to move at all; and if $\Delta\mu$ is negative, the ions will tend to move from side B to side A.

Electrochemical Equilibrium and the Nernst Equation

$\Delta\mu$ may be thought of as the *net* force on the ion, whereas $RT\ln [x]_A/[x]_B$ is the force caused by the concentration difference, and $zF(E_A - E_B)$ is the force caused by the electrical potential difference. When the two forces are equal and opposite, $\Delta\mu = 0$, and there is no net force on the ion. When there is no net force on the ion, there will be no net movement of the ion, and the ion is said to be in *electrochemical equilibrium* across the membrane. At equilibrium $\Delta\mu = 0$. From equation (3), therefore

$$RT\ln\frac{[x]_A}{[x]_B} + zF(E_A - E_B) = 0 \quad (4)$$

Solving for $E_A - E_B$, we obtain

$$E_A - E_B = \frac{-RT}{zF}\ln\frac{[x]_A}{[x]_B} = \frac{RT}{zF}\ln\frac{[x]_B}{[x]_A} \quad (5)$$

Equation 5 is called the *Nernst equation*. The condition of equilibrium was assumed in its derivation, and *the Nernst equation is valid only for ions at equilibrium*. It allows one to compute the electrical potential difference, $E_A - E_B$, required to produce an electrical force, $zF(E_A - E_B)$, that is equal and opposite to the concentration force, $RT\ln ([x]_A/[x]_B)$, that tends to move the ion from A to B.

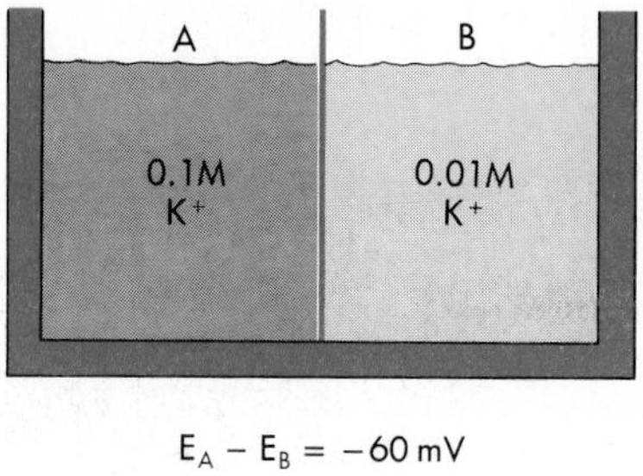

Fig. 2-1. A membrane separates chambers containing different K^+ concentrations. At an electrical potential difference ($E_A - E_B$) of -60 mV, K^+ is in electrochemical equilibrium across the membrane.

Use of the Nernst equation. It is often convenient to convert the Nernst equation to a form involving $\log_{10}$ rather than natural logarithms ($\ln x = 2.303 \log_{10} x$). Since biological potentials are usually expressed in millivolts (mV), the units of R are selected so RT comes out in millivolts. At 29.2° C the quantity 2.303 RT/F is equal to 60 mV. Since this quantity is proportional to the absolute temperature, it changes roughly by only 1/273 for each centigrade degree. Thus the value of 60 mV for 2.303 RT/F holds approximately for most experimental conditions, and a useful form of the Nernst equation is

$$E_A - E_B = \frac{-60\text{ mV}}{z}\log_{10}\frac{[x]_A}{[x]_B} = \frac{60\text{ mV}}{z}\log_{10}\frac{[x]_B}{[x]_A} \quad (6)$$

Examples of uses of the Nernst equation

Example 1. In Fig. 2-1 K^+ is 10 times more concentrated in chamber A than in chamber B. Following is a calculation of the electrical potential difference that must exist between the chambers for K^+ to be in equilibrium across the membrane.

Since we have specified that K^+ should be *in equilibrium*, the Nernst equation will hold.

$$E_A - E_B = \frac{-60\text{ mV}}{+1}\log_{10}\frac{[K^+]_A}{[K^+]_B}$$

$$= -60\text{ mV}\log_{10}\frac{0.1}{0.01} \quad (7)$$

$$= -60\text{ mV}\log(10) = -60\text{ mV}$$

The Nernst equation tells us that at equilibrium, side A must be 60 mV negative relative to side B. We can see that this polarity is correct, since K^+ will tend to move from B to A due to the electrical force, which will counteract the tendency for it to move from A to B because of the concentration difference.

This example shows that an electrical potential difference of about 60 mV is required to balance a tenfold concentration difference of a univalent ion. This is a useful rule of thumb.

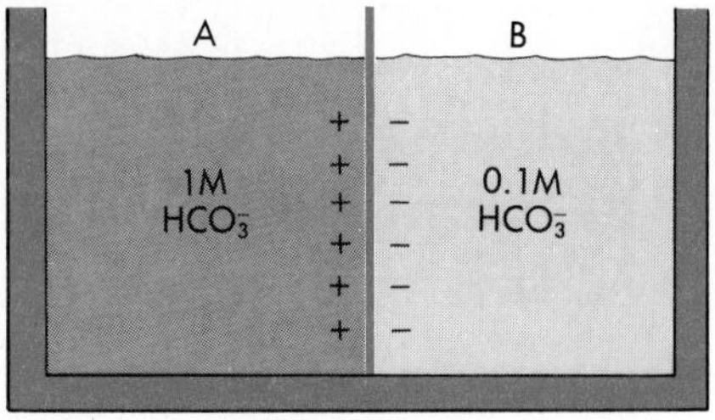

■ **Fig. 2-2.** A membrane separates chambers that contain different HCO_3^- concentrations. $E_A - E_B = +100$ mV. HCO_3^- is not in electrochemical equilibrium. If $E_A - E_B$ were +60 mV, HCO_3^- would be in equilibrium. $E_A - E_B$ is stronger than it needs to be to just balance the tendency for HCO_3^- to move from *A* to *B* because of its concentration difference. Thus net movement of HCO_3^- from *B* to *A* will occur.

Example 2. In Fig. 2-2 the Nernst equation can help decide whether HCO_3^- is in equilibrium. If HCO_3^- is not in equilibrium, the Nernst equation determines the direction of net flow of HCO_3^-.

The Nernst equation tells us the electrical potential difference, $E_A - E_B$, that will *just balance* the concentration difference of HCO_3^- across the membrane.

$$E_A - E_B = \frac{-60 \text{ mV}}{z} \log_{10} \frac{[HCO_3^-]_A}{[HCO_3^-]_B]}$$

$$= \frac{-60 \text{ mV}}{-1} \log_{10} \frac{1}{0.1} \qquad (8)$$

$$= 60 \text{ mV} \log_{10} (10) = 60 \text{ mV}$$

Thus a potential difference of +60 mV between A and B would just balance the tendency of HCO_3^- to move from A to B because of its concentration difference. Since $E_A - E_B$ is actually +100 mV, the electrical potential is of the right sign to oppose the concentration force, but it is 40 mV larger than it needs to be to just balance the concentration force. Since the electrical force on HCO_3^- is larger than the concentration force, it will determine the net direction of HCO_3^- movement. Net HCO_3^- flow will occur from B to A.

In brief, the Nernst equation can be used to predict the direction that ions will tend to flow:

1. If the potential difference *measured* across a membrane is equal to the potential difference *calculated* from the Nernst equation for a particular ion, then *that particular ion* is in electrochemical equilibrium across the membrane, and there will be no net flow of that ion across the membrane.

2. If the measured electrical potential is of the same sign as that calculated from the Nernst equation for a particular ion but is larger in magnitude, then the electrical force is larger than the concentration force, and net movement of that particular ion will tend to occur in the direction determined by the electrical force.

3. When the electrical potential difference is of the same sign but is numerically less than that calculated from the Nernst equation for a particular ion, then the concentration force is larger than the electrical force, and net movement of that ion tends to occur in the direction determined by the concentration difference.

4. If the electrical potential difference measured across the membrane is of the sign opposite to that predicted by the Nernst equation for a particular ion, then the electrical and concentration forces are in the same direction. Thus that ion cannot be in equilibrium, and it will tend to flow in the direction determined by both electrical and concentration forces.

■ *Ionic Mechanisms*

In the discussion on electrical activity of nerve and muscle cells that follows, the principles of ionic equilibrium are used to explain the ionic mechanisms of the resting membrane potential and the action potential.

■ *The Gibbs-Donnan Equilibrium*

Cytoplasm typically contains proteins, organic polyphosphates, and other ionized substances that cannot permeate the plasma membrane. Cytoplasm also contains Na^+, K^+, Cl^-, and other ions to which the plasma membrane is somewhat permeable. The steady-state properties of this mixture of permeant and impermeant ions are described by the *Gibbs-Donnan equilibrium*.

Consider a membrane separating a solution of KCl from a solution of KY, where Y^- is an anion to which the plasma membrane is completely impermeable (Fig. 2-3). The membrane is permeable to water, K^+, and Cl^-. Suppose that *initially* on side A there is a 0.1M solution of KY, and an equal volume of 0.1M KCl is on side B. Because $[Cl^-]_B$ exceeds $[Cl^-]_A$, there will be a net flow of Cl^- from chamber B to chamber A. Negatively charged Cl^- ions flowing from side B to side A will create an electrical potential difference (side A negative) that will cause K^+ also to flow from side B to side A. Essentially the same number of K^+ ions as Cl^- ions will flow from side B to side A to preserve *electroneutrality. The principle of electroneutrality* states that any macroscopic region of a solution must have an equal number of positive and negative charges. In reality slight separation of charges does occur in certain situations, but in chemical terms the imbalance between positive and negative charges is very small and essentially impossible to measure by chemical techniques.

If enough time passes, those components of the system that can permeate the membrane—K^+ and Cl^- in this example—will come to equilibrium. At equilibrium

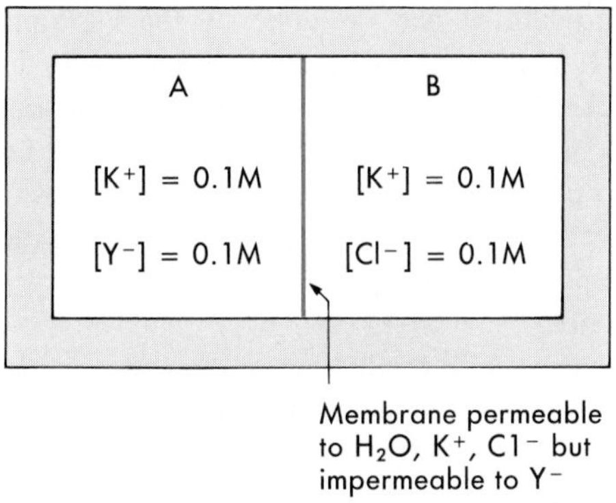

Fig. 2-3. Before a Gibbs-Donnan equilibrium is established, a membrane separates two aqueous compartments. The membrane is permeable to H_2O, K^+, and Cl^- but impermeable to Y^-.

both $\Delta\mu_{K^+}$ and $\Delta\mu_{Cl^-}$ must equal zero. From equation 3

$$\Delta\mu_{K^+} = RT\ln\frac{[K^+]_A}{[K^+]_B} + F(E_A - E_B) = 0$$
$$\Delta\mu_{Cl^-} = RT\ln\frac{[Cl^-]_A}{[Cl^-]_B} - F(E_A - E_B) = 0 \quad (9)$$

Adding these two equations and dividing the result by RT gives

$$\ln\frac{[K^+]_A}{[K^+]_B} + \ln\frac{[Cl^-]_A}{[Cl^-]_B} = 0 \quad (10)$$

This gives

$$\ln\frac{[K^+]_A}{[K^+]_B} = -\ln\frac{[Cl^-]_A}{[Cl^-]_B} = \ln\frac{[Cl^-]_B}{[Cl^-]_A} \quad (11)$$

Thus

$$\frac{[K^+]_A}{[K^+]_B} = \frac{[Cl^-]_B}{[Cl^-]_A} \quad (12)$$

Cross multiplying gives

$$[K^+]_A\,[Cl^-]_A = [K^+]_B[Cl^-]_B \quad (13)$$

Equation 13 is called the *Donnan relationship* (or the Gibbs-Donnan equation) and holds for any pair of univalent cation and anion in equilibrium between the two chambers. If other ions that could attain an equilibrium distribution were present, the same reasoning and an equation similar to equation 13 would apply to them as well.

Example of a Gibbs-Donnan equilibrium. The Donnan relationship and the principle of electroneutrality make it possible to determine the equilibrium concentrations of the components in the problem posed at the beginning of this section. The initial situation is shown in Fig. 2-3. If b represents the change in $[Cl^-]$ when Cl^- moves from B to A, the equilibrium value of $[Cl^-]_B$ can be denoted as $0.1 - b$. By the electroneutrality principle $[K^+]_B = [Cl^-]_B = 0.1 - b$. If the volumes of A and B are kept the same, at equilibrium, $[Cl^-]_A = b$, and $[K^+]_A = 0.1 + b$. Substituting these concentrations into the Donnan relationship:

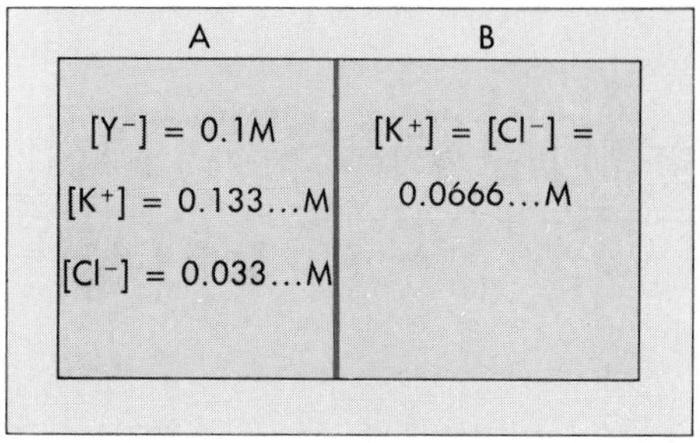

Fig. 2-4. Ion concentrations after Gibbs-Donnan equilibrium has been attained. The initial ion concentrations were as shown in Fig. 2-3.

$$[K^+]_A\,[Cl^-]_A = [K^+]_B[Cl^-]_B$$
$$(0.1 + b)\,(b) = (0.1 - b)\,(0.1 - b) \quad (14)$$

Solving this equation for b gives $b = 0.0333$ so that at equilibrium we obtain the concentrations shown in Fig. 2-4.

In this Gibbs-Donnan equilibrium both K^+ and Cl^- (but *not* Y^-) are in electrochemical equilibrium. This means that both K^+ and Cl^- must satisfy the Nernst equation, so that the equilibrium transmembrane potential difference can be computed from the Nernst equation for *either* K^+ or Cl^-.

$$E_A - E_B = \frac{-60\text{ mV}}{+1}\log_{10}\frac{[K^+]_A}{[K^+]_B}$$
$$= -60\text{ mV}\log_{10}\frac{0.1333}{0.0666}$$
$$= \frac{-60\text{ mV}}{-1}\log_{10}\frac{[Cl^-]_A}{[Cl^-]_B} = \quad (15)$$
$$60\text{ mV}\log_{10}\frac{0.0333}{0.0666}$$
$$= -60\text{ mV}\log_{10}2 = -60\text{ mV}\,(0.3) = -18\text{ mV}$$

Note that only the permeant ions attain equilibrium. The impermeant ion, Y^-, *cannot* reach an equilibrium distribution. What may not yet be evident is that water also will not achieve equilibrium, unless provision is made for that to occur. The total number of K^+ and Cl^- ions on side A in the preceding example exceeds that on side B. This is a general property of Gibbs-Donnan equilibria.

Taking the impermeant Y into account as well, the total concentration of osmotically active ions is considerably greater on side A than on side B. Because K^+ and Cl^- are present at equilibrium, they will have their full osmotic force even though the membrane is quite permeable to them. Water will tend to flow by osmosis

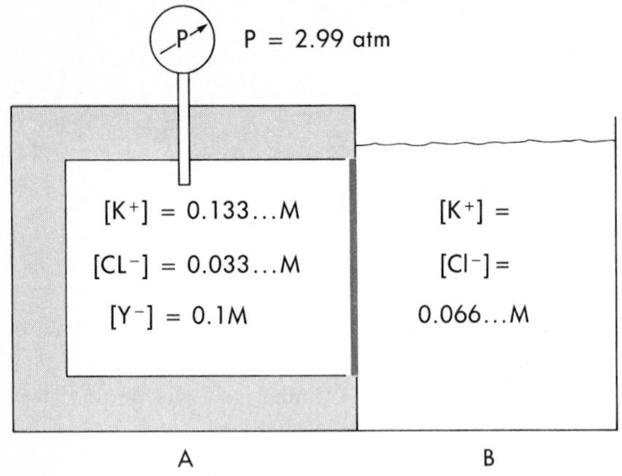

Fig. 2-5. A hydrostatic pressure of 2.99 atm is required to prevent water from flowing from *B* to *A* in the Gibbs-Donnan equilibrium in Fig. 2-4.

from side B to side A until the total osmotic pressure of the two solutions is equal. But then ions will flow to set up a new Gibbs-Donnan equilibrium, and that requires there be more osmotically active ions on the side with Y^-. All the water from side B will end up on side A unless water is restrained from moving.

This can be done by enclosing the solution on side A in a rigid container (Fig. 2-5). Then, as fluid flows from side B to side A, pressure will build up in A and that pressure will oppose further osmotic water flow. The pressure in chamber A at equilibrium is equal to the difference between the total osmotic pressures of the solutions in chambers A and B. In this example the approximate hydrostatic pressure (P) in chamber A at equilibrium (at 0° C) is

$$\begin{aligned} P &= \Delta\pi_{K^+} + \Delta\pi_{Cl^-} + \Delta\pi_{Y^-} \\ &= RT\,(\Delta[K^+] + \Delta[Cl^-] + \Delta[Y^-]) \\ &= RT\,(0.06667 - 0.03333 + 0.1) \\ &= (22.4 \text{ atm})\, 0.13333) = 2.99 \text{ atm} \end{aligned} \tag{16}$$

Regulation of Cell Volume

K^+ and Cl^- are nearly in equilibrium across many plasma membranes, and their distribution is influenced by the predominantly negatively charged impermeant ions, such as proteins and nucleotides, in the cytoplasm. K^+ and Cl^- approximately satisfy the Donnan relationship. This being the case, why does the osmotic imbalance discussed previously not cause the cell to swell and finally burst? One reason is that cells actively pump Na^+ out of the cytoplasm to the extracellular fluid, decreasing the osmotic pressure of the cytoplasm and increasing that of the extracellular fluid. Much of the pumping of Na^+ is done by the Na^+ pump, the Na^+, K^+-ATPase, in the plasma membrane. The Na^+, K^+-ATPase (p. 17) splits an ATP and uses some of the energy released to extrude 3 Na^+ from the cytoplasm and to pump 2 K^+ into the cell. Whereas K^+ is only slightly removed from an equilibrium distribution, Na^+ is pumped out against a large electrochemical potential difference.

When the ATP production of the cell is compromised (in the presence of metabolic inhibitors or low O_2 levels), or when the Na^+, K^+-ATPase is specifically inhibited (by cardiac glycosides), cells swell. Some investigators believe that the irreversible brain damage that occurs after acute oxygen deprivation is caused by the injury that results from inhibition of Na^+ pumping and osmotic swelling of brain cells. Consistent with this hypothesis, when brain cells are "preshrunk" with hypertonic solutions, they may have a greater ability to survive periods of low oxygen.

While the Na^+, K^+-ATPase is involved in volume regulation, other mechanisms play important roles in controlling cell volume. When cells are osmotically shrunk or osmotically swollen, ion transport processes are activated that return cell volume toward normal. In response to an initial osmotic shrinkage, cells gain Na^+ and Cl^-, water enters the cells osmotically, and the cells swell back toward normal volume. In response to an initial osmotic swelling, K^+ and Cl^- exit from the cells, and consequently the cells lose water.

Our knowledge of these volume-regulating ion fluxes is still incomplete. Some volume-regulating ion fluxes may be modulated by volume-dependent alterations in intracellular Ca^{++} concentration by calmodulin-dependent processes.

Resting Membrane Potentials

Communication between nerve cells depends on an electrical disturbance that is propagated in the plasma membrane, and is called an *action potential*. In striated muscle an action potential propagates rapidly over the entire cell surface, allowing the cell to contract synchronously. The action potential and the ionic mechanisms that account for its properties are discussed in Chapter 3. All cells that are able to produce action potentials have sizable *resting membrane potentials* across their plasma membranes. Most inexcitable cells also have a resting membrane potential.

The resting membrane potential of many cells can be measured using glass microelectrodes that have tip diameters of about 0.1 μm and can puncture the plasma membrane of some cells without greatly injuring the cell. The electrical potential difference between the tip of a microelectrode inside a skeletal muscle cell and a reference electrode in the extracellular fluid is about

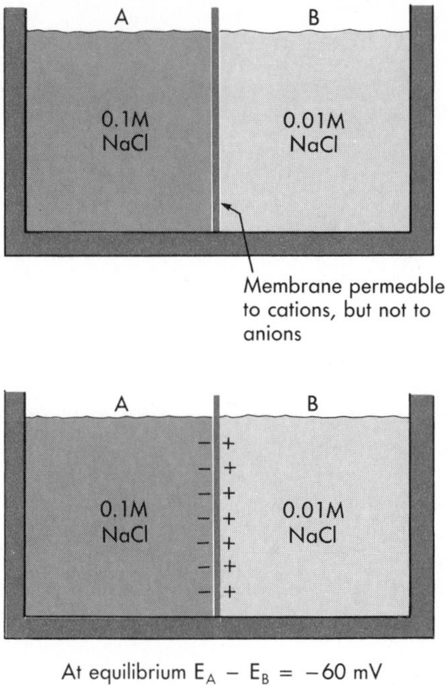

■ **Fig. 2-6.** *Top,* A concentration cell. The membrane, which is permeable to cations but not to anions, separates NaCl solutions of different concentrations. *Bottom,* The concentration cell after electrochemical equilibrium has been established. The flow of an infinitesimal amount of Na^+ generated an electrical potential difference across the membrane.

−90 mV. The resting membrane potential is necessary for the cell to fire an action potential. If the resting membrane potential is decreased to −50 mV or less, the cell is no longer able to produce an action potential.

Ions that are actively transported are not in electrochemical equilibrium across the plasma membrane. It is shown later that the flow of ions across the plasma membrane, down their electrochemical potential gradients, is directly responsible for generating most of the resting membrane potential. To understand how the electrochemical potential gradient of an ion can give rise to a transmembrane difference in electrical potential, let us first consider a model system known as a concentration cell.

■ *The Concentration Cell*

In Fig. 2-6, *A*, the membrane that separates chambers A and B is permeable to cations but not to anions. Initially there is no electrical potential difference across the membrane. Na^+ will flow from A to B because of the concentration force acting on it. Cl^- has the same force on it but cannot flow from A to B because the membrane is impermeable to anions. The flow of Na^+ from A to B will transfer net positive charge to side B and leave a very slight excess of negative charges behind on side A, causing A to become electrically negative to side B (Fig. 2-6, *B*). This electrical force is oppositely directed to the concentration force on Na^+. The more Na^+ that flows, the larger the opposing electrical force. Net Na^+ flow will stop when the electrical force just balances the concentration force, that is, when the electrical potential difference is equal to the equilibrium (Nernst) potential for Na^+. Algebraically,

$$E_A - E_B = \frac{-60 \text{ mV}}{+1} \log_{10} \frac{0.1}{0.01} =$$

$$-60 \text{ mV} \log (10) = -60 \text{ mV} \quad (17)$$

Note that *only a very small amount of Na^+ flows* from A to B before equilibrium is reached. This is because the separation of positive and negative charges requires a large amount of work. The potential difference that builds up to oppose further Na^+ movement is a manifestation of that work.

The Na^+ concentration difference in this example acts much like a *battery*. The natural tendency for any ion that can flow is to seek equilibrium; thus Na^+ tends to flow until its equilibrium potential difference is established. As is explained later, in a system such as a cell membrane, with more than one permeant ion, *each* ion ''strives'' to make the transmembrane potential difference equal to its equilibrium potential. The more permeant the ion, the greater its ability to force the electrical potential difference toward its equilibrium potential.

■ *The Distribution of Ions across Plasma Membranes*

In most tissues a number of ions are not in equilibrium between the extracellular fluid and the cytoplasm. Table 2-1 gives the concentrations of Na^+, K^+, and Cl^- in the extracellular fluid and in the cytoplasmic water of frog skeletal muscle and squid giant axon. Intracellular ion concentrations for mammalian muscle are similar to those for frog muscle.

Chloride is nearly in equilibrium across the plasma membrane of both frog muscle and squid axon. This is known because chloride's potential difference for equilibrium, as calculated from the Nernst equation, is about equal to the measured transmembrane potential difference. In both tissues K^+ has a concentration force tending to make it flow out of the cell. The electrical force on K^+ is oppositely directed to the concentration force. If the $E_{in} - E_{out}$ in frog muscle were −105 mV, electrical and concentration forces on K^+ would exactly balance. Since $E_{in} - E_{out}$ is only −90 mV, the concentration force is greater than the electrical force, and K^+ therefore has a net tendency to flow out of the cell. In frog muscle and in squid axon *both* the concentration and electrical forces on Na^+ tend to cause it to flow

Table 2-1. Distribution of Na^+, K^+, and Cl^- across the plasma membranes of frog muscle and squid axon

	Extracellular fluid (mM)	*Cytoplasm (mM)*	*Approximate equilibrium potential (mV)*	*Actual resting potential (mV)*
Frog muscle				
$[Na^+]$	120	9.2	+67	
$[K^+]$	2.5	140	−105	
$[Cl^-]$	120	3 to 4	−89 to −96	−90
Squid axon				
$[Na^+]$	460	50	+58	
$[K^+]$	10	400	−96	
$[Cl^-]$	540	About 40	About −68	−70

Data from Katz, B.: Nerve, muscle, and synapse, New York, 1966, McGraw-Hill Book Co. Copyright © 1966 by McGraw-Hill Book Co. Used with permission of McGraw-Hill Book Co.

into the cell. Na^+ is the ion furthest from an equilibrium distribution. The larger the difference between the measured membrane potential and the equilibrium potential for an ion, the larger the net force tending to make that ion flow.

Active Ion Pumping and the Resting Potential

The Na^+, K^+-ATPase, located in the plasma membrane, uses the energy of the terminal phosphate ester bond of ATP to actively extrude Na^+ from the cell and to actively take K^+ into the cell. The Na^+, K^+ pump is responsible for the high intracellular K^+ concentration and the low intracellular Na^+ concentration. Since the pump moves a larger number of Na^+ ions out than K^+ ions in (3 Na^+ to 2 K^+), it causes a net transfer of positive charge out of the cell and thus contributes to the resting membrane potential. Because it brings about net movement of charge across the membrane, the pump is termed *electrogenic*.

The size of the pump's contribution to the resting potential can be estimated by completely inhibiting the pump with a cardiac glycoside, such as ouabain. Such studies show that in some cells the electrogenic Na^+, K^+ pump is responsible for a large fraction of the resting potential. In most vertebrate nerve and skeletal muscle cells, however, the direct contribution of the pump to the resting potential is small under most circumstances—less than 5 mV. The resting membrane potential in nerve and skeletal muscle results primarily from the diffusion of ions down their electrochemical potential gradients. The ionic gradients are maintained by active ion pumping. In other types of excitable cells electrogenic pumping of ions may contribute more to the resting membrane potential. In certain vertebrate smooth muscle cells, for example, the electrogenic effect of the Na^+, K^+ pump may be responsible for 20 mV or more of the resting membrane potential.

Generation of the Resting Membrane Potential by the Ion Gradients

The earlier section on concentration cells shows how an ion gradient can act as a battery. When a number of ions are distributed across a membrane, all being removed from electrochemical equilibrium, *each* ion will tend to force the transmembrane potential toward *its own* equilibrium potential, as calculated from the Nernst equation. The more permeable the membrane to a particular ion, the greater strength that ion will have in forcing the membrane potential toward its equilibrium potential. In frog muscle (Table 2-1) the Na^+ concentration difference can be regarded as a battery that tries to make $E_{in} - E_{out}$ equal to +67 mV. The K^+ concentration difference is like a battery that attempts to make $E_{in} - E_{out}$ equal to −105 mV. The Cl^- concentration difference resembles a battery trying to make $E_{in} - E_{out}$ equal to −90 mV.

We can draw an equivalent electrical circuit for the plasma membrane. In Fig. 2-7 each ion gradient is represented by a battery of the appropriate polarity. The resistor in series with each battery represents the resistance *(R)* to the passage of that ion through the membrane. The reciprocal of each resistance is the conductance *(g)* of the membrane for that ion. C_m represents the membrane capacitance that stores the transmembrane potential difference. In this circuit, if the transmembrane resistance to a particular ion is decreased, the total transmembrane electrical potential difference will move toward the battery potential of that ion.

The Chord Conductance Equation

The way in which the interplay of ion gradients creates the resting membrane potential is also illustrated by a simple mathematical model. If the transmembrane electrical potential difference is equal to the equilibrium potential for a particular ion, there is no net force on that

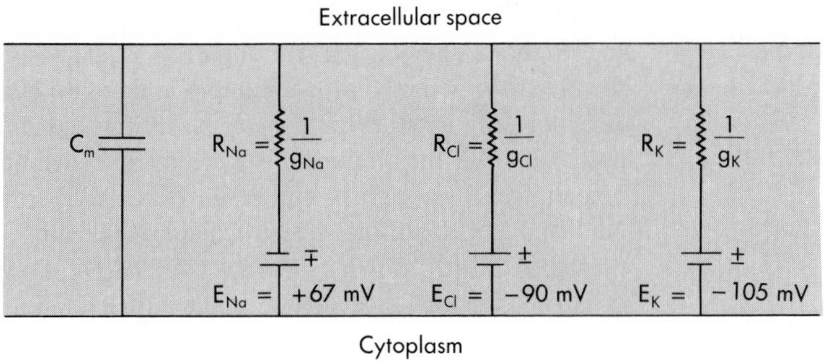

Fig. 2-7. An electrical equivalent circuit model of the plasma membrane of a skeletal muscle cell. The equilibrium potentials of Na^+, K^+, and Cl^- are represented as batteries. The resistances *(R)* and conductances (*g*, reciprocal of resistance) to each ion are shown. C_m represents the membrane capacitance.

ion, and there will be no net flow of that ion. However, if the membrane potential is *not* equal to the equilibrium potential for a given ion, then the difference between the membrane potential and the ion's equilibrium potential can be regarded as the driving force for that ion. Since ions bear charge, ionic flow is equivalent to electrical current. The net current (I) of an ion across the membrane is equal to the driving force on the ion times the conductance of the membrane for that ion. For Na^+, K^+, and Cl^-

$$I_K = g_K(E_m - E_K)$$
$$I_{Na} = g_{Na}(E_m - E_{Na}) \qquad (18)$$
$$I_{Cl} = g_{Cl}(E_m - E_{Cl})$$

where E_K, and E_{Na}, and E_{Cl} are the equilibrium (Nernst) potentials for the indicated ion.

In the steady state, when the transmembrane electrical potential difference is constant, the sum of all the ionic currents across the membrane must be zero. This is so because, if there *were* net current flowing across the membrane, the membrane capacitor would be charging or discharging, and the transmembrane potential difference would be changing. Any net charge transfer across the membrane leads to a change in the degree of charge separation across the membrane and hence in the membrane potential. If K^+, Na^+, and Cl^- are the only important ions, then the algebraic sum of their currents must be zero in the steady state.

$$I_K + I_{Na} + I_{Cl} = 0 \qquad (19)$$

Substituting from equation 18 for the currents, we obtain

$$g_K (E_m - E_K) + g_{Na} (E_m - E_{Na}) + g_{Cl} (E_m - E_{Cl}) = 0 \qquad (20)$$

For a cell membrane across which chloride is in equilibrium, $E_m = E_{Cl}$, and $E_m - E_{Cl} = 0$, so the third term can be dropped. However, it should be kept in mind that Cl^- is not in equilibrium for all excitable cells. Also, whenever E_m moves away from E_{Cl}, Cl^- will exert a restoring force to bring E_m back toward E_{Cl} just as any ion tries to force E_m toward its equilibrium potential. Without the chloride term equation 10 becomes

$$g_K (E_m - E_K) + g_{Na} (E_m - E_{Na}) = 0 \qquad (21)$$

Solving for E_m yields

$$E_m = \frac{g_K}{g_K + g_{Na}} E_K + \frac{g_{Na}}{g_K + G_{Na}} E_{Na} \qquad (22)$$

Equation 22 is one form of the *chord conductance equation*. It expresses the transmembrane electrical potential difference as a weighted average of the equilibrium potentials of K^+ and Na^+. The weighting factor for each ion is that fraction of the total ionic conductance caused by that particular ion. To consider more ions, we need only add appropriate terms to the chord conductance equation. In a cell in which Cl^- and Ca^{++} play important roles, the chord conductance equation becomes

$$E_m = \frac{g_K}{g_T} E_K + \frac{g_{Na}}{g_T} E_{Na} + \frac{g_{Cl}}{g_T} E_{Cl} + \frac{g_{Ca}}{g_T} E_{Ca} \qquad (23)$$

where

$$g_T = g_K + g_{Na} + g_{Cl} + g_{Ca}$$

For the frog muscle fiber discussed earlier, $E_{in} - E_{out} = -90$ mV. The membrane potential is much closer to E_K (-105 mV) than to E_{Na} ($+67$ mV). The chord conductance equation predicts that in resting muscle g_K is 10.5 times larger than g_{Na}. This has been confirmed by ion flux measurements with radioactive tracers. In resting frog sartorius muscle, g_K is about 10 times g_{Na}. In resting squid axon ($E_m = -70$ mV), the chord conductance equation predicts that g_K is about five times larger than g_{Na}. In other types of excitable cells the relationship between g_K and g_{Na} is somewhat different. Other ions also may play a role in generating the resting membrane potential. Resting membrane potentials vary from about -7 mV or so in human erythrocytes to -30 mV in some types of smooth muscle and up to -90 mV in vertebrate skeletal muscle and cardiac ventricular cells.

The chord conductance equation shows that, if g_{Na} were suddenly increased, the membrane potential would move toward E_{Na} (toward $+67$ mV in frog muscle). This is what occurs during an action potential when g_{Na} transiently increases. The ionic mechanism of the action potential is discussed in Chapter 3.

The Constant Field Equation: an Alternative to the Chord Conductance Equation

The chord conductance equation is derived by considering the current of each individual ion as the product of its conductance and its driving force, where the driving force is equal to the difference between the membrane potential and the equilibrium potential for the ion. By setting the sum of the ionic currents equal to zero (steady state assumption), the chord conductance equation can be obtained.

Another way of dealing with the fluxes of individual ions across the membrane treats the flux of each ion as the product of its permeability coefficient (p. 11) and a driving force that is a function of the membrane potential and the intracellular and extracellular concentrations of the ion. By assuming that, in the steady state, the sum of the fluxes of K^+, Na^+, and Cl^- is zero, the following expression for the membrane potential is derived:

$$E_m = \frac{RT}{F} \ln \frac{k_p^{K^+}[K^+]_o + k_p^{Na^+}[Na^+]_o + k_p^{Cl^-}[Cl^-]_i}{k_p^{K^+}[K^+]_i + k_p^{Na^+}[Na^+]_i + k_p^{Cl^-}[Cl^-]_o} \qquad (24)$$

where the k_p's are permeability coefficients and the subscripts *o* and *i* represent extracellular and intracellular concentrations, respectively. Equation 24 is known as the *constant field equation*. This is because the expressions for the driving forces of the individual ions are derived by assuming that the electric field within the membrane is of constant strength.

If one of the permeability coefficients in equation 24 becomes very large relative to the other two, then the predicted membrane potential approaches the equilibrium potential for the highly permeable ion. The more permeable a particular ion, the larger its contribution to E_m. Thus the constant field equation and the chord conductance equation lead to qualitatively similar explanations of the membrane potential.

Summary of the Ionic Mechanism of the Resting Membrane Potential

The Na^+, K^+ pump establishes gradients of Na^+ and K^+ across the plasma membranes of the cells. Since the amount of Na^+ pumped out is larger than the amount of K^+ pumped in, the pump transfers net charge across the membrane and contributes to the resting membrane potential. The pump is therefore said to be electrogenic. In vertebrate skeletal and cardiac muscle and in nerve the electrogenic activity of the pump is directly responsible for only a small fraction of the resting membrane potential. The major portion of the resting membrane potential in these tissues is a result of the diffusion of Na^+ and K^+ down their electrochemical potential gradients, with Na^+ flowing into the cell and K^+ flowing out. The principal role of the pump in these tissues is to maintain the ion gradients it has established. K^+ and Na^+ each tend to force the transmembrane potential toward their own equilibrium potential. The resulting E_m is a weighted average of E_K and E_{Na}, with the weighting factor for each ion being the fraction of the total membrane conductance caused by that ion. In these cells the resting membrane potential is *directly* due to the diffusion of K^+ and Na^+ down their respective electrochemical potential gradients. The resting membrane potential is *indirectly* due to the Na^+, K^+ pump, which maintains the gradients. In mammalian smooth muscle cells and in certain other cell types the electrogenic effect of the active pumping of Na^+ and K^+ may contribute a substantial fraction of the resting membrane potential.

Bibliography

Aidley, D.J.: The physiology of excitable cells, ed. 2, Cambridge, 1978, Cambridge University Press.

Davson, H.: A textbook of general physiology, ed. 4, Baltimore, 1970, Williams & Wilkins.

Dowben, R.M.: General physiology: a molecular approach, New York, 1969, Harper & Row, Publishers, Inc.

Junge, D.: Nerve and muscle excitation, ed. 2, Sunderland, Mass., 1981, Sinauer Associates, Inc.

Kandel, E.R., and Schwartz, J.H.: Principles of neural science, ed. 2, New York, 1985, Elsevier Science Publishing Co., Inc.

Katz, B.: Nerve, muscle, and synapse, New York, 1966, McGraw-Hill Book Co.

Kuffler, S.W., Nicholls, J.G., and Martin, A.R.: From neuron to brain, ed. 2, Sunderland, Mass., 1984, Sinauer Associates, Inc.

CHAPTER

3

Generation and Conduction of Action Potentials

An *action potential* is a rapid change in the membrane potential followed by a return to the resting membrane potential (Fig. 3-1). The size and shape of action potentials differ considerably from one excitable tissue to another. An action potential is propagated with the same shape and size along the whole length of a nerve axon or muscle cell. The action potential is the basis of the signal-carrying ability of nerve cells. It allows all parts of a long muscle cell to contract almost simultaneously. This chapter discusses the ionic currents that generate action potentials and the ways in which action potentials are propagated and conducted.

■ *Experimental Observations of Membrane Potentials*

Our knowledge of the ionic mechanism of an action potential was first obtained from experiments on the squid giant axon. The large diameter (up to 0.5 mm) of the squid giant axon makes it a convenient object for electrophysiological research with intracellular electrodes. The frog sartorius muscle is another useful preparation. The sartorius muscle can be removed from the frog without damage to the muscle. The surface muscle cells can be visualized with a microscope and penetrated with one or more microelectrodes.

The following experiment illustrates some basic aspects of the resting membrane potential. A sartorius muscle is removed from a frog and put in a dish of fluid with composition similar to the frog's extracellular fluid. Two microelectrodes (tip diameter less than 0.5 μm) are placed in the extracellular fluid. When both microelectrodes are placed in the extracellular fluid, no electrical potential difference between them is observed. One electrode then is moved slowly toward a muscle cell until it penetrates the plasma membrane. At the instant the electrode pops through the membrane, an abrupt change of the potential difference between the

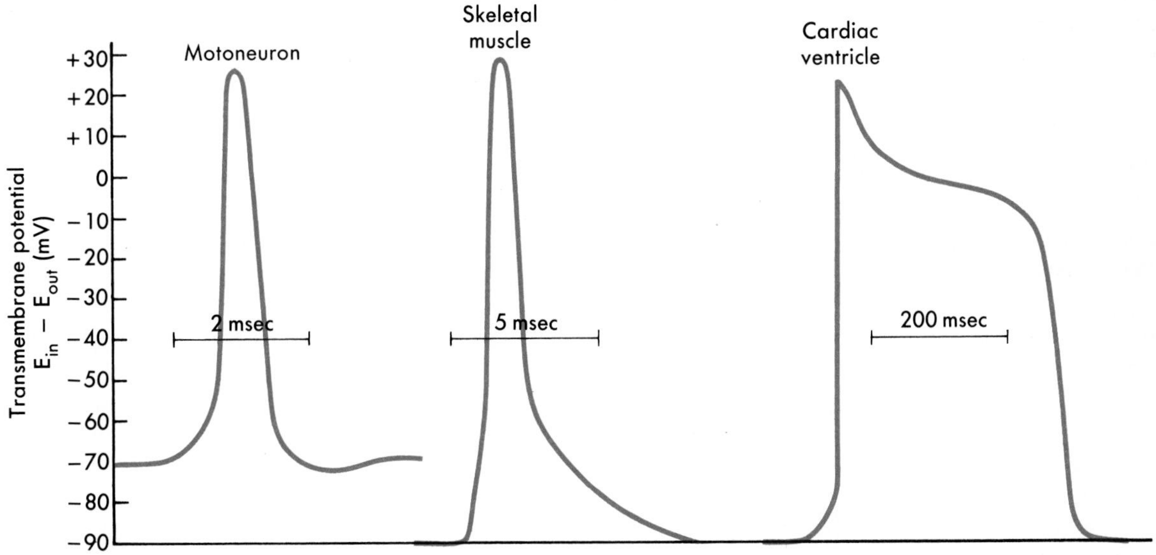

■ **Fig. 3-1.** Action potentials (note the different times scales) from three vertebrate cell types. (Redrawn from Flickinger, C.J., et al.: Medical cell biology, Philadelphia, 1979, W.B. Saunders Co.)

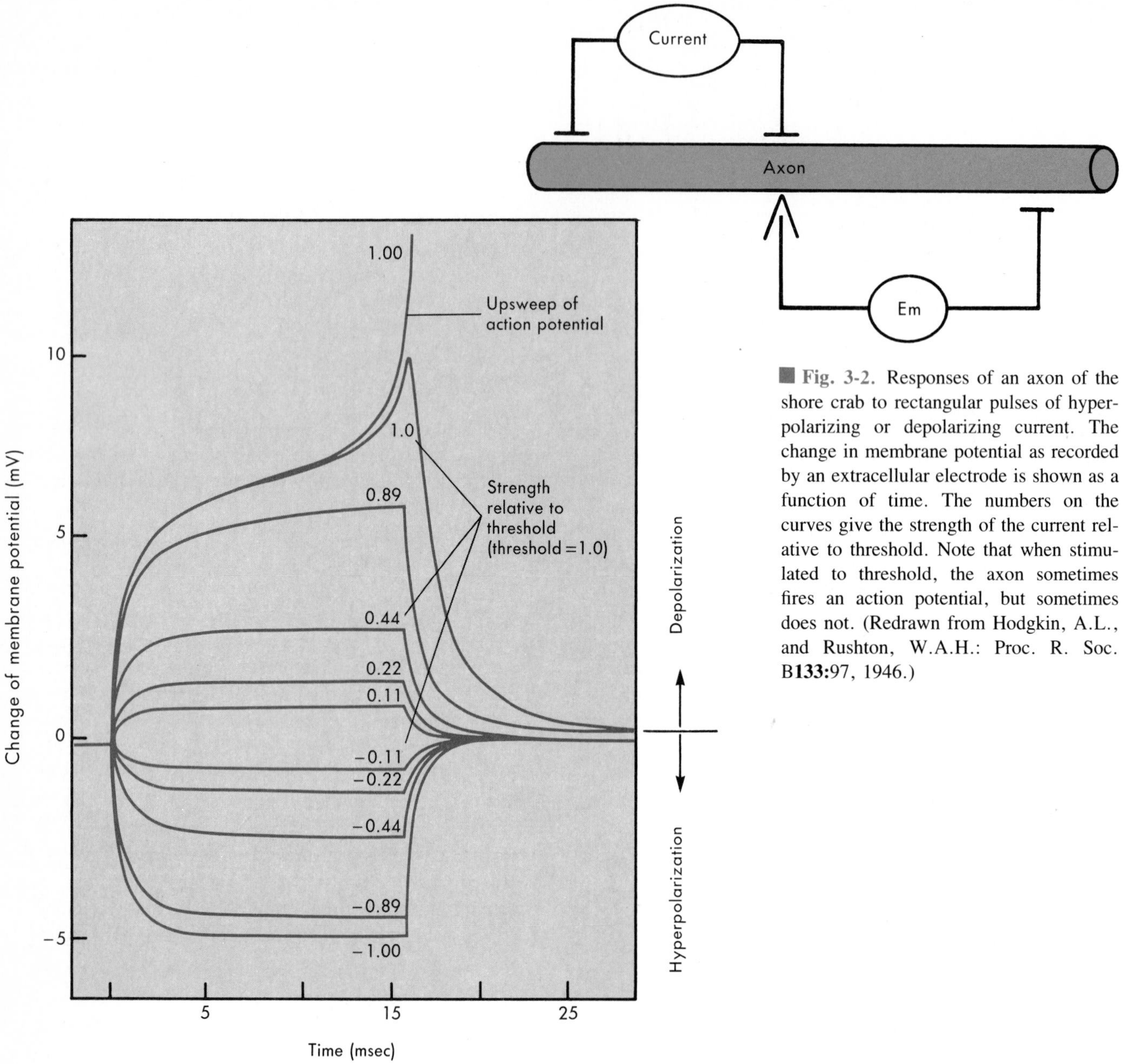

Fig. 3-2. Responses of an axon of the shore crab to rectangular pulses of hyperpolarizing or depolarizing current. The change in membrane potential as recorded by an extracellular electrode is shown as a function of time. The numbers on the curves give the strength of the current relative to threshold. Note that when stimulated to threshold, the axon sometimes fires an action potential, but sometimes does not. (Redrawn from Hodgkin, A.L., and Rushton, W.A.H.: Proc. R. Soc. B**133**:97, 1946.)

two electrodes is observed. The intracellular electrode suddenly becomes about 90 mV negative with respect to the external electrode. This −90 mV potential difference is the *resting membrane potential* of the muscle fiber. A third microelectrode placed in the same cell also will register −90 mV relative to the external solution. At rest there is no net internal current in the muscle cell and thus no potential difference between these two intracellular electrodes. In the absence of perturbing influences the resting membrane potential remains at −90 mV.

Subthreshold Responses: The Local Response

Fig. 3-2 illustrates the results of an experiment in which the membrane potential of an axon of a shore crab is perturbed by passing rectangular pulses of current across the plasma membrane. Current pulses are depolarizing or hyperpolarizing depending on the direction of current flow. The terms *depolarizing* and *hyperpolarizing* may be confusing. A change of the membrane potential from −90 mV to −70 mV is a depolarization, since it is a decrease of the potential difference, or po-

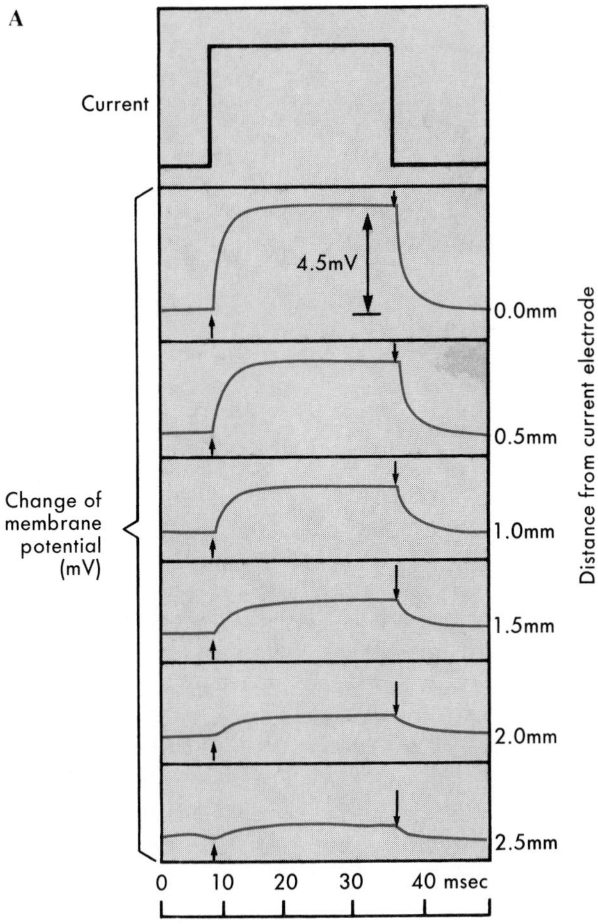

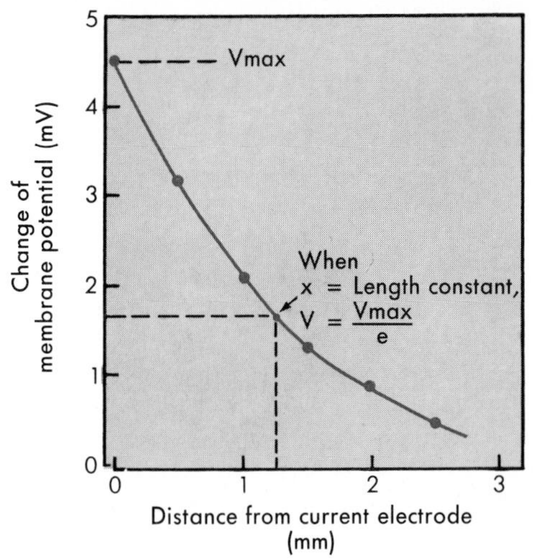

■ **Fig. 3-3. A,** Responses of an axon of a shore crab to a rectangular pulse of current recorded extracellularly by an electrode located different distances from the current-passing electrode. As the recording electrode is moved farther from the point of stimulation, the response of the membrane potential is slower and smaller. **B,** The maximal change in membrane potential from A plotted versus distance from the point of current passage. The distance over which the response falls to 1/e (37%) of the maximal response is called the length constant. (Part A is redrawn from Hodgkin, A.L., and Rushton, W.A.H.: Proc. R. Soc. B**133:**97, 1946.)

larization, across the cell membrane. If the membrane potential changes from −90 mV to −100 mV, the polarization across the membrane has increased; this is hyperpolarization.

The larger the current passed, the larger the perturbation of the membrane potential. As shown in Fig. 3-2, in response to depolarizing current pulses above a certain threshold strength, the cell fires an action potential.

When subthreshold current pulses are passed, the size of the potential change observed depends on the *distance* of the recording microelectrode from the point of current passage (Fig. 3-3, *A*). The closer the recording electrode to the site of current passage, the larger the potential change observed. The size of the potential change is found to decrease exponentially with distance from the site of current passage (Fig. 3-3, *B*). The distance over which the potential change decreases to 1/e (37%) of its maximal value is called the *length constant* (or *space constant*). A length constant of 1 to 3 mm is typical for mammalian nerve or muscle cells. Because these potential changes are observed primarily near the site of current passage and the changes are not propagated along the length of the cell (as are action potentials), they are called *local responses*.

■ *Action Potentials*

If progressively larger depolarizing current pulses are applied, a point is reached at which a different sort of response, the *action potential,* occurs (Figs. 3-2 and 3-4). An action potential is triggered when the depolarization is sufficient for the membrane potential to reach a *threshold* value, which is near −60 mV for frog satorius muscle. The action potential differs from the local depolarizing response in two important ways: (1) it is a much larger response, with the polarity of the membrane potential actually reversing (the cell interior becoming positive with respect to the exterior), and (2) the action potential is *propagated without decrement* down the entire length of the muscle fiber. The size and shape of an action potential remain the same as it travels along the muscle fiber; it does not decrease in size with distance as does the local response. When a stimulus larger than the threshold stimulus is applied, the size and shape of the action potential do not change; the size of the action potential does not increase with increased stimulus strength. A stimulus either fails to elicit an action potential (a subthreshold stimulus), or it produces a full-size action potential. For this reason the action potential is an *all-or-none response*.

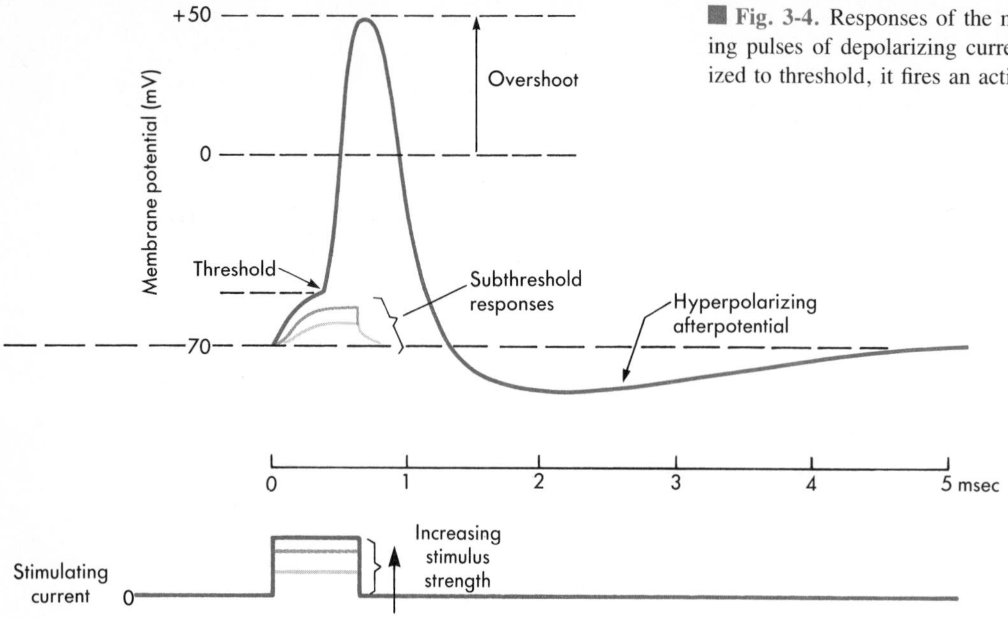

Fig. 3-4. Responses of the membrane potential to increasing pulses of depolarizing current. When the cell is depolarized to threshold, it fires an action potential.

The Shape of the Action Potential

The form of an action potential of a squid giant axon is shown in Fig. 3-4. Once the membrane is depolarized to the threshold, an explosive depolarization occurs, which completely depolarizes the membrane and even *overshoots* so that the membrane becomes polarized in the reverse direction. The peak of the action potential reaches about +50 mV. The membrane potential then returns toward the resting membrane potential almost as rapidly as it was depolarized. After repolarization a transient hyperpolarization occurs that is known as the *hyperpolarizing afterpotential*. It persists for about 4 msec. The following section discusses the ionic currents that cause the various phases of the action potential.

Ionic Mechanisms of the Action Potential

In Chapter 2 the resting membrane potential was seen to be a weighted sum of the equilibrium potentials for Na^+, K^+, Cl^-, etc., with the weighting factor for each ion being the fraction it contributes to the total ionic conductance of the membrane (the chord conductance equation). In squid giant axon the resting membrane potential (E_m) is about −70 mV. E_K is about −100 mV in squid axon, so an increase in g_K would hyperpolarize the membrane. E_{Cl} is about −70 mV, so an increase in g_{Cl} would stabilize E_m at −70 mV. An increase in g_{Na} of sufficient magnitude would cause depolarization and reversal of the membrane polarity, since E_{Na} is about +60 mV in squid axon. A decrease in g_K also would have the effect of depolarizing the membrane.

It has been known for a long time that Na^+ is involved in the action potential. It was shown in 1902 that Na^+ in the extracellular fluid is required for excitability. Some 50 years later Hodgkin and Huxley showed that the action potential of squid giant axon is caused by successive conductance increases to sodium and potassium ions. They found that the conductance to Na^+, g_{Na}, increases very rapidly during the early part of the action potential (Fig. 3-5). The sodium conductance reaches a peak about the same time as the peak of the action potential, then it decreases rather rapidly. The K^+ conductance, g_K, increases more slowly, reaches a peak at about the middle of the repolarization phase, and then returns more slowly to resting levels.

As described in Chapter 2, the chord conductance equation shows that the membrane potential is a result of the opposing tendencies of the K^+ gradient to bring E_m toward the equilibrium potential for K^+ and the Na^+ gradient to bring E_m toward the equilibrium potential for Na^+. Increasing the conductance of either ion will increase its ability to pull E_m toward its equilibrium potential. The rapid increase in g_{Na} during the early part of the action potential causes the membrane potential to move toward the equilibrium potential for Na^+ (+60 mV). The peak of the action potential reaches only about +50 mV. This is because the conductance of K^+ also increases, albeit more slowly, providing an opposing tendency, and also because g_{Na} quickly decreases toward resting levels. The rapid return of the membrane potential toward the resting potential is caused by the rapid decrease of g_{Na} and the continued increase in g_K, both of which decrease the size of the Na^+ term in the chord conductance equation and increase the size of the K^+ term. During the hyperpolarizing afterpotential,

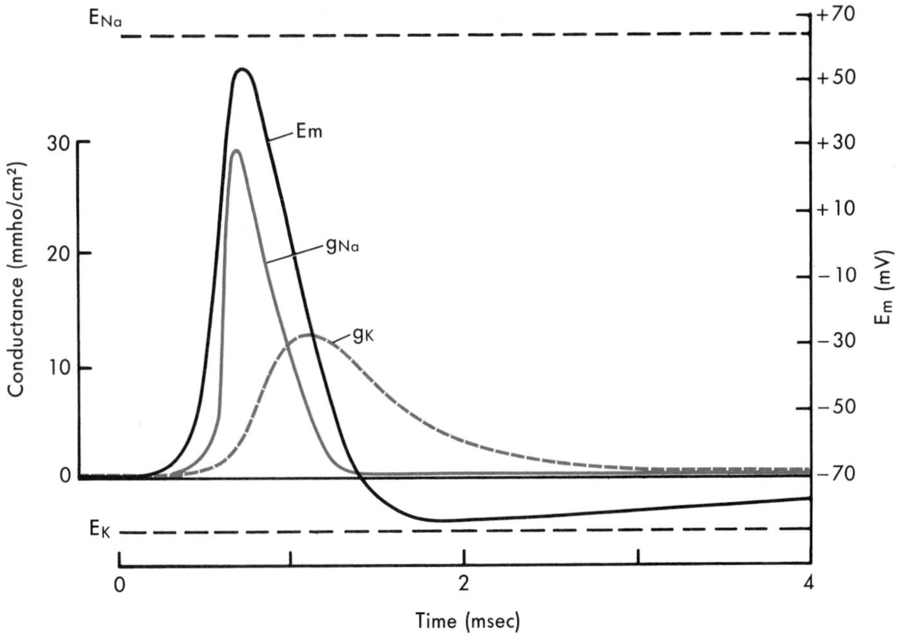

■ **Fig. 3-5.** The action potential (E_m) of a squid giant axon is shown on the same time scale with the changes in the conductance of the axon membrane to sodium and potassium ions that occur during the action potential. (Redrawn from Hodgkin, A.L., and Huxley, A.F.: J. Physiol. **117:**500, 1952.)

when the membrane potential is actually more negative than the resting potential (more polarized), g_{Na} has returned to baseline levels, but g_K remains elevated above resting levels. Thus E_m is pulled closer to the K^+ equilibrium potential (-100 mV) for a short time.

■ *The Basis for Knowledge of the Action Potential*

Early in this century it was hypothesized that the action potential was caused by a transient increase of the permeability of the plasma membrane to all ions. Later evidence indicated that the entry of Na^+ into the cell plays a central role in generating the action potential. Removal of Na^+ from the extracellular fluid was found to abolish the action potential. Both the height of the action potential and the maximum rate of depolarization during the action potential depend on the extracellular concentration of Na^+, increasing with increasing external Na^+.

By repetitively stimulating the squid giant axon in a bath containing radioactive $^{22}Na^+$, it was found that 3 to 4×10^{-12} mole of Na^+ enter the axon through each square centimeter of surface area during one action potential. This amount of Na^+ is in good agreement with calculations of the amount of Na^+ that should enter the axon based on electrical considerations. The capacitance (C) of the membrane is defined as dq/dv, the amount of charge that must flow to cause a 1 V change in potential difference. It is known that the capacitance of 1 cm^2 of squid axon membrane is about 1 microfarad. The amount of charge that must flow to discharge each cm^2 of the membrane capacitor by 100 mV (the approximate height of the action potential) is C times V, or 10^{-6} farad $\times$ 10^{-1} V $=$ 10^{-7} coulombs. Dividing this by the amount of charge on 1 mole of Na^+ ions (96,500 coulombs/mole) gives about 10^{-12} mole of Na^+ ions that must enter each square centimeter of axonal membrane to depolarize the membrane by 100 mV.

If only Na^+ and K^+ are involved, the amount of K^+ that leaves the cell during the repolarization phase must be equal to the amount of Na^+ that enters during depolarization. The Na^+, K^+-ATPase pumps out the Na^+ that enters and reaccumulates the lost K^+. Because of the small number of ions that cross the membrane during each impulse, the Na^+, K^+ pump is not required in the short run. A squid axon that has been poisoned with cyanide or ouabain so that ion pumping is abolished can fire nearly 100,000 action potentials before failing. Smaller axons, with a much smaller ratio of volume to surface, will fail after fewer impulses if the Na^+ pump is poisoned.

■ *The Voltage Clamp Technique*

Much of the current knowledge of the ionic mechanism of the action potential comes from experiments using the *voltage clamp* technique. This technique uses electronic feedback to rapidly set the transmembrane potential difference to some chosen value. The voltage clamp circuit holds the membrane potential at this level and measures the net ionic current that flows across the membrane during the clamp (Fig. 3-6).

When the membrane of the squid giant axon is depolarized rapidly to some point beyond the threshold, there is an almost instantaneous current that is caused by discharge of the membrane capacitance. This is followed by currents that flow over the same time course as a normal action potential (Fig. 3-6). The inward phase of the current results from Na^+ entering the axon, and the outward current results from K^+ leaving the axon. The Na^+ current can be separated from the K^+ current as follows (Fig. 3-7).

1. Enough of the external Na^+ is replaced by choline$^+$ (an impermeant cation), so that the Na^+ concentration inside the axon equals that in the extracellular fluid. If the membrane potential then is clamped to zero, there will be no net force causing Na^+ to flow, since there is no concentration difference across the membrane and no electrical potential difference. When this is done, the inward current component disappears (Fig. 3-7, curve B), indicating that the inward current is normally caused by Na^+. Since curve A is the sum of I_K and I_{Na} and curve B is I_K alone, subtracting B from A yields curve C in Fig. 3-7, which represents I_{Na}.

2. The involvement of Na^+ in the inward current during the action potential also can be shown without changing the external $[Na^+]$. If E_m is clamped to +58 mV, which is the Na^+ equilibrium potential, no net flow of Na^+ occurs, and the inward current disappears. Clamping the axon to positive E_m greater than +58 mV causes a net outward Na^+ current. The potential at which the Na^+ current reverses direction (+58 mV) is called the *reversal potential* for the Na^+ current. The fact that the reversal potential for the inward current phase of the action potential is equal to the equilibrium potential for Na^+ supports the contention that Na^+ carries most of the current in the inward phase.

In their pioneering studies of action potentials in squid giant axons Hodgkin and Huxley found that rapidly depolarizing the membrane caused a rapid increase in the flow of Na^+ into the axon and a slower increase in the flow of K^+ out of the axon. By separating the Na^+ current from the K^+ current, they were able to determine the conductances, g, for Na^+ and K^+ as a function of time during voltage clamp. [If I and E_m are known, g can be calculated, since $I_{Na} = g_{Na} (E_m - E_{Na})$, and $I_K = g_K (E_m - E_K)$.] When the membrane potential was clamped to zero, the conductances shown in Fig. 3-8 were obtained. The increase in g_{Na} shuts off

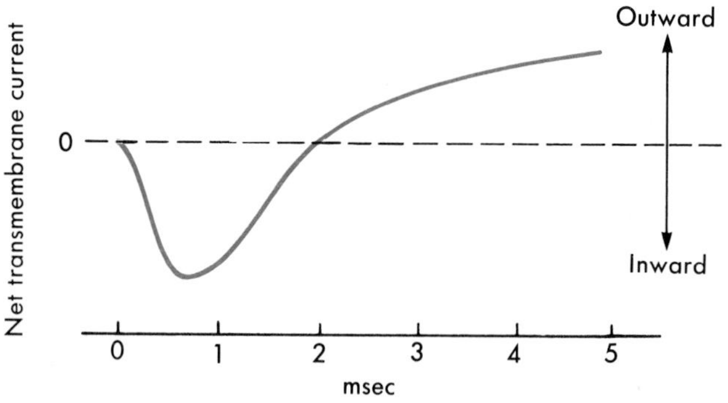

Fig. 3-6. Net transmembrane current flow in a squid giant axon after clamping the membrane potential to zero.

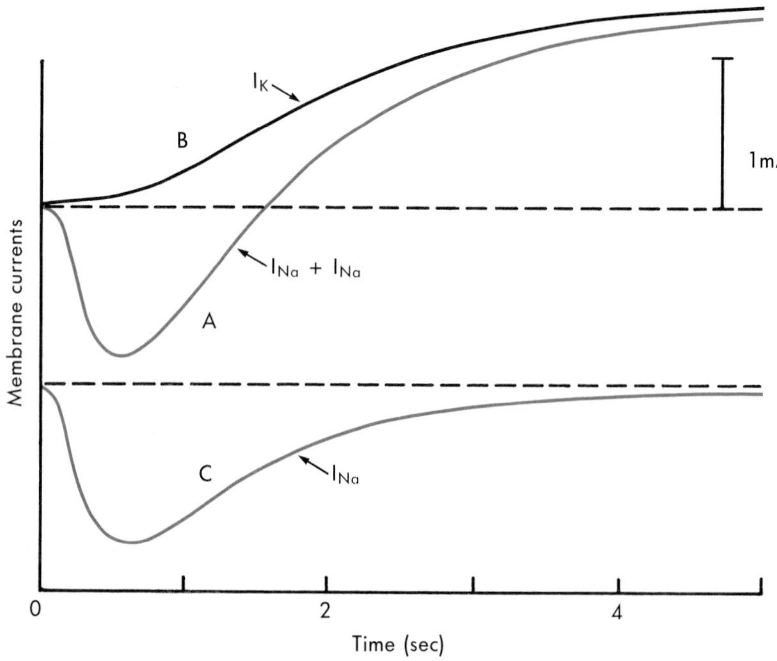

Fig. 3-7. A, The net ionic current that flows when a squid giant axon in sea water is clamped from its resting potential to a transmembrane potential of 0 mV. This current consists primarily of sodium and potassium ions flowing across the membrane. **B,** The net ionic current when the axon is placed in artificial sea water with most of the sodium replaced by choline (an impermeant cation), so that the intracellular and extracellular sodium concentrations are equal. This current is assumed to be caused by potassium only. **C,** Curve A minus curve B. This curve is taken to represent the sodium current. (Redrawn from Hodgkin, A.L., and Huxley, A.F.: The conduction of the nervous impulse, J. Physiol. **116:**449, 1952. Courtesy of Charles C. Thomas, Publisher, Springfield, Illinois.)

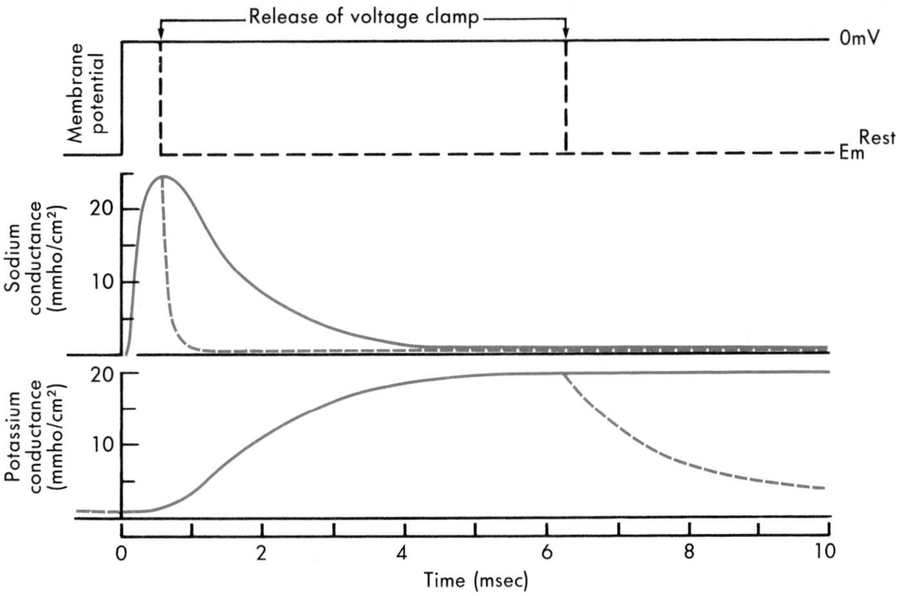

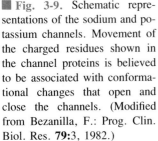

Fig. 3-8. Sodium and potassium conductance changes are shown in response to voltage clamp of the squid giant axon from the resting potential to a membrane potential of zero. Note that the sodium conductance inactivates after a slight delay even if the voltage clamp is maintained, but the potassium conductance remains elevated until the clamp is released. (Redrawn from Hodgkin, A.L.: Proc. Roy. Soc. B**148:**1, 1958.)

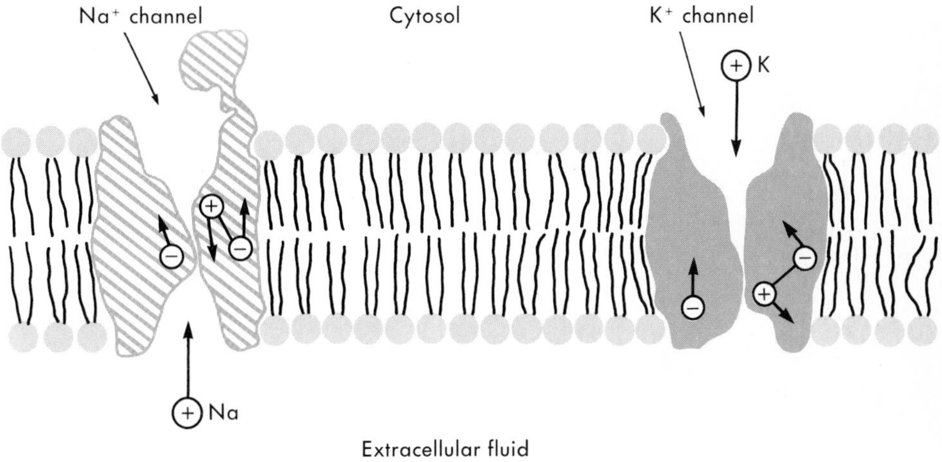

Fig. 3-9. Schematic representations of the sodium and potassium channels. Movement of the charged residues shown in the channel proteins is believed to be associated with conformational changes that open and close the channels. (Modified from Bezanilla, F.: Prog. Clin. Biol. Res. **79:**3, 1982.)

after a short delay even though the clamp is maintained. The increase in g_K, however, remains and does not decrease to resting levels until the clamp is released. Smaller depolarizations cause smaller changes in g_K and g_{Na}.

In brief, there are three basic events in the action potential: the increase of g_{Na}, the return of g_{Na} to resting levels (called inactivation of Na^+ channels), and the increase of g_K. The rate and extent of each of these depends on the level of the membrane potential. By doing voltage clamp experiments at different E_m levels, the changes in g_{Na} and g_K that must occur during the normal action potential can be reconstructed.

Ion Channels and Gates

Hodgkin and Huxley proposed that ion currents pass through separate Na^+ and K^+ channels, each with distinct characteristics, in the plasma membrane. Recent research supports this interpretation and has determined some of the properties of the channels (Fig. 3-9).

To enter the narrowest part of the channel known as the selectivity filter, it is believed that K^+ and Na^+ must shed most of their water of hydration. This allows Na^+ and K^+ to form coordination bonds with oxygen atoms in the interior of the selectivity filter. The configuration of the oxygen atoms in the selectivity filter determines the specificity of the Na^+ and K^+ channels. Tetrodotoxin (TTX) is a specific blocker of the Na^+ channel, and tetraethylammonium ion (TEA^+) blocks the K^+ channel. Tetrodotoxin binds to the extracellular side of the sodium channel. TEA^+ blocks the K^+ channel from the inside surface of the membrane.

The Na^+ channel appears to have both an *activation gate* and an *inactivation gate* that account for the changes in g_{Na} during voltage clamp. Digestion of the interior of the squid axon with the proteolytic enzyme pronase does away with the self-inactivation of g_{Na} by destroying the inactivation gate.

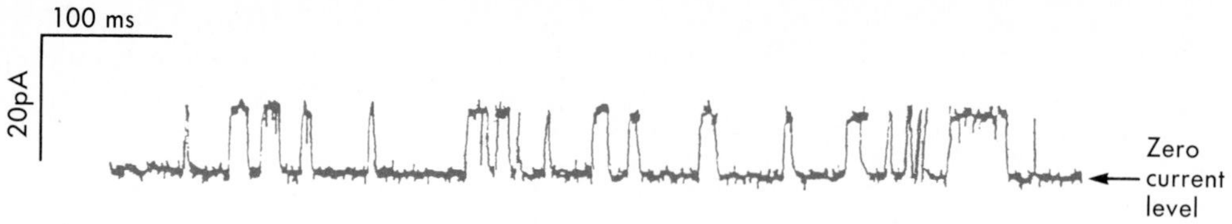

■ **Fig. 3-10.** Ionic current through a single ion channel from rat muscle incorporated into a planar lipid bilayer membrane. The channel opens and closes spontaneously. The fraction of time this channel spends in the open state is a function of calcium ion concentration and membrane potential. (Reproduced from Moczydlowski, E., and Latorre, R.: *The Journal of General Physiology,* 1983, vol. 82, pp. 511-542 by copyright permission of the Rockefeller University Press.)

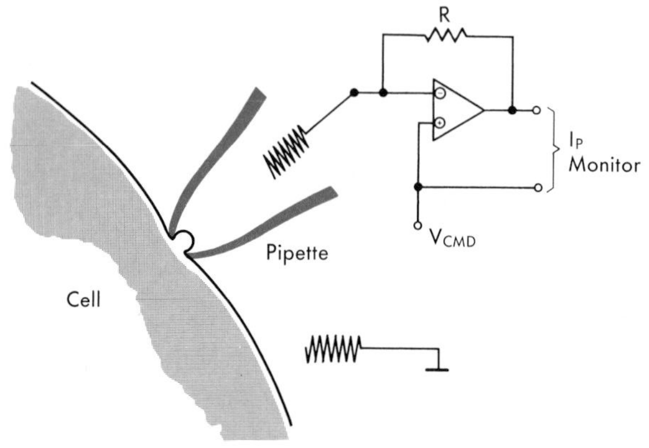

■ **Fig. 3-11.** A patch electrode and circuitry required to record the currents that flow through the small number of ion channels isolated in the electrode. (Redrawn by permission from Sigworth, F.J., and Neher, E.: *Nature,* vol. 287, pp. 447-449. Copyright © 1980 Macmillan Journals Limited.)

The gates of the channels may be charged and probably are peptide chains. The opening and closing of the gates cause redistributions of charges across the membrane. Such redistributions are measurable as small *gating currents.*

■ *Behavior of Individual Ion Channels*

Recently it has been possible to study the behavior of individual ion channels. One way to do this is to incorporate either purified ion channel proteins or bits of membrane into planar lipid bilayers that separate two aqueous compartments. Then electrodes placed in the aqueous compartments can be used to monitor or impose currents and voltages across the membrane. Under some conditions only one, or a few, ion channels of a particular type may be present in the planar membrane. The ion channels spontaneously oscillate between two conductance states, an open state and a closed state (Fig. 3-10).

Another way to study individual ion channels involves the use of so-called patch electrodes. A fire-polished microelectrode is placed against the surface of a cell and suction is applied to the electrode (Fig. 3-11). A high-resistance seal is formed around the tip of the electrode. The sealed patch electrode can then be used to monitor the activity of whatever channels happen to be trapped inside the seal. The patch of membrane can either be studied *in situ,* or removed from the cell so that the composition of the solution in contact with the intracellular face of the membrane may be manipulated, or even turned inside out if desired! Sometimes the patch trapped inside the electrode contains more than one functional ion channel of a particular type (Fig. 3-12).

Patch electrodes have been used to study Na^+ channels in rat muscle cells. In response to a step depolarization of the muscle cell membrane, some of the Na^+ channels show an opening event, some do not open at all, and some open more than once (Fig. 3-13). When the currents of a large number of channels are averaged (Fig. 3-13, *B*), the resulting behavior resembles the macroscopic behavior of Na^+ channels discussed earlier. That is to say the "average channel" opens promptly in response to depolarization, then after a short time delay, the channel closes (inactivates), even though the depolarization is maintained.

Our knowledge of ion channels is rapidly expanding. The primary amino acid sequence of the Na^+ channel has recently been elucidated by determining the sequence of the DNA that encodes the channel protein. While the 3-dimensional structure of the Na^+ channel remains to be determined, its intramembrane domain is known to consist of a number of α-helices that span the membrane and probably surround the ion channel. Groups of charged amino acid residues that may form the activation and inactivation gates have been tentatively identified. A detailed structural model of how the Na^+ channel operates should be forthcoming in the next few years.

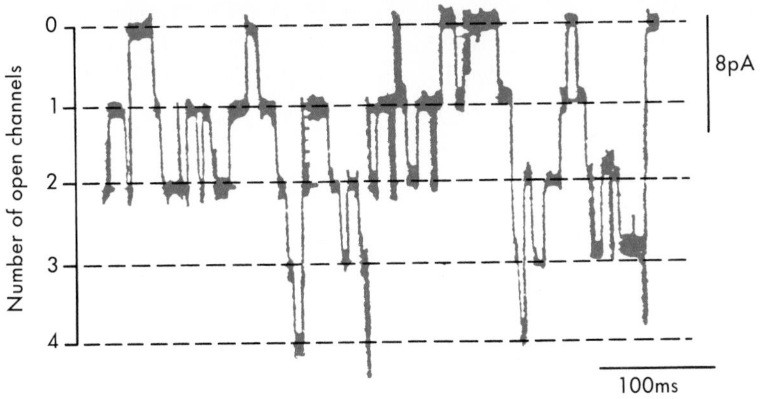

■ **Fig. 3-12.** A current recording from a patch electrode on a muscle cell plasma membrane. The four different current levels show that this particular patch of membrane contains four different ion channels. (Redrawn from Hammill, O.P., et al.: Pflugers Arch. **391:**85, 1981.)

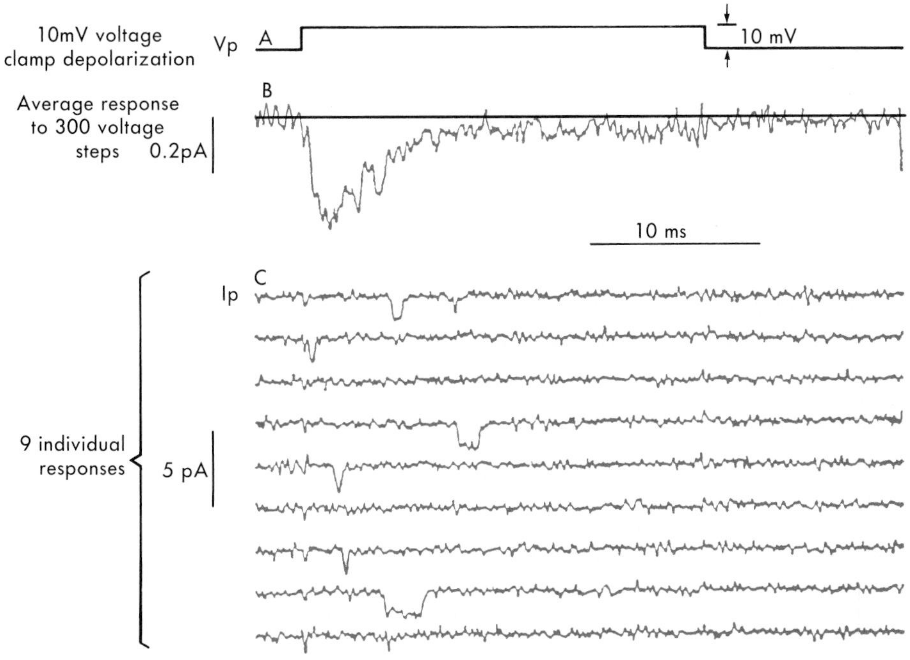

■ **Fig. 3-13.** A patch electrode records the currents that flow in a small patch of rat muscle membrane in response to a 10-mV depolarization (trace *A*). Tetraethylammonium is used to block potassium channels that may be present in the patch. The traces in curves *C* show responses to nine individual 10 mV depolarizations. The tracing in *B* is the average of 300 individual responses. Note that this average response resembles the response of large numbers of sodium channels as seen in conventional voltage clamp experiments (Fig. 3-7). (Redrawn by permission from Sigworth, F.J., and Neher, E.: *Nature,* vol. 287, pp. 447-449. Copyright © 1980 Macmillan Journals Limited.)

■ *Action Potentials in Cardiac and Smooth Muscle*

■ *Cardiac Muscle*

An action potential in a cardiac ventricular cell is schematically shown in Fig. 3-1. The initial rapid depolarization and overshoot is caused by the rapid entry of Na^+ through channels that are very similar to the Na^+ channels of nerve and skeletal muscle. Because of the rapid kinetics of opening and closing of these channels, they are called *fast* Na^+ channels.

After the initial depolarization and overshoot the cardiac ventricular action potential has a *plateau phase.* The plateau is due to another set of channels that are distinct from the fast Na^+ channels. These channels open and close much more slowly than the fast Na^+ channels and are called *slow channels.* The slow channels conduct both Na^+ and Ca^{++} ions, and the Ca^{++} that enters the cell during the plateau phase helps to initiate contraction of the ventricular cell. The repolarization of the ventricular cell is brought about by the closing of the slow channels and by a much delayed opening of K^+ channels. The ionic mechanism of the cardiac action potential is discussed in more detail in Chapter 27.

■ *Smooth Muscle*

Action potentials vary considerably among different types of smooth muscle. Characteristically action potentials in smooth muscle have slower rates of depolarization and repolarization and less overshoot than skeletal muscle action potentials. Smooth muscle cells lack fast Na^+ channels. The depolarizing phase of smooth muscle action potentials is caused primarily by channels that resemble the cardiac slow channels in their kinetics and in conducting both Na^+ and Ca^{++}. The Ca^{++} that enters via the slow channels is often vital for excitation-contraction coupling in smooth muscle. Repolarization is caused by the closing of the slow Na^+/Ca^{++} channels and a simultaneous opening of K^+ channels.

Properties of Action Potentials

Voltage Inactivation of the Action Potential

If a neuron or skeletal muscle cell is depolarized, for example, by increasing the concentration of K^+ in the extracellular fluid, its action potential has a slower rate of rise and a smaller overshoot. This is a result of two factors: (1) a smaller electrical force driving Na^+ into the depolarized cell, and (2) *voltage inactivation* of the Na^+ channels. The increase in g_{Na} during the action potential is self-inactivating. Once the Na^+ channels are inactivated, the membrane must be repolarized toward the normal resting membrane potential before the channels can be reopened. As the membrane potential is restored toward normal resting levels, more and more of the Na^+ channels again become capable of being activated. Since the action potential mechanism requires a critical density of open Na^+ channels, an action potential may not be generated in response to stimulation when a considerable fraction of Na^+ channels is inactivated because of partial depolarization. This is called *voltage inactivation* of the action potential because of voltage inactivation of the Na^+ channels. Voltage inactivation of the Na^+ channels is involved in important properties of excitable cells, such as refractoriness and accommodation.

The Refractory Periods

During much of the action potential the membrane is completely refractory to further stimulation. This means that, no matter how strongly the cell is stimulated, it is unable to fire a second action potential. This is called the *absolute refractory period* (Fig. 3-14). The cell is refractory because a large fraction of its Na^+ channels is voltage inactivated and cannot be reopened until the membrane is repolarized.

During the last part of the action potential the cell is able to fire a second action potential, but a stronger than normal stimulus is required. This is the *relative refractory period*. Early in the relative refractory period, before the membrane potential has returned to the resting potential level, some Na^+ channels are voltage inactivated, so a stronger than normal stimulus is needed to open the critical number of Na^+ channels needed to trigger an action potential. Throughout the relative refractory period the conductance to K^+ is elevated, which results in increased opposition to depolarization of the membrane. This also contributes to the refractoriness.

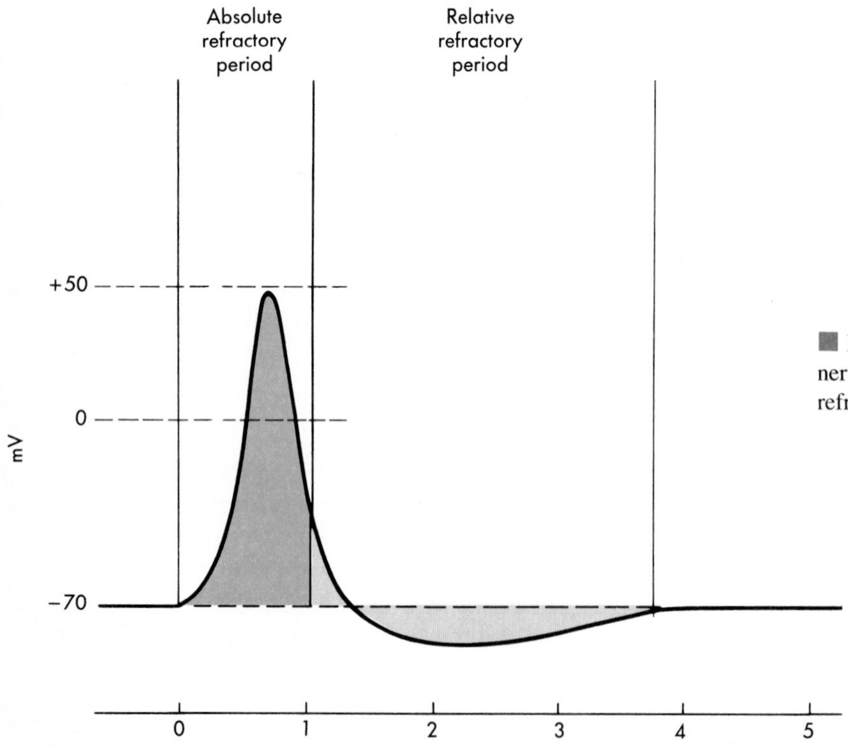

Fig. 3-14. The action potential of nerve showing the absolute and relative refractory periods.

Accommodation to Slow Depolarization

When a nerve or muscle cell is depolarized slowly, the threshold may be passed without an action potential being fired. This is called *accommodation*. Na^+ and K^+ channels are both involved in accommodation. During slow depolarization some of the Na^+ channels that are opened by depolarization have enough time to become voltage inactivated before the threshold potential is attained. If depolarization is slow enough, the critical number of open Na^+ channels required to trigger the action potential may never be achieved. In addition, K^+ channels open in response to the depolarization. The increased g_K tends to repolarize the membrane, making it still more refractory to depolarization.

The Strength-Duration Curve

A stimulus depolarizes the cell by causing charge to flow across the plasma membrane. The relevant quantity is the *total amount* of charge that flows across the membrane (current × time). A strong stimulus depolarizes the membrane to threshold quickly. A weaker stimulus must be applied longer for the critical amount of charge to flow across the membrane. This often is depicted in the *strength-duration curve,* which is a plot of the stimulus strength versus the minimum time the stimulus must be applied to cause an action potential (Fig. 3-15).

A very weak stimulus will not cause an action potential even if applied for a very long time (accommodation). The smallest stimulus strength that can elicit an action potential from a particular preparation is called the *rheobase*. The time needed by a stimulus *twice* as strong as the rheobase to elicit an action potential is called the *chronaxie* of the preparation. Chronaxie is a useful index of the excitability of a preparation. The larger the chronaxie, the less excitable the preparation.

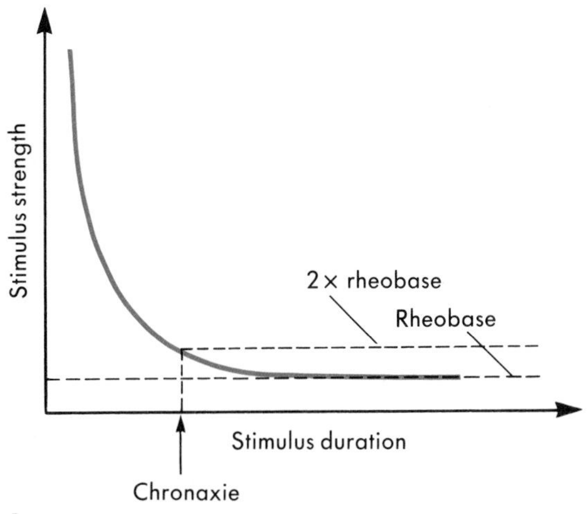

Fig. 3-15. The strength-duration curve. The ordinate shows stimulus strength, and the abscissa shows the minimal time a stimulus of that strength must be applied to produce an action potential.

Conduction of the Action Potential

A principal function of neurons is to transfer information by the conduction of action potentials. The axons of the motor neurons of the ventral horn of the spinal cord conduct action potentials from the cell body of the neuron in the spinal cord to a number of skeletal muscle fibers. The distance from the motor neuron to one of the muscle fibers it innervates may be longer than 1 m. The mechanism of action potential conduction and the factors that determine the speed of conduction are considered next.

Action potentials are conducted along a nerve or muscle fiber by local current flow, just as occurs in electrotonic conduction of subthreshold potential changes. Thus the same factors that govern the velocity of electrotonic conduction also determine the speed of action potential propagation.

The Local Response

Fig. 3-16, *A,* shows the membrane of an axon or muscle fiber that has been depolarized in a small area. In

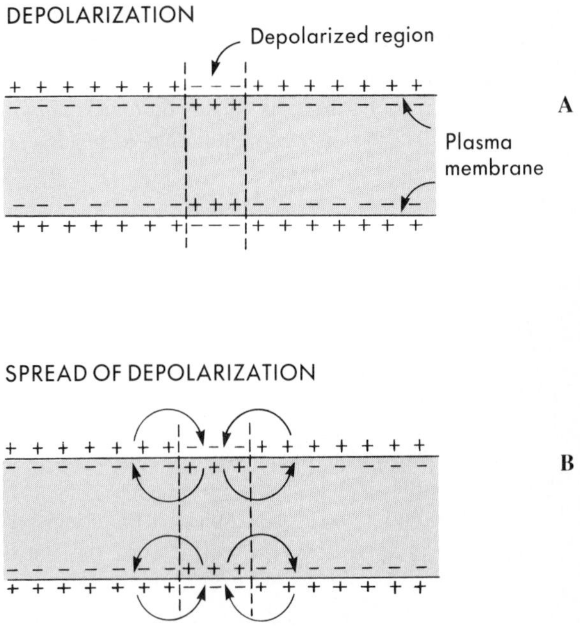

Fig. 3-16. Mechanism of electrotonic spread of depolarization. **A,** The reversal of membrane polarity that occurs with local depolarization. **B,** The local currents that flow to depolarize adjacent areas of the membrane and allow spreading of the depolarization.

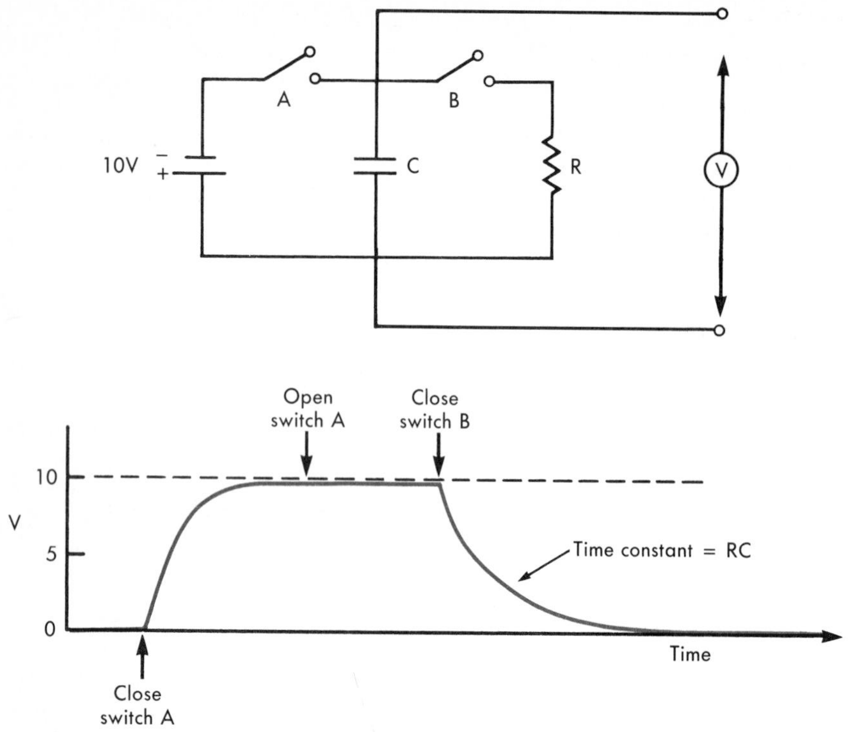

■ **Fig. 3-17.** An electrical circuit that illustrates the charging and discharging of a capacitor (C). When switch A is closed, currents flow to charge the capacitor to the battery voltage (10 v). When switch A is opened, the charge remains on the capacitor since there is no path for current to flow to discharge it. When switch B is closed, current flows through resistor R to discharge the capacitor.

the depolarized region the external aspect of the membrane is negative relative to the adjacent membrane, and the internal face of the depolarized membrane is positively charged relative to neighboring internal areas. The potential differences cause local currents to flow (Fig. 3-16, *B*), which depolarize the membrane adjacent to the initial site of depolarization. These newly depolarized areas then cause current flows which depolarize other segments of the membrane that are still further removed from the initial site of depolarization. This depolarization spread is called the *local response*. This mechanism of conduction is known as *electrotonic conduction*.

■ *RC Circuits*

The speed of electrotonic conduction is determined by the passive electrical properties of the cell involved. In particular, the *capacitance* of the plasma membrane and the *resistance* through which charge must flow to charge or discharge the membrane capacitor help to determine the conduction velocity.

The hydrophobic core of the plasma membrane is a good electrical insulator. The surfaces of the plasma membrane and the electrolyte solutions of the cytosol and extracellular fluid are better conductors of electrical current. Capacitors, devices that can store electrical charge, are constructed from a sheet of insulating material sandwiched between two conductors. The plasma membrane has a structure analagous to a capacitor and plasma membranes have significant capacitance (about 1 microfarad/cm^2). Before considering how the capacitance of the plasma membrane helps to determine the velocity of electrotonic conduction, let us consider the electrical circuits shown in Fig. 3-17.

The circuit in Fig. 3-17 consists of a 10-volt battery, a capacitor, and a resistor. When switch A is closed, the currents that flow will bring the voltage difference between the two plates of the capacitor to the battery voltage (10 volts). If switch A is now opened, the voltage across the capacitor will remain constant because there is no conductive path that will allow the capacitor to discharge. When switch B is closed, current will flow through the resistor and discharge the capacitor. As shown in the figure, the voltage across the capacitor will decay exponentially to zero.

What determines how rapidly the current will decay? The capacitance of the capacitor is equal to the amount of charge stored per unit voltage difference across the capacitor. The capacitance (C) times the initial voltage difference across the capacitor is then equal to the amount of charge (in coulombs) that must flow to totally discharge the capacitor. The larger the capacitance, the more charge must flow, and the more time will be required to discharge the capacitor. The resistance (R) in the circuit (and the voltage difference V_C) will determine the current flow that discharges the capacitor. The higher the resistance, the smaller the current, and the slower the discharge of the capacitor. Large values of R and large values of C both increase the time required to discharge the capacitor. The time required for the capacitor to discharge from its original voltage (V_o) to V_o/e (37% of V_o) is equal to R times C;

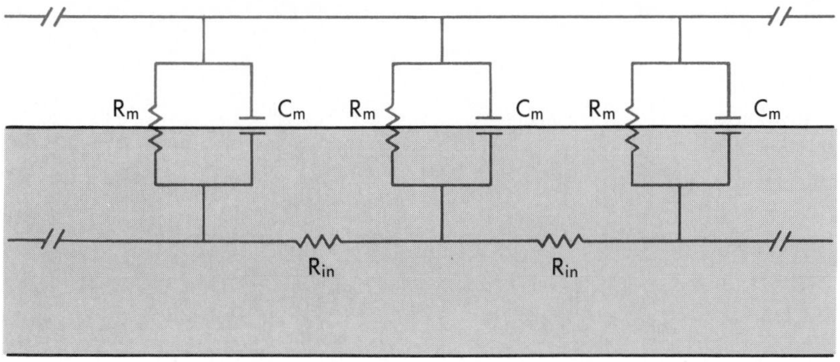

■ **Fig. 3-18.** The equivalent electrical circuit for electrotonic conduction. The membrane capacitor (C_m) can be discharged by currents flowing through the membrane resistance (R_m) and the internal longitudinal resistance (R_{in}). The effective time constant for this circuit is $\sqrt{R_m R_{in}}\,C_m$

RC is called the *time constant* for the circuit. The smaller the time constant, the more rapidly can the capacitor be discharged.

■ *Conduction Velocity*

The speed of electrotonic conduction along a nerve or muscle fiber is determined by the membrane capacitance and the electrical resistance to the flow of current. The membrane potential is a measurable manifestation of the charge stored by the membrane capacitor. The amount of charge that must flow to depolarize 1 cm^2 of membrane is proportional to the membrane capacitance (the coulombs of charge stored per volt of potential difference across the membrane) per unit area. A typical value of membrane capacitance (C_m) is about 10^{-6} farad/cm^2 membrane. To depolarize the membrane from -100 mV (-0.1 V) to 0 mV, 10^{-7} coulombs of charge must flow across each square centimeter of rmembrane:

$$\text{Charge flow/cm}^2 = \text{Capacitance/cm}^2 \times \text{Voltage change}$$
$$= 10^{-6}\ \text{farad/cm}^2 \times 0.1\ \text{V} \qquad (1)$$
$$= 10^{-7}\ \text{coulombs/cm}^2$$

The membrane capacitance thus determines *how much charge* must flow to depolarize the membrane. The larger the membrane capacitance, the greater the amount of charge that must flow, and the slower the rate of electrotonic spread.

The resistance to electrotonic current flow determines *how rapidly* charge can flow. The resistance to electrotonic current flow depends on the resistance to current flow across the membrane (R_m) and the resistance to longitudinal current flow in the cytoplasm (R_{in}) (Fig. 3-18). The effective resistance is proportional to the geometric mean of R_m and R_{in} ($\sqrt{R_m R_{in}}$). The larger $\sqrt{R_m R_{in}}$, the slower will electrotonic current flow, and the slower will be the rate of electrotonic conduction.

The product $\sqrt{R_m R_{in}}\,C_m$ is the time constant for electrotonic conduction. The smaller $\sqrt{R_m R_{in}}\,C_m$, the more rapidly electrotonic conduction can occur, and vice versa.

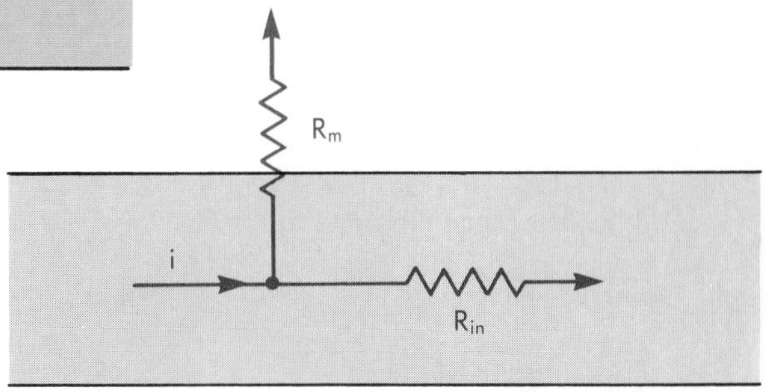

■ **Fig. 3-19.** An axon or a muscle fiber resembles an electrical cable. Currents that flow across the membrane resistance (R_m) are lost from the cable. Currents that flow through the longitudinal resistance (R_{in}) carry the electrical signal along the cable.

Effect of fiber size on conduction velocity. Consider a cylindrical nerve or muscle cell. The surface area of the cell increases with increasing radius ($A_s = 2\pi rl$). The cross-sectional area of the cell increases with the *square* of the radius ($A_x = \pi r^2$). The membrane capacitance increases in direct proportion to membrane area. The membrane resistance decreases in proportion to an increase in membrane area. The internal resistance is inversely proportional to the cross-sectional area. The effect of doubling the radius of the cell thus will increase C_m by a factor of 2, decrease R_m by a factor of 2, and decrease R_{in} by a factor of 4. Thus C_m will increase twofold, but $\sqrt{R_m R_{in}}$ will decrease by a factor of $\sqrt{2 \cdot 4}$, or $2\sqrt{2}$. The product $\sqrt{R_m R_{in}}\,C_m$ thus will decrease to $1/\sqrt{2}$ of its former value with a doubling of cell radius, and the velocity of electrotonic conduction will increase to $\sqrt{2}$ times the conduction velocity of the smaller fiber. Thus larger fibers have larger conduction velocities.

■ *Electrotonic Conduction Involves Decrement*

Earlier in this chapter it was noted that the local response dies away to almost nothing over the course of several millimeters (see Fig. 3-3). A nerve or muscle fiber has some of the properties of an electrical cable. In a perfect cable the insulation surrounding the core conductor prevents all current loss so that a signal is transmitted along the cable with undiminished strength (Fig. 3-19). The plasma membrane of an unmyelinated nerve or muscle fiber serves as the insulation. The membrane has a resistance much higher than the resistance of the cytoplasm, but (partly because of its thinness) the plasma membrane is not a perfect insulator. The higher the ratio of R_m to R_{in}, the better the cell can function as a cable, and the longer the distance that a signal can be transmitted electrotonically without significant decrement. $\sqrt{R_m/R_{in}}$ determines the *length constant* (see Fig. 3-3) of a cell. The length constant is the distance over which an electrotonically conducted signal falls to 37% (1/e) of its initial strength. A typical length constant for unmyelinated mammalian nerve and muscle fibers is about 1 to 3 mm. Some axons in the human body are about 1 m long, so it is clear that the local response cannot conduct a signal over so great a distance.

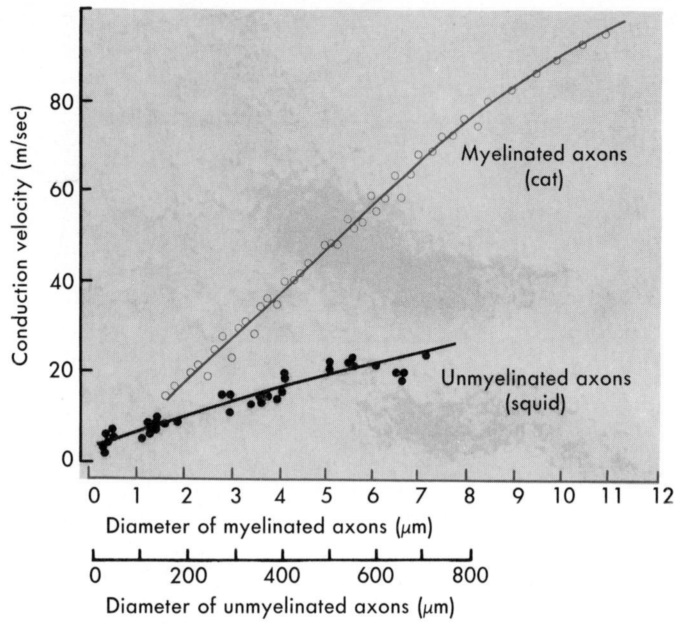

Fig. 3-20. Conduction velocities of myelinated and unmyelinated axons as functions of axon diameter. Myelinated axons *(colored curve* and *abscissa)* are from cat saphenous nerve at 38 C. Unmyelinated axons *(black curve* and *abscissa)* are from squid and cuttlefish at 20 to 22 C. Note that the myelinated axons have faster conduction velocities than unmyelinated axons 100 times greater in diameter. (Data for colored curve and abscissa from Gasser, H.S., and Grundfest, H.: Am. J. Physiol. **127:**393, 1939; data from black curve and abscissa from Pumphrey, R.J., and Young, J.Z.: J. Exp. Biol. **15:**453, 1938.)

The Action Potential as a Self-Reinforcing Signal

Many nerve and muscle fibers are much longer than their length constants. The action potential serves to conduct an electrical impulse with undiminished strength along the full length of those fibers. To do this, the action potential *reinforces itself* as it is propagated along the fiber. The propagation of the action potential occurs by the mechanism depicted in Fig. 3-16. When the areas on either side of the depolarized region reach threshold, these areas also fire action potentials, which locally reverses the polarity of the membrane potential. By local current flow the areas of the fiber adjacent to these areas are brought to threshold, and then these areas in turn fire action potentials. There is a cycle of depolarization by local current flow followed by generation of an action potential in a restricted region that then travels along the length of the fiber, with "new" action potentials being generated as they spread. In this way the action potential propagates over long distances, keeping the same size and shape.

Since the shape and size of the action potential are ordinarily invariant, only variations in the frequency of the action potentials can be used in the code for information transmission along axons. The maximum frequency is limited by the duration of the absolute refractory period (about 1 msec) to about 1000 impulses/second in large mammalian nerves.

Effect of Myelination on Conduction Velocity

A squid giant axon with 500 μm diameter has a conduction velocity of 25 m/second and is unmyelinated. If conduction velocity were directly proportional to fiber radius, a human nerve fiber with a 10μm diameter would conduct at 0.5 m/second. With this conduction velocity a reflex withdrawal of the foot from a hot coal would take about 4 seconds. Even though our nerve fibers are much smaller in diameter than squid giant axons, our reflexes are much faster than this. The myelin sheath that surrounds certain vertebrate nerve fibers results in a much greater conduction velocity than that of unmyelinated fibers of similar diameters. A 10 μm *myelinated* fiber has a conduction velocity about 50 m/second, which is twice that of the 500 μm squid giant axon. The high conduction velocity permits reflexes that are fast enough to allow us to avoid dangerous stimuli. Fig. 3-20 shows the large increase in conduction velocity caused by myelination. A myelinated axon has a greater conduction velocity than an unmyelinated fiber that is 100 times larger in diameter. As discussed below, the myelin sheath increases the velocity of action potential conduction by decreasing the capacitance of the axon and by allowing action potentials to be generated only at the *nodes of Ranvier*.

The evolution of myelinated fibers in vertebrates is salutary. If each of our peripheral nerve fibers were as large as a squid giant axon, then the nerve trunks (each containing hundreds of nerve fibers) would be so large that the peripheral nerves would alter the human form.

Myelination is caused by the wrapping of Schwann cell plasma membrane around a nerve axon (Fig. 3-21).

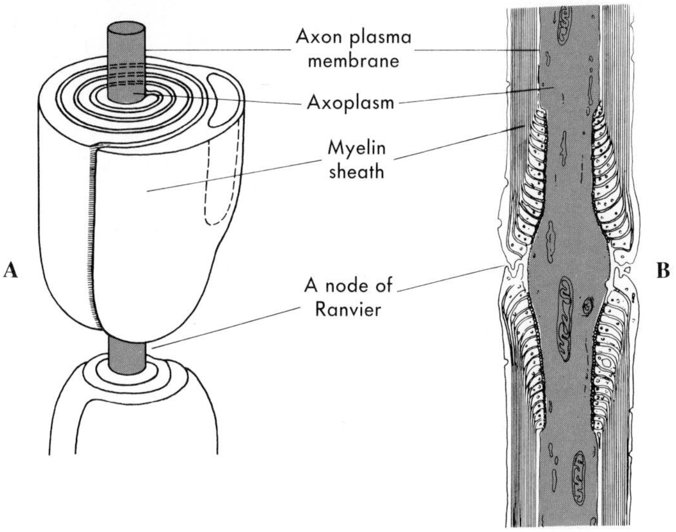

■ **Fig. 3-21.** The myelin sheath. **A,** Schematic drawing of Schwann cells wrapping around an axon to form a myelin sheath. **B,** Drawing of a cross-section through a myelinated axon near a node of Ranvier. (Redrawn from Elias, H., et al.: Histology and human microanatomy, ed. 4, New York, 1978, John Wiley & Sons, Inc.)

Thus the myelin sheath consists of several to over 100 layers of plasma membrane. Gaps that occur in the sheath every 1 to 2 mm are known as nodes of Ranvier. Nodes of Ranvier are about 1 μm wide and are the lateral spaces between different Schwann cells along the axon.

Myelination greatly alters the electrical properties of the axon. The many wrappings of membrane around the axon increase the effective *membrane resistance,* so that R_m/R_{in} and thus the length constant is much greater. Less of a conducted signal is lost through the electrical insulation of the myelin sheath, so that the amplitude of a conducted signal declines less with distance along the axon.

Each Schwann cell membrane has a capacitance similar to that of the plasma membrane of the axon. The capacitances of the membranes in the myelin sheath act as though they were connected in series. Capacitors in series are added according to the equation $1/C_t = 1/C_1 + 1/C_2 + 1/C_3 + \ldots$ etc., where C_t is the effective overall capacitance, and C_1, C_2, C_3, etc., are the individual capacitances. If there are 50 identical capacitances in series (25 Schwann cell wraps), then $1/C_t = 50/C$, and $C_t = C/50$. A myelin sheath with 50 membranes thus *lowers* the membrane capacitance by a factor of 50. Resistances in series, however, add directly, so the myelination will increase the membrane resistance by fiftyfold. Earlier in this chapter we saw that the time constant $\sqrt{R_m R_{in}}\ C_m$ determines the electrotonic conduction velocity. The smaller this product, the greater the conduction velocity. The 25 Schwann cell wraps should have no effect on R_{in} but will lower C_m by a factor of 50 and increase R_m by a factor of 50. Thus $\sqrt{R_m R_{in}}\ C_m$ will decrease by $50/\sqrt{50}$, resulting in a sevenfold increase in electrotonic conduction velocity because of myelination.

Another property of conduction in myelinated fibers that enhances conduction velocity is called *saltatory conduction,* because the impulse "jumps" from one node of Ranvier to the next. Saltatory conduction occurs because the action potential is regenerated only at the nodes (1 to 2 mm apart). The action potential in myelinated fibers is not regenerated at each place along the axon as the impulse is propagated. The internodal plasma membrane cannot produce action potentials, because the depolarization of the internodal membrane is divided among 50 or so membranes of the myelin sheath. The resulting depolarization of the internodal plasma membrane is only 1 or 2 mV—not nearly sufficient to reach threshold.

In brief, conduction in myelinated axons is characterized by rapid electrotonic conduction (because of the decreased time constant for conduction) with little decrement (because of the increased length constant) between the nodes of Ranvier. Only at the nodes is the action potential regenerated.

Myelinated axons are also more efficient metabolically than nonmyelinated axons. The sodium-potassium pump extrudes the sodium that enters and reaccumulates the potassium that leaves the cell during action potentials. In a myelinated axon, ionic currents are restricted to the small fraction of the membrane surface at the nodes of Ranvier. For this reason fewer Na^+ and K^+ ions traverse a unit area of membrane, and less ion pumping is required to maintain Na^+ and K^+ gradients.

■ *Bibliography*

Aidley, D.J.: The physiology of excitable cells, ed. 2, Cambridge, 1978, Cambridge University Press.

Davson, H.: A textbook of general physiology, ed. 4, Baltimore, 1970, Williams & Williams.

Hodgkin, A.L.: The conduction of the nervous impulse, Springfield, Ill., 1964, Charles C Thomas, Publisher.

Junge, D.: Nerve and muscle excitation, ed. 2, Sunderland, Mass., 1981, Sinauer Associates, Inc.

Kandel, E.R., and Schwartz, J.H.: Principles of neural science, ed. 2, New York, 1985, Elsevier Science Publishing Co., Inc.

Katz, B.: Nerve muscle, and synapse, New York, 1966, McGraw-Hill Book Co.

Keynes, R.D.: Ion channels in the nerve-cell membrane, Sci. Am. **240**:126, 1979.

Stevens, C.F.: The neuron, Sci. Am. **241**(3):54, 1979.

Stevens, C.F.: Studying just one molecule: single channel recording, Trends Pharm. Sci. **5**:131, 1984.

CHAPTER 4

Synaptic Transmission

A synapse is a site at which an impulse is transmitted from one cell to another. There are two types of synapses: electrical synapses and chemical synapses. At an *electrical synapse* two excitable cells communicate by the direct passage of electrical current between them. This is called *ephaptic* or *electrotonic* transmission. *Gap junctions* link electrotonically coupled cells and are thought to be low-resistance pathways for current flow directly between the cells. There are few well-studied examples of ephaptic transmission in the vertebrate central nervous system.

Information more often is transferred between excitable cells by means of *chemical synapses*. Chemical synapses may be better suited for the complex modulation of synaptic activity and the integration that occurs at synapses in vertebrate central nervous systems. At a chemical synapse an action potential causes a transmitter substance to be released from the presynaptic neuron. The transmitter diffuses across the extracellular synaptic cleft and binds to receptors on the membrane of the postsynaptic cell to cause a change in its electrical properties. Chemical synapses have *synaptic delay*—the time required for these events to occur. The neuromuscular (or myoneural) junction is a particularly well-studied vertebrate chemical synapse.

Although the nature of the presynaptic and postsynaptic cells, the structure of the synapse, and the transmitter substance vary, there are certain characteristics that chemical synapses have in common.

Neuromuscular Junctions

The synapses between the axons of motoneurons and skeletal muscle fibers are called neuromuscular junctions, myoneural junctions, or motor endplates. The neuromuscular junction was the first vertebrate synapse to be well characterized. The neuromuscular junction serves as a model chemical synapse that provides a basis for understanding more complex synaptic interactions among neurons in the central nervous system.

General characteristics of transmission at chemical synapses

Action potential in presynaptic cell
↓
Depolarization of the plasma membrane of the presynaptic axon terminal
↓
Entry of Ca^{++} into presynaptic terminal
↓
Release of the transmitter by the presynaptic terminal
↓
Chemical combination of the transmitter with specific receptors on the plasma membrane of the postsynaptic cell
↓
Transient change in the conductance of the postsynaptic plasma membrane to specific ions
↓
Transient change in the membrane potential of the postsynaptic cell

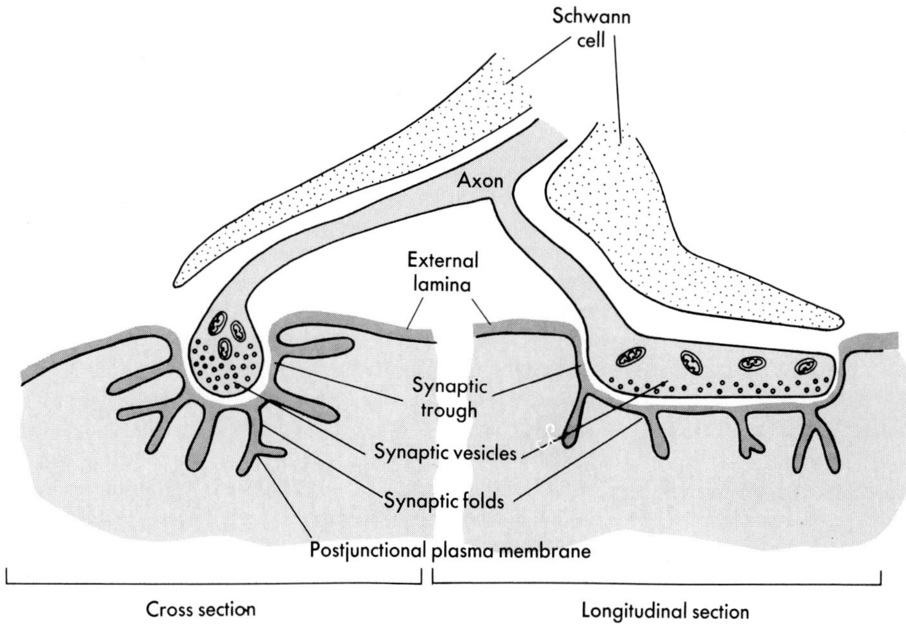

Fig. 4-1. The ultrastructure of the neuromuscular junction in skeletal muscle. (Redrawn from Flickinger, C.J., et al.: Medical cell biology, Philadelphia, 1979, W.B. Saunders Co.)

Structure of the Neuromuscular Junction

Near the neuromuscular junction the motor nerve loses its myelin sheath and divides into fine terminal branches (Fig. 4-1). The terminal branches of the axon lie in synaptic troughs on the surfaces of the muscle cells (Fig. 4-1). The plasma membrane of the muscle cell lining the trough is thrown into numerous junctional folds. The axon terminals contain many 400 Å smooth-surfaced synaptic vesicles that contain acetylcholine. The axon terminal and the muscle cell are separated by the synaptic cleft, which contains a carbohydrate-rich amorphous material.

Acetylcholine receptor molecules are concentrated near the mouths of the junctional folds. Acetylcholinesterase appears to be evenly distributed on the external surface of the postsynaptic membrane. The synaptic vesicles in the nerve terminals and specialized release sites on the presynaptic membrane are concentrated opposite the mouths of the junctional folds.

Overview of Neuromuscular Transmission

The action potential is conducted down the motor axon to the presynaptic axon terminals. Depolarization of the plasma membrane of the axon terminal brings about a transient increase in its calcium conductance. Ca^{++} flows down its electrochemical potential gradient into the axon terminal. In ways that are not yet well understood, the influx of Ca^{++} causes synaptic vesicles to fuse with the plasma membrane and to empty their acetylcholine into the synaptic cleft by exocytosis. Acetylcholine diffuses across the synaptic cleft and combines with a specific acetylcholine receptor protein on the external surface of the muscle plasma membrane of the motor endplate. The combination of acetylcholine with the receptor protein causes a transient increase in the conductance of the postjunctional membrane to Na^+ and K^+. Ionic currents (Na^+ and K^+) result in a transient depolarization of the endplate region. The transient depolarization is called the *endplate potential* or *EPP* (Fig. 4-2). The EPP is transient because the action of acetylcholine is ended by the hydrolysis of acetylcholine to form choline and acetate. The hydrolysis is catalyzed by the enzyme acetylcholinesterase, which is present in high concentration on the postjunctional membrane.

The postjunctional plasma membrane of the neuromuscular junction is not electrically excitable and does not fire action potentials. After it is depolarized, adjacent regions of the muscle cell membrane are depolarized by electrotonic conduction (Fig. 4-2). When those regions reach threshold, action potentials are generated. Action potentials are propagated along the muscle fiber at high velocity and induce the muscle cells to contract. The steps involved in neuromuscular transmission are listed below, and some are considered next in more detail.

Synthesis of Acetylcholine

Motoneurons and their axons synthesize acetylcholine. Most other cells are not able to make acetylcholine. The enzyme choline-*O*-acetyltransferase in the motoneuron catalyzes the condensation of acetyl coenzyme A (acetyl CoA) and choline. Acetyl CoA is produced by the neuron (as well as by most cells). Choline cannot be synthesized by the motoneuron and is obtained by ac-

A

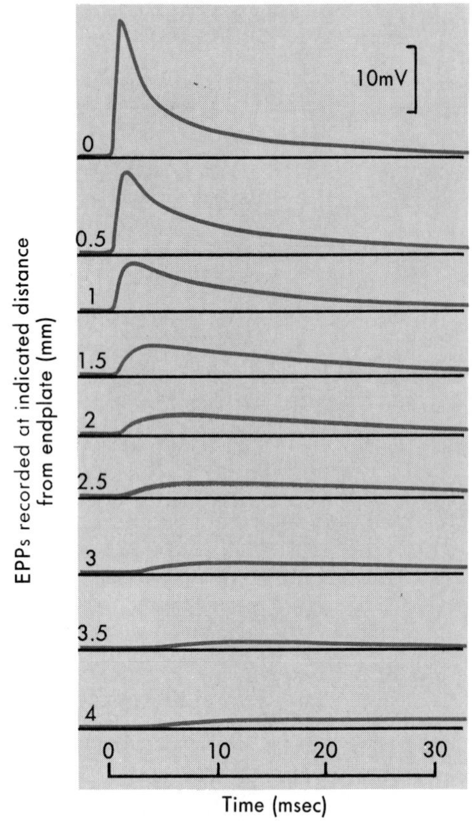

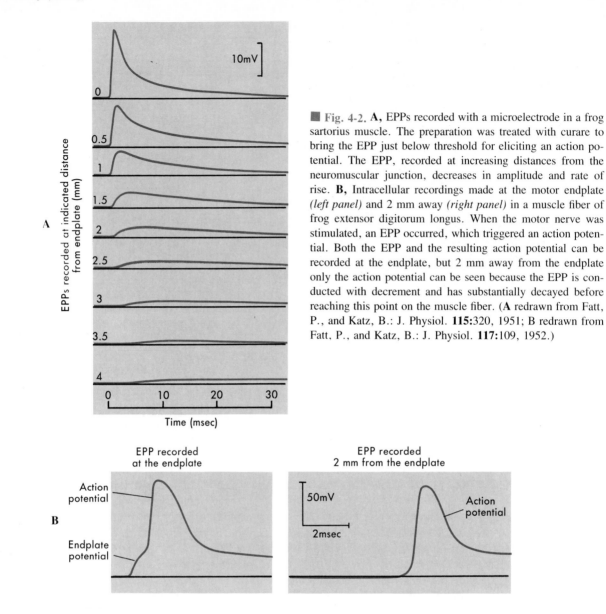

■ Fig. 4-2. **A,** EPPs recorded with a microelectrode in a frog sartorius muscle. The preparation was treated with curare to bring the EPP just below threshold for eliciting an action potential. The EPP, recorded at increasing distances from the neuromuscular junction, decreases in amplitude and rate of rise. **B,** Intracellular recordings made at the motor endplate *(left panel)* and 2 mm away *(right panel)* in a muscle fiber of frog extensor digitorum longus. When the motor nerve was stimulated, an EPP occurred, which triggered an action potential. Both the EPP and the resulting action potential can be recorded at the endplate, but 2 mm away from the endplate only the action potential can be seen because the EPP is conducted with decrement and has substantially decayed before reaching this point on the muscle fiber. (**A** redrawn from Fatt, P., and Katz, B.: J. Physiol. **115:**320, 1951; B redrawn from Fatt, P., and Katz, B.: J. Physiol. **117:**109, 1952.)

B

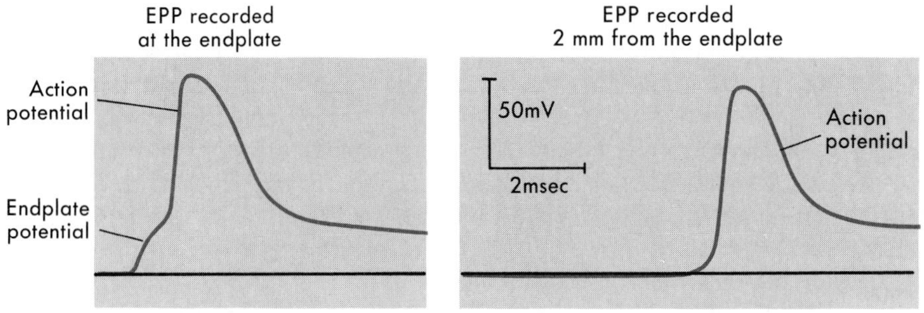

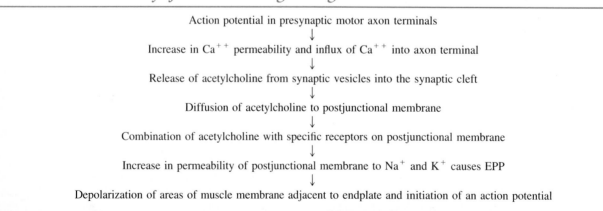

Summary of events occurring during neuromuscular transmission

Action potential in presynaptic motor axon terminals

↓

Increase in Ca^{++} permeability and influx of Ca^{++} into axon terminal

↓

Release of acetylcholine from synaptic vesicles into the synaptic cleft

↓

Diffusion of acetylcholine to postjunctional membrane

↓

Combination of acetylcholine with specific receptors on postjunctional membrane

↓

Increase in permeability of postjunctional membrane to Na^+ and K^+ causes EPP

↓

Depolarization of areas of muscle membrane adjacent to endplate and initiation of an action potential

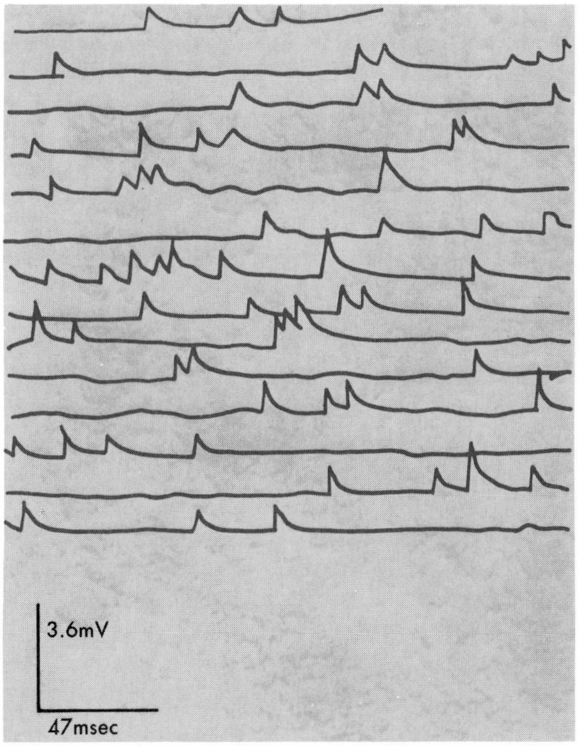

■ Fig. 4-3. Spontaneous MEPPs recorded at a neuromuscular junction in a fiber of frog extensor digitorum longus. (Redrawn from Fatt, P., and Katz, B.: Nature **166:**597, 1950.)

tive uptake from the extracellular fluid. The plasma membrane of the motoneuron has a transport system that can accumulate choline against a large electrochemical potential gradient. About half the choline that is freed in the synaptic cleft when acetylcholine is hydrolyzed is actively taken back up into the motoneuron to be used in the resynthesis of acetylcholine.

Quantal Release of Transmitter

Even if the motoneuron is not stimulated, small depolarizations of the postjunctional muscle cell occur spontaneously. These small spontaneous depolarizations are known as *miniature endplate potentials,* or *MEPPs* (Fig. 4-3). They occur at random times with a frequency that averages about 1 per second. Each MEPP depolarizes the postjunctional membrane by only about 0.4 mV on average. The MEPP has the same time course as an EPP that is evoked by an action potential in the nerve terminal. The MEPP is similar to the EPP in its response to most drugs. The EPP and MEPP are both prolonged by drugs that inhibit acetylcholinesterase, and both are similarly depressed by compounds that compete with acetylcholine for binding to the receptor protein. The frequency of MEPPs varies in time, but their amplitudes are within a relatively narrow range (Figs. 4-3 and 4-4). A MEPP is caused by the spontaneous release of a small number (probably only one) of vesicles of transmitter into the synaptic cleft.

The quantal nature of transmitter release has been shown in another way. Extracellular Ca^{++} is required for transmitter release. If the extracellular Ca^{++} is reduced to low levels, the sizes of the EPPs evoked by stimulation of the motoneuron are greatly reduced. Under these conditions spontaneous variations occur in the size of the stimulation-evoked EPPs. The size of the EPP does not vary continuously, but in small steps that correspond to the size of a single MEPP (Fig. 4-4).

Acetylcholine can also be released via a pathway that does not involve membrane-bounded vesicles. The relative roles of transmitter release from synaptic vesicles and via the vesicle-independent pathway remain to be determined.

Action of Cholinesterase and Reuptake of Choline

Acetylcholinesterase is concentrated on the external surface of the postjunctional membrane and in the basal lamina. The EPP is terminated by the hydrolysis of acetylcholine. Eserine and edrophonium are inhibitors of the enzyme and are called *anticholinesterases.* In the presence of an anticholinesterase, the EPP is larger and dramatically prolonged.

Approximately half of the choline released by the hydrolysis of acetylcholine is taken back up into the prejunctional nerve terminal by a Na^{+}-mediated secondary active transport system. The motoneuron cannot synthesize choline, so the reuptake provides choline needed for the resynthesis of acetylcholine. Hemicholiniums are drugs that block the choline transport system and inhibit choline uptake. Prolonged treatment with hemicholiniums results in depletion of the store of transmitter and ultimately causes a decrease in the acetylcholine content of the quanta.

The Ionic Mechanism of the EPP

The cation channels that acetylcholine causes to open in the postjunctional membrane differ from the cation channels of nerve and muscle by being independent of the membrane potential. The postjunctional channels are gated by the action of acetylcholine, rather than by the transmembrane potential. The acetylcholine-dependent cation channels of the postjunctional membrane are permeable only to cations, but the channels are not very selective among small cations. Na^{+}, K^{+}, Rb^{+}, and NH_4^{+} pass through these channels with roughly equal ease.

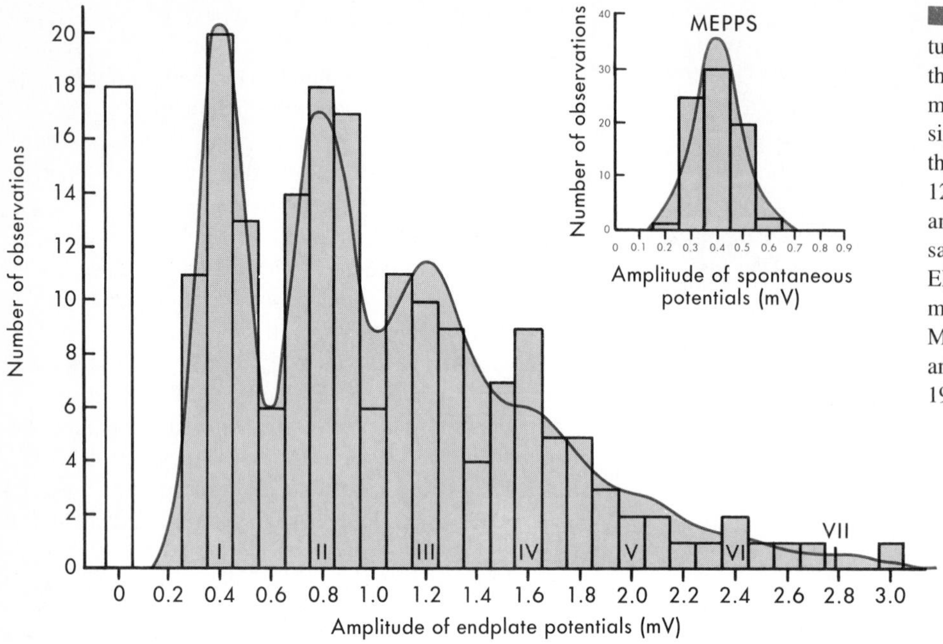

Fig. 4-4. Histogram of the amplitudes of EPPs elicited by stimulating the motor nerve to a cat tenuissimus muscle fiber. Neuromuscular transmission was severely inhibited by bathing the muscle in a solution containing 12.5 mM Mg^{++}. The insert shows an amplitude histogram for MEPPs in the same fiber. Note that the distribution of EPPs has peaks that occur at integral multiples of the mean amplitude of the MEPPs. (Redrawn from Boyd, I.A., and Martin, A.R.: J. Physiol. **132:**74, 1956.)

The membrane potential may be determined primarily by the membrane conductances to K^+ and Na^+, as shown by the chord conductance equation (short form):

$$E_m = \frac{g_K}{g_K + g_{Na}} E_K = \frac{g_{Na}}{g_K + g_{Na}} E_{Na} \qquad (1)$$

Because acetylcholine causes the postsynaptic membrane to become permeable to both Na^+ and K^+, the membrane potential of the endplate in the presence of acetylcholine should tend toward a value in between E_K and E_{Na}. This has been shown by experiments in which the membrane potential of the postjunctional muscle cell is voltage clamped to various values. The prejunctional motor axon is then stimulated, and the resulting endplate currents are measured (Fig. 4-5, *A*). As the postjunctional cell is depolarized by the voltage clamp, the endplate currents diminish. For positive membrane potentials the net endplate current becomes outward. The potential of the postjunctional membrane at which no net current flows (about 0 mV) is called the *reversal potential* of the neuromuscular junction (Fig. 4-5, *B*).

The Acetylcholine Receptor Protein

Recently the acetylcholine receptor protein has been studied intensively. Development of methods for isolating and purifying hydrophobic membrane proteins and the availability of snake venom neurotoxins that bind very tightly to the acetylcholine receptor have been essential in these studies. α-Bungarotoxin, from the venom of the Formosan krait (a relative of the cobra), is a useful toxin that binds to the acetylcholine receptor almost irreversibly.

Binding of radioactively labeled α-bungarotoxin by neuromuscular junctions suggests that there are 10^7 to 10^8 binding sites per motor endplate. At mouse diaphragm neuromuscular junctions the acetylcholine binding sites concentrated near the mouths of the postjunctional folds have a density of about 20,000 per μm^2. This suggests that the receptor molecules are quite tightly packed, since the maximum density possible has been estimated to be about 50,000 per μm^2.

The acetylcholine receptor protein is an integral membrane protein and is deeply embedded in the hydrophobic lipid matrix of the postjunctional membrane. Cholinesterase, on the other hand, is loosely associated with the surface of the postjunctional membrane by hydrophilic interactions.

Acetylcholine receptor protein has been isolated from certain neuromuscular junctions and from the electroplax of electrical fish, where the receptor concentration is especially high. The receptor protein has been extensively purified by affinity chromatography. Purified receptor protein has been reconstituted into a lipid bilayer model membrane. In such systems binding of acetylcholine by the receptor protein leads to an increase in the Na^+ and K^+ conductance of the lipid bilayer.

The acetylcholine receptor protein consists of five subunits, two of which are identical, so that there are four different polypeptide chains. The duplicated subunit is the α subunit (molecular weight 40,000). The acetylcholine binding sites are located on the α subunits. The other subunits are β (50,000), γ (60,000), and δ (65,000). In some species the pentamer $\alpha_2\beta\gamma\delta$ is the predominant form of the acetylcholine receptor. In

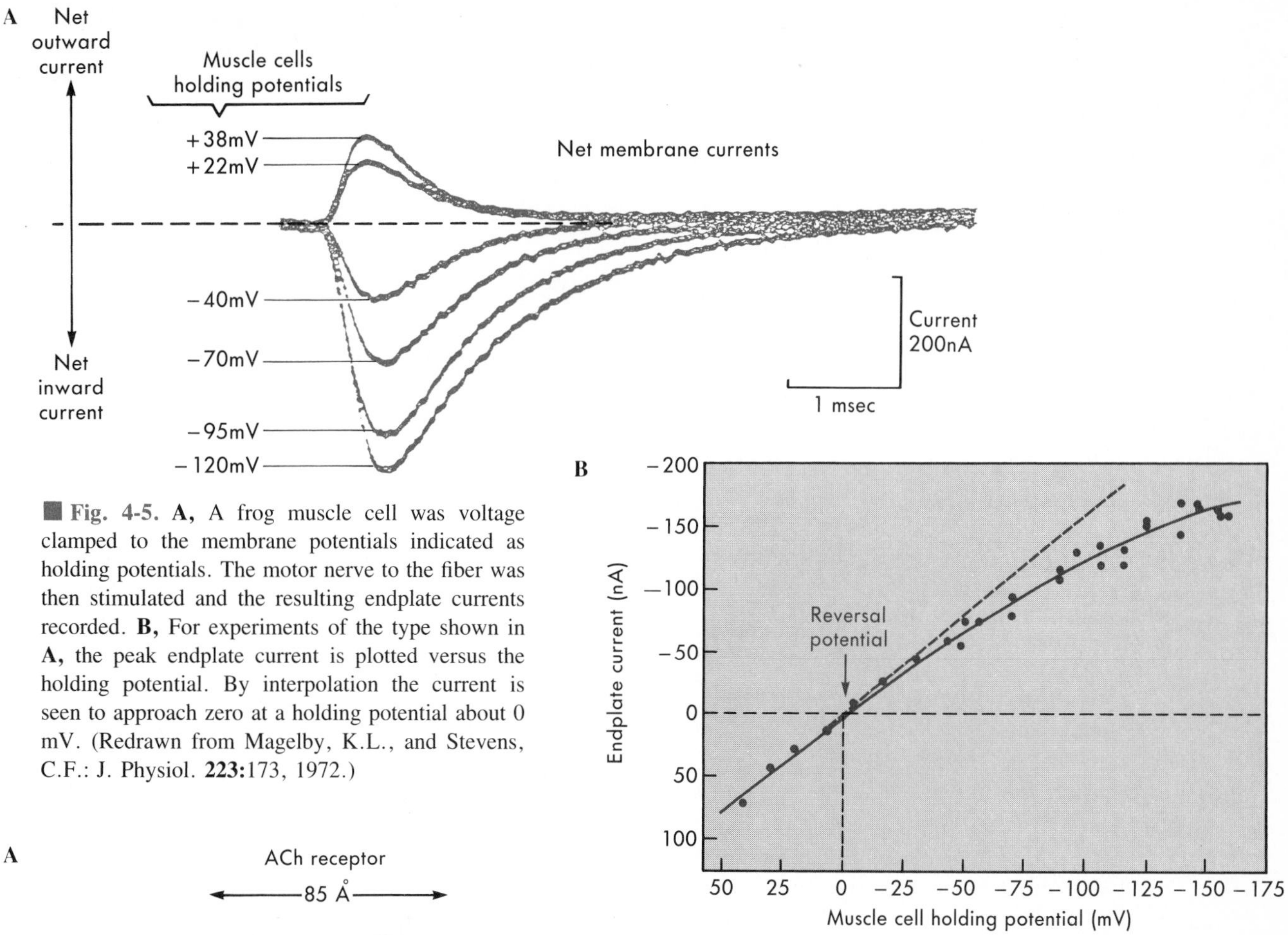

■ **Fig. 4-5. A,** A frog muscle cell was voltage clamped to the membrane potentials indicated as holding potentials. The motor nerve to the fiber was then stimulated and the resulting endplate currents recorded. **B,** For experiments of the type shown in **A,** the peak endplate current is plotted versus the holding potential. By interpolation the current is seen to approach zero at a holding potential about 0 mV. (Redrawn from Magelby, K.L., and Stevens, C.F.: J. Physiol. **223:**173, 1972.)

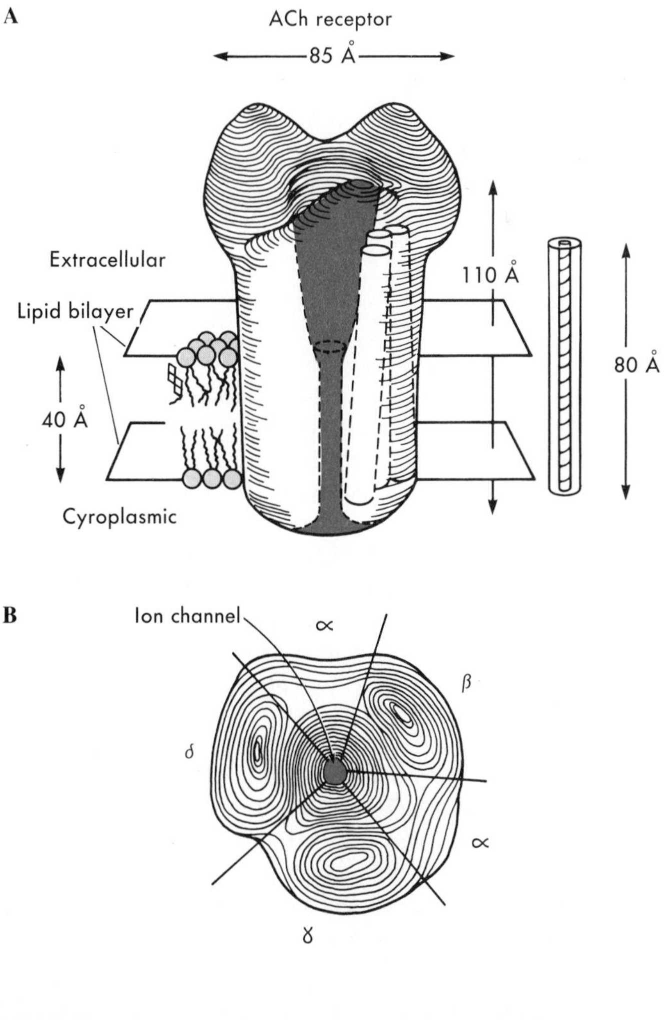

■ **Fig. 4-6.** A model of the structure of the acetylcholine receptor protein, **A,** Viewed from the side and, **B,** viewed looking down on the acetylcholine receptor from the cytoplasmic surface. The closed curves are electron density profiles. The five subunits surround a central ion channel. (Redrawn from Kistler, J., et al.: Biophys. J. **37:**371, 1982. Reproduced from the *Biophysical Journal* by copyright permission of the Biophysical Society.)

other species two pentamers are linked by a disulfide bridge between their γ subunits. The function of these two forms is apparently identical. The genes for the α,β,γ, and δ subunits have been cloned and sequenced. The amino acid sequences of the α,β,γ, and δ subunits have extensive sequence homology, and thus the four subunits have probably evolved from a common precursor.

X-ray diffraction and image analysis of electron micrographs of negatively stained preparations have revealed a good deal about the three-dimensional structure of the acetylcholine receptor in the membrane. The five subunits surround a central ion channel (Fig. 4-6). All of the subunits span the membrane, with more of their mass protruding from the extracellular face of the

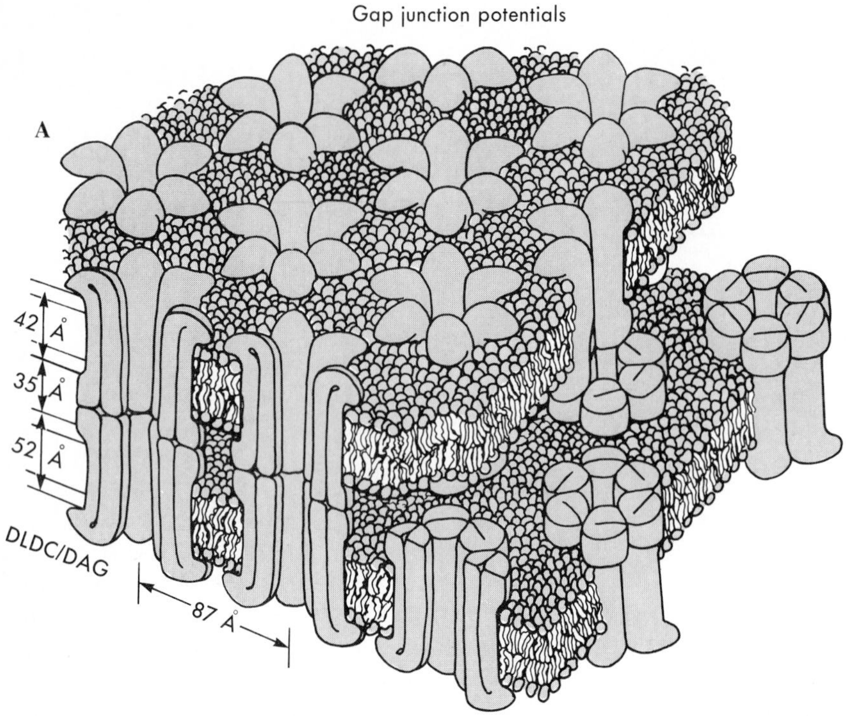

■ **Fig. 4-7. A,** A model for the structure of gap junction channels. Each plasma membrane contains connexons, each of which consists of an hexagonal array of six connexin polypeptides. The connexons of the two membranes are aligned at the gap junction to form channels between the cytosolic compartments of the two cells. **B,** A model of the opening and closing of the gap junction channel. The individual subunits of the connexon are proposed to twist relative to one another to open and close the central channel. (**A** redrawn from Makowski, L., et al.: J. Cell Biol. **74:**629, 1977. Reproduced from *The Journal of Cell Biology* by copyright permission of The Rockefeller University Press; **B** redrawn from Unwin, P.N.T., and Zampighi, G.: Nature **283:**545, 1980. Reprinted by permission from *Nature*. Copyright © 1980 Macmillan Journals Limited.)

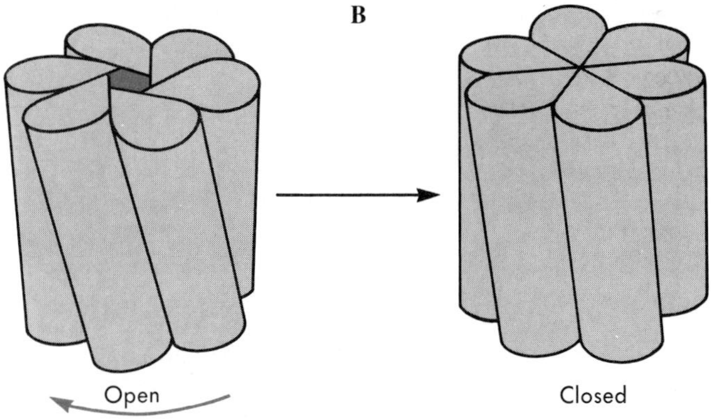

membrane than from the intracellular face. The subunits form a funnel around the mouth of the ion channel on the external side of the membrane. The binding of acetylcholine is postulated to cause a conformational change in the α subunit that results in alterations in the conformations of the other subunits and opens the central ion channel.

Defects in neuromuscular transmission are involved in myasthenia gravis. Experiments involving tritiated bungarotoxin binding suggest that in this disease there is decreased density of acetylcholine receptors on the postjunctional membrane. Most myasthenic patients have circulating antibodies to acetylcholine receptor protein. Whether these antibodies are the cause of myasthenia gravis or are produced secondarily as the result of endplate degeneration remains to be determined.

■ *Synapses between Neurons*

Chemical synapses between neurons are more prevalent than electrical synapses and are better understood. Chemical transmission between neurons has many of the same properties that characterize the neuromuscular junction. Electrical synapses have been described in the central nervous systems of animals from invertebrates to mammals. The prevalence of electrical synapses in the nervous systems of higher animals and their physiological roles are still poorly defined.

■ *Electrical Synapses*

At an electrical synapse a change in the membrane potential of one cell is transmitted to the other cell by the direct flow of current. Because current flows directly between the two cells that make an electrical synapse, there is essentially no synaptic delay. Chemical synapses typically have a synaptic delay of about 0.5 msec. In general, electrical synapses allow conduction in both directions. In this respect they differ from chemical synapses, which are obligatorily unidirectional. Certain electrical synapses conduct more readily in one direction than in the other; this property is called *rectification*.

Cells that form electrical synapses typically are joined by *gap junctions*. Gap junctions are plaquelike structures in which the plasma membranes of the coupled cells are very close (less than 30 Å). In freeze-fracture electron micrographs of gap junctions there are regular arrays of intramembrane protein particles. The intramembrane particles consist of six subunits surrounding a central channel that is accessible to water. The hexagonal array is called a *connexon*. Each of the six subunits is a single protein (one polypeptide chain) called connexin (molecular weight 25,000). At the gap junction the connexons of the coupled cells are aligned to form channels (Fig. 4-7, *A*). The channels allow the passage of water-soluble molecules up to molecular weights of 1200 to 1500 from once cell to the other. Such channels are the pathways for electrical current flow between the cells.

Cells that are electrically coupled may become uncoupled by closing of the connexon channels. The channels may close in response to increased intracellular Ca^{++} or H^{+} in one of the cells or in response to depolarization of one or both of the cells. A model for the mechanism of closing the channels is shown in Fig. 4-7, *B*.

Numerous electrical synapses have been described in the peripheral and central nervous systems of invertebrates and vertebrates. Electrical synapses appear to be particularly useful in reflex pathways in which rapid transmission between cells (little synaptic delay) is necessary or when the synchronous response of a number of neurons is required. Among the many nonneuronal cells that are coupled by gap junctions are hepatocytes, myocardial cells, intestinal smooth muscle cells, and the epithelial cells of the lens.

Chemical Synapses

When one neuron makes a chemical synapse with another, the presynaptic nerve terminal characteristically broadens to form a *terminal bouton*. At the synapse itself the presynaptic and postsynaptic membranes come more or less into close apposition and lie parallel to one another. Substantial structures stabilize the synaptic association, so that when nervous tissue is disrupted, the relationship of the presynaptic and postsynaptic membranes at the synapse often is preserved.

Electron micrographs of synapses in the central nervous system often show areas of high electron density subjacent to the plasma membranes in the region of synaptic contact. Synapses where this electron-dense region is of roughly similar extent and intensity on both presynaptic and postsynaptic sides of the synapse are called *symmetrical synapses* (Fig. 4-8, *A*). Synapses characterized by a postsynaptic electron density that is greater in extent and intensity than the presynaptic electron density are called *asymmetrical synapses* (Fig. 4-8, *B*). These classifications may indicate the two ends of a spectrum of synaptic structure rather than two distinct classes. The presynaptic nerve terminals at asymmetrical synapses often contain spherical synaptic vesicles, whereas symmetrical synapses are characterized by flattened or ellipsoidal vesicles. The asymmetrical synapses (round vesicles) may involve excitatory transmitters, whereas the symmetrical synapses (flattened vesicles) may involve inhibitory transmitters. This interpretation is, however, controversial.

Because of the structure and organization of chemical synapses, conduction is necessarily one way. *One-way conduction* of chemical synapses contributes to the organization of central nervous systems of vertebrates. A *synaptic delay* of about 0.5 msec is a property of transmission at chemical synapses. Synaptic delay is primarily caused by the time required for the release of transmitter. The time required for calcium channels to open in response to depolarization of the presynaptic terminal appears to be a major component of synaptic delay. In polysynaptic pathways synaptic delay accounts for a significant fraction of the total conduction time.

At chemical synapses the transmitter released by the presynaptic neurons alters the conductance of the postsynaptic plasma membrane to one or more ions. A change of the conductance of the postsynaptic membrane to an ion that is not in equilibrium across the membrane brings about a flow of that ion, which causes a change in the membrane potential of the postsynaptic cell. In most cases transmitters produce their effect by increasing the conductance of the postsynaptic membrane to one or more ions, but some invertebrate transmitters (and perhaps vertebrate transmitters as well) act by decreasing the postsynaptic conductances to specific ions.

The part of the membrane of the postsynaptic neuron that forms the synapse is specialized for *chemical sensitivity* rather than electrical sensitivity. Action potentials are not produced at the synapse. The change in membrane potential, be it depolarization or hyperpolarization, that occurs at the synapse is conducted electrotonically over the membrane of the postsynaptic neuron to the *axon hillock–initial segment* region (Fig. 4-9). The axon hillock–initial segment has a lower threshold than the rest of the plasma membrane of the postsynaptic cell, and an action potential will be generated here if the sum of all the inputs to the cell exceeds threshold. Once the action potential has been generated, it is conducted back over the surface of the soma of the postsynaptic cell and is propagated along its axon.

Input-Output Relations

The neuromuscular junction is representative of a particularly simple type of synapse in which one action potential in the presynaptic cell (the input) results in a single action potential in the postsynaptic cell (the output). In other types of synapses the output may differ from the input. Synapses can be classified as one-to-one, one-to-many, or many-to-one, based on the relationship between input and output.

In a *one-to-one synapse,* like the neuromuscular junc-

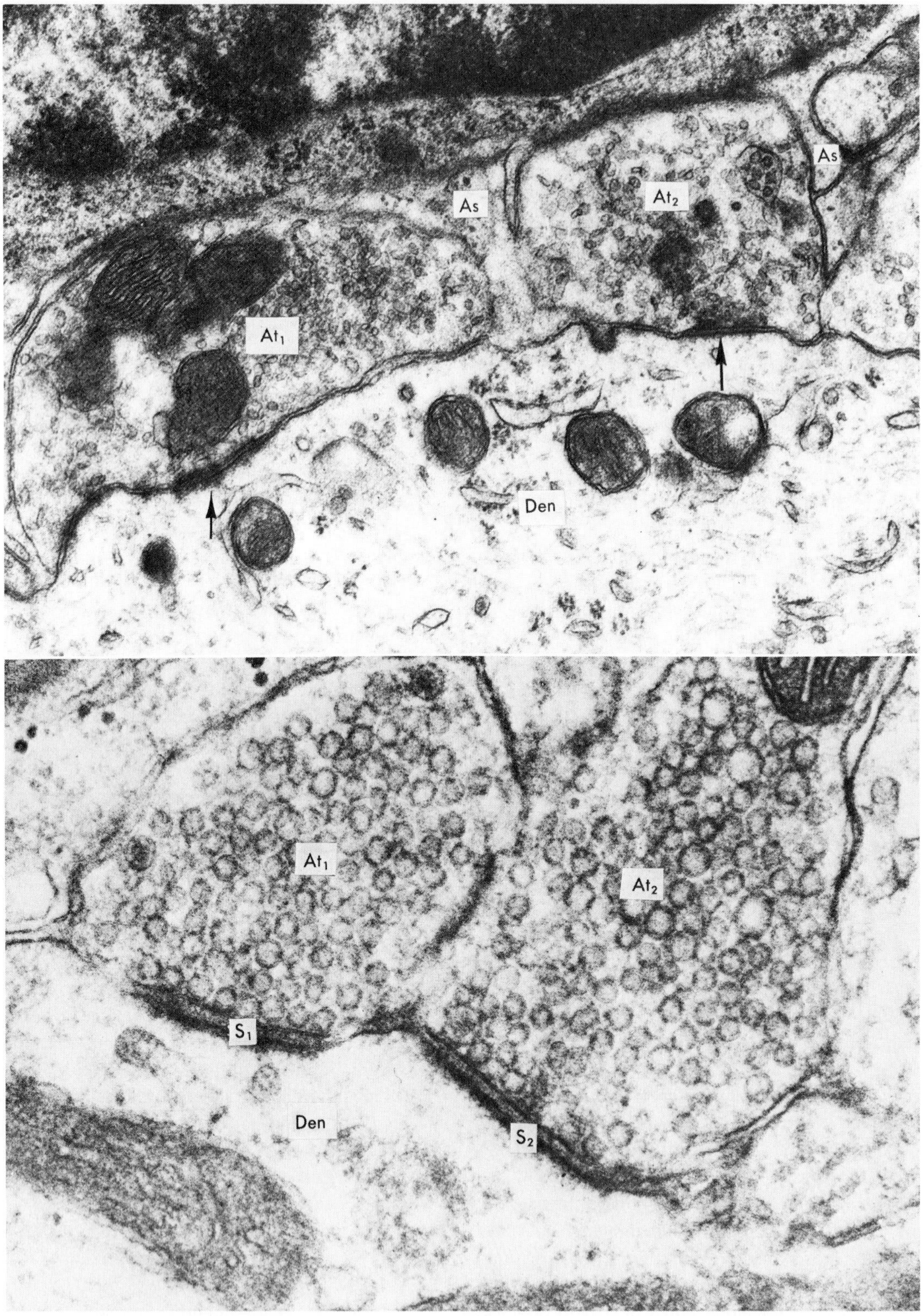

■ Fig. 4-8. **A,** Symmetrical synapses. Two axon terminals (At_1 and At_2) synapse with a large dendrite *(Den)* in the anterior horn of the spinal column. Note that the presynaptic and postsynaptic electron densities *(arrow)* are similar in extent. Astroglial cells *(As)* invest the axon terminals. **B,** Asymmetrical synapses (S_1 and S_2) in the cerebral cortex. Two axon terminals synapse with dendrite of a stellate cell. Note the greater extents of the electron densities on the postsynaptic (dendrite) side of the two synapses. (From Peters, A., Palay, S.L., and Webster, H. de F.: The fine structure of the nervous system, Philadelphia, 1976, W.B. Saunders Co.)

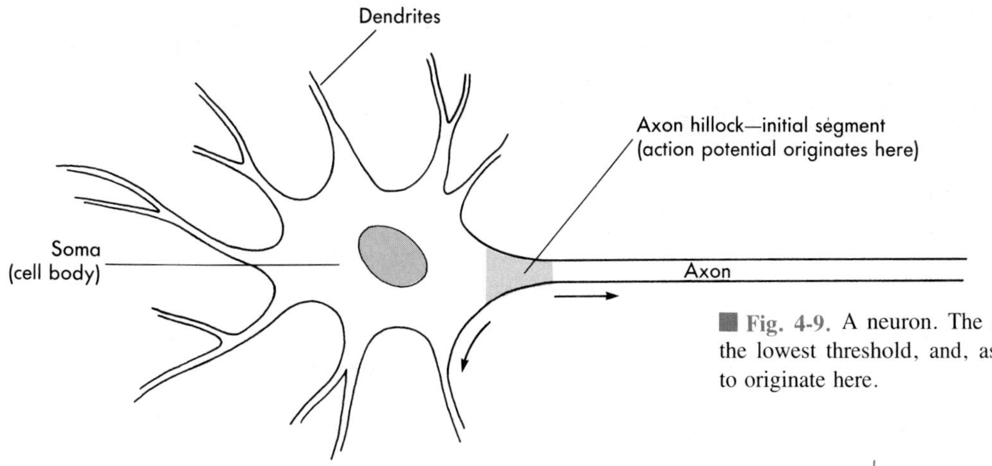

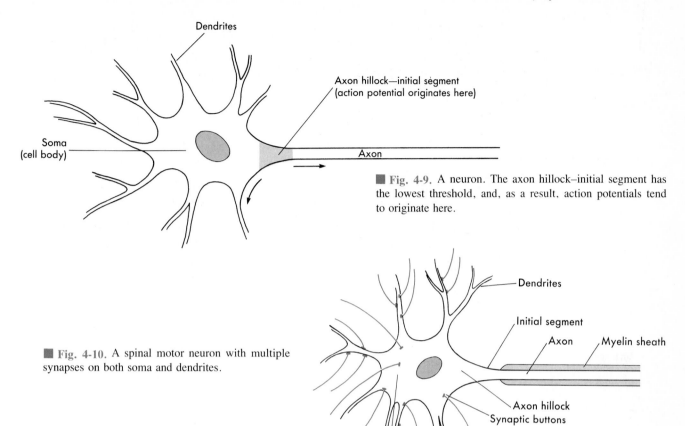

■ **Fig. 4-9.** A neuron. The axon hillock–initial segment has the lowest threshold, and, as a result, action potentials tend to originate here.

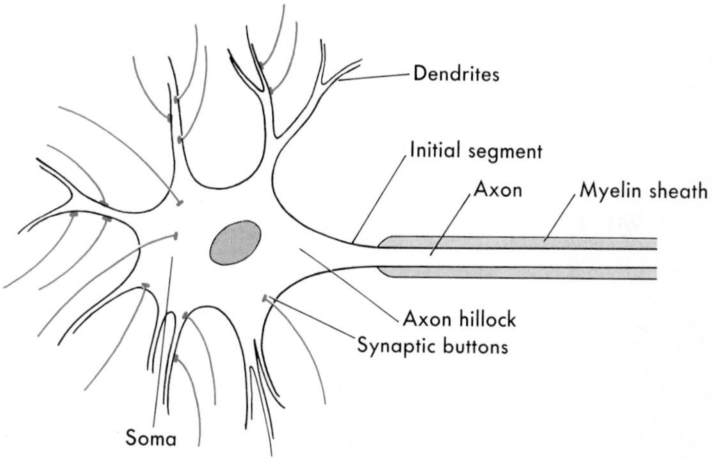

■ **Fig. 4-10.** A spinal motor neuron with multiple synapses on both soma and dendrites.

tion, the input and output are the same. A single action potential in the presynaptic cell evokes a single action potential in the postsynaptic cell. Because the output is the same as the input, no integration can occur at this type of snyapse.

In a *one-to-many synapse* a single action potential in the presynaptic cell elicits many action potentials in the postsynaptic cell. One-to-many synapses are not common, one example being the synapse of motoneurons on Renshaw cells in the spinal cord. One action potential in the motoneuron induces the Renshaw cell to fire a burst of action potentials.

In a *many-to-one synaptic arrangement* one action potential in the presynaptic cell is not enough to make the postsynaptic cell fire an action potential. The nearly simultaneous arrival of presynaptic action potentials in several input neurons that synapse on the postsynaptic cell is necessary to depolarize the postsynaptic cell to threshold. The spinal motoneuron has this type of synaptic organization. One hundred or more presynaptic axons synapse on each spinal motoneuron (Fig. 4-10). Some of these are excitatory inputs that *depolarize* the postsynaptic cell and bring it closer to its threshold. Other inputs are inhibitory and *hyperpolarize* the motoneuron, taking it farther away from threshold. The changes in postsynaptic potential caused by an action potential in a single input are about 1 to 2 mV. Thus no one excitatory input is capable of bringing the motoneuron to threshold. A transient depolarization of the postsynaptic neuron as the result of an action potential in the presynaptic cell is called an *excitatory postsynaptic potential (EPSP)* (Fig. 4-11). The transient hyperpolarization caused by an action potential in an inhibitory input is called an *inhibitory postsynaptic potential (IPSP)* (Fig. 4-11). At any instant the postsynaptic cell *integrates* the various inputs. If the momentary sum of the inputs depolarizes the postsynaptic cell to its threshold, it will fire an action potential. This is integration at the level of a single postsynaptic neuron.

■ *Summation of Synaptic Inputs*

The summation (or integration) of inputs can occur by either *spatial summation* or *temporal summation* (Fig. 4-12). *Spatial* summation occurs when two separate inputs arrive simultaneously. The two postsynaptic potentials are added so that two simultaneous excitatory inputs will depolarize the postsynaptic cell about twice as much as either input alone. One EPSP and one IPSP that occur simultaneously will tend to cancel one another. Even inputs that synapse at opposite ends of the postsynaptic body cell will act in this way. The postsynaptic potentials (EPSPs and IPSPs) are conducted

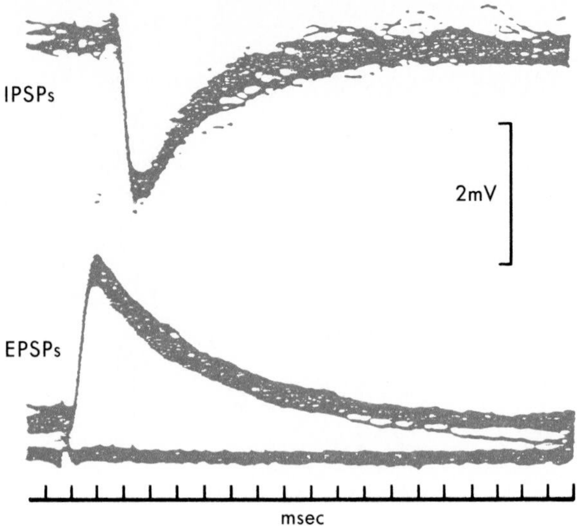

■ **Fig. 4-11.** Inhibitory postsynaptic potentials *(IPSPs)* and excitatory postsynaptic potentials *(EPSPs)* recorded with a microelectrode in a cat spinal motor neuron in response to stimulation of an appropriate peripheral afferent fiber. Forty traces are superimposed. (Redrawn from Curtis, D.R., and Eccles, J.C.: J. Physiol. **145:**529, 1959.)

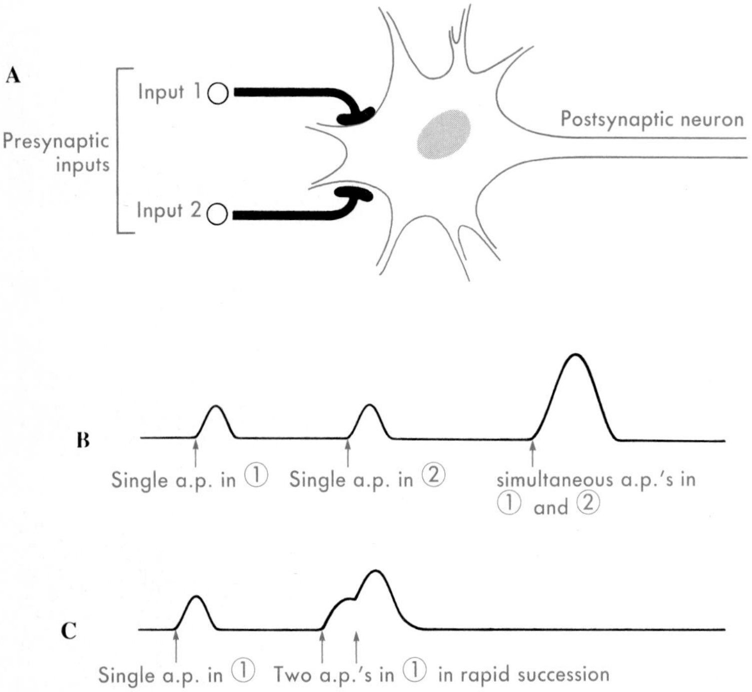

■ **Fig. 4-12. A,** Spatial and temporal summation at a postsynaptic neuron with two synaptic inputs *(1* and *2).* **B,** Spatial summation. The postsynaptic potential in response to single action potentials in inputs *1* and *2* occurring separately and simultaneously. **C,** Temporal summation. The postsynaptic response to two impulses in rapid succession in the same input.

rapidly over the entire cell membrane of the postsynaptic cell body with almost no decrement. This is because cellular dimensions (about 100 μm) are much smaller than the length constant (about 1 to 2 mm) for electrotonic conduction. Synaptic potentials that originate in fine dendritic branches decrease in magnitude as they are conducted to the cell body.

Temporal summation occurs when two or more action potentials in a single presynaptic neuron are fired in rapid succession, so that the resulting postsynaptic potentials overlap in time. A train of impulses in a single presynaptic neuron can cause the potential of the postsynaptic cell to change in a stepwise manner, each step caused by one of the presynaptic impulses.

Integration at the spinal motoneuron takes place because many positive and negative inputs impinge on a single motoneuron. This permits fine control of the firing pattern of the spinal motoneuron.

■ *Modulation of Synaptic Activity*

The responses of a postsynaptic neuron to individual stimulations of a particular presynaptic neuron are relatively constant in magnitude and time course. However, when a presynaptic cell is stimulated repeatedly at relatively high frequency, the postsynaptic response may depend on the frequency and duration of the presynaptic stimulation. When a presynaptic axon is stimulated repeatedly, the postsynaptic response may grow with each stimulation. This phenomenon is called *facilitation* (Fig. 4-13, *A*). As shown in Fig. 4-13, *B*, the extent of facilitation depends on the frequency of presynaptic impulses. Facilitation dies away rapidly, within tens to hundreds of milliseconds.

When a presynaptic neuron is stimulated tetanically (many stimuli at high frequency) for several seconds, a longer-lived enhancement of postsynaptic response oc-

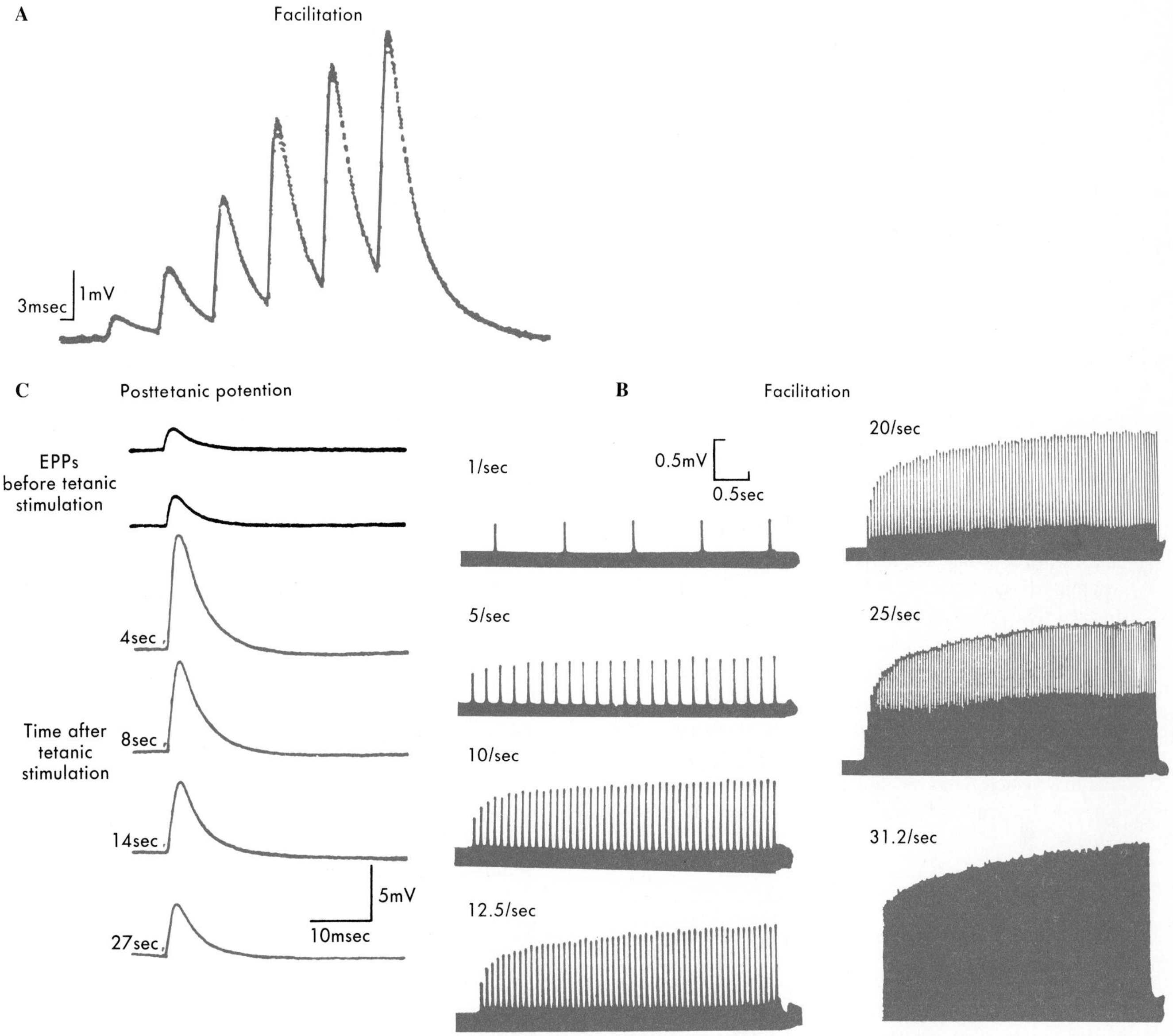

■ **Fig. 4-13. A,** Facilitation at a neuromuscular junction. EPPs at a neuromuscular junction in toad sartorius muscle were elicited by successive action potentials in the motor axon. Neuromuscular transmission is depressed by 5 mM Mg^{++} and 2.1 μM curare, so that action potentials do not occur. **B,** EPPs at a frog neuromuscular junction elicited by repetitively stimulating the motor axon at different frequencies. Note that facilitation fails to occur at the lowest frequency of stimulation (1 per second) and that the degree of facilitation increases with increasing frequency of stimulation in the range of frequency employed. Neuromuscular transmission was inhibited by bathing the preparation in 12 to 20 mM Mg^{++}. **C,** Posttetanic potentiation at a frog neuromuscular junction. The top two traces indicate control EPPs in response to single action potentials in the motor nerve. The subsequent traces indicate the responses to single action potentials following tetanic stimulation (50 impulses/sec for 20 seconds) of the motor nerve. The time interval between the end of tetanic stimulation and the single action potential is shown on each trace. The muscle was treated with tetrodotoxin to prevent generation of action potentials. (**A** redrawn from Belnave, R.J., and Gage, P.W.: J. Physiol. **266:**435, 1977; **B** redrawn from Magelby, K.L.: J. Physiol. **234:**327, 1973; **C** redrawn from Weinrich, D.: J. Physiol. **212:**431, 1971.)

curs called *posttetanic potentiation* (Fig. 4-13, *C*). Posttetanic potentiation persists much longer than facilitation; it lasts tens of seconds to several minutes after cessation of tetanic stimultion.

An enhancement of synaptic efficacy may occur that is intermediate in time course between facilitation and posttetanic potentiation. This enhancement is called *augmentation,* and it persists for about 10 seconds after repetitive stimulation is ended.

Facilitation, augmentation, and posttetanic potentiation are the result of the effects of repeated stimulation on the presynaptic neuron. These phenomena do not involve a change in the sensitivity of the postsynaptic cell to transmitter. With repeated stimulation, an increased number of quanta of transmitter is released. Increased levels of intracellular calcium may play a role in the enhancement of transmitter release with repetitive stimulation, but other intracellular events are involved as well. The details of the mechanisms that bring about facilitation, augmentation, and posttetanic potentiation are unknown.

When a synapse is repetitively stimulated for a long time, a point is reached at which each successive presynaptic stimulation elicits a smaller postsynaptic response. This phenomenon is called *synaptic fatigue* (neuromuscular depression at the motor endplate). The postsynaptic cell at a fatigued synapse responds normally to transmitter applied from a micropipette, so the defect appears to be presynaptic. In some cases a decrease in quantal content (the amount of transmitter per synaptic vesicle) has been implicated in synaptic fatigue. A fatigued synapse typically recovers in a few seconds.

Ionic Mechanisms of Postsynaptic Potentials in Spinal Motoneurons

Much of our current knowledge of synaptic mechanisms in the mammalian central nervous system is derived from studies of cat spinal motoneurons.

The EPSP. The EPSP (see Fig. 4-11) of the cat spinal motoneuron is caused by a transient increase of the conductance of the postsynaptic membrane to both Na^+ and K^+. One way in which this was demonstrated was by injecting Na^+ or K^+ into the cell to raise the intracellular concentration of that particular ion. Injection of either Na^+ or K^+ results in a smaller EPSP because, when the conductance increase occurs, there is a smaller tendency for Na^+ to flow in and depolarize the cell after Na^+ injection and a greater tendency for K^+ to flow out and oppose depolarization after K^+ injection. If the postsynaptic membrane is progressively depolarized, the EPSP progressively decreases in size because of a decreased tendency for Na^+ to enter and an increased tendency for K^+ to leave the cell. When E_m reaches about zero, the EPSP disappears, and if E_m is made positive, the EPSP changes direction. Thus zero is the *reversal potential* for EPSP. If Na^+ were the only ion flowing during the EPSP, the reversal potential for the EPSP should equal the equilibrium potential for Na^+ (about +65 mV). If K^+ were the only ion involved in the EPSP, the reversal potential should equal the K^+ equilibrium potential (about −100 mV). The reversal potential for the EPSP is a weighted average of E_{Na} and E_K. Injection of Cl^- into the cell does not alter the EPSP, so Cl^- apparently is not involved in the EPSP.

The IPSP. The IPSP (see Fig. 4-11) of cat spinal motoneurons is caused by an increased chloride conductance of the postjunctional membrane. The reversal potential for the IPSP is more negative than the normal resting membrane potential of the postsynaptic neuron (Fig. 4-14). The reversal potential is near the equilibrium potential for Cl^-. At rest there is a net tendency for Cl^- to enter the cell. The increase in chloride conductance, as the result of transmitter release at the inhibitory synapse, allows Cl^- to enter the postsynaptic cell and hyperpolarize it. Injecting Cl^- into the cell or hyperpolarizing it decreases the net tendency for Cl^- to enter the cell and decreases the size of the IPSP. Injection of Na^+ or K^+ produces no change in the IPSP, suggesting that neither Na^+ nor K^+ is involved in the IPSP of spinal motoneurons. In certain other cell types the IPSP appears to be caused by an increased K^+ conductance.

Presynaptic inhibition. Inhibitory interactions are vital in stabilizing the central nervous system. Another type of inhibition is called *presynaptic inhibition.* If an inhibitory input to a spinal motoneuron is stimulated tetanically and then an excitatory input is stimulated, the EPSP elicited by stimulating the excitatory input may be reduced in magnitude after the inhibitory volley. This is believed to occur by a mechanism in which axon collaterals of the inhibitory axons synapse on the excitatory nerve terminals (Fig. 4-15). An action potential in the inhibitory nerve produces a depolarization (rather long lived) of the excitatory nerve terminal. This brings the excitatory nerve terminal closer to threshold. What is important here is that the partly depolarized excitatory terminal *will release less transmitter in response to an action potential.* The smaller release of transmitter results in a smaller EPSP. The phenomenon of decreased transmitter release from a partially depolarized nerve terminal is well known at the neuromuscular junction.

The presumed anatomical arrangement that underlies presynaptic inhibition is diagramed in Fig. 4-15. The mechanisms described for presynaptic inhibition have been worked out in invertebrates. They are believed to apply in mammalian nervous systems, but they have not been as fully characterized in mammals.

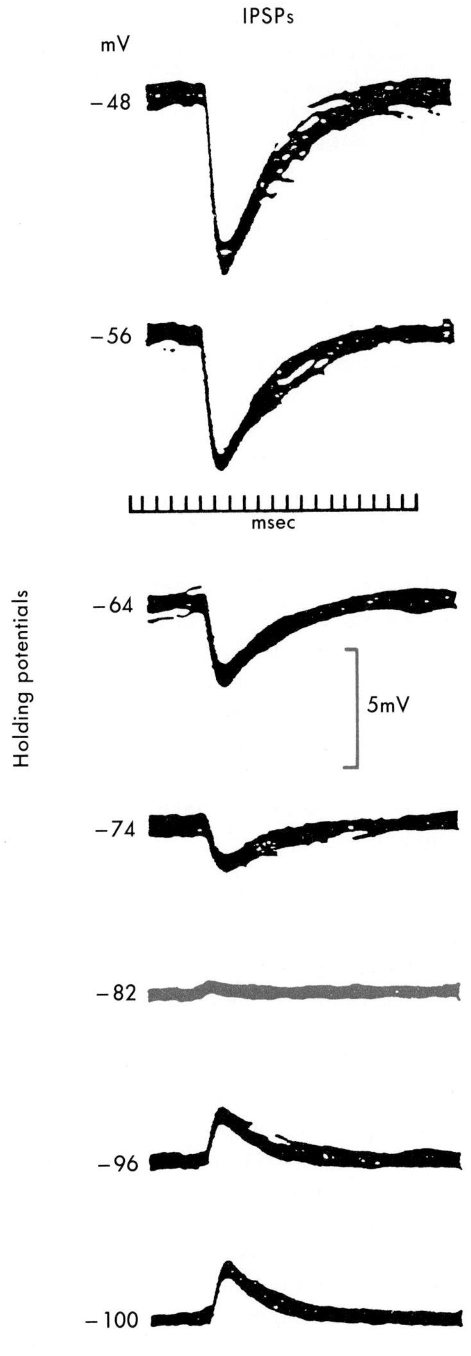

■ Fig. 4-14. A cat spinal motor neuron was voltage-clamped to the membrane potentials shown and then IPSPs were elicited by stimulating a peripheral afferent input to the spinal motor neuron. Each record was formed by superimposing 40 individual records. As the ''resting'' potential was made more negative, the IPSP first diminished in amplitude and, finally, at about −80 mV, reversed direction. The reversal potential (about −80 mV) of the IPSP is close to the value of the equilibrium potential for chloride. This supports the hypothesis that chloride currents are responsible for the IPSP in cat spinal motor neurons. (Redrawn from Coombs, J.S., et al.: J. Physiol. **130:**326, 1955.)

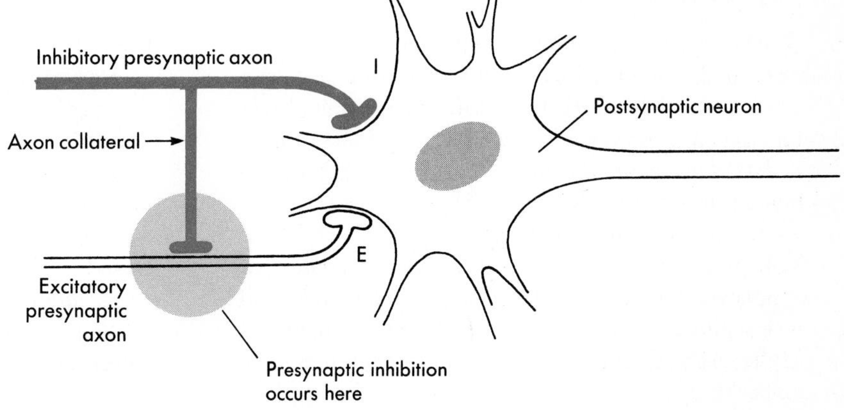

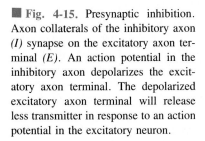

■ Fig. 4-15. Presynaptic inhibition. Axon collaterals of the inhibitory axon *(I)* synapse on the excitatory axon terminal *(E)*. An action potential in the inhibitory axon depolarizes the excitatory axon terminal. The depolarized excitatory axon terminal will release less transmitter in response to an action potential in the excitatory neuron.

Transmitters in the Nervous System

Identification of Transmitter Substances

A number of compounds have been proposed to function as neurotransmitters. Such compounds are called *candidate neurotransmitters*, or *putative neurotransmitters*. Candidate neurotransmitters usually are concentrated in specific neurons or in specific neuronal pathways. Microapplication of the putative transmitter to particular areas of the central nervous system may evoke specific responses. Correlation of information about the localization of the putative transmitter with knowledge of the location of neurons that respond to the candidate transmitter and the ways in which they respond allows intelligent speculation about the functions of a putative neurotransmitter substance.

It is often difficult to *prove* that a substance is the transmitter at a particular synapse. A putative transmitter (X) must satisfy the following criteria before it is accepted as a proven transmitter at a particular synapse:

1. The presynaptic neurons must contain X and must be able to synthesize it.
2. X must be released by the presynaptic neurons on appropriate stimulation.
3. Microapplication of X to the postsynaptic membrane must mimic the effects of stimulation of the presynaptic neuron.
4. The effects of presynaptic stimulation and of microapplication of X should be altered in the same way by pharmacological agents.

Our knowledge of neurotransmitters has increased greatly in recent years. Some transmitters have rapid and transient effects on the postsynaptic cell. Other transmitters have effects that are much slower in onset and may last for minutes or even hours. Most of the candidate neurotransmitters that have been discovered so far fall into three major chemical classes: amines, amino acids, and oligopeptides.

Transmitters and Putative Transmitters in the Central Nervous System

Acetylcholine. As discussed previously, acetylcholine is the transmitter used by all motor axons that arise from the spinal cord. Acetylcholine plays a central role in the autonomic nervous system, being the transmitter for all preganglionic neurons and also for postganglionic parasympathetic fibers. The Betz cells of the motor cortex use acetylcholine as their transmitter. The basal ganglia, which are involved in the control of movement, contain high levels of acetylcholine, and acetylcholine is believed to be a transmitter in the basal ganglia. Acetylcholine may be the transmitter in a large number of central pathways.

Biogenic Amine Transmitters

Among the amines that may serve as neurotransmitters are norepinephrine, epinephrine, dopamine, serotonin, and histamine.

Dopamine, norepinephrine, and *epinephrine* are catecholamines and share a common biosynthetic pathway that starts with the amino acid tyrosine. Tyrosine is converted to L-dopa by tyrosine hydroxylase. L-Dopa is converted to dopamine by a specific decarboxylase. In dopaminergic neurons the pathway stops here. Noradrenergic neurons have another enzyme, dopamine β-hydroxylase, that converts dopamine to norepinephrine. Other cells add a methyl group to norepinephrine to produce epinephrine. S-Adenosylmethionine is the methyl donor, and the reaction is catalyzed by phenylethanolamine-*N*-methyltransferase.

As discussed earlier, norepinephrine is the primary transmitter for postganglionic sympathetic neurons. In the brain norepinephrine-containing cell bodies are found in the locus ceruleus. The neurons of the locus ceruleus project to the cortex, hypothalamus, cerebellum, and spinal cord. The noradrenergic neurons in the brain may be involved in arousal, in regulation of mood, and in dreaming.

Neurons that contain high levels of dopamine are prominent in the midbrain regions known as the substantia nigra and the ventral tegmentum. Some of the axons of these neurons travel to the forebrain, where they may play a role in emotional responses. Other dopaminergic axons terminate in the corpus striatum, where they are believed to play an important role in control of complex movements. The degeneration of these dopaminergic synapses in the corpus striatum occurs in Parkinson's disease and is believed to be a major cause of the muscular tremors and rigidity that characterize this disease.

Some evidence suggests that hypersecretion of dopamine by neurons in the limbic system and/or increased sensitivity to dopamine may be involved in schizophrenia. Antipsychotic drugs such as chlorpromazine and related compounds are dopamine-receptor antagonists and thus tend to diminish the effects of endogenous dopamine.

Serotonin (5-hydroxytryptamine)-containing neurons are present in high concentration in the raphe nuclei located in the midline of the brainstem. These nuclei project widely to brain and spinal cord. Serotonergic neurons may be involved in regulating temperature, sensory perception, onset of sleep, and control of mood.

Histamine is present in certain neurons in the hypothalamus. The functions of these presumably histaminergic neurons are not yet known.

Amino Acid Transmitters

Glycine, the simplest amino acid, is an inhibitory neurotransmitter released by certain spinal interneurons.

Glutamate and aspartate, dicarboxylic amino acids, have strong excitatory effects on many neurons in the brain. It may be that glutamate and aspartate are the most prevalent excitatory transmitters in the brain.

γ-Aminobutyric acid (GABA) is not incorporated into proteins, nor is it present in all cells (as are the other naturally occurring amino acids). GABA is produced from glutamate by a specific decarboxylase present only in the central nervous system. Among the cells that contain GABA are some cells in the basal ganglia, the cerebellar Purkinje cells, and certain spinal interneurons. In all known cases GABA functions as an inhibitory transmitter. It is the most common transmitter in the brain, and it may be that as many as a third of the synapses in the brain have GABA as their neurotransmitter.

GABA is believed to be important in many different central control pathways. A deficit in the level of GABA and other neurotransmitters in the corpus striatum occurs in patients with Huntington's chorea. The uncontrolled movements that characterize this disease may be partly due to diminished effectiveness of GABA-mediated inhibition in central motor pathways. GABA may also be involved in control of mood and emotion, since some evidence suggests that benzodiazepines (such as diazepam), which are effective antianxiety drugs, may act to enhance the effectivenss of GABA as a transmitter.

Neuroactive peptides

Gut-brain peptides

- Vasoactive intestinal polypeptide (VIP)
- Cholecystokinin octapeptide (CCK-8)
- Substance P
- Neurotensin
- Methionine enkephalin
- Leucine enkephalin
- Motilin
- Insulin
- Glucagon

Hypothalamic-releasing hormones

- Thyrotropin-releasing hormone (TRH)
- Luteinizing hormone-releasing hormone (LHRH)
- Somatostatin (growth hormone releasing-inhibiting factor, or SRIF)

Pituitary peptides

- Adrenocorticotropin (ACTH)
- β-Endorphin
- α-Melanocyte-stimulating hormone (α-MSH)

Others

- Dynorphin
- Angiotensin II
- Bradykinin
- Vasopressin
- Oxytocin
- Carnosine
- Bombesin

Modified from Snyder, S.H.: Science **209**:976, 1980. Copyright 1980 by American Association for the Advancement of Science.

Neuroactive Peptides

Relatively recently it has been found that certain cells release peptides that act at very low concentrations to excite or inhibit neurons. To date, about 25 of these so-called neuropeptides, ranging from 2 amino acids to about 40 amino acids long, have been identified. Most of the neuropeptides that have been discovered so far are listed above. It is likely that more neuropeptides will be added to this list in the near future.

Neuropeptides typically affect their target neurons at lower concentrations than the classical neurotransmitters discussed previously, and the action of neuropeptides usually lasts longer. A number of the neuropeptides listed in the box are more familiar as hormones. Neuropeptides may act as hormones, as neurotransmitters, or as neuromodulators. A hormone is a substance that is released into the blood and that reaches its target cell via the circulation. A neurotransmitter or neuromodulator is released near the surface of its target cell and simply diffuses to the target cell. Neurotransmitters, as discussed earlier, act to change the conductance of the target cell to one or more ions, and in that way they change the membrane potential of the target cell. A neuromodulator modulates synaptic transmission. The neuromodulator may act presynaptically to change the amount of transmitter released in response to an action potential or it may act on the postsynaptic cell to modify its response to transmitter. A number of neuropeptides are suspected to act as true transmitters at particular synapses, but no neuropeptide has yet satisfied all four of the criteria required of an established neurotransmitter. Whether a neuropeptide acts as a true synaptic transmitter or as a synaptic modulator is often difficult to determine.

In a number of instances, some of which are listed in Table 4-1, neuropeptides coexist in the same nerve terminals with classical transmitters. In some of these cases, the neuropeptide is released along with the transmitter in response to nerve stimulation. Whether the transmitter and the neuropeptide co-exist in the same

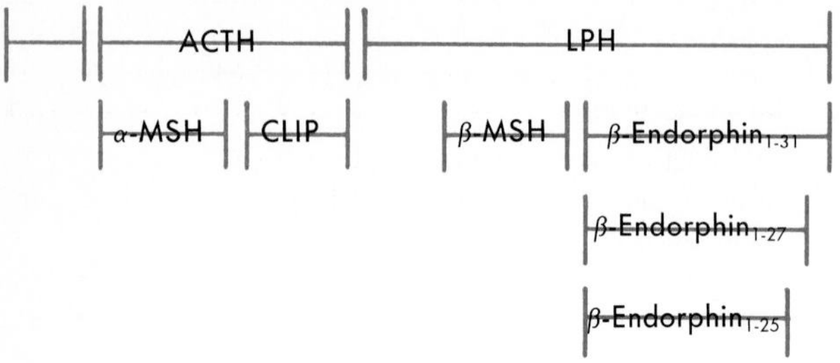

Fig. 4-16. Endorphins are secreted in an inactive prohormone form that contains other important peptides. Specific proteases are presumed to release the active peptides. ACTH: adrenocorticotrophic hormone, LPH: β-lipotropin, MSH: melanocyte-stimulating hormone, CLIP, corticotrophin-like intermediate lobe peptide. (Redrawn from Smyth, D.G.: Br. Med. Bull. **39:**25-30, 1983. By permission of Churchill Livingstone, Edinburgh.)

Table 4-1. Examples of the coexistence within the same nerve terminal of a classical transmitter and a neuropeptide*

Transmitter	*Peptide*
Acetylcholine	Vasoactive intestinal peptide (VIP)
Norepinephrine	Somatostatin
	Enkephalin
	Neurotensin
Dopamine	Cholecystokinin (CCK)
	Enkephalin
Adrenalin	Enkephalin
Serotonin	Substance P
	Thyrotropic-releasing hormone (TRH)

Reprinted by permission of the publisher from *Chemical Messengers: small molecules and peptides* by Schwartz, J.H. in Kandel, E.R., and Schwartz, J.H., editors: Principles of neural science. Copyright © 1981 by Elsevier Science Publishing Co., Inc.

*Evidence for the coexistence of a classical transmitter substance with a neuroactive peptide has been reported for these combinations. With the information thus far available, it is not yet possible to determine the specificity of the pairs and their physiological significance.

synaptic vesicles, or are packaged in separate vesicles, is not yet clear.

Neuropeptides are very important physiologically. They are being intensively investigated, but our knowledge of the function of neuropeptides is still fragmentary. This section briefly summarizes our current knowledge of neuropeptides; additional information is presented in Chapter 52.

Synthesis of neuropeptides. Most classical neurotransmitters are synthesized in nerve terminals by pathways that involve soluble enzymes and simple precursors. Neuropeptides are synthesized in the cell body. They are encoded by the cell's DNA and transcribed into messenger RNA, which is translated on polyribosomes bound to the endoplasmic reticulum. As is the case for other secretory proteins, the nascent peptide of neuropeptides begins with a hydrophobic signal sequence of 18 to 25 amino acids. The signal sequence leads the nascent chain into the lumen of the endoplasmic reticulum, so that the completely translated polypeptide chain ends up in a cistern of the endoplasmic reticulum. The signal sequence is cleaved off and smooth vesicles containing the polypeptide bud off the endoplasmic reticulum. These vesicles fuse with the forming face of the Golgi complex and are moved toward the mature face of the Golgi complex. Secretory vesicles, containing the neuropeptides, emerge from the mature face of the Golgi complex. Secretory vesicles are moved by fast axonal transport to the axon terminals, where they are known as synaptic vesicles.

A number of neuropeptides are synthesized as preprohormones. Cleavage of the signal sequence converts the preprohormone to a prohormone. Proteolytic cleavage of the prohormone may then release one or more active peptides. In some cases one prohormone may contain several active peptide sequences. As shown in Fig. 4-16, the prohormone of the opioid peptide, β-endorphin, is a 31,000-dalton polypeptide that contains a number of active sequences. One cleavage of the prohormone releases adrenocorticotrophic hormone (ACTH) and β-lipotropin. Cleavage of ACTH releases a hormone, melanocyte-stimulating hormone (α-MSH). Cleavages of β-lipotropin release β-MSH and a number of active β-endorphins.

Opioid peptides. *Opiates* are drugs that are derived from the juice of the opium poppy. Opiates are useful therapeutically as powerful analgesics. They exert their analgesic effect by binding to specific opiate receptors. The binding of opiates to their receptors is stereospecifically inhibited by a morphine derivative called naloxone. Compounds that do not derive from the opium poppy but that exert direct effects by binding to opiate receptors are called *opioids*. Operationally, opioids are defined as directly acting compounds whose effects are stereospecifically antagonized by naloxone. Table 4-2 lists some of the major effects of opioid compounds and the locations of the opiate receptors believed to mediate these effects.

The three major classes of endogenous opioid peptides in mammals are enkephalins, endorphins, and dynorphin. Enkephalins are the simplest opioids; they are pentapeptides. Met-enkephalin is Tyr-Gly-Gly-Phe-Met. Leu-enkephalin (Tyr-Gly-Gly-Phe-Leu) has leucine in place of methionine. Dynorphin and the endorphins are somewhat longer peptides that share one or the other of the enkephalin sequences at their *N*-terminal ends.

Opioid peptides are widely distributed in neurons of the central nervous system and intrinsic neurons of the gastrointestinal tract. Opioid peptides are found in vesicles that resemble synaptic vesicles. The endorphins are discretely localized in particular structures of the central nervous system, while the enkephalins and dynorphins are more widely distributed. Enkephalins are

Table 4-2. Some of the effects that have been attributed to opioid peptides and the locations of the receptors proposed to mediate the effects.

Opioid effect	*Location of opioid receptors*
Analgesia	
Spinal (body)	Laminae I and II of dorsal horn
Trigeminal (face)	Substantia gelatinosa of trigeminal nerve
Supraspinal	Periaqueductal gray matter, medial thalamic nuclei, intralaminar thalamic nuclei, (?) striatum
Autonomic reflexes	
Suppression of cough, orthostatic hypotension, inhibition of gastric secretion	Nuclei tractus solitarius, commisuralis, ambiguous, and locus ceruleus
Respiratory depression	Nucleus tractus solitarius, parabrachial nuclei
Nausea and vomiting	Area postrema
Meiosis	Superior colliculus, pretectal nuclei
Endocrine effects	
Inhibition of vasopressin secretion	Posterior pituitary
Hormonal effects	Hypothalamic infundibulum, hypothalamic nuclei, accessory optic system, (?) amygdala
Behavioral and mood effects	Amygdala, nucleus stria terminalis, hippocampus, cortex, medial thalamic, nuclei, nucleus accumbens, (?) basal ganglia
Motor rigidity	Striatum

From Atweh, S.F., and Kuhar, M.F.: Br. Med. Bull. **39**:47, 1983.

present at high levels in corpus striatum (particularly globus pallidus). Enkephalin-containing nerve endings are located in the trigeminal complex and in the dorsal horn of the spinal column. Enkephalin-containing and substance P–containing synapses overlap in the spinal cord.

Calcium-dependent, depolarization-induced release of enkephalins from brain slices has been demonstrated, which suggests that enkephalins serve as neurotransmitters in the brain. When opioids are directly applied to particular groups of neurons, the effects may resemble those of transmitters (altered membrane conductance to particular ions) or those of peptide hormones (changes in intracellular levels of cyclic nucleotides or calcium). Opioids hyperpolarize certain neurons of sympathetic ganglia and locus ceruleus by increasing the K^+ conductance of the plasma membrane. Opioids may be inhibitory transmitters in these cells. In some cases opioids appear to act presynaptically to diminish the release of synaptic transmitter (perhaps by decreasing calcium influx in response to stimulation). Opioids may act presynaptically to decrease the release of acetylcholine by preganglionic nerve endings in sympathetic ganglia. Substance P is the suspected transmitter at synapses made by primary sensory neurons (their cell bodies are in the dorsal root ganglia) with spinal interneurons in the dorsal horn of the spinal column. Enkephalins act to decrease the release of substance P at these synapses, thereby inhibiting the pathway for pain sensation at the first synapse in the pathway (Fig. 4-17).

Opioids appear to have inhibitory effects in structures in the brain involved in the perception of pain. They also increase the firing rate of certain neurons, for example,

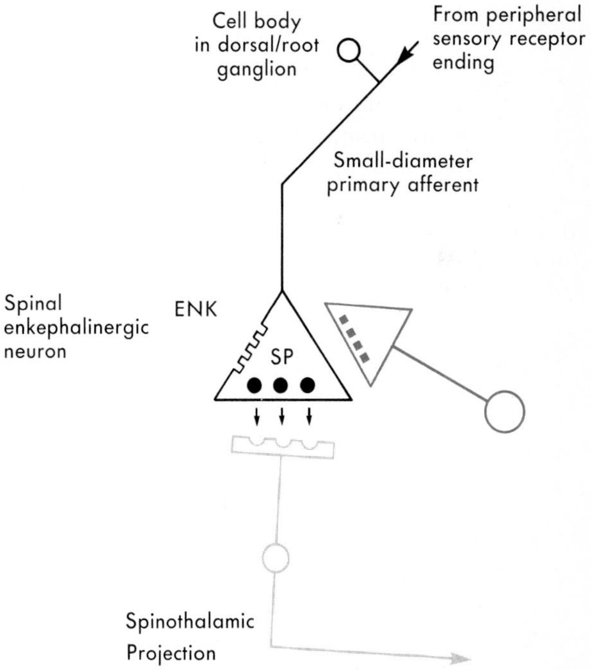

Fig. 4-17. A schematic model of the hypothesis that enkephalinergic synapses in the dorsal column of the spinal column inhibit the release of substance P *(SP),* the putative transmitter at the first synapse made by primary sensory neurons in the spinal column. *ENK,* Enkephalins. (Redrawn from Iversen, L.L.: Br. Med. Bull. **38:**277, 1982. By permission of Churchill Livingstone, Edinburgh.)

Renshaw cells. The output of both Renshaw cells and cerebellar pyramidal cells inhibits motor activity.

In certain cell types intracellular levels of cyclic AMP promptly decrease in response to opioids. In such cases the opioid receptor may be linked in an inhibitory fashion to the adenylate cyclase of the plasma membrane. In other cases responses that are triggered by increases in intracellular calcium are inhibited by opioids. In these instances opioids are believed to block calcium influx or calcium release from internal stores.

Opioids also affect certain neurons that secrete hormones. In the anterior pituitary, opioids have both direct and indirect effects. They stimulate anterior pituitary cells that produce ACTH, prolactin, and growth hormone, but inhibit the cells that produce thyroid-stimulating hormone (TSH). Opioids suppress the release of follicle-stimulating hormone (FSH) and luteinizing hormone (LH) by the anterior pituitary by suppressing the secretion of LHRH (luteinizing hormone–releasing hormone) from the hypothalamus. Opioids inhibit release of vasopressin (antidiuretic hormone) from the posterior pituitary. The physiological roles of opioids in modulating the release of pituitary hormones remain to be elucidated.

Endogenous opioids may be involved in a variety of physiological control mechanisms. They probably play a role in the response to stress and the perception of pain. The inhibitory effect of opioids on primary sensory neurons was mentioned earlier. They also diminish the perception of painful stimuli, probably by inhibiting other neurons in the spinal column and in the brain that are involved in processing signals from pain receptors. Injury stress causes the parallel release of β-endorphins and ACTH from the anterior pituitary, as well as secretion of enkephalins and of epinephrine by the adrenal medulla. The functions of β-endorphins and enkephalins in modulating the physiological responses to stress remain to be determined.

Nonopioid neuropeptides. As shown in the list of neuroactive peptides on p. 60, most of the known neuropeptides are not opioids. This section briefly summarizes current knowledge of nonopioid neuropeptides.

Substance P. *Substance P,* a peptide of 11 amino acids, is present in specific neurons in the brain, in primary sensory neurons, and in plexus neurons in the wall of the gastrointestinal tract. Substance P is a member of a group of neuropeptides called tachykinins (most of the other known tachykinins are present in amphibians). Substance P was the first so-called gut-brain peptide to be discovered. The wall of the gastrointestinal tract is richly innervated with neurons that form networks or plexuses. The intrinsic plexuses of the gastrointestinal tract exert primary control over its motor and secretory activities (Chapters 43 to 45). These enteric neurons contain many of the neuropeptides, including substance P, that are found in the brain and spinal column.

The postulated role of substance P as the neurotransmitter at the synapses between primary sensory neurons and dorsal horn neurons was discussed earlier (see Fig. 4-17). Substance P may also be a transmitter in the brain. Calcium-dependent release of substance P in response to depolarization has been found in brain slices or in synaptosomes from hypothalamus, substantia nigra, and other regions of the brain. Substance P, applied to certain neurons in substantia nigra, causes these neurons to release dopamine at their terminals in corpus striatum.

Vasoactive intestinal peptide (VIP). VIP is a member of a family of neuropeptides related to secretin, which was first discovered as a gastrointestinal hormone but is now known to be a neuropeptide as well. Secretin (27 amino acids) and glucagon (29 amino acids) have 14 amino acids in common at similar positions. VIP (28 amino acids) and gastric inhibitory peptide (GIP, 43 amino acids) have extensive sequence homology with secretin and glucagon. These four neuropeptides probably have a common ancestor peptide and arose in evolution by gene duplication.

VIP is widely distributed in the central nervous system and in the intrinsic neurons of the gastrointestinal tract. In neurons in the brain it has been localized in synaptic vesicles. It is found in particularly high concentrations in cerebral cortex, hippocampus, amygdala, and hypothalamus. VIP appears to function as an inhibitory "transmitter" to vascular and nonvascular smooth muscle and as an excitatory "transmitter" to glandular epithelial cells. VIP and acetylcholine coexist in parasympathetic nerve terminals that innervate pancreatic acinar cells. VIP and acetylcholine potentiate one another in their effects on pancreatic acinar cells. When applied locally, VIP excites a wide variety of cells in the central nervous system. VIP is especially potent when applied to hippocampal pyramidal cells, being excitatory at 0.1 femtomole/sec (femto $= 10^{15}$). Small amounts of VIP excite or inhibit neurons in preoptic, septal, and midbrain central gray matter. VIP is present at high levels in the anterior hypothalamus and in the median eminence. VIP is found in hypophyseal portal blood, and it may modulate pituitary function. VIP may enhance prolactin secretion by the anterior pituitary.

Secretin, glucagon, and GIP are molecules whose function as hormones has been well characterized. These peptides have also been found in particular neurons in the central nervous system, but their functions in the CNS remain to be determined.

Cholecystokinin (CCK). Cholecystokinin is a member of a group of neuropeptides that includes gastrin and cerulein, which have similar C-terminal sequences. CCK is a well-known gastrointestinal hormone that elicits contraction of the gallbladder (Chapter 43). CCK(39) is cleaved near its N-terminus to produce CCK(33), the physiological form in the gastrointestinal tract. The N-terminal octapeptide, CCK(8), is present in particular neurons of the central nervous system.

CCK(8) has been found in synaptic vesicles in the brain and is present in high levels in amygdala, cortex, frontal cortex, striatum, and hippocampus. CCK(8) probably functions as a true synaptic transmitter in some of these regions. CCK and dopamine apparently coexist in certain nerve terminals in the mesencephalon. CCK(8) is excitatory when applied directly, at femtomolar levels, to pyramidal cells of the hippocampus and to certain neurons in the spinal column. CCK may be involved in the neural pathways that mediate satiety, because systemic injection of the peptide causes satiety. Application of CCK to certain areas of the brain lessens the perception of injurious stimuli.

Neurotensin. Neurotensin is one of the most recently discovered gut-brain neuropeptides. It is present in enteric neurons and in the brain, particularly in median eminence, substantia nigra, periaqueductal gray matter, locus ceruleus, raphe nuclei, and superior and inferior colliculi. When injected into cerebrospinal fluid at low concentrations, neurotensin elicits hypothermia. Thus neurotensin may be involved in the regulation of body temperature. Systemic injections of neurotensin cause profound hyperglycemia and elevated glucagon levels in blood. Neurotensin may also be involved in the control of pituitary function. Injection of neurotensin in cerebrospinal fluid increases somatostatin secretion (and lowered growth hormone secretion as a result) and inhibits prolactin and LH secretion from the anterior pituitary. Neurotensin, when applied locally, inhibits cerebellar purkinje cells.

Effects of other neuropeptides. Neuropeptides affect many specific neurons of the central nervous system. ACTH depolarizes hippocampal pyramidal cells. Angiotensin II increases the firing rate of supraoptic neurosecretory cells. Angiogensin II applied near the subfornical organ produces drinking behavior. Bombesin excites Al pyramidal neurons. Bradykinin excites terminals of cutaneous and visceral afferent sensory fibers. Motilin excites most corticospinal neurons. Oxytocin excites hippocampal interneurons. LHRH and angiotensin II depolarize neurons in frog sympathetic ganglia by closing K^+ channels.

Neuropeptides are being investigated intensively. Information about their localization, effects, and mechanisms of action is accumulating rapidly.

Other Neuromodulators

There are important neuromodulators that are not peptides. Purines and purine nucleotides and nucleosides function as neuromodulators in the central, autonomic, and peripheral nervous systems. Substances that serve as classical neurotransmitters may also act as neuromodulators. In some cases the transmitter binds to receptors on the presynaptic neuron that released it, with the result that the transmitter acts as a modulator to regulate its own release. The study of neuromodulators is at an early stage of development. Other neuromodulators remain to be discovered and their mechanisms of action elucidated.

Bibliography

Aidley, D.J.: The physiology of excitable cells, ed. 2, Cambridge, 1978, Cambridge University Press.

Bloom, F.E.: Neuropeptides, Sci. Am. **245**:148, 1981.

Cooper, J.R., Bloom, F.E., and Roth, R.H.: The biochemical basis of neuropharmacology, ed. 4, New York, 1982, Oxford University Press, Inc.

Davson, H.: A textbook of general physiology, ed. 4, Baltimore, 1970, Williams & Wilkins.

Eccles, J.C.: The physiology of synapses, Berlin, 1964, Springer Verlag.

Gregory, R.A., editor: Regulatory peptides of gut and brain, Br. Med. Bull. **38**:219, 1982.

Hughes, J., editor: Opioid peptides, Br. Med. Bull. **39**:1, 1983

Iversen, L.L.: The chemistry of the brain, Sci. Am. **241**:134, 1979.

Kandel, E.R., and Schwartz, J.H.: Principles of neural science, ed. 2, New York, 1985, McGraw-Hill Book Co.

Katz, B.: Nerve, muscle, and synapse, New York, 1966, McGraw-Hill Book Co.

Kuffler, S.W., Nicholls, J.G., and Martin, A.R.: From neuron to brain, ed. 2, Sunderland, Mass., 1984, Sinauer Associates, Inc.

Llinas, R.K.: Calcium in synaptic transmission, Sci. Am. **247**:56, 1982.

SECTION

II

THE NERVOUS SYSTEM

David H. Cohen

S. Murray Sherman

CHAPTER

5

The Nervous System and Its Components

The contemporary era of the neurosciences began in the mid-1950s with the introduction of neurophysiological techniques for studying the dynamics of neural activity. Concomitantly important developments in neuroanatomical methods helped specify the detailed connectivity of neural pathways. Together these advances initiated a rigorous cellular approach to understanding brain function. In the succeeding three decades gains in knowledge of the nervous system have been explosive. This exciting field is generating a true cell biology of the nervous system and its ultimate output—behavior.

When the nervous system is discussed in a physiology textbook, selection of material is a critical factor. Space constraints prevent comprehensive coverage, and to preserve some historical continuity it is thus necessary to bias the presentation toward the inclusion of traditional material. Selectivity is particularly difficult in a field that is advancing as rapidly as neuroscience.

Although the subject matter is indeed traditional, we have attempted to treat it in a contemporary manner and present the nervous system with a functional neuroanatomical orientation. It is assumed that the student has been exposed to fundamental neuroanatomy as background for this text.

Throughout we have stressed the analysis of neural function. In particular, we have emphasized the importance of analyzing the brain from the sensory or motor peripheries centrally, allowing one to relate central activity to either sensory stimulus patterns or motor behavior. Such an approach relies first on establishing the relevant neural circuitry. This, in turn, provides the substrate for applying contemporary analytical techniques to study at a cellular level how the brain organizes its sensory inputs, motor outputs, and their coupling. Through this approach remarkable achievements have been realized in describing how the brain processes sensory information to establish a basis for perception and how goal-directed motor behavior is organized. Understanding of sensorimotor integration and more complex functions has not advanced as rapidly. In fact, the level of understanding of any given structure and its function varies inversely with its proximity to the periphery.

Glial Cells and Neurons

The cellular elements of the nervous system are *glial cells* and *neurons*. Although attention will be focused almost exclusively on neurons, glial cells actually outnumber neurons by an order of magnitude. Roughly speaking, there are 10^{13} glial cells versus 10^{12} neurons. In this section, we shall briefly review some salient features of these cellular elements; details of cellular biology are found elsewhere in this book (Section I).

Glial Cells

Glial cells neither conduct action potentials nor form functional synapses with other cells, although they can be passively polarized in response to nearby neural activity. Their functions are complex and not completely understood. Glial cells generally provide support for neurons. They form a mechanical matrix in which neurons are embedded; they probably play metabolic and nutritive roles; they may help to regulate blood flow through the brain; they may act as a sink or source of ions; they can electrically insulate axons and synapses from one another; and they phagocytose neural debris in response to damage.

Five broad types of glial cells have been described: *astrocytes (astroglia), oligodendrocytes (oligodendrog-*

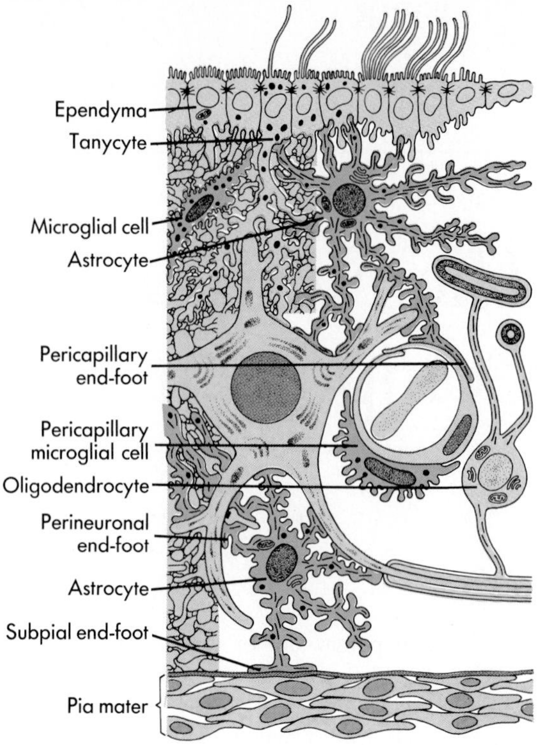

Fig. 5-1. Schematic representation of nonneural elements in the central nervous system. Two astrocytes *(darker color)* are shown ending on a neuron's soma and dendrites. They also contact the pial surface and/or capillaries. An oligodendrocyte *(lighter color)* provides the myelin sheaths for axons. Also shown are microglia *(darker color)* and ependymal cells *(lighter color)*. (Redrawn from Williams, P.L., and Warwick, R.: Functional neuroanatomy of man, Edinburgh, 1975, Churchill Livingstone.)

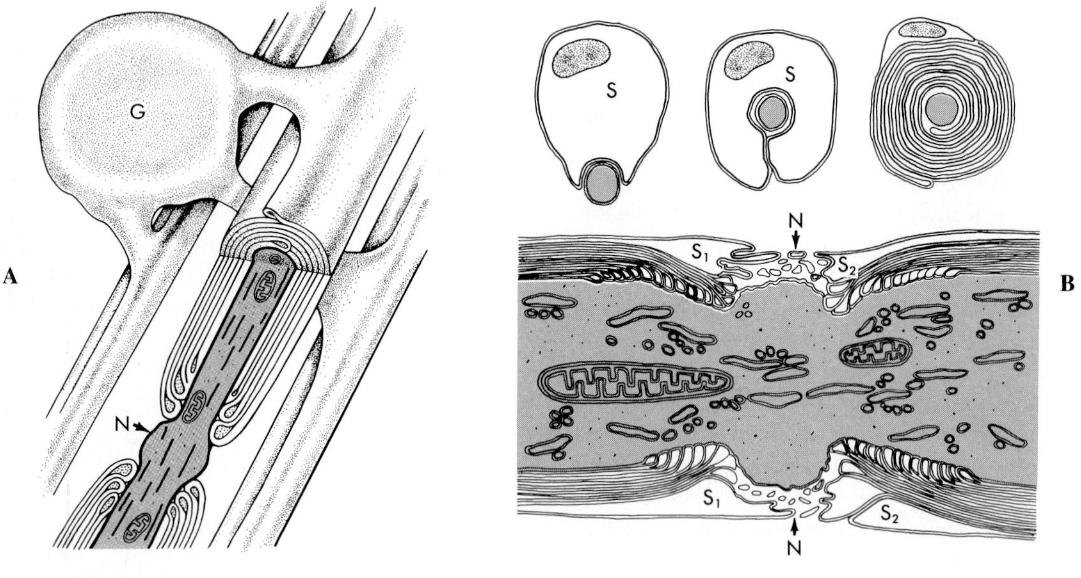

Fig. 5-2. Myelin sheaths of axons. **A,** Myelinated axons in the central nervous system. A single oligodendrocyte *(G)* emits several processes, each of which winds in a spiral fashion around an axon to form the myelin sheath. The axon *(color)* is shown in cutaway. The myelin from a single oligodendrocyte ends before the next wrapping from another oligodendrocyte. The bare axon between sheaths is the node of Ranvier *(N)*. Conduction of action potentials is saltatory down the axon, skipping from node to node. **B,** Myelinated axon in the peripheral nervous system. A Schwann cell forms a myelinated sheath for peripheral axons in much the same fashion as oligodendrocytes do for central ones, except that each Schwann cell myelinates a single axon. The top shows a cross-sectional view of progressive stages in myelin sheath formation by a Schwann cell *(S)* around an axon *(color)*. The bottom shows a longitudinal view of a myelinated axon *(color)*. The node of Ranvier *(N)* is shown between adjacent sheaths formed by two Schwann cells *(S_1 and S_2)*. (Redrawn from Patton, H.D., et al.: Introduction to basic neurology, Philadelphia, 1976, W.B. Saunders Co.)

lia), ependymal cells, microglia, and *Schwann cells* (Figs. 5-1 and 5-2). The first four are associated with the central nervous system and the last with peripheral nerves.

Astrocytes, so named because they are star shaped, are found abundantly throughout the brain and spinal cord. Their many processes surround neurons and their axons often end on the walls of blood vessels. They probably serve metabolic and nutritive functions for the neurons as well as provide a mechanical matrix. Furthermore, many synapses are surrounded by astrocytic processes that separate and perhaps electrically insulate these synapses from one another.

Oligodendrocytes are found chiefly among myelinated axons in the brain and spinal cord. In fact, processes from these cells wrap many times around an axon to form the myelin sheath (Fig. 5-2). This sheath not only insulates axons from one another, but it also limits current flow across the axon membrane (axolemma), except at its discontinuities or *nodes of Ranvier*. Since current flows across the membrane effectively only at these nodes, the action potential is conducted in a saltatory fashion (Chapter 3). Such conduction is much more rapid than would result if myelin were absent. Schwann cells provide the myelin sheaths, in much the same fashion, for peripheral nerves.

Ependymal cells line the surfaces of the brain's ventricles and the central canal of the spinal cord. Their function is unclear.

Finally, microglia are the smallest cells in the central nervous system. Under normal conditions, they have no obvious function. However, if nervous tissue is damaged, these cells enlarge and act as phagocytes to eliminate debris.

Neurons

Neurons are the cells responsible for information processing and transfer in the nervous system. With the exception of certain interneurons that lack long axons, neurons possess axons that conduct regenerative action potentials over considerable distances. Neurons communicate with one another via synapses.

Fig. 5-3 illustrates a "typical" neuron, but many variations in morphology will be described in later chapters. The neuron has several distinct regions. The *soma (or cell body)* and *dendritic processes* comprise the major input zone of the neuron; that is, axons from other neurons form synapses onto the dendrites and soma. The soma is also the metabolic factory for the neuron. Many dendrites issue from the soma, but only a single *axon* does so. The *myelin sheath,* if present, begins near the soma. Although not shown in Fig. 5-3, many axons branch repeatedly at nodes of Ranvier to provide collateral fibers. Finally, the myelin sheath ends, and the axon exhibits unmyelinated, preterminal branches that form synapses onto other neurons, muscle fibers, etc.

In the traditional view of the neuron, synaptic afferent fibers are located on the dendritic arbor and soma, which thus represent the cell's input zone. The axon represents the cell's sole output via the synapses formed from its arbor (Fig. 5-3). Also, the soma, dendrites, and perhaps the preterminal portions of the axon are not electrically excitable. In other words, the membranes of these cellular components do not display voltage-dependent conductance changes caused by the opening or closing of various ion channels and thus do not exhibit action potentials or other active spikes. Electrical excitability and the capacity to propagate an action potential are properties limited to the axon (Chapter 3). In this view, the dendrites and soma behave as passive electri-

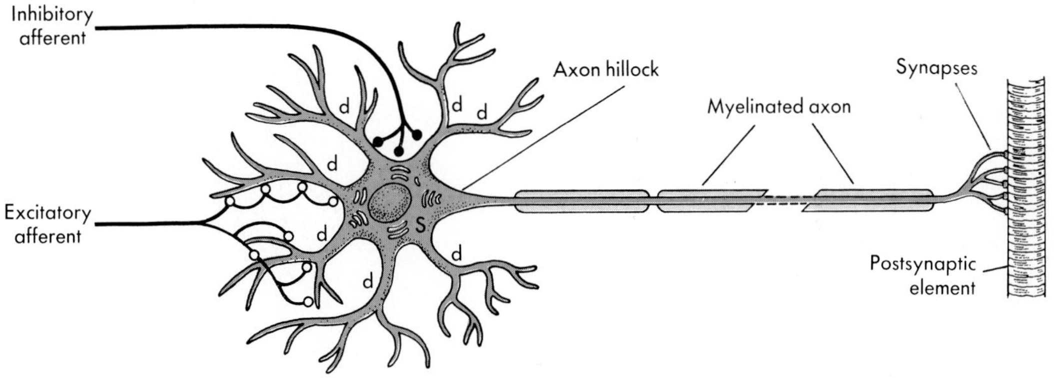

Fig. 5-3. Schematic diagram of an idealized neuron and its major components. Most afferent input from axons of other cells terminates in synapses on the dendrites *(d),* although some may terminate on the soma *(S)*. Excitatory terminals tend to terminate more distally on dendrites than do inhibitory ones, which often terminate on the soma. (Redrawn from Williams, P.L., and Warwick, R.: Functional neuroanatomy of man, Edinburgh, 1975, Churchill Livingstone.)

cal cables, characterized by a low internal resistance (a result of the cytoplasm) with an insulation of high resistance plus some capacitance (a function of the outer membrane). Chapter 3 describes certain cable properties of fibers, such as length and time constants, that determine how electrical signals are transmitted in these cablelike processes.

The dendrites and soma can be modeled as a branched cable. Such analysis requires a number of assumptions that may not always pertain, but it offers a means of determining the efficacy of signal integration and transfer from synapses located on the dendrites or soma to the most proximal portion of the electrically excitable axon (the *initial segment*). The analysis indicates that the most peripheral dendritic location of a synaptic input is only about one length constant from the axon initial segment. Thus all synapses can influence the firing rate of the postsynaptic cell. If the electrotonic conduction of excitatory and inhibitory postsynaptic potentials (EPSPs and IPSPs, respectively) sum at the initial segment to produce a sufficient depolarization, an action potential will be produced and be propagated down the axon. Although the action potential might cease at the preterminal axon branches, the resultant depolarization will travel electrotonically to the synapses and thereby release transmitter.

However, contemporary research has raised some important qualifications and limitations to this traditional view of the neuron. Synapses are not limited to those from axon to dendrite (axodendritic) or from axon to soma (axosomatic), although these are the most common. Axon to axon (axoaxonic) and dendrite to dendrite (dendrodendritic) synapses exist in many regions of the mammalian central nervous system. Also, neurons with somata and dendrites that exhibit strictly linear cable properties may be rare. Instead, many neurons possess somata and dendrites with voltage- and time-dependent conductance changes to specific ions; this represents an active and nonlinear response to electrical signals quite unlike the passive and linear response of a cable.

An example of this is the *low threshold calcium spike* exhibited by some neurons. At normal levels of resting membrane potential (−60 mV), deplorization from an EPSP is conducted electrotonically to the axon initial segment. If the cell is hyperpolarized slightly (to perhaps −65 mV) for at least 100 msec, the same EPSP can increase conductance to calcium. This leads to a large calcium influx that strongly depolarizes the cell, a response that is in fact a calcium spike, as opposed to the sodium spike of the traditional action potential. The calcium spike can lead to a complex series of other voltage-dependent conductance changes. Relative depolarization (at −60 mV) *inactivates* the conductance change underlying the calcium spike, so that an EPSP from such a level will not trigger it. However, hypo-

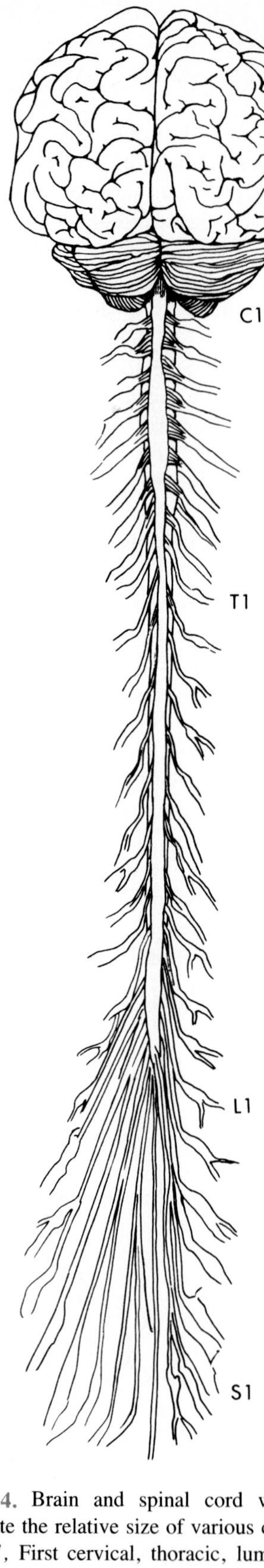

Fig. 5-4. Brain and spinal cord with attached spinal nerves. Note the relative size of various components. *C1, T1, L1,* and *S1,* First cervical, thoracic, lumbar, and sacral segments, respectively. (Redrawn from Williams, P.L., and Warwick, R.: Functional neuroanatomy of man, Edinburgh, 1975, Churchill Livingstone.)

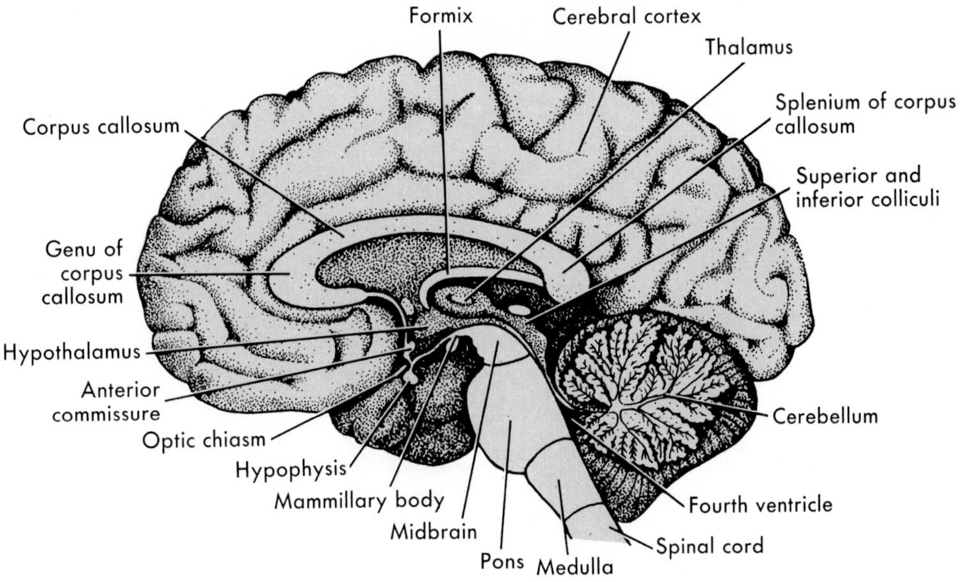

■ **Fig. 5-5.** Midsagittal section of the brain. Note the relationships among the cerebral cortex, cerebellum, thalamus, and brainstem plus the location of various commissures. (Redrawn from Kandel, E.R., and Schwartz, J.H.: Principles of neuroscience, New York, 1981, Elsevier North-Holland, Inc.)

polarization of the membrane for a sufficiently long time (5 mV for 150 msec) will *deinactivate* the calcium conductance, which thus enables a small EPSP to trigger the calcium spike. Shorter hyperpolarizations are insufficient to deinactivate the calcium conductance, which is why the conductance is said to have a time dependency along with its voltage dependency. This is an example of a nonlinear, active response to electrical signals that is most unlike cable behavior. It is now clear that not all information transfer between synapses and the axon's initial segment need be conducted electrotonically.

■ *Gross Topography of the Brain and Spinal Cord*

The central nervous system of vertebrates consists of five major components: the *spinal cord,* the *brainstem,* the *cerebellum,* the *diencephalon,* and the *telencephalon* (or *cerebral hemispheres*) (Figs. 5-4 to 5-6). Because these regions will be considered in more detail in later chapters, the purpose here is merely to indicate their gross relationships with one another and to identify certain prominent structures.

The brainstem is further subdivided into the *medulla, pons,* and *midbrain*. The cerebellum is prominent in humans and is located between the brainstem and posterior cortex. The *thalamus* and *hypothalamus* are located in the diencephalon. The massive telencephalon includes the *corpus striatum* and *cerebral cortex*. The cortex is divided into the *occipital, temporal, parietal,* and *frontal lobes,* and its convolutions form numerous *sulci* and *gyri* (Fig. 5-6). Joining the cerebral hemispheres is a massive commissure, the *corpus callosum.* Its posterior end is called the *splenium,* and anteriorly it bends downward at the *genu*. The *optic chiasm,* formed by the junction of the optic nerves, is found just anterior to the *hypophysis* and ventral to the hypothalamus.

Fig. 5-7 shows the *ventricles,* which are continuous with the *central canal* of the spinal cord and contain *cerebrospinal fluid*. The *fourth ventricle* is enclosed by the cerebellum, pons, and anterior portion of the medulla, and the *third ventricle* is located in the diencephalon. Between these ventricles is the *cerebral aqueduct*. The paired *lateral ventricles* communicate with the third ventricle via the *intraventricular foramen* on each side. The relationship of the extensive lateral ventricles to the hemispheres is shown in Fig. 5-8.

This information serves as a framework for later chapters. The reader should consult standard neuroanatomy books for more detailed anatomical information.

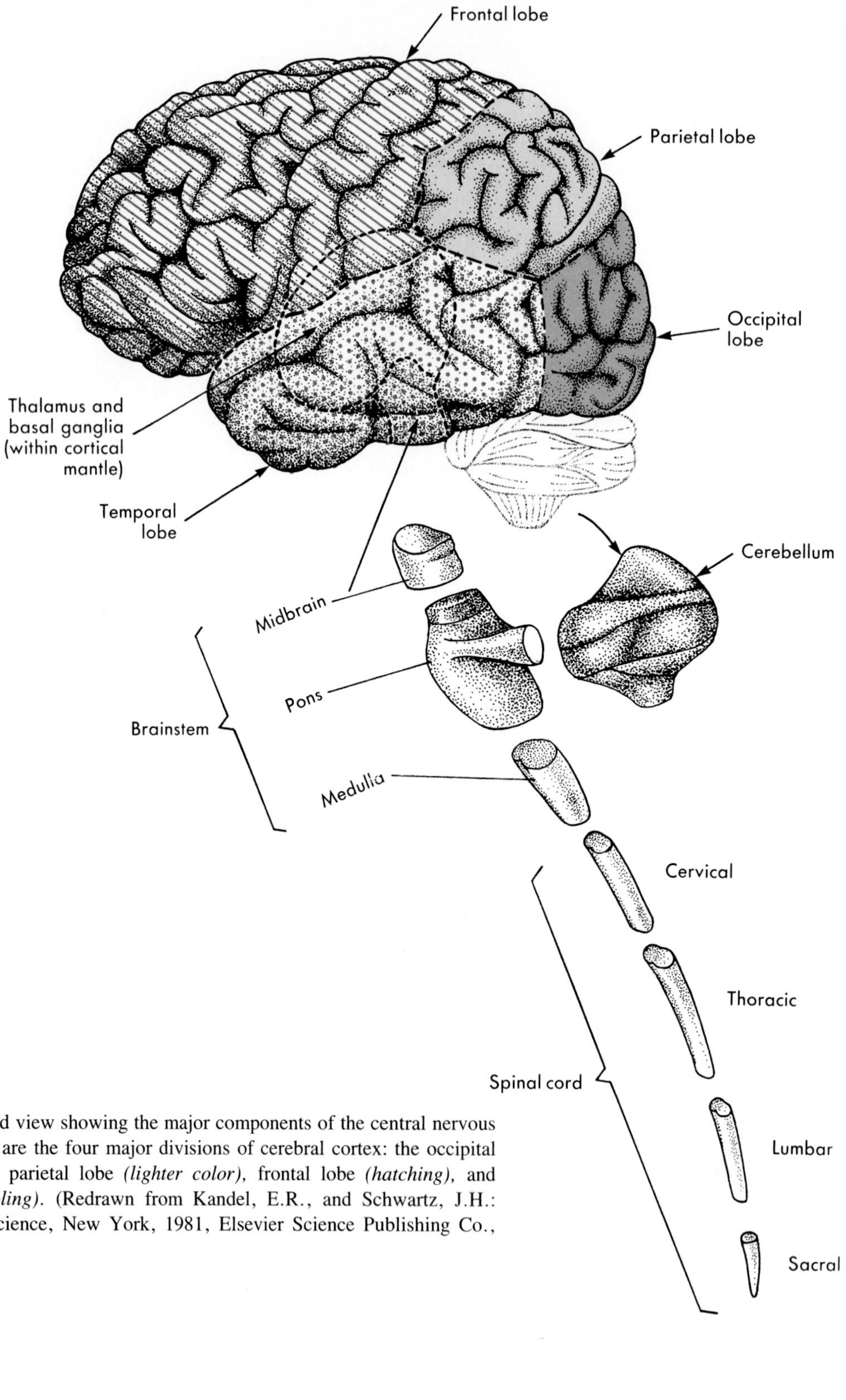

■ **Fig. 5-6.** Exploded view showing the major components of the central nervous system. Also shown are the four major divisions of cerebral cortex: the occipital lobe *(darker color),* parietal lobe *(lighter color),* frontal lobe *(hatching),* and temporal lobe *(stippling).* (Redrawn from Kandel, E.R., and Schwartz, J.H.: Principles of neuroscience, New York, 1981, Elsevier Science Publishing Co., Inc.)

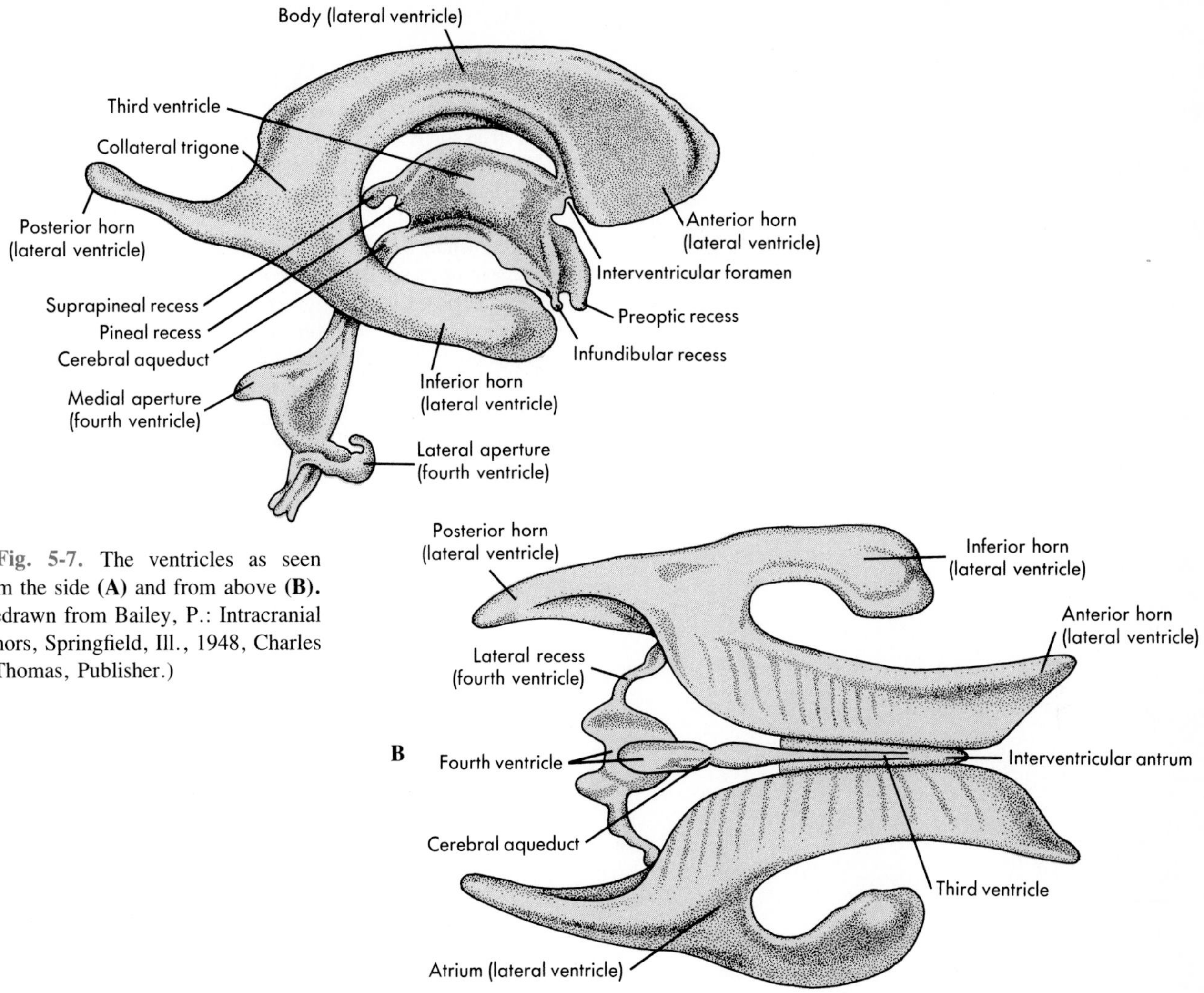

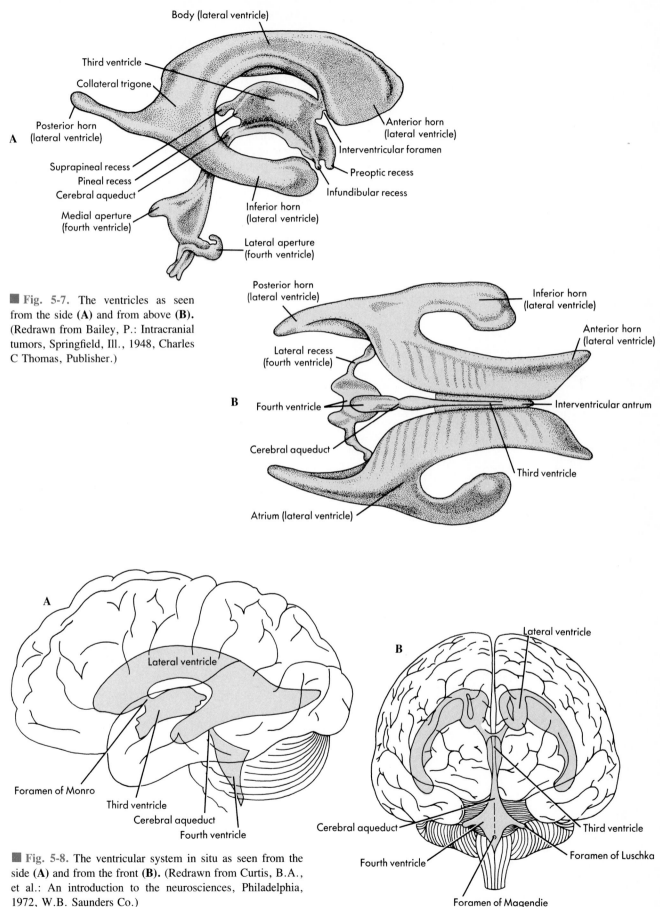

■ **Fig. 5-7.** The ventricles as seen from the side **(A)** and from above **(B).** (Redrawn from Bailey, P.: Intracranial tumors, Springfield, Ill., 1948, Charles C Thomas, Publisher.)

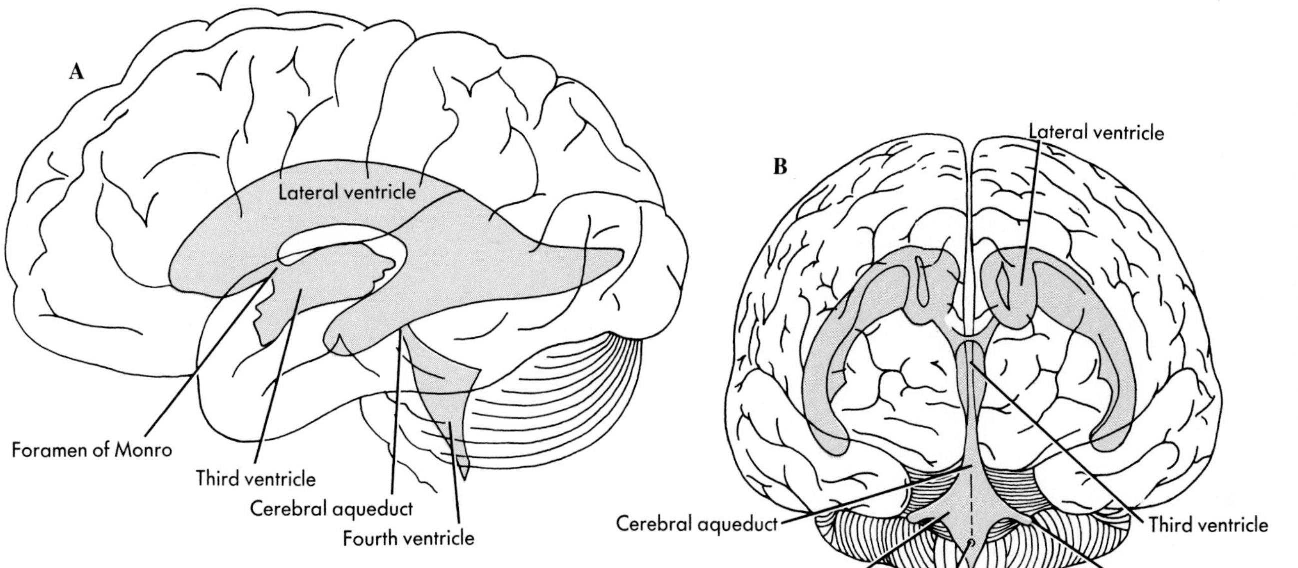

■ **Fig. 5-8.** The ventricular system in situ as seen from the side **(A)** and from the front **(B).** (Redrawn from Curtis, B.A., et al.: An introduction to the neurosciences, Philadelphia, 1972, W.B. Saunders Co.)

Bibliography

Journal articles

Jahnsen, H., and Llinas, R.: Electrophysiological properties of guinea-pig thalamic neureons: an in vitro study, J. Physiol. (Lond.) **349:**205, 1984.

Jahnsen, H., and Llinas, R.: Ionic basis for the electroresponsiveness and oscillatory properties of guinea-pig thalamic neurons in vitro, J. Physiol. (Lond.) **349:**227, 1984.

Llinas, R., and Sugimori, M.: Calcium conductances in Purkinje cell dendrites: their role in development and integration, Prog. Brain Res. **51:**323, 1979.

Wong, R.K.S., Prince, D.A., and Basbaum, A.I.: Intradendritic recordings from hippocampal neurons, Proc. Natl. Acad. Sci. USA **76:**986, 1979.

Books and monographs

Carpenter, M.B.: Core text of neuroanatomy, Baltimore, 1972, Williams & Wilkins.

Eccles, J.C.: The physiology of nerve cells, Baltimore, 1957, The Johns Hopkins University Press.

Kuffler, S.W., and Nicholls, J.G.: From neuron to brain, Sunderland, Mass., 1976, Sinauer Associates, Inc.

Peters, A., Palay, S.L., and de F. Webster, H.: The fine structure of the nervous system: the nervous system: the neurons and supporting cells, Philadelphia, 1976, W.B. Saunders Co.

Rall, W.: Core conductor theory and cable properties of neurons. In Kandel, E., and Geiger, S., editors: Handbook of sensory physiology: the nervous system I, Bethesda, Md., 1977, American Physiological Society.

Williams, P.L., and Warwick, R.: Functional neuroanatomy of man, Philadelphia, 1975, W.B. Saunders Co.

CHAPTER

6

Peripheral Units of the Nervous System

The central nervous system can be viewed as an integrator that analyzes signals from the sensory pathways and uses this information to generate command signals along the motor pathways to muscles and other effectors. The nervous system thus senses and analyzes the environment in order to generate behavior appropriate to that environment. All its interactions with the environment must pass through the peripheral units, which are sensory receptors or transducers and motor units, since these are the only interfaces between the nervous system and the environment. Because it is simpler to understand the functional organization of the nervous system by starting at the periphery and proceeding centrally, the peripheral units are considered before detailed descriptions of specific sensory or motor pathways.

Sensory Receptors or Transducers

Sensory systems start with receptors or transducers. (For reasons given later, *receptor* and *transducer* are terms that will be used interchangeably.) These receptors initiate neural signals in response to sensory stimuli, and these signals are transmitted to the central nervous system to be integrated with other neural signals at a number of synaptic sites. Each sensory system has its own unique population of transducers and neural pathways.

General Considerations

To encode the sensory environment, the energy of environmental signals (light, heat, touch, etc.) must be converted into electrochemical energy that can be used to generate neural signals. This is accomplished by the transducers. Each transducer is able to convert a particular form of stimulus energy (light, heat, etc.) into a graded, slow potential called a *generator potential*. The generator potential is similar in many ways to the excitatory postsynaptic potential seen in the central nervous system and will be considered in more detail later.

Transducers thus exhibit a high degree of modality specificity; that is, they are relatively sensitive to one stimulus modality and not to others. However, this specificity is relative and not absolute. A photoreceptor in the retina has a low generator potential threshold for photic energy but will also exhibit a generator potential to sufficiently intense stimuli of other modalities. Thus pressure against the eyeball, if great enough, excites these photoreceptors. Nonetheless, the sensation perceived is one of flashes or spots of light, not pressure, and the significance of this is considered later in the description of sensory systems. The various classes of transducers found in mammals include mechanoreceptors involved in the somatosensory, auditory, and vestibular systems; photoreceptors for vision; thermoreceptors involved in the somatosensory system; chemoreceptors used for the senses of smell and taste as well as for signaling the chemical composition of the blood; and nociceptors involved in pain sensation (Table 6-1).

Although nociceptors are separately indicated in Table 6-1, they appear to be special subgroups of mechanical, thermal, and chemical transducers with such high thresholds that they respond to tissue-damaging stimuli; also, other nociceptors are chemical transducers sensitive to metabolic by-products of cell damage caused by nociceptive stimuli.

Sensory receptors are typically located at or near the periphery, often in specialized structures, such as the retina or organ of Corti. Some, however, are located deep within the body in connective tissue, joints and major blood vessels. These receptors can be either specialized portions of the distal endings of primary afferent neurons or specialized nonneural cells that form

■ Table 6-1. Classification of neural transducers

Stimulus energy	*Transducer types*
Mechanical	1. Mechanoreceptors of skin and deep tissues, including free nerve endings and specialized structures 2. Mechanoreceptors of joints 3. Stretch receptors of muscle and tendon 4. Hair cells, including vestibular and cochlear 5. Visceral pressure receptors
Photic	Photoreceptors of retina, including rods and cones
Thermal	Thermoreceptors for "cool" and "warm" stimuli
Chemical	1. Chemoreceptors for taste 2. Olfactory receptors 3. Osmoreceptors 4. Carotid and aortic body receptors
Extremes of mechanical, thermal, or chemical energy	Nociceptors

synapses on the primary afferents. Each primary afferent is associated with only one or a very few transducers of the same type. The generator potential initiated by the transducer, if sufficiently large, typically leads to action potentials in the primary afferents (an exception noted later is the retina, in which action potentials first occur several synapses central to the photoreceptors). To explain more completely the properties of transducers, specific detailed examples of these kinds of receptors are considered next.

■ *Pacinian Corpuscle*

Pacinian corpuscles are mechanoreceptors located in the skin and deep tissues. They mediate the sensation of pressure or vibration. Fig. 6-1 illustrates their general appearance and functional properties. The pacinian corpuscle consists of a body composed of concentric layers, much like an onion, into which the distal end of a primary afferent fiber penetrates. The soma of this afferent fiber lies in a dorsal root ganglion of the spinal cord, and its axon enters the spinal cord via the dorsal root. Pressure applied to the "onion" leads to a graded generator potential in the axon (Fig. 6-1, *A*). Its size increases monotonically with increasing pressure, and if it is large enough to surpass the axon's threshold, a regenerative action potential occurs. It should be noted that a subthreshold generator potential, which is electrotonically conducted by the axon, decays to an insignificant size long before the spinal cord is reached. Consequently, the only message transmitted to the central nervous system results from the action potentials.

The distal end of the axon, and not the onion, is the actual transducer; that is, when the onion is stripped away, force applied to the nerve ending (Fig. 6-1, *B*), but not more centrally along the axon (Fig. 6-1, *C*), leads to a generator potential. Evidently such a mechanical distortion of the membrane opens up ionic channels, particularly for sodium. The passage of such ions down their electrochemical gradients depolarizes the axon, which elicits the generator potential.

Although not the actual transducer, the onion nonetheless serves an important mechanical function. Fig. 6-2 shows that the pacinian corpuscle is a *phasic* or rapidly adapting transducer; that is, the generator potential rapidly returns to the baseline (Fig. 6-2, *left.*) This potential occurs only during rapid pressure changes such as the onset or termination of the stimulus. However, if the onion is removed and the force is applied directly to the nerve ending, the generator potential becomes *tonic* or very slowly adapting; it is sustained with slow temporal decay as long as the stimulus is present (Fig. 6-2, *right.*) Apparently the onion, because of its viscoelastic properties, acts as a high-pass temporal filter for mechanical stimuli, thereby transforming these stimuli in such a way that only transient pressure changes reach the nerve ending. However, even with the onion removed so that a stimulus evokes a sustained generator potential, the responses actually reaching the spinal cord (action potentials) reflect further significant adaptation at the impulse-initiation site. Action potentials are generated only briefly during rapid changes in the generator potential, even though the site remains very depolarized. The phasic responses of the pacinian corpuscle, then, are only partially explained by the mechanical properties of the onion.

Phasic responses, such as those exemplified by the pacinian corpuscle, subserve an important role in sensory processing. Phasic activity is the most sensitive indicator of a temporal change in a stimulus. Because a tonic response or generator potential would increase or decrease monotonically with stimulus strength, small changes in the stimulus would be difficult to detect in such a response. Phasic transducers generate a potential only when such changes occur and thus more dramatically signal these changes. However, phasic cells are ill suited to signal quantitatively the magnitude of such changes. They are even ill suited to distinguish increases from decreases in stimulus intensity. For this, tonic or slowly adapting transducers are needed.

■ *Tonic Mechanoreceptors*

A tonic receptor is one from which the evoked generator potential is maintained for the duration of the stimulus with only slight adaptation; that is, the generator potential decays very slowly during a maintained stim-

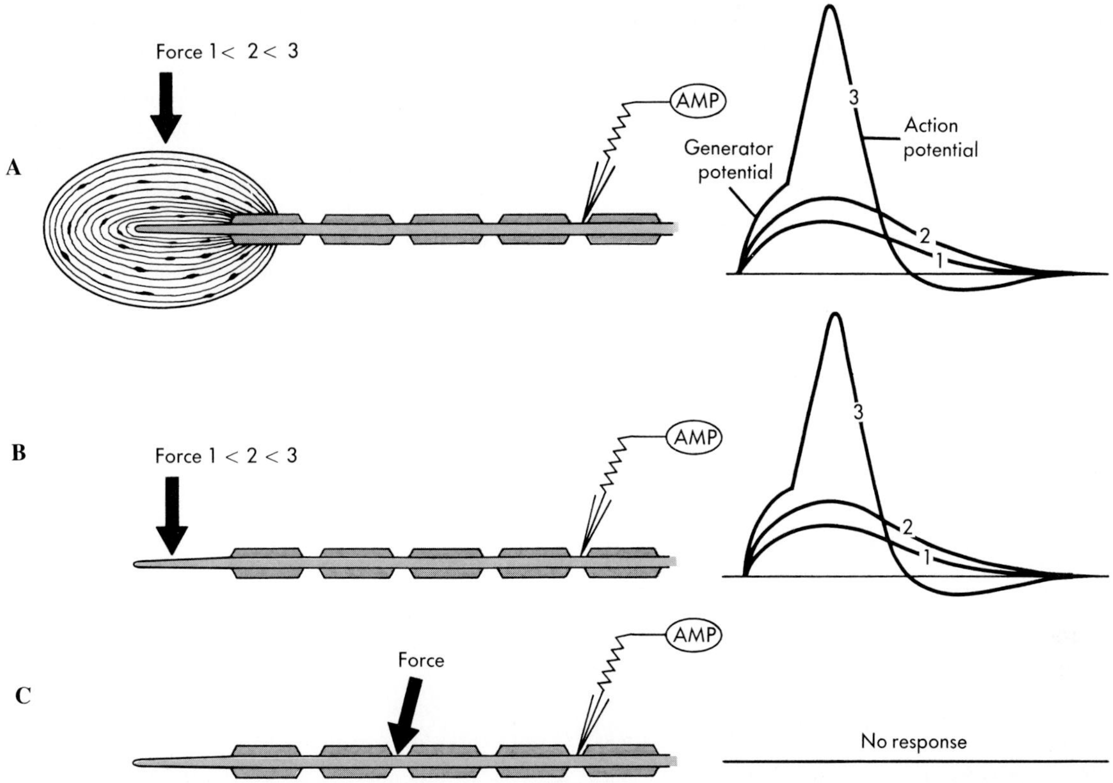

■ **Fig. 6-1.** Schematic illustration of response properties of the pacinian corpuscle. A myelinated axon *(color)* penetrates an onionlike structure that forms the pacinian corpuscle; the axon is the distal end of a primary afferent fiber that enters the spinal cord centrally via a dorsal root. In all drawings the electrophysiological responses of the axon (as recorded through an electrode and amplified by appropriate electronic equipment) are shown to three forces applied to the receptor. Force *1* is smaller than force *2,* which is smaller than force *3,* and the responses are numbered accordingly. **A,** Intact corpuscle. Forces applied to the surface of the corpuscle result in electrotonically transmitted generator potentials that demonstrate temporal and spatial summation (not illustrated). The amplitude of each potential monotonically reflects the size of the applied force. If the force and resultant potential are sufficiently large, a regenerative action potential results and travels to the spinal cord. Smaller generator potentials that do not trigger an action potential are functionally irrelevant because they decay to undetectable size long before reaching the spinal cord. **B,** Experiment to demonstrate that the onionlike structure is not necessary for transduction. Forces applied directly to the distal tip of the axon result in responses as in **A. C,** Experiment to demonstrate that the actual transducer is the distal tip of the axon, since forces applied elsewhere evoke no electrophysiological response. (Redrawn from Loewenstein, W.R.: Biological transducers. Sci. Am. **203:**98, 1960.)

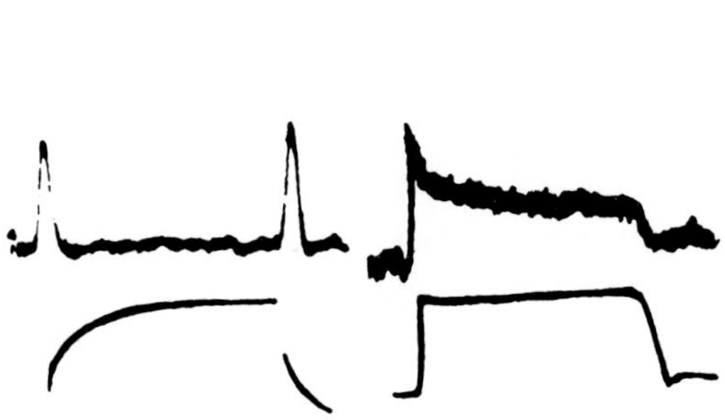

■ **Fig. 6-2.** Stimulus-response records of a pacinian corpuscle. The top traces represent the generator potentials recorded in the axon in response to the application of forces 50 msec in duration as shown in the lower traces. On the left is the stimulus-response pattern of an intact pacinian corpuscle. Note the extremely phasic responses that occur only when the stimulus changes, that is, at onset and termination. The right half shows the pattern of stimuli applied directly to the distal tip of the axon after removal of the onionlike structure. Here the response is relatively tonic. (From Lowenstein, W.R., and Mendelsohn, M.: J. Physiol. [Lond.] **177:**377, 1965.)

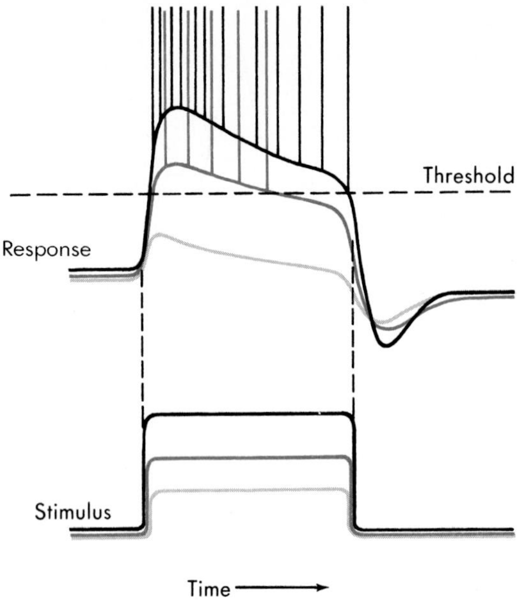

■ **Fig. 6-3.** Relationships among stimulus intensity, generator potential amplitude, and frequency of generated action potentials for a tonic transducer. A monotonic relationship exists between stimulus intensity and generator potential amplitude. If this amplitude surpasses some threshold value, action potentials result with a frequency that is monotonically related to the generator potential's amplitude because the greater this amplitude, the sooner the cell's relative refractory period can be overcome. Thus a code is established for transmission to the central nervous system whereby the frequency of action potentials relates to stimulus intensity in a reliable monotonic fashion.

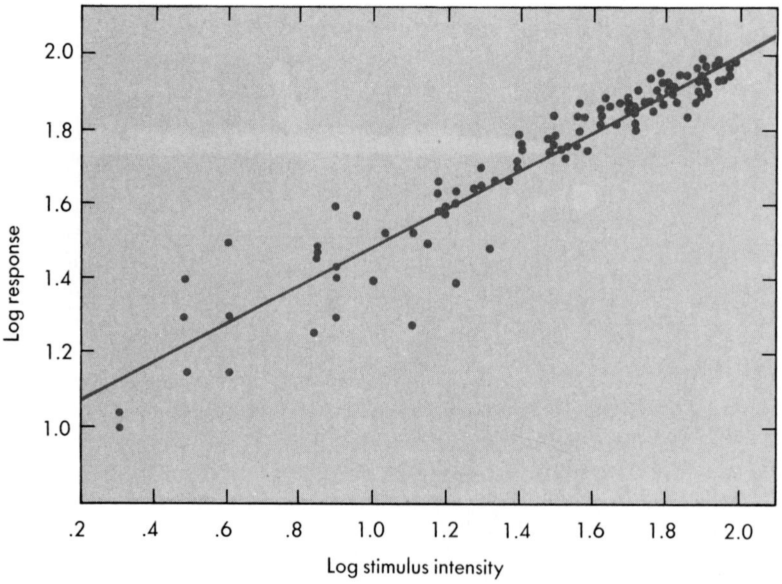

■ **Fig. 6-4.** Recordings of responses from a single cutaneous fiber are shown as a function of stimulus intensity applied to the skin. The log of the response monotonically encodes the log of the stimulus intensity. These data verify the relationships shown in Fig. 6-3. (From Werner, G., and Mountcastle, V.B.: J. Neurophysiol. **28:**359, 1965.)

ulus in a fashion similar to that illustrated in Fig. 6-2, *right.* A tonic transducer and afferent fiber, unlike the pacinian corpuscle, exhibits no adaptation at its spike-initiation region. Thus the presence of action potentials closely corresponds to the maintained level of depolarization. If the generator potential exceeds the sensory axon's threshold, action potentials ensue. The nature of a neuron's membrane properties are such that the more the maintained generator potential exceeds this threshold, the sooner threshold can be reached again after an action potential. Thus the greater the generator potential is, the higher the rate of action potentials or firing frequency (Fig. 6-3). From this series of processes, a neural code that signals stimulus strength reaches the central nervous system; that is, firing frequency is monotonically related to stimulus intensity (Fig. 6-3). This is an important property of tonic transducers and their associated axons.

Recordings from primary afferent fibers from glabrous skin in the monkey hand have shown that these fibers respond tonically in such a way that the firing frequency accurately encodes stimulus strength. This is illustrated by Fig. 6-4. A phasic transducer, such as an intact pacinian corpuscle, could never encode stimulus strength in this fashion. The actual transducers that provide the generator potentials for the fibers shown in Fig. 6-4 have not been identified. Excellent candidates are specialized cells that form synaptic contacts with the distal end of primary afferent fibers. The generator potential might be evoked in these cells and synaptically transmitted to the afferent nerve fiber.

A 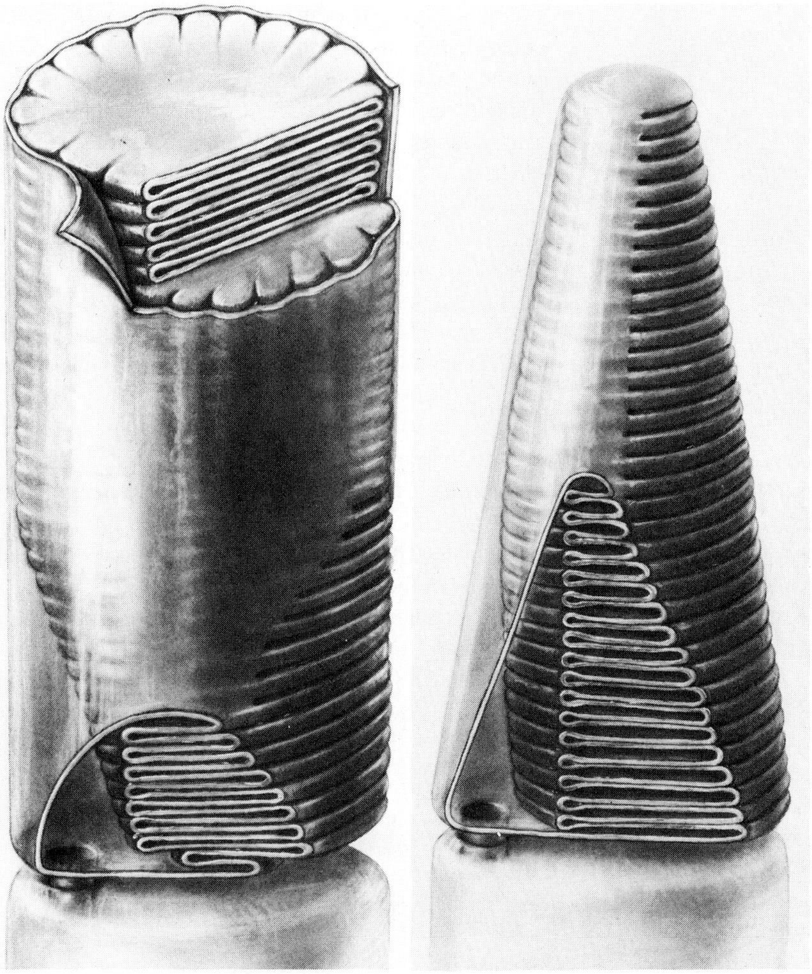 B

Fig. 6-5. Schematic drawings of the outer segments of a vertebrate rod **(A)** and cone **(B).** Because of its much greater length, only the proximal portion of the rod outer segment is shown. (Courtesy R.W. Young.)

Photoreceptors: Rods and Cones

The preceding sections illustrate the differences between fast-adapting (phasic) and slowly adapting (tonic) transducers, as well as the functional significance of these differences. It is important to remember, however, that tonic transducers do adapt, albeit slowly. The photoreceptors in the retina illustrate this point.

Humans have two broad types of photoreceptors, more sensitive rods for black-and-white vision and less sensitive cones for color vision. At low light levels *(scotopic illumination),* cones do not respond, and only the rods function; thus colors are not appreciated in very dim light. At higher light levels *(photopic illumination),* cones function and rods are saturated; color vision is now possible. At middle brightness levels *(mesopic illumination),* rods and cones function together. These photoreceptors transduce photic energy into generator potentials that are synaptically transmitted to retinal interneurons (Chapter 8).

Rods and cones have similar morphological features (Fig. 6-5). An outer segment (cylindrical in rods and conical in cones) is attached to an inner segment and the rest of the cell by a thin stalk. The outer segment is stacked with parallel membranes into which the photosensitive chemicals (or photopigments) are embedded. In rods the photopigment is called *rhodopsin,* which consists of a small vitamin A derivative attached to a large protein. The vitamin A derivative is *retinene,* and the protein is known as an *opsin.* The photochemistry and transduction processes of rods are described later; those of cones are thought to be similar.

When a photon strikes rhodopsin, it changes the isomerization of the retinene portion of the molecule, and this eventually leads to a breakdown of rhodopsin into retinene or vitamin A and the opsin. This breakdown is known as *bleaching* and is reversible (Fig. 6-6). Bleaching is the first in a series of events that ultimately lead to the generator potential. A particularly unusual feature of rods and cones is the polarity of their generator potentials: these neurons are hyperpolarized by light. In the dark, sodium ions flow into photoreceptors. This is the ''dark current,'' and it depolarizes the cell. In response to light, a series of incompletely understood steps ultimately blocks many sodium channels in the membrane; the dark current is reduced, and the

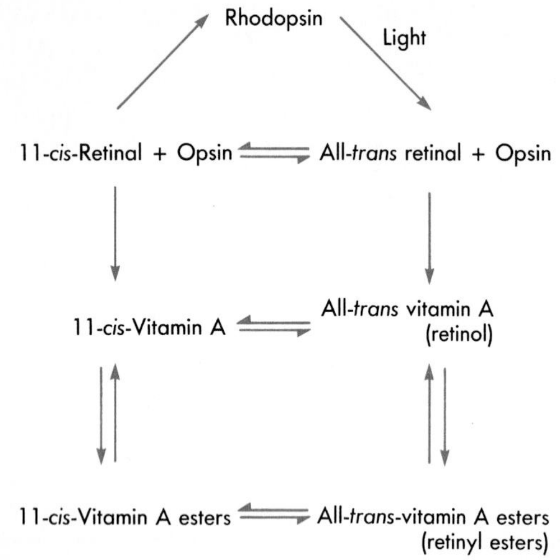

■ Fig. 6-6. Chemical reactions involved in synthesis and breakdown of rhodopsin. For simplicity enzymes and metabolic prerequisites are not shown. (Redrawn from Wald, G.: Science **162:**230, 1968. © The Nobel Foundation 1968.)

cell becomes hyperpolarized. This hyperpolarization signals the presence of light quite effectively to other retinal neurons. Indeed, in a cell without action potentials (i.e., a photoreceptor cell), there is no reason to suppose that the more conventional depolarization is a better means of signaling than is hyperpolarization.

This hyperpolarization, or generator potential, slowly adapts in the presence of constant light. A small portion of this adaptation can be related to the photochemistry of bleaching, although even this may be significant only at high levels of illumination. The magnitude of the generator potential is related to the rate of the bleaching reaction. For a given number of photons, the rate of bleaching (and thus the generator potential size) is monotonically related to the concentration of intact rhodopsin still available. When a constant light is first applied to the retina, more intact rhodopsin is present, and the potential is maximal. As rhodopsin bleaches, the reaction slows down, and the potential diminishes. In other words, it slowly adapts. However, both in terms of rate and extent, the amount of photoreceptor adaptation actually measured far exceeds that predicted on the basis of bleaching alone. Other unknown factors contribute more significantly to this process.

This slow adaptation serves a useful purpose at the expense of slightly degrading the stimulus intensity/frequency code described earlier. Each photoreceptor at any one time has a dynamic response range that can encode only about a 4 log unit intensity range. However, the human visual system can operate over roughly a 10 log unit range. These values for response range of receptors and the system depend on the actual method of experimental measurement. Gradual adaptation is the key to this difference, and in vision most such adaptation has its origin in the transduction process. Each receptor can gradually adjust its limited dynamic range to suit the light level present. For instance, when a person remains in a dark room, the photoreceptors become *dark adapted* for maximal sensitivity. If the person then emerges into a bright, sunlit area, he is "dazzled," and everything seems equally bright. As the retina *light adapts* to become less sensitive, shades of gray are appreciated, and shapes can be perceived. The converse is true when a person goes from a bright to a dark environment. Therefore slow adaptation increases the dynamic range of a sensory system beyond that of the individual transducers. An alternative strategy for such an increase would require transducers to maintain fixed sensitivity ranges that differ from one another. To a very limited extent, the sensitivity difference between rods and cones achieves this, but not nearly enough to account for the extensive dynamic range of human vision, given the individual ranges of rods or cones. This alternate strategy of different, fixed ranges for transducers would result in fewer responsive receptors and neural elements to analyze a given stimulus than is the case with slowly adapting cells. Again, this important feature of sensory processing is a property of the transducers.

■ *Effectors and Their Control*

In the previous section the peripheral units at the input or sensory end of the nervous system were discussed. We shall now focus on the peripheral units of the output or motor end of the nervous system. These outputs are ultimately expressed by three classes of effector action: contraction of striated or skeletal muscle, contraction of smooth muscle of the viscera and eye, and glandular secretion. The neural control of glandular secretion is discussed elsewhere in this text (Chapter 43), and the neural (autonomic) control of the viscera is treated in Chapter 20. Here we will focus on the peripheral units of movement.

■ *Basic Unit of Movement*

With respect to the contraction of skeletal muscle, which induces movement, it is most appropriate to consider first the motoneuron. Muscles receive their innervation, and thus commands, from only one source, the motoneurons of the spinal cord and cranial nerve motor nuclei. These motoneurons are polygonal, multipolar cells that vary in size up to 70 μm. They give rise to large myelinated axons, 9 to 20 μm in diameter, that leave the central nervous system to reach their target muscle through either cranial or peripheral nerves.

In the spinal cord the motoneurons are located in the

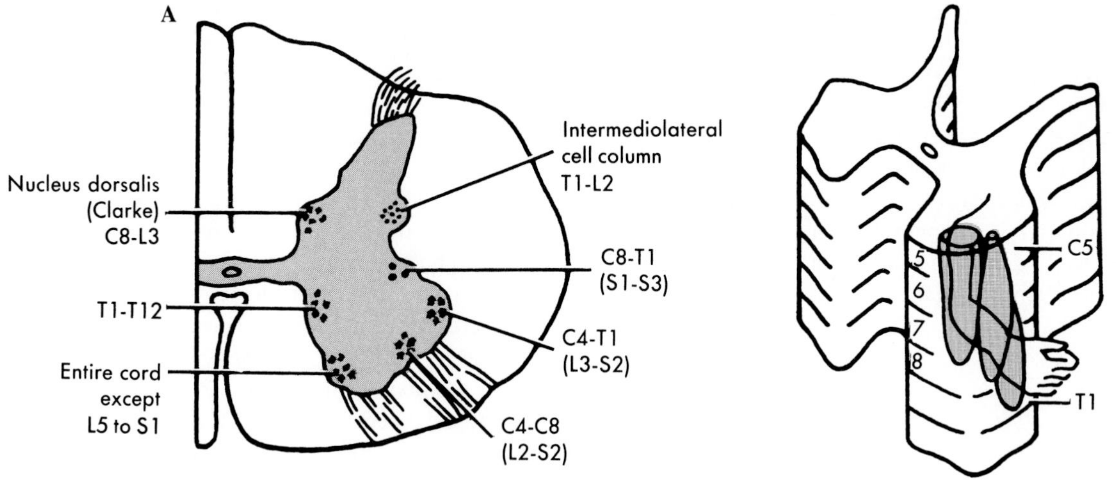

■ **Fig. 6-7.** Diagrammatic illustration of the topographical organization of the motoneurons and the muscles they innervate. **A** shows a transverse section of the spinal cord with the positions of various cellular columns and their longitudinal extents. **B** shows three longitudinal columns of motoneurons *(color)* at cervical levels in relation to the parts of the arm they innervate. It should be appreciated that the more proximal muscles of the arm are innervated by more ventromedially situated motoneurons and the more distal muscles by more dorsolaterally situated motoneurons. (Redrawn from Brodal, A.: Neurological anatomy, ed. 3. Copyright © 1981 by Oxford University Press, Inc. Reprinted by permission.)

ventral horn. In a transverse section through the spinal cord different groups of motoneurons varying in mediolateral location can clearly be identified. In sagittal or horizontal section these groups are arranged in longitudinal columns (Fig. 6-7). As described in greater detail in Chapter 13, this cytoarchitectural arrangement of the spinal motoneurons reflects a highly topographic organization with respect to target musculature. Indeed, within each column are subsets of spatially contiguous motoneurons that innervate a single muscle. The complement of motoneurons innervating a given muscle is defined as the *motoneuron pool* of that muscle. (Fig. 6-8).

In mammals each muscle fiber is innervated by a single motoneuron. However, any given motoneuron will innervate a group of fibers within a single muscle. A motoneuron and the fibers it innervates are referred to as a *motor unit* (Fig. 6-8), and this can be viewed as the basic unit of movement. The innervation of a given muscle fiber by only one motoneuron leads to some important clinical applications. It is possible in humans to record the electrical activity of muscles during contraction *(electromyography)*. Also, the action potentials of a single muscle fiber can be studied with needle electrodes implanted in muscles. This activity directly reflects the discharge of a single motoneuron. Electromyography is useful in various clinical contexts such as the differential diagnosis of atrophies and dystrophies and the assessment of regeneration after peripheral nerve injury.

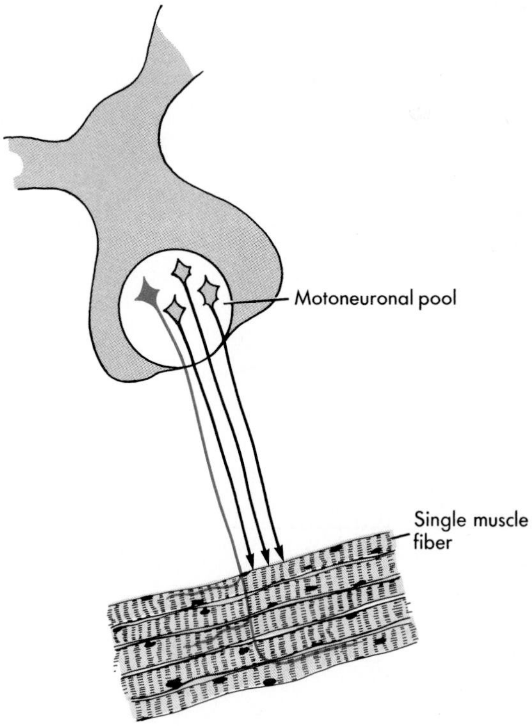

■ **Fig 6-8.** Schematic illustation of a motoneuron pool, that is, the group of motoneurons innervating a given muscle. The motoneuron in color illustrates the concept of the motor unit, that is, a single motoneuron and all the muscle fibers it innervates.

The *size of motor units* varies considerably. Some muscles, such as the extrinsic eye muscles and the intrinsic muscles of the digits, are used for highly differentiated movements. In such muscles a few fibers are innervated by a single motoneuron. In contrast, the *innervation ratio* (the number of muscle fibers innervated by each motoneuron) may be as high as 2000:1 in large postural muscles. The muscle fibers of a motor unit are rather evenly distributed throughout the muscle, as opposed to being spatially restricted in a contiguous group.

Control of Contraction

Because a single motoneuron innervates many muscle fibers, it is clear that on reaching a muscle the axon of the motoneuron must ramify extensively. At each ramifying branch the termination forms the *motor endplate,* which is a specialized ending that constitutes the neural component of the neuromuscular junction. The morphology and function of the neuromuscular junction are discussed in Chapter 4. However, it is important to reemphasize that motoneurons only excite the muscle; that is, activation of a motoneuron results in contraction of the fibers it innervates. Relaxation or lengthening of muscle fibers can only be achieved by a reduction of the motoneuron discharge. The tension generated in a motor unit will thus be a function of the discharge rate

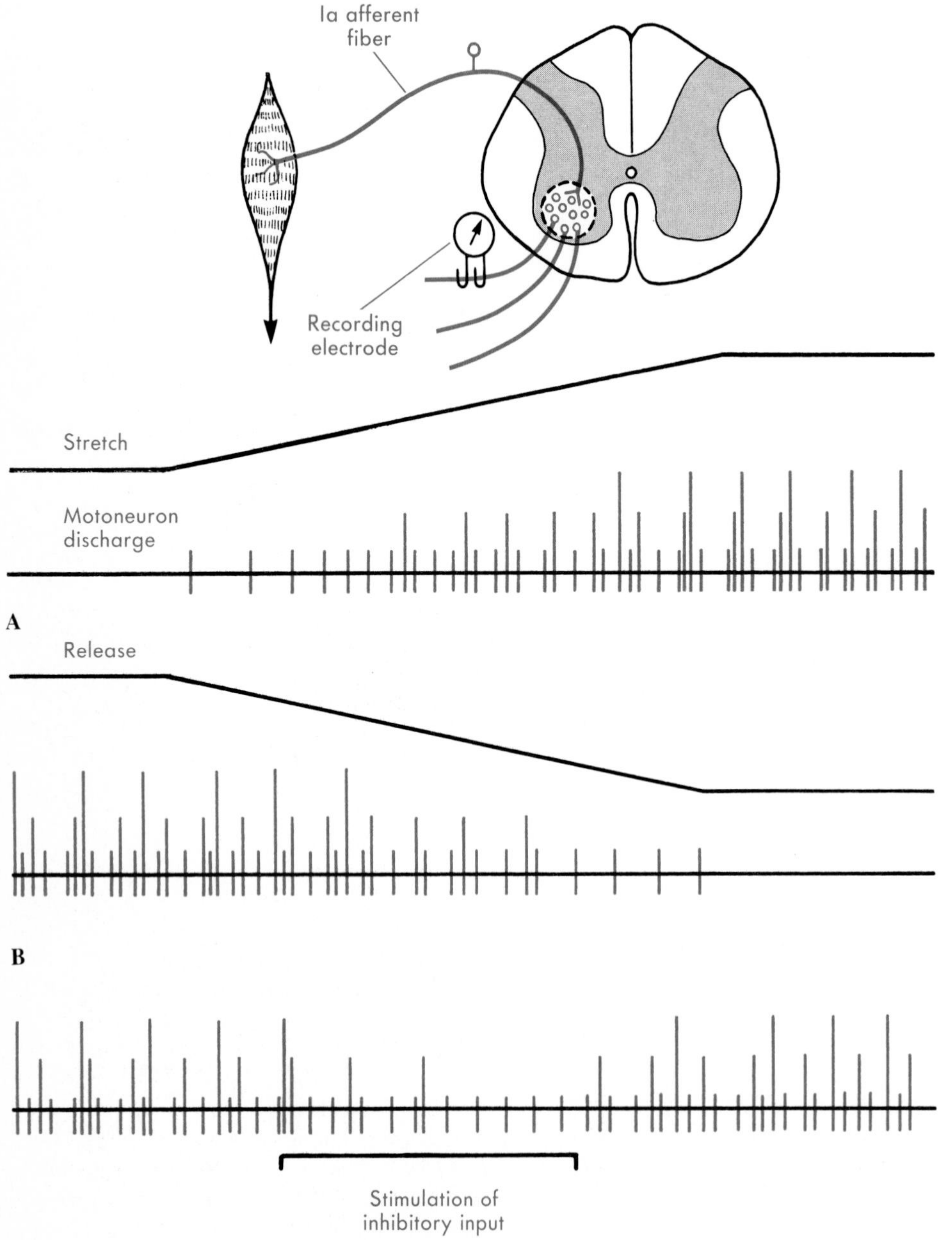

Fig. 6-9. The size principle in the recruitment of motoneurons. The schematic at the top shows a Ia afferent fiber from a muscle spindle and recording electrodes on a dissected ventral root filament arising from an homonymous motoneuron. **A,** Excitation. As a muscle is stretched, the increased activity of the Ia afferent fibers first recruits the smaller motoneurons. As the stretch increases, successively larger motoneurons are recruited. On release from stretch, the larger motoneurons stop discharging first and the smaller motoneurons last. **B,** Inhibition. Activating an inhibitory input to the motoneuron pool first silences the larger motoneurons and then successively smaller motoneurons. (Reproduced with permission from Eyzaguirre, C., and Fidone, S.J.: Physiology of the nervous system: an introductory text, 2nd edition. Copyright © 1975 by Year Book Medical Publishers, Inc., Chicago.)

of its motoneuron, and the tension generated in an entire muscle will be the sum of the tensions of all the active motor units of that muscle.

Graded muscle contraction is achieved in part by activating an increasingly larger number of motor units *(recruitment)*. This recruitment occurs in an orderly fashion. For some time, orderly recruitment was thought to be governed by a *size principle,* in which the excitability threshold of a motoneuron is a function of its size (Fig. 6-9). According to this principle, the first motor units recruited are those of the smaller motoneurons, which generate smaller contractile forces than do the larger motor units. As successively larger motor units are recruited, larger and larger increments of contractile forces are added. However, recent experiments suggest that the order of recruitment may be more closely related to the contraction strength of the motor unit, with weaker motor units being recruited before stronger ones (Fig. 6-10).

The motoneuronal discharge frequency also contributes to the gradation of contraction. Once a motor unit is recruited, increased force of contraction is associated with an increased discharge frequency of its motoneuron. Furthermore, the different motor units of a given muscle discharge asynchronously with respect to each other because of their different thresholds, and this assures smoothly graded contraction.

The Final Common Path

It is clear from the previous discussion that considerable central integration is required for the smooth onset of contraction in just a single muscle. The coordinated action of a group of muscles around a joint requires still more integration. Complex sequences of muscle contractions, to produce spatially and temporally organized movement, demand truly impressive neuronal interactions. This is reflected in the large number of pathways involved in the control of movement and the thousands of synaptic contacts distributed over the soma and dendrites of each motoneuron.

Despite the complexity of the integration of motor activity that occurs at many levels of the neuraxis, muscle contraction and therefore movement must ultimately be expressed through the motoneuron. It is for this reason that Sherrington described the motoneuron as the *final common path.* Any lesion compromising the final

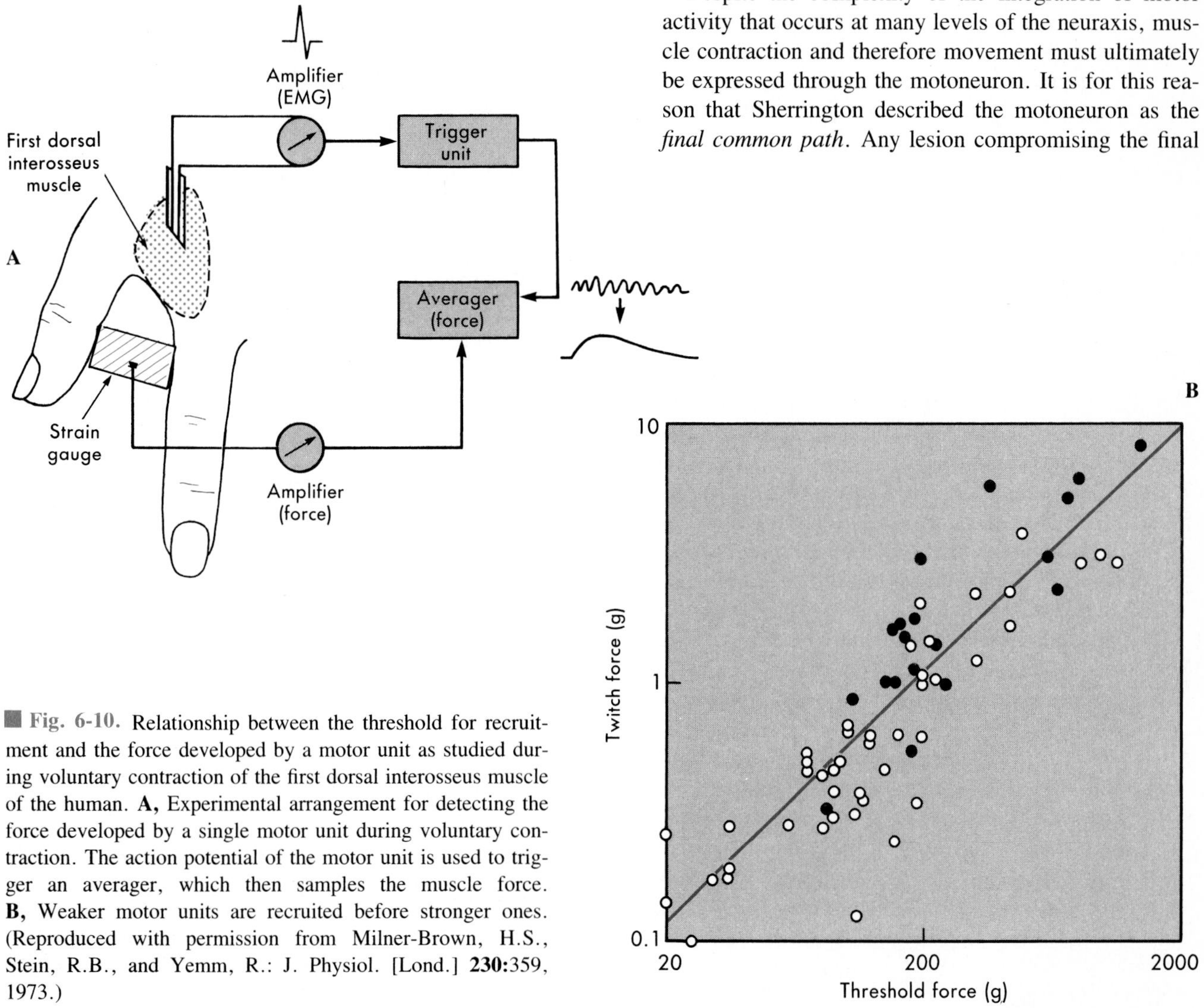

Fig. 6-10. Relationship between the threshold for recruitment and the force developed by a motor unit as studied during voluntary contraction of the first dorsal interosseus muscle of the human. **A,** Experimental arrangement for detecting the force developed by a single motor unit during voluntary contraction. The action potential of the motor unit is used to trigger an averager, which then samples the muscle force. **B,** Weaker motor units are recruited before stronger ones. (Reproduced with permission from Milner-Brown, H.S., Stein, R.B., and Yemm, R.: J. Physiol. [Lond.] **230:**359, 1973.)

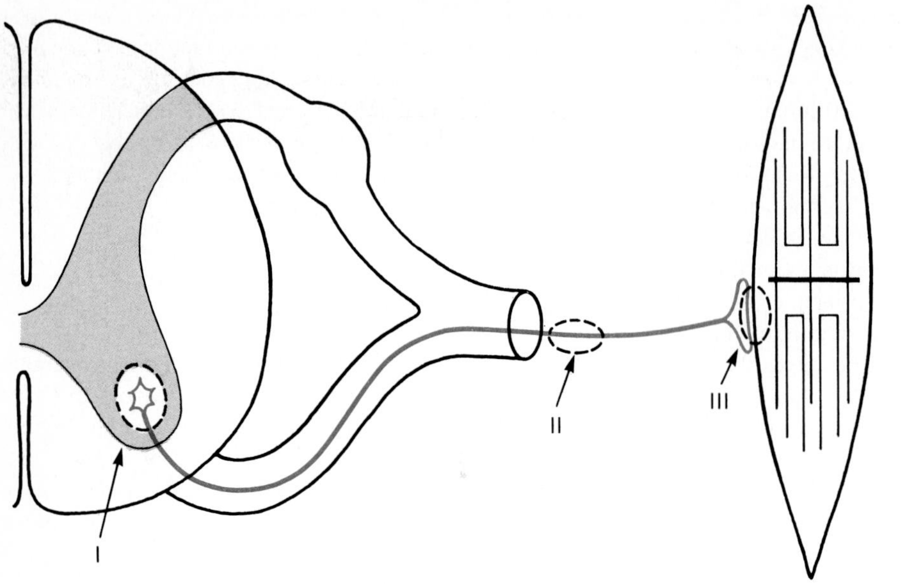

■ **Fig. 6-11.** Sites involved in diseases that affect the final common path for skeletomotor activity. *I,* The motoneuron cell body that can be directly affected by such diseases as poliomyelitis and amyotrophic lateral sclerosis. *II,* The axon of the motoneuron that can be affected by injury and various diseases such as the demyelinating Guillain-Barré syndrome. *III,* The neuromuscular junction where transmission can be compromised by a disease such as myasthenia gravis or by a toxin such as botulinus toxin. Compromise of the motoneuron or its axon produces lower motoneuron disease. (Modified from Ruch, T., and Patton, H.D.: Physiology and biophysics, vol. IV, Excitable tissues and reflex control of muscle, Philadelphia, 1982, W.B. Saunders Co.)

common path of a given muscle will preclude the ability of that muscle to contract and thereby result in its paralysis. Such lesions are sometimes referred to as *lower motoneuron lesions,* and they may result from compromise of different segments of the final common path (Fig. 6-11). For example, diseases such as poliomyelitis and amyotrophic lateral sclerosis destroy the motoneuronal cell body, whereas the various peripheral neuropathies, such as the Guillain-Barré syndrome, can destroy the axon of the motoneuron. If the lesion destroys the motoneuron or interrupts its connection with a muscle, then the muscle fibers of the motor unit will degenerate. This leads to muscle atrophy and reflects the trophic relationship that exists between the motoneuron and muscle fibers of a motor unit (Chapter 23).

Beyond lower motoneuron lesions, the function of the motor unit can be compromised by any disease process affecting transmission at the neuromuscular junction, such as myasthenia gravis. Still more peripherally, the motor unit can be affected by a number of diseases involving the muscle itself—the myopathies, which include diseases such as the muscular dystrophies.

■ Bibliography

Journal articles

Attwell, D., and Wilson, M.: Behavior of the rod network in the tiger salamander retina mediated by membrane properties of individual rods, J. Physiol. (Lond.) **309:**287, 1980.

Baylor, D.A., Lamb, T.D., and Yau, K.W.: The membrane current of single rod outer segments, J. Physiol. (Lond.) **288:**589, 1979.

Baylor, D.A., Matthews, G., and Yau, K.W.: Two components of electrical dark noise in toad retinal rod outer segments, J. Physiol. (Lond.) **309:**591, 1980.

Baylor, D.A., Nunn, B.J., and Schrapf, J.L.: The photocurrent, noise, and spectral sensitivity of rods of the monkey *Macaca fascicularis,* J. Physiol. (Lond.) **357:**575, 1984.

Buchtal, F.: The electromyogram: its value in the diagnosis of neuromuscular disorders, World Neurol. **3:**16, 1962.

Detwiler, P.B., Hodgkin, A.L., and McNaughton, P.: Temporal and spatial characteristics of the voltage response of rods in the retina of the snapping turtle, J. Physiol. **300:**213, 1980.

Fain, G.L., and Lisman, J.E.: Membrane conductances of photoreceptors, Prog. Biophys. Mol. Biol. **37:**91, 1981.

Griff, E.R., and Pinto, L.H.: Interactions among rods in the isolated retina of *Bufo marinus,* J. Physiol. (Lond.) **314:**237, 1981.

Hubbell, W.L., and Bownds, M.D.: Visual transduction in vertebrate photoreceptors, Annu. Rev. Neurosci. **2:**17, 1979.

Iggo, A.: Is the physiology of cutaneous receptors determined by morphology? Prog. Brain Res. **43:**15, 1976.

Lasanskey, A.: Synaptic actions mediating cone responses to annular stimulation in the retina of the larval tiger salamander, J. Physiol. (Lond.) **310:**205, 1981.

Schwartz, E.A.: Phototransduction in vertebrate rods, Annu. Rev. Neurosci. **8:**339, 1985.

Tomita, T.: Electrical activity of vertebrate photoreceptors. Q. Rev. Biophys. **3:**179, 1970.

Wald, G.: The receptors of human color vision, Science **145:**1007, 1964.

Wald, G.: Molecular basis of visual excitation, Science **162:**230, 1968.

Zajac, F.E., and Faden, J.S.: Relationship among recruitment order, axonal conduction velocity, and muscle-unit properties of type-identified motor units in cat plantaris muscle, J. Neurophysiol. **53:**1303, 1985.

Books and monographs

Brodal, A.: Neurological anatomy, ed. 3, New York, 1981, Oxford University Press, Inc.

Burke, R.E.: Motor units: anatomy, physiology and functional organization. In Brookhart, J.M., and Mountcastle, V.B., editors: Handbook of physiology: the nervous system, section 1, volume II, part 1, chapter 10, Bethesda, Md., 1981, American Physiological Society.

Cohen, A.I.: Rods and cones. In Fuortes, M.G.F., editor: Handbook of sensory physiology, vol. VII, Physiology of photoreception organs, New York, 1972, Springer-Verlag New York, Inc.

Henneman, E.: Skeletal muscle, the servant of the nervous system. In Mountcastle, V.B., editor: Medical physiology, ed. 14, vol. I, St. Louis, 1980, The C.V. Mosby Co.

Henneman, E., and Mendell, L.M.: Functional organization of motoneuronal pool and its inputs. In Brookhart, J.M., and Mountcastle, V.B., editors: Handbook of physiology: the nervous system, section 1, volume II, part 1, chapter 10, Bethesda, Md., 1981, American Physiological Society.

Miller, W.H., editor: Current topics in membranes and transport, vol. 15, Molecular mechanisms of photoreceptor transduction, New York, 1981, Academic Press, Inc.

Rowland, L.P.: Diseases of the motor unit: the motor neuron, peripheral nerve, and muscle. In Kandel, E.R., and Schwartz, J.H., editors: Principles of neuroscience, New York, 1981, Elsevier Science Publishing Co., Inc.

Sherrington, C.S.: The integrative action of the nervous system, ed. 2, New Haven, Conn., 1947, Yale University Press.

Teorell, T.: A biophysical analysis of mechano-electrical transduction. In Lowenstein, W.R., editor: Handbook of sensory physiology, vol. I, Principles of receptor physiology, New York, 1971, Springer-Verlag New York, Inc.

CHAPTER 7

General Principles of Sensory Systems

Mammalian sensory systems have evolved to provide the organism with information about its environment so that these systems can guide its behavior accordingly. These systems must tell the organism *what* is in the environment, or the precise form of stimulus energy present, be it photic, mechanical, thermal, or chemical. They must specify *when* changes in the environment occur, or the sudden presence or absence of a stimulus. In addition to these qualitative considerations, mammalian sensory systems have evolved the capability to quantify stimulus parameters in terms of *how much* (intensity) and *where* (location).

How the various sensory systems accomplish this is of fundamental importance. These systems share many basic organizational features. Thus once a fundamental framework is understood, the details unique to each specific modality are easy to elaborate. Consequently, the visual system will be described in detail because an in-depth understanding of this system establishes the framework common to other sensory systems. Also, more is known about the neural basis of vision than about any other modality.

Mammalian sensory systems are complex combinations of interacting pathways. A great deal of the complexity of these pathways derives from their evolutionary history, and their organization can more readily be grasped with some understanding of this evolution. Although the details of such an evolutionary history are not known, the general processes can be suggested. An imaginary but plausible process by which sensory systems might have evolved is outlined in the following sections. Although this is largely based on guesswork, it should facilitate understanding the enormous complexity of mammalian sensory systems. Finally, the concept of a receptive field will be discussed because this has proven important in analyzing sensory pathways.

■ *General Evolution of Sensory Systems*

Imagine a simple, multicellular organism swimming in the primordial seas (Fig. 7-1, *A*). Two behaviors, among others, are essential to this organism's survival; it must find food and must avoid serving as another animal's meal. The evolution of these behaviors involves the development of a rudimentary analogue of a nervous system, including various sensory systems.

Sensory receptors or transducers sensitive only to specific nourishing chemicals in the sea might evolve as laterally paired cells on or near the organism's surface. These could be functionally connected to paired contractile units (analogous to muscles) in an ipsilateral fashion. This arrangement would cause the organism to move along an appropriate chemical gradient toward the source of food (Fig. 7-1, *A*). Note that this scheme requires both that the receptors induce contraction only in the presence of the appropriate chemicals and that information from these receptors be conducted along specific lines or channels. Thus activity along these lines can uniquely reflect the presence or absence of appropriate chemicals.

If the same contractile elements are also used in the behavior to avoid predators, however, confusion can result. For instance, the organism might have evolved the strategy of avoiding light in order to avoid predators that require vision. This could be accomplished by the evolution of laterally paired photoreceptors connected with the contractile units in a contralateral fashion. Again, activity in the photoreceptors and their "labeled lines" specifies the signal as visual in origin. This activity would cause the organism to swim away from sources of light (Fig. 7-1, *A*).

A problem for the organism in Fig. 7-1, *A* arises each time light and food appear simultaneously from the same direction. Evolution might favor the development of some sort of sensory integrator that could control the contractile elements based on the nature of all sensory

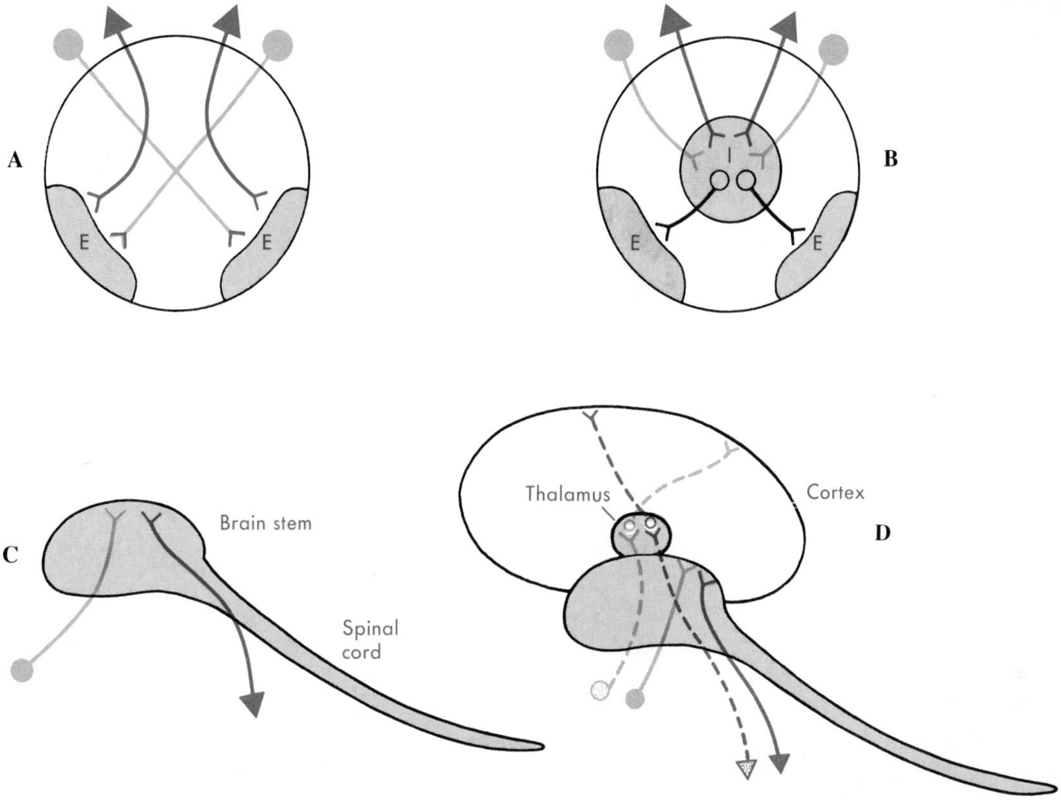

■ **Fig. 7-1.** Schema for a plausible but imaginary course of evolution for sensory systems. **A,** A primitive animal with bilateral symmetry and swimming in primordial seas must find food and avoid predators in order to survive. For food acquisition, a pair of specialized chemosensory elements *(colored triangles)* have evolved sensitivity to chemical nutrients in the environment; these are analogous to neural transducers. By virtue of their ipsilateral activation of contractile effector *(E)* elements (analogous to muscles), their activation causes the organism to orient toward the food source and swim down its chemical gradient to feed. To avoid predators it might be useful to keep to dark places to avoid visual detection. For this purpose a pair of photosensitive elements *(colored circles)* exists that activates the contralateral effector elements, thereby establishing a negative phototactic response for the organism. **B,** A problem with the primitive design in **A** is that when food and light appear from the same direction, the effector elements receive contradictory signals from the chemoreceptors and photoreceptors. Thus an integrator *(I)* has evolved to analyze the raw sensory signals and send a coordinated command to the effectors. **C,** The integrator may be viewed as a primitive brain that performs the same basic function as the spinal cord and brainstem as it may have appeared at the beginning of vertebrate evolution. For simplicity only two transducers are shown (analogous to the two pairs in **A** and **B**) that represent an early photoreceptor *(colored circle)* and pressure-sensitive receptor *(colored triangle).* **D,** Finally, as mammals appear, a great expansion of thalamus and cortex appears (although homologous structures seem to exist in nonmammalian vertebrates). Corresponding to this is the evolution of newer transducers and their central connections through thalamocortical pathways. However, note that the older spinal cord and brainstem pathways do not disappear as the newer portions of the sensory systems evolve. Indeed, these older pathways continue to contribute significantly to sensory processing.

signals received (Fig. 7-1, *B*). This integrator can be viewed as a primordial central nervous system.

The organism represented in Fig. 7-1, *B*, has the basic analogue of a nervous system. It is relatively easy to convert this to the plan evident in primitive vertebrates (Fig. 7-1, *C*); that is, the integrator can be represented by a dorsal neuraxis that consists of the spinal cord and brainstem. This neuraxis receives sensory input that originates with the receptors, and it controls the muscles (and other effector organs) through various motoneurons. Each receptor is specifically sensitive to a particular sensory quality, such as light, pressure, or certain chemicals, and each has its own labeled line to the neuraxis via centripetal axon projections. These enter either the spinal cord by way of dorsal roots or the brainstem by way of cranial nerves.

Basic Plan of Mammalian Sensory Systems

Evolution of the mammalian brain, which might have derived from such a hypothetical ancestral plan, involved a tremendous proliferation of the telencephalon, particularly with regard to thalamocortical pathways (Fig. 7-1, *D*). Phylogenetically newer portions of the sensory system reached the sensory cortex through appropriate thalamocortical pathways. However, evolution is a conservative process, and older sensory pathways generally are not discarded as the newer thalamocortical pathways are evolved. Mammalian sensory systems are consequently a combination of older (subcortical and extrathalamic) and newer (thalamocortical) pathways that cooperate and interact in analyzing the sensory environment. Unfortunately it is never possible in existing mammalian species to identify unambiguously the relative phylogenetic age of these components in any sensory system. Such tentative identifications are inferential rather than factual. Also, the properties and functions of "older" pathways or components may change as "newer" ones evolve. The important lesson here is that older subcortical sensory pathways may be important to perception. Too often in the past these older pathways have been swept under the rug like some evolutionary debris, but recently much deserved attention has been directed toward these subcortical pathways.

As an example of this organizational principle, the visual system will be considered briefly. In nonmammalian vertebrates, the optic tectum is the largest and most important central structure for vision. A similar condition probably held true for vertebrates before the evolution of mammalian geniculocortical pathways, although even the most primitive existing vertebrates, for example, elasmobranchs, exhibit small but distinct thalamic-telencephalic visual pathways. In any case, as these newer pathways evolved, the presumably older retinotectal pathways that involve the superior colliculus (the mammalian homologue of the optic tectum) did not disappear. Indeed, substantial recent evidence, considered in more detail later, emphasizes the importance of the superior colliculus and other subcortical structures to mammalian vision. It is thus insufficient to consider only the newer (retinogeniculocortical) part of the visual system. The same principle applies to other sensory systems such as the somatosensory and auditory systems. The following sections are organized around this concept.

How mammalian sensory systems are able to analyze the nature of stimuli in the environment can now be considered in a general way. Because transducers are relatively sensitive to one specific form of stimulus energy and because they are connected with little or no convergence or divergence to axons projecting toward the central nervous system, these pathways are *labeled lines* and their activity specifies *what* the stimulus is. Thus if a transducer is excited by extreme forms of inappropriate stimulus energy (remember that their sensitivity is relative), the percept is as if the appropriate stimulus energy were present. For instance, extreme pressure applied to the retina is perceived as flashes of light and not as a mechanical stimulus.

Fast-adapting or phasic cells are ideally suited to signal *when* a change occurs in the environment. Many transducers are phasic and thus can serve this function. As described later, central synaptic integration can convert tonic input signals into phasic signals. For example, the visual system, which has only tonic photoreceptors, can transform these within the brain to phasic responses that reliably signal temporal changes.

As indicated in the previous description of sensory receptors, however, tonic receptors and neurons are needed to signal stimulus intensity, or *how much* is present. This is accomplished with monotonic relationships among intensity, the size of the generator potential, and the resultant firing frequency of the central neurons involved. The slow adaptation of these cells sacrifices some precision of the intensity code in favor of an increase dynamic range.

Finally, the ability to determine precisely *where* a stimulus is located, or its spatial distribution, is also largely a function of the transducers. In the various sensory systems, these transducers are spatially arranged in peripheral tissue (retina, skin, etc.), and the spatial pattern of active transducers conveys the basic information required to determine the stimulus location. The central nervous system has mechanisms that then sharpen this spatial image.

Receptive Fields

Receptive field analysis is a useful tool in analyzing the functional organization of sensory systems. For any sensory neuron in the central nervous system, a *receptive field* can be described in terms of that area of transducer-containing tissue which, when properly stimulated, causes a change in the neuron's electrochemical properties. This change is usually revealed as a change in the firing rate. The precise region of transducer-containing tissue involved in a receptive field often depends on the nature of the stimulus used. For instance, a visual neuron has a receptive field related to a region of retina. If the retina is explored with small spots of light, flashed on and off, a map of the receptive field can be constructed with separate regions excited by light turned off, excited by light turned on, inhibited by light turned off, and/or inhibited by light turned on. If the

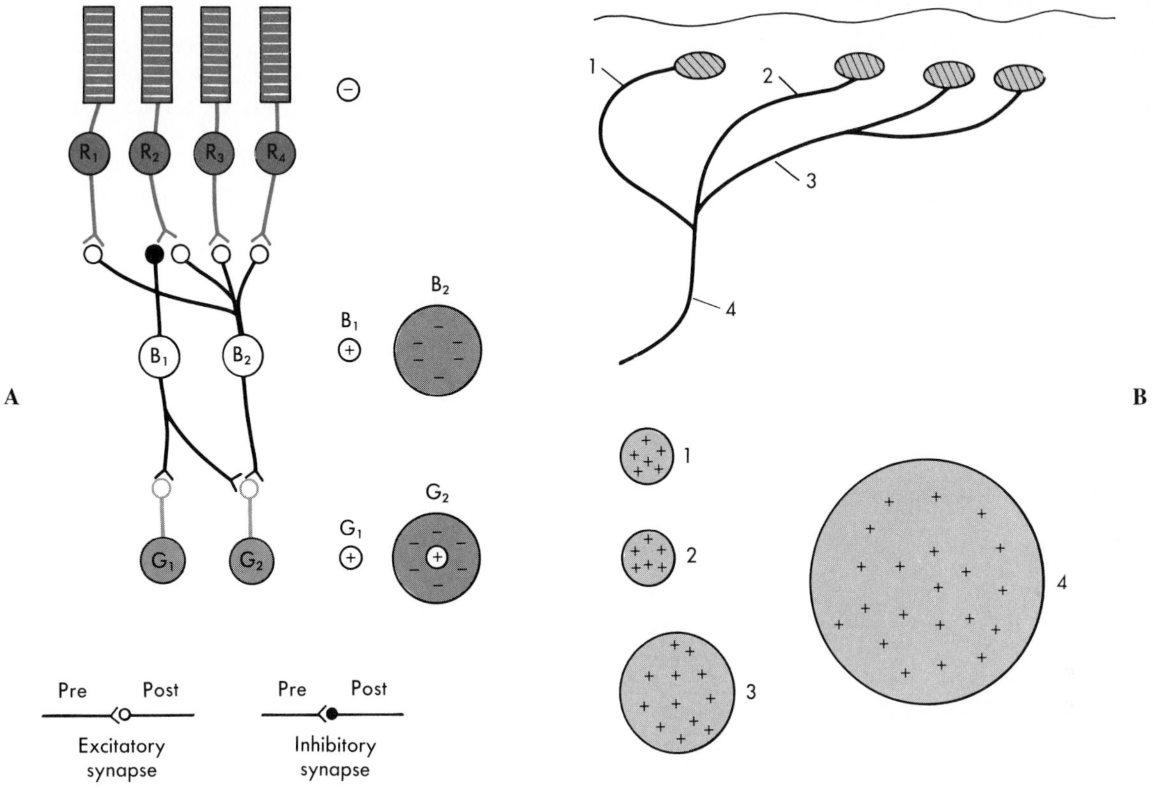

Fig. 7-2. Examples of neuronal receptive fields based on functional connections between sensory transducers and neuronal element under study. **A,** Visual receptive fields. The receptive field of each photoreceptor (R_1, R_2, R_3, or R_4) is equal to the patch of retina occupied by the photoreceptor. Because photoreceptors hyperpolarize to photic stimulation, the photoreceptor receptive field by convention is drawn as inhibitory and indicated with minus signs. For bipolar cell B_1, which receives an inhibitory input from photoreceptor R_2, the receptive field is equal in size but opposite in sign to that of R_2. For bipolar cell B_2, which receives excitatory input from each photoreceptor, the receptive field is equal to the spatial sum of those of R_1, R_2, R_3, and R_4. Likewise, the receptive fields of ganglion cells G_1 and G_2 can be inferred from the circuitry as shown. **B,** Somatosensory receptive fields. For this example assume that the action potential is generated near each distal tip of the axonal branches by the transducer mechanism and that these potentials are not antidromically (retrogradely) conducted toward a distal tip from a branch point. Thus receptive fields of distal branches *1* or *2* are simply equivalent to the skin area innervated by the single transducer attached to each. The receptive field recorded at point *3* is equal to the skin area innervated by the right pair of transducers, and the receptive field recorded more proximally at point *4* is the skin area innervated by all four transducers.

same receptive field is plotted by a spot of light moving across the retina, the map might look somewhat different from the map based on flashing spots, and it might also vary with changes in the speed or direction of movement. Indeed, some visual cells might exhibit strong responses to some stimulus parameters applied to a proscribed region of retina and none to other parameters.

Receptive fields, then, can be complicated to describe, but they are always related to areas of transducers that have been stimulated. Fig. 7-2 shows several simple examples of different receptive fields for visual and somatosensory cells. The receptive field for a neuron could be readily inferred if one could know the detailed functional connections between transducers, intervening neurons, and the neuron in question. Such detailed information on connectivity is rarely available. The value of a receptive field analysis is the converse. A knowledge of receptive field properties of many neurons at various levels in a sensory system allows inferences regarding the functional interactions among these neurons.

Bibliography

Journal article

Campbell, L.B.G., and Hodos, W.: The concept of homology and the evolution of the nervous system, Brain Behav. Evol. **3**:353, 1970.

Northcutt, R.G.: Evolution of the telencephalon in nonmammals, Annu. Rev. Neurosci. **4**:301, 1981.

Books and monographs

Ariens Kappers, C.U., Huber, G.C., and Crosby, E.C.: The comparative anatomy of the nervous system of vertebrates, including man, New York, 1967, Hafner Press. (Originally published in 1936.)

Nauta, W.J.H., and Karten, H.J.: A general profile of the vertebrate brain, with sidelights on ancestry of cerebral cortex. In Schmitt, F.O., editor: The neurosciences second study program, New York, 1970, Rockefeller University Press.

CHAPTER

8

The Visual System

The Eye and Physiological Optics

Although the visual system begins with the transduction process in the retina, a complete understanding of vision must take into account the transformations in photic stimuli as they pass through the eye's optics to the retina. Thus this section discusses the role of the eye as an optical instrument.*

General Structure

Fig. 8-1 shows a horizontal section through the middle of the human right eye as viewed from above. The eye has three coats. Most of the outside of the eye is covered by the *sclera,* a tough connective tissue that reflects most light striking it. It thus appears white in normal illumination. At the front of the eye, the sclera is replaced by a transparent outer coat known as the *cornea*. The circular junction between sclera and cornea is called the *limbus*. The middle and inner coats occupy only the posterior two thirds of the eye. The middle coat is the *choroid,* which contains a rich blood supply as well as considerable melanin. This pigment absorbs most light striking it. The inner coat is the retina.

The posterior portion of the eye is occupied by the *vitreous body* or *vitreous humor,* a clear, jellylike substance. The vitreous body is bounded at its anterior end by the *ciliary body, zonule fibers,* and *lens;* these structures will be discussed more thoroughly later. The *iris* lies anterior to the lens. It reflects and absorbs most of the light that strikes it, but some light passes through a circular hole, the *pupil,* in the center of the iris. The remaining space between the lens and cornea is occupied by a watery, clear substance called the *aqueous humor*. This fluid is segregated into two compartments: an *anterior chamber* in front of the iris and a *posterior chamber* between the iris and lens.

The *optic disc* and *fovea* appear as depressions in the retina. These form two obvious and important landmarks that can be seen with an ophthalmoscope. The fovea is the portion of retina specialized for most acute vision. The eye is typically directed so that the most important point in the visual world, the *fixation point,* is imaged onto the fovea.

The fovea serves as the center of a frame of reference for locations and directions on the retina. All retina between the fovea and nose is *nasal* retina, and that between the fovea and temple is *temporal* retina. The view from above the right eye (Fig. 8-1) thus depicts nasal retina to the left of the fovea and temporal retina to the right. If the left eye were also drawn in this figure, the left portion of its retina would be temporal and the right, nasal. The optic disc lies in each nasal retina. The retina above the fovea is *superior,* and below is *inferior* retina. Superior and inferior retinas are roughly equal in size, but the nasal retina is somewhat greater in extent than the temporal retina. Finally, nasal, temporal, superior, and inferior directions on the retina refer to relative positions in this coordinate system. For instance, the fovea is temporal to the optic disc on the retina; similarly, points *A* and *B* in Fig. 8-1 are both in the temporal retina, although point *A* is nasal to point *B*.

Fig. 8-1 also shows the *visual* and *optic axes*. The visual axis is a line that passes through the fovea, nodal point of the eye, and fixation point. The *nodal point* of an optical system can be defined in terms of the relationship between an object and the reversed, inverted image formed by the optics. Straight lines can be drawn between each point on the object and its corresponding point on the image. All of these lines intersect at the nodal point (Figs. 8-1 and 8-2). The visual axis is the axis along which the eye is directed. It is different from the eye's optic axis, which is the axis of optical symmetry, or the line that passes through the center of the cornea, pupil, and lens. It, too, passes through the

*The reader is expected to have a basic knowledge of optics that includes an understanding of (1) how lenses form images and (2) the meanings of terms such as *refraction, focal length, diopter,* and *spherical* or *chromatic aberration*.

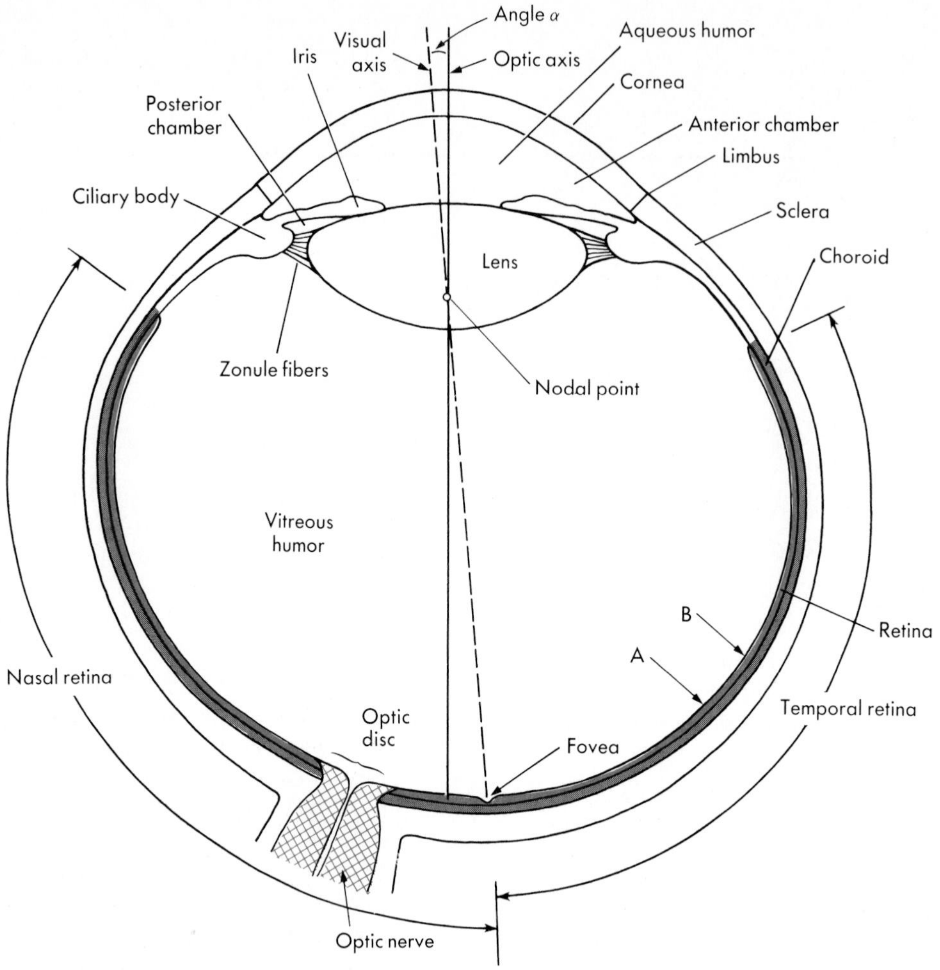

■ **Fig. 8-1.** Schematic diagram of a horizontal section through the human right eye as viewed from above. (Redrawn from Walls, G.L.: The vertebrate, eye and its adaptive radiation, Bloomfield Hills, Mich., 1942, Cranbrook Institute of Science.)

■ **Fig. 8-2.** Image formation by the eye. If straight lines are drawn between equivalent points on the object *(O)* and image *(I)*, they intersect at the nodal point. The relationship between *D* (approximately the distance between the object and the eye), *d* (roughly the eye's focal length), *L* (object size), *l* (image size), and θ (the angular subtense of the object and image) is given. Because *d* is a fixed constant, *l* and θ bear a constant trigonometric relationship to each other. Note, however, that the approximation that tan θ/2 = 1/2d breaks down for large values of θ or *l*, since *l* represents an arc rather than a chord of a circle.

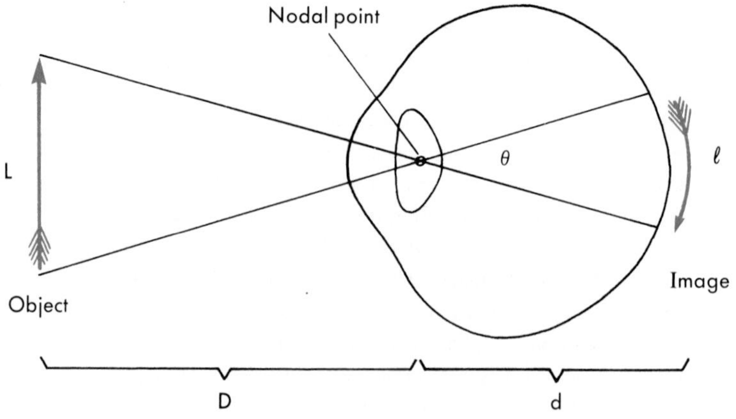

$$\frac{\ell}{L} = \frac{d}{D} \qquad \tan\frac{\Theta}{2} = \frac{L}{2D} \cong \frac{\ell}{2d}$$

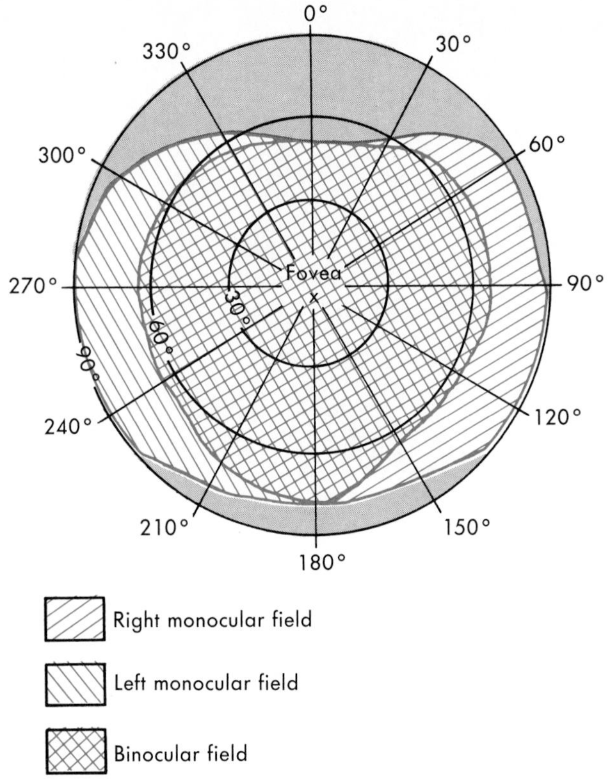

Fig. 8-3. Human visual fields shown in polar coordinates. The fields for the right and left eye are shown separately. The binocular segment of visual field is the overlap region of the two monocular fields. The two monocular segments are the peripheral crescents that can be seen by only one eye.

nodal point. Typically, the optic axis intersects the retina nasal to the fovea (Fig. 8-1).

The angle between the optic and visual axes is called the angle α (Fig. 8-1). In humans the angle α averages roughly 5 degrees in each eye. Consequently, fixation of a distant object, which leads to parallel visual axes, also leads to about a 10-degree angle between the optic and visual axes. Ignorance of the angle α can lead to misdiagnosis of the clinical condition of *strabismus,* which is defined as the inability to direct both foveas to the same fixation point. When one informally judges the direction in which an eye is aimed, one usually estimates the direction the center of the pupil seems to assume. This refers to the optic and not the visual axis. If the angle between these axes is large enough, one might not diagnose eyes that are pathologically crossed, because the optic axes might be parallel. Conversely, eyes might be misdiagnosed as pathologically diverged because the optic axes are divergent, when in fact the visual axes are parallel. Although the angle α averages roughly 5 degrees, it is quite variable among humans. The angle may be much larger or smaller than 5 degrees and can even be reversed in sign; that is, in rare cases, usually involving a high degree of myopia (p. 100), the visual axis intersects the retina temporal to the fovea.

Fig. 8-2 demonstrates a final point regarding the eye's general structure. The size of an image on the retina can be specified in two equivalent fashions: (1) the linear extent of the image or (2) the angle formed by the nodal point and limits of the object or image. For the human retina each millimeter of retina equals slightly more than 3 degrees of visual space. The optic disc is roughly 5 degrees in diameter, and the size of the visual field seen by one eye is approximately 150 degrees horizontally by 120 degrees vertically (Fig. 8-3).

Images Formed by the Eye

A major role of the nonneural parts of the eye is to form an optically ideal image on the retina. The refractive power of the cornea and lens focuses images on the retina. The transparency of the cornea and lens, as well as that of the aqueous and vitreous humors, preserves the quality of these focused images with little intensity loss or light scatter. The reflective sclera and absorbant choroid serve to prevent improperly focused light from reaching the retina and degrading the image.

Camera optics. In many ways the eye can be compared to a camera (Fig. 8-4). A good camera has a compound lens system (two or more elements to minimize optical distortions) that forms an inverted and reversed image of an object on the photosensitive film plane. The focus can be adjusted by changing the distance between the lens and the film. The image formed can be thought of as a precise, point-to-point map of the visual space within the camera's field of view; each of the points in space can be precisely located in the image once the direction and orientation of the camera are known.

Stray light (i.e., photons that are not imaged by the optics) reduces contrast in the image. Stray light is prevented from reaching the film in two ways. The camera body is opaque, so light can enter only through the lens. Also, internal reflections of stray light are minimized by the camera's interior, which is painted black to absorb such light.

Finally, the camera has an iris/diaphragm system in the lens to control the size of the entrance aperture of the optical system. This serves two purposes. First, the amount of light striking the film is controlled, because it is proportional to the cross-sectional area of the aperture. This is relatively unimportant because the shutter speed also controls the amount of light. Second and more importantly, *depth of field* and overall optical quality are controlled by the aperture size. Fig. 8-5 shows how a small aperture produces a greater depth of field than does a large aperture. Each optical system has an aperture size that optimizes the optics. The size is usually intermediate; it is a compromise between

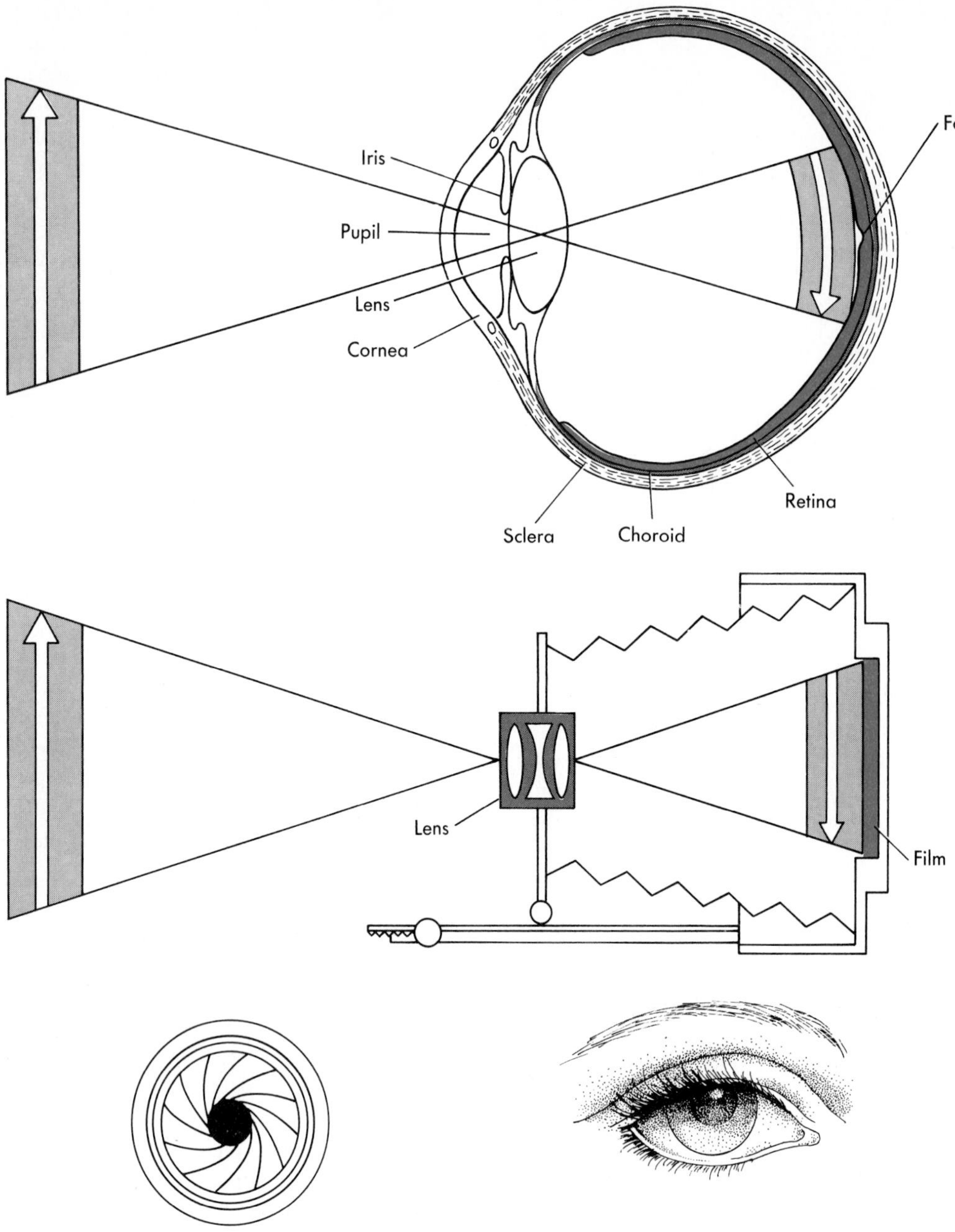

■ **Fig. 8-4.** Comparison of eye and camera. Many close parallels can be seen in the optical properties of eyes and cameras, only some of which are illustrated here. See text for details. (Redrawn from Wald, G.: Sci. Am. **183:**32, 1950.)

smaller sizes, which reduce spherical aberrations and increase depth of field, and larger sizes, which minimize diffraction of the image.

Physiological optics. Each of the optical features considered in the previous section has a counterpart in the eye, which has a compound lens system that forms an inverted and reversed image on the photosensitive retina. The compound lens system has two chief components: the cornea and lens. The cornea is actually responsible for most of the refractive power of the eye, but its power of approximately 43 diopters (D) is fixed. The lens has less power but can be continuously adjusted between approximately 13 and 26 D to change the focus of the eye. Most of this change in power, or *accommodation,* results from changing the shape of the lens, although there is also some displacement of the lens relative to the retina.

The retinal image can be specified in terms of the fixation point imaged on the fovea; that is, once an observer in a normal, upright position fixates a point, images of other points can be precisely specified in terms of their retinal locations.

Stray light stimulation of the retina is minimized in two ways. First, light is prevented from reaching the

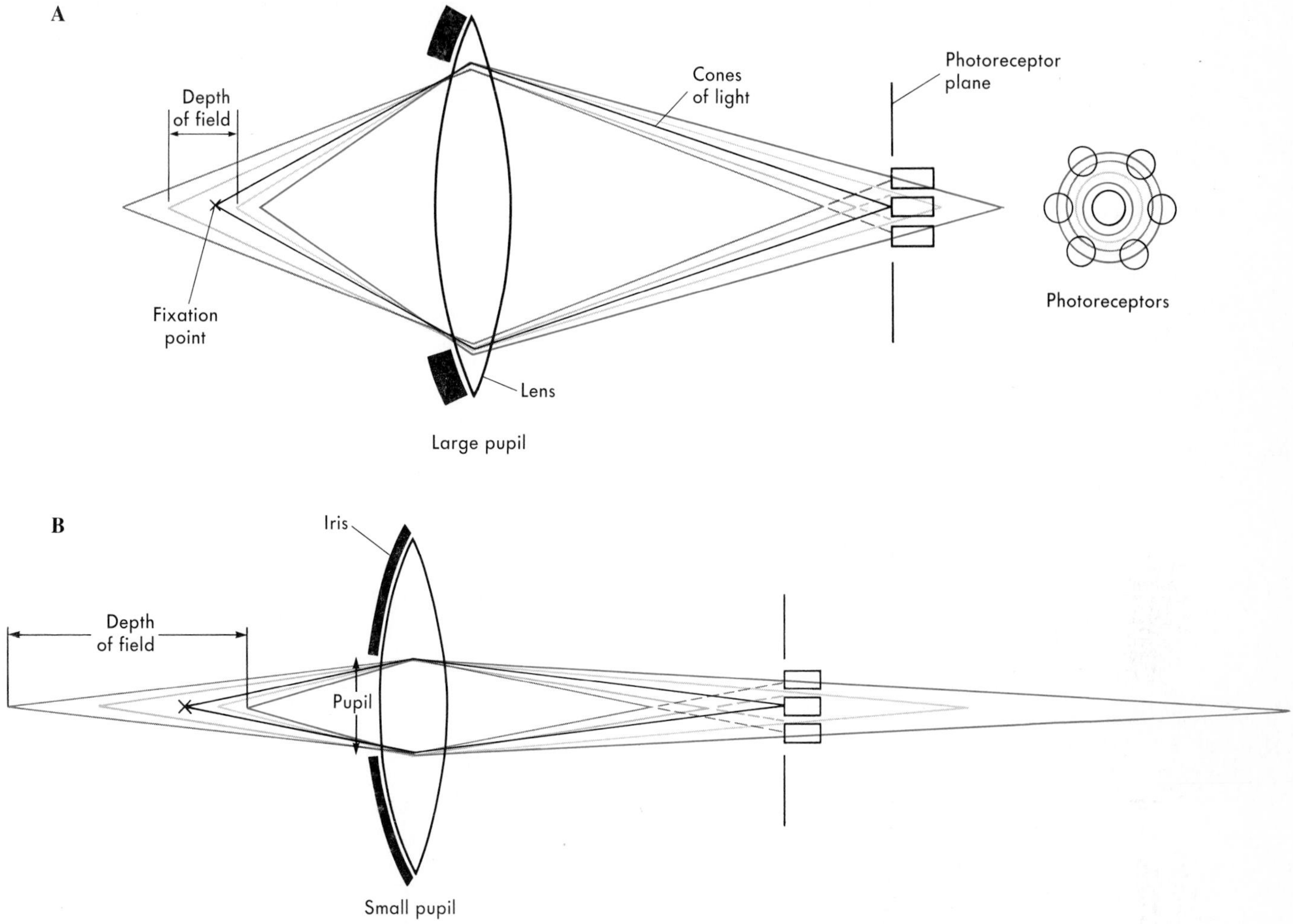

■ **Fig. 8-5.** The relationship between pupil size and depth of field. The imaging for a large pupil is drawn above **(A),** and that for a small pupil is drawn below **(B)**. As rays of light from a single object point become focused, they describe a cone of light, and the intersection of these cones with the image plane (i.e., the retina) defines "blur circles." The inset at the far right of each drawing shows a view of the photoreceptors and blur circles as might be seen from looking onto the retinal surface. These blur circles become smaller as the object points become better focused, and they are single points in the image plane for object points that are perfectly focused. Imprecise focus thus causes some object points to be imaged as blur circles. However, as long as these circles impinge on no more than a single photoreceptor, no blur will be perceived, and in practical terms, excellent focus results. Thus if the eye is accurately focused onto a single point *(x),* there is a range in front of and behind that point within which other points are effectively focused because their blur circles fall onto a single photoreceptor. This range is called the *depth of field* or *depth of focus*. Note that the rays from the object in focus for each drawing (black rays) converge to a point on the middle photoreceptor; the rays from both the "near point" (i.e., the midcolor rays from the edge of the depth of focus closest to the eye) and the "far point" (i.e., the lightest color rays from the edge of the depth of focus farthest from the eye) form blur circles that cover the middle photoreceptor but do not impinge on neighboring photoreceptors; and the rays outside the depth of focus (darkest color and gray rays) form blur circles that impinge on these neighboring photoreceptors. A larger pupil **(A)** creates cones with larger vertex angles than does a smaller pupil **(B)**. Consequently, larger pupils result in smaller depths of focus.

retina along paths that do not pass through the cornea and lens. The sclera reflects most of this light, and the choroid absorbs most of the remainder. Second, light that does enter the eye and is not absorbed by photopigment in the photoreceptors is prevented from reflecting to inappropriate photoreceptors by absorption in the *pigment epithelium,* a layer of retina that lies just beyond the photoreceptors. Pigment in the choroid and pigment epithelium are thus important in the prevention of stray light and reduced contrast. Albinos lack such pigment, and the image of the world they see is thus rather poor. Albinos have other visual deficits as well that seem to involve abnormal visual pathways, but much of their visual deficit can be attributed to poor optics.

Finally, the eye has an iris that creates an adjustable diaphragm or pupil. Part of the adjustment serves to control the light levels striking the retina, but this function is relatively unimportant. The human pupil can be altered in area by approximately a factor of 30 (1.5 to 8mm in diameter). Since at any one time photoreceptors can respond to roughly a 4 log unit range of intensities, whereas the human visual system can function over roughly a 10 log unit range, it is clear that photoreceptor adaptation is much more important than is pupil size in the visual system's reaction to different illumination levels.

Much more important is the pupil's role in the optical quality of the eye, particularly with regard to depth of field (Fig. 8-5). Control of depth of field becomes more important for near than for distant vision, because the depth of field is always larger the further away the optical system is focused. Thus pupil size is not of great concern for distance vision. For near vision, such as during threading a needle, the pupil becomes smaller to ensure adequate depth of field. Otherwise too much of the relevant field of view would always be blurred.

The *near reaction* is a *synkinesis* that involves three related actions for transfer of fixation from a more distant to a nearer object: the lens accommodates, the visual axes converge, and the pupil constricts. This synkinetic pupillary constriction should be distinguished from the reflex constriction to light. The neural pathways involved in the synkinetic constriction are not known. The reflex constriction involves a pathway from the retina to cells in the pretectum, from these cells to preganglionic cells in the Edinger-Westphal portion of the oculomotor nucleus, and from these to postganglionic cells that innervate the pupillary constrictor muscle. Argyll Robertson described patients who lack the pupillary reflex to light but who still demonstrate pupillary constriction when asked to shift fixation from a distant to a near object. This abnormality is still known as the *Argyll Robertson pupil.* It is, however, rarely seen, and its pathophysiology is not well understood.

Dynamic Properties of the Iris

Pupillary changes are brought about by changes in the configuration of the iris. Two dominant smooth muscle groups are found in the iris. One is circumferentially organized and receives parasympathetic innervation; its contraction causes pupillary constriction. The other is radially organized and receives sympathetic innervation; its contraction causes pupillary dilation.

Dynamic Properties of the Lens and Accommodation

Normal accommodation. Accommodation is a process whereby the eye changes its focal length, primarily by changes in lens shape and partially by a shift in lens position. The eye or lens is *relaxed* when the lens is most flat and in its most posterior position. Different *degrees of accommodation* imply a more curved lens that shifts forward slightly. Thus the relaxed eye has a longer focal length and is focused furthest away; the accommodated eye has a shorter focal length that causes the eye to be focused for closer objects.

Fig. 8-6 shows details of the structures responsible for accommodation. The lens itself is a plastic, jellylike substance that is contained within a tough, elastic *lens capsule.* This is analogous to petrolatum placed under pressure inside a balloon. The shape of the lens inside its capsule tends to be spherical. However, tough *zonular fibers* are attached at one end around the equator of the lens capsule, and at their other end they attach to the *ciliary body.* The ciliary body, in turn, is firmly attached at its other end to the sclera. Tension from the zonular fibers pulls the lens and its capsule into a relatively flat shape. The amount of tension in the zonular fibers thus determines the extent of the flattening. The tension is controlled by action of smooth muscles that are located in the ciliary body and receive parasympathetic innervation. These muscles run predominantly in a circumferential fashion and thus act as a sphincter. Contraction of the ciliary muscle reduces tension in the zonular fibers. This allows the lens to assume a more spherical shape as a result of the passive elastic properties of the lens capsule. The entire lens also moves slightly forward during ciliary muscle contraction. Tension on the zonular fibers is maximal and lens curvature minimal during relaxation of the muscle. The lens then is in its most posterior position. Accommodation thus implies ciliary muscle contraction and near vision; this muscle is relaxed to permit distance vision.

Presbyopia. Accommodation achieves nearer focus largely because of the lens capsule's elasticity. Imagine a lens capsule stretched out of shape and no longer elastic. Ciliary muscle contraction would have a negligible

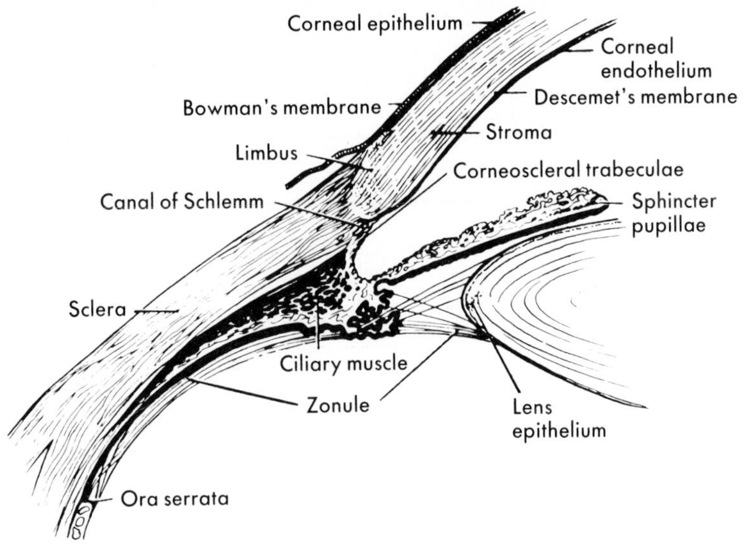

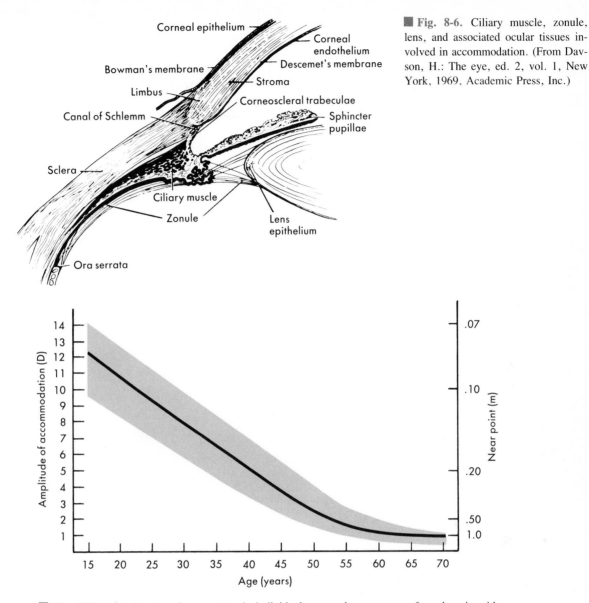

■ **Fig. 8-6.** Ciliary muscle, zonule, lens, and associated ocular tissues involved in accommodation. (From Davson, H.: The eye, ed. 2, vol. 1, New York, 1969, Academic Press, Inc.)

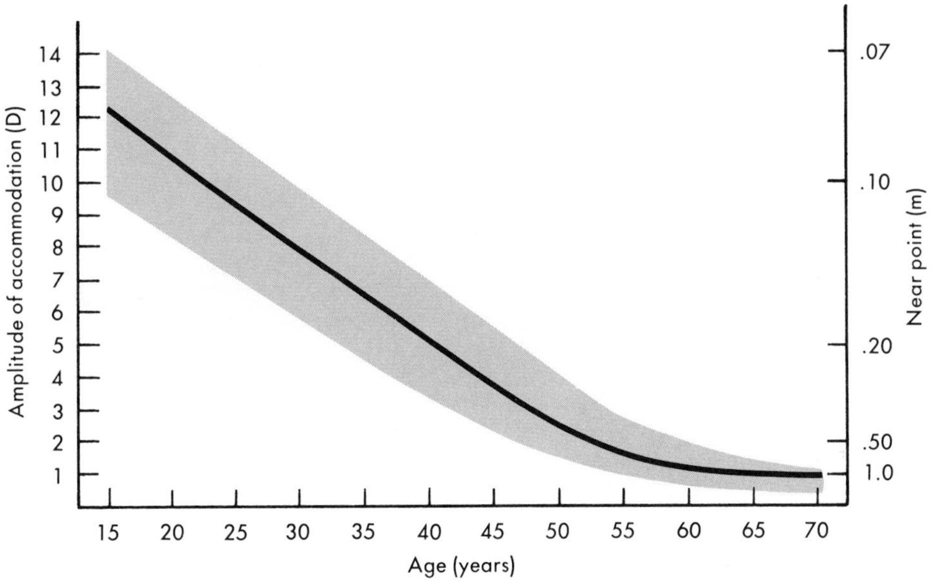

■ **Fig. 8-7.** Typical values for emmetropic individuals; normal appearance of presbyopia with age. As a normal consequence of aging, the amplitude of accommodation decreases and the near point (i.e., the closest point that can be focused) increases. *D*, Diopters. (Redrawn from Mountcastle, V.B., editor: Medical physiology, ed. 14, St. Louis, 1980, The C.V. Mosby Co.)

effect on lens shape, and the ability to focus near objects would be severely compromised. This is precisely the condition of *presbyopia,* a normal concomitant of the aging process during which tissue elasticity generally declines throughout the body. Consequently, focusing ability is lost with age (Fig. 8-7), and older people often wear "reading glasses" or bifocals for close work.

■ *Emmetropia and Ametropia*

Probably the most common and easily treated disturbance of vision is improper focus of the eyes. Normal focus is called *emmetropia*. Improper focus is called *ametropia,* and several forms exist. *These conditions are always defined with respect to the relaxed eye.* This is the reason that tests for emmetropia or ametropia usually involve pharmacological paralysis of accommodation (achieved by appropriate eye drops) to ensure that the relaxed eye is being tested. Because the relaxed eye is tested and is focused to its furthest point, that is, its longest focal length, the test measures its ability to focus distant objects. In practice, anything further than approximately 6 m from the human eye can be treated as infinitely distant for these tests.

The relaxed, emmetropic eye sharply focuses distant objects (Fig. 8-8, *A*). Nearer objects can be focused through accommodation. A relaxed ametropic eye cannot focus distant objects. Three major forms of ametropia exist.

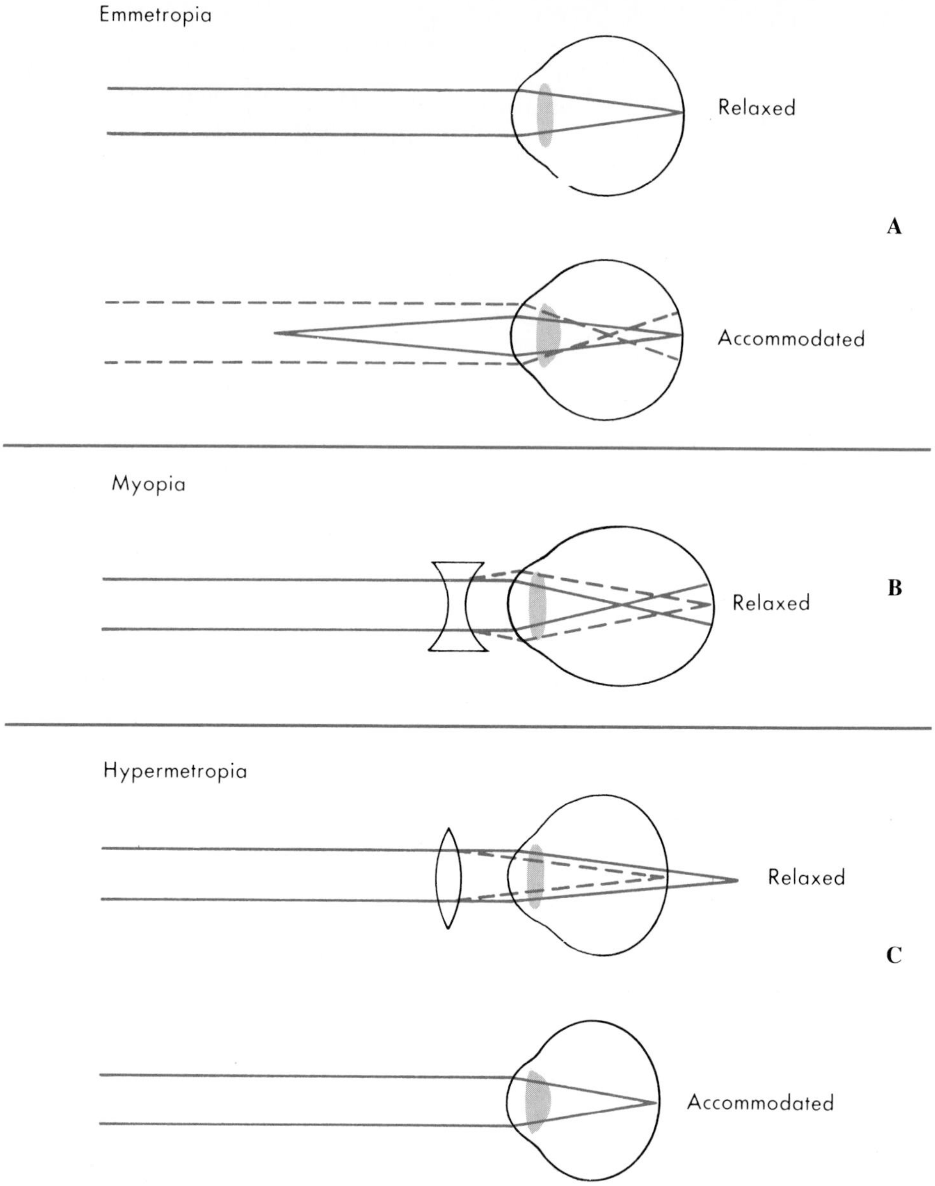

Fig. 8-8. Emmetropia and spherical ametropias. **A,** Emmetropia. When the eye is relaxed *(upper),* rays from an infinitely distant point source, which are effectively parallel, converge at the retina where they are focused. For practical purposes any distance greater than 6 m (or 20 feet) is considered ''infinitely distant'' and leads to parallel rays from a point source. Accommodation of the lens *(lower)* permits points from closer objects to be focused onto the retina *(solid lines),* while distant objects are focused in front of the retina *(dashed lines).* **B,** Myopia. Parallel rays are focused in front of the retina *(solid lines),* typically because the eyeball is too long. This is corrected by the addition of a negative lens *(dashed lines).* **C,** Hypermetropia. In the relaxed eye *(upper)* parallel rays are focused behind the retina *(solid lines),* typically because the eyeball is too short. This is corrected by the addition of a positive lens *(dashed lines).* Note that for limited hypermetropia (i.e., within the range of accommodation) lens accommodation can bring distant objects to focus on the retina *(lower).* Thus the eye should always be relaxed to test for hypermetropia.

Myopia (nearsightedness) is illustrated in Fig. 8-8, *B.* Distant objects are focused in front of the retina because the length of the eye is too great for its refractive power. Myopia can be corrected by a simple spectacle or contact lens of appropriate negative power.

Hypermetropia (hyperopia or farsightedness) is illustrated in Fig. 8-8, *C.* Distant objects are brought to focus behind the retina because the eye is too short for its refractive power. The correction in this case is a positive lens. Hypermetropia is often confused with presbyopia for two reasons. First, constant accommodation can compensate for mild hypermetropia so that distant objects are always in focus, and no problem may be evident. Second, less accommodation is thus available for closer focusing, with the result that the closest point that can be focused is rather distant. Constant accommodation, which an uncorrected hypermetropic individual might employ, is not desirable. Instead, such an individual should wear an optical corrective device constantly. In contrast, presbyopic persons limit such correction to close work only.

Astigmatism is a more complex form of ametropia. An emmetropic eye or well-designed optical system must be able to image objects clearly regardless of the object orientation in visual space; that is, an emmetropic eye can readily focus lines oriented vertically, obliquely or horizontally. In practice, such an eye is radially symmetrical because the radii of curvature of the cornea, lens, and retina are constant for all meridians. An *astigmatic eye* lacks such symmetry; thus not all orientations at a given distance can be focused (Fig. 8-9). The usual cause of this ametropia is a cornea that has different radii of curvature for different meridians. Thus the corneal surface does not approximate a spherical surface. More rarely, astigmatism can be caused by a lens or retinal shape that is similarly distorted. Astigmatism can be corrected by a spectacle lens with appropriately different radii of curvature for different orien-

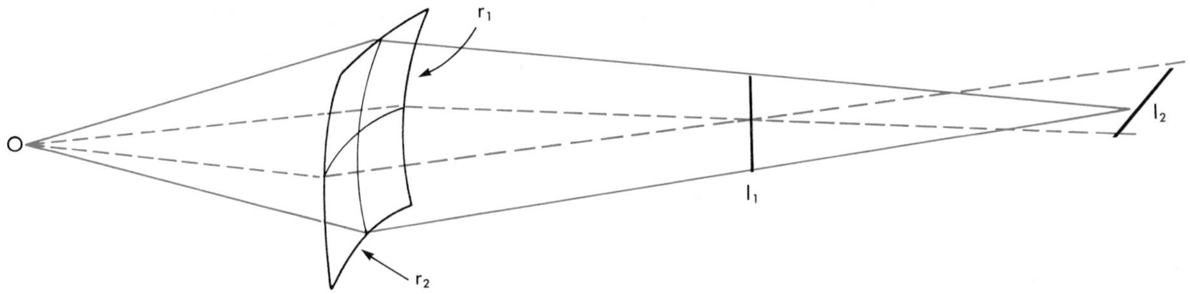

■ **Fig. 8-9.** Astigmatism caused by a nonspherical corneal surface. A section of the cornea is shown with different radii of curvature for the vertical and horizontal axes (respectively, r_1 and r_2). In this example, $r_1 > r_2$; thus the horizontal axis of the cornea has more dioptric power and a shorter focal length than does the vertical axis. Because of this, rays (colored lines) from a single point object *(O)* cannot be imaged in a single point. Rays that span the vertical axis of the cornea *(solid colored line)* are focused at I_2, but the rays spanning the horizontal axis *(dashed colored line)* are focused closer at I_1. As a consequence, a vertical line is imaged at I_1 and a horizontal line at I_2. In between, the object point is imaged as an oval blur. Because the object point is nowhere imaged as a point, blur is always present. Such astigmatism can be corrected by a nonspherical (or "cylindrical") spectacle lens that creates an opposite astigmatism to cancel out the cornea's astigmatism. Alternatively, a contact lens can optically replace the corneal surface with a spherical refractive surface and eliminate the astigmatism.

tations. Contact lenses can compensate for most forms of astigmatism that result from corneal distortion, since the spherical surface of the contact lens optically replaces the distorted corneal surface.

Finally, it should be reemphasized that presbyopia is not a form of ametropia, and it has a different etiology. Indeed, it is common for an individual to have presbyopia and ametropia, and these conditions independently contribute to image formation and need to be corrected separately. Also, astigmatism can occur in addition to myopia or hypermetropia. Finally, although it is most common for an individual to have roughly equivalent emmetropia or ametropia in each eye, exceptions to this occur. When the two eyes differ in this regard, the condition is known as *anisometropia*.

■ Retina

■ Layering and Regional Variations

General layering scheme. Fig. 8-10 shows a cross section through the human retina with ten layers indicated. This layering pattern is common to all vertebrates. Layer 1 is the *pigment epithelium*. It is adjacent to the choroid on one side and the photoreceptors on the other. Cells in this layer are nonneural and serve at least two functions. Because they contain pigment that absorbs photons not captured by the photoreceptors, they reduce scattered light. The pigment epithelium also serves a nutritive and metabolic function by acting as an intermediary between the choroidal blood supply and the photoreceptors. Layer 2 is the *receptor layer*. It is divided into layers comprising the *outer* (2a) and *inner* (2b) segments of rods and cones. Layer 3 is the *external limiting membrane*. This is not a true membrane, but rather it is composed of closely apposed processes from glial elements, known as Müller cells, with somata in layer 6. Layer 4 is the *outer nuclear layer*, and it contains the somata of the photoreceptors. Those of the cones are in the outer half, and those of the rods are in the inner half. Layer 5 is the *outer plexiform layer*. It represents a zone of synaptic interactions among the photoreceptors and certain of the retinal interneurons. Layer 6 is the *inner nuclear layer*. It contains the somata of Müller cells and the interneurons. These interneurons include horizontal cells, bipolar cells, amacrine cells, and interplexiform cells. Layer 7 is the *inner plexiform layer*. It represents a zone of synaptic interactions among certain of the interneurons and retinal ganglion cells. Layer 8 is the *ganglion cell layer*, and it contains the somata of these neurons. Layer 9 is the *optic fiber layer*, and it contains unmyelinated ganglion cell axons that course toward the optic disc. These axons become myelinated as they penetrate the optic disc, and they represent the only output from the retina. Finally, layer 10 is the *inner limiting membrane*, which is not a true membrane but is formed by apposed processes of the same Müller cells that form the outer limiting membrane. Below layer 10 is the vitreous humor.

Neural input to the retina from the rest of the brain has never been demonstrated unequivocally in any mammal, although such a pathway clearly exists in many species of nonmammalian vertebrates. Details of functional interactions among the retinal neurons are provided later, but first we shall consider variations in the general plan as illustrated in Fig. 8-10.

Blood supply. The blood supply of the retina has two general divisions. The ciliary artery penetrates the sclera at the back of the eye to distribute in the choroid as the *choriocapillaris,* which is a rich network of

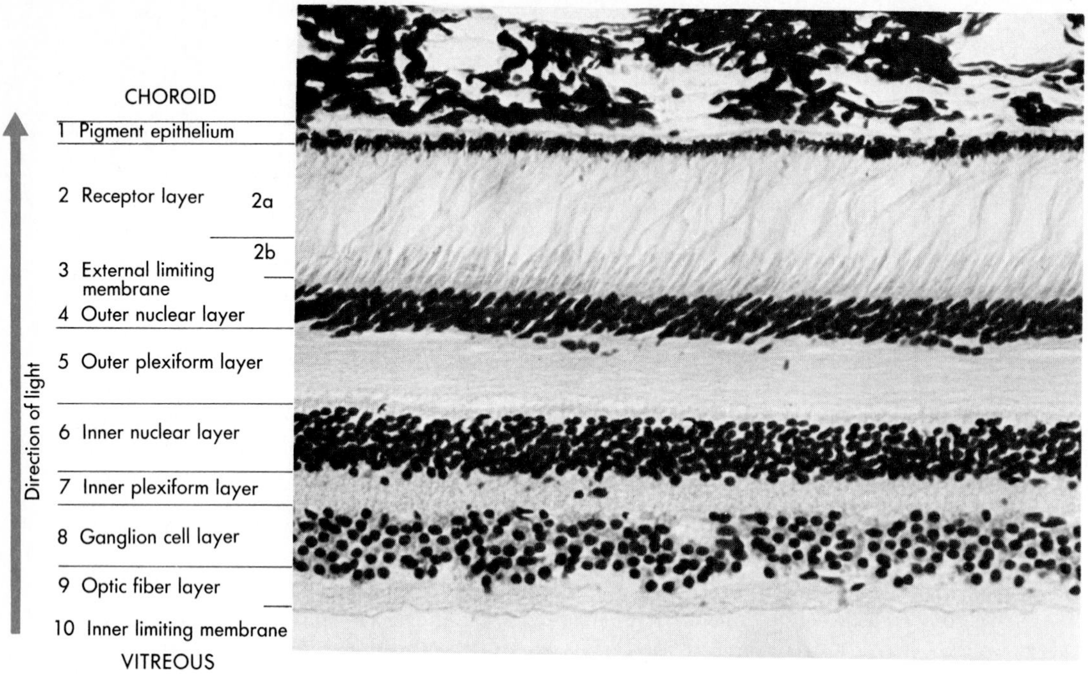

■ Fig. 8-10. Cross section of a macaque monkey's retina. The arrow denotes the direction of light as it passes through the retinal layers to reach the photoreceptors. (Nissl stain; courtesy R.E. Weller.)

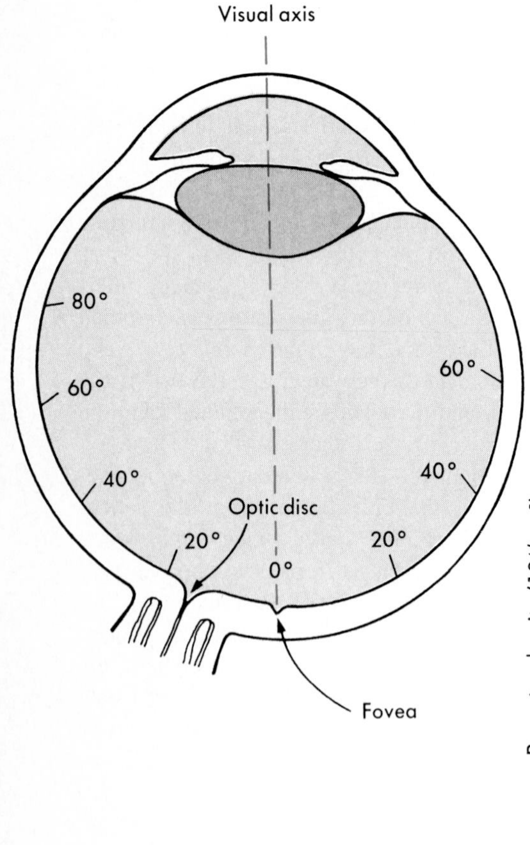

■ Fig. 8-11. Change with eccentricity in density of rods *(colored curve)* and cones *(black curve)*. Cone density peaks at the fovea, and that of rods peaks parafoveally, but both density functions decrease with eccentricity. (Redrawn from Cornsweet, T.N.: Visual perception, New York, 1970, Academic Press, Inc.)

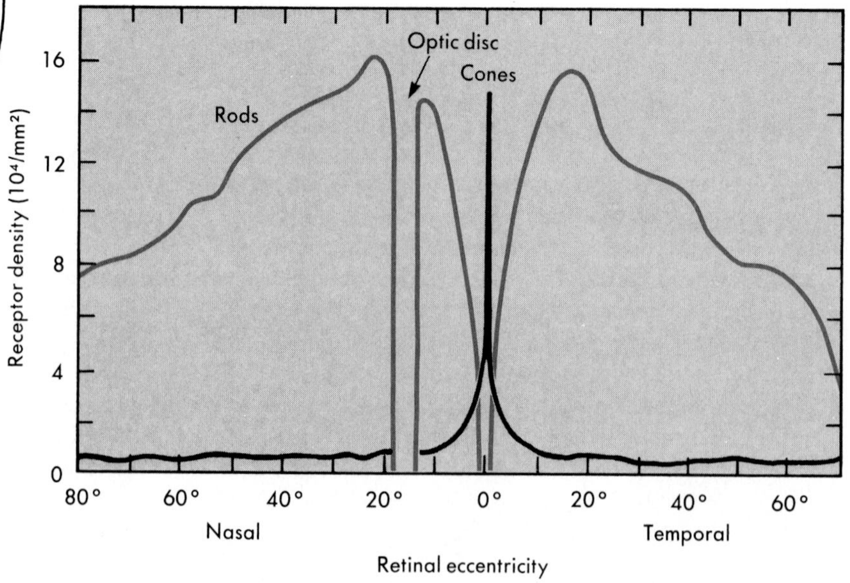

blood vessels that feeds the photoreceptors. As noted earlier, the pigment epithelium serves as an important link in this nutritive function. Several *vorticose veins* collect at the back of the eye to drain the choriocapillaris. The remaining retinal neurons are fed from blood vessels lying on the vitreal surface of the retina. These vessels are the *central retinal artery* and *central retinal vein*. They penetrate the globe at the optic disc, branch repeatedly, and fan out to cover the retina (see Fig. 8-13).

Changes with eccentricity. In general, the density of neurons in all layers decreases monotonically as the distance from the fovea, or *eccentricity,* increases. Retinal thickness consequently decreases with increasing eccentricity. This is illustrated for rods and cones in Fig. 8-11, but similar functions apply to the other retinal cell types as well (Fig. 8-12). As a result, visual performance declines monotonically as the test stimulus is imaged further from the fovea. For example, spatial resolution, or the finest detail that can be resolved, declines with eccentricity from the fovea.

Fovea. As seen in Fig. 8-12, the fovea is a depression, or pit, in the retina. The purpose of this structure probably relates to the curious placement of photoreceptors as seen in Fig. 8-10. Because light enters the retina from the vitreous humor, photons must pass through most of the retina to reach the rods and cones. Although fairly transparent, these retinal layers are not optically ideal, and the resultant image at the photoreceptors is somewhat degraded. The effect would be noticeable only for species with particularly acute vision such as certain birds, reptiles, fish, and mammals, including humans and other primate species. Even in these species the reduced visual performance with reduced retinal cell density away from the fovea suggests that optical degradation by the retinal layers becomes important only where cell density is sufficiently high.

The fovea provides a solution for this problem by sweeping away from the path of light most unnecessary tissue in order to optimize the image at the photoreceptors. As shown in Fig. 8-12 for the human fovea, this excavation leaves only layers 1 to 4, and thus only the photoreceptors are present among the retina's neural components. Because of the excavation, image quality is best at the fovea, which is the only region that can benefit from it.

The fovea of humans contains only cones; rods do not appear until the nearby parafoveal region is reached. Thus cone density peaks in the fovea, whereas rod density peaks in an annulus surrounding the fovea (Fig. 8-11). A simple experiment on a starry evening will illustrate this arrangement. Outside in the darkness the illumination is so dim that only the rods function; the less sensitive cones do not respond. In such illumi-

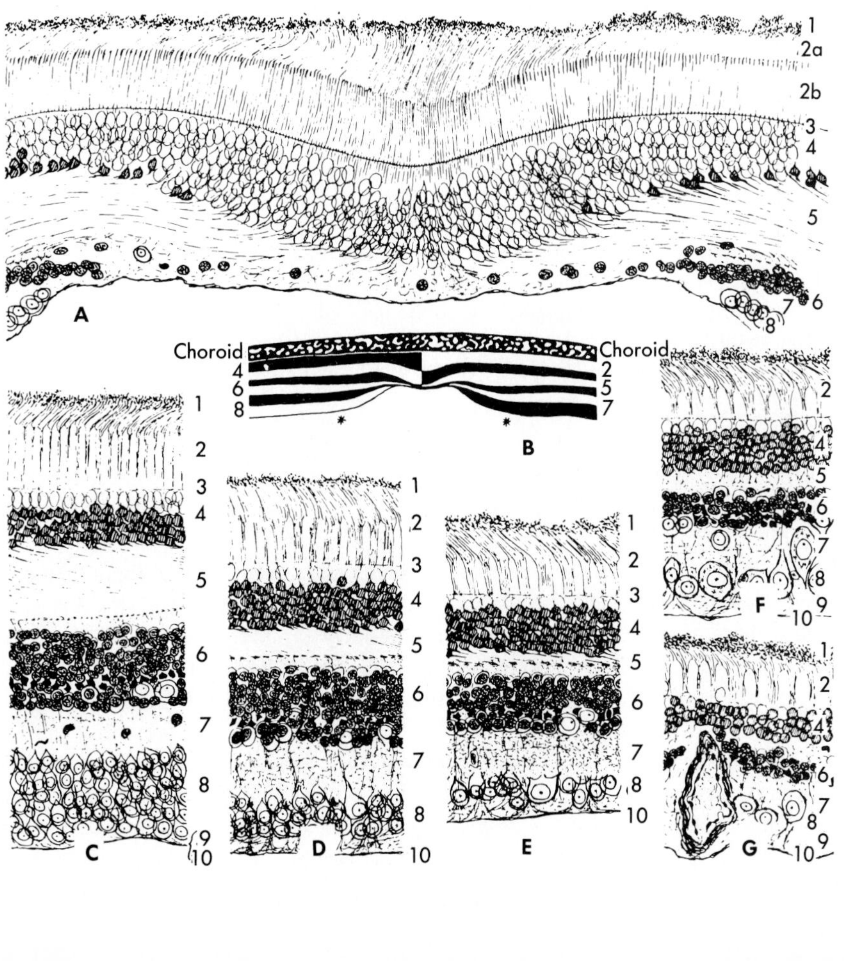

Fig. 8-12. Cross sections of macaque monkey's retina at different locations show changes with eccentricity. Numbers identify the various retinal layers. **A,** Fovea. Note that nearly all elements other than the photoreceptors and their cell bodies are swept aside to create a pit. In this monkey (as in humans), the fovea contains only cones (with open cell bodies in layer *4*) and no rods (with filled cell bodies in layer *4*). Receptor density is increased by virtue of the long, thin cone outer segments. **B,** More schematic view of fovea and layering. The asterisks in the parafoveal region mark the limits of the fovea, and no rods are found between the asterisks. **C** to **G,** Increasingly peripheral retina. Note decrease in retinal thickness and cell numbers with increasing eccentricity. (From Polyak, S.: The vertebrate visual system, Chicago, 1957, University of Chicago Press. Copyright © 1957 by University of Chicago Press.)

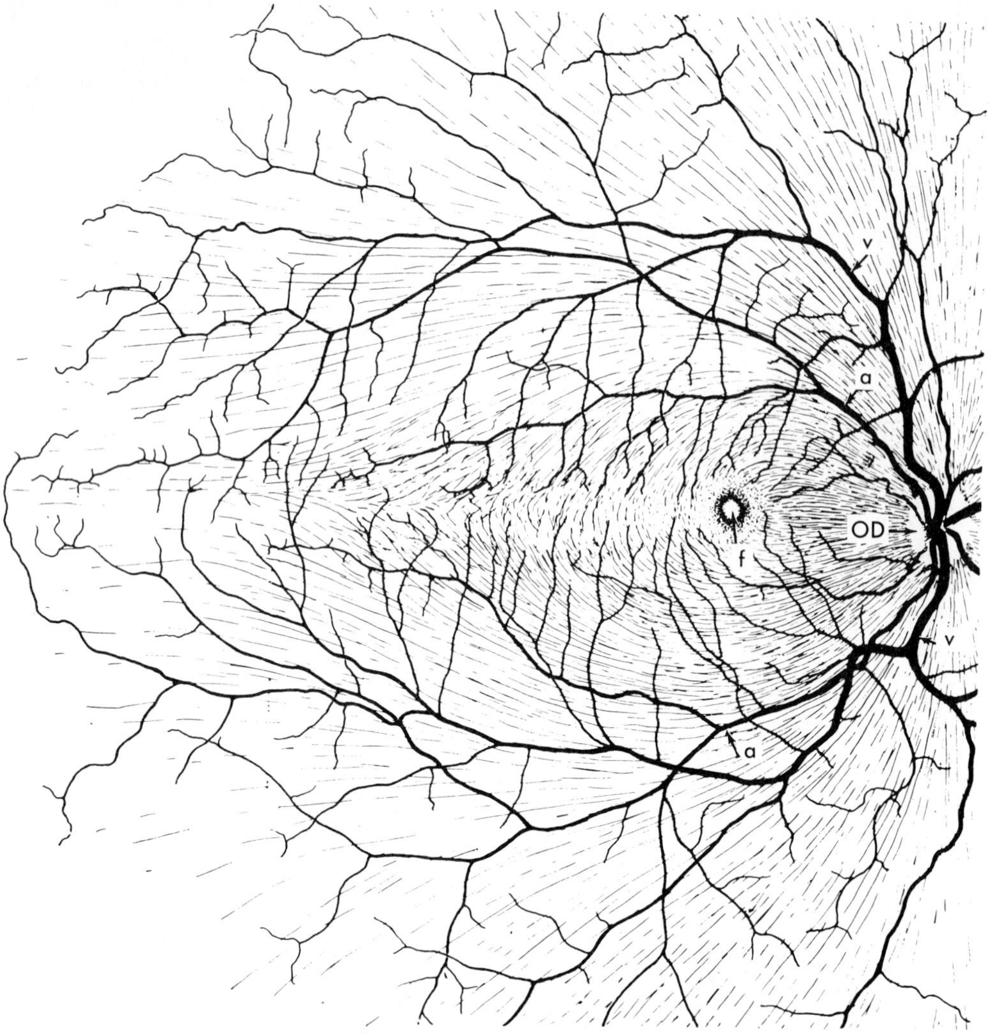

■ **Fig. 8-13.** Part of a human ocular fundus or that part of the retina and associated structures visible during ophthalmoscopic examination. The pattern of ganglion cell axons *(fine lines)* and veins *(v)* coursing toward and arteries *(a)* radiating from the optic disc *(OD)* can be seen. The key to this pattern is that the axons and blood vessels avoid the fovea *(f)*. The axons from the temporal retina take looping courses toward the optic disc, looping first away from and then toward the horizontal raphe (a straight line extending from the optic disc through the fovea). In nasal retina, fibers take a more direct, essentially straight route to the optic disc, since no foveal region need be avoided. The blood vessel pattern is also more cruciate in nasal retina. (From Polyak, S.: The vertebrate visual system. Chicago, 1957, University of Chicago Press. Copyright © 1957 by University of Chicago Press.)

nation the fovea will not function, and the most sensitive retinal region becomes the parafoveal annulus of peak rod density. Thus a very dim, barely detectable star can be seen just beside the fixation point (parafoveally) but not at the fixation point (foveally). In other words, the star disappears when one looks directly at it and reappears when one looks just to one side of it.

A final consequence of the fovea concerns the trajectory of ganglion cell axons toward the optic disc, where they become myelinated and form the optic nerve. As noted earlier, these fibers travel in layer 9, and as a result of the foveal structure, they must detour around the fovea. Fig. 8-13 illustrates the course taken by these fibers. Generally, nasal to the optic disc the course is straight because there is no fovea to be avoided. Also, between the fovea and optic disc the fibers take a direct route. In the rest of the retina all of the nerve fibers must arch superiorly or inferiorly around the fovea in long, curved courses. A horizontal raphe is formed so that ganglion cells superior to this raphe arch their axons superior to the fovea. The converse is true for inferior ganglion cells. Lesions restricted to a small area of retina but involving all layers produce blind areas, or *scotomata* (singular, *scotoma*), of complicated shape.

These scotomata can be mapped by *perimetry*, which is a process whereby the visibility is determined for small targets placed throughout the visual field with reference to the fixation point. Fig. 8-14 shows that such a lesion nasal to the disc produces a wedge-shaped scotoma because the bundle of fibers interrupted by the lesion originates from the corresponding retinal region. A lesion temporal to the disc, however, produces a curious scimitar-shaped scotoma that rests on the horizontal meridian of the visual field. The shapes of these scotomata are readily explicable from a knowledge of the course of retinal ganglion cell axons as described earlier.

Optic disc. Because the ganglion cell fibers run along the inside of the retina, they must penetrate all of the retinal layers to exit from the eye (another consequence of the inverted vertebrate retina). They do this at the optic disc (Fig. 8-15). This is also called the *blind spot* because, with no retina at this location, any image formed here is unseen.

Fig. 8-16 offers an experiment to demonstrate and map the blind spot. Close the left eye; fixate on the circle with the right eye, and hold the page so that the cross is horizontal to the circle. Now move the page back and forth at a distance of roughly 30 cm from the eye. The cross disappears and reappears as its image moves in and out of the optic disc. This technique is a rough perimetry test and can be used to map the blind spot fairly accurately. The process can be repeated for the left eye by closing the right eye, fixating on the cross, and noting the disappearance and reappearance of the circle.

Note that with both eyes open no blind spot appears. This is simply because when a target is imaged on one eye's optic disc, it is also imaged in the other eye's temporal retina where no blind spot exists. More interesting is the common observation that viewing the world through one eye does not generally provide the percept of a blind spot; that is, with only a single eye open, one is not aware of a white, gray, or black spot that corresponds to the blind spot in the visual field.

This exemplifies a perceptual phenomenon known as *filling in of the scotoma*. A small enough scotoma is usually undetected unless objective tests, such as visual perimetry, are used. Thus the optic disc creates a blind spot that is rather obvious with perimetry testing and that is ignored or undetected otherwise. The same principle applies to pathological scotomata resulting from lesions of the retina or central visual pathways. Surprisingly large scotomata can be ignored due to the filling-in phenomenon. Perceptual problems usually result from sufficiently large scotomata, but these problems are often not associated with a blind area. Thus objective tests must always be relied on rather than subjective impressions when scotomata are suspected.

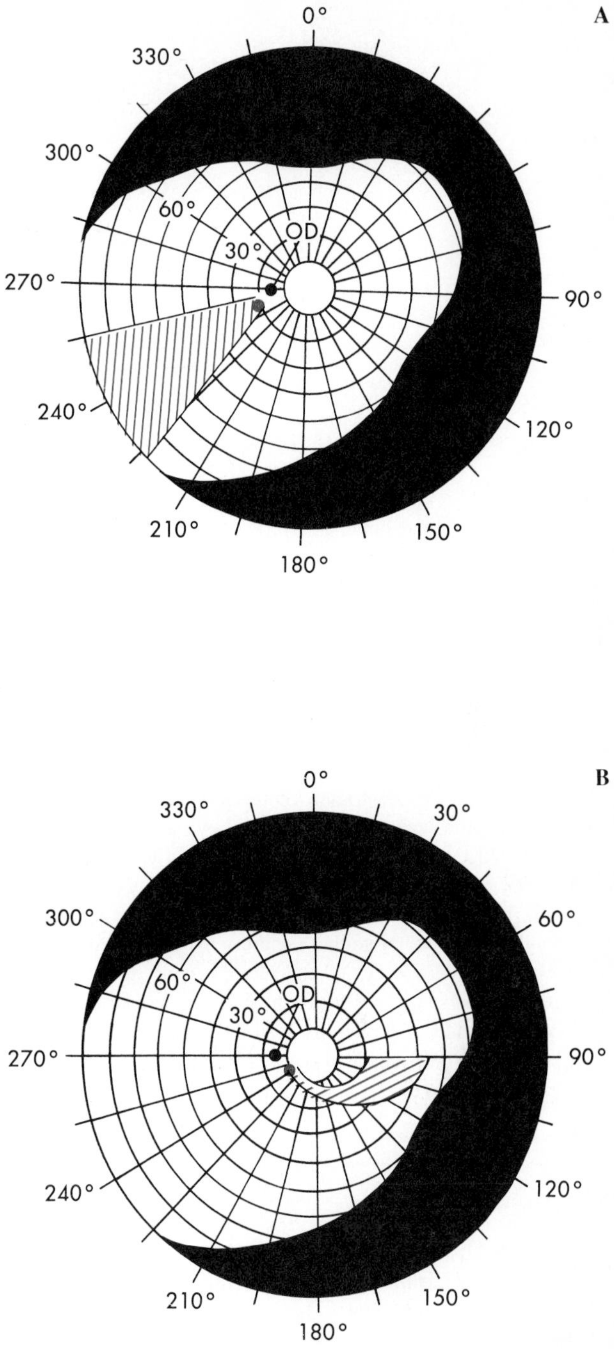

Fig. 8-14. Scotomata caused by various retinal lesions shown in polar coordinates as in Fig. 8-3. The lesions are drawn as colored circles and affect all retinal layers. Thus not only are ganglion cells and photoreceptors damaged at the site of the lesion, but also axons from more distant ganglion cells are destroyed, thereby increasing the scotoma *(colored hatching)*. The shape of the scotoma is determined by the course of ganglion cell fibers through the lesion (see Fig. 8-13). **A,** Lesion nasal to the optic disc *(OD)*. This creates a wedge-shaped lesion that fans out symmetrically toward the periphery. **B,** Lesion temporal to the optic disc. This creates a scimitar-shaped lesion that abuts the horizontal midline of the visual field. (Redrawn from Moses, R.A.: Adler's physiology of the eye, St. Louis, 1970, The C.V. Mosby Co.)

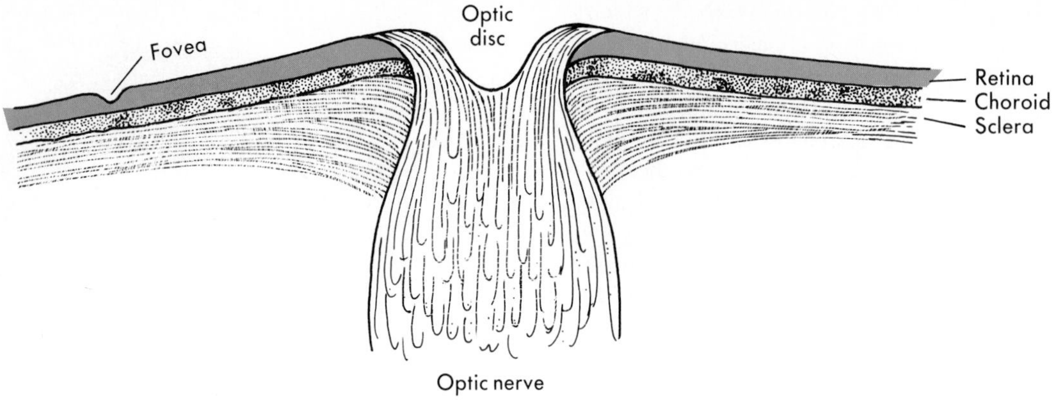

Fig. 8-15. Cross section through the eye at the optic disc. Ganglion cell fibers converge at the disc and penetrate all layers of the eye to enter the cranium and form the optic nerve. Consequently the retina is interrupted here, leading to the physiological blind spot. Note the shape of the optic disc profile. It is sensitive to pressure and becomes more excavated during increased intraocular pressure (e.g., glaucoma) and less excavated during increased intracranial pressure.

Fig. 8-16. Demonstration of the physiological blind spot created by the optic disc. Close one eye and hold the page roughly 30 cm from the open eye. If the right eye is used, fixate on the spot, keep the cross oriented directly to the right of the spot, and move the page slightly closer and further from the open eye. Note that the cross appears and disappears as the image moves out of and into the optic disc. With this technique the shape and location of the optic disc can be accurately plotted. To plot the left optic disc, close the right eye, open the left, and fixate the cross. The spot can be used to ploy the optic disc.

Organization

Circuitry. Fig. 8-17 summarizes a simplified version of the microcircuitry of the vertebrate retina. In the outer plexiform layer, photoreceptors form synapses onto bipolar and horizontal cells. Horizontal cells send processes along the outer plexiform layer to innervate bipolar cells. Horizontal cells also receive input from interplexiform cells. In the inner plexiform layer, bipolar cells synapse onto amacrine and/or ganglion cells, and ganglion cells are also innervated by amacrine cells. Synapses are found as well between amacrine cells and between amacrine and interplexiform cells. Information is thus carried from the outer to the inner plexiform layer by bipolar cells and in the reverse direction by interplexiform cells.

Ganglion cells integrate the synaptic processing that cascades through the retina. The efferent axons of these cells form the optic nerve, which is the *only* route from the retina to the rest of the brain. Fig. 8-17 illustrates variations of neural paths that can be taken to reach ganglion cells from the photoreceptors. One extreme shown by the ganglion cell on the right involves a more direct circuit with ganglion cells dominated by bipolar cell input. The other extreme shown by the ganglion cell on the left is a less direct circuit and involves ganglion cells dominated by amacrine cell input. The central ganglion cell represents probably the most common example of convergent input from both bipolar and amacrine cells. Comparative studies of many vertebrate retinas have shown that the ratio of bipolar-to-amacrine synapses onto ganglion cells increases in a progression from nonmammalian species through mammals to primates. Thus circuits involving ganglion cell innervation by amacrine cells are probably phylogenetically older than those involving bipolar cell innervation.

Functional interactions. To date there is little direct evidence regarding functional properties of the interplexiform cell, but the other retinal neurons have been extensively studied. Consequently, this section focuses on these other cell types.

Two curious, related properties of retinal neurons will be described before their functional properties are discussed. First, only amacrine and ganglion cells generate action potentials. The other cell types communicate via graded, electrotonically conducted potentials. Such electrotonic conduction is sufficient for the short distances of the intraretinal circuitry. Action potentials are essential for the retinal ganglion cells, because they must communicate with the rest of the brain via long axons. Why amacrine cells also produce action potentials when other retinal interneurons do not is a mystery. Furthermore, when a threshold process such as an action potential is not involved, hyperpolarizing and depolarizing potentials can convey equal information. By convention a depolarizing response will be referred to as *excitatory* and a hyperpolarizing one as *inhibitory,* even though these designations are somewhat arbitrary. Because photoreceptors hyperpolarize to photic stimu-

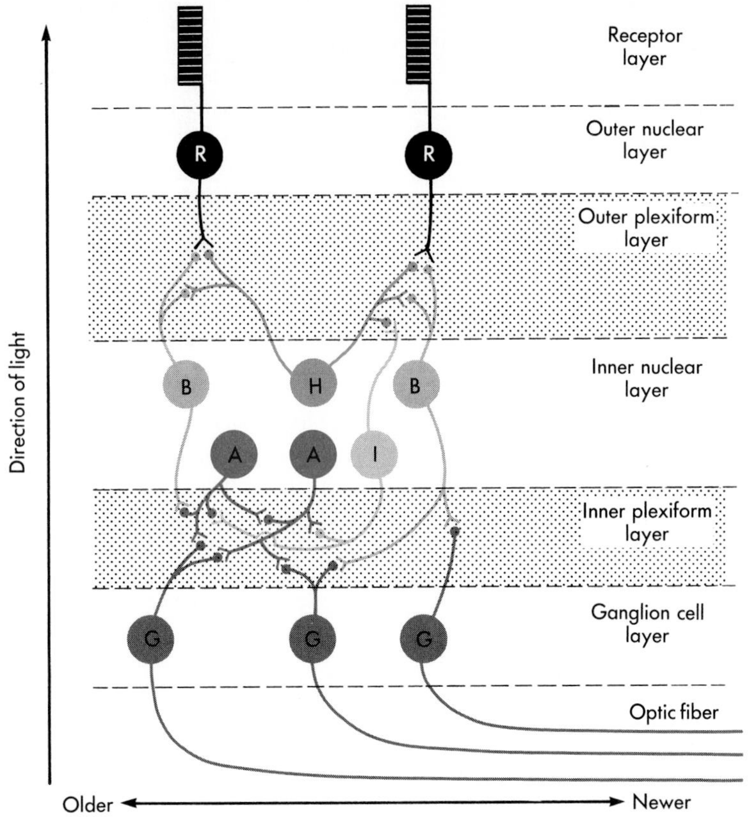

Fig. 8-17. The connections among retinal neurons and the significance of prominent layers. The neurons shown are photoreceptors *(R)*, horizontal cells *(H)*, bipolar cells *(B)*, interplexiform cells *(I)*, amacrine cells *(A)*, and ganglion cells *(G)*. It has been suggested that ganglion cells dominated by amacrine cell inputs represent phylogenetically older circuitry and that ganglion cells dominated by bipolar cell inputs represent newer circuitry. The arrow indicates the direction of light as it passes through the retina to reach the photoreceptors.

lation (Chapter 6), they are inhibited by light according to this convention. Indeed, photic stimulation seems to reduce neurotransmitter release from the photoreceptors.

Most of our understanding of and hypotheses about functional circuitry within the retina derives from receptive field studies and pharmacological manipulations of retinal neurons. The receptive field approach for visual neurons involves the determination of how they are affected by visual stimuli that vary in contrast, shape, temporal parameters, and retinal location. Fig. 8-18 presents a greatly simplified summary of receptive field properties for some of the retinal neurons as well as the possible synaptic interactions that can be inferred from an analysis of these properties. This exemplifies the value of the receptive field approach in understanding functional circuitry. However, the speculative and incomplete nature of Fig. 8-18 must be emphasized.

As noted in Chapter 6, photoreceptors are relatively depolarized in the dark because of an inward flow of sodium ions (the "dark current"), and their synapses are maximally active in the dark. Photoreceptors probably use glutamate as their transmitter. Light inhibits photoreceptors by reducing the inward sodium current, thereby hyperpolarizing the cell and reducing transmitter release. The receptive field of a photoreceptor is small and circular, being coextensive with its retinal location (Fig. 8-18, *A*). Some weak response to light applied outside this region can be detected, due mostly to electrical coupling between neighboring photoreceptors. However, cones receive a synaptic feedback inhibitory input from nearby rods via horizontal cells (these functionally weak pathways are not shown in Figs. 8-17 and 8-18).

Horizontal cells have larger, more uniform receptive fields and are tonically inhibited by photic stimuli (i.e., they hyperpolarize). They seem to receive convergent excitatory input from several adjacent photoreceptors (Fig. 8-18, *B*). This input from photoreceptors is considered excitatory, because transmitter release from the photoreceptors depolarizes the horizontal cell. The horizontal cell hyperpolarizes to light because photoreceptors do.

Bipolar cells also respond tonically to photic stimulation, and there are two complementary types. One is the "depolarizing" or "on-center" type, and the other is the "hyperpolarizing" or "off-center" type. For the former (the left example in Fig. 8-18, *C*), light shone anywhere within a small central region of the receptive field (called the *center*) depolarizes or excites the cell. Light applied to an annular region surrounding the center (called the *surround*) hyperpolarizes or inhibits the cell. This represents an antagonistic organization of the center-surround receptive field, since the response to a stimulus placed in the center of the cell's receptive field is antagonistic to the response elicited by a stimulus placed in the surround. As might be predicted from such an arrangement, a small light spot that just fills the

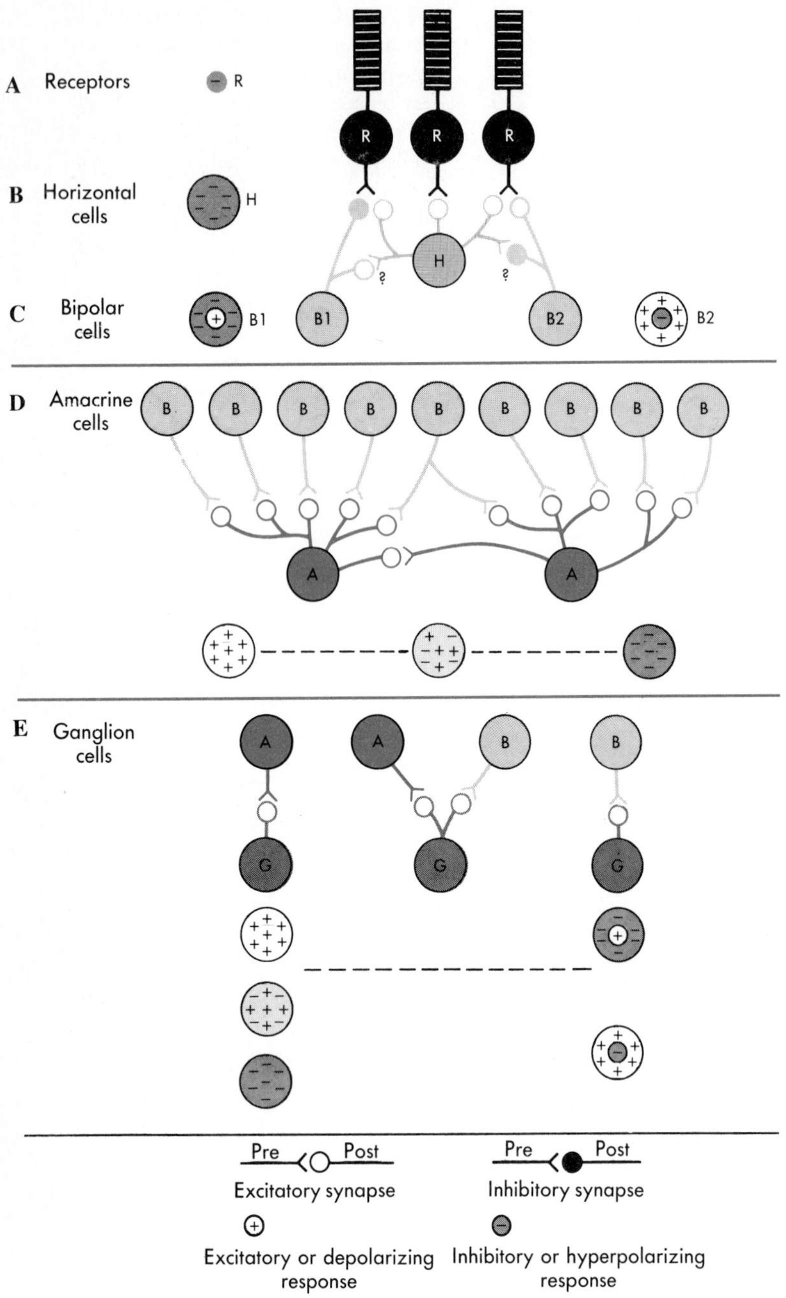

■ **Fig. 8-18.** Receptive fields and proposed neural circuitry for many retinal neurons (**A** to **E**), including photoreceptors *(R)*, horizontal cells *(H)*, bipolar cells *(B)*, amacrine cells *(A)*, and ganglion cells *(G)*. Interplexiform cells are not included due to lack of receptive field information specific to these cells. Photoreceptors have small receptive fields equal in size to the retinal area these cells occupy. By convention these fields are considered inhibitory (shown by *minus signs*) because photoreceptors are hyperpolarized by light. Horizontal cells receive convergent excitatory input from several photoreceptors and thus have larger inhibitory fields. Bipolar cells exhibit a center-surround receptive field organization. On-center cells *(B1)* receive inhibitory input from photoreceptors and excitatory input from horizontal cells; the converse holds for off-center cells *(B2)*. Many varieties of amacrine cell receptive fields have been described. These include some that are mostly excitatory due to convergent input from many on-center bipolar cells, some that exhibit excitatory and inhibitory regions diffusely intermixed due to mixed input from on- and off-center bipolar cells, and many that are mostly inhibitory due to convergent input from many off-center bipolar cells. A variety of ganglion cell types has also been described. Some seem to have receptive fields quite similar to those of amacrine cells and presumably reflect nearly exclusive amacrine cell inputs. Some have receptive fields with center-surround organization similar to those of bipolar cells, presumably resulting from nearly exclusive bipolar cell input. Some have more complicated receptive fields that may reflect a combination of amacrine and bipolar cell inputs. Clearly our understanding of how circuitry of the outer plexiform layer results in neuronal receptive fields is more detailed than that of the inner plexiform layer.

center will maximally depolarize the cell. A larger spot that simultaneously falls on the surround will depolarize the cell less, if at all, because the depolarization from central stimulation sums more or less linearly with the hyperpolarization from surround stimulation. This cell type is called *on center* because light turned on in the center excites the cell. Typically the size of the receptive field center of these bipolar cells is equal to or a little larger than the size of the receptive field of a photoreceptor; the surround is roughly as large as a horizontal cell's receptive field. Thus the circuitry illustrated in Fig. 8-18, *A* to *C*, can be proposed for these on-center, bipolar cells. The central response is subserved by inhibitory synapses onto the bipolar cell from one or a small number of photoreceptors. Such an inhibitory synapse can "invert" the photoreceptor hyperpolarization to postsynaptic depolarization. The surround response is provided by excitatory input from a horizontal cell. The evidence for this explanation for the surround response is weak; hence the question marks in Fig. 8-18.

The complementary type of bipolar cell, or *off-center* type, is almost the negative image of the on-center type. The off-center cell (shown to the right of Fig. 8-18, *C*) has a similar center-surround receptive field organization, but light in the center hyperpolarizes the cell, whereas light in the surround depolarizes it. Such a cell is called off-center because light turned off in the center excites the cell (or more precisely, disinhibits the cell). Fig. 8-18, *A* to *C*, also summarizes the circuitry likely to subserve the responses of these off-center cells. The center response is provided by excitatory input from one or a few photoreceptors and the surround response by inhibitory input from a horizontal cell. Again, inhibition is needed to "invert" a hyperpolarizing presynaptic response (horizontal cell) to a depolarizing postsynaptic response (bipolar cell). The evidence for circuitry subserving the surround response is inferential and is indicated by a question mark in Fig. 8-18.

On- and off-center bipolar cells could conceivably respond in opposite polarity to the same neurotransmitter. Consequently the same photoreceptors (or horizontal cells) could excite one type of bipolar cell and inhibit the other by using one synaptic neurotransmitter. Evidence for this has been obtained for synapses between photoreceptors and bipolar cells. As noted previously, all photoreceptors probably use the same neurotransmitter, which is glutamate. Applied to the off-center bipolar cell this transmitter seems to *increase* conductance to sodium, which allows sodium ions to enter the cell and thus depolarize it. This is an excitatory synaptic effect. However, the same transmitter applied to the on-center bipolar cell *decreases* conductance to sodium. The entry of sodium ions is reduced, thereby hyperpolarizing the cell, which is consistent with an inhibitory synaptic effect. Off- and on-center bipolar cells seem to have different postsynaptic receptors* associated with photoreceptor synapses. These different synaptic receptors, when combined with glutamate, have different effects on opening and closing of their associated sodium channels. It is not known whether analogous differences apply to postsynaptic receptors for inputs from horizontal cells.

Amacrine cells (Fig. 8-18, *D*) form a much more varied and complicated group. They differ in response properties, morphology, and the transmitters they use. Amacrine cells are the most distal retinal cells to exhibit action potentials, and thus excitation and inhibition are no longer arbitrary definitions. Amacrine cells can respond either tonically or physically to visual stimuli. Most seem to have relatively large and diffuse receptive fields and can be excited by light turned on, by light turned off, or by both light onset and termination. Amacrine cells receive convergent input from a number of bipolar cells as well as from other amacrine cells, and it is difficult to specify the circuitry responsible for their receptive field properties.

Finally, ganglion cells receive input from amacrine cells, bipolar cells, or both (Fig. 8-18, *E*). Many ganglion cells have large, diffuse receptive fields similar to those of some amacrine cells, and these ganglion cells probably receive their predominant input from amacrine cells. They generally respond best to stimuli in the center of the receptive field, and the responsiveness diminishes as the stimulus is placed more eccentrically in the receptive field. Such ganglion cells were only recently discovered in mammals, and now they appear to represent roughly one third of all ganglion cells in cats and monkeys. At the other extreme are ganglion cells with center-surround receptive field organization much like that of bipolar cells, which probably represents their predominant input. These ganglion cells are either excited by photic stimulation in the center and inhibited by stimulation of the surround (on-center cells) or are excited by stimulation of the surround and inhibited by stimulation of the center (off-center cells). Also, many ganglion cells have a center-surround configuration that seems superimposed on a more diffusely organized receptive field component, and this may represent ganglion cells with more balanced amacrine and bipolar cell input. The different ganglion cell types are considered in more detail below.

The bipolar-to-ganglion cell circuit is thought to be phylogenetically more recent than the amacrine-to-ganglion cell circuit. This, in turn, implies that the center-surround receptive field organization evolved relatively recently. Although this seems a likely speculation, a weakness in this inference stems from the fact that center-surround receptive fields or visual neurons were described first for the primitive horseshoe crab *(Limulus)*.

Surround or Lateral Inhibition

The center-surround receptive field organization seen most distally in bipolar cells and represented by many or most ganglion cells is also called *surround* or *lateral inhibition*. Analogous lateral inhibition is seen in other sensory systems and is generally associated with their more recently evolved components. Fig. 8-19 shows how such receptive field organization can enhance two-point discrimination or spatial resolution compared with that possible without lateral inhibition. This example illustrates on-center cells, but a similar result is derived from off-center cells.

Fig. 8-19, *A*, shows how cells with overlapping center-surround receptive fields respond to a small light spot. The spot is in the center of one cell and excites it; it lies in the surrounds of neighboring cells and inhibits them; it lies completely outside the receptive fields of more distantly located cells and does not affect them.

*These synaptic receptors are not to be confused with sensory receptors, which are the transducers described in Chapter 6 and include photoreceptors. This is unfortunate terminology. Synaptic receptors are specialized molecules located in the postsynaptic membranes of neurons. The combination of a receptor with the appropriate neurotransmitter molecule alters the state of the receptor, perhaps because of a conformational change. This in turn opens or closes an adjacent ionic channel. Opening or closing of these channels represents a conductance increase or decrease to specific ions and leads to depolarization or hyperpolarization of the cell. This event is otherwise known as an excitatory postsynaptic potential (EPSP) or inhibitory postsynaptic potential (IPSP).

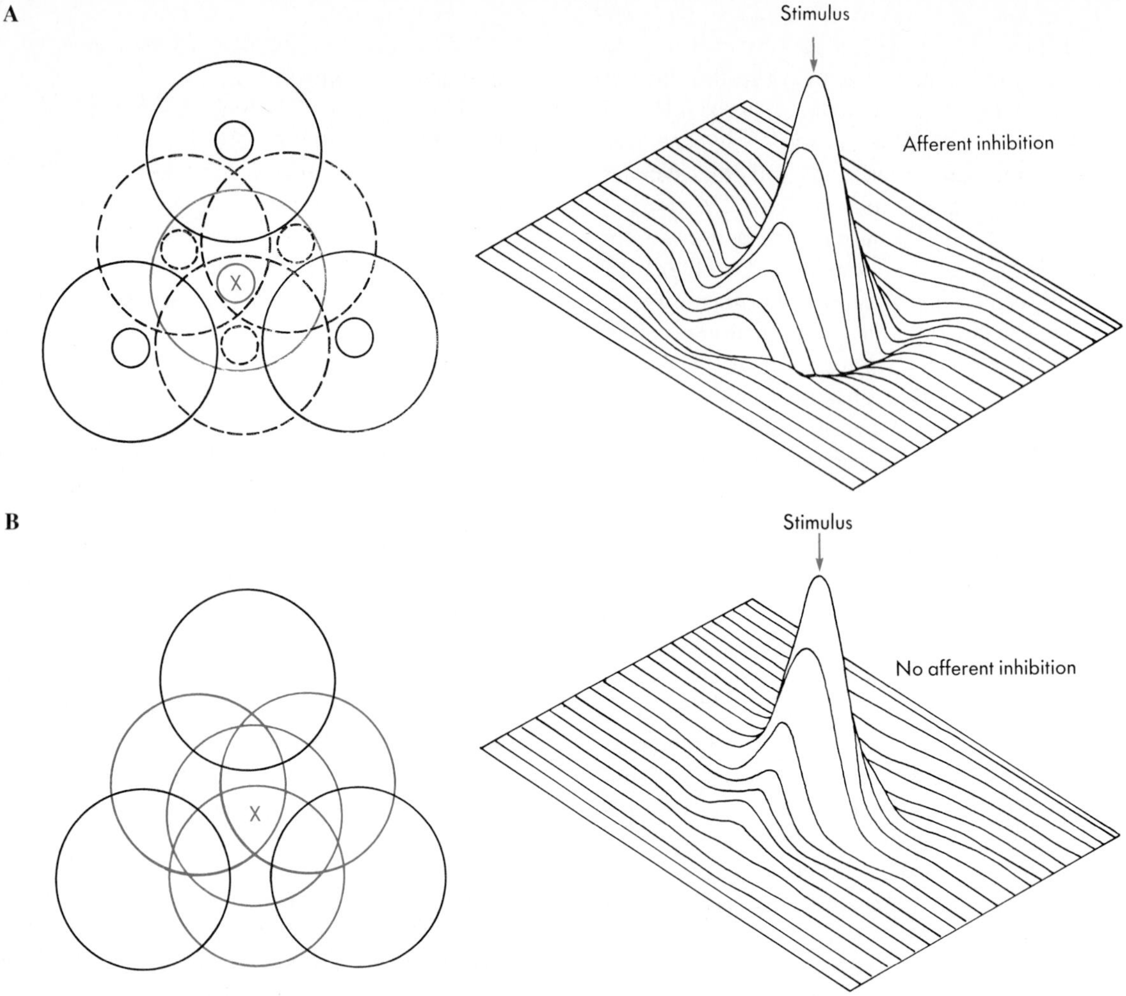

■ **Fig. 8-19.** Significance of afferent inhibition to discrimination of two stimuli from a single stimulus. **A,** With afferent inhibition. On the left are shown seven partially overlapping receptive fields, each with a central excitatory zone and an inhibitory surround. A stimulus placed at the point indicated by the colored *X* would fall in the excitatory center of one cell *(solid color),* in the inhibitory surround of three cells *(dashed black),* and outside the field of three cells *(solid black).* The pattern of neuronal activity evoked by such a single stimulus is shown on the right. Neurons just beneath the stimulus location are maximally excited; a surrounding annulus of neurons is inhibited, and beyond this is no neuronal response. **B,** Without afferent inhibition. On the left are shown seven partially overlapping receptive fields. A stimulus placed at the point indicated by the colored *X* would fall within four fields to excite their cells and would fall outside the fields of the other three cells. No cells are inhibited by the stimulus. Thus the pattern of neuronal activity evoked by such a stimulus *(right)* includes a single peak with no surrounding inhibitory trough.

The three-dimensional graph on the right depicts the response profile to a single photic stimulus of a patch of retinal ganglion cells with center-surround stimulation. As indicated, a central zone of active cells is surrounded by a zone of inhibited cells.

Fig. 8-19, *B,* shows that for cells with overlapping but diffusely organized fields, no surround zone of inhibited cells exists to a single stimulus. As noted earlier, each of these cells responds best to a stimulus applied to the center of its receptive field and less so as the stimulus is placed more eccentrically in the receptive field. Thus in the presence of a single stimulus the active neuronal zone peaks at the stimulus location and gradually diminishes, without a surrounding zone of inhibited cells.

Fig. 8-19, *C,* shows the consequence of these types of receptive fields for two-point discrimination. With afferent inhibition two nearby stimuli produce two separate zones of activity, a pattern that clearly signals the presence of two stimuli. Without afferent inhibition a single broad zone of activity results, and this can easily be confused with a more intense stimulus placed in the middle.

C

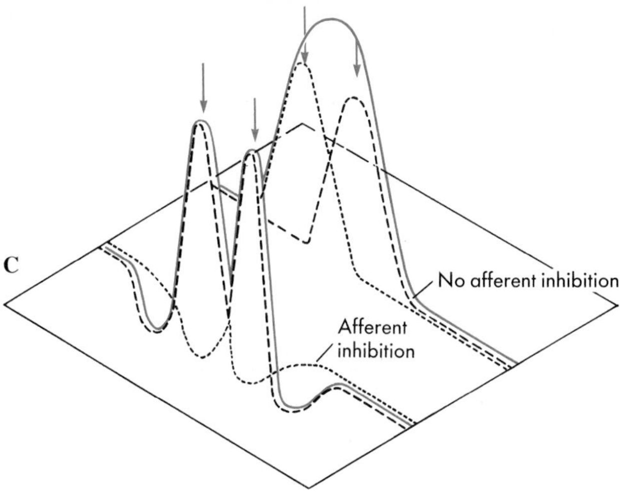

■ Fig. 8-19. **cont'd C,** Comparison of neuronal activity patterns to paired stimuli with and without afferent inhibition. From the activity patterns shown in **A** and **B,** the pattern of neuronal activity to each stimulus alone is indicated by the dashed black curves. The colored curves indicate the summed activity to the pair of stimuli. With afferent inhibition two peaks of active neurons are evident, and because a single stimulus could not evoke such an activity pattern, a pair of stimuli is unambiguously signaled. Without afferent inhibition, a single broad peak of active neurons results, and such activity could result from a single more intense stimulus. Thus afferent inhibition more clearly encodes fine spatial details of the stimulus than is possible without afferent inhibition.

■ *W-, X-, and Y-cells*

Recent studies of cats have demonstrated a division of retinal ganglion cells into physiologically and morphologically distinct classes called *W-, X-, and Y-cells.* These neuronal classes represent, respectively, about 40%, 55%, and 5% of all ganglion cells. More limited data from other mammalian species, including monkeys and probably humans, suggest a similar classification of retinal ganglion cells. We shall refer to these as W-, X-, and Y-cells in primates, although more detailed understanding may emphasize differences between the cat and primates and may require another terminology. Most of the comments concerning W-, X-, and Y-cells derive from experiments on cats, and their application to other species, including monkeys and humans, is as yet somewhat tenuous. Indeed, we already know that W-, X-, and Y-cells of the cat are different in certain morphological and physiological details from their presumed counterparts in monkeys.

In the cat retina, X- and Y-cells seem to be internally homogeneous classes, but W-cells are not and may well include several further distinct neuronal classes. With this important proviso, we shall refer to these neurons simply as W-cells. X- and Y-cells (and most W-cells) have center-surround receptive field configurations that suggest strong bipolar cell input. Most W-cells have more uniform, diffuse receptive field organization indicative of strong amacrine cell input. Generally, however, we lack precise knowledge concerning the relative strength of amacrine and bipolar cell inputs to these ganglion cell classes.

Many morphological and physiological differences can be briefly noted for W-, X-, and Y-cells in cats. W-cells have small to medium-sized somata and small axons; somata and axons are medium-sized in X-cells; and in Y-cells they are large. Each of these cell classes seems to have distinctive dendritic morphology, as is illustrated for the cat's retina in Fig. 8-20, *A*. Compared with X- and Y-cells, W-cells have large, diffuse receptive fields, respond poorly to visual stimuli, and possess very slowly conducting axons. For these reasons it has been speculated that W-cells are phylogenetically older than X- or Y-cells. The many differences between X- and Y-cells are more subtle, and some of the more obvious differences follow: compared with Y-cells, X-cells generally have smaller receptive fields, more tonic responses to visual stimuli, better responses to small stimuli or fine detail but poorer responses to larger or cruder forms, more slowly conducting axons, and more linear summation of responses to visual stimuli. Regarding this last difference, an X-cell generally responds to the combination of two or more stimuli in a fashion that can be predicted from the linear addition of the responses to each of the stimuli alone. Y-cell responses to complex stimuli often exhibit nonlinear summation and thus cannot be readily predicted from knowledge of responses to simpler components of the stimulus.

In monkeys, too, retinal X- and Y-cells have distinctive physiology and morphology (see Fig. 8-20, *B*). At present, very little is known of the detailed morphology and physiology of W-cells in primates. In addition to most of the X- and Y-cell properties noted for cats, two further differences have been noted for monkeys. First, X-cells respond differently to different wavelengths or colors of light, whereas Y-cells are relatively insensitive to such wavelength differences. Second, X-cells are much less sensitive to luminance variations in visual stimuli than are Y-cells.

One final point of distinction can be made for the retinal W-, X-, and Y-cells in both cats and monkeys. These cell types display different central projections and represent the retinal point of departure for at least three parallel and functionally independent pathways involved in processing of visual information. It seems reasonable to assume that all of these differences signify different roles for W-, X-, and Y-cells in visual perception; these roles are considered later in this chapter.

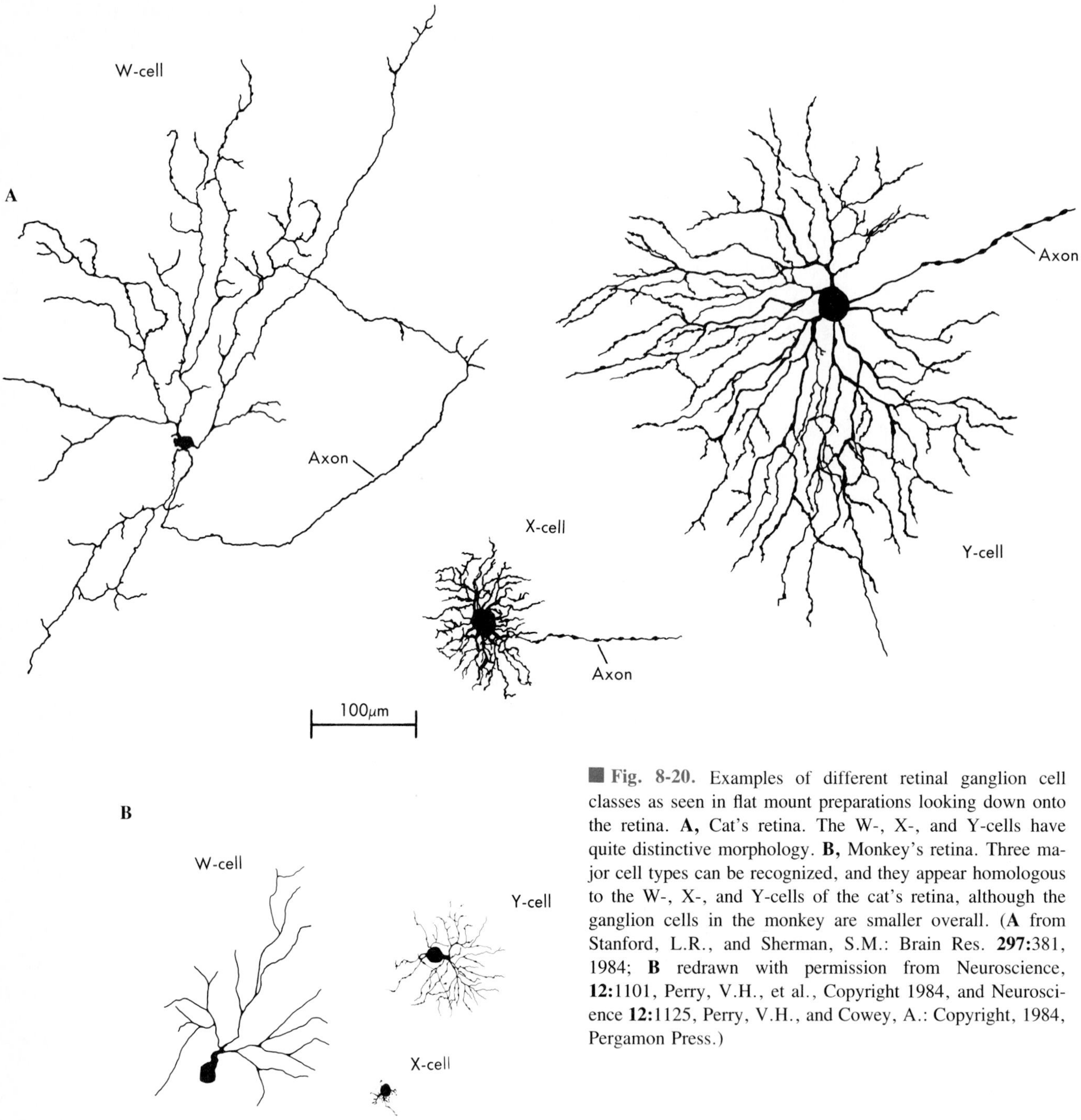

■ **Fig. 8-20.** Examples of different retinal ganglion cell classes as seen in flat mount preparations looking down onto the retina. **A,** Cat's retina. The W-, X-, and Y-cells have quite distinctive morphology. **B,** Monkey's retina. Three major cell types can be recognized, and they appear homologous to the W-, X-, and Y-cells of the cat's retina, although the ganglion cells in the monkey are smaller overall. (**A** from Stanford, L.R., and Sherman, S.M.: Brain Res. **297:**381, 1984; **B** redrawn with permission from Neuroscience, **12:**1101, Perry, V.H., et al., Copyright 1984, and Neuroscience **12:**1125, Perry, V.H., and Cowey, A.: Copyright, 1984, Pergamon Press.)

■ *Retinofugal Projections*

■ *Route of Retinofugal Fibers*

Axons of retinal ganglion cells exit from the eye at the *optic nerve* and course medially and posteriorly to the *optic chiasm*. A partial decussation of axons occurs at the chiasm; some cross the midline and project contralaterally, while others do not and project ipsilaterally. The axons from each eye then form the *optic tract,* in which these fibers continue posteriorly toward their targets in the brainstem (Fig. 8-21). No ganglion cell has a bifurcating axon that innervates both hemispheres; each of these axons enters only one optic tract.

Fig. 8-22 illustrates the simple rules that govern whether or not an axon crosses in the optic chiasm. To a first approximation, axons of ganglion cells in the nasal retina cross to enter the contralateral optic tract, and those of cells in the temporal retina enter the ipsilateral optic tract without crossing. This feature, combined with the optically reversed image of visual objects, leads to a continuous map of visual space in the brain

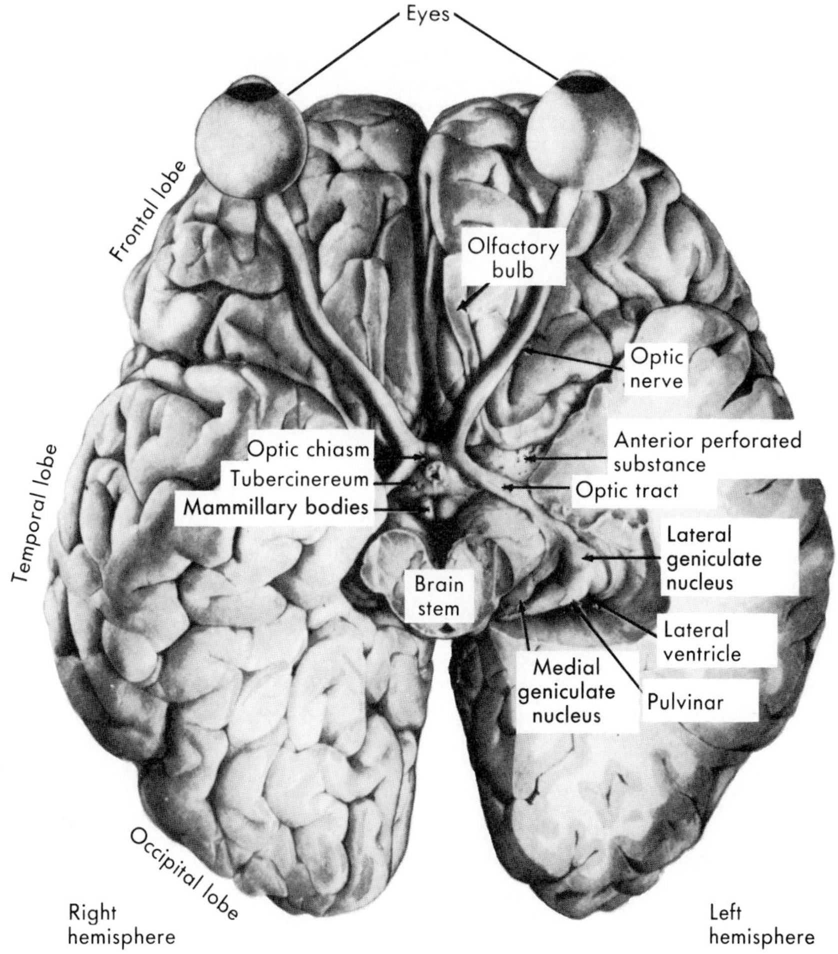

■ **Fig. 8-21.** Ventral surface of the human brain. Among various landmarks can be seen the eyes, the optic nerves, the optic chiasm, the optic tracts, and the lateral geniculate nucleus. (From Polyak, S.: The vertebrate visual system, Chicago, 1957, University of Chicago Press. Copyright © 1957 by University of Chicago Press.)

so that each hemifield is mapped in the contralateral hemisphere. Furthermore, the optic nerves, chiasm, and tracts contain axons of retinal ganglion cells; no synapses from these retinal neurons have yet occurred in the visual pathway. Central to the optic chiasm, axons representing both eyes are intermingled.

One detail must be added to this description of the decussation at the optic chiasm. The retinal division between ganglion cells that project to one or the other optic tract is not a precise vertical line. Rather, there is a vertical strip of retina, approximately 1 to 2 degrees across and passing through the fovea, in which ganglion cells that project to each hemisphere are intermingled. This has been termed the *strip of nasotemporal overlap*. The significance of this overlap probably relates to stereopsis, which is described later.

■ *Terminations of Retinofugal Axons*

Retinal ganglion cells project to several brainstem sites, but the vast majority of the optic tract fibers terminate in the dorsal lateral geniculate nucleus (LGN) of the thalamus and the superior colliculus of the midbrain. Other sites include the suprachiasmatic nucleus of the hypothalamus, the ventral LGN* of the thalamus, several pretectal nuclei (including the nucleus of the optic tract) in the midbrain, and the accessory optic nucleus in the midbrain tegmentum (Fig. 8-23).

The retinofugal projections differ significantly among W-, X-, and Y-cells (Fig. 8-23). The dorsal LGN receives input from all retinal X- and Y-cells. Limited W-cell input to the dorsal LGN has been described for the cat and probably exists as well in primates. The superior colliculus receives a mixture of W- and Y-cell inputs. Each retinofugal Y-cell axon in the cat bifurcates to innervate both the dorsal LGN

*The ventral LGN is a thalamic structure quite distinct from the dorsal LGN in most mammals. Among other differences, dorsal LGN neurons project exclusively to the visual cortex, whereas ventral LGN neurons project only to subcortical loci. In many monkey species and in humans, however, the ventral LGN homologue is unclear. It probably includes part or all of a band of neurons called the *pregeniculate nucleus,* which lies immediately dorsomedial to the dorsal LGN (Fig. 8-23).

■ **Fig. 8-22.** Rules governing optical image formation and the partial decussation of optic nerve fibers at the optic chiasm. In the example the object is an arrow that spans the entire horizontal extent of visual field, and the fixation point is imaged onto each fovea as indicated. Optical considerations lead to a reversal of the image onto the retina so that the right half of the arrow *(color)* is mapped onto ganglion cells *(triangles)* of the left temporal and right nasal retinas, while the left half *(black)* is mapped onto ganglion cells *(circles)* of the right temporal and left nasal retinas. Also, the extreme limits of the arrow (monocular segments) can be viewed by only the nasal retina on that side, and the binocular segments are imaged in each eye. Ganglion cells from the nasal retina *(open symbols)* decussate in the optic chiasm, and ganglion cells from the temporal retina *(filled symbols)* project ipsilaterally. Thus the right half of visual space is mapped onto the left hemisphere and vise versa. Although not shown, ganglion cells within 1 to 2 degrees of a vertical line running through the fovea project to one or the other optic tract regardless of their nasal or temporal location so that portions of the visual field near the vertical midline *(colored stippled circles)* are represented in both hemispheres.

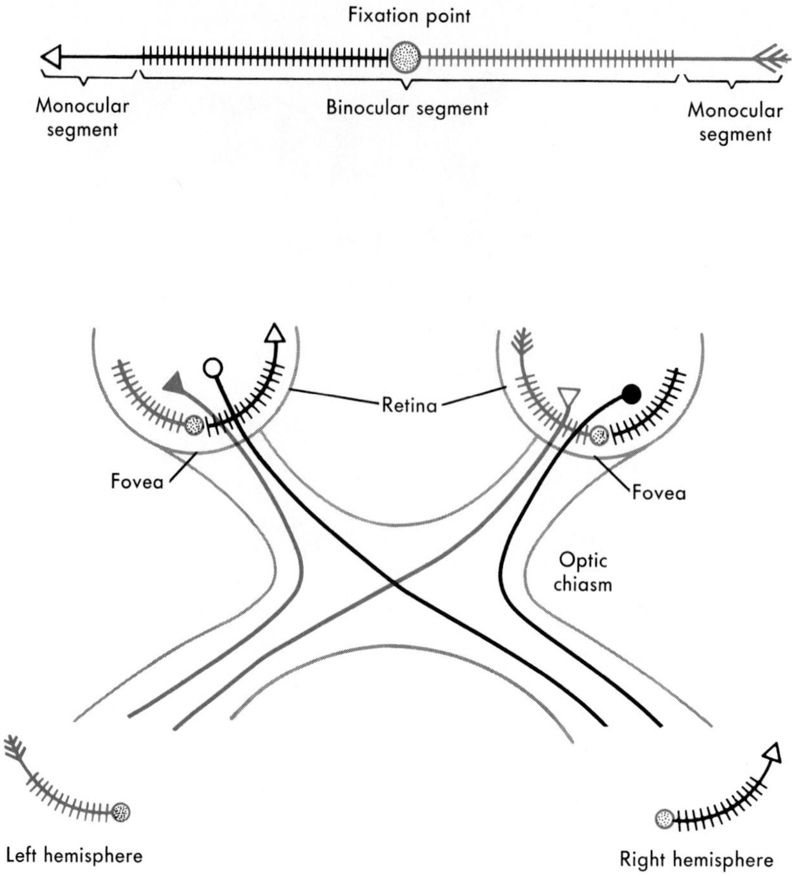

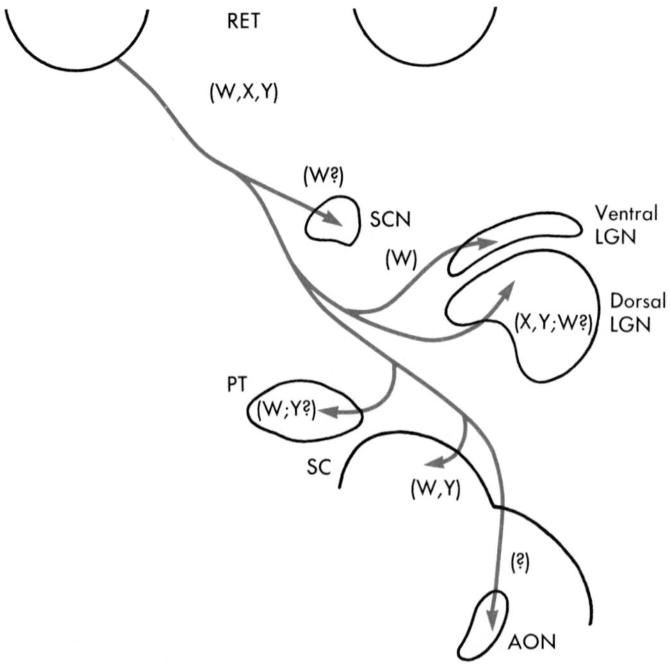

■ **Fig. 8-23.** Schematic diagram of retinofugal projections. Terminal zones include the suprachiasmatic nucleus *(SCN)*, the ventral lateral geniculate nucleus (ventral LGN), the dorsal lateral geniculate nucleus *(dorsal LGN)*, the pretectal nuclei *(PT)*, the superior colliculus *(SC)*, and the accessory optic nucleus *(AON)*. Letters in parentheses indicate the source of projections from W-, X-, or Y-cells in the retina. Question marks refer to uncertain or unknown sources.

and the superior colliculus; similar data for primates are not yet available.

Although detailed information is lacking for most of the other sites of retinofugal projections, they probably receive W-cell input and little, if any, X-cell or Y-cell input. In other words, X-cells project to the dorsal LGN, Y-cells project primarily to the dorsal LGN and superior colliculus, and W-cells probably project to all of the retinofugal sites. The dorsal LGN is associated with the phylogenetically newest portion of the visual system; the superior colliculus probably represents an intermediate stage of evolution; and the remaining retinofugal sites are much older.

Geniculostriate System

Geniculate Lamination

Retinal X-cells and Y-cells from both eyes project to the dorsal LGN, and their precise termination depends largely on the geniculate laminar patterns. Fig. 8-24 illustrates the lamination of the human infant dorsal LGN

A

B

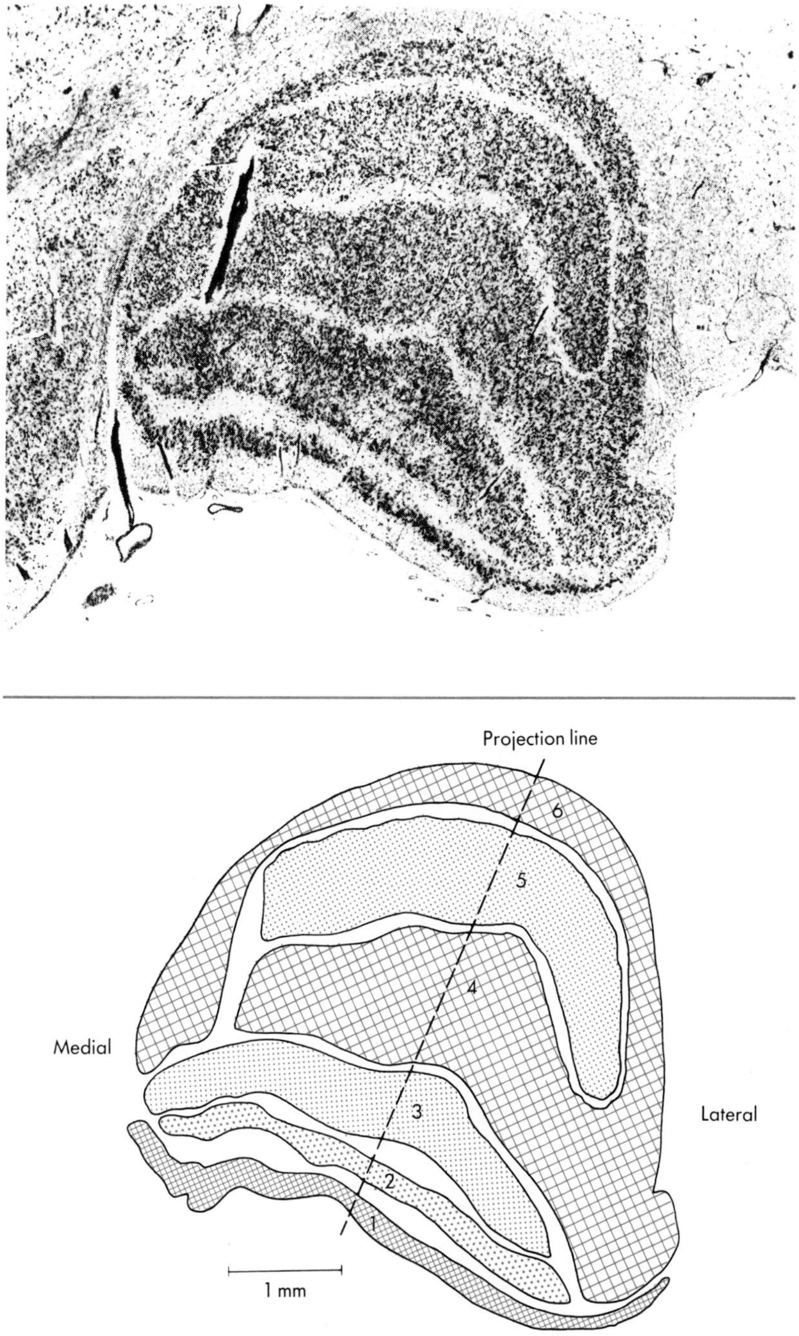

Fig. 8-24. Coronal section through the lateral geniculate nucleus of a human infant. **A,** Photomicrograph of Nissl-stained material. **B,** Explanation of photomicrograph in **A.** The magnocellular laminae are *1* and *2,* and the parvocellular laminae are *3, 4, 5,* and *6.* The contralateral eye innervates laminae *1, 4,* and *6,* and the ipsilateral eye innervates laminae *2, 3,* and *5.* Lines running approximately at right angles to the laminae represent projection lines (one example is shown). Projection lines roughly represent the locus of geniculate neurons that map a single point in visual space. (**A** courtesy T.L. Hickey and R.W. Guillery.)

as seen in a coronal section. The dorsal LGN is kidney shaped with its hilus oriented ventromedially. Typically, six laminae can be distinguished. These are more or less horizontally oriented, but they curve according to the shape of the dorsal LGN. The laminae are numbered 1 to 6 from the ventral to the dorsal surface (Fig. 8-24, *B*). Laminae 1 and 2 tend to have larger cells than do laminae 3 to 6. Thus laminae 1 and 2 are called the *magnocellular laminae,* and laminae 3 to 6 are called the *parvocellular laminae*.

The significance of these laminae lies in their ocular input, or "ocular dominance" (Fig. 8-24, *B*), and probably also in the differential X- and Y-cell inputs. Although fibers from each eye are intermingled in the optic tract, they segregate before terminating in the dorsal LGN. Thus each lamina receives input exclusively from one or the other eye. Laminae 1, 4, and 6 are innervated by the contralateral eye and laminae 2, 3, and 5 by the ipsilateral eye. Y-cell axons innervate laminae 1 and 2, and X-cell axons innervate laminae 3 to 6. Finally, on-center X-cells tend to innervate laminae 5 and 6 and off-center X-cells tend to terminate in laminae 3 and 4. On- and off-center Y-cell axons are intermixed in their innervation of laminae 1 and 2. Thus the magnocellular laminae represent a matched pair of Y-cell layers innervated by each eye, and the parvocellular laminae represent two pairs of X-cell layers (mostly separated into on- and off-center) innervated by each eye.

Functional Organization and Possible Role of the LGN

Practically every dorsal LGN neuron has an axon that terminates in the cerebral cortex, particularly in the *striate cortex* (area 17). These cells innervate no known subcortical site, although many of their projection axons exhibit local collaterals. Some neurons in the dorsal LGN, perhaps 20% to 30%, do not have projection axons and seem to be strictly involved in local circuits; these are the *interneurons*. Physiological evidence suggests that each neuron of the dorsal LGN receives a dominant excitatory input from one or a very few retinal ganglion cells of the same type (that is, X- or Y-cell, on- or off-center). This seems to be equally true for the projection neurons and the interneurons. Consequently, with minor and subtle exceptions, each dorsal LGN neuron exhibits response properties to visual stimuli that are virtually indistinguishable from those of its retinal input. Because projection cells of the dorsal LGN convey the same on- or off-center receptive field information to cortex that they receive from retina, they are said to be *relay cells*. This receptive field correspondence between the retinal ganglion cells and neurons of the dorsal LGN also means that the terms "X-cell" and "Y-cell" are equally applicable to both cell groups.

Furthermore, this equivalence of receptive field properties implies that the synaptic circuitry of the dorsal LGN is unique among regions of the visual system. For every other hierarchical level of the visual system, such as the retina, superior colliculus, and various areas of visual cortex (the superior colliculus and visual cortex are considered later in this chapter), synaptic interactions among visual neurons lead to increasingly complex elaborations of receptive field properties. An example is the change in receptive field properties between photoreceptors and bipolar cells shown in Fig. 8-18. These changes underlie the basic ability of the visual system to extract meaningful information from complex visual stimuli. If the dorsal LGN does not alter receptive field information arriving from retina before transmitting it to cortex, then it does not participate in the further neuronal elaboration of visual perception. What, then, is the functional significance of the dorsal LGN? Before answering this question it is necessary to consider the nonretinal inputs to the dorsal LGN as well as certain features of its intrinsic organization.

Nonretinal inputs to the dorsal LGN. Three structures other than the retina provide significant input to neurons of the dorsal LGN. These are summarized in Fig. 8-25. First and most prominent, neurons in layer VI* of several areas of visual cortex, including the striate cortex, originate the corticogeniculate pathway. Second, a diffuse input derives from various cell groups in the brainstem reticular formation, cell groups that also innervate most of the remaining thalamus as well as the neocortex. These brainstem regions include (a) the raphe nuclei, which are a loose aggregation of neurons that extend along the midline of the brainstem from the medulla to the midbrain; (b) the locus ceruleus, which is a small nucleus in the pons at the level of the isthmus; and (c) the parabrachial nucleus, which is a small cell group located in the midbrain tegmentum. Third, cells in the visual portion of the reticular nucleus of the thalamus (RNT), which is a portion of the ventral thalamus, innervate the dorsal LGN. Interestingly, these RNT cells receive input from axon collaterals of dorsal LGN relay cells as these axons pass through en route to cortex. The connection between the RNT and dorsal LGN is thus reciprocal.†

Synaptology of the dorsal LGN. A survey of the synaptic inputs onto the dendrites and somata of LGN

*Chapter 19 includes a more complete description of cortical layering.

†Much or most of the NRT can be divided into regions specific for a particular sensory modality, such as vision, somesthesis, and audition, and the connections of these other regions are similar to those for the visual portion of the NRT. For example, relay cell axons from the ventral posterolateral nucleus innervate the appropriate portion of the NRT with collaterals as they pass through en route to cortex, and this region of the NRT sends an axon back to the ventral posterolateral nucleus.

relay cells reveals that only 10% to 20% of these inputs derive from retina. This is astonishing in view of the receptive field physiology of these relay cells, because only this minority of retinal inputs is expressed in their receptive field properties. The receptive field technique clearly has its limitations. The nonretinal inputs onto these relay cells are almost equally divided between excitatory and inhibitory synapses. The former derive from corticogeniculate axons, and the latter derive partly from the few interneurons described earlier and partly from axons of RNT cells.

Functional "gating" of dorsal LGN cells. Both the interneurons and RNT cells are *GABAergic,* which means that they employ γ *aminobutyric acid* as their inhibitory transmitter. Maximum action of these GABAergic synapses can prevent the relay cells from transmitting their retinal inputs to cortex; less than maximum action has graded effects. Transmission at the retinogeniculate synapse can be regulated in this manner over a large range from practically nil to highly reliable.

To understand how the activity of the inhibitory interneurons and RNT cells is controlled is of considerable interest, because such activity determines the efficacy of retinogeniculate transmission. A simplified scheme is represented in Fig. 8-25, which is based largely on work done in cats. Both types of inhibitory neuron receive a fairly direct visual input, which derives from retinal axons for the interneurons and from relay cell axon collaterals for the RNT cells. These inhibitory cells also receive a prominent input from the corticogeniculate pathway. Corticogeniculate cells also monosynaptically innervate the relay cells. This implies that the cortex can simultaneously depolarize the relay cells directly and inhibit them or other relay cells indirectly through the inhibitory interneurons and RNT cells. Furthermore, many inputs from the brainstem reticular formation to the dorsal LGN mentioned earlier do not actually impinge directly on relay cells. These axons from brainstem direct their synapses instead onto the interneurons within the dorsal LGN, and they also innervate the RNT cells. Recent evidence also indicates a direct pathway from the brainstem reticular formation to relay cells (not shown in Fig. 8-25), a pathway that is mostly excitatory.

The X- and Y-cells probably fit into this circuitry somewhat differently (not illustrated in Fig. 8-25): the interneurons, which receive X-cell rather than Y-cell input from retina, dominate the inhibition of the relay X-cells; the RNT cells, which appear to receive more input from collaterals of relay Y-cell axons than from relay X-cell axons, dominate the inhibition of the relay Y-cells. In any case, these circuits suggest that both the cortex and the brainstem reticular formation gate retinogeniculate transmission.

The ability to shift attention both between vision and other sensory modalities as well as between different visual targets is probably dependent on neuronal circuits such as those illustrated in Fig. 8-25. This might be regarded as a means of protecting cortex from needless or unimportant visual information, on the one hand, and of directing cortex to attend to interesting or important new visual objects on the other. In this view, the dorsal LGN is a gateway to cortex for otherwise undigested information from retina, a gateway whose ease of passage is under continuous control from the cortex itself and from the brainstem reticular formation. The dorsal LGN does not exist for further analysis of visual signals per se. As shall be noted in later chapters, other sensory modalities probably use their thalamic "relays" for the same purpose.

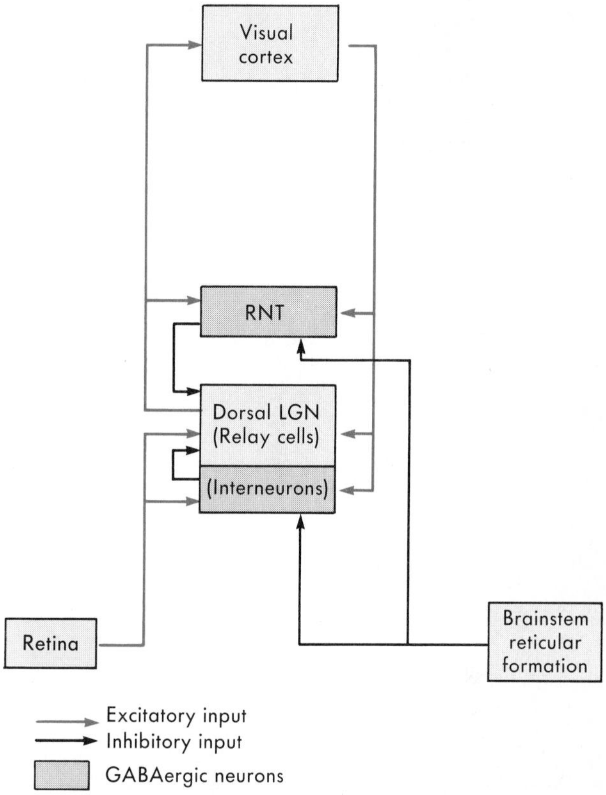

Fig. 8-25. Inputs to the dorsal LGN. Inhibitory afferents are shown in black, and excitatory ones are colored. Note the different inputs to the relay cells and interneurons of the dorsal LGN. (Although not shown, recent evidence suggests a direct pathway from the brainstem reticular formation to relay cells. This pathway is mostly excitatory.) (Redrawn from Sherman, S.M., and Koch, C.: Artificial Intelligence Memo 825, Cambridge, Mass., 1985, The MIT Press.)

Geniculostriate Projection

As in other areas of neocortex, the striate cortex (area 17) has six distinct layers of cells and fibers that run

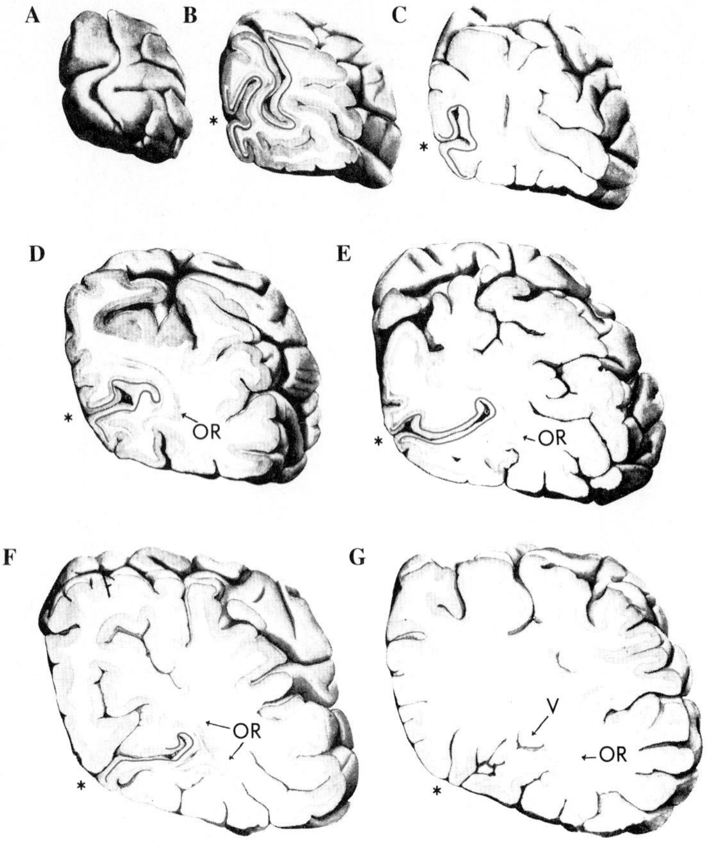

■ Fig. 8-26. Sections through the right occipital lobe of a human. The sections begin in **A** at the posterior pole and ending in **G** at the level of the lateral horn of the lateral ventricle *(V)*. Asterisks denote the calcarine fissure, and the colored lines indicate the stripe of Gennari that delineates the striate cortex. Note how much of the striate cortex lies buried in the calcarine fissure. Also shown are fibers of the optic radiation *(OR)* as they approach the striate cortex. (From Polyak, S.: The vertebrate visual system, Chicago, 1957, University of Chicago Press. Copyright © 1957 by University of Chicago Press.)

parallel to the pial surface. Layer I is nearest the pia, and layer VI is next to the white matter (see Chapter 19 for details). Essentially every X- and Y-cell from the dorsal LGN, which involves the magnocellular (laminae 1 and 2) and parvocellular (laminae 3 to 6) laminae, innervates the striate cortex. A tiny minority of dorsal LGN cells appear to innervate cortex beyond area 17. These are probably W-cells, about which so little is presently known in primates that the following account is largely limited to the geniculostriate projection of X- and Y-cells. In cats and many other infraprimate mammalian species, the geniculocortical projection densely innervates many cortical areas outside striate cortex. This obviously differs from the situation in primates, for which only a sparse extrastriate geniculocortical projection exists. The significance of this interspecies difference remains unclear. Nonetheless, it raises a note of caution, because much of the evidence on which our understanding of the central visual pathways rests is derived from studies of cats.

Course of geniculostriate axons. The striate cortex is located at the posterior pole of the occipital cortex and extends anteriorly from the pole along the medial surface of the hemisphere. Much of the striate cortex lies buried in a deep sulcus called the *calcarine fissure*. This fissure runs rostrocaudally and divides the striate cortex into roughly equal portions lying above and below the fundus of this sulcus (Fig. 8-26). However, geniculostriate axons do not take a direct route from the dorsal LGN to the striate cortex but rather loop around the lateral ventricles before coursing toward their terminal sites. Those which terminate in the cortex inferior to the calcarine fissure take an exaggerated detour well into the temporal lobe to skirt the ventricle. This detour is known as *Meyer's loop* (Fig. 8-27), and it explains why lesions of the temporal lobe can interfere with the geniculostriate projection.

Termination in striate cortex. The geniculocortical fibers terminate primarily in layer IV of the striate cortex, although a sparse terminal field has also been reported for layer VI. The axon terminal fields in layer IV are particularly dense and can be seen macroscopically as a glistening white band in gross specimens. This band is often called the *stria of Gennari,* and its stripe-like appearance accounts for the term *striate cortex*.

Layer IV of the striate cortex in primates is not uniformly innervated by dorsal LGN axons. The precise termination patterns depend on the dorsal LGN laminae from which the axons derive (Fig. 8-28). A thin sublayer at the dorsal tier of layer IV, called layer IVa, receives input from the parvocellular laminae. A slightly larger zone just beneath this, layer IVb, receives no geniculate innervation. The ventral region, layer IVc, is approximately half the width of layer IV,

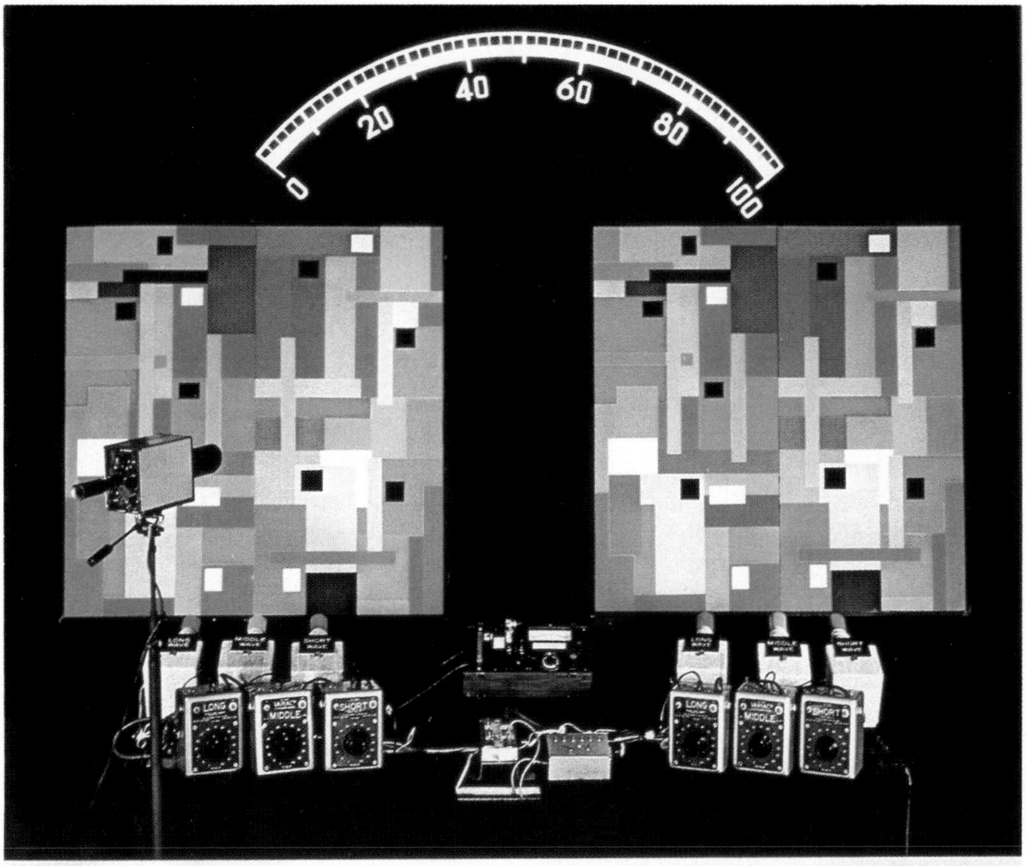

"COLOR MONDRIAN" EXPERIMENT employs two identical displays of sheets of colored paper mounted on boards four and a half feet square. The colored papers have a matte finish to minimize specular reflection. Each "Mondrian" is illuminated with its own set of three projector illuminators equipped with band-pass filters and independent brightness controls so that the long-wave ("red"), middle-wave ("green") and short-wave ("blue") illumination can be mixed in any desired ratio. A telescopic photometer can be pointed at any area to measure the flux, one wave band at a time, coming to the eye from that area. The photometer reading is projected onto the scale above the two displays. In a typical experiment the illuminators can be adjusted so that the white area in the Mondrian at the left and the green area (or some other area) in the Mondrian at the right are both sending the same triplet of radiant energies to the eye. The actual radiant-energy fluxes cannot be re-created here because of the limitations of color reproduction. Under actual viewing conditions white area continues to look white and green area continues to look green even though the eye is receiving the same flux triplet from both areas.

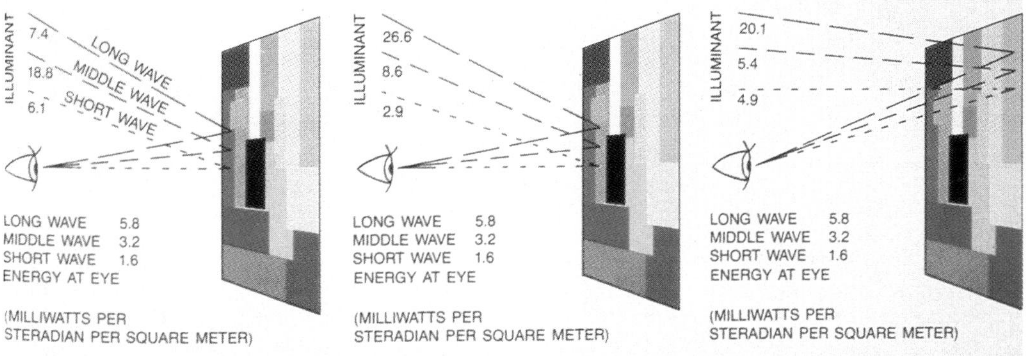

IDENTICAL ENERGY FLUXES AT THE EYE provide different color sensations in the Mondrian experiments. In this example, with the illuminants from the long-wave, middle-wave and short-wave illuminators adjusted as indicated, an area that looks red continues to look red (*left*), an area that looks blue continues to look blue (*middle*) and an area that looks green continues to look green (*right*), even though all three are sending to the eye the same triplet of long-, middle- and short-wave energies. The same triplet can be made to come from any other area: if the area is white, it remains white; if the area is gray, it remains gray; if it is yellow, it remains yellow, and so on.

■ **Plate I.** Color Mondrian experiments. (From Land, E.: Sci. Am. **237:**105, 1977.)

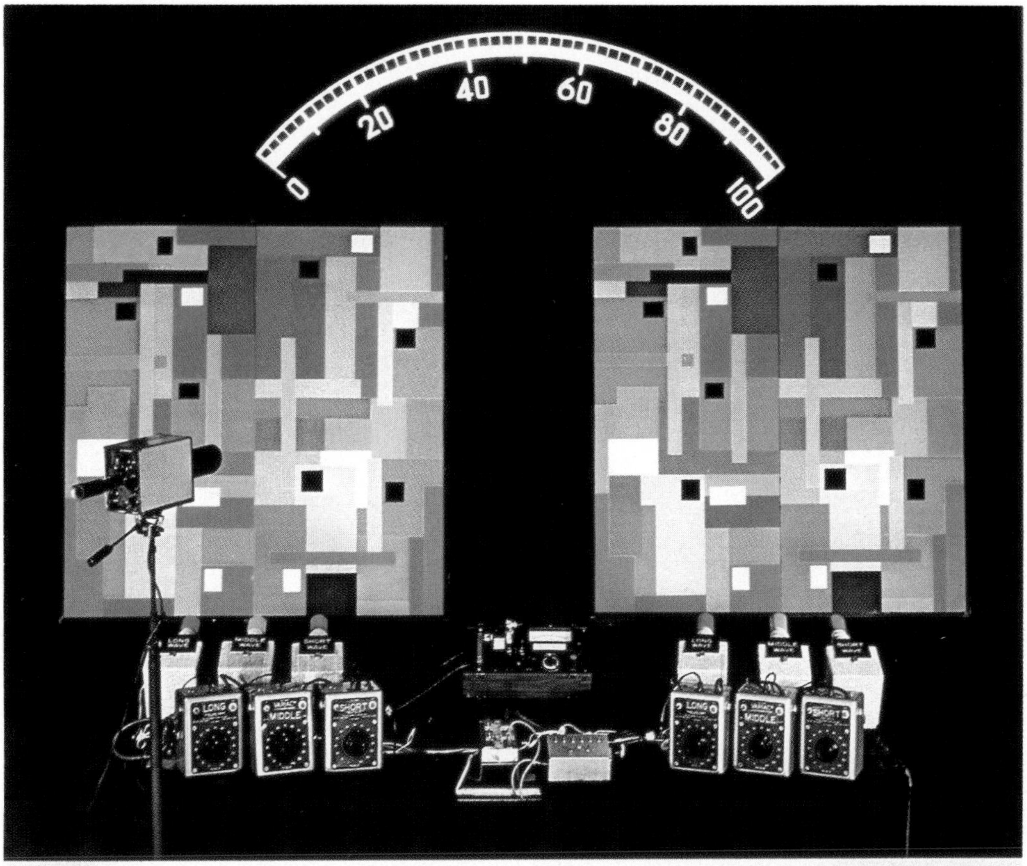

"COLOR MONDRIAN" EXPERIMENT employs two identical displays of sheets of colored paper mounted on boards four and a half feet square. The colored papers have a matte finish to minimize specular reflection. Each "Mondrian" is illuminated with its own set of three projector illuminators equipped with band-pass filters and independent brightness controls so that the long-wave ("red"), middle-wave ("green") and short-wave ("blue") illumination can be mixed in any desired ratio. A telescopic photometer can be pointed at any area to measure the flux, one wave band at a time, coming to the eye from that area. The photometer reading is projected onto the scale above the two displays. In a typical experiment the illuminators can be adjusted so that the white area in the Mondrian at the left and the green area (or some other area) in the Mondrian at the right are both sending the same triplet of radiant energies to the eye. The actual radiant-energy fluxes cannot be re-created here because of the limitations of color reproduction. Under actual viewing conditions white area continues to look white and green area continues to look green even though the eye is receiving the same flux triplet from both areas.

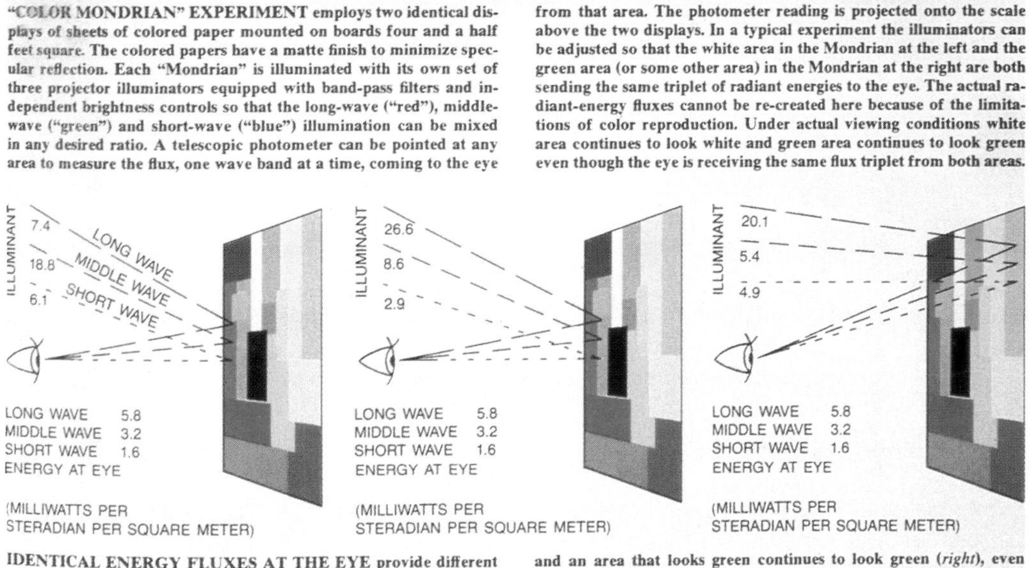

IDENTICAL ENERGY FLUXES AT THE EYE provide different color sensations in the Mondrian experiments. In this example, with the illuminants from the long-wave, middle-wave and short-wave illuminators adjusted as indicated, an area that looks red continues to look red (*left*), an area that looks blue continues to look blue (*middle*) and an area that looks green continues to look green (*right*), even though all three are sending to the eye the same triplet of long-, middle- and short-wave energies. The same triplet can be made to come from any other area: if the area is white, it remains white; if the area is gray, it remains gray; if it is yellow, it remains yellow, and so on.

■ **Plate I.** Color Mondrian experiments. (From Land, E.: Sci. Am. **237:**105, 1977.)

A

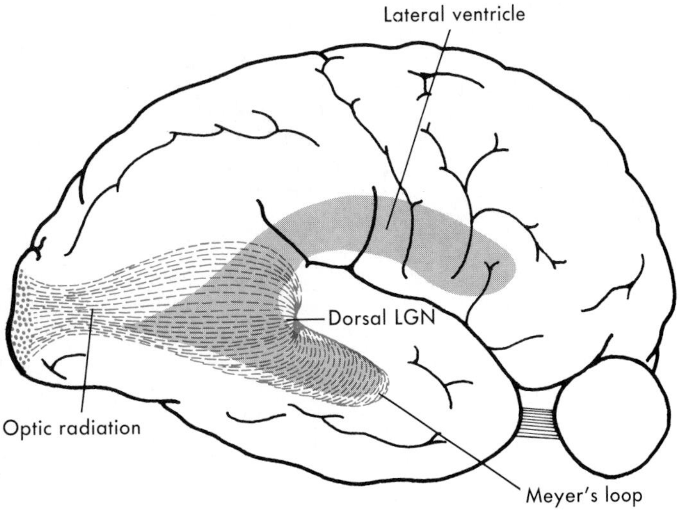

B

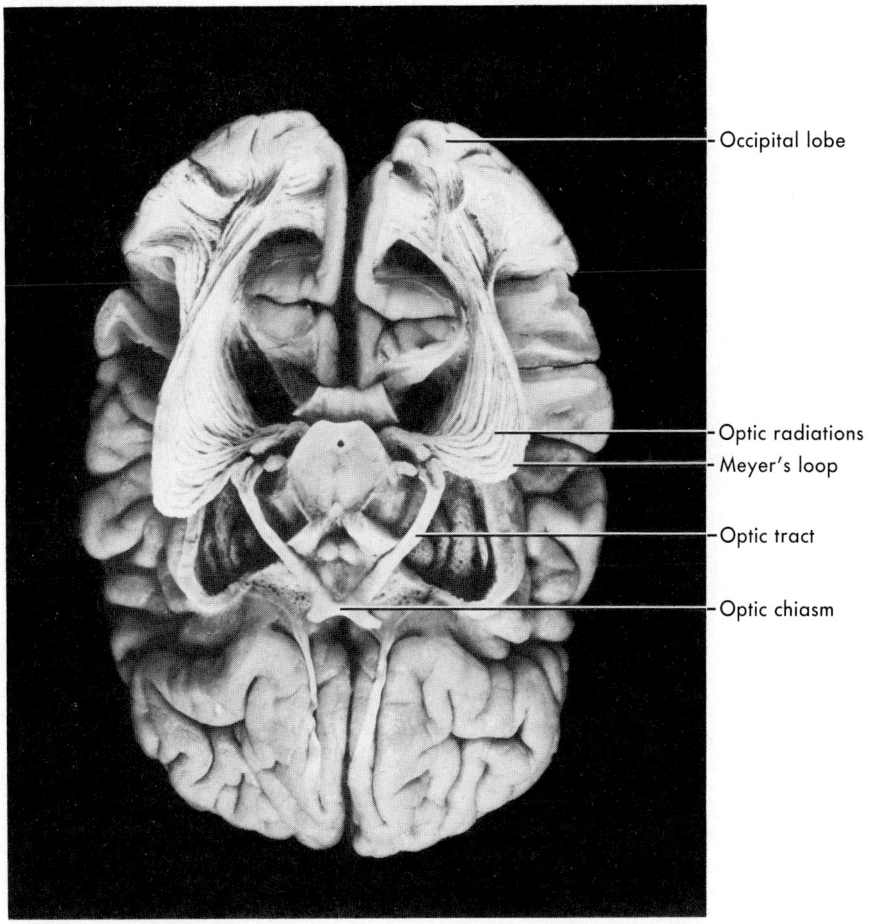

■ Fig. 8-27. Course of the geniculocortical fibers in the optic radiation. **A,** Schematic drawing of the optic radiation *(color)*. The geniculocortical fibers start among cell bodies of the dorsal LGN, which lies medial to the lateral ventricle *(gray)*. The fibers sweep to the lateral side of the ventricle before turning caudally to reach the striate cortex. This turning is known as *Meyer's loop*. Because of the shape of the lateral ventricle, some of the optic radiation extends well into the temporal lobe. **B,** Photograph of dissection of the ventral surface of the human brain showing the optic chiasm, optic tracts, and optic radiations. (**A** redrawn from Sandford, H.S., and Bair, H.L.: Arch Neurol. Psychiatry **42:**21, 1939; **B** from Gluhbegovic, N., and Williams, T.H.: The human brain: a photographic guide, New York, 1980, Harper & Row, Publishers, Inc.)

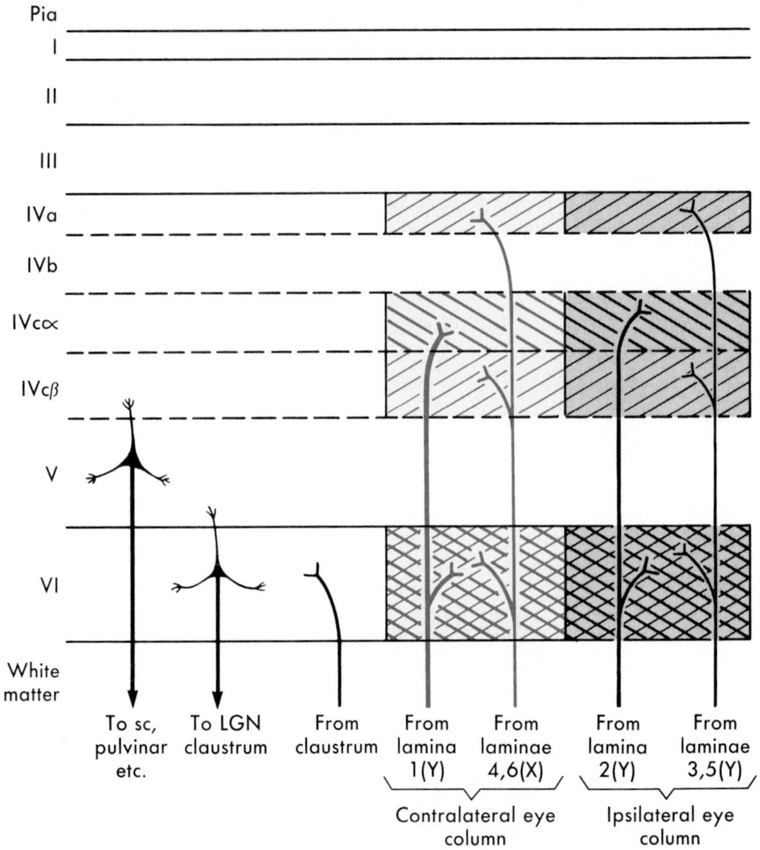

Fig. 8-28. The subcortical connections involving the striate cortex. Projections to most subcortical loci (e.g., the superior colliculus and pulvinar) derive from layer V pyramidal cells. Although not shown, pulvinar projections to the striate cortex probably exist, and they may terminate largely in layer I. Layer VI pyramidal and fusiform cells innervate the lateral geniculate nucleus and parts of the claustrum, and this region of claustrum projects to cortical layer VI. Geniculate innervation from the magnocellular laminae terminates in layers IV cα and VI, whereas that from the parvocellular laminae terminates in layers IVa, IVcβ, and VI. Note also that input representing one or the other eye is segregated into separate "ocular dominance columns" that are roughly 500 μm across. Left and right eye columns (or more precisely, vertically oriented slabs 500 μm wide and several millimeters long) lie next to one another in a fairly regular array.

and it is divided into two zones of rougly equal width. Layer IVcα is innervated by magnocellular laminae. Immediately ventral to this is a layer, IVcβ, that receives terminals from the parvocellular laminae. Little or no overlap exists between magnocellular and parvocellular terminal zones.

Another pattern of dorsal LGN input relates to ocular dominance. Adjacent and alternating patches of layer IV, approximately 500 μm wide, are innervated by dorsal LGN laminae 1, 4, and 6 or 2, 3, and 5. In other words, these alternate patches represent the ipsilateral or contralateral eye (Fig. 8-28).

Retinotopic Organization

Both the dorsal LGN and the striate cortex exhibit a precise point-to-point, but slightly distorted, map of the retina and thus of visual space. That is, neurons in these structures have well-defined receptive fields, and as one proceeds in an orderly sequence among neurons across the dorsal LGN, striate cortex, etc., the positions of the receptive fields gradually shift in a predictable fashion. This reflects the precise and orderly arrangment of connections along the entire retinogeniculostriate pathway. A general term for this arrangement is *neurotopic organization,* and it is a general feature of newer portions of sensory systems. Older portions often lack such organization. In the visual system the more specific term is *retinotopic organization.*

There are actually two retinotopic maps in the dorsal LGN and striate cortex. There is one for each eye, and these are in register with one another. For instance, the dorsal LGN laminae are stacked in such manner that a line passing orthogonally through them corresponds to a small region of visual space (Fig. 8-24). This same region is separately mapped for one or the other eye as the appropriate laminae are considered. Likewise, a line passing perpendicularly through the cortical layers represents a similar region of visual space. As these lines move across the dorsal surface of the dorsal LGN or striate cortex, there is a shift in the small region mapped, as predicted by the retinotopic map.

Fig. 8-29 illustrates details of the map in striate cortex. A distortion in these maps is evident. Areas of retina nearer the fovea are represented by more neural tissue in the striate cortex than are areas more distant from the fovea. This is called the *magnification factor* and is also evident in the dorsal LGN. To an extent, the magnification factor is a consequence of the ganglion cell density at various retinal locations. For a given retinotopic location, a more or less constant ratio exists among numbers of cortical, dorsal LGN, and retinal

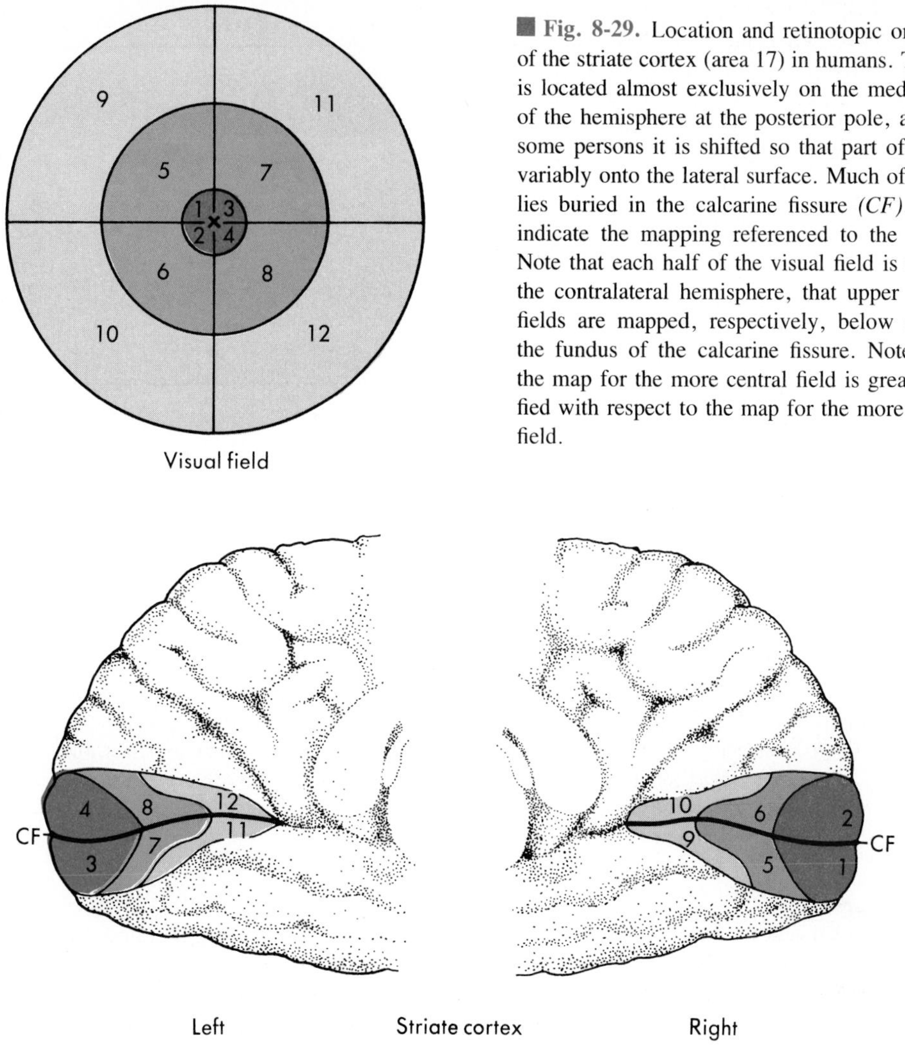

Fig. 8-29. Location and retinotopic organization of the striate cortex (area 17) in humans. This cortex is located almost exclusively on the medial surface of the hemisphere at the posterior pole, although in some persons it is shifted so that part of it extends variably onto the lateral surface. Much of the cortex lies buried in the calcarine fissure *(CF)*. Numbers indicate the mapping referenced to the fovea *(x)*. Note that each half of the visual field is mapped to the contralateral hemisphere, that upper and lower fields are mapped, respectively, below and above the fundus of the calcarine fissure. Note also that the map for the more central field is greatly magnified with respect to the map for the more peripheral field.

ganglion cells. In order to devote more neurons to the analysis of visual targets closer to the fovea, more retinal ganglion cells must obtain their inputs from foveally located photoreceptors via the retinal interneurons. However, a combination of optical constraints and the fact that most functional connections within the retina are organized vertically across the plexiform layers require that these ganglion cells lie physically close to their respective photoreceptors. This leads to ganglion cell density changes across the retina. For instance, this density increases as the fovea is approached. Magnification factor in the retina is thus expressed as a change in the density of retinal neurons with retinal eccentricity rather than as a change of the volume of ganglion cells arranged in a fixed density. No such optical and physical constraints exist in the dorsal LGN or striate cortex, and these structures maintain a fairly homogeneous neuron density throughout. Magnification factor here is thus expressed as a change in the volume of neural tissue for different retinotopic locations. A concomitant of retinotopic organization and magnification factor is the change with eccentricity of spatial resolution and receptive field size. Fields get smaller and resolution consequently improves as one nears the foveal representation, because more neurons are available to map a given region of visual space.

A detailed knowledge of the retinotopic organization of the central visual pathways, particularly of the striate cortex, can be very useful clinically. A small lesion results in a proscribed scotoma. From the location and size of this scotoma one can often localize a lesion precisely.

Cortical Receptive Field Properties

The receptive field properties of striate cortical cells are quite complicated, particularly with respect to those of the geniculate afferent neurons. As noted earlier, each geniculate neuron has a monocular receptive field (i.e.,

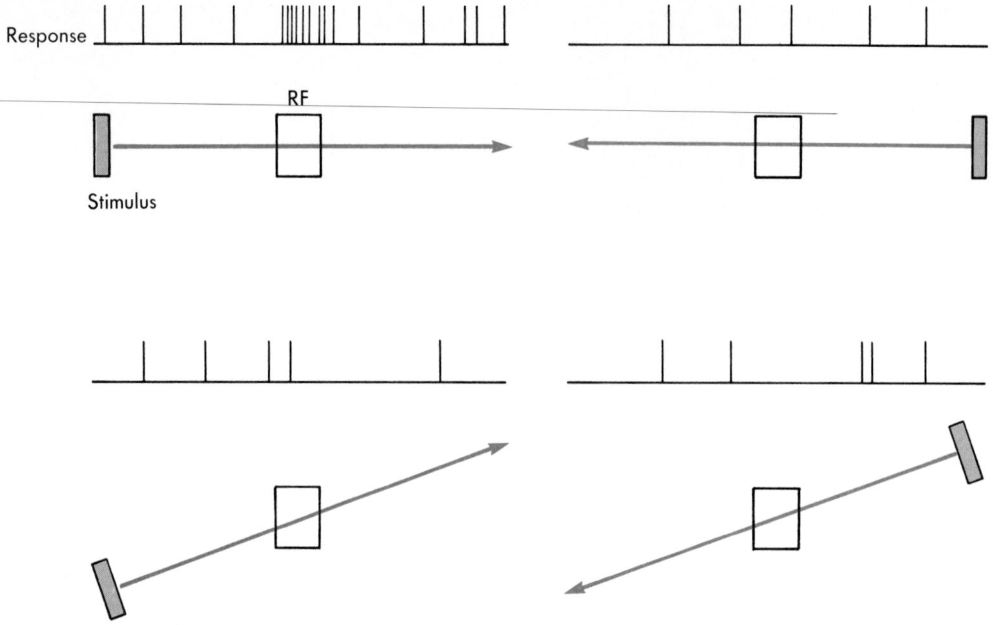

■ **Fig. 8-30.** Orientation and direction selectivity for a neuron in the visual cortex. The response of the cell is shown above the drawing of the stimulus *(colored bar)* moving as indicated through the receptive field *(RF; open rectangle)*. The vertically oriented stimulus moving left to right *(upper left)* maximally excites the cell. The same stimulus moving right to left with vertical orientation *(upper right)* or in either direction with an orientation other than vertical *(lower left and right)* fails to discharge the cell.

related to only one eye), and one need specify only the position of a bright or dark spot within the center or surround in order to excite or inhibit the cell in a predictable fashion.

Three additional stimulus parameters must be specified, however, to predict the response properties of most neurons in the striate cortex. One is known as *orientation selectivity* (Fig. 8-30). The stimulus must be elongated in one dimension, such as a rectangle or straight border between light and dark regions, and the cells are quite sensitive to the orientation of the long axis of this stimulus. For instance, a given neuron in the striate cortex might respond to a vertically oriented stimulus, but a tilt from vertical by as little as 5 to 10 degrees is often sufficient to render the cell unresponsive, and such a slight orientation shift may even inhibit the cell. Different cortical cells are ''tuned'' for different stimulus orientations so that orientations of every angle are represented in the cortex. The manner by which the geniculate inputs that are not orientation sensitive are formed into orientation-sensitive cortical neurons is unclear.

In addition to orientation selectivity, most cortical neurons display *direction selectivity*. If a properly oriented target is moved back and forth through the receptive field, responses for one direction of motion are larger than those for the other, and frequently only one direction of motion evokes a response (Fig. 8-30). As with orientation selectivity, the manner by which direction selectivity is generated remains a mystery.

Finally, most cortical neurons have binocular receptive fields, which means that stimulation of either eye can activate the cell. The receptive field in each eye maps roughly the same region of visual space (but see facing page). A distinct minority of cortical neurons are monocular, but these tend to be concentrated in layer IV, where the dorsal LGN input remains largely segregated in terms of ocularity. Evidently intracortical circuitry beyond layer IV includes convergence onto single cells of inputs related to each eye. This binocular convergence is crucial to stereopsis and binocular single vision: one is aware of a single, integrated view of the world and not separate views for each eye.

■ *Serial Versus Parallel Processing*

As noted earlier, it is far from clear how the geniculostriate pathways and cortical circuitry are organized to produce the functional types of neurons seen in the cortex. Two different hypotheses, referred to as *serial* and *parallel processing,* have been suggested. These hypotheses are not mutually exclusive, and it is quite likely that some synthesis of these ideas is closer to the truth.

Hubel and Wiesel, who proposed the hypothesis of serial processing, argued that retinal ganglion and geniculate cells are fairly homogeneous (the W-, X-, and

Y-cell classification was not known at the time their hypothesis was formulated) and that cells in the striate cortex could be classified in groups that reflect receptive field complexity and thus hierarchial order. More complex receptive fields require more specification of the stimulus to discharge the cell. *Simple cells* thus are the first-order cortical cells that receive geniculate input; *complex cells* are innervated by simple cells; *hypercomplex cells* are innervated by complex cells; and so on. Thus visual processing proceeds along a functional neuronal chain. As the chain ascends, more stimulus detail is encoded. The pattern of activity at the pinnacle of each hierarchy can specify the stimulus. These hierarchial neural chains represent a set of homogeneous building blocks for stimulus analysis. Identical neuronal chains are used repeatedly in a point-by-point analysis of the visual scene, and thus each point is analyzed in only this manner. Although there is little direct evidence for or against this hypothesis, it has the virtue of providing a simplifying theoretical framework.

Because it was formulated before knowledge of W-, X-, and Y-cells, however, the hypothesis of serial processing does not account for these cell classes. Proponents of the parallel processing hypothesis have most forcefully emphasized the importance of W-, X-, and Y-cells by suggesting that they represent three distinct, functionally independent pathways to and through the cortex. It is further suggested that these pathways analyze different aspects of the visual scene in parallel.

It is presently not clear how W-, X-, and Y-cell innervation from the dorsal LGN relates to the simple, complex, and hypercomplex cell classes in the striate cortex. This is an active area of research that is replete with new ideas and considerable controversy. The functional implications of the geniculostriate pathway and the W-, X-, and Y-cell pathways are considered later in this chapter.

Stereopsis

Stereopsis can be defined as binocular depth perception. It is distinguished from the many monocular cues to depth, such as different retinal image sizes for objects at different distances and different rates of object movement at different distances (motion parallax). The cues for stereopsis are based on the slightly different retinal image each eye forms of a given visual scene because each eye has a slightly different perspective. This is the same principle used to calculate distances in rangefinder cameras and to generate stereoscopic photographs by cameras with two lenses offset from one another.

The useful range of stereopsis is thus determined by interocular distance; in humans it is best suited for visually guided work with the hands. Human stereopsis is useful only up to a distance of 125 m. *Stereoacuity* (the minimum separation of two objects necessary to be perceived as such) improves dramatically as object distance lessens, and this is a straightforward consequence of trigonometry. Therefore steropsis is more useful for threading a needle than for landing an airplane.

A possible neural basis for stereopsis has been described in the cat and extended to monkeys. Binocular neurons in the striate cortex do not have receptive fields in precisely corresponding locations for each retina. Instead, they usually exhibit *receptive field disparities,* which means disparate receptive field locations in each retina. Fig. 8-31, *A,* illustrates this phenomenon for three cortical neurons with receptive fields near the fovea. Cell *1* has a receptive field in each fovea; cell *2,* in each nasal retina; and cell *3,* in each temporal retina. Cell *1* has no "disparity" because its receptive fields are in precisely corresponding retinal locations, but cells *2* and *3* exhibit disparity. When object *1* is fixated, cell *1* responds best to object *1;* cell *2* responds best to a more distant object (object *2*); and cell *3* responds best to a nearer object (object *3*). The largest disparities for cortical cells in monkeys are approximately 1 to 2 degrees. That is, the retinal location of one receptive field may be that far from precisely corresponding position for the other receptive field.

Receptive field disparities thus provide these cells with sensitivity to the third dimension relative to a fixation point. The range of disparities exhibited by these cortical cells closely matches that predicted by psychophysical studies of stereopsis. Given information about the alignment of the visual axes or the amount of convergence between the eyes, the brain can deduce three-dimensional information from knowledge of the pattern of activity of cells with receptive field disparities. Unfortunately little is known concerning how the brain encodes the critical information of visual axis alignment.

If all ganglion cells temporal to a vertical line passing through the fovea projected ipsilaterally and all cells nasal to this line projected contralaterally, then the presence of cortical cells with binocularly disparate receptive fields from each nasal or temporal retina (e.g., cells *2* and *3,* respectively, in Fig. 8-31, *A*) would require complicated interhemisphere pathways. This is avoided because of the "nasotemporal overlap" seen in the retina. A 1- to 2-degree wide vertical strip of retina includes intermixed ganglion cells that project ipsilaterally or contralaterally. This is the strip of nasotemporal overlap described earlier in this chapter. The width of this strip roughly matches the range of disparities seen in cortical cells. This feature thereby provides disparate input to cortical cells with receptive fields near the vertical meridian or midline of the visual field, and complicated interhemisphere pathways are not needed. It should finally be noted that neither stereopsis nor receptive field disparities are limited to foveal regions. The entire binocular field of view exhibits these features (Fig. 8-31, *B*).

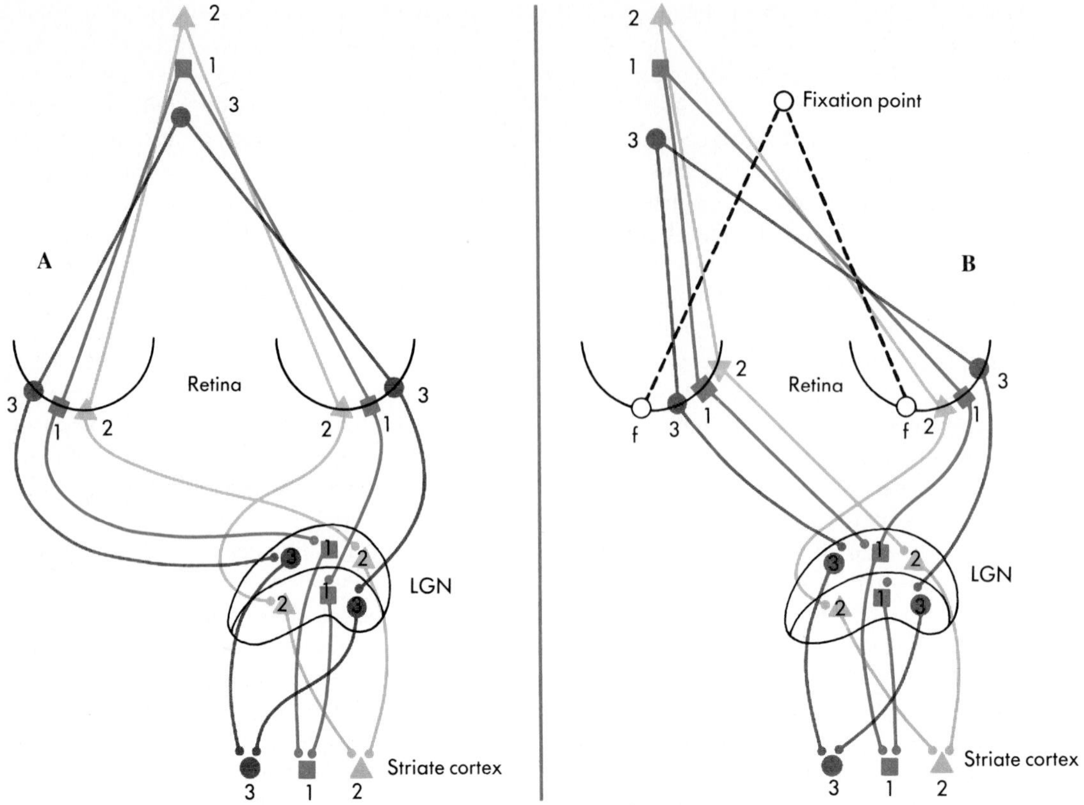

■ **Fig. 8-31.** Schematic diagrams illustrating how receptive field disparities might provide the neural basis for stereopsis. A cortical neuron with nondisparate receptive fields in each retina means that the distance and direction (i.e., left, right, up, or down) of each field from the fovea are identical. Receptive field disparities are caused by differences in the distance and/or direction of the fields in each retina. In each drawing the neural circuitry provides cortical cell *1* with nondisparate receptive fields, whereas cells *2* and *3* have disparities in opposite directions. Thus cell *2* encodes farther objects than does cell *1*, and cell *3* encodes nearer objects. The maximum disparities seen in studies of cats and monkeys are roughly 2 degrees. **A,** Objects near the fixation point. The fixation point *(1)* is imaged onto each fovea. Thus the nearer point *(3)* falls onto each temporal retina, and the farther point *(2)* falls onto each nasal retina. **B,** Objects far from the fixation point. The same principles apply as in **A,** except that all of the disparities exist to one side of the fovea.

■ *Color*

How the visual pathways are organized to subserve wavelength or color discrimination is not well understood. The geniculostriate pathways play a key role in this process. Actually, wavelength and color discrimination are not equivalent, a point soon to be taken up in more detail. Wavelength discrimination is an easier concept to explain, and thus it will be considered first.

Wavelength discrimination. Visible light includes a range of wavelengths, and our appreciation of different colors is somehow bound up in the wavelengths of light associated with visible objects. We can define wavelength discrimination simply as the ability to distinguish among various wavelengths or their combinations in visual stimuli. Clearly, wavelength discrimination begins with the cone photoreceptors; in humans there are three types. Fig. 8-32 illustrates the *spectral sensitivity* (the sensitivity to different wavelengths) of each type. However, physiological data suggest that most retinal ganglion cells receive both rod and cone input via the retinal interneurons. This implies that these ganglion cells can function at photopic (i.e., cone only) and scotopic (i.e., rod only) levels of illumination. Thus wavelength discrimination is not determined by unique cone inputs to various cells; rather the nature and interactions of these inputs lead to receptive fields of certain neurons (e.g., X-cells) that respond differently to different patterns of cone stimulation.

Retinal ganglion and geniculate cells can be divided into two general types based on their responses to different wavelengths. One cell type produces responses that do not depend significantly on the spectral content, or wavelength combinations, of the stimulus. This cell

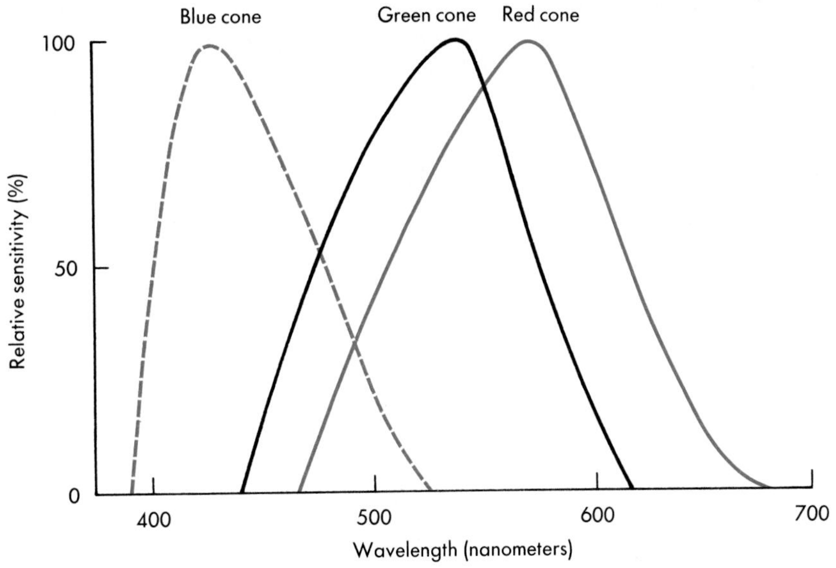

Fig. 8-32. Relative spectral sensitivity of the three cone types in humans.

type instead responds to luminance contrast and is called *broadband,* because it responds to a broad band of wavelengths. The other cell type is known as *spectrally opponent,* because wavelength is a key stimulus parameter. Most spectrally opponent cells are excited by one spectral band turned on (or off) in the receptive field center and turned off (or on) by another in the surround. For instance, many of these cells are excited by red light turned on in the center or by green light turned off in the surround. Other more complicated patterns of spectral opponency have also been described. Finally, spectrally opponent cells have been described in the striate cortex and in certain extrastriate cortical areas, but details of their spectral sensitivity are currently uncertain.

Within the monkey's geniculostriate pathways, the broadband neurons have been associated with Y-cells and the magnocellular laminae of the dorsal LGN. Spectrally opponent neurons have been associated with X-cells and the parvocellular laminae. Perhaps another functional significance of the parallel processing in X- and Y-cell geniculostriate pathways concerns wavelength discrimination, which might be a unique function of the X-cell pathway.

Color discrimination. Wavelength sensitivity is obviously important to color discrimination. The appreciation of *color constancy,* which means the capacity in humans to recognize the same colors in spite of vastly different spectral constituents of these colors, suggests that color discrimination is complicated.

Perhaps the clearest and most dramatic demonstration of color constancy and its functional significance derives from psychophysical experiments. Observers can correctly identify the colors in a complexly colored scene, such as a Mondrian painting, despite quite different conditions of illumination. The basic experiment is outlined in Plate I. The wavelengths actually present for each colored rectangle in the scene are a product of the wavelengths in the illuminating light and the relative reflectance of each wavelength by the colored patch. Therefore it is possible to equalize the actual *spectral* components of different colored patches, such as the red, green, and blue patches, by appropriately adjusting the spectral quality of the illuminating light. However, even though these patches are spectrally identical, one can still perceive their true colors.

In retrospect, it seems obvious why mammalian evolution would have favored color constancy. Natural lighting conditions exhibit enormous spectral variations, including the greens of forest shade, the reds of dusk, the blues of an overcast day, etc. Nonetheless, true colors must be recognized despite this variation: a monkey must recognize an orange by its color regardless of its spectral composition.

In the experiment illustrated by Plate I, the brain is able to extract the true colors in a visual scene over a wide range of changing spectral qualities. The underlying hypothesis for this process has been called the *retinex theory,* because it may be a function of retina, cortex, or both. The visual system may accomplish this by parallel visual pathways, each of which is most sensitive to a limited range of wavelengths. Each of these pathways thus acts as a spectral filter and perceives a set of relative brightnesses for each colored patch that differs from the relative brightnesses perceived by the other spectrally limited pathways. From a comparison of these different relative brightnesses, true or natural color can be deduced. Note that this requires several colors in a scene for comparison. Once a stimulus is limited to a single color or too few colors for compari-

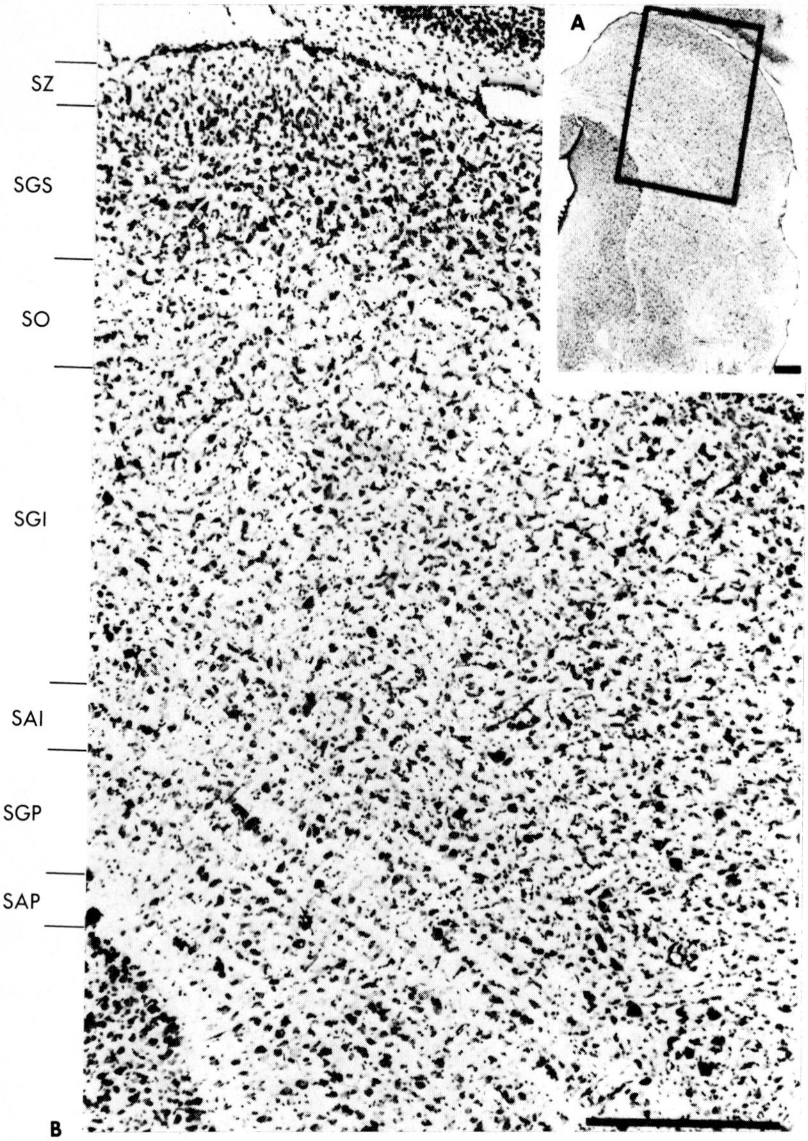

■ Fig. 8-33. Coronal section through the superior colliculus in a prosimian primate *(Galago senegalensis)*. **A,** Lower power photomicrograph of midbrain region. **B,** Higher power photomicrograph of area demarcated in **A.** Collicular lamination is poorly differentiated in primates, but the layers are indicated. *SZ,* Stratum zonale; *SGS,* stratum griseum superficiale; *SO,* stratum opticum; *SGI,* stratum griseum intermedium; *SAI,* stratum album intermedium; *SGP,* stratum griseum profundum; *SAP,* stratum album profundum. Scale bars in **A** and **B** are 0.5 mm. (Courtesy D. Raczkowski.)

son, color constancy fails. Perceived colors can then be predicted on the basis of the actual spectral composition of the stimulus.

Such a theory suggests that somewhere in the visual system some cells may exist that demonstrate true color discrimination rather than just wavelength discrimination. Preliminary evidence from monkeys suggests that the spectrally opponent cells of the retina, dorsal LGN, and striate cortex are strictly sensitive to wavelength, and their responses can be predicted simply on the basis of the spectral composition of the stimulus. However, one specialized area of extrastriate cortex, called V4 (see Fig. 8-34, *C*), may be specialized for color vision. It may possess cells that respond to the true colors of stimuli rather than their spectral composition.

■ Superior Colliculus Pathways

■ Lamination and Connections

A large portion of the visual system involves pathways through the superior colliculi, which are bilaterally paired, prominently laminated structures that form the dorsal portion of the rostral midbrain (Fig. 8-33). Seven collicular layers can be distinguished. From the dorsal surface ventrally, these layers are the *stratum zonale, stratum griseum superficiale, stratum opticum, stratum griseum intermedium, stratum album intermedium, stratum griseum profundum,* and *stratum album profundum.* Below this last layer is the periaqueductal gray matter. The three dorsal layers are exclusively involved in vision, and the four ventral layers are involved in more complex multimodal processing. Therefore only the three dorsal layers are considered here.

Two major visual inputs of roughly equal extent innervate these dorsal layers. One is represented by retinocollicular fibers that pass into the brachium of the superior colliculus, enter the stratum opticum, and terminate in the upper three layers (Fig. 8-33). These axons arise from retinal W- and Y-cells, predominantly from the contralateral nasal retina. The other input is a corticocollicular pathway predominantly from the striate cortex (area 17) and nearby cortex (areas 18 and 19) but also involving other cortical zones as well. Much of this pathway involves the following neuronal chain: retinal Y-cell to geniculate Y-cell to complex cell in layer V of the striate cortex; these complex cells comprise the source of the corticocollicular pathway. There is some evidence that these cells in layer V of striate cortex receive direct geniculocortical input from Y-cells via their dendrites that extend into layer IV and/or layer VI. It follows then that the W- and Y-cell pathways, but not the X-cell pathway, innervate the superior colliculus.

Neurons in the upper layers of the superior colliculus project to the thalamus. Most of this projection terminates in the pulvinar, and some terminates in the dorsal LGN. The pulvinar and dorsal LGN innervate virtually all known cortical areas involved in visual processing (see ''Extrastriate Visual Cortex''), so the superior colliculus is indirectly involved in the innervation of visual cortex. The possible functional significance of this is considered later in this chapter.

Retinotopic Organization

As befits a relatively new portion of the visual system, the superior colliculus is organized retinotopically. The fovea is represented slightly lateral to the rostral pole of the superior colliculus. The line that divides the visual fields into superior and inferior halves runs mediocaudally. Rostromedial to this line is represented the superior visual field, and the inferior visual field is represented caudolaterally.

Receptive Field Properties

Collicular neurons have fairly large receptive fields that are most sensitive to rapid stimulus motion in a particular direction. For most cells, the preferred direction of stimulus motion is away from the fovea. These cells seem less concerned with spatial detail than with stimulus motion. In addition to direction selectivity, most possess binocular receptive fields. However, unlike cortical cells, these collicular neurons lack clear orientation selectivity.

In cats, removal of the visual cortex and thus the corticocollicular input dramatically alters the collicular receptive field properties. Postoperatively, collicular neurons lose their direction selectively, respond poorly to rapidly moving stimuli, and tend to respond almost exclusively to stimulation of the contralateral eye. These properties presumably reflect the retinocollicular input. Corticocollicular input is thus needed for binocularity, direction selectivity, and responsiveness to rapidly moving targets. Nonetheless, collicular cells still respond well to visual stimuli after cortical removal. Curiously, effects of cortical removal or inactivation on collicular receptive fields in monkeys are somewhat different; cells dorsal to the upper half of layer III are relatively unaffected, whereas more ventral cells lose responsiveness to visual stimuli. The reasons for these differences between cat and monkey are not understood.

Extrastriate Visual Cortex

In the cat, at least 13 separate visual areas have been mapped in the cerebral cortex. In primates, a large number of separate visual cortical areas also exists; the actual number perhaps varies among species, and contemporary research continues to uncover new visual areas. The actual number of these separate visual areas thus continues to increase with the more complete mapping of cortex. Nearly all of these areas are retinotopically organized, but receptive field data are generally unavailable for most of these regions. Fig. 8-34 illustrates these areas for the cat, the owl monkey (a New World monkey), and the macaque monkey (an Old World monkey).

These visual areas have many complex interconnections with each other and with the thalamus, particularly the pulvinar. Indeed the pulvinar innervates most or all of these areas. Although the functional significance of these multiple extrastriate areas is generally unknown, recent research in macaque monkeys offers clues of the role played by certain areas. As noted earlier in this chapter, V4 may be involved in the processing of color information. Likewise, area MT may be specialized to analyze visual motion. Finally, a broad generalization has emerged regarding the relationship of these areas to the parallel X- and Y-cell inputs to cortex (the W-cell input is too small and weak to be detected by present methodology). This is illustrated schematically in Fig. 8-35. Both pathways enter cortex via the striate cortex (area 17 or VI). As noted earlier, they innervate separate sublayers there, and they are routed differently to areas of extrastriate cortex. The X-cell pathway flows mainly through the inferotemporal areas of visual cortex and is involved in the recognition and

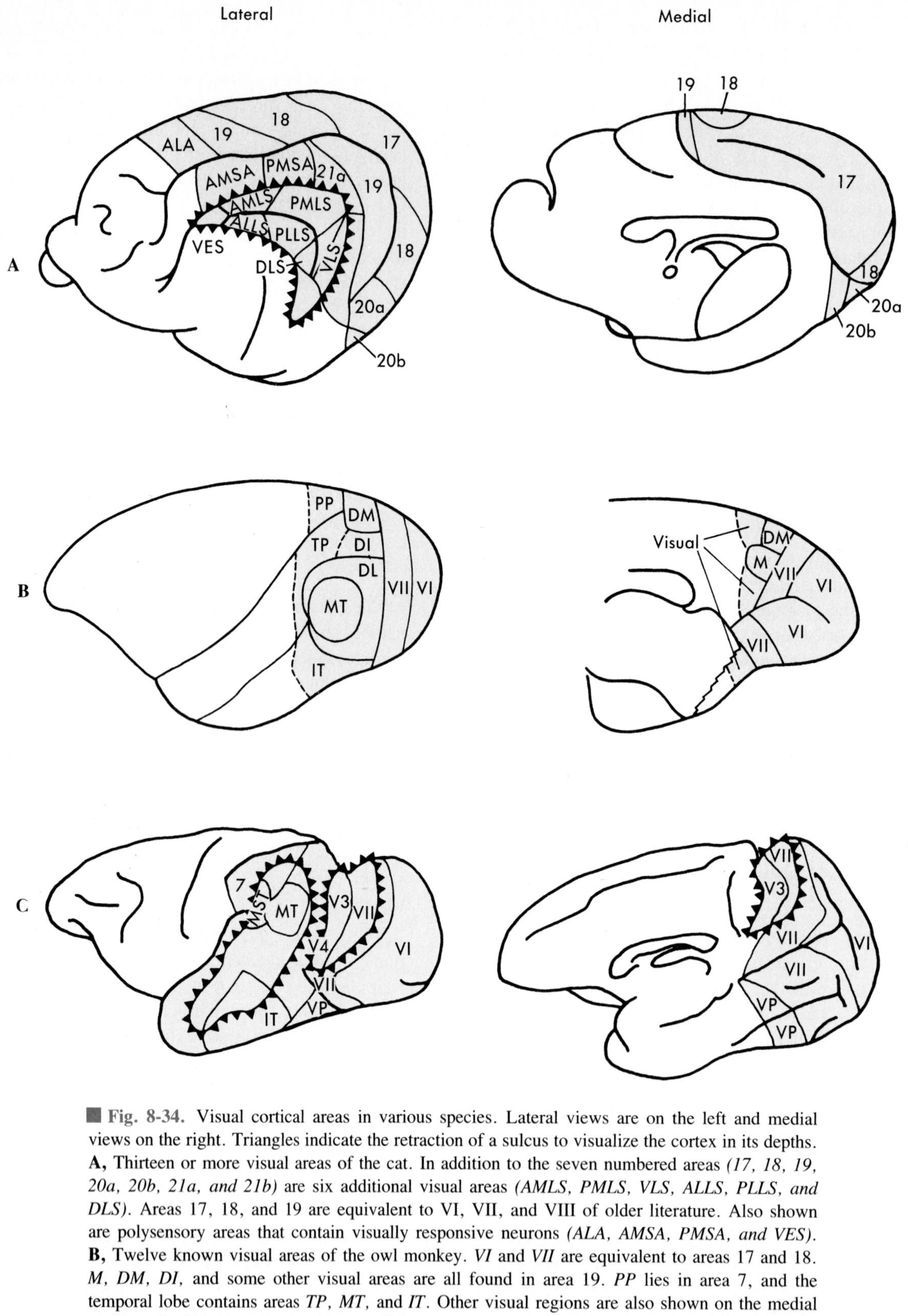

Fig. 8-34. Visual cortical areas in various species. Lateral views are on the left and medial views on the right. Triangles indicate the retraction of a sulcus to visualize the cortex in its depths. **A,** Thirteen or more visual areas of the cat. In addition to the seven numbered areas *(17, 18, 19, 20a, 20b, 21a, and 21b)* are six additional visual areas *(AMLS, PMLS, VLS, ALLS, PLLS, and DLS)*. Areas 17, 18, and 19 are equivalent to VI, VII, and VIII of older literature. Also shown are polysensory areas that contain visually responsive neurons *(ALA, AMSA, PMSA, and VES)*. **B,** Twelve known visual areas of the owl monkey. *VI* and *VII* are equivalent to areas 17 and 18. *M, DM, DI,* and some other visual areas are all found in area 19. *PP* lies in area 7, and the temporal lobe contains areas *TP, MT,* and *IT*. Other visual regions are also shown on the medial surface. **C,** Eight known visual areas of the macaque monkey. *VI* is equivalent to area 17. Area 18 contains *VII, V3, V4,* and *VP*. *MT, MST,* and *IT* are located in the temporal lobe. It should be noted that studies of multiple visual areas of macaque monkeys are recent compared with those of cats and owl monkeys, and more areas may emerge as more detailed maps become available. (Redrawn from Tusa, R.J.: Visual cortex: multiple areas and multiple functions. In Morrison, A.R., and Strick, P.L., editors: Changing concepts of the nervous system, New York, 1982, Academic Press, Inc.)

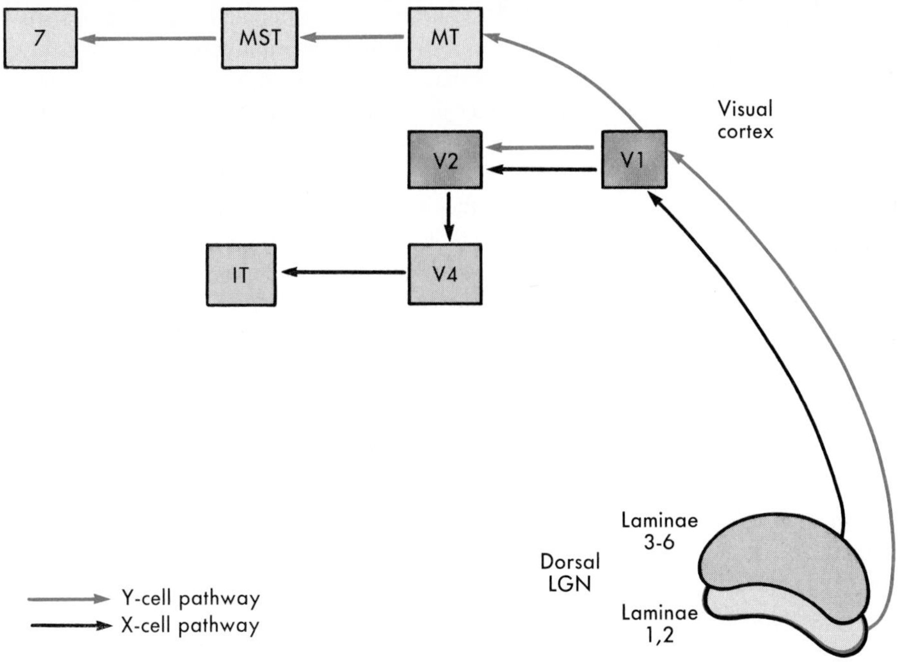

Fig. 8-35. Highly speculative schema of parallel pathways emanating from the magnocellular (colored) and parvocellular (black) laminae of the dorsal LGN in primates. Both of these pathways are routed through V1 (striate cortex), and both also seem to be present in V2, but they separate thereafter into a dorsal, magnocellular tier of areas and a ventral, parvocellular tier.

analysis of fine spatial details. The Y-cell pathway flows mainly through the parietal areas of visual cortex and is involved in motion analysis and in analysis of the relative location of objects.

These multiple areas have been clarified only in the last decade. A similar, but presently less complete, picture is emerging for the representation of other sensory systems in the cortex. Previously no specific sensory function was attributed to most of the posterior half of the cortex, and it was designated as *association cortex* by default. The contemporary view is that little or no such association cortex exists in this region and that all of this posterior cortical area is devoted to specific sensory analysis. This is considered more fully in Chapter 19.

Phylogenetically Older Visual Pathways

The previous discussion describes some of the complex interactions among certain of the phylogenetically newer visual pathways. These are characterized by the presence of retinotopic organization, and often the neurons of these pathways respond tonically to visual stimuli and display lateral inhibition in their receptive fields. As noted earlier, the retina projects sparsely to other brainstem sites that are regarded here as phylogenetically older. These include retinal projections to the suprachiasmatic nucleus, ventral LGN, pretectum, and accessory optic nucleus.

Where function can be suggested for these structures, it is generally reflexive and subconscious in nature. The suprachiasmatic nucleus is involved in circadian rhythms involving the light/dark diurnal cycle. The nucleus of the optic tract and accessory optic nucleus participate in certain reflexive eye movements, such as optokinetic nystagmus and the vestibulo-ocular reflex. The nucleus of the optic tract is also involved in the pupillary light reflex. No obvious function can yet be suggested for the ventral LGN.

Role of Various Visual Pathways in Spatial Vision

The previous sections deal with the functional organization of the various central visual pathways. This section deals with the possible functional role each of these pathways plays in conscious visual perception. Most attention in this context has been focused on the geniculostriate pathways and the extrastriate pathways involving the superior colliculus.

Cortical Versus Collicular Pathways

Until recently most investigators emphasized the importance of the geniculostriate pathways to visual perception and relegated the superior colliculus to vestigial or reflex roles such as crude brightness analysis or reflex eye movement. However, the importance of the superior colliculus has been emphasized from recent behavioral studies of animals and humans with brain lesions. Lesions of the visual cortex commonly lead to severe visual dysfunction in many mammalian species, including humans, and this has led to the concept that this cortex and not the superior colliculus is crucial to visual perception. However, there are two flaws in this conclusion. The first is logical. Lesions of the visual cortex interrupt the corticocollicular pathways, and the superior colliculus thus becomes extensively denervated. Visual capacity following such lesions hardly elucidates the functional role of the superior colliculus in normal individuals. Second, more recent and complete behavioral studies have revealed surprising visual capacity in animals and humans with lesions of the visual cortex. Most of this work has been done with cats, but a limited amount has involved monkeys and human patients.

Studies of Visual Function in Cats

Complete bilateral destruction of the striate cortex in the cat produces remarkably little impairment in the animal's visual performance. There is moderate loss of acuity, but otherwise the cats are indistinguishable before and after such an ablation on tests of visual performance. (As noted later in this chapter, good form vision does not depend on acuity per se.) However, lesions that involve extensive extrastriate areas in the occipitotemporal cortex do dramatically impair form vision.

Because these areas are innervated by both the dorsal LGN and pulvinar (which in turn receives collicular input), it is not clear to what extent geniculocortical or colliculothalmocortical pathways contribute to the cat's spatial vision. Similar data exist for other infraprimate mammalian species. Generally, it seems clear for these species that the geniculostriate pathways are not essential for good form vision but may be needed for high acuity vision and perhaps stereopsis.

Behavior-ablation data cannot be interpreted decisively, however, because of the complex effects of neural lesions throughout the brain. Fig. 8-36, which illustrates a phenomenon known as the *Sprague effect,* exemplifies the underestimated role of the superior colliculus in normal form vision. Large unilateral ablations of the occipitotemporal cortex that destroy all known visual areas produce a profound and permanent *hemianopsia* (blindness in a half-field vision) for the hemifield contralateral to the lesion (Fig. 8-36, *B*). In the past such data were used to emphasize the importance of the cortex in vision and to deemphasize subcortical pathways. However, if the superior colliculus contralateral to the earlier cortical lesion is then ablated or if the colliculi are disconnected from one another by splitting their commissure, there is a dramatic restoration of vision for the previously blind hemifield (Fig. 8-36, *C*). We still lack an adequate explanation for the partial restoration of vision after blindness had been produced by an earlier lesion.

Such "collicular vision" is by no means normal. It allows the cat to move about visually and to locate visual objects of interest. However, detailed pattern vision and the ability to discriminate among various spatial patterns are absent or rudimentary. Conversely, removal of the superior colliculus but with the cortex intact results in rather subtle deficits that do not seem to involve spatial vision. This evidence and similar data from other mammals have led to the concept that the superior colliculus is involved primarily in determining *where* objects are, whereas the cortex is needed to determine *what* objects are. However, it should be reemphasized that the functioning of the superior colliculus after removal of its massive corticocollicular innervation can hardly be expected to be normal.

Studies of Visual Function in Monkeys and Humans

Analogous ablation-behavior data are much less complete for monkeys and humans. Many older studies have repeatedly emphasized the nearly total loss of spatial vision in monkeys and humans after damage to the striate cortex. Although it certainly seems clear that such damage reduces visual performance more drastically for primates than for cats, two recent series of studies have emphasized certain similarities between primates and cats regarding these ablation-behavior data.

First, it appears that many or all of the earlier experimental studies of monkeys involved lesions that extended well beyond the striate cortex. Thus much of the cortical innervation from the pulvinar, as well as from the dorsal LGN, was interrupted. More recent studies with monkeys that had lesions limited to the striate cortex demonstrated significant remaining spatial vision, although vision in these monkeys was drastically impaired.

Second, in studies of human patients with cortical damage resulting from vascular accidents or penetrating missile wounds, the extents of these lesions are difficult to assess. Nonetheless, surprising visual function has been described in such patients when appropriate measures are used. For example, on a standard perimetry test for which the patient must verbally indicate visualization of the target, a large scotoma might be mapped that can be related to the cortical injury. If nonverbal responses are then employed, such as requiring the patient to point a finger at the verbally denied target, it is clear that visual stimuli can be crudely localized. The cortical lesion seems to interfere with language related to visual stimuli in addition to its effect on visual capacity per se. Even in humans some visual capacity thus seems to survive interruption of the geniculostriate pathways. Although of great and obvious importance to vision in primates, these pathways are not the sole substrate of conscious vision as was once thought.

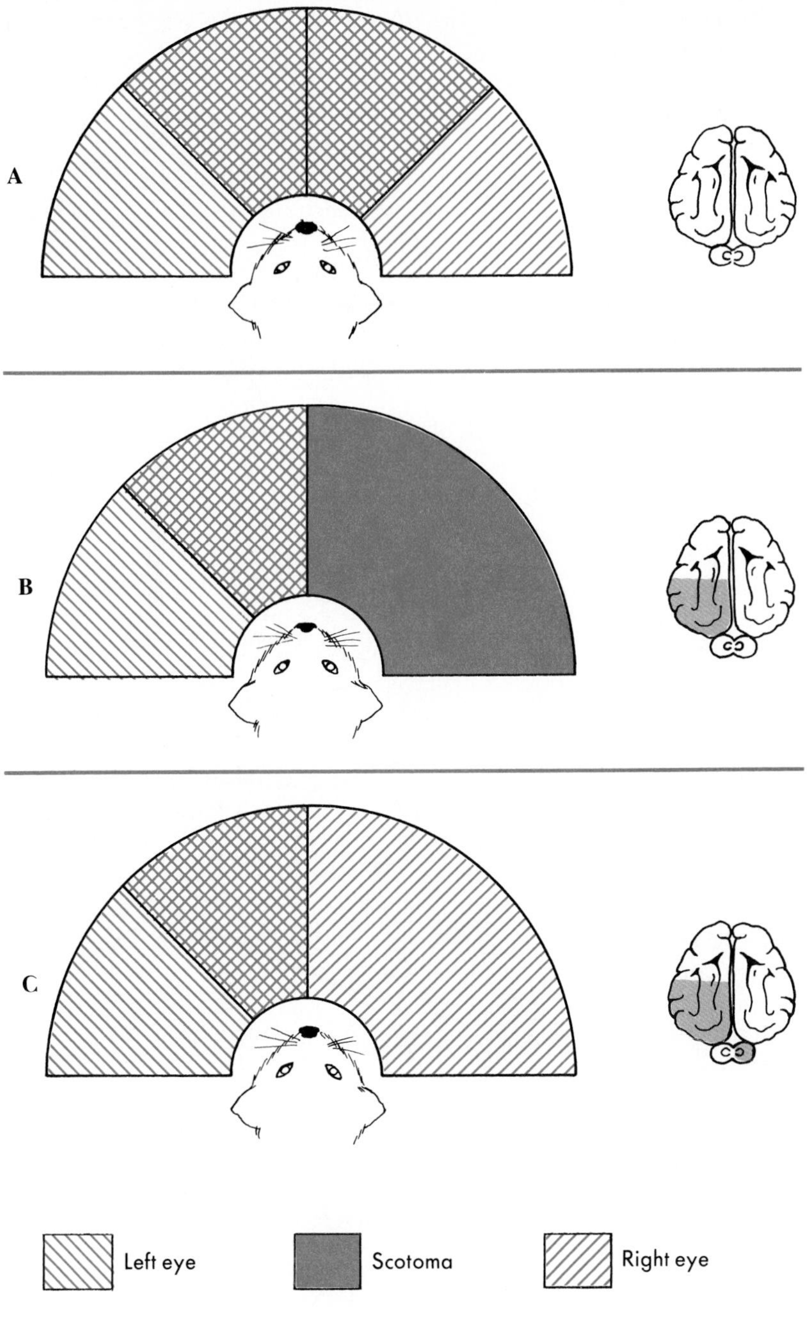

Fig. 8-36. The Sprague effect from studies of cats. **A,** Visual fields in a normal cat. Each eye sees from about 45 degrees across the midline to 90 degrees ipsilaterally. Thus each temporal retina sees from the midline to 45 degrees into the opposite hemifield, and each nasal retina sees the entire ipsilateral hemifield. **B,** Visual fields after left visual cortex removal. The cat now exhibits a right hemianopia, and vision is restricted to the left nasal and right temporal retinas. **C,** Visual fields after subsequent removal of the right superior colliculus. Vision is restored to the right hemifield, although visual capacity here remains subnormal. Note, however, that the left temporal retina remains blind. (Redrawn from Sherman, S.M.: J. Comp. Neurol. **172:** 231, 1977.)

W-, X-, and Y-cell Pathways

Because W-, X-, and Y-cells have been appreciated relatively recently, our understanding of the significance of their differences is far from complete. Studies of these cells and their pathways represent an active research area, replete with speculation and controversy. Nonetheless, because an understanding of these pathways will undoubtedly prove crucial to appreciation of visual processing, it is worth considering them in greater detail with emphasis on the tentative and speculative nature of many of the conclusions.

We know a great deal more about these pathways in the cat than in any other species, and even in the cat we know far more about X- and Y-cells than about W-cells. However, because W-cells respond poorly to most visual stimuli and because they seem to play a relatively minor role in geniculocortical innervation (particularly for their homologues in monkeys), most relevant hypotheses have focused on X- and Y-cells.

An appealing hypothesis (among many) is derived from a consideration of both X- and Y-cell response properties, the different geniculocortical innervation patterns of the cells, and the behavioral consequences of some of the ablation-behavior data noted earlier. Y-cells are most responsive to large targets or cruder forms. X-cells are relatively unresponsive to such stimuli but respond better to smaller targets having finer spatial details. Evidence from behavioral studies in which the cruder and finer forms in a visual scene were separated suggests that basic form analysis is performed on the cruder forms and that the finer ones add details to this basic analysis.* The Y-cell pathway may thus be involved with basic form analysis, whereas the X-cell pathway may be involved in high acuity vision that analyzes the minute details in a visual scene. Also, the X-cell pathway in many primate species may be used to analyze colors.

Behavioral studies of cortically lesioned animals are consistent with this notion. In cats, the geniculocortical projection of X-cells is limited nearly exclusively to the striate cortex, whereas the Y-cells directly innervate striate and extrastriate cortical areas, often via branching axons. Thus a bilateral ablation of the striate cortex effectively eliminates the X-cell pathway but leaves much of the Y-cell pathway intact. As noted earlier, such a lesion has remarkably little effect on basic form vision in a cat, except for reduced acuity. Such a lesion in a monkey, in contrast, is much more devastating to form vision. Perhaps this results not from the essential nature of the striate cortex per se but from evidence that in monkeys all geniculocortical homologues of Y-cells, as well as of X-cells, terminate only in the striate cortex. It may be the course of the Y-cell pathway that determines the extent of visual impairment consequent to striate cortex damage.

Anatomical evidence from the cat further supports this notion of X- and Y-cell roles (Fig. 8-37). Compared with X-cells and the X-cell pathway, relatively few Y-cells exist in the retina, but the Y-cell pathway diverges extensively to the dorsal LGN and visual cortex. As a result, a small minority of retinal ganglion cells seems to dominate the geniculocortical innervation patterns. Few Y-cells may be needed in retina because coding of the important crude forms requires no more

*In much the same way that complex sounds can be broken down into their pure tones or frequency components (Chapter 10), so can complex visual scenes be analyzed in terms of similar components. These visual components are *sine wave gratings,* which are sinusoidal variations of brightness with position. These are analogous to tones, which are sinusoidal variations of sound level with time, except that the longitudinal dimension is spatial rather than temporal. Thus a visual scene can be described in terms of its component spatial frequencies (i.e., cycles of a sine wave grating per spatial unit). Lower spatial frequencies correspond to the cruder shapes in a visual scene, and higher spatial frequencies correspond to the finer details. Behavioral evidence indicates that basic form information is carried by the lower spatial frequencies and that the higher ones add details and raise acuity. If the higher frequencies are removed from a typical scene (photograph of a landscape, a face, etc.), details are lost, but the basic subject can be recognized.

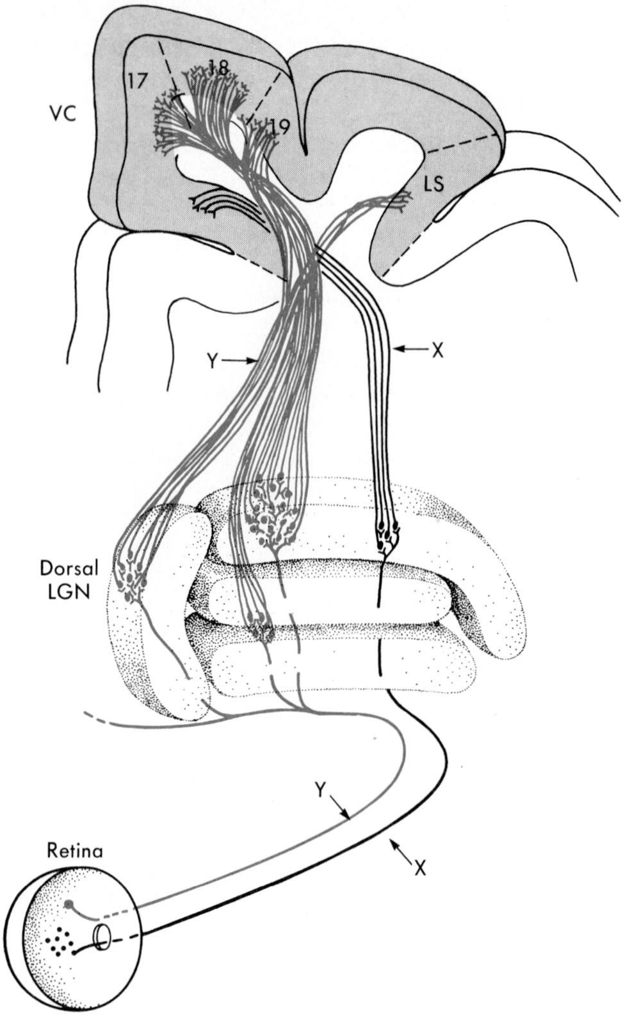

Fig. 8-37. Hypothetical illustration of the cat's X- and Y-cell pathways, drawn for simplicity only from the contralateral eye. The X-cell pathway is shown in black and the Y-cell pathway in color. X-cells outnumber Y-cells in the retina by roughly 10:1, but Y-cell axons of both the optic tract and optic radiation exhibit much more divergence and numbers of synapses than do X-cell axons. Consequently, many more postsynaptic cells are innervated by each Y-cell axon than by each X-cell axon. Single Y-cell retinogeniculate axons innervate all subregions of the dorsal LGN. The geniculocortical Y-cell axons innervate nearly all known areas of visual cortex (see Fig. 8-34), and although single axons do not innervate all areas, they commonly branch to innervate two or more. Retinogeniculate X-cell axons, by contrast, innervate only one laminar region of the dorsal LGN, and their geniculocortical axons innervate only the striate cortex. The result is that the 10:1 X-to-Y-cell retinal ratio is roughly 1:1 in the dorsal LGN and probably becomes dominated by Y-cells in the cortex. (Redrawn from Sherman, S.M.: Functional organization of the W-, X-, and Y-cell pathways in the cat: a review and hypothesis, Prog. Psychobiol. Physiol. Psychol. **11:**233, 1985.)

than a coarse retinal "grain." The importance of this information to basic form vision, however, requires that much of the cortex need be devoted to its analysis. In contrast, many X-cells may exist in retina because a fine grain is needed to code the smallest details in a visual stimulus. Relatively little cortex need be devoted to this analysis performed via the X-cells. To support such a relative divergence in the Y-cell pathway, individual Y-cell axons branch more extensively and issue larger terminal arbors than do those of X-cells, both at the retinogeniculate and geniculocortical levels. Whether or not a similar pattern of connectivity exists for the homologue of X- and Y-cells in primates remains to be determined.

The hypothesis just stated, as noted, applies to the X- and Y-cell pathways in the cat. If true, it should at least provide a useful starting point for understanding the functional significance of X- and Y-cells in monkeys and humans. However, perhaps evolution has led to functional specializations for these pathways that differ markedly between species. This concept is similar to the belief that our fingers, while homologous to a bird's wings, have evolved to peform quite a different function. The great sensitivity of Y-cells to spatial patterns in monkeys is consistent with their playing a similar role to the basic form analysis ascribed to the Y-cell pathway in the cat. In contrast, the unique wavelength sensitivity of monkey X-cells has no obvious counterpart in the cat.

■ *Bibliography*

Journal articles

Ahlsen, G., Lindstrom, S., and Lo, F.-S. Inhibition from the brain stem of inhibitory interneurones of the cat's dorsal lateral geniculate nucleus, J. Physiol. (Lond.) **347:**593, 1984.

Bowling, D.B., and Michael, C.R. Terminal patterns of single physiologically characterized optic tract fibers in the cat's lateral geniculate nucleus, J. Neurosci. **4:**198, 1984.

Barbur, J.L., Ruddock, K.H., and Waterfield, V.A.: Human visual responses in the absence of a geniculo-calcarine projection, Brain **103:**905, 1980.

Benevento, L.A., and Yoshida, K.: The afferent and efferent organization of the lateral geniculo-prestriate pathways in the macaque monkey, J. Comp. Neurol. **203:**455, 1981.

Daw, N.W.: The psychology and physiology of colour vision, Trends Neurosci. **7:**330, 1984.

Derrington, A.M., and Lennie, P.: Spatial and temporal contrast sensitivities of neurones in lateral geniculate nucleus of macaque, J. Physiol. (Lond.) **357:**219, 1984.

Ferster, D.: A comparison of depth mechanisms in areas 17 and 18 of the cat visual cortex, J. Physiol. (Lond.) **311:**623, 1981.

Fitzpatrick, D., Itoh, K., and Diamond, I.T.: The laminar organization of the lateral geniculate body and the striate cortex in the Squirrel monkey *(Saimiri sciureus),* J. Neurosci. **3:**673, 1983.

Fitzpatrick, D., Penny, G.R., and Schmechel, D.E.: Glutamic acid decarboxylase-immunocytoreactive neurons and terminals in the lateral geniculate nucleus of the cat, J. Neurosci. **4:**1809, 1984.

Friedlander, M.J., et al.: Morphology of functionally identified X- and Y-cells in the cat's lateral geniculate nucleus, J. Neurophysiol. **46:**80, 1981.

Hubel, D.H., and Wiesel, T.N.: Functional architecture of macaque monkey visual cortex (Ferrier Lecture), Proc. R. Soc. Lond. Biol. **198:**1, 1977.

Kaplan, E., and Shapley, R.M.: X and Y cells in the lateral geniculate nucleus of macaque monkeys, J. Physiol. (Lond.) **330:**125, 1982.

Land, E.H.: The retinex theory of color vision, Sci. Am. **237:**108, 1977.

Lehmkuhle, S., Kratz, K.E., and Sherman, S.M.: Spatial and temporal sensitivity of normal and amblyopic cats, J. Neurophysiol. **48:**372, 1982.

Perry, V.H., and Cowey, A.: Retinal ganglion cells that project to the superior colliculus and pretectum in the macaque monkey, Neuroscience **12:**1125, 1984.

Perry, V.H., Oehler, R., and Cowey, A.: Retinal ganglion cells that project to the dorsal lateral geniculate nucleus in the macaque monkey, Neuroscience **12:**1101, 1984.

Poggio, G.F., and Talbot, W.H.: Mechanisms of static and dynamic stereopsis in foveal cortex of the rhesus monkey, J. Physiol. (Lond.) **315:**469, 1981.

Rodieck, R.W., Binmoeller, K.F., and Dineen, J.: Parasol and midget ganglion cells of the human retina, J. Comp. Neurol. **233:**115, 1984.

Rodieck, R.W., and Brening, R.K.: Retinal ganglion cells: properties, types, genera, pathways and trans-species comparisons, Brain Behav. Evol. **23:**121, 1983.

Schiller, P.H., et al.: Response characteristics of single cells in the monkey superior colliculus following ablation or cooling of visual cortex, J. Neurophysiol. **37:**181, 1974.

Sherman, S.M., and Koch, C.: The control of retinogeniculate transmission in the mammalian lateral geniculate nucleus, Exp. Brain Res. **63:**1, 1986.

Sherman, S.M., and Spear, P.D.: Organization of visual pathways in normal and visually deprived cats, Physiol. Rev. **62:**738, 1982.

Slaughter, M.M., and Miller, R.F.: Characterization of an extended glutamate receptor of the ON bipolar neuron in the vertebrate retina, J. Neurosci. **5:**224, 1985.

Sur, M., and Sherman, S.M.: Retinogeniculate terminations in cats: morphological differences between X- and Y-cell axons, Science **218:**389, 1982.

Van Essen, D.C., and Maunsell, J.H.R.: Hierarchical organization and functional streams in the visual cortex, Trends Neurosci. **6:**370, 1983.

Yukie, M., and Iwai, E.: Direct projection from the dorsal lateral geniculate nucleus to the prestriate cortex in macaque monkeys, J. Comp. Neurol. **187:**679, 1979.

Books and monographs

Boynton, R.M.: Human color vision, New York, 1979, Holt, Rinehart, & Winston.

Diamond, I.T.: Changing views of the organization and evolution of the visual pathways. In Morrison, A.R., and Strick, P.L., editors: Changing concepts of the nervous system, New York, 1982, Academic Press, Inc.

Dowling, J.E.: Information processing by local circuits: the vertebrate retina as a model system. In Schmitt, F.O., and Worden, F.G., editor: The neurosciences: fourth study program, Cambridge, Mass., 1979, The MIT Press.

Graybiel, A.M., and Berson, D.M.: On the relationship between transthalamic and transcortical pathways in the visual system. In Schmitt, F.O., et al., editors: The organization of the cerebral cortex, Cambridge, Mass., 1981, The MIT Press.

Lund, J.S.: Intrinsic organization of the primate visual cortex, area 17, as seen in Golgi preparations. In Schmitt, F.O., et al., editors: The organization of the cerebral cortex, Cambridge, Mass., 1981, The MIT Press.

Rosenquist, A.C., Raczkowski, D., and Symonds, L.: The functional organization of the lateral posterior-pulvinar complex in the cat. In Morrison, A.R., and Strick, P.L., editors: Changing concepts of the nervous system, New York, 1982, Academic Press, Inc.

Sherman, S.M.: Functional organization of the W-, X-, and Y-cell pathways in the cat: a review and hypothesis. In: Sprague, J.M., and Epstein, A.N., editor: Progress in psychobiology and physiological psychology, vol 11, New York, 1985, Academic Press, Inc.

Sherman, S.M.: Parallel pathways in the cat's geniculocortical system: W-, X-, and Y-cells. In Morrison, A.R., and Strick, P.L., editors: Changing concepts of the nervous system, New York, 1982, Academic Press Inc.

Stone, J.: Parallel Processing in the visual system, New York, 1983, Plenum Publishing Corp.

Tusa, R.J.: Visual cortex: multiple areas and multiple functions. In Morrison, A.R., and Strick, P.L., editors: Changing concepts of the nervous system, New York, 1982, Academic Press, Inc.

Woolsey, C.N., editor: Cortical sensory organization, vol. 2, Multiple visual areas, Clifton, N.J., 1981, Humana Press.

CHAPTER

9

The Somatosensory System

Humans perceive at least four distinct body sensations. These are touch (also known as touch-pressure), kinesthesia (or joint sensation), temperature, and nociception (or pain). Most schemata also include tactile perception of vibration as a submodality distinct from touch. The following sections discuss the organization of somatosensory pathways represented via the dorsal root ganglia, and a subsequent section deals with the analogous representation of the cranial component, mostly via the trigeminal nerve.

Transducers

In general, the transducers, or sensory receptors, for somesthesia are poorly understood. Two examples, the pacinian corpuscle and a tonically responsive transducer, are covered in Chapter 6, and they are involved in touch or pressure sensation. In most cases of a tonically responsive transducer we cannot identify particular types of transducers with particular somesthetic properties. A variety of specialized corpuscular and disk-shaped structures have been found near endings of primary afferent fibers in the skin and deeper tissue (joints, fascia, etc.). These presumably act as transducers (e.g., the tonically responsive transducer), or they assist in the transduction process (e.g., the "onion" of the pacinian corpuscle). Many free nerve endings without specialized structures have also been found, and most or all of the transducers for nociception or pain are thought to be free nerve endings. Although the details have yet to be elucidated, there is no reason to doubt that a variety of specialized and uniquely sensitive transducers exist for touch, kinesthesia, temperature, and nociception.

Afferent Fibers

Table 9-1 summarizes the classification of afferent fiber types that enter the spinal cord. From largest to smallest, they are designated group I through group IV, and an alternate designation frequently used is given in brackets. All of these fibers have their somata in the dorsal root ganglia. The largest fibers, group I (also known as A-α), originate in muscle spindles and tendon organs. Most seem to be involved strictly in spinal reflexes, such as the knee-jerk reflex, but recent evidence suggests that some group I muscle afferent fibers may be involved in kinesthesia. Group II (also known as A-β, γ) includes large myelinated fibers that convey information about touch and kinesthesia. Group III (also known as A-δ) includes small myelinated fibers that convey information about touch, temperature, and no-

Table 9-1. Somatosensory afferent fibers

Modality	*Group I** *(A-α; new)*	*Group II** *(A-B, γ; new)*	*Group III†* *(A-δ; old)*	*Group IV†* *(C; oldest)*
Touch		X	X	X
Kinesthesia	?	X		
Temperature			X	?
Nociception			X	X‡

*Dorsal column system
†Anterolateral system
‡Mostly anterolateral system, but some contribution to dorsal column system via dorsal column postsynaptic cells

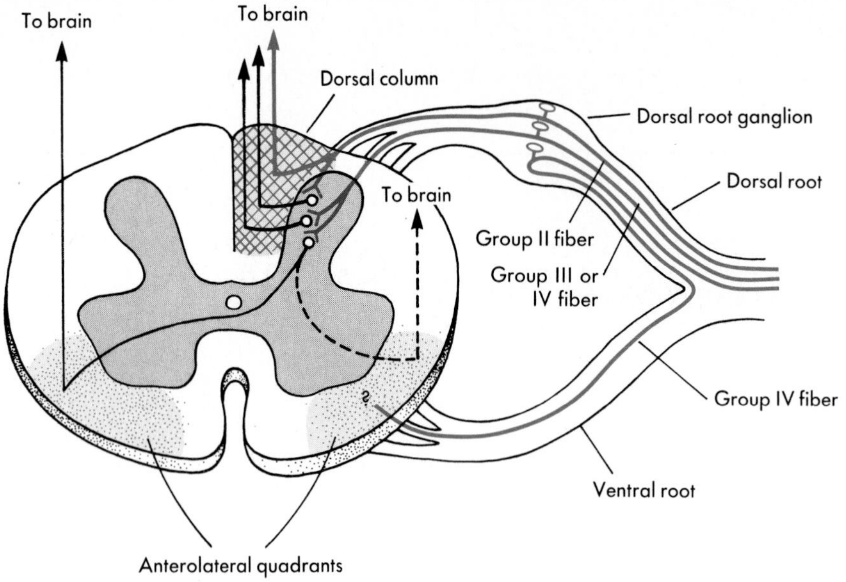

■ **Fig. 9-1.** Spinal cord organization of dorsal column and anterolateral systems. Somata of afferent fibers are located in the dorsal root ganglion. Group II afferent fibers enter the cord via the dorsal root. Their branches ascend in the ipsilateral dorsal column to terminate in the dorsal column nuclei of the medulla. Other branches synapse onto neurons of the spinal gray matter, mostly in the dorsal horn (see Fig. 9-2, *A,* for the distribution of the cell bodies of these second-order cells). These neurons are known as *dorsal column postsynaptic cells,* and their axons also ascend in the ipsilateral dorsal columns. Not shown are the few group I muscle afferents that probably have central connections much like the group II afferents, and thus they contribute to the axons ascending in the dorsal columns. Group III and most of group IV afferent fibers also enter the cord via the dorsal root and terminate on second-order neurons with cell bodies in the spinal gray matter. A few of the group IV nociceptive input seem to reach dorsal column post-synaptic cells, but the majority of group IV and all group III fibers innervate second-order neurons with cell bodies distributed as shown in Fig. 9-2, *B*. Most of these cells project axons across the midline into the contralateral anterolateral quadrant of the spinal cord, although some appear to enter the ipsilateral quadrant. These secondary fibers ascend in the anterolateral quadrants to terminate at many levels of the neuraxis, including brainstem reticular formation and thalamus (see Fig. 9-9). Some group IV afferent fibers enter the cord via the ventral root, and their central connections are presently unknown.

ciception. Group IV (also known as C) represents small unmyelinated fibers that convey information about touch, nociception, and perhaps temperature. The pathways originating from muscle spindles and tendon organs (e.g., all group I and some group II fibers) are discussed in Chapter 14 and will not be considered further in this chapter.

Recent studies have shown that many group IV fibers enter the spinal cord via the *ventral* root. Therefore the classical notion that the dorsal and ventral roots are strictly afferent and efferent, respectively, must be revised. The central pathways of these ventral root afferent fibers and the location of their somata are presently unclear, although it is thought that their somata lie in the dorsal root ganglia. The possible functional significance of this is considered later in the discussion of nociception or pain.

Within the spinal cord, two main ascending fiber tracts are involved in the flow of somatosensory signals to the brain. These tracts are the dorsal columns, which participate in the *dorsal column system,* and the anterolateral columns, which participate in the *anterolateral system*. Although most of the fibers in the dorsal columns are primary, with cell bodies in a dorsal root ganglion, perhaps as many as one third are secondary. The latter derive from neurons known as *dorsal column postsynaptic cells;* their cell bodies are concentrated in laminae III to V of the dorsal horn, although many can be found in the ventral horn. (Fig. 9-1, *A*, illustrates the distribution of these cells.) They are innervated by the primary afferents, and their ascending axons project into the ipsilateral dorsal columns (Fig. 9-2). Fibers in the anterolateral system are derived from neurons with cell bodies located mostly in the spinal gray matter (Fig. 9-1, *B*); these cells typically receive innervation from primary afferents via the dorsal root and are thus second order.

Table 9-1 also indicates the relationships among modalities, fiber type, and participation in the dorsal column or anterolateral system. Some Group I fibers from

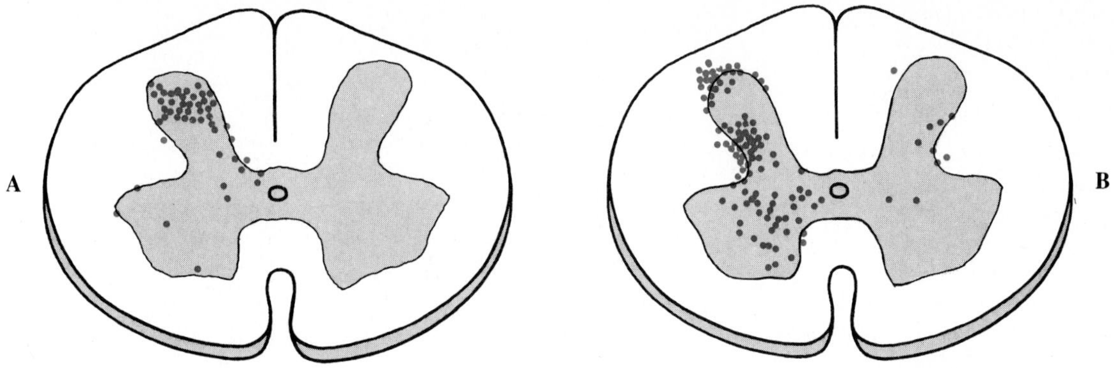

■ Fig. 9-2. Location of cell bodies *(colored dots)* that give rise to second-order axons ascending in the spinal cord. **A,** Location of dorsal column postsynaptic cells, the axons of which ascend ipsilaterally in the dorsal columns of the spinal cord. **B,** Location of cells whose axons ascend contralaterally in the anterolateral quadrant of the spinal cord.

muscle spindles may ascend in the dorsal columns. Group II fibers are related to the dorsal column system. Fibers of Groups III and IV relate mainly to the anterolateral system. However, some axons of the dorsal column postsynaptic cells that ascend in the dorsal columns seem to convey nociceptive information from Group IV afferents. Consequently, touch and nociception are the only modalities represented by both systems. Kinesthesia is processed almost completely via the dorsal column system, and temperature is processed only via the anterolateral system.

Like the visual system, the somatosensory system can be roughly divided into phylogenetically newer and older components. The dorsal column system is newer than is the anterolateral system. Furthermore, the anterolateral system itself contains older and newer portions. There appears to be a good correlation between fiber size and phylogenetic age such that smaller fibers are older. Touch, then, is represented by older and newer components; kinesthesia is represented only by the newer component, and temperature and nociception are represented only by the older one. However, it is important to recognize the incomplete and sometimes implausible nature of these divisions. For instance, kinesthesia was probably present fairly early in mammalian evolution, yet it is currently seen almost exclusively in the newest dorsal column system.

■ Dorsal Column System

■ General Anatomical Considerations

Ascending pathways. Fig. 9-3 illustrates the ascending pathways involved in the dorsal column system. A large branch of the primary afferent fiber ascends through the spinal cord in the dorsal columns; other branches are involved in local spinal circuits. For simplicity, Fig. 9-3 does not include fibers of the dorsal column postsynaptic cells, because projection patterns do not appear to differ from those illustrated for the dorsal columns.

Fibers entering the dorsal columns at higher spinal segments occupy a progressively lateral position so that there is a precise topographical order to these fibers. At higher levels a division of the dorsal columns into two funiculi can be discerned (Fig. 9-3). The more medial *funiculus gracilis* includes fibers from T7 and below; the more lateral *funiculus cuneatus* includes fibers from T6 and above.

These dorsal column fibers, both primary and secondary, terminate at the level of the lower medulla in either the *nucleus gracilis* or *nucleus cuneatus,* depending on the funiculus in which they ascend (Fig. 9-3). Together these nuclei are called the *dorsal column nuclei.* The secondary and tertiary fibers issue forth from these nuclei as the *internal arcuate fibers,* which soon cross the midline at the *decussation of the medial lemniscus.* These fibers continue to ascend in the medial lemniscus to their terminus in the *ventralis posterolateralis* (VPL) nucleus of the thalamus. Tertiary and quaternary fibers then issue from the VPL nucleus to terminate in the somatosensory cortex, mostly in layer IV but also in layer VI. This thalamocortical termination pattern closely resembles that in the visual system.

Historically, two somatosensory areas were defined: SI, or "primary" somatosensory cortex, and SII, or "secondary" somatosensory cortex (Fig. 9-4). SI is located in the postcentral gyrus, and it was thought to be the exclusive terminus of VPL nucleus afferents; SII is located in the parietal cortex just above the sylvian fissure, and it was thought to receive secondary VPL nucleus input from SI. This terminology is outdated by contemporary research for two reasons. First, we know now that SII receives direct VPL nucleus input. Second, recent studies of owl monkeys have established a

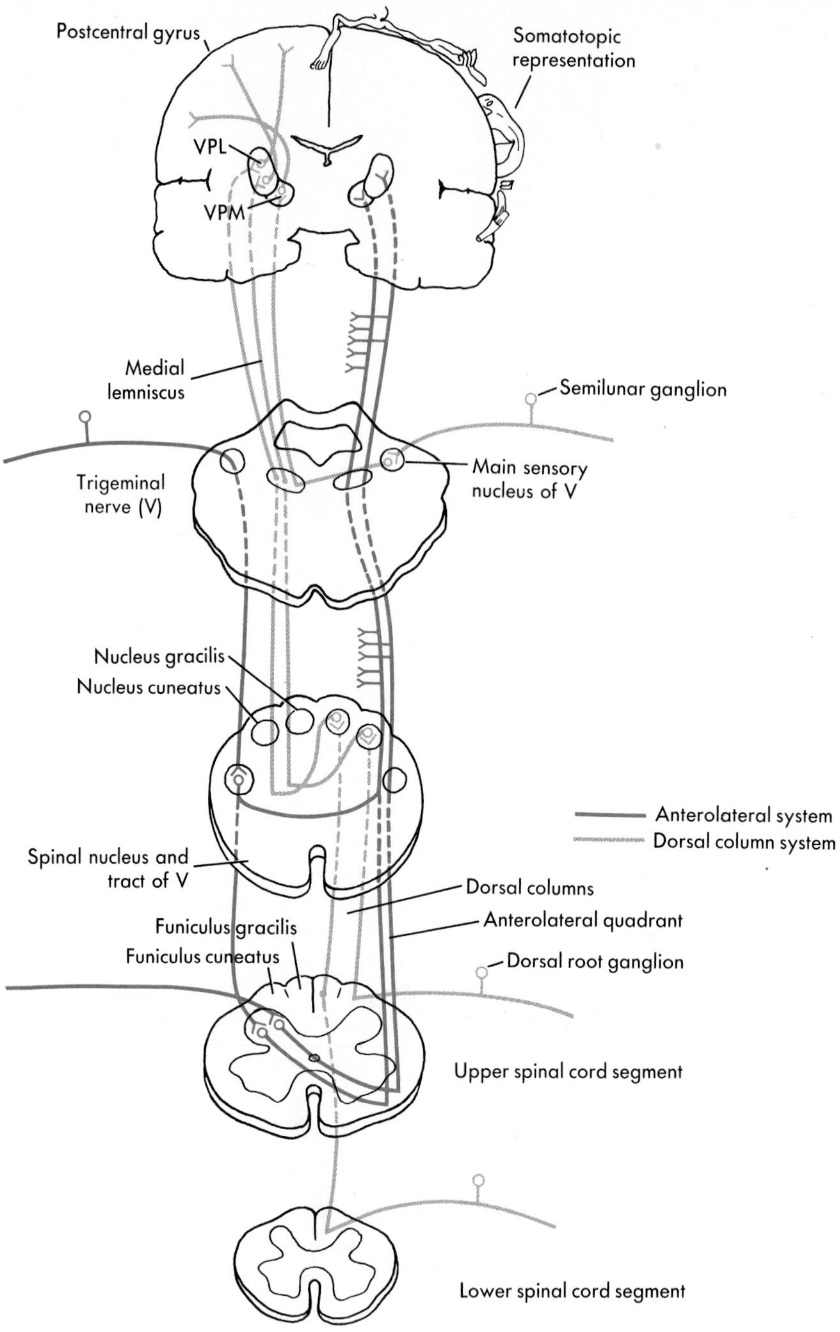

Fig. 9-3. Schematic representation of some central connections in the dorsal column and anterolateral systems. For simplicity, only the primary fibers are shown in the dorsal columns. No obvious difference has yet emerged between the central connections of the primary and secondary fibers that ascend. Both trigeminal and spinal components are shown. The dorsal column system is shown in lighter color, and the anterolateral system is shown in darker color. Note the somatotopic arrangement throughout the neuraxis for the dorsal column system. Fibers that enter dorsal roots at higher spinal segments ascend in the dorsal columns more laterally, and this relationship is preserved throughout; furthermore, the trigeminal component merges with the spinal component in the medial lemniscus next to fibers that represent the highest cervical spinal segments. Consequently an accurate map of the body surface exists throughout, including the cortical areas of the postcentral gyrus. *VPL,* Ventralis posterolateralis nucleus; *VPM,* ventralis posteromedialis nucleus.

multiplicity of somatosensory cortical areas, again as in the visual system. SI is actually composed of at least four different zones and SII of at least two. Other separate cortical representations might also be described.

Anatomical and electrophysiological studies in cats and monkeys suggest a refinement in the functional organization of the somatosensory thalamocortical pathways (Fig. 9-5). A central core region of the VPL nucleus represents cutaneous tactile information for fine discriminations: the surrounding shell represents kinesthesia plus less precise information about temperature and nociception in the skin and deeper tissues. A correlate to this core and shell arrangement is also evident in the thalamocortical projections (see Fig. 9-5). The core region of the VPL nucleus innervates chiefly cortical areas 1 and 3B, whereas the shell region innervates chiefly areas 2 and 3a. Finally, different cell types within the VPL nucleus project to different zones within layer IV of SI cortex, a relationship analogous to the parallel X- and Y-cell projections of the geniculostriate pathway.

Descending pathways. As in the visual system, the newer somatosensory pathways include a large descending component; that is, corticofugal fibers, originating mostly from the somatosensory cortex, project directly to the thalamic and medullary relay nuclei of the dorsal

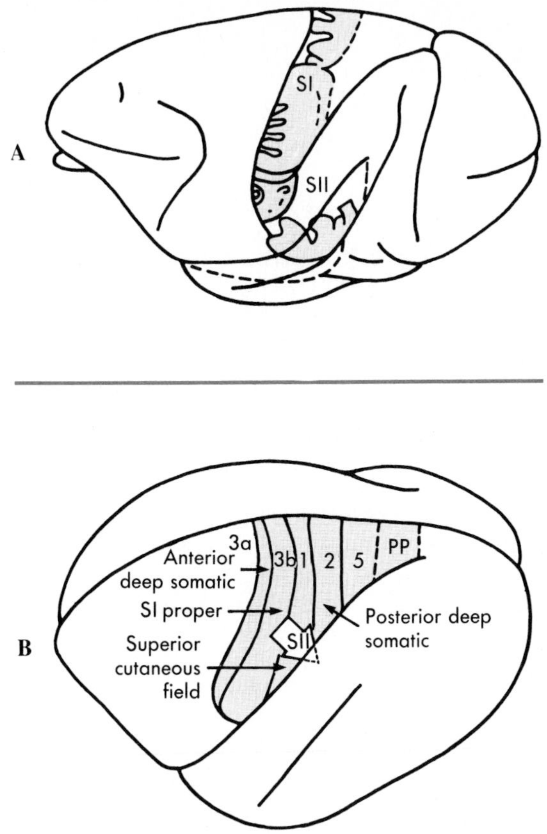

■ **Fig. 9-4.** Organization of somatosensory cortical areas in primates. This representation is derived from studies of the owl monkey. **A,** Older, traditional view with a primary *(SI)* and secondary *(SII)* area. **B,** Contemporary view with at least four and probably more separate areas. The four definite areas are the anterior deep somatic area (area *3a*), SI (area *3b*), the posterior cutaneous area (area *1*), and the posterior deep somatic area (area 2). Areas *SII, 5,* and the posterior parietal *(PP)* cortex probably each represent more than one somatosensory area. (Redrawn from Merzenich, M.M., and Kaas, J.H.: Prog. Psychobiol. Physiol. Psychol. **9:**1, 1980.)

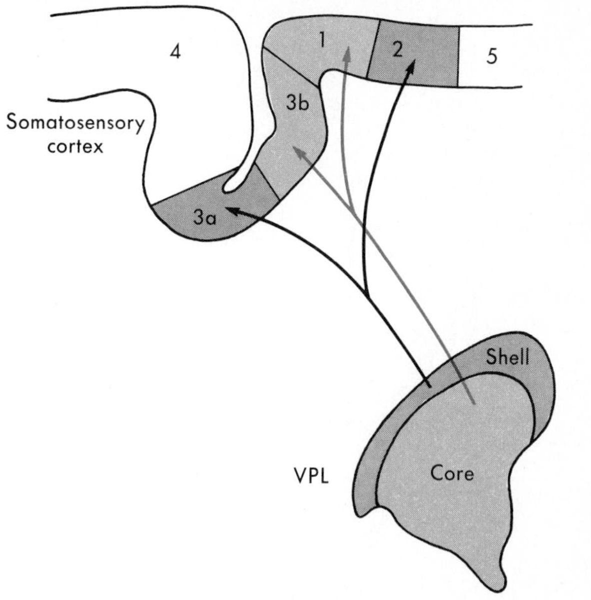

■ **Fig. 9-5.** Schematic relationships in thalamocortical projection from VPL to areas of somatosensory cortex. A core region *(color)* of VPL projects primarily to cortical areas 1 and 3b, and a shell region projects primarily to areas 2 and 3a. (Redrawn from Jones, E.G., and Friedman, D.P.: J. Neurophysiol. **48:**521, 1982.

column system. There is probably also an indirect projection to these nuclei via corticobulbar fibers from the somatosensory cortex to the brainstem reticular formation. The functional significance of this descending component has yet to be elucidated but is likely to gate transmission in the dorsal column and VPL nuclei. This is analogous to the function of similar connections described for the visual system in Chapter 8. Other similarities are noted below.

Other pathways. Other connections of the VPL nucleus are remarkably similar to those of the dorsal LGN (see also Chapter 8). A region of the RNT (the reticular nucleus of the thalamus) is devoted to somesthesia and is reciprocally connected with the VPL nucleus. Input to the RNT derives from VPL axons as they pass through en route to the cortex. RNT cells are GABAergic, as are the few intrinsic neurons in the VPL nucleus. Finally, neurons of the brainstem reticular formation innervate the VPL nucleus and the RNT. These pathways may gate the VPL inputs to cortex in much the same way that their counterparts in the visual system control retinogeniculate transmission to the cortex.

■ *Touch Sensation*

Many neurons throughout the dorsal column system are sensitive to light touch applied to the skin or to pressure at certain deep tissues, such as fascia, joint capsules, periosteum, and ligaments. We know most about those related to the skin because it is easier to study receptive fields on the body surface than to study those related to deep tissues.

Spatial resolution for touch sensitivity can be defined as the minimal separation needed for two stimuli to be distinguished from a single, more intense stimulus. This refers, for example, to how closely applied to the skin two pinpricks can be to one another and still be distinguished from a single prick. Two-point discrimination varies with position on the body's surface from roughly 2 mm for the fingertips to roughly 40 mm for the back. As expected from knowledge of the visual system, receptive field size and afferent inhibition both play roles in such discrimination. Cutaneous receptive fields are smaller where discrimination is more acute (e.g., the

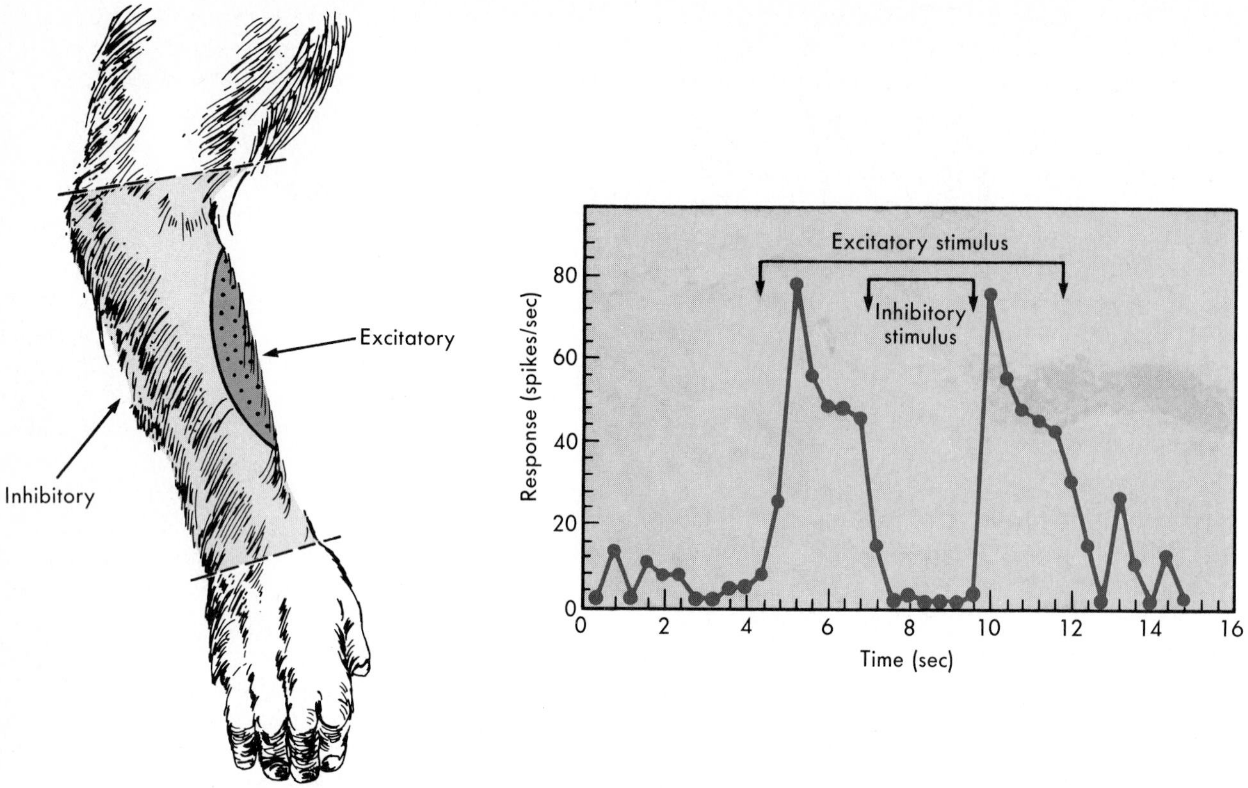

■ **Fig. 9-6.** Afferent inhibition in a somatosensory neuron of the postcentral gyrus in a monkey. On the left is the receptive field with an excitatory center *(darker color)* surrounded by an inhibitory region *(lighter color)*. Responses are shown on the right to cutaneous stimuli applied to the receptive field. When a stimulus is applied to the center, the cell discharges. Stimulation of the surround silences the cell even in the presence of an excitatory stimulus. (Redrawn from Mountcastle, V.B., and Powell, T.P.S.: Johns Hopkins Med. J. **105:**201, 1959. Copyright 1959 by The Johns Hopkins University Press.)

■ **Fig. 9-7.** Direction selectivity in a somatosensory neuron of the postcentral gyrus in a monkey. The neuron is most responsive to cutaneous stimuli moving along the axis *UW* to *RF* and least responsive along the axis *RW* to *UF*. (From Costanzo, R.M., and Gardner, EP.: J. Neurophysiol. **43:**1319, 1980.)

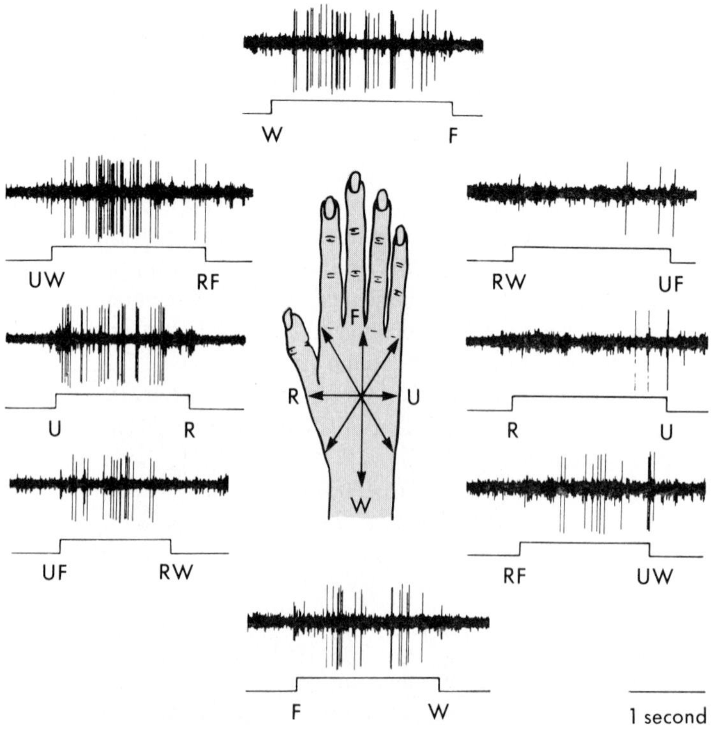

fingertips) and larger where discrimination is poorer (e.g., the surface of the trunk); this is analogous to the smaller visual receptive fields that occur as acuity increases nearer the foveal representation. Furthermore, whereas primary cutaneous afferent fibers do not display lateral inhibition, neurons of the dorsal column nuclei, VPL nucleus, and somatosensory cortex do exhibit such inhibition (Fig. 9-6); that is, stroking the cutaneous receptive field excites such a neuron for some central cutaneous zone and inhibits it for a surrounding annular zone. This arrangement seems analogous to the center-surround receptive fields of retinal ganglion and lateral geniculate cells.

Another analogy between the geniculocortical and dorsal column systems concerns the specificity of adequate stimuli to excite the neuron. For primary afferent fibers, cells of the dorsal column nuclei, and VPL neurons, only the position of the stimulus is crucial to the cell's response, and this is similar to the limited specificity of retinal and geniculate cells. However, many cells in the somatosensory cortex exhibit a dramatic direction selectivity (Fig. 9-7), that is, they respond when a cutaneous stimulus, such as a brush, is moved in one direction across the skin but not for the reverse or orthogonal directions. This is analogous to the direction and orientation selectivity seen first in the geniculocortical system among cortical neurons.

Many touch-sensitive neurons in the dorsal column system respond tonically to touch or pressure, much like the responses of many neurons in the newer, geniculocortical portion of the visual system. These tonic responses are characterized by a monotonic relationship between firing rate and touch intensity. This applies to neurons throughout the system, including primary afferent neurons (Fig. 6-4) and cortical neurons.

Kinesthesia

Neurons sensitive to joint position or angle have also been studied throughout the dorsal column system. They generally exhibit tonic responses that encode joint angle in the frequency of firing (Fig. 9-8, *A*). As Fig. 9-8, *B*, shows, some of these cells increase firing with increased flexion, and others increase firing with increased extension.

The plots in Fig. 9-8, *B*, should not be mistaken for receptive fields. They merely represent the relationship between stimulus intensity and response, and they are strictly analogous to the example in Fig. 6-4. A change in joint angle changes the pressure on related tissue, and thus it changes the intensity of the stimulus at the transducer. The receptors for kinesthesia were first thought to be located in the joint capsules or ligaments, but these receptors seem only to signal the extremes of joint position. Fibers such as those illustrated in Fig. 9-8 are probably muscle afferents that signal the change in muscle length consequent to joint rotation. These muscle afferents are now thought to be the peripheral source of kinesthesia.

Joint angle thus has little to do with the concept of a receptive field for such a cell. Rather, as defined in

A

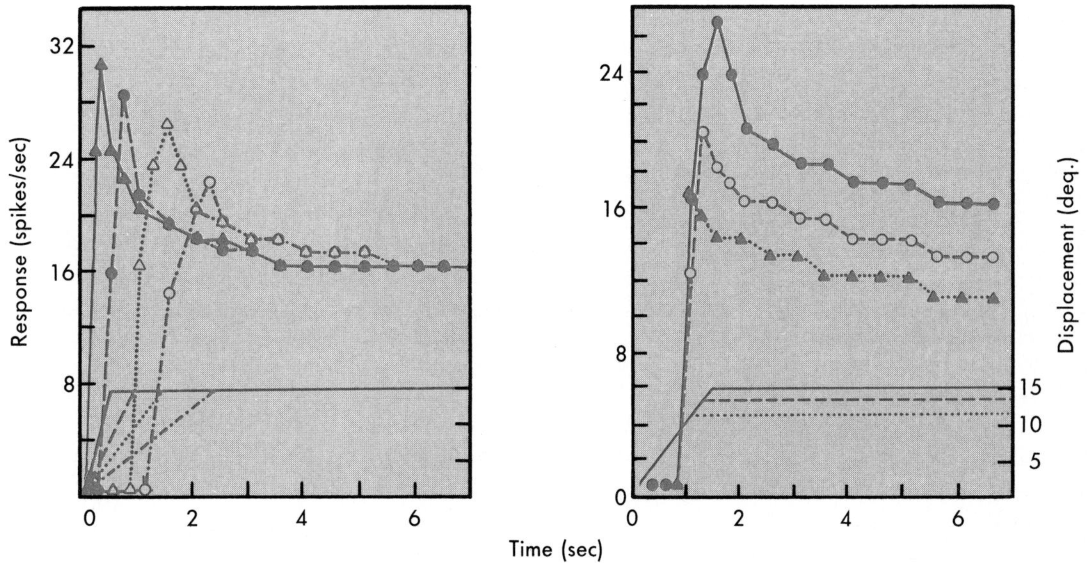

Fig. 9-8. Response properties of kinesthesia-sensitive neurons. **A,** Responses of primary afferent fibers to flexion of the knee of a cat. As shown, the magnitude of the sustained response encodes the extent of flexion, whereas the rate of flexion determines the rate of rise of the response. *Continued.*

B

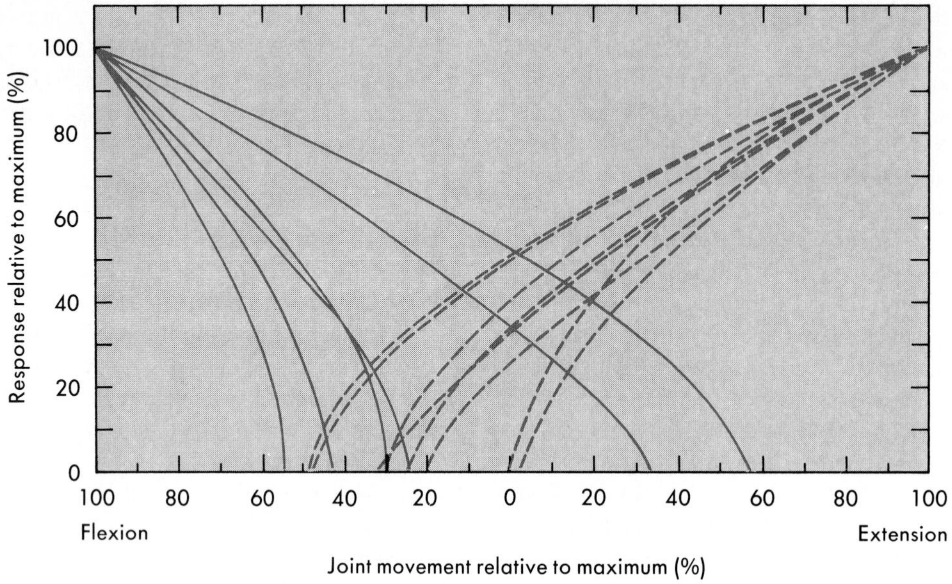

■ Fig. 9-8, cont'd. **B,** Response ranges of 14 separate neurons in the monkey's thalamic VPL nucleus. Each neuron responds monotonically either to increasing flexion or extension of the appropriate joint, but never to both. Note that most neurons respond to stimuli that correspond to half or more of the total range of joint movements. These curves illustrate the frequency code by which neuronal discharges represent joint position. (**A** redrawn from Boyd, I.A., and Roberts, T.D.M.: J. Physiol. [Lond.] **122:**38, 1953; **B** redrawn from Mountcastle, V.B., et al.: J. Neurophysiol. **26:**807, 1963.)

Chapter 7, the receptive field of such a cell is determined by the area of appropriate transducer-containing tissue. One might refer to the receptive field of such a cell as "elbow" as opposed to "wrist" or "shoulder." Afferent inhibition, if it existed in such a cell, would be seen not as inhibition of the cell by certain joint angles, but rather it would be seen as inhibition caused by stimulation of neighboring joints. To our knowledge such an experiment has yet to be described, but afferent inhibition for these neurons may be unnecessary. The brain presumably has little difficulty determining which joint has been stimulated, and spatial resolution is of less concern in kinesthesia than is the frequency code that is used to analyze joint angle.

■ *Somatotopic Organization*

Throughout the dorsal column system a precise neurotopic or *somatotopic organization* exists (Fig. 9-3). This results from two factors. First, each primary afferent fiber entering the dorsal funiculi of the spinal cord ascends lateral to those arising from lower segments. This also applies to the secondary fibers ascending from the dorsal column postsynaptic cells. Thus caudal regions of the body are mapped more medially in the dorsal columns. Second, the primary, secondary, and tertiary fibers maintain a precise and regular spatial relationship among themselves (Fig. 9-3). An orderly map of the body, or *homunculus* (i.e., "little human"), thus exists for the somatosensory cortex. As is the case with vision, this map is distorted by differing magnification factors that relatively enlarge the neural representation of those somatic regions, such as the fingertips, that are capable of high spatial resolution (Fig. 9-3).

Despite the somatotopic organization, there is no modality mixing within single cells at any level of the system from the dorsal funiculi of the spinal cord to the somatosensory cortex; that is, the cells respond either to touch or to joint angle changes but never to both. Because of the somatotopy, kinesthesia and touch-sensitive cells that map similar body parts intermingle. For instance, a neuron that maps the skin over the elbow would be found near one that responds to elbow extension or flexion. This is another example of cells being responsive to a single modality, a feature that seems essential to the ability of the central nervous system to identify the nature of the stimulus.

■ *Cortical Columnar Organization*

Evidence for columnar organization of cortex similar to that already described for visual cortex, was derived from receptive field studies of cells in the postcentral gyrus. These cells are organized into columns from pia to white matter. Cells in each column share many functional features with one another, including modality.

■ Fig. 9-9. Schematic representation of terminal zones of fibers ascending in the anterolateral quadrant of the spinal cord. Terminal zones include the medullary, pontine, and mesencephalic reticular formation plus certain areas of the thalamus. Most thalamic terminals are found in midline nuclei, but some reach the VPL and VPM nuclei. Roman numerals, cranial nerve nuclei; *IC,* inferior colliculus; *SC,* superior colliculus; *RF,* reticular formation; *CL,* central lateral nucleus of the thalamus; *VB,* ventrobasal complex (VPL and VPM) of the thalamus. (Redrawn from Mountcastle, V.B., editor: Medical physiology, ed. 14, St. Louis, 1980, The C.V. Mosby Co.; courtesy W.R. Mehler.)

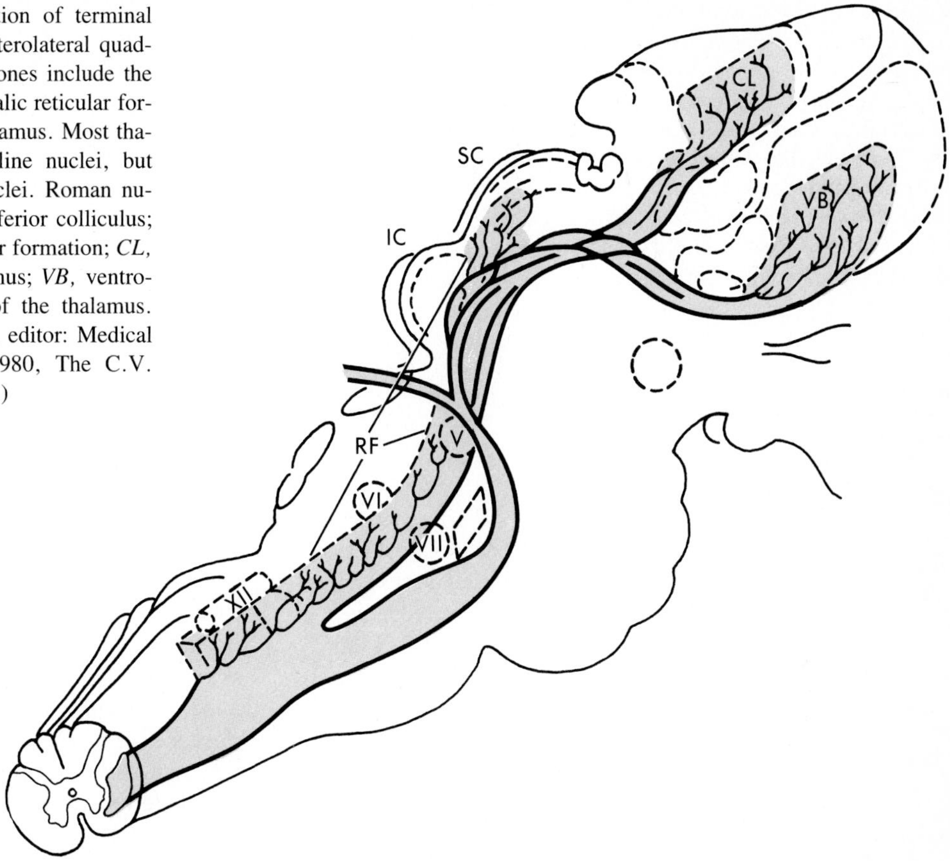

Thus columns of cells sensitive to the angle of a particular joint are found intermingled among cell columns sensitive to touch of skin near that joint. However, more recent evidence suggests that submodalities are represented by adjoining slabs of neurons running for several millimeters across the cortex rather than by much smaller discrete columns.

■ *Anterolateral System: Touch and Temperature*

The anterolateral system is phylogenetically older than is the dorsal column system. However, the former can be subdivided into newer and older portions that relate to innervation by group III and group IV afferent fibers, respectively. The anterolateral system processes information concerning touch, temperature, and pain. Because of its unique importance, pain is treated separately later in this chapter.

■ *General Anatomical Considerations*

Fig. 9-1 illustrates the intraspinal circuitry of the anterolateral system. Primary afferent fibers enter the dorsal horn of the spinal cord and course towards the ventral horn. These fibers contact the second-order cells of the anterolateral system (see Fig. 9-2, *B,* for the locations of the cell bodies of these neurons). In the newer portions of the system, the second-order cells are located more dorsally in the spinal gray matter and usually project an axon across the midline to the anterolateral quadrant of the spinal cord where it ascends to brainstem levels. In older portions there may be interneurons in the dorsal horn so that the ascending fiber is of a higher order than secondary, or in rare cases the fiber may ascend in the anterolateral column ipsilateral to the dorsal horn synapses (Fig. 9-1). Furthermore, for the older portions of the system, the cells of origin of the ascending fibers in the anterolateral quadrant are located more ventrally in the spinal gray matter.

Fig. 9-9 illustrates the diverse brainstem sites innervated by these ascending fibers. These sites include the medullary and pontine reticular formation, the superior colliculus, the central gray matter of the mesencephalon, some midline thalamic nuclei (such as the nucleus centralis lateralis), and the VPL nucleus of the thalamus. This suggests that the system can be divided into *spinobulbar* (spinoreticular, spinotectal, etc.), *paleospinothalamic* (e.g., involving midline thalamic nuclei), and *neospinothalamic* (involving VPL nucleus) tracts, from oldest to newest. This may be a useful functional division, but there seems to be little evidence for the frequent claim that these tracts are anatomically separated within the anterolateral quadrants of the spinal cord.

■ **Fig. 9-10.** Examples of response of various primary afferent fibers to stimuli of different temperatures. The solid curves represent responses of fibers sensitive to cool *(color)* or warm *(black)* stimuli. The dashed curves represent responses of nociceptive fibers to cold *(color)* or hot *(black)* stimuli. (Modified from Zotterman Y.: Annu. Rev. Physiol. **15**:357, 1953.)

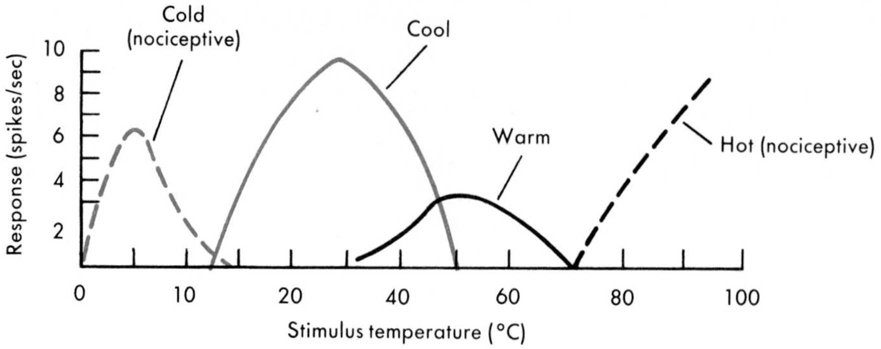

■ *Touch Sensation*

The neospinothalamic pathway that ascends to the VPL nucleus joins the fibers of the medial lemniscus en route. Because all VPL neurons share basic receptive field features (such as small fields, afferent inhibition, and precise somatotopic organization), touch-sensitive fibers of the neospinothalamic pathway seem to have properties identical to those of fibers emerging from the dorsal column nuclei. Central to their merger in the medial lemniscus, the two populations of fibers from the dorsal column and anterolateral systems are no longer distinguishable.

Little is known about the functional organization of the older portions of the anterolateral system. Limited data, mostly from recordings of cutaneous receptive fields in the midline thalamic nuclei, suggest detailed similarities with older portions of the visual system. These cutaneous receptive fields are large and diffusely organized and lack lateral inhibition. Furthermore, the neurons respond phasically to tactile stimulation and are not arranged in any obvious somatotopic order.

■ *Temperature Sensation*

Little is known about the neural substrate of temperature sensation. Recordings of peripheral nerves indicate that many group III afferent fibers have small cutaneous receptive fields that are insensitive to mechanical deformation but respond tonically to small temperature changes in the skin. Fig. 9-10 illustrates the responses of typical thermosensitive fibers. Within a range of roughly 20° to 40° C, these fibers increase their firing rate monotonically with either increasing temperature (''warm'' fibers) or decreasing temperature (''cool'' fibers). There are roughly 5 to 10 times as many cool fibers as warm fibers. Interestingly, at temperatures below 20° and above 40° C most of these fibers cease responding. At such extreme temperatures, tissue damage will begin, and a different set of neurons that uniquely signal nociception become responsive.

Reports of temperature-sensitive neurons in the brain have been rare. Occasional temperature-sensitive units have been found in the anterolateral quadrants of the spinal cord, VPL nucleus of the thalamus, and somatosensory cortex. Temperature sensation thus seems to be at least partly processed via the neospinothalamic pathways, but a great deal more data are needed. Finally, some indirect evidence, based on differential nerve blockades of myelinated and unmyelinated peripheral fibers, suggests that some group IV afferent fibers might contribute to temperature sensation.

■ *Anterolateral System: Pain or Nociception*

Pain is obviously an important subject to understand clinically. Unfortunately we know surprisingly little about the neural substrates for pain sensation. This is partly a result of the complex and poorly understood central pathways that mediate pain and partly a result of the difficulty in defining pain. Pain is an emotional and changeable perception. What is painful for one person at one time may not be at another time, or it may never be painful for another person. The battlefield has provided many dramatic examples of the difficulty in defining pain. Many soldiers have experienced horrible wounds, but in the excitement of battle, many claim awareness of the injury but experience no great pain. Pain typically begins much later when the excitement associated with the battle has ceased.

For these reasons it is easier to define and describe the concept of *nociception* than to deal with the qualitative nature of pain. Nociception is the sensation associated with tissue damage, and it is the sensation that leads to the percept of pain. A precise relationship between the nociception and pain is illustrated by Fig. 9-11, which plots the rate of heat transfer to the skin necessary to elicit pain as a function of initial skin temperature. The linear relationship intercepts the abscissa at approximately 45° C. If the skin were maintained at or above this temperature, irreversible tissue damage

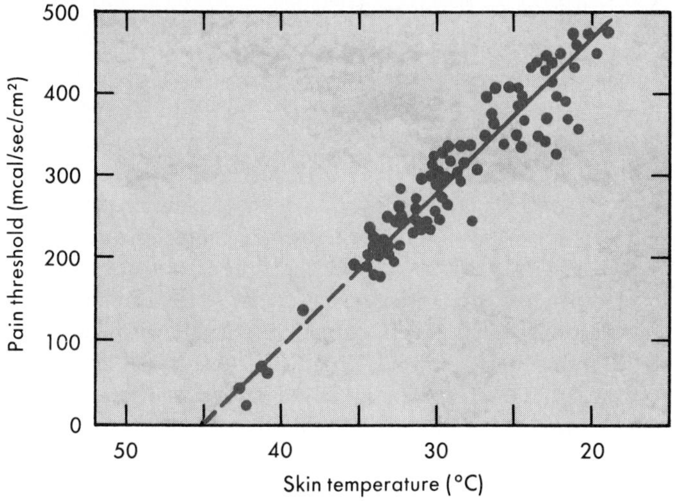

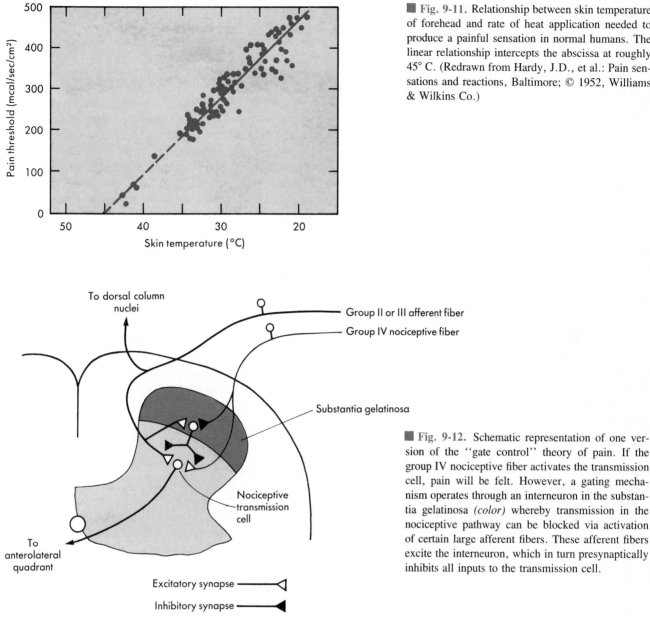

■ **Fig. 9-11.** Relationship between skin temperature of forehead and rate of heat application needed to produce a painful sensation in normal humans. The linear relationship intercepts the abscissa at roughly 45° C. (Redrawn from Hardy, J.D., et al.: Pain sensations and reactions, Baltimore; © 1952, Williams & Wilkins Co.)

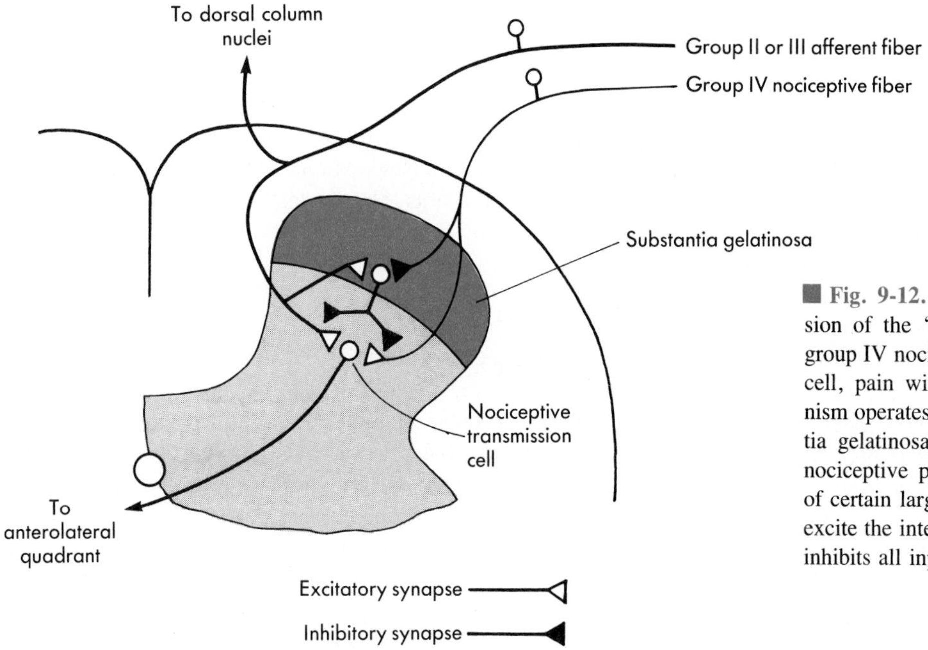

■ **Fig. 9-12.** Schematic representation of one version of the "gate control" theory of pain. If the group IV nociceptive fiber activates the transmission cell, pain will be felt. However, a gating mechanism operates through an interneuron in the substantia gelatinosa *(color)* whereby transmission in the nociceptive pathway can be blocked via activation of certain large afferent fibers. These afferent fibers excite the interneuron, which in turn presynaptically inhibits all inputs to the transmission cell.

would soon result. The sensation of pain thus relates rather precisely to actual or threatened tissue damage. In the discussion that follows we shall use the terms *nociception* and *pain* interchangeably.

■ *Specificity and Pattern Theories of Pain*

Before what is known of the central nociceptive pathways is described in detail, it is useful to consider two general theories regarding the neural basis of pain. The first is the *specificity theory* that nociception is processed as any other submodality of somesthesia; that is, specific nociceptive transducers exist that respond only to nociceptive stimuli, and they are functionally connected to specific neural pathways. Nociception thus has its own transducers and "labeled lines," much like kinesthesia or touch sensation.

The second hypothesis is the *pattern theory*. It argues that nociception shares transducers and/or pathways with other submodalities, but that the different pattern of activity in the same neuron can be used to signal either nociception or a nonpainful stimulus. For example, light touch applied to the skin might cause transducers and their central pathways to fire at some low frequency, but intense pressure that damages the skin might cause the same cells to discharge at a much higher frequency that signals pain.

Fig. 9-12 summarizes a popular theory that incorporates features of specificity and pattern. The theory is known as the *gate control* theory of pain, and it states that synaptic transmission of nociceptive information can be "gated" in the dorsal horn of the spinal cord by activity from other pathways. Specifically, the activity of large somatosensory afferent fibers discharges an interneuron in the substantia gelatinosa, which, in turn, can cause presynaptic inhibition of smaller, nociceptive afferent fibers (Fig. 9-12). Activity in the second-order neuron of the nociceptive pathway can thus be modified by activity in other somatosensory pathways. Theories such as this one have been influential in our thoughts about pain mechanisms. They also serve as plausible explanations for certain phenomena. For instance, minor cutaneous pain often can be abolished by rubbing the affected skin area; that is, rubbing activates the large somatosensory afferent fibers to inhibit transmission of the nociceptive afferent fibers. Also, some have attempted to explain the effects of acupuncture by these theories. Although there is little direct evidence to support such a gate control theory, we now know that the transmission of nociceptive signals can be gated effectively at several levels in the central nociceptive pathways, including the first synapse in the dorsal horn of the spinal cord. This is considered more fully below in our discussion of the pharmacology related to *algesia* (the presence of pain) and *analgesia* (the absence of or relief from pain).

Most relevant data point rather convincingly toward the simpler specificity theory. Neurophysiological studies have demonstrated afferent fibers and central neurons that are specifically activated by nociceptive stimuli, although some others can be activated by nociceptive and non-nociceptive stimuli as if they have a wide dynamic range of responsiveness to a particular submodality, such as pressure or temperature. Furthermore, many other somatosensory neurons, for example, those sensitive to touch and temperature, actually cease responding when stimulus levels become intense enough to cause tissue damage (Fig. 9-10). Thus our ensuing description of nociceptive pathways will treat these as submodalities, like those for touch or temperature, with unique transducers and labeled lines to the brain.

Peripheral Mechanisms

Afferent fiber types. Nerve block experiments and neurophysiological studies of single neurons both demonstrate that nociception is conveyed centrally by group III and IV afferent fibers. Roughly one fifth of all group III and half of all group IV afferent fibers are nociceptive. All of the group III afferent fibers enter the spinal cord via the dorsal root, but many of the unmyelinated nociceptive afferent fibers are found in the ventral root and probably enter the spinal cord there. In the framework of our description of sensory systems, it is likely that the nociceptive pathways include phylogenetically very old (group IV afferent fibers) and moderately old (group III afferent fibers) components.

Fast and Slow Pain

Two qualitatively different types of pain can be readily appreciated. These are termed *fast* and *slow pain*. Fast pain is a short, well-localized sensation that is well matched to the stimulus. Examples are a pinprick and strong pinch. Such stimuli are usually distinguishable from one another. The sensation starts and stops abruptly when the stimulus is applied and removed, respectively. Slow pain is a throbbing, burning, or aching sensation that is poorly localized and less specifically related to the stimulus. One cannot usually ascribe a specific nociceptive stimulus (heat, cutting, pinching, etc.) to this sensation. Also, the onset of the sensation has a long latency following application of the stimulus, and this pain continues for hours or days after removal of the stimulus. Fast pain is conveyed by group III afferent fibers and is associated strictly with the skin. Slow pain is conveyed by group IV afferent fibers and involves both cutaneous and deep tissues such as internal organs, joints, and muscles.

Neuropeptides and Pain

One of the most exciting advances in neuroscience research during the past decade has been the discovery of neuropeptides. Many such peptides have been described, and the list continues to grow. The roles of the neuropeptides in synaptic transmission and in endocrine physiology are discussed also in Chapters 4 and 49, respectively.

Neuropeptides are found in neurons throughout the central nervous sytem and are generally thought to play a role in synaptic transmission. Some are thought to be simple neurotransmitters in the same way that acetylcholine and γ-aminobutyric acid serve this function. Others are thought to serve as *neuromodulators* by coexisting with more conventional neurotransmitters in the presynaptic terminal. Neuromodulators do not themselves cause any substantive voltage or conductance change in the postsynaptic cell, but they can amplify, prolong, reduce, or shorten the postsynaptic response to the neurotransmitter. In cerebral cortex, in fact, when neuropeptides are found in synaptic terminals, they are always co-localized with GABA. Finally, some neuropeptides may function as blood-borne agents much like hormones. Although it is beyond the scope of this book to provide an account of neuropeptides, we shall briefly describe three main types that seem to play a prominent role in nociception. They are *substance P, bradykinin,* and the *opioids*. Table 9-2 shows the amino acid con-

Table 9-2. Constituents of neuropeptides

Neuropeptide	*Sequence*
Substance P	Arg-Pro-Lys-Pro-Gln-Gln-Phe-Phe-Gly-Leu-Met-NH_2
Bradykinin	Arg-Pro-Pro-Gly-Phe-Ser-Pro-Phe-Arg
[Leu]enkephalin	Tyr-Gly-Gly-Phe-Leu
[Met]enkephalin	Tyr-Gly-Gly-Phe-Met

stituents of the first two of these substances and the active portions of the opioids.

Substance P. Substance P, which is comprised of 11 amino acids (see Table 9-2), is widely distributed throughout the central nervous sytem. Substance P is the probable transmitter for most group IV nociceptive afferents. However, the discovery of substance P in certain areas, such as the retina, that have no plausible role in pain sensation demonstrates that this neuropeptide is not an exclusive agent of nociception.

Bradykinin. Bradykinin consists of 9 amino acids (Table 9-2). It seems to play a specific role in the transduction of nociceptive events at the periphery. This role is further considered below.

The opioids. The opioids are morphine-like substances that are manufactured in many regions of the central nervous sytem, including the pituitary gland. They are called ''morphine-like'' because they produce similar responses to those elicited by morphine and certain other narcotic substances; these responses include analgesia and euphoria. The endogenous opioids seem to underlie many forms of complex and changeable behavior, such as different moods, responses to pain, and stress.

Many endogenous opioids have been described and undoubtedly more await discovery. All have *[leu]enkephalin, [met]enkephalin,* or both as their active cores. Each core is a short peptide containing five amino acids (Table 9-2). The neuroactive opioids are generally produced by cleaving larger peptides or proteins. Three main classes are known, grouped according to the three precursor peptides. The first class derives from *pro-opiomelanocortin* (or *POMC*). POMC gives rise to the opioid *β-endorphin* (among other substances), which includes [met]enkephalin. The second class derives from *proenkephalin*. This is the precursor for numerous opioids, including [leu]enkephalin, [met]enkephalin, or longer peptides that include these enkephalins. The third class results from cleavage of *prodynorphin,* which also yields several endogenous opioids that contain [leu]enkephalin as the neuroactive core; most of these are the various *dynorphins*.

These opioids are produced in widely scattered cell groups distributed from the cerebral cortex to the spinal cord, and they are found in many axons and synaptic terminals. Fig. 9-13 summarizes the locations of cell bodies and fibers that contain the various classes of endogenous opioids. Although many specific types of opioids exist and must interact with equally specific receptors on the target cells to produce their neuroactive result, it is not yet possible to ascribe specific functions, analgesic or otherwise, to each of the opioids. We shall thus refer only to the general term ''opioids'' in the remainder of this chapter.

Transduction Mechanisms

Our understanding of the transduction process in nociception is far from complete or satisfactory, and the following should be considered as mostly speculative and a basis for further research.

Fast pain. The transducers for fast pain can be considered to be mechanical, thermal, or chemical. These transducers employ the same mechanisms as their non-nociceptive counterparts, but with higher thresholds that require nociceptive stimuli. This would explain the relatively close relationship between the fast pain perceived and the nature of the stimulus.

Slow pain. Fig. 9-14, *A,* outlines one version of our current understanding of the less direct transduction process for slow pain. A nociceptive stimulus causes tissue damage that leads to cell death. As some of these cells die, they release various proteolytic enzymes into the interstitial fluid, and these enzymes cleave certain γ-globulins circulating in the blood. The result is a variety of short-chain neuropeptides, including bradykinin and substance P, in the vicinity of the injury. Some of these polypeptides then activate the group IV nociceptive afferent fibers, which can thus be considered chemical transducers, in the same manner that neurotransmitters lead to excitatory postsynaptic potentials (EPSPs) in central neurons. A variety of indirect data is consistent with this hypothesis. For instance, injection of bradykinin or the contents of blister fluid into healthy skin evokes the sensation of slow pain. Analysis of blister fluid and certain insect bite venoms has revealed the presence of bradykinin-like polypeptides. Finally, bradykinin has also been identified in the superfusate of tooth pulp exposed to noxious stimuli. This perfusate proves to be a powerful nociceptive stimulus when injected into the skin.

Fig. 9-14, *B,* illustrates a related and more contemporary version of the sequelae that underlie slow pain transduction. Here, the group IV nociceptive afferent fiber is depicted as a branching process that innervates a considerable area of tissue. A nociceptive stimulus applied to the skin activates the subjacent ending or endings in the traditional manner. However, after the orthodromically traveling (i.e., in the forward direction toward the spinal cord) action potential reaches a branch point, the collaterals distal to the branch are in-

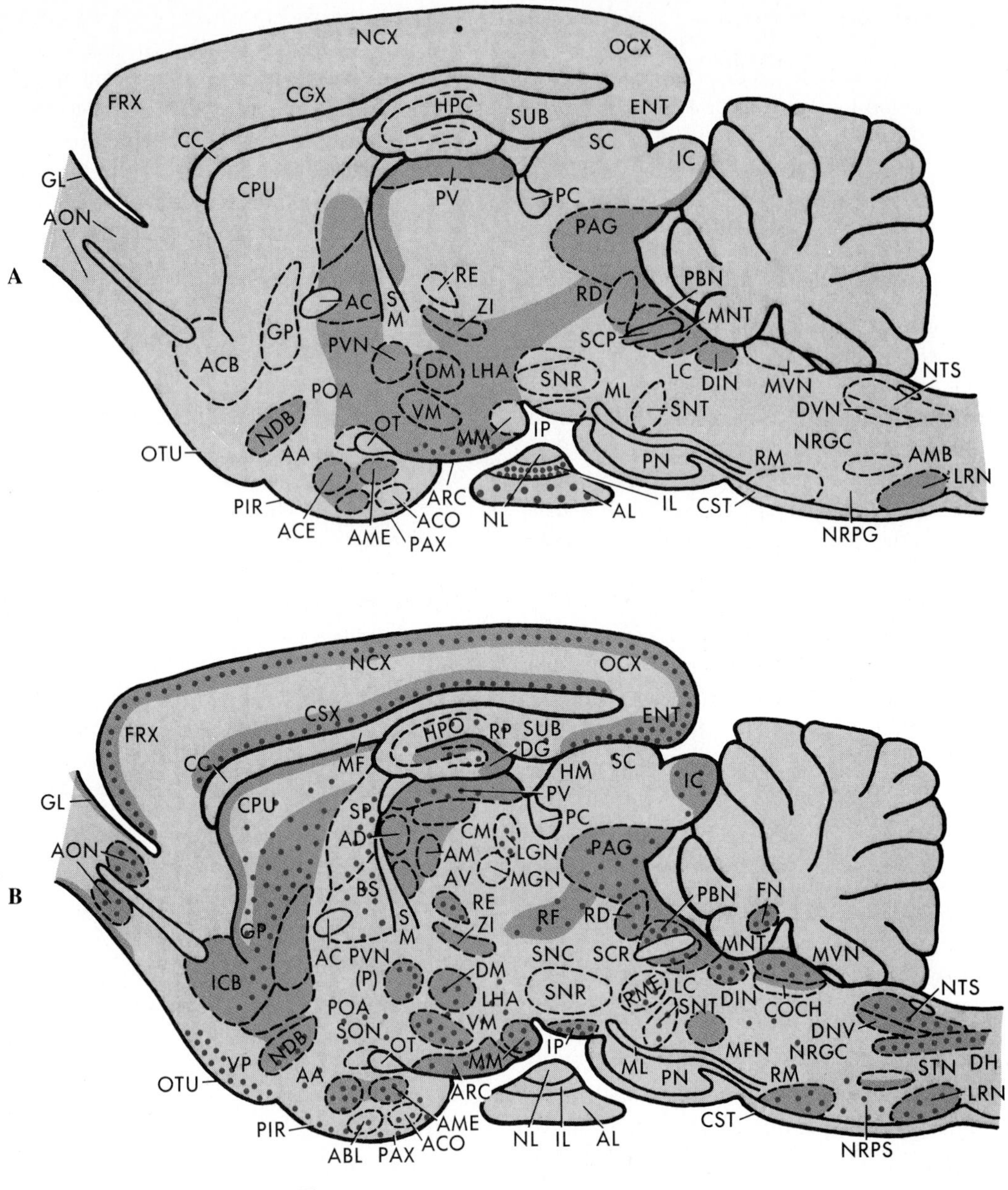

■ **Fig. 9-13.** For legend see opposite page.

vaded antidromically (i.e., in the backward direction). This so-called *axon reflex* signifies that an action potential can travel backward down the axon to its terminals in the innervated tissue. These Group IV afferents appear to discharge a chemical, perhaps substance P, at their peripheral terminals in much the same manner as neurotransmitter is discharged from synaptic terminals in the spinal cord. The chemical release causes cells in the nearby tissue to discharge various substances, such as bradykinin, histamine and prostaglandins, which further excite the nociceptive Group IV afferents.

The events diagrammed by Fig. 9-14 account for many of the phenomena peculiar to slow pain. The long latency of the sensation to a nociceptive stimulus can be explained by the time needed for cell death or for the production of such neuroactive substances as bradykinin and histamine. The prolonged sensation that follows removal of the offending stimulus is caused by the continued presence of these neuroactive substances. Finally, the lack of differential sensations for different nociceptive stimuli (heat, mechanical, etc.) would result from the indirect response of group IV afferent fibers to these neuroactive substances, regardless of which specific nociceptive stimulus produced them.

■ *Central Pathways*

We know distressingly little about the central pathways that represent nociception, although the advances made

C

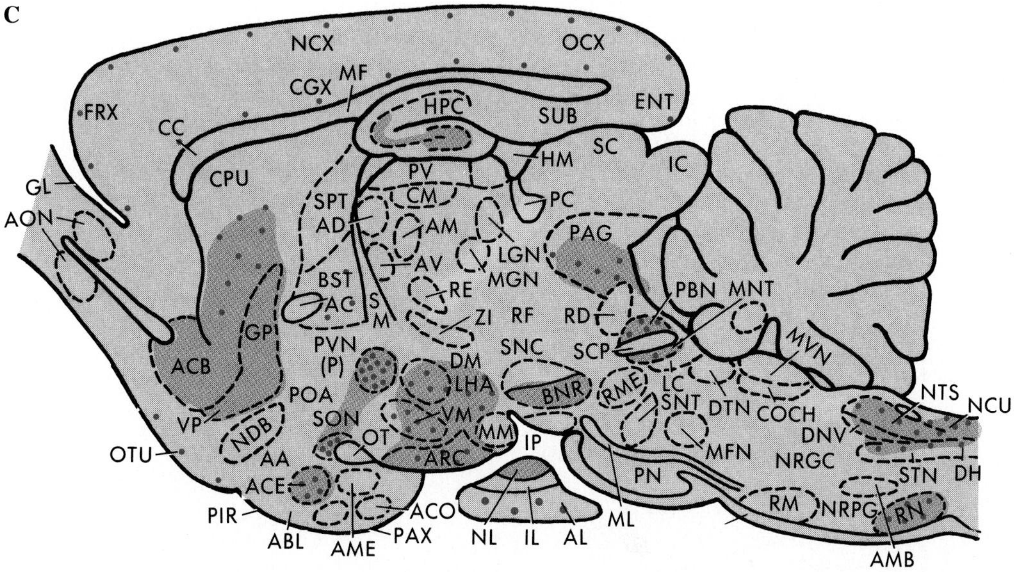

■ Fig. 9-13. Location of opioid-containing neurons in the rat's central nervous system. The colored dots represent concentrations of opioid-containing cells and the lightly colored background indicates regions in which opioid-containing fibers and terminals are concentrated. The drawings are broken into neural elements with opioids deriving from pro-opiomelanocortin **(A),** proenkephalin **(B),** and prodynorphin **(C).** *AA,* anterior amygdala; *ABL,* basolateral nucleus of amygdala; *AC,* anterior commissure; *ACB,* nucleus acumbens; *ACE,* central nucleus of amygdala; *ACO,* cortical nucleus of amygdala; *AD,* anterodorsal nucleus of thalamus; *AL,* anterior lobe of pituitary; *AM,* anteromedial nucleus of thalamus; *AMB,* nucleus ambiguus; *AME,* medial nucleus of amygdala; *AON,* anterior olfactory nucleus; *ARC,* arcuate nucleus; *AV,* anteroventral nucleus of thalamus; *BST,* bed nucleus of stria terminalis; *CC,* corpus callosum; *CGX,* cingulate cortex; *CM,* central-medial nucleus of thalamus; *COCH,* cochlear nuclear complex; *CPU,* caudate-putamen; *CST,* corticospinal tract; *DG,* dentate gyrus; *DH,* dorsal horn of spinal cord; *DM,* dorsomedial nucleus of hypothalamus; *DNV,* dorsal motor nucleus of vagus; *DTN,* dorsal tegmental nucleus; *ENT,* entorhinal cortex; *FN,* fastigial nucleus of cerebellum; *FRX,* frontal cortex; *GL,* glomerular layer of olfactory bulb; *GP,* globus pallidus; *HM,* medial habenular nucleus; *HPC,* hippocampus; *IC,* inferior colliculus; *IL,* intermediate lobe of pituitary; *IP,* interpeduncular nuclear complex; *LC,* locus coeruleus; *LGN,* lateral geniculate nucleus; *LHA,* lateral hypothalamic area; *LRN,* lateral reticular nucleus; *MF,* mossy fibers of hippocampus; *MFN* motor facial nucleus; *MGN,* medial geniculate nucleus; *ML,* medial lemniscus; *MM,* medial mammillary nucleus; *MNT,* mesencephalic nucleus of trigeminal; *MVN,* medial vestibular nucleus; *NCU,* cuneate nucleus; *NCX,* neocortex; *NBD,* nucleus of diagonal band; *NL,* neural lobe of pituitary; *NRGC,* nucleus gigantocellularis reticularis; *NRPG,* nucleus reticularis paragigantocellularis; *NTS,* nucleus tractus solitarius; *OCX,* occipital cortex; *OT,* optic tract; *OTU,* olfactory tubercle; *PAG,* periaqueductal gray; *PAX,* periamygdaloid cortex; *PBN,* parabrachial nucleus; *PC,* posterior commissure; *PIR,* piriform cortex; *PN,* pons; *POA,* preoptic area; *PP,* perforant path; *PV,* periventricular nucleus of thalamus; *PVN(M),* paraventricular nucleus (pars magnocellularis); *PVN(P)* paraventricular nucleus (pars parvocellularis); *RD,* nucleus raphé dorsalis; *RE,* nucleus reuniens of thalamus; *RF,* reticular formation; *RM,* nucleus raphé magnus; *RME,* nucleus raphé medianus; *SC,* superior colliculus; *SCP,* superior cerebellar peduncle; *SM,* stria medullaris thalami; *SNC,* substantia nigra (pars compacta); *SNR,* substantia nigra (pars reticulata); *SNT,* sensory nucleus of trigeminal (main); *SON,* supraoptic nucleus; *SPT,* septal nuclei; *STN,* spinal nucleus of trigeminal; *SUB,* subiculum; *VM,* ventromedial nucleus of hypothalamus; *VP,* ventral pallidum; *ZI,* zona incerta. (Redrawn from Khachaturian, et al.: Trends in Neurosci. **8:**111, 1985.)

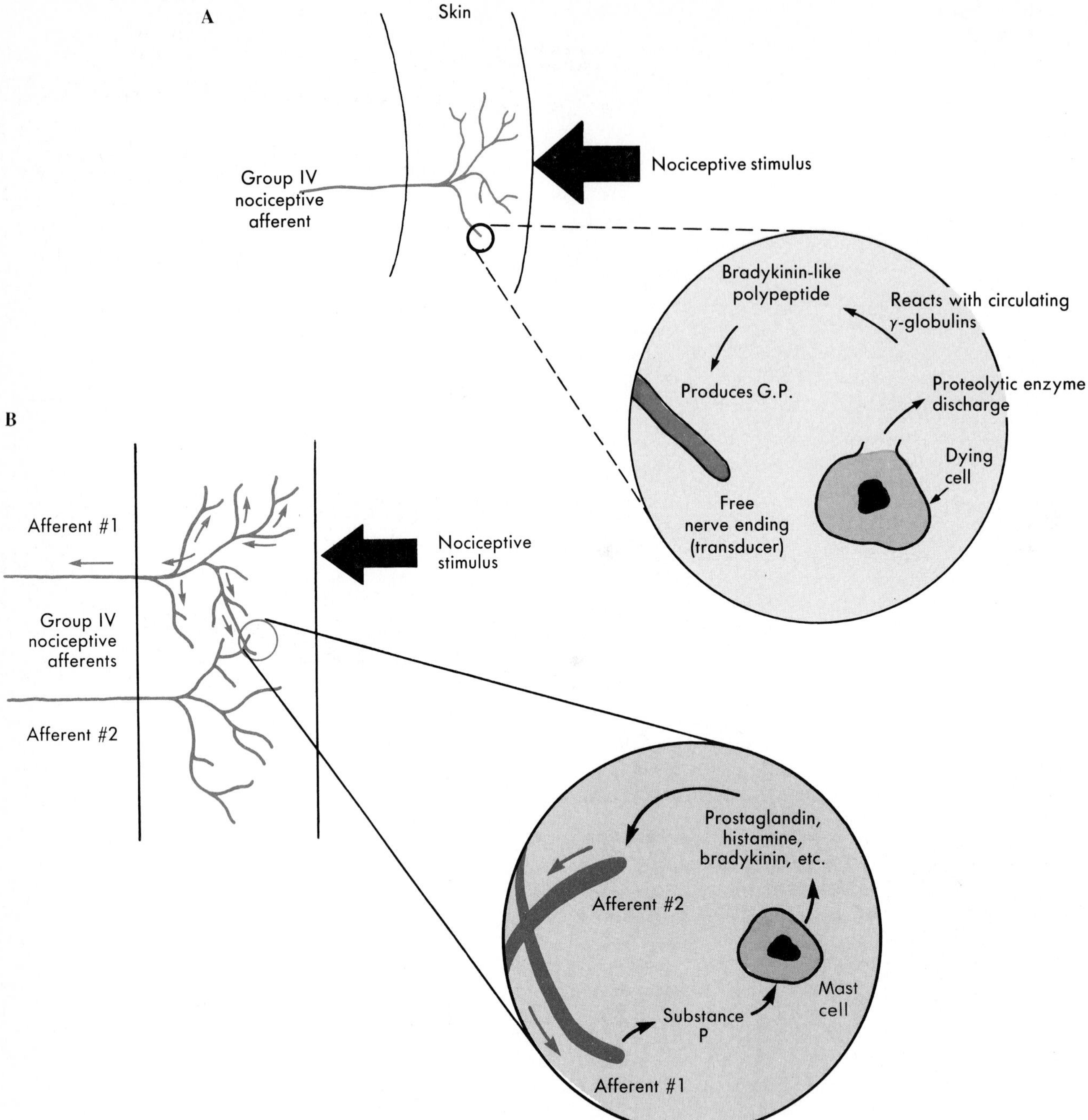

Fig. 9-14. Two hypothetical schemas to suggest transduction mechanisms for slow pain. In both examples, the nociceptive afferent conveying slow pain information responds to the presence of a neuropeptide, such as bradykinin, histamine, etc. **A,** Release of bradykinin from damaged tissue. A nociceptive stimulus applied near the free nerve endings of a group IV nociceptive afferent fiber *(color)* causes tissue damage and cell death. As some cells die, they discharge proteolytic enzymes that react with circulating γ-globulins to produce bradykinin-like polypeptides. The free nerve ending is a chemical transducer sensitive to these polypeptides; consequently a generator potential *(GP)* is evoked. **B,** Release of bradykinin from mast cells in response to the presence of substance P. The nociceptive stimulus activates terminal of afferent no. 1, which conveys fast pain. Because only part of the terminal field is activated by the stimulus, the propagation of impulses antidromically invade other terminals *(see arrows)*. This antidromic invasion causes release of substance P, which in turn causes release of bradykinin-like agents from nearby mast cells. As in **A,** these secondary substances activate the terminals of the axon conveying slow pain information (afferent no. 2).

during the last decade have been impressive. The diffuse connections, small neurons and fibers, and difficulty in defining painful or nociceptive stimuli render this a particularly difficult problem for experimental analysis. Indeed, much of our understanding stems from clinical observations, particularly from surgical attempts to alleviate intractable pain by interruption of various neuronal pathways.

Ascending pathways. After group II and group IV afferents enter the spinal gray matter, nociceptive information is relayed from spinal cord neurons to the brain. The most important ascending nociceptive tracts lie in the anterolateral quadrants of the cord, and are thus part of the anterolateral system (Fig. 9-9). Most of the ascending axons cross from the spinal gray to enter the contralateral anterolateral column, but some ascend ipsilaterally. Finally, axons from some of the nociceptive spinal cord neurons ascend in the ipsilateral dorsal column along with other axons of dorsal column postsynaptic cells (see above).

Nociceptive neurons are difficult to study. Limited data from animals suggest that such neurons occur in the following sites: the dorsal horn and the anterolateral and dorsal columns of the spinal cord; various brainstem reticular nuclei; various thalamic nuclei, including the VPL nucleus and certain midline thalamic nuclei; and the somatosensory cortex. Those in the VPL nucleus and somatosensory cortex are probably involved in analysis of fast pain; slow pain is probably analyzed in the other regions. Little is known about cortical involvement in slow pain.

Unfortunately there may be many ascending pathways other than the anterolateral and dorsal column systems involved in the processing of nociceptive information. For example, many group IV afferent fibers enter the spinal cord via the ventral root, and we presently know virtually nothing regarding the central connections related to this afferent pathway. It is likely that many of these afferent fibers are nociceptive.

The ubiquity of the ascending nociceptive pathways is evident from attempts to interrupt them in the treatment of intractable pain such as often occurs in terminal cancer patients. If the suffering is intense and drugs no longer provide relief, often an attempt will be made to destroy the presumptive nociceptive cells or fibers. The most common procedure is to cut through the anterolateral quadrant of the spinal cord at the appropriate spinal level, thereby interrupting fibers ascending in the anterolateral funiculus as well as causing other damage. Virtually every level of the neuraxis shown in Fig. 9-9 has been interrupted in some patient, and overall the results have been disappointing. Relief, if it occurs at all, is usually incomplete and temporary. This is true even when the appropriate dorsal roots have been sectioned, presumably because of the course taken by many group IV afferent fibers through the ventral roots. Apparently, ascending nociceptive pathways are so diffuse and redundant that it is impractical to achieve permanent analgesia surgically.

Referred pain. Often pain is sensed in a region of the body that is inappropriate to the nociceptive stimulus. This is known as *referred pain* because the pain seems to be referred from the damaged region to healthy tissue. Most commonly such pain may be sensed in the skin when internal organs become damaged.

The explanation for pain referral is not understood, but it is an important clinical phenomenon that generally follows a basic rule called the *dermatome rule;* that is, the area of skin in which pain is felt is usually innervated by the same spinal segment as is the affected organ. Thus heart attack victims often feel cutaneous pain radiating from the left shoulder down the arm because sensory innervation from these skin regions and the affected cardiac tissue pass through the same spinal segments. One should always be suspicious of an internal pathological condition when cutaneous pain is inexplicable. The dermatome rule can be used as a guide to locate the source.

The dermatome rule offers two possible explanations (among many) for referred pain. One is illustrated in Fig. 9-1, *A*. Although most nociceptive afferent fibers have unique pathways to the brain (represented by afferent fibers *1* and *4* and neurons *a* and *c* in Fig. 9-15, *A*), perhaps some cutaneous and visceral afferent fibers (*2* and *3*) converge on the same second-order neuron (*b*). Activity of neuron *b* cannot distinguish between cutaneous and visceral pain. However, in most persons cutaneous pain is more often experienced than is visceral pain, and the brain might simply have learned to interpret activity in neuron *b* as due to activation of the cutaneous afferent fiber *(2)*. When neuron *b* is activated by the visceral afferent fiber *(3)*, the brain would therefore misinterpret this as cutaneous in origin. This is a plausible explanation of referred pain and of the dermatome rule, since such convergence of afferent fibers is most probable for fibers innervating the same spinal segment. Nonetheless, there is no direct evidence to support this hypothetical neural basis of referred pain.

Another possible explanation for referred pain is illustrated in Fig. 9-15, *B*. Some primary afferents that innervate visceral structures branch to innervate cutaneous regions as well. Activation of the visceral afferents could be misinterpreted just as in the convergence theory illustrated in Fig. 9-15, *A*. However, this activation would also lead to antidromic activation of the branch leading to the skin. This in turn could release substance into the skin, and create the necessary conditions for slow pain in the skin. This process has been illustrated in Fig. 9-14, *B*.

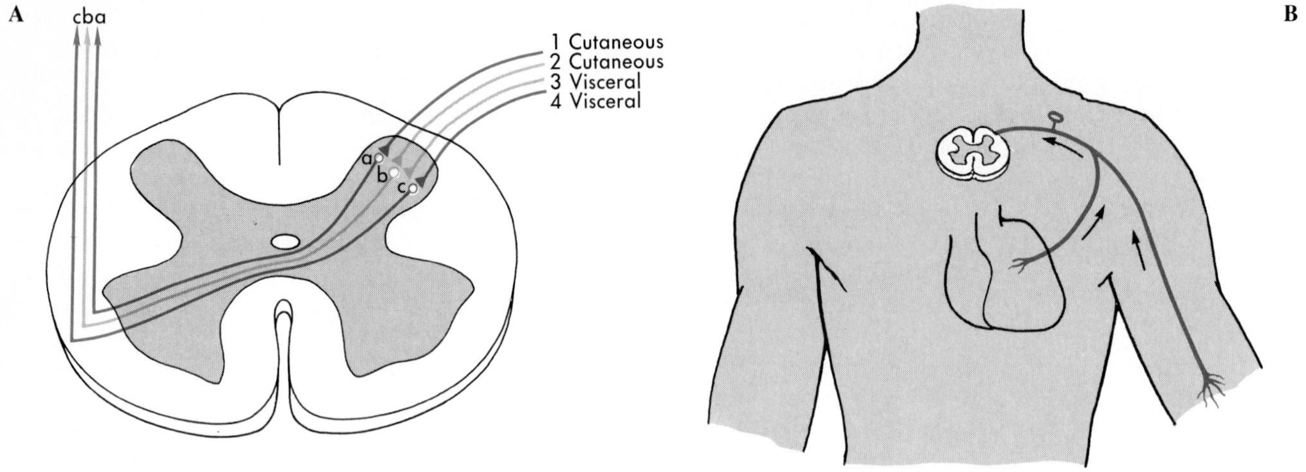

■ **Fig. 9-15.** Two plausible explanations for referred pain. **A,** Convergence of primary afferents. Four nociceptive afferent fibers *(1, 2, 3,* and *4)* innervate three secondary neurons (*a, b,* and *c*) in the spinal cord. Fibers *1* and *2* innervate the skin, and fibers *3* and *4* innervate visceral structures. Fibers *1* and *4* each innervate a single neuron (*a* and *c,* respectively). Activity along these pathways *(darker colors)* unambiguously reflects cutaneous stimulation for neuron *a* and visceral stimulation for neuron *c*. However, fibers *2* and *3* converge onto neuron *b (lighter colors)* so that activity of neuron *b* cannot be unambiguously interpreted by the brain. Frequently even if fiber *3* (visceral) excites neuron *b,* the perception is one of pain as if the peripheral signals were transmitted along fiber *2* (cutaneous). Thus pain of visceral origin is often referred to a cutaneous site in a pattern that obeys the dermatome rule. (Modified from Ruch, T.C., et al.: Neurophysiology, ed. 2, Philadelphia, 1965, W.B. Saunders Co.) **B,** Antidromic activation of skin afferents. An afferent nociceptive axon (*color),* which branches according to the dermatome rule to innervate both the skin and an internal organ (the heart in this example), is activated by a nociceptive stimulus at its terminations in the heart. The propagation of impulses *(colored arrows)* proceeds orthodromically toward the spinal cord and antidromically down the cutaneous branch. The antidromic activity in the skin could result in the local release of substance P, which is a nociceptive agent and would in turn activate nociceptive axons from the skin (see Fig. 9-14, *B*). Thus a painful stimulus at the heart can cause the sensation of pain in the skin within the constraints of the dermatome rule.

Descending and pharmacological control of nociception. As noted above, the changeable nature of our responses to nociceptive stimuli may be dramatic. Pain thresholds vary enormously and are quite strongly affected by many drugs, such as aspirin and morphine. Aspirin seems to act peripherally, probably at the level of transduction, to minimize or eradicate the afferent nociceptive signal. It is thus a true analgesic, because it affects the entire sensation of pain. Morphine acts centrally at many sites to reduce the suffering caused by nociceptive stimuli; it may even block completely the central transmission of nociceptive stimuli. Morphine and other narcotics appear to mimic the effects of the endogenous opioids, the action of which probably underlies the changeable reactions to nociceptive stimuli.

Fig. 9-16, *A,* summarizes a circuit that regulates the transmission of nociceptive signals. Nociceptive cells of the spinal gray matter and their ascending axons in the anterolateral columns can be effectively inhibited by stimulating various sites in the brainstem reticular formation. Furthermore, such stimulation often produces dramatic analgesia. Two particularly active regions for this are the *periaqueductal gray* of the midbrain and the *nucleus raphe magnus* and nearby cell groups of the rostral ventral medulla. Bulbospinal fibers from both regions descend to the spinal cord, and the periaqueductal gray also innervates the medullary neurons. It is not yet clear if the bulbospinal fibers inhibit the nociceptive cells directly or inhibit them indirectly by controlling the activity of interneurons (see Fig. 9-16, *A*).

The CNS regions, shown in Fig. 9-16, *A,* are especially rich in endogenous opioids. Injection of opioids directly into any of these regions typically produces a profound analgesia. The general effect of opioids on neurons is inhibitory, yet the bulbospinal neurons must be activated to block nociception. Thus opioids (and opioid containing neurons) may act by inhibiting interneurons that, in turn, inhibit the bulbospinal fibers (Fig. 9-16, *A*). In other words, the opioids activate the descending bulbospinal neurons by disinhibition. As might be expected, application of opioids to the spinal gray matter strongly inhibits the nociceptive neurons in that region. Whether this effect results from a presyn-

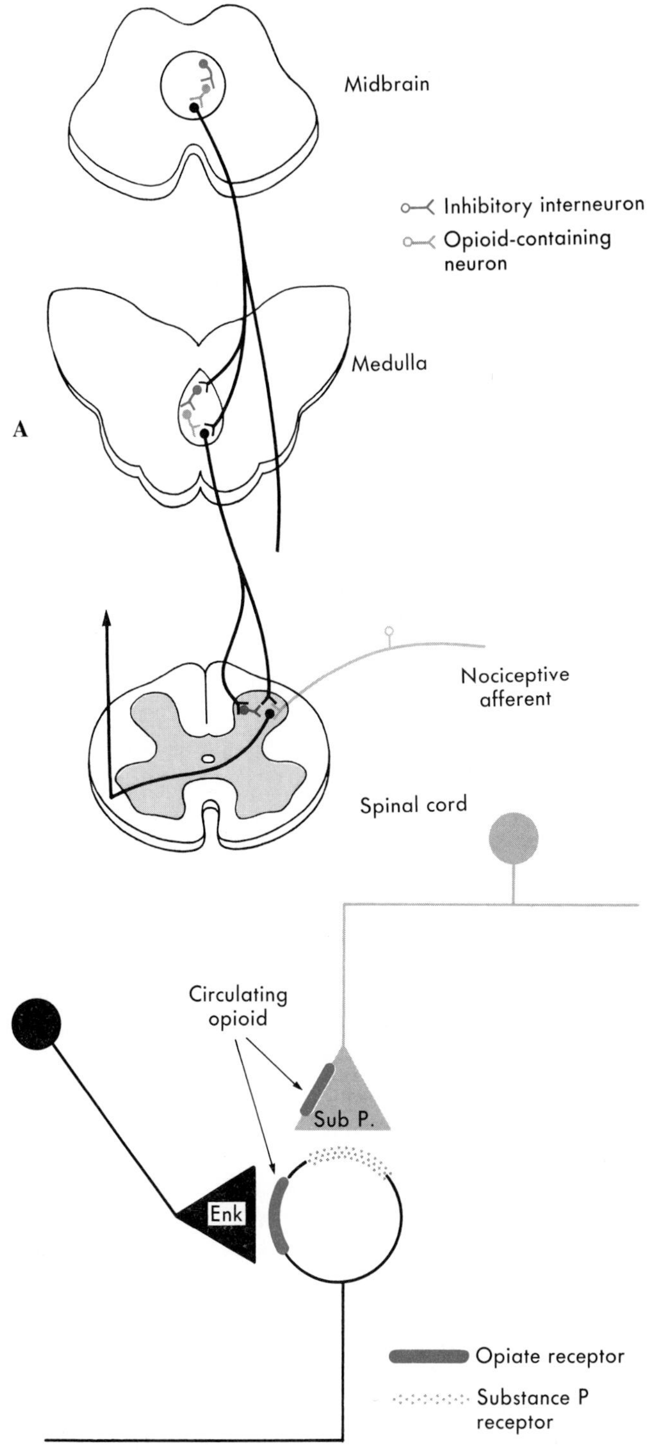

Fig. 9-16. Descending control of nociception. **A,** Schema of major pathways involved. Opioid-containing neurons *(darker color)* seem to act by inhibition, and their activity can block transmission of nociceptive information. At brain stem levels, the opioid-containing neurons seem to inhibit interneurons *(gray),* thereby disinhibiting certain descending pathways. Activity in these descending pathways causes activation of opioid-containing neurons in the dorsal horn of the spinal cord to block transmission from nociceptive afferents *(lighter color)* to the second order neurons. **B,** Detail of organization in the dorsal horn of the spinal cord. Primary nociceptive afferents *(lighter color),* which may use substance P as a neurotransmitter, have opioid receptors *(darker color)* on their terminals, although no axo-axonic synapses have been described there. The second order neuron also contains opioid receptors *(darker color)* as well as receptors for substance P *(gray stippling).* Activation of the opioid receptors, either by local opioid-containing neurons via conventional synapses or by circulating opioids, can disrupt nociceptive transmission both pre- and postsynaptically. (Redrawn from Henry, J.L. In Porter, R., and O'Connor, M., editor, Substance P in the Nervous System, London, 1982, Ciba Foundation Symposium 91, Pitman, p. 206.)

aptic inhibitory action on the primary nociceptive afferents that contain substance P, or to a more conventional postsynaptic inhibition of the nociceptive spinal cord cells, or to both is debatable. Substance P–containing terminals do have opioid receptors, but synapses have not been found on the primary afferent fibers that could explain true presynaptic inhibition. The presynaptic effect might be modulated by opioids that have diffused or circulated to the site. Fig. 9-16, *B,* illustrates these possibilities.

Trigeminal Pathways

The functional organization of somesthesia for the head is virtually the same as that for the body with the exception that the afferent fibers have cell bodies in the semilunar ganglion, and most enter the neuraxis via the trigeminal nerve; a few enter via the facial and vagus nerves. The trigeminal system is organized into analogues of the dorsal column and anterolateral systems. These pathways from the head and body come together to form a continuous and complete somatotopic map (Fig. 9-3).

The *main sensory nucleus* of the trigeminal nerve is located in the pons and is analogous to the dorsal column nuclei of the medulla. This nucleus receives afferent fibers from the trigeminal nerve, and its neurons project axons across the midline where they merge with the medial lemniscus to innervate the *ventralis posteromedialis* (VPM) nucleus of the thalamus. The VPL and VPM nuclei are contiguous, and since they appear to be functionally equivalent counterparts for the head and body, they are often considered together as the *ventrobasal* (VB) *complex* (VB = VPL + VPM). One minor difference between the main trigeminal and dorsal column systems is that a very small projection from the main nucleus to the VPM nucleus exists ipsilaterally.

The analogue of the anterolateral system is represented by the *spinal nucleus and tract* of the trigeminal nerve. This is an elongated column of cells and fibers that is continuous caudally with the dorsal horn at upper cervical levels and nearly merges rostrally with the main nucleus. Some fibers entering the trigeminal nerve are directed caudally and innervate neurons of the spinal nucleus. Most second-order fibers from the spinal nucleus cross to ascend the neuraxis with the anterolateral system to end in the same brainstem sites (Fig. 9-9). Some of these fibers from the spinal nucleus ascend ipsilaterally.

The functional divisions in the trigeminal system are thought to mimic those of the ascending spinal pathways. The main nucleus is innervated mainly by large myelinated fibers that convey information about touch and kinesthesia. As yet, no analogue to the dorsal column postsynaptic cells of the spinal cord has been described for the trigeminal pathways; likewise, no analogue to the nociceptive fibers ascending in the dorsal columns has yet been described in relationship to the main sensory nucleus. The spinal nucleus receives smaller myelinated and unmyelinated fibers that signal touch, temperature, and nociception. Nociceptive cells in the VPM nucleus are located in an outer shell of this nucleus, an arrangement similar to that described earlier in this chapter for nociceptive representation in the VPL nucleus.

As is the case with the spinal components of the somatosensory system, the somatosensory cortex projects fibers to the trigeminal components. Heavy direct projections have been described to the VPM nucleus and to the main and spinal trigeminal nuclei. Indirect projections via brainstem reticular pathways probably exist as well. The principles described above for control of pain transmission via descending pathways and release of endogenous opioids apply equally well to nociceptive transmission within the trigeminal pathways.

Functional Significance of Old and New Pathways

In the visual system the newer geniculostriate portion may not be the sine qua non for vision that is often claimed. The same principle applies to the somatosensory pathways. For instance, lesions of the dorsal funiculi of the spinal cord (and thus the newest position of the pathways) lead to only a transient and partial loss of touch sensitivity. (Kinesthesia is permanently and nearly totally lost after such a lesion because this submodality is poorly represented in older portions of the system.) Likewise, lesions of the somatosensory cortex produce a temporary anesthesia with a gradual return of sensation that is only subtly abnormal. Thus one must keep in mind the principle that somesthesia results from complex interactions among the various somatosensory pathways, old and new.

The somatosensory submodalities of touch sensation, kinesthesia, temperature sensation, nociception or pain, and perhaps vibration sensation can be distinguished. Each has its own transducers and labeled lines or neuronal pathways. These submodalities are conveyed to the central nervous system by large myelinated group II fibers (touch and kinesthesia), small myelinated group III fibers (touch, temperature, and pain), and small unmyelinated group IV fibers (touch, pain, and perhaps temperature). The central pathways can be divided roughly into phylogenetically newer and older portions. The former are represented by the dorsal column system and main sensory trigeminal nucleus; the latter are represented by the anterolateral system and spinal trigeminal nucleus. The newer pathways begin with group II afferent fibers, whereas the older ones receive smaller afferent fibers. Generally the newer portions display a high degree of somatotopic organization, small receptive fields with lateral inhibition, and many neurons that respond tonically to sensory stimuli. The older pathways lack these features. Older and newer pathways combine in the processing of somesthesia, and the newer pathways appear to be necessary only for the processing of kinesthesia.

In general, the functional organization of the somatosensory pathways shares much in common with that of the visual pathways. Both systems have older and newer components with similar distinguishing characteristics. It is interesting to compare the newer portions: the retinogeniculostriate pathways for vision and the dorsal column system and neospinothalamic pathways for somesthesia. Fig. 9-17 outlines this comparison. Because neurons analogous to retinal horizontal and amacrine cells are not evident in somesthesia, these cells are omitted from Fig. 9-17. The somatosensory transducers are both represented by specialized cells that synapse onto the primary afferent fiber (e.g., the tonically responsive transducer described in Chapter 6). In all three pathways, *cell 1* is a transducer (photoreceptor or tonically responsive transducer) that synapses onto a bipolar neuron *(cell 2)*. The bipolar neuron then synapses onto *cell 3* (ganglion cell, dorsal column nucleus cell, or dorsal horn cell), which projects an axon to the thalamus (LGN or VPL nucleus). This projection is such that the sensory environment is mapped onto the contralateral thalamus. Thus the axon from each somatosensory *cell 3* crosses the midline, and that from the retinal ganglion cell does or does not cross in the optic chiasm so that each hemifield is contralaterally mapped in the LGN. The thalamocortical projection from *cell 4* terminates mostly in layer IV but also in layer VI of the appropriate cortex. The cortical neurons *(cell 5)* are functionally arranged in columns perpendicular to the layering.

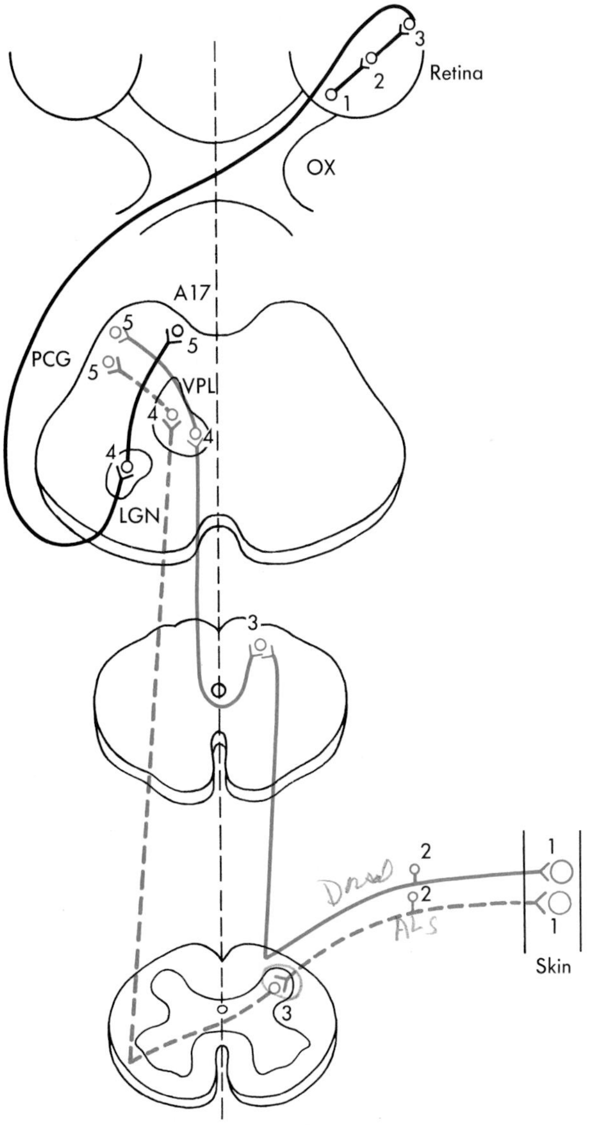

■ **Fig. 9-17.** Some similarities among the visual pathways *(solid black),* dorsal column somatosensory pathways *(solid color),* and newer portion of the anterolateral somatosensory pathways *(dashed color).* In all pathways transduction begins with a specialized cell (cell *1,* a rod or cone for vision or a specialized mechanoreceptor in the skin) that synapses onto a bipolar-type neuron (cell *2,* a bipolar cell in retina or primary afferent fiber in the skin). The fiber from cell *2* synapses onto cell *3* (retinal ganglion cell, dorsal column nucleus cell, or dorsal horn cell). In all systems the fiber from cell *3* crosses the midline if needed to map sensory space onto the contralateral half of the neuraxis. Thus both somatosensory fibers decussate, and the ganglion cell axon decussates in the optic chiasm *(OX)* if it derives from nasal but not temporal retina. The fibers from cell *3* terminate onto cell *4* in the thalamus *(VPL* or *LGN),* and cell *4* projects to the appropriate area of cortex, either area 17 *(A17)* or the postcentral gyrus *(PCG).* Cell *5* in cortex represents the first stage of direction or orientation specificity in response to stimulus parameters.

Throughout each system neurotopic organization is evident, and the sensory maps are distorted to magnify areas of higher acuity. Neurons display lateral inhibition at the most peripheral site possible (i.e., the first site of synaptic interactions); this occurs in *cell 2* of the visual system (because of the action of horizontal cells that have no somatosensory counterpart) but not until *cell 3* of the somatosensory system. More specific stimulus requirements, such as the shape and direction of movement, are first seen among the cortical neurons *(cell 5).*

It should be clear that, except for certain anatomical details, the basic organization of the central pathways that underlie vision are remarkably similar to those related to somesthesia. It is these common principles that are most fundamental to sensory processing and, therefore, should be understood most clearly.

■ *Bibliography*

Journal articles

Adriaensen, H., et al.: Response properties of thin myelinated (A-δ) fibers in human skin nerves, J. Neurophysiol. **49:**111, 1983.

Akil, H.: Endogenous opioids: biology and function, Ann. Rev. Neurosci. **7:**223, 1984.

Basbaum, A.I. and Fields, H.L.: Endogenous pain control systems: brainstem spinal pathways and endorphin circuitry, Ann. Rev. Neurosci. **7:**309, 1984.

Bennett, G.J. et al.: (1983) The cells of origin of the dorsal column postsynaptic projection in the lumbosacral enlargements of cats and monkeys, Somatosensory Res. **1:**131, 1983.

Berkley, K.J.: Spatial relationships between the terminations of somatic sensory motor pathways in the rostral brainstem of cats and monkeys. II. Cerebellar projections compared with those of the ascending somatic sensory pathways in lateral diencephalon, J. Comp. Neurol. **220:**229, 1983.

Casey, K.L., and Morrow, T.J.: Ventral posterior thalamic neurons differentially responsive to noxious stimulation of the awake monkey, Science **221:**675, 1983.

Constanzo, R.M., and Gardner, E.P.: A quantitative analysis of responses of direction-sensitive neurons in somatosensory cortex of awake monkeys, J. Neurophysiol. **43:**1319, 1980.

Duclaux, R., Schafer, K., and Hensel, E.: Response of cold receptors to low skin temperatures in nose of the cat, J. Neurophysiol. **43:**1571, 1980.

Dunn, R.C., Jr., and Tolbert, D.L.: The corticotrigeminal projection in the cat: a study of the organization of cortical projections to the spinal trigeminal nucleus, Brain Res. **240:**13, 1982.

Dykes, R.W.: Parallel processing of somatosensory information: a theory, Brain Res. Rev. **6:**47, 1982.

Fields, H.L.: Anatomy and physiology of a nociceptive modulatory system, Phil. Trans. R. Soc. Lond. B **308:**361, 1985.

Gebhart, G.F., et al.: Quantitative comparison of inhibition in spinal cord of nociceptive information by stimulation in periaqueductal gray or nucleus raphe magnus of the cat, J. Neurophysiol. **50:**1433, 1983.

Gebhart, G.F., et al.: Inhibition in spinal cord of nociceptive information by electrical stimulation and morphine microinjection at identical sites in midbrain of the cat, J. Neurophysiol. **51:**75, 1984.

Hoffman, D.S., et al.: Neuronal activity in medullary dorsal horn of awake monkeys trained in a thermal discrimination task. I. Responses to innocuous and noxious thermal stimuli. J. Neurophysiol. **46:**409, 1981.

Honda, C.N., et al.: Neurons in ventrobasal region of cat thalamus selectively responsive to noxious mechanical stimulation, J. Neurophysiol. **49:**662, 1983.

Huerta, M.F., et al.: Studies of the principal sensory and spinal trigeminal nuclei of the rat: projections to the superior colliculus, inferior olive, and cerebellum, *J. Comp. Neurol.* **220:**147, 1983.

Kaas, J.H., et al.: The somatotopic organization of the ventroposterior thalamus of the squirrel monkey, *Saimiri sciureus, J. Comp. Neurol.* **226:**111, 1984.

Khachaturian, H., et al.: Anatomy of the CNS opioid systems, Trends Neurosci. **8:**111, 1985.

Kenshalo, D.R., Jr., and Isensee, O.: Responses of primate SI cortical neurons to noxious stimuli, J. Neurophysiol. **50:**1479, 1983.

Kniffki, K.-D., and Mizumura, K.: Responses of neurons in VPL and VPL-VL region of the cat to algesic stimulation of muscle and tendon, J. Neurophysiol. **49:**649, 1983.

Kosar, E., and Hand, P.J.: First somatosensory cortical columns and associated neuronal clusters of nucleus ventralis posterolateralis of the cat: an anatomical demonstration, J. Comp. Neurol. **198:**515, 1981.

Snyder, S.H., and Childers, S.R.: Opiate receptors and opioid peptides, Annu. Rev. Physiol. **34:**315, 1979.

Snyder, S.H.: Drug and neurotransmitter receptors in the brain, Science **224:**22, 1984.

Spreafico, R., et al.: Cortical relay neurons and interneurons in the N. Ventralis Posterolateralis of cats: a horseradish peroxidase, electron-microscopic, Golgi and immunocytochemical study, Neuroscience **9:**491, 1983.

Steriade, M., and Deschenes, M.: The thalamus as a neuronal oscillator, Brain Res. Rev. **8:**1, 1984.

Sur, M., et al.: Modular distribution of neurons with slowly adapting responses in area 3b of somatosensory cortex in monkeys, J. Neurophysiol. **51:**724, 1984.

Tanji, J., and Wise, S.P.: Submodality distribution in sensorimotor cortex of the unanesthetized monkey, J. Neurophysiol. **45:**467, 1981.

Woolf, C.J.: Evidence for a central component of post-injury pain hypersensitivity, Nature **306:**686, 1983.

Yokota, T., and Matsumoto, N.: Somatotopic distribution of trigeminal nociceptive specific neurons within the caudal somatosensory thalamus of cat, Neuroscience Letters **39:**125, 1983.

Books and monographs

Bonica, J.J., editor: Pain, research publications: Association for Research in Nervous and Mental Disorders, vol. 58, New York, 1980, Raven Press.

Brown, A.G.: Organization in the spinal cord: the anatomy and physiology of identified neurones, Berlin, 1981, Springer.

Burgess, P.R., and Perl, E.R.: Cutaneous mechanoreceptors and nociceptors. In Iggo, A., editor: Handbook of sensory physiology, somatosensory system, vol. II, New York, 1973, Springer-Verlag New York, Inc.

Darian-Smith, I.: The trigeminal system. In Iggo, A., editor: Handbook of sensory physiology, somatosensory system, vol. II, New York, 1973, Springer-Verlag New York, Inc.

Hardy, J.D., Wolff, H.G., and Goodell, H.: Pain sensations and reactions, Baltimore, 1952, Williams & Wilkins.

Kaas, J.H., et al.: Organization of somatosensory cortex in primates. In Schmitt, F.O., et al.: editors: The organization of the cerebral cortex, Cambridge, Mass., 1981, The MIT Press.

Luttinger, D., et al.: Peptides and nociception. In Smythies, J.R., and Bradley, R.J. editors: International Review of Neurobiology, vol. 25, New York, 1984, Academic Press, Inc., p. 185.

Oliverio, A., et al.: Psychobiology of opioids. In Smythies, J.R., and Bradley, R.J., editors: International review of neurobiology, vol. 25, New York, 1984, Academic Press, Inc., p. 277.

Porter, R., and O'Connor, M., editors: (1982) Substance P in the nervous system, Ciba Foundation Symposium 91, Pitman, 1982, London.

Willis, W.D.: The pain system: the neural basis of nociceptive transmission in the mammalian nervous system, vol. 8, Pain and headache, Gildenberg, P.L., (series editor), Basel, 1985, Karger.

Woolsey, C.N., editor: Cortical sensory organization, vol. I, Multiple somatic areas, Clifton, N.J., 1981, Humana Press.

Zieglgansberger, W.: Opioid actions on mammalian spinal neurons. In Smythies, J.R., and Bradley, R.J., editors: International review of neurobiology, vol. 25, New York, 1984, Academic Press, Inc., p. 243.

CHAPTER

10

The Auditory System

The sense of hearing enables us to detect and analyze minute pressure changes that propagate as waves through air (or other media). By a complex but elegant series of events, these pressure changes are transduced into neural signals. Before this process is described, however, some salient features of the physical nature of sound will be reviewed briefly.

Sound Waves

Sounds can be characterized as changes in pressure with time. These changes can be described by waveforms that are often complex. Each waveform uniquely describes a particular sound. For example, Fig. 10-1 depicts a waveform for a note (the C below middle C) played on a piano. For reasons given later, we consider the basic components of sound, called *pure tones,* to be sinusoidal changes in pressure with time (Fig. 10-2). Each pure tone can be completely characterized by three parameters: *amplitude, temporal frequency,* and *temporal phase* (Fig. 10-2).* The amplitude is a measure of the peak-to-trough pressure change. The frequency is the number of cycles of the sine wave that occurs per unit time and is usually expressed in cycles per second, or hertz (Hz). The phase refers to the absolute temporal position of the sine wave; two otherwise equal but out-of-phase tones are shown in Fig. 10-2. Phase is expressed in fractions of a cycle, or in degrees, where 360 degrees equals one full cycle. Thus the two tones of equal frequency and amplitude in Fig. 10-2 are 90 degrees out of phase with one another.

*In a more general sense, a fourth parameter is needed to characterize sine waves; this is the mean or "d.c." level, which is the mean level of the sine wave, or more simply, the average value of a peak and trough of the sinusoid. In sound waves, however, the d.c. level is the ambient air pressure and is thus not a crucial variable under most conditions.

Fourier produced a general theorem that can be applied directly to sound and explains why pure tones are considered to be the basic components of sound. He stated that *any* complex waveform can be synthesized by the linear addition of sine waves that have been appropriately selected for amplitude, frequency, and phase. *Fourier synthesis* is the construction of complex waveforms from sine waves. The converse, *Fourier analysis* is the description of complex waveforms in terms of their component sine waves. Fig. 10-3 illustrates the principles of Fourier analysis and synthesis. A relatively simple waveform (a triangle wave) can be created by the linear addition of suitably chosen sine waves (Fig. 10-3, *A*). The same set of sine waves, if shifted in phase, can be used to create a square wave (Fig. 10-3, *B*). For square waves the peaks of one component sine wave are aligned with the troughs of the next higher frequency component, whereas for triangle waves the peaks of all components are aligned. This illustrates the importance of phase in Fourier synthesis. Finally, Fig. 10-3, *C* depicts the synthesis of a more arbitrary waveform from a more complicated set of constituent sine waves. (For a more complete description of Fourier's theorem and its application, suitable mathematical texts should be consulted.) Fourier thus emphasized the elemental nature of sine waves.

From the preceding description it follows that any sound wave, no matter how complicated a function of pressure versus time, can be synthesized and analyzed in terms of pure tones, which are sinusoidal functions of pressure versus time. Fig. 10-4 demonstrates this for the sound illustrated in Fig. 10-1. Fig. 10-4, *B,* shows its "Fourier spectrum," or the relative amplitude of the pure tones that make up this sound. When phase information is added, the Fourier spectrum is a useful and concise means of characterizing a sound.

This analytical approach allows us to determine how a sound will be changed by adding or subtracting certain tones. For instance, an audio amplifier typically

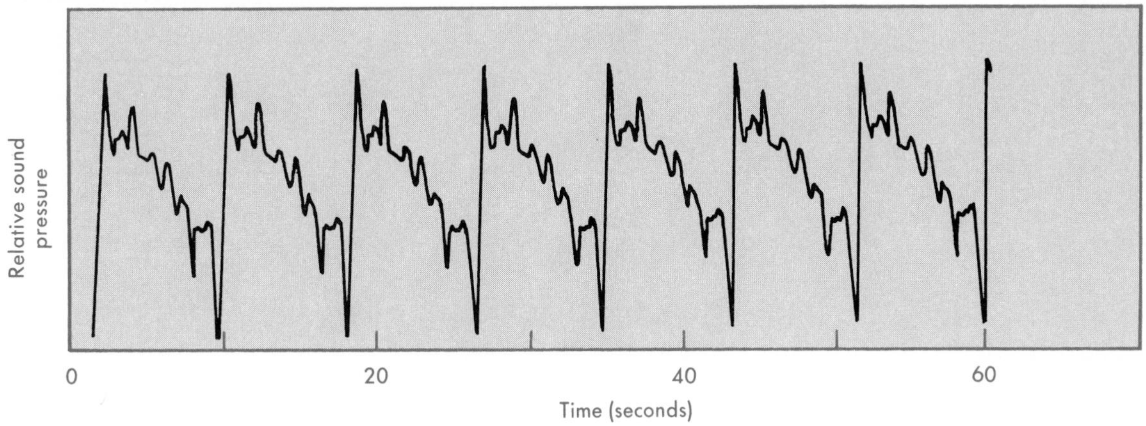

Fig. 10-1. Pressure versus time waveform representing the sound created by striking the C below the middle C on a piano. (Redrawn from Cornsweet, T.N.: Visual perception, New York, 1970, Academic Press, Inc.)

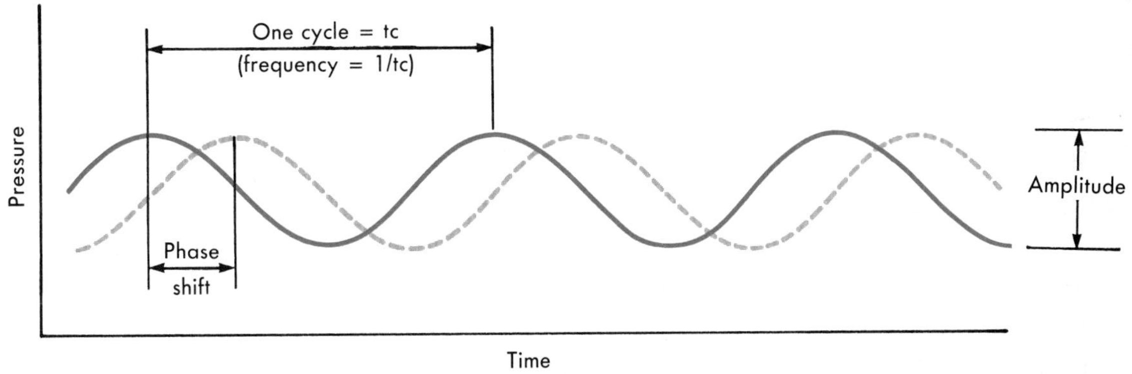

Fig. 10-2. Representation of two pure tones 90 degrees out of phase with one another. A pure tone is a sinusoidal change in pressure with time. The parameters needed to characterize such a tone are the temporal frequency (the inverse of the time needed to encompass one full cycle), the amplitude of the pressure change, and the temporal phase.

does not reproduce all frequencies equally, and a particular amplifier might be incapable of reproducing tones of less than 20 Hz or more than 20,000 Hz. If sounds (e.g., music) containing frequencies under 20 or over 20,000 Hz were to be played through such an amplifier, the following could be done. Fourier analysis of the original would show the component tones lower than 20 Hz and higher than 20,000 Hz; these could be removed. Fourier synthesis of the remaining tones would then precisely match the sound reproduced by the amplifier.

In human hearing, all frequencies are not equally perceived. The unit of sound pressure commonly used to measure hearing is the decibel (dB). The number of decibels for a given sound pressure (P) equals 20 log P/P_R, where P_R is a reference pressure. P_R most often is 0.002 dyne/cm^2 (the threshold for human hearing), but it is sometimes considered to be 1 dyne/cm^2. Fig. 10-5 shows the approximate range of human hearing as a function of frequency. The darkly colored region represents sounds that are heard. Below this *(gray)* sounds are too faint to be detected; and above this *(lightly colored)* the pressure changes are so powerful that they are felt and thus sensed by somesthesia.

Human hearing is most sensitive at roughly 3,000 Hz, and it rapidly deteriorates for higher and lower frequencies. Fourier's theorem as applied to Fig. 10-5 enables one to predict which components of a complex sound can actually be heard and how they would sound. For instance, the predominant frequency for the tone shown in Fig. 10-1 is slightly above 100 Hz, and its amplitude is nearly five times greater than that of components near 3,000 Hz. However, humans detect tones near 100 Hz nearly 40 dB (or 100 times) worse than they do those near 3,000 Hz. Thus this major component of the note C below middle C is not perceived nearly as well as components nearer 3,000 Hz. Human auditory perception distorts actual sounds in a predictable fashion. One can predict this distortion simply by

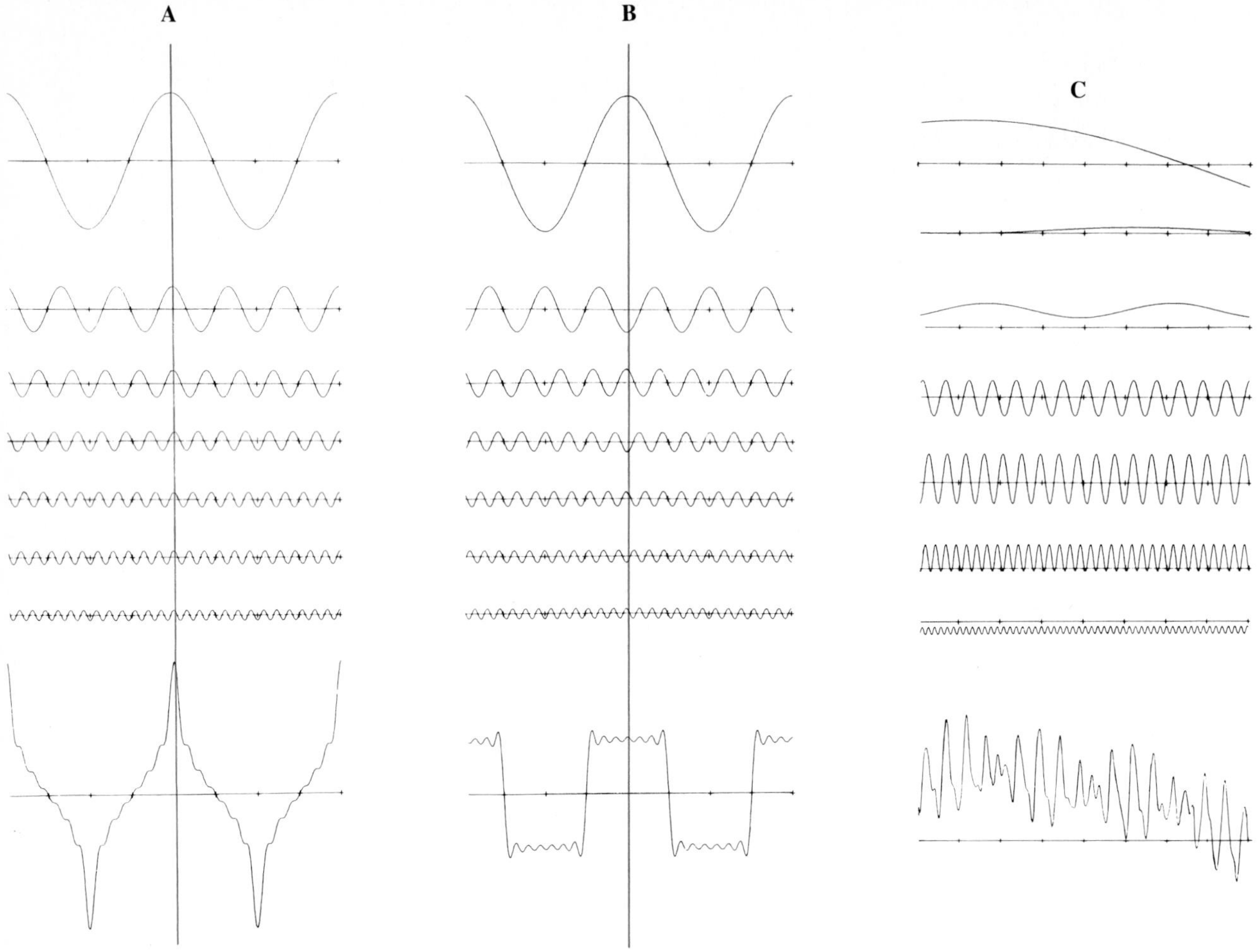

■ **Fig. 10-3.** Examples of Fourier analysis and synthesis. At the bottom of each of the three columns is a complex waveform synthesized from the linear addition of seven different sine waves. Modulation amplitude (a measure of the vertical peak-to-trough distance), mean amplitude (the mean of the peak and trough amplitudes), and frequency (number of cycles per unit along the abscissa) of the waveforms are in arbitrary units. **A,** Synthesis of an approximately triangular waveform from sine waves. All of these sine waves have the same mean amplitude. If the top sine wave has a modulation amplitude of *A* and a frequency of *F*, then the succeeding ones have respective modulation amplitudes and frequencies of A/3 and 3F, A/5 and 5F, A/7 and 7F, A/9 and 9F, A/11 and 11F, and A/13 and 13F. A precise triangular waveform would be formed from the continued addition of odd sine wave components with the modulation amplitude and frequency of the nth component being A/n and nF. Note that the peaks and troughs of the top (lowest frequency) sine wave are aligned with peaks and troughs of all of the higher frequency components *(colored line)*. This establishes the phase relationships among the component sine waves. **B,** Synthesis of an approximately square waveform. Note that the only difference between this and the triangular waveform in **A** lies in the phase relationships among the component sine waves. Here, the peaks (and troughs) of each sine wave are aligned with troughs (and peaks) of the next higher frequency sine wave component *(colored line)*. Phase relationships are thus important to Fourier synthesis and analysis. **C,** Synthesis of an arbitrary waveform from component sine waves that differ in modulation amplitude, mean amplitude, frequency, and phase. Given enough sine waves with appropriate values of amplitude, frequency, and phase, any arbitrary waveform can be synthesized. (From Sherman, S.M.: In Sprague, J.M., and Epstein, A.N., editors: Progress in psychobiology and physiological psychology, vol. 2, New York, 1985, Academic Press, Inc.)

A
B

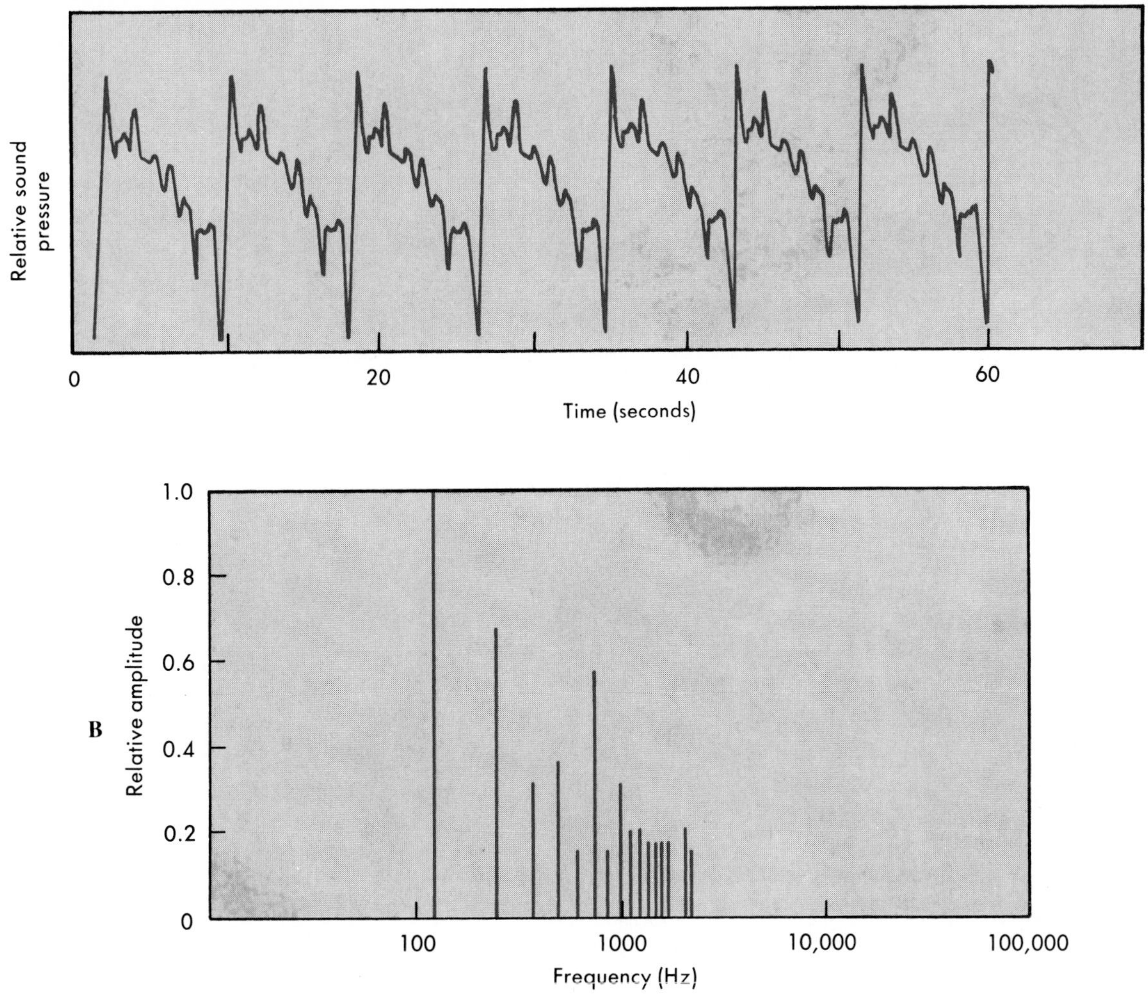

■ **Fig. 10-4.** A complex sound and its Fourier spectrum. **A,** Waveform of same sound as depicted in Fig. 10-1. **B,** Fourier spectrum of the sound showing the relative strength of each of the component pure tones. (From Cornsweet, T.N.: Visual perception, New York, 1970. Academic Press, Inc.)

multiplying a sound's Fourier spectrum (Fig. 10-4, *B*) by the relative sensitivity curve of Fig. 10-5 and then applying Fourier's theorem to synthesize the resultant product.

■ *The Auditory Periphery*

The three major divisions of the ear are the *external* or *outer ear,* the *middle ear,* and the *internal* or *inner ear* (Fig. 10-6). Actual neural transduction occurs in the inner ear, but the nature of the stimulus is significantly transformed before reaching the transducers.

■ *External or Outer Ear*

The external or outer ear consists of the *pinna* and *external meatus,* or *auditory canal*. The pinna has a complicated shape that probably directs sound waves toward the auditory canal. The pinna may also be important in localizing the source of sounds. The human pinna is relatively immobile, but most animals can move the pinna toward the source of sound and thereby improve sensitivity. The auditory canal is an air-filled tube down which sound waves must travel to reach the middle ear. The shape of the canal provides it with a resonant frequency of approximately 3,500 Hz, but this is broadly tuned (roughly 800 to 6,000 Hz) because the canal surfaces are elastic. Frequencies nearer the resonant frequency arrive at the middle ear with less attenuation than do higher and lower ones. Much of the differential sensitivity of human hearing is thus derived from the shape and physical characteristics of the auditory canal.

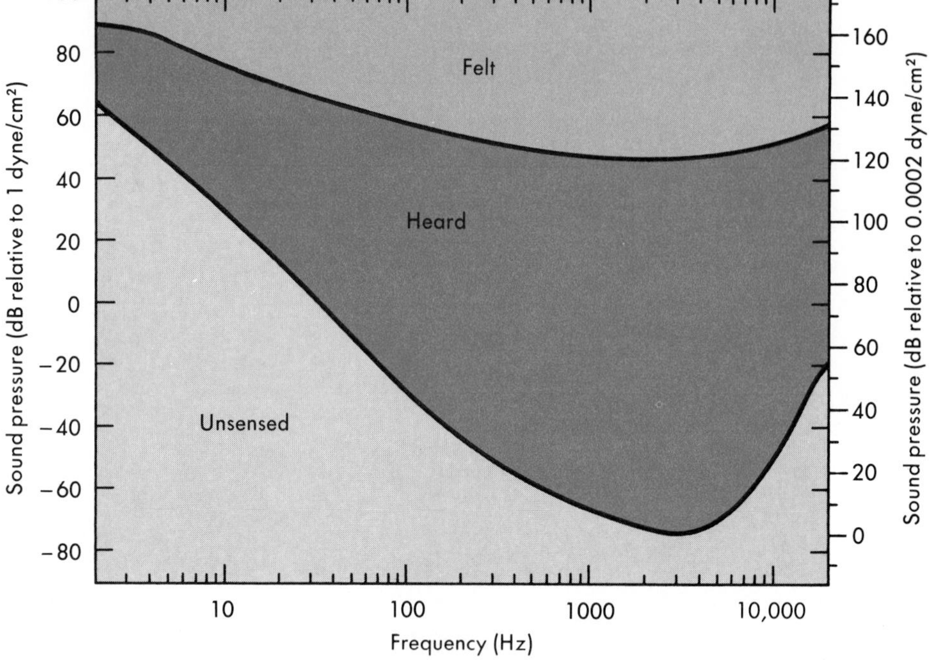

Fig. 10-5. Range of human hearing *(darker color)*. Above this range *(lighter color)* sound pressure waves are sufficiently intense to be felt with somethesia. Below this range sound pressure waves are too weak to be sensed.

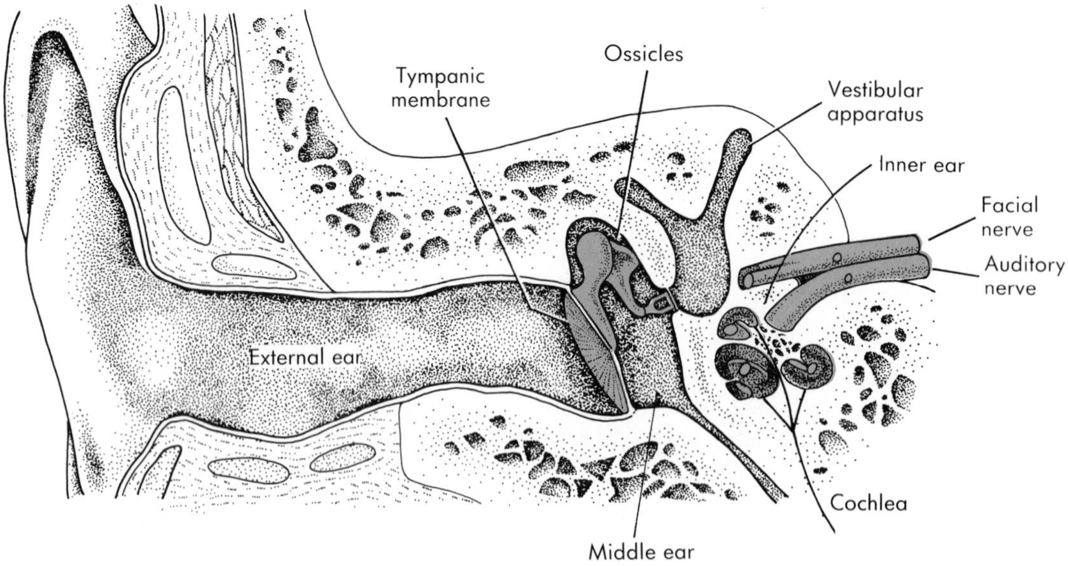

Fig. 10-6. Major divisions of the human ear. Shown in color are the tympanic membrane and ossicles of the middle ear plus the cochlea and auditory nerve of the inner ear. (Redrawn from von Békésy, G.: Sci. Am. **197:**66, 1957.)

A

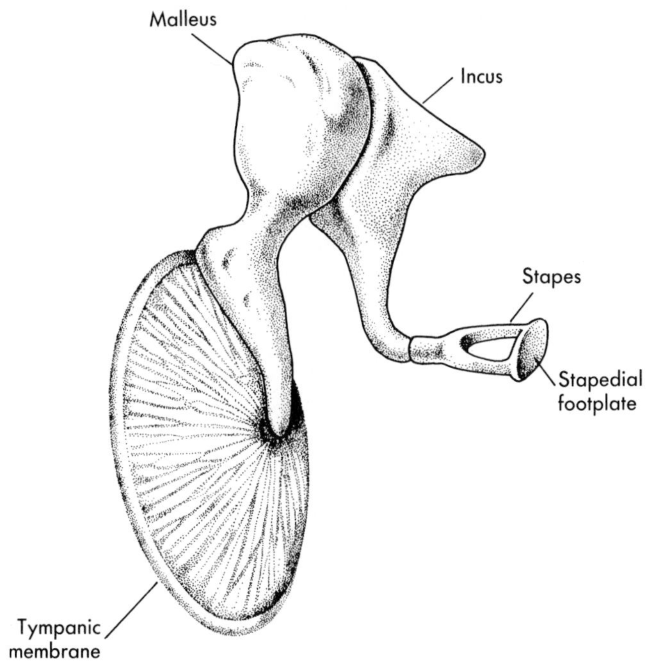

B

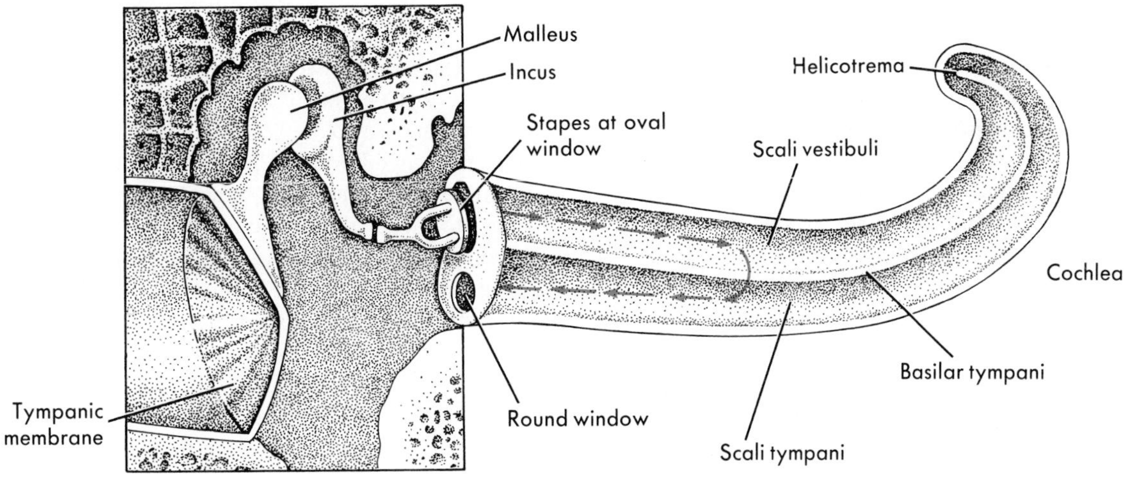

Fig. 10-7. Middle ear and its relationship to the cochlea. **A,** Tympanic membrane (eardrum), malleus (hammer), incus (anvil), and stapes (stirrup). The stapedial footplate rests on the membrane covering the oval window. **B,** Relationship of ossicles and cochlea. The liquid-filled, bony cochlea has elastic membranes covering two apertures, the round and oval windows. Movement of the stapedial footplate causes displacement of the oval window, resulting in displacement of the liquids *(arrows)* and round window. In the process the basilar membrane is displaced. (**A** redrawn from von Békésy, G.: Sci. Am. **197:**66, 1957; **B** redrawn from Kandel, E., and Schwartz, J.H.: Principles of neural science, New York, 1981, Elsevier North-Holland, Inc.)

Middle Ear

The middle ear consists of the *tympanic membrane* (or *eardrum*) and three bones, or ossicles, known as the *malleus* (or *hammer*), the *incus* (or *anvil*), and the *stapes* (or *stirrup*) (Fig. 10-7, *A*). The rest of the middle ear is air filled. The ossicles are firmly attached to one another and move as a unit. The base of the malleus is attached to the tympanic membrane, and the footplate of the stapes is attached to a membrane covering an opening in the bony shell of the inner ear. This opening is called the *oval window*. Its membranous covering separates the air-filled middle ear from the liquid-filled cochlea (Fig. 10-7, *B*).

Sound waves that travel down the outer ear impinge on the tympanic membrane and cause it to vibrate at an amplitude and frequency that correspond to the sound itself. These vibrations are transmitted to the ossicles, which vibrate about an axis running near the juncture of the malleus and incus. This, in turn, causes the footplate of the stapes to vibrate against the oval window, and the sound waves are thereby transmitted to the liquids of the inner ear.

What function does the middle ear serve? What

would happen if the airborne sound waves bypassed the middle ear and directly struck the membrane covering the oval window? The answers to these questions lie in the fact that the *airborne* sound waves in the auditory canal must pass to the *watery* liquids of the inner ear. The sound-conducting properties of air and water are quite different. A complicated parameter known as *acoustic impedance* describes much of a medium's sound-conducting qualities. (Acoustic impedance is determined by the amplitude of the sound pressure variations divided by the volume velocity of the sound waves; volume velocity is the product of the velocity of the sound wave and the area through which it passes.) The acoustic impedance of water is considerably higher than that of air. When such an impedance mismatch occurs, most of the sound traveling from air to water would be reflected off the surface back into the air; very little would enter the water. More specifically, the ratio of transmitted energy (E_t) to incident energy (E_i) can be specified as

$$\frac{E_t}{E_i} = \frac{4r}{(r + 1)^2}$$

where *r* is the ratio of impedances. In the case of an air-to-water interface, the impedance mismatch would result in reflection of 99.9% of the sound energy. Because sound energy is proportional to the square of the sound pressure, this represents a 30 dB loss. Indeed such a loss would result if there were no middle ear and if the tympanic membrane were applied directly to the fluids of the inner ear.

The middle ear provides a much better impedance match between the outer and inner ears so that much more of the sound energy is transmitted. In humans only about 10 to 15 dB are lost. The ossicles of the middle ear perform this impedance-matching function. They act as a transformer that increases the pressures from the tympanic membrane to the oval window. Two factors subserve this function. First, the area of the tympanic membrane is much greater than that of the oval window. Hence a force applied to the malleus and transmitted to the stapedial footplate exerts more pressure at the oval window than that which originated at the tympanic membrane. Second, the ossicular chain moves as a lever that rotates around the malleus-incus junction. The lever action is such that the stapedial footplate moves less and, consequently, exerts more force than do movements of the malleus caused by the tympanic membrane. Again this amplifies sound pressure between the tympanic membrane and oval window.

The treatment of *otosclerosis* in the recent past illustrates the function of the ossicles. Otosclerosis is a condition in which pathological bone growth around the ossicles tends to cement or freeze them against the skull (Fig. 10-8). In such a condition no vibrations or sound can be transmitted from the outer to the inner ear. This form of deafness was previously alleviated by a *fenestration operation* (Fig. 10-8, *C*). The ossicles were removed or disconnected from the tympanic membrane, and an artificial membrane-covered opening in the cochlea was made to replace the immobile oval window. Sounds could then be transmitted to the cochlear fluids, but without the benefit of the ossicular impedance matching. The restored hearing carried a 15 to 20 dB loss. However, given the large dynamic range of human hearing (Fig. 10-5), this loss is compatible with useful hearing. Fenestration operations are now rarely needed. Instead otosclerosis is typically treated with artificial ossicular replacement and hearing is restored to almost normal levels.

Internal or Inner Ear

The inner ear consists of the cochlea as well as organs for vestibular sensation. The vestibular system is treated in Chapter 11, and only the cochlea is considered here.

Cochlear structure. Fig. 10-9 shows the complicated structure of the cochlea. It is a bony labyrinth filled with liquids and various structural elements. (See Fig. 11-1 for the relationship between the cochlear and vestibular labyrinths in the inner ear.) The *oval* and *round windows* are two membrane-covered openings in the bone that permit displacement of the cochlear liquids in response to motion of the stapedial footplate. Like most liquids, those in the cochlea are practically incompressible. Therefore they could not move in response to sound stimuli without these windows. Among the structures inside the cochlea are the *basilar membrane* and *organ of Corti* (Fig. 10-9). The organ of Corti rests on the basilar membrane and contains the transducers, which are known as *hair cells*. The cochlea is tightly coiled into a spiral. The fluids in the cochlea are divided among three compartments (Fig. 10-9): the *scala vestibuli* is separated from the scala media by *Reissner's membrane;* the *scala media* lies between the basilar and Reissner's membranes, and it contains the organ of Corti; the final chamber is the *scala tympani*.

Before proceeding with the events leading to transduction we will consider a simple model of the cochlear plan (Fig. 10-10). The cochlear model begins with a liquid-filled glass container covered at each end by an elastic membrane (Fig. 10-10, *A*). Downward or upward movement of the top membrane (the "oval window") causes displacement of the liquid and a similar movement of the bottom membrane (the "round window"). Fig. 10-10, *B* adds the stapes (to vibrate against the oval window) and the basilar membrane. The basilar membrane moves in phase with the oval and round windows, and it divides the fluid into two compartments, the scala vestibuli and scala tympani.

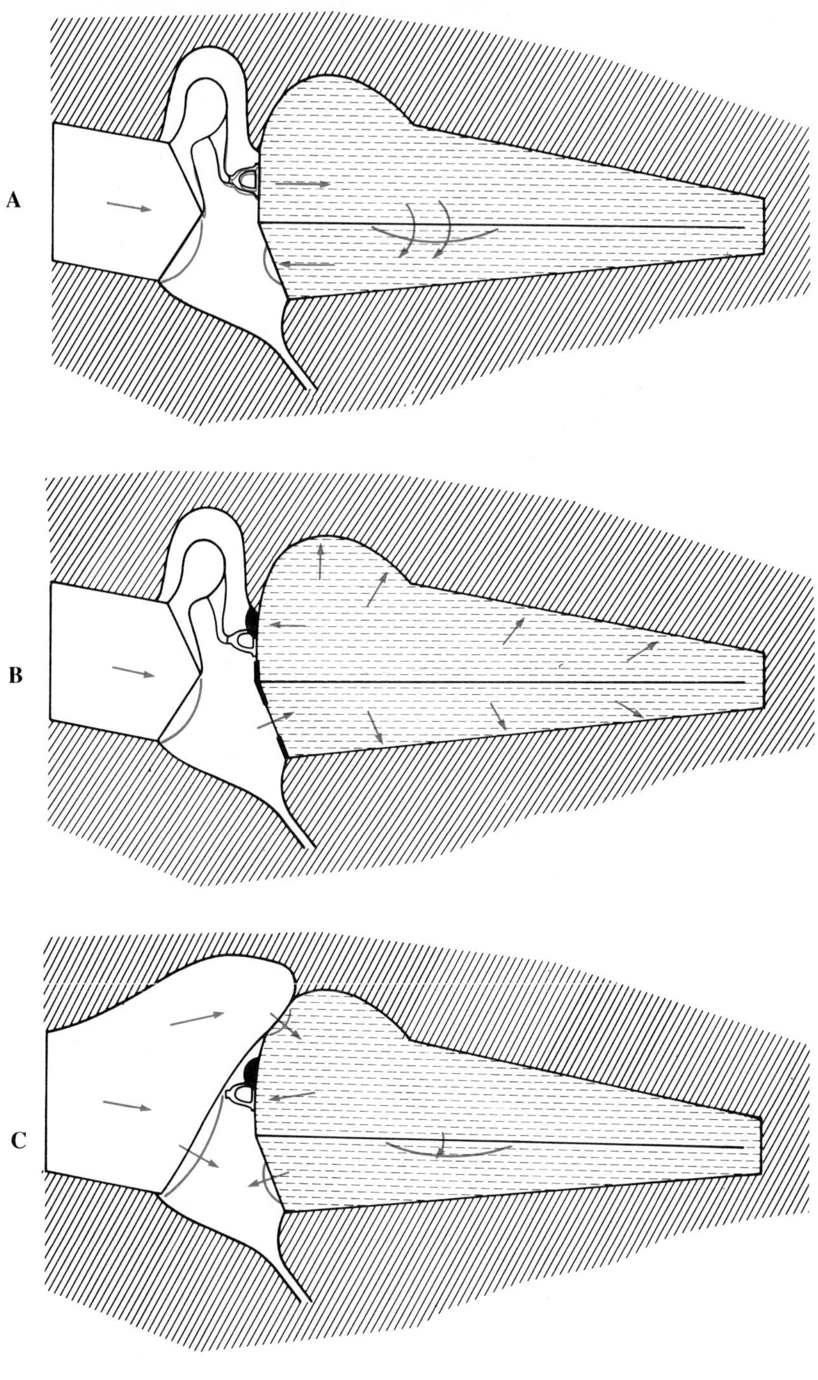

■ **Fig. 10-8.** Otosclerosis and its alleviation by a fenestration operation. **A,** Normal ear. Displacements *(color)* caused by sound waves are transmitted via the middle ear to the basilar membrane in the cochlea. **B,** Otosclerosis. Pathological bone formation *(black)* freezes the stapedial footplate and oval window so that they cannot move in response to tympanic membrane displacement. Because the cochlear liquids are incompressible, no round window or basilar membrane movement is possible either. **C,** Fenestration operation. An extra aperture is created in the cochlear wall exposed to the outer ear, and the aperture is covered with a flexible membrane. Sound waves can now cause displacement of the basilar membrane, although the impedance-matching quality of the middle ear is lost. (Redrawn from von Békésy, G.: Sci. Am. **197:**66, 1957.)

The hair cells and fibers are added to the basilar membrane in Fig. 10-10, *C.* Later we shall see how vibrations of the basilar membrane affect the hair cells. Also shown is the *helicotrema,* which is a small gap in the basilar membrane. It permits the liquids in the scala vestibuli and scala tympani to be continuous. Furthermore, the helicotrema allows the basilar membrane to return to its resting position in the presence of a constant displacement of the oval and round windows. In other words, it dampens out low-frequency movements of the liquids so that only high-frequency displacements (i.e., in the range of sound frequencies) cause appreciable vibration of the basilar membrane. A static change in air pressure is an example of a low-frequency stimulus that the helicotrema filters out as a potential auditory stimulus. In Fig. 10-10, *D,* the cochlea is stretched along a dimension parallel to the basilar membrane, and the cochlea is coiled in Fig. 10-10, *E.* The human cochlea is approximately 35 mm in length and contains slightly more than 2½ turns. For simplicity the scala media is not shown in Fig. 10-10. The scala media is a self-contained, membrane-bound chamber in which the organ of Corti is located (Fig. 10-9). Fig. 10-10, *E,* is a fairly accurate model of the mammalian cochlea, and

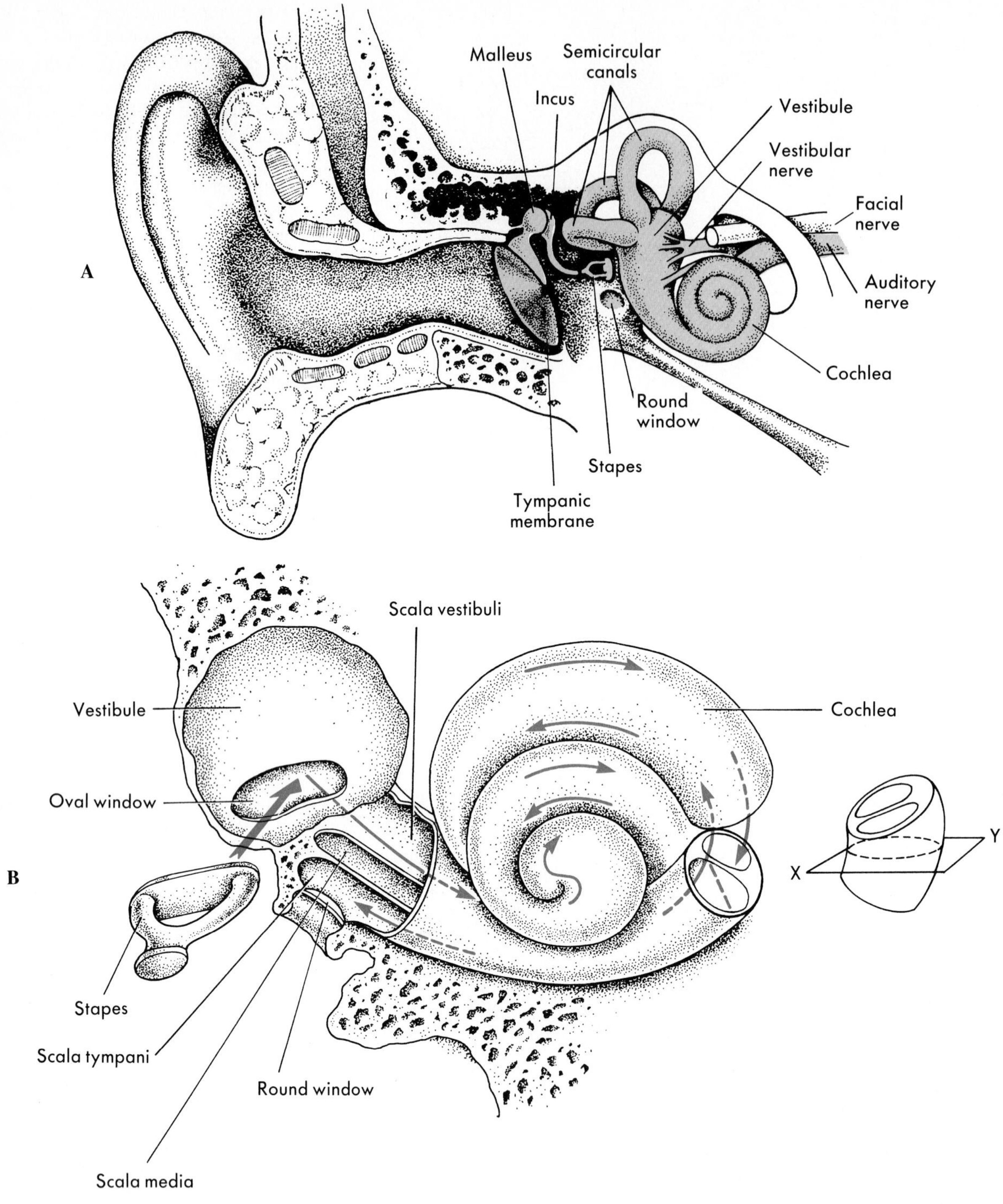

Fig. 10-9. Structure of the right cochlea. **A,** Location of the cochlea within the head. **B,** Coiled structure of cochlea showing the relationships among the oval window, scala vestibuli, scala media, and scala tympani. The arrows indicate the path of continuous fluid flow from the scala vestibuli through the helicotrema to the scala tympani. The small drawing at the right illustrates the plane of section for **C. C,** Cross section through the cochlea. The *X* and *Y* indicate the orientation vis-a-vis the drawing on the right in **B.** The hair cells and sensory fibers are shown in color. The dashed rectangle outlines the area shown in **D. D,** Organ of Corti, including the hair cells and sensory fibers *(color)*. The hair cells have fine hairlike processes embedded in the tectorial membrane. A single inner hair cell and several outer hair cells are separated by the rods of Corti. (**B** redrawn from Gulick, W.L.: Hearing: physiology and psychophysics, New York, 1971, Oxford University Press; after Maloney.)

C

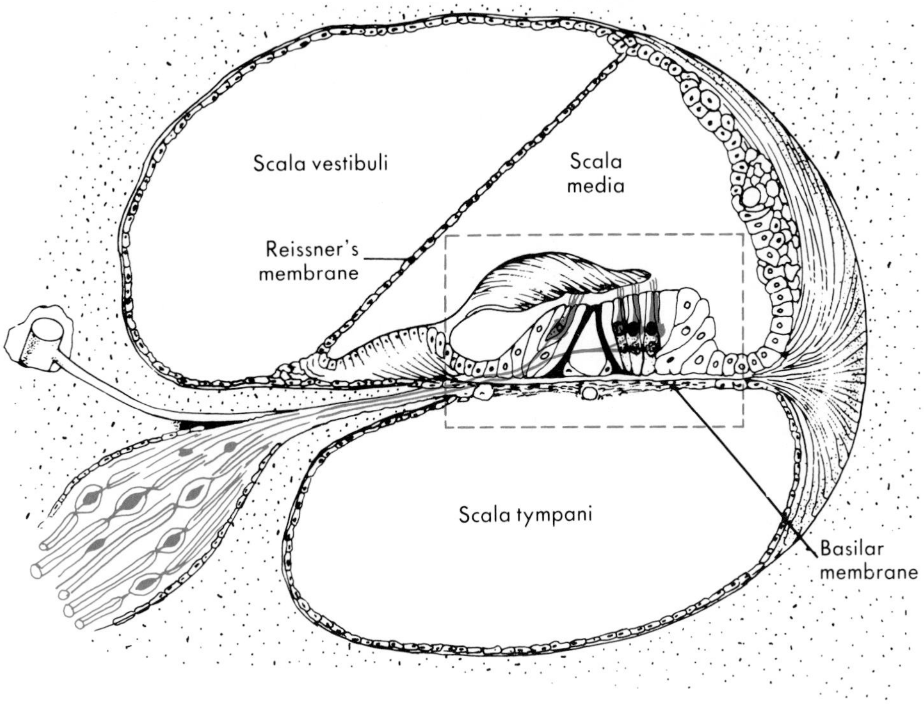
Y
Scala vestibuli
Scala media
Reissner's membrane
Scala tympani
Basilar membrane

D

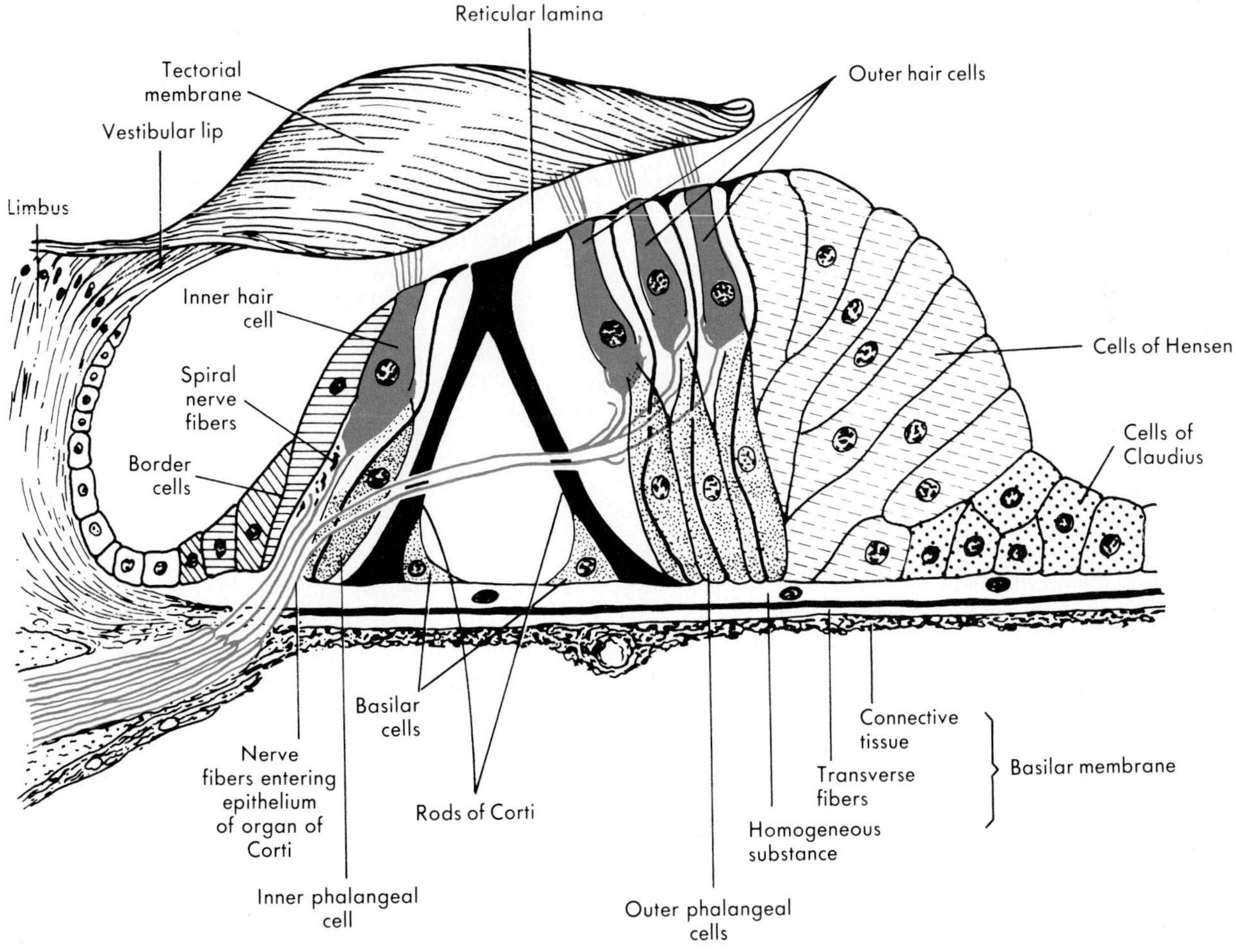
Reticular lamina
Tectorial membrane
Outer hair cells
Vestibular lip
Limbus
Inner hair cell
Cells of Hensen
Spiral nerve fibers
Cells of Claudius
Border cells
Basilar cells
Connective tissue
Nerve fibers entering epithelium of organ of Corti
Transverse fibers
Basilar membrane
Rods of Corti
Homogeneous substance
Inner phalangeal cell
Outer phalangeal cells

Fig. 10-9, cont'd. For legend see opposite page.

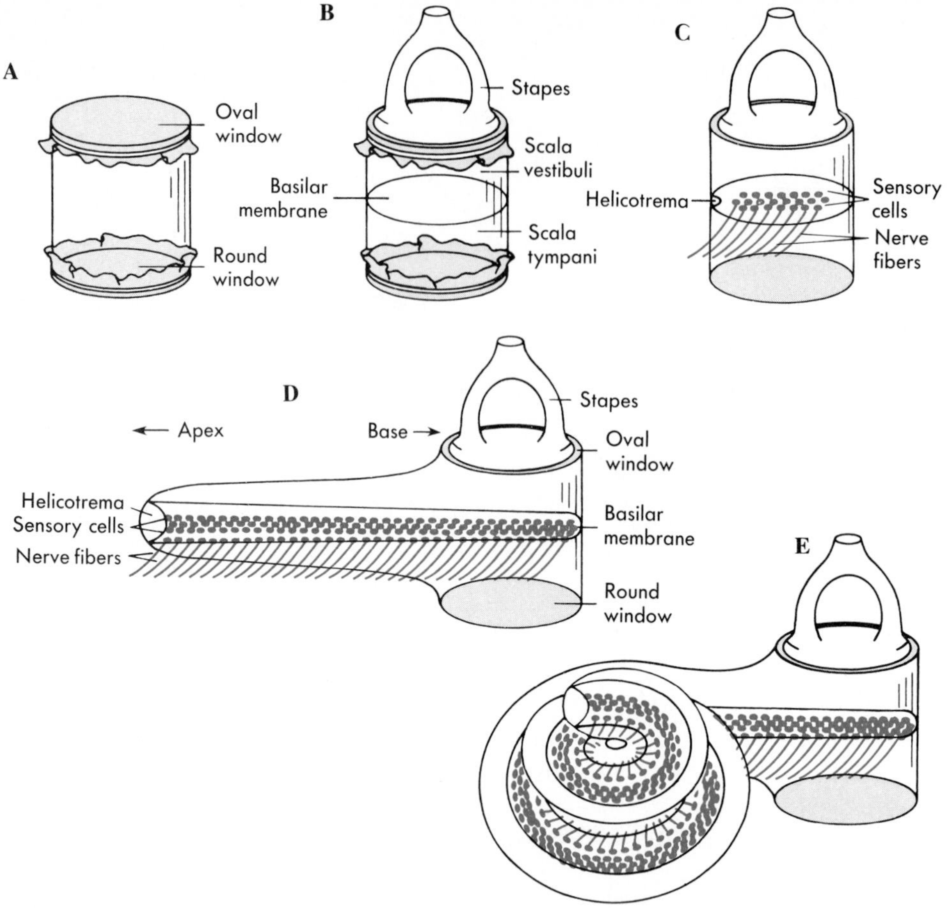

■ **Fig. 10-10.** Simplified schema for the construction of a model cochlea. **A,** Glass cylinder filled with liquid and covered at each end with an elastic membrane. The top covering models the oval window, and the bottom covering models the round window. **B,** Addition of the stapes (to serve as a piston against the oval window) and the basilar membrane stretched across the cylinder. The latter divides the fluids into the scala vestibuli and scala tympani (for simplicity the scala media is not represented in this figure). **C,** Addition of sensory hair cells and auditory nerve fibers *(color)*. Also, a gap in the basilar membrane known as the *helicotrema* is added. **D,** Elongation of the cochlea along axis parallel to the basilar membrane. **E,** Coiling of the cochlea. (Redrawn from Kiang, N.Y.S.: Stimulus representation in the discharge patterns of auditory neurons. In Tower, D.B., editor: The human nervous system, vol. 3, Human communication and its disorders, New York, 1975, Raven Press. © 1975.)

from it one should readily comprehend the manner whereby sound vibrations are transmitted through the ossicles and cause the basilar membrane to vibrate.

Organ of Corti and transduction. Fig. 10-9, *D,* shows the key structural features of the organ of Corti. The hair cells are supported in a rigid structure, the *reticular lamina,* which in turn is supported on stiff triangular structures called the *rods of Corti.* All of this rests on the basilar membrane. The rods of Corti divide the hair cells into a single row of inner hair cells and three rows of outer hair cells. Although the inner and outer hair cells differ morphologically and physiologically, the functional significance of these differences is unknown.

The hair cell (Fig. 10-11) is so named because hairlike cilia protrude from the top of the cell. These are called *stereocilia,* and they gradually increase in length from rows of shorter cilia to rows of longer ones. These cochlear hair cells are similar to the hair cells in the vestibular apparatus, with one conspicuous difference. Vestibular hair cells (described more fully in Chapter 11) also have a single *kinocilium,* which is structurally different from a stereocilium. The kinocilium lies next to the longest stereocilia and is longer than any of the hair cell cilia.

The stereocilia can be deflected by external forces. However, they are fairly rigid and are attached to one another. They thus move as a unit, with a pivot at their basal attachment at the top of the hair cell. Many of the stereocilia are embedded at their distal ends into the *tec-*

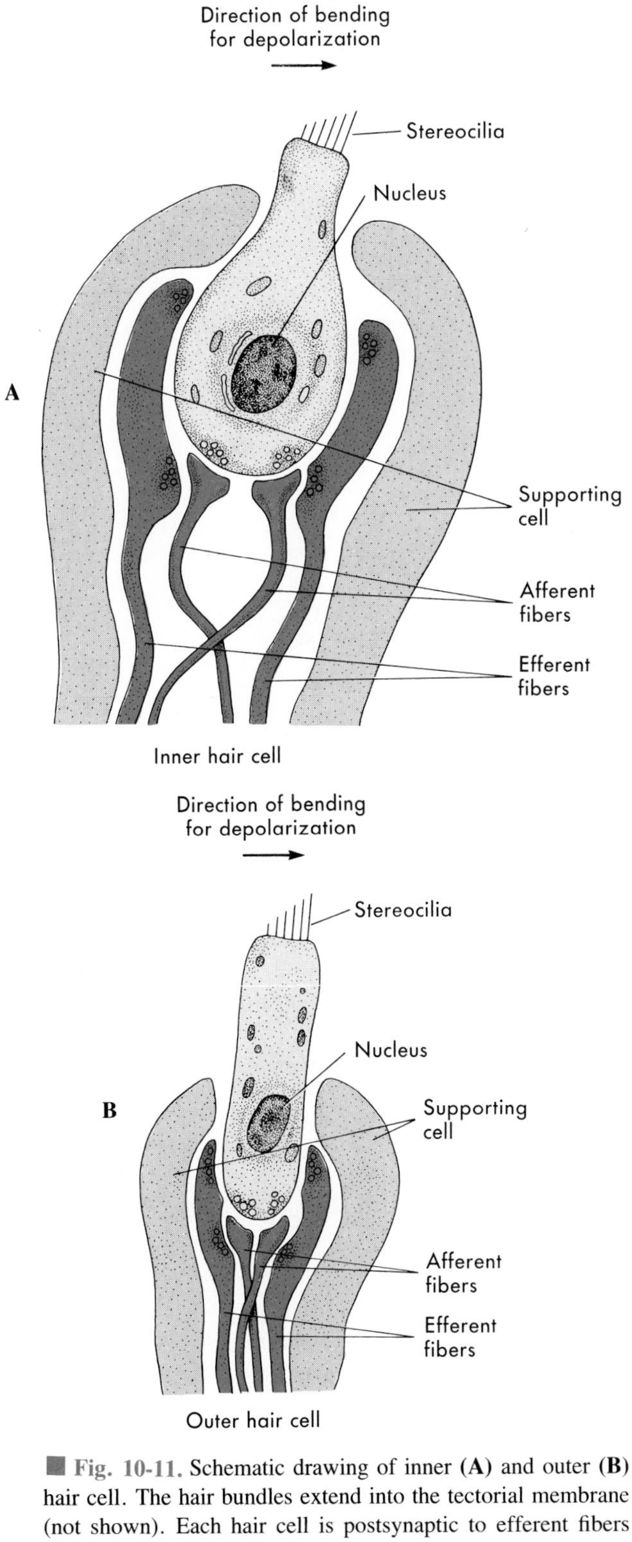

■ **Fig. 10-11.** Schematic drawing of inner **(A)** and outer **(B)** hair cell. The hair bundles extend into the tectorial membrane (not shown). Each hair cell is postsynaptic to efferent fibers *(light color)* from the olivocochlear bundle and presynaptic to afferent fibers *(dark color)* that comprise the auditory nerve. The efferent fibers may also form synapses onto the terminals of the afferent fibers. Note that the hair bundles become gradually longer from left to right, which sets up an axis of polarization *(arrows)*. Bending of the hairs in this direction depolarizes the hair cell, and bending in the opposite direction hyperpolarizes the cell.

torial membrane, which lies just dorsal to the reticular lamina (Fig. 10-9). As the oval and round windows vibrate and thereby displace the basilar membrane, the rods of Corti and reticular lamina pivot as a unit around the base of the inner rod. This creates a shearing force on the stereocilia and that force is transformed to a mechanical deformation of the top of the hair cell (Fig. 10-12). This somehow leads to the generator potential (GP), presumably by altering one or more ionic conductances near the top of the hair cell. Deflections of the stereocilia along the axis from shorter to longer stereocilia depolarize the hair cell, whereas deflections in the opposite direction hyperpolarize the hair cell. Deflections at right angles have little net effect on hair cell polarization, but the anatomy of the organ of Corti greatly minimizes such deflections. For a given displacement of the stereocilia, the depolarization is greater than the hyperpolarization. Thus at the frequencies that exist in most sounds the GP consists of a net depolarization.

At the base of the hair cell can be found a number of synapses formed between the cell and axons of the eighth nerve (Fig. 10-11). The hair cell is presynaptic to many of these axons, which are thus afferent fibers that innervate the brain, and it is postsynaptic to others. The efferent fibers also seem to synapse onto the afferent ones. The hair cell transmits the GP synaptically to these afferent fibers, and if the GP is sufficiently large, action potentials travel toward the brain. As in other systems, the amplitude of the GP, and thus frequency of action potentials, varies directly with the basilar membrane displacement. This displacement, in turn, relates to the amplitude or intensity of the sound waves. In this way sound stimuli produce the coded signals that reach the brain.

A unique feature of the auditory system is the fact that the transducers are under efferent control from the brain. As noted earlier, some of the axons in the organ of Corti form presynaptic terminals onto the hair cells and afferent fibers. These efferent axons run in the *olivocochlear bundle* that orginates with cells in the *superior olivary nucleus* of the pons (see Fig. 10-15). This pathway is predominantly crossed, but it has an ipsilateral component. The function of this efferent pathway may be to affect the hair cell threshold or to contribute to afferent inhibition in auditory nerve fibers.

Tonotopic organization of the basilar membrane. A final important structural feature of the inner ear is its *tonotopic map*. The basilar membrane is not structurally homogeneous from the stapes to the helicotrema; it gradually changes in width, tension, and viscoelastic properties. Consequently each portion of the basilar membrane vibrates with maximum amplitude only to one specific resonant frequency. In other words, a standing wave is set up for each sound frequency within the range of human hearing (Fig. 10-13, *A*). Those

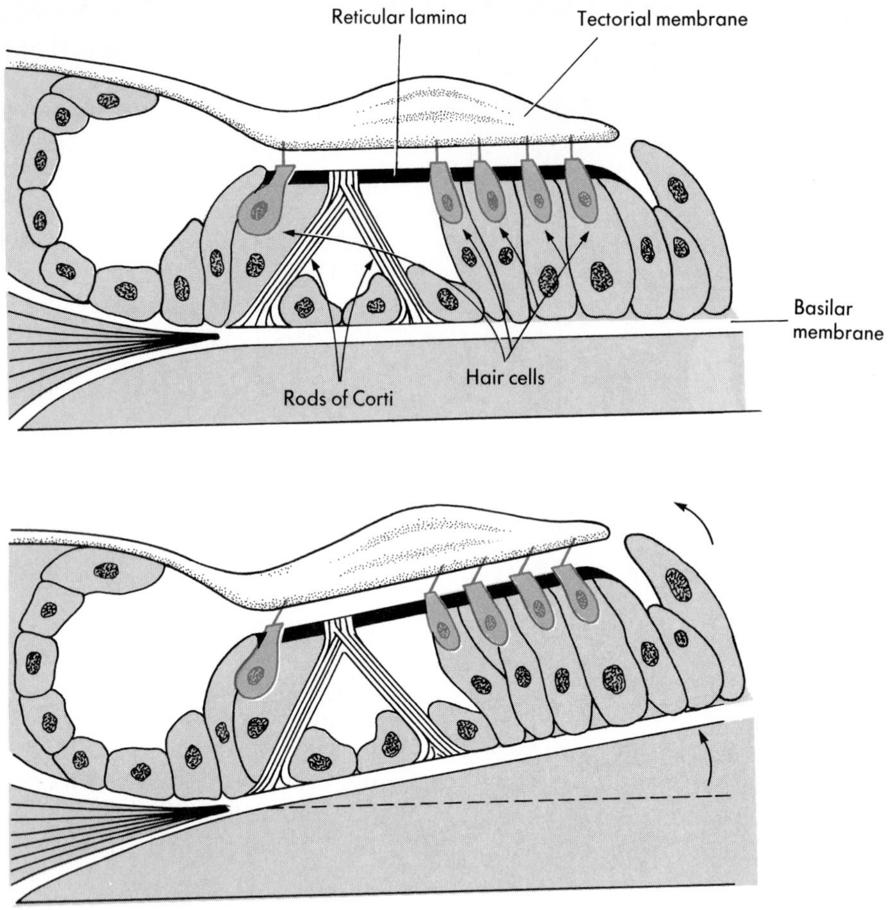

■ **Fig. 10.12.** Effect of basilar membrane displacement on hair cells. Note how upward movement of the membrane causes a shearing force to be exerted on the hairs.

waves for higher frequencies reach peak amplitude closer to the stapes (Fig. 10-13, *B*).

The consequence of this is that each tone causes a limited patch of basilar membrane to vibrate sufficiently to produce large GPs and subsequent action potentials. A tonotopic map of the cochlea is shown in Fig. 10-14. Because of the placement of each hair cell in a limited patch of basilar membrane, each of these cells is "tuned" to a particular sound frequency or tone. It is not clear whether all of the tuning seen in the hair cells can be explained by the tuning of the basilar membrane. Nonmammalian vertebrates, such as turtles, have no appreciable tuning in the basilar membrane but do exhibit considerable tuning in the mechanical and electrical properties of individual hair cells. Likewise mammalian hair cells may possess a degree of mechanical and electrical tuning independent of that in the basilar membrane.

Fig. 10-14 also illustrates a fundamental difference between the auditory system and the visual or somatosensory systems. Receptive fields are defined in terms of the spatial distribution of relevant transducers in the periphery. The functional significance of this for vision and somesthesia is that receptive fields in these systems relate to space on the retina or skin. Auditory receptive fields, however, are *not* organized along a spatial dimension; rather they are organized in terms of tones or frequencies, since the frequency domain is what is mapped by the auditory periphery. Thus while vision and somesthesia are elegant spatial senses, the auditory system is basically organized to discriminate tones and not their source locations. Note that this concept of the auditory receptive field follows from the same definition and principle as applies to visual or somesthetic receptive fields. The spatial distribution of transducers along the cochlea characterizes the frequency distribution of a sound rather than its spatial location, whereas the spatial distributions of transducers along the retina or cutaneous surface characterize the spatial location of the stimulus. Auditory spatial localization is also performed, but the underlying neural processes are more complicated and less basic than those for tonotopic analysis.

A

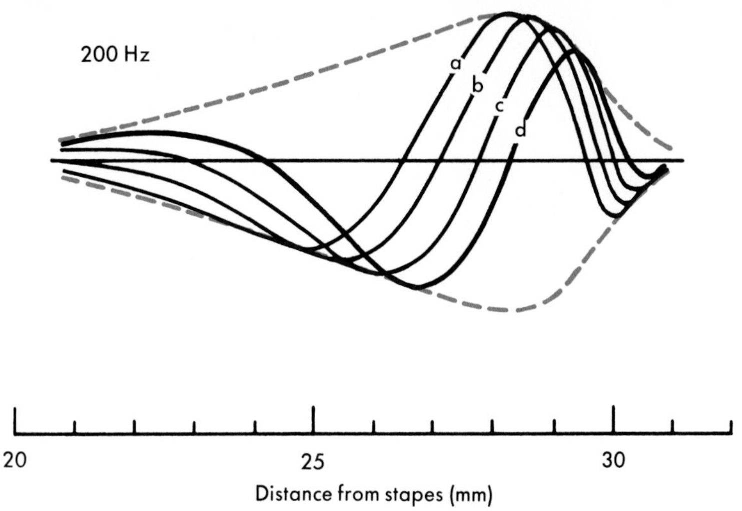

Fig. 10-13. Differential frequency responses of the basilar membrane. **A,** Traveling wave evoked along the basilar membrane by a 200 Hz sound. The wave is shown at four successive intervals *(a, b, c,* and *d).* and the dashed line describes the envelope of the peaks. Note how the maximum amplitude varies with distance along the basilar membrane, thereby resulting in the maximum response to a 200 Hz stimulus evoked from hair cells located approximately 29 mm from the stapes or oval window. **B,** Envelopes of traveling waves for a variety of frequencies. Note that the maximum response varies regularly with frequency so that regions closer to the stapes or oval window are more sensitive to higher frequencies. (Redrawn from von Békésy. G.: Experiments in hearing [research articles from 1928 to 1958], New York, 1960, McGraw-Hill, Inc. Copyright © 1960 by McGraw-Hill Book Co. Used with permission.)

B

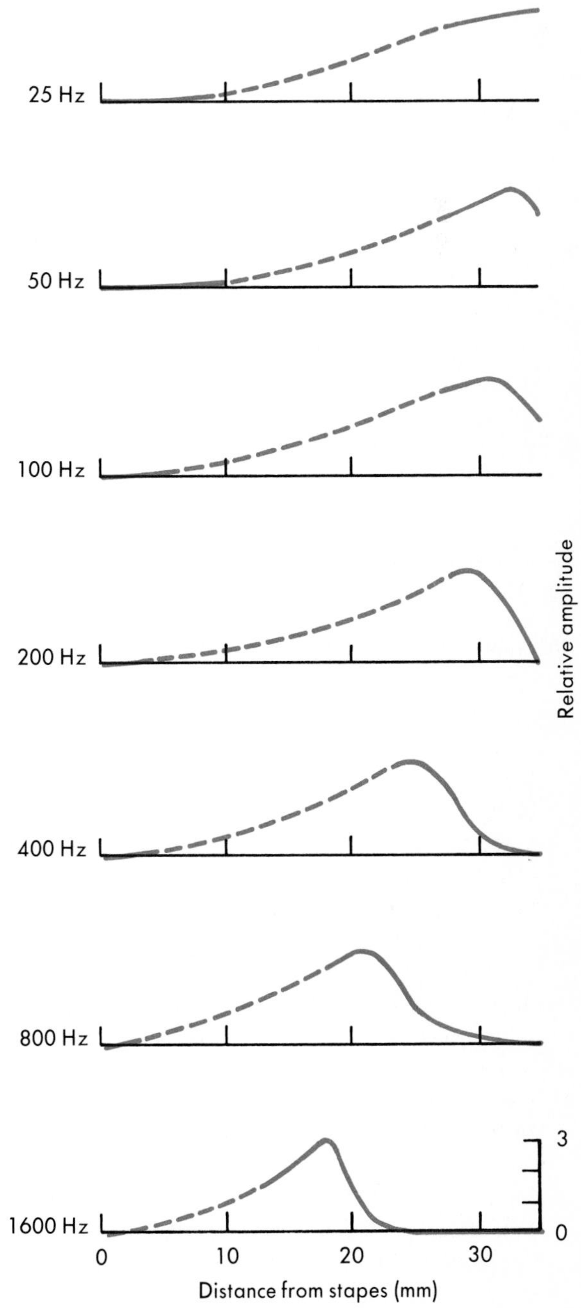

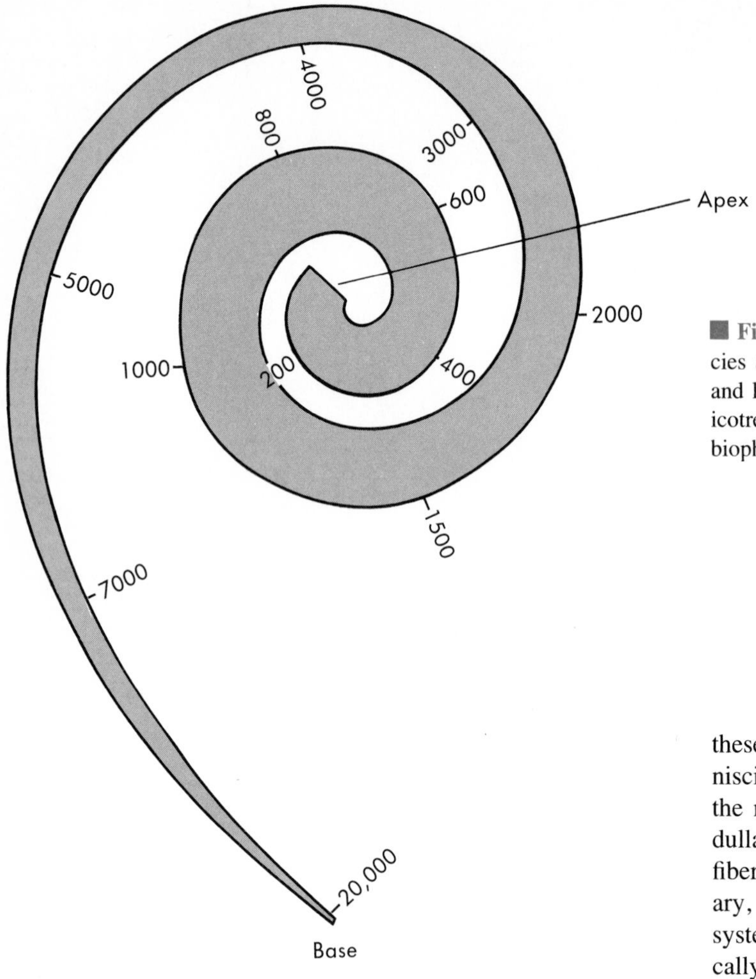

■ **Fig. 10-14.** Tonotopic map of the cochlea. Higher frequencies are represented nearer the base (i.e., the oval window), and lower ones are represented nearer the apex (i.e., the helicotrema). (Redrawn from Stuhlman, O.: An introduction to biophysics, New York, 1943, John Wiley & Sons, Inc.)

■ *The Central Auditory Pathways*

■ *Anatomical Features*

As for other sensory systems, the central auditory pathways include both ascending and descending pathways. These pathways also seem to be organized into phylogenetically older and newer portions, but sufficient data to elucidate these fully are not available.

Ascending pathways. Fig. 10-15 is a simplified version of some of the main features of the ascending pathways, which are complex. Auditory fibers from the auditory or cochlear nerve (part of the eighth nerve) terminate ipsilaterally in the *dorsal* and *ventral cochlear nuclei*. Many fibers from these nuclei cross the midline to innervate the *superior olivary complex,* and others ascend in the contralateral *lateral lemniscus*. The superior olivary complex is composed of a number of distinct nuclei, the most prominent of which are the *lateral* and *medial superior olivary nuclei*. Many of the crossing fibers from the cochlear nuclei decussate in the *trapezoid body,* although some innervate the ipsilateral superior olivary complex. Each superior olivary complex innervates the other across the trapezoid body, and these nuclei also contribute fibers to the lateral lemnisci. Therefore the lateral lemniscus, which represents the major auditory component ascending from the medulla, contains fibers that represent both cochleae; these fibers are a mixture of fibers of different orders (secondary, tertiary, etc.) Unlike the visual and somatosensory systems, in which various levels of the pathways typically map only the contralateral representation of space and include units of the same order (e.g., secondary, tertiary), the auditory system is much more complicated even at such a peripheral level as the lateral lemniscus.

The lateral lemniscus ascends to the midbrain, where most of its fibers terminate in the *inferior colliculus*. At this level, too, considerable auditory information is relayed across the midline in the *commissure of the inferior colliculus*. Fibers leave the inferior colluculus via the *brachium of the inferior colliculus* to terminate in the *medial geniculate nucleus*. Most of these fibers travel ipsilaterally, but some innervate the contralateral medial geniculate nucleus after crossing the collicular commissure. Neurons of the medial geniculate nucleus project fibers via the auditory radiations to the auditory cortex. Finally, as is the case for the visual and somatosensory thalamocortical projections, axon collaterals from cells of the medial geniculate nucleus innervate an exclusively auditory portion of the reticular nucleus of the thalamus (RNT) *en route* to the cortex. Cells of this auditory portion of the RNT provide GABAergic inhibitory input to the medial geniculate nucleus. Interneurons that are GABAergic (and thus presumably inhibitory) can also be found among the relay cells of the medial geniculate nucleus.

The colliculogeniculocortical projection is even more complex than described in the above paragraph. It has

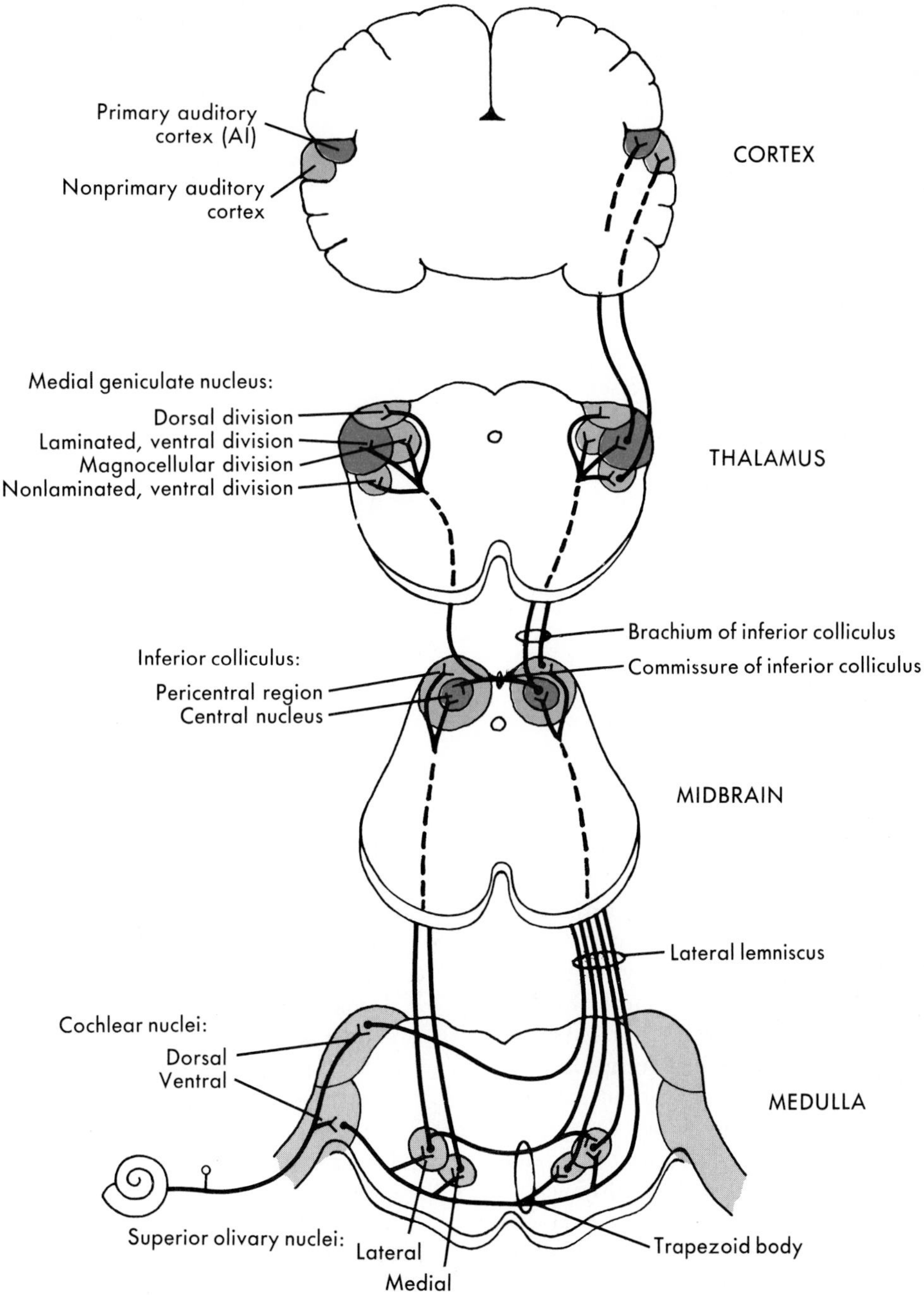

Fig. 10-15. Diagram of some of the central auditory pathways. The partial division of these pathways into core *(darker color)* and belt *(lighter color)* regions is shown at the levels of the inferior colliculus, medial geniculate nucleus, and auditory cortex. Note that as peripherally as the lateral lemniscus the pathways include a mixture of secondary, tertiary, and other higher level fibers as well as a representation of both cochleae.

been divided very generally into *core* and *belt* portions that may relate to phylogenetic age. Fig. 10-15 shows these divisions. The presumably newer *core* portion includes the central nucleus of the inferior colliculus, the laminated portion of the ventral division of the medial geniculate nucleus, and the primary auditory cortex. This cortex has been designated AI, and it corresponds to Brodmann's areas 41 and 42 (Chapter 13). The core geniculocortical projection terminates mainly in layer IV, although a sparse projection to layer VI has also been described.

The presumably older belt portion includes the pericentral nucleus and tegmentum of the inferior colliculus, the nonlaminated ventral, dorsal, and magnocellular divisions of the medial geniculate nucleus, and cortical auditory areas that include AI and neighboring

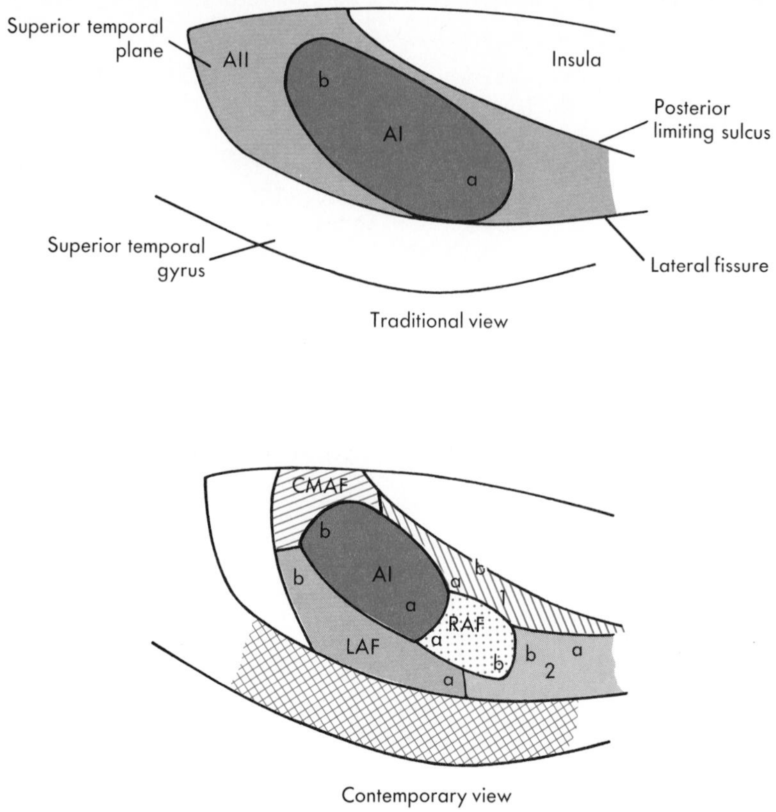

Fig. 10-16. Organization of auditory cortex in the macaque monkey; view of the surface of the superior temporal plane and dorsolateral surface of the superior temporal gyrus. The representation of the cochlear apex *(a)* and base *(b)*, or low and high frequencies, respectively, are designated for those areas in which tonotopic organization has been verified. Drawing at top is the older traditional view with the primary *(AI)* and secondary *(AII)* auditory areas. Drawing at bottom is the contemporary view that illustrates the five or more auditory areas. These include the primary auditory cortex *(AI)*, the nontonotopic caudiomedial auditory field *(CMAF)*, the lateral auditory field *(LAF)*, plus two as yet unnamed areas (designated *1* and *2*) with tonotopic organization. Furthermore, the superior temporal gyrus seems to contain one or more additional auditory areas. (Redrawn from Merzenich, M.M., and Kass, J.H.: Progr. Psychobiol. Physiol. Psychol. **9**:1, 1980.)

regions. As in the visual and somatosensory systems, there are numerous separate cortical representations of sensory space in the auditory system (at least six are presently known). They all receive geniculocortical input via the belt portion of the pathway. Fig. 10-16 shows these known auditory cortical areas. Here the projection is substantially to layer I of the cortex. In many ways the core and belt portions seem analogous, respectively, to the geniculocortical and pulvinar-cortical portions of the visual system (Chapter 8).

In addition to the newer core and older belt portions of the auditory pathways shown in Figs. 10-15 and 10-16, there are more diffuse and presumably even older portions that for simplicity are not diagrammed. For instance, fibers passing through the medulla at the level of the superior olivary complex emit fine collaterals among neurons of the tegmentum reticular formation and among cells scattered along the trapezoid body. Fibers ascending in the lateral lemniscus terminate among neurons of the midbrain reticular formation and among cells scattered along the lateral lemniscus. These latter cells are usually referred to as the *dorsal and ventral nuclei of the lateral lemniscus*. These diffuse pathways are probably functionally analogous to the oldest parts of the visual and somatosensory systems.

The complexity of the ascending auditory pathways makes it difficult to localize lesions in humans on the basis of the resultant hearing deficits. Unilateral lesions produce a unilateral deafness only if they occur in or peripherally to the cochlear nuclei (Fig. 10-15). At all other levels unilateral lesions produce bilateral hearing disorders that are only subtly different from one another. An example of this is an animal's ability to locate the source of a sound stimulus. This ability can be degraded by a lesion in one of many locations throughout the auditory system. Only the subtlest differences in deficits provide any clue to the location of the lesion. Furthermore, the cerebral cortex is not essential to hearing, at least in most experimental animals. Complete, bilateral removal of the auditory cortical areas only slightly diminishes auditory sensitivity. These lesions however, do create difficulties for the animals in the discrimination of tonal patterns (e.g., tones given in a different sequence such as AABA or ABAA) that are readily discriminated by intact animals. Such bilateral cortical lesions rarely occur in humans. However, evidence from experiments with animals indicates that considerable sensory perception occurs subcortically.

Descending pathways. As in older sensory systems, the auditory pathways include a significant descending component that probably interacts with the ascending pathways at many levels to create numerous, complex feedback loops. All of the known auditory cortical areas ipsilaterally innervate all portions of the medial geniculate nucleus, and they probably also innervate the auditory portion of the RNT. However, no descending fibers from this nucleus to other auditory structures have been described. The auditory cortex also ipsilaterally innervates the inferior colliculus, which in turn ipsilaterally innervates the superior olivary nucleus and the dorsal cochlear nucleus. Finally, as noted earlier, hair cells in the cochlea are innervated bilaterally from the superior olivary nuclei via the olivocochlear bundle.

Functional Organization

As for other sensory systems, the study of individual neuronal receptive fields has done much to elucidate the functional organization of the auditory system. Most of this analysis has been limited to the core or newer portion of the auditory system. Thus afferent inhibition and neurotopic organization are key features that have been described in experimental animals.

Receptive fields. Because receptive fields refer to regions of transducer-containing tissue, the tonotopic organization of the cochlea dictates that auditory receptive fields be described in terms of frequency and not space (pp. 169-170).

Fig. 10-17, *A,* shows a receptive field for a typical auditory neuron. This is the conventional manner in which to describe auditory receptive fields, and it corresponds to a ''tuning curve.'' This is a curve that depicts the minimum sound intensity needed to evoke a neural discharge as a function of sound frequency. The dark colored region indicates intensity/frequency combinations that discharge the cell; the light colored region indicates combinations that inhibit the cell; other combinations do not affect the cell. Two important features of these receptive fields are the *characteristic frequency,* which is found at the lowest point of the curve, and the *selectivity for frequency,* which is a function of how steeply the curve rises from the characteristic frequency. To illustrate these two characteristics, Fig. 10-17, *B,* shows auditory receptive fields for cochlear nerve fibers. These fields include a range of characteristic frequencies. Fig. 10-17, *B,* shows relatively highly selective or tightly tuned cells and relatively poorly selective or broadly tuned ones. A tightly tuned cell is analogous to a visual or somatosensory neuron with a small receptive field. Presumably, tonal discriminations depend on tightly tuned neurons.

Note that curves for most auditory neurons are asymmetrical (Fig. 10-17, *B*), although this is largely an artifact of the logarithmic scale for the abscissa. The curves rise from the characteristic frequency more steeply toward higher than toward lower frequencies. This is a straightforward consequence of the tuning of the basilar membrane itself. As shown in Fig. 10-13, traveling waves in this membrane set up by pure tones possess a similar asymmetry.

Afferent inhibition. As is discussed in Chapters 8 and 9, *afferent inhibition* means that a stimulus applied to transducers just beyond a neuron's receptive field will inhibit that neuron. Afferent inhibition has the same connotation for auditory neurons; i.e., stimuli of intensity/frequency combinations just outside the excitatory receptive field will inhibit an auditory neuron that possesses afferent inhibition. Fig. 10-17, *A,* shows such an inhibitory receptive field *(light colored region).* As is the case in vision and somesthesia, such afferent inhibition enhances the system's ability to discriminate stimuli applied to slightly different regions of the transducer surface. In other words, afferent inhibition in the auditory system sharpens frequency discriminations. Afferent inhibition occurs at all levels of the auditory pathways, including fibers of the cochlear nerve, but it is enhanced at each higher level of the system.

Tonotopic organization. The quality of neurotopic organization for newer portions of a sensory system appears in the newer portions of the auditory system as *tonotopic organization.* Neighboring neurons in a structure exhibit characteristic frequencies that differ only slightly. These frequencies change systematically as one moves across the structure. This is precisely equivalent to the retinotopic and somatotopic organizations that are found in the visual and somatosensory systems, respectively. Tonotopic organization and neurons that are sharply tuned for sound frequency are general features of the newer portions of the auditory system, including the auditory nerve, cochlear nuclei, superior olivary complex, lateral lemniscus, inferior colliculus, medial geniculate nucleus, and auditory cortex. Fig. 10-16 shows the tonotopic arrangements of the auditory cortex.

Binaural interactions. Input from each cochlea is present at every level of the auditory system central to the cochlear nuclei (Fig. 10-15). Most cells of the superior olivary complex, inferior colliculus, medial geniculate nucleus, and auditory cortex have binaural receptive fields; that is, a neural response can be generated from stimulation of either cochlea, and the two receptive fields generally have similar tuning characteristics.

Such binaural interactions are important in sound localization. Humans can localize the direction of sounds with an accuracy that depends on frequency, but the accuracy can approach 1 degree. Two cues are important. A sound source closer to one ear will reach that ear sooner and with greater intensity than it will for the other ear. Thus interaural differences in temporal phase and intensity are key factors in sound localization. Such a difference actually specifies a locus of directions that fall within a vertical plane passing through the head.

Presumably, binaural neurons sensitive to differences in phase and firing rate (since firing rate encodes intensity) of their monaural inputs could encode certain features of the direction of the sound source. Most attention has been focused on the superior olivary complex

A

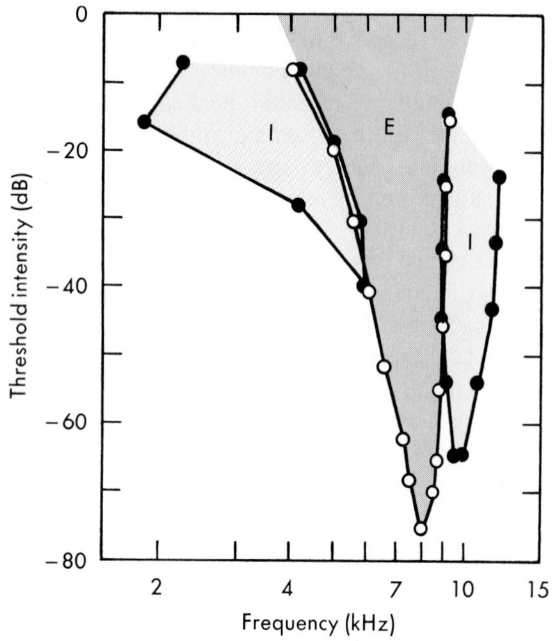

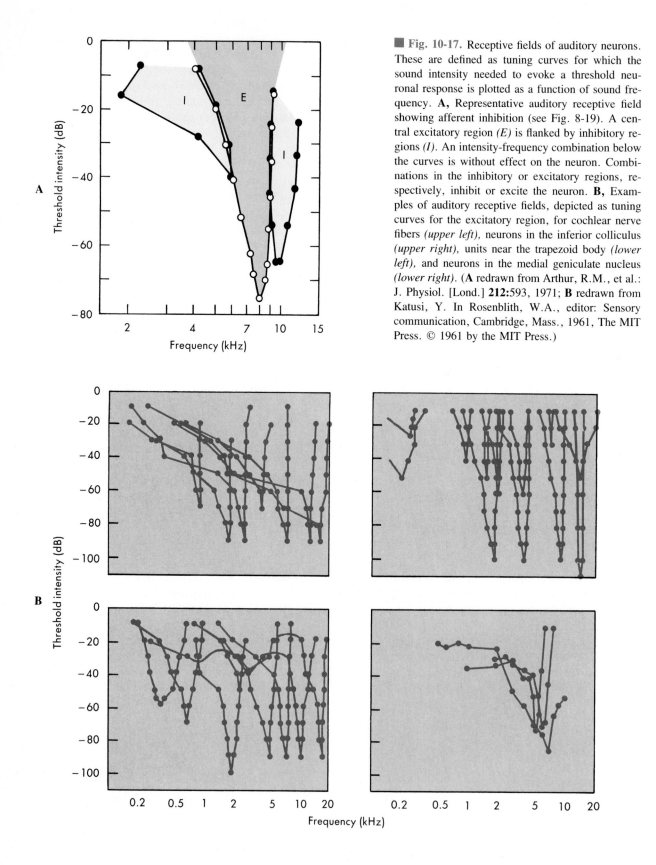

■ **Fig. 10-17.** Receptive fields of auditory neurons. These are defined as tuning curves for which the sound intensity needed to evoke a threshold neuronal response is plotted as a function of sound frequency. **A,** Representative auditory receptive field showing afferent inhibition (see Fig. 8-19). A central excitatory region *(E)* is flanked by inhibitory regions *(I)*. An intensity-frequency combination below the curves is without effect on the neuron. Combinations in the inhibitory or excitatory regions, respectively, inhibit or excite the neuron. **B,** Examples of auditory receptive fields, depicted as tuning curves for the excitatory region, for cochlear nerve fibers *(upper left),* neurons in the inferior colliculus *(upper right),* units near the trapezoid body *(lower left),* and neurons in the medial geniculate nucleus *(lower right)*. (**A** redrawn from Arthur, R.M., et al.: J. Physiol. [Lond.] **212:**593, 1971; **B** redrawn from Katusi, Y. In Rosenblith, W.A., editor: Sensory communication, Cambridge, Mass., 1961, The MIT Press. © 1961 by the MIT Press.)

B

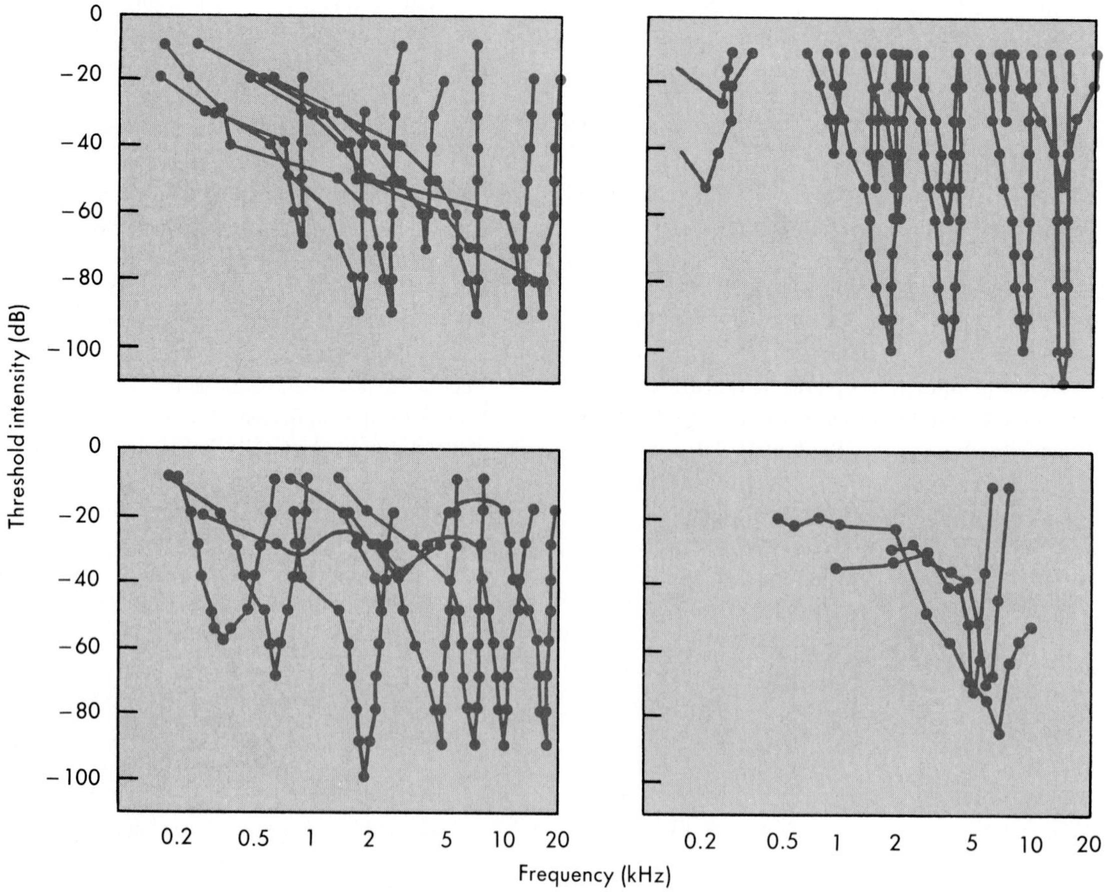

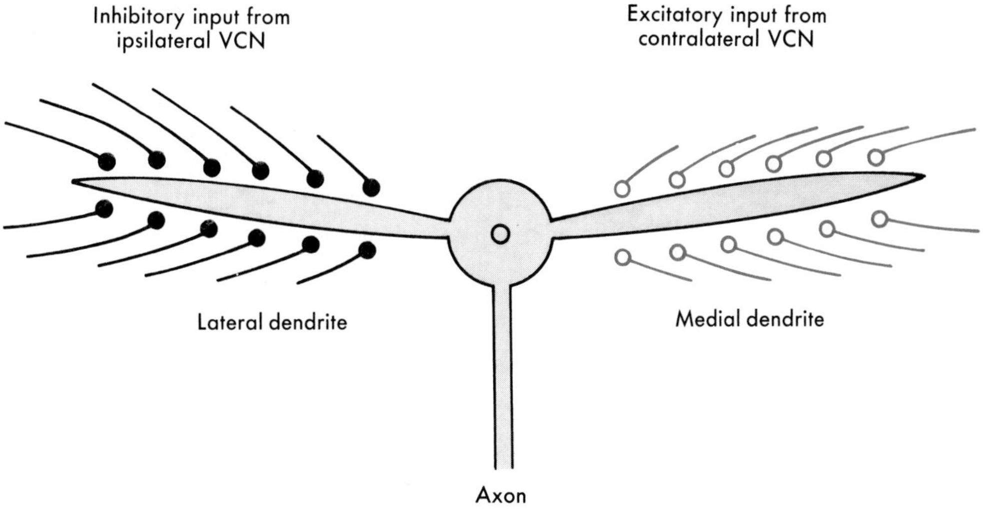

Fig. 10-18. Schematic diagram of binaural interactions into neuron of the medial superior olivary nucleus on the left side. The cell's medial dendrite receives excitatory input from the contralateral ventral cochlear nucleus, whereas the lateral dendrite receives inhibitory input from the ipsilateral ventral cochlear nucleus.

because this is the first stage of effective binaural integration. (However, the olivocochlear pathways raise the possibility that even hair cells might exhibit binaural interactions, and some evidence of such limited interactions has been found for cells of the dorsal cochlear nucleus.) In particular, many neurons of the medial superior olivary nucleus have two major dendrites, one directed medially and the other directed laterally (Fig. 10-18). Most of the inputs to the medial dendrites are excitatory, and such inputs derive from the contralateral ventral cochlear nucleus. Most inputs to the lateral dendrites are inhibitory, and they derive from the ipsilateral ventral cochlear nucleus. It has been suggested that, in order for these cells to fire, the timing of arrival of excitatory and inhibitory inputs must be precise. One of the factors that governs this timing is the phase difference of the sound arriving at the two ears. Activity transmitted through such a neuron of the medial superior olivary nucleus might thus signal a particular interaural time difference and thus a locus of directions for the sound source.

Our ability to use interaural differences to detect the locations of sounds suggests that some portions of our auditory system might be mapped in actual spatial rather than frequency coordinates. Studies of experimental animals have so far failed to detect such mapping; all areas studied to date indicate either tonotopic organization or no obvious mapping of any sort. However, lesions occurring in many auditory areas, including the auditory cortex, can result in a pronounced deficit in the ability to localize sources of brief (but not prolonged) sounds. Because of the tonotopic arrangement of auditory areas, a lesion typically produces a deficit only to the appropriate frequency. Such a deficit, which is analogous to a visual scotoma, has been termed a *sigoma*.

Cortical architecture and specificity. From the descriptions of the visual and somatosensory systems, one might predict two features that distinguish the auditory cortex from lower levels within the system. First, the cortical neurons are arranged in functional columns of cells perpendicular to the pial surface. Second, cortical neurons may display certain forms of stimulus specificity, such as the direction that a frequency-modulated stimulus is swept (i.e., from high to low frequencies or the reverse) through the auditory receptive field. This is precisely analogous to direction selectivity for visual or somatosensory receptive fields.

Bibliography

Journal articles

Allen, J.B.: Cochlear micromechanics—a physical model of transduction, J. Acoust. Soc. Am. Caps. **68:**1660, 1981.

Allon, N., et al.: Responses of single cells in the medial geniculate body of awake squirrel monkeys, Exp. Brain Res. **41:**222, 1981.

Brunso-Bechtold, J.K., et al.: HRP study of the organization of auditory afferents ascending to central nucleus of inferior colliculus in cat, J. Comp. Neurol. **197:**705, 1981.

Calford, M.B., and Webster, W.R.: Auditory representation within principal division of cat medial geniculate body: an electrophysiological study, J. Neurophysiol. **45:**1013, 1981.

Calford, M.B.: The parcellation of the medial geniculate body of the cat defined by the auditory response properties of single units, J. Neurosci. **3:**2350, 1983.

Calford, M.B., and Aitkin, L.M.: Ascending projections to the medial geniculate body of the cat: evidence for multiple, parallel auditory pathways through thalamus, J. Neurosci. **3**:2365, 1983.

Flock, A., and Strelioff, D.: Graded and nonlinear mechanical properties of sensory hairs in the mammalian hearing organ, Nature **310:**597, 1984.

Galaburda, A., and Sanides, F.: Cytoarchitectonic organization of the human auditory cortex, J. Comp. Neurol. **190:**597, 1980.

Glendenning, K.K., et al.: Ascending auditory afferents to the nuclei of the lateral lemniscus, J. Comp. Neurol. **197:**673, 1981.

Hudspeth, A.J.: Mechanoelectrical transduction by hair cells in the acousticolateralis sensory system, Annu. Rev. Neurosci. **6:**187, 1983.

Imig, T.J., and Morel, A.: Organization of the thalamocortical auditory system in the cat, Annu. Rev. Neurosci. **6:**95, 1983.

Imig, T.J., and Morel, A.: Topographic and cytoarchitectonic organization of thalamic neurons related to their targets in low-, middle-, and high-frequency representations in cat auditory cortex, J. Comp. Neurol. **227:**511, 1984.

Imig, T.J., and Morel, A.: Tonotopic organization in ventral nucleus of medial geniculate body in the cat, J. Neurophysiol. **53:**309, 1985.

Jenkins, W.M., and Merzenich, M.M.: Role of cat primary cortex for sound-localization behavior, J. Neurophysiol. **52:**819, 1984.

Phillips, D.P., and Irvine, D.R.F.: Responses of single neurons in physiologically defined primary auditory cortex (AI) of the cat: frequency tuning and responses to intensity, J. Neurophysiol. **45:**48, 1981.

Russell, I.J., and Sellick, P.M.: Low-frequency characteristics of intracellularly recorded receptor potentials in guineapig cochlear hair cells, J. Physiol. (Lond.) **338:**179, 1983.

Shosaku, A., and Sumitomo, I.: Auditory neurons in the rat thalamic reticular nucleus, Exp. Brain Res. **49:**432, 1983.

Books and monographs

Brodal, A.: Neurological anatomy in relation to clinical medicine, ed. 3, New York, 1981, Oxford University Press, Inc.

Kiang, N.Y.S.: Stimulus representation in the discharge patterns of auditory neurons. In Tower, D.B., editor: The nervous system, vol. 3, Human communication and its disorders, New York, 1975, Raven Press.

Neff, W.D., et al.: Behavioral studies of auditory discrimination: central nervous system. In Keidel, W.D., and Neff, W.D., editors: Handbook of sensory physiology, auditory system: physiology, central nervous system, behavioral studies, and psychoacoustics, vol. V/2, New York, 1975, Springer-Verlag New York, Inc.

Von Békésky, G.: Experiments in hearing, New York, 1960, McGraw-Hill, Inc.

Yin, T.C.T. and Kuwada, S.: Neuronal mechanisms of binaural interaction. In Edelman, G.M., and Cowan, W.M., editors: Dynamic aspects of neocortical function, New York, 1984 John Wiley & Sons.

Woolsey, C.N., editor: Cortical sensory organization, vol. 3, Multiple auditory areas, Clifton, N.J., 1981, Humana Press.

CHAPTER

11

The Vestibular System

The function of the vestibular system is to sense forces of acceleration, both linear and rotational. The most prominent of these forces is the linear force of gravity. Because the vestibular apparatus is part of the inner ear and is thus located inside the head, it is head acceleration that is sensed.

Two general and related functions are served by the vestibular system. In the presence of forces acting on the head, body balance is maintained through postural reflexes that are largely determined from vestibular input. Also, because of vestibuloocular reflexes, stable eye positions are maintained with respect to visual objects in spite of pronounced head movements. This allows the maintenance of ocular fixation and a stable visual field during activities such as running and jumping.

Discussion of the vestibular system is more superficial than that of vision, somesthesia, and hearing for two reasons. First, the vestibular pathways have nothing analogous to the "lemniscal" thalamocortical system that is associated with these other sensory systems. Second, although we are conscious of acceleration, most of the vestibular information is processed subconsciously to serve postural and oculomotor reflexes (see also Chapter 18). This discussion concerns the peripheral vestibular apparatus and some of the related central pathways.

Vestibular Apparatus

Gross Structure of the Labyrinth

The peripheral apparatus of the vestibular system lies in the inner ear. The vestibular apparatus is adjacent to and continuous with the cochlea (Fig. 11-1), and it consists of *bony labyrinth* into which is fitted the *membranous labyrinth*. Between the bony and membranous labyrinths is the *perilymph;* viscous *endolymph* fills the membranous labyrinth. The perilymph and endolymph are continuous with the fluids contained within the cochlea (Chapter 10).

The labyrinths are functionally divided into five prominent divisions. These include two bulbous enlargements, the *utricle* and *saccule,* and three *semicircular ducts* or *canals,* the *superior, posterior,* and *horizontal.* Fig. 11-2 shows the relationship of these structures to one another and their orientation within the head. Note that the semicircular canals of one side are in planes that are roughly perpendicular to one another. Also, the planes of the two horizontal canals are approximately parallel with one another. This is also true for both the planes of the right superior and left posterior canals as well as those of the left superior and right posterior canals.

Near one end of each canal is an enlargement called the *ampulla* (Fig. 11-1). The epithelium inside each ampulla is thickened in a region called the *crista;* the crista contains the vestibular transducers. These transducers are *hair cells,* which are quite similar to those in the cochlea (Fig. 11-3). The saccule and utricle also have a region of epithelium called the *macula,* which contains a dense aggregate of hair cells. The utricular macula is oriented in the horizontal plane and the saccular macula is oriented vertically, approximately in a parasagittal plane.

Hair Cells

Fig. 11-3 shows the main features of the vestibular hair cells, which form synapses onto sensory fibers of the eighth nerve. Also, as is the case for cochlear hair cells, synapses are formed onto these receptors from efferent fibers. At the top of each hair cell are a number of *stereocilia* of variable length, plus a single *kinocilium* at one edge of the bundle of cilia. Bending of the stereocilia leads to a generator potential that synaptically affects the firing rate of the eighth nerve fiber. The longer

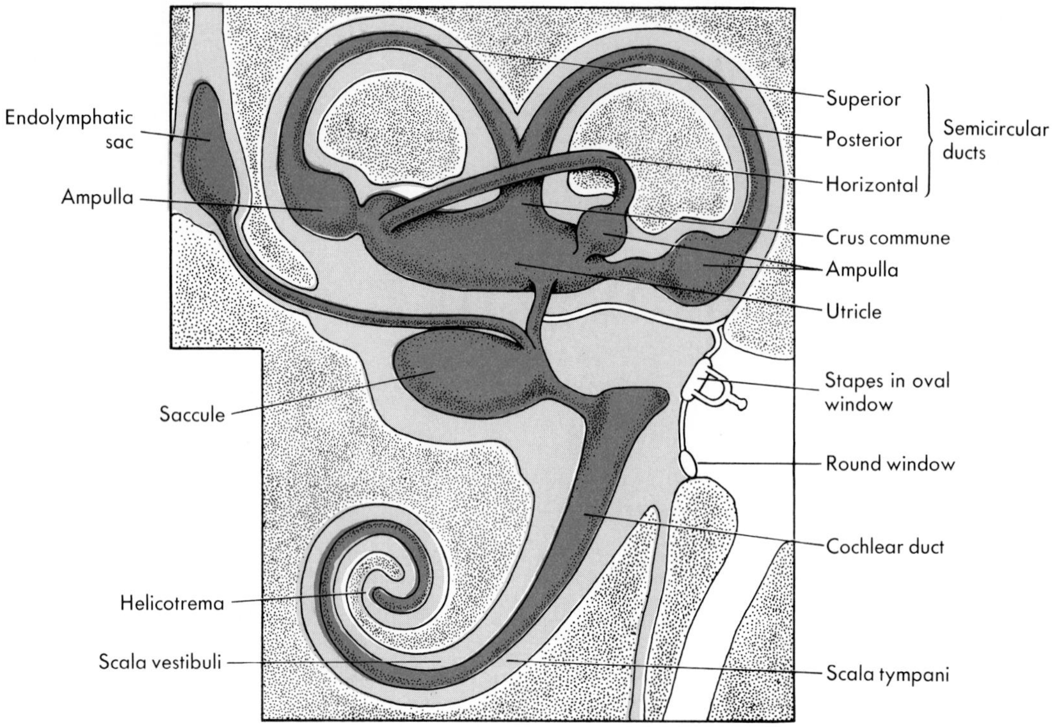

■ **Fig. 11-1.** Relationships between the membranous labyrinths *(darker color),* perilymph *(lighter color),* and bone *(stippling)* in the inner ear. (Redrawn from Kandel, E.R., and Schwartz, J.H.: Principles of neural science, New York, 1981, Elsevier.)

A

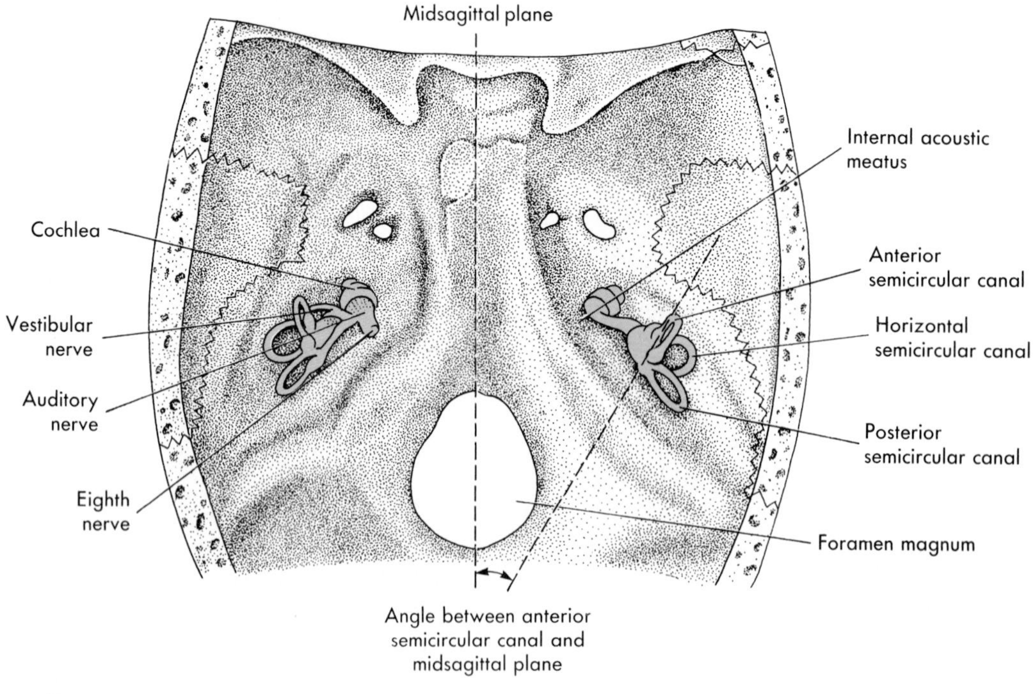

■ **Fig. 11-2.** Structure of the inner ear. **A,** View of cranium from above, showing relative location and orientation of the inner ear *(color).* Note that the horizontal semicircular canals are roughly coplanar, and that the anterior canal and contralateral posterior canal lie in roughly parallel planes.

Fig. 11-2, cont'd. B, Labyrinths and nerves of the inner ear. (**A** redrawn from Kandel, E.R., and Schwartz, J.H.: Principles of neural science, 1981, Elsevier. **B** redrawn from Best, C.H., and Taylor, N.B.: Physiological basis of medical practice, Baltimore, © 1966, The Williams & Wilkins Co.)

B

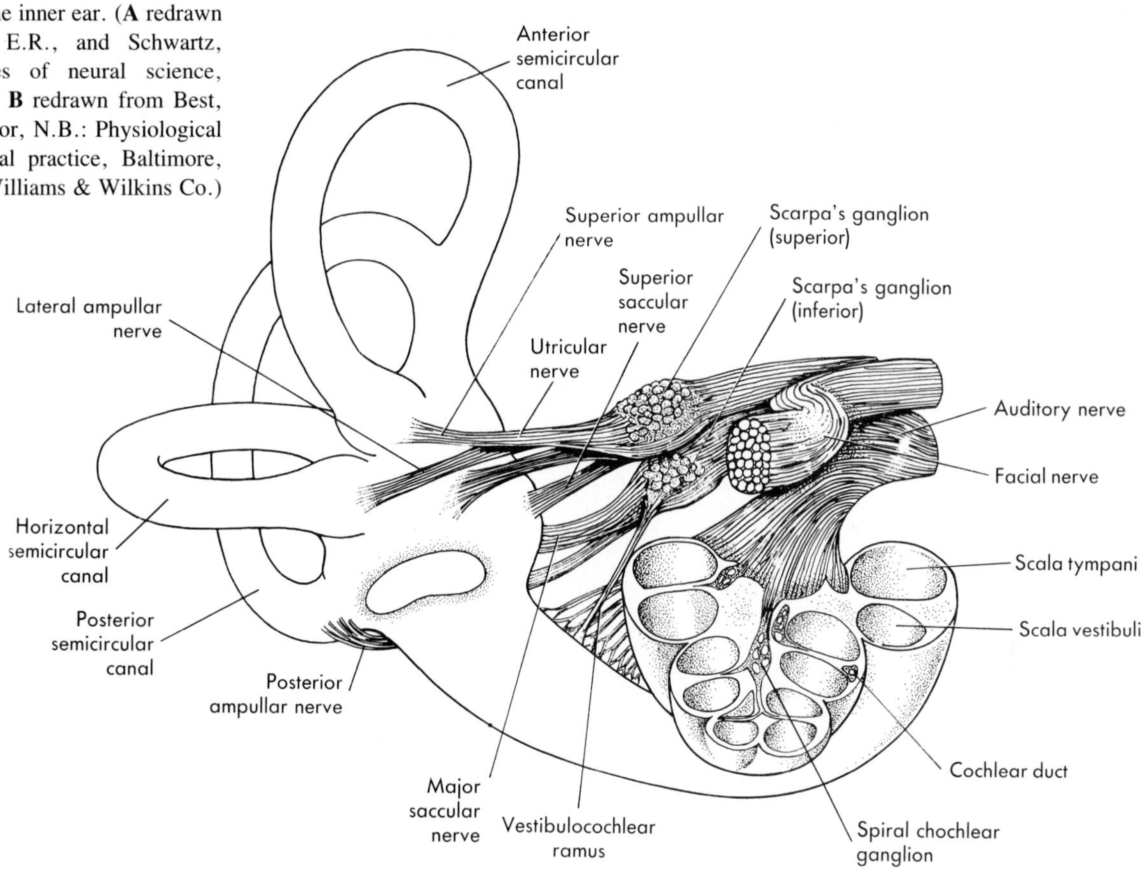

stereocilia are closer to the kinocilium; this provides a functional axis of polarization similar to that of cochlear hair cells (Fig. 11-4). Bending of stereocilia toward the kinocilium depolarizes the hair cell and increases the firing rate of the afferent vestibular nerve fiber; bending in the reverse direction hyperpolarizes the hair cell and decreases the afferent fiber's firing rate. The axis and direction of a hair cell are denoted by the line from the shortest stereocilium to the kinocilium.

Semicircular canals. The cilia of the hair cells in each ampulla are embedded in a structure called the *cupula,* which is a gelatinous capsule that hangs down from the roof of the ampulla (Fig. 11-5). Movements of the viscous endolymph with respect to the walls of the semicircular canal push on and displace the cupula, thereby bending the stereocilia and effecting neural transduction. Because the endolymph has considerable inertia, it lags behind in response to head rotations or angular accelerations. This lagging provides the necessary stimulus for a generator potential. Thus the direction of bending of the stereocilia is opposite to the direction of head rotation (Fig. 11-6).

Within an ampulla the hair cells are all oriented (with regard to their functional axis as just defined) in the same direction. This direction points toward the utricle for the ampulla of the horizontal canal and away from the utricle for the ampullae of the other semicircular canals. This arrangement allows the complementary pairs of semicircular canals on each side (left and right horizontal, left superior and right posterior, left posterior and right superior) (Fig. 11-2) to synchronize their outputs in a heteronymous fashion. Fig. 11-6 shows how this works for the horizontal canals. In this example a clockwise head rotation (as viewed from above) causes a counterclockwise bending of the hair cells. This depolarizes hair cells of the right ampulla and increases the firing rate of vestibular nerve afferents on the right side; hyperpolarization and decreased firing result on the left. The functional connections illustrated in Fig. 11-6 convert this asymmetrical pattern of activity between the right and left vestibular nerves into an appropriate leftward eye movement (see also below). An analogous result can be predicted for head rotations in other planes that involve the superior or posterior canals. Because the semicircular canals respond in this manner to head movements, their responses are most important in the control of eye movements to maintain fixation.

■ **Fig. 11-3.** Vestibular hair cells, including a type I cell *(left)* and a type II cell *(right)*. The hair cells are presynaptic to afferent fibers *(lighter color)*. Efferent fibers *(darker color)* synapse directly onto type I hair cells and onto the afferent fibers innervating type II hair cells. Extending from the top of each hair cell is a single kinocilium and many stereocilia.

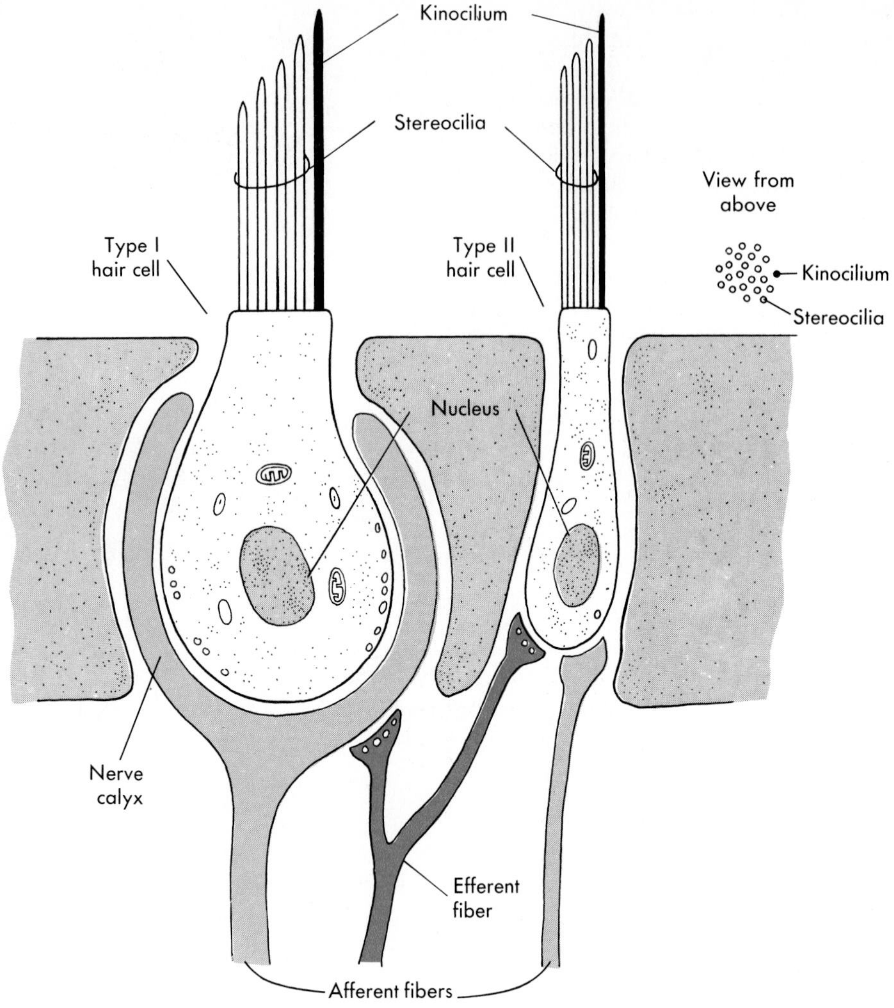

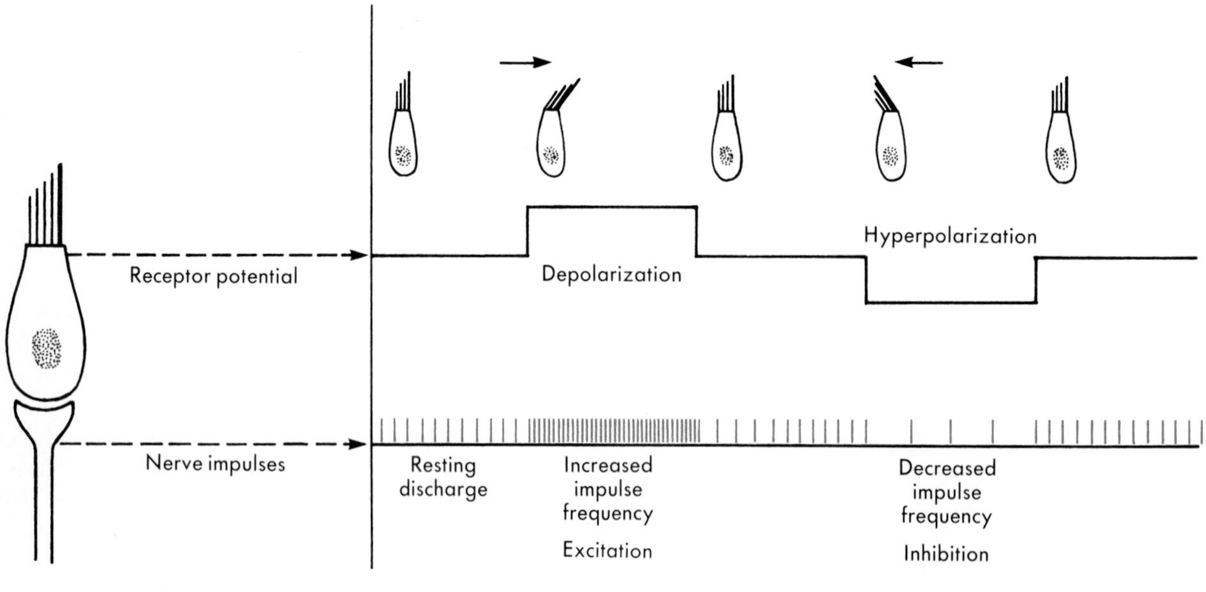

■ **Fig. 11-4.** Directional selectivity of hair cells. Bending of the kinocilium away from the stereocilia depolarizes the hair cell and increases the firing rate in its afferent fiber. Bending of the kinocilium toward the stereocilia hyperpolarizes the hair cell and decreases the firing rate in its afferent fiber. (Redrawn from Kandel, E.R., and Schwartz, J.H.: Principles of neural science, New York, 1981, Elsevier.)

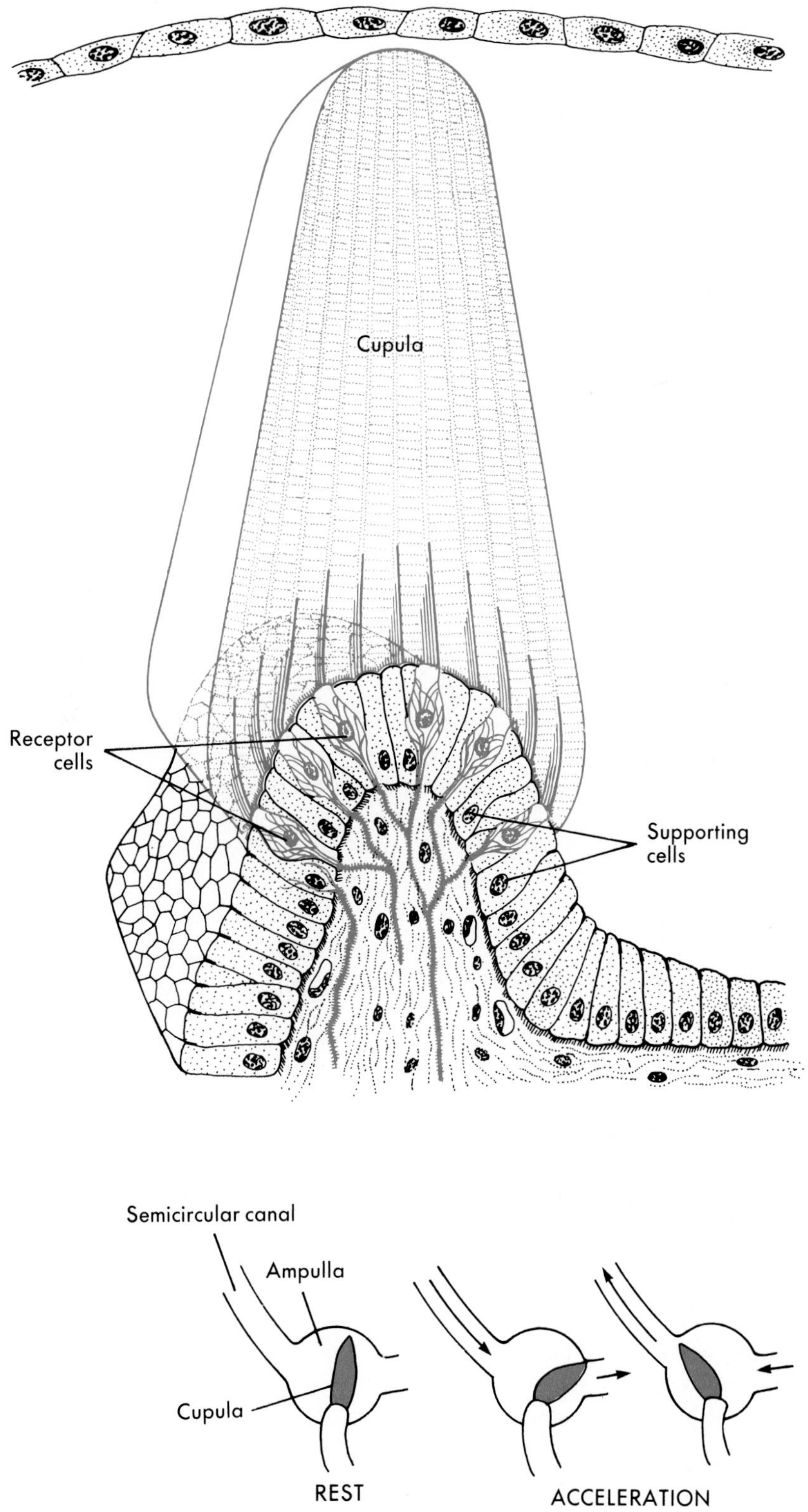

Fig. 11-5. Response of the cupula in the ampulla to angular acceleration. *Above,* Note the hair cells with their cilia embedded in the ampulla. *Below,* Ampullar displacement (and thus bending of the stereocilia and kinocilia) in response to acceleration. (Redrawn from Wersäll, J.: Acta Otolaryngol. [Stock.] Suppl. **126:**1, 1956.)

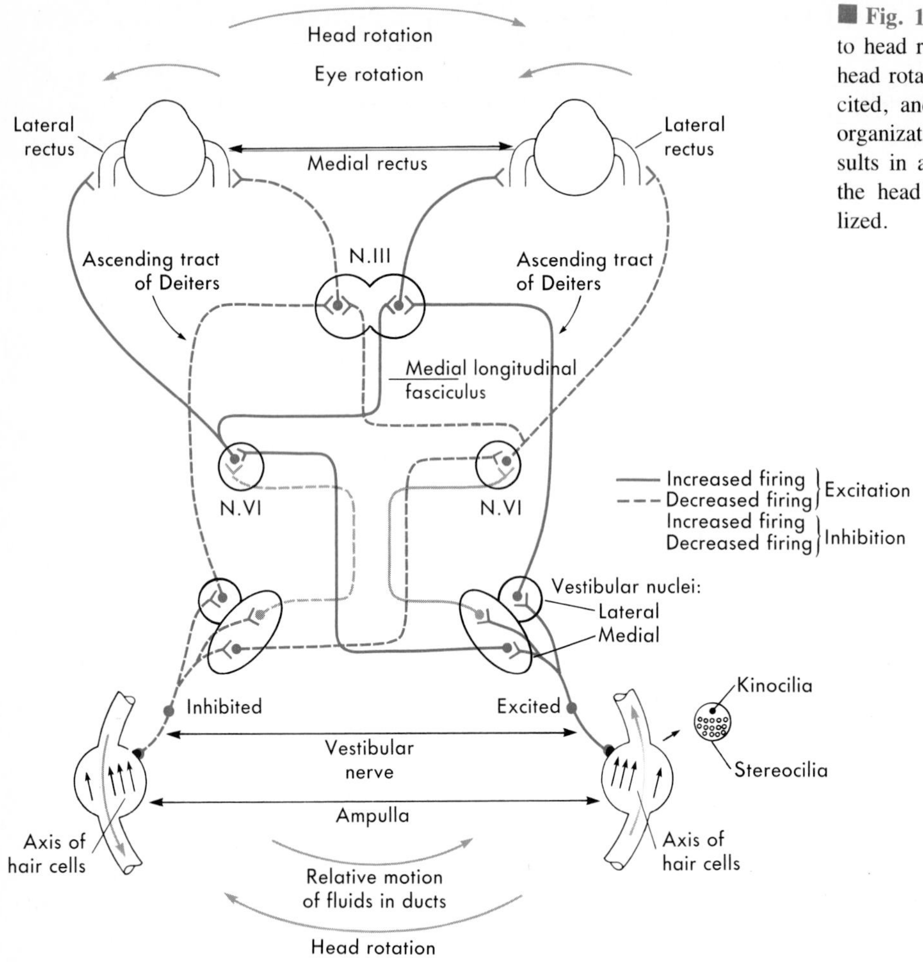

Fig. 11-6. Vestibuloocular reflex in response to head rotation in the horizontal plane. For the head rotation as shown, the right ampulla is excited, and the left is inhibited. Because of the organization of the vestibular pathways, this results in an eye movement that compensates for the head rotation, so retinal images are stabilized.

Utricle and saccule. The utricle and saccule respond principally to changes in linear acceleration, particularly to that caused by gravity. Their responses are most important for the control of posture and maintenance of balance. Hair cells in the macula of the utricle and saccule are stimulated by forces similar to those that affect the canals, although some of the details differ. The cilia of macular hair cells extend from the cells into a dense *otolithic membrane,* which is a matrix floating within the utricle or saccule. The utricular macula is oriented horizontally within the head, so head tilt causes the otolithic membrane to deform the stereocilia. The saccular macula is oriented vertically in the sagittal plane. Hence a change in upward or downward acceleration (e.g., jumping or falling) results in an analogous deformation of the stereocilia by the otolithic membrane. Both macular structures are sensitive to linear accelerations in the horizontal plane.

The hair cell polarization axes are orderly but complex in both the utricle and saccule. Fig. 11-7 shows the axes for each structure. From these patterns of polarity it is clear that each direction of linear acceleration results in a particular pattern of increased and decreased responsiveness of hair cells and their related vestibular afferent nerve fibers.

Vestibular Pathways

Fibers of the eighth nerve carry the vestibular information from the hair cells to the medulla and lower pons. Somata of these fibers are located in Scarpa's ganglion (Fig. 11-2). Most of their axons terminate among the four nuclei of the vestibular nuclear complex: the *superior vestibular nucleus* (Bechterew's nucleus), the *lateral vestibular nucleus* (Deiters' nucleus), the *medial vestibular nucleus* (Schwalbe's nucleus), and the *inferior vestibular nucleus* (Roller's nucleus; also known as the *descending vestibular nucleus*). The remaining few fibers of the vestibular nerve bypass the vestibular nuclei and directly innervate the cerebellum.

All of the vestibular nuclei receive some input from the cerebellum as well as from the vestibular nerve, but further details of their inputs and outputs differ (Fig. 11-8). The semicircular canals innervate all vestibular nuclei, although their densest input is to the superior and medial vestibular nuclei. The utricle innervates mainly the lateral and inferior vestibular nuclei, and the saccule innervates only the inferior vestibular nucleus.

The major descending output from the vestibular nuclei is the *lateral vestibulospinal tract,* which arises

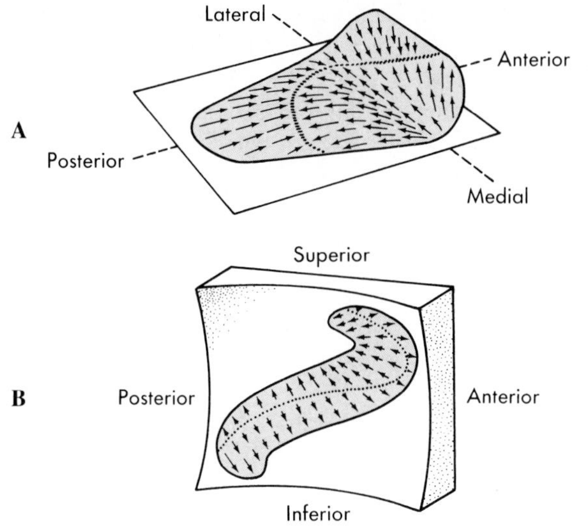

■ **Fig. 11-7.** Axes of polarization of macular hair cells. The arrows represent the direction from the stereocilia to the kinocilium. **A,** Utricle. **B,** Saccule. (Redrawn from Spoendlin, H.H. In Wolfson, R.J., editor: The vestibular system and its diseases, Philadelphia, 1966, University of Pennsylvania Press.)

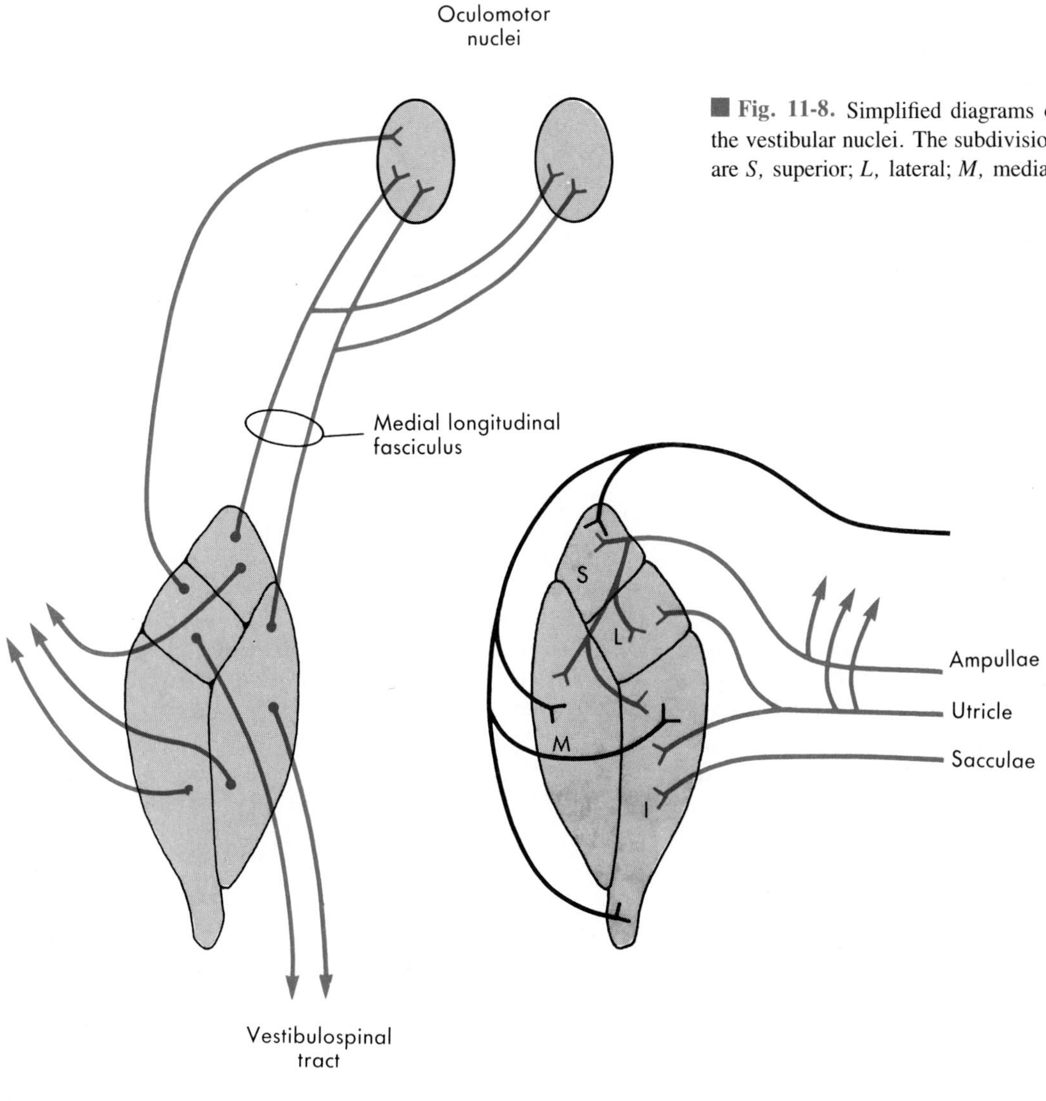

■ **Fig. 11-8.** Simplified diagrams of the major pathways of the vestibular nuclei. The subdivisions of the vestibular nuclei are *S,* superior; *L,* lateral; *M,* medial; and *I,* inferior.

from the lateral vestibular nucleus and reaches all spinal levels. Another prominent output of the vestibular nuclei is the *medial longitudinal fasciculus,* which has both descending and ascending components. The former arises from the medial vestibular nucleus and becomes the *medial vestibulospinal tract,* which innervates cervical and upper thoracic levels of the spinal cord. The ascending portion of the medial longitudinal fasciculus derives mostly from the medial and superior vestibular nuclei and innervates oculomotor nuclei. A less prominent ascending pathway is the *ascending tract of Deiter,* which originates from the lateral vestibular nucleus, lies lateral to the medial longitudinal fasciculus, and innervates the oculomotor nucleus (Figs. 11-6 and 11-8). Finally, a large input to the cerebellum arises from the inferior vestibular nucleus, and some cells of the superior and medial vestibular nuclei also project axons to the cerebellum.

The main role of the ascending vestibular pathways to the oculomotor nuclei is reflex control of eye movements, otherwise known as the *vestibuloocular reflex.* This reflex maintains visual fixation by stabilizing the position of the eyes in space during head movements. Fig. 11-6 illustrates how this works for a horizontal clockwise rotation of the head. Such a rotation causes the horizontal ducts to move clockwise relative to the endolymph because of inertia of the endolymph within the ducts. This, in turn, increases firing in the right vestibular nerve and decreases firing in the left nerve. Such head rotation ultimately evokes appropriate changes in firing rates for neurons in the abducens and oculomotor nuclei so that conjugate movement of the eyes compensates precisely for the head movement. If the head rotation continues, eventually each eye moves to its extreme position in the orbit and cannot be deviated further. At this point a rapid eye movement in the direction of the head rotation is made, a new fixation point is chosen, and a slow drift to compensate for the further head rotation begins again. These rhythmic eye movements are called *vestibular nystagmus* (nystagmus refers to periodic, conjugate eye movements) and are involuntary. The slow phase of vestibular nystagmus is the eye movement opposite to the direction of head rotation; the fast phase is in the direction of head rotation. Eventually friction between the ducts and endolymph reduces the movement between them. The nystagmus slows down and stops after about 20 seconds of head rotation. However, on abrupt termination of head rotation the endolymph continues to move in the same direction as the prior head rotation. Thus relative movement between the endolymph and duct begins, and this movement is opposite in relative direction to that present during the head rotation. This leads to *postrotary nystagmus,* which is characterized by fast and slow phases opposite in direction to those of the nystagmus observed during the head rotation. Spontaneous nystagmus or a failure to elicit normal nystagmus is often associated with brainstem lesions that interrupt the vestibular pathways.

The descending vestibulospinal tracts are concerned with postural reflexes and contribute to medial descending motor pathways (see Chapters 15 and 18 for details). The functional significance of connections between the cerebellum and the vestibular nuclei is described in Chapter 16.

Bibliography

Journal articles

Anastosopoulos, D., and Mergner, T.: Canal-neck interaction in vestibular nuclear neurons of the cat, Exp. Brain Res. **46:**269, 1982.

Boyle, R., and Pompeiano, O.: Responses of vestibulospinal neurons to sinusoidal rotation of neck and macular vestibular inputs on vestibulospinal neurons, J. Neurophysiol. **45:**852, 1981.

Büttner, U., and Waespe, W.: Vestibular nerve activity in the alert monkey during vestibular and optokinetic nystagmus, Exp. Brain Res. **41:**310, 1981.

Carleton, S.C., and Carpenter, M.B.: Afferent and efferent connections of the medial, inferior, and lateral vestibular nuclei, Brain Res. **278:**29, 1983.

Chubb, M.C., and Fuchs, A.F.: Contribution of y group of vestibular nuclei and dentate nucleus of cerebellum to generation of vertical smooth eye movements, J. Neurophysiol. **48:**75, 1982.

Corey, D.P., and Hudspeth, A.J.: Kinetics of the receptor current in bullfrog saccular hair cells, J. Neurosci. **3:**962, 1983.

Goldberg, J.M., and Fernández, C.: Efferent vestibular system in the squirrel monkey: anatomical location and influence on afferent activity, J. Neurophysiol. **43:**986, 1980.

Hudspeth, A.J.: Extracellular current flow and the site of transduction by vertebrate hair cells, J. Neurosci. **2:**1, 1982.

Ito, J., et al.: Vestibular efferent fibers to ampulla of anterior, lateral, and posterior semicircular canals in cats, Brain Res. **259:**293, 1983.

Kubin, L., et al.: Responses of lateral reticular neurons to convergent neck and macular vestibular inputs, J. Neurophysiol **46:**48, 1981.

Maciewicz, R., et al.: The vestibulothalamic pathway: contribution of the ascending tract of Dieter's, Brain Res. **252:**1, 1982.

Mergner, T., et al.: Vestibulo-thalamic projection to the anterior suprasylvian cortex of the cat, Exp. Brain Res. **44:**455, 1981.

Shotwell, S.L., et al.: Direction sensitivity of individual vertebrate hair cells to controlled deflection of their hair bundles, Ann. N.Y. Acad. Sci. **374:**1, 1981.

Tomko, D.L., et al.: Responses to head tilt in cat eighth nerve afferents, Exp. Brain Res. **41:**216, 1981.

Uchino, Y., et al.: Branching patterns and properties of vertical- and horizontal-related excitatory vestibuloocular neurons in the cat, J. Neurophysiol. **48:**891, 1982.

Books and monographs

Brodal, A.: Neurological anatomy in relation to clinical medicine, ed. 3, New York, 1981, Oxford University Press.

Carpenter, M.C., and Sutin, J.: Human neuroanatomy, ed. 8, Baltimore, 1983, Williams and Wilkins, Co.

Naunton, R.F., editor: The vestibular system, New York, 1975, Academic Press, Inc.

Wilson, V.J., and Melvill Jones, G.: Mammalian vestibular physiology, New York, 1979, Plenum Press.

CHAPTER

12

Chemical Senses

Gustation (taste) and olfaction (smell) are the only senses that can identify the chemical nature of stimuli in the external environment. In many animals these important senses not only help to define the environment but also affect social behavior, such as mating, territoriality, and feeding. During the course of primate evolution, however, these chemical senses have become less important to conscious perception, whereas vision, touch, and hearing have dramatically evolved. For this reason, and because we know little about the functional organization of the chemical senses, less attention will be devoted to this subject.

Taste

The elemental taste qualities are salty, sweet, sour, and bitter. More complex tastes are combinations of these four qualities, in much the same way that three primary colors mixed in various combinations can provide a full spectrum of colors.

Receptors

The chemoreceptors involved in taste are the *taste buds*. They are located throughout the mouth but are most concentrated on the dorsal surface of the tongue. They are also found on the palate, faucial pillars, pharynx, and larynx. Taste buds typically occur in groups of one to five on raised *papillae*. Fig. 12-1 includes drawings of a papilla and a taste bud with the transducer cells. The taste bud has 40 to 50 cells arranged in the shape of a bottle; the opening at the top is the *taste pore*.

The transducer cells have microvilli that extend into the taste pore. These microvilli are thought to be chemosensitive, and somehow their contact with the appropriate molecule or molecules leads to a generator potential. Each transducer cell is innervated at its base by fine nerve fibers. Although direct evidence is lacking, these fibers probably include afferent axons that conduct taste information toward the brain as well as efferent fibers from the brain. Also, conventional synapses probably exist between these fibers and the transducer cells.

Recordings from individual sensory fibers indicate that individual taste buds and fibers are not usually selective for any of the four elemental taste stimuli. However, these fibers usually are selective for some of these stimuli but not others. The combined activity of many of these partially selective fibers provides the information needed to extract the relative content of the four elemental stimuli.

Pathways

Fig. 12-2 summarizes the taste pathways. The innervation of the tongue is supplied by fibers from the *facial* (seventh), *glossopharyngeal* (ninth), and *vagus* (tenth) nerves (Fig. 12-2, *B*). The somata of these fibers are located, respectively, in the *geniculate, superior petrosal,* and *nodose ganglia*. These fibers enter the medulla and terminate in the *nucleus of the solitary tract*. Fibers from each half of the tongue innervate the ipsilateral nucleus.

Our knowledge of the central taste pathways is sketchy at best. In the rat, neurons from the nucleus of the solitary tract project ipsilaterally to a pontine nucleus just lateral to the brachium conjunctivum; this pontine structure is located in the caudal region of the *parabrachial nucleus*.* Limited evidence from hamsters, cats, and fish suggests a homologous pathway. In rats, fibers from this region of the parabrachial nucleus

*The more anterior regions of the parabrachial nucleus are functionally distinct from this gustatory region. As noted in Chapter 8, the more rostral cell group innervates the thalamus and cortex.

A

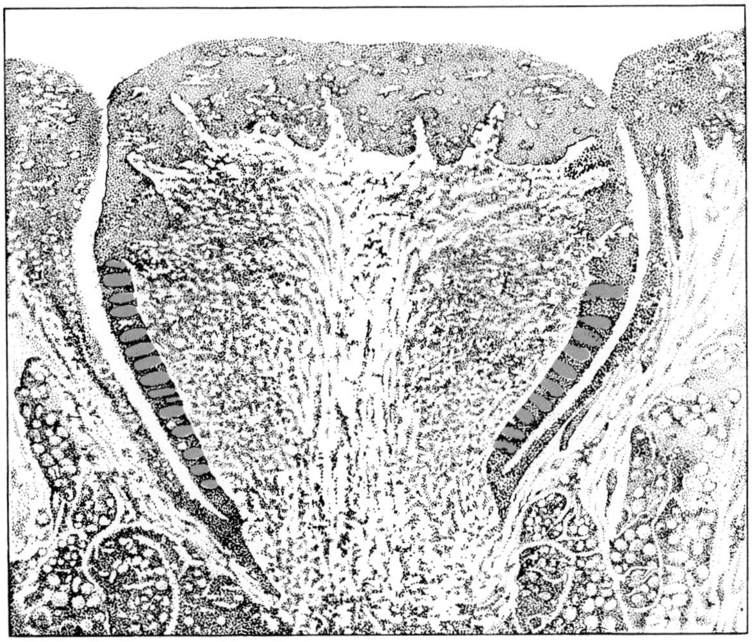

B

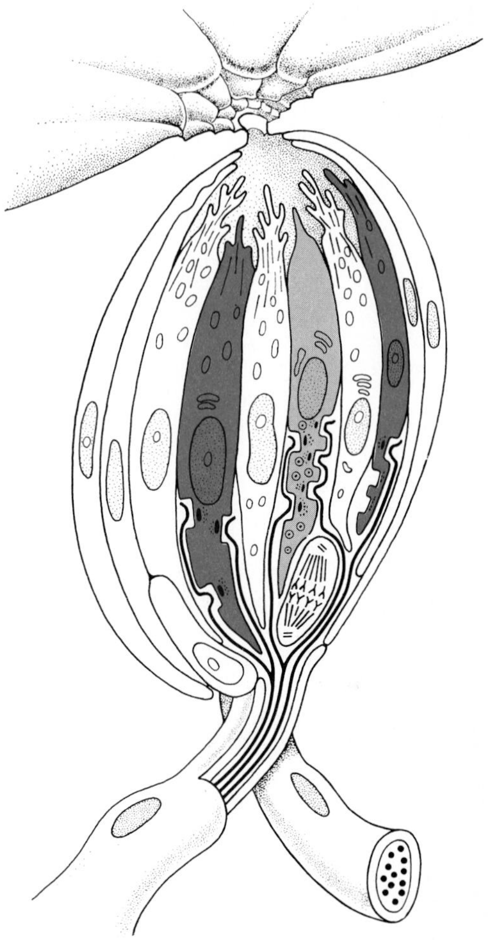

■ **Fig. 12-1.** Taste buds and receptors. **A,** Section through the circumvillate papilla of a macaque monkey. The taste buds are shown in color. The top surface of the papilla is roughly 1.5 mm across. **B,** Taste bud, cut away to expose two types of receptor cells. One *(lighter color)* has presynaptic vesicles with dense cores, and the other *(darker color)* has standard vesicles. Supporting cells are not colored. The receptor cell innervation is shown in black. (**A** redrawn from Bloom, W., and Fawcett, D.W.: A textbook of histology, ed. 8, Philadelphia, 1962, W.B. Saunders Co.; **B** redrawn from Williams, P.L., and Warwick, R.: Functional neuroanatomy of man, Philadelphia, 1975, W.B. Saunders Co.)

terminate in a small-celled region of the thalamic VPM nucleus (Fig. 12-2, *C*). Fibers from this nucleus project to a region of the cerebral cortex immediately ventral to the somatic representation of the tongue. Some of the fibers from these neurons of the parabrachial nucleus bypass the thalamus and directly innervate the cerebral cortex. The details of the central pathways differ in rats and monkeys (and thus probably humans). In monkeys, some axons from the nucleus of the solitary tract bypass the parabrachial nucleus to innervate the VPM nucleus directly. No innervation from this caudal region of the parabrachial nucleus to the thalamus has been documented (see footnote, p. 188). Also, no direct cortical projection from the parabrachial nucleus has been described for monkeys.

The gustatory thalamic relay is part of the ventrobasal (VB) complex described in Chapter 9 (see also Fig. 12-2, *C*). The pontothalamic projection is predominantly ipsilateral, although a small crossed component seems also to be present. The thalamocortical gustatory fibers innervate well-defined cortical taste areas that have been described in many mammalian species (Fig. 12-2, *D*). At least two have been found in primates—one associated within the somatosensory region on the lateral convexity and another in the opercular insular cortex. Note that, unlike the predominantly crossed somatosensory pathway, each half of the tongue is mapped primarily within the ipsilateral hemisphere, at least in primates.

Finally, gustatory information reaches many other brainstem areas. These include the hypothalamus, the amygdaloid complex, the substantia innominata, and the bed nucleus of the stria terminalis.

■ *Smell*

The sense of smell is the least understood of all the major senses, for at least two reasons. First, it is difficult to define adequately either an olfactory stimulus or the response to such a stimulus. There are no "elemental" odors analogous to salty, sweet, sour, or bitter tastes. Thus, even though we have remarkable abilities to detect and identify molecules based on odor (and among mammals, primates are quite poor at this), we lack a detailed, quantitative understanding of these abilities. Hence, investigations of the olfactory pathways are often carried out without a theoretical frame-

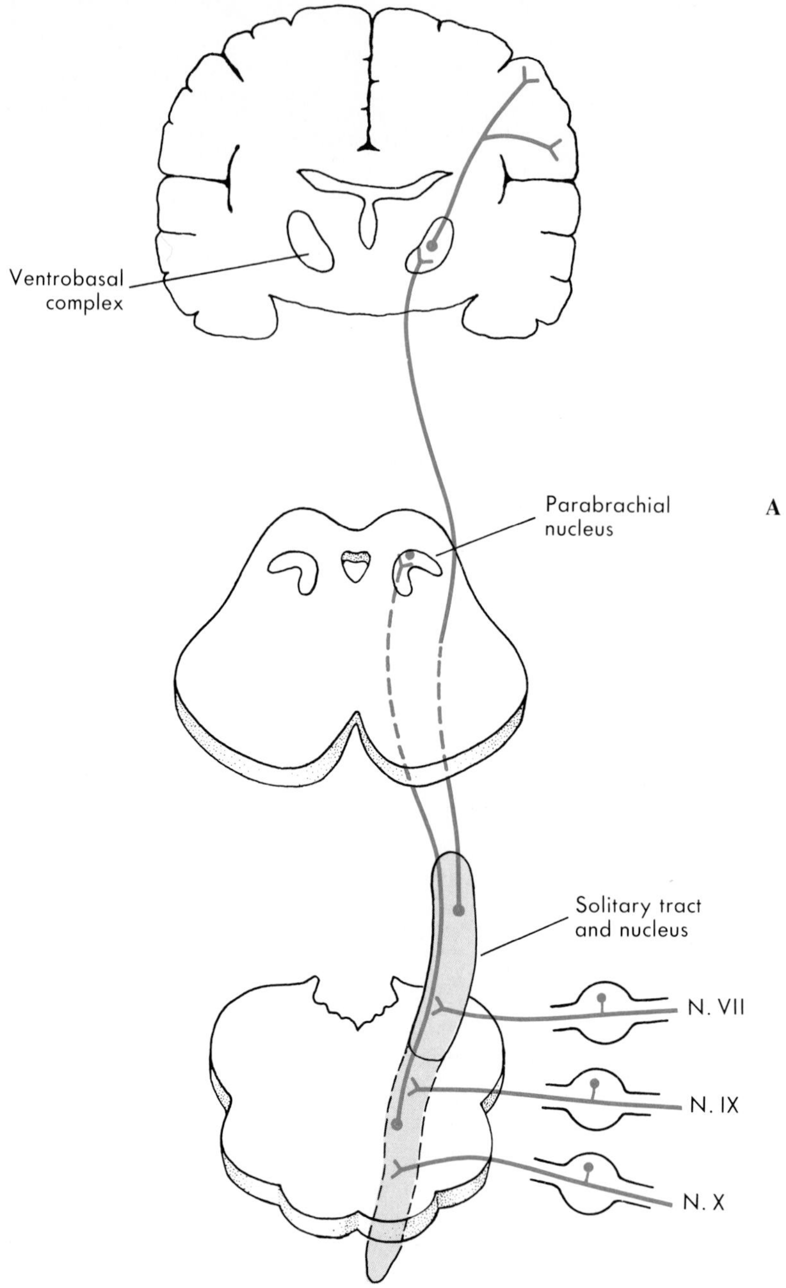

■ **Fig. 12-2.** Summary of known taste pathways in monkeys. **A,** Organization of central pathways. The primary afferent fibers enter the medulla via the facial *(seventh),* glossopharyngeal *(ninth),* and vagus *(tenth)* nerves. Synapses are formed in the nucleus of the solitary tract *(lighter color).* Some second-order fibers ascend to terminate in the parabrachial nucleus of the pons, whereas others terminate in the ventrobasal complex of the thalamus. A projection from the parabrachial nucleus to the thalamus has not yet been described for the monkey, although it appears to exist in many other vertebrates (e.g., rats, hamsters, cats, and fish). Neurons in the thalamus project to the cortex. Finally, in rats, many parabrachial neurons project to both the ventrobasal complex of the thalamus and also directly to cerebral cortex, bypassing the thalamic relay. Such a direct extrathalamic gustatory projection to cortex has not yet been described in primates. **B,** Details of the gustatory innervation of the tongue. **C,** Coronal section through the thalamus of a monkey at the level of the gustatory thalamic relay, the parvicellular portion of the ventral posteromedial nucleus *(VPM(PC)).* Other nuclei shown are ventral posterolateral *(VPL),* ventral posteromedial *(VPM),* centromedial *(CM),* ventral posteroinferior *(VPI),* and parafascicular *(P).* **D,** Two separate cortical representations of taste, termed TNI *(lighter color)* and TNII *(darker color),* in the monkey. Neurons in TNI are responsive to somatosensory as well as gustatory stimuli, but TNII seems to be a purely gustatory area. (**B,** Redrawn from Beidler, L.M.: In Mountcastle, V.B., editor: Medical physiology, ed. 14, St. Louis, 1979, The C.V. Mosby Co. **D,** Redrawn from Benjamin, R.M., and Burton, H.: Brain Res. **7:**221, 1968.)

work built on specific functional questions. Second, the olfactory pathways themselves are diffuse and difficult to characterize. These pathways are probably the most primitive of the major sensory systems, and this might explain the diffuse nature of the central connections.

■ *Receptors*

The olfactory receptor cells lie in the *olfactory mucosa,* which is located on the upper posterior portion of the nasal septum and the opposite part of the lateral wall (Fig. 12-3). The olfactory mucosa occupies only about

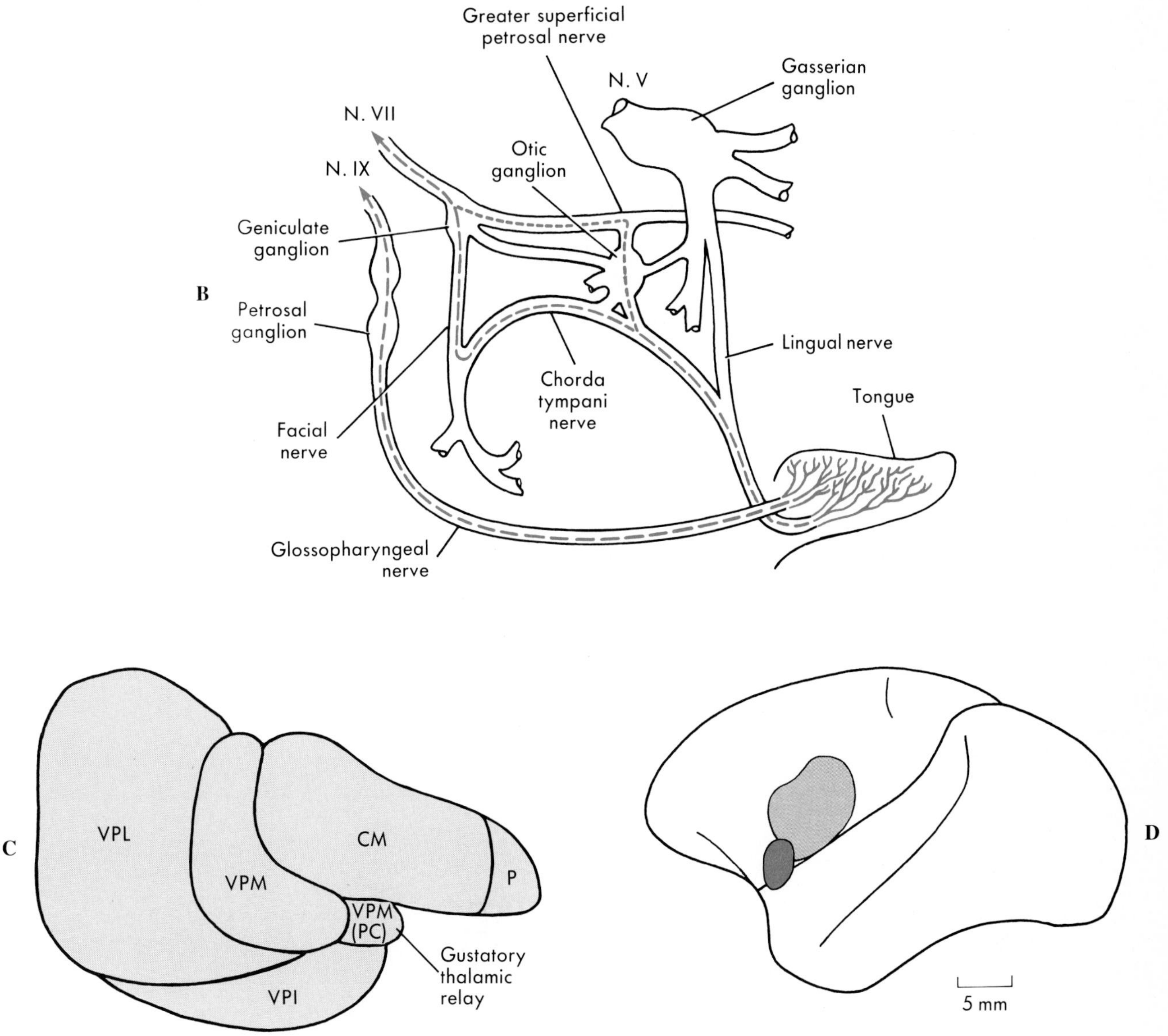

■ **Fig. 12-2, cont'd.** For legend see opposite page.

2.5 square centimeters in each nostril, and it consists of a number of supporting cells as well as the receptor cells.

The receptor cells act as both transducer and ganglion cells (Fig. 12-4, *A*). At one end of the cell a number of microvilli project into the nasal cavity; at the other end the soma thins to a single, fine unmyelinated axon. These axons pass through fenestrations in the cribriform plate of the ethmoid bone and enter the *olfactory bulbs* of the brain. The axons entering the olfactory bulbs form the *olfactory (first) nerve*.

■ *Pathways*

Historically, two general names have been used to describe the olfactory system: the *limbic system* and the *rhinencephalon*. However, the structures conventionally associated with the limbic system are not equivalent to those of the rhinencephalon, and neither is isomorphic with known olfactory pathways. This section deals only with the olfactory pathways.

The olfactory bulbs are located at the anterior ventral surface of the brain. They contain three major neuron types: *tufted* and *mitral cells,* which are second-order projection neurons, and *granule cells,* which are interneurons (Fig. 12-4, *B*). Synapses formed between the olfactory nerve afferent fibers and tufted or mitral cells occur in pronounced morphological specializations known as *glomeruli*. The olfactory bulb also receives inputs from other brain regions, most of unknown origin, but many derive from the contralateral olfactory bulb via the *anterior commissure*.

Fig. 12-5 summarizes some of the known connec-

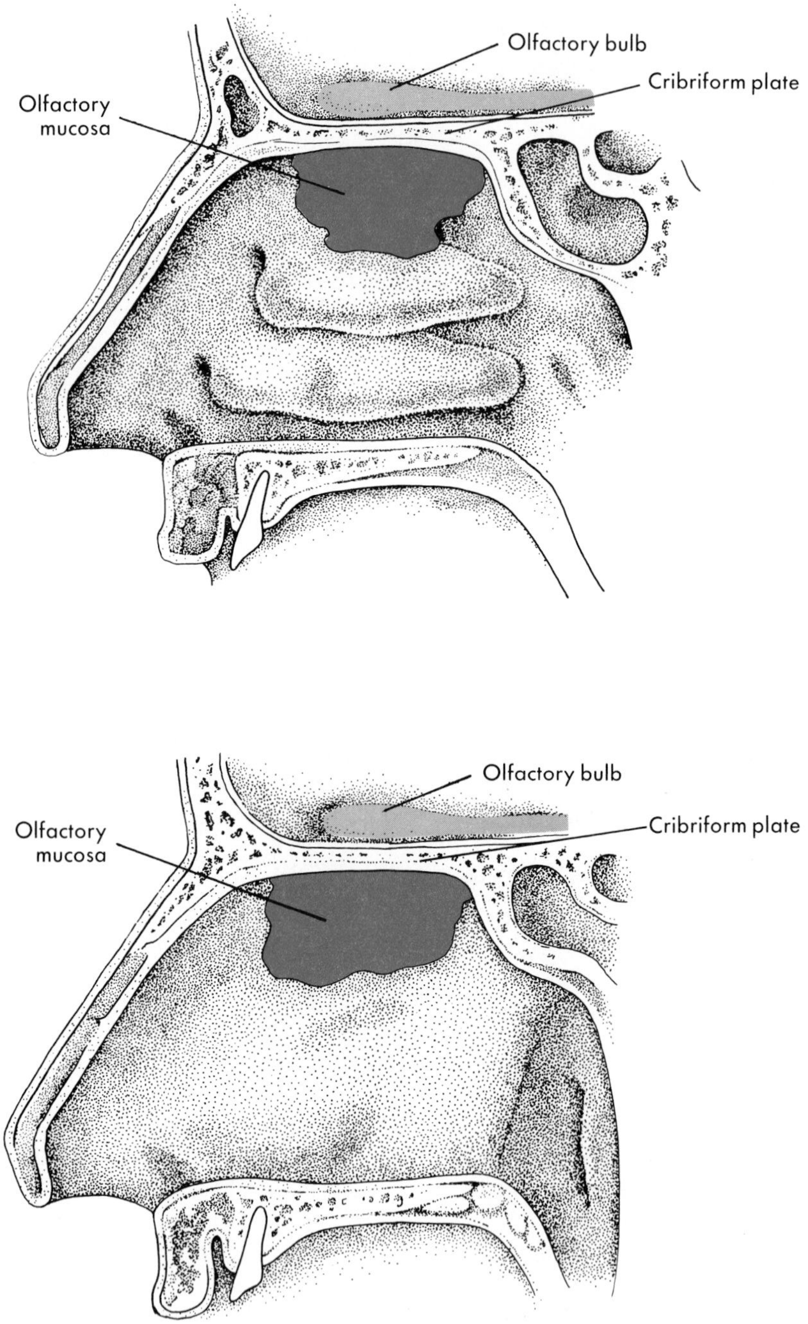

■ **Fig. 12-3.** Location of olfactory mucosa *(darker color)* relative to the olfactory bulb *(lighter color)*. Note the separation of mucosa and bulb by the bony cribriform plate. *Top,* Distribution on lateral nasal wall; *bottom,* distribution on septum.

tions of the olfactory bulb. Fibers of tufted and mitral cells leave the olfactory bulb posteriorly and form the *olfactory tract*. The tract splits into *lateral* and *medial olfactory striae,* and the enclosed triangular zone is known as the *olfactory trigone*. The olfactory tract innervates a number of related structures that collectively define the *olfactory cortex*. These include the *anterior olfactory nucleus,* the *olfactory tubercle* (or *anterior perforated substance*), the *prepiriform cortex,* parts of the *amygdaloid nucleus,* and *transitional entorhinal cortex*. The prepiriform cortex is the largest terminus of the olfactory tract. Finally, as noted above, some fibers of the olfactory tract innervate the contralateral olfactory bulb.

The neocortex is probably involved in olfaction. A pathway has been described from several regions of olfactory cortex to the *mediodorsal* and *submedial* thalamic nuclei. These thalamic nuclei in turn project to prefrontal and orbital regions of neocortex, particularly in the dorsal bank and fundus of the rhinal sulcus. Cells in these cortical areas respond to olfactory bulb stimulation. The functional significance of this newly discovered thalamocortical pathway for olfaction is not yet known.

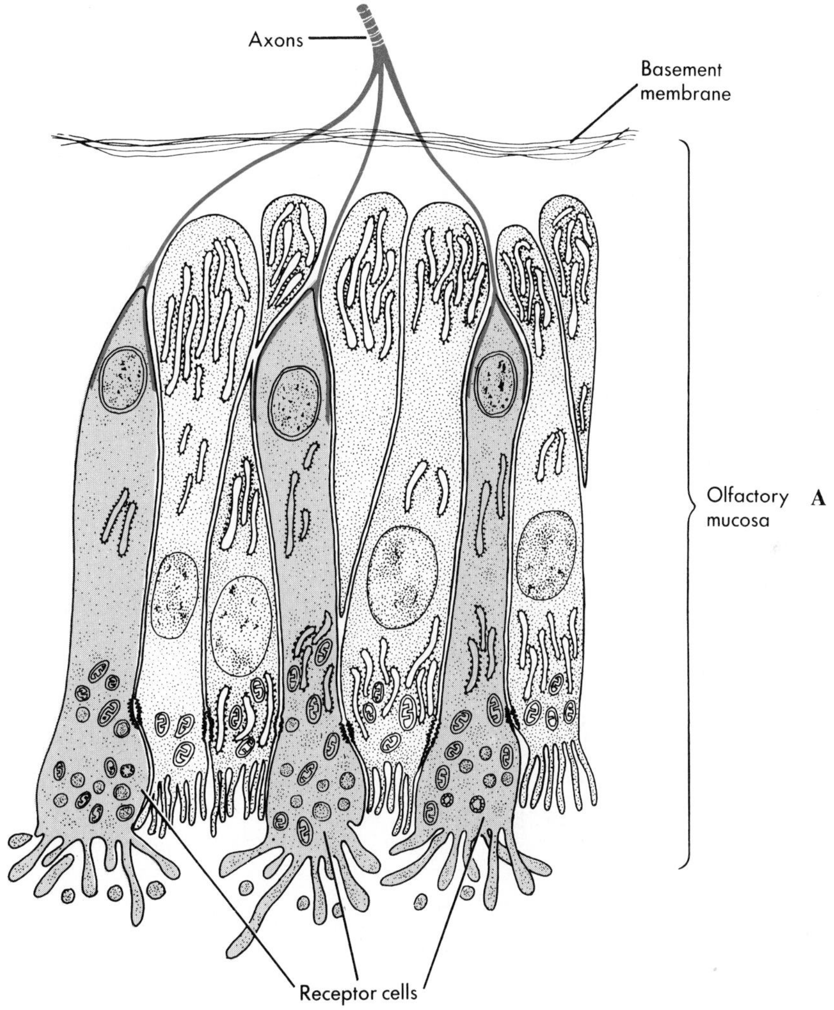

■ **Fig. 12-4. A,** Olfactory receptors. Each receptor (lighter color) is continuous with an axon (darker color) that penetrates the cribriform plate and forms the olfactory nerve. Supporting, or sustentacular, cells are also present in the olfactory mucosa. **B,** Section through the olfactory mucosa and bulb. Axons of olfactory receptors *(darker color)* penetrate the cribriform plate and enter the bulb, where they terminate in glomeruli, ending on processes of mitral and tufted cells *(both black)*. These latter cells send axons out of the bulb. Also shown are granule cells *(lighter color),* which are involved in intrinsic circuitry of the bulb. (**A** redrawn from de Lorenzo, A.J.D. In Zotterman, Y., editor: Olfaction and taste, Elmsford, N.Y., 1963, The Pergamon Press: **B** redrawn from House, E.L., and Pansky, B.: A functional approach to neuroanatomy, ed. 2, New York, 1967, McGraw-Hill Book Co. Copyright © 1967 by McGraw-Hill Book Co. Used with permission.)

Continued.

B

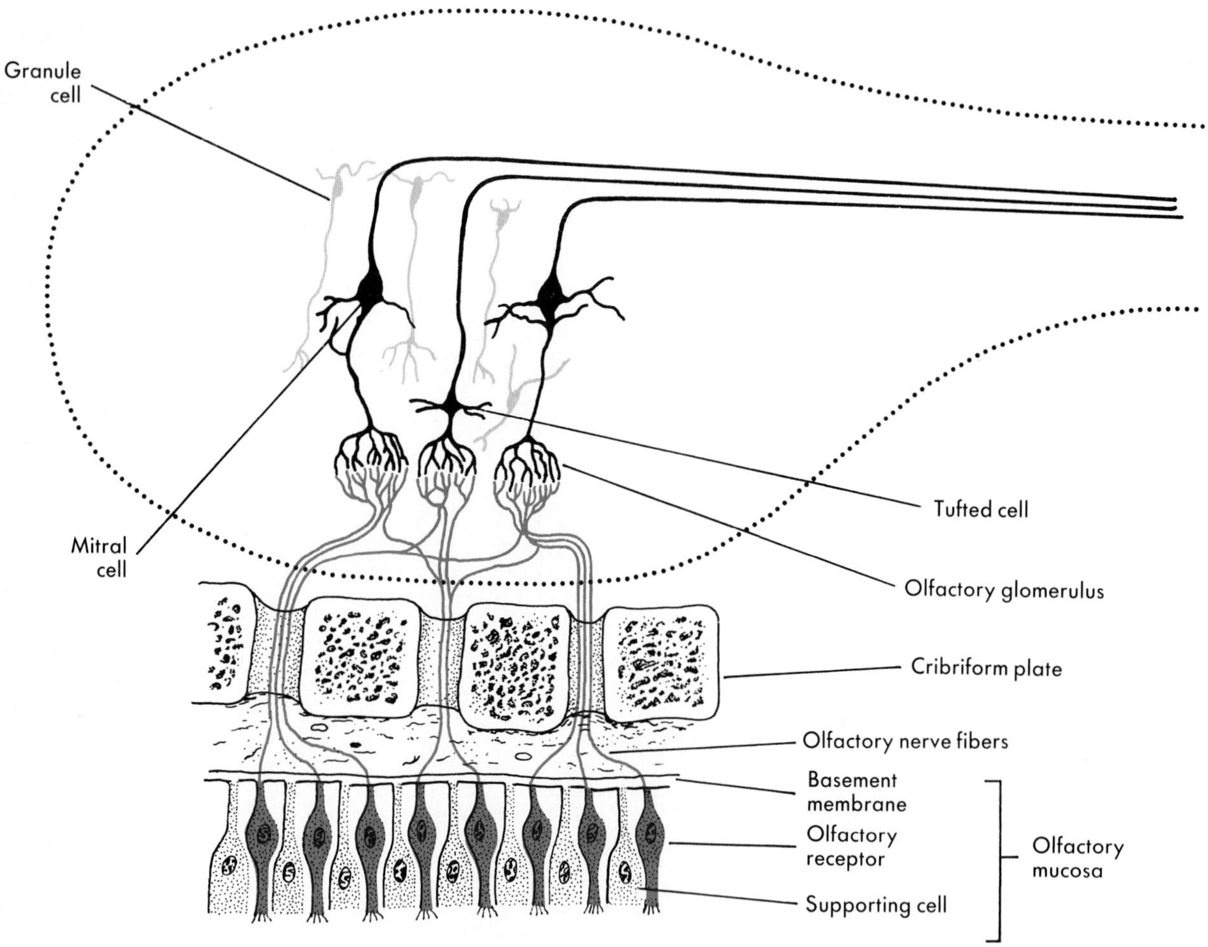

■ **Fig. 12-4, cont'd.** For legend see previous page.

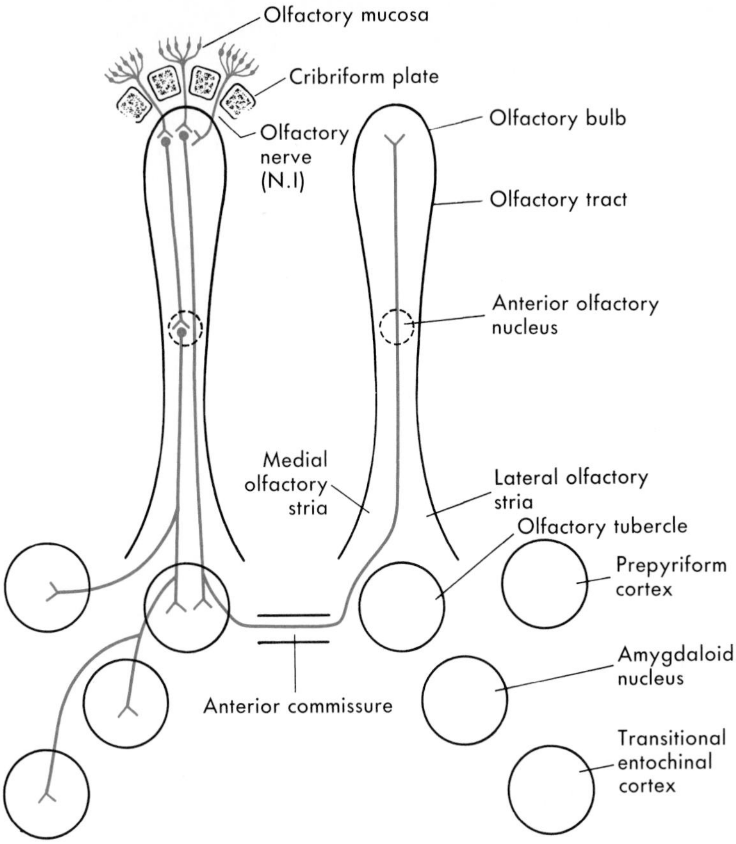

■ **Fig. 12-5** Some of the major olfactory pathways.

■ *Bibliography*

Journal articles

Beckstead, R.M., et al.: The nucleus of the solitary tract in the monkey: projections to the thalamus and brainstem nuclei, J. Comp. Neurol. **190:**259, 1980.

Hamilton, R.B., and Norgren, R.: Central projections of gustatory nerves in the rat, J. Comp. Neurol. **222:**560, 1984.

Lasiter, P.S., and Glanzman, D.L.: Axon collaterals of pontine taste area neurons project to the posterior ventromedial thalamic nucleus and to the gustatory neocortex, Brain Res. **258:**299, 1983.

Macrides, F., and Schneider, S.P.: Laminar organization of mitral and tufted cells in the main olfactory bulb of the adult hamster, J. Comp. Neurol. **208:**419, 1982.

Pfaffmann, C., et al.: Neural mechanisms and behavioral aspects of taste, Annu. Rev. Psychol. **30:**283, 1979.

Price, J.L., and Slotnick, B.N.: Dual olfactory representation in the rat thalamus: an anatomical and electrophysiological study, J. Comp. Neurol. **205:**63, 1983.

Smith, D.V., et al.: Gustatory neuron types in hamster brain stem, J. Neurophysiol. **50:**522, 1983.

Whitehead, M.C., and Frank, M.E.: Anatomy of the gustatory system in the hamster: central projections of the chorda tympani and the lingual nerve, J. Comp. Neurol. **220:**378, 1983.

Yamamoto, T., et al.: Gustatory responses of cortical neurons in rats. I. Response characteristics, J. Neurophysiol. **51:**616, 1984.

Books and monographs

Beidler, L.M.: The chemical senses: gustation and olfaction. In Mountcastle, V.B., editor: Medical physiology, ed. 14, St. Louis, 1979, The C.V. Mosby Co.

Heimer, L.: The olfactory cortex and the ventral striatum. In Livingston, K.E. and Hornykiewicz, O., editors: Limbic mechanisms: the continuing evolution of the limbic system concept, New York, 1978, Plenum Press.

Shepherd, G.M.: The synaptic organization of the brain, ed. 2, New York, 1979, Oxford University Press.

Shepherd, G.M.: Neurobiology, New York, 1983, Oxford University Press.

CHAPTER

13

A Functional Neuroanatomical Framework for Motor Systems

The preceding discussion of sensory systems illustrates the remarkable advances in our understanding of how inputs to the brain are processed. In analyzing sensory processing one is able to present precisely controlled stimuli to a peripheral array of receptors. One can then describe the receptive fields of successively more central neurons along a sensory pathway. When such data are combined with information on the detailed neuronal connectivity of the pathway, one can infer how the pathway is functionally organized to analyze stimulus information.

This capability of analyzing sensory pathways centrally from the periphery has facilitated progress in sensory neurobiology. Unfortunately our understanding of motor systems has not developed as rapidly, in part because this powerful peripheral-to-central approach has been applied only recently. This is not surprising because for the motor systems this requires analysis in a "retrograde" direction, that is, from the muscles centrally. Until recently the techniques available for such an analysis have been limited. However, developments over the past few decades have established a foundation for an organized, functional neuroanatomical treatment of the motor systems beginning, appropriately, at the motor periphery.

A Functional Classification of Muscles

Muscles can be classified in various ways. For example, anatomically one can define flexors and extensors, abductors and adductors, and pronators and supinators. Physiologically one can classify most muscles as flexors or extensors. *Extensors* are defined as muscles that increase the angle at the joint, and *flexors* are muscles that decrease the angle at the joint. In the legs, extensors resist gravity, whereas flexors assist gravity. In all extremities flexors participate in withdrawal reflexes.

A somewhat different approach, however, has considerable heuristic value in the analysis of the central control of movement. This approach is based on a medial-lateral or proximal-distal distinction, in which a class of medial or proximal muscles and a class of lateral or distal muscles are defined. The medial musculature would include, for example, the axial and girdle muscles, be they flexors or extensors. The actions of these muscles involve the axis and proximal limbs. Thus they subserve functions that require body and whole limb movement, such as posture, progression, and equilibration. Although the medial flexor muscles do not perform an antigravity function, they do act in an organized manner with the medial extensors to control posture. The lateral or distal musculature would include, for example, the intrinsic muscles of the digits and the distal muscles of the extremities. Such muscles, flexor or extensor, do not serve a primary postural function, but they mediate manipulatory activity. This is not a rigorous anatomical formulation because the two classes are not necessarily mutually exclusive. However, this formulation helps in understanding the functional neuroanatomy of movement.

Topographical Organization of Motoneuronal Pools

As previously described, the spinal motoneurons, or final common path, are located in the ventral horn and are arrayed in longitudinal columns with different mediolateral positions (see Fig. 6-7). These different positions reflect a topographical organization with respect to target muscles. The most medially situated motoneurons innervate the most proximal muscle groups, and the most laterally situated motoneurons innervate the most distal muscles of the extremities; for example, the intrinsic muscles of the digits are innervated by the

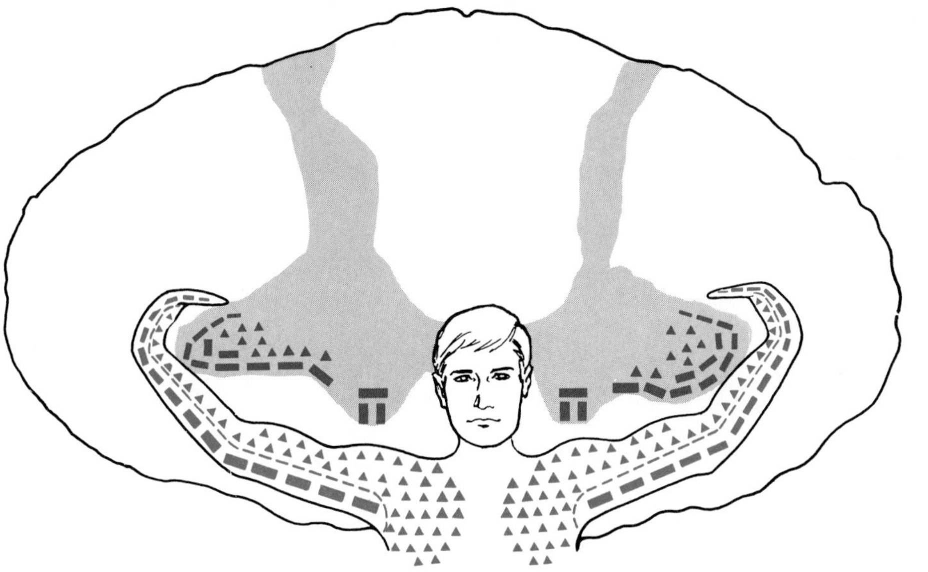

■ Fig. 13-1. Upper body superimposed on a cross section of the spinal cord with corresponding symbols in the spinal cord and on the upper body. The positions of the motoneurons that innervate the muscles of different portions of the upper extremity are shown. More proximal muscles are indicated by larger symbols and more distal muscles by smaller symbols. Correspondingly, the motoneuronal locations supplying the innervation to more proximal muscles are indicated by larger symbols and the locations supplying the innervation to more distal muscles by smaller symbols. This shows the shift from ventromedial to dorsolateral motoneurons as one moves from more proximal to distal muscles of the extremity. Triangles indicate flexor muscles, and rectangles indicate extensor muscles. (Redrawn from Crosby, E.C., Humphrey, T., and Lauer, E.W.: Correlative anatomy of the nervous system, New York, 1962, The Macmillan Co. Copyright © 1962 by the Macmillan Publishing Co., Inc.)

most dorsolaterally located motoneurons (Fig. 13-1). Consequently, this motoneuronal cell column is restricted to the brachial enlargement, and it extends only over the lower cervical and first thoracic segments. As one moves ventromedially in the ventral horn, the motoneuronal pools for successively more proximal limb muscles are encountered. For example, near the base of the ventral horn at middle to lower cervical levels is a column of motoneurons innervating the muscles that act at the shoulder and elbow. Located at lumbosacral levels are analogous columns that innervate the muscles of the lower limbs. Situated more medially are the motoneurons that innervate the muscles of the trunk and those that innervate the muscles of the neck. Thus this cell column extends throughout most of the spinal cord.

This topographical organization, taken in conjunction with the functional muscle classification developed earlier, allows one to consider medial and lateral groups of motoneurons that innervate, respectively, the proximal muscles involved in postural control and the distal muscles involved in manipulatory activity. From this it can be inferred that inputs that preferentially distribute to the more medial motoneurons will be primarily involved in the control of posture. Conversely, inputs that preferentially distribute to the more lateral motoneuronal cell groups will be principally concerned with manipulatory activity.

■ *Topographical Organization of Spinal Interneurons*

Most influences on the motoneurons are not exerted via direct projections on these cells, but rather through spinal interneurons. Thus, whether the mediolateral topography of the motoneurons is preserved in the interneuronal cell groups that project on these motoneurons is critical. Anatomical studies have established that this appears to be the case, although the precision of the topography has not been determined. The more dorsolaterally situated interneurons of the spinal gray matter project preferentially on the more dorsolaterally situated motoneurons (Fig. 13-2). Analogously, more ventromedially situated interneurons project perferentially on the more medially located motoneuronal cell groups. Consequently, it can be concluded that any pathway that preferentially projects on the more medial motoneurons or their associated medial interneurons will be primarily involved in controlling posture. Conversely, any pathway that projects preferentially on the more lat-

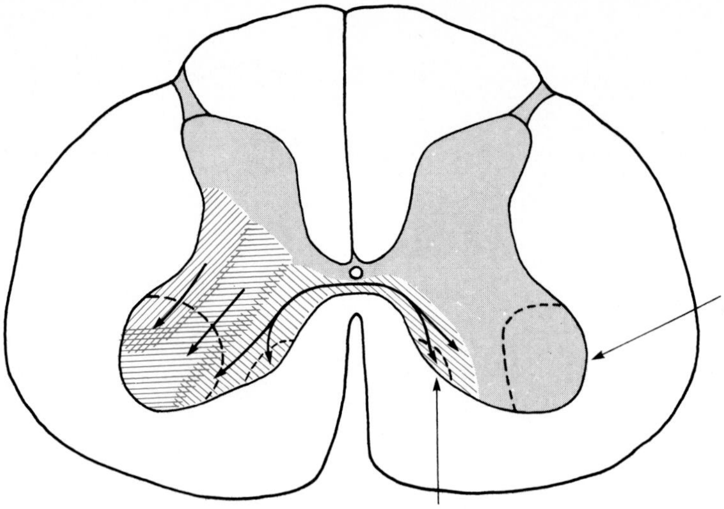

■ Fig. 13-2. Summary of the relationship of interneurons in the intermediate and ventral spinal gray matter to the motoneuronal cell groups. Note that more medially situated interneurons project preferentially to more medially situated motoneurons (bilaterally), whereas more laterally situated interneurons project preferentially on more laterally situated motoneurons. (Redrawn from Sterling, P., and Kuypers, H.G.J.M.: Brain Res. **7**:419, 1968.)

eral motoneurons or their associated lateral interneurons will be more concerned with manipulatory activity involving the distal limb muscles.

■ Bibliography

Journal articles

Sterling, P., and Kuypers, H.G.J.M.: Anatomical organization of the brachial spinal cord of the cat. II. The motoneuron plexus, Brain Res., **4**:16, 1967.

Sterling, P., and Kuypers, H.G.J.M.: Anatomical organization of the brachial spinal cord of the cat. IV. The propriospinal connections, Brain Res. **7**:419, 1968.

Books and monographs

Brodal, A.: Neurological anatomy, ed. 3, New York, 1981, Oxford University Press.

Kuypers, H.G.J.M.: The anatomical organization of the descending pathways and their contributions to motor control especially in primates. In Desmedt, J.E., editor: New developments in electromyography and clinical neurophysiology, vol. 3, Basel, 1973, S. Karger.

Kuypers, H.G.J.M.: A new look at the organization of the motor system. In Kuypers, H.G.J.M., and Martin, G.F., editors: Descending pathways to the spinal cord, Prog. Brain Res. **57**:390, 1982.

CHAPTER

14

Spinal Organization of Motor Function

The various descending pathways that control motor behavior must ultimately act on spinal circuitry (or brainstem circuitry in the case of the cranial nerve motor nuclei). As mentioned previously, only a small proportion of suprasegmental (from above the spinal cord) and segmental (at the level of the spinal cord) inputs that affect the motoneurons terminate monosynaptically on them. Consequently, the neuronal circuitry of the spinal cord must be understood to appreciate how movements are organized. Indeed, a certain amount of motor organization can be mediated by spinal circuitry, and descending influences can elicit patterns of motor behavior by accessing these circuits via appropriate spinal interneurons.

The motor organization at spinal levels has been investigated by studying animals after spinal cord transection. In such an experimental preparation, one can explore the patterns of motor behavior elicited by activating various segmental inputs. One can then delineate the intraspinal circuitry that mediates these patterns. In this chapter we will consider (1) the effects of spinal transection, (2) the classes of segmental inputs that influence the motoneurons, (3) the specific patterns of reflex behavior that are elicited by such inputs, and (4) some general principles of spinal organization.

Effects of Spinal Transection

Immediately on transection of the spinal cord all muscles innervated by spinal segments below the lesion are *paralyzed*. Voluntary contraction of these muscles is permanently abolished. Also, there is a permanent *anesthesia* (loss of sensation) in all parts of the body innervated by segments below the transection. Reflexes mediated at spinal levels are abolished as well, but only temporarily. This *areflexia* is often referred to as *spinal shock*. It includes the loss both of somatic reflexes, such as those mediating muscle tone and the protective withdrawal responses, and of autonomic reflexes, such as those involved in peristalsis, sweat secretion, maintenance of arterial blood pressure, and emptying the bladder and bowels. The level of the transection will, of course, determine the specific pattern of autonomic areflexia.

Although the anesthesia and loss of voluntary muscle contraction are permanent, the areflexia is transient. The spinally mediated reflexes show varying courses of recovery. The evidence is rather compelling that spinal shock results from the loss of suprasegmental influences on the spinal cord. First, the extent and duration of the areflexia are much greater in species that have more extensive suprasegmental control of the spinal cord. For example, in nonmammalian vertebrates, such as amphibia, spinal shock is minimal, and the spinal reflexes reappear in a few minutes. In contrast, recovery in humans requires months. Second, after recovery from spinal shock after an initial transection, the areflexia of the previously affected muscles does not reappear after a second transection caudal to the first. Because this second transection cannot further compromise suprasegmental influences, this experimental result seems to implicate the loss of descending pathways in the genesis of spinal shock. Third, neurophysiological evidence indicates that motoneuronal excitability is depressed after spinal transection.

In humans the time course of recovery from spinal shock is highly variable. Infrequently reflex activity appears as soon as 24 hours. However, most often no reflex activity appears for 2 to 6 weeks. The first reflexes to reappear are flexion responses to tactile stimulation and reflexes involving the genitals and the anal sphincter. With further recovery the flexor responses generally become hyperactive so that an innocuous tactile stimu-

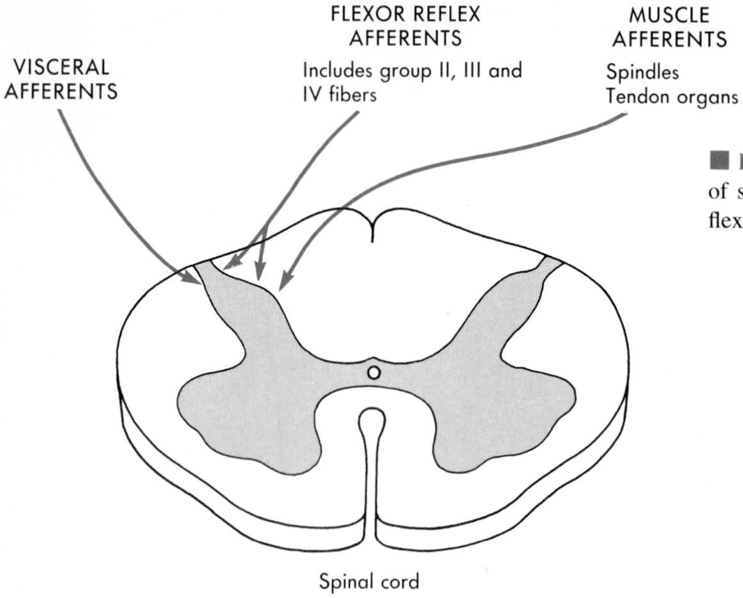

■ Fig. 14-1. Schematic illustration of the three major classes of segmental input to the spinal cord: muscle afferent fibers, flexor reflex afferent fibers, and visceral afferent fibers.

lus can elicit a mass flexion reflex involving many joints. Finally, tendon reflexes recover slowly and may become hyperactive and accompanied by clonus.

The mechanisms of recovery from spinal shock are not known; however, segmental influences probably become more effective. It has been shown that the activity of the motoneurons and the fusimotor neurons increases, and this parallels recovery in spinal animals. A possible mechanism that might contribute to recovery is axonal sprouting (Chapter 21); that is, sensory fibers that enter the cord or spinal interneurons might sprout additional terminals. These terminals may occupy the synaptic space vacated by the terminal degeneration of the suprasegmental projections to the spinal gray matter.

Segmental Inputs, Including Muscle Afferent Fibers, That Elicit Reflex Activity

General Considerations

As a foundation for describing the spinal reflexes, it is first necessary to review the major classes of input from the periphery to the spinal cord (Fig. 14-1). One such class originates in muscles and it transmits to the nervous system information *(proprioception)* on the length or tension of muscles. One type of muscle afferent derives from the *primary (annulospiral) endings* of the muscle spindle. These endings are stretch receptors that transmit information on the length and rate of change of length of muscles. The fibers from these receptors are large (12 to 20 μm) and myelinated and conduct impulses at 70 to 120 m/second; these fibers are in the Ia group. Another proprioceptive type of fibers derives from the *secondary (flower-spray) endings* of the muscle spindle. These, too, are stretch receptors, and they provide information primarily on muscle length. They do so over somewhat smaller myelinated fibers (5 to 12 μm) that conduct impulses at 30 to 70 m/second and are in the group II category. The third type of proprioceptive inputs is derived from the *Golgi tendon organs*. These are specialized stretch receptors located in the region of the tendon, and they transmit information on muscle tension. Their afferent fibers are in the group Ib category; such fibers have axonal diameters and conduction velocities that approximate those of the Ia afferent fibers.

The second major class of segmental inputs germane to the spinal reflexes consists of somatosensory receptors. This class includes a wide variety of receptors that are located throughout the body and contribute to the group II, III, and IV afferent fiber classes. However, from the perspective of the spinal reflexes these fibers can be aggregated into a single functional class designated *flexor reflex afferent fibers* because they exert a common reflex action, namely flexor patterns.

The final major class of segmental inputs consists of the *visceral afferent fibers*. This complex and highly differentiated class of inputs forms the afferent limb of the spinal autonomic reflexes. Included, for example, are stretch receptors that are located in the urinary bladder; the activation of these receptors elicits reflexes that empty the bladder.

Muscle Afferent Fibers

Three of the just mentioned classes of segmental afferent fibers originate in muscle. It is important to elaborate on these before proceeding to the reflexes. These

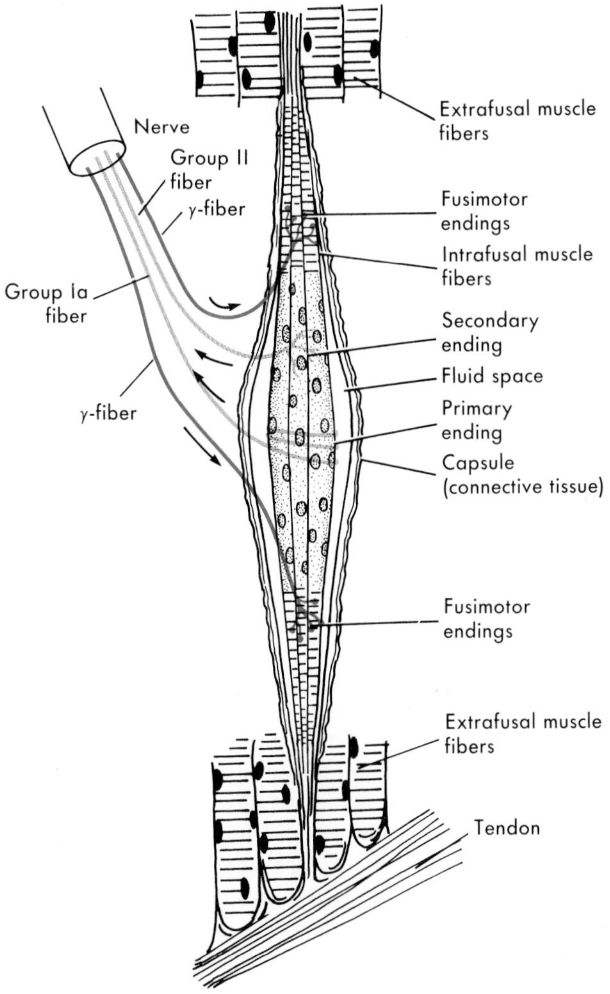

■ **Fig. 14-2.** Schematic illustration of the muscle spindle and its principal components. (Redrawn from Brodal, A.: Neurological anatomy, ed. 3. Copyright © 1981 by Oxford University Press, Inc. Reprinted by permission.)

proprioceptors consist of the muscle spindle receptors and the Golgi tendon organ; almost all muscles contain this receptor complement. The *muscle spindle* is a fusiform- or spindle-shaped structure (Fig. 14-2) that is several millimeters in length and just a few hundred microns in width. The spindle consists of a connective tissue sheath that encloses 2 to 12 *intrafusal muscle fibers;* each end of the spindle is attached to extrafusal muscle fibers. The spindles are distributed throughout the muscle. Their density appears to vary with the degree of control required by a given muscle. The intrinsic muscles of the digits have a considerably greater density of spindles per muscle mass than, for example, do the large muscles involved in postural control.

There are at least two types of intrafusal muscle fibers, the *nuclear bag fibers* and the *nuclear chain fibers* (Fig. 14-3). The nuclear bag fibers are longer and wider than the nuclear chain fibers. They have an expanded central or *equatorial* region with densely aggregated nuclei; hence, the name nuclear bag. Beyond the equatorial region and most prominently at the distal poles, the nuclear bag fiber contains contractile elements. The smaller nuclear chain fibers have fewer nuclei that are serially arrayed within the central region. These fibers too, have contractile material distal to the central region.

The primary endings of the muscle spindle coil around the central regions of both the nuclear bag and nuclear chain fibers, whereas the secondary endings arise from pericentral regions of the nuclear chain fiber with minimal termination on the nuclear bag fiber (Fig. 14-3). A typical spindle contains two nuclear bag fibers and about five nuclear chain fibers. The spindle is innervated by a single group Ia afferent fiber and a single group II afferent fiber. These fibers ramify within the spindle to supply the various primary and secondary endings.

The intrafusal fibers, in addition to their sensory innervation, receive motor innervation from the central nervous system; they are therefore under *centrifugal* control (Fig. 14-3). This innervation derives from smaller neurons, *γ-motoneurons* or *fusimotor neurons,* that are located in the ventral horn among the α-motoneurons that innervate the extrafusal muscle fibers. The axon diameter of the γ-motoneurons is smaller than that of the α-motoneuron. The γ-motoneuron axons travel in muscle nerves, and they consist of two prominent types (Fig. 14-3). One terminates in discrete motor endplates *(plate endings)* on the distal (contractile) poles of the nuclear bag fibers. The other terminates as an extensive network of *trail endings,* primarily on the nuclear chain fibers. One class of fibers, Aβ fibers, innervates both intrafusal and extrafusal muscle fibers via plate endings. The Aβ fibers are sparse, however, and will not be discussed further.

The third source of muscle afferent fibers, the Golgi tendon organ, is not located in the muscle spindle but at the musculotendinous junction (Fig. 14-4). The tendon organ consists of a rich arborization of small unmyelinated fibers that extend for perhaps 500 μm and are enclosed in a fine capsule. Tendon organs are less numerous than spindles in any given muscle, and they are more abundant in muscles that contract slowly.

The muscle spindles are arranged *in parallel* to the extrafusal fibers, whereas the Golgi tendon organs are situated *in series* with the muscle (Fig. 14-5). Because the primary endings, secondary endings, and Golgi tendon organs are stretch receptors, this arrangement is critical for their adequate stimulation. Being in parallel with the muscle, the spindle will stretch as the muscle lengthens (relaxes). This will deform the primary and secondary endings on the central and pericentral regions of the intrafusal fibers. The deformation initiates a generator potential and ultimately an action potential in the group Ia or II fibers that arise from the spindle. Conversely, contraction of a muscle will shorten the muscle spindle and reduce the stretch on the primary and sec-

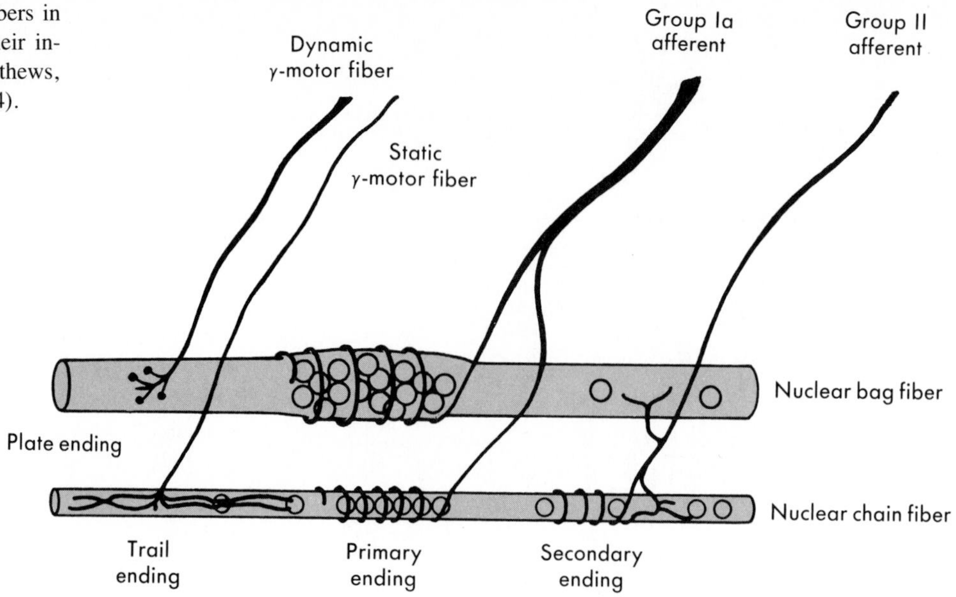

■ **Fig. 14-3.** Types of intrafusal fibers in mammalian muscle spindles and their innervation. (Redrawn from Matthews, P.B.C.: Physiol. Rev. **44:**219, 1964).

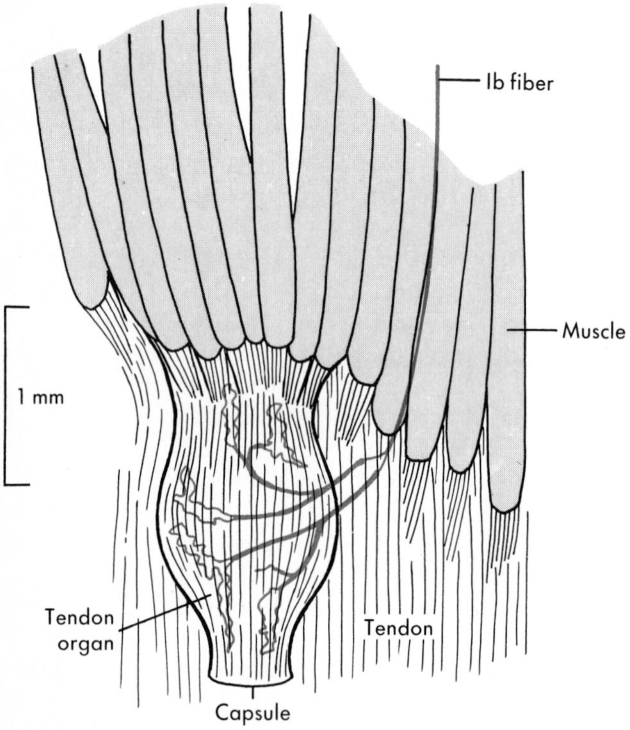

■ **Fig. 14-4.** Schematic illustration of a Golgi tendon organ. (Redrawn from Barker, D.: Muscle receptors, Hong Kong, 1962, Hong Kong University Press.)

ondary endings, thus decreasing their discharge. In contrast to this behavior of spindle receptors, the Golgi tendon organ, which is in series with the muscle, is most effectively stretched, and hence activated, by muscle contraction (Fig. 14-6). Although stretch of a muscle can generate enough tension at the musculotendinous junction to activate the Golgi tendon organ, this is a much less effective stimulus than muscle contraction because of the relative elasticities of the muscle fibers and the tendon.

The Golgi tendon organ is a highly sensitive stretch receptor that transmits information regarding the *force* generated at the musculotendinous junction; indeed, its threshold is less than 0.1 g. Thus the discharge of the tendon organ does not signal muscle length because the force generated during a muscle contraction will vary with the load against which the muscle is working; for example, considerable tension can be generated during isometric contraction. It should also be appreciated that a tendon organ will reflect the forces of the specific muscle fibers inserting on it. Therefore these receptors do not signal the average force of the entire muscle, but they contribute information regarding local force. They also show both static and dynamic components in their discharge; that is, they are sensitive to both the velocity of force development as well as to the steady-state force.

In contrast to the tendon organs, spindle receptors are activated by lengthening of muscle. However, in the discussion thus far, no functional distinction has been made between the primary and secondary endings. Both receptor types signal muscle length (Fig. 14-7). If the discharges of primary (Ia) and secondary (II) endings at different fixed muscle lengths are recorded, their discharge rates show an increase as a function of length. Moreover, at any given length these receptors show a *static response* that is maintained as long as the muscle is maintained at that length; that is, little adaptation occurs. If the activity of these receptors is recorded *during the lengthening* of muscle, a distinction can be made between the primary and secondary endings (Fig. 14-7). In such an experiment the primary ending discharge is velocity sensitive during lengthening; that is, it has a *dynamic response component* so that during muscle

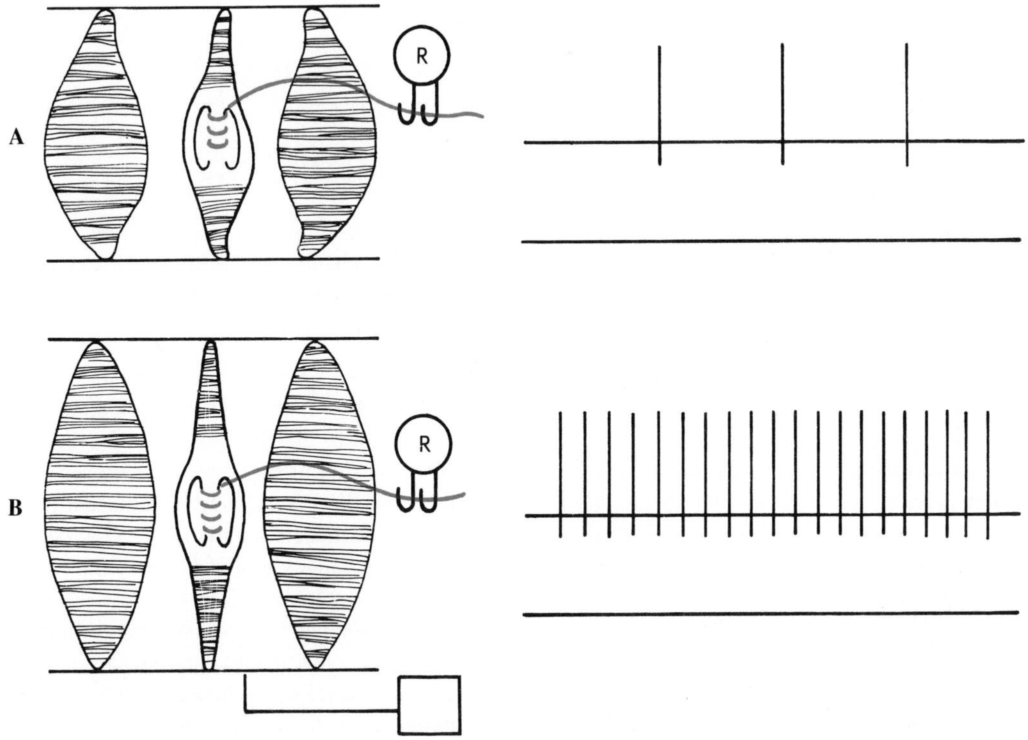

Fig. 14-5. The effects of muscle shortening (contraction) **(A)** and muscle lengthening (stretch) **(B)** on the discharge rate of an afferent fiber from a spindle of the muscle. Contraction of the muscle shortens the spindle and reduces the discharge rate of the afferent fiber. Stretch of the muscle stretches the spindle and increases the discharge rate. *R,* Recording electrode. (Redrawn from Eyzaguirre, C., and Fidone, S.J.: Physiology of the nervous system, ed. 2, Chicago, 1975, Year Book Medical Publishers, Inc.; modified from Ruch, T.C., and Patton, H.D.: Physiology and biophysics, ed. 19, Philadelphia, 1965, W.B. Saunders Co.: & Hunt, C.C., and Kuffler, S.W.: J. Physiol. **113:** 298, 1951.)

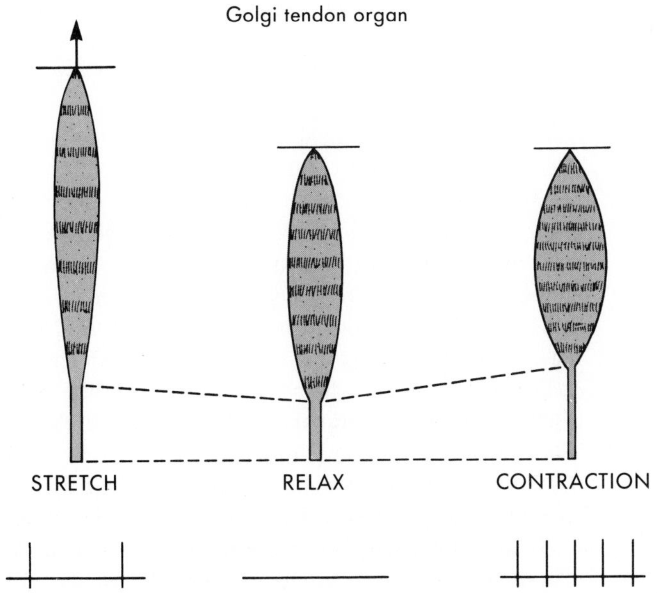

Fig. 14-6. Tendon elongation during muscle stretch, relaxation, and active contraction. Contraction is more effective in elongating the tendon than externally applied stretch, where some of the force is dissipated in the elongation of the muscle. Consequently, active contraction is more effective in eliciting an increase in discharge from the Golgi tendon organ. (Redrawn with permission from Eyzaguirre, C., and Fidone, S.J.: Physiology of the nervous system, ed. 2. Copyright © 1975 by Year Book Medical Publishers, Inc., Chicago.)

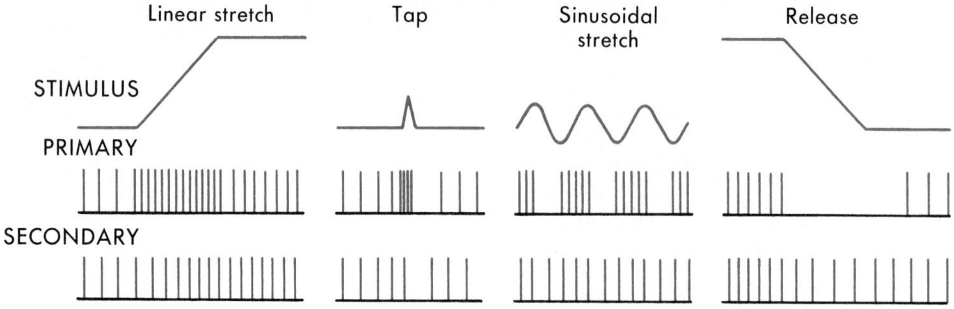

■ **Fig. 14-7.** Idealized responses of primary and secondary endings to various manipulations of muscle length. The responses are shown as if the muscle were initially under moderate stretch and with no fusimotor activity. (From Matthews, P.B.C.: Physiol. Rev. **44:** 219, 1964.)

lengthening its discharge is a function of both the length and the velocity of lengthening. In contrast, the secondary ending has little or no dynamic response, and it continues to signal principally length during the active phase of stretch. Again, under static or steady-state conditions (at fixed length) the behaviors of the two types of receptors are very similar. During release from stretch (as during contraction), the muscle spindle shortens and "unloads" the force at the receptor regions (Fig. 14-7). During this unloading the secondary ending shows a gradual decrease in discharge proportional to muscle length. However, the primary ending shows a dynamic response during such shortening, providing information on the velocity of shortening. In fact, the response may even show a cessation of discharge or *silent period* during the release from stretch (Fig. 14-7).

The mechanisms of the different discharge properties of the primary and secondary endings are not fully understood. They may reflect differences in the viscoelastic properties of the nuclear bag and nuclear chain fibers. The secondary endings terminate primarily on the nuclear chain fibers (Fig. 14-3). Thus these fibers may have properties that give rise to the static response of the secondary endings. If this is correct, then the static component of the primary ending discharge could derive from its termination on the nuclear chain fiber, whereas the dynamic component would be mediated by its termination on the nuclear bag fiber (Fig. 14-3).

The final aspect of the muscle afferent fibers to be considered is the motor innervation of the muscle spindle (Fig. 14-3). This innervation originates from γ-motoneurons in the ventral horn. The fusimotor innervation is confined to the intrafusal fibers, and it causes the more distal regions of these fibers to contract. If an intrafusal fiber is maintained at a constant length, then contraction of its distal poles by fusimotor activation would result in a lengthening of its more central regions where the receptors are located (Fig. 14-8). Thus activation of the fusimotor fibers increases the tension in the receptor regions. This central control of the spindle receptor sensitivity allows these receptors to respond with high resolution over a wide range of muscle lengths. For example, if the receptors can discharge at rates of up to 500 per second, then the fusimotor control of tension at the receptor region can be set so that this entire discharge range is dedicated to but a few millimeters of muscle lengthening. This permits resolution of length changes on the order of hundreds of microns. If the fusimotor discharge is then readjusted, this "sensitivity window" can be shifted to another portion of the muscle's possible excursion. During movements initiated by descending neural pathways, the α- and γ-motoneurons to a given muscle are probably *coactivated* (Fig. 14-9). With such an arrangement the spindle-receptor sensitivity is maintained during muscle contraction because the parallel activation of the fusimotor fibers of that muscle prevents unloading of the spindle.

Thus, the spindle receptors do not signal absolute muscle length. A given discharge rate of a spindle receptor may correspond to various muscle lengths, depending on the level of fusimotor discharge. The spindle receptors actually transmit information about the length of the spindle relative to the muscle length. Both the fusimotor discharge and the spindle afferent discharge must be known to infer absolute muscle length. How this integrative computation is achieved centrally is not yet understood.

The two morphological types of fusimotor fibers, plate endings and trail endings, may correspond to two functional fusimotor classes (Fig. 14-10). One such class, the *static fusimotor fiber,* serves to increase the static responses from both the primary and secondary endings. It has been suggested that this functional class may correspond to the fusimotor fibers with trail endings. (The trail endings distribute preferentially to the nuclear chain fibers, which may give rise to the static discharge from the primary and secondary endings.) The other functional type, the *dynamic fusimotor fiber,* has little effect on the static response from the spindle receptors, but it increases its velocity sensitivity. These fibers may correspond to the fibers with plate endings. (These are distributed preferentially to the nuclear bag

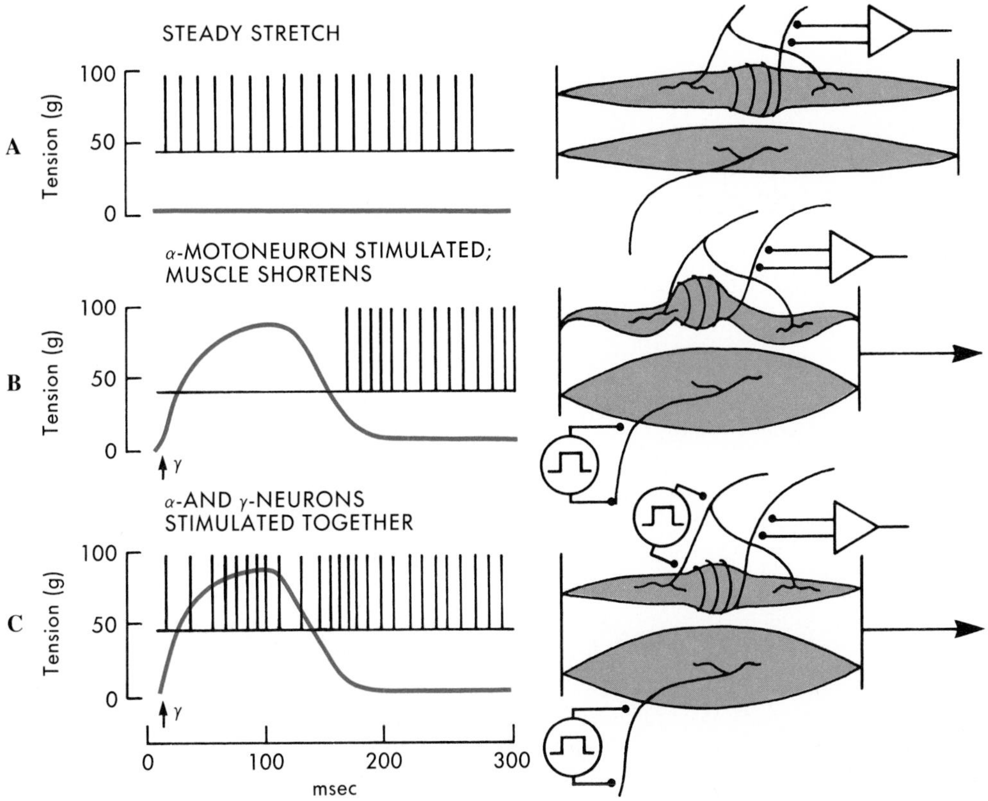

■ **Fig. 14-8.** Effect of fusimotor stimulation during muscle contraction. **A,** During steady stretch the discharge of the spindle afferent fiber is maintained. **B,** With muscle contraction the spindle shortens, and the discharge of the afferent fiber ceases. **C,** If the fusimotor innervation is activated together with the innervation of the extrafusal fibers, then the discharge is maintained during shortening of the muscle. (Redrawn from Kuffler, S.W., and Nicholls, J.G.: From neuron to brain, Sunderland, Mass., 1976, Sinauer Associates, Inc.)

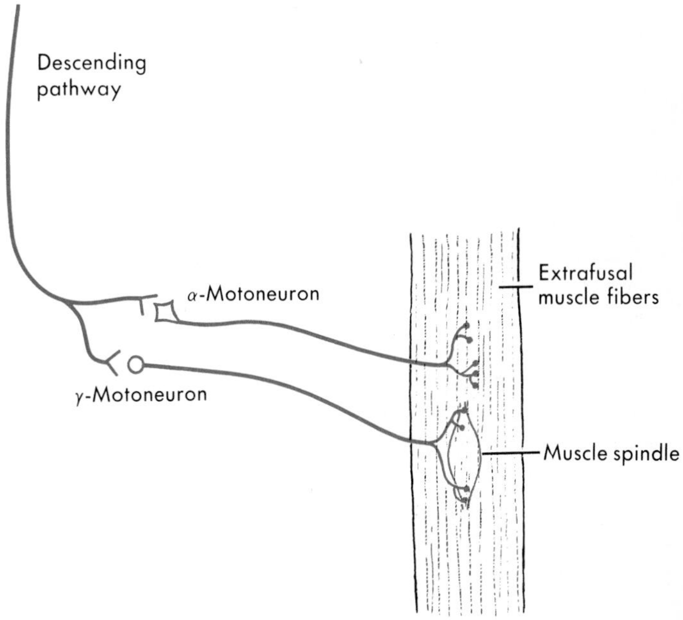

■ **Fig. 14-9.** Schematic illustration of "coactivation" of α- and γ-motoneurons. A descending fiber of a motor pathway activates in parallel an α-motoneuron and a γ-motoneuron innervating the same muscle. In this way the shortening of the spindle during contraction of the muscle is prevented; thus the discharge of the spindle sensory fibers is maintained. See Fig. 14-8.

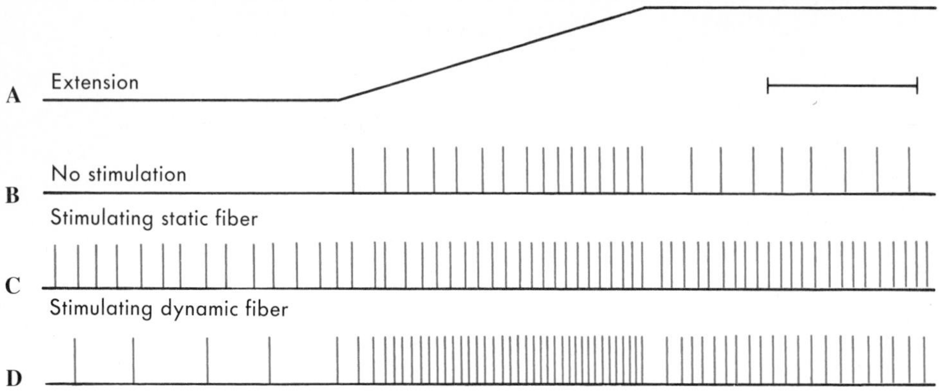

Fig. 14-10. Effects of stimulating single fusimotor fibers on the response of a primary ending to muscle stretch. **A,** Extension of a muscle. **B,** Absence of stimulation. **C,** The effects of stimulating a static fusimotor fiber. **D,** The effects of stimulating a dynamic fusimotor fiber. (From Crowe, A., and Matthews, P.B.C.: J. Physiol. **174:**109, 1964.)

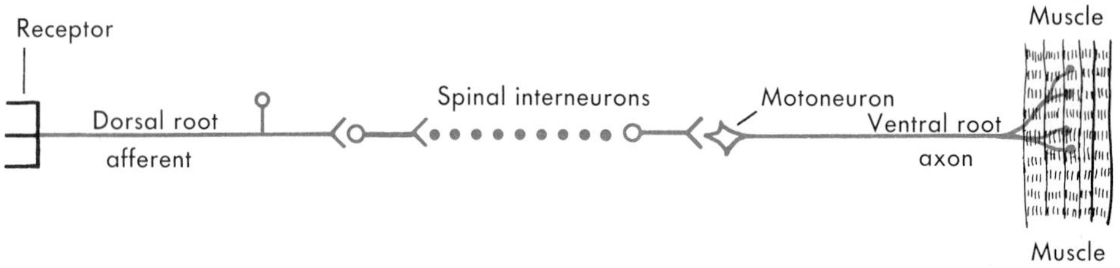

Fig. 14-11. Schematic illustration of a reflex pathway consisting of a receptor and its afferent fiber, spinal interneurons, and a motoneuron and the effector it innervates.

fibers, which presumably give rise to the dynamic response of the primary ending.) Some nuclear bag fibers apparently have trail endings. These fibers are a subtype of nuclear bag fibers with some static properties.

In brief, the muscle spindle is a complex receptor organ that has a dual afferentation and at least a dual efferentation. The receptors respond to muscle stretch or lengthening and provide information to the nervous system regarding muscle length (primary and secondary endings) and rate of change of length (primary endings). The sensitivity of these receptors can be adjusted by the central nervous system via the γ-motoneurons. The static and dynamic fusimotor innervations allow the static and dynamic receptor discharges to be differentially controlled. Such a complex proprioceptive system is important because the sequence of muscle contractions required to move an arm to a given position is a function of the starting position of the limb. Information about muscle length is sufficient to infer that starting position. Given this, the appropriate sequence of muscle contractions can be programed. However, in much of our motor behavior, the limb will not be static, but moving. Therefore the system must also be able to predict the limb's position at any given moment; this requires information on the velocity of change in muscle length. The muscle spindle system with its receptors and their centrifugal control is clearly designed to provide the information needed to program goal-directed movements.

Spinal Reflexes

A foundation has now been established for discussing the spinal reflexes elicited by each of the afferent classes described earlier. By way of introduction, Fig. 14-11 illustrates a basic spinal reflex circuit. It consists of (1) a receptor and its afferent fiber, (2) one or more spinal interneurons (with the exception of the myotatic reflex), (3) a motoneuron and its axon, and (4) an effector or muscle. The pathways for autonomic reflexes are more complex and will be described later. The properties of any reflex depend on the characteristics of the receptors initiating the reflex, the nature of the spinal circuitry, and the anatomical distribution of the involved motoneurons.

The Myotatic Reflex

The *myotatic (tendon) reflex (myotatic* means muscle stretch) is one of the most important postural reflexes. It is spinally mediated and is elicited by activation of the primary (Ia) endings of the muscle spindle (Fig. 14-12).

The Ia afferent fiber, on entering the spinal cord, bifurcates to give rise to a branch that ascends in the dorsal columns and a segmental branch that terminates largely within the segment of entry. The ascending Ia fibers ultimately provide information on muscle length

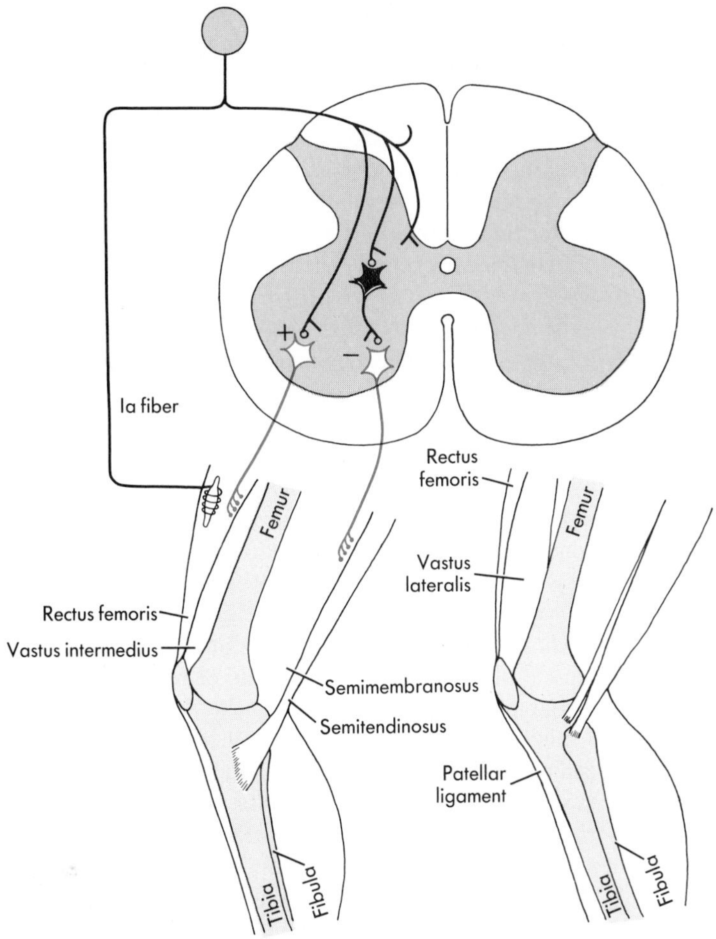

■ **Fig. 14-12.** The neural circuitry for the myotatic reflex. The afferent segment of the reflex pathway consists of the group Ia fibers from a spindle of the homonymous muscle. The afferent segment then monosynaptically innervates motoneurons of that muscle (*plus sign*). Through an inhibitory interneuron, shown in black, it disynaptically inhibits *(minus sign)* the motoneurons of the antagonistic muscle (reciprocal innervation).

and velocity of lengthening to various central structures, such as the cerebellum and cerebral cortex. The segmental branch terminates not only on various interneurons, but also directly *(monosynaptically)* on α-motoneurons. The pattern of this monosynaptic termination is specific, because it is confined to motoneurons *(homonymous motoneurons),* innervating the muscle from which the Ia afferent fiber arose. In fact, a single Ia afferent fiber terminates on most, if not all, cells of the homonymous motoneuron pool. To a lesser extent the Ia afferent fibers also terminate on motoneurons innervating muscles that act synergistically with their muscles of origin.

Given the properties of the primary endings and this simple monosynaptic circuit, the nature of the myotatic reflex can readily be inferred. The adequate stimulus for the reflex is muscle stretch. When this occurs, the primary endings are activated, and the resulting discharge in the rapidly conducting Ia afferent fibers provides a monosynaptic excitatory input to the homonymous motoneurons. Activation of these motoneurons then elicits contraction of the stretched muscle. Consequently, the myotatic reflex is a *stretch-evoked reflex,* and it produces muscle contraction in response to stretch. Because the primary ending discharge has both dynamic and static components, the myotatic reflex can be activated by either of these discharge components. Thus in the case of a rapid muscle stretch, as with a tap of the patellar tendon, the resulting myotatic reflex (contraction of the quadriceps muscles and hence extension around the knee) will be driven principally by the dynamic component of the primary ending discharge. With sustained stretch of a muscle, the resulting myotatic reflex will be driven by the static component of the primary ending discharge (Fig. 14-7).

The myotatic reflex circuitry illustrates a general principle of spinal organization called *reciprocal innervation*. The Ia afferent fibers, in addition to their monosynaptic terminations on homonymous motoneurons, project on a population of spinal interneurons that inhibit the motoneurons of muscles antagonistic to the homonymous muscle (Fig. 14-12). This reciprocal innervation functions to relax the muscles that would otherwise oppose the myotatic reflex action. This feature is prominent in spinal organization, and it integrates the action of agonists and antagonists at a given joint.

The myotatic reflex, particularly its static component, is prominent in the medial extensor muscles that serve a critical antigravity function. Thus this reflex is a fundamental postural mechanism. For example, when a person is upright, gravity tends to stretch the quadriceps muscles. This stretching force elicits a sustained reflex contraction of the quadriceps. That reflex contraction maintains extension around the knee joint and thereby an upright posture. Because the primary endings show little adaptation, this reflex contraction can be sustained as long as the stretch is imposed. Furthermore, the finely graded nature of the primary ending discharge, its faithful transmission over rapidly conducting afferent fibers, and the involvement of but a single synapse in the reflex circuit all combine to permit rapid and finely graded reflex contractions in response to small increments in muscle length. The secondary endings of the muscle spindle may also contribute to the myotatic reflex (static component) by means of a monosynaptic connection to the homonymous motoneurons.

The myotatic reflex is also critical for muscle tone. *Muscle tone* is defined as the resistance of a muscle to active or passive stretch, and this is precisely the action of the myotatic reflex. Alterations in tone frequently occur clinically with lesions at many levels of the neuraxis. In most instances of altered tone, the lesion affects the myotatic reflex by perturbing the fusimotor control of spindle sensitivity. For example, the atonia after spinal transection largely results from a severe reduction of γ-motoneuronal discharge consequent to the loss of suprasegmental inputs to the γ-motoneurons. The fusimotor system maintains a background level of discharge that is modulated up or down. Elimination of this discharge shifts the spindle to one extreme of its dynamic range. Hence, even large muscle stretches may be insufficient to generate enough force on the intrafusal fibers to activate their receptors. At the other extreme, any lesion that removes major inhibitory influences on the γ-motoneurons will increase the background discharge and raise the sensitivity of the spindle receptors. The myotatic reflexes may then become abnormally active. This results in *hypertonia* or *spasticity*. *Spastic paralysis* would then refer to the situation in which interruption of descending pathways has eliminated the ability to contract a muscle voluntarily via the α-motoneurons (paralysis), but the spinally mediated myotatic reflexes involving that muscle are hyperactive (spasticity). Interrupting the myotatic reflex circuit by cutting the dorsal roots (dorsal rhizotomy), and therefore the Ia afferent fibers, will generally eliminate the hypertonia and transform it to hypotonia.

A common clinical sign of hypertonia is *clonus*, which is a sequence of alternating muscle contractions and relaxations. In the hypertonia resulting from enhanced fusimotor discharge, a tendon tap will elicit an abnormally brisk myotatic reflex. For example, exaggerated extension of the leg occurs in response to a patellar tendon tap. Because the tendon tap is a transient stimulus and the myotatic reflex ends abruptly with stimulus termination, the extension is immediately followed by relaxation or lengthening. Under normal fusimotor control, the leg would return to a resting position. However, with enhanced fusimotor discharge the stretch imposed by the return to a resting position elicits another myotatic reflex contraction, although somewhat weaker. This second contraction could then elicit a third contraction which in turn could elicit a fourth contraction. Thus, a damped sequence of a number of contractions could arise from a single tendon tap.

Although tendon stretch (tap) is an effective clinical means of evaluating the myotatic reflexes of various muscles, the monosynaptic myotatic circuit can also be monitored electrophysiologically in humans. Superficial electrical stimulation of the skin behind the knee (the popliteal fossa) spreads sufficient current to excite the group Ia afferent fibers of the medial popliteal nerve, which carries sensory information from the gastrocnemius muscle. This elicits in that muscle a myotatic response called *Hoffmann's reflex* (H reflex). With recording electrodes on the surface of the gastrocnemius muscle, this myotatic response can be recorded electromyographically.

The dynamic and static components of the primary ending discharge can be modulated independently by the dynamic and static fusimotor fibers, respectively. Hence it is possible to affect differentially myotatic responses to rapid muscle stretch and sustained stretch. Because the response to tendon tap is primarily driven by the dynamic component of the Ia fiber discharge, deep tendon reflexes could be augmented by any lesion that selectively increases the activity of the dynamic fusimotor fibers. In contrast, a lesion that selectively increases the activity of the static fusimotor fibers would not enhance deep tendon reflexes, but it would augment myotatic responses to sustained stretch. This kind of hypertonia, as in the plastic rigidity of parkinsonism, would not be velocity sensitive.

In brief, the myotatic reflex is elicited by stretch of a muscle; it is a reflex contraction in response to such stretch. This reflex is the basic mechanism of muscle tone and serves a critical postural function. The afferent segment of the reflex circuit consists of the primary (and possibly secondary) endings of the muscle spindle and their afferent fibers. The spinal circuitry is simple, consisting of a monosynaptic excitatory connection to the homonymous motoneurons. Thus the reflex is local because its effects are largely restricted to the muscle from which the activating afferent fiber arises. Moreover, the reflex action is rapid (short latency), stimulus-locked, finely graded, and capable of being sustained (particularly in medial extensor muscles). The myotatic circuitry also includes a disynaptically mediated inhibition of the antagonistic muscles, producing relaxation of the antagonist muscles. This constitutes reciprocal innervation.

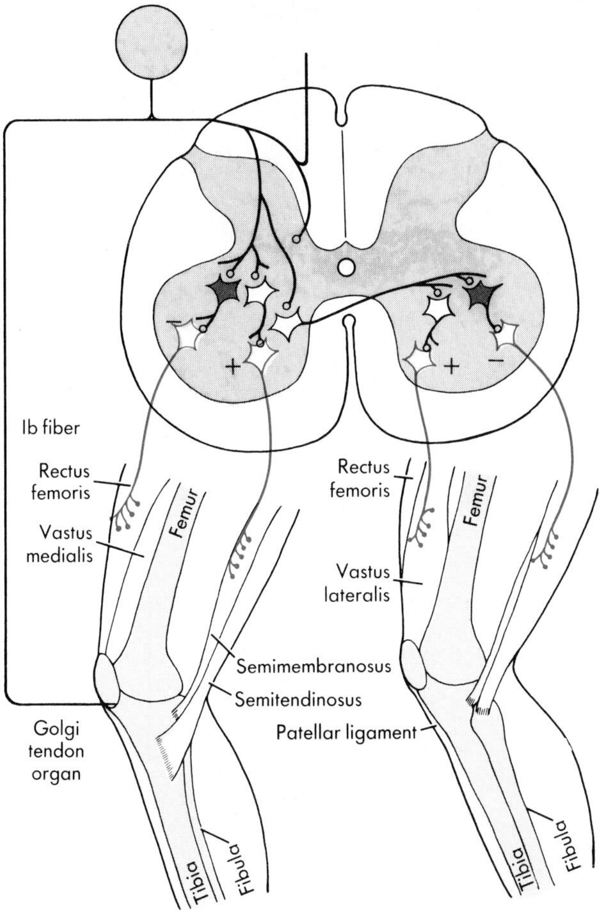

■ **Fig. 14-13.** The neural circuitry for the inverse myotatic reflex. The afferent segment of the reflex pathway consists of the group Ib fibers from Golgi tendon organs. These afferent fibers disynaptically inhibit motoneurons innervating the contracting muscle. They disynaptically excite the motoneurons innervating the antagonistic muscle to produce reciprocal innervation. The afferent fibers also excite a commissural neuron that disynaptically mediates a crossed-extensor response that also shows reciprocal innervation. Black interneurons and minus signs are inhibitory, and clear interneurons and plus signs are excitatory.

■ *The Inverse Myotatic Reflex*

Another important spinal reflex is initiated by the activation of the Golgi tendon organs. In this circuit (Fig. 14-13) the Ib afferent fibers from the tendon organs terminate on spinal interneurons that inhibit the homonymous motoneurons and synergistic motoneurons, including those acting at other joints. Thus the reflex action relaxes a muscle in response to its contraction. The distribution of this reflex effect is most prominent in antigravity muscles. Because the myotatic reflex is a contraction of a muscle in response to lengthening, this tendon organ–mediated reflex has been called the *inverse myotatic reflex*. However, the adequate stimulus for the myotatic reflex is muscle length, whereas that for the inverse myotatic reflex is muscle tension. Therefore, the designation inverse myotatic reflex is not truly appropriate.

Like the myotatic circuit, the inverse myotatic reflex includes a disynaptic reciprocal innervation (Fig. 14-13). However, in contrast to the myotatic reflex, it also includes a crossed-component (Fig. 14-13). If, for example, the inverse myotatic action is to relax the quadriceps muscles of one limb in response to their contraction, the crossed-component excites the motoneurons of the contralateral quadriceps muscles. This crossed-component, called *Phillipson's reflex,* itself includes the circuitry for reciprocal innervation so that the antagonistic hamstring muscles of the contralateral limb are relaxed. This reciprocal innervation in the crossed-component of a reflex is sometimes referred to as *double reciprocal innervation*. A teleological view of the crossed-component is to consider it as postural in function. If the inverse myotatic action is on the antigravity muscles of a limb, leading to their relaxation, then the crossed-component will be extensor in nature. This crossed extension establishes postural support.

Unlike the myotatic reflex, the inverse myotatic reflex cannot be demonstrated behaviorally, except in a hypertonic limb. If, for example, one begins to flex a spastic lower limb at the knee, increasing resistance to the stretch of the extensor muscles will be encountered. This resistance will continue to increase as more force is imposed until at some point the extensor contraction abruptly terminates, and the limb passively flexes. This behavior inspired the descriptive name *clasp-knife reaction;* another frequently used designation is the *lengthening reaction*.

This reflex is mediated via the Golgi tendon organs. It was originally thought that these tendon organs were high-threshold receptors and that the inverse myotatic reflex protected the muscle from overloading. However, we now know that tendon organs are, in fact, highly sensitive receptors if the appropriate stimulus, muscle contraction, is applied. The inverse myotatic reflex is involved in spinally organized motor patterns. It may, for instance, contribute to the smooth onset and termination of contraction, the rotation among motor units during sustained contraction, the rapid switching between flexion and extension required for such behavior as running, and as a tension feedback system.

■ *Stretch-Evoked Flexion*

In spinally transected animals, activating the group II spindle afferent fibers results in an excitation of flexor motoneurons that resembles that seen in the flexor withdrawal reflex; extensor motoneurons are inhibited. This finding generated the hypothesis that the secondary endings mediate a stretch-evoked flexion reflex that might contribute to motor patterns such as stepping.

However, in intact or decerebrate preparations, activating the group II spindle afferent fibers produces an effect that is opposite to that in the spinal animal. In the intact or decerebrate animal the secondary endings appear to excite extensor and inhibit flexor motoneurons. This effect is more consistent with the recently discovered role of the secondary endings in contributing to the myotatic reflex.

The secondary endings may contribute to the clasp-knife effect (inverse myotatic reflex). This would again implicate the group II spindle afferent fibers in a stretch-evoked flexion. How this relates to their excitatory action on the extensor motoneurons is unclear.

The Flexor Withdrawal Reflex

The final spinal reflex to be considered is elicited by activation of a diverse group of receptors that signal somatosensory information, particularly pain. These receptors give rise to afferent fibers in the group II, III, and IV classes. On entering the spinal cord, the fibers course both rostrally and caudally, terminating extensively among the spinal interneurons of many segments (Fig. 14-14). Through complex and incompletely understood spinal circuitry, these afferent fibers ultimately excite flexor motoneurons that innervate flexor muscles (Fig. 14-15).

If only the group II somatic afferent fibers (not to be confused with the group II spindle afferent fibers that arise from the secondary endings) are activated, a weak and rather restricted flexor pattern results. With inclusion of the group III afferent fibers, which include fibers from nociceptors that mediate fast pain, a much more active flexion response occurs that involves more joints. The maximal reflex response is evoked when the group IV afferent fibers, which mediate slow pain, are included as well. Because this diverse class of somatic afferent fibers elicits a common reflex pattern of flexion, they have been aggregated for functional purposes as the *flexor reflex afferent fibers*. It is clear from the afferent information that elicits the reflex that the reflex protects by withdrawing the stimulated body part from potentially damaging stimuli *(nociceptive stimuli)*. This contrasts with the other spinal reflexes that subserve postural functions.

As with the other spinal reflexes, the general principle of reciprocal innervation applies to the organization of the flexor withdrawal reflex (Fig. 14-15). In this case the reciprocal innervation consists of a polysynaptic inhibition of the extensor motoneurons antagonistic to the excited flexor motoneurons. Moreover, the flexor reflex includes a crossed-extensor component (Fig. 14-15) that serves a postural function (as in the inverse myotatic

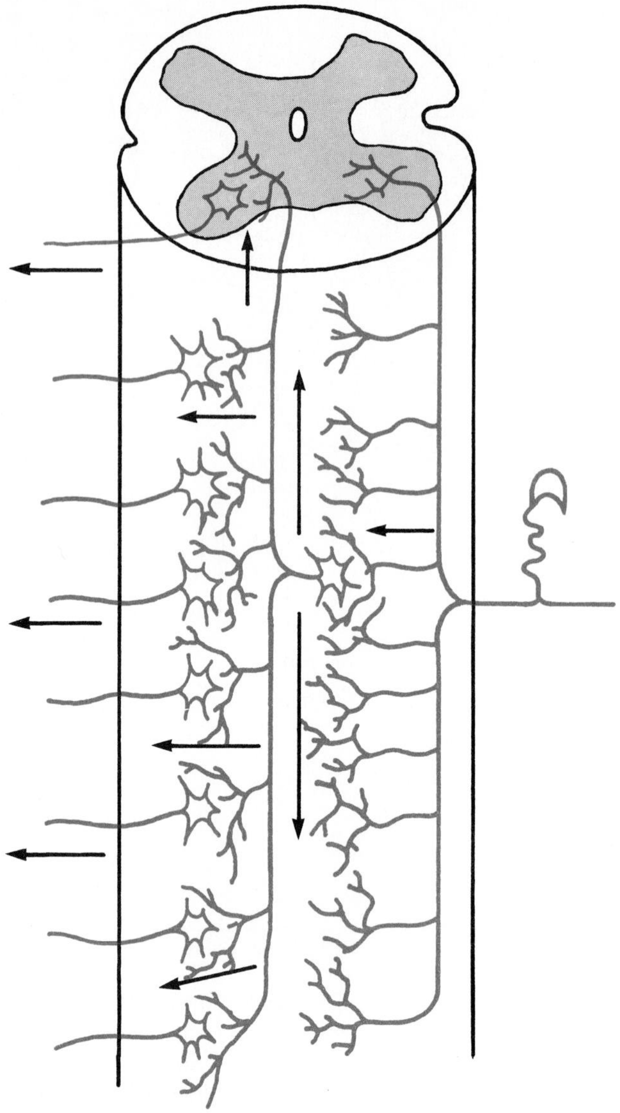

Fig. 14-14. Schematic illustration of the divergence that occurs with the flexor reflex afferent fibers that constitute the afferent segment of the flexor withdrawal reflex. (Redrawn with permission from Eyzaguirre, C., and Fidone, S.J.: Physiology of the nervous system, ed. 2. Copyright © 1975 by Year Book Medical Publishers, Inc., Chicago. Slightly modified from Cajal, S.R.: Histologie du système nerveux, Paris, 1909, Maloine.)

reflex), and the crossed-component includes reciprocal innervation. This constitutes a complex neural network that mediates a basic motor pattern consisting of flexion of one limb, for example, and extension of the contralateral limb.

The properties of the flexor withdrawal reflex are different from those of the myotatic reflex. These differences are derived from the nature of the involved receptors, the degree of divergence of the afferent projections to the spinal cord, the complexity of the intraspinal circuitry, and the distribution of the reflex effects to target muscles. First, the more slowly conducting fibers of the

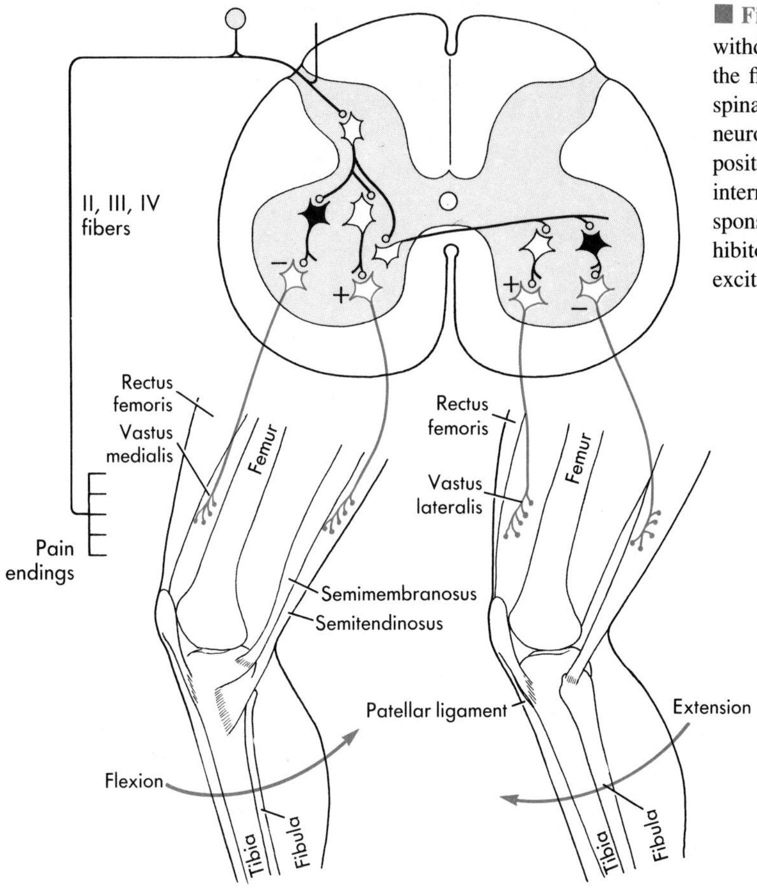

■ **Fig. 14-15.** The neural circuitry of the flexor withdrawal reflex. The afferent segment consists of the flexor reflex afferent fibers. Through a chain of spinal interneurons these fibers excite flexor motoneurons and inhibit extensor motoneurons. The opposite pattern occurs contralaterally via commissural interneurons to produce a crossed-extensor response. Black interneurons and minus signs are inhibitory, and clear interneurons and plus signs are excitatory.

afferent segment and the more extensive intraspinal circuitry increase the latency of the flexor withdrawal reflex. Second, whereas the myotatic reflex is finely graded as a function of muscle stretch, the flexor withdrawal reflex has a much more nonlinear input-output relationship. At low stimulus intensities the response, is small. However, as the nociceptive afferent fibers are activated, a full response develops that is augmented little by further increases in stimulus intensity. Third, whereas the myotatic reflex is stimulus locked and terminates abruptly with stimulus termination, the flexor response persists beyond stimulus termination. This reflects the persistence of neuronal activity within the complex spinal circuitry that mediates the reflex, a phenomenon designated *afterdischarge*. Recurrent circuits probably are responsible for such afterdischarge, which serves to maintain withdrawal from a tissue-damaging stimulus for a sufficient time to allow adjustments that prevent recontacting that stimulus. Fourth, the reflex effect is less specific than that of the myotatic reflex, reflecting the considerable divergence of the sensory inputs, the complexity of the spinal circuitry, and the wide distribution to different motoneuronal pools.

Despite the extensive divergence in the flexor withdrawal circuitry (Figs. 14-14 and 14-15), the output is well organized. The extent to which flexors around a given joint are activated is a function of the location of the stimulus, a property called *local sign* (Fig. 14-16). For example, if the nociceptive stimulus is delivered to the plantar surface of the foot, then the resulting reflex action will include considerable flexion at the ankle, somewhat less at the knee, and still less at the hip. Stimulation of the skin of the popliteal fossa will evoke a reflex response that includes considerable flexion at the hip and knee, with much less at the ankle. Thus the final limb position will be a function of the site of stimulation. The flexor pattern removes the stimulated site from the threatening stimulus.

The preceding discussion describes the general flexor withdrawal response. Specific reflexes of a different nature can be evoked from various localized regions. For example, in the spinal animal, tactile stimulation of the footpads can elicit an extension of the entire limb, ostensibly a postural response (the *positive supporting reaction)*. The most effective stimulus for producing this effect is not activation of the cutaneous afferent fibers but stretch of the interosseus muscles.

Some Principles of Spinal Organization

Two important properties of spinal organization are *convergence* and *divergence*. Although these can apply to the connections between any two cell groups, they are readily illustrated by the segmental afferent fibers to the spinal cord. Convergence refers to a many-to-one projection, where axons from a number of neurons converge to terminate on a single neuron. An excellent example of this is the convergence of many Ia afferent fibers from a given muscle on each motoneuron of the motoneuronal pool of that muscle. An example of divergence, a one-to-many projection, is the distribution of group III and IV cutaneous afferent fibers that ramify extensively on entering the spinal cord (Fig. 14-14).

Convergence is the anatomical foundation for *spatial summation*. In this phenomenon a neuron receives multiple inputs, none of which alone may be capable of discharging it. However, activity in more than one of these convergent inputs can summate to discharge the target neuron. A related phenomenon is *temporal summation*. In this case a given level of discharge of a single input may be insufficient to discharge the target neuron. However, increasing the frequency of that input may permit sufficient summation of postsynaptic potentials to activate the target cell.

Both spatial and temporal summation are examples of *facilitation,* where an input to a neuron increases the probability of its discharging in response to an input from another source (spatial) or to a greater number of inputs from the same source (temporal). A basis for this phenomenon is that the distribution of terminals of a given fiber, for example, to a motoneuronal pool, may involve considerable divergence. Some of the motoneurons contacted by that fiber may be insufficiently excited to discharge, whereas others may be activated by that fiber. Thus one can define a *subliminal fringe* and a *discharge zone* for the set of motoneurons contacted by a given fiber (Fig. 14-17). Another fiber projecting onto the motoneuronal pool will also diverge to contact a number of cells; some of these will be in its subliminal fringe and some in its discharge zone. The population of motoneurons contacted by this second fiber may not be identical to that contacted by the first fiber. Thus these inputs *fractionate* the motoneuronal pool. However, the neurons contacted by each fiber may overlap, so that convergence occurs on certain of them. If this overlap involves their subliminal fringes, then facilitation by spatial summation can occur. The effect of stimulating the two fibers simultaneously will then be greater than the sum of the effects of stimulating each individually. In contrast, if their discharge zones overlap, the effects of stimulating the two fibers simultaneously may be less than the sum of the effects of stimulating them individually. This phenomenon is called *occlusion*.

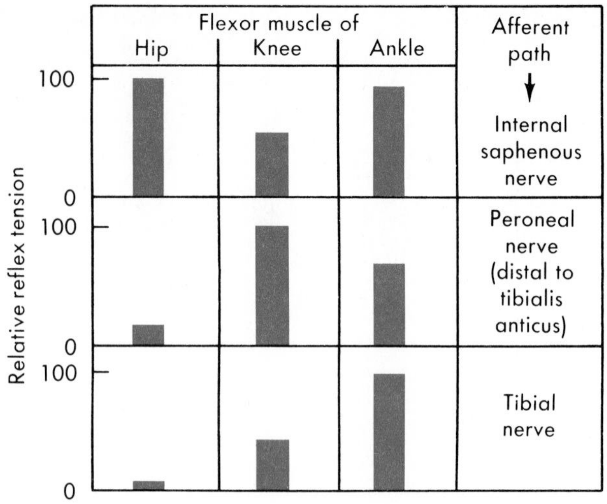

Fig. 14-16. Local sign in the flexor withdrawal reflex. The bars indicate relative reflex tensions developed in three flexor muscles as a result of stimulation of each of three afferent nerves innervating different parts of the hindlimb. Each path activates all three muscles, but the relative participation of each muscle in the reflex movement varies as a function of the afferent path that is stimulated. (Redrawn from Patton, H.D.: Reflex regulation of movement and posture. In Ruch, T., and Patton, H.D., editors; Physiology and biophysics, vol. IV, Philadelphia, 1982, W.B. Saunders Co.)

Another principle that can be illustrated with the motoneuronal pool is that of feedback circuits (Fig. 14-18). To this point we have described the ventral horn as containing α- and γ-motoneurons. However, also located in the ventral horn is an intrinsic neuron (interneuron), designated the *Renshaw cell,* that is inhibitory to local motoneurons. The Renshaw cell receives inputs from various sources; one important source is the collaterals of motoneuron axons. The Renshaw cell may then project back onto a motoneuron that innervates it, and this establishes a negative feedback circuit *(recurrent inhibition)* that limits the discharge of the motoneuron. The spinal cord also contains positive feedback circuits *(recurrent facilitation)*. An example is a collateral of a motoneuron axon that terminates on an inhibitory interneuron, which in turn terminates on a neuron that inhibits the motoneuron. Such a circuit "turns off" inhibition of the motoneuron *(disinhibition)* and thereby facilitates it.

Finally, some principles, discussed previously, are worth recalling in this context. These include reciprocal innervation (Fig. 14-12), crossed-components involving a spinal commissural neuron (Fig. 14-13), and double reciprocal innervation (Fig. 14-13). These are characteristics of the spinal circuitry that form the basis for organizing simple motor patterns at spinal levels. In a sense they provide simple "motor programs" that can be activated by descending pathways.

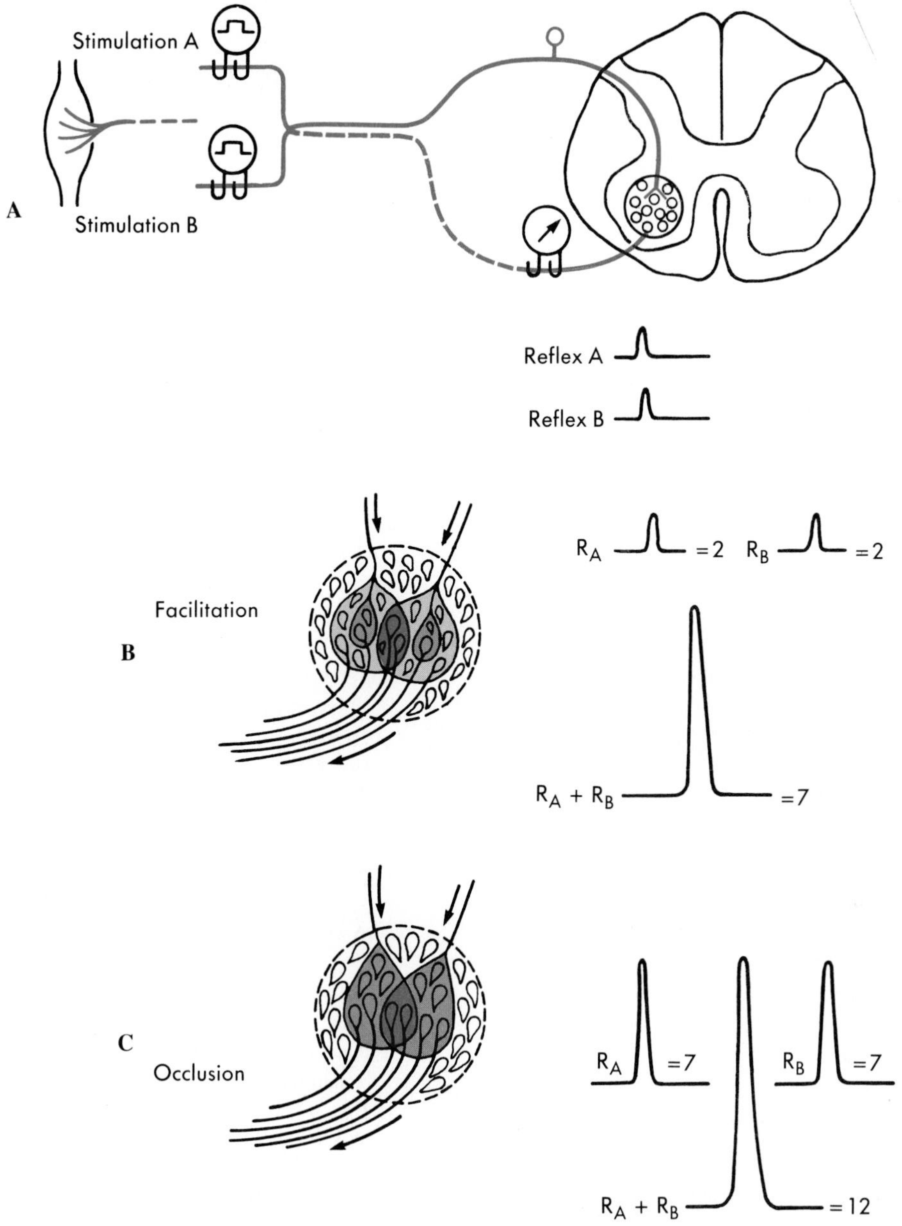

Fig. 14-17. The phenomena of facilitation and occlusion. The monosynaptic reflex is used as an example. **A** shows the preparation, where a muscle nerve is dissected into two branches of equal size (*A* and *B*), which are placed on stimulating electrodes. Stimulating each of these branches will evoke monosynaptic reflexes of equal size (*reflex A* and *reflex B*). The reflexes are recorded from the ventral root, which is cut *(broken line)* distal to the recording electrode. **B** shows facilitation. The discharge zone of each afferent nerve branch *(A* and *B)* is included in the dark oval. Stimulating each branch will activate the motoneurons in its discharge zone. The subliminal fringe of each branch includes those motoneurons within the lighter oval. When both nerves are simultaneously stimulated, the neurons in their overlapping subliminal fringes will be activated as well as the neurons in each discharge zone. Consequently, the number of neurons activated with simultaneous stimulation of both branches will be greater than the sum of the neurons activated by stimulating each branch individually. **C** shows occlusion. In this case the discharge zones overlap. Consequently, the number of motoneurons activated by simultaneously stimulating the two branches will be less than the sum of the motoneurons activated by stimulating each branch individually. (Redrawn with permission from Eyzaguirre, C., and Fidone, S.J.: Physiology of the nervous system, ed. 2. Copyright © 1975 by Year Book Medical Publishers, Inc., Chicago.)

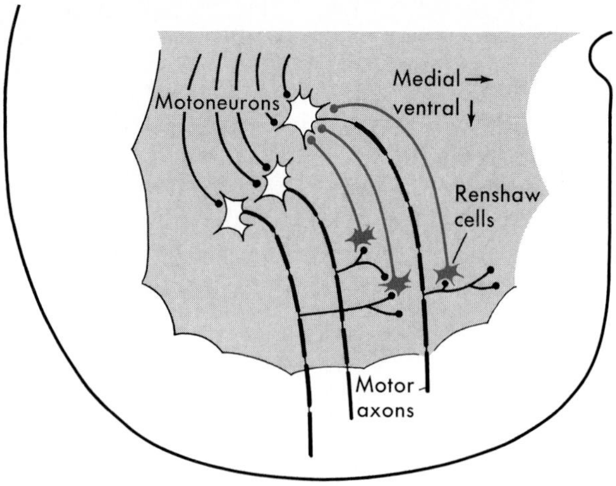

■ **Fig. 14-18.** Schematic illustration of recurrent inhibition of motoneurons by inhibitory interneurons *(color),* the Renshaw cells. (Redrawn from Eccles, J.C.: The physiology of synapses. New York, 1964, Academic Press, Inc.)

■ *Bibliography*

Journal articles

Burke, D., et al.: Spasticity, decerebrate rigidity and the clasp-knife phenomenon: an experimental study in the cat, Brain **95:**31, 1972.

Crowe, A., and Matthews, P.B.C.: The effects of stimulation of static and dynamic fusimotor fibres on the response to stretching of the primary endings of muscle spindles, J. Physiol. **174:**109, 1964.

Houk, J., and Henneman, E.: Responses of Golgi tendon organs to active contractions of the soleus muscle of the cat, J. Neurophysiol. **30:**466, 1967.

Hunt, C.C., and Perl, E.R.: Spinal reflex mechanisms concerned with skeletal muscle, Physiol. Rev. **40:**538, 1960.

Kuffler, S.W., and Hunt, C.C.: The mammalian small-nerve fibers: a system for efferent nervous regulation of muscle spindle discharge, Res. Publ. Assoc. Nerv. Ment. Dis. **30:**24, 1952.

Kuhn, R.A.: Functional capacity of the isolated human spinal cord, Brain **73:**1, 1950.

Landau, W.M., and Clare, M.H.: Fusimotor function, VI. H reflex, tendon jerk, and reinforcement in hemiplegia, Arch. Neurol. **10:**128, 1964.

Laporte, Y., and Lloyd, D.P.C.: Nature and significance of the reflex connections established by large afferent fibers of muscular origin, Am. J. Physiol. **169:**609, 1952.

Liddell, E.G.T.: Spinal shock and some features in isolation-alteration of the spinal cord in cats, Brain **57:**386, 1934.

Liddell, E.G.T., and Sherrington, C.S.: Reflexes in response to stretch (myotatic reflexes), Proc. R. Soc. Lond. (Biol.) **96:**212, 1924.

Lloyd, D.P.C.: Facilitation and inhibition of spinal motoneurons, J. Neurophysiol. **9:**421, 1946.

Lloyd, D.P.C.: Integrative pattern of excitation and inhibition in two-neuron reflex arcs, J. Neurophysiol. **9:**439, 1946.

Mendell, L.M., and Henneman, E.: Terminals of single Ia fibers: location, density, and distribution within a pool of 400 homonymous motoneurons, J. Neurophysiol. **34:**171, 1971.

Renshaw, B.: Central effects of centripetal impulses in axons of spinal ventral roots, J. Neurophysiol. **9:**191, 1946.

Sahs, A.L. and Fulton, J.F.: Somatic and autonomic reflexes in spinal monkeys, J. Neurophysiol. **3:**258, 1940.

Sherrington, C.S.: Flexion-reflex of the limb, crossed extension-reflex, and reflex stepping and standing, J. Physiol. **40:**28, 1910.

Stein, R.B.: Peripheral control of movement, Physiol. Rev. **54:**215, 1974.

Stuart, D.G., et al.: Stretch responsiveness of Golgi tendon organs, Exp. Brain Res. **10:**463, 1970.

Books and monographs

Creed, R.S., et al.: Reflex activity of the spinal cord, London, 1932, Oxford University Press.

Eccles, J.C.: The physiology of synapses, New York, 1964, Academic Press, Inc.

Harris, D.A., and Henneman, E.: Feedback signals from muscle and their efferent control. In Mountcastle, V.B., editor: Medical physiology, ed. 14, St. Louis, 1980, The C.V. Mosby Co.

Henneman, E.: Organization of the spinal cord and its reflexes. In Mountcastle, V.B., editor: Medical physiology, ed. 14, St. Louis, 1980, The C.V. Mosby Co.

Matthews, P.B.C.: Muscle spindles: their messages and their fusimotor supply. In Brookhart, J.M., and Mountcastle, V.B., editors: Handbook of physiology, the nervous system, section 1, vol. II, pt. 1, Bethesda, 1981, American Physiological Society.

Mendell, L.M., and Henneman, E.: Input to motoneuron pools and its effects. In Mountcastle, V.B., editor: Medical physiology, ed. 14, St. Louis, 1980, The C.V. Mosby Co.

Patton, H.D.: Reflex regulation of movement and posture. In Ruch, T., and Patton, H.D., editors: Physiology and biophysics, vol. IV, Philadelphia, 1982, W.B. Saunders Co.

Sherrington, C.S.: The integrative action of the nervous system, ed. 2, New Haven, Conn., 1947, Yale University Press.

CHAPTER

15

Descending Pathways Involved in Motor Control

An axon that descends from the brain to the spinal cord can influence the local spinal circuitry in a number of ways. Its specific influence will be a function of its discharge properties, the nature of its physiological effects (e.g., excitatory or inhibitory), and the distribution of its terminals. Fig. 15-1 illustrates some of the different possibilities of terminal distribution within a spinal reflex circuit. Each distribution will have a different influence. The axon can terminate on cells that influence the receptor component of the circuit, as in the centrifugal control of muscle spindle receptor sensitivity. The axon can terminate on one or more interneurons of the circuit, or it can bypass the intrinsic spinal circuitry and terminate on motoneurons directly to gain greater control over muscle action. Also important is how a fiber system distributes among its population of target neurons. For example, if the fiber projects to motoneurons, its terminal field might be restricted to certain motoneurons, such as those innervating the intrinsic muscles of the digits. This would restrict the musculature that the fiber can influence, which in turn shapes its functional role in motor control. Finally, how a descending fiber terminates on a postsynaptic neuron contributes to its effect. For example, if the terminals are restricted to the distal dendrites of the postsynaptic cell, the fiber will have less influence on the neuron than if its terminals were close to the spike initiation zone. The various descending pathways use all of these possibilities in controlling motor behavior.

Before the actual pathways involved in motor control are discussed, a definitional issue should be raised. We tend to describe tracts that descend from higher centers as "motor pathways." However, only the final common path should properly be defined as motor. To appreciate this definitional problem one need only consider that the Ia afferent fibers from the muscle spindles terminate monosynaptically on the motoneurons, yet these are dorsal root fibers that are clearly sensory. Consequently, from a morphological viewpoint the basis for defining any descending pathway as motor is problematic. Nevertheless, we take the license to do so, based on the functional effects of certain suprasegmental systems.

■ *A Classification of Descending Pathways*

Traditionally, the motor pathways have been designated as *pyramidal or extrapyramidal*. This classification is derived from the early clinical literature but has little anatomical foundation. Strictly speaking, the pyramidal system consists of all fibers traversing the medullary pyramids, regardless of their cells of origin or sites of termination. Extrapyramidal systems are then defined, largely by exclusion, to include "motor systems" that contain fibers that do not travel in the pyramids. Strikingly different motor deficits consequent to lesions of the pyramidal and extrapyramidal systems have reinforced this classification for many years.

The pyramidal-extrapyramidal classification has not contributed to an understanding of the descending pathways involved in motor control. To abandon this traditional view of the motor pathways is compelling, because a more useful classification has become available. The basis for this alternative is developed in Chapter 13; it focuses on the *sites of termination* of a given fiber system. In brief, the spinal cord includes topographically organized columns of motoneurons. The more ventromedially situated motoneuronal cell groups innervate the axial and girdle muscles, and the more dorsolaterally located motoneurons innervate the muscles of the distal extremities. This mediolateral topography is maintained in the interneuronal projections on the mo-

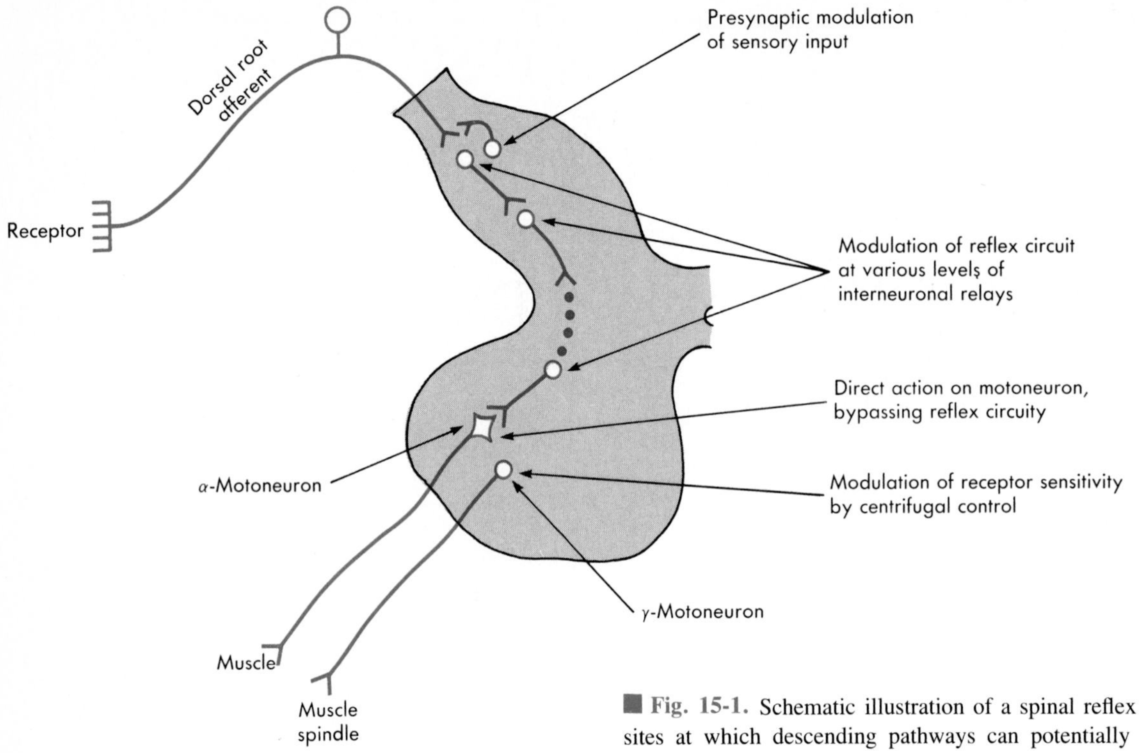

■ **Fig. 15-1.** Schematic illustration of a spinal reflex pathway and the sites at which descending pathways can potentially influence such a pathway.

toneurons. Thus from a functional perspective one can consider a *lateral system* that governs fine manipulatory activity of the extremities and a *medial system* that controls body and whole limb movement.

If the descending pathways are now considered with respect to their spinal sites of termination, then each can be classified with respect to the extent to which it participates in lateral as opposed to medial system types of functions. Pathways that terminate primarily on the ventromedial motoneurons or their associated interneurons will be more concerned with controlling posture, equilibration, and progression. Conversely, those that terminate on the lateral motoneurons or their associated interneurons will be more concerned with controlling finer manipulatory activity of the extremities.

■ *Lateral System Pathways*

Two major pathways terminate on the more laterally situated motoneurons or their associated interneurons: the *lateral corticospinal tract* and the *rubrospinal tract* (Figs. 15-2 and 15-3).

The *corticospinal tract* arises from both the precentral gyrus (including the motor and premotor cortices) and the postcentral gyrus (the somatosensory cortex), deriving from both small and large neurons, primarily in layer V. The tract then descends in the posterior limb of the internal capsule, the medial two thirds of the cerebral peduncle, and the medullary pyramids. Most of the fibers then decussate at the medullary-spinal junction. The crossed pathway then descends in the lateral funiculus of the spinal cord, where it constitutes the lateral corticospinal tract. A small contingent of fibers does not decussate, but remains ipsilateral and descends in the ventral funiculus. This ventral corticospinal tract is a minor pathway that does not exist in all persons. When present, it does not descend beyond thoracic levels. (It should be noted that along its descending brainstem trajectory the descending fibers from the precentral gyrus also project onto the motor cranial nerve nuclei.)

By definition the pyramidal tract consists of all fibers coursing longitudinally in the medullary pyramids, regardless of their sites of origin or termination. The corticospinal tract thus consists of a subset of these fibers that has its cells of origin in the cortex and sites of termination in the spinal cord. The corticospinal tract component that arises from the postcentral gyrus terminates in the dorsal horn and functions to modulate afferent input (Fig. 15-3). The motor component arises from the precentral gyrus, and it terminates more ventrally in the spinal gray matter (Figs. 15-2 and 15-4). As they descend, the sensory and motor components of the corticospinal tract issue collaterals at numerous levels of the neuraxis, including the basal ganglia, thalamus, and brainstem.

The motor component of the corticospinal tract will now be considered in more detail. The precentral gyrus, from which the tract arises, has an orderly topographi-

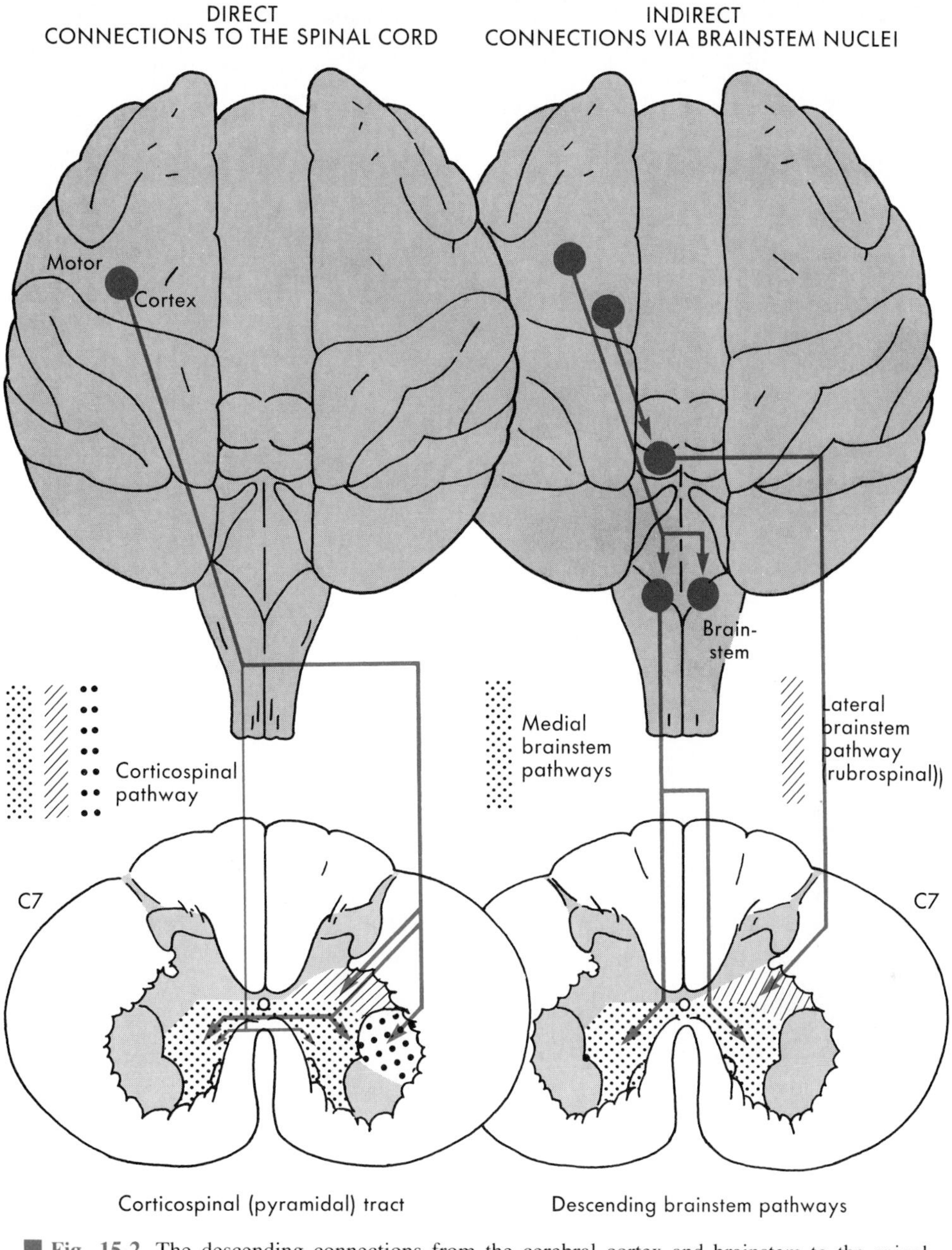

■ **Fig. 15-2.** The descending connections from the cerebral cortex and brainstem to the spinal cord of the monkey. The dorsolateral motoneurons innervating the muscles of the distal extremities are indicated by dots. The dorsolateral interneurons that project onto these motoneurons are shown by hatching, and the ventromedial interneurons that project onto motoneurons innervating more medial muscles are shown by stippling. (Redrawn from Brinkman, C.: Split-brain monkeys: cerebral control of contralateral and ipsilateral arm, hand, and finger movements, doctoral dissertation, Rotterdam, 1974, Erasmus University Rotterdam.)

cal representation (homunculus) of the body musculature (Fig. 15-5). This representation was discovered by electrical stimulation experiments in which it was found that localized stimulation of the motor cortex will elicit discrete movements. For example, stimulation of the "wrist area" will evoke a rapid flexion at the wrist. Noteworthy is the disproportionate representation of certain musculatures, such as those of the face and limbs (Fig. 15-5). The fibers of the corticospinal tract that originate more rostrally in the precentral gyrus, where the more proximal muscles are represented, terminate bilaterally in the more ventral spinal gray matter. Here they influence the interneurons and motoneurons that innervate the more proximal muscles. However, the greatest proportion of corticospinal fibers influence the face and extremities (Fig. 15-5). Consequently, the tract terminates prominently in the more lateral spinal gray matter, where it influences the inter-

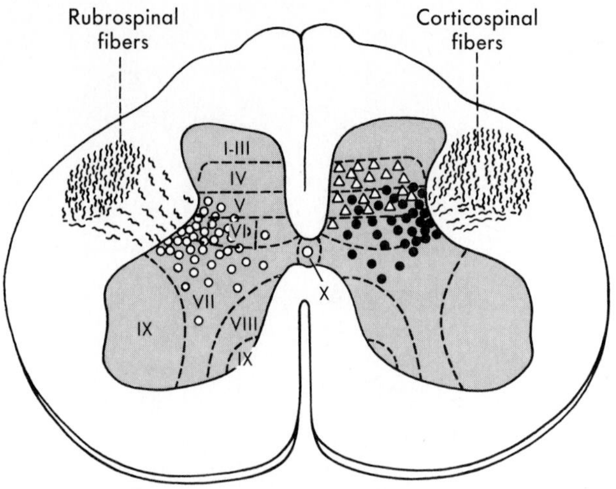

■ **Fig. 15-3.** A transverse section of the cervical spinal cord of the cat. The diagram shows the location in the lateral funiculus and the sites of termination in the spinal gray matter of rubrospinal and corticospinal fibers from "motor" and "sensory" regions of the sensorimotor cortex. Note the similarities in the areas of termination of rubrospinal and corticospinal fibers from the motor cortex. Because the section is from the cat, there are no direct corticospinal projections to mononeurons. *Open circles,* Sites of termination of rubrospinal fibers; *solid circles,* sites of termination of corticospinal fibers from motor cortex; *open triangles,* sites of termination of corticospinal fibers from sensory cortex. (Redrawn from Brodal, A.: Neurological anatomy, ed. 3. Copyright © 1981 by Oxford University Press, Inc. Reprinted by permission.)

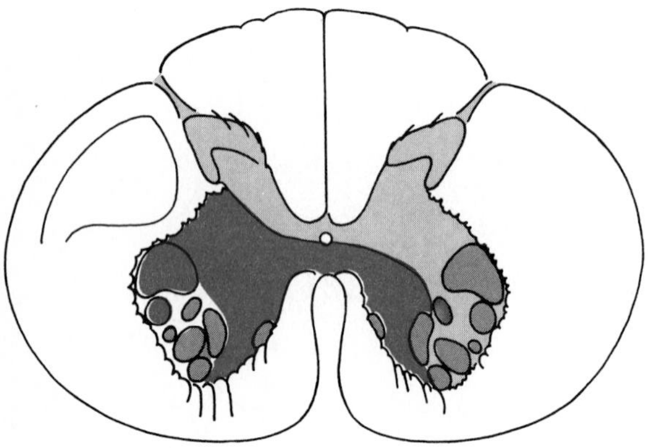

■ **Fig 15-4.** Schematic representation of the termination *(color)* of the precentral (motor) component of the corticospinal tract in the monkey. These fibers terminate unilaterally in the dorsolateral intermediate zone but bilaterally in the ventromedial intermediate zone. In addition, fibers terminate unilaterally on the dorsolateral motoneurons. (Redrawn from Brinkman, C.: Split-brain monkeys: cerebral control of contralateral and ipsilateral arm, hand and finger movements, doctoral dissertation, Rotterdam, 1974, Erasmus University Rotterdam.)

■ **Fig. 15-5. A,** The motor homunculus of the precentral gyrus. In this schematic coronal section the location of the motor cortical representation of different parts of the body is shown, and size of the various parts is proportional to the amount of cortical surface area serving them. **B,** The somatotopic organization of primary, supplementary, and secondary motor areas in the cerebral cortex of the monkey. The central and longitudinal fissures are shown opened out, with the broken line indicating the floor of the fissure and the solid line the lip of the fissure on the brain's surface. At the bottom is an ipsilateral face area (secondary motor area). (Redrawn with permission from Eyzaguirre, C., and Fidone, S.J.: Physiology of the nervous system, ed. 2. Copyright © 1975 by Year Book Medical Publishers, Inc., Chicago. **A** slightly modified from Penfield, W., and Rasmussen, T.: The cerebral cortex of man, New York, 1950, The Macmillan Co.; **B** slightly modified from Woolsey, C.N., et al.: Res. Publ. Assoc. Res. Nerv. Ment. Dis. **30:**238, 1952.)

A

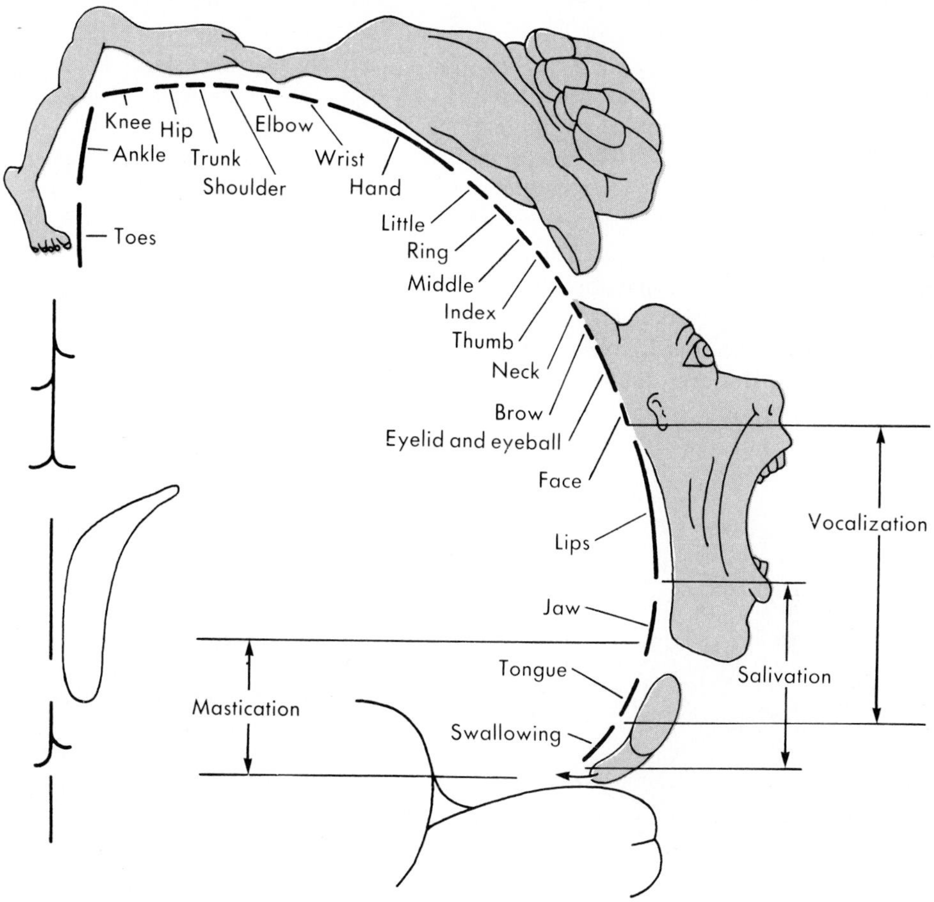

B

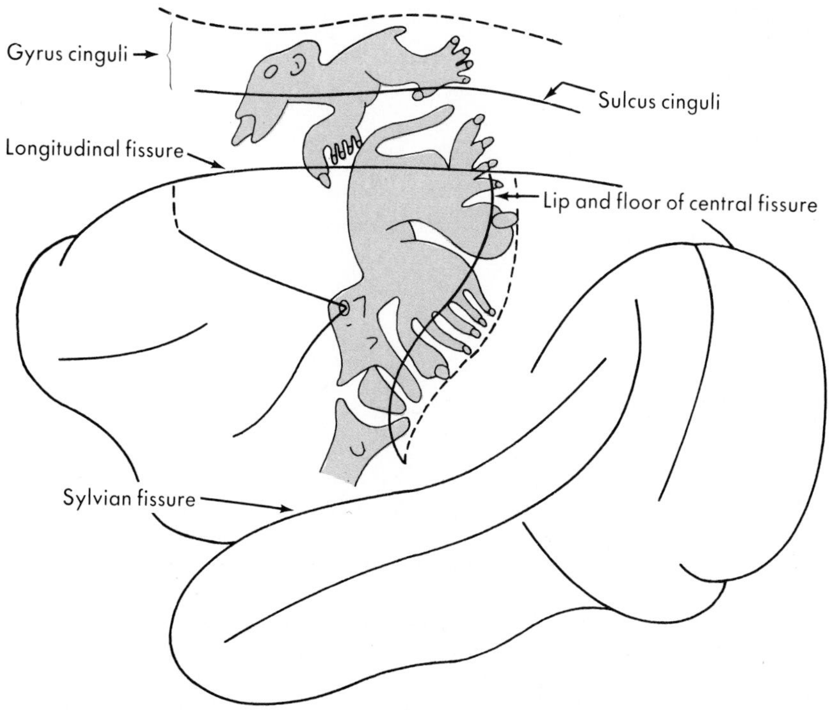

■ **Fig. 15-5.** For legend see opposite page.

neurons and motoneurons associated with the muscles of the extremities. On the basis of this prominent termination in the lateral spinal gray matter, the corticospinal tract is classified as a component of the lateral system.

An important feature of the tract is its direct termination on the motoneurons that innervate the intrinsic muscles of the digits (Figs. 15-2 and 15-4). Indeed, the motor component of the tract is most highly developed in species such as the chimpanzee and man that have evolved considerable capacity for independent movement of the digits. In this context, independent finger movement in primates does not occur ontogenetically until development of the corticospinal projection on the motoneurons of the digits. However, the size of the entire corticospinal tract does not necessarily correlate with the development of the digits, because the size of the sensory component can vary independently of digital development. Thus cetaceans, such as whales, have prominent corticospinal tracts, but most of the fibers originate in the postcentral gyrus and terminate in the dorsal horn.

The second major component of the lateral system is the *rubrospinal tract*. It arises from both large and small cells of the caudal part of the red nucleus (Fig. 15-6), a midbrain structure. Although not organized as discretely as in the motor cortex, the body musculature is represented topographically in the red nucleus. Immediately on leaving the nucleus, the descending fibers cross the midline and ultimately assume a trajec-

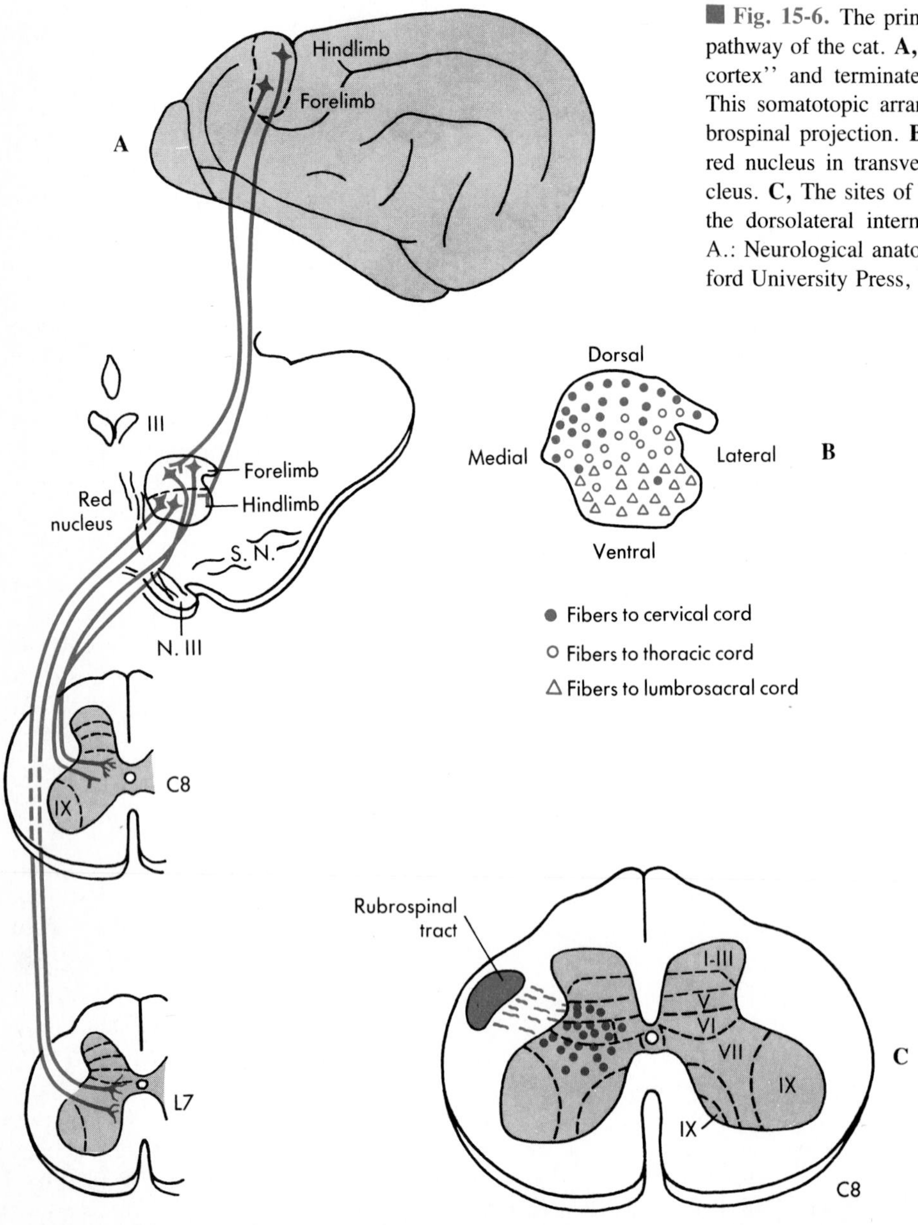

■ **Fig. 15-6.** The principal features of the corticorubrospinal pathway of the cat. **A,** This pathway originates in the "motor cortex" and terminates somatotopically in the red nucleus. This somatotopic arrangement is then maintained in the rubrospinal projection. **B,** The somatotopic organization of the red nucleus in transverse section at the midlevel of the nucleus. **C,** The sites of termination of the rubrospinal fibers in the dorsolateral intermediate zone. (Redrawn from Brodal, A.: Neurological anatomy, ed. 3. Copyright, © 1981 by Oxford University Press, Inc. Reprinted by permission.)

tory in the lateral funiculus of the spinal cord just ventrolateral to that of the corticospinal tract (Fig. 15-3). The terminal field of the rubrospinal tract overlaps to a large extent with that of the corticospinal tract. However, it has few, if any, direct terminations on motoneurons. As in the corticospinal tract, the rubrospinal fibers terminate preferentially on interneurons associated with the more laterally situated motoneurons (Figs. 15-3 and 15-6).

The rubrospinal tract is a component of a corticorubrospinal system, since the motor cortex projects in a topographical fashion onto the cells of origin of the rubrospinal tract (Fig. 15-6). However, the corticorubral projection apparently arises largely from a population of cortical neurons different from that giving rise to the corticospinal tract. This projection is excitatory, but weakly so because it is restricted primarily to the distal dendrites of the rubrospinal neurons. A more powerful excitatory input to the rubrospinal neurons originates from the intermediate or interposed cerebellar nuclei, the axons of which terminate on the proximal dendrites and somata (Chapter 16).

Thus, the lateral system pathways consist primarily of the corticospinal and rubrospinal tracts. (The lateral system also includes a crossed pontospinal tract from the ventrolateral pontine tegmentum, but little is known about this pathway.) These pathways terminate most prominently in the interneuronal zone that is associated with the motoneurons that innervate the muscles of the extremities. The corticospinal tract is unique in having substantial monosynaptic connections with the most dorsolaterally located motoneurons that innervate the intrinsic muscles of the digits. This tract is, in fact, the only descending pathway with obvious direct access to the final common path for independent movement of the digits.

■ *Medial System Pathways*

A number of descending pathways travel in the ventral funiculus of the spinal cord and terminate preferentially on the more ventromedially situated interneurons and motoneurons (Fig. 15-2). These motoneurons innervate the axial and girdle musculatures, the principal functions of which are postural. Most prominent among these medial system pathways are the *lateral vestibulospinal tract* (Fig. 15-7) and the *pontine reticulospinal tract* (Fig. 15-8).

The *lateral vestibulospinal tract* arises from the lateral vestibular, or Deiters', nucleus. The pathway descends, without crossing, in the ventral funiculus, and it terminates ventromedially in the spinal gray matter throughout the rostrocaudal extent of the spinal cord (Fig. 15-7). This terminal distribution is the substrate for its influence on the extensor motoneurons that innervate the proximal postural muscles. In fact, most medial extensor motoneurons show monosynaptic excitatory postsynaptic potentials (EPSPs) in response to activation of the lateral vestibulospinal tract, whereas the medial flexor motoneurons are disynaptically inhibited. Moreover, the ventromedially situated interneurons involved in the disynaptic connections of lateral vestibulospinal fibers to motoneurons show substantial recruitment. Because the most prominent source of input to the lateral vestibular nucleus originates in the utricles of the labyrinth (Fig. 15-8), this pathway is probably principally concerned with adjustment of the postural muscles to linear acceleratory displacements of the body.

The *pontine reticulospinal tract* arises from cells of the medial two thirds of the pons, including the entire nucleus reticularis pontis caudalis and the caudal segment of nucleus reticularis pontis oralis. Similar to the lateral vestibulospinal tract, it descends ipsilaterally in the ventral funiculus, and it terminates in the more ventromedial spinal gray matter throughout the rostrocaudal extent of the spinal cord (Fig. 15-9). At the level of termination, some fibers cross in the anterior commissure to innervate the contralateral ventromedial gray matter. Also, as in the lateral vestibulospinal tract, the pontine reticulospinal tract is excitatory to the more medial motoneurons to the proximal extensor muscles, primarily through interneurons.

Other pathways that contribute to the medial system are the *medial vestibulospinal tract, interstitiospinal tract, tectospinal tract,* and *medullary reticulospinal tract* (Fig. 15-10). The medial vestibulospinal tract (Figs. 15-7 and 15-8) arises from the medial vestibular nucleus, which receives its primary input from the semicircular canals. The tract descends ipsilaterally in the ventral funiculus to midthoracic levels, and it adjusts the neck and upper limbs to angular acceleration of the body. Ascending fibers from the medial vestibular nucleus mediate adjustments of eye position to angular acceleration. The interstitiospinal tract arises from the interstitial nucleus of Cajal in the rostral midbrain (Fig. 15-10). It descends without crossing to travel in the ventral funiculus, and it terminates in the ventromedial spinal gray matter throughout the rostrocaudal extent of the spinal cord. This tract also helps mediate the rotation of the head and body about the longitudinal axis. The tectospinal tract has its cells of origin in the superior colliculus (Fig. 15-10). In contrast to other pathways of the medial system, the tectospinal tract crosses the midline before descending in the ventral funiculus. However, it distributes only to upper cervical levels and mediates visually guided head movement.

The medullary reticulospinal tract arises from cells in the medial tegmental field of the medulla. The fibers of this tract descend in both the ipsilateral and contralat-

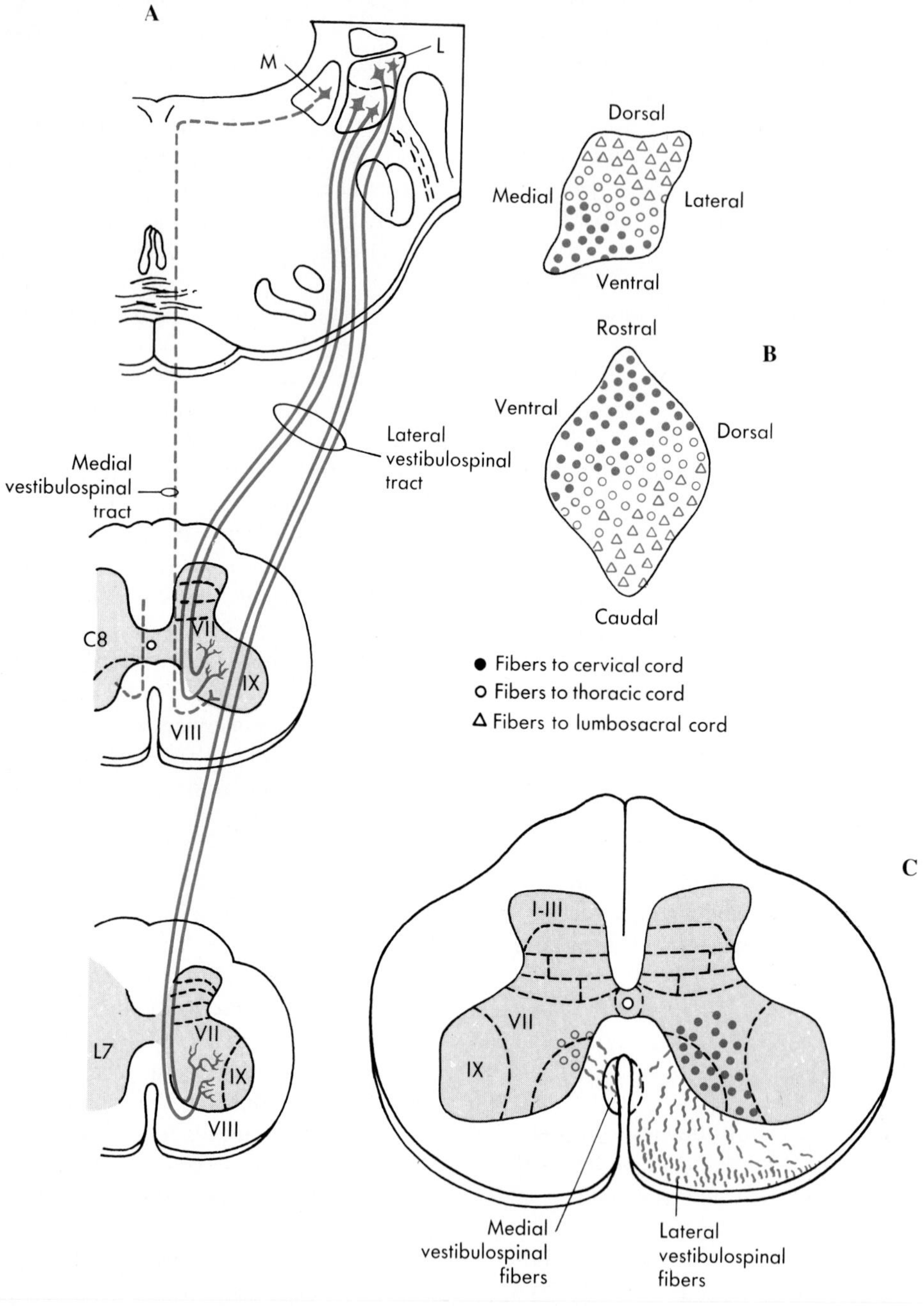

■ **Fig. 15-7. A,** Principal features of the vestibulospinal pathways of the cat. The lateral vestibulospinal tract arises from the lateral vestibular nucleus of Deiters *(L),* which is somatotopically organized. The medial vestibulospinal tract arises from the medial vestibular nucleus *(M)* and does not project beyond the thoracic spinal cord, **B,** The somatotopic reconstructions. **C,** The spinal termination patterns of the vestibulospinal pathways. (Redrawn from Brodal, A.: Neurological anatomy, ed. 3. Copyright © 1981 by Oxford University Press, Inc. Reprinted by permission.)

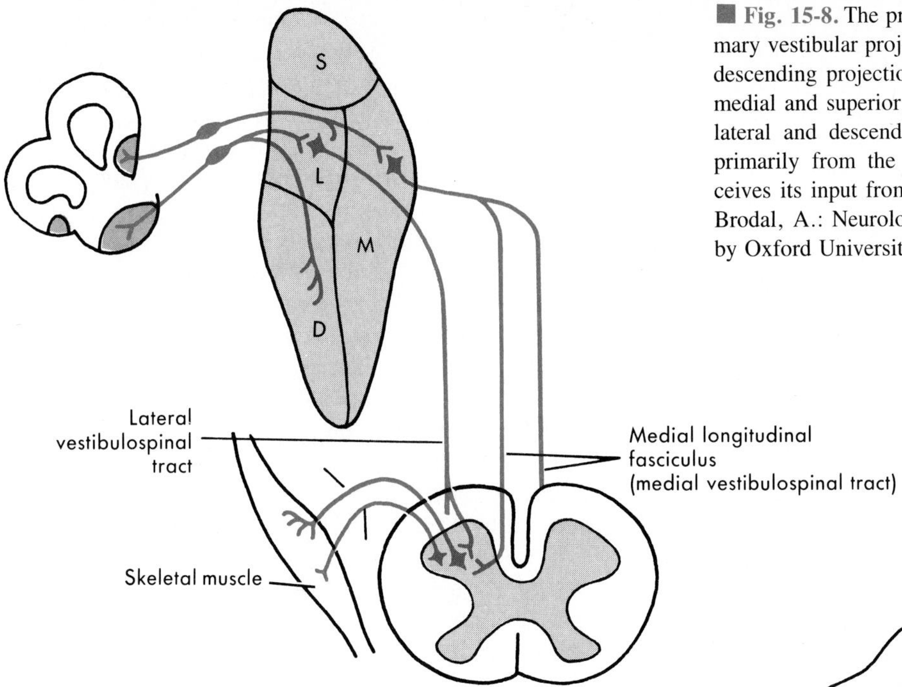

Fig. 15-8. The principal features of the organization of primary vestibular projections on the vestibular nuclei and of the descending projections of these nuclei. *M* and *S* refer to the medial and superior vestibular nuclei, respectively. Note that lateral and descending vestibular nuclei receive projections primarily from the utricle, whereas the medial nucleus receives its input from the semicircular canals. (Redrawn from Brodal, A.: Neurological anatomy, ed. 3. Copyright © 1981 by Oxford University Press, Inc. Reprinted by permission.)

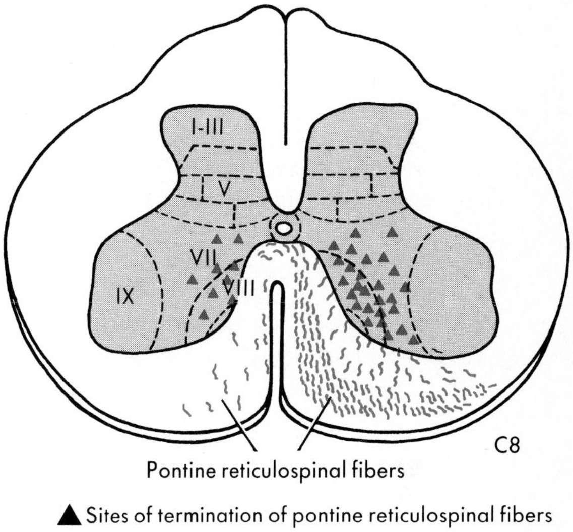

Fig. 15-9. A transverse section of the cervical spinal cord of the cat. The diagram shows the position and terminal distribution of the pontine reticulospinal tract. (Redrawn from Brodal, A.: Neurological anatomy, ed. 3. Copyright © 1981 by Oxford University Press, Inc. Reprinted by permission.)

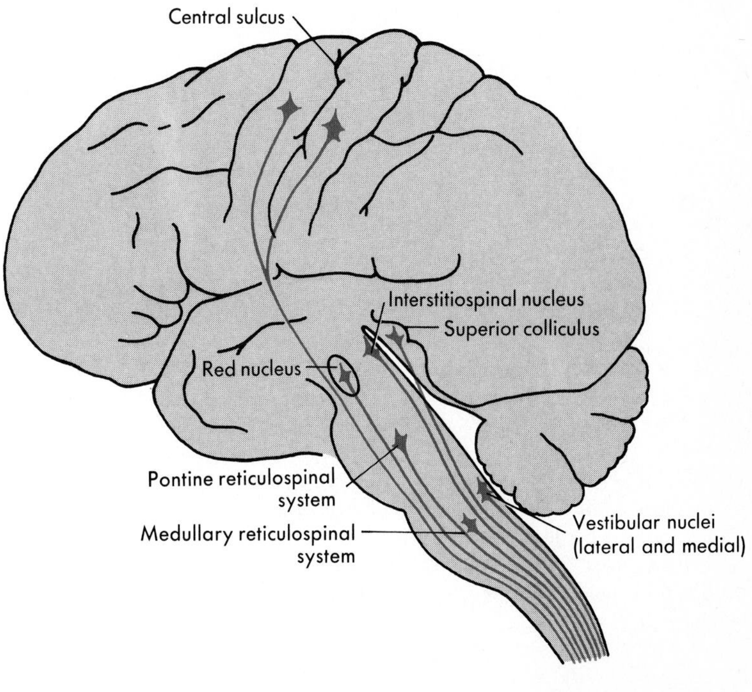

Fig. 15-10. The major fiber systems descending to the spinal cord. Shown are the corticospinal, rubrospinal, tectospinal, interstitiospinal, pontine and medullary reticulospinal, and medial and lateral vestibulospinal tracts. (Redrawn from Brodal, A.: Neurological anatomy, ed. 3. Copyright © by Oxford University Press, Inc. Reprinted by permission.)

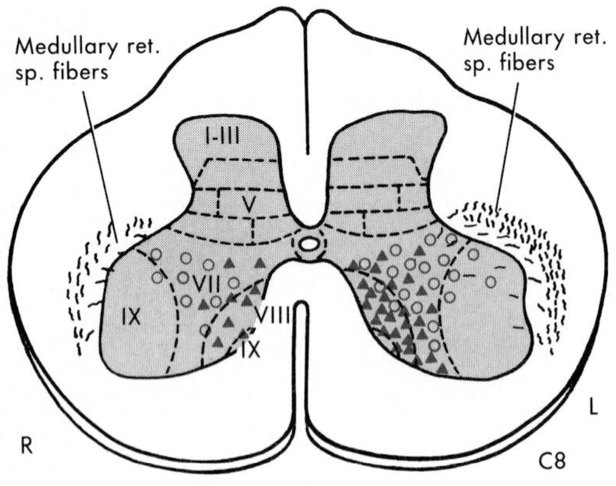

■ **Fig. 15-11.** A transverse section of the cervical spinal cord of the cat. The diagram shows the position and terminal distribution of the medullary reticulospinal tract. (Redrawn from Brodal, A.: Neurological anatomy, ed. 3. Copyright © 1981 by Oxford University Press, Inc. Reprinted by permission.)

eral ventrolateral funiculus and terminate in the spinal gray matter with a wider and more lateral distribution than the pontine reticulospinal tract (Fig. 15-11). This tract inhibits proximal limb motoneurons, in part through direct motoneuronal connections.

Thus, in contrast to the lateral system pathways, the tracts constituting the medial system originate primarily from the brainstem. With the exception of the medullary reticulospinal pathway, their spinal trajectories are through the ventral, rather than the lateral, funiculus, and they distribute preferentially to the ventromedial spinal gray matter. (The tectospinal and medullary reticulospinal tracts, although considered medial system pathways, share some properties with the lateral system pathways in that their fibers terminate in more lateral, as well as medial, parts of the spinal gray matter.) The terminal fields of the medial pathways in the spinal cord thus determine that their influences will be more restricted to motoneurons that innervate the proximal muscle groups. Functionally this implicates these pathways in the control of posture, equilibration, and progression, in contrast to lateral system pathways, the principal function of which is the control of spatially organized and fractionated movements involving the extremities.

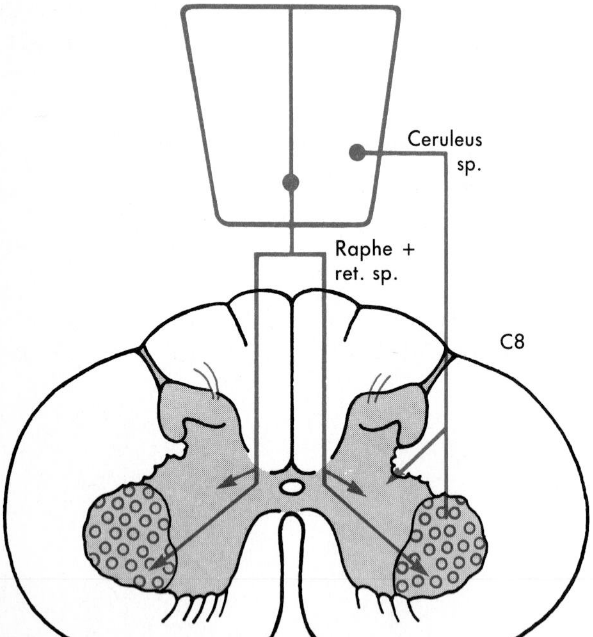

■ **Fig. 15-12.** The descending monoaminergic pathways that terminate in the ventral horn. (Redrawn from Kuypers, H.G.J.M.: A new look at the organization of the motor system. In Kuypers, H.G.J.M., and Martin, G.F., editors: Descending pathways to the spinal cord, Prog. Brain Res. **57**:390, 1982.)

■ *Monoaminergic Pathways*

In recent years two monoaminergic pathways have been identified that influence motoneurons directly and indirectly (Fig. 15-12). A noradrenergic pathway arises from neurons in the area of the locus ceruleus and nucleus subceruleus. This pathway descends primarily ipsilaterally in the ventrolateral funiculus, and its fibers distribute widely throughout the ventral horn. The terminal field includes the motoneuronal cell groups, particularly at the cervical and lumbar enlargements.

A serotonergic pathway arises from cells in the more caudal parts of the medullary raphe nuclei. This pathway descends in the lateral, and possibly ventral, funiculus. It distributes widely throughout the ventral horn, and includes synapses with the motoneuronal cell groups. In all likelihood, one or more neuropeptides, such as substance P, are colocalized with serotonin in many terminals of this pathway.

The projections of these pathways to the ventral horn are extensive, but their role in motor control is not fully understood. The monoaminergic pathways probably do not participate in organizing movements. Rather, they are more likely to modulate the responsiveness of the final common pathway, either directly or through interneurons. In a sense, these pathways may constitute a third component of the motor system, a component that would contribute a feature such as motivational drive.

Interruption of the Descending Pathways

Interruption of lateral system pathways would be expected to compromise control of the extremities and of the muscles of the digits in particular. In contrast, interruption of medial system pathways would be expected to compromise the axial and girdle musculatures. Considerable experimental and clinical evidence supports these predictions.

The *corticospinal tract* can be *interrupted* in monkeys without damage to other ascending or descending systems by a lesion at the medullary pyramids. Such a lesion has surprisingly minimal effect, and the few lasting signs are confined to the distal limb muscles. A mild flexor hypotonia may ensue, consistent with evidence that the corticospinal tract primarily excites flexor motoneurons and inhibits extensor motoneurons of the extremities. The primary persistent deficit is a loss of the ability to move the digits independently, that is, a *loss of independent or fractionated finger movement*. Because the lateral corticospinal tract is the only descending pathway that has direct access to the dorsolateral motoneurons that innervate the intrinsic muscles of the digits, this deficit is not unexpected. There are no deficits involving the more proximal muscles; thus posture, equilibration, and progression are not compromised. Indeed, the monkey can sit, walk, and climb quite normally. Although a deficit restricted to the digits may not appear to be critical, it seriously compromises the ability for fine manipulatory activity, a most important aspect of the human's motor repertoire. With respect to the lower extremities, the toes are affected as well. In a normal adult, stroking the sole of the foot will elicit a plantar flexion of the big toe. Frequently the other toes will flex and adduct as well. In the infant, where the corticospinal innervation of the dorsolateral lumbar motoneurons has not yet developed, such a stimulus elicits instead a dorsiflexion (i.e., an inverted plantar reflex) with fanning of the toes. This is called *Babinski's sign*. In an adult in which the corticospinal tract has been interrupted, Babinski's sign also occurs. Because the plantar response is generally considered a flexor response, the shift to an inverted plantar, or extensor, response reflects the loss of flexor motoneuronal innervation consequent to interruption of the lateral corticospinal tract.

Interruption of the other lateral system pathway, the *rubrospinal tract,* has almost no evident effect, and the role of this pathway in humans remains moot. Perhaps the only sequela to interrupting this pathway is a reduction in limb movement, that is, *limb hypokinesia*. However, if one surgically *interrupts both the corticospinal and rubrospinal tracts,* then striking deficits involving movements of the extremities ensue. The monkey has difficulty in flexing its arm and closing its hand. Independent finger movement is, of course, lost, because this follows interruption of the corticospinal tract alone. After a period of recovery, the capacity for grasping or hand closure is regained, but it occurs only as part of a total arm movement. Thus in attempting to guide the affected arm to an object and grasp it, the animal must implement a sweeping arm movement that involves flexion at the elbow and wrist, with a grasp in which the fingers flex in concert. Although predictable, it is still striking that such combined interruption of the two lateral system pathways does not affect the axial and girdle muscles. Thus posture and equilibration are totally unaffected.

The effects of experimentally *interrupting the medial system pathways* by a brainstem lesion are quite different. Such a procedure immediately reduces tone in the proximal extensor muscles, so that there is a flexor bias of the trunk. Several weeks elapse before the righting reflexes are sufficiently recovered to allow the animal to maintain an upright posture. After recovery of the righting reflexes, the animal tends to slump forward and frequently falls. This results from impairment of correcting movements that involve the axial and girdle muscles and the proximal muscles of the limbs. Walking is extremely difficult, and when walking does occur, directing the course of progression is very difficult. However, if the animal is supported properly, it can demonstrate perfectly intact control of the distal extremity muscles, allowing fine manipulatory activity of the fingers.

Pyramidal Tract Signs Compared with Corticospinal Tract Interruption

The traditional pyramidal-extrapyramidal classification of the motor pathways is in part based on clinical phenomenology that includes a group of deficits that are presumed to reflect damage to the corticospinal tract. This set of observations is referred to as the *pyramidal tract syndrome, corticospinal signs,* or sometimes *upper motoneuron disease*. The constellation of abnormalities includes (1) increased deep tendon reflexes, (2) hypertonia (spasticity), (3) paresis, (4) Babinski's sign, and (5) decreased superficial (flexor) reflexes. However, when a lesion restricted to the corticospinal tract is produced experimentally, the consequent abnormalities consist primarily of the loss of independent finger movement. Although it is not generally included among the set of pyramidal tract signs, this deficit occurs in human patients. Similarly, Babinski's sign is shared by both the experimental and clinical lesions, yet the increased deep tendon reflexes and spasticity do not follow experimental interruption of the corticospinal tract. If muscle tone is altered, it is reflected more in the flexor than in the extensor muscles.

The resolution of this discrepancy is reasonably straightforward. The clinician rarely, if ever, sees a case of pure corticospinal tract damage. Corticospinal tract signs constitute a syndrome that generally involves damage to the cortex or to the internal capsule, where systems beyond the corticospinal tract are compromised as well. For example, a cerebrovascular accident involving the middle cerebral artery characteristically produces pyramidal tract signs. However, such a lesion clearly involves other cortical projections, such as corticoreticular ones (Fig. 15-2). On this basis Brodal has emphasized that the pyramidal tract syndrome is indeed a misnomer. He has suggested substituting the designation *internal capsule syndrome* or *pure motor hemiplegia*.

Even lesions restricted to the motor cortex (area 4) may not provide a reasonable model of the pyramidal tract syndrome. The literature on experimental lesions in primates suggests that ablating the motor cortex can produce the deficits that follow experimental interruption of the corticospinal tract by pyramidotomy; this would be expected. Additionally, at least a transient paresis ensues, as with stroke in humans. However, producing the spasticity, a sine qua non of the pyramidal tract syndrome, requires a lesion that extends well beyond the motor cortex and includes at least the premotor cortex (area 6) as well. With a lesion restricted to the motor cortex the experimental animal shows a flaccid paralysis that is most severe in the more distal musculature, as one would anticipate from the functional neuroanatomy. Some recovery occurs with time, but such improvement is primarily seen in the more proximal muscles. Although this discrepancy between the experimental and human clinical literature may reflect species differences, it is more likely that a lesion restricted to the motor cortex rarely occurs in humans.

Although the genesis of the spasticity that follows cortical lesions in humans is not fully understood, one possible explanation is that it results from destruction of corticoreticular projections. For example, interruption of inhibitory cortical influences on the pontine reticulospinal tract would release it and thus increase medial extensor tone. Similarly, interruption of excitatory cortical influences on the medullary reticulospinal tract would reduce its inhibition of proximal motoneurons and thus increase medial extensor tone. This subject is discussed in more detail in Chapter 18.

Bibliography

Journal articles

Asanuma, H.: Cerebral cortical control of movement, Physiologist **16**:143, 1973.

Coulter, J.D., Ewing, L., and Carter, C.: Origin of primary sensorimotor cortical projections to lumbar spinal cord of cat and monkey, Brain Res. **403**:366, 1976.

Kuypers, H.G.J.M., and Brinkman, J.: Precentral projections to different parts of the spinal intermediate zone in the rhesus monkey, Brain Res. **24**:29, 1970.

Kuypers, H.G.J.M., and Lawrence, D.G.: Cortical projections to the red nucleus and the brainstem in the rhesus monkey, Brain Res. **4**:151, 1967.

Landau, W.M., and Clare, M.H.: The plantar reflex in man, with special reference to some conditions where the extensor response is unexpectedly absent, Brain **82**:321, 1959.

Lawrence, D.G., and Hopkins, D.A.: The development of motor control in the rhesus monkey: evidence concerning the role of corticomotoneuronal connections, Brain **99**:235, 1976.

Lawrence, D.G., and Kuypers, H.G.J.M.: The functional organization of the motor system in the monkey. I. The effects of bilateral pyramidal lesions, Brain **91**:1, 1968.

Lawrence, D.G., and Kuypers, H.G.J.M.: The functional organization of the motor system in the monkey. II. The effects of lesions of the descending brainstem pathways, Brain **91**:15, 1968.

Nyberg-Hansen, R.: Functional organization of descending supraspinal fibre systems to the spinal cord: anatomical observations and physiological correlations, Ergeb. Anat. Entwickl. Gesch. **39**(2):1, 1966.

Sterling, P., and Kuypers, H.G.J.M.: Anatomical organization of the brachial spinal cord of the cat. III. The propriospinal connections, Brain Res. **7**:419, 1968.

Books and monographs

Bowker, R.M., et al.: Organization of descending serotonergic projections to the spinal cord. In Kuypers, H.G.J.M., and Martin, G.F., editors: Descending pathways to the spinal cord, Prog. Brain Res. **57**:390, 1982.

Brodal, A.: Neurological anatomy, ed. 3, Oxford, 1981, Oxford University Press.

Ghez, C.: Introduction to the motor systems. In Kandel, E.R., and Schwartz, J.H.: Principles of neural science, New York, 1981, Elsevier Science Publishing Co., Inc.

Kuypers, H.G.J.M.: Anatomy of the descending pathways. In Brookhart, J.M., and Mountcastle, V.B., editors: Handbook of physiology, the nervous system, section 1, vol. II, pt. 2, Bethesda, 1981, American Physiological Society.

Kuypers, H.G.J.M.: A new look at the organization of the motor system. In Kuypers, H.G.J.M., and Martin, G.F., editors: Descending pathways to the spinal cord, Prog. Brain Res. **57**:390, 1982.

Westlund, K.M., et al.: Descending noradrenergic projections and their spinal terminations. In Kuypers, H.G.J.M., and Martin, G.F., editors: Descending pathways to the spinal cord, Prog. Brain Res. **57**:390, 1982.

CHAPTER

16

The Cerebellum

The cerebellum has been viewed traditionally as a motor structure involved in the regulation rather than in the execution of movements. It is situated above the brainstem in the posterior fossa, and its afferent and efferent connections are via three pairs of fiber tracts: the inferior cerebellar peduncle or restiform body, the middle cerebellar peduncle or brachium pontis, and the superior cerebellar peduncle or brachium conjunctivum cerebelli (Fig. 16-1). The cerebellum receives a complex array of sensory information from most, if not all, modalities. In fact, it is the principal central target for proprioceptive information. For this reason, it was described by Sherrington as the "head ganglion of the proprioceptive system." The moment-by-moment input of detailed sensory information, particularly concerning body position, muscle length, and muscle tension, is integrated by the cerebellar cortex, and the resultant output is transmitted to the descending motor pathways to modulate ongoing movements. In broad descriptive terms the cerebellum coordinates and smooths muscular activity on the basis of such continuous sensory inputs. More specifically it regulates the rate, range, force, and direction of movements.

The preceding description of the role of the cerebellum in motor control is clearly imprecise. Unfortunately it reflects our present understanding of how this structure integrates its inputs and then interacts with descending motor pathways. This inadequacy is ironic because the microanatomy and microphysiology of the cerebellum are perhaps better characterized than are those for any other structure of the mammalian central nervous system.

Anatomical Considerations

Topography

A superficial view of the cerebellum (Fig. 16-2) shows its complexity. Immediately striking are the extensive transverse convolutions of the surface and the pattern of fissures that differentiates the cerebellum into lobes and lobules. Indeed, many traditional descriptions of cerebellar structure are long scholarly treatments of the various lobules, the names of which defy recall. Briefly, the most prominent divisions are the anterior, posterior, and flocculonodular lobes, and these are defined by two large fissures (Fig. 16-3). The anterior and posterior lobes are divided by the primary fissure, and the posterior and flocculonodular lobes are separated by the posterolateral fissure. Each of these lobes is then further differentiated into a number of lobules by smaller fissures. Although this lobular organization is meaningful, as will become evident in the discussion of the afferent pathways to the cerebellum, a more effective initial approach is to consider the cerebellum's longitudinal organization.

Longitudinal Corticonuclear Zones

The cerebellum consists of a three-layered cortex and deep nuclear groups (Fig. 16-4). The anatomy will be reviewed in more detail later. However, it is helpful to consider first the general relationship between the cortex and the deep nuclei.

The flow of information through the cerebellum begins with inputs arriving at the cerebellar cortex. The information is then processed within the cortical circuitry; its output element is the Purkinje cell (Fig. 16-4). The Purkinje cells project onto the deep cerebellar nuclei, which relay the cerebellar outflow to various areas of the central nervous system. (There are some exceptions, however, where the Purkinje cell axons exit the cerebellum directly without relaying through the deep nuclei.) An orderly topography prevails in this corticonuclear projection (Fig. 16-5). The most medial or *vermal zone* of the cerebellar cortex projects onto the most medial deep nucleus, the *fastigial nucleus*. The most lateral or *hemispheric zone* projects onto the most lateral deep nucleus, the *dentate nucleus*. Between

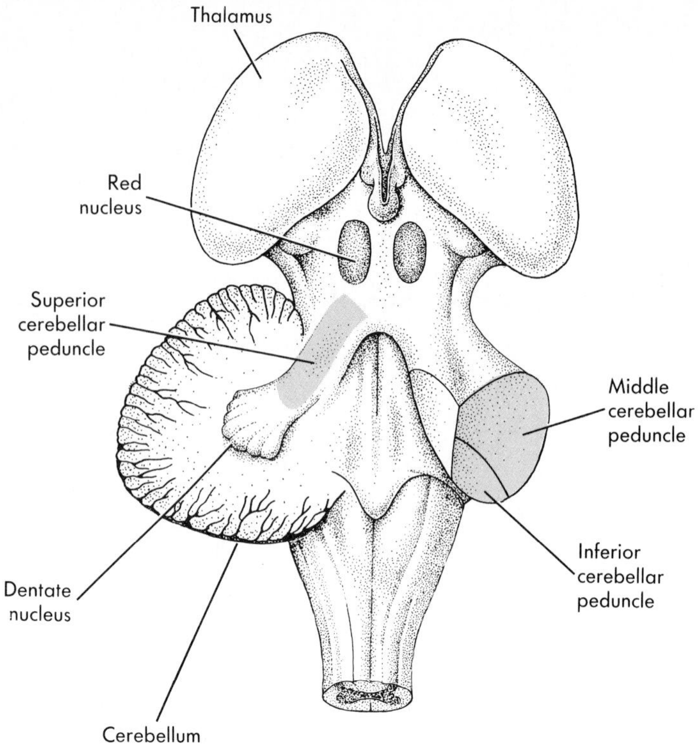

■ **Fig. 16-1.** Schematic illustration showing the cerebellar peduncles. (Redrawn from Carpenter, M.B.: Human neuroanatomy, ed. 7. © 1976, The Williams & Wilkins Co., Baltimore.)

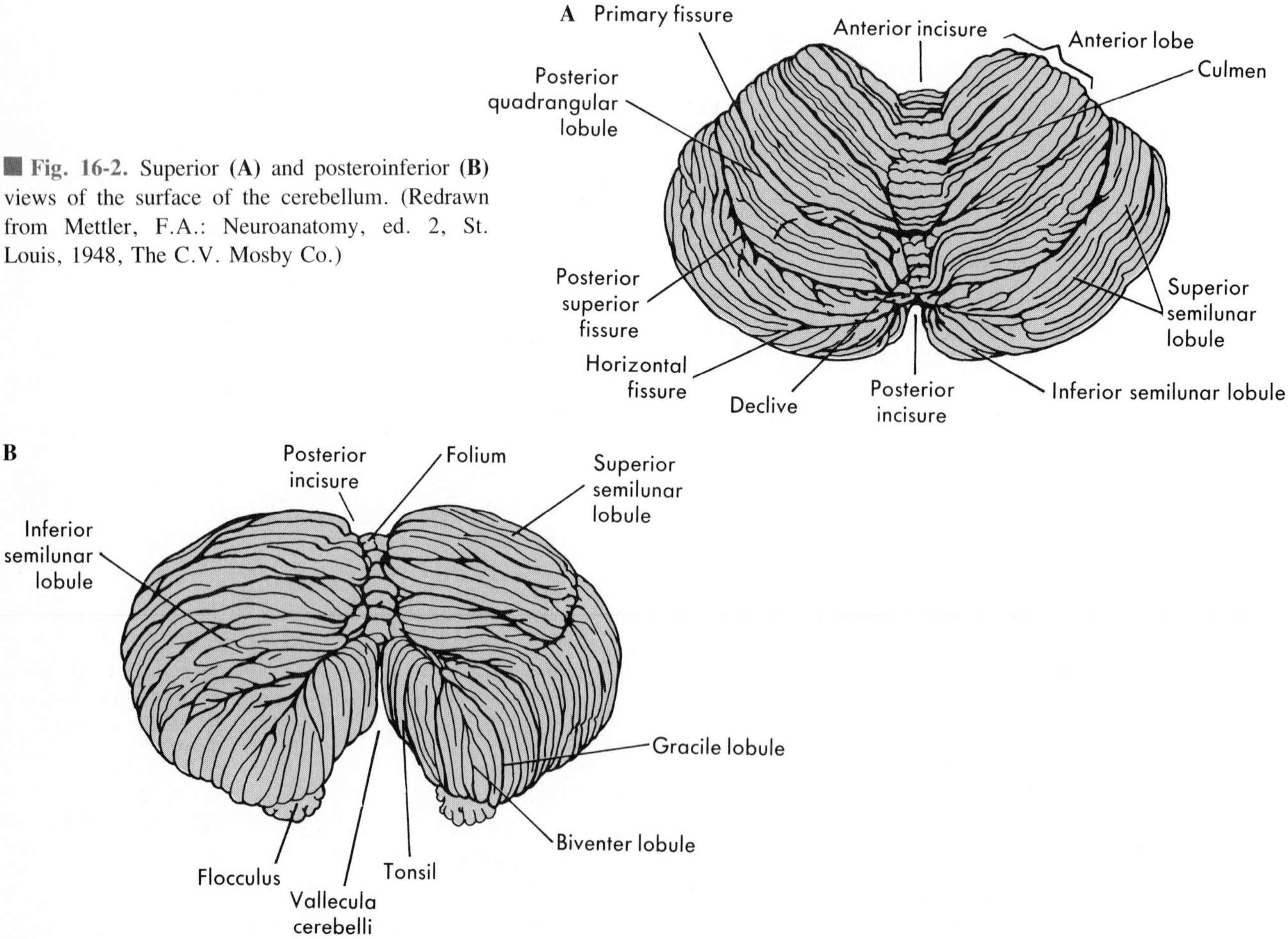

■ **Fig. 16-2.** Superior **(A)** and posteroinferior **(B)** views of the surface of the cerebellum. (Redrawn from Mettler, F.A.: Neuroanatomy, ed. 2, St. Louis, 1948, The C.V. Mosby Co.)

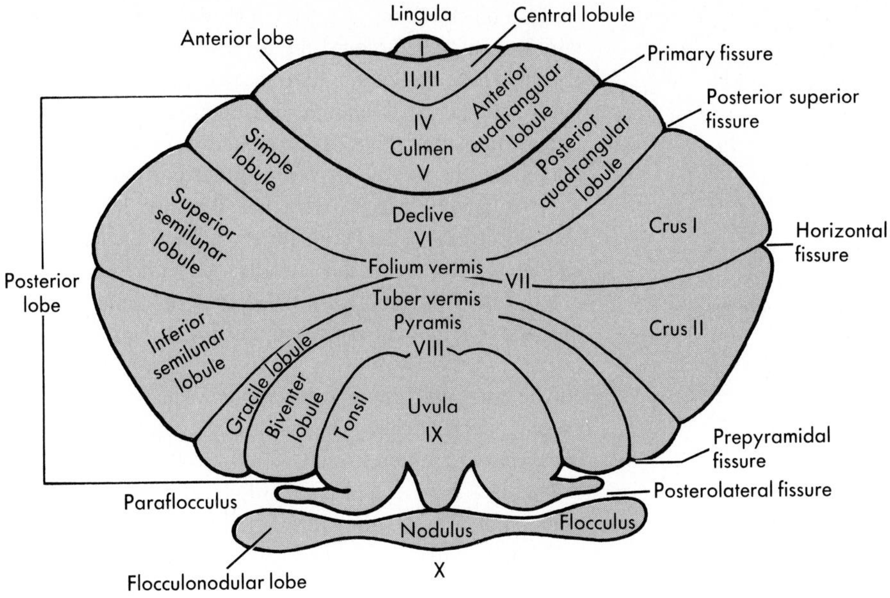

■ **Fig. 16-3.** Fissures and lobules of the cerebellum. Portions of the cerebellum caudal to the posterolateral fissure constitute the flocculonodular lobe (archicerebellum). Portions rostral to the primary fissure constitute the anterior lobe (paleocerebellum). The neocerebellum lies between the primary and posterolateral fissures. Roman numerals refer to portions of the cerebellar vermis only. (Redrawn from Carpenter, M.B.: Human neuroanatomy, ed. 7. © 1976, The Williams & Wilkins Co., Baltimore; modified from Larsell, Jansen and Brodal [1958], and Angevine et al.)

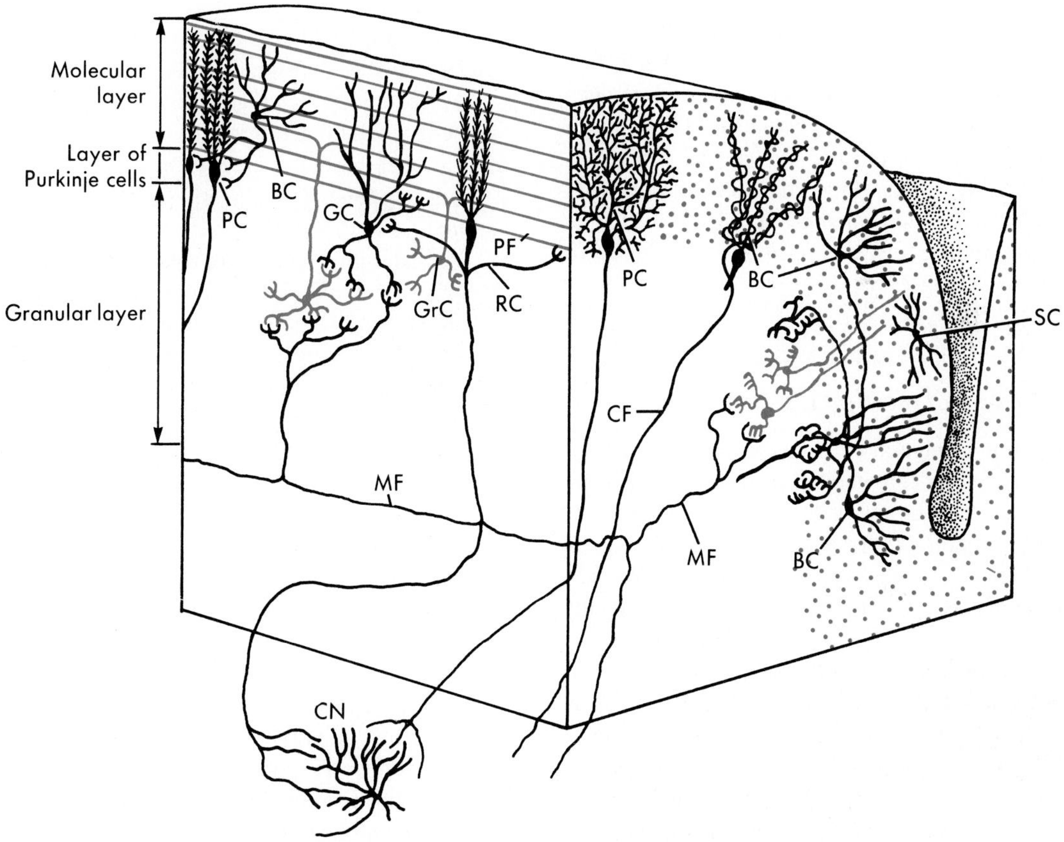

■ **Fig. 16-4.** Organization of the cerebellar cortex shown diagrammatically in a wedge-shaped section of the cortex along the longitudinal axis of a folium. *PC,* Purkinje cell; *BC,* basket cell; *GC,* Golgi cell; *GrC,* granule cell; *PF,* parallel fiber; *RC,* recurrent collateral; *MF,* mossy fiber; *CF,* climbing fiber; *CN,* deep cerebellar nuclear cell; *SC,* stellate cell. (Redrawn from Fox, C.A.: The structure of the cerebellar cortex. In Crosby, E.C., Humphrey, T.H., and Lauer, E.W., editors: Correlative anatomy of the nervous system, New York, 1962, The Macmillan Co. Copyright © 1962 by Macmillan Publishing Co., Inc.)

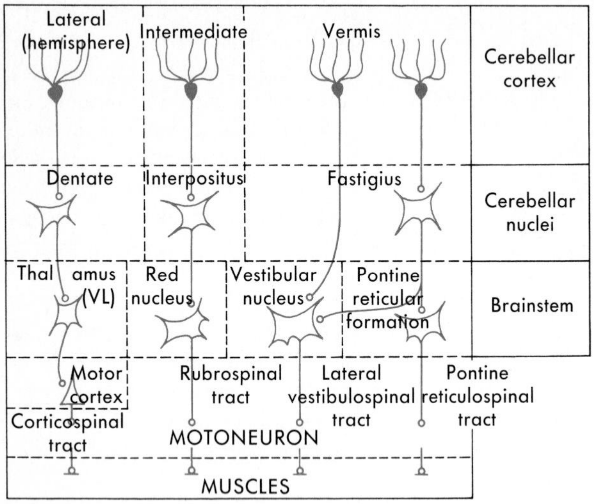

■ Fig. 16-5. Pattern of outflow from the cerebellar cortex. Inhibitory neurons are in solid color. (Redrawn from Bell, C.C., and Dow, R.S.: Cerebellar circuitry. In Schmitt, F.O., et al., editors: Neurosciences Research Symposium Summaries, vol. 2, Cambridge, Mass., 1967, The MIT Press. Copyright © 1967 by The MIT Press.)

these medial and lateral longitudinal zones is an intermediate or interposed zone, the *paravermal zone,* which projects onto the interposed or intermediate deep nuclei, the *globus* and *emboliform nuclei.* (These nuclei correspond to the *interpositus nucleus* in commonly studied carnivores.) Thus the cerebellum is organized into three *longitudinal corticonuclear zones:* medial or vermal, intermediate or paravermal, and lateral or hemispheric.

■ Distribution of Cerebellar Efferent Fibers

A significant advantage in viewing the cerebellum from the perspective of its longitudinal or mediolateral organization is that the outputs of these zones distribute in a predictable fashion with respect to the medial and lateral descending motor pathways (Fig. 16-5). The medial or vermal zone, via the fastigial nucleus, projects prominently onto the pontine reticular formation to access the cells of origin of the pontine reticulospinal tract. This zone also projects onto the lateral vestibular nucleus, which contains the cells of origin of the lateral vestibulospinal tract. This projection is of particular interest because, in addition to fibers from the fastigial nucleus, it includes a complement of Purkinje cell axons that do not relay through the fastigial nucleus (Fig. 16-4). Thus the medial cerebellar zone primarily influences the cells of origin of the principal pathways of the medial descending system: the pontine reticulospinal tract and the lateral vestibulospinal tract.

In contrast to the outflow from the medial zone, the lateral or hemispheric zone projects, via the dentate nucleus, onto the ventral lateral nucleus of the thalamus, which then projects onto the motor cortex (Fig. 16-5). By this route the lateral zone influences the most prominent pathway of the lateral descending system, the lateral corticospinal tract. The projection from the dentate nucleus to the rostral or parvocellular portion of the red nucleus is much less extensive. Because most of the cells of origin of the rubrospinal tract are located in the caudal two thirds of the nucleus, it is unclear to what extent this projection from the dentate nucleus provides the lateral cerebellar zone with access to the rubrospinal tract.

The intermediate or paravermal zone, via the intermediate deep nuclei, projects primarily onto the red nucleus and to a much lesser extent onto the ventral lateral thalamus. Thus its principal influence is on the rubrospinal tract, associating it with the lateral descending system.

■ Distribution of Cerebellar Afferent Fibers

The pattern of cerebellar afferentation is complex. Vestibular information influences a restricted area of the cerebellum, namely, the flocculonodular lobe, with a very small extension into the adjacent area (uvula) of the posterior lobe (Fig. 16-6). Therefore this region of the cerebellum is referred to as the *vestibulocerebellum.* It is also the oldest area of the cerebellum phylogenetically, and thus it is also referred to as the *archicerebellum.*

A later phylogenetic development consists of the anterior lobe and the posterior segment of the vermis. Consequently, these areas are referred to as the *paleocerebellum.* It includes segments of both the vermal and paravermal longitudinal zones. Sensory information from the spinal cord distributes over the paleocerebellum and is relayed via a number of direct and indirect pathways. In view of its spinal input, this cerebellar region is also referred to as the *spinocerebellum.*

The body surface is topographically mapped on the paleocerebellum in multiple fashion (Fig. 16-7). One sensory map is found in the anterior lobe. The axial body surface is represented medially, and the limb representation extends laterally. This map is inverted, and the head representation is situated posteriorly. A second representation is found in the posterior lobe. Again, the axial surface is represented closest to the midline, and the limb representation extends laterally. This posterior map is longitudinally oriented in a direction opposite to the anterior map so that the head representations of each map are found more centrally on the cerebellar surface. Visual and auditory information is relayed to the cerebellum. Such information influences areas within the head representations in a zone between the anterior and posterior components of the spinocerebellum (Fig. 16-7). Furthermore, these sensory maps coincide with mo-

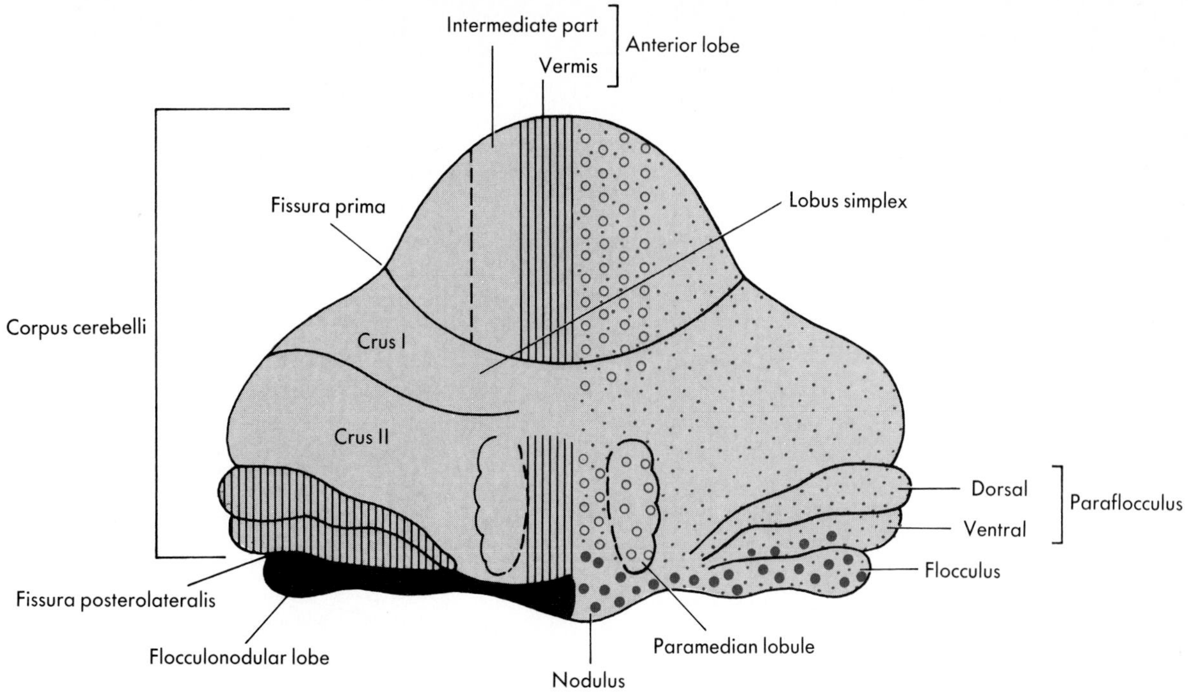

■ **Fig. 16-6.** Cerebellar afferentation. On the left are the archicerebellum *(black),* paleocerebellum *(hatched),* and neocerebellum *(grey)*. On the right are the distributions of vestibulocerebellar fibers *(closed circles),* spinocerebellar fibers *(open circles),* and pontocerebellar fibers *(dots)*. (Redrawn from Brodal, A.: Neurological anatomy, ed. 3. Copyright © 1981 by Oxford University Press, Inc. Reprinted by permission.)

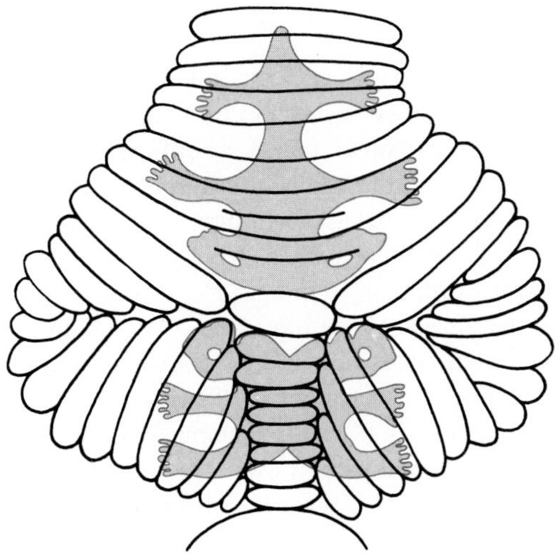

■ **Fig. 16-7.** Somatotopic representations of the body on the cerebellar cortex. (Redrawn from Snider, R.: The cerebellum, Sci. Am. **199**(6):4, 1958.)

tor maps that have been derived by electrical stimulation of the cerebellar surface.

The most prominent direct spinocerebellar pathways are the dorsal and ventral spinocerebellar pathways that relay information from thoracic and upper lumbar levels (Fig. 16-8). The pathways that provide information from cervical levels are the cuneocerebellar and rostral spinocerebellar pathways. The dorsal spinocerebellar tract arises from cells in Clarke's column, and it projects ipsilaterally to the vermal and paravermal zones. The cuneocerebellar pathway arises from the external cuneate nucleus, and it also terminates ipsilaterally in the vermal and paravermal zones. Both of these pathways provide detailed cutaneous and proprioceptive information from very circumscribed regions, and they distribute in a highly localized fashion on the cerebellum. The ventral and rostral spinocerebellar pathways arise from cells that are distributed more diffusely in the spinal gray matter and that project bilaterally onto the spinocerebellum. They relay cutaneous and proprioceptive information from wider regions than the dorsal and cuneocerebellar pathways, and they terminate in a less discrete pattern on the cerebellum. The indirect pathways are complex, but the most prominent pathways include relays in the inferior olive (spino-olivocerebellar pathways) and in the lateral reticular nucleus (spinoreticulocerebellar pathways).

Such spinal input is consistent with the longitudinal pattern of cerebellar organization that was described in the context of the cerebellar efferent pathways (Fig. 16-8). Vestibular inputs, which are associated with the medial descending system, distribute to the flocculonodular lobe and uvula. These regions constitute the part of the medial longitudinal zone that influences the lateral vestibulospinal tract. The representation of the head, as

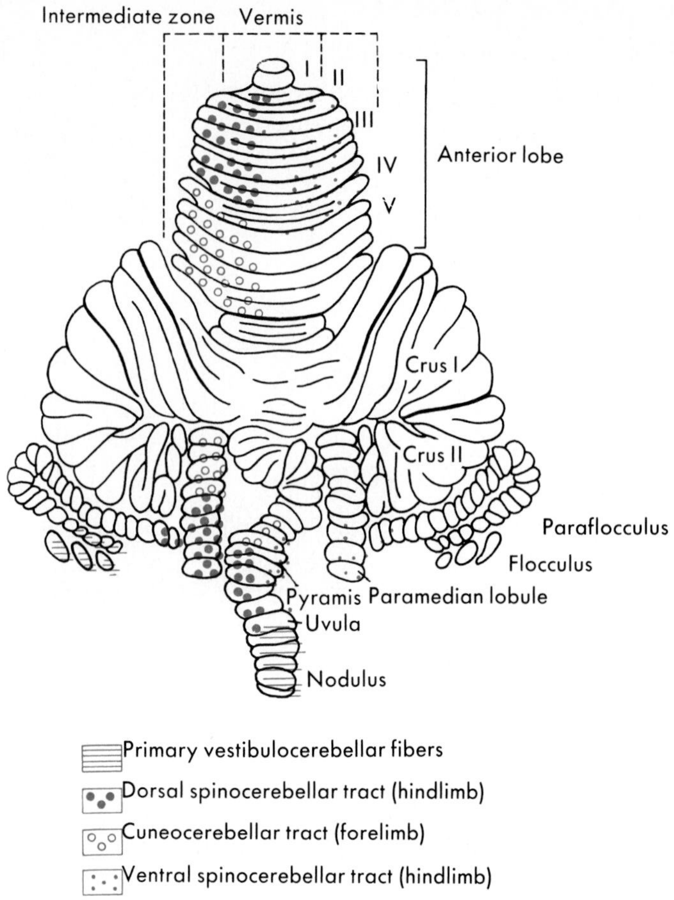

Fig. 16-8. The cerebellar surface showing the distribution of fibers of the dorsal and ventral spinocerebellar tracts and fibers from the external cuneate nucleus. Note the somatotopic pattern. The terminal distribution of primary vestibular fibers is also shown. (Redrawn from Brodal, A.: Neurological anatomy, ed. 3. Copyright © 1981 by Oxford University Press, Inc. Reprinted by permission.)

well as of visual and auditory inputs, is in the medial zone, and head and eye movements are associated with medial system functions. The axial body surface is represented within the vermal zone, which influences the pontine reticulospinal tract. Thus the cerebellar zone that influences the medial musculature appropriately receives information about the status of the medial muscles, joints, and body surface. The more distal limb representation extends into the paravermal zone, the outflow of which most prominently influences the rubrospinal tract, which in turn controls the limb muscles.

The phylogenetically newest portion of the cerebellum is the lateral or hemispheric zone; it is referred to as *neocerebellum*. This zone has no direct sensory representation. Its principal source of afferent fibers is from the cortex, and this corticocerebellar influence is relayed via the pontine nuclei (Fig. 16-6). Thus the hemispheric zone is also referred to as *pontocerebellum*. This corticopontocerebellar projection arises from widespread areas of the neocortex, including the association cortex. The projection includes collaterals of corticospinal and corticobulbar pathways, which provide information to the cerebellum regarding commands to motoneurons.

The Position of the Cerebellum in the Chain of Motor Command

The cerebellum occupies an interesting position among the structures that control movement. It receives a wealth of sensory information, particularly of a proprioceptive nature. Such information is important for the regulation of motor activity. In addition, the cerebellum receives information regarding the commands being delivered to motoneurons, and this information is provided from many levels of the chain of motor command. For example, collaterals of corticospinal tract fibers influence the cerebellum through the medial and lateral pontine nuclei and inform the cerebellum of the output of the motor cortex. Also, certain spinocerebellar inputs originate from spinal interneurons that influence motoneurons, either directly or indirectly, and these interneurons also receive messages from various descending motor pathways. Thus the cerebellum receives inputs about motor commands from cells near the final common path.

The cerebellum is thus one of the best-informed motor structures of the nervous system. Consequently, it can process this mass of information and relay its processed output to the various descending systems. Thus the cerebellum may be viewed as a principal component of many major loops involving the descending pathways. For example, the neocerebellum receives extensive cortical input via the corticopontocerebellar pathway. In turn, it relays information back to the cortex via the ventral thalamus, where it also interacts with the outflow of the basal ganglia. The flocculonodular lobe is the primary cerebellar target of vestibular information, and its outflow then plays back on the vestibulospinal pathways.

Functional Neuroanatomy

Functional Correlates of the Mediolateral Longitudinal Organization

On the basis of the efferent distribution of the longitudinal zones the distribution of motor deficits consequent to cerebellar lesions may be predicted. Indeed, it follows the mediolateral organization. Lesions involving the medial zone of the cerebellum produce ipsilateral truncal deficits. If the flocculonodular region is included, then deficits in equilibration will result as well. These deficits appear as disturbances of tone and synergy, where synergy refers to the temporal properties of movements. Thus *asynergia* would involve disturbances in the timing of the onset and termination of contractions, the relationship between the contractions of agonistic and antagonistic muscles, and the temporal sequences of muscular contractions.

Lesions confined to the intermediate zone affect principally the ipsilateral proximal limbs, impairing skilled movement of the limb and its participation in postural adjustments. Lesions confined to the lateral zone compromise spatially organized movements of the limbs, but they have no effect on posture, progression, or equilibrium. Although the dentate projection to the thalamus is crossed, the corticospinal tract then decussates to return the cerebellar influence to the ipsilateral side. Hence, the effects of lesions are ipsilateral.

Thus the distribution of deficits caused by restricted cerebellar lesions is entirely understandable, given the longitudinal organization and its efferent pathway distribution with respect to the medial and lateral descending systems. More of the specific nature of the deficits will be considered later.

Cerebellar Disorders

The longitudinal organization of the cerebellum also provides a predictive model for understanding cerebellar deficits in disease. Although human cerebellar disease frequently does not respect the mediolateral boundaries of this organization, some disorders illustrate well the basic principles of longitudinal organization.

Damage to the cerebellum or its direct connections causes unique motor disorders. Although the classification is somewhat oversimplified, these disorders can be grouped into three classes: disturbances of *synergy, equilibrium,* or *tone*. Synergy describes coordination, and it refers to the regulation of the rate, range, force, and direction of movements. Cerebellar lesions that disturb synergy are frequently referred to clinically as producing *ataxia*. The asynergia can manifest itself in a number of ways. For example, a common form of asynergia is *dysmetria,* in which errors in the direction and force of a movement compromise an individual's ability to bring a limb smoothly to a desired position. The limb may either overshoot the desired position *(hypermetria)* or undershoot it *(hypometria)*. When the lower limbs are involved, a cerebellar lesion produces an unsteady gait that is frequently broad based. If the desired movement is more complex, the asynergia is expressed as a *decomposition of movement,* where the required sequence of muscle contractions is executed in discrete steps rather than as a smooth movement. The *intention tremor* associated with certain cerebellar lesions is a tremor that is perpendicular to the direction of movement and that increases in magnitude toward the end of the movement. It is a disturbance of synergy that involves the temporal relationship between the contraction of agonistic and antagonistic muscles. The ways in which asynergia is expressed are probably as numerous as the clinical tests that can be devised. Depending on the locus of the cerebellar damage, asynergia can involve any muscle group, including the extrinsic eye muscles (producing *nystagmus*) and the muscles of speech (producing *dysarthria*). However, the underlying theme of the disturbance is a loss of the precise temporal regulation of sequences of contractions.

Disorders of equilibration, that is, impairment of the ability to maintain an upright posture, can occur if the lesion involves the vestibulocerebellum. With respect to muscle tone, cerebellar lesions in humans most often produce a hypotonia, which can result in a *pendular limb*. After elicitation of a patellar tendon reflex, for example, the limb will continue to swing back and forth in a pendular fashion.

It is also instructive to consider specific lesions in the context of longitudinal cerebellar organization. Two prominent disorders affect the medial longitudinal zone. One is a childhood tumor, a medulloblastoma, that most commonly involves the flocculonodular lobe. The principal deficit is impaired equilibration, with no ataxia or alterations in muscle tone. Another medial disorder is the cerebellar degeneration in alcoholism. This degeneration is largely restricted to the paleocerebellum. The motor deficits are principally postural, and they include a truncal ataxia and broad-based gait. Both of these diseases involve mainly the vermal cerebellum, and the motor abnormalities are thus confined to the more proximal muscles. In contrast to this, various degenerative disorders and tumors of the cerebellum involve principally the neocerebellum or lateral zone. In these disturbances the deficits involve the ipsilateral limbs, and hypotonia, limb ataxia, and tremor are common.

Thus the distribution of motor deficits in cerebellar disease is predictable from the anatomical organization of the cerebellar outflow. The deficits are ipsilateral to the lesion, and the mediolateral position and extent of the lesion will determine the proximodistal distribution of involved muscles.

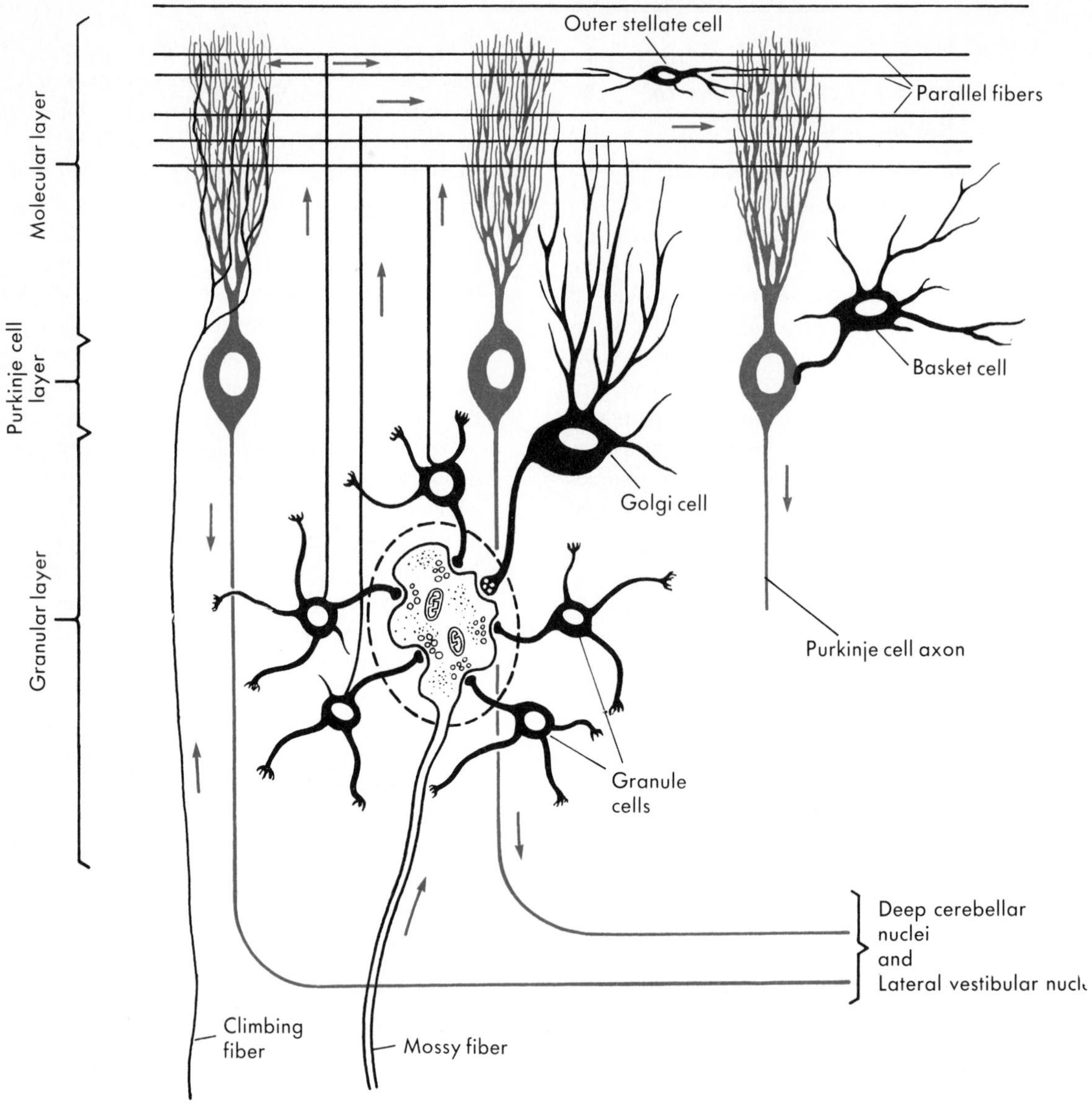

■ **Fig. 16-9.** Elements of the cerebellar cortex in the longitudinal axis of a folium. (Redrawn from Carpenter, M.B.: Human neuroanatomy, ed. 7. © 1976, The Williams & Wilkins Co., Baltimore; based on Gray and Eccles et al.)

A final point is that the patient can frequently compensate well for cerebellar damage, particularly if the lesion occurs in childhood. The mechanism of this compensation is not understood.

■ Cerebellar Function

■ Organization of the Cerebellar Cortex

The cerebellar cortex consists of three layers (Fig. 16-9). The deepest, or *granular,* layer contains mainly densely packed *granule cells,* which receive most of the input to the cerebellar cortex. Also in the granular layer are the *Golgi neurons,* one of three types of interneurons of the cerebellar cortex. Above the granular layer is the *Purkinje layer,* which consists of a single layer of *Purkinje cells,* the output element of the cerebellar cortex. The most superficial layer is the *molecular layer.* The two other interneuronal classes, the *stellate* and *basket cells,* are situated in this layer. In addition, the molecular layer contains the dendrites of the Purkinje and Golgi cells. The final important constituents of this layer are the axons of the granule cells. These axons ascend from the granular layer, and on reaching the molecular layer, they bifurcate to run longitudinally for a few millimeters in the orientation of the folium. These axons constitute the *parallel fiber system.*

Most input to the cerebellar cortex is transmitted via *mossy fibers,* which arise from many different cell

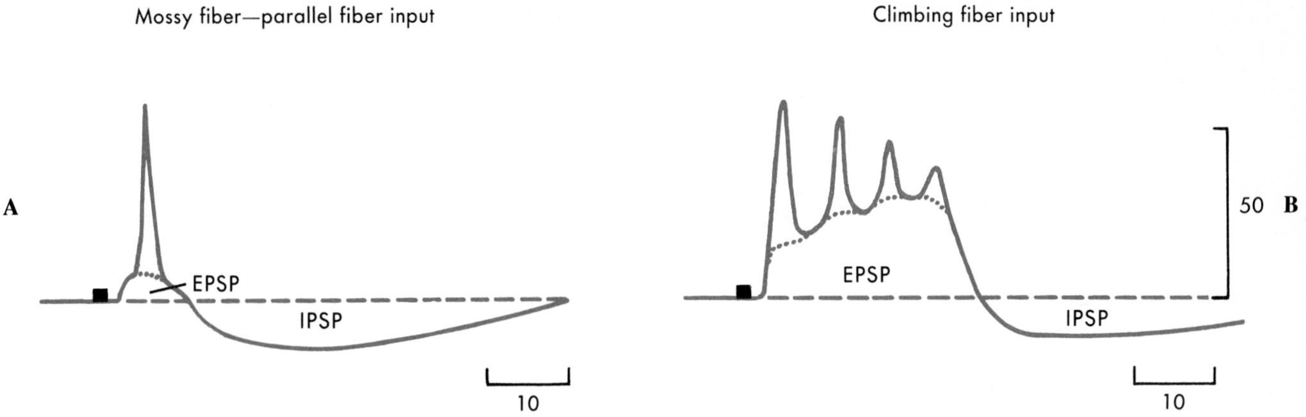

Fig. 16-10. Comparison of responses of Purkinje cells to inputs from the mossy fiber (**A**) and climbing fiber (**B**) systems. Note the burst, or complex spike, in response to a climbing fiber input.

groups of the spinal cord and brainstem. The mossy fibers terminate on and excite the granule cells in the deepest layer of the cerebellar cortex. Therefore they activate the parallel fiber system, which consists of the axons of the granule cells (Fig. 16-9). This afferent fiber system shows massive convergence and divergence; each granule cell receives a large number of mossy fiber inputs and each mossy fiber synapses with a large number of granule cells. The contacts are axodendritic and form distinctive glomeruli (Fig. 16-9). In addition, the mossy fibers innervate the deep cerebellar nuclei *en passage* to the cerebellar cortex.

A second source of afferent fibers originates from cells of the contralateral inferior olive. These neurons give rise to the *climbing fiber* system, which differs strikingly from the mossy fiber system, both anatomically and physiologically. The climbing fiber system shows little convergence or divergence. Each Purkinje cell receives input from only a single climbing fiber, and each climbing fiber contacts no more than a dozen Purkinje cells. Furthermore, the climbing fibers make numerous excitatory contacts with the somas and proximal dendrites of their target Purkinje cells (Fig. 16-9) and innervate the deep nuclei via collaterals. Such multiple contacts lead to a powerful excitatory action, in which a single discharge along a climbing fiber can initiate a burst of impulses from the Purkinje cell (Fig. 16-10). The organization of the inferior olive also has unusual properties, such as electrotonic coupling of its neurons. The possible functions of this unique climbing fiber input to the cerebellar cortex are unclear.

A third source of input to the cerebellar cortex is aminergic. This includes noradrenergic afferents from the locus ceruleus and serotonergic afferents from the raphe. These inputs appear to regulate the excitability of cerebellar cortical neurons.

Whereas a climbing fiber input will briefly activate a restricted ensemble of Purkinje cells, mossy fiber inputs have a more divergent influence. Mossy fiber inputs activate a large number of granule cells and thereby many parallel fibers (Fig. 16-9). Those parallel fibers will then excite a band of Purkinje cells over wide areas of the cerebellum. The Purkinje cell dendrites are oriented perpendicularly to the longitudinal axis of the folium, and the parallel fibers course parallel to this axis (Fig. 16-4). Hence each parallel fiber runs perpendicularly through a row of Purkinje cell dendrites, contacting perhaps 50 Purkinje cells. Each Purkinje cell is contacted by approximately 200,000 parallel fibers.

The parallel fibers also activate the three classes of cerebellar cortical interneurons, all of which have their dendrites in the molecular layer (Figs. 16-9 and 16-11). These interneurons are all inhibitory. The Golgi cell feeds back to inhibit the granule cells, thus shutting off the input via the mossy fiber system. The basket and stellate cells both feed forward to inhibit the Purkinje cells. The stellate cells inhibit more distal dendrites, and the basket cells inhibit more proximal dendrites and the soma (Fig. 16-11).

The geometry of these intracortical connections is precise. Activation of a narrow band of parallel fibers will activate a narrow beam of Purkinje cells. After a short delay the inhibitory feedback from the Golgi cells will turn off the parallel fiber input over that narrow band. The feed-forward inhibition via the basket and stellate cells is directed toward the bands of Purkinje cells that are arrayed parallel to and on either side of the activated band. The net effect is that (1) mossy fiber activation of a small cluster of granule cells excites a well-defined beam of Purkinje cells via the parallel fibers; (2) these same parallel fibers excite basket and stellate cells that inhibit the surrounding beams of Purkinje cells; and (3) after a short delay the Golgi cell feedback inhibition shuts off the activating input. Consequently, mossy fiber activation of a cluster of granule cells produces a well-focused beam of Purkinje cell excitation for a brief period, and concomitantly the off-beam Purkinje cells are inhibited. This is a sophisti-

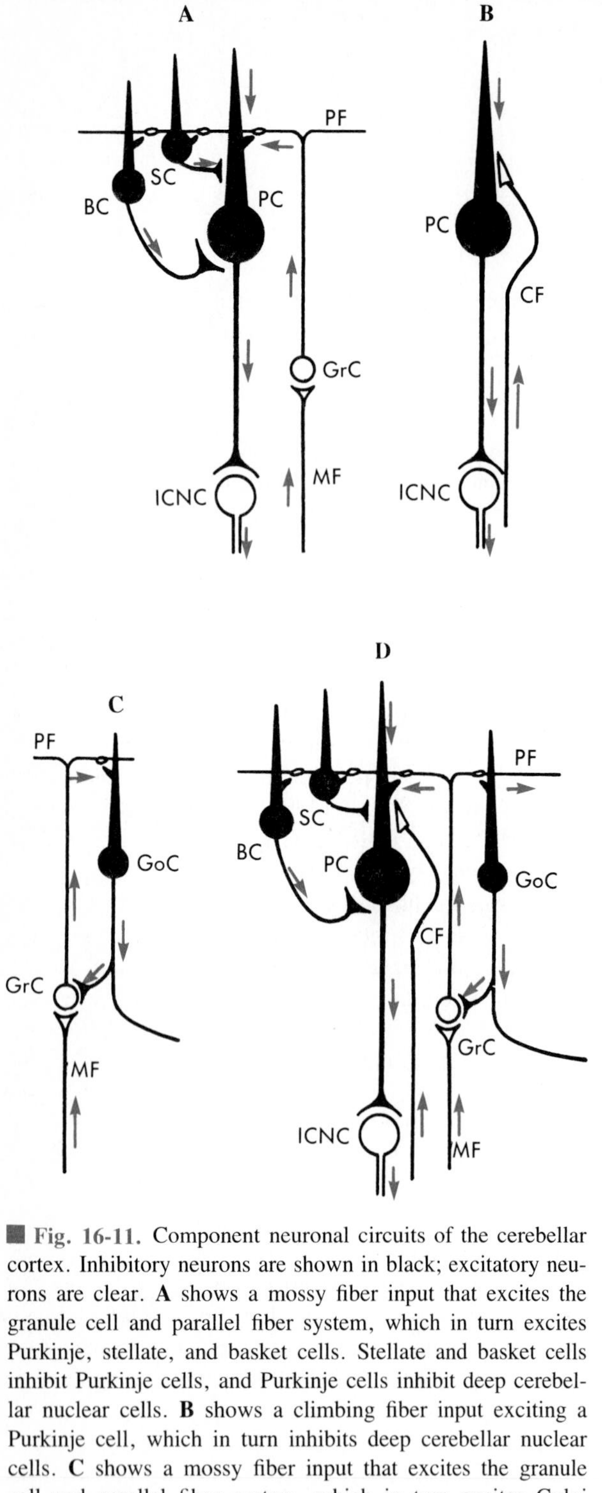

Fig. 16-11. Component neuronal circuits of the cerebellar cortex. Inhibitory neurons are shown in black; excitatory neurons are clear. **A** shows a mossy fiber input that excites the granule cell and parallel fiber system, which in turn excites Purkinje, stellate, and basket cells. Stellate and basket cells inhibit Purkinje cells, and Purkinje cells inhibit deep cerebellar nuclear cells. **B** shows a climbing fiber input exciting a Purkinje cell, which in turn inhibits deep cerebellar nuclear cells. **C** shows a mossy fiber input that excites the granule cell and parallel fiber system, which in turn excites Golgi cells. The Golgi cell axon returns to inhibit the granule cell. **D** shows a combination of the circuits shown in **A** to **C.** *PC,* Purkinje cell; *SC,* stellate cell; *BC,* basket cell; *PF,* parallel fiber; *GrC,* granule cell; *MF,* mossy fiber; *ICNC,* deep cerebellar nuclear cell; *CF,* climbing fiber; *GoC,* Golgi cell. (Redrawn from Eccles, J.C.: Functional organization of the cerebellum in relation to its role in motor control. In Nobel Symposium I: Muscular afferents and motor control, New York, 1966, John Wiley & Sons, Inc. © The Nobel Foundation 1966.)

cated form of lateral inhibition, similar to that seen in sensory systems, and it contributes to a high degree of spatial localization of the cerebellar output. With natural stimuli many such beams become activated in a highly organized spatial pattern. The pattern changes constantly as inputs from the periphery continuously impinge on the cerebellum.

Intracellular recordings from the Purkinje cells that project directly to the lateral vestibular nucleus have demonstrated that activation of the Purkinje cells produces inhibition. This finding has since been generalized, establishing that the output of the cerebellar cortex is inhibitory. Indeed, the only excitatory cell type in the cerebellar cortex is the granule cell. Apparently the deep cerebellar nuclei maintain a high level of background activity, most likely via collateral activation by climbing and mossy fibers. The output of the cerebellar cortex then inhibits this ongoing excitatory activity in a precise geometrical fashion. Thus the maintained excitatory influences of the deep cerebellar nuclei on the various targets of the cerebellum are modulated up or down by the output of the cerebellar cortex. For example, an increase in Purkinje cell activity of the paravermal cortex would inhibit cells of the nucleus interpositus and thus *disfacilitate* the rubrospinal neurons. This regulatory action occurs with a temporal and spatial precision appropriate for a structure that governs the rate, range, direction, and force of movements.

Cellular Studies of Cerebellar Function

With respect to the receptive fields of cerebellar cells, little transformation occurs from the mossy fiber to the granule cell, although considerable transformation of sensory information can occur before mossy fiber input. Most granule cells respond to a single modality, most frequently to changes in muscle length or tension. Responses to tactile stimulation are also common, whereas joint movement is a less effective stimulus for the granule cells. The receptive fields of the granule cells vary considerably in size and laterality and in the directionality of their responses (excitation or inhibition). Moreover, there is no apparent clustering of receptive fields in neighboring granule cells.

In contrast to the granule cell receptive fields, the Purkinje cells generally respond to inputs from an entire ipsilateral limb, and there is extensive overlap of fields among neighboring Purkinje cells. These neurons do not show the modality specificity of the granule cells, which is not surprising in view of the fact that perhaps 200,000 granule cells contact a single Purkinje cell. However, as in the granule cells, the preferred stimulus is muscle stretch. It is likely that the muscle spindle afferent fibers, and particularly the secondary endings, predominate. Most Purkinje cells respond to displacement around only a single joint, and the direction of their response can differ as a function of the direction of joint displacement. Although knowledge of the receptive field characteristics of cerebellar cortical neurons is primitive, the data do suggest that information arising from the muscle spindles may be the most important influence on Purkinje cell discharge. The general distribution of receptive fields is consistent with the mediolateral organization of the cerebellar output and with participation of the cerebellum in regulating movements in response to sensory feedback.

The nature of the sensory information that reaches the cerebellum and the transformation of this information within the cerebellar cortex are prerequisite to understanding cerebellar function. Relevant information may be derived by studying the responses of cerebellar neurons during movement. Although such an approach has not yet yielded a definitive answer, it has generated some provocative suggestions. Such studies have shown that the discharge of neurons in the interpositus nucleus is altered after the start of a movement, and this discharge is related to such variables as muscle force and velocity. In contrast, neurons in the dentate nucleus frequently discharge before a movement, sometimes even before the movement-related activity of neurons in the motor cortex. Moreover, the discharge of these cerebellar neurons is more closely related to joint position and the direction of movement.

Such data suggest that the association areas of the neocortex act in concert with the neocerebellum to organize and initiate movements, with the final command being transmitted by the motor cortex. The intermediate zone in this scheme may be regulatory through the feedback of sensory information.

Bibliography

Journal articles

Allen, G.L., and Tsukahara, N.: Cerebrocerebellar communication systems, Physiol. Rev. **54:**957, 1974.

Chambers, W.W., and Sprague, J.M.: Functional localization in the cerebellum. I. Organization in longitudinal corticonuclear zones and their contribution to the control of posture, both extrapyramidal and pyramidal, J. Comp. Neurol. **103:**105, 1955.

Evarts, E.V., and Thach, W.T.: Motor mechanisms of the CNS: cerebrocerebellar interrelations, Annu. Rev. Physiol. **31:**451, 1969.

Holmes, G.: The cerebellum of man, Brain **62:**1, 1939.

Jansen, J., and Brodal, A.: Experimental studies on the intrinsic fibers of the cerebellum. II. The corticonuclear projection, J. Comp. Neurol. **73:**267, 1940.

Snider, R.S., and Stowell, A.: Receiving areas of the tactile, auditory and visual systems in the cerebellum, J. Neurophysiol. **7:**331, 1944.

Thach, W.T.: Correlation of neural discharge with pattern and force of muscular activity, joint position, and direction of intended next movement in motor cortex and cerebellum, J. Neurophysiol. **41:**654, 1978.

Books and monographs

Bloedel, J.R., and Courville, J.: Cerebellar afferent systems. In Brookhart, J.M., and Mountcastle, V.B., editors: Handbook of physiology, the nervous system, section 1, vol. II, pt. 2, Bethesda, 1981, American Physiological Society.

Bloedel, J.R., Dichgans, J., and Precht, W.: Cerebellar functions, Berlin, 1985, Springer-Verlag.

Brodal, A.: Neurological anatomy, ed. 3, Oxford, 1981, Oxford University Press.

Brooks, V.B., and Thach, W.T.: Cerebellar control of posture and movement. In Brookhart, J.M., and Mountcastle, V.B., editors: Handbook of physiology, the nervous system, section 1, vol. II, pt. 2, Bethesda, 1981, American Physiological Society.

Eccles, J.C., Ito, M., and Szentágothai, J.: The cerebellum as a neuronal machine, New York, 1967, Springer-Verlag New York, Inc.

Ghez, C., and Fahn, S.: The cerebellum. In Kandel, E.R., and Schwartz, J.H., editors; Principles of neural science, New York, 1981, Elsevier Science Publishing Co., Inc.

Llinas, R.: Electrophysiology of the cerebellar networks. In Brookhart, J.M., and Mountcastle, V.B., editors: Handbook of physiology, the nervous system, section 1, vol. II, pt. 2, Bethesda, 1981, American Physiological Society.

CHAPTER

17

The Basal Ganglia

■ *Anatomical Overview*

■ *Definition*

Full consensus has not yet been achieved regarding a precise anatomical definition of the basal ganglia. However, there is agreement that the *caudate nucleus, putamen,* and *globus pallidus* constitute its main mass (Fig. 17-1). The amygdaloid nucleus is generally excluded, and the claustrum is generally included. Other anatomical terms relating to the basal ganglia are the *corpus striatum* (caudate nucleus, putamen, globus pallidus, and claustrum), *neostriatum* (caudate nucleus and putamen), *paleostriatum* (globus pallidus), and *lentiform nucleus* (putamen and globus pallidus). As used here, basal ganglia refers to the caudate nucleus, putamen, and globus pallidus, the major subcortical cell masses of the telencephalon.

Also associated with the basal ganglia are certain subtelencephalic cell groups, the most prominent of which are the *substantia nigra* of the mesencephalon and the *subthalamic nucleus* of the basal caudal diencephalon (Fig. 17-1). The substantia nigra consists of a ventral pars reticulata, which appears to be a caudal continuation of the internal segment of the globus pallidus, and a dorsally located pars compacta, which contains highly pigmented (melanin-rich) neurons. The basal ganglia and their associated diencephalic and mesencephalic structures have been traditionally referred to as the *extrapyramidal system,* a term with little significance in contemporary treatments of the motor system.

■ *Afferentation*

The structures that constitute the basal ganglia, in addition to being extensively interconnected, receive inputs from four major sources: the cortex, thalamus, substantia nigra, and raphe (Fig. 17-2). The neocortex projects topographically onto the caudate nucleus and putamen. A prominent component of these corticostriatal projections is an input from the precentral gyrus (the motor and premotor cortices).

The fibers from the cortical motor areas apparently arise from a population of small pyramidal cells in the upper half of layer V. These cells are different from the cells of origin of other descending cortical projections such as the corticospinal tract. The cortical input to the caudate nucleus and putamen is excitatory, and glutamate may be the transmitter.

Perhaps the next most prominent source of striatal afferentation is the substantia nigra (Fig. 17-2). This *nigrostriatal pathway* derives primarily from neurons in the pars compacta, and it terminates topographically in the caudate nucleus and putamen. The neurotransmitter of this nigrostriatal projection is dopamine.

A third source of afferent fibers to the basal ganglia is the thalamus; this thalamostriatal projection issues principally from the intralaminar nuclei (Fig. 17-2). Like the corticostriatal projection, the fibers from the intralaminar nuclei distribute topographically on the caudate nucleus and putamen and are excitatory. More recently a thalamic projection has also been identified to the globus pallidus and to the pars reticulata of the substantia nigra, the output structures of the basal ganglia. However, this originates in the subthalamic nucleus rather than in the intralaminar thalamus (Fig. 17-2).

Finally, the dorsal raphe nucleus projects extensively onto the neostriatum (Fig. 17-2). Indeed most, if not all, of the serotonin in the basal ganglia is derived from this pathway.

■ *Efferentation*

The major afferents to the basal ganglia from the cortex and thalamus project to the neostriatum (caudate nucleus and putamen) and are excitatory. The neostriatum then projects topographically onto the output elements of the basal ganglia, namely, the internal segment of

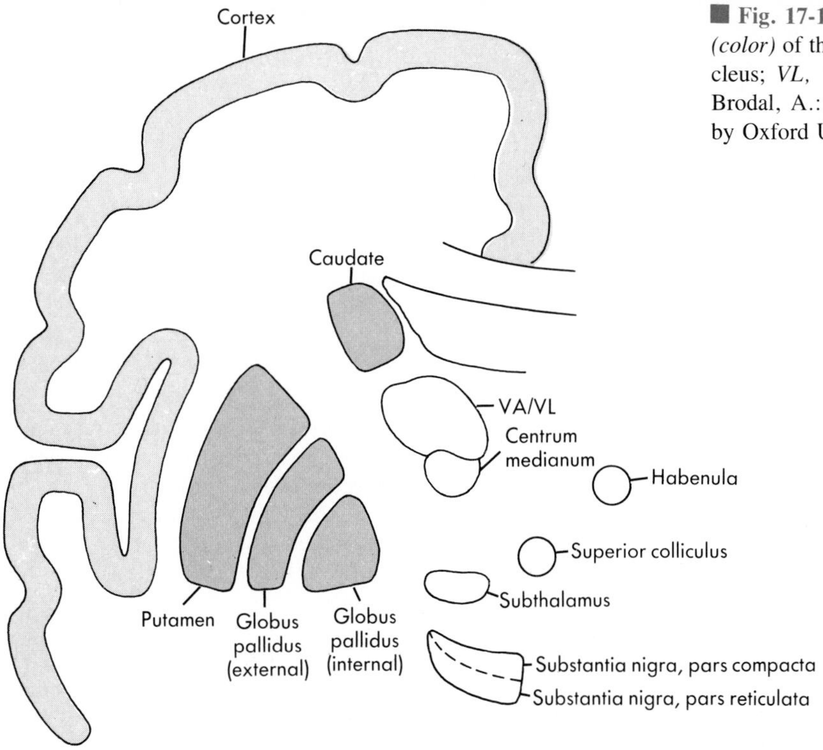

■ **Fig. 17-1.** A coronal section showing the main cell masses *(color)* of the basal ganglia. *VA,* Ventral anterior thalamic nucleus; *VL,* ventral lateral thalamic nucleus. (Redrawn from Brodal, A.: Neurological anatomy, ed. 3. Copyright © 1981 by Oxford University Press, Inc. Reprinted by permission.)

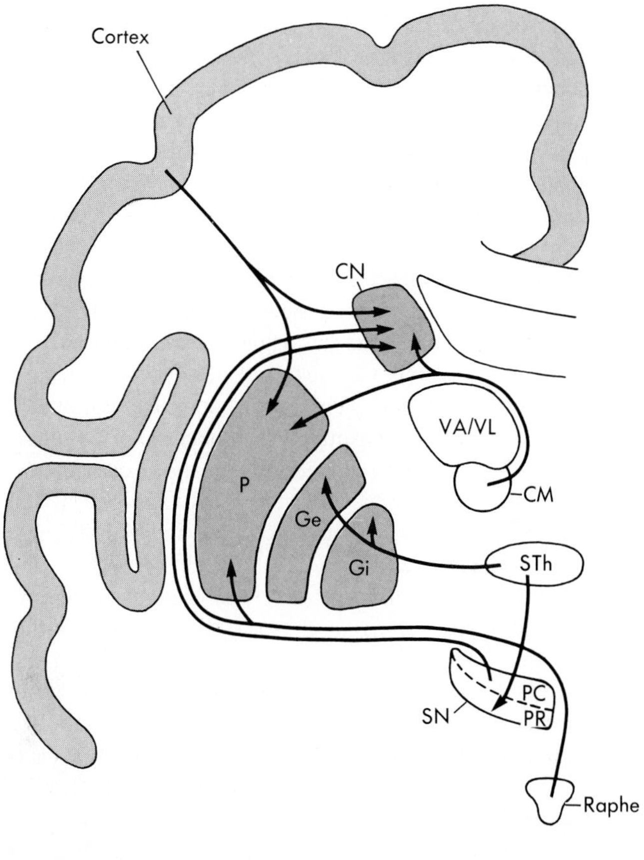

■ **Fig. 17-2.** A coronal section showing the major afferent fibers to the basal ganglia. See Fig. 17-1 for identification of structures. (Redrawn from Brodal, A.: Neurologic anatomy, ed. 3. Copyright © 1981 by Oxford University Press, Inc. Reprinted by permission.)

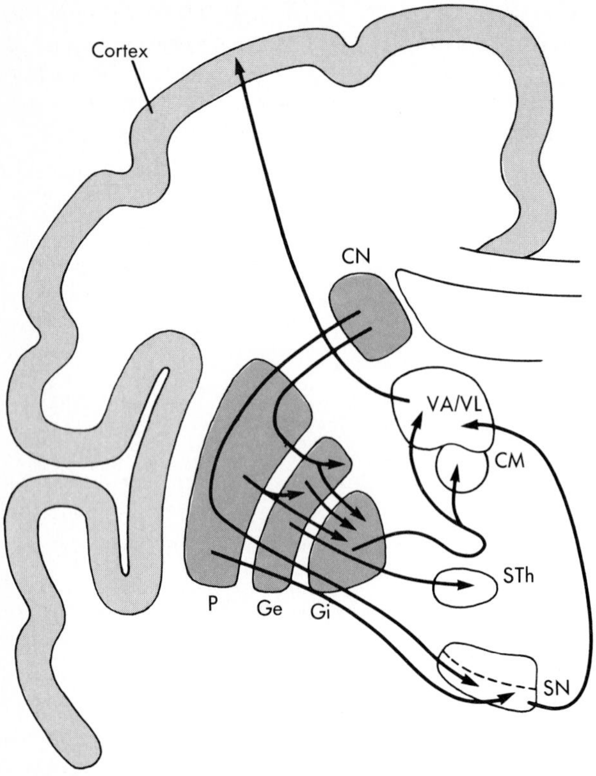

■ **Fig. 17-3.** A coronal section showing the intrinsic connections and major efferent projections of the basal ganglia. See Fig. 17-1 for identification of structures. (Redrawn from Brodal, A.: Neurological anatomy, ed. 3. Copyright © 1981 by Oxford University Press, Inc. Reprinted by permission.)

the globus pallidus and the pars reticulata of the substantia nigra (Fig. 17-3). These projections are inhibitory, and the transmitter is thought to be γ-aminobutyric acid (GABA). The cells of the internal pallidal segment give rise to two distinct fiber systems: the ansa lenticularis and the fasciculus lenticularis. These merge in the field of Forel to form the fasciculus thalamicus, which terminates topographically in the ventral thalamus, primarily in the ventral anterior (VA) and ventral lateral (VL) nuclei (Fig. 17-3). These projections from the globus pallidus and the pars reticulata of the substantia nigra to the ventral thalamus are also thought to be inhibitory.

The VA/VL complex in turn projects onto the motor cortex (Fig. 17-3) and is presumed to be excitatory. Because the motor cortex is a major source of afferent fibers to the caudate nucleus and putamen, this establishes a loop from the motor cortex to the basal ganglia and back to the motor cortex via the ventral thalamus (Figs. 17-2 and 17-3).

In brief, the principal output structures of the basal ganglia are the internal segment of the globus pallidus and the pars reticulata of the substantia nigra. Both project onto the VA/VL complex of the thalamus, which in turn projects onto the motor cortex.

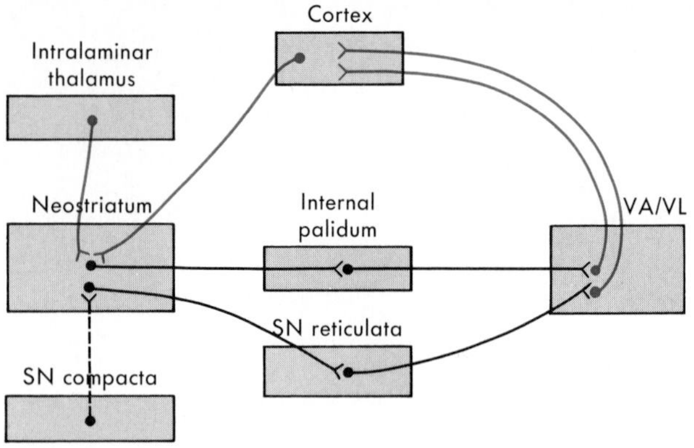

■ **Fig. 17-4.** Schematic diagram of some major functional neuronal connections of the basal ganglia. Colored connections are excitatory; black are inhibitory. The solid black inhibitory connections use GABA as a neurotransmitter. The solid colored excitatory connections are presumed to use glutamate as a transmitter. The broken black connection (the nigrostriatal pathway) uses dopamine; whereas the pathway is generally viewed as inhibitory, its influence is not firmly established and could in fact be excitatory. Other connections, such as from the subthalamic nucleus and raphe, are shown in Figs. 17-2 and 17-3.

■ *Functional Organization of the Basal Ganglia Circuitry*

The structure and function of the basal ganglia circuitry are very complex and many important questions must be resolved. Fig. 17-4 presents a simplified scheme of that circuitry, because the circuitry is fundamental to understanding diseases of the basal ganglia.

The cortex and intralaminar thalamus provide the major sources of phasic, excitatory input to the basal ganglia. They project in topographic fashion onto the neostriatum (caudate nucleus and putamen), whose neurons have little or no maintained activity. They then activate the striatal neurons during movement.

The striatal neurons project topographically to the output structures of the basal ganglia, the internal segment of the globus pallidus, and the pars reticulata of the substantia nigra. These projections are inhibitory, and GABA is the transmitter. The neurons of the internal pallidal segment and pars reticulata discharge continuously at high rates. Therefore activation of striatal neurons by cortical or thalamic inputs inhibits this high rate of background discharge.

The output elements of the basal ganglia then send inhibitory projections to the VA/VL complex of the thalamus. Because of their high rate of background discharge, they tonically inhibit the VA/VL. Thus when cortical or thalamic inputs to the neostriatum inhibit the pallidum and pars reticulata, the inhibition of VA/VL is reduced. Consequently, excitation of the striatum increases VA/VL activity by disinhibition.

At least three systems modulate the activity through this circuitry. The subthalamic nucleus regulates the output elements, the globus pallidus and the pars reticulata of the substantia nigra. The influence of the subthalamus is probably inhibitory, but precisely how the subthalamus affects information flow from the basal ganglia is not known. The nigrostriatal projection from the pars compacta of the substantia nigra to the caudate nucleus and putamen uses dopamine as a transmitter. Whereas its action is generally viewed as inhibitory, evidence also exists for an excitatory influence. For the present purposes we will treat it as an inhibitory pathway, but the reader should be aware that this conclusion is not firm. If one assumes an inhibitory action, then increasing the discharge along this dopaminergic pathway would decrease VA/VL activity by reducing inhibitory input to the output elements of the basal ganglia. Because neurons of the pars compacta do not discharge in specific relationship to movements and because they maintain low continuous rates of firing, the nigrostriatal pathway may have a tonic, modulatory influence. Finally, the serotonergic pathway from the raphe also regulates the striatum in a nonspecific manner, but a clear picture is not yet available as to the nature of that regulation.

Lesions of the Basal Ganglia

Experimental lesions of the basal ganglia have not been very informative. Such lesions have not reproduced the motor abnormalities of basal ganglia diseases. These diseases frequently involve disturbances of a particular transmitter system. Traumatic experimental lesions provide an inappropriate approach because they generally are not confined to a single transmitter system.

Clinical disorders involving the basal ganglia generally cause three types of deficits: (1) abnormal involuntary movements, or *dyskinesias,* (2) *changes in muscle tone,* and (3) slowness in initiating and executing movements, or *bradykinesia*. The abnormal movements are largely characteristic of the basal ganglia disease. These movements include regular involuntary *tremor* at rest, slow writhing movements most prominently involving the distal extremities *(athetosis),* flicklike movements involving the extremity and facial muscles *(chorea),* and violent flailing movements most commonly involving the proximal limb musculature *(ballismus)*. Changes in muscle tone can result in rigidity, spasticity, or hypotonus, depending on the particular disease process and the site of the lesion.

Parkinson's Disease

Perhaps the most common disorder involving the basal ganglia is Parkinson's disease, which is characterized by a resting tremor, cogwheel or plastic rigidity, and bradykinesia. The primary lesion site in this disease is the substantia nigra, particularly the dopaminergic neurons of the pars compacta. Most of the dopamine in the central nervous system is found in the basal ganglia, and brain dopamine levels are severely reduced in persons with Parkinson's disease. This would seem to implicate the nigrostriatal pathway. It is presently thought that in Parkinson's disease the dopaminergic neurons of the substantia nigra are destroyed, resulting in degeneration of the nigrostriatal fibers. If it is assumed that the nigrostriatal pathway inhibits its target neurons in the caudate nucleus and putamen, then degeneration of this dopaminergic pathway would remove an inhibitory influence on the striatal neurons and thereby increase the activity of the VA/VL neurons that project to the motor cortex (Fig. 17-4). The symptoms of Parkinson's disease would presumably reflect this change in activity of VA/VL neurons.

This reasoning led to the treatment of parkinsonian patients with L-dopa, a precursor of dopamine that crosses the blood-brain barrier. The rationale of this approach is to restore brain dopamine levels and thus to reverse the loss of the dopaminergic input from the nigrostriatal pathway. This treatment has, in fact, relieved some of the distressing symptoms. The positive results of L-dopa therapy support the argument that the dopaminergic influence on striatal neurons is nonspecific, since it involves nonspecific transmitter replacement. Unfortunately, however, the L-dopa treatment is not presently viewed with the same enthusiasm that prevailed when it was introduced. (More recently, transplantation of dopamine-secreting neurons is being explored as a therapeutic modality in Parkinson's disease.)

With respect to the functional anatomy of the disease, the tremor and rigidity, *but not the bradykinesia,* have been alleviated by surgical lesions that interrupt the outflow of the basal ganglia either at the globus pallidus or ventral thalamus. Surgical lesions of the motor cortex also alleviate these symptoms. Furthermore, if a person with Parkinson's disease sustains a stroke involving the internal capsule, then the tremor disappears on the hemiplegic side. It is significant that these lesions, which are all on the output side of the basal ganglia, do not alleviate the bradykinesia. This is consistent with the hypothesis that bradykinesia is a primary deficit of certain lesions of the basal ganglia, whereas the tremor and rigidity are release phenomena. It has been suggested that the tremor may result from loss of the serotonergic input from the raphe to the basal ganglia. Whether or not this is correct, it emphasizes the importance of determining the roles of the different afferent systems to the basal ganglia in producing the various symptoms of Parkinson's disease.

A final point regarding this disorder relates to rigidity, which is more evident in flexor than in extensor muscles. In some patients the rigidity is cogwheel in nature, where passive movement of the limb results in an alternation between increased and decreased resistance to the movement. Other persons show a plastic rigidity, in which the resistance to passive movement is independent of the velocity of the movement. In either case neither clonus nor a clasp-knife reaction occurs. Although agreement has not been reached with respect to the mechanism of parkinsonian rigidity, enhanced fusimotor discharge probably participates. Consistent with this hypothesis is that rigidity can be abolished by dorsal root section. Furthermore, the insensitivity to the velocity of passive movement and the absence of clonus suggest that such enhanced discharge may occur preferentially in the static fusimotor system. Thus the rigidity may be viewed as a release phenomenon, in which inhibitory influences that restrain the static fusimotor activity are interrupted.

Huntington's Disease

Huntington's disease is an autosomal dominant disorder characterized by dementia, chorea, and sometimes hypotonia. (Recently the application of recombinant DNA techniques has led to the discovery of a genetic marker for Huntington's disease.) The chorea results from involvement of the basal ganglia. This involvement includes degeneration of the GABAergic neurons that give rise to the striatal projections to the internal pallidum and the pars reticulata of the substantia nigra (Fig. 17-4); the cholinergic interneurons of the striatum degenerate as well. Loss of these GABAergic striatal neurons remove the inhibitory input to the output elements of the basal ganglia (Fig. 17-4). This, of course, reduces the activity of the VA/VL complex, since its tonic inhibitory input from the pallidum and pars reticulata would be disinhibited (Fig. 17-4). The choreiform movements then presumably result from the reduced activity of the VA/VL complex. This contrasts with Parkinson's disease, where VA/VL activity is increased consequent to degeneration of the nigrostriatal pathway. It is of interest that excessive L-dopa administration in individuals with Parkinson's disease can produce chorea; presumably this decreases VA/VL activity (Fig. 17-4), as occurs in Huntington's disease. Although the identification of the transmitter systems involved in Parkinson's and Huntington's diseases represents a significant advance in understanding the pathophysiology of these disorders, the specific patterns of abnormal motor activity are still not explained.

Other Disorders

Various other diseases affect the basal ganglia, but only two will be mentioned. Athetosis can follow damage to the basal ganglia at birth (cerebral palsy), and the lesion probably involves both the corpus striatum and the globus pallidus. As in Parkinson's disese, some relief from the motor abnormality can be achieved with lesions of the pallidal outflow or of the motor cortex. Ballismus is a striking abnormality consequent to damage, usually vascular, of the subthalamic nucleus. The ballistic movements can be abolished by lesions of the motor cortex or the corticospinal tract.

Role of the Basal Ganglia in Motor Control

Relationship to Medial and Lateral Systems

Although the connections of the basal ganglia are complex, the principal extrinsic influence of these structures is on the motor cortex. The most prominent source of input to the basal ganglia is the neocortex, which projects onto the neostriatum. The caudate nucleus and putamen then project onto the internal segment of the globus pallidus and the pars reticulata of the substantia nigra, the output structures of the basal ganglia that relay through the ventral thalamus to influence the precentral gyrus (motor cortex). Thus the basal ganglia influence motor behavior primarily through pathways, such as the corticospinal tract, that arise in the motor cortex. On the basis of these connections, it can be inferred that the basal ganglia act as part of the lateral system and are thus preferentially concerned with organized activity of the limb musculature.

Various observations are consistent with this hypothesis. Most importantly, neurons in the globus pallidus generally discharge during movement of a contralateral limb, and the motor deficits in disorders of the basal ganglia are more prominent in the extremities. For example, the resting tremor in Parkinson's disease is most evident in the fingers. Abnormal orofacial movements also occur in Parkinson's disease, and neurons in the pars reticulata of the substantia nigra discharge preferentially in relation to orofacial movement. Athetotic movements principally involve the fingers and hands, sometimes the toes and feet, and less often the more proximal muscle groups. Choreiform movements in Huntington's disease mainly involve the limbs, as in ballismus, and the rigidity in parkinsonism is more prominent in flexor muscles than in the antigravity extensors. Moreover, many of the motor abnormalities of disease of the basal ganglia are attenuated or abolished by destruction of the pallidal outflow, its thalamic relay to the cortex, the motor cortex itself, or the primary descending output of the motor cortex—the corticospinal tract.

The flow of information from the basal ganglia to the cortex converges with cerebellar influences at the VA and VL nuclei of the thalamus. The lateral or hemispheric cerebellar zone accesses the motor cortex and the corticospinal tract via this thalamic region. Consequently, the VA and VL nuclei represent critical relays for the lateral system because both cerebellar and basal ganglia influences are integrated there before being relayed to the cells of origin of the corticospinal tract. Unfortunately information about the nature of this interaction is minimal. However, it may be significant that tremor can result from disease of either the cerebellum or the basal ganglia. Although cerebellar "intention tremor" is considered to be distinct from the "resting tremor" of basal ganglial disease, they may share a common pathological basis. The disorder may be expressed differently because the background tone in cerebellar disease differs from that in disease of the basal ganglia. Surgical lesions of the VL nucleus can abolish either tremor.

Contribution to Motor Behavior

The role of the basal ganglia in motor behavior is still unclear; however, some suggestions can be ventured. Although it is not exclusively associated with the distal musculature, this motor complex appears to act preferentially through the lateral system via its influence on the motor cortex. Strongly supportive of this contention is that disease of the basal ganglia is most prominently expressed in abnormal movements of the extremities, and lesions of the motor cortex can abolish such movements. Furthermore, recording of cellular activity in the globus pallidus during movement indicates that the basal ganglia are concerned with movement of the contralateral extremities and that they often discharge prior to neurons of the motor cortex. This suggests that the basal ganglia play a role early in the initiation of movement and do not merely have a feedback relationship with the motor cortex.

The basal ganglia may assist in setting the limb and girdle musculatures as the background for spatially and temporally organized fine manipulatory activity. For example, they might appropriately position the shoulder and elbow as the basis for a goal-directed movement of the wrist and fingers. Furthermore, they may set the "postural background" for limb movements. In this sense they would constitute an important bridge between the medial and lateral systems. Viewed in this way, many of the motor abnormalities consequent to disease of the basal ganglia might be interpreted as loss of control in setting the appropriate muscle groups to support organized, voluntary, manipulatory activity.

Bibliography

Journal articles

Coyle, J.T., and Schwarcz, R.: Lesion of striatal neurons with kainic acid provides a model for Huntington's chorea, Nature **263**:244, 1976.

Hore, J., Meyer-Lohmann, J., and Brooks, V.B.: Basal ganglia cooling disables learned arm movements of monkeys in the absence of visual guidance, Science **195**:584, 1977.

Kennard, M.: Experimental analysis of functions of the basal ganglia in monkeys and chimpanzees, J. Neurophysiol. **7**:127, 1944.

Books and monographs

Brodal, A.: Neurological anatomy, ed. 3, Oxford, 1981. Oxford University Press.

Carpenter, M.B.: Anatomy of the corpus striatum and brain stem integrating systems. In Brookhart, J.M., and Mountcastle, V.B., editors: Handbook of physiology, the nervous system, section 1, vol. II, pt. 2, Bethesda, 1981, American Physiological Society.

Côté, L.: Basal ganglia, the extrapyramidal motor system, and diseases of transmitter metabolism. In Kandel, E.R., and Schwartz, J.H., editors: Principles of neural science, New York, 1981, Elsevier Science Publishing Co., Inc.

DeLong, M., and Georgopoulos, A.P.: Motor functions of the basal ganglia. In Brookhart, J.M., and Mountcastle, V.B., editors: Handbook of physiology, the nervous system, section 1, vol. II, pt. 2, Bethesda, 1981, American Physiological Society.

Hornykiewicz, O.: Neurochemical pathology and pharmacology of brain dopamine and acetylcholine: rational basis for the current drug treatment of parkinsonism. In McDowell, F.H., and Markham, C.H., editors: Recent advances in Parkinson's disease. Philadelphia, 1971, F.A. Davis Co.

Kitai, S.T.: Electrophysiology of the corpus striatum and brain stem integrating systems. In Brookhart, J.M., and Mountcastle, V.B., editors: Handbook of physiology, the nervous system, section 1, vol. II, pt. 2, Bethesda, 1981, American Physiological Society.

Penney, J.B., Jr., and Young, A.B.: Speculations on the functional anatomy of basal ganglia disorders, Ann. Rev. Neurosci. **6**:73, 1983.

CHAPTER

18

Control of Movement and Posture

In the preceding chapters on the motor system, motor control was analyzed by beginning at the periphery and proceeding centrally. Analysis of the central nervous system is usually facilitated when the structure under study can be linked to the sensory or motor peripheries. Indeed the remarkable progress in our understanding of the sensory systems in large measure reflects the ability to study successive transformations in receptive fields as one proceeds systematically from the receptors to the central nervous system. Advances in understanding the motor system have been realized from efforts to study the control of motor behavior by beginning at the muscles and their innervation. An understanding of the functional organization of the final common path then establishes a basis for the rational classification of the major descending pathways involved in motor control; Chapters 13 to 15 develop these ideas. Chapters 16 and 17 introduce the cerebellum and basal ganglia within this framework.

The objectives of this chapter, which concludes the discussion of the motor system, are (1) to introduce additional material on the control of posture and movement and (2) to present some examples of sensorimotor integration. These examples will illustrate the directions that are required to approach the question of how motor outputs are guided by sensory inputs.

■ Control of Posture

Posture is defined as the *active muscular resistance to displacement of the body by gravity or acceleration*. The maintenance of an upright position is a critical substrate for the performance of phasic goal-directed movements. This is achieved largely through reflex adjustments of the proximal extensor muscles in response to perturbations that displace the body. For this reason the proximal extensor muscles are often referred to as the *antigravity muscles*. Because the proximal musculature mediates these important functions, the suprasegmental pathways that constitute the output segments of these reflexes are components of the medial system. Therefore an analysis of postural control is equivalent to a further analysis of the pathways that comprise the medial system and of the inputs that govern their activities.

■ The Postural Reflexes

There are two broad classes of postural reflexes: the *statokinetic reflexes,* which are elicited by acceleratory displacement of the body, and the *static reflexes,* which are elicited by gravitational displacement. Thus the first step in categorizing the postural reflexes is to segregate them with respect to the origin of their initiating stimuli. Further subdivision within each class can then be made on the basis of the target muscles of the reflex.

By definition the *statokinetic reflexes* are elicited by acceleratory stimuli. Therefore the afferent limb of all such reflex circuits must involve the labyrinths. Our knowledge of labyrinthine receptors indicates that reflexes evoked by linear acceleration, as occur in falling, originate with signals from the utricle. Moreover, such reflexes are mediated by the lateral vestibulospinal tract, because this is the primary target of utricular information.

An example of a statokinetic reflex evoked by linear acceleration is the *vestibular placing reaction*. For example, if a cat is blindfolded and held by the pelvis with its head oriented downward, a sudden lowering of the animal will stimulate the utricular maculae. This will elicit extension of the forelimbs. This reflex response is an adaptive reaction that prepares the animal for appropriate support by the limbs on surface contact. Destruction of the utricles abolishes this response. However, if the blindfold is removed, the response can be initiated by visual cues; in this case it is a *visual placing reaction*.

Many postural reflexes that are mediated by the vestibular system can also be elicited by visual stimuli.

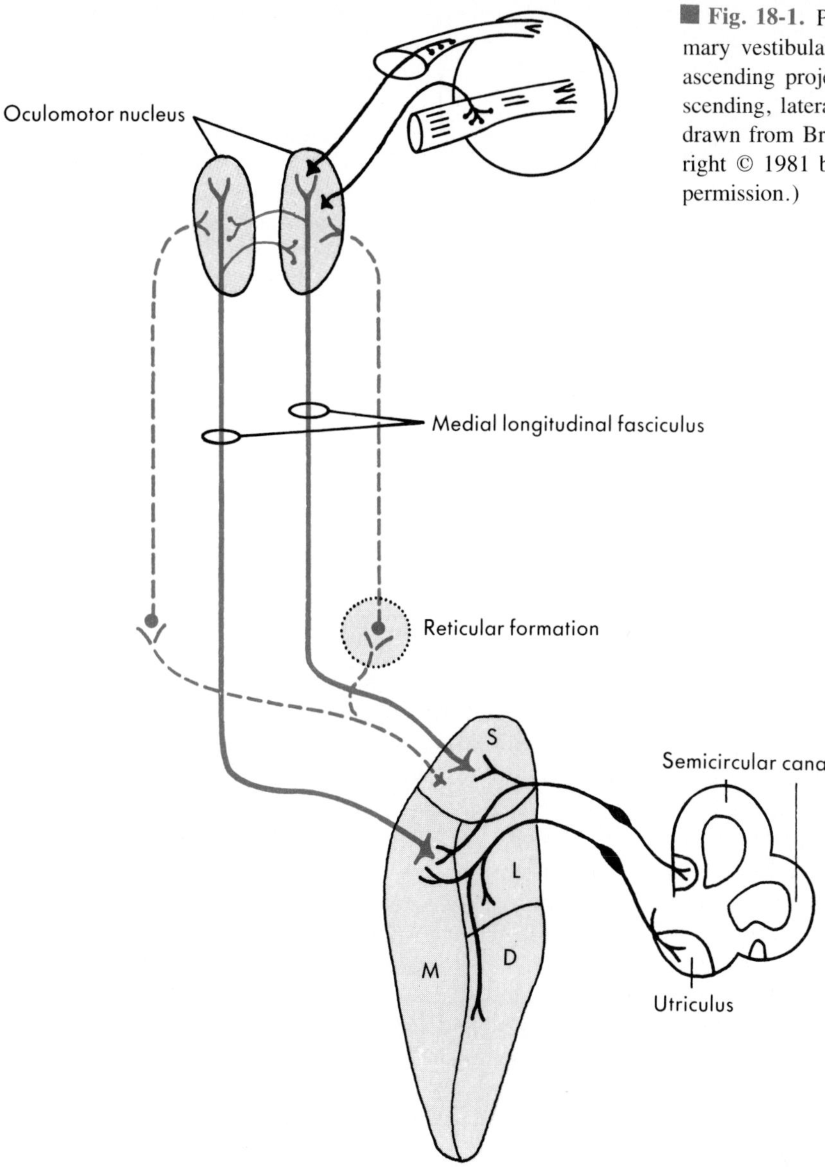

Fig. 18-1. Principal features of the organization of the primary vestibular projections on the vestibular nuclei and the ascending projections of these nuclei. *D, L, M,* and *S,* Descending, lateral, medial, and superior vestibular nuclei. (Redrawn from Brodal, A.: Neurological anatomy, ed. 3. Copyright © 1981 by Oxford University Press, Inc. Reprinted by permission.)

Thus the visual system frequently compensates for lesions of the vestibular apparatus or its central pathways. In fact, an animal with a bilateral labyrinthectomy can appear to be normal unless it is deprived of visual cues, in which case the ability to maintain an upright posture is seriously impaired. Indeed, humans are more compromised by abnormal signals from the labyrinths, as in Ménière's disease, than by the loss of labyrinthine information—provided visual cues are available.

Statokinetic reflexes evoked by angular acceleration, as during rotation, originate with signals from the semicircular canals. Information from the canals mainly influences the medial vestibular nucleus. Hence, the evoked reflex effects are primarily mediated by the ascending (medial longitudinal fasciculus) and descending (medial vestibulospinal tract) pathways from this nucleus (Fig. 18-1). Consider the response in an individual who is rotated to the right. The endolymph of the horizontal canals will flow to the left until the inertia of the endolymph is overcome. This flow deflects the hair cells of the cristae and changes the discharge of the vestibular fibers from the horizontal canals (Fig. 18-2) (see Chapter 11). The change in discharge then informs the brain of an angular acceleration to the left in the horizontal plane. This signal will elicit important reflex effects on the eyes, neck, and upper limbs.

The reflex effects on the eyes are called *vestibular nystagmus*. They are mediated by projections from the medial vestibular nucleus, through interneurons of the brainstem, to the cranial nerve nuclei that innervate the extrinsic eye muscles (Fig. 18-1). As the head rotates to the right, the eyes will undergo a slow conjugate deviation to the left, that is, in the direction of the endolymph flow. This maintains the center of gaze. As the rotation continues, the fixation point can no longer be maintained, and the eyes move rapidly to the right to

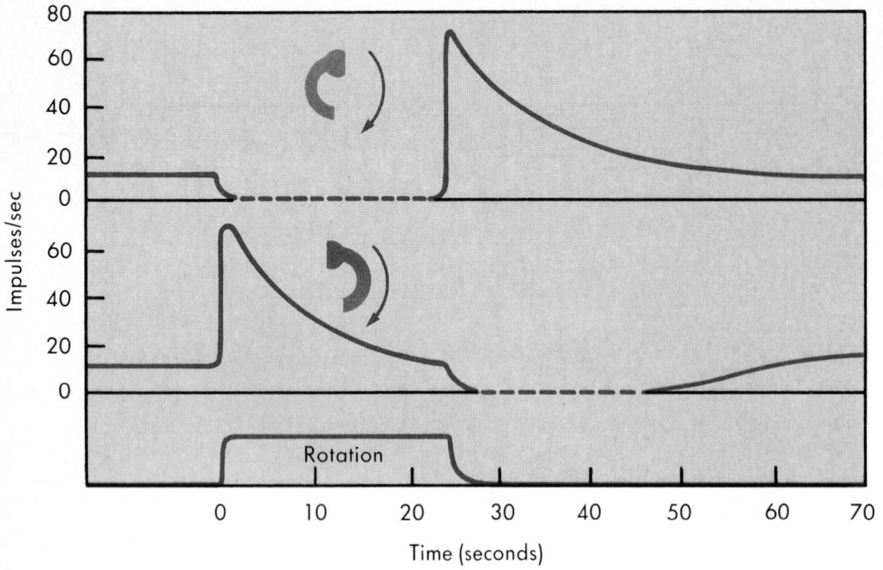

■ **Fig. 18-2.** Effects of head rotation of the discharge of the vestibular nerve. There is a steady discharge at rest, which is suppressed by rotation in one direction and increased by rotation in the opposite direction. Cessation of rotation produces an afterdischarge in one case and a silent period in the other. (Redrawn from Adrian, E.D.: J. Physiol. **100:**389, 1943.)

establish a new fixation point. This slow conjugate deviation in the direction of the endolymph flow and the rapid return in the opposite direction occur repeatedly until the inertia of the endolymph is overcome and the endolymph follows the rotation of the canal (Fig. 18-2). The slow deviation to the left is called the *slow component* of the nystagmus, and the rapid return to the right is the *fast component*. The direction of nystagmus is designated as that of the fast component.

When the endolymph is stationary within the canal, that is, when there is no longer an accelerative stimulus and the head is rotating at a constant velocity, vestibular nystagmus ceases. However, if visual cues are available, reflex eye movements will continue. This *optokinetic nystagmus* is independent of vestibular input. To return to the earlier example of vestibular nystagmus, if the rotation is now terminated, a deceleratory stimulus is imposed. The inertia of the endolymph will cause it to continue to move to the right in the direction of the previous rotation (Fig. 18-2). This will evoke a *postrotatory nystagmus* in which the slow deviation of the eyes is to the right, again in the direction of the endolymph flow, and the rapid return is to the left.

Vestibular nystagmus can also be evoked by *caloric stimulation* of the semicircular canals. This provides a convenient clinical means of assessing the integrity of the brainstem circuitry that mediates vestibuloocular reflexes. In this test the head is tilted 60 degrees from the vertical plane to orient the horizontal canal vertically. If the external auditory canal is then irrigated with warm water, a convection current is established, and the endolymph rises toward the ampulla. This evokes a nystagmus with the slow component toward the contralateral side. Irrigating the canal with cold water will establish a convection current in the opposite direction. It will elicit a nystagmus with the slow component toward the side of irrigation.

Certain diseases, particularly those involving the brainstem, produce nystagmus. Thus nystagmus in the absence of an appropriate stimulus is a clear pathological sign, which can be useful in the localization of brainstem lesions.

In the earlier example of body rotation, the rotation to the right will elicit extension of the right arm and relaxation or flexion of the left arm. The effect on the neck muscles is to turn the head to the left. These reflex responses are mediated via the medial vestibulospinal tract. They constitute head and upper limb adjustments that tend to resist the rotation (displacement) of the body. In general, postural reflexes can be understood intuitively as adjustments that resist displacement. This concept can assist in remembering the positions that will be assumed by the head and limbs in the various reflexes. Changes in eye position will generally be in a direction such that the starting fixation point is maintained.

The other major class of postural reactions, the *static reflexes,* produce adjustments to displacements by gravity. This large class of reactions may be further subdivided with respect to the target of the reflex action. Conventionally the static reflexes are designated as *local, segmental,* or *general static reflexes*. Local static reflexes are those in which the reflex effect is on the same limb from which the stimulus was initiated. The most important of these, the myotatic reflex, is described in detail in Chapter 14. To review briefly, the myotatic reflex is the basis of muscle tone. Its importance as a postural mechanism can be easily appreciated by considering the quadriceps muscles in an individual who is standing in a normal upright posture. In this position, gravity tends to displace the body downward, stretching the quadriceps muscles as the legs flex at the knees. The muscle stretch evokes discharge from the muscle spindles of the quadriceps, and this initiates a myotatic reflex contraction that extends the leg at the knee. This maintains the leg as a pillar of support and thus counteracts the gravitational displacement of the body.

Segmental static reactions are those reflexes in which the stimulus originates in one limb and the reflex effect is on the contralateral limb. The crossed-extensor components of reflexes, such as the crossed-component of the flexor withdrawal response, fall into this category (Chapter 14).

In the largest group of postural reflexes, the general static reactions, a stimulus that arises at one site of the body affects many muscle groups. For example, numerous postural adjustments occur in response to changes in head position. A change in head position relative to body position stimulates the stretch receptors of the neck muscles. Such proprioceptive information from the neck muscles can reflexly affect the limbs *(tonic neck reflexes)* and the eyes *(head-on-eyes reflexes),* and it helps evoke the complex ensemble of *righting reactions* as well. The tonic neck reflexes are most easily illustrated in a quadruped such as the cat (Fig. 18-3). Turning the head to the right, relative to the body, will extend the right limbs and relax the left limbs; rotation to the left will elicit an opposite limb response. Dorsiflexion of the head will elicit extension of the forelimbs and flex the hindlimbs, whereas ventroflexion will flex the forelimbs and extend the hindlimbs. Again, these limb responses compensate for changes in the animal's center of gravity.

The neck proprioceptors also contribute to the reflex adjustments of eye position that help maintain an invariant visual field. In this reaction a rotation of the head to the right will result in an eye movement to the left; rotation to the left yields an eye movement to the right. Dorsiflexion elicits a downward eye movement, whereas ventroflexion produces an upward movement. This head-on-eyes reflex is demonstrable in a comatose patient whose appropriate brainstem circuitry is still intact; the reflex is referred to as the *doll's eye phenomenon.*

The topic of the postural reflexes has not been treated comprehensively. The postural reactions are numerous and often complex, and in most instances their precise circuitry remains unspecified. These reflexes involve virtually every level of the neuraxis, and they occur continuously. They interact to adjust for displacements of the body and to maintain a constant visual field. This is accomplished through the various pathways of the medial system. The appropriate occurrence of these reflexes constitutes an essential substrate for phasic, goal-directed motor activity.

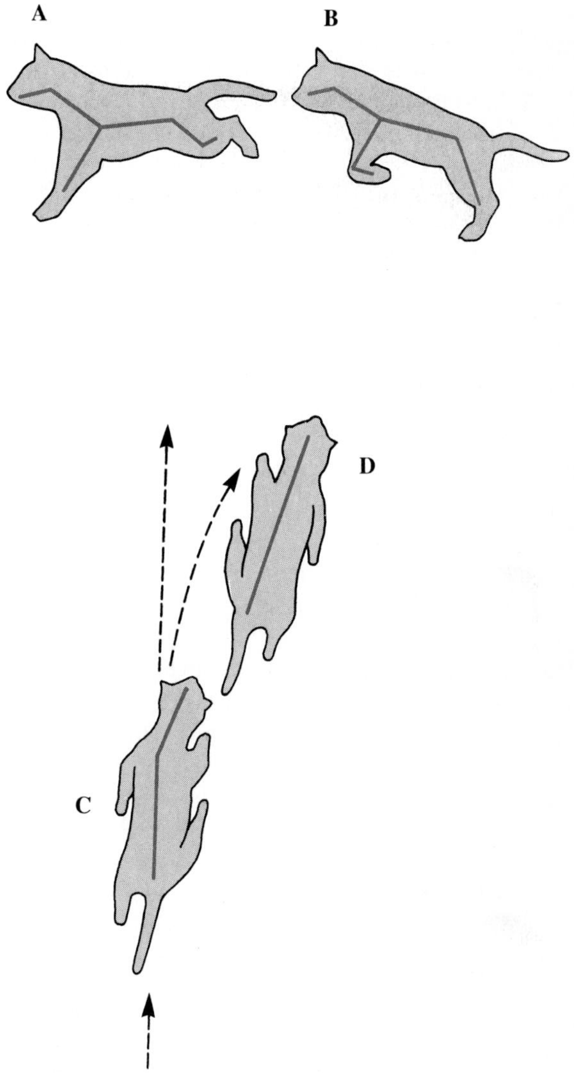

Fig. 18-3. Tonic neck reflexes during locomotion in the cat. During running or leaping movements (**A** and **B**) the animal's head remains in a fixed attitude in relation to the ground. **A,** Dorsiflexion of the head and neck in relation to the back produces forelimb extension and hindlimb flexion. **B,** Ventroflexion of the head and neck in relation to the back produces inhibition of the forelimb extensors and facilitation of the hindlimb extensors. **C** and **D,** The animal is running; turning the head to the right reflexly increases stability during lateral movement to the right. Facilitation of forelimb and hindlimb extensors on the right is accompanied by inhibition of extensors on the left side. The extensor inhibition and flexor facilitation prepare the animal for the first step in the new direction. (Redrawn with permission from Eyzaguirre, C., and Fidone, S.J.: Physiology of the nervous system, ed. 2. Copyright © 1975 by Year Book Medical Publishers, Inc., Chicago.)

Control of Tone in Proximal Extensor Muscles

The maintenance of tone in the proximal extensor muscles is an important element of postural control. Indeed, the continuous inputs to the medial motoneurons originate with stimuli that elicit the postural reflexes. Such inputs contribute significantly to the maintenance of tone. This is produced by activity in the various medial system pathways that descend to excite both the α- and γ-motoneurons that innervate the proximal muscles and their spindles. Spinal transection abolishes the tone of the antigravity muscles that are innervated by segments below the transection, because these important suprasegmental contributions to the maintenance of tone are interrupted. Many aspects of spinal shock, such as the loss of myotatic reflexes, are direct manifestations of this loss of medial system input to the spinal cord (see Figs. 15-2 and 15-10).

The two pathways of the medial system that are most important in the maintenance of tone are the lateral vestibulospinal and pontine reticulospinal tracts. Thus spinal shock can be produced with any transection caudal to the lateral vestibular nucleus. However, if the transection is just rostral to this nucleus, then in contrast to spinal shock, the tone of the antigravity muscles is actually enhanced and an exaggerated antigravity posture obtains. This phenomenon is called *decerebrate rigidity*. When this occurs in humans (Fig. 18-4), the arms and legs are extended, the back arched, and the head dorsiflexed. The feet are ventroflexed and the arms pronated. However, there is flexion at the wrists and minimal involvement of the digits, indicating that the rigidity is, indeed, a medial system phenomenon. Clinically this can occur after trauma or with compression of the brain by a space-occupying lesion, such as a tumor. A possible explanation for this antigravity posture is that the lateral vestibular nucleus, which exerts a powerful excitatory action on the proximal extensor motoneurons, is released from inhibitory inputs from more rostral structures. Another contributing factor may be the loss of descending excitatory inputs to the cells of origin of the medullary reticulospinal tract, a tract that inhibits proximal extensor motoneurons.

The decerebrate rigidity can be enhanced by transecting more rostrally so that the cells of origin of the pontine reticulospinal tract remain in continuity with the spinal cord. This tract also provides a powerful tonic excitatory influence on the proximal extensor motoneurons, and the transection releases it from inhibitory control, possibly by corticoreticular projections. Thus the released pontine reticulospinal tract acts synergistically with the lateral vestibulospinal tract to enhance the activity of the α- and γ-motoneurons that innervate the proximal extensor muscles.

Any experimental procedure that reduces these descending excitatory influences from the vestibulospinal and pontine reticulospinal tracts will attenuate the rigidity. Thus placing a lesion in the ventral funiculus of the spinal cord or destroying the cells of origin of the pontine reticulospinal or lateral vestibulospinal tract will attenuate or abolish the rigidity. Because the cerebellum influences the cells of origin of these two pathways, experimental manipulations involving these cerebellar influences will also affect the rigidity. For instance, lesions of the fastigial nucleus will eliminate an excitatory drive on the pontine reticulospinal and lateral vestibulospinal tracts and will therefore reduce rigidity. Electrical stimulation of the fastigial nucleus will have the opposite effect. In contrast, a lesion of the vermal cortex will eliminate the Purkinje cell inhibition of the fastigial nucleus and the lateral vestibular nucleus. Thus a vermal lesion will enhance the rigidity; vermal stimulation will, of course, attenuate it. Yet another exercise is to remove the labyrinths or transect the vestibular portion of the eighth nerve. The utricles provide the major source of input to the lateral vestibular nucleus. Therefore either of these procedures will remove that excitatory drive and thus reduce the rigidity.

Cutting the appropriate dorsal roots will also abolish decerebrate rigidity. This interrupts the myotatic reflex loop but leaves the α-motoneuron innervation intact. The abolition of the rigidity suggests that it is mediated primarily by enhanced drive on γ-motoneurons to spindles of the antigravity muscles. On this basis decerebrate rigidity is considered a *γ-rigidity* because it reflects hyperactivity of the myotatic reflexes that involve the antigravity muscles.

Although many rigidities are γ-mediated, *α-rigidity* also exists. It can be demonstrated experimentally by a puzzling phenomenon that involves the anterior cerebellum. In a preparation in which decerebrate rigidity has been established and then has been eliminated by dorsal rhizotomy, a subsequent lesion of the anterior cerebellum will reinstate the rigidity. Because the dorsal rhizotomy precludes any involvement of the γ-motoneurons, the reestablished rigidity must be mediated by the α-motoneurons. Indeed it has been shown that anterior cerebellectomy increases the activity of the α-motoneurons. This observation has generated the hypothesis that in the intact animal the cerebellum helps to determine the relative influence of descending systems on the α- and γ-motoneurons. Furthermore, this α-rigidity appears to be mediated primarily by the lateral vestibulospinal tract because it is abolished by section of the eighth cranial nerve. This is in contrast to γ-rigidity, in which the pontine reticulospinal pathway plays the predominant role. The exaggerated antigravity posture of decerebrate rigidity is mediated by the interruption of descending influences on medial system pathways. Therefore, if more such inputs are left intact and if the lateral system pathways are preserved, the capability for phasic movements to be superimposed on the rigidity will be increased.

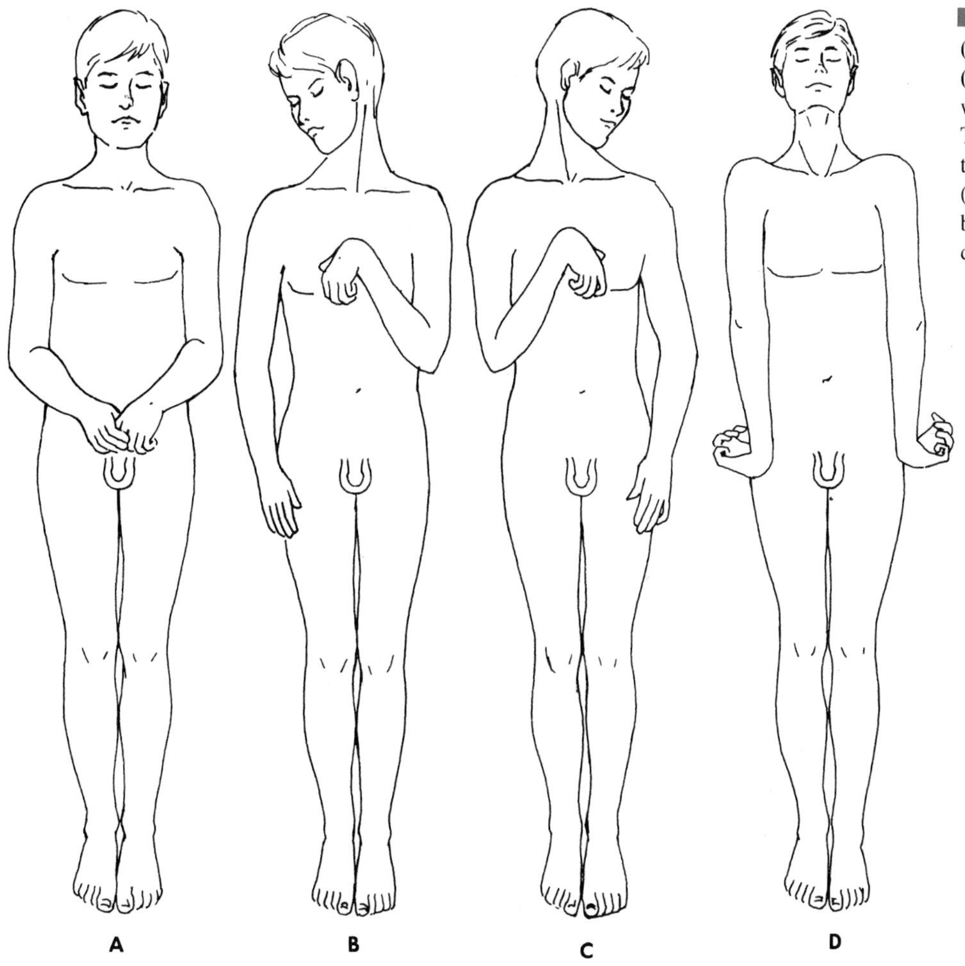

Fig. 18-4. Decorticate rigidity (**A** to **C**) and decerebrate rigidity (**D**). **A,** The patient is lying supine with the head unturned. **B** and **C,** The tonic reflexes are shown by turning the head to the right or left. (Redrawn from Fulton, J.F: Textbook of physiology, ed. 17, Philadelphia, 1955, W.B. Saunders Co.)

As transections are made more rostrally, the rigidity can be modified by applying stimuli that elicit various reflexes. For example, in humans if only the cortex is compromised, then a rigidity ensues that is characterized by extension of the lower limbs. However, the position of the upper limbs will depend on head position (Fig. 18-4). If the head is rotated to the right, the left arm will be flexed and the right arm extended; the converse occurs if the head is rotated to the left. This reflects the tonic neck reflexes, and it suggests that such reflexes are interacting with the *decorticate rigidity* to modify the classic posture. In many mammalian species decorticate rigidity is not prominent. In such animals transections rostral to the midbrain may permit walking.

Control of Movement

The control of phasic, goal-directed movement is now considered further to provide an opportunity for a more detailed treatment of the lateral system.

Motor Areas of the Cortex

Movements can be elicited by stimulating many areas of the cortex. However, if only those areas from which reasonably localized responses can be evoked at low stimulation thresholds are considered, the "motor regions" of the cortex become more limited (Fig. 18-5). These are generally identified as the primary motor, premotor, and supplementary motor areas. A secondary motor area and frontal eye fields are also often identified. The cortical motor representation can perhaps be viewed as two regions. One region is encompassed by the representation of the musculature on the dorsal surface of the precentral gyrus (Fig. 15-5). Area 4, the primary motor area, represents the extremities and is closest to the central sulcus. Area 6, the premotor area, represents the axial musculature and much of the head. Area 8, the frontal eye fields, represents the eyes within the head. The second motor region is the supplementary motor cortex, which is located on the mesial surface of the precentral cortex; it, too, includes a full representation of the body muscles (see Fig. 15-5).

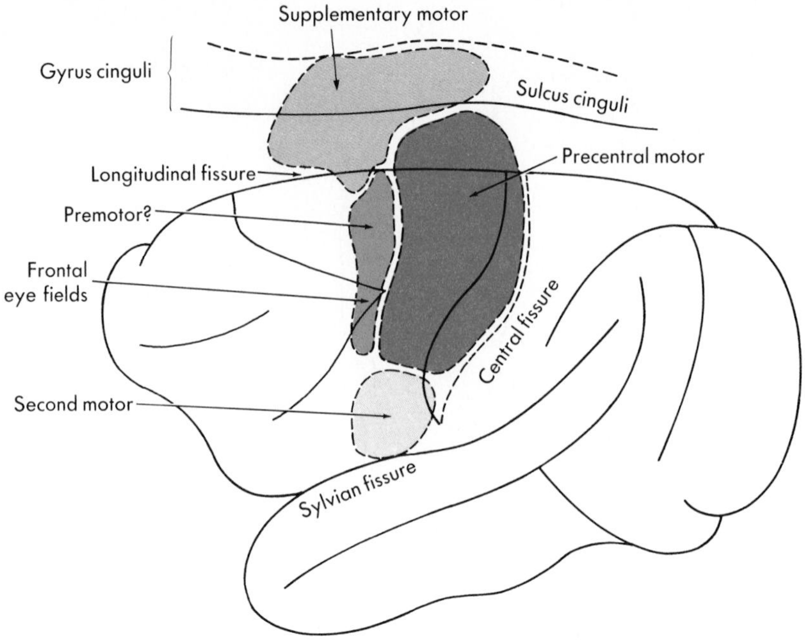

Fig. 18-5. Principal motor areas of the cerebral cortex of the monkey. Shown are the precentral or primary motor area, the supplementary motor area, the second motor area, and the premotor area, each with a different shade. The central and longitudinal fissures are shown opened out, with the broken line indicating the floor of the fissure and the solid line the lip of the fissure on the brain surface. (Redrawn from Eyzaguirre, C., and Fidone, S.J.: Physiology of the nervous system, ed. 2. Copyright © 1975 by Year Book Medical Publishers, Inc., Chicago.)

Stimulation of the primary motor area is described in Chapter 15. Discrete surface stimulation elicits low-threshold, highly localized movements such as wrist flexion. Stimulation with microelectrodes (microstimulation) deeper in the motor cortex can evoke contraction of single muscles. Moreover, stimulation of localized *efferent cortical zones* consistently contracts the same muscle or muscles. These zones may be analogous to the columnar organizations of the sensory cortices. Of course, an efferent cortical zone for any given muscle or muscle group will be within the appropriate area of the topographical map of the musculature. Stimulation of the supplementary motor cortex requires higher currents and yields less localized and more complex responses. Stimulating the primary motor area elicits discrete contralateral responses, but activating the supplementary motor cortex evokes postural adjustments that involve many muscles, often bilaterally.

As stated earlier, the corticospinal tract is not the only motor pathway that originates from the cortical motor areas. This can be demonstrated readily for the primary motor cortex by interrupting the corticospinal tract. Stimulating the motor cortex still elicits movements, and a topographical map of the musculature remains. However, the character of the evoked movements changes strikingly. The rapid flicklike movements that primarily involve the contralateral flexor muscles are now replaced by higher threshold, slower movements that are long lasting and frequently involve the entire limb. Fractionated movements of the digits do not occur.

The movements elicited by stimulating the supplementary motor cortex do not require an intact corticospinal tract. Moreover, they are not affected by lesions of the primary motor area. Thus the supplementary motor cortex must influence motor activity through cortical projections to the brainstem independently of those derived from the primary motor area.

The multiple cortical representation of the musculature is reminiscent of the multiple sensory representations. What function is served by such multiple representation in either sensory or motor systems is not yet clear.

Activity of Pyramidal Tract Neurons During Movement

The aspects of movement that are controlled by the corticospinal tract have been clarified by experiments on monkeys. Pyramidal tract neurons can be identified in the awake animal by implanting stimulating electrodes in the medullary pyramids. If a cortical neuron responds antidromically to pyramidal stimulation, it can be assumed it is a cell of origin of a pyramidal tract fiber. Such cortical cells can then be studied during various controlled movements, and their discharge characteristics can be correlated with the various features of the movements.

If a monkey is trained to respond to a stimulus cue by displacing a lever in such a way that the required movement involves flexion of the wrist, one can record from pyramidal tract neurons in a cortical efferent zone associated with the wrist muscles. One finds that cells associated with wrist flexors discharge before the movement. Some will generate a transient high-frequency burst *(dynamic neurons);* others will continue discharging at increased frequency throughout the wrist flexion *(static neurons);* still others have mixed properties *(mixed neurons).* If the amount of force required to dis-

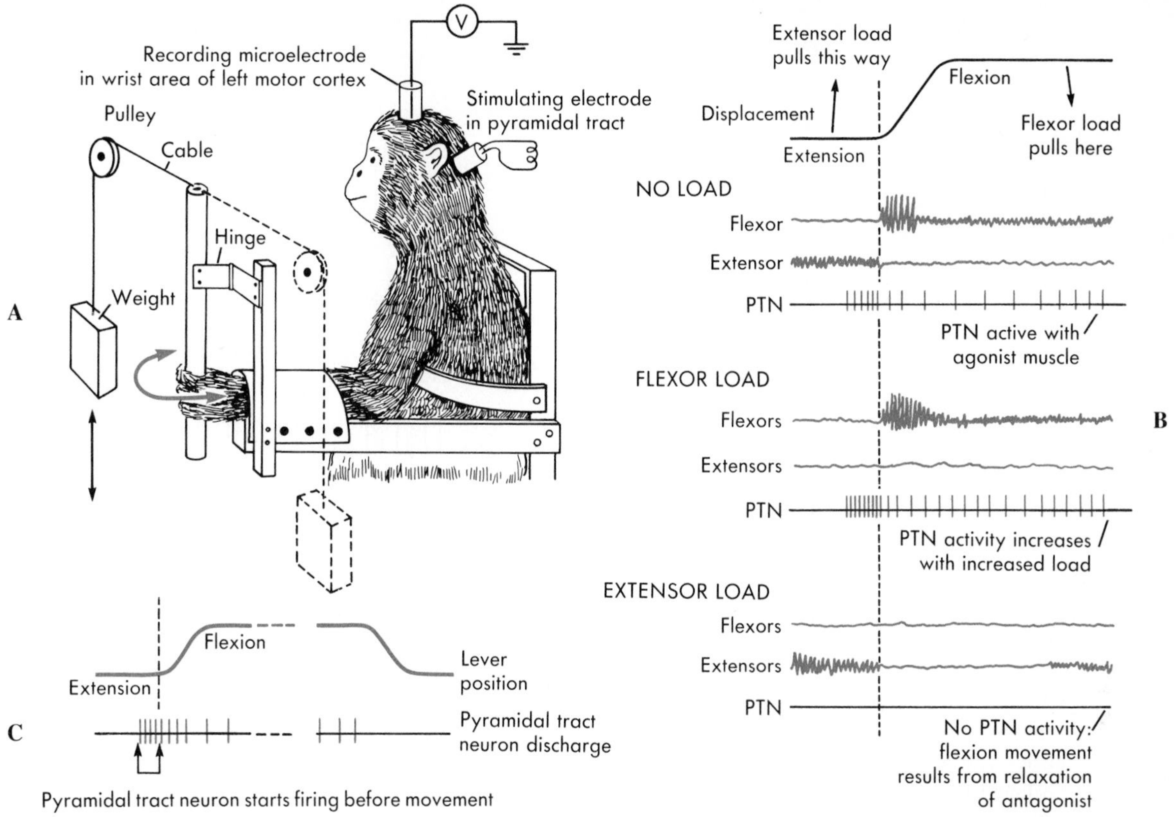

Fig. 18-6. The activity of corticospinal neurons reflects the direction of force exerted. **A** shows the experimental arrangement for recording the discharge of a pyramidal tract neuron while the awake monkey alternately flexes and extends its wrist. **B** demonstrates that the cortical neuron begins firing before the movement *(arrows)*. **C** shows a pyramidal tract neuron *(PTN)* that increases its activity with flexion of the wrist. Flexor and extensor electromyograms and PTN discharge records are shown under different load conditions. The absence of neuronal activity with an extensor load indicates that the neuronal output codes for force rather than displacement. (Redrawn from Kandel, E.R., and Schwartz, J.H.: Principles of neural science, New York, 1981, Elsevier Science Publishing Co., Inc.)

place the lever is now varied, a class of pyramidal tract neurons discharges at a frequency that varies with the force of contraction (or rate of change of force) rather than with the intended displacement (Fig. 18-6). Other pyramidal tract neurons influence the wrist flexors, but their discharge properties relative to the movement are more complex and are not yet sufficiently understood. The rubrospinal pathway has a similar physiological organization but a higher proportion of dynamic neurons (Fig. 18-7). From such investigations it is clear that the corticospinal neurons of the primary motor area and the rubrospinal neurons discharge shortly before a movement and that some of these neurons regulate the force of contraction.

The extent to which the corticospinal tract excites a given motoneuron is determined by the extent to which the innervated muscle participates in (1) antigravity functions and (2) phasic movements. If the muscle is primarily involved in antigravity or postural function, then its motoneurons are inhibited by the corticospinal tract. If, in contrast, the muscle serves no postural role but is involved in spatially organized movements, then the corticospinal tract will excite its motoneurons (Fig. 18-8). Thus the intrinsic muscles of the digits, both flexor and extensor, are excited by the corticospinal tract. However, for the muscles of the proximal limb, the extensors will be inhibited, and the flexors will be excited. This reinforces the concept that the corticospinal and rubrospinal tracts, the principal components of the lateral system, are mainly concerned with spatially organized, goal-directed movements. To perform this function, they inhibit the activity of antigravity muscles, presumably to reduce the excitation of postural muscles that might compete with the desired phasic movement. Because adjustments of the axial and proximal limb muscles are usually necessary for a spatially organized movement involving the distal muscles, pathways other than the corticospinal tract also participate.

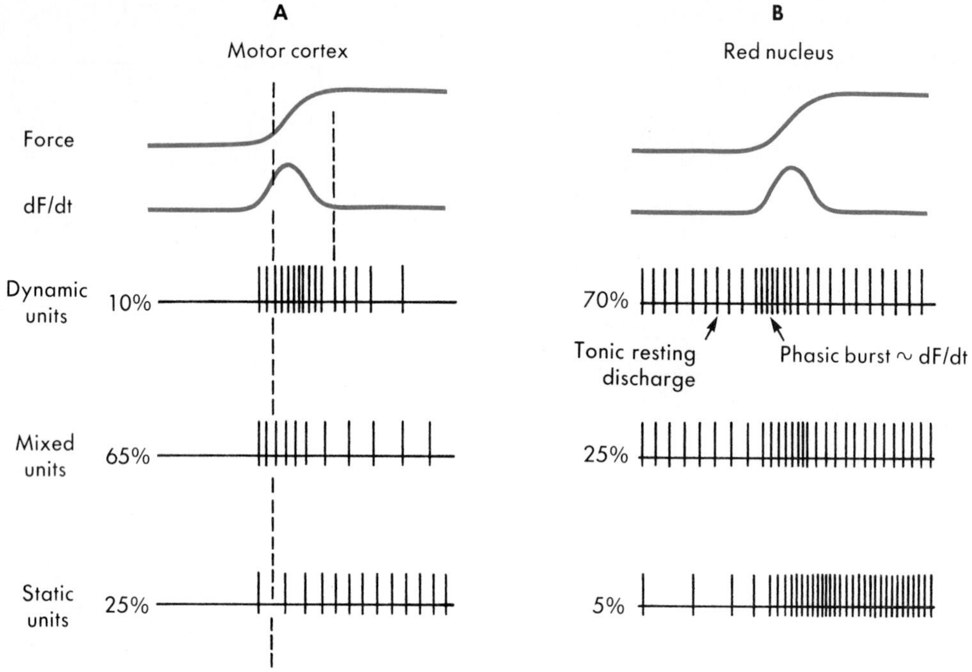

■ **Fig. 18-7.** Patterns of activity of dynamic, static, and mixed neurons in the motor cortex **(A)** and red nucleus **(B)** of the cat during voluntary isometric contraction. (Redrawn from Kandel, E.R., and Schwartz, J.H.: Principles of neural science, New York, 1981, Elsevier Science Publishing Co., Inc.)

■ **Fig. 18-8.** The actions of the corticospinal tract on the muscles controlling the joints of the forelimbs and hindlimbs of the cat and baboon. These animals are shown schematically in their habitual standing postures. Inhibitory *(I)* and facilitative *(F)* effects of the corticospinal tract are shown for the motoneurons innervating the flexor and extensor muscles at each joint. (Redrawn from Preston, J.B., Shende, M.C., and Uemura, K.: The motor cortex-pyramidal system: patterns of facilitation and inhibition on motoneurons innervating limb musculature of cat and baboon and their possible adaptive significance. In Yahr, M.D., and Purpura, D.P., editors: Neurophysiological basis of normal and abnormal motor activities, New York, 1967, Raven Press.)

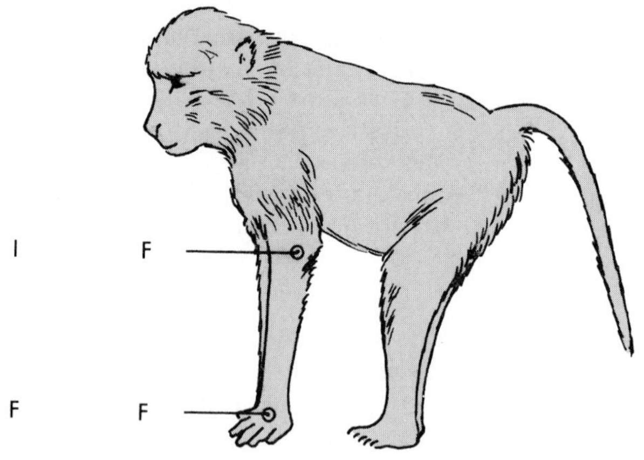

■ *Interaction of the Corticospinal Tract with Spinal Circuitry*

The corticospinal tract has prominent monosynaptic connections with the motoneurons that innervate the intrinsic muscles of the digits. These connections allow it to bypass the intrinsic spinal circuitry in controlling independent movement of the digits. However, much of the influence of the corticospinal tract on motoneurons is exerted through spinal interneurons. This raises the issue of whether this tract, and descending motor pathways in general, uses independent populations of spinal interneurons or converges on interneurons that participate in spinal reflexes, such as flexor withdrawal responses. There is evidence that the corticospinal projections onto interneurons generally use the existing spinal organization. For example, a corticospinal fiber that excites a flexor motoneuron is most likely to terminate on an interneuron that also receives input from flexor reflex afferent fibers and from the Ib fibers. Thus the fiber would use the existing circuitry for flexor withdrawal and inverse myotatic reflexes, thereby using the intrinsic spinal connections for inhibiting antagonistic extensor motoneurons (reciprocal innervation) and perhaps even for crossed-components (see Fig. 14-15).

Sensorimotor Integration

The descending motor pathways terminate at spinal cord sites that are not far removed from the final common path. Moreover, evidence indicates that the corticospinal and rubrospinal tracts discharge shortly before movement. What distinguishes these pathways from spinal interneurons? Is the corticospinal tract merely a glorified interneuron with a long axon derived from cells of origin in the cortex? These pathways do, indeed, differ from spinal interneurons in an exceedingly important respect. Because their cells of origin are located suprasegmentally, they have access to a rich variety of information that is not available to the spinal circuitry. Thus one of the important challenges in analyzing motor control is to describe how sensory stimuli initiate and guide motor behavior, that is, the problem of sensorimotor integration.

The spinal reflexes described in Chapter 14 represent simpler models of such integration. In the myotatic reflex, for example, much of the circuitry is specified, and the characteristics of the sensory input and motor output are generally understood. The flexor withdrawal reflex illustrates a more complex and less understood circuit for sensorimotor integration. Yet, these spinal reflexes are much simpler than the organization of the complex movements that involve suprasegmental pathways. In the following sections are two examples of an investigative approach to this important problem.

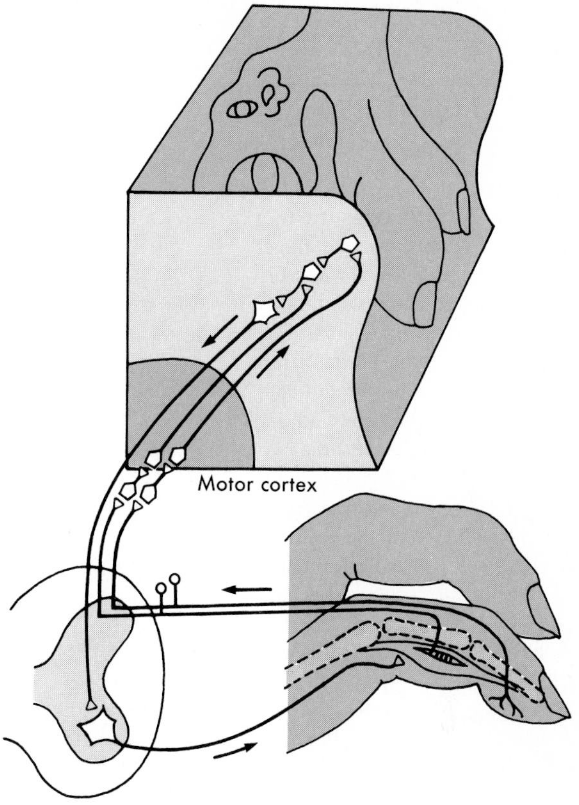

Fig. 18-9. The input-output organization of the cortical efferent zone controlling the flexor muscle of a digit. See text for further explanation. (Redrawn from Asanuma, H.: Physiologist **16:**143, 1973.)

Input-Output Relationships Across the Motor Cortex

In the motor cortex, pyramidal tract neurons respond to cutaneous input as well as to active and passive movements of a limb. For a specific cortical efferent zone, the cutaneous fields of its neurons lie in the path of the movement elicited by stimulating that zone. For example, if stimulation of the cortical zone produces flexion of a digit, then the cutaneous fields of neurons in that zone will be on the ventral surface of the digit, and the cutaneous input will excite the cortical neurons polysynaptically (Fig. 18-9). Those joint receptors activated by the flexion also will excite the neurons of the zone and the spindle receptors of the digit flexor will excite the cortical neurons (Fig. 18-9). Because these receptors are activated when the flexor muscle lengthens (relaxes), a slowing of the flexion will act to excite the cortical neurons that drive the flexion.

As a general rule these sensory stimuli influence the motor cortex in a manner that facilitates the movement. This feedback contributes to serial motor acts and assists the motor cortex in tracking objects. For example, the flexor contractions of the muscles involved in grasping an object are facilitated by the cutaneous and proprioceptive stimuli that are generated by the movement and by its consequent contact with the object.

Another example involves the *tactile placing reaction,* which is mediated by the motor cortex. If a cat is held and moved toward a surface, it will flex its limb and then extend it to establish the leg as a source of support when the dorsal aspect of its paw contacts that surface. In this placing response, the cutaneous stimuli generated by contact of the dorsum of the paw with the surface will facilitate the discharge of pyramidal tract neurons involved in flexing the paw. The cutaneous stimuli subsequently generated by the contact of the ventral aspect of the paw with the surface would then facilitate pyramidal tract neurons that activate extension.

Visually Guided Finger Movement

The preceding section illustrates an approach to sensorimotor integration based on receptive field analysis of pyramidal tract neurons. Although far from fully specifying the sensory control of these cells, it represents an exciting start with respect to cutaneous and propriocep-

Fig. 18-10. Drawings from a film showing the hand and finger movements of a split-brain monkey with a complete commissurotomy. The monkey is taking a small food pellet (shown in color) from a test board. Under guidance of the contralateral eye **(A)** the hand and fingers in reaching out assume the precision grip posture *(top drawing)*, and the index finger and thumb dislodge the pellet from the well. In this situation the visual information can influence the corticospinal neurons controlling the hand and fingers being used. Under the guidance of the ipsilateral eye **(B)** the hand and fingers do not assume the precision grip posture until the hand has touched the board. The hand is brought to the proper place, but the pellet is not taken from the well. Instead the hand and fingers explore the surface of the board as if "blind." In this case visual information cannot gain access to the corticospinal neurons controlling the hand. (Redrawn from Brinkman, C.: Split-brain monkeys: cerebral control of contralateral and ipsilateral arm, hand and finger movements, doctoral dissertation, Rotterdam, 1974, Erasmus University Rotterdam.)

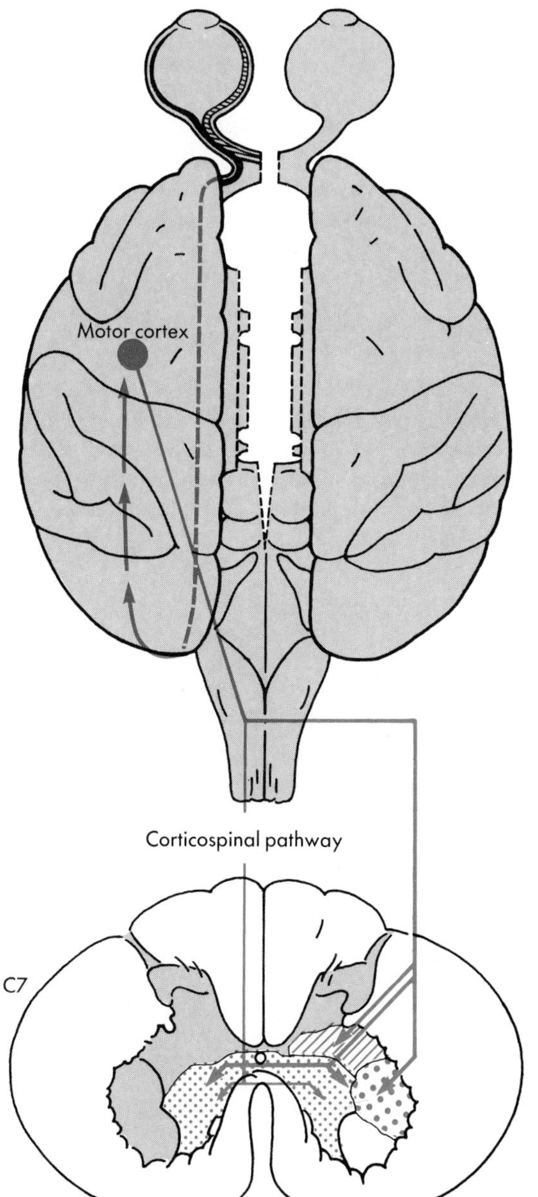

Fig. 18-11. The corticospinal projections in the monkey (see Fig. 19-2). Note that in the split-brain preparation visual information presented to the left eye reaches only the left hemisphere. It can therefore influence the corticospinal fibers controlling the right distal extremity but not those controlling the left. (Redrawn from Brinkman, C.: Split-brain monkeys: cerebral control of contralateral and ipsilateral arm, hand and finger movements, doctoral dissertation, Rotterdam, 1974, Erasmus University Rotterdam.)

tive inputs. Another approach is based on a functional neuroanatomical analysis of the pathways that mediate visual guidance of independent finger movements. A monkey is required to remove a small piece of food from a well in a board, a task that demands fractionated movement of the digits (Fig. 18-10, *A*). This task can be accomplished only via the corticospinal tract that arises from the motor cortex contralateral to the hand that is used (Fig. 18-11). As described more fully in Chapter 19, if the optic chiasm and certain of the transverse commissures of the brain are sectioned, visual information presented to an eye will reach only the ipsilateral hemisphere (Fig. 18-11). In such a preparation, if the right eye is covered, the monkey is able to retrieve food from the board with its right hand but not its left hand (Fig. 18-10); that is, the visual information from the food board is available only to the left hemisphere via the left eye. The left hemisphere can then control the fingers of the right hand because the corticospinal tract decussates at the pyramids. Of course, if the left eye is covered and the visual information is available to the right eye, then retrieval of the food can only be negotiated by the left hand.

This finding suggests that visual information reaching a given hemisphere can access the motor cortex of that hemisphere via intracortical circuitry. If a lesion is made in the parietal lobe between the visual and motor cortices, visually guided finger movement is eliminated in this task. (It is known that somatosensory information also reaches the motor cortex via intracortical pathways, in this case involving the somatosensory cortex of the parietal lobe.) This suggests the outline of a pathway whereby visual stimuli access the cells of origin of the corticospinal tract that control fractionated finger movement.

Bibliography

Journal articles

Asanuma, H., and Rosén, I.: Topographical organization of cortical efferent zones projecting to distal forelimb muscles in the monkey, Exp. Brain Res. **14:**243, 1972.

Asanuma, H., Stoney, S.D., and Abzug, C.: Relationship between afferent input and motor outflow in cat motorsensory cortex, J. Neurophysiol. **31:**670, 1968.

Evarts, E.V.: et al.: Central control of movement, Neurosci. Res. Program Bull. **9:**1, 1971.

Haaxma, R., and Kuypers, H.G.J.M.: Role of the occipitofrontal cortico-cortical connections in visual guidance of relatively independent hand and finger movements in rhesus monkeys, Brain Res. **71:**361, 1974.

Lawrence, D.G., and Kuypers, H.G.J.M.: The functional organization of the motor system in the monkey. II. The effects of lesions of the descending brain-stem pathways, Brain **91:**15, 1968.

Sherrington, C.S.: Decerebrate rigidity and reflex coordination of movements, J. Physiol. (Lond.) **22:**319, 1898.

Sprague, J.M., and Chambers, W.W.: Regulation of posture in intact and decerebrate cat. I. Cerebellum, reticular formation, vestibular nuclei, J. Neurophysiol. **16:**451, 1953.

Sprague, J.M., and Chambers, W.W.: Control of posture by reticular formation and cerebellum in the intact, anesthetized and unanesthetized and in the decerebrated cat, Am. J. Physiol. **176:**52, 1954.

Tanji, J., and Evarts, E.V.: Anticipatory activity of motor cortex neurons in relation to direction of an intended movement, J. Neurophysiol. **39:**1062, 1976.

Books and monographs

Asanuma, H.: The pyramidal tract. In Brookhart, J.M., and Mountcastle, V.B., editors: Handbook of physiology, the nervous system, section 1, vol. II, pt. 1, Bethesda, 1981, American Physiological Society.

Brodal, A.: Neurological anatomy, ed. 3, Oxford, 1981, Oxford University Press.

Evarts, E.V.: Role of motor cortex in voluntary movements in primates. In Brookhart, J.M., and Mountcastle, V.B., editors: Handbook of physiology, the nervous system, section 1, vol. II, pt. 2, Bethesda, 1981, American Physiological Society.

Ghez, C.: Cortical control of voluntary movement. In Kandel, E.R., and Schwartz, J.H., editors: Principles of neural science, New York, 1981, Elsevier Science Publishing Co., Inc.

Lundberg, A.: Control of spinal mechanisms from the brain. In Tower, D.B., editor: The nervous system, vol. 1, The basic neurosciences, New York, 1975, Raven Press.

Penfield, W., and Rasmussen, T.: The cerebral cortex of man, New York, 1950, The Macmillan Co.

Phillips, C.G., and Porter, R.: Corticospinal neurones: their role in movement, London, 1977, Academic Press, Inc.

Ruch, T.C.: Brain stem control of posture and orientation in space. In Ruch, T., and Patton, H.D., editors: Physiology and biophysics, vol. 1, Philadelphia, 1979, W.B. Saunders Co.

Sherrington, C.S.: The integrative action of the nervous system, ed. 2, New Haven, Conn., 1947, Yale University Press.

Weisendanger, M.: Organization of secondary motor areas of cerebral cortex. In Brookhart, J.M., and Mountcastle, V.B., editors: Handbook of physiology, the nervous system, section 1, vol. II, pt. 2, Bethesda, 1981, American Physiological Society.

CHAPTER

19

The Cerebral Cortex

Viewed from an evolutionary perspective, the expansion of the mammalian *cerebral cortex* is striking. In all vertebrates, however, the evolutionary trend can be discerned from three analogous cortical divisions, namely, the *archicortex,* the *paleocortex,* and the *neocortex.* Neocortex is often called *isocortex,* and *allocortex* indicates the contiguous expanse of archicortex and paleocortex. The *cingulate gyrus* may represent a transitional form between allocortex and neocortex. Among these cortical divisions, the neocortex has enlarged most dramatically during mammalian evolution. Its large surface area (2500 cm^2) is folded among numerous *sulci* and *gyri,* and its overall volume approximates 600 cm^3. Equally impressive is the evolutionary development of the *corpus callosum,* which in humans is a massive cerebral commissure interconnecting neocortical areas between the two hemispheres.

Neocortex

The human neocortex represents roughly 90% of cortical tissue. Neocortex can be distinguished from archicortex and paleocortex on the basis of lamination. During at least some stage of ontogeny, six layers parallel to the surface can be clearly discerned in neocortex. Subsequent development and differentiation obscure some of these layers in certain areas of adult neocortex.

Cell Types

Nearly all neocortical neurons can be placed into one of three broad categories: *pyramidal cells, stellate cells,* and *fusiform cells* (Fig. 19-1). Other rare cell types have also been described.

Pyramidal cells are the largest, with somata typically 30 to 50 μm across and occasionally over 100 μm in diameter. Each cell has a pyramid-shaped soma. The apex is directed toward the pial surface, a large *apical dendrite* branches and often extends to the pial surface, and a set of branched *basal dendrites* fan out from the base of the soma for considerable distances. The dendrites are usually covered with spiny protrusions that are thought to be postsynaptic specializations. A single axon emerges from the base of the soma and projects into white matter to other cortical or subcortical areas. The axon also emits collaterals that contribute to intrinsic cortical circuitry.

Stellate cells, also known as *granule cells,* have star-shaped or spherical somata that are roughly 10 μm in diameter. Their short dendritic processes are richly branched, and they emanate in all directions from the soma. Some of these cells have spines on their dendrites, but others do not (Fig. 19-1); these cells are respectively called spiny and smooth stellate cells. Stellate cell axons branch and ramify locally. Consequently these cells are regarded as cortical interneurons, although some may project axons into the white matter. Most of the spiny stellate cells are probably excitatory interneurons, whereas the smooth stellate cells are inhibitory and tend to be GABAergic (i.e., employing γ-aminobutyric acid as a neurotransmitter).

Fusiform cells are relatively rare. They have elongated somata, typically 10 to 15 μm wide and up to 50 μm long. Their dendrites emerge as two tufts from either end of the soma, and the cell thus has a bipolar appearance. The cell is oriented perpendicular to the pial surface. Axons from fusiform cells project into white matter, locally, and to other cortical or subcortical sites.

In brief, pyramidal and fusiform cells are the major efferent neurons of the neocortex, although they also contribute significantly to intrinsic circuitry. Stellate cells are the main cortical interneurons.

Functional Organization

Cortical layers. The six main neocortical layers that lie parallel to the pial surface are designated layers I to VI, from the pial surface to white matter (Fig. 19-2).

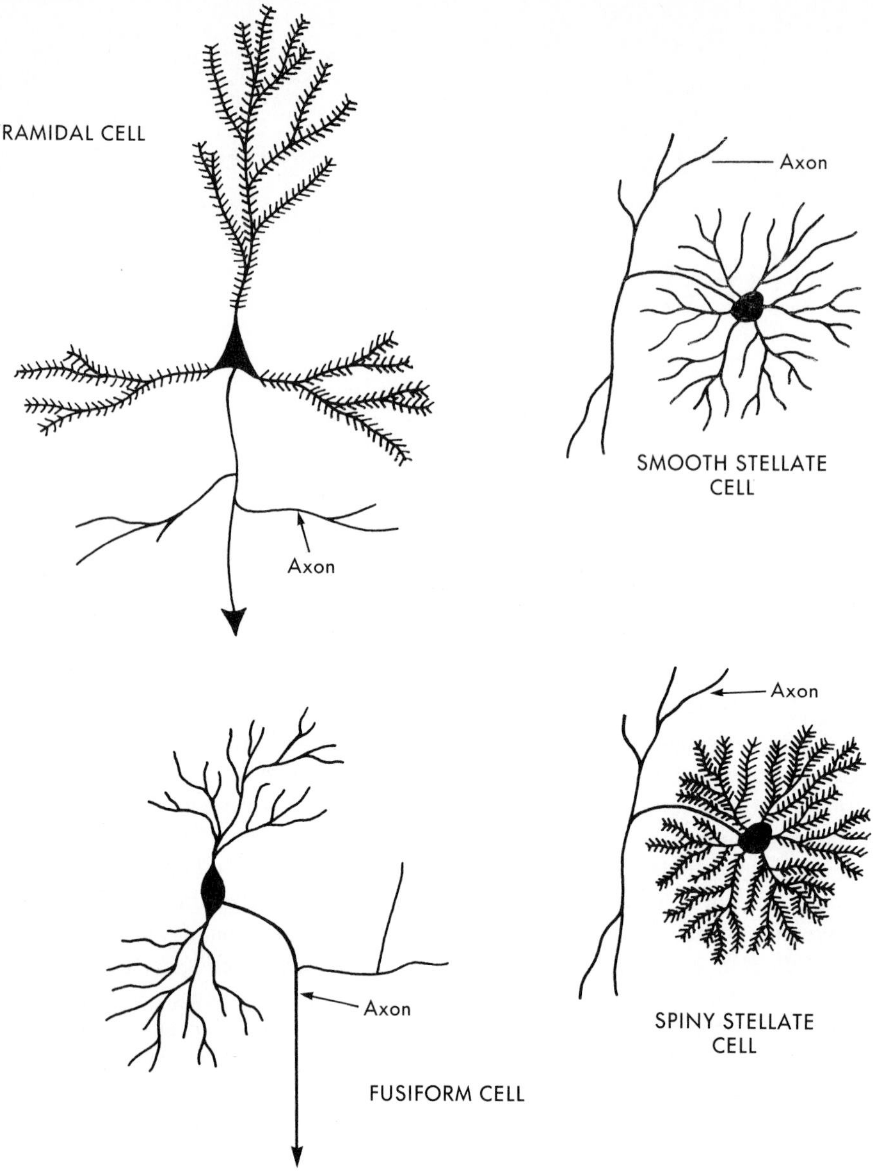

■ **Fig. 19-1.** Major neuron types within the cerebral cortex. These include the pyramidal cell, fusiform cell, smooth stellate cell, and spiny stellate cell.

Layer I is the thin *molecular, plexiform,* or *zonal layer*. It has few somata and is basically a synaptic plexiform layer comprising axonal and dendritic processes. Layer II is the *external granular layer,* comprising mostly stellate cells, but also some pyramidal cells. Layer III is the *pyramidal* or (or *external pyramidal*) *layer,* and its neurons are predominantly pyramidal cells, with some stellate cells. Layer IV, the *internal granular layer,* consists mostly of stellate cells with some pyramidal cells. Layer V, the *ganglionic (internal pyramidal) layer,* is dominated by large pyramidal cells but also includes some stellate cells. Finally, layer VI, the *multiform layer,* contains fusiform cells and a variety of other neuron types, including some stellate and pyramidal cells.

Pyramidal and stellate cell bodies can be found in any layer, but pyramidal cell bodies are concentrated in layers III and IV, and stellate cell bodies in layers II and IV. Fusiform cell bodies are associated nearly exclusively with layer VI. The dendrites of stellate and fusiform cells are contained mostly or completely within a single layer. In contrast, pyramidal cell dendrites cross layers freely and can receive synaptic inputs through much or all of the cortical depth. For instance, many pyramidal cells with somata in layer V or VI have apical dendrites that reach layer I, and basal dendrites of pyramidal cells commonly descend toward the white matter.

Cortical afferents and efferents. The various afferent fibers to neocortex and sources of corticofugal pathways have distinctly layered arrangements (Fig. 19-3). To a first approximation, the supragranular layers (i.e.,

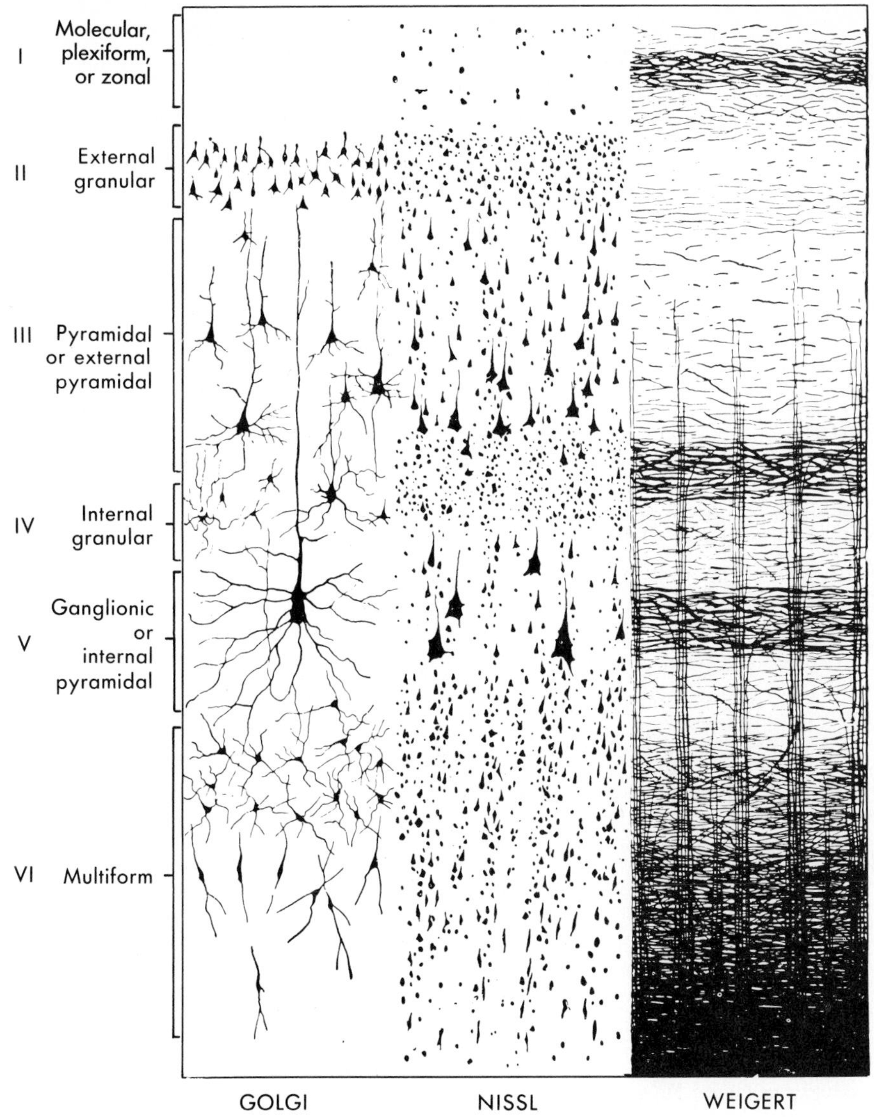

Fig. 19-2 Layers of cerebral cortex as shown by the different staining methods of Golgi (entire structure of a few isolated neurons), Nissl (somata), and Weigert (myelinated axons). (From Brodal, A.: Neurological anatomy, ed. 3. Copyright © 1981 by Oxford University Press, Inc. Reprinted by permission.)

layers I to III) are involved in corticocortical pathways, and the granular and infragranular layers (i.e., layers IV to VI) participate in subcortical pathways. Exceptions to this generalization exist.

Corticocortical axons enter the gray matter from below and ramify extensively in layer II, although some of these also terminate in layers I and III and may issue collaterals in layer VI. These axons arise from other cortical areas, either in the same hemisphere or contralaterally via commissures, chiefly the *corpus callosum*. Thalamocortical axons terminate chiefly in layer IV, frequently with collaterals in layer VI; however, some thalamic cell groups innervate cortex more diffusely, with terminals in many layers, including layer I. The claustrum innervates all layers, but particularly layers IV and VI of many cortical areas. This pathway is predominantly ipsilateral, but a small crossed component has been demonstrated. Finally, cholinergic, noradrenergic, and serotonergic inputs that diffusely innervate all areas and layers of cortex have been described. The cholinergic inputs derive both from the *basal nucleus of Meynert* located in the substantia innominata and from the *parabrachial nucleus* of the brainstem reticular formation; the noradrenergic inputs derive from the *locus ceruleus* of the brainstem reticular formation and the serotonergic inputs derive from the *dorsal raphe nucleus* of the brainstem reticular formation. As noted in Chapter 8, these regions of the brainstem reticular formation also innervate many other brain areas, including the thalamus.

Although many of the afferent fibers to cortex have specific laminar terminations, their target cells are not

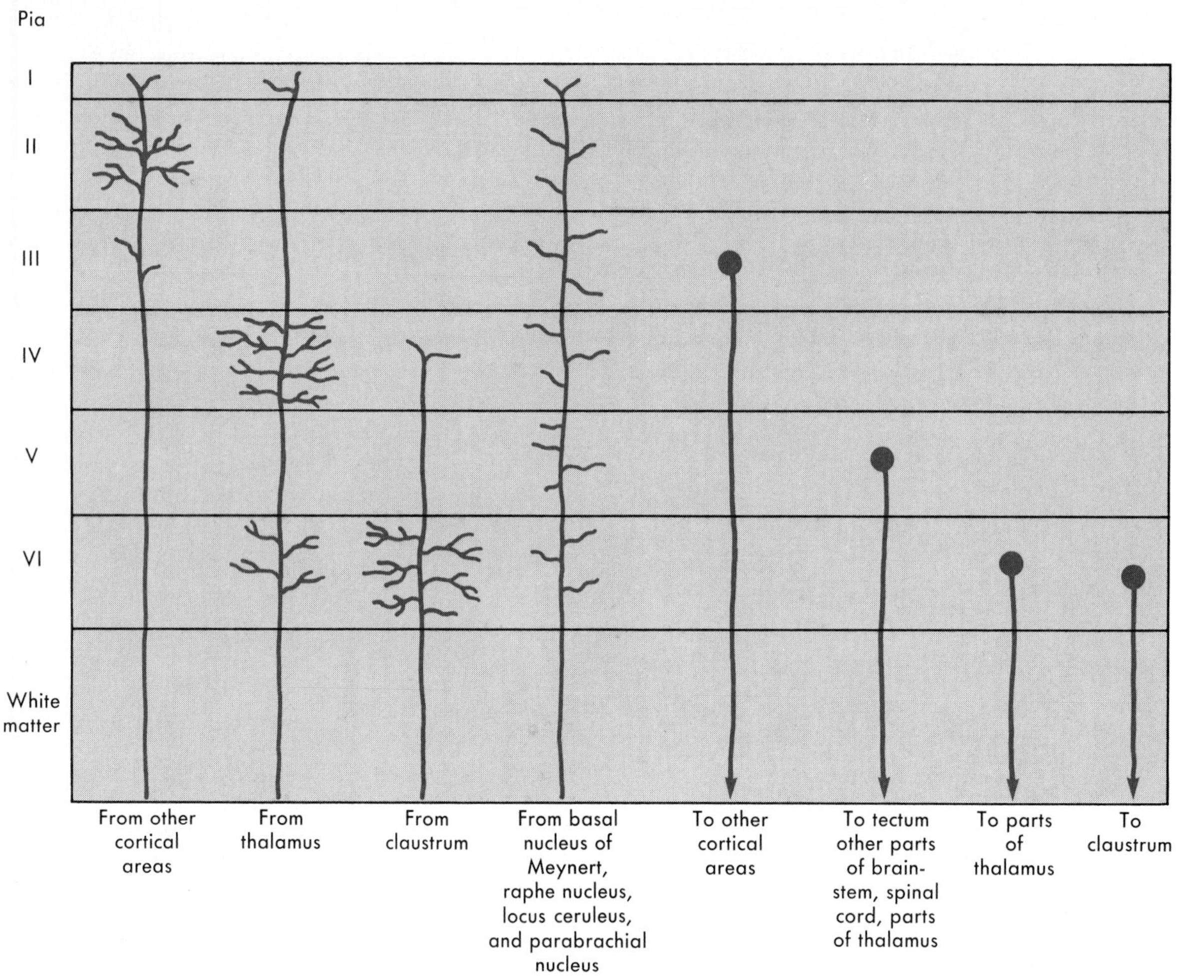

Fig. 19-3. Major inputs and outputs of different layers of cerebral cortex.

easy to define. For instance, the lateral geniculate nucleus projects extensively to layers IV and VI of striate cortex, with no appreciable input to layer V. Yet pyramidal cells with somata in layer V typically receive direct geniculate innervation, by virtue of either the apical dendrite that ascends through layer IV or basal dendrites that descend into layer VI.

The layering pattern of cortical efferent fibers depends on the projection target. Corticocortical projections (both ipsilateral and contralateral) originate predominantly with pyramidal cells of layer III, although a few such cells in layers II and V may participate in this pathway. Also, a sparse projection from supragranular stellate cells may contribute to this pathway. Layer V pyramidal cells are the major source of descending corticofugal pathways, including the corticotectal, corticobulbar, corticospinal, and certain corticothalamic (e.g., to pulvinar) pathways. Layer VI pyramidal cells and fusiform cells innervate certain thalamic targets (e.g., the lateral geniculate nucleus) and the claustrum. Different layer VI cell populations are involved in each pathway. Corticothalamic cells are located throughout layer VI, whereas corticoclaustral cells concentrate in the middle of the layer. These two cell populations differ pharmacologically and morphologically. In general, the connections between cortex and thalamus are reciprocal, and many more corticothalamic axons exist than do thalamocortical axons. As an example, the layer VI cells of striate cortex that innervate the dorsal LGN are roughly 10 times more numerous than the geniculate cells that innervate striate cortex.

Columnar organization. The functional unit of neocortex appears to be a column of interconnected cells that runs vertically through all the layers. Ample anatomical and physiological evidence for this can be found in visual cortex (Chapter 8), somatosensory cortex (Chapter 9), auditory cortex (Chapter 10), and motor cortex (Chapter 18). This was evident to early anatomists, who noted that cortical cells and fibers seemed to be organized into discrete vertical columns (Fig. 19-2). Although the rich complexity of neuronal connectivity within a column is still incompletely defined, the broad outline of these interconnections is beginning to emerge. A schematized and greatly simplified summary of these intracolumnar connections is provided in Fig. 19-4, *A*.

Presumably, inputs to cortex are analyzed by a column of cells. This information is integrated, and appro-

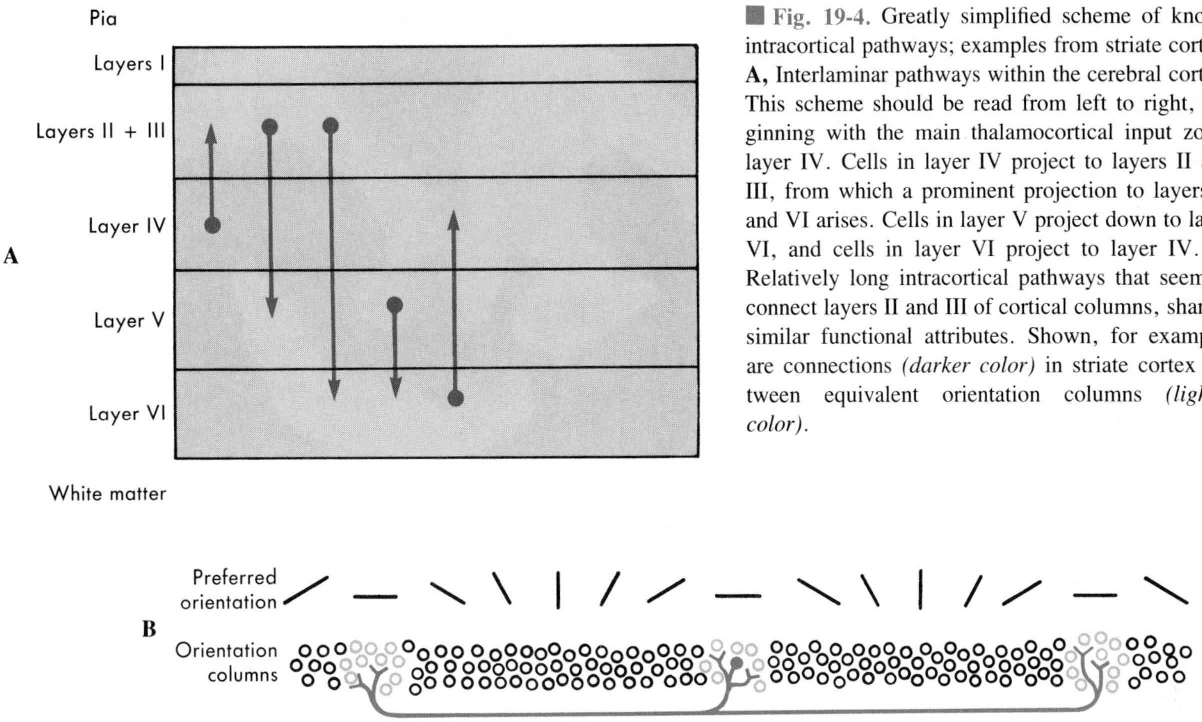

Fig. 19-4. Greatly simplified scheme of known intracortical pathways; examples from striate cortex. **A,** Interlaminar pathways within the cerebral cortex. This scheme should be read from left to right, beginning with the main thalamocortical input zone, layer IV. Cells in layer IV project to layers II and III, from which a prominent projection to layers V and VI arises. Cells in layer V project down to layer VI, and cells in layer VI project to layer IV. **B,** Relatively long intracortical pathways that seem to connect layers II and III of cortical columns, sharing similar functional attributes. Shown, for example, are connections *(darker color)* in striate cortex between equivalent orientation columns *(lighter color)*.

priate efferent cells in the column discharge accordingly. Also, functional connections within a column are more numerous than those across columns. Neocortex resembles the retina in this regard. Both structures have a laminar arrangement, but most functional connections and the major direction of information flow occur within cell columns organized perpendicular to the layering. Indeed, the layering emerges because the cell columns are stacked next to one another in register, so that homologous cells in each column form the different layers across columns.

The column is thus the functional module of cortex, and its size is relatively constant throughout neocortex and across species. Those mammalian species with larger volumes of cortex have more of these modular units of a fairly fixed size, rather than a similar number of larger columns. Also long, horizontal connections extend several millimeters within cortex to link together columns of similar function. These horizontal connections are not as plentiful as the within-column connections. In striate cortex, for example, columns that map the same orientation of an elongated visual stimulus are interconnected, whereas the columns that map dissimilar orientations are not interconnected (Fig. 19-4, *B*). This results in a regular array of interdigitating connections across cortex.

Regional Variations

The cortex is divided into many discrete zones that subserve different brain functions. The evidence for such functionally discrete zones consists of (1) histologically distinct areas of cortex called *cytoarchitectonic* divisions, (2) physiological studies of various cortical areas, and (3) neuropsychological investigations of the behavioral effects of discrete cortical lesions.

Cytoarchitectonic divisions. The basic layering scheme varies for different cortical areas. Fig. 19-5 shows five different types of cortex and their locations on the surface of the hemisphere. Types *2* to *4* represent variations on the basic laminar scheme, because six layers with appropriate cell types can be clearly seen. These types are thus called *homotypic* cortex. Types *1* and *5* are *heterotypic cortex* because six layers are not clearly present. In fetal life these heterotypical areas also pass through a six-layer stage and are thus neocortical. Type *1* is called *agranular cortex,* because pyramidal cells abound and few stellate cells exist. Layers II and IV are poorly developed, so that the efferent cells in layers III and V are the dominant cell type. Agranular cortex is thus often associated with motor areas, and indeed motor cortex of the precentral gyrus is agranular (Fig. 19-5). Type *5, granular cortex (koniocortex),* has few pyramidal but many stellate cells. The afferent layers II and IV are well developed, and granular cortex is often associated with sensory areas. Indeed, the primary visual, somatosensory, and auditory areas are all granular cortex (Fig. 19-5).

These five cortical types can often be further subdivided. A detailed cytoarchitectonic map that is based on these subtle differences was produced by Brodmann (Fig. 19-6). Brodmann's map includes roughly 50 discrete areas. The common designations include area 17,

Fig. 19-5. Five major types of cerebral cortex as described by C. von Economo. *1,* Agranular cortex; *2,* frontal type of cortex; **3,** parietal type of cortex; *4,* polar type of cortex; and *5,* granular cortex of koniocortex. **A,** Distribution in the cerebral hemisphere. **B,** Schematic appearance in cross section. (From Kornmüller, A.E., and Janzen, R.: Arch. Psychiatr. Nervenkr. **110:**224, 1939.)

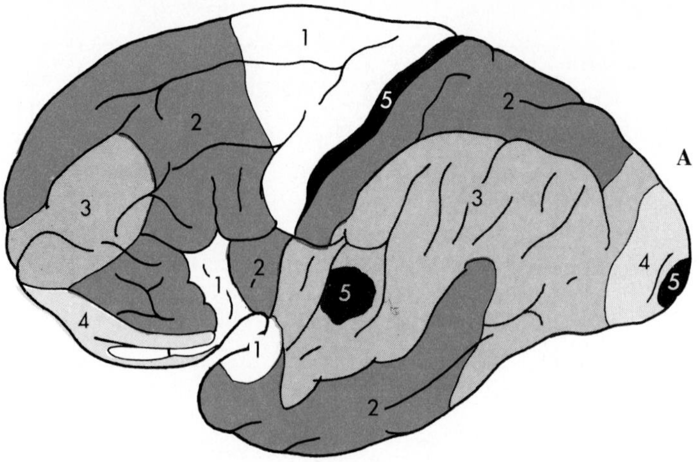

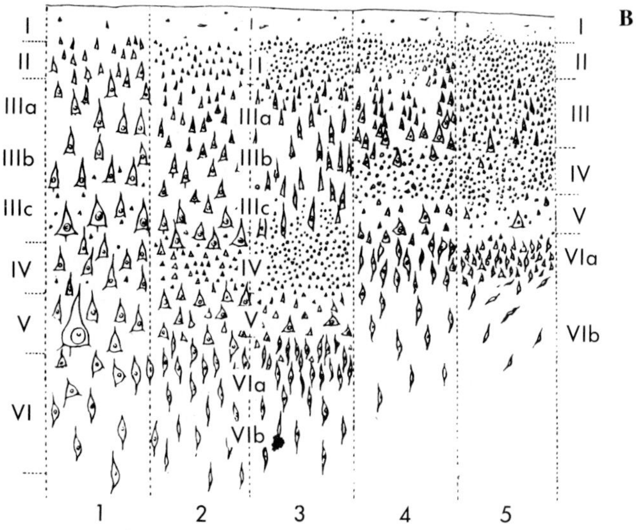

which is striate cortex, or VI; areas 1, 2, and 3, which are the primary somatosensory cortex, or SI; areas 41 and 42, which are the primary auditory cortex, or AI; area 4, which is the motor cortex; and area 8, which is the frontal eye field involved in saccadic eye movements. In general, these maps can be correlated well with cortical areas that subserve different functions.

Sensory and motor areas of cortex. Concepts of the functional organization of sensory and motor areas have changed markedly in the past 20 years. Previously it was thought that only areas 17, 18, and 19 were strictly visual, 41 and 42 auditory, and 1, 2, and 3 somatosensory. All cortex in between these areas was designated *association cortex,* which was presumed to be the locus at which all sensory information was analyzed, integrated, and passed on to motor cortex.

Recent studies, however, have shown that much or all of what was presumed to be association cortex actually contains multiple, specific sensory representations. For instance, functional mapping studies in cats and many species of monkey show that little room is left for association cortex. Fig. 19-7 summarizes the contemporary mapping data for the owl monkey's cerebral cortex. Further mapping studies will probably all but eliminate association cortex and its theoretical consequences from notions of cortical function. Instead, sensory and motor processing involve a bewildering multiplicity of areas that are richly interconnected with each other and with the thalamus. The significance of these multiple representations is uncertain. For instance, some of the visual areas are thought to specialize in the analysis of color, movement, stereopsis, or forms. This problem remains an active area of research.

Frontal lobes. The specific sensory and motor areas of cortex are located in the occipital, parietal, and temporal lobes, which are relatively well understood functionally. By contrast, the functional significance of the frontal lobes is poorly understood. This is an important gap in our knowledge; the frontal lobes are larger, proportionally and absolutely, in humans than in any other animal.

Afferent fibers to the frontal lobe arrive directly from other cortical areas, from the thalamus (chiefly the dorsomedial nucleus, or MD), from the brainstem reticular formation and basal nucleus of Meynert (see above), from the amygdala, from the cingulate gyrus, and from the septum. The MD relays further information from some of these areas and from the hypothalamus and olfactory areas. Frontal cortex, in turn, innervates these areas plus the pulvinar and intralaminar thalamic nuclei. Considering that the frontal lobe is extensively connected with the amygdala, cingulate gyrus, septum, and

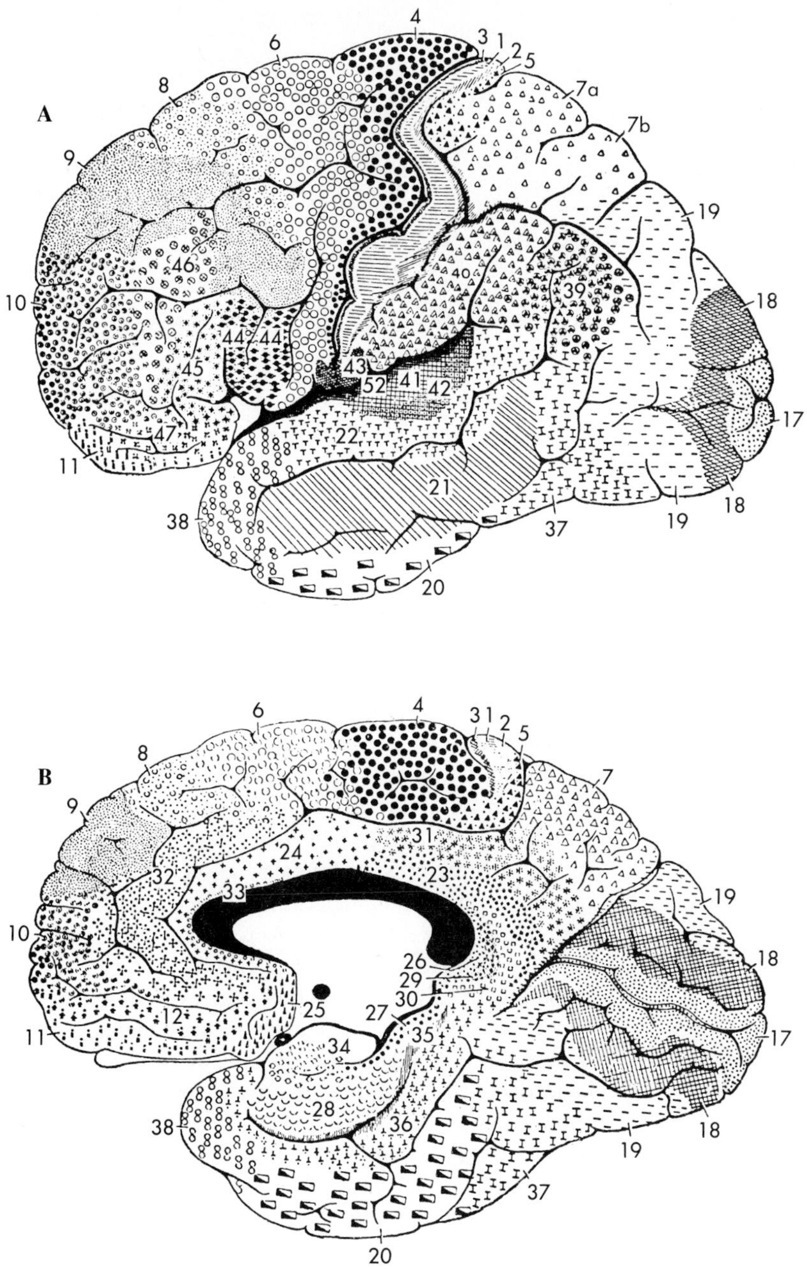

■ **Fig. 19-6.** Cytoarchitectonic divisions of human cerebral cortex according to Brodmann. **A,** Lateral surface of left hemisphere. **B,** Medial surface of right hemisphere. (From Crosby, E.C., et al.: Correlative anatomy of the nervous system, New York, 1962, Macmillan Publishing Co., Inc.)

hypothalamus, this cortical region probably has important personality and emotional functions.

Behavioral studies of frontal lobe damage in humans and animals generally suggest that this region of cortex has emotional and intellectual functions, but a precise and detailed consensus as to its function and functional subdivisions is only just emerging. In animal studies, bilateral frontal lobe lesions do not affect sensory abilities, motor abilities, or the ability to learn simple tasks. However, learning of more sophisticated sensorimotor activities is devastated. Monkeys with such lesions have trouble concentrating and are easily distracted. They cannot learn to respond to one stimulus (perhaps a sound from above) with a particular behavior (perhaps reaching for the left of two food wells) while responding to a second stimulus (perhaps a sound from below) with another behavior (perhaps reaching for the right of two food wells). Unilateral lesions produce much milder deficits in animals.

A similar pattern occurs in humans, although the effects of unilateral ablations of frontal cortex is more pronounced than might be expected from the animal studies. Such ablations are performed to remove tumors or to control epilepsy. Simple intelligence tests do not generally reveal deficits in human patients with such ablations. However, more complicated behavior is greatly affected. As in the experimental animals, these patients are easily distracted and have trouble concentrating. Complex problem-solving abilities are greatly impaired. Response to external stimuli is often inappro-

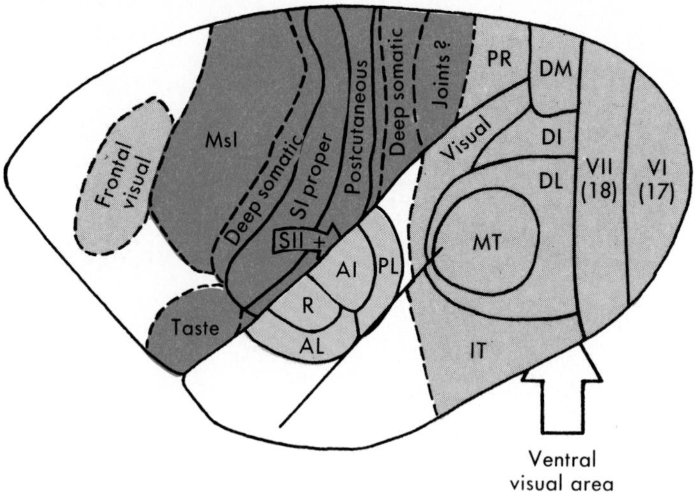

Fig. 19-7. Known sensory areas of cerebral cortex in a New World monkey. After delineation of visual areas *(gray)*, somatosensory areas *(darker color)*, and auditory areas *(lighter color)*, little cortex is unaccounted for. Much of this replaces the outdated notion of "association cortex." Not shown is a medial visual area contiguous with *DM* on the medial surface. *MsI*, Motor-sensory area I; *SI* primary somatosensory area; *SII* +, other somatosensory areas; *AI*, primary auditory area; *R*, rostral auditory area; *AL*, anterior lateral auditory area; *PL*, posterior lateral auditory area; *VI*, primary visual area; *VII*, secondary visual area; *DM*, dorsal medial visual area; *DI*, dorsal intermediary visual area; *DL*, dorsal lateral visual area; *MT*, middle temporal visual area; and *IT*, inferotemporal visual area. (Redrawn from Merzenich, M.M., and Kaas, J.H. In Sprague, J.M., and Epstein, A.N., editors: Progress in psychobiology and physiological psychology, vol. 9, New York, 1980, Academic Press, Inc.)

priate. For instance, social situations are frequently misread, and following instructions is often difficult. These patients often cannot string together simple tasks to reach straighforward goals. For example, an experienced housewife with frontal lobe damage could no longer prepare entire meals, even though she could easily prepare the individual dishes. Finally, the effects of damage to the left or right frontal lobe are subtly different.

Frontal lobe damage greatly reduces aggressive behavior. Because of this, certain emotional disorders, such as schizophrenia, were commonly treated with bilateral frontal lobotomies. This form of psychosurgery is highly controversial, since it often fails to produce the desired effect. The resultant intellectual, personality, and emotional changes are sometimes worse than the original disorders. Frontal lobotomies are currently rarely performed, particularly because powerful drugs are available to achieve more safely, predictably, and reversibly most of the desired personality changes.

Cerebral Dominance and Language

Human cortex is divided into two more or less equivalent hemispheres. However, these hemispheres are not strict mirror images of one another. Most of the known hemispheric asymmetry is a result of language functions, which are typically concentrated in a single hemisphere. This language function defines the *dominant hemisphere*. With the possible exceptions of apes and perhaps cetaceans, no nonhuman mammals are believed to have obvious language ability, although subtle hemispheric asymmetries have been described for a variety of vertebrate species.

Evidence supporting the concept of a dominant hemisphere comes from processes that affect one hemisphere and interfere with language. Such processes include unilateral trauma (e.g., stroke, tumor, physical injury) or unilateral, temporary anesthesia. Unilateral anesthesia can be induced by injection into one carotid artery of a short-acting anesthetic, such as amobarbital. In the vast majority of humans, the left hemisphere is dominant. In the remainder, either the right hemisphere is dominant or neither hemisphere is dominant, because each hemisphere subserves language.

Although language function defines cerebral dominance, a strong correlation exists with handedness. Left hemisphere dominance nearly always results in right-handedness. Right dominance less clearly results in left-handedness because substantial numbers of right-handed or ambidextrous individuals are found with such dominance. Handedness is another example of interhemispheric asymmetry, because each hand is represented predominantly in the contralateral hemisphere. Other examples of functional interhemispheric differences are considered in the discussion of the corpus callosum.

In addition to these examples of functional differences between the hemispheres, striking anatomical differences also exist. For instance, careful measurements of the *planum temporale*, which is located at the posterosuperior surface of the temporal lobe within the region associated with language function (Fig. 19-8), reveal interhemispheric differences. This region is consistently larger in the dominant hemisphere. This difference has been found in the great apes but not in monkeys. It has also been found in infants, which suggests a predilection for a specified dominant hemisphere before the development of language. This is interesting, because the effects of a unilateral lesion to the language areas in an adult differ dramatically from these effects

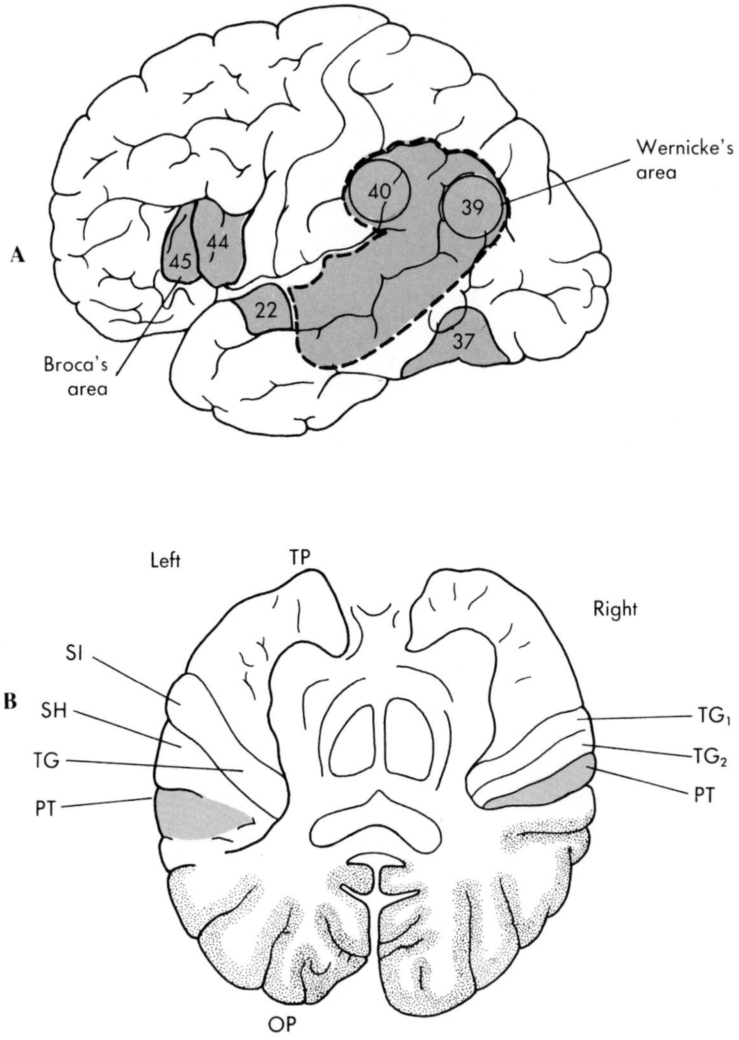

Fig. 19-8. Language areas of cerebral cortex. **A,** Regions of left hemisphere associated with language *(color)*. The numbers represent Brodmann areas (see Fig. 19-6). Wernicke's area is approximately delineated by the dashed line, and Broca's area is roughly coextensive with areas 44 and 45. **B,** Horizontal section through the brain showing interhemispheric asymmetry in the region of the upper temporal lobe. Note the larger planum temporale *(PT)* on the left compared with that on the right. This region is associated with Wernicke's area, which is larger in the dominant, left hemisphere. *TP,* Temporal pole; *OP,* occipital pole; *TG,* transverse gyrus (including primary auditory area); two transverse gyri *(TG_1,TG_2)* are commonly found in the right hemisphere; *SI,* sulcus intermedius; *SH,* sulcus of Heschl. (**A** redrawn from Williams, P.L., and Warwick, R.: Functional neuroanatomy of man, Philadelphia, 1975, W.B. Saunders Co.; **B** redrawn from Geschwind, N. In Mountcastle, V.B., editor: Medical physiology, ed. 14, St. Louis, 1979, The C.V. Mosby Co.)

in an infant. In an adult such a lesion produces a permanent interference with language ability; this disturbance is known as *aphasia*. In infancy such a lesion usually does not result in persistent aphasia, presumably because the immature, nondominant hemisphere retains the ability to develop language capacity. The greater flexibility, plasticity, and adaptability of the young brain is reconsidered in Chapter 21.

From cortical lesions that produce aphasia in humans, those areas involved in language can be delineated. Results of electrical stimulation of the cortex in humans undergoing neurosurgical procedures confirm the location and extent of these areas. However, these methods are imprecise and difficult to control. For instance, aphasic patients suffer from a wide range of cortical lesions that typically produce many deficits in addition to aphasia. Therefore the precise extent and location of the lesions are difficult to assess. For these reasons, the following discussion of cortical language areas and aphasia is tentative.

Fig. 19-8 shows the two major language areas of the dominant hemisphere. *Wernicke's area* is located in the posterior portion of the superior temporal gyrus, near the auditory cortex. The anterior region, *Broca's area,* lies just anterior to the representation of the face area in the motor cortex. Lesions of either area produce language difficulties. However, lesions of Wernicke's area cause more difficulty with understanding written or spoken language than with coherent speech or writing. Conversely, lesions of Broca's area interfere more with speech and writing than with comprehension of written or spoken language. Wernicke's area is thus more or less associated with *sensory aphasia* (difficulty in understanding language) and Broca's area with *motor aphasia* (difficulty in producing coherent language). However, rarely, if ever, does a lesion produce a purely sensory or purely motor aphasia.

A final point must be emphasized regarding aphasia. An aphasic patient may be unable to comprehend or produce language yet typically have no purely sensory or motor disorder. For instance, despite motor aphasia, nonverbal vocalization and humming of complicated melodies is often normal, as is the ability to draw complex pictures or diagrams. Similarly, sensory aphasia

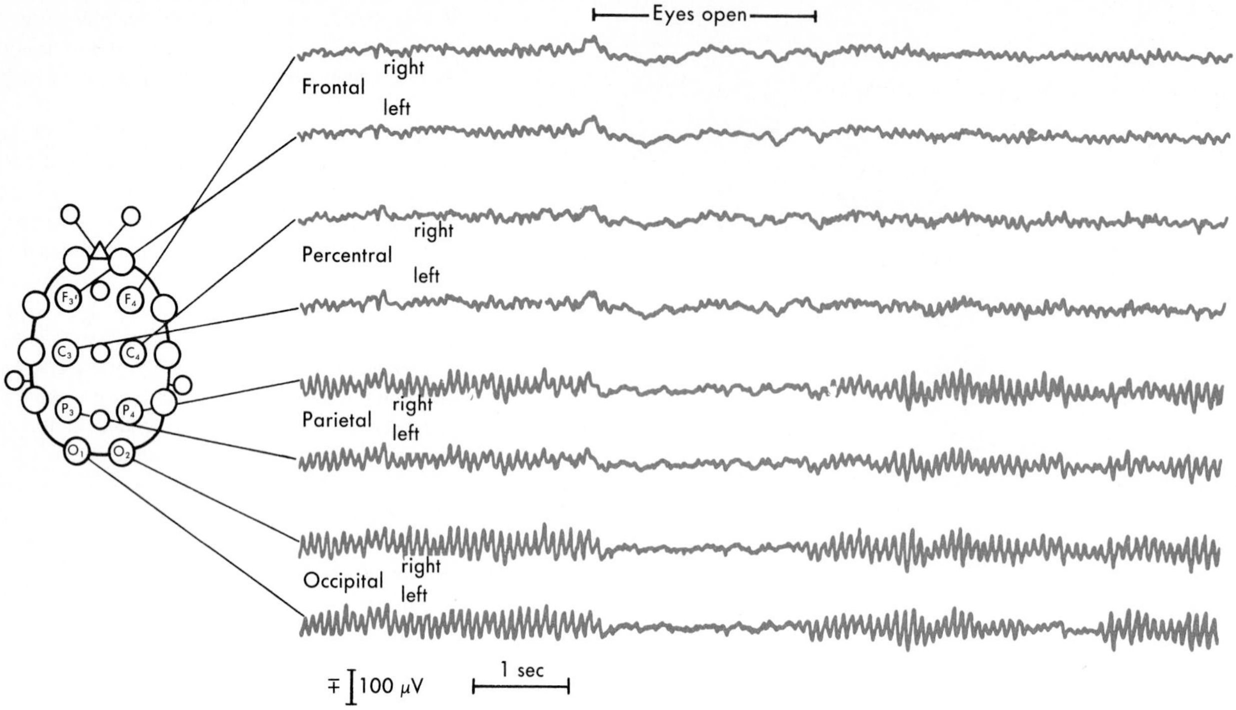

■ Fig. 19-9. Normal EEG of a resting, awake human. Simultaneous eight-channel unipolar recording from the sites on the skull that are indicated at left. An electrode was attached to each earlobe, and the two together constituted the indifferent electrode. Opening of the eyes *(middle of record)* blocks the alpha rhythm. (From Schmidt, R.F., editor: Fundamentals of neurophysiology, ed. 2, New York, 1978, Springer-Verlag.)

may appear with no other auditory or visual symptoms; the identification of complex sounds or tunes and comprehension of complex visual scenes may be completely normal. As a final example, deaf mutes who have been trained to communicate fluently with sign language may suffer severe disability in understanding or producing sign language after lesions to their language areas, even though their manual dexterity and vision are unaffected by lesions. Thus aphasia is a deficit in language ability quite separate from sensory or motor performance.

■ The Electroencephalogram

By placing a set of disk electrodes at various standardized locations on the scalp, the activity of underlying neural tissue can be recorded as waves that vary in frequency from 1 to over 30 cycles per second (hertz). The record of these waveforms, taken under highly standardized conditions, is called the *electroencephalogram* (EEG). It is a conventional tool in clinical neurology, particularly in dealing with epileptic patients.

In a normal individual the electroencephalogram varies as a function of a number of factors, including the recording site, age, and the state of wakefulness. In an awake relaxed individual with the eyes closed, slow waves that average 10 hertz (Hz) in frequency occur most prominantly over the occipital lobe (Fig. 19-9). This *alpha rhythm* represents synchronized neuronal activity in the underlying cortex. During alpha activity any change that makes the individual more alert, such as opening the eyes, will block the alpha rhythm and produce less synchronized activity of lower amplitude and higher frequency (Fig. 19-9). This is characterized by *beta waves,* which have an average frequency of 20 Hz. Such waves are characteristic of an alert or aroused state. These potentials recorded in the EEG reflect the activity, both postsynaptic potentials and action potentials, of many hundreds of thousands of neurons in the underlying cortex.

■ Sleep Stages

The EEG has been particularly useful in the study of sleep, because the different stages of sleep are reflected in different EEG waveforms. Four stages of slow-wave sleep can be distinguished, in which there is a progressive slowing of the EEG and an increase in wave amplitude (Fig. 19-10). The EEG in the relaxed, awake state is dominated by alpha activity, and the transition to the first stage of slow-wave sleep is characterized by

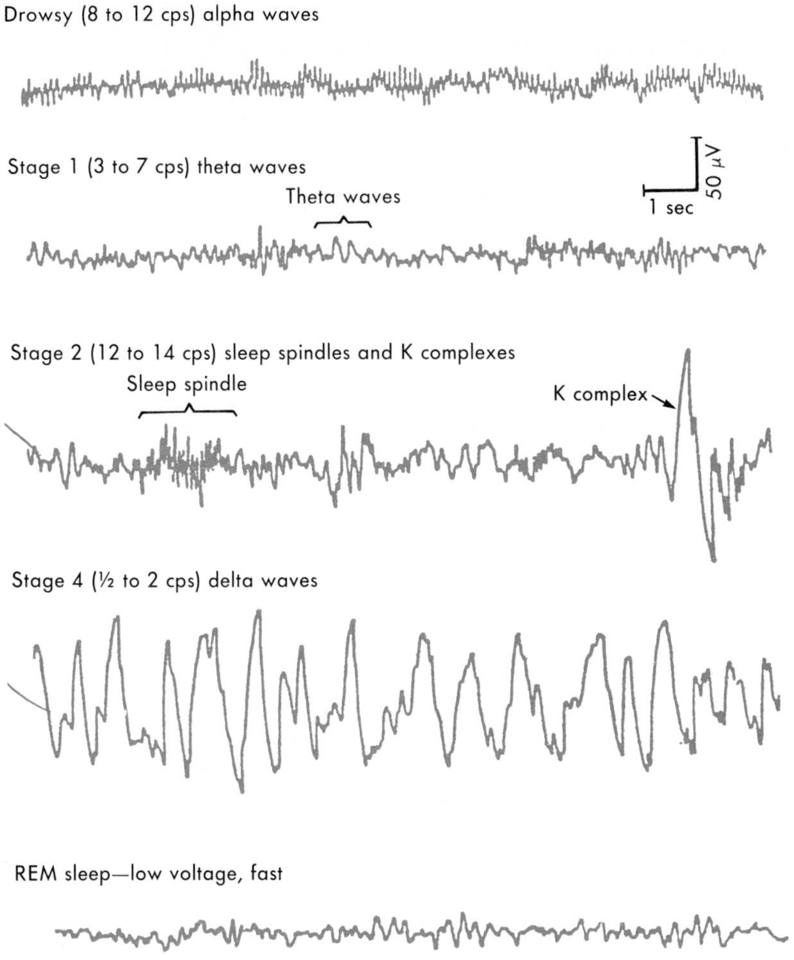

■ Fig. 19-10. Human electroencephalographic records during the progression from drowsiness through slow-wave sleep. Stage 3 is similar to stage 4 and is not shown. The bottom panel shows the EEG during REM sleep. (Modified from Shepherd, G.M.: Neurobiology, London, 1983, Oxford University Press.)

fluctuations between alpha waves and low-amplitude *theta waves,* which have a frequency of 3 to 7 Hz. As sleep deepens into stage 2, further EEG slowing occurs with bursts of activity at 12 to 14 Hz, sleep spindles, and periodic K complexes. In moderately deep sleep, stage 3 the EEG consists primarily of large *delta waves* at frequencies of 0.5 to 2 Hz, with infrequent sleep spindles. The EEG during the deepest stage of sleep, stage 4, consists almost entirely of delta waves. During slow-wave sleep the muscles are relaxed, sympathetic activity diminishes, and heart rate and blood pressure decrease. The transition through the four stages of slow-wave sleep is smooth, and the ease with which an individual can be wakened varies inversely with the stage. On waking, there is a smooth transition through the stages in the reverse order and over a similar time span.

During the course of sleep, epochs of EEG desynchronization occur (Fig. 19-10), accompanied by sympathetic activation and a loss of muscle tone. Phasic bursts of rapid eye movement (REM) are a prominent feature of these epochs. These eye movements were one of the first observations in the study of this phase of sleep, and thus it is generally referred to as REM sleep. Because the EEG is desynchronized during REM sleep and thus resembles the alert state, the epochs are sometimes referred to as *paradoxical sleep*. The first REM epoch generally occurs 1 to 2 hours after the onset of sleep. Several such epochs occur at regular intervals throughout the night. In a typical night's sleep, REM epochs constitute approximately 20% to 25% of the sleep time, although this varies with age. Of the different phases of sleep, it is most difficult to awaken an individual from REM sleep, although spontaneous waking occurs most often in this phase. Moreover, waking an individual during REM sleep leads to the highest frequency of dream recall.

The various theories of the neural mechanisms underlying the sleep-wake cycle concern the involvement of certain neural circuits and different neurotransmitters. However, no single theory has yet gained overwhelming support.

■ *Abnormal EEGs*

Various diseases of the nervous system, such as tumors, can produce an abnormal EEG. However, most such

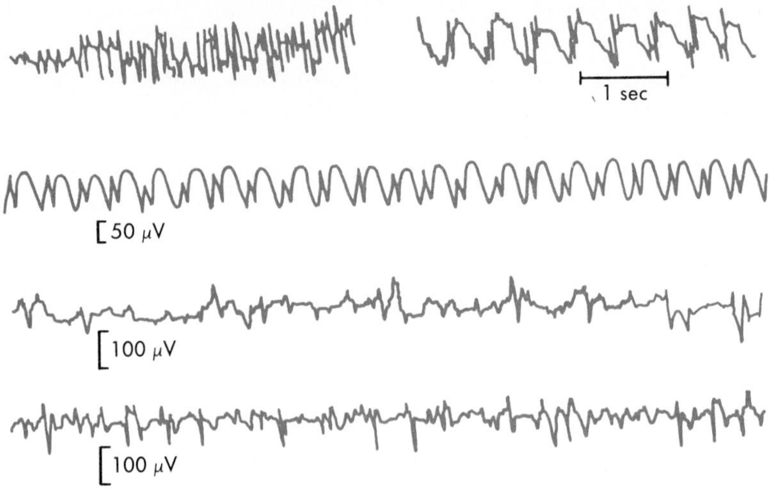

■ **Fig. 19-11.** Examples of abnormal EEG activity. **A,** Tonic (left) and clonic (right) phases of grand mal seizure. **B,** Petit mal seizure. **C,** Temporal lobe epilepsy. **D,** Focal seizure. (Redrawn from Eyzaguirre, C., and Fidone, S.J.: Physiology of the nervous system, ed. 2. Copyright 1975 by Yearbook Medical Publishers, Inc., Chicago.)

diseases are more effectively diagnosed by other procedures. Where the EEG has remained most useful is in the diagnosis of epilepsy and sleep disorders. Epilepsy, one of the most common neurological disorders, involves the synchronous discharge of large aggregates of neurons. Since cortical neurons are frequently involved, the EEG is an obvious and effective instrument in diagnosing epileptiform activity (Fig. 19-11). With regard to sleep disorders the EEG is particularly effective, since the stages of sleep are defined on the basis of EEG activity.

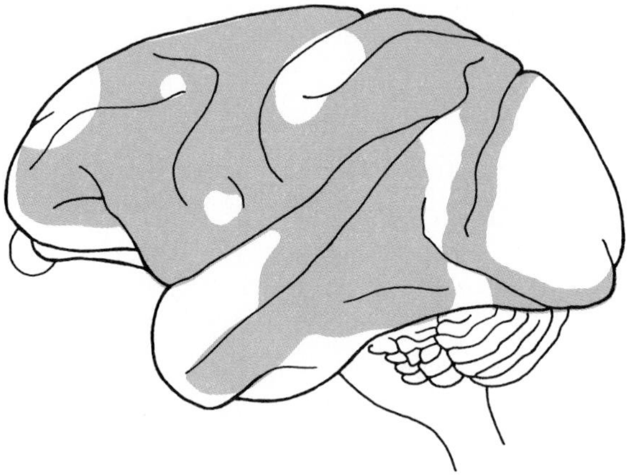

■ **Fig. 19-12.** Distribution of fibers of corpus callosum in the cerebral hemisphere of a macaque monkey. The zones of distribution are shown in color. (Redrawn from Ettlinger, E.G., de Reuck, A.V.S., and Porter, R., editors: Functions of the corpus callosum, CIBA Foundation Study Group no. 20, London, 1965, J. & A. Churchill, Ltd.)

■ *Corpus Callosum*

The great size and cortical distribution of the corpus callosum suggested functional importance (Fig. 19-12) and its role in interhemispheric integration was elucidated in the early 1950s.

■ *Interhemispheric Transfer*

Animal experiments. Fig. 19-13 outlines the basic experiment that uncovered the function of the corpus callosum. An experimental animal (e.g., a cat or monkey) can be forced to learn a visual discrimination through one eye, while the other eye, which is occluded, never views the test stimuli. For instance, the animal may learn that it can obtain food by pushing a door with a cross on it but pushing a door with a circle leads to no reward. Once the animal has learned this task through the experienced eye and then is tested monocularly through the naive eye, *it* continues to perform as if the naive eye had participated in the training (Fig. 19-13, *A*). This is called *interocular transfer* (an unfortunate term, because the transfer of information occurs between the hemispheres and not between the eyes). Because of the partial decussation of the optic chiasm, which results in both retinas projecting to each hemisphere, such transfer is not surprising. However, if after midsagittal transection of the optic chiasm, a cat or monkey is put through a training regimen similar to that just described, interocular transfer remains excellent even though each eye now projects only to the ipsilateral hemisphere (Fig. 19-13, *B*). Finally, if an animal has both the optic chiasm and corpus callosum transected, no interocular transfer is seen, and, the naive eye must relearn the discrimination from the very beginning (Fig. 19-13, *C*). This simple experiment elegantly illustrates the function of the corpus callosum in interhemispheric transfer of information.

Such transfer is by no means limited to the visual domain. For instance, if a normal cat is taught to discriminate between two somatosensory stimuli applied to one limb, it immediately demonstrates familiarity with the discrimination when tested on a contralateral limb. If the corpus callosum is transected, no such transfer is evident, and the cat must relearn the discrimination for the contralateral limb. Because each limb is represented predominantly in the contralateral hemisphere, information available to contralateral limbs can be compared only in the presence of a corpus callosum.

Two details can be added to this simplified view of corpus callosum function, based on studies of experimental animals. First, information that can be processed subcortically can be transferred between the hemispheres subcortically without a corpus callosum. For instance, a cat with transections of the optic chiasm and corpus callosum exhibits no interocular transfer of visual pattern discriminations. However, the animal does demonstrate excellent transfer of discriminations based on crude brightness differences. Visual patterns are processed in cortex, but brightness can be appreciated without cortex. If, however, midbrain commissures (the *habenular, inferior collicular, superior collicular,* and *posterior commissures*) are transected along with the optic chiasm and corpus callosum, interocular transfer of brightness discrimination is blocked. Second, under some conditions the *anterior commissure* seems capable of subserving limited interhemispheric transfer in the absence of a corpus callosum.

Finally, one can ask whether the corpus callosum immediately involves both hemispheres in newly acquired memories, or whether one hemisphere forms the memory and the other can reach it via the commissure. The answer to this seems to depend on the species involved. Cats seem to form memories bilaterally and monkeys do not. For instance, suppose a cat or monkey has only its optic chiasm sectioned and learns a visual task monocularly. Following this, and before the task is presented monocularly to the other eye, the corpus callosum is sectioned. The cat exhibits familiarity with the task when it uses the naive eye, but the monkey does not. Before the callosal section, the cat had already formed the memory bilaterally, but the monkey had not (Fig. 19-14).

Human studies. Studies in humans show that the same general principles that apply to animals apply as well to the role of the human corpus callosum in interhemispheric transfer. Extensive testing has been carried out in human patients with therapeutic transection of this commissure.*

Because the optic chiasm was intact in these patients, special procedures were needed to ensure that visual information was provided to only one hemisphere (Fig.

*In the 1950s several patients were discovered to have an unusual form of epilepsy. They had an epileptiform focus in one hemisphere that initially produced mild seizures, but when the epileptiform activity was transferred via the corpus callosum to the other hemisphere, serious grand mal seizures resulted. To prevent this, each patient had his or her corpus callosum transected, and epileptic symptoms were greatly reduced. Such a procedure is still occasionally performed to treat certain forms of epilepsy.

A

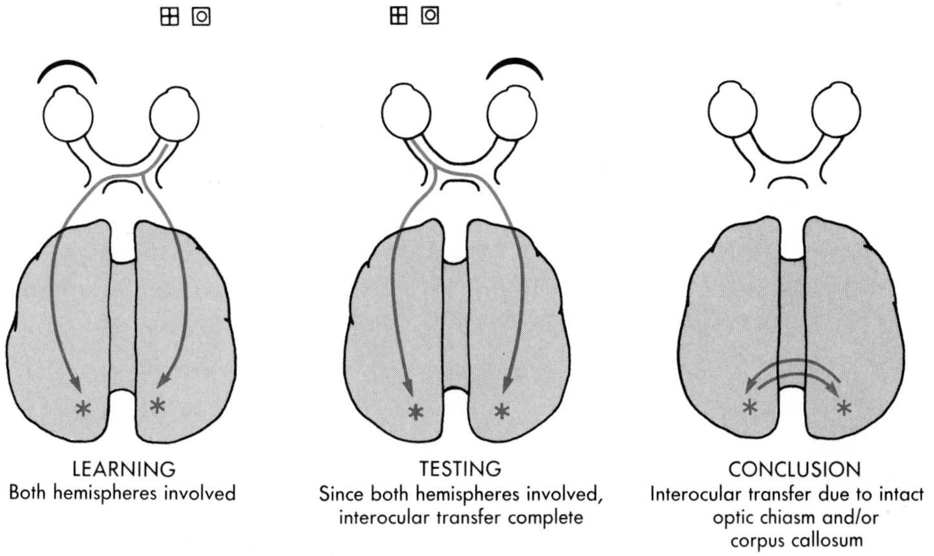

Fig. 19-13. Experiments demonstrating role of corpus callosum in interhemispheric transfer of visual information. In all cases a visual discrimination (e.g., whether a cross of circle indicates the correct door to obtain a food reward) is learned through only one eye. After learning is achieved, the animal is retested through only the naive eye. **A,** Intact animal. Because of the intact optic chiasm, the original learning involves both hemispheres. During testing the naive eye can access both hemispheres that "know" the task, and thus the animal performs as well as with the former eye. Interocular transfer is complete because of the original bilateral learning and/or interhemispheric transfer through the corpus callosum. *Continued.*

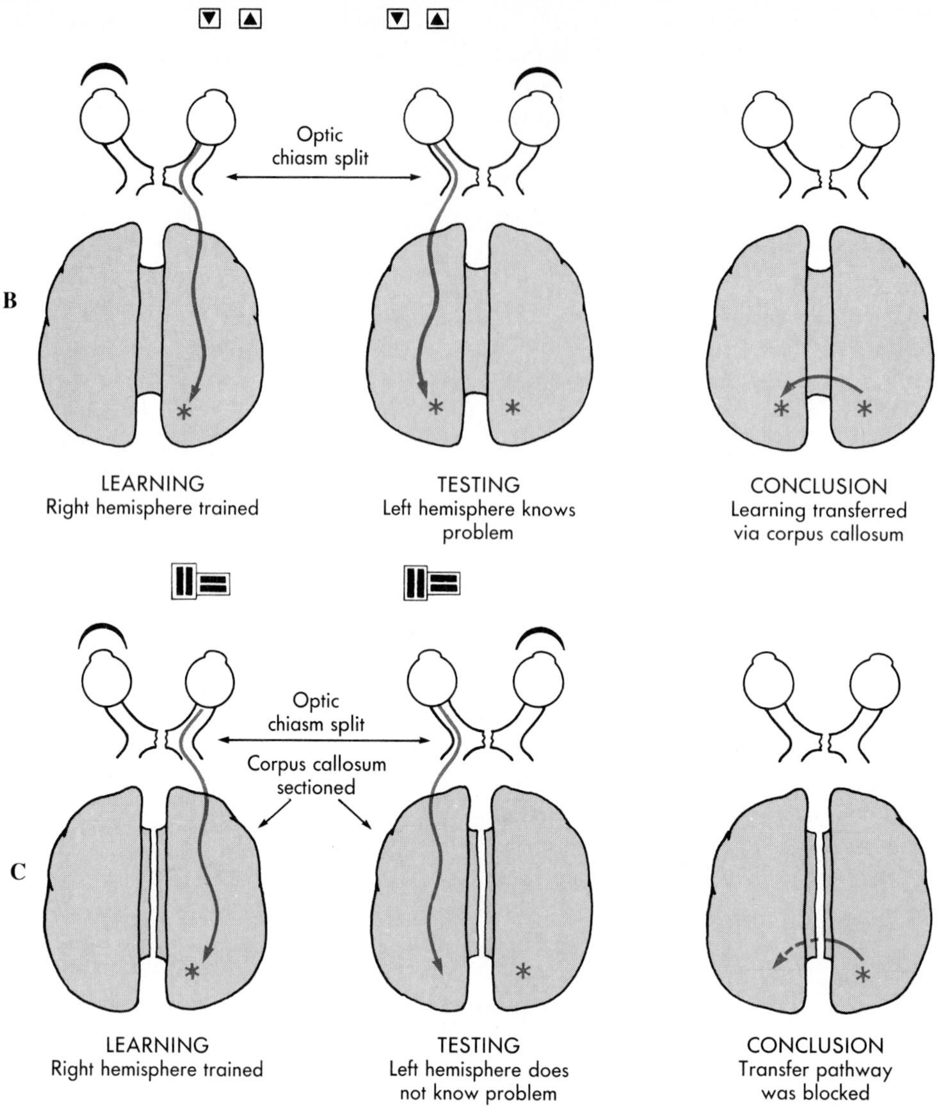

Fig. 19-13, cont'd. B, Animal with transection of optic chiasm. Now the original learning is limited to the hemisphere ipsilateral to the open eye. Nonetheless, testing with the naive eye shows complete interocular transfer as if that eye knew the task all along. This occurs because information is transferred through the corpus callosum from the originally involved hemisphere to that related to the naive eye. **C,** Animal with transection of optic chiasm and corpus callosum. Now interhemispheric transfer is blocked, so no interocular transfer occurs. The naive eye must relearn the task as if it were completely new to the animal.

19-15). The subject faced a projection screen and fixated on a small cross in the center. A photograph of a common object was projected to one side of this fixation point. It thus fell within one hemifield and was represented only in the contralateral hemisphere. The subject was then instructed to reach under the screen with one hand to palpate a number of unseen objects and pick out the object projected on the screen. If the hand was ipsilateral to the hemifield exposed, the correct object was consistently chosen, but with the other hand, objects were chosen at random. (A normal subject successfully locates the correct object under these conditions equally well with either hand.)

The explanation for this result in the callosum-sectioned patients is straightforward. If the visual and somesthetic information was contained within the same hemisphere (as occurred if the left hand explored for objects after its picture was projected to the left hemifield), the information was successfully integrated. If the information was contained within opposite hemi-

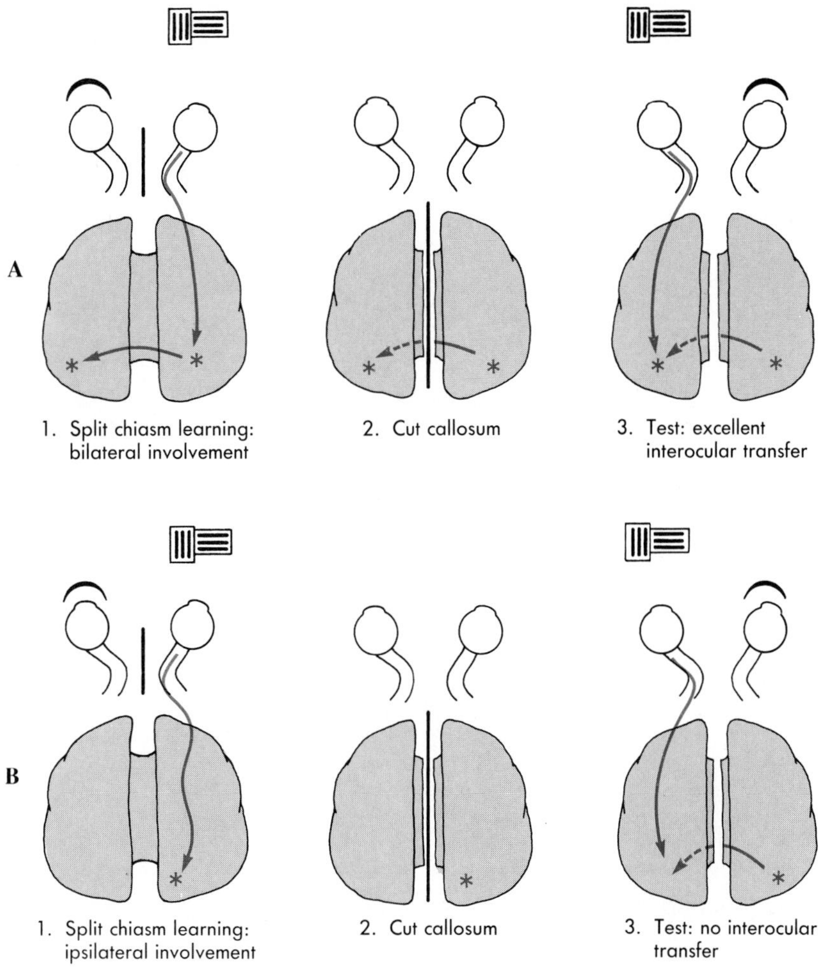

■ **Fig. 19-14.** Experiments demonstrating different role of corpus callosum during learning tasks in cats and monkeys with optic chiasm transections. Because of the transections, the visual information is primarily limited to one hemisphere during monocular learning. **A,** Cat. With an intact corpus callosum the visual information is immediately transferred to the other hemisphere. If the corpus callosum is transected between learning and testing of the naive eye, the naive eye thus shows evidence of complete interocular transfer. **B,** Monkey. The original learning is limited to the ipsilateral hemisphere. Thus no interocular transfer is evident when the corpus callosum is transected between learning and testing of the naive eye.

spheres (as occurred if the right hand was used after projection of the picture to the left hemifield), it could not be integrated without a corpus callosum. Finally, if a separate picture was projected to each hemifield and one hand was used, an object appropriate for that hand was chosen; if two hands were used simultaneously, an appropriate and different object was chosen by each hand.

Because language abilities are typically limited to one dominant hemisphere, predictable language deficits are seen in patients with corpus callosum transection. If a patient with a dominant left hemisphere is commanded verbally to raise his right hand, this is readily accomplished because the motor control for that hand is located primarily in the dominant and thus communicative hemisphere. The patient cannot, however, raise his left hand on such a simple language command. Likewise, the patient can verbally describe details of somesthetic stimuli applied to the right side of his body but not to the left side.

Visual testing dramatically underscores the language asymmetries in the hemispheres of these patients. When the testing condition of Fig. 19-15 is used, the patient can correctly name any object seen in his right hemifield and palpated by his right hand. When the left hemifield is stimulated, although the patient can correctly choose the object with his left hand, he verbally denies having seen or palpated anything. Most interestingly, if a different object is projected to each hemifield, such as a comb to the left and a ball to the right, and the patient is allowed to use only his left hand to locate an object, that hand unerringly locates the comb. When asked

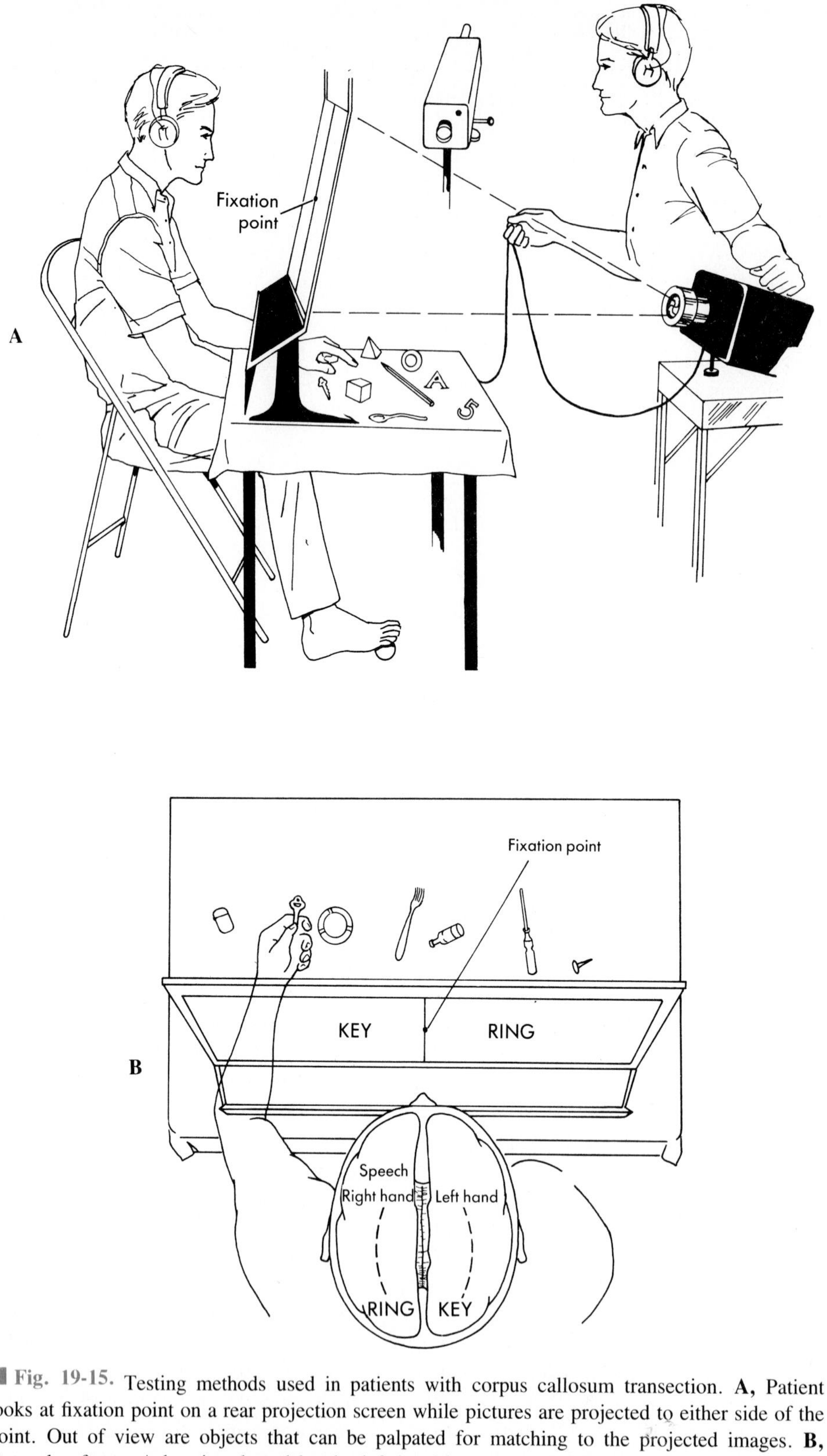

■ **Fig. 19-15.** Testing methods used in patients with corpus callosum transection. **A,** Patient looks at fixation point on a rear projection screen while pictures are projected to either side of the point. Out of view are objects that can be palpated for matching to the projected images. **B,** Example of test. A key is selected by the left hand because the picture of a key is represented only in the appropriate (right) hemisphere; the left hand, without an intact corpus callosum, is ignorant of the picture of a ring. Verbally, the patient would deny seeing the key, because the right hemisphere is verbally uncommunicative. Instead, despite reaching with the left hand for a key, the patient would report seeing only a ring. (Redrawn from Sperry, R.W. In Schmitt, F.O., and Worden, F.G., editors: The neurosciences third study program, Cambridge, Mass., 1974, M.I.T. Press. Copyright © 1974 by the M.I.T. Press.)

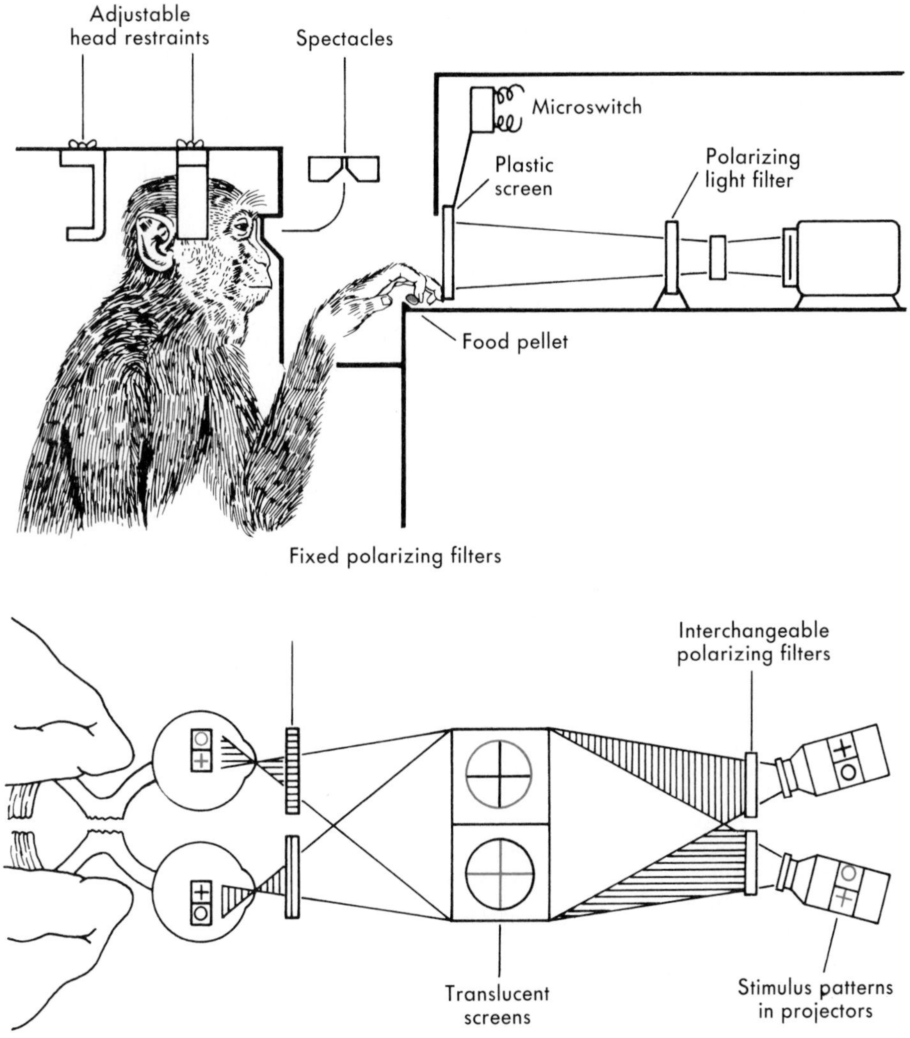

■ **Fig. 19-16.** Experimental set up to demonstrate role of corpus callosum in interhemispheric coordination of behavior. *Above,* Monkey trained to push one of two doors to obtain food reward. The correct door is signaled by visual pattern projected onto it. As the food reward is randomly switched from door to door, the same visual pattern switches with it. *Below,* Optical arrangement to train simultaneously each eye on the opposite task. By the appropriate use of polarizing filters, one eye sees a cross on one door and a circle on the other while the second eye sees the reverse relationship. Thus one eye learns to associate the circle with the correct choice while the other eye learns to attribute the correct choice to the cross. An intact monkey or a monkey with a transected optic chiasm but intact corpus callosum cannot deal with this paradoxical information, shows signs of frustration, and soon abandons the task. In contrast, a monkey with transections of both structures blissfully performs the task, presumably because no paradox is evident to it. (Redrawn from Sperry, R.W.: Sci. Am. **210:**42, 1964.)

what he saw, the patient consistently claims to have seen only a ball and denies having viewed a comb.

■ *Integration of Different Hemispheric Functions*

Coordination of the differentially specialized hemispheres. Because language abilities are typically limited to one hemisphere, what special function is performed by areas of the other hemisphere located in the mirror image of language areas? The callosum-sectioned patients offered a unique opportunity to explore this problem. When asked to solve complicated spatial tasks, such as three-dimensional puzzles, these subjects performed much better with the left hand than with the right. This suggests that the right, nondominant hemisphere specializes for spatial tasks in much the same sense that the other hemisphere specializes for verbal ones. When the corpus callosum is intact, of course, the hands exhibit no such consistent difference in spatial abilities. In addition to its role in spatial abilities, the right nondominant hemisphere deals with nonverbal communication, such as facial expressions, body language, and intonations in speech. Perhaps neural substrates of many higher functions in addition to those mentioned here exhibit interhemispheric asymmetries in humans.

A consequence of interhemispheric asymmetries is that the same function need not require two cortical areas. An intact corpus callosum, then, obviates the need for unnecessary redundancy. For instance, the experiment outlined in Fig. 19-14 shows that cats engage bilateral cortical areas during a visual learning task that involves only one hemisphere in monkeys. Monkeys do not suffer from the consequences of this lateralization as long as the corpus callosum is intact. In a sense the corpus callosum is a prerequisite for the release of cortical areas from bilateral redundancy. This release might be the first step in an evolutionary process that led to the development of higher functions such as language.

Uniting of two hemispheres into a single "mind." Not only is the corpus callosum a vital communication link between the hemispheres, but it also coordinates the hemispheres in the promulgation of behavior. That is, although each of us has two hemispheres capable of neural processing, we normally act as if we have but one mind instead of two. Several studies in animals and humans illustrate this point.

Fig. 19-16 illustrates one outline of an experiment carried out in monkeys. One group of monkeys was normal and the other underwent a transection of the optic chiasm and corpus callosum. The monkeys were trained to discriminate between simple targets (a circle and a cross) to select the appropriate panel for an appetitive reward. By a clever optical design involving polarizing filters, one eye was exposed to stimuli that indicated the cross was the correct stimulus, and the other eye simultaneously learned that the circle was correct. The intact monkeys, for which this confusing information reached both hemispheres, were very frustrated and soon ceased working on the problem. The monkeys with optic chiasm and corpus callosum transections were not confused or frustrated. They quickly learned the task and contentedly selected the correct panel. However, one hemisphere learned to associate "cross" with this panel, whereas the other simultaneously learned to associate "circle" with it. The two hemispheres, without a corpus callosum, are capable of learning diametrically opposite tasks. An intact corpus callosum prevents this.

Other examples come from observations of patients with therapeutic transection of the corpus callosum. These patients frequently display evidence of poor coordination between the hemispheres in everyday tasks involving the hands. For instance, while one of their hands buttons a shirt during dressing, the other hand not only fails to cooperate, it even attempts to unbutton the shirt at the same time. Such dramatic examples of lack of cooperation occurred randomly and intermittently, but they underscore the function of the corpus callosum in ensuring interhemispheric coordination.

The corpus callosum was transected in a young man who happened to have bilateral representation of certain language skills, such as limited reading and writing abilities. Each hemisphere could thus communicate separately. When each was asked to provide a simple list, such as a favorite food, movie star, color, and so forth, different answers often appeared from each hemisphere, as if two different minds were set free by destruction of the corpus callosum. Also, when instructed via the right hemisphere to perform an action, the left (conversant) hemisphere, which was ignorant of the motivation for what the body was now doing, would often fabricate elaborate explanations for the act. For instance, the patient can be shown two pictures, one lateralized to each hemisphere, and must select a match for each from a separate group of pictures set before him. With his right hand, he selects a picture of a chicken to match one of a chicken claw flashed to his left hemisphere; with his left hand, he picks a picture of a snow shovel to match that of a snow scene flashed to his right hemisphere. When his left hemisphere was asked to explain the choices, the patient claimed that the chicken fit with a chicken claw and that the shovel was needed to clean out the chicken shed. Obviously the left hemisphere had no knowledge of why the picture of the shovel was selected, since it was done by the incommunicado left hemisphere—yet it did not hesitate to fabricate an elaborate, plausible explanation for the behavior. Gazzaniga* suggests the following from these findings:

> . . . the human is composed of a variety of mental systems, most of them non-verbal in nature and most only able to communicate by actually carrying out an overt behavior. Once the behavior is carried out, the speech and language system must give the activity an interpretation.

The function of the corpus callosum as derived from the study of callosum-sectioned patients has been summarized as follows†:

> Everything we have seen so far indicates that the surgery has left these people with two separate minds, that is, two separate spheres of consciousness. What is experienced in the right hemisphere seems to be entirely outside the realm of awareness of the left hemisphere. This mental division has been demonstrated in regard to perception, cognition, volition, learning, and memory. One of the hemispheres, the left, dominant or major hemisphere, has speech and is normally talkative and conversant. The other, the minor hemisphere, however, is mute and dumb, being able to express itself only through nonverbal reactions.

■ *Allocortex*

Allocortex represents about 10% of human cortical volume. It can be distinguished from neocortex largely on the basis of its layering: typically no more than three layers can be distinguished in allocortex at any stage of ontogeny as opposed to the six-layered neocortex. Because the significance and functional organization of the allocortex are not well understood, our description of these areas is brief. As noted, allocortex includes both archicortex and paleocortex.

*From Gazzaniga, M.S.: Cornell University Med. Coll. Alumni Q. **45**:2, 1982.

†From Sperry, R.W.: Brain bisection and mechanisms of consciousness. In Eccles, J.C., editor: Brain and conscious experience, New York, 1966, Springer-Verlag, p. 299.

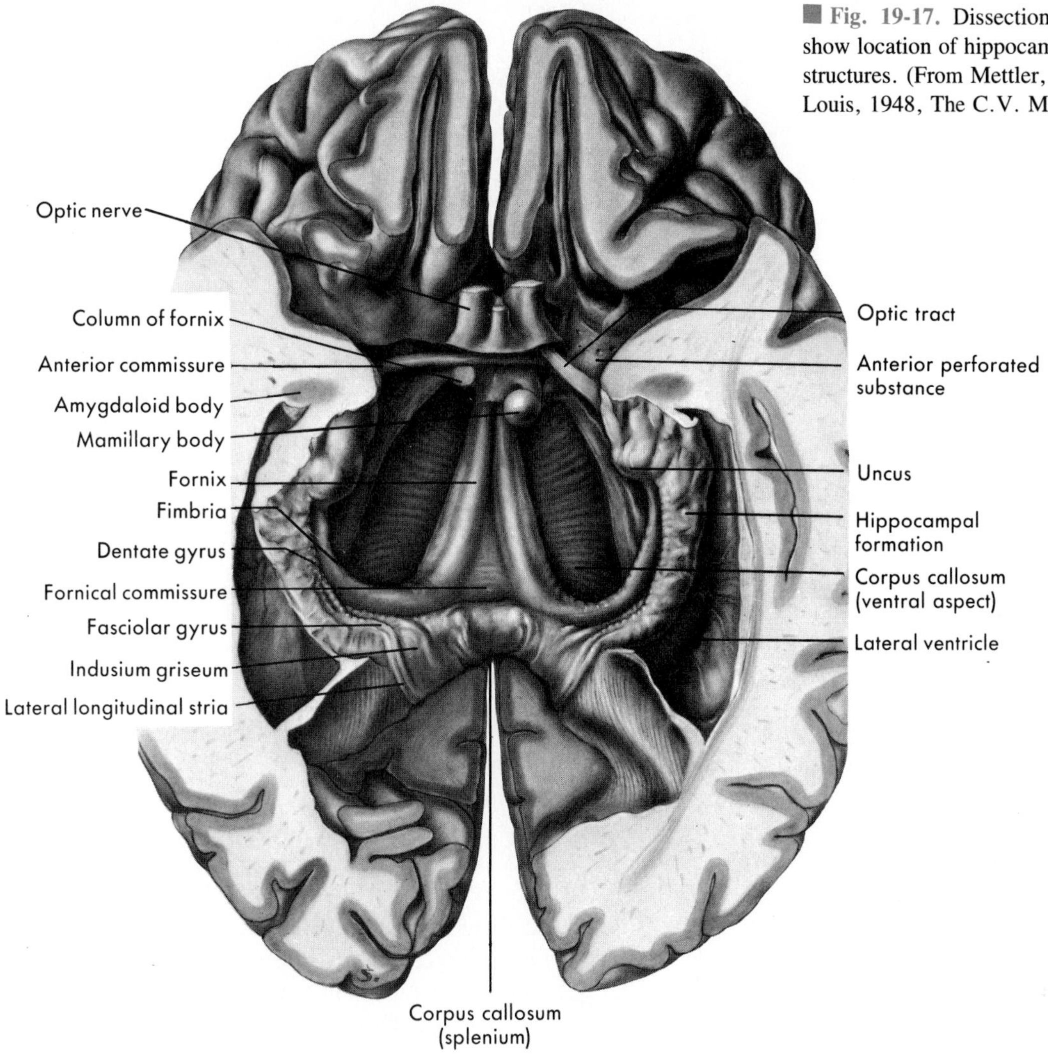

Fig. 19-17. Dissection of base of the brain to show location of hippocampal formation and related structures. (From Mettler, F.A.: Neuroanatomy, St. Louis, 1948, The C.V. Mosby Co.)

Archicortex: Hippocampal Formation

The archicortex, which is equivalent to a group of structures known as the *hippocampal formation,* is located in the medial and ventral portion of each hemisphere (Fig. 19-17). The hippocampal formation consists of the *dentate gyrus, hippocampus,* and the *subiculum* (sometimes divided into the *prosubiculum,* the *subiculum,* the *presubiculum,* and the *parasubiculum*) (Figs. 19-18 and 19-19).

Subiculum. The subiculum lies in the *parahippocampal gyrus* adjacent to the *hippocampal sulcus* and merges into the neocortex of the parahippocampal gyrus near the *collateral sulcus.* The *entorhinal area* (Brodmann's area 28), which is located adjacent to the subiculum anteriorly, may be a part of paleocortex. Also, because the layering patterns in the subiculum are more variable than those of the remaining hippocampal formation, the subiculum may be transitional between allocortex and neocortex.

Hippocampus. The hippocampus derives its name from its resemblance, in coronal sections, to a sea horse. It also resembles a ram's horn and thus is often called *cornu ammonis* or *Ammon's horn* (Ammon was an ancient Egyptian god with a ram's head). Thus separate regions of the hippocampus are designated CA1 through CA4 (see Fig. 19-8, *A*).

The hippocampus bulges into the inferior horn of the lateral ventricle. The *alveus* is a subependymal sheath of afferent and efferent fibers, and the latter collect at the *fimbria* and enter the *fornix.* The three main layers of the hippocampus in a direction from the brain surface to the alveus are the *molecular layer,* which is a neuropil similar to layer I of neocortex; the *pyramidal cell layer,* which is a collection of pyramidal cells oriented with their bases toward the alveus; and the *polymorphic cell layer,* which resembles layer VI of neocortex.

Dentate gyrus. The dentate gyrus is trilayered like the hippocampus, except that instead of the pyramidal cell layer there is the *granule cell layer* of small cells.

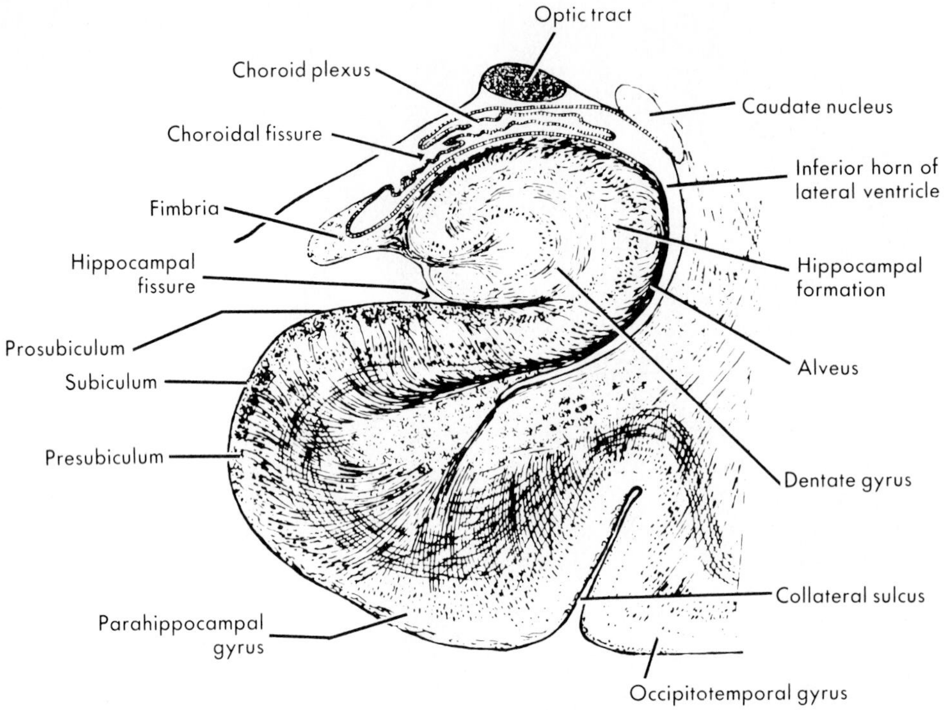

■ **Fig. 19-18.** Cross section through the human hippocampus. (From Carpenter, M.B.: Human neuroanatomy, ed. 7, Baltimore; © 1976, The Williams & Wilkins Co.)

The dentate gyrus is C shaped, with the open side directed toward the fimbria, whereas the other side forms the dorsal surface of the hippocampal sulcus. The *molecular layer* is located nearest this sulcus; the *polymorphic cell layer* is located nearest the fimbria.

Hippocampal pathway. Fig. 19-19, *B*, summarizes many of the connections of the hippocampal formation. Inputs derive largely from the parahippocampal gyrus beyond the subiculum, and these derive largely from entorhinal cortex. These inputs enter the hippocampus via either the *perforant path* or the *alvear path*. Other afferent pathways reach all parts of the hippocampal formation via the fornix and arise in the *cingulate gyrus,* the *septal nucleus,* the *induseum griseum,* and the contralateral *hippocampal formation* (via a commissure in the fornix). Because the fornix contains predominantly efferent fibers from the hippocampus, relatively few afferent fibers reach the hippocampus via this route. Most functional input arrives indirectly via the perforant and alvear paths. In addition, a major intrinsic pathway arises from granule cells of the dentate gyrus, and it terminates on pyramidal cells of the hippocampus.

Nearly all efferent fibers from the hippocampal formation are axons of pyramidal cells in the hippocampus, although some efferent fibers arise from polymorphic cells and from the subiculum. These fibers course along the alveus, collect at the fimbria, and enter the *fornix* (Fig. 19-20) at the posterior tip of the hippocampus formation adjacent to the *splenium of the corpus callosum*. These fibers form the *crus* of the fornix on each side. They pass over the thalamus, where the two crura merge to form the body of the fornix. Some fibers enter the contralateral crus via a commissure. The body then divides into two ventrally directed columns of fibers that pass through the hypothalamus. Some fibers exit the columns in the *precommissural portion* (relative to the anterior commissure), and the remaining columns of fibers constitute the *postcommissural portion*. Axons of this latter portion innervate the *mammillary body,* the *anterior thalamic nucleus,* and the *periaqueductal gray* of the midbrain. Precommissural fibers innervate parts of the *hypothalamus* and the *septal nucleus,* as well as the mammillary body and anterior thalamic nucleus.

Functional considerations. Although the hippocampal formation has received considerable experimental attention, its functional role is not clear. Given its widespread, diffuse connections, it seems likely that no single function can be attributed to it, but that it rather participates in many complex behaviors. Among the roles suggested for the hippocampal formation are memory, many emotional responses, and spatial mapping of the environment. However, it should be noted that birds have only a rudimentary hippocampus, yet

A

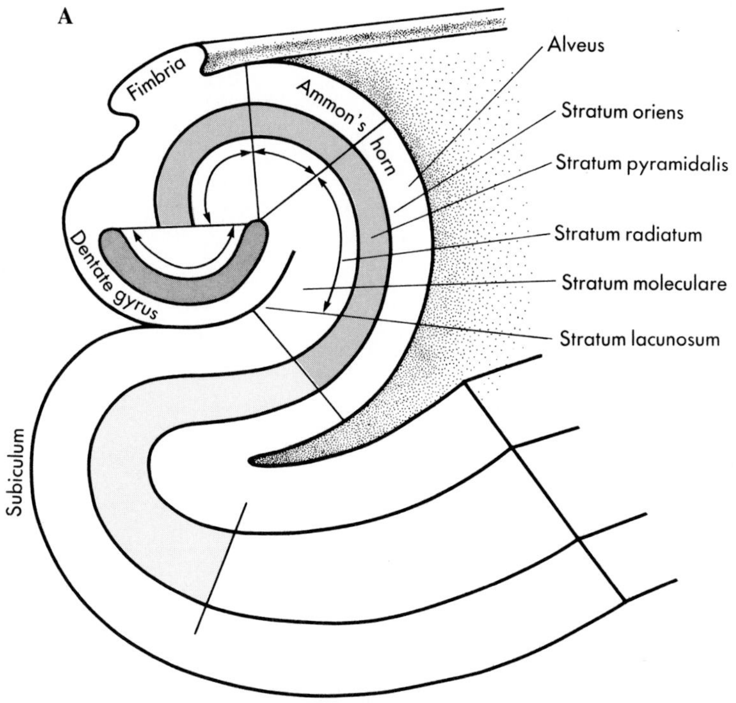

B

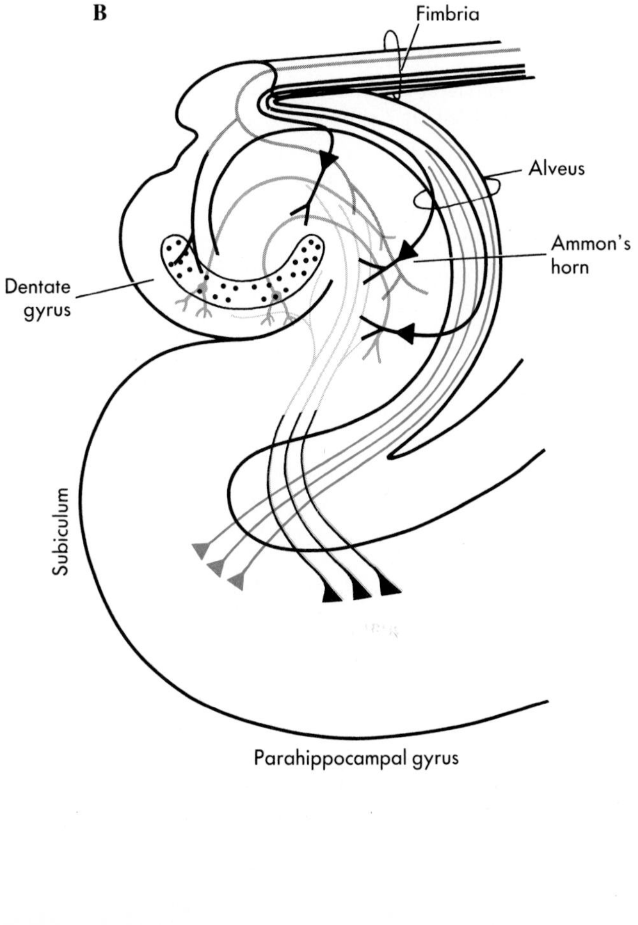

■ **Fig. 19-19. A,** Major regions of hippocampus. Cell bodies of the dentate gyrus are in gray. The prominent pyramidal cell bodies of Ammon's horn are in darker color, and this layer merges gradually with neurons of the subiculum *(lighter color)*. Approximate locations of other layers are indicated. **B,** Some connections of the hippocampus. Cells of the dentate gyrus *(medium color)* form the mossy fibers of the hippocampus and innervate the pyramidal cells *(black)*. Pyramidal cell axons course along the alveus and enter the fimbria. Afferent fibers *(gray)* also enter from the fimbria and innervate the dentate gyrus and/or Ammon's horn. Inputs from the parahippocampal gyrus arrive via the alvear path *(dark color)* or perforant path *(light color)*. (Redrawn from Williams, P.L., and Warwick, R.: Functional neuroanatomy of man, Philadelphia, 1975, W.B. Saunders Co.)

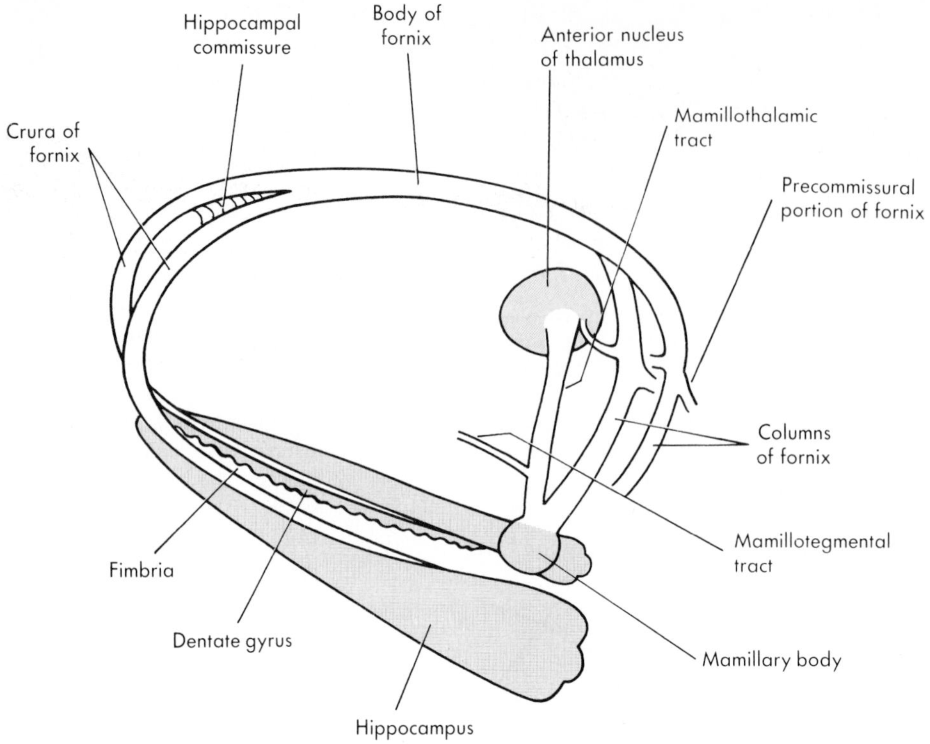

■ **Fig. 19-20.** The fornix and related pathways. (Redrawn from Barr, M.: The human nervous system: an anatomic viewpoint, ed. 3, New York, 1979, Harper & Row, Publishers.)

none of the functions suggested for this structure are absent from the behavioral repertoire of birds. The hippocampal formation is certainly not essential to olfaction, as was once thought.

■ *Paleocortex: Rhinencephalon*

Which structures form mammalian paleocortex are not known precisely. One common classification considers the paleocortex and rhinencephalon to be equivalent. In other words, those structures identified with olfaction, including the *olfactory bulb, tract, tubercle,* and *striae,* the *anterior olfactory nucleus,* parts of the *amygdala,* and parts of the *pyriform cortex,* form the paleocortex (Chapter 12).

■ *Bibliography*

Journal articles

Amaral, D.G., and Cowan, W.M.: Subcortical afferents to the hippocampal formation in the monkey, J. Comp. Neurol. **189:**573, 1980.

Amaral, D.G., et al.: The commissural connections of the monkey hippocampal formation, J. Comp. Neurol. **224:**307, 1984.

Azmitia, E.C.: Bilateral serotonergic projections to the dorsal hippocampus of the rat: simultaneous localization of ^{3}H-5HT and HRP after retrograde transport, J. Comp. Neurol. **203:**737, 1981.

Benowitz, L.I., et al.: Hemispheric specialization in nonverbal communication, Cortex **19:**5, 1983.

Damasio, A.R., and Geschwind, N.: The neural basis of language, Ann. Rev. Neurosci. **7:**127, 1984.

Deacon, T.W., et al.: Afferent connections of the perirhinal cortex in the rat, J. Comp. Neurol. **220:**168, 1983.

Gaarskjaeb, F.B.: The hippocampal mossy fiber system of the rate studied with retrograde tracing techniques: correlation between topographic organization and neurogenetic gradients, J. Comp. Neurol. **203:**717, 1981.

Gazzaniga, M.S.: Split-brain research: a personal history, Cornell University Med. Coll. Alumni Q. **45:**2, 1982.

Gilbert, C.D.: Horizontal integration in the neocortex, Trends Neurosci. **8:**160, 1985.

Hubel, D.H., and Wiesel, T.N.: Functional architecture of macaque monkey visual cortex (Ferrier lecture), Proc. R. Soc. Lond. Biol. **198:**1, 1977.

Loy, R., et al.: Noradrenergic innervation of the adult rat hippocampal formation, J. Comp. Neurol. **189:**699, 1980.

Milner, B.: Some cognitive effects of frontal-lobe lesions in man, Philos. Trans. R. Soc. Lond. [Biol.] **298:**211, 1982.

Moruzzi, G.: The sleep-waking cycle, Rev. Physiol. Biochem. Exp. Pharmacol. **64:**1, 1972.

Moruzzi, G., and Magoun, H.W.: Brain stem reticular formation and activation of the EEG, Electroenceph. Clin. Neurophysiol. **1**:455, 1949.

Phillips, C.G., et al.: Localization of function in the cerebral cortex: past, present and future, Brain **107**:327, 1984.

Prince, D.A.: Neurophysiology of epilepsy, Annu. Rev. Neurosci. **1**:395, 1978.

Reep, R.: Relationship between prefrontal and limbic cortex: a comparative anatomical review, Brain Behav. Evol. **25**:5, 1984.

Rockel, A.J., et al.: The basic uniformity in structure of the neocortex, Brain **103**:221, 1980.

Rockland, K.S., and Pandya, D.N.: Laminar origins and terminations of cortical connections of the occipital lobe in the rhesus monkey, Brain Res. **179**:3, 1979.

Rosenkilde, C.E.: Functional heterogeneity of the prefrontal cortex in the monkey: a review, Behav. Neurol. Biol. **25**:301, 1979.

Sidtis, J.J., et al.: Variability in right hemisphere language function after callosal section: evidence for a continuum of generative capacity, J. Neurosci. **1**:323, 1981.

Sperry, R.W.: Mental unity following surgical disconnection of the cerebral hemispheres, Harvey Lecture **62**:293, 1964.

Sperry, R.W.: Some effects of disconnecting the cerebral hemispheres, Science **217**:1223, 1982.

Van Essen, D.C., et al.: The pattern of interhemispheric connections and its relationship to extrastriate visual areas in the macaque monkey, J. Neurosci. **2**:265, 1982.

Wyss, J.M., et al.: A study of subcortical afferents to the hippocampal formation in the rat, Neuroscience **4**:463, 1979.

Wyss, J.M., et al.: The organization of the fimbria, dorsal fornix and ventral hippocampal formation in the rat, Anat. Embryol. **158**:303, 1980.

Zola-Morgan, S., et al.: The neuroanatomy of amnesia: amygdala-hippocampus versus temporal stem, Science **218**:1337, 1982.

Books and monographs

Benson, D.F.: Aphasia alexia, and agraphia, New York, 1979, Churchill Livingstone.

Brodal, A.: Neurological anatomy in relation to clinical medicine ed. 3, New York, 1981, Oxford University Press.

Diamond, I.T.: The subdivisions of neocortex: a proposal to revise the traditional view of sensory, motor, and association areas. In Sprague, J.M., and Epstein, A.N., editors: Progress in psychobiology and physiological psychology, vol. 8, New York, 1979, Academic Press, Inc.

Geschwind, N.: Some special functions of the human brain: dominance, language, apraxia, memory, and attention. In Mountcastle, V.B., editor: Medical physiology, ed. 14, vol. 1, St. Louis, 1979, The C.V. Mosby Co.

Kleitman, N.: Sleep and wakefulness, ed. 2, Chicago, 1963, University of Chicago Press.

Merzenick, M.M., and Kaas, J.H.: Principles of organization of sensory-perceptual systems of mammals. In Sprague, J.M., and Epstein, A.N., editors: Progress in psychobiology and physiological psychology, vol. 9, New York, 1980, Academic Press, Inc.

O'Keef, J., and Nadel, L.: The hippocampus as a cognitive map, Oxford, 1978, Clarendon Press.

Schmitt, F.O., et al., editors: The organization of the cerebral cortex, Cambridge, 1981, M.I.T. Press.

Sperry, R.W.: Lateral specialization in the surgically separated hemispheres. In Schmitt, F.O., and Worden, F.G., editors: The neurosciences third study program, Cambridge, Mass., 1974, M.I.T. Press.

Woolsey, C.N.: Cortical sensory organization, vols. 1 to 3, Clifton, N.J., 1981, Humana Press.

CHAPTER

20

The Autonomic Nervous System and Its Central Control

The autonomic (vegetative, visceral) nervous system innervates the internal organs to modulate and control the internal environment; in this context it contributes to maintaining homeostasis. However, it also participates in adjustments to environmental stimuli. For example, a fall of the ambient temperature requires heat production and the prevention of heat loss; exercise necessitates an increase in cardiac output and shunting of blood to the active muscle beds. Such adjustments are mediated in part through the autonomic nervous system, but they also involve complex central circuitry. Thus an important component of central nervous system activity is directed toward the control or modulation of the autonomic outflow.

The significance of the autonomic nervous system and its central control for overall behavior should not be underestimated. By way of illustration, consider the regulation of body temperature. In poikilothermic (cold-blooded) animals much of the regulation of body temperature is behavioral, such as seeking warmer sites when the ambient temperature falls. This places restrictions on both behavior and acceptable environments. Homeothermic (warm-blooded) animals, however, have a greater freedom of movement in environments of varying temperature. This can be strikingly demonstrated by depriving a homeothermic animal of its temperature-regulating capability and then placing it in a cold room with a source of heat in one corner. With thermoregulatory mechanisms intact, the animal will roam the entire room. However, with these mechanisms compromised it will remain close to the heat source, restricting its ambulatory freedom substantially. Thus autonomic regulation of the internal milieu, particularly in response to environmental challenges, provides considerable adaptive advantage.

Organization of the Autonomic Innervation

Most rigorously defined, the autonomic nervous system consists only of the efferent innervation of the viscera, which has two major subdivisions: the *sympathetic* and *parasympathetic systems* (Fig. 20-1). The final common path of the autonomic innervation is conventionally viewed as involving a two-neuron chain. A *preganglionic neuron* is located in the central nervous system, and this projects onto a *postganglionic neuron* located in the periphery (Fig. 20-1). For the sympathetic system the preganglionic neurons are situated in the thoracic and upper lumbar spinal cord. Thus this division of the autonomic nervous system is often referred to as the *thoracolumbar* division. The parasympathetic preganglionic neurons are found in certain cranial nerve nuclei and in the sacral spinal cord; therefore the alternate designation of the parasympathetic system is the *craniosacral* division. The locations of the postganglionic neurons of the two divisions also differ (Fig. 20-1). In the sympathetic system the postganglionic neurons are generally situated in either the *paravertebral (sympathetic) chains* that parallel much of the vertebral column or in *prevertebral ganglia* such as the celiac ganglion (Fig. 20-1). In contrast, the postganglionic neurons of the parasympathetic system are located in close proximity to their target organ. (At this point it is important to emphasize that in the peripheral autonomic innervation exceptions are the rule.)

Many generalities have emerged regarding the structure and function of the autonomic nervous system and its central control. For example, the sympathetic outflow is traditionally viewed as diffuse, whereas the parasympathetic system is described as being more localized. Also, the two divisions are generally conceived as being antagonistic in function. However, as knowledge of autonomic control has advanced, it has become clear that such generalities are oversimplified. Indeed sympathetic control is more specific than originally envisioned, and the relationship between the two divisions is better characterized as a cooperative integrated action than as adversative. In situations where the sympathetic and parasympathetic innervations of an organ produce apparently opposite responses, it is more productive to view these innervations as acting reciprocally than antagonistically. Yet another generality is that there is no voluntary control of the autonomic innervation, and this, too, has been challenged in recent years.

The Sympathetic System

The preganglionic neurons of the sympathetic, or thoracolumbar, division of the autonomic nervous system are located primarily in the *lateral horn* or *intermediolateral cell column* of the thoracic and first two or three lumbar segments of the spinal cord (Figs. 20-1 and 20-2). However, sympathetic preganglionic neurons are

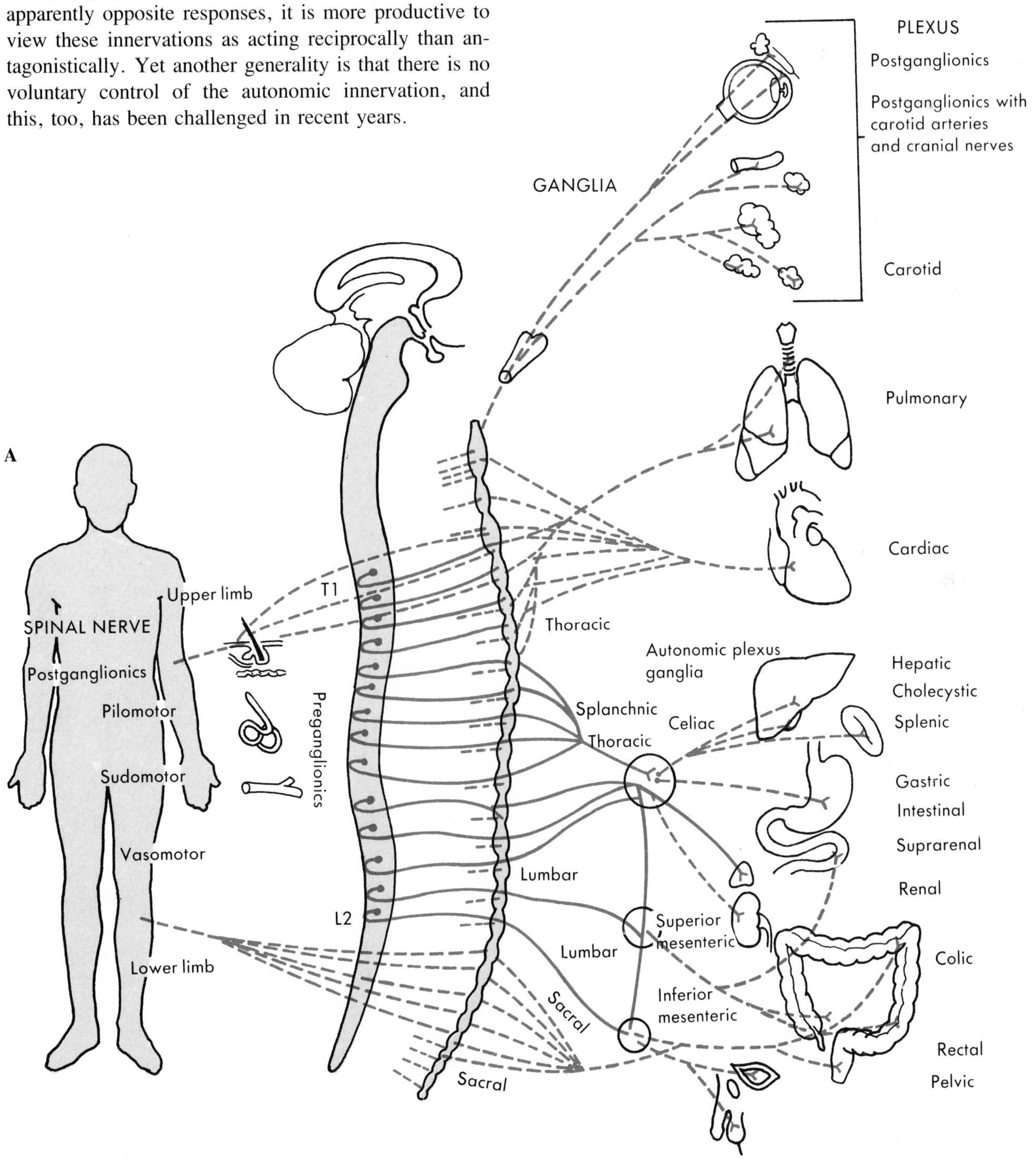

■ **Fig. 20-1. A,** Schematic representation of the sympathetic system.

Continued.

■ **Fig. 20-1, cont'd. B,** Schematic representation of the parasympathetic system. *Solid lines,* preganglionics; *broken lines,* postganglionics. (Redrawn from Bhagat, B.D., Young, P.A., and Biggerstaff, D.E.: Fundamentals of visceral innervation, 1977. Courtesy Charles C Thomas, Publisher, Springfield, Ill.)

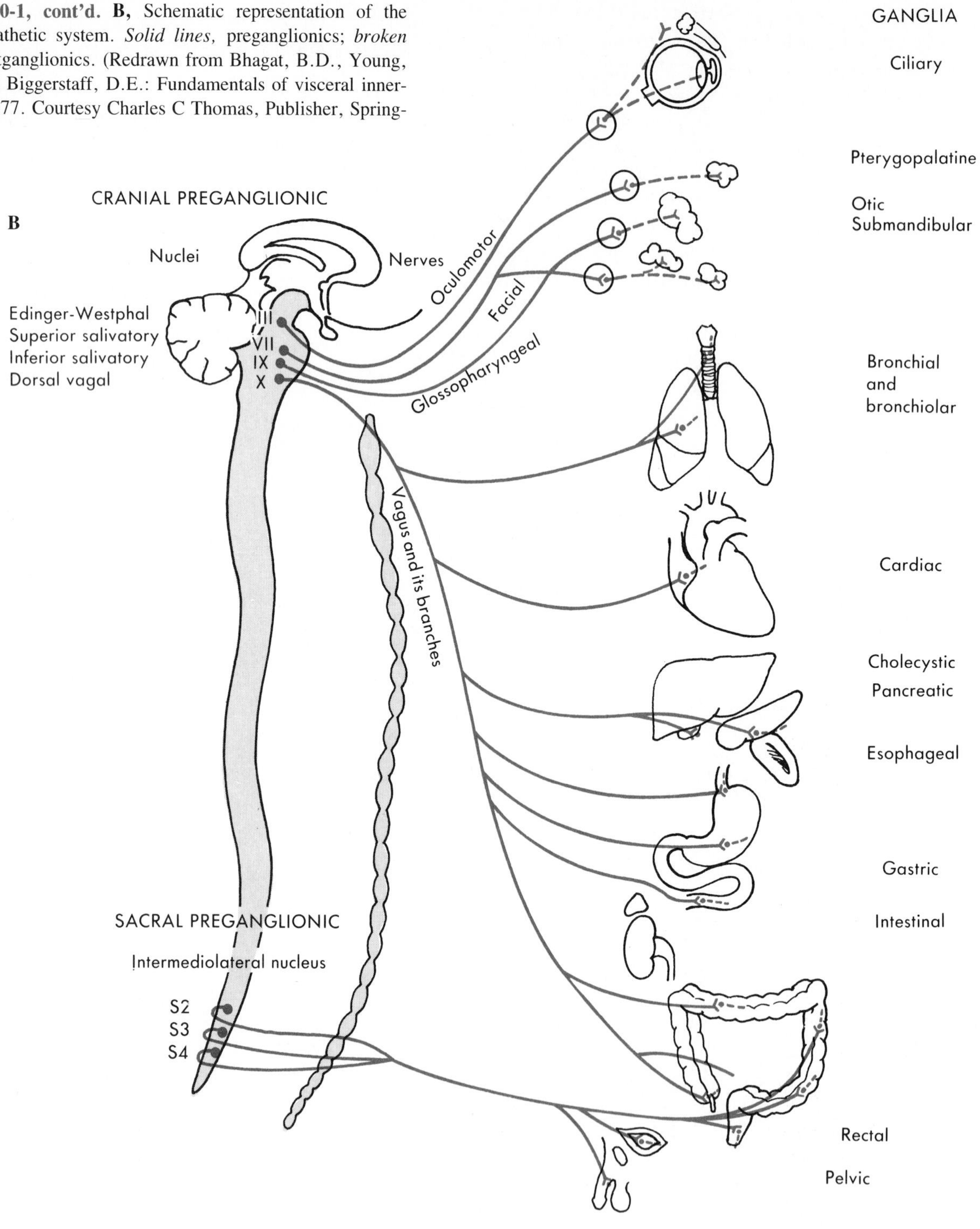

also found in lesser numbers more medially, dorsolateral to the central canal. In comparison with the α-motoneurons, the preganglionic neurons are small and distributed in clusters; their axons are of small diameter and are myelineated (group B). These axons exit the spinal cord through the ventral roots along with the axons of α- and γ-motoneurons (Fig. 20-2). However, a preganglionic axon may course for a number of segments within the paravertebral chain before synapsing on postganglionic neurons. Furthermore, the lumbar preganglionic neurons can distribute their axons bilaterally.

The most common trajectory of a sympathetic preganglionic fiber is to leave the ventral root and course through a *white ramus* (Fig. 20-2). The white ramus, so named because the sympathetic preganglionic axons

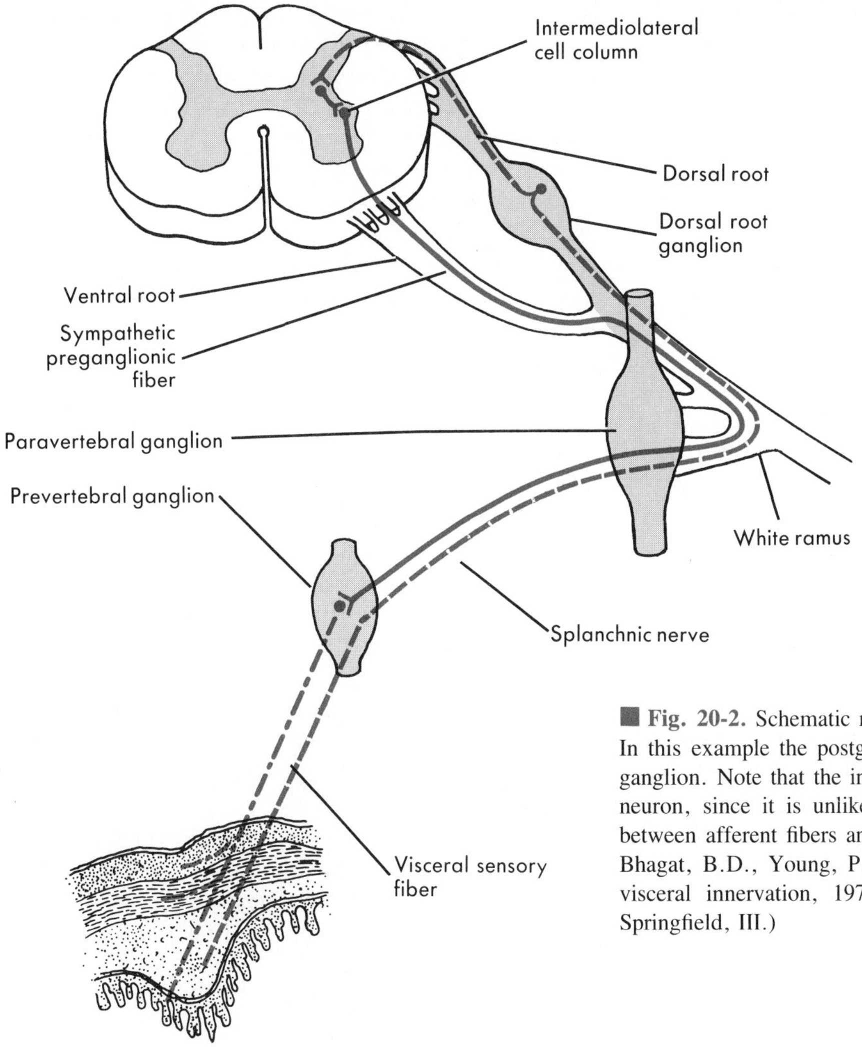

■ **Fig. 20-2.** Schematic representation of a simple visceral reflex arc. In this example the postganglionic neuron is located in a prevertebral ganglion. Note that the intraspinal circuitry includes at least one interneuron, since it is unlikely that any monosynaptic connections exist between afferent fibers and the preganglionic neurons. (Redrawn from Bhagat, B.D., Young, P.A., and Biggerstaff, D.E.: Fundamentals of visceral innervation, 1977, Courtesy Charles C Thomas, Publisher Springfield, Ill.)

are myelinated, constitutes the pathway to the paravertebral or sympathetic ganglia. This chain of ganglia extends from cervical through sacral levels. However, white rami are found only at segments that contain preganglionic cell bodies, i.e., the thoracic and upper lumbar segments. The preganglionic axons ramify extensively in the paravertebral chain and synapse within a number of ganglia. It is via such ramification that the ganglia at cervical, lower lumbar, and sacral levels receive their inputs. At cervical levels there are only three ganglia, each rather large (Fig. 20-1). These are (in descending order) the *superior, middle,* and *inferior* cervical sympathetic ganglia. Often the inferior cervical ganglion merges with the first thoracic ganglion to form the *stellate ganglion,* an important structure in the innervation of the heart. The postganglionic neurons of the sympathetic ganglia give rise to small unmyelinated axons (group C) that exit the ganglia to join visceral or spinal nerves via the *gray ramus,* so named because these axons are unmyelinated.

Although a preganglionic-postganglionic synapse in the paravertebral chain is the most common organization of the sympathetic outflow, there are two other prominent arrangements. Certain preganglionic axons continue peripherally, with or without synapsing in the paravertebral chain, to terminate on postganglionic neurons in *prevertebral ganglia,* such as the celiac ganglion or mesenteric ganglia (Figs. 20-1 and 20-2). The other exception involves the innervation of the adrenal medulla by the splanchnic nerve. In this case the target cells, the chromaffin cells, are innervated directly by preganglionic axons without relay through a postganglionic neuron.

The sympathetic preganglionic neurons all use *acetylcholine* as a neurotransmitter, and preganglionic-postganglionic synapses in the sympathetic system have been studied as models of cholinergic transmission. These are *nicotinic cholinergic synapses* because the drug nicotine acts as an agonist at low doses (it blocks transmission at higher doses). Neuromuscular transmission is also nicotinic, although the cholinergic receptors on muscle and on postganglionic neurons do not behave identically with respect to blocking agents such as curare or α-bungarotoxin. The postganglionic sympathetic neurons most commonly use *norepinephrine* as a neurotransmitter. In certain cases, however, such as the innervation of the sweat glands and of certain blood vessels, the junctions are cholinergic. Of interest in this context is the direct innervation of the adrenal medulla by preganglionic axons. The chromaffin cells release catecholamines (in humans approximately 80% is epinephrine and 20% norepinephrine), and hence they may perhaps be viewed as analogous to postganglionic neurons.

Advances in understanding of synaptic transmission in the autonomic nervous system have been explosive in recent years, particularly with respect to the pharmacology of autonomic synapses (Table 20-1), and a contemporary pharmacology text should be consulted for further information.

The Parasympathetic System

For the cranial portion of the parasympathetic system the preganglionic neurons are located in the motor nuclei of cranial nerves III, VII, IX, and X (Fig. 20-1). The sacral portion arises from preganglionic neurons in sacral segments 2 to 4. Their location in the spinal gray matter is analogous to that of the intermediolateral column of the thoracolumbar system. The cranial preganglionic neurons issue myelinated axons that travel in the respective cranial nerves to terminate on postganglionic neurons within the target organs. The sacral preganglionic fibers also issue myelinated axons that emerge via the ventral roots to travel in the pelvic nerves and ultimately terminate on postganglionic cells within the target organs (Fig. 20-1).

As in the sympathetic system, the parasympathetic preganglionic fibers release *acetylcholine* as their neurotransmitter, and the receptors are *nicotinic*. However, unlike the sympathetic innervation, the parasympathetic postganglionic fibers are also cholinergic, and these are *muscarinic* synapses; that is, the drug muscarine acts as an agonist, and transmission is blocked by atropine.

With but a few exceptions (such as the afferent fibers from the carotid sinus and carotid body) our knowledge of visceral afferent fibers is limited. Much research is required before we have a comprehensive view of the afferent innervation of the viscera. These afferent fibers are extremely important. In addition to mediating visceral sensation, including pain, they participate in basic regulatory reflexes and provide the central nervous system with important information about the state of the internal environment.

One route by which such information can reach the brain is through sensory components of cranial nerves. For example, information from the carotid sinus regarding arterial blood pressure is transmitted over fibers that are located in the glossopharyngeal nerve (IX) and that synapse in the nucleus of the solitary tract. Information from aortic arch receptors is transmitted over vagal afferent fibers that also terminate in the nucleus of the solitary tract. Most visceral afferent fibers, however, have their cell bodies in dorsal root ganglia and enter the spinal cord through dorsal *or* ventral roots. They appear to terminate primarily in laminae I and V of the dorsal horn. In these laminae, as well as in the intermediate spinal gray, there is considerable convergence of visceral and somatic afferent fibers. This is consistent with the hypothesis that referred visceral sensation is in part a function of viscerosomatic convergence at the spinal cord. Visceral afferent fibers from different organs also converge upon single neurons of the dorsal horn. For example, many spinal neurons that respond to nociceptive stimulation of the heart also respond to such stimulation of the gallbladder. This viscerovisceral convergence may explain the difficulty in discriminating sensations from different organs, such as the heart and gallbladder. Some visceral afferent fibers may synapse directly in autonomic ganglia without entering the spinal cord, providing the possibility for peripheral mediation of autonomic reflexes.

A final point is that some visceral receptors, such as the chemoreceptors of the carotid body, receive an efferent innervation from the central nervous system that can modulate their sensitivity; i.e., the receptors are under centrifugal control.

Autonomic Ganglia

The autonomic ganglia are conventionally viewed as simple relays in which preganglionic neurons synapse on postganglionic neurons with varying degrees of convergence and divergence. However, as the peripheral autonomic innervation is studied in greater detail, this generalization has faltered. For example, in the paravertebral chain ganglionic interneurons may exist. More compelling evidence has been gathered with respect to certain more peripheral ganglia, such as the enteric nervous system innervating the gastrointestinal tract. The enteric nervous system includes more than 100 million neurons, about as many as the number of neurons in the spinal cord. The enteric system includes a number of neuronal types that are organized into complex circuits that are capable of substantial local processing. Moreover, in addition to acetylcholine and norepinephrine, serotonin and an array of neuropeptides are probably involved in synaptic transmission in the enteric nervous system.

Functional Considerations Regarding the Autonomic Innervation

Simplified characterizations of the autonomic innervation describe sympathetic function as concerned with mobilizing the organism for ''fight or flight'' and parasympathetic functions as ''vegetative'' and ''emptying.'' However, inspection of Fig. 20-1 and Table 20-1 suggests that this is not realistic. It now seems more appropriate to abandon such generalizations and to attend to more specific effects at target organs.

Distribution and Specificity of the Peripheral Autonomic Innervation

The distribution of the autonomic efferent nerves is summarized in Fig. 20-1 and Table 20-1. Some structures, such as the adrenal medulla, pilomotor muscles, and most blood vessels, receive only sympathetic innervation. Others, such as the sublingual gland, receive only parasympathetic innervation. However, most structures are *dually innervated*. Although in many instances the sympathetic and parasympathetic innervations have opposite effects on such dually innervated organs, this should not dominate our view of their interaction. For example, in the male sex organs the parasympathetic innervation mediates erection and the sympathetic innervation induces ejaculation; at the salivary glands both innervations produce secretion but with different properties. Such examples do not illustrate antagonistic action.

In other dually innervated organs the two subdivisions of the autonomic nervous system do appear to have opposite effects. For example, the sympathetic cardiac innervation increases heart rate and contractility, whereas the parasympathetic innervation decreases both. Sympathetic system activation decreases intestinal motility, whereas parasympathetic system activation increases it. However, such antagonistic action would occur only if both innervations were activated concomitantly. In fact, they often behave synergistically. By way of illustration, in situations that demand increased cardiac output, there is an increase in the sympathetic outflow to the heart to increase both rate and contractility. However, the vagal outflow to the heart does not compete with the sympathetic innervation. Instead vagal activity is inhibited, and this also contributes to the increase in rate and contractility. Consequently when one considers (1) singly innervated organs, (2) dually innervated organs without antagonistic action, and (3) dually innervated organs with antagonistic action but where physiologically the two innervations act synergistically, the concept of antagonistic action loses much of its conceptual utility.

With regard to specificity, the traditional view of the parasympathetic innervation is that it exerts restricted local actions. This view is based in part on the low innervation ratio of preganglionic to postganglionic neurons in the parasympathetic system. However, there are exceptions. In contrast, the sympathetic system has been traditionally viewed as discharging diffusely to elicit an ensemble of mobilizing responses. These would include, for instance, increased cardiac output, increased arterial blood pressure, pupillary dilation (mydriasis), decreased gastrointestinal activity, pilomotor responses, and release of catecholamines from the adrenal medulla.

Although these mass responses of the sympathetic system can indeed occur in concert in emergency or stress situations, it is now known that the sympathetic system can mediate highly localized responses as well. For example, blood flow can be increased in a single muscle. More often patterned responses occur that are generally specific to the stimulus situation. Thus the sympathetic system is more appropriately viewed as having a specificity that is perhaps analogous to local sign in flexor withdrawal reflexes. For example, the cardiovascular adjustments to different metabolic or somatic demands are quite specific. In exercise cardiac output increases and blood is shunted to the active muscles; arterial blood pressure may not change at all. During temperature changes the alterations in peripheral resistance are largely confined to the cutaneous bed; cutaneous vasodilation occurs with thermal stimulation, and cutaneous vasoconstriction occurs with cold. During orthostasis (moving from a prone to a standing position) cardiac output may not change, but carotid blood flow may increase, which maintains perfusion of the brain.

Autonomic Reflexes

As in the somatomotor system, some important autonomic regulatory actions involve spinally mediated reflexes. For instance, certain important components of control of the urinary bladder, gastrointestinal system, and sexual organs are spinally mediated. The afferent limb of such reflexes consists mainly of visceral afferent fibers, although somatic afferent fibers initiate certain autonomic reflexes. In some cases both visceral and somatic afferent fibers participate. The literature increasingly documents convergence of somatic and visceral afferent nerves onto neurons of the dorsal horn. The afferent fibers that evoke autonomic reflexes generally involve one or more spinal interneurons, and the intraspinal reflex pathway ultimately influences the preganglionic cell column.

Table 20-1. Responses of effector organs to autonomic nerve impulses

Effector organs	*Receptor type*	*Adrenergic impulses*[1] *Responses*[2]	*Cholinergic impulses*[1] *Responses*[2]
Eye			
Radial muscle, iris	α	Contraction (mydriasis) + +	—
Sphincter muscle, iris		—	Contraction (miosis) + + +
Ciliary muscle	β	Relaxation for far vision +	Contraction for near vision + + +
Heart			
SA node	β_1	Increase in heart rate + +	Decrease in heart rate; vagal arrest + + +
Atria	β_1	Increase in contractility and conduction velocity + +	Decrease in contractility, and (usually) increase in conduction velocity + +
AV node	β_1	Increase in automaticity and conduction velocity + +	Decrease in conduction velocity; AV block + + +
His-Purkinje system	β_1	Increase in automaticity and conduction velocity + + +	Little effect
Ventricles	β_1	Increase in contractility, conduction velocity, automaticity, and rate of idioventricular pacemakers + + +	Slight decrease in contactility
Arterioles			
Coronary	α,β_2	Constriction +; dilation[3] + +	Dilation ±
Skin and mucosa	α	Constriction + + +	Dilation[4]
Skeletal muscle	α,β_2	Constriction + +; dilation[3,5] + +	Dilation[6] +
Cerebral	α	Constriction (slight)	Dilation[4]
Pulmonary	α,β_2	Constriction +; dilation[3]	Dilation[4]
Abdominal viscera; renal	α,β_2	Constriction + + +; dilation[5] +	—
Salivary glands	α	Constriction + + +	Dilation + +
Veins (systemic)	α,β_2	Constriction + +; dilation + +	—
Lung			
Bronchial muscle	β_2	Relaxation +	Contraction + +
Bronchial glands	?	Inhibition (?)	Stimulation + + +
Stomach			
Motility and tone	α_2,β_2	Decrease (usually)[7] +	Increase + + +
Sphincters	α	Contraction (usually) +	Relaxation (usually) +
Secretion		Inhibition (?)	Stimulation + + +
Intestine			
Motility and tone	α_2,β_2	Decrease[7] +	Increase + + +
Sphincters	α	Contraction (usually) +	Relaxation (usually) +
Secretion		Inhibition (?)	Stimulation + +
Gallbladder and ducts		Relaxation +	Contraction +
Kidney	β_2	Renin secretion + +	—
Urinary bladder			
Detrusor	β	Relaxation (usually) +	Contraction + + +
Trigone and sphincter	α	Contraction + +	Relaxation + +

From Goodman, L.S., and Gilman, A.: The pharmacological basis of therapeutics, ed. 6, New York, 1980, Macmillan Publishing Co., Inc. Copyright © 1980 by Macmillan Publishing Co., Inc.

[1]A long dash signifies no known functional innervation.

[2]Responses are designated 1+ to 3+ to provide an approximate indication of the importance of adrenergic and cholinergic nerve activity in the control of the various organs and functions listed.

[3]Dilation predominates *in situ* due to metabolic autoregulatory phenomena.

[4]Cholinergic vasodilatation at these sites is of questionable physiological significance.

[5]Over the usual concentration range of physiologically released, circulating epinephrine, β-receptor response (vasodilatation) predominates in blood vessels of skeletal muscle and liver, α-receptor response (vasoconstriction), in blood vessels of other abdominal viscera. The renal and mesenteric vessels also contain specific dopaminergic receptors, activation of which causes dilatation, but their physiological significance has not been established.

[6]Sympathetic cholinergic system causes vasodilatation in skeletal muscle, but this is not involved in most physiological responses.

[7]It has been proposed that adrenergic fibers terminate at inhibitory β-receptors on smooth muscle fibers, and at inhibitory α-receptors on parasympathetic cholinergic (excitatory) ganglion cells of Auerbach's plexus.

[8]Depends on stage of menstrual cycle, amount of circulating estrogen and progesterone, and other factors.

[9]Palms of hands and some other sites (''adrenergic sweating'').

[10]There is significant variation among species in the type of receptor that mediates certain metabolic responses.

Table 20-1. Responses of effector organs to autonomic nerve impulses—cont'd

Effector organs	*Receptor type*	*Adrenergic impulses*[1] *Responses*[2]	*Cholinergic impulses*[1] *Responses*[2]
Ureter			
Motility and tone	α	Increase (usually)	Increase (?)
Uterus	α,β_2	Pregnant: contraction (α); nonpregnant: relaxation (β)	Variable[8]
Sex organs, male	α	Ejaculation + + +	Erection + + +
Skin			
Pilomotor muscles	α	Contraction + +	—
Sweat glands	α	Localized secretion[9] +	Generalized secretion + + +
Spleen capsule	α,β_2	Contraction + + +; relaxation +	—
Adrenal medulla		—	Secretion of epinephrine and norepinephrine
Liver	α,β_2	Glycogenolysis, gluconeogenesis[10] + + +	Glycogen synthesis +
Pancreas			
Acini	α	Decreased secretion +	Secretion + +
Islets (β cells)	α	Decreased secretion + + +	—
	β_2	Increased secretion +	—
Fat cells	α,β_1	Lipolysis[10] + + +	—
Salivary glands	α	Potassium and water secretion +	Potassium and water secretion + + +
	β	Amylase secretion +	—
Lacrimal glands		—	Secretion + + +
Nasopharyngeal glands		—	Secretion + +
Pineal gland	β	Melatonin synthesis	—

An excellent example of segmental reflex control in the autonomic nervous system is provided by the urinary bladder (Fig. 20-3). As the fluid volume of the bladder increases, the intravesical pressure rises. Stretch receptors in the bladder wall transmit information about bladder distension over the pelvic nerves to the sacral spinal cord. This information, via intraspinal circuitry, can activate parasympathetic preganglionic neurons in the second, third, and fourth sacral segments. The axons of these neurons travel in the pelvic nerve to terminate on local postganglionic parasympathetic neurons in the vesical plexuses at the urinary bladder. Activation of these neurons then contracts the detrusor muscle and hence the bladder, contributing to emptying. The afferent information from the bladder also influences sympathetic preganglionic neurons at upper lumbar levels. The axons of these cells relay through postganglionic neurons in prevertebral ganglia (the inferior mesenteric and inferior hypogastric plexuses). The pathway through the inferior mesenteric ganglion inhibits contraction of the internal sphincter, thereby contributing to bladder emptying. As voiding is initiated, urine flow through the urethra stimulates an afferent discharge that elicits parasympathetic reflex contraction of the detrusor muscle and contributes to the continuation of micturition.

Spinally mediated reflexes, such as the ones just mentioned, permit a certain degree of bladder control after spinal transection. However, in the intact animal micturition critically involves supraspinal structures, particularly in the medulla and pons. Spinally mediated control of the bladder is normally under massive control by descending pathways. This descending control brings micturition under voluntary control, independent of the autonomic nervous system. The external sphincter muscles are striated and are innervated by α-motoneurons in the third and fourth sacral segments. Activation of these motoneurons by descending pathways can contract the sphincter and prevent voiding.

Many important reflex adjustments of the internal organs are mediated by suprasegmental reflex pathways. A particularly prominent example is the carotid sinus reflex. Increased pressure in the carotid sinus increases the discharge of afferent fibers that travel in the ninth cranial nerve and synapse in the nucleus of the solitary tract. (This nucleus may be viewed as a brainstem equivalent of the dorsal horn of the spinal cord.) Through local medullary circuitry, descending fibers are activated that inhibit the activity of sympathetic preganglionic neurons that control peripheral resistance. The action of this circuitry is to decrease peripheral resistance in response to increased arterial blood pressure (Chapter 32).

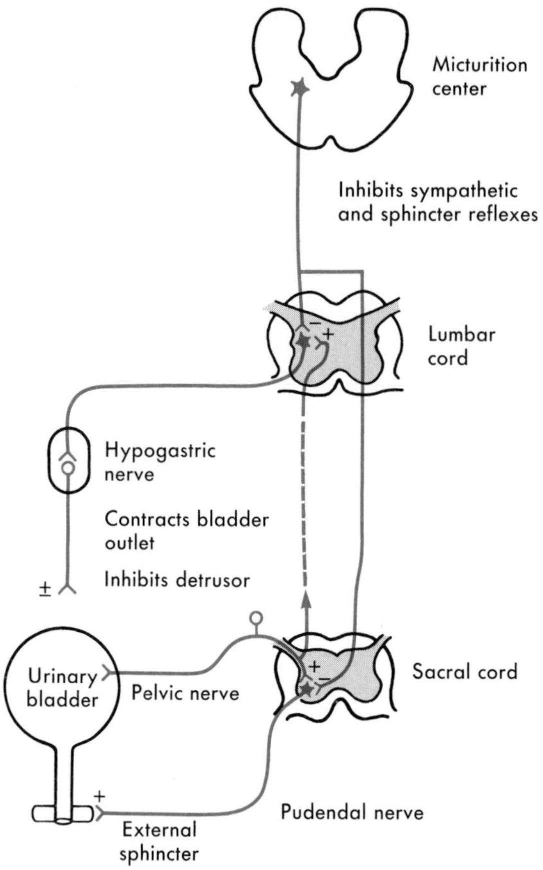

■ **Fig. 20-3.** Schematic representation of detrusor-sphincter reflexes. (Redrawn from de Groat, W.C., and Booth, A.M.: Autonomic systems to bladder and sex organs. In Dyck, P.J., Thomas, P.K., Lambert, E., et al., editors: Peripheral neuropathy, ed. 2, Philadelphia, 1984, W.B. Saunders Co.)

■ *Higher Autonomic Control*

The autonomic reflexes are relatively simple examples of central control of autonomic activity. However, many critical functions, such as feeding, drinking, reproduction, and temperature regulation, have autonomic components that involve complex central circuitry. Unfortunately such circuitry has not been specified in sufficient detail.

As in the somatomotor system, central autonomic control is most productively analyzed by investigating the functional neuroanatomy of the major pathways that influence the autonomic outflow (the autonomic final common path). Analogous to those that control movement, these pathways are primarily organized longitudinally, and it is most effective to analyze them from the periphery centrally. For the autonomic nervous system this entails first specifying the primary descending pathways that terminate on the sympathetic and parasympathetic preganglionic cell groups. One can then approach the problem of ''sensorimotor integration'' in order to specify how various stimuli (internal and external) guide the motor (autonomic) outflow. Again, as in the somatomotor system, the cells of origin of the major descending pathways are located from the lower brainstem to the neocortex. However, unlike the pathways that control movement, the major confluence of such pathways in autonomic control is at the hypothalamus. Thus it is appropriate at this point to review briefly the organization of the hypothalamus.

■ *The Hypothalamus*

The hypothalamus is a diencephalic structure bordering the third ventricle (Fig. 20-4). Rostrocaudally it can be divided into *supraoptic, tuberal,* and *mammillary regions,* each consisting of aggregates of nuclear groups. The rostral continuation of the hypothalamus is the *preoptic region* (Fig. 20-4), which continues forward as the *septum*. Both the preoptic and septal regions participate in autonomic control. Major fiber tracts course longitudinally through the hypothalamus, the most prominent being the *medial forebrain bundle* and the *fornix*. The latter tract divides the hypothalamus into *medial* and *lateral regions* (Fig. 20-5). Many of the hypothalamic cell groups are named on the basis of their location with respect to this mediolateral division.

Although various well-defined nuclear groups are located in the hypothalamus, it is difficult to delineate nuclear boundaries in some regions. Thus, a cytoarchitectonic description of the hypothalamus does not have the precision it does when applied to other areas of the nervous system, such as the cortex and cerebellum. This and other factors have rendered it more difficult to study afferent and efferent connections. A highly schematic summary of these connections is presented in Fig. 20-6, but it should be emphasized that important intrinsic hypothalamic connections establish communication between the medial and lateral divisions.

The hypothalamus is very important in autonomic control. Indeed certain hypothalamic nuclei, such as the supraoptic nucleus, constitute final common paths in neuroendocrine control (Chapter 52). However, extrahypothalamic structures are also involved in the control of autonomic function. For example, in respiration, micturition, and emesis most processing involves central structures other than the hypothalamus.

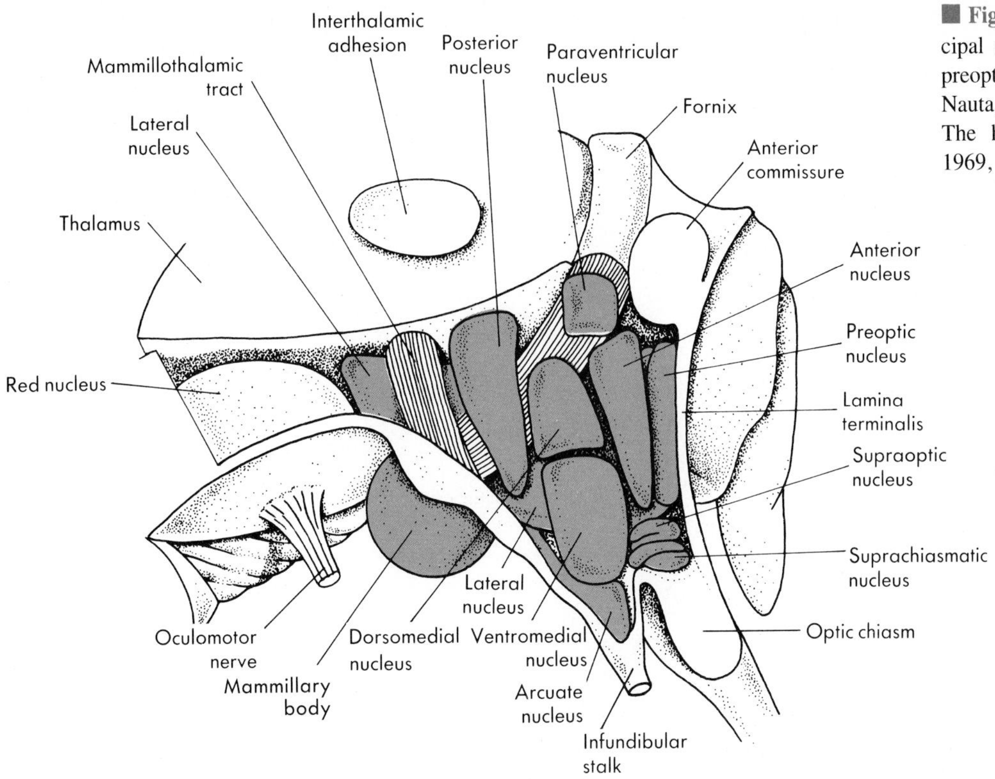

■ **Fig. 20-4.** Medial view of the principal nuclei of the hypothalamus and preoptic area *(color)*. (Redrawn from Nauta, W.J.H., and Haymaker, W.: The hypothalamus, Springfield, Ill., 1969, Charles C Thomas, Publisher.)

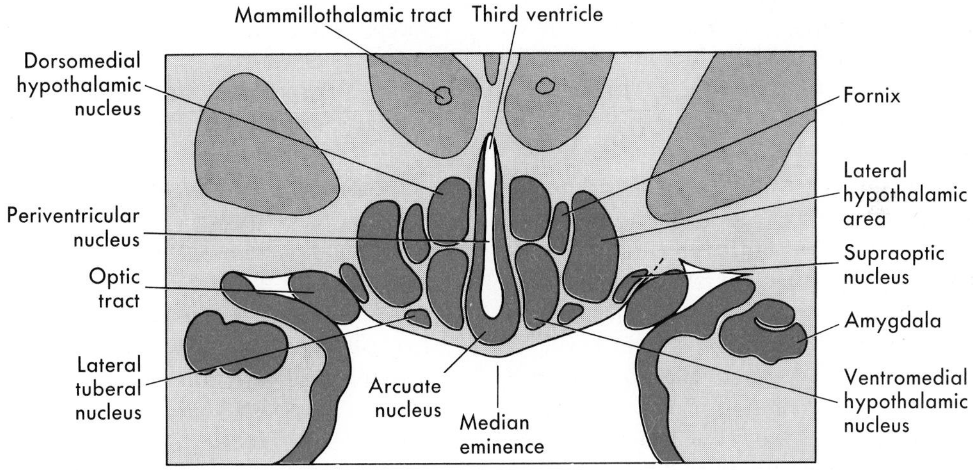

■ **Fig. 20-5.** Frontal view of the hypothalamus showing the fornix, which divides the hypothalamus into medial and lateral regions. (Redrawn by permission of the publisher from Chapter 46, by Kandel, E.R., and Schwartz, J.H., Principles of neural science, ed. 2, p. 616. Copyright 1985 by Elsevier Science Publishing Co., Inc.)

■ *The Center Concept*

The concept of "centers" that mediate particular autonomic functions has dominated tne literature on central autonomic control. Examples are the vasomotor and vasodilator centers of the medulla, the heat production and heat loss centers of the diencephalon, and the hypothalamic feeding and satiety centers. This notion has some heuristic value, and indeed it reflects a certain localization of autonomic function. However, the concept is somewhat misleading with respect to the functional neuroanatomy of central autonomic control because it turns attention away from the longitudinal organization. It is perhaps equivalent to describing the motor cortex and the red nucleus as "movement centers" while ignoring the corticospinal and rubrospinal tracts and their connections.

The evolution and dominance of the center concept

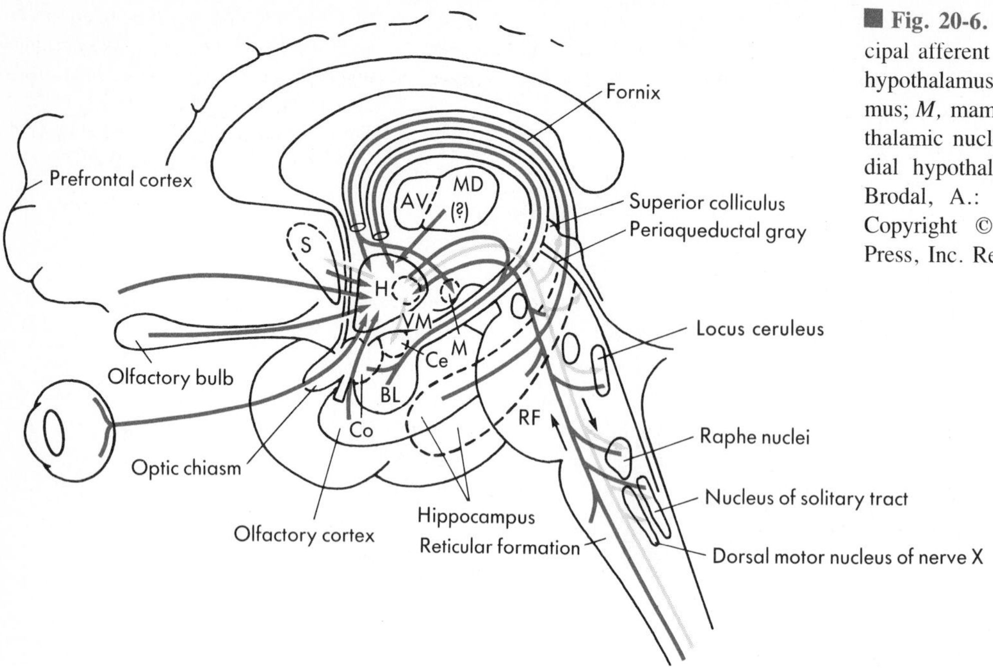

■ **Fig. 20-6.** Schematic diagram of the principal afferent and efferent connections of the hypothalamus. *A,* amygdala; *H,* hypothalamus; *M,* mammillary body; *MD,* dorsomedial thalamic nucleus; *S,* septum; *VM,* ventromedial hypothalamic nucleus. (Redrawn from Brodal, A.: Neurological anatomy, ed. 3, Copyright © 1981 by Oxford University Press, Inc. Reprinted by permission.)

largely reflect two properties of the organization of central autonomic control. First, the hypothalamus is a major conduit for pathways involved in such control. Thus in early investigations, which primarily involved the lesion approach, major fiber bundles *en passage* through the hypothalamus were destroyed in addition to local cell groups. This compromised both hypothalamic and extrahypothalamic autonomic control and generated an exaggerated view of the hypothalamic involvement, thereby reinforcing the center concept. Second, many of the pathways involved in central autonomic control are composed of fine fibers that do not appear as prominent tracts. Thus they were not easily identified by older anatomical methods. This did not encourage the concept of longitudinally organized central autonomic control but again reinforced the center concept.

These traditional views are now being modified as the central pathways involved in different autonomic functions are being characterized. As will be seen in the sections to follow, we are in a transitional period where a strong residual of the center concept remains for some autonomic activities, but a more contemporary description is available for others.

■ *Temperature Regulation*

In homeothermic animals, which regulate their body temperatures, a fall in the ambient temperature requires additional heat production and reduction of heat loss. In contrast, an increase in ambient temperature requires greater heat loss and a reduction of heat production. Such regulation has the properties of a servomechanism. Information on body temperature is provided by *thermoreceptors* distributed throughout the body, as well as by temperature-sensitive neurons in the anterior hypothalamus. The firing rates of these temperature-sensitive cells reflect the temperature of the regional blood supply. Normal body temperature is the *set point* of the system, and the inputs from the various thermoreceptors signal deviations (error signals) from that set point. Maintenance of the set point is then accomplished by a complex set of responses involving autonomic, somatic, and endocrine systems.

By way of illustration, cooling of the body elicits shivering (the asynchronous contraction of muscle fibers), which is a primary mechanism of thermogenesis. In addition to this somatic response, the output of the thyroid gland increases, which represents a hormonal contribution to heat production through increased metabolism. The autonomic nervous system contributes to thermogenesis by releasing norepinephrine from sympathetic nerve terminals, and this stimulates the release and oxidation of fatty acids. Heat loss is restricted mainly by two autonomic responses: piloerection, which establishes a dead space, and constriction of the cutaneous vessels, which reduces cooling of the blood. In response to warming of the body, heat loss is enhanced by autonomic responses such as sweating, which increases evaporation, and dilation of the cutaneous vessels, which promotes cooling of the blood. Hormonal participation is reflected by a decrease in thyroxine release, which lessens heat production by decreasing metabolic activity.

In the servomechanism control of body temperature the thermoreceptors are the sensors, and the responses described above are the effector actions. The central integrator or controller critically involves the hypothalamus. The preoptic region and the anterior hypothalamus mediate the responses that result in heat loss. Lesions of this region prevent sweating and cutaneous vasodilation in response to increased ambient temperature and therefore produce hyperthermia. Electrical stimulation of the anterior hypothalamus elicits cutaneous vasodilation and suppresses shivering. Such experimental results led to the designation of the anterior hypothalamus and preoptic regions as a *heat loss center*. Cells within the posterior hypothalamus appear to be involved in heat production and conservation. Posterior hypothalamic lesions dorsolateral to the mammillary body eliminate these responses and result in a chronic hypothermia. Electrical stimulation elicits responses such as shivering. Thus this region of the posterior hypothalamus has been viewed as containing a *heat production and conservation center*.

Heating or cooling of the hypothalamus also evokes thermoregulatory responses consistent with the existence of temperature-sensitive neurons in certain regions of the hypothalamus. However, warming of the heat production and conservation center of the posterior hypothalamus does not abolish the shivering that is elicited by cooling of the skin. Moreover, the center concept of thermoregulation does not help explain how the posterior hypothalamus mediates shivering. It does not explain how preferential constriction of the cutaneous vasculature is effected. Stated more broadly, it discourages the investigation of extrahypothalamic involvement and thus the delineation of the neuroanatomical pathways that mediate the various thermoregulatory responses. The disturbance of thermoregulatory responses produced by hypothalamic and preoptic lesions is certainly striking. However, the extent to which the hypothalamus initiates and integrates such responses has not been resolved.

With regard to the pathophysiology of thermoregulation, fever reflects an increase in the set point for body temperature. Since deviation from the set point evokes thermoregulatory responses, an increase in the set point activates mechanisms for heat production (shivering) and conservation (cutaneous vasoconstriction). The membranes of certain bacteria contain a pyrogenic lipopolysaccharide that stimulates leukocytes to produce an endogenous pyrogen that is released into the circulation. Injection of this pyrogen directly into the hypothalamus produces fever, possibly by altering the sensitivity of central thermoreceptors.

Regulation of Food Intake

The regulation of body weight is much more complex than the regulation of body temperature. As with temperature regulation, the concept of a set point seems to apply. However, in body weight control, the set point varies substantially among individuals and is affected more readily by a number of factors.

The relevant sensory signals are complex because they include short-term inputs that control food intake and long-term inputs that control body weight. Moreover, the effects of these two kinds of inputs are interactive in that long-term inputs can affect responsiveness to short-term signals. It had been proposed that hypothalamic *glucoreceptors* sense changes in blood glucose levels and on that basis control food intake. Whereas hypothalamic glucoreceptors are involved, they appear to influence food intake only in response to large decreases in blood glucose levels. More recent research has implicated a variety of neuropeptides that affect feeding. For example, the opioid peptides and the pancreatic polypeptide family stimulate food intake. Other peptides (e.g., cholecystokinin) inhibit food intake. *Cholecystokinin* is found in the hypothalamus and is released from the duodenum and upper intestine, providing the possibility for both peripheral and central signals. The effects of the peptides on food intake are convincing, but their specificity has not been established. Note also that hormones such as insulin and adrenal glucocorticoids can profoundly influence food intake.

Concerning central neural mechanisms, lesions of the lateral hypothalamus produce *aphagia,* a decrease in food intake that may lead to death. Electrical stimulation of this region induces eating, even after satiety. In contrast, destruction of the ventromedial nucleus of the hypothalamus increases food intake *(hyperphagia),* which can result in obesity. Electrical stimulation of this hypothalamic area will terminate feeding. Such results generated the hypothesis that the lateral hypothalamus contains a *feeding center* and the ventromedial hypothalamus contains a *satiety center*. Moreover, these two centers are thought to interact reciprocally. For example, if feeding is induced by lateral hypothalamic stimulation, stimulation of the ventromedial nucleus will terminate the ongoing feeding abruptly.

As with other functions discussed previously, hypothalamic involvement is unquestionable. However, once again it is important to consider the extent to which the hypothalamus is involved in regulation of food intake because of the apparent involvement of other structures in this function. By way of illustration, the lateral hypothalamic lesions that produce aphagia interrupt many fibers *en passage,* including the nigrostriatal pathway. Destruction of this dopaminergic tract outside the hypothalamus may produce aphagia. Similarly, lateral hypothalamic lesions can interrupt the trigeminal lemniscus and abolish ascending sensory information that may stimulate ingestion and deglutition; this too can produce aphagia. Indeed, many extrahypothalamic structures have now been implicated in the control of feeding (Fig. 20-7). Thus it is more productive to consider neural circuits that mediate the regulation of food intake and body weight than to consider hypothalamic centers.

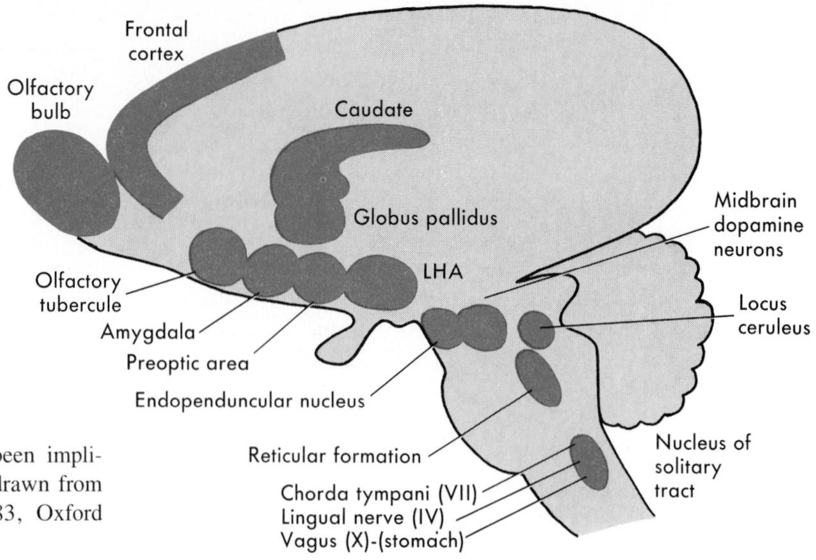

■ **Fig. 20-7.** Some neural structures that have been implicated in the control of food intake in the rat. (Redrawn from Shepherd, G.M.: Neurobiology, New York, 1983, Oxford University Press.)

■ *Regulation of Water Intake*

In certain respects the control of fluid intake is similar to that for the regulation of food intake. The concepts of set point and negative feedback apply, but the process is simpler and better understood than the regulation of feeding and body weight. An important difference is that for drinking, the amount of fluid intake is not critical as long as it is sufficient. Excess water is rapidly eliminated by inhibiting the release of antidiuretic hormone (ADH), a peptide contained in neurons of the supraoptic nucleus and secreted by the posterior pituitary gland. Such factors as increased blood volume and decreased osmolality of extracellular fluid inhibit ADH release (Chapter 52).

A variety of stimuli influence fluid intake, but the two primary inputs for the control of drinking are osmolality and blood volume (Fig. 20-8, *A*). During water deprivation the extracellular fluid becomes hyperosmotic. Because the extracellular and intracellular compartments are in equilibrium, the cell cytoplasm becomes hyperosmotic. Certain cells, which are probably localized to the hypothalamus, function as *osmoreceptors* and constitute one afferent limb of the neural circuitry that regulates drinking.

Water deprivation also diminishes vascular volume, which is signalled by receptors in the low pressure side of the vasculature, such as the right atrium. The reduced blood volume also induces the release of renin by the kidney. Renin cleaves circulating angiotensinogen to make *angiotensin I,* which is then hydrolyzed to *angiotensin II*. Angiotensin II, an octapeptide, stimulates drinking. Angiotensin II receptors are prominent in the hypothalamus, in the walls of the cerebral ventricles, and in structures such as the circumventricular organ, where there is no blood-brain barrier. In addition to eliciting drinking, angiotensin II also mediates water

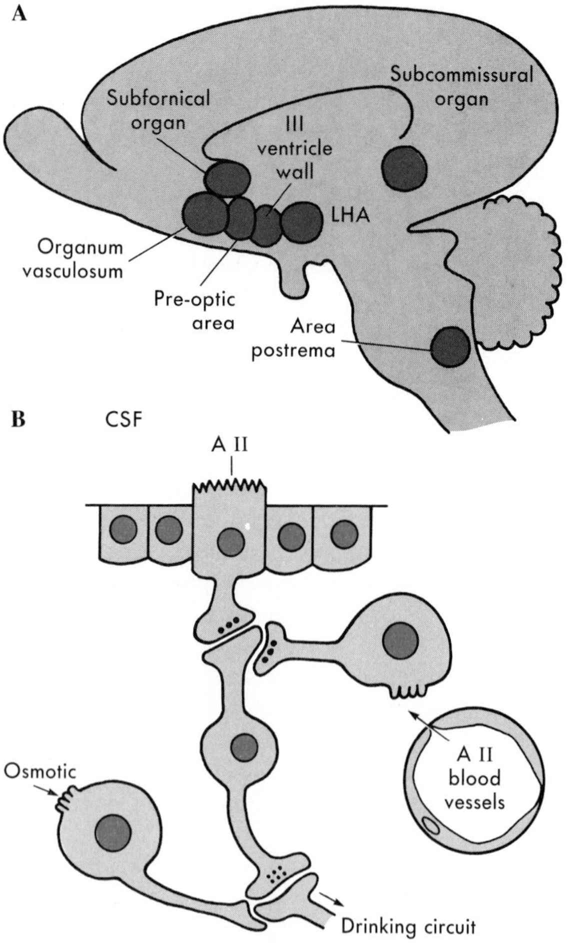

■ **Fig. 20-8. A,** Afferent circuitry for signalling changes in osmolality and vascular volume. Site 1, angiotensin II receptor in the wall of the cerebral ventricle. Site 2, angiotensin II receptor in the circumventricular organ. Site 3, osmoreceptor in the hypothalamus. **B,** Structures implicated in the regulation of drinking in the rat. (Redrawn from Shepherd, G.M.: Neurobiology, New York, 1983, Oxford University Press.)

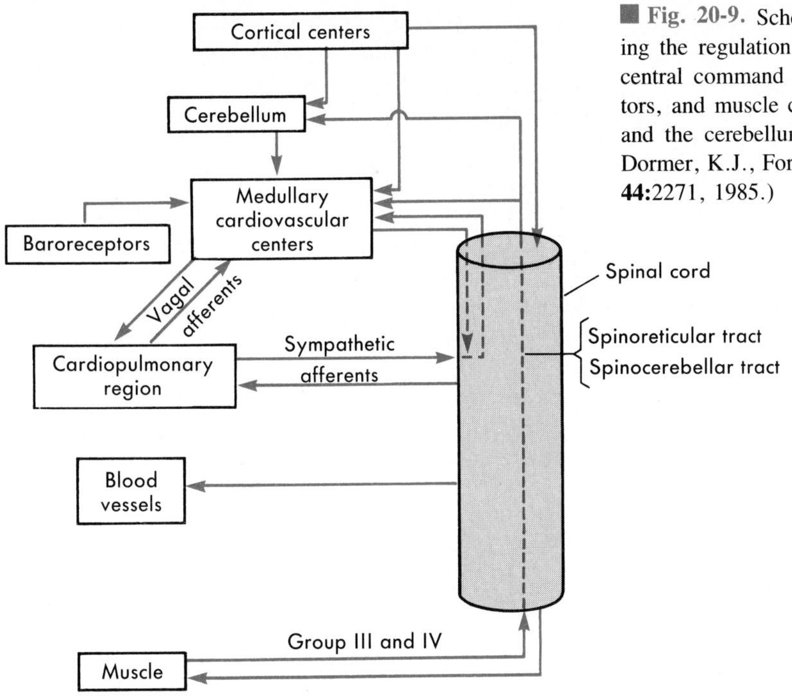

Fig. 20-9. Schematic representation of current knowledge concerning the regulation of the cardiovascular system during exercise. Both central command and feedback loops are included. Heart, baroreceptors, and muscle compose the feedback loops in the peripheral system and the cerebellum for a central feedback loop. (From Stone, H.L., Dormer, K.J., Foreman, R.D., Thies, R., and Blair, R.W.,: Fed. Proc. **44:**2271, 1985.)

conservation by eliciting vasoconstriction, aldosterone release, and ADH release. Of the two primary stimuli that signal water deprivation, osmolality is more powerful than vascular volume.

With respect to central structures that regulate drinking, reciprocal hypothalamic centers, analogous to those believed to regulate food intake, have been proposed. Particularly prominent in these formulations are the preoptic region and the lateral hypothalamus. However, more recently a number of extrahypothalamic structures have also been implicated in the control of drinking (Fig. 20-8 *B*).

Central Cardiovascular Control

As a final illustration of central autonomic control, cardiovascular function will be considered. It provides a particularly useful case study because it demonstrates (1) the transition from the center concept to longitudinal organization, (2) patterned sympathetic activity appropriate to environmental demands, and (3) the integration of reflex activity with such phasic adjustments.

Historically, central cardiovascular control has been approached from two directions. One focuses principally on cardiovascular adjustments in response to various peripheral inputs, that is, primarily a description of reflex activity. Changes in cardiac dynamics and peripheral resistance evoked by activation of the baroreceptors are characteristic of this approach. The other approach focuses on the nervous system and is directed primarily toward describing ''centers'' for the homeostatic regulation of cardiovascular activity. Constructs such as the medullary *vasopressor* and *vasodepressor centers* were derived from this approach, and they generated a hierarchical view of the organization of central cardiovascular control.

The concept now prevails that descending functional pathways mediate patterns of cardiovascular adjustment specific to behavioral demands. The adjustments required by exercise provide an excellent example. In this case there is a need for increased cardiac output and increased blood flow in the active muscle beds. A pathway originating in the motor cortex selectively increases the blood flow in muscles that are represented in the activated region. The precise trajectory and connectivity of this pathway have yet to be specified. However, the important point is that viewing the cardiovascular adjustments in exercise in the context of functional neuroanatomical pathways has been considerably more productive than using the center concept (Fig. 20-9).

The so-called exercise pathway illustrates other important points regarding central autonomic control. First, it demonstrates that the sympathetic system can behave in a selective and patterned manner. Second, it increases our understanding of the interaction between the sympathetic and parasympathetic innervations in a physiological context. Specifically, the increased cardiac output during exercise involves the synergistic action of the sympathetic and parasympathetic cardiac innervations, with the sympathetic drive on the heart increasing and vagal cardiac inhibition decreasing during exercise. Third, the exercise pathway has allowed investigation of the interaction between patterned responses evoked by environmental demands and reflex activity. During the increase in cardiac output the baroreceptor reflex is inhibited, preventing competition via its negative feedback effect. Fourth, a central organization originating in the cortex allows cardiovascular adjustments to occur in anticipation of exercise so that the central control can prepare the organism for impending demands. This is seen, for example, in a runner at the starting line, and it clearly demonstrates that learned control of autonomic outflow plays a functional role, a fifth point. Finally, it supports the view of central cardiovascular control in terms of both phasic and tonic functions; i.e., certain central pathways mediate phasic adjustments in response to specific classes of transient demands, such as exercise, and other central pathways mediate more tonic or homeostatic functions, such as the moment-by-moment regulation of arterial blood pressure.

We continue to learn more of the pathways responsible for specific patterns of cardiovascular adjustment and how they interact. For example, during a shift from a prone to a standing position, it is necessary to effect adjustments that maintain blood flow to the brain, i.e., to oppose gravitationally induced pooling of blood in the lower body *(orthostasis)*. This is mediated by a reflex pathway, the afferent limb of which involves the vestibular system. This labyrinthine information is then relayed to a localized region of the cerebellum, the ros-

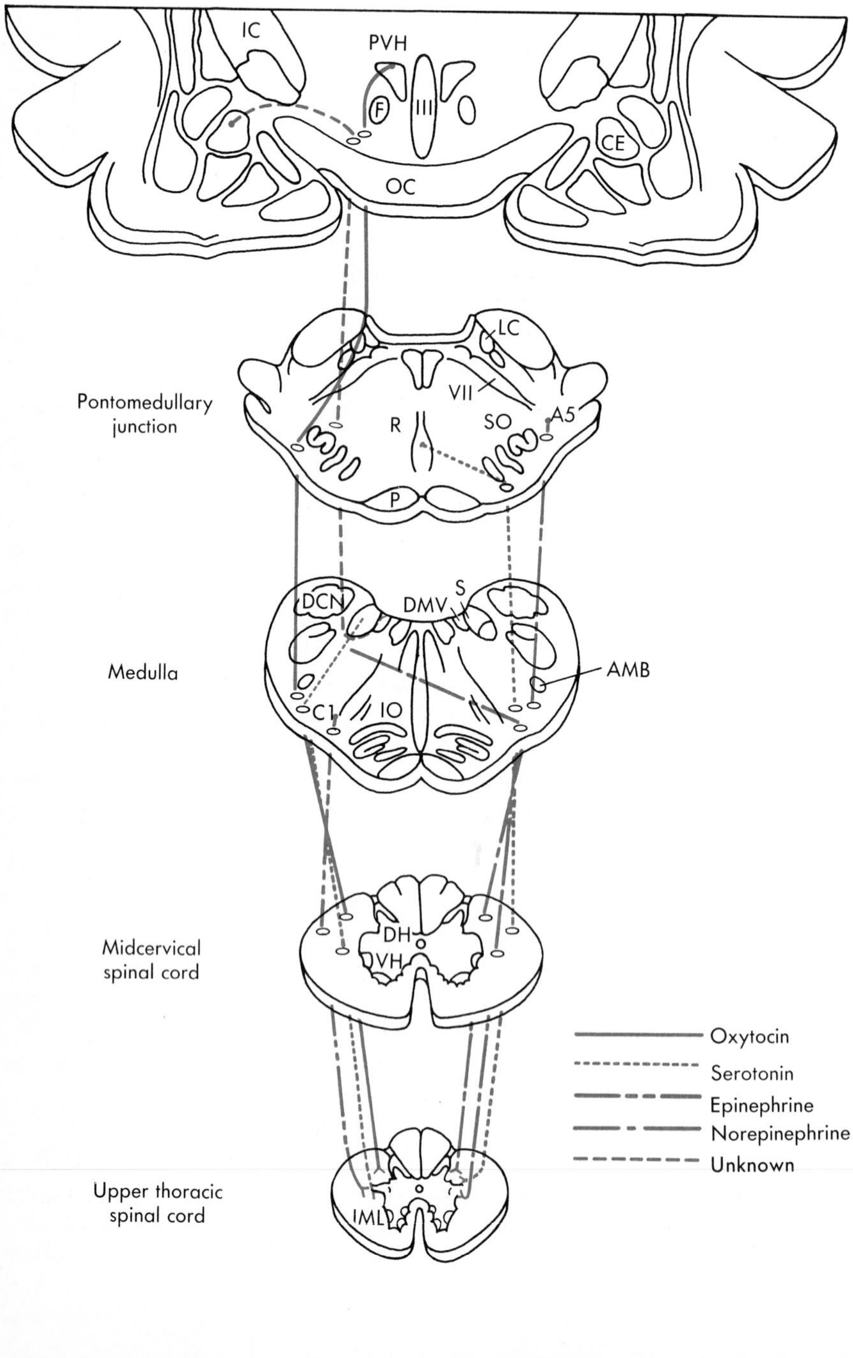

■ **Fig. 20-10.** Schematic summary of descending pathways that project directly to the dorsal motor nucleus of the vagus and/or the sympathetic preganglionic cell column. Hypothesized transmitters are indicated. *A5,* A5 catecholamine cell group of Dahlstrom and Fuxe; *AMB,* nucleus ambiguus; *C1,* epinephrine cell group; *CE,* central nucleus of the amygdala; *DCN,* dorsal column nuclei; *DH,* dorsal horn; *DMV,* dorsal motor nucleus of the vagus; *F,* fornix; *IC,* internal capsule; *III,* third ventricle; *IML,* intermediolateral (sympathetic preganglionic) cell column; *IO,* inferior olive; *LC,* locus ceruleus; *OC,* optic chiasm; *P,* pyramid; *PVH,* paraventricular nucleus of the hypothalamus; *R,* raphe nucleus; *S,* solitary nucleus; *SO,* superior olive; *VH,* ventral horn; *VII,* facial nerve rootlets. (Redrawn from Cohen, D.H., and Cabot, J.B.: Trends Neurosci. **2:**273, 1979.)

tral third of the medial zone, which then projects back on the brainstem to influence cardiovascular structures that project to the spinal cord. Interruption of this pathway can produce a postural hypotension.

An important contemporary thrust is to describe rigorously the various descending pathways that impinge on autonomic preganglionic neurons (Fig. 20-10). This establishes a foundation for analyzing the different functional "cardiovascular pathways" from the periphery to the central nervous system. Moreover, as the neurotransmitters of these pathways are discovered, pharmacological manipulation becomes possible. This is important in dealing with cardiovascular diseases that have a neurogenic etiology, and there is now increasing acceptance of the idea that these may, indeed, be quite common. For example, many now believe that sudden cardiac death may often be initiated neurogenically and that a significant proportion of cases of essential hypertension may involve the central nervous system. It is noteworthy in this regard that many of the most widely used antihypertensive drugs, such as clonidine, have neural sites of action.

Thus more recent investigations of central cardiovascular control are teaching us the merits of reconceptualizing our view of autonomic control. In the case of cardiovascular control this has diminished the historically dominant concept of medullary centers. It has placed them in a more comprehensive context that permits rational investigation of longitudinally organized pathways that mediate specific classes of adjustment, be they homeostatic or in response to environmental demands.

Neural Influences on the Immune System

It has long been suspected that psychological stress can affect the immune system and alter susceptibility to disease, but direct evidence was lacking. Recent studies have shown that various environmental stresses do have an immunosuppressive effect, including a reduction in helper T cells and reduced activity of natural killer cells. Furthermore, rigorous behavioral experiments have established that one can classically condition immunosuppression.

One relevant pathway for such effects is hormonal and involves the hypothalamus. Stressful stimuli elicit the release of corticotropin-releasing factor from the hypothalamus. This increases ACTH release from the pituitary gland, which then stimulates secretion of adrenal corticosteroids. Among the effects of these hormones is immunosuppression. However, neural influences may also affect immune responses directly. For example, the thymus, spleen, lymph nodes, and bone marrow are apparently innervated by nerve fibers, in addition to those supplying the vasculature. Furthermore, this innervation is generally well-developed in regions of T cells and absent in regions where B cells are concentrated. The function of this innervation is not yet known.

The immune system may also influence neuronal activity. Experiments indicate that the brain receives signals, either directly from the immune system or via substances released by infected cells. Also, changes in neuronal discharge have been recorded during immune responses.

Emotional Behavior

Just as the hypothalamus was historically designated as the principal suprasegmental structure for autonomic control, the limbic system was assigned the mediation of emotional behavior. Presumed to be in a control position over the hypothalamus, limbic structures then performed higher-order integrative functions involving autonomic control. Again, this reflected the earlier tendency to view the nervous system as being hierarchically organized.

The limbic lobe is phylogenetically the oldest part of the cortex. It was subsequently proposed that a complex circuit involving the limbic lobe (Papez circuit) participated with the hypothalamus in mediating emotional behavior; this became known as the *limbic system* (Fig. 20-11). Neocortical areas subserving various higher functions then communicated with the hypothalamus through the limbic system. The circuit involved a flow of information from the cingulate gyrus, via the hippo-

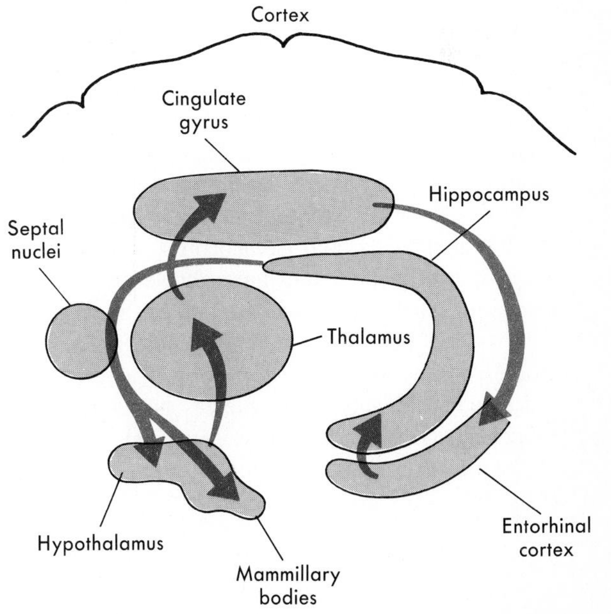

Fig. 20-11. Schematic representation of the Papez circuit linking the hypothalamus and limbic system. (From Groves, Phillip M., and Schlesinger, Kurt, Introduction to biological psychology, ed. 2, © 1979, 1982, Wm. C. Brown Publishers, Dubuque, Iowa. All Rights Reserved. Reprinted by permission.

campus, to the mammillary bodies of the hypothalamus. A loop was closed by information returning from the hypothalamus to the cingulate gyrus via the anterior thalamic nuclei. The concept of this circuit gained popularity for some time, and as more was learned about autonomic areas of the forebrain, the basic circuit was expanded to include structures such as the amygdala, which is also connected to a loop with the hypothalamus.

The data supporting this approach were derived largely from studying the results of lesions and electrical stimulation because various emotional behaviors could be evoked or perturbed by experimental manipulations of limbic structures. For example, stimulation of certain regions of the amygdala can evoke aggressive behavior, and amygdalar lesions can produce docility. Hypothalamic stimulation can also elicit defensive behavior, and this is modulated by concomitant stimulation of the amygdala. This reinforced the concept of cortical limbic structures that modulate the hypothalamic integration of somatic, autonomic, and endocrine components of emotional behavior. An early finding that supported this concept was that surgically separating the hypothalamus from its cortical influences produced an animal that showed *sham rage*. Such an animal responds to mild stimulation with the extreme aggressive responses that are characteristic of rage.

Another important contribution to the limbic concept is a syndrome caused by temporal lobe lesions. The *Klüver-Bucy syndrome* includes (1) loss of the ability to detect and recognize the meaning of objects on the basis of visual cues (visual agnosia), (2) a tendency to examine objects orally (oral tendencies), (3) an inability to suppress attention to irrelevant stimuli (hypermetamorphosis), (4) hypersexuality with a loss of discriminability of appropriate sexual objects, (5) dramatic changes in dietary habits that include a loss of discriminability of appropriate food objects, and (6) a flatness of emotional responsiveness (taming). These temporal lobe lesions include both the neocortex and limbic cortex, and subsequent studies have allowed separation of the components of the syndrome that are attributable to each of these areas. Thus it is now appreciated that the changes in emotional behavior are primarily the result of damage to the amygdala, whereas such symptoms as visual agnosia reflect damage to visual areas of the temporal neocortex.

Bibliography

Journal articles

Anand, B.K., and Brobeck, J.R.: Localization of a "feeding center" in the hypothalamus of the rat, Proc. Soc. Exp. Biol. Med. **77:**323, 1951.

Andersson, B.: Regulation of water intake, Physiol. Rev. **58:**582, 1978.

Cabanac, M.; Temperature regulation, Annu. Rev. Physiol. **37:**415, 1975.

Castonguay, T.W.: The neural and metabolic basis of feeding, Brain Res. Bull. **14:**Whole issue, 1985.

Cohen,D.H., and Cabot, J.B.: Toward a cardiovascular neurobiology, Trends Neurosci. **2:**273, 1979.

Gershon, M.D.: The enteric nervous system, Annu. Rev. Neurosci. **4:**227, 1981.

Hilton, S.M.: Ways of viewing the central nervous control of the circulation—old and new, Brain Res. **87:**213, 1975.

Kirchheim, H.R.: Systemic arterial baroreceptor reflexes, Physiol. Rev. **56:**100, 1976.

Koizumi, K., and Brooks, C.M.: The integration of autonomic system reactions: a discussion of autonomic reflexes, their control and their association with somatic reactions, Ergeb. Physiol. **67:**1, 1972.

Papez, J.W.: A proposed mechanism of emotion, Arch. Neurol. Psychiatry **38:**725, 1937.

Petras, J.M., and Faden, A.I.: The origin of sympathetic preganglionic neurons in the dog, Brain Res. **144:**353, 1978.

Smith, O.A.: Reflex and central mechanisms involved in the control of the heart and circulation, Annu. Rev. Physiol. **36:**93, 1974.

Smith, O.A., and DeVito, J.L.: Central neural integration for the control of autonomic responses associated with emotion, Annu. Rev. Neurosci. **7:**43, 1984.

Books and monographs

Brodal, A.: Neurological anatomy, ed. 3, New York, 1981, Oxford University Press Inc.

Cannon, W.B.: The wisdom of the body, ed. 2, New York, 1939, W.W. Norton & Co., Inc.

de Groat, W.C., and Booth, A.M.: Autonomic systems to bladder and sex organs. In Dyck, P.J., Thomas, P.K., Lambert, E., and Bunge, R., editors: Peripheral neuropathy, ed. 2, Philadelphia, 1984, W.B. Saunders Co.

Jänig, W.: The autonomic nervous system. In Schmidt, R.F., editor: Fundamentals of neurophysiology, ed. 2, New York, 1978, Springer-Verlag New York, Inc.

Pick, J.: The autonomic nervous system: morphological, comparative, clinical and surgical aspects, Philadelphia, 1970, J.B. Lippincott Co.

CHAPTER

21

Neural Plasticity

The integrative functioning of the nervous system is based primarily on the connectivity among its basic elements—the neurons. To a large extent these connections are determined genetically and, once established, remain stable. However, it is becoming increasingly apparent that, under certain circumstances, some synaptic connections can be modified, and this modifiability is referred to as *neural plasticity*.

Considerable contemporary research is focused on this phenomenon. The central questions are: What situations can produce synaptic modification? Is modifiability a property of all neurons or only certain classes of nerve cells? What are the mechanisms of this plasticity, and do the mechanisms that mediate neuronal modification in different situations share common properties?

It is beyond the scope of this chapter to treat this captivating and important topic comprehensively. Rather, we present examples from three major classes of plastic change: (1) postnatal modifications of connectivity that occur as a result of interactions with the environment, (2) the changes in connectivity that occur as a consequence of injury to the brain, and (3) the plasticity during learning or experience. As yet, scientists lack a detailed understanding of the mechanisms that mediate these various modifications of the nervous system, but research is providing rapid progress toward such understanding.

Postnatal Development

A few decades ago it was thought that the development of neural connections was essentially under genetic control and that environmental influences had a negligible effect on synapse formation. Thus neural development was thought to be essentially immutable. Scientists now know that the development of many neural connections can be affected by the neuronal environment and are thus *plastic*. Two major environmental manipulations have been used to study these neuroplastic processes: brain lesions and sensory deprivation. The former directly alters the neuronal environment and the latter does so indirectly by changing the levels of evoked activity among the relevant neurons. As we shall describe later with regard to the visual system, neonatal lesions or deprivation cause the development of certain pathways to be abnormal. Many others develop normally under equivalent conditions, presumably because they are under strict genetic control.

Normal Development

Before considering the effect of the environment on neuronal development, we shall first briefly describe a phenomenon that seems common to the development of most pathways in many vertebrate species. With few exceptions*, neuron generation in the mammalian brain is complete near the time of birth. However, development passes through a phase whereby an excess of neurons and pathways is created. The unnecessary or redundant elements are eventually pruned as development proceeds. In some cases extraneous neurons and their processes simply die; in others some of the exuberant axonal branches of a neuron are retracted as the cell develops, which stabilizes its other efferent connections.

Numerous examples of these processes can be cited. For instance, the adult avian retina (unlike the mammalian retina) receives massive input from the contralateral isthmo-optic nucleus of the midbrain. Neurons of this nucleus are significantly more numerous in the immature brain than in the adult, and the immature pathway is bilateral rather than contralateral, as it is in the adult (Fig. 21-1). Presumably those neurons that fail to make appropriate connections in the retina (i.e., mostly the ipsilaterally projecting neurons) die and dis-

*In mammals, the only known exception to the dogma that no new neurons are produced postnatally is in the olfactory mucosa. Here, a constant turnover of the sensory neurons occurs throughout life.

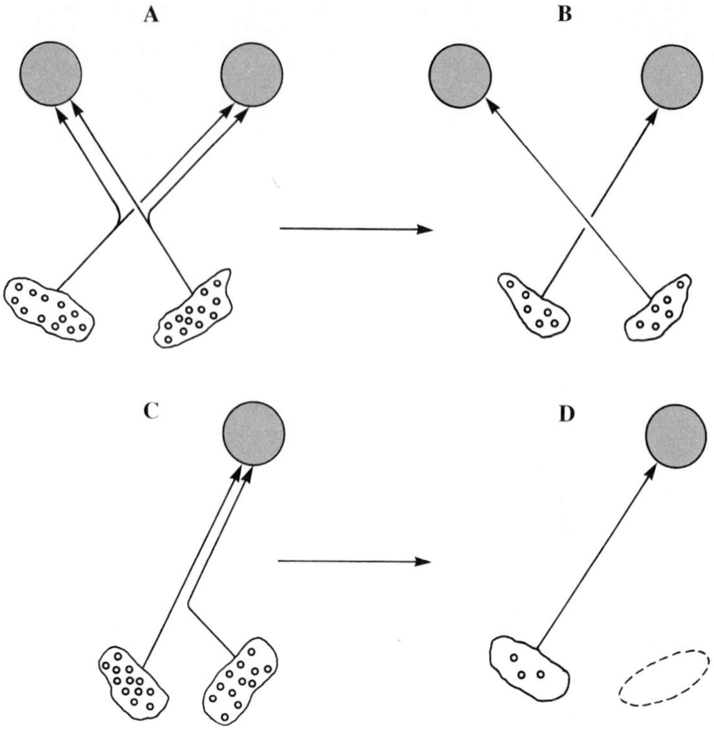

■ **Fig. 21-1.** Cell death in the development of the avian isthmooptic nucleus. Cells of this nucleus project to the retina (color). **A,** Normal early development. A bilateral projection with many cells is present. **B,** Normal adult. The projection is entirely crossed with fewer cells. Presumably, cells not able to sustain a projection (i.e., those projecting ipsilaterally) die and disappear. **C,** Early development with one eye removed. Both nuclei project to the remaining eye. **D,** Later development after early removal of one eye, as in **C.** Only the contralateral projection remains, and cells of the other isthmooptic nucleus die and disappear. Note, however, that fewer cells remain in the intact nucleus than is normally the case, as in **B,** possibly because some cells of the ipsilateral nucleus successfully competed for synaptic sites before dying. (Redrawn from Lund, R.D.: Development and plasticity of the brain, New York, 1978, Oxford University Press.)

appear. As another example, the corpus callosum in immature rodents and cats includes fibers from neurons that do not contribute to this same structure in adults. Most of these cells survive during development, but they retract their exuberant callosal axons during maturation. Also, the immature callosal fibers innervate the entire hemisphere in immature animals, whereas a more limited pattern of innervation is seen in adults (see Fig. 19-12).

The development of cellular and axonal elements is not monotonic (i.e., excessive development of these elements is followed by removal of many). This is not true, however, for synaptic development (Fig. 21-2). That is, the absolute number of synapses increases monotonically during development until maximum levels are gradually reached in adulthood. Apparently the excessive axonal pathways during early development do not signify excessive synaptic numbers, since few synapses have been formed anywhere by this stage. It is not yet clear if the pruning of excessive projections and cells is a prelude to or a result of the dramatic increase in synaptic numbers.

■ *Critical Period*

The examples of developmental neuroplasticity considered here all exhibit an early *critical period of development* during which neuronal connections are sensitive, or plastic, to environmental influences. After this period they are not. For instance, a kitten raised with visual deprivation, induced by closing or patching an eye, develops many abnormal connections related to that eye if, and only if, the deprivation occurs at least partly during the first 3 or 4 postnatal months. After this time, deprivation will not produce abnormal connections, and a normal sensory environment will not correct abnormalities that developed during deprivation. Even a few days of visual deprivation during the critical period can produce permanent developmental changes. There also seems to be a critical period for abnormal development caused by brain injury because neonatal lesions can lead to abnormal connections that are not seen after adult lesions. Adult lesions can lead to plastic changes, but these involve different pathways and processes than those that are involved in developmental neuroplasticity.

■ *Neonatal Lesions*

Animal studies. In a variety of vertebrate species and neuronal pathways, neonatal lesions result in the development of aberrant pathways and cellular development that cannot be produced by later lesions. Thus timing is critical. The plastic effects of two complementary types of lesions have been studied separately and in combination. In one, a set of afferent fibers to a structure is removed to determine what anomalous afferent fibers might develop in its place. In another, a termination site of an afferent pathway is removed to determine if the pathway will develop anomalous connections to another site.

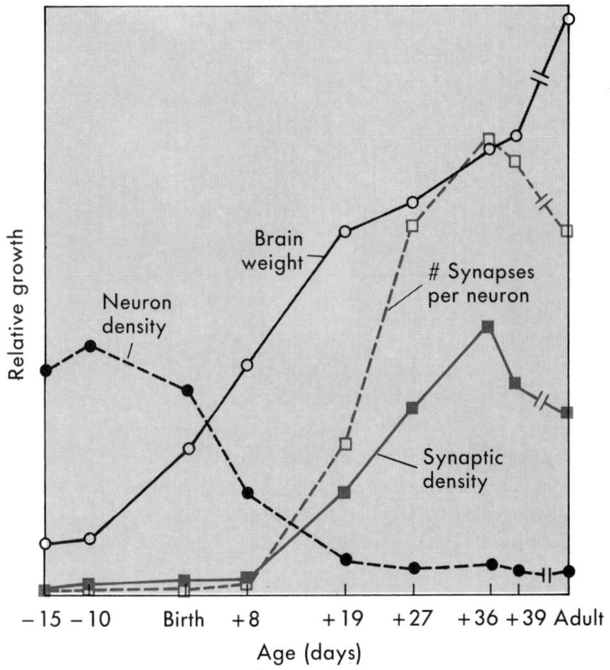

■ **Fig. 21-2.** Neuronal development of the kitten's visual cortex. A large postnatal increase in brain weight is mostly due to growth of the synaptic neuropil. Because no new neurons are added postnatally as the brain grows, neuronal density actually decreases. (Redrawn from Cragg, B.G.: Invest. Ophthalmol. **11**:377, 1972.)

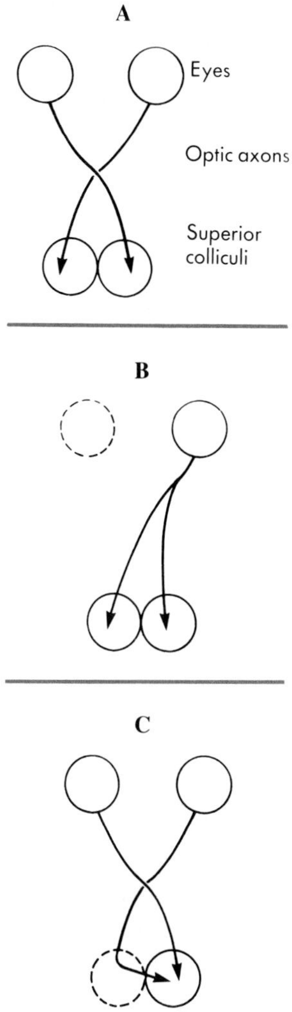

■ **Fig. 21-3.** Effect of neonatal lesions on development of the rat's retinocollicular pathway. **A,** normal development. The pathway is almost entirely crossed. **B,** Removal of one source of afferent input (left eye removed). The remaining eye develops bilateral projections. **C,** Removal of one target (left colliculus removed). The remaining superior colliculus receives input from both eyes.

Examples of the former process often involve removal of one eye neonatally to determine the effects on development of the other eye's connections. For instance, the normal adult rat's retinotectal pathway is nearly exclusively crossed (Fig. 21-3, *A*). However, if the rat is raised with one eye removed, the remaining eye develops a bilateral retinotectal projection (Fig. 21-3, *B*). Similarly, early monocular enucleation in a kitten or monkey leads to an expanded retinogeniculate projection from the remaining eye into "inappropriate" laminae. Plasticity also develops in response to neonatal removal of a normal target zone for a pathway. In rodents, for instance, neonatal removal of one superior colliculus results in the development of binocular input to the remaining colliculus (Fig. 21-3, *C*) instead of the normally crossed retinotectal pattern (Fig. 21-3, *A*).

As might be expected, the most dramatic examples of developmental plasticity result from lesions that remove both a normal input to a structure and a normal termination site of an afferent pathway. For instance, the superior colliculus projects to the pulvinar, which normally receives no retinal afferents (Fig. 21-4, *A*; see also Chapter 8). Neonatal removal of the superior colliculus in rodents leads to growth of retinal afferent fibers into the denervated zone of the pulvinar (Fig. 21-4, *B*). Even more dramatic, if the neonatal tectal lesions involve the inferior colliculus or its brachium, then a major auditory pathway to the medial geniculate nucleus is eliminated (Fig. 21-4, *A*; see also Chapter 11), and aberrant retinal fibers will grow into this auditory structure and form synapses (Fig. 21-4, *C*).

Finally, several unusual examples of cell death often result from neonatal lesions. Early removal of one eye in a bird leads to death and disappearance of cells of the contralateral isthmo-optic nucleus (Fig. 21-1, *C* and *D*), presumably because these cells never formed functional connections. Also, prenatal removal of one eye in mammals leads to slightly more ganglion cells in the remaining retina, because fewer cells die. Part of the explanation may be the greater opportunity for these

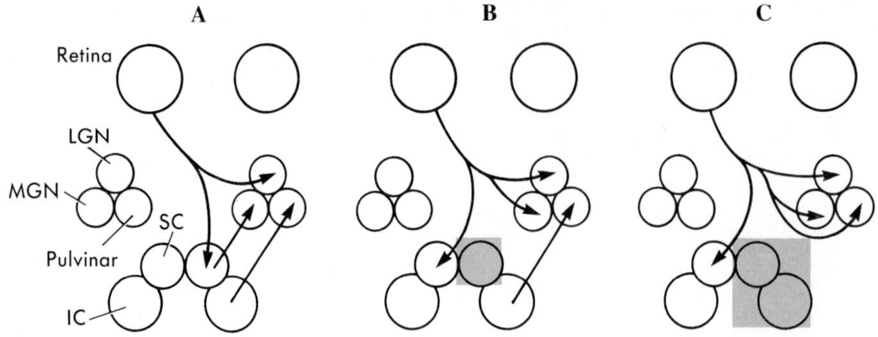

■ **Fig. 21-4.** Effect of neonatal lesions on development of the hamster's retinofugal pathways. **A,** Normal development. The retinofugal pathways are practically completely crossed and innervate the lateral geniculate nucleus *(LGN)* and superior colliculus *(SC)*. Also shown is the ipsilateral projection from the SC to the pulvinar and that from the inferior colliculus *(IC)* to the medial geniculate nucleus *(MGN)*. **B,** Removal of the SC. The contralateral retina not only innervates the remaining SC (Fig. 21-3, *C*), but also innervates the denervated pulvinar. **C,** Removal of the SC and IC. The contralateral retina not only innervates the other SC and the denervated pulvinar, but it also innervates the MGN, even though the MGN is normally an auditory structure.

cells to maintain central connections because more postsynaptic targets are available (Fig. 21-3, *B*).

More interesting is the result of early visual cortex removal in cats or monkeys. Not only do most geniculate neurons die and disappear, but so do a class of retinal ganglion cells. These are the X-cells; W- and Y-cells remain. As noted in Chapter 8, retinal X-cells project only to the lateral geniculate nucleus, and failure of their normal target cells to develop causes these retinal neurons to die and disappear. Retinal W- and Y-cells survive because they project to additional neuronal targets, such as the superior colliculus, that survive the cortical removal. These observations support the notion that the cell death that occurs normally during development eliminates neurons that fail to form functional connections.

We emphasize that these examples of neuroplasticity result from abnormal development after neonatal lesions. Plastic changes after adult lesions are much smaller or nonexistent. Also, not all pathways show such plasticity. These others may be more strictly under genetic control. Furthermore, even within a single structural pathway that demonstrates developmental plasticity, it is not clear whether all cells or only some subpopulation of cells forms aberrant connections. For instance, it was noted earlier that neonatal monocular enucleation in kittens leads to an expanded projection from the remaining eye into ''inappropriate laminae'' of the dorsal LGN. However, this expanded projection is limited to the retinogeniculate axons of Y-cells; the retinogeniculate connections from X-cells continue to obey their proper laminar boundaries. Conversely, as noted in the prior paragraph, retinal X-cells, but not Y-cells, are uniquely affected by early removal of the cortex.

The mechanisms for the formation of aberrant connections after early lesions are obscure. How much is caused by failure of retraction of neonatally exuberant pathways and how much is caused by lesion-induced abnormal growth, or ''sprouting,'' is not clear. Both factors may be involved. Whether by failure to retract or by sprouting, two other processes seem to combine to determine the final pattern of connections. First, axons seem to compete with one another for synaptic targets. A denervated zone thus becomes a particularly attractive target for a developing axon because competition there is reduced. Second, a neuron tends to conserve its axonal terminal arbor, so that if it is prevented from growing at one point it tends to grow more at another. This is called the pruning hypothesis because of its analogous horticultural connotation. Removal of one eye results in the development of an aberrant pathway from the remaining eye to the superior colliculus ipsilateral to that eye. This occurs because less competition for synaptic space is associated with that colliculus. Removal of one colliculus serves to ''prune'' axons from the contralateral eye, and they then tend to grow aberrant connections (e.g., to the ipsilateral, remaining colliculus). When denervation and pruning coincide, as in aberrant retinal input to the pulvinar, the developmental plasticity is particularly evident. Finally, if neonatal lesions prevent neurons from forming functional connections, those neurons may die and disappear. This can even be a retrograde transneuronal process, as it is for retinal X-cells after neonatal cortex removal.

Clinical implications. Detailed studies of human brains that developed after neonatal lesions are generally unavailable or technically not feasible. However, the same neuroplastic changes probably occur. This

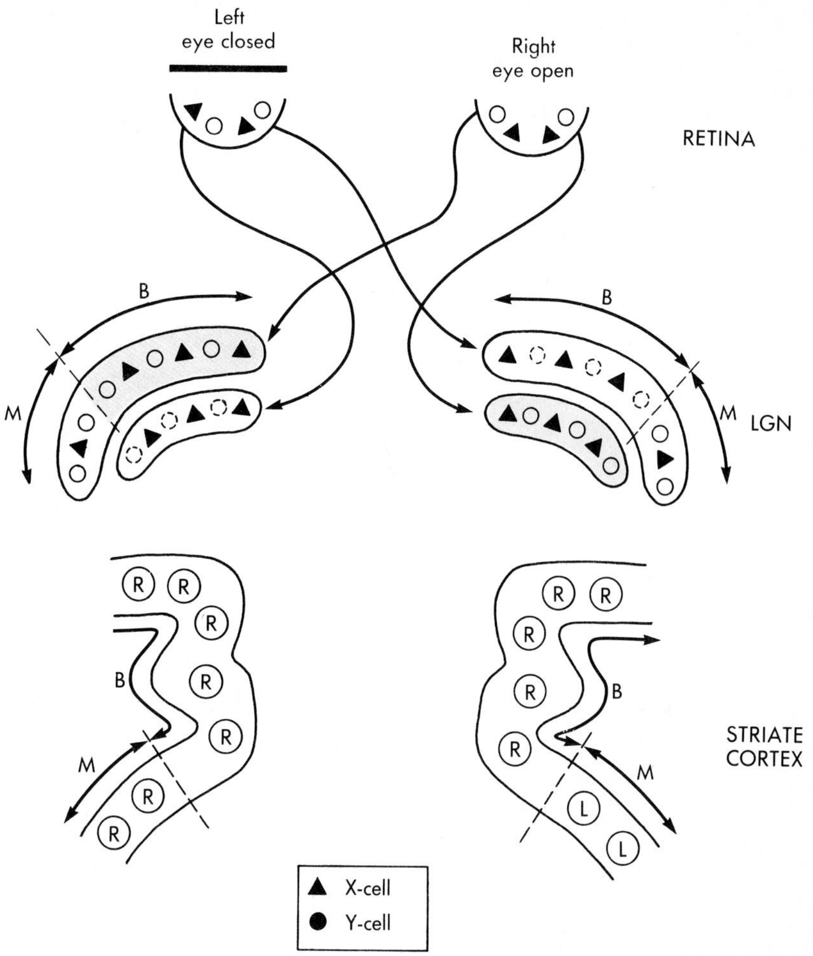

■ **Fig. 21-5.** Summary of some of the abnormalities seen in the visual system of cats raised with monocular deprivation due to maintained closure of one eye (the left eye in this example). Ganglion cells in the deprived retina develop normally. In geniculate laminae receiving input from the deprived eye, the X-cells develop fairly normally, but the Y-cells do not. Y-cell abnormalities, however, are limited to the deprived, binocular segment of the nucleus *(color)*. In the binocular segment of striate cortex nearly all cells have purely monocular receptive fields for the right eye (instead of the binocular fields seen in normal cats), but many normal fields can be seen for the deprived eye in the deprived, monocular segment. *LGN,* dorsal lateral geniculate nucleus; *B,* binocular segment; *M,* monocular segment; *R,* cortical cell with receptive field for only the right eye; *L,* cortical cell with receptive field for only the left eye.

may explain the common observation that lesions in young children often lead to different symptoms than similar lesions in adults. To review an example cited in Chapter 19, a lesion to language areas of cortex in the adult typically leads to permanent aphasia. The same lesion in a child often results in little or no aphasia. Perhaps after the neonatal damage, aberrant pathways can form that subserve language functions. Obviously, such examples of developmental neuroplasticity may be important clinically, and we need to know a great deal more about the underlying mechanisms.

■ *Monocularly Deprived Cats*

Sensory deprivation offers a more physiological and clinically relevant means of studying developmental neuroplasticity than does neural ablation. The deprivation-induced changes are more subtle and do not involve direct physical trauma. The primary change is limited to different levels of neural activity, and these can have profound effects on the developing brain.

The most thoroughly studied and best understood example of developmental neuroplasticity is probably the monocularly deprived kitten. These animals are raised to adulthood with monocular deprivation by suturing closed the lids over one cornea. The eye can be subsequently reopened either to study its ability to influence neural activity or to test the cat's ability to see with that eye. Just a few examples of the affects of such deprivation on visual development will serve to underscore the profound influence of the sensory environment on the developing nervous system.

Geniculostriate abnormalities. Although the deprived retinal ganglion cells seem to develop normally, geniculate and cortical neurons do not (Fig. 21-5). Deprived geniculate cells (i.e., those in laminae innervated by the sutured eye) grow less than in the normal state and exhibit many abnormal responses. Interestingly, these abnormalities are largely limited to Y-cells; the W- and X-cells are relatively unaffected by the deprivation. Geniculate cells in nondeprived laminae appear to develop normally.

The effects seen in striate cortex are dramatic (Fig. 21-6, *B*). Here nearly all cells can be normally driven through the nondeprived eye, but they fail to respond to stimulation of the deprived eye. Thus, instead of the normal binocular activation of these cells, it seems as if the deprived geniculate cells, even the fairly normal W- and X-cells, fail to develop functional geniculostriate connections.

The dimensions of the ocular dominance columns de-

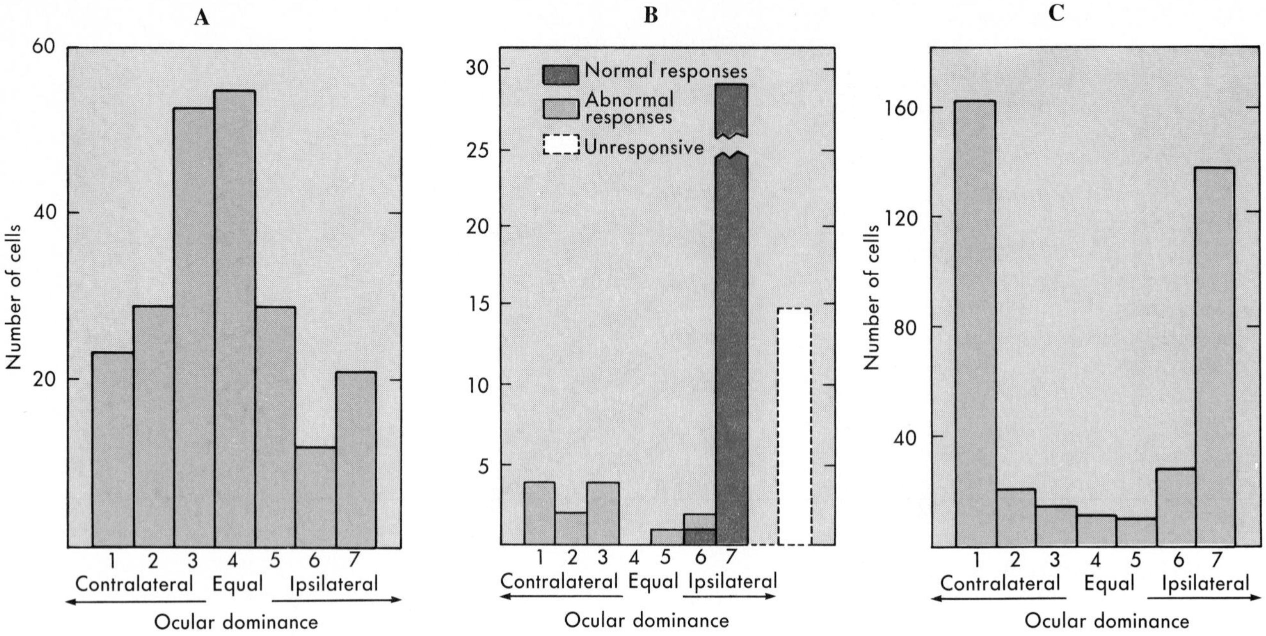

■ **Fig. 21-6.** Ocular dominance histograms of cells in the binocular segment of the cat's striate cortex. Seven classes of ocular dominance are shown: *1,* responds only to stimulation of contralateral eye; *2,* responds well to contralateral eye and poorly to ipsilateral eye; *3,* responds somewhat better to contralateral than to ipsilateral eye; *4,* responds equally well to either eye; *5,* responds somewhat better to ipsilateral than to contralateral eye; *6,* responds well to ipsilateral eye, and poorly to contralateral eye; and *7,* responds only to ipsilateral eye. **A,** Normal cat. All cells respond to visual stimuli and most have binocular receptive fields. **B,** Cat raised with monocular suture of contralateral eye. Most cells have normal receptive fields only for the (nondeprived) ipsilateral eye. The few with fields for the (deprived) contralateral eye respond abnormally, and many neurons are unresponsive. In the other hemisphere, a mirror-image pattern is evident, since the normal cells are class 1 (i.e., with receptive fields for the nondeprived eye). **C,** Cats raised with exotropia (divergent strabismus). All cells are normally responsive, but they tend to have monocular receptive fields for one or the other eye. (Data from Wiesel, T.N., and Hubel, D.H.: J. Neurophysiol. **26:**1003, 1963; Hubel, D.H., and Wiesel, T.N.: J. Neurophysiol **28:**1041, 1965.)

scribed in Chapter 8 are an anatomical correlate. Fig. 21-7 illustrates this for normal and monocularly deprived monkeys. (The results from cats and monkeys are essentially the same, but the monkey data are described because they more clearly illustrate the point.) Normally, the dimensions of the ocular dominance columns from each eye are equal. However, during early monocular deprivation those related to the nondeprived eye grow excessively at the expense of those from the deprived eye. Because these columns reflect the pattern of geniculocortical afferents related to each eye, this implies that the deprived geniculate neurons are unable to establish the normal complement of connections in the cortex, whereas the complement from the nondeprived geniculate cells increases.

Critical period. These abnormalities show a distinct critical period. For instance, if one eye is sutured closed at birth and then opened at 4 to 6 months of age, at which time the other eye is sutured closed for an additional several years, the deficits are as just described and are limited to the initially deprived eye. Also, eye closure for just a few days during the critical period can impair vision permanently.

Binocular competition. Much of the neuroplasticity caused by monocular deprivation results from competition among developing axons for synaptic connections. In some form of *binocular competition,* pathways from the open eye develop at the expense of those from the closed eye. This is indicated by a comparison of the development that occurs in the binocular segment with that in the monocular segment (Figs. 21-5 and 21-8).

Because the visual field is retinotopically mapped onto the visual parts of the brain, one can define the *binocular segment* of the striate cortex, dorsal LGN, superior colliculus, etc., as that portion onto which the binocularly viewed, central region of the field is mapped. The remaining *monocular segment* corresponds to that extreme lateral portion of the visual field

A

B

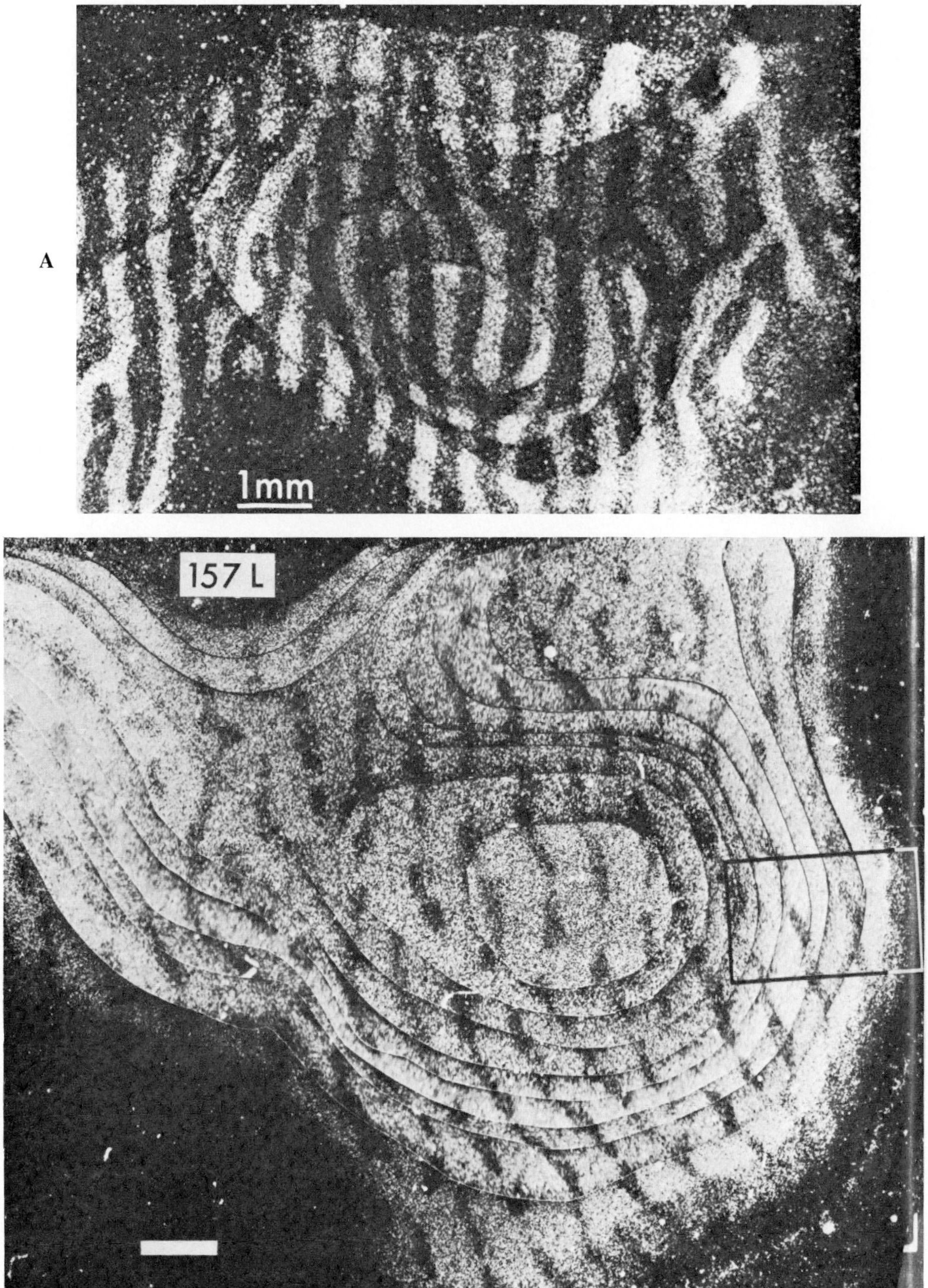

■ **Fig. 21-7.** Photographic montages showing ocular dominance columns in striate cortex of monkeys; view of a plane parallel to the cortical layering. The bright bands represent ocular dominance columns related to one eye and its appropriate set of geniculate laminae, and the intervening dark areas are the columns for the other eye and geniculate laminae. The bands were created by autoradiography after injecting a radioactive tracer into one eye: the tracer is first transported by retinogeniculate axons to geniculate neurons, and it is then transneuronally transported to striate cortex. The labeled cortical columns are seen as bright bands in these dark-field views. **A,** Normal monkey. Notice that the bright and dark bands relating to each eye are comparable in size. **B,** Monkey raised with monocular deprivation. The nondeprived eye was injected with tracer, so that the labeled (bright) columns reflect this eye. These columns are much larger than the unlabeled (dark) ones from the deprived eye.

Continued.

C

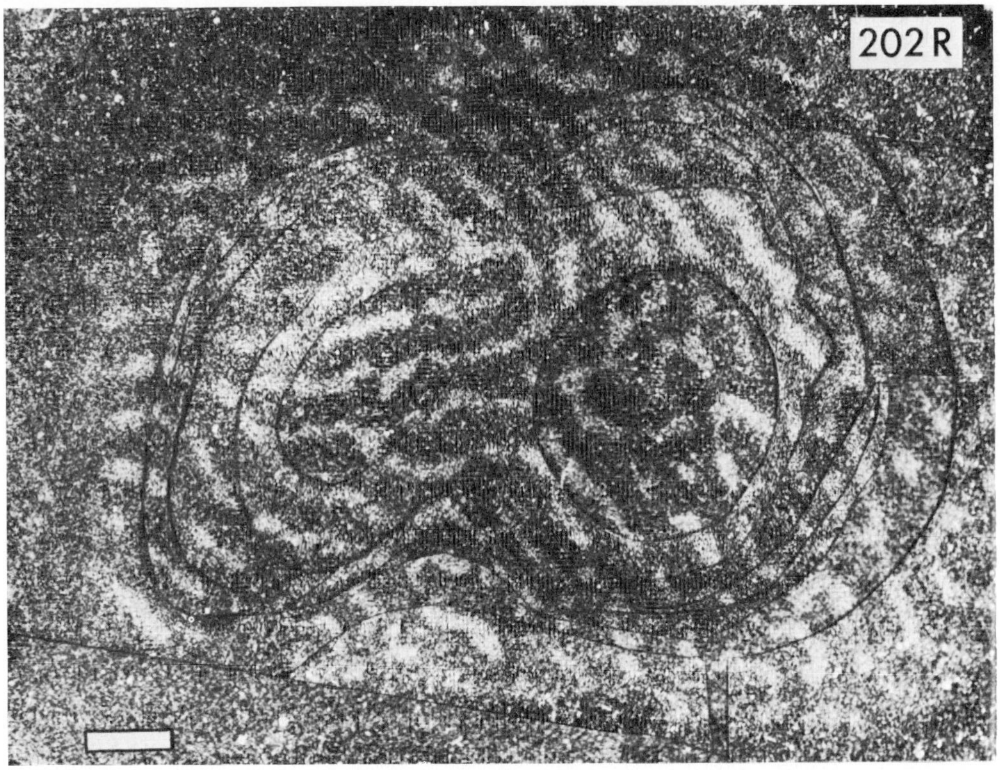

■ **Fig. 21-7, cont'd. C,** Another monkey raised with monocular deprivation. The deprived eye was injected with tracer, providing the complementary picture to **B.** Here, the deprived columns are much smaller than are the nondeprived ones. (**A** reproduced from Hubel, D.H., and Wiesel, T.N.: Proc. Roy. Soc. (B) **198:**1, 1977; **B** and **C** reproduced from LeVay, S., Hubel, D.H., and Wiesel, T.N.: J. Comp. Neurol. 191:1-51, 1980.)

on each side that can be viewed only by the eye on that side. Fig. 21-8 shows that, in addition to this natural monocular segment *(NMS),* an artificial monocular segment *(AMS)* can be created for one eye by placing a retinal lesion *(L)* in the other.

This division is most interesting because of the different effects of early monocular deprivation on the two segments. Nearly all of the abnormalities just noted for the deprived eye among geniculate and striate cortex neurons are limited to the binocular segment (Figs. 21-5 and 21-8). The natural or artificial monocular segment innervated by the deprived eye is just as deprived of pattern vision as the binocular segment. In these monocular segments, the geniculate cells grow to normal size; the response properties even for Y-cells develop fairly normally, and many striate cortex neurons develop normal receptive field properties for the deprived eye. Subtle abnormalities do develop in the deprived monocular segment, but they are smaller than those in the binocular segment. A similar pattern can be found in the superior colliculus and in extrastriate cortical areas that have been studied in these cats. Furthermore, the behavioral capacity with the deprived eye is closer to normal in response to stimuli placed in the monocular segment than to those in the binocular segment.

Fig. 21-9 illustrates the main conclusion that can be derived from these observations. That is, central pathways related to each eye compete with one another for synaptic connections during development. We shall use the geniculostriate connections as an example, although we are not certain about the specific sites of this competitive development. During normal development, pathways from one or the other eye gain no systematic competitive advantage; a balance is struck, and geniculate neurons develop normally with convergent binocular afferentation of cortical cells (Fig. 21-9, *A*). Monocular deprivation upsets the balance (Fig. 21-9, *B*). Deprived geniculate neurons develop at a competitive disadvantage because of the reduction in evoked activity. This allows their antagonists from normally innervated laminae to capture the vast majority of synaptic space in the cortex at the expense of the deprived neurons. The latter neurons fail to develop normally, and cortical cells become effectively innervated only by the nondeprived geniculate neurons. One anatomical consequence of this is the distorted pattern of ocular dominance columns illustrated in Fig. 21-7.

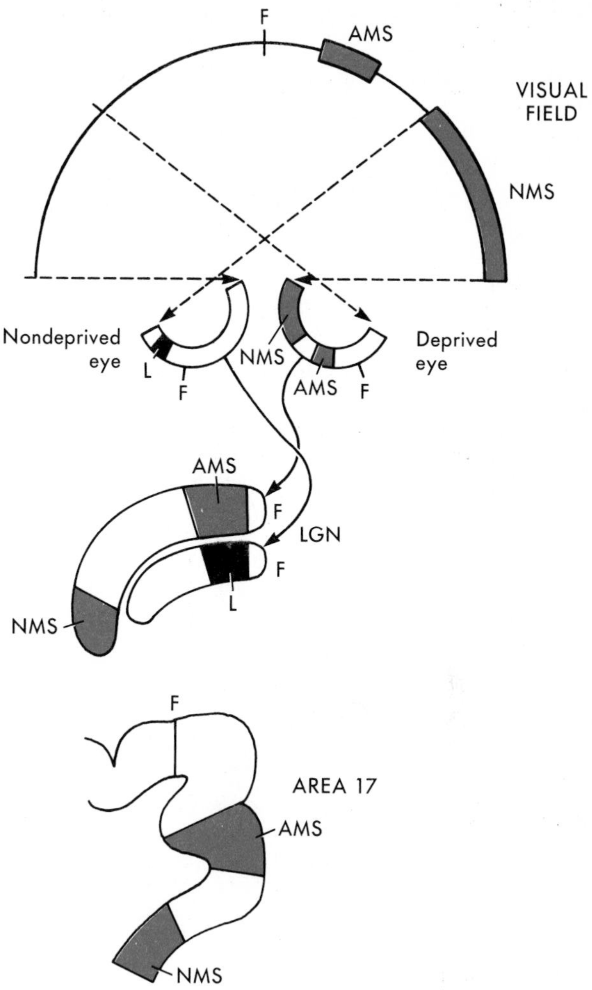

■ **Fig. 21-8.** Effects of monocular deprivation on natural and artificial monocular segments *(NMS* and *AMS)* in cats. If a lesion *(L)* is placed in the retina of the nondeprived eye of a neonatal cat, two monocular segments exist for the deprived eye. The NMS corresponds to the extreme nasal retina that views a portion of peripheral visual field beyond the view of the other eye. The AMS corresponds to the retinal region that maps the same portion of visual space as the region lesioned in the other eye. Because of the retinotopic map in the dorsal lateral geniculate nucleus *(LGN)* and area 17, the NMS and ALS can be found in these structures. As summarized in Fig. 21-5, the deprived NMS develops relatively normally in the dorsal LGN and area 17. Likewise, the deprived AMS develops equally normally, and perimetry tests indicate relatively normal vision for the deprived eye only in the NMS and AMS regions of the visual field. Since the deprived AMS develops fairly normally even though it represents a fairly central portion of visual field, this implies that the deprived monocular segment is spared many of the deleterious consequences of monocular deprivation, not because it represents peripheral visual field but rather because it represents monocular visual field, *F,* Fixation point and its representation in the retina, dorsal LGN, and area 17. (Redrawn from Sherman, S.M., and Spear, P.D.: Physiol. Rev. **62:**738, 1982.)

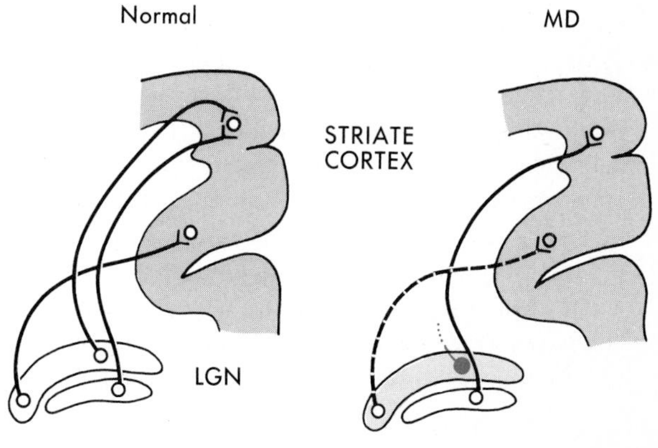

■ **Fig. 21-9.** Hypothesis of binocular competition used to explain observations summarized in Figs. 21-5 and 21-8. The hypothesis suggests that development largely proceeds via a process of competitive interactions between the pathways related to each eye. For the sake of clarity, the hypothesis is described as if these interactions occur among geniculostriate synapses, although the actual site of the interactions is currently unknown. ***Normal development.*** Normally, there is an enormous postsynaptic proliferation of geniculocortical synapses, and they compete with one another for space onto and control of the cortical cell. With a normal binocular environment, a competitive advantage in this process is conferred to neither geniculate lamina's neurons, a balance is struck, and binocular cortical cells develop. ***Monocular deprivation (MD).*** The cell in the binocular segment of the deprived lamina *(colored circle)* is somehow placed at a competitive disadvantage with respect to its nondeprived rival. Perhaps its reduced level of activity leads to this disadvantage. The result is that the nondeprived geniculate neuron develops exclusive synaptic control of the cortical cell. However, by definition, the deprived cell in the monocular segment *(open circle)* cannot be placed at a competitive disadvantage, since no nondeprived cells are available to innervate its target neuron. Thus, although deprived as much as its neighbor in the binocular segment, the monocular segment cell is free to develop synaptic control of the cortical cell. To the extent that development in the deprived monocular segment is less affected than that in the binocular segment, a mechanism of binocular competition is implicated. If, on the other hand, equal abnormalities were seen in both segments, there would be no reason to infer such a competitive process of synaptic development. (Redrawn from Sherman, S.M. In Freeman, R.D., editor: Developmental neurobiology of vision, New York, 1979, Plenum Press.)

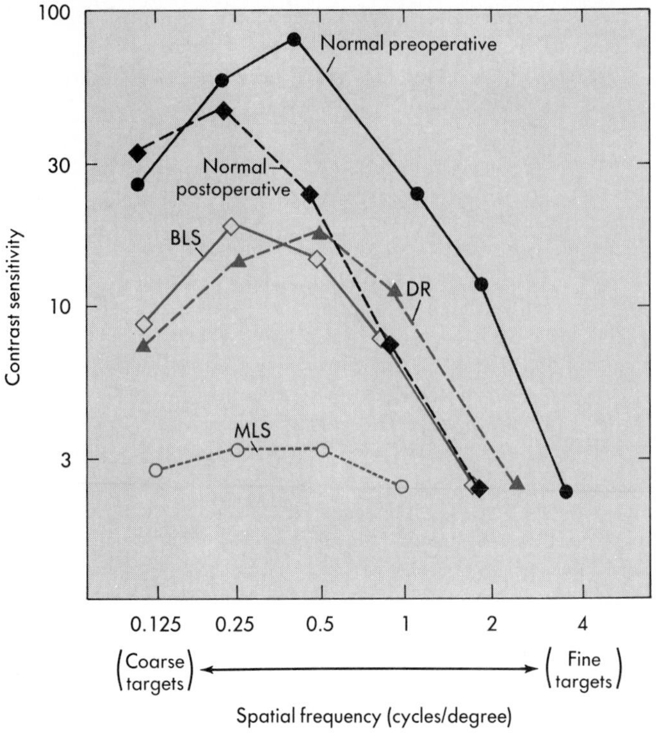

■ **Fig. 21-10.** Behavioral contrast sensitivity functions of cats. These functions plot the inverse of the minimum contrast necessary for the animal to detect a sine wave grating of various spatial frequencies. Low spatial frequencies correspond to coarse or large targets, and high frequencies represent fine or small targets. The function for normally reared cats is shown in black before *(solid line)* and after *(dashed line)* bilateral removal of striate cortex, which also effectively eliminates the X-cell pathway. Functions for cats raised with visual deprivation are shown in color and include binocular lid suture *(BLS),* total dark rearing *(DR),* and monocular lid suture *(MLS).* The MLS data are for the deprived eye only; behavior with the nondeprived is normal. Where sensitivity to low spatial frequencies is good (normal and destriate cats), spatial vision is good; where sensitivity to these frequencies is poor (BLS, DR, and MLS cats), spatial vision is poor. (Redrawn from Lehmkuhle, S., et al.: J. Neurophysiol. **48:**372, 1982.)

However, a deprived cell can be placed at such a competitive disadvantage only in the binocular segment, where such competitive interactions are possible. The deprived cell in the monocular segment may develop more slowly and less completely than normal, but it need not compete against a favored antagonist. Thus many normal connections can be formed in this segment despite the deprivation. These normal connections in the deprived monocular segment amply illustrate that the deficits that develop in the binocular segment are not caused by deprivation per se. Rather they are the result of the effects of competitive interactions that become unbalanced during monocular deprivation.

Neural basis of amblyopia. The serious disruption in the development of the Y-cell pathway during deprivation may explain some of the profound amblyopia (i.e., reduced visual capacity) exhibited by the deprived eye. In Chapter 8 we suggested that the Y-cell pathway, because of its sensitivity to lower spatial frequencies, seems sufficient and perhaps even necessary for normal pattern vision. Its disruption during deprivation results in poor sensitivity to low spatial frequencies and thus results in amblyopia.

Fig. 21-10 compares the spatial contrast sensitivity among several groups of cats. These include normally reared cats before and after bilateral removal of striate cortex, monocularly deprived cats, and binocularly deprived cats. (Binocular deprivation also leads to somewhat abnormal development of the Y-cell pathway.) Removal of striate cortex completely interrupts the X-cell pathway but not the Y-cell pathway. The destriate cats exhibit far better spatial vision and better low-frequency sensitivity than do the deprived cats, even though there may be no obvious differences in spatial acuity. This suggests that abnormal development of the striate cortex is not sufficient to explain the amblyopia of visually deprived cats because its total removal in normally raised cats leads to less visual deficit than does deprivation rearing. A more likely explanation for the amblyopia seems to be the Y-cell abnormalities because geniculate Y-cells in the cat innervate many cortical areas (see Chapter 8). Consequently, a vast expanse of cortex develops abnormal geniculate input. In any case, the neuroplastic changes in the visual system caused by visual deprivation dramatically affect visual capacity.

■ *Clinical Significance*

The message for human pediatrics from these animal studies is clear: the infant's sensory environment is a crucial factor in normal brain development. This is most evident from the many forms of amblyopia that develop from visual deprivation. Numerous parallels with visually deprived cats are apparent, although by no means is it clear that cats and humans share the same developmental mechanisms. Nonetheless, if an infant is raised with any ocular abnormality that distorts the visual environment, a permanent amblyopia is likely to result. Such distortions can be caused by corneal scarring, cataracts, ametropia, strabismus, ptosis of the eyelid, and even therapeutic patching.

The deficits caused by these forms of deprivation exhibit a critical period that extends through the first 5 to 10 years of life. One example of this comes from studies of astigmatic patients. Astigmatism means that all target orientations cannot be simultaneously focused (see Chapter 8). Because only one meridian is properly focused during astigmatism, the deprivation is similar to that in the kittens whose visual experience was limited to stripes of a particular orientation. The patients who were studied focused one meridian and failed to focus the orthogonal one. Their astigmatism was corrected optically with an appropriate cylindrical lens. If the correction was made in infancy or early childhood, no substantive amblyopia developed. If the correction was not made until adulthood, an amblyopia specific to the previously defocused meridian was evident despite optical correction. Astigmatism of late onset (i.e., after the critical period) did not lead to such a meridional amblyopia, and vision was normal after appropriate optical correction. These results indicate that a critical period of development exists and also that the resultant amblyopia is specifically related to the precise nature of the visual deprivation.

These and similar observations emphasize that the causes of visual distortion should be removed as soon as possible in infants to avoid permanent amblyopia. That is, a cataract should be removed, strabismus corrected, and myopia eliminated as soon as diagnosed. This current view is in stark contrast to the prevailing opinion of 20 or 30 years ago, when children were given a chance to outgrow these symptoms, and attempts to correct them were postponed until late in the critical period.

The importance of the critical period also dictates caution in the early treatment of visual disorders. For instance, when many young children begin to show the first signs of a weak or amblyopic eye, the weakness can often be reversed through forced usage of that eye by patching the good or dominant eye. This is still excellent therapy, but it must be used carefully with regard to the effects of even a few days' deprivation of a healthy eye. (Remember that a few days' monocular occlusion in a kitten, or slightly longer in an infant monkey, produces permanent defects in the central pathways related to that eye.)

Finally, these observations in the visual system can probably be extrapolated in a general sense to neural development. Anything that affects the developing neuronal environment, including sensory stimuli, motor activity, learning, motivational experience, and nutrition, may influence the final product of neural development.

■ Response to Injury of the Adult Brain

If axons in the central nervous system of poikilothermic vertebrates are damaged, they have the capacity to regenerate. For instance, if the optic nerve of a fish or frog is transected, the distal fragments will degenerate, as in homeothermic vertebrates. However, the proximal axonal stumps will then regenerate and reestablish connections with the appropriate retinofugal targets, such as the optic tectum. These regenerated connections eventually function and a nearly normal pathway is reestablished.

In contrast, homeothermic vertebrates show such regeneration only in the peripheral nervous system. With injury of central pathways, regeneration is incomplete. Why successful regeneration does not occur centrally in homeothermic vertebrates is under extensive investigation.

Although regeneration does not occur in the adult mammal, plastic changes do follow injury to the central nervous system. The most thoroughly studied example of such injury-induced modification is the phenomenon of *axonal sprouting* (Fig. 21-11). Such sprouting occurs in adult mammals when the dorsal roots are unilaterally transected for several segments above and below a dorsal root that was left intact. After several months the projection pattern of the fibers from the remaining dorsal root on that side was compared with the projection pattern of the corresponding dorsal root on the control side of the spinal cord. The projections of the dorsal root on the experimental side were found to have expanded dramatically, presumably reflecting the sprouting of the axons into nearby denervated zones that had been left vacant by the previous dorsal root transections.

Such axonal sprouting occurs in many structures of the mammalian central nervous system, including retinofugal targets, the hippocampus, the septum, and the red nucleus. In some cases the new connections have indeed formed functional connections. For example, the red nucleus receives prominent projections from both the cerebellum and motor cortex. The cerebellar projection is preferentially distributed to the soma and proximal dendrites of the rubral neurons, whereas the cortical projection is primarily on more distal dendrites. If the cerebellar projection is destroyed, then the cortical projection sprouts to establish more proximal connections at the synaptic sites that have been vacated by degeneration of the cerebellorubral pathway. These new connections are functional, and they increase the influence of the motor cortex on the red nucleus. Such injury-induced plasticity may contribute to the recovery of motor behavior.

The phenomenon of axonal sprouting implies that denervated portions of dendrites and somata can attract collateral inputs from nearby intact axons. Moreover, such plasticity occurs in many areas of the adult mammalian brain. However, not all pathways are equally capable of sprouting, and studies of several pathways have failed to demonstrate sprouting.

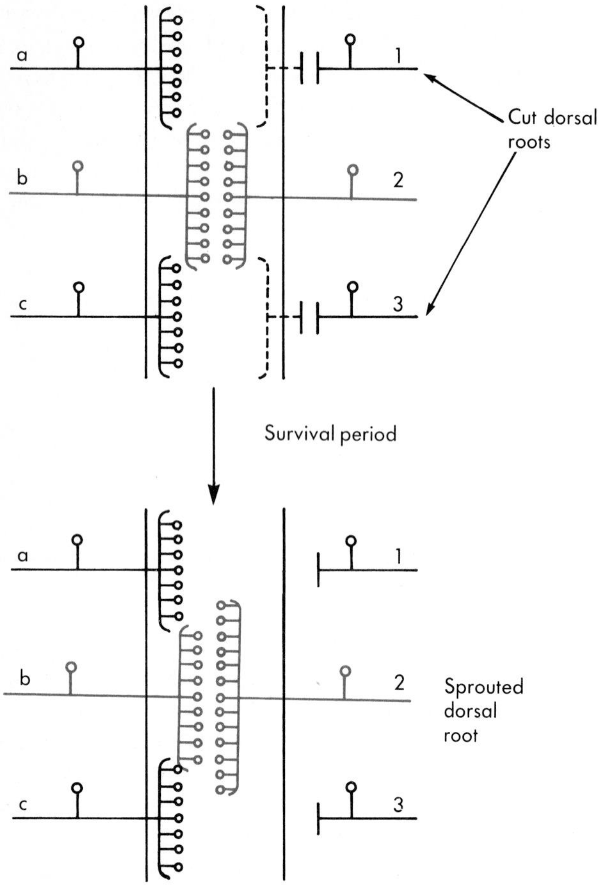

■ **Fig. 21-11.** Lesion-induced sprouting in the mammalian spinal cord. Normally fibers from dorsal roots *b* and *2* have similar central projections *(upper)*. If fibers from dorsal roots *1* and *3* (neighboring segments to 2) are cut, their central connections will degenerate. After a survival period, fibers from dorsal root *2* "sprout" to occupy some of the denervated space so that their terminal extent is greater than that of fibers from dorsal root *b (lower)*.

Learning and Memory

The capacity of at least certain classes of neurons to change their behavior as a function of environmental inputs is a very important property of the nervous system because it subserves learning and adaptive behavior. In the broadest sense *learning* refers to a process that mediates a change in behavior as a result of experience. *Memory* is the end point of the process and refers to the actual storage of information representing the experience-induced modification. Such storage of information involves actual physical changes in neurons. The mechanisms of such change have commanded intense interest for many decades, and our understanding of the cellular changes that mediate information storage has advanced substantially.

Classes of Learning

Controversies have raged over the years among psychologists as to whether there are different kinds of learning and, if so, what the different classes of learned behavior might be. We will not review such arguments here. But two very broad classes of learned behavior—*nonassociative* and *associative*—can be distinguished. Nonassociative learning does not require that an organism learn a predictive relationship between stimuli. The simplest example of such learning is *habituation* during which a particular stimulus elicits some unlearned response that declines in magnitude (habituates) with repeated presentation of that stimulus. In teleological terms the organism is learning that the stimulus is not biologically meaningful in a given context. Therefore the organism becomes increasingly unresponsive to it (i.e., it learns not to respond). A commonplace example is a dog or a cat that orients its head and ears toward a novel auditory stimulus. With repetition of that stimulus, such orienting behavior will attenuate or habituate if no meaningful event is associated with the occurrence of the stimulus.

Another kind of nonassociative learning is *sensitization*. For example, some noxious event may increase an organism's responsiveness to other stimuli that would normally be neutral. If a dog or cat experiences a painful stimulus of some sort, then the subsequent occurrence of the auditory stimulus might elicit more vigorous orienting behavior than normally occurs. Teleologically one could view this kind of learning as an adaptive behavior that prepares the animal for responding to a potentially threatening environment.

In associative learning an organism acquires information about the relationship among stimuli. In *classical conditioning,* described extensively by Pavlov, the temporal pairing of a neutral stimulus, the *conditioned stimulus,* with a stimulus that elicits an unlearned response, the *unconditioned stimulus,* can endow the conditioned stimulus with the novel capacity to elicit a response. Often this conditioned response is similar to that evoked by the unconditioned stimulus, although that need not be the case. A traditional example of such learning is provided by Pavlov's observation that his experimental dogs salivated on the appearance of the caretaker who fed them. Salivation is an unconditioned response to ingestion of food, the unconditioned stimulus. The visual and auditory stimuli associated with the caretaker normally would not elicit salivation, but these became conditioned stimuli because they were associated with the delivery of food. Consequently, these stimuli gained the capacity to elicit a conditioned response, salivation, which in this example resembled the unconditioned response to food intake.

The other major form of associative learning is *instrumental* or *operant conditioning*. In this case responses that are immediately followed by an event that has reinforcing value will show an altered probability of occurrence. If the reinforcing event is positive, such as presentation of food, then the probability of response will increase. If the reinforcing event is negative, such as pain, then the probability of response will decrease. As in classical conditioning, a predictive relationship is being learned.

Stages of Memory

The storage of information, or memory, may involve different stages. For instance, early in learning the information may be represented in short-term memory and over time this information is then transferred in some way to long-term memory. The number of stages, their relationships to each other, and the mechanisms underlying each remain an area of debate. Indeed, the usefulness of distinguishing among different phases of memory has been questioned by some. However, a distinction between short-term and long-term memory retains some heuristic value in understanding a broad range of experimental and clinical findings. For example, head trauma often results in a *retrograde amnesia* in which the memory of recently learned events is lost, but memory for older events remains unimpaired. A similar phenomenon occurs after electroconvulsive shock therapy, a treatment for severe depression disorders.

Such observations suggest a more fragile short-term memory that ultimately is transformed into a more robust long-term form of storage. It is generally agreed that long-term storage must be mediated by physical changes in the involved neurons. However, whether there is a qualitative difference between these forms of storage has not been resolved.

Localization of Information Storage

Neuroscientists have been concerned for many years with what parts of the brain are involved in storing information. It is now clear that for any given learning task there are specific neural pathways that participate in the performance of that task. Moreover, these pathways may differ, depending on the nature of the specific learning task. Thus storage may be localized, since in a given learning situation not all areas of the brain are necessarily involved. In any specific situation changes occur along much of the involved pathways, although not necessarily in all segments of those pathways. Thus it is unlikely that memory is stored in only a few structures. Related questions that remain to be resolved are whether all neurons can store information on a long-term basis and whether specific classes of neurons are specialized for this function.

The neuropathology of learning and memory disorders in humans has not contributed substantially to our understanding of these phenomena. Lesions of the temporal lobe, hippocampus, and certain midline diencephalic structures, such as the dorsomedial nucleus of the thalamus, result in a generalized *anterograde amnesia*. That is, previously stored information is preserved, but the ability to store new information is impaired. In Korsakoff's syndrome these lesions result from a thiamine deficiency caused by inappropriate nutrition in chronic alcoholism. Although experimental lesions of these structures in animals can produce similar effects, what remains puzzling is why these are the only structures that have been persistently associated with memory deficits in humans.

Cellular Mechanisms of Learning

For many years investigators were seriously frustrated in their efforts to describe the cellular mechanisms of learning. However, significant advances have been made. These have derived mainly from the development of experimental animal models in which the relevant circuitry for some well-specified learned response has been delineated. Describing such circuitry in the complex brains of vertebrates is difficult, and only a few experimental models involve vertebrate species. However, the use of simpler invertebrate systems reduces the complexity of mapping the relevant pathways, and progress has been significant when such systems have been used.

The available invertebrate and vertebrate models have generated exciting information. For example, results from vertebrate models suggest that training-induced modification in a given learning task is probably distributed extensively over the involved neural pathways, including the sensory pathways. The invertebrate models are progressing to a stage of development where a true cell biology of learning is at hand. Some of these models involve simple systems consisting of a monosynaptic connection between a single sensory neuron and a single motoneuron (Fig. 21-12). In such systems short-term habituation and sensitization have now been studied intensively. Such investigations have shown that the neuronal changes are presynaptic, involving modification of transmitter release at the terminals of the sensory neuron. In habituation, transmitter release is decreased, whereas in sensitization it is increased. Most recently, studies of nonassociative learning in *Aplysia*, a gastropod mollusc, have permitted a biophysical analysis of the presynaptic depression and facilitation that mediate habituation and sensitization, respectively. These studies have implicated changes in the Ca^{++} current at the presynaptic terminals (Fig. 21-13).

■ **Fig. 21-12.** Simplified neuronal circuit for the gill-withdrawal reflex in the marine snail *Aplysia*. The site of plasticity in this reflex that underlies its habituation is at the terminals of the sensory neurons on the central target cells *(color)*—the interneurons and motoneurons. (Redrawn from Kandel, E.R., and Schwartz, J.H.: Principles of neural science, New York, 1981, Elsevier/North Holland.)

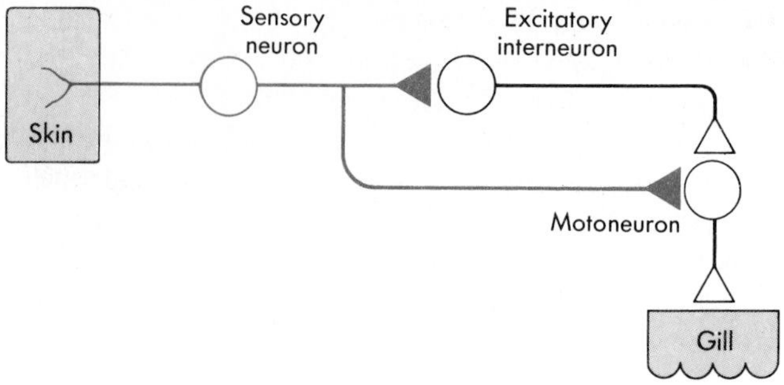

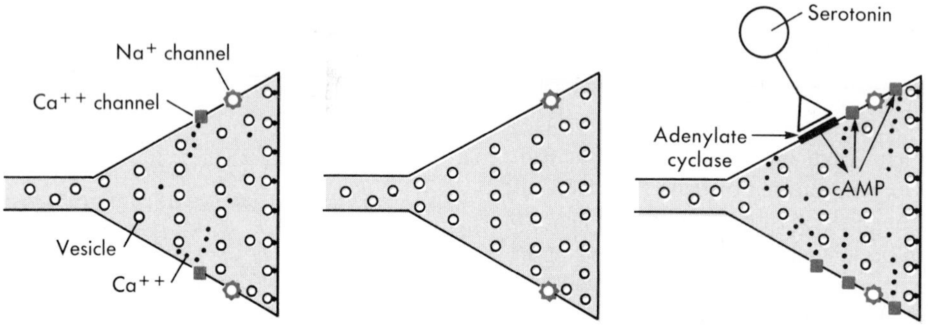

■ **Fig. 21-13.** Model of short term habituation and sensitization. **A,** In the control state an action potential in the terminal membranes of the sensory neurons opens a number of Ca^{++} channels *(black squares)* in parallel with the Na^{+} channels *(hexagons)*. As a result some Ca^{++} flows into the terminals and allows a certain number of synaptic vesicles to bind to release sites and be released. **B,** Repeated action potentials in the terminals decrease the number of open Ca^{++} channels in the sensory terminal and, in the limit, may shut them down altogether. The resulting depression in Ca^{++} influx functionally inactivates the synapse by preventing synaptic vesicles from binding, C, Sensitization is produced by cells thought to be serotonergic. In the terminals serotonin acts on adenylate cyclase, which stimulates the synthesis of cAMP. cAMP in turn acts on a cAMP-dependent protein kinase to phosphorylate a membrane protein thought to be the K^{+} channel, and this leads to a decrease in the repolarizing K^{+} current and a broadening of the action potential. The increase in the duration of the action potential increases the time during which Ca^{++} channels can open, leading to a greater influx of Ca^{++} and greater binding of vesicles to release sites, and therefore to an increased release. (Redrawn from Kandel, E.R., and Schwartz, J.H.: Principles of neural science, New York, 1981, Elsevier/North Holland.)

■ *Bibliography*

Journal articles

Blakemore, C., et al.: Recovery from monocular deprivation in the monkey. I. Reversal of physiological effects in the visual cortex, Proc. R. Soc. Lond. [Biol.] **213:**399, 1981.

Crawford, M.L.J., et al.: Binocular neurons and binocular function in monkeys and children, Invest. Ophthal. Vis. Sci. **24:**491, 1983.

Easter, Jr., S.S.: Birth of olfactory neurons: lifelong neurogenesis, Trends Neurosci. **7:**105, 1984.

Finlay, B.A., et al.: Cell death in the mammalian visual system during normal development. II. Superior colliculus, J. Comp. Neurol. **204:**318, 1982.

Garey, L.J., and Vital-Durand, F.: Recovery from monocular deprivation in the monkey. II. Reversal of morphological effects in the lateral geniculate nucleus, Proc. R. Soc. Lond. [Biol.] **213:**425, 1981.

Harwerth, R.S., et al.: Behavioral studies of stimulus deprivation amblyopia in monkeys, Vision Res. **21:**779, 1981.

Hess, R.F., et al.: Residual vision in humans who have been monocularly deprived of pattern stimulation in early life, Exp. Brain Res. **44:**295, 1981.

Innocenti, G.M.: Growth and reshaping of axons in the establishment of visual callosal connections, Science **212:**824, 1981.

Jeffery, G.: Retinal ganglion cell death and terminal field retraction in the developing rodent visual system, Dev. Brain Res. **13:**81, 1984.

Jeffery, G., et al.: Does the early exuberant retinal projection to the superior colliculus in the neonatal rat develop synaptic connections? Dev. Brain Res. **14:**135, 1984.

Jen, L.S., and Lund, R.D.: Experimentally induced enlargement of the uncrossed retino-tectal pathway in rats, Brain res. **211:**37, 1981.

Kaas, J.H., et al.: The reorganization of the somatosensory

cortex following peripheral nerve damage in adult and developing mammals, Annu. Rev. Neurosci. **6:**325, 1983.

Knudsen, E.I., et al.: Monaural occlusion alters sound localization during a sensitive period in the barn owl, J. Neurosci. **4:**1001, 1984.

Landmesser, L.T.: The generation of neuromuscular specificity, Annu. Rev. Neurosci. **3:**279, 1980.

Lehmkuhle, S., et al.: Spatial and temporal sensitivity of normal and amblyopic cats, J. Neurophysiol. **48:**372, 1982.

Manny, R.E., and Levi D.M.: Pyschophysical investigations of the temporal modulation sensitivity function in amblyopia: spatiotemporal interactions, Invest. Ophthalmol. Vis. Sci. **22:**425, 1982.

Mendell, L.M.: Modifiability of spinal synapses, Physiol. Rev. **64:**260, 1984.

Purves, D., and Lichtman, J.W.: Elimination of synapses in the developing nervous system, Science **210:**153, 1980.

Rakic, P.: Limits of neurogenesis in primates, Science **227:**1054, 1985.

Sengelaub, D.R., and Finlay, B.L.: Early removal of one eye reduces normally occurring cell death in the remaining eye, Science **212:**573, 1981.

Sherman, S.M.: Development of retinal projections to the cat's lateral geniculate nucleus, Trends Neurosci. **8:**350, 1985.

Sherman S.M., and Spear, P.D.: Organization of visual pathways in normal and visually deprived cats, Physiol. Rev. **62:**738, 1982.

Sur, M., et al.: Monocular deprivation affects X- and Y-cell retinogeniculate terminations in cats, Nature **300:**183, 1982.

Swindale, N.V., et al.: Recovery from monocular deprivation in the monkey. III. Reversal of anatomical affects in the visual cortex, Proc. R. Soc. Lond. [Biol.] **213:**435, 1981.

Tsukahara, N.: Synaptic plasticity in the mammalian central nervous system, Annu. Rev. Neurosci. **4:**351, 1981.

Tsukahara, N., et al.: Specificity of the newly-formed corticorubral synapses in the kitten red nucleus, Exp. Brain Res. **51:**45, 1983.

Books and monographs

Blakemore, C.: Maturation and modification in the developing visual system. In Held, R., Leibowitz, H.W., and Teuber, H.L., editors: Handbook of sensory physiology, vol. 8, New York, 1978, Springer-Verlag New York, Inc.

Cohen, D.H.: Some organizational principles of a vertebrate conditioning pathway: is memory a distributed property? In Weinberger, N.M., McGaugh, J.L., and Lynch, G., editors: Memory systems of the brain: animal and human cognitive processes, New York, 1985, Guilford Publications, Inc.

Cotman, C.W., editor: Neuronal plasticity, New York, 1978, Raven Press.

Freeman, R.D., editor: Developmental neurobiology of vision, New York, 1979, Plenum Press.

Hirsch, H.V.B., and Leventhal, A.G.: Functional modification of the developing visual system. In Jacobson, M., editor: Handbook of sensory physiology, vol. 9. New York, 1978, Springer-Verlag New York, Inc.

Innocenti, G.M.: Two types of brain plasticity? In Cuénod, M., Kreutzberg, G.W., and Bloom, F.E., editors: Progress in brain research, vol. 51, Amsterdam, 1979, Elsevier/North-Holland.

Kupferman, I.: Learning. In Kandel, E.R., and Schwartz, J.H., editors: Principles of neural science, New York, 1981, Elsevier-North-Holland.

LeVay, S., et al.: The postnatal development and plasticity of ocular-dominance columns in the monkey. In Schmitt, F.O., et al., editors: The organization of the cerebral cortex, Cambridge, Mass., 1981, The MIT Press.

Lund, R.D.: Development and plasticity of the brain, New York, 1978, Oxford University Press Inc.

Purves, D., and Lichtman, J.W.: Principles of neural development, Sunderland, Massachusetts, 1985, Sinauer Associates, Inc.

Sherman, S.M.: Functional organization of the W-, X- and Y-cell pathways of the cat: a review and hypothesis. In Sprague, J.M., and Epstein, A.N., editors: Progress in psychobiology and physiological psychology, vol. II, New York, 1985, Academic Press, Inc.

Steward, O.: Events within the sprouting neuron and the denervated neuropil during lesion-induced synaptogenesis. In Morrison, A.R., and Strick, P.L., editors: Changing concepts of the nervous system, New York, 1982, Academic Press, Inc.

SECTION

III

MUSCLE

Richard A. Murphy

CHAPTER

22

Contraction of Muscle Cells

Functional Classification of Muscle

Muscles are composed of highly specialized cells that can generate force and produce movement. Skeletal muscle cells are effectors responsible for voluntary actions that range from speaking to running. Cardiac and smooth muscle cells subserve the functions of the cardiovascular, respiratory, digestive, gastrointestinal, and genitourinary systems. Muscle comprises about 45% to 50% of the total body mass, and the enormous increase in energy expenditure during sustained exercise dictates many of the characteristics of the cardiovascular and respiratory systems.

Muscles are traditionally classified in terms of their anatomy (striated versus smooth). Another distinction is made between voluntary muscles under conscious control and involuntary muscles in the internal organ systems with autonomic innervation. However, the properties of different muscle cells are more readily understood in terms of their functional roles. The basic distinction is between muscle cells attached to the skeleton and those in the walls of hollow organs (Fig. 22-1).

Cells attached to a skeleton (these are all voluntary and striated) are often very long and bridge the attachment points of the muscle (Fig. 22-1, *A*). Thus the individual cells are anatomically and mechanically arranged in parallel. This means that the cells function independently, and the total force equals the sum of the forces generated by the cells. The musculoskeletal system is arranged so that most gravitational loads are borne by the skeleton and ligaments. Skeletal muscle cells are normally relaxed and are usually recruited to generate force and movement.

Muscle cells in the walls of hollow organs cannot function independently (Fig. 22-1, *B*). In a continuous sheet of muscle, cells must be connected in series with each other as well as in parallel. Cells mechanically linked in series are like links in a chain: all must bear the same stress and contract uniformly. Furthermore, all cells are interdependent because pressures are transmitted throughout hollow organs, and contraction of some cells necessarily alters the load on the others. Cells in hollow organs (typically involuntary and smooth) have two functional roles. They must be capable not only of generating force and movement, like skeletal muscle cells, but also of maintaining organ dimensions against applied loads. For example, vascular smooth muscle must bear the load imposed by the blood pressure to regulate blood flow. These cells have more complex contractile and regulatory systems to carry out this latter function economically.

This chapter describes the cellular processes of contraction and the mechanisms involved in control of contraction. The basic apparatus for conversion of chemical energy into mechanical energy (chemomechanical transduction) in the form of force development or movement is considered first in a discussion of skeletal muscle cells. The additional elements involved in smooth muscle function are then introduced. Chapter 23 is devoted to integrated tissue function and the capacity of various muscles to adapt to changing requirements. Muscle function exhibits striking diversity among tissues. These specific characteristics may be found in the chapters describing the various organ systems.

Structure of the Contractile Apparatus in Skeletal Muscle

A skeletal muscle is composed of bundles of enormous, multinucleated cells up to 80 μm in diameter and sometimes many centimeters long (Fig. 22-2). These cells, like those in cardiac muscle, have a striking banding pattern responsible for their classification as *striated* muscles. The striations arise from a highly organized arrangement of subcellular structures. There are few instances in biology in which ultrastructure provides a

Fig. 22-1. Muscle function in the body. **A,** Individual skeletal muscle cells are long and directly connect two points on the skeleton. Each cell receives a branch of a motor nerve. These anatomical arrangements are associated with the following functional characteristics: each cell acts independently; normally relaxed (skeleton and ligaments bear most gravitational loads); contraction usually produces movement and work (= force × distance); small length changes. **B,** Cardiac and smooth muscle cells are small and connected together in layers around hollow organs. A blood vessel is diagrammed because of the simplicity of its concentric rings of smooth muscle cells. Muscle in hollow organs has a distinctive set of functional characteristics and roles. Cells are interdependent: cells attached in series bear equal stress; pressure (load) is transmitted throughout organ. Smooth muscle has two functional roles: to contract with shortening and work; to maintain organ dimensions against imposed loads (internal pressure).

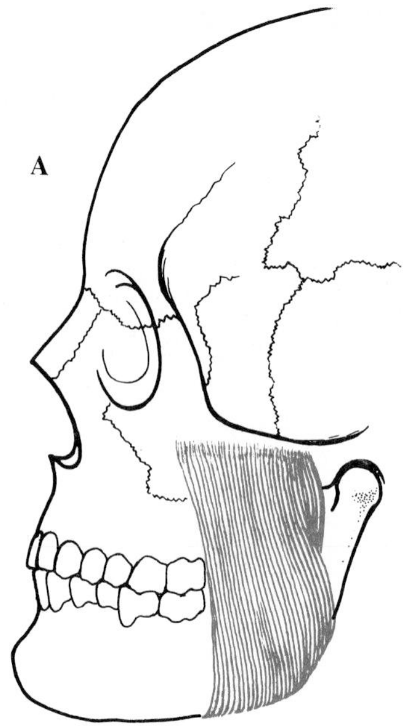

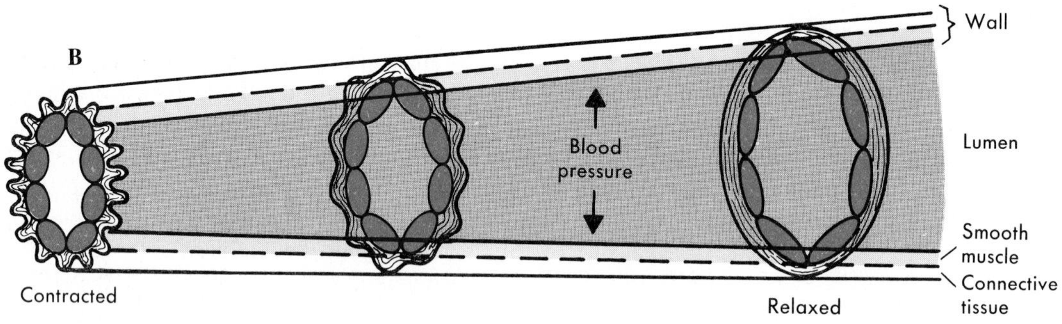

clearer basis for understanding cell function than in striated muscle. Therefore careful study of Fig. 22-2 is warranted. Electron micrographs reveal bundles of filaments running along the axis of the cell. These bundles are termed *myofibrils*. The gross striation pattern of the cell arises from a repeating pattern in the myofibrils that is in transverse register across the whole cell (Fig. 22-2, *C*).

The banding pattern arises from two sets of filaments in the myofibrils. Their organization is most clearly seen in a drawing that exaggerates the cross-sectional dimensions (Fig. 22-3). The dark striations are a region containing a lattice of *thick filaments*. This region of the myofibril is termed the *A band* because it appears dark (anisotropic) when viewed in a microscope using polarized light. A second lattice consists of *thin filaments* that are attached to a transverse, darkly staining structure termed the *Z line* or Z disk. The thin filaments extend from two adjacent Z lines to interdigitate with the thick filaments. Areas of the myofibril or cell containing only thin filaments and Z disks are termed *I bands* since they are light (isotropic) when viewed under polarized light. The thick and thin filament arrays form the contractile system, and the repeating unit in each myofibril delimited by the Z disks is the basic contractile unit called a *sarcomere*. Each sarcomere contains half of two I bands with a central A band. The latter has a less dense central region (H zone) where there is no overlap of thin filaments. The H zone is bisected by a darkly staining M line containing proteins that link the thick filaments together. Cross sections of the myofibril reveal the relationship between thin and thick filaments (Fig. 22-3). The thin filaments form a hexagonal array around each thick filament, whereas each thin filament is equidistant from three thick filaments. This arrangement reflects the presence of two thin filaments for each thick filament per half sarcomere in vertebrate striated muscle. The complex membrane systems associated with each sarcomere are considered later.

The Thin Filament

Composition and structure. Thin filaments are ubiquitous constituents of all nucleated cells and are a dominant feature in muscle. All thin filaments contain two major proteins. A globular protein called *actin*,

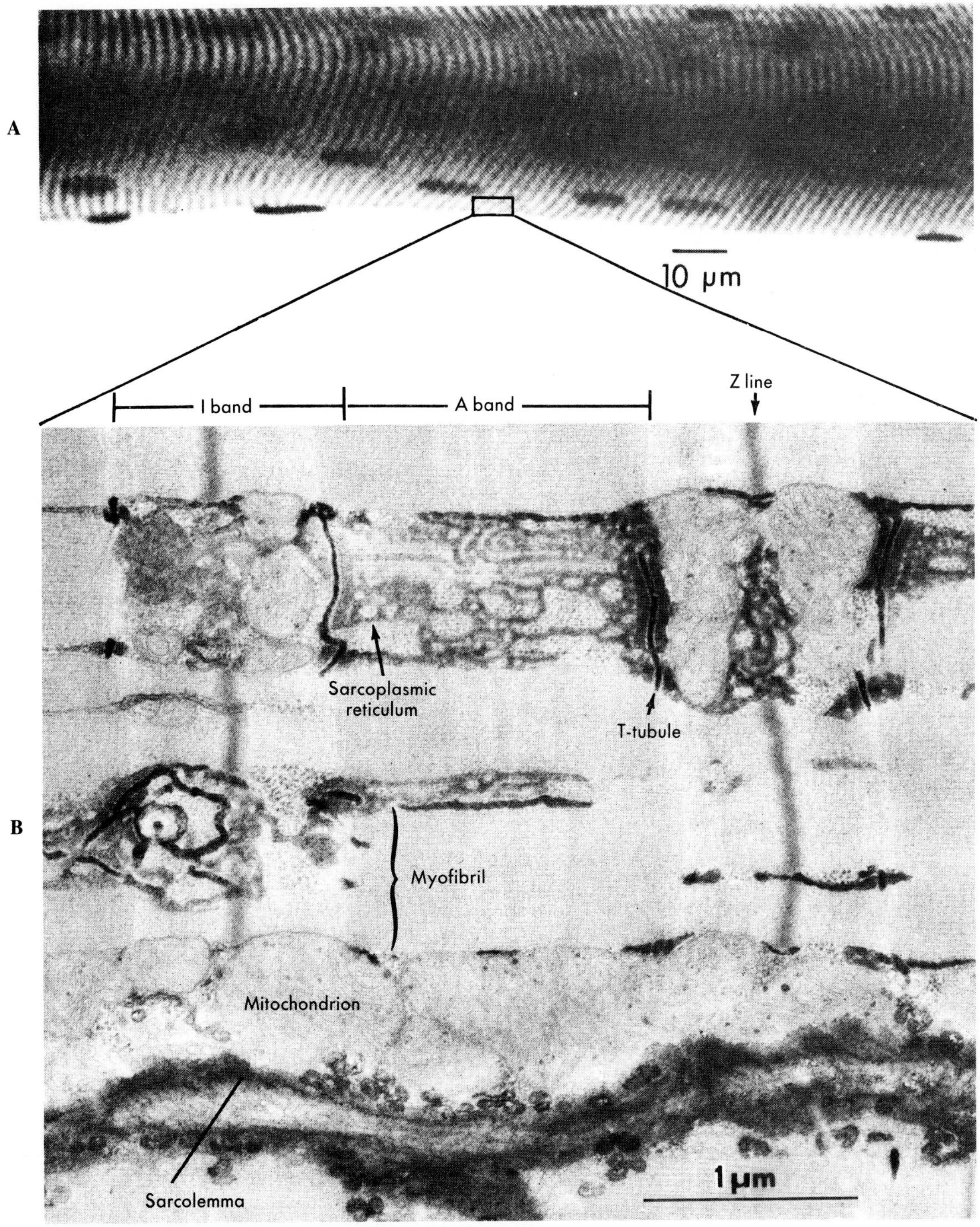

■ Fig. 22-2. **A,** Segment of a single cell from human gastrocnemius muscle. Cross striations of alternating dark *(A)* and light *(I)* bands are visible along with nuclei located peripherally. **B,** Electron micrograph of a longitudinal section through a mouse skeletal muscle cell. The tissue is stained to show the internal membrane system of the sarcoplasmic reticulum as a very dark network of interconnecting tubules. Invaginations of the cell membranes (T-tubules) are visible as dark elements extending into the cell at the level of the junction between the A and I bands. (**A** and **C** redrawn from Leeson, C.R., and Leeson, T.S.: Histology, ed. 3, Philadelphia, 1976, W.B. Saunders Co.; **B** courtesy Dr. Michael S. Forbes.)

Continued.

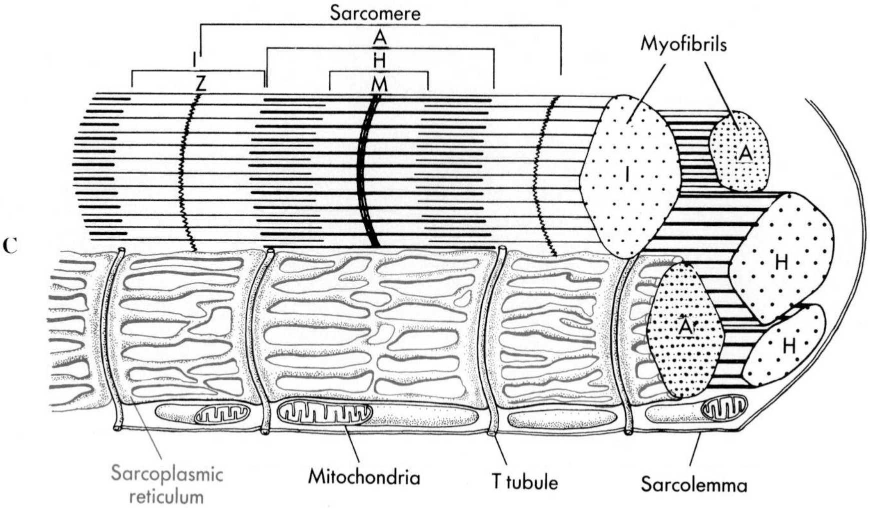

■ **Fig. 22-2, cont'd. C,** Drawing illustrating the three-dimensional relationships between membrane elements and the filament lattice.

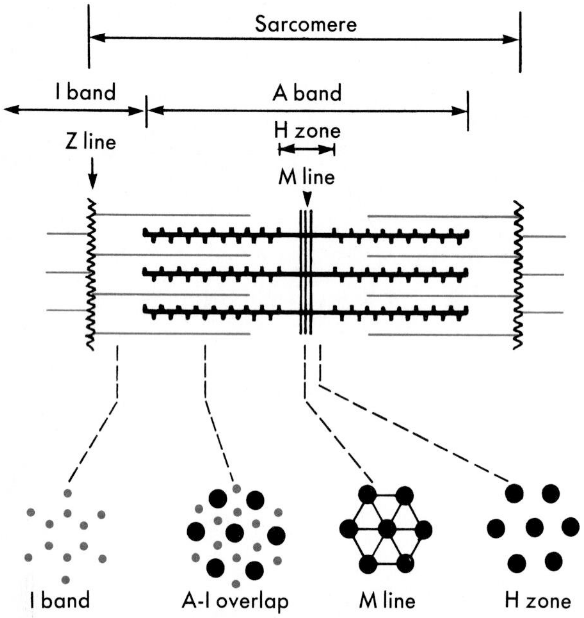

■ **Fig. 22-3.** Longitudinal *(top)* and cross-sectional *(bottom)* diagrams showing the relationships between thick *(black)* and thin *(color)* filaments of a sarcomere. (Redrawn from Squire, J.M.: The structural basis of muscular contraction, New York, 1981, Plenum Press.)

with a molecular weight of 45,000 daltons, polymerizes under conditions existing in the cytoplasm to form twisted, two-stranded filaments (Fig. 22-4). Rod-shaped molecules of *tropomyosin* stretch along each strand of the thin filament and each molecule of tropomyosin is associated with six or seven actins in one strand. Tropomyosin is composed of two separate polypeptide chains. The individual polypeptides have a basic α-helical structure, and the two helical peptides are wound around each other to form a supercoil. Molecules of this type are long, rigid, and insoluble. Other proteins, present in smaller amounts, are associated with the thin filaments. These minor proteins are involved in the attachment of thin filaments to Z disks and probably contribute to the remarkably uniform thin filament length of 1 μm in vertebrate striated muscles. Finally, additional thin-filament proteins are involved in regulation of the interaction of actin with the thick filament. The most important regulatory protein in striated muscles is troponin, which is linked to tropomyosin (Fig. 22-4). The regulatory proteins are considered later.

Lattice organization and polarity. Thin filaments are anchored at one end. The filaments have a polarity, although this is not apparent in electron micrographs. Thus thin filaments on each side of the Z disk "point" in opposite directions. Fig. 22-4 indicates one view of how this may arise from a splitting of the two strands of the thin filament at the Z disk so that each strand forms half a thin filament of opposite polarity. In cross section the thin filaments can be seen to form a hexagonal lattice around each thick filament (Fig. 22-3). Each thin filament lies equidistant from three thick filaments.

■ *The Thick Filament*

A very large protein (about 470,000 daltons) called *myosin* forms the thick filament, although small amounts of other proteins are present. The myosin molecule is formed by the association of six different polypeptides. These peptides, which are not covalently linked, can be dissociated by detergents or denaturing agents (Fig. 22-5) and separated into three pairs: one set of large *heavy chains* and two sets of *light chains*. In the intact molecule most of the heavy chain has an α-helical structure, and the two strands are twisted around each other in a supercoil that forms a long, rigid, insoluble "tail." However, one end of the

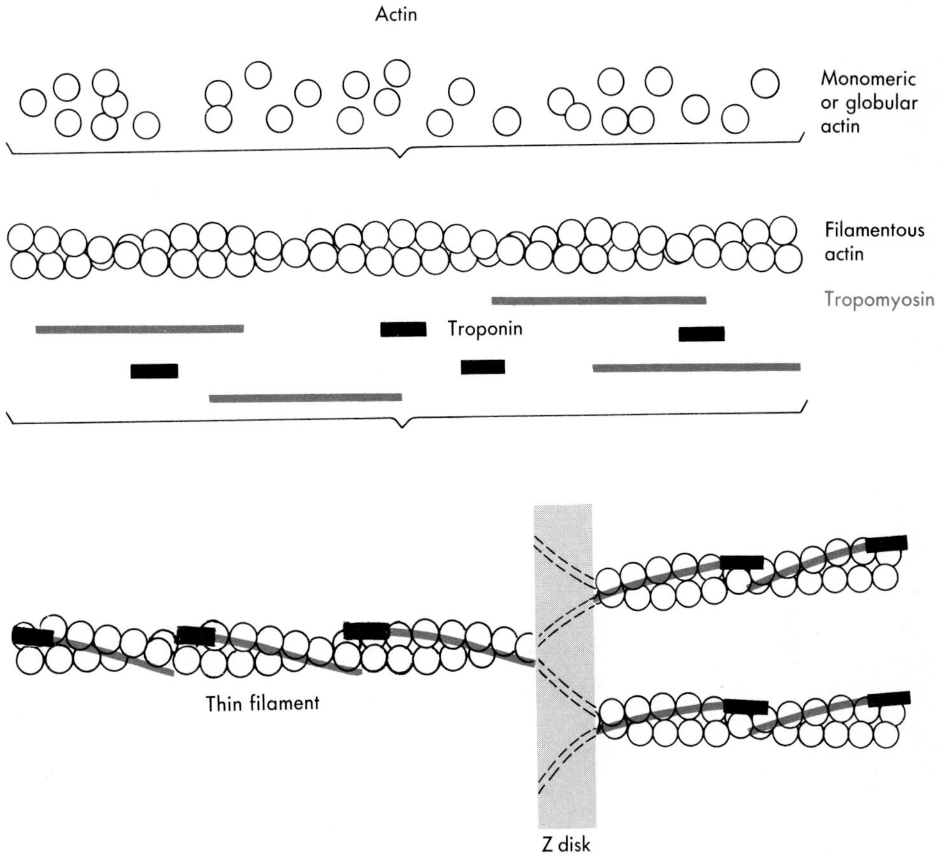

Fig. 22-4. Composition and structure of thin filaments in muscle. Globular actin monomers *(top)* polymerize into a two-stranded helical filament. The thin filament structure is completed with the addition of stiff, rod-shaped tropomyosin molecules *(colored area in center)*. Troponin *(black rectangles)* is a regulatory protein bound to the tropomyosin component of the thin filament in vertebrate striated muscles *(bottom)*. For clarity the tropomyosin-troponin complexes associated with only one strand of the actin helix are illustrated. Thin filaments are anchored to Z lines in striated muscles. Filaments on each side of a Z line have opposite polarities. (Redrawn from Murray, J.M., and Weber, A.: Sci. Am. **230:**58, 1974.)

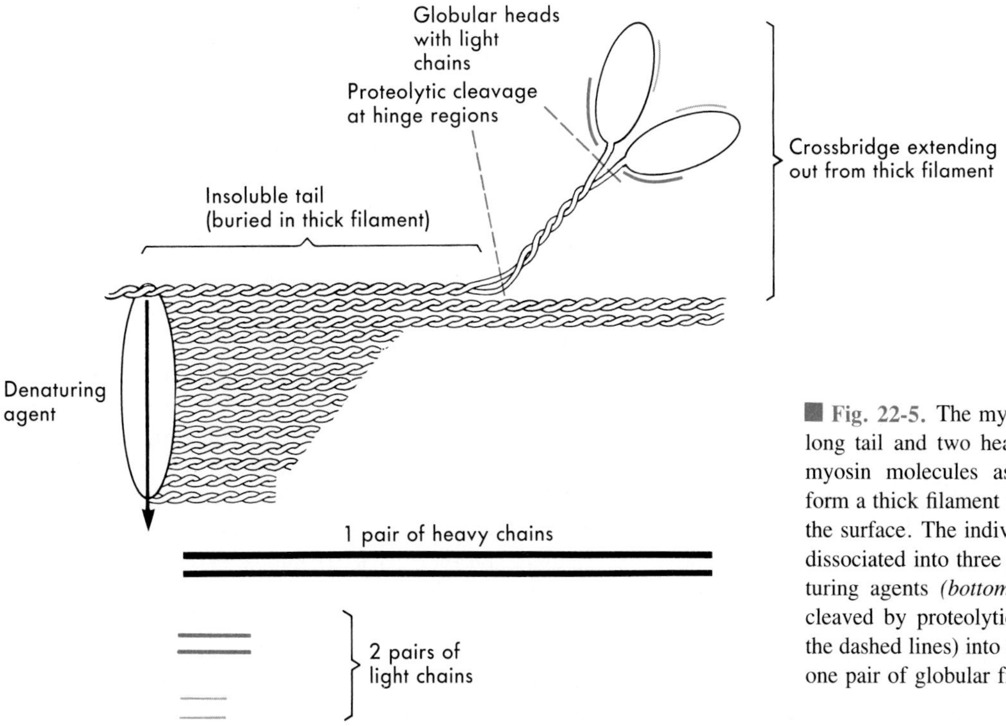

Fig. 22-5. The myosin molecule *(top)* contains a long tail and two heads. The tail regions of many myosin molecules associate in the cytoplasm to form a thick filament with the heads projecting from the surface. The individual myosin molecule can be dissociated into three pairs of polypeptides by denaturing agents *(bottom)*. The molecule can also be cleaved by proteolytic enzymes (at sites shown by the dashed lines) into two rod-shaped fragments plus one pair of globular fragments.

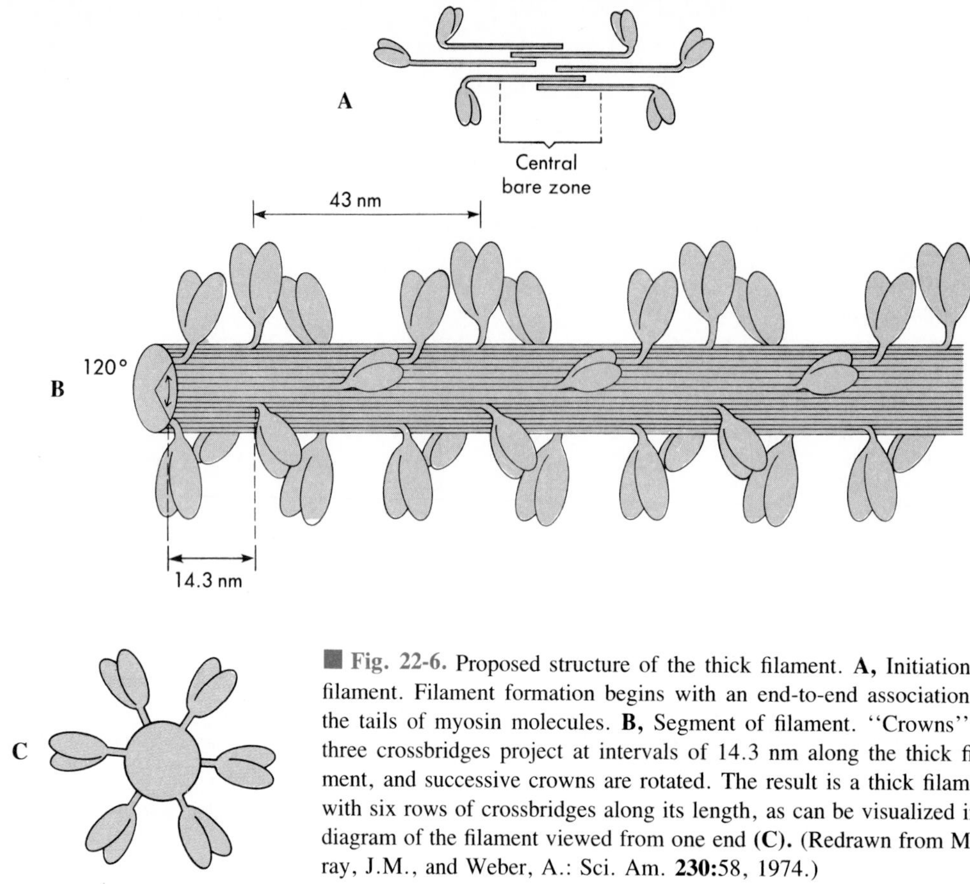

■ **Fig. 22-6.** Proposed structure of the thick filament. **A,** Initiation of filament. Filament formation begins with an end-to-end association of the tails of myosin molecules. **B,** Segment of filament. ''Crowns'' of three crossbridges project at intervals of 14.3 nm along the thick filament, and successive crowns are rotated. The result is a thick filament with six rows of crossbridges along its length, as can be visualized in a diagram of the filament viewed from one end **(C).** (Redrawn from Murray, J.M., and Weber, A.: Sci. Am. **230:**58, 1974.)

heavy chains has a globular tertiary structure. Thus each myosin molecule has two ''heads'' attached to one end of the long tail (Fig. 22-5). One polypeptide of each set of light chains is associated with each head of the molecule.

If myosin is briefly exposed to proteolytic enzymes such as trypsin or papain, it is preferentially cleaved into one long and one short rod segment and two globular fragments (Fig. 22-5). The points of cleavage occur at regions where the basic supercoiled structure is perturbed, exposing the peptide backbone to enzymatic attack. Cleavage points indicate flexible regions that serve as hinges for the molecule. Studies of these proteolytic fragments show the functions of each part of the myosin molecule, as described later.

Filament formation and structure. Myosin aggregates in the cytoplasm to form thick filaments. The basic filament represents aggregates of the tail segment of the molecules, and smoother filaments can form from the tail segments obtained by proteolytic cleavage. The remainder of the molecule, including the globular heads and the rod portion between the hinges, projects laterally from the filaments. These projections are visible in electron micrographs and are termed *crossbridges* because they can link adjacent thick and thin filaments. The process of filamentogenesis is extraordinarily specific. Each filament begins with an end-to-end association of the tails of the myosin molecules (Fig. 22-6, *A*). This produces a central thick filament segment lacking crossbridge projections. The crossbridges in each half of the filament are consequently oriented in the opposite direction. Therefore the molecules of the thick and thin filaments in each half of the sarcomere have the same relative orientation as a result of the thin filament polarity. The crossbridges project in groups of three from the filament (Fig. 22-6, *B*). Successive crowns of crossbridges are rotated, producing a helical arrangement of crossbridges along the thick filament. In a striated muscle the crossbridges project toward the six thin filaments surrounding each thick filament. Thick filaments are 1.6 μm long and contain an estimated 300 to 400 crossbridges. Minor protein constituents probably contribute to the highly ordered and regular structure of thick filaments and their basic triangular lattices (Fig. 22-3) in cells. Adjacent thick filaments in the sarcomere are linked by protein connections between the central bare zones, contributing to the stability of the filament lattice.

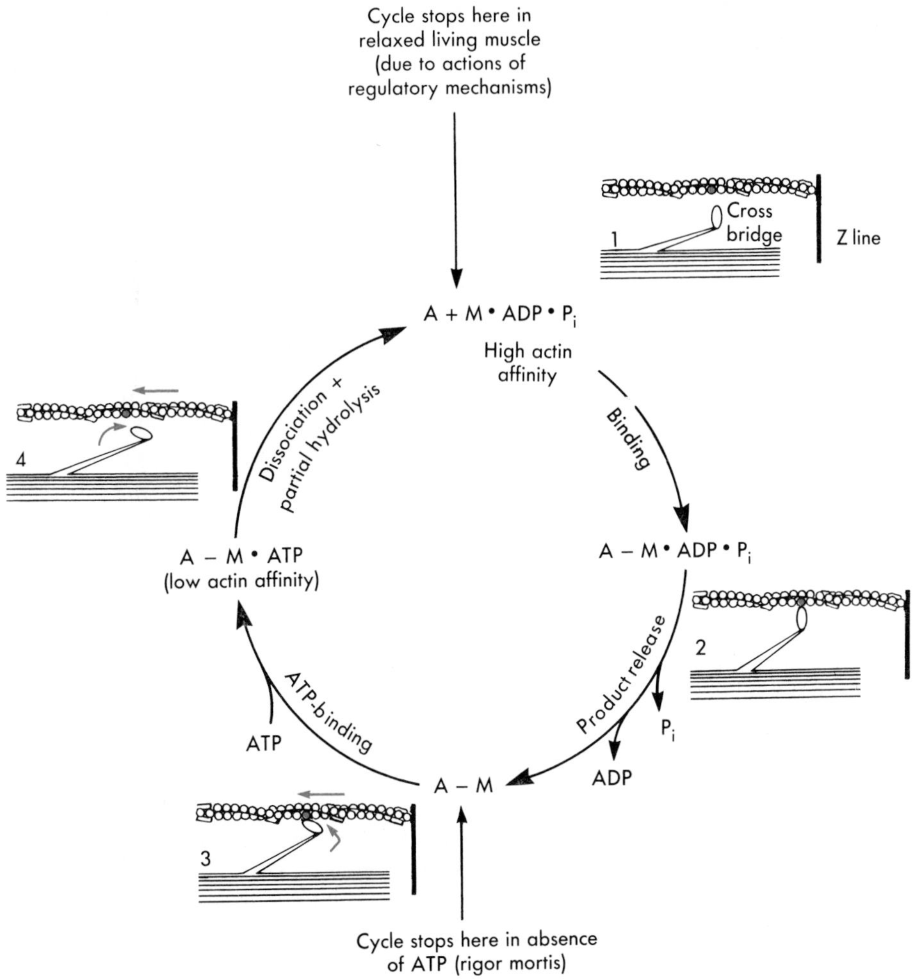

■ **Fig. 22-7.** The mechanism of ATP hydrolysis by actomyosin and the major steps in the crossbridge cycle. Actin in the thin filament *(A)* and the myosin crossbridge projecting from the thick filament *(M)* interact cyclically. This interaction involves a number of steps during which ATP is hydrolyzed, and the energy released is harnessed to induce conformational changes in the crossbridge. Each cycle causes the thick and thin filaments to interdigitate by about 10 nm (note position of colored actin monomer).

■ *Crossbridge Interactions with the Thin Filament*

■ *Crossbridge Properties*

Each crossbridge consists of two identical heads that exhibit a remarkable set of properties. Most studies suggest that the two heads act largely independently in the reactions discussed in this section. However, the behavior of one head may be somewhat influenced by reactions involving the other (cooperativity). The properties of the myosin head depend on the intact globular part of the heavy chain and the two associated light chains.

Myosin can catalyze the hydrolysis of adenosine triphosphate (ATP), producing adenosine diphosphate (ADP) and inorganic phosphate (P_i). However, the ATPase activity of myosin is inhibited by the high magnesium concentrations existing in cells. Myosin can also bind to actin, forming an *actomyosin* complex. Actomyosin is a very active ATPase in the muscle cells. The interaction between actin and myosin, associated with ATP hydrolysis, represents the fundamental *chemomechanical transduction* process, considered later, in which chemical energy is converted into mechanical energy by the muscle.

The splitting of ATP by actomyosin is a complex cycle involving many steps. The more important steps are illustrated in Fig. 22-7. In a relaxed muscle, regulatory systems prevent actin-myosin interactions (Fig. 22-7, *top*). This inhibition is overcome on stimulation by Ca^{++} ions and is considered later. In the presence of ATP, myosin has ADP and P_i bound to each head and, in this state, exhibits a high affinity for actin. The ADP and P_i are released when the myosin heads bind to actin. Product release allows an ATP molecule to bind to myosin, and the affinity of myosin for actin (as A—M) is greatly reduced. The bound ATP is hydrolyzed with dissociation of myosin and actin, although the products (ADP and P_i) remain bound to myosin in the succeeding step. The energy released by splitting of ATP is stored in the myosin molecule, which is now in a high-energy state and has a renewed high affinity for actin.

The Crossbridge Cycle

The cycle associated with ATP hydrolysis by isolated actomyosin releases the energy in ATP without any conversion into mechanical work. Chemomechanical transduction involves conformational changes in the crossbridges, which lead to filament movements with shortening and force development (Fig. 22-7). The details of this process are uncertain, but considerable evidence favors the general model illustrated for the crossbridge interacting with the thin filament. In a resting muscle the crossbridge is not attached to the thin filament and is oriented perpendicularly to the myosin filament (step 1). When a muscle is stimulated a rise in myoplasmic Ca^{++} concentration produces changes in the myofilament structure, allowing crossbridge binding to the thin filament (considered in the section on control mechanisms; step 2). The hinge regions in the crossbridge permit the head to swing toward the thin filament. After attachment, myosin heads change their conformation (tilt), using the energy stored in the high-energy myosin-ADP-P_i complex (step 3). This conformational change in the crossbridge generates a force moving the thin filament relative to the thick filament. It is likely that the conformational change also leads to release of ADP and P_i, setting the stage for crossbridge detachment when another ATP is bound (step 4). Each cycle can move the filaments about 10 nm relative to each other. The way in which enormous numbers of such cycles generate muscular contraction is considered in the following section. The illustrated cycle will continue until interrupted in the detached state by control systems (which remove Ca^{++} from the myoplasm and produce relaxation) or until the ATP is exhausted. ATP depletion is an abnormal situation in muscle cells and arrests the cycle with the formation of permanent actomyosin complexes as shown in Fig. 22-7. Death causes muscular rigidity *(rigor mortis)* because ATP depletion leads to permanent crossbridge attachment.

Biophysics of the Contractile System

Quantitative estimates of the mechanical output of the muscles have provided important inferences about the mechanism of chemomechanical transduction and its control mechanisms. The measurement of muscle contraction also provides a way of assessing the effects of neurotransmitters, drugs, and hormones and of quantifying pathological changes. The important mechanical variables are force, length, and the derived variable of shortening velocity (Δ length/Δ time). The basic approach in muscle mechanics is to control all of these factors except the measured dependent variable. Such measurements should employ muscle preparations in which the cells are aligned along the axis in which force and length are determined.

Force-Length Relationships and the Sliding Filament Mechanism

The dependence of force generation on the length of the muscle is shown in Fig. 22-8. Points on these curves are determined under conditions in which the length of the muscle is fixed, and force is measured during a contraction. Such a contraction at constant length is termed *isometric*. Force will vary with the size of the muscle and is best expressed as a stress (force/cross-sectional area of the muscle) to allow comparisons with other tissues. A relaxed muscle is elastic and it requires a force to stretch it to an increased length. The resulting *passive force-length or stress-length relationship* primarily reflects the properties of connective tissue in intact muscles. Connective tissue is a more important component in the walls of hollow organs, such as the heart and gastrointestinal tract, than in striated muscle, where the in vivo length is constrained by the skeletal attachments. When a muscle is stimulated, the total isometric stress is greater than the passive stress at any length. The difference between the stress-length curves for contracting and relaxed muscles is the *active stress-length relationship* that characterizes the contractile system (Fig. 22-8, *B*). There is a family of such curves for any muscle when it is not maximally activated (Fig. 22-8, *C*).

The correlation of structural and mechanical information reveals that active stress is proportional to the overlap between thick and thin filaments in a sarcomere (Fig. 22-8, *D*). As the muscle is stretched beyond the length (L_o) at which maximum active force (F_o) is developed, active stress decreases linearly with the overlap between thick and thin filaments—showing that stress is proportional to the number of active crossbridges that can interact with the thin filament. This number includes all the interacting crossbridges in each half sarcomere. There is a small segment at the peak of the stress-length curve at which stress does not change with length. At these lengths the thin filaments move past the central thick filament bare zones that lack crossbridges. Force declines at sarcomere lengths less than L_o. This is partially a result of disturbances in the filament lattice geometry as diagrammed in Fig. 22-8, *D*, when thick filaments collide with Z disks and thin filaments overlap. However, the control mechanisms for the contractile machinery become less sensitive to the stimulus at short muscle lengths, and partial inactivation contributes to the low stress at short lengths. Slight variations in sarcomere lengths in whole muscles obscure the inflection points detected in the stress-length curves for sarcomeres.

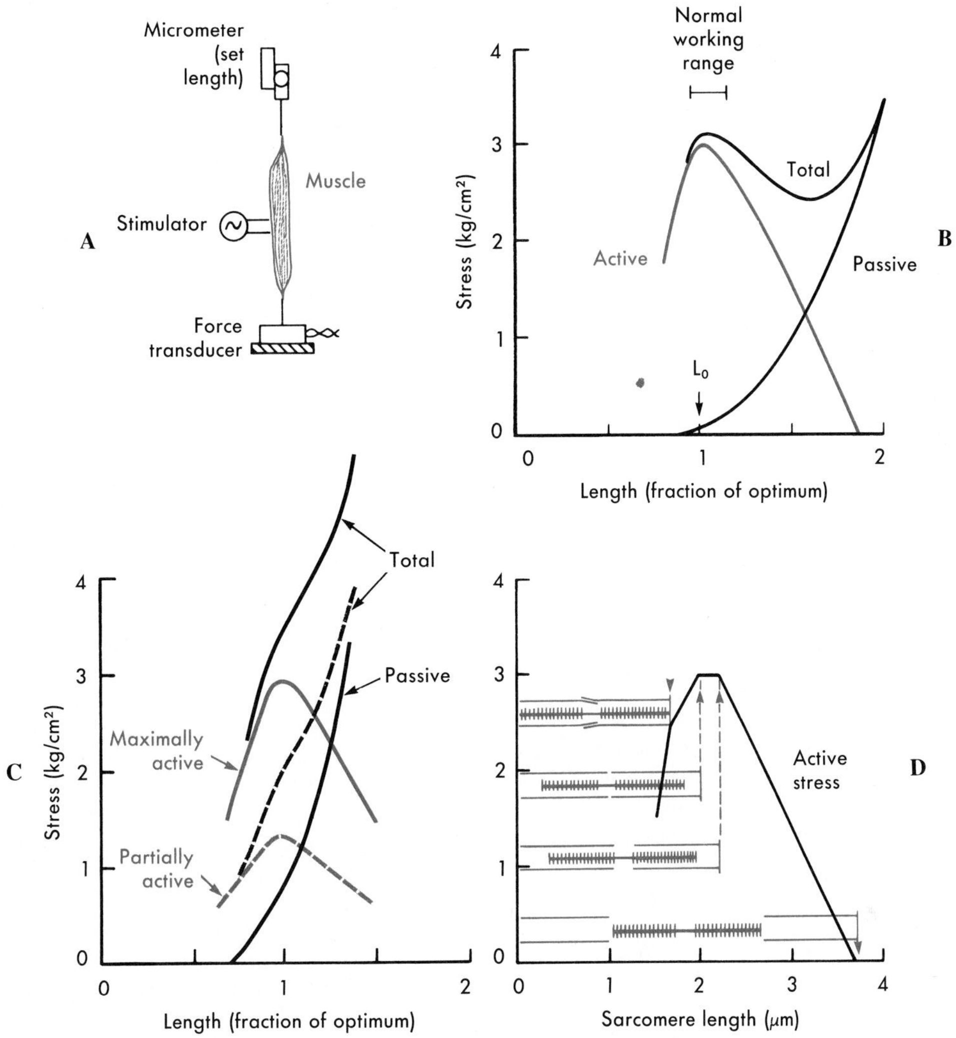

■ **Fig. 22-8.** Stress-length relationships in muscle. **A,** Schematic diagram of experimental apparatus in which the tissue is attached to a micrometer to set the length and a transducer to measure force. Force is normalized as a stress, to allow comparisons of muscles of differing size, that is, force/cross-sectional area of the muscle cells. Values of stress at different lengths are obtained and plotted as shown in **B** to **D. B,** Three stress-length curves are shown for skeletal muscle: (1) the passive stress exerted as a function of the length of the relaxed muscle, (2) the total stress exerted by the maximally stimulated muscle (passive + active), and (3) the active stress-length curve for the contractile machinery obtained as the difference between the total and passive stresses at any length (colored curve). All muscles have an inherent maximal force-generating capacity, which is obtained at the optimal length (L_o). **C,** In cardiac and smooth muscles, which are not attached to the skeleton, connective tissue limits how far the tissue can be stretched. The characteristics of the connective tissue alter the passive, but not the active, stress-length relationship. Active stress will vary with the extent to which the contractile system is activated. **D,** Precise studies of the stress-length behavior of the sarcomeres in single skeletal muscle cells reveal the dependence of stress on the overlap of thick and thin filaments. Diagrams of four sarcomeres at lengths where the slope of the stress-length curve changes show how filament interactions and active stress depend on sarcomere length. (Data from Gordon, A.M., Huxley, A.F., and Julian, F.J.: J. Physiol. [London] **184:**170, 1966.)

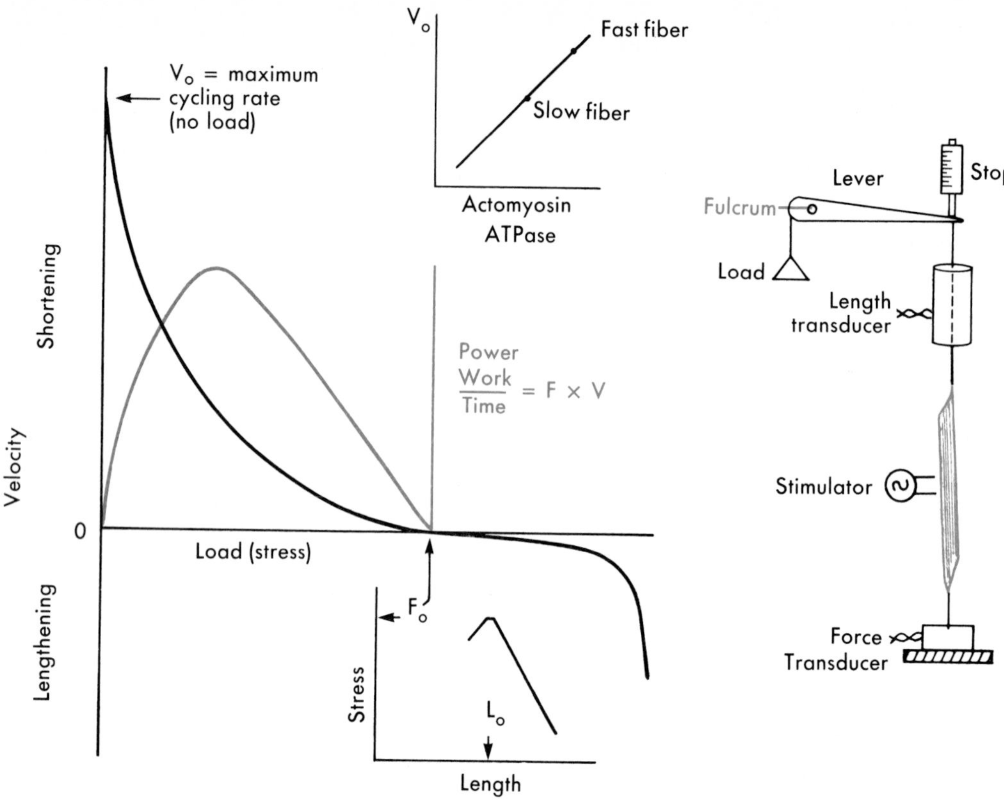

■ **Fig 22-9.** The dependence of shortening velocity on the load (stress) on a muscle. Shortening velocity may be measured using a lever system, which permits a muscle to shorten against a constant load. Velocity is measured using a length transducer (velocity = Δ length/Δ time). The velocity-load curve is constructed from points obtained in a series of contractions against different loads. (The initial length in these experiments is at L_o, and only a slight shortening is permitted.) If the load placed on the muscle is greater than the load that the active crossbridges can bear, the muscle will lengthen. Consequently, the velocity-load curve can be extended to describe this situation. Insets emphasize that the maximal stress (F_o) a muscle can develop depends on the number of interacting crossbridges, whereas the maximal shortening velocity (V_o) is limited by the rate at which a particular isoenzymatic form of myosin synthesized in a cell can interact with actin and release the energy stored in ATP. The power output of a muscle *(colored curve)* is the mechanical work (force times distance shortened) per unit of time and can be calculated as the product of load times shortening velocity.

Force-generation and muscle type. The maximal stress generated depends on muscle length (filament overlap), the number of crossbridges that are turned on (the level of *activation*), and the type of muscle. Because the lengths of the thick and thin filaments and their packing density are similar in vertebrate striated muscles, they all generate similar maximal stresses at L_o. The maximal stress is about 3 kg/cm^2 of cell cross-sectional area.

Velocity-stress relationships and the crossbridge cycle. Another basic relationship characterizing the output of a muscle is obtained when the load or stress on a muscle is held constant and the shortening velocity is measured (Fig. 22-9). A contraction at constant load is termed *isotonic*. Shortening velocity is determined by the load on the muscle as illustrated by the hyperbolic velocity-stress curve. Muscles lift a heavy load slowly but can shorten rapidly when lightly loaded. Fig. 22-9 also shows that a contracting muscle can withstand (briefly) a heavier stress when forcibly stretched than it can develop isometrically. The strength of the crossbridge attachment to the thin filament is greater than the force generated by its movement. Consequently, muscles can bear a load about 1.6 times F_o before the crossbridge attachment is mechanically broken and rapid lengthening occurs. This situation is important physiologically when a muscle is contracted to decelerate the body, as when a person is walking downhill.

The contractile system operates most efficiently (i.e., the greatest mechanical work output for the chemical energy used) with optimal loading. This is illustrated by the power curve. Power, or work/time, is simply the product of force and velocity, and power can be calculated by multiplying these values at any point on the velocity-force curve (Fig. 22-9). Maximal power is obtained with a load of about 0.3 of the maximal force that can be developed. A muscle contracting isometrically does no work (force × distance = 0), nor does a muscle shortening with no load. The mechanical efficiency of such contractions is necessarily zero. Inefficient contractions are often physiologically important when either high speed or maximal force is appropriate. The optimal efficiency of the contractile system in converting chemical energy into mechanical work is about 40% to 45%.

If a sarcomere is to shorten more than the 10 nm associated with a single crossbridge cycle, each crossbridge must detach and then reattach at a new site on the thin filament closer to the Z line. The crossbridges must cycle asynchronously to maintain a constant force and permit continuous shortening. This is probably facilitated by the fact that the distances between successive crossbridges along a thick filament are different from the repeat distances on the thin filament.

The number of active crossbridges is the primary variable subject to control. Crossbridge cycling rates depend primarily on the load. A partially activated skeletal muscle developing a low maximum force (F_o) will have the same maximal shortening rate (V_o) as that of a fully activated muscle.

Shortening velocities in a muscle depend on several factors. First, the velocity varies with the number of sarcomeres in a cell. Total shortening and shortening velocity are the sum of the movements of thin filaments past thick filaments times the number of half sarcomeres in the cell. Velocities can be calculated in terms of μm per second per half sarcomere to allow comparisons between different muscles. However, it is more common to normalize shortening velocities by reporting values in terms of optimal muscle lengths (L_o) per second, which accounts for the number of sarcomeres in a cell. Shortening velocities for smooth muscle can be normalized only this way. The velocity also depends on the load on the muscle; the velocity-stress relationship shows that crossbridge cycling rates fall as the stress on the crossbridges increases. This effect can be understood by referring to Fig. 22-9. The conformational change in the crossbridge, which causes shortening between attachment and detachment, is opposed by the load. Heavier loads progressively increase the average time for this to occur and allow a new cycle to take place. An unloaded crossbridge can cycle at a maximal rate, indicated by V_o. This maximal rate depends on the molecular properties of the myosin synthesized within a cell. The direct proportionality between the ATPase activity of myosin isolated from a cell and V_o for that cell (inset, Fig. 22-9) illustrates this molecular diversity, which is responsible for physiological differences in the speed of contraction of muscle cells from different sources.

Intracellular Control Mechanisms for the Contractile System

Two elements are involved in regulation of the contraction-relaxation cycle. The first involves cell membrane events that alter the cellular concentration of a *second messenger,* that is, an ion or molecule that diffuses to the fibrillar contractile apparatus and regulates the crossbridge cycle. The second messenger in the myoplasm of all types of muscle is Ca^{++}. The second aspect of regulation concerns the mechanisms whereby Ca^{++} produces its effects on the contractile machinery.

Troponin and Thin Filament Regulation in Striated Muscle

Vertebrate skeletal and cardiac muscles contain a regulatory protein, termed *troponin,* in the thin filament. One molecule of troponin is bound to the end of each tropomyosin molecule (Figs. 22-4 and 22-10). Troponin binds up to four Ca^{++} ions in a cooperative manner. In the presence of micromolar concentrations of Ca^{++} the binding sites on troponin C are occupied, and the conformation of the thin filament changes (Fig. 22-10). This shift allows crossbridges adjacent to that portion of the thin filament to attach and cycle. In effect, Ca^{++} binding to troponin acts as a switch that regulates the number of crossbridges that are interacting with the thin filament.

In vertebrate skeletal and cardiac muscles the contraction-relaxation cycle involves five steps: (1) a stimulus at the cell membrane causes the myoplasmic Ca^{++} to rise above 0.1 μM; (2) binding of Ca^{++} to troponin exposes the sites at which crossbridges attach to actin; (3) crossbridges now bind and cycle at all exposed sites; (4) cessation of stimulation is followed by Ca^{++} removal from the myoplasm and dissociation of Ca^{++} from troponin; (5) the thin filament returns to a configuration in which further crossbridge interactions are blocked.

The binding of Ca^{++} ions to troponin exhibits a steep response curve so that relatively small changes in Ca^{++} produce large changes in the number of active crossbridges and force development (Fig. 22-10). Each crossbridge cycle is associated with ATP hydrolysis; therefore the myofibrillar ATPase activity is also proportional to the Ca^{++} concentration.

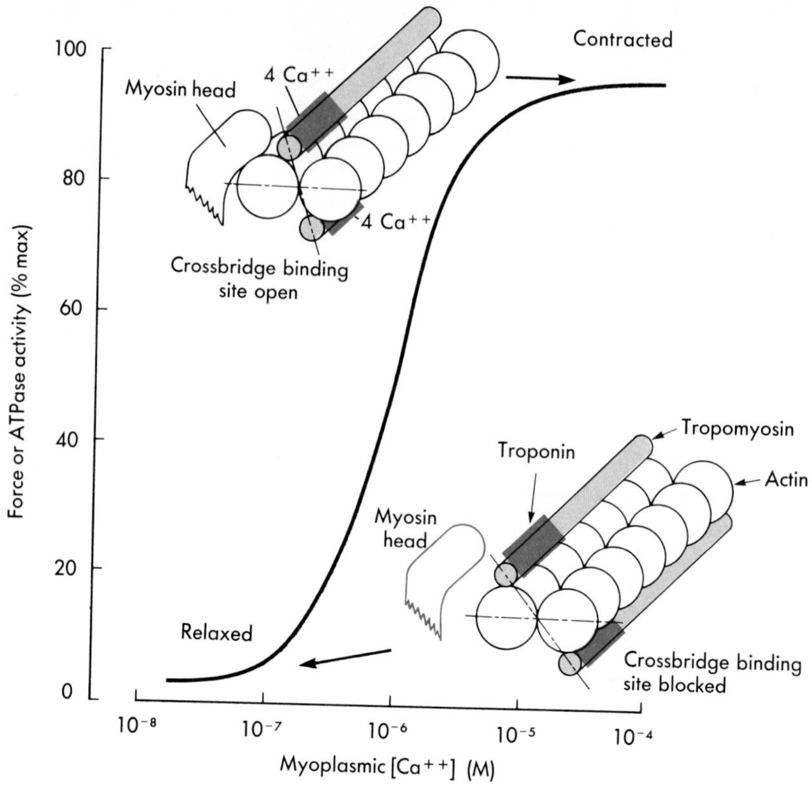

Fig. 22-10. Regulation of the crossbridge cycle by Ca^{++} and troponin. Force development by muscle cells or ATP splitting by actomyosin containing troponin and tropomyosin requires very small concentrations of Ca^{++} in the myoplasm. Ca^{++} activates contraction by binding to the regulatory protein troponin in the thin filament of striated muscle. Troponin is associated with tropomyosin as a 1:1 complex in the thin filament. A conformational change occurs in the thin filament when Ca^{++} binds to troponin. This change may involve movement of the tropomyosin rods into the groove of the two-stranded actin helix (here shown as though it were straight for clarity). This shift exposes the myosin-binding site and allows the crossbridges to interact with all seven actins associated with each Ca^{++}-troponin-tropomyosin complex. (Redrawn from Hartshorne, D.J.: In Lapedes, D.N., editor: Yearbook of science and technology, New York, 1976, McGraw-Hill, Inc. Copyright 1976 by McGraw-Hill Book Co. Used with the permission of McGraw-Hill Book Co.)

The Contractile System in Smooth Muscle

Smooth muscle is far more important to health care professionals than is striated muscle. Inappropriate (pathological) behavior of smooth muscle is involved in most illnesses (e.g., hypertension, atherosclerosis, coronary artery disease, asthma, gastrointestinal disorders, stroke, and so forth). Nevertheless, smooth muscle often receives cursory attention because it has been considered to be a primitive contractile cell type that is slower, weaker, and more specialized for a large shortening capacity than striated muscle is. This characterization is inaccurate. Unlike striated muscle, smooth muscle in vertebrates has no counterpart in the "primitive" invertebrate phyla. Emerging evidence shows that smooth muscle has a contractile system capable of more flexible behavior than striated muscle, consistent with its functional role noted in the introduction to this chapter.

Smooth Muscle Structure

Cells. Smooth muscle cells vary considerably in size but are typically 100 to 300 μm long and 2 to 5 μm in maximal width. Cells characteristically taper toward their ends but may be irregular in cross-section. Juxtaposed cells exhibit a variety of junctions that serve as sites of communication and mechanical linkages (see Chapter 23). Smooth muscle cells are also embedded in a connective tissue matrix, which is a product of the cell's synthetic and secretory activities. This matrix limits distension of the tissue to lengths near the optimum for force development.

Cell cytoskeleton. The cytoskeleton in muscle cells serves as attachment points for the thin filaments and permits force transmission to the ends of the cell. The contractile apparatus in smooth muscle is not organized into myofibrils, and Z-disks are lacking. The functional equivalents of the Z-disks are centrally located ellipsoidal *dense bodies* and *dense areas* that form bands along the sarcolemma (Fig. 22-11). These structures serve as attachment points for the thin filaments and contain α-actinin, a protein also found in the Z-disks of striated muscle.

Intermediate filaments with diameters between those of thin filaments (7 nm) and thick filaments (15 nm) are prominent in smooth muscle and link the dense bodies and areas into a cytoskeletal network (Fig. 22-11). All smooth muscles contain intermediate filaments consisting of polymers of a protein termed *desmin*. Smooth muscle cells in some tissues also contain intermediate filaments composed of *vimentin*.

Myofilaments. Thick and thin filaments are extremely long relative to their diameter. Consequently, complete filaments are almost never found in thin sec-

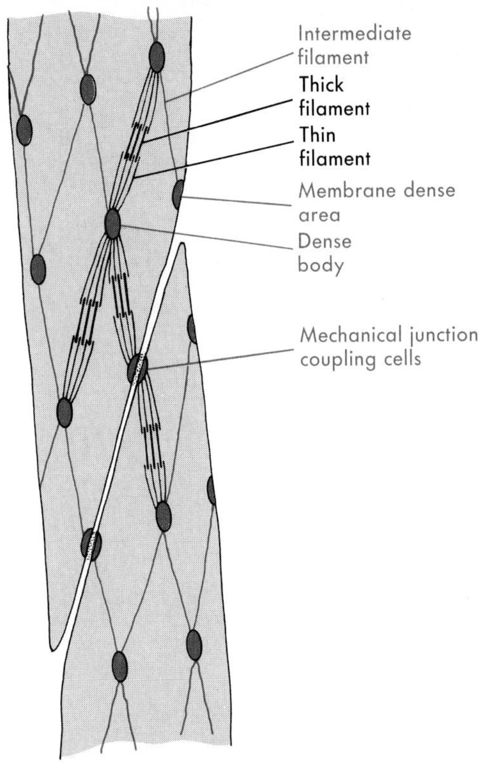

Fig. 22-11. Apparent organization of the cytoskeleton *(color)* and myofilaments in smooth muscles. Small contractile elements functionally equivalent to a sarcomere presumably underlie the similarities in mechanics between smooth and skeletal muscles. Linkages consisting of specialized junctions or interstitial fibrillar material functionally couple the contractile apparatus of adjacent cells.

tions prepared for electron microscopy. Therefore the relationships between thick and thin filaments cannot be determined unless very large numbers are arranged in parallel, as they are in striated muscles. The filaments in smooth muscle are not all aligned, and little is known about the organization of the contractile apparatus into a contractile unit functionally equivalent to a sarcomere. Fig. 22-11 synthesizes some information about myofilament and cytoskeletal arrangements; important details are lacking about filament lengths and overlap at various points on the force-length relationship.

The thin filaments of smooth and striated muscle have the same basic structure (Fig. 22-4), except that the regulatory protein, troponin, is not present in smooth muscle. However, the cellular content of actin and tropomyosin in smooth muscle is about twice that of striated muscle, and most of the myoplasm is filled with thin filaments approximately aligned along the long axis of the cell. The thin filaments exhibit the same polarity with respect to their attachments as those in striated muscle. In contrast, the myosin content of smooth muscle is only one-fourth that of striated muscle. Small groups of three to five thick filaments are aligned, surrounded by many thin filaments. Such groups of thick filaments with interdigitating thin filaments connected to dense bodies or areas (Fig. 22-11) may be the equivalent of the sarcomere. The relative sparcity of thick filaments and the absence of transverse registration leads to a ''smooth'' (i.e., not striated) appearance, which accounts for the common name of this muscle type. The contractile apparatus of adjacent cells is mechanically coupled via the links between membrane dense areas. This is an anatomical correlate of the fact that the cells are not functionally independent contractile units (Fig. 22-1, *B*).

Contractile Function in Smooth Muscle

The evidence that the sliding filament/crossbridge mechanism underlies contraction largely rests on similarities in the mechanics of smooth and striated mucles.

The force-length relationship. Smooth muscle tissues contain large amounts of connective tissue containing extensible *elastin* fibrils and inextensible *collagen* fibrils. This matrix is responsible for the passive force-length curve measured in relaxed tissues (Fig. 22-8, *C*). Highly distending forces or loads are withstood by this extracellular matrix, which limits organ volume. The active force developed on stimulation depends on tissue length. When lengths are normalized to L_o (the optimal length for force development), the force-length curves for smooth and skeletal muscle are very similar. This similarity provides strong support for a sliding filament mechanism in smooth muscle. The force-length curves differ quantitatively, however. Smooth muscle cells often shorten more in vivo than do striated muscle cells, although this is not caused by differences in their contractile systems. Smooth muscles are characteristically only partially activated, and the peak isometric forces attained vary with the stimulus (Fig. 22-8, *C*). However, smooth muscles can generate active stresses of about 6 kg/cm^2 of cell cross-sectional area, a value twice that of striated muscle. How this is accomplished with only one fourth of the myosin (and crossbridge) content of striated muscle is unknown.

Velocity-stress relationships and the crossbridge cycle. Smooth and striated muscles exhibit the same hyperbolic dependence of shortening velocity on load, the same optimal power output at a load of 0.3 F_o, and the same ability to bear (transiently) an applied load greater than the developed active stress (Fig. 22-9). Contraction velocities are far slower in smooth muscle than in striated muscle. One factor underlying these slow velocities is a myosin isoenzyme with a low ATPase activity.

Smooth muscles exhibit many different velocity-stress curves. Peak force, F_o, varies with the stimulus and the level of activation of the contractile apparatus (Fig. 22-12, *A*). This is consistent with control of the

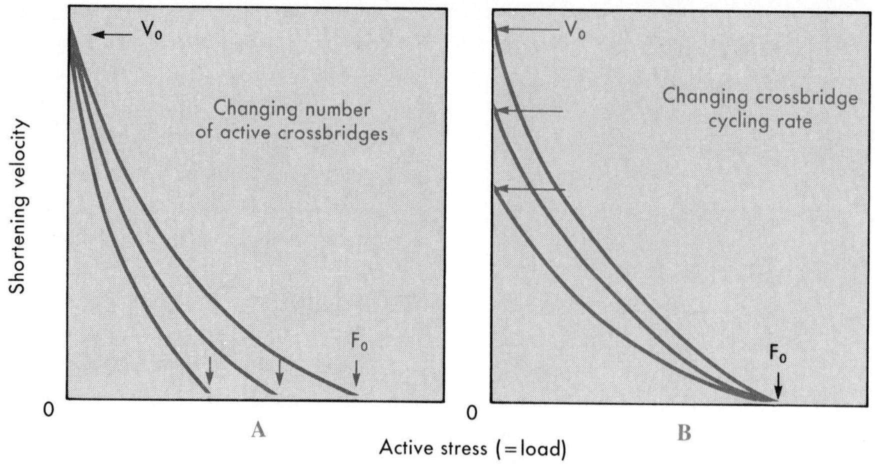

Fig. 22-12. Velocity-stress relationships in smooth muscle can vary considerably in the same preparation. **A,** Different stimuli elicit varying maximal isometric forces. A family of curves reflect differences in activation. **B,** Unlike striated muscles, the maximal unloaded shortening velocities in smooth muscle can vary, even when the maximal isometric stress is the same. This reflects the operation of mechanisms that regulate crossbridge cycling rates.

numbers of crossbridges that are turned on and that interact with the thin filament. The values of V_o also differ (Fig. 22-12, *B*), depending on the level of activation of the muscle. Such results imply that both crossbridge cycling rates and the number of active crossbridges are regulated, in marked contrast to striated muscle. This behavior probably reflects properties of a complex regulatory system for the crossbridge rather than basic differences between crossbridges in smooth and striated muscle.

Latch and crossbridge physiology. Many smooth muscles normally contract tonically, i.e., exhibit sustained partial activation. Such *tone* is characteristic of the smooth muscle in blood vessels or in sphincters. These sustained tonic contractions are characterized by reduced rates of crossbridge cycling. Although average crossbridge cycling rates in a striated muscle fall during an isometric contraction as force rises to a maintained level, the unloaded shortening velocity, V_o, remains constant. Fig. 22-13 illustrates how V_o can fall within seconds after stimulation in a smooth muscle. The resulting maintenance of force with low crossbridge cycling rates and ATP consumption has been termed *latch*. Latch appears to reflect a transition from a population of rapidly cycling crossbridges to a population of attached, noncycling, or slowly cycling crossbridges.

The capacity to arrest a crossbridge cycle in the attached state (latch) as well as in the detached state (relaxation) is functionally important for a muscle type that often acts to withstand imposed loads and maintain organ dimensions. The metabolic savings are considerable. The heart and upper esophagus are the only hollow organs that contain striated muscle. This is functionally appropriate for ''pumps'' that must shorten rhythmically.

Ca^{++}-Dependent Regulatory Mechanisms in Smooth Muscle

Smooth muscle cells lack troponin, the Ca^{++}-binding regulatory protein on the thin filaments of skeletal and cardiac muscles. Our current understanding of the link between Ca^{++} and initiation of crossbridge cycling stems from the discovery that all muscles contain a pair of enzymes, *myosin kinase* and *myosin phosphatase,* which phosphorylate and dephosphorylate the crossbridges, respectively. These reactions take place at a specific site on the regulatory light chains, one site per crossbridge head. Crossbridge phosphorylation depends on micromolar concentrations of Ca^{++} (Fig. 22-14), because the active form of myosin kinase is a complex of kinase-calmodulin-Ca^{++}. *Calmodulin* is a small cytoplasmic protein that binds four Ca^{++} ions. The calmodulin-$Ca^{++}{}_4$ complex can combine with several enzymes besides myosin kinase to confer Ca^{++}-dependent activity. Crossbridge phosphorylation occurs when the myoplasmic Ca^{++} increases on stimulation of a muscle. During relaxation, when Ca^{++} decreases, myosin kinase is inactivated, and net dephosphorylation occurs. No mechanisms regulating the activity of myosin phosphatase have been discovered. These enzymatic reactions are slow; almost a second is required to phosphorylate the crossbridges in a muscle. Force development occurs much faster than phosphorylation in striated muscles, in which the physiological signifi-

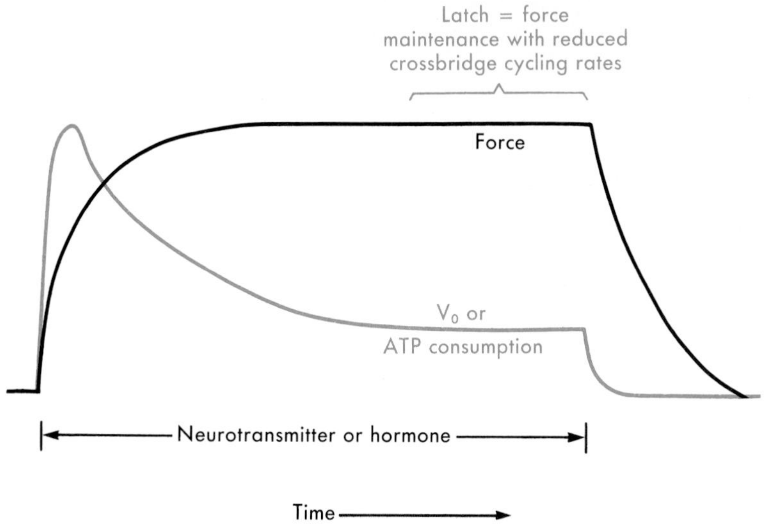

■ **Fig. 22-13.** The maximal unloaded rates for crossbridge cycling in smooth muscle characteristically fall with time from peak values produced on stimulation. The result is maintained force with reduced ATP consumption (latch). Tonic contractions in smooth muscle are associated with an ATP consumption that may be only 1⁄300 of that required to maintain the same force by a striated muscle.

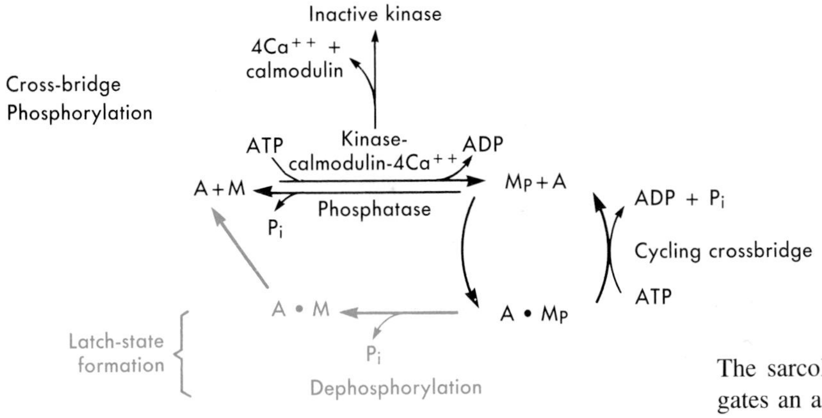

■ **Fig. 22-14.** Hypothesis for regulation of crossbridge interactions in smooth muscle. Ca^{++}-stimulated myosin phosphorylation (M_p) is obligatory for the initiation of crossbridge cycling with actin *(A)*. A noncycling or slowly cycling complex (A·M) is postulated to explain latch. Unloaded shortening velocities, V_o, vary with the relative sizes of the populations of phosphorylated cycling crossbridges and "latchbridges."

cance of crossbridge phosphorylation is uncertain.

In contrast, crossbridge phosphorylation is an obligatory initial step in activation of smooth muscle, in which phosphorylated crossbridges are "switched on" and cycle rapidly with the thin filament (Fig. 22-14). However, when the Ca^{++} concentration falls, dephosphorylated crossbridges remain attached to the thin filament to maintain force in the latch state. The latch state requires low concentrations of Ca^{++} and a small pool of phosphorylated crossbridges.

■ *Regulation of Cellular Ca^{++}*

The myoplasmic Ca ion concentration in relaxed muscle is very low (less than 0.1 μM). This is maintained by active transport mechanisms against very large concentration gradients. Contraction is initiated by release of Ca^{++} into the myoplasm, and relaxation follows Ca^{++} removal. The Ca^{++} involved in initiating contraction is localized in an internal compartment called the *sarcoplasmic reticulum* (Fig. 22-2).

The *sarcolemma,* or plasma membrane, separates the extracellular space from the *myoplasm,* or the intracellular space, which contains the contractile apparatus. The sarcolemma is an excitable membrane and propagates an action potential, which is the first event in the excitation of striated muscle cells (Chapter 3). Small *transverse tubules* (T-tubules) open to the extracellular space at the sarcolemma and form a reticulum in the interior of the cell (Fig. 22-15). This network of T-tubules around the myofibrils forms a grid across the cell at the level of the junction between the A and the I bands in each sarcomere of mammalian striated muscles (Fig. 22-2 and 22-15). Despite the extent of T-tubule development, the volume of this system that is continuous with the extracellular fluid is only 0.1% to 0.5% of the cell volume. Depolarization of the sarcolemma spreads down the T-tubule into the interior of the cell. The T-tubules are in close association with the fenestrated sheath of sarcoplasmic reticular membranes surrounding the myofibrils (Fig. 22-16). This sheath consists of repeating units with (1) expanded elements (or terminal cisternae) that surround the T-tubules and with (2) narrower elements that run along the myofibrils.

■ *Ca^{++} Regulation by the Sarcoplasmic Reticulum*

The membranes of the sarcoplasmic reticulum form an intracellular compartment occupying about 1% to 5% of the volume of mammalian skeletal muscle cells, which

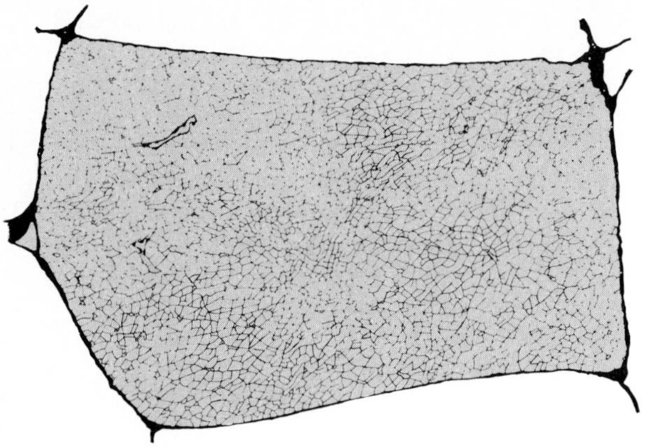

■ **Fig. 22-15.** The T tubular system in a skeletal muscle, reconstructed from high-voltage electron micrographs of serial transverse sections. The extensive network of T tubules across the fiber has many openings to the extracellular space. (From Peachey, L.D., and Eisenberg, B.R. Reproduced from the Biophysical journal, 1978, **22:**145, by copyright permission of the Biophysical Society.)

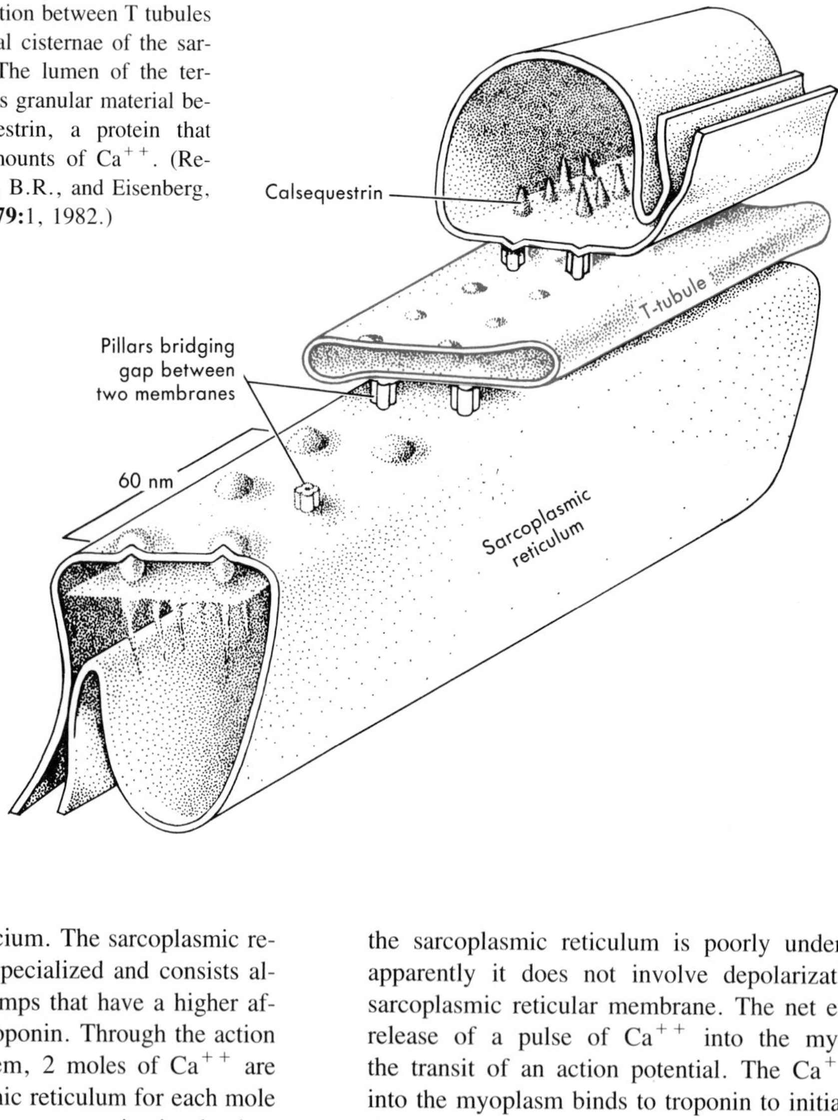

■ **Fig. 22-16.** Association between T tubules *(color)* and the terminal cisternae of the sarcoplasmic reticulum. The lumen of the terminal cisternae contains granular material believed to be calsequestrin, a protein that weakly binds large amounts of Ca^{++}. (Redrawn from Eisenberg, B.R., and Eisenberg, R.S.: J. Gen Physiol. **79:**1, 1982.)

contain large amounts of calcium. The sarcoplasmic reticular membrane is highly specialized and consists almost entirely of transport pumps that have a higher affinity for Ca^{++} than does troponin. Through the action of this active transport system, 2 moles of Ca^{++} are sequestered in the sarcoplasmic reticulum for each mole of ATP hydrolyzed. These pumps maintain the low resting myoplasmic Ca^{++} concentration. The transit of an action potential along the sarcolemma causes Ca^{++} release from the sarcoplasmic reticulum into the myoplasm. The nature of the coupling between the sarcolemma T-tubular system to induce Ca^{++} release from the sarcoplasmic reticulum is poorly understood, but apparently it does not involve depolarization of the sarcoplasmic reticular membrane. The net effect is the release of a pulse of Ca^{++} into the myoplasm by the transit of an action potential. The Ca^{++} released into the myoplasm binds to troponin to initiate contraction.

The interior of the sarcoplasmic reticulum contains a protein called *calsequestrin* (Fig. 22-16). Each molecule of calsequestrin can bind about 43 Ca^{++} ions. The binding affinity is low. Calsequestrin plus another protein with similar properties acts to reduce the sarco-

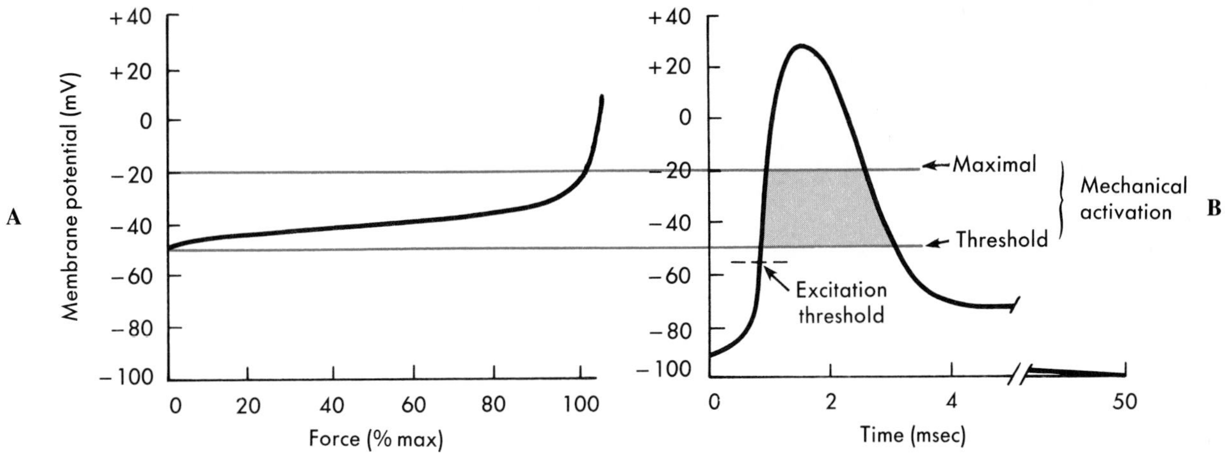

■ Fig. 22-17. **A,** The relationship between steady state force development and membrane potential (E_m) obtained by increasing the K^+ concentration in the bathing solution. The mechanical threshold for force development occurs at an E_m of about −50 mV when Ca^{++} release from the sarcoplasmic reticulum leads to myoplasmic concentrations exceeding threshold values of about 0.1 μM. At an E_m of −20 mV, myoplasmic Ca^{++} concentrations are maximal for activation. **B,** Under physiological conditions the transit of an action potential along a muscle fiber leads to release of a Ca^{++} pulse. Ca^{++} release is proportional to the time the membrane is depolarized within the values producing mechanical activation, as indicated by the colored area in the illustrated action potential. (Redrawn from Sandow, A.: Arch. Phys. Med. Rehabil. **45:**62, 1964.)

plasmic reticular concentration of Ca^{++} from perhaps 20 mM, if all were free, to an estimated 0.5 mM. This action greatly reduces the concentration gradient against which the pump must act.

■ *Excitation-Contraction Coupling in Skeletal Muscle*

The overall process by which depolarization of the sarcolemma causes Ca^{++} release into the myoplasm and Ca^{++} binding to regulatory sites to initiate crossbridge cycling is termed *excitation-contraction coupling* (E-C coupling). The response of the contractile system depends on the amount of Ca^{++} released into the myoplasm. Ca^{++} release from the sarcoplasmic reticulum depends on the value of the membrane potential, and it occurs when the potential becomes less negative than about −50 mV. Maximal Ca^{++} release with full activation of the cell occurs when the cell is depolarized to a level of about −20 mV (Fig. 22-17). The more negative value is termed the *mechanical threshold* and is similar whether it is achieved during a normal action potential or when the membrane is experimentally depolarized (Fig. 22-17). Agents that lower the mechanical threshold or prolong the action potential increase the magnitude of the Ca^{++} pulse. Release of a greater amount of Ca^{++} potentiates the contractile response to a single action potential. Caffeine is a drug that elicits both of these effects.

A single action potential may release sufficient Ca^{++} to activate the contractile machinery in skeletal muscles fully. However, the Ca^{++} is very rapidly pumped back into the sarcoplasmic reticulum before the muscle has time to develop its maximal force (Fig. 22-18). The resulting submaximal response to a single action potential is termed a *twitch*. The time relationships among the action potential, the Ca^{++} pulse, and isometric force development are illustrated in Fig. 22-18 for a skeletal muscle fiber. Repetitive action potentials can cause summation of twitches, producing a partial or complete *tetanus* as the Ca^{++} pulses summate to maintain saturating Ca^{++} concentrations in the myoplasm (Figs. 22-18 and 22-19).

Force development is much slower than Ca^{++} release and binding to troponin, because a number of crossbridge cycles must produce some sarcomere shortening before maximal force is registered at the tendons. This behavior is a result of the fact that the structural elements of the muscle, including the crossbridges, are somewhat elastic and lengthen when a force is generated. This elasticity (termed *series elasticity*) is seen only in a contracting muscle and is distinct from the *passive (parallel) elasticity* observed in a relaxed muscle (Fig. 22-8), which is largely caused by the presence of connective tissue.

■ *Excitation-Contraction Coupling in Cardiac Muscle*

The contractile apparatus of cardiac muscle is organized like that of skeletal muscle. Excitation-contraction coupling involves action potential–induced release of

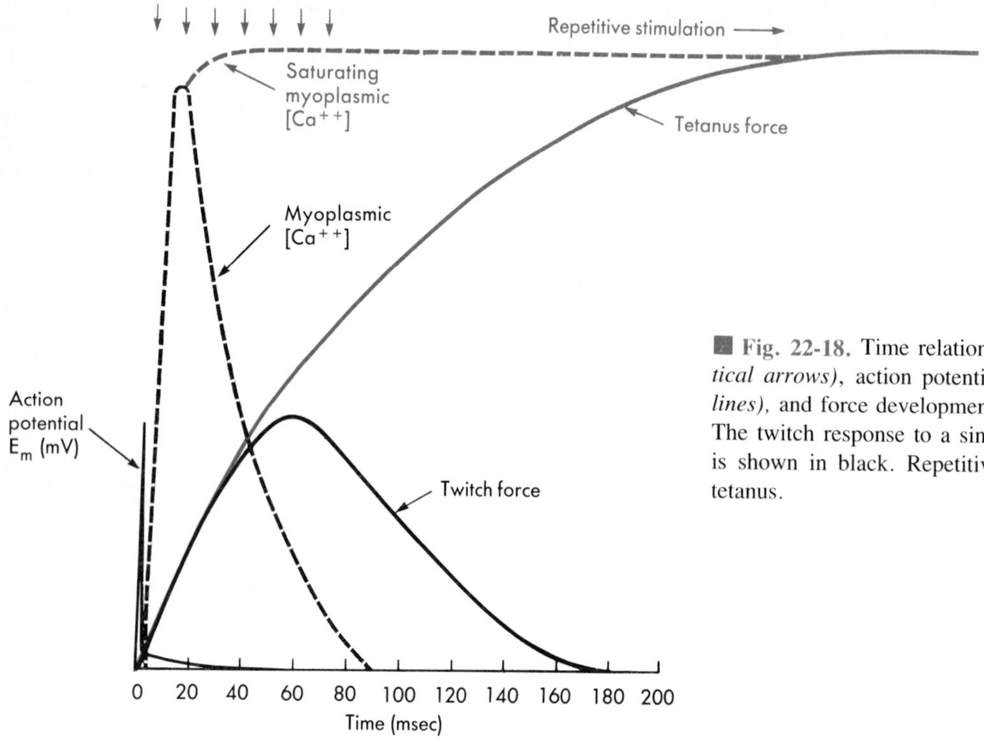

■ **Fig. 22-18.** Time relationships between the stimulus *(vertical arrows)*, action potential, myoplasmic $[Ca^{++}]$ *(dashed lines)*, and force development *(solid lines)* in skeletal muscle. The twitch response to a single stimulus and action potential is shown in black. Repetitive stimulation *(color)* leads to a tetanus.

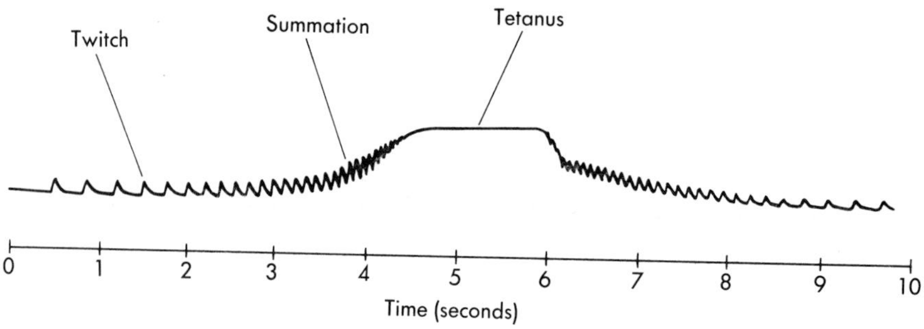

■ **Fig. 22-19.** The force developed by a skeletal muscle fiber in response to repetitive stimulation at gradually increasing frequencies, followed by gradually decreasing frequencies. Individual twitch responses start to fuse and sum as the frequency increases, leading to a complete or fused tetanus. (From Buchthal, F.: Dan. Biol. Med. **17**:1, 1942.)

Ca^{++} from the sarcoplasmic reticulum, diffusion to the thin filament (where it combines with troponin), and a conformational change in the thin filament that permits crossbridges to cycle. However, crossbridge regulation has several special characteristics associated with the heart's physiological function as a pump. These characteristics are briefly described in this section, and are considered in detail in Chapters 27 and 28.

Electrical and mechanical relationships. Action potentials in cardiac muscle cells vary, but they are characteristically very long (Fig. 22-20). The mechanical response is a twitch that is largely completed within the refractory period of the cell membrane. The heart beat is simply the sum of the nearly simultaneous twitches of all the cardiac cells. Three mechanisms are involved in physiological regulation of the force of contraction: (1) changes in Ca^{++} influx from the extracellular space, (2) modulation of the Ca^{++} sensitivity of the contractile apparatus, and (3) modulation of Ca^{++} extrusion out of the cell. Fig. 22-21 illustrates how each of these can potentiate the twitch.

Extracellular Ca^{++} and potential-sensitive Ca^{++} channels. The action potential in skeletal muscles is caused by the opening of *fast channels* in the membrane. This process permits an initial rapid inward Na^+ current, followed by an outward K^+ movement and repolarization (see Chapter 3). There is no significant movement of Ca^{++} ions through fast channels from the extracellular space to the myoplasm. In addition to the fast channels, cardiac muscle has membrane potential–dependent *slow channels,* which open during the transit of an action potential (Fig. 22-21, *B*). Ca^{++} ions are the major inward current-carrying ions in the slow channels. The inward Ca^{++} current is responsible for the maintained depolarization characterized by the plateau phase of the action potential in cardiac muscle (Fig. 22-20). Although only a small amount of Ca^{++} enters a cell during a cardiac action potential, this influx

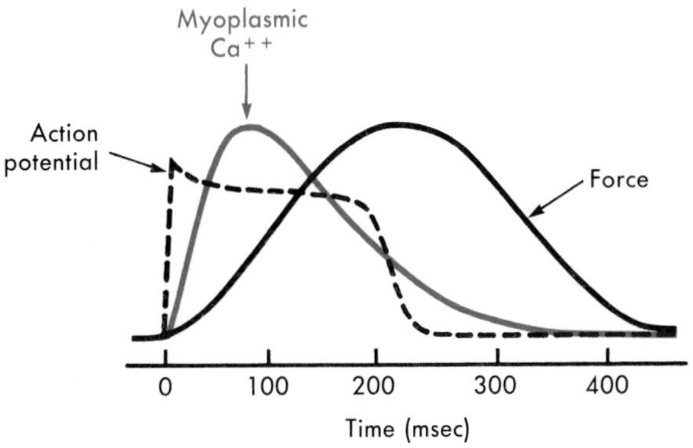

Fig. 22-20. Approximate time course of changes in membrane potential, the myoplasmic Ca^{++} concentration, and the force generated in a cardiac muscle cell. Cardiac muscle cannot normally be tetanized because relaxation occurs during the refractory period.

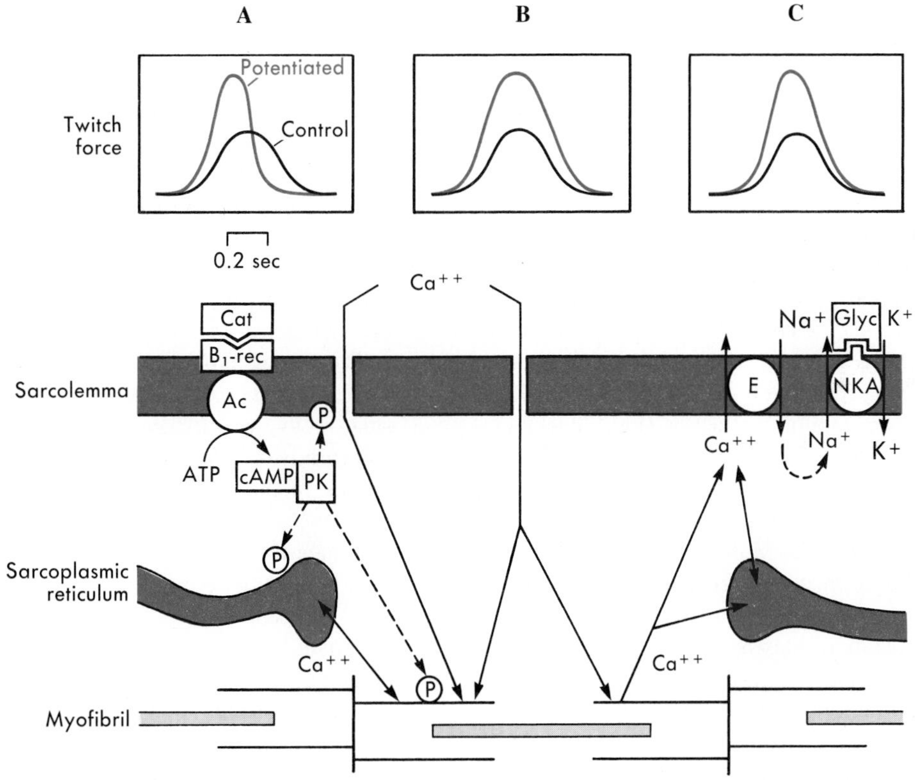

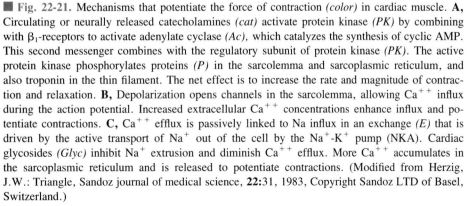

Fig. 22-21. Mechanisms that potentiate the force of contraction *(color)* in cardiac muscle. **A,** Circulating or neurally released catecholamines *(cat)* activate protein kinase *(PK)* by combining with β_1-receptors to activate adenylate cyclase *(Ac),* which catalyzes the synthesis of cyclic AMP. This second messenger combines with the regulatory subunit of protein kinase *(PK).* The active protein kinase phosphorylates proteins *(P)* in the sarcolemma and sarcoplasmic reticulum, and also troponin in the thin filament. The net effect is to increase the rate and magnitude of contraction and relaxation. **B,** Depolarization opens channels in the sarcolemma, allowing Ca^{++} influx during the action potential. Increased extracellular Ca^{++} concentrations enhance influx and potentiate contractions. **C,** Ca^{++} efflux is passively linked to Na influx in an exchange *(E)* that is driven by the active transport of Na^+ out of the cell by the Na^+-K^+ pump (NKA). Cardiac glycosides *(Glyc)* inhibit Na^+ extrusion and diminish Ca^{++} efflux. More Ca^{++} accumulates in the sarcoplasmic reticulum and is released to potentiate contractions. (Modified from Herzig, J.W.: Triangle, Sandoz journal of medical science, **22:**31, 1983, Copyright Sandoz LTD of Basel, Switzerland.)

contributes significantly to the maintenance of the normal levels of Ca^{++} that are handled by the sarcoplasmic reticulum. The force of the rhythmical cardiac contractions rapidly declines if extracellular Ca^{++} is reduced, and it is enhanced when extracellular Ca^{++} is increased. Similarly, higher frequencies of cardiac contraction increase the magnitude of the Ca^{++} pool available for E-C coupling over the course of several beats. This produces the *staircase phenomenon* or *Treppe* characterized by progressive increases in the force of contraction over several beats (see Chapter 29). Thus both the sarcoplasmic reticulum and the extracellular Ca^{++} pools are involved in E-C coupling in the heart.

Modulation of activation by Ca^{++}. During exercise, both the force and the frequency of the twitch increase. This response is caused by increased activity of the cardiac sympathetic-adrenergic nerves and high plasma catecholamine levels. The enhanced cardiac output reflects several regulatory processes that occur when the catecholamines combine with β-adrenergic receptors on the cardiac cell membrane (Fig. 22-21, *A*). β-receptor occupancy activates adenylate cyclase, which produces cyclic adenosine monophosphate (AMP) in the cells. Like Ca^{++}, cyclic AMP is a second messenger that mediates cellular processes. Cyclic AMP exerts its effects by binding to the regulatory subunit of a protein kinase that phosphorylates a number of proteins involved in Ca^{++} handling in cardiac muscle. Phosphorylation of a membrane protein leads to the activation of potential-sensitive Ca^{++} channels in the plasma membrane. Thus each action potential is associated with a greater Ca^{++} influx. This in turn potentiates the twitch and leads to a faster rate of contraction (Fig. 22-21, *A*). The maximal response occurs after a number of contractions, with a buildup of the amount of Ca^{++} that is stored in the sarcoplasmic reticulum and that is released with each action potential. More frequent and forceful contractions require enhanced relaxation rates, so that the heart can more effectively refill with blood between beats. Faster relaxation rates occur because of two additional actions of protein kinase. One is to phosphorylate proteins of the Ca^{++} pumps of the sarcoplasmic reticulum; this enhances the activity of those pumps. The second action is to phosphorylate troponin. Phosphorylated troponin has a lower affinity for Ca^{++}. The net effect is to shorten the twitch through faster removal of Ca^{++} from the troponin and the myoplasm. Thus contraction is graded in cardiac muscle by changing the amount of Ca^{++} mobilized, the rates of release and uptake, and the sensitivity of the contractile apparatus to Ca^{++}.

Modulation of Ca^{++} extrusion. Ca^{++} influx during the cardiac action potential must be balanced over time by efflux. An important mechanism for Ca^{++} efflux in cardiac muscle cells is a passive transport that is coupled with the inward movement of Na^+ down its diffusion gradient (Fig. 22-21, *C*). This passive Ca^{++} extrusion is maintained through an active outward transport of Na^+ by the Na^+-K^+ pump in the plasma membrane. The balance of factors that determine Na^+-K^+ pump activity is not well understood. However, drugs such as the cardiac glycosides inhibit Na^+-K^+ pump activity. This in turn reduces Ca^{++} efflux by slowing the passive Na^+-Ca^{++} exchange, because less Na^+ enters the cell as the myoplasmic Na^+ concentrations rise. Such glycosides enhance the twitch without affecting its time course (Fig. 22-21, *C*). Digitalis is a cardiac glycoside that has been used for over a century to treat heart failure.

Excitation-Contraction Coupling in Smooth Muscle

Smooth muscles contain a sarcoplasmic reticulum. The sarcoplasmic reticulum consists of a tubular network in the periphery of the cells adjacent to the plasma membrane or its invaginations, termed *caveoli* (Fig. 22-22). The volume of the sarcoplasmic reticulum varies among smooth muscles, and in some it equals the volume of the cell occupied by the sarcoplasmic reticulum in striated muscles. However, the many neurotransmitters, hormones, and drugs that stimulate contraction cause only transient responses in tissues bathed in Ca^{++}-free solutions. Most responses also involve Ca^{++} influx into the cell from the extracellular space.

The sarcolemma and extracellular space function as the equivalent of a sarcoplasmic reticulum in these small cells, which have short diffusion distances and comparatively slow rates of contraction and relaxation. Pumps in the smooth muscle cell membrane actively extrude Ca^{++} from the myoplasm. Passive Na^+-Ca^{++} exchange does not play a significant role in Ca^{++} extrusion in smooth muscle. Ca^{++} influx depends on two types of channels (Fig. 22-22). The first is the potential-sensitive slow channel, which responds to the membrane potential. Smooth muscle cell membranes do not contain fast channels. Enough Ca^{++} crosses the membrane during an action potential to initiate a small contraction, which can be summed to produce larger contractions with trains of action potentials (Fig. 22-23, *A* and *B*).

Not all smooth muscles generate slow action potentials when depolarized. This is probably the result of a simultaneous rise in K^+ permeability, which prevents a regenerative response. Variations in Ca^{++} influx through potential-sensitive slow channels in the absence of action potentials produce graded contractions (Fig. 22-23, *C*).

The "resting" membrane potential (E_m) reflects ionic concentration gradients and conductances as in striated muscle. However, there is a significant and variable

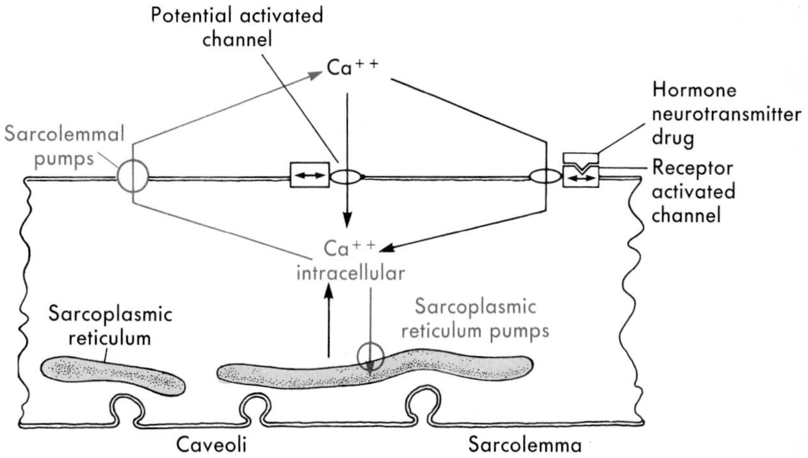

■ **Fig. 22-22.** Cellular compartments and Ca^{++} pools in smooth muscle. The sarcolemma and its invaginations (caveoli) separate the myoplasm from the extracellular space. The sarcoplasmic reticulum is a simple membrane system delineating an intracellular compartment in which Ca^{++} is present in high concentrations. Ca^{++} can enter the myoplasm to activate the contractile apparatus from the sarcoplasmic reticulum or through the sarcolemma. Relaxation follows Ca^{++} movement back into the sarcoplasmic reticulum or extracellular space by active transport processes.

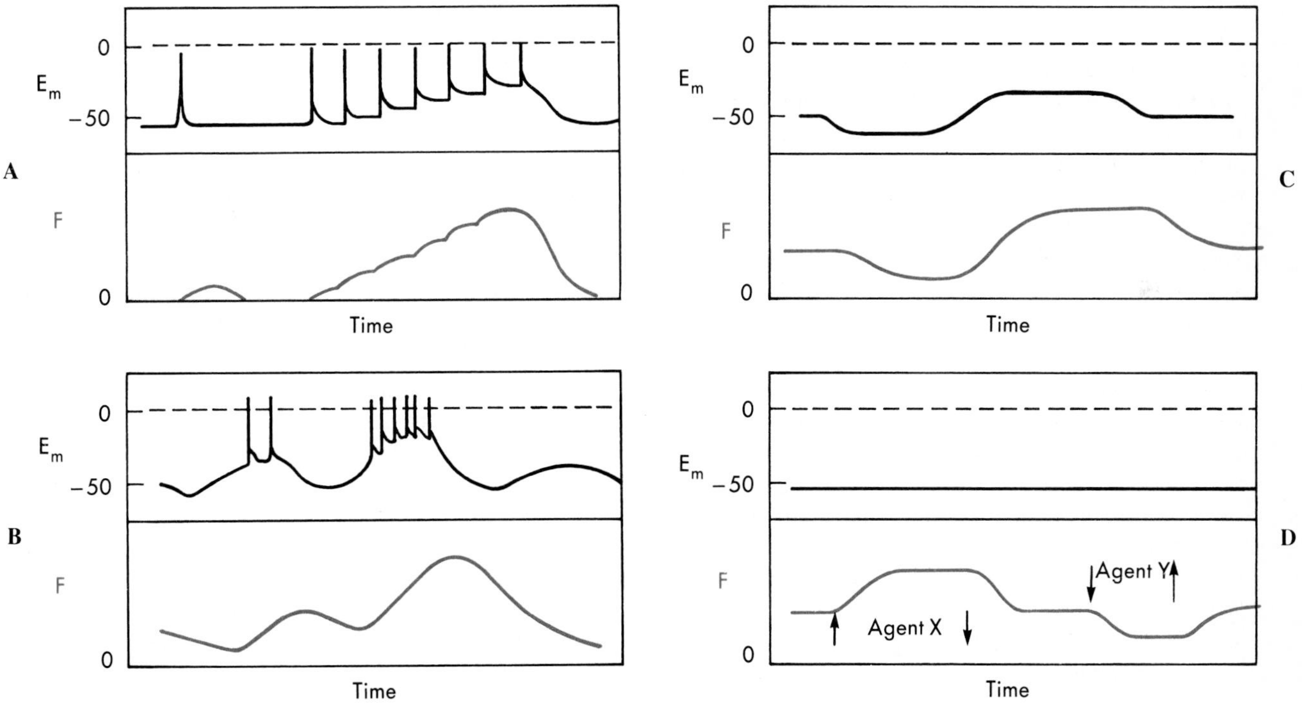

■ **Fig. 22-23.** Relationships between membrane potential *(E_m)* and force generation *(F)* characteristic of different types of smooth muscle. **A,** Action potentials may be generated and lead to a twitch or larger summed mechanical responses. Action potentials are characteristic of single-unit smooth muscles (many visceral). Gap junctions permit the spread of action potentials throughout the tissue. **B,** Rhythmical activity produced by slow waves that trigger action potentials. The contractions are usually associated with a burst of action potentials. Slow oscillations in membrane potential usually reflect the activity of electrogenic pumps in the cell membrane. **C,** Tonic contractile activity may be related to the value of the membrane potential in the absence of action potentials. Graded changes in E_m are common in multiunit smooth muscles (many vascular), where electrical activity is not propagated from cell to cell. **D,** Pharmacomechanical coupling; changes in force produced by the addition or removal *(arrows)* of drugs or hormones that have no significant effect on the membrane potential.

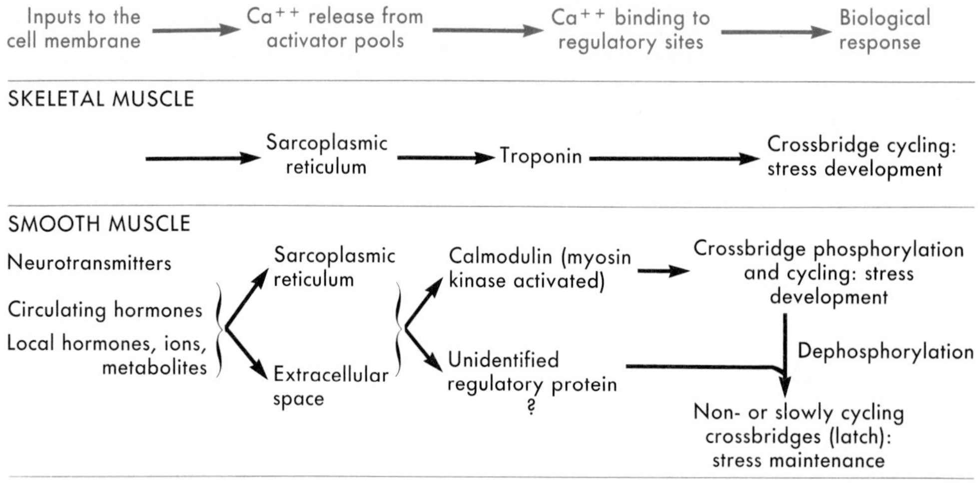

■ **Fig. 22-24.** Regulation of contraction involves four elements *(color)*. The sequence is comparatively simple in skeletal muscle, involving a single event in each step. Cardiac muscle is similar, although somewhat more complicated because the sarcolemma is involved in determining myoplasmic Ca^{++} and the force of contraction. The steps involved in the coupling of activation and contraction in smooth muscle are only partially understood and involve multiple events at each step, reflecting the diverse functional roles of this muscle type.

contribution to E_m by electrogenic pumps that extrude 3 Na^+ ions in exchange for 2 K^+ ions (Fig. 22-21, *C*; see Chapter 2). Changes in pump activity alter the membrane potential (Fig. 22-23, *C*). Sufficient depolarization may produce trains of action potentials, most typically in smooth muscles that contract rhythmically, as in the gastrointestinal tract (Fig. 22-23, *B*).

Receptor-activated Ca^{++} channels. Many neurotransmitters, hormones, and drugs can combine with highly specific receptors on smooth muscle cell membranes (Fig. 22-22). Such agents can elicit contractions in some cases without changing the membrane potential (Fig. 22-23, *D*). The Ca^{++} sources can involve both extracellular or membrane-bound Ca^{++} and Ca^{++} released internally from the sarcoplasmic reticulum. This process is termed *pharmacomechanical coupling*. Depolarization may also occur in response to the binding of neurotransmitters, hormones, or drugs to the cell membrane. Additional Ca^{++} influx through potential-activated Ca^{++} channels may occur in such cells during depolarization.

The complexity of the processes involved in E-C coupling or pharmacomechanical coupling in smooth muscle reflects the diverse and complex physiology of these tissues. In addition, tissue function can be modified by a wide range of neural and hormonal influences. A broad comparison of how contractile output is graded in the different muscle types is given in the next section.

Other second messengers and crossbridge regulation. A variety of drugs relax smooth muscles by increasing the cellular concentrations of cyclic AMP or cyclic guanosine monophosphate (GMP). The protein kinases activated by these second messengers can phosphorylate a variety of enzymes and modify their activities. Relaxation can result from enhanced Ca^{++} extrusion or sequestration, presumably by phosphorylation of Ca^{++} pumps.

■ *Grading Contractile Force*

The mechanisms described earlier for regulating the myoplasmic Ca^{++} concentration and the crossbridge interaction with the thin filament are summarized in Fig. 22-24. The physiologically important ways in which various muscle tissues vary their output are summarized in Table 22-1.

In skeletal muscle, contraction occurs only in response to activity in the motor nerves, and Ca^{++} release from the sarcoplasmic reticulum is produced by the action potential. Contractile force can be graded by two mechanisms. One mechanism is to recruit more cells in a muscle. Recruitment involves increments of groups of cells, each group consisting of all the muscle fibers innervated by a single motor nerve *(motor unit)*. The second mechanism is to sum the twitch responses by increasing the frequency of stimulation to produce a tetanus. The gradation of contraction is considered in greater detail in the following chapter and in Chapter 14. Force will vary somewhat with shortening according to the stress-length relationship. However, skeletal muscles are attached to the skeleton in arrangements that usually limit the normal working range to lengths near the peak of the stress-length curve (Figs. 22-8, *B*, and 23-3).

Cardiac muscle cells are all electrically coupled, and the cells contract almost synchronously as the action potential propagates rapidly from cell to cell. Furthermore, the prolonged action potentials and subsequent

Table 22-1. Mechanisms for grading contractile force

General mechanism	Occurrence		
	Skeletal	*Cardiac*	*Smooth**
Recruit more cells (motor units)	+	−	+
Sum twitches by increasing stimulation frequency (tetanus)	+	−	+
Alter filament overlap by stretch	(+)	+	+
Vary twitch via potential-activated Ca^{++} channels	−	+	+
Tonic depolarization and activation of potential-dependent channels without action potentials	−	−	+
Receptor-activated channels (pharmacomechanical coupling)	−	−	+

*The relative importance of these mechanisms varies greatly with the type of smooth muscle.

refractory period of cardiac cells prevent tetanization. Thus different mechanisms are used to grade the force of the twitch. One mechanism involves the action of neurotransmitters or hormones to alter Ca^{++} influx from the extracellular space during the action potential. Other mechanisms change Ca^{++} sequestration or extrusion or the sensitivity of the regulatory proteins to the myoplasmic Ca^{++}. The other factor grading contraction involves stretching the cells to lengths more satisfactory for force generation when the cardiac filling pressure increases (the Starling mechanism, Chapter 28).

Some smooth muscles resemble skeletal muscle, with extensive innervation and little electrical coupling between smooth muscle cells. Such tissues grade contractile force in ways that are analogous to those described for skeletal muscle: recruitment of more cells and summation of twitch responses. Other smooth muscles are more like the heart, where action potentials are propagated between electrically coupled muscle cells. Gradation of output is a function of (1) action potential frequency and mechanical summation, (2) the level of the membrane potential and opening of potential-dependent Ca^{++} channels, and (3) alterations in length of smooth muscle cells by variations in the volume of the hollow organs. However, all smooth muscle cells are sensitive to a variety of neurotransmitters and hormones, and the responses can be further modified by receptor-activated Ca^{++} channels. Force development by most smooth muscles reflects a balance of neural, hormonal, mechanical, and local metabolic factors, some of which may be inhibitory and others excitatory.

Energy Use and Supply

Muscular contraction demands a constant supply of ATP at a rate proportional to consumption. In this section the ATP requirements for various types of contractions (muscle energetics) are considered. The ways in which cells can adequately provide ATP are then reviewed. Finally, the matching of ATP production and use is described for particular fiber types. Muscle cells are specialized for specific contractile activities and a knowledge of fiber types is needed to understand many physiological aspects of muscle function in the body.

Energetics

The mechanical output of muscle can be readily determined in terms of work performed (force × distance shortened) or the force-time integral (force × time) in isometric contractions. Measurement of energy use by the contractile system is more difficult. Heat production can be determined very accurately and reflects ATP use. On the basis of assumptions concerning the total energy released during ATP hydrolysis and the efficiency of its conversion to mechanical work, energy consumption can be calculated from heat production. There are potential errors in these assumptions, and the method is not well suited for estimating rapid events. The same is true for metabolic measurements in which oxygen consumption (or lactate production when oxidative metabolism is blocked) is used to estimate ATP use. However, more difficult direct chemical measurements of ATP and creatine phosphate use (in muscles where metabolic resynthesis of these compounds is inhibited) have confirmed the basic relationships in muscular energetics.

Muscle has an ATP consumption or metabolism associated with basal cellular activities, such as maintaining ion gradients and synthesizing and degrading cellular constituents. Muscles are unusual in that this *resting metabolism* represents only a small fraction of the maximal ATP use that is associated with contraction. The energy requirements for activation, associated with action potentials and Ca^{++} release into the myoplasm, are also comparatively small.

Most energy consumption is a consequence of the fact that each complete cycle of the enormous numbers of crossbridges in a muscle cell requires one ATP molecule (Fig. 22-7). Cycling rates in skeletal muscle depend on two factors. One is the load on the muscle (Fig. 22-9). A muscle contracting isometrically has a low crossbridge cycling rate and a comparatively moderate rate of ATP use. This rate is reduced even more in a muscle subjected to an imposed stretch. This situation is termed *negative work,* since it is characterized by the product of force times the distance stretched.

Such an imposed stretch occurs when muscles are used to decelerate the body. When a muscle can shorten, crossbridge cycling rates increase, and ATP consumption rises with the mechanical work done. Low forces and high velocities are associated with extremely high rates of ATP consumption. Speed is costly, as in other aspects of life. Shortening velocity is ultimately limited by the myosin isoenzyme present. Fast fibers, which contain a form of myosin that is capable of high rates of ATP hydrolysis, have much greater requirements for rapid ATP synthesis than do slow fibers. Thus the type of myosin is the second factor that dictates metabolic requirements.

Relaxation is also associated with above-resting levels of ATP use. In part this increased ATP use provides energy for the Ca^{++} pumps in the sarcoplasmic reticulum or the sarcolemma. The creatine pool is also rephorphorylated if it were depleted during the contraction, and glycogenesis contributes to ATP use in resynthesis of glycogen stores.

The energetics of crossbridge cycling and the efficiency of work performance during shortening in smooth muscles are similar to those described for striated muscle. Nevertheless, mammalian smooth muscles can maintain force tonically (steadily) with a rate of ATP use that is several hundredfold less than that in striated muscles. Part of this performance may be attributed to a form of myosin present in smooth muscles that has intrinsically lower rates of crossbridge cycling. However, part of the economy in maintaining a steady force is a consequence of the regulatory system that can greatly slow the average crossbridge cycle (latch). This allows a maintained contraction in which the muscle can withstand distending pressure and yet expend relatively little energy.

Metabolism

Muscle shares the same ATP-generating mechanisms that are found in all nucleated cells (Fig. 22-25), although the relative importance of the different mechanisms varies in different muscle cell types.

1. *Direct phosphorylation* of ADP to regenerate ATP from creatine phosphate is an extremely rapid reaction. This pathway usually functions as a buffer to maintain the normal ATP levels of 3 to 5 mM at the beginning of contraction, while other systems for regenerating ATP are being turned on. Myoplasmic creatine phosphate concentrations are 5 (smooth) to 20 (skeletal) mM, which are only sufficient to provide the energy for a few twitches. In the second direct phosphorylation reaction, *adenylate kinase* (often termed *myokinase* in muscle) transfers a phosphate group from one ADP to another to form ATP and AMP. This reaction has little metabolic significance, but it plays an important regulatory role in glycolysis (*colored arrows,* Fig. 22-25). Phosphofructokinase, the rate-limiting enzyme in glycolysis, is inhibited by ATP and stimulated by ADP and AMP. Thus ADP and AMP formed on activation of the contractile apparatus will stimulate glycolysis.

2. *Anaerobic glycolysis* is very rapid and readily meets the ATP demands, even of very fast muscle cells. Consequently it is important in cells of this type and in all muscle cells when the oxygen supply is inadequate. However, this pathway has a net yield of only 2 moles of ATP per mole of glucose (or 3 if the glucose is derived from cell glycogen), and it is comparatively inefficient. In the presence of oxygen, pyruvate is converted to CO_2 instead of lactate (aerobic glycolysis), and the net yield of ATP is improved threefold. ATP production by glycolysis may also be limited by the cellular stores of glycogen, which can be rapidly depleted.

3. *Oxidative phosphorylation* of fatty acids is the primary source of energy in muscles that are frequently active. Oxidative phosphorylation is not only efficient (36 moles ATP/mole glucose), but it can operate continuously when the circulation is adequate. However, oxidative phosphorylation is a slow process. It cannot meet the maximal ATP consumption rates of rapidly contracting skeletal muscle fibers unless a major fraction of the fibers consists of mitochondria in close proximity to a capillary.

Matching of ATP Production and Consumption

One of the two isoenzymatic forms of myosin characteristic of skeletal muscle cells has a higher rate of ATP hydrolysis than the other. Cells in which the "fast" myosin is synthesized have faster shortening velocities. With appropriate histochemical methods, thin sections of frozen muscles can be cut and incubated with ATP under conditions where only the slow myosin is enzymatically active. The phosphate released is trapped on the section, "staining" the slow fibers. Fig. 22-26, *A*, illustrates how fast fibers (types IIA and IIB) and slow fibers (type I) are typically intermixed in most mammalian skeletal muscles. Table 22-2 provides a summary of the fiber types and the nomenclature in general use.

Similar histochemical reactions can also be applied to serial sections cut from the same muscle to estimate the activities of enzymes in the oxidative (Fig. 22-26, *B*) and glycolytic (Fig. 22-26, *C*) metabolic pathways. The metabolic capacities in muscle fibers can vary continuously over a wide range. However, most fast (type II) fibers show high activities of glycolytic enzymes and low activities of oxidative enzymes, a fact confirmed by the comparatively few mitochondria that can be ob-

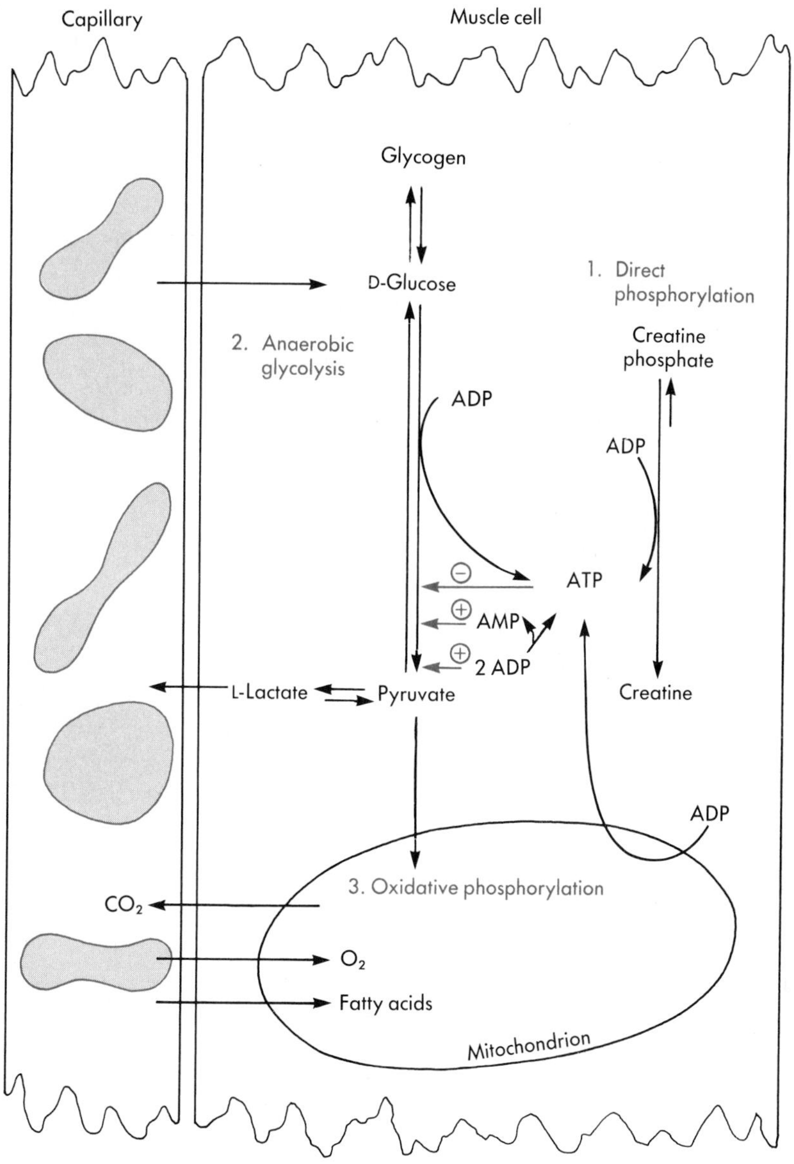

■ **Fig. 22-25.** Metabolic pathways in muscle. ATP is supplied via direct phosphorylation of ADP *(1)*, glycolysis *(2)*, and oxidative phosphorylation *(3)*.

served in electron micrographs. Fast fibers with high contraction velocities have a much more extensive sarcoplasmic reticulum with high pumping rates to quickly activate and inactivate the contractile machinery than do slow fibers, cardiac muscle, and smooth muscle. Because the ATP consumption rates in the myofibrils and sarcoplasmic reticulum of such fast fibers can most readily be matched by glycolysis, a metabolism based primarily on glycolysis is appropriate, provided that the fibers are recruited only occasionally for brief efforts. This is true, as is shown in the next chapter. Theoretically a fiber of small diameter with very high mitochondrial and surrounding capillary densities could oxidatively synthesize ATP at rates used by the fast myosin isoenzyme. Some fast fibers with high glycolytic and oxidative capacities are found in mammals and have led to the subclassification of fast, type II fibers shown in Table 22-2.

Slow, type I fibers can meet relatively modest metabolic demands with oxidative phosphorylation. Molecules associated with oxygen binding (e.g., hemoglobin, myoglobin, cytochromes) contain iron and are red. Therefore the red color of an oxidative muscle cell or a tissue containing mostly type I (or IIA) fibers has led to the designation of such types as *red fibers*.

Cardiac muscle is slow and contains myosin isoenzymes with low ATPase activities. As might be expected, this continuously active muscle is almost en-

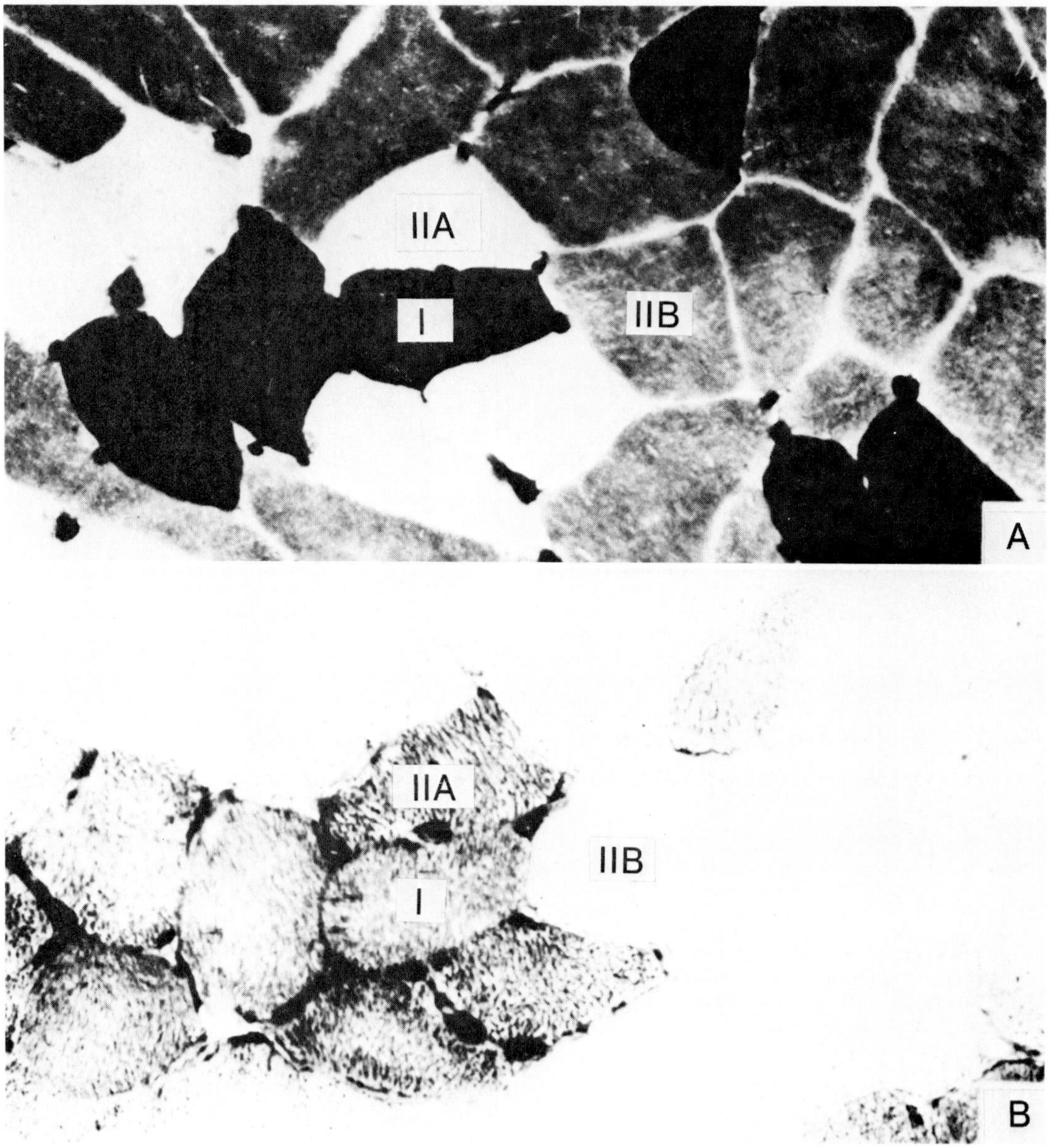

■ Fig. 22-26. Histochemical staining of cross sections of the feline semitendinosus skeletal muscle. **A,** Staining for the ATPase activity of the slow myosin isoenzyme of type I fibers. Type IIB fibers stain more than type IIA fibers in this species, and the myosins in these two types of fast fibers may differ. **B,** Staining for succinic dehydrogenase activity, an enzyme associated with oxidative phosphorylation.

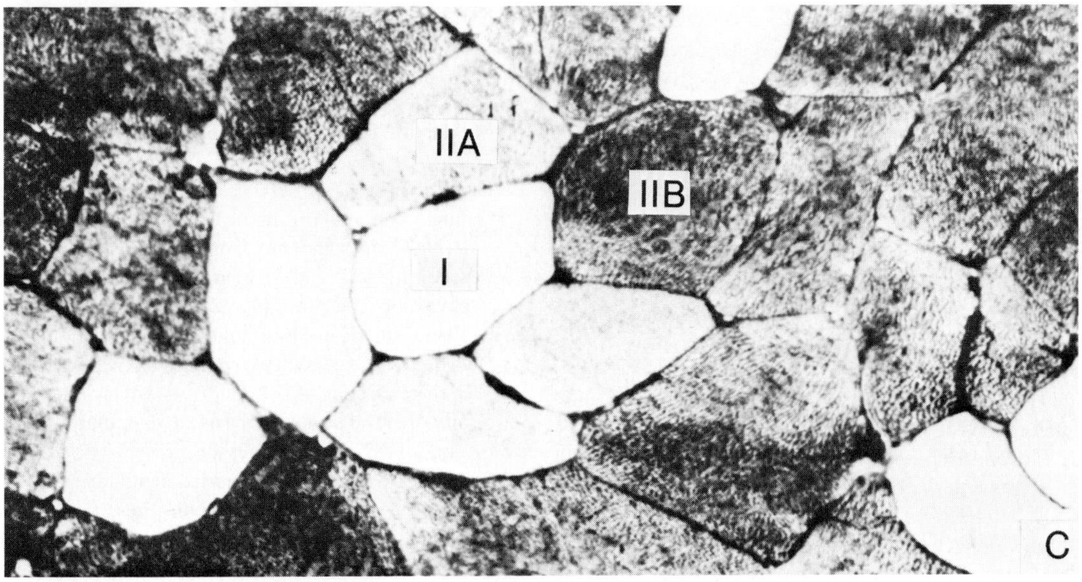

Fig. 22-26, cont'd. C, PAS stain indicating glycolytic capacity. Three distinct fiber types can be differentiated in these serial sections: slow oxidative or type I, fast glycolytic or type IIB, and fast oxidative (and glycolytic) or type IIA fibers. (×450.) (From Hoppeler, H., et al.: Respir. Physiol. **44:**94, 1981.)

Table 22-2. Basic classification of skeletal muscle fiber types

	Type I: slow oxidative (red)	*Type IIB: fast glycolytic (white)*	*Type IIA*: fast oxidative (red)*
Myosin isoenzyme (ATPase rate)	Slow	Fast	Fast
Sarcoplasmic reticular Ca^{++} pumping capacity	Moderate	High	High
Diameter (diffusion distance)	Moderate	Large	Small
Oxidative capacity: mitochondrial content, capillary density, myoglobin	High	Low	Very high
Glycolytic capacity	Moderate	High	High

*Comparatively infrequent in humans and other primates. In the text a simple designation of type II fiber refers to a fast-glycolytic (type IIB) fiber.

tirely oxidative and is highly sensitive to interruptions in its blood supply. Total ATP consumption rates in smooth muscle are much lower than in striated muscles.

The oxidation of fatty acids provides most of the ATP used by the muscles in the body. However, all muscles have a significant capacity to use other substrates, including carbohydrates, certain amino acids, and ketone bodies.

Bibliography

Journal articles

Harrington, W.F., and Rogers, M.E.: Myosin, Ann. Rev. Biochem. **53:**35, 1984.

Kamm, K.E., and Stull, J.T.: The function of myosin and myosin light chain kinase phosphorylation in smooth muscle, Ann. Rev. Pharmacol. Toxicol. **25:**593, 1985.

Kodama, T.: Thermodynamic analysis of muscle ATPase mechanisms, Physiol. Rev. **65:**467, 1985.

Martonosi, A.N.: Mechanisms of Ca^{++} release from sarcoplasmic reticulum of skeletal muscle, Physiol. Rev. **64:**1240, 1984.

Morgan, D.L., and Proske, U.: Vertebrate slow muscle: its structure pattern of innervation, and mechanical properties, Physiol. Rev. **64:**103, 1984.

Pollack, G.H.: The crossbridge theory, Physiol. Rev. **63:**1049, 1983.

Rasmussen, H., and Barrett, P.Q.: Calcium messenger system: an integrated view, Physiol. Rev. **64:**938, 1984.

Staehelin, L.A., and Hull, B.E.: Junctions between cells, Sci. Am. **238:**140, 1978.

Tada, M., and Katz, A.M.: Phosphorylation of the sarcoplasmic reticulum and sarcolemma, Ann. Rev. Physiol. **44:**401, 1982.

Books and monographs

Berne, R.M., and Sperelakis, N., editors: Handbook of physiology, section 2, the cardiovascular system, vol. 1, The heart, Bethesda, Md., 1979, American Physiological Society.

Bohr, D.F., et al.: editors: Handbook of physiology, section 2, The cardiovascular system, vol. 2, Vascular smooth muscle, Bethesda, Md., 1980, American Physiological Society.

Bülbring, E., et al.: Smooth muscle, an assessment of current knowledge, Austin, Tex., 1981, University of Texas Press.

Hoyle, G.: Muscles and their neural control, New York, 1983, John Wiley & Sons.

Huddart, H.: The comparative structure and function of muscle, Oxford, 1975, Pergamon Press, Ltd.

Huddart, H., and Hunt, S.: Visceral muscle: its structure and function, New York, 1975, Halsted Press.

Huxley, A.: Reflections on muscle, Princeton, N.J., 1980, Princeton University Press.

Katz, A.M.: Regulation of cardiac contractile activity by cyclic nucleotides. In Kebabian, J.W., and Nathanson, J.A., editors: Cyclic nucleotides. II. Physiology and pharmacology, Berlin, 1982, Springer-Verlag.

McMahon, T.A.: Muscles, reflexes, and locomotion, Princeton, N.J., 1984, Princeton University Press.

Needham, D.M.: *Machina carnis:* the biochemistry of muscular contraction in its historical development, Cambridge, 1971, Cambridge University Press.

Peachey, L.D., and Adrian, R.H., editors: Handbook of physiology, section 10, Skeletal muscle, Bethesda, Md., 1983, American Physiological Society.

Rüegg, J.C.: Calcium in muscle activation, Berlin, 1986, Springer-Verlag.

Squire, J.: The structural basis of muscular contraction, New York, 1981, Plenum Press.

Stein, R.B.: Nerve and muscle: membranes, cells, and systems, New York, 1980, Plenum Press.

Woledge, R.C., et al.: Energetic aspects of muscle contraction, London, 1985, Academic Press.

CHAPTER

23

Muscle as a Tissue

■ *Voluntary Muscle: Physiological Control and Function*

Individual muscle cells are encased in a connective tissue layer called the *endomysium*. Groups of skeletal muscle cells form *fascicles* that are bounded by the *perimysium*. These fascicles are grouped into the definitive muscle, which is covered by the *epimysium*. The three connective tissue layers are composed primarily of elastin and collagen fibrils. An estimated 250 million cells are found in the more than 400 skeletal muscles in humans. Each of these muscles exerts specific movements via tendons that are attached to the skeleton in most cases.

■ *Musculoskeletal Relationships*

The output of a muscle depends on the size of the muscle's cells and their anatomical arrangement. Increasing the diameter of a fiber by synthesis of new myofibrils *(hypertrophy)* will increase the force-generating capacity of a cell (Fig. 23-1). The formation of more cells *(hyperplasia)* also increases tissue force output. However, differentiated skeletal muscle has only a limited capacity to form new cells. The force-generating capacity is unchanged if the length of the cells is increased by adding more sarcomeres in series without increasing the cross-sectional area of the cells. However, the absolute velocity of contraction and the total shortening capacity of the cell increase with the addition of more sarcomeres (Fig. 23-1).

The output of muscles not only depends on the number of sarcomeres and myofibrils in the cells but also on the orientation of the cells within the tissue. Measurements on intact muscles directly reflect cellular properties only when the cells are arranged parallel with each other and also with the axes of the tendons that transmit the force (Fig. 23-2). This parallel arrangement maximizes shortening capacity and velocity for a muscle. However, force-generating capacity can be enhanced at the expense of shortening capacity and velocity by arrangements placing the muscle fibers at an angle to the tendons. Examples of such *pennate* (featherlike) arrangements are illustrated in Fig. 23-2.

Some skeletal muscles, including those surrounding the mouth and anus, serve as sphincters. Striated muscles are also found in the upper portions of the esophagus, and they are active during swallowing. Others may be attached to the skin. However, most skeletal muscles are attached to the skeleton. The attachments involve the highly inextensible protein *collagen,* which forms *tendons* or flattened sheets termed *aponeuroses*. Muscles are described in terms of an origin at the proximal or relatively fixed point and an insertion on the bone that is moved. (The distinction between origin and insertion is sometimes arbitrary.) Muscles bridge one or, more frequently, two joints. The contraction of an individual muscle can consequently lead to movement of more than one bone.

Specific coordinated movements require the actions of two or more muscles. These may be *synergists,* when the muscles act together, or *antagonists,* when the actions of the muscles are opposed. *Kinesiology* is the study of the interactions of groups of muscles. The complex nature of coordinated movements has been revealed by electromyographic techniques, in which the summed electrical activity of a muscle is detected from electrodes placed on the overlying skin. Another technique can detect the activity of a single cell with needle electrodes inserted into the muscle. These techniques reveal that a specific movement may involve contractions of antagonistic muscles and may not involve contractions of some synergistic muscles. The activity patterns are often unpredictable and can only be determined by direct recordings. In general, the interactions of the muscles serve not only to move a specific bone but also to fix or stabilize another bone or joint.

Fig. 23-1. Effects of growth on the mechanical output of a muscle cell. Growth may consist of adding new myofibrils (depicted as a series of model sarcomeres) within a cell (hypertrophy), formation of new cells (hyperplasia), or adding more sarcomeres in series as the muscle cells lengthen along with skeletal growth. The effects of the illustrated cell growth on the absolute force (kg), shortening velocity (m/sec), and shortening capacity (m) of the muscle are summarized in the table.

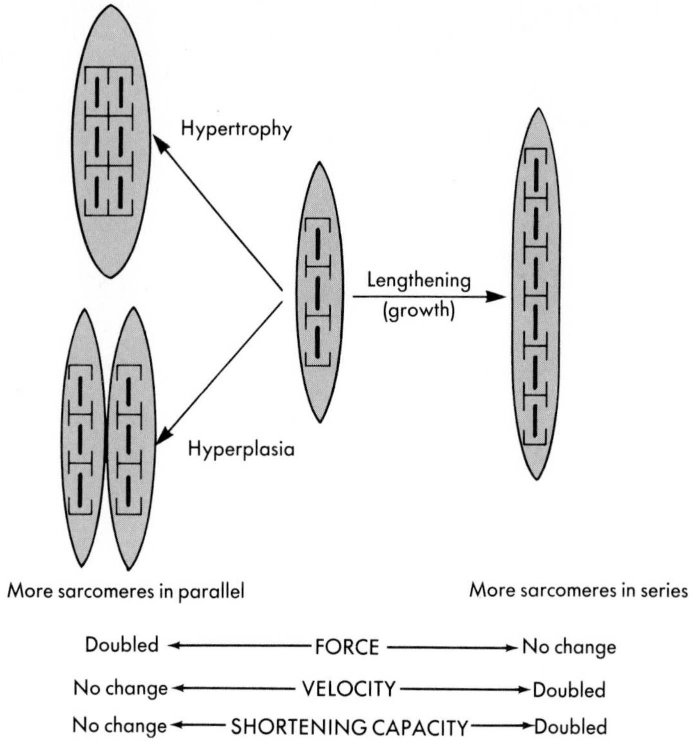

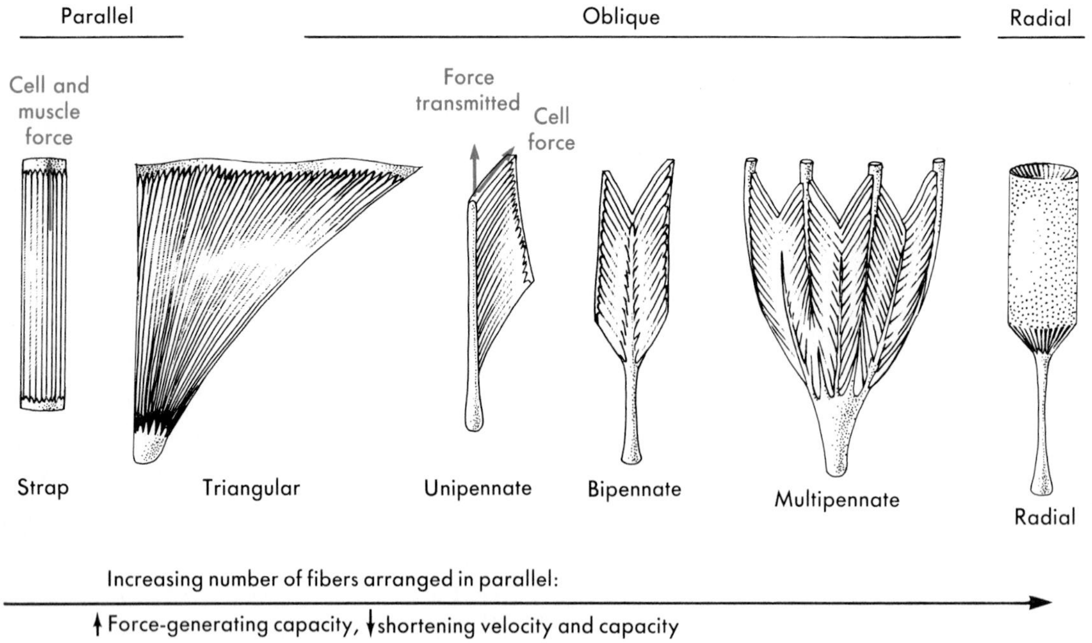

Fig. 23-2. Some arrangements of skeletal muscle fibers. The force generated by all of the individual cells is fully transmitted to the skeleton in only a few straplike muscles. The cells in most muscles are arranged at an angle to the axis of the muscle. This allows more fibers to be attached to a tendon and increases the total force-generating capacity. However, not all of the force generated by each cell is usefully transmitted to the tendon with an oblique geometry, and the overall shortening velocity and shortening capacity of the muscle are less than that of the individual cells. (Redrawn from Gray's anatomy, ed. 35 [British], Philadelphia, 1973, W.B. Saunders Co.)

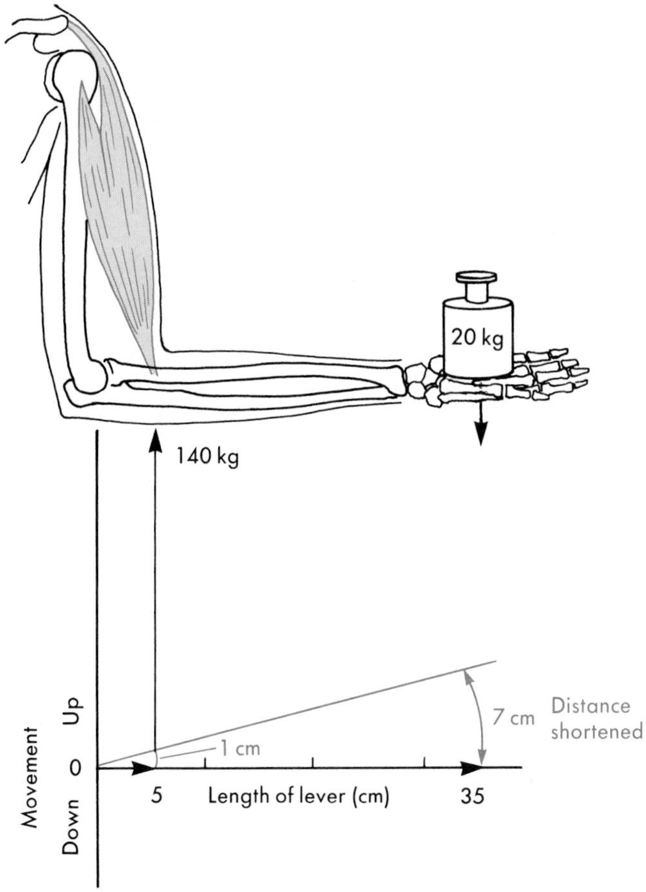

Fig. 23-3. Example of the musculoskeletal lever system. The biceps muscle operates at a 1:7 mechanical disadvantage in this situation and must generate high forces to support a weight on the hand. However, little shortening of the biceps muscle is required to produce large displacements of the hand. (Redrawn from Guyton, A.C.: Textbook of medical physiology, ed. 6, Philadelphia, 1981, W.B. Saunders Co.)

The skeleton also acts as a lever system with significant mechanical consequences. As illustrated in Fig. 23-3, the biceps muscle of a person holding a 20 kg weight at approximately right angles must develop an isometric force of 140 kg. Enormous forces can be placed on tendons by large muscles such as the gastrocnemius. This can lead to rupture of the tendon. The greatest stress on a tendon occurs upon gravitational loading of a contracting muscle because the cross-bridges can bear more force than they can develop. Injuries usually occur as the result of a fall, which subjects a contracting muscle to a sudden large increase in stress. Another consequence of the skeletal lever system is that large movements of the limbs can occur with far less shortening of the muscles (Fig. 23-3). With this arrangement sarcomere lengths remain close to their optimum for force development in most movements.

Coordination of Muscular Activity

Motor nerves and motor units. The cell bodies of the *motor nerves* (α-motor axons) are in the ventral horn of the spinal cord (Fig. 23-4). The axon exits via the ventral root and reaches the muscle through a mixed peripheral nerve. The motor nerves branch in the muscle, with each terminal innervating a single muscle cell in mammals. The specialized cholinergic synapse that forms the neuromuscular junction and the neuromuscular transmission process that generates an action potential in the muscle fiber are described in Chapter 4. A *motor unit* consists of the motor nerve and all of the muscle fibers innervated by that nerve. The motor unit (and not the individual cell) is the functional contractile unit because all of the cells within a motor unit contract synchronously when the motor nerve fires. The muscle cells of a motor unit are not segregated anatomically into distinct groups, and considerable intermixture of cells occurs among neighboring motor units.

Motor units exhibit considerable specialization. They may consist of only two or three muscle fibers, or there may be over a thousand cells in some motor units in large muscles. The cell bodies and axons of the motor nerve increase in size with the number of muscle fibers in the unit. This relationship is understandable in terms of the metabolic requirements for synthesis and release of acetylcholine. A distinction between slow oxidative (type I) and fast glycolytic (type II) muscle fibers is shown in Table 22-2. This classification also applies to motor units because all the fibers in one motor unit are of the same type (Table 23-1). The smaller motor units normally consist of type I cells (Table 23-1) (Fig. 23-4). An important point is that only the contraction velocity or myosin ATPase activity clearly distinguishes the fiber types. These two parameters reflect the presence of different isoenzymatic forms of myosin in the cells. Metabolic differences arise from different cellular

Table 23-1. Properties of motor units

	Motor unit classification	
Characteristics	***Type I***	***Type II***
Properties of nerve		
Cell diameter	Small	Large
Conduction velocity	Fast	Very fast
Excitability	High	Low
Properties of muscle cells		
Number of fibers	Few	Many
Fiber diameter	Moderate	Large
Force of unit	Low	High
Metabolic profile	Oxidative	Glycolytic
Contraction velocity	Moderate	Fast
Fatigability	Low	High

Fig. 23-4. The motor unit and some inputs. Large and small motor units are mixed within a single muscle. An example of each is illustrated in a pair of contralateral muscles. The motor unit contracts in response to an action potential in the motor axon. Contraction is elicited when the sum of the inputs from synapses on the cell body depolarizes the motor neuron to its critical firing potential.

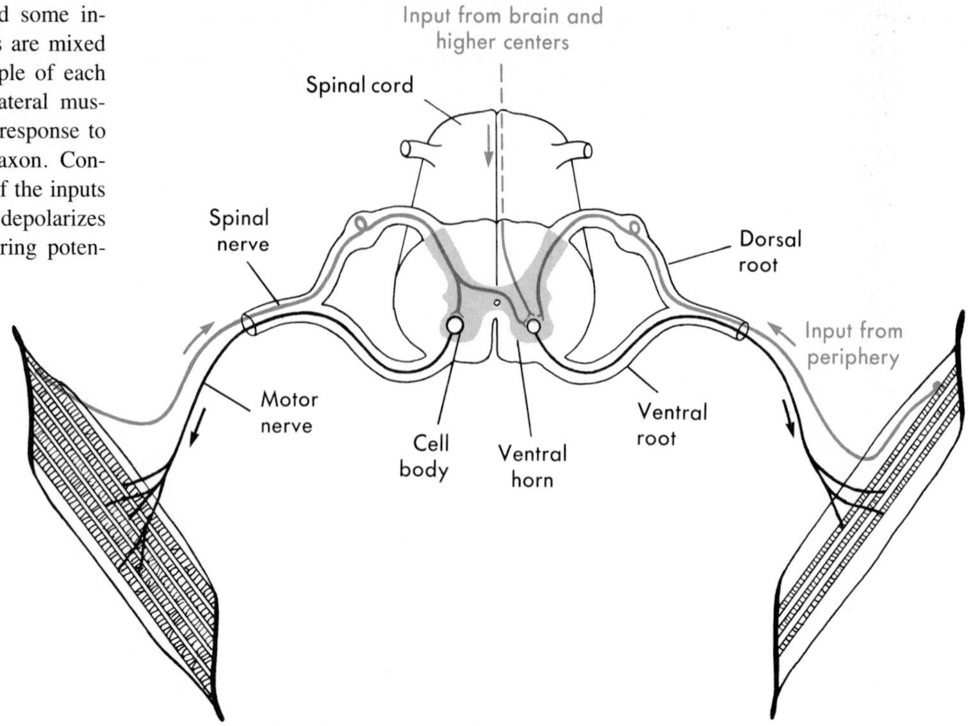

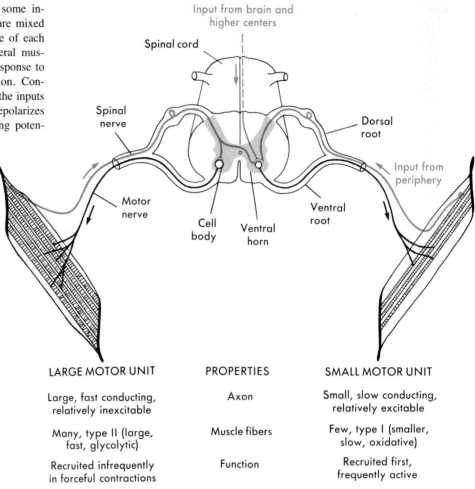

LARGE MOTOR UNIT	PROPERTIES	SMALL MOTOR UNIT
Large, fast conducting, relatively inexcitable	Axon	Small, slow conducting, relatively excitable
Many, type II (large, fast, glycolytic)	Muscle fibers	Few, type I (smaller, slow, oxidative)
Recruited infrequently in forceful contractions	Function	Recruited first, frequently active

contents of mitochondria, for example, and may vary considerably between cells of the same type.

The inputs to the motor nerve are both excitatory and inhibitory. The inputs involve (1) neurons from the brain, (2) neurons from elsewhere in the spinal cord, and (3) neurons originating from a variety of receptors within a muscle, from its antagonists and synergists, and from the same group of contralateral muscles (Fig. 23-4). These inputs to the motor neuron and their integrated activity are described in Chapter 14. A motor neuron fires when the sum of the excitatory and inhibitory inputs depolarizes the cell to its critical membrane potential.

Recruitment of motor units. The functional importance of the variations among motor units can be illustrated by considering how the motor units in a muscle are progressively activated in graded contractions. Increasing excitatory or decreasing inhibitory input to the motor neuron pool in the ventral horn will depolarize the cell bodies. However, a given level of excitatory input will produce more depolarization of the smallest neurons because of their smaller membrane areas. Thus the first axons to fire are those of the smallest motor units. The conduction velocity of the action potential will be relatively low, reflecting the cable properties of the axons of small diameter. The total force developed by the muscle will be small because there are only a few cells of moderate diameter in these units.

Fig. 23-5 shows the basic relationships between motor unit recruitment and total force generated by a muscle. In the left column *31* indicates the number of motor units among those sampled in the muscle that developed up to 10 g force (average, 5 g) when tetanized. If all 31 motor units fired together, the total force generated would be about 0.15 kg, assuming an average force of 5 g/unit. The important point is that these units are recruited first, and they remain active as long as any part of the muscle is contracting. Many such small motor units permit fine gradations of delicate movements.

With increasing excitatory input somewhat larger axons fire. In this sample there were fewer motor units that could generate between 10 and 20 g force, but their summed contributions to the total force in the muscle were comparable, i.e., 10 units $\times$ 15 g (average force) $=$ 0.15 kg, shown by the height of the colored column. The total force generated by the muscle with the increased level of excitatory input is now the sum of 31

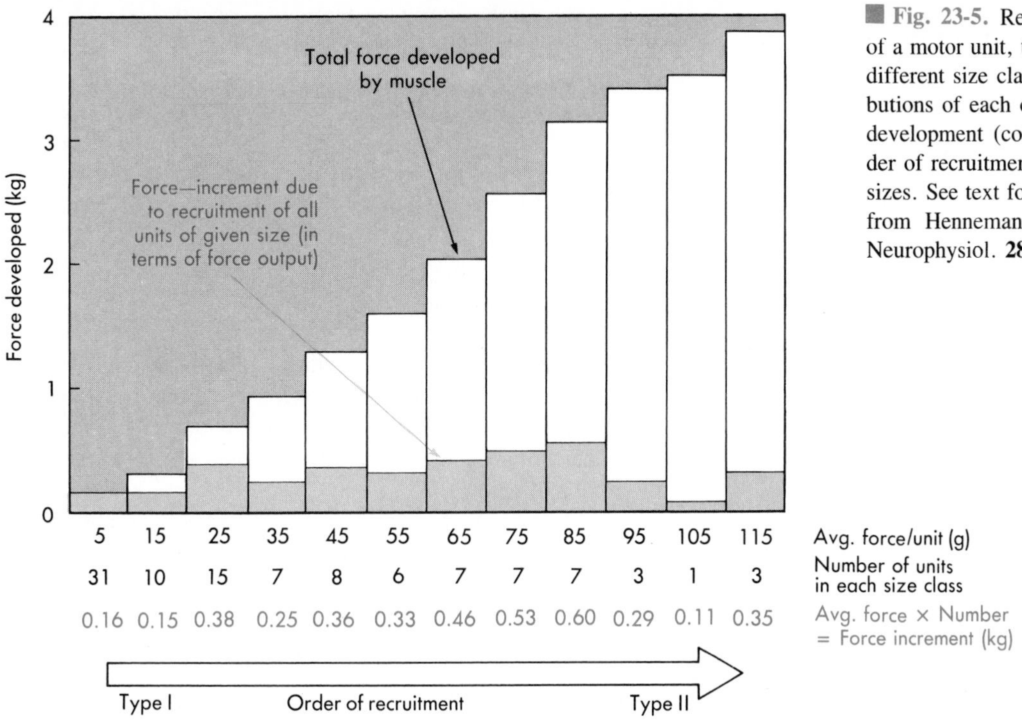

Fig. 23-5. Relationships between the size of a motor unit, the number of motor units of different size classes in a muscle, the contributions of each class of motor units to force development (colored numbers), and the order of recruitment of motor units of different sizes. See text for further explanations. (Data from Henneman, E., and Olson, C.B.: J. Neurophysiol. **28:**581, 1965.)

units averaging 5 g force plus 10 units averaging 15 g force, or 0.31 kg. A smooth gradation in contractile force is still maintained because the percent increase in force produced by a larger motor unit remains small when added to the force already being generated. Fig. 23-5 shows that smaller numbers of larger motor units are successively recruited until all motor units are contracting. There is some evidence that the largest motor units are so inexcitable that most persons cannot recruit them voluntarily. They may account for the exceptional displays of strength exhibited by persons under stress when increased excitatory activity occurs in the central nervous system.

The pattern of increasing force illustrated in Fig. 23-5 is a consequence of recruitment by the *size principle*. Recruitment by size and the simultaneous increase in firing rates, which allows each unit to increase its force by tetanization, are responsible for gradations in contractile force. Because the larger motor units are also the faster units, whole muscle contraction velocities can be increased by recruiting more motor units. Recruiting more units also contributes to the increase in speed by reducing the effective load on each cell, allowing faster crossbridge cycling rates.

The size principle also explains the metabolic profile pattern of the motor units. Highly oxidative units are those which are used most. Maximal efforts, in which fast motor units are recruited, cannot be sustained because of the rapid depletion of glycogen. This intracellular source of energy is needed to supply the glycolytic pathway, which is the principal way in which the high rates of ATP consumption can be met.

Slow or tonic muscle fibers. In amphibians, reptiles, and birds the function of the slow, type I fiber is served by an entirely different type of skeletal muscle cell. This type is confusingly termed a *slow* or sometimes a *tonic* fiber. It differs from the slow, type I fibers described earlier in that it has many motor endplates that receive branches from one motor neuron. These slow muscle fibers generally do not generate action potentials, and they are depolarized only by the decremental spread of the closely spaced endplate potentials. The slow or tonic fibers of lower vertebrates exhibit only a small response to a single neural stimulus, but the force output can be continuously graded by increasing the frequency of motor nerve firing. Tonic fibers are very rare in mammals, but they are found in the extraocular muscles that control eye movement.

Tone in skeletal muscle. The skeletal system supports the body mass efficiently when the posture is normal. The amount of energy expended for the muscular contraction that is required to maintain a standing posture is remarkably small. However, muscles normally exhibit some level of contractile activity. Isolated, unstimulated muscles are relaxed and quite flaccid. ''Relaxed'' muscles in the body are comparatively firm. This *tone* is apparently a result of low levels of contractile activity in some motor units driven by reflex arcs from receptors in the muscles; tone is abolished by dorsal root section. Tone in skeletal muscles should be distinguished from ''tone'' in vascular smooth muscles. The latter refers to the normal tonic or continuous contraction that maintains vascular resistance.

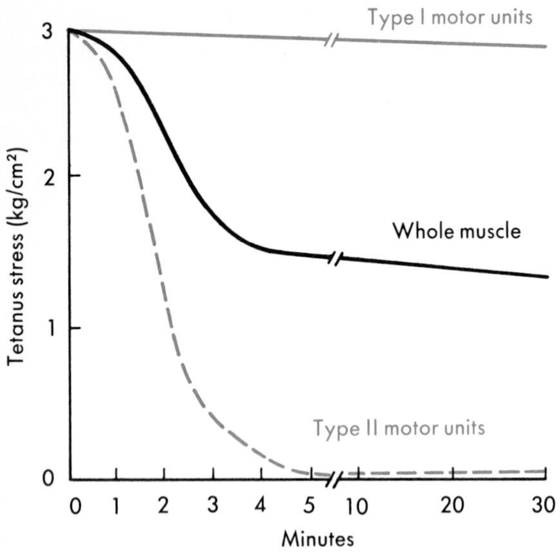

■ Fig. 23-6. Fatigue of a skeletal muscle in which half of the cross-sectional area is composed of slow, oxidative type I motor units and half of fast, glycolytic type II motor units. The muscle was briefly tetanized once every second by stimulation of its motor nerve in situ. With this regimen, type II motor units exhibit rapid cellular fatigue and failure to contract, while type I motor units maintain almost normal contractile responses.

■ *Fatigue*

Remarkably little is known about the factors responsible for fatigue. Fatigue may potentially occur in any of the steps involved in muscular contraction, from the brain to the muscle cells, as well as in the systems involved in maintaining energy supplies, including cardiovascular and respiratory functions.

Cellular fatigue. Fatigue of motor units can be assessed experimentally by recording the maximal stress maintained during prolonged contraction or during a series of brief tetani elicited by direct stimulation of the motor nerve to the muscle. The latter regimen allows adequate perfusion when the circulation is intact. Tetanic stress decays rapidly to a level that can be maintained for long periods (Fig. 23-6). This decay represents the rapid and almost total failure of fast motor units. The decline in tetanic stress is paralleled by glycogen and creatine phosphate depletion and by lactic acid production. This implies that ATP depletion leads to the failure of contraction. However, the decline in stress occurs when the ATP pool is not greatly reduced, and the muscle fibers do not go into rigor. Slow motor units, which can meet the energy demands of the stimulus regimen, do not exhibit significant fatigue for many hours. Evidently some factor associated with energy metabolism can inhibit contraction, but this factor has not been clearly identified. There is some evidence that *neuromuscular fatigue* can occur in the largest fast motor units, in which the ability of the motor nerve to synthesize and release acetylcholine may be limiting.

General fatigue. Most persons tire and cease exercise long before there is any motor unit fatigue of the type illustrated in Fig. 23-6. *General physical fatigue* may be defined as a state of disturbed homeostasis produced by work. The basis for the perceived discomfort or even pain probably involves many factors. These factors may include a lowering of plasma glucose levels and accumulation of metabolites. Motor system function in the central nervous system is not impaired. Highly motivated and trained athletes are able to bear the discomfort and will exercise to the point where some motor unit fatigue occurs. Part of the enhanced performance observed after training involves motivational factors.

Recovery. Muscle blood flow and oxygen uptake remain elevated for some time after exercise. *Oxygen debt* (Fig. 23-7) is the excess amount of oxygen consumed over that required for resting metabolism when energy use by the contractile system has ceased. Some oxygen debt occurs even at low levels of exercise, because slow oxidative motor units consume considerable ATP (derived from creatine phosphate or glycolysis) before oxidative metabolism can increase ATP production to the steady-state requirements. The oxygen debt is much greater in strenuous exercise, during which fast glycolytic motor units are recruited. The oxygen debt is approximately equal to the energy consumed during exercise minus that supplied by oxidative metabolism (i.e., the dark- and light-colored areas in Fig. 23-7 are roughly equal). The additional oxygen used during recovery represents the energy requirements for restoring normal cellular metabolite levels.

■ *Trophic Responses of Skeletal Muscle*

Trophic factors are responsible for the long-term development and maintenance of specific characteristics of a tissue. Skeletal muscle exhibits considerable plasticity (variability in the phenotype or properties of the cells) as shown by a variety of experimental studies.

Growth and development. Skeletal muscle fibers differentiate before innervation (neuromuscular junctions may be formed well after birth). Before innervation the fibers physiologically resemble slow, type I cells. These uninnervated fibers have acetylcholine receptors distributed throughout the sarcolemma and are supersensitive to that neurotransmitter. An endplate is formed when the first growing nerve terminal establishes contact. The fiber forms no further association with nerves, and the receptors to acetylcholine become concentrated in the endplate membranes. Fibers that are innervated by a small motor neuron form slow oxidative motor units. Fibers innervated by large motor nerves

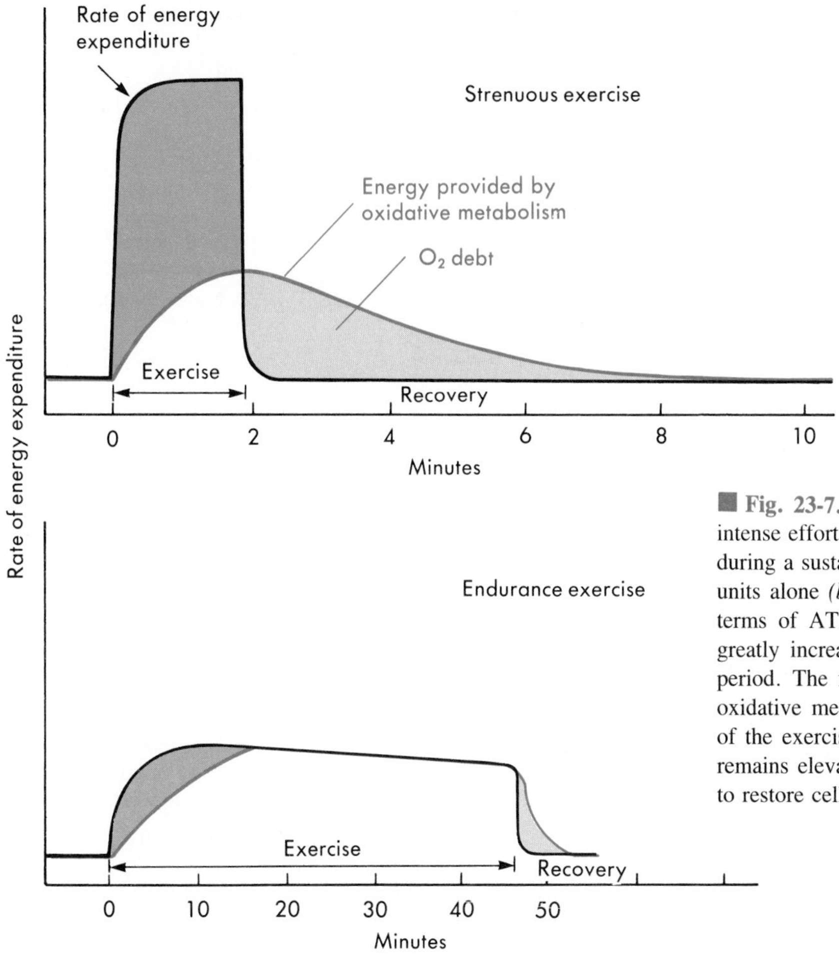

Fig. 23-7. The rate of energy expenditure during a brief, intense effort involving both type I and II motor units *(top)* or during a sustained, moderate effort generated by type I motor units alone *(bottom)*. The total rate of energy expenditure in terms of ATP *(black)* has a low basal resting level that is greatly increased by crossbridge cycling during the exercise period. The fraction of this energy expenditure provided by oxidative metabolism *(colored line)* varies with the duration of the exercise and the level of effort. Oxygen consumption remains elevated during recovery to provide the energy used to restore cellular homeostasis.

develop all the characteristics of fast, type II motor units. Thus innervation produces major changes in the cells, including the synthesis of the fast and slow myosin isoenzymes, which replace the embryonic variant.

An increase in muscle strength and size occurs during maturation. The cells must lengthen with skeletal growth. This is accomplished by formation of additional sarcomeres at the ends of the cells. This process is reversible. The cells shorten by destruction of terminal sarcomeres if a limb is immobilized in a shortened position or if an improperly set fracture leads to a shortened limb segment. The gradual increase in strength and diameter of a muscle during growth is achieved mainly by hypertrophy (Fig. 23-1). Skeletal muscles have a limited capacity to form new fibers (hyperplasia) by differentiation of satellite cells that are present in the tissues. Injured cell segments that contain nuclei can grow and fuse with other segments to regenerate a cell. However, major cellular destruction leads to replacement with scar tissue. Overall, skeletal muscle cells exhibit remarkable dynamic adjustments of their morphology to the demands of the organism.

Denervation, reinnervation, and cross-innervation. A variety of studies have shown the importance of innervation for the skeletal muscle phenotype (Fig. 23-8). If the motor nerve is cut, muscle *fasciculation* occurs. This term describes small, irregular contractions caused by the release of acetylcholine from the terminals of the degenerating distal portion of the axon. Several days after denervation, muscle *fibrillation* begins. Fibrillation is characterized by spontaneous, repetitive contractions. These originate after the spread of cholinergic receptors occurs over the entire cell membrane (a return to the preinnervation embryonic characteristic) and reflect a supersensitivity to acetylcholine. *Atrophy* also occurs, with a decrease in the size of the muscle and its cells. Atrophy is progressive in humans, with degeneration of some cells after 3 or 4 months. Most of the muscle fibers are replaced by fat and connective tissue after 1 to 2 years. These changes are all reversed if reinnervation occurs within a few months. Reinnervation is normally achieved by growth of the peripheral stump of the motor nerve axons along the old nerve sheath.

Reinnervation of a formerly fast, type II fiber by a

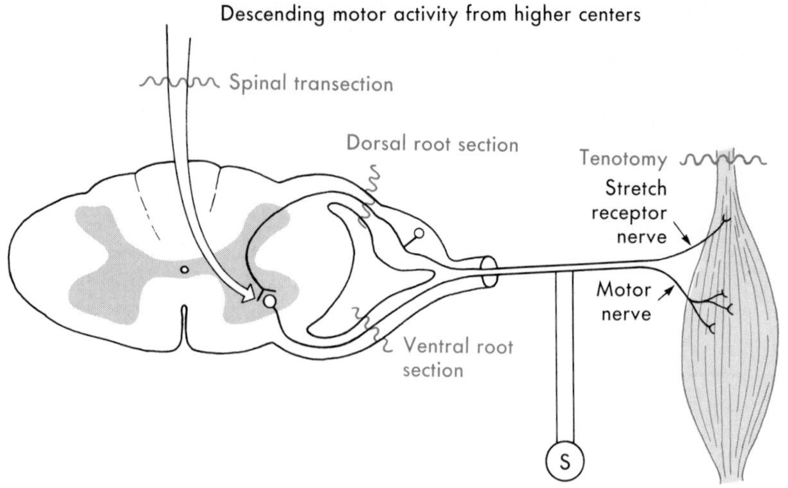

Fig. 23-8. Experimental or pathological situations that modify the phenotype of skeletal muscles. Normal contractile activity in the muscle can be abolished by ventral root section that cuts the motor nerves (peripheral denervation). In a muscle with normal innervation, contractile activity can be increased by pacemakers *(S)* implanted on the motor nerves. Activity can be decreased by blocking excitatory pathways from higher centers (spinal transection) or by interrupting the reflex arcs that modify muscle activity through dorsal root section. If the muscle tendon is severed (tenotomy), stretch receptors within the tendon and muscle become inactive. Tenotomy also decreases excitatory input to the motor nerves.

Table 23-2. Effects of exercise

Type of training	*Example*	*Major adaptive response*
Learning/coordination	Typing	Increases the rate and accuracy of motor skills (central nervous system)
Endurance (submaximal, sustained efforts)	Marathon running	Increased oxidative capacity in all involved motor units with limited cellular hypertrophy
Strength (brief, maximal efforts)	Weight lifting	Hypertrophy and enhanced glycolytic capacity of motor units employed

small motor axon causes that cell to redifferentiate into a slow, type I fiber and vice versa. Such observations suggest that there are qualitative differences in large and small motor nerves and that the nerves have specific ''trophic'' effects on the muscle fibers. Although it is increasingly recognized that nerve terminals release substances in addition to the primary neurotransmitter at synapses, an alternative explanation of the trophic effects of innervation fits most observations. Simply, it is the frequency of contraction that determines fiber development and phenotype. Electrical stimulation via electrodes implanted in the muscle or its motor nerves can ameliorate denervation atrophy. More strikingly, chronic low-frequency stimulation of fast motor units (by a pacemaker with electrodes placed on their motor nerves) causes fast motor units to be converted to slow units. Some shifts toward a typical fast-fiber phenotype can be observed when the frequency of contraction in slow units is greatly decreased by reducing the excitatory input in the ventral horn. This may be achieved through appropriate spinal or dorsal root section or by severing the tendon, which functionally inactivates peripheral mechanoreceptors (Fig. 23-8) (Chapter 14). In brief, fibers that undergo frequent contractile activity form many mitochondria and synthesize the slow myosin isoenzyme. Fibers innervated by large, inexcitable axons contract infrequently. Such relatively inactive fibers form few mitochondria and have large concentrations of glycolytic enzymes. The fast myosin isoenzyme is synthesized in such cells.

Response to exercise. Exercise physiologists identify three categories of training regimens and responses (Table 23-2). In practice, most athletic endeavors involve elements of all three. The learning aspect can involve motivational factors as well as neuromuscular coordination. This aspect of training does not involve adaptive changes in the muscle fibers per se. However, motor skills can persist for years without regular training, unlike the responses of muscle cells to exercise.

All healthy persons can maintain some level of continuous muscular activity that is supported by oxidative metabolism. This level can be greatly increased by a regular exercise regimen that is sufficient to induce adaptive responses. The adaptive response of skeletal muscle fibers to endurance exercise is mainly an increase in the metabolic capacity of the motor units involved. This demand places an increased load on the cardiovascular and respiratory systems and increases the capacity of the heart and respiratory muscles. The latter effects are responsible for the principal physiological benefits associated with endurance exercise.

Muscle strength can be increased by regular massive efforts that involve all motor units. Such efforts recruit fast glycolytic motor units and are brief. The blood supply may be interrupted as the tissue pressures rise above the intravascular pressures during maximal contractions, further limiting the duration of the contraction. Regular maximal strength exercise, such as weight lifting, induces synthesis of more myofibrils and hypertrophy of the active muscle cells. The increased stress also induces growth of tendons and bones.

The precise cellular adaptive responses to exercise are difficult to determine. However, most studies indicate that the effects of exercise on muscle are quantitative. Endurance exercise does not cause fast motor units to become slow, with the synthesis of the slow myosin isoenzyme, nor do maximal muscular efforts produce a shift from slow to fast motor units. Experiments involving cross-innervation and other techniques show that such shifts are possible. It seems likely that any practical exercise regimen, when superimposed on normal daily activities, does not provide sufficient changes in the pattern of activation of motor units to shift the myosin isoenzyme distribution.

Two additional factors that affect the motor unit phenotype and athletic performance are hormones and the genetic endowment of the person. Androgens stimulate muscle hypertrophy and may account for the greater average muscle mass in males.

Involuntary Muscles: Physiological Control and Function

Involuntary muscles include a highly diverse group of tissue ranging from striated muscle in the heart to many types of smooth muscles. Some general functional characteristics of muscles surrounding hollow organs are outlined in this section. Specific characteristics of various smooth muscle types are given in the chapters devoted to the different organ systems.

Cell-to-Cell Contacts

A variety of specialized contacts between involuntary muscle cells serves two functions: mechanical linkages and communication. Smooth and cardiac muscle cells are connected to each other rather than spanning the distance between two tendons. Thus cells anatomically arranged in series should not only be linked mechanically but should also be activated simultaneously and to the same degree. If this were not true, contraction in one region would simply stretch another region without a substantial decrease in radius or increase in pressure. The mechanical connections are provided by sheaths of connective tissue and by specific junctions between muscle cells. One example is the intercalated disk in cardiac muscle (Chapter 28).

In smooth muscle there are several types of junctions (Fig. 23-9). One is the *gap junction* (or nexus) in which adjacent plasma membranes are separated by only 2 or 3 nm. Gap junctions often exhibit a typical five-layered appearance in electron micrographs. Other specialized contacts are the *intermediate junction* and the *attachment plaque* or *desmosome-like attachment*. The membranes of these attachments are separated by a space of 20 to 60 nm, often marked by a central dense line. A variety of simple appositions between areas of cell membranes that are separated more widely are also common between smooth muscle cells. Some of these are quite elaborate with protrusions of one cell into another.

All these junctions may subserve the roles of both mechanical linkage and cell-to-cell communication. However, gap junctions are of particular importance in forming low-resistance pathways through which currents generated by action potentials in one cell can fire adjacent cells. In certain tissues, such as in the heart or the outer longitudinal layer of smooth muscle in the intestine, there are large numbers of such junctions. A wave of depolarization is readily propagated from cell to cell through such tissues. Intermediate junctions and desmosomes are presumed to subserve mechanical functions.

Neuromuscular Relationships

Some smooth muscles show little or no electrical coupling via junctions, and these cells may not normally fire action potentials. Synchrony of cellular activation occurs by diffusion of hormones or neurotransmitters through the tissue. Neural control of contraction in involuntary muscle is more complex than in skeletal muscle. Three factors must be considered: (1) the types of innervation and neurotransmitters, (2) the proximity of the nerves to the muscle cells, and (3) the type and distribution of the receptors for neurotransmitters in the muscle cell membranes (Fig. 23-10).

The types of innervation can be divided into three categories. *Extrinsic innervation* is derived from axons of the autonomic nervous system. In arteries this is usually limited to sympathetic nerves, but both sympathetic and parasympathetic innervation commonly are present in other tissues. *Intrinsic nerves* contained in plexuses may occur within the smooth muscle tissue, particularly in the gastrointestinal tract. Finally, *afferent sensory neurons* that mediate various reflexes may be found in the plexuses. The nervous system for the gastrointestinal tract alone is larger than the total motor system for

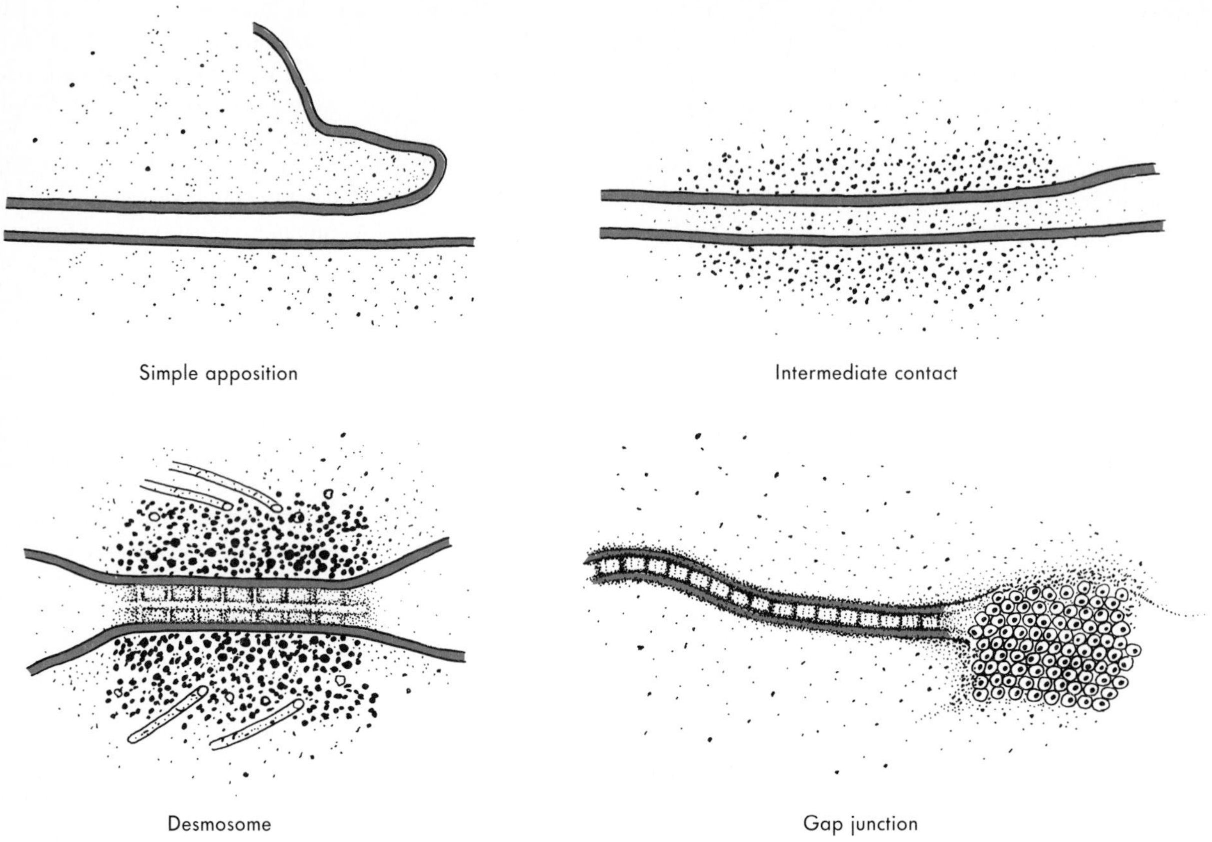

■ **Fig. 23-9.** Junctions between smooth muscle cells. The types and frequency of junctions vary widely among different types of smooth muscles. (Courtesy of Dr. Michael S. Forbes.)

all the skeletal muscles. A few smooth muscle tissues have no innervation.

Neuromuscular junctions in involuntary muscle are functionally comparable to those in skeletal muscle: presynaptic transmitter release, diffusion across the "junction," and combination with a postsynaptic receptor. However, elaborate neuromuscular contacts at axon terminals are not found. Autonomic nerves have a series of swollen areas or varicosities that are spaced at intervals along the axon. These varicosities contain the vesicles in which the neurotransmitters are found (Fig. 23-10). Each varicosity functions as a neuromuscular junction, although the adjacent muscle membranes exhibit little specialization. The varicosities are closely apposed to the muscle cell membranes with a gap of 6 to 20 nm in tissues with a rich neural regulation. The average gap may be 80 to 120 nm and, occasionally, considerably more in tissues in which neural control is less extensive. In arteries the nerves tend to be concentrated in the outer layer of smooth muscle cells just under the adventitial sheath of connective tissue. In such cases neurally mediated activation will be greatest at the periphery, and it may be nonexistent for smooth muscle cells in the luminal layers.

The neurotransmitters that are released and the presynaptic and postsynaptic effects of the neurohormones are highly variable. Marked individuality of the responses of different tissues that contain involuntary muscle is achieved by differences in their innervation, the types of transmitter released, and the nature of the receptors for each transmitter.

Relationships with other cell types. Contractile function in smooth muscle can be modulated by cells other than nerves. For example, specialized junctional regions can be observed between the membranes of the smooth muscle cells and the endothelial cells that line blood vessels. The actions of some circulating hormones or drugs can be mediated by the endothelial cells. For example, acetylcholine, acting on endothelial cell receptors, releases an unidentified substance that subsequently produces arterial smooth muscle relaxation. Acetylcholine causes contraction when it combines with cholinergic receptors in the arterial smooth muscle cell membrane.

■ *Patterns of Function*

In a classic grouping of involuntary muscles, a distinction is made between *multiunit* and *single-unit* tissues

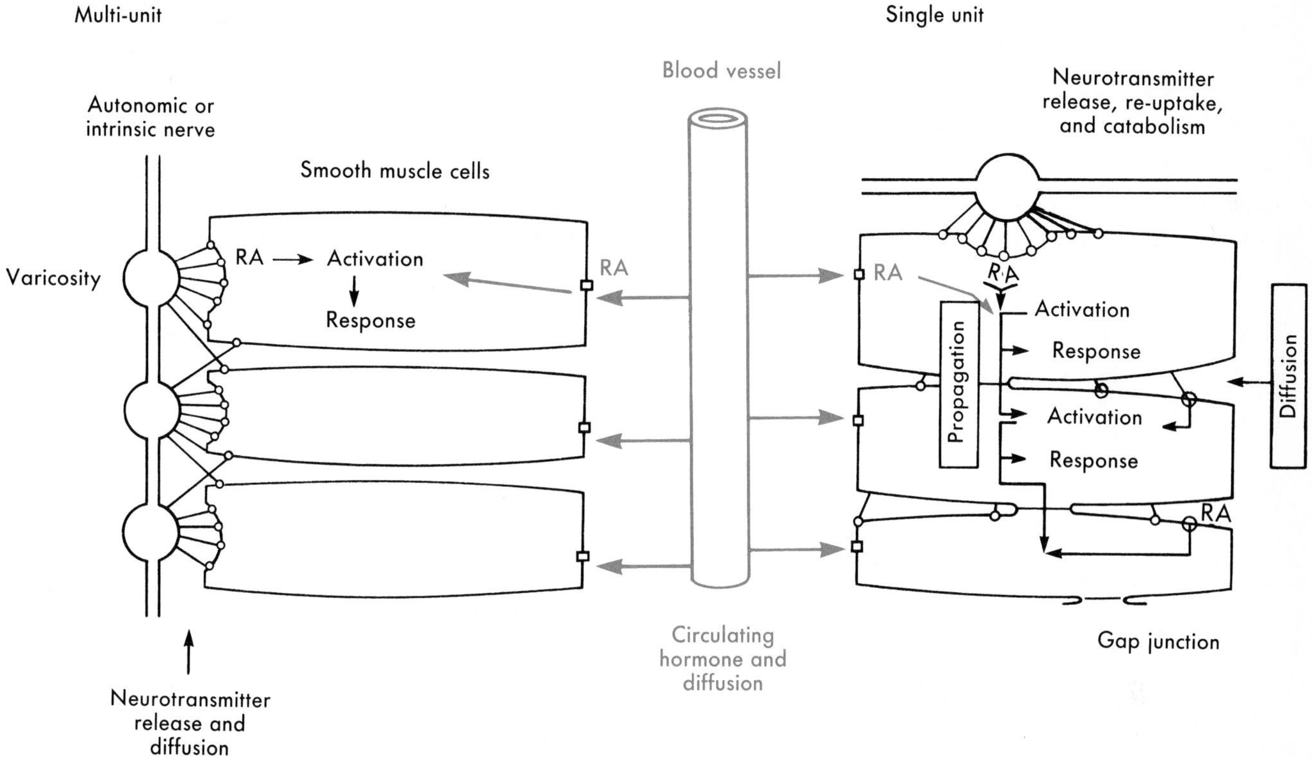

Fig. 23-10. Control of smooth muscle. Contraction (or inhibition of contraction) of smooth muscles can be initiated by (1) intrinsic activity of pacemaker cells, (2) neurally released transmitters, or (3) circulating hormones. The combination of a neurotransmitter, hormone, or drug with specific receptors activates contraction *(RA)* by increasing cell Ca^{++}. The response of the cells depends on the concentration of the transmitters or hormones at the cell membrane and the nature of the receptors present. Hormone concentrations depend on diffusion distance, release, reuptake, and catabolism. Consequently, cells lacking close neuromuscular contacts will have a limited response to neural activity unless they are electrically coupled so that depolarization is transmitted from cell-to-cell. **A,** Multiunit smooth muscles resemble striated muscles in that there is no electrical coupling, and neural regulation is important. **B,** Single-unit smooth muscles are like cardiac muscle, and electrical activity is propagated throughout the tissue. Most smooth muscles probably lie between the two ends of the single-unit–multiunit spectrum. (Redrawn from Ljung, B.: Physiological patterns of neuroeffector control mechanisms. In Bevan, J.A., et al., editors: Vascular neuroeffector mechanisms, Basel, 1976, S. Karger.)

(Fig. 23-10). Multiunit tissues are those in which each cell does not communicate with other muscle cells through junctions. Contraction of multiunit smooth muscle is controlled by extrinsic innervation or hormonal diffusion. Skeletal muscles are an example of this pattern, which is also typical of a few smooth muscles (i.e., those in the vas deferens and iris). However, all smooth muscle cells that are anatomically arranged in series must be part of the same functional motor unit, although many nerves may be involved. Single-unit tissues (Fig. 23-10, *B*) are exemplified by the heart and by a number of smooth muscles that undergo rhythmical contractile activity. All the cells in such tissues are electrically coupled through cell-to-cell junctions. Single-unit tissues can maintain fairly normal contractile activity without extrinsic innervation. The activity pattern in denervated single-unit muscles is caused by pacemaker cells and intrinsic reflex pathways.

The distinction between single-unit and multiunit tissues is overly simplified. Most smooth muscles are controlled and coordinated by a combination of neural elements and some degree of coupling. One generalization is that tonic muscles, such as arteries and sphincters, which maintain more or less continuous levels of tone, approach the multiunit end of the spectrum. Characteristically, such tissues do not exhibit action potentials when stimulated. Tissues that undergo phasic (rhythmical) activity, such as peristalsis, usually generate action potentials that are propagated from cell to cell. Such tissues more fully meet the criteria characterizing single-unit muscles.

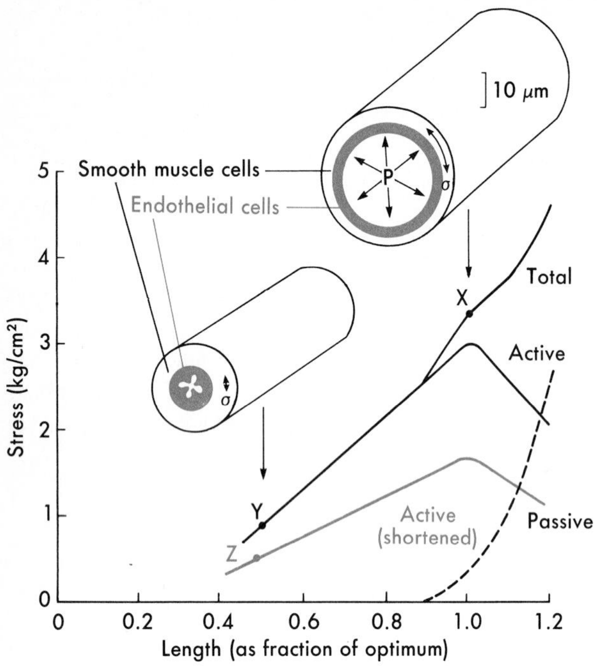

Fig. 23-11. The mechanical properties of smooth muscle in hollow organs. In this simple example the smooth muscle of a small arteriole shortens by 50% from its optimum length for force generation. The stress-length relationships for the arteriolar wall before constriction are illustrated in black. The actual wall stress (σ) in the arteriole is governed by the Laplace relationship (text). If the pressure remained constant, σ would fall from a value proportional to *X* to a value proportional to *Z* during the constriction. Thus the smooth muscle needs to develop considerably less force to maintain the constriction. This geometrical factor more than compensates for the fall in active stress *(Y)* due to the length-stress relationship of the smooth muscle cells.

Biophysics of Hollow Organs

The mechanical output of a skeletal muscle is readily calculated from the stress-length and velocity-stress relationships of the fibers (Figs. 22-8 and 22-9) and the angle formed between the fibers and the axis of the tendon. However, additional factors affect the mechanics of hollow organs.

Structural factors. Many arteries have a comparatively simple geometry in which the smooth muscle cells run circumferentially around the vessel. However, the heart and many organs that contain smooth muscle have multiple layers of cells with different orientations. Such structural arrangements can have a profound effect on the pressure-volume relationships of the tissue. A moderate shortening of smooth muscle can produce a disproportionately large volume change. This is illustrated for an arteriole with a lumen diameter of 18 μm and a wall composed of a 1 μm thick endothelium surrounded by a 2 μm thick smooth muscle cell (Fig. 23-11). A 50% shortening will halve the mean radius of the smooth muscle cylinder. However, the cells are not compressible so wall thickening reduces the inner radius of the smooth muscle layer to 7 μm. Contraction of the smooth muscle cells produces bulging of the endothelial cell layer. The overall volume reduction is 36-fold for an arteriolar segment that undergoes a cell shortening of 50%. This reduction can stop red cell flow.

Physical factors. The stress that is borne by the organ wall and that results from transmural pressure (e.g., the internal pressure minus the pressure outside a hollow organ) can be calculated from a derivation of the *law of Laplace* (considered further in Chapter 32). It states that the stress (σ) in the wall of the vessel (double-headed arrows, Fig. 23-11) produced by the transmural pressure (P) is equal to the product of the pressure and the mean radius (r) of the wall divided by the wall thickness (w), or $\sigma = Pr/w$, with units of force per unit area. Obviously a given pressure in the lumen produces a smaller stress on shortened smooth muscle cells because of the reduced vessel radius and the increased wall thickness. At larger radii the stress on the arterial wall produced by a given transmural pressure is withstood in part by the passive elasticity of the connective tissue and in part by active contraction of the arterial smooth muscle. This is illlustrated by point *X* in Fig. 23-11, where the muscle cells are at their optimal length for force development. After the muscle shortens by 50%, all the load must be borne by the smooth muscle cells. Their ability to withstand the distending pressure is reduced (point *Y*) because of their force-length properties. However, the actual wall stress on the smooth muscle cells (neglecting endothelial cell compression) is now significantly lower because of the Laplace relationship (point *Z*). The smooth muscle cells would have to "relax" to maintain the reduced diameter if the compressive forces were insignificant and would be operating on a lower stress-length curve (colored curve, Fig. 23-11).

The relatively simple example illustrated in Fig. 23-11 shows the general relationship between muscle function and hollow organ mechanics. Pressure-volume relationships rather than stress-length relationships are normally used to describe the mechanical properties of hollow organs. The pressure-volume curve of an organ depends on the stress-length properties of the smooth muscle. However, structural factors and the Laplace relationship also influence pressure-volume behavior.

Functional Adaptations in Involuntary Muscle

Neurotrophic relationships. Unlike skeletal muscles, involuntary muscles show little dependence on their extrinsic innervation for maintaining a normal phenotype. This may be related to the fact that the heart

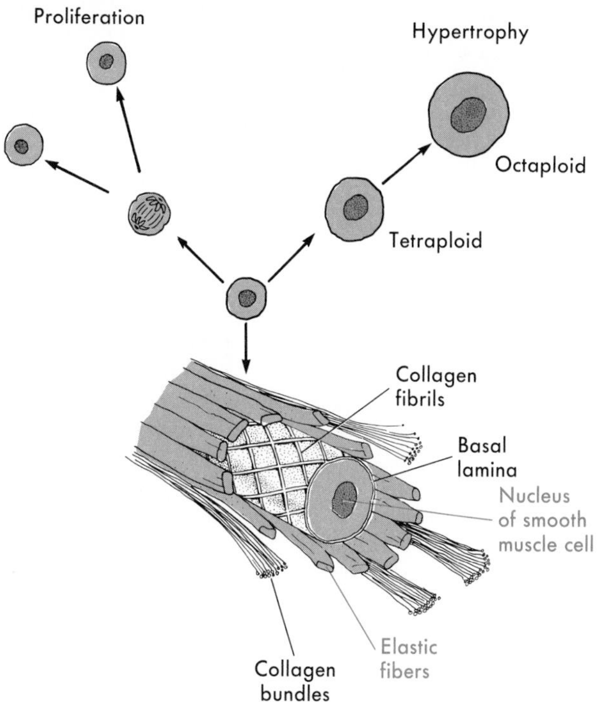

Fig. 23-12. Smooth muscle cells carry out many activities. **A,** They retain the capacity to divide during normal growth or in certain pathological responses such as formation of atherosclerotic plaques. **B,** Cells also may hypertrophy in response to increased loads. Chromosomal replication, not followed by cell division, yields cells with a greater content of contractile proteins. **C,** Smooth muscle cells also synthesize and secrete the constituents of the connective tissue matrix.

and smooth muscles can maintain contractile activity by other mechanisms when denervated. However, supersensitivity to neurotransmitters occurs in the heart and smooth muscle after denervation.

Development and hypertrophy. During development and growth the number of smooth and cardiac muscle cells increases (hyperplasia) (Fig. 23-12, *A*). In adults tissue mass increases if an organ is subjected to a sustained increase in mechanical work. This is termed *compensatory hypertrophy*. A striking example of this occurs in the ventricles and arterial media in hypertension. The increased mechanical load on the muscle cells appears to be the common factor inducing hypertrophy in involuntary muscles. Recent studies of myocardial and arterial smooth muscle cells indicate that tissue hypertrophy is sometimes due to cellular hypertrophy. Chromosomal replication (which may or may not be followed by nuclear replication) results in significant numbers of polyploid muscle cells. The polyploid cells, with multiples of the normal sets of chromosomes, synthesize more contractile proteins and increase in size with each chromosomal replication (Fig. 23-12, *B*).

The capacity of skeletal muscles to synthesize different myosin isoenzymes is demonstrated by cross-innervation or reinnervation experiments. However, such changes have only been ''physiologically'' demonstrated in cardiac muscle after hypertrophy resulting from disease or from experimental procedures to induce cardiac overload. The myosin isoenzyme synthesized in the hypertrophied cells has a lower specific ATPase activity, which decreases contraction velocity of the cells and changes the myocardial energetics. The overloaded heart is subjected to a continuous stress, unlike the activity pattern in skeletal muscle associated with an exercise regimen. This may be responsible for the appearance of new myosin isoenzymes in cardiac muscle.

The myometrium, which is the smooth muscle component of the uterus, exhibits another striking example of hypertrophy as parturition approaches. Hormones play an important role in this response. During pregnancy (when the hormone progesterone predominates) the smooth muscle is quiescent, and there are few gap junctions electrically coupling the smooth muscle cells. At term, under the dominant influence of estrogen, there is a marked hypertrophy of the myometrium. A great increase in the number of gap junctions occurs just before birth, converting the myometrium into a single-unit tissue to coordinate contraction during parturition.

Synthetic and Secretory Functions

Growth and development of tissues containing smooth muscles are associated with increases in the connective tissue matrix. Smooth muscle cells can synthesize and secrete the materials that make up this matrix. These constituents include collagen, elastin, and proteoglycans (Fig. 23-12, *C*). The synthetic and secretory capacities are evident when smooth muscle cells with extensive contractile filament arrays are isolated and placed in tissue culture. The cells rapidly modulate and lose thick myosin filaments and much of the thin filament lattice. Their places are filled by a greatly expanded rough endoplasmic reticulum and Golgi apparatus (cellular structures that are associated with protein synthesis and secretion). The modulated cells multiply and lay down connective tissues in the culture plate. This process is reversible, and some degree of redifferentiation with formation of thick filaments is possible after cessation of cell replication. Such studies suggest that the synthetic, secretory, and proliferative smooth muscle cell is phenotypically and functionally distinguishable from the differentiated contractile cell. However, these functions may all be present in varying degrees. The determinants of the smooth muscle cell phenotype are largely unknown, but hormones and growth factors in the blood, as well as the mechanical loads on the cells, are implicated in the control of phenotypic modulation.

Bibliography

Journal articles

Buchthal, F., and Schmalbruch, H.: Motor unit of mammalian muscle, Physiol. Rev. **60**:90, 1980.

Freund, H.-J.: Motor unit and muscle activity in voluntary motor control, Physiol. Rev. **63**:387, 1983.

Books and monographs

Alexander, R.M.: Animal mechanics, Seattle, 1968, University of Washington Press.

Åstrand, P.O., and Rodahl, K.: Textbook of work physiology, ed. 2, New York, 1977, McGraw-Hill Book Co.

Bertaccini, G., editor: Mediators and drugs in gastrointestinal motility, I, morphological basis and neurophysiological control, Berlin, 1982, Springer-Verlag.

Bohr, D.F., Somlyo, A.P., and Sparks, H.V., Jr., editors: Handbook of physiology, section 2, The cardiovascular system, vol. 2, Vascular smooth muscle, Bethesda, Md., 1980, American Physiological Society.

Brooks, V.R., editor: Handbook of physiology, section 1, The nervous system, vol. 2, Motor control, parts 1 and 2, Bethesda, Md., 1981, American Physiological Society.

Bülbring, E., et al., editors: Smooth muscle: an assessment of current knowledge, Austin, 1981, University of Texas Press.

Crass, M.F., and Barnes, C.D., editors: Vascular smooth muscle: metabolic, ionic, and contractile mechanisms, New York, 1982, Academic Press, Inc.

Garfield, R.E.: Cell-to-cell communication in smooth muscle, Chapter 6. In Grover, A.K., and Daniel, E.E., editors: Calcium and contractility: smooth muscle, Clifton, N.J., 1985, Humana Press.

Goldspink, D.F., editor: Development and specialization of skeletal muscle, Cambridge, England, 1980, Cambridge University Press.

Grundy, D.: Gastrointestinal motility: the integration of physiological mechanisms, Lancaster, England, 1985, MTP Press, Ltd.

Guba, F., Maréchal, G., and Takács, Ö., editors: Mechanism of muscle adaptation to functional requirements, Budapest, 1981, Akadémiai Kiadó.

Johnson, L.R., editor: Physiology of the gastrointestinal tract, vol. 1, New York, 1981, Raven Press.

McMahon, T.A.: Muscles, reflexes, and locomotion, Princeton, N.J., 1984, Princeton University Press.

Peachey, L.D., and Adrian, R.H., editors: Handbook of physiology: skeletal muscle, section 10, Bethesda, Md., 1983, American Physiological Society.

Pette, D., editor: Plasticity of muscle, Berlin, 1980, Walter de Gruyter & Co.

Stephens, N.L., editor: Smooth muscle contraction, New York, 1984, Marcel Dekker, Inc.

SECTION

IV

BLOOD

Oscar D. Ratnoff

CHAPTER 24

Blood Components

Blood is the vehicle of transportation that makes possible the specialization of structure and function that is characteristic of all but the lowest organisms. Blood, which makes up about 6% of total body weight, is a suspension of various types of cells in a complex aqueous medium, the plasma. The elements of blood serve multiple functions essential for metabolism and the defense of the body against injury.

■ *Plasma*

In the study of *plasma* and its constituents, venous or arterial blood must be mixed with an anticoagulant to inhibit coagulation. Two types of anticoagulant are in general use: (1) heparin, a complex sulfated mucopolysaccaride that is usually derived from lung or intestinal mucosa and that inhibits the enzymes that induce coagulation (see Chapter 25), or (2) agents, such as sodium citrate or salts of ethylenediamine tetraacetic acid (EDTA), that reduce the concentration of calcium ions, which are needed for coagulation. Plasma is then separated from the blood cells by centrifugation. If blood is drawn without adding an anticoagulant, it clots. After coagulation, the cell-depleted, fluid phase of blood, devoid of fibrinogen, is called *serum*.

The normal adult has an average of 25 ml to 45 ml of plasma per kg of body weight, that is, a total volume of about 3L. Many substances are dissolved in plasma, including electrolytes, proteins, lipids, carbohydrates (particularly glucose), amino acids, vitamins, hormones, nitrogenous breakdown products of metabolism (such as urea and uric acid), and gaseous oxygen, carbon dioxide, and nitrogen (Tables 24-1 and 24-2). The concentrations of these constituents may vary as the result of such diverse influences as diet, metabolic demand, and the levels of hormones and vitamins. Normally the composition of blood is maintained at biologically safe and useful levels by a variety of ''homeostatic'' mechanisms. The balance may be upset by impaired function in a multitude of disorders, particularly those involving the lungs, cardiovascular system, kidneys, liver, and endocrine organs.

Atmospheric oxygen, needed for the oxidative metabolic processes of the body, diffuses from the pulmonary alveoli into the circulating plasma in the pulmonary capillaries and thence into red blood cells, where it combines with hemoglobin, the major carrier of oxygen in blood. Similarly, carbon dioxide, elaborated by the tissues through the oxidation of carbon-containing compounds, diffuses into peripheral capillaries and is then carried by the blood to the lung, where it is excreted. Carbon dioxide is transported in several ways: in solution, as bicarbonate ions, and as carbaminohemoglobin.

The ionic constituents of plasma maintain the pH of blood within physiological limits. Along with other, nonionic solutes, the ions also maintain the osmolarity of the plasma; the normal osmolarity is 280 to 300 mOsm/kg water (see Chapters 45 and 48). The chief inorganic cation of plasma is sodium, present normally at an average concentration of 142 mEq/L (Table 24-1). Plasma also contains small amounts of ionic potassium, calcium, magnesium, and hydrogen. The principal anion of plasma is chloride, present at an average concentration of 103 mEq/L. Ionic equilibrium is maintained by the presence of other anions, including bicarbonate, plasma protein, and (to a lesser degree) phosphate, sulfate, and organic acids. Elaborate mechanisms preserve the normal ionic composition and pH of plasma.

Literally hundreds of different proteins are dissolved in plasma. In all, plasma normally contains about 7 g of protein/dl. The bulk of protein belongs to two groups, albumin and the various immunoglobulins. Plasma (or serum) proteins have been characterized by their migration in an electric field at pH 8.6. Albumin migrates most rapidly toward the anode; globulin species described as alpha-1, alpha-2, and beta migrate successively less rapidly; and gamma globulin migrates very slowly.

Table 24-1. Ionic constituents of plasma in adults

Constituent	*Concentration*
Cations	
Sodium (mEq/L)	135-145
Potassium (mEq/L)	3.5-5.0
Calcium* (mEq/L)	2.2-2.5
Magnesium (mEq/L)	1.5-2.0
Hydrogen (pH)	7.35-7.45
Anions	
Chloride (mEq/L)	95-107
Bicarbonate (mEq/L)	22-26
Lactate (mEq/L)	1.0-1.8
Sulfate (mEq/L)	1.0
Phosphate† (mEq/L)	2.0

*Total plasma calcium is 8.5 to 10.5 mg/dl.
†Total inorganic phosphorus is 2.5 to 4.5 mg/dl.

Table 24-2. Some constituents of plasma in adults

Constituent	*Concentration*
Proteins	6.0-8.0
Albumin (g/dl)	3.4-5.0
Total globulin (g/dl)	2.2-4.0
Transferrin (mg/dl)	250
Haptoglobin (mg/dl)	30-205
Hemopexin (mg/dl)	50-100
Ceruloplasmin (mg/dl)	25-45
Ferritin (μg/l)	15-300
Nonproteins	
Cholesterol (mg/dl)	140-250
Glucose (mg/dl)	70-110
Urea nitrogen (mg/dl)	6-23
Uric acid (mg/dl)	4.1-8.5
Creatinine (mg/dl)	0.7-1.4
Iron (μg/dl)	50-150

Albumin, synthesized by the parenchymal cells of the liver, is nomally present at an average concentration of about 4 g/dl. Because albumin diffuses poorly through intact vascular endothelium, it provides the critical *colloid osmotic* or *oncotic* pressure that regulates the passage of water and diffusible solutes through the capillaries. When the concentration of albumin is severely reduced, excess extracellular fluid may accumulate. In extravascular tissues, this fluid accumulation is described as *edema,* whereas in closed body cavities, it is described as either *ascites* (in the peritoneal cavity) or *effusion* (in the pleural or pericardial cavities). A decreased concentration of albumin may be the result of impaired synthesis in hepatic disease, urinary loss in renal disorders, or gastrointestinal loss in protein-losing enteropathy.

Albumin is also the carrier for substances that are adsorbed to it. These substances include certain normal components of blood, such as bilirubin and fatty acids, and certain exogenous agents, such as drugs. Albumin also furnishes some of the anions needed to balance the cations of plasma.

Among the proteins, the antibodies *(immunoglobulins)* are second to albumin in concentration. The antibodies arise on stimulation of lymphocytes in response to their exposure to antigens, which are agents that are foreign to the normal body and that evoke the formation of specific antibodies. Immunoglobulins are synthesized by plasma cells in the lymphoid organs and are critical as a defense against infection. The immunoglobulins constitute the bulk of gamma globulin in plasma. Other plasma proteins include (1) the "clotting factors" needed for the coagulation of blood, of which the most plentiful is *fibrinogen;* (2) the components of *complement,* a group of proteins that mediate the biological effects of immune reactions; (3) many enzymes or their precursors; (4) enzyme inhibitors; (5) specific carriers of such constituents as iron *(transferrin),* copper *(ceruloplasmin),* hormones, and vitamins; and (6) scavengers of agents inadvertently released into plasma—for example, *haptoglobin,* which binds free hemoglobin, and *hemopexin,* which binds heme. Plasma lipids, the chief of which are triglycerides, cholesterol, and phospholipids, are transported as complexes with plasma proteins, the *apolipoproteins.*

Blood Cells

The cellular constituents of blood include red blood cells (erythrocytes), a variety of white blood cells (leukocytes), and platelets. Five classes of leukocytes are recognized: neutrophils, eosinphils, basophils, monocytes, and lymphocytes. The classes are distinguished in blood smears by their morphologic and tinctorial characteristics on staining with a mixture of dyes (Wright-Romanowsky stain) (Fig. 24-1, Plate II, and Table 24-3).

Erythrocytes

The mature *erythrocyte* is an anuclear cell surrounded by a deformable membrane that is well adapted to the need to traverse narrow capillaries (Fig. 24-1 and Plate II). The red cells are biconcave disks, each with a diameter of about 8 μm, a thickness of 2 μm at its edge, and a volume of about 87 μm^3. In normal adults, the red cells occupy on the average about 48% of the volume of blood in men and about 42% in women; the percentage of the volume of blood made up by erythrocytes is defined as the *hematocrit*. The red blood cell

■ Table 24-3. Peripheral blood cells in adults

Blood cells	***Content***
Erythrocytes (red blood cells/μL)	
Men	4.3-5.9 × 10^6
Women	3.5-5.5 × 10^6
Hematocrit (%)	
Men	39-55
Women	36-48
Hemoglobin (g/dl)	
Men	13.9-16.3
Women	12.0-15.0
Leukocytes (white blood cells/μL)	4.8-10.8 × 10^3
Neutrophils (% leukocytes)	50-70
Bands (% leukocytes)	0.6
Lymphocytes (% leukocytes)	20-40
T cells (% lymphocytes)	70
B cells (% lymphocytes)	10-20
Monocytes (% leukocytes)	1-10
Eosinophils (% leukocytes)	0-3
Basophils (% leukocytes)	0-1
Platelets (/μL)	150-350 × 10^3

count—that is, the concentration of red cells in blood—normally averages about 5.2 million/μL in men and 4.8 million/μL in women.

The principal protein constituent of the cytoplasm of the mature erythrocyte is hemoglobin. Normal blood has about 15 g of hemoglobin/dl in adult men and about 13.5 g/dl in adult women. The cytoplasm also contains enzymes that are needed to provide enough energy to preserve the integrity of the cell's membrane, to maintain the intracellular concentrations of potassium above and of sodium below those in the surrounding plasma, to convert carbon dioxide to bicarbonate ion, and to prevent the oxidative transformation of hemoglobin to a nonfunctional protein, methemoglobin.

Hemoglobin, a protein synthesized in the marrow by the nucleated precursors of the erythrocytes, is a complex molecule with a molecular weight of 68,000 daltons. It is composed of two dissimilar pairs of polypeptide "globin" subunits—two alpha chains and two nonalpha chains, either beta, gamma, or delta chains. In the fetus and newborn, most of the hemoglobin is composed of two alpha and two gamma chains, a configuration designated as fetal hemoglobin or hemoglobin F; the remainder consists of two alpha and two beta chains, hemoglobin A. The proportion of hemoglobin F declines in early childhood, so that when the child reaches the age of 4 years, the hemoglobin F has virtually disappeared, and more than 95% of the hemoglobin is hemoglobin A. A small proportion, up to 3.5% of hemoglobin, is made up of two alpha and two gamma chains, hemoglobin A_2.

Each globin subunit is attached covalently to a prosthetic group consisting of a tetrapyrrole, heme. Heme is synthesized from glycine and succinyl-coenzyme A by a sequence of steps that leads to the formation of a pyrrole, porphobilinogen, and then to a tetrapyrrole ring compound, protoporphyrin IX. A mitochondrial enzyme, heme synthetase, inserts an atom of ferrous iron into protoporphyrin IX. Oxygen binds reversibly to the iron incorporated into the heme unit. The combination of oxygen with hemoglobin in the pulmonary capillaries and the release of oxygen from hemoglobin in the capillaries of other tissues are described in Chapter 39.

The average life span of the normal erythrocyte in the circulation is 120 days. Perhaps 10% to 20% of senescent red cells break up within the blood stream, where the liberated hemoglobin is bound to a specific carrier protein, *haptoglobin*. Some plasma hemoglobin is cleaved intravascularly into globin and heme; the latter binds to another carrier protein, *hemopexin*. Both the hemoglobin-haptoglobin and heme-hemopexin complexes are cleared from the circulation by the liver and catabolized by the hepatic parenchymal cells.

The great bulk of senescent red cells are engulfed by the macrophages of the reticuloendothelial system, particularly in the liver and spleen. In the macrophages, the hemoglobin is freed from the cells and catabolized into globin and heme. The globin is broken down by cellular proteases into its constituent amino acids, which join the pool of amino acids in the plasma and are reutilized for protein synthesis. The heme tetrapyrrole is split enzymatically, releasing its iron atom and forming a linear tetrapyrrole, *biliverdin*. The released iron is reutilized for the most part for the formation of heme in erythroblasts. The biliverdin is reduced to *bilirubin,* which is liberated into the plasma, where it is bound to albumin and transported to the parenchymal cells of the liver. There it is coupled to glucuronic acid, forming a water-soluble conjugate that is excreted into the bile. The bilirubin loses its glucuronic acid in the process. Some of the bilirubin is reabsorbed as such into the blood stream, to be reexcreted by the liver. Most of it, however, is reduced by bacterial enzymes in the gut, forming, successively, tetrapyrroles, the *urobilinogens* (which are colorless), and *stercobilin* and *urobilin* (which provide the brown color of the stool). In turn, some urobilinogen and urobilin are reabsorbed from the gut. They are then either reexcreted by the liver into the biliary ducts or excreted into the urine.

Nearly all the iron needed for the synthesis of hemoglobin comes from this catabolism of heme-containing compounds. The iron is transported from the macrophages into the plasma, where it is bound in the ferric state to a carrier protein, *transferrin* (iron-binding globulin). Transferrin is normally about one third saturated with iron, and its concentration in plasma is about 250 mg/dl. Transferrin carries the iron to the cells that need it for heme synthesis. These cells are almost entirely bone marrow erythroid precursors, the erythroblasts and reticulocytes. The transferrin binds to specific membrane receptors on these cells. The iron, reduced to the ferrous state, is either incorporated into heme or stored as ferritin. Ferritin is a complex of a water-soluble protein (apoferritin), and ferrous hydroxide. The macrophages of the reticuloendothelial system also store iron as ferritin, providing a reserve of this essential metal. Some of the ferritin is degraded there into an insoluble form, hemosiderin, which also can furnish iron when needed. Ferritin is found in normal plasma as well as in such organs as the liver, spleen, and heart.

Although the recycling of iron is highly efficient, small amounts are continually lost, largely through the desquamation of intestinal mucosa. A normal adult must ingest about 1 mg of iron daily to compensate for this loss. In women, additional iron must be absorbed to make up for the blood loss during menstruation and for utilization of iron by the developing fetus. Most of the iron in the diet is derived from the heme moiety of meat, but a small amount is also provided by inorganic iron that is reduced by the acid of gastric juice.

Absorption of iron takes place in the mucosal cells of the duodenum and jejunum. Through mechanisms that are disputed, just enough iron is absorbed to make up for that which has been lost to the body. In the mucosal cells, iron that is attached to heme is liberated and reduced to the ferrous form, joining the inorganic ferrous iron that has also been absorbed. The ferrous iron is then transported through the cytoplasm of the mucosal cells to the underlying small blood vessels. Transport through the cells is facilitated by its binding to an intracellular transferrin-like molecule. Other iron molecules are bound to a mucosal cell ferritin and may be lost to the body during the desquamation of these cells. The iron that reaches the plasma is oxidized to the ferric state and bound to transferrin, joining the plasma pool of iron that is utilized for the formation of heme.

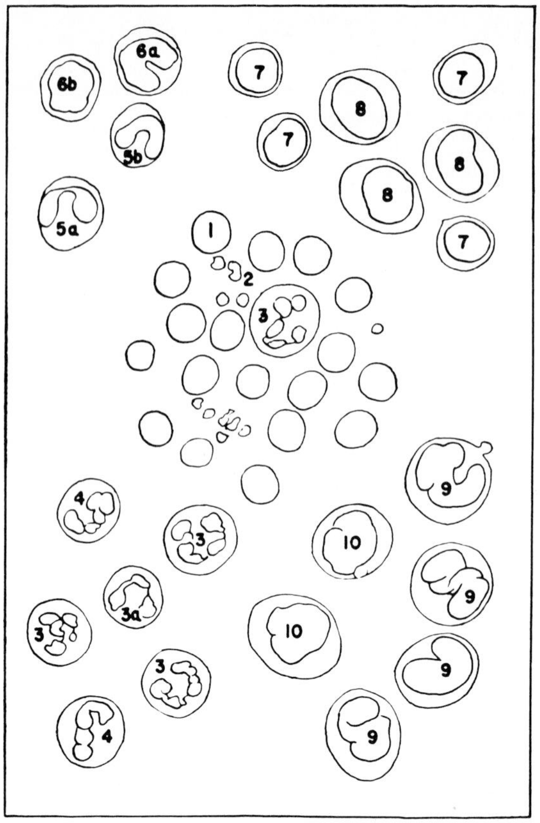

Fig. 24-1. The morphology of blood cells. *1,* Normal red cells. *2,* Platelets. *3,* Neutrophil, adult. *3a,* Neutrophil, adult (two lobes). *4,* Neutrophil, band form. *5a,* Eosinophil, two lobes. *5b,* Eosinophil, band form. *6a,* Basophil, band form. *6b,* Metamyelocyte, basophilic, *7,* Lymphocyte, small. *8,* Lymphocyte, large. *9,* Monocyte, mature. *10,* Monocyte, young. (From Daland, G.A.: A color atlas of morphologic hematology, Cambridge, Mass., 1951, With permission. Harvard University Press.)

Leukocytes

Normal blood contains between 4000 and 10,000 *leukocytes*/ μL (Table 24-3). Of these cells, about 40% to 75% are neutrophils, 20% to 45% are lymphocytes, 2% to 10% are monocytes, 1% to 6% are eosinophils, and less than 1% are basophils.

Neutrophils, eosinophils and basophils, described collectively as granulocytes, are distinguished by the nature of the granules in their cytoplasm (Fig. 24-1 and Plate II). These cells average about 12 to 15 μm in diameter. The mature forms have small, multilobed nuclei and abundant cytoplasm.

Neutrophils have nuclei with 2 to 5 lobes and, in stained blood smears, fine purple granules in a pink cytoplasm. A small fraction of the neutrophils in peripheral blood have a single, sausage-shaped nonlobulated nucleus, and are described as band or juvenile cells. Primary granules, which make up a minority of the neutrophils' granules, contain a variety of enzymes, among them lysozyme, an enzyme that digests the walls of certain bacteria, and a peroxidase that reduces hydrogen peroxide. Much more numerous in stained preparations are secondary or specific granules that contain an iron-binding protein, lactoferrin, a cationic bactericidal protein, and a vitamin B_{12}-binding protein. Besides the neutrophils detected in samples of peripheral blood, about an equal number are distributed along the margins of the smallest blood vessels, particularly in the spleen and liver; this "marginal pool" of neutrophils can be mobilized into the circulation by various stimuli, e.g., the injection of epinephrine.

Within 12 hours or less after the neutrophils have been discharged from the marrow into the blood stream, they migrate into extravascular tissues where they sur-

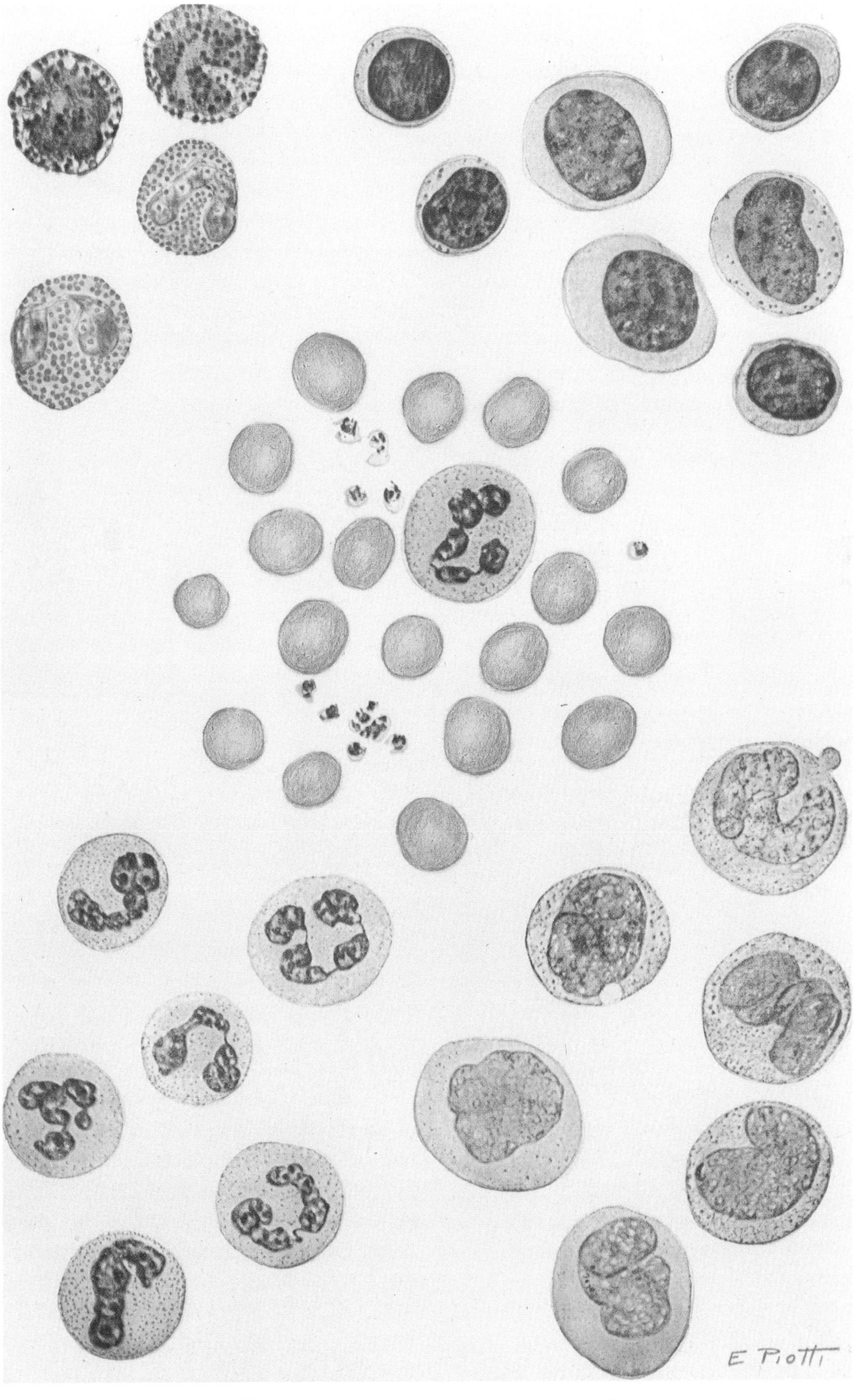

■ **Plate II.** Typical cells of normal human blood.

vive 4 or 5 days. There they are attracted to sites of injury in extravascular tissues by "chemotactic" agents, among them agents released by microorganisms or injured tissues, components of the clotting system, and components of complement. Neutrophils provide a major defense against infection by bacteria, which they can ingest and destroy.

The mechanisms for the destruction of bacteria begin with attachment of the bacteria to the neutrophil surface. This process is mediated by fibronectin and by antibody and complement *opsonins* that are fixed to the bacterial surface. Pseudopodia extruded by the neutrophils encircle the adherent bacteria and enclose them in newly formed vacuoles, the *phagosomes;* the process is called *phagocytosis*. The bacteria are destroyed in these vacuoles by enzymes released from the neutrophilic granules that fuse to the membranes lining the phagosomes. Among these enzymes are lysozyme, which disrupts the outer membrane of certain microorganisms, and enzymes that bring a sudden increase in the consumption of oxygen by the neutrophil. This "respiratory burst" results in the elaboration of hydrogen peroxide, superoxide ions (O_2^-), hydroxyl radicals (OH), and singlet oxygen ($^1O_2$). Of these species, the principal bactericidal agent is hydrogen peroxide, which oxidizes components of the bacteria by the generation of hypochlorous acid (HOCl). This acid is generated through the interaction of chloride ions and myeloperoxidase, an enzyme discharged from the primary granules. Lactoferrin, a component of the specific granules, contributes to bacterial killing by binding iron that would be utilized by the microorganisms and by catalyzing the generation of hydroxyl radicals. The processes through which bacteria are phagocytized and killed may lead to spillage of the neutrophil's enzymes and oxygen metabolites into the surrounding milieu. There they induce changes associated with the inflammatory reaction and may bring about proteolytic injury of tissue cells and of the neutrophil itself. In a similar manner, the engulfment of particulate matter by the neutrophil leads to its disruption and the release of agents that induce local inflammatory changes.

A sharp increase in the number of circulating neutrophils is a characteristic response to infection with certain microorganisms, such as pneumococcal, streptococcal, or staphylococcal infection. The increase is caused both by mobilization of marginated neutrophils and by stimulation of production of these cells. A similar *leukocytosis* accompanies nonbacterial tissue injury, for example, myocardial infarction.

Eosinophils most often have bilobed nuclei; in stained preparations their cytoplasm contains large red granules (Fig. 24-1). Once released into the blood stream, most eosinophils migrate within 30 minutes into extravascular tissues, where they survive for 8 to 12 days. Like neutrophils, the eosinophils are mobile cells whose movement is directed by chemotactic agents derived from a variety of sources, including mast cells and lymphocytes. The eosinophils are phagocytic and destroy organisms through oxidative mechanisms similar but not identical to those of neutrophils. Eosinophils are increased in number in the peripheral blood in many situations, among them, such parasitic infestations as trichinosis and schistosomiasis, in which they appear to serve an important defense function. Eosinophils are also increased in numbers in patients with such allergic or hypersensitivity states as asthma, in which exposure to abnormal exogenous or endogenous antigens leads to an immediate immunologic reaction. In this situation, the eosinophils dampen the host's response by limiting the antigen-induced release of mediators of inflammation.

Basophils have a multilobulated nucleus and, in stained blood smears, large, deep-blue cytoplasmic granules (Fig. 24-1). Like other granulocytes, basophils are motile cells with phagocytic properties. Basophils may migrate into extravascular tissues. There they may be stimulated, e.g., by complexes of antigens that are bound to immunoglobulin E (IgE) and that react with specific IgE receptors on the surface of basophils. The stimulated cells may discharge histamine from their granules into the surrounding tissues. This brings about either an explosive systemic anaphylactic response or local dilation and increased permeability of blood vessels, resulting in local edema. The relationship between blood basophils and the morphologically similar "*mast cells*" that are found scattered throughout extravascular tissues is not certain, but they probably have different origins. Like the basophils, the mast cells may be responsible for some of the phenomena associated with localized immunologic reactions. On appropriate stimulation, mast cells also release histamine from their granules, bringing about acute immunologic reactions or localized wheals.

A fourth class of leukocytes, the *monocytes,* are larger than other leukocytes, having an average diameter of about 15 to 20 μm (Fig. 24-1 and Plate II). In stained smears they have an indented, often kidney-shaped nucleus and fine pink cytoplasmic granules that are readily differentiated from those of the granulocytes. The monocytes are released into the blood stream from the marrow, where they mature. After a day or two, they migrate into the tissues, particularly the liver, spleen, lymph nodes, and lungs. There they comprise the macrophages of the reticuloendothelial system, cells with multiple functions that can replicate themselves in situ. The monocytes and macrophages are motile cells that are actively phagocytic, ingesting particulate matter, including microorganisms, injured or dead cells, and denatured proteins. The killing of microorganisms is brought about by devices similar to those used by neutrophils.

Monocytes also participate in immune responses. They appear to do this by "processing" antigens so that they can be recognized by T and B lymphocytes (described later). The monocytes are stimulated to secrete an agent, *interleukin-1,* that promotes the proliferation and maturation of T lymphocytes. Interleukin-1 is also a pyrogen, i.e., an agent that induces fever. The T helper cells are stimulated to secrete *interleukin-2,* which activates the T cells. Either this agent or an agent designated as B cell growth factor signals B lymphocytes to proliferate and differentiate into antibody-producing plasma cells. Monocytes have many other biological properties, including the elaboration of tissue thromboplastin, activators of plasminogen, proteolytic enzymes, and other biologically active agents.

The fifth class of leukocytes, the *lymphocytes,* is a heterogeneous group of cells with large nuclei (Fig. 24-1 and Plate II). The amount of cytoplasm depends on their size, which varies from 6 to 20 μm in diameter. The cytoplasm of most lymphocytes is devoid of granules in stained preparations; an exception is "natural killer cells" (described later) that have azurophilic cytoplasmic granules. Some lymphocytes, called *B cells,* are recognized by the presence of immunoglobulins on their surface. B cells are transformed on stimulation with antigen into plasma cells that synthesize and secrete the specific immunoglobulin antibodies. Other lymphocytes, called *T cells,* can be identifed because their surfaces have receptors for sheep erythrocytes. They participate in cell-mediated immune responses, such as sensitivity to tuberculin, that are not dependent on the presence of circulating antibodies. Certain T cells, designated as *helper* or *suppressor cells,* respectively abet or inhibit the transformation of B lymphocytes to antibody-producing cells. When sensitized by antigens on the surface of tissue cells, other T cells, designated as *cytotoxic lymphocytes,* bring about lysis of the tissue cells. In this way they participate in such phenomena as the rejection of grafted tissues that are incompatible with tissues of the host. Some lymphocytes with characteristics of neither T or B lymphocytes are described as *null cells.* Some null cells can destroy certain tumor cells, virus-infected cells, or tissue cells that have been coated with antibody; these are called *natural killer cells.*

Platelets

Platelets are anuclear cytoplasmic fragments of *megakaryocytes,* large polyploid cells found in the bone marrow (Fig. 24-1 and Plate II). They have an important role in the control of bleeding and the genesis of thrombosis, i.e., the formation of clots within blood vessels. The platelets are discussed in detail in Chapter 25.

Hematopoiesis

Our understanding of the origin of peripheral blood cells derives from observations in animals, fortified by knowledge gained from patients with disorders of hematopoiesis (the process of generating blood cells). Current evidence suggests that blood cells arise through a succession of steps from primitive *totipotent stem* cells. The totipotent stem cell is derived, ultimately, from blood islands in the embryonal yolk sac, whose cells colonize in the liver, spleen, and marrow. In the second trimester of fetal life, the generation of erythrocytes, granulocytes, monocytes, and megakaryocytes (the precursors of platelets) is centered largely in the liver and spleen. Blood cell production gradually shifts to the marrow, where cellular development takes place in endothelium-lined sinuses in the extravascular spaces, the marrow stroma. By the time of birth, the development of these cells is confined to marrow tissues. During childhood the generation of all these cell lines gradually becomes restricted to the marrow of the calvaria, pelvis, ribs, sternum, vertebrae, and ends of the long bones. Lymphocytes too are derived from the totipotent stem cells, first in the liver and spleen and later in the marrow. Their subsequent generation and differentiation occur not only in the marrow but also in the thymus and peripheral lymphatic organs.

The most primitive totipotent stem cell in the marrow can reproduce itself indefinitely. Under conditions that are not yet clear it can differentiate into stem cells that are the precursors of lymphocytes or into a *pluripotential stem cell* (Fig. 24-2). The pluripotential stem cell is also able to perpetuate itself. Additionally it may undergo further differentiation into cells that are committed to the development of erythrocytes, megakaryocytes, granulocytes, or monocytes; the last two are probably derived from a common precursor. The differentiation of the pluripotential stem cells into one or another *committed cell* (Fig. 24-2) is apparently determined by environmental conditions local to the growing cell colony.

The evolution of the mature blood cells that will enter the blood stream takes place through the successive division and differentiation of the committed or progenitor cells. These steps represent the *mitotic compartment* (Fig. 24-2) of hematopoiesis. The earliest forms recognized, the *blast* cells, have large nuclei with prominent nucleoli and relatively scant cytoplasm. Further division and differentiation of these cells is accompanied by shrinkage of the nucleus, loss of visible nucleoli, and expansion of the cytoplasm, which takes on the characteristics of the peripheral blood cells. Ultimately cell division ceases. Maturation to the functional cells that are to be released into the blood continues in what is described as the *postmitotic compartment* (Fig. 24-2) of hematopoiesis.

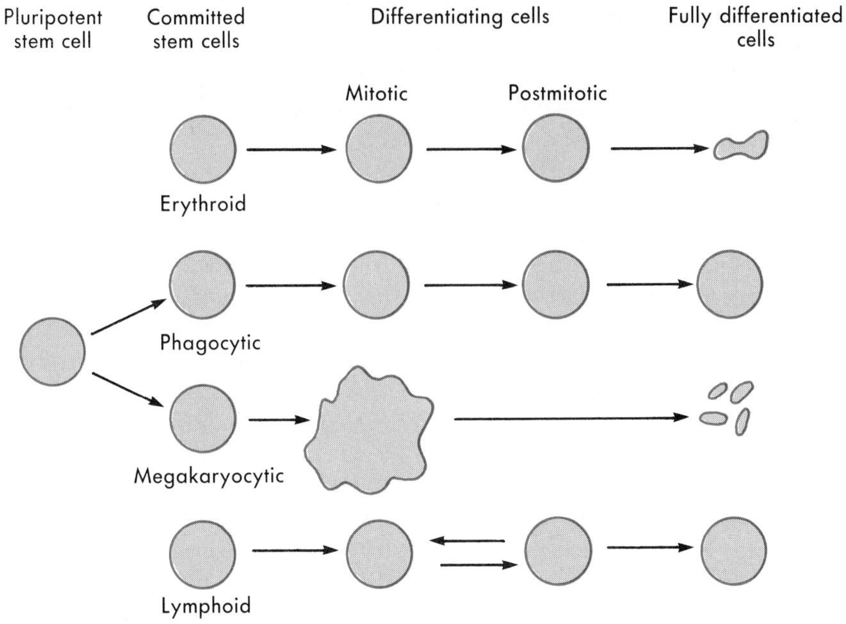

■ **Fig. 24-2.** An overall look at hematopoiesis. (Modified from Babior, B.M., and Stossel, T.P.: Hematology: a pathophysiological approach, New York, 1984, Churchill Livingstone.)

Following this general pattern, the earliest committed erythroid precursor cells differentiate into nucleated *erythroblasts* or *pronormoblasts,* which begin the synthesis of the oxygen-carrying protein, hemoglobin. Further cell division and differentiation bring about the formation of the most mature nucleated cells, the *normoblasts,* in which the nuclear chromatin is condensed. Eventually the normoblasts extrude their nuclei and are released as erythrocytes through the capillaries of the marrow into the blood stream.

Maturation of red blood cells from the earliest committed precursor cell takes place over a period of 3 to 4 days. For the first 2 days of their life in the blood stream, the anuclear erythrocytes contain remnants of RNA and other intracellular organelles and continue to synthesize hemoglobin; such cells are termed *reticulocytes.* For the remainder of their life span, the erythrocytes cannot synthesize any protein and can no longer synthesize hemoglobin. The proportion of reticulocytes in peripheral blood may rise when the rate of erythropoesis is increased in response to severe blood loss or to destruction of red cells by hemolysis (that is, disruption of the outer membrane of the erythrocyte with loss of its intracellular contents).

Differentiation of the erythroid precursors is under the influence of a polypeptide hormone, *erythropoietin,* that is elaborated chiefly by the parenchymal cells of the kidney. The synthesis of erythropoietin is stimulated by tissue hypoxia, which may be brought about by anemia, residence at high altitudes, impaired oxygenation of hemoglobin as the result of cardiac or pulmonary disease, or other stimuli.

Other committed cells differentiate through similar steps from blast cells to mature cells with the characteristics of neutrophils, eosinophils, basophils, or monocytes. Differentiation is under the influence of one or more agents described collectively as *colony-stimulating activity* (CSA). CSA is produced by many tissues, among them macrophages or monocytes, T lymphocytes, and endothelial cells within the stroma of the marrow. CSA is found in the plasma and in lung and placental tissues.

The normal differentiation and maturation of marrow cells depends on adequate supplies of *vitamin* B_{12} and *folic acid.* These agents are required for the formation of DNA, a requisite for cell division.

Vitamin B_{12} (cyanocobalamin) is a cobalt-containing compound that is synthesized by bacteria and molds. This vitamin cannot be synthesized by man but is furnished in the diet by foods of animal origin; approximately 5 μg of the vitamin is needed daily. In the stomach, vitamin B_{12} is bound to a specific protein, intrinsic factor, that is secreted by the parietal cells in the body of the stomach and fundus. The complex of vitamin B_{12} and intrinsic factor attaches to and is incorporated into mucosal cells in the lower portion of the ileum. In these cells vitamin B_{12} is split from the complex. Vitamin B_{12} is thence carried to the blood-forming tissues, bound to a specific transport protein, transcobalamin II.

Folic acid is also furnished in the diet, particularly by vegetables and dairy products; the daily requirement is about 50 μg. Dietary folate is conjugated to a polyglutamate chain, from which it is cleaved as it is absorbed from the intestine, and is reduced in intestinal epithelial and other cells to its biologically active form, tetrahydrofolate.

Lymphocytes are generated from the primitive hematopoietic stem cells in marrow. There they enter the blood stream and colonize in the lymphoid organs, principally the thymus, spleen, lymph nodes, tonsillar tissue, and submucosal Peyer's patches of the intestines. In these peripheral lymph organs, the lymphocytes can reproduce themselves and undergo further differentiation, sustaining the population of these cells. The lymphocytes in these peripheral sites can reenter the circulating blood, from which they can again return to the lymphoid tissues by way of the lymphatic channels.

On the basis of animal experiments, the T lymphocytes are thought to be derived from lymphocytes that have migrated from the marrow to the cortex of the thymus gland. There they proliferate and migrate to the medulla of the thymus, where they mature and then migrate to other lymphoid tissues. The origin of human B cells is less certain; their designation derives from the observation that in birds antibodies are synthesized by lymphocytes derived from the bursa of Fabricius. In mammals, B cells probably mature in bone marrow and then migrate to the various lymphoid tissues, where they can reproduce themselves. Synthesis of immunoglobulins takes place in the lymphocytes that have migrated to the lymphoid organs, where they are transformed on antigenic stimulation to antibody-producing plasma cells. In the lymphoid organs the T and B cells segregate into separate zones. Both types of cells freely recirculate between the lymphoid organs and peripheral blood.

Cells committed to the development of megakaryocytes, the precursors of platelets, are stimulated to mature under the influence of a hormone-like substance, thrombopoietin; the platelets are shed from the cytoplasm of mature megakaryocytes.

■ Anemia

Anemia is a decrease in the circulating mass of erythrocytes. It may result from decreased generation of these cells or their premature destruction or loss through hemorrhage. Failure of synthesis can arise in many ways: (1) as a consequence of hypocellularity of the marrow; (2) from replacement of the marrow by tumor tissue; (3) by suppression of hematopoiesis—as occurs in renal failure or from deficiencies of agents, such as vitamin B_{12} or folic acid, that are essential for the maturation of red cells; or (4) from a deficiency of iron, needed for the formation of heme. Premature destruction of red cells can come about from such diverse causes as hereditary defects in their outer membranes or direct chemical, physical, or immunologic injury. Of the innumerable causes of anemia, only a few are discussed here.

Iron-deficiency anemia is undoubtedly the most common form of anemia in Western countries. It is almost invariably caused either by blood loss, such as occurs with menstruation or from an ulcer or tumor in the gastrointestinal tract, or by the utilization of iron by the fetus during pregnancy. Anemia occurs when the iron stores of the body are depleted and, reflecting this, the concentrations of serum ferritin and of transferrin-bound iron are reduced. The red cells are characteristically smaller than normal (microcytosis) and are pallid from the resultant deficiency of hemoglobin (hypochromia).

Deficiencies of vitamin B_{12} and folic acid impair DNA synthesis. Such deficiencies bring about a characteristic anemia in which the erythrocytes are larger than normal (macrocytosis). In the marrow the erythroblasts are abnormally large, and the nuclei of these "megaloblasts" are both larger and less mature in appearance than they are normally. Similarly the precursors of the granulocytes may be abnormally large. The granulocytes that reach the circulation may be "hypersegmented," i.e., they have an increased number of nuclear lobules. Thrombocytopenia, a decrease in the number of platelets in circulating blood, may be present—the result of ineffective maturation of megakaryocytes. The prototypic megaloblastic anemias are (1) *pernicious anemia,* in which gastric atrophy results in deficient secretion of intrinsic factor, so that vitamin B_{12} cannot be absorbed from the gut; and (2) *folate deficiency,* most often from inadequate dietary intake, such as occurs in alcoholics, but also from increased demand for folate when the rate of erythropoiesis is greatly increased, as in some hemolytic anemias. The deficiency of vitamin B_{12} also leads to the characteristic neurologic damage observed in patients with pernicious anemia. The posterior and lateral columns of the spinal cord and peripheral nerves are demyelinated in this neuropathy. This neurologic abnormality is not seen in patients with folate deficiency.

Sickle cell anemia is particularly instructive, since it represents a host of defects of hemoglobin synthesis, in which an alteration in a single base pair of the DNA that directs synthesis results in the replacement of a specific amino acid in the globin molecule. Thus in sickle cell anemia a valine is substituted for the glutamic acid residue that is the sixth amino acid from the aminoterminal end of the globin beta chain. This genetically determined abnormal hemoglobin is designated as hemoglobin S. In individuals who are homozygotes for the sickle hemoglobin abnormality, this single amino acid substitution brings about the aggregation of hemoglobin molecules into polymers when the oxygen tension is low. This aggregation distorts the normally pliable discoid red cells into a characteristic sickled shape. In this configuration the cells are no longer flexible and may occlude small blood vessels, with resultant tissue damage. The distorted red cells have a shortened life span, bringing about a hemolytic anemia, that is, an anemia caused by premature destruction of red cells. Numerous other abnormal hemoglobins have been described.

Another instructive group of hereditary hemoglobin disorders are the *thalassemias*, which are characterized by deficient synthesis of the alpha or beta chains. The thalassemias are inherited as autosomal recessive traits; the hemoglobin of homozygotes lacks either alpha or beta chains, whereas heterozygotes synthesize varying proportions of the two chains. The thalassemias form a heterogeneous group of disorders in which the specific defects are genetically determined. For example, in alpha-thalassemia, deficient synthesis of alpha chains may be accompanied by increased synthesis of hemoglobin H, which has four beta chains, or hemoglobin Bart's, which has four gamma chains. Similarly, deficient synthesis of beta chains may be associated with increased synthesis of hemoglobin F and A_2. Numerous other forms of thalassemia have been described; usually the homozygotes are anemic and have decreased numbers of circulating red cells and decreased concentrations of hemoglobin within each cell.

The therapy of anemia depends on its pathogenesis. For example, iron may be prescribed to correct iron-deficiency anemia, or vitamin B_{12} to treat pernicious anemia. In some cases replacement of red cells by transfusion may be needed.

Blood Groups

Blood transfusion has been made feasible by the recognition that there are hereditary differences in the chemical structure of the red cell membranes. For example, individuals can be separated into "blood groups" O, A, B, or AB, depending on the genetically determined saccharide groups of the membrane glycoprotein. Furthermore those with blood group O have in their plasma antibodies directed against group A and B cells, those of group A have antibodies directed against B cells and vice versa, whereas those of group AB do not have antibodies directed at any of these blood groups. These "natural antibodies" are for the most part IgM molecules. For a transfusion to be successful, the recipient must be lacking in antibodies directed against the infused cells. Thus individuals of blood group O are "universal donors," and those of blood group AB are "universal recipients." In common practice, however, blood of the donor is carefully matched to that of the recipient.

The ABO system is only one of many inherited blood group systems. Of special note are the complex Rh blood groups. Ordinarily individuals whose red cells lack the Rh factor do not have antibodies against this substance in their plasma. Women who are Rh negative, however, may develop such antibodies if they carry a fetus who has inherited the Rh factor from its father. The maternal antibodies can cross the placenta to the fetus, resulting in a devastating destruction of its erythrocytes. This disorder is described as hemolytic disease of the newborn. Sensitization to the Rh antigen can also occur if an Rh-negative individual is transfused with Rh-positive blood cells. In such individuals a subsequent transfusion of Rh-positive blood may lead to premature destruction of the infused erythrocytes.

Immunoglobulins

An *antigen* is a substance that is foreign to the normal body and immunogenic—that is, an antigen can induce the formation of an *antibody* (or immunoglobulin) that can combine in a specific way with that antigen. Antigens are usually proteins, but some antigens are polysaccharides, lipopolysaccharides (such as the endotoxin produced by certain gram-negative bacteria), or nucleic acids. Much smaller molecules, called *haptens,* may bring about antibody formation, but only when they are combined with larger carrier molecules that are immunogenic. Subsequently haptens alone can bind to the specific immunoglobulins that they have evoked. Antibodies are proteins synthesized by plasma cells (derived from B lymphocytes), in conjunction with monocytes and T lymphocytes.

The combination of antigen and antibody, an *immune complex,* is held together by noncovalent bonds. Both antigens and antibodies may have more than one combining site, i.e., they are polyvalent, and hence they may form a latticework of proteins. In the test tube the lattice of soluble antigens and antibodies may precipitate out of solution, an event called the *precipitin reaction.* If the antigen is a component of a particle, such as a bacterium, the polyvalent antibodies serve as bridges that make the particles stick together: the phenomenon of *agglutination.* Antibodies may also enhance the phagocytosis of particulate antigens, a process known as *opsonization.* In the presence of a group of serum proteins known collectively as *complement,* bacteria or blood cells that have combined with their specific antibodies may be disrupted, a process called *immune lysis,* or *cytolysis.*

Antibodies provide a major defense against infectious agents, and they may neutralize toxic agents, such as the toxins elaborated by certain organisms. The reaction between antigen and antibody is, however, mindless, and under some circumstances it may injure the host. For example, tissue damage may be evoked by immune complexes, by the destruction of blood or tissue cells, or by immediate reactions that may threaten life by severe bronchoconstriction or a fall in blood pressure that results in shock (*anaphylaxis*).

The formation of antibodies may be induced experimentally by the parenteral administration of antigen. If the host has not previously been exposed to the antigen, antibodies appear in peripheral blood only after several days have elapsed. The titer of antibodies then rises, reaching a peak after about 2 or 3 weeks, and then it usually gradually decreases. Subsequent exposure to the same antigen induces a much more rapid increase in titer, the so-called anamnestic response (*anamnesis* means ''memory''). This secondary rise in titer, which usually lasts longer than the primary response, comes about by the stimulation of lymphocytes that have previously been exposed to the antigen, so-called memory cells.

Unstimulated B lymphocytes synthesize immunoglobulins that adhere to their surfaces but are not secreted. The B lymphocytes are highly diverse, synthesizing one or another of innumerable immunoglobins, each with unique variable regions. The steps leading to the secretion of antibody begin with the transport of antigen to the lymph nodes and spleen via lymphatic and blood vascular channels, respectively. Antigen is engulfed there by monocytes or macrophages that ''present'' the antigen, which is attached to the surface of these cells, to B lymphocytes in the lymphoid organs. Antigens adhere only to those lymphocytes that have on their surfaces antibodies specific for the presented antigen. This signal is ordinarily not sufficient in itself to induce the production of antibodies. An additional signal is required. It is not yet certain whether this is furnished by a protein hormone, interleukin-2, or by a specific B cell growth factor; these agents are probably released by T helper lymphocytes that have been similarly exposed to macrophages coated with the specific antigen. Thus stimulated by their matching antigen, the B lymphocytes proliferate and transform into plasma cells that secrete immunoglobulins specific for the particular antigen into the surrounding milieu. Inherent in this concept is that immunogenic substances in some way stimulate proliferation of only those clones of B lymphocytes that are committed to the synthesis of the antibody proteins matching the specific antigen.

Five classes of immunoglobulins have been recognized, designated IgG, IgA, IgM, IgD, and IgE (Table 24-4). Common to all is a basic unit with a molecular weight of about 150,000 daltons (Fig. 24-3). Each unit is composed of two large (or heavy) and two small (or light) polypeptide chains, held together by disulfide bonds and by noncovalent forces. Digestion of the basic unit by the proteolytic enzyme (papain) splits the basic unit into two ''Fab'' fragments and one ''Fc'' fragment. Cleavage takes place at a portion of the heavy chain, known as the hinge region, an area thought to provide a flexibility that allows room for contact of the molecule with antigen. Each Fab fragment contains one of the light chains and the aminoterminal portion of one of the heavy chains. The Fc fragment is a dimer of the carboxyterminal portion of the two heavy chains.

Table 24-4. Immunoglobulins

Physical characteristics	*IgG*	*IgA*	*IgM*	*IgD*	*IgE*
Molecular weight ($\times 10^3$)	150	150; 400	900	180	190
Basic units/molecule	1	1, 2*	5*	1	1
Heavy chain†	gamma	alpha	mu	delta	epsilon
Concentration in serum (mg/dl)	600-1500	85-380	50-400	< 15	0.01-0.03
Biological half-disappearance time (days)	21-23	6	5	2-8	1-5
Crosses placenta	+	−	−	−	−
Principal Ig in secretions	−	+	−	−	−
Functional characteristics					
Toxin neutralization	+	+	+	−	−
Agglutination of particulate antigens	+	+	+	−	−
Opsonization	+	?	−	−	−
Bacterial lysis	+	−	+	−	−
Virus inactivation	+	+	+	−	−
Macrophage (monocyte) and neutrophil binding	+	±	−	−	−
Binding to and degranulation of mast cells and basophils with histamine release	−	−	−	−	+
Damage to host tissues	+	−	+	−	+
Complement fixation	+‡	−	+	−	−

Data are from various sources.
*Polymeric form joined by a J chain.
†All immunoglobulins may have kappa or lambda chains but not both.
‡Complement fixed by IgG of subclasses 1, 2, and 3, but not 4.

The striking ability of antibodies to react with specific antigens depends on the genetically determined structural diversity of the aminoterminal ends of the light and heavy chains of the Fab fragment. These portions of the Fab fragment are therefore designated as the "variable" regions of the immunoglobulin; that of the light chain is V_L, and that of the heavy chain is V_H. The variable regions of the Fab fragments contain the sequence of amino acids that bind to antigen. Each of the two Fab fragments thus has one combining site for antigen. The remainder of the Fab fragments and the Fc fragments do not display this variability and are therefore referred to as the "constant" regions of the immunoglobulin. The Fc fragment of the immunoglobulins does not combine with antigens, but it is the site of receptors for complement and for such cells as neutrophils, monocytes, and mast cells.

The Fab and Fc fragments of the immunoglobulins are heterogeneous. The light chains can be divided into two classes, kappa (κ) and lambda (λ). They are distinguished from each other immunologically, i.e., by the reaction of the light chains with antibodies specific to either kappa or lambda light chains. About two thirds of the immunoglobulins of normal plasma contain kappa chains, and the other third, lambda chains. Evidence suggests that any one plasma cell can secrete immunoglobulins with either kappa or lambda chains but not both.

The IgG, IgD, and IgE molecules are each composed of a single basic unit. The IgA molecules may be monomeric or dimeric. Those of the IgM class have five units that are joined at the carboxyterminal ends of the Fc fragments. The dimeric IgA and pentameric IgM molecules are each joined by an additional polypeptide, the J chain. Each class of immunoglobulins has a distinctive heavy chain, denoted by the corresponding lower case Greek letter.

At different stages in the immune response, lymphocytes that have been stimulated by antigens to differentiate into plasma cells secrete different classes of immunoglobulin. For example, some clones of plasma cells first synthesize IgM and later IgG antibodies directed against the same antigen; in this switch the proteins retain the same variable regions.

The different composition of the heavy chains accounts for the different biological properties of the five classes of immunoglobulins. *IgG* is the dominant antibody that reacts with bacteria and probably viruses. It can cross the placenta to enter the fetal circulation. It is also secreted into the colostrum, the initial milk fed to the newborn infant, and may thereby serve as a source of immunity against infection. Four subclasses of the IgG molecule are known, differing in their heavy chains. Of these, the Fc regions of the IgG of subclasses 1, 2, and 3 react with complement.

IgA is elaborated by plasma cells localized to secretory tissues, where two IgA molecules, joined by a J chain, may combine with *secretory piece*, a protein that is synthesized by epithelial cells. Combined with secretory piece, IgA is secreted into the tears, saliva, and milk and into respiratory, intestinal, and cervical secretions. The complex forms a barrier to infection by or-

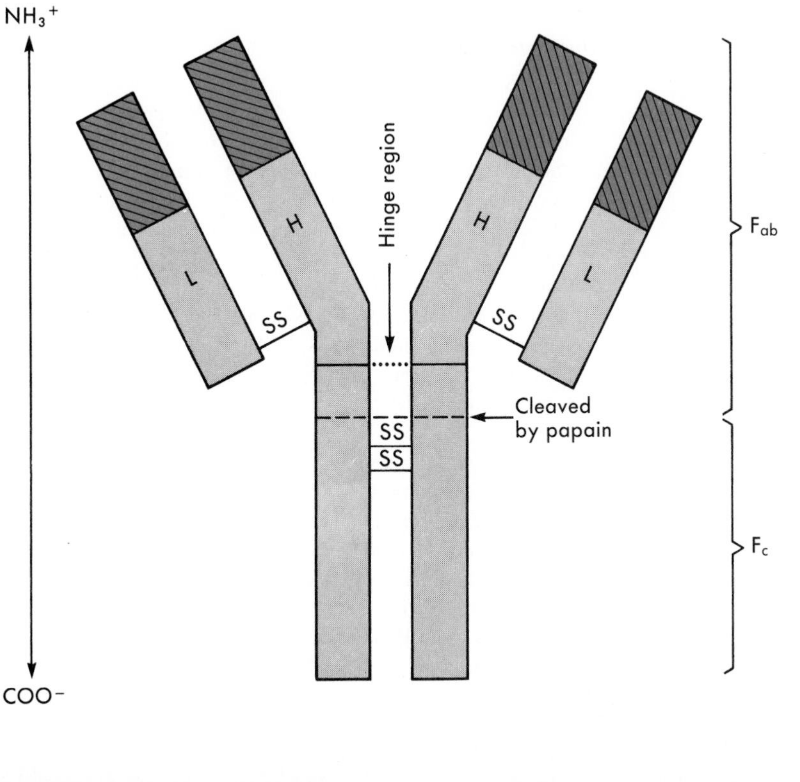

Fig. 24-3. The basic unit of the immunoglobulins is a tetramer of 2 light *(L)* and two heavy *(H)* chains, joined by disulfide *(SS)* bonds. The amino terminals of the light and heavy chains *(crosshatching)* are the variable regions of the molecule that combine with specific antigen. The remainder of the molecule is designated as the constant region. Papain splits the molecule into three fragments: two Fab (antigen-binding) fragments and one Fc (crystallizable) fragment. Neutrophils, monocytes, and the Clq component of complement bind to the Fc region. The hinge region is indicated by dotted line.

ganisms to which the IgA molecules are specifically directed.

IgM is the most prominent immunoglobulin on the surface of lymphocytes. It is the antibody first synthesized by plasma cells after an initial exposure to antigen. The natural antibodies against blood cell antigens are of the IgM class.

IgE binds avidly to mast cells. Exposure of the mast cells to specific antigens may then trigger reactions that lead to discharge of the contents of intracellular granules, including histamine, leukotrienes (oxidation derivatives of arachidonic acid), and other agents. These substances induce vasodilation, increase vascular permeability, contract certain smooth muscles, such as those of the respiratory tree, and thus may bring about attacks of bronchial asthma. In instances that are fortunately rare, the reaction between antigen and IgE may be so explosive that there is a life-threatening fall in blood pressure (from vasodilation) and bronchoconstriction (from contraction of bronchial smooth muscle). These *anaphylactic reactions* may be triggered in previously sensitized individuals by the sting of venomous insects, such as bees or wasps, or by the injection of animal serum or drugs. The more banal allergic reactions, such as hives (urticaria), hay fever, and some types of bronchial asthma, are more localized. The mast cells are degranulated in the tissues exposed to ingested or inhaled antigens.

The biological function of *IgD* is not yet clarified.

Complement

The term *complement* encompasses a group of plasma proteins that, when appropriately activated, bring about responses that defend the body against injury. Among the functions of complement are (1) *chemotaxis*, the attraction of leukocytes to the site of injury; (2) *opsonization*, the presentation of insoluble immune complexes for phagocytosis by granulocytes or monocytes; (3) the *release of anaphylatoxins* (which increase vascular permeability) from mast cells; and (4) the *disruption of the exterior membranes* of certain microorganisms and blood cells.

The complement system is complex, involving the participation of many different plasma proteins (Table 24-5). Two convergent chains of chemical reactions have been delineated: the classic and alternative (or properdin) pathways. The *classic complement pathway* involves the interaction of nine distinct proteins, C1 (i.e., the first component of complement) to C9. The components are not numbered in their order of action but rather on historical grounds. The agents participating in the *alternative complement pathway* include proteins designated B, D, H, I, and P (for properdin), as well as components C3 to C9 of the classic pathway.

Table 24-5. Complement components in serum

Factor	*Molecular weight* × 10^3	*Concentration (mg/dl)*
Classical pathway		
C1	± 1,000	
C1q	410	7
C1r	190	5
C1s	88	4
C2	117	2.5
C3	185	83-177
C4	206	15-45
Alternative pathway		
Factor B	93	17-42
Factor D	24	0.1-0.2
Factor H	150	40-50
Factor I	88	3-5
Properdin (P)	185	2.5
Membrane attack pathway		
C5	190	8.5
C6	128	7.5
C7	121	5.5
C8	163	5.5
C9	79	5-20

Data are from various sources.

Bibliography

Journal articles

Klebanoff, S.J.: Oxygen metabolism and the toxic properties of phagocytes, Ann. Intern. Med. **93:**480, 1980.

Mayer, M.M.: Complement: historical perspectives and some current issues, Complement **1:**2, 1984.

Books and monographs

Babior, B.M., and Stossel, T.P.: Hematology: a pathophysiological approach, New York, 1984, Churchill Livingston.

Donaldson, V.H.: Complement. In Frohlich, E.D., editor: Pathophysiology: altered regulation mechanisms in diseases, ed. 3, Philadelphia, 1984, J.B. Lippincott Co.

Muller-Eberhard, H.J.: Complement: molecular mechanisms, regulation and biologic function. In Berlin, R.D., et al., editors: Molecular basis of biological degradative processes, New York, 1978, Academic Press.

Reichlin, M., and Harley, J.B.: Adaptive immunity. In Frohlich, E.D., editor: Pathophysiology: altered regulation, mechanisms in disease, ed. 3, Philadelphia, 1984, J.B. Lippincott Co.

Roitt, I.: Essential immunology, ed. 4, Oxford, U.K., 1980, Blackwell Scientific Publications.

Spivak, J.L., editor: Fundamentals of clinical hematology, Hagerstown, Md., 1980, Harper & Row.

Stites, D.P., et al.: Basic and clinical immunology, ed. 5, Los Altos, Calif. 1984, Lange Medical Publications.

Unanue, E.R., and Benacerraf, B.: Textbook of immunology, ed. 2, Baltimore, 1984, Williams and Wilkins.

CHAPTER

25

Hemostasis and Blood Coagulation

The life of every organism depends on the preservation of its internal environment. Even one-cell creatures can seal disruptions of their external membranes. Invertebrates protect themselves from loss of their vital hemolymph by processes analogous to vertebrate clotting (in that they use circulating blood cells, hemolymph, or both).

Vertebrates have evolved complex mechanisms to stem hemorrhage after injury. Failure of the hemostatic mechanisms may lead to fatal exsanguination. In mammals, an early event after trauma is transitory contraction of damaged blood vessels, but this furnishes little toward *hemostasis,* i.e., the arrest of bleeding. Within seconds after vascular injury, however, *platelets,* small anuclear circulating blood cells, adhere to the site of damage and pile up, one on another, to provide a mechanical plug that effectively stops bleeding from minor injuries. Hemorrhage from more formidable wounds is stanched by the coagulation of blood. When the skin is unbroken, bleeding may be checked by external compression of the injured vessels by the extravasated blood. This back pressure is highly effective where the skin is tightly bound to the underlying tissues, as in the fingertip, but it is essentially useless where the skin is distensible. For example, hemorrhage around the eye is unchecked by the counterpressure of extravasated blood, the pathogenesis of a "black eye." After childbirth, hemostasis in the uterus is aided by contraction of the uterine musculature, which compresses the blood vessels that feed the site from which the placenta was wrenched.

Incredibly, that clots may provide a mechanical barrier to loss of blood after vascular injury was not appreciated until the beginning of the eighteenth century. A clot is a network of protein fibers whose meshes trap blood cells and *serum,* that is, plasma that has undergone coagulation. Speculation about the source of the fibers of the blood (now called *fibrin*) goes back at least as far as Plato and his pupil, Aristotle, in the fourth century BC. About 200 years ago William Hewson, an English physician, proposed that fibrin was derived during clotting from plasma, the liquid portion of blood. Over the succeeding years, it became clear that this came about through the action of a proteolytic enzyme, *thrombin,* that catalyzed the transformation of a plasma protein, *fibrinogen* (factor I) to fibrin.

Thrombin is not normally present in circulating blood, but when blood is shed, thrombin is cleaved from its precursor in plasma, *prothrombin* (factor II). The release of thrombin is the final step in one or both of two convergent and intertwined chains of chemical reactions, the *extrinsic* and *intrinsic* pathways. When trauma disrupts the vascular endothelial lining, blood comes into contact with subendothelial structures and other exposed injured tissues. This sets in motion a succession of catalytic events through either or both pathways. At each step, a proenzyme *clotting factor* is transformed to its enzymatic form in which it can activate the next proenzyme in the chain. The series of enzymatic steps magnifies the original disturbance of the blood until, in the end, thrombin is released explosively.

The enzymes that are activated via the extrinsic or intrinsic pathways are endopeptidases (i.e., they cleave specific peptide bonds that are not located at the extreme ends of the substrate molecules). Cleavage exposes the site in the substrate clotting factor that is responsible for its biologic function. In most cases, the broken bond is located within a folded portion of the polypeptide chain that is held together by internal disulfide cross-links. Thus with disruption of the peptide bond the polypeptide chain is split into two parts that are held together by a disulfide bridge. The clotting enzymes of the intrinsic and extrinsic pathways are serine proteases (i.e., the catalytic site contains a serine residue). Three clotting factors, *high molecular weight kininogen, antihemophilic factor* (AHF, factor VIII), and *proaccelerin* (factor V), have not yet been shown to have catalytic activity and probably act as nonenzymatic cofactors.

Reactions of the extrinsic pathway of thrombin formation are initiated by contact of blood with injured tissues. The damaged cells furnish a clot-promoting agent, *tissue thromboplastin* (tissue factor, factor III). This agent was first studied extensively by a German-Estonian physiologist, Alexander Schmidt, who thought that clotting took place in the following two steps. First, tissue thromboplastin transformed prothrombin to thrombin, and then this enzyme converted fibrinogen into a fibrin clot. We now know that generation of thrombin is much more complex.

Studies of the intrinsic pathway of thromboplastin, whereby blood clots in the absence of tissue thromboplastin, were pioneered by John Lister, the great British surgeon and a contemporary of Schmidt. Lister observed that blood clotted rapidly when it was drawn into a cup, but much more slowly when it was placed in an india-rubber tube. He concluded that clotting came about by contact of blood with a "foreign surface" that differed from the normal vascular endothelial lining. The cup, in this regard, was more "foreign" than the india-rubber lining.

The successive steps of the intrinsic pathway are now known to begin with activation of a plasma proenzyme, *Hageman factor* (factor XII), that is brought about by contact with a suitable foreign surface. Activated Hageman factor (factor XIIa) then initiates a succession of enzymatic events that involve at least eight other plasma proteins and that lead to the elaboration of thrombin. The final steps of the intrinsic and extrinsic pathways are identical.

Clotting is modulated by inhibitory agents within the plasma. The first such substance to be delineated was an inhibitor of thrombin, now called *antithrombin III,* whose action is greatly potentiated by *heparin,* a mucopolysaccharide that is now widely used therapeutically (p. 384).

Blood also has the capacity to redissolve clots that might inadvertently form within blood vessels. This property is attributed for the most part to the elaboration of a plasma proteolytic enzyme, *plasmin.* Plasmin can be generated from its precursor, *plasminogen,* in many ways. One of the most interesting is the process of coagulation itself.

The integrity of blood vessel walls is also critical in hemostasis. When the supporting structures surrounding the blood vessels are defective, as may occur, for example, in scurvy (vitamin C deficiency), a hemorrhagic tendency may ensue. Our understanding of the role of vessels in the control of bleeding is still primitive. Blood flow itself affects hemostasis. The formation of an intravascular clot is fostered by stasis of flow and impeded by rapid blood flow.

Blood Coagulation

Nomenclature

As each new clotting factor was discovered, it acquired a trivial name that reflected its supposed function or was derived from the name of a patient whose plasma appeared to be deficient in the particular agent. Often, several names were used to designate the same substance. To bring order out of this chaos, the International Committee for the Nomenclature of Clotting Factors has assigned a Roman numeral to most of the clotting factors and has urged that each author refer to this number along with the trivial name he or she prefers. Currently, most of the scientific literature uses only the numerical nomenclature. Inevitably, this has led to numerous errors that are of little consequence in published articles but of great moment if patients or physicians are confused. Table 25-1 provides a key to commonly used synonyms. This chapter will avoid the pitfalls of pied type by retaining the admittedly archaic trivial nomenclature, followed, where clarity is needed, by the appropriate Roman numeral. Familiarity with both systems of nomenclature will aid the student in understanding the relationships between basic knowledge and clinical application.

The Synthesis of Clotting Factors

With the exception of antihemophilic factor (factor VIII), the clotting factors in plasma are synthesized mainly in the liver. Antihemophilic factor is a complex molecule that can be dissociated into two subcomponents of unequal size. The larger subcomponent, designated von Willebrand factor or factor VIII:VWF, is synthesized in vascular endothelial cells and megakaryocytes. The site of synthesis of the smaller subcomponent, factor VIII:C, is unknown, although one likely candidate is the liver. Megakaryocytes also synthesize the a (or α) chains of *fibrin-stabilizing factor* (factor XIII) and a form of fibrinogen whose identity with plasma fibrinogen is disputed. Proaccelerin (factor V) appears to be synthesized by the liver and by vascular endothelial cells and megakaryocytes.

Vitamin K. Hepatocytes synthesize four clotting factors, Christmas factor (factor IX), factor VII, Stuart factor (factor X), and prothrombin (factor II). Synthesis occurs only if vitamin K is present. Vitamin K is the generic name for a group of fat-soluble quinone derivatives that are plentiful in leafy vegetables. In the newborn infant, vitamin K is provided by milk; cow's milk is a much richer source than human milk. Within a few days after birth, bacteria that have begun to grow within the lumen of the gut become an important source of the

Table 25-1 Blood clotting factors

Roman numeral	*Trivial names*	*Activated or altered state*
Factor I	Fibrinogen	Fibrin
Factor II	Prothrombin	Thrombin
Factor III	Tissue thromboplastin; tissue factor	—
Factor IV	Calcium ions	—
Factor V	Proaccelerin; Ac-globulin	Altered proaccelerin (factor V_a)
Factor VII	—	Factor α-VII_a
Factor VIII	Antihemophilic factor (AHF); antihemophilic factor complex	
(Factor VIII/VWF)*		
Factor VIII:C*	Coagulant subcomponent of AHF	Altered antihemophilic factor (factor VIII:C_a)
Factor VIII:VWF	von Willebrand factor subcomponent of AHF, von Willebrand factor	
Factor IX	Christmas factor	Factor IX_a
Factor X	Stuart factor	Factor X_a
Factor XI	Plasma thromboplastin anticedent (PTA)	Factor XI_a
Factor XII	Hageman factor (HF)	Factor XII_a
Factor XIII	Fibrin-stabilizing factor (FSF)	Factor XIII_a (fibrinoligase)
—	Plasma prekallikrein (Fletcher factor)	Plasma kallikrein
—	High molecular weight (HMW) kininogen (Fitzgerald, Williams, or Flaujeac factor)	

*The terminology of factor VIII is now under review. Historically, the term was first applied to the coagulant property of antihemophilic factor (factor VIII:C) and later to the noncovalent complex of coagulant AHF and what is now called von Willebrand factor. The von Willebrand factor has been termed factor VIII:VWF, factor VIIIR:Ag, or ristocetin cofactor (factor VIII:RCo), depending on the context (see text). Some authors now restrict the meaning of the term factor VIII to the coagulant part of the AHF complex and describe von Willebrand factor as VWF.

vitamin. Because vitamin K is fat-soluble, its absorption from the gut depends on (1) the presence of bile salts that are excreted by the liver into the duodenum and (2) normal digestive and absorptive mechanisms for fat. Synthesis of the vitamin K–dependent clotting factors appears to be controlled by humoral factors, the coagulopoietins.

These physiologic considerations help to explain the commonplace occurrence of combined deficiencies of the vitamin K–dependent clotting factors. In *hemorrhagic disease of the newborn,* e.g., the infant ingests inadequate amounts of vitamin K before bacterial flora are established. Vitamin K deficiency also results from *suppression of intestinal flora* by orally administered antibiotics, particularly if little or none of the vitamin is ingested. Deficiency of the four clotting factors may result from its malabsorption, as may occur if bile salts are excluded from the gut by *obstruction of the common bile duct* or if *pancreatic* or *bowel disease* impair absorption of lipids. The hemorrhagic symptoms and laboratory abnormalities in all these states can be corrected by parenteral administration of vitamin K. Deficiencies of the vitamin K–dependent factors may also complicate *hepatic disease,* but in this situation the administration of vitamin K is usually without benefit.

Recent studies have elucidated the long puzzling function of vitamin K. Synthesis of the vitamin K–dependent clotting factors proceeds in two stages. First, the hepatocytes synthesize polypeptide progenitors of each factor, a process that does not require the vitamin. In the second step, vitamin K, reduced to its hydroquinone form in the liver, acts as a cofactor for a specific microsomal carboxylase (Fig. 25-1). This carboxylase inserts a second carboxyl group into the γ-carbon of certain glutamic acid residues in the polypeptide chains. These unique tricarboxylic glutamic acid residues serve as points of attachment for the calcium ions that are needed for transformation of the vitamin K–dependent factors to their enzymatically active states. During carboxylation the reduced vitamin K is oxidized to an epoxide form from which vitamin K is then regenerated by enzymatic reduction.

Synthesis of the vitamin K–dependent factors is inhibited by dicumarol and its congeners, notably warfarin. These agents have had long usage as "anticoagulants" in the prevention and treatment of thrombosis. Dicumarol was first extracted from spoiled sweet clover, which induced a mysterious hemorrhagic disorder of cattle. Dicumarol and similar agents competitively inhibit the enzymes that reduce vitamin K and its epoxide (Fig. 25-1). In this way they suppress the carboxylation of the progenitors of the vitamin K–dependent factors so that the synthesis of these factors is not completed.

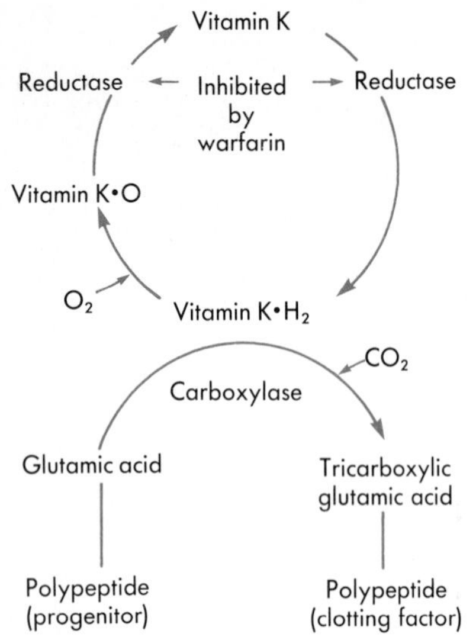

■ **Fig. 25-1.** The role of vitamin K is the insertion of a second carboxyl group into the γ-carbon of glutamic acid residues of the vitamin K–dependent clotting factors. *Vitamin K·O,* vitamin K epoxide.

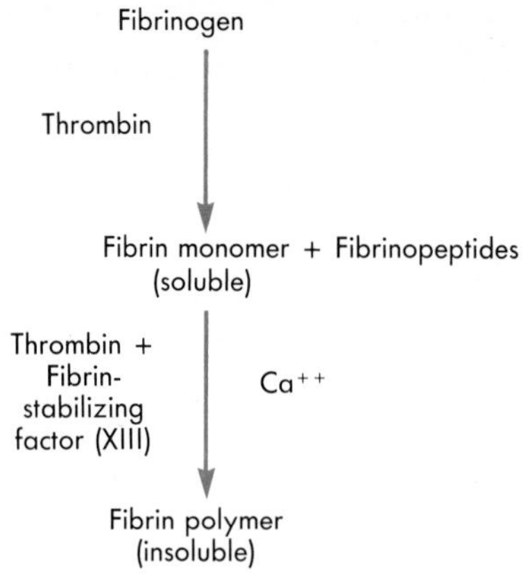

■ **Fig. 25-2.** The formation of fibrin. (Redrawn from Ratnoff, O.D.: Hemorrhagic disorders: coagulation defects. In Beeson, P.B., McDermott, W., and Wyngaarden, J.B., editors: Cecil textbook of medicine, ed. 15, Philadelphia, 1979, W.B. Saunders Co.)

That four distinct clotting factors require vitamin K to complete their synthesis suggests that they have a common evolutionary origin. In agreement with this, extensive similarities in amino acid sequence have been found. A reasonable explanation is that the different factors arose as the result of the duplication and subsequent mutation of the genes responsible for the synthesis of an ancestral protein.

For many years, the common wisdom held that vitamin K was needed only for the synthesis of clotting factors. Now we have learned that it is also required for synthesis of other proteins found, for example, in plasma, bone, kidney, lung, spleen, and placenta. Presumably in each case the functioning of these proteins requires the attachment of calcium ions to the tricarboxylic glutamic acid residues formed through the action of vitamin K. Of pertinent interest are two vitamin K–dependent plasma agents, *protein C* (autoprothrombin II-A) and *protein S*. Protein C, when activated by thrombin, inhibits the coagulant properties of antihemophilic factor (factor VIII:C) and proaccelerin (factor V) whereas protein S enhances this property of protein C (p. 385).

■ *Plasma Fibrinogen and the Formation of Fibrin*

Blood clotting is the visible result of the conversion of a soluble plasma protein, fibrinogen (factor I), into an insoluble meshwork of fibrin. Fibrinogen is a dimeric glycoprotein with a molecular weight of 340,000. Each half of the dimer is composed of three polypeptide chains, designated Aα, Bβ, and γ, respectively. The six chains are held together by disulfide bonds.

The conversion of fibrinogen to fibrin takes place in three stages (Fig. 25-2). First, fibrinogen undergoes limited proteolysis by thrombin that has evolved during the earlier steps of the coagulation process. The partially digested fibrinogen molecules, called *fibrin monomers,* polymerize by electrostatic forces into insoluble strands of fibrin. Finally, the constituent fibrin monomers of fibrin are cross-linked covalently by a plasma enzyme, *activated fibrin-stabilizing factor* (fibrinoligase, factor XIII_a).

In the first step of fibrin formation, thrombin cleaves four small polypeptide fragments, each with a molecular weight of about 1500, from each fibrinogen molecule, reducing its weight by about 2%. One fragment, *fibrinopeptide A,* is released from the amino-terminal end of each Aα chain, and another, *fibrinopeptide B,* from the amino-terminal end of each Bβ chain. The residue, fibrin monomer, retains its dimeric structure, each half having three chains, designated α, β, and γ. Fibrinopeptide A is released by thrombin more rapidly than fibrinopeptide B, and with its separation polymerization begins. Polymerization depends primarily on the separation of fibrinopeptide A.

As the first fibrin monomers are generated, they are surrounded by unaltered fibrinogen molecules with which they form loose complexes. As more fibrin monomers accumulate, the equilibrium shifts so that fibrin monomers now polymerize with each other. At the same time the complexes of fibrinogen and fibrin monomer dissociate, making more fibrin monomers available for polymerization. The monomers aggregate both side-to-side and end-to-end, gradually building insoluble polymers that thicken and lengthen to form visible fibrous strands. Some of the polymers appear to be

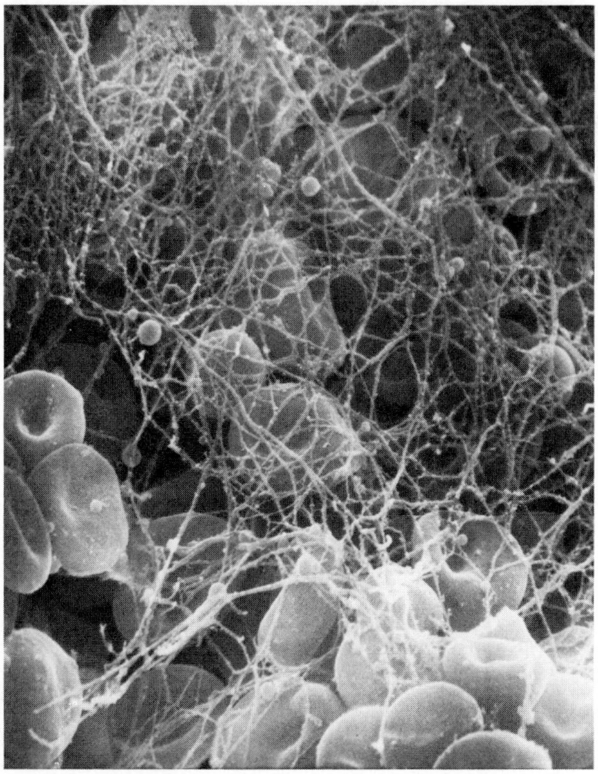

■ Fig. 25-3. Human blood clot showing red blood cells immobilized within a network of fibrin threads. The small spheres are platelets. Scanning electron micrograph (×9000). (Reproduced with permission from Shelly, W.B.: Of red cell bondage, JAMA **249:**3089, Copyright 1983, American Medical Association.)

branched so that the fibrin strands build into a complicated meshwork that traps blood cells and serum. Additionally, platelets bind to the polymerizing strands of fibrin (Fig. 25-3).

Calcium ions are not required for the transformation of fibrinogen to fibrin monomer. However, at the concentration present in plasma, calcium ions greatly accelerate the polymerization process.

Clots formed in purified mixtures of thrombin and fibrinogen are held together by noncovalent forces and have low tensile strength. Such clots dissolve readily in solutions of urea or monochloroacetic acid, which dissociate noncovalent bonds. In contrast, clots formed in normal human plasma are insoluble in these dispersing agents and have high tensile strength. Plasma contains a proenzyme, fibrin-stabilizing factor (factor XIII). When activated this factor acts as a transamidase to forge covalent links between the γ-carboxyl groups of glutamic acid residues of one fibrin monomer and the ϵ amino groups of lysyl residues of another. Fibrin-stabilizing factor is a tetramer composed of two a (or α) chains and two b (or β) chains; one half of the fibrin-stabilizing factor in blood is found in platelets, where it is composed only of a chains.

Activation of fibrin-stabilizing factor is brought about by thrombin, which cleaves a small polypeptide from each a chain. In the presence of calcium ions, fibrin-stabilizing factor then induces covalent links among fibrin monomers. It creates dimers between the γ chains of two adjacent monomers, and, at a slower rate, polymers among several α-chains. Resistance to dissolution by urea or monochloroacetic acid occurs only when these α-chain polymers are formed.

Fibrin-stabilizing factor has other actions besides covalent cross-linking of fibrin. For example, it forges links between molecules of contractile proteins of muscles or platelets; it bonds fibronectin (cold insoluble globulin) to itself and to fibrin or collagen (p. 389); and it induces binding of α_2-plasmin inhibitor to fibrin (p. 384). In the absence of fibrin-stabilizing factor experimental wound healing is retarded. This phenomenon may be related to stimulation of the growth of fibroblasts by activated fibrin-stabilizing factor.

■ *Disorders of Fibrin Formation*

The concentration of fibrinogen in normal human plasma averages 270 to 300 mg/dl. Thus there is about 10 g of fibrinogen in the circulating blood; perhaps an additional 5 g is present in extravascular fluid. An increased concentration of fibrinogen is commonplace in normal pregnancy and in innumerable disease states, particularly those associated with inflammation, tissue damage, or neoplasm. In these disease states, fibrinogen is thought to be an ''acute phase reactant,'' i.e., a plasma component whose concentration is increased under the stress of disease.

An increased concentration of fibrinogen accelerates the settling of red blood cells when blood is allowed to stand. This increased ''erythrocyte sedimentation rate'' was described by Hippocrates, who interpreted the rapid settling of red cells as a basic cause of disease. Settling is fostered by the higher density of red cells than that of the plasma and is resisted by the surface area that the cells present to the surrounding medium. Normally these forces are almost balanced so that red cells sediment only very slowly. Rapid settling comes about because the increased amounts of fibrinogen neutralize the electrostatic forces that exist on the surface of red cells and normally keep these cells from sticking to each other. Under these conditions the red cells form stacks of cells that resemble piles of coins. The stacked cells, in contrast to individual erythrocytes, present relatively less surface area to the surrounding plasma to oppose the downward pull of gravity, and therefore the rate at which the cells sediment is accelerated.

The biological half-life of fibrinogen under normal conditions is about 3 to 4 days, i.e., half of the fibrin-

Table 25-2. Molecular weight, concentration, and biological half-life of plasma hemostatic factors

Factor	*Molecular weight (× 1000)*	*Approximate plasma concentration (mg/dl)*	*Biological half-life**
Fibrinogen (I)	340	200-400	3-4 days
Prothrombin (II)	72	12	3 days
Proaccelerin (V)	330	0.4-1.4	15-30 hrs.
Factor VII	50	0.05-0.06	4-5 hrs.
Antihemophilic factor (VIII)	1,000-12,000	0.5-1	10-12 hrs.
(VIII:C)	265	0.005-0.010	
(VIII:VWF)	450-12,000	0.5-1.0	
Christmas factor (IX)	57	0.4-0.5	18-24 hrs.
Stuart factor (X)	59	1.2	24-36 hrs.
Plasma thromboplastin antecedent (XI)	124	0.4-0.6	2-3 days
Hageman factor (XII)	80	1.5-4.5	2 days
Plasma prekallikrein	88	3.5-4.5	35 hrs.
HMW kininogen	110-120	8-9	6 days
Fibrin-stabilizing factor (XIII)	320	1-2	11-12 days
Plasminogen	92	20	2-2½ days

*The biological half-life of a clotting factor is measured by infusing it into a patient with a congenital deficiency of the factor or by infusing radiolabeled factor into a normal individual. Ordinarily the curve describing the rate of disappearance of the factor from plasma has two components. First, the concentration of the infused substance decreases relatively rapidly as it diffuses into extravascular spaces. Thereafter the concentration decreases more slowly, reflecting the catabolism of the factor. The biological half-life, i.e., the length of time until the titer of the infused factor decreases by 50%, is calculated from the second component of the disappearance curve.

ogen present in the plasma at any given time will have disappeared in this interval (Table 25-2). Thus about 15% of plasma fibrinogen must be replaced daily by its continual synthesis, which takes place in the parenchymal cells of the liver. Among the agents proposed as stimuli to synthesis are growth hormone, thyroxin, corticotropin, hormones released from inflammatory lesions, and the degradation products released by the digestion of fibrinogen and fibrin by plasmin, a plasma proteolytic enzyme (p. 382).

The normal site and mechanism of the catabolism of fibrinogen are unknown. Fibrinogen is utilized through its conversion to fibrin both in hemostasis and in the formation of inflammatory lesions. Perhaps some fibrinogen is digested by plasmin, particularly when this enzyme has been activated by severe stress. But these mechanisms participate only marginally in the normal catabolism of fibrinogen. The hypothesis that fibrinogen catabolism is the result of continual slow intravascular clotting has little to support it.

Both hypofibrinogenemia (an abnormally low concentration of fibrinogen in plasma) and afibrinogenemia (the absence of detectable fibrinogen) may be hereditary or acquired. *Congenital afibrinogenemia* is a rare autosomal recessive disorder in which affected individuals do not synthesize fibrinogen; the blood does not clot even when large amounts of thrombin are added. Although patients with this disorder bleed uncontrollably from sites of injury, they have surprisingly little difficulty if unchallenged. The lesson learned from these patients is that hemostasis from trivial wounds is apparently provided by platelets when the generation of thrombin is normal. Impaired synthesis of fibrinogen resulting in hypofibrinogenemia may occur in catastrophic liver disease, but this is an uncommon phenomenon.

Hypofibrinogenemia and afibrinogenemia must be distinguished from *dysfibrinogenemia,* a group of rare autosomal dominant disorders in which plasma contains abnormal variants of fibrinogen that clot unusually slowly upon addition of thrombin.

The most frequent disorder of fibrinogen is *disseminated intravascular coagulation,* a serious and sometimes lethal complication of many disease processes. Widespread thrombosis (i.e., intravascular clotting) within small blood vessels occurs because agents that can bring about the elaboration of thrombin have gained access to the blood stream. Much or all of the plasma fibrinogen may be consumed in the formation of thrombi, resulting in hypofibrinogenemia or afibrinogenemia. Other clotting factors may be depleted and the platelet count may be reduced, while the proteolytic enzyme plasmin (p. 382) may become active and may dissolve the intravascular clots. Among the incitants of disseminated intravascular clotting are such complications of pregnancy and childbirth as retention of a dead fetus within the uterus for several weeks, premature separation of the placenta, and amniotic fluid embolism in which amniotic fluid and its contents enter the maternal blood stream during parturition. Other causes of disseminated intravascular coagulation include such diverse events as envenoming by the bite of snakes whose venom contains clot-promoting agents; sepsis with a va-

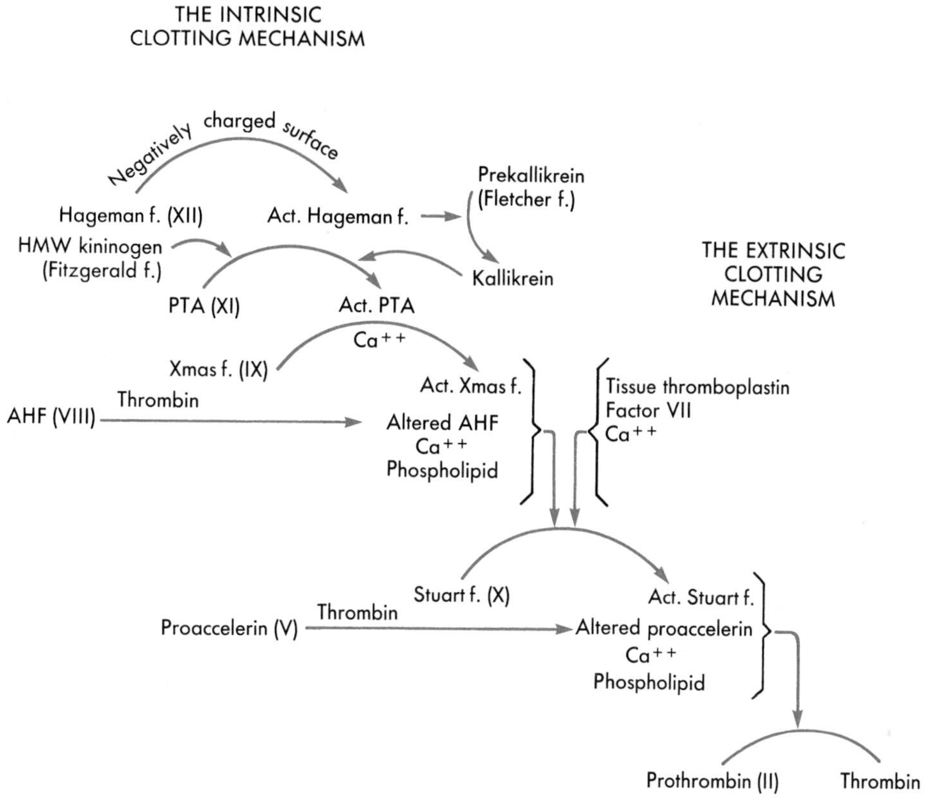

■ **Fig. 25-4.** The intrinsic and extrinsic pathways for the formation of thrombin. Omitted from the diagram are inhibitors of the various steps. The phospholipid portion of tissue thromboplastin may function in the activation and action of Stuart factor (factor X). The phospholipid for the intrinsic pathway is furnished by platelets and by the plasma itself. Augmentation of the action of factor VII by thrombin and by the activated forms of Hageman factor (factor XII_a), Christmas factor (factor IX_a), and Stuart factor (factor X_a), and the activation of Christmas factor by the factor VII–tissue thromboplastin complex are not depicted. *Act,* activated; *HMW,* high molecular weight; *PTA,* plasma thromboplastin antecedent; *X-mas,* Christmas; *AHF,* antihemophilic factor. (Reprinted, with permission, from Ratnoff, O.D.: Hemorrhagic disorders: coagulation defects. In Beeson, P.B., McDermott, W., and Wyngaarden, J.B., editors: Cecil textbook of medicine, ed. 15, Philadelphia, 1979, W.B. Saunders Co., pp. 1881-1897.)

riety of organisms (an all too frequent cause of death when abortion was illegal); transfusion of incompatible blood in which damaged red blood cells appear to foster the evolution of thrombin; and the presence of certain tumors in which clot-promoting products of the neoplasm gain entrance into the blood stream.

■ *The Formation of Thrombin*

Thrombin, the enzyme ultimately responsible for the formation of fibrin monomers, is generated by the catalytic scission of prothrombin by activated Stuart factor (factor X_a). Under physiological conditions, activation of Stuart factor (factor X) takes place through two series of enzymatic steps, the intrinsic and extrinsic pathways (Fig. 25-4). These reactions are so intertwined that the reader may find it helpful to read the next few sections twice to appreciate the nature of the feed-back mechanisms that are involved.

■ *Surface-Mediated Reactions Initiating the Intrinsic Pathway*

Activation of Stuart factor (factor X) by the intrinsic pathway is the end result of a series of reactions that begins when blood comes into contact with negatively charged surfaces. When venous blood is drawn into a polystyrene or silicone-coated tube and centrifuged to

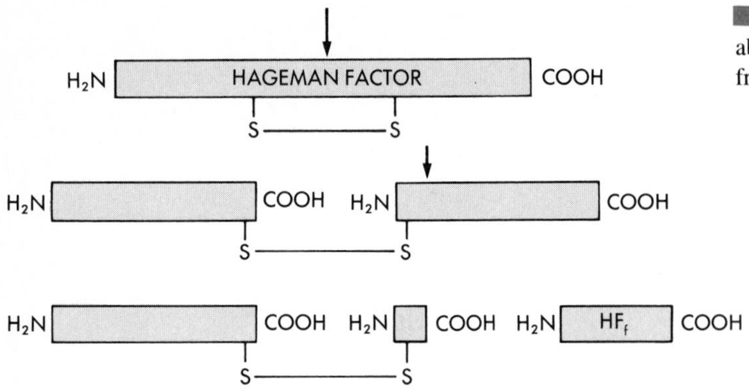

■ **Fig. 25-5.** The activation of Hageman factor. What brings about scission *(arrows)* is uncertain. *HF*$_f$, Carboxy-terminal fragment.

sediment its cells, the plasma thus separated clots readily when placed in glass tubes, but much more slowly in polystyrene or silicone-coated tubes. Such experiments suggest that glass has qualities that induce a change in plasma leading to the elaboration of thrombin. The clot-promoting properties of glass have been related to its negative surface charge, and many similarly charged insoluble substances, such as kaolin (clay), diatomaceous earth (Celite), and talc have a similar effect. These substances are foreign to the body, but addicts may inject themselves with drugs adulterated with kaolin or talc and in this way may self-induce thrombosis.

Uncertainty exists about the natural analogues of the clot-promoting solids. Sebum (the oily secretion of skin), some forms of collagen or basement membrane proteins, sulfatides (such as are found in brain tissue), disrupted vascular endothelial cells, and endotoxins derived from gram-negative bacteria have all been implicated.

The events triggered by negatively charged surfaces include not only the generation of thrombin but also the mediation of other defense mechanisms of the body against injury, including inflammation, the immune response, and fibrinolysis, i.e., the dissolution of clots. Four plasma proteins are involved in the initiation of one or another of these "surface-mediated defense reactions": *Hageman factor* (factor XII), *plasma thromboplastin antecedent* (PTA, factor XI), *plasma prekallikrein* (Fletcher factor), and *high molecular weight kininogen* (Fitzgerald, Williams, or Flaujeac factor). All four factors are readily adsorbed from normal plasma to negatively charged surfaces such as glass. It is apparently in this adsorbed state that they exert their clot-promoting functions.

The first step in the various surface-mediated defense reactions, including the initiation of events of the intrinsic pathway, is conversion of Hageman factor to an enzymatic form that can activate PTA. When Hageman factor is adsorbed to clot-promoting negatively charged surfaces, it undergoes a change in shape that exposes amino acid residues that are buried within the native molecule. Whether this shape change is enough to induce clot-promoting activity is disputed. Adsorbed to such surfaces from normal plasma, Hageman factor is split internally within a disulfide loop (Fig. 25-5). The two-chain species that results can transform PTA to its enzymatically active state. Hageman factor is then further cleaved into two polypeptide fragments. The carboxy-terminal fragment, designated HF_f, includes the sequence of amino acids that is responsible for Hageman factor's enzymatic activity, but this fragment is a much weaker activator of PTA than the two-chain species.

The conversion of PTA to its activated state by activated Hageman factor requires the presence of high molecular weight kininogen. The conversion takes place on the surface of clot-promoting, negatively charged surfaces to which the three clotting factors are adsorbed. Activation is accelerated by similarly adsorbed plasma prekallikrein, but this factor is not an absolute requirement for this step.

Long before their role as clotting factors was discovered, both plasma prekallikrein and high molecular kininogen had been identified in plasma by their participation in experimental inflammatory reactions. In the presence of high molecular weight kininogen, activated Hageman factor (factor XII_a) changes plasma prekallikrein to its enzymatic form, plasma kallikrein. In turn, plasma kallikrein releases small polypeptide *kinins,* notably a nonapeptide, bradykinin, from plasma proteins described as *kininogens*. Plasma contains two groups of kininogens, distinguished by their molecular weight. The high molecular weight kininogens are much more avid substrates for plasma kallikrein than those of lower molecular weight; only the high molecular weight kininogens promote clotting. Bradykinin, released through the action of plasma kallikrein, can dilate small blood vessels, increase their permeability, and induce pain. In this way bradykinin can bring about the warmth, redness, swelling, and pain of inflammatory lesions. Bradykinin also contracts certain smooth muscles, a property useful in its biological assay. Human (but not bovine) high molecular weight kininogen still retains its clot-promoting properties after the bradykinin sequence has been removed.

Activated Hageman factor participates in at least three other surface-mediated defense reactions. Directly or indirectly, it augments the clot-promoting properties of factor VII (p. 380), it converts plasminogen to plasmin (p. 383), and it transforms C1, the first component of complement (a group of proteins participating in immune reactions), to its enzymatically active state ($C\overline{1}$) (p. 370). In the test tube, activated Hageman factor also converts prorenin to renin, an enzyme that brings about the formation of angiotensin, a polypeptide that contracts arterioles.

Some individuals have a hereditary deficiency of one or another of the four plasma proteins that participate in surface-mediated defense reactions. In each case, the affected individuals have impaired blood coagulation *in vitro*. Paradoxically those lacking Hageman factor (said to have Hageman trait), plasma prekallikrein (Fletcher trait), or high molecular weight kininogen (Fitzgerald, Williams, or Flaujeac trait) are curiously free of symptoms of a bleeding disorder, and those with PTA deficiency have only a mild bleeding tendency. Each of these disorders is ordinarily inherited as an autosomal recessive trait.

Activation of Christmas factor (factor IX) via the intrinsic pathway. The function of activated PTA (factor XI_a) in the intrinsic pathway is the activation of Christmas factor (factor IX), the first step that requires calcium ions, Christmas factor is a vitamin K–dependent protein that is synthesized by the liver under the direction of a gene on the X chromosome (p. 373). Activated PTA cleaves Christmas factor at two points (Fig. 25-6). First, the single chain of Christmas factor is split within an internal disulfide loop. The two-chain molecule that results acquires enzymatic properties only after activated PTA goes on to sever a small "activation" polypeptide from the longer of the two chains.

In the test tube, activation of Christmas factor can also be brought about by plasma kallikrein, activated Stuart factor (factor X_a), and factor VII. Two of these agents, activated Stuart factor and factor VII, participate in the extrinsic pathway of thrombin formation (p. 381). Perhaps this explains the benign nature of deficiencies of Hageman factor, PTA, plasma prekallikrein, or high molecular weight kininogen. Activation of the extrinsic pathway may circumvent the steps of the intrinsic pathway before the participation of Christmas factor.

Christmas factor is functionally deficient in the plasma of patients with Christmas disease (hemophilia B). This bleeding syndrome mimics classic hemophilia both in its mode of inheritance and its symptomatology, and it is distinguished from classic hemophilia only by specific assays (p. 382). The affected males in some families synthesize reduced amounts of Christmas factor, and in others, functionally incompetent variants of this protein.

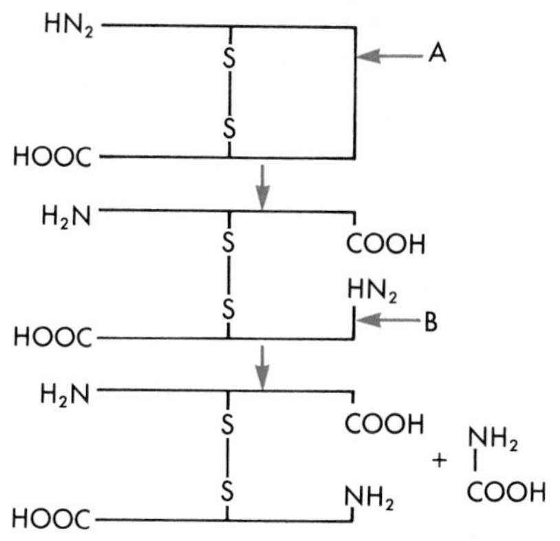

Fig. 25-6. The activation of Christmas factor (factor IX). Activated plasma thromboplastin antecedent (factor XI_a) splits Christmas factor successively at points *A* and *B*. Christmas factor acquires enzymatic properties only with the release of the "activation polypeptide."

Activation of Stuart factor (factor X) via the intrinsic pathway. The role of activated Christmas factor (factor IX_a) is the conversion of Stuart factor to its activated state. Under physiological conditions the activation of Stuart factor depends on the interaction of activated Christmas factor with *antihemophilic factor* (AHF, factor VIII), *phospholipids* (derived principally from platelets but also from plasma), and *calcium ions*. Antihemophilic factor is a complex molecule with a variety of defined properties, and is sometimes designated as factor VIII/VWF (see below). It is functionally deficient in several hereditary bleeding disorders, the most common of which are *classic hemophilia* and *von Willebrand's disease*.

Antihemophilic factor is a complex molecule that is readily dissociated into two subcomponents of unequal size. The larger, high molecular weight subcomponent, synthesized in vascular endothelial cells and megakaryocytes, contains the bulk of the protein of antihemophilic factor. Its molecular weight varies from about 860,000 to 12,000,000 daltons. It is composed of multimers of subunits, each with a molecular weight of about 240,000 daltons. The high molecular weight subcomponent forms precipitates when it is mixed with antiserum raised in goats or rabbits that have been immunized with the antihemophilic factor complex. For this reason it has been described as factor VIIIR:Ag, i.e., an antigen (Ag) related (R) to antihemophilic factor (factor VIII). An antigen is an agent that induces the formation of specific antibodies (p. 367).

The larger subcomponent of antihemophilic factor enhances the adhesion of platelets to subendothelial structures. In this way the larger subcomponent may be important for hemostasis when the subendothelial structures are exposed to the circulating blood by vascular injury. *In vitro,* higher molecular weight species of the larger subcomponent bind to platelets in the presence of the antibiotic ristocetin and cause the platelets to clump ("agglutinate"). This property of the high molecular weight subcomponent of antihemophilic factor has therefore been designated as factor VIII:RCo (RCo representing ristocetin-cofactor). Since this property is deficient in von Willebrand's disease but not in classic hemophilia, it has been also designated as factor VIII:VWF or von Willebrand factor (VWF). Additionally, when normal blood is filtered through a column of glass beads, most of its platelets are retained within the column, a phenomenon that depends on the presence of the large subcomponent of antihemophilic factor.

The second, smaller subcomponent of antihemophilic factor contains the amino acid sequences needed for blood coagulation and has therefore been designated factor VIII:C or, less clearly, as factor VIII. This subcomponent has a molecular weight of about 270,000 daltons, and is probably synthesized mainly in the liver.

The titer of antihemophilic factor is under hormonal control. Exercise, stress, pregnancy and the administration of estrogens, epinephrine, vasopressin or its synthetic derivative, 1-desamino-8-D-arginine vasopressin (DDAVP), all increase the titers both of the lower molecular weight coagulant part of the antihemophilic factor complex (factor VIII:C) and of the higher molecular weight subcomponent (factor VIII:VWF).

Classic hemophilia (hemophilia A) is the most common hereditary disorder of the intrinsic pathway. In this disease the coagulant titer of antihemophilic factor (factor VIII:C) is reduced, but the concentration of antigens related to antihemophilic factor complex (factor VIIIR:Ag) is normal. Similarly, the capacity to agglutinate platelets in the presence of ristocetin (factor VIII:VWF or factor VIII:RCo) is normal.

The severity of classic hemophilia parallels the degree of the defect measured in clotting tests. Within a family, affected individuals are closely similar. In severe hemophilia, the titer of coagulant antihemophilic factor is less than 1% of the titer in the average normal individual. The patient may have repeated bleeding into joints (hemarthrosis), and this process may lead to crippling. Apparently spontaneous bleeding into the skin or soft tissues is frequent, and hematuria (blood in the urine) is an important clinical feature. Injuries and surgical procedures, including dental extraction or tonsillectomy, may result in devastating hemorrhage. Death from exsanguination is unusual and is more likely to come from bleeding into a vital area or infection at the site of bleeding. In milder cases significant bleeding may occur only after injury or surgical procedures. To control bleeding the patient is transfused with fractions of normal plasma rich in antihemophilic factor. Alternatively, in patients with mild disease, hemostasis may be fostered by the intravenous injection of 1-desamino-8-D-arginine vasopressin (DDAVP), which causes a transient increase in the titer of antihemophilic factor (factor VIII).

Classic hemophilia is the prototype of X chromosome–linked disorders. It affects only males, who pass the abnormal gene to their daughters, all of whom are carriers. In turn, half of a carrier's sons have hemophilia, and half of her daughters are carriers. Women who carry the abnormal gene are usually asymptomatic. However, they can be recognized in about 95% of cases because their plasma contains relatively less coagulant antihemophilic factor (factor VIII:C) than antihemophilic factor–related antigen (factor VIIIR:Ag or von Willebrand factor) as detected by an antiserum against the antihemophilic factor complex.

Von Willebrand's disease is a hereditary disorder of both sexes. It is inherited as an autosomal dominant trait in which affected individuals have symptoms suggestive of mild classic hemophilia. In women menorrhagia and bleeding after childbirth may be troublesome. The plasma is deficient in antihemophilic factor.

Stuart factor (factor X) is a vitamin K–dependent plasma proenzyme (p. 372) that is the precursor of activated Stuart factor, the enzyme immediately responsible for the release of thrombin from prothrombin. Stuart factor is activated by proteolytic separation of a polypeptide fragment from one of its two chains. In the intrinsic pathway this cleavage is brought about by a complex of activated Christmas factor (factor IX_a), antihemophilic factor (factor VIII), and calcium ions, all adsorbed to phospholipid micelles. The protease responsible for cleavage of Stuart factor is activated Christmas factor, whereas antihemophilic factor appears to act as a nonenzymatic cofactor.

Activation of Stuart factor (factor X) via the extrinsic pathway. The early events of the intrinsic pathway are short-circuited when plasma is exposed to injured tissues. Such tissues contain one or more powerful agents known generically as tissue thromboplastin (tissue factor). Tissue thromboplastin is found principally in cell membranes of almost every cell, including peripheral blood monocytes; platelets are an important exception. Tissue thromboplastin is a lipoprotein complex composed of heat-stable phospholipids and a heat-labile glycoprotein with a molecular weight that varies from 50,000 to 330,000 daltons, depending on the source of the thromboplastin. Removal of the phospholipid from tissue thromboplastin inactivates its clot-promoting properties, which can be restored by readdition of the phospholipid portion.

The clot-promoting properties of tissue thromboplastin are mediated through its action on a trace plasma protein, *factor VII,* a single-chain protein that is synthesized when vitamin K is available (p. 372). By itself,

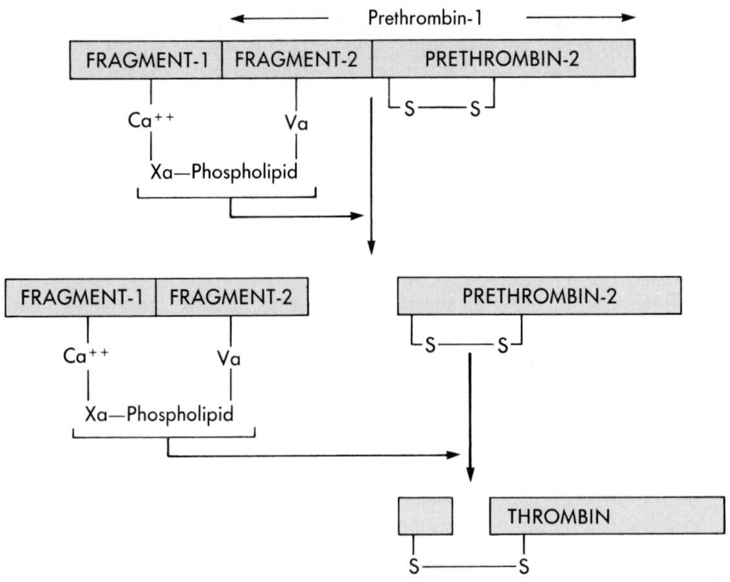

■ **Fig. 25-7.** The release of thrombin from prothrombin.

factor VII has no activity. Complexed stoichiometrically with tissue thromboplastin, it cleaves Stuart factor enzymatically, converting it to its activated state (factor X_a). The subsequent steps of the extrinsic and intrinsic pathways are identical.

As was noted earlier, the factor VII–tissue thromboplastin complex also activates Christmas factor. *Thus reactions of the extrinsic pathway affect the intrinsic pathway and vice versa.*

The generation of thrombin. Thrombin, the protease that is ultimately responsible for the formation of a fibrin clot, is separated from its parent molecule, prothrombin, by activated Stuart factor (factor X_a) through reactions that are augmented by proaccelerin (factor V), calcium ions, and phospholipid.

Prothrombin (factor II) is a vitamin K–dependent glycoprotein (p. 372). The molecule has been dissected enzymatically into several distinct parts (Fig. 25-7). Prothrombin is split by thrombin into an amino-terminal segment, fragment-1, and a carboxy-terminal portion, prethrombin-1. Fragment-1 contains the tricarboxylic glutamic acid residues that serve as points of attachment for calcium ions. The prethrombin-1 segment can be further cleaved by activated Stuart factor into an amino-terminal portion, fragment-2, and a carboxy-terminal fragment, prethrombin-2, that contains an internal disulfide loop.

Thrombin formation proceeds through several stages. The first activated Stuart factor that forms via reactions of the intrinsic or extrinsic pathways binds via calcium links to tricarboxylic glutamic acid residues in fragment-1 of prothrombin (Fig. 25-7). There it slowly cleaves prothrombin in two places, between fragment-2 and prethrombin-2 and within the internal disulfide loop in prethrombin-2. This second split converts prethrombin-2 to thrombin, a molecule with a molecular weight of 39,000 daltons.

As thrombin is liberated it alters proaccelerin (factor V), a glycoprotein that in its native state has no clot-promoting properties, by cleavage of its polypeptide chains. This alteration is also brought about by the activated form of Stuart factor (factor X_a). In this altered form, described as factor V_a, proaccelerin attaches to the fragment-2 portion of as yet unaltered prothrombin molecules. There it is a nonenzymatic accelerator of the action of activated Stuart factor, bringing about a rapid release of thrombin. Both activated Stuart factor and proaccelerin are adsorbed to phospholipid, which appears to augment their activity. Phospholipid for the intrinsic pathway is furnished by platelets and (to a lesser extent) plasma, and for the extrinsic pathway it is furnished by tissue thromboplastin. Thus the events that lead to the activation of Stuart factor and to the formation of thrombin are parallel. Indeed, there are homologies in the amino acid sequences of the coagulant portion of the antihemophilic factor complex (factor VIII:C) and proaccelerin.

Thrombin converts fibrinogen to fibrin monomer (p. 374), activates fibrin-stabilizing factor (factor XIII) (p. 375) and factor VII (p. 380), augments the coagulant actions of antihemophilic factor (factor VIII) (p. 380) and proaccelerin (factor V) (see above), and aggregates platelets (p. 387). Thrombin also binds to fibrin, and in this way fibrin contributes in a minor way to the inactivation of this enzyme.

The role of vascular endothelium in blood coagulation. The endothelium that lines blood vessels has long been thought to provide a surface that lacks the capacity to initiate blood clotting. Recent studies, however, suggest that endothelial cells may under appropriate circumstances influence the clotting process. Thus activated PTA (factor XI_a), Christmas factor (factor IX), activated Christmas factor (factor IX_a), Stuart factor (factor X), thrombin, and antithrombin III (p. 385) can bind to the surface of endothelial cells grown in tissue culture. Furthermore, endothelial cells synthesize the high molecular weight subcomponent of antihemophilic factor (von Willebrand factor or factor VIII:VWF) and proaccelerin (factor V).

The significance of these observations is only now beginning to be understood. When thrombin is added to endothelial cells grown in culture to confluence (i.e., to the point that each cell is in contact with the next) the endothelial cells appear to shrink, inducing gaps between them. If this phenomenon were to occur *in vivo* one might anticipate that the circulating blood could come into contact with subendothelial structures, fostering local thrombosis.

When activated PTA binds to endothelial cells it can activate Christmas factor that has been similarly bound. This can then lead to the formation of fibrin via the reactions of the intrinsic pathway. If instead the endothelial surface is perturbed, as can be brought about by thrombin or by endotoxin (a lipopolysaccharide released by gram-negative bacteria), the endothelial cells acquire tissue thromboplastin-like activity, and they can then interact with factor VII. In this way, endothelial cells bring about formation of fibrin both by reactions of the extrinsic pathway and by activation of Christmas factor that is adsorbed to the endothelial surface. Factor V for these various reactions is furnished by the plasma, by the endothelial cells themselves, and much more effectively by its release from platelets. Thus under appropriate conditions *in vivo* endothelial cells may provide a template for the assembly of clotting factors and in this way promote localized coagulation.

The role of endothelial cells in the initiation of fibrinolysis is described on pp. 382-383.

Laboratory Studies of the Clotting Mechanism

Delineation of the nature of defects of the clotting mechanism is largely a laboratory exercise. The tests in everyday use in clinical laboratories depend on the physiological principles that have been reviewed.

The *whole blood clotting time* measures the period elapsing until blood withdrawn from a vein clots in glass test tubes under standardized conditions. The clotting time reflects the integrity of the intrinsic pathway of thrombin formation. It is abnormally long for patients with deficiencies or qualitative defects of any clotting factor except factor VII and fibrin-stabilizing factor (factor XIII). It is also long if abnormal inhibitors of clotting are present. The clotting time is insensitive to partial deficiencies of clotting factors, and it is now used almost exclusively to measure the anticoagulant effect of therapeutically administered heparin (p. 384).

The *partial thromboplastin time* (PTT) has largely supplanted the whole blood clotting time as a measure of the intrinsic pathway. Blood is drawn into citrate solution to reduce the concentration of the calcium ions needed for coagulation. Plasma separated from citrated blood is mixed with phospholipid in glass tubes. The mixture is then recalcified and the clotting time is measured. The PTT is prolonged under the same conditions as the whole blood clotting time but is much more sensitive to minor abnormalities. Usually kaolin, diatomaceous earth (Celite), or solutions of ellagic acid are added to the mixture to ensure brisk activation of Hageman factor (factor XII). Like the whole blood clotting time, the PTT is normal in deficiencies of factor VII and fibrin-stabilizing factor.

The *prothrombin time* assesses the integrity of the extrinsic pathway. Tissue thromboplastin, usually in the form of rabbit brain tissue, is added to citrated plasma. The mixture is recalcified and the clotting time is measured. The prothrombin time is abnormally long if any of the plasma clotting factors of the extrinsic pathway are deficient or qualitatively defective or if inhibitors of this pathway are present. The prothrombin time is normal in deficiencies of fibrin-stabilizing factor or of the intrinsic pathway components that act before the participation of Stuart factor (factor X).

The *thrombin time* is the clotting time of a mixture of thrombin and plasma. It is abnormally long when the concentration of fibrinogen is abnormally low or when this protein is qualitatively defective. The thrombin time is also lengthened by the presence of endogenous or exogenous inhibitors of the formation of thrombin, e.g., therapeutically administered heparin.

Specific assays for each of the clotting factors are available. Uniquely, the concentration of fibrinogen can be assayed chemically by converting it to fibrin and measuring the protein content of the fibrin. The clotting factors of the extrinsic and intrinsic pathways are assayed by measuring the ability of small amounts of the sample to be tested to correct the abnormal prothrombin time or PTT of plasma that is known to be deficient solely in the specific factor under study.

Fibrinolysis and Related Phenomena

Under appropriate conditions, clotted blood may reliquefy. The agent responsible for fibrinolysis, the dissolution of fibrin, is *plasmin,* a typical serine protease that is the activated form of a plasma glycoprotein, *plasminogen*. Plasminogen is a single-chain polypeptide that is synthesized in the liver and perhaps elsewhere as well. It can be changed to plasmin by many activators, not all of which are physiological.

Streptokinase, a protein elaborated by certain β-hemolytic streptococci, is of particular interest as an activator of plasminogen because it has been used clinically to dissolve thrombi and emboli. Streptokinase combines stoichiometrically with plasminogen to form a complex that transforms plasminogen to plasmin by cleavage within an internal disulfide loop. Staphylococci synthesize a similar activator of plasminogen, staphylokinase.

Streptokinase and staphylokinase are foreign to the body, but many physiological activators also exist. Normal urine, for example, contains a serine protease, *urokinase,* that is elaborated and excreted by the kidney. This enzyme converts plasminogen to plasmin through the same cleavages as the streptokinase-plasminogen complex. Like streptokinase, it has been introduced for the clinical dissolution of thrombi and emboli.

Activators of plasminogen have also been identified in human plasma, milk, tears, saliva, seminal fluid, and many tissues, including vascular endothelium, blood monocytes, and tumor cells. The *tissue plasminogen activators* are serine proteases, and the generation of plasmin is the result of their proteolytic cleavage of plasminogen. The tissue activators of plasminogen fall into two groups, depending on whether or not they resemble urokinase by immunological and functional tests. Those tissue plasminogen activators that are not urokinase-like bind more avidly to fibrin and they lyse clots much more effectively than they hydrolyze fibrinogen. Tissue plasminogen activators are now undergoing clinical trials in the expectation that they will effectively dissolve thrombi.

A tissue plasminogen activator of considerable physiological interest can be demonstrated in blood drawn from a vein distal to a tourniquet that has been applied to the upper arm for several minutes. Clots formed from the plasma of such blood dissolve at an increased rate. Activation of plasminogen under these conditions appears to be initiated by an agent that is released from venous endothelium and that is similar to the nonurokinase-like tissue plasminogen activator.

Fibrinolysis, as measured *in vitro,* is greatly enhanced by emotional stress, strenuous physical activity, or the injection of epinephrine, pyrogens, or vasopressin or its synthetic analogue, 1-desamino-8-D-arginine vasopressin (DDAVP). This phenomenon is probably the consequence of the release of the venous endothelial plasminogen activator. Vascular plasminogen activator can also be released by the vitamin K–dependent plasma protease, activated protein C (p. 385).

Plasma itself can convert plasminogen to plasmin "spontaneously" through one of several pathways. Three enzymes activated through surface-mediated reactions (p. 377), plasma kallikrein, activated PTA (factor XI_a), and (much more weakly) activated Hageman factor (factor XII_a), can activate plasminogen. Thus activation of the intrinsic pathway of thrombin formation initiates reactions that lead to fibrinolysis.

Plasmin is not very specific. It digests not only fibrin but numerous other substances as well, including fibrinogen, antihemophilic factor (factor VIII), proaccelerin (factor V), and other clotting factors. Plasmin may participate in the immune defenses of the body, for it converts the first component of complement (C1) to its enzymatically active state ($C\overline{1}$). It also separates a peptide fragment from the fifth component of complement (C5) that has chemotactic properties (i.e., the capacity to attract leukocytes). It fragments Hageman factor, liberating its enzymatically active carboxy-terminal fragment (HF_f), and it releases kinins from kininogens, both directly and possibly by converting prekallikrein to kallikrein (p. 378). Additionally, plasmin digests casein and certain synthetic amino acid esters and small polypeptide amides, all of which have been used to examine and assay its function.

The most notable property of plasmin is the digestion of fibrinogen and fibrin. In purified systems, plasmin attacks these two proteins with equal avidity. In plasma, however, generation of plasmin occurs more readily on the surface of a clot. One explanation for this phenomenon is that plasminogen and its activators adhere to fibrin. In this situation, plasmin can digest fibrin relatively unhampered by the inhibitors of this enzyme in plasma (p. 384). The practical result is that the therapeutic injection of streptokinase, urokinase, or tissue plasminogen activator may bring about dissolution of intravascular clots without a major effect on circulating fibrinogen. Any plasmin that leaks into the circulation after its activation on the surface of a clot is readily inactivated by its inhibitors in plasma, notably α_2-plasmin inhibitor (p. 384).

The digestion of fibrinogen takes place in steps (Fig. 25-8). First, plasmin separates fragments sequentially from the carboxy-terminal ends of the α-chains and the amino-terminal ends of the β-chains of fibrinogen. The principal residue, fragment X, is still coagulable by thrombin. Fragment X is then further digested by plasmin into fragments designated Y, D, and E. The degradation products of fibrinogen are inhibitors of clotting. These products interfere with the formation and action of thrombin and with the polymerization of fibrin monomers. In part, this inhibition of clotting may be caused by the formation of soluble complexes of fragments X and Y with fibrinogen or fibrin monomer. The fragmentation of fibrin proceeds along similar lines, but fragment X is not coagulable.

Fibrinogen and fibrin can be digested not only by plasmin but also by proteases released by leukocytes.

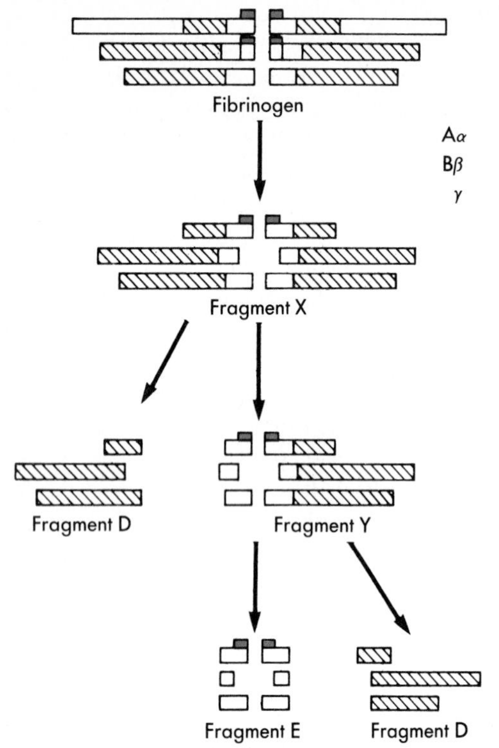

■ **Fig. 25-8.** The digestion of fibrinogen by plasmin. The three chains of fibrinogen are Aα, Bβ, and γ; the solid black areas represent fibrinopeptides A and B. The first cleavage of fibrinogen removes a carboxy-terminal segment of the Aα chain; the residue (fragment X) is still coagulable. Fragment X is then cleaved successively to fragment Y, two fragments D, and fragment E. (Redrawn from Sherry, S.: Mechanisms of fibrinolysis. In Williams, W.J., et al.: editors: Hematology, ed. 2, New York, 1977, McGraw-Hill, Inc. Copyright 1977 by McGraw-Hill Book Co. Used with the permission of McGraw-Hill Book Co.)

The proteases responsible have been identified as elastases, collagenases, and chymotrypsin-like enzymes.

■ *Laboratory Examination of the Fibrinolytic System*

Spontaneous activity of the *fibrinolytic system* of plasma can be assessed by measuring the time required for dissolution of clots formed from whole blood, plasma, or the euglobulin fraction of plasma (i.e., the fraction that is insoluble when plasma is diluted in acidified water). The rate of fibrinolysis reflects the combined effects of plasminogen, activators of plasminogen, and inhibitors of the activation and action of plasmin (p. 383). Tests of fibrinolysis measure what takes place after fibrinogen has been converted to fibrin *in vitro;* their significance *in vivo* is not clear. The plasma content of plasminogen can be estimated by converting this proenzyme to plasmin by addition of urokinase or streptokinase. The enzymatic activity of plasmin can then be assayed on protein or synthetic ester or amide substrates. Inhibitors of plasmin, particularly α_2-plasmin inhibitor (see below), are readily estimated by testing the inhibitory effect of plasma on the fibrinolytic or amidolytic activity of plasmin.

■ *Inhibitors of Clotting and Fibrinolysis*

Plasma is rich in substances that can inhibit *activated* clotting factors and plasmin (Table 25-3). Presumably these serve to restrict the growth of intravascular clots and to minimize the deleterious effects of unbridled fibrinolysis.

Antithrombin III is the principal inhibitor of thrombin and activated Stuart factor (factor X_a). To a lesser extent, it blocks the action of the enzymatically active forms of each of the other plasma serine proteases involved in the formation and dissolution of clots; it is without effect on fibrin-stabilizing factor (factor XIII). Inhibition results from the formation of a stoichiometric complex between antithrombin III and the activated enzymes. The importance of antithrombin III is evident from the frequency of thrombosis in individuals with a partial deficiency of this inhibitor.

The inhibitory properties of antithrombin III are greatly potentiated by the addition of heparin, with which it forms a complex. Heparin is a negatively charged sulfated polysaccharide that is synthesized principally in mast cells. It is found in many tissues, including lung, liver, and intestinal mucosa, but it is not a normal constituent of blood. Without antithrombin III, heparin is not a significant anticoagulant. Heparin is widely used in the prevention and treatment of thrombotic states, in which its enhancement of the action of antithrombin III limits the growth of intravascular clots. Overdosage of heparin is readily controlled by administration of *protamine sulfate,* a highly positively charged polypeptide derived from fish sperm.

Recently a second plasma cofactor for the inhibition of thrombin by heparin has been detected. This agent, variously called *heparin cofactor II* and *heparin cofactor A,* does not react with other proteases of the clotting mechanism.

The major inhibitor of plasmin is *α_2-plasmin inhibitor* (α_2-antiplasmin) with which plasmin combines stoichiometrically and probably covalently in an irreversible link. α_2-Plasmin inhibitor also inhibits the adsorption of plasminogen to fibrin and in this additional way inhibits fibrinolysis. Inhibition of fibrinolysis is enhanced by cross-linkage of α_2-plasmin inhibitor to fibrin by activated fibrin-stabilizing factor (factor

Table 25-3. Molecular weight, concentration, and biological half-life of plasma inhibitors of clotting and fibrinogen

Factor	*Molecular weight (× 1000)*	*Approximate plasma concentration (mg/dl)*	*Biological half-life**	*Deficiency state*
Antithrombin III	62	15-30	45-60 hrs.	Thrombotic tendency
Heparin cofactor II	66	6-12		
α_1-Antitrypsin (α_1 proteinase inhibitor)	55	130	5-6 days	Chronic obstructive pulmonary disease
α_2-Macroglobulin	725	285-350	5 days	
α_2-Plasmin inhibitor	67	5-8	2½ days	Bleeding tendency
C1-esterase inhibitor ($C\bar{1}$-INH)	105	18-25	3-4 days	Hereditary angioneurotic edema
Protein C	62	0.4-0.5	16-18 hrs.	Thrombotic tendency
Protein S	69	1-4	11-21 hrs.	Thrombotic tendency
Histidine-rich glycoprotein	75	9-17	3 days	
Activated protein C inhibitor	57	0.4-0.6		

*See footnote to Table 25-2.

XIII_a). Patients with a hereditary deficiency of this inhibitor have a severe hemorrhagic disorder in which the rapid dissolution of clots by the unchecked action of plasmin impedes hemostasis. α_2-Plasmin inhibitor inactivates the activated forms of Hageman factor, plasma prekallikrein, activated Stuart factor (factor X_a), and thrombin, but these reactions are probably not significant.

Plasma also contains less well-defined *inhibitors of the activation of plasminogen* by tissue plasminogen activators and urokinase. *Histidine-rich glycoprotein* is a plasma protein that inhibits the activation of plasminogen by reducing its binding to fibrin; it also neutralizes the anticoagulant properties of heparin.

α_2-Macroglobulin is a plasma protein that slowly inactivates the proteolytic properties of plasmin, thrombin, and plasma kallikrein. It is a major inhibitor of plasma kallikrein and the chief back-up system for the inhibition of plasmin when supplies of α_2-plasmin inhibitor are exhausted. Its mode of action is unusual. The enzyme binds to and partially digests the inhibitor. As a consequence, the shape of the enzyme is so changed that its ability to attack protein substrates is greatly reduced.

α_1-Antitrypsin (α_1-proteinase inhibitor) is a glycoprotein that has broad specificity and that is synthesized in the liver. Among enzymes that it inactivates in the test tube are activated PTA (factor XI_a), plasmin, and probably thrombin and plasma kallikrein.

C1 esterase inhibitor ($C\bar{1}$-INH) is a plasma protein that inhibits the activated form of the first component of complement ($C\bar{1}$). In *hereditary angioneurotic edema,* a disorder in which the affected individuals have recurrent bouts of soft tissue swelling without pain or itching, the concentration of $C\bar{1}$-INH is very low. $C\bar{1}$-INH also inhibits activated HF, plasma kallikrein, activated PTA, and plasmin. These properties are probably of no clinical importance since patients with hereditary angioneurotic edema do not have defective hemostasis or fibrinolysis.

Protein C (autoprothrombin II-A) is a two-chain vitamin K–dependent plasma proenzyme (p. 373) that can be changed to an enzymatically active state by thrombin and by the thrombin-altered form of proaccelerin (factor V_a). Activation is greatly accelerated on the surface of endothelial cells, which furnish a cofactor, *thrombomodulin,* that binds thrombin to the cell surface (Fig. 25-9). In its activated form protein C inhibits the coagulant properties of antihemophilic factor (factor VIII:C) and proaccelerin (factor V), particularly after their alteration or activation by thrombin. This may explain in part the low titers of factors VIII:C and V in some cases of widespread intravascular coagulation. Activated protein C also enhances fibrinolysis *in vivo* by releasing an activator of plasminogen from endothelial cells. A plasma inhibitor of activated protein C has been described.

Protein S is another vitamin K–dependent protein that forms a complex with activated protein C. This complex enhances inactivation of thrombin-altered proaccelerin (factor V_a).

Platelets

The *platelets* are small, disc-shaped anuclear cells with an average diameter of about 2 to 4 μm. The platelets serve multiple functions in the hemostatic and defense mechanisms of the body. They arise by budding from the cytoplasm of their progenitors, the *megakaryocytes,* which are large polyploid cells with 4, 8, or 16 nuclei and which are derived from the primitive hematopoietic stem cells. In the normal adult, megakaryocytes are found almost exclusively in bone marrow. In the newborn and under certain pathologic conditions in the adult, megakaryocytes are also found in the liver,

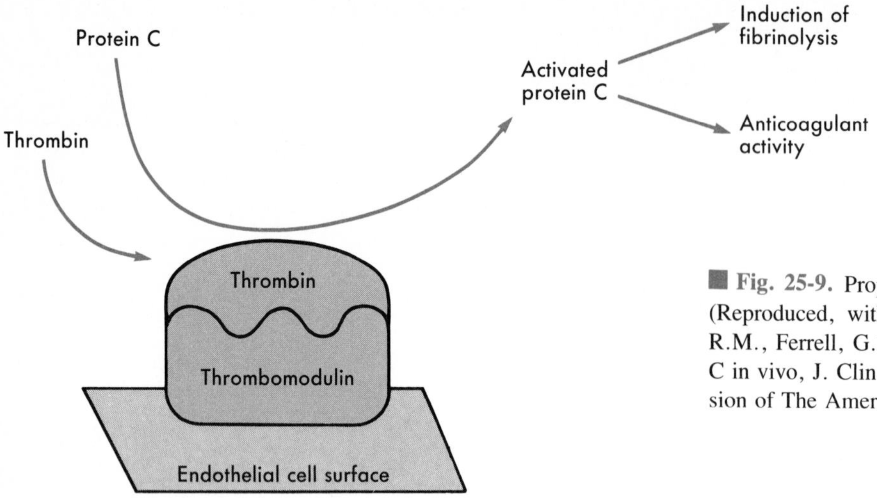

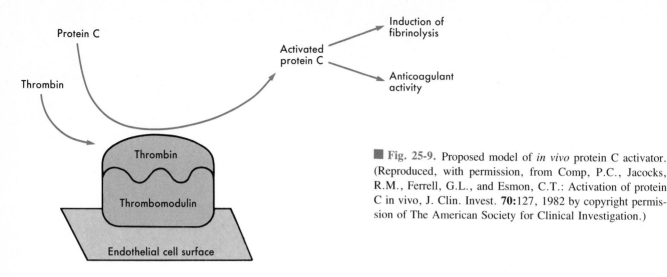

Fig. 25-9. Proposed model of *in vivo* protein C activator. (Reproduced, with permission, from Comp, P.C., Jacocks, R.M., Ferrell, G.L., and Esmon, C.T.: Activation of protein C in vivo, J. Clin. Invest. **70:**127, 1982 by copyright permission of The American Society for Clinical Investigation.)

lungs, and other organs. Maturation of megakaryocytes to the point at which they shed their platelets takes 4 or 5 days; it has been estimated that each megakaryocyte gives rise to 1,000 to 1,500 platelets.

Normal blood contains between 150,000 and 350,000 platelets/μl. Not all platelets are in the circulating blood; as many as a third are sequestered elsewhere, principally in the spleen. The life of an individual platelet is about 8 to 12 days; their destruction appears to be a consequence of senescence. Aged or damaged platelets are removed from the circulation by the reticuloendothelial system. The spleen is an important site of removal, and in individuals whose spleens have been removed the platelet count may be modestly elevated.

Platelet production is probably regulated at least partially by the total number of platelets destroyed rather than by their concentration in circulating blood, as though the metabolites of destroyed platelets stimulate thrombopoiesis. Nonetheless, the signal for platelet production is uncertain. Maturation of megakaryocytes and shedding of their platelets is stimulated by a poorly defined hormone, *thrombopoietin,* but this substance is probably not required for platelet production.

Platelet Structure

The anatomy of the platelets is complex (Fig. 25-10). The exterior coat or glycocalyx is rich in glycoproteins that may be responsible for adhesion of platelets to subendothelial structures and for aggregation of platelets one to another when these cells are appropriately stimulated. Separated from plasma, the platelets have a loose covering of plasma proteins. Among these proteins are fibrinogen, antihemophilic factor (factor VIII), and proaccelerin (factor V). The surface of the platelets has receptors for many agents, including fibrinogen, thrombin, adenosine diphosphate (ADP), catecholamines, serotonin (5-hydroxytryptamine), collagen, and immune complexes (i.e., complexes of antigen and antibody).

Just within the exterior coat is a phospholipid-rich "unit membrane" that is the principal source of clot-promoting phospholipids. This layer is the site of many enzymes, among them, phospholipases, glycogen synthetase, and adenylate cyclase. The innermost layer of the platelet membrane contains fibrillar microtubular structures and microfilaments that contain actin and myosin and that maintain the shape of the platelet.

The cytoplasm of the platelets is equally complex. At its periphery is an "open canalicular system" whose channels appear to be invaginations of the external platelet membranes. Through these tubular structures the contents of platelet granules can be discharged to the exterior of the cell. Scattered through the cytoplasm is a second series of channels, the dense tubular system, which does not connect with the open canalicular system. These structures, which represent smooth endoplasmic reticulum, are probably the sites of prostaglandin synthesis and of calcium ion sequestration. The cytoplasm also contains a potent contractile system that includes an actin-binding protein and proteins that resemble muscle actin and myosin. Within the cytoplasm are the a chains of fibrin-stabilizing factor (factor XIII) and *a* metabolic pool of ATP and ADP that furnishes energy for the cell's metabolism.

Dispersed throughout the cytoplasm are numerous organelles, including small electron-dense bodies, lysosomes, α-granules, mitochondria, glycogen granules, and, inconstantly, a Golgi apparatus (see the box on p. 388).

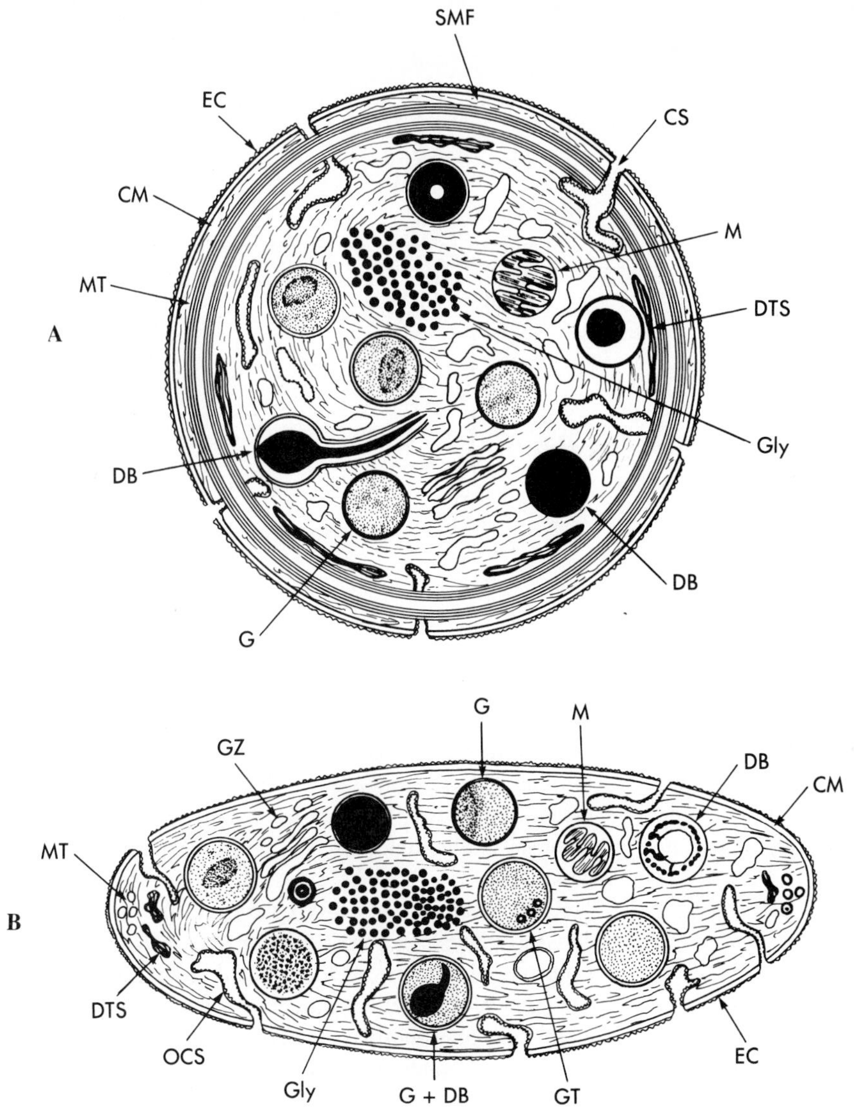

■ **Fig. 25-10.** Diagrammatic representation of a platelet in an equatorial plane **(A)** and in cross section **(B).** Components of the peripheral zone include the exterior coat *(EC),* trilaminar unit membrane *(CM),* and submembrane area containing specialized filaments *(SMF),* which form the platelet wall and line channels of the surface-connected canalicular system *(CS).* The matrix of platelet cytoplasm contains active microfilaments, structured filaments, the circumferential band of microtubules *(MT),* and glycogen *(Gly).* Organelles embedded in the sol-gel zone include mitrochondria *(M),* granules *(G),* and electron-dense bodies *(DB);* some α-granules appear to contain tubular structures *(GT).* The membrane systems include the surface-connected open canalicular system *(CS* or *OCS)* and the dense tubular system *(DIS).* An occasional Golgi apparatus *(GZ)* is found in some platelets. (Redrawn from White, J.G.: Am. J. Clin. Pathol. **71:**363, 1979.)

■ *Platelet Adhesion, Aggregation, and the Release Reaction (Fig. 25-11)*

Within seconds after the endothelial lining of blood vessels is disrupted, some circulating platelets adhere to the exposed subendothelial structures, such as collagen. Factor VIII:VWF (p. 379), which is synthesized in endothelial cells and megakaryocytes, appears to enhance platelet adhesion. The adherent platelets become more spherical. By active contraction that involves polymerization of microfibrillar actin, the platelets send out pseudopods or spicules that spread out like the legs of a spider along the subendothelial fibrils. Simultaneously, contraction of the platelet's microfibrils forces the cytoplasmic granules to migrate toward the center of the cell and to discharge the contents of the granules into the surrounding blood via the open canalicular system. This process is called the *release reaction.* Substances extruded from the granules, including the nonmetabolic pool of ADP in the dense bodies, attract other platelets to those adherent to the injured vessel walls. At the same time, the clotting mechanisms are activated as blood comes into contact with the damaged vascular surfaces. Thrombin, generated locally in this way, promotes further platelet aggregation and release. Thrombin also brings about the formation of strands of fibrin that bind the platelet aggregates into a firm *hemostatic plug* that may control bleeding from small vascular injuries. The plug may also serve as a nidus for a more formidable intravascular clot or thrombus.

How platelets stick to subendothelial structures and to each other is under intensive study. Platelet aggre-

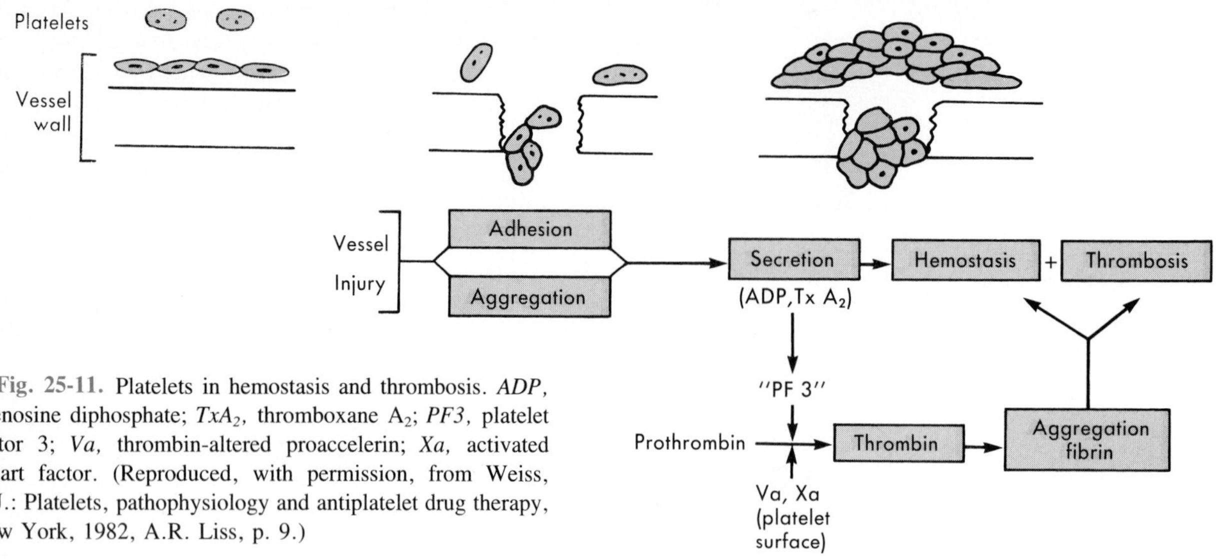

■ **Fig. 25-11.** Platelets in hemostasis and thrombosis. *ADP,* adenosine diphosphate; *TxA*$_2$, thromboxane A_2; *PF3,* platelet factor 3; *Va,* thrombin-altered proaccelerin; *Xa,* activated Stuart factor. (Reproduced, with permission, from Weiss, H.J.: Platelets, pathophysiology and antiplatelet drug therapy, New York, 1982, A.R. Liss, p. 9.)

*Platelet organelles**

Electron dense bodies (dense granules)
- Serotonin
- Catecholamines
- Nonmetabolic "storage pool" of ATP, ADP
- Calcium ions
- Pyrophosphate

α-Granules
- Albumin
- Platelet factor 4
- β-thromboglobulin
- Fibronectin
- Plasminogen
- Platelet fibrinogen
- Platelet-derived growth factor
- High molecular weight kininogen
- α_2-plasmin inhibitor
- Proaccelerin (factor V)
- Antihemophilic factor–related antigen (factor VIIIR:Ag; von Willebrand factor; factor VIII:VWF)
- Thrombospondin
- C1-esterase inhibitor ($C\bar{1}$-INH)

Lysosomes
- Acid hydrolases

Mitochondria

Golgi apparatus (inconstant)

Glycogen granules

*The localization of certain of the substances in the specific granules is only tentative.

gation and the release reaction can be induced in the test tube by addition of such diverse agents as collagen, thrombin, ADP, catecholamines, serotonin, arachidonic acid, aggregated immunoglobulins, antigen-antibody complexes, and antibodies directed against platelets.

Platelet aggregation takes place only if the surrounding milieu contains fibrinogen, which adheres to specific glycoprotein receptors on the surface of platelets that have been stimulated by the aggregation agent. The platelets of subjects who have ingested aspirin within the preceding few days do not undergo the release reaction. Additionally, platelet aggregation is incomplete and reversible. In normal individuals, the effect of aspirin is reflected only by a slight prolongation of the bleeding time (p. 390), but in patients with disorders of hemostasis the tendency to hemorrhage may be enhanced.

Recent studies partially explain the mechanisms underlying platelet aggregation and release (Fig. 25-12). Collagen, thrombin, ADP, or catecholamines bind to receptors in the platelet membrane where they activate membrane phospholipases, perhaps by releasing membrane-bound calcium ions. The activated phospholipases hydrolyze membrane phosphatidylcholine and phosphatidylinositol. Arachidonic acid released from these phosphatides is converted to cyclic endoperoxides (prostaglandins G_2 and H_2) by a cyclo-oxygenase that utilizes molecular oxygen. The endoperoxides are transformed by a microsomal enzyme, thromboxane synthetase, to a short-lived compound, thromboxane A_2, that can aggregate platelets, bring about the release reaction, and constrict small blood vessels. The release of arachidonic acid from membrane phospholipids is blocked by elevated concentrations of cyclic AMP (cAMP). Aspirin inhibits platelet aggregation and release by inactivating the cyclo-oxygenase, which is irreversibly acetylated, preventing the formation of thromboxane A_2. Not yet clear is the relationship between thromboxane A_2 and the release of ADP from platelet granules.

Thromboxane A_2 is one of a family of arachidonate derivatives known collectively as *eicosanoids*. Other eicosanoids of biological importance are *prostacyclin* (prostaglandin I_2) and the *leukotrienes*. Prostacyclin is synthesized by endothelial cells which, like platelets, convert arachidonic acid to cyclic endoperoxides through the action of a cyclo-oxygenase. But in the en-

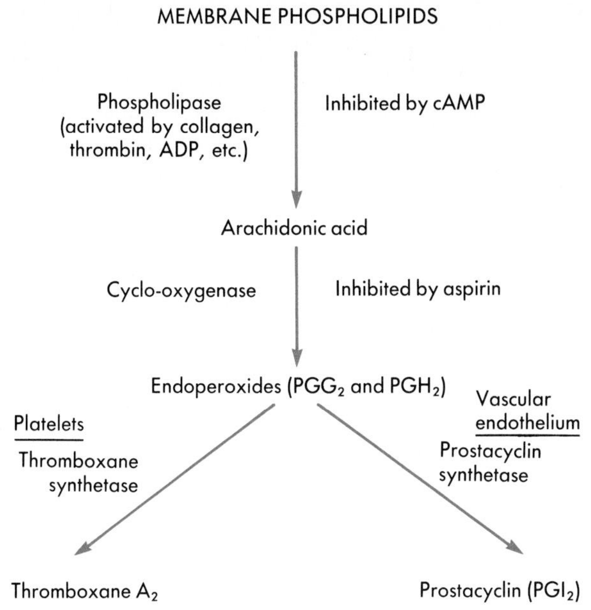

Fig. 25-12. Pathways to the formation of thromboxane A_2 in platelets, and of prostacyclin in vascular endothelial cells.

dothelial cell the cyclic endoperoxides are changed by a microsomal enzyme, prostacyclin synthetase, to prostacyclin (prostaglandin I_2) rather than to thromboxane A_2. One stimulus for prostacyclin synthesis is thrombin. Prostacyclin inhibits platelet aggregation, perhaps by increasing the level of cAMP, and dilates blood vessels. Endothelial cells can use endoperoxides released from stimulated platelets to synthesize prostacyclin. Perhaps this phenomenon limits platelet aggregation at sites of vascular damage. Aspirin, at somewhat higher doses than are needed to block the synthesis of thromboxane A_2 in platelets, inhibits prostacyclin formation, suggesting that this drug may have a double-edged effect.

The third class of eicosanoids, the leukotrienes, are synthesized by granulocytes through the action of cytoplasmic enzymes described as lipoxygenases. The leukotrienes play no direct role in hemostasis, but they are chemotactic (i.e., they bring about migration of leukocytes to the site of their release), they increase vascular permeability locally, leading to the formation of wheals, and they constrict the smallest bronchioles, inducing the bronchoconstriction that is characteristic of bronchial asthma and anaphylaxis (Chapter 24).

Through the thromboxane pathway thrombin brings about aggregation of platelets and discharge of their granular contents. But thrombin also aggregates platelets and induces the release reaction when this pathway is blocked, suggesting that some additional mechanism may be involved, perhaps by the elaboration of another mediator. One candidate for this role is a protein component of the α-granules, *thrombospondin,* which upon stimulation of the platelets by thrombin migrates to the surface of the cells. There thrombospondin can form a complex with fibrinogen, and this complex may serve as a bridge between adjacent platelets. Recent studies suggest, too, that fibrinogen bound to thrombospondin on the platelet membrane may stabilize platelet aggregates that are formed by stimulation of these cells by thrombin or ADP. Thrombospondin is synthesized not only by platelets but also by fibroblasts, monocytes, and endothelial cells. Hence, this protein may play a more general role in the adhesion of cell to cell and of cells to the extracellular matrix.

A minor constituent of the α-granules that is discharged during the release reaction is *fibronectin,* which is also synthesized by fibroblasts, hepatocytes, endothelial cells, and other tissues. Fibronectin is localized to many cellular surfaces where it may serve as a glue that fosters the adhesion of cells to one another. It also serves as an "opsonin" that enhances the engulfment of microorganisms and particulate matter by phagocytes. It may also play an as yet undefined role in promoting platelet aggregation by thrombin. Fibronectin is an important component of the matrix of connective tissue where it is bound to collagen by an enzyme similar to activated fibrin-stabilizing factor (factor XIII_a).

Fibronectin is also present in plasma, where it has been called *cold-insoluble globulin.* Plasma fibronectin may participate during disseminated intravascular coagulation in reactions that inhibit the polymerization of fibrin monomers. During *in vitro* clotting it may be covalently linked to fibrin by activated fibrin-stabilizing factor. Perhaps this property enhances the adhesion of cells to fibrin and in this way promotes wound healing.

Platelets and Blood Coagulation

In the test tube stimulation of platelets by ADP or other agents (or contact of platelets with such negatively charged substances as kaolin) makes available membrane phospholipids (platelet factor 3, PF3) for the intrinsic pathway of thrombin formation. Presumably similar stimulation of platelets occurs *in vivo.* Additionally, activated Stuart factor (factor X_a) binds to receptor sites on the surface of thrombin-stimulated platelets. In this situation activated Stuart factor converts prothrombin to thrombin much more effectively. The receptor for activated Stuart factor has been identified as proaccelerin (factor V) that has been altered to its clot-promoting state by thrombin. The function of platelet fibrinogen is not clear, although this clotting factor makes up as much as 15% of platelet protein. Both proaccelerin and fibrinogen are found in the platelet's α-granules and are secreted during the release reaction.

Platelets may alter coagulation and fibrinolysis through yet another mechanism. As noted above, when platelets are stimulated by thrombin, the α-granules release a protein, thrombospondin. This protein migrates to the surface of these platelets where it can bind heparin, fibrinogen, fibronectin, histidine-rich glycoprotein, plasminogen, and collagen. These properties of thrombospondin may inhibit the anticoagulant properties of heparin and the activation of plasminogen by tissue plasminogen activator.

Clot Retraction

After normal blood coagulates, the clot gradually shrinks, extruding clear serum. This phenomenon of clot retraction depends on the presence of viable, metabolically active platelets. Retraction is initiated by an action of thrombin on these platelets, but the subsequent steps are only vaguely appreciated. Perhaps 15% or more of platelet protein consists of actin and myosin, which are similar but not identical to that in muscle. One construction of the mechanism of clot retraction is that thrombin releases calcium ions from intracellular storage sites. The calcium ions then inactivate regulatory proteins that are similar to muscle troponin. When this takes place, platelet myosin, an adenosine triphosphatase (ATPase), is phosphorylated, and it combines with actin through cross-bridges. The complex then contracts in a manner similar to the contraction of muscle; the energy required comes from the splitting of ATP by myosin ATPase. The contractile process causes the extrusion of pseudopods and migration of granules to the center of the cell. The platelet spicules stick to polymerizing fibrin in the surrounding blood, whereupon the contractile process pulls the fibrin strands toward the platelet. This decreases the volume of the clot so that the entrapped serum is extruded; the fibrin strands themselves are not shortened. Clot retraction draws the platelets together and these cells gradually assume an amorphous appearance that has been described as *platelet metamorphosis,* a term now seldom used. Clot retraction is not inhibited by aspirin and is thus not dependent on the release reaction.

The utility of clot retraction has been puzzling. Perhaps the retraction of fibrin strands can bring together the edges of a wound filled with clot or the shrinkage of the clot allows room for vascular walls to approximate.

Laboratory Studies of Platelet Function

Platelets can be counted with reasonable accuracy in a hemocytometer (counting chamber), but currently they are more likely to be enumerated with an electron particle counter. The latter technique is suitable for most purposes, but it gives erratic results when the platelet count is very low, e.g., 10,000 to 20,000 platelets/μl. Accuracy in this range is critical, however, in patients with severe thrombocytopenia.

The *bleeding time* measures the duration of bleeding from a deliberately incised wound. Currently popular is the use of a template to provide a standardized incision of the forearm, but this technique sometimes leaves permanent scars. This complication can be avoided by measuring the bleeding time in the ball of the finger. The bleeding time is long when the platelet count is abnormally low (< 80,000 platelets/μl) or high (> 800,000 platelets/μl) or the platelets are qualitatively abnormal; the paradox of a long bleeding time in association with high platelet counts is poorly understood. The bleeding time is also long in von Willebrand's disease (p. 380) and in the presence of appreciable amounts of abnormal plasma proteins, as in multiple myeloma.

Clot retraction can be observed qualitatively in blood allowed to clot in glass tubes and incubated for 24 hours. Normally, the clot shrinks, expressing clear serum. Impaired retraction is observed when the platelet count is less than about 80,000 platelets/μl or when the platelets have the qualitative abnormality described as thrombasthenia or Glanzmann's disease.

Platelet aggregation is usually measured by adding the substances to be tested to platelet-rich plasma that is stirred continuously. Aggregation is recognized turbidimetrically by the increase in light transmission that occurs as platelets clump. The aggregating agents usually include collagen, ADP, epinephrine, and thrombin. Aggregation is impaired in many hereditary or acquired qualitative platelet disorders. The same technique is used to assay *platelet agglutination* by ristocetin, a measure of the titer of the von Willebrand factor subcomponent of antihemophilic factor (factor VIII:VWF).

Platelet Disorders

Thrombocytopenia, a reduced number of circulating platelets, may reflect their impaired production, increased destruction or utilization, or sequestration into a space, such as the spleen, outside the effective circulation. The causes of thrombocytopenia are legion. The hemorrhagic tendency is proportional to the degree of thrombocytopenia and is characterized by petechiae, ecchymoses, menorrhagia, and bleeding from mucous membranes and into the central nervous system. Besides the low platelet count, the patients may have a long bleeding time and impaired clot retraction. Differentiation among the causes of thrombocytopenia requires careful examination of the bone marrow to see if megakaryocytes are present; their absence implies a failure of platelet production.

Thrombocytosis, an abnormally high number of circulating platelets, occurs in a wide variety of disorders, among them, iron deficiency anemia, chronic infection, and malignancy. Thrombocytosis is also a feature of some disorders of blood production, such as polycythemia vera, a disease in which the numbers of red and white blood cells are also increased. Paradoxically, when the platelet count is above about 800,000 platelets/μl the patients may experience a bleeding tendency. Under these conditions, described as *thrombocythemia,* the platelets are often qualitatively abnormal.

Many qualitative abnormalities of platelets have been described. In different cases clot retraction is reduced, the release of clot-promoting activity (PF3) from platelets is inadequate, aggregation of platelets by collagen or ADP is impaired, or the electron dense bodies are absent or discharge their ADP poorly. Often combinations of these or other defects are detected. In some instances the platelet abnormality is inherited while in others it is a complication of some other disorder, notably renal failure. The bleeding time is nearly always prolonged.

Thrombosis

A thrombus is a clot within a blood vessel. The conditions that foster the formation of thrombi include damage to the vascular wall, impaired blood flow (stasis), and alterations in the blood composition that render it more readily coagulable. The initial event is probably a disruption of the normal vascular surface that induces localized platelet adhesion and aggregation and localized blood coagulation; such lesions are more evident in arterial than venous thrombosis. Thereafter a blood clot extends from this nidus into the lumen of the vessel. The anatomy of the thrombus is conditioned by its location. In arteries, in which blood flow is swift, the adherent portion of the thrombus is rich in platelets, whereas in veins, in which flow is sluggish, the base of the thrombus may resemble a typical blood clot. In veins, thrombi are particularly likely to begin at the site of valves (a region of stagnant flow).

The degree to which changes in the blood itself foster thrombosis is much debated. Experimentally, the injection of activated clotting factors induces intravascular clots but only in areas in which blood flow has been halted by the subsequent application of ligatures. Thus stasis and the introduction of activated clotting factors foster this type of experimental thrombosis. The role of increased concentrations of clotting factors is uncertain. A familial tendency to recurrent thrombosis has been described in individuals with partial deficiencies of antithrombin III, heparin cofactor II, protein C, or protein S, as if intravascular clotting were fostered by impairment of normal mechanisms that limit the generation or action of thrombin.

Embolism is the process through which thrombi break off from their vascular attachment and are carried by the flowing blood until they lodge in a blood vessel too small to allow their passage. The most frequent example is the often lethal disorder, pulmonary embolism, in which a venous thrombus that originates in the veins of the legs or the pelvis becomes wedged in the pulmonary arteries.

Bibliography

Journal articles

Aoki, N., and Harpel, P.C.: Inhibitors of the fibrinolytic enzyme system, Semin. Thromb. Hemost. **10:**24, 1984.

Bachman, F., and Kruithof, I.E.K.O.: Tissue plasminogen activator: clearance and physiological aspects, Semin. Thromb. Hemost. **10:**6, 1984.

Davie, E.W., et al.: The role of serine proteases in the blood coagulation cascade, Adv. Enzymol. **48:**277, 1979.

Esmon, C.T., editor: Protein C, Semin. Thromb. Hemost. **10:**109, 1984.

Jackson, C.M., and Nemerson, Y.: Blood coagulation, Annu. Rev. Biochem. **49:**765, 1980.

Leung, L.L.K., and Nachman, R.L.: Complex formation of platelet thrombospondin with fibrinogen, J. Clin. Invest. **70:**542, 1982.

Marcus, A.J.: The eicosanoids in biology and medicine, J. Lipid Res. **25:**1511, 1984.

Shattil, S.J., and Bennett, J.S.: Platelets and their membranes in hemostasis: physiology and pathophysiology, Ann. Intern. Med. **94:**108, 1981.

Suttie, J.W.: Vitamin K–dependent carboxylase, Am. Rev. Biochemistry **54:**459, 1985.

Suttie, J.W., and Jackson, C.M.: Prothrombin structure, activation and biosynthesis, Physiol. Rev. **57:**1, 1977.

White, J.G.: Current concepts of platelet structure, Am. J. Clin. Pathol. **71:**373, 1979.

Books and monographs

Gordon, J.L., editor: Platelets in biology and pathology 2, New York, 1981, Elsevier/Science Publishing Co., Inc.

Kline, D.A., and Reddy, K.N.N.: Fibrinolysis, Boca Raton, Florida, 1980, CRC Press, Inc.

Mosher, D.F., and Mosesson, M.W.: Plasma fibronectin: roles in hemostasis, macrophage function and related processes. In Bowie, E.J.W., and Sharp, A.A., editors: Hemostasis and thrombosis, London, 1985, Butterworth & Co., Publishers.

Ogston, D.: The physiology of hemostasis, Cambridge, 1983, Harvard University Press.

Ratnoff, O.D.: The significance of contact activation. In Bowie, E.J.W., and Sharp, A.A., editors: Hemostasis and thrombosis, London, 1985, Butterworth & Co., Publishers.

Ratnoff, O.D., and Forbes, C.D., editors: Disorders of hemostasis, Orlando, Florida, 1984, Grune & Stratton, Inc.

Weiss, H.J.: Platelets: pathophysiology and antiplatelet drug therapy, New York, 1982, Alan R. Liss, Inc.

Zimmerman, T.S., et al.: Factor VIII/von Willebrand factor. In Brown, E.B., editor: Progress in hematology, vol. 13, Orlando, Florida, 1983, Grune & Stratton, Inc.

SECTION

V

THE CARDIOVASCULAR SYSTEM

Robert M. Berne
Matthew N. Levy

CHAPTER

26

The Circuitry

The circulatory, endocrine, and nervous systems constitute the principal coordinating and integrating systems of the body. Whereas the nervous system is primarily concerned with communications and the endocrine glands with regulation of certain body functions, the circulatory system serves to transport and distribute essential substances to the tissues and to remove by-products of metabolism. The circulatory system also shares in such homeostatic mechanisms as regulation of body temperature, humoral communication throughout the body, and adjustments of oxygen and nutrient supply in different physiological states.

The cardiovascular system that accomplishes these chores is made up of a pump, a series of distributing and collecting tubes, and an extensive system of thin vessels that permit rapid exchange between the tissues and the vascular channels. The primary purpose of this section is to discuss the function of the components of the vascular system and the control mechanisms (with their checks and balances) that are responsible for alteration of blood distribution necessary to meet the changing requirements of different tissues in response to a wide spectrum of physiological and pathological conditions.

Before considering the function of the parts of the circulatory system in detail, it is useful to consider it as a whole in a purely descriptive sense. The heart consists of two pumps in series: one to propel blood through the lungs for exchange of oxygen and carbon dioxide (the *pulmonary circulation*) and the other to propel blood to all other tissues of the body (the *systemic circulation*). Unidirectional flow through the heart is achieved by the appropriate arrangement of effective flap valves. Although the cardiac output is intermittent, continuous flow to the periphery occurs by distension of the aorta and its branches during ventricular contraction (*systole*), and elastic recoil of the walls of the large arteries with forward propulsion of the blood during ventricular relaxation (*diastole*). Blood moves rapidly through the aorta and its arterial branches. The branches become narrower in the more peripheral arteries, and their walls become thinner and change histologically. From a predominantly elastic structure, the aorta, the peripheral arteries become more muscular until at the arterioles the muscular layer predominates (Fig. 26-1).

As far out as the beginning of the arterioles, frictional resistance to blood flow is relatively small, and, despite a rapid flow in the arteries, the pressure drop from the root of the aorta to the point of origin of the arterioles is relatively small (Fig. 26-2). The arterioles, the stopcocks of the vascular tree, are the principal points of resistance to blood flow in the circulatory system. The large resistance offered by the arterioles is reflected by the considerable fall in pressure from arterioles to capillaries. Adjustment in the degree of contraction of the circular muscle of these small vessels permits regulation of tissue blood flow and aids in the control of arterial blood pressure.

In addition to a sharp reduction in pressure across the arterioles, a change from pulsatile to steady flow also occurs. The pulsatile arterial blood flow, caused by the intermittency of cardiac ejection, is damped at the capillary level by the combination of distensibility of the large arteries and frictional resistance in the arterioles. Many capillaries arise from each arteriole so that the total cross-sectional area of the capillary bed is very large, despite the fact that the cross-sectional area of each capillary is less than that of each arteriole. As a result, blood flow becomes quite slow in the capillaries, analogous to the decrease in flow rate seen at the wide regions of a river. Since the capillaries consist of short tubes whose walls are only one cell thick, and since flow rate is slow, conditions in the capillaries are ideal for the exchange of diffusible substances between blood and tissue.

On its return to the heart from the capillaries, blood passes through venules and then through veins of increasing size. As the heart is approached, the number

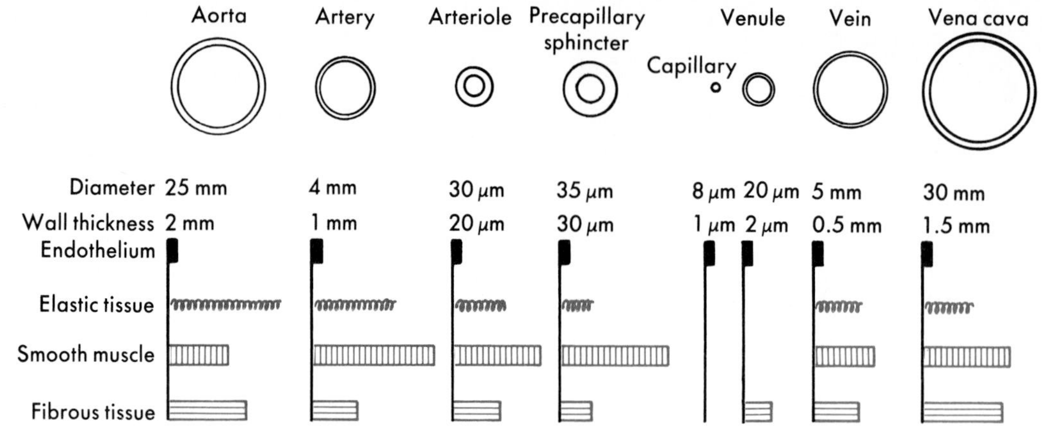

■ **Fig. 26-1.** Internal diameter, wall thickness, and relative amounts of the principal components of the vessel walls of the various blood vessels that compose the circulatory system. Cross sections of the vessels are not drawn to scale because of the huge range from aorta and venae cavae to capillaries. (Redrawn from Burton, A.C.: Physiol. Rev. **34:**619, 1954.)

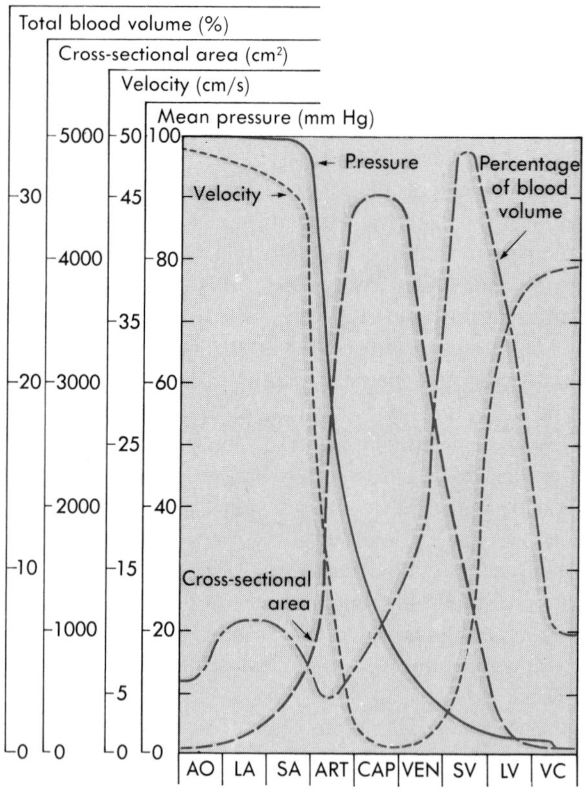

■ **Fig. 26-2.** Pressure, velocity of flow, cross-sectional area, and capacity of the blood vessels of the systemic circulation. The important features are the inverse relationship between velocity and cross-sectional area, the major pressure drop across the arterioles, the maximal cross-sectional area and minimal flow rate in the capillaries, and the large capacity of the venous system. The small but abrupt drop in pressure in the venae cavae indicates the point of entrance of these vessels into the thoracic cavity and reflects the effect of the negative intrathoracic pressure. To permit schematic representation of velocity and cross-sectional area on a single linear scale, only approximations are possible at the lower values. *AO,* Aorta; *LA,* large arteries; *SA,* small arteries; *ART,* arterioles; *CAP,* capillaries; *VEN,* venules; *SV,* small veins; *LV,* large veins; *VC,* venae cavae.

of veins decreases, the thickness and composition of the vein walls change (Fig. 26-1), the total cross-sectional area of the venous channels diminishes, and the velocity of blood flow increases (Fig. 26-2). Also note that most of the circulating blood is located in the venous vessels (Fig. 26-2).

Data from a 20 kg dog (Table 26-1) indicate that the number of vessels increases about 3 billion-fold, and the total cross-sectional area increases about 500-fold from the aorta to the capillaries. The volume of blood in the capillaries is only 5% of the total blood volume, compared to 11% in the aorta, arteries, and arterioles and 67% in the veins and venules. In contrast, blood volume in the pulmonary vascular bed is about equally divided among the arterial, capillary, and venous vessels. The cross-sectional area of the venae cavae is larger than that of the aorta (although this is not evident from Fig. 26-2 because cross-sectional areas of venae

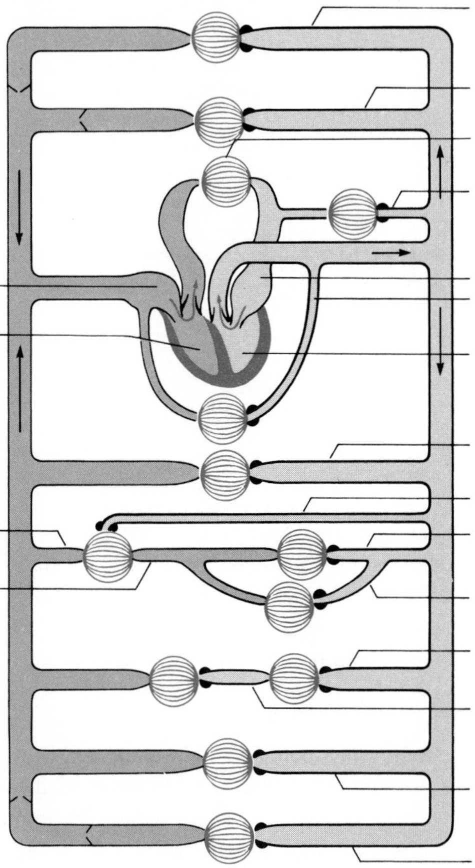

■ **Fig. 26-3.** Schematic diagram of the parallel and series arrangement of the vessels composing the circulatory system. The capillary beds are represented by thin lines connecting the arteries (on the right) with the veins (on the left). The crescent-shaped thickenings proximal to the capillary beds represent the arterioles (resistance vessels). (Redrawn from Green, H.D.: In Glasser, O., editor: Medical physics, vol. 1, Chicago, 1944, Year Book Medical Publishers, Inc.)

■ **Table 26-1.** Vascular dimensions in a 20 kg dog

Vessels	*Number*	*Total cross-sectional area (cm^2)*	*Total blood volume (%)*
Systemic			
Aorta	1	2.8	
Arteries	40-110,000	40	11
Arterioles	2.8×10^6	55	
Capillaries	2.7×10^9	1357	5
Venules	1×10^7	785	
Veins	660,000-110	631	67
Venae cavae	2	3.1	
Pulmonary			
Arteries and arterioles	$1\text{-}1.5 \times 10^6$	137	3
Capillaries	2.7×10^9	1357	4
Venules and veins	$2 \times 10^6\text{-}4$	210	5
Heart			
Atria	2		5
Ventricles	2		

Data from Milnor, W.R.: Hemodynamics, Baltimore, 1982, Williams & Wilkins.

cavae and aorta are so close to zero with a scale that includes the capillaries), and hence the flow is slower than that in the aorta.

The circulatory system is composed of conduits arranged in series and in parallel (Fig. 26-3). Blood entering the right ventricle via the right atrium is pumped through the pulmonary arterial system at a mean pressure of about one seventh of that in the systemic arteries. The blood then passes through the lung capillaries, where carbon dioxide is released and oxygen is taken up. The oxygen-rich blood returns via the pulmonary veins to the left atrium and ventricle to complete the cycle.

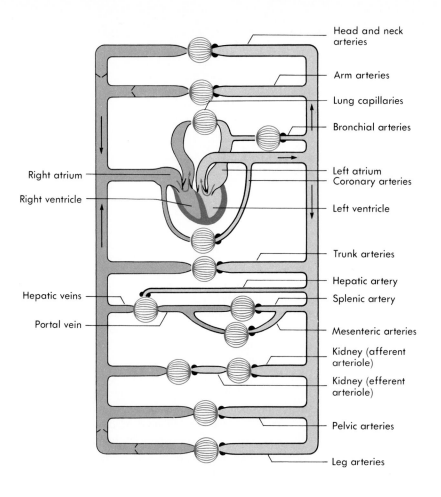

CHAPTER

27

Electrical Activity of the Heart

The experiments on "animal electricity" conducted by Galvani and Volta two centuries ago led to the discovery that electrical phenomena were involved in the spontaneous contractions of the heart. In 1855 Kölliker and Müller observed that when they placed the nerve of a nerve-muscle preparation in contact with the surface of a frog's heart, the muscle twitched with each cardiac contraction. This phenomenon may be observed in the laboratory by allowing the phrenic nerve of an anesthetized dog to lie across the exposed surface of the heart; the diaphragm will contract with each heartbeat. At the end of the last century, the construction of sensitive, high-fidelity galvanometers permitted registration of the changes in electrical potential during the various phases of the cardiac cycle, which led to the science of electrocardiography. Furthermore, progress in electronics and cardiac electrophysiology has led to the development of devices for converting various abnormal cardiac rhythms to normal rhythms and for maintaining normal heart rates in patients with severe conduction blocks (artificial pacemakers).

■ *Transmembrane Potentials*

The electrical behavior of single cardiac muscle cells has been investigated by inserting microelectrodes into the interior of cells from various regions of the heart. The potential changes recorded from a typical ventricular muscle fiber are illustrated schematically in Fig. 27-1. When two electrodes are situated in an electrolyte solution near a strip of quiescent cardiac muscle, no measurable potential difference exists between the two electrodes (from point *A* to point *B*, Fig. 27-1). At point *B* one of the electrodes, a microelectrode with a tip diameter less than 1 μm was inserted into the interior of a cardiac muscle fiber. Immediately the galvanometer recorded a potential difference across the cell membrane, indicating that the potential of the interior of the cell was about 90 mV lower than that of the surrounding medium. Such electronegativity of the interior of the resting cell with respect to the exterior is also characteristic of skeletal and smooth muscle, of nerve, and indeed of most cells within the body (see also Chapter 2). At point *C* a propagated action potential was transmitted to the cell impaled with the microelectrode. Very rapidly the cell membrane became depolarized; actually, the potential difference was reversed (positive overshoot), so that the potential of the interior of the cell exceeded that of the exterior by about 20 mV. The rapid upstroke of the action potential is designated phase 0. Immediately after the upstroke, there was a brief period of partial repolarization (phase 1), followed by a *plateau* (phase 2) that persisted for about 0.1 to 0.2 s. The potential then became progressively more negative (phase 3), until the resting state of polarization was again attained (at point *E*). The process of rapid repolarization (phase 3) proceeds at a much slower rate of change than does the process of depolarization (phase 0). The interval from the completion of repolarization until the beginning of the next action potential is designated phase 4.

The time relationships between the electrical events and the actual mechanical contraction are shown in Fig. 27-2. It can be seen that rapid depolarization (phase 0) precedes tension development and that completion of repolarization coincides approximately with peak tension development. The duration of contraction tends to parallel the duration of the action potential. Also as the frequency of cardiac contraction is increased, the duration of both the action potential and the mechanical contraction decreases.

■ *Principal Types of Cardiac Action Potentials*

Two main types of action potentials are observed in the heart, as shown in Fig. 27-3. One type, the *fast response,* occurs in the normal myocardial fibers in the atria and ventricles and in the specialized conducting

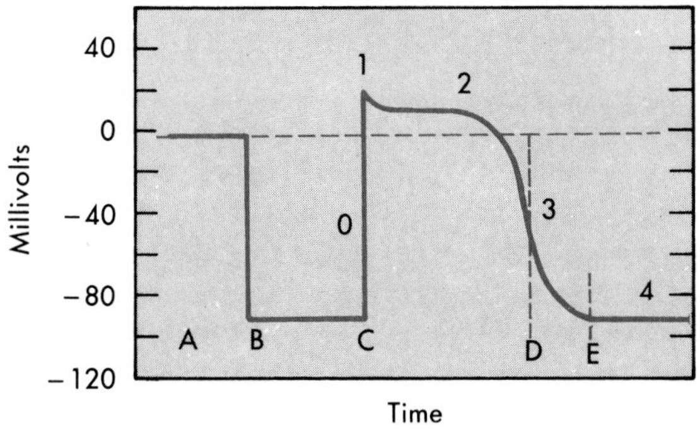

■ Fig. 27-1. Changes in potential recorded by an intracellular microelectrode. From time *A* to *B* the microelectrode was outside the fiber; at *B* the fiber was impaled by the electode. At time *C* an action potential began in the impaled fiber, Time *C* to *D* represents the effective refractory period, and *D* to *E* represents the relative refractory period.

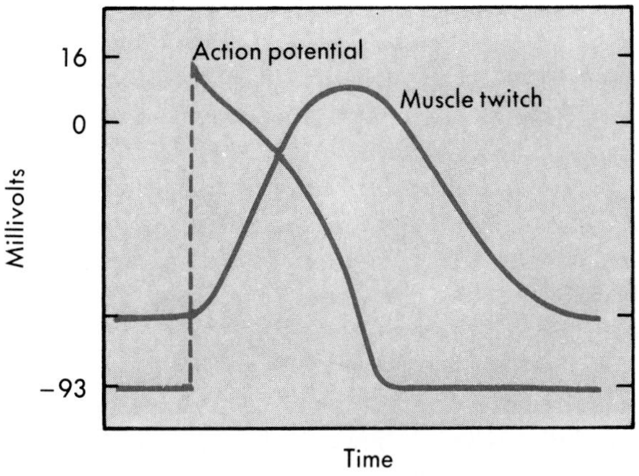

■ Fig. 27-2. Time relationships between the mechanical force developed by a thin strip of ventricular muscle and the changes in transmembrane potential. (Redrawn from Kavaler, F., Fisher, V.J., and Stuckey, J.H.: Bull. N.Y. Acad. Med. **41**:592, 1965.)

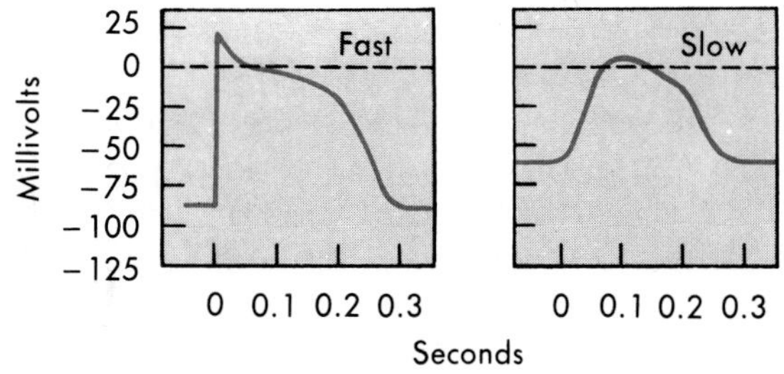

■ Fig 27-3. A fast- and slow-response action potential recorded from the same canine Purkinje fiber. In the left panel the Purkinje fiber bundle was perfused with a solution containing K^+ at a concentration of 4mM. In the right panel epinephrine was added and the K^+ concentration was raised to 16 mM. (Redrawn from Wit, A.L., Rosen, M.R., and Hoffman, B.F.: Am. Heart J. **88**:515. 1974.)

fibers *(Purkinje fibers)* in these chambers. The action potentials shown in Figs. 27-1 and 27-2 are also typical fast responses. The other type of action potential, the *slow response,* is found in the *sinoatrial (SA) node,* the natural pacemaker region of the heart, and the *atrioventricular (AV) node,* the specialized tissue involved in conducting the cardiac impulse from atria to ventricles. Furthermore, fast responses may be converted to slow responses either spontaneously or under certain experimental conditions. For example, in a myocardial fiber, a gradual shift of the resting membrane potential from its normal level of about −90 mV to a value of about −60 mV will convert the subsequent action potentials to the slow response. Such conversions may occur spontaneously in patients with severe coronary artery disease, in those regions of the heart in which the blood supply has been severely curtailed.

As shown in Fig. 27-3, not only is the resting membrane potential of the fast response considerably more negative than that of the slow response, but also the slope of the upstroke (phase 0), the amplitude of the action potential, and the extent of the overshoot of the fast response are greater than in the slow response. It will be explained later that the magnitude of the resting potential is largely responsible for these other distinctions between the fast and slow responses. Furthermore, the amplitude of the action potential and the rate of rise of the upstroke are important determinants of propagation velocity. Hence, in cardiac tissue characterized by the slow response, conduction velocity is much slower, and there is a much greater tendency for impulses to be blocked than in tissues displaying the fast response. Slow conduction and a tendency toward blocking increase the likelihood of certain rhythm disturbances.

■ *Ionic Basis of the Resting Potential*

The various phases of the cardiac action potential are associated with changes in the permeability of the cell membrane, mainly to sodium (Na^+), potassium (K^+), and calcium (Ca^{++}) ions. Changes in cell membrane permeability alter the rate of ion passage across the membrane. The permeability of the membrane to a given ion defines the net quantity of the ion that will diffuse across each unit area of the membrane per unit concentration difference across the membrane.

Just as with all other cells in the body (see also Chapter 2), the concentration of potassium ions inside a cardiac muscle cell, $[K^+]_i$, greatly exceeds the concentration outside the cell, $[K^+]_o$, as shown in Fig. 27-4. The reverse concentration gradient exists for Na ions and for unbound Ca ions. Estimates of the extra-cellular and intracellular concentrations of Na^+, K^+, and Ca^{++}, and of the equilibrium potentials (defined below) for these ions, are compiled in Table 27-1. The

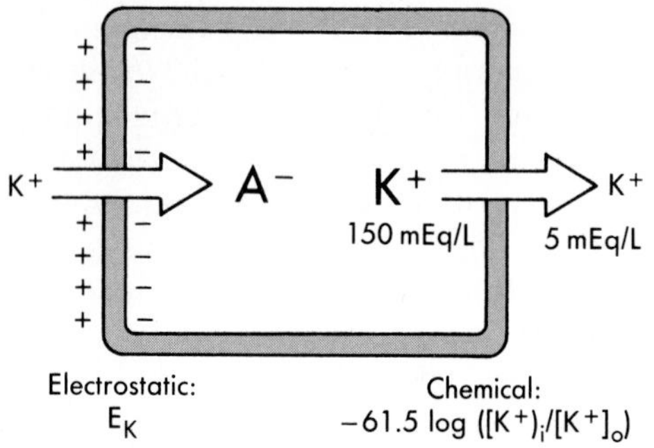

Fig. 27-4. The balance of chemical and electrostatic forces acting on a resting cardiac cell membrane, based on a 30:1 ratio of the intracellular to extracellular K^+ concentrations, and the existence of a nondiffusible anion (A^-) inside but not outside the cell.

Table 27-1. Intracellular and extracellular ion concentrations and equilibrium potentials in cardiac muscle cells

Ion	*Extracellular concentrations (mM)*	*Intracellular concentrations (mM)**	*Equilibrium potential (mV)*
Na^+	145	10	70
K^+	4	135	−94
Ca^{++}	2	10^{-4}	132

Modified from Ten Eick, R.E., Baumgarten, C.M., and Singer, D.H.: Prog. Cardiovasc. Dis. **24:**157, 1981.

*The intracellular concentrations are estimates of the free concentrations in the cytoplasm.

resting cell membrane is relatively permeable to K^+, but much less so to Na^+ and Ca^{++}. Because of the high permeability to K^+, there tends to be a net diffusion of K^+ from the inside to the outside of the cell, in the direction of the concentration gradient, as shown on the right side of the cell in Fig. 27-4. Many of the anions (labeled A^-) inside the cell, such as the proteins, are not free to diffuse out with the K^+. Therefore as the K^+ diffuses out of the cell and leaves the A^- behind, the deficiency of cations causes the interior of the cell to become electronegative, as shown on the left side of the cell in Fig. 27-4.

Therefore two opposing forces are involved in the movement of K^+ across the cell membrane. A chemical force, based on the concentration gradient, results in the net outward diffusion of K^+. The counterforce is an electrostatic one; the positively charged K ions are attracted to the interior of the cell by the negative potential that exists there. If the system came into equilibrium, the chemical and the electrostatic forces would be equal. This equilibrium is expressed by the Nernst equation (Chapter 2) for potassium:

$$E_K = -61.5 \log ([K^+]_i/[K^+]_o) \qquad (1)$$

The right-hand term represents the chemical potential difference at the body temperature of 37° C. The left-hand term, E_K, represents the electrostatic potential difference that would exist across the cell membrane if K^+ were the only diffusible ion. E_K is called the *potassium equilibrium potential*. When the measured concentrations of $[K^+]_i$ and $[K^+]_o$ for mammalian myocardial cells are substituted into the Nernst equation, the calculated value of E_K equals about −94 mV (Table 27-1). This value is close to, but slightly more negative than, the resting potential actually measured in myocardial cells. Therefore a small potential of about 10 to 15 mV tends to drive K^+ out of the resting cell.

The balance of forces acting on the Na ions is entirely different from that acting on the K ions in resting cardiac cells. The intracellular Na^+ concentration, $[Na^+]_i$, is much lower than the extracellular concentration, $[Na^+]_o$. At 37° C, the *sodium equilibrium potential,* E_{Na}, expressed by the Nernst equation, is:

$$-61.5 \log ([Na^+]_i/ [Na^+]_o) \qquad (2)$$

For cardiac cells, E_{Na} is about 40 to 70 mV (Table 27-1). At equilibrium, therefore, an electrostatic force of 40 to 70 mV, oriented with the inside of the cell more positive than the outside, would be necessary to counterbalance the chemical potential for Na^+. However, the actual polarization of the resting cell membrane is just the opposite. The resting membrane potential of myocardial fibers is about −90 mV. Hence both chemical and electrostatic forces act to pull extracellular Na^+ into the cell. The influx of Na^+ through the cell membrane is small, however, because the permeability of the resting membrane to Na^+ is very low. Nevertheless, it is mainly this small inward current of positively charged Na ions that causes the potential on the inside of the resting cell membrane to be slightly less negative than the value predicted by the Nernst equation for K^+.

The steady inward leak of Na^+ would cause a progressive depolarization of the resting cell membrane were it not for the metabolic pump that continuously extrudes Na^+ from the cell interior and pumps in K^+. The metabolic pump involves the enzyme, Na^+, K^+-activated ATPase, which is located in the cell membrane itself (Chapter 1). Because the pump must move Na^+ against both a chemical and an electrostatic gradient, operation of the pump requires the expenditure of metabolic energy. Increases in $[Na^+]_i$ or in $[K^+]_o$ accelerate the activity of the pump. The quantity of Na^+ extruded by the pump slightly exceeds the quantity of K^+ transferred into the cell. Therefore the pump itself tends to create a potential difference across the cell membrane, and thus it is termed an *electrogenic pump*. If the pump is partially inhibited, as by large doses of digitalis, the concentration gradients for Na^+ and K^+

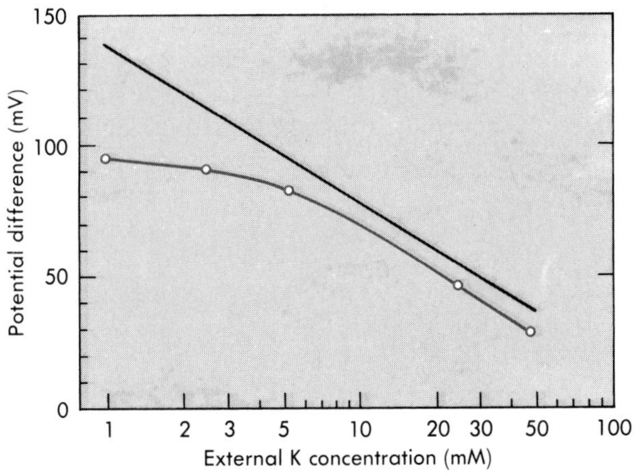

■ Fig. 27-5. Transmembrane potential of a cardiac muscle fiber varies inversely with the potassium concentration of the external medium *(colored curve)*. The straight black line represents the change in transmembrane potential predicted by the Nernst equation for E_K. (Redrawn from Page, E.: Circulation **26**:582, 1962. By permission of the American Heart Association, Inc.)

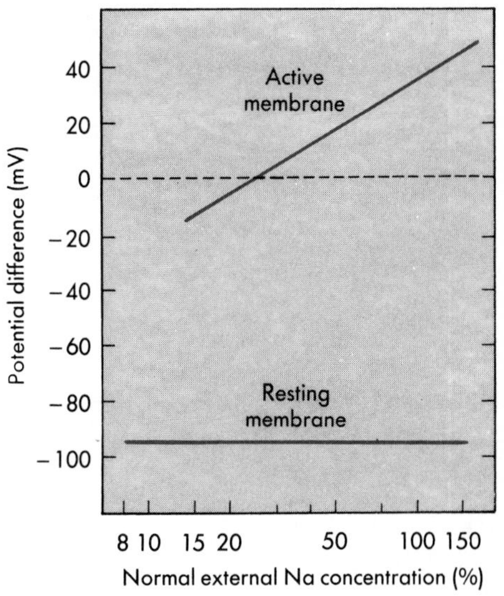

■ Fig. 27-6. Concentration of sodium in the external medium is a critical determinant of the amplitude of the action potential in cardiac muscle *(upper curve)* but has relatively little influence on the resting potential *lower curve)*. (Redrawn from Weidmann, S.: Elektrophysiologie der Herzmuskelfaser, Bern, 1956, Verlag Hans Huber.)

are partially dissipated, and the resting membrane potential becomes less negative than normal.

The dependence of the transmembrane potential, V_m, on the intracellular and extracellular concentrations of K^+ and Na^+ and on the conductances (g_K and g_{Na}) of these ions is described by the chord conductance equation:

$$V_m = \frac{g_K}{g_K + g_{Na}} E_K + \frac{g_{Na}}{g_K + g_{Na}} E_{Na} \qquad (3)$$

For a given ion (X), the conductance (g_X) is defined as the ratio of the current (i_X) carried by that ion to the difference between the V_m and the Nernst equilibrium potential (E_X) for that ion; i.e.,

$$g_X = \frac{i_X}{V_m - E_X} \qquad (4)$$

The chord conductance equation reveals that the relative, not the absolute, conductances to Na^+ and K^+ determine the resting potential. In the resting cardiac cell, g_K is about 100 times greater than g_{Na}. Therefore the chord conductance equation for the resting cell essentially reduces to the Nernst equation for K^+.

When the ratio $[K^+]_i/[K^+]_o$ is decreased experimentally by raising $[K^+]_o$, the measured value of V_m (dashed line, Fig. 27-5) approximates that predicted by the Nernst equation for K^+ (continuous line). For extracellular K^+ concentrations of about 5 mM and above, the measured values correspond closely with the predicted values. The measured levels are slightly less than those predicted by the Nernst equation because of the small but finite value of g_{Na}. For values of $[K^+]_o$ below about 5 mM, g_K decreases as $[K^+]_o$ is diminished. As g_K decreases, the effect of the Na^+ gradient on the transmembrane potential becomes relatively more important, as predicted by the chord conductance equation. This change in g_K accounts for the greater deviation of the measured V_m from that predicted by the Nernst equation for K^+ at low levels of $[K^+]_o$ (Fig. 27-5). Also, in accordance with the chord conductance equation, changes in $[Na^+]_o$ have relatively little effect on resting V_m (Fig. 27-6) because of the low value of g_{Na}.

■ *Ionic Basis of the Fast Response*

Genesis of the upstroke. Any process that abruptly changes the resting membrane potential to a critical value (called the *threshold*) will result in a propagated action potential. The characteristics of fast-response action potentials resemble those shown on the left side of Fig. 27-3. The rapid depolarization that takes place during phase 0 is related almost exclusively to the inrush of Na^+ by virtue of a sudden increase in g_{Na}. The amplitude of the action potential (the magnitude of the potential change during phase 0) varies linearly with the logarithm of $[Na^+]_o$, as shown in Fig. 27-6. When $[Na^+]_o$ is reduced from its normal value of about 140 mM to about 20 mM, the cell is no longer excitable.

The physical and chemical forces responsible for these transmembrane movements of Na^+ are explained

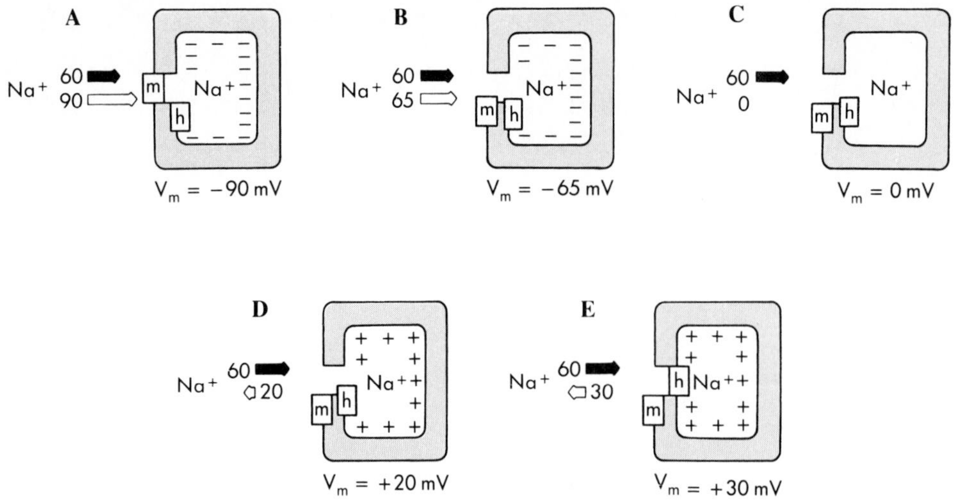

Fig. 27-7. The gating of a sodium channel in a cardiac cell membrane during phase 4 (panel A) and during various stages of phase 0, when the transmembrane potential (V_m) had attained values of −65 mV (panel **B**), 0 mV (panel **C**), 20 mV (panel **D**), and 30 mV (panel **E**). The positions of the *m* and *h* gates in the fast Na^+ channels are shown at the various levels of V_m. The electrostatic forces are represented by the white arrows, and the chemical (diffusional) forces by the black arrows.

in Fig. 27-7. When the resting membrane potential, V_m, is suddenly changed to the threshold level of about −60 to −70 mV, the properties of the cell membrane change dramatically; it becomes much more permeable to Na^+.

Na^+ moves through *fast* Na^+ *channels* in a manner which suggests that the flux is controlled by two types of "gates." One of these gates, the *m* gate, tends to open the channel as V_m becomes less negative and is therefore called an *activation gate*. The other, the *h* gate, tends to close the channel as V_m becomes less negative and hence is called an *inactivation gate*. The *m* and *h* designations were originally employed by Hodgkin and Huxley in their mathematical model of conduction in nerve fibers (Chapter 3).

With the cell at rest, V_m is about −90 mV. At this level, the *m* gates are closed and the *h* gates are wide open, as shown in Fig. 27-7, *A*. The concentration of Na^+ is much greater outside than inside the cell, and the interior of the cell is electrically negative with respect to the exterior. Hence, both chemical and electrostatic forces are oriented to draw Na^+ into the cell. The electrostatic force in Fig. 27-7, *A*, is a potential difference of 90 mV, and it is represented by the white arrow. The chemical force, based on the difference in Na^+ concentration between the outside and inside of the cell, is represented by the black arrow. For a Na^+ concentration difference of about 130 mM, a potential difference of 60 mV (inside more positive than outside) would be necessary to counterbalance the chemical, or diffusional, force, according to the Nernst equation for Na^+. Therefore we may represent the net chemical force favoring the inward movement of Na^+ in Fig. 27-7 (black arrows) as being equivalent to a potential of 60 mV. With the cell at rest the total electrochemical force favoring the inward movement of Na^+ is 150 mV (panel *A*). The *m* gates are closed, however, and the conductance of the resting cell membrane to Na^+ is very low. Hence, virtually no Na^+ moves into the cell; that is, the *inward* Na^+ *current* is negligible.

Any process that makes V_m less negative tends to open the *m* gates, thereby "activating" the fast Na^+ channels. The activation of the fast channels is therefore called a *voltage-dependent* phenomenon. The potential at which the *m* gates swing open varies somewhat from channel to channel in the cell membrane. As V_m becomes progressively less negative, more and more *m* gates will open. As the *m* gates open, Na^+ enters the cell (Fig. 27-7, *B*) by virtue of the chemical and electrostatic forces referred to before.

The entry of positively charged Na^+ into the interior of the cell tends to neutralize some of the negative charges inside the cell and thereby tends to diminish further the transmembrane potential, V_m. The resultant reduction in V_m, in turn, opens more *m* gates, thereby further increasing the inward Na^+ current. This is called a *regenerative process*. As V_m approaches the threshold value of about −65 mV, the remaining *m* gates rapidly swing open in the fast Na^+ channels until virtually all of the *m* gates are open (Fig. 27-7, *B*).

The rapid opening of the *m* gates in the fast Na^+ channels is responsible for the large and abrupt increase in Na^+ conductance, g_{Na}, coincident with phase 0 of the action potential (Fig. 27-8). The rapid influx of

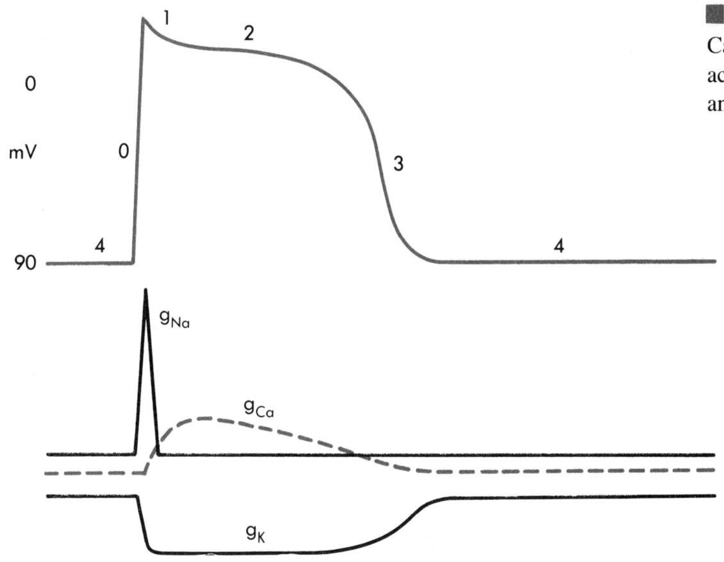

Fig. 27-8. Changes in the conductances of Na^+ (g_{Na}), Ca^{++} (g_{Ca}), and K^+ (g_k) during the various phases of the action potential of a fast-response cardiac cell. The conductance diagram shows directional changes only.

Na^+ accounts for the rapid rate of change of V_m during phase 0. The maximum rate of change of V_m (that is, the maximum dV_m/dt) varies from 100 to 200 V/s in myocardial cells and from 500 to 1000 V/s in Purkinje fibers. Although the quantity of Na^+ that enters the cell during one action potential alters V_m by over 100 mV, that quantity is too small to change the intracellular Na^+ concentration measurably. The chemical force remains virtually constant, and only the electrostatic force changes appreciably throughout the action potential. Hence, the lengths of the black arrows in Fig. 27-7 remain constant at 60 mV, whereas the white arrows change in magnitude and direction.

As the Na^+ rushes into the cardiac cell during phase 0, the negative charges inside the cell are neutralized, and V_m becomes progressively less negative. When V_m becomes zero (Fig. 27-7, *C*), there is no longer an electrostatic force pulling Na^+ into the cell. As long as the fast Na^+ channels are open, however, Na^+ continues to enter the cell because of the large concentration gradient that still exists. This continuation of the inward Na^+ current causes the inside of the cell to become positively charged with respect to the exterior of the cell (Fig. 27-7, *D*). This reversal of the membrane polarity is the so-called overshoot of the cardiac action potential, which is evident in Fig. 27-1. Such a reversal of electrostatic gradient would, of course, tend to repel the entry of Na^+ (Fig. 27-7, *D*). However, as long as the inwardly directed chemical forces exceed these outwardly directed electrostatic forces, the net flux of Na^+ will still be inward, although the rate of influx will be diminished.

The inward Na^+ current finally ceases when the *h* (inactivation) gates close (Fig. 27-7, *E*). The activity of the *h* gates is governed by the value of V_m just as is that of the *m* gates. However, whereas the *m* gates tend to open as V_m becomes less negative, the *h* gates tend to close under this same influence. The opening of the *m* gates occurs very rapidly (in about 0.1 to 0.2 ms), whereas the closure of the *h* gates is relatively slow, requiring 1 ms or more. Phase 0 is finally terminated when the *h* gates have closed and have thereby "inactivated" the fast Na^+ channels. The closure of the *h* gates so soon after the opening of the *m* gates accounts for the quick return of g_{Na} to near its resting value (Fig. 27-8).

The *h* gates then remain closed until the cell has partially repolarized during phase 3 (at about point *D* in Fig. 27-1). Until these gates do reopen partially, the cell is refractory to further excitation. This mechanism prevents a sustained, tetanic contraction of cardiac muscle, which would of course be inimical to the intermittent pumping action of the heart.

Statistical characteristics of the "gate" concept. The recent development of the patch-clamping technique has made it possible to measure ionic currents through single membrane channels. The individual channels have been observed to open and close repeatedly in a quasirandom sequence. This process is illustrated in Fig. 27-9, which shows the current flow through single Na^+ channels in a cultured myocardial cell. To the left of the arrow, the membrane potential was clamped at −85 mV. At the arrow, the potential was suddenly changed to −45 mV, at which value it was held for the remainder of the record.

Fig. 27-9 indicates that immediately after the membrane potential was made less negative, one Na^+ channel opened three times in sequence. It remained open for about 2 or 3 ms each time and was closed for about 4 or 5 ms between openings. In the open state, it allowed 1.5 pA of current to pass. During the first and second openings of this channel, a second channel also

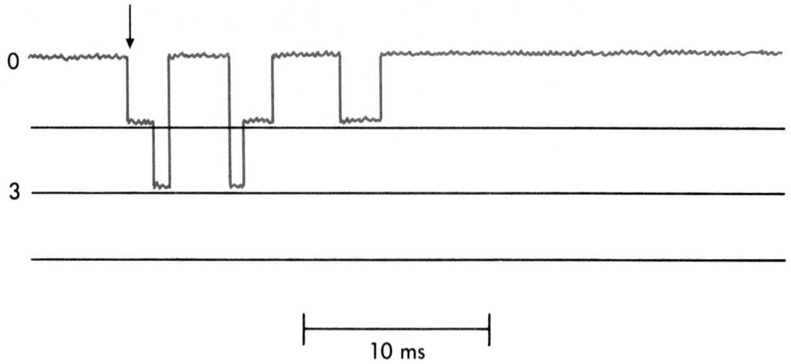

■ **Fig. 27-9.** The current flow (in picoamperes) through two individual Na^+ channels in a cultured cardiac cell, recorded by the patch-clamping technique. The membrane potential had been held at −85 mV, but was suddenly changed to −45 mV at the arrow and held at this potential for the remainder of the record. One channel opened three times for about 3 ms each time, which allowed about 1.5 pA of current to flow through the membrane. During the first and second openings, a second channel opened for about 1 ms each time, permitting an additional 1.5 pA of current to flow. (Redrawn from Cachelin, A.B., DePeyer, J.E., Kokubun, S., and Reuter, H.: J. Physiol. **340:**389, 1983.)

opened, but for periods of only 1 ms. During the brief times that both channels were open simultaneously, the total current was 3 pA. After the first channel closed for the third time, both channels remained closed for the rest of the recording, even though the membrane was held at a constant level of relative depolarization.

The overall change in ionic conductance of the entire cell membrane at any given time reflects the number of channels that happen to be open at that time. Since the individual channels open and close in an irregular pattern, the overall membrane conductance represents the statistical probability of the open or closed state of the individual channels. The temporal characteristics of the activation process then represent the time course of the increasing probability that the specific channels will be open, rather than the kinetic characteristics of the activation gates in the individual channels. Similarly, the temporal characteristics of inactivation reflect the time course of the decreasing probability that the channels will be open and not the kinetic characteristics of the inactivation gates in the individual channels.

Genesis of the plateau. In cardiac cells that have a prominent plateau and especially in Purkinje fibers, phase 1 constitutes an early, brief period of limited repolarization between the end of the upstroke and the beginning of the plateau. Phase 1 mainly reflects the initial influence of the inactivation of the fast sodium channels.

During the plateau (phase 2) of the action potential, there is an influx mainly of Ca^{++}, but also of Na^+, into the cell through *slow channels*. The entry of these ions into the cell through the slow channels has been called the *slow inward current*. The slow channels appear to be entirely different from the fast Na^+ channels that are open during phase 0. The activation, inactivation, and recovery processes are much slower for the slow than for the fast channels. The fast channels may be blocked by tetrodotoxin, whereas the slow channels may be blocked by manganese ions (Mn^{++}) or verapamil, agents that are known to impede the movement of Ca^{++} into the cell.

The slow channels are activated when V_m reaches their threshold voltage of about −30 to −40 mV. Opening of these channels is reflected by an increase in Ca^{++} conductance (g_{Ca}), which begins shortly after the upstroke of the action potential (Fig. 27-8). At the beginning of the action potential, the intracellular Ca^{++} concentration is much less than the extracellular concentration (Table 27-1). Consequently, with the increase in g_{Ca}, there is an influx of Ca^{++} into the cell throughout the plateau. The Ca^{++} that enters the myocardial cell during the plateau is involved in *excitation-contraction coupling,* as described in Chapters 22 and 28.

Various factors may influence the slow inward current. This current may be increased by catecholamines, such as epinephrine and norepinephrine. This is probably the principal mechanism by which catecholamines enhance cardiac muscle contractility. Conversely, drugs such as verapamil, nifedipine, and diltiazem impede the slow inward current; such drugs are known as *calcium channel blocking agents*. By reducing the amount of Ca^{++} that enters the myocardial cells during phase 2, these drugs diminish the strength of the cardiac contraction (Fig. 27-10).

During the plateau of the action potential, the concentration gradient for K^+ between the inside and outside of the cell is virtually the same as it is during phase 4, but the V_m is close to 0 mV. Therefore the chemical forces acting on K^+ greatly exceed the electrostatic forces during the plateau, and K^+ tends to diffuse out of the cell. The efflux of the positively charged ion

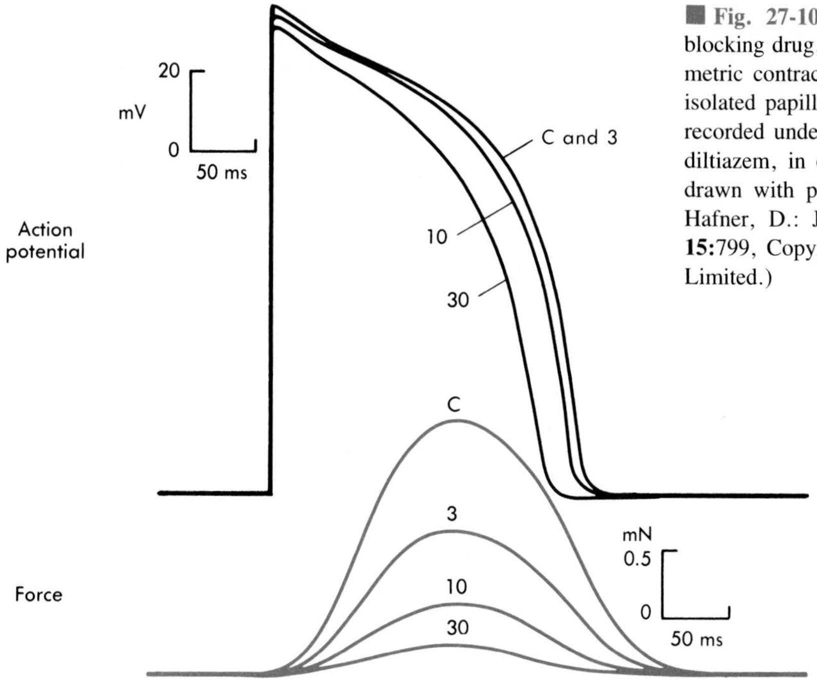

Fig. 27-10. The effects of diltiazem, a Ca^{++} channel blocking drug, on the action potentials (in millivolts) and isometric contractile forces (in millinewtons) recorded from an isolated papillary muscle of a guinea pig. The tracings were recorded under control conditions *(C)* and in the presence of diltiazem, in concentrations of 3, 10, and 30 μmol/L. (Redrawn with permission from Hirth, C., Borchard, U., and Hafner, D.: Journal of Molecular and Cellular Cardiology **15:**799, Copyright 1983 by Academic Press, Inc. [London], Limited.)

tends to make the interior of the cell membrane more negative, i.e., it tends to repolarize the cell membrane, thereby terminating the plateau. In nerve fibers, g_K increases when the neuron is depolarized, and the resultant outward current of K^+ promotes rapid repolarization (Chapter 3). Conversely, in cardiac cells the $g_K{}^+$ of the cell membrane in the outward direction diminishes during phase 2 (Fig. 27-8), although the inward K^+ conductance is considerably greater. This unidirectional decrease in g_K during the plateau has been called *anomalous rectification.*

As a consequence of the reduction in g_K in the outward direction, there is only a small outward current of K^+ during the plateau. It tends to balance the slow inward currents of Ca^{++} and Na^+ and thereby helps maintain a prolonged plateau at a V_m close to 0 mV. The effects of altering this balance between the inward currents of Ca^{++} and Na^+ and the outward current of K^+ are demonstrated by the administration of a calcium channel blocking drug. Fig. 27-10 shows that with increasing concentrations of diltiazem, the voltage of the plateau becomes less positive, and the plateau does not persist as long, i.e., the duration of the action potential diminishes.

Genesis of repolarization. The process of final repolarization (phase 3) depends on two principal processes: (1) an increase in g_K (Fig. 27-8) and (2) inactivation of the slow inward Ca^{++} and Na^+ currents. The increase in g_K is induced in part by the elevation in intracellular Ca^{++}, consequent to the inward Ca^{++} current during the plateau. The enhancement of g_K leads to an efflux of K^+ from the cell; the outward current of K^+ is no longer balanced by the slow inward currents of Ca^{++} and Na^+. This efflux of K^+ therefore causes the charge on the inside of the cell membrane to become progressively more negative. The increase in g_K is voltage dependent; as the inside of the cell becomes more negative, g_K increases and the outward flux of K^+ is accelerated. Hence, this rapid phase of repolarization (phase 3) can be considered to be a *regenerative process,* just as is the inward current of Na^+ during phase 0. The efflux of K^+ during phase 3 rapidly restores the resting level of the membrane potential. The excess Na^+ that had entered the cell mainly during phases 0 and 2 is eliminated by the Na^+/K^+ pump, in exchange for the K^+ that had exited chiefly during phases 2 and 3.

Ionic Basis of the Slow Response

Fast-response action potentials may be considered to consist of two principal components, a spike (phases 0 and 1) and a plateau (phase 2). In the slow response, the first component is absent or inoperative, and the second component accounts for the entire action potential. In the fast response, the spike is produced by the inward Na^+ current through the fast channels. These channels can be blocked by certain interventions, such as the administration of tetrodotoxin. When the fast Na^+ channels are blocked, slow responses may be generated in the same fibers under appropriate conditions.

The Purkinje fiber action potentials shown in Fig. 27-11 clearly exhibit the two components. In the control

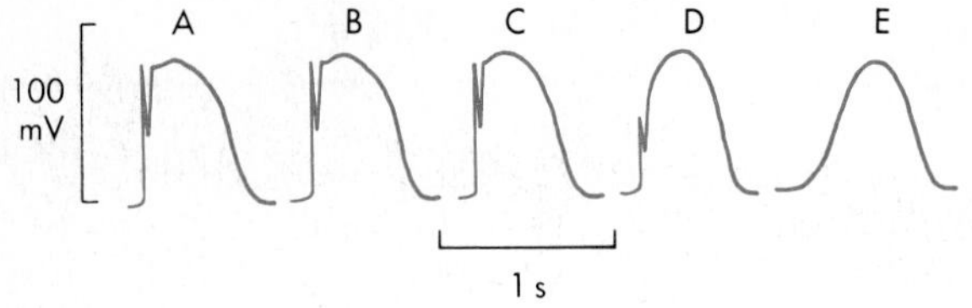

■ Fig. 27-11. Effect of tetrodotoxin on the action potential recorded in a calf Purkinje fiber perfused with a solution containing epinephrine and 10.8 mM K^+. The concentration of tetrodotoxin was 0 M in **A,** 3×10^{-8} M in **B,** 3×10^{-7}**M** in **C,** and 3×10^{-6} M in **D** and **E; E** was recorded later than **D.** (Redrawn from Carmeliet, E., and Vereecke, J.: Pflügers Arch. **313:**300, 1969.)

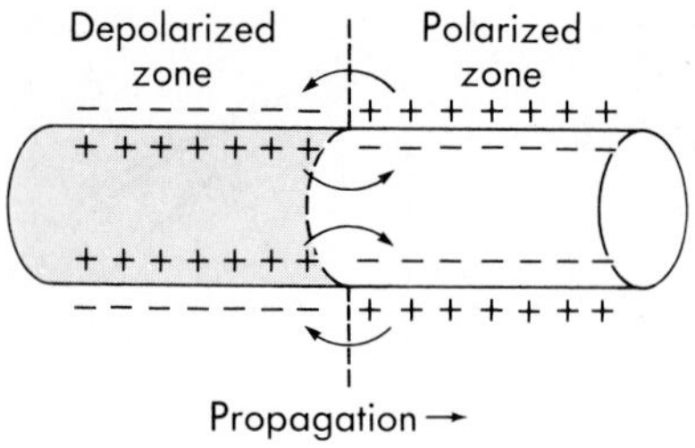

■ Fig. 27-12. The role of local currents in the propagation of a wave of excitation down a cardiac fiber.

tracing (panel *A*), a prominent notch separates the spike from the plateau. In panels *B* to *E*, progressively larger quantities of tetrodotoxin were added to the bathing solution to produce a graded blockade of a larger and larger fraction of the fast Na^+ channels. It is evident that the spike becomes progressively less prominent in panels *B* to *D*, and it disappears entirely in *E*. Thus the tetrodotoxin had a pronounced effect on the spike and only a negligible influence on the plateau. With elimination of the spike (panel *E*), the action potential resembles a typical slow response.

Certain cells in the heart, notably those in the SA and AV nodes, are normally slow response fibers. In such fibers, depolarization is achieved by the inward currents of Ca^{++} and Na^+ through the slow channels. These ionic events closely resemble those that occur during the second component of fast-response action potentials. The slow channels in nodal cells can be blocked by Mn^{++} or verapamil, just as in fast-response fibers.

Conduction in Cardiac Fibers

An action potential traveling down a cardiac muscle fiber is propagated by local circuit currents, similar to the process that occurs in nerve and skeletal muscle fibers (Chapter 3). In Fig. 27-12, consider that the left half of the cardiac fiber already has been depolarized, whereas the right half is still in the resting state. The fluids normally in contact with the external and internal surfaces of the membrane are essentially solutions of electrolytes and thus are good conductors of electricity. Hence, current (in the abstract sense) will flow from regions of higher to those of lower potential, denoted by the plus and minus signs, respectively. In the external fluid, current will flow from right to left between the active and resting zones, and it will flow in the reverse direction intracellularly. In electrolyte solutions, the true current is carried by a movement of cations in one direction and anions in the opposite direction. At the cell exterior, for example, cations will flow from right to left, and anions from left to right (Fig. 27-12). In the cell interior, the opposite migrations will occur. These local currents at the border between the depolarized and polarized sections of the fiber will tend to depolarize the region of the resting fiber adjacent to the border.

Conduction of the Fast Response

In the fast response, the fast Na^+ channels will be activated when the transmembrane potential is suddenly brought to the threshold value of about −70 mV. The inward Na^+ current will then depolarize the cell very rapidly at that site. This portion of the fiber will become part of the depolarized zone, and the border will be displaced accordingly (to the right in Fig. 27-12). The same process will then begin at the new border. This process will be repeated over and over, and the border will move continuously down the fiber as a wave of depolarization.

At any given point on the fiber, the greater the amplitude of the action potential and the greater the rate of change of potential (dV_m/dt) during phase 0, the more rapid the conduction down the fiber. The amplitude of the action potential equals the difference in potential between the fully depolarized and the fully polarized regions of the cell interior (Fig. 27-12). The magnitude of the local currents is proportional to this potential difference. Since these local currents shift the potential of the resting zone toward the threshold value, they are the local stimuli that depolarize the adjacent resting portion of the fiber to its threshold potential. The greater the potential difference between the depolarized and polarized regions (i.e., the greater the amplitude of the action potential), the more efficacious the local stimuli, and the more rapidly the wave of depolarization is propagated down the fiber.

The dV_m/dt during phase 0 is also an important determinant of the conduction velocity. The reason can be appreciated by referring again to Fig. 27-12. If the active portion of the fiber depolarizes very gradually, the

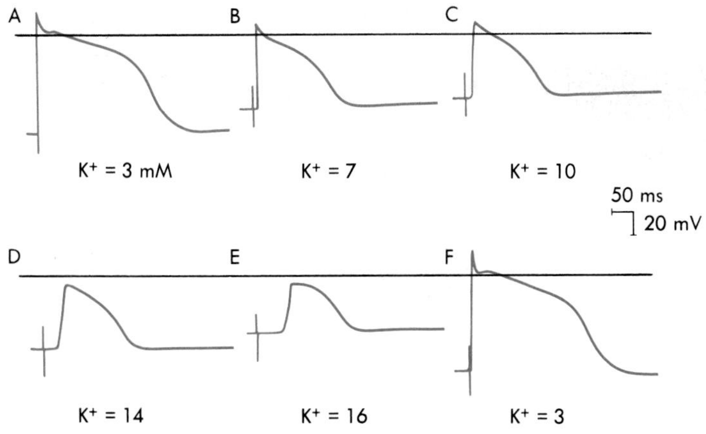

Fig. 27-13. The effect of changes in external potassium concentration on the transmembrane action potentials recorded from a Purkinje fiber. The stimulus artifact appears as a biphasic spike to the left of the upstroke of the action potential. Horizontal calibration, 50 ms; vertical calibration, 20 mV. The horizontal lines near the peaks of the action potentials denote 0 mV. (From Myerburg, R.J, and Lazzara, R. In Fisch, E., editor: Complex electrocardiography, Philadelphia, 1973, F.A. Davis Co.)

local currents across the border between the depolarized and polarized regions would be very small. Thus the resting region adjacent to the active zone would be depolarized very slowly, and consequently each new section of the fiber would require more time to reach threshold.

The level of the resting membrane potential is also an important determinant of conduction velocity. This factor operates through its influence on the amplitude and maximum slope of the action potential. The resting potential may vary for several reasons: (1) it can be altered experimentally by varying $[K^+]_o$ (Fig. 27-5); (2) in cardiac fibers that are intrinsically automatic, V_m becomes progressively less negative during phase 4 (Fig. 27-17, *B*); and (3) during a premature depolarization, repolarization may not have been completed before the beginning of the next excitation (Fig. 27-14). In general, the less negative the level of V_m, the less the velocity of impulse propagation, regardless of the reason for alteration of the V_m level.

The V_m level affects conduction velocity because the inactivation, or *h,* gates (Fig. 27-7) in the fast Na^+ channels are voltage dependent. The less negative the V_m, the greater is the number of *h* gates that tend to close. During the normal process of excitation, depolarization proceeds so rapidly during phase 0 that the comparatively slow *h* gates do not close until the end of that phase. If partial depolarization is produced by a more gradual process, however, such as by elevating the level of external K^+, then the *h* gates do have ample time to close. When the cell is in a partial state of depolarization, many of the fast Na^+ channels will already be inactivated, and only a fraction of these channels will be available to conduct the inward Na^+ current during phase 0.

The results of an experiment in which the resting V_m of a bundle of Purkinje fibers was varied by altering the value of $[K^+]_o$ are shown in Fig. 27-13. When $[K^+]_o$ was 3 mM (panels *A* and *F*), the resting V_m was -82 mV, and the slope of phase 0 was great. At the end of phase 0, the overshoot attained a value of 30 mV. Hence, the amplitude of the action potential was 112 mV. The tissue was stimulated at some distance from the impaled cell, and the stimulus artifact appears as a diphasic deflection just before phase 0. The time separating this artifact from the beginning of phase 0 is inversely proportional to the conduction velocity.

When $[K^+]_o$ was increased to 7 and 10 mM (panels *B* and *C*), the resting V_m became -63 and -58 mV, respectively. There were concomitant reductions in the amplitude and duration of the action potential and in the slope of phase 0. Also, the stimulus artifact appears at a greater and greater distance in front of phase 0, demonstrating that conduction velocity was reduced accordingly. As the resting V_m was made less and less negative by elevating $[K^+]_o$, many of the *h* gates closed, thereby reducing the number of available fast Na^+ channels. When the wave of excitation reached the impaled cell in panels *B* and *C,* the *m* gates rapidly swung open as soon as the threshold voltage for activating the fast Na^+ channels was reached in that cell. However, because many of the *h* gates were already closed by virtue of the elevated $[K^+]_o$, some of the fast Na^+ channels were no longer available. Therefore the magnitude of the inward Na^+ current was considerably curtailed, and so dV_m/dt was diminished. It had been estimated that, on the average, about half of the fast Na^+ channels are inactivated when the resting V_m is about -70 mV, and all channels are inactivated when the resting V_m is about -55 mV.

When $[K^+]_o$ was raised to levels of 14 and 16 mM (panels *D* and *E*), the resting V_m became less negative than the critical value of about -55 mV, and therefore the fast Na^+ channels were completely inactivated. The action potentials in panels *D* and *E* are characteristic slow responses, presumably mediated by inward Ca^{++} and Na^+ currents through the slow channels exclusively. The slow conduction is reflected in the tracings by the long time intervals between the stimulus artifacts and the upstrokes of the action potentials.

Conduction of the Slow Response

Local circuits are also responsible for propagation of the slow response (Fig. 27-12). However, the characteristics of the conduction process are entirely different from those of the fast response. The threshold potential is −35 mV for the slow response, and the conduction velocity is very much less than that for the fast response. The conduction velocities of the slow responses in the SA and AV nodes are about 0.02 to 0.1 m/s. The fast response conduction velocities are about 0.3 to 1 m/s for myocardial cells and 1 to 4 m/s for the specialized conducting fibers in the atria and ventricles. Slow responses are more likely to be blocked than are fast responses. Also, the former cannot be conducted at as rapid repetition rates. Fast responses are easily conducted in either an antegrade or a retrograde direction. The velocity of conduction is virtually the same in both directions. The slow response conduction velocity in one direction may be much greater than in the opposite direction. Not uncommonly, the slow response will be conducted in one direction only, but it will be blocked in the opposite direction. This condition of *undirectional* block is a sine qua non for reentry and is the basis for many arrhythmias.

Cardiac Excitability

Currently, knowledge of cardiac excitability is important because of the use of artificial pacemakers and other electrical devices for correcting serious disturbances of rhythm. The excitability characteristics of cardiac cells differ considerably, depending on whether the action potentials are fast or slow responses.

Fast Response

Once the fast response has been initiated, the depolarized cell will no longer be excitable until about the middle of the period of final repolarization (phase 3). The interval from the beginning of the action potential until the fiber is able to conduct another action potential is called the *effective refractory period*. In the fast response, this period extends from the beginning of phase 0 to a point in phase 3 where repolarization has reached about −50 mV (period *C* to *D* in Fig. 27-1). At about this value of V_m the electrochemical gates (*m* and *h*) for many of the fast Na^+ channels have been reset.

Full excitability is not regained until the cardiac fiber has been fully repolarized (point *E* in Fig. 27-1). During period *D* to *E* in the figure, an action potential may be evoked, but only when the stimulus is stronger than that which would be capable of eliciting a response during phase 4. Period *D* to *E* is called the *relative refractory period*.

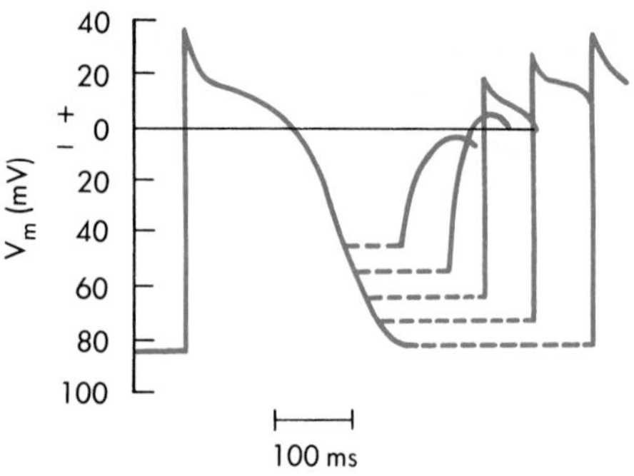

Fig. 27-14. The changes in action potential amplitude and slope of the upstroke as action potentials are initiated at different stages of the relative refractory period of the preceding excitation. (Redrawn from Rosen, M.R., Wit, A.L., and Hoffman, B.F.: Am. Heart J. **88:**380, 1974.)

When a fast response is evoked during the relative refractory period of an antecedent excitation, its characteristics vary with the membrane potential that exists at the time of stimulation. The nature of this voltage dependency is illustrated in Fig. 27-14. It is evident that as the fiber is stimulated later and later in the relative refractory period, the amplitude of the response and the rate of rise of the upstroke increase progressively. Presumably, the number of fast Na^+ channels that have recovered from inactivation increases as repolarization proceeds during phase 3. As a consequence of the increase in amplitude and slope of the evoked response, the propagation velocity becomes greater the later the cell is stimulated during the relative refractory period. Once the fiber is fully repolarized, the response is constant no matter what time in phase 4 the stimulus is applied. By the end of phase 3, the *m* and *h* gates of all channels are in their final positions and no further change in excitability occurs.

Slow Response

The effective refractory period during the slow response frequently extends well beyond phase 3. Even after the cell has completely repolarized, it may not be possible to evoke a propagated response for some time. The relative refractory period then extends into phase 4, during which V_m is virtually constant in nonpacemaker cells. During the relative refractory period, excitability progressively improves despite the constant level of V_m. The recovery of full excitability is much slower than for the fast response. The process might involve a few seconds, as compared to a few tenths of a second for re-

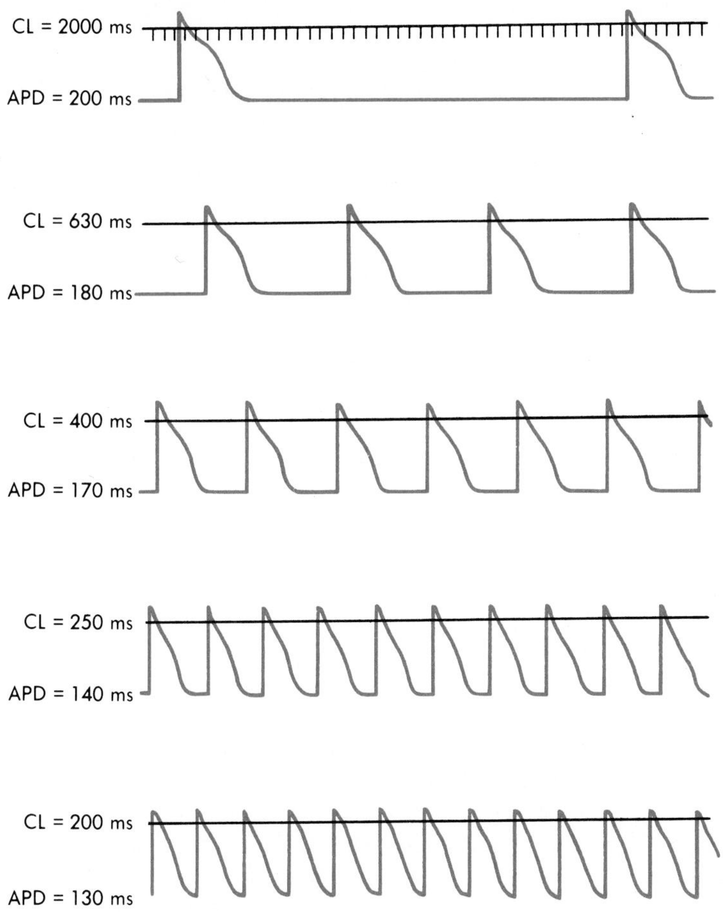

Fig. 27-15. The effect of changes in cycle length *(CL)* on the action potential duration *(APD)* of canine Purkinje fibers. (Modified from Singer, D., and Ten Eick, R.E.: Am. J. Cardiol. **28:**381, 1971.)

covery in the fast response. Until full recovery of excitability is achieved, conduction velocity varies with excitability. Impulses arriving at a slow response fiber early in its relative refractory period are conducted much more slowly than those arriving late in that period. The lengthy refractory periods also account for the observed tendency toward conduction blocks. Even when slow response fibers are excited at a relatively low repetition rate, the fiber may be able to conduct only a fraction of those impulses.

Effects of Cycle Length

Changes in cycle length alter the duration of the action potentials of cardiac cells and thus change their refractory periods. Consequently, the changes in cycle length are often important factors in the initiation or termination of certain arrhythmias. The changes in action potential durations produced by stepwise reductions in cycle length from 2000 to 200 ms in a Purkinje fiber are shown in Fig. 27-15. Note that as the cycle length is diminished, the action potential duration decreases.

This direct correlation between action potential duration and cycle length is ascribable to the relationship between the intracellular Ca^{++} concentration and g_K that was described on p. 405. Ca^{++} enters the cardiac cells mainly during the plateau of the action potential. As the cycle length is diminished, the number of plateaus per minute increases, augmenting the influx of Ca^{++} into the cell per minute. The resultant elevation of intracellular Ca^{++} concentration increases g_K, which terminates phase 2 more quickly and thereby shortens the action potential.

Natural Excitation of the Heart

The nervous system controls various aspects of the behavior of the heart, including the frequency at which it beats and the vigor of each contraction. However, cardiac function certainly does not completely depend on intact nervous pathways.

The properties of *automaticity* (the ability to initiate its own beat) and of *rhythmicity* (the regularity of such pacemaking activity) are intrinsic to cardiac tissue. The heart will continue to beat even when it is completely removed from the body. If the coronary vasculature is artificially perfused, rhythmic cardiac contraction will

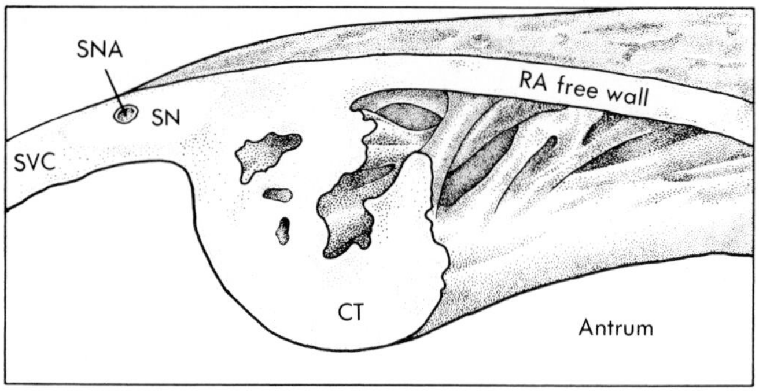

■ **Fig. 27-16.** The location of the SA node near the junction between the superior vena cava and the right atrium. *SN,* SA node; *SNA,* sinoatrial artery; *SVC,* superior vena cava; *RA,* right atrium; *CT,* crista terminalis. (Redrawn from James, T.N.: Am. J. Cardiol. **40:**965, 1977.)

persist for considerable periods of time. Apparently, at least some cells in the walls of all four cardiac chambers are capable of initiating beats; such cells probably reside in the nodal tissues or specialized conducting fibers of the heart.

The region of the mammalian heart that ordinarily displays the highest order of rhythmicity is the SA node; it is called the *natural pacemaker* of the heart. Recently, excitation of the right atrium has been mapped in detail by means of a large number of closely spaced electrodes. The excitation maps indicate that two or three sites of automaticity, located 1 or 2 cm from the SA node itself, may serve along with the SA node as an *atrial pacemaker complex.* At times, all of these loci initiate impulses simultaneously. At other times, the site of earliest excitation shifts from locus to locus, depending on such conditions as the level of autonomic neural activity.

Other regions of the heart that initiate beats under special circumstances are called *ectopic foci,* or *ectopic pacemakers.* Ectopic foci may become pacemakers when (1) their own rhythmicity becomes enhanced, (2) the rhythmicity of the higher order pacemakers becomes depressed, or (3) all conduction pathways between the ectopic focus and those regions with a higher degree of rhythmicity become blocked.

When the SA node and the other components of the atrial pacemaker complex are excised or destroyed, pacemaker cells in the AV junction usually have the next highest order of rhythmicity, and they become the pacemakers for the entire heart. After some time, which may vary from minutes to days, automatic cells in the atria usually become dominant. In the dog, the most common site for the ectopic atrial pacemaking region is at the junction between the inferior vena cava and the right atrium. Purkinje fibers in the specialized conduction system of the ventricles also possess automaticity. Characteristically, they fire at a very slow rate. When the AV junction is unable to conduct the cardiac impulse from the atria to the ventricles, such *idioventricular pacemakers* in the Purkinje fiber network initiate the ventricular contractions. Such ventricular contractions occur at a frequency of only 30 to 40 beats per minute.

■ *Sinoatrial Node*

The SA node, which is the phylogenetic remnant of the sinus venosus of lower vertebrate hearts, was first described for mammalian hearts in 1906. In humans it is 15 mm long, 5 mm wide, and 2 mm thick. It lies in the terminal sulcus on the posterior aspect of the heart, where the superior vena cava joins the right atrium (Fig. 27-16). The sinus node artery runs lengthwise through the center of the node.

The SA node contains two principal types of cells: (1) small, round cells, which have few organelles and myofibrils, and (2) slender, elongated cells, which are intermediate in appearance between the round and the ordinary myocardial cells. The round cells are probably the pacemaker cells, whereas the transitional cells probably conduct the impulses within the node and to the nodal margins.

A typical transmembrane action potential recorded from a cell in the SA node is depicted in Fig. 27-17, *B*. Compared with the transmembrane potential recorded from a ventricular myocardial cell (Fig. 27-17, *A*), the resting potential of the SA node cell is usually less, the upstroke of the action potential (phase 0) has a much slower velocity, a plateau is absent, and repolarization (phase 3) is more gradual. These are all characteristic of the slow response. Under ordinary conditions, tetrodotoxin has no influence on the SA nodal action potential. This indicates that the upstroke of the action potential is not produced by an inward current of Na^+ through the fast channels. By special techniques, it has been shown that such fast Na^+ channels do exist in the nodal pacemaker cells. However, they are much more sparse in nodal cells than in myocardial or Purkinje fibers. Also, those fast Na^+ channels are ordinarily inactivated (i.e., the *h* gates are closed [Fig. 27-7]) because the maximum negativity of the transmembrane potential is only about −65 mV in SA node cells.

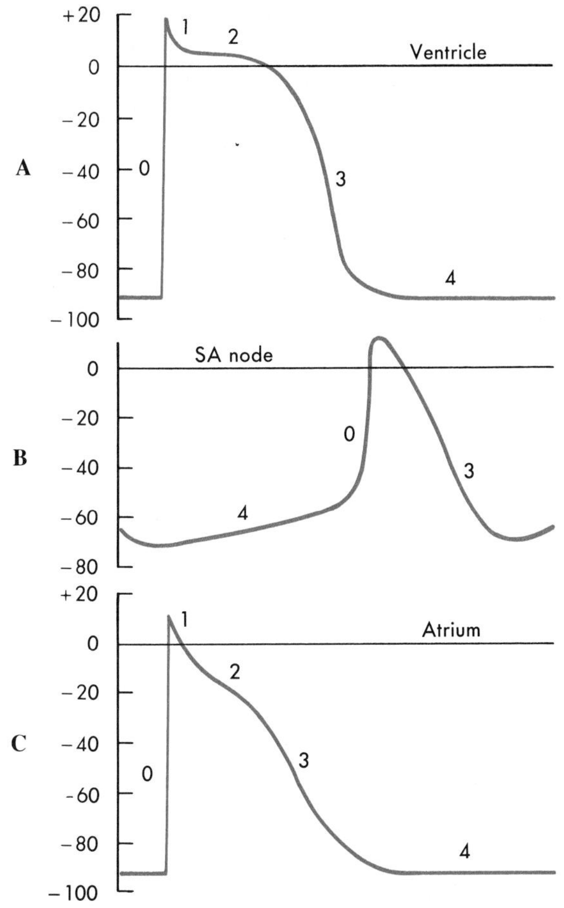

■ **Fig. 27-17.** Typical action potentials (in millivolts) recorded from cells in the ventricle, **A,** SA node, **B,** and atrium, **C.** Sweep velocity in **B** is one half that in **A** or **C.** (From Hoffman, B.F., and Cranefield, P.F.: Electrophysiology of the heart, New York, 1960. Used by permission of McGraw-Hill Book co.)

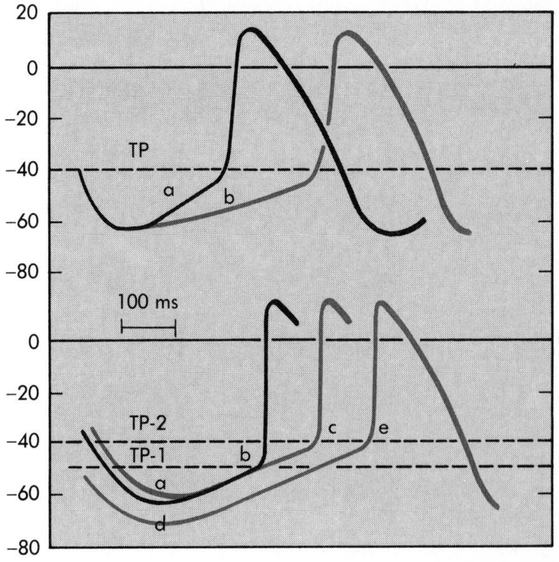

■ **Fig. 27-18.** Mechanisms involved in changes of frequency of pacemaker firing. In section **A** a reduction in the slope of the pacemaker potential from *a* to *b* will diminish the frequency. In section **B** an increase in the threshold (from *TP-1* to *TP-2*) or an increase in the magnitude of the resting potential (from *a* to *d*) will also diminish the frequency. (Redrawn from Hoffman, B.F., and Cranefield, P.F.: Electrophysiology of the heart, New York, 1960. Used by permission of McGraw-Hill Book Co.)

However, the principal distinguishing feature of a pacemaker fiber resides in phase 4. In nonautomatic cells the potential remains constant during this phase, whereas in a pacemaker fiber there is a slow depolarization, called the *pacemaker potential.* Depolarization proceeds at a steady rate until the threshold is attained, and then an action potential is triggered.

The frequency of discharge of pacemaker cells may be varied by a change in (1) the rate of depolarization during phase 4, (2) the threshold potential, or (3) the resting potential (Fig. 27-18). When the rate of depolarization increases (*b* to *a* in Fig. 27-18, *A*) the threshold potential will be attained earlier, and the heart rate will increase. A rise in the threshold potential (from *TP-1* to *TP-2* in Fig. 27-18, *B*) will delay the onset of phase 0 (from time *b* to time *c*), and the heart rate will be reduced accordingly. Similarly, when the magnitude of the resting potential is increased (from *a* to *d*), more time will be required to reach threshold *TP-2*, and the heart rate will diminish.

Ordinarily, the frequency of pacemaker firing is controlled by the activity of both divisions of the autonomic nervous system. Increased sympathetic nervous activity, through the release of norepinephrine, raises the heart rate principally by increasing the slope of the pacemaker potential. Increased vagal activity, through the release of acetylcholine, diminishes the heart rate by hyperpolarizing the pacemaker cell membrane and by reducing the slope of the pacemaker potential. Fig. 27-19 shows the changes in transmembrane potential in an SA nodal cell in response to a brief vagal stimulus *(arrow).* The immediately ensuing hyperpolarization caused the cardiac cycle length to increase to 1250 ms from a basic cycle length of 700 ms. The next several cycles were also lengthened. However, these longer cycles were associated with a decreased slope of the pacemaker potential but no significant hyperpolarization. It is likely that the acetylcholine released at the vagal nerve endings persists only as long as the hyperpolarization. The mechanism responsible for the later phase of bradycardia, associated with the reduced rate of diastolic depolarization, is not known. Vagal stimulation frequently also evokes a *pacemaker shift,* wherein the true pacemaker cells are inhibited more than some of

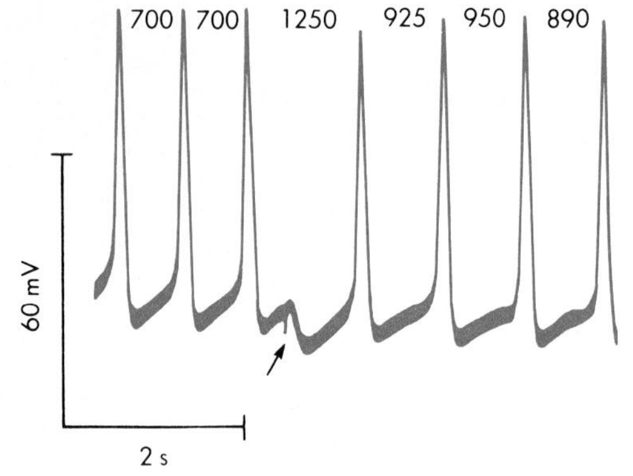

■ **Fig. 27-19.** Effect of a brief vagal stimulus *(arrow)* on the transmembrane potential recorded from an SA node pacemaker cell in an isolated cat atrium preparation. The cardiac cycle lengths, in milliseconds, are denoted by the numbers at the top of the figure. (Modified from Jalife, J., and Moe, G.K.: Circ. Res. **45**:595, 1979.)

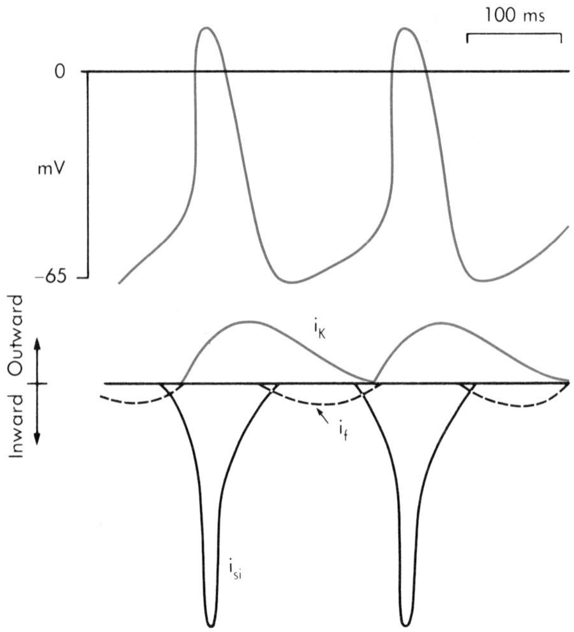

■ **Fig. 27-20.** The transmembrane potential changes *(top half)* that occur in SA node cells are produced by three principal currents *(bottom half)*: *(1)* the slow inward current, i_{si}; *(2)* a hyperpolarization-induced inward current, i_f; and *(3)* an outward K^+ current, i_k. (Redrawn from Brown, H.F.: Physiol. Rev. **61**:644, 1981.)

the latent pacemakers within the node or atrial pacemaker complex, and these latent pacemaker cells then become the true pacemakers.

■ *Ionic Basis of Automaticity*

Our understanding of the ionic basis of automaticity in cardiac cells has improved recently as a result of advances in voltage-clamping techniques. Several ionic currents have been shown to contribute to the slow depolarization that occurs during phase 4 in automatic cells in the heart. In the pacemaker cells of the SA node, the diastolic depolarization is ascribable to at least three ionic currents: (1) an inward current, i_f, induced by hyperpolarization; (2) the slow inward current, i_{si}; and (3) an outward K^+ current, i_K (Fig. 27-20).

The inward current, i_f, is carried mainly by Na^+. This current becomes activated during the repolarization phase of the action potential, as the membrane potential becomes more negative than about -50 mV. The more negative the membrane potential becomes at the end of repolarization, the greater will be the activation of the i_f current.

The second current responsible for diastolic depolarization is the slow inward current, i_{si}. This current, comprised of Ca^{++} and Na^+, becomes activated toward the end of phase 4, as the transmembrane potential reaches a value of about -55 mV (Fig. 27-20). Once the slow channels become activated, the influx of Ca^{++} and Na^+ into the cell increases. The influx of cations accelerates the rate of diastolic depolarization, which then leads to the upstroke of the action potential. A decrease in the external Ca^{++} concentration (Fig. 27-21) or the addition of calcium channel blocking agents (Fig. 27-22) diminishes the amplitude of the action potential and the slope of the pacemaker potential in SA node cells.

The progressive diastolic depolarization mediated by the two inward currents, i_f and i_{si}, is opposed by a third current, an outward K^+ current, i_K. The efflux of K^+ tends to repolarize the cell. However, the outward K^+ current decays steadily throughout phase 4 (Fig. 27-20), and the opposition to the depolarizing effects of the two inward currents gradually decreases.

The ionic basis for automaticity in the AV node pacemaker cells is probably identical to that in the SA node cells. Similar mechanisms probably also account for automaticity in Purkinje fibers, except that the slow inward current is not involved. Hence, the slow diastolic depolarization is mediated principally by the balance between the hyperpolarization-induced inward current, i_f, and the outward K^+ current, i_K.

The autonomic neurotransmitters affect automaticity by altering the ionic currents across the cell membranes. The adrenergic transmitters increase all three currents that are involved in SA nodal automaticity. The adrenergically mediated increase in the slope of diastolic depolarization indicates that the augmentations of i_f and i_{si} must exceed the enhancement of i_K. The hyperpolarization (Fig. 27-19) induced by the acetyl-

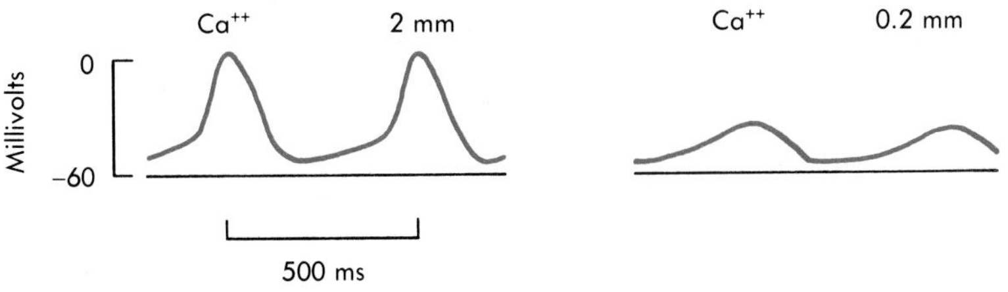

Fig. 27-21. Transmembrane action potentials recorded from an SA node pacemaker cell in an isolated rabbit atrium preparation. The concentration of Ca^{++} in the bath was changed from 2 to 0.2 mM. (Modified from Kohlhardt, M., Figulla, H.-R., and Tripathi, O.: Basic Res. Cardiol. **71:**17, 1976.)

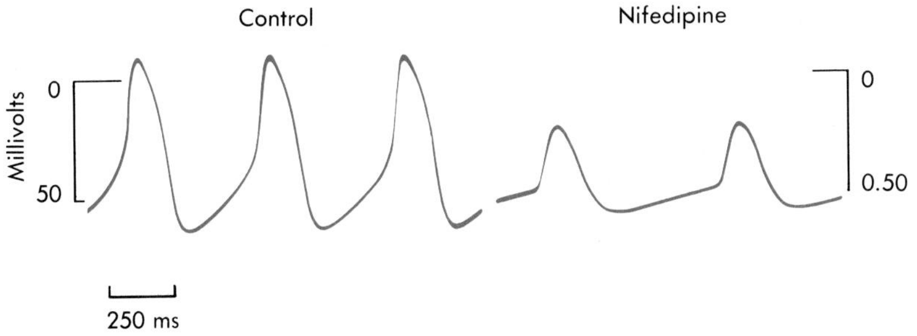

Fig. 27-22. The effects of nifedipine (5.6×10^{-7}M), a Ca^{++} channel blocking drug, on the transmembrane potentials recorded from a rabbit's **SA** node cell. (From Ning, W., and Wit, A.L.: Am. Heart J. **106:**345, 1983.

choline released at the vagus endings in the heart is achieved by an increase in g_K. This change in conductance is mediated through activation of specific ionic channels that are controlled by the cholinergic receptors, rather than through the activation of the regular i_K channels.

Overdrive Suppression

The automaticity of pacemaker cells becomes depressed after a period of excitation at a high frequency. This phenomenon is known as *overdrive suppression*. Because of the greater intrinsic rhythmicity of the SA node than that of the other latent pacemaking sites in the heart, the firing of the SA node tends to suppress the automaticity in the other loci. If an ectopic focus in one of the atria suddenly began to fire at a rate of 150 impulses per minute in an individual with a normal heart rate of 70 beats per minute, the ectopic center would become the pacemaker for the entire heart. When that rapid ectopic focus suddenly stopped firing, the SA node might remain quiescent briefly because of overdrive suppression. The interval from the end of the period of overdrive until the SA node resumes firing is called the *sinus node recovery time*. In patients with the so-called *sick sinus syndrome,* the sinus node recovery time may be markedly prolonged. The resultant period of asystole might result in syncope.

The mechanism responsible for overdrive suppression is uncertain. A reasonable hypothesis is based on the activity of the membrane pump that actively extrudes Na^+ from the cell, in partial exchange for K^+ (p. 405). During depolarization, a certain quantity of Na^+ enters a given cell. The more frequently it is depolarized, therefore, the more Na^+ that enters the cell per minute. Under conditions of overdrive, the Na^+ pump becomes more active in the process of extruding this larger quantity of Na^+ from the cell interior. The quantity of Na^+ extruded by the pump exceeds the quantity of K^+ that enters the cell. This enhanced activity of the pump results in some hyperpolarization of the cell, since there is a net loss of cations from the cell interior. Because of the hyperpolarization, the pacemaker potential requires more time to reach the threshold, as shown in Fig. 27-18, *B*. It has also been postulated that, when the overdrive suddenly ceases, the Na^+ pump does not decelerate instantaneously but continues to operate at an accelerated rate for some time. This excessive extrusion of Na^+ tends to oppose the gradual depolarization of the pacemaker cell during phase 4, thereby suppressing its intrinsic automaticity temporarily.

Atrial Conduction

From the SA node, the cardiac impulse spreads radially throughout the right atrium (Fig. 27-23) along ordinary atrial myocardial fibers, at a conduction velocity of approximately 1 m/s. A special pathway, the *anterior interatrial myocardial band* (or *Bachmann's bundle*), conducts the impulse from the SA node directly to the left atrium. Three tracts, the *anterior, middle,* and *posterior internodal pathways,* have been described. These tracts consist of ordinary myocardial cells and specialized conducting fibers. Some authorities assert that these pathways constitute the principal routes for conduction of the cardiac impulse from the SA to the AV node.

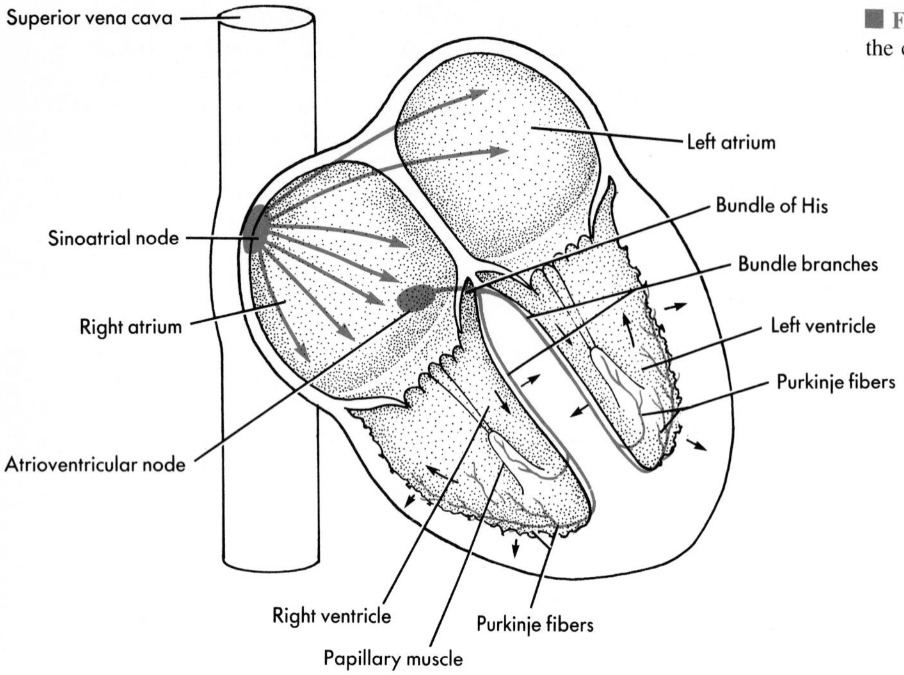

Fig. 27-23. Schematic representation of the conduction system of the heart.

The configuration of the atrial transmembrane potential is depicted in Fig. 27-17, *C*. Compared with the potential recorded from a typical ventricular fiber (Fig. 27-17, *A*), the plateau (phase 2) is not as well developed, and repolarization (phase 3) occurs at a slower rate.

Atrioventricular Conduction

The cardiac action potential proceeds along the internodal pathways in the atrium and ultimately reaches the AV node. This node is approximately 22 mm long, 10 mm wide, and 3 mm thick. It is situated posteriorly on the right side of the interatrial septum near the ostium of the coronary sinus. The AV node contains the same two cell types as the SA node, but the round cells are more sparse and the elongated cells preponderate. The internodal pathways join the AV node at its crest and along the convex surface facing the right atrium.

The AV node has been divided into three functional regions: (1) the A-N region, the transitional zone between the atrium and the remainder of the node; (2) the N region, the midportion of the AV node; and (3) the N-H region, the zone in which nodal fibers gradually merge with the *bundle of His,* which is the upper portion of the specialized conducting system for the ventricles. Normally, the AV node and bundle of His constitute the only pathways for conduction from atria to ventricles.

Several features of AV conduction are of physiological and clinical significance. It is in the A-N region of the AV node that the principal delay occurs in the passage of the impulse from the atrial to the ventricular myocardial cells. The conduction velocity is actually less in the N region than in the A-N region. However, the path length is substantially greater in the latter than in the former zone, which accounts for the difference in the total conduction time through the two regions. The conduction times through the A-N and N zones account for a considerable fraction of the time interval between the onsets of the atrial and ventricular systoles and between the onsets of the *P wave* (the electrical manifestation of the spread of atrial excitation) and the *QRS complex* (spread of ventricular excitation) in the electrocardiogram. The period of time between the initiation of the P wave and the QRS complex is called the *P-R interval* (Fig. 27-33). Functionally, this delay between atrial and ventricular excitation permits optimal ventricular filling during atrial contraction.

In the N region, the action potentials display many of the characteristics of the slow response. The resting potential is about −60 mV, the upstroke velocity is very low (about 5 V/s), and the conduction velocity is about 0.05 m/s. Tetrodotoxin, which blocks the fast Na^+ channels, has virtually no effect on the action potentials in this region. Conversely, the slow channel blocking agents decrease the amplitude and duration of the action potentials (Fig. 27-24) and depress AV conduction. The action potentials in the A-N region are intermediate in configuration between those in the N region and atria. Similarly, the action potentials in the N-H region are transitional between those in the N region and bundle of His.

The relative refractory period of the cells in the N region extends well beyond the period of complete re-

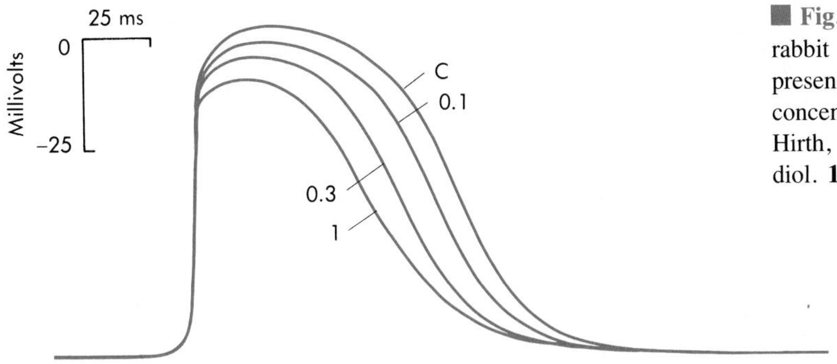

■ **Fig. 27-24.** Transmembrane potentials recorded from a rabbit AV node cell under control conditions *(C)* and in the presence of the calcium channel blocking drug, diltiazem, in concentrations of 0.1, 0.3, and 1 μmol/L. (Redrawn from Hirth, C., Borchard, U., and Hafner, D.: J. Mol. Cell Cardiol. **15:**799, 1983.)

polarization. Such refractoriness is said to be *time dependent,* in contrast to the *voltage-dependent* refractoriness that characterizes most of the other types of cells in the heart (Fig. 27-14). As the repetition rate of atrial depolarizations is increased, conduction through the AV junction tends to be prolonged. Most of that prolongation takes place in the N region of the AV node. Impulses tend to be blocked at stimulus repetition rates that are easily conducted in other regions of the heart. If the atria are paced at a high frequency, only one half or one third of the impulses might be conducted through the AV junction to the ventricles. This tends to protect the ventricles from excessive contraction frequencies, wherein the filling time between contractions might be inadequate. Retrograde conduction, from ventricles to atria, can usually occur through the AV node. However, the propagation time is significantly longer and the impulse is blocked at lower repetition rates in the retrograde than in the antegrade direction. Finally, the AV node is a common site for reentry; the underlying mechanisms are explained on p. 417.

The autonomic nervous system regulates AV conduction. Weak vagal activity may simply prolong AV conduction time. Stronger vagal activity may cause some or all of the impulses arriving from the atria to be blocked in the node. The delayed conduction or block tends to occur largely in the N region of the node. The cardiac sympathetic nerves, on the other hand, have a facilitative effect. They act to decrease the AV conduction time and to enhance the rhythmicity of the latent pacemakers in the AV junction. The norepinephrine released at the sympathetic nerve terminals increases the amplitude and slope of the upstroke of the AV nodal action potentials, principally in the N region of the node.

■ *Ventricular Conduction*

The bundle of His passes subendocardially down the right side of the interventricular septum for approximately 12 mm and then divides into the right and left *bundle branches* (Figs. 27-23 and 27-25). The right bundle branch is a direct continuation of the bundle of His and proceeds down the right side of the interventricular septum. The left bundle branch, which is considerably thicker than the right, arises almost perpendicularly from the bundle of His and perforates the interventricular septum. On the subendocardial surface of the left side of the interventricular septum, the main left bundle branch splits into a thin *anterior division* and a thick *posterior division.* Clinically, impulse conduction in the right bundle branch, the main left bundle branch, or either division of the left bundle branch may be impaired. Conduction blocks in one or more of these pathways give rise to characteristic electrocardiographic patterns. A block of either of the main bundle branches is known as right or left *bundle branch block.* Block of either division of the left bundle branch is called *left anterior hemiblock* or *left posterior hemiblock.*

The right bundle branch of the two divisions of the left bundle branch ultimately subdivide into a complex network of conducting fibers called *Purkinje fibers,* which ramify over the subendocardial surfaces of both ventricles. In certain mammalian species, such as cattle, the Purkinje fiber network is arranged in discrete, encapsulated bundles (Fig. 27-25).

Purkinje fibers are the broadest cells in the heart, 70 to 80 μm in diameter, compared with 10 to 15 μm for ventricular myocardial cells. The large diameter accounts in part for the greater conduction velocity in Purkinje than in myocardial fibers. Purkinje cells have abundant, linearly arranged sarcomeres, just as do myocardial cells. However, the T-tubular system is absent in the Purkinje cells of many species but is well developed in the myocardial cells.

The conduction velocity of the action potential over the Purkinje fiber system is the fastest of any tissue within the heart; estimates vary from 1 to 4 m/s. This permits a rapid activation of the entire endocardial surface of the ventricles.

The configuration of the action potentials recorded from Purkinje fibers is quite similar to that from ordinary ventricular myocardial fibers (Fig. 27-17, *A*). In general, phase 1 is more prominent in Purkinje fiber action potentials than in those recorded from ventricular

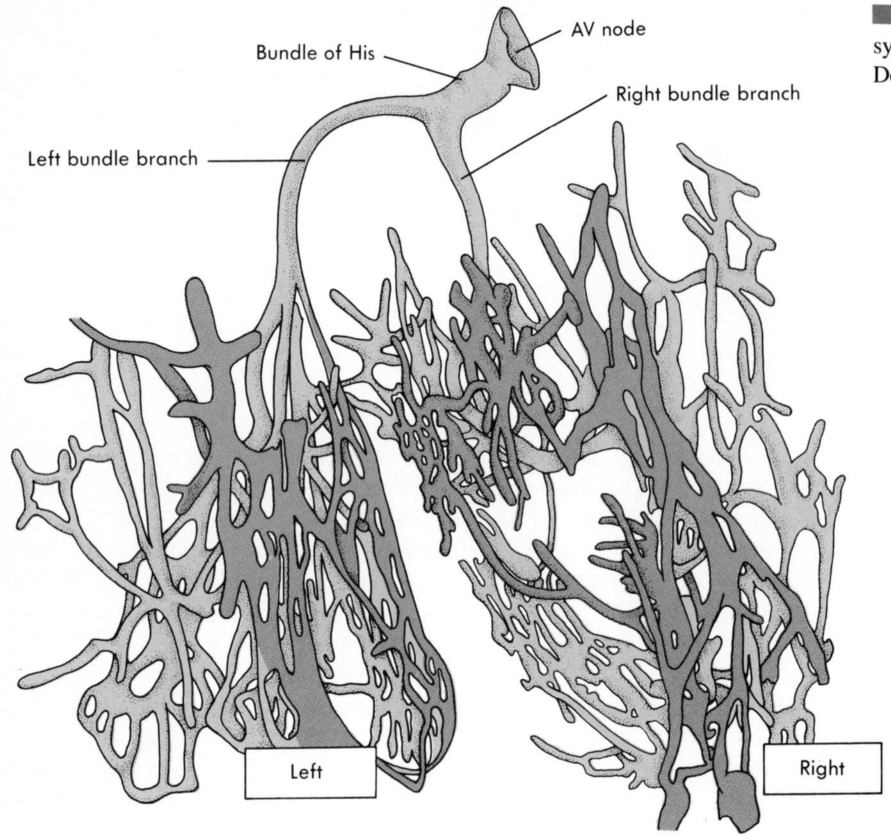

■ **Fig. 27-25.** AV and ventricular conduction system of the calf heart. (Redrawn from DeWitt, L.M.: Anat. Rec. **3**:475, 1909.)

fibers, and the duration of the plateau (phase 2) is longer. In certain regions of the Purkinje fiber network, the action potential durations, and hence the effective refractory periods, are especially long. The Purkinje fibers in these regions are called *gate cells;* the comparative durations of the action potentials of the ordinary Purkinje cells *(P)* and of the gate cells *(G)* are shown in Fig. 27-26.

Because of the long refractory period of the gate cells, many premature activations of the atria are conducted through the AV junction but are blocked by the gate cells. Therefore they fail to evoke a premature contraction of the ventricles. This function of protecting the ventricles against the effects of premature atrial depolarizations is especially pronounced at slow heart rates because the action potential duration and hence the effective refractory period of the gate cells vary inversely with the heart rate (Fig. 27-15). Similar directional changes in the refractory period occur in most of the other cells in the heart with changes in rate. However, in the AV node, the effective refractory period does not change appreciably over the normal range of heart rates, and it actually increases at very rapid heart rates. Therefore at high rates, it is the AV node, rather than the gate cells, that protects the ventricles when impulses arrive at excessive repetition rates.

The spread of the action potential over the ventricles is of major concern in clinical cardiology. Numerous

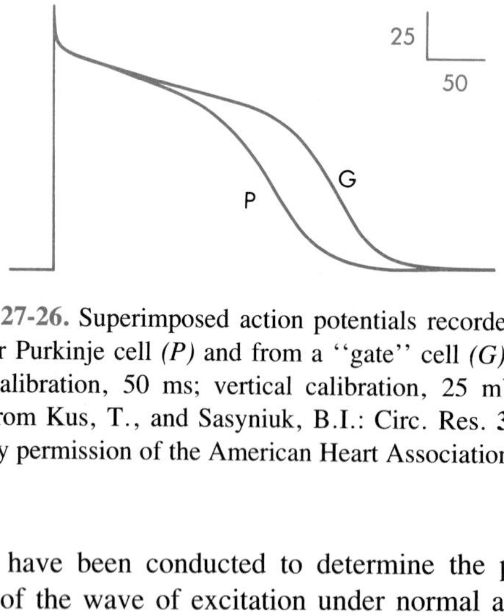

■ **Fig. 27-26.** Superimposed action potentials recorded from a regular Purkinje cell *(P)* and from a ''gate'' cell *(G)*. Horizontal calibration, 50 ms; vertical calibration, 25 mV (Redrawn from Kus, T., and Sasyniuk, B.I.: Circ. Res. **37**:844, 1975. By permission of the American Heart Association, Inc.)

studies have been conducted to determine the precise course of the wave of excitation under normal and abnormal conditions. Such knowledge serves as a basis for the interpretation of the electrocardiogram. However, only the elementary, salient features of ventricular activation are considered here.

The first portions of the ventricles to be excited are the interventricular septum (except the basal portion) and the papillary muscles. The wave of activation spreads into the substance of the septum from both its left and its right endocardial surfaces. Early contraction of the septum tends to make it more rigid and allows it to serve as an anchor point for the contraction of the remaining ventricular myocardium. Also, early contraction of the papillary muscles prevents eversion of the AV valves during ventricular systole.

The endocardial surfaces of both ventricles are activated rapidly, but the wave of excitation spreads from endocardium to epicardium at a slower velocity (about 0.3 to 0.4 m/s). Because the right ventricular wall is appreciably thinner than the left, the epicardial surface of the right ventricle is activated earlier than that of the left ventricle. Also, apical and central epicardial regions of both ventricles are activated somewhat earlier than their respective basal regions. The last portions of the ventricles to be excited are the posterior basal epicardial regions and a small zone in the basal portion of the interventricular septum.

Reentry

Under appropriate conditions, a cardiac impulse may reexcite some region through which it had passed previously. This phenomenon, known as *reentry,* is responsible for many clinical disturbances of cardiac rhythm. The reentry may be *ordered* or *random*. In the ordered variety, the impulse traverses a fixed anatomical path, whereas in the random type the path continues to change. The principal example of random reenty is *fibrillation* (p. 427).

The conditions necessary for reentry are illustrated in Fig. 27-27. In each of the four panels, a single bundle *(S)* of cardiac fibers splits into a left *(L)* and a right *(R)* branch. A connecting bundle *(C)* runs between the two branches. Normally, the impulse coming down bundle *S* is conducted along the *L* and *R* branches (panel *A*). As the impulse reaches connecting link *C,* it enters from both sides and becomes extinguished at the point of collision. The impulse from the left side cannot proceed further because the tissue beyond is absolutely refractory, since it had just undergone depolarization from the other direction. The impulse cannot pass through bundle *C* from the right either, for the same reason.

It is obvious from panel *B* that the impulse cannot make a complete circuit if antegrade block exists in the two branches *(L* and *R)* of the fiber bundle. Furthermore, if bidirectional block exists at any point in the loop (for example, branch *R* in panel *C*), the impulse will not be able to reenter.

A necessary condition for reenty is that at some point in the loop, the impulse is able to pass in one direction but not in the other. This phenomenon is called *unidirectional block.* As shown in panel *D,* the impulse may travel down branch *L* normally. However, the impulse traveling down branch R may be blocked in the antegrade direction. The impulse that had been conducted down branch *L* and through the connecting branch *C*

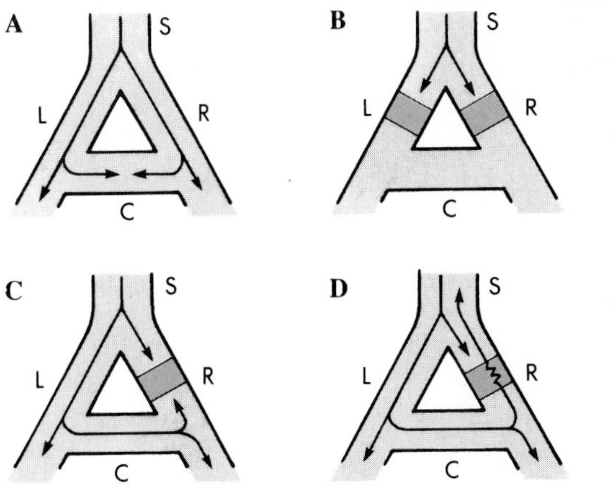

Fig. 27-27. The role of unidirectional block in reentry. In panel **A,** an excitation wave traveling down a single bundle *(S)* of fibers continues down the left *(L)* and right *(R)* branches. The depolarization wave enters the connecting branch *(C)* from both ends and is extinguished at the zone of collision. In panel **B,** the wave is blocked in the *L* and *R* branches. In panel **C,** bidirectional block exists in branch *R*. In panel **D,** unidirectional block exists in branch *R*. The antegrade impulse is blocked, but the retrograde impulse is conducted through and reenters bundle *S*.

may be able to penetrate the depressed region in branch *R* from the retrograde direction, even though the antegrade impulse had been blocked previously at this same site. The antegrade impulse in branch R will arrive at the depressed region earlier than the impulse that traverses a longer path and enters branch *R* from the opposite direction. The antegrade impulse may be blocked simply because it arrives at the depressed region during its effective refractory period. If the retrograde impulse is delayed sufficiently because of the longer path, the refractory period may have ended, and the retrograde impulse will then be conducted back into bundle *S*.

Unidirectional block is a necessary condition for reentry but not a sufficient one. It is also essential that the effective refractory period of the reentered region be less than the propagation time around the loop. In panel *D,* if the retrograde impulse is conducted through the depressed zone in branch *R* and if the tissue just beyond is still refractory from the antegrade depolarization, branch *S* will not be reexcited. Therefore the conditions that promote reentry are those that prolong conduction time or shorten the effective refractory period.

An example of sustained reentry in a frog atrial preparation is shown in Fig. 27-28. The central portion of the preparation, which contained the automatic cells, was excised, making the preparation quiescent. The electrical activity of the ring-shaped tissue was monitored at eight sites by means of a voltage-sensitive dye.

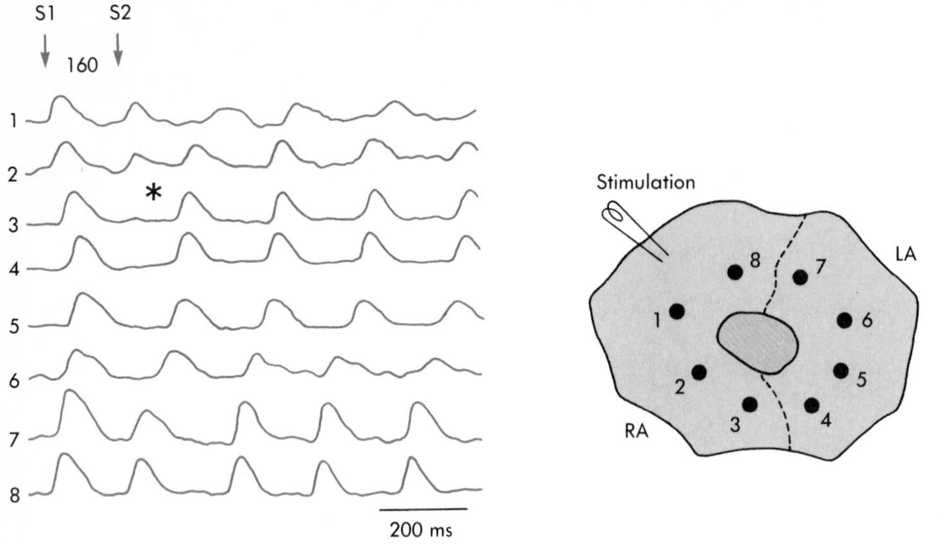

■ **Fig. 27-28.** The potential changes recorded from eight sites in an isolated atrium preparation from a frog. The potential changes were obtained by staining the tissue with a voltage-sensitive dye. The central region, which contained the automatic cells, was excised. Stimulating electrodes were attached at a point equidistant from sites *1* and *8*. When two stimuli *(S1* and *S2)* were given 160 ms apart, a continuous, clockwise reentrant rhythm was induced. *RA,* Right arm; *LA* left arm. (Modified from Sawanobori, T., Hirano, Y., Hirota, A., and Fujii, S.: Am. J. Physiol. **247:**H185, 1984.)

Electrical stimuli could be applied at a locus equidistant from sites 1 and 8. When stimulus *S1* was given, sites 1 and 8 were depolarized first. The onsets of the initial potential changes at various sites indicate that the wave of depolarization traveled in two directions around the ring of tissue; it proceeded counterclockwise from site 1 and clockwise from site 8. Site 5 was the last site to be depolarized. Therefore the two waves must have met at or near site 5 and extinguished each other. This resembles the hypothetical situation illustrated in branch *C* of Fig. 27-27, *A*.

When a second stimulus *(S2)* was given after sites 1 and 8 had just regained excitability (Fig. 27-28), sustained reentrant activity could be induced. Conduction proceeded counterclockwise from site 1 to site 2, but it was blocked somewhere between sites 2 and 3, as indicated by the asterisk. Presumably, the effective refractory period of the tissue near site 3 was longer than that of sites 1 and 2.

Stimulus *S2* was also conducted in a clockwise direction past site 8, but it continued around the entire loop. It was not blocked at site 3 when it approached from the clockwise direction, undoubtedly because it arrived sufficiently late, which allowed the tissue to regain its excitability. Thereafter the reentrant excitation continued around and around the loop in a clockwise direction. This experiment on the frog atrium verifies the validity of the hypothetical reentry loop diagrammed in Fig. 27-27, *D*.

The functional components of reentry loops responsible for specific arrhythmias in intact hearts are protean. Some loops are very large and involve entire specialized conduction bundles; others are microscopic. The loop may include myocardial fibers, specialized conducting fibers, nodal cells, and junctional tissues, in almost any conceivable arrangement. Also, the cardiac cells in the loop may be normal or deranged.

■ *Triggered Activity*

Considerable attention has been directed recently toward the possible role played by *triggered activity* in the genesis of certain types of arrhythmias. Such a mechanism probably accounts for some of the extrasystoles and paroxysmal tachycardias that previously had been ascribed to reentry. Certain critical characteristics of arrhythmias evoked by triggered activity closely resemble those induced by reentry. Notably, depolarizations ascribable to triggered activity, just as those ascribable to reentry, are always coupled to a preceding beat.

The principal electrophysiologic characteristics of triggered activity are shown in Fig. 27-29. The transmembrane potentials were recorded from Purkinje fibers that were exposed to a high concentration of acetylstrophanthidin, a rapidly acting digitalis-like substance. In the absence of driving stimuli these fibers were quiescent. In each panel, a sequence of six driven depolarizations were induced at various basic cycle lengths.

When the cycle length was 800 ms (panel *A*), the last driven depolarization was followed by a brief *afterdepolarization* that did not reach threshold. Such an afterdepolarization is the hallmark of triggered activity. Once the last afterdepolarization in panel *A* had subsided, the transmembrane potential remained constant until another driving stimulus was given. The upstroke of an afterdepolarization (positive slope of phase 4) can be detected after each of the first five driven depolarizations. Note that the slope of this afterdepolarization

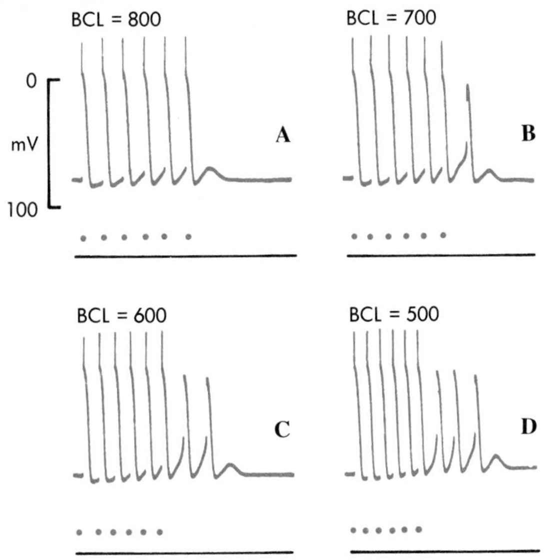

■ **Fig. 27-29.** Transmembrane action potentials recorded from isolated canine Purkinje fibers. Acetylstrophanthidin was added to the bath, and sequences of six driven beats (denoted by the dots) were produced at basic cycle lengths *(BCL)* of 800, 700, 600, and 500 ms. Note that afterpotentials occurred after the driven beats, and that these afterpotentials reached threshold after the last driven beat in panels **B** to **D.** (From Ferrier, G.R., Saunders, J.H., and Mendez, C.: Circ. Res. **32:**600, 1973. By permission of the American Heart Association, Inc.)

increased progressively after each driven depolarization in the sequence.

When the basic cycle length was diminished to 700 ms (panel *B*), the afterdepolarization that followed the last driven beat did reach threshold, and a nondriven depolarization (or *extrasystole*) ensued. This extrasystole was itself followed by an afterpotential that was subthreshold. Diminution of the basic cycle length to 600 ms (panel *C*) also evoked an extrasystole after the last driven depolarization. The afterpotential that followed the extrasystole did reach threshold, however, and a second extrasystole occurred. A sequence of three extrasystoles followed the six driven depolarizations that were separated by intervals of 500 ms (panel *D*). Slightly shorter basic cycle lengths or slightly greater concentrations of acetylstrophanthidin evoked a continuous sequence of nondriven beats, resembling a *paroxysmal tachycardia* (described on p. 427).

The afterdepolarizations responsible for triggered activity are associated with elevated intracellular Ca^{++} concentrations. The amplitude of the afterdepolarizations is increased by interventions that raise intracellular Ca^{++} concentrations, such as elevated extracellular Ca^{++} concentrations and toxic levels of digitalis glycosides. The elevated levels of intracellular Ca^{++} provoke the oscillatory release of Ca^{++} from the sarcoplasmic reticulum. Hence, in myocardial cells the afterdepolarizations are accompanied by small changes in developed tension. The high intracellular Ca^{++} concentrations also activate certain membrane channels that permit the passage of Na^+ and K^+. The net flux of these cations constitutes a transient inward current that is responsible for the afterdepolarization of the cell membrane.

■ *Basis of Electrocardiography*

The electrocardiograph is a valuable instrument because it enables the physician to infer the course of the cardiac impulse simply by recording the variations in electrical potential at various loci on the surface of the body. By analyzing the details of these potential fluctuations, the physician gains valuable insight concerning (1) the anatomical orientation of the heart, (2) the relative sizes of its chambers, (3) a variety of disturbances of rhythm and of conduction, (4) the extent, location, and progress of ischemic damage to the myocardium, (5) the effects of altered electrolyte concentrations, and (6) the influence of certain drugs (notably digitalis and its derivatives). It should be emphasized, however, that the electrocardiogram gives no direct information concerning the mechanical performance of the heart. The science of electrocardiography is extensive and complex, but only the elementary basis of electrocardiography is considered here.

■ *Direct Leads*

Electrocardiograms are usually recorded from *indirect leads* (recording electrodes located on the skin), which are at some distance from the heart, the source of potential. However, *direct leads* have been applied to the surface of the heart in experimental animals and in humans during thoracic surgical procedures. Electrocardiograms recorded by direct leads are discussed first because they are simpler to comprehend.

The principles are illustrated by describing the potential changes recorded directly from the surface of a single, long myocardial fiber. In Fig. 27-30 the changes in potential in one localized region (in contact with electrode *A*) are first considered. Electrode *A* is connected to the lower vertical deflecting plate of a cathode ray oscilloscope; electrode *C*, far to the right of *A*, is connected to the upper deflecting plate. If the strip of muscle is stimulated at its left end, the action potential travels from left to right along the strip. Before excitation reaches region *A*, however, the external surface of the fiber at *A* is at the same potential as the surface at *C*, and no difference in potential is recorded between *A* and *C* (section *1*). When the action potential reaches

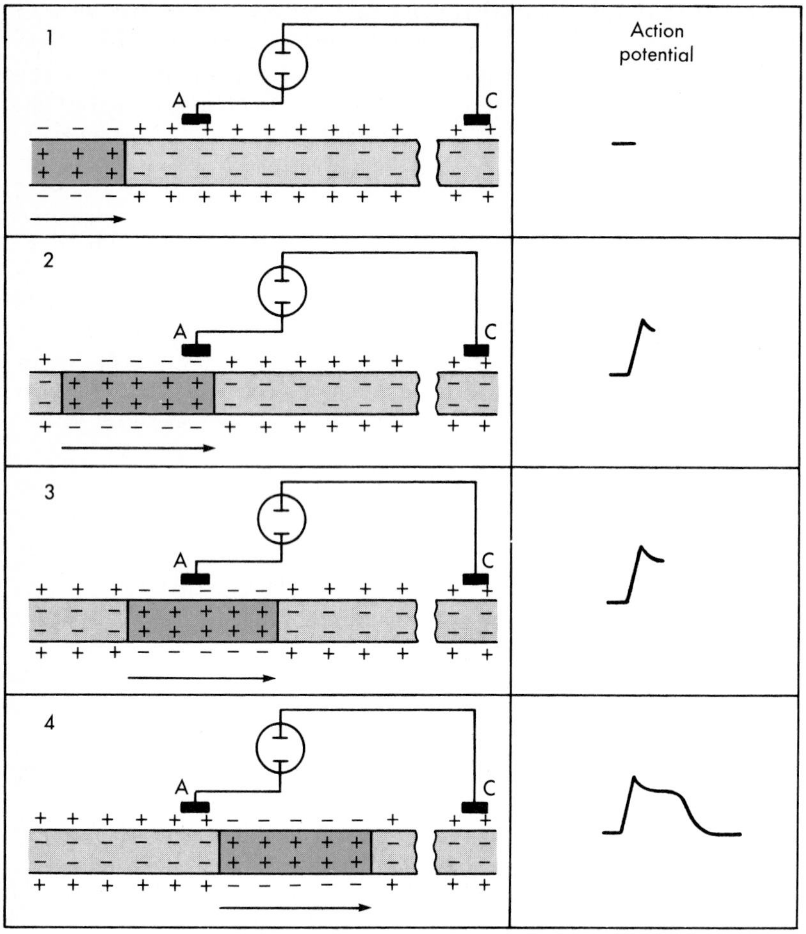

■ **Fig. 27-30.** Sequential changes of potential recorded from an external electrode, *A*, as an action potential travels along a strip of cardiac muscle. Electrode *C* is located far to the right of *A;* the entire cycle of depolarization and repolarization is completed under *A* before the action potential reaches *C*.

region *A*, the transmembrane potential in that region rapidly reverses, and the external surface at *A* becomes electronegative with respect to the surface at *C* (section 2). This difference in potential persists as long as region *A* remains depolarized (section *3*). As region *A* repolarizes, the potential at *A* again becomes equal to the potential at *C* (section *4*). Repolarization proceeds much more slowly than depolarization; hence, the slope of phase 3 is more gradual than that of phase 0.

Fig. 27-31 illustrates the potential differences that would be recorded when the wave of activation is recorded simultaneously from two active regions of the same fiber. In section *1* of Fig. 27-31 the lead locations and oscilloscope connections are identical to those represented in Fig. 27-30. Hence, the recorded action potential is the same. Section *2* of Fig. 27-31 shows the record that is obtained under identical conditions, except that the potential difference between electrodes *B* and *C* is registered. In this case *B* is connected to the upper and *C* to the lower vertical deflecting plate. These connections were chosen because electrode *B* is also connected to the upper deflecting plate in section *3*, in which the potential difference between *A* and *B* is recorded. Relative to the action potential recorded in section *1* of Fig. 27-31, the action potential generated in section *2* is inverted and delayed. The delay depends on the distance between electrodes *A* and *B* and on the propagation velocity of the action potential.

Finally, if electrodes *A* and *B* are connected to the lower and upper deflecting plates, respectively, then the potential difference recorded between them has the configuration shown at the right of section *3*. This action potential represents the algebraic sum of the potentials recorded separately under *A* and *B* (with respect to resting region *C*). The externally recorded action potential consists of an initial upright spike, the *R wave,* followed by a small downward deflection, the *S wave*. The second major wave, termed the *T wave,* is inverted and occurs during repolarization. In electrocardiograms the analogous deflection occurring during ventricular depolarization is often triphasic and therefore is designated the *QRS complex*. The T wave is smaller but of greater duration than the QRS complex because the rate of repolarization is slower than the rate of depolarization (as is evident in the monophasic tracings shown in sections *1* and *2*). The portion of the tracing between the end of the QRS complex and the start of the T wave is called the *ST segment*. During this interval the regions under

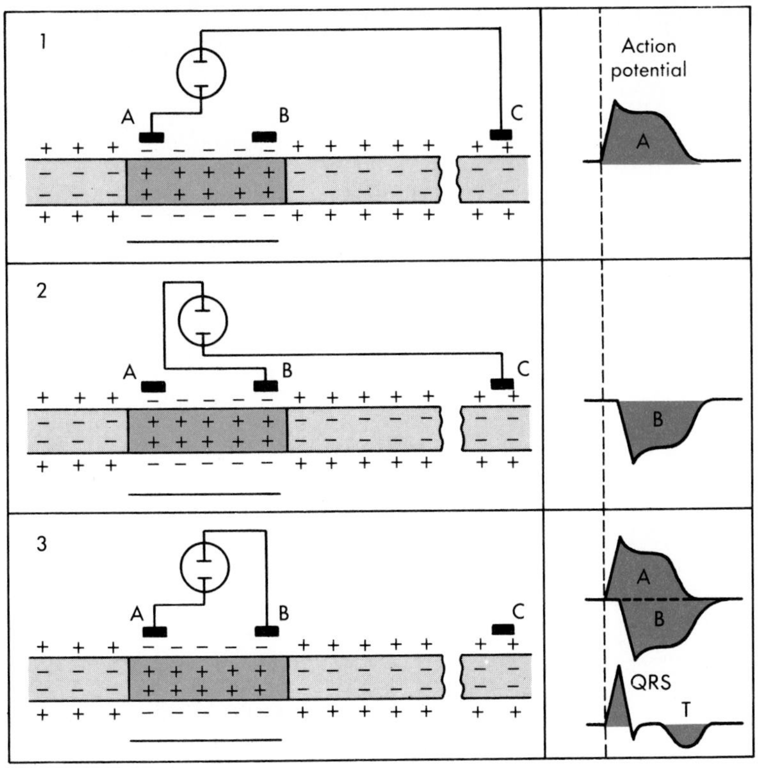

■ **Fig. 27-31.** Monophasic and biphasic action potentials recorded from the surface of a strip of cardiac muscle. In section **1,** electrodes *A* and *C* are connected to the oscilloscope as in the preceding figure, and the recorded action potential is therefore identical with that shown in Section **4** of Fig. 27-30. In section **2,** electrodes *B* and *C* are connected to the oscilloscope as shown, and the resulting deflection, *B,* will be inverted and displaced in time relative to deflection *A* in section **1.** In section **3,** electrodes *A* and *B* are connected to the oscilloscope, and the resulting QRS and T waves represent the algebraic sum of the individual deflections, *A* and *B*.

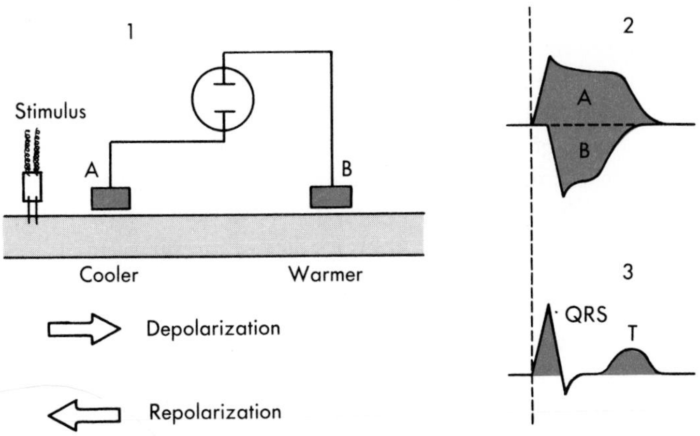

■ **Fig. 27-32.** When the wave of depolarization progresses in one direction along a cardiac muscle strip and the wave of repolarization travels in the opposite direction, then the QRS and T waves will be deflected in the same direction relative to the isoelectric line.

both electrodes are depolarized. Since they are of equal negativity, the ST segment lies along the *isoelectric line,* or line of zero potential difference.

In most normal electrocardiograms the T wave is deflected in the same direction as the QRS complex from the isoelectric line, contrary to the oppositely directed waves seen in section *3* of Fig. 27-31. The explanation may be illustrated by subjecting a strip of cardiac muscle to a temperature gradient (temperature increasing from left to right), as shown in Fig. 27-32. If a stimulus is applied to the left end, then the wave of depolarization is, of course, propagated from left to right. However, because the right end of the strip is warmer than the left end, the fibers in the right end may begin to repolarize before the fibers in the left end. Hence, the repolarization process may actually proceed from right to left, that is, the reverse of the direction of propagation of the wave of depolarization. Stated in another way, the action potential duration under electrode *A* (near the cooler end) will exceed that under *B* (near the warmer end). Therefore the T wave will be deflected in the same direction as the QRS complex under such conditions, as shown in Fig. 27-32. When depolarization and repolarization proceed in the same direction, the T wave is inverted with respect to the QRS complex; when these processes proceed in opposite directions, the deflections occur in the same direction.

As stated on p. 417, the wave of depolarization in the ventricles proceeds from endocardium to epicardium. However, the wave of repolarization normally travels in the opposite direction across the ventricular walls, producing the concordant QRS and T deflections. Stated in another way, the duration of the action potentials in the subendocardial cells is greater than that in the subepicardial cells (analogous to action potentials *A* and *B* in section *2* of Fig. 27-32). It has been postulated that the pressure developed in the ventricular chambers during systole retards the repolarization process more in the subendocardial than in the subepicardial region.

Scalar Electrocardiography

The systems of leads used to record routine electrocardiograms are oriented in certain planes of the body. The diverse electromotive forces that exist in the heart at any moment can be represented by a three-dimensional vector. A system of recording leads oriented in a given plane detects only the projection of the three-dimensional vector on that plane. Furthermore, the potential difference between two recording electrodes represents the projection of the vector on the line between the two leads. Components of vectors projected on such lines are not vectors but are *scalar quantities* (having magnitude but not direction). Hence, a recording of the changes over time of the differences of potential between two points on the surface of the skin is called a *scalar electrocardiogram*

Configuration of the scalar electrocardiogram. The scalar electrocardiogram reflects the changes over time of the electrical potential between pairs of points on the skin surface. The cardiac impulse progresses through the heart in an extremely complex three-dimensional pattern. Hence, the precise configuration of the electrocardiogram varies from individual to individual, and in any given individual the pattern varies with the anatomical location of the leads.

In general, the pattern consists of P, QRS, and T waves (Fig. 27-33). The P-R interval is a measure of the time from the onset of atrial activation to the onset of ventricular activation; it normally ranges from 0.12 to 0.20 s. A considerable fraction of this time involves passage of the impulse through the AV conduction system. Pathological prolongations of this interval are associated with disturbances of AV conduction produced by inflammatory, circulatory, pharmacological, or nervous mechanisms.

The configuration and amplitude of the QRS complex vary considerably among individuals. The duration is usually between 0.06 and 0.10 s. Abnormal prolongation may indicate a block in the normal conduction pathways through the ventricles (such as a block of the left or right bundle branch). During the ST interval the entire ventricular myocardium is depolarized. Therefore the ST segment lies on the isoelectric line under normal conditions. Any appreciable deviation from the isoelectric line is noteworthy and may indicate ischemic damage of the myocardium. The Q-T interval is sometimes referred to as the period of ''electrical systole'' of the ventricles. Its duration is about 0.4 s, but it varies inversely with the heart rate, mainly because the myocardial cell action potential duration varies inversely with the heart rate (see Fig. 27-15).

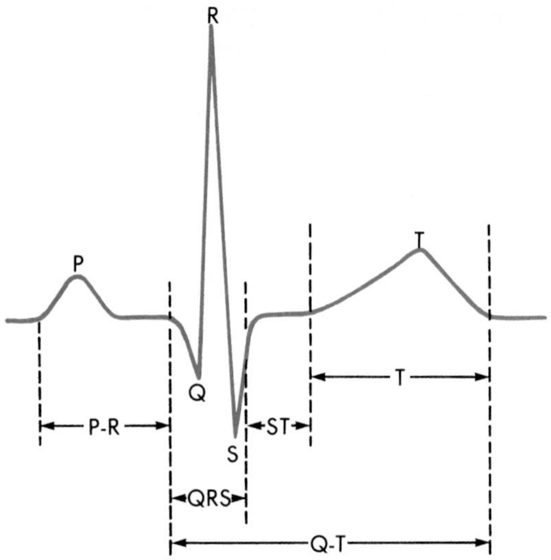

Fig. 27-33. Configuration of a typical scalar electrocardiogram, illustrating the important deflections and intervals.

In most leads the T wave is deflected in the same direction from the isoelectric line as the major component of the QRS complex, although biphasic or oppositely directed T waves are perfectly normal in certain leads. When the T wave and QRS complex deviate in the same direction from the isoelectric line, it indicates that the repolarization process does not follow the same route as the depolarization process, as explained previously (Fig. 27-32). T waves that are abnormal either in direction or amplitude may indicate myocardial damage, electrolyte disturbances, or cardiac hypertrophy.

Standard limb leads. The original electrocardiographic lead system was devised by Einthoven. In his lead system the *resultant cardiac vector* (the vector sum of all electrical activity occurring in the heart at any given moment) was considered to lie in the center of a triangle (assumed to be equilateral) formed by the left and right shoulders and the pubic region (Fig. 27-34). This triangle, called the *Einthoven triangle,* is oriented in the frontal plane of the body. Hence, only the projection of the resultant cardiac vector on the frontal plane will be detected by this system of leads. For convenience, the electrodes are connected to the right and left forearms rather than to the corresponding shoulders, since the arms are considered to represent simple extensions of the leads from the shoulders; this assumption has been validated experimentally. Similarly, the leg (the left leg, by convention) is taken as an extension of

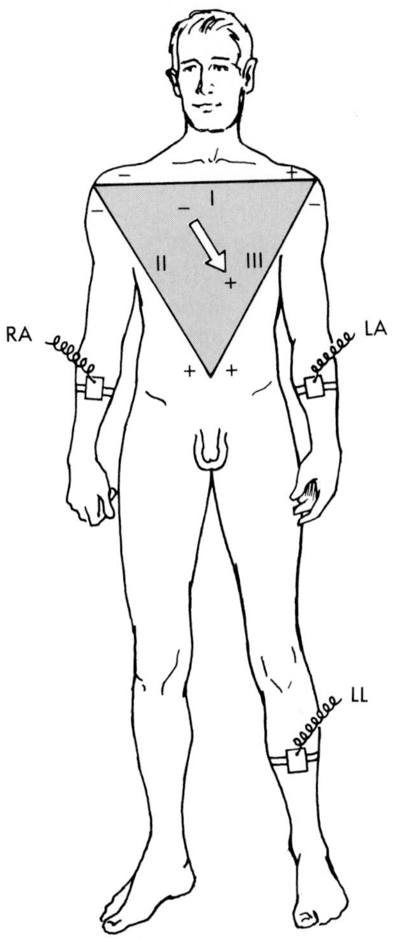

■ **Fig. 27-34.** Einthoven triangle, illustrating the galvanometer connections for standard limb leads I, II, and III.

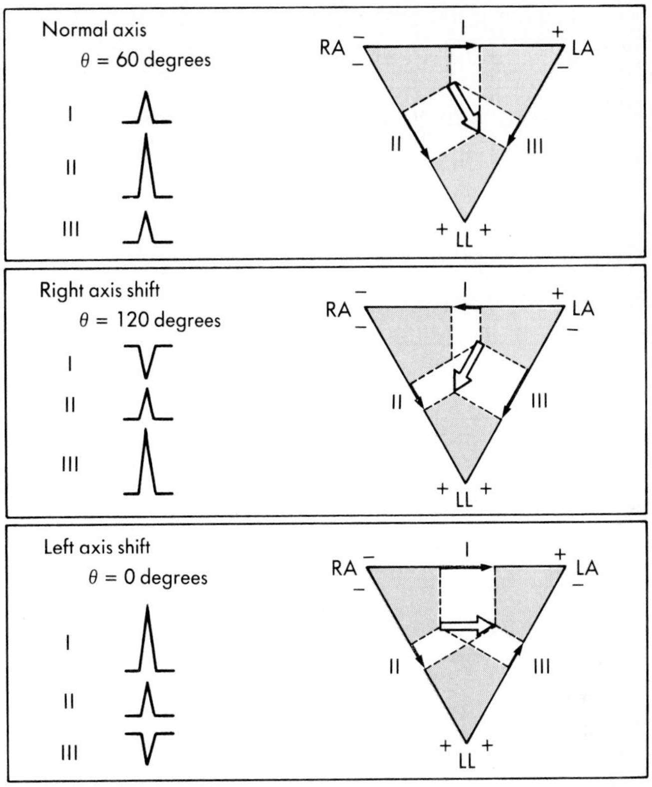

■ **Fig. 27-35.** Magnitude and direction of the QRS complexes in limb leads I, II, and III, when the mean electrical axis (θ) is 60 degrees *(top section),* 120 degrees *(middle section),* and 0 degrees *(bottom section).*

the lead system from the pubis, and thus the third electrode is connected to the left leg.

Certain conventions dictate the manner in which these *standard limb leads* are connected to the galvanometer. Lead I records the potential difference between the left arm (LA) and the right arm (RA). The galvanometer connections are such that when the potential at LA (V_{LA}) exceeds the potential at RA (V_{RA}), the galvanometer will be deflected upward from the isoelectric line. In Figs. 27-34 and 27-35 this arrangement of the galvanometer connections for lead I is designated by a (+) at LA and by a (−) at RA. Lead II records the potential difference between RA and LL (left leg) and yields an upward deflection when V_{LL} exceeds V_{RA}. Finally, lead III registers the potential difference between LA and LL and yields an upward deflection when V_{LL} exceeds V_{LA}. These galvanometer connections were arbitrarily chosen so that the QRS complexes will be upright in all three standard limb leads in the majority of normal individuals.

Let the frontal projection of the resultant cardiac vector at some moment be represented by an arrow (tail negative, head positive), as in Fig. 27-34. Then the potential difference, $V_{LA} - V_{RA}$, recorded in lead I will be represented by the component of the vector projected along the horizontal line between LA and RA, as shown in Fig. 27-35. If the vector makes an angle, θ, of 60 degrees with the horizontal (as in the top section of Fig. 27-35), the magnitude of the potential recorded by lead I will equal the vector magnitude times cosine 60 degrees. The deflection recorded in lead I will be upward, since the positive arrowhead lies closer to LA than to RA. The deflection in lead II also will be upright, since the arrowhead lies closer to LL than to RA. The magnitude of the lead II deflection will be greater than that in lead I because in this example the direction of the vector parallels that of lead II; therefore the magnitude of the projection on lead II exceeds that on lead I. Similarly, in lead III the deflection will be upright, and in this example, where θ = 60 degrees, its magnitude will equal that in lead I.

If the vector in the top section of Fig. 27-35 happens to represent the resultant of the electrical events occurring during the peak of the QRS complex, then the ori-

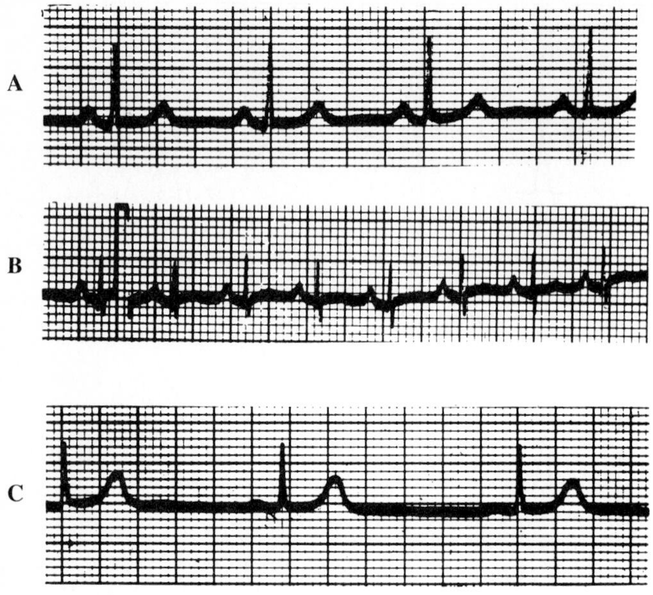

■ **Fig. 27-36.** Sinoatrial rhythms. **A,** Normal sinus rhythm. **B,** Sinus tachycardia. **C,** Sinus bradycardia.

entation of this vector is said to represent the *mean electrical axis* of the heart in the frontal plane. The positive direction of this axis is taken in the clockwise direction from the horizontal plane (contrary to the usual mathematical convention). For normal individuals the average mean electrical axis is approximately + 60 degrees (as in the top section of Fig. 27-35). Therefore the QRS complexes are usually upright in all three leads and largest in lead II.

Changes in the mean electrical axis may occur with alterations in the anatomical position of the heart or with changes in the relative preponderance of the right and left ventricles. For example, the axis tends to shift toward the left (more horizontal) in short, stocky individuals and toward the right (more vertical) in tall, thin persons. Also, with left or right ventricular hypertrophy (increased myocardial mass), the axis will shift toward the hypertrophied side.

With appreciable shift of the mean electrical axis to the right (middle section of Fig. 27-35, where $\theta = 120$ degrees), the displacements of the QRS complexes in the standard leads will change considerably. In this case the largest upright deflection will be in lead III and the deflection in lead I will be inverted, since the arrowhead will be closer to RA than to LA. With left axis shift (bottom section of Fig. 27-35, where $\theta = 0$ degrees), the largest upright deflection will be in lead I, and the QRS complex in lead III will be inverted.

As is evident from this discussion, the standard limb leads, I, II, and III, are oriented in the frontal plane at 0, 60, and 120 degrees, respectively, from the horizontal plane. Other limb leads, which are also oriented in the frontal plane, are usually recorded in addition to the standard leads. These "unipolar limb leads" lie along axes at angles of +90, −30, and −150 degrees from the horizontal plane. Such lead systems are described in all textbooks on electrocardiography and will not be considered further here.

To obtain information concerning the projections of the cardiac vector on the sagittal and transverse planes of the body in scalar electrocardiography, the *precordial leads* are usually recorded. Most commonly, each of six selected points on the anterior and lateral surfaces of the chest in the vicinity of the heart is connected in turn to the galvanometer. The other galvanometer terminal is usually connected to a *central terminal,* which is composed of a junction of three leads from LA, RA, and LL, each in series with a 5000 ohm resistor. It can be shown that the voltage of this central terminal remains at a theoretical zero potential throughout the cardiac cycle.

■ *Arrhythmias*

Cardiac arrhythmias reflect disturbances of either *impulse propagation* or *impulse initiation.* The principal disturbances of impulse propagation are conduction blocks and reentrant rhythms. Disturbances of impulse initiation include those that arise from the SA node and those that originate from various ectopic foci.

In the past it was believed that disturbances of impulse initiation, as a class, could be distinguished from reentry disturbances. The former are independent of any preceding depolarization, whereas the latter do depend on a preceding depolarization. This generalization no longer applies, however, because triggered activity (p. 418) constitutes a disturbance of impulse initiation, even though it depends on a preceding depolarization. For this and various other reasons, it is often difficult to determine the mechanism responsible for many of the rhythm disturbances that are observed clinically.

■ *Altered Sinoatrial Rhythms*

The frequency of pacemaker discharge varies by the mechanisms described earlier in this chapter (Fig. 27-18). Changes in SA nodal discharge frequency are usually produced by the cardiac autonomic nerves. Examples of electrocardiograms of sinus tachycardia and sinus bradycardia are shown in Fig. 27-36. The P, QRS, and T deflections are all normal, but the duration of the cardiac cycle (the *P-P interval*) is altered. Characteristically, when sinus bradycardia or tachycardia occurs, the cardiac frequency changes gradually and requires several beats to attain its new steady-state value. Electrocardiographic evidence of *respiratory cardiac arrhythmia* is common and is manifested as a rhythmic

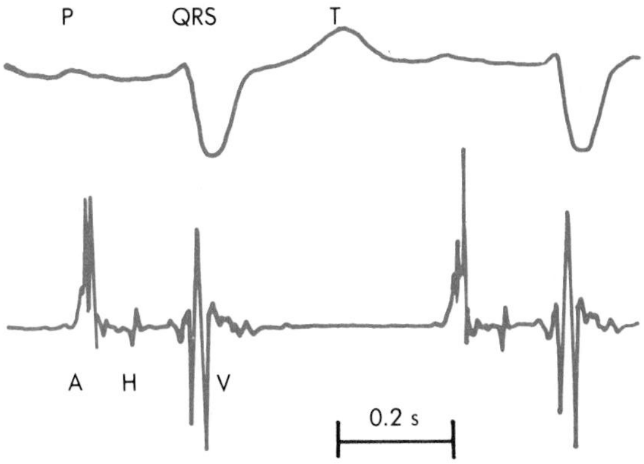

■ Fig. 27-37. His bundle electrogam (lower tracing, retouched) and lead II of the scalar electrocardiogram (upper tracing). The deflection, *H*, which represents the impulse conduction over the bundle of His, is clearly visible between the atrial, *A*, and ventricular, *V*, deflections. The conduction time from the atria to the bundle of His is denoted by the A-H interval; that from the bundle of His to the ventricles, by the H-V interval. (Courtesy Dr. J. Edelstein.)

A

B

C

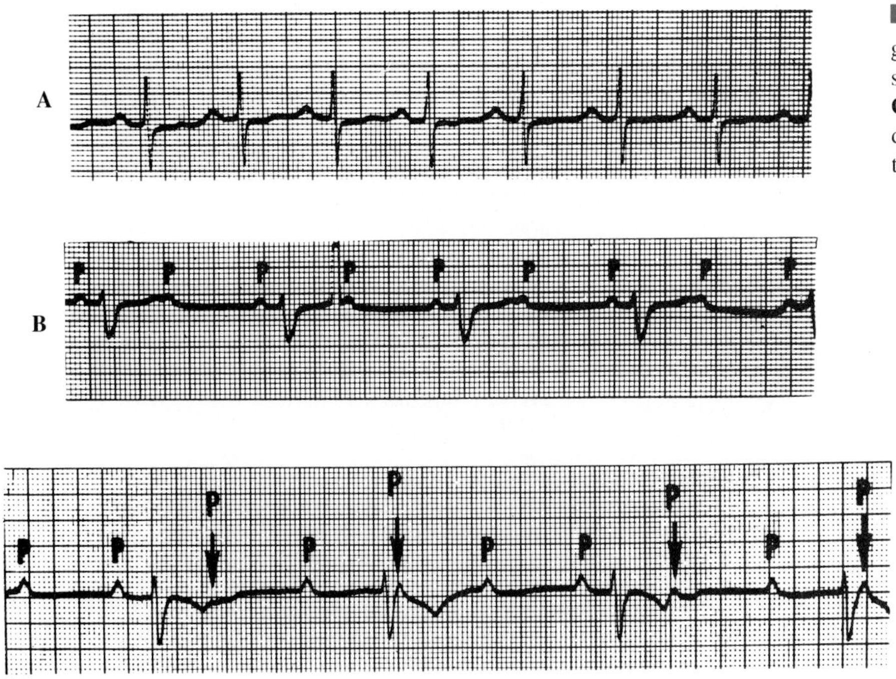

■ Fig. 27-38. AV blocks. **A,** First-degree heart block; P-R interval is 0.28 s. **B,** Second-degree heart block (2:1). **C,** Third-degree heart block; note the dissociation between the P waves and the QRS complexes.

variation in the P-P interval at the respiratory frequency (p. 456).

Atrioventricular Transmission Blocks

Various physiological, pharmacological, and pathological processes can impede impulse transmission through the AV conduction tissue. The site of block can be localized more precisely by recording the *His bundle electrogram* (Fig. 27-37). To obtain such tracings, an electrode catheter is introduced into a peripheral vein and is threaded centrally until the tip containing the electrodes lies in the junctional region between the right atrium and ventricle. When the electrodes are properly positioned, a distinct deflection (Fig. 27-37, *H*) is registered, which represents the passage of the cardiac action potential down the bundle of His. The time intervals required for propagation from the atrium to the bundle of His *(A-H interval)* and from the bundle of His to the ventricles *(H-V interval)* may be measured accurately. Abnormal prolongation of the former or latter interval indicates block above or below the bundle of His, respectively.

Three degrees of AV block can be distinguished, as shown in Fig. 27-38. *First-degree AV block* is characterized by a prolonged P-R interval. In Fig. 27-38, *A,* the P-R interval is 0.28 s; an interval greater than 0.2 s is abnormal. In most cases of first-degree block the A-H interval of the His bundle electrogram is prolonged, and the H-V interval is normal. Hence, the delay is usually located above the bundle of His, that is, in the AV node.

In *second-degree AV block* all QRS complexes are preceded by P waves, but not all P waves are followed by QRS complexes. The ratio of P waves to QRS complexes is usually the ratio of two small integers (such as 2:1, 3:1, 3:2). Fig. 27-38, *B*, illustrates a typical 2:1 block. His bundle electrograms have demonstrated that the site of block may be above or below the bundle of His. When the block arises above the bundle, the H deflections are absent after the A waves of the blocked beats. When the block occurs below the bundle, the H deflection is readily apparent after each A wave, but the V wave is absent during the blocked beats. This type of block implies a graver prognosis than when the block exists above the bundle, and an artificial pacemaker is frequently required.

Third-degree AV block is often referred to as *complete heart block* because the impulse is unable to traverse the AV conduction pathway from atria to ventricles. His bundle electrograms reveal that the most common sites of complete block are distal to the bundle of His. In complete heart block the atrial and ventricular rhythms are entirely independent. A classical example is displayed in Fig. 27-38, *C*, where the QRS complexes bear no fixed relationship to the P waves. Because of the slow ventricular rhythm (32 beats per minute in this example), circulation is often inadequate, especially during muscular activity. Third-degree block is often associated with syncope (so-called Stokes-Adams attacks) caused principally by insufficient cerebral blood flow. Third-degree block is one of the most common conditions requiring treatment by artificial pacemakers.

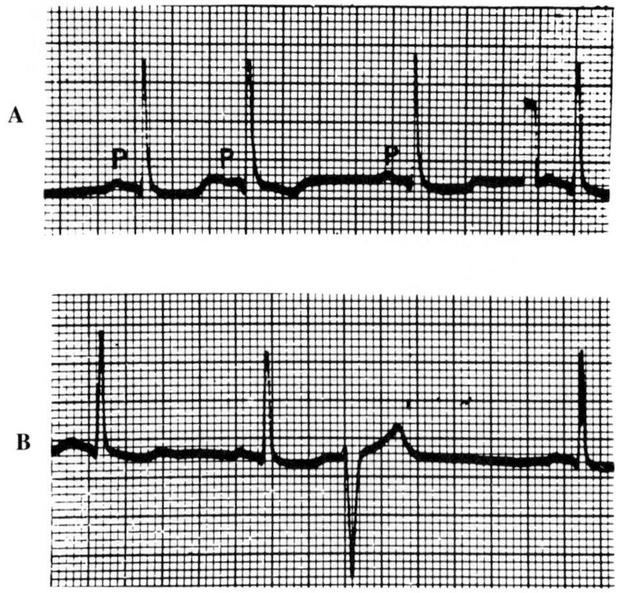

Fig. 27-39. A premature atrial systole, **A,** and a premature ventricular systole, **B,** recorded from the same patient. The premature atrial systole (the second beat in the top tracing) is characterized by an inverted P wave (superimposed on the preceding T wave) and normal QRS and T waves. The interval following the premature beat is not much longer than the usual interval between beats. The brief rectangular deflection just before the last beat is a standardization signal. The premature ventricular systole, **B,** is characterized by bizarre QRS and T waves and is followed by a compensatory pause.

Premature Systoles

Premature contractions occur at times in most normal individuals but are more common under certain abnormal conditions. They may originate in the atria, AV junction, or ventricles. One type of premature beat is coupled to a normally conducted beat. If the normal beat is suppressed in some way (for example, by vagal stimulation), the premature beat also will be abolished. Such premature beats are called *coupled extrasystoles,* or simply *extrasystoles,* and they probably reflect a reentry phenomenon (Fig. 27-27). A second type of premature beat occurs as the result of enhanced automaticity in some ectopic focus. This ectopic center may fire regularly and be protected in some way from depolarization by the normal cardiac impulse. If this premature beat occurs at a regular interval or at a simple multiple of that interval, the disturbance is called *parasystole*.

A *premature atrial systole* is shown in the electrocardiogram in Fig. 27-39, *A*. The normal interval between beats was 0.89 s (heart rate, 68 beats per minute). The premature atrial contraction (second P wave in the figure) followed the preceding P wave by only 0.56 s. The configuration of the premature P wave differs from the configuration of the other, normal P waves because the course of atrial excitation, originating at some ectopic focus in the atrium, is different from the normal spread of excitation originating at the SA node. The QRS complex of the premature beat is usually normal in configuration because the spread of ventricular excitation occurs over the usual pathways. However, the QRS complex of the premature atrial beat is sometimes bizarre, revealing *aberrant intraventricular conduction,* especially with very premature contractions. This is because the action potential duration of the ventricular Purkinje fibers is so long. Premature atrial impulses, traversing the AV conduction system, may find some of the Purkinje fibers still refractory from the preceding ventricular depolarization. In this event the sequence of ventricular activation may be abnormal, and therefore the configuration of the QRS may be aberrant.

The duration of the interval following a premature atrial systole depends, in part, on the location of the ectopic activity. The impulse is conducted from the ectopic site to all regions of the atria and to the SA node. If the impulse reaches the SA node before it has spontaneously generated its next natural impulse, the SA node is depolarized and then begins a completely new cycle of activity. The interval between the premature atrial beat and the next normal beat is therefore equal to the duration of a normal cardiac cycle plus the time required for the ectopic impulse to be conducted to the SA node. One additional factor that can also prolong this interval is that a premature depolarization of pacemaker cells may transiently depress their rhythmicity. In Fig. 27-39, *A*, the interval from the premature depolarization to the succeeding normal P wave is virtually equal to the normal cardiac cycle duration for that patient. Therefore the ectopic focus must have been near the SA node. If the ectopic focus is sufficiently far from the SA node, the SA node will fire naturally before the premature excitation wave will have reached it. In this event the natural cycle of activity of the SA node will not have been disturbed; the interval from the P wave of the beat just preceding the premature depolarizaiton to the P wave of the beat just following it will be equal to the duration of two normal cycles. However, if the SA node is depolarized by the ectopic excitation, then the interval from the beat preceding the premature depolarization to the beat following it will be significantly less than the duration of two normal cycles (such as in the example illustrated in Fig. 27-39, *A*).

A *premature ventricular systole,* recorded from the same patient, appears as tracing *B* of Fig. 27-39. Since the premature excitation originated at some ectopic focus in the ventricles, the impulse spread was aberrant and the configurations of the QRS and T waves were entirely different from the normal deflections. The premature QRS complex followed the preceding normal QRS complex by only 0.47 s. The interval following the premature excitation was 1.28 s, considerably longer than the normal interval between beats (0.89 s). The interval (1.75 s) from the QRS complex just before the premature excitation to the QRS complex just after it, was virtually equal to the duration of two normal cardiac cycles (1.78 s).

The prolonged interval that usually follows a premature ventricular systole is called a *compensatory pause*. The reason for the compensatory pause after a premature ventricular systole is that, contrary to most premature atrial systoles, the ectopic ventricular impulse does not disturb the natural rhythm of the SA node. Either the ectopic ventricular impulse is not conducted retrograde through the AV conduction system, or, if it is, the time required is such that the SA node has already fired at its natural interval before the ectopic impulse could have reached it. Likewise, the SA nodal impulse usually does not affect the ventricle, because the AV junction and perhaps also the ventricles are still refractory from the premature ventricular excitation. In Fig. 27-39, *B,* the P wave originating in the SA node at the time of the premature systole occurred at the same time as the T wave of the premature cycle and therefore cannot be identified in the tracing.

Ectopic Tachycardias

When a tachycardia originates from some ectopic site in the heart, the onset and termination are typically abrupt, in contrast to the more gradual changes in heart rate in sinus tachycardia. Because of the sudden appearance and abrupt reversion to normal, such ectopic tachycardias are usually referred to as *paroxysmal tachycardias*. Episodes of ectopic tachycardia may persist for only a few beats or for many hours or days, and the episodes often recur. Paroxysmal tachycardias may occur as the result of either (1) the rapid firing of an ectopic pacemaker, (2) triggered activity secondary to afterpotentials that reach threshold, or (3) an impulse circling a reentry loop repetitively.

Paroxysmal tachycardias originating either in the atria or in the AV junctional tissues are usually indistinguishable, and therefore both are included in the term *paroxysmal supraventricular tachycardia*. The tachycardia often results from an impulse repetitively circling a reentry loop that includes atrial tissue and the AV junction. An electrocardiogram illustrating this arrhythmia is shown in Fig. 27-40, *A*. The QRS complexes are normal, because ventricular activation proceeds over the normal pathways. When the supraventricular frequency is excessively rapid, the AV conduction tissue may be incapable of conducting all impulses, and second-degree AV blocks (for example, 2:1 block) may be a concomitant of the paroxysmal tachycardia. Not infrequently the supraventricular impulses may be conducted through the junctional tissues at a frequency that is too high for some of the ventricular Purkinje fibers (because of their long refractory periods). In this event the QRS complexes will reflect the resultant aberrant intraventricular conduction, and the arrhythmia may be very difficult to distinguish from a paroxysmal tachycardia of ventricular origin.

Paroxysmal ventricular tachycardia originates from an ectopic focus in the ventricles. The electrocardiogram is characterized by repeated, bizarre QRS complexes that reflect the aberrant intraventricular impulse conduction (Fig. 27-40, *B*). Paroxysmal ventricular tachycardia is much more ominous than supraventricular tachycardia because it is frequently a precursor of ventricular fibrillation, a lethal arrhythmia that will be described in the next section. The paroxysmal ventricular tachycardia illustrated in Fig. 27-40, *B,* was recorded from a patient immediately after resuscitation from ventricular fibrillation.

Fibrillation

Under certain conditions cardiac muscle undergoes an irregular type of contraction that is entirely ineffectual in propelling blood. Such an arrhythmia is termed *fi-*

A

B

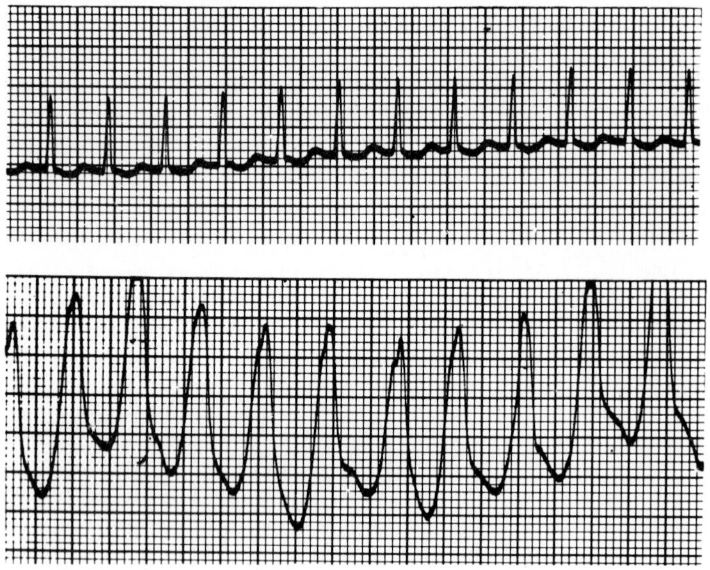

■ **Fig. 27-40.** Paroxysmal supraventricular **(A)** and ventricular **(B)** tachycardia.

A

B

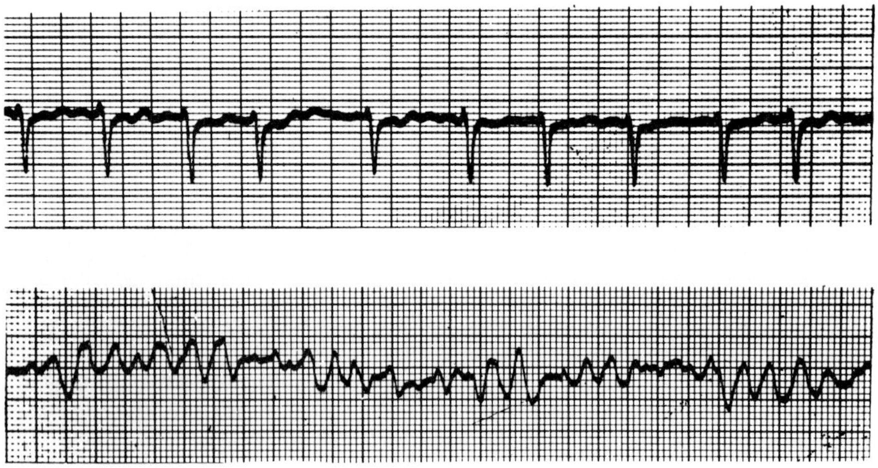

■ **Fig. 27-41.** Atrial **(A)** and ventricular **(B)** fibrillation.

brillation and may involve either the atria or ventricles. Fibrillation probably represents a reentry phenomenon, in which the reentry loop fragments into multiple, irregular circuits.

The tracing in Fig. 27-41, *A,* illustrates the electrocardiographic changes in *atrial fibrillation*. This condition occurs commonly in various types of chronic heart disease. The atria do not contract and relax sequentially during each cardiac cycle and, hence, do not contribute to ventricular filling. Instead, the atria undergo a continuous, uncoordinated, rippling type of activity. In the electrocardiogram, there are no P waves; they are replaced by continuous irregular fluctuations of potential, called *f* waves. The AV node is activated at intervals that may vary considerably from cycle to cycle. Hence, there is no constant interval between QRS complexes and therefore between ventricular contractions. Because the strength of ventricular contraction depends on the interval between beats (because it determines the time available for ventricular filling, and for other reasons as well, as explained on p. 464), the pulse is extremely irregular with regard to both rhythm and force. In many patients the atrial reentry loop and the pattern of AV conduction are more regular. The rhythm is then referred to as *atrial flutter*.

Although atrial fibrillation and flutter are compatible with life and even with full activity, the onset of *ventricular fibrillation* leads to loss of consciousness within a few seconds. The irregular, continuous, uncoordinated twitchings of the ventricular muscle fibers pump no blood. Death ensues unless immediate, effective resuscitation is achieved or unless the rhythm reverts to normal spontaneously, which rarely occurs. Ventricular fibrillation may supervene when the entire ventricle, or some portion of it, is deprived of its normal blood supply. It may also occur as a result of electrocution or in response to certain drugs and anesthetics. In the electrocardiogram (Fig. 27-41, *B*) irregular fluctuations of potential are manifest.

Fibrillation is often initiated when a premature impulse arrives during the *vulnerable period*. In the ventricles, this period coincides with the downslope of the T wave. During this period, the excitability of the cardiac cells varies. Some fibers are still in their effective refractory periods, others have almost fully recovered their excitability, and still others are able to conduct impulses, but only at very slow conduction velocities. As a consequence, the action potentials are propagated over the chambers in multiple wavelets that travel along circuitous paths and at various conduction velocities. As a region of cardiac cells becomes excitable again, it will ultimately be reentered by one of the wave fronts traveling about the chamber. The process tends to be self-sustaining.

Atrial fibrillation may be reverted to a normal sinus rhythm by drugs that prolong the refractory period. As the cardiac impulse completes the reentry loop, it may then find the myocardial fibers no longer excitable. However, much more dramatic therapy is required in ventricular fibrillation. Conversion to a normal sinus rhythm is accomplished by means of a strong electric current that places the entire myocardium briefly in a refractory state. Techniques have been developed to safely administer the current through the intact chest wall. In successful cases the SA node again takes over as the normal pacemaker for the entire heart. Direct current shock has been found to be more effective than alternating current shock. Such shocks are now widely used to treat not only ventricular fibrillation, but atrial fibrillation and certain other arrhythmias as well. Electrical abolition of such disturbances of rhythm is often called *cardioversion*.

Bibliography

Journal articles

Abildskov, J.A.: The sequence of normal recovery of excitability in the dog heart, Circulation **52:**442, 1975.

Armstrong, C.W.: Sodium channels and gating currents, Physiol. Rev. **61:**644, 1981.

Brown, H.F.: Electrophysiology of the sinoatrial node, Physiol. Rev. **62:**505, 1982.

Childers, R.: The AV node: normal and abnormal physiology, Prog. Cardiovasc. Dis. **19:**361, 1977.

De Mello, W.C.: Modulation of junctional permeability, Fed. Proc. **43:**2692, 1984.

Glitsch, H.G.: Characteristics of active Na transport in intact cardiac cells, Am. J. Physiol. **236:**H189, 1979.

Hauswirth, O., and Singh, B.N.: Ionic mechanisms in heart muscle in relation to the genesis and the pathological control of cardiac arrhythmias, Pharmacol. Rev. **30:**5, 1979.

Hoffman, B.F., and Rosen, M.R.: Cellular mechanisms for cardiac arrhythmias, Circ. Res. **49:**69, 1981.

Horackova, M.: Transmembrane calcium transport and the activation of cardiac contraction, Can. J. Physiol. Pharmacol. **62:**874, 1984.

Jalife, J., and Moe, G.K.: Phasic effects of vagal stimulation on pacemaker activity of the isolated sinus node of the young cat, Circ. Res. **45:**595, 1979.

Kohlhardt, M., and Mnich, Z.: Studies on the inhibitory effect of verapamil on the slow inward currents in mammalian ventricular myocardium, J. Mol. Cell. Cardiol. **10:**1037, 1978.

Langer, G.A.: Sodium-calcium exchange in the heart, Annu. Rev. Physiol. **44:**435, 1982.

Lee, C.O.: Ionic activities in cardiac muscle cells and application of ion-selective microelectrodes, Am. J. Physiol. **241:**H459, 1981.

Maylie, J., and Morad, M.: Ionic currents responsible for the generation of pacemaker current in the rabbit sinoatrial node, J. Physiol. **355:**215, 1984.

Pappano, A.J.: Ontogenetic development of autonomic neuroeffector transmission and transmitter reactivity in embryonic and fetal hearts, Pharmacol. Rev. **29:**3, 1977.

Shamoo, A.E., and Ambudkar, I.S.: Regulation of calcium transport in cardiac cells, Can. J. Physiol. Pharmacol. **62:**9, 1984.

Singer, D.H., Baumgarten, C.M., and Ten Eick, R.E.: Cellular electrophysiology of ventricular and other dysrhythmias: studies on diseased and ischemic heart, Prog. Cardiovasc. Dis. **24:**97, 1981.

Spach, M.S., and Kootsey, J.M.: The nature of electrical propagation in cardiac muscle, Am. J. Physiol. **244:**H3, 1983.

Sperelakis, N.: Hormonal and neurotransmitter regulation of Ca^{++} influx through voltage-dependent slow channels in cardiac muscle membrane, Membr. Biochem. **5:**131, 1984.

Strauss, H.C., Prystowsky, R.N., and Scheinman, M.M.: Sinoatrial and atrial electrogenesis, Prog. Cardiovasc. Dis. **19:**385, 1977.

Ten Eick, R.E., Baumgarten, C.M., and Singer, D.H.: Ventricular dysrhythmia: membrane basis, or of currents, channels, gates, and cables, Prog. Cardiovasc. Dis. **24:**157, 1981.

Wit, A.L., and Cranefield, P.F.: Reentrant excitation as a cause of cardiac arrhythmias, Am. J. Physiol. **235:**H1, 1978.

Books and monographs

Blaustein, M.P., and Lieberman, M.: Electrogenic transport: fundamental principles and physiological implications, New York, 1984, Raven Press.

Bouman L.N., and Jongsma, H.J.: Cardiac rate and rhythm, The Hague, 1982, Martinus Nijhoff Publishers.

Fozzard, H.A.: Conduction of the action potential. In Handbook of physiology; Section 2: The cardiovascular system—the heart, vol. I, Bethesda, Md., 1979, American Physiological Society.

Levy, M.N., and Vassalle, M.: Excitation and neural control of the heart, Bethesda, Md., 1982, American Physiological Society.

Mullins, L.J.: Ion transport in heart, New York, 1981, Raven Press.

Nelson, C.V., and Geselowitz, D.B.: Theoretical basis of electrocardiology, Oxford, 1976, Clarendon Press.

Noble, D.: The initiation of the heartbeat, ed. 2, Oxford, 1979, Oxford University Press.

Paes de Carvalho, A., Hoffman, B.F., and Lieberman, M.: Normal and abnormal conduction in the heart, Mt. Kisco, N.Y., 1982, Futura Publishing Co., Inc.

Sakmann, B., and Neher, E.: Single channel recording, New York, 1983, Plenum Press.

Scher, A.M., and Spach, M.S.: Cardiac depolarization and repolarization and electrocardiogram. In Handbook of physiology; Section 2: The cardiovascular system—the heart, vol. I, Bethesda, Md., 1979, American Physiological Society.

Sperelakis, N.: Origin of the cardiac resting potential. In Handbook of physiology; Section 2: The cardiovascular system—the heart, vol. I, Bethesda, Md., 1979, American Physiological Society.

Sperelakis, N.: Physiology and pathophysiology of the heart, Hingham, Mass., 1984, Martinus Nijhoff Publishers.

Stein, W.D.: Ion channels: molecular and physiological aspects, Orlando, Fla., 1985, Academic Press, Inc.

Zipes, D.P., and Jalife J.: Cardiac electrophysiology and arrhythmias, Orlando, Fla., 1985, Grune & Stratton, Inc.

CHAPTER

28

The Cardiac Pump

It is nearly impossible to contemplate the pumping action of the heart without being struck by its simplicity of design, its wide range of activity and functional capacity, and the staggering amount of work it performs relentlessly over the lifetime of an individual. To understand how the heart accomplishes its important task, it is first necessary to consider the relationships between the structure and function of its components.

■ *Structure of the Heart in Relation to Function*

■ *Myocardial Cell*

A number of important morphological and functional differences exist between myocardial and skeletal muscle cells (see also Chapter 22). However, the contractile elements within the two types of cells are quite similar; each skeletal and cardiac muscle cell is made up of *sarcomeres* (from Z line to Z line) containing thick filaments composed of myosin (in the A band) and thin filaments containing actin. The thin filaments extend from the point where they are anchored to the Z line (through the I band) to interdigitate with the thick filaments. As in the case of skeletal muscle, shortening occurs by the sliding filament mechanism. Actin filaments slide along adjacent myosin filaments by cycling of the intervening crossbridges, thereby bringing the Z lines closer together.

Skeletal and cardiac muscle show similar length-force relationships. The sarcomere length has been determined with electron microscopy in papillary muscles and in intact ventricles rapidly fixed during systole or diastole. Maximal developed force is observed at resting sarcomere lengths of 2 to 2.4 μm for cardiac muscle. At such lengths, there is overlap of thick and thin filaments, and a maximal number of crossbridge attachments. Developed force of cardiac muscle is less than the maximum value when the sarcomeres are stretched beyond the optimum length, because of less overlap of the filaments, and hence less cycling of the crossbridges. At resting sarcomere lengths shorter than the optimum value, the thin filaments overlap each other, which diminishes contractile force.

In general, the length-force relationship for fibers in the papillary muscle also holds true for fibers in the intact heart. This relationship may be expressed graphically, as in Fig. 28-1, by substituting ventricular systolic pressure for force and end-diastolic ventricular volume for myocardial resting fiber (and hence sarcomere) length. The lower curve in Fig. 28-1 represents the increment in pressure produced by each increment in volume when the heart is in diastole. The upper curve represents the peak pressure developed by the ventricle during systole at each degree of filling and illustrates the *Frank-Starling relationship* of initial myocardial fiber length (or initial volume) to force (or pressure) development by the ventricle. Note that the pressure-volume curve for the ventricle in diastole is initially quite flat, indicating that large increases in volume can be accommodated with only small increases in pressure. Nevertheless, systolic pressure development is considerable at the lower filling pressures. However, the ventricle becomes much less distensible with greater filling, as evidenced by the sharp rise of the diastolic curve at large intraventricular volumes. In the normal intact heart, peak force may be attained at a filling pressure of 12 mm Hg. At this intraventricular diastolic pressure, which is about the upper limit observed in the normal heart, the sarcomere length is 2.2 μm. However, developed force peaks at filling pressures as high as 30 mm Hg in the isolated heart. Even at higher diastolic pressure (>50 mm Hg) the sarcomere length is not greater than 2.6 μm in cardiac muscle. This resistance to stretch of the myocardium at high filling pressures probably resides in the noncontractile constituents of the tissue (connective tissue) and may serve as a safety factor against overloading of the heart in diastole. Usually, ventricular diastolic pressure is about 0 to 7 mm Hg, and the average diastolic sarcomere length is about 2.2 μm. Thus the normal heart operates on the

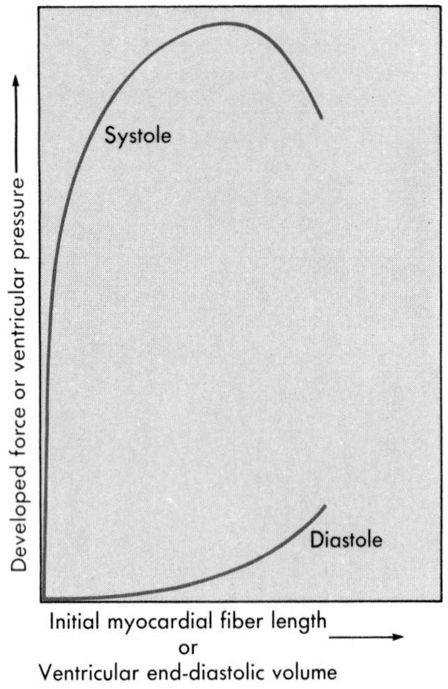

Fig. 28-1. Relationship of myocardial resting fiber length (sarcomere length) or end-diastolic volume to developed force or peak systolic ventricular pressure during ventricular concentration in the intact dog heart. (Redrawn from Patterson, S.W., Piper, H., and Starling, E.H.: J. Physiol. **48:**465, 1914.)

ascending portion of the Frank-Starling curve depicted in Fig. 28-1.

A striking difference in the appearance of cardiac and skeletal muscle is the semblance of a syncytium in cardiac muscle with branching interconnecting fibers (Figs. 28-2 and 28-3). However, the myocardium is not a true anatomical syncytium, since laterally the myocardial fibers are separated from adjacent fibers by their respective sarcolemmas. The end of each fiber is separated from its neighbor by dense structures, *intercalated disks,* that are continuous with the sarcolemma (Figs. 28-2 to 28-4). Nevertheless, cardiac muscle functions as a syncytium, since a wave of depolarization followed by contraction of the entire myocardium (an all-or-none response) occurs when a suprathreshold stimulus is applied to any one focus.

As the wave of excitation approaches the end of a cardiac cell, the spread of excitation to the next cell depends on the electrical conductance of the boundary

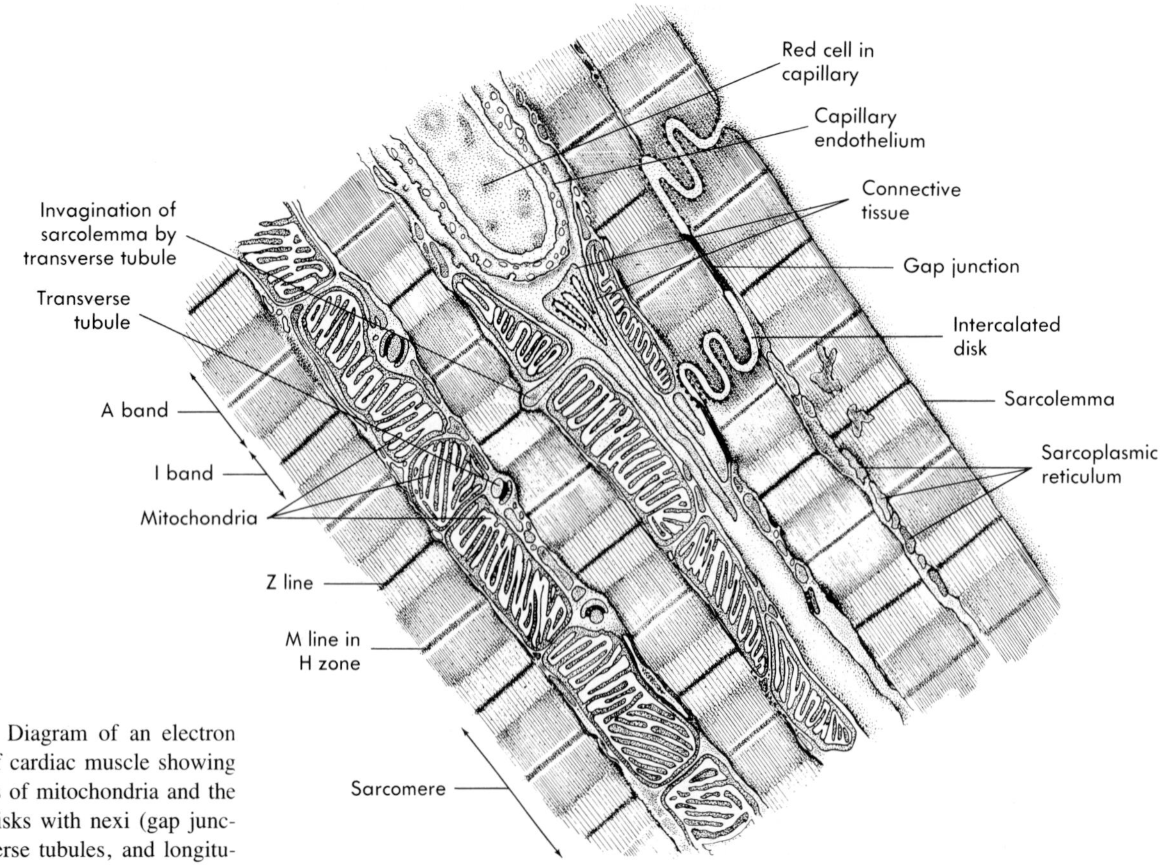

Fig. 28-2. Diagram of an electron micrograph of cardiac muscle showing large numbers of mitochondria and the intercalated disks with nexi (gap junctions), transverse tubules, and longitudinal tubules.

between the two cells. Gap junctions (nexi) with high conductances are present in the intercalated disks between adjacent cells (Figs. 28-2 to 28-4). These gap junctions facilitate the conduction of the cardiac impulse from one cell to the next. Impulse conduction in cardiac tissues progresses more rapidly in a direction parallel to the long axes of the constituent fibers than in a direction perpendicular to the long axes of those fibers. Gap junctions exist in the borders between myocardial fibers that are in contact with each other longitudinally; they are very sparse or absent in the borders between myocardial fibers that lie side by side.

Another difference between cardiac and fast skeletal muscle fibers is in the abundance of mitochondria *(sarcosomes)* in the two tissues. Fast skeletal muscle, which is called on for relatively short periods of repetitive or sustained contraction and which can metabolize anaerobically and build up a substantial oxygen debt, has relatively few mitochondria in the muscle fibers (Chapter 23). In contrast, cardiac muscle, which is required

A

B

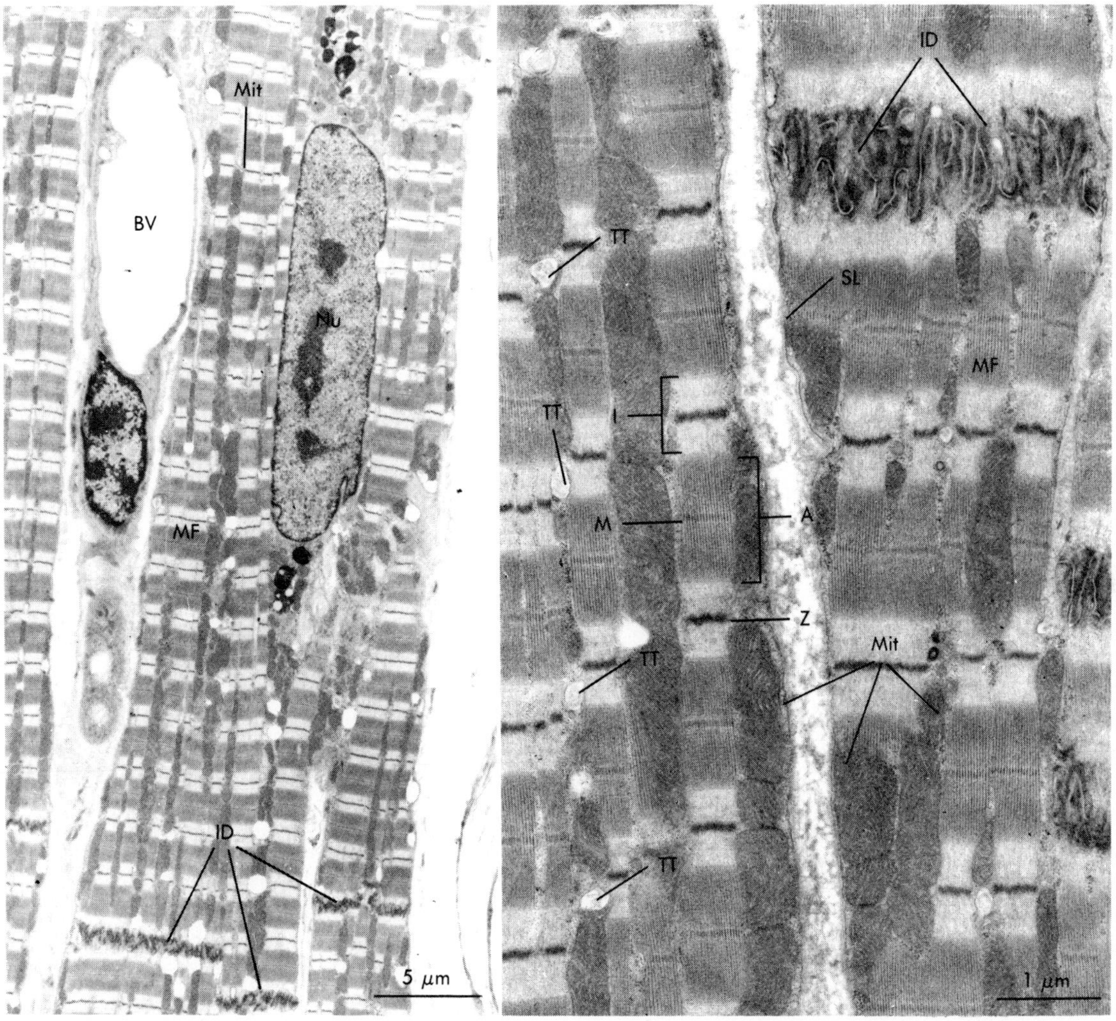

■ **Fig. 28-3. A,** Low-magnification electron micrograph of a monkey heart (ventricle). Typical features of myocardial cells include the elongated nucleus *(Nu),* striated myofibrils *(MF)* with columns of mitochondria *(Mit)* between the myofibrils, and intercellular junctions (intercalated disks, *ID*). A blood vessel *(BV)* is located between two myocardial cells. **B,** Medium-magnification electron micrograph of monkey ventricular cells, showing details of untrastructure. The sarcolemma *(SL)* is the boundary of the muscle cells and is thrown into multiple folds where the cells meet at the intercalated disk region *(ID).* The prominent myofibrils *(MF)* show distinct banding patterns, including the A band *(A),* dark Z lines *(Z),* I band regions *(I),* and M lines *(M)* at the center of each sarcomere unit. Mitochondria *(Mit)* occur either in rows between myofibrils or masses just underneath the sarcolemma. Regularly spaced transverse tubules *(TT)* appear at the Z line levels of the myofibrils.

Continued.

C 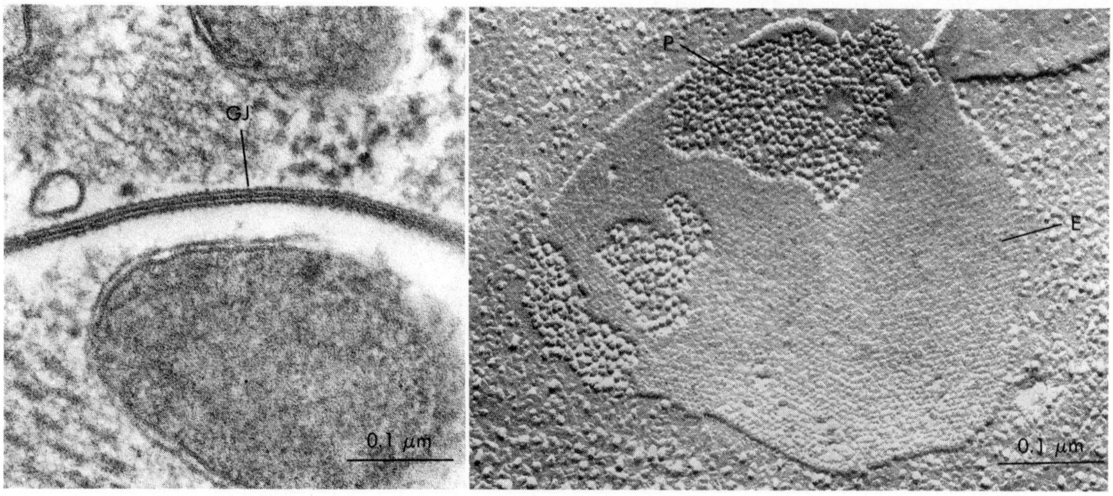 D

■ **Fig. 28-3, cont'd. C,** High-magnification electron micrograph of a specialized intercellular junction between two myocardial cells of the mouse. Called a *gap junction (GJ)* or nexus, this attachment consists of very close apposition of the sarcolemmal membranes of the two cells and appears in thin section to consist of seven layers. **D,** Freeze-fracture replica of mouse myocardial gap junction, showing distinct arrays of characteristic intramembranous particles. Large particles *(P)* belong to the inner half of the sarcolemma of one myocardial cell, whereas the "pitted" membrane face *(E)* is formed by the outer half of the sarcolemma of the cell above. (Electron micrographs courtesy of Dr. Michael S. Forbes.)

to contract repetitively for a lifetime and which is incapable of developing a significant oxygen debt, is very rich in mitochondria (Figs. 28-2 to 28-4). Rapid oxidation of substrates with the synthesis of adenosine triphosphate (ATP) can keep pace with the myocardial energy requirements because of the large numbers of mitochondria containing the respiratory enzymes necessary for oxidative phosphorylation.

To provide adequate oxygen and substrate for its metabolic machinery, the myocardium is also endowed with a rich capillary supply, about one capillary per fiber. Thus diffusion distances are short, and oxygen, carbon dioxide, substrates, and waste material can move rapidly between myocardial cell and capillary. With respect to exchange of substances between the capillary blood and the myocardial cells, electron micrographs of myocardium show deep invaginations of the sarcolemma into the fiber at the Z lines (Figs. 28-2 to 28-4). These sarcolemmal invaginations constitute the transverse-tubular, or *T-tubular system*. The lumina of these T-tubules are continuous with the bulk interstitial fluid, and they play a key role in excitation-contraction coupling.

In mammalian ventricular cells, adjacent T-tubules are interconnected by longitudinally running or axial tubules, thus forming an extensively interconnected lattice of "intracellular" tubules (Fig. 28-4). This T-tubule system is open to the interstitial fluid, is lined with a basement membrane continuous with that of the surface sarcolemma, and contains micropinocytotic-like vesicles (Fig. 28-5). Thus in ventricular cells the myofibrils and mitochondria have ready access to a space that is continuous with the interstitial fluid. The T-tubular system is absent or poorly developed in atrial cells of many mammalian hearts.

A network of sarcoplasmic reticulum (Figs. 28-4 and 28-5) consisting of small-diameter sarcotubules is also present surrounding the myofibrils; these sarcotubules are believed to be "closed," since colloidal tracer particles (2 to 10 nm in diameter) do not enter them. They do not contain basement membrane. Flattened elements of the sarcoplasmic reticulum are often found very close to the T-tubular system, as well as to the surface sarcolemma, forming "diads" (Fig. 28-5).

Excitation-contraction coupling. The earliest studies on isolated hearts perfused with isotonic saline solutions indicated the need for optimum concentrations of Na^+, K^+, and Ca^{++}. In the absence of Na^+ the heart is not excitable and will not beat, because the action potential depends on extracellular Na ions. In contrast, the resting membrane potential is independent of the Na ion gradient across the membrane (Fig. 28-6). Under normal conditions the extracellular K^+ concentration is about 4 mM. A reduction in extracellular K^+ has little effect on myocardial excitation and contraction. However, increases in extracellular K^+, if great enough, produce depolarization, loss of excitability of the myocardial cells, and cardiac arrest in diastole. Ca^{++} is also essential for cardiac contraction; removal of Ca^{++} from the extracellular fluid results in de-

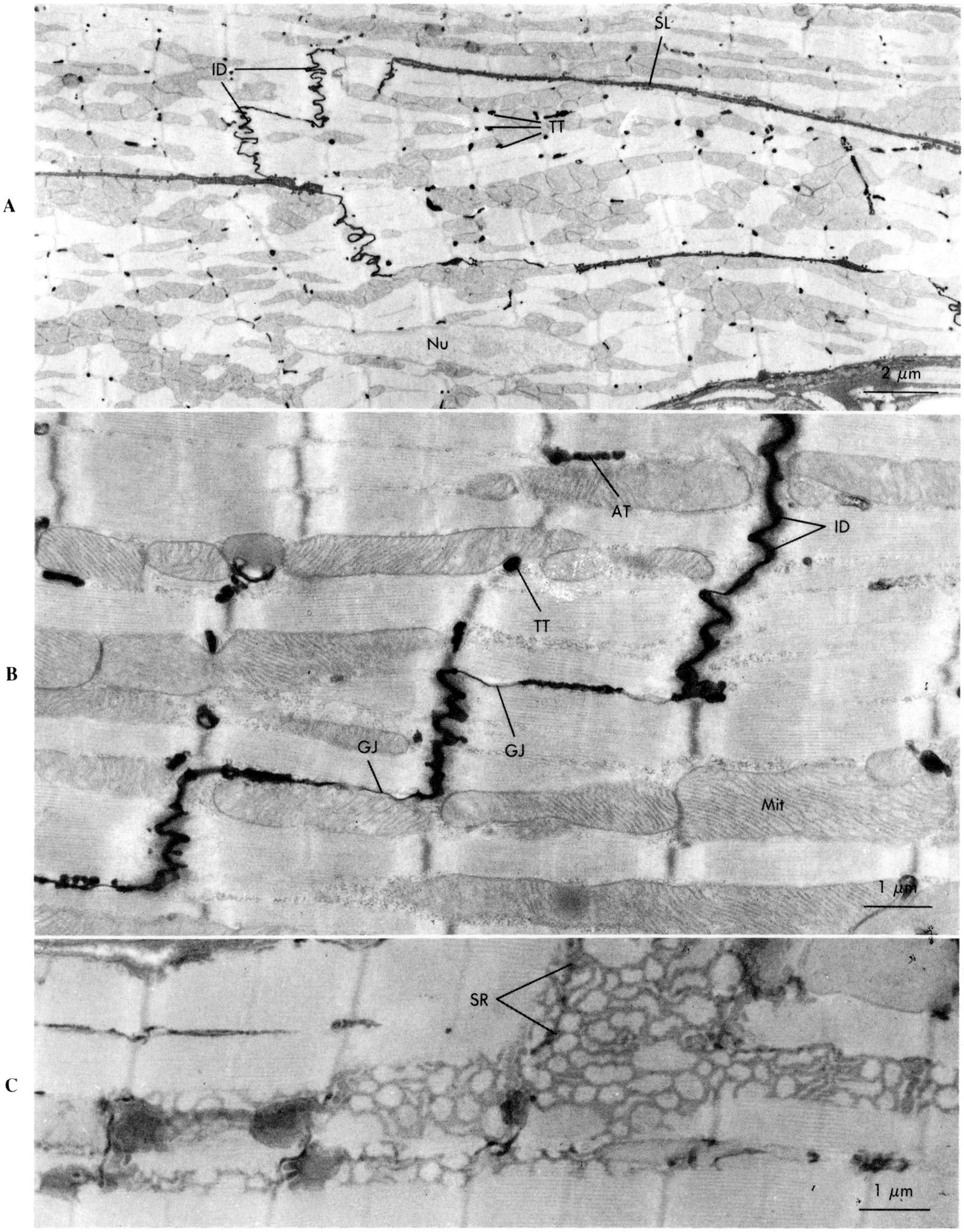

■ **Fig. 28-4. A,** Low-magnification electron micrograph of the right ventricular wall of a mouse heart. Tissue was fixed in a phosphate-buffered glutaraldehyde solution and postfixed in ferrocyanide-reduced osmium tetroxide. This procedure has resulted in the deposition of electronopaque precipitate in the extracellular space, thus outlining the sarcolemmal borders *(SL)* of the muscle cells and delineating the intercalated disks *(ID)* and transvere tubules *(TT)*. *Nu,* Nucleus of the myocardial cell. **B,** Mouse cardiac muscle in longitudinal section, treated as in panel **A.** The path of the extracellular space is traced through the intercalated disk region *(ID),* and sarcolemmal invaginations that are oriented transverse to the cell axis (transverse tubules, *TT*) or parallel to it (axial tubules, *AT*) are clearly identified. Gap junctions *(GJ)* are associated with the intercalated disk. Mitochondria are large and elongated and lie between the myofibrils. **C,** Mouse cardiac muscle. Tissue treated with ferrocyanide-reduced osmium tetroxide so as to identify the internal membrane system (sarcoplasmic reticulum, *SR*). Specific staining of the SR reveals its architecture as a complex network of small-diameter tubules that are closely associated with the myofibrils and mitochondria. (Electron micrographs courtesy Dr. Michael S. Forbes.)

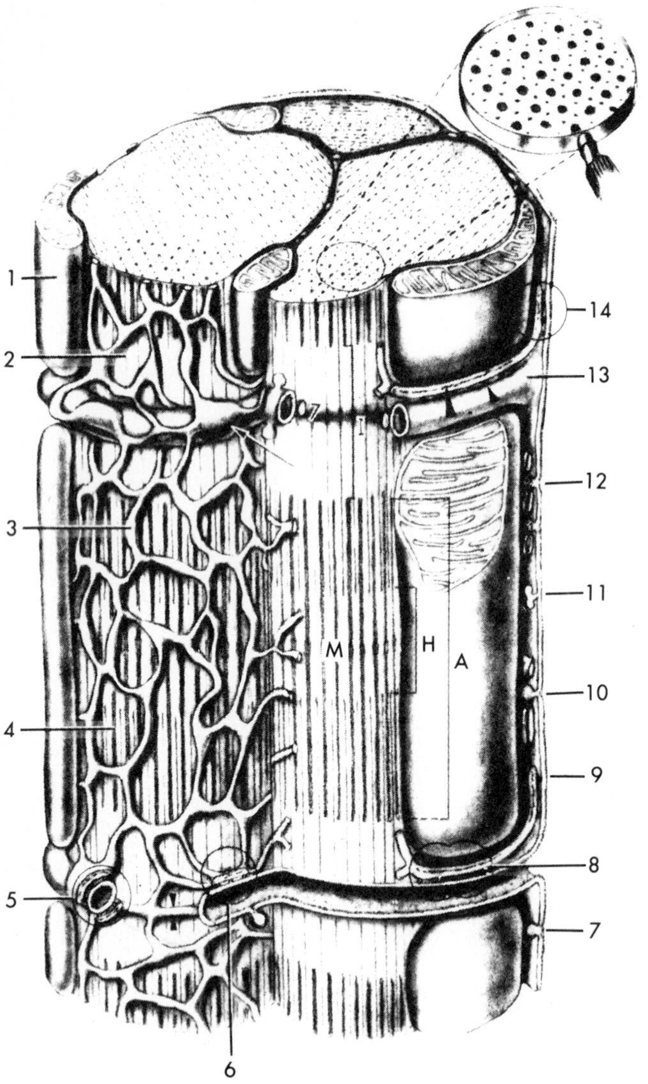

■ **Fig. 28-5.** Model of mammalian cardiac muscle showing a portion of a muscle fiber; *1,* mitochondrion; *2,* actin (thin) filament; *3,* sarcoplasmic reticulum; *4,* myosin (thick) filament; *5,* triad, made up of two couplings at a transverse tubule; *6,* diad, made up of one coupling at a transverse tubule. The couplings in *5,6,* and *8* are interior couplings. *7,* Pit (or coated vesicle) also commonly seen at transverse tubules; *8,* interior coupling; *9,*sarcolemma; *10,* pinocytotic vesicle or caveola; *11,* branched caveola; *12,* caveola containing dense granules; *13,* transverse tubule; *14,* peripheral coupling. *Large arrowhead,* Junctional granules (or central membrane), only within junctional sarcoplasmic reticulum. *Small arrowhead,* Junctional processes. *Arrow,* Junctional sarcoplasmic reticulum with connections of the sarcosplasmic reticulum across the Z line. Also shown are the M line, A band, H band, and I band. Note the narrow junction of the sarcoplasmic reticulum of the couplings at 5, 6, 8, and 14. (Reproduced by permission from Z. Zellforsch. **98:**437, 1969.)

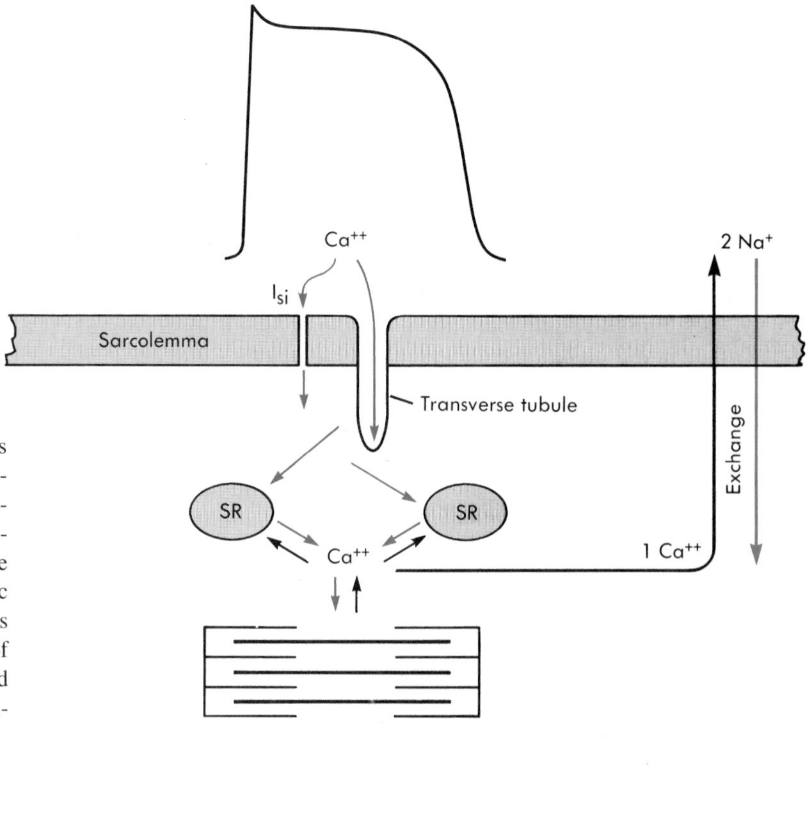

■ **Fig. 28-6.** Schematic diagram of the movements of calcium in excitation-contraction coupling in cardiac muscle. The influx of Ca^{++} from the interstitial fluid during excitation (plateau of action potential) triggers the release of Ca^{++} from the sarcoplasmic reticulum *(SR).* The free cytosolic Ca^{++} activates contraction of the myofilaments (systole). Relaxation (diastole) occurs as a result of uptake of Ca^{++} by the sarcoplasmic reticulum and extrusion of intracellular Ca^{++} by Na^+-Ca^{++} exchange.

creased contractile force and eventually arrest in diastole. Conversely, an increase in extracellular Ca^{++} enhances contractile force, and at very high Ca^{++} concentrations cardiac arrest occurs in systole (rigor). It is now well documented that free intracellular Ca^{++} is the agent responsible for the contractile state of the myocardium.

Initially a wave of excitation spreads rapidly along the myocardial sarcolemma from cell to cell via gap junctions. Excitation also spreads into the interior of the cells via the T-tubules (Figs. 28-2 to 28-5), which invaginate the cardiac fibers at the Z lines. (Electrical stimulation at the Z line or the application of ionized Ca to the Z lines in the skinned [sarcolemma removed] cardiac fiber elicits a localized contraction of adjacent myofibrils.) During the plateau (phase 2) of the action potential, Ca^{++} permeability of the sarcolemma increases. Ca^{++} flows down its electrochemical gradient and is largely responsible for the slow inward current (Chapter 27). Ca^{++} enters the cell through Ca^{++} channels in the sarcolemma and in the invaginations of the sarcolemma, the T-tubules (Fig. 28-6). Opening of the Ca^{++} channels is believed to be caused by phosphorylation of the channel proteins by a cyclic AMP (cAMP)-dependent protein kinase. The primary source of extracellular Ca^{++} is the interstitial fluid (10^{-3}M of Ca^{++}). Some Ca^{++} also may be bound to the sarcolemma and to the *glycocalyx,* a mucopolysaccharide that covers the sarcolemma. The amount of calcium entering the cell interior from the extracellular space is not sufficient to induce contraction of the myofibrils, but it serves as a trigger *(trigger Ca^{++})* to release Ca^{++} from the intracellular Ca^{++} stores, the sarcoplasmic reticulum (Fig. 28-6). The cytosolic free Ca^{++} increases from a resting level of about 10^{-7}M to levels of 10^{-6} to 10^{-5}M during excitation, and the Ca^{++} binds to the protein troponin C. The Ca^{++}-troponin complex interacts with tropomyosin to unblock active sites between the actin and myosin filaments. This interaction allows crossbridge cycling and hence contraction of the myofibrils (systole).

Mechanisms that raise cytosolic Ca^{++} increase the developed force, and those that lower Ca^{++} decrease the developed force. For example, catecholamines increase the movement of Ca^{++} into the cell by phosphorylation of the Ca^{++} channels by a cAMP-dependent protein kinase (Fig. 22-21). An increase in cytosolic Ca^{++} is also achieved by increasing extracellular Ca^{++} or decreasing the Na^{+} gradient across the sarcolemma. With a decrease in the Na^{+} gradient (e.g., lowering extracellular Na^{+}) less Na^{+} enters the cell and consequently less Ca^{++} leaves the cell by the Na^{+}-Ca^{++} exchange mechanism. Developed force is diminished by a reduction in extracellular Ca^{++}, by an increase in the Na^{+} gradient across the sarcolemma, or by the administration of Ca^{++} blockers that prevent Ca^{++} from entering the myocardial cell (Fig. 27-10).

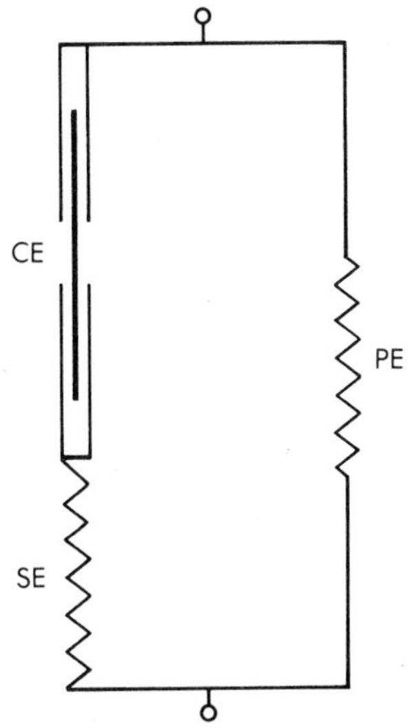

■ **Fig. 28-7.** A simplified model for contraction of striated muscle. *CE,* Contractile element; SE, series elastic element; *PE,* parallel elastic element.

At the end of systole the Ca^{++} influx ceases and the sarcoplasmic reticulum is no longer stimulated to release Ca^{++}. In fact, the sarcoplasmic reticulum avidly takes up Ca^{++} by means of an ATP-energized calcium pump that is activated by phosphorylation. The resultant decrease in cytosolic Ca^{++} reverses the binding of Ca^{++} by troponin C. This permits tropomyosin to again block the sites for interaction between the actin and myosin filaments, and relaxation (diastole) occurs. A sarcoplasmic reticular protein, *phospholamban,* accelerates Ca^{++} sequestration and may help regulate the Ca^{++} pump of the sarcoplasmic reticulum. Cardiac contraction and relaxation are both accelerated by catecholamines and adenylate cyclase activation, but the precise chemical mechanisms responsible for the transition from systole to diastole have not been elucidated. Mitochondria also take up and release Ca^{++}, but the process is too slow to be involved in excitation-contraction coupling.

The Ca^{++} that enters the cell to initiate contraction must be removed during diastole. The removal is primarily accomplished by an electroneutral exchange of 2 Na^{+} for 1 Ca^{++} (Fig. 28-6). Ca^{++} also may be removed from the cell by an electrogenic pump that utilizes energy to transport three or more Na^{+} for each Ca^{++} across the sarcolemma. A similar electrogenic pump may transport Ca^{++} into the cell. The role of Ca^{++} electrogenic pumps and their contributions to the movement of Ca^{++} across the sarcolemma is controversial; hence this transport mechanism has not been included in Fig. 28-6.

Myocardial contractile machinery and contractility. A simple mechanical model of striated muscle is shown in Fig. 28-7. The *contractile element* (the inter-

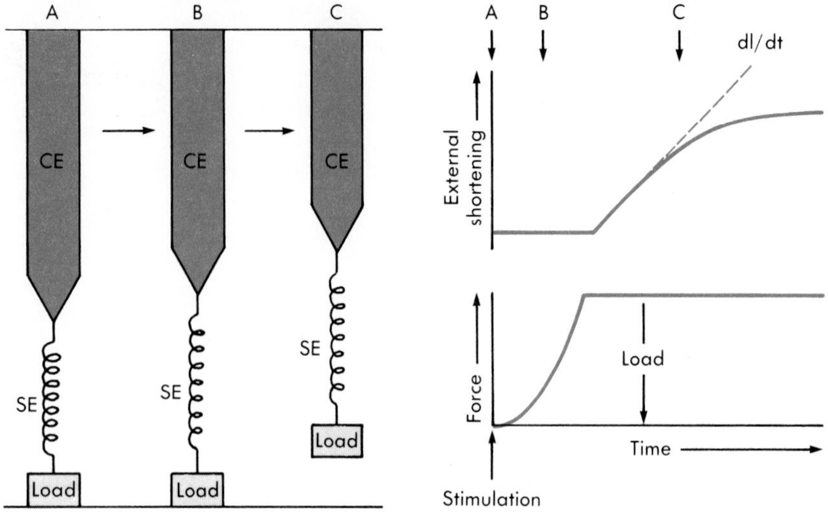

■ **Fig. 28-8.** Model for a preloaded and afterloaded isotonic contraction of papillary muscle. **A,** Muscle at rest. Preload is represented by partial stretch of the series elastic element *(SE)*. **B,** Partial contraction of the contractile element *(CE)* with stretch of the series elastic element and no external shortening (the isometric phase of the contraction). **C,** Further contraction of the contractile element with external shortening and lifting of the afterload. The tangent *(dl/dt)* to the initial slope of the shortening curve on the right is the velocity of initial shortening. (Redrawn from Sonnenblick, E.H.: The myocardial cell, Philadelphia, 1966, University of Pennsylvania Press.)

digitating actin and myosin filaments) is connected in series with an elastic element, the *series elastic element*. As the contractile element shortens, the series elastic element is stretched; the degree of stretch is proportional, within limits, to the force developed. The anatomical counterpart of the series elastic element is controversial. Also shown in Fig. 28-7 is an elastic element that is parallel to the contractile element. This *parallel elastic element* is important in the stretch of the resting muscle, the initial length. With stretch of a papillary muscle or with an increase in diastolic filling of the intact ventricles, the nonactivated actin and myosin filaments slide past each other with little resistance. However, the parallel elastic element, which anatomically is probably connective tissue, offers increasing counterforce at greater lengths. As mentioned on p. 431, this counterforce protects against overstretch of the cardiac muscle during diastole.

Velocity and force of contraction are a function of the intracellular concentration of free Ca ions. Force and velocity are inversely related, so that with no load, force is negligible and velocity is maximal. In an isometric contraction, where no external shortening occurs, force is maximal and velocity is zero.

The sequence of events in a preloaded and afterloaded isotonic contraction of a papillary muscle is illustrated in Fig. 28-8. Point *A* represents the resting state in which the preload is responsible for the existing initial stretch. With stimulation the contractile element begins to shorten, and at point *B* the series elastic element has been stretched, but the load has not yet been lifted (the isometric phase of the contraction). This stretch of the series elastic element (an expression of the muscle extensibility) consumes a certain amount of energy. Therefore the energy used for shortening of the muscle is actually less than the total energy expenditure in a single contraction. Stretch of the series elastic element is represented in the diagram at the lower right as a progressive rise in force with no external shortening. At point *C* the force developed by the contractile element has equaled the load, and the load has been raised without further stretch of the series elastic element. This is represented in the diagram on the right as external shortening of the muscle without a further increase in force.

The initial slope (dashed tangent) of the shortening curve (Fig. 28-8, upper right section) depicts the initial rate of shortening (change in length with change in time, dl/dt). Since the initial velocity depends on the magnitude of the afterload, a series of initial velocities can be obtained from a papillary muscle by varying the afterload. This is portrayed in Fig. 28-9 where the slope of the initial velocity curve and the degree of shortening can be seen to decrease with increasing loads. Furthermore, the onset of shortening is delayed (longer isometric phase), but the time from stimulation to maximal shortening (peaks of shortening curves) is unchanged (Fig. 28-9).

If the initial velocity of shortening is plotted against the afterload, the force-velocity curves shown in Fig. 28-10, *A*, are obtained. The maximum velocity (V_o) may be estimated by extrapolation of the force-velocity curve back to zero load (as indicated by the dotted lines in Fig. 28-10, *A*) and represents the maximum rate of cycling of the crossbridges.

Contractility can be defined as a change in developed force at a given resting fiber length. However, it also may be defined in terms of a change in V_o. Augmentation of contractility is observed with certain drugs, such as norepinephrine or digitalis, and with an increase in contraction frequency (*tachycardia,* when applied to the whole heart). The increase in contractility *(positive*

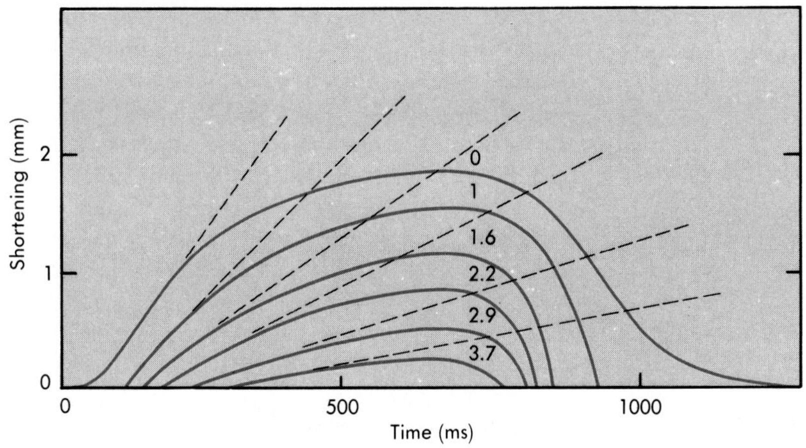

■ **Fig. 28-9.** Series of superimposed afterloaded contractions of a papillary muscle at a constant initial length. The numbers refer to the magnitude of the afterload in grams and the dashed lines represent the initial velocity of shortening. (Redrawn from Sonnenblick, E.H.: The myocardial cell, Philadelphia, 1966, University of Pennsylvania Press.)

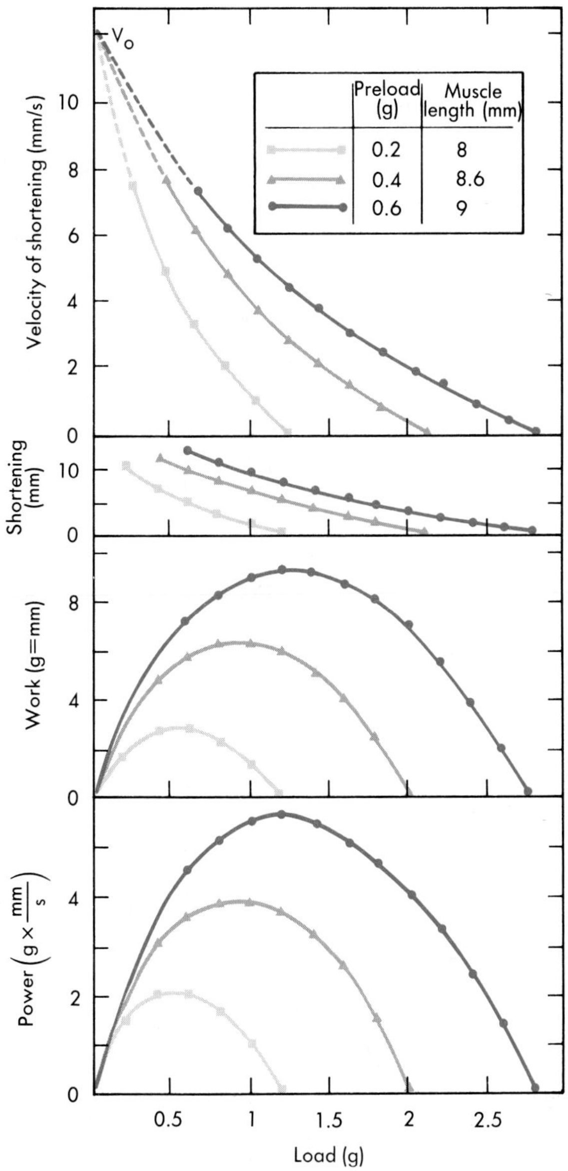

■ **Fig. 28-10.** The effect of increasing initial length of a cat papillary muscle on the force-velocity relationship, degree of shortening, muscle work, and muscle power. (Redrawn and reproduced by permission from Am. J. Physiol. **202:**931, 1962.)

inotropic effect) produced by any of the preceding interventions is reflected by increments in developed force and V_O.

An increase in initial fiber length produces a more forceful contraction, as shown in Fig. 28-10. However, this greater force development is not associated with any change in contractility, as estimated by V_o (Fig. 28-10, *A*). Fig. 28-10 also illustrates that at any given load, the degree of shortening, the work (shortening × load, or distance × force), and the power (velocity × load, or work/time) all increase with the initial length of the papillary muscle. It is apparent that with an increase in initial fiber length, greater force may be developed, but the estimated V_o is the same for all three initial lengths. Hence changes in resting length may alter force development but not contractility. This conclusion is, of course, based on the assumption that the displayed extrapolation of the force-velocity curves back to the verticle axis provide the true value for V_o. This assumption has recently been challenged; some investigators have failed to obtain a hyperbolic force-velocity relationship and have observed changes in V_o with changes in initial length (that is, V_o is *length-dependent*). For this and several other reasons, estimates of V_o may not be a reliable index of contractility.

Although the experiments from which the length-force and force-velocity relationships have been derived were carried out on papillary muscles (essentially one-dimensional), the findings are to some extent applicable

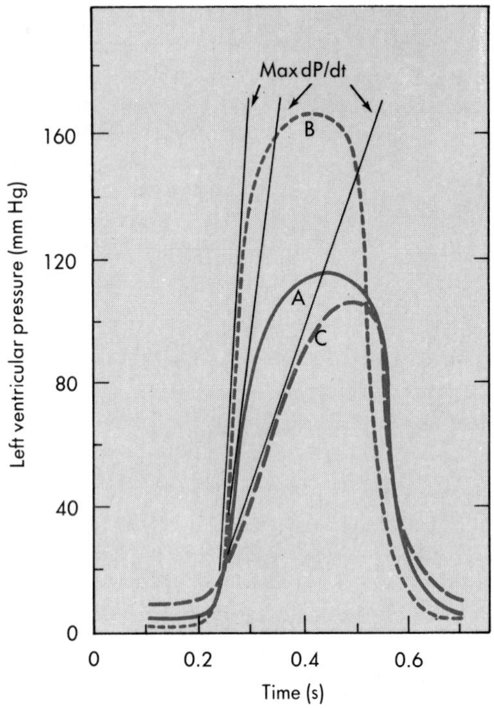

■ **Fig. 28-11.** Left ventricular pressure curves with tangents drawn to the steepest portions of the ascending limbs to indicate maximum dP/dt values. *A,* Control; *B,* hyperdynamic heart, as with norepinephrine administration; *C,* hypodynamic heart, as in cardiac failure.

to the intact heart (three-dimensional). For the left ventricle the preload is the ventricular pressure at the end of diastole (end-disatolic pressure) and the afterload is the aortic pressure. However, the complex changes in ventricular shape that occur in systole and the fact that fibers branch at different angles and therefore do not all contract in their optimum longitudinal axis make accurate determinations of contractility in the whole heart very difficult.

A reasonable index of myocardial contractility can be obtained from the contour of ventricular pressure curves (Fig. 28-11). A hypodynamic heart is characterized by an elevated end-diastolic pressure, a slowly rising ventricular pressure, and a somewhat reduced ejection phase (curve *C,* Fig. 28-11). A normal ventricle under adrenergic stimulation shows a reduced end-diastolic pressure, a fast-rising ventricular pressure, and a brief ejection phase (curve *B,* Fig. 28-11). The slope of the ascending limb of the ventricular pressure curve indicates the maximum rate of force development by the ventricle (maximum rate of change in pressure with time; maximum dP/dt, as illustrated by the tangents to the steepest portion of the ascending limbs of the ventricular pressure curves in Fig. 28-11). The slope is maximal during the isovolumic phase of systole and, at any given degree of ventricular filling, provides an index of the initial contraction velocity and hence of contractility. Similarly, one can obtain an indication of the contractile state of the myocardium from the initial velocity of blood flow in the ascending aorta (the initial slope of the aortic flow curve). The *ejection fraction,* which is the ratio of the volume of blood ejected from the left ventricle per beat *(stroke volume)* to the volume of blood in the left ventricle at the end of diastole is widely used clinically as an index of contractility. Other measurements or combination of measurements that in general are concerned with the magnitude or velocity of the ventricular contraction also have been used to assess the contractile state of the cardiac muscle. There is no index that is entirely satisfactory at present, and this undoubtedly accounts for the large number of indices that are currently in use.

■ *Cardiac Chambers*

The atria are thin-walled, low-pressure chambers that function more as large reservoir conduits of blood for their respective ventricles than as important pumps for the forward propulsion of blood. The ventricles were once thought to be made up of bands of muscle. However, it now appears that they are formed by a continuum of muscle fibers that take origin from the fibrous skeleton at the base of the heart (chiefly around the aortic orifice). These fibers sweep toward the apex at the epicardial surface and also pass toward the endocardium as they gradually undergo a 180-degree change in direction to lie parallel to the epicardial fibers and form the endocardium and papillary muscles (Fig. 28-12). At the apex of the heart the fibers twist and turn inward to form papillary muscles, whereas at the base and around the valve orifices they form a thick powerful muscle that not only decreases ventricular circumference for ejection of blood but also narrows the AV valve orifices as an aid to valve closure. In addition to a reduction in circumference, ventricular ejection is accomplished by a decrease in the longitudinal axis with descent of the base of the heart. The earlier contraction of the apical part of the ventricles coupled with approximation of the ventricular walls propels the blood toward the outflow tracts. The right ventricle, which develops a mean pressure about one seventh that developed by the left ventricle, is considerably thinner than the left.

■ *Cardiac Valves*

The cardiac valves consist of thin flaps of flexible, tough endothelium-covered fibrous tissue firmly attached at the base to the fibrous valve rings. Movements of the valve leaflets are essentially passive, and the orientation of the cardiac valves is responsible for

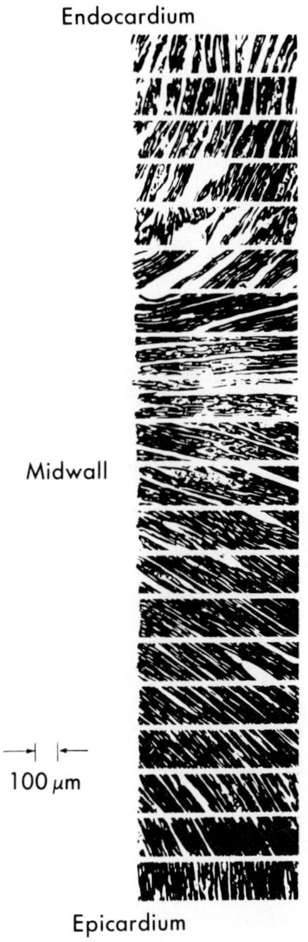

■ **Fig. 28-12.** Sequence of photomicrographs showing fiber angles in successive sections taken from the middle of the free wall of the left ventricle from a heart in systole. The sections are parallel to the epicardial plane. Fiber angle is 90 degrees at the endocardium, running through 0 degrees at the midwall to −90 degrees at the epicardium. (From Streeter, D.D., Jr., Spotnitz, H.M., Patel, D.P., Ross, J., Jr., and Sonnenblick, E.H.: Circ. Res. **24:**339, 1969. By permission of the American Heart Association, Inc.)

unidirectional flow of blood through the heart. There are two types of valves in the heart—the *atrioventricular valves,* or *AV valves,* and the *semilunar valves* (Figs. 28-13 and 28-14).

Atrioventricular valves. The valve between the right atrium and right ventricle is made up of three cusps *(tricuspid valve),* whereas that between the left atrium and left ventricle has two cusps *(mitral valve).* The total area of the cusps of each AV valve is approximately twice that of the respective AV orifice so that there is considerable overlap of the leaflets in the closed position (Figs. 28-13 and 28-14). Attached to the free edges of these valves are fine, strong filaments *(chordae tendineae),* which arise from the powerful papillary muscles of the respective ventricles and prevent eversion of the valves during ventricular systole.

The mechanism of closure of the AV valves has been the subject of considerable investigation, and a number of factors are thought to play a role in approximating the valve leaflets. In the normal heart the valve leaflets are relatively close during ventricular filling and provide a funnel for the transfer of blood from atrium to ventricle. This partial approximation of the valve surfaces during diastole is believed to be caused by eddy currents behind the leaflets and possibly also by some tension on the free edges of the valves, exerted by the chordae tendineae and papillary muscles that are stretched by the filling ventricle. Movements of the mitral valve leaflets throughout the cardiac cycle are shown in an *echocardiogram* (Fig. 28-15). Echocardiography consists of sending short pulses of high-frequency sound waves (ultrasound) through the chest tissues and the heart and recording the echoes reflected from the various structures. The timing and the pattern of the reflected waves provide such information as the diameter of the heart, the ventricular wall thickness, and the magnitude and direction of the movements of various components of the heart.

In Fig. 28-15 the echocardiograph is positioned to depict movement of the anterior leaflet of the mitral valve. The posterior leaflet moves in a pattern that is a mirror image of the anterior leaflet, but in the projection shown in Fig. 28-15 the excursions appear much smaller. At point *D* the mitral valve opens, and during rapid filling *(D* to *E)* the anterior leaflet moves toward the ventricular septum. During the reduced filling phase *(E* to *F)* the valve leaflets float toward each other but the valve does not close. The ventricular filling contributed by atrial contraction *(F* to *A)* forces the leaflets apart, and a second approximation of the leaflets follows *(A* to *C).* At point *C* the valve is closed by ventricular contraction. The valve leaflets, which bulge toward the atrium, stay pressed together during ventricular systole *(C* to *D).*

Semilunar valves. The valves between the right ventricle and the pulmonary artery and between the left ventricle and the aorta consist of three cuplike cusps attached to the valve rings (Figs. 28-13 and 28-14). At the end of the reduced ejection phase of ventricular systole, there is a brief reversal of blood flow toward the ventricles (shown as a negative flow in the phasic aortic flow curve in Fig. 28-16) that snaps the cusps together and prevents regurgitation of blood into the ventricles. During ventricular systole the cusps do not lie back against the walls of the pulmonary artery and aorta but float in the bloodstream approximately midway between the vessel walls and their closed position. Behind the semilunar valves are small outpocketings of the pulmonary artery and aorta *(sinuses of Valsalva),* where eddy currents develop that tend to keep the valve cusps away from the vessel walls. The orifices of the right and left coronary arteries are located behind the right and the left cusps, respectively, of the aortic valve. Were it not

■ **Fig. 28-13.** Drawing of a heart split perpendicular to the interventricular septum to illustrate the anatomical relationships of the leaflets of the AV and aortic valves.

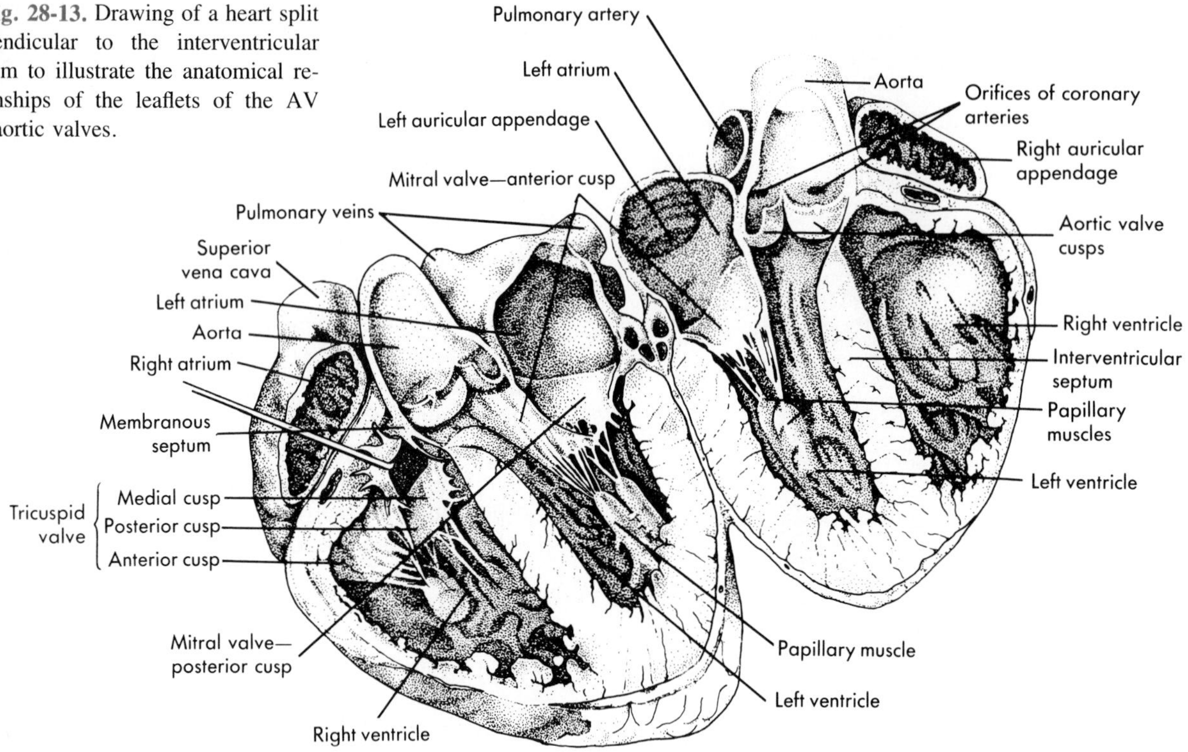

■ **Fig. 28-14.** Four cardiac valves as viewed from the base of the heart. Note how the leaflets overlap in the closed valves.

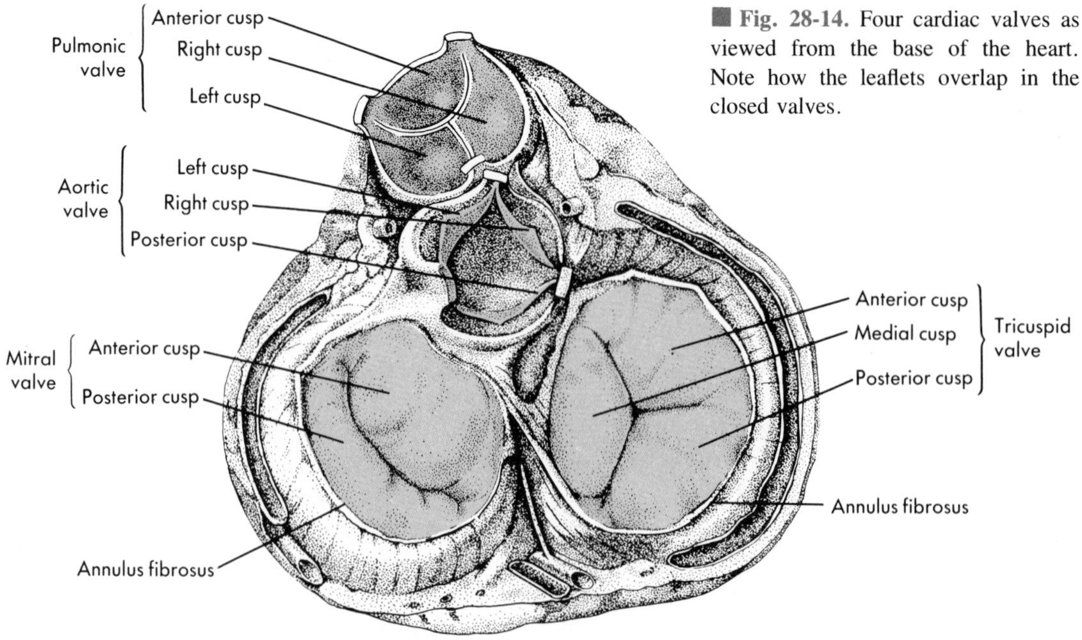

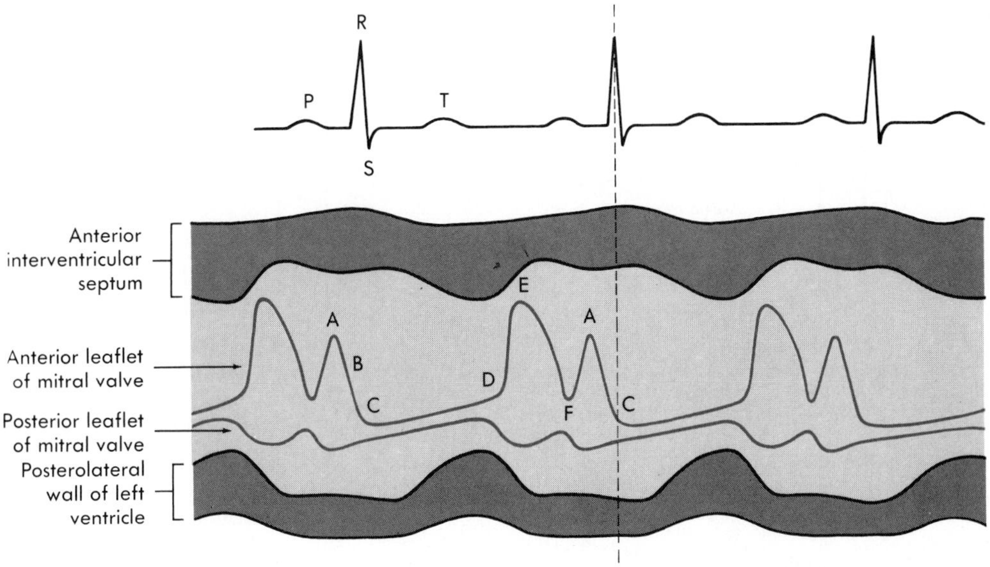

■ **Fig. 28-15.** Drawing made from an echocardiogram showing movements of the mitral valve leaflets (particularly the anterior leaflet) and the changes in the diameter of the left ventricular cavity and the thickness of the left ventricular walls during cardiac cycles in a normal person. *D* to *C*, Ventricular diastole; *C* to *D*, ventricular systole; *D* to *E*, rapid filling; *E* to *F*, reduced filling (diastasis); *F* to *A*, atrial contraction. Mitral valve closes at *C* and opens at *D*. Simultaneously recorded electrocardiogram at top. (Original echocardiogram courtesy of Dr. Sanjiv Kaul.)

for the presence of the sinuses of Valsalva and the eddy currents developed therein, the coronary ostia could be blocked by the valve cusps.

■ *The Pericardium*

The pericardium is an epithelialized fibrous sac. It closely invests the entire heart and the cardiac portion of the great vessels and is reflected onto the cardiac surface as the epicardium. The sac normally contains a small amount of fluid, which provides lubrication for the continuous movement of the enclosed heart. The distensibility of the pericardium is small, so that it strongly resists a large, rapid increase in cardiac size. Because of this characteristic, the pericardium plays a role in preventing sudden overdistension of the chambers of the heart. However, in congenital absence of the pericardium or after its surgical removal, cardiac function is within physiological limits. Nevertheless, with the pericardium intact, an increase in diastolic pressure in one ventricle increases the pressure and decreases the compliance of the other ventricle. In contrast to an acute change in intracardiac pressure, progressive and sustained distension of the heart (as occurs in cardiac hypertrophy) or a slow progressive increase in pericardial fluid (as occurs in pericarditis with pericardial effusion) gradually stretches the intact pericardium.

■ *Heart Sounds*

There are usually four sounds produced by the heart, but only two are ordinarily audible through a stethoscope. With electronic amplification the less intense sounds can be detected and recorded graphically as a *phonocardiogram*. This means of registering heart sounds that may be inaudible to the human ear aids in delineating the precise timing of the heart sounds relative to other events in the cardiac cycle.

The first heart sound is initiated at the onset of ventricular systole (Fig. 28-16) and consists of a series of vibrations of mixed, unrelated, low frequencies (a noise). It is the loudest and longest of the heart sounds, has a crescendo-decrescendo quality, and is heard best over the apical region of the heart. The tricuspid valve sounds are heard best in the fifth intercostal space just to the left of the sternum, and the mitral sounds are heard best in the fifth intercostal space at the cardiac apex.

The first heart sound is chiefly caused by the oscillation of blood in the ventricular chambers and vibration of the chamber walls. The vibrations are engendered in part by the abrupt rise of ventricular pressure with acceleration of blood back toward the atria, but primarily by sudden tension and recoil of the AV valves and adjacent structures with deceleration of the blood by closure of the AV valves. The vibrations of the ventricles

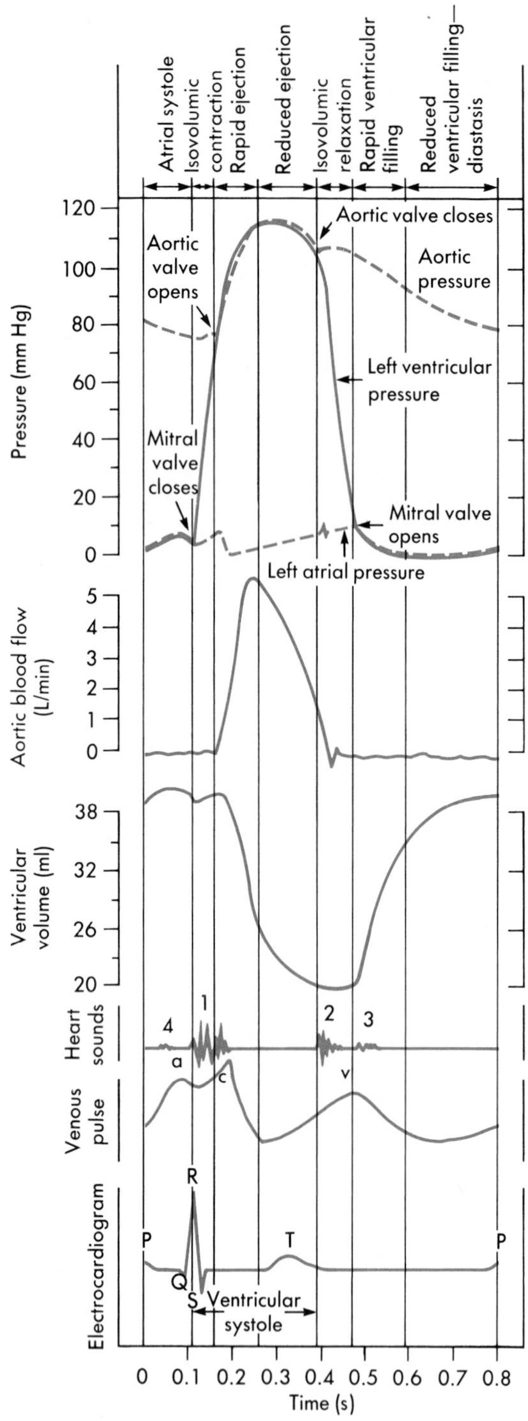

Fig. 28-16. Left atrial, aortic, and left ventricular pressure pulses correlated in time with aortic flow, ventricular volume, heart sounds, venous pulse, and the electrocardiogram for a complete cardiac cycle in the dog.

and the contained blood are transmitted through surrounding tissues and reach the chest wall where they may be heard or recorded. The intensity of the first sound is a function of the force of ventricular contraction and of the distance between the valve leaflets. When the leaflets are farthest apart, either when the in-

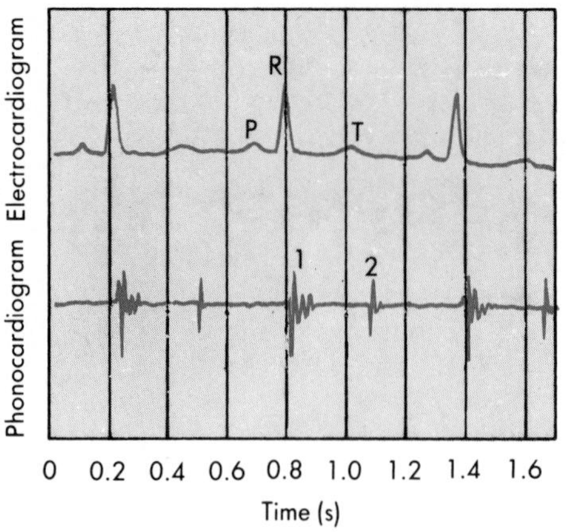

Fig. 28-17. Phonocardiogram illustrating the first and second heart sounds and their relationship to the P, R, and T waves of the electrocardiogram. (Time lines = 0.04 s.)

terval between atrial and ventricular systoles is prolonged and AV valve leaflets float apart or when ventricular systole immediately follows atrial systole, the first sound is loudest.

The second heart sound, which occurs with closure of the semilunar valves (Fig. 28-16), is composed of higher frequency vibrations (higher pitch), is of shorter duration and lower intensity, and has a more snapping quality than the first heart sound. The second sound is caused by abrupt closure of the semilunar valves, which initiates oscillations of the columns of blood and the tensed vessel walls by the stretch and recoil of the closed valve. The second sound caused by closure of the pulmonic valve is heard best in the second thoracic interspace just to the left of the sternum, whereas that caused by closure of the aortic valve is heard best in the same intercostal space but to the right of the sternum. Conditions that bring about a more rapid closure of the semilunar valves, such as increases in pulmonary artery or aortic pressure (for example, pulmonary or systemic hypertension), will increase the intensity of the second heart sound. In the adult the aortic valve sound is usually louder than the pulmonic, but in cases of pulmonary hypertension the reverse is often true.

A normal phonocardiogram taken simultaneously with an electrocardiogram is illustrated in Fig. 28-17. Note that the first sound, which starts just beyond the peak of the R wave, is composed of irregular waves and is of greater intensity and duration than the second sound, which appears at the end of the T wave. A third and fourth heart sound do not appear on this record.

The third heart sound, which is sometimes heard in children with thin chest walls or in patients with left ventricular failure, consists of a few low-intensity, low-frequency vibrations heard best in the region of the apex. It occurs in early diastole and is believed to be the result of vibrations of the ventricular walls caused by abrupt cessation of ventricular distension and deceleration of blood entering the ventricles. This occurs in overloaded hearts when the ventricular volume is very large and the ventricular walls are stretched to the point where distensibility abruptly decreases. A third heart sound in patients with heart disease is usually a grave sign.

A fourth, or atrial, sound, consisting of a few low-frequency oscillations, is occasionally heard in normal individuals. It is caused by oscillation of blood and cardiac chambers created by atrial contraction (Fig. 28-16).

Since the onset and termination of right and left ventricular systoles are not precisely synchronous, differences in the time of vibration of the two AV valves or two semilunar valves can sometimes be detected with the stethoscope. Such asynchrony of valve vibrations, which may sometimes indicate abnormal cardiac function, is manifest as a *split sound* over the apex of the heart for the AV valves and over the base for the semilunar valves. The heart sounds also may be altered by deformities of the valves; *murmurs* may be produced, and the character of the murmur serves as an important guide in the diagnosis of valvular disease. When the third and fourth (atrial) sounds are accentuated, as occurs in certain abnormal conditions, triplets of sounds may occur, resembling the sounds of a galloping horse. These *gallop rhythms* are essentially of two types—*presystolic gallop* caused by accentuation of the atrial sound, and *protodiastolic gallop* caused by accentuation of the third heart sound.

Cardiac Cycle

Ventricular Systole

Isovolumic contraction. The onset of ventricular contraction coincides with the peak of the R wave of the electrocardiogram and the initial vibration of the first heart sound. It is indicated on the ventricular pressure curve as the earliest rise in ventricular pressure after atrial contraction. The interval of time between the start of ventricular systole and the opening of the semilunar valves (when ventricular pressure rises abruptly) is termed *isovolumic contraction,* since ventricular volume is constant during this brief period (Fig. 28-16).

The increment in ventricular pressure during isovolumic contraction is transmitted across the closed valves and is evident in Fig. 28-16 as a small oscillation on the aortic pressure curve. Isovolumic contraction also has been referred to as isometric contraction. However, some fibers shorten and others lengthen, as evidenced by changes in ventricular shape; it is therefore not a true isometric contraction.

Ejection. Opening of the semilunar valves marks the onset of the ejection phase, which may be subdivided into an earlier, shorter phase *(rapid ejection)* and a later, longer phase *(reduced ejection).* The rapid ejection phase is distinguished from the reduced ejection phase by (1) the sharp rise in ventricular and aortic pressures that terminates at the peak ventricular and aortic pressures, (2) a more abrupt decrease in ventricular volume, and (3) a greater aortic blood flow (Fig. 28-16). The sharp decrease in the left atrial pressure curve at the onset of ejection results from the descent of the base of the heart and stretch of the atria. During the reduced ejection period, runoff of blood from the aorta to the periphery exceeds ventricular output and therefore aortic pressure declines. Throughout ventricular systole the blood returning to the atria produces a progressive increase in atrial pressure. Note that during approximately the first third of the ejection period left ventricular pressure slightly exceeds aortic pressure and flow accelerates (continues to increase), whereas during the last two thirds of ventricular ejection the reverse holds true. This reversal of the ventricular/aortic pressure gradient in the presence of continued flow of blood from the left ventricle to the aorta (caused by the momentum of the forward blood flow) is the result of the storage of potential energy in the stretched arterial walls, which produces a deceleration of blood flow into the aorta. The peak of the flow curve coincides in time with the point at which the left ventricular pressure curve intersects the aortic pressure curve during ejection. Thereafter flow decelerates (continues to decrease) because the pressure gradient has been reversed.

With right ventricular ejection there is shortening of the free wall of the right ventricle (descent of the tricuspid valve ring) in addition to lateral compression of the chamber. However, with left ventricular ejection there is very little shortening of the base-to-apex axis, and ejection is accomplished chiefly by compression of the left ventricular chamber.

The effect of ventricular systole on left ventricular diameter is shown in an echocardiogram (Fig. 28-15). During ventricular systole (Fig. 28-15, *C* to *D*) the septum and the free wall of the left ventricle become thicker and move closer to each other.

The venous pulse curve shown in Fig. 28-16 has been taken from a jugular vein and the *c* wave is caused by impact of the adjacent common carotid artery. Note that except for the *c* wave, the venous pulse closely follows the atrial pressure curve.

At the end of ejection, a volume of blood approximately equal to that ejected during systole remains in the ventricular cavities. This *residual volume* is fairly constant in normal hearts but is smaller with increased heart rate or reduced outflow resistance and larger when the opposite conditions prevail. An increase in myocardial contractility may decrease residual volume (or increase stroke volume and ejection fraction), especially in the depressed heart. With severely hypodynamic and dilated hearts, as in *heart failure*, the residual volume can become many times greater than the stroke volume. In addition to serving as a small adjustable blood reservoir, the residual volume to a limited degree can permit transient disparities between the outputs of the two ventricles.

Ventricular Diastole

Isovolumic relaxation. Closure of the aortic valve produces the incisura on the descending limb of the aortic pressure curve and the second heart sound (with some vibrations evident on the atrial pressure curve) and marks the end of ventricular systole. The period between closure of the semilunar valves and opening of the AV valves is termed *isovolumic* relaxation and is characterized by a precipitous fall in ventricular pressure without a change in ventricular volume.

Rapid filling phase. The major part of ventricular filling occurs immediately on opening of the AV valves when the blood that had returned to the atria during the previous ventricular systole is abruptly released into the relaxing ventricles. This period of ventricular filling is called the *rapid filling phase*. In Fig. 28-16 the onset of the rapid filling phase is indicated by the decrease in left ventricular pressure below left atrial pressure, resulting in the opening of the mitral valve. The rapid flow of blood from atria to relaxing ventricles produces a decrease in atrial and ventricular pressures and a sharp increase in ventricular volume.

The decrease in pressure from the peak of the *v* wave of the venous pulse is caused by transmission of the pressure decrease incident to the abrupt transfer of blood from the right atrium to the right ventricle with opening of the tricuspid valve. Elastic recoil of the previous ventricular contraction may aid in drawing blood into the relaxing ventricle when residual volume is small, but it probably does not play a significant role in ventricular filling under most normal conditions.

Diastasis. The rapid filling phase is followed by a phase of slow filling, called *diastasis*. During diastasis, blood returning from the periphery flows into the right ventricle and blood from the lungs into the left ventricle. This small, slow addition to ventricular filling is indicated by a gradual rise in atrial, ventricular, and venous pressures and in ventricular volume.

Pressure-volume relationship. The changes in left ventricular pressure and volume throughout the cardiac cycle are summarized in Fig. 28-18. The element of time is not considered in this pressure-volume loop. Diastolic filling starts at *A* and terminates at *C*, when the mitral valve closes. The initial decrease in left ventricular pressure *(A* to *B)*, despite the rapid inflow of blood from the atrium, is due to progressive ventricular relaxation and distensibility. During the remainder of diastole *(B* and *C)* the increase in ventricular pressure reflects ventricular filling and the passive elastic characteristics of the ventricle. Note that only a small increase in pressure occurs with the increase in ventricular volume during diastole *(B* to *C)*. With isovolumic contraction *(C* to *D)* there is a steep rise in pressure and no change in ventricular volume. At *D* the aortic valve opens and during the first phase of ejection (rapid ejection, *D* to *E*), the large reduction in volume is associated with a continued but lesser increase in ventricular pressure than that which occurred during isovolumic contraction. This volume reduction is followed by reduced ejection *(E* to *F)* and a small decrease in ventricular pressure. The aortic valve closes at *F*, and this event is followed by isovolumic relaxation *(F* to *A)*, which is characterized by a sharp drop in pressure and no change in volume. The mitral valve opens at *A* to complete one cardiac cycle.

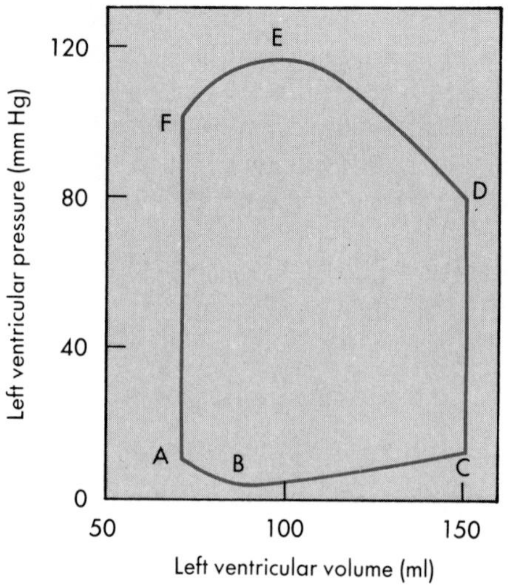

Fig. 28-18. Pressure-volume loop of the left ventricle for a single cardiac cycle *(ABCDEF)*.

Atrial systole. The onset of atrial systole occurs soon after the beginning of the P wave of the electrocardiogram (curve of atrial depolarization), and the transfer of blood from atrium to ventricle made by the peristalsis-like wave of atrial contraction completes the period of ventricular filling. Atrial systole is responsible for the small increases in atrial, ventricular, and venous (*a* wave) pressures, as well as in ventricular volume shown in Fig. 28-16. Throughout ventricular diastole, atrial pressure barely exceeds ventricular pressure, indicating a low-resistance pathway across the open AV valves during ventricular filling. A few small vibrations produced by atrial systole constitute the fourth, or atrial, heart sound.

Since there are no valves at the junctions of the venae cavae and right atrium or of the pulmonary veins and left atrium, atrial contraction can force blood in both

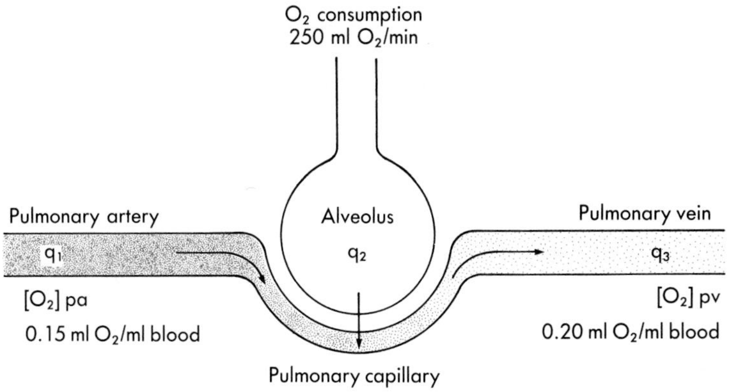

■ Fig. 28-19. Schema illustrating the Fick principle for measuring cardiac output. The change in color intensity from pulmonary artery to pulmonary vein represents the change in color of the blood as venous blood becomes fully oxygenated.

directions. Actually, little blood is pumped back into the venous tributaries during the brief atrial contraction, mainly because of the inertia of the inflowing blood.

Atrial contraction is not essential for ventricular filling, as can be observed in atrial fibrillation or complete heart block. However, its contribution is governed to a great extent by the heart rate and the structure of the AV valves. At slow heart rates, filling practically ceases toward the end of diastasis, and atrial contraction contributes little additional filling. During tachycardia diastasis is abbreviated and the atrial contribution can become substantial, especially if it occurs immediately after the rapid filling phase when the AV pressure gradient is maximal. Should tachycardia become so great that the rapid filling phase is encroached on, atrial contraction assumes great importance in rapidly propelling blood into the ventricle during this brief period of the cardiac cycle. Of course, if the period of ventricular relaxation is so brief that filling is seriously impaired, even atrial contraction cannot prevent inadequate ventricular filling. The consequent reduction in cardiac outkkput may result in syncope. Obviously, if atrial contraction occurs simultaneously with ventricular contraction, no atrial contribution to ventricular filling can occur. In certain disease states the AV valves may be markedly narrowed *(stenotic)*. Under such conditions atrial contraction may play a much more important role in ventricular filling than it does in the normal heart.

Ventricular contraction has been shown to aid indirectly in right ventricular filling by its effect on the right atrium. Descent of the base of the heart stretches the right atrium downward, and pressure measurements indicate a sharp reduction in right atrial pressure associated with acceleration of blood flow in the venae cavae toward the heart. Enhancement of venous return by ventricular systole provides an additional supply of atrial blood for ventricular filling during the subsequent rapid filling phase of diastole. However, this mechanism is probably of little physiological importance, except possibly at rapid heart rates.

■ *Measurement of Cardiac Output*

■ *Fick Principle*

In 1870 the German physiologist, Adolph Fick, contrived the first method for measuring cardiac output in intact animals and people. The basis for this method, called the *Fick principle,* is simply an application of the law of conservation of mass. It is derived from the fact that the quantity of oxygen (O_2) delivered to the pulmonary capillaries via the pulmonary artery plus the quantity of O_2 that enters the pulmonary capillaries from the alveoli must equal the quantity of O_2 that is carried away by the pulmonary veins.

This is depicted schematically in Fig. 28-19. The rate, q_1, of O_2 delivery to the lungs equals the O_2 concentration in the pulmonary arterial blood, $[O_2]_{pa}$, times the pulmonary arterial blood flow, Q, which equals the cardiac output; that is,

$$q_1 = Q[O_2]_{pa} \quad (1)$$

Let q_2 be the net rate of O_2 uptake by the pulmonary capillaries from the alveoli. At equilibrium, q_2 equals the *O_2 consumption* of the body. The rate, q_3, at which O_2 is carried away by the pulmonary veins equals the O_2 concentration in the pulmonary venous blood, $[O_2]_{pv}$, times the total pulmonary venous flow, which is virtually equal to the pulmonary arterial blood flow, Q; that is,

$$q_3 = Q[O_2]_{pv} \quad (2)$$

From conservation of mass,

$$q_1 + q_2 = q_3 \quad (3)$$

Therefore

$$Q[O_2]_{pa} + q_2 = Q[O_2]_{pv} \quad (4)$$

Solving for cardiac output,

$$Q = q_2/([O_2]_{pv} - [O_2]_{pa}) \quad (5)$$

Equation 5 is the statement of the Fick principle.

In the clinical determination of cardiac output, O_2 consumption is computed from measurements of the volume and O_2 content of expired air over a given interval of time. Since the O_2 concentration of peripheral arterial blood is essentially identical to that in the pulmonary veins, $[O_2]_{pv}$ is determined on a sample of peripheral arterial blood withdrawn by needle puncture. Pulmonary arterial blood actually represents mixed systemic venous blood. Samples for O_2 analysis are obtained from the pulmonary artery or right ventricle through a catheter. In the past a stiff catheter was used, and it had to be introduced into the pulmonary artery under fluoroscopic guidance. Now, a very flexible catheter with a small balloon near the tip can be inserted into a peripheral vein. As the tube is advanced, it is carried by the flowing blood toward the heart. By following the pressure changes, the physician is able to advance the catheter tip into the pulmonary artery without the aid of fluoroscopy.

An example of the calculation of cardiac output in a normal, resting adult is illustrated in Fig. 28-19. With an O_2 consumption of 250 ml/min, an arterial (pulmonary venous) O_2 content of 0.20 ml O_2/ml blood, and a mixed venous (pulmonary arterial) O_2 content of 0.15 ml O_2/ml blood, the cardiac output would equal 250 ÷ (0.20 − 0.15) = 5000 ml/min.

The Fick principle is also used for estimating the O_2 consumption of organs in situ, when blood flow and the O_2 contents of the arterial and venous blood can be determined. Algebraic rearrangement reveals that O_2 consumption equals the blood flow times the arteriovenous O_2 concentration difference. For example, if the blood flow through one kidney is 700 ml/min, arterial O_2 content is 0.20 ml O_2/ml blood, and renal venous O_2 content is 0.18 ml O_2/ml blood, then the rate of O_2 consumption by that kidney must be 700 (0.20 − 0.18) = 14 ml O_2/min.

Indicator Dilution Techniques

The indicator dilution technique for measuring cardiac output is also based on the law of conservation of mass and is illustrated by the model in Fig. 28-20. Let a liquid flow through a tube at a rate of Q ml/s, and let q mg of dye be injected as a slug into the stream at point *A*. Let mixing occur at some point downstream. If a small sample of liquid is continually withdrawn from point *B* farther downstream and passed through a densitometer, a curve of the dye concentration, c, may be recorded as a function of time, t, as shown in the lower half of the figure.

If no dye is lost between points *A* and *B*, the amount of dye, q, passing point *B* between times t_1 and t_2 will be

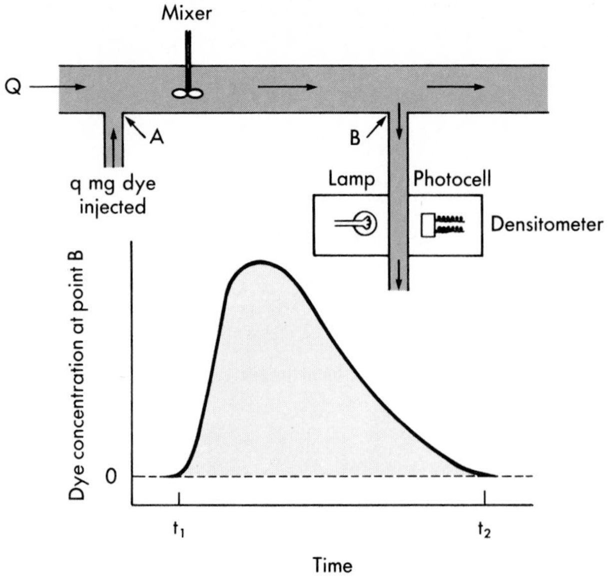

Fig. 28-20. The indicator dilution technique for measuring cardiac output. In this model, in which there is no recirculation, q mg of dye are injected instantaneously at point *A* into a stream flowing at Q ml/min. A mixed sample of the fluid flowing past point *B* is withdrawn at a constant rate through a densitometer. The resulting dye concentration curve at point *B* has the configuration shown in the lower section of the figure.

$$q = \bar{c}Q\,(t_2 - t_1) \qquad (6)$$

where $\bar{c}$ is the mean concentration of dye. The value of $\bar{c}$ may be computed by dividing the area of the dye concentration by the duration $(t_2 - t_1)$ of that curve; that is

$$\bar{c} = \int_{t_1}^{t_2} c\,dt/(t_2 - t_1) \qquad (7)$$

Substituting this value of $\bar{c}$ in to equation 6, and solving for Q yields

$$Q = \frac{q}{\int_{t_1}^{t_2} c\,dt} \qquad (8)$$

Thus flow may be measured by dividing the amount of indicator injected upstream by the area under the downstream concentration curve.

This technique has been widely used to estimate cardiac output in humans. A measured quantity of some indicator (a dye or isotope that remains within the circulation) is injected rapidly into a large central vein or into the right side of the heart through a catheter. Arterial blood is continuously drawn through a detector (densitometer or isotope rate counter), and a curve of indicator concentration is recorded as a function of time.

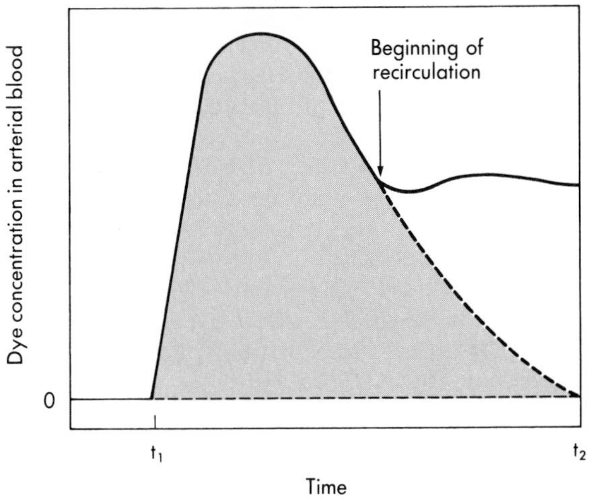

■ **Fig. 28-21.** Typical dye concentration curve recorded from a human. Because of recirculation of the dye, the concentration does not return to 0, as in the model in Fig. 28-20. The dashed line on the decending limb represents the semilogarithmic extrapolation of the upper portion of the descending limb, prior to the beginning of recirculation.

Because some of the indicator recirculates and reappears at the site of arterial withdrawal before the entire curve is inscribed, the concentration curve is not as simple as that shown in Fig. 28-20. Instead, on the downstroke of the curve a secondary increase in concentration (Fig. 28-21) appears as the recirculated dye becomes mixed with the last portions of dye still undergoing its primary passage past the site of withdrawal. To compute the area under the concentration curve, the downslope of the curve beyond the beginning of recirculation is extrapolated to zero concentration (dashed line, Fig. 28-21). The extrapolation, of course, introduces some error into the estimation of cardiac output.

Presently the most popular indicator dilution technique is *thermodilution*. The indicator is cold saline solution. The temperature and volume of the saline are measured accurately before injection. A flexible catheter is introduced into a peripheral vein and advanced so that the tip lies in the pulmonary artery. A small thermistor at the catheter tip records the changes in temperature. The opening in the catheter lies a few inches proximal to the tip. When the tip is in the pulmonary artery, the opening lies in or near the right atrium. The cold saline solution is injected rapidly into the right atrium through the catheter. The resultant change in temperature downstream is recorded by the thermistor in the pulmonary artery.

The thermodilution technique has the following advantages: (1) an arterial puncture is not necessary; (2) the small volumes of saline solution used in each determination are innocuous, allowing repeated determinations to be made; and (3) recirculation is negligible. Temperature equilibration takes place as the cooled blood flows through the pulmonary and systemic capillary beds, before it flows by the thermistor in the pulmonary artery the second time. Therefore the curve of temperature change resembles that shown in Fig. 28-20, and the extrapolation errors are averted.

■ *Bibliography*

Journal articles

Alpert, N.R., Hamrell, B.B., and Mulieri, L.A.: Heart muscle mechanics, Annu. Rev. Physiol. **41:**521, 1979.

Brutsaert, D.L., Claes, V.A., and Sonnenblick, E.H.: The velocity of shortening of unloaded heart muscle relative to the length-tension relation, Circ. Res. **29:**63, 1971.

Chapman, R.A.: Control of cardiac contractility at the cellular level, Am. J. Physiol. **245:**H535, 1983.

Fabiato, A., and Fabiato, F.: Calcium and cardiac excitation—contraction coupling, Annu. Rev. Physiol. **41:**473, 1979.

Horackova, M.: Transmembrane calcium transport and the activation of cardiac contraction, Can. J. Physiol. Pharmacol. **62:**874, 1984.

Jewell, B.R.: A reexamination of the influence of muscle length on myocardial performance, Circ. Res. **40:**221, 1977.

Langer, G.A.: Sodium-calcium exchange in the heart, Annu. Rev. Physiol. **44:**435, 1982.

Little, R.C.: The mechanism of closure of the mitral valve: a continuing controversy, Circulation **59:**615, 1979.

Molaug, M., et al.: Myocardial function of the interventricular septum: effects of right and left ventricular pressure loading before and after pericardiotomy in dogs, Circ. Res. **49:**521, 1981.

Noble, M.I.M.: The contribution of blood momentum to left ventricular ejection in the dog, Circ. Res. **23:**663, 1968.

Ross, J., Jr., and Sobel, B.E.: Regulation of cardiac contraction, Annu. Rev. Physiol. **34:**47, 1972.

Sagawa, K.: The ventricular pressure-volume diagram revisited, Circ. Res. **43:**677, 1978.

Sonnenblick, E.: Force-velocity relations in mammalian heart muscle, Am. J. Physiol. **202:**931, 1962.

Streeter, D.D., Jr., et al.: Fiber orientation in the canine left ventricle during diastole and systole, Circ. Res. **24:**339, 1969.

Weber, K.T., et al.: The contractile behavior of the heart and its functional coupling to the circulation, Prog. Cardiovasc. Dis. **24:**375, 1982.

Wexler, L.F., et al.: The relationship of the first heart sound to mitral valve closure in dogs, Circulation **66:**235, 1982.

Winegrad, S.: Calcium release from cardiac sarcoplasmic reticulum, Annu. Rev. Physiol. **44:**451, 1982.

Wohlfart, B., and Nobel, M.I.M.: The cardiac excitation-contraction cycle, Pharmacol. Ther. **16:**1, 1982.

Books and monographs

Brady, A.J.: Mechanical properties of cardiac fibers. In Handbook of physiology; Section 2: The cardiovascular system—the heart, vol. I, Bethesda, Md., 1979, American Physiological Society.

Braunwald, E., Ross, J., Jr., and Sonnenblick, E.H.: Mechanisms of contraction of the normal and failing heart, ed. 2, Boston, 1976, Little, Brown & Co.

Katz, A.M.: Physiology of the heart, New York, 1977, Raven Press.

Katz, A.M.: In Stone, P.H., and Antman, E.M., editors: Role of calcium in contraction of cardiac muscle in calcium channel blocking agents in the treatment of cardiovascular disorders, Mt. Kisco, N.Y., 1983, Future Publishing Co., Inc.

Langer, G.A., and Brady, A.J.: The mammalian myocardium, New York, 1974, John Wiley & Sons, Inc.

Lassen, N.A.: Indicator methods for measurement of organ and tissue blood flow. In Handbook of physiology; Section 2: The cardiovascular system—peripheral circulation and organ blood flow, vol. III, Bethesda, Md., 1983, American Physiological Society.

Parmley, W.W., and Talbot, L.: Heart as a pump. In Handbook of physiology; Section 2: The cardiovascular system—the heart, vol. I, Bethesda, Md., 1979, American Physiological Society.

Sommer, J.R., and Johnson, E.A.: Ultrastructure of cardiac muscle. In Handbook of physiology; Section 2: The cardiovascular system—the heart, vol. I, Bethesda, Md., 1979, American Physiological Society.

CHAPTER

29

Regulation of the Heartbeat

The quantity of blood pumped by the heart may be varied by changing the frequency of its beats or the volume ejected per stroke. A discussion of the control of cardiac activity may therefore be subdivided into a consideration of the regulation of pacemaker activity and the regulation of myocardial performance. However, in the intact organism, a change in the behavior of one of these features of cardiac activity almost invariably alters the other.

Experiments have shown that certain local factors, such as temperature changes and tissue stretch, can affect the discharge frequency of the SA node. Under natural conditions, however, the principal control of heart rate is relegated to the autonomic nervous system, and discussion will be restricted to this aspect of heart rate control. Relative to myocardial performance, intrinsic and extrinsic factors warrant consideration.

Nervous Control of Heart Rate

In normal adults the average heart rate at rest is approximately 70 beats per minute, but the rate is significantly greater in children. During sleep the heart rate diminishes by 10 to 20 beats per minute, but during emotional excitement or muscular activity it may accelerate to rates considerably above 100. In well-trained athletes at rest the rate is usually only 50 to 60 beats per minute.

The SA node is usually under the tonic influence of both divisions of the autonomic nervous system. The sympathetic system enhances automaticity, whereas the parasympathetic system inhibits it. Changes in heart rate usually involve a reciprocal action of the two divisions of the autonomic nervous system. Thus an increased heart rate is produced by a diminution of parasympathetic activity and concomitant increase in sympathetic activity; deceleration is usually achieved by the opposite mechanisms. Under certain conditions the heart rate may change by selective action of just one division of the autonomic nervous system, rather than by reciprocal changes in both divisions.

Ordinarily, in healthy, resting individuals parasympathetic tone is predominant. Abolition of parasympathetic influences by the administration of atropine usually elicits a pronounced tachycardia, whereas abrogation of sympathetic effects by the administration of propranolol usually slows the heart only slightly (Fig. 29-1). When both divisions of the autonomic nervous system are blocked, the heart rate of young adults averages about 100 beats per minute. The rate that prevails after complete autonomic blockade is called the *intrinsic heart rate*.

Parasympathetic Pathways

The cardiac parasympathetic fibers originate in the medulla oblongata, in cells that lie in the dorsal motor nucleus or the nucleus ambiguus. The precise location varies from species to species. Centrifugal vagal fibers pass inferiorly through the neck close to the common carotid arteries and then through the mediastinum to synapse with postganglionic cells located within the heart itself. Most of the cardiac ganglion cells are located near the SA node and AV conduction tissue.

The right and left vagi are usually distributed differentially to the various cardiac structures. The right vagus nerve affects the SA node predominantly. Stimulation produces sinus bradycardia or even complete cessation of SA nodal activity for several seconds. The left vagus nerve mainly influences AV conduction tissue, to produce various degrees of AV block. However, there is a considerable overlap, such that left vagal stimulation also depresses the SA node and right vagal stimulation impedes AV conduction.

The SA and AV nodes are rich in cholinesterase. Hence the effects of any given vagal impulse are ephemeral because the acetylcholine released at the nerve terminals is rapidly hydrolyzed. Also, parasympathetic influences preponderate over sympathetic effects at the SA node, as shown in Fig. 29-2. As the frequency of vagal stimulation in an anesthetized dog

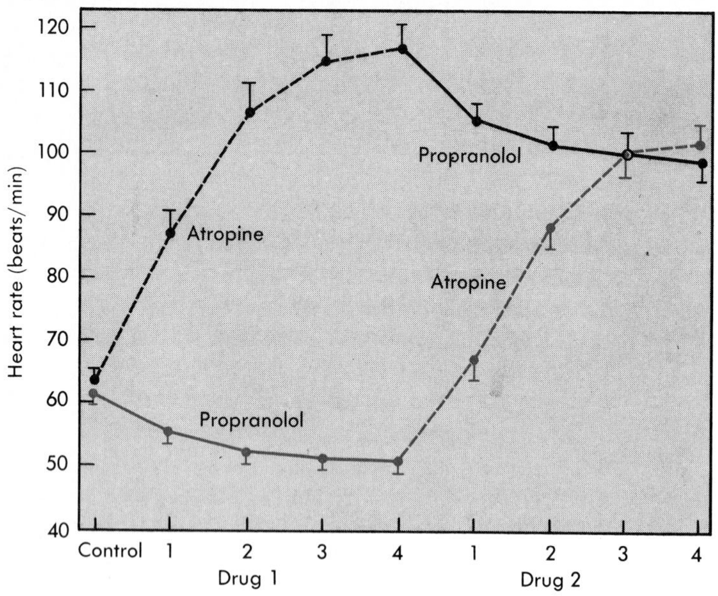

■ **Fig. 29-1.** The effects of four equal doses of atropine (0.04 mg/kg total) and of propranolol (0.2 mg/kg total) on the heart rate of 10 healthy young men (mean age, 21.9 years). In half of the trials, atropine was given first *(top curve);* in the other half, propranolol was given first *(bottom curve).* (Redrawn from Katona, P.G., McLean, M., Dighton, D.H., and Guz, A.: J. Appl. Physiol. **52:**1652, 1982.)

was increased from 0 to 8 pulses per second (left panel), the heart rate decreased by 75 beats per minute in the absence of concomitant cardiac sympathetic stimulation (S = 0). Sympathetic stimulation at 4 pulses per second (S = 4) increased the heart rate by 80 beats per minute in the absence of vagal stimulation (i.e., vagal frequency = 0). However, when vagal stimulation at 8 pulses per second was combined with sympathetic stimulation at 4 pulses per second (S = 4), the heart rate declined by 155 beats per minute. The right panel shows that increasing the sympathetic stimulation frequency from 0 to 4 pulses per second produced a substantial cardioacceleration in the absence of vagal stimulation (V = 0). However, when the vagi were stimulated at 8 pulses per second (V = 8), increasing the sympathetic stimulation frequency from 0 to 4 pulses per second had only a negligible influence on heart rate.

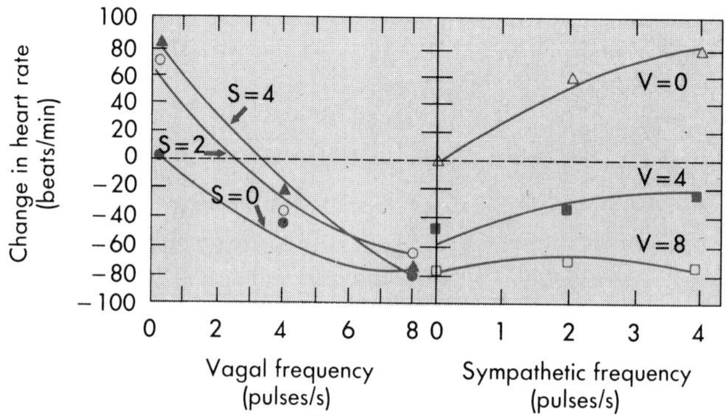

■ **Fig. 29-2.** The changes in heart rate in an anesthetized dog when the vagus and cardiac sympathetic nerves were stimulated simultaneously. The sympathetic nerves were stimulated at 0, 2, and 4 pulses/second; the vagus nerves at 0, 4 and 8 pulses/second. The symbols represent the observed changes in heart rate; the curves represent a derived regression equation. (Modified from Levy, M.N., and Zieske, H.: J. Appl. Physiol. **27:**465, 1969.)

■ *Sympathetic Pathways*

The cardiac sympathetic fibers originate in the intermediolateral columns of the upper five or six thoracic and lower one or two cervical segments of the spinal cord (see also Chaper 20). The fibers emerge from the spinal column through the white communicating branches and enter the paravertebral chain of ganglia. The anatomical details of the sympathetic innervation of the heart vary among mammalian species; the innervation has been elaborated in greatest detail in the dog (Fig. 29-3). In that species, virtually all of the preganglionic neurons ascend in the paravertebral chain and funnel through the stellate ganglia. In many species, including the cat, the preganglionic and postganglionic neurons synapse mainly in the stellate ganglia. In other species, such as the dog, the preganglionic neurons traverse the two limbs of the ansa subclavia and then synapse with the postganglionic neurons in the caudal cervical ganglia. These latter ganglia lie close to the vagus nerves in the superior portion of the mediastinum. Sympathetic and parasympathetic fibers then join to form a complex network of mixed efferent nerves to the heart (Fig. 29-3). Other sympathetic fibers ascend to the head and neck—as the cervical sympathetic chain in some species and in a common bundle with the vagus nerve (as the vagosympathetic trunk) in other species.

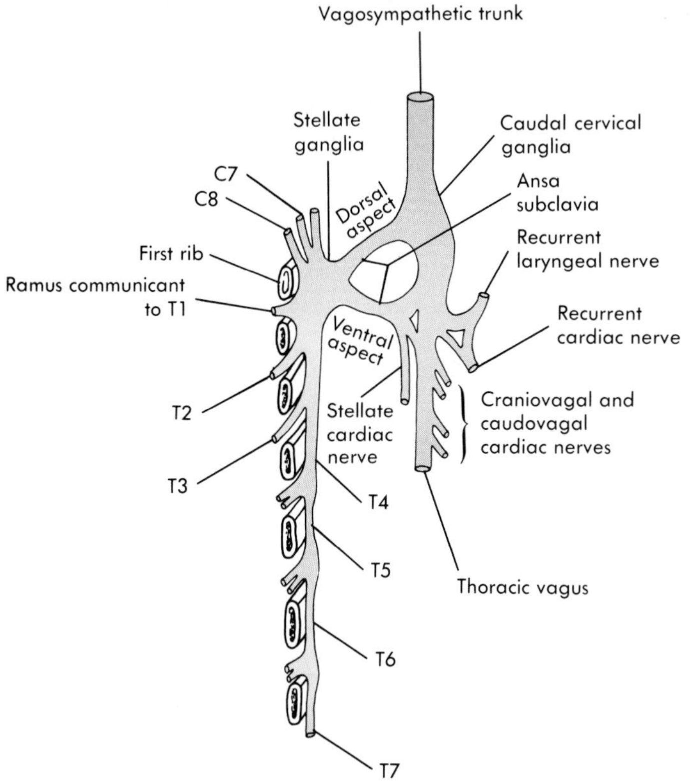

■ **Fig. 29-3.** Upper thoracic sympathetic chain and the cardiac autonomic nerves on the right side in the dog. (Modified from Mizeres, N.J.: Anat. Rec. **132**:261, 1958.)

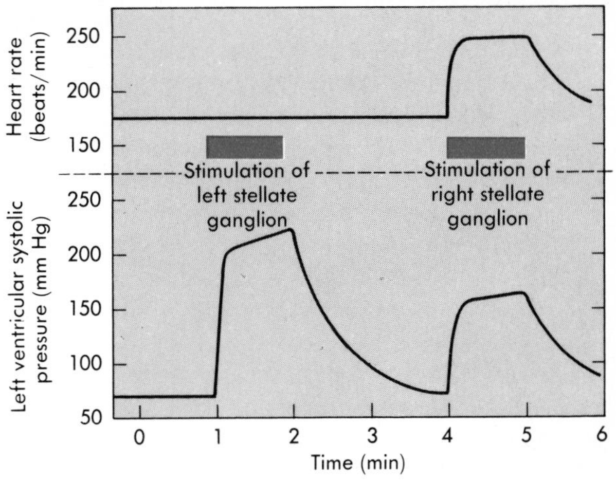

■ **Fig. 29-4.** In the dog, stimulation of the left stellate ganglion has a greater effect on ventricular contractility than does right-sided stimulation, but it has a lesser effect on heart rate. In this example, traced from an original record, left stellate ganglion stimulation had no detectable effect at all on heart rate but had a considerable effect on ventricular performance in an isovolumic left ventricle preparation.

The postganglionic cardiac sympathetic fibers approach the base of the heart along the adventitial surface of the great vessels. On reaching the base of the heart, these fibers are distributed to the various chambers as an extensive epicardial plexus. They then penetrate the myocardium, usually accompanying the branches of the coronary vessels. The adrenergic receptors in the nodal regions and in the myocardium are predominantly of the beta type; that is, they are responsive to beta-adrenergic agonists, such as isoproterenol, and are inhibited by specific beta-adrenergic blocking agents, such as propranolol.

As with the vagus nerves, there is a differential distribution of the left and right sympathetic fibers. In the dog, for example, the fibers on the left side have more pronounced effects on myocardial contractility than on heart rate. In some dogs left cardiac sympathetic nerve stimulation may not affect heart rate, even though it may exert pronounced facilitatory effects on ventricular performance, as shown in Fig. 29-4. Conversely, right cardiac sympathetic nerve stimulation elicits considerably greater cardiac acceleration and less augmentation of contractile force than does equivalent stimulation of sympathetic fibers on the left side. This bilateral asymmetry probably also exists in humans. In a group of patients right stellate ganglion blockade caused a mean reduction in heart rate of 14 beats per minute, whereas left-sided blockade elicited an average reduction of only 2 beats per minute.

It is evident from Fig. 29-4 that the effects of sympathetic stimulation decay very gradually after the cessation of stimulation, in contrast to the abrupt termination of the response after vagal activity (not shown). Most of the norepinephrine released during sympathetic stimulation is taken up again by the nerve terminals, and much of the remainder is carried away by the bloodstream. Relatively little of the released norepinephrine is degraded in the tissues, in contrast to the fate of the acetylcholine released at the vagus nerve endings.

■ *Control by Higher Centers*

Dramatic alterations in cardiac rate, rhythm, and contractility have been induced experimentally by stimulation of various regions of the brain. In the cerebral cortex the centers regulating cardiac function are located mostly in the anterior half of the brain, principally in the frontal lobe, the orbital cortex, the motor and premotor cortex, the anterior part of the temporal lobe, the insula, and the cingulate gyrus. In the thalamus, tachycardia may be induced by stimulation of the midline, ventral, and medial groups of nuclei. Variations in heart rate also may be evoked by stimulating the posterior and posterolateral regions of the hypothalamus. Stimuli applied to the H_2 fields of Forel in the diencephalon elicit a variety of cardiovascular responses, including tachycardia; such changes simulate closely those observed during muscular exercise. Undoubtedly the cortical and diencephalic centers are responsible for initiating the cardiac reactions that occur during excitement, anxiety, and other emotional states. The hypothalamic centers are also involved in the cardiac response to alterations in environmental temperature. Recent studies have shown that localized temperature changes in the preoptic anterior hypothalamus evoke pronounced changes in heart rate and peripheral resistance.

Stimulation of the parahypoglossal area of the medulla produces a reciprocal activation of cardiac sympathetic and inhibition of cardiac parasympathetic pathways. In certain dorsal regions of the medulla, distinct cardiac accelerator and augmentor sites have been detected in animals with transected vagi. Stimulation of accelerator sites increases heart rate, whereas stimulation of augmentor sites increases cardiac contractility. The accelerator regions are more abundant on the right and the augmentor sites more prevalent on the left. A similar distribution also exists in the hypothalamus. It appears, therefore, that for the most part the sympathetic fibers descend the brainstem ipsilaterally.

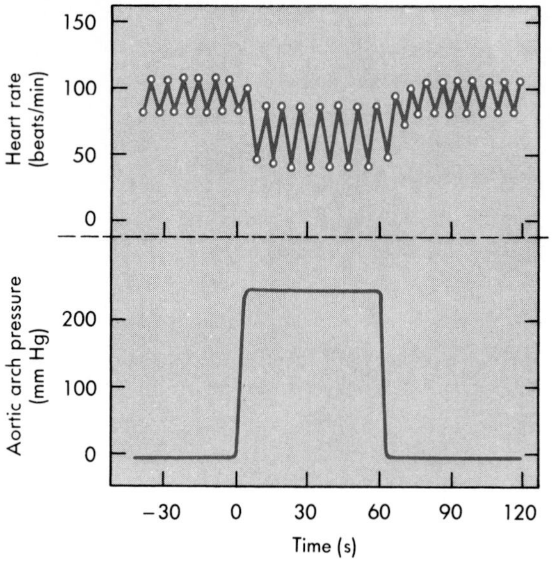

Fig. 29-5. Effect of a stepwise pressure change in the isolated aortic arch on heart rate. When pressure is raised, the mean heart rate decreases and the rhythmic fluctuations of heart rate increase at the frequency of respiratory movements. (Redrawn from Levy, M.N., Ng, M.L., and Zieske, H.: Circ. Res. **19**:930, 1966.)

Baroreceptor Reflex

An inverse relationship between arterial blood pressure and heart rate was first described in 1859 by Marey and is often referred to as *Marey's law of the heart*. It was demonstrated subsequently that the alterations in heart rate evoked by changes in blood pressure depended on baroreceptors located in the aortic arch and carotid sinuses (p. 517). These two sets of baroreceptors are about equally potent in the regulation of heart rate. An example of the effects on heart rate elicited by a stepwise elevation of pressure in an isolated aortic arch preparation is illustrated in Fig. 29-5. An abrupt rise in pressure in the aortic arch from 0 to 250 mm Hg resulted in bradycardia and an exaggeration of the respiratory cardiac arrhythmia (defined on p. 456). The subsequent decline of pressure was followed by return of the heart rate to control levels.

When the arterial blood pressure is in the normal range, the changes in heart rate that occur in response to moderate alterations in baroreceptor stimulation are mediated by reciprocal changes in activity in the two autonomic divisions. However, when blood pressure is increased by more than 20 or 30 mm Hg, cardiac sympathetic tone is completely suppressed. Thereafter the additional reduction in heart rate produced by any further rise in blood pressure is evoked entirely by increased vagal activity. The converse applies during the development of severe hypotension. Vagal tone virtually disappears after a relatively small drop in blood pressure. As the pressure continues to decline, further acceleration of the heart is ascribable solely to a progressive increase in sympathetic activity.

Participation of both divisions of the autonomic nervous system in the heart rate response to a moderate hypotension is shown in Fig. 29-6. In a group of 13 normal unanesthetized dogs, a 30 mm Hg decrease in mean arterial blood pressure was produced by nitroglycerin, which dilates peripheral arterioles. With efferent vagal and sympathetic pathways both intact, this reduction in pressure reflexly increased the heart rate by about 100 beats per minute. Blockade of the beta-adrenergic receptors alone with propranolol had a negligible effect on heart rate in these animals; the heart rate decreased from 77 to 75 beats/minute. After administration of propranolol, the 30 mm Hg reduction in blood pressure elicited by nitroglycerin increased the heart rate by only 50 beats/minute.

When the effects of propranolol had disappeared, atropine was given to block the cholinergic receptors. The basal heart rate increased to 180 beats/minute (Fig. 29-6). Hypotension increased the heart rate by about 30 beats/minute. When both autonomic divisions were blocked, heart rate was not affected by a 30 mm Hg decline in mean arterial pressure.

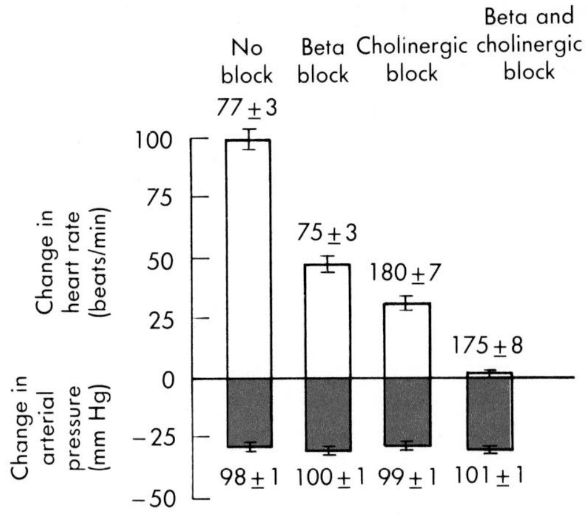

■ **Fig. 29-6.** The effects of beta-adrenergic receptor blockade, cholinergic blockade, and combined blockade on the change in heart rate in a group of 13 resting dogs when the mean arterial blood pressure was decreased by about 30 mm Hg by means of nitroglycerin. The control values of heart rate and blood pressure prior to hypotension are shown for each column. (From Vatner, S.F., Higgins, C.B., and Braunwald, E.: Cardiovasc. Res. **8**:53, 1974.)

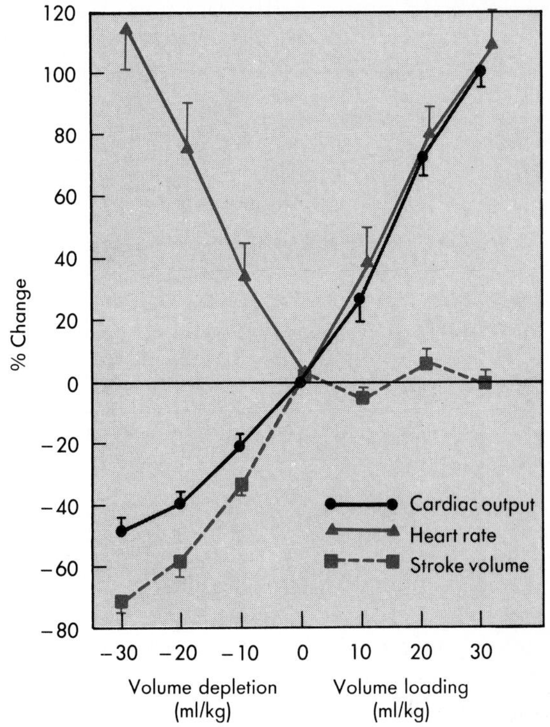

■ **Fig. 29-7.** Effects of blood transfusion and of bleeding on cardiac output, heart rate, and stroke volume in unanesthetized dogs. (From Vatner, S.F., and Boettcher, D.H.: Circ. Res. **42**:557, 1978. By permission of the American Heart Association, Inc.)

■ *Bainbridge Reflex and Atrial Receptors*

In 1915 Bainbridge reported that infusions of blood or saline solution accelerated the heart rate. This increase in heart rate occurred whether arterial pressure was unaffected or increased. Acceleration was observed whenever central venous pressure rose sufficiently to distend the right side of the heart, and the effect was abolished by bilateral transection of the vagi. Bainbridge postulated that increased cardiac filling elicited tachycardia reflexly and that the afferent impulses were conducted by the vagi.

Many investigators have confirmed the acceleration of the heart in response to the intravenous administration of fluid. However, the magnitude and direction of the response depends on the prevailing heart rate. At relatively slow rates, intravenous infusions usually accelerate the heart. At more rapid initial rates, however, infusions will ordinarily slow the heart. Increases in blood volume not only evoke the Bainbridge reflex, but they also activate other reflexes (notably the baroreceptor reflex) that tend to elicit oppositely directed heart rate changes. The actual change in heart rate evoked by an alteration of blood volume under a given set of conditions is therefore the resultant of these antagonistic reflex effects.

In unanesthetized dogs, volume loading with blood increased heart rate and cardiac output proportionately (Fig. 29-7). Consequently, stroke volume remained virtually constant. Conversely, reductions in blood volume diminished the cardiac output but also evoked a tachycardia. Undoubtedly, the Bainbridge reflex was prepotent over the baroreceptor reflex when the blood volume was raised, but the baroreceptor reflex prevailed over the Bainbridge reflex under conditions of hypovolemia.

Receptors that influence heart rate exist in both atria. The receptors are located principally in the venoatrial junctions—in the right atrium at its junctions with the venae cavae and in the left atrium at its junctions with the pulmonary veins. Distension of these receptors sends impulses centripetally in the vagi. The efferent impulses are carried by fibers from both autonomic divisions to the SA node. The cardiac response is highly selective. Even when the reflex increase in heart rate is large, changes in ventricular contractility have been negligible. Furthermore, the increase in heart rate is unattended by any augmentation of sympathetic activity to the peripheral arterioles.

Stimulation of the atrial receptors causes an increase in urine volume, in addition to the cardiac acceleration. The reduction in activity in the renal sympathetic nerve fibers might be partially responsible for this diuresis. However, the principal mechanism appears to be a neurally mediated reduction in the secretion of vasopressin (antidiuretic hormone) by the posterior pituitary

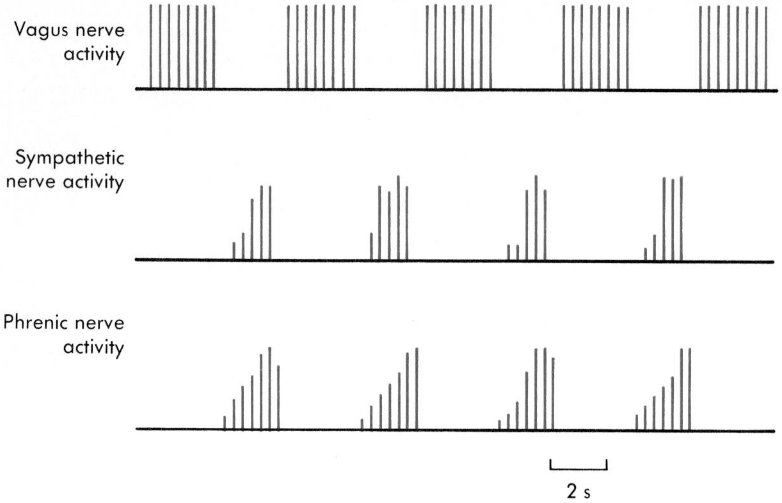

Fig. 29-8. The respiratory fluctuations in efferent neural activity in the cardiac nerves of an anesthetized dog. Note that the sympathetic nerve activity occurs synchronously with the phrenic nerve discharges (which initiate diaphragmatic contraction), whereas the vagus nerve activity occurs between the phrenic nerve discharges. (From Kollai, M., and Koizumi, K.: J. Auton. Nerv. Syst. **1**:33, 1979.)

gland. A peptide with potent diuretic and natriuretic properties is present in atrial tissue. This peptide is called *atrial natriuretic factor,* and its role in the regulation of water and electrolyte balance is being investigated intensively (see also Chapters 48 and 52).

Respiratory Cardiac Arrhythmia

Rhythmic variations in heart rate, occurring at the frequency of respiration, are detectable in most individuals and tend to be more pronounced in children. Typically the cardiac rate accelerates during inspiration and decelerates during expiration.

Recordings from the autonomic nerves to the heart reveal that the neural activity increases in the sympathetic fibers during inspiration, whereas the neural activity in the vagal fibers increases during expiration (Fig. 29-8). The acetylcholine released at the vagal endings is removed so rapidly that the rhythmic changes in activity can elicit rhythmic variations in heart rate. Conversely, the norepinephrine released at the sympathetic endings is removed more slowly, thus damping out the effects of rhythmic variations in release of this transmitter on heart rate. Hence rhythmic changes in heart rate are ascribable almost entirely to the oscillations in vagal activity.

Respiratory sinus arrhythmia is exaggerated when vagal tone is enhanced. An example is shown in Fig. 29-5, where vagal tone was increased by elevating pressure in the aortic arch. Under these conditions the amplitude of the cyclic variations in heart rate was twice as great as when the pressure in the aortic arch was low (Fig. 29-5).

Both reflex and central factors contribute to the genesis of the respiratory cardiac arrhythmia. During inspiration the intrathoracic pressure decreases and therefore venous return to the right side of the heart is accelerated, which elicits the Bainbridge reflex. After the time delay required for the increased venous return to reach the left side of the heart, left ventricular output increases and raises arterial blood pressure. This in turn reduces heart rate reflexly through baroreceptor stimulation.

Fluctuations in sympathetic activity to the arterioles cause peripheral resistance to vary at the respiratory frequency. Consequently, arterial blood pressure fluctuates rhythmically, which affects heart rate via the baroreceptor reflex. Stretch receptors in the lungs may also affect heart rate. Moderate pulmonary inflation may increase heart rate reflexly. The afferent and efferent limbs of this reflex are located in the vagus nerves.

Central factors are also responsible for respiratory cardiac arrhythmia. The respiratory center in the medulla influences the cardiac autonomic centers. In the experiment displayed in Fig. 29-5 a total heart-lung bypass was employed to eliminate variations in respiration, venous return, or arterial blood pressure. Yet in that experiment, a respiratory cardiac arrhythmia was still prominent. The changes in heart rate that occurred at the frequency of the rib-cage movements were almost certainly induced by an interaction between the respiratory and cardiac centers in the medulla.

Chemoreceptor Reflex

The cardiac response to peripheral chemoreceptor stimulation merits special consideration because it illustrates the complexity that may be introduced when one stimulus excites two organ systems simultaneously. In intact animals, stimulation of the carotid chemoreceptors consistently increases ventilatory rate and depth but ordinarily evokes only slight increases or decreases in heart

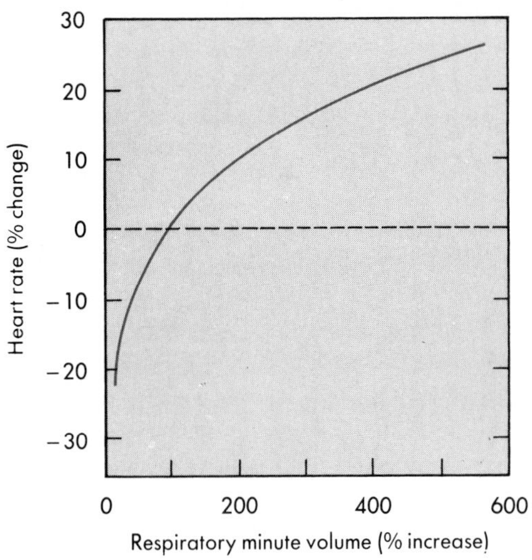

■ **Fig. 29-9.** Relationship between the change in heart rate and the change in respiratory minute volume during carotid chemoreceptor stimulation in spontaneously breathing cats and dogs. When respiratory stimulation was relatively slight, heart rate usually diminished; when respiratory stimulation was more pronounced, heart rate usually increased. (Modified from Daly, M.deB., and Scott, M.J.:J. Physiol. **144:**148, 1958.)

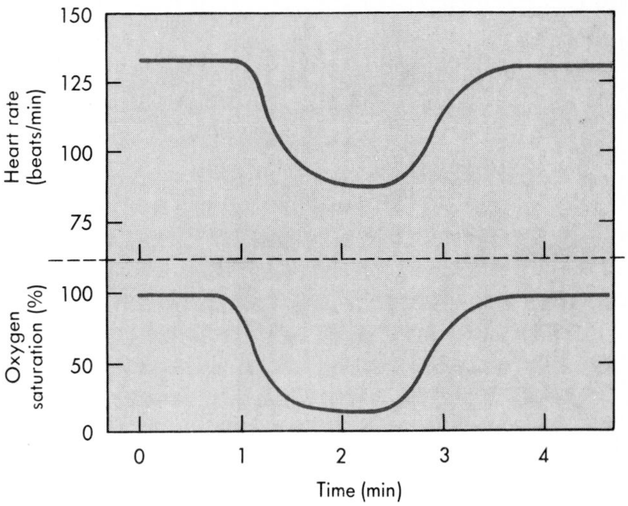

■ **Fig. 29-10.** Changes in heart rate during carotid chemoreceptor stimulation when the lungs remain deflated and respiratory gas exchange is accomplished by an artificial oxygenator. The lower tracing represents the oxygen saturation of the blood perfusing the carotid chemoreceptors. The blood perfusing the remainder of the animal, including the myocardium, was fully saturated with oxygen throughout the experiment. (Modified from Levy, M.N., DeGeest, H., and Zieske, H.: Circ. Res. **18:**67, 1966.)

rate. The directional change in heart rate is related to the enhancement of pulmonary ventilation, as shown in Fig. 29-9. When respiratory stimulation is relatively mild, heart rate usually diminishes; when the increase in pulmonary ventilation is more pronounced, heart rate usually accelerates.

The cardiac response to peripheral chemoreceptor stimulation is the resultant of primary and secondary reflex mechanisms. The primary reflex effect of carotid chemoreceptor excitation on the SA node is inhibitory. Secondary effects are facilitatory, and they vary with the concomitant stimulation of respiration.

An example of the primary inhibitory influence is displayed in Fig. 29-10. In this experiment on an anesthetized dog, the lungs were completely collapsed and blood oxygenation was accomplished by an artificial oxygenator. When the carotid chemoreceptors were stimulated, an intense bradycardia ensued that was usually accompanied by some degree of AV block. Such effects are mediated primarily by efferent vagal fibers.

The identical primary inhibitory effect also operates in humans. The electrocardiogram in Fig. 29-11 was recorded from a quadriplegic patient who could not breathe spontaneously, but required tracheal intubation and artificial respiration. When the tracheal catheter was briefly disconnected to permit nursing care, the patient quickly developed a profound bradycardia. His heart rate was 65 beats per minute just before the tracheal catheter was disconnected. In less than 10 seconds after cessation of artificial respiration, his heart rate dropped to about 20 beats per minute. This bradycardia could be prevented with atropine, and its onset could be delayed considerably by hyperventilating the patient before disconnecting the tracheal catheter.

The pulmonary hyperventilation that is ordinarily evoked by carotid chemoreceptor stimulation influences heart rate secondarily, both by initiating more pronounced pulmonary inflation reflexes and by producing hypocapnia. Each of these influences accelerates the heart and depresses the primary cardiac response to chemoreceptor stimulation. Hence when pulmonary hyperventilation is not prevented, carotid chemoreceptor stimulation affects heart rate minimally.

■ *Ventricular Receptor Reflexes*

Sensory receptors located near the endocardial surfaces of the ventricular walls initiate reflex effects similar to those elicited by the arterial baroreceptors. Excitation of these endocardial receptors diminishes the heart rate and peripheral resistance. The receptors discharge in a pattern that parallels the changes in ventricular pressure. Impulses that originate in these receptors are transmitted to the medulla by myelinated vagal fibers.

Other sensory receptors have been identified in the epicardial regions of the ventricles. These receptors discharge in patterns that are not related to the changes in

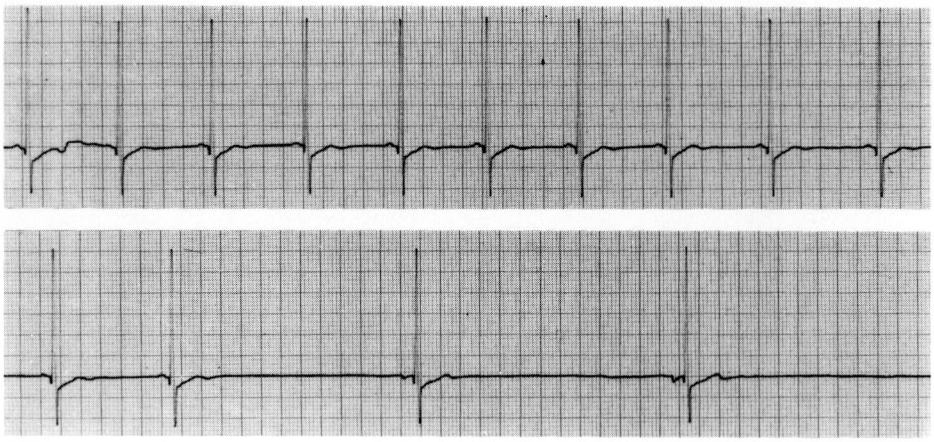

■ **Fig. 29-11.** Electrocardiogram of a 30-year-old quadriplegic man who could not breathe spontaneously and required tracheal intubation and artificial respiration. The two strips are continuous. The tracheal catheter was temporarily disconnected from the respirator at the beginning of the top strip, at which time his heart rate was 65 beats per minute. In less than 10 seconds, his heart rate decreased to about 20 beats/minute. (Modified from Berk, J.L., and Levy, M.N.: Eur. Surg. Res. **9**:75, 1977.)

ventricular pressure. Ventricular receptors are excited by a variety of mechanical and chemical stimuli, and their physiological functions are not clear.

■ *Intrinsic Regulation of Myocardial Performance*

Just as the heart can initiate its own beat in the absence of any nervous or hormonal control, so also can the myocardium adapt to changing hemodynamic conditions by mechanisms that are intrinsic to cardiac muscle itself. Experiments on denervated hearts reveal that this organ adjusts remarkably well to stress. For example, racing greyhounds with denervated hearts perform almost as well as those with intact innervation. Their maximal running speed is only 5% less after complete cardiac denervation. In these dogs, the threefold to fourfold increase in cardiac output that occurs when they run is achieved principally by an increase in stroke volume. In normal dogs the increase of cardiac output with exercise is accompanied by a proportionate increase of heart rate; stroke volume does not change much. It is unlikely that the cardiac adaptation in the denervated animals is achieved entirely by intrinsic mechanisms; circulating catecholamines undoubtedly contribute. If the β-adrenergic receptors are blocked by drugs in greyhounds with denervated hearts, their racing performance is severely impaired.

The heart is partially or completely denervated in a variety of clinical situations: (1) the surgically transplanted heart is totally decentralized, although the intrinsic, postganglionic parasympathetic fibers persist; (2) atropine blocks vagal effects on the heart, and propranolol blocks sympathetic influences; (3) certain drugs, such as reserpine, deplete cardiac norepinephrine stores and thereby restrict or abolish sympathetic control; and (4) in chronic congestive heart failure, cardiac norepinephrine stores are often severely diminished, thereby attenuating any sympathetic influences.

The intrinsic cardiac adaptation that has received the greatest attention involves changes in the resting length of the myocardial fibers. This adaptation is designated *Starling's law of the heart* or the *Frank-Starling mechanism*. The mechanical and structural bases for this mechanism have been explained in Chapters 22 and 28. The term *heterometric autoregulation* also has been applied to those adaptive mechanisms that involve changes in myocardial fiber length, whereas the term *homeometric autoregulation* is applied to those other intrinsic adjustments of cardiac performance that are independent of changes in myocardial fiber length.

Heterometric Autoregulation

Studies on isolated hearts. In 1895 Frank described the response of the isolated heart of the frog to alterations in the load on the myocardial fibers just prior to contraction—the *preload*. He observed that as the preload increased, the heart responded with a more forceful contraction. Frank recognized that cardiac muscle behavior was similar to that of skeletal muscle when it is stretched to progressively greater initial lengths prior to contraction.

In 1914 Starling described the intrinsic response of the heart to changes in right atrial and aortic pressure in the canine heart-lung preparation, which is depicted in Fig. 29-12. In this preparation the right atrium is filled with blood from a reservoir. Right atrial pressure is varied either by altering the height of the reservoir or by adjusting a screw clamp on the connecting tube. From the right atrium, blood enters the right ventricle, which then pumps it through the pulmonary vessels. The trachea is cannulated and the lungs are artificially ventilated.

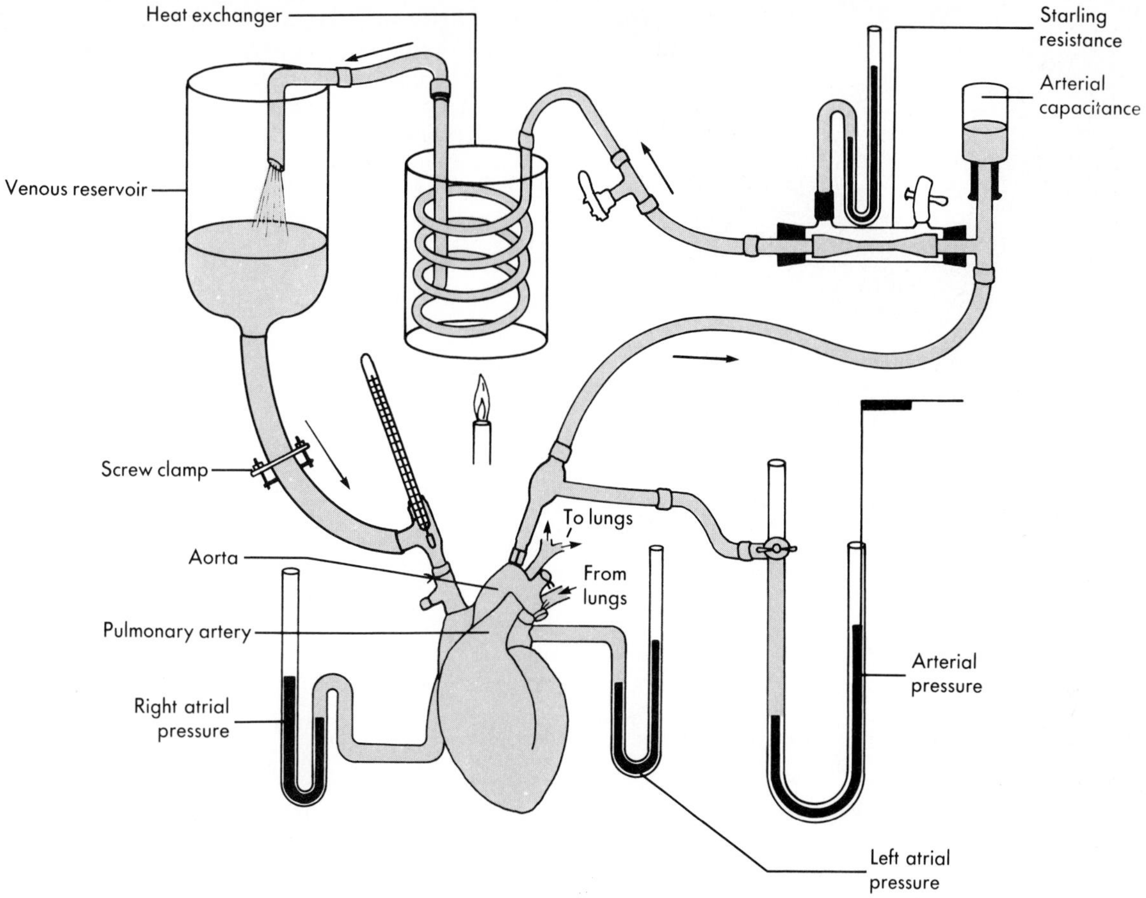

■ **Fig. 29-12.** Heart-lung preparation. (Redrawn from Patterson, S.W., and Starling, E.H.: J. Physiol. **48:**357, 1914.)

The pulmonary venous return enters the left atrium. The aorta is ligated distal to the arch, and a cannula is inserted into the brachiocephalic artery. Blood is pumped by the left ventricle through this cannula and through artificial tubing that ultimately conducts the blood back through a heating coil to the right atrial reservoir. The volume of both ventricles is recorded by a special device.

Peripheral resistance is adjusted by a pressure-limiting system, which has become known as a *Starling resistance*. This device consists of a piece of collapsible tubing in a rigid chamber. Any desired air pressure is applied to the collapsible tubing through an inlet. The cardiac output is measured by temporarily diverting the flow, which is returning to the venous reservoir, into a graduated cylinder for a measured interval of time.

The response of the isolated heart to a sudden augmentation of right atrial pressure is shown in Fig. 29-13. Aortic pressure was permitted to increase only slightly. In the top tracing, an increase of ventricular volume is registered as a downward deflection. Hence the upper border of the tracing represents the systolic volume, the lower border indicates the diastolic volume, and the amplitude of the deflections reflects the stroke volume.

For several beats after the augmentation of right atrial pressure, the ventricular volume progressively increased. This indicates that a disparity must have existed between ventricular inflow during diastole and ventricular output during systole; that is, during systole the ventricles did not expel an amount of blood equal to that which entered during diastole, until a new equilibrium had been attained. This progressive accumulation of blood dilated the ventricles and lengthened the individual myocardial fibers that comprised the walls of the ventricles.

The increased diastolic fiber length somehow facilitates ventricular contraction and enables the ventricles to pump a greater stroke volume, so that at equilibrium cardiac output exactly matches the augmented venous return. An optimum fiber length apparently exists, beyond which contraction is actually impaired (Chapter 22). Therefore excessively high filling pressures may depress rather than enhance the pumping capacity of the ventricles by overstretching the myocardial fibers.

Changes in diastolic fiber length also permit the iso-

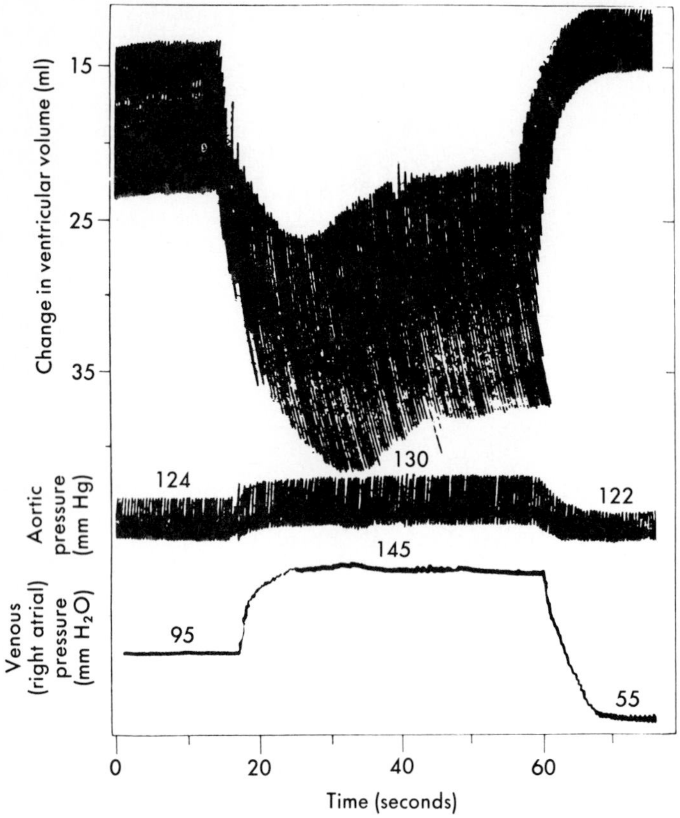

■ Fig. 29-13. Changes in ventricular volume in a heart-lung preparation when the venous reservoir was suddenly raised (right atrial pressure increased from 95 to 145 mm H_2O) and subsequently lowered (right atrial pressure decreased from 145 to 55 mm H_2O). Note that an increase in ventricular volume is registered as a downward shift in the volume tracing. (Redrawn from Patterson, S.W., Piper, H., and Starling, E.H.: J. Physiol. **48**:465. 1914.)

lated heart to compensate for an increase of peripheral resistance. In the experiment depicted in Fig. 29-14 the arterial resistance was abruptly raised in three steps, whereas venous inflow was held constant. Each rise in resistance increased the arterial pressure and ventricular volume. With each abrupt elevation of arterial pressure, which reflects the afterload, the left ventricle was at first unable to pump a normal stroke volume. Since venous return was held constant, the diminution of stroke volume was attended by a rise in ventricular diastolic volume and therefore in the length of the myocardial fibers. This change in end-diastolic fiber length finally enabled the ventricle to pump a given stroke volume against a greater peripheral resistance.

The external work performed per stroke by the left ventricle is approximately equal to the product of the mean arterial pressure and stroke volume (p. 486). Therefore the increased diastolic length of the cardiac muscle fiber increases the work production by the left ventricle. However, at an excessively high peripheral resistance, further augmentation of resistance will reduce stroke volume and stroke work.

Changes in ventricular volume are also involved in the cardiac adaptation to alterations in heart rate. During bradycardia, for example, the increased duration of diastole permits greater ventricular filling. The consequent augmentation of myocardial fiber length increases stroke volume. Therefore the reduction in heart rate may be fully compensated by the increase in stroke volume, such that cardiac output may remain constant.

When cardiac compensation involves ventricular dilation, the force required by each myocardial fiber to generate a given intraventricular systolic pressure must be appreciably greater than that developed by the fibers in a ventricle of normal size. The relationship between wall tension and cavity pressure resembles that for cylindrical tubes in that for a constant internal pressure, wall tension varies directly with the radius. As a consequence, the dilated heart requires considerably more oxygen to perform a given amount of external work than does the normal heart.

In the intact animal, the heart is enclosed in the pericardial sac. The relatively rigid pericardium determines the pressure-volume relationship at high levels of pressure and volume. The pericardium exerts this limitation of volume even under normal conditions, when an individual is at rest and the heart rate is slow. In the cardiac dilation and hypertrophy that accompanies chronic heart failure, the pericardium is stretched considerably (Fig. 29-15). The pericardial limitation of cardiac filling is exerted at pressures and volumes that are entirely different from those in normal individuals.

Studies on more intact preparations. The major problem of assessing the role of the Frank-Starling mechanism in intact animals and humans is the difficulty of measuring end-diastolic myocardial fiber

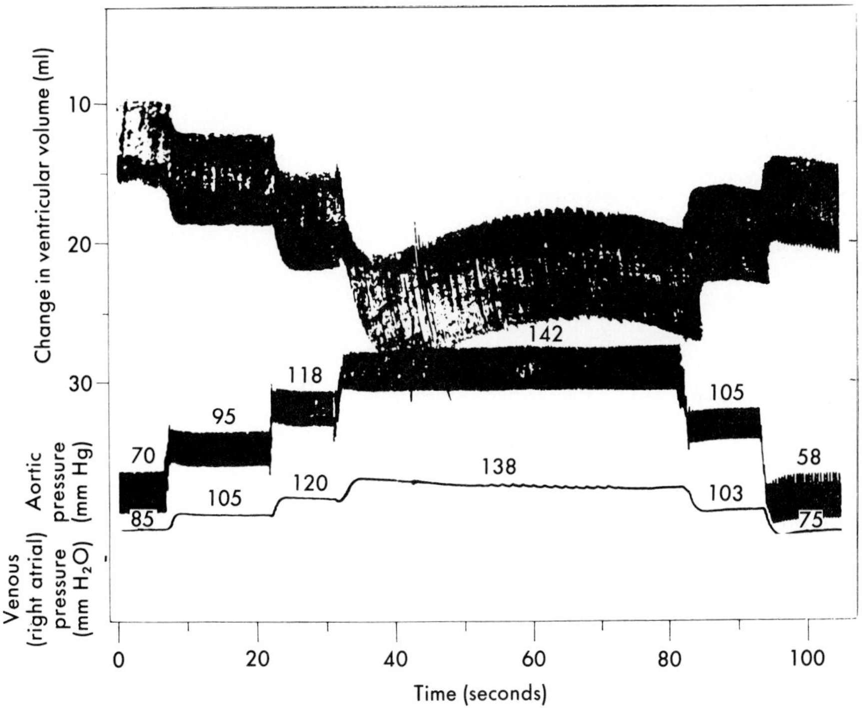

■ **Fig. 29-14.** Changes in ventricular volume, aortic pressure, and right atrial pressure in a heart-lung preparation when peripheral resistance was raised and subsequently lowered in several steps. Note that an increase in ventricular volume is registered as a downward shift in the volume tracing. (Redrawn from Patterson, S.W., Piper, H., and Starling, E.H.: J. Physiol. **48**:465, 1914.)

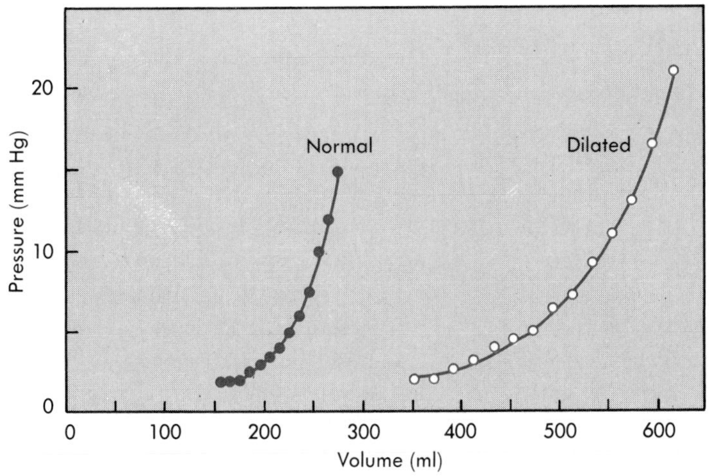

■ **Fig. 29-15.** Pericardial pressure-volume relations in a normal dog and in a dog with experimentally induced chronic cardiac hypertrophy. (Modified from Freeman, G.L., and Le Winter, M.M.: Circ. Res. **54**:294, 1984. By permission of the American Heart Association, Inc.)

length. The Frank-Starling mechanism has been represented graphically by plotting some index of ventricular performance along the ordinate and some index of fiber length along the abscissa. The most commonly used indices of ventricular performance are cardiac output, stroke volume, and stroke work. The indices of fiber length include ventricular end-diastolic volume, ventricular end-diastolic pressure, ventricular circumference, and mean atrial pressure.

The Frank-Starling mechanism is better represented by a family of so-called ventricular function curves, rather than by a single curve. To construct a given ventricular function curve, blood volume is altered over a wide range of values, and stroke work and end-diastolic pressure are measured at each step. Similar observations are then made during the desired experimental intervention. For example, the ventricular function curve obtained during a norepinephrine infusion lies above and to the left of a control ventricular function curve (Fig. 29-16). It is evident that, for a given level of left ventricular end-diastolic pressure, the left ventricle performs more work during a norepinephrine infusion than during control conditions. Hence a shift of the ventricular function curve to the left usually signifies an improvement of ventricular contractility; a shift to the right usually indicates an impairment of contractility and a consequent tendency toward *cardiac failure*. A shift in a ventricular function curve does not uniformly indicate a change in contractility, however. Contractility is a measure of cardiac performance at a given level

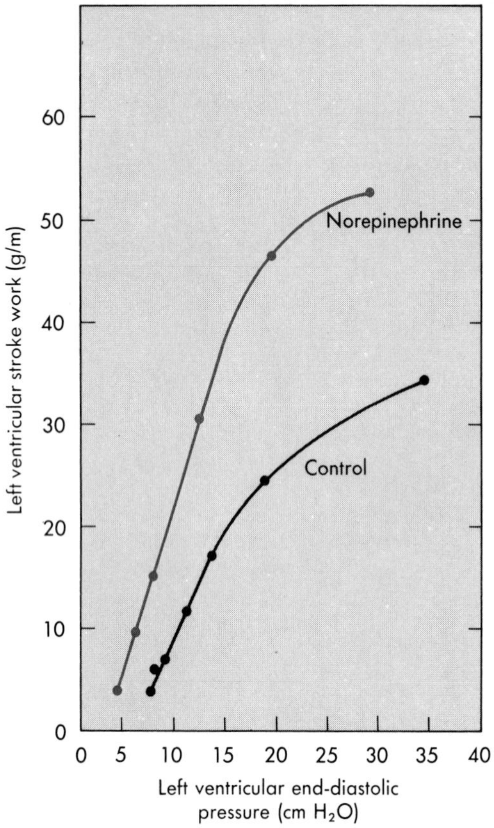

■ **Fig. 29-16.** A constant infusion of norepinephrine in a dog shifts the ventricular function curve to the left. This shift signifies an enhancement of ventricular contractility. (Redrawn from Sarnoff, S.J., et al.: Circ. Res. **8:**1108, 1960.)

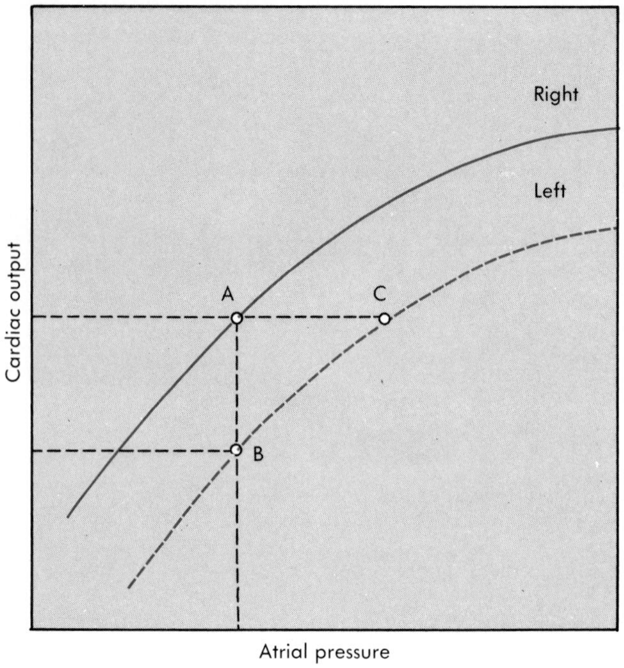

■ **Fig. 29-17.** Curves relating the outputs of right and left ventricles to mean right and left atrial pressure, respectively. At a given level of cardiac output, mean left atrial pressure (for example, point *C*) exceeds mean right atrial pressure (point *A*).

of preload and afterload. The end-diastolic pressure is ordinarily a good index of preload, whereas the aortic systolic pressure is a good index of afterload. In assessing myocardial contractility, the cardiac afterload must be held constant as the end-diastolic pressure is varied over a range of values.

It is difficult to determine the precise position on a Frank-Starling curve at which the heart of an intact, conscious person or animal may be operating. When a dog is reclining and relaxed, the left ventricle seems to be at nearly its maximum size at the end of diastole; it is probably limited by the pericardium. When blood volume is reduced, the end-diastolic volume diminishes, and the Frank-Starling mechanism undoubtedly helps regulate cardiac output. Conversely, when blood volume is acutely expanded, end-diastolic volume apparently cannot increase appreciably. Hence extrinsic mechanisms probably play a more crucial role in cardiac output regulation. In the experiments illustrated in Fig. 29-7, for example, the transfusion of blood caused a parallel increase in heart rate and cardiac output; stroke volume did not change appreciably. The Bainbridge reflex and other extrinsic mechanisms probably play the dominant role in this adaptation to an increased blood volume.

The Frank-Starling mechanism is ideally suited for matching the cardiac output to the venous return. Any sudden, excessive output by one ventricle soon results in a greater venous return to the other ventricle. The consequent increase in diastolic fiber length serves as the stimulus to increase the output of the second ventricle to correspond with that of its mate. For this reason it is the Frank-Starling mechanism that maintains a precise balance between the outputs of the right and left ventricles. Because the two ventricles are arranged in series in a closed circuit, even a small, but maintained, imbalance in the outputs of the two ventricles would be catastrophic.

The curves relating cardiac output to mean atrial pressure for the two ventricles are not coincident; the curve for the left ventricle usually lies below that for the right, as shown in Fig. 29-17. At equal right and left atrial pressures (points *A* and *B*) right ventricular output would exceed left ventricular output. Hence venous return to the left ventricle (a function of right ventricular output) would exceed left ventricular output, and left ventricular diastolic volume and pressure would rise. By the Frank-Starling mechanism, left ventricular output would therefore increase (from *B* toward *C*). Only when the outputs of both ventricles are identical

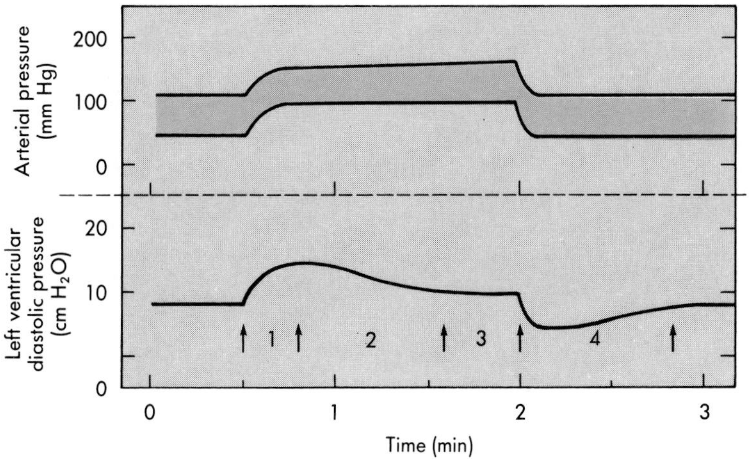

■ **Fig. 29-18.** Homeometric autoregulation in the dog heart in response to a sustained increase in peripheral resistance. Initially left ventricular end-diastolic pressure increased (phase *1*), but then returned toward the control level (phases *2* and *3*), even though peripheral resistance was still augmented. (Redrawn from Sarnoff, S.J., and Mitchell, J.H.: Am. J. Med. **34:**440, 1963.)

(points *A* and *C*) would the equilibrium be stable. Under such conditions, however, left atrial pressure *(C)* would exceed right atrial pressure *(A),* and this is precisely the relationship that prevails in both humans and dogs. This difference in atrial pressures accounts for the observation that in *congenital atrial septal defects,* where a communication exists between the two atria, the direction of the shunt flow is ordinarily from left to right.

■ *Homeometric Autoregulation*

Afterload-induced regulation. An increase in afterload may evoke a type of homeometric autoregulation, which is sometimes referred to as the *Anrep effect.* In the experiment shown in Fig. 29-18, arterial and left ventricular diastolic pressures rose when peripheral resistance was increased abruptly. However, the rise (phase *1*) in left ventricular diastolic pressure was only temporary. After the peak diastolic pressure was reached, the diastolic pressure diminished progressively toward the control level (phase *2*) and aortic pressure remained elevated. Finally, a new equilibrium level of ventricular pressure was reached (phase *3*); in many cases this pressure was actually at or slightly below the control value. The ventricular volume during diastole paralleled the changes in ventricular diastolic pressure. With abrupt return of peripheral resistance to the control level, left ventricular diastolic pressure declined to a value appreciably below control and then gradually returned to the control level (phase *4*). Coronary blood flow was controlled in these experiments, so that the adaptation of the heart cannot be attributed to an improved myocardial circulation.

Thus the mammalian ventricle has the intrinsic capability of adapting to changes in filling pressure and arterial resistance without a continued increase in resting fiber length. The conditions under which homeometric autoregulation, rather than maintained heterometric autoregulation, would prevail are not known.

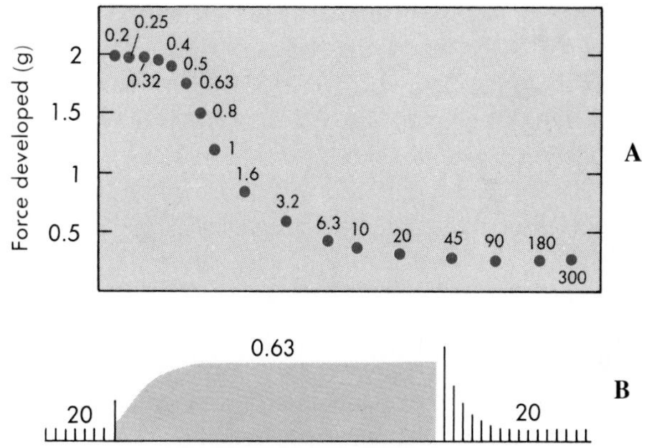

■ **Fig. 29-19.** Changes in force development in an isolated papillary muscle from a cat as the interval between contractions is varied. The numbers in both sections of the record denote the interval (in seconds) between beats. In section **A** the points represent the steady-state forces developed at the intervals indicated. (Redrawn from Koch-Weser, J., and Blinks, J.R.: Pharmacol. Rev. **15:**601, 1963).

Rate-induced regulation. The effects of contraction frequency on the force developed in an isometrically contracting cat papillary muscle are shown in Fig. 29-19, *B.* Initially the strip of cardiac muscle was stimulated to contract only once every 20 seconds. When the muscle was made to contract once every 0.63 second, the developed force increased progressively over the next several beats. This progressive increase in developed force induced by a change in contraction frequency is known as the *staircase,* or *Treppe, phenomenon.*

At the new steady state, the developed force was more than five times as great as it was at the lower contraction frequency. A return to the slower rate had the opposite influence on developed force, with an ultimate return to the initial value.

The effect of the interval between contractions on the

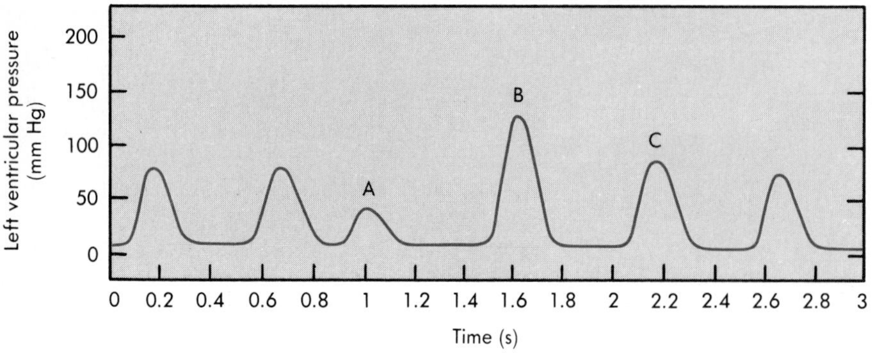

Fig. 29-20. In an isovolumic canine left ventricle preparation a premature ventricular systole (beat *A*) is typically feeble, whereas the postextrasystolic contraction (beat *B*) is characteristically strong, and the enhanced contractility may persist to a diminishing degree over a few beats (for example, contraction *C*).

steady-state level of developed force is shown in panel *A* (Fig. 29-19) for a wide range of intervals. As the interval is diminished from 300 seconds down to about 10 to 20 seconds, there is little change in developed force. As the interval is reduced further, to a value of about 0.5 second, force increases sharply. Further reduction of the interval to 0.2 second has little additional effect on developed force.

The progressive rise in developed force observed in panel *B* as the contraction frequency is increased is ascribable to a gradual rise in intracellular calcium. During the plateau of the action potential, Ca^{++} enters the cell (Chapter 27). As the contraction frequency is increased, there is a greater influx of Ca^{++} into the cell. The duration of the cardiac action potential diminishes as contraction frequency is raised (Fig. 27-15), and so less Ca^{++} actually enters per contraction. The influx of Ca^{++} per minute equals the influx per beat times the number of beats per minute. As the frequency of contraction is increased, the increment in the number of beats per minute exceeds the decrement in Ca^{++} influx per beat. Hence the Ca^{++} influx per minute increases. The intracellular content of Ca^{++} increases, and as a consequence, developed force rises as the interval between contractions is diminished, as shown in Fig. 29-19, *A*.

Another influence of the time interval between beats has been termed *postextrasystolic potentiation*. When a premature ventricular systole occurs, the premature contraction itself is feeble, whereas the beat after the compensatory pause is very strong. This response is partly ascribable to the Frank-Starling mechanism. Inadequate ventricular filling just before the premature beat accounts partly for the feeble premature contraction. Subsequently, the exaggerated degree of filling associated with the compensatory pause explains in part the vigorous postextrasystolic contraction.

Although heterometric autoregulation is certainly involved in the usual ventricular adaptation to a premature beat, it is not the exclusive mechanism. For example, in the ventricular pressure curves recorded from an isovolumic left ventricle preparation (Fig. 29-20), in which neither filling nor ejection takes place during the cardiac cycle, the premature beat *(A)* is feeble and the succeeding contraction *(B)* is supernormal. Such enhanced contractility in contraction B is an example of postextrasystolic potentiation, and it may persist for one or more additional beats (for example, contraction *C*).

Postextrasystolic potentiation probably depends on the time course of the intracellular circulation of Ca^{++} during the contraction and relaxation process. The Ca^{++} that activates contraction is discharged from some "release compartment" in the sarcoplasmic reticulum. This compartment probably contains only a little more Ca^{++} than is required for one beat. The Ca^{++} during excitation is then taken up at some other site in the sarcotubular network to induce relaxation. The Ca^{++} is translocated from this "uptake compartment" back to the "release compartment," where it will be released again with the next excitation. The process of translocation requires about 500 to 800 milliseconds.

The premature beat itself (beat *A*, Fig. 29-20) is feeble because it occurs before much of the Ca^{++} has moved from the uptake site in the sarcotubular network, where it is unavailable for interaction with the contractile proteins, to the release compartment, where the discharge of Ca^{++} to the myofibrils does occur. The postextrasystolic beat (Fig. 29-20, beat *B*), conversely, is considerably stronger than normal. A plausible reason is that after the compensatory pause (the pause between beats *A* and *B*) the release compartment will have available for discharge the Ca^{++} that had been taken up at the site in the sarcotubular network after two contractions—the extrasystole (beat *A*) and the normal beat that had preceded it.

Extrinsic Regulation of Myocardial Performance

Although the completely isolated heart can adapt well to changes in preload and afterload, in the intact animal various extrinsic factors also influence the myocardium. Under many natural conditions these extrinsic mechanisms dominate the intrinsic mechanisms. These extrin-

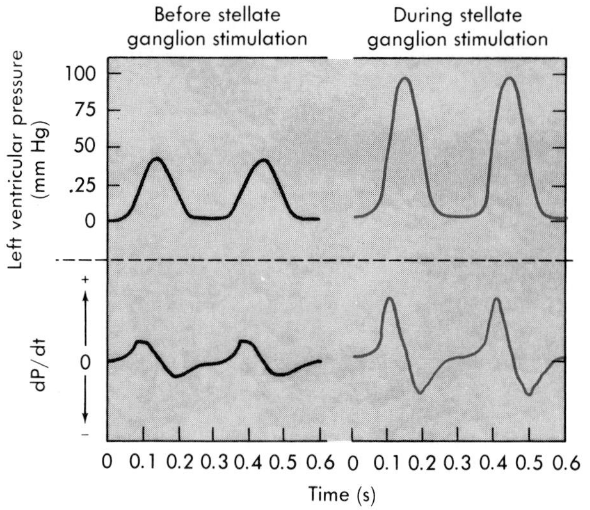

■ Fig. 29-21. In an isovolumic left ventricle preparation, stimulation of cardiac sympathetic nerves evokes a substantial rise in peak left ventricular pressure and in the maximum rates of intraventricular pressure rise and fall *(dP/dt)*.

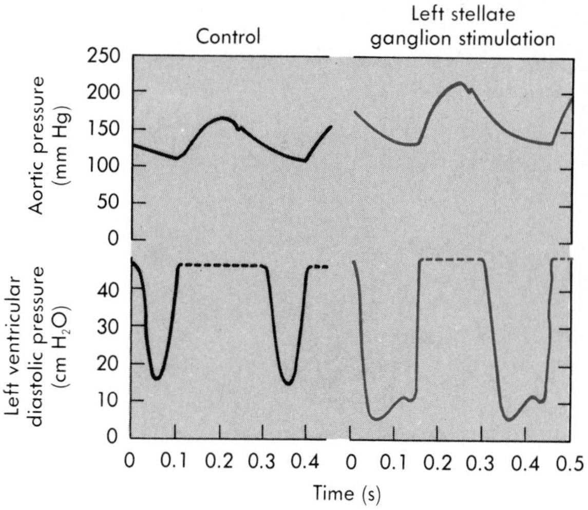

■ Fig. 29-22. Stimulation of the left stellate ganglion of a dog increases arterial pressure, stroke volume, and stroke work despite a concomitant reduction in ventricular end-diastolic pressure. Note also the abridgment of systole, thereby allowing more time for ventricular filling; the heart was paced at a constant rate. In the ventricular pressure tracings the pen excursion is limited at 45 mm Hg; acutal ventricular pressures during systole can be estimated from the aortic pressure tracings. (Redrawn from Mitchell, J.H., Linden, R.J., and Sarnoff, S.J.: Circ. Res. **8**:1100, 1960.)

sic regulatory factors may be subdivided into nervous and chemical components.

■ *Nervous Control*

Sympathetic influences. The sympathetic division of the autonomic nervous system enhances atrial and ventricular contractility. The concentration of norepinephrine in the various regions of the heart reflects the relative density of the sympathetic innervation to those areas. In the normal heart the norepinephrine concentration in the atria is about three times that in the ventricles. The norepinephrine concentration in the SA and AV nodes is similar to that in the surrounding atrial regions. When the heart is denervated, the tissue concentration of norepinephrine approaches zero.

The effects of increased cardiac sympathetic activity on the ventricular myocardium are asymmetrical. The cardiac sympathetic nerves on the left side of the body usually have a much greater effect on ventricular contraction than do those on the right side (Fig. 29-4). The alterations in ventricular contraction evoked by electrical stimulation of the left stellate ganglion in a canine isovolumic left ventricle preparation are shown in Fig. 29-21. The peak pressure and the maximum rate of pressure rise (dP/dt) during systole are markedly increased. Also, the duration of systole is reduced and the rate of ventricular relaxation is increased during the early phases of diastole. The shortening of systole and more rapid ventricular relaxation assist ventricular filling. The more rapid relaxation permits a greater filling for a given filling pressure. For a given cardiac cycle length, the abbreviation of systole allows more time for diastole and hence for ventricular filling. In the experiment shown in Fig. 29-22, for example, the animal's heart was paced at a constant rapid rate. Sympathetic stimulation (right panel) shortened systole, which allowed substantially more time for ventricular filling. These factors gain importance at fast heart rates, which of course prevail when sympathetic activity is increased.

Sympathetic nervous activity enhances myocardial performance. Neurally released norepinephrine or circulating catecholamines interact with β-adrenergic receptors on the cardiac cell membranes (Fig. 22-21). This reaction activates adenylate cyclase, which raises the intracellular levels of cyclic AMP (cAMP). As a consequence, protein kinases are activated that promote the phosphorylation of various proteins within the myocardial cells. Phosphorylation of specific sarcolemmal proteins augments the opening of the calcium channels in the myocardial cell membranes. Hence Ca^{++} influx increases during each action potential plateau, and more Ca^{++} is released from the sarcoplasmic reticulum in response to each cardiac excitation. The contractile strength of the heart is thereby increased.

The overall effect of increased cardiac sympathetic activity in intact animals can best be appreciated in terms of families of ventricular function curves. When

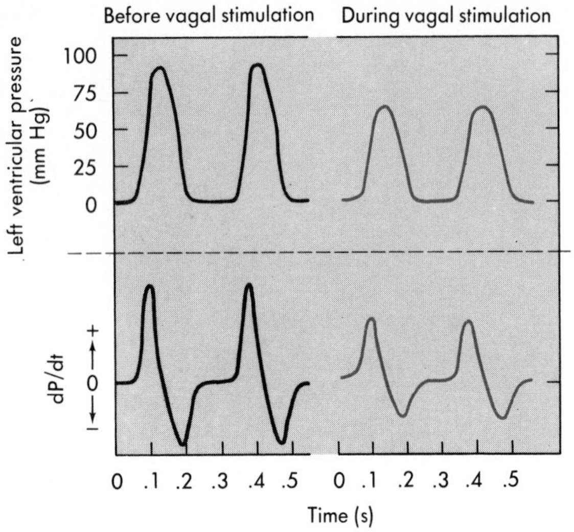

■ **Fig. 29-23.** In an isovolumic left ventricle preparation, when the ventricle is paced at a constant frequency, vagal stimulation decreases the peak left ventricular pressure and diminishes the maximum rates of pressure rise and fall *(dP/dt)*.

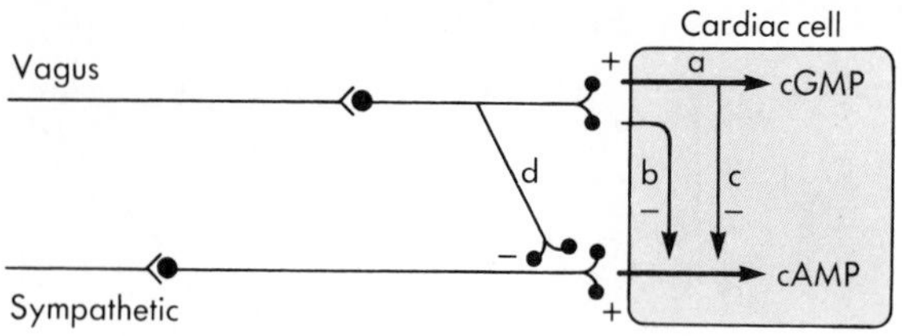

■ **Fig. 29-24.** The interneuronal and intracellular mechanisms responsible for the vagal-sympathetic interactions. (Modified from Levy, M.N.: In Baan, J., Noordergraaf, A., and Raines, J., editors: Proceedings of the International Conference on Cardiovascular System Dynamics, Cambridge, Mass., 1976, The MIT Press.)

stepwise increases in the frequency of electrical stimulation are applied to the left stellate ganglion, the ventricular function curves shift progressively to the left. The changes parallel those produced by catecholamine infusions (Fig. 29-16). Hence, for any given left ventricular end-diastolic pressure, the ventricle is capable of performing more work as the level of sympathetic nervous activity is raised. During cardiac sympathetic stimulation the increase in work is usually accompanied by a reduction in left ventricular end-diastolic pressure. An example of the response to stellate ganglion stimulation in a paced heart is shown in Fig. 29-22. In this experiment, stroke work increased by about 50%, despite a 7 cm H_2O reduction in the left ventricular end-diastolic pressure. Note also the pronounced abridgment of the duration of ventricular systole, with the consequent lengthening of the filling period. The reason for the reduction in ventricular end-diastolic pressure is explained on p. 532.

Parasympathetic influences. The vagus nerves inhibit the cardiac pacemaker, atrial myocardium, and AV conduction tissue. The vagus nerves also depress the ventricular myocardium, but the effects are less pronounced. In the isovolumic left ventricle preparation, vagal stimulation decreases the peak left ventricular pressure, maximum rate of pressure development (dP/dt), and maximum rate of pressure decline during diastole (Fig. 29-23). In pumping heart preparations the ventricular function curve shifts to the right during vagal stimulation.

The vagal effects on the ventricular myocardium are achieved by at least four mechanisms, as shown in Fig. 33-24. One process is direct, and the others involve an interaction with the sympathetic nervous system. With respect to the direct mechanism, vagal stimulation or acetylcholine infusions raise the intracellular levels of cyclic GMP (cGMP) (arrow *A*). This nucleotide in turn may depress myocardial contractility through some unknown process.

With respect to the indirect mechanisms, when the existing level of cardiac sympathetic nervous activity is low, the depressant effect of increased parasympathetic activity on the ventricular myocardium is relatively feeble. However, against a background of tonic sympathetic activity, the negative inotropic effect of vagal stimulation is considerably greater. The chronotropic responses to autonomic neural activity are similar to the inotropic responses. In the left half of Fig. 29-2, for example, vagal stimulation reduced the heart rate much more during simultaneous cardiac sympathetic stimulation at 4 pulses per second (S = 4) than it did in the absence of sympathetic stimulation (S = 0).

This accentuated antagonism between the parasympathetic and sympathetic systems may be accomplished in three different ways (Fig. 29-24). First, the acetylcholine released at the vagal endings directly reduces the intracellular levels of cAMP (arrow *B*). As stated in the preceding section, the enhancement of contractility by sympathetic neural activity is mediated by cAMP. Second, the elevated intracellular levels of cGMP accelerate the hydrolysis of cAMP (arrow *C*), thereby lowering its concentration in the myocardial cell.

The third mechanism responsible for the accentuated antagonism involves interneuronal processes. As shown in Fig. 29-24, some postganglionic vagal terminals *(D)* end near the postganglionic sympathetic terminals in the heart. The acetylcholine released at these vagal endings inhibits the release of norepinephrine from the sympathetic fibers. Fig. 29-25 demonstrates that stimulation of the cardiac sympathetic nerves *(S)* results in the overflow of substantial amounts of norepinephrine into the coronary sinus blood. Concomitant vagal stimula-

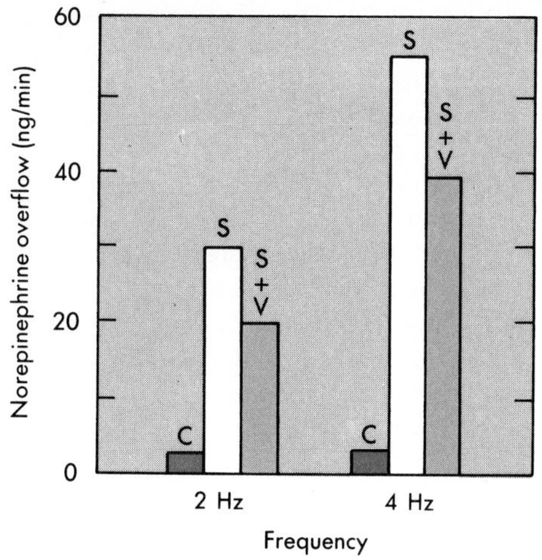

■ **Fig. 29-25.** The mean rates of overflow of norepinephrine into the coronary sinus blood in a group of seven dogs under control conditions *(C)*, during cardiac sympathetic stimulation *(S)* at 2 or 4 Hz and during combined sympathetic and vagal stimulation *(S + V)*. The combined stimulus consisted of sympathetic stimulation at 2 or 4 Hz, and vagal stimulation at 15 Hz. (Redrawn from Levy, M.N., and Blattberg, B.: Circ. Res. **38**:81, 1976. By permission of the American Heart Association, Inc.)

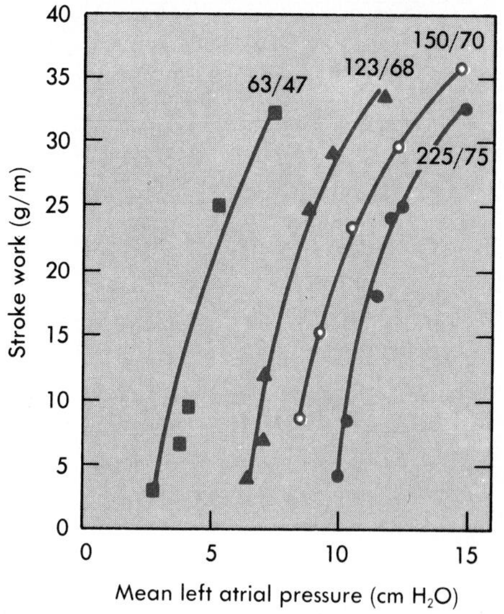

■ **Fig. 29-26.** As the pressure in the isolated carotid sinus is progressively raised, the ventricular function curves shift to the right. The numbers at the tops of each curve represent the systolic/diastolic perfusion pressures (in mm Hg) in the carotid sinus regions of the dog. (Redrawn from Sarnoff, S.J., et al.: Circ. Res. **8**:1123, 1960.)

tion (S + V) reduces the overflow of norepinephrine by about 30%. The amount of norepinephrine overflowing into the coronary sinus blood probably parallels the amount released at the sympathetic terminals.

Baroreceptor reflex. Just as stimulation of the carotid sinus and aortic arch baroreceptors may change heart rate (p. 454), so also may it alter myocardial performance. Evidence of reflex alterations of ventricular contractility is presented in Fig. 29-26. Ventricular function curves were obtained at four levels of carotid sinus pressure. With each successive rise in pressure, the ventricular function curves were displaced farther and farther to the right, denoting a progressively greater reflex depression of ventricular performance.

In normal, resting individuals and animals the tonic level of sympathetic activity is usually very low. Under such conditions, moderate changes in baroreceptor activity may have little reflex influence on myocardial contractility. In states of augmented sympathetic neural activity, however, the effects of the baroreceptor reflex on contractility may be substantial. In the adaptation to blood loss, for example, a reflex change in myocardial contractility may help to compensate. Blood loss diminishes cardiac output. The associated reduction in arterial blood pressure alters the intensity of baroreceptor stimulation, thereby not only increasing heart rate but also enhancing myocardial contractility.

When blood loss is substantial, however, severe generalized arteriolar vasoconstriction ensues. The increased peripheral resistance may diminish the cardiac preload, as explained in Chapter 34 (p. 533). The reduction in preload will limit the ability of the increased cardiac sympathetic activity to improve cardiac output.

Chemoreceptor reflex. When pulmonary ventilation is controlled, carotid chemoreceptor stimulation results in profound bradycardia (Figs. 29-10 and 29-11), often with some degree of AV conduction block. This indicates a considerable increase of vagal activity. Increased vagal activity may depress atrial contractility substantially and ventricular contractility moderately.

■ *Chemical Control*

Hormones

Adrenomedullary hormones. The adrenal medulla is essentially a component of the autonomic nervous system. The principal hormone secreted by the adrenal medulla is epinephrine, although some norepinephrine is also released. The rate of secretion of catecholamines by the adrenal medulla is largely regulated by the same mechanisms that control the activity of the sympathetic nervous system. The concentrations of catecholamines in the blood rise under the same conditions that activate the sympathoadrenal system. However, the cardiovascular effects of circulating catecholamines are probably minimal under normal conditions.

The changes in myocardial contractility induced by norepinephrine infusions have been tested in resting,

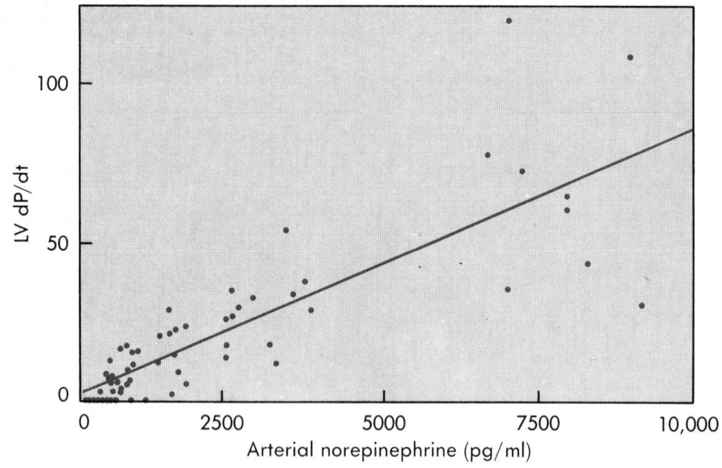

■ **Fig. 29-27.** The effect of norepinephrine infusions on ventricular contractility in a group of resting, unanesthetized dogs. The plasma concentrations of norepinephrine *(pg/ml)* plotted along the abscissa are the increments above the control values. The maximum rate of rise of left ventricular pressure *(LV dP/dt)* is an index of contractility. (Redrawn from Young, M.A., Hintze, T.H., and Vatner, S.F.: Am. J. Physiol. **248:**H82, 1985.)

unanesthetized dogs. The maximum rate of rise of left ventricular pressure (dP/dt), an index of myocardial contractility, was found to be proportional to the norepinephrine concentration in the blood (Fig. 29-27). In these same animals, moderate exercise increased the maximum dP/dt by almost 100%, but it raised the circulating catecholamines by only 0.5 ng/ml. Such a rise in blood norepinephrine concentration, by itself, would have had only a negligible effect on left ventricular dP/dt (Fig. 29-27). Hence the pronounced change in dP/dt observed during exercise must have been ascribable mainly to the catecholamines released from the cardiac sympathetic nerve fibers rather than from the adrenal medulla. In most other conditions in which sympathoadrenal activity is increased, the cardiac effects are also mainly ascribable to the catecholamines released at the nerve endings in the heart rather than to those circulating in the blood.

Adrenocortical hormones. Cardiovascular problems are common in adrenocortical insufficiency *(Addison's disease)*. The blood volume tends to fall, which may lead to severe hypotension and cardiovascular collapse, the so-called addisonian crisis.

The influence of adrenocortical steroids on the myocardium is controversial. Cardiac muscle removed from adrenalectomized animals and placed in a tissue bath is more likely to fatigue than that obtained from normal animals. In some species the adrenocortical hormones enhance contractility. Furthermore, hydrocortisone potentiates the cardiotonic effects of the catecholamines. This potentiation may be mediated in part by an inhibition of the uptake mechanisms for the catecholamines.

Thyroid hormones. Cardiac activity is sluggish in patients with inadequate thyroid function *(hypothyroidism);* that is, the heart rate is slow and cardiac output is diminished. The converse is true in patients with overactive thyroid glands *(hyperthyroidism)*. Characteristically, such patients exhibit tachycardia, high cardiac output, palpitations, and arrhythmias (such as atrial fibrillation). In experimental animals the cardiovascular manifestations of hyperthyroidism may be simulated by the administration of thyroxine.

Numerous studies on intact animals and humans have demonstrated that thyroid hormones enhance myocardial contractility. The rates of Ca^{++} uptake and of ATP hydrolysis by the sarcoplasmic reticulum are increased in experimental hyperthyroidism, and the opposite effects occur in hypothyroidism. Thyroid hormones increase protein synthesis in the heart, which leads to cardiac hypertrophy. These hormones also affect the composition of myosin isoenzymes in cardiac muscle. They increase principally those isoenzymes with the greatest ATPase activity, which thereby enhances myocardial contractility.

The cardiovascular changes in thyroid dysfunction also depend on indirect mechanisms. Thyroid hyperactivity increases the body's metabolic rate, and this in turn dilates the arterioles. The consequent reduction in the total peripheral resistance increases cardiac output, as explained on p. 533. Substantial evidence indicates that with hyperthyroidism either sympathetic neural activity is increased or the sensitivity of the heart to such activity is enhanced. However, other evidence contradicts these conclusions. There is general agreement that thyroid hormone does increase the density of β-adrenergic receptors in cardiac tissue.

Insulin. Insulin has a prominent, direct, positive inotropic effect on the heart of several mammalian species. The effect of insulin is evident even when hypoglycemia is prevented by glucose infusions and when the β-adrenergic receptors are blocked. In fact, the positive inotropic effect of insulin is potentiated by β-adrenergic receptor blockade. The enhancement of contractility cannot be explained satisfactorily by the concomitant augmentation of glucose transport into the myocardial cells. The inotropic effect of insulin may be mediated by inhibition of intracellular Ca^{++} binding, thereby increasing the intracellular concentration of Ca^{++} available for interaction with the myofilaments.

Glucagon. Glucagon has potent positive inotropic and chronotropic effects on the heart. The endogenous hormone probably plays no significant role in the normal regulation of the cardiovascular system, but it has been used to treat a variety of cardiac conditions. The effects of glucagon on the heart closely resemble those of the catecholamines, and certain of the metabolic effects are similar. Both glucagon and catecholamines activate adenylate cyclase to increase the myocardial tissue levels of cAMP. The catecholamines activate adenylate cyclase by interacting with β-adrenergic receptors, but glucagon activates this enzyme through a different mechanism. Nevertheless, the consequent rise in cAMP increases Ca^{++} influx through the Ca^{++} channels in the sarcolemma and facilitates Ca^{++} release and reuptake by the sarcoplasmic reticulum, just as do the catecholamines.

Anterior pituitary hormones. The cardiovascular derangements in hypopituitarism are related principally to the associated deficiencies in adrenocortical and thyroid function. Growth hormone probably affects the myocardium, at least in combination with thyroxine. In hypophysectomized experimental animals growth hormone alone has little effect on the depressed heart, whereas thyroxine by itself restores adequate cardiac performance under basal conditions. However, with hypervolemia or increased peripheral resistance, thyroxine alone does not restore adequate cardiac function, but the combination of growth hormone and thyroxine does reestablish normal cardiac performance.

Blood gases

Oxygen. Changes in oxygen tension (Pa_{O_2}) of the blood perfusing the brain and the peripheral chemoreceptors affect the heart through nervous mechanisms, as described earlier in this chapter. These indirect effects of hypoxia are usually prepotent. Moderate degrees of hypoxia characteristically increase heart rate, cardiac output, and myocardial contractility. These changes are largely abolished by β-adrenergic receptor blockade. The Pa_{O_2} of the blood perfusing the myocardium also influences myocardial performance directly. The effect of hypoxia is biphasic; moderate degrees are stimulatory and more severe degrees are depressant. As shown in Fig. 29-28, when the O_2 saturation is reduced to levels below 50% in isolated hearts, the peak left ventricular pressures are less than the control levels. However, with less severe degrees of hypoxia (O_2 saturation > 50%), the peak pressures exceed the control level.

Carbon dioxide. Changes in Pa_{CO_2} may also affect the myocardium directly and indirectly. The indirect, neurally mediated effects produced by increased Pa_{CO_2} are similar to those evoked by a decrease in Pa_{O_2}. With respect to the direct effects on the heart, alterations in

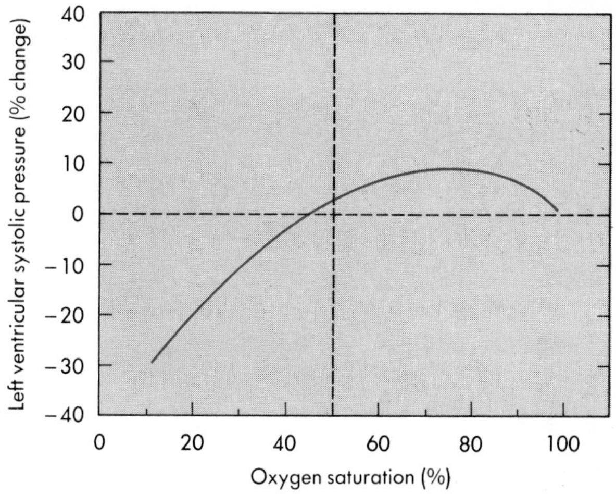

■ **Fig. 29-28.** In the isovolumic left ventricle preparation, a reduction in the O_2 saturation of coronary arterial blood to between 45% and 100% stimulates ventricular contractility, wheras an O_2 saturation below 45% depresses ventricular contractility. (Redrawn from Ng, M.L., et al.: Am. J. Physiol. **211**:43, 1966.)

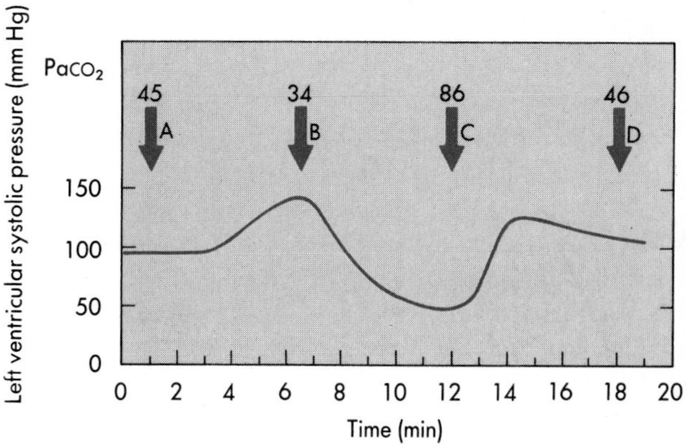

■ **Fig. 29-29.** Decrease in Pa_{CO_2} increases left ventricular systolic pressure (arrow *B*) in an isovolumic left ventricle preparation; a rise in Pa_{CO_2} (arrow *C*) has the reverse effect. When the Pa_{CO_2} is returned to the control level (arrow *D*), left ventricular systolic pressure returns to its original value (arrow *A*).

myocardial performance elicited by changes of Pa_{CO_2} in the coronary arterial blood are illustrated in Fig. 29-29. In this experiment on an isolated left ventricle preparation, the control Pa_{CO_2} was 45 mm Hg (arrow *A*). Decreasing the Pa_{CO_2} to 34 mm Hg (arrow *B*) was stimulatory, whereas increasing Pa_{CO_2} to 86 mm (arrow *C*) was depressant. In intact animals, systemic hypercapnia activates the sympathoadrenal system, which tends to compensate for the direct depressant effect of the increased Pa_{CO_2} on the heart.

Neither the Pa_{CO_2} nor the blood pH are primary determinants of myocardial behavior. The resultant change in intracellular pH is the critical factor. The reduced intracellular pH diminishes the amount of Ca^{++} released from the sarcoplasmic reticulum in response to excitation. The diminished pH also depresses the myofilaments directly. When they are exposed to a given concentration of Ca^{++}, the lower the prevailing pH, the less force the myofibrils develop.

Bibliography

Journal articles

Dampney, R.A.L.: Functional organization of central cardiovascular pathways, Clin. Exp. Pharmacol. Physiol. **8:**241, 1981.

Farah, A.E.: Glucagon and the circulation, Pharmacol. Rev. **35:**181, 1983.

Farah, A.E., and Alousi, A.A.: The actions of insulin on cardiac contractility, Life Sci. **29:**975, 1981.

Geis, G.S., and Wurster, R.D.: Cardiac responses during stimulation of the dorsal motor nucleus and nucleus ambiguus in the cat, Circ. Res. **46:**606, 1980.

Katona, P.G., et al.: Sympathetic and parasympathetic cardiac control in athletes and nonathletes at rest, J. Appl. Physiol. **52:**1652, 1982.

Katz, A.M.: Cyclic adenosine monophosphate effects on the myocardium: a man who blows hot and cold with one breath, J. Am. Coll. Cardiol. **2:**143, 1983.

Katz, A.M.: Inotropic drugs and their mechanisms of action, J. Am. Coll. Cardiol. **4:**389, 1984.

Kirchheim, H.R.: Systemic arterial baroreceptor reflexes, Physiol. Rev. **56:**100, 1976.

Klein, I., and Levey, G.S.: New perspectives on thyroid hormone, catecholamines, and the heart, Am. J. Med. **76:**167, 1984.

Kollai, M., and Koizumi, K.: Cardiovascular reflexes and interrelationships between sympathetic and parasympathetic activity, J. Auton. Nerv. Syst. **4:**135, 1981.

Kunze, D.L.: Regulation of activity of cardiac vagal motoneurons, Fed. Proc. **39:**2513, 1980.

Löffelholz, K., and Pappano, A.J.: The parasympathetic neuroeffector junction of the heart, Pharmacol. Rev. **37:**1, 1985.

Levy, M.N.: Cardiac sympathetic-parasympathetic interactions, Fed. Proc. **43:**2598, 1984.

Loewy, A.D., and McKellar, S.: Neuroanatomical basis of central cardiovascular control, Fed. Proc. **39:**2495, 1980.

Mark, A.L.: The Bezold-Jarisch reflex revisited: clinical implications of inhibitory reflexes originating in the heart, J. Am. Coll. Cardiol. **1:**90, 1983.

Morkin, E., Flink, I.L., and Goldman, S.: Biochemical and physiologic effects of thyroid hormone on cardiac performance, Prog. Cardiovasc. Dis. **25:**435, 1983.

Sagawa, K.: The ventricular pressure-volume diagram revisited, Circ. Res. **43:**677, 1978.

Scott, E.M.: Reflex control of the cardiovascular system and its modification—some implications for pharmacologists, J. Auton. Pharmacol. **3:**113, 1983.

Shamoo, A.E., and Ambudkar, I.S.: Regulation of calcium transport in cardiac cells, Can. J. Physiol. Pharmacol. **62:**9, 1984.

Spyer, K.M.: Neural organisation and control of the baroreceptor reflex, Rev. Physiol. Biochem. Pharmacol. **88:**23, 1981.

Vanhoutte, P.M., and Levy, M.N.: Prejunctional cholinergic modulation of adrenergic neurotransmission in the cardiovascular system, Am. J. Physiol. **238:**H275, 1980.

Vatner, S.F., and Boettcher, D.H.: Regulation of cardiac output by stroke volume and heart rate in conscious dogs, Circ. Res. **42:**557, 1978.

Watanabe, A.M., et al.: Cardiac autonomic receptors, Circ. Res. **50:**161, 1982.

Winegrad, S.: Regulation of cardiac contractile proteins, Circ. Res. **55:**565, 1984.

Wohlfart, B., and Noble, M.I.M.: The cardiac excitation-contraction cycle, Pharmacol. Ther. **16:**1, 1982.

Young, M.A., Hintze, T.H., and Vatner, S.F.: Correlation between cardiac performance and plasma catecholamine levels in conscious dogs, Am. J. Physiol. **248:**H82, 1985.

Books and monographs

Abboud, F.M., et al., editors: Disturbances in neurogenic control of the circulation, Bethesda, Md., 1981, American Physiological Society.

Bishop, V.S., Malliani, A., and Thorén, P.: Cardiac mechanoreceptors. In Handbook of physiology; Section 2: The cardiovascular system—peripheral circulation and organ blood flow, vol. III, Bethesda, Md., 1983, American Physiological Society.

Braunwald, E., and Ross, J., Jr.: Control of cardiac performance. In Handbook of physiology; Section 2: Cardiovascular system—the heart, vol. I, Washington, D.C., 1979, American Physiological Society.

Brown, A.M.: Cardiac reflexes. In Handbook of physiology; Section 2: Cardiovascular system—the heart, vol. I, Washington, D.C., 1979, American Physiological Society.

Coleridge, J.C.G., and Coleridge, H.M.: Chemoreflex regulation of the heart. In Handbook of physiology; Section 2: Cardiovascular system—the heart, vol. I, Washington, D.C., 1979, American Physiological Society.

Downing, S.E.: Baroreceptor regulation of the heart. In Handbook of physiology; Section 2: Cardiovascular system—the heart, vol. I, Washington, D.C., 1979, American Physiologic Society.

Fowler, N.O.: Pericardium in health and disease, Mount Kisco, N.Y., 1985, Futura Publishing Co., Inc.

Hainsworth, R., Kidd, C., and Linden, R.J., editors: Cardiac receptors, Cambridge, 1979, Cambridge University Press.

Korner, P.I.: Central nervous control of autonomic cardiovascular function. In Handbook of physiology; Section 2: Cardiovascular system—the heart, vol. I, Washington, D.C., 1979, American Physiological Society.

Levine, H.J., and Gaasch, W.H., editors: The ventricle: basic and clinical aspects, Hingham, Mass., 1985, Martinus Nijhoff Publishers.

Levy, M.N., and Martin, P.J.: Neural control of the heart. In Handbook of physiology; Section 2: Cardiovascular system—the heart, vol. I, Washington, D.C., 1979, American Physiological Society.

Opie, L., editor: The heart, Orlando Fla., 1984, Grune & Stratton, Inc.

Randall, W.C., editor: Nervous control of cardiovascular function, New York, 1984, Oxford University Press.

Sperelakis, N., editor: Physiology and pathophysiology of the heart, Hingham, Mass., 1984, Martinus Nijhoff Publishers.

Stull, J.T., and Mayer, S.E.: Biochemical mechanisms of adrenergic and cholinergic regulation of myocardial contractility. In Handbook of physiology; Section 2: Cardiovascular system—the heart, vol. I, Washington, D.C., 1979, American Physiological Society.

CHAPTER

30

Hemodynamics

The problem of treating the pulsatile flow of blood through the cardiovascular system in precise mathematical terms is insuperable. The heart is a complicated pump, and its behavior is affected by a variety of physical and chemical factors. The blood vessels are multibranched, elastic conduits of continuously varying dimensions. The blood itself is a suspension of red and white corpuscles, platelets, and lipid globules suspended in a colloidal solution of proteins. Despite these complicating factors, considerable insight may be gained from understanding the elementary principles of fluid mechanics as they pertain to simpler physical systems. Such principles will be expounded in this chapter to explain the interrelationships among velocity of blood flow, blood pressure, and the dimensions of the various components of the systemic circulation.

■ *Velocity of the Bloodstream*

In describing the variations in blood flow in different vessels it is first essential to distinguish between the terms *velocity* and *flow*. The former term, sometimes designated as linear velocity, refers to the rate of displacement with respect to time and has the dimensions of distance per unit time, for example, cm/s. The latter term is frequently designated as volume flow and has the dimensions of volume per unit time, for example, cm^3/s. In a conduit of varying cross-sectional dimensions, velocity, v, flow, Q, and cross-sectional area, A, are related by the equation:

$$v = Q/A \qquad (1)$$

The interrelationships among velocity, flow, and area are portrayed in Fig. 30-1. In the case of an incompressible fluid flowing through rigid tubes, the flow past successive cross sections must be the same. For a given constant flow the velocity varies inversely as the cross-sectional area. Thus for the same volume of fluid per second passing from section *a* into section *b*, where the cross-sectional area is five times greater, the velocity of flow diminishes to one fifth of its previous value. Conversely, when the fluid proceeds from section *b* to section *c*, where the cross-sectional area is one tenth as great, the velocity of each particle of fluid must increase tenfold.

The velocity at any point in the system depends not only on area, but also on the flow, Q. This, in turn, depends on the pressure gradient, properties of the fluid, and dimensions of the entire hydraulic system, as discussed in the following section. For any given flow, however, the ratio of the velocity past one cross section relative to that past a second cross section depends only on the inverse ratio of the respective areas; that is,

$$v_1/v_2 = A_2/A_1 \qquad (2)$$

This rule pertains regardless of whether a given cross-sectional area applies to a single large tube or to several smaller tubes in parallel.

As shown in Fig. 26-2, velocity decreases progressively as the blood traverses the aorta, its larger primary branches, the smaller secondary branches, and the arterioles. Finally, a minimum value is reached in the capillaries. As the blood then passes through the venules and continues centrally toward the venae cavae, the velocity progressively increases again. The relative velocities in the various components of the circulatory system are related only to the cross-sectional area. Thus each point on the cross-sectional area curve is inversely proportional to the corresponding point on the velocity curve (see Fig. 26-2).

■ *Relationship Between Velocity and Pressure*

In that portion of a hydraulic system in which the total energy remains virtually constant, changes in velocity may be accompanied by appreciable alterations in the measured pressure. Consider three sections *(A, B,* and

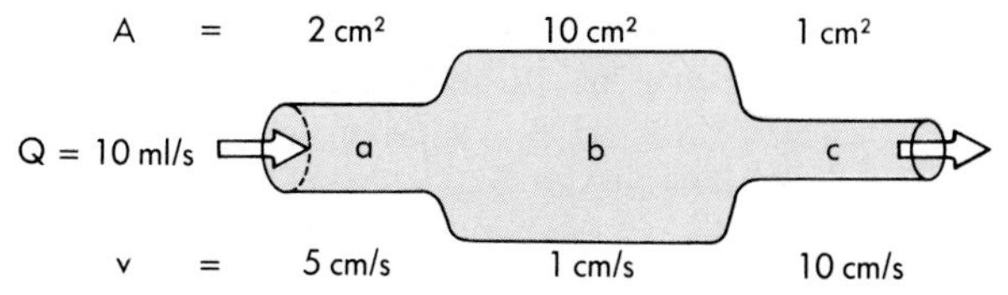

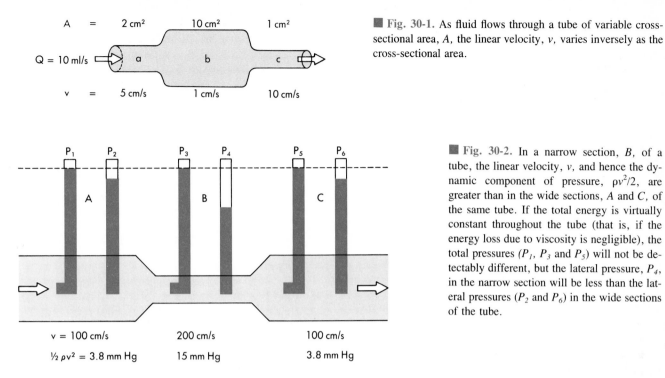

Fig. 30-1. As fluid flows through a tube of variable cross-sectional area, *A*, the linear velocity, *v*, varies inversely as the cross-sectional area.

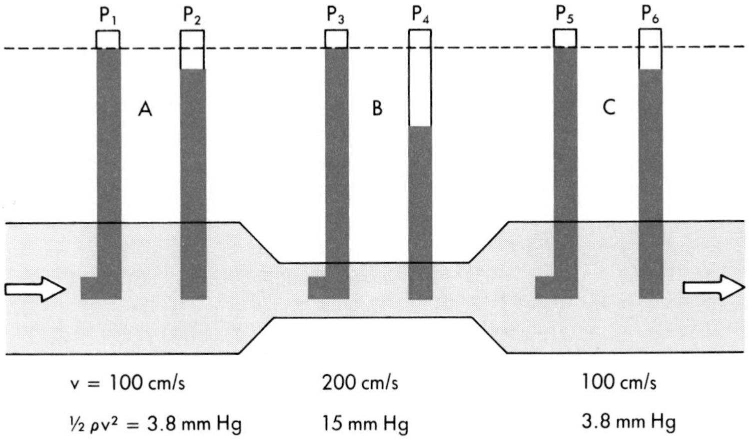

Fig. 30-2. In a narrow section, *B*, of a tube, the linear velocity, *v*, and hence the dynamic component of pressure, $\rho v^2/2$, are greater than in the wide sections, *A* and *C*, of the same tube. If the total energy is virtually constant throughout the tube (that is, if the energy loss due to viscosity is negligible), the total pressures (P_1, P_3 and P_5) will not be detectably different, but the lateral pressure, P_4, in the narrow section will be less than the lateral pressures (P_2 and P_6) in the wide sections of the tube.

C) of such a hydraulic system, as depicted in Fig. 30-2. Six pressure probes, or *pitot tubes,* have been inserted. The openings of three of these *(2, 4,* and *6)* are tangential to the direction of flow and hence measure the *lateral,* or *static,* pressure within the tube. The openings of the remaining three pitot tubes *(1, 3,* and *5)* face upstream. Therefore they detect the *total pressure,* which is the lateral pressure plus a pressure component ascribable to the kinetic energy of the flowing fluid. This dynamic component, P_d, of the total pressure may be calculated from the following equation:

$$P_d = \rho v^2/2 \quad (3)$$

where ρ is the density of the fluid, and v is the velocity. If the midpoints of segments *A, B,* and *C* are at the same hydrostatic level, then the corresponding total pressure, P_1, P_3, and P_5, will be equal, if the energy loss from viscosity in these segments is negligible. However, with the changes in cross-sectional area, the concomitant velocity changes alter the dynamic component.

In sections *A* and *C*, let $\rho = 1\ \text{g/cm}^3$, and $v = 100$ cm/s. Therefore,

$$\begin{aligned} P_d &= 5000\ \text{dynes/cm}^2 \\ &= 3.8\ \text{mm Hg} \end{aligned} \quad (4)$$

since $1330\ \text{dynes/cm}^2 = 1$ mm Hg. In the narrow section, *B*, let the velocity be twice as great as in sections *A* and *C*. Therefore,

$$\begin{aligned} P_d &= 20{,}000\ \text{dynes/cm}^2 \\ &= 15\ \text{mm Hg} \end{aligned} \quad (5)$$

Hence, in the wide sections of the conduit, the lateral pressures (P_2 and P_6) will be only 3.8 mm Hg less than the respective total pressures (P_1 and P_5), whereas in the narrow section, the lateral pressure (P_4) is 15 mm Hg less than the total pressure (P_3).

The peak velocity of flow in the ascending aorta of normal dogs is about 100 to 200 cm/s. Therefore the measured pressure may vary significantly, depending on the orientation of the pressure probe. In the descending thoracic aorta the peak velocity is about half as great as in the ascending aorta (Fig. 30-3), and lesser velocities have been recorded in more distal arterial sites. In most arterial locations the dynamic component will be a negligible fraction of the total pressure, and the orientation of the pressure probe will not materially influence the magnitude of the pressure recorded. At the site of a constriction, however, the dynamic pressure component may attain substantial values. In *aortic stenosis,* for example, the entire output of the left ventricle is ejected through a narrow valve orifice. The high flow velocity is associated with a large kinetic energy, and therefore the lateral pressure is correspondingly reduced.

The pressure tracings shown in Fig. 30-4 were obtained from two pressure transducers inserted into the left ventricle of a patient with aortic stenosis. The transducers were located on the same catheter and were 5 cm apart. When both transducers were well within the left ventricular cavity (Fig. 30-4, *A*), they both recorded the same pressures. However, when the proximal transducer was positioned in the aortic valve orifice (Fig. 30-4, *B*), the lateral pressure recorded during ejection was much less than that recorded by the transducer in the ventricular cavity. This pressure difference was ascrib-

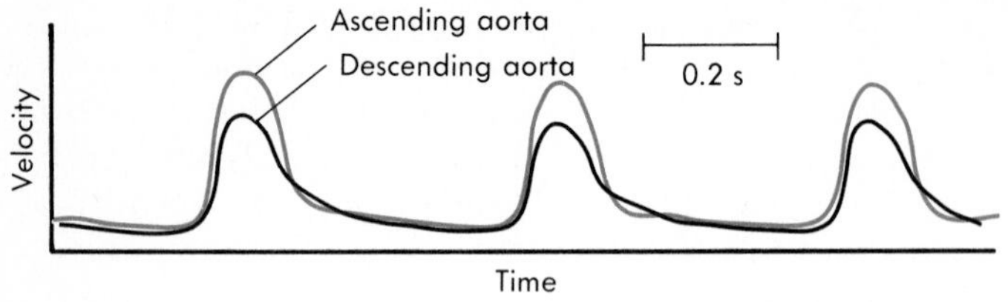

■ **Fig. 30-3.** Velocity of the blood in the ascending and descending aorta of a dog. (Redrawn from Falsetti, H.L., et al.: Circ. Res. **31**:328, 1972. By permission of the American Heart Association, Inc.)

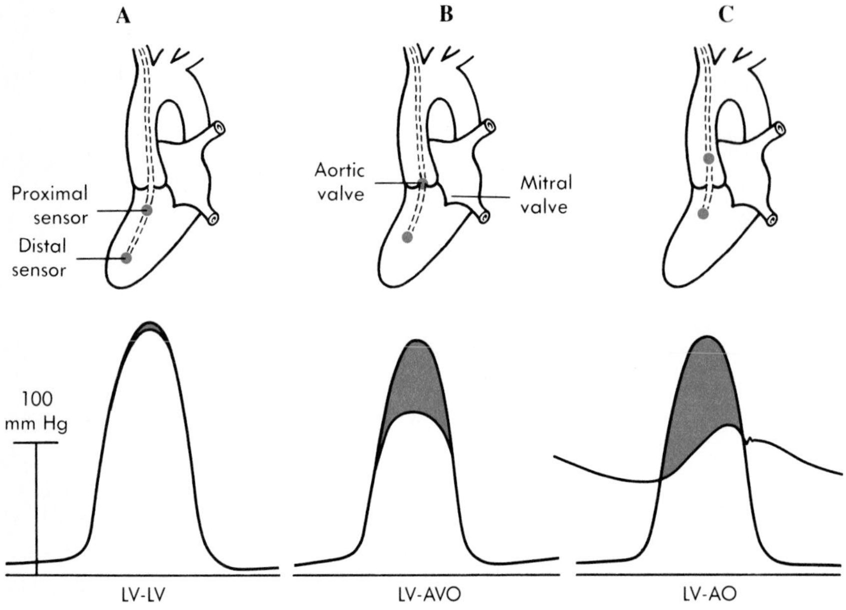

■ **Fig. 30-4.** Pressures recorded by two transducers in a patient with aortic stenosis. **A,** Both transducers were in the left ventricle *(LV-LV)*. **B,** One transducer was in the left ventricle, and the other was in the aortic valve orifice *(LV-AVO)*. **C,** One transducer was in the left ventricle, and the other was in the ascending aorta *(LV-AO)*. (Redrawn from Pasipoularides, A., et al.: Am. J. Physiol. **246:**H542, 1984.)

able almost entirely to the much greater velocity of flow in the narrowed valve orifice than in the ventricular cavity. The pressure difference reflects mainly the conversion of some potential energy to kinetic energy. When the catheter was withdrawn still farther, so that the proximal transducer was in the aorta (Fig. 30-4, *C*), the pressure difference was even more pronounced, because substantial energy was lost through friction as blood flowed rapidly through the narrow orifice.

The reduction of lateral pressure in the region of the stenotic valve orifice influences the coronary blood flow in patients with aortic stenosis. The orifices of the right and left coronary arteries are located in the sinuses of Valsalva, just behind the valve leaflets. Hence the initial segments of these vessels are oriented at right angles to the direction of blood flow through the aortic valves. Therefore the lateral pressure is that component of the total pressure that propels the blood through the two major coronary arteries. During the ejection phase of the cardiac cycle, the lateral pressure is diminished by the conversion of potential energy to kinetic energy.

This process is grossly exaggerated in aortic stenosis, because of the high flow velocities. Angiographic studies in patients with aortic stenosis have revealed that the direction of flow often reverses in the large coronary arteries toward the end of the ejection phase of systole.

The decreased lateral pressure in the aorta in aortic stenosis is undoubtedly an important factor in causing this reversal of coronary blood flow. An important aggravating feature in this condition is that the demands of the heart muscle for oxygen are greatly increased. Therefore the pronounced drop in lateral pressure during cardiac ejection may be a contributory factor in the tendency for patients with severe aortic stenosis to experience *angina pectoris* (anterior chest pain associated with inadequate blood supply to the heart muscle) and to die suddenly.

■ *Relationship Between Pressure and Flow*

The most fundamental law governing the flow of fluids through cylindrical tubes was derived empirically over a century ago by Poiseuille, a French physician. He was primarily interested in the physical determinants of blood flow, but substituted simpler liquids for blood in his measurements of flow through glass capillary tubes. His work was so precise and important that his observations have been designated *Poiseuille's law*. Subsequently, this same law has been derived theoretically.

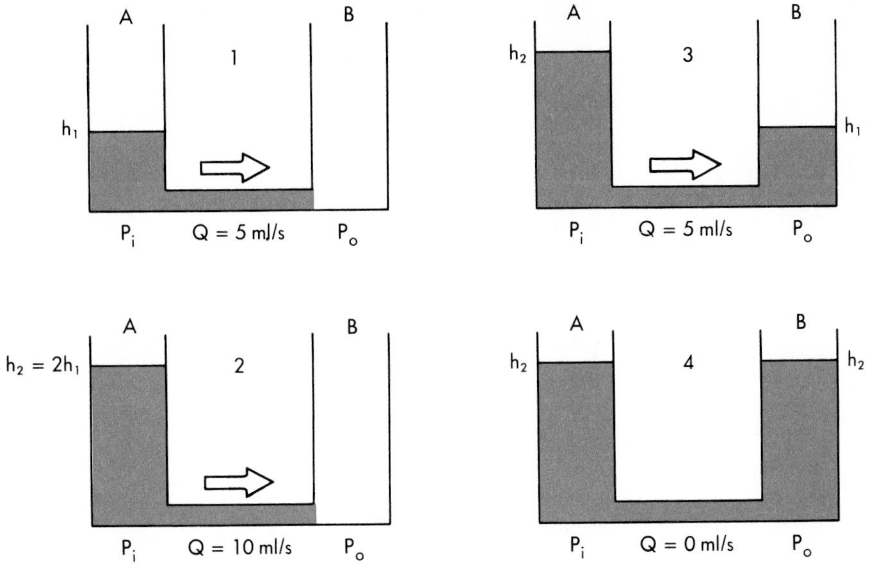

Fig. 30-5. The flow, Q, of fluid through a tube connecting two reservoices, A and B, is proportional to the difference between the pressure, P_i, at the inflow end and the pressure, P_o, at the outflow end of the tube.

Poiseuille's law is applicable to the flow of fluids through cylindrical tubes only under special conditions. It applies to the case of steady, laminar flow of newtonian fluids. The term *steady flow* signifies the absence of variations of flow in time, that is, a nonpulsatile flow. *Laminar flow* is the type of motion in which the fluid moves as a series of individual layers, with each stratum moving at a different velocity from its neighboring layers. In the case of flow through a tube, the fluid consists of a series of infinitesimally thin concentric tubes sliding past one another. Laminar flow will be described in greater detail below, where it will be distinguished from turbulent flow. Also a *newtonian fluid* will be defined more precisely. For the present discussion it will suffice to consider it as a homogeneous fluid, such as air or water, in contradistinction to a suspension, such as blood.

Pressure is one of the principal determinants of the rate of flow. The pressure, P, in dynes/cm^2, at a distance h centimeters below the surface of a liquid is

$$P = h\rho g \tag{6}$$

where ρ is the density of the liquid in g/cm^3 and g is the acceleration of gravity in cm/s^2. For convenience, however, pressure is frequently expressed simply in terms of height, h, of the column of liquid above some arbitrary reference point.

Consider the tube connecting reservoirs *A* and *B* in Fig. 30-5. Let reservoir *A* be filled with liquid to height h_1, and let reservoir *B* be empty, as in section *1* of Fig. 30-5. The outflow pressure, P_o, is therefore equal to the atmospheric pressure, which shall be designated as the zero, or reference, level. The inflow pressure, P_i, is then equal to the same reference level plus the height, h_1, of the column of liquid in reservoir *A*. Under these conditions let the flow, Q, through the tube be 5 ml/s. If reservoir *A* is filled to height h_2, which is twice h_1, and reservoir *B* is again empty (as in section *2*), the flow will be twice as great, that is, 10 ml/s. Thus with reservoir *B* empty, the flow will be directly proportional to the inflow pressure, P_i. If reservoir *B* is now allowed to fill to height h_1, and the fluid level in *A* is maintained at h_2 (as in section *3*), the flow will again become 5 ml/s. Thus flow is directly proportional to the difference between inflow and outflow pressures;

$$Q \propto P_i - P_o \tag{7}$$

If the fluid level in *B* attains the same height as in *A*, flow will cease (section *4*).

For any given pressure difference between the two ends of a tube, the flow will depend on the dimensions of the tube. Consider the tube connected to reservoir *1* in Fig. 30-6. With length l_1 and radius r_1, the flow Q_1 is observed to be 10 ml/s. The tube connected to reservoir *2* has the same radius, but is twice as long. Under these conditions the flow Q_2 is found to be *5* ml/s, or only half as great as Q_1. Conversely, for a tube half as long as l_1 the flow would be twice as great as Q_1. In other words, flow is inversely proportional to the length of the tube:

$$Q \propto 1/l \tag{8}$$

The tube connected to reservoir *3* in Fig. 30-6 is the same length as l_1, but the radius is twice as great. Under these conditions, the flow Q_3 is found to increase to a value of 160 ml/s, which is sixteen times greater than Q_1. The precise measurements of Poiseuille revealed that flow varies directly as the fourth power of the radius:

$$Q \propto r^4 \tag{9}$$

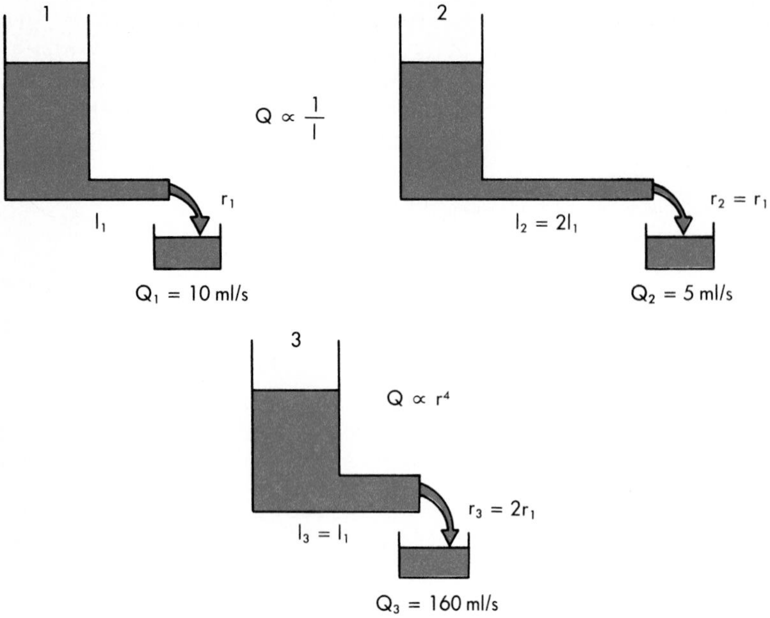

■ **Fig. 30-6.** The flow, Q, of fluid through a tube is inversely proportional to the length, l, and directly proportional to the fourth power of the radius, r, of the tube.

Thus, in the example above, since $r_3 = 2r_1$, Q_3 will be proportional to $(2r_1)^4$, or $16r_1^4$; therefore, Q_3 will equal $16Q_1$.

Finally, for a given pressure difference and for a cylindrical tube of given dimensions, the flow will vary, depending on the nature of the fluid itself. This flow-determining property of fluids is termed *viscosity,* η, which has been defined by Newton as the ratio of *shear stress* to the *shear rate* of the fluid. These terms may be comprehended most clearly by considering the flow of a homogeneous fluid between parallel plates. In Fig. 30-7 let the bottom plate (the bottom of a large basin) be stationary, and let the upper plate move at a constant velocity along the upper surface of the fluid. The *shear stress,* τ, is defined as the ratio of F:A, where F is the force applied to the upper plate in the direction of its motion along the upper surface of the fluid and A is the area of the upper plate in contact with the fluid. The shear rate is du/dy, where u is the velocity in the direction parallel to the motion of the upper plate for any minute fluid element contained between the plates and y is the distance of that element above the bottom, stationary plate.

For a movable plate traveling with constant velocity across the surface of a homogeneous fluid, the velocity profile of the fluid will be linear. The fluid layer in contact with the upper plate will adhere to it and therefore will move at the same velocity, U, as the plate. Each minute element of fluid between the plates will move at a velocity, u, proportional to its distance, y, from the lower plate. Therefore the shear rate will be U/Y, where Y is the total distance between the two plates. Since viscosity, η, is defined as the ratio of shear stress, τ, to the shear rate du/dy, in the case illustrated in Fig. 28-7,

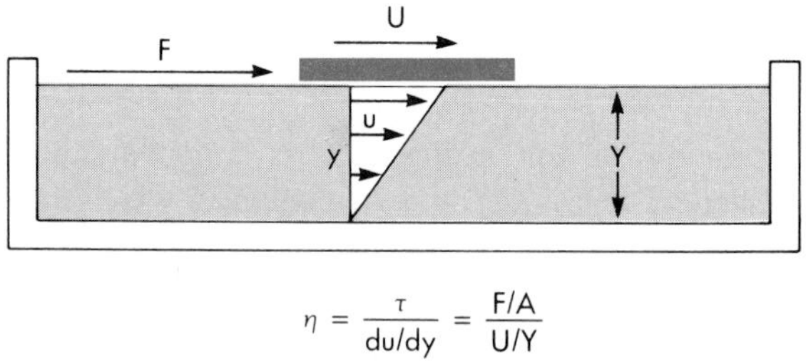

■ **Fig. 30-7.** For a newtonian fluid, the viscosity, η, is defined as the ratio of shear stress, τ, to shear rate, *du/dy*. For a plate of contact area, A, moving across the surface of a liquid, τ equals the ratio of the force, F (applied in the direction of motion) to the contact area, a, and *du/dy* equals the ratio of the velocity of the plate, U, to the depth of the liquid, Y.

$$\eta = (F/A)/(U/Y) \tag{10}$$

Thus the dimensions of viscosity are dynes/cm^2 divided by (cm/s)/cm, or dyne-s-cm^{-2}. In honor of Poiseuille, 1 dyne-s-cm^{-2} has been termed a *poise*. The viscosity of water at 20° C is approximately 0.01 *poise,* or 1 centipoise. In the case of certain nonhomogeneous fluids, notably suspensions such as blood, the ratio of the shear stress to the shear rate is not constant; such fluids are said to be *nonnewtonian*.

With regard to the flow of fluids through cylindrical tubes, the flow will vary inversely as the viscosity. Thus in the example of flow from reservoir *1* in Fig. 30-6, if the viscosity of the fluid in the reservoir were doubled, then the flow would be halved (5 ml/s instead of 10 ml/s).

In summary, for the steady, laminar flow of a newtonian fluid through a cylindrical tube, the flow, Q, varies directly as the pressure difference, $P_i - P_o$, and the

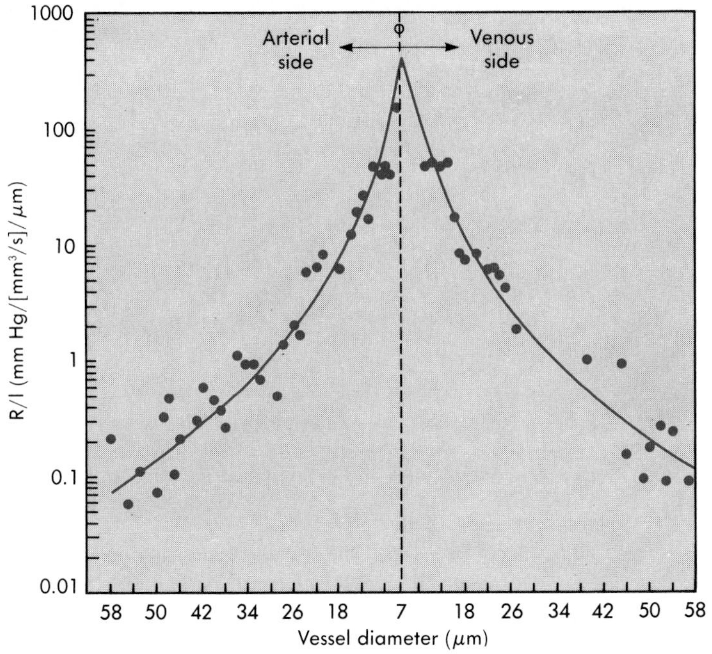

■ **Fig. 30-8.** The resistance per unit length *(R/l)* for individual small blood vessels in the cat mesentery. The capillaries, diameter 7 μ, are denoted by the vertical dashed line. Resistances of the arterioles are plotted to the left and resistances of the venules are plotted to the right of the vertical dashed line. The solid circles represent the actual data. The two curves through the data represent the following regression equations for the arteriole and venule data, respectively: (a) arterioles, $R/l = 1.02 \times 10^6 D^{-4.04}$, and (b) venules, $R/l = 1.07 \times 10^6 D^{3.94}$. Note that for both types of vessels, the resistance per unit length is universely proportional to the fourth power (within 1%) of the vessel diameter (D). (Redrawn from Lipowsky, H.H., Kovalcheck, S., and Zweifach, B.W.: Circ. Res. **43:**738. By permission of the American Heart Association, Inc.)

fourth power of the radius, r, of the tube, and it varies inversely as the length, l, of the tube and the viscosity, η, of the fluid. The full statement of Poiseuille's law is

$$Q = \frac{\pi(P_i - P_o)r^4}{8\,\eta\,l} \quad (11)$$

where π/8 is the constant of proportionality.

■ *Resistance to Flow*

In electrical theory the concept of resistance is useful. Resistance, R, is defined as the ratio of voltage drop, E, to current flow, I. Similarly, in fluid mechanics the analogous hydraulic resistance, R, may be defined as the ratio of pressure drop, $P_i - P_o$, to flow, Q. For the steady, laminar flow of a newtonian fluid through a cylindrical tube the physical components of hydraulic resistance may be perceived readily by rearranging Poiseuille's law to give the hydraulic resistance equation

$$R = \frac{P_i - P_o}{Q} = \frac{8\,\eta\,I}{\pi\,r^4} \quad (12)$$

Thus, when Poiseuille's law applies, the resistance to flow depends only on the dimensions of the tube and on the characteristics of the fluid.

The principal determinant of the resistance to blood flow through any individual vessel within the circulatory system is its caliber. The resistance to flow through small blood vessels in cat mesentery has been measured, and the resistance per unit length of vessel (R/l) is plotted against the vessel diameter in Fig. 30-8. The resistance is highest in the capillaries (diameter 7 μm), and it diminishes as the vessels increase in diameter on the arterial and venous sides of the capillaries. The values of R/l were found to be virtually proportional to the fourth power of the diameter for the larger vessels on both sides of the capillaries.

Changes in vascular resistance induced by natural stimuli occur by changes in radius. The principal changes are achieved by alterations in the contraction of the circular smooth muscle cells in the vessel wall. However, changes in internal pressure also alter the caliber of the blood vessels and therefore alter the resistance to blood flow through those vessels.

From Fig. 26-2 it may be noted that the greatest pressure drop across the various types of vessels in series with each other occurs across the arterioles. Since the total flow is the same through the various series components of the circulatory system, it follows that the greatest resistance to flow resides in the arterioles. The reason why the highest resistance is not located in the capillaries (as might otherwise be suspected from Fig. 30-8) is related to the relative numbers of parallel capillaries and parallel arterioles, as explained later. The arterioles are vested with a thick coat of circularly arranged smooth muscle fibers, by means of which the lumen radius may be varied. From the hydraulic resistance equation, wherein R varies inversely as r^4, it is clear that small changes in radius will alter resistance greatly.

In the cardiovascular system the various types of vessels listed along the horizontal axis in Fig. 26-2 lie in series with one another. Furthermore, the individual members within each category of vessels are ordinarily arranged in parallel with one another (see Fig. 26-3). For

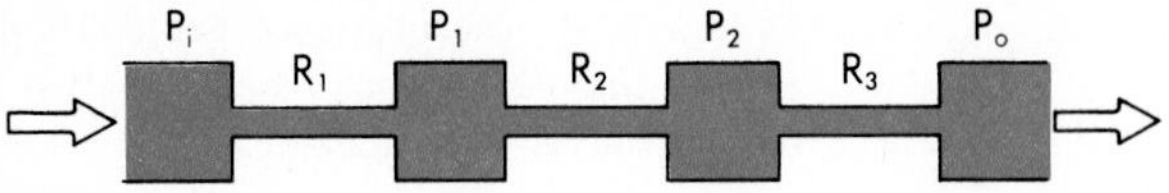

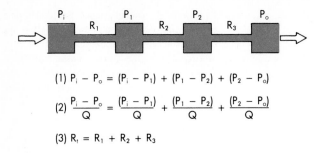

(1) $P_i - P_o = (P_i - P_1) + (P_1 - P_2) + (P_2 - P_o)$

(2) $\frac{P_i - P_o}{Q} = \frac{(P_i - P_1)}{Q} + \frac{(P_1 - P_2)}{Q} + \frac{(P_2 - P_o)}{Q}$

(3) $R_t = R_1 + R_2 + R_3$

■ **Fig. 30-9.** For resistances (R_1, R_2, and R_3) arranged in series, the total resistance R_t, equals the sum of the individual resistances.

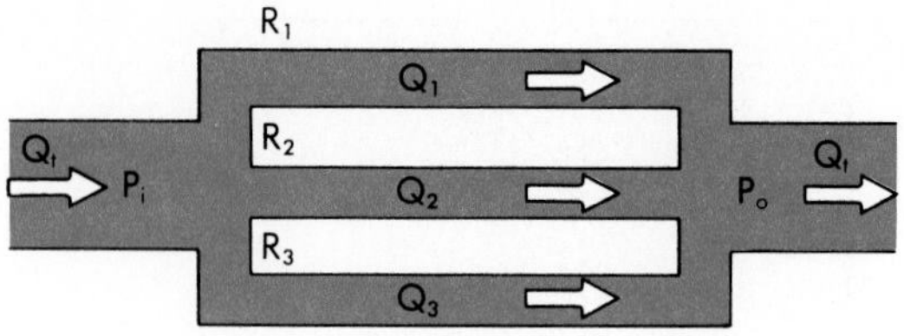

(1) $Q_t = Q_1 + Q_2 + Q_3$

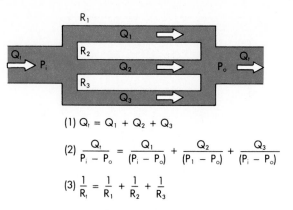

(2) $\frac{Q_t}{P_i - P_o} = \frac{Q_1}{(P_i - P_o)} + \frac{Q_2}{(P_1 - P_o)} + \frac{Q_3}{(P_i - P_o)}$

(3) $\frac{1}{R_t} = \frac{1}{R_1} + \frac{1}{R_2} + \frac{1}{R_3}$

■ **Fig. 30-10.** For resistances (R_1, R_2, and R_3) arranged in parallel, the reciprocal of the total resistance, R_t, equals the sum of the reciprocals of the individual resistances.

example, the capillaries throughout the body are in most instances parallel elements, with the notable exceptions of the renal vasculature (wherein the peritubular capillaries are in series with the glomerular capillaries) and the splanchnic vasculature (wherein the intestinal and hepatic capillaries are aligned in series). Formulas for the total hydraulic resistance of components arranged in series and in parallel have been derived in the same manner as those for similar combinations of electrical resistances.

Three hydraulic resistances, R_1, R_2, and R_3, are arranged in series in the schema depicted in Fig. 30-9. The pressure drop across the entire system—that is, the difference between inflow pressure, P_i, and outflow pressure, P_o—consists of the sum of the pressure drops across each of the individual resistances (equation *1*). Under steady-state conditions, the flow, Q, through any given cross section must equal the flow through any other cross section. By dividing each component in equation *1* by Q (equation *2*), it becomes evident from the definition of resistance that the total resistance, R_t, of the entire system equals the sum of the individual resistances, that is,

$$R_t = R_1 + R_2 + R_3 \tag{13}$$

For resistances in parallel, as illustrated in Fig. 30-10, the inflow and outflow pressures are the same for all tubes. Under steady-state conditions, the total flow, Q_t, through the system equals the sum of the flows through the individual parallel elements (equation *1*). Because the pressure gradient ($P_i - P_o$) is identical for all parallel elements, each term in equation *1* may be divided by that pressure gradient to yield equation *2*. From the definition of resistance, equation *3* may be derived. This states that the reciprocal of the total resistance, R_t, equals the sum of the reciprocals of the individual resistances, that is,

$$\frac{1}{R_t} = \frac{1}{R_1} + \frac{1}{R_2} + \frac{1}{R_3} \tag{14}$$

Stated in another way, if we define hydraulic *conductance* as the reciprocal of resistance, it becomes evident that, for tubes in parallel, the total conductance is the sum of the individual conductances.

By considering a few simple illustrations, some of the fundamental properties of parallel hydraulic systems become apparent. For example, if the resistances of the three parallel elements in Fig. 30-10 were all equal, then

$$R_1 = R_2 = R_3 \tag{15}$$

Therefore

$$1/R_t = 3/R_1$$

and

$$R_t = R_1/3$$

Thus the total resistance is less than any of the individual resistances. After further consideration, it becomes evident that for any parallel arrangement, the total resistance must be less than that of any individual component. For example, consider a system in which a very high-resistance tube is added in parallel to a low-resistance tube. The total resistance must be less than that of the low-resistance component by itself, because the high-resistance component affords an additional pathway, or conductance, for fluid flow.

As a physiological illustration of these principles, consider the relationship between the *total peripheral resistance (TPR)* of the entire systemic vascular bed and the resistance of one of its components, such as the renal vasculature. In an individual with a cardiac output of 5000 ml/min and an arterial pressure of 100 mm Hg, the TPR will be 0.02 mm Hg/ml/min, or 0.02 PRU (peripheral resistance units). Blood flow through one kidney would be approximately 600 ml/min. Renal resistance would therefore be 100 mm Hg/600 ml/min, or 0.17 PRU, which is 8.5 times as great as the total resistance of the entire systemic vascular bed.

From Fig. 26-2 it seems paradoxical that the resis-

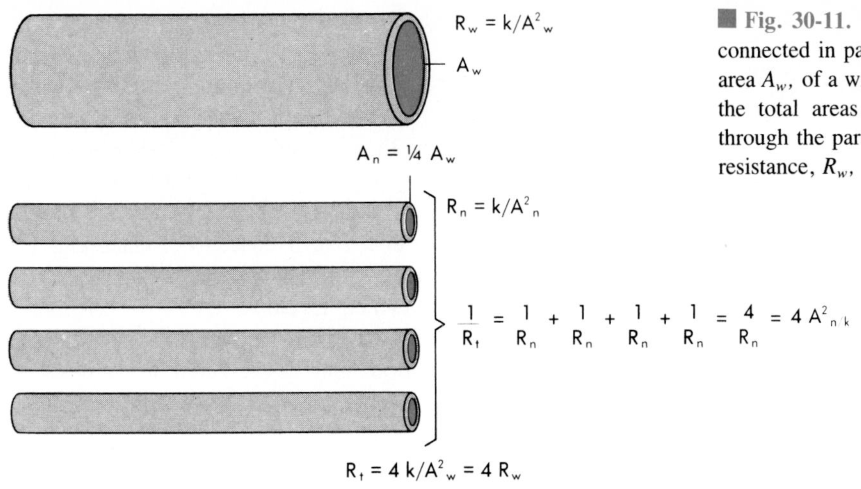

■ **Fig. 30-11.** When four narrow tubes, each of area A_n, are connected in parallel, the total cross-sectional area equals the area A_w, of a wide tube of area such that $A_w = 4A_n$. Although the total areas are equal, the total resistance, R_t, to flow through the parallel narrow tubes is four times as great as the resistance, R_w, through the single wide tube.

tance to flow through the arterioles (as manifested by the pressure drop between the arterial and capillary ends of these vessels) is considerably greater than that through certain other vascular components, despite the fact that the total cross-sectional area of the arterioles exceeds that for these same vascular components. For example, the total cross-sectional area of the arterioles greatly exceeds that of the large arteries, yet the resistance to flow through the arterioles exceeds that through the large arteries.

Consideration of simple models of tubes in parallel will help resolve this apparent paradox. In Fig. 30-11 the resistance to flow through one wide tube of cross-sectional area A_w is compared with that through four narrower tubes in parallel, each of area A_n. The total cross-sectional area of the parallel system of four narrow tubes equals the area of the wide tube; that is,

$$A_w = 4\ A_n \tag{16}$$

Since resistance, R, is inversely proportional to the fourth power of the radius, r, and since

$$A = \pi r^2 \tag{17}$$

for cylindrical tubes, it follows that

$$R = k/A^2 \tag{18}$$

The proportionality constant, k, is related to tube length and fluid viscosity, both of which will be held constant in the example under consideration. From equation *18*, the resistances of the wide tube, R_w, and a single narrow tube, R_n, are

$$R_w = k/A_w^2 \tag{19}$$

$$R_n = k/A_n^2 \tag{20}$$

From equation *3* in Fig. 30-10,

$$\frac{1}{R_t} = \frac{1}{R_n} + \frac{1}{R_n} + \frac{1}{R_n} + \frac{1}{R_n} = \frac{4}{R_n} \tag{21}$$

Substituting the value of R_n in equation *20*,

$$1/R_t = 4A_n^2/k \tag{22}$$

Rearranging,

$$R_t = k/4A_n^2 \tag{23}$$

From equations *16* and *19*,

$$R_t = 4k/A_w^2 = 4R_w \tag{24}$$

Hence the resistance of four such tubes in parallel is four times as great as that of a single tube of equal total cross-sectional area.

If a similar calculation is made for eight such tubes in parallel, with each tube having one fourth the cross-sectional area of the single wide tube, it will be found that the total resistance will equal $2R_w$. In this circumstance the resistance to flow through eight such narrow tubes in parallel will still be twice as great as that through the single tube, despite the fact that the total cross-sectional area for the eight narrow tubes is twice as great as that for the single wide tube. This is analogous to the relationship that exists between resistance and area in the circulatory system when comparing the arterioles with the large arteries. Despite the fact that the total cross-sectional area of all the arterioles greatly exceeds that of all the large arteries, the resistance to flow through the arterioles is considerably greater than that through the large arteries.

If the example is carried still further, it will be found that 16 such narrow tubes in parallel, now with four times the total cross-sectional area of the single wide tube, will exert a resistance to flow just equal to the resistance through the wide tube. Any number of these narrow tubes in excess of 16, then, will have a lower resistance than that of the single wide tube. This is analogous to the situation for the arterioles and capillaries. The resistance to flow through a single capillary is much greater than that through a single arteriole (Fig.

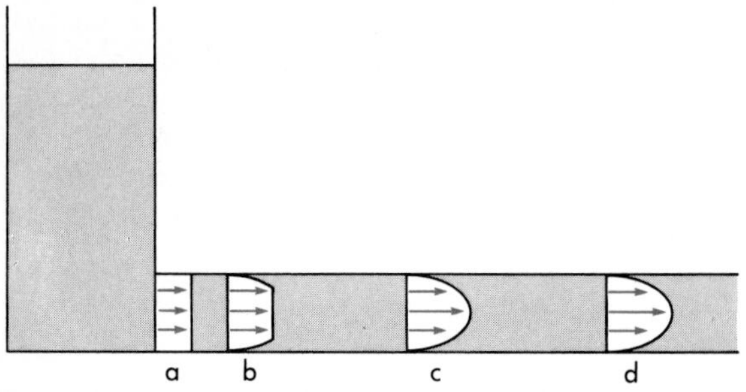

■ **Fig. 30-12.** Laminar flow in a cylindrical tube. At the inlet, *a,* the velocities are equal at all radial distances from the center of the tube. Near the inlet, *b,* the velocity profile is flat near the center of the tube, but a velocity gradient is established near the wall. When flow becomes fully developed, *c* and *d,* the velocity profile is parabolic.

30-8). Yet the number of capillaries so greatly exceeds the number of arterioles, as reflected by the relative difference in total cross-sectional areas, that the pressure drop across the arterioles is considerably greater than the pressure drop across the capillaries (see Fig. 26-2).

■ *Laminar and Turbulent Flow*

Under certain conditions, the flow of a fluid in a cylindrical tube will be *laminar* (sometimes called *streamlined*), as illustrated in Fig. 30-12. At the entrance of the tube all the fluid elements will have the same linear velocities, regardless of their radial positions. In progressing along the tube, however, the thin layer of fluid in contact with the wall of the tube adheres to the wall and hence is motionless. The layer of fluid just central to this external lamina must shear against this motionless layer and therefore moves slowly, but with a finite velocity. Similarly, the adjacent, more central layer travels still more rapidly. Close to the tube inlet the fluid layers near the axis of the tube still move with the same velocity, and do not shear against one another. However, at a distance from the tube inlet equal to several tube diameters, laminar flow becomes *fully developed;* that is, the velocity profiles do not change with longitudinal distance along the tube. In fully developed laminar flow the longitudinal velocity profile is that of a paraboloid. The velocity of the fluid adjacent to the wall is zero, whereas the velocity at the center of the stream is maximum and equal to twice the mean velocity of flow across the entire cross section of the tube. In laminar flow, fluid elements remain in one lamina, or streamline, as the fluid progresses longitudinally along the tube.

Irregular motions of the fluid elements may develop in the flow of fluid through a tube; this flow is called *turbulent*. Under such conditions fluid elements do not remain confined to definite laminae but rapid, radial mixing occurs. A considerably greater pressure is required to force a given flow of fluid through the same tube when the flow is turbulent than when it is laminar. In turbulent flow the pressure drop is approximately proportional to the square of the flow rate, whereas in laminar flow, the pressure drop is proportional to the first power of the flow rate. Hence, to produce a given flow, a pump such as the heart must do considerably more work if turbulence develops.

Whether turbulent or laminar flow will exist in a tube under given conditions may be predicted on the basis of a dimensionless number called *Reynold's number,* N_R. This number represents the ratio of inertial to viscous forces and equals $\rho D\bar{v}/\eta$, where D is the tube diameter, $\bar{v}$ is the mean velocity, ρ is the density, and η is the viscosity. For $N_R < 2000$, the flow will usually be laminar; for $N_R > 3000$, turbulence will usually exist. Various possible conditions may develop in the transition range of N_R between 2000 and 3000. Since flow tends to be laminar at low N_R and turbulent at high N_R, it is evident from the definition of N_R that large diameters, high velocities, and low viscosities predispose to the development of turbulence. In addition to these factors, abrupt variations in tube dimensions or irregularities in the tube walls will produce turbulence.

Turbulence is usually accompanied by vibrations in the auditory frequency ranges. When turbulent flow exists within the cardiovascular system, it is usually detected as a *murmur*. The factors listed earlier that predispose to turbulence may account for murmurs heard clinically. In severe anemia *functional cardiac murmurs* (murmurs not caused by structural abnormalities) are frequently detectable. The physical basis for such murmurs resides in (1) the reduced viscosity of blood caused by the low red cell content and (2) the high flow velocities associated with the marked augmentation of cardiac output that usually occurs in anemic patients.

Blood clots, or *thrombi,* are much more likely to develop with turbulent than with laminar flow. One of the problems with the use of artificial valves in the surgical treatment of valvular heart disease is that thrombi tend to occur in association with the prosthetic valve. The thrombi then may be dislodged, often occluding a crucial vessel in the peripheral circulation. It is thus important to design such valves to avert development of turbulent flow.

■ *Shear Stress on the Vessel Wall*

In Fig. 30-7 an external force was applied to a plate floating on the surface of a volume of liquid contained in a large basin. This force, exerted parallel to the surface, caused a shearing stress on the liquid below, thereby producing a differential motion of each layer of liquid relative to the adjacent layers. At the bottom of the basin, the flowing liquid produced a shearing stress on the surface of the basin in contact with the liquid. By rearranging the formula for viscosity stated in Fig. 30-7, it is apparent that the shear stress, τ, equals $\eta(du/dy)$; that is, the shear stress equals the product of the viscosity and the shear rate. Hence the greater the rate of flow, the greater the shear stress that the liquid exerts on the walls of the container or tube in which it flows.

For precisely the same reasons, the rapidly flowing blood in a large artery tends to pull the endothelial lining of the artery along with it. This force *(viscous drag)* is proportional to the shear rate (du/dy) of the layers of blood near the wall. For a flow regimen that obeys Poiseuille's law,

$$\tau = 4\eta Q/\pi r^3 \quad (25)$$

The greater the rate of blood flow (Q) in the artery, the greater will be du/dy near the arterial wall, and the greater will be the viscous drag (τ).

In certain types of arterial disease, particularly with hypertension, the subendothelial layers tend to degenerate locally, and small regions of the endothelium may lose their normal support. The viscous drag on the arterial wall may cause a tear between a normally supported and an unsupported region of the endothelial lining. Blood may then flow from the vessel lumen, through the rift in the lining, and dissect between the various layers of the artery. Such a lesion is called a *dissecting aneurysm*. It occurs most commonly in the proximal portions of the aorta and is extremely serious. One reason for its predilection for this site is the very high velocity of blood flow, with the associated high values of du/dy at the endothelial wall.

Rheological Properties of Blood

The viscosity of a newtonian fluid, such as water, may be determined by measuring the rate of flow of the fluid at a given pressure gradient through a cylindrical tube of known length and radius. As long as the fluid flow is laminar, the viscosity may be computed by substituting these values into Poiseuille's equation. The viscosity of a given newtonian fluid at a specified temperature will be constant over a wide range of tube dimensions and flows. However, for a nonnewtonian fluid, the viscosity calculated by substituting into Poiseuille's equation may vary considerably as a function of tube dimensions and flows. Therefore, in considering the rheological properties of a suspension such as blood, the term *viscosity* does not have a unique meaning. The terms *anomalous viscosity* and *apparent viscosity* are frequently applied to the value of viscosity obtained for blood under the particular conditions of measurement.

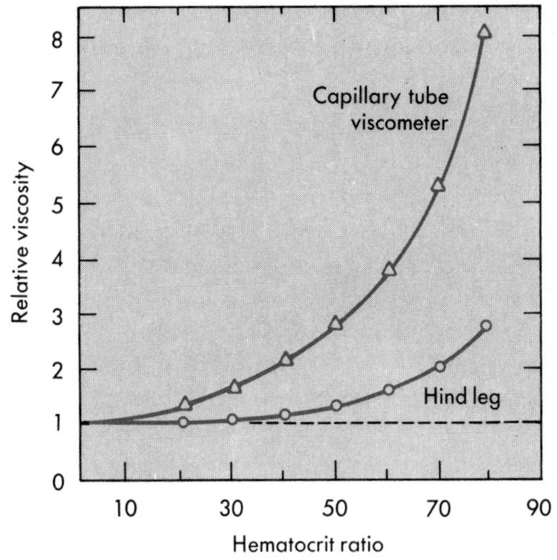

Fig. 30-13. Viscosity of whole blood, relative to that of plasma, increases at a progressively greater rate as the hematocrit ratio increases. For any given hematocrit ratio the apparent viscosity of blood is less when measured in a biological viscometer (such as the hind leg of a dog) than in a conventional capillary tube viscometer. (Redrawn from Levy, M.N., and Share, L.: Circ. Res. **1**:247, 1953.)

Rheologically, blood is a suspension, principally a suspension of erythrocytes in a relatively homogeneous liquid, the blood plasma. For this reason the apparent viscosity of blood varies as a function of the *hematocrit ratio* (ratio of volume of red blood cells to volume of whole blood). In Fig. 30-13 the upper curve represents the ratio of the apparent viscosity of whole blood to that of plasma over a range of hematocrit ratios from 0% to 80%, measured in a tube 1 mm in diameter. The viscosity of plasma is 1.2 to 1.3 times that of water. Fig. 30-13 (upper curve) shows that blood, with a normal hematocrit ratio of 45%, has an apparent viscosity 2.4 times that of plasma. In severe anemia, blood viscosity is low. With increasing hematocrit ratios the slope of the curve increases progressively; it is especially steep at the upper range of erythrocyte concentrations. A rise in hematocrit ratio from 45% to 70%, which occurs in *polycythemia,* increases the apparent viscosity more than twofold, with a proportionate effect on the resistance to blood flow. The effect of such a change in hematocrit ratio on peripheral resistance may be appreciated when it is recognized that even in the most severe cases of essential hypertension, the total peripheral resistance rarely increases by more than a factor of two. In hypertension, the increase in peripheral resistance is achieved by arteriolar vasoconstriction.

For any given hematocrit ratio the apparent viscosity of blood depends on the dimensions of the tube employed in estimating the viscosity. Fig. 30-14 demonstrates that the apparent viscosity of blood diminishes

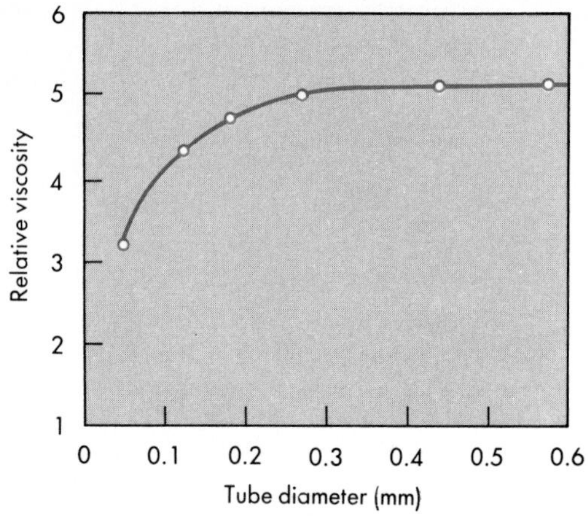

Fig. 30-14. Viscosity of blood, relative to that of water, increases as a function of tube diameter up to a diameter of about 0.3 mm. (Redrawn from Fåhraeus, R., and Lindqvist, T.: Am. J. Physiol. **96**:562, 1931.)

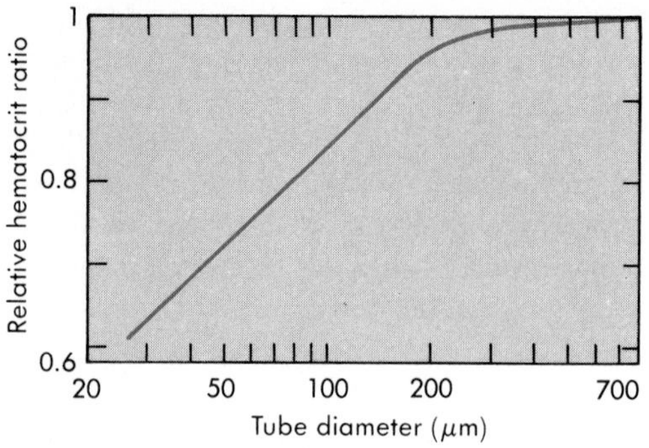

Fig. 30-15. The "relative hematocrit ratio" of blood flowing from a feed reservoir through capillary tubes of various calibers, as a function of the tube diameter. The relative hematocrit is the ratio of the hematocrit of the blood in the tubes to that of the blood in the feed reservoir. (Redrawn from Barbee, J.H., and Cokelet, G.R.: Microvasc. Res. **3**:6, 1971.)

progressively as tube diameter decreases below a value of about 0.3 mm. The diameters of the highest resistance blood vessels, the arterioles, are considerably less than this critical value. This phenomenon therefore reduces the resistance to flow in the blood vessels that possess the greatest resistance. The apparent viscosity of blood, when measured in living tissues, is considerably less than when measured in a conventional capillary tube viscometer with a diameter greater than 0.3 mm. In the lower curve of Fig. 30-13, the apparent relative viscosity of blood was assessed by using the hind leg of an anesthetized dog as a biological viscometer. Over the entire range of hematocrit ratios, the apparent viscosity was less as measured in the living tissue than in the capillary tube viscometer (upper curve), and the disparity was greater the higher the hematocrit ratio.

The influence of tube diameter on apparent viscosity is ascribable in part to the change in actual composition of the blood as it flows through small tubes. The composition changes because the red blood cells tend to accumulate in the faster axial stream, whereas it is largely plasma that flows in the slower marginal layers. To illustrate this phenomenon, a reservoir such as *A* in Fig. 30-5 has been filled with blood possessing a given hematocrit ratio. The blood in *A* was constantly agitated to prevent settling and was permitted to flow through a narrow capillary tube into reservoir *B*. As long as the tube diameter was substantially greater than the diameter of the red blood cells, the hematocrit ratio of the blood in *B* was not detectably different from that in *A*. Surprisingly, however, the hematocrit ratio of the blood contained within the tube was found to be considerably lower than the hematocrit ratio of the blood in either reservoir. The smaller the tube, the more pronounced the reduction in hematocrit ratio in the tube. In Fig. 30-15, the relative hematocrit is the ratio of the hematocrit in the tube to that in the reservoir at either end of the tube. For tubes of 500 μm diameter or greater, the relative hematocrit ratio was close to 1. However, as the tube diameter was diminished below 500 μm, the relative hematocrit ratio progressively diminished; for a tube diameter of 30 μm, the relative hematocrit ratio was only 0.6.

That this situation results from a disparity in the relative velocities of the red cells and plasma can be appreciated on the basis of the following analogy. Consider the flow of traffic across a bridge that is 3 miles long. Let the cars move in one lane at a speed of 60 miles per hour and the trucks in another lane at 20 miles per hour, as illustrated in Fig. 30-16. If one car and one truck start out across the bridge each minute, then except for the initial few minutes of traffic flow across the bridge one car and one truck will arrive at the other end each minute. Yet if one counts the actual number of cars and trucks on the bridge at any moment, there will be three times more of the slower moving trucks than of the more rapidly traveling cars.

Since the axial portions of the bloodstream contain a greater proportion of red cells and move with a greater velocity, the red cells tend to traverse the tube in a shorter period of time than the plasma. Therefore the red cells correspond to the rapidly moving cars in the analogy, and the plasma corresponds to the slowly moving trucks. Measurement of transit times through various organs have shown that red cells do travel faster than the plasma. Furthermore, the hematocrit ratios of the blood contained in various tissues are lower than those in blood samples withdrawn from large arteries or veins in the same animal.

The physical forces responsible for the drift of the

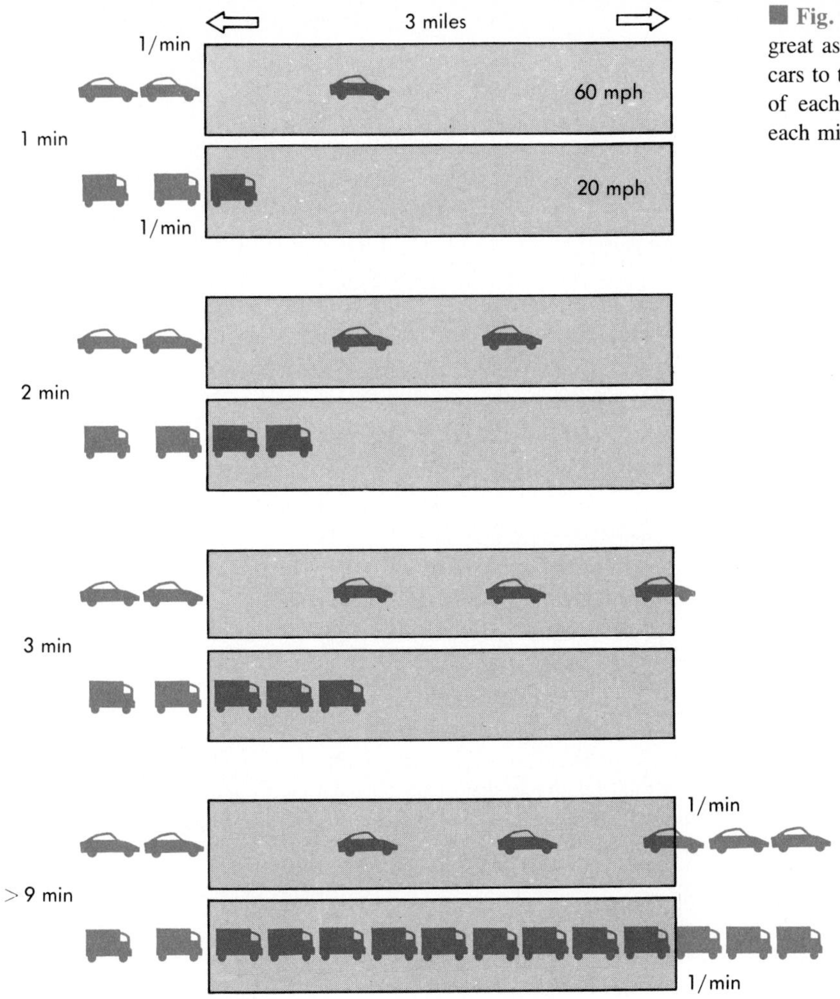

■ **Fig. 30-16.** When the car velocity is three times as great as the truck velocity, the ratio of the numbers of cars to trucks on a bridge will be 1:3, even though one of each type of vehicle enters and leaves the bridge each minute.

erythrocytes toward the axial stream and away from the vessel walls are not fully understood. One factor is the great flexibility of the red blood cells. At low flow (or shear) rates, comparable to those rates that prevail in the microcirculation, rigid particles do not migrate toward the axis of a tube, whereas flexible particles do migrate. The concentration of flexible particles near the tube axis is enhanced by increasing the shear rate.

The apparent viscosity of blood diminishes as the shear rate is increased (Fig. 30-17), a phenomenon called *shear thinning*. The greater tendency of the erythrocytes to accumulate in the axial laminae at higher flow rates is partly responsible for this nonnewtonian behavior. However, a more important factor is that at very slow rates of shear, the suspended cells tend to form aggregates, which would increase viscosity. This tendency toward aggregation decreases as the flow is augmented, producing the diminution in apparent viscosity that is exhibited in Fig. 30-17. The tendency toward aggregation at low flows depends on the concentration in the plasma of the larger protein molecules, especially fibrinogen. For this reason, the changes in blood viscosity with shear rate are much more pronounced when the concentration of fibrinogen is high. Also, at low flow rates, the leukocytes tend to adhere to the endothelial lining of the microvessels, thereby increasing the apparent viscosity.

The deformability of the erythrocytes is also a factor in shear thinning, especially at high hematocrit ratios. The mean diameter of human red blood cells is about 8 μm, yet they are able to pass through openings with a diameter of only 3 μm. As blood that is densely packed with erythrocytes is caused to flow at progressively greater rates, the erythrocytes become more and more deformed, diminishing the apparent viscosity of the blood. The flexibility of human erythrocytes is enhanced as the concentration of fibrinogen in the plasma increases (Fig. 30-18). If the red blood cells become hardened, as they are in certain spherocytic anemias, shear thinning may become much less prominent.

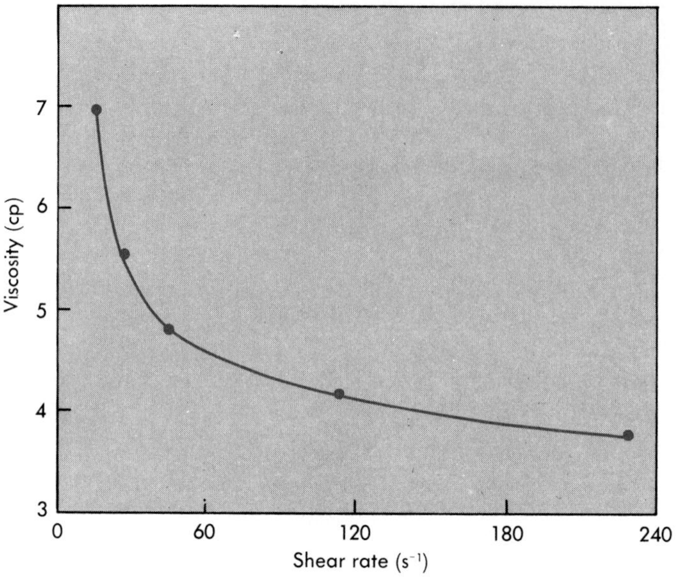

■ **Fig. 30-17.** Decrease in the viscosity of blood (centipoise) at increasing rates of shear. The shear rate refers to the velocity of one layer of fluid relative to that of the adjacent layers and is directionally related to the rate of flow. (Redrawn from Amin, T.M., and Sirs, J.A.: Q.J. Exp. Physiol. **70:**37, 1985.)

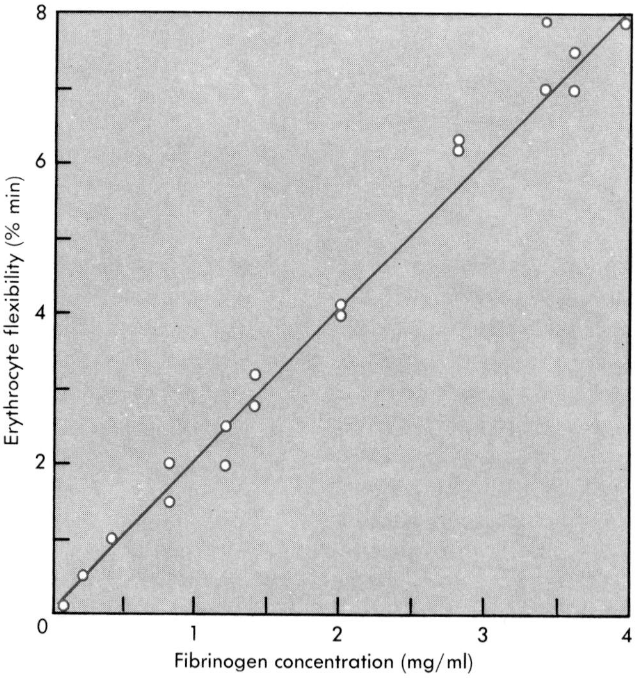

■ **Fig. 30-18.** The effect of the plasma fibrinogen concentration on the flexibility of human erythrocytes. (Redrawn from Amin, T.M., and Sirs, J.A.: Q.J. Exp. Physiol. **70:**37, 1985.)

Bibliography

Journal articles

Amin, T.M., and Sirs, J.A.: The blood rheology of man and various animal species, Q. J. Exp. Physiol. **70**:37, 1985.

Barbee, J.H., and Cokelet, G.R.: The Fåhraeus effect, Microvasc. Res. **3**:6, 1971.

Fåhraeus, R., and Lindqvist, T.: The viscosity of blood in narrow capillary tubes, Am. J. Physiol. **96**:562, 1931.

Lindgard, P.S.: Capillary pore rheology of erythrocytes. IV. Effect of pore diameter and haematocrit, Microvasc. Res. **13**:59, 1977.

Prokop, E.K., Palmer, R.F., and Wheat, M.W., Jr.: Hydrodynamic forces in dissecting aneurysms, Circ. Res. **27**:121, 1970.

Sarelius, I.H., and Duling, B.R.: Direct measurement of microvessel hematocrit, red cell flux, velocity, and transit time, Am. J. Physiol. **243**:H1018, 1982.

Talbot, L., and Berger, S.A.: Fluid-mechanical aspects of the human circulation, Am. Scientist **62**:671, 1974.

Winter, D.C., and Nerem, R.M.: Turbulence in pulsatile flows, Ann. Biomed. Eng. **12**:357, 1984.

Zamir, M.: The role of shear forces in arterial branching, J. Gen. Physiol. **67**:213, 1976.

Books and monographs

Charm, S.E., and Kurland, G.S.: Blood flow and microcirculation, New York, 1974, John Wiley & Sons, Inc.

Chien, S., Usami, S., and Skalak, R.: Blood flow in small tubes. In Renkin, E.M., and Michel, C.C., editors: Handbook of physiology; Section 2: The cardiovascular system—Microcirculation, vol. IV, Bethesda, Md., 1984, American Physiological Society.

Cokelet, G.R., Meiselman, H.J., and Brooks, D.E., editors: Erythrocyte mechanics and blood flow, New York, 1980, Alan R. Liss, Inc.

Fung, Y.C.: Biodynamics: circulation, New York, 1984, Springer-Verlag New York, Inc.

Milnor, W.R.: Hemodynamics, Baltimore, 1982, Williams & Wilkins.

Noordergraaf, A.: Circulatory system dynamics, New York, 1979, Academic Press, Inc.

Taylor, D.E.M., and Stevens, A.L., editors: Blood flow: theory and practice, New York, 1983, Academic Press, Inc.

CHAPTER

31

The Arterial System

Hydraulic Filter

The principal function of the systemic and pulmonary arterial systems is to distribute blood to the capillary beds throughout the body. The arterioles, the terminal components of this system, regulate the distribution to the various capillary beds. Between the heart and the arterioles, the aorta and pulmonary artery and their major branches constitute a system of conduits of considerable volume and distensibility. An arterial system composed of elastic conduits and high-resistance terminals constitutes a *hydraulic filter* analogous to the resistance-capacitance filters of electrical circuits.

Hydraulic filtering converts the intermittent output of the heart to a steady flow through the capillaries (Fig. 31-1).

The entire ventricular stroke volume is discharged into the arterial system during systole, which usually occupies approximately one third of the duration of the cardiac cycle. In fact, most of the stroke volume is pumped during the rapid ejection phase, which constitutes about half of total systole. Part of the energy of cardiac contraction is dissipated as forward capillary flow during systole; the remainder is stored as potential energy, in that much of the stroke volume is retained by the distensible arteries. During diastole the elastic recoil of the arterial walls converts this potential energy into capillary blood flow. If the arterial walls were rigid, then capillary flow would cease during diastole.

Hydraulic filtering minimizes the work load of the heart. More work is required to pump a given flow intermittently than steadily; the more effective the filtering, the less the excess work. A simple example will illustrate this point.

Consider first the steady flow of a fluid at a rate of 100 ml/s through a hydraulic system with a resistance of 1 mm Hg/ml/s. This combination of flow and resistance would result in a constant pressure of 100 mm Hg, as shown in Fig. 31-2, *A*. Neglecting any inertial effect, hydraulic work (W) may be defined as

$$W = \int_{t_1}^{t_2} P dV \qquad (1)$$

that is, each small increment of volume, dV, pumped is multiplied by the pressure, P, existing at the time, and the products are integrated over the time interval of interest, $t_2 - t_1$, to give the total work, W. For steady flow, W = PV. In the example in Fig. 31-2, *A*, the work done in pumping the fluid for 1 second would be 10,000 mm Hg-ml (or 1.33×10^7 dyne-cm).

Next, consider an intermittent pump that ejects the same volume per second, but pumps the entire volume at a steady rate over 0.5 second and then pumps nothing during the next 0.5 second. Hence, it pumps at the rate of 200 ml/s for 0.5, as shown in Fig. 31-2, *B* and *C*. In *B* the conduit is rigid and the fluid is incompressible, but the system has the same resistance as in *A*. During the pumping phase of the cycle (systole) the flow of 200 ml/s through a resistance of 1 mm Hg/ml/s would produce a pressure of 200 mm Hg. During the filling phase of the cycle (diastole) the pressure would be 0 mm Hg in this rigid system. The work done during systole would be 20,000 mm Hg-ml, which is twice that required in the example shown in Fig. 31-2, *A*.

If the system were very distensible, hydraulic filtering would be very effective, and the pressure would remain virtually constant throughout the entire cycle (Fig. 31-2, *C*). Of the 100 ml of fluid pumped during the 0.5 second of systole, only 50 ml would be emitted through the high-resistance outflow end of the system during systole. The remaining 50 ml would be stored by the distensible conduit during systole and would flow out during diastole. Hence the pressure would be virtually constant at 100 mm Hg throughout the cycle. The fluid pumped during systole would be ejected at only half the pressure that prevailed in Fig. 31-2, *B*, and therefore, the work would be only half as great. With nearly perfect filtering, as in Fig. 31-2, *C*, the work would be identical to that for steady flow (Fig. 31-2, *A*).

Naturally, the filtering accomplished by the systemic

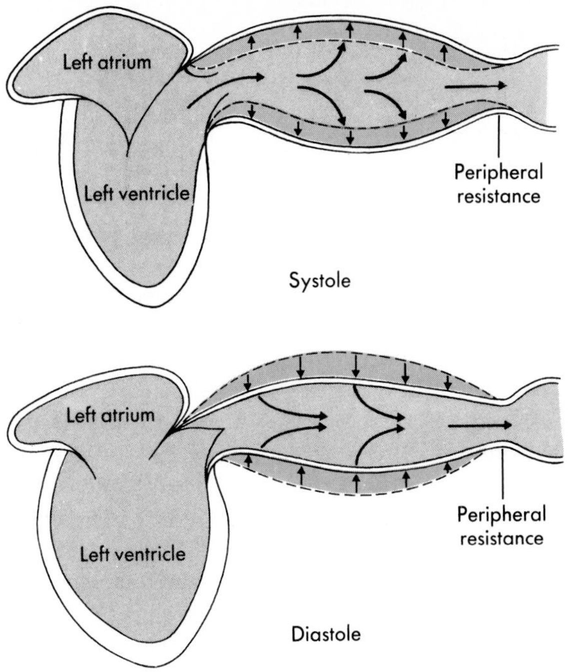

■ **Fig. 31-1.** During ventricular systole the stroke volume ejected by the ventricle results in some forward capillary flow, but most of the ejected volume is stored in the elastic arteries. During ventricular diastole the elastic recoil of the arterial walls maintains capillary flow throughout the remainder of the cardiac cycle.

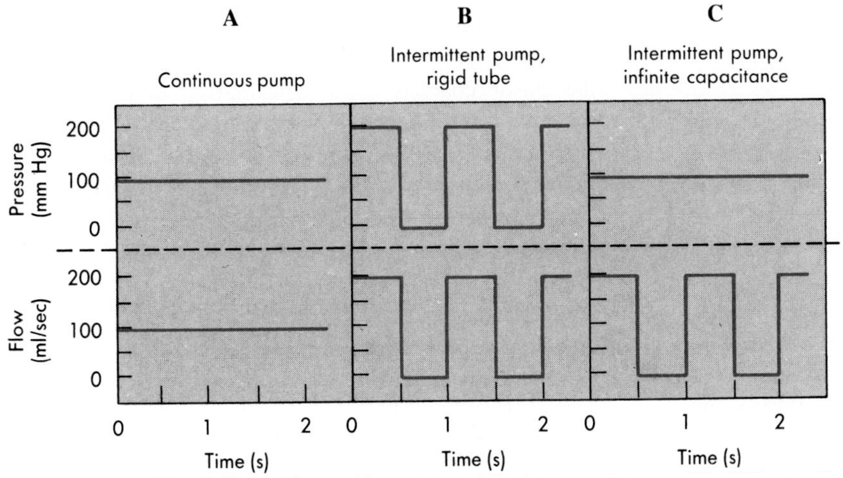

■ **Fig. 31-2.** The relationships between pressure and flow for three hydraulic systems, in each of which the flow is 100 ml/second and the resistance is 1 mm Hg/ml/second. In system **A** the flow is steady and the distensibility of the conduit is immaterial. In systems **B** and **C** the flow is intermittent; it is steady for half the cycle and ceases for the remainder of the cycle. In system **B** the conduit is rigid, whereas in system **C** the conduit is infinitely distensible, resulting in perfect filtering of the pressure. In systems **A** and **C** the work per second is 10,000 mm Hg-ml (or 1.33 × 10^7 dyne-cm); in system **B** the work per second is twice as great.

and pulmonic arterial systems is intermediate between the examples in Fig. 31-2, *B* and *C*. Under average normal conditions, the additional work imposed by intermittency of pumping, in excess of that for the steady flow case, is about 35% for the right ventricle and about 10% for the left ventricle. These fractions change, however, with variations in heart rate, peripheral resistance, and arterial distensibility.

■ *Arterial Elasticity*

The elastic properties of the arterial wall may be appreciated by considering first the *static pressure–volume relationship* for the aorta. To obtain the curves shown in Fig. 31-3, aortas were obtained at autopsy from individuals in different age groups. All branches of the aorta were ligated and successive volumes of liquid were injected into this closed elastic system in the same manner that successive increments of water might be introduced into a balloon. After each increment of volume, the internal pressure was measured. In Fig. 31-3 the curve relating pressure to volume for the youngest age group (curve *A*) is sigmoidal. Although the curve is quite linear over most of its extent, the slope decreases at the upper and lower ends. At any given point, the slope (dV/dP) represents the aortic *capacitance* (or *compliance*). Thus in normal individuals the aortic capacitance is least at extremely high and low pressures and greatest over the usual range of pressure variations.

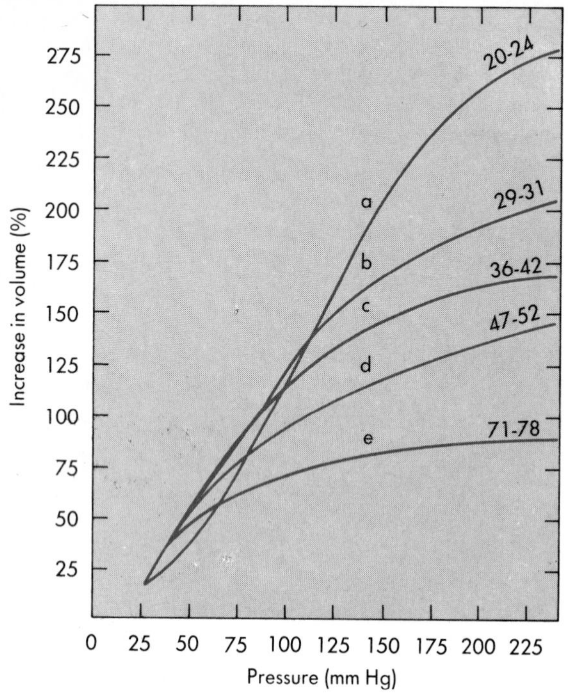

Fig. 31-3. Pressure-volume relationships for aortas obtained at autopsy from humans in different age groups (denoted by the numbers at the right end of each of the curves). (Redrawn from Hallock, P., and Benson, I.C.: J. Clin. Invest. **16:**595, 1937.)

This sequence of capacitance changes resembles the familiar capacitance changes encountered in inflating a balloon. The greatest difficulty in introducing air into the balloon is experienced at the beginning of inflation and again at near-maximum volume, just prior to rupture of the balloon. At intermediate volumes, the balloon is relatively easy to inflate.

It is also apparent from Fig. 31-3 that the curves become displaced downward and the slopes diminish as a function of advancing age. Thus for any given pressure above about 80 mm Hg, the capacitance decreases with age, a manifestation of increased rigidity caused by progressive changes in the collagen and elastin contents of the arterial walls. The heart is unable to eject its stroke volume into a rigid arterial system as rapidly as into a more compliant system. As capacitance diminishes, peak arterial pressure occurs progressively later in systole. Hence the rapid ejection phase of systole is significantly prolonged as aortic capacitance decreases.

A complete representation of arterial elasticity cannot be derived from such static pressure-volume curves. When *dynamic curves* are obtained, that is, when pressure is recorded during continuous injection or withdrawal of fluid, the pressure is a function not only of volume but also of the rate of change of volume. Therefore the arterial wall has *viscoelastic,* rather than purely elastic, properties. In a purely elastic system, pressure would be a function of volume alone.

Determinants of the Arterial Blood Pressure

The determinants of the pressure that may exist at any moment within the arterial system cannot be evaluated with great precision. Yet the arterial blood pressure is a quantitative measurement routinely obtained for diagnosis in most patients, and it provides a useful clue to their cardiovascular status. A simplified approach therefore will be undertaken in an attempt to gain a general understanding of the principal determinants of the arterial blood pressure. To accomplish this, the determinants of the *mean arterial pressure* (defined in the following section) will first be analyzed. The *systolic* and *diastolic arterial pressures* will then be considered as the upper and lower limits of periodic oscillations about this mean pressure. Finally, the changes in arterial pressure as the pulse wave progresses from the origin of the aorta toward the capillaries will be considered.

The determinants of the arterial blood pressure will be arbitrarily subdivided into "physical" and "physiological" factors. For the sake of simplicity, the arterial system will be assumed to be a static, elastic system, and the only two "physical" factors to be considered will be the blood volume within the arterial system and the elastic characteristics (capacitance) of the system. Several "physiological" factors will be considered, such as heart rate, stroke volume, cardiac output, and peripheral resistance. Such physiological factors will be shown to operate through one or both of the physical factors, however.

Mean Arterial Pressure

The *mean arterial pressure* is the pressure in the arteries, averaged over time. It may be obtained from an arterial pressure tracing by measuring the area under the curve and dividing this area by the time interval involved, as shown in Fig. 31-4. The mean arterial pressure, P_a, usually can be approximated satisfactorily from the measured values of the systolic (P_s) and diastolic (P_d) pressures by means of the following formula:

$$\overline{P}_a \cong P_d + \frac{1}{3}(P_s - P_d) \qquad (2)$$

The mean pressure will be considered to depend only on the mean volume of blood in the arterial system and the elastic properties of the arterial walls. The arterial volume, V_a, in turn, depends on the rate of inflow, Q_i, from the heart into the arteries *(cardiac output)* and the rate of outflow, Q_o, from the arteries through the capillaries *(peripheral runoff);* expressed mathematically,

$$dV_a/dt = Q_i - Q_o \qquad (3)$$

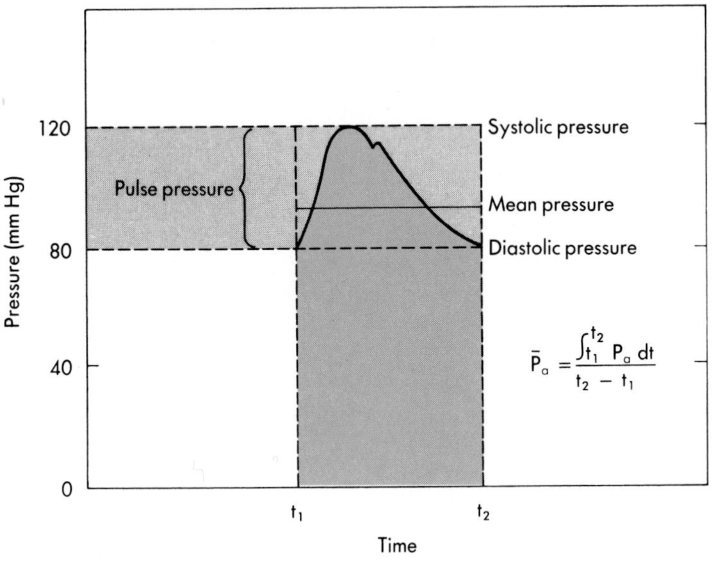

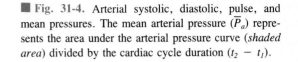

Fig. 31-4. Arterial systolic, diastolic, pulse, and mean pressures. The mean arterial pressure ($\bar{P}_a$) represents the area under the arterial pressure curve (*shaded area*) divided by the cardiac cycle duration ($t_2 - t_1$).

If arterial inflow exceeds outflow, then arterial volume increases, the arterial walls are stretched more, and pressure rises. The converse happens when arterial outflow exceeds inflow. When inflow equals outflow, arterial pressure remains constant.

The change in pressure in response to an alteration of cardiac output can be better appreciated by considering some simple examples. Under control conditions, let cardiac output be 5 L/min and mean arterial pressure ($\bar{P}_a$) be 100 mm Hg. From the definition of total peripheral resistance

$$R = (\bar{P}_a - \bar{P}_{ra})/Q_o \qquad (4)$$

If $\bar{P}_{ra}$ (right atrial pressure) is negligible compared with $\bar{P}_a$,

$$R \cong \bar{P}_a/Q_o \qquad (5)$$

Therefore, in the example, R is 100/5, or 20 mm Hg/L/min.

Now let cardiac output, Q_i, suddenly increase to 10 L/min. Instantaneously $\bar{P}_a$ will be unchanged. Since the outflow, Q_o, from the arteries depends on $\bar{P}_a$ and R, Q_o also will remain unchanged at first. Therefore Q_i, now 10 L/min, will exceed Q_o, still only 5 L/min. This will increase the mean arterial blood volume ($\bar{V}_a$). From equation 2, when $Q_i > Q_o$, then $d\bar{V}_a/dt > 0$; that is, volume is increasing.

Since $\bar{P}_a$, depends on the mean arterial blood volume, V_a, and the arterial capacitance, C_a, an increase in V_a, will raise the $\bar{P}_a$. By definition

$$C_a = d\bar{V}_a/d\bar{P}_a \qquad (6)$$

Therefore

$$d\bar{V}_a = C_a d\bar{P}_a \qquad (7)$$

and

$$\frac{d\bar{V}_a}{dt} = C_a\frac{d\bar{P}_a}{dt} \qquad (8)$$

From equation 2,

$$\frac{d\bar{P}_a}{dt} = \frac{Q_i - Q_o}{C_a} \qquad (9)$$

Hence P_a will rise when $Q_i > Q_o$, will fall when $Q_i < Q_o$ and will remain constant when $Q_i = Q_o$.

In this example, in which Q_i is suddenly increased to 10 L/min, $\bar{P}_a$ will continue to rise as long as Q_i exceeds Q_o. It is evident from equation 5 that Q_o will not attain a value of 10 L/min until $\bar{P}_a$ reaches a level of 200 mm Hg, as long as R remains constant at 20 mm Hg/L/min. Hence as $\bar{P}_a$ approaches 200, Q_o will almost equal Q_i and $\bar{P}_a$ will rise very slowly. When Q_i is first raised, however, Q_i is greatly in excess of Q_o, and therefore $\bar{P}_a$ will rise sharply. The pressure-time tracing in Fig. 31-5 indicates that, regardless of the value of C_a, the slope gradually diminishes as pressure rises, to approach a final value asymptotically.

Furthermore, the height to which $\bar{P}_a$ will rise is independent of the elastic characteristics of the arterial walls. $\bar{P}_a$ must rise to a level such that $Q_o = Q_i$. It is apparent from equation 5 that Q_o depends only on pressure gradient and resistance to flow. Hence C_a determines only the rate at which the new equilibrium value of $\bar{P}_a$ will be attained, as illustrated in Fig. 31-5. When C_a is small (rigid vessels), a relatively slight increment in $\bar{V}_a$ (caused by a transient excess of Q_i over Q_o) increases $\bar{P}_a$ greatly. Hence $\bar{P}_a$ attains its new equilibrium level quickly. Conversely, when C_a is large, then considerable volumes can be accommodated with relatively small pressure changes. Therefore the new equilibrium

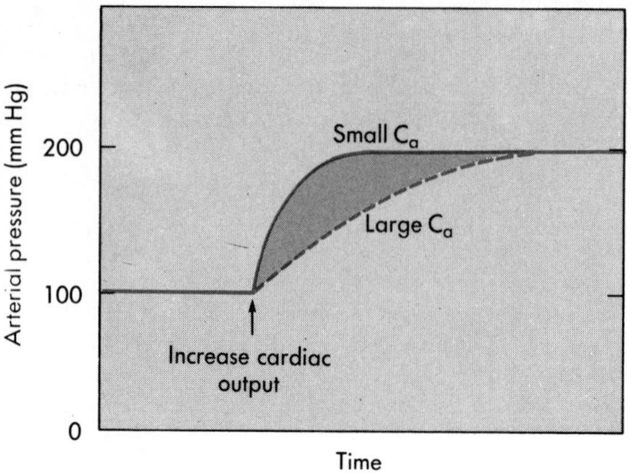

■ **Fig. 31-5.** When cardiac output is suddenly increased, the arterial capacitance (C_a) determines the *rate* at which the mean arterial pressure will attain its new, elevated value, but will not determine the *magnitude* of the new pressure.

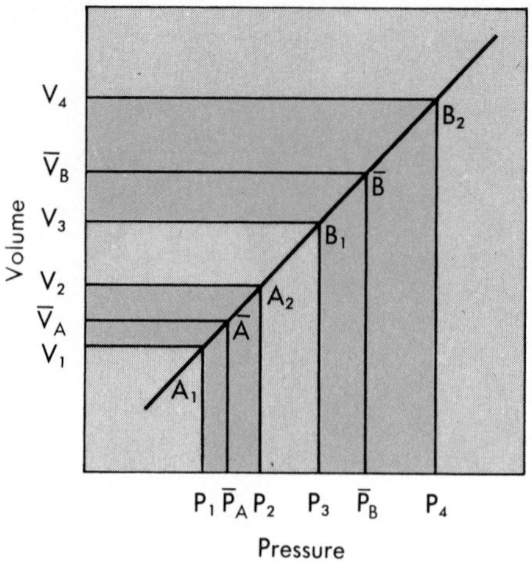

■ **Fig. 31-6.** Effect of a change in stroke volume on pulse pressure in a system in which arterial capacitance is constant over the range of pressures and volumes involved. A larger volume increment ($V_4 - V_3$ as compared to $V_2 - V_1$) results in a greater mean pressure ($\overline{P}_B$ as compared to $\overline{P}_A$) and a greater pulse pressure ($P_4 - P_3$ as compared to $P_2 - P_1$).

value of $\overline{P}_a$ is reached at a slower rate.

Similar reasoning may now be applied to explain the changes in $\overline{P}_a$ that accompany alterations in peripheral resistance. Let the control conditions be identical with those of the preceding example, that is, $Q_i = 5$, $\overline{P}_a = 100$, and $R = 20$. Then let R suddenly be increased to 40. Instantaneously $\overline{P}_a$ will be unchanged. With $\overline{P}_a = 100$ and $R = 40$, $Q_o = \overline{P}_a/R = 2.5$ L/min. If Q_i remains constant at 5 L/min, $Q_i \geqslant Q_o$, and $\overline{V}_a$ will increase; hence $\overline{P}_a$ will rise. $\overline{P}_a$ will continue to rise until it reaches 200 mm Hg. At this level $Q_o = 200/40 = 5$ L/min, which equals Q_i. $\overline{P}_a$ will then remain at this new elevated equilibrium level as long as Q_i and R do not change again.

It is clear, therefore that *the level of the mean arterial pressure depends only on cardiac output and peripheral resistance*. It is immaterial whether any change in cardiac output is accomplished by an alteration of heart rate, of stroke volume, or of both. Any change in heart rate that is balanced by a concomitant, oppositely directed change in stroke volume will not alter Q_i. Hence $\overline{P}_a$ will not be affected.

■ *Pulse Pressure*

If we assume that the arterial pressure, P_a, at any moment depends primarily on arterial blood volume, V_a, and arterial capacitance, C_a, it can be shown that the arterial *pulse pressure* (difference between systolic and diastolic pressures) is principally a function of stroke volume and arterial capacitance.

Stroke volume. The effect of a change in stroke volume on pulse pressure may be analyzed under conditions in which C_a remains virtually constant over the range of pressures under consideration. C_a is constant over any linear region of the pressure-volume curve (Fig. 31-6). Volume is plotted along the vertical axis, and pressure is plotted along the horizontal axis; the slope, dV/dP, is equivalent to the capacitance, C_a.

In an individual with such a linear P_a:V_a curve, the arterial pressure would oscillate about some mean value ($\overline{P}_A$ in Fig. 31-6) that depends entirely on cardiac output and peripheral resistance, as explained previously. This mean pressure corresponds to some mean arterial blood volume, $\overline{V}_A$. The coordinates $\overline{P}_A$, $\overline{V}_A$ define point $\overline{A}$ on the graph. During diastole, peripheral runoff occurs in the absence of ventricular ejection of blood, and P_a and V_a diminish to minimum values, P_1 and V_1, just prior to the next ventricular ejection. P_1 is then, by definition, the *diastolic pressure*.

During the rapid ejection phase of systole, the volume of blood that is introduced into the arterial system exceeds the volume that exits through the arterioles. Arterial pressure and volume therefore rise from point A_1 toward point A_2 in Fig. 31-6. The maximum arterial volume, V_2, is reached at the end of the rapid ejection phase, and this volume corresponds to a peak pressure, P_2, which is the *systolic pressure*.

The *pulse pressure* is the difference between systolic and diastolic pressures ($P_2 - P_1$ in Fig. 31-6), and it corresponds to some *volume increment*, $V_2 - V_1$. This increment equals the volume of blood discharged by the left ventricle during the rapid ejection phase minus the volume that has run off to the periphery during this same phase of the cardiac cycle. When a normal heart beats at a normal frequency, the volume increment during the rapid ejection phase is a large fraction of the stroke volume (about 80%). It is this increment that will

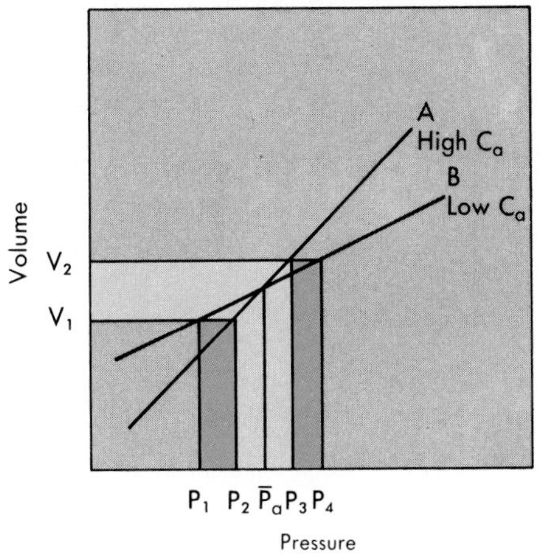

Fig. 31-7. For a given volume increment ($V_2 - V_1$) a reduced arterial capacitance (curve *B* as compared to curve *A*) results in an increased pulse pressure ($P_4 - P_1$ as compared to $P_3 - P_2$).

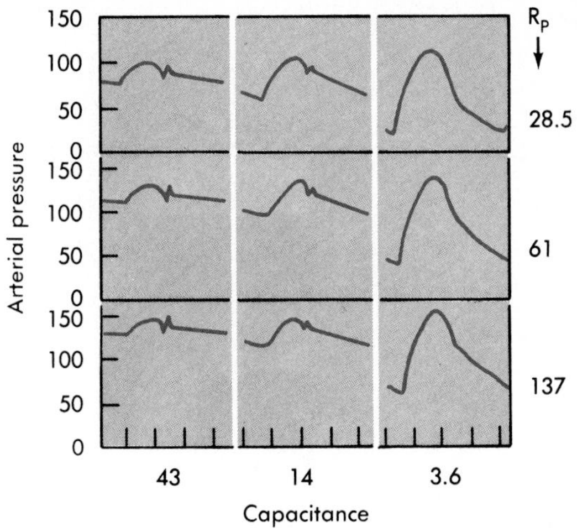

Fig. 31-8. The changes in aortic pressure with changes in arterial capacitance and peripheral resistance (R_p) in an isolated cat heart preparation. (Modified from Elizinga, G., and Westerhof, N.: Pressure and flow generated by the left ventricle against different impedances, Circ. Res. **32:**178, 1973. By permission of the American Heart Association, Inc.)

raise arterial volume rapidly from V_1 to V_2 and hence will cause the arterial pressure to rise from the diastolic to the systolic level (P_1 to P_2 in Fig. 31-6). During the remainder of the cardiac cycle, peripheral runoff will greatly exceed cardiac ejection. During diastole, of course, cardiac ejection equals zero. The resultant arterial blood volume decrement will cause volumes and pressures to fall from point A_2 back to point A_1.

If stroke volume is now doubled, while heart rate and peripheral resistance remain constant, the mean arterial pressure will be doubled, to $\bar{P}_B$ in Fig. 31-6. Thus the arterial pressure will now oscillate each heartbeat about this new value of the mean arterial pressure. A normal, vigorous heart will eject this greater stroke volume during a fraction of the cardiac cycle approximately equal to the fraction observed at the lower stroke volume. Therefore the volume increment, $V_4 - V_3$, will be a large fraction of the new stroke volume and hence will be approximately twice as great as the previous volume increment ($V_2 - V_1$). With a linear P_a:V_a curve, the greater volume increment will be reflected by a pulse pressure ($P_4 - P_3$) that will be approximately twice as great as the original pulse pressure ($P_2 - P_1$). With a rise in both mean and pulse pressures, it is evident from inspection of Fig. 31-6 that the rise in systolic pressure (from P_2 to P_4) must exceed the rise in diastolic pressure (from P_1 to P_3).

Arterial capacitance. To assess arterial capacitance as a determinant of pulse pressure, the relative effects of the same volume increment ($V_2 - V_1$ in Fig. 31-7) in a young person (curve *A*) and in an elderly person (curve *B*) will be compared. Let cardiac output and total peripheral resistance be the same in both cases; therefore $\bar{P}_a$ will be the same. It is apparent from Fig. 31-7 that the same volume increment ($V_2 - V_1$) will result in a greater pulse pressure ($P_4 - P_1$) in the less distensible arteries of the elderly individual than in the more compliant arteries of the young one ($P_3 - P_2$). For the reasons enunciated on p. 486, this will impose a greater work load on the left ventricle of the elderly than of the young person, even if the stroke volumes, total peripheral resistances (TPR), and mean arterial pressures are equivalent.

Fig. 31-8 displays the effects of changes in arterial capacitance and in peripheral resistance, R_p, on the arterial pressure in an isolated cat heart preparation. As the capacitance was reduced from 43 to 14 to 3.6 units, the pulse pressure increased significantly. In this preparation, the stroke volume decreased as the capacitance was diminished. This accounts for the failure of the mean arterial pressure to remain constant at the different levels of arterial capacitance. The effects of changes in peripheral resistance in this same preparation are described in the next section.

Total peripheral resistance and arterial diastolic pressure. It is often stated that increased TPR affects primarily the level of the diastolic arterial pressure. The validity of such an assertion deserves close scrutiny. First, let TPR be increased in an individual with a linear P_a:V_a curve, as depicted in Fig. 31-9, *A*. If heart rate and stroke volume remain constant, then an increase in TPR will evoke a proportionate increase in $\bar{P}_a$ (from P_2 to P_5). If the volume increments ($V_2 - V_1$ and $V_4 - V_3$) are equal at both levels of TPR, then the pulse pressure ($P_3 - P_1$ and $P_6 - P_4$) will also be equal. Hence systolic (P_6) and diastolic (P_4) pressures will have been

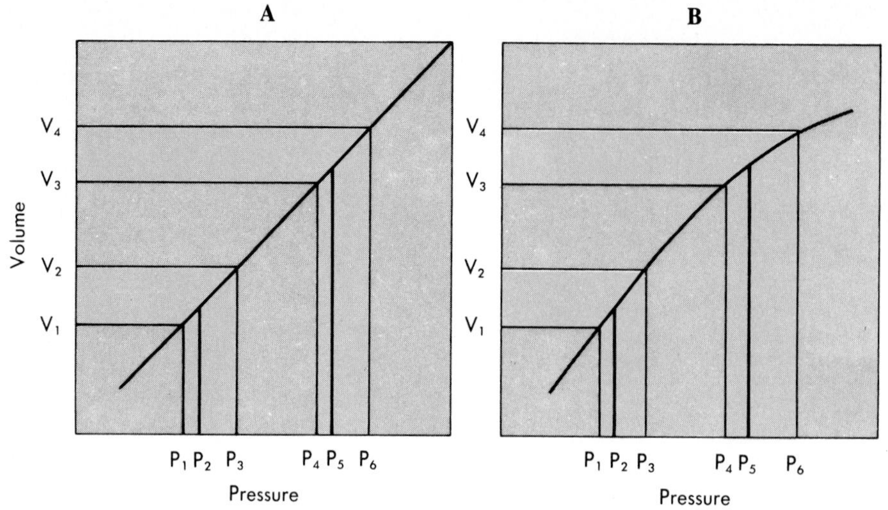

Fig. 31-9. Effect of a change in TPR (volume increment remaining constant) on pulse pressure when the pressure-volume curve for the arterial system is rectilinear, **A,** or curvilinear, **B.**

elevated by exactly the same amounts from their respective control levels (P_3 and P_1).

Chronic hypertension, a condition characterized by a persistent elevation of TPR, occurs more commonly in middle-aged and elderly individuals than in younger persons. The $P_a:V_a$ curve for a hypertensive patient would therefore possess the configuration shown in Fig. 31-9, *B*, which resembles the curves in Fig. 31-3 for older individuals.

The type of curve in Fig. 31-9, *B,* reveals that C_a is less at higher than at lower pressures. As before, if cardiac output remains constant, an increase in TPR would cause a proportionate rise in $\overline{P}_a$ (from P_2 to P_5). For equivalent increases in TPR, the elevation of pressure from P_2 to P_5 will be the same in Fig. 31-9, *A* and *B*, for reasons discussed on p. 489. Assuming the volume increment ($V_4 - V_3$ in Fig. 31-9, *B*) at elevated TPR to be equal to the control increment ($V_2 - V_1$), it is evident that the pulse pressure ($P_6 - P_4$) in the hypertensive range will greatly exceed that ($P_3 - P_1$) which prevails at normal pressure levels. In other words, a given volume increment will produce a greater pressure increment when the arteries are more rigid than when they are more compliant. Hence the rise in systolic pressure ($P_6 - P_3$) will greatly exceed the increase in diastolic pressure ($P_4 - P_1$).

These hypothetical changes in arterial pressure closely resemble those actually seen in patients with hypertension. Diastolic pressure is indeed elevated in such individuals, but ordinarily not more than 10 to 40 mm Hg above the average normal level of 80 mm Hg, whereas it is not uncommon for systolic pressures to be elevated by 50 to 150 mm Hg above the average normal level of 120 mm Hg. The combination of an increased resistance and diminished arterial capacitance would be represented in Fig. 31-8 by a shift in direction from the top left panel to the bottom right panel; that is, both the mean pressure and the pulse pressure would be increased significantly. These results also coincide with the changes predicted by Fig. 31-9, *B*.

Peripheral Arterial Pressure Curves

The radial stretch of the ascending aorta brought about by left ventricular ejection initiates a pressure wave that is propagated down the aorta and its branches. The pressure wave travels with a finite velocity that is considerably faster than the actual forward movement of the blood itself. It is this propagated pressure wave that one perceives by palpating the radial or any other peripheral artery.

The velocity of transmission of the pressure wave varies inversely with the vascular capacitance. Accurate measurement of the transmission velocity has provided valuable information about the elastic characteristics of the arterial tree. In general, transmission velocity increases with age, confirming the observation that the arteries become less compliant with advancing age (Fig. 31-3). Also, velocity increases progressively as the pulse wave travels from the ascending aorta toward the periphery. This indicates that vascular capacitance is less in the more distal than in the more proximal portions of the arterial system, a fact that also has been confirmed by direct measurement.

The arterial pressure contour becomes distorted as the wave is transmitted down the arterial system; the changes in configuration of the pulse with distance are shown in Fig. 31-10. Aside from the increasing delay in the time of onset of the initial pressure rise, three

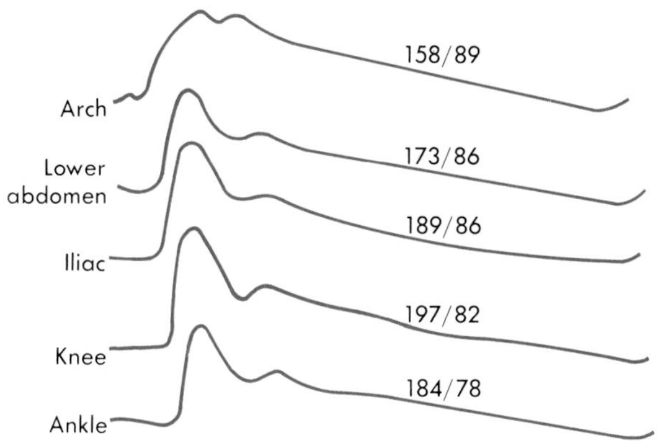

■ **Fig. 31-10.** Arterial pressure curves recorded from various sites in an anesthetized dog. (From Remington, J.W., and O'Brien, L.J.: Am. J. Physiol. **218:**437, 1970.)

major changes occur in the arterial pulse contour as the pressure wave travels away from the aortic arch. First, the high-frequency components of the pulse, such as the incisura (that is, the notch that appears at the end of ventricular ejection), are damped out and soon disappear. Second, the systolic portions of the pressure wave become narrowed and elevated. In the curves shown in Fig. 31-10, the systolic pressure at the level of the knee is 39 mm Hg greater than that recorded in the aortic arch. Third, a hump may become prominent on the diastolic portion of the pressure wave. These changes in contour of the pulse wave are pronounced in young individuals, but the alterations diminish with age. In elderly patients with less compliant arteries the pulse wave may be transmitted virtually unchanged from the ascending aorta to the periphery.

The damping of the high-frequency components of the arterial pulse is largely ascribable to the viscoelastic properties of the arterial walls. The precise mechanism for the peaking of the pressure wave is controversial. Probably several factors contribute, including (1) reflection, (2) tapering, (3) resonance, and (4) changes in transmission velocity with pressure level. Relative to the first of these mechanisms, whenever significant changes in configuration or in dimensions occur (such as at points of branching), pressure waves are reflected backward. Hence pressure at any time and location is determined by the algebraic summation of an antegrade incident wave and retrograde reflected waves. About 80% of the incident wave is reflected back from the peripheral bed at normal levels of peripheral resistance. The second factor, tapering, alters the pulse contour because in proceeding distally along the aorta or any other large artery, the lumen progressively narrows beyond each successive branch. A pressure wave becomes amplified as it progresses down a tapered tube. With respect to resonance, the arterial tree resonates at certain frequencies, and other frequencies are effectively damped. Finally, as stated previously, transmission velocity varies inversely with arterial capacitance. Furthermore, capacitance varies inversely with pressure level (Fig. 31-3). Hence the points on the pressure curve at the higher levels of pressure tend to travel faster than those at lower pressure levels. Thus the peak of the arterial pressure curve tends to catch up with the beginning, or ''foot,'' of the same curve. This contributes to the peaking and narrowing of the curve in more distal vessels such as the femoral artery. Reflection and resonance also account for the diastolic humps on these same peripheral curves.

■ *Blood Pressure Measurement in Humans*

In hospital intensive care units, needles or catheters may be introduced into peripheral arteries of patients, and arterial blood pressure can then be measured *directly* by means of strain gauges. Ordinarily, however, the blood pressure is estimated *indirectly* by means of a *sphygmomanometer*. This instrument consists of an inextensible cuff containing an inflatable bag. The cuff is wrapped around the extremity (usually the arm, occasionally the thigh) so that the inflatable bag lies between the cuff and the skin, directly over the artery to be compressed. The artery is occluded by inflating the bag, by means of a rubber squeeze bulb, to a pressure in excess of arterial systolic pressure. The pressure in the bag is measured by means of a mercury manometer or an aneroid manometer. Pressure is released from the bag at a rate of 2 or 3 mm Hg per second by means of a needle valve in the inflating bulb.

The practitioner listens with a stethoscope applied to the skin of the antecubital space over the brachial artery. While the pressure in the bag exceeds the systolic pressure, the brachial artery is occluded and no sounds are heard. When the inflation pressure falls just below the systolic level (point *A* in Fig. 31-11), the small spurt of blood escapes through the cuff and slight tapping sounds (called *Korotkoff sounds*) are heard with each heartbeat. The pressure at which the first sound is detected represents the systolic pressure. It usually corresponds closely with the directly measured systolic pressure and it exceeds by a few millimeters of mercury the pressure. As inflation pressure continues to fall, more blood escapes under the cuff per beat and the sounds become louder thuds. As the inflation pressure approaches the diastolic level, the Korotkoff sounds become muffled. As they fall just below the diastolic level (point *B* in Fig. 31-11), the sounds disappear; this indicates the diastolic pressure. The origin of the Korotkoff sounds is related to the spurt of blood passing un-

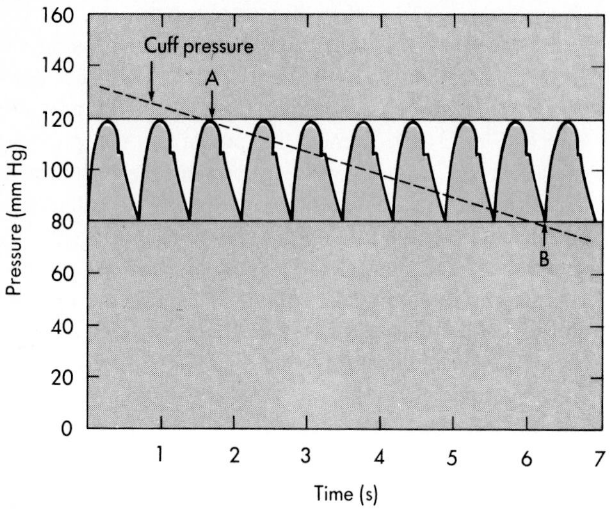

■ **Fig. 31-11.** Principle of measurement of arterial blood pressure with a sphygmomanometer. The oblique line represents pressure in the inflatable bag in the cuff. At cuff pressures greater than the systolic pressure (to the left of *A*), no blood progresses beyond the cuff and no sounds can be detected below the cuff. At cuff pressures between the systolic and diastolic levels (between *A* and *B*), spurts of blood traverse the arteries under the cuff and produce the Korotkoff sounds. At cuff pressures below the diastolic pressure (to the right of *B*), arterial flow past the region of the cuff is continuous, and no sounds are audible.

der the cuff and meeting a static column of blood; the impact and turbulence generate audible vibrations. Once the inflation pressure is less than the diastolic pressure, flow is continuous in the brachial artery and sounds are no longer heard.

■ Bibliography

Journal articles

Farrar, D.J., et al.: Aortic pulse wave velocity, elasticity, and composition in a nonhuman primate model of atherosclerosis, Circ. Res. **43:**52, 1978.

Finkelstein, S.M., and Collins, V.R.: Vascular hemodynamic impedance measurement, Prog. Cardiovasc. Dis. **24:**401, 1982.

Murgo, J.P., et al.: Aortic input impedance in normal man: relationship to pressure wave forms, Circulation **62:**105, 1980.

Nichols, W.W., and Pepine, C.J.: Left ventricular afterload and aortic input impedance: implications of pulsatile blood flow, Prog. Cardiovasc. Dis. **24:**293, 1982.

O'Rourke, M.F.: Vascular impedance in studies of arterial and cardiac function, Physiol. Rev. **62:**570, 1982.

Pepine, C.J., and Nichols, W.W.: Aortic input impedance in cardiovascular disease, Prog. Cardiovasc. Dis. **24:**307, 1982.

Simon, A.C., et al.: An evaluation of large arteries' compliance in man, Am. J. Physiol. **237:**H550, 1979.

Stettler, J.C., Niederer, P., and Anliker, M.: Theoretical analysis of arterial hemodynamics including the influence of bifurcations, Ann. Biomed. Eng. **9:**145, 1981.

Van den Bos, G.C., et al.: Reflection in the systemic arterial system: effects of aortic and carotid occlusion, Cardiovasc. Res. **10:**565, 1976.

Books and monographs

Bauer, R.D., and Busse, R., editors: Arterial system: dynamics, control theory and regulation, Heidelberg, 1978, Springer-Verlag.

Dobrin, P.B.: Vascular mechanics. In Handbook of physiology; Section 2: The cardiovascular system—peripheral circulation and organ blood flow, vol. III, Bethesda, Md., 1983, American Physiological Society.

Milnor, W.R.: Hemodynamics, Baltimore, 1982, Williams & Wilkins.

Noordergraaf, A.: Circulatory system dynamics, New York, 1978, Academic Press, Inc.

Taylor, D.E.M., and Stevens, A.L., editors: Blood flow: theory and practice, New York, 1983, Academic Press, Inc.

CHAPTER

32

The Microcirculation and Lymphatics

The entire circulatory system is geared to supply the body tissues with blood in amounts commensurate with their requirements for oxygen and nutrients. The capillaries, consisting of a single layer of endothelial cells, permit rapid exchange of water and solutes with interstitial fluid. The muscular arterioles, which are the major *resistance vessels,* regulate regional blood flow to the capillary beds, and the venules and veins serve primarily as collecting channels and storage, or *capacitance, vessels.*

The arterioles, which range in diameter from about 5 to 100 μm, have a thick smooth muscle layer, a thin adventitial layer, and an endothelial lining (see Fig. 26-1). The arterioles give rise directly to the capillaries (5 to 10 μm diameter) or in some tissues to *metarterioles* (10 to 20 μm diameter), which then give rise to capillaries (Fig. 32-1). The metarterioles can serve either as thoroughfare channels to the venules, bypassing the capillary bed, or as conduits to supply the capillary bed. There are often cross connections between arterioles and between venules, as well as in the capillary network. Arterioles that give rise directly to capillaries regulate flow through their cognate capillaries by constriction or dilation. At the points of origin of the capillaries in some tissues there is a small cuff of smooth muscle called the *precapillary sphincter* that controls blood flow through the cognate capillaries (Fig. 32-1). The capillaries form an interconnecting network of tubes of different lengths with an average length of 0.5 to 1 mm.

Capillary distribution varies from tissue to tissue. In metabolically active tissues, such as cardiac and skeletal muscle and glandular structures, capillaries are numerous, whereas in less active tissues, such as subcutaneous tissue or cartilage, *capillary density* is low. Also, all capillaries are not of the same diameter, and since some capillaries have diameters less than that of the erythrocytes, it is necessary for the cells to become temporarily deformed in their passage through these capillaries. Fortunately, the normal red cells are quite flexible and readily change their shape to conform with that of the small capillaries.

Blood flow in the capillaries is not uniform and depends chiefly on the contractile state of the arterioles and, where present, precapillary sphincters. The average velocity of blood flow in the capillaries is approximately 1 mm/s; however, it can vary from zero to several millimeters per second in the same vessel within a brief period. These changes in capillary blood flow may be of random type or may show rhythmical oscillatory behavior of different frequencies that are caused by contraction and relaxation *(vasomotion)* of the precapillary vessels. This vasomotion is to some extent an intrinsic contractile behavior of the vascular smooth muscle and is independent of external input. Furthermore, changes in *transmural pressure* (intravascular minus extravascular pressure) influence the contractile state of the precapillary vessels. An increase in transmural pressure, whether produced by an increase in venous pressure or by dilation of arterioles, results in contraction of the terminal arterioles at the points of origin of the capillaries, whereas a decrease in transmural pressure elicits precapillary vessel relaxation. In addition, humoral and possibly neural factors also affect vasomotion. For example, when the precapillary sphincters contract in response to increased transmural pressure, the contractile response can be overridden and vasomotion abolished. This effect is accomplished by metabolic (humoral) factors (p. 511) when the oxygen supply becomes too low for the requirements of the parenchymal tissue, as occurs in muscle during exercise.

Although reduction of transmural pressure will induce relaxation of the terminal arterioles, blood flow through the capillaries obviously cannot increase if the reduction in intravascular pressure is caused by severe constriction of the parent arterioles or metarterioles. Large arterioles and metarterioles also exhibit vasomo-

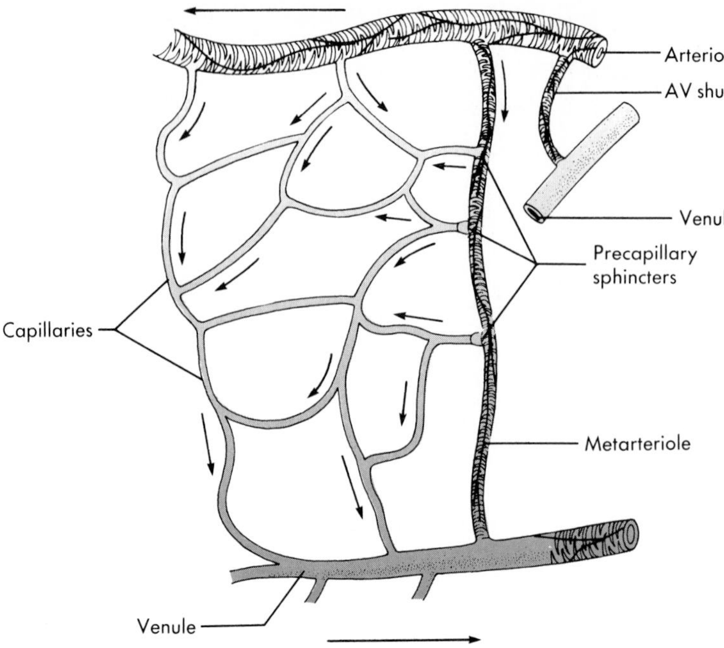

■ **Fig. 32-1.** Schematic drawing of the microcirculation. The circular structures on the arteriole and venule represent smooth muscle fibers, and the branching solid lines represent sympathetic nerve fibers. The arrows indicate the direction of blood flow.

tion, but in the contraction phase they usually do not completely occlude the lumen of the vessel and arrest blood flow as may occur with contraction of the terminal arterioles and precapillary sphincters. Thus flow rate may be altered by arteriolar and metarteriolar vasomotion.

Since blood flow through the capillaries provides for exchange of gases and solutes between blood and tissue, it has been termed *nutritional flow,* whereas blood flow that bypasses the capillaries in traveling from the arterial to the venous side of the circulation has been termed *nonnutritional,* or *shunt flow* (Fig. 32-1). In some areas of the body (fingertips) true arteriovenous shunts exist (p. 544). However, in many tissues, such as muscle, evidence of anatomical shunts is lacking. Nevertheless, nonnutritional flow does occur and can be demonstrated in muscle by stimulation of the adrenergic sympathetic fibers to the muscle vessels during constant flow perfusion or by stimulation of the sympathetic cholinergic fibers to muscle (p. 516). With adrenergic sympathetic fiber stimulation at constant flow, the capillary surface area available for solute exchange is reduced, as measured by a decrease in intraarterially administered ^{86}Rb uptake by the tissue. With sympathetic cholinergic fiber stimulation, blood flow increases but tissue clearance (washout) of radioactive isotope deposited in the tissue is unchanged. With this technique for the estimation of nutritional blood flow, a radioactive substance such as $^{22}Na^{+}$ is injected into the tissue in a small volume of fluid to minimize tissue damage. The rapidity with which it is removed by the perfusing blood is indicative of the rate of nutritional blood flow. Increased flow through open capillaries in the absence of morphological arteriovenous shunts has been termed a *physiological shunt.* It is the result of a greater flow of blood through open capillaries (shunts) with either no change or an increase in the number of closed capillaries. Adrenergic sympathetic fiber stimulation elicits arteriolar constriction, but when flow is set constant by a pump, all the perfusing blood passes through the decreased number of open channels during the sympathetic nerve stimulation. Cholinergic sympathetic fiber stimulation dilates arterioles, and the increased volume flow of blood caused by the decrease in vascular resistance occurs through the same capillaries that were open before nerve stimulation. In tissues that have metarterioles, shunt flow may be continuous from arteriole to venule during low metabolic activity when many precapillary vessels are closed. When metabolic activity increases in such tissues and precapillary vessels open, blood passing through the metarterioles is readily available for capillary perfusion.

The true capillaries are devoid of smooth muscle and are therefore incapable of active constriction. Nevertheless, the endothelial cells that form the capillary wall contain actin and myosin and can alter their shape in response to certain chemical stimuli. Evidence is lacking, however, that changes in endothelial cell shape regulate blood flow through the capillaries. Hence changes in capillary diameter are passive and are caused by alterations in precapillary and postcapillary resistance.

For many years it was believed that capillaries were inert and merely served as barriers to blood cells and large molecules such as plasma proteins. Recently, however, the metabolic activity of the endothelial cells and the interaction between blood components (particularly platelets) and the endothelial cells has been recognized. For example, stimulation of endothelium with acetylcholine releases a vascular smooth muscle relaxant, *endothelial-derived relaxing factor (EDRF)*. In an intact perfused arterial vessel, acetylcholine produces vasodilation (indirect effect by EDRF release), but in a vessel that is stripped of endothelium, acetylcholine elicits vasoconstriction (direct effect on vascular smooth muscle). The identity of EDRF is not yet known, but evidence suggests a nitric oxide–like substance.

The thin-walled capillaries can withstand high internal pressures without bursting because of their narrow lumen. This can be explained in terms of the law of Laplace and is illustrated in the following comparison of wall tension of a capillary with that of the aorta:

	Aorta	*Capillary*
Radius (r)	1.5 cm	5×10^{-4} cm
Height of Hg column (h)	10 cm Hg	2.5 cm Hg
ρ	13.6 g/cm^3	13.6 g/cm^3
g	980 cm/s^2	980 cm/s^2
P	$10 \times 13.6 \times 980 = 1.33 \times 10^5$ $dyne/cm^2$	$2.5 \times 13.6 \times 980 = 3.33 \times 10^4$ $dyne/cm^2$
w	0.2 cm	1×10^{-4} cm
T = Pr	$(1.33 \times 10^5)(1.5) = 2 \times 10^5$ dyne/cm	$(3.33 \times 10^4)(5 \times 10^{-4}) =$ 16.7 dyne/cm
$\sigma = \frac{Pr}{w}$	$\frac{2 \times 10^5}{0.2} = 1 \times 10^6$ $dyne/cm^2$	$\frac{16.7}{1 \times 10^{-4}} = 1.67 \times 10^5$ $dyne/cm^2$

The Laplace equation is

$$T = Pr \quad (1)$$

where

T = Tension in the vessel wall
P = Transmural pressure
r = Radius of the vessel

Wall tension is the force per unit length tangential to the vessel wall that opposes the distending force (Pr) that tends to pull apart a theoretical longitudinal slit in the vessel (Fig. 32-2). Transmural pressure is essentially equal to intraluminal pressure, since extravascular pressure is negligible. The Laplace equation applies to very thin wall vessels, such as capillaries. Wall thickness must be taken into consideration when the equation is applied to thick wall vessels, such as the aorta. This is done by dividing Pr (pressure × radius) by wall thickness (w). The equation now becomes

$$\sigma \text{ (wall stress)} = Pr/w \quad (2)$$

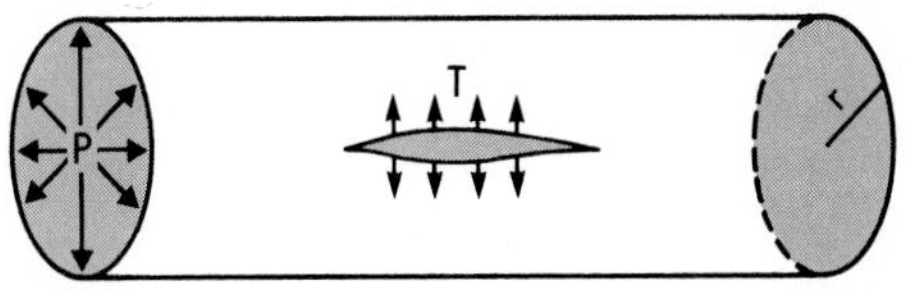

Fig. 32-2. Diagram of a small blood vessel to illustrate the law of Laplace: T = Pr, where *P* = intraluminal pressure, *r* = radius of the vessel, and *T* = wall tension as the force per unit length tangential to the vessel wall, tending to pull apart a theoretical longitudinal slit in the vessel.

To convert pressure in mm Hg (height of Hg column) to dynes per square centimeter, P = hpg, where h = the height of Hg column in centimeters, p = the density of Hg in g/cm^3, g = gravitational acceleration in cm/s^2, σ = force per unit area, and w = wall thickness.

Thus at normal aortic and capillary pressures the wall tension of the aorta is about 12,000 times greater than that of the capillary. In a person standing quietly, capillary pressure in the feet may reach 100 mm Hg. Under such conditions capillary wall tension increases to 66.5 dynes/cm, a value that is still only one three-thousandth that of the wall tension in the aorta at the same internal pressure. However, σ (stress), which takes wall thickness into consideration, is only about tenfold greater in the aorta than in the capillary.

In addition to providing an explanation for the ability of capillaries to withstand large internal pressures, the preceding calculations also point out that in dilated vessels, wall tension increases even when internal pressure remains constant and may, under certain circumstances (for example, aneurysm of the aorta), be an important factor in rupture of the vessel. The foregoing equation also indicates that as the wall of the vessel becomes thicker, the wall stress decreases. In *hypertension* (high blood pressure) the arterial vessel walls thicken (hypertrophy of the vascular smooth muscle), thereby minimizing the arterial wall stress and hence the possibility of vessel rupture.

The diameter of the resistance vessels is determined by the balance between the contractile force of the vascular smooth muscle and the distending force produced by the intraluminal pressure. The greater the contractile activity of the vascular smooth muscle of an arteriole, the smaller its diameter, until a point is reached, in the case of small arterioles, when complete occlusion of the vessel occurs, in part because of infolding of the endothelium, and the cells trapped in the vessel. With progressive reduction in the intravascular pressure, vessel diameter decreases, as does tension in the vessel wall (law of Laplace). When perfusion pressure is reduced, a point is reached where blood flow ceases even though there is still a positive pressure gradient. This phenomenon has been referred to as the "critical closing pressure," and its mechanism is still controversial. This critical closing pressure is low when vasomotor activity is reduced by inhibition of sympathetic nerve activity to the vessel, and it is increased when vasomotor tone is enhanced by activation of the vascular sympathetic nerve fibers. It has been suggested that flow stops because of vessel collapse when vascular smooth muscle contractile stress exceeds the stress associated with vessel radius, wall thickness, and intraluminal pressure (law of Laplace).

Transcapillary Exchange

Solvent and solute move across the capillary endothelial wall by three processes: diffusion, filtration, and via endothelial vesicles (pinocytosis). By far the greatest number of molecules traverse the capillary endothelium by diffusion.

Capillary Pores

The permeability of the capillary endothelial membrane is not the same in all body tissues. For example, the liver capillaries are quite permeable, and albumin escapes at a rate severalfold greater than from the less permeable muscle capillaries. Also, there is not uniform permeability along the whole capillary; the venous ends are more permeable than the arterial ends, and permeability is greatest in the venules. The greater permeability at the venous end of the capillaries and in the venules is due to the greater number of pores in these regions of the microvessels.

The sites where filtration occurs have been a controversial subject for a number of years. A fraction of the water flows through the capillary endothelial cell membranes. However, most investigators believe that water flows mainly through apertures in the endothelial wall

A

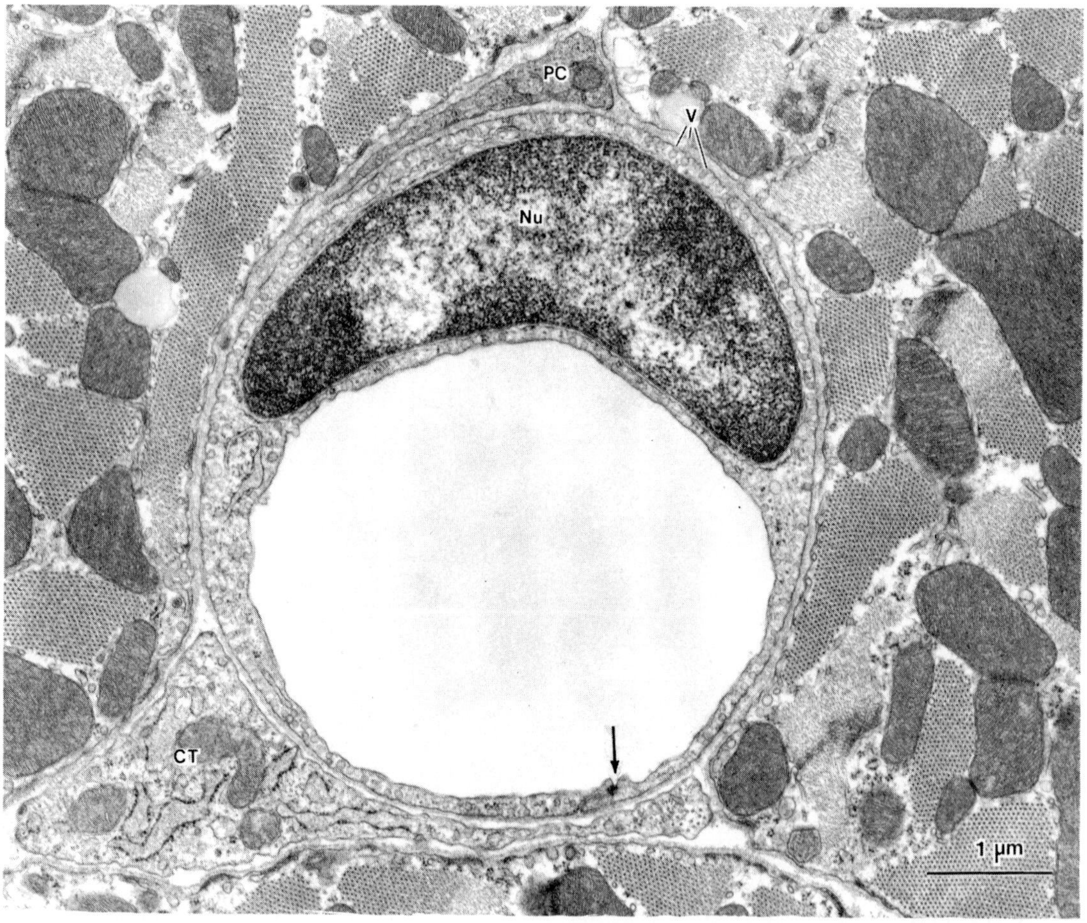

■ Fig. 32-3. **A,** Cross-sectioned capillary in mouse ventricular wall. Luminal diameter is approximately 4 μm. In this thin section, the capillary wall is formed by a single endothelial cell (*Nu,* endothelial nucleus), which forms a junctional complex *(arrow)* with itself. The thin pericapillary space is occupied by a pericyte *(PC)*—believed to be equivalent to vascular smooth muscle cells of larger vessels—and a connective tissue *(CT)* cell ("fibroblast"). Note the numerous endothelial vesicles *(V)*. (Micrographs courtesy Dr. Michael S. Forbes.) *Continued.*

of the capillaries (Figs. 32-3 and 32-4). Calculations based on the transcapillary movement of solutes of small molecular size led to the prediction of pore diameters of about 4 nm. However, electron microscopy failed to reveal pores, and the clefts at the junctions of endothelial cells appeared to be fused at the tight junctions (Figs. 32-3 and 32-4). However, studies on cardiac and skeletal muscle with horseradish peroxidase, a protein with a molecular weight of 40,000, have demonstrated that many of the clefts between adjacent endothelial cells are open. Electron microscopy revealed filling of the clefts with peroxidase from the lumen side of the capillaries with a gap at the narrowest point of about 4 nm providing morphological support of the physiological evidence for the existence of capillary pores. The clefts (pores) are sparse and represent only about 0.02% of the capillary surface area. In cerebral capillaries, where a blood-brain barrier to many small molecules exists, peroxidase studies do not reveal any pores. Some studies have failed to reveal the presence of interendothelial pores, even with microperoxidase (molecular weight of 1900), and transcapillary movement of solute (large and small molecules) is thought to occur through channels formed by the fusion of vesicles (vesicular channels) across the endothelial cells (Fig. 32-4). The basement membrane, a layer of fine fibrillar material around the capillaries, retards the passage of large molecules (greater than 10 nm radius).

In addition to clefts, some of the more porous capillaries (for example, in kidney, intestine) contain fenestrations (Fig. 32-4) 20 to 100 nm wide, whereas others (such as in the liver) have a discontinuous endothelium (Fig. 32-4). The fenestrations that appear to be sealed by a thin diaphragm are quite permeable to horseradish peroxidase and a number of other tracers. Hence larger molecules can penetrate capillaries with fenestrations or gaps caused by discontinuous endothelium that can pass through the intercellular clefts of the endothelium.

Diffusion

Under normal conditions only about 0.06 ml of water per minute moves back and forth across the capillary wall per 100 g of tissue as a result of filtration and absorption, whereas 300 ml of water per minute transfers across the endothelium per 100 g of tissue by diffusion, a 5000-fold difference. Relating filtration and diffusion to blood flow, we find that about 2% of the plasma passing through the capillaries is filtered, whereas the diffusion of water is 40 times greater than the rate that it is brought to the capillaries by blood flow. The trans-

B

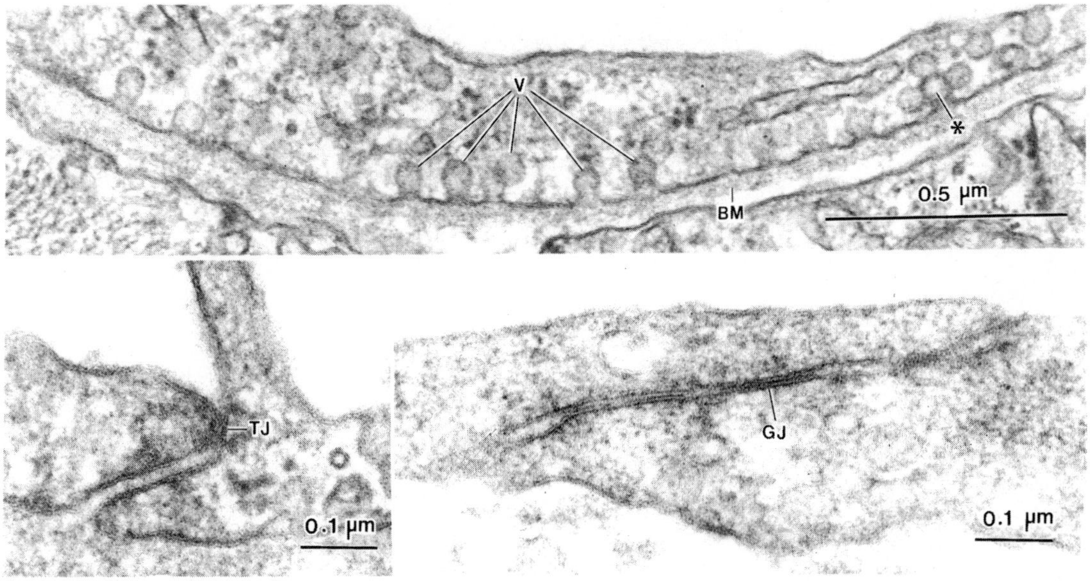

C D

Fig. 32-3, cont'd. **B,** Detail of endothelial cell in panel **A,** showing plasmalemmal vesicles *(V)* that are attached to the endothelial cell surface. These vesicles are especially prominent in vascular endothelium and are believed to be involved in transport of substances across the blood vessel wall. Note the complex alveolar vesicle (*). *BM,* Basement membrane. **C,** Junctional complex in a capillary of mouse heart. "Tight" junctions *(TJ)* typically form in these small blood vessels and appear to consist of fusions between apposed endothelial cell surface membranes. **D,** Interendothelial junction in a muscular artery of monkey papillary muscle. Although tight junctions similar to those of capillaries are found in these large blood vessels, extensive junctions that resemble gap junctions in the intercalated disks between myocardial cells (Figs. 3-3 and 3-4) often appear in arterial endothelium (example shown at *GJ*).

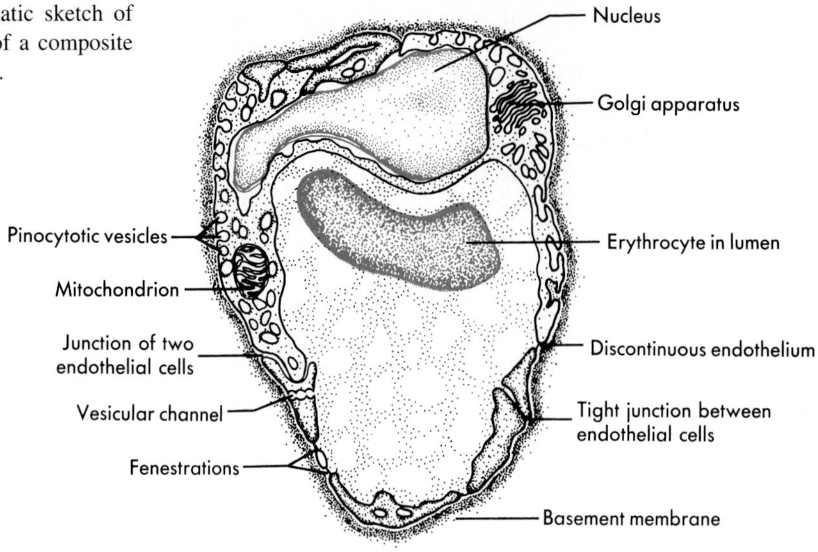

Fig. 32-4. Diagrammatic sketch of an electron micrograph of a composite capillary in cross section.

capillary exchange of solutes is also primarily governed by diffusion. Thus diffusion is the key factor in providing exchange of gases, substrates, and waste products between the capillaries and the tissue cells.

The process of diffusion is described by Fick's law:

$$J = -DA\frac{dc}{dx} \qquad (3)$$

where

- J = Quantity of a substance moved per unit time (t)
- D = Free diffusion coefficient for a particular molecule (the value is inversely related to the square root of the molecular weight)
- A = Cross-sectional area of the diffusion pathway
- $\frac{dc}{dx}$ = Concentration gradient of the solute

Fick's law is also expressed as:

$$J = -PS(C_o - C_i) \qquad (4)$$

Where

- P = Capillary permeability of the substance
- S = Capillary surface area
- C_i = Concentration of the substance inside the capillary
- C_o = Concentration of the substance outside the capillary

Hence the PS produce provides a convenient expression of available capillary surface, since permeability is rarely altered under physiological conditions

In the capillaries, diffusion of lipid-insoluble molecules is not free but is restricted to the pores whose mean size can be calculated by measurement of the diffusion rate of an uncharged molecule whose free diffusion coefficient is known. Movement of solutes across the endothelium is quite complex and involves corrections for attractions between solute and solvent molecules, interactions between solute molecules, pore configuration, and charge on the molecules relative to charge on the endothelial cells. It is not simply a question of random thermal movements of molecules down a concentration gradient.

For small molecules, such as water, NaCl, urea, and glucose, the capillary pores offer little restriction to diffusion (low *reflection coefficient*), and diffusion is so rapid that the mean concentration gradient across the capillary endothelium is extremely small. Water passes directly through the endothelial cells and through the capillary pores between endothelial cells. With lipid-insoluble molecules of increasing size, diffusion through muscle capillaries becomes progressively more restricted, until diffusion becomes minimal with molecules of a molecular weight above about 60,000. With small molecules the only limitation to net movement across the capillary wall is the rate at which blood flow transports the molecules to the capillary *(flow limited)*.

When transport across the capillary is flow limited, the concentration of a small molecular solute in the blood reaches equilibrium with its concentration in the interstitial fluid near the origin of the capillary from the cognate arteriole. If a small molecular inert tracer is infused intraarterially, its concentration falls to negligible levels near the arterial end of the capillary (Fig. 32-5, *A*). A somewhat larger molecule moves further along the capillary before reaching an insignificant concentration in the blood, and the number of still larger molecules that enter the arterial end of the capillary and cannot pass through the capillary pores is the same as the number leaving the venous end of the capillary (Fig. 32-5, *A*). An increase in blood flow extends the detectable concentration farther down the capillary and increases the capillary diffusion capacity (rate of tissue uptake of the solute).

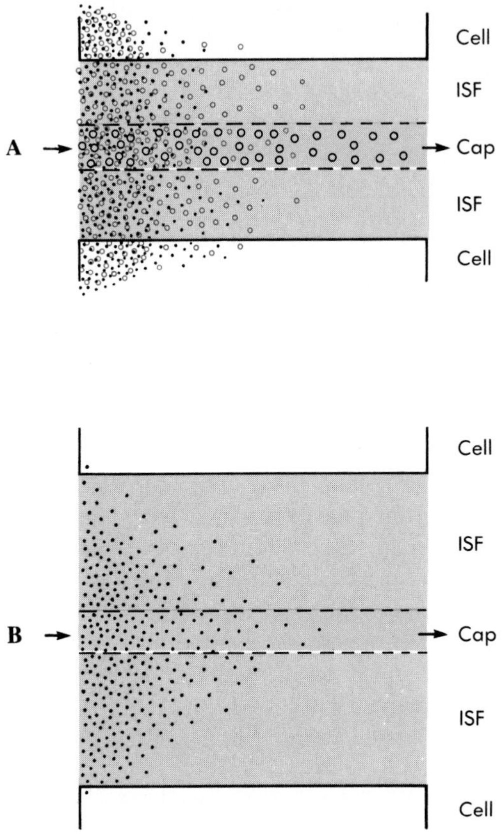

■ Fig. 32-5. Flow and diffusion-limited transport from capillaries *(Cap)* to tissue. **A,** Flow-limited transport. The smallest water-soluble inert tracer particles *(black dots)* reach negligible concentrations after passing only a short distance down the capillary. Larger particles *(colored circles)* with similar properties travel farther along the capillary before reaching insignificant intracapillary concentrations. Both substances cross the interstitial fluid *(ISF)* and reach the parenchymal tissue (cell). Because of their size, more of the smaller particles are taken up by the tissue cells. The largest particles *(black circles)* cannot penetrate the capillary pores and hence do not escape from the capillary pores and hence do not escape from the capillary lumen except by pinocytotic vesicle transport. An increase in the volume of blood flow or an increase in capillary density will increase tissue supply for the diffusible solutes. Note that capillary permeability is greater at the venous end of the capillary (also in the venule, not shown) because of the larger number of pores in this region. **B,** Diffusion-limited transport. When the distance between the capillaries and the parenchymal tissue is large, as a result of edema or low capillary density, diffusion becomes a limiting factor in the transport of solutes from capillary to tissue even at high rates of capillary blood flow.

With large molecules diffusion across the capillaries becomes the limiting factor *(diffusion limited)*. In other words, capillary permeability to a large molecular solute limits its transport across the capillary wall (Fig. 32-5, *A*). The rate of diffusion of small lipid-insoluble molecules is so rapid that the rate-limiting factor in blood-tissue exchange only can become important when long distances exist between capillaries (for example, tissue edema or very low capillary density) (Fig. 32-5, *B*). Furthermore, the rate of diffusion is uninfluenced by filtration in the direction opposite to the concentration gradient of the diffusible substance. In fact, filtration *or* absorption accelerates the movement of tracer ions from interstitial fluid to blood (tissue clearance). The reason for this enhanced tissue clearance is not known, but it may be the result of a stirring effect on the interstitial fluid or to changes in its structure (for example, gel to sol transformation or "canals" in a gel matrix).

Movement of lipid-soluble molecules across the capillary wall is not limited to capillary pores (only about 0.02% of the capillary surface), since such molecules can pass directly through the lipid membranes of the entire capillary endothelium. Consequently, lipid-soluble molecules move with great rapidity between blood and tissue. The degree of lipid solubility (oil-to-water partition coefficient) proves a good index of the ease of transfer to lipid molecules through the capillary endothelium.

Oxygen and carbon dioxide are both lipid soluble and readily pass through the endothelial cells. Calculations based on (1) the diffusion coefficient for O_2, (2) capillary density and diffusion distances, (3) blood flow, and (4) tissue O_2 consumption indicate that the O_2 supply of normal tissue at rest and during activity is not limited by diffusion or the number of open capillaries. Recent measurements of Po_2 and oxygen saturation of blood in the microvessels indicate that in many tissues O_2 saturation at the entrance of the capillaries has already decreased to a saturation of about 80% as a result of diffusion of O_2 from arterioles. Such studies also have shown that CO_2 loading and the resultant intravascular shifts in the oxyhemoglobin dissociation curve occur in the precapillary vessels. These findings reflect not only the movement of gas to respiring tissue at the precapillary level, but also the direct flux of O_2 and CO_2 between adjacent arterioles, venules, and possibly arteries and veins (countercurrent exchange). This exchange of gas represents a diffusional shunt of gas around the capillaries, and, at low blood flow rates, it may limit the supply of O_2 to the tissue.

■ *Capillary Filtration*

The direction and the magnitude of the movement of water across the capillary wall are determined by the algebraic sum of the hydrostatic and osmotic pressures existing across the membrane. An increase in intracapillary hydrostatic pressure favors movement of fluid from the vessel to the interstitial space, whereas an increase in the concentration of osmotically active parti-

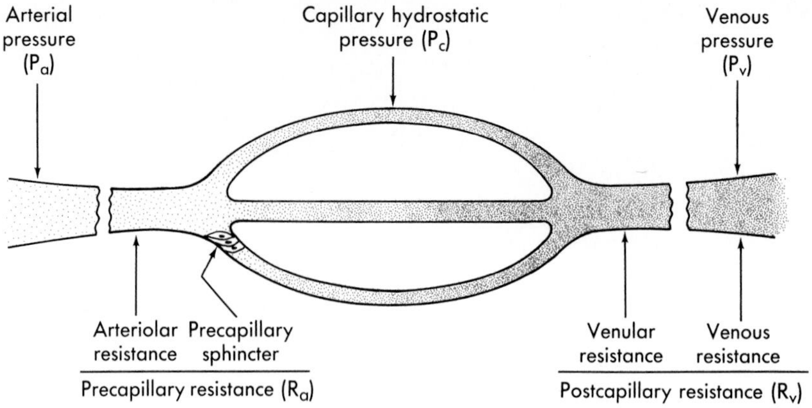

Fig. 32-6. Diagram of the terminal vascular bed, illustrating precapillary and postcapillary pressures and resistances used in the calculation of capillary hydrostatic pressure.

cles within the vessels favors movement of fluid into the vessels from the interstitial space.

Hydrostatic Forces

The hydrostatic pressure (blood pressure) within the capillaries is not constant and depends on the arterial pressure, the venous pressure, and the precapillary (arteriolar and precapillary sphincter) and postcapillary (venules and small veins) resistances. A gain in arterial or venous pressure increases capillary hydrostatic pressure, whereas a reduction in each has the opposite effect. An increase in arteriolar resistance or closure of precapillary sphincters reduces capillary pressure, whereas greater venous resistance (venules and veins) increases capillary pressure (Fig. 32-6).

The relationship of precapillary and postcapillary pressure and resistance and their effect on capillary pressure can be expressed by the equation:

$$P_c = \frac{(R_v/R_a)\,P_a + P_v}{1 + (R_v/R_a)} \qquad (5)$$

where

P_c = Capillary hydrostatic pressure
P_a = Arterial pressure
P_v = Venous pressure
R_a = Precapillary resistance (on the arterial side)
R_v = Postcapillary resistance (on the venous side)

A given increment in venous pressure produces a greater effect on capillary hydrostatic pressure than the same increment in arterial pressure, and about 80% of the increase in venous pressure is transmitted back to the capillaries. The important variable in the preceding equation is the ratio of postcapillary resistance to precapillary resistance (R_v/R_a), where R_a is greater than R_v (about 4 to 1). Hence an increase in one resistance (R_a) may be offset by a proportional increase in the other (R_v).

Despite the fact that capillary hydrostatic pressure is variable from tissue to tissue (even within the same tissue), average values, obtained from many direct measurements in human skin, are about 32 mm Hg at the arterial end of the capillaries and 15 mm Hg at the venous end of the capillaries at the level of the heart (Fig. 32-7). The hydrostatic pressure in capillaries of the lower extremities will be higher and that of capillaries in the head will be lower in the standing position. Measurements of pressure at the arteriolar and venous ends of the capillaries in different tissues in several mammals give values that are reasonably close to the ones found in human skin capillaries. This hydrostatic pressure is the principal force in filtration across the capillary wall.

Tissue pressure, or more specifically, interstitial fluid pressure (P_i) outside the capillaries, opposes capillary filtration, and it is $P_c - P_i$ that constitutes the driving force for filtration. The true value of P_i is still controversial. For years it was assumed to be close to zero in the normal (nonedematous) state. However, recent studies using perforated, plastic capsules implanted in the subcutaneous tissue or wicks inserted through the skin, indicate a negative P_i of from -1 to -7 mm Hg. If the pressures recorded by these techniques are representative of interstitial fluid pressure in undisturbed tissue, then the hydrostatic driving force for capillary filtration is greater than the value of P_c.

Osmotic Forces

The key factor that restrains fluid loss from the capillaries is the osmotic pressure of the plasma proteins—usually termed the *colloid osmotic pressure* or *oncotic pressure* (π_p). The total osmotic pressure of plasma is about 6000 mm Hg, whereas the oncotic pressure is only about 25 mm Hg. However, this small oncotic pressure plays an important role in fluid exchange across the capillary wall, because the plasma proteins are essentially confined to the intravascular space, whereas the electrolytes that are responsible for the ma-

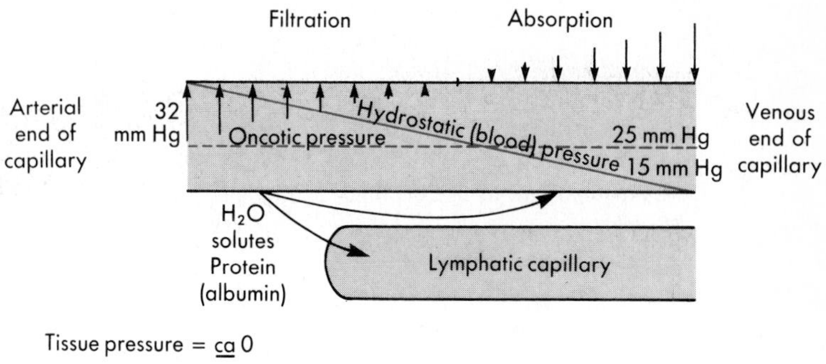

■ **Fig. 32-7.** Schematic representation of the factors responsible for filtration and absorption across the capillary wall and the formation of lymph.

jor fraction of plasma osmotic pressure are practically equal in concentration on both sides of the capillary endothelium. The relative permeability of solute to water influences the actual magnitude of the osmotic pressure. The *reflection coefficient* (σ) is the relative impediment to the passage of a substance through the capillary membrane. The reflection coefficient of water is zero and that of albumin (to which the endothelium is essentially impermeable) is 1. Filterable solutes have reflection coefficients between 0 and 1. Also, different tissues have different reflection coefficients for the same molecule, and therefore movement of a given solute across the endothelial wall varies with the tissue. The true oncotic pressure is defined by

$$\pi = \sigma RT\,(C_i - C_o) \qquad (6)$$

where

σ = Reflection coefficient
R = Gas constant
T = Absolute temperature
C_i and C_o = Solute (essentially albumin) concentration inside and outside the capillary

Of the plasma proteins, albumin is preponderate in determining oncotic pressure. The average albumin molecule (molecular weight 69,000) is approximately half the size of the average globulin molecule (molecular weight 150,000) and is present in almost twice the concentration as the globulins (4.5 vs. 2.5 g/100 ml of plasma). Albumin also exerts a greater osmotic force than can be accounted for solely on the basis of the number of molecules dissolved in the plasma, and therefore it cannot be completely replaced by inert substances of appropriate molecular size such as dextran. This additional osmotic force becomes disproportionately greater at high concentrations of albumin (as in plasma) and is weak to absent in dilute solutions of albumin (as in interstitial fluid). One reason for this behavior of albumin is its negative charge at the normal blood pH and the attraction and retention of cations (principally Na^+) in the vascular compartment (the *Gibbs-Donnan effect*). Furthermore, albumin binds a small number of chloride ions, which increases its negative charge and, hence, its ability to retain more sodium ions inside the capillaries. The small increase in electrolyte concentration of the plasma over that of the interstitial fluid produced by the negatively charged albumin enhances its osmotic force to that of an ideal solution containing a solute of molecular weight of 37,000. If albumin did indeed have a molecular weight of 37,000, it would not be retained by the capillary endothelium because of its small size, and obviously could not function as a counterforce to capillary hydrostatic pressure. If, however, albumin did not have an enhanced osmotic force, it would require a concentration of about 12 g of albumin/100 ml of plasma to achieve a plasma oncotic pressure of 25 mm Hg. Such a high albumin concentration would greatly increase blood viscosity and the resistance to blood flow through the vascular system. The other factors that contribute to the nonlinearity of the relationship of albumin concentration to osmotic force are not known. About 65% of plasma oncotic pressure is attributable to albumin, about 15% to the globulins, and the remainder to other ill-defined components of the plasma.

Small amounts of albumin escape from the capillaries and enter the interstitial fluid, where they exert a very small osmotic force (0.1 to 5 mm Hg). This force, π_i, is small because of the low concentration of albumin in the interstitial fluid and because at low concentrations the osmotic force of albumin becomes simply a function of the number of albumin molecules per unit volume of interstitial fluid.

■ *Balance of Hydrostatic and Osmotic Forces—Starling Hypothesis*

The relationship between hydrostatic pressure and oncotic pressure and the role of these forces in regulating fluid passage across the capillary endothelium were expounded by Starling in 1896 and constitute the *Starling hypothesis*. It can be expressed by the equation:

$$\text{Fluid movement} = k[P_c + \pi_i) - (P_i + \pi_p)] \qquad (7)$$

where

P_c = Capillary hydrostatic pressure
P_i = Interstitial fluid hydrostatic pressure
π_p = Plasma protein oncotic pressure
π_i = Interstitial fluid oncotic pressure
k = Filtration constant for the capillary membrane

Filtration occurs when the algebraic sum is positive, and absorption occurs when the algebraic sum is negative.

Classically, it has been thought that filtration occurs at the arterial end of the capillary and absorption at its venous end because of the gradient of hydrostatic pressure along the capillary. This is true for the idealized capillary as depicted in Fig. 32-7, but direct observations have revealed that many capillaries show filtration for their entire length, whereas others show only absorption. In some vascular beds (for example, the renal glomerulus) hydrostatic pressure in the capillary is high enough to result in filtration along the entire length of the capillary. In other vascular beds, such as in the intestinal mucosa, the hydrostatic and oncotic forces are such that absorption occurs along the whole capillary. As discussed earlier in this chapter, capillary pressure is quite variable and depends on several factors, the principal one being the contractile state of the precapillary vessel. In the normal steady state, arterial pressure, venous pressure, postcapillary resistance, interstitial fluid hydrostatic and oncotic pressures, and plasma oncotic pressure are relatively constant, and change in precapillary resistance is the determining factor with respect to fluid movement across the wall for any given capillary. Since water moves so quickly across the capillary endothelium, the hydrostatic and osmotic forces are nearly in equilibrium along the entire capillary. Hence filtration and absorption in the normal state occur at very small degrees of imbalance of pressure across the capillary wall. Only a small percentage (2%) of the plasma flowing through the vascular system is filtered, and of this about 85% is absorbed in the capillaries and venules. The remainder returns to the vascular system as lymph fluid along with the albumin that escapes from the capillaries.

In the lungs the mean capillary hydrostatic pressure is only about 8 mm Hg. Since the plasma oncotic pressure is 25 mm Hg, and interstitial fluid oncotic pressure has been estimated to be about 15 mm Hg, the net force slightly favors reabsorption. Only in pathological conditions, such as left ventricular failure or stenosis of the mitral valve, does pulmonary capillary hydrostatic pressure exceed plasma oncotic pressure. When this occurs, it may lead to pulmonary edema, a condition that can seriously interfere with gas exchange in the lungs.

Capillary Filtration Coefficient

The rate of movement of fluid across the capillary membrane (Q_f) depends not only on the algebraic sum of the hydrostatic and osmotic forces across the endothelium (ΔP), but also on the area of the capillary wall available for filtration (A_m), the distance across the capillary wall (Δx), the viscosity of the filtrate (η), and the filtration constant of the membrane (k). These factors may be expressed by the equation:

$$Q_f = \frac{kA_m\Delta P}{\eta\Delta x} \qquad (8)$$

The dimensions are units of flow per unit of pressure gradient across the capillary wall per unit of capillary surface area. This expression, which describes the flow of fluid through a membrane (pores), is essentially Poiseuille's law for flow through tubes (p. 477).

Since the thickness of the capillary wall and the viscosity of the filtrate are relatively constant, they can be included in the filtration constant, k, and if the area of the capillary membrane is not known, the rate of filtration can be expressed per unit weight of tissue. Hence the equation can be simplified to

$$Q_f = k_t\Delta P \qquad (9)$$

where k_t is the capillary filtration coefficient for a given tissue, and the units for Q_f are milliliters per minute per 100 g of tissue per millimeter of mercury.

The rates of filtration and absorption are determined by the rate of change in tissue weight or volume at different mean capillary hydrostatic pressures that are altered by adjustment of arterial and venous pressures. At the isogravimetric or isovolumic point (constant weight or constant volume, respectively, as continuously measured with an appropriate scale or volume recorder), the hydrostatic and osmotic forces are balanced across the capillary wall and there is neither net filtration nor absorption. An abrupt increase in arterial pressure will increase capillary hydrostatic pressure and fluid will move from the capillaries to the interstitial fluid compartment. Since the pressure increment, the weight increase of the tissue per unit time, and the total weight of the tissue are known, the capillary filtration coefficient (or k_t) in milliliters per minute per 100 g of tissue per millimeter of mercury can be calculated. With the isogravimetric and isovolumic techniques, it is assumed that 80% of the increments in venous pressure are transmitted back to the capillaries, that precapillary and postcapillary resistances are constant when venous pressure is changed, and that the weight or volume change that occurs immediately after raising venous pressure is the result of vascular distension and not filtration. These assumptions may not always be correct; nevertheless, the filtration coefficient constitutes a useful index of capillary permeability and surface area.

In any given tissue the filtration coefficient per unit area of capillary surface, and hence capillary permeability, is not changed by different physiological conditions, such as arteriolar dilation and capillary distension, or by such adverse conditions as hypoxia, hypercapnia, or reduced pH. With capillary injury (toxins, severe burns) capillary permeability increases greatly, as indicated by the filtration coefficient, and significant amounts of fluid and protein leak out of the capillaries into the interstitial space.

Since capillary permeability is constant under normal conditions, the filtration coefficient can be used to determine the relative number of open capillaries (total capillary surface area available for filtration in tissue). For example, increased metabolic activity of contracting skeletal muscle induces relaxation of precapillary resistance vessels with opening of more capillaries (*capillary recruitment,* resulting in an increased filtering space).

Some protein is apparently required to maintain the integrity of the endothelial membrane. If the plasma proteins are replaced by nonprotein colloids so as to give the same oncotic pressure, the filtration coefficient is doubled and edema occurs. However, if as little as 0.2% albumin is added, normal permeability is restored. One reasonable explanation is that the albumin binds to certain sites on the endothelial membrane and alters pore dimensions.

Disturbances in Hydrostatic-Osmotic Balance

Changes in arterial pressure per se may have little effect on filtration, since the change in pressure may be countered by adjustments of the precapillary resistance vessels (termed *autoregulation* [p. 510]), so that hydrostatic pressure in the open capillaries remains the same. However, with severe reduction in arterial pressure, as may occur in hemorrhage, there may be arteriolar constriction mediated by the sympathetic nervous system and a fall in venous pressure resulting from the blood loss. These changes will lead to a decrease in the capillary hydrostatic pressure. Furthermore, the low blood pressure in hemorrhage causes a decrease in blood flow (and hence in oxygen supply) to the tissue with the result that vasodilator metabolites accumulate and induce relaxation of arterioles. Precapillary vessel relaxation is also engendered by the reduced transmural pressure. As a consequence of these several factors, absorption predominates over filtration and occurs at a larger capillary surface area. This is one of the compensatory mechanisms employed by the body to restore blood volume (p. 568).

An increase in venous pressure alone, as occurs in the feet when one changes from the lying to the standing position, would elevate capillary pressure and enhance filtration. However, the increase in transmural pressure causes precapillary vessel closure (myogenic mechanism [p. 511]) so that the capillary filtration coefficient actually decreases. This reduction in capillary surface available for filtration protects against the extravasation of large amounts of fluid into the interstitial space (edema). With prolonged standing, particularly when associated with some elevation of venous pressure in the legs (such as that caused by tight garters or pregnancy) or with sustained increases in venous pressure as seen in congestive heart failure, filtration is greatly enhanced and exceeds the capacity of the lymphatic system to remove the capillary filtrate from the interstitial space.

A large amount of fluid can move across the capillary wall in a relatively short time. In a normal individual the filtration coefficient (k_t) for the whole body is about 0.0061 ml/min/100 g of tissue/mm Hg. For a 70 kg man, elevation of venous pressure of 10 mm Hg for 10 minutes would increase filtration from capillaries by 342 ml. This would not lead to edema formation, since the fluid is returned to the vascular compartment by the lymphatic vessels. When edema does develop, it usually appears in the dependent parts of the body, where the hydrostatic pressure is greatest, but its location and magnitude are also determined by the type of tissue. Loose tissues, such as the subcutaneous tissue around the eyes or in the scrotum, are more prone to collect larger quantities of interstitial fluid than are firm tissues, such as muscle, or encapsulated structures, such as the kidney.

The concentration of the plasma proteins may also change in different pathological states and, hence, alter the osmotic force and movement of fluid across the capillary membrane. The plasma protein concentration is increased in dehydration (for example, water deprivation, prolonged sweating, severe vomiting, and diarrhea), and water moves by osmotic forces from the tissues to the vascular compartment. In contrast, the plasma protein concentration is reduced in nephrosis (a renal disease in which there is loss of protein in the urine) and edema may occur. When there is extensive capillary injury, as happens in burns, leaks occur and plasma protein escapes into the interstitial space along with fluid and increases the oncotic pressure of the interstitial fluid. This greater osmotic force outside the capillaries leads to additional fluid loss and possibly may lead to severe dehydration of the patient.

Pinocytosis

Some transfer of substances across the capillary wall can occur in tiny pinocytotic vesicles (pinocytosis). These vesicles (Figs. 32-3 and 32-4), formed by a pinching off of the surface membrane, can take up substances on one side of the capillary wall, move by thermal kinetic energy across the cell, and deposit their contents at the other side. The amount of material that can be transported in this way is very small relative to that moved by diffusion. However, pinocytosis may be responsible for the movement of large lipid-insoluble molecules (30 nm) between blood and interstitial fluid. The number of pinocytotic vesicles in endothelium varies with the tissue (muscle>lung>brain) and increases from the arterial to the venous end of the capillary.

Lymphatics

The terminal vessels of the lymphatic system consist of a widely distributed closed-end network of highly permeable lymph capillaries that are similar in appearance to blood capillaries. However, they are generally lacking in tight junctions between endothelial cells and possess fine filaments that anchor them to the surrounding connective tissue. With muscular contraction these fine strands may distort the lymphatic vessel to open spaces between the endothelial cells and permit the entrance of protein and large particles and cells present in the interstitial fluid. The lymph capillaries drain into larger vessels that finally enter the right and left subclavian veins at their junctions with the respective internal jugular veins. Only cartilage, bone, epithelium, and tissues of the central nervous system are devoid of lymphatic vessels. The plasma capillary filtrate is returned to the circulation by virtue of tissue pressure, facilitated by intermittent skeletal muscle activity, contractions of the lymphatic vessels, and an extensive system of one-way valves. In this respect they resemble the veins, although even the larger lymphatic vessels have thinner walls than the corresponding veins and contain only a small amount of elastic tissue and smooth muscle.

The volume of fluid transported through the lymphatics in 24 hours is about equal to an animal's total plasma volume and the protein returned by the lymphatics to the blood in a day is about one fourth to one half of the circulating plasma proteins. This is the only means whereby protein (albumin) that leaves the vascular compartment can be returned to the blood, since back diffusion into the capillaries cannot occur against the large albumin concentration gradient. Were the protein not removed by the lymph vessels, it would accumulate in the interstitial fluid and act as an oncotic force to draw fluid from the blood capillaries to produce increasing severe edema. In addition to returning fluid and protein to the vascular bed, the lymphatic system filters the lymph at the lymph nodes and removes foreign particles, such as bacteria. The largest lymphatic vessel, the *thoracic duct,* in addition to draining the lower extremities, returns protein lost through the permeable liver capillaries and carries substances absorbed from the gastrointestinal tract, principally fat in the form of chylomicrons, to the circulating blood.

Lymph flow varies considerably, being almost nil from resting skeletal muscle and increasing during exercise in proportion to the degree of muscular activity. It is increased by any mechanism that enhances the rate of blood capillary filtration, for example, increased capillary pressure or permeability or decreased plasma oncotic pressure. When either the volume of interstitial fluid exceeds the drainage capacity of the lymphatics or the lymphatic vessels become blocked, as may occur in certain disease states, interstitial fluid accumulates, chiefly in the more compliant tissues (for example, subcutaneous tissue) and gives rise to clinical edema.

Bibliography

Journal articles

Bundgaard, M.: Transport pathways in capillaries—in search of pores, Annu. Rev. Physiol. **42:**325, 1980.

Duling, B.R., and Klitzman, B.: Local control of microvascular function: role in tissue oxygen supply, Annu. Rev. Physiol. **42:**373, 1980.

Furchgott, R.: Role of endothelium in responses of vascular smooth muscle, Circ. Res. **53:**557, 1983.

Gore, R.W., and McDonagh, P.F.: Fluid exchange across single capillaries, Annu. Rev. Physiol. **42:**337, 1980.

Karnovsky, M.J.: The ultrastructural basis of transcapillary exchanges, J. Gen. Physiol. **52:**645, 1968.

Krogh, A.: The number and distribution of capillaries in muscles with calculation of the oxygen pressure head necessary for supplying the tissue, J. Physiol. **52:**409, 1919.

Lewis, D.H., editor: Symposium on lymph circulation, Acta Physiol. Scand. Suppl. **463:**9, 1979.

Rosell, S.: Neuronal control of microvessels, Annu. Rev. Physiol. **42:**359, 1980.

Starling, E.H.: On the absorption of fluids from the connective tissue spaces, J. Physiol. **19:**312, 1896.

Books and monographs

Bert, J.L., and Pearce, R.H.: The interstitium and microvascular exchange. In Handbook of physiology; Section 2: The cardiovascular system—microcirculation, vol. IV, Bethesda, Md., 1984, American Physiological Society.

Crone, C., and Levitt, D.G.: Capillary permeability to small solutes. In Handbook of physiology; Section 2: The cardiovascular system—microcirculation, vol. IV, Bethesda, Md., 1984, American Physiological Society.

Hudlicka, O.: Development of microcirculation: capillary growth and adaptation. In Handbook of physiology; Section 2: The cardiovascular system—microcirculation, vol. IV, Bethesda, Md., 1984, American Physiological Society.

Kaley, G., and Altura, A., editors: Microcirculation, vols. 1 and 2, Baltimore, 1977, 1978, University Park Press.

Krogh, A.: The anatomy and physiology of capillaries, New York, 1959, Hafner Co.

Michel, C.C.: Fluid movements through capillary walls. In Handbook of physiology, Section 2: The cardiovascular system—microcirculation, vol. IV, Bethesda, Md., 1984, American Physiological Society.

Mortillaro, N.A.: Physiology and pharmacology of the microcirculation, vol. 1, New York, 1983, Academic Press, Inc.

Renkin, E.M.: Control microcirculation and blood-tissue exchange. In Handbook of physiology; Section 2: The cardiovascular system—microcirculation, vol. IV, Bethesda, Md., 1984, American Physiological Society.

Shepro, D., and D'Amore, P.A.: Physiology and biochemistry of the vascular wall endothelium. In Handbook of physiology; Section 2: The cardiovascular system—microcirculation, vol. IV, Bethesda, Md., 1984, American Physiological Society.

Simionescu, M., and Simionescu, N.: Ultrastructure of the microvascular wall: functional correlations. In Handbook of physiology; Section 2: The cardiovascular system—microcirculation, vol. IV, Bethesda, Md., 1984, American Physiological Society.

Taylor, A.E., and Granger, D.N.: Exchange of macromolecules across the microcirculation. In Handbook of physiology; Section 2: The cardiovascular system—microcirculation, vol. IV, Bethesda, Md., 1984, American Physiological Society.

Wiedeman, M.P.: Architecture. In Handbook of physiology. Section 2: The cardiovascular system—microcirculation, vol. IV, Bethesda, Md., 1984, American Physiological Society.

Yoffey, J.M., and Courtice, F.C.: Lymphatics, lymph and the lymphomyeloid complex, London, 1970, Academic Press.

Zweifach, B.W., and Lipowsky, H.H.: Pressure-flow relations in blood and lymph microcirculation. In Handbook of physiology; Section 2: The cardiovascular system—microcirculation, vol. IV, Bethesda, Md., 1984, American Physiological Society.

CHAPTER

33

The Peripheral Circulation and Its Control

The peripheral circulation is essentially under dual control, centrally through the nervous system and locally in the tissues by the environmental conditions in the immediate vicinity of the blood vessels. The relative importance of these two control mechanisms is not the same in all tissues. In some areas of the body, such as the skin and the splanchnic regions, neural regulation of blood flow predominates, whereas in others, such as the heart and brain, this mechanism plays a minor role.

The vessels chiefly involved in regulating the rate of blood flow throughout the body are called the *resistance vessels* (arterioles). These vessels offer the greatest resistance to the flow of blood pumped to the tissues by the heart and thereby are important in the maintenance of arterial blood pressure. Smooth muscle fibers constitute a large percentage of the composition of the walls of the resistance vessels (see Fig. 26-1). Therefore the vessel lumen can be varied from one that is completely obliterated by strong contraction of the smooth muscle, with infolding of the endothelial lining, to one that is maximally dilated as a result of full relaxation of the vascular smooth muscle. Some resistance vessels are closed at any given moment in time and partial contraction (or *tone*) of the vascular smooth muscle exists in the arterioles. Were all the resistance vessels in the body to dilate simultaneously, blood pressure would fall precipitously to very low levels.

Vascular Smooth Muscle

Vascular smooth muscle is the tissue responsible for the control of total peripheral resistance, arterial and venous tone, and the distribution of blood flow throughout the body. The smooth muscle cells are small, mononucleate, and spindle shaped. They are generally arranged in helical or circular layers around the larger blood vessels and in a single circular layer around arterioles (Fig. 33-1, *A* and *B*). Also, parts of endothelial cells project into the vascular smooth muscle layer *(myoendothelial junctions)* at various points along the arterioles. These projections suggest a functional interaction between endothelium and adjacent vascular smooth muscle (Fig. 33-1, *C*). In general, the close association between action potentials and contraction observed in skeletal and cardiac muscle cells cannot be demonstrated in vascular smooth muscle, and vascular smooth muscle lacks transverse tubules.

Graded changes in membrane potential are often associated with increases or decreases in force. Contractile activity is generally elicited by neural or humoral stimuli. The behavior of smooth muscle in different vessels varies. For example, some vessels, particularly in the portal or mesenteric circulation, contain longitudinally oriented smooth muscle that is spontaneously active and that shows action potentials which correlate with the contractions (electrical-mechanical coupling).

The vascular smooth muscle cells contain large numbers of thin, actin filaments and comparatively small numbers of thick, myosin filaments. These filaments are aligned in the long axis of the cell but do not form visible sarcomeres with striations. Nevertheless, the sliding filament mechanism is believed to operate in this tissue, and phosphorylation of crossbridges regulates their rate of cycling. Compared to skeletal muscle, the smooth muscle contracts very slowly, develops high forces, maintains force for long periods of time with low ATP utilization, and operates over a considerable range of lengths under physiological conditions.

The interaction between myosin and actin, leading to contraction, is controlled by the myoplasmic Ca^{++} concentration as in other muscles, but the molecular mechanism whereby Ca^{++} regulates contraction differs. For example, smooth muscle lacks troponin and fast sodium channels. The increased myoplasmic Ca^{++} that initiates contraction can come from intracellular stores in the sarcoplasmic reticulum, be displaced from the plasma membrane, or pass into the cell following an increase in membrane Ca^{++} permeability. Calcium is extruded from the cell by a Ca^{++} pump in the cell membrane. The relative importance of intracellular and extracellular Ca^{++} for activation varies with different vascular smooth muscles and different agonists.

Most of the arteries and veins of the body are supplied to different degrees solely by fibers of the sympathetic nervous system (Fig. 33-1, *A* and *B*). These nerve fibers exert a tonic effect on the blood vessels, as evidenced by the fact that cutting or freezing the sympathetic nerves to a vascular bed (such as muscle) results in an increase in blood flow. Activation of the sympathetic nerves either directly or reflexly (p. 513) enhances vascular resistance. In contrast to the sympathetic nerves the parasympathetic nerves tend to decrease vascular resistance, but they innervate only a small fraction of the blood vessels in the body, mainly in certain viscera and pelvic organs. Vascular smooth muscle also responds to humoral stimulation (hormones and drugs) without evidence of electrical excitation. This has been referred to as *pharmacomechanical coupling* and is mediated by Ca^{++} influx or release. In the category of pharmacological stimuli are such substances as catecholamines, histamine, acetylcholine, serotonin, angiotensin, adenosine, and prostaglandins. Local environmental changes alter the contractile state of vascular smooth muscle and such alterations as increased temperature or increased carbon dioxide levels induce relaxation of this tissue.

A

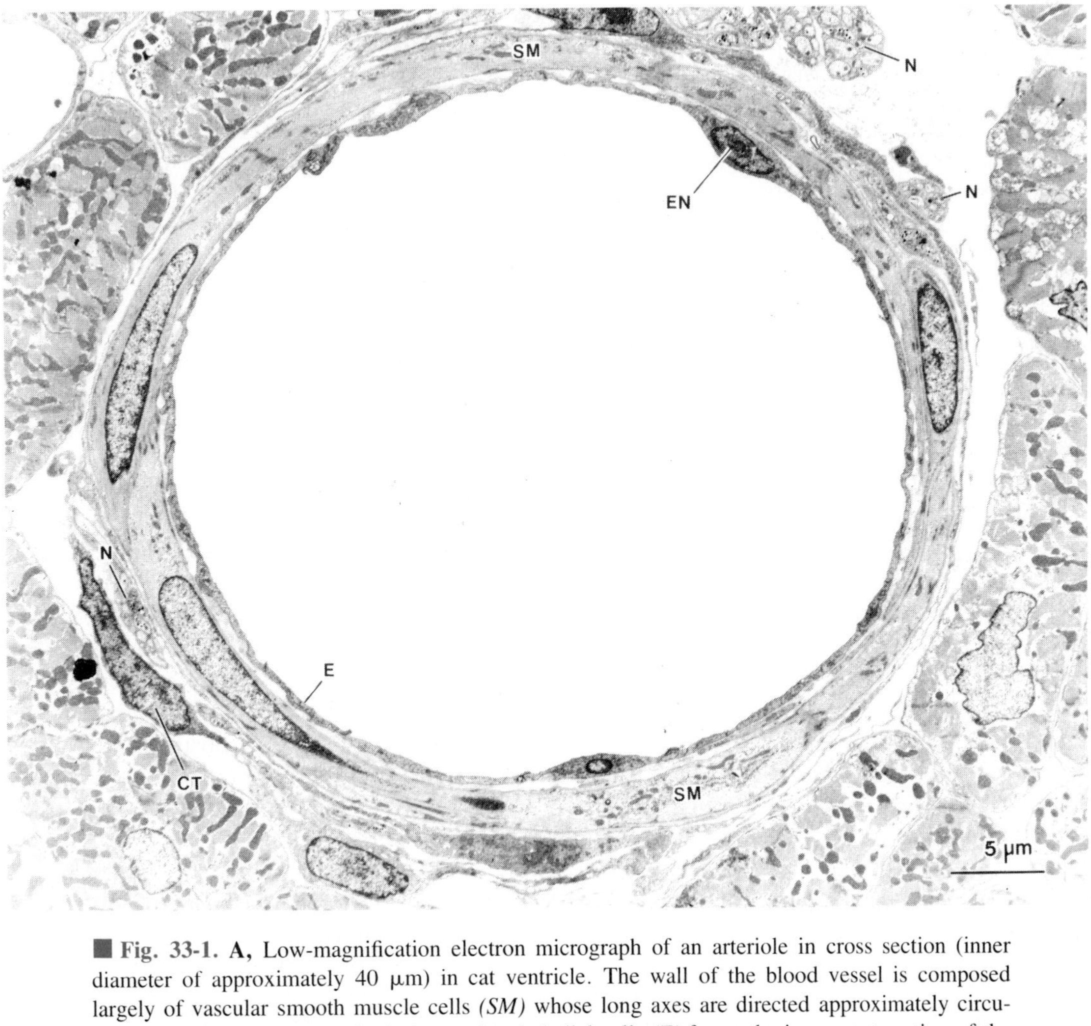

■ **Fig. 33-1. A,** Low-magnification electron micrograph of an arteriole in cross section (inner diameter of approximately 40 μm) in cat ventricle. The wall of the blood vessel is composed largely of vascular smooth muscle cells *(SM)* whose long axes are directed approximately circularly around the vessel. A single layer of endothelial cells *(E)* forms the innermost portion of the blood vessel. Connective tissue elements *(CT)* such as fibroblasts and collagen make up the adventitial layer at the periphery of the vessel; nerve bundles also appear in this layer *(N)*. *EN*, Endothelial cell nucleus. *Continued.*

B

C

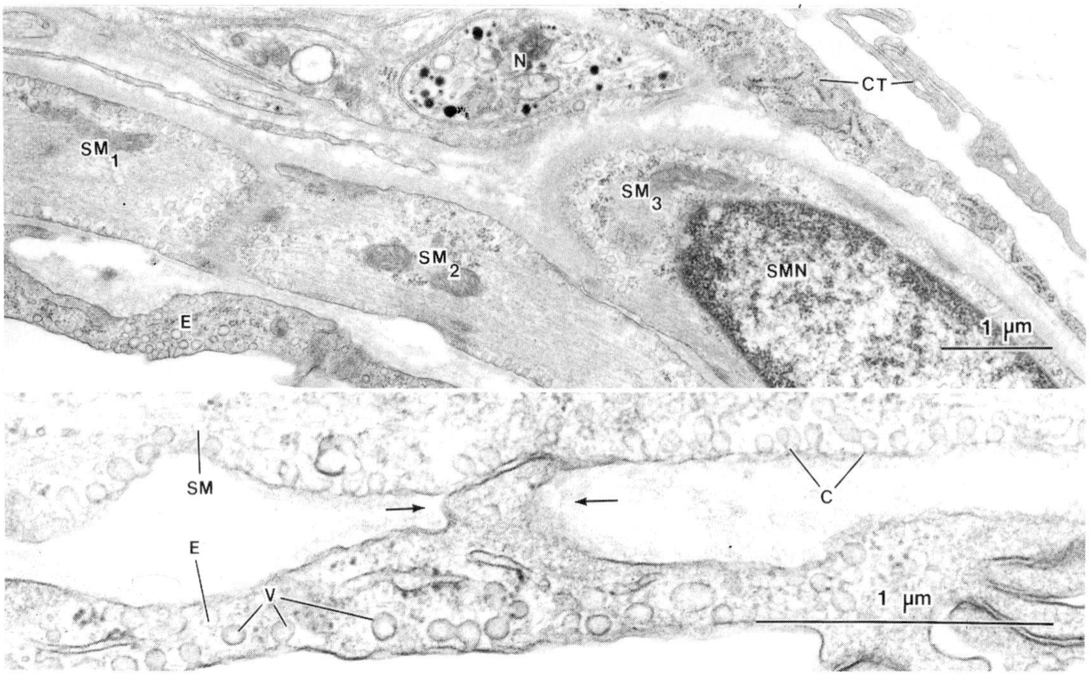

■ **Fig. 33-1, cont'd. B,** Detail of the wall of the blood vessel in panel **A.** This field contains a single endothelial layer *(E),* the medial smooth muscle layer (three smooth muscle cell profiles: SM_1, SM_2, SM_3), and the adventitial layer (containing nerves [*N*] and connective tissue [*CT*]. *SMN,* Smooth muscle nucleus. **C,** Another region of the arteriole, showing the area in which the endothelial *(E)* and smooth muscle *(SM)* layers are apposed. A projection of an endothelial cell (between arrows) is closely applied to the surface of the overlying smooth muscle, forming a "myoendothelial junction." Plasmalemmal vesicles are prominent in both the endothelium *(V)* and the smooth muscle cell (where such vesicles are known as "caveolae," *C*).

■ *Intrinsic or Local Control of Peripheral Blood Flow*

In a number of different tissues the blood flow appears to be adjusted to the existing metabolic activity of the tissue. Furthermore, imposed changes in the perfusion pressure (arterial blood pressure) at constant levels of tissue metabolism, as measured by oxygen consumption, are met with vascular resistance changes that tend to maintain a constant blood flow. This mechanism is commonly referred to as *autoregulation* of blood flow and is illustrated graphically in Fig. 33-2. In the skeletal muscle preparation from which these data were gathered, the muscle was completely isolated from the rest of the animal and was in a resting state. From a control pressure of 100 mm Hg, the pressure was abruptly increased or decreased, and the blood flows observed immediately after changing the perfusion pressure are represented by the closed circles. Maintenance of the altered pressure at each new level was followed within 30 to 60 seconds by a return of flow to or toward the control levels; the open circles represent these steady-state flows. Over the pressure range of 20 to 120 mm Hg, the steady-state flow is relatively constant. Calculation of resistance across the vascular bed (pressure/flow) during steady-state conditions indicates that with elevation of perfusion pressure the resistance vessels constricted, whereas with reduction of perfusion pressure, dilation occurred.

The mechanism responsible for this constancy of blood flow in the presence of an altered perfusion pressure is not known. However, three explanations have been suggested—the *tissue pressure hypothesis,* the *myogenic hypothesis,* and the *metabolic hypothesis.* According to the tissue pressure concept, an increase in perfusion pressure produces an increase in blood volume of the tissue and a net transfer of fluid from the intravascular to the extravascular compartments. The resultant increase in tissue pressure (turgor) is believed to compress the very thin-walled vessels such as the capillaries, venules, and veins, and thereby to reduce the flow of blood into the tissue. A reduction in perfusion pressure would elicit the opposite response. Such a mechanism can operate only in an encapsulated structure where expansion of the tissue is restricted. However, even in the kidney, which possesses a fairly rigid connective tissue capsule and shows a high degree of autoregulation of blood flow, conclusive evidence for a tissue pressure mechanism is lacking.

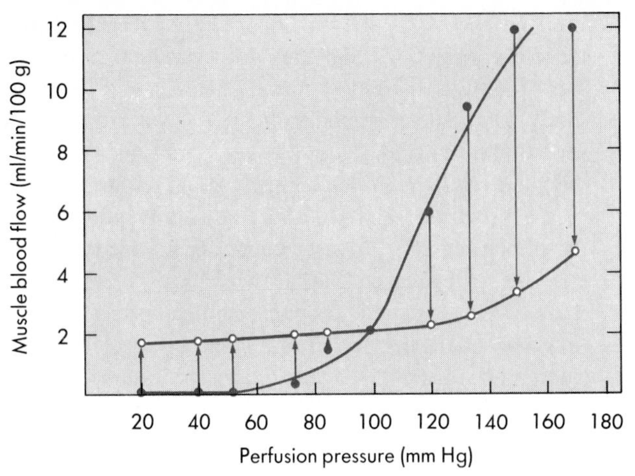

■ **Fig. 33-2.** Pressure-flow relationship in the skeletal muscle vascular bed of the dog. The closed circles represent the flows obtained immediately after abrupt changes in perfusion pressure from the control level (point where lines cross). The open circle represent the steady-state flows obtained at the new perfusion pressure. (Redrawn from Jones, R.D., and Berne, R.M.: Circ. Res. **14:**126, 1964.)

The myogenic hypothesis states that the vascular smooth muscle contracts in response to stretch and relaxes with a reduction in tension. Therefore the initial flow increment produced by an abrupt increase in perfusion pressure that passively distends the blood vessels would be followed by a return of flow to the previous control level by contraction of the smooth muscles of the resistance vessels. One difficulty with the myogenic hypothesis is that for flow to remain constant following an increase in perfusion pressure, it is necessary for the caliber of the resistance vessels to be less than it was prior to the elevation of pressure. Should this occur, this stretch stimulus for the maintenance of increase resistance would be removed.

However, if an increase in vessel wall tension, and not stretch, triggers a contractile (constrictor) response, then it is possible, by consideration of the Laplace equation ($T = Pr$) to explain autoregulation in terms of the myogenic mechanism. Raising perfusion pressure increases the pressure across the vessel wall and by passive stretch increases the radius, resulting in a large increment in wall tension. Constriction of the vessel in response to the increase in tension results in a reduction of its diameter to less than the original value so that the product of pressure (increased) and radius (decreased) is restored to the control level. The resistance vessels also show intermittent contraction and relaxation, and it is possible to avoid the paradox encountered when a simple maintained contractile response to stretch is postulated. It has been proposed that an increase in the frequency of contractile responses (via stretch-induced increase in frequency of action potentials) occurs with elevation of perfusion pressure. If such a mechanism were operative, the resistance vessels as a whole would spend more time in the contracted state when pressure was raised than they did prior to elevation of perfusion pressure.

The myogenic mechanism has been demonstrated in certain tissues and in isolated arterioles. Since blood pressure is reflexly maintained at a fairly constant level under normal conditions, operation of a myogenic mechanism would be expected to be minimized. However, when one changes position (from lying to standing) a large change in transmural pressure occurs in the lower extremities. The precapillary vessels constrict in response to this imposed stretch, which results in cessation of flow in most capillaries. After flow stops, capillary filtration diminishes until the increase in plasma oncotic pressure and the increase in interstitial fluid pressure balance the elevated capillary hydrostatic pressure produced by changing from a horizontal to a vertical position. If arteriolar resistance did not increase with standing, the hydrostatic pressure in the lower parts of the legs would reach such high levels that large volumes of fluid would pass from the capillaries into the interstitial fluid compartment and produce edema.

According to the metabolic hypothesis, the blood flow is governed by the metabolic activity of the tissue, and any intervention that results in an O_2 supply that is inadequate for the requirements of the tissue gives rise to the formation of vasodilator metabolites. These metabolites are released from the tissue and act locally to dilate the resistance vessels. When the metabolic rate of the tissue increases or the O_2 delivery to the tissue decreases, more vasodilator substance is released and the metabolite concentration in the tissue increases. When metabolic activity at constant perfusion pressure decreases or perfusion pressure at constant metabolic activity increases, the tissue concentration of the vasodilator agent falls. A decrease in metabolite production or an increase in washout and/or inactivation of the metabolite elicits an increase in precapillary resistance. An attractive feature of the metabolic hypothesis is that, in most tissues, blood flow closely parallels metabolic activity. Hence, although blood pressure is kept fairly constant, metabolic activity and blood flow in the different body tissues vary together under physiological conditions (for example, exercise).

Many substances have been proposed as mediators of metabolic vasodilation. Some of the earliest ones suggested are lactic acid, CO_2, and hydrogen ions. However, the decrease in vascular resistance induced by supernormal concentrations of these dilator agents falls considerably short of the dilation observed under physiological conditions of increased metabolic activity.

Changes in O_2 tension can evoke changes in the contractile state of vascular smooth muscle; an increase in Po_2 elicits contraction, and a decrease in Po_2, relaxation. If significant reductions in the intravascular Po_2

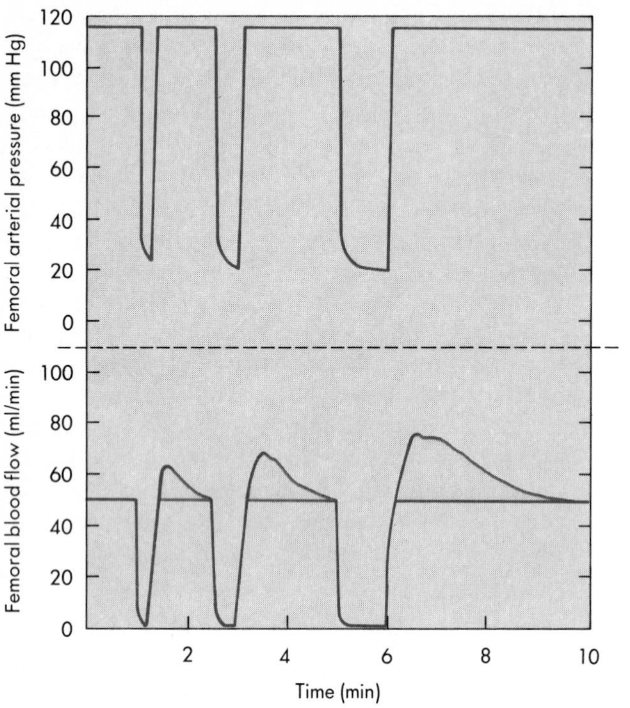

■ **Fig. 33-3.** Reactive hyperemia in the hind limb of the dog after 15-, 30-, and 60-second occlusions of the femoral artery.

occur before the arterial blood reaches the resistance vessels (diffusion through the arterial and arteriolar walls [p. 501]), small changes in O_2 supply or consumption could elicit contraction or relaxation of the resistance vessels. However, direct measurements of Po_2 at the resistance vessels indicate that over a wide range of Po_2 (11 to 343 mm Hg) there is no correlation between O_2 tension and arteriolar diameter. Furthermore, if Po_2 were directly responsible for vascular smooth muscle tension, one would not expect to find a parallelism between the duration of arterial occlusion and the duration of the reactive hyperemia (flow above control level on release of an arterial occlusion) (Fig. 33-3). With either short occlusions (5 to 10 seconds) or long occlusions (1 to 3 minutes) the venous blood becomes bright red (well oxygenated) within 1 or 2 seconds after release of the arterial occlusion and hence the smooth muscle of the resistance vessels must be exposed to a high Po_2 in each instance. Nevertheless, the longer occlusions result in longer periods of reactive hyperemia. These observations are more compatible with the release of a vasodilator metabolite from the tissue than with a direct effect of Po_2 on the vascular smooth muscle.

Potassium ions, inorganic phosphate, and interstitial fluid osmolarity can also induce vasodilation. Since K^+ and phosphate are released and osmolarity is increased during skeletal muscle contraction, it has been proposed that these factors contribute to *active hyperemia* (increased blood flow caused by enhanced tissue activity). However, significant increases of phosphate concentration and osmolarity are not consistently observed during muscle contraction, and they may produce only transient increases in blood flow. Therefore they are not likely candidates as mediators of the vasodilation observed with muscular activity. Potassium release occurs with the onset of skeletal muscle contraction or an increase in cardiac activity and could be responsible for the initial decrease in vascular resistance observed with exercise or increased cardiac work. However, K^+ release is not sustained, despite continued arteriolar dilation throughout the period of enhanced muscle activity. Therefore some other agent must serve as mediator of the vasodilation associated with the greater metabolic activity of the tissue. Reoxygenated venous blood obtained from active cardiac and skeletal muscles under steady-state conditions of exercise does not elicit vasodilation when infused into a test vascular bed. It is difficult to see how oxygenation of the venous blood could alter its K^+ or phosphate content or its osmolarity and thereby destroy its vasodilator effect.

Recent evidence indicates that adenosine, which is involved in the regulation of coronary blood flow, may also participate in the control of the resistance vessels in skeletal muscle; also, some of the prostaglandins have been proposed as important vasodilator mediators in certain vascular beds.

Thus there are a number of candidates for the mediator of metabolic vasodilation, and the relative contribution of each of the various factors remains the subject for future investigation. Several factors may be involved in any given vascular bed, and different factors preponderate in different tissues.

Metabolic control of vascular resistance via the release of a vasodilator substance is predicated on the existence of basal vessel tone. This tonic activity, or *basal tone,* of the vascular smooth muscle is readily demonstrable, but in contrast to tone in skeletal muscle it is independent of the nervous system. The factor responsible for basal tone in blood vessels is not known, but one or more of the following factors may be involved: (1) an expression of myogenic activity in response to the stretch imposed by the blood pressure, (2) the high O_2 tension of arterial blood, (3) the presence of calcium ions, or (4) some unknown factor in plasma, since addition of plasma to the bathing solution of isolated vessel segments evokes partial contraction of the smooth muscle.

If arterial inflow to a vascular bed is stopped for a few seconds to several minutes, the blood flow, on release of the occlusion, immediately exceeds the flow before occlusion and only gradually returns to the control level *(reactive hyperemia).* This is illustrated in Fig. 33-3 where blood flow to the leg was stopped by clamping the femoral artery for 15, 30, and 60 seconds. Release of the 60-second occlusion resulted in a peak blood flow 70% greater than the control flow, with a return to control flow within about 110 seconds. When this same experiment is done in humans by inflating a blood pressure cuff on the upper arm, dilation of the resistance vessels of the hand and forearm, immediately after release of the cuff, is evident from the bright red color of the skin and the fullness of the veins. Within limits the peak flow and particularly the duration of the reactive hyperemia are proportional to the duration of the occlusion (Fig. 33-3). If the extremity is exercised during the occlusion period, reactive hyperemia is increased. These observations and the close relationship that exists between metabolic activity and blood flow in the unoccluded limb are consonant with a metabolic mechanism in the local regulation of tissue blood flow.

Extrinsic Control of Peripheral Blood Flow

Neural Sympathetic Vasoconstriction

A number of regions in the medulla influence cardiovascular activity. Some of the effects of stimulation of the dorsal lateral medulla are vasoconstriction, cardiac acceleration, and enhanced myocardial contractility. Caudal and ventromedial to the *pressor* region is a zone that produces a decrease in blood pressure on stimulation. This *depressor area* exerts its effect by direct spinal inhibition and by inhibition of the medullary pressor region. However, the precise mechanism of its depressor actions is still unknown. These areas comprise a center not in an anatomical sense in that a discrete group of cells is discernible, but in a physiological sense in that stimulation of the pressor region produces the responses mentioned previously. From the vasoconstrictor regions, fibers descend in the spinal cord and synapse at different levels of the thoracolumbar region (T1 to L2 or L3). Fibers from the intermediolateral gray matter of the cord emerge with the ventral roots, but leave the motor fibers to join the paravertebral sympathetic chains through the white communicating branches. These preganglionic white (myelinated) fibers may pass up or down the sympathetic chains to synapse in the various ganglia within the chains or in certain outlying ganglia. Postganglionic gray branches (unmyelinated) then join the corresponding segmental spinal nerves and accompany them to the periphery to innervate the arteries and veins. Postganglionic sympathetic fibers from the various ganglia join the large arteries and accompany them as an investing network of fibers to the resistance and capacitance vessels (veins).

The vasoconstrictor regions are tonically active, and reflexes or humoral stimuli that enhance this activity result in an increase in frequency of impulses reaching the terminal branches to the vessels, where a constrictor neurohumor (norepinephrine) is released and elicits constriction (alpha-adrenergic effect) of the resistance vessels. Inhibition of the vasoconstrictor areas reduces their tonic activity and hence diminishes the frequency of impulses in the efferent nerve fibers, resulting in vasodilation. In this manner, neural regulation of the peripheral circulation is accomplished primarily by alteration of the number of impulses passing down the vasoconstrictor fibers of the sympathetic nerves to the blood vessels. The vasomotor regions may show rhythmic changes in tonic activity manifested as oscillations of arterial pressure. Some occur at the frequency of respiration *(Traube-Hering waves)* and are due to an increase in sympathetic impulses to the resistance vessels coincident with inspiration. Others are independent of and at a lower frequency than respiration *(Mayer waves)*.

Sympathetic Constrictor Influence on Resistance and Capacitance Vessels

The vasoconstrictor fibers of the sympathetic nervous system supply the arteries, arterioles, and veins, but neural influence on the larger vessels is of far less functional importance than it is on the microcirculation. Capacitance vessels are apparently more responsive to sympathetic nerve stimulation than are resistance vessels, since they reach maximum constriction at a lower frequency of stimulation than do the resistance vessels. However, capacitance vessels do not possess beta-adrenergic receptors nor do they respond to vasodilator metabolites. Norepinephrine is the neurotransmitter released at the sympathetic nerve terminals at the blood vessels, and many factors, such as circulating hormones and particularly locally released substances, modify the liberation of norepinephrine from the vesicles of the nerve terminals.

The response of the resistance and capacitance vessels to stimulation of the sympathetic fibers is illustrated in Fig. 33-4. At constant arterial pressure, sympathetic fiber stimulation evoked a reduction of blood flow (constriction of the resistance vessels) and a decrease in blood volume of the tissue (constriction of the capacitance vessels). The abrupt decrease in tissue volume is caused by movement of blood out of the capacitance vessels and out of the hindquarters of the cat, whereas the late, slow, progressive decline in volume (to the right of the arrow) is caused by movement of extravascular fluid into the capillaries and hence away from the tissue. The loss of tissue fluid is a consequence of the lowered capillary hydrostatic pressure brought about by constriction of the resistance vessels, with establishment of a new equilibrium of the forces responsible for filtration and absorption across the capillary wall.

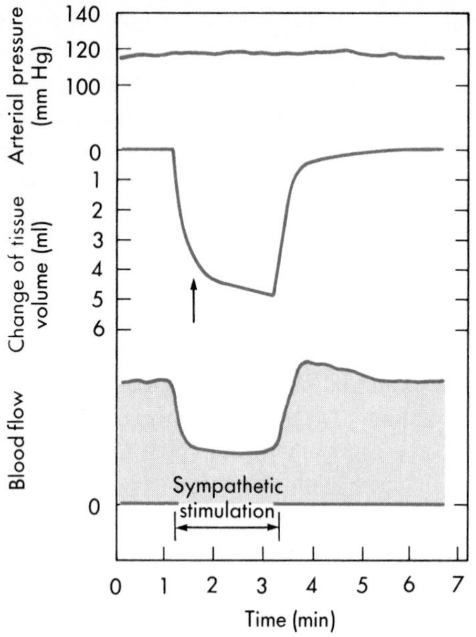

■ **Fig. 33-4.** Effect of sympathetic nerve stimulation (2 Hz) on blood flow and tissue volume in the hindquarters of the cat. The arrow denotes the change in slope of the tissue volume curve where the volume decrease due to emptying of capacitance vessels ceases and loss of extravascular fluid becomes evident. (Redrawn from Mellander, S.: Acta Physiol. Scand. **50**[Suppl. 176]:1-86, 1960.)

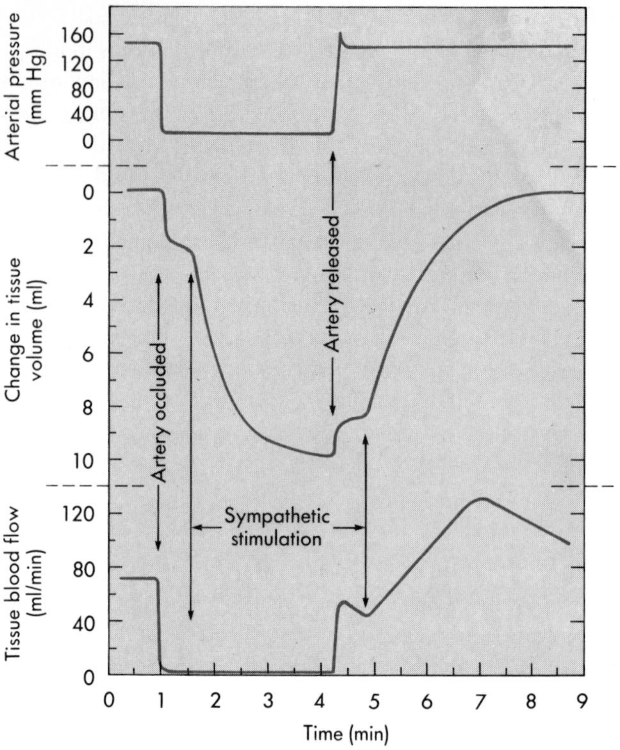

■ **Fig. 33-5.** Effect of arterial occlusion and sympathetic nerve stimulation (6 impulses/second) on the tissue volume in the hindquarters of the cat. Note the small changes in tissue volume with arterial occlusion and release compared to the large volume change obtained with sympathetic nerve stimulation. (Redrawn from Mellander, S.: Acta Physiol. Scand. **50**[Suppl. 176]:1-86, 1960.)

In addition to active changes (contraction and relaxation of the vascular smooth muscle) in vessel caliber, there are also passive changes caused solely by alteration in intraluminal pressure; an increase in intraluminal pressure produces distension of the vessels, and a decrease produces a reduction in caliber by recoil of the elastic components of the vessel walls. The relative effects of the passive and active forces on the volume changes of the tissues are depicted in Fig. 33-5. With occlusion of the arterial blood supply to the tissue, perfusion pressure dropped almost to zero, blood flow became nil, and there was an abrupt 2 ml decrease in tissue volume. This rapid reduction in tissue volume is a passive response to the reduction in arterial pressure, with expulsion of blood from the tissue. The slow decrease in tissue volume (between the first two arrows) is quite small, relative to that shown in Fig. 33-4, because of the brevity of the period between arterial occlusion and the start of sympathetic stimulation. This slight reduction in volume is caused by the decrease in capillary hydrostatic pressure incident to the reduction of the perfusion pressure. It represents loss of extravascular fluid via absorption by the capillaries and movement out through the venous channels. When the sympathetic nerve fibers were stimulated during the period of arterial occlusion, a large decrease in tissue volume was observed. Since blood flow already had been stopped by arterial occlusion, this change in tissue volume cannot be a passive response secondary to arteriolar constriction but must represent active constriction of the capacitance vessels. Note that a slow decrease in tissue volume (2.5 to 4.2 minutes, Fig. 33-5) followed the rapid change and indicates a continuation of the loss of extravascular fluid as a result of the reduction in capillary hydrostatic pressure. If the capillary pressure is held at low levels for 2 to 3 minutes, either by arterial occlusion or by constriction of the resistance vessels, a new pressure equilibrium is established across the capillary wall and further loss of tissue fluid becomes negligible. The protein concentration of the blood leaving the tissue during the period of slow decrease in tissue volume was found to be reduced, indicating dilution of the plasma proteins by the absorption of low-protein tissue fluid by the capillaries. When the arterial occlusion was released during continuous sympathetic nerve stimulation, a small abrupt increase in intravascular pressure occurred because of a passive stretch of the capacitance vessels by the increased intraluminal pressure. This rapid increment in tissue volume was followed by a small gradual increase (between pair of arrows to the right, Fig. 33-5) that represents movement of fluid from the capillaries into the interstitial spaces in response to

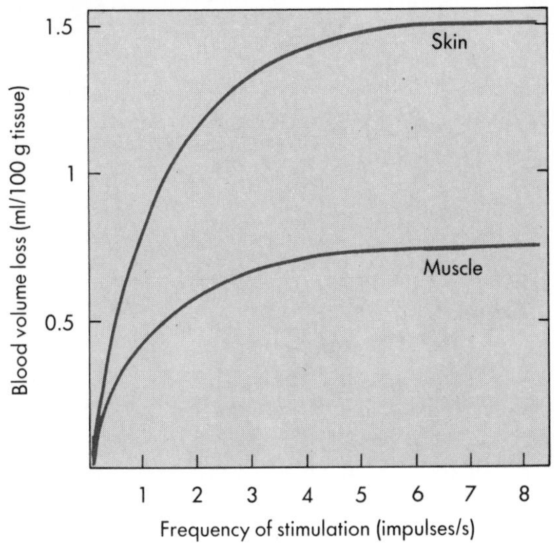

■ **Fig. 33-6.** Comparison of the blood volume loss from skin and muscle with sympathetic nerve stimulation. (Redrawn from Mellander, S.: Acta Physiol. Scand. **50**[Suppl. 176]:1-86, 1960.)

the elevation of capillary hydrostatic pressure. Cessation of sympathetic stimulation resulted in relaxation of the smooth muscle of the capacitance vessels and a restoration of their preexisting blood volume. Note that blood flow increased above, and gradually returned toward, the control level, illustrating a typical reactive hyperemia response to the period of ischemia. From these observations it can be seen that the passive changes of the capacitance vessels are small, relative to those induced by active contraction and relaxation of their vascular smooth muscle. Under physiological conditions, in which such drastic reduction in perfusion pressure would not occur, the contribution of a passive factor to changes in volume of the capacitance vessels would be even less than shown in Fig. 33-5.

At basal tone approximately one third of the blood volume of a tissue can be mobilized on stimulation of the sympathetic nerves at physiological frequencies. The basal tone is very low in capacitance vessels; with veins denervated, only small increases in volume are obtained with maximum doses of the vasodilator, acetylcholine. Therefore the blood volume at basal tone is close to the maximum blood volume of the tissue. The amount of blood that can be mobilized from skin and muscle at different frequencies of stimulation of the sympathetic nerves is depicted in Fig. 33-6. The greater mobilization of blood from the skin than from the muscle capacitance vessels may in part be caused by a greater sensitivity of these vessels to sympathetic stimulation, but it is also caused by the fact that basal tone is lower in skin than in muscle. Therefore, in the absence of neural influence, the skin capacitance vessels contain more blood than do the muscle capacitance vessels.

Blood is mobilized from capacitance vessels in response to physiological stimuli. In exercise, activation of the sympathetic nerve fibers produces constriction of veins and hence augments the cardiac filling pressure. Also, in arterial hypotension (as in hemorrhage), the capacitance vessels constrict to aid in overcoming the decreased central venous pressure associated with this condition. In addition, the resistance vessels constrict in shock, thereby assisting in the maintenance or restoration of arterial pressure. With arterial hypotension the enhanced arteriolar constriction also leads to a small mobilization of blood from the tissue by virtue of recoil of the postarteriolar vessels when intraluminal pressure is reduced. Furthermore, there is mobilization of extravascular fluid because of greater absorption into the capillaries in response to the lowered capillary hydrostatic pressure.

Clear dissociation between responses of the resistance and capacitance vessels can be demonstrated with nerve stimulation or with the use of epinephrine and acetylcholine. In Fig. 33-7, *A*, a small intravenous dose of epinephrine elicited an increase in blood flow (active dilation of resistance vessels) and a small abrupt increase in tissue volume (passive expansion of the capacitance vessels by the increased intraluminal pressure). Following the passive increase in tissue volume (to the left of the first arrow) a steady gradual tissue volume increase occurred (between arrows), attributable to movement of fluid from the capillaries to the interstitial space as a result of the elevated capillary hydrostatic pressure. Fig. 33-7, *B*, represents the effect of a dose of acetylcholine that elicited the same degree of resistance vessel dilation as that obtained with epinephrine in Fig. 33-7, *A*. Note that the acetylcholine produced a large increment in tissue volume (blood volume) and only a small increase in tissue volume attributable to increased capillary filtration (between arrows). In this case the concomitant dilation of the capacitance and resistance vessels resulted in a small increment in capillary pressure, hence a slower rate of capillary filtration. In Fig. 33-7, *C*, the dose of epinephrine was about twice that given in *A;* it elicited dilation of the resistance vessels but constriction of the capacitance vessels, resulting in a small decrease in tissue blood volume. These opposite effects on the resistance and capacitance vessels increased capillary hydrostatic pressure and enhanced the rate of capillary filtration, as evidenced from the slowly rising portion of the tissue volume curve (between arrows). A still larger dose of epinephrine (Fig. 33-7, *D*) produced constriction of both resistance and capacitance vessels with a decrease in tissue blood volume (to left of first arrow) and movement of fluid from the extravascular to the intravascular compartment (between arrows) because of a decrease in net capillary hydrostatic pressure.

From these observations it becomes apparent that

■ Fig. 33-7. Effect of acetylcholine and different doses of epinephrine on resistance and capacitance vessels in skeletal muscle, *A,* Epinephrine 0.3 μg/kg/min. *B,* Acetylcholine 1.7 μg/kg/minute (to give same degree of dilation of resistance vessels as in *A*). *C,* Epinephrine 0.7 μg/kg/minute. *D,* Epinephrine 3 μg/kg/minute. Pairs of arrows indicate periods of filtration and absorption. (Redrawn from Mellander, S.: Acta Physiol. Scand. **50**[Suppl. 176]:1-86, 1960.)

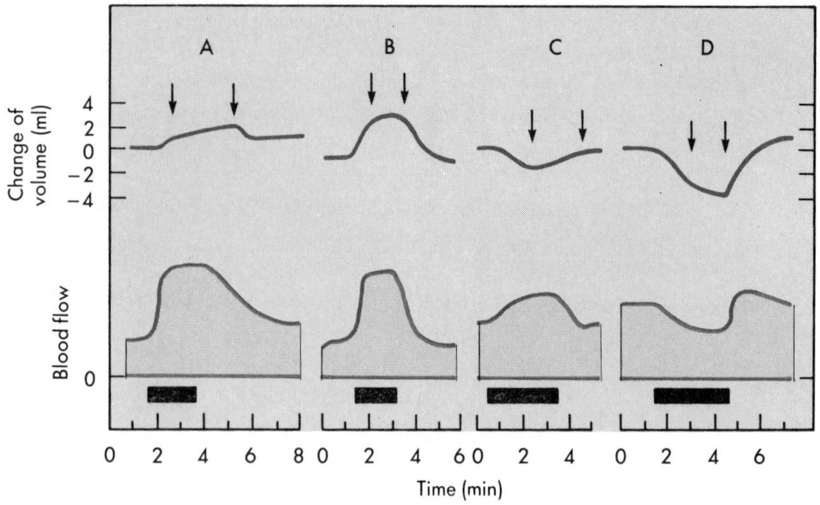

neural and humoral stimuli can exert similar or dissimilar effects on different segments of the vascular tree, and in so doing can alter blood flow, tissue blood volume, and extravascular volume to meet the physiological requirements of the organism.

■ *Active Sympathetic Vasodilation*

Although major neural control of the peripheral vessels is provided by the adrenergic vasoconstrictor fibers of the sympathetic nervous system, there are also *sympathetic cholinergic fibers* innervating the resistance vessels in skeletal muscle and skin. Electrical stimulation of the sympathetic nerves to blood vessels produces vasoconstriction. However, if the adrenergic constrictor effect is blocked by a suitable adrenergic receptor blocking agent or if the neural stores of norepinephrine are depleted by prior treatment with reserpine, the stimulation results in *active dilation,* which can be blocked by the administration of atropine. There is no evidence that these cholinergic sympathetic dilator fibers innervate the capacitance vessels. However, a small fraction of the vasodilation observed with activation of the baroreceptors may occur through the sympathetic cholinergic neurons.

Fibers of the cholinergic sympathetic dilator system arise in the motor cortex of the cerebrum and pass through the hypothalamus and the ventral medulla before joining the other sympathetic outflow in the spinal cord. Activation of the cholinergic sympathetic dilator system produces a relatively large transient initial increase in blood flow followed by a smaller sustained increment in flow during the period of nerve stimulation. There is no evidence for any tonic activity of these fibers. Since excitement or apprehension seems to activate this system and induce vasodilation in skeletal muscle, it has been suggested that the role of the sympathetic vasodilators is to provide the muscles with an increased blood flow in anticipation of the use of the muscles (for example, flight or fight). These cholinergic sympathetic fibers have been demonstrated in the dog and cat, but their existence in primates is questionable.

In addition to the cholinergic sympathetic vasodilator system, there are beta-adrenergic receptors on the resistance vessels that can be demonstrated by sympathetic nerve stimulation after the administration of atropine and an alpha-adrenergic receptor blocker. The vasodilation observed in muscle with intraarterial injection of small doses of epinephrine or with isoproterenol is caused by stimulation of these beta-adrenergic receptors.

■ *Parasympathetic Neural Influence*

The efferent fibers of the cranial division of the parasympathetic nervous system supply blood vessels of the head and viscera, whereas fibers of the sacral division supply blood vessels of the genitalia, bladder, and large bowel. Skeletal muscle and skin do not receive parasympathetic innervation. Because only a small proportion of the resistance vessels of the body receives parasympathetic fibers, the effect of these cholinergic fibers on total vascular resistance is small.

Stimulation of the parasympathetic fibers to the salivary glands induces marked vasodilation. However, the increase in submaxillary gland blood flow seen with chorda tympani stimulation is believed by some investigators to be secondary to the increase in metabolic activity of the gland and by others to be a primary action on the arterioles of the gland. A vasodilator polypeptide, *bradykinin,* formed locally from the action of enzyme on a plasma protein substrate present in the glandular lymphatics has been considered to be the metabolic mediator of the vasodilation produced by

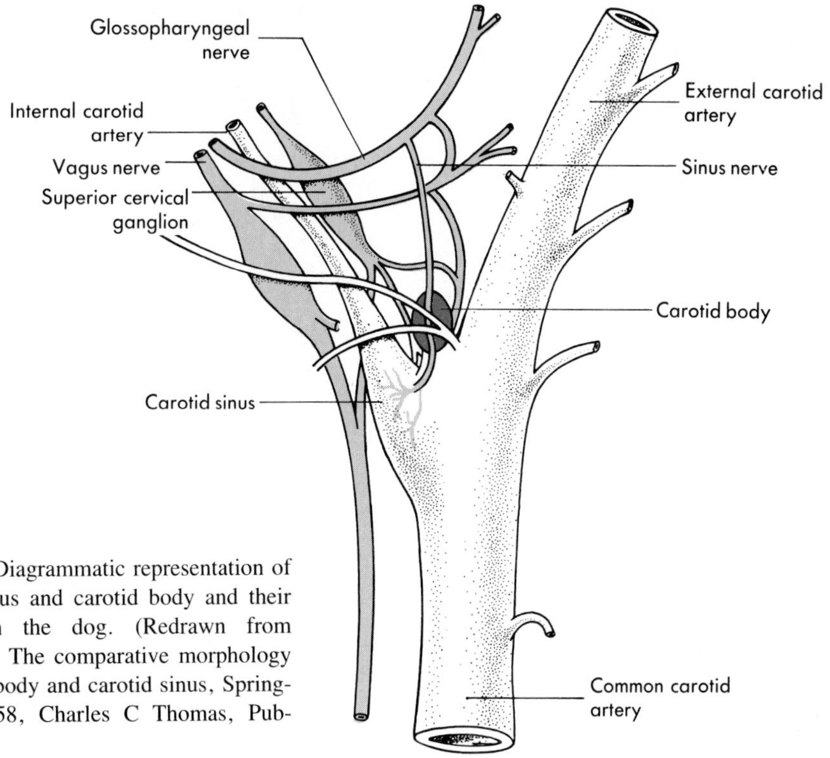

Fig. 33-8. Diagrammatic representation of the carotid sinus and carotid body and their innervation in the dog. (Redrawn from Adams, W.E.: The comparative morphology of the carotid body and carotid sinus, Springfield, Ill., 1958, Charles C Thomas, Publisher.)

chorda tympani stimulation. Whether vasodilation in salivary glands results from the release of a cholinergic neurohumor from nerve endings, from the formation and release of bradykinin, or from both is unsettled. Bradykinin also has been reported to be formed in other exocrine glands, such as the lacrimal glands and the sweat glands. Its presence in sweat is thought to be partly responsible for the dilation of cutaneous blood vessels with sweating.

Dorsal Root Dilators

Vasomotor impulses were at one time thought to travel antidromically in the spinal sensory nerves. Other than the antidromic impulses seen with the axon reflex (p. 549), there is no good evidence that the spinal nerves transmit impulses from the spinal cord to the peripheral blood vessels.

Humoral Factors

Epinephrine and norepinephrine exert a profound effect on the peripheral blood vessels. In skeletal muscle, epinephrine in low concentrations dilates resistance vessels (beta-adrenergic effect) and in high concentrations produces constriction (alpha-adrenergic effect). In skin only vasoconstriction is obtained with epinephrine, whereas in all vascular beds the primary effect of norepinephrine is vasoconstriction. When stimulated, the adrenal gland can release epinephrine and norepinephrine into the systemic circulation. However, under physiological conditions, the effect of catecholamine release from the adrenal medulla is of lesser importance than norepinephrine release produced by sympathetic nerve activation.

Vascular Reflexes

Areas of the medulla that mediate sympathetic and vagal influences are under the influence of neural impulses arising in the baroreceptors, chemoreceptors, hypothalamus, cerebral cortex, and skin and also can be altered by changes in the blood concentrations of CO_2 and O_2.

Baroreceptors. The *baroreceptors* (or *pressoreceptors*) are stretch receptors located in the carotid sinuses (slightly widened areas of the internal carotid arteries at their points of origin from the common carotid arteries) and in the aortic arch (Figs. 33-8 and 33-9). Impulses arising in the carotid sinus travel up the sinus nerve (nerve of Hering) to the glossopharyngeal nerve and, via the latter, to the nucleus of the tractus solitarius (NTS) in the medulla. The NTS is the site of central projection of the chemoreceptors and baroreceptors. Stimulation of the NTS inhibits sympathetic nerve impulses to the peripheral blood vessels (depressor), whereas lesions of the NTS produce vasoconstriction (pressor). Impulses arising in the pressoreceptors of the

Fig. 33-9. **A,** Anterior view of the aortic arch showing the innervation of the aortic bodies and pressoreceptors in the dog. **B,** Posterior view of the aortic arch showing the innervation of the aortic bodies and pressoreceptors. (Modified from Nonidez, J.F.: Anat. Rec. **69:**299, 1937.)

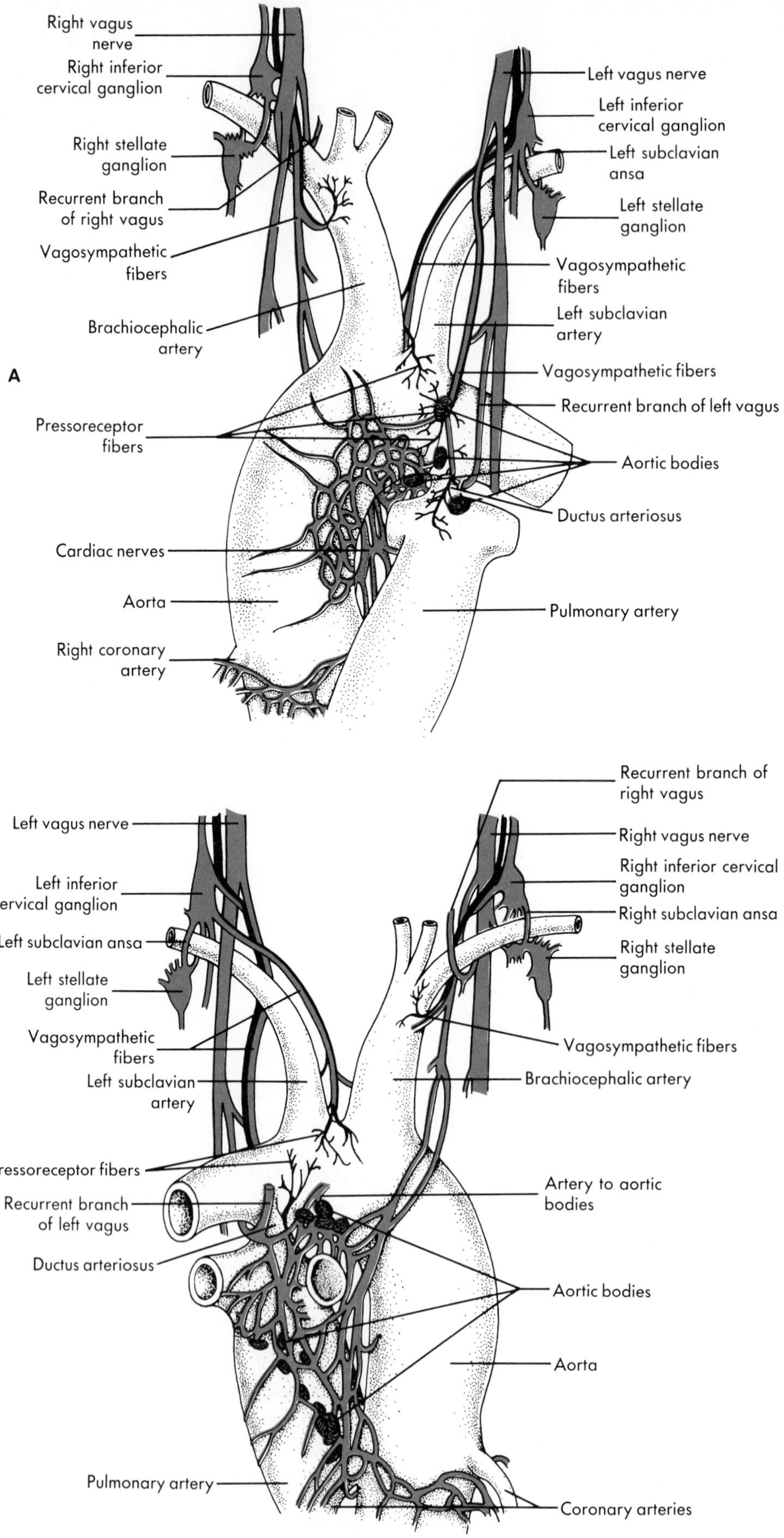

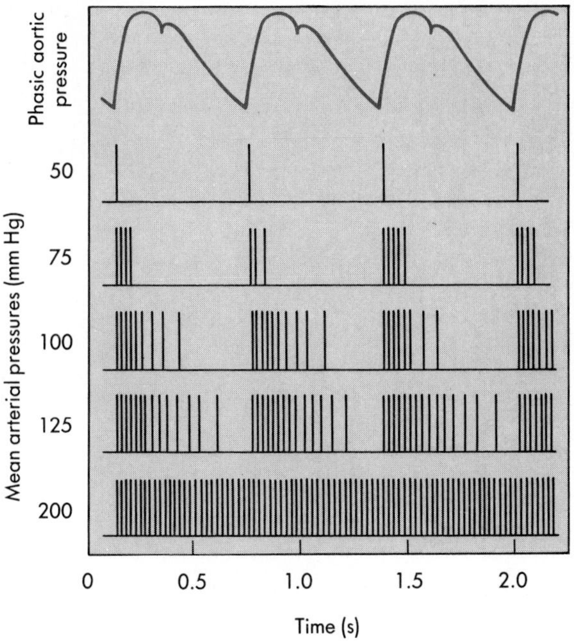

Fig. 33-10. Relationship of phasic aortic blood pressure in the firing of a single afferent nerve fiber from the carotid sinus at different levels of mean arterial pressure.

aortic arch reach the NTS via afferent fibers in the vagus nerves. The pressoreceptor nerve terminals in the walls of the carotid sinus and aortic arch respond to the stretch and deformation of the vessel induced by the arterial pressure. The frequency of firing is enhanced with an increase in blood pressure and diminished with a decrease in blood pressure. An increase in impulse frequency, as occurs with a rise in arterial pressure, inhibits the vasoconstrictor regions, resulting in peripheral vasodilation and a lowering of blood pressure. Contributing to a lowering of the blood pressure is a bradycardia brought about by stimulation of the vagal regions. The carotid sinus and aortic baroreceptors are not equipotent in their effects on peripheral resistance in response to nonpulsatile alterations in blood pressure. The carotid sinus baroreceptors are more sensitive than those in the aortic arch. Changes in pressure in the carotid sinus evoke greater alterations in perfusion pressure, and hence in resistance, than do equivalent changes in aortic arch pressure. However, with pulsatile changes in blood pressure the two sets of baroreceptors respond similarly.

The carotid sinus with the sinus nerve intact can be isolated from the rest of the circulation and perfused by either a donor dog or an artificial perfusion system. Under these conditions changes in the pressure within the carotid sinus are associated with reciprocal changes in the blood pressure of the experimental animal. The receptors in the walls of the carotid sinus show some adaptation and therefore are more responsive to constantly changing pressures than to sustained constant pressures. This is illustrated in Fig. 33-10, where at normal levels of blood pressure a barrage of impulses from a single fiber of the sinus nerve is initiated in early systole by the pressure rise and only a few spikes are observed during late systole and early diastole. At lower pressures these phasic changes are even more evident, but the overall frequency of discharge is reduced. The blood pressure threshold for eliciting sinus nerve impulses is about 50 mm Hg, and a maximum sustained firing is reached at around 200 mm Hg. Since the pressoreceptors show some degree of adaptation, their response at any level of mean arterial pressure is greater with a large than with a small pulse pressure. This is illustrated in Fig. 33-11, which shows the effects of damping pulsations in the carotid sinus on the frequency of firing in a fiber of the sinus nerve and on the systemic arterial pressure. When the pulse pressure in the carotid sinuses is reduced with an air chamber (Windkessel), but mean pressure remains constant, the rate of electrical impulses recorded from a sinus nerve fiber decreases and the systemic arterial pressure increases. Restoration of the pulse pressure in the carotid sinus restores the frequency of sinus nerve discharge and systemic arterial pressure to control levels.

The resistance increases that occur in the peripheral vascular beds in response to a reduced pressure in the carotid sinus vary from one vascular bed to another and thereby produce a redistribution of blood flow. For example, in the dog the resistance changes elicited by altering carotid sinus pressure around the normal operating sinus pressure are greatest in the femoral vessels, less in the renal, and least in the mesenteric and celiac vessels. Furthermore, the sensitivity of the carotid sinus reflex can be altered. Local application of norepinephrine or stimulation of sympathetic nerve fibers to the carotid sinuses enhances the sensitivity of the receptors in the sinus so that a given increase in intrasinus pressure produces a greater depressor response. A decrease in baroreceptor sensitivity occurs in hypertension when the carotid sinus becomes stiffer and less deformable as a result of the high intraarterial pressure. Under these conditions a given increase in carotid sinus pressure elicits a smaller decrement in systemic arterial pressure than it does at normal levels of blood pressure. In other words, the set point of the baroreceptors is raised in hypertension so that the threshold is increased and the receptors are less sensitive to change in transmural pressure.

In some individuals the carotid sinus is quite sensitive to pressure. Hence tight collars or other forms of external pressure over the region of the carotid sinus may elicit marked hypotension and fainting. In some patients with severe coronary artery disease and chest pain *(angina pectoris),* symptoms have been temporarily relieved by stimulation of the sinus nerve by means

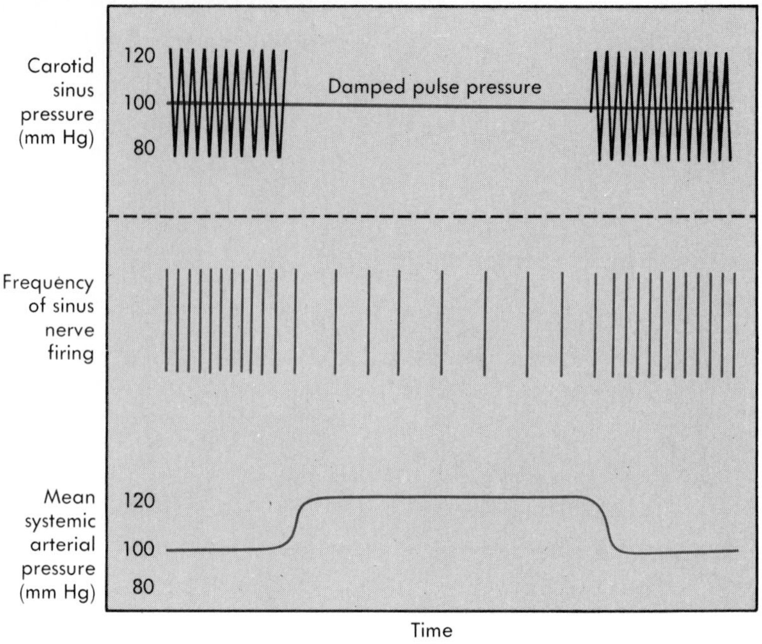

■ **Fig. 33-11.** Effect of reducing pulse pressure in the vascularly isolated perfused carotid sinuses *(top record)* on impulses recorded from a fiber of a sinus nerve *(middle record)* and on mean systemic arterial pressure *(bottom record)*. Mean pressure in the carotid sinuses *(colored line, top record)* is held constant when pulse pressure is damped.

of a chronically implanted stimulator that can be activated externally. The reduction in blood pressure achieved by sinus nerve stimulation decreases the pressure work of the heart and hence the myocardial ischemia responsible for the pain. As would be expected, denervation of the carotid sinus can produce temporary, and in some instances prolonged, hypertension.

The baroreceptors play a key role in short-term adjustments of blood pressure when relatively abrupt changes in blood volume, cardiac output, or peripheral resistance (as in exercise) occur. However, long-term control of blood pressure—that is, over days, weeks, and longer—is determined by the fluid balance of the individual, namely, the balance between fluid intake and fluid output. By far the single most important organ in the control of body fluid volume, and hence blood pressure, is the kidney. With overhydration, excessive fluid intake is excreted, whereas with dehydration there is a marked reduction in urine output.

Cardiopulmonary baroreceptors. In addition to the carotid sinus and aortic baroreceptors, there are also cardiopulmonary receptors with vagal and sympathetic afferent and efferent nerves. These cardiopulmonary reflexes are tonically active and can alter peripheral resistance with changes in intracardiac, venous, or pulmonary vascular pressures. The receptors are located in the atria, ventricles, and pulmonary vessels; when stimulated they reflexly inhibit sympathetic vasoconstrictor tone of the resistance vessels, thereby lowering blood pressure. The afferent limb of this depressor reflex consists of vagus nerve fibers. In contrast, the afferent sympathetic nerve fibers of the cardiopulmonary baroreceptors carry impulses centrally that primarily augment sympathetic tone of the resistance vessels (pressor effect). The atrial receptors are of two types: those stimulated by atrial contraction (A receptors) and those stimulated by atrial filling and distension (B receptors).

The cardiopulmonary baroreceptors also play a role in the regulation of vasomotor tone and urine output. Stimulation of the receptors inhibits angiotensin, aldosterone, and vasopressin (antidiuretic hormone) release; interruption of the reflex pathway has the opposite effects. Changes in urine volume elicited by changes in cardiopulmonary baroreceptor activation are important in the regulation of blood volume. For example, a decrease in blood volume (hypovolemia), as occurs in hemorrhage, enhances sympathetic vasoconstriction in the kidney and increases secretions of renin, angiotensin, aldosterone, and antidiuretic hormone. The renal vasoconstriction (primarily afferent arteriolar) reduces glomerular filtration and increases renin release from the kidney. Renin acts on a plasma substrate to form angiotensin, which increases aldosterone release from the adrenal cortex. The enhanced release of antidiuretic hormone increases water reabsorption. The net result is retention of salt and water by the kidney and a sensation of thirst. Angiotensin also raises systemic arteriolar tone.

The cardiopulmonary receptors and carotid and aortic baroreceptors are necessary for the full expression of blood pressure regulation. The major responses elicited by stimulation of these receptors differ in different vascular beds. For example, activation of the carotid pressoreceptors has a large effect on the splanchnic vascular bed but only a little influence on forearm vascular resistance. The reverse is true for the cardiopulmonary receptors.

Peripheral chemoreceptors. The chemoreceptors consist of small, highly vascular bodies in the region of the aortic arch and just medial to the carotid sinuses (Figs. 33-8 and 33-9). They are sensitive to changes in the Po_2, Pco_2, and pH of the blood. Although they are primarily concerned with the regulation of respiration, they reflexly influence the vasomotor regions to a minor degree. A reduction in arterial blood O_2 tension (Pao_2) stimulates the chemoreceptors, and the increase in the number of impulses in the afferent nerve fibers from the carotid and aortic bodies stimulates the vasoconstrictor regions, resulting in increased tone of the resistance and capacitance vessels. The chemoreceptors are also stimulated by increased arterial blood CO_2 tension ($Paco_2$) and reduced pH, but the reflex effect induced is quite small compared to the direct effect of hypercapnia and hydrogen ions on the vasomotor regions in the medulla. When hypoxia and hypercapnia occur at the same time, the stimulation of the chemoreceptors is greater than the sum of the two stimuli when they act alone. When the chemoreceptors are stimulated simultaneously with a reduction in pressure in the baroreceptors, the chemoreceptors potentiate the vasoconstriction observed in the peripheral vessels. However, when the baroreceptors and chemoreceptors are both stimulated (for example, high carotid sinus pressure and low Pao_2), the effects of the baroreceptors predominate.

There are also chemoreceptors with sympathetic afferent fibers in the heart. These cardiac chemoreceptors are activated by ischemia and transmit the precordial pain (angina pectoris) associated with an inadequate blood supply to the myocardium.

Hypothalamus. Optimum function of the cardiovascular reflexes requires the integrity of pontine and hypothalamic structures. Furthermore, these structures are responsible for behavioral and emotional control of the cardiovascular system. Stimulation of the anterior hypothalamus produces a fall in blood pressure and bradycardia, whereas stimulation of the posterolateral region of the hypothalamus produces a rise in blood pressure and tachycardia. The hypothalamus also contains a temperature-regulating center that affects the skin vessels. Stimulation by cold applications to the skin or by cooling of the blood perfusing the hypothalamus results in constriction of the skin vessels and heat conservation, whereas warm stimuli result in cutaneous vasodilation and enhanced heat loss.

Cerebrum. The cerebral cortex can also exert a significant effect on blood flow distribution in the body. Stimulation of the motor and premotor areas can affect blood pressure; usually a pressor response is obtained. However, vasodilation and depressor responses may be evoked, as in blushing or fainting, in response to an emotional stimulus.

Skin and viscera. Painful stimuli can elicit either pressor or depressor responses, depending on the magnitude and location of the stimulus. Distension of the viscera often evokes a depressor response, whereas painful stimuli on the body surface usually evoke a pressor response. In the anesthetized animal, strong electrical stimulation of a sensory nerve will produce a strong pressor response. However, it is sometimes possible to obtain a depressor response with low-intensity and low-frequency stimulation. Furthermore, all vascular beds do not exhibit the same response; in some, resistance increases, whereas in others it decreases. In addition, muscle contractions can elicit reflex changes in the magnitude of vasoactivity in the muscle. For the most part, these reflexes are mediated through the vasomotor areas in the medulla, but there are also spinal areas that can aid in the regulation of peripheral resistance.

Pulmonary Reflexes

Inflation of the lungs reflexly induces systemic vasodilation and a decrease in arterial blood pressure. Conversely, collapse of the lungs evokes systemic vasoconstriction. Afferent fibers mediating this reflex run in the vagus nerves and possibly to a limited extent in the sympathetic nerves. Their stimulation by stretch of the lungs inhibits the vasomotor areas. The magnitude of the depressor response to lung inflation is directly related to the degree of inflation and to the existing level of vasoconstrictor tone.

Chemosensitive Regions of the Medulla

Increases of $Paco_2$ stimulate the vasoconstrictor regions, thereby increasing peripheral resistance. Reduction in $Paco_2$ below normal levels (as with hyperventilation) decreases the degree of tonic activity of these areas, thereby decreasing peripheral resistance. The chemosensitive regions are also affected by changes in pH. A lowering of blood pH stimulates and a rise in blood pH inhibits these areas. These effects of changes in $Paco_2$ and blood pH possibly operate through changes in cerebrospinal fluid pH, as appears to be the case for the respiratory center. Whether there are special hydrogen ion chemoreceptors mediating pH-induced vasomotor effects has not been established.

Oxygen tension has relatively little direct effect on the vasomotor region. The primary effect of hypoxia is reflexly mediated via the carotid and aortic chemoreceptors. Moderate reduction of Pao_2 will stimulate the vasomotor region, but severe reduction will depress vasomotor activity in the same manner that other areas of the brain are depressed by very low O_2 tensions.

Cerebral ischemia, which may occur because of excessive pressure exerted by an expanding intracranial tumor, results in a marked increase in vasoconstriction. The stimulation is probably caused by a local accumulation of CO_2 and reduction of O_2 and possibly by excitation of intracranial baroreceptors. With prolonged, severe ischemia, central depression eventually supervenes, and the blood pressure falls.

Balance Between Extrinsic and Intrinsic Factors in Regulation of Peripheral Blood Flow

Dual control of the peripheral vessels by intrinsic and extrinsic mechanisms makes possible a number of vascular adjustments that enable the body to direct blood flow to areas where it is needed in greater supply and away from areas whose immediate requirements are less. In some tissues a more or less fixed relative potency of extrinsic and intrinsic mechanisms exists, and in other tissues the ratio is changeable, depending on the state of activity of that tissue.

In the brain and the heart, both vital structures with very limited tolerance for a reduced blood supply, intrinsic flow-regulating mechanisms are dominant. For instance, massive discharge of the vasoconstrictor region over the sympathetic nerves, which might occur in severe, acute hemorrhage, has negligible effects on the cerebral and cardiac resistance vessels, whereas skin, renal, and splanchnic blood vessels become greatly constricted.

In the skin the extrinsic vascular control is dominant. Not only do the cutaneous vessels participate strongly in a general vasoconstrictor discharge, but they also respond selectively through hypothalamic pathways to subserve the heat loss and heat conservation function

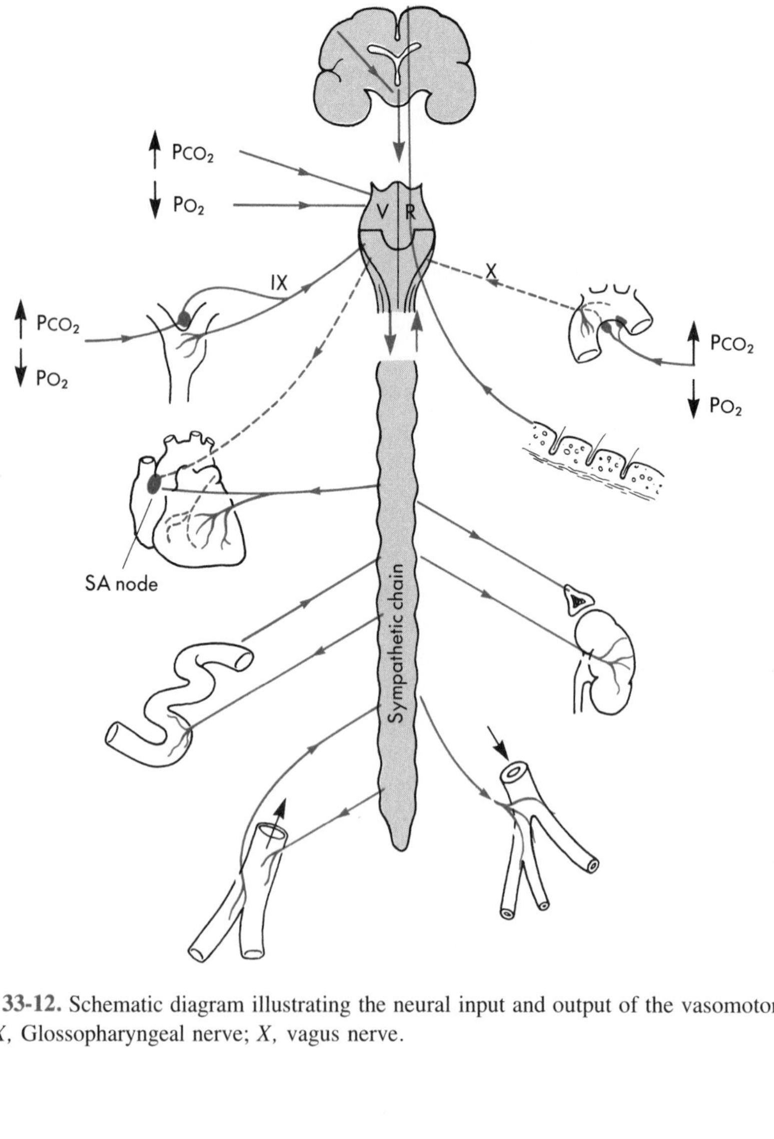

Fig. 33-12. Schematic diagram illustrating the neural input and output of the vasomotor region *(VR)*. *IX,* Glossopharyngeal nerve; *X,* vagus nerve.

required in body temperature regulation. However, intrinsic control can be demonstrated by local changes of temperature that can modify or override the central influence on resistance and capacitance vessels.

In skeletal muscle the interplay and changing balance between extrinsic and intrinsic mechanisms can be clearly seen. In resting skeletal muscle, neural control (vasoconstrictor tone) is dominant, as can be demonstrated by the large increment in blood flow that occurs immediately after section of the sympathetic nerves to the tissue. In anticipation of and at the start of exercise, such as running, blood flow increases in the leg muscles, possibly mediated by activation of the cholinergic sympathetic dilator system. After the onset of exercise the intrinsic flow-regulating mechanism assumes control, and because of the local increase in metabolites, vasodilation occurs in the active muscles. Vasoconstriction occurs in the inactive tissues as a manifestation of the general sympathetic discharge, but constrictor impulses reaching the resistance vessels of the active muscles are overridden by the local metabolic effect. Operation of this dual control mechanism thus provides increased blood where it is required and shunts it away from relatively inactive areas. Similar effects may be achieved with an increase in Pa_{CO_2}. Normally the hyperventilation associated with exercise keeps Pa_{CO_2} at normal levels. However, were Pa_{CO_2} to increase, a generalized vasoconstriction would occur because of stimulation of the vasoconstrictor region by CO_2. In the active muscles, where the CO_2 concentration is highest, the smooth muscle of the arterioles would relax in response to the local P_{CO_2}. Factors affecting and affected by the vasomotor region are summarized in Fig. 33-12.

Bibliography

Journal articles

Abboud, F.M., Eckberg, D.L., Johannsen, J.H., and Mark, A.L.: Carotid and cardiopulmonary baroreceptor control of splanchnic and forearm vascular resistance during venous pooling in man, J. Physiol. **286:**173, 1979.

Belloni, F.L., and Sparks, H.V.: The peripheral circulation, Annu. Rev. Physiol. **40:**67, 1978.

Berne, R.M., Knabb, R.M., Ely, S.W., and Rubio, R.: Adenosine in the local regulation of blood flow: a brief overview, Fed. Proc. **42:**3136, 1983.

Brown, A.M.: Receptors under pressure—an update on baroreceptors, Circ. Res. **46:**1, 1980.

Coleridge, H.M., and Coleridge, J.C.G.: Cardiovascular afferents involved in regulation of peripheral vessels, Annu. Rev. Physiol. **42:**413, 1980.

Donald, D.E., and Shepherd, J.T.: Reflexes from the heart and lungs: physiological curiosities or important regulatory mechanisms, Cardiovasc. Res. **12:**449, 1978.

Donald D.E., and Shepherd J.T.: Autonomic regulation of the peripheral circulation, Annu. Rev. Physiol. **42:**429, 1980.

Hilton, S.M., and Spyer, K.M.: Central nervous regulation of vascular resistance, Annu. Rev. Physiol. **42:**399, 1980.

Kunze, D.L., Krauhs, J.M., and Orlea, C.J.: Direct action of norepinephrine on aortic baroreceptors of rat adventitia, Am. J. Physiol. **247:**H811, 1984.

Shepherd, J.T.: Reflex control of arterial blood pressure, Cardiovasc. Res. **16:**357, 1982.

Books and monographs

Abboud, F.M., and Thames M.D.: Interaction of cardiovascular reflexes in circulatory control. In Handbook of physiology; Section 2: The cardiovascular system—peripheral circulation and organ blood flow, vol. III, Bethesda, Md., 1983, American Physiological Society.

Bevan, J.A., Bevan R.D., and Duckles, S.P.: Adrenergic regulation of vascular smooth muscle. In Handbook of physiology; Section 2: The cardiovascular system—vascular smooth muscle, vol. II, Bethesda, Md., 1980, American Physiological Society.

Bishop, V.S., Malliani, A., and Thoren, P.: Cardiac mechanoreceptors. In Handbook of physiology; Section 2: The cardiovascular system, vol. III, Bethesda, Md., 1983, American Physiological Society.

Brown, A.M.: Cardiac reflexes. In Handbook of physiology; Section 2: The cardiovascular system—the heart, vol. I, Bethesda, Md., 1979, American Physiological Society.

Crass, M.F., and Barnes, D.C., editors: Vascular smooth muscle—metabolic, ionic and contractile mechanisms, New York, 1982, Academic Press, Inc.

Eyzaguirre, C., Fitzgerald, R.S., Lahiri, S., and Zapata, P.: Arterial chemoreceptors. In Handbook of physiology; Section 2: The cardiovascular system—peripheral circulation and organ blood flow, vol. III, Bethesda, Md., 1983, American Physiological Society.

Johnson, P.C.: The myogenic response. In Handbook of physiology; Section 2: The cardiovascular system—vascular smooth muscle, vol. II, Bethesda, Md., 1980, American Physiological Society.

Korner, P.I.: Central nervous control of autonomic cardiovascular function. In Handbook of physiology; Section 2: The cardiovascular system—the heart, vol. I, Bethesda, Md., 1979, American Physiological Society.

Kovach, A.G.B., Sandos, P., and Kollii, M., editors: Cardiovascular physiology: neural control mechanisms, New York, 1981, Academic Press, Inc.

Mancia, G., and Mark, A.L.: Arterial baroreflexes in humans. In Handbook of physiology; Section 2: The cardiovascular system—peripheral circulation and organ blood flow, vol. III, Bethesda, Md., 1983, American Physiological Society.

Mark, A.L., and Mancia, G.: Cardiopulmonary baroreflexes in humans. In Handbook of physiology; Section 2: The cardiovascular system—peripheral circulation and organ blood flow, vol. III, Bethesda, Md., 1983, American Physiological Society.

Mulvany, M.J., Strandgaard, S., and Hammersen, F., editors: Resistance vessels: physiology, pharmacology and hypertensive pathology, Basel, Switzerland, 1985, S. Karger.

Rhodin, J.A.G.: Architecture of the vessel wall. In Handbook of physiology; Section 2: The cardiovascular system—vascular smooth muscle, vol. II, Bethesda, Md., 1980, American Physiological Society.

Rothe, C.F.: Venous system: physiology of the capacitance vessels. In Handbook of physiology; Section 2: The cardiovascular system—peripheral circulation and organ blood flow, vol. III, Bethesda, Md., 1983, American Physiological Society.

Sagawa, K.: Baroreflex control of systemic arterial pressure and vascular bed. In Handbook of physiology; Section 2: The cardiovascular system—peripheral circulation and organ blood flow, vol. III, Bethesda, Md., 1983, American Physiological Society.

Sparks, H.V., Jr.: Effect of local metabolic factors on vascular smooth muscle. In Handbook of physiology; Section 2: The cardiovascular system—vascular smooth muscle, vol. II., Bethesda, Md., 1980, American Physiological Society.

CHAPTER

34

Control of Cardiac Output: Coupling of Heart and Blood Vessels

The following four factors (Fig. 34-1) are usually considered to control cardiac output: heart rate, myocardial contractility, preload, and afterload. Heart rate and myocardial contractility are strictly "cardiac factors." They are characteristics of the cardiac tissues, although they are subject to modulation by various neural and humoral mechanisms. Preload and afterload, however, depend on the characteristics of both the heart and the vascular system. On the one hand, preload and afterload are important determinants of cardiac output. On the other hand, preload and afterload are themselves determined by the cardiac output and certain vascular characteristics. Preload and afterload may be designated "coupling factors," because they constitute a functional coupling between the heart and blood vessels. The heart pumps the blood around the vascular system. Concomitantly, the vessels partly determine the preload and afterload and hence regulate the quantity of blood that the heart will pump around the circuit per unit time.

To understand the regulation of cardiac output, therefore, it is important to appreciate the nature of the coupling between the heart and the vascular system. Guyton and his colleagues have developed graphic techniques that we shall use in modified form to analyze the interactions between the cardiac and vascular components of the circulatory system.

The graphic analysis involves two simultaneous functional relationships between the *cardiac output* and the *central venous pressure* (that is, the pressure in the right atrium and thoracic venae cavae). The curve defining one of these relationships will be called the *cardiac function curve*. It is an expression of the well-known Frank-Starling relationship (p. 458) and reflects the fact that the cardiac output depends, in part, on the preload (this is, the central venous, or right atrial, pressure). The cardiac function curve is a characteristic of the heart itself and has been studied in hearts completely isolated from the rest of the circulatory system (Fig. 29-12).

The second functional relationship between the central venous pressure and the cardiac output is defined by a second curve, which we shall call the *vascular function curve*. This relationship depends only on certain characteristics of the vascular system, namely, the peripheral resistance, the arterial and venous capacitances, and the blood volume. The vascular function curve is entirely independent of the characteristics of the heart, and it can be studied even if the heart were replaced by a mechanical pump.

■ *Vascular Function Curve*

The vascular function curve defines the change in central venous pressure that occurs as a consequence of a change in cardiac output; that is, central venous pressure is the dependent variable (or response), and cardiac output is the independent variable (or stimulus). This contrasts with the cardiac function curve, for which the central venous pressure is the independent variable and the cardiac output is the dependent variable.

The simplified model of the circulation illustrated in Fig. 34-2 will help explain how the cardiac output determines the level of the central venous pressure. The essential components of the cardiovascular system have been lumped into four elements. The right and left sides of the heart, as well as the pulmonary vascular bed, are considered simply as a pump-oxygenator, much as that employed during open heart surgery. The high-resistance microcirculation is designated the peripheral resistance. Finally, the entire *capacitance* of the system is subdivided into two components, the total arterial capacitance, C_a, and the total venous capacitance, C_v. As

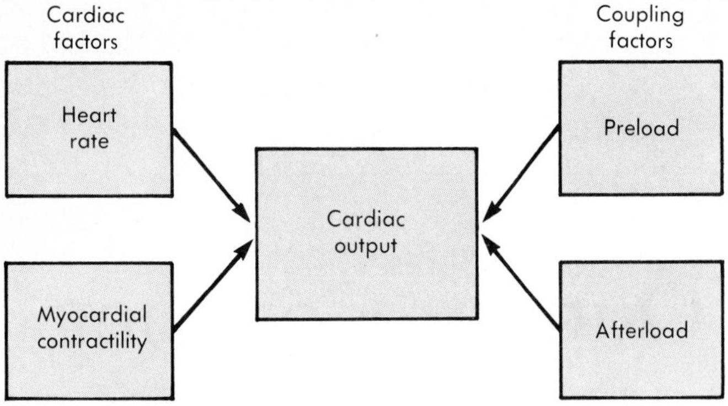

■ **Fig. 34-1.** The four factors that determine cardiac output.

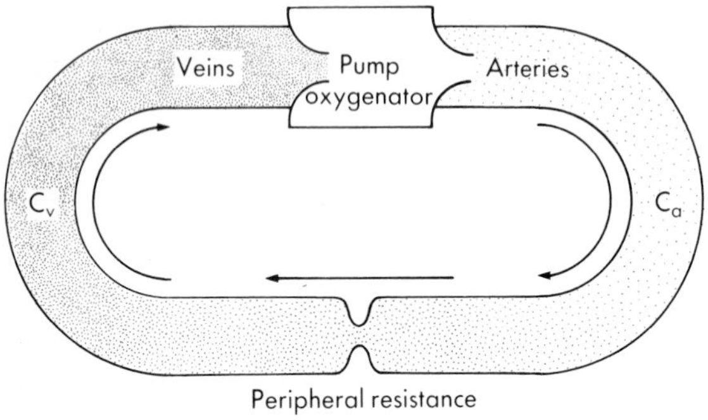

■ **Fig. 34-2.** Simplified model of the cardiovascular system, consisting of a pump-oxygenator, an arterial capacitance *(C_a)*, a peripheral resistance, and a venous capacitance *(C_v)*.

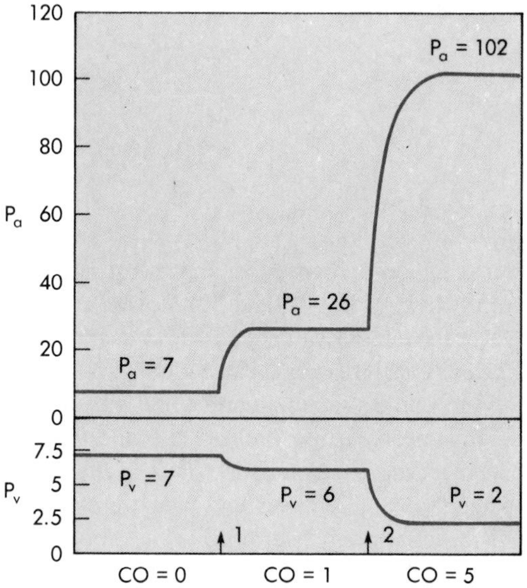

■ **Fig. 34-3.** The changes in arterial *(P_a)* and venous *(P_v)* pressures in the circulatory model shown in the preceding figure. The total peripheral resistance is 20 mm Hg/L/minute, and the ratio of C_v to C_a is 19:1. The cardiac output *(CO)* is 0 to the left of arrow *1*. It is increased to 1 L/minute at arrow *1*, and to 5 L/minute at arrow 2.

defined on p. 487, *capacitance* (C) is the increment of volume (dV) accommodated per unit change of pressure (dP); that is,

$$C = dV/dP$$

The venous capacitance is about 20 times as great as the arterial capacitance. In the example to follow, the ratio of C_v to C_a will be set at 19:1 to simplify certain calculations. Thus, if it were necessary to add *x* ml of blood to the arterial system to produce a 1 mm Hg increment in arterial pressure, then it would be necessary to add 19*x* ml of blood to the venous system to raise venous pressure by the same amount.

With the model system at rest, pressures are the same throughout the entire circuit. The pressure that exists at rest is a function of only the total volume of blood in the system and the elastic characteristics of the walls (that is, the overall capacitance of the system). This equilibrium pressure has been termed the *mean circulatory pressure*. At normal blood volumes and with normal vessels, the magnitude of the mean circulatory pressure has been estimated to be about 7 mm Hg. To the left of arrow *1* in Fig. 34-3, the arterial pressure (P_a) and the venous pressure (P_v) are both equal to 7 mm Hg when the cardiac output (CO) is zero.

Let the pump-oxygenator (or simply the pump) in Fig. 34-2 start suddenly to deliver a constant flow of 1 L/min (at arrow *1*, Fig. 34-3), and let peripheral resistance remain constant at 20 mm Hg/L/min. The direction of transfer of blood will be from the venous to the arterial side of the circuit. Because of the blood volume shift, pressure will begin to fall on the venous side and rise on the arterial side. The arterial pressure, P_a, will continue to rise and the venous pressure, P_v, will continue to fall until the pressure gradient, $P_a - P_v$, is great enough to force a flow, Q, of 1 L/min through the peripheral resistance, R. The pressure gradient across the peripheral resistance is often referred to as the *vis a tergo,* or force from behind. *It is the single most important factor responsible for the venous return to the heart and is directly ascribable to the pumping action of the heart itself.*

At equilibrium, the flow through the peripheral resistance must equal the flow being pumped by the heart. A pressure gradient of 20 mm Hg is necessary to force a flow of 1 L/min through a resistance of 20 mm Hg/L/min. When the ratio of C_v to C_a equals 19:1, this pressure gradient will be achieved by a 19 mm Hg rise in P_a and a 1 mm Hg fall in P_v. For readers interested in the derivation of these results, the equations are presented in the next section.

Mathematical Analysis

From the definition of peripheral resistance (p. 478):

$$R = (P_a - P_v)/Q \tag{1}$$

Given that Q = 1 and R = 20 and solving equation 1 for P_a:

$$P_a = P_v + QR = P_v + (1 \times 20) \tag{2}$$

Thus P_a will increase to a value 20 mm Hg greater than P_v. It will continue to be 20 mm Hg above P_v, as long as the pump output is maintained at 1 L/min and the peripheral resistance remains at 20 mm Hg/L/min.

We can calculate what the actual changes in P_a and P_v will be under these conditions. The arterial volume increment needed to achieve the required level of P_a depends entirely on the arterial capacitance, C_a. For a rigid arterial system (low capacitance) this volume will be small; for a distensible system the volume will be large. Whatever the magnitude, however, the change in volume represents the translocation of some quantity of blood from the venous to the arterial side of the circuit.

For a given total blood volume, any increment in arterial volume (ΔV_a) must equal the decrement in venous volume (ΔV_v); that is,

$$\Delta V_a = -\Delta V_v \tag{3}$$

From the definition of capacitance,

$$C_a = \Delta V_a/\Delta P_a \tag{4}$$

and

$$C_v = \Delta V_v/\Delta P_v \tag{5}$$

By substitution into equation 3,

$$\frac{\Delta P_v}{\Delta P_a} = \frac{C_a}{C_v} \tag{6}$$

Given that C_v is 19 times as great as C_a, then the increment in P_a will be 19 times as great as the decrement in P_v; that is,

$$\Delta P_a = -19\Delta P_v \tag{7}$$

To calculate the absolute values of P_a and P_v, let ΔP_a represent the difference between the prevailing P_a and the mean circulatory pressure (P_{mc}); that is, let

$$\Delta P_a = P_a - P_{mc} \tag{8}$$

and let ΔP_v represent the difference between the prevailing P_v and the mean circulatory pressure:

$$\Delta P_v = P_v - P_{mc} \tag{9}$$

Substituting these values for ΔP_a and ΔP_v into equation 7:

$$P_a - P_{mc} = -19(P_v - P_{mc}) \tag{10}$$

By solving equations 2 and 10 simultaneously:

$$P_a = P_{mc} + 19 \tag{11}$$

and

$$P_v = P_{mc} - 1 \tag{12}$$

Hence for a the mean circulatory pressure of 7 mm Hg, P_a increases to 26 mm Hg and P_v decreases to 6 mm Hg (Fig. 34-3). These pressure changes provide the required arteriovenous pressure gradient of 20 mm Hg.

If the pump output is abruptly increased to a constant level of 5 L/min (Fig. 34-3, arrow *2*) and peripheral resistance remains constant at 20 mm Hg/L/min, an additional volume of blood again will be translocated from the venous to the arterial side of the circuit. It will progressively accumulate in the arteries until P_a reaches a level of 100 mm Hg above P_v, as shown by substitution into equation 2:

$$P_a = P_v + QR = P_v + (5 \times 20) \tag{13}$$

By solving equations 10 and 13 simultaneously, we find that P_a rises to a value of 95 mm Hg above P_{mc}, and P_v falls to a value 5/mm Hg below P_{mc}. In Fig. 34-3, therefore P_v declines to 2 mm Hg and P_a rises to 102 mm Hg. The resultant pressure gradient of 100 mm Hg will force a cardiac output of 5 L/min through a constant peripheral resistance of 20 mm Hg/L/min.

Venous Pressure Dependence on Cardiac Output

Experimental and clinical observations have shown that alterations in cardiac output do indeed evoke the directional changes in P_a and P_v that have been predicted earlier for our simplified model. In an experiment on an anesthesized dog, a mechanical pump was substituted for the right ventricle (Fig. 34-4). As the cardiac output, Q, was diminished in a series of small steps, P_a fell and P_v rose. Similarly, a major coronary artery may suddenly become occluded in a human patient. The resultant *acute myocardial infarction* often leads to a substantial reduction in cardiac output, which is attended

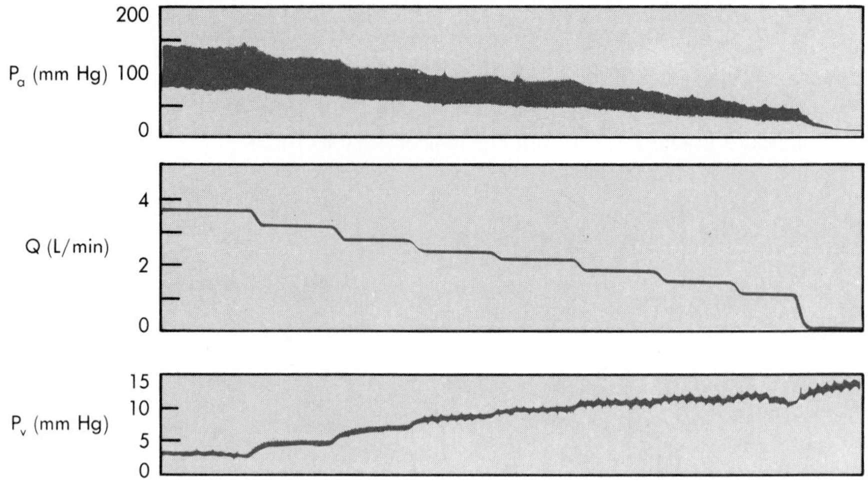

Fig. 34-4. The changes in arterial *(P_a)* and central venous *(P_v)* pressures produced by changes in systemic blood flow *(Q)* in a canine right-heart bypass preparation. Stepwise changes in Q were produced by altering the rate at which blood was mechanically pumped from the right atrium to the pulmonary artery. (From Levy, M.N.: Circ. Res. **44**:739, 1979. By permission of the American Heart Association, Inc.)

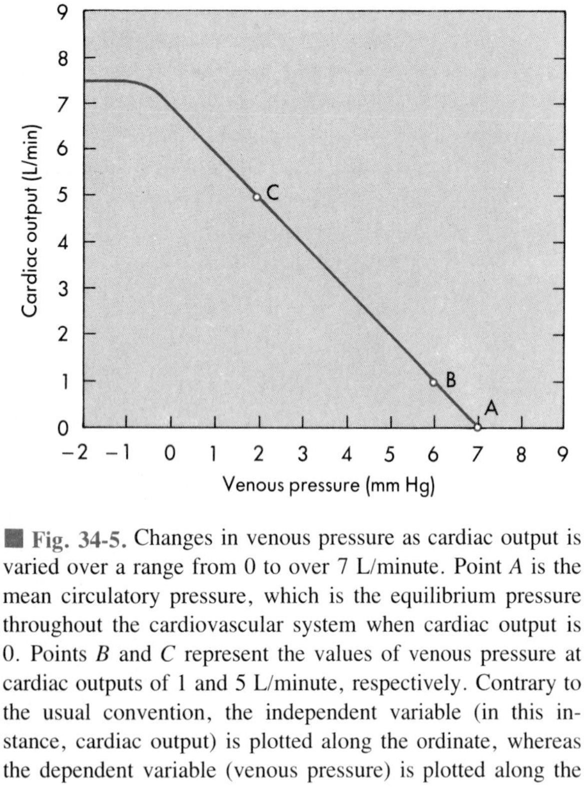

Fig. 34-5. Changes in venous pressure as cardiac output is varied over a range from 0 to over 7 L/minute. Point *A* is the mean circulatory pressure, which is the equilibrium pressure throughout the cardiovascular system when cardiac output is 0. Points *B* and *C* represent the values of venous pressure at cardiac outputs of 1 and 5 L/minute, respectively. Contrary to the usual convention, the independent variable (in this instance, cardiac output) is plotted along the ordinate, whereas the dependent variable (venous pressure) is plotted along the abscissa.

by a fall in the arterial pressure and a rise in the central venous pressure.

In animal experiments, if the output of the heart or of a substitute mechanical pump were varied and the resultant values of P_v were recorded, a *vascular function curve* could be constructed. Such a curve is shown in Fig. 34-5 and is derived from the theoretical data shown in Fig. 34-3. Contrary to the usual convention, the independent variable, cardiac output in this instance, is plotted along the ordinate, and the dependent variable, venous pressure, is plotted along the abscissa. The customary arrangement of axes is reversed for reasons that will become apparent later in this chapter. In Fig. 34-5, point *A* represents the mean circulatory pressure, point *B*, the value of P_v at a cardiac output of 1 L/min, and point *C*, the value of P_v at a cardiac output of 5 L/min. From equations 1, 6, 8, and 9, the following equation for the linear portion of the vascular function curve can be derived:

$$P_v = -\frac{RC_a}{C_a + C_v} Q + P_{mc} \qquad (14)$$

Note that the slope depends only on R, C_a, and C_v. Note also that when Q = O, then $P_v = P_{mc}$; that is, at zero flow, P_v equals the mean circulatory pressure.

The reduction of P_v that can be produced by an increase in cardiac output is limited. At some critical maximum value of cardiac output, sufficient fluid will be translocated from the venous to the arterial side of the circuit such that P_v will drop below the ambient pressure. In a system of distensible vessels, such as the venous system, the vessels will be collapsed by this greater external pressure. This venous collapse will, of course, limit the maximum value of cardiac output, regardless of the capabilities of the pump. Note that, in Fig. 34-5, cardiac output remains constant as P_v decreases below zero.

In the simplified schema in Fig. 34-2, the venous system was considered to be without resistance. In the body, however, there is a continuous pressure gradient from the venules to the right side of the heart. In an animal whose chest is open, the ambient pressure about the great central veins will be the atmospheric pressure. If an artificial pump is substituted for the heart and if the pump output is progressively increased, the site of collapse will be at the junction of the venae cavae with the inflow side of the pump. In the normal animal with an intact chest, if venous collapse occurs, it will take place at the points of entry of the veins into the chest, for reasons to be described later in this chapter.

Blood Volume

The vascular function curve is affected by variations in total blood volume. During circulatory standstill (zero cardiac output), the mean circulatory pressure depends only on vascular capacitance and blood volume, as stated previously. Thus for a given vascular capacitance the mean circulatory pressure will be increased when the blood volume is expanded *(hypervolemia)* and decreased when the blood volume is diminished *(hypovolemia)*. This is illustrated by the x-axis intercepts in Fig. 34-6, where the mean circulatory pressure is 5 mm Hg with hemorrhage and 9 mm Hg with transfusion, as compared to a value of 7 mm Hg when the blood volume is normal *(normovolemia)*.

Furthermore, the differences in P_v during hypervolemia, normovolemia, and hypovolemia in the static system are preserved at each level of cardiac output, so that the vascular function curves parallel each other (Fig. 34-6). To illustrate, consider the example of hypervolemia, in which the mean circulatory pressure is 9 mm Hg. In Fig. 34-6 both P_a and P_v would be 9 mm Hg, instead of 7 mm Hg, when the cardiac output is zero. With a sudden increase in cardiac output to 1 L/min (at arrow *1*, Fig. 34-3), if the peripheral resistance were still 20 mm Hg/L/min, an arteriovenous pressure gradient of 20 mm Hg would still be necessary for 1 L/min to flow through the resistance vessels. This does not differ from the example for normovolemia. Assuming the same ratio of C_v to C_a of 19:1, the pressure gradient would be achieved by a 1 mm Hg decline in P_v and a 19 mm Hg rise in P_a. Hence a change in cardiac output from 0 to 1 L/min would evoke the same 1 mm Hg reduction in P_v irrespective of the blood volume, as long as the C_v/C_a ratio and the peripheral resistance were independent of blood volume. Equation 14 also discloses (1) that the slope of the vascular function curve remains constant as long as R, C_v, and C_a do not change, and (2) that the slope does not depend on P_{mc}.

From Fig. 34-6 it is also apparent that the cardiac output at which P_v becomes 0 varies directly with the blood volume. Therefore the maximum value of cardiac output becomes progressively more limited as the total blood volume is reduced. However, the pressure at which the veins collapse (sharp change in slope of the vascular function curve) is not altered appreciably by changes in blood volume. This pressure depends only on the ambient pressure.

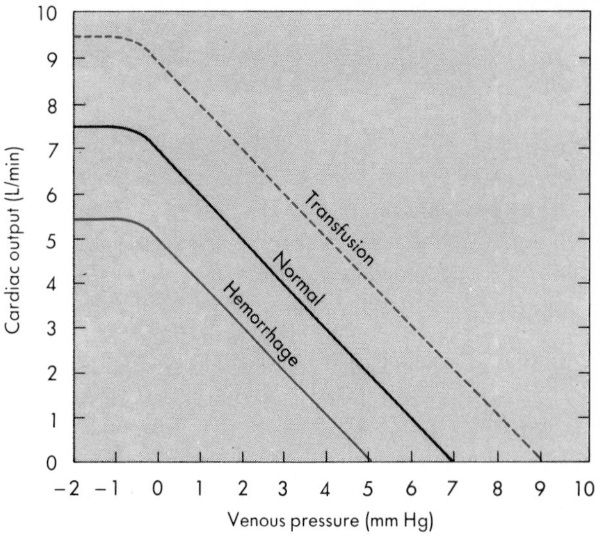

Fig. 34-6. Effects of increased blood volume (*transfusion* curve) and of decreased blood volume (*hemorrhage* curve) on the vascular function curve. Similar shifts in the vascular function curve are produced by increases and decreases, respectively, in venomotor tone.

Venomotor Tone

The effects of changes in venomotor tone on the vascular function curve closely resemble those for changes in blood volume. In Fig. 34-6, for example, the transfusion curve could just as well represent increased venomotor tone, whereas the hemorrhage curve could represent decreased tone. During circulatory standstill, for a given blood volume, the pressure within the vascular system will rise as the tension exerted by the smooth muscle within the vascular walls increases. The arteriolar and venous smooth muscle are under substantial nervous and humoral control. The fraction of the blood volume located within the arterioles is small, whereas the blood volume in the veins is large (see Fig. 26-2). Therefore changes in venous tone can alter the mean circulatory pressure appreciably, whereas changes in arteriolar constriction do not affect mean circulatory pressure substantially. Hence mean circulatory pressure rises with increased venomotor tone and falls with diminished tone.

Experimentally, the pressure attained shortly after abrupt circulatory standstill is usually above 7 mm Hg, even when blood volume is normal. This is attributable to the generalized venoconstriction elicited by cerebral ischemia, activation of the chemoreceptors, and reduced stimulation of the baroreceptors. If resuscitation is not successful, this reflex response subsides as central nervous activity ceases. At normal blood volume the mean circulatory pressure usually then approaches a value close to 7 mm Hg.

Blood Reservoirs

The extent of venoconstriction is considerably greater in certain regions of the body than in others. In effect, vascular beds that undergo appreciable venoconstriction constitute blood reservoirs. The vascular bed of the skin is one of the major blood reservoirs in people. During blood loss, profound subcutaneous venoconstriction occurs, giving rise to the characteristic pale appearance of the skin. The resultant redistribution of blood away from the skin liberates several hundred milliliters of blood to be perfused through more vital regions. The vascular beds of the liver, lungs, and spleen are important blood reservoirs. In the dog the spleen is packed with red blood cells and can constrict to a small fraction of its normal size. During hemorrhage this mechanism autotransfuses blood of high erythrocyte content into the general circulation. However, in humans the volume changes of the spleen are considerably smaller.

Peripheral Resistance

The modification of the vascular function curve introduced by changes in arteriolar tone is shown in Fig. 34-7. The arterioles contain only about 3% of the total blood volume (see Fig. 26-2). Hence changes in the contractile state of these vessels do not significantly alter the mean circulatory pressure, as stated previously. Thus the family of venous pressure curves representing a range of peripheral resistances converges at a common point on the abscissa.

At any given cardiac output, P_v varies inversely with the arteriolar tone, all other factors remaining constant. Arteriolar constriction sufficient to double the peripheral resistance will cause a twofold rise in P_a. In the example shown in Fig. 34-3, a change in the cardiac output from 0 to 1 L/min caused P_a to rise from 7 to 26 mm Hg, an increment of 19 mm Hg. If peripheral resistance had been twice as great, the same change in cardiac output would have evoked twice as great an increment in P_a. To achieve this greater rise in P_a, twice as great an increment in blood volume would be required on the arterial side of the circulation, assuming a constant arterial capacitance. Given a constant total blood volume, this larger arterial volume signifies a corresponding reduction in venous volume. Hence the decrement in venous volume would be twice as great when the peripheral resistance is doubled. With a constant venous capacitance, a twofold reduction in venous volume would be reflected by a twofold decline in P_v. Therefore, in Fig. 34-3, an increase in cardiac output to 1 L/min (arrow *1*) would have caused a 2 mm Hg decrement in P_v, to a level of 5 mm Hg, instead of the 1

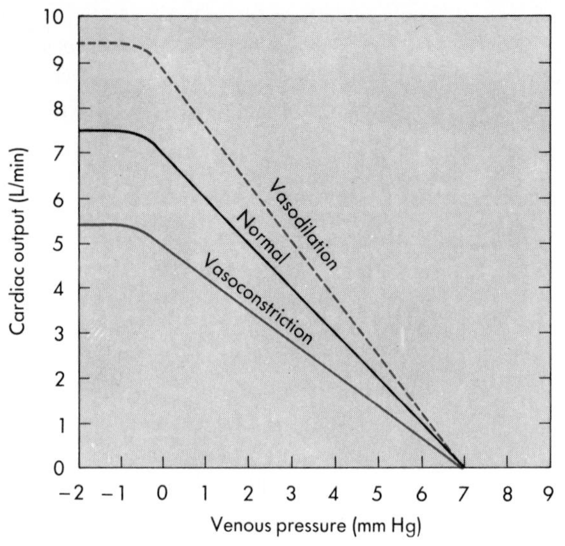

Fig. 34-7. Effects of arteriolar vasodilation and vasoconstriction on the vascular function curve.

mm Hg decrement that occurred with the normal peripheral resistance. Similarly, greater increases in cardiac output would have evoked proportionately greater decrements in P_v under conditions of increased peripheral resistance than with normal levels of resistance. This inverse relationship between the peripheral resistance and the decrement in P_v, together with the failure of peripheral resistance to affect the mean circulatory pressure, accounts for the counterclockwise rotation of the venous pressure curves with increased peripheral resistance (Fig. 34-7). Conversely. arteriolar vasodilation produces a clockwise rotation from the same horizontal axis intercept. A higher maximum level of cardiac output is attainable with vasodilation than with normal or increased arteriolar tone (Fig. 34-7).

Interrelationships Between Cardiac Output and Venous Return

Cardiac output and venous return are inextricably interdependent. Clearly, except for small, transient disparities, the heart is unable to pump any more blood than is delivered to it through the venous system. Similarly, since the circulatory system is a close circuit, the rate of venous return must equal the cardiac output over any appreciable time interval. The flow around the entire closed circuit depends on the capability of the pump, the characteristics of the circuit, and the total volume of fluid in the system. Cardiac output and venous return are simply two terms for the flow around the closed circuit. Cardiac output is the volume of blood being pumped by the heart per unit time. Venous return is the volume of blood returning to the heart per unit time. At equilibrium, these two flows are equal.

The techniques of circuit analysis will be applied in an effort to gain some insight into the control of flow around the circuit. Acute changes in cardiac contractility, peripheral resistance, or blood volume may transiently exert disparate effects on cardiac output and venous return. Except for such brief disparities, however, such factors simply alter flow around the entire circuit, and it is irrelevant whether one thinks of that flow as ''cardiac output'' or ''venous return.'' Commonly, authors have ascribed the reduction in cardiac output during hemorrhage, for example, to a decrease in venous return. It will become clear that such an explanation is a blatant example of circular reasoning. Hemorrhage reduces flow around the entire circuit, for reasons to be elucidated. *To attribute the reduction in cardiac output to a curtailment of venous return is equivalent to ascribing the decrease in total flow to a decrease in total flow!*

Coupling Between the Heart and the Vasculature

In accordance with Starling's law of the heart, or so-called heterometric autoregulation, cardiac output is intimately dependent on P_v, since right atrial pressure is virtually identical to central venous pressure. Furthermore, the right atrial pressure is approximately equal to the right ventricular end-diastolic pressure because the normal tricuspid valve constitutes a low resistance junction between the right atrium and ventricle. In the discussion to follow, graphs of cardiac output as a function P_v will be called *cardiac function curves*. Supervention of other types of intrinsic regulatory mechanisms (for example, homeometric autoregulation, p. 463) will modify such curves appreciably. For completeness such changes must be considered, but for simplicity the cardiac function curves will be drawn to represent only the Frank-Starling relationship. Extrinsic regulatory influences may be expressed as shifts in such curves, as indicated previously (p. 461).

A typical cardiac function curve is plotted on the same coordinates as a normal vascular function curve in Fig. 34-8. Contrary to the vascular function curve, the cardiac function curve is plotted according to the usual convention; that is, the variable (P_v) plotted along the abscissa is the independent variable (stimulus), and the variable (cardiac output) plotted along the ordinate is the dependent variable (response). In accordance with the Frank-Starling mechanism, the cardiac function curve reveals that a rise in P_v causes an *increase* in cardiac output. Conversely, the vascular function curve describes an inverse relationship between cardiac output and P_v; that is, a rise in cardiac output causes a *reduction* in P_v. The *equilibrium point* of such a system is defined by the intersection of these two curves. The coordinates of this equilibrium point represent the values of cardiac output and P_v at which such a system tends to operate. Only transient deviations from such values for cardiac output and P_v are possible, as long as the given cardiac and vascular function curves accurately describe the system.

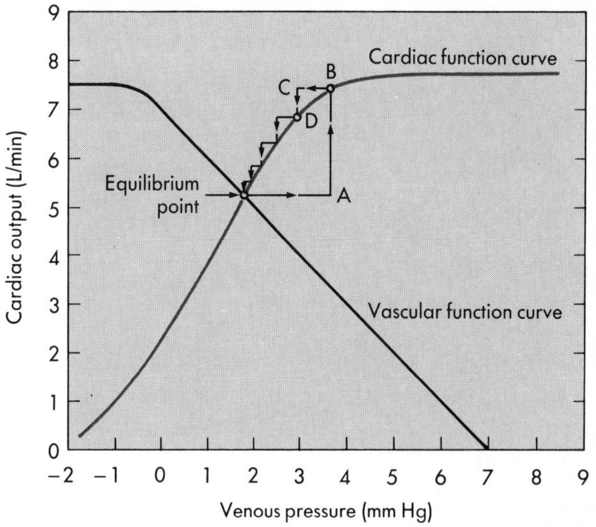

Fig. 34-8. Typical vascular and cardiac function curves plotted on the same coordinate axes. The coordinates of the equilibrium point, at the intersection of these curves, represent the stable values of cardiac output and central venous pressure at which the system tends to operate. Any perturbation (such as when venous pressure is suddenly increased to point *A*) institutes a sequence of changes in cardiac output and venous pressure such that these variables are returned to their equilibrium values.

The tendency for the cardiovascular system to operate about such an equilibrium point may best be illustrated by examining its response to a sudden perturbation. Consider the changes elicited by a sudden rise in P_v from the equilibrium point to point *A* in Fig. 34-8. Such a change might be induced by the rapid injection, during ventricular systole, of a given volume of blood on the venous side of the circuit, accompanied by the withdrawal of an equal volume from the arterial side so that total blood volume would remain constant. Because of the Frank-Starling mechanism, this elevated P_v would increase cardiac output (point *B*) during the very next ventricular systole. The increased cardiac output, in turn, would result in the net transfer of blood from the venous to the arterial side of the circuit, with a consequent reduction in P_v (to point *C*). During the next systole, cardiac output would therefore be less (point *D*), although still above the equilibrium point. This process would continue in ever-diminishing steps until the equilibrium values for P_v and cardiac output were reestablished (at the same point of intersection).

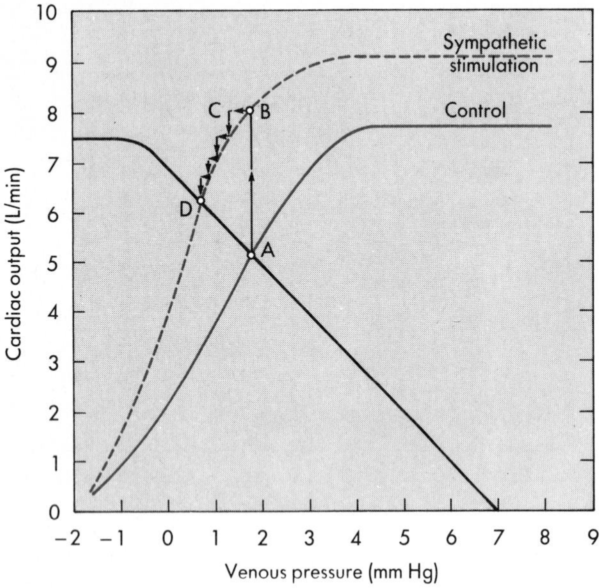

■ **Fig. 34-9.** Enhancement of myocardial contractility, as by cardiac sympathetic nerve stimulation, causes the equilibrium values of cardiac output and P_v to shift from the intersection (point *A*) of the control vascular and cardiac function curves *(continuous lines)* to the intersection (point *D*) of the same vascular function curve with that cardiac function curve *(dashed line)* representing enhanced myocardial contractility.

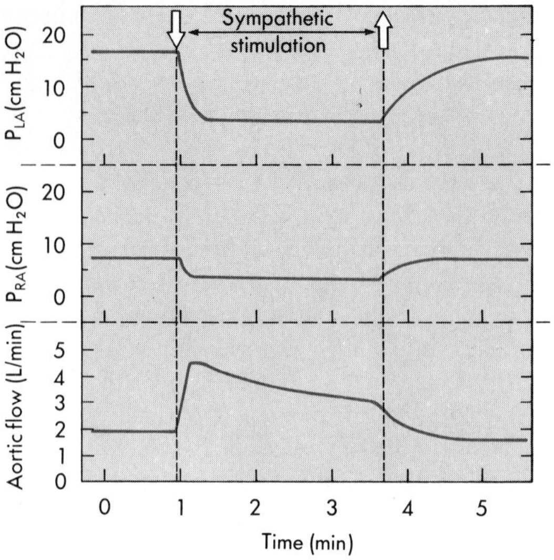

■ **Fig. 34-10.** During electrical stimulation of the left stellate ganglion (containing cardiac sympathetic nerve fibers), aortic blood flow increased while pressures in the left atrium *(P_{LA})* and right atrium *(P_{RA})* diminished. These data conform with the conclusions derived from Fig. 34-9, in which the equilibrium values of cardiac output and venous pressure are observed to shift from point *A* to point *D* during cardiac sympathetic nerve stimulation. (Redrawn from Sarnoff, S.J., et al.: Circ. Res. **8:**1108, 1960.)

■ *Enhanced Myocardial Contractility*

Graphs of cardiac and vascular function curves help explain the effects of alterations in ventricular contractility. In Fig. 34-9 the lower cardiac function curve represents the control state, whereas the upper curve reflects an improved contractility. This pair of curves is analogous to the family of ventricular function curves described on p. 461. Such enhancement of ventricular contractility might be achieved by electrical stimulation of the cardiac sympathetic nerves. Since the effects of such stimulation are restricted to the heart, the vascular function curve would be unaffected. Therefore one vascular function curve would suffice, as shown in Fig. 34-9.

During the control state the equilibrium values for cardiac output and P_v are designated by point *A*. With the onset of cardiac sympathetic nerve stimulation (assuming the effects to be instantaneous and constant), the prevailing level of P_v would abruptly raise cardiac output to point *B* because of the enhanced contractility. However, this high cardiac output would increase the net transfer of blood from the venous to the arterial side of the circuit, and consequently P_v will fall (to point *C*). Cardiac output will continue to fall until a new equilibrium point *(D)* is reached, which is located at the intersection of the vascular function curve with the new cardiac function curve. The new equilibrium point *(D)* lies above and to the left of the control equilibrium point *(A)*. This shift reveals that sympathetic stimulation evokes a greater cardiac output at a lower level of P_v. Such a change accurately describes the true response. In the experiment represented in Fig. 34-10, the left stellate ganglion was stimulated between the two arrows. During stimulation aortic flow rose substantially, accompanied by reductions in right and left atrial pressures (P_{RA} and P_{LA}).

■ *Blood Volume*

Changes in blood volume do not directly affect myocardial contractility but they do influence the vascular function curve in the manner shown in Fig. 34-6. Therefore, to understand the circulatory alterations evoked by a given change in blood volume, it is necessary to plot the appropriate cardiac function curve along with the vascular function curves that represent the control and experimental states.

Fig. 34-11 illustrates the response to a blood transfusion. Equilibrium point *B*, which denotes the values for cardiac output and P_v after transfusion, lies above and to the right of the control equilibrium point *A*. Thus transfusion increases both cardiac output and P_v. Hemorrhage has the opposite effect. Pure increases or decreases in venomotor tone elicit responses that are anal-

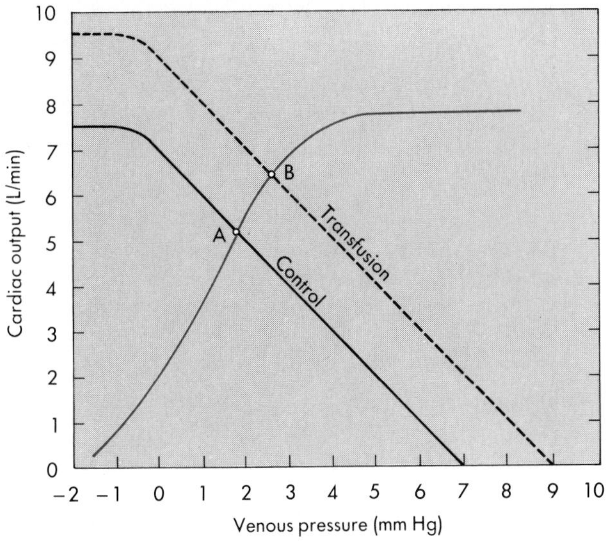

■ **Fig. 34-11.** After a blood transfusion the vascular function curve is shifted to the right *(dashed curve)*. Therefore cardiac output and venous pressure are both increased, as denoted by the translocation of the equilibrium point from *A* to *B*.

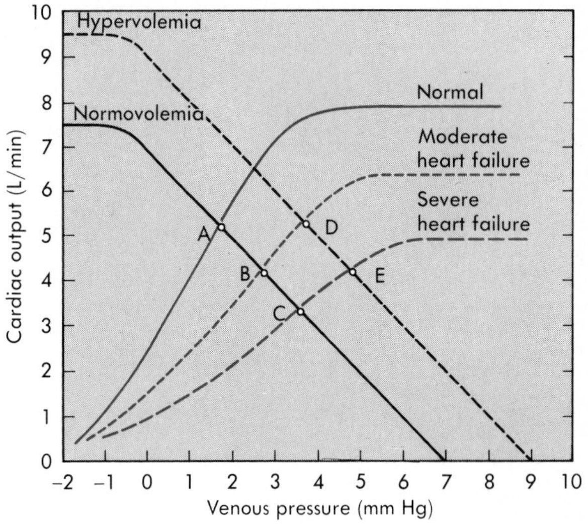

■ **Fig. 34-12.** With moderate or severe heart failure, the cardiac function curves are shifted to the right. With no change in blood volume, cardiac output decreases and venous pressure rises (from control equilibrium point *A* to point *B* or point *C*). With the increase in blood volume that usually occurs in heart failure, the vascular function curve is shifted to the right. Hence venous pressure may be elevated with no reduction in cardiac output (point *D*) or (in severe heart failure) with some diminution in cardiac output (point *E*).

ogous to those evoked by augmentations or reductions, respectively, of the total blood volume, for reasons that were discussed on p. 529.

■ *Heart Failure*

Heart failure may be acute or chronic. Acute heart failure may be caused by toxic quantities of drugs and anesthetics or by certain pathological conditions, such as coronary artery occlusion. Chronic heart failure may occur in such conditions as essential hypertension or ischemic heart disease. In these various forms of heart failure, myocardial contractility is impaired. Consequently, the cardiac function curve is shifted to the right, as depicted in Fig. 34-12.

In acute heart failure there may not have been sufficient time for blood volume to change. Therefore the equilibrium point will shift from the intersection of the normal curves (Fig. 34-12, point *A*) to the intersection of the normal vascular function curve with a depressed cardiac function curve (point *B* or *C*).

In chronic heart failure there is a shift not only in the cardiac function curve but also in the vascular function curve, by virtue of an increase in blood volume caused in part by fluid retention by the kidneys. The fluid retention is related to the concomitant reduction in glomerular filtration rate and to the increased secretion of aldosterone by the adrenal cortex. The resultant hypervolemia is reflected by a rightward shift of the vascular function curve, as shown in Fig. 34-12. Hence, with moderate degrees of heart failure, P_v will be elevated but cardiac output will be approximately normal (point *D*). With more severe degrees of heart failure, P_v will be still higher and cardiac output will be subnormal (point *E*).

■ *Peripheral Resistance*

Predictions concerning the effects of changes in peripheral resistance are complex because both the cardiac and vascular function curves shift. With increased peripheral resistance (Fig. 34-13) the vascular function curve is displaced downward but converges to the same x-axis intercept as the control curve, as described on p. 530. The cardiac function curve is also shifted downward because at any given P_v the heart is able to pump less blood against a greater resistive load. Because both curves are displaced downward, the new equilibrium point, *B*, will fall below the control point, *A*.

Whether point *B* will fall directly below point *A* or will lie to the right or left of it depends on the magnitude of the shift in each curve. For example, if a given increase in peripheral resistance shifts the vascular function curve more than the cardiac function curve, equilibrium point *B* would fall below and to the left of *A;* that is, both cardiac and P_v would diminish. Conversely, if the cardiac function curve is displaced more than the vascular function curve, point *B* will fall below and to the right of point *A;* that is, cardiac output would decrease but P_v would rise.

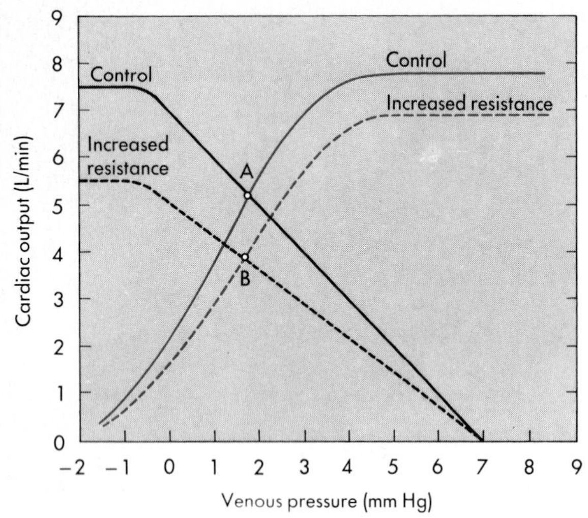

■ **Fig. 34-13.** With increased peripheral resistance, both the cardiac and vascular function curves are displaced downward.

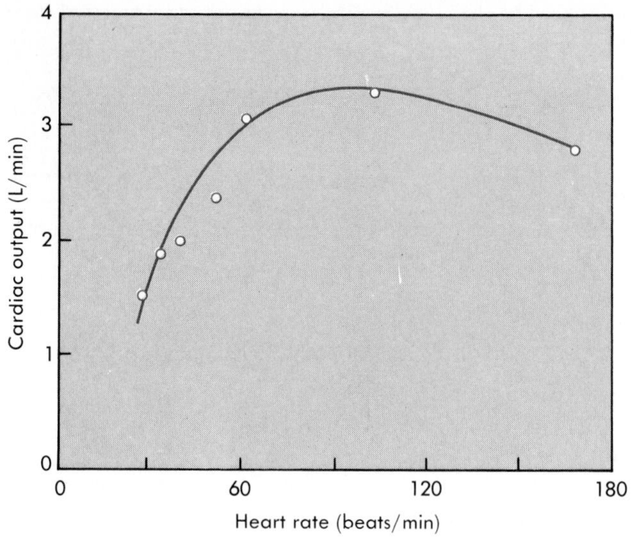

■ **Fig. 34-14.** The changes in cardiac output induced by changing the rate of ventricular pacing in a dog with complete heart block. (Redrawn from Miller, D.E., et al.: Circ. Res. **10:**658, 1962. By permission of the American Heart Association, Inc.)

■ *Role of Heart Rate*

Cardiac output is the product of stroke volume and heart rate. The preceding analysis of the control of cardiac output was, in reality, restricted to the control of stroke volume, and the role of heart rate was neglected.

The effect of changes in heart rate on cardiac output will now be considered. The analysis is complex, because a change in heart rate will alter the other three factors (preload, afterload, and contractility) that determine stroke volume (Fig. 34-1). An increase in heart rate, for example, would decrease the duration of diastole. Hence ventricular filling would be diminished; that is, preload would be reduced. If the proposed increase in heart rate did alter cardiac output, the arterial pressure would change; that is, afterload would be altered. Finally, the rise in heart rate would increase the net influx of Ca^{++} into the myocardial cells, and this would enhance myocardial contractility (p. 464).

Heart rate has been varied by artificial pacing in many types of experimental preparations and in humans. The effects on cardiac output have usually resembled the experimental results shown in Fig. 34-14. In that experiment on a dog with a third-degree heart block, contraction frequency was varied by ventricular pacing. As the ventricular rate was increased from 30 to 60 beats/minute, the cardiac output increased substantially. Presumably, at the slowest heart rates within this range, the greater filling per cardiac cycle is not adequate to compensate for the decreased number of contractions per minute.

Over the frequency range from 60 to 170 beats/minute, however, cardiac output did not change very much (Fig. 34-14). Hence, as heart rate was increased, the stroke volume was proportionately reduced. The decreased time for filling at the faster rates partly accounts for the observed proportionality between heart rate increase and stroke volume decrease. Also, vascular autoregulation tends to hold tissue blood flow constant. This adaptation leads to changes in preload and afterload that maintain cardiac output nearly constant.

During certain activities, notably physical exercise, the cardiac output increases and this change in output is attended by a proportionate increase in heart rate and a negligible change in stroke volume. The temptation is great to conclude that the increase in cardiac output must be caused by the observed increase in heart rate, because of the striking correlation between cardiac output and heart rate. Yet Fig. 34-14 emphasizes that, over a wide range of heart rates, a change in heart rate has little influence on cardiac output.

The principal increase in cardiac output during exercise must therefore be ascribed to the pronounced reduction in peripheral vascular resistance (Chapter 36). The attendant changes in heart rate are not inconsequential, however. If the heart rate cannot increase normally during exercise, the augmentation of cardiac output and the capacity for exercise may be severely limited. The increase in heart rate does play a permissive role in augmenting cardiac output, even if it is not proper to assign it a primary, causative role. The mechanisms responsible for raising heart rate in precise proportion to the increase in cardiac output are undoubtedly neural in origin, but the specific components of the reflex arcs remain to be elucidated.

Ancillary Factors That Affect the Venous System and Cardiac Output

The interrelationships between central venous pressure and cardiac output described above have been oversimplified. The effects elicited by changes in single variables have been invoked. However, because many feedback control loops regulate the cardiovascular system, an isolated change in a single variable rarely occurs. A change in blood volume, for example, reflexly alters cardiac function, peripheral resistance, and venomotor tone. Several auxiliary factors also regulate cardiac output. Such ancillary factors may be considered to modulate the more basic factors that have been considered earlier.

Gravity

Gravitational forces may affect cardiac output profoundly. Among soldiers standing at attention for long periods, particularly in warm weather, it is not unusual for some individuals to faint because of reduced cardiac output. Gravitational effects are exaggerated in airplane pilots during pullouts from dives. The centrifugal force in the footward direction may be several times greater than the force of gravity. Such individuals characteristically black out during the maneuver, as blood is drained from the cephalic regions and pooled in the lower parts of the body.

The explanation for the reduction in cardiac output under such conditions is often specious. It is argued that when an individual is standing, the forces of gravity impede venous return from the dependent regions of the body. This statement is incomplete, however, because it ignores the facilitative counterforce on the arterial side of the same circuit.

In this sense the vascular system resembles a U tube. To comprehend the action of gravity on flow through such a system, the models depicted in Figs. 34-15 and 34-16 will be analyzed. In Fig. 34-15 all the U tubes represent rigid cylinders of constant diameter. With both limbs of the U tube oriented horizontally *(A)* the flow depends only on the pressures at the inflow and outflow ends of the tube (P_i and P_o, respectively), the viscosity of the fluid, and the length and radius of the tube, in accordance with Poiseuille's equation (p. 477). With a constant cross section the pressure gradient will be uniform; hence the pressure midway down the tube (P_m) will equal the average of the inflow and outflow pressures.

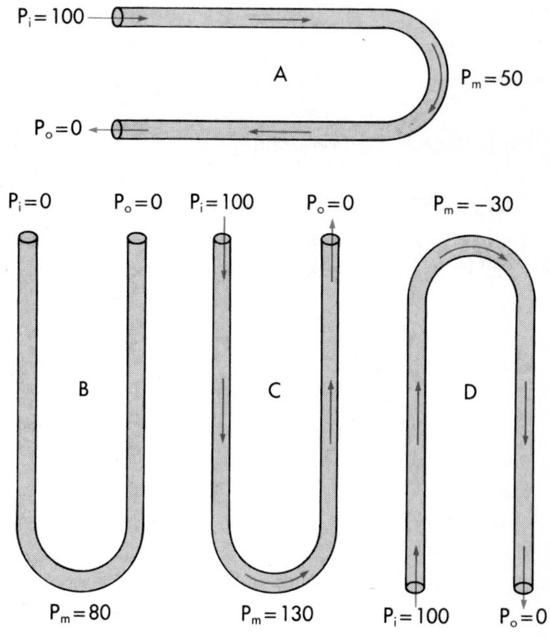

■ Fig. 34-15. Pressure distributions in rigid U tubes with constant internal diameters, all with the same dimensions. For a given inflow pressure (P_i = 100) and outflow pressure (P_o = 0), the pressure at the midpoint (P_m) depends on the orientation of the U tube, but the flow through the tube is independent of the orientation.

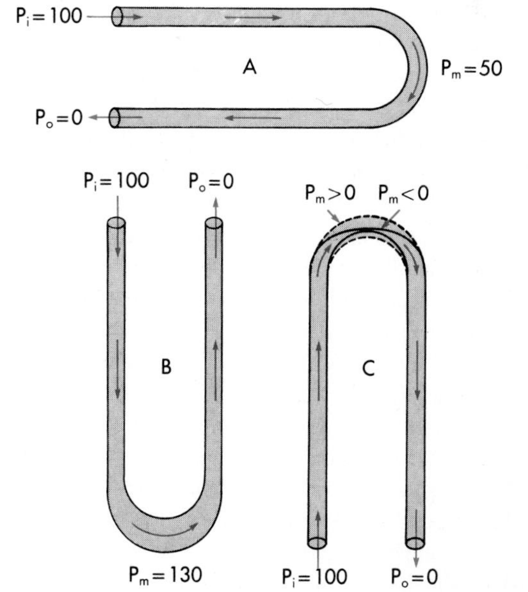

■ Fig. 34-16. In U tubes with a distensible section at the bend, even when inflow (P_i) and outflow (P_o) pressures are the same, the resistance to flow, and the fluid volume contained within each tube vary with the orientation of the tube. P_m, Pressure at the midpoint of the tube.

When the U tube is oriented vertically *(B* to *D),* hydrostatic forces must be taken into consideration. In tube *B* both limbs are open to atmospheric pressure and both ends are located at the same hydrostatic level; hence there is no flow. The pressure at the midpoint of the tube will simply be ρhg. It will depend on the density of the fluid, ρ; the height of the U tube, h; and the acceleration of gravity, g. In the example, the length of the U tube is such that the midpoint pressure is 80 mm Hg.

Now consider tube *C,* where the tube is oriented the same as tube *B,* but where a 100 mm Hg pressure difference is applied across the two ends. The flow will precisely equal that in *A,* because the pressure gradient, tube dimensions, and fluid viscosity are all the same. Gravitational forces are precisely equal in magnitude but opposite in direction in the two limbs of the U tube. Because the flow will be the same as that in *A,* the pressure drop will be 50 mm Hg at the midpoint because of the viscous losses resulting from flow. Furthermore, gravity will tend to increase pressure by 80 mm Hg at the midpoint, just as in tube *B.* The actual pressure at the midpoint of tube *C* will be the resultant of the viscous loss and hydrostatic gain, or 130 mm Hg in this example.

In *D* a pressure gradient of 100 mm Hg is applied to the same U tube, but the tube is oriented in the opposite direction. Gravitational forces will be so directed that the pressure at the midpoint will tend to be 80 mm Hg less than that at the end of the U tube. Viscous losses will still produce a 50 mm Hg pressure drop at the midpoint relative to P_i. Hence, with this orientation, pressure at the midpoint of the U tube will be 30 mm Hg below ambient pressure. Flow will, of course, be the same as in tubes *A* and *C,* for the reasons stated in relation to *C.*

In a system of rigid tubes gravitational effects will not alter the rate of blood flow. However, experience shows that gravity does affect the cardiovascular system. The reason is that the vessels are distensible rather than rigid. To explain this phenomenon, the pressures in a set of U tubes with distensible components (at the bends in the tubes of Fig. 34-16) will be examined. In tubes *A* and *B* the pressure distributions will resemble those in tubes *A* and *C,* respectively, of Fig. 34-15. Because the pressure is higher at the bend of tube *B* than at the bend of tube *A,* and the segment is distensible in this region, the distension at bend *B* will exceed that at bend *A.* The extent of the distension will depend on the elastic characteristics of these tube segments. Because flow is directly related to the tube diameter, the flow through *B* will exceed the flow through *A* for a given pressure difference applied at the ends.

Because orienting a U tube with its bend downward actually increases rather than diminishes flow, how then is the observed impairment of cardiovascular function explained when the body is similarly oriented? The explanation is that the cardiovascular system is a closed circuit of constant fluid (blood) volume, whereas the U tube is an open conduit supplied by a fluid source of unlimited volume. In the dependent regions of the cardiovascular system, the distension will occur more on the venous than on the arterial side of the circuit, because the venous capacitance is so much greater than the arterial capacitance. Such venous distension is readily observed on the back of the hands when the arms are allowed to hang down. The hemodynamic effects of such venous distension *(venous pooling)* resemble those caused by the hemorrhage of an equivalent volume of blood from the body. When a person shifts from a supine position to a relaxed standing position, from 300 to 800 ml of blood are pooled in the legs. This may reduce cardiac output by about 2 L/minute.

The compensatory adjustments to the erect position are similar to the adjustments to blood loss. For example, the diminished baroreceptor excitation reflexly speeds the heart, strengthens the cardiac contraction, and constricts the arterioles and veins. The baroreceptor reflex has a greater effect on the resistance than on the capacitance vessels. Warm ambient temperatures tend to interfere with the compensatory vasomotor reactions, and the absence of muscular activity exaggerates the effects.

Many of the drugs used to treat hypertension also interfere with the reflex adaptation to standing. Similarly, astronauts exposed to weightlessness lose their adaptations after a few days, and they experience difficulties when they first return to a normal gravitational field. When individuals with impaired reflex adaptations stand, their blood pressures may drop substantially. This response is called *orthostatic hypotension,* which may cause lightheadedness or fainting.

When the U tube is rotated so that the bend is directed upward, the opposite effects prevail. In Fig. 34-16 the pressure at the bend of tube *C* would tend to be −30 mm Hg, just as in tube *D* of Fig. 34-15. Because the ambient pressure exceeds the internal pressure, however, the tube will collapse. Flow will then cease, and there will no longer be any viscous decline of pressure associated with flow. In U tube *C,* when flow stops, the pressure at the top of each limb will be 80 mm Hg less than at the bottom (the hydrostatic pressure difference). Hence, near the top of the left, or inflow, limb the pressure will approach 20 mm Hg. As soon as this pressure exceeds ambient pressure, the collapsed tubing will be forced open and flow will begin. With the onset of flow, however, pressure will again drop below the ambient pressure. Thus the tubing at the bend will flutter; that is, it will fluctuate between the open and closed states.

When an arm is raised, the cutaneous veins in the

hand and forearm collapse, for the reasons described previously. Fluttering does not occur here because the deeper veins are protected from collapse by being tethered to surrounding structures. This protection allows these deeper veins to accomodate the flow ordinarily carried by the collapsed superficial veins. The analogy would be to add a rigid tube (representing the deeper veins) in parallel with the collapsible tube (representing the superficial veins) at the bend of tube *C* in Fig. 34-16. The collapsible tube would no longer flutter, but would remain closed. All flow would occur through the rigid tube, just as in tube *D* in Fig. 34-15.

The superficial veins in the neck are ordinarily partially collapsed when a normal individual is upright. Venous return from the head is conducted largely through the deeper cervical veins. However, when *central venous pressure* is abnormally elevated, the superficial neck veins are distended and they do not collapse even when the subject sits or stands. Such cervical venous distension is an important clinical sign of congestive heart failure.

Muscular Activity and Venous Valves

When an individual becomes upright but remains at rest, the pressure rises in the veins in the dependent regions of the body. The venous pressure in the legs increases gradually and does not reach an equilibrium value until almost 1 minute after standing. The gradual rise in P_v is attributable to the venous valves, which permit flow only toward the heart. When a person stands, the valves prevent blood in veins from actually falling toward the feet. Hence the column of venous blood is supported at numerous levels by these valves; temporarily the venous column consists of many separate segments. However, blood continues to enter the column from many venules and small tributary veins, and pressure continues to rise. As soon as the pressure in one segment exceeds that in the segment just above it, the valve is forced open. Ultimately, all the valves are open, and the column is continuous, similar to the outflow limbs of the U tubes shown in Figs. 34-15 and 34-16.

Precise measurement reveals that the final level of P_v in the feet during quiet standing is only slightly greater than that in a static column of blood extending from the right atrium to the feet. This indicates that the pressure drop caused by flow from foot veins to the right atrium is very small. This very low resistance justifies lumping the veins as a common venous capacitance in the model illustrated in Fig. 34-2.

When an individual who has been standing quietly begins to walk, the venous pressure in the legs decreases appreciably. Because of the intermittent venous compression produced by the contracting muscles and because of the orientation of the venous valves, blood is forced from the veins toward the heart. Hence muscular contraction lowers the mean venous pressure and serves as an auxiliary pump. Furthermore, it prevents venous pooling and lowers capillary hydrostatic pressure, thereby reducing the tendency for edema fluid to collect in the feet during standing.

Respiratory Activity

The normal, periodic activity of the respiratory muscles causes rhythmic variations in vena caval flow, and it constitutes an auxiliary pump to promote venous return. Coughing, straining at stool, and other activities that require the respiratory muscles may affect cardiac output substantially.

The changes in blood flow in the superior vena cava during the normal respiratory cycle are shown in Fig. 34-17. During respiration the reduction in intrathoracic pressure is transmitted to the lumina of the thoracic blood vessels. The reduction in central venous pressure during inspiration increases the pressure gradient between extrathoracic and intrathoracic veins. The consequent acceleration of venous return to the right atrium is displayed in Fig. 34-17 as an increase in superior vena caval blood flow from 5.2 ml/second during expiration to 11 ml/second during inspiration.

An exaggerated reduction in intrathoracic pressure achieved by a strong inspiratory effort against a closed glottis (called Müller's maneuver) does not increase venous return proportionately. The extrathoracic veins collapse near their entry into the chest when their pressures fall below the ambient level. As the veins collapse, flow into the chest momentarily stops. The cessation of flow raises pressure upstream, forcing the collapsed segment to open again. The process is repetitive; the venous segments adjacent to the chest alternately open and close.

During expiration, flow into the central veins decelerates. However, the mean rate of venous return during normal respiration exceeds the flow rate in the temporary absence of respiration. Hence normal inspiration apparently facilitates venous return more than normal expiration impedes it. In part, this must be attributable to the valves in the veins of the extremities and neck. These valves prevent any reversal of flow during expiration. Thus the respiratory muscles and venous valves constitute an auxiliary pump for venous return.

Sustained expiratory efforts increase intrathoracic pressure and thereby impede venous return. Straining against a closed glottis (termed *Valsalva's maneuver*) regularly occurs during coughing, defecation, and weight lifting. Intrathoracic pressures in excess of 100 mm Hg have been recorded in trumpet players and pressures over 400 mm Hg have been observed during par-

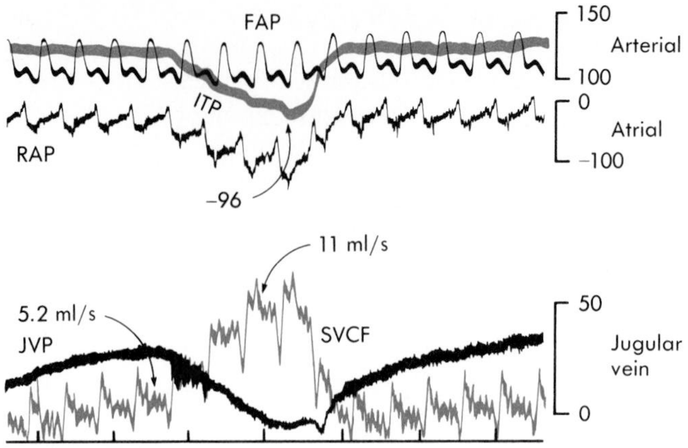

Fig. 34-17. During a normal inspiration, intrathoracic *(ITP),* right atrial *(RAP),* and jugular venous *(JVP)* pressures decrease, and flow in the superior vena cava *(SVCF)* increases in this case from 5.2 to 11 ml/second. All pressures are in millimeters of water, except for femoral arterial pressure *(FAP),* which is in millimeters of mercury. (Modified from Brecher, G.A.: Venous return, New York, 1956. Grune & Stratton, Inc.)

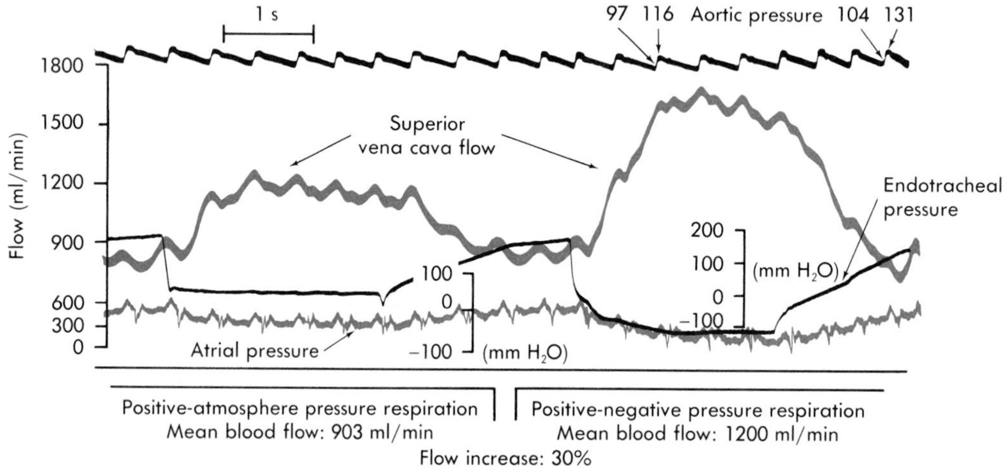

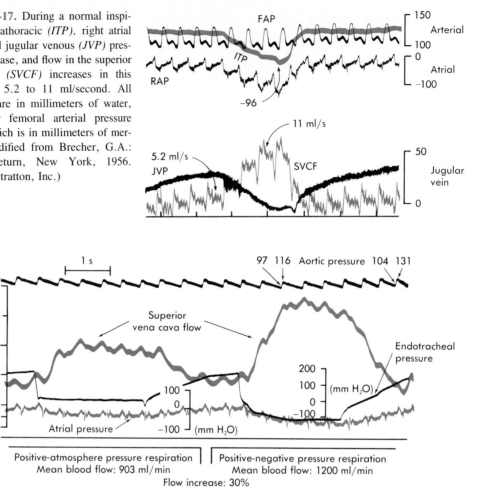

Fig. 34-18. During intermittent positive-pressure respiration, the flow in the superior vena cava is approximately 30% greater when the lungs are deflated actively by applying negative endotracheal pressure *(right side)* than when they are allowed to deflate passively against atmospheric pressure *(left side).* (Modified from Brecher, G.A.: Venous return, New york, 1956, Grune & Stratton, Inc.)

oxysms of coughing. Such pressure increases are transmitted directly to the lumina of the intrathoracic arteries. After cessation of coughing the arterial blood pressure may fall precipitously because of the preceding impediment to venous return.

The dramatic increase in intrathoracic pressure induced by coughing constitutes an auxiliary pumping mechanism for the blood, despite its concurrent tendency to impede venous return. During certain diagnostic procedures, such as coronary angiography or electrophysiological testing, patients are at increased risk for ventricular fibrillation. Such patients have been trained to cough rhythmically on command. If ventricular fibrillation does occur, substantial arterial blood pressure pulses are generated by each cough, and loss of consciousness is delayed considerably. The cough raises the intravascular pressure equally in intrathoracic arteries and veins. Blood is propelled through the tissues, however, because the increased pressure is transmitted to the extrathoracic arteries but not to the extrathoracic veins, because of the venous valves.

Artificial Respiration

In most forms of artificial respiration (mouth-to-mouth resuscitation, mechanical respiration), lung inflation is achieved by applying endotracheal pressures above atmospheric pressure,and expiration occurs by passive recoil of the thoracic cage. Thus lung inflation is attended by an appreciable rise in intrathoracic pressure. Vena cava flow decreases sharply during the phase of positive-pressure lung inflation (indicated by the progressive rise in endotracheal pressure in the central portion of Fig. 34-18). When negative endotracheal pressure (indicated by the abrupt decrease in endotracheal pressure in the right half of Fig. 34-18) is used to facilitate deflation, vena cava flow accelerates more than when the lungs are allowed to deflate passively (near the left border of Fig. 34-18).

Bibliography

Journal articles

Bromberger-Barnea, B.: Mechanical effects of inspiration on heart functions: a review, Fed. Proc. **40:**2172, 1981.

Carneiro, J.J., and Donald, D.E.: Blood reservoir function of dog spleen, liver, and intestine, Am. J. Physiol. **232:**H67, 1977.

Cassidy, S.S., and Mitchell, J.H.: Effects of positive pressure breathing on right and left ventricular preload and afterload, Fed. Proc. **40:**2178, 1981.

Levy, M.N.: The cardiac and vascular factors that determine systemic blood flow, Circ. Res. **44:**739, 1979.

Ludbrook, J.: The musculovenous pumps of the human lower limb, Am. Heart J. **71:**635, 1966.

Melbin, J., et al.: Coherence of cardiac output with rate changes, Am. J. Physiol. **243:**H499, 1982.

Miller, R.R., et al.: Differential systemic arterial and venous actions and consequent cardiac effects of vasodilator drugs, Prog. Cardiovasc. Dis. **24:**353, 1982.

Narahara, K.A., and Blettel, M.L.: Effect of rate on left ventricular volumes and ejection fraction during chronic ventricular pacing, Circulation **67:**323, 1983.

Niemann, J.T., et al.: Mechanical "cough" cardiopulmonary resuscitation during cardiac arrest in dogs, Am. J. Cardiol. **55:**199, 1985.

Rothe, C.F.: Reflex control of veins and vascular capacitance, Physiol. Rev. **63:**1281, 1983.

Sagawa, K.: Closed-loop physiological control of the heart, Ann. Biomed. Eng. **8:**415, 1980.

Shoukas, A.A., and Sagawa, K.: Control of total systemic vascular capacity by the carotid sinus baroreceptor reflex, Circ. Res. **33:**22, 1973.

Weisel, R.D., Berger, R.L., and Hechtman, H.B.: Measurement of cardiac output by thermodilution, N. Engl. J. Med. **292:**682, 1975.

Books and monographs

Green J.F.: Determinants of systemic blood flow. International review of physiology III: Cardiovascular physiology, vol. 18, Baltimore, 1979, University Park Press.

Guyton, A.C., Jones, C.E., and Coleman, T.G.: Circulatory physiology: cardiac output and its regulation, ed. 2, Philadelphia, 1973, W.B. Saunders Co.

Lassen, N.A.: Indicator methods for measurement of organ and tissue blood flow. In Handbook of physiology; Section 2: The cardiovascular system—peripheral circulation and organ blood flow, vol. III, Bethesda, Md., 1983, American Physiological Society.

Rothe, C.F.: Venous system: physiology of the capacitance vessels. In Handbook of physiology; Section 2: The cardiovascular system—peripheral circulation and organ blood flow, vol. III, Bethesda, Md., 1983. American Physiological Society.

Shepherd, J.T., and Vanhoutte, P.M.: Veins and their control, Philadelphia, 1975, W.B. Saunders Co.

CHAPTER

35

Special Circulations

Coronary Circulation

Anatomy of Coronary Vessels

The right and left coronary arteries, which arise at the root of the aorta behind the right and left cusps of the aortic valve, respectively, provide the entire blood supply to the myocardium. The right coronary artery supplies principally the right ventricle and atrium; the left coronary artery, which divides near its origin into the anterior descendens and the circumflex branches, supplies principally the left ventricle and atrium, but there is some overlap. In the dog the left coronary artery supplies about 85% of the myocardium, whereas in humans the right coronary artery is dominant in 50% of individuals, the left coronary artery is dominant in another 20%, and the flow delivered by each main artery is about equal in the remaining 30%.

Coronary blood flow is most commonly measured in humans by thermodilution. Thermodilution is the same procedure used for the measurement of cardiac output (p. 449), except that the indicator (cold saline solution) is ejected at the tip of a catheter inserted into the coronary sinus via a peripheral vein. The thermal sensor (thermistor) is located on the catheter a few centimeters from the catheter tip. The greater the coronary sinus outflow, the less is the temperature decrease produced by the cold saline injection. This method does not measure total coronary blood flow, because only about two thirds of coronary arterial inflow returns to the venous circulation through the coronary sinus. However, almost all of the blood flow that does empty into the coronary sinus comes from the left ventricle. Hence the thermodilution method provides a good estimate of left ventricular coronary blood flow. Right and left coronary artery inflow, as well as inflow of the major branches of the left coronary artery, can be measured with reasonable accuracy by injection of a radioactive tracer (for example, ^{133}Xe) through a catheter threaded into the coronary artery via a peripheral artery. Myocardial clearance of the isotope is monitored with a detector appropriately placed over the precordium. Total coronary blood flow also can be measured by the nitrous oxide method, in a similar manner to that described for cerebral blood flow (p. 552). However, about 10 minutes are required for each measurement, which obviates the use of this method for following rapid changes in coronary blood flow.

After passage through capillary beds, most of the venous blood returns to the right atrium through the coronary sinus, but some reaches the right atrium by way of the anterior coronary veins. There are also vascular communications directly between the vessels of the myocardium and the cardiac chambers; these comprise the *arteriosinusoidal,* the *arterioluminal,* and the *thebesian vessels.* The arteriosinusoidal channels consist of small arteries or arterioles that lose their arterial structure as they penetrate the chamber walls and divide into irregular, endothelium-lined sinuses (50 to 250 μm). These sinuses anastomose with other sinuses and with capillaries and communicate with the cardiac chambers. The arterioluminal vessels are small arteries or arterioles that open directly into the atria and ventricles. The thebesian vessels are small veins that connect capillary beds directly with the cardiac chambers and also communicate with cardiac veins and other thebesian veins. On the basis of anatomical studies, intercommunication appears to exist among all the minute vessels of the myocardium in the form of an extensive plexus of subendocardial vessels. It has been suggested that some myocardial nutrition can be derived from the cardiac cavities through those channels. Isotope-labeled blood in the cardiac chambers does penetrate a short distance into the endocardium, but does not constitute a significant source of oxygen and nutrients to the myocardium. In the dog a major fraction of the left coronary inflow returns to the right atrium via the coronary sinus, and a small fraction supplying the interventricular septum returns directly to the right ventricular cavity. Right cor-

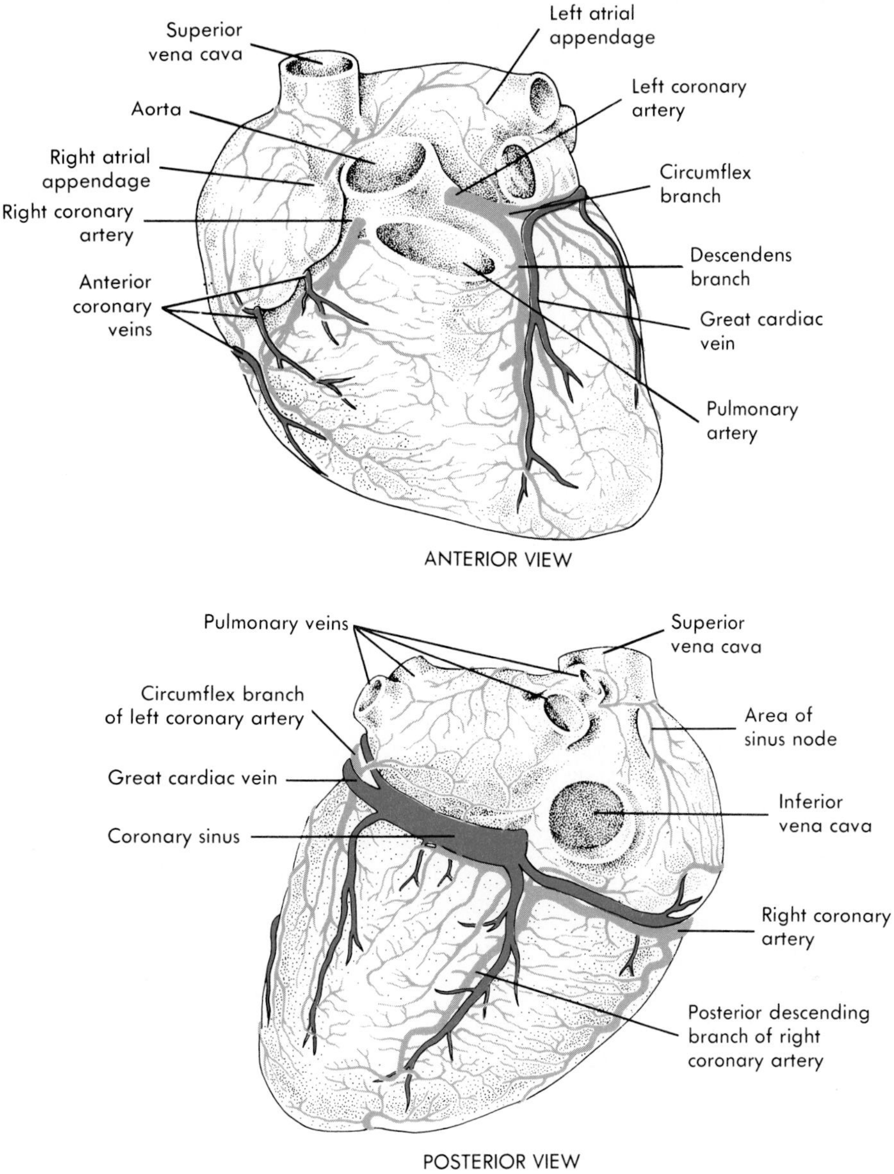

■ **Fig. 35-1.** Anterior and posterior surfaces of the heart, illustrating the location and distribution of the principal coronary vessels.

onary artery drainage is primarily via the anterior cardiac veins to the right atrium. The epicardial distribution of the coronary arteries and veins is illustrated in Fig. 35-1.

■ *Factors That Influence Coronary Blood Flow*

Physical factors. The primary factor responsible for perfusion of the myocardium is the aortic pressure, which is, of course, generated by the heart itself. Changes in aortic pressure generally evoke parallel directional changes in coronary blood flow. However, alterations of cardiac work, produced by an increase or decrease in aortic pressure, have a considerable effect on coronary resistance. By means of mechanisms that have not yet been fully elucidated, increased metabolic activity of the heart results in a decrease in coronary resistance, and a reduction in cardiac metabolism produces an increase in coronary resistance. If a cannulated coronary artery is perfused by blood from a pressure-controlled reservoir, perfusion pressure can be altered without changing aortic pressure and cardiac work. Under these conditions abrupt variations in perfusion pressure produce equally abrupt changes in coronary blood flow in the same direction. However, maintenance of the perfusion pressure at the new level is associated with a return of blood flow toward the level observed

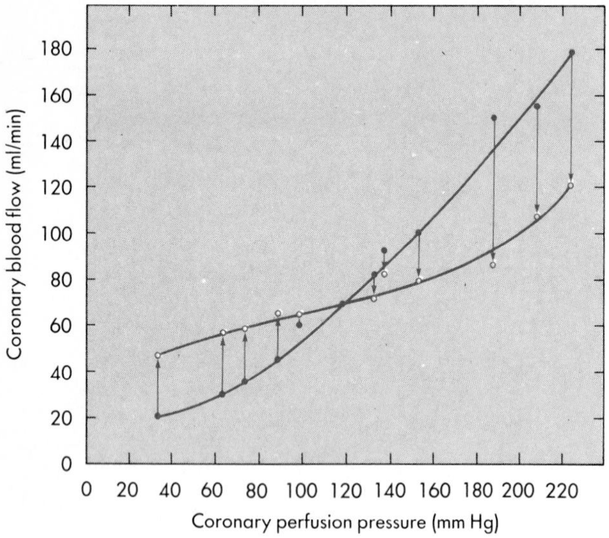

■ **Fig. 35-2.** Pressure-flow relationships in the coronary vascular bed. At constant aortic pressure, cardiac output, and heart rate, coronary artery perfusion pressure was abruptly increased or decreased from the control level indicated by the point where the two lines cross. The closed circles represent the flows that were obtained immediately after the change in perfusion pressure, and the open circles represent the steady-state flows at the new pressures. There is a tendency for flow to return toward the control level (autoregulation of blood flow), and this is most prominent over the intermediate pressure range (about 60 to 180 mm Hg).

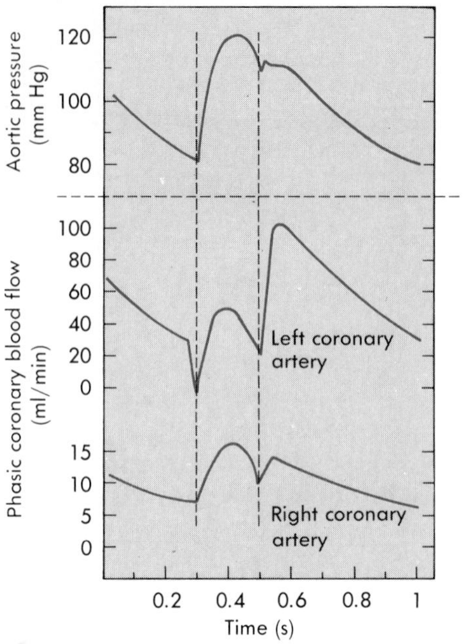

■ **Fig. 35-3.** Comparison of phasic coronary blood flow in the left and right coronary arteries.

before the induced change in perfusion pressure (Fig. 35-2). This phenomenon is an example of autoregulation of blood flow and is discussed in Chapter 33. Under normal conditions blood pressure is kept within relatively narrow limits by the baroreceptor reflex mechanisms so that changes in coronary blood flow are primarily caused by caliber changes in the coronary resistance vessels in response to metabolic demands of the heart.

In addition to providing the head of pressure to drive blood through the coronary vessels, the heart also influences its blood supply by the squeezing effect of the contracting myocardium on the blood vessels that course through it *(extravascular compression* or *extracoronary resistance)*. This force is so great during early ventricular systole that blood flow, as measured in a large coronary artery supplying the left ventricle, is briefly reversed. Maximum left coronary inflow occurs in early diastole, when the ventricles have relaxed and extravascular compression of the coronary vessels is virtually absent. This flow pattern is seen in the phasic coronary flow curve for the left coronary artery (Fig. 35-3). After an initial reversal in early systole, left coronary blood flow follows the aortic pressure until early diastole, when it rises abruptly and then declines slowly as aortic pressure falls during the remainder of diastole.

The high extravascular resistance and the large amount of work done by the left ventricle during systole, as opposed to the minimum extravascular resistance and absence of left ventricular work during diastole, are used to advantage clinically to improve myocardial perfusion in patients with damaged myocardium and low blood pressure. The method is called *counterpulsation* and consists of the insertion of an inflatable balloon into the thoracic aorta through a femoral artery. The balloon is inflated during ventricular diastole and deflated during systole. This procedure enhances coronary blood flow during diastole by raising diastolic pressure at a time when coronary extravascular resistance is lowest. Furthermore, it reduces cardiac energy requirements by lowering aortic pressure during ventricular ejection.

Left ventricular myocardial pressure (pressure within the wall of the left ventricle) is greatest near the endocardium and lowest near the epicardium. However, under normal conditions this pressure gradient does not impair endocardial blood flow, since a greater blood flow to the endocardium during diastole compensates for the greater blood flow to the epicardium during systole. In fact, when 10 μm diameter radioactive microspheres are injected into the coronary arteries, their distribution indicates that the blood flow to the epicardial and endocardial halves of the left ventricle are approximately equal (slightly higher in the endocardium) under normal conditions. Since extravascular compression is greatest at the endocardial surface of the ventricle, equality of epicardial and endocardial blood flow must

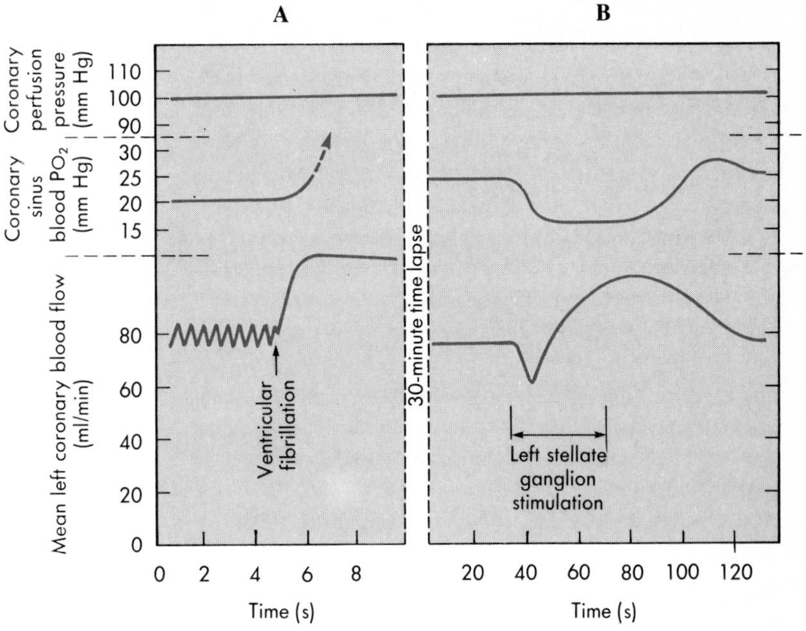

■ **Fig. 35-4. A,** Unmasking of the restricting effect of ventricular systole on mean coronary blood flow by induction of ventricular fibrillation during constant pressure perfusion of the left coronary artery. **B,** Effect of cardiac sympathetic nerve stimulation on coronary blood flow and coronary sinus blood O_2 tension in the fibrillating heart during constant pressure perfusion of the left coronary artery.

mean that the tone of the endocardial resistance vessels is less than that of the epicardial vessels.

Under abnormal conditions, when diastolic pressure in the coronary arteries is low, such as in severe hypotension, partial coronary artery occlusion, or severe aortic stenosis, the ratio of endocardial to epicardial blood flow falls below a value of 1. This indicates that the blood flow to the endocardial regions is more severely impaired than that to the epicardial regions of the ventricle. This redistribution of coronary blood flow is also reflected in an increase in the gradient of myocardial lactic acid and adenosine concentrations from epicardium to endocardium. For this reason, the myocardial damage observed in arteriosclerotic heart disease (for example, following coronary occlusion) is greatest in the inner wall of the left ventricle.

Flow in the right coronary artery shows a similar pattern (Fig. 35-3), but because of the lower pressure developed during systole by the thin right ventricle, reversal of blood flow does not occur in early systole and systolic blood flow constitutes a much greater proportion of total coronary inflow than it does in the left coronary artery. The extent to which extravascular compression restricts coronary inflow can be readily seen when the heart is suddenly arrested in diastole or with the induction of ventricular fibrillation. Fig. 35-4 depicts mean left coronary flow when the vessel was perfused with blood at a constant pressure from a reservoir. At the arrow in record *A,* ventricular fibrillation was electrically induced and an immediate and substantial increase in blood flow occurred. Subsequent increase in coronary resistance over a period of many minutes reduced myocardial blood flow to below the level existing before induction of ventricular fibrillation (record *B,* before stellate ganglion stimulation).

Tachycardia and bradycardia have dual effects on coronary blood flow. A change in heart rate is accomplished chiefly by shortening or lengthening of diastole. With tachycardia the proportion of time spent in systole, and consequently the period of restricted inflow, increases. However, this mechanical reduction in mean coronary flow is overridden by the coronary dilation associated with the increased metabolic activity of the more rapidly beating heart. With bradycardia the opposite is true; restriction of the coronary inflow is less (more time in diastole) but so are the metabolic (O_2) requirements of the myocardium.

Neural and neurohumoral factors. Stimulation of the sympathetic nerves to the heart elicits a marked increase in coronary blood flow. However, the increase in flow is associated with cardiac acceleration and a more forceful systole. The stronger myocardial contractions and the tachycardia (with the consequence that a greater proportion of time is spent in systole) tend to restrict coronary flow, whereas the increase in myocardial metabolic activity, as evidenced by the rate and contractility changes, tends to evoke dilation of the coronary resistance vessels. The increase in coronary blood flow observed with cardiac sympathetic nerve stimulation is the algebraic sum of these factors. In perfused hearts in which the mechanical effect of extravascular compression is eliminated by cardiac arrest or ventricular fibrillation, an initial coronary vasoconstriction is often observed with cardiac sympathetic nerve stimulation before the vasodilation attributable to the metabolic effect comes into play (Fig. 35-4, *B*). Such observations suggest that the primary action of the sympathetic nerve fibers on the coronary resistance vessels is vasoconstriction.

In contrast to skeletal muscle (p. 516), sympathetic cholinergic innervation of the coronary vessels does not exist, and whether there are sympathetid fibers to beta-adrenergic receptors on the coronary arterioles is questionable. However, experiments with the use of alpha- and beta-adrenergic drugs and their respective blocking agents reveal the presence of alpha-receptors (constrictors) and beta-receptors (dilators) on the coronary vessels. Furthermore, the coronary resistance vessels participate in the baroreceptor and chemoreceptor reflexes and there is sympathetic constrictor tone of the coronary arterioles than can be reflexly modulated. Nevertheless, coronary resistance is predominantly under local nonneural control.

Vagus nerve stimulation has little direct effect on the caliber of the coronary arterioles. In the fibrillating heart with the coronary arteries perfused at a constant pressure, or in the beating paced heart with flow measured in late diastole when there is essentially no extravascular compression, small increments in coronary blood flow can be observed with stimulation of the peripheral ends of the vagi. In addition, activation of the carotid and aortic chemoreceptors can elicit a decrease in coronary resistance via the vagus nerves to the heart. However, it is likely that the vagi exert less effect on coronary resistance in the normal animal, because of the overriding influence of metabolic mechanisms. The failure of strong vagal stimulation to evoke a large increase in coronary blood flow is not because of insensitivity of the coronary resistance vessels to acetylcholine, since intracoronary administration of this agent elicits marked vasodilation.

Reflexes originating in the myocardium and altering vascular resistance in peripheral systemic vessels, including the coronary vessels, have been conclusively demonstrated. However, the existence of extracardiac reflexes, with the coronary resistance vessels as the effector sites, has not been established.

Metabolic factors. One of the most striking characteristics of the coronary circulation is the close parallelism between the level of myocardial metabolic activity and the magnitude of the coronary blood flow. This relationship is also found in the denervated heart or the completely isolated heart, whether in the beating or the fibrillating state. The ventricles will continue to fibrillate for many hours when the coronary arteries are perfused with arterial blood from some external source. With the onset of ventricular fibrillation, an abrupt increase in coronary blood flow occurs because of the removal of extravascular compression (Fig. 35-4). Flow then gradually returns toward, and often decreases below, the prefibrillation level. The increase in coronary resistance that occurs despite the elimination of extravascular compression is a manifestation of the heart's ability to adjust its blood flow to meet its energy requirements. The fibrillating heart utilizes less O_2 than the pumping heart, and blood flow to the myocardium is reduced accordingly.

The link between cardiac metabolic rate and coronary blood flow remains unsettled. Numerous agents, generally referred to as metabolites, have been suggested as mediators of the vasodilation observed with increased cardiac work. Accumulation of vasoactive metabolites also may be responsible for reactive hyperemia, since the duration of coronary flow following release of the briefly occluded vessel is, within certain limits, proportional to the duration of the period of occlusion. Among the substances implicated are CO_2, O_2 (reduced O_2 tension), lactic acid, hydrogen ions, histamine, potassium ions, increased osmolarity, polypeptides, prostaglandins, and adenine nucleotides. None of these has satisfied all the criteria for the physiological mediator. Although potassium release from the myocardium can account for about half of the initial decrease in coronary resistance, it cannot be responsible for the increased coronary flow observed with prolonged enhancement of cardiac metabolic activity, since its release from the cardiac muscle is transitory.

Recent evidence suggests that adenosine plays a role as a metabolic vasodilator. According to the adenosine hypothesis, a reduction in myocardial O_2 tension produced by low coronary blood flow, hypoxemia, or increased metabolic activity of the heart leads to the myocardial formation of adenosine, which crosses the interstitial fluid space to reach the coronary resistance vessels and induces vasodilation by activating an adenosine receptor.

Coronary dilation results in an increase in coronary blood flow that hastens the washout of adenosine and reduces its formation. Under normal resting conditions, only small amounts of adenosine are released by the heart and exert a minimum dilating effect. Factors that alter coronary vascular resistance are schematized in Fig. 35-5.

Coronary Collateral Circulation and Vasodilators

In the normal human heart there are virtually no functional intercoronary channels, whereas in the dog there are a few small vessels that link branches of the major coronary arteries. Abrupt occlusion of a coronary artery or one of its branches in a human or dog leads to ischemic necrosis and eventual fibrosis of the areas of myocardium supplied by the occluded vessel. However, if narrowing of a coronary artery occurs slowly and progressively over a period of days, weeks, or longer, collateral vessels develop and may furnish sufficient blood to the ischemic myocardium to prevent or reduce the

■ **Fig. 35-5.** Schematic representation of factors that increase (+) or decrease (−) coronary vascular resistance. The intravascular pressure (arterial blood pressure) stretches the vessel wall.

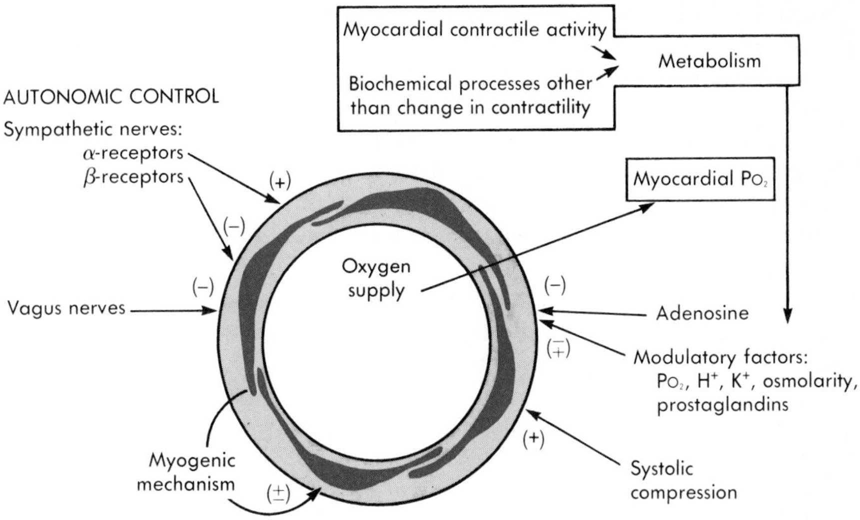

extent of necrosis. The development of collateral coronary vessels has been extensively studied in dogs, and the clinical picture of coronary atherosclerosis, as it occurs in humans, can be simulated by gradual narrowing of the normal dog's coronary arteries. Collateral vessels develop between branches of occluded and nonoccluded arteries. They originate from preexisting small vessels that undergo proliferative changes of the endothelium and smooth muscle, possibly in response to wall stress and chemical agents released by the ischemic tissue.

Numerous surgical attempts have been made to enhance the development of coronary collateral vessels. However, the techniques used do not increase the collateral circulation over and above that produced by coronary artery narrowing alone. When discrete occlusions or severe narrowing occurs in coronary arteries (even vessels as small as 1 mm in diameter), the lesions can be bypassed with a vein graft or the narrow segment can be dilated by inserting a balloon-tipped catheter into the diseased vessel via a peripheral artery and inflating the balloon. Distension of the vessel by balloon inflation *(angioplasty)* can produce a lasting dilation of a narrowed coronary artery.

A number of drugs are available that induce coronary vasodilation, and they are used in patients with coronary artery disease to relieve *angina pectoris,* the chest pain associated with myocardial ischemia. Many of these compounds are nitrites. They are not selective dilators of the coronary vessels, and the mechanism whereby they accomplish their beneficial effects has not been established. The arterioles distal to the stenosed artery are undoubtedly already maximally dilated by the ischemia responsible for the symptoms and cannot dilate more in response to the drugs. It has been suggested that the relief of angina pectoris by nitrites is brought about by a reduction in cardiac work and myocardial O_2 requirement caused by the moderate hypotension these drugs produce. In short, the reduction in pressure work and O_2 requirement must be greater than the reduction in coronary blood flow and O_2 supply consequent to the lowered coronary perfusion pressure. It has also been demonstrated that nitrites dilate large coronary arteries and coronary collateral vessels, thus increasing blood flow to ischemic myocardium and alleviating precordial pain.

■ *Cardiac Oxygen Consumption and Work*

The volume of O_2 consumed by the heart is determined by the amount and the type of activity the heart performs. Under basal conditions, myocardial O_2 consumption is about 8 to 10 ml/min/100 g of heart. It can increase severalfold with exercise and decrease moderately under conditions such as hypotension and hypothermia. The cardiac venous blood is normally quite low in O_2 (about 5 ml/dl), and the myocardium can receive little additional O_2 by further O_2 extraction from the coronary blood, a situation that also exists in contracting skeletal muscle. Therefore increased O_2 demands of the heart must be met primarily by an increase in coronary blood flow. When the heartbeat is arrested, as with administration of potassium, but coronary perfusion is maintained experimentally, the O_2 consumption falls to 2 ml/min/100 g or less, which is still six to seven times greater than that for resting skeletal muscle.

Left ventricular work per beat (*stroke work,* p. 486) is generally considered to be equal to the product of the stroke volume and the mean aortic pressure against which the blood is ejected by the left ventricle. At resting levels of cardiac output the kinetic energy component is negligible (p. 473). However, at high cardiac outputs as in severe exercise, the kinetic component can account for up to 50% of total cardiac work. One can simultaneously halve the aortic pressure and double the cardiac output, or vice versa, and still arrive at the same value for cardiac work. However, the O_2 requirements are greater for any given amount of cardiac work when a major fraction is pressure work as opposed to volume work. An increase in cardiac output at a constant aortic pressure (volume work) is accomplished with a small increase in left ventricular O_2 consumption, whereas increased arterial pressure at constant cardiac output (pressure work) is accompanied by a large increment in myocardial O_2 consumption. Thus myocardial O_2 consumption may not correlate well with overall cardiac work. The magnitude and duration of left ventricular pressure do correlate with left ventricular O_2 consumption.

The area under the systolic portion of the left ventricular pressure curve has been termed the *tension-time index,* and this index correlates reasonably well with myocardial O_2 consumption in many different physiological states. Since this index does not take into consideration the velocity of contraction, and since the velocity factor influences myocardial O_2 consumption, there are conditions (for example, exercise, sympathetic activation, epinephrine administration) in which the tension-time index fails to reflect accurately the O_2 consumption. The tension development (ventricular wall tension), velocity of shortening, and, to a lesser extent, degree of shortening of the myocardial fibers constitute the chief determinants of myocardial O_2 consumption.

The greater energy demand of pressure work over volume work can be readily demonstrated in the heart-lung preparation and may be associated with a decrease in myocardial creatine phosphate levels, but the reason for this difference in energy requirements is unknown. Nevertheless, it is also of great clinical importance, especially in aortic stenosis, in which left ventricular O_2 consumption is increased because of the high intraventricular pressures developed during systole, but coronary perfusion pressure is normal or reduced because of the pressure drop across the narrowed orifice of the diseased aortic valve.

Since mean pulmonary artery pressure is about one seventh that of aortic pressure, and since the outputs of the two ventricles are equal, work of the right ventricle is one seventh that of the left ventricle.

Cardiac Efficiency

As with an engine, the efficiency of the heart is the ratio of the work accomplished to the total energy utilized. Assuming an average O_2 consumption of 9 ml/min/100 g for the two ventricles, a 300 g heart consumes 27 ml O_2/min, which is equivalent to 130 small calories at a respiratory quotient of 0.82. Together the two ventricles do about 8 kg-m of work per minute, which is equivalent to 18.7 small calories. Therefore the gross efficiency is 14%

$$\frac{18.7}{130} \times 100 = 14\% \qquad (1)$$

The net efficiency is slightly higher (18%) and is obtained by subtracting the O_2 consumption of the nonbeating (asystolic) heart (about 2 ml/min/100 g) from the total cardiac O_2 consumption in the calculation of efficiency. It is thus evident that the efficiency of the heart as a pump is relatively low and is comparable to the efficiency of many mechanical devices used in everyday life. With exercise, efficiency improves, since mean blood pressure shows little change, whereas cardiac output and work increase considerably without a proportional increase in myocardial O_2 consumption. The energy expended in cardiac metabolism that does not contribute to the propulsion of blood through the body appears in the form of heat. The energy of the flowing blood is also dissipated as heat, chiefly in passage through the arterioles.

Substrate Utilization

The heart is quite versatile in its use of substrates, and within certain limits the uptake of a particular substrate is directly proportional to its arterial concentration. The utilization of one substrate is also influenced by the presence or absence of other substrates. For example, the addition of lactate to the blood perfusing a heart metabolizing glucose leads to a reduction in glucose uptake, and vice versa. At normal blood concentrations, glucose and lactate are consumed at about equal rates, whereas pyruvate uptake is very low, but so is its arterial concentration. For glucose, the threshold concentration is about 4 mM and below this blood level no myocardial glucose uptake occurs. Insulin reduces this threshold and increases the rate of glucose uptake by the heart. A very low threshold exists for cardiac utilization of lactate; insulin does not affect its uptake by the myocardium. With hypoxia, glucose utilization is facilitated by an increase in the rate of transport across the myocardial cell wall, whereas lactate cannot be metabolized by the hypoxic heart and is in fact produced by the heart under anaerobic conditions. Associated with lactate production by the hypoxic heart is the breakdown of cardiac glycogen.

Of the total cardiac O_2 consumption, only 35% to 40% can be accounted for by the oxidation of carbohydrate. Thus the heart derives the major part of its energy from oxidation of noncarbohydrate sources. The chief noncarbohydrate fuel used by the heart is esterified and nonesterified fatty acid, which accounts for about 60% of myocardial O_2 consumption in the postabsorptive state. The various fatty acids show different thresholds for myocardial uptake but are generally utilized in direct proportion to their arterial concentration. Ketone bodies, especially acetoacetate, are readily oxidized by the heart and contribute a major source of energy in diabetic acidosis. As is true of carbohydrate substrates, use of a specific noncarbohydrate substrate is influenced by the presence of other substrates, both noncarbohydrate and carbohydrate. Therefore, within certain limits, the heart uses preferentially that substrate which is available in the largest concentration. Most evidence indicates that the contribution to myocardial energy expenditure provided by the oxidation of amino acids is small.

Normally the heart derives its energy by oxidative phosphorylation, in which each mole of glucose yields 36 moles of ATP. However, during hypoxia, glycolysis supervenes and 2 moles of ATP are provided by each mole of glucose. Beta-oxidation of fatty acids is also curtailed. If hypoxia is prolonged, cellular creatine phosphate and eventually ATP are depleted.

In ischemia lactic acid accumulates (lack of washout) and causes a decrease in intracellular pH. This condition inhibits glycolysis, fatty acid use, and protein synthesis, which results in impairment of membrane function, cellular damage, and eventually necrosis of myocardial cells.

Detailed analysis of myocardial metabolism and its alteration in hypoxia and ischemia is an important subject, but it lies beyond the scope of this book. A number of procedures, some specific and some nonspecific, have been extensively studied in an effort to reduce myocardial infarct size after coronary occlusion and to restore and preserve marginally affected cardiac muscle.

Cutaneous Circulation

The oxygen and nutrient requirements of the skin are relatively small and, in contrast to most other body tissues, the supply of these essential materials is not the chief governing factor in the regulation of cutaneous blood flow. The primary function of the cutaneous circulation is maintenance of a constant body temperature. Consequently, the skin shows wide fluctuations in blood flow, depending on the need for loss or conservation of body heat. Mechanisms responsible for alterations in skin blood flow are primarily activated by changes in ambient and internal body temperatures.

Regulation of Skin Blood Flow

There are essentially two types of resistance vessels in skin: arterioles and *arteriovenous (AV) anastomoses.* The arterioles are similar to those found elsewhere in the body, AV anastomoses shunt blood from arterioles to venules and venous plexuses; hence they bypass the capillary bed. They are found primarily in the fingertips, palms of the hands, toes, soles of the feet, ears, nose, and lips. AV anastomoses differ morphologically from the arterioles in that they are either short, straight, or long, coiled vessels about 20 to 40 μm in lumen diameter, with thick muscular walls richly supplied with nerve fibers. These vessels are almost exclusively under sympathetic neural control and become maximally dilated when their nerve supply is interrupted. Conversely, reflex stimulation of the sympathetic fibers to these vessels may produce constriction to the point of complete obliteration of the vascular lumen. Although AV anastomoses do not exhibit *basal tone* (tonic activity of the vascular smooth muscle independent of innervation), they are highly sensitive to vasoconstrictor agents like epinephrine and norepinephrine. Furthermore, AV anastomoses do not appear to be under metabolic control, and they fail to show reactive hyperemia or autoregulation of blood flow. Thus the regulation of blood flow through these anastomotic channels is governed principally by the nervous system in response to reflex activation by temperature receptors or from higher centers of the central nervous system.

The bulk of the skin resistance vessels exhibits some basal tone and is under dual control of the sympathetic nervous system and local regulatory factors, in much the same manner as are other vascular beds. However, in the case of skin, neural control plays a more important role than local factors. Stimulation of sympathetic nerve fibers to skin blood vessels (arteries and veins, as well as arterioles) induces vasoconstriction, and severance of the sympathetic nerves induces vasodilation. With chronic denervation of the cutaneous blood vessels, the degree of tone that existed before denervation is gradually regained over a period of several weeks. This is accomplished by an enhancement of basal tone that compensates for the degree of tone previously contributed by sympathetic nerve fiber activity. Epinephrine and norepinephrine elicit only vasoconstriction in cutaneous vessels. Whether the increased basal tone following denervation of the skin vessels is the result of their enhanced sensitivity to circulating catecholamines *(denervation hypersensitivity)* has not been estabished.

Parasympathetic vasodilator nerve fibers do not supply the cutaneous blood vessels. However, stimulation of the sweat glands, which are innervated by cholinergic fibers of the sympathetic nervous system, results in dilation of the skin resistance vessels. Sweat contains an enzyme that acts on a protein moiety in the tissue fluid to produce *bradykinin,* a polypeptide with potent vasodilator properties. Bradykinin formed in the tissue can act locally to dilate the arterioles and increase blood flow to the skin. The skin vessels of certain regions, particularly the head, neck, shoulders, and upper chest, are under the influence of the higher centers of the central nervous system. Blushing, as with embarrassment or anger, and blanching, as with fear or anxiety, are examples of cerebral inhibition and stimulation, respectively, of the sympathetic nerve fibers to the affected regions. Whether blushing is caused in part by activation of the sweat glands and the local formation of bradykinin is unknown.

In contrast to AV anastomoses in the skin, the cutaneous resistance vessels show autoregulation of blood flow and reactive hyperemia. If the arterial inflow to a limb is stopped with an inflated blood pressure cuff for a brief period of time, the skin shows marked reddening below the point of vascular occlusion when the cuff is deflated. This increased cutaneous blood flow (reactive hyperemia) is also manifested by the distension of the superficial veins in the erythematous extremity. With respect to autoregulation of blood flow in the skin, the relatively low metabolic activity of the tissue favors a myogenic rather than a metabolic mechanism (p. 511).

Ambient and body temperature in regulation of skin blood flow. Since the primary function of the skin is to preserve the internal milieu and protect it from adverse changes in the environment and since ambient temperature is one of the most important external variables the body must contend with, it is not surprising that the vasculature of the skin is chiefly influenced by environmental temperature. Exposure to cold elicits a generalized cutaneous vasoconstriction that is most pronounced in the hands and feet. This response is chiefly mediated by the nervous system, since arrest of the circulation of a hand with a pressure cuff and immersion of that hand in cold water results in vasoconstriction in the skin of the other extremities that are exposed to room temperature. With the circulation of the chilled hand unoccluded, the reflex vasoconstriction is caused in part by the cooled blood returning to the general circulation and stimulating the temperature-regulating center in the anterior hypothalamus. Direct application of cold to this region of the brain produces cutaneous vasoconstriction.

The skin vessels of the cooled hand also show a direct response to cold. Moderate cooling or exposure for brief periods to severe cold (0° to 15° C) results in constriction of the resistance and capacitance vessels, including AV anastomoses. However, prolonged exposure of the hand to severe cold has a secondary vasodilator effect. Prompt vasoconstriction and severe pain are elicited by immersion of the hand in water near 0° C, but are soon followed by dilation of the skin vessels with reddening of the immersed part and alleviation of the pain. With continued immersion of the hand, alternating periods of constriction and dilation occur, but the skin temperature rarely drops to as low a degree as it did with the initial vasoconstriction. Prolonged severe cold, of course, results in tissue damage. The rosy faces of people working or playing in a cold environment are examples of cold vasodilation. However, the blood flow through the skin of the face may be greatly reduced despite the flushed appearance. The red color of the slowly flowing blood is in large measure the result of the reduced oxygen uptake by the cold skin and the cold-induced shift to the left of the oxyhemoglobin dissociation curve.

Direct application of heat produces not only local vasodilation of resistance and capacitance vessels and AV anastomoses but also reflex dilation in other parts of the body. The local effect is independent of the vascular nerve supply, whereas the reflex vasodilation is a combination of anterior hypothalamic stimulation by the returning warmed blood and of stimulation of receptors in the heated part. However, evidence for a reflex from peripheral temperature receptors is not as definitive for warm stimulation as it is for cold stimulation.

The close proximity of the major arteries and veins to each other permits considerable heat exchange (countercurrent) between artery and vein. Cold blood that flows in veins from a cooled hand toward the heart takes up heat from adjacent arteries, resulting in warming of the venous blood and cooling of the arterial blood. Heat exchange is of course in the opposite direction with exposure of the extremity to heat. Thus heat conservation is enhanced and heat gain is minimized during exposure of extremities to cold and warm environments, respectively.

Skin Color and Special Reactions of the Skin Vessels

The color of the skin is of course caused in large part by pigment; however, in all but very dark skin, the degree of pallor or ruddiness is primarily a function of the amount of blood in the skin. With little blood in the venous plexuses the skin appears pale, whereas with moderate to large quantities of blood in the venous plexuses, the skin shows color. Whether this color is bright red, blue, or some shade between is determined by the degree of oxygenation of the blood in the subcutaneous vessels. For example, a combination of vasoconstriction and reduced hemoglobin can produce an ashen gray color of the skin, whereas a combination of venous engorgement and reduced hemoglobin can result in a dark purple hue. Skin color provides little information about the rate of cutaneous blood flow. There

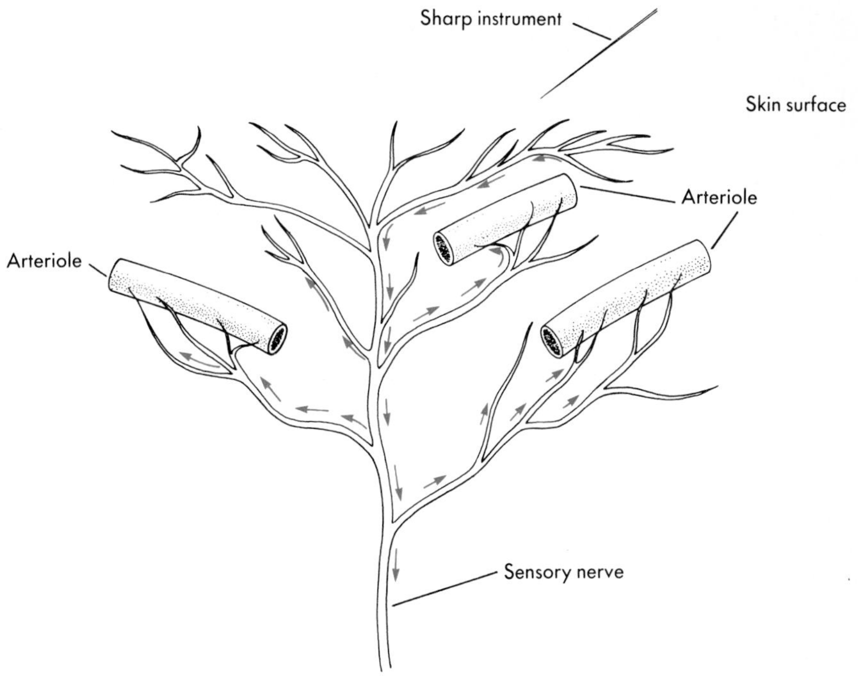

Fig. 35-6. Schematic representation of the axon reflex in response to a scratch on the skin surface with a sharp instrument. Arrows indicate the pathways of impulses in a sensory nerve from the site of stimulation to adjacent arterioles to produce local vasodilation (flare).

may coexist rapid blood flow and pale skin when the AV anastomoses are open, and slow blood flow and red skin when the extremity is exposed to cold.

White reaction and triple response. If the skin of the forearm of many individuals is lightly stroked with a blunt instrument, a white line appears at the site of the stroke within 20 seconds. The blanching becomes maximum in about 30 to 40 seconds and then gradually disappears within 3 to 5 minutes. This response is known as a *white reaction* and has been attributed to capillary contraction, since it occurs in the denervated limb and is unaffected by arrest of the limb circulation. Since all direct evidence indicates that capillaries do not contract and since skin color is primarily a result of blood content of venous plexuses, venules, and small veins, it seems logical to attribute the white reaction to venous constriction induced by mechanical stimulation.

If the skin is stroked more strongly with a sharp pointed instrument, a *triple response* is elicited. Within 3 to 15 seconds a thin *red line* appears at the site of the stroke, followed in about 15 to 30 seconds by a red blush, or *flare,* extending out 1 to 2 cm from either side of the red line. This in turn is followed in 3 to 5 minutes by an elevation of the skin along the red line, with gradual fading of the red line as the elevation, a *wheal,* becomes more prominent. The red line is probably caused by dilation of the vessels because of mechanical stimulation. The flare, however, is the result of dilation of neighboring arterioles caused to relax by an *axon reflex* originating at the site of mechanical stimulation. In an axon reflex, the nerve impulse travels centripetally in the cutaneous sensory nerve fiber and then antidromically down the small branches of the afferent nerve to adjacent arterioles to elicit vasodilation (Fig. 35-6). The flare is not affected by acute section or anesthetic block of the sensory nerve central to the point of branching, whereas it is abolished when the nerve degenerates after section. The wheal is caused by increased capillary permeability induced by the trauma. Fluid containing protein leaks out of the capillaries locally and produces edema at the site of injury. Since the triple response can be elicited by an intradermal injection of histamine, it has been attributed to histamine or a histamine-like substance *(H-substance).* Whether it is histamine, ATP, a vasoactive polypeptide like bradykinin, or some yet unidentified substance is unknown.

Skeletal Muscle Circulation

The rate of blood flow in skeletal muscle varies directly with the contractile activity of the tissue and the type of muscle. Blood flow and capillary density in red (slow-twitch, high-oxidative) muscle are greater than in white (fast-twitch, low-oxidative) muscle. In resting muscle the precapillary arterioles exhibit asynchronous intermittent contractions and relaxations, so that at any given moment, a very large precentage of the capillary bed is not perfused. Consequently, total blood flow through quiescent skeletal muscle is low (1.4 to 4.5 ml/min/100 g). With exercise the resistance vessels relax and the muscle blood flow may increase manyfold (up to 15 to 20 times the resting level), the magnitude of the increase depending largely on the severity of the exercise.

Regulation of Skeletal Muscle Blood Flow

Control of muscle circulation is achieved by neural and local factors; the relative contribution of these factors is dictated by muscle activity. At rest neural and myogenic regulations are predominant, whereas in exercise metabolic control supervenes. As with all tissues, physical factors such as arterial pressure, tissue pressure, and blood viscosity influence muscle blood flow. However, another physical factor comes into play during exercise—the squeezing effect of the active muscle on the vessels. With intermittent contractions, inflow is restricted and venous outflow is enhanced during each brief contraction. The presence of the venous valves prevents backflow of blood in the veins between contractions, thereby aiding in the forward propulsion of the blood. With strong sustained contractions the vascular bed can be compressed to the point where blood flow actually ceases temporarily.

Neural factors. Although the resistance vessels of muscle possess a high degree of basal tone, they also display tone attributable to continuous low frequency activity in the sympathetic vasoconstrictor nerve fibers. Evidence for this sympathetic tone is the increase in blood flow observed with local anesthetic block of the sympathetic fibers.

The basal frequency of firing in the sympathetic vasoconstrictor fibers is quite low (about 1 to 2 per second), and maximum vasoconstriction is observed at frequencies as low as 8 to 10 per second. Stimulation of the sympathetic nerve fibers to skeletal muscle elicits vasoconstriction that is caused by the release of norepinephrine at the fiber endings. Intraarterial injection of norepinephrine elicits only vasoconstriction, whereas low doses of epinephrine produce vasodilation in muscle and large doses cause vasoconstriction. The vasodilator effect of epinephrine is the result of stimulation of beta-adrenergic receptors on the resistance vessels.

The tonic activity of the sympathetic nerves is greatly influenced by reflexes from the baroreceptors. An increase in carotid sinus pressure results in dilation of the vascular bed of the muscle, and a decrease in carotid sinus pressure elicits vasoconstriction (Fig. 35-7). When the existing sympathetic constrictor tone is high, as in the experiment illustrated in Fig. 35-7, the decrease in blood flow associated with common carotid artery occlusion is small, but the increase following the release of occlusion is large. The vasodilation produced by baroreceptor stimulation is caused by inhibition of sympathetic vasoconstrictor activity. Since muscle is the major body component on the basis of mass and thereby represents the largest vascular bed, the participation of its resistance vessels in vascular reflexes plays an important role in maintaining a constant arterial blood pressure.

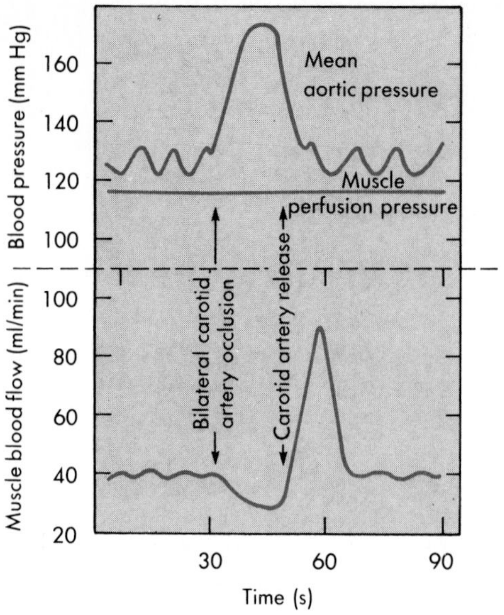

Fig. 35-7. Evidence for participation of the muscle vascular bed in vasoconstriction and vasodilation mediated by the carotid sinus baroreceptors after common carotid artery occlusion and release. In this preparation the sciatic and femoral nerves constituted the only direct connection between the hind leg muscle mass and the rest of the dog. The muscle was perfused by blood at a constant pressure that was completely independent of the animal's arterial pressure. (Redrawn from Jones, R.D., and Berne, R.M.: Am. J. Physiol. **204:**461, 1963.)

Reference has previously been made to the sympathetic cholinergic vasodilator pathway from the cortex and hypothalamus to the muscle resistance vessels (p. 516). These nerve fibers are believed to induce vasodilation of the muscle vessels in anticipation of exercise. Whether these nerve fibers in truth serve this function and whether they exist in humans has not been established.

A comparison of the vasoconstrictor and vasodilator effects of the sympathetic nerves to blood vessels of muscle and skin is summarized in Fig. 35-8. Note the lower basal tone of the skin vessels, their greater constrictor response, and the absence of active cutaneous vasodilation.

Local factors. It already has been stressed that neural regulation of muscle blood flow is superseded by metabolic regulation (p. 523) when the muscle changes from the resting to the contracting state. However, local control is also demonstrable in innervated resting skeletal muscle of the dog's hind leg when the reflex stimulation of the vasomotor nerves is minimum and local thermal or mechanical stimulation of the muscle is eliminated. Such preparations show autoregulation of blood flow (a prime example of local control) to the same degree as do actively contracting muscle and de-

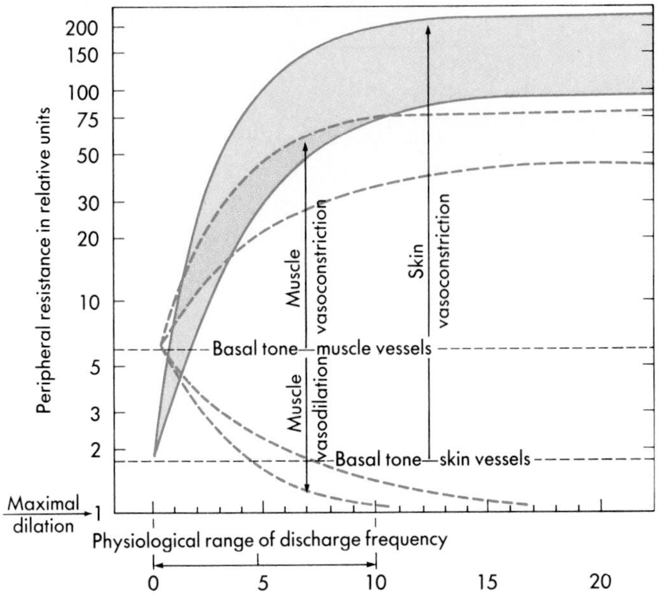

■ Fig. 35-8. Basal tone and the range of response of the resistance vessels in muscle and skin to sympathetic nerve stimulation. Peripheral resistance plotted on a logarithmic scale. (Redrawn from Celander, O., and Folkow, B.: Acta Physiol. Scand. **29**:241, 1953.)

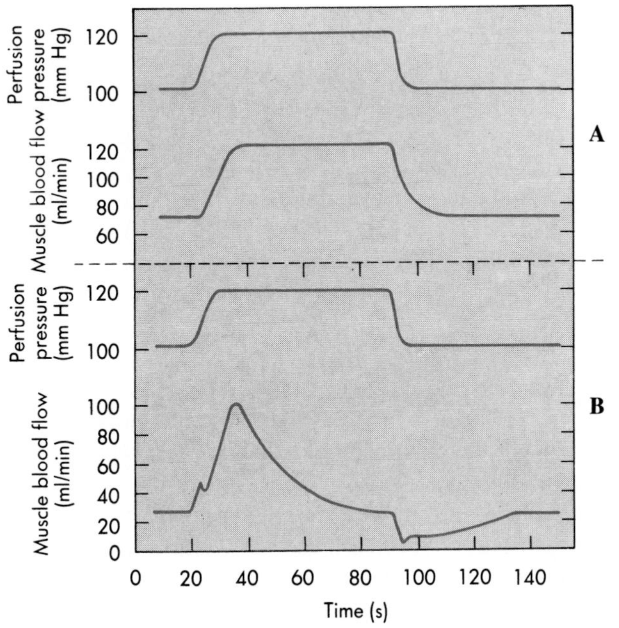

■ Fig. 35-9. **A,** Absence of autoregulation of blood flow. Skeletal muscle blood flow follows changes in perfusion pressure. **B,** Autoregulation of skeletal muscle blood flow. Blood flow returns to control level despite maintenance of elevated perfusion pressure.

nervated muscle. If, however, vascular reflexes are activated when the vasomotor nerves to the muscle are intact or if the muscle receives mechanical or thermal stimulation (with or without the nerves intact), blood flow increases and autoregulation is either lost or diminished (Fig. 35-9). In resting muscle an autoregulating preparation is characterized by low blood flow and low venous blood O_2 saturation, whereas in contracting muscle venous blood O_2 saturation is low, but the flow is high. Thus the common denominator for autoregulation of blood flow in skeletal muscle is a low venous blood O_2 level and presumably a low muscle O_2 tension. Direct measurements of Po_2 in resting skeletal muscle cells yield a range of values from 0 to 10 mm Hg with a mean Po_2 of 4 mm Hg. Studies on humans also reveal that at rest blood flow through muscle is low and O_2 saturation of blood draining the muscles is frequently less than 50%. These observations indicate that at rest muscle blood flow is under either local or neural control, depending on the number and the nature of the stimuli impinging on the vasomotor regions of the medulla. During exercise local metabolic factors take over blood flow regulation, regardless of the degree of sympathetic nerve activity.

■ *Cerebral Circulation*

Blood reaches the brain through the internal carotid and vertebral arteries. The latter join to form the basilar artery, which, in conjunction with branches of the internal carotid arteries, forms the *circle of Willis*. A unique feature of the cerebral circulation is that it all lies within a rigid structure, the cranium. Since intracranial contents are incompressible, any increase in arterial inflow, as with arteriolar dilation, must be associated with a comparable increase in venous outflow. The volume of blood and of extravascular fluid can show considerable variations in most tissues. In brain the volume of blood and extravascular fluid is relatively constant; changes in either of these fluid volumes must be accompanied by a reciprocal change in the other. In contrast to most other organs, the rate of total cerebral blood flow is held within a relatively narrow range and in humans averages 55 ml/min/100 g of brain.

Estimation of Cerebral Blood Flow

Total cerebral blood flow can be measured in humans by the nitrous oxide (N_2O) method, which is based on the Fick principle (p. 447). The subject breathes a gas mixture of 15% N_2O, 21% O_2, and 64% N_2 for 10 minutes, which is sufficient time to permit equilibration of the N_2O between the brain tissue and the blood leaving the brain. Simultaneous samples of arterial (any artery) blood and mixed cerebral venous (internal jugular vein) blood are taken at the start of N_2O inhalation and at 1-minute intervals throughout the 10-minute period of N_2O administration. From these data, the cerebral blood flow can be calculated by the Fick equation:

$$\text{Cerebral blood flow} = \frac{\text{Amount of } N_2O \text{ taken up by brain during time } (t_2 - t_1)}{\text{AV difference of } N_2O \text{ across brain during time } (t_2 - t_1)} \quad (2)$$

Since the arterial and venous concentrations are continuously changing with time, it is necessary to get the true AV difference during the period of N_2O inhalation by integration of the AV difference over the 10-minute period. This is represented in Fig. 35-10, *A*, by the shaded area between the arterial and venous N_2O concentration curves constructed from the blood concentrations observed at successive 1-minute intervals during N_2O administration. The amount of N_2O removed by the brain, as well as the concentration of N_2O in the brain tissue, are unknown. Since the partition coefficient between brain and blood is about 1 and since equilibrium of N_2O between the brain and the blood leaving the brain is reached by the end of 10 minutes, the concentration of N_2O in the brain tissue closely approximates that of the cerebral venous blood in the 10-minute sample. The total weight of the brain is not known, and so for convenience the concentration in brain tissue is multiplied by 100 to express the cerebral blood flow (CBF) in ml/min/100 g brain tissue. The equation is:

$$\text{CBF} = \frac{V_{10} \cdot S \cdot 100}{\int_0^{10} (A - V)dt} \quad (3)$$

V_{10} = Venous concentration of N_2O at equilibrium (at 10 minutes)
S = Partition coefficient of N_2O between brain and blood = 1
A − V = Arteriovenous difference of N_2O

Cerebral blood flow also can be calculated from the desaturation A − V curves, which are constructed from N_2O concentrations of simultaneously drawn arterial

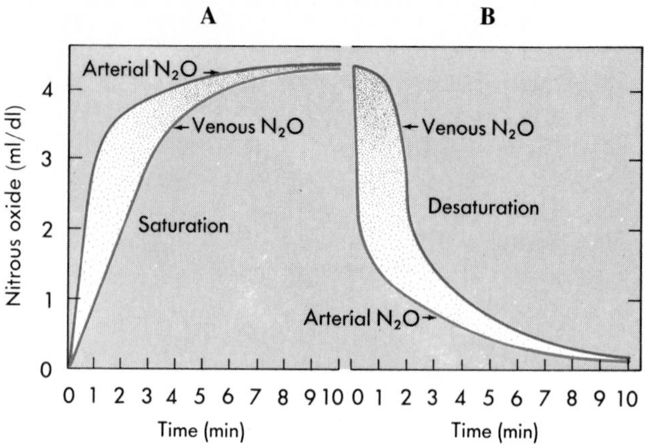

Fig. 35-10. Concentrations of N_2O in arterial and cerebral venous blood during saturation, **A**, and desaturation, **B**. The shaded areas represent the arteriovenous differences of N_2O during the 10-minute period of N_2O inhalation and the 10 minutes after discontinuing the N_2O administration.

and venous blood samples taken each minute for 10 minutes, starting when equilibrium is reached between brain tissue and cerebral venous blood (Fig. 35-10, *B*). In this procedure the subject breathes the N_2O mixture for 10 minutes, and sampling starts at the moment N_2O inhalation is stopped. The only difference from the preceding equation is that the denominator becomes

$$\int_{10}^{20} (V - A)dt \quad (4)$$

Cerebral blood flow and its distribution to different areas of the brain can be measured in animals by injection into the internal carotid artery of microspheres (about 15 μm) labeled with radioactive substances. The microspheres become lodged in the arterioles and capillaries, the brain tissue is sampled, and the radioactivity of the tissue is determined. Blood flow to each tissue sample is proportional to the radioactivity in that sample. By the use of microspheres labeled with different radioactive isotopes, several measurements of cerebral blood flow can be made. A gamma counter is used to measure each isotope independently of the other isotopes in the sample. This method is also used for measurements of blood flow in other tissues, such as the myocardium. One can also measure cerebral blood flow in animals with the use of ^{14}C-antipyrine, which is taken up by the brain in proportion to the blood flow. The brain is then sliced and the radioactivity of the slice is determined by radioautography. The advantage of these methods over the N_2O method or the direct measurement of venous outflow from the brain is that blood flow to different regions of the brain *(regional blood flow)* can be determined. The obvious disadvantage is that the animals must be sacrificed to obtain the samples of brain tissue.

Recently, the development of multiple collimated scintillation detectors built into a helmet that fits over the cranium has made possible the measurement of regional blood flow (cortical blood flow) in animals and humans. An inert radioactive gas (such as ^{133}Xe) is injected into an internal carotid artery, and from its rate of washout from the brain, regional cerebral blood flow can be determined. The radioactive gas also may be given by inhalation, but more sophisticated techniques are required to eliminate noncerebral blood flow and to distinguish between blood flow to cortical (gray matter) and deep cerebral (white matter) tissue.

Regulation of Cerebral Blood Flow

Of the various body tissues, the brain is the least tolerant of ischemia. Interruption of cerebral bloodflow for as little as 5 seconds results in loss of consciousness, and ischemia lasting just a few minutes results in irreversible tissue damage. Fortunately, regulation of the cerebral circulation is primarily under direction of the brain itself. Local regulatory mechanisms and reflexes originating in the brain tend to maintain cerebral circulation relatively constant in the presence of possible adverse extrinsic effects such as sympathetic vasomotor nerve activity, circulating humoral vasoactive agents, and changes in arterial blood pressure. Under certain conditions the brain also regulates its blood flow by initiating changes in systemic blood pressure. For example, elevation of intracranial pressure results in an increase in systemic blood pressure. This response, called *Cushing's phenomenon,* is apparently caused by ischemic stimulation of vasomotor regions of the medulla. It aids in maintaining cerebral blood flow in such conditions as expanding intracranial tumors.

Neural factors. The cerebral vessels receive innervation from the cervical sympathetic nerve fibers that accompany the internal carotid and vertebral arteries into the cranial cavity. The importance of neural regulation of the cerebral circulation is controversial, but the prevalent belief is that relative to other vascular beds sympathetic control of the cerebral vessels is weak and that the contractile state of the cerebral vascular smooth muscle depends primarily on local metabolic factors. There are no known sympathetic vasodilator nerves to the cerebral vessels, but the vessels do receive parasympathetic fibers from the facial nerve, which produce a slight vasodilation on stimulation.

Local factors. Generally total cerebral blood flow is constant. However, regional cortical blood flow varies with and is tightly coupled with regional metabolic activity. For example, movement of one hand results in increased blood flow only in the hand area of the contralateral sensory-motor and premotor cortex. Stimulation of the retina with flashes of light increases blood flow only in the visual cortex. Glucose uptake also corresponds with regional cortical neuronal activity. For example, when the retina is stimulated by light, uptake of ^{14}C-2-deoxyglucose is enhanced in the visual cortex. This analogue of glucose is taken up and phosphorylated by cerebral neurons but cannot be metabolized further. The magnitude of its uptake is determined from radioautographs of slices of the brain. The mediator of the link between cerebral metabolism and blood flow has not been established, but the three principal candidates are pH, potassium, and adenosine.

It is well known that the cerebral vessels are very sensitive to carbon dioxide tension. Increases in arterial blood CO_2 tension (Pa_{CO_2}) elicit marked cerebral vasodilation; inhalation of 7% CO_2 results in a twofold increment in cerebral blood flow. By the same token decreases in Pa_{CO_2}, such as elicited by hyperventilation, produce a decrease in cerebral blood flow. CO_2 produces changes in arteriolar resistance by altering perivascular (and probably intracellular vascular smooth muscle) pH. By independently changing P_{CO_2} and bicarbonate concentration, it has been demonstrated that pial vessel diameter (and presumably blood flow) and pH are inversely related, regardless of the level of the P_{CO_2}.

Carbon dioxide can diffuse to the vascular smooth muscle from the brain tissue or from the lumen of the vessels, whereas hydrogen ions in the blood are prevented from reaching the arteriolar smooth muscle by the *blood-brain barrier.* Hence the cerebral vessels dilate when the hydrogen ion concentration of the cerebrospinal fluid is increased, but they show only minimal dilation in response to an increase in the hydrogen ion concentration of the arterial blood. Despite the responsiveness of the cerebral vessels to pH changes, the precise role of hydrogen ions in the regulation of cerebral blood flow remains obscure. The initiation of increases in cerebral blood flow that is produced by seizures has been reported to be associated with transient increases rather than decreases in perivascular pH. Also, the intracellular and extracellular decreases in pH that occur with electrical stimulation of the brain or hypoxia often occur after cerebral blood flow has increased in response to the stimulus. Furthermore, with prolonged hypocapnia, cerebrospinal fluid pH may return to control in the face of a persistent reduction in cerebral blood flow. Therefore pH probably plays no significant role in the normal regulation of cerebral blood flow.

With respect to K^+, such stimuli as hypoxia, electrical stimulation of the brain, and seizures elicit rapid increases in cerebral blood flow and are associated with increases in perivascular K^+. The increments in K^+ are similar to those which produce pial arteriolar dilation when K^+ is applied topically to these vessels. However, the increase in K^+ is not sustained throughout the period of stimulation. Hence only the initial increment in cerebral blood flow can be attributed to the release of K^+.

Adenosine levels of the brain increase with ischemia, hypoxemia, hypotension, hypocapnia, electrical stimulation of the brain, or induced seizures. When it is applied topically, adenosine is a potent dilator of the pial arterioles. In short, any intervention that either reduces the O_2 supply to the brain or increases the O_2 need of the brain results in rapid (within 5 seconds) formation of adenosine in the cerebral tissue. Unlike pH or K^+, the adenosine concentration of the brain increases with initiation of the stimulus and remains elevated throughout the period of O_2 imbalance. The adenosine released into the cerebrospinal fluid during conditions associated with inadequate brain O_2 supply is available to the brain tissue for reincorporation into cerebral tissue adenine nucleotides.

All three factors, pH, K^+, and adenosine may act in concert to adjust the cerebral blood flow to the metabolic activity of the brain, but how these factors interact to accomplish this regulation of cerebral blood flow remains to be elucidated.

The cerebral circulation shows reactive hyperemia and excellent autoregulation between pressures of about 60 and 160 mm Hg. Mean arterial pressures below 60 mm Hg result in reduced cerebral blood flow and syncope, whereas mean pressures above 160 may lead to increased permeability of the blood-brain barrier and cerebral edema. Autoregulation of cerebral blood flow is abolished by hypercapnia or any other potent vasodilator, and none of the candidates for metabolic regulation of cerebral blood flow has been shown to be responsible for this phenomenon. To what extent autoregulation of the cerebral vessels is attributable to a myogenic mechanism has not been established.

Splanchnic Circulation

The splanchnic circulation consists of the blood supply to the gastrointestinal tract, liver, spleen, and pancreas. Several features distinguish the splanchnic circulation, the most noteworthy of which is that two large capillary beds are partially in series with one another. The small splanchnic arterial branches supply the capillary beds in the gastrointestinal tract, spleen, and pancreas. From these capillary beds, the venous blood ultimately flows into the portal vein, which normally provides most of the blood supply to the liver. However, the hepatic artery also supplies blood to the liver.

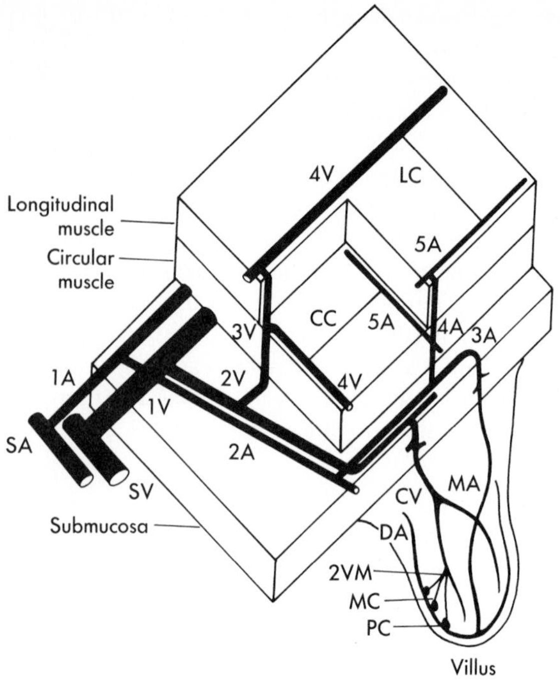

Fig. 35-11. The distribution of small blood vessels to the rat intestinal wall. *SA*, Small artery; *SV*, small vein; *1A* to *5A*, first- to fifth-order arterioles; *1V* to *4V*, first- to fourth-order venules; *CC* and *LC*, capillaries in circular and longitudinal muscle layers; *MA* and *CV*, main arteriole and collecting venule of a villus; *DA*, distribution arteriole; *2VM*, second-order mucosal venule; *PC*, precapillary sphincter; *MC*, mucosal capillary. (From Gore, R.W., and Bohlen, H.G.: Am. J. Physiol. **233:**H685, 1977.)

Intestinal Circulation

Anatomy. The gastrointestinal tract is supplied by the celiac, superior mesenteric, and inferior mesenteric arteries. The superior mesenteric artery is the largest of all the aortic branches and carries over 10% of the cardiac output. Small mesenteric arteries form an extensive vascular network in the submucosa (Fig. 35-11). Their branches penetrate the longitudinal and circular muscle layers and give rise to third- and fourth-order arterioles. Some third-order arterioles in the submucosa become the main arterioles to the tips of the villi.

The direction of the blood flow in the capillaries and venules in a villus is opposite to that in the main arteriole (Fig. 35-12). This arrangement constitutes a countercurrent exchange system. An effective countercurrent multiplier in the villus facilitates the absorption of sodium and water. The countercurrent exchange permits diffusion of O_2 from arterioles to venules. At low flow rates, a substantial fraction of the O_2 may be shunted from arterioles to venules near the base of the villus, thereby curtailing the supply of O_2 to the mucosal cells

Fig. 35-12. Scanning electron micrographs of rabbit intestinal villi *(left panel)* and corrosion cast of the microcirculation in the villus *(right panel). A,* Arteriole; *V,* venule. (From Gannon, B.J., Gore, R.W., and Rogers, P.A.W.: Biomed. Res. **2**[Suppl.]:235, 1981.)

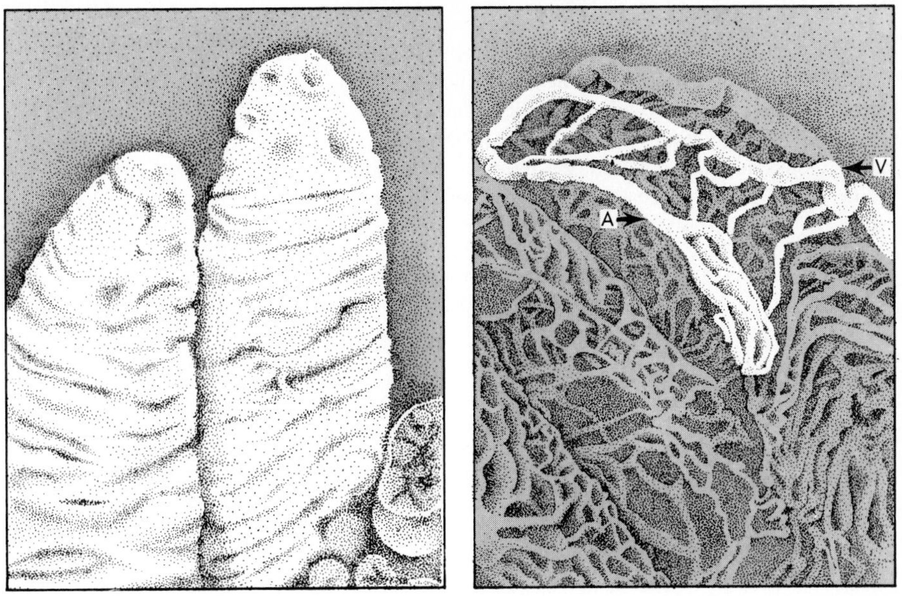

at the tip of the villus. When intestinal blood flow is reduced, the shunting of O_2 is exaggerated, which may cause extensive necrosis of the intestinal villi.

Neural regulation. The neural control of the mesenteric circulation is almost exclusively sympathetic. Increased sympathetic activity constricts the mesenteric arterioles, papillary sphincters, and capacitance vessels. These responses are mediated by alpha-receptors, which are prepotent in the mesenteric circulation; however, beta-receptors are also present. Infusion of a beta-receptor agonist, such as isoproterenol, causes vasodilation.

During fighting or in response to artificial stimulation of the hypothalamic "defense" area pronounced vasoconstriction occurs in the mesenteric vascular bed. This shifts blood flow from the temporarily less important intestinal circulation to the more crucial skeletal muscles, heart, and brain.

Autoregulation. Autoregulation in the intestinal circulation is not as well developed as it is in certain other vascular beds, such as those in the brain and kidney. The principal mechanism responsible for autoregulation is metabolic, although a myogenic mechanism probably also participates. The adenosine concentration in the mesenteric venous blood rises fourfold after brief arterial occlusion. Adenosine is a potent vasodilator in the mesenteric vascular bed and may be the principal metabolic mediator of autoregulation. However, potassium and altered osmolality may also contribute to the overall response.

The O_2 consumption of the small intestine is more rigorously controlled than is the blood flow. In one series of experiments, the O_2 uptake of the small intestine remained constant when the arterial perfusion pressure was varied between 30 and 125 mm Hg.

Functional hyperemia. Food ingestion increases intestinal blood flow. The secretion of certain gastrointestinal hormones contributes to this hyperemia. Gastrin and cholecystokinin augment intestinal blood flow, and they are secreted when food in ingested. The absorption of food affects the intestinal blood flow. Undigested food has no vasoactive influence, whereas several products of digestion are potent vasodilators. Among the various constituents of chyme, the principal mediators of mesenteric hyperemia are glucose and fatty acids.

Hepatic Circulation

Anatomy. The blood flow to the liver normally is about 25% of the cardiac output. The flow is derived from two sources, the portal vein and the hepatic artery. Ordinarily, the portal vein provides about three fourths of the blood flow. The portal venous blood already has passed through the gastrointestinal capillary bed, and therefore much of the O_2 already has been extracted. The hepatic artery delivers the remaining one fourth of the blood, which is fully saturated with O_2. Hence about three fourths of the O_2 used by the liver is derived from the hepatic arterial blood.

The small branches of the portal vein and hepatic artery give rise to terminal portal venules and hepatic arterioles (Fig. 35-13). These terminal vessels enter the hepatic acinus (the functional unit of the liver) at its center. Blood flows from these terminal vessels into the sinusoids, which constitute the capillary network of the liver. The sinusoids radiate toward the periphery of the acinus, where they connect with the terminal hepatic venules. Blood from these terminal venules drains into

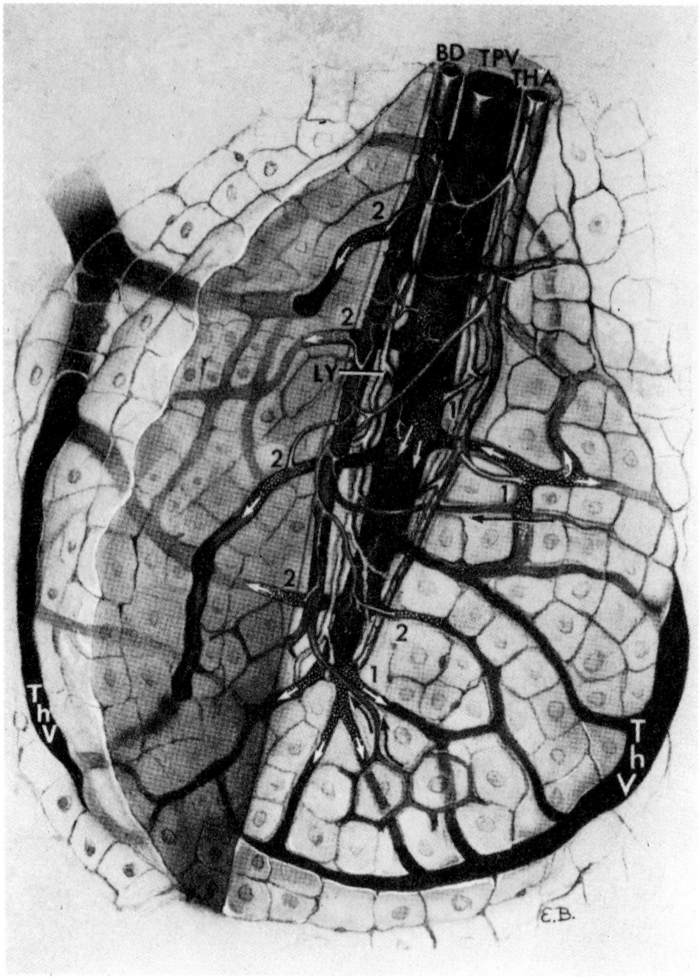

■ **Fig. 35-13.** Microcirculation to a hepatic acinus. *THA,* Terminal hepatic arteriole; *TPV,* terminal portal venule; *BD,* bile ductule; *THV,* terminal hepatic venule; *LY,* lymphatic. The hepatic arterioles empty either directly *(1)* or through the peribiliary plexus *(2)* into the sinusoids that run from the terminal portal venule to the terminal hepatic venules. (From Rappaport, A.M.: Microvasc. Res. **6**:212, 1973.)

progressively larger branches of the hepatic veins, which are tributaries of the inferior vena cava.

Hemodynamics. The mean blood pressure in the portal vein is about 10 mm Hg and that in the hepatic artery is about 90 mm Hg. The resistance of the vessels upstream of the hepatic sinusoids is considerably greater than that of the downstream vessels. Consequently, the pressure in the sinusoids is only 2 or 3 mm Hg greater than that in the hepatic veins and inferior vena cava. The ratio of presinusoidal to postsinusoidal resistance in the liver is much greater than is the ratio of precapillary to postcapillary resistance for almost any other vascular bed. Hence drugs and other interventions that alter the presinusoidal resistance usually affect the pressure in the sinusoids only slightly. Such changes in presinusoidal resistance have little effect on the fluid exchange across the sinusoidal wall. Conversely, changes in hepatic venous (and in central venous) pressure are transmitted almost quantitatively to the hepatic sinusoids and profoundly affect the transsinusoidal exchange of fluids. When central venous pressure is elevated, as in congestive heart failure, plasma water transudes from the liver into the peritoneal cavity, leading to *ascites*.

Regulation of flow. Blood flows in the portal venous and hepatic arterial systems vary reciprocally. When blood flow is curtailed in one system, the flow increases in the other system. However, the resultant increase in flow in one system usually does not fully compensate for the initiating reduction in flow in the other system.

The portal venous system does not autoregulate. As the portal venous pressure and flow are raised, resistance either remains constant or it decreases. The hepatic arterial system does autoregulate, however.

The liver tends to maintain a constant O_2 consumption, because the extraction of O_2 from the hepatic blood is very efficient. As the rate of O_2 delivery to the liver is varied, the liver compensates by an appropriate change in the fraction of O_2 extracted from the blood. This extraction is facilitated by the distinct separation of the presinusoidal vessels at the acinar center from the postsinusoidal vessels at the periphery of the acinus (Fig. 35-11). The substantial distance between these types of vessels prevents the countercurrent exchange

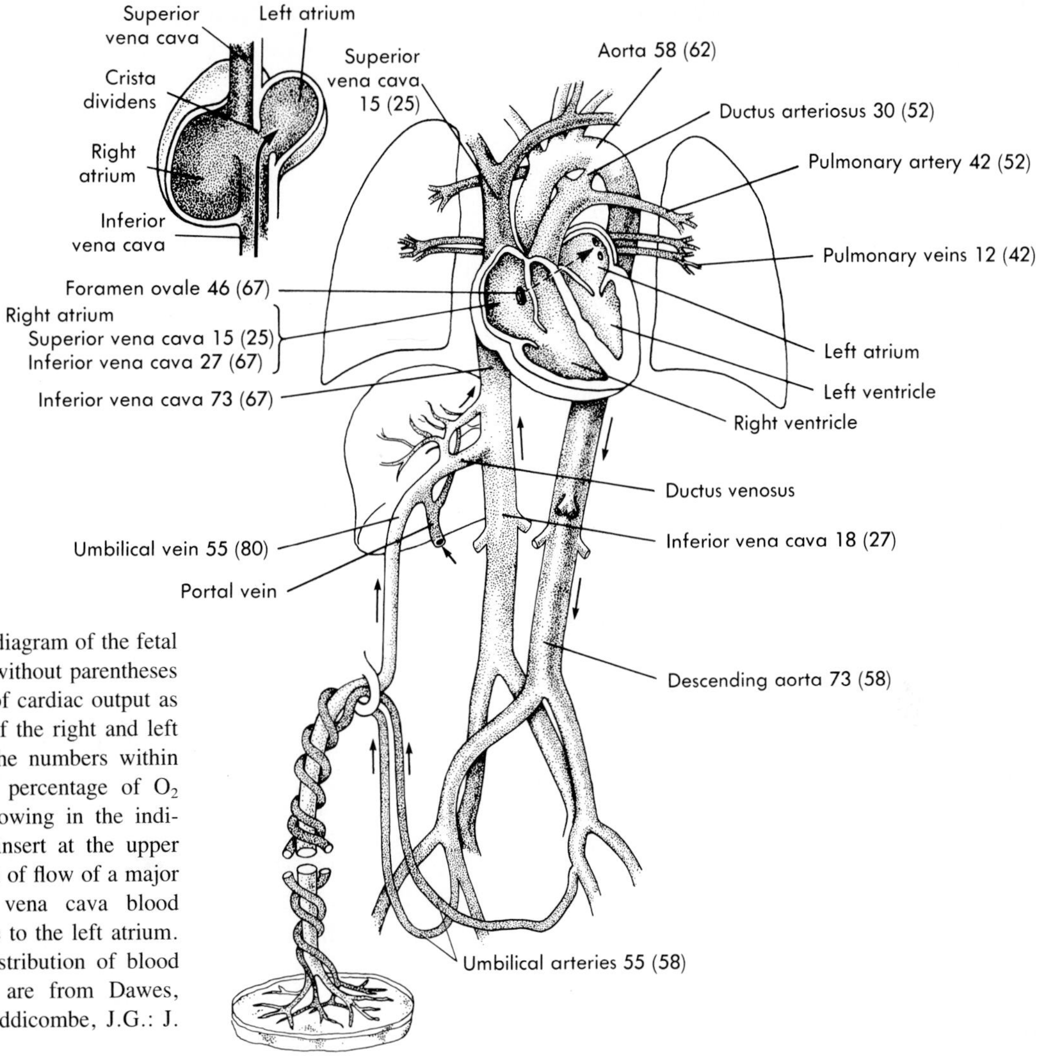

■ **Fig. 35-14.** Schematic diagram of the fetal circulation. The numbers without parentheses represent the distribution of cardiac output as a percentage of the sum of the right and left ventricular outputs, and the numbers within parentheses represent the percentage of O_2 saturation of the blood flowing in the indicated blood vessel. The insert at the upper left illustrates the direction of flow of a major portion of the inferior vena cava blood through the foramen ovale to the left atrium. (Values for percentage distribution of blood flow and O_2 saturations are from Dawes, G.S., Mott, J.C., and Widdicombe, J.G.: J. Physiol. **126:**563, 1954.)

of O_2, contrary to the condition that exists in an intestinal villus (Fig. 35-11).

The sympathetic nerves constrict the presinusoidal resistance vessels in the portal venous and hepatic arterial systems. Neural effects on the capacitance vessels are more important, however. The effects are mediated mainly by alpha-receptors.

Capacitance vessels. The liver contains about 15% of the total blood volume of the body. Under appropriate conditions, such as in response to hemorrhage, about half of the hepatic blood volume can be rapidly expelled. Hence the liver constitutes an important blood reservoir in humans. In certain other species, such as the dog, the spleen is a more important blood reservoir. Smooth muscle in the capsule and trabeculae of the spleen contract in response to increased sympathetic neural activity, such as occurs during exercise or hemorrhage. However, this mechanism does not exist in humans.

■ *Fetal Circulation*

The circulation of the fetus shows a number of differences from that of the postnatal infant. The fetal lungs are functionally inactive, and the fetus depends completely on the placenta for O_2 and nutrient supply. Oxygenated fetal blood from the placenta passes through the umbilical vein to the liver. A major fraction passes through the liver and a small fraction bypasses the liver to the inferior vena cava through the *ductus venosus* (Fig. 35-14). In the inferior vena cava, blood from the ductus venosus joints blood returning from the lower trunk and extremities, and this combined stream is in turn joined by blood from the liver through the hepatic veins. The streams of blood tend to maintain their identity in the inferior vena cava and are divided into two streams of unequal size by the edge of the interatrial septum *(crista dividens)*. The larger stream, which is primarily blood from the umbilical vein, is shunted to the left atrium through the *foramen ovale,* which lies between the inferior vena cava and the left atrium (*inset,* Fig. 35-14). The other stream passes into the right atrium, where it is joined by superior vena caval blood returning from the upper parts of the body and by blood from the myocardium. In contrast to the adult, in whom the right and left ventricles pump in series, in the fetus the ventricles operate essentially in parallel. Because of the large pulmonary resistance, less than one third of the right ventricular output goes through the lungs. The remainder passes through the *ductus arteriosus* from the pulmonary artery to the aorta at a point distal to the origins of the arteries to the head and upper extremities. Flow from pulmonary artery to aorta occurs because pulmonary artery pressure is about 5 mm Hg higher than aortic pressure in the fetus. The large volume of blood coming through the foramen ovale into the left atrium is joined by blood returning from the lungs and is pumped out by the left ventricle into the aorta. About one third of the aortic blood goes to the head, upper thorax, and arms and the remaining two thirds go to the rest of the body and the placenta. The amount of blood pumped by the left ventricle is about 20% greater than that pumped by the right ventricle, and the major fraction of the blood that passes down the descending aorta flows by way of the two umbilical arteries to the placenta.

In Fig. 35-14 the distribution of fetal blood flow is given in a percentage of the combined right and left ventricular outputs. Note that over half of the combined cardiac output is returned directly to the placenta without passing through any capillary bed. Also indicated in Fig. 35-14 are the O_2 saturations of the blood (numbers in parentheses) at various points of the fetal circulation. Fetal blood leaving the placenta is 80% saturated, but the saturation of the blood passing through the foramen ovale is reduced to 67% by mixing with desaturated blood returning from the lower part of the body and the liver. Addition of the desaturated blood from the lungs reduces the O_2 saturation of left ventricular blood to 62%, which is the level of saturation of the blood reaching the head and upper extremities. The blood in the right ventricle, a mixture of desaturated superior vena caval blood, coronary venous blood, and inferior vena caval blood, is only 52% saturated with O_2. When the major portion of this blood traverses the ductus arteriosus and joins that pumped out by the left ventricle, the resultant O_2 saturation of blood traveling to the lower part of the body and back to the placenta is 58% saturated. Thus it is apparent that the tissues receiving blood of the highest O_2 saturation are the liver, heart, and upper parts of the body, including the head.

At the placenta the chorionic villi dip into the maternal sinuses and O_2, CO_2, nutrients, and metabolic waste products exchange across the membranes. The barrier to exchange is quite large, and the equilibrium of O_2 tension between the two circulations is not reached at normal rates of blood flow. Therefore the O_2 tension of the fetal blood leaving the placenta is very low. Were it not for the fact that fetal hemoglobin has a greater affinity for O_2 than does adult hemoglobin, the fetus would not receive an adequate O_2 supply. The fetal oxyhemoglobin dissociation curve is shifted to the left so that at equal pressures of O_2 fetal blood will carry significantly more O_2 than will maternal blood. If the mother is subjected to hypoxia, the reduced blood O_2 tension is reflected in the fetus by tachycardia and an increase in blood flow through the umbilical vessels. If the hypoxia persists or if flow through the umbilical vessels is impaired, fetal distress occurs and is first manifested as bradycardia. In early fetal life the high cardiac glycogen levels that prevail (which gradually decrease to adult levels by term) may protect the heart from acute periods of hypoxia.

Circulatory Changes That Occur at Birth

The umbilical vessels have thick muscular walls that are very reactive to trauma, tension, sympathomimetic amines, bradykinin, angiotensin, and changes in O_2 tension. In animals in which the umbilical cord is not tied, hemorrhage of the newborn is prevented by constriction of these large vessels in response to one or more of those stimuli. Closure of the umbilical vessels produces an increase in total peripheral resistance and of blood pressure. When blood flow through the umbilical vein ceases, the ductus venosus, a thick-walled vessel with a muscular sphincter, closes. What initiates closure of the ductus venosus is still unknown. The asphyxia, which starts with constriction or clamping of the umbilical vessels, plus the cooling of the body activate the respiratory center of the newborn infant. With the filling of the lungs with air, pulmonary vascular resistance decreases to about one tenth of the value existing prior to lung expansion. This resistance change is not caused by the presence of O_2 in the lungs, since the change is just as great if the lungs are filled with nitrogen. However, filling the lungs with liquid does not reduce pulmonary vascular resistance.

The left atrial pressure is raised above that in the inferior vena cava and right atrium by (1) the decrease in pulmonary resistance, with the resulting large flow of blood through the lungs to the left atrium, (2) the reduction of flow to the right atrium caused by occlusion of the umbilical vein, and (3) the increased resistance to left ventricular output produced by occlusion of the umbilical arteries. This reversal of the pressure gradient across the atria abruptly closes the valve over the foramen ovale, and fusion of the septal leaflets occurs over a period of several days.

With the decrease in pulmonary vascular resistance, the pressure in the pulmonary artery falls to about one half its previous level (to about 35 mm Hg), and this change in pressure, coupled with a slight increase in aortic pressure, reverses the flow of blood through the ductus arteriosus. However, within several minutes the large ductus arteriosus begins to constrict, producing turbulent flow, which is manifest as a murmur in the newborn. Constriction of the ductus arteriosus is progressive and usually is complete within 1 to 2 days after birth. Closure of the ductus arteriosus appears to be initiated by the high O_2 tension of the arterial blood passing through it, since pulmonary ventilation with O_2 or with air low in O_2 induces, respectively, closure and opening of this shunt vessel. Whether O_2 acts directly on the ductus or through the release of a vasoconstrictor substance is not known. Similarly, in a heart-lung preparation made from a newborn lamb, the ductus arteriosus may be made to close with high PaO_2. The mechanism whereby increases in PaO_2 produce closure of the ductus arteriosus is unknown, but changes in the concentrations of bradykinin, prostaglandins, and adenosine in blood or ductal tissue may be involved.

At birth the walls of the two ventricles are approximately of the same thickness, with a possibly slight preponderance of the right ventricle. Also present in the newborn is thickening of the muscular levels of the pulmonary arterioles, which is apparently responsible in part for the high pulmonary vascular resistance of the fetus. After birth the thickness of the walls of the right ventricle diminishes, as does the muscle layer of the pulmonary arterioles; the left ventricular walls increase in thickness. These changes are progressive over a period of weeks after birth.

Failure of the foramen ovale or ductus arteriosus to close after birth is occasionally observed and constitutes two of the more common congenital cardiac abnormalities that are now amenable to surgical correction.

Bibliography

Journal articles

Abboud, F.M., editor: Regulation of the cerebral circulation (symposium), Fed. Proc. **40**:2296, 1981.

Bacchus, A.N., et al.: Adenosine and coronary blood flow in conscious dogs during normal physiological stimuli, Am. J. Physiol. **243**:H629, 1982.

Belloni, F.L.: The local control of coronary blood flow, Cardiovasc. Res. **13**:63, 1979.

Berne, R.M.: Regulation of coronary blood flow, Physiol. Rev. **44**:1, 1964.

Berne, R.M.: Role of adenosine of the regulation of coronary blood flow, Circ. Res. **47**:807, 1980.

Berne, R.M., Winn, H.R., and Rubio, R.: The local regulation of cerebral blood flow, Prog. Cardiovasc. Dis. **24**:243, 1981.

Buckberg, G.D., et al.: Variable effects of heart rate on phasic and regional left ventricular muscle blood flow in anesthetized dogs, Cardiovasc. Res. **9**:1, 1975.

Chou, C.C.: Contribution of splanchnic circulation to overall cardiovascular and metabolic homeostasis (symposium), Fed. Proc. **42**:1656, 1983.

Feigl, E.O.: Sympathetic control of coronary circulation, Circ. Res. **20**:262, 1967.

Feigl, E.O.: Parasympathetic control of coronary blood flow in dogs, Circ. Res. **25**:509, 1969.

Feigl, E.O.: Coronary physiology, Physiol. Rev. **63**:1, 1983.

Granger, D.N., and Kvietys, P.R.: The splanchnic circulation: intrinsic regulation, Annu. Rev. Physiol. **43**:409, 1981.

Greenway, C.V.: The role of the splanchnic venous system in overall cardiovascular homeostasis, Fed. Proc. **42**:1678, 1983.

Greenway, C.V., Seaman, K.L., and Innes, I.R: Norepinephrine on venous compliance and unstressed volume in cat liver, Am. J. Physiol. **248**:H468, 1985.

Gregg, D.E.: The natural history of coronary collateral development, Circ. Res. **35**:335, 1974.

Heymann, M.A., Iwamoto, H.S., and Rudolf, A.M.: Factors affecting changes in the neonatal systemic circulation. Annu. Rev. Physiol. **43**:371, 1981.

Heymann, M.A., and Rudolph, A.M.: Control of the ductus arteriosus, Physiol. Rev. **55**:62, 1975.

Kalsner, S., editor: The coronary artery, London, 1982, Croom Helm, Ltd.

Klocke, F.J., and Ellis, A.K.: Control of coronary blood flow, Annu. Rev. Med. **31**:489, 1980.

Klocke, F.J., Mates, R.E., Canty, J.M., Jr., and Ellis, A.K.: Coronary pressure-flow relationships—controversial issues and probable implications, Circ. Res. **56**:310, 1985.

Kontos, H.A.: Regulation of the cerebral circulation, Annu. Rev. Physiol. **43**:397, 1981.

Lautt, W.W.: Hepatic vasculature: a conceptual review, Gastroenterology **73**:1163, 1977.

Lautt, W.W.: Hepatic nerves: a review of their functions and effects, Can. J. Physiol. Pharmacol. **58**:105, 1980.

Murray, P.A., Belloni, F.L., and Sparks, H.V.: The role of potassium in the metabolic control of coronary vascular resistance in the dog, Circ. Res. **44**:767, 1979.

Olsson, R.A.: Local factors regulating cardiac and skeletal muscle blood flow, Annu. Rev. Physiol. **43**:385, 1981.

Olsson, R.A., and Patterson, R.E.: Adenosine as a physiological regulator of coronary blood flow, Prog. Molec. Subcel. Biol. **4**:227, 1976.

Rappaport, A.M.: The microcirculatory hepatic unit, Microvasc. Res. **6**:212, 1973.

Rappaport, A.M., and Schneidermann, J.H.: The function of the hepatic artery, Rev. Physiol. Biochem. Pharmacol. **76**:129, 1976.

Rubio, R., and Berne, R.M.: Regulation of coronary blood flow, Prog. Cardiovasc. Dis. **18**:105, 1975.

Thompson, C.I., Rubio, R., and Berne, R.M.: Changes in adenosine and glycogen phosphorylase activity during the cardiac cycle, Am. J. Physiol. **238**:H389, 1980.

Wearn, J.T., et al.: The nature of the vascular communications between the coronary arteries and the chambers of the heart, Am. Heart J. **9**:143, 1933.

Books and monographs

Berne, R.M., and Rubio, R.: Coronary circulation. In Handbook of physiology; Section 2: The cardiovascular system—the heart, vol. I, Bethesda, Md., 1979, American Physiological Society.

Berne, R.M., Winn, H.R., and Rubio, R.: Metabolic regulation of cerebral blood flow. In Mechanisms of vasodilation—Second Symposium, New York, 1981, Raven Press.

Donald, D.E.: Splanchnic circulation. In Handbook of physiology; Section 2: The cardiovascular system—peripheral circulation and organ blood flow, vol. III, Bethesda, Md., 1983, American Physiological Society.

Faber, J.J., and Thornburg, K.L.: Placental physiology, New York, 1983, Raven Press.

Gootman, N., and Gootman, P.M., editors: Perinatal cardiovascular function, New York, 1983, Marcel Dekker, Inc.

Gregg, D.E.: Coronary circulation in health and disease, Philadelphia, 1950, Lea & Febiger.

Heistad, D.D., and Kontos, H.A.: Cerebral circulation. In Handbook of physiology; Section 2: The cardiovascular system—peripheral circulation and organ blood flow, vol. III, Bethesda, Md., 1983, American Physiological Society.

Hellon, R.: Thermoreceptors. In Handbook of physiology; Section 2: The cardiovascular system—peripheral circulation and organ blood flow, vol. III, Bethesda, Md., 1983, American Physiological Society.

Lautt, W.W., editor: Hepatic circulation in health and disease, New York, 1981, Raven Press.

Lewis, T.: Blood vessels of the human skin and their responses, London, 1927, Shaw & Son, Ltd.

Longo, L.D., and Reneau, D.D., editors: Fetal and newborn cardiovascular physiology, vol. 1, Developmental aspects, New York, 1978, Garland STPM Press.

Marcus, M.L.: The coronary circulation in health and disease, New York, 1983, McGraw-Hill Book Co.

Morgan, H.E., Rannels, D.E., and McKee, E.E.: Protein metabolism of the heart. In Handbook of physiology; Section 2: The cardiovascular system—the heart, vol. I, Bethesda, Md., 1979, American Physiological Society.

Mott, J.C., and Walker, D.W.: Neural and endocrine regulation of circulation in the fetus and newborn. In Handbook of physiology; Section 2: The cardiovascular system—peripheral circulation and organ blood flow, vol. III, Bethesda, Md., 1983, American Physiological Society.

Owman, C., and Edvinsson, L., editors: Neurogenic control of the brain circulation, Oxford, 1977, Pergamon Press.

Randle, P.J., and Tubbs, P.K.: Carbohydrate and fatty acid metabolism. In Handbook of physiology; Section 2: The cardiovascular system—the heart, vol. I., Bethesda, Md., 1979, American Physiological Society.

Roddie, E.C.: Circulation to skin and adipose tissue. In Handbook of physiology; Section 2: The cardiovascular system—peripheral circulation and organ blood flow, vol. III, Bethesda, Md., 1983, American Physiological Society.

Schaper, W.: The collateral circulation of the heart, New York, 1971, North-Holland Publishing Co.

Shepherd, A.P., and Granger, D.N.: Physiology of the intestinal circulation, New York, 1984, Raven Press.

Shepherd, J.T.: Circulation of skeletal muscle. In Handbook of physiology; Section 2: The cardiovascular system—peripheral circulation and organ blood flow, vol. III, Bethesda, Md., 1983, American Physiological Society.

Sparks, H.V., Jr., Wangler, R.D., and DeWitt, D.F.: Control of the coronary circulation in physiology and pathophysiology of the heart. In Sperelakis, N., editor: Physiology and pathophysiology of the heart, Hingham, Mass., 1984, Martinus Nijhoff Publishers.

CHAPTER 36

Interplay of Central and Peripheral Factors in the Control of the Circulation

The primary function of the circulatory system is to deliver the supplies needed for tissue metabolism and growth and to remove the products of metabolism. To explain how the heart and blood vessels serve this function, it has been necessary to analyze the system morphologically and functionally and to discuss the mechanisms of action of the component parts in their contribution to maintaining adequate tissue perfusion under different physiological conditions. Once the functions of the various components are understood, it is essential that their interrelationships in the overall role of the circulatory system be considered. Tissue perfusion depends on arterial pressure and local vascular resistance, and arterial pressure in turn depends on cardiac output and total peripheral resistance (TPR). Arterial pressure is maintained within a relatively narrow range in the normal individual, a feat that is accomplished by reciprocal changes in cardiac output and TPR. However, cardiac output and peripheral resistance are each influenced by a number of factors, and it is the interplay among these factors that determines the level of these two variables. The autonomic nervous system and the baroreceptors play the key role in regulating blood pressure. However, from the long-range point of view the control of fluid balance by the kidney, adrenal cortex, and central nervous system, with maintenance of a constant blood volume, is of the greatest importance.

In a well-regulated system one way to study the extent and sensitivity of the regulatory mechanism is to disturb the system and observe its response to restore the preexisting steady state. Disturbances in the form of physical exercise and hemorrhage will be used to illustrate the effects of the various factors that go into regulation of the circulatory system.

Exercise

The cardiovascular adjustments in exercise comprise a combination and integration of neural and local (chemical) factors. The neural factors consist of (1) *central command,* (2) reflexes originating in the contracting muscle, and (3) the baroreceptor reflex. Central command is the cerebrocortical activation of the sympathetic nervous system that produces cardiac acceleration, increased myocardial contractile force, and peripheral vasoconstriction. Reflexes can be activated intramuscularly by stimulation of mechanoreceptors (stretch, tension) and "chemoreceptors" (products of metabolism) in response to muscle contraction. Impulses from these receptors travel centrally via small myelinated (group III) and unmyelinated (group IV) afferent nerve fibers. The group IV unmyelinated fibers may represent the muscle chemoreceptors; no morphological chemoreceptor has been identified. The central connections of this reflex are unknown, but the efferent limb is the sympathetic nerve fibers to the heart and peripheral blood vessels. The baroreceptor reflex has been described on p. 454, and the local factors that influence skeletal muscle blood flow (metabolic vasodilators) are described on p. 511. Vascular chemoreceptors do not play a significant role in regulation of the cardiovascular system in exercise. The pH, P_{CO_2} and P_{O_2} of arterial blood are normal during exercise, and the vascular chemoreceptors are located on the arterial side of the circulatory system.

Mild to Moderate Exercise

In humans or in trained animals, anticipation of physical activity inhibits the vagus nerve impulses to the heart and increases sympathetic discharge. The concerted inhibition of parasympathetic areas and activation of sympathetic areas of the medulla on the heart result in an increase in heart rate and myocardial contractility. The tachycardia and the enhanced contractility increase cardiac output.

Peripheral resistance. At the same time that cardiac stimulation occurs, the sympathetic nervous system also elicits vascular resistance changes in the periphery. In cats and dogs, and possibly in humans, the cholinergic sympathetic vasodilator system is activated and dilates the resistance vessels in the muscles. In skin, kidneys, splanchnic regions, and active muscle, sympathetic-mediated vasoconstriction increases vascular resistance, which diverts blood away from these areas (Fig. 36-1). This increased resistance in vascular beds of inactive tissues persists throughout the period of exercise.

As cardiac output and blood flow to active muscles increase with progressive increments in the intensity of exercise, blood flow to the splanchnic and renal vasculatures decreases. Blood flow to the myocardium increases, whereas that to the brain is unchanged. The net effect is an increase in the total blood flow to these two key organs. Skin blood flow initially decreases during exercise and then increases as body temperature rises with increments in duration and intensity of exercise. Skin blood flow finally decreases when the skin vessels constrict as the total body O_2 consumption nears maximum (Fig. 36-1).

The major circulatory adjustment to prolonged exercise involves the vasculature of the active muscles. Local formation of vasoactive metabolites induces marked dilation of the resistance vessels, which progresses with increases in the intensity level of exercise. Potassium is one of the vasodilator substances released by contracting muscle, and it may be in part responsible for the initial decrease in vascular resistance in the active muscles. Other contributing factors may be an increase in interstitial fluid osmolarity during the initial phase of exercise and the release of adenosine and a decrease in pH during sustained exercise. The local accumulation of metabolites relaxes the terminal arterioles. Blood flow through the muscle may increase 15 to 20 times above the resting level. This metabolic vasodilation of the precapillary vessels in active muscles occurs very soon after the onset of exercise, and the decrease in TPR enables the heart to pump more blood at a lesser load and more efficiently (less pressure work, p. 546) than if TPR were unchanged. Only a small percentage of the capillaries is perfused at rest, whereas in actively contracting muscle all or nearly all of the capillaries contain flowing blood *(capillary recruitment)*. The surface available for exchange of gases, water, and solutes is increased manyfold. Furthermore, the hydrostatic pressure in the capillaries is increased because of the relaxation of the resistance vessels and the increased osmolarity of the interstitial fluid. Hence there is a net movement of water and solutes into the muscle tissue. Tissue pressure rises and remains elevated during exercise as fluid continues to move out of the capillaries and is carried away by the lymphatics. Lymph flow is increased as a result of the increase in capillary hydrostatic pressure and the massaging effect of the contracting muscles on the valve-containing lymphatic vessels.

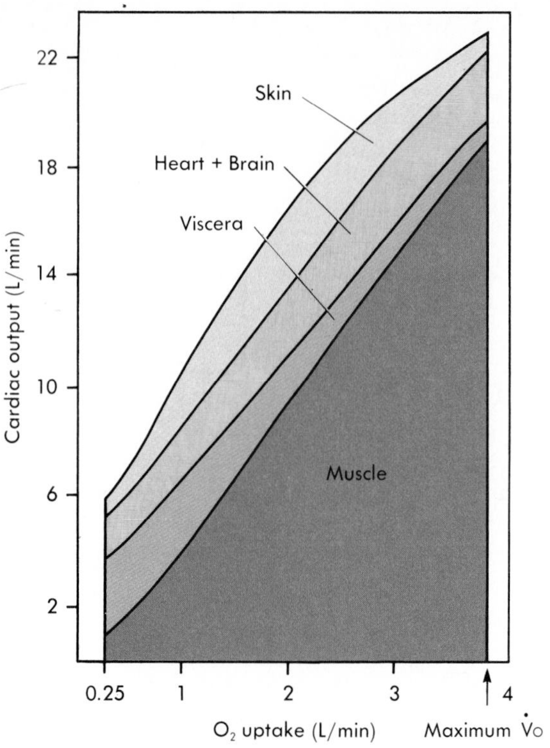

Fig. 36-1. Approximate distribution of cardiac output at rest and at different levels of exercise up to the maximum O_2 consumption ($\dot{V}o_{2max}$) in a normal young man. Viscera refers to splanchnic and renal blood flow. (Redrawn from Ruch, H.P., and Patton, T.C.: Physiology and biophysics, ed. 12, 1974, W.B. Saunders Co.)

The contracting muscle avidly extracts O_2 from the perfusing blood (increased AV-O_2 difference, Fig. 36-2), and the release of O_2 from the blood is facilitated by the nature of oxyhemoglobin dissociation. The reduction in pH caused by the high concentration of CO_2 and the formation of lactic acid and the increase in temperature in the contracting muscle contribute to shifting the oxyhemoglobin dissociation curve to the right. At any given partial pressure of O_2, less O_2 is held by the hemoglobin in the red cells, and consequently there is a more effective O_2 removal from the blood. Oxygen consumption may increase as much as sixty-fold with only a fifteen-fold increase in muscle blood flow. Muscle myoglobin may serve as a limited O_2 store in exercise and can release attached O_2 at very low partial pressures. However, it can facilitate O_2 transport from capillaries to mitochondria by serving as an O_2 carrier.

Cardiac output. The enhanced sympathetic drive and the reduced parasympathetic inhibition of the sinoatrial node continue during exercise, and consequently tachycardia persists. If the work load is moderate and constant, the heart rate will reach a certain level and remain there throughout the period of exercise. However, if the work load increases, a concomitant increase in heart rate occurs until a plateau is reached in severe exercise at about 180 beats per minute (Fig. 36-2). In contrast to the large increment in heart rate, the increase in stroke volume is only about 10% to 35% (Fig. 36-2), the larger values occurring in trained individuals. (In very well-trained distance runners, whose cardiac outputs can reach six to seven times the resting level, stroke volume reaches about twice the resting value.)

Thus it is apparent that the increase in cardiac output observed with exercise is achieved principally by an increase in heart rate. If the baroreceptors are denervated, the cardiac output and heart rate responses to exercise are sluggish when compared to the changes in animals with normally innervated baroreceptors. However, in the absence of autonomic innervation of the heart, as produced experimentally in dogs by total cardiac denervation, exercise still elicits an increment in cardiac output comparable to that observed in normal animals, but chiefly by means of an elevated stroke volume. However, if a beta-adrenergic receptor blocking agent is given to dogs with denervated hearts, exercise performance is impaired. The beta-adrenergic receptor blocker apparently prevents the cardiac acceleration and enhanced contractility caused by increased amounts of circulating catecholamines and hence limits the increase in cardiac output necessary for maximum exercise performance.

Venous system. In addition to the contribution made by sympathetically mediated constriction of the capacitance vessels in both exercising and nonexercising parts of the body, venous return is aided by the working skeletal muscles and the muscles of respiration. As pointed out in Chapter 34, the intermittently contracting muscles compress the vessels that course through them and, in the case of veins with their valves oriented toward the heart, pump blood back toward the right atrium. The flow of venous blood to the heart is also aided by the increase in the pressure gradient developed by the more negative intrathoracic pressure produced by deeper and more frequent respirations. In humans, with the exception of the skin, lungs, and liver, there is little evidence that blood reservoirs contribute much to the circulating blood volume. In fact, blood volume is usually reduced slightly during exercise, as evidenced by a rise in the hematocrit ratio, because of water loss externally by sweating and enhanced ventilation, and fluid movement into the contracting muscle. The fluid loss from the vascular compartment into contracting muscle reaches a plateau as interstitial fluid pressure rises and opposes the increased hydrostatic pressure in the capillaries of the active muscle. The fluid loss is partially offset by movement of fluid from the splanchnic regions and inactive muscle into the bloodstream. This influx of fluid occurs as a result of a decrease of hydrostatic pressure in the capillaries of these tissues and of an increase in plasma osmolarity due to movement of osmotically active particles into the blood from the contracting muscle. In addition, reduced urine formation by the kidneys helps to conserve body water.

The large volume of blood returning to the heart is so rapidly pumped through the lungs and out into the aorta that central venous pressure remains essentially constant. Thus the Frank-Starling mechanism of a greater initial fiber length does not account for the greater stroke volume in moderate exercise. Chest x-ray films of individuals at rest and during exercise reveal a decrease in heart size in exercise, which is in harmony with the observations of a constant ventricular filling pressure. However, in maximum or near-maximum exercise, right atrial pressure and end-diastolic ventricular volume do increase. Thus the Frank-Starling mechanism contributes to the enhanced stroke volume in very vigorous exercise.

Arterial pressure. If the exercise involves a large proportion of the body musculature, such as in running or swimming, the reduction in total vascular resistance can be considerable (Fig. 36-2). Nevertheless, arterial pressure starts to rise with the onset of exercise, and the increase in blood pressure roughly parallels the severity of the exercise performed (Fig. 36-2). Therefore the increase in cardiac output is proportionally greater than the decrease in TPR. The vasoconstriction produced in the inactive tissues by the sympathetic nervous system (and to some extent by the release of catecholamines from the adrenal medulla) is important for maintaining normal or increased blood pressure, since sympathectomy or drug-induced block of the adrenergic sympathetic nerve fibers results in a decrease in arterial pressure *(hypotension)* during exercise.

Sympathetic-mediated vasoconstriction also occurs in active muscle when additional muscles are activated after about half of the total skeletal musculature is contracting. In experiments in which one leg is working at maximum levels and then the other leg starts to work, blood flow decreases in the first working leg. Further-

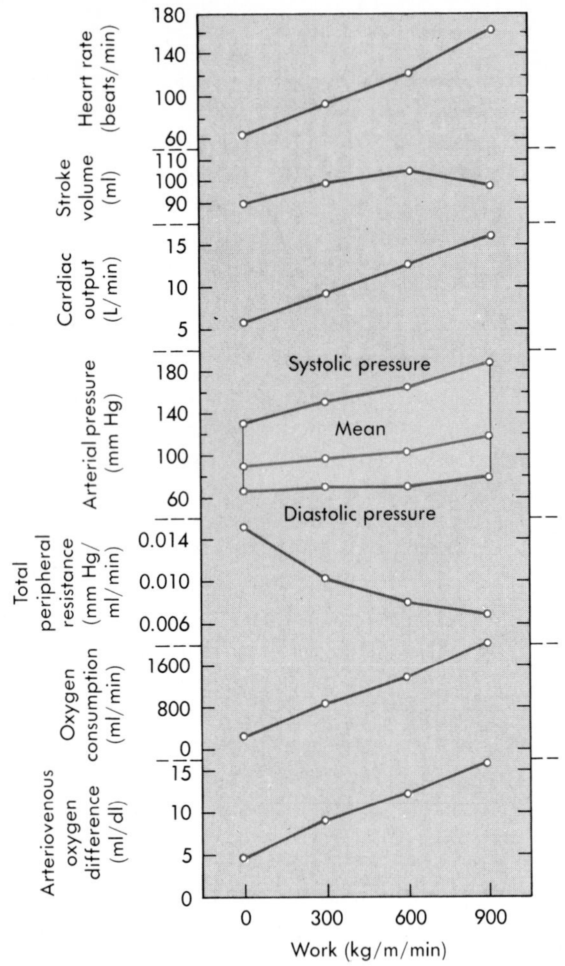

■ **Fig. 36-2.** Effect of different levels of exercise on several cardiovascular variables. (Data from Carlsten, A., and Grimby, G.: The circulatory response to muscular exercise in man, Springfield, Ill., 1966, Charles C Thomas, Publisher.)

more, blood levels of norepinephrine rise significantly in exercise, and most of it comes from sympathetic nerves in the active muscles.

As body temperature rises during exercise, the skin vessels dilate in response to thermal stimulation of the heat-regulating center in the hypothalamus, and TPR decreases further. This would result in a decline in blood pressure were it not for the increasing cardiac output and constriction of arterioles in the renal, splanchnic, and other tissues.

In general, mean arterial pressure rises during exercise as a result of the increase in cardiac output. However, the effect of enhanced cardiac output is offset by the overall decrease in TPR so that the mean blood pressure increase is relatively small. Vasoconstriction in the inactive vascular beds contributes to the maintenance of a normal arterial blood pressure for adequate perfusion of the active tissues. The actual pressure attained represents the product of cardiac output and TPR (p. 527). Systolic pressure usually increases more than diastolic pressure, resulting in an increase in pulse pressure (Fig. 36-2). The larger pulse pressure is primarily attributable to a greater stroke volume and to a lesser degree to a more rapid ejection of blood by the left ventricle with less peripheral runoff during the brief ventricular ejection period.

■ *Severe Exercise*

In severe exercise taken to the point of exhaustion, the compensatory mechanisms begin to fail. Heart rate attains a maximum level of about 180 beats per minute, and stroke volume reaches a plateau and often decreases, resulting in a fall in blood pressure. Dehydration occurs. Sympathetic vasoconstrictor activity supersedes the vasodilator influence on the cutaneous vessels and has the hemodynamic effect of a slight increase in effective blood volume. However, cutaneous vasoconstriction also results in a decrease in the rate of heat loss. Body temperature is normally elevated in exercise, and reduction in heat loss through cutaneous vasoconstriction can, under these conditions, lead to very high body temperatures with associated feelings of acute distress. The tissue and blood pH decrease, as a result of increased lactic acid and CO_2 production, and the reduced pH is probably the key factor in determining the maximum amount of exercise a given individual can tolerate because of muscle pain, subjective feeling of exhaustion, and inability or loss of will to continue. A summary of the neural and local effects of exercise on the cardiovascular system is schematized in Fig. 36-3.

■ *Postexercise Recovery*

When exercise stops, there is an abrupt decrease in heart rate and cardiac output—the sympathetic drive to the heart is essentially removed. In contrast, TPR remains low for some time after the exercise is ended, presumably because of the accumulation of vasodilator metabolites in the muscles during the exercise period. As a result of the reduced cardiac output and persistence of vasodilation in the muscles, arterial pressure falls, often below preexercise levels, for brief periods. Blood pressure is then stabilized at normal levels by the baroreceptor reflexes.

■ *Limits of Exercise Performance*

The two main forces that could limit skeletal muscle performance in the human body are the rate of O_2 utilization by the muscles and the O_2 supply to the mus-

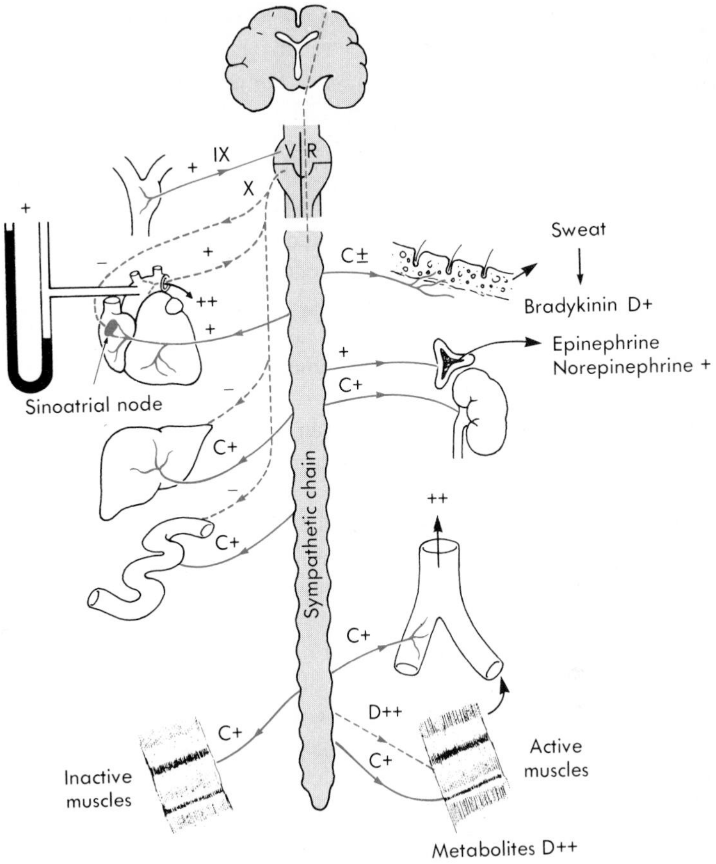

Fig. 36-3. Cardiovascular adjustments in exercise. *VR* Vasomotor region; *C,* vasoconstrictor activity; *D,* vasodilator activity; *IX,* glossopharyngeal nerve; *X,* vagus nerve; ---, sympathetic cholinergic system; +, increased activity; −, decreased activity.

cles. Muscle O_2 usage is probably not critical, since during exercise maximum O_2 consumption ($\dot{V}O_{2max}$) by a large percentage of the body muscle mass is not increased when additional muscles are activated. If muscle O_2 utilization were limiting, recruitment of more contracting muscle would use additional O_2 to meet the enhanced O_2 requirements and would thereby increase total body O_2 consumption. Limitation of O_2 supply could be caused by inadequate oxygenation of blood in the lungs or limitation of the supply of O_2-laden blood to the muscles. Failure to fully oxygenate blood by the lungs can be excluded, since even with the most strenuous exercise at sea level, arterial blood is fully saturated with O_2. Therefore O_2 delivery (or blood flow, since arterial blood O_2 content is normal) to the active muscles appears to be the limiting factor in muscle performance. This limitation could be caused by the inability to increase cardiac output beyond a certain level as a result of a limitation of stroke volume, since heart rate reaches maximum levels before $\dot{V}O_{2max}$ is reached. However, blood pressure provides the energy for muscle perfusion, and blood pressure depends on peripheral resistance, as well as on cardiac output. With increasing levels of exercise at peak $\dot{V}O_2$ and peak cardiac output, blood pressure would fall as more muscle vascular beds dilate in response to locally released vasodilator metabolites if some degree of centrally mediated vasoconstriction (baroreceptor reflex) did not occur in the resistance vessels of the active muscles. Hence the adjustment of resistance in the active muscles appears to be the critical limitation of whole body exercise, although a maximum pumping capacity of the heart has not been ruled out. With exercise of a small group of muscles, such as the hand, when the cardiovascular system is not severely taxed, the limiting factor is unknown but lies within the muscle.

Physical Training and Conditioning

The response of the cardiovascular system to regular exercise is to increase its capacity to deliver O_2 to the active muscles and to improve the ability of the muscle to utilize O_2. The $\dot{V}O_{2max}$ is quite reproducible in a given individual and varies with the level of physical conditioning. Training progressively increases the $\dot{V}O_{2max}$, which reaches a plateau at the highest level of conditioning. Highly trained athletes have a lower resting heart rate, greater stroke volume, and lower peripheral resistance than they had before training or after deconditioning (becoming sedentary). The low resting heart rate is due to a higher vagal tone and a lower sympathetic tone. With exercise, the maximum heart rate of the trained individual is the same as that in the untrained, but is attained at a higher level of exercise. The trained person also exhibits a low vascular resistance that is inherent in the muscle. For example, if an individual exercises one leg regularly over an extended period and does not exercise the other leg, the vascular resistance is lower and the $\dot{V}O_{2max}$ is higher in the "trained" leg than in the "untrained" leg. Also, the well-trained athlete has a lower resting sympathetic outflow to the viscera than does a sedentary counterpart. Physical conditioning is also associated with greater extraction of O_2 from the blood (increased $A\text{–}VO_2$) by the muscles but not with an improvement in cardiac contractility. With long-term training capillary density in skeletal muscle increases. One can speculate that the number of arterioles also increases and can account for the decrease in muscle vascular resistance. The number of mitochondria increases, as do the oxidative enzymes in the mitochondria. Also, it appears that ATPase activity, myoglobin, and enzymes involved in lipid metabolism increase with physical conditioning. Endurance training, such as running or swimming, produces an increase in left ventricular volume without an increase in left ventricular wall thickness. In contrast, strength exercises, such as weight lifting, appear to produce some increase in left ventricular wall thickness (hypertrophy) with little effect on ventricular volume. However, this increase in wall thickness is small relative to that observed in hypertension in which there is a persistent elevation of afterload due to the high peripheral resistance.

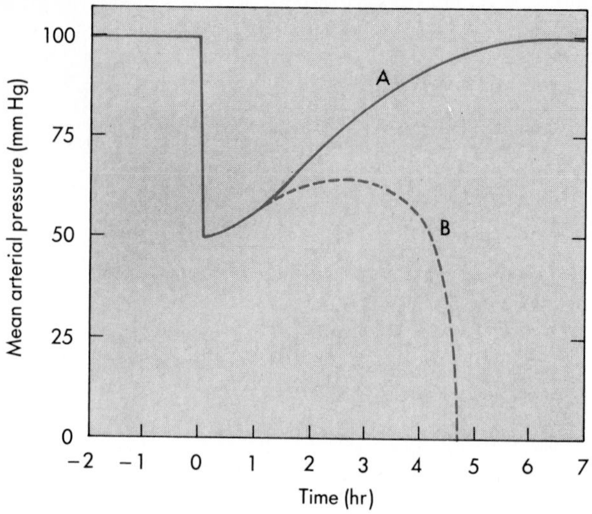

Fig. 36-4. The changes in mean arterial pressure after a rapid hemorrhage. At time zero, the animal is bled rapidly to a mean arterial pressure of 50 mm Hg. After a period in which the pressure returns toward the control level, some animals will continue to improve until the control pressure is attained (curve *A*). However, in other animals the pressure will begin to decline until death ensues (curve *B*).

Hemorrhage

In an individual who has lost a large quantity of blood, the principal findings are related to the cardiovascular system. The arterial systolic, diastolic, and pulse pressures diminish and the pulse is rapid and feeble. The cutaneous veins are collapsed and fill slowly when compressed centrally. The skin is pale, moist, and slightly cyanotic. Respiration is rapid, but the depth of respiration may be shallow or deep.

Course of Arterial Blood Pressure Changes

Cardiac output decreases as a result of blood loss (p. 532). The changes in mean arterial pressure evoked by an acute hemorrhage in experimental animals are illustrated in Fig. 36-4. If sufficient blood is withdrawn rapidly to bring mean arterial pressure to 50 mm Hg, the pressure tends to rise spontaneously toward control over the subsequent 20 or 30 minutes. In some animals (curve *A*, Fig. 36-4) this trend continues, and normal pressures are regained within a few hours. In other animals (curve *B*), after the initial rise of pressure above 50 mm Hg, the pressure begins to decline and continues to fall at an accelerating rate until death ensues.

In other experiments, animals are bled to a given hypotensive level, for example, to 35 mm Hg, by connecting a peripheral artery to a reservoir elevated to an appropriate height (Fig. 36-5). The arterial blood runs rapidly into the reservoir until the pressures become equilibrated and then continues to flow into the reservoir at a progressively slower rate for about 2 hours. This gradual increase in shed blood volume is a manifestation of the same compensatory mechanisms that produced the rise in arterial blood pressure after hemorrhage in the experiment depicted in Fig. 36-4. However, in the experiment represented in Fig. 36-5, as the arterial pressure tends to rise higher than the pressure in the reservoir, blood flows from the cannulated vessel into the reservoir.

About 2 hours after the beginning of hemorrhage, the arterial pressure tends to fall below the established level, and blood begins to flow from the reservoir to the animal. Once 40% to 50% of the maximum shed volume has returned to the animal, rapid reinfusion of the remaining shed blood improves arterial pressure only transiently. The arterial pressure then falls progressively until the animal dies (Fig. 36-5). This progressive deterioration of cardiovascular function is termed *shock*. At some point the deterioration becomes irreversible; a lethal outcome can be retarded only temporarily by any known therapy, including massive transfusions of donor blood.

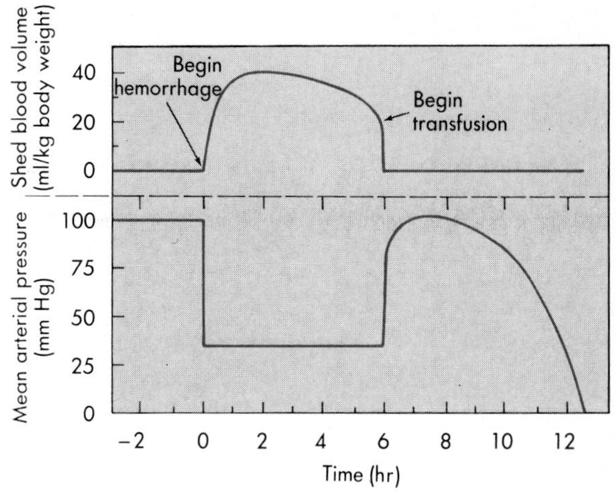

■ **Fig. 36-5.** Changes in shed blood volume and mean arterial pressure during and after a 6-hour period of hemorrhage sufficient to hold mean arterial pressure at 35 mm Hg. The shed blood was reinfused after 6 hours.

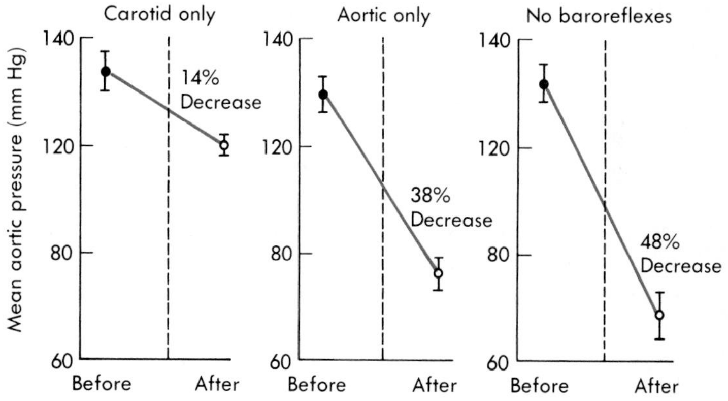

■ **Fig. 36-6.** The changes in mean aortic pressure in response to an 8% blood loss in a group of eight dogs. *Left panel.* The carotid sinus baroreceptor reflexes were intact and the aortic reflexes were interrupted. *Middle panel.* The aortic reflexes were intact and the carotid sinus reflexes were interrupted. *Right panel,* All sinoaortic reflexes were abrogated. (From Shepherd, J.T.: Circulation **50:**418, 1974. By permission of the American Heart Association, Inc.; derived from the data of Edis, A.J.: Am. J. Physiol. **221:**1352, 1971.)

■ *Compensatory Mechanisms*

The changes in arterial pressure immediately after an acute blood loss (Fig. 36-4) and in shed blood volume during the initial stages of sustained hemorrhage (Fig. 36-5) indicate that certain compensatory mechanisms must be operating. Any mechanism that raises the arterial pressure toward normal in response to the reduction in pressure may be designated a *negative feedback mechanism.* It is termed "negative" because the secondary change in pressure is opposite to the initiating change. The following negative feedback responses are evoked: (1) the baroreceptor reflexes, (2) the chemoreceptor reflexes, (3) cerebral ischemia responses, (4) reabsorption of tissue fluids, (5) release of endogenous vasoconstrictor substances, and (6) renal conservation of salt and water.

Baroreceptor reflexes. The reduction in mean arterial pressure and in pulse pressure during hemorrhage decreases the stimulation of the baroreceptors in the carotid sinuses and aortic arch. Several cardiovascular responses are thus evoked, all of which tend to return the arterial pressure toward normal. Reduction of vagal tone and enhancement of sympathetic tone elicit tachycardia and a positive inotropic effect on the atrial and ventricular myocardium. The increased sympathetic discharge also produces generalized venoconstriction, which has the same hemodynamic consequences as a transfusion of blood (p. 532). Sympathetic activation constricts certain blood reservoirs, which provides an autotransfusion of blood into the circulating bloodstream. In the dog, considerable quantities of blood are mobilized by contraction of the spleen. In humans the spleen is not an important blood reservoir. Instead, the cutaneous, pulmonary, and hepatic vasculatures probably constitute the principal blood reservoirs.

Generalized arteriolar vasoconstriction is a prominent response to the diminished baroreceptor stimulation during hemorrhage. The reflex increase in TPR minimizes the fall in arterial pressure resulting from the reduction of cardiac output. Fig. 36-6 shows the effect of an 8% blood loss on mean aortic pressure in a group of dogs. With both vagi cut and only the carotid sinus baroreceptors operative (left panel), this hemorrhage decreases mean aortic pressure by 14%, which was not significantly different from the pressure reduction

(12%) that occurred with all baroreceptor reflexes intact. When the carotid sinuses were denervated and the aortic baroreceptor reflexes were intact, the same percentage reduction in blood volume decreased mean aortic pressure by 38% (middle panel). Hence the carotid sinus baroreceptors are more effective than the aortic baroreceptors in attenuating the fall in pressure. The aortic baroreceptors were operative, however, because when both sets of afferent baroreceptor pathways were interrupted, an 8% blood loss reduced arterial pressure by 48%.

Although the arteriolar vasoconstriction is widespread during hemorrhage, it is by no means uniform. Vasoconstriction is most severe in the cutaneous, skeletal muscle, and splanchnic vascular beds and is slight or absent in the cerebral and coronary circulations. In many instances the cerebral and coronary vascular resistances are diminished. Thus the reduced cardiac output is redistributed to favor flow through the brain and the heart.

The severe cutaneous vasoconstriction accounts for the characteristic pale, cold skin of patients suffering from blood loss. Warming the skin of such patients improves their appearance considerably, much to the satisfaction of well-meaning individuals rendering first aid. However, it also inactivates an effective, natural compensatory mechanism—to the detriment of the patient.

In the early stages of mild to moderate hemorrhage, the changes in renal resistance are usually slight. The tendency for increased sympathetic activity to constrict the renal vessels is counteracted by autoregulatory mechanisms. With more-prolonged and severe hemorrhages, however, renal vasoconstriction becomes intense. The reductions in renal circulation are most severe in the outer layers of the renal cortex. The inner zones of the cortex and outer zones of the medulla are spared.

The severe renal and splanchnic vasoconstriction during hemorrhage favors the heart and brain. However, if such constriction persists too long, it may be harmful. Frequently patients survive the acute hypotensive period only to die several days later from kidney failure resulting from renal ischemia. Intestinal ischemia also may have dire effects. In the dog, for example, intestinal bleeding and extensive sloughing of the mucosa occur after only a few hours of hemorrhagic hypotension. Furthermore, the low splanchnic flow swells the centrilobular cells in the liver. The resultant obstruction of the hepatic sinusoids raises the portal venous pressure, which intensifies the intestinal blood loss. Fortunately, the pathological changes in the liver and intestine are usually much less severe in humans than in dogs.

Chemoreceptor reflexes. Reductions in arterial pressure below about 60 mm Hg do not evoke any additional responses through the baroreceptor reflexes, because this pressure level constitutes the threshold for stimulation. However, low arterial pressure may stimulate peripheral chemoreceptors because of anoxia of the chemoreceptor tissue consequent to inadequate local blood flow. Chemoreceptor excitation enhances the already existent peripheral vasoconstriction associated with the baroreceptor reflexes. Also, respiratory stimulation assists venous return by the auxiliary pumping mechanism described on p. 537.

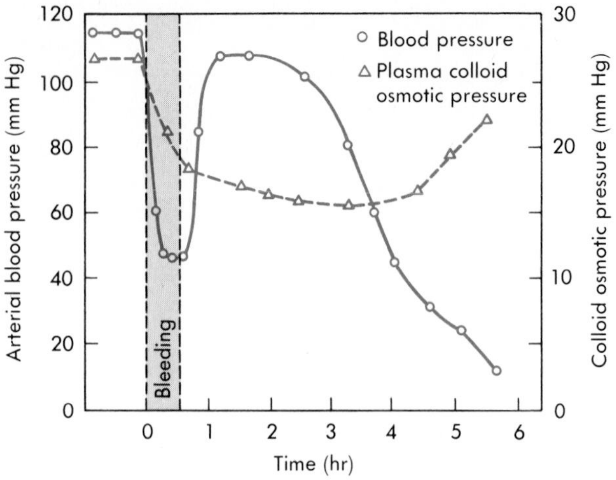

Fig. 36-7. The changes in arterial blood pressure and plasma colloid osmotic pressure in response to withdrawal of 45% of the estimated blood volume over a 30-minute period, beginning at time zero. The data are the average values for 23 cats. (Redrawn from Zweifach, B.W.: Anesthesiology **41:**157, 1974.)

Cerebral ischemia. At low arterial pressures (below 40 mm Hg) inadequate cerebral blood flow activates the sympathoadrenal system. The sympathetic nervous discharge is several times greater than the maximum that occurs when the baroreceptors cease to be stimulated. Therefore the vasoconstriction and facilitation of myocardial contractility may be pronounced. With more severe degrees of cerebral ischemia, however, the vagal centers become activated. The resulting bradycardia may aggravate the hypotension that initiated the cerebral ischemia.

Reabsorption of tissue fluids. The arterial hypotension, arteriolar constriction, and reduced venous pressure during hemorrhagic hypotension lower the hydrostatic pressure in the capillaries. The balance of these forces promotes the net reabsorption of interstitial fluid into the vascular compartment. The rapidity of this response is displayed in Fig. 36-7. In a group of cats, 45% of the estimated blood volume was removed over a 30-minute period. The mean arterial blood pressure declined rapidly to about 45 mm Hg. The pressure then returned rapidly, but only temporarily, to near the control level. The plasma colloid osmotic pressure declined markedly during the bleeding and continued to decrease at a more gradual rate for several hours. The reduction in colloid osmotic pressure reflects the dilution of the blood by tissue fluids.

Considerable quantities of fluid thus may be drawn into the circulation during hemorrhage. About 0.25 ml of fluid per minute per kilogram of body weight may be reabsorbed. Approximately 1 liter of fluid per hour might be autoinfused into the circulatory system of an average individual from the interstitial spaces after an acute blood loss.

Considerable quantities of fluid also may be slowly shifted from intracellular to extracellular spaces. This fluid exchange is probably mediated by the secretion of cortisol by the adrenal cortex in response to hemorrhage. Cortisol appears to be essential for a full restoration of the plasma volume after hemorrhage.

Endogenous vasoconstrictors. The *catecholamines,* epinephrine and norepinephrine, are released from the adrenal medulla in response to the same stimuli that evoke widespread sympathetic nervous discharge. Blood levels of catecholamines are high during and after hemorrhage. When animals were bled to an arterial pressure level of 40 mm Hg, the catecholamines increased as much as fifty times.

Epinephrine comes almost exclusively from the adrenal medulla, whereas norepinephrine is derived both from the adrenal medulla and the peripheral sympathetic nerve endings. These humoral substances reinforce the effects of sympathetic nervous activity listed previously.

Vasopressin, a potent vasoconstrictor, is actively secreted by the posterior pituitary gland in response to hemorrhage. Removal of about 20% of the blood volume in experimental animals increases the vasopressin secretion to about 40 times the normal rate. The receptors responsible for the accelerated release are the sinoaortic baroreceptors and receptors in the left atrium.

The diminished renal perfusion during hemorrhagic hypotension leads to the secretion of *renin* from the juxtaglomerular apparatus. This enzyme acts on a plasma protein, *angiotensinogen,* to form *angiotensin,* a very powerful vasoactive substance.

Renal conservation of water. Fluid and electrolytes are conserved by the kidneys during hemorrhage in response to various stimuli, including the increased secretion of vasopressin (antidiuretic hormone) noted previously. The lower arterial pressure decreases the glomerular filtration rate, and thus curtails the excretion of water and electrolytes. Also the diminished renal blood flow raises the blood levels of angiotensin, as described previously. This polypeptide accelerates the release of *aldosterone* from the adrenal cortex. Aldosterone in turn stimulates sodium reabsorption by the renal tubules, and water accompanies the sodium that is actively reabsorbed.

Decompensatory Mechanisms

In contrast to the negative feedback mechanisms just described, latent *positive feedback mechanisms* are also evoked by hemorrhage. Such mechanisms exaggerate any change that occurs. Specifically, positive feedback mechanisms aggravate the hypotension induced by blood loss and tend to initiate *vicious cycles,* which may lead to death. The operations of positive feedback mechanisms are manifest in curve *B* of Fig. 36-4, and in the uptake of blood from the reservoir in Fig. 36-5.

Whether a positive feedback mechanism will lead to a vicious cycle depends on the *gain* of that mechanism. Gain is defined as the ratio of the secondary change evoked by a given mechanism to the initiating change itself. A gain greater than 1 induces a vicious cycle; a gain less than 1 does not. For example, consider a positive feedback mechanism with a gain of 2. If, for any reason, mean arterial pressure decreases by 10 mm Hg, the positive feedback mechanism would then evoke a secondary reduction of pressure of 20 mm Hg, which in turn would cause a further decrement of 40 mm Hg; that is, each change would induce a subsequent change that was twice as great. Hence mean arterial pressure would decline at an ever-increasing rate until death supervened, much as is depicted by curve *B* in Fig. 36-4.

Conversely, a positive feedback mechanism with a gain of 0.5 would indeed exaggerate any change in mean arterial pressure but would not necessarily lead to death. For example, if arterial pressure suddenly decreased by 10 mm Hg, the positive feedback mechanism would initiate a secondary, additional fall of 5 mm Hg. This in turn would provoke a further decrease of 2.5 mm Hg. The process would continue in ever-diminishing steps, with the arterial pressure approaching an equilibrium value asymptotically.

Some of the more important positive feedback mechanisms include (1) cardiac failure, (2) acidosis, (3) inadequate cerebral blood flow, (4) aberrations of blood clotting, and (5) depression of the reticuloendothelial system.

Cardiac failure. The role of cardiac failure in the progression of shock during hemorrhage is controversial. All investigators agree that the heart fails terminally, but opinions differ concerning the importance of cardiac failure during earlier stages of hemorrhagic hypotension. Shifts to the right in ventricular function curves (Fig. 36-8) constitute experimental evidence of a progressive depression of myocardial contractility during hemorrhage.

The hypotension induced by hemorrhage reduces the coronary blood flow and therefore depresses ventricular function. The consequent reduction in cardiac output leads to a further decline in arterial pressure, a classical example of a positive feedback mechanism. Furthermore, the reduced tissue blood flow leads to an accu-

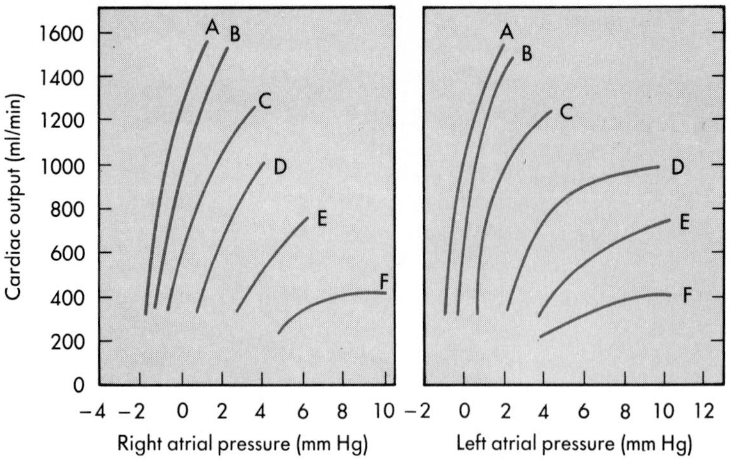

Fig. 36-8. Ventricular function curves for the right and left ventricles during the course of hemorrhagic shock. Curves *A* represent the control function curve; curves *B*, 117 min; curves *C*, 247 mm; curves *D*, 380 min; curves *E*, 295 min; and curves *F*, 310 min after the initial hemorrhage. (Redrawn from Crowell, J.W., and Guyton, A.C.: Am. J. Physiol. **203:**248, 1962.)

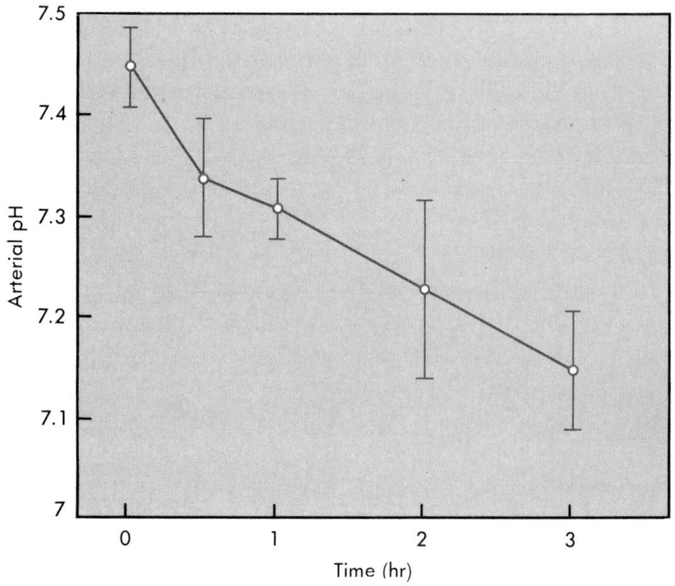

Fig. 36-9. The reduction in arterial blood pH (mean ± SD) in a group of 11 dogs whose blood pressure had been held at a level of 35 mm Hg by bleeding into a reservoir, beginning at time zero. (Modified from Markov, A.K., et al.: Circ. Shock **8:**9, 1981.)

mulation of vasodilator metabolites, which decreases peripheral resistance and therefore aggravates the fall in arterial pressure.

In addition to the curtailment of coronary blood flow, other mechanisms contribute to the development of cardiac failure during hemorrhagic hypotension. As described in the following section, acidosis develops during hemorrhagic shock. This may depress the myocardium directly, and also diminish the responsiveness of the heart to sympathetic stimulation and to circulating catecholamines.

Subendocardial necrosis and hemorrhage frequently occur in hemorrhagic shock. Such pathological changes result from inadequate coronary perfusion and excessive sympathoadrenal activity. These anatomical alterations accentuate the functional impairment of the heart.

Acidosis. The inadequate blood flow during hemorrhage affects the metabolism of all cells in the body. The resultant stagnant anoxia accelerates the production of lactic acid and other acid metabolites by the tissues. Furthermore, impaired kidney function prevents adequate excretion of the excess H^+, and generalized metabolic acidosis ensues (Fig. 36-9). The resulting depressant effect of acidosis on the heart further reduces tissue perfusion and thus aggravates the metabolic acidosis. Acidosis also diminishes the reactivity of the resistance vessels to neurally released and circulating catecholamines, thereby intensifying the hypotension.

Central nervous system depression. The hypotension in shock reduces cerebral blood flow. Moderate degrees of cerebral ischemia induce a pronounced sympathetic nervous stimulation of the heart, arterioles, and veins. With severe degrees of hypotension, however, the cardiovascular centers in the brainstem eventually become depressed because of inadequate blood flow to the brain. The resultant loss of sympathetic tone then reduces cardiac output and peripheral resistance. The resulting reduction in mean arterial pressure intensifies the inadequate cerebral perfusion.

Various endogenous *opioids,* such as *enkephalins* and *beta-endorphin,* may be released into the brain substance or into the circulation in response to the same stresses that provoke circulatory shock. Enkephalins exist along with catecholamines in secretory granules in the adrenal medulla, and they are released together in response to stress. Similar stimuli release beta-endorphin and adrenocorticotrophic hormone (ACTH) from the anterior pituitary gland. These opioids depress the centers in the brainstem that mediate some of the compensatory autonomic adaptations to blood loss, endotoxemia, and other shock-provoking stresses. Furthermore, the opioid antagonist, *naloxone,* improves cardiovascular function and survival in various forms of shock.

Aberrations of blood clotting. The alterations of blood clotting after hemorrhage are typically biphasic—an initial phase of hypercoagulability followed by a secondary phase of hypocoagulability and fibrinolysis. In the initial phase intravascular clots, or *thrombi,* develop within a few minutes of the onset of severe hemorrhage, and coagulation may be extensive throughout the minute blood vessels.

Thromboxane A_2 may be released from various ischemic tissues. It aggregates platelets, and more thromboxane A_2 is released from the trapped platelets, which serves to trap additional platelets. This form of positive feedback intensifies and prolongs the clotting tendency.

The mortality from certain standard shock-provoking procedures has been reduced considerably by anticoagulants such as heparin.

In the later stages of hemorrhagic hypotension, the clotting time is prolonged and fibrinolysis is prominent. It was mentioned previously that in the dog, hemorrhage into the intestinal lumen is common after several hours of hemorrhagic hypotension. Blood loss into the intestinal lumen would, of course, aggravate the effects of the original hemorrhage.

Reticuloendothelial system. During the course of hemorrhagic hypotension, reticuloendothelial system (RES) function becomes depressed. The phagocytic activity of the RES is modulated by an opsonic protein. The opsonic activity in plasma diminishes during shock, which may account in part for the depression of RES function. As a consequence, the antibacterial and antitoxic defense mechanisms are impaired. Endotoxins from the normal bacterial flora of the intestine constantly enter the circulation. Ordinarily they are inactivated by the RES, principally in the liver. When the RES is depressed, these endotoxins invade the general circulation. Endotoxins produce a form of shock that resembles in many respects that produced by hemorrhage. Therefore depression of the RES aggravates the hemodynamic changes caused by blood loss.

In addition to their role in inactivating endotoxin, the macrophages release many of the mediators that are associated with shock: acid hydrolases, neutral proteases, certain coagulation factors, and the arachidonic acid derivatives—prostaglandins, thromboxanes, and leukotrienes. Macrophages also release certain regulatory proteins, called *monokines,* that modulate temperature regulation, intermediary metabolism, hormone secretion, and the immune system.

Interactions of Positive and Negative Feedback Mechanisms

Hemorrhage provokes a multitude of circulatory and metabolic derangements. Some of these changes are compensatory; others are decompensatory. Some of these feedback mechanisms possess a high gain, others, a low gain. Furthermore, the gain of any specific mechanism varies with the severity of the hemorrhage. For example, with only a slight loss of blood, mean arterial pressure is within the range of normal and the gain of the baroreceptor reflexes is appreciable. With greater losses of blood, when mean arterial pressure is below about 60 mm Hg (that is, below the threshold for the baroreceptors), further reductions of pressure will have no additional influence through the baroreceptor reflexes. Hence below this critical pressure the baroreceptor reflex gain will be zero or near zero.

As a general rule, with minor degrees of blood loss, the gains of the negative feedback mechanisms are high, whereas those of the positive feedback mechanisms are low. The converse is true with more severe hemorrhages. The gains of the various mechanisms are additive algebraically. Therefore, whether a vicious cycle develops depends on whether the sum of the various gains exceeds 1. Total gains in excess of 1 are, of course, more likely with severe losses of blood. Therefore, to avert a vicious cycle, serious hemorrhages must be treated quickly and intensively, preferably by whole blood transfusions, before the process becomes irreversible.

Bibliography

Journal articles

Averill, D.B., Scher, A.M., and Feigl, E.O.: Angiotensin causes vasoconstriction during hemorrhage in baroreceptor-denervated dogs. Am. J. Physiol. **245:**H667, 1983.

Bernton, E.W., Long, J.B., and Holaday, J.W.: Opioids and neuropeptides: mechanisms in circulatory shock, Fed. Proc. **44:**290, 1985.

Bevegård, B.S., and Shepherd, J.T.: Regulation of the circulation during exercise in man, Physiol. Rev. **47:**178, 1967.

Blomqvist, C.G., and Saltin, B.: Cardiovascular adaptations to physical training, Annu. Rev. Physiol. **45**:169, 1983.

Bond, R.F., and Johnson, G., III: Vascular adrenergic interactions during hemorrhagic shock, Fed. Proc. **44**:281, 1985.

Brengelmann, G.L.: Circulatory adjustments to exercise and heat stress, Annu. Rev. Physiol. **45**:191, 1983.

Chaudry, I.H.: Cellular mechanisms in shock and ischemia and their correction, Am. J. Physiol. **245**:R117, 1983.

Chen, H.I., and Bishop, V.S.: Baroreflex open-loop gain and arterial pressure compensation in hemorrhagic hypotension, Am. J. Physiol. **245**:H54, 1983.

Christensen, N.J., and Galbo, H.: Sympathetic nervous activity during exercise, Annu. Rev. Physiol. **45**:139, 1983.

Clausen, J.P.: Effect of physical training on cardiovascular adjustments to exercise in man, Physiol. Rev. **57**:779, 1977.

Curtis, M.T., and Lefer, A.M.: Actions of opiate antagonists with selective receptor interactions in hemorrhagic shock, Circ. Shock **10**:131, 1983.

Filkins, J.P.: Monokines and the metabolic pathophysiology of septic shock, Fed. Proc. **44**:300, 1985.

Fredholm, B.B., Farnebo, L.O., and Hamberger, B.: Plasma catecholamines, cyclic AMP and metabolic substrates in hemorrhagic shock of the rat: the effect of adrenal demedullation and 6-OH-dopamine treatment, Acta Physiol. Scand. **105**:481, 1979.

Goldfarb, R.D.: Cardiac mechanical performance in circulatory shock: a critical review of methods and results, Circ. Shock **9**:633, 1982.

Hock, C.E., Su, J.-Y., and Lefer, A.M.: Role of AVP in maintenance of circulatory homeostasis during hemorrhagic shock, Am. J. Physiol. **246**:H174, 1984.

Jones, S.B., and Romano, F.D.: Plasma catecholamines in the conscious rat during endotoxicosis, Circ. Shock **14**;189, 1984.

Kaufman, M.P., Longhurst, J.C., Rybicki, K.J., Wallach, J.H., and Mitchell, J.H.: Effects of static muscular contraction on impulse activity of groups III and IV afferents in cats, J. Appl. Physiol. **55**:105, 1983.

Laughlin, M.H., and Armstrong, R.B.: Muscle blood flow during locomotory exercise, Exerc. Sport Sci. Rev. **13**:95, 1985.

Lefer, A.M.: Eicosanoids as mediators of ischemia and shock, Fed. Proc. **44**:275, 1985.

Liang, C., and Hood, W.B., Jr.: Afferent neural pathway in the regulation of cardiopulmonary responses to tissue hypermetabolism. Circ. Res. **38**:209, 1976.

Liard, J.F.: Vasopressin in cardiovascular control: role of circulating vasopressin, Clin. Sci. **67**:473, 1984.

Loegering, D.J.: Humoral factor depletion and reticuloendothelial depression during hemorrhagic shock, Am. J. Physiol. **232**:H283, 1977.

Ludbrook, J.: Reflex control of blood pressure during exercise, Ann. Rev. Physiol. **45**:155, 1983.

Ludbrook, J., and Graham, W.F.: The role of cardiac receptor and arterial baroreceptor reflexes in control of the circulation during acute change of blood volume in the conscious rabbit, Circ. Res. **54**:424,1984.

Markov, A.K., Oglethorpe, N., Young, D.B., and Hellems, H.K.: Irreversible hemorrhagic shock: treatment and cardiac pathophysiology, Circ. Shock **8**:9, 1981.

Mitchell, J.H., Kaufman, M.P., and Iwamoto, G.A.: The exercise pressor reflex: its cardiovascular effects, afferent mechanisms, and central pathways, Ann. Rev. Physiol. **45**:229, 1983.

Nightingale, L.M., Tambolini, W.P., and Kish, P., Weber, P., and Goldfarb, R.D.: Depression of left ventricular performance during canine splanchnic artery occlusion shock, Circ. Shock **14**:93, 1984.

Pinardi, G., Talmaciu, R.K., Santiago, E., and Cubeddu, L.X.: Contribution of adrenal medulla, spleen and lymph to the plasma levels of dopamine β-hydroxylase and catecholamines induced by hemorrhage hypotension in dogs. J. Pharmocol. Exp. Ther. **209**:176, 1979.

Rowell, L.B.: Human cardiovascular adjustments to exercise and thermal stress, Physiol. Rev. **54**:75, 1974.

Saltin, B., and Rowell, L.B.: Functional adaptations to physical activity and inactivity, Fed. Proc. **39**:1506, 1980.

Scheuer, J., Penpargkul, S., and Bhan, A.K.: Experimental observations on the effects of physical training upon intrinsic cardiac physiology and biochemistry, Am. J. Cardiol. **33**:744, 1974.

Smith, E.E., Guyton, A.C., Manning, R.D., and White, R.J.: Integrated mechanisms of cardiovascular response and control during exercise in the normal human, Prog. Cardiovasc. Dis. **18**:421, 1976.

Stone, H.L. Control of the coronary circulation during exercise, Ann. Rev. Physiol. **45**:213, 1983.

Vatner, S.F., and Pagani, M.: Cardiovascular adjustments to exercise: hemodynamics and mechanisms, Prog. Cardiovasc. Dis. **19**:91, 1976.

Wilson, M.F., and Brackett, D.J.: Release of vasoactive hormones and circulartory changes in shock, Circ. Shock **11**:225, 1983.

Zweifach, B.W., and Fronek, A.: The interplay of central and peripheral factors in irreversible hemorrhagic shock, Prog. Cardiovasc. Dis. **18**:147, 1975.

Books and monographs

Altura, B.M., Lefer, A.M., and Schumer, W.: Handbook of shock and trauma, vol. 1, Basic science, New York, 1983, Raven Press.

Brooks, G.A., and Fahey, T.D.: Exercise physiology—human bioenergetics and its applications, New York, 1984, John Wiley & Sons, Inc.

Carlsten, A., and Grimby, G.: The circulatory response to mscular exercise in man, Springfield, Ill., 1966, Charles C Thomas, Publisher.

Lefer, A.M., editor: Advances in shock research, vol. 5, New York, 1981, Alan R. Liss, Inc.

Lind, A.R.: Cardiovascular adjustments to isometric contractions: static effort. In Handbook in physiology; Section 2: The cardiovascular system—peripheral circulation and organ blook flow, vol. III, Bethesda, Md., 1983, American Physiological Society.

Mitchell, J.H., and Schmidt, R.F.: Cardiovascular reflex control by afferent fibers from skeletal muscle receptors. In Handbook of physiology; Section 2: The cardiovascular system—peripheral circulation and organ blood flow, vol. III, Bethesda, Md., 1983, American Physiological Society.

Reichard, S.M., Reynolds, D.G., and Adams, H.R., editors: Advances in shock research, vol. 10, New York, 1983, Alan R. Liss, Inc.

SECTION

VI

THE RESPIRATORY SYSTEM

Neil S. Cherniack
Murray D. Altose
Steven G. Kelsen

CHAPTER

37

Organization and Mechanics of the Respiratory System

■ *Organization of the Respiratory System*

Normal cellular metabolism requires a continuous supply of oxygen (O_2) and a continuous disposal of carbon dioxide (CO_2). The major responsibility of the respiratory system is to ventilate the lungs. This provides an ongoing source of fresh air to the gas-exchanging surfaces to allow the addition of O_2 and the removal of CO_2 from the blood passing through the lungs. From a functional standpoint, the respiratory system can be considered to be a control system (Fig. 37-1). The system adjusts the rate of ventilation as environmental conditions vary, as metabolic demands are altered, or as the physical characteristics of the ventilatory apparatus are modified by growth, senescence, or disease.

The system controller consists of a network of neurons in the medulla and pons (Chapter 40). The discharges of the neurons set the cyclic inspiratory and expiratory pattern of breathing. The activity of brain stem respiratory neurons, however, may be influenced by inputs from other brain centers, such as the cerebral cortex and cerebellum. Examples of the cerebral cortical influence on breathing are the cessation of respiratory activity that accompanies voluntary breathholding and the modulation of respiration during speech. Inputs that influence the automatic behavior of brain stem respiratory neurons arise from chemoreceptors that sense the level of O_2 and CO_2 in the blood and brain and from mechanoreceptors in the lung and chest wall that monitor lung inflation and chest expansion. Signals from central medullary chemoreceptors and peripheral chemoreceptors in the carotid and aortic bodies increase neuronal inspiratory activity in the brain stem. Also, afferent impulses from lung stretch receptors are relayed back to the brain stem via the vagus nerves; these are important determinants of breathing frequency.

Respiratory activity emanating from medullary neurons is transmitted along the spinal cord to cervical and thoracic spinal motor neurons that innervate the muscles of respiration. During inspiration, contractions of the major inspiratory muscles, namely, the diaphragm and the external intercostal muscles, expand the chest and inflate the lungs. Whereas inspiration is an active event involving muscle contractions, expiration during quiet breathing occurs passively by an elastic recoil of the lungs and the chest wall. However, at increased levels of breathing, the muscles of expiration, including the internal intercostal muscles and the muscles of the anterior abdominal wall, assist in expelling air from the lungs and in returning the chest bellows to its end-expiratory position.

At the end of a normal exhalation, the respiratory muscles are essentially at rest. The ventilatory apparatus, consisting of the lungs and the surrounding chest wall, is in an equilibrium position; that is, in the absence of any active contraction of the respiratory muscles, air neither moves into or out of the lungs and thoracic volume is constant (Fig. 37-2). This position of equilibrium is achieved by the opposing elastic recoil forces of the lungs and the chest wall. At end expiration, the recoil forces of the lung are directed inward, favoring collapse, whereas the chest wall recoils outwardly, favoring expansion. The lung and chest wall recoil forces at end expiration are equal in magnitude, but opposite in direction. The lungs actually fill the chest cavity, so that the visceral pleura of the lungs is in contact with the parietal pleura of the chest cage. The two pleural surfaces are separated by only a thin liquid film that provides an adhesive bond holding the lung and the chest wall together. The opposing lung and chest wall recoil forces produce a subatmospheric pressure of about 5 cm H_2O in the potential space between the visceral and parietal pleurae. The pressure in the

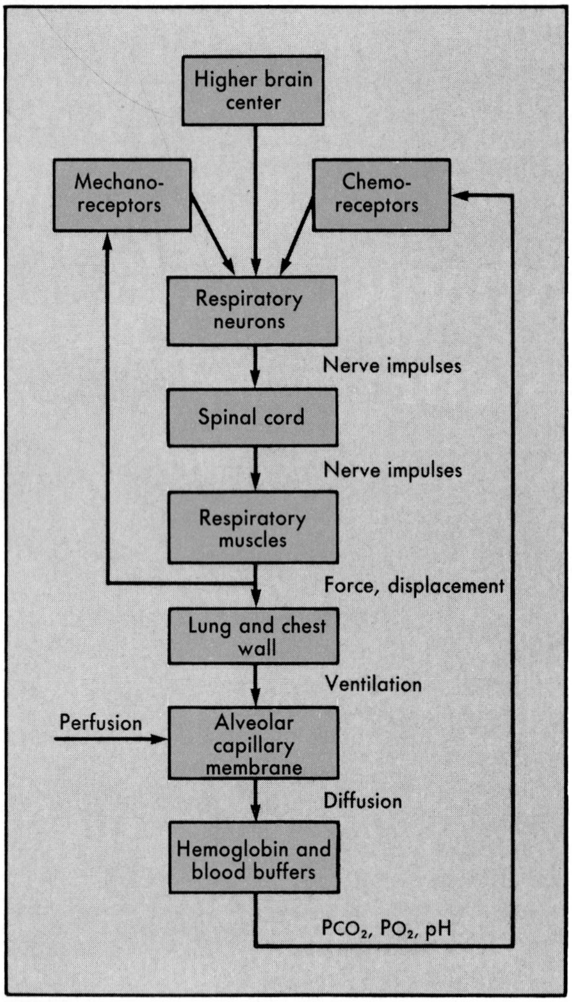

■ **Fig. 37-1.** Diagram of the respiratory control system.

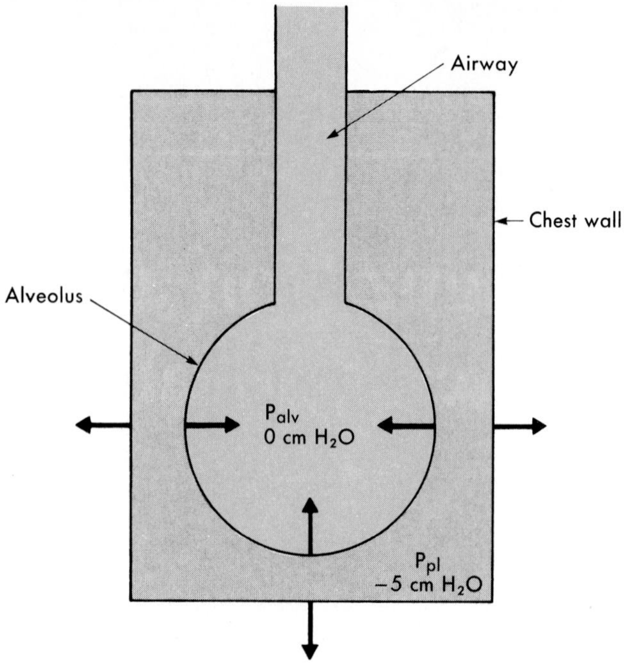

■ **Fig. 37-2.** An airway and an alveolus (representing the tracheobronchial tree and lungs) are enclosed within the chest wall. The pleural pressure (P_{pl}) is the pressure in the potential space between the lungs and chest wall. The alveolar pressure (P_{alv}) is the pressure within the alveoli. The transpulmonary pressure is the difference between P_{alv} and P_{pl}. The pressure difference across the chest wall is the difference between the pleural pressure and the pressure at the body surface. At the end of a normal exhalation, the elastic recoil forces of the lung are directed inward, while the elastic recoil forces of the chest wall are directed outward.

alveolar spaces within the lungs at end expiration, however, equals atmospheric pressure. Because no pressure gradient exists between the alveoli and the airway opening at end expiration and end inspiration, air does not move along the tracheobronchial tree during these periods of the respiratory cycle.

Contraction of the muscles of inspiration expands the chest cavity outward and causes the pleural pressure to become more subatmospheric. This pressure change is transmitted to the interior of the lungs during inspiration, so that alveolar pressure also becomes subatmospheric. The pressure difference between the alveoli and the airway opening induces air to flow into the lung from the atmosphere. The lung continues to expand until its recoil force equals the opposing force of the chest wall and the contracting inspiratory muscles.

Inspired air passes through the pharynx and larynx to enter the tracheobronchial tree. The trachea divides into right and left main stem bronchi to the respective lungs. Bronchi continue to divide repeatedly in an irregular dichotomous pattern into daughter branches of differing sizes and angulations (Fig. 37-3). With each successive branching, individual airways are narrower but the total cross-sectional area of all of the airways in that generation of branches exceeds that of the preceding generation. With this increase in cross-sectional area at successive levels, the resistance to airflow progressively decreases. The number of branchings from the trachea to the terminal bronchioles varies in different parts of the lung from about 10 to over 20.

Beyond the terminal bronchioles are the respiratory bronchioles, which are airways with alveoli in their walls. Gas exchange occurs in the alveoli. There are three generations of respiratory bronchioles, each with a greater number of alveoli, then an alveolar duct completely surrounded by alveoli, and finally alveolar sacs that arise from the alveolar ducts both terminally and laterally. The total number of alveoli in a lung varies from 2×10^8 to 6×10^8. Inspired air moves through the proximal conducting airways largely by convection (bulk movement of gas). Distally, as the velocity of airflow falls in conducting and respiratory airways, gas is distributed primarily by diffusion (a physical process that depends on the kinetic activity of the gas molecules).

Mixed venous blood from the systemic circulation is

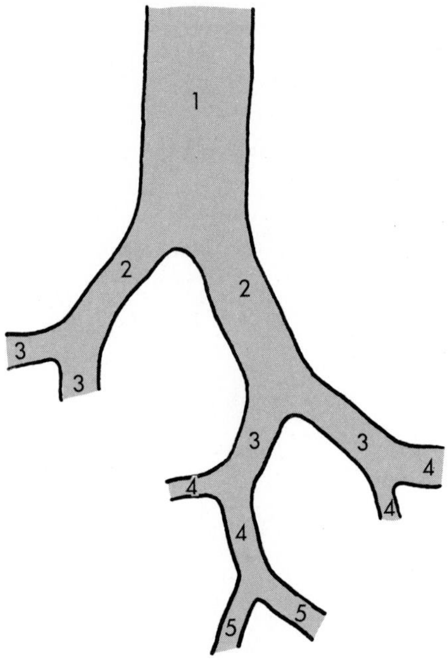

■ Fig. 37-3. Irregular dichotomous branching of the tracheobronchial tree. The numbers denote the generation of the branches.

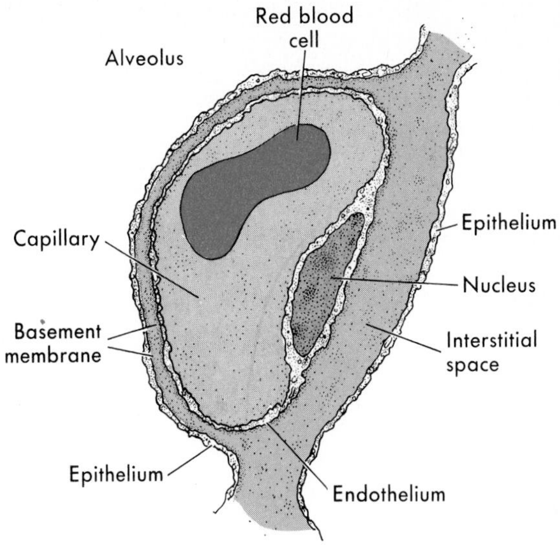

■ Fig. 37-4. Alveolar-capillary membrane.

pumped by the right ventricle through the pulmonary arteries to the pulmonary capillary bed (Chapter 38). Branches of the bronchial tree and pulmonary arterial systems course together through the lung and achieve their most intimate contact with one another in the gas-exchanging regions at the alveolar-capillary membrane (Fig. 37-4). Within the interalveolar system, the alveolar epithelium and the capillary endothelium rest on separate basement membranes. In some places the septum is thin and the two basement membranes appear to be fused. The diffusion of gas occurs mainly across this thin portion. In other regions the alveolar and capillary basement membranes are separated by an interstitial space containing some cells and connective tissue. Water and nongaseous solutes are transported mainly across this thick portion of the alveolar wall.

After passage through the pulmonary capillary bed, oxygen-rich blood is returned to the left atrium and is then delivered to the body tissues by the contraction of the left ventricle. Oxygen is transported in the blood primarily in chemical combination with hemoglobin and to a small extent in physical solution in plasma (Chapter 39). Carbon dioxide is carried in the blood in physical solution in plasma, in chemical combination with hemoglobin (carbaminohemoglobin), but mainly in the form of bicarbonate.

Rates of O_2 delivery to and CO_2 removal from body tissues depend on the activity of both the respiratory and circulatory systems. The levels of respiratory and circulatory activity are shaped by the respective system responses to changes in metabolic activity, as during exercise. Important interactions between the two systems coordinate their responses (Fig. 37-5). The chest bellows and the circulatory pump interact mechanically. An example of this is the churning action of cardiac contractions, which promotes the distribution of gas in peripheral air spaces of the lung. Also, changes in intrathoracic pressure during the breathing cycle influence venous return and cardiac output (Chapter 34). Certain interactions among sensory receptors also regulate respiration and circulation. The activation of chemoreceptors not only stimulates respiration but also affects heart rate and blood pressure (Chapters 29 and 33). Similarly, stimulation of arterial baroreceptors affects not only the circulation but the level of respiratory activity as well.

Thus, the respiratory system is but one element in a larger integrated network that maintains O_2 and CO_2 homeostasis in the body. Although one component of the network can partially compensate for defects in another, optimal operation depends on the coordinated activity of all parts of the network.

■ *Respiratory Muscles*

The respiratory muscles act together in a coordinated fashion to move the chest bellows and produce ventilation. Normally, the forces required to overcome the resistance and elastance of the chest bellows are only a small fraction of the maximum force-generating capacity of the respiratory muscles. As a result, breathing is almost effortless. In patients with lung disease, however, abnormalities in the mechanical properties of the lung and airways may require the respiratory muscles to contract strenuously even when the patient is at rest.

Because of intricate anatomical and mechanical inter-

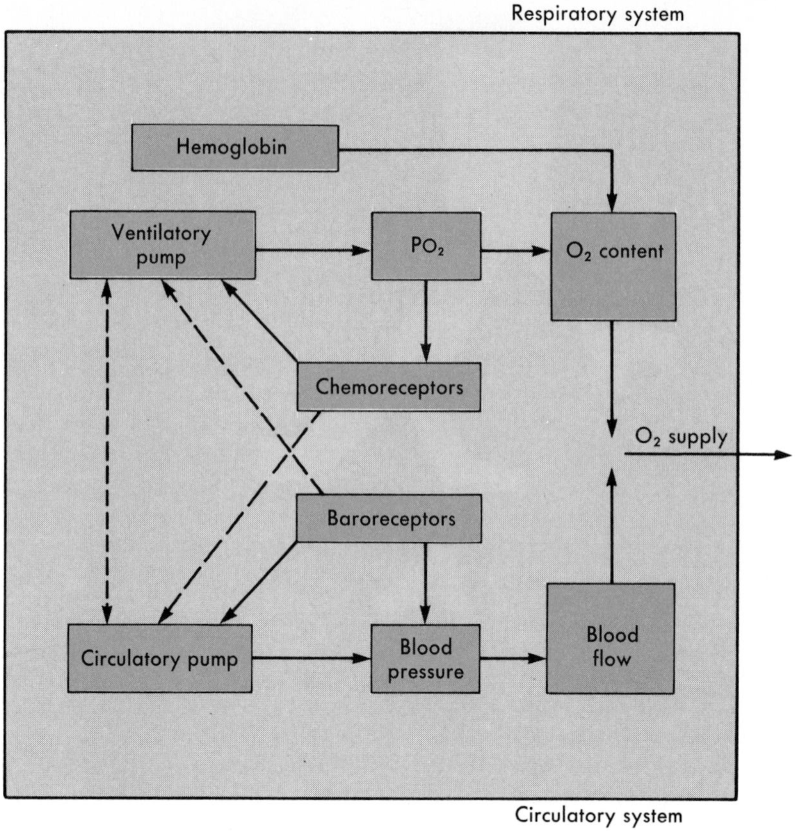

Fig. 37-5. Oxygen delivery to body tissues depends on the activity of the respiratory and circulatory systems. Important neural and mechanical interactions coordinate the responses of the two systems.

relationships, the actions of the individual muscles are complicated. Muscles that are active during inspiration include the diaphragm, the parasternal muscles, the external intercostal muscles, and the scalene and sternocleidomastoid muscles. During quiet breathing, expiration is generally passive, resulting from the elastic recoil of the lungs. However, at increased levels of ventilation and in the presence of airway obstruction, the muscles of expiration become active. These muscles include the internal intercostal muscles and certain abdominal muscles, such as the external and internal oblique muscles and the transverse and rectus muscles.

The Diaphragm

The diaphragm, which is supplied by the phrenic nerves, is a thin dome-shaped sheet of muscle that is inserted into the lower ribs. It is the principal muscle of inspiration, and it accounts for more than two thirds of the air that enters the lungs during quiet breathing. During anesthesia, the diaphragm usually accounts for an even greater proportion of the inspiratory air volume. Two thirds of the fibers making up the human diaphragm are slow-twitch fibers, which are resistant to fatigue.

During inspiration, the dome of the diaphragm flattens as the myofibrils shorten. The diaphragm increases the vertical, anteroposterior, and lateral dimensions of the thoracic cage (Fig. 37-6). During diaphragm contraction, pressure within the abdomen becomes positive. The lower ribs remain caudal to the diaphragm, even as it descends, and the positive abdominal pressure causes the lower rib cage to be displaced outward.

External Intercostal Muscles

The external intercostal muscles extend downward and forward from the lower surface of the rib above to the upper surface of the rib below. These muscles are innervated by the intercostal nerves from the first to the eleventh thoracic spinal cord segments. Contraction of the external intercostal muscles raises the ribs during inspiration. The upward displacement of the upper ribs increases the anteroposterior dimensions of the chest, whereas elevation of the lower ribs increases the transverse dimension of the thorax. Intercostal muscle contraction tenses the intercostal spaces, thereby stabilizing the chest and preventing retraction of those spaces as the pleural pressure becomes more negative consequent to contraction of the diaphragm during inspiration.

During quiet breathing, the contribution of intercostal muscle contraction to the movement of air into the lungs is relatively small; the intercostal muscles mainly optimize the action of the diaphragm. When the dia-

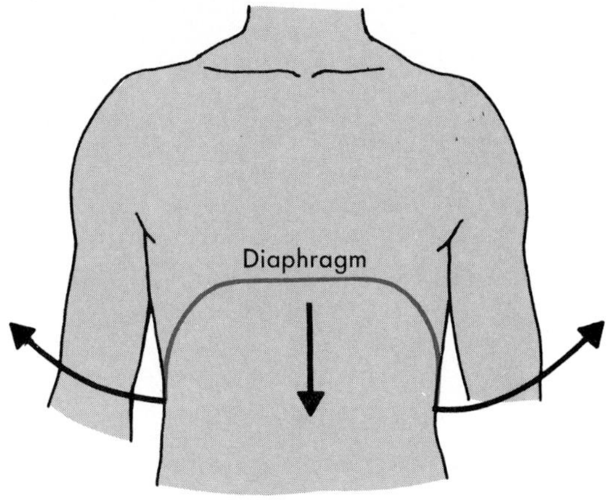

■ Fig. 37-6. Contraction and descent of the diaphragm during inspiration increase the vertical, anteroposterior, and lateral dimensions of the thorax.

phragm is paralyzed or inactivated, the external intercostal muscles become essential to breathing.

■ *Accessory Muscles of Inspiration*

At high levels of ventilation or when movement of air into the lungs is obstructed, the scalene or sternocleidomastoid muscles contribute to inspiration. The scalene muscles arise from the transverse processes of the lower five cervical vertebrae and insert into the upper aspects of the first and second ribs. Contraction of these muscles elevates and enlarges the upper parts of the chest. Similarly, the sternocleidomastoid muscles elevate the sternum and enlarge slightly the anteroposterior and longitudinal dimensions of the chest.

■ *Muscles of Expiration*

A variety of muscles may assist expiration. The most important muscles of expiration are the abdominal muscles, which are innervated by nerves from the lower six thoracic and first lumbar segments of the spinal cord. Abdominal muscle contraction depresses the lower ribs, pulls down the anterior part of the lower chest, and compresses the abdominal contents during expiration. The increased abdominal pressure forces the relaxed diaphragm upward, causing its fibers lengthen.

The abdominal muscles are assisted during expiration by the internal intercostal muscles, which depress the ribs and move them downward and inward. Contraction of the internal intercostal muscles also stabilizes the rib cage and prevents bulging of the intercostal spaces during forceful expiratory maneuvers.

The muscles of expiration are critical in forceful expiratory maneuvers, such as coughing. The abdominal muscles also contribute to maximum inspiratory maneuvers. At lung volumes near total lung capacity, activation of the abdominal muscles augments the action of the diaphragm in expanding the rib cage.

■ *Oxygen Cost of Breathing*

The energy expenditure during breathing can be determined from the oxygen cost of breathing. To determine the oxygen cost of breathing, the oxygen consumption is measured at rest and at an increased level of ventilation produced by voluntary hyperventilation or CO_2 breathing. Provided no other factors act to increase oxygen consumption, the added oxygen uptake at the higher level of ventilation represents that oxygen used by the muscles of respiration. The oxygen cost of breathing is normally about 1 ml/L of ventilation. Ordinarily, this cost constitutes less than 5% of the total oxygen consumption of the body. At high levels of ventilation, however, the oxygen cost of breathing increases.

The metabolic demands on the respiratory muscles are reflected in their oxygen consumption. When the resistance or elastance of the chest bellows is increased by disease, greater forces are required of the respiratory muscles to achieve a given level of ventilation. Accordingly, respiratory muscle oxygen consumption increases. The high oxygen cost of breathing may limit exercise capacity and the maximum level of ventilation in some individuals with severe lung disease.

■ *Lung Volumes*

Lung volumes can be subdivided into a number of compartments (Fig. 37-7). The volume of air in the lungs at the normal, resting, end-expiratory position is termed the *functional residual capacity (FRC)*.

The volume of air that is inhaled during inspiration and that leaves the lungs during expiration is the *tidal volume (TV)*. The maximum volume of air that can be drawn into the lungs from functional residual capacity is termed the *inspiratory capacity (IC)*. The inspiratory capacity is made up of the tidal volume plus the *inspiratory reserve volume (IRV)*. The volume of air contained in the lungs at the end of a maximum inspiration is the *total lung capacity (TLC)*.

The maximum volume of air that can be forcibly exhaled from functional residual capacity (i.e., after the completion of a quiet expiration) is the *expiratory reserve volume (ERV)*. The volume of air remaining in the lungs after such a maximum expiratory effort is the *residual volume (RV)*.

The *vital capacity (VC)* is the maximum volume of air that can be exhaled after a maximum inspiration; it

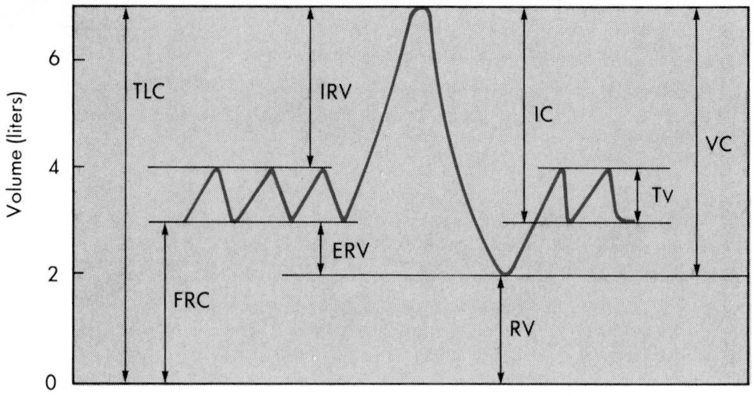

■ Fig. 37-7. Lung volume and subdivisions. *TLC,* Total lung capacity; *FRC,* functional residual capacity; *IRV,* inspiratory reserve volume; *ERV,* expiratory reserve volume; *RV,* residual volume; *IC,* inspiratory capacity; *TV,* tidal volume; *VC,* vital capacity.

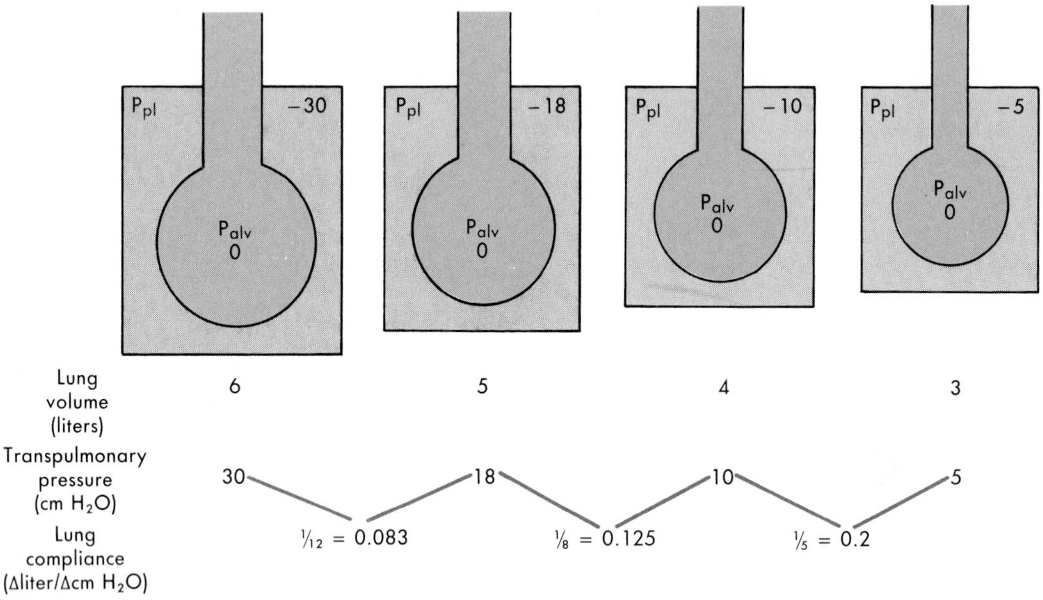

■ Fig. 37-8. Volume-pressure relationships of the lung. At functional residual capacity *(far right),* the transpulmonary pressure is 5 cm H_20. As lung volume increases *(right to left),* the transpulmonary pressure progressively increases. Lung compliance, that is, the ratio of the change in lung volume to the change in transpulmonary pressure, falls progressively.

is the difference between total lung capacity and residual volume.

■ *Elastic Properties of the Lung: Volume-Pressure Relationships*

Volume-pressure relationships describe the elastic properties of the lung. In practice, measurements of volume-pressure relationships in human subjects are made under static conditions, when airflow is temporarily arrested at successive lung volumes during the course of expiration from total lung capacity (Fig. 37-8). At each lung volume, transpulmonary pressure is determined from the difference between alveolar and pleural pressure. Pleural pressure is determined indirectly by measuring the pressure in the esophagus with a loose, thin-walled balloon attached to a fine catheter. Because the wall of the esophagus is flaccid (except during swallowing), the pressure in the interior of the esophagus approximates pleural pressure. When airflow along the tracheobronchial tree has ceased, alveolar pressure equals the pressure at the mouth, because resistive pressure losses do not occur.

At lung volumes near functional residual capacity (about 40% of total lung capacity), lung distensibility, or compliance, is relatively great; compliance is measured from the slope of the volume-pressure curve (Fig. 37-9). However, the volume-pressure characteristics of the lung are nonlinear. Compliance progressively falls as the total lung capacity is approached, and hence greater changes in pressures are required to produce a given volume change (Fig. 37-9).

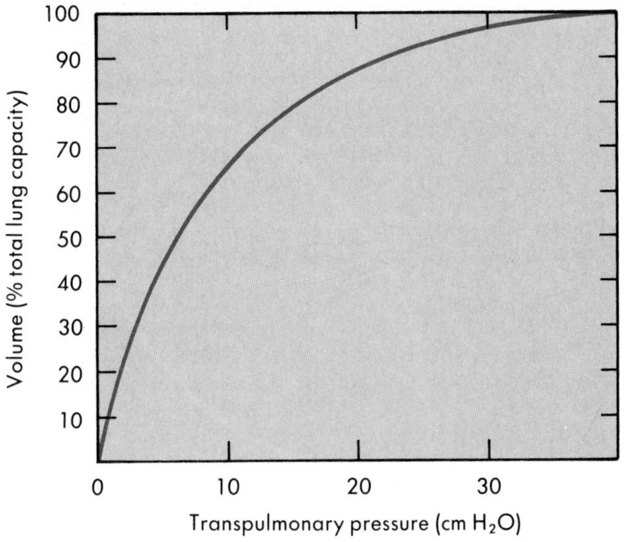

■ Fig. 37-9. Relationship between lung volume and transpulmonary pressure. Lung compliance is maximum at low lung volumes, but it decreases progressively as total lung capacity is approached.

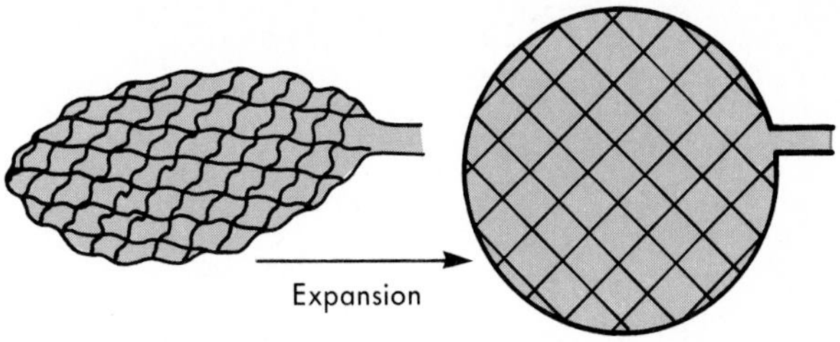

■ Fig. 37-10. Lung expansion involves an unfolding of elastin and collagen fibers in the alveolar walls. The actual lengths of the individual fibers change little.

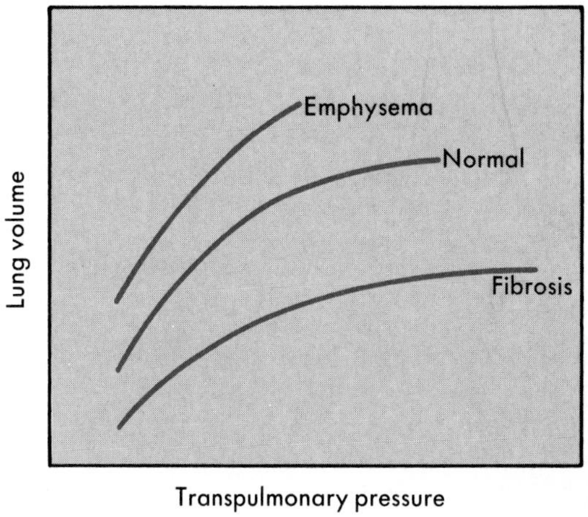

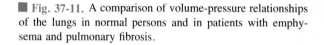

■ Fig. 37-11. A comparison of volume-pressure relationships of the lungs in normal persons and in patients with emphysema and pulmonary fibrosis.

■ *Tissue Forces*

The elastic properties of the lungs depend on the physical characteristics of the lung tissues and on the surface tension of the film lining the alveolar walls. Lung tissue elasticity arises from the elastin and collagen fibers in the alveolar walls, the surrounding bronchioles, and the pulmonary capillaries. Whereas the elastin fibers can be stretched to approximately double their resting length, the collagen fibers resist stretch and act mainly to limit further expansion at large lung volumes. The lungs expand during breathing through an unfolding and geometric rearrangement of fibers in the alveolar walls. This process is analogous to the manner in which a nylon stocking is stretched, without much change in the length of individual fibers (Fig. 37-10). Changes in the arrangement and physicochemical properties of the elastin and collagen fibers in the lungs account for the increasing distensibility of the lungs with advancing age.

Disease processes can also alter lung distensibility. *Emphysema,* which is characterized by a degradation of the elastin and collagen framework and a destruction of alveolar walls, markedly increases the distensibility of the lungs. At any lung volume the elastic recoil of the lung, as determined from the transpulmonary pressure, is subnormal, and thus a given change in transpulmonary pressure evokes a large change in lung volume. Conversely, *pulmonary fibrosis,* which involves an increase in the interstitial tissues of the lung, causes the lung to stiffen. At any lung volume the elastic recoil exceeds that of the normal lung. Thus, a given change in transpulmonary pressure produces a smaller than normal change in lung volume (Fig. 37-11).

■ *Surface Forces*

At an air-liquid interface in a spherical structure, the strong intermolecular forces in the liquid lining cause the area of the lining to shrink. The lung is filled with air, but the inner surface of the alveoli is covered with a thin liquid film. Accordingly, an air-liquid interface exists in the alveoli. Surface forces acting at the air-liquid interface contribute significantly to the elastic recoil of the lung. The effects of surface forces on lung elasticity become evident when the volume-pressure relationships of air-filled and saline-filled lungs of experimental animals are compared (Fig. 37-12). Filling the lung with saline eliminates the air-liquid interface and abolishes the surface forces without affecting the elasticity of the pulmonary tissues. At any given volume, the transpulmonary pressure of the liquid-distended lung is only about one half of that of the lung inflated with air. Thus, when the air-liquid interface is removed, the lung becomes more compliant. The contribution of surface forces to the elastic recoil of the lung is greater at low than at high lung volumes.

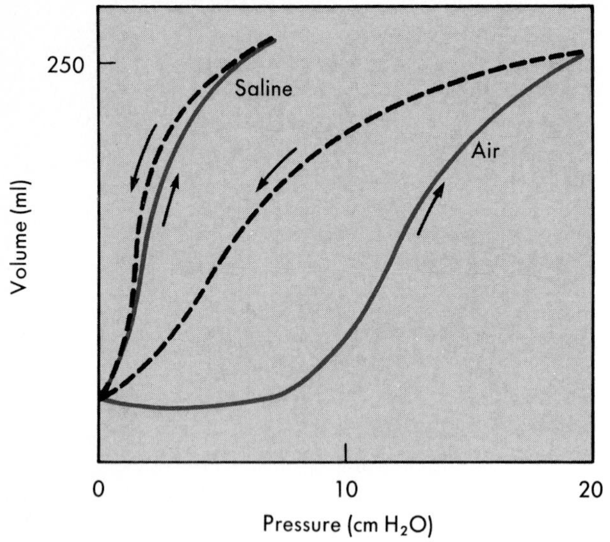

■ Fig. 37-12. Pressure-volume relationships of an air-filled and a saline-filled lung. In the saline-filled lung, an air-liquid interface is absent, and surface forces are abolished. The compliance of the saline-filled lung is considerably greater than that of the air-filled lung. Hysteresis, which is considerable in the air-filled lung, is greatly reduced in the saline-filled lung.

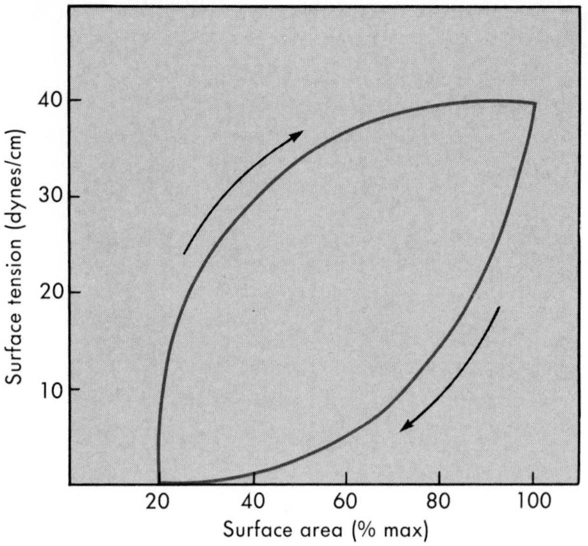

■ Fig. 37-13. Surface tension–surface area relationships of lung extract containing surfactant. The surface tension of surfactant is very low. As the surface area of the film is reduced, surface tension is further reduced. The surface tension–surface area relationship is characterized by marked hysteresis; that is, relationship varies with the direction *(arrows)* of the change in surface area.

■ *Surfactant*

The liquid film lining the interior of alveolar walls is termed *surfactant*. This material is produced by the type II granular pneumocyte, a specialized epithelial cell in the alveolar wall. Surfactant consists of dipalmitoyl lecithin, which is a phospholipid with detergent-like properties and which is conjugated to protein. Surfactant has a number of important characteristics. First, the surface tension of surfactant is very low. This minimizes the surface forces. Second, as the surface area of the film is reduced, the surface tension decreases further (Fig. 37-13). This is critical in maintaining the stability of alveoli and preventing their collapse. The alveoli are essentially spheric structures and thus behave according to Laplace's equation (Chapter 32) for a thin-walled sphere:

$$P = T/r \qquad (1)$$

This equation states that the pressure (P) inside the alveolus is directly proportional to the tension (T) in the wall and is inversely proportional to the radius (r). If the surface tension was the same in all alveoli, the pressure in small alveoli would exceed the pressure in large alveoli. Therefore, smaller alveoli would empty into larger alveoli and collapse (Fig. 37-14). Because the surface tension of surfactant decreases as the alveoli shrink, however, the alveolar pressures of parallel large and small alveoli are virtually the same. Therefore, small alveoli that would otherwise empty into larger alveoli maintain their volume.

When the surface area of a surfactant film is kept small, a rearrangement of its molecules causes the surface tension to increase progressively with time. Therefore, the peripheral air spaces tend to collapse and lung compliance tends to decrease during prolonged periods of shallow breathing. A single large breath or sigh reopens alveoli and expands the surface, thus lowering the surface tension and restoring the elastic properties of the lung.

The surface tension–surface area relationship of surfactant displays marked hysteresis (Fig. 37-13); that is, the relationship between tension and area is different during increases and decreases of surface area. The surface tension–surface area curve forms a loop. The hysteresis of surfactant largely accounts for the differences in the volume-pressure relationships during inflation and deflation of the air-filled lung (Fig. 37-12).

Type II granular pneumocytes first appear in the alveolar epithelium of the fetal lung at about 21 weeks of gestation. They begin to produce surfactant between 28 and 32 weeks of gestation. The phospholipids, including lecithin, that comprise surfactant pass into the amniotic fluid. After about 35 weeks of gestation, the concentration of lecithin in the amniotic fluid begins to rise.

If the production of surfactant by the fetus is delayed or if an infant is born prematurely, before adequate

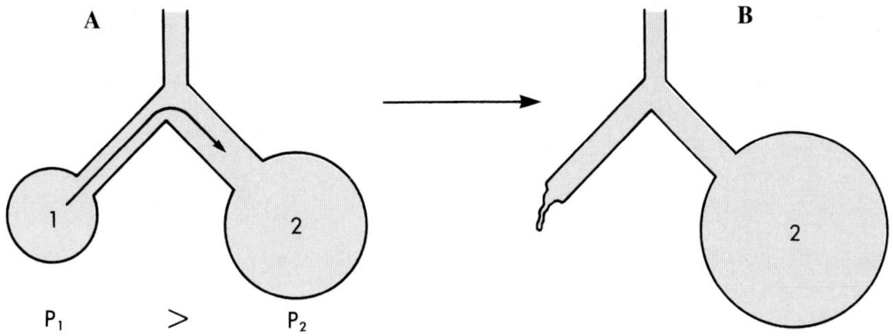

■ **Fig. 37-14.** Communicating alveoli of different sizes. If the surface tension in both alveoli were the same *(A)*, the transpulmonary pressure of the smaller alveolus *(1)* would exceed that of the larger alveolus *(2)*. The smaller alveolus would collapse *(B)* and empty into the larger alveolus.

amounts of surfactant have been synthesized, the infant is prone to develop the *respiratory distress syndrome*. In this condition the alveoli are very unstable and the lung is less compliant. The occurrence of the respiratory distress syndrome can be predicted from the ratio of lecithin to sphingomyelin in the amniotic fluid. If the ratio is less than 2:1, the syndrome is likely to occur.

Surfactant synthesis in the fetus depends mainly on gestational age, but it is also under hormonal influences. The administration of a glucocorticoid to the mother prior to delivery may accelerate surfactant synthesis in the fetus.

■ *Elastic Properties of the Chest Wall*

When the chest wall is at functional residual capacity, the elastic recoil of the chest wall is directed outward, which serves to assist inspiration. The elastic recoil of the chest wall is such that the chest, were it unopposed by the recoil of the lungs, would expand to about 70% of the total lung capacity (Fig. 37-15). This volume represents the equilibrium, or resting position, of the chest wall unopposed by the lung. In this position, with the respiratory muscles at rest, the pressure gradient across the chest wall (that is, the difference between the pleural pressure and the pressure at the body surface) is zero. If the thorax is expanded beyond its equilibrium position, the chest wall, like the lung, would tend to recoil inward. This inward recoil resists further expansion and favors a return to the equilibrium position.

At volumes less than 70% of the total lung capacity, the recoil of the chest is directed outward and is opposite to that of the lung. As residual volume is approached during lung deflation, not only is the outward recoil of the chest wall large, but the compliance of the chest wall falls to very low levels. The increased stiffness of the chest wall at low lung volumes is a major determinant of residual volume.

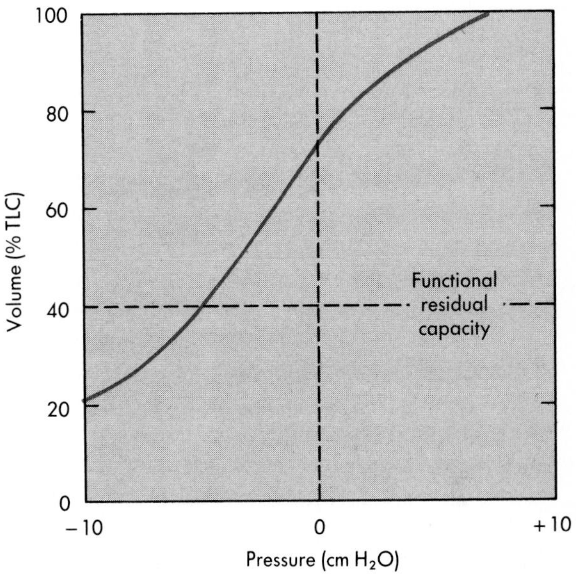

■ **Fig. 37-15.** Pressure-volume relationships of the chest wall. The pressure difference across the chest wall is the difference between pleural pressure and pressure at the body surface. The equilibrium position of the chest wall, that is, the lung volume at which the pressure difference across the chest wall is zero, is approximately 70% of total lung capacity. At volumes above the equilibrium position, the chest wall tends to recoil inward, whereas at volumes below the equilibrium position, the chest wall tends to recoil outward.

The compliance of the chest wall is deranged by abnormalities of the bones and joints of the thorax. Such abnormalities include kyphoscoliosis (curvature of the spinal column), ankylosing spondylitis (arthritis of the spine, producing immobility), and extreme obesity, which imposes a mass load on the chest wall.

Elastic Properties of the Lung–Chest Wall System

Mechanically, the lung and the chest wall operate in series with one another. Therefore, the separate pressures exerted by the passive recoil of the lung (P_L) and the passive recoil of the chest wall (P_{CW}) can be summed algebraically to yield the overall recoil pressure of the total respiratory system (P_{RS}):

$$P_{RS} = P_L + P_{CW} \quad (2)$$

At functional residual capacity the elastic recoil pressures of the lung and chest wall are equal in magnitude but opposite in direction. The algebraic sum of the two, the overall recoil pressure of the total respiratory system, equals zero. The functional residual capacity represents the equilibrium or relaxation position of the respiratory system (Fig. 37-16).

At any volume above functional residual capacity, the net recoil pressure of the respiratory system exceeds atmospheric pressure; the net recoil pressure at such volumes thus favors a decrease in lung volume. With the airway open to the atmosphere, lung volumes above functional residual capacity can be maintained only by the action of the muscles of inspiration. Total lung capacity is achieved when the inward passive elastic recoil pressure of the respiratory system reaches the maximum force that the muscles of inspiration can generate.

At lung volumes below functional residual capacity, the respiratory system recoil pressure is less than atmospheric pressure, and the net effect is directed toward increasing lung volume. Lung volumes below functional residual capacity can be maintained with the airway open only by the opposing action of the muscles of expiration. Even at residual volume, the lung maintains its tendency to collapse and the chest wall is solely responsible for the outward recoil of the respiratory system.

Nonelastic Properties of the Lungs and Resistance to Airflow

Whereas elastic forces of the lung oppose lung expansion, viscous and frictional forces impede airflow into and out of the lung. The total resistance of the ventilatory apparatus consists of the flow resistance of the airways and the frictional resistance to the displacement of lung and chest wall tissues during breathing. Tissue resistance normally comprises only a small fraction (about 10%) of the total resistance, but it may increase considerably with diseases of the lung parenchyma.

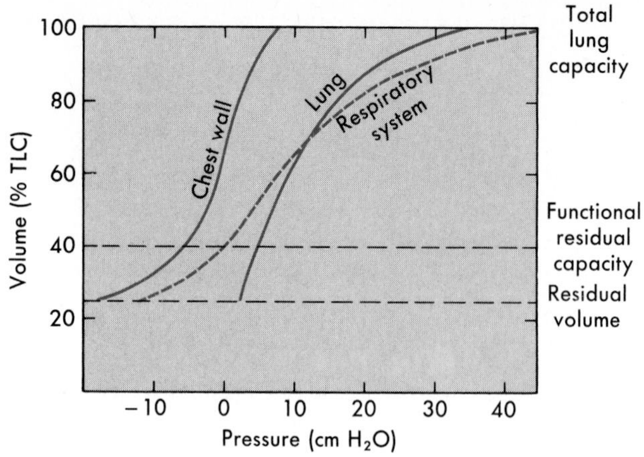

Fig. 37-16. Pressure-volume relationships of the lung, chest wall, and combined system. At any lung volume the elastic recoil of the total system is the algebraic sum of the recoil pressures of the lung and chest wall.

Airflow

The action of the muscles of inspiration and expiration, acting in concert with the recoil forces of the lung, produce the pressures that drive gas into and out of the lungs. Pressure-flow relationships in the lung are particularly complicated because the airways comprise a system of dichotomous branching tubes. To illustrate the range and variety of the pressure-flow relationships that may occur along the airways of the lung, it is convenient to consider patterns of flow through rigid circular tubes. In many respects, the fluid mechanical features resemble those of the cardiovascular system (Chapter 30).

Laminar flow is characterized by organized streamlines that run parallel to the sides of the tube (Fig. 37-17). The streamlines slide over one another, and those at the center of the tube move faster than those closest to the walls. Consequently, the flow profile is parabolic. The rate of airflow ($\dot{V}$) is directly proportional to the driving pressure (ΔP) and to the fourth power of the tube radius (r) and is inversely proportional to the length (l) of the tube and the viscosity (η) of the gas, according to Poiseuille's equation:

$$\dot{V} = \frac{\pi \Delta P\ r^4}{8 \eta l} \quad (3)$$

Thus, for a given driving pressure the rate of laminar flow critically depends on tube radius. If the radius of the tube is reduced to half, the driving pressure must increase sixteenfold to maintain a given flow rate.

Turbulent flow is characterized by a complete disorganization of streamlines (Chapter 30). Molecules of gas move laterally, change their velocities, and collide

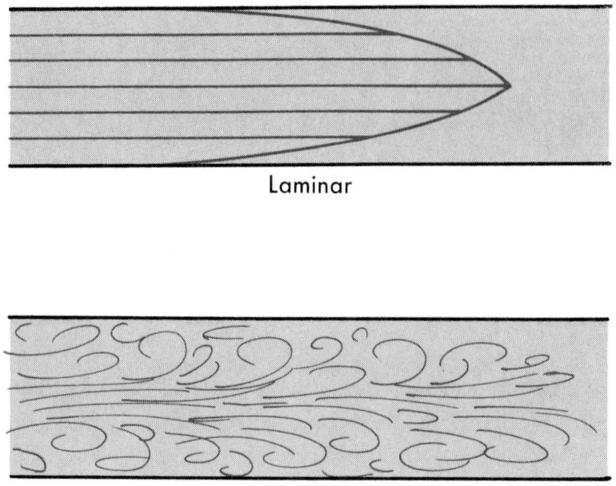

■ Fig. 37-17. Patterns of fluid flow through a tube.

with one another to produce eddies and swirls (Fig. 37-17). The flow rate is not proportional to the driving pressure, but rather to the square root of the driving pressure, and it also varies with gas density.

At branches in the tracheobronchial tree, where air from two separate channels comes together, the parabolic profile of laminar flow may become blunted. The streamlines may separate from the walls of the tube and minor eddy formation may develop to produce a mixed or transitional flow pattern. Under these conditions, flow is a complex function of the driving pressure and the gas density and viscosity.

Reynold's number (N_R), an empirically derived formulation, can be used to predict the pattern of airflow (Chapter 30).

When Reynold's number is less than 2000, flow is fully laminar. However, when gas density and flow velocity are high and viscosity is low, Reynold's number is large, and flow may become turbulent.

In the normal lung, laminar flow occurs only in very small peripheral airways. The overall cross-sectional area of these airways is great; hence, the mean velocity of flow through any given airway is very slow. In general, airflow through the trachea is turbulent, but flow is transitional through most of the remainder of the branching tracheobronchial tree.

The driving pressure that produces flow overcomes viscous forces but also accelerates the air. Acceleration is important during expiration, because as air moves from the alveoli toward the airway opening, the total cross-sectional area of the airways decreases. Hence, the molecules of air must accelerate through the converging channels, even though the overall flow rate does not change. The pressure required to produce this acceleration is proportional to the gas density and to the square of the flow velocity.

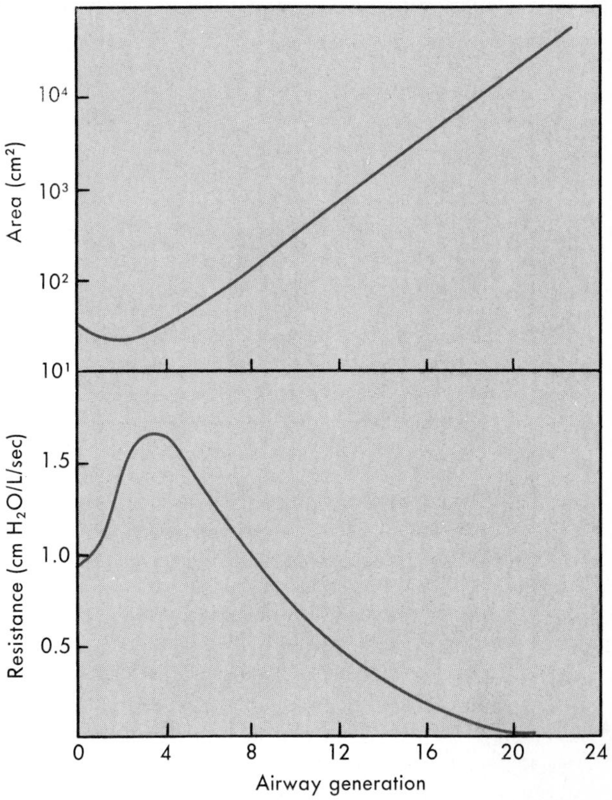

■ Fig. 37-18. Beyond the segmental bronchi, the total cross-sectional area of all airways in a given airway generation increases progressively toward the alveoli. Correspondingly, airway resistance falls with each successive airway generation.

■ *Distribution of Airway Resistance*

The different portions of the tracheobronchial tree do not contribute equally to the total airway resistance. Much of the resistance to airflow is provided by the airways of the upper respiratory tract. The resistance of the nasal passages is very high; it may comprise 50% of the total airway resistance during nose breathing. The mouth, pharynx, larynx, and trachea account for about 20% to 30% of airway resistance during quiet mouth breathing. However, this fraction may increase to about 50% at increased levels of ventilation, such as those encountered during exercise.

The tracheobronchial tree is a system of branching tubes. Along the tracheobronchial tree toward the alveoli, the total cross-sectional area of the airways progressively increases with each successive generation of branching (Fig. 37-18). Accordingly, airway resistance falls dramatically as the alveoli are approached. The major sites of resistance in intrapulmonary airways are in medium-sized lobar, segmental, and subsegmental airways up to about the tenth generation. The smaller peripheral airways, less than 2 mm in diameter, contrib-

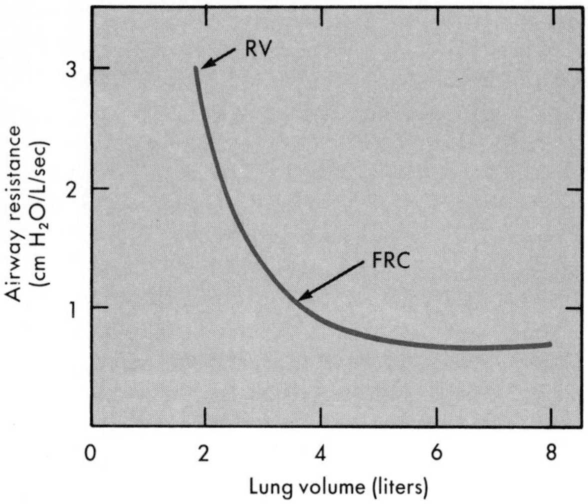

■ Fig. 37-19. Relationship between lung volume and airway resistance. *RV*, Residual volume; *FRC*, functional residual capacity.

ute less than 40% of the total airway resistance of the normal lung. The resistance of airways beyond the eighteenth to twentieth generation approaches zero. Certain pulmonary disorders significantly reduce the caliber of small bronchi and bronchioles. Because these airways contribute relatively little to the overall airway resistance, significant disease at these sites may not be detected by routine tests of pulmonary function.

■ *Effects of Lung Volume on Airway Caliber*

The airways are elastic and can be compressed or distended. The caliber of the airways depends on the difference between the pressures within and surrounding the airways, that is, the transmural airway pressure. The pressure surrounding intrathoracic airways approximates the pleural pressure. As lung volumes are increased, the pleural pressure becomes more subatmospheric, the transmural airway pressure increases, and airway calibers enlarge. Thus, the airways are essentially tethered to the parenchymal lung tissue; the lung tissue pulls outward on the airways and helps to keep them open.

The relationship between lung volume and airway resistance is hyperbolic (Fig. 37-19). At large volumes, lung elastic recoil is high and the traction applied to the walls of the intrathoracic airways is great. The airways widen because of the large transmural airway pressure, and the resistance to airflow falls. At small lung volumes, transmural airway pressure is low, the airways become narrow, and airflow resistance increases.

If the elastic recoil of the lung is reduced by destruction of alveolar walls, as in pulmonary emphysema, the transmural airway pressure at any given lung volume will be correspondingly less and the airways will be narrowed. Even though the disease might not affect the airways directly, the resistance to airflow will increase.

The effects of changes in transmural pressure on airway caliber depend on airway compliance, which is determined by structural composition. The trachea is almost completely encircled by cartilaginous rings, which prevent complete collapse even when the pressure surrounding the trachea exceeds the intraluminal pressure. The bronchi are less well supported by incomplete cartilaginous rings and cartilaginous plates, and the bronchioles lack any cartilaginous support. These structures are much more compliant, and they may narrow considerably when the extraluminal pressure exceeds that within the airways. Additionally, all airways can be stiffened, although to different degrees, by contraction of the smooth muscle in their walls.

■ Table 37-1. Control of airway smooth muscle

Stimulus	*Contraction*	*Relaxation*
Nervous	Cholinergic	Adrenergic
Neurohumoral	Acetylcholine	Norepinephrine
Chemical	Histamine	Prostaglandin E
	SRS-A	
	Prostaglandin F-2α	
Physical	Smoke, dust, SO_2	

■ *Control of Airway Smooth Muscle*

The tone of the smooth muscle cells that encircle the airways affects their caliber and hence airflow resistance. Airway smooth muscle reacts to autonomic nervous activity, circulating hormones, inhaled particles, and chemicals released by cells located near the tracheobronchial tree (Table 37-1). The reactivity of the airways can be assessed in humans by determining the changes in resistance as graded doses of a constricting agent, for example, methacholine, are inspired. These bronchial challenge tests indicate that the airways of asthmatic patients are hyperreactive; that is, in these patients smaller doses of the provocative drug are required to increase airway resistance than in normal individuals. Airway smooth muscle tone depends on the balance between the mechanisms that tend to contract and relax the smooth muscles. In patients with hyperreactive airways, this balance is altered, so that the constricting influences dominate.

Nerve fibers from the parasympathetic and sympathetic nervous systems converge on airway smooth muscle. Stimulation of parasympathetic nerve fibers, carried in the vagus nerves, releases acetylcholine, which contracts airway smooth muscle. These effects can be blocked by atropine and other cholinergic antagonists. In humans the parasympathetic nervous system probably has the most important neural influence on air-

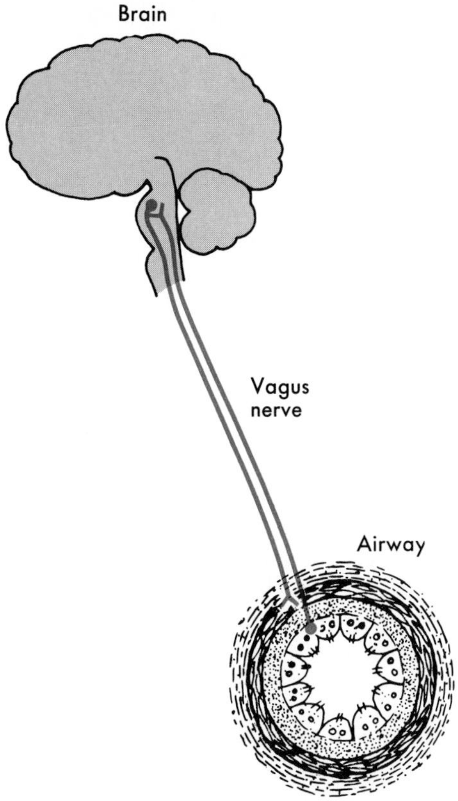

■ **Fig. 37-20.** Vagal control of airway smooth muscle. Afferent impulses from submucosal irritant receptors are carried to the brain via the vagus nerves. These impulses elicit a reflex contraction of airway smooth muscle. The efferent signals also travel in the vagus nerve.

way smooth muscle tone. Airway pollutants, such as SO_2, incite coughing and bronchial narrowing. These reactions are initiated by stimulating irritant receptors that are located in the airway submucosa. The afferent and efferent fibers are all carried in the vagi. Irritant receptor stimulation triggers reflex bronchoconstriction by releasing acetylcholine from efferent vagal nerve endings (Fig. 37-20).

In experimental animals sympathetic stimulation releases norepinephrine and dilates the bronchioles. Norepinephrine reacts with β-adrenergic receptors in the smooth muscle cell membrane to dilate the bronchioles and also with α-receptors in the cell membrane to constrict the airways. Its net effect is bronchodilation, because there are many more β-receptors than α-receptors in the airways. Histochemical studies have demonstrated that few, if any, sympathetic fibers exist in the airways of humans. However, the β-receptors present in airway smooth muscle membranes can be stimulated by circulating catecholamines released from the adrenal medulla or exogenously administered.

Chemicals released from mast cells and immunological reactions can substantially alter airway caliber. Mast cells are located in the connective tissue underlying the smooth muscle and, in lesser numbers, in the walls of the airways. Histamine and slow-reacting substance of anaphylaxis (SRS-A) are the most important bronchoconstrictor mediators released from mast cells. Histamine constricts airway smooth muscle by binding to H_1 receptors. However, histamine also relaxes airway muscle by binding to H_2 receptors, which are located more peripherally in the lungs. In humans the H_1-mediated bronchoconstriction is the predominant effect of histamine.

Histamine and SRS-A increase the production of prostaglandins. Not all prostaglandins affect smooth muscle tone in the same way. Certain prostaglandins, such as prostaglandin E, dilate the bronchioles. Other prostaglandins, such as $F_2\alpha$, constrict the bronchioles.

Smooth muscle tone also depends on membrane potential and membrane permeability to calcium. Contraction of airway smooth muscle is initiated by calcium influx, and uptake of intracellular calcium by the sarcoplasmic reticulum brings about relaxation.

The airway smooth muscle cells, the irritant receptors, and the majority of intrapulmonary mast cells lie beneath the airway epithelium. The epithelium acts as a barrier between noxious agents in the airstream and these cellular components. This barrier helps stabilize airway smooth muscle tone.

Cell-to-cell connections within the epithelial layer are "tight junctions." Agents such as cigarette smoke or infectious viruses can disrupt the epithelial barrier by loosening these junctions, thereby permitting reactions that constrict the airway.

■ *Flow-Volume Relationships*

A maximum expiratory flow-volume curve (Fig. 37-21) is constructed by plotting airflow against lung volume when an individual, after inspiring to total lung capacity, exhales as forcefully, rapidly, and completely as possible to residual volume (a vital capacity maneuver). The airflow reaches a peak near the beginning of the forced expiration, at a lung volume slightly less than total lung capacity. As lung volume decreases, lung elastic recoil diminishes, the transmural airway pressure decreases, the intrathoracic airways narrow, and the airway resistance increases. Consequently, the rate of airflow progressively falls throughout the remainder of the forced expiration.

In a similar fashion, a maximum inspiratory flow-volume curve is generated during a rapid, forceful inspiratory maneuver from residual volume to total lung capacity (Fig. 37-21). The changes in airflow with changes in lung volume differ during maximum inspiratory and expiratory maneuvers. During a maximum inspiratory maneuver, flow reaches a high level while lung volume is still low, but the flow remains high over a wide range of lung volumes. During a maximum in-

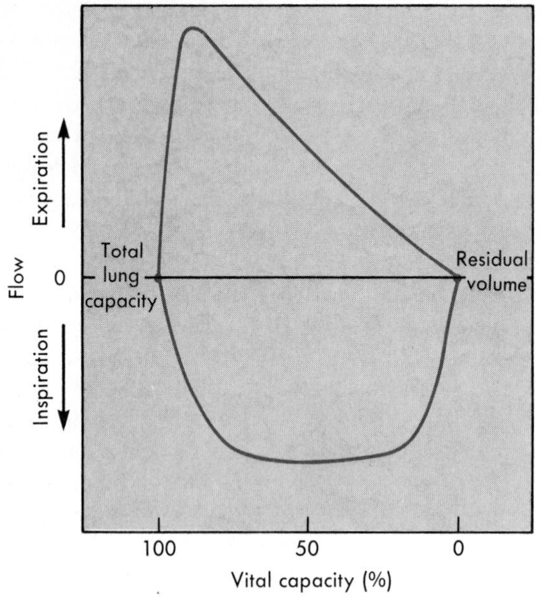

Fig. 37-21. A maximum flow-volume loop.

spiratory effort, the pleural pressure is markedly subatmospheric, the transmural airway pressure is large, and the intrathoracic airways are held open. As lung volume increases in the course of the inspiratory maneuver, the muscles of inspiration shorten. According to the length-tension properties of skeletal muscle, the forces generated by those muscles fall. The decrease in inspiratory muscle force with increasing lung volume favors a progressive reduction in inspiratory flow. However, as the lung expands, the airways widen, so that airway resistance diminishes. Consequently, inspiratory flow remains relatively constant until total lung capacity is approached.

A series of flow-volume loops may be obtained by repeating the expiratory and inspiratory maneuvers for a full vital capacity at different degrees of effort (Fig. 37-22). The greater the inspiratory effort, the greater the rate of airflow over the entire vital capacity. Hence, at any lung volume from residual volume to total lung capacity, inspiratory flow rates directly relate to the forces generated by the muscles of inspiration.

Fig. 37-22. A series of flow-volume loops. Each pair of curves represents inspiratory and expiratory vital capacity maneuvers performed at a specific level of effort. The range from minimum to maximum effort is designated by the numbers *1* to *5*.

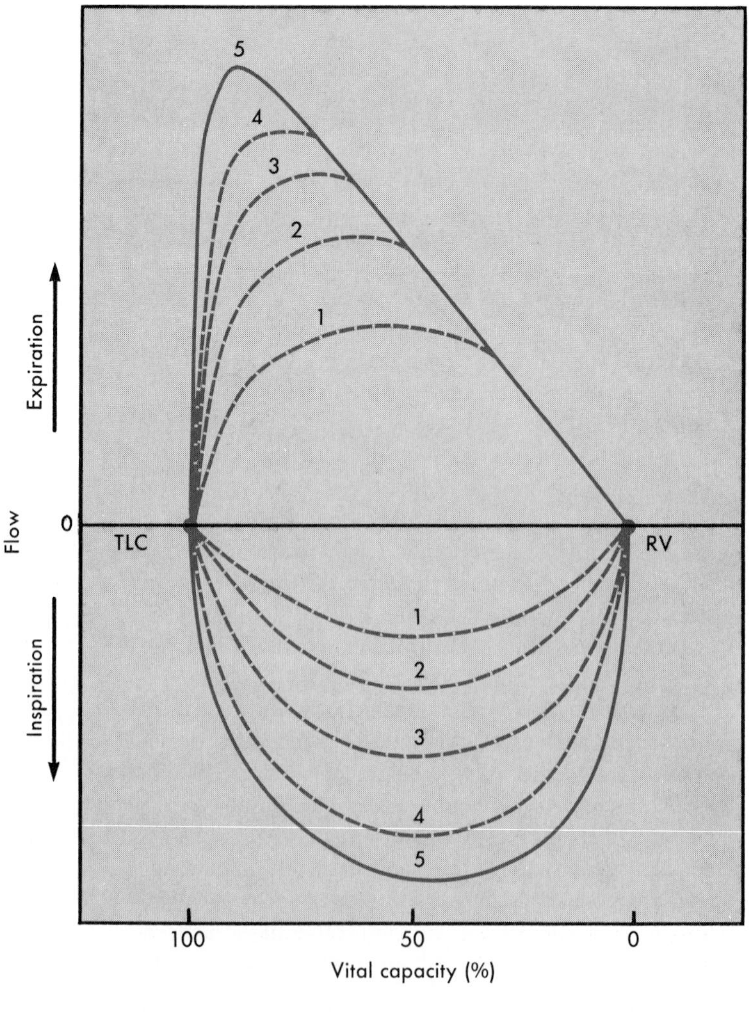

With increasing expiratory efforts, the expiratory airflow rates progressively increase. At low lung volumes, near residual volume, maximum airflow rates are achieved with relatively small effort. Progressively greater efforts are required to achieve maximum airflows at larger and larger volumes. However, at all lung volumes below 40% to 50% of the vital capacity, greater airflows are not achieved with increases in expiratory effort. Thus, at volumes less than 40% to 50% of the vital capacity, expiratory flow rates are relatively independent of effort. In contrast, at lung volumes greater than 40% to 50% of vital capacity, expiratory airflows continue to increase with increasing effort, until a fully maximum effort is expended.

Isovolume Pressure-Flow Curves

The influence of lung volume on the pressure-flow relationships during expiration is further illustrated by isovolume pressure-flow curves. Fig. 37-23 shows the relationship between pleural pressure and expiratory flow at comparable lung volumes during different expiratory efforts. Repeated expiratory vital capacity maneuvers were performed with different degrees of effort, as shown in Fig. 37-22; airflow, lung volume, and pleural pressure are measured simultaneously. At various lung volumes (80%, 50%, and 30% of the vital capacity are shown in Fig. 37-23), the instantaneous airflow is plotted against the instantaneous pleural pressure.

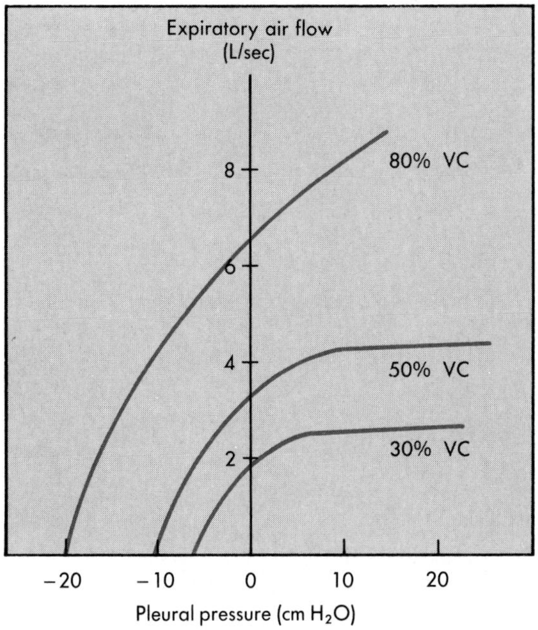

Fig. 37-23. Isovolume pressure-flow curves. During repeated expiratory vital capacity maneuvers, each performed with a different degree of effort, simultaneous measurements are made of airflow rates, lung volume changes, and pleural pressure. The instantaneous expiratory flow rate is plotted against the instantaneous pleural pressure during the various maneuvers at lung volumes of 80%, 50%, and 30% of the vital capacity *(VC)*.

Expiratory airflow is zero during breath-holding; lung volume is held constant with the glottis open by the actions of the respiratory muscles. Under these conditions, the pleural pressure is subatmospheric, because of the elastic recoil of the lung. The greater the lung volume, the more negative is the pleural pressure. At any lung volume above functional residual capacity, expiration is first produced by relaxation of the inspiratory muscles. This causes the pleural pressure to become less negative; the alveolar pressure becomes positive and the expiratory airflow increases. Expiration is aided by contraction of the muscles of expiration. With increasing expiratory effort, the pleural pressure may exceed atmospheric pressure.

At a lung volume of 80% of the vital capacity, airflow increases progressively as pleural pressure becomes more positive; flow is effort dependent. In contrast, at lung volumes of 30% and 50% of the vital capacity, the airflow reaches a plateau as pleural pressure becomes positive. Thereafter, further increases in effort and in pleural pressure fail to increase the airflow. Over this range of pleural pressures, expiratory flow is virtually independent of effort. Because airflows remain constant, despite increases in expulsive force, it follows that the resistance to airflow increases in direct proportion to the increase in the pressure gradient along the entire airway. This occurs because of compression of large intrathoracic airways by the positive pleural and extraluminal airway pressure.

Equal Pressure Point Theory

The equal pressure point theory helps explain the maximum expiratory flow rates that occur during forced expiratory maneuvers. In the model of the lung shown in Fig. 37-24, the alveoli are represented by an elastic sac and the intrathoracic airways by a compressible tube, both enclosed within the chest cage. When, for example, a lung volume of 50% of the vital capacity is maintained constant by the action of the inspiratory muscles and when there is no airflow, pleural pressure is subatmospheric, and it counterbalances the elastic recoil pressure of the lung (Fig. 37-24, *A*). If the airway is open to the atmosphere, and if there is no airflow, the pressure along the entire airway, including the alveoli, is equal to atmospheric pressure.

At the same lung volume (50% of vital capacity), but during quiet expiration, pleural pressure is less subatmospheric. Because the elastic recoil pressure of the lung is unchanged, alveolar pressure is positive and air flows out of the lung. Pressure is gradually dissipated along the airway in overcoming resistance, and the

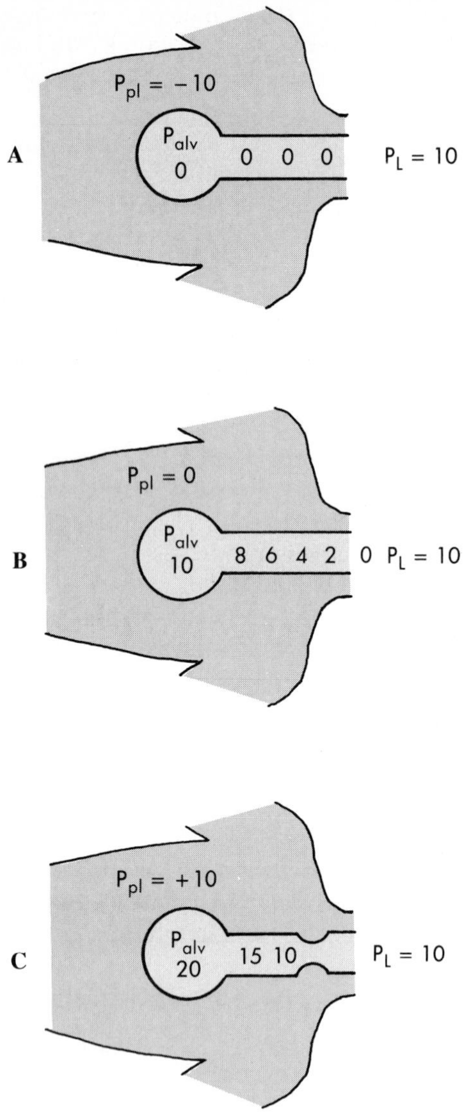

■ **Fig. 37-24.** A lung model in which the alveoli are represented by an elastic sac and the intrathoracic airways by a compressible tube. In the three conditions (**A,** no airflow; **B** and **C,** expiration) the lung volumes are the same; consequently, the elastic recoil pressures of the lung (P_L) are the same. Pleural pressure is designated P_{pl}, and alveolar pressure is designated P_{alv}. In each condition, the pressures at various points along the airway are shown. In **C** the intrathoracic airways downstream from the equal pressure point are compressed because the extraluminal pressure exceeds that within the airway lumen.

pressure at the airway opening is zero. However, all along the intrathoracic airway, airway pressure exceeds pleural pressure. The transmural airway pressure is positive and the airways remain open (Fig. 37-24, *B*).

Pleural pressure exceeds atmospheric pressure during a forceful expiratory effort, and alveolar pressure is yet more positive (Fig. 37-24, *C*). Airway pressure falls more steeply from the alveolus to the airway opening because of the greater airflow. At some point along the airways, the drop in airway pressure equals the elastic recoil pressure of the lung. At that point, referred to as *the equal pressure point,* the intraluminal pressure equals the pressure surrounding the airway; that is, it equals the pleural pressure. Downstream from the equal pressure point, toward the airway opening, the transmural airway pressure is negative; that is, the intraluminal airway pressure falls below the pleural pressure. Consequently, the airways are compressed and the resistance to airflow increases. Once expiratory flow is maximum, any further increase in pleural pressure, achieved by increasing expiratory force, simply compresses the downstream segment more. The resultant increase in airflow resistance is proportional to the increase in pleural pressure, and therefore airflow does not change.

At maximum expiratory airflows, the equal pressure point is located at the level of segmental bronchi. The equal pressure point divides the airways into two components aligned in series: an upstream segment from the alveoli to the equal pressure point and a downstream segment from the equal pressure point to the airway opening. The driving pressure of the upstream segment is the difference between alveolar pressure (P_{alv}) and the pressure at the equal pressure point; the latter corresponds to the pleural pressure (P_{pl}). This difference between P_{alv} and P_{pl} equals the elastic recoil pressure of the lung. Accordingly, maximum expiratory flow ($\dot{V}_{max}$) can be expressed in terms of the elastic recoil pressure of the lung (P_L) and the resistance (R_{US}) of the upstream segment; that is,

$$\dot{V}_{max} = P_L/R_{US} \quad (5)$$

This equation indicates that lung elasticity and the resistance of the upstream airways are important determinants of maximum expiratory flow.

Changes in the airflow velocity that occur in the tracheobronchial tree contribute to dynamic compression of the airways. During forced expiration, as air moves downstream from the alveoli toward the airway opening, the molecules of air accelerate and their velocity increases as the total cross-sectional airway diameter decreases. According to Bernoulli's principle, the intraluminal pressure diminishes inversely as the square of the velocity. When the intraluminal airway pressure falls below the pleural pressure, the airways just downstream from the equal pressure points are compressed and their caliber is reduced. This increases linear velocity more, and hence decreases intraluminal pressure further. These changes in pressure produce a flow-limiting segment that prevents a rise in flow rate, despite an increase in expiratory effort.

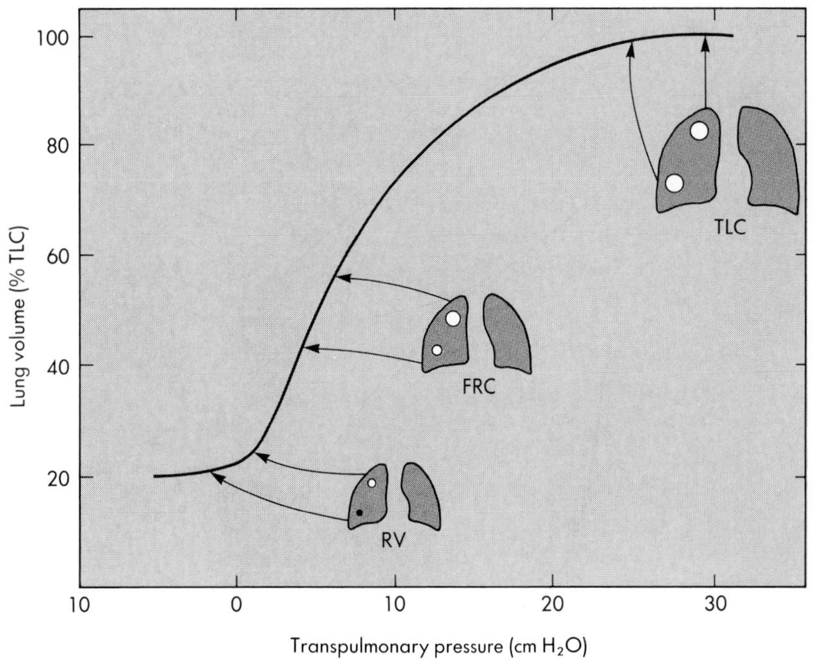

■ **Fig. 37-25.** Regional distribution of lung volumes. Because of the differences in pleural pressure from the apex to the base of the lung, the transpulmonary pressures of the alveoli at the lung apex will differ from those of alveoli at the lung base. At a given lung volume, alveoli at the apex and at the base will fall on different locations of the same pressure-volume curve.

■ *Regional Distribution of Lung Volume and Ventilation*

■ *Alveolar Size*

Alveolar volume is not uniform throughout the lungs. The size of alveoli may vary from the top to the bottom of the lung because of differences in the distending pressure to which they are subjected.

In the upright posture, there is a pleural pressure gradient of approximately 0.25 cm H_2O per centimeter of vertical distance along the lung. Because of the weight of the lung and the effects of gravity, the pleural pressure is most negative at the apex of the lung, and it becomes progressively less negative toward the lung base. Consequently, the transpulmonary pressure, that is, alveolar minus pleural pressure, is greater at the apex than at the base of the lung. The elastic properties of the lung are uniform throughout the lung; hence, the pressure-volume relationships in all areas of the lung are almost identical.

At any overall lung volume, alveoli at the apex and at the base will fall on different coordinates of the same pressure-volume curve, by virtue of their differing transpulmonary pressures (Fig. 37-25). Near total lung capacity, the pressure-volume curve is flat (Fig. 37-9). Thus, despite the differences in transpulmonary pressure between the top and bottom of the lung, alveoli at the lung apex and base are approximately the same size (Fig. 37-25). At intermediate lung volumes (e.g., functional residual capacity) the pressure-volume curve is steep. Because of the regional differences in transpulmonary pressure, alveoli at the lung apices are more expanded than those at the lung bases. At low lung volumes, near residual volume, the pleural pressure at the bottom of the lung may actually exceed the pressure inside the airways. The negative transmural pressure closes airways in the gravity-dependent zones of the lung base. Airway closure prevents the alveoli from emptying completely as the transpulmonary pressure falls to zero.

During a slow inspiration from residual volume, the alveoli of the upper lung zones start to fill first. Trapped units at the lung bases receive no ventilation until the transpulmonary pressure exceeds a critical level and the airways reopen. In contrast, during an inspiration from functional residual capacity, ventilation to alveoli in the lung base is greater than to alveoli at the apex of the lung. At functional residual capacity, the alveoli at the lung base are smaller than those at the lung apex (Fig. 37-25). Hence, for a given change in transpulmonary pressure, alveoli at the lung base will enlarge more than those at the lung apex.

■ *Distribution of Ventilation*

The distribution of ventilation in the lungs and the volume at which the airways in the lung bases begin to close can be assessed by the single-breath nitrogen washout test. The subject first takes a single full inspiration of 100% oxygen from residual volume to total lung capacity. The oxygen enters the lungs and dilutes the nitrogen contained in the lung gas. At the beginning of the inspiration, the nitrogen-rich gas remaining in the

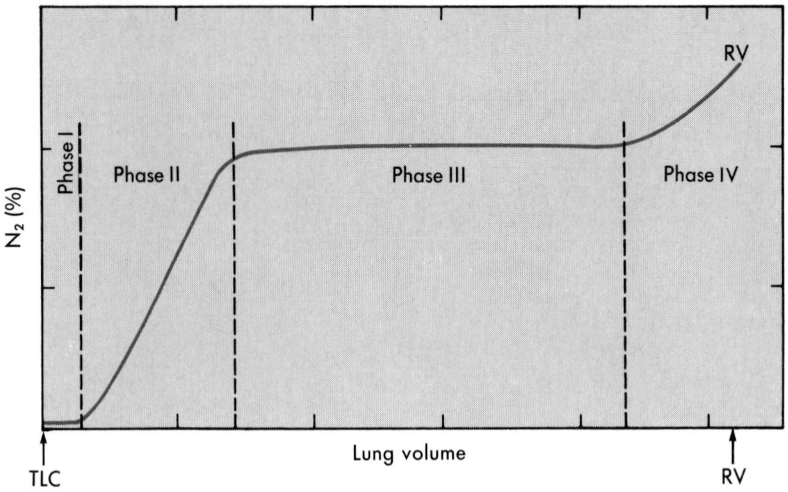

■ **Fig. 37-26.** Single-breath nitrogen washout curve.

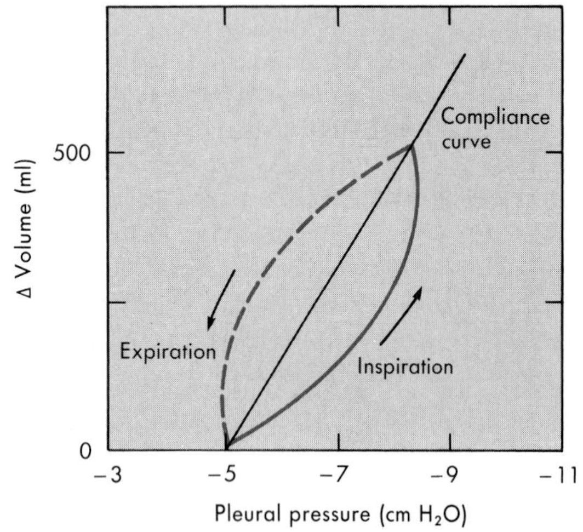

■ **Fig. 37-27.** Dynamic pressure-volume relationship during a single breath.

proximal airways goes to the alveoli in the upper lung zones. However, the fresh breath, containing only oxygen, is preferentially distributed to the lung bases. The pattern of distribution of the inspired gas is such that at the end of the maximum inspiration of oxygen, there will be a greater dilution with oxygen and a lower concentration of nitrogen in the alveoli of the lung bases than in the alveoli of the upper lung zones.

During the subsequent slow but complete exhalation from total lung capacity to residual volume, the concentration of nitrogen in the exhaled air at the mouth is plotted against the expired volume (Fig. 37-26). The initial portion of the exhaled air consists only of the oxygen that filled the larger airways; this portion of the exhaled air contains no nitrogen (phase I). As the alveolar air that does contain nitrogen begins to be exhaled, the concentration of nitrogen in the expired air rises steeply (phase II) and then reaches a plateau (phase III). If ventilation is distributed evenly and if the alveolar gas leaves all regions of the lung synchronously during expiration, phase III will be flat. In lung units that fill poorly during the inspiration of pure oxygen, the nitrogen concentration will be relatively high. These units will also empty more slowly and later than will the well-ventilated units. Consequently, if the distribution of ventilation is not uniform, the nitrogen concentration will rise progressively during the course of expiration in phase III. At low lung volumes near the end of expiration, when the airways at the lung bases close, only alveoli at the top of the lung will continue to empty. Because the concentration of nitrogen in the alveoli of the upper lung zones is higher, the slope of the nitrogen–lung volume curve will increase abruptly (phase IV). The volume at which this deflection occurs is known as the *closing volume*.

■ *Dynamic Pressure-Volume Relationships*

The changes in lung volume and pleural pressure during a normal breathing cycle may be displayed as a pressure-volume loop. Such changes in pressure and volume are affected by the elastic properties of the lung and by the resistance of the airways (Fig. 37-27).

Airflow ceases momentarily at the ends of expiration and inspiration. The change in pleural pressure from end expiration to end inspiration reflects the increased elastic recoil of the lungs. The ratio of the change in volume to the change in pressure (i.e., the slope of the line connecting the end inspiratory and end expiratory points on the pressure-volume loop) is the dynamic compliance.

During inspiration, when air is flowing into the lung, the difference between the actual pleural pressure at any given lung volume and that at end expiration reflects the forces required to overcome not only the elastic recoil of the lungs but also the airway and tissue resistance.

Normally, the dynamic compliance closely approxi-

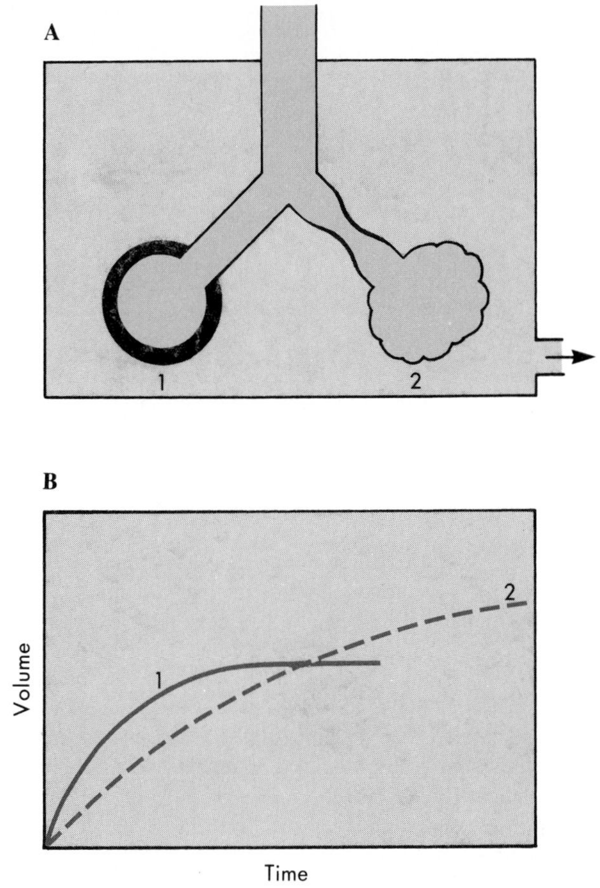

■ Fig. 37-28. Two lung units are arranged in parallel **(A).** Unit 1, with a low resistance and low compliance has a short time constant and fills quickly **(B).** Unit 2, with a large resistance and a high compliance, has a long time constant. It fills more slowly and reaches a greater volume on application of the same distending force **(B).**

mates the static compliance. It remains essentially unchanged even when breathing frequency is increased up to 60 breaths per minute. This indicates that air spaces normally fill evenly and empty synchronously even when the rates of airflow are high.

The functional units in the lung are arranged in parallel with each other (Fig. 37-28). The distribution of ventilation to parallel lung units depends on the resistance and compliance of the respective airways and alveoli. For ventilation to be distributed evenly at high breathing frequencies and rates of airflow, the resistance and compliance of parallel units must be distributed such that the products of resistance and compliance are approximately the same for each unit. The product of resistance and compliance constitutes the *time constant,* which is a measure of the time required for a system to go from one equilibrium state to another.

If the airway resistance of one unit (e.g., unit 2 in Fig. 37-28) is relatively great, it will not fill as quickly, nor will it empty as rapidly as other parallel units (e.g., unit 1 in Fig. 37-28). Similarly, lung units that are more compliant (unit 2) than other parallel units (unit 1) may not fill to their capacity if the time for inspiration is limited. Also, high-compliance alveoli will fill more than low-compliance alveoli, but empty more slowly during expiration.

Other mechanisms also promote uniformity in alveolar filling and emptying. Fenestrations in the lung tissue between alveoli (pores of Kohn) and between alveoli and respiratory bronchioles (canals of Lambert) allow air to reach alveoli even if the bronchioles that supply them are blocked. In healthy individuals, these fenestrations minimize the effects of unequal time constants of ventilation within the lung. Also, the structural organization of the lung tissues promotes an even distribution of ventilation by an effect known as *interdependence*. Pleural pressure is transmitted along the walls common to adjacent alveoli and thereby maintains a relatively constant distending force on alveoli in all regions of the lung. The movements of the lung and of the chest wall are also interdependent. Failure of a region of the lung to enlarge when the chest wall expands will make the local pleural pressure more negative. This will increase the transpulmonary pressure across the region and thereby promote expansion.

Work of Breathing

The respiratory muscles perform mechanical work in overcoming the elastic recoil of the lung and the nonelastic resistance of the airways and tissues during breathing. Work (W) is defined as the product of pressure (P) and volume (V), according to the equation:

$$W = \int PdV \qquad (6)$$

The work of breathing is represented as an area on a pressure-volume diagram. Fig. 37-29 shows a dynamic pressure-volume loop (ABEDA) of the lung during the course of a breath starting from functional residual capacity. The dark-colored area (ACEFA) represents the work performed during inspiration in overcoming the elastic forces of the lung, and the light-colored area (ABECA) represents the work required of the respiratory muscles to overcome airflow and tissue resistance.

The mechanical work used to overcome the elastic recoil and to expand the lung during inspiration is partially stored as potential energy. That energy is released during expiration and is utilized to overcome airway and tissue resistance. Accordingly, expiration is passive during quiet breathing. However, at high levels of ventilation and when airway resistance is increased, additional mechanical work may be required during expiration to overcome these nonelastic forces.

Mechanical work should be distinguished from energy expenditure. For a given change in pressure the thorax may not expand much if the various respiratory

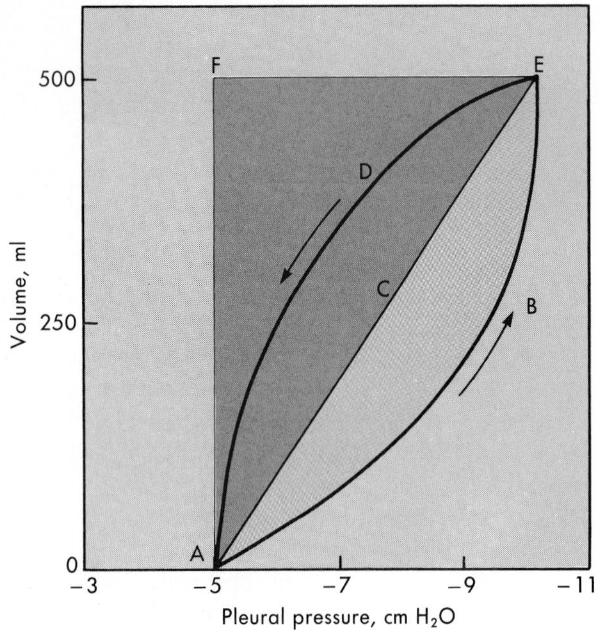

■ **Fig. 37-29.** Dynamic pressure-volume loop of the lung *(ABEFA)* illustrating the mechanical work during inspiration required to overcome lung elastic forces *(dark color, ACEFA)* and flow and tissue resistance *(light color, ABECA)*.

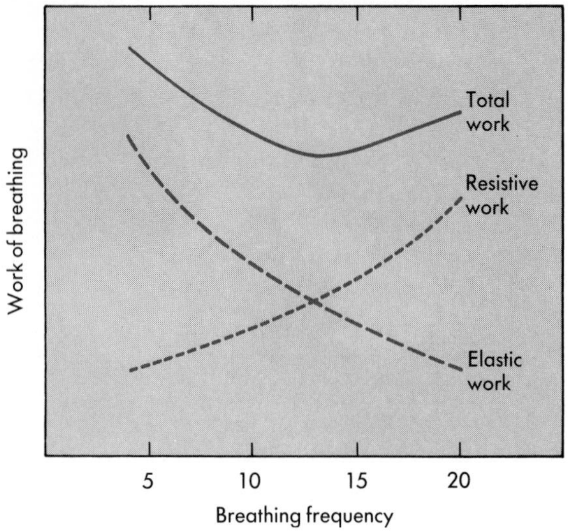

■ **Fig. 37-30.** Relationship of work of breathing to breathing frequency at a constant level of ventilation. During ventilation at low breathing frequencies with large tidal volumes, the resistive work of breathing is low, but the elastic work of breathing is high. In contrast, rapid, shallow breathing is characterized by higher resistive work and lower elastic work. During spontaneous breathing, tidal volume and breathing frequency are usually regulated to minimize the total work of breathing.

muscles fail to act in a coordinated manner. Similarly, if agonist and antagonist muscle groups contract simultaneously, considerable energy may be expended, but little mechanical work will be performed.

At any level of ventilation, the work of breathing depends on the breathing pattern. Large tidal volumes increase the elastic work of breathing, whereas rapid breathing results in high airflows and increases the work against resistive forces (Fig. 37-30). Tidal volume and breathing frequency tend to be set at levels that minimize the work of breathing.

■ *Assessment of the Mechanical Function of the Lung*

■ *Lung Volumes*

The subdivisions of thoracic gas volume can be determined with a spirometer, which is a gas volume recorder. Many different types of spirometers exist. The simplest and one of the most commonly used is the water spirometer, which consists of a double-walled drum into which is fitted a bell. The bell is attached by a pulley to a pen that writes on a rotating cylinder (Fig. 37-31).

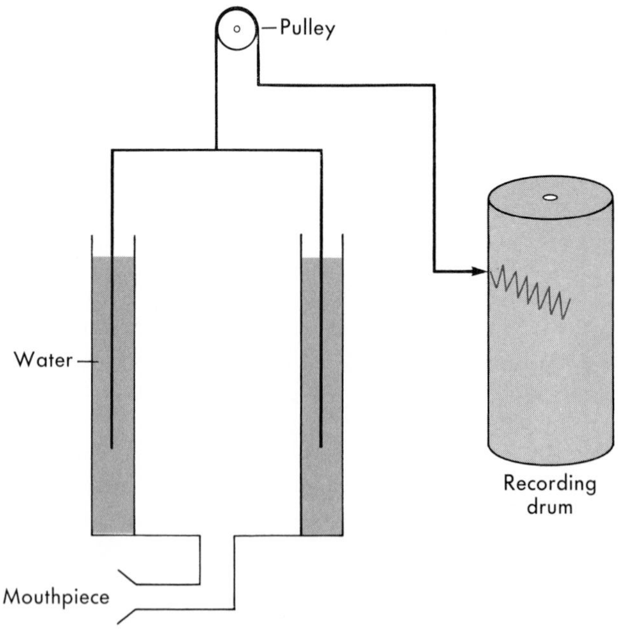

■ **Fig. 37-31.** Diagram of a spirometer.

The residual volume, the air that remains in the lung after a maximum expiration, cannot be determined directly by a spirometer, and alternative methods must be employed. In practice, functional residual capacity is measured by some technique, such as body plethysmography. The expiratory reserve volume (Fig. 37-7) can be determined by a spirometer. The residual volume is then ascertained by subtracting the expiratory reserve volume from the functional residual capacity.

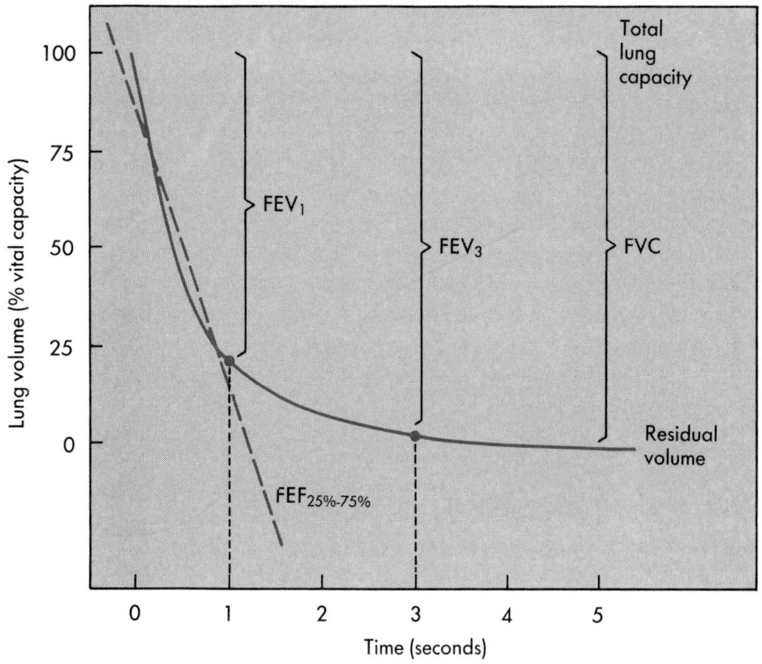

Fig. 37-32. Spirometric tracing of a maximum expiratory vital capacity maneuver. Lung volume, expressed as a percentage of the vital capacity, is plotted as a function of the time from the onset of the expiratory maneuver. The forced vital capacity *(FVC)* and the forced expiratory volumes in 1 second *(FEV_1)* and 3 seconds *(FEV_3)* are shown. The forced midexpiratory flow rate *($FEF_{25\%-75\%}$)* is the average airflow rate over the middle half of the vital capacity.

The body plethysmograph is a sealed enclosure within which an individual may be seated. The subject makes gentle inspiratory and expiratory efforts against a closed shutter at the mouth. During inspiratory efforts against the closed airway, the air in the lungs is decompressed, lung volume increases slightly, and the alveolar pressure becomes subatmospheric. Because the plethysmograph is sealed, the increase in lung volume compresses the air in the plethysmograph. The change in lung volume therefore can be calculated from the increase in pressure within the plethysmograph. Expiratory efforts against an obstructed airway produce the opposite effects.

The volume of air in the lungs is determined by applying Boyle's law, which states that the product of pressure and volume is constant for a given quantity of gas at a constant temperature. Boyle's law also can be expressed as follows:

$$P_i \times V_i = (P_i + \Delta P) \times (V_i + \Delta V), \qquad (7)$$

where

P_i = initial pressure
V_i = initial volume
ΔP = change in pressure
ΔV = corresponding change in volume

In the determination of the functional residual capacity, V_i is the lung volume at the end of a normal expiration and P_i is the pressure in the airways and alveoli (equal to atmospheric pressure at the end of expiration). Changes in intrapulmonary pressure (ΔP) during breathing efforts against the closed shutter are determined from changes in pressure at the mouth. In the absence of airflow, pressures are equal throughout the system. Changes in lung volume (ΔV) consequent to gas expansion or compression are computed from the measured changes in pressure within the plethysmograph.

Airflow Rates

The resistive properties of the airways are commonly assessed from measurements of airflow rates out of the lungs during a rapid, forceful expiratory maneuver from total lung capacity to residual volume (Fig. 37-32). When a spirometer is used, the rates of airflow are determined from the volume of air exhaled during particular time intervals. Measurements are made of the volume exhaled during the first second (the forced expiratory volume in 1 second, FEV_1) and over the first 3 seconds (the forced expiratory volume in 3 seconds, FEV_3). The FEV_1 and FEV_3 are also expressed as a percentage of the forced vital capacity. The mean airflow rate is also measured over the middle half of the forced vital capacity, that is, between 25% and 75% of the vital capacity. This is called the maximum midexpiratory flow rate, or the forced midexpiratory flow rate ($FEF_{25\%-75\%}$).

Respiratory disorders, such as asthma, that obstruct the airways decrease the expiratory airflow rates. The reduction is proportional to the severity of the obstruction. The $FEF_{25\%-75\%}$ is a sensitive index of such obstruction, and the index may be decreased when airway obstruction is mild. With more severe obstruction, the FEV_1 and the rate of airflow throughout expiration are significantly impaired and the time required to expel the vital capacity is prolonged (Fig. 37-33).

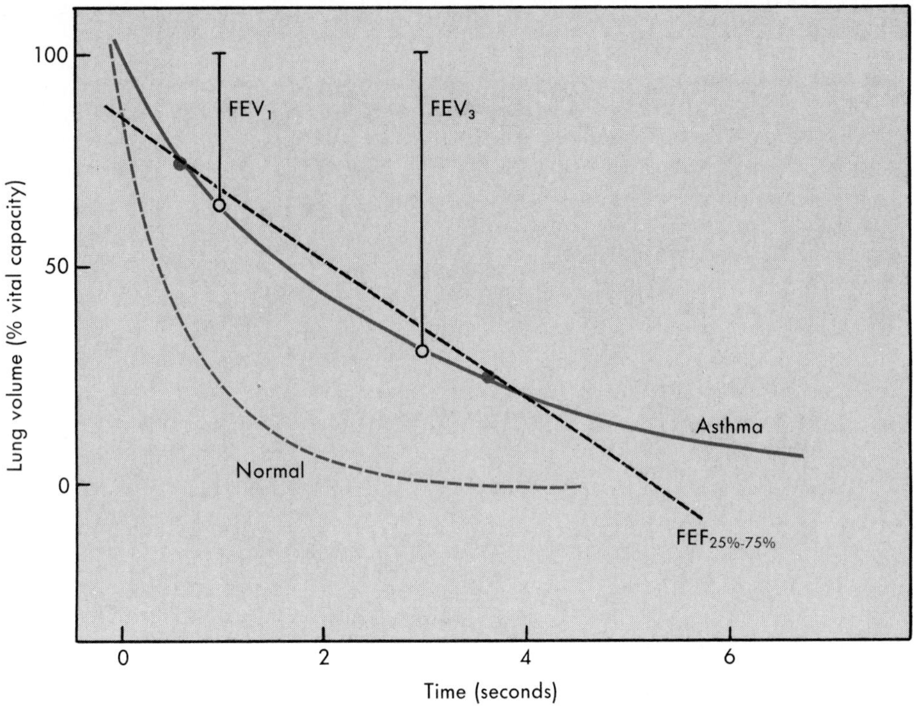

■ **Fig. 37-33.** Spirometric tracing of a maximum expiratory vital capacity maneuver in a patient with airway obstruction *(solid curve)*. The FEV_1, FEV_3 and $FEF_{25\%-75\%}$ are less than obtained in a normal person.

■ *Assessment of Respiratory Muscle Function*

The strength of the respiratory muscles can be determined from the maximum pressures generated during inspiratory or expiratory efforts against a closed airway (Fig. 37-34). As with other skeletal muscles, the contractile force of respiratory muscles depends on the velocity of shortening. The slower the shortening, the greater the force exerted. During maneuvers against a closed airway, the respiratory muscles cannot shorten much, and hence they contract almost isometrically. Muscle length does change slightly, consequent to gas expansion or compression. The forces that can be developed during an isometric contraction depend on the precontraction length of the muscles. The muscles of expiration, for example, are lengthened at total lung capacity. With the lungs fully expanded, they are at their most advantageous mechanical position and can exert their greatest force. At smaller lung volumes, maximum expiratory airway pressures progressively decrease. The muscles of inspiration are longer and the maximum inspiratory force is greater at lung volumes near residual volume. The resting length of the muscles of inspiration decreases as thoracic volume increases, and consequently the maximum inspiratory airway pressure falls.

The earliest abnormalities resulting from respiratory muscle weakness are reductions in maximum inspiratory and expiratory pressures. These changes occur even before lung volumes are significantly altered. More marked weakness of the inspiratory muscles reduces the inspiratory capacity and lowers the vital capacity and total lung capacity. Expiratory muscle weakness, on the other hand, results in elevation of the residual volume.

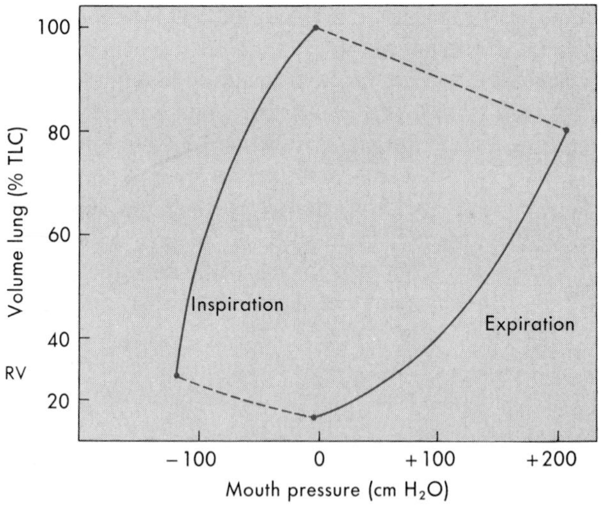

■ **Fig. 37-34.** Effects of lung volume on maximum inspiratory and expiratory pressures. *RV*, Residual volume.

Neuromuscular disorders may not affect all muscle groups uniformly. Because the diaphragm is the major muscle of inspiration, disorders that mainly involve the diaphragm are the most serious. Diaphragm function can be evaluated by measuring transdiaphragmatic pressure, either during normal breathing or during maximum inspiratory maneuvers against a closed airway. Transdiaphragmatic pressure, the pressure gradient

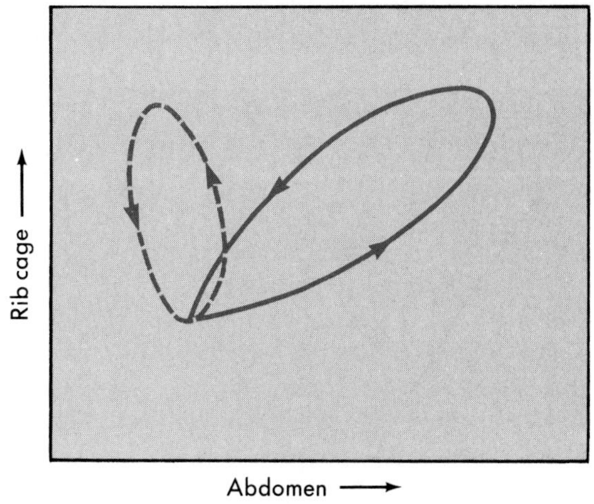

■ Fig. 37-35. A plot of rib cage and abdomen displacement during a breath. Normally the outward displacements of the rib cage and abdomen are synchronous during inspiration, and the inward movements of the rib cage and abdomen are synchronous during expiration *(solid loop)*. When the diaphragm is paralyzed, the abdomen moves inward during inspiration as the rib cage expands *(dashed loop)*.

across the diaphragm, is the difference between intrathoracic and intraabdominal pressure. Catheters are positioned in the esophagus to measure intrathoracic pressure and in the stomach to measure intraabdominal pressure.

A less direct, noninvasive method of evaluating the actions of the diaphragm and intercostal muscles is to record the movements of the rib cage and abdomen during breathing. As the diaphragm descends during inspiration, the abdomen is displaced outward. The extent of abdominal movement measures the contribution of the diaphragm to tidal volume. Although the diaphragm can move the lower portion of the rib cage outward, displacement of the upper rib cage is achieved by contractions of the intercostal muscles and perhaps also the accessory muscles of respiration. Normally, the rib cage and abdomen move nearly synchronously during inspiration. When the diaphragm is paralyzed, the abdomen moves inward as the rib cage expands during inspiration (Fig. 37-35). When the intercostal muscles are paralyzed, the rib cage moves inward as the abdomen expands during inspiration.

■ Bibliography

Journal articles

Agostoni, E.: Mechanics of the pleural space, Physiol. Rev. **42:**57, 1972.

Bolton, T.B.: Mechanisms of action of transmitters and other substances on smooth muscle, Physiol. Rev. **59:**606, 1979.

Chang, H.K., and El Masry, O.A.: A model study of flow dynamics in human central airways. I. Axial velocity profiles, Respir. Physiol. **49:**75, 1982.

Engel, L.A., and Paiva, M.: Analyses of sequential filling and emptying of the lung, Respir. Physiol. **45:**309, 1981.

Haber, P.S., et al.: Alveolar size as a determinant of pulmonary distensibility in mammalian lungs, J. Appl. Physiol. **54:**837, 1983.

Hyatt, R.E., and Black, L.F.: The flow-volume curve: a current perspective, Am. Rev. Respir. Dis. **107:**191, 1973.

Isabey, D., and Chang, H.K.: Steady and unsteady pressure-flow relationships in central airways, J. Appl. Physiol. **51:**1338, 1981.

Knudson, R.J., and Kallenborn, W.T.: Evaluation of lung elastic recoil by exponential curve analysis, Respir. Physiol. **46:**29, 1981.

Macklem, P.T.: Airway obstruction and collateral ventilation, Physiol. Rev. **51:**365, 1971.

Melissinos, C.G., et al.: Chest wall shape during forced expiratory maneuvers, J. Appl. Physiol. **50:**84, 1981.

Morgan, T.E.: Pulmonary surfactant, N. Engl. J. Med. **284:**1185, 1971.

Oldmixon, E.H., and Hoppin, F.G., Jr.: Comparison of amounts of collagen and elastin in pleura and parenchyma of dog lung, J. Appl. Physiol. **56:**1383, 1984.

Sharp, J.T., et al.: Relative contributions of rib cage and abdomen to breathing in normal subjects, J. Appl. Physiol. **39:**608, 1975.

Wilson, T.A.: Relations among recoil pressure, surface area, and surface tension in the lung, J. Appl. Physiol. **50:**921, 1981.

Books and monographs

Campbell, E.J.M., et al.: The respiratory muscles: mechanics and neural control, ed. 2, Philadelphia, 1970, W.B. Saunders Co.

Clements, J.A., and King, R.J.: Composition of the surface active material. In Crystal, R.G., editor: The biochemical basis of pulmonary function, New York, 1976, Marcel Dekker, Inc.

Macklem, P.T., and Mead, J., editors: Handbook of physiology. Section 3: Respiratory system—mechanics of breathing, vol. III, Bethesda, Md., 1986, American Physiological Society.

CHAPTER

38

The Pulmonary Circulation

The pulmonary circulation delivers the entire cardiac output to the lungs and distributes it to the pulmonary capillaries as a very thin film of blood in close proximity to the alveolar gas. The gas-exchanging surfaces of the lungs constitute a maze of interdigitating sheets of capillaries separated only by air-filled spaces. In fact, capillaries make up 85% to 95% of the alveolar surface area.

Differences in function of the pulmonary and systemic circulations are associated with differences in structure of the vessels in the two systems. The pulmonary vascular system is a low-resistance network of highly distensible vessels. The main pulmonary artery is much shorter than the aorta. The walls of the pulmonary artery and its branches are much thinner than the walls of the aorta, and they contain less smooth muscle and elastin. Contrary to systemic arterioles, which have very thick walls composed mainly of circularly arranged smooth muscle, the pulmonary arterioles are thin and contain little smooth muscle. The pulmonary arterioles therefore do not have the same capacity for vasoconstriction as do their counterparts in the systemic circulation. The pulmonary veins are also thin and possess little smooth muscle.

The pulmonary capillaries also differ markedly from the systemic capillaries. Whereas the systemic capillaries are usually arranged as a network of tubular vessels with some interconnections, the pulmonary capillaries are sandwiched between adjacent alveoli in such a manner that the blood flows as if in thin sheets. This provides for maximal exposure of the capillary blood to the alveolar gases. The total surface area for exchange between alveoli and blood has been estimated to be about 50 to 100 square meters. Only thin layers of vascular endothelium and alveolar epithelium separate the blood and alveolar gas. The thickness of the sheets of blood between adjacent alveoli depends on the intravascular and intraalveolar pressures. During pulmonary vascular congestion, as when the left atrial pressure becomes elevated, the width of the sheet may increase severalfold. Conversely, when the local alveolar pressure exceeds the adjacent capillary pressure, the capillaries are compressed and may collapse. Hydrostatic factors are crucial in this phenomenon, particularly with respect to the distribution of blood flow to the various regions of the lungs, as described later.

The total cross-sectional area of the pulmonary vascular bed markedly increases from the level of the pulmonary artery to the level of the capillary bed. Despite this increase in cross-sectional area, the resistance of all of the smaller blood vessels, arranged in parallel, is greater than it is in the major pulmonary arteries, for the same reason that prevails in the systemic circulation (Chapter 30). However, the relative increase in resistance with diminishing vessel size is far less than in the systemic circulation.

In addition to functioning in gas exchange, the pulmonary capillaries also act as filters of systemic venous blood. These filters cleanse the blood of small clots and other particulate materials before they can reach the capillary beds of other vital organs such as the brain. Finally, the capillary endothelium influences the circulating levels of certain vasoactive substances.

The lungs also receive blood through bronchial vessels from the systemic circulation. These vessels nourish the walls of the tracheobronchial tree down to the terminal bronchioles, the supporting vessels, and the outer layers of the pulmonary arteries and veins.

Pulmonary Pressures and Their Effects on Regional Pulmonary Blood Flow

In normal persons the average systolic and diastolic pressures in the pulmonary artery are about 25 and 10 mm Hg, respectively, and the mean pressure is about 15 mm Hg. These pressures are much lower than those in the aorta and reflect the much lower resistance of the pulmonary than of the systemic vascular bed (Fig. 38-1). The mean pressure in the left atrium is normally

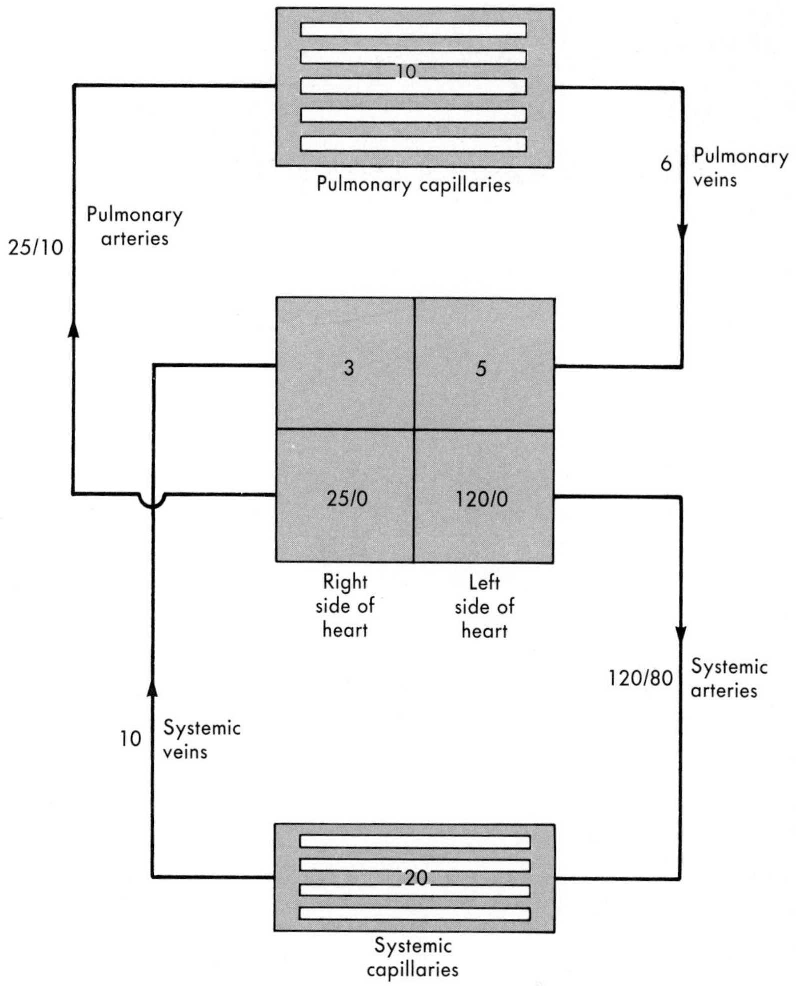

■ **Fig. 38-1.** Systolic and diastolic pressures (in mm Hg) in the systemic and pulmonary arteries and mean pressures in the systemic and pulmonary capillaries and veins.

about 5 to 8 mm Hg, so the total pulmonary arteriovenous pressure gradient is only about 10 mm Hg. The mean hydrostatic pressure in the pulmonary capillaries lies between the pulmonary arterial and pulmonary venous values, but it is somewhat closer to the latter. Mean pulmonary capillary pressure is therefore about 10 mm Hg.

Vascular pressure, the difference between pressure in the lumen of a blood vessel and atmospheric pressure, varies in similar-sized pulmonary vessels, depending on their positions in the lung. In a person standing erect, vascular pressure is greater at the base of the lung than at the apex. The pressure differences in the lung vessels are caused by gravitational effects. The pressure at a point at the top of the lung (the apex of the lung in the erect posture) will be equal to the inlet pressure of the pulmonary artery less the height of a column of blood extending from the inlet to that point. The pressure in capillaries at the bottom of the lung will be greater than inlet pressure by the height of such a column extending from the inlet to that point. When an individual is recumbent, the effects of gravity on vascular pressure are less than when that person is in the erect position because the lungs are not as wide as they are long. Because arterial pressures are low in the pulmonary circulation, these gravitational effects substantially influence the driving pressure.

In accordance with Poiseuille's law (see Chapter 30), for a vessel of a given radius and length, the flow between two points in a system of vessels depends on the driving pressure (the difference between the vascular pressure at these points). If the pulmonary vessels had rigid walls, there would be no effect of gravity on the total driving pressure (the difference between arterial and venous pressures) because gravity would affect both arterial and venous pressures equally (see Chapter 34). The smaller blood flow at the top of the lung than at the bottom occurs because (1) the walls of the pulmonary capillaries are collapsible and not rigid, and (2) alveolar pressure can be greater than capillary vascular pressure at the top of the lung.

As shown in Fig. 38-2, there is a potential region at the top of the lung *(zone 1)* where pulmonary arterial pressure may fall so low that it is less than alveolar pressure (normally close to atmospheric pressure). In the erect person, capillaries in this zone would be completely collapsed, and there would be no blood flow. *Zone 1* does not exist under normal conditions, but it

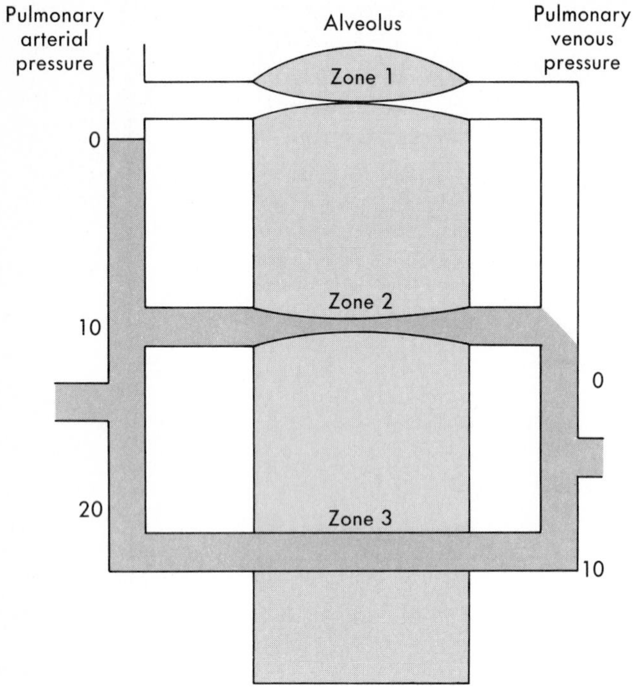

■ Fig. 38-2. Distribution of pulmonary blood flow at different levels in the lung.

may be present if pulmonary arterial pressure is reduced or if alveolar pressure is increased (by positive-pressure artificial ventilation).

Pulmonary arterial pressure becomes greater down the lung because of the effects of gravity. Pulmonary artery pressure may exceed alveolar pressure, but venous pressure, which is always lower than arterial pressure, may be less than alveolar pressure *(zone 2)*. Blood flow in *zone 2* depends on the difference between arterial and alveolar pressures and not the arterial-venous pressure difference. This condition corresponds to that of a waterfall and can be simulated by a Starling resistor, which is a flexible tube inside a rigid chamber (see Chapter 29). When chamber pressure is greater than outflow pressure, the tube collapses at its downstream end, and the pressure in the tube at this point limits flow. The pulmonary capillary bed behaves similarly to the Starling resistor. The lower the position in the lung within *zone 2,* the greater will be the arterial pressure. However, alveolar pressure throughout the lung is unaffected by gravity. Because of the increasing gradient between arterial and alveolar pressure, blood flow progressively increases down this zone from top to bottom.

In *zone 3* venous pressure exceeds alveolar pressure, and flow is regulated by the arterial-venous pressure difference. The capillary pressure (which is between arterial and venous pressure) increases down the zone, whereas the alveolar pressure outside the vessels remains constant. The increased intravascular pressure distends the capillaries and reduces resistance, increasing blood flow in the apex to base direction within this region.

The resistance of larger pulmonary arteries and veins can also influence the distribution of blood flow. The pressure surrounding these vessels approximates pleural pressure. Because pleural pressure is less subatmospheric at the very bottom of the lung than it is at the top of the lung, the transmural vascular pressure is smaller, and the blood vessels narrow. This diminishes regional blood flow at the bottom of the lung. When the lung is at residual volume, the distribution of perfusion is much more uniform than when the lung is at total lung capacity. The distribution is determined by the large vessel resistance, which increases from the top to the bottom of the lung, thereby counteracting the gravitational effect on flow through the pulmonary capillaries.

The driving pressure across the entire pulmonary vascular bed is the difference between the pressures in the pulmonary artery and the left atrium. Catheters can be inserted to measure pressure at these specific points in the pulmonary circulation. In normal humans the driving pressure across the pulmonary vascular bed is 10 to 15 mm Hg. Obstruction to flow into the left ventricle, as occurs in mitral stenosis, can increase all the pulmonary vascular pressures without affecting the total driving pressure. However, this can raise capillary pressure enough to induce *pulmonary edema,* which is a condition of excessive fluid in the pulmonary interstitium and alveoli.

Capillary pressures cannot be measured directly, nor can they be reliably deduced from measurements of pulmonary arterial or left atrial pressures. For example, obstruction to pulmonary artery flow raises mean pulmonary artery pressure proximal to the obstruction, but pulmonary capillary pressure and left atrial pressure may be normal or subnormal.

■ *Passive Effects on Pulmonary Vascular Resistance*

The resistance of a system of blood vessels is the ratio of driving pressure to blood flow. The high resistance of the systemic circulation is largely caused by the muscular arterioles (see Chapter 33), which control the distribution of blood flow to various organs of the body. The pulmonary circulation does not have such thick-walled small vessels, and the total resistance of the pulmonary circulation in healthy subjects is only one tenth that in the systemic circulation. Pulmonary resistance, however, is not constant, but it varies passively with changes in vascular pressure and lung volume.

The relationship between pulmonary artery pressure

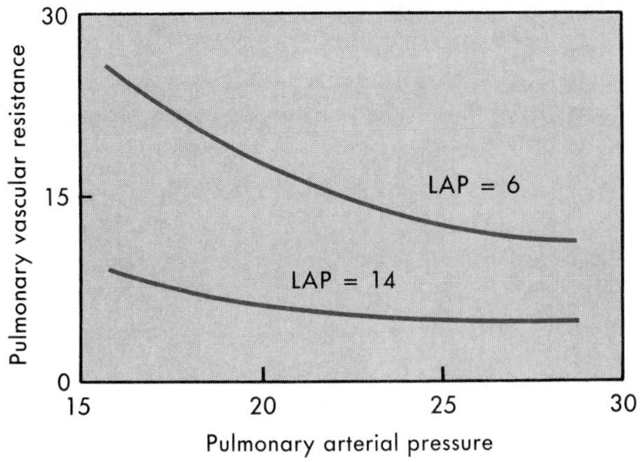

■ Fig. 38-3. The effects of changes in pulmonary artery pressure on pulmonary vascular resistance. As pulmonary artery pressure increases, vascular resistance falls. This effect is attenuated when left atrial pressure is high.

and pulmonary vascular resistance is shown in Fig. 38-3. Pulmonary vascular resistance decreases as the pressure within the vessels rises. As the left atrial or pulmonary artery pressure rises, some capillaries that are usually closed begin to conduct blood, thus lowering the overall resistance. In addition, as vascular pressures rise, more and more capillary segments become distended. Because the distension of the pulmonary vascular bed has a limit, the effects of pulmonary arterial pressure changes on resistance are greater when left atrial pressure is low.

Changes in lung volume affect pulmonary vascular resistance by altering transmural pressure in the blood vessels. Transmural pressure is the difference in pressure inside and outside the vessel wall. Increased transmural pressure tends to enlarge blood vessel diameter. Because vascular pressures are so low in the pulmonary circulation, small changes in the pressure surrounding the vessel can produce relatively large percentage changes in the transmural pressure.

From a functional standpoint, it is useful to think of two categories of pulmonary vessels: the extraalveolar vessels (e.g., the major arteries and veins) and the alveolar vessels (e.g., the capillaries and some arterioles and venules), which behave as if they are surrounded by alveolar pressure (Fig. 38-4).

At large lung volumes the pressure surrounding extraalveolar blood vessels is highly subatmospheric. The large transmural vascular pressure that exists when the lung is distended increases the caliber of the pulmonary arteries and veins. This tends to lower pulmonary vascular resistance.

In contrast, the alveolar vessels are compressed as the lung expands and the alveoli enlarge (Fig. 38-4). This tends to increase pulmonary vascular resistance. Because of these opposing effects of lung volume on

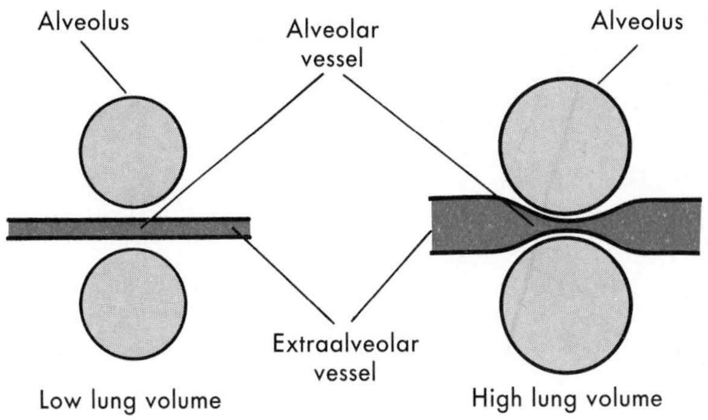

■ Fig. 38-4. Effects of lung volume on the caliber of alveolar and extraalveolar vessels.

the caliber of alveolar and extraalveolar vessels, total pulmonary vascular resistance usually is minimal at functional residual capacity. As lung volume decreases, the resistance of the extraalveolar vessels increases. If the lung is completely collapsed, the tone of the smooth muscle and the elastic tissue in the walls of extraalveolar vessels tend to obstruct blood flow. Pulmonary artery pressure must be raised before blood flow begins; this pressure is called the *critical opening pressure*. When vascular pressures are high, the effects of lung volume changes on transmural pressure are less than at normal pressures.

■ *Active Regulation of Pulmonary Vascular Resistance*

Although changes in pulmonary resistance are achieved mainly by passive factors, resistance can be actively modified by neural, chemical, and humoral influences. The pulmonary blood vessels are innervated by sympathetic fibers, but more sparsely than systemic vessels. Unlike most systemic vessels, the pulmonary blood vessels receive some parasympathetic innervation. The density of nerve fibers is greatest in the largest vessels and decreases with vessel size. Vessels less than 3 mm in diameter have few nerve fibers. In the adult, stimulation of the stellate ganglion decreases the compliance of the large blood vessels but produces a less consistent increase in vascular resistance. The fetal pulmonary circulation responds to both parasympathetic and sympathetic nerve stimulation. The influence of baroreceptor and chemoreceptor stimulation on resistance is weak, although systemic hypoxia does increase the resistance of the large vessels in the lung.

Local alveolar hypoxia produces vasoconstriction, which increases vascular resistance even in the denervated lung. This reaction shifts blood away from poorly ventilated alveoli. The reaction is accentuated by hyper-

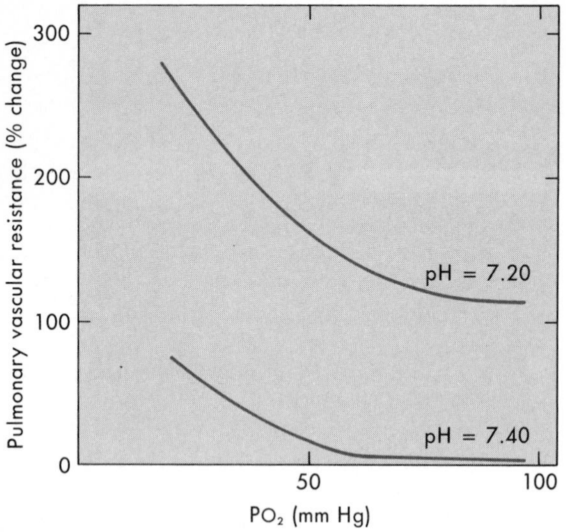

■ **Fig. 38-5.** Effects of hypoxia on pulmonary vascular resistance. Hypoxia causes pulmonary vasoconstriction and an increase in pulmonary vascular resistance. This effect is accentuated by acidosis.

capnia, acidosis, and hypertrophy of the vascular smooth muscle (Fig. 38-5). The mechanism by which local hypoxia produces its effects is still under investigation. Hypoxia does not appreciably constrict isolated vessels but has a much greater effect when the vessel exposed to hypoxia is surrounded by a cuff of lung tissue. This suggests that hypoxia releases some mediator from the tissue cells, and the mediator then produces the vasoconstriction. Mast cells, which are found in the interstitial space, degranulate when exposed to hypoxia. They contain vasoactive substances that, when released, can produce vasoconstriction. Prostaglandins, angiotensin, epinephrine, norepinephrine, and serotonin are pulmonary vasoconstrictors that have been implicated in the hypoxic vascular response. Histamine is the pulmonary vasoconstrictor that has been most extensively studied. Although considerable evidence supports the idea that histamine released by mast cells mediates the hypoxic response or at least modulates that response, not all experimental data support this idea.

■ *Fluid Movement Through the Lung*

Efficient gas exchange requires that the alveolocapillary membrane be kept free of excess water. The alveolocapillary membrane has thick and thin portions. In the thin part, the barrier consists of the alveolar epithelium, the capillary endothelium, and their fused basement membranes. In the thick parts, a layer of interstitial tissue separates the two basement membranes. The alveolar epithelium is made of membranous pneumocytes (type I), which have few organelles, and granular pneumocytes (type II), which contain vesicles and multilaminar bodies that are believed to contain the precursors of surfactant. Type I and type II cells are held tightly together so that molecules can pass into the alveoli only by going through cells and not between them. The pulmonary capillary endothelium is continuous and nonfenestrated. The junctions between endothelial cells are only partially obliterated, and between some endothelial cells there is a gap through which molecules can pass.

Fluid transfer across pulmonary capillaries and systemic capillaries depends on hydrostatic and osmotic forces and obeys the Starling hypothesis (see Chapter 32). The capillary hydrostatic pressure favors movement of fluid out of the capillary, and it is opposed by the hydrostatic pressure in the interstitial fluid. The colloid osmotic pressure of the plasma proteins tends to draw fluid into the capillary, but it is opposed by the osmotic pressure of the proteins in the interstitial fluid.

The colloid osmotic pressure within the capillary is about 25 mm Hg. The capillary hydrostatic pressure is about 10 mm Hg, and it is higher at the bottom of the lung than at the top. The colloid osmotic pressure of the interstitial fluid is not known precisely but has been estimated to be about 15 mm Hg. The interstitial hydrostatic pressure is unknown, but it is thought to be between mean alveolar and mean pleural pressure (slightly negative).

Based on these estimated values, the net force favors a small continuous leak of fluid out of the capillaries into the thick portion of the interstitial space. This fluid travels through the interstitial space to the perivascular and peribronchial spaces within the lung; then it passes into lymphatic channels. Lymphatic channels do not extend into the alveolar septum, but instead end near the terminal bronchioles. However, they are strategically placed to drain fluid that has passed into the interstitial space. They are held open by tethers that connect their external surfaces to the surrounding connective tissue. Pulmonary lymph is propelled toward the hilum of the lung by the active contraction of smooth muscle in the walls of lymphatics as well as by pulsations arising from ventilatory movement and circulatory contractions. Unicuspid, funnel-shaped valves in the small lymphatics prevent backward flow. Lymph flow from the lung, which is normally about 20 ml/minute, greatly increases if the capillary pressure is raised over a long period. This helps to prevent pulmonary edema.

Pulmonary edema may result from an increase in hydrostatic pressure or in capillary permeability, such that both water and proteins can easily cross the capillary wall. For example, release of endotoxin by bacteria infecting the lung can cause pulmonary edema by increasing capillary permeability. Greater hydrostatic pressure may also make the capillary more permeable by widening the endothelial cell pores through which water and protein exit. Increased left atrial pressure can produce pulmonary edema by increasing capillary hydrostatic pressure as in the case of mitral valve stenosis.

Interstitial edema occurs before alveolar edema. It has been suggested that the pressure-volume relationship in the interstitial space is such that even a small amount of excess fluid makes interstitial pressure much more positive, which would oppose fluid flow from the capillary. When large amounts of fluid are present in the interstitium, the compliance of the interstitial space becomes much larger and therefore more fluid can be accepted with only small increases in pressure. This allows the interstitial space to serve as a fluid reservoir that protects the alveolus. The alveolus is further protected by the tight junctions between the epithelial lining cells. Surface-active forces in the alveoli, which would tend to draw fluid into the airspace, are diminished by the layer of surfactant that coats the alveolus. Only when edema is severe does fluid enter the alveolus and interfere significantly with gas exchange.

Water accidentally entering the alveoli through the airway can be reabsorbed because capillary oncotic pressure is much greater than capillary hydrostatic pressure. For example, large quantities of water may be absorbed from the alveoli in drowning.

Metabolic Function of the Pulmonary Endothelium

The pulmonary circulation has a unique position in the vascular system because all of the venous blood from the body tissues must pass through the lung before it is recycled. This allows the lung to play an important role in regulating the level of vasoactive substances in the circulation. Attention has been focused on the capillary endothelial cell as a possible site of this metabolic activity. The surface area of the endothelium is large because of the dense network of capillaries in the lung and because endothelial cells have many projections and indentations (caveolae) that contain many enzyme-rich sites. Some prostaglandins of the E and F types, serotonin, norepinephrine, and histamine, are removed from the blood while passing through the lungs. Uptake of some substances, such as serotonin, depends on a specific enzymatic carrier that can be saturated.

Angiotensin I is converted to the more powerful vasoconstrictor, angiotensin II, on passage through the lungs. Angiotensin I, a decapeptide, is formed by the action of renin (secreted by the kidney) on angiotensinogen (produced by the liver). In the lung two amino acids are enzymatically cleaved from angiotensin I to form the octapeptide, angiotensin II. Conversion of angiotensin I to angiotensin II is delayed by hypoxia.

The Bronchial Vessels

The bronchial arteries usually originate from the midthoracic aorta, or they may arise from the branches of the intercostal, internal mammary, or subclavian arteries. Two or three arteries accompany the bronchi and form a peribronchial plexus. Venous blood from the large bronchi and pleural space drains into the azygos and hemiazygos systems. Blood from small bronchi enters the pulmonary veins, adding some poorly oxygenated blood to the blood leaving the lungs. A network of arterioles, venules, and capillaries, which form on either side of the muscular layer of the bronchi, is interposed between the arteries and veins. The submucosal network intermingles with pulmonary capillaries, and therefore changes in the permeability of the small bronchial vessels may affect fluid exchange across the pulmonary capillaries.

Under normal conditions, the blood passing through the bronchial artery does not reach gas-exchanging surfaces. Because the pressure in the bronchial arteries is the same as systemic pressure, it is sufficiently great to perfuse bronchi, bronchioles, and supporting tissue everywhere in the lungs, regardless of body position. Although in humans bronchial blood flow is only 1% to 2% of pulmonary blood flow, this is appropriate to the mass of the lung that is nourished (about 475 g). If the pulmonary circulation becomes obstructed, the bronchial arteries enlarge and establish connections with precapillary vessels in the pulmonary circuit. In these circumstances, little or no O_2 may be extracted from the alveoli that receive bronchial blood because the bronchial arterial blood is already well oxygenated. However, this bronchial blood supplies nutritive substrates and maintains the viability of the alveolar ducts and alveoli.

Bibliography

Journal articles

Bhattacharya, J., et al.: Factors affecting lung microvascular pressure, Ann. N.Y. Acad. Sci. **384:**107, 1982.

Culver, B.H., and Butler, J.: Mechanical influences on the pulmonary microcirculation, Annu. Rev. Physiol. **42:**187, 1980.

Graham, R., et al.: Critical closure in the canine pulmonary vasculature, Circ. Res. **50:**566, 1982.

Koyama, T., and Horimoto, M.: Pulmonary microcirculatory response to localized hypercapnia, J. Appl. Physiol. **53:**1556, 1982.

Lucas, C.L.: Fluid mechanics of the pulmonary circulation. CRC Crit. Rev. Biomed. Eng. **10:**317, 1984.

Marshall, C., and Marshall, B.: Site and sensitivity for stimulation of hypoxic pulmonary vasoconstriciton, J. Appl. Physiol. **55:**711, 1983.

McMurtry, I.F., et al.: Studies of the mechanism of hypoxic pulmonary vasoconstriction, Adv. Shock Res. **8:**21, 1982.

Nandiwada, P.A., et al.: Pulmonary vasodilator responses to vagal stimulation and acetylcholine in the cat, Circ. Res. **53:**86, 1983.

Orchard, C.H., et al.: The relationship between hypoxic pulmonary vasoconstriction and arterial oxygen tension in the intact dog, J. Physiol. [London] **338:**64, 1983.

Rabinovitch, M., et al.: Changes in pulmonary blood flow affect vascular response to chronic hypoxia in rat, Circ. Res. **52:**432, 1983.

Sobin, S.S., et al.: Changes in arteriole in acute and chronic hypoxic pulmonary hypertension and recovery in rat, J. Appl. Physiol. **55:**1445, 1983.

Sylvester, J.T., et al: Hypoxic constriction of alveolar and extraalveolar vessels in isolated pig lungs, J. Appl. Physiol. **54:**1660, 1983.

Zhuang, F.Y., et al.: Analysis of blood flow in cat's lung with detailed anatomical and elasticity data, J. Appl. Physiol. **55:**1341, 1983.

Books and monographs

Fishman, A.P.: Pulmonary circulation. In Fishman, A.P., and Fisher, A.B., editors: Handbook of physiology, Section 3. The respiratory system—circulation and nonrespiratory functions, vol. 1, Bethesda, Md., 1985, American Physiological Society.

Grover, R.F., et al.: Pulmonary circulation. In Shepard, J.T., and Abboud, F.M., editors: Handbook of physiology, Section 2. The cardiovascular system—peripheral circulation and organ blood flow, vol. 3, Bethesda, Md., 1983, American Physiological Society.

Parker, J.C., et al.: Pulmonary transcapillary exchange and pulmonary edema. In Guyton, A.C., and Young, D.B., editors: International review of physiology: cardiovascular physiology III, vol. 18, Baltimore, 1979, University Park Press.

Taylor, A.E., and Parker, J.C.: Pulmonary interstitial spaces and lymphatics. In Fishman, A.P., and Fisher, A.B., editors: Handbook of physiology, Section 3. The respiratory system—circulation and nonrespiratory functions, vol. 1, Bethesda, Md., 1985, American Physiological Society.

CHAPTER

39

Gas Exchange and Gas Transport

Properties of Gases

An appreciation of the gas-exchanging function of the lung requires an understanding of some of the fundamental properties of gases. A gas consists of molecules that are in a state of random motion. The molecules fill any container in which they are enclosed and exert a pressure that is caused by the collision of molecules with one another and with the container walls.

The respiratory gases, which include oxygen, carbon dioxide, and nitrogen, obey the law of perfect gases, which is expressed as follows:

$$PV = nRT \qquad (1)$$

where

P = Pressure
V = Volume
n = Number of gas molecules
R = Gas constant
T = Absolute temperature

This expression indicates that if temperature is kept constant, the volume occupied by an aliquot of gas varies inversely with the pressure to which it is subjected (Boyle's law). Also, at constant pressure the volume of a gas is proportional to the absolute temperature (Charles' law). Similarly, at constant volume the pressure exerted by a gas varies directly with the absolute temperature (Gay-Lussac's law).

At sea level the total pressure of atmospheric air is about 1000 cm H_2O or 760 mm Hg (torr). More and more commonly pressure is expressed in kilopascals. One kilopascal is equal to 10 cm H_2O.

Air is a mixture of gases including nitrogen, oxygen, carbon dioxide, and certain inert gases, such as argon and neon. Each gas in the mixture exerts a partial pressure that is proportional to its concentration (Dalton's law). The partial pressure of each gas in the mixture is the same as it would be if that gas alone occupied the entire volume. The sum of the partial pressures of each of the gases in the mixture equals the total pressure.

As the inspired air enters the body, it is heated and humidified as it passes over the nasal turbinates and mucous membranes of the airways. At body temperature, the partial pressure of water vapor is 47 mm Hg, regardless of the barometric pressure. The total pressure of dry gases in the airways is consequently equal to the barometric pressure (760 mm Hg at sea level) minus water vapor pressure (47 mm Hg), or 713 mm Hg. The partial pressures of the gases in inspired air are proportional to their concentrations. The concentration of O_2 is about 21%, and the partial pressure of O_2 (Po_2) is 149 mm Hg. The concentration of CO_2 is very low, 0.03%, and its partial pressure (Pco_2) is less than 1 mm Hg. The concentration of N_2 is 79%, and its partial pressure ($P{N_2}$) is 563 mm Hg.

A gas such as O_2 or CO_2, when exposed to a liquid, will dissolve in that liquid. According to Henry's law, the volume of gas dissolved is proportional to its partial pressure. The greater the concentration of O_2 or CO_2 in the gas phase, the greater the volume that will go into solution. At equilibrium the partial pressures in the gas and liquid phases will be the same. If, however, the partial pressure in the gas phase is then reduced, some of the dissolved molecules of gas would leave solution and reenter the gas phase.

The volume of gas dissolved in a liquid also depends on the solubility of the gas in the solvent and on the temperature. CO_2 is more soluble in water than is O_2 so that at a given partial pressure, greater quantities of CO_2 than of O_2 will be dissolved.

Gas Exchange

Ventilation, O_2 Uptake, and CO_2 Output

Ventilation is measured as the volume of gas that is inspired or expired in a unit of time. Minute ventilation is the product of tidal volume and the breathing frequency.

In the lung, gas is exchanged in the alveoli. Blood passing through the pulmonary capillary bed removes O_2 from the alveolar air and adds CO_2. Ventilation of the alveoli, on the other hand, adds O_2 to the alveolar air and removes CO_2. The composition of alveolar gas thus depends on the balance between the levels of alveolar ventilation and pulmonary capillary blood flow. Consequently, in alveoli to which blood flow is excessive relative to ventilation, the P_{O_2} will be low and the P_{CO_2} will be high. Conversely, in alveoli to which blood flow is deficient relative to ventilation, the P_{O_2} will be high and the P_{CO_2} will be low.

The volume of O_2 that is taken up by the lungs may be calculated from the difference in the amount of O_2 in the inspired and expired air, according to the equation:

$$\dot{V}_{O_2} = (\dot{V}_I \times F_{I_{O_2}}) - (\dot{V}_E \times F_{E_{O_2}}) \quad (2)$$

where

$\dot{V}_{O_2}$ = O_2 uptake per minute
$\dot{V}_I$ = Inspired volume of ventilation per minute
$F_{I_{O_2}}$ = Concentration of O_2 in inspired air
$\dot{V}_E$ = Expired volume of ventilation per minute
$F_{E_{O_2}}$ = Concentration of O_2 in expired gas

The volume of CO_2 that is eliminated is determined in a similar fashion. Because the concentration of CO_2 in the inspired air is negligible, CO_2 ouput per minute ($\dot{V}_{CO_2}$) is calculated simply as the product of the expired volume of ventilation ($\dot{V}_E$) and the concentration of CO_2 in the expired gas ($F_{E_{CO_2}}$):

$$\dot{V}_{CO_2} = \dot{V}_E \times F_{E_{CO_2}} \quad (3)$$

In the steady state, when whole body metabolism is constant, the relationship between tissue CO_2 production and O_2 consumption is fixed. The *respiratory quotient* is the ratio of CO_2 production to O_2 consumption in the tissues. The quotient depends on whether carbohydrate, protein, or fat is being metabolized; it may range from 0.7 to 1.0, indicating that O_2 consumption equals or exceeds CO_2 production.

The ratio of the amount of CO_2 eliminated to the amount of O_2 taken up by the lungs is called the *respiratory exchange ratio;* i.e., $\dot{V}_{CO_2}/\dot{V}_{O_2}$. Under steady-state conditions, the respiratory exchange ratio equals the respiratory quotient. Ordinarily, the amount of O_2 taken up by the capillary blood is greater than the amount of CO_2 added by the blood to the alveolar gas. As a result, the expired volume is somewhat less than the inspired volume. Nitrogen is not exchanged in the lung. Hence, the volume of nitrogen in the inspired air (i.e., the product of the inspired volume of ventilation [$\dot{V}_I$] and the concentration of nitrogen in the inspired air [$F_{I_{N_2}}$]) will equal the volume of nitrogen in the expired gas (i.e., the product of the expired volume of ventilation [$\dot{V}_E$] and the concentration of nitrogen in the expired gas [$F_{E_{N_2}}$]):

$$\dot{V}_I \times F_{I_{N_2}} = \dot{V}_E \times F_{E_{N_2}} \quad (4)$$

Differences between the inspired and the expired volumes of ventilation will be reflected in differences in the concentrations of nitrogen in the inspired and expired gas.

Dead Space

A portion of each breath does not participate in O_2 and CO_2 exchange. The volume of inspired air that does not participate in gas exchange has been termed the *dead space*. The lung components that contain nonexchanging air are the pharynx, larynx, conducting airways, and those alveoli that receive little or no perfusion. The volume of each breath that fills the first three of these components constitutes the *anatomical dead space*. The volume in underperfused alveoli is the *alveolar dead space;* normally, this space is much smaller than the anatomical dead space. The sum of the anatomical and alveolar dead spaces is the *physiological dead space*.

The anatomical dead space (in milliliters) during normal breathing is about equal to a person's body weight (in pounds). In an adult male the anatomical dead space is about 150 ml. The alveolar dead space in normal subjects is about 5 to 10 ml. In patients with lung disease the alveolar dead space may become very large and may exceed the anatomical dead space.

Although the dead space of the conducting airways does not exchange gas, it does exchange heat and water. In the dead space the inspired air is humidified and brought to body temperature. Also, harmful constituents, such as inhaled bacteria, are removed in the dead space before they reach the more delicate gas-exchanging surfaces.

The volume of the anatomical dead space can be determined from the nitrogen washout method. A single breath of pure O_2 is inspired to total lung capacity, and the concentration of N_2 is then determined during the next expiration (Fig. 39-1). The initial portion of the exhaled air consists of dead space containing no nitrogen (phase I). The next part of the exhalation, which is characterized by a steep rise in nitrogen concentration (phase II), is made up of a mixture of dead space and alveolar gas. Because some mixing occurs, the boundary between the dead space and the alveolar gas is not sharp. Therefore the rise in N_2 concentration in the expired air is not instantaneous; i.e., it is not represented by the vertical line in Fig. 39-1. Phase II in this figure can be divided so that the areas of the colored segments are equal. The total anatomical dead space is the volume of phase I plus the volume of phase II up to the point of division.

Each tidal volume (V_T) consists of a dead space volume (V_D) and an alveolar volume (V_A), i.e.:

$$V_T = V_D + V_A \quad (5)$$

The concentration of CO_2 in the anatomical dead space is the same as that in the inspired air; it is virtually zero. During expiration the alveolar gas mixes with the dead space air in the airways. Thus the concentration of CO_2 in the mixed expired gas is lower than that in alveolar gas.

The physiological dead space can be determined by means of the Bohr equation, which states that the volume of CO_2 in mixed expired gas equals the volume of CO_2 in the dead space plus the volume of CO_2 in the alveolar component of the tidal volume. The volume of CO_2 in mixed expired gas is the product of the expired tidal volume (V_T) and the CO_2 concentration in the expired gas (F_{ECO_2}). The volume of CO_2 in the anatomical dead space is the product of the anatomical dead space volume (V_D) and the CO_2 concentration in the inspired air (F_{ICO_2}). The volume of CO_2 in the alveolar component of the tidal volume is the product of the alveolar volume (V_A) and the CO_2 concentration in alveolar gas (F_{ACO_2}). Hence

$$V_T \times F_{ECO_2} = (V_D \times F_{ICO_2}) + (V_A \times F_{ACO_2}) \quad (6)$$

The amount of air entering the alveolar space is equal to the tidal volume minus the dead space volume. Hence

$$V_A = V_T - V_D \quad (7)$$

Combining equations (6) and (7) and assuming that $F_{ICO_2} \cong 0$,

$$V_T \times F_{ECO_2} = (V_T - V_D) \times F_{ACO_2} \quad (8)$$

Solving for V_D, the expression becomes

$$V_D = V_T \frac{(F_{ACO_2} - F_{ECO_2})}{F_{ACO_2}} \quad (9)$$

When F_{ICO_2} is not zero, the equation becomes

$$V_D = V_T \frac{(F_{ACO_2} - F_{ECO_2})}{(F_{ACO_2} - F_{ICO_2}} \quad (10)$$

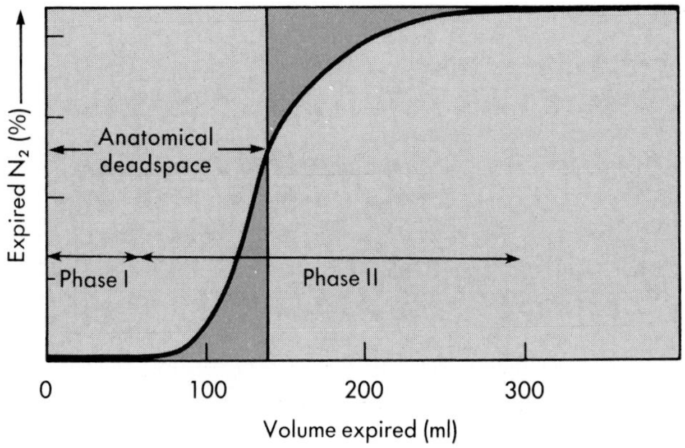

■ **Fig. 39-1.** Single-breath nitrogen washout curve.

The dead space volume does not remain constant, but it enlarges with inspiration as the conducting airways widen because of the traction exerted on the air passages by the surrounding lung parenchyma. This increase in dead space during inspiration is not totally detrimental because it lowers airway resistance.

Even when the tidal volume is less than the anatomical dead space (during very shallow respiration), some of the inspired gas may reach the gas-exchanging surfaces of the alveoli, especially when the airflow is high. Air tends to travel through smaller airways with a parabolic profile (Fig. 37-15). Hence the more central streamlines will travel farther than those nearer the walls. When the breathing frequencies and hence the airflows are very high, the alveoli can be ventilated even with very low tidal volumes. Thus alveolar ventilation depends on the pattern of breathing as well as on the total ventilation. Shallow, rapid breathing (small tidal volumes and high breathing frequencies) provides less alveolar ventilation than the same minute ventilation produced by slow and deep breathing. The pattern of breathing also affects the work performed by the respiratory muscles. Therefore the breathing pattern actually used, particularly in patients with lung disease, may reflect an attempt by the respiratory controller to maintain adequate ventilation at a minimal energy cost.

■ *Alveolar Ventilation*

Alveolar ventilation ($\dot{V}_A$) is total (or minute) ventilation minus dead space ventilation. Because dead space air does not participate in gas exchange, the entire output of CO_2 in expired gas comes from alveolar gas. Accordingly, CO_2 output can be expressed in terms of alveolar ventilation ($\dot{V}_A$), i.e., the product of the alveolar component of the tidal volume and the breathing frequency, and the partial pressure of CO_2 in alveolar air (P_{ACO_2}):

$$\dot{V}_{CO_2} = K \times \dot{V}_A \times P_{ACO_2} \quad (11)$$

where K is a constant of proportionality. Rearranging the equation, it becomes

$$P_{ACO_2} = \frac{\dot{V}_{CO_2}}{K \times \dot{V}_A} \quad (12)$$

This equation indicates that, at any given level of metabolic activity, the P_{ACO_2} varies inversely with the level of alveolar ventilation. This equation is graphically represented in Fig. 39-2.

Similarly, the O_2 uptake, which occurs only in the alveoli, can also be considered in terms of alveolar ventilation ($\dot{V}_A$) and the difference in the partial pressures of O_2 between the inspired air and the alveolar gas:

$$\dot{V}_{O_2} = K \times \dot{V}_A(P_{IO_2} - P_{AO_2}) \quad (13)$$

Rearranging the equation, it becomes

$$P_{AO_2} = P_{IO_2} - \frac{\dot{V}_{O_2}}{K \times \dot{V}_A} \quad (14)$$

This equation is graphed in Fig. 39-3. When equations (11) and (14) are combined and $\dot{V}_A$ is eliminated, P_{AO_2} can be expressed

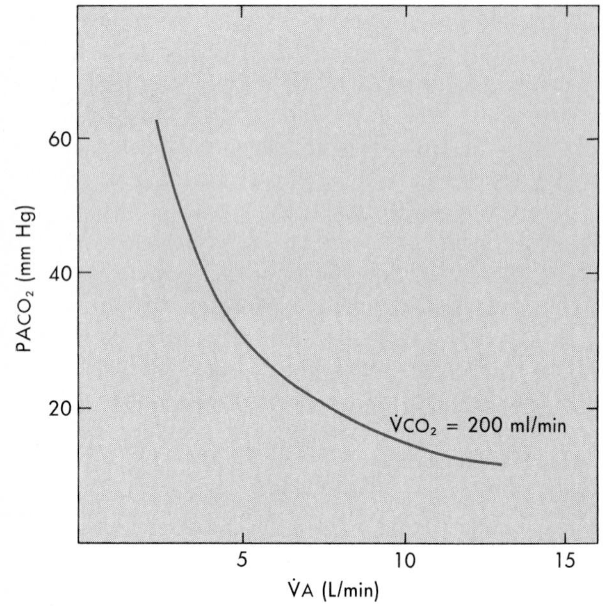

■ Fig. 39-2. At a constant level of CO_2 production, the alveolar P_{CO_2} varies inversely with alveolar ventilation.

$$P_{AO_2} = P_{IO_2} - \frac{P_{ACO_2}}{R} \quad (15)$$

since $\frac{\dot{V}_{O_2}}{\dot{V}_{CO_2}} = \frac{1}{R}$, where R is the respiratory exchange ratio.

This equation is a simplified form of the alveolar air equation; it assumes no CO_2 in the inspired air. The equation indicates that the alveolar P_{O_2} is dependent on the partial pressure of O_2 in the inspired air (P_{IO_2}) and on the alveolar P_{CO_2}, with a correction (R) for the respiratory exchange ratio. Also, at a given inspired O_2 concentration and a constant value of R, P_{AO_2} and P_{ACO_2} are inversely related to one another.

A correction can be introduced for the small amount of CO_2 that is present in the inspired air. In its complete form the alveolar air equation is:

$$P_{AO_2} = P_{IO_2} - \frac{P_{ACO_2}}{R} + \frac{P_{ACO_2} \times F_{ICO_2} \times (1 - R)}{R} \quad (16)$$

The relationship between P_{AO_2} and P_{ACO_2} can be illustrated by the $O_2 - CO_2$ diagram (Fig. 39-4). During room air breathing (P_{IO_2} = 150 mm Hg; P_{ICO_2} = 0) and when *R* is 1, all possible values for alveolar P_{O_2} and P_{CO_2} will fall on a line that has a slope of −1 and intersects the horizontal axis at a P_{O_2} of 150 mm Hg (point *I*). Point *I* represents the P_{O_2} and P_{CO_2} of inspired air. The location on the line of point *A*, repre-

■ Fig. 39-3. At a constant level of O_2 consumption and a fixed inspired O_2 concentration, there is a direct but nonlinear relationship between alveolar P_{O_2} and the level of alveolar ventilation.

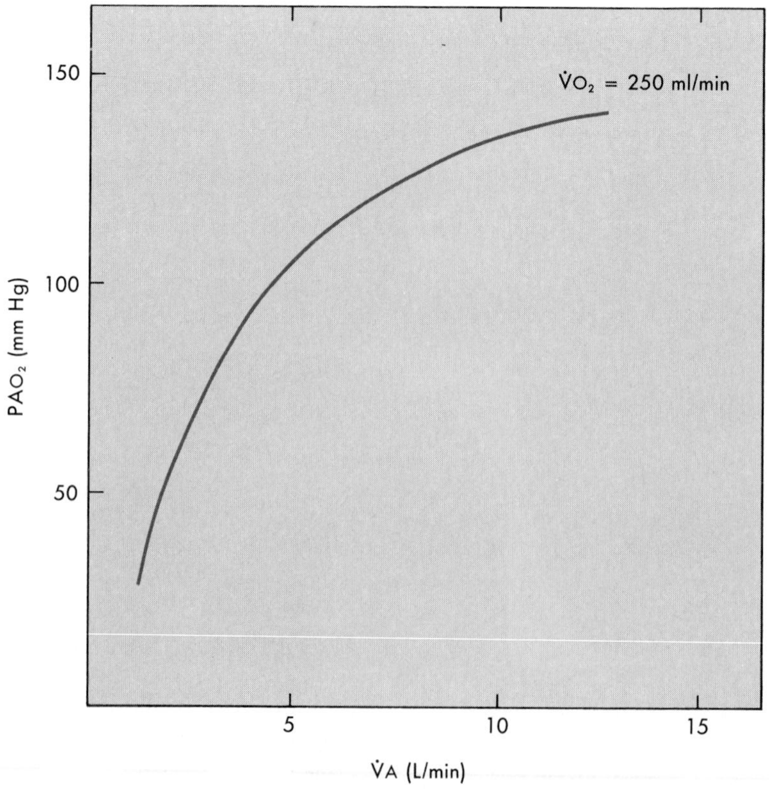

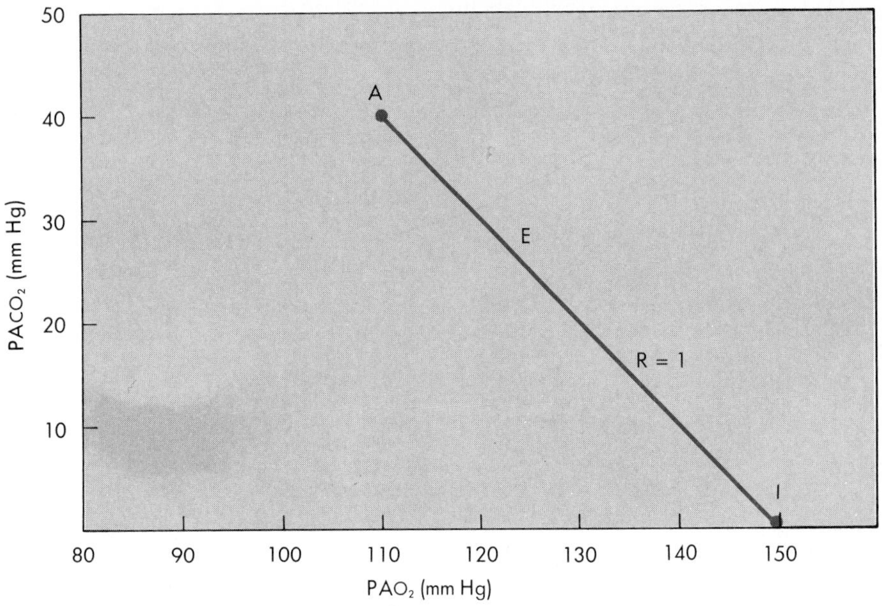

Fig. 39-4. Oxygen–carbon dioxide diagram. The coordinates of points *I*, *E*, and *A* represent the O_2 and CO_2 tensions of inspired, expired, and alveolar gas, respectively.

senting alveolar P_{O_2} and P_{CO_2}, will depend on the level of alveolar ventilation. The higher the alveolar ventilation, the higher will be the P_{AO_2} and the lower will be the P_{ACO_2}. Conversely, the lower the alveolar ventilation, the lower will be the P_{AO_2} and the higher will be the P_{ACO_2}. The location of point *E*, representing the composition of mixed expired gas, will depend on the size of the dead space. The larger the dead space, the greater the difference between points *A* and *E*.

Table 39-1. Forms of O_2 transport in the blood

	Arterial (P_{O_2} = 100 mm Hg)	*Venous (P_{O_2} = 40 mm Hg)*	*A-V difference*
Physically dissolved (vol%*)	0.3	0.1	0.2
Bound to hemoglobin (vol%)	19.5	14.4	5.1
Total (vol%)	19.8	14.5	5.3

*Vol% = Milliliters of gas per deciliter of blood.

Gas Transport by the Blood

Less than 5% of the O_2 and CO_2 carried by the blood is present in physical solution. Primarily, the ability of the blood to carry O_2 and CO_2 depends on reversible chemical reactions. These reactions require specialized compounds and catalysts so that these gases can be transported between the lungs and tissues within the time constraint imposed by the circulation.

The amount of O_2 carried in the blood in physical solution and as O_2 chemically bound to hemoglobin in the erythrocytes is shown in Table 39-1.

The volume of O_2 in solution in a given volume of plasma follows Henry's law and is proportional to its partial pressure. At a P_{O_2} of 100 mm Hg, 3 ml of O_2 is dissolved in each liter of blood. In resting humans, with an O_2 requirement of about 250 ml/minute and a normal cardiac output of about 6 L/minute, extraction of dissolved O_2 by the body tissues supplies only about 1.2% of the body's O_2 requirement. The O_2 requirements of the body at rest could be satisfied by complete extraction by the tissues of dissolved O_2 from the blood only if the cardiac output were to exceed 80 L/minute.

O_2 combines rapidly and reversibly with hemoglobin in the red blood cell to form *oxyhemoglobin* (HbO_2). Hemoglobin that is not combined with O_2 is called *reduced hemoglobin* (Hb). Most of the O_2 (about 98%) present in the blood is bound to hemoglobin.

The hemoglobin molecule consists of *heme* (a pigment) and *globin* (a protein) (Chapter 24). The globin portion of the molecule is composed of four chains; two are identical α-chains, each with 141 amino acids, whereas the other two are identical β-chains, each with 146 amino acids. In normal human hemoglobin (hemoglobin A), the component amino acids, the sequence in which they are arranged, and their spatial relationships are precisely known. In addition, the different amino acid compositions that characterize more than 100 variants of normal hemoglobin have also been determined. Changes in the amino acid sequence may affect the ability of hemoglobin to combine with O_2. Each of the four globin chains in hemoglobin contains one heme unit. Iron in the ferrous state is present at the center of each heme group; it can combine with one molecule of O_2. Only when heme, iron, and globin are together in their proper spatial relationship can the combination with O_2 occur.

The amount of O_2 in chemical combination with he-

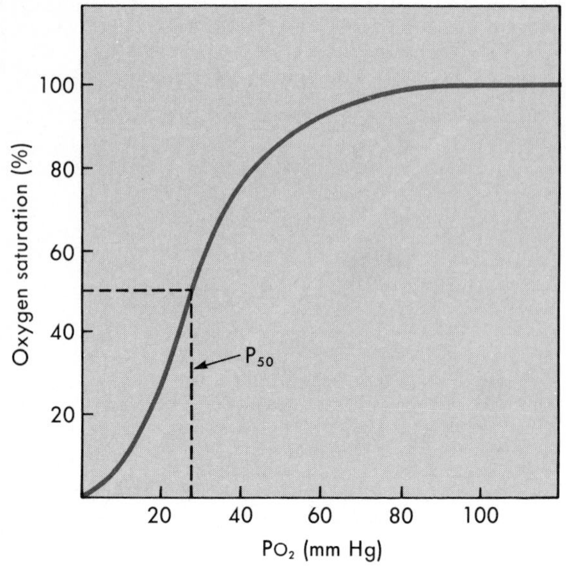

Fig. 39-5. Oxyhemoglobin (HbO_2) dissociation curve. P_{50} is the value of Po_2 at which the blood is 50% saturated with O_2.

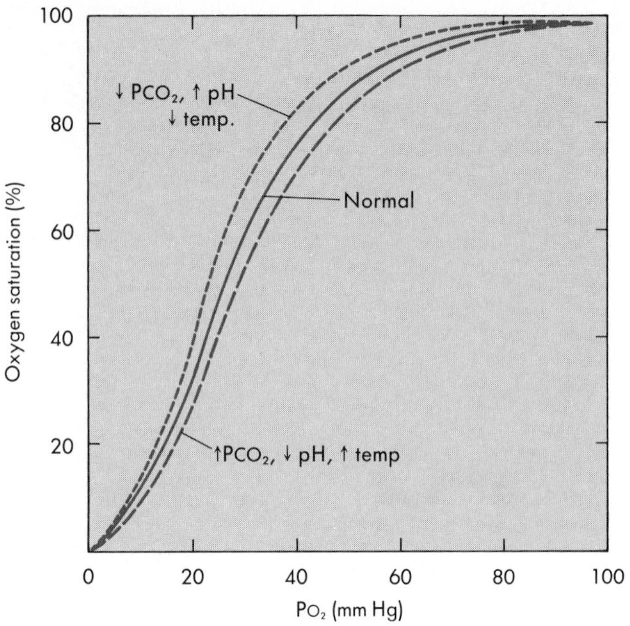

Fig. 39-6. Effect of changes in temperature and pH on O_2 binding to hemoglobin. An increase in temperature or a decrease in pH produces a shift to the right of the HbO_2 dissociation curve. The HbO_2 dissociation curve is shifted to the left by a fall in temperature or an increase in pH.

moglobin depends on the partial pressure of O_2. Each hemoglobin molecule can bind up to four molecules of O_2. Oxygenation of the iron atoms in one of the chains in hemoglobin accelerates oxygenation in the remaining chains. Similarly, release of O_2 by any of the iron atoms enhances the release of O_2 molecules from the remaining iron atoms.

One gram of hemoglobin can combine with 1.34 ml of O_2. At a normal hemoglobin concentration of 15 g/dl of blood, the blood can carry about 20 vol%, i.e., 20 ml O_2/dl blood. This maximal amount is referred to as the *oxygen capacity*. Normally, arterial Po_2 is not sufficiently high to saturate all the hemoglobin with O_2. The actual amount of O_2 in combination with hemoglobin is less than the capacity; the ratio of the O_2 content to the O_2 capacity, expressed as percent, is called the *oxygen saturation*.

The O_2 content and saturation of blood increase progressively with increasing Po_2. However, the relationship between Po_2 and O_2 saturation (the HbO_2 dissociation curve) is not linear, but rather it is sigmoid in shape (Fig. 39-5). The upper flat region of the HbO_2 dissociation curve indicates that the amount of O_2 bound to hemoglobin is relatively constant over a fairly wide range of high O_2 tensions. However, the changes in O_2 saturation are much more pronounced for a given change in Po_2 at the lower ranges of O_2 tension. In a normal person breathing room air, the O_2 saturation of arterial blood is 97%. Small changes in Po_2 produced by alterations in alveolar ventilation will ordinarily have little effect on the amount of O_2 in the arterial blood.

For the tissues to take up O_2 from peripheral capillary blood, the Po_2 in the tissues must be lower than the Po_2 of the blood perfusing them. Because of the steep slope of the HbO_2 curve when Po_2 is below 55 mm Hg, large quantities of O_2 can be unloaded from blood to the tissues with relatively small decreases in blood Po_2. Hence, increased metabolic demands for O_2 can be satisfied by small reductions in tissue Po_2.

When healthy humans with an O_2 saturation of 97% breathe gas mixtures containing more than 21% O_2, the alveolar Po_2 rises above 100 mm Hg, but the arterial O_2 content will increase very little. However, when lung disease decreases the arterial Po_2 and O_2 saturation, even a slight increase in the O_2 concentration in the inspired air may raise the arterial O_2 content substantially.

Factors That Alter the Combination of Oxygen with Hemoglobin

The amount of O_2 bound to hemoglobin depends not only on the O_2 tension but also on the partial pressure of CO_2, the pH, and the temperature of the blood (Fig. 39-6). The HbO_2 dissociation curve is shifted to the right by an increase in temperature, a rise in the CO_2 tension, or a decrease in pH (termed the *Bohr shift*). Therefore, at any given partial pressure of O_2, hemoglobin is less saturated with oxygen.

This shift in oxyhemoglobin dissociation aids in the delivery of oxygen from the blood to the tissues, as shown in Fig. 39-7. The Po_2 of the tissues is much less

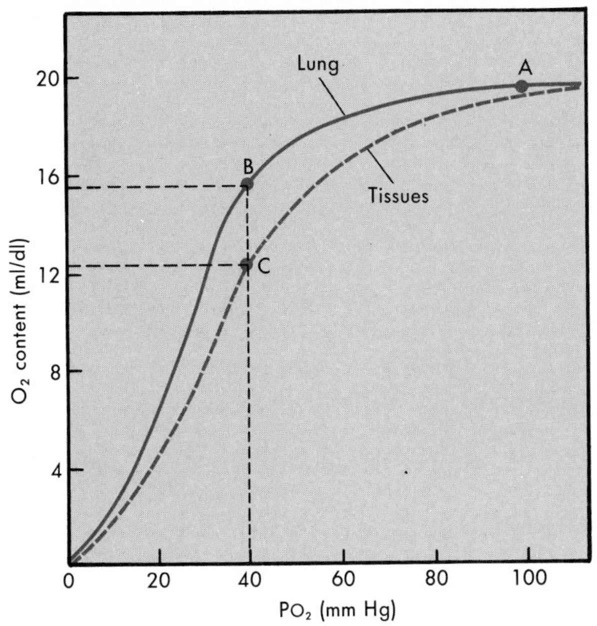

Fig. 39-7. Effect of tissue P_{CO_2} on O_2 delivery to the tissues. If the tissue P_{O_2} were 40 mm Hg, the O_2 content of arterial blood would decrease from 20 ml/dl **(A)** to 15 ml/dl **(B)** in the tissues based on O_2 pressure gradient alone. The high P_{CO_2} in the tissues causes a rightward shift *(dashed curve)* in the oxyhemoglobin dissociation curve. Thus the O_2 content of the capillary blood **(C)** decreases to 12 ml/dl because of the increase in P_{CO_2} in the capillary blood.

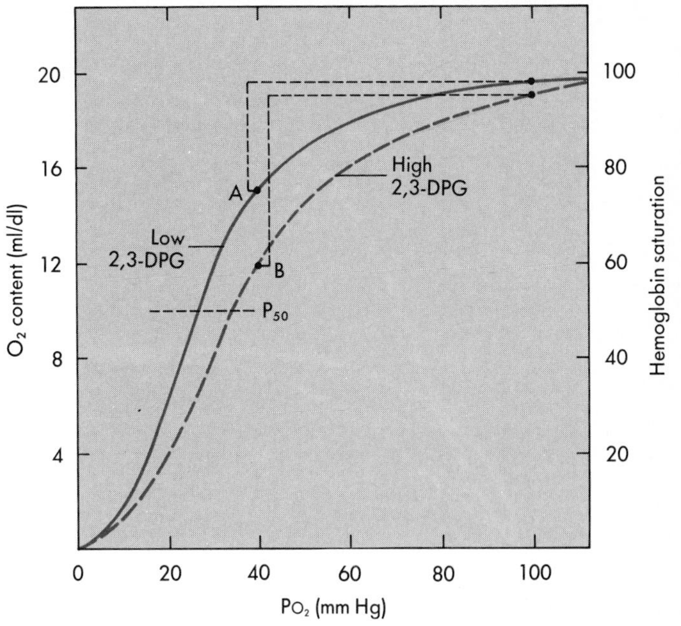

Fig. 39-8. Effect of 2,3-DPG on O_2 binding to hemoglobin. An increased concentration of 2,3-DPG *(dashed curve)* shifts the HbO_2 dissociation curve to the right. The P_{50}, that is, the P_{O_2} required for 50% hemoglobin saturation, is increased. Also, the amount of O_2 released at the tissues from the arterial blood as the P_{O_2} falls from 100 mm Hg to 40 mm Hg is greater (*B* versus *A*).

than the P_{O_2} of arterial blood. As the blood reaches the capillaries, O_2 diffuses from the blood to the tissues because of the difference in partial pressures. The P_{O_2} of capillary blood approaches the P_{O_2} that prevails in the surrounding tissues.

In the example shown in Fig. 39-7, the tissue P_{O_2} is 40 mm Hg. Based on the P_{O_2} gradient alone, each deciliter of blood would deliver to the tissues about 5 ml of O_2; i.e., the O_2 content would decrease from 20 ml/dl in arterial blood *(A)* to about 15 ml/dl in the capillary blood *(B)*.

The P_{CO_2} is higher in the tissues than in arterial blood. When the blood arrives at the tissue capillaries, the P_{CO_2} of the blood approaches the P_{CO_2} of the tissues. The increased P_{CO_2} causes a shift in the oxyhemoglobin dissociation curve (*dashed line,* Fig. 39-7); the shift reflects the diminished ability of hemoglobin to bind O_2. Consequently, at a P_{O_2} of 40 mm Hg, the capillary blood binds only 12 ml/dl of O_2 *(C)* instead of 15 ml/dl *(B)*. Thus, in this example, the rise in P_{CO_2} as the blood traverses the tissue capillaries makes available to the tissues an additional 3 ml of O_2 from each deciliter of blood.

Because of the effect of P_{CO_2} on the ability of hemoglobin to bind O_2, the ability of the blood to take on O_2 as the blood flows through the pulmonary capillaries is also improved. The alveolar P_{CO_2} is less than the P_{CO_2} of the systemic venous blood that flows to the lungs in the pulmonary arteries. The diminished P_{CO_2} in the pulmonary capillaries causes a leftward shift in the oxyhemoglobin dissociation curve, which thereby allows hemoglobin to bind more O_2 for a given P_{O_2}.

An increase in metabolism decreases the P_{O_2} and increases the P_{CO_2} and the hydrogen ion (H^+) concentration in the tissues and capillary blood. The shift in the HbO_2 dissociation curve resulting from the changes in P_{CO_2} and pH facilitates the unloading of O_2 from the blood to the tissues for the reasons just outlined.

A specific compound (2,3-diphosphoglycerate, or 2,3-DPG) that is present in high concentrations in red blood cells regulates the release of O_2 from HbO_2. This substance is formed in the erythrocyte by anaerobic glycolysis from 1,3-DPG. When added to red cells, 2,3-DPG shifts the HbO_2 curve to the right. The effect on the dissociation curve can be measured by determining the P_{O_2} that is required for a hemoglobin-O_2 saturation of 50% at a temperature of 37° C and a pH of 7.40; that P_{O_2} is called the P_{50} (Figs. 39-5 and 39-8). Increased concentrations of 2,3-DPG raise the P_{50}. At any given P_{O_2}, the O_2 content of blood is less as the 2,3-DPG concentration is raised, but the availability of O_2 to the tissues is increased (Fig. 39-8). Conversely, lowered concentrations of 2,3-DPG reduce the P_{50}. Hence the O_2 saturation is higher at a given P_{O_2} and less O_2 is

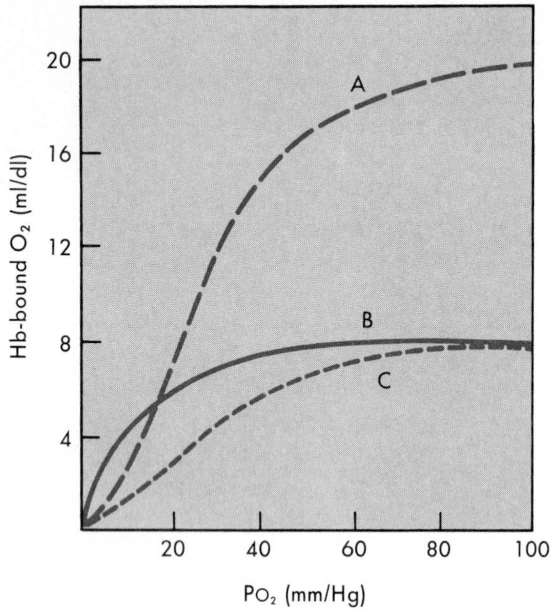

Fig. 39-9. Oxyhemoglobin dissociation curves. *A,* A normal HbO_2 dissociation curve with a hemoglobin concentration of about 15 g/dl. *B,* An HbO_2 dissociation curve after exposure to CO. The hemoglobin concentration is 15 g/dl, but the O_2-carrying capacity is markedly reduced, and the dissociation curve is shifted to the left. *C,* An HbO_2 dissociation curve in a person with anemia and a hemoglobin concentration of about 6 g/dl.

available to the tissues. Increased amounts of 2,3-DPG acidify red blood cells, and this acidification increases the P_{50}. Also, 2,3-DPG binds more tightly to reduced hemoglobin and interferes with the changes in the shape of the molecule that are required for its combination with O_2. An excess of 2,3-DPG is not always advantageous, especially when a person breathes O_2-poor gas mixtures or resides at high altitude. Excess 2,3-DPG reduces the O_2 binding to hemoglobin in the lung and thus reduces O_2 uptake.

Certain drugs and chemicals, such as nitrates and sulfonamides, oxidize the iron in hemoglobin from the ferrous to the ferric form. Hemoglobin with iron in ferric form *(methemoglobin)* cannot take up oxygen.

Carbon monoxide (CO) has more than 200 times the affinity for hemoglobin than does O_2. Because of this greater affinity for hemoglobin, even very small concentrations of CO may produce appreciable levels of carboxyhemoglobin (HbCO). Because CO competes with O_2 for binding sites on the hemoglobin molecule, the amount of HbO_2 will consequently be reduced. Combination of CO with hemoglobin also affects O_2 binding and causes the O_2 dissociation curve to shift to the left and become more hyperbolic. This causes a greater deficit in O_2 uptake than might be expected just from the decrease in carrying capacity of hemoglobin for O_2 (Fig. 39-9). Even when HbCO levels are high and consequently HbO_2 levels are reduced, the Po_2 in arterial blood remains normal because the Po_2 is determined by the amount of O_2 in physical solution. Hence, the chemoreceptors, which respond to low Po_2, are not stimulated and ventilation fails to increase. CO poisoning is treated by administering high concentrations of inspired O_2 to displace some of the CO that had combined with hemoglobin.

Variations in the globin chains may affect the P_{50}. Certain variants of hemoglobin are designated by letters of the alphabet, such as Hb A for normal adult hemoglobin and Hb F for fetal hemoglobin. Other variants bear the name of the location where they were first described, such as Hb Seattle. Usually, in these variants one amino acid is substituted. In only one, Hb M, does the structural change involve the heme group.

The globin chains in Hb F and Hb A differ, which causes Hb F to have a greater affinity for O_2 than does Hb A. Fetal red blood cells contain Hb F; therefore they are able to combine with more O_2 as they flow through the capillaries in the placenta, where the Po_2 is low. Shortly after birth, Hb F begins to disappear from the red blood cells and is replaced by Hb A.

Tissue Oxygenation

Respiration must be coordinated with circulatory activity to provide the body tissues with enough O_2 to prevent tissue hypoxia. The amount of O_2 carried to the tissues by the capillaries depends on the O_2 content of the arterial blood and the cardiac output. The O_2 content of the blood in turn depends on the level of Po_2 and the amount of hemoglobin in the blood. Normally, the total amount of O_2 carried in the blood is less than a liter, so the O_2 reserve is small if breathing should stop. Tissue hypoxia may result from anemia or from excessive reductions in arterial Po_2 or blood flow.

O_2 is transported from the capillaries to the tissues by diffusion. Hence, tissue Po_2 must be less than capillary Po_2. The rate of diffusion of O_2 from capillary to tissue depends on the surface area available for diffusion and the thickness of the tissue layer through which O_2 must diffuse. Under resting conditions, as shown in Fig. 39-10, not all tissue capillaries are open at any one instant. More capillaries open when the metabolic rate increases (e.g., during exercise). This facilitates diffusion and allows more O_2 to be extracted by the tissues. Arterial hypoxemia also opens more tissue capillaries, thereby acting to improve tissue perfusion.

The metabolism of O_2 in tissue involves a series of coupled chemical processes that take place in the mitochondria and require enzymatic facilitation. Various cytoplasmic poisons can interfere with enzymatic function and reduce the O_2 uptake by cells. If the cell cannot use O_2, the capillary and venous Po_2 will be abnormally high in the face of tissue hypoxia.

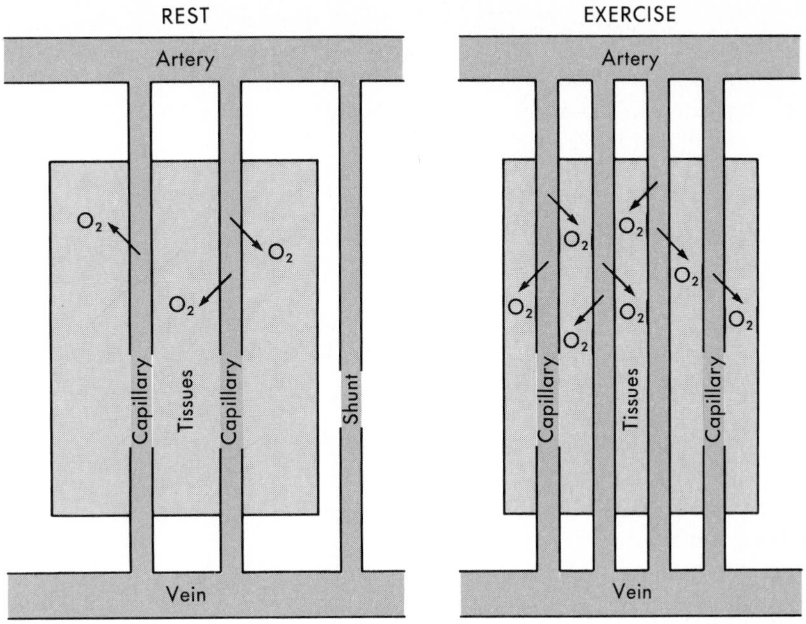

Fig. 39-10. Tissue oxygenation. At rest not all capillaries are open. During exercise more capillaries open, which facilitates diffusion and allows more O_2 to be extracted by the tissues.

Table 39-2. Forms of CO_2 transport in blood

	Venous (P_{CO_2} = 46 mm Hg)	*Arterial (P_{CO_2} = 40 mm Hg)*	*Difference*
Dissolved CO_2 (vol%*)	3.1	2.7	0.4
HCO_3^- (vol%)	47.0	43.9	3.1
Carbamino CO_2 (vol%)	3.9	2.4	1.5
Total (vol%)	54.0	49.0	5.0

*Vol% = Milliliters of gas per deciliter of blood.

Types of Hypoxia

Hypoxia has been classified into four main types: hypoxic, anemic, circulatory, and histotoxic.

In *hypoxic hypoxia* the tissues are supplied by blood with an abnormally low O_2 tension, and thus the hemoglobin is poorly saturated with O_2. This condition occurs when the partial pressure of O_2 in the inspired air is lower than normal (e.g., at altitude) or in patients with respiratory diseases that interfere with gas exchange in the lung.

In *anemic hypoxia* the tissues are supplied by arterial blood with a normal O_2 tension but a low hemoglobin concentration. Hence, the O_2 content and the O_2 capacity of the blood are both lower than normal. This variety of hypoxia can occur in anemia as well as when toxic substances (such as CO) combine with hemoglobin and interfere with its capacity to combine with O_2.

Circulatory hypoxia may occur when cardiac output (e.g., in shock or congestive heart failure) or when there is a local obstruction to arterial blood flow.

Histotoxic hypoxia occurs when a toxic substance such as cyanide interferes with the ability of tissues to use O_2.

Carbon Dioxide Transport and Its Effects on Acid-Base Balance

The CO_2 produced in body tissues is carried in the blood to the lungs in three forms: in physical solution, as bicarbonate (HCO_3^-), and as a carbamino compound combined with blood proteins (Table 39-2).

Physically dissolved carbon dioxide. The volume of CO_2 dissolved in blood depends on the partial pressure of CO_2. Because CO_2 is about 20 times more soluble than O_2, CO_2 in the dissolved form is more important in CO_2 transport than is dissolved O_2 in O_2 transport. As with most other gases, an increase in body temperature decreases the volume of CO_2 in physical solution. At normal body temperature, an additional 0.6 ml of CO_2 is dissolved in each liter of plasma for every 1 mm Hg increase in P_{CO_2}. Dissolved CO_2 accounts for about 5% of the CO_2 in the blood.

Bicarbonate. Some of the dissolved CO_2 reacts with water to form carbonic acid (H_2CO_3). This hydration of CO_2 takes place according to the equation:

$$CO_2 + H_2O \rightleftarrows H_2CO_3 \qquad (17)$$

In the plasma this reaction takes place very slowly, and the concentration of dissolved CO_2 is about 1,000 times greater than the concentration of H_2CO_3. A small portion of the H_2CO_3 so formed subsequently ionizes, according to the equation:

$$H_2CO_3 \rightleftarrows H^+ + HCO_3^- \qquad (18)$$

Only a minute amount of bicarbonate (HCO_3^-) is produced in plasma, but much larger amounts are formed in red blood cells. Red blood cells contain the enzyme *carbonic anhydrase,* which catalyzes the combination of H_2O with CO_2.

$$CO_2 + H_2O \xrightleftharpoons{\text{carbonic anhydrase}} H_2CO_3 \rightleftharpoons HCO_3^- + H^+ \quad (19)$$

The increase in red cell HCO_3^- ultimately leads to increased plasma HCO_3^- by diffusion. HCO_3^- is the major form in which CO_2 is transported in the blood.

Also, hemoglobin in the red blood cell provides basic groups, which combine with the hydrogen ions formed by the dissociation of H_2CO_3. This increases the formation of HCO_3^- by shifting the following reaction to the right:

$$H_2CO_3 + Hb^- \rightleftharpoons Hb + H^+$$
$$HCO_3^- \rightleftharpoons HHb + HCO_3^- \quad (20)$$

Reduced hemoglobin is a better H^+ acceptor than is HbO_2.

Carbamino compounds. CO_2 can also combine with amine groups in proteins to form carbamino compounds, as shown in the following reaction:

$$Hb - N\begin{matrix} H \\ H \end{matrix} + CO_2 \rightleftharpoons Hb - N\begin{matrix} H \\ COO^- \end{matrix} + H^+ \quad (21)$$

Because more amine groups in the blood are available in hemoglobin than in plasma proteins, more CO_2 combines with hemoglobin than with plasma proteins to form carbamino compounds. Reduced hemoglobin binds CO_2 more readily than does HbO_2. About 30% of the arteriovenous difference in CO_2 is accounted for by the difference in the carbamino CO_2 contents in the arterial and venous blood.

Carbon Dioxide Dissociation Curve

The CO_2 dissociation curve relates the CO_2 content of the blood to the blood P_{CO_2} (Fig. 39-11). Note that the relationship between CO_2 content and P_{CO_2} is steeper than the relationship between O_2 content and P_{O_2}. The total quantity of CO_2 in the blood is more than twice that of O_2. In the physiological range of CO_2 content and tension, the CO_2 dissociation curve is almost linear, and there is no plateau.

The degree of oxygenation affects the CO_2 dissociation curve. The greater the saturation of hemoglobin with O_2, the less will be the CO_2 content for a given P_{CO_2}. This effect, called the *Haldane effect,* is caused by the greater ability of reduced hemoglobin to buffer H^+ and to form carbamino hemoglobin. The Haldane effect facilitates CO_2 elimination in the lungs and CO_2 transfer from tissues to blood.

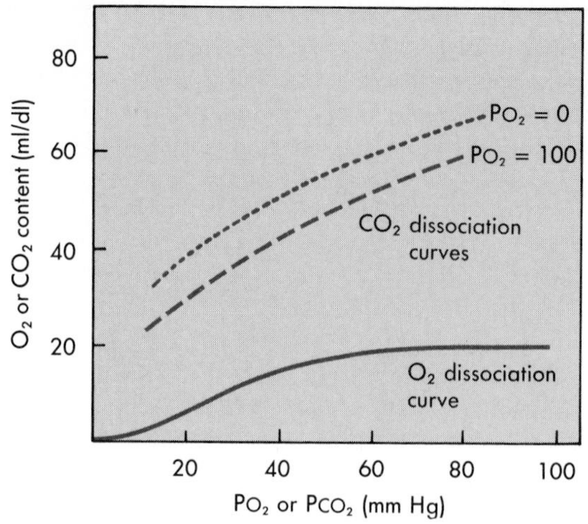

Fig. 39-11. Comparison of HbO_2 and CO_2 dissociation curves. The CO_2 content of blood is more than twice the O_2 content of blood. Also, the CO_2 dissociation curves are much steeper than the O_2 dissociation curve. The CO_2 content at any given P_{CO_2} depends on the level of oxygenation (P_{O_2} of 0 versus P_{O_2} of 100). As the O_2 saturation of hemoglobin is reduced, the CO_2 content at a given P_{CO_2} will increase.

Carbon Dioxide Exchange in the Lung

CO_2 exchange in the lungs is summarized in Fig. 39-12. Because systemic venous P_{CO_2} is higher than alveolar P_{CO_2}, the physically dissolved CO_2 diffuses from the pulmonary capillary plasma into the alveoli. By mass action,CO_2 is also released from carbamino binding and from HCO_3^- in the plasma. Because of the absence of carbonic anhydrase in plasma, this last reaction is slow. As CO_2 leaves plasma, it is replaced by CO_2 from red cells. This CO_2 comes mainly from HCO_3^- in red cells but it also comes from CO_2 in the carbamino form. Carbonic anhydrase is present within red cells and as a result, CO_2 formation from both reactions proceeds rapidly. As blood passes through the pulmonary capillaries, HCO_3^- in red blood cells decreases more rapidly than it does in plasma. Hence, HCO_3^- moves into red blood cells from the plasma as the HCO_3^- in the red blood cells decreases. This increases the negative charge inside the red cell. Consequently, some chloride exits from the red cells into the plasma as a result of electrostatic forces. This transfer of chloride ions between red blood cells and plasma is referred to as the *chloride* (or *Hamburger*) *shift*.

In the pulmonary capillary bed, hemoglobin is oxygenated at the same time that CO_2 is eliminated. HbO_2 is less able to combine with CO_2 and is more acidic.

■ **Fig. 39-12.** CO_2 exchange in the lungs. CO_2 is carried in solution in the plasma, in combination with hemoglobin in the red cell ($HHbCO_2$), in the form of carbonic acid (H_2CO_3) in the red cell, and as bicarbonate (HCO_3^-) in the red cell and plasma.

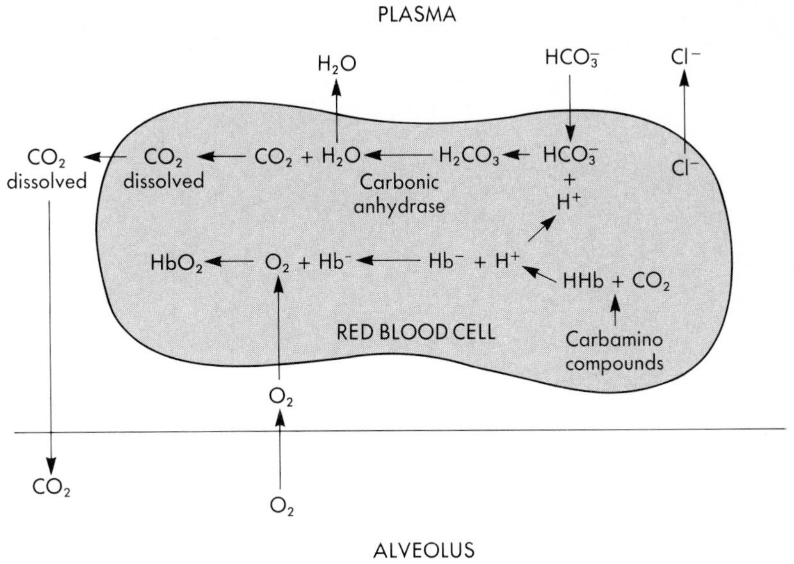

The consequent decrease in carbamino hemoglobin and increase in the concentration of H^+ as HbO_2 is formed (Haldane effect) facilitates the conversion of HCO_3^- to CO_2. The Haldane effect accounts for almost half of the transfer of CO_2 in the lung (Table 39-2). The change in HCO_3^- and H^+ in the red blood cell decreases the osmotic pressure. Therefore water leaves the red cells, and the red cells shrink.

None of the chemical and physical processes involved in CO_2 transfer occur instantaneously. For example, 0.16 second is required for completion of the chloride shift. The rate of pulmonary blood flow limits the time the erythrocytes are present in the lungs, and hence the time available for CO_2 liberation from the venous blood is also limited. Normally the time required for chemical reactions to reach completion does not affect CO_2 transfer. However, if carbonic anhydrase were not available in the red cells, the uncatalyzed hydration of CO_2 would be so slow that only 10% of the available CO_2 would be removed from the blood during its passage through the pulmonary capillaries.

Changes in H^+ and HCO_3^- concentrations in the plasma are slow because of the absence of carbonic anhydrase. Moreover, because the red cell does not allow the ready passage of cations across its membrane (although it allows free passage of anions), the H^+ in the plasma cannot equilibrate with the H^+ in the red cell directly by diffusion. Hence, H^+ concentrations and P_{CO_2} may continue to change even after the red cell has left the pulmonary capillary.

■ *Exchange of Carbon Dioxide in Tissue*

Processes that take place in the tissue are the reverse of those that occur in the lung. In the tissue, uptake of CO_2 by red blood cells occurs because the tissue P_{CO_2} is greater than the P_{CO_2} in capillary blood. CO_2 uptake is facilitated by reduction of hemoglobin (Haldane effect).

CO_2 is transported across the tissue-capillary membrane by diffusion, as is O_2. Tissue and capillary P_{CO_2} rapidly reach equilibrium. CO_2 transfer from tissues to blood is therefore limited by perfusion.

The ratio of tissue perfusion to tissue volume affects the speed at which the P_{CO_2} in different tissues reaches equilibrium with P_{CO_2} in the blood. The greater the perfusion, the faster the equilibration. In some poorly perfused tissues, such as resting muscle, tissue P_{CO_2} will continue to change for very long periods (minutes and even hours) after a change in Pa_{CO_2} produced, for instance, by breathing air with added CO_2.

The rapidity with which changes in whole body metabolism can change P_{CO_2} also depends on the ratio of cardiac output to tissue volume. Because chemoreceptor responses are influenced by the level of Pa_{CO_2}, cardiac output can affect the responses to environmental or metabolic disturbances in CO_2 balance. Large increases in cardiac output can therefore change Pa_{CO_2} abruptly.

■ *Ventilation-Perfusion Relationships*

Gas exchange is maximally efficient when both ventilation and blood flow are appropriately matched in all regions of the lung. However, this ideal situation does not exist even in the normal lung. The amount of CO_2 delivered to, and the amount of O_2 extracted from, each alveolus is in proportion to its blood flow. If blood flow increases to an alveolus but ventilation remains constant, more CO_2 is delivered and more O_2 is removed; consequently alveolar P_{CO_2} rises, and alveolar P_{O_2} falls. If perfusion remains the same but alveolar ventilation is increased, more CO_2 will be expelled from the alveolus, and alveolar P_{CO_2} will fall. P_{O_2} in the alveolus will be increased because more O_2 is carried into

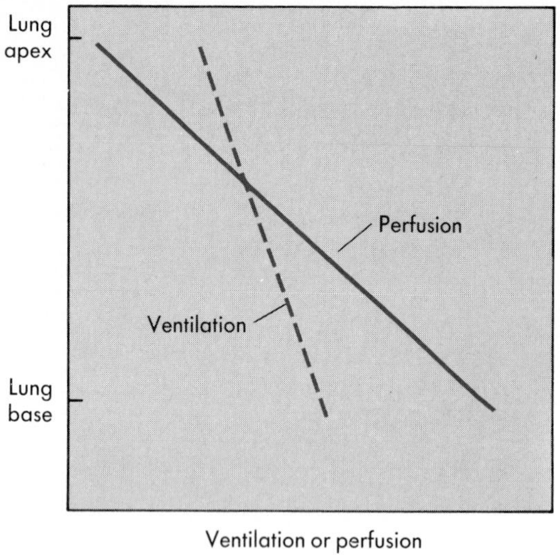

■ **Fig. 39-13.** Distribution of ventilation and perfusion in the lungs. In a person in the upright position, ventilation is greater at the bottom than at the top of the lung. Similarly, pulmonary blood flow is greater at the bottom than at the top of the lung, but the difference in blood flow exceeds the difference in ventilation between the lung apex and lung base.

the alveolus as a result of the increased ventilation. Thus the Po_2 and Pco_2 in each alveolus depend on the ratio of *ventilation to perfusion*.

When a subject is in the upright posture, more ventilation is distributed to the bottom than to the top of the lung because of the effects of gravity (Chapter 38). Similarly, the distribution of pulmonary blood flow decreases from the bottom to the top of the lung. The degree of change in blood flow with vertical distance along the lung is greater than the corresponding change in ventilation. Therefore the ventilation/perfusion ratio progressively increases from the bottom to the top of the lung (Fig. 39-13). Overall, the ratio ($\dot{V}_A/\dot{Q}$) of alveolar ventilation to pulmonary blood flow, both in liters per minute, is about 0.8.

In alveoli at the top of the lung, ventilation is about three times greater than blood flow. O_2 uptake is limited by the relatively low pulmonary blood flow, but the blood passing through the pulmonary capillary bed is well oxygenated. CO_2 is readily extracted from each unit of blood flowing through the capillary bed, and ventilation is high so that the quantity of CO_2 exhaled from this apical region of the lung is relatively high and alveolar Pco_2 is low.

At the bottom of the lung, alveolar ventilation is less than pulmonary capillary blood flow; the $\dot{V}_A/\dot{Q}$ is approximately 0.6. Because of the relative hypoventilation of alveoli at the bottom of the lung, the alveolar Po_2 is lower and the Pco_2 is higher at the lung bases than at the top of the lungs. The greater blood flow at the bottom than at the top of the lung causes most of the O_2 uptake to take place near the bottom of the lung.

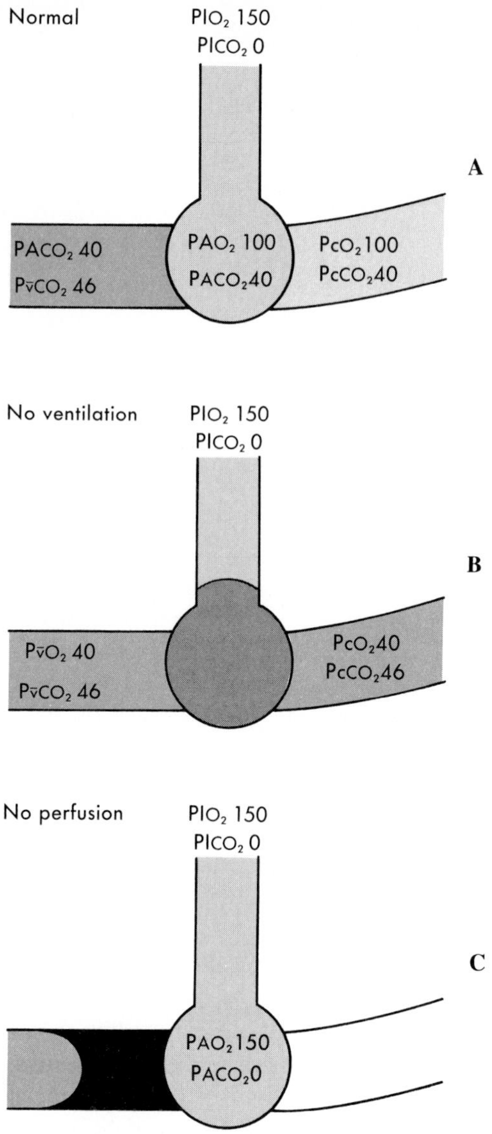

■ **Fig. 39-14.** Schematic diagram of gas exchange in different lung units. **A,** Well-ventilated, well-perfused lung unit. Blood leaving this unit in the pulmonary capillaries has equilibrated with the alveolar Pco_2 and Pco_2. **B,** Lung unit with normal perfusion, but no ventilation. Blood leaving the unit in the pulmonary capillaries has the same Po_2 and Pco_2 values as in pulmonary artery (or mixed systemic venous) blood. **C,** Lung unit with normal ventilation but no perfusion. Alveolar Po_2 and Pco_2 are the same as those in inspired gas. *P,* Partial pressure; *I,* inspired air; *A,* alveolar gas; *a,* arterial blood; *v̄,* mixed systemic venous blood; *c,* pulmonary capillary blood.

The range of ventilation/perfusion ratios throughout the normal lung is relatively narrow so that regional differences in arterial blood gas tensions are small. However, in pulmonary diseases, ventilation/perfusion ratios vary markedly in the lung because of patchy abnormalities in airway resistance, lung compliance, and blood vessel caliber. Ventilation/perfusion ratios in the lung may range from zero to infinity.

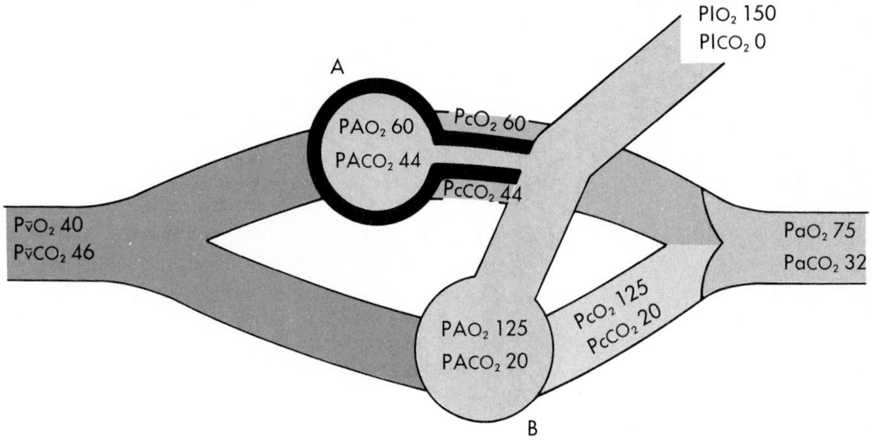

■ **Fig. 39-15.** Schematic diagram of two parallel lung units. Both lung units are equally well perfused. Lung unit *A* has a reduced ventilation and a low $\dot{V}_A/\dot{Q}$ ratio. Lung unit *B* has excessive ventilation and a high $\dot{V}_A/\dot{Q}$ ratio. *P*, Partial pressure; *I*, inspired air; *A*, alveolar gas; *c*, end-capillary; *a*, arterial blood; $\bar{v}$, mixed venous blood.

In the normal lung, the P_{O_2} and P_{CO_2} of the pulmonary capillary blood come into equilibrium with the respective gas tensions in the alveoli, as shown in Fig. 39-14, *A*. Some alveoli in a diseased lung may receive no ventilation at all because of airway obstruction; an example is shown in Fig. 39-14, *B*. The $\dot{V}_A/\dot{Q}$ of this unit is zero. Under these conditions the end-capillary P_{O_2} and P_{CO_2} will equal those in mixed venous blood (P_{O_2} = 40 mm Hg; P_{CO_2} = 46 mm Hg).

The ventilation/perfusion ratio in a lung unit increases as a result of constriction, obstruction, or obliteration of pulmonary blood vessels, which reduces pulmonary blood flow. The air in poorly perfused units has a relatively high P_{O_2} and a low P_{CO_2}. With complete obstruction to blood flow, the $\dot{V}_A/\dot{Q}$ of the unit rises to infinity. The P_{O_2} and P_{CO_2} in the alveoli of that unit approach those of inspired air; that is, P_{O_2} = 150 mm Hg, and P_{CO_2} = 0 mm Hg (Fig. 39-14, *C*).

The consequences of uneven ventilation/perfusion ratios on arterial gas tensions can be illustrated by a model consisting of two lung units (Fig. 39-15). Both units are assumed to be equally well perfused, but the ventilation rates differ. In this example the inspired gas has a P_{O_2} of 150 mm Hg and a P_{CO_2} of zero. Alveolar P_{CO_2} will be lower in the unit with a greater ventilation, and alveolar P_{O_2} will be higher. Assume that the ventilation rates are such that the PA_{CO_2} in the poorly ventilated unit *(A)* is 44 mm Hg and the PA_{O_2} is only 60 mm Hg, whereas in the well-ventilated unit *(B)* PA_{CO_2} is 20 mm Hg, and the PA_{O_2} is 125 mm Hg. The mean arterial P_{O_2} must be ascertained by averaging the O_2 contents of the mixed end-capillary blood rather than by averaging the end-capillary O_2 tensions of the two units. The blood from poorly ventilated unit with a P_{O_2} of 60 mm Hg is about 90% saturated, whereas the blood from the normal unit with a P_{O_2} of 125 mm Hg is about 99% saturated. The mixture of blood from the two units will be 94.5% saturated, and according to the HbO_2 dissociation curve, the partial pressure of O_2 in the arterial blood will be about 75 mm Hg.

The CO_2 dissociation curve between the arterial and mixed venous points is essentially linear. In contrast to the effects of the P_{O_2}, the P_{CO_2} of mixed arterial blood will be equal to the average of the CO_2 tensions of the end-capillary blood from the two units, or about 32 mm Hg.

Because the HbO_2 dissociation curve is almost flat at high P_{O_2} levels, increasing the ventilation and thus the PA_{O_2} of a lung unit with an already high $\dot{V}_A/\dot{Q}$ will not appreciably alter the O_2 content of blood leaving that unit. Consequently, it will have little effect on the overall P_{O_2} of the mixed arterial blood. However, increasing the ventilation to high $\dot{V}_A/\dot{Q}$ units will significantly decrease arterial P_{CO_2}, because the CO_2 dissociation curve is not flat.

In the normal lung, the gas tensions in the pulmonary capillary blood come almost into equilibrium with the alveolar gas tensions. Systemic arterial blood gas tensions are almost identical to the tensions in the pulmonary capillaries. Hence the systemic arterial gas tensions normally are very similar to the mixed alveolar gas tensions, which can be estimated from expired gas samples obtained at the end of expiration. However, when the $\dot{V}_A/\dot{Q}$ ratios vary widely, the gas tensions in end-expiratory air and in systemic arterial blood may be very disparate, as shown in Fig. 39-16. The P_{O_2} of the expired air will depend on the P_{O_2} in each alveolus and the relative amount that each alveolus contributes to the total alveolar ventilation. Alveoli with a high $\dot{V}_A/\dot{Q}$ ratio have higher levels of P_{O_2} and less perfusion, whereas alveoli with a low $\dot{V}_A/\dot{Q}$ ratio have lower O_2 tensions and relatively more perfusion. As the figure shows, a wide distribution of $\dot{V}_A/\dot{Q}$ ratios causes the P_{O_2} of expired gas to be higher than the average P_{O_2} of the blood in the pulmonary veins and systemic arteries. Likewise, a wide range of $\dot{V}_A/\dot{Q}$ ratios causes the systemic arterial P_{CO_2} to be higher than the P_{CO_2} in the expired air. The normal difference between end-expiratory and arterial O_2 tensions is about 5 mm Hg. The difference between end-expiratory and

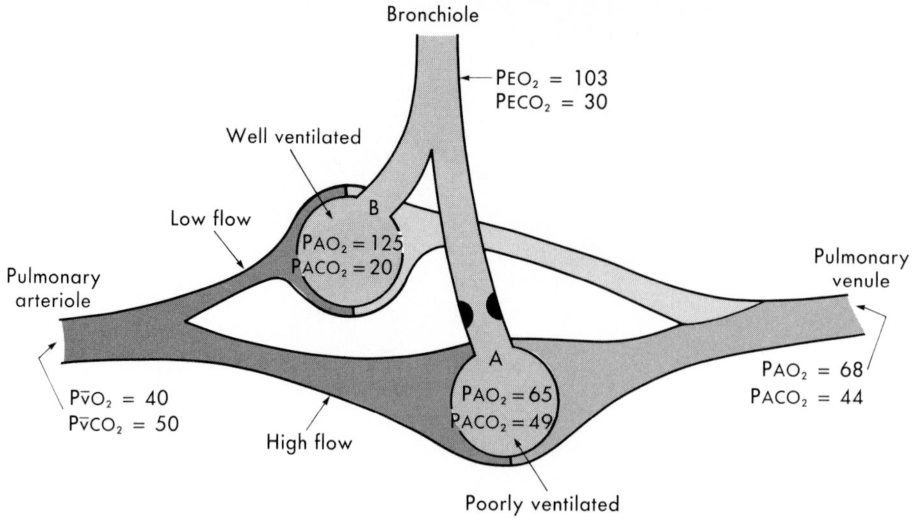

Fig. 39-16. Effect of ventilation-perfusion mismatch on the gas tensions in systemic arterial (or pulmonary venous) blood and in mixed expired alveolar gas. Blood flow to alveolus *A* (a low $\dot{V}_A/\dot{Q}$ unit) is twice the flow to alveolus *B* (a high $\dot{V}_A/\dot{Q}$ unit). Ventilation of alveolus *A*, on the other hand, is half that of alveolus *B*. Note that the P_{O_2} and P_{CO_2} of mixed expired alveolar gas (P_{EO_2}, P_{ECO_2}) more closely resemble the values in alveolus *B*, whereas the systemic arterial P_{O_2} (Pa_{O_2}) more closely resembles the value (P_{AO_2}) in alveolus *A*. Note that partial pressure gradients exist between mixed expired alveolar gas and systemic arterial blood for both O_2 and CO_2.

arterial CO_2 tensions is normally almost indiscernible.

Ventilation-perfusion inequalities interfere with the efficiency of gas exchange. For the same level of ventilation, arterial P_{CO_2} is higher and arterial P_{O_2} is lower the greater the spread of $\dot{V}_A/\dot{Q}$ ratios.

The lowering of arterial P_{O_2} produced by low $\dot{V}_A/\dot{Q}$ alveoli cannot be overcome by increasing ventilation to normal regions of the lung. Low $\dot{V}_A/\dot{Q}$ regions are the most common cause of hypoxemia in lung disease. Elevations in arterial P_{CO_2} caused by $\dot{V}_A/\dot{Q}$ abnormalities give rise, by stimulating chemoreceptors, to a compensatory increase in ventilation and hence to a fall in alveolar P_{CO_2}. Because the CO_2 dissociation curve has no plateau, increased ventilation can restore arterial P_{CO_2} to nearly normal levels. Ultimately, when $\dot{V}_A/\dot{Q}$ abnormalities become very severe, ventilation may not increase sufficiently to prevent chronic elevation of arterial P_{CO_2}.

The ventilation/perfusion ratio of an alveolus can be calculated if the alveolar P_{CO_2}, alveolar P_{O_2}, mixed venous P_{O_2}, and respiratory exchange ratio, R, are known:

$$\dot{V}_A/\dot{Q} \cong R \times \frac{(P_{AO_2} - P_{\bar{v}O_2})}{P_{ACO_2}} \qquad (22)$$

All the possible values for alveolar P_{CO_2} and P_{O_2} that occur in lung alveoli with different $\dot{V}_A/\dot{Q}$ ratios can be incorporated in an O_2/CO_2 diagram (Fig. 39-17). It is evident from the shape of the $\dot{V}_A/\dot{Q}$ curve in Fig. 39-17 that low $\dot{V}_A/\dot{Q}$ ratios depress alveolar P_{O_2} more than they raise arterial P_{CO_2}, whereas high $\dot{V}_A/\dot{Q}$ ratios lower alveolar P_{CO_2} proportionally more than they increase alveolar P_{O_2}.

A wide range of ventilation/perfusion ratios in the lung produces a disparity between mixed alveolar P_{O_2} and the P_{O_2} of the mixed pulmonary venous and systemic arterial blood. Arterial P_{O_2} can be further decreased by systemic venous blood, which, for anatomical reasons, fails to come into contact with alveoli but then mixes with blood in the pulmonary veins.

Normally about 5% of the blood ejected by the left ventricle has not been oxygenated. Flow of blood from bronchial veins, which empty into pulmonary veins, and blood from thebesian veins of the left ventricular myocardium, which empty into the left ventricle, constitutes a normal right-to-left shunt of underoxygenated blood. In some congenital heart diseases, blood may be shunted directly from the right to the left side of the heart through intracardiac defects or through pulmonary arteriovenous communications.

In normal subjects the P_{O_2} gradient between alveolar gas and arterial blood is 5 to 10 mm Hg. This gradient results mainly from anatomical shunting of mixed venous blood into the pulmonary veins, with failure of the blood to come in contact with alveolar gas. This effect is functionally equivalent to having alveoli with a $\dot{V}_A/\dot{Q}$ ratio of zero.

The fraction of the cardiac output that fails to reach gas-exchanging surfaces in the lung can be calculated

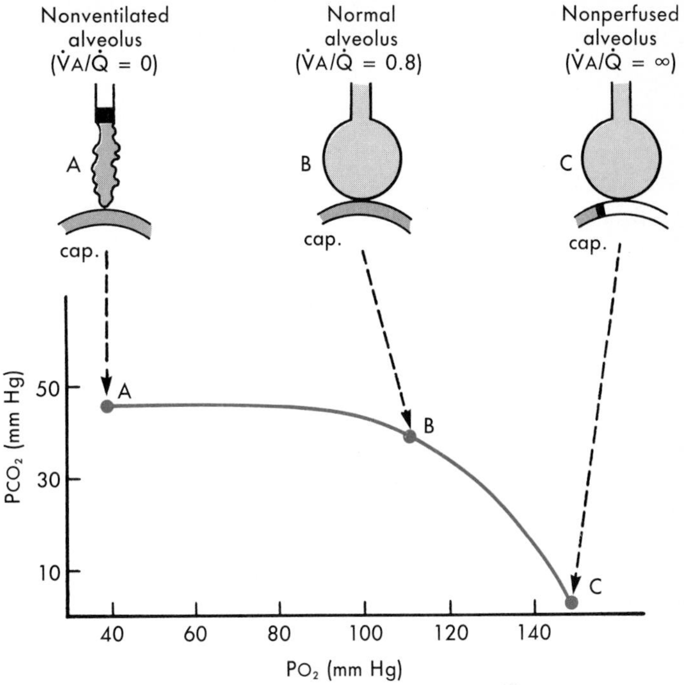

■ **Fig. 39-17.** Effects of variations in the $\dot{V}A/\dot{Q}$ ratio on the PO_2 and PCO_2 of the alveolar gas (and pulmonary capillary blood) in a nonventilated alveolus *(A)*, a normal alveolus *(B)*, and a nonperfused alveolus *(C)*. The gas tensions in the nonventilated alveolus *(A)* will equal that of mixed systemic venous blood (PO_2 = 40, PCO_2 = 46). In the normal alveolus *(B)*, PO_2 will equal 110 and PCO_2 will equal 40. The gas tensions in the nonperfused alveolus *(C)* will equal that of inspired air (PO_2 = 150, PCO_2 = 0).

from the *shunt equation*. The total cardiac output ($\dot{Q}T$) consists of the shunt flow ($\dot{Q}S$) plus the blood flow past the ventilated alveoli ($\dot{Q}T - \dot{Q}S$). The amount of O_2 in the blood pumped each minute by the left ventricle is the product of the cardiac output and the O_2 content of the arterial blood (CaO_2). This total quantity of O_2 is made up of the contribution from the shunted blood, the O_2 content of which is the same as that of mixed systemic venous blood ($C\bar{v}O_2$), and the contribution from the blood that has perfused ventilated alveoli ($\dot{Q}T - \dot{Q}S$) and the O_2 content of the end-capillary blood (CcO_2). This can be expressed as follows:

$$\dot{Q}T \times CaO_2 = (\dot{Q}S \times C\bar{v}O_2) + (\dot{Q}T - \dot{Q}S) \times CcO_2 \quad (23)$$

By rearranging this equation,

$$\dot{Q}S/\dot{Q}T = \frac{CcO_2 - CaO_2}{CcO_2 - C\bar{v}O_2} \quad (24)$$

The O_2 content of the arterial and mixed venous blood can be measured directly. The O_2 content of the end-capillary blood must be estimated by measuring the mixed alveolar PO_2 and considering it to equal the end-capillary PO_2. If there are many low $\dot{V}A/\dot{Q}$ alveoli in the lung, this assumption will be incorrect. Therefore the usual procedure is to have the subject breathe 100% O_2. After a sufficient period of O_2 breathing, nitrogen is washed out of the lung, and even the most poorly ventilated alveoli will contain only O_2 and CO_2. The end-capillary PO_2 of even poorly ventilated lung units will be the same as the PO_2 in the alveolar gas of those units.

■ *Diffusion of Gases in the Lung*

■ *Diffusion in the Alveolus*

In the gas-exchanging units of the lung, diffusion is the principal process that accounts for the transfer of O_2 and CO_2 within the alveoli and from the alveoli, through the intervening interstitial tissue and into the capillary blood. The rates of diffusion of molecules in a gaseous phase are inversely proportional to the square root of their densities (Graham's law). Therefore the diffusions of two gases are not much different unless the disparities in molecular weights are substantial. For O_2 (molecular weight, 32) and CO_2 (molecular weight, 44) the relative rates of diffusion in the alveolus are not too different.

$$\frac{O_2 \text{ rate}}{CO_2 \text{ rate}} = \frac{\sqrt{44}}{\sqrt{32}} = \frac{6.6}{5.7} \quad (25)$$

In normal alveoli (diameter, approximately 20 μm) CO_2 and O_2 appear to mix evenly in less than 10 msec. Consequently, diffusion in the alveolus itself does not limit gas exchange. In severe emphysema, a disease in which alveolar walls are destroyed, the air sacs may become very large. Under such conditions, gas tensions within the terminal lung units are not uniform and gas exchange is impaired.

Diffusion Through the Alveolar-Capillary Membrane

Gas transfer through tissue sheets, such as the alveolar-capillary membrane, into a liquid medium, such as the blood, depends on the gas tension gradient, the solubility of the gas in the liquid, the molecular weight of the gas, and the properties of the membrane. This relationship is expressed by Fick's law (Chapter 1):

$$J = -DA\frac{\Delta c}{\Delta x} \quad (26)$$

where

J = Flux of the gas across the membrane
D = Diffusion coefficient of the membrane
A = Area of the membrane
Δx = Thickness of the membrane
Δc = Concentration gradient across the membrane

According to Henry's law:

$$\Delta c = \alpha \Delta P \quad (27)$$

where

α = Solubility of the gas in the tissue
ΔP = Gas tension gradient across the membrane

Therefore

$$J = -DA\alpha\frac{\Delta P}{\Delta x} \quad (28)$$

The alveolar capillary membrane averages about 0.5 μm in thickness (Fig. 39-18) and consists of the following layers from the alveolus to the capillary side: surfactant, alveolar epithelium, its basement membrane, a connective tissue layer, and the basement membrane of the capillary and its endothelium.

Capillaries are situated in the lung interstitium between alveoli. One side of each capillary is thin and the other thick. On the thin side, the alveolar and capillary basement membranes are fused, and no connective tissue layer intervenes. Gas exchange probably takes place mainly through this thin side. The thick side seems to be important in fluid exchange (Chapter 38). The area of the blood-gas barrier in the lung is enormous, 50 to 100 m^2. Over 80% of this area is covered by a thin layer of blood. The pulmonary capillaries and associated alveolar-capillary membranes constitute a thin layer of blood with interposed connective tissue struts that separate two alveolar sheets.

Although the potential area for gas exchange is large, the effective surface may be much more limited, particularly in the presence of lung disease. Areas of the lung that are neither ventilated nor perfused will not participate in gas exchange. As a consequence, diseases (e.g., emphysema) that affect the distribution of inspired air or of pulmonary blood flow may influence gas diffusion.

In a resting individual the red blood cells remain within the pulmonary capillary for only about 0.75 second, which imposes a limit on the time available for gas transfer. Depending on the specific gas in the alveoli, transfer of the gas from alveoli to blood may be limited by the rates of perfusion or diffusion (Fig. 39-19). Transfer of a gas is said to be *diffusion-limited*

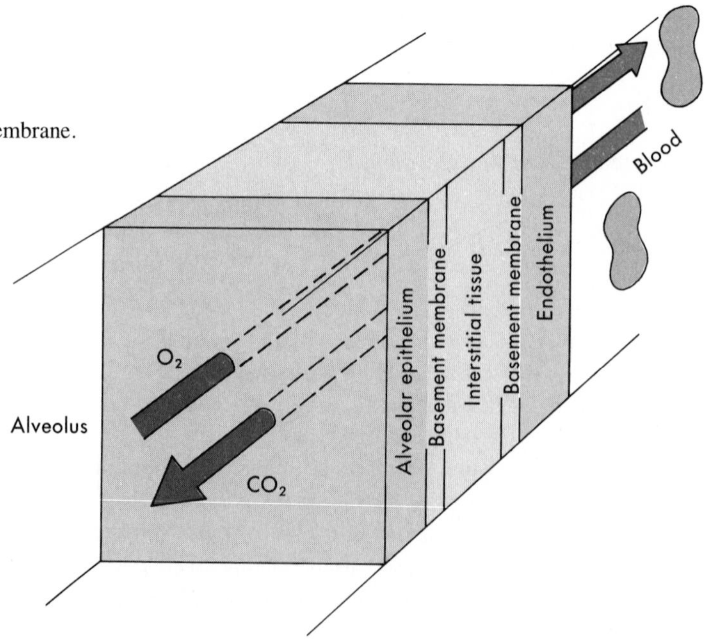

Fig. 39-18. Schematic diagram of the alveolar-capillary membrane.

when the gas tensions in the alveoli and the capillary fail to reach equilibrium during the time of red cell transit through the capillary. For example, the transfer of CO is diffusion-limited. CO moves rapidly across the extremely thin blood-gas barrier from the alveolar gas into the red blood cell. However, because of the high affinity of CO for hemoglobin in the red blood cell, a large amount of CO can be taken up by the cell, but its partial pressure in the blood scarcely increases. Thus no appreciable back pressure of CO develops because of its chemical combination with hemoglobin. Gas continues to cross the alveolar wall rapidly, and equilibration is not achieved even if flow through the capillaries is very low. Therefore the amount of CO that enters the blood is limited by the diffusion properties of the blood-gas barrier and not by the rate of perfusion.

On the other hand, the transfer of an inert gas, such as nitrous oxide, is *perfusion-limited*. When nitrous oxide moves across the alveolar wall, it dissolves in the blood, but it does not combine with any substance in the blood. As a result, the partial pressure of nitrous oxide in the blood increases rapidly. Hence, by the time the red cell is only one tenth of the way along the capillary, the partial pressures of nitrous oxide on both sides of the membrane are the same, preventing further gas transfer. Continued transfer of nitrous oxide can occur only if new blood is supplied. Thus the amount of nitrous oxide taken up by the blood depends almost entirely on the rate of perfusion.

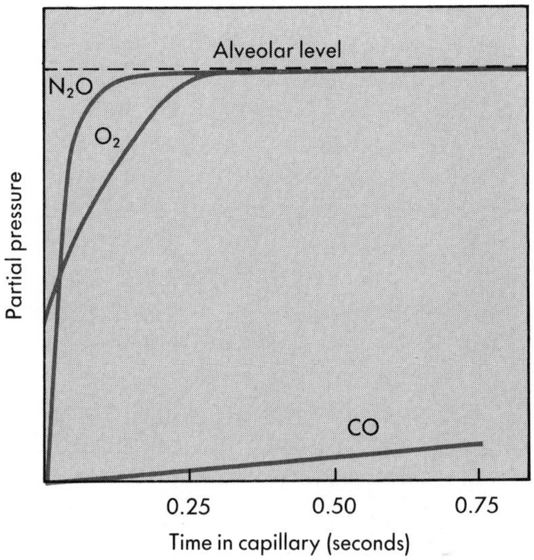

Fig. 39-19. Changes in the partial pressure of carbon monoxide (CO), nitrous oxide (N_2O), and oxygen (O_2) in blood during transit through the pulmonary capillary bed. During transit through the pulmonary capillaries, blood P_{CO} increases only slightly. P_{N_2O} of blood reaches equilibrium with that in the alveolar gas very rapidly. The P_{O_2} of blood and alveolar gas become virtually the same after the blood has travelled only about one third of the way along the capillary bed.

In healthy persons breathing quietly, the transfer of O_2 is mainly perfusion-limited. O_2, like CO, combines with the hemoglobin, but it does so less avidly. Therefore, when a given volume of O_2 enters a red blood cell, the rise in partial pressure is much greater than when the same volume of CO is transferred. Under resting conditions, the capillary P_{O_2} is nearly the same as the alveolar P_{O_2} by the time the red cell is about one third of the way along the capillary (Fig. 39-20, *A*).

When cardiac output is increased during exercise, the time spent by the red blood cell in the capillary is reduced to one third of normal (approximately 0.25 second). Even under these conditions, the P_{O_2} in the blood will become equal to that in the alveoli before the end of the capillary is reached.

Breathing a hypoxic gas mixture at sea level or ambient air at altitude will decrease alveolar P_{O_2} and reduce the partial pressure difference that drives O_2 across the alveolar-capillary membrane. Hence, P_{O_2} will increase more slowly in the capillary blood. Also, because of the shape of the HbO_2 dissociation curve, a given increase in O_2 content increases the blood P_{O_2}

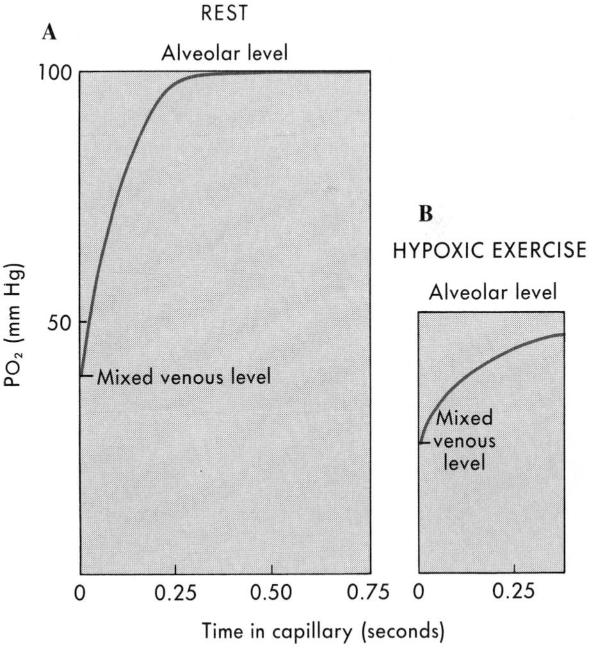

Fig. 39-20. Change in partial pressure of O_2 in blood during transit through the pulmonary capillary bed. **A,** At rest the transit time of blood through the pulmonary capillary bed is about 0.75 second. The P_{O_2} gradient between alveolar gas and mixed venous blood is large. Diffusion is rapid, and O_2 equilibrates early. **B,** Exercise increases blood flow and shortens the transit time of blood through the capillary bed. When a hypoxic gas mixture is breathed, the alveolar-mixed venous P_{O_2} difference is smaller and hence diffusion is slower. Under these conditions blood and alveolar P_{O_2} may not reach equilibrium by the end of the capillary.

less when the Po_2 is relatively low than when the Po_2 is relatively high. For both these reasons, hypoxia decreases the likelihood that blood Po_2 and alveolar Po_2 will reach equilibrium by the end of the capillary. But in healthy subjects, a limitation of O_2 transfer by diffusion probably only occurs when exhausting exercise is performed under hypoxic conditions (Fig. 39-20, *B*). In some lung diseases (e.g., pulmonary fibrosis), the pulmonary interstitium may be so thickened that diffusion limits the transfer of O_2, even under less severe conditions.

Although CO_2 has a greater molecular weight than O_2, it is much more soluble in water. Therefore, for a given partial pressure difference, the flux of CO_2 in water is 20 times greater than that of O_2. However, because so much more CO_2 than O_2 is dissolved in blood, it takes nearly as long for Pco_2 as for Po_2 to reach equilibrium between the capillary blood and the alveolus. This means that any abnormality in diffusion that affects O_2 equilibration may also affect the equilibration of CO_2. In addition, the chemical reactions between O_2 and hemoglobin and those involved in CO_2 transfer (the carbamino reaction and CO_2 buffering in the red cell) take time, which could affect the equilibration time in the capillary. Nonetheless, diffusion abnormalities rarely cause CO_2 retention. When arterial Pco_2 rises, ventilation increases, and restores the normal level of arterial Pco_2.

Measurement of Diffusing Capacity of the Lung

The amount of gas transferred across the alveolar-capillary membrane is directly proportional to the area of the membrane and is inversely proportional to the thickness of the membrane. It is not possible to measure during life either the total area (A) or thickness of the alveolar-capillary membrane (Δx). Instead the membrane characteristics can be estimated from a simplified version of equation 28:

$$D_L = \frac{DA\alpha}{\Delta x} = -\frac{J}{\Delta P} = \frac{J}{P_A - P_C} \qquad (29)$$

where

D_L = Diffusing capacity of the lung
J = Volume of gas transferred per unit of time
P_A = Alveolar tension of the gas
P_C = Capillary tension of the gas

D_L includes area, thickness, and diffusion properties of the lung.

The adaptation of equation 29 for the diffusing capacity of the lung for oxygen is

$$D_{L}O_2 = \frac{\dot{V}O_2}{P_{A}O_2 - P_{C}O_2} \qquad (30)$$

where

D_LO_2 = Diffusing capacity in ml O_2/minute/mm Hg Po_2
$\dot{V}O_2$ = Volume of O_2 (in ml) transferred per minute from alveolar gas to the blood
P_AO_2 = Mean alveolar Po_2 in mm Hg
P_CO_2 = Mean pulmonary capillary Po_2 in mm Hg

A decrease in D_LO_2 may be the result of a decreased area or an increased thickness of the alveolar-capillary membrane, or both. D_LO_2 may be unaffected when the thickness is increased if the surface area available for diffusion is simultaneously increased.

O_2 diffusing capacity is difficult to determine because (a) the uptake of O_2 depends mainly on blood flow and (b) the instantaneous partial pressure gradient of O_2 from the alveolus to the capillary blood cannot be measured. It is much simpler to use CO than O_2 to measure the diffusing capacity.

The diffusing capacity for CO (D_LCO) is given by

$$D_LCO = \frac{\dot{V}CO}{P_ACO - P_CCO} \qquad (31)$$

where

P_ACO = Partial pressure of CO in alveolar gas
P_CCO = Partial pressure of CO in capillary blood
$\dot{V}CO$ = Volume of CO transferred (ml/min)

Except in habitual cigarette smokers, the partial pressure of CO in capillary blood is so small that it can be neglected. The equation then becomes

$$D_LCO = \frac{\dot{V}CO}{P_ACO} \qquad (32)$$

The binding of O_2 or CO to hemoglobin is not instantaneous. The time required for the chemical reactions with hemoglobin appreciably delays the uptake of O_2 and CO by the red cell. Hence, the transfer of O_2 (or CO) across the alveolar-capillary membrane can be considered as occurring in two stages: (1) diffusion through the blood-gas barrier (including the plasma and red cell interior) and (2) reaction of the gas with hemoglobin. Each stage offers a resistance to gas uptake by the blood, and these resistances can be summed to produce an overall "diffusion" resistance. The reciprocal of D_L is the overall resistance to uptake of gas by diffusion. The resistance of the blood-gas barrier to diffusion can be considered to be $1/D_M$, where M signifies membrane. The reciprocal of the product of the rate of reaction (θ) of O_2 (or CO) with hemoglobin in milliliters of gas/minute/mm Hg/milliliters of blood and the volume of capillary blood (V_C), i.e., $1/(\theta \times V_C)$, describes the "resistance" of this chemical reaction to the uptake of gas. The two resistances to uptake offered by diffusion across the membrane and the uptake of O_2 or CO by hemoglobin are added to obtain the total diffusion resistance. Thus the complete expression is

$$\frac{1}{D_L} = \frac{1}{D_M} + \frac{1}{\theta \times V_C} \qquad (33)$$

The resistances offered by the membrane and blood components are approximately equal. The equation indicates that a reduction of capillary blood volume or a reduction in the hemoglobin content of the blood lowers the diffusing capacity.

Normal values for $D_{L}CO$ are 25 ml/minute/mm Hg in resting subjects. $D_{L}CO$ rises during exercise when pulmonary blood flow and pressure increase, probably because the area available for diffusion increases as more pulmonary capillaries open.

CO diffusing capacity also varies with body size, probably because of differences in the area available for gas exchange. The diffusing capacity is about 25% greater in the supine than in the standing position, because more capillaries are open in the upper lobe when the subject is supine. The CO diffusing capacity also depends on the PO_2. When the arterial PO_2 is reduced, the $D_{L}CO$ is increased because of changes in the rate of reaction of CO with hemoglobin.

In smokers a considerable portion of the hemoglobin may be combined with CO (as much as 12%); hence, the back pressure of CO in the blood is appreciable. The values obtained for diffusing capacity for CO in smokers should be corrected by measurements of actual levels of HbCO.

Effects of Ventilation on Acid-Base Balance

The pH of blood and H^+ homeostasis is regulated by the HCO_3^-/H_2CO_3 buffer system. According to the Henderson-Hasselbalch equation:

$$pH = pK + \log\frac{[HCO_3^-]}{[H_2CO_3]} \qquad (34)$$

It is important to note that H_2CO_3 is in equilibrium with dissolved CO_2:

$$CO_2 + H_2O \rightleftarrows H_2CO_3 \qquad (35)$$

The equilibrium position of this reaction is far to the left; that is, for every molecule of H_2CO_3, there are about 500 molecules of CO_2 in solution. The denominator in the Henderson-Hasselbalch equation is in fact the sum of the dissolved CO_2 and H_2CO_3, but this H_2CO_3 pool is overwhelmingly made up of dissolved CO_2. According to Henry's law, the dissolved CO_2 is proportional to the PCO_2 so that the Henderson-Hasselbalch equation can be rewritten:

$$pH = pK + \log\frac{HCO_3^-}{0.03\ PCO_2} \qquad (36)$$

where 0.03 is the solubility constant.

The HCO_3^- concentration in the blood is regulated chiefly by the kidneys, and the blood PCO_2 is regulated by the lungs.

Blood buffers other than the HCO_3^-/H_2CO_3 system prevent wide swings in blood pH when PCO_2 changes. These blood buffers include the $HPO_4^-/H_2PO_4^-$ system and numerous proteins, particularly hemoglobin, as explained in Chapter 48. The effects of various respiratory and metabolic disturbances on acid-base balance are also discussed in Chapter 48.

Bibliography

Journal articles

Allen, C.J., et al.: Alveolar gas exchange during exercise: a single-breath analysis, J. Appl. Physiol. **57:**1704, 1984.

Clark, J.S., et al.: Gas-blood PCO_2 and PO_2 equilibration in a steady-state rebreathing dog preparation, J. Appl. Physiol. **56:**1229, 1984.

Forster, R.E., and Crandall, E.D.: Pulmonary gas exchange, Annu. Rev. Physiol. **38:**69, 1976.

Heaf, D.P., et al.: Postural effects on gas exchange in infants, N. Engl. J. Med. **308:**1505, 1983.

Hornstein, D., and Ledsome, J.R.: Alveolar gas pressure changes with resistive loads in man, Respiration **45:**347, 1984.

Paiva, M., and Engel, L.A.: Model analysis of intra-acinar gas exchange, Respir. Physiol. **62:**257, 1985.

Riley, R.L., and Permutt, S.: Venous mixture component of the A-aPO_2 gradient, J. Appl. Physiol. **35:**430, 1973.

Thews, G.: Theoretical analysis of the pulmonary gas exchange at rest and during exercise, Int. J. Sports Med. **5:**113, 1984.

Werlen, C., et al.: Alveolar-arterial equilibration in the lung of sheep, Respir. Physiol. **55:**205, 1984.

West, J.B.: Ventilation-perfusion relationships, Am. Rev. Respir. Dis. **116:**919, 1977.

West, J.B., et al.: Pulmonary gas exchange on the summit of Mount Everest, J. Appl. Physiol. **55:**678, 1983.

Books and monographs

Filley, G.F.: Acid-base and blood gas regulation, Philadelphia, 1971, Lea & Febiger.

Rahn, H., and Fenn, W.O.: A graphical analysis of the respiratory gas exchange: the O_2-CO_2 diagram, Bethesda, Md., 1955, American Physiological Society.

West, J.B., editor: Pulmonary gas exchange. Vol. I. Ventilation, blood flow, and diffusion, New York, 1980, Academic Press.

West, J.B., editor: Pulmonary gas exchange. Vol. II. Organism and environment, New York, 1980, Academic Press.

CHAPTER

40

Control of Respiration

The effects of breathing are continuously monitored by *chemoreceptors,* which detect changes in P_{CO_2} and P_{O_2}, and by *mechanoreceptors,* which survey changes in thoracic mechanics. The activity of these receptors allows ventilation to be adjusted automatically so that arterial blood gases are kept within acceptable limits despite changing internal and external conditions.

Ventilation is also affected by signals from thermal receptors, vascular baroreceptors, and other mechanoreceptors situated throughout the body and by the activity of higher brain centers. The multiplicity of inputs to the respiratory neurons permits satisfactory levels of ventilation to be maintained even when disease affects certain afferent pathways or when the response to some sensory cue is blunted by a depressed state of consciousness.

Respiratory motor activity is transmitted from the brain along the phrenic nerve to the diaphragm, along the tenth nerve to the laryngeal muscles, and along spinal nerves to the intercostal and abdominal muscles. Certain accessory muscles, such as the sternocleidomastoid, and muscles in the upper airways, such as the tongue and the dilator muscles of the nose, may be activated, particularly when respiration is stimulated. Although the actual movement of some of the upper airway muscles with respiration may be small, their activation affects the resistance to airflow through the upper airway passages and maintains upper airway patency. Finally, rhythmic changes in the parasympathetic nerves with respiration may alter tracheal tone, tracheal resistance, and anatomical dead space.

The nerve pathways that connect the sensory input from chemoreceptors and mechanoreceptors to the respiratory motor neurons are complex and involve one or more interneurons. This allows flexibility in respiratory responses.

Central Organization of Respiratory Neurons

Supraspinal Organization

Various ablation, stimulation, and recording techniques have been used to explore the brain and to locate respiratory areas. Two types of areas have been found: (1) primary respiratory areas whose major function is respiration and (2) subsidiary respiratory areas that are concerned primarily with other functions but that project to the primary respiratory areas and modify respiration.

At least two regions in the brainstem function as intrinsic respiratory areas: (1) the medullary respiratory area and (2) the pneumotaxic center in the pons. Subsidiary respiratory areas of the brain, with projections to one or more of the primary centers, are located in the cortex (which provides volitional inputs to the respiratory centers), the cerebellum, and the medullary vasomotor areas.

The behavior of the medullary neurons can be strongly modified by influences arising in the brain rostral to the medulla. Nevertheless, the pattern of breathing is virtually normal even when the medulla is separated from the rest of the brain.

The precise anatomical and functional organization of the respiratory medullary neurons is under investigation. Substantial evidence indicates that two separate neuronal networks in the medulla are crucial (Fig. 40-1). These networks include neuronal groups in the nucleus tractus solitarius (NTS), located dorsally near the exit of the ninth cranial nerve, and a ventrally located group of neurons that extends rostrally from the first cervical spinal segment to the caudal border of the pons.

The dorsal group of neurons discharges mainly in inspiration. The ventral group (the nucleus retroambiguus, paraambiguus, and retrofacialis) contains both inspiratory and expiratory neurons. Neurons from both

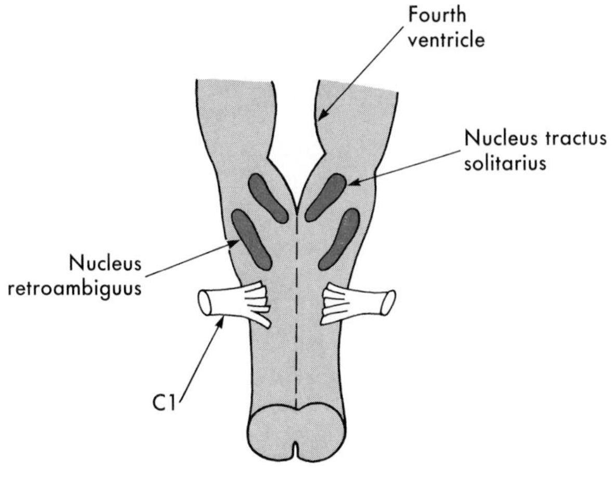

■ Fig. 40-1. Brainstem respiratory neuronal groups.

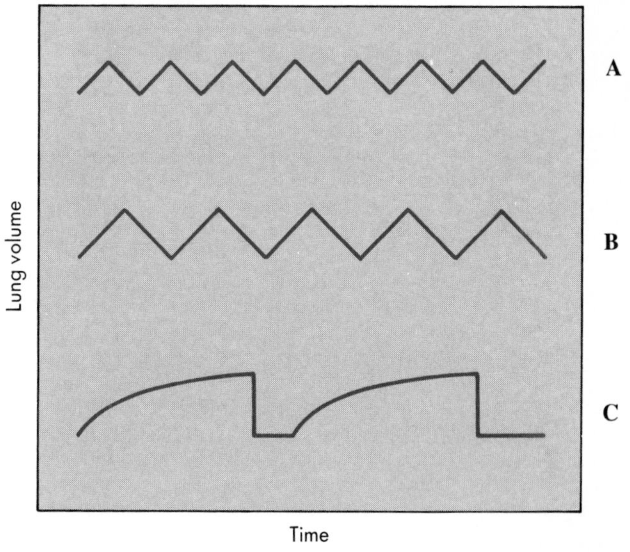

■ Fig. 40-2. Patterns of breathing. **A,** Normal breathing. **B,** Prolonged duration of inspiration and expiration caused by vagal section. **C,** Apneustic breathing characterized by deep and prolonged inspiration produced by lesions in the rostral pons.

the ventral and dorsal groups direct the respiratory activity of the muscles of the chest wall and abdomen.

Efferent fibers from the dorsal and ventral group decussate at the level of the obex, and they travel contralaterally in the ventrolateral portion of the spinal cord to reach the spinal motor neurons. The fiber tracts to the expiratory and inspiratory neurons separate within the ventrolateral white matter, where the fibers to the expiratory muscles lie ventral and medial to the fibers to the inspiratory muscles. Afferent fibers from the peripheral chemoreceptors, baroreceptors, and pulmonary mechanoreceptors synapse in the NTS.

The mechanism by which these networks of medullary neurons cause switching between inspiration and expiration is unclear. Studies to date have failed to demonstrate pacemaker cells, that is, neurons that demonstrate oscillation of membrane potential in the absence of excitatory and inhibitory inputs. Most of the evidence supports the concepts that rhythmic respiratory activity requires continuous (tonic) or intermittent (phasic) excitatory inputs and that respiratory oscillations are the result of reciprocal inhibition of interconnected neuronal networks.

Studies of anesthetized animals indicate that thoracic mechanoreceptors, chiefly stretch receptors in the walls of the airways, are critical in setting the breathing frequency. Interruption of pulmonary stretch receptor impulses by section of the vagus nerves markedly prolongs the duration of inspiration and expiration (Fig. 40-2). Inputs from pulmonary stretch receptors have less influence in conscious than in anesthetized animals, and their effects can be abolished by visual or auditory stimulation.

Networks of neurons in the pons influence the switching between inspiration and expiration. When lesions are made in the pneumotaxic center in the rostral pons, inspiration becomes even more prolonged, and *apneusis* ensues (i.e., deep and prolonged inspirations that last tens of seconds).

Afferent impulses from *lung stretch receptors* traveling in the vagi affect the duration of inspiration and expiration. Stretch receptors are stimulated by lung inflation. During spontaneous breathing, lung inflation occurs during inspiration. By artificially inflating the lung at different times in the respiratory cycle, the effect of lung volume changes on the timing of inspiration and expiration can be demonstrated. In anesthetized animals with intact vagi, artificial *inflation of the lung* during the *inspiratory phase* of phrenic nerve activity can terminate inspiration. The effect of lung inflation on the respiratory pattern depends on the volume of gas introduced (Fig. 40-3). Early in inspiration, large volume changes are needed to terminate inspiration, whereas late in inspiration, smaller volumes are needed. After vagotomy, lung inflation has no effect on the duration of inspiration. Neurons whose firing is excited by lung inflation have been found near the NTS.

The regulation of time in expiration (Te) is less well understood. In either spontaneously breathing animals or animals in which inspiratory time (Ti) is manipulated experimentally by lung inflation, Te and Ti are closely and directly correlated. As one time becomes shorter, so does the other. Hence, those mechanisms that determine Ti probably also influence Te. Other factors also operate, however, because lung inflation during the first 75% of expiration prolongs Te without affecting the subsequent Ti.

These reflex effects on respiratory timing may be beneficial. During spontaneous breathing, termination of inspiration in response to large lung inflations prevents overinflation of the lung. By shortening inspira-

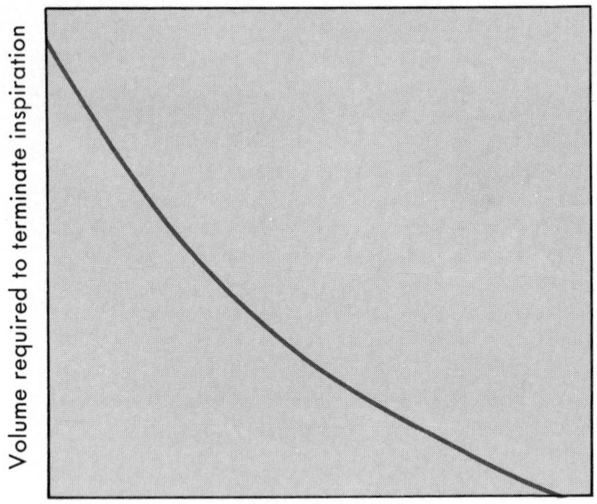

■ Fig. 40-3. In anesthetized animals with intact vagi, lung inflation during the phase of inspiration can terminate inspiration by stimulating pulmonary stretch receptors. Early in inspiration a large volume inflation is required to terminate inspiration. Late in the inspiratory phase inspiration is terminated by small inflations.

tory time, it also facilitates the increases in respiratory frequency needed for adequate increases in ventilation when breathing is stimulated (e.g., by breathing CO_2-enriched gases and exercise). When the resistance to inspiratory airflow is increased, the decreased rate of lung inflation prolongs inspiration. Hence more time is available for gas to enter the lung and a constant tidal volume can be maintained. On the other hand, increased lung volume or obstruction to airflow during expiration reflexly increases the time available for lung emptying by forestalling the next inspiratory effort.

These fundamental observations can be synthesized by an operational model of the brainstem mechanisms that generate the respiratory rhythm (Fig. 40-4). In the model, signals from the *central chemoreceptors* (in the medulla) and *peripheral chemoreceptors* (aortic and carotid bodies) impinge on a pool of inspiratory neurons (A pool), probably in the NTS. These neurons project to the respiratory spinal motor neurons and increase their activity *(central inspiratory activity)*. The time course of the increase in activity resembles that of the changes in tidal volume. Increased input from chemoreceptors (excited by hypoxia or hypercapnia) increases the neural activity. Central inspiratory activity stimulates another pool of neurons (B pool, Fig. 40–4), probably also in the NTS. In addition, the B pool receives signals from pulmonary stretch receptors. These stretch receptor signals, which increase with lung expansion, are summed with the A pool input at the B pool. The B pool, in turn, activates the C pool of ''inspiratory off-switch neurons.'' These neurons inhibit the A pool neurons. When a critical level of excitation occurs in the C pool, the activity of the A pool is extinguished and expiration occurs. During this inhibitory period, the pulmonary stretch receptors further excite the C pool neurons, thereby prolonging expiration.

An increase in chemoreceptor activity raises the ''off-switch'' threshold and enhances central inspiratory activity. Thus, tidal volume can increase even without any change in inspiratory duration (Fig. 40–4, *dashed lines*). Ablation of the pneumotaxic center in the vagotomized animal increases the ''off-switch'' threshold without affecting central inspiratory activity, further lengthening inspiration.

This model of respiratory pattern generation has been challenged. Instead, inspiratory and expiratory neurons may directly inhibit each other's activities. Nonetheless the aforementioned ideas may help the reader to integrate the effects of chemical and reflex stimuli on neuronal activity and to relate them to changes in tidal volume and frequency.

Higher brain centers modulate the function of the respiratory controller. Projections from the cerebral motor cortex subserve the volitional control (speaking, breath-holding, and voluntary hyperventilation) of the respiratory system. These cortical fibers descend to the brainstem respiratory neurons via the corticobulbar tracts and to the spinal motor neurons via the corticospinal tracts, which are located in the dorsolateral columns of the cord. Discrete lesions in the spinal cord may eliminate rhythmic activity in the respiratory muscles, even though these muscles still can be contracted voluntarily.

■ *Spinal Integration*

Intracellular recordings from intercostal motor neurons in the ventral horn of the spinal cord reveal the presence of descending excitatory and inhibitory activity. Descending respiratory impulses from the brain reach various segments of the spinal cord, where they are integrated with intrasegmental and intersegmental information. This descending neural activity alternately depolarizes and hyperpolarizes the motor cells, leading to rhythmic increases and decreases of their excitability.

Excitation and inhibition of respiratory muscles involve segmental interneuronal networks and descending influences. Inhibition of antagonist muscles takes place through interneurons that connect the motor neurons of inspiratory and expiratory muscles. Excitatory and inhibitory intersegmental reflexes that modulate intercostal and phrenic motor neuron output have been demonstrated in anesthetized animals and conscious humans. For example, stretch of intercostal muscles or electrical stimulation of dorsal roots in the lowermost thoracic segments (T_9 to T_{12}) reflexly excites intercostal and phrenic motor neurons. In contrast, similar stimuli to more rostral segments (T_1 to T_8) inhibit phrenic motor neuron activity and terminate inspiration prematurely by means of a supraspinal reflex arc.

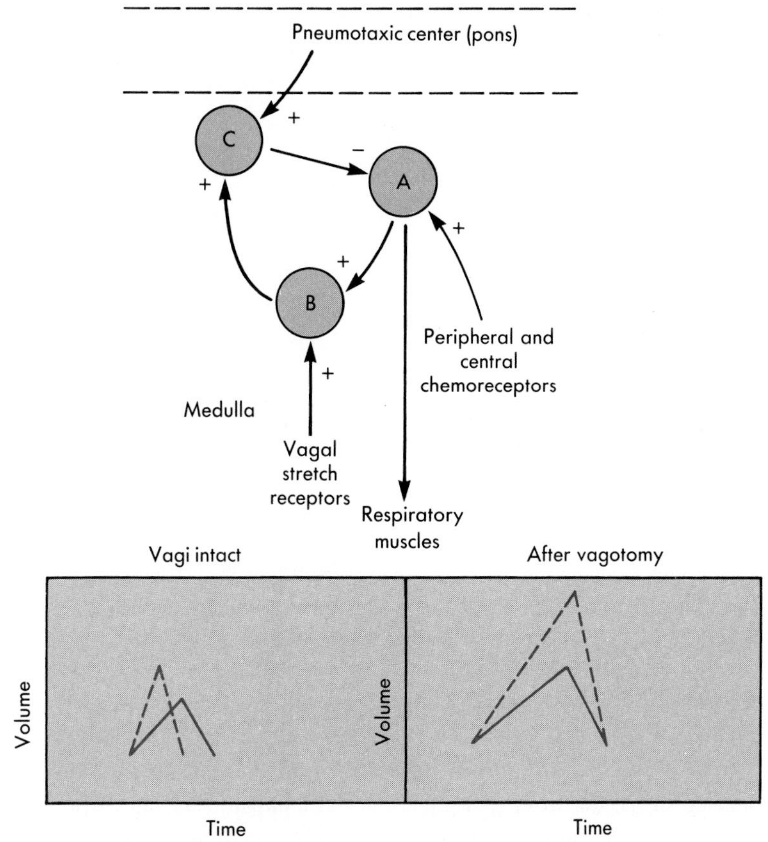

■ **Fig. 40-4.** Brainstem mechanisms that regulate respiratory rhythm (see text).

Spinal reflexes are important because they stabilize the rib cage and prevent its collapse when intrathoracic pressure is markedly subatmospheric. They also augment muscle force when respiratory resistance is increased or compliance is decreased. Intercostal to phrenic reflexes, which cut short inspiratory efforts, may be beneficial in the newborn. The rib cage of the newborn is very compliant. Hence the intercostal spaces retract when subatmospheric pressures are exaggerated in the thorax by forceful diaphragm contractions. The intercostal retraction reduces tidal volume; the inhibitory reflex abbreviates the ineffectual contractions of the diaphragm.

■ Chemical Respiratory Control

■ Effects of Carbon Dioxide

When CO_2-enriched air is inspired, ventilation increases. The greater the augmentation of ventilation caused by a given increment in inspired CO_2 (i.e., the greater the sensitivity of the chemoreceptors), the less is the increase in arterial P_{CO_2}. In conscious humans the *central chemoreceptors,* located within the medulla, account for 70% to 80% of the CO_2-induced increases in ventilation. *Peripheral chemoreceptors,* located in the carotid and aortic bodies, account for the remainder. Both central and peripheral chemoreceptors respond in proportion to the level of Pa_{CO_2}. The carotid body may also be influenced by the rate of change of Pa_{CO_2}. When Pa_{CO_2} increases rapidly, the excitation of the carotid body is greater than when Pa_{CO_2} increases slowly.

In man and animals, elevations of Pa_{CO_2} over a wide range cause a virtually linear increase in ventilation (Fig. 40-5). At arterial P_{CO_2} levels above about 100 mm Hg, the ventilatory response to hypercapnia reaches a plateau and may even diminish. In anesthetized animals and man, artificial hyperventilation with room air reduces Pa_{CO_2} and may produce apnea (cessation of breathing). However, in normal awake man, hyperventilation with 100% O_2 or even room air rarely causes apnea. This has been attributed to a "wakefulness drive," which stems mainly from nonspecific environmental stimuli (e.g., noise, light, or touch) that impinge on the cerebral cortex.

Most of the excitatory effects of CO_2 on breathing are mediated by specialized chemosensitive cells, the central chemoreceptors, in the medulla. The exact location of the central chemoreceptors in the medulla is still disputed. Substantial evidence indicates that (1)

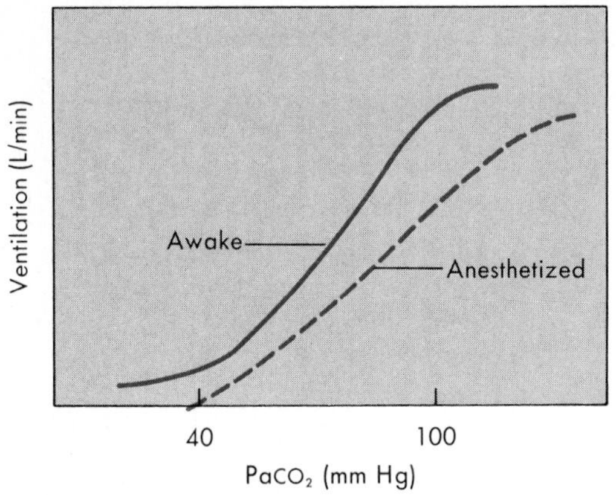

■ **Fig. 40-5.** Ventilatory responses to increases in Pa_{CO_2}. Ventilation increases linearly with Pa_{CO_2} between Pa_{CO_2} levels of 40 to 100 mm Hg. At higher values, ventilation levels off and may decrease as Pa_{CO_2} increases further. During wakefulness ventilation persists even at low Pa_{CO_2} levels. In anesthetized animals and humans apnea results as the Pa_{CO_2} is reduced below a threshold level.

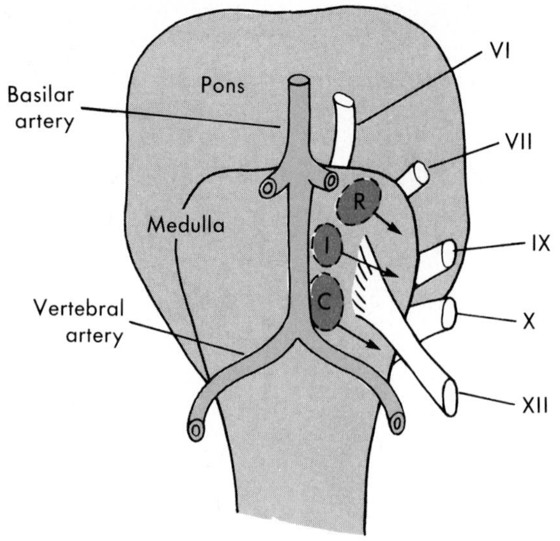

■ **Fig. 40-6.** Chemoreceptors on the ventrolateral surface of the medulla. *R, I,* and *C* are the rostral, intermediate, and caudal central chemoreceptor areas.

they are distinct from the respiratory neurons themselves, (2) they are not located in the dorsal or ventral groups of respiratory neurons described earlier, and (3) they respond to changes in the hydrogen ion (H^+) concentration of either the brain interstitial fluid or the intracellular fluid of the receptors.

Studies with drugs and temperature probes suggest that the central chemoreceptors are located near the ventrolateral surface of the medulla. Striking ventilatory effects can be produced from three superficial areas that lie between the origins of the seventh and twelfth cranial nerves (Fig. 40-6).

However, the central chemoreceptors have not been definitely identified; consequently, central chemoreceptor activity cannot be measured directly, even in animals. The function of the central chemoreceptors is assessed indirectly by enriching the content of the CO_2 in the inspired air and relating the increased ventilation or phrenic nerve activity to the level of Pa_{CO_2}. Central chemoreceptor activity is therefore inferred from the effect of a CO_2 stimulus on the output of the respiratory neurons. Ventilation reaches its steady state 10 to 20 minutes after the onset of breathing a CO_2-enriched gas mixture. When a steady state of ventilation is achieved, changes in Pa_{CO_2} are assumed to reflect changes in the interstitial hydrogen ion concentration of the brain.

Unlike ventilation, arterial P_{CO_2} achieves its steady state value within 2 to 3 minutes. The explanation for the slower changes in ventilation is not entirely clear. In response to CO_2 inhalation the H^+ concentration at the ventral surface of the medulla rises rapidly, almost as quickly as in the arterial blood. Therefore, if the central chemoreceptors are near the ventral surface, it is surprising that ventilation does not reach a steady state at a much faster rate. Appreciable time may be required before central chemoreceptor activity is translated into ventilation because the processing by the brain neuronal circuits may be slow. Whether such neuronal delays actually occur is uncertain. However, it has been shown that when electrical stimulation of peripheral chemoreceptors is abruptly stopped, ventilation declines gradually. This slow decline in ventilation has been attributed to reverberations in neuronal circuits set up by the stimulation.

The assumption that changes in Pa_{CO_2} reflect a change in brain pH, even when a steady state has been reached, may not be valid. In metabolic acidosis or alkalosis, neither changes in arterial P_{CO_2} nor in arterial pH reliably indicate changes in the acid-base status of the brain interstitial fluid. Bicarbonate (HCO_3^-) swings in the cerebrospinal fluid occur much more slowly than in the blood. As a consequence, changes in blood bicarbonate levels are not mirrored immediately in the brain fluids. The effects of chronic metabolic acidosis and alkalosis on the ventilatory response to CO_2 are shown in Fig. 40-7. At any level of P_{CO_2}, the ventilation is greater in metabolic acidosis (Fig. 40-7, left side) and less in metabolic alkalosis, probably reflecting the altered level of bicarbonate in brain fluids. If the same ventilation results are plotted, for example, as a function of the H^+ concentration in the brain interstitial fluid, the response lines become identical.

Acid injected directly into the blood immediately stimulates the peripheral chemoreceptors. Hence, ventilation increases and the Pa_{CO_2} falls. The transfer of CO_2 is faster than the transfer of H^+ or HCO_3^- between

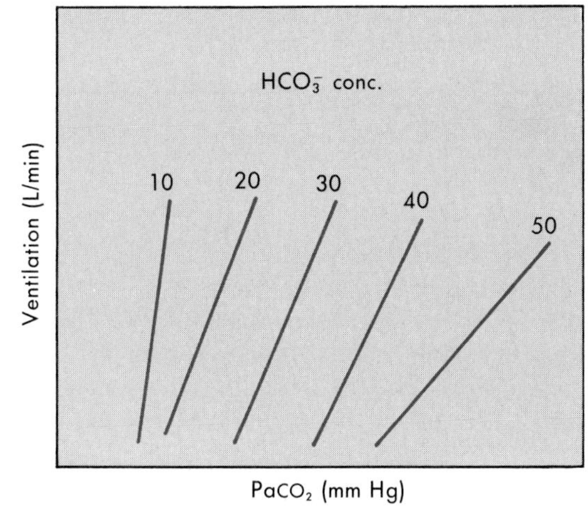

■ **Fig. 40-7.** Effects of HCO_3^- ion concentrations (mEq/L) on the ventilatory response to hypercapnia. An increase in HCO_3 (metabolic acidosis) decreases the response to Pa_{CO_2}.

the brain interstitial fluid and the blood. In the acute phase of metabolic acidosis, therefore, the brain interstitial fluid may transiently become alkaline, thereby diminishing central chemoreceptor activity. In chronic acid-base disturbances, H^+ changes in the cerebrospinal fluid usually occur in the same direction as those in the blood, but they are less pronounced. Over many hours, therefore, ventilation will steadily rise in metabolic acidosis as the brain interstitial pH decreases. The converse is true when the blood pH is acutely increased during a metabolic alkalosis.

The relationship between the arterial P_{CO_2} and the brain interstitial fluid P_{CO_2} also depends on the cerebral blood flow. The greater the cerebral blood flow, the less the difference between the arterial and interstitial fluid P_{CO_2}. High levels of cerebral blood flow wash acid out of the brain and decrease the level of brain P_{CO_2} for a given arterial value. Hence, elevations in cerebral blood flow reduce the change in ventilation produced by a given rise in arterial P_{CO_2}. Because hypercapnia increases cerebral blood flow, the effects of CO_2 on the cerebral vessels influences the relationship between ventilation and Pa_{CO_2}.

■ *Effects of Hypoxia*

Hypoxia increases ventilation primarily by exciting sensors in the carotid body (innervated by the ninth cranial nerve) and, to a lesser extent, in the aortic body (innervated by the tenth cranial nerve). The relationship between Pa_{O_2} and ventilation is hyperbolic as shown in Fig. 40-8, *A*. The nonlinear respiratory response to hypoxia is therefore different from the linear respiratory response to hypercapnia. If the carotid and aortic bodies are removed or nonfunctional, hypoxia depresses breathing because reduction in brain P_{O_2} may directly depress the respiratory cells in the brain. In addition, ventilation may be reduced because hypoxia increases brain blood flow, so that brain P_{CO_2} and H^+ concentration fall as P_{O_2} is lowered.

The carotid and aortic bodies also respond to changes in arterial P_{CO_2} and H^+ concentration. Such chemoreceptor responses are particularly important in the immediate increase in ventilation that occurs when blood pH is acutely reduced. Hypoxia accentuates the effects of hypercapnia and acidosis on peripheral chemoreceptor activity (Fig. 40-8, *B*).

Similarly, increases in Pa_{CO_2} enhance the ventilatory response to hypoxia (Fig. 40-8, *A*). This synergism occurs both at the level of the carotid body itself and probably in the central nervous system as well, where the inputs from the central and peripheral chemoreceptors converge. If isocapnic conditions are not maintained and the Pa_{CO_2} is allowed to fall when hypoxic gases are inhaled, the response to hypoxia is markedly attenuated.

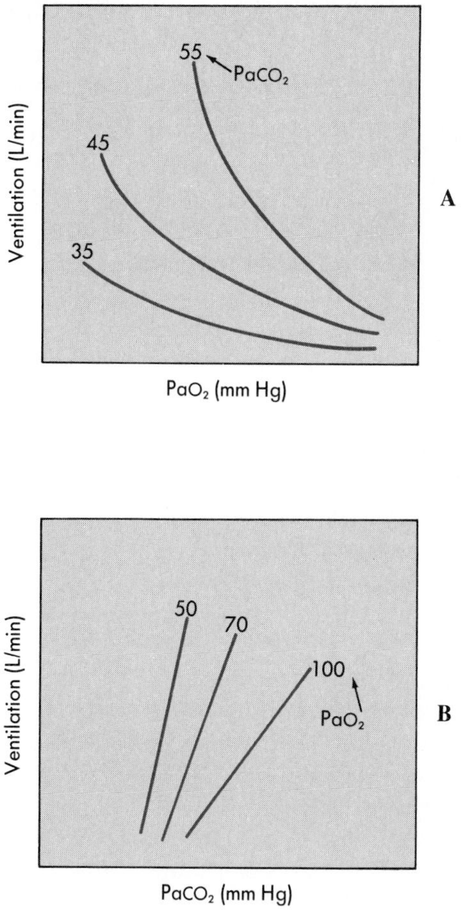

■ **Fig.. 40-8.** The effects of hypoxia and hypercapnia on ventilation. **A,** At a given P_{CO_2}, ventilation increases in a hyperbolic fashion as the P_{O_2} is reduced. The ventilatory response to hypoxia is greater the higher the P_{CO_2}. **B,** The ventilatory response to hypercapnia is accentuated by hypoxia.

Arterial Po_2 and Pco_2 are altered by inspiration and expiration so that carotid body discharge varies during the breathing cycle. Because the carotid body also is excited by increases in the rate of change of Pco_2, carotid body discharges are enhanced by increasing the amplitude of the respiratory oscillations in arterial Pco_2, even when the mean level of Pco_2 is unchanged. By contrast, the amplitude of the oscillations in Po_2 does not affect the carotid body.

The ventilatory response to a given level of carotid body activity is also related to the respiratory phase during which information from this receptor is received by the central nervous system. Carotid body activity during inspiration increases ventilation more effectively than during expiration. The relationship between the phases of respiration and the cyclic variations in carotid body firing depend on the circulation time. Therefore changes in cardiac output may affect breathing, but such effects are probably small.

Carotid Body Structure and Function

At least two types of cells occur in the carotid body: type I cells, which have large vesicles that contain catecholamines (such as dopamine) and polypeptides (such as substance P), and type II cells, which do not have these vesicles. Either the type I cells or the afferent nerve endings are probably the actual hypoxia sensors.

Neural activity in chemoreceptor afferent fibers increases as hypoxia becomes more severe. The partial pressure of O_2, rather than the O_2 content, of the arterial blood is the stimulus. The mechanism that allows the carotid body to respond even to mild hypoxia has not been elucidated completely, but some details are known. Although the carotid body blood flow is unusually high, so also is its metabolic rate. Vascular shunts within the carotid body, in combination with its high metabolic rate, may produce local areas of hypoxia within the carotid body, even when the arterial blood is fully saturated with O_2. Measurements of carotid body Po_2 with microelectrodes have shown some regions with extremely low O_2 tensions, but the O_2 tensions in different regions vary widely.

Carotid body excitation may depend on the receptor phosphate potential, that is, on the ratio of adenosine triphosphate (ATP) to adenosine diphosphate (ADP) and P_i. Even mild hypoxia may limit ATP production in the carotid body. Within the carotid body, the cytochrome enzymes that produce ATP may have an especially low affinity for O_2, thereby accounting for the great sensitivity of the carotid body to changes in Po_2.

Perfusion of the carotid body with hypoxemic blood releases acetylcholine from this structure. Thus acetylcholine may be the transmitter that mediates the increased neural activity during hypoxia. However, antagonists of acetylcholine do not prevent hypoxia from increasing chemoreceptor discharge. Certain catecholamines, such as dopamine, also are released during hypoxia, but dopamine probably inhibits rather than augments the carotid body responses to hypoxia. Substance P and vasoactive intestinal peptide stimulate carotid body activity, and they may help mediate the hypoxic response.

Both the sensitivity of the peripheral chemoreceptors to alterations in Po_2 (nerve discharge rate/ΔPo_2) and the range of O_2 tensions over which they respond may be modified by the central nervous system. For example, changes in sympathetic nervous activity alter carotid body blood flow, and this may affect the sensitivity of the carotid body to hypoxia. Decreases in carotid body blood flow increase its sensitivity to hypoxia. Also, the carotid body responses to hypoxia can be inhibited by direct efferent discharge from the central nervous system via fibers coursing through the carotid body nerve.

The mechanisms that allow the aortic bodies to respond to hypoxia are probably similar to those in the carotid body. The aortic body is more affected by changes in blood flow than is the carotid body.

Pulmonary Mechanoreceptors

Receptors in the Airways and Lungs

Sensory receptors in the lungs and airways, as in other hollow viscera, are stimulated by irritation of the lining layers and changes in distending forces. The afferent fibers of these receptors travel to the brain in the tenth cranial nerve, the vagus.

There are three types of pulmonary receptors: *stretch receptors,* located within the smooth muscle layer of the extrapulmonary airways; *irritant receptors,* which ramify among airway epithelial cells and have a distribution similar to the stretch receptors; and *J (juxtacapillary) receptors,* situated in the lung interstitium near alveolar capillaries.

The stretch receptors were discussed earlier in the section on the organization of respiratory neurons. These receptors are excited by an increase in bronchial transmural pressure, and they adapt slowly to a sustained stimulus. As the lung is inflated, they reflexly inhibit inspiration and promote expiration. They are responsible for the *Hering-Breuer reflex,* which produces apnea in response to large lung inflations and augments expiratory muscle contraction. The Hering-Breuer reflex is weaker in humans than in animals. Newborn infants, however, may display a prominent Hering-Breuer reflex, presumably because descending inhibitory influences from the immature cortex have not yet developed.

Irritant receptors and J receptors are served by vagal fibers that are smaller than are those from the stretch receptors. Unlike the stretch receptors, these other receptors rapidly adapt when subjected to a sustained stimulus.

Irritant receptors are stimulated chemically by noxious agents, such as sulfur dioxide, ammonia, nitrogen dioxide, or antigens. They are also stimulated mechanically by lung inflation, by increases in airflow, by particulate matter, and by changes in bronchial smooth muscle tone. Irritant receptor stimulation reflexly augments the activity of inspiratory motor neurons. It constricts the airways and it interacts with the stretch receptors to promote rapid, shallow breathing. This pattern of breathing, in combination with airway constriction, may limit penetration of dangerous agents into the lung and may prevent such agents from reacting with the gas exchanging surfaces. The irritant receptors may also stabilize lung compliance by initiating the periodic sighs that occur sporadically during normal breathing. These sighs reexpand collapsed areas of the lung. The chemical mediators released in the lung during allergic reactions (e.g., histamine, slow reacting substance of anaphylaxis, and bradykinin) also stimulate irritant receptors. The augmentation of inspiratory activity and increases in breathing frequency produced by irritant receptor excitation may enhance ventilation during asthmatic attacks when the work of breathing is severely increased.

J receptors are supplied by unmyelinated fibers. They are excited by interstitial edema in the lung and by certain chemicals, such as histamine, halothane, and phenyldiguanide. Activation of the J receptors causes laryngeal closure and apnea, followed by rapid, shallow breathing. J receptors, together with the irritant receptors, may be responsible for the tachypnea (rapid breathing) seen in patients with a pulmonary embolus, pulmonary edema, or pneumonia. Receptors that are similar to J receptors and that are innervated by unmyelinated fibers can also be found in the airways.

Chest Wall Proprioceptors

Like all skeletal muscles, the respiratory muscles of the chest wall develop forces that depend on their starting length (preload) and their afterload. The preload varies with body posture, and the afterload varies with chest wall expansion and the resistance to airflow. Receptors in the chest wall reflexly modify motor nerve discharge to the respiratory muscles in such a manner that ventilation changes are minimized, despite varying preloads and afterloads.

Three types of receptors in the chest wall are the *joint, tendon,* and *spindle receptors*. They signal to the respiratory neurons information about the forces exerted by the respiratory muscles and the movements of the chest wall. Specialized Ruffini receptors and pacinian and Golgi organs are present in the joints within the chest wall. Joint receptor activity varies with the degree

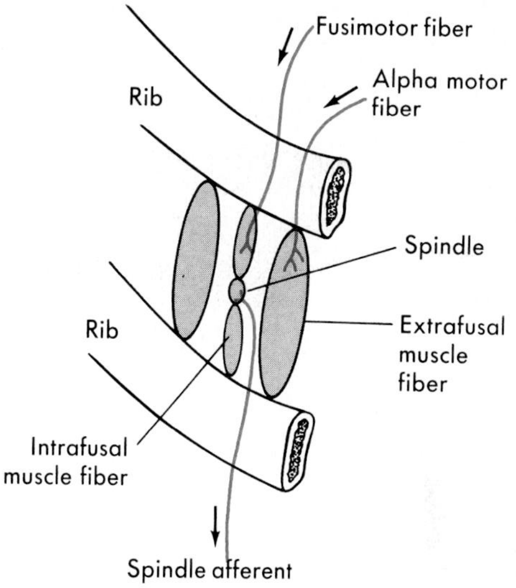

Fig. 40-9. Intercostal muscle spindle and its neural connections.

and rate of change of rib movement. Tendon organs in the intercostal muscles and the diaphragm monitor the force of muscle contraction and tend to inhibit inspiration. Muscle spindles are abundant in the intercostal muscles, but scarce in the diaphragm. Spindles help coordinate breathing during changes in posture and speech. They also help stabilize the rib cage when breathing is impeded by increases in airway resistance or decreases in lung compliance.

Fig. 40-9 shows the operation of the spindle and its neural connections. Spindles are located on intrafusal muscle fibers, which are aligned in parallel with the extrafusal fibers that elevate the ribs. Motor innervation of the extrafusal fibers originates in α motor neurons. The intrafusal fibers, on the other hand, are innervated by γ (fusimotor) motor neurons.

Passive stretch of an intercostal spindle, as occurs during lateral flexion of the trunk, increases afferent spindle activity. Through a monosynaptic segmental reflex, such spindle excitation causes contraction of the parent extrafusal fiber, which helps restore the upright position of the trunk. The spindles can also be stretched by efferent fusimotor discharge, which causes contraction of the intrafusal fiber itself. Some fusimotor fibers fire phasically, such that their discharge rises during inspiration and falls during expiration; other fusimotor fibers are tonically active. Without phasic fusimotor activity, spindle discharge would decrease when the extrafusal fibers in the external intercostal muscles contract during inspiration. Simultaneous activation of fusimotor and α motor neurons causes the spindles to be under continuous stretch during inspiration, thereby enhancing the contribution made by the intercostal muscles to respiration. If inspiratory movements are impeded, afferent activity heightens in a spindle innervated by a phasically active fusimotor fiber. This increases inspiratory muscle force and helps to preserve tidal volume. Spindle afferent fibers project all the way to the cerebral cortex. Such connections may provide the information that allows respiratory movements to be perceived consciously. The sensation of breathlessness may result from an imbalance in the demand for muscle shortening and the actual degree of shortening, as reflected in spindle afferent activity.

Tests of Chemoreceptor Sensitivity

Steady State Hypercapnia

When lung and respiratory muscle function is normal, the sensitivity of the peripheral and central chemoreceptors to CO_2 can be evaluated by measuring the ventilatory response to inspired CO_2. In the conventional "*steady state test,*" the inspired CO_2 is increased in steps. The ventilation at each step is related to the change in arterial P_{CO_2}. Sensitivity to CO_2 is determined from the slope of the line relating ventilation to Pa_{CO_2} (Fig. 40-5). Although inspired CO_2 has easy access to the brain, considerable time is required for ventilation to reach its final steady state. Usually the inspired CO_2 concentrations at each step must be maintained constant for 10 to 20 minutes to ensure that the steady state is reached. Ventilation and the cerebral venous P_{CO_2} reach steady state values at about the same time. Relative rates of equilibration of P_{CO_2} in arterial and cerebral venous blood indicate that arterial P_{CO_2} reaches its steady state level long before the venous P_{CO_2} (Fig. 40-10). Hence, if ventilation is measured too soon, chemosensitivity will be underestimated.

In normal individuals the average ventilatory response to inspired CO_2 is about 2.5 L/minute for each millimeter of mercury rise in Pa_{CO_2}. The response is somewhat less in women than in men, and it may decline with advanced age. The CO_2 response varies considerably among individuals. Some of this variability is caused by differences in body size, genetic makeup, and even personality. The variability among individuals is reduced when the CO_2 response is corrected for differences in vital capacity.

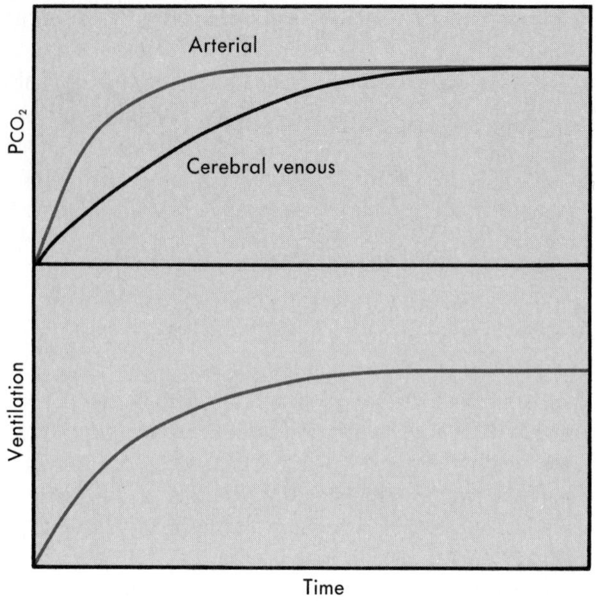

Fig. 40-10. Relative equilibration times for P_{CO_2} in the cerebral arterial and venous blood and for ventilation in response to a stepwise increase in inspired CO_2 concentration.

Tests of Peripheral Chemoreceptors

The peripheral chemoreceptors (the carotid and aortic bodies) contribute about 25% to the total ventilatory increase observed when CO_2 is inhaled. Because the peripheral chemoreceptors react rapidly to changes in inspired CO_2, their response to hypercapnia has been evaluated by measuring the immediate change in ventilation that occurs in the first few breaths after an abrupt change in inspired CO_2 concentration.

The response to hypoxia can be measured either by rebreathing or by steady state techniques. Because of the prominent effects of CO_2 on breathing, the arterial P_{CO_2} must be held constant while the hypoxic response is being measured. Peripheral chemoreceptor responses to O_2 can also be evaluated by measuring the effect of a few breaths of 100% N_2 or 100% O_2 on ventilation.

The normal ventilatory response to hypoxia is more variable than that to CO_2. The ventilatory response to hypoxia is influenced by the metabolic rate, the CO_2 response, prior chronic exposure to hypoxia, and genetic factors.

Chronic hypoxemia, as occurs at high altitude, depresses the ventilatory response to hypoxia, presumably because of adaptation of the chemoreceptors or of neurons in the central nervous system. This depressant effect of chronic hypoxemia is especially pronounced if it begins at birth.

Effects of Thoracic Disease on Ventilatory Responses to Hypercapnia and Hypoxia

Lung disease frequently depresses the ventilatory responses to chemical stimuli. At least four factors contribute to the reduction in ventilatory responses: (1) increased work of breathing, (2) decreased efficiency of gas exchange, (3) impaired performance of the respiratory muscles, and (4) reduced responses to chemical or mechanical stimuli.

Experimentally produced, moderate increases in airway resistance in normal subjects may depress resting ventilation or the CO_2 response very slightly, or they may even heighten ventilation. The mechanisms that preserve ventilation in these circumstances include (1) increased stimulation of mechanoreceptors in the lung and chest wall, (2) improvement of the mechanical advantage and coordination of the respiratory muscle, (3) conscious respiratory efforts, and (4) the intrinsic properties of the muscles themselves, which allow them to increase their contractile force when the velocity of shortening is reduced.

Large increases in inspiratory resistance, however, depress resting ventilation and reduce ventilatory responses to CO_2. Hence, arterial P_{CO_2} may rise and arterial P_{O_2} may fall. This suggests that at any given level of P_{CO_2} the respiratory system is able to perform some maximal amount of work.

Lung disease alters the efficiency of gas exchange by increasing ventilation/perfusion mismatching, increasing dead space ventilation, and increasing the amount of blood shunted through the lung. Increases in dead space ventilation are particularly important when the tidal volume is small, as it tends to be when lung or chest wall compliance diminishes. All these impairments of pulmonary gas exchange heighten the amount of ventilation, and hence the work, that is required to maintain normal levels of arterial gas tensions.

Enlargement of the functional residual capacity (FRC) decreases the length of the inspiratory muscles. Because they are less stretched, they are able to exert less force for a given amount of activation. If enlarged lung volumes are maintained for long periods in experimental animals, the number of sarcomeres per muscle fiber decreases. Therefore sarcomere length is eventually restored to normal levels. Whether this adaptive process occurs in humans is unknown. Furthermore, chronic respiratory disease may lead to poor nutrition, which will reduce the strength and endurance of the respiratory muscles, further impairing ventilation.

Some patients with severe impairment of pulmonary function develop respiratory failure. Those patients who have the poorest chemosensitivity seem to be the most likely to develop CO_2 retention when the performance of the chest bellows is reduced. Considerable indirect evidence supports this hypothesis. For example, offspring of patients with severe obstructive lung disease respond more poorly to changes in inspired CO_2 than do offspring of normal individuals. In children who have retained CO_2 because of upper airway obstruction caused by hypertrophy of the adenoids and tonsils, the CO_2 sensitivity is depressed, and it remains depressed even after the tonsils and adenoids have been removed. Patients who have retained CO_2 during an asthmatic attack also respond poorly to CO_2 after recovery from the asthmatic episode.

Some people retain CO_2 even though lung function is normal. Often the cause of the depressed CO_2 sensitivity is unknown, but some central nervous system abnormality is suspected. CO_2 retention may also be associated with certain metabolic abnormalities, such as alkalosis, or with the chronic administration of drugs, such as methadone, that depress respiration.

Altered responses of mechanoreceptors can also impair the ventilatory responses to hypercapnia and hypoxia and alter the blood gas tensions. Increased sensitivity of irritant receptors in patients with bronchitis or asthma may lead to rapid shallow breathing and a decreased effectiveness of gas exchange. Conversely, blunted sensitivity of chest wall mechanoreceptors may impair the ability of the respiratory system to compensate for changes in airway resistance and lung compliance.

Indices of Respiratory Efferent Nerve Activity

Ventilation

Ventilation is usually used to assess the respiratory responses to alterations in the inspired concentrations of CO_2 and O_2. If the performance of the chest bellows severely limits the ventilatory responses to chemical stimuli, other methods must be used to assess the response of respiratory motor neurons.

Occlusion Pressure

In the *occlusion pressure technique* the isometric force of contraction of the inspiratory muscles is measured. This test is based on the principle that under nearly isometric conditions, the contractile force of skeletal muscle correlates closely with its electrical activity. This relationship has been confirmed experimentally for the respiratory muscles. To perform the test, the airway is momentarily occluded at the beginning of inspiration, and the negative pressure developed at the mouth during inspiration is measured. In conscious subjects the response is more reproducible when the occlusion is brief, about one tenth of a second. The occlusion pressure becomes more negative during hypercapnia and hypoxia. These pressure changes can be related to the change in Pa_{CO_2} and Pa_{O_2} to estimate chemosensitivity. Because the occlusion pressure is measured in the absence of airflow, it is not affected by changes in airway resistance.

The tensions developed by the inspiratory muscles depend on their initial length. Hence, occlusion pressure measurements in patients with lung disease may be influenced by the alterations in inspiratory muscle length caused by changes in FRC. Increased FRC in animals reduces the occlusion pressure. However, in conscious man even large changes in FRC (about 1 L) have little effect on occlusion pressure.

Measurement of the electrical activity of the diaphragm is another way of evaluating respiratory motor output in humans. The activity can be measured by passing two electrodes down the esophagus and positioning them so that they straddle the diaphragm.

Abnormal Breathing Patterns

Breathing is usually a smoothly recurring, continuous cycle of inspiration and expiration. However, in some diseases of the central nervous system, episodes of respiratory arrest (apnea) recur.

In *Cheyne-Stokes breathing* tidal volume waxes and wanes cyclically in association with recurrent periods of apnea (Fig. 40-11). This form of breathing occurs in disorders of the central nervous system and in congestive heart failure. However, in some humans without an obvious neurological defect, this breathing pattern may appear on exposure to hypoxia, during sleep, and immediately after voluntary hyperventilation. During these regular cycles, blood concentrations of oxygen and carbon dioxide also fluctuate. In the cerebral arterial blood, $Paco_2$ is higher and Pao_2 is lower during the hyperventilatory portion of the cycle (Fig. 40-11).

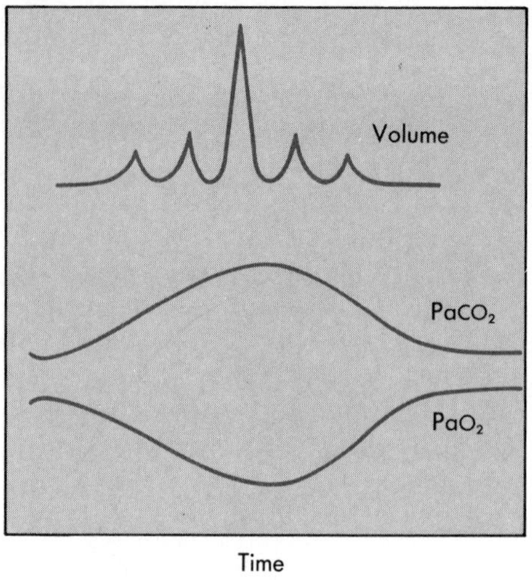

■ **Fig. 40-11.** Cheyne-Stokes breathing. Tidal volume waxes and wanes rhythmically. The Pao_2 is lowest and the $Paco_2$ highest in the cerebral arterial blood when ventilation volume is the greatest.

Cheyne-Stokes breathing is a manifestation of instability in the ventilatory control system. Similar oscillatory changes in output occur in man-made feedback control systems in predictable circumstances. Delayed information transfer around the feedback system and excessive controller gains (an increased ratio of the change in response to the change in stimulus level) are two important conditions that lead to instability in a physical feedback system.

The sensitivity of the respiratory controller can be assessed by measuring the slopes of the lines that relate ventilation to $Paco_2$ and Pao_2. Over a broad range of $Paco_2$ values, controller sensitivity is constant (Fig. 40-8, *B*); the ventilation changes are linear. By contrast, ventilation increases nonlinearly with hypoxia (Fig. 40-8, *A*). As hypoxia becomes more severe, controller sensitivity is more pronounced in the presence of hypercapnia. Controller sensitivity may also increase in certain neurological diseases that affect the cortex.

Controller sensitivity can become so great that the controller overacts and increases ventilation too much. For example, hyperventilation sufficient to drop arterial carbon dioxide tension below the threshold value for the chemoreceptors will induce apnea. During apnea, the arterial carbon dioxide tension rises, and the arterial oxygen tension drops. When the carbon dioxide tension rises to the threshold value, spontaneous ventilation begins and increases rapidly because of the hypoxia. Although the oxygen tension is quickly restored to normal values, the excessive ventilation again lowers the carbon dioxide tension below the threshold, apnea ensues, and the cycle continues.

Delays in information transfer within the respiratory system occur when the transit time around the circulation is prolonged, i.e., when cardiac output is reduced, as in congestive heart failure. The respiratory controller must receive information about the prevailing alveolar Pco_2 and Po_2 if it is to adjust ventilation correctly. When the time required for the blood to circulate from the lungs to the central and peripheral chemoreceptors is prolonged, the controller may react inappropriately. Thus, if cyclic changes in blood gas composition occur in the pulmonary capillaries because of fluctuations in pulmonary ventilation, the same cyclic changes in blood gases will occur in the arteries perfusing the medullary respiratory center, but the timing will be substantially delayed. In Fig. 40-11, e.g., as the ventilatory volume increases, the pulmonary capillary Pco_2 will fall and the Po_2 will rise. After a substantial delay, these changes in blood gases will occur in the cerebral arterial blood and they will inhibit ventilation. Consequently, the pulmonary capillary Pco_2 will rise and Po_2 will fall. After a similar delay, these changes in blood gas concentration will appear in the arterial blood perfusing the medullary respiratory center, and ventilation will increase again. Congestive heart failure is a frequent cause of Cheyne-Stokes breathing.

Increased levels of $Paco_2$ during quiet respiration may also induce Cheyne-Stokes breathing. This circumstance often occurs during sleep, and it may account for some forms of recurrent apnea during sleep. Other factors that predispose to Cheyne-Stokes breathing are (1) decreases in the total amount of O_2 in the lungs (e.g., as a result of a diminished FRC); (2) decreased buffering capacity of the blood for CO_2, so that changes in Pa_{CO_2} occur more rapidly during breathing; and (3) changes in cerebral blood flow that allow exaggerated changes in H^+ concentration to occur at central chemoreceptors.

Other less common forms of abnormal breathing patterns also occur. *Apneustic breathing,* with its prolonged inspiratory pauses, has already been discussed. In *Biots breathing,* as in Cheyne-Stokes breathing, apnea recurs, but during the period of ventilation, tidal volume and frequency remain fixed. The mechanism for this pattern is unclear; it may be a variant of Cheyne-Strokes breathing. It occurs in patients with central nervous system diseases, especially meningitis.

During sleep, periods of absent airflow may occur because of upper airway obstruction despite continued respiratory activity (see p. 637). Grossly irregular, or ataxic, breathing can be observed in some patients with medullary lesions and occasionally in persons with absent responses to chemical stimuli *(primary alveolar hypoventilation)*. Increased breathing and hypocapnia occur in diseases that excite irritant and lung J receptors (e.g. , asthma and pulmonary embolism) or that cause metabolic acidosis. In diabetes this hyperventilation is sometimes termed *Kussmaul respiration*.

Bibliography

Journal articles

Altose, M., et al.: Respiratory sensations and dyspnea, J. Appl. Physiol. **58:**1051, 1985.

Asmussen, E.: Control of ventilation in exercise, Exercise Sports Sci. Rev. **11:**24, 1983.

Cherniack, N.S.: Sleep apnea and its causes, J. Clin. Invest. **73:**1501, 1984.

Dempsey, J.A., and Forster, H.V.: Mediation of ventilatory adaptations, Physiol. Rev. **62:**262, 1982.

Euler, C. von: On the central pattern generator for the basic breathing rhythmicity, J. Appl. Physiol. **55:**1647, 1983.

Eyzaguirre, C., and Zapata, P.: Perspectives in carotid body research, J. Appl. Physiol. **57:**931, 1984.

Lagercrantz, H.: Classical and ''new'' neurotransmitters during development: some examples from control of respiration, J. Dev. Physiol. **6:**195, 1984.

Long, S., and Duffin, J.: The neuronal determinants of respiratory rhythm, Prog. Neurobiol. **27:**101, 1986.

Pack, A.I. Sensory inputs to the medulla, Annu. Rev. Physiol. **43:**73, 1981.

Richter, D.W.: Generation and maintenance of the respiratory rhythm, J. Exp. Biol. **100:**93, 1982.

Santiago, T.V., and Edelman, N.H.: Opioids and breathing, J. Appl. Physiol. **59:**1675, 1985.

Schlaefke, M.E.: Central chemosensitivity: a respiratory drive, Rev. Physiol. Biochem. Pharmacol. **90:**171, 1981.

Whipp, B.J., and Ward, S.A.: Cardiopulmonary coupling during exercise, J. Exp. Biol. **100:**175, 1982.

Books and monographs

Fishman, A.P., et al., editors: Handbook of physiology. Section 3: Respiratory system. Vol. II. Control of breathing. Bethesda, Md., 1986, American Physiological Society.

CHAPTER

41

Environmental and Developmental Aspects of Respiration

■ *Effects of Sleep on Respiration*

In alert, conscious humans, stimuli from the environment act reflexly via brain centers to excite breathing. Such stimuli sustain breathing even at very low levels of chemical drive.

Fluctuations in the state of alertness occur regularly, even in awake subjects. However, they are most pronounced during sleep and significantly affect respiration. The neural and biochemical mechanisms that produce sleep involve an excitatory and inhibitory interplay among various areas of the brain. They alter the balance of activity among noradrenergic, serotonergic, and cholinergic neurons. Sleep is not a homogeneous phenomenon. It can be divided into a slow-wave (SW) stage and a rapid eye movement (REM) stage, during which dreaming occurs. Each stage is associated with fairly characteristic changes in muscular, cardiovascular, and respiratory activity.

■ *Regulation of Breathing in Sleep*

A reduction in the level of environmental stimulation and the withdrawal of cerebral influences on respiration decrease ventilation and thereby increase the arterial P_{CO_2} during SW sleep. Systemic blood pressure also falls, and heart rate is reduced. Hypercapnic ventilatory responses are attenuated in SW sleep, and the CO_2 response curve is shifted to the right (Fig. 41-1).

REM sleep is divided into phasic and tonic stages. In tonic REM sleep breathing maintains its regularity, but tidal volume may decrease. In addition, the ventilatory responses to inspired CO_2 are further reduced. The ventilatory responses to hypoxia are better maintained in sleep than are the responses to CO_2.

External stimuli and changes in blood gas tensions are less effective in producing arousal in REM sleep than in SW sleep. Phasic REM sleep is associated with irregular breathing patterns because the intrinsic activity of higher brain centers dominates respiratory neuron activity.

People with depressed ventilatory responses to hypercapnia and hypoxia while awake breathe even less during sleep than do those that show normal responses. In some patients with lung or respiratory muscle disease, the hypoxemia that develops during sleep seems disproportionate to the degree of hypoventilation. In these patients ventilation/perfusion ratios may be abnormal in some regions of the lung because of the recumbent position and discoordinated respiratory muscle movement. Such abnormal ventilation/perfusion ratios may contribute to the arterial oxygen desaturation during sleep.

The level of ventilation and the adequacy of gas exchange also depend on the caliber of the upper airways. The upper airways are convoluted, semirigid, and include movable structures. During inspiration the pressure within the upper airways is subatmospheric, and the transmural pressure of the extrathoracic upper airways is negative. This gradient compresses the airway and displaces structures such as the tongue. Consequently, upper airway caliber is reduced, and resistance to air flow increases.

These effects are counterbalanced by the actions of the upper airway muscles, including the alae nasi (the dilator muscles of the external nares), the genioglossus (the tongue protrussor muscle), the posterior cricoarytenoids (vocal cord abductors), and the geniohyoid and sternohyoid muscles (which displace the hyoid arch anteriorly). These muscles enlarge and stiffen the upper airways.

During wakefulness all of these muscles are tonically active. They also undergo phasic changes in synchrony

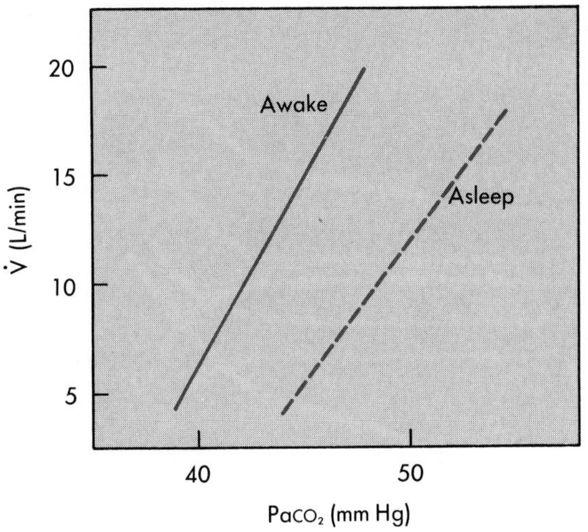

Fig. 41-1. Ventilatory responses to hypercapnia during wakefulness and during sleep. With sleep the response line is shifted to the right, and the change in ventilation for a given change in Pa_{CO_2} (the slope of the response line) is reduced.

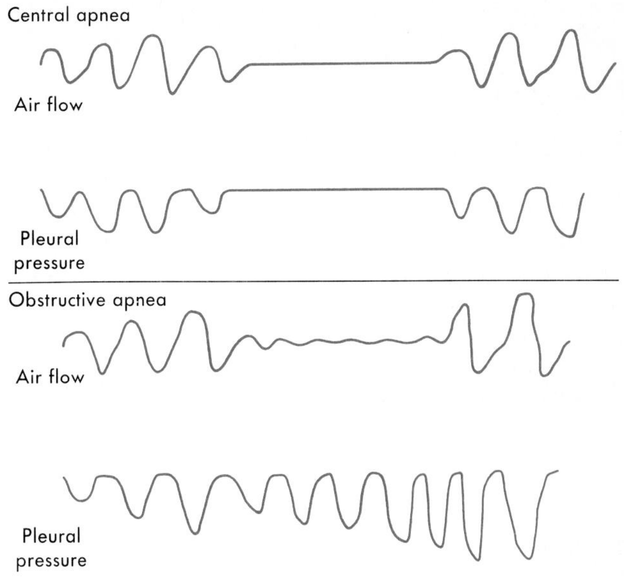

Fig. 41-2. Sleep apneas. **A,** Central apneas are characterized by a cessation of air flow as a result of a temporary suspension of all respiratory efforts. The cessation of respiratory efforts is reflected in absence of any pleural pressure deflections. **B,** In obstructive apneas air flow ceases despite persistent respiratory efforts and pleural pressure swings.

with breathing. As with the chest wall muscles hypoxia and hypercapnia also increase upper airway muscle activity. In sleep, however, the activity of upper airway muscles is markedly reduced.

A variety of neural reflexes modulate respiration during both wakefulness and sleep. Mechanical stimulation of the airways in animals in either SW or REM sleep elicits reflex responses that differ from those observed in the awake state. For example, in experimental animals laryngeal stimulation during wakefulness produces coughing, but it causes apnea in REM sleep. Pulmonary vagal irritant and stretch reflexes not only affect the activity of chest wall muscles but also influence the muscles of the upper airway. The excitation of pulmonary stretch receptors by lung inflation terminates inspiration. Even small increases in lung volume can markedly reduce the level of activity of the vocal cord abductor and tongue protrussor muscles. Changes in pressure and flow within the upper airway also reflexly affect upper airway muscle activation and may reduce the caliber and increase the resistance.

The phasic and tonic activity of skeletal muscle decreases during sleep, particularly in the REM stage. The loss of activity in upper airway muscles during sleep is far greater than that of the diaphragm. Because of the relative loss of tone in the upper airway, the negative airway pressure created by the diaphragm during inspiration may be sufficient to occlude the upper airway. Brief periods of upper airway obstruction occur even in normal people during sleep. Such obstruction may be caused by changes in the level of activity of these muscles, in their activation time, or in their responses to chemical and mechanical stimuli.

Apneic Pauses During Sleep

Apneic periods occur during sleep in a third of normal individuals and are particularly frequent among older men. Apnea may last for more than 10 seconds and may be associated with falls in arterial O_2 saturation to 75% or less. These apneas appear during all sleep stages but are more common in the lighter stages of SW and REM sleep.

Sleep apneas have been classified into two distinct categories: central and obstructive (Fig. 41-2). Central apnea is characterized by a cessation of respiratory efforts. In obstructive apnea, despite persistent respiratory efforts, air flow ceases because of total obstruction of the upper airway. Snoring is a manifestation of partial obstruction of the upper airway. The sites of upper airway obstruction during sleep include the larynx, the pharynx, and the oropharynx. Arousal may be an important element in terminating sleep apnea. Arousal may result from chemoreceptor excitation by hypoxia and hypercapnia.

Respiratory disturbances occurring during sleep may be primary factors in certain diseases. In patients with such disturbances prolonged and frequent obstructive apneas occur. The recurrent periods of hypoxia and hypercapnia may lead to polycythemia, rightsided heart failure, and pulmonary hypertension.

Respiratory Adjustments During Exercise

The ability to exercise depends on the capacity of the cardiovascular and respiratory systems, acting in coordination, to increase O_2 delivery to the tissues and to remove the excess CO_2. The circulatory adjustments are covered in Chapter 36, and only the respiratory adaptations will be considered here.

Cardiac output increases during exercise, for the reasons outlined in Chapter 36. Because of the low resistance and great distensibility of the pulmonary vascular bed, this increase in pulmonary blood flow is accompanied by only a small rise in pulmonary vascular pressure. More capillaries are opened in the lung, and the area available for gas diffusion increases; the diffusing capacity for both O_2 and CO_2 rises. The alveolar-arterial P_{O_2} difference (A-a P_{O_2}) decreases slightly from the resting value in moderate exercise, reflecting the more even distribution of $\dot{V}_A/\dot{Q}$ ratios in the lung during exercise. However, as exercise intensifies, the A-a P_{O_2} widens because mixed venous P_{O_2} is decreased.

The airways distend slightly during exercise, increasing the anatomic dead space. However, the alveolar dead space is decreased because of the improvement in ventilation-perfusion matching. Thus the physiological dead space is nearly the same during exercise and rest. Because tidal volume is larger during exercise the dead space/tidal volume ratio falls as exercise becomes more severe.

The ventilatory adjustments that take place are geared to the intensity of exercise and to its duration. In very brief, intense exercise the breath is frequently held until the end of the exercise. With more prolonged exercise, however, ventilation is elevated above the resting level, and it increases even more as exercise becomes more strenuous.

Acid-base balance is normal during moderate exercise, when O_2 delivery to the mitochondria is adequate to meet energy requirements aerobically. However, with sufficiently severe exercise, total energy requirements can be satisfied only by a combination of aerobic metabolism and anaerobic glycolysis. The lactic acid formed during glycolysis diffuses into the blood and increases the H^+ ion concentration. The chemical reaction of the acid with blood bicarbonate forms CO_2, which adds to the CO_2 produced in the tissues. The highest level of work that can be performed without inducing a sustained metabolic acidosis is called the *anaerobic threshold*. The level of the anaerobic threshold is higher in trained, physically fit than in untrained subjects.

When the level of exercise is below the anaerobic threshold, ventilation is linearly related to both CO_2 production and O_2 consumption (Fig. 41–3). Arterial P_{O_2}, P_{CO_2}, and pH are virtually unchanged, although venous values may be altered. The decrease in arterial pH that occurs when the level of exercise is above the anaerobic threshold stimulates the carotid body. Ventilation increases out of proportion to the rise in O_2 consumption. Because of the hyperventilation that results from the increasing blood lactic acid at very high work intensities, Pa_{CO_2} falls and Pa_{O_2} rises.

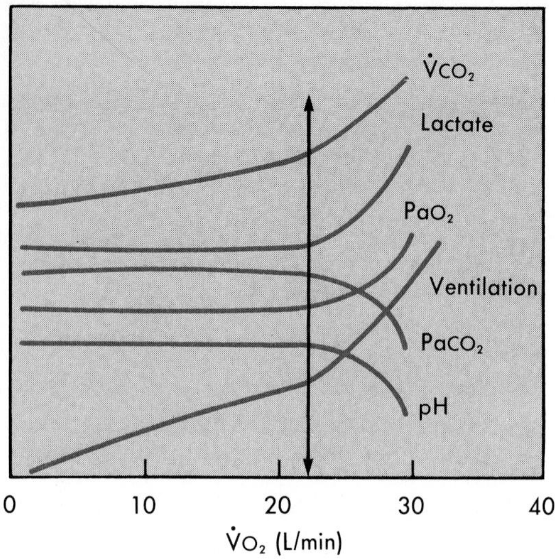

Fig. 41-3. Changes in ventilation, CO_2 production ($\dot{V}_{CO_2}$), Pa_{CO_2}, Pa_{O_2}, pH, and blood lactate during progressive exercise. The *vertical line* represents the anaerobic threshold.

Mechanism of Exercise Hyperpnea

The increase in ventilation is the most obvious and important respiratory adjustment to exercise. It takes 4 to 6 minutes after the beginning of moderate exercise before a steady rate of ventilation is reached; with severe exercise it may take even longer. Exercise increases ventilation in fairly distinct stages (Fig. 41-4). In stage I ventilation increases abruptly; in stage II ventilation increases more gradually; and in stage III ventilation remains constant.

Both neural and chemical factors regulate ventilation during exercise, just as they do during rest. Some of the mechanisms whereby ventilation can be increased during exercise are by activation of:

1. Cardiovascular mechanoreceptors in the systemic or the pulmonary circulation
2. Thermal receptors
3. Central or peripheral chemoreceptors
4. Central nervous system
5. Mechanoreceptors in muscle, such as joint receptors of muscle spindles

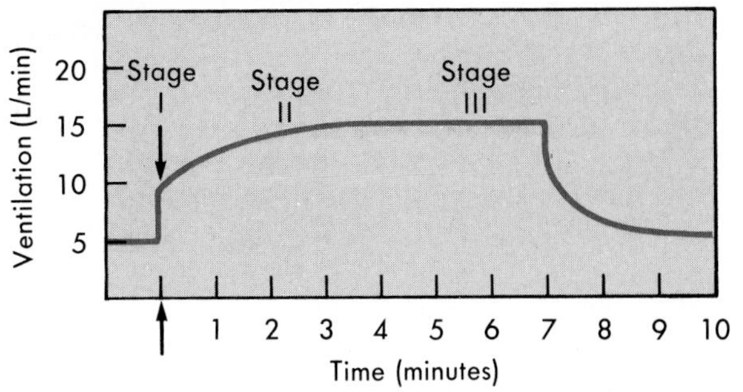

■ Fig. 41-4. Three stages in the time course of changes in ventilation during exercise.

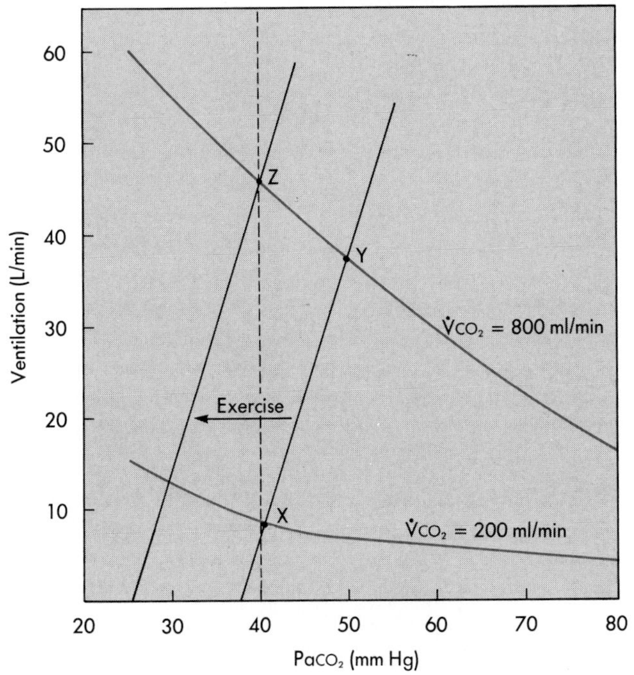

■ Fig. 41-5. At rest the level of ventilation and the Pa_{CO_2} are designated by the point *X*, the intersection of the hyperbola describing the effect of ventilation on Pa_{CO_2} and the line representing the ventilatory response to changes in Pa_{CO_2}. $\dot{V}_{CO_2}$ increases during exercise and the hyperbola is shifted upward and to the right. If the ventilatory response to CO_2 were unchanged, the level of ventilation and the Pa_{CO_2} during exercise would be at point *Y*. In fact, the ventilatory response line is shifted to the left during exercise and intersects the exercise hyperbola at point *Z*. The adjustments in the ventilatory reponse are such that the Pa_{CO_2} during exercise is virtually identical to that at rest.

6. Receptors that monitor metabolic activity in muscle or blood gas concentrations in the mixed venous blood

It is not at all obvious which of the above-mentioned signals mediates the tight coupling of ventilation to metabolic rate.

Despite very large changes in CO_2 production and O_2 consumption during exercise, the steady-state arterial P_{O_2} and P_{CO_2} levels remain remarkably constant in humans. Hence stimulation of chemoreceptors alone by changes in arterial blood gases cannot account for exercise hyperpnea. Some other factor must regulate ventilation during exercise.

Fig. 41-5 illustrates the response of the central controller to increasing Pa_{CO_2}, along with the two hyperbolas that show the effect of ventilation on Pa_{CO_2} at two different levels of metabolic activity (Chapter 40). If the exercise-induced change in ventilation were caused by the ventilatory drive from the newly produced CO_2 accumulating in the bloodstream, the response line would intersect the curve for the higher $\dot{V}_{CO_2}$ at point *Y*. This point defines the arterial P_{CO_2} that would occur during exercise if CO_2 sensitivity were the only factor responsible for the hyperpnea. The fact that during exercise (below the anaerobic threshold) Pa_{CO_2} is unchanged indicates that some factor sensitive to metabolic rate has increased breathing. This factor shifts the CO_2 response line by some as yet unknown mechanism.

The H^+ concentration and the P_{CO_2} in the arterial blood fluctuate with inspiration and expiration. These changes produce corresponding fluctuations in peripheral chemoreceptor discharge. The oscillations in chemoreceptor activity are greater during exercise because of the increased level of metabolic activity and the larger tidal volume. This may represent the additional drive that maintains ventilation in proportion to metabolic rate.

■ *Respiratory Effects of Breathing Low and High Oxygen Gas Mixtures*

Barometric pressure decreases with increasing altitude (Fig. 41-6); the relation is approximately exponential. The inspired air contains virtually the same percent of O_2 at high and low altitudes. Thus inspired P_{O_2} and hence alveolar P_{O_2} fall as an individual ascends from sea level to high altitudes. The resulting hypoxemia elicits a variety of compensatory responses. Some of these occur acutely, but others develop gradually during prolonged exposure.

The initial hyperventilation that occurs during exposure to high altitude is caused by stimulation of the peripheral chemoreceptors, particularly the carotid bodies. The increase in ventilation reduces the arterial P_{CO_2} and increases the pH. This decreases the excitation of central chemoreceptors and thus limits the degree of the ventilatory increase. After 2 or 3 days at high altitude ventilation increases further. This increase is part of the process of acclimatization, a poorly understood phe-

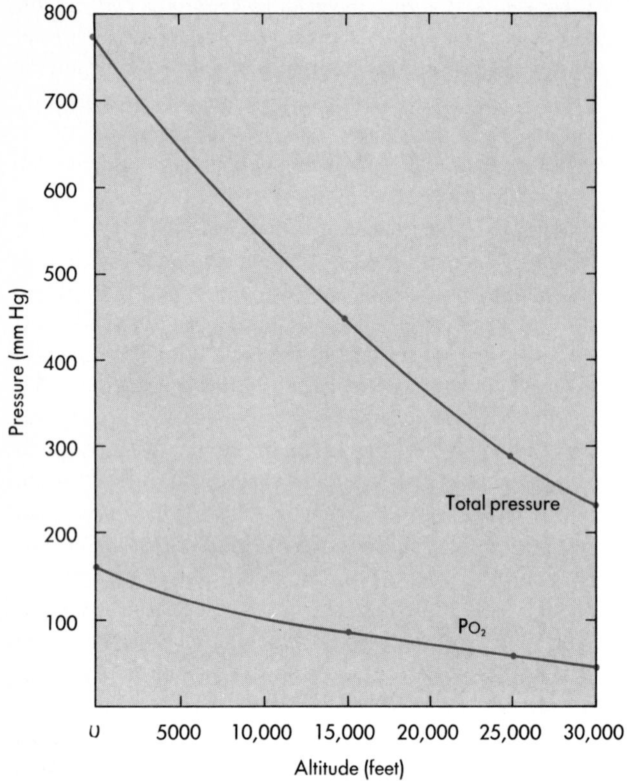

■ **Fig. 41-6.** Changes in barometric pressure and the P_{O_2} of inspired air with altitude.

nomenon. This secondary increase in ventilation occurs, in part, by the following mechanisms: (1) the renal excretion of bicarbonate reduces plasma HCO_3^- and returns blood pH toward normal; (2) the HCO_3^- concentration is decreased in brain interstitial fluid, perhaps by some active process that expels HCO_3^- ions from the interstitial fluid into the blood; and (3) anaerobic metabolism in the hypoxic brain allows lactic acid to accumulate, which lowers bicarbonate levels. Another important alteration that occurs is a shift to the left of the ventilatory response to CO_2 and a steepening of the slope of the response curve. Consequently, the threshold for a stimulatory effect occurs at a lower P_{CO_2}.

Persons who live at high altitudes and who are chronically hypoxic have a blunted ventilatory response to hypoxia. They breathe less at a given level of arterial P_{O_2} than do those who live at sea level. The response to hypercapnia, on the other hand, is the same in people who live at high and low altitudes. Hypoxic desensitization develops over years and is more likely to occur when chronic exposure to hypoxia begins in infancy. A similar blunted response to hypoxia is found in patients who are hypoxemic because they have a type of congenital heart disease in which blood is shunted from the right to the left side of the heart. It is not clear whether the blunted ventilatory response is caused by some depressant effect of chronic hypoxia on the central nervous system or whether it originates at the peripheral chemoreceptors. Although the structure of the carotid bodies changes in natives of a high-altitude environment, the functional significance of such changes is uncertain. The blunted response to hypoxia ultimately disappears when such natives stay at sea level for years or when hypoxemia is relieved by surgical correction of cardiac defects.

In addition to these alterations in the respiratory controller, other respiratory changes also occur. Lung volume is the same in infants born at high altitude and at sea level. However, adult natives of high altitudes have disproportionately large lungs. The increase in size does not involve the airways, but it occurs mainly in the gas exchanging units of the lung. This increase in lung size may be responsible for the increase in pulmonary diffusing capacity observed in high-altitude natives.

Exposure to high altitude also increases the hemoglobin concentration in the blood and consequently augments the oxygen-carrying capacity. The increase in hemoglobin is caused by the release of *erythropoietin* from the kidney and possibly from other organs. This substance accelerates red cell production in the bone marrow.

During exposure to high altitude the concentration of 2,3-diphosphoglycerate also rises in the red cell. This decreases the affinity of hemoglobin for O_2 (Chapter 39). The P_{50} of the blood is decreased. Although this change enhances the unloading of O_2 from hemoglobin in the tissues, it also interferes with the uptake by hemoglobin in the lungs. Hence its adaptative value is uncertain.

Circulatory changes also occur with exposure to high altitude. Cardiac output, heart rate, and pulmonary artery pressure all increase. With prolonged exposure the density of capillaries may increase in many tissues. These changes improve O_2 transfer in the lung and O_2 delivery to tissue mitochondria.

Occasionally, tolerance at high altitude disappears and serious symptoms develop, such as ventilatory depression, polycythemia, and heart failure. This intolerance to altitude, called *Monge's disease,* is relieved by descent to lower altitude or by administration of oxygen.

■ Effects of Breathing High-Oxygen Mixtures

Breathing O_2-enriched gas mixtures for prolonged periods of time can be injurious. Damage to the capillary endothelium results in the exudation of fluid into the interstitial spaces of the lung. Pulmonary edema can develop. Larger alveolar cells proliferate and form a continuous lining over the alveoli, replacing the membranous pneumocytes. In newborn infants prolonged breathing of O_2 may cause severe inflammation and obliteration of distal bronchioles. High-oxygen breathing can also cause alveolar collapse. Alveoli that are served by obstructed airways tend to collapse because the total pressure of the gases in mixed venous blood is less than that in the alveoli. Hence the alveolar gas is absorbed into the blood. The N_2 in alveoli is more slowly absorbed than O_2. Inhalation of 100% O_2 eliminates the N_2 in the alveoli and accelerates collapse. In addition, oxygen breathing for even a few hours depresses epithelial ciliary function and mucus transport.

In humans the pulmonary function abnormalities caused by breathing O_2-enriched gas mixtures include a decreased vital capacity, a reduced lung compliance, a widened A-a Po_2 gradient, and a diminished diffusing capacity.

Animals become more resistant to O_2-induced lung injury during graded exposure to elevated concentrations of O_2 in the air. Tolerance to the effects of O_2 is achieved by certain enzymatic adaptations.

Oxygen breathing may also injure nonpulmonary tissues. In infants, breathing of 100% O_2 may produce blindness by causing fibrous tissue to form behind the lens. The condition is known as *retrolental fibroplasia*.

Brief exposures to hyperbaric O_2 at several atmospheres of pressure may produce convulsions. However, with care, hyperbaric O_2 therapy carried out in specially constructed chambers is sometimes useful in the treatment of gas gangrene and severe CO poisoning.

Divers breathe air at high pressure, which increases the partial pressure of O_2 and N_2 in the alveoli. The high partial pressure of N_2 forces more N_2 into solution in tissues. If the diver is decompressed too rapidly, bubbles of N_2 may form in the tissues and blood, producing pain and neurological damage (the "bends") because the N_2 bubbles obstruct blood flow. Very high levels of N_2 also may depress the central nervous system.

Respiratory Changes With Age

The Newborn

One of the principal functions of the first few breaths after birth is to transform the fluid-filled fetal lung to one containing air. High distending forces are needed in the first few breaths to move liquid out of the lung and to overcome surface forces that oppose the movement of air. With the first few breaths the end-expiratory lung volume (functional residual capacity) increases, and breathing becomes easier. The ability of the newborn to maintain air in the lungs depends on how well the chest wall can resist the collapsing forces produced by contraction of the diaphragm and how much surfactant has been produced to reduce surface tension (Chapter 37). In the premature infant the chest is very pliable, and surfactant production may be poor, increasing the danger of lung collapse. Resistance to air flow is higher in the infant than in the adult. The recoil forces of both the lung and chest wall are low in the infant. New alveoli are added until the child is about 8 years of age. Then the increase in lung size is achieved mainly by enlargement of existing alveoli.

Pulmonary vascular resistance is very high in the fetus, and the pulmonary blood vessels are very thick because of a medial layer of smooth muscle. After birth the vessel walls become thinner so that, by age 4 months, the ratio of wall thickness to external diameter is about the same as in the adult.

Diffusing capacity of the lungs increases during infancy as the alveolar surface area enlarges. By 6 years of age the diffusing capacity per square meter of body surface area is about the same as in the adult.

Breathing may be irregular at birth, particularly in the preterm infant. The patterns of breathing range from regular respiration to frequent pauses or apneic episodes. In preterm infants apnea is predominantly central; that is, respiratory movements are absent. In infants pure obstructive apnea may occur after the first month of life, and it may predispose to the *sudden infant death syndrome*.

Both full-term and preterm infants increase minute ventilation in response to small increases in inspired CO_2 concentration. Responsiveness to CO_2 is less well developed the more immature the infant.

Preterm infants respond to a reduction in inspired O_2 concentrations with a transient increase in ventilation for approximately 1 minute, followed by a sustained depression of ventilation. In term infants the response is similar during the first week of life. This biphasic response to hypoxemia in the neonate markedly differs from the ventilatory response in adults, in whom a low Pao_2 elicits a sustained increase in ventilation. The characteristic response to low Pao_2 in infants is explained by initial peripheral chemoreceptor stimulation, followed by an overriding depression of the brain stem respiratory center.

Changes in lung volume reflexly alter the timing of respiration much more in the newborn than in the adult. A small but sustained increase in lung volume in the newborn significantly prolongs the expiratory time and decreases the respiratory rate. The lung deflation reflexly increases the respiratory rate. The increase in breathing rate is due primarily to a shortening of expiratory time. This may help the infant maintain an adequate functional residual capacity. Pulmonary irritant reflexes have also been elicited in the neonate. Direct stimulation of the lining of the tracheal wall augments respiratory efforts.

Respiration in the Elderly

Pulmonary performance declines with advancing age. The changes proceed at a variable rate that is dependent

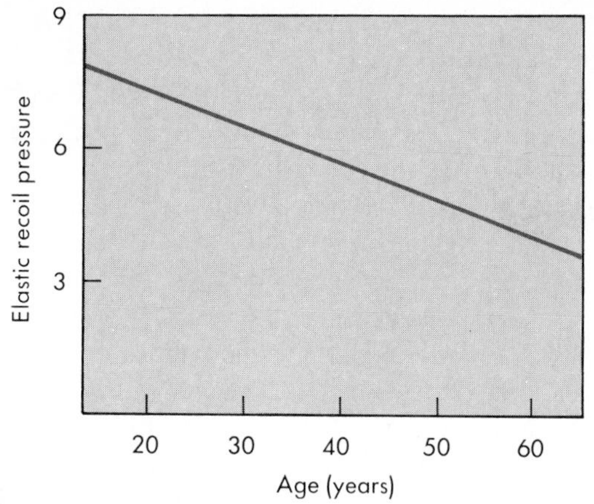

■ Fig. 41-7. Effects of age on lung elastic recoil pressure (at 60% of total lung capacity).

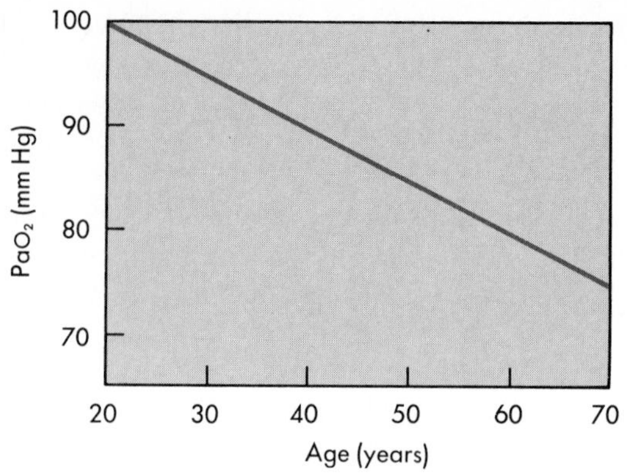

■ Fig. 41-8. Effects of age on the partial pressure of oxygen in the arterial blood (PaO_2).

both on the ''aging process'' and on the extent of exposure to noxious agents in the environment.

Lung elasticity is diminished in the elderly, and thus the transpulmonary pressure at a given lung volume is less than in young adults (Fig. 41-7). Because of the loss of lung elastic recoil, airways, particularly in dependent lung zones, will close at higher lung volumes than will those in younger individuals. This accounts for the elevated residual volume in the elderly.

With advancing age the chest wall stiffens because of structural changes in the bones and joints of the rib cage. The compliances in lung and chest wall change in opposite directions, and therefore the total lung capacity or functional residual capacity changes little with senescence. Respiratory muscle strength decreases in the elderly. This diminishes vital capacity and forced expiratory flow rates.

In the elderly the internal surface area of the lung decreases. The alveoli become wider and more shallow. The dead space enlarges, and the diffusing capacity decreases. The closing volume of airways increases progressively with advancing age, again reflecting a greater unevenness in the distribution of ventilation.

Ventilation/perfusion ratios become more variable with advancing age, and as a result the A-a PO_2 increases. With increasing age in adults, the arterial PO_2 falls about 0.3 mm Hg per year (Fig. 41-8). The arterial PCO_2 is similar in young and elderly adults. However, in the elderly the ventilatory responses to both hypercapnia and hypoxia are reduced.

■ *Pulmonary Defenses*

A normal 70 kg man inspires approximately 6 L of air from the environment each minute, exposing the entire respiratory tree to any toxic or pathogenic agents in the atmosphere. Maintenance of the lung function depends on the presence of effective defense mechanisms.

In humans the defenses consist of (1) mechanical barriers (the nose, nasal hair, and turbinates), (2) the mucociliary escalator, (3) immunologic systems that operate both in the conducting airways as well as in the alveoli, and (4) phagocytic cells in the alveoli.

■ *Fate of Inhaled Particles*

The nostrils and turbinates effectively trap any particles with diameters greater than 10 μm. Thus particles in this range never reach the conducting airways. Particles between 2 and 10 μm become trapped on the mucus layers of the trachea, bronchi, and terminal bronchioles and are deposited by sedimentation onto the airway walls. This mucus layer normally moves the particles upward, toward the epiglottis and posterior pharynx, by the action of cilia (see below). When the particles entrapped in the mucus layer reach the posterior pharynx, they are swallowed. The swallowed material passes through the gastrointestinal system and either is metabolized in the stomach and small intestine or is excreted.

■ *Mucociliary Escalator*

Between 10 and 100 ml of mucus secretions is produced by the tracheobronchial epithelium each day. These secretions spread out in a very thin (2 to 5 μm) layer (usually called the gel layer) over the ciliated surface and are propelled toward the glottis by the beating cilia that line the airways (Fig. 41-9). Beneath the gel layer is an aqueous (sol) layer (also called the periciliary fluid) in which the cilia beat freely. The nature and source of the periciliary fluid are unknown.

Tracheobronchial mucus secretions come from two

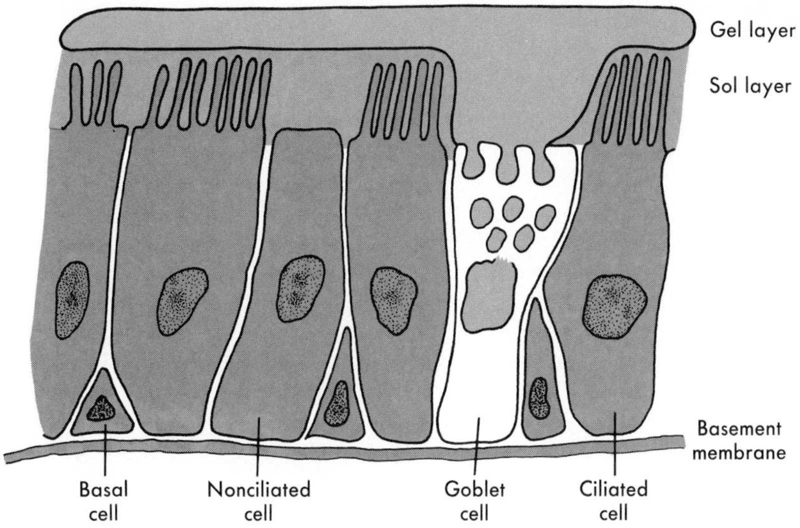

■ **Fig. 41-9.** The airway epithelium and lining material.

sources: from unicellular glands (goblet cells) in the surface epithelium and from submucosal glands, mainly in the smaller airways. Those submucosal glands, which contain both mucus-secreting and serous cells, are distributed from the larynx to the smallest bronchi.

Vagal stimulation increases mucus secretion from the submucosal glands. The goblet cells respond mainly to local irritants, either physical or chemical, with an outpouring of mucus.

Most of the cells lining the tracheobronchial tract from the larynx to the terminal bronchioles are ciliated. Cilia beat from 12 to 20 times per second. Adjacent rows of cilia are in different phases of motion so that wavelike patterns are discernible on the surface of the epithelium.

Cilia are about 6 μm long and 0.3 μm in diameter, and they are anchored by basal bodies and rootlets in the apical cytoplasm. Each ciliated cell has approximately 200 cilia. Cross sections of these cilia demonstrate a series of microtubules, nine peripheral microtubule doublets arranged in a circle around two centrally positioned single tubules (Fig. 41-10). This microtubular system is called the axoneme. The microtubules contain a special ATPase called dynein (from the term *dynamic protein*). This ATPase catalyzes the release of energy from adenosine triphosphate, which diffuses into the cilia from the body of the cell. The peripheral microtubular doublets are loosely connected to a central tubule apparatus by radial spokes. When energy is supplied each peripheral doublet can move longitudinally, relative to the central tubules. The differential movement of peripheral microtubules bends the cilia.

Effective beating of the cilia and proper amounts of mucus with appropriate physical properties are needed for normal mucociliary transport. The mucus film is propelled at 0.5 to 1.5 cm/minute in the human trachea.

■ *Other Defense Mechanisms*

Lung secretions contain immunoglobulin, which inhibits viral infection, promotes clumping of some bacteria, and, in conjunction with lysozyme and complement, promotes phagocytosis.

Cellular elements in the lung include lymphocytes, which are important mediators of the immune reaction, and alveolar macrophages, which ingest and digest bacteria that reach the alveolar surfaces (Fig. 41-11).

The alveolar macrophage normally resides on the alveolar surface. It is free to move about, migrating from alveolus to alveolus. Its activity can be random, phagocytosing whatever debris, particles, and organisms it encounters. The phagocytic activity of the macrophage probably also can become activated by chemical messengers released by the T-lymphocytes.

During prolonged, heavy cigarette smoking, white blood cells and alveolar macrophages release proteolytic enzymes, particularly elastase, into the lung. Elastase destroys the normal elastic fibers that support the lungs' alveolar structure, and this destruction may lead to emphysema. Certain inhibitors, such as α_1-antitrypsin, inactivate elastase. In an autosomal recessive hereditary disorder characterized by a deficiency of α_1-antitrypsin, the inactivation of elastase is impaired and emphysema is even more likely to occur.

■ *Respiratory Sensation*

Dyspnea is a cardinal symptom of respiratory disease. The term refers to the unpleasant or distressing sensation of labored or difficult breathing. The sensation arises through a complex series of steps involving the activation of sensory receptors, the transmission of sensory signals to the central nervous system, and the processing of those signals by higher brain centers. A va-

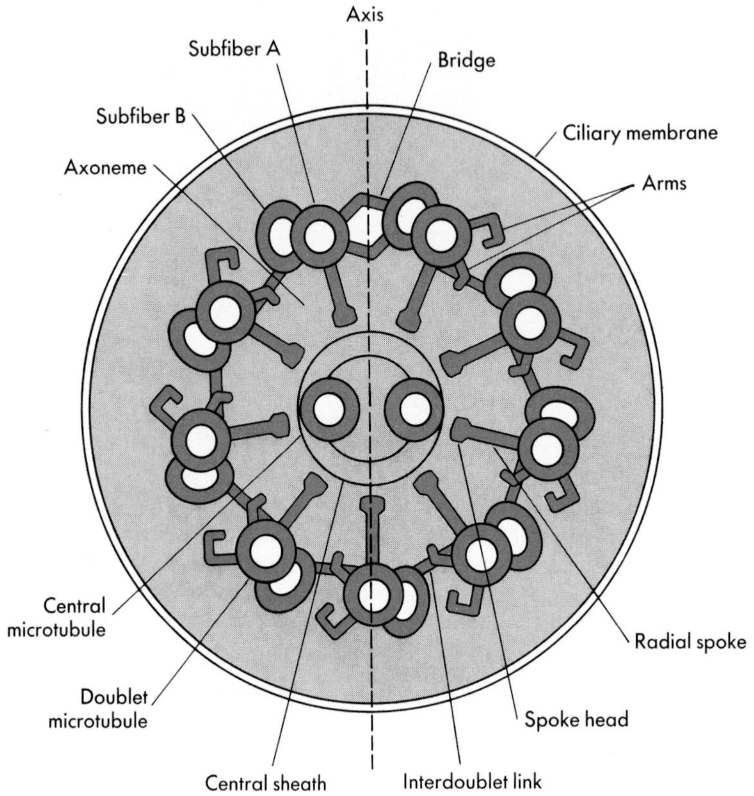

Fig. 41-10. Cross section of a cilium.

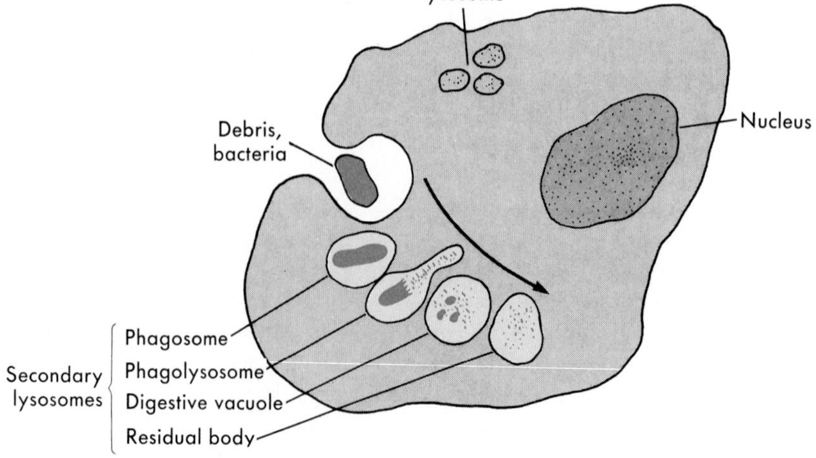

Fig. 41-11. The structure and function of the alveolar macrophage.

riety of signals arising from (a) chemoreceptors in the blood and brain, (b) mechanoreceptors in the airways, lungs, chest wall, or respiratory muscles, or (c) outgoing respiratory motor command signals from the central nervous system may mediate the sensation of dyspnea.

Breath-Holding

The sensation of dyspnea may resemble that experienced during breath-holding. Hypoxia and hypercapnia contribute to the unpleasant sensation and to eventual termination of breath-holding. However, alterations in blood gas tensions or H^+ concentration are not the sole factors limiting breath-holding time. If an individual at the breaking point of a breath-hold exhales and then inspires a mixture of gas with an even lower P_{O_2} and a higher P_{CO_2} than in the alveolar gas just expired, a further period of breath-holding can be accomplished. Thus certain nonchemical or mechanical factors appear to influence breath-holding time. Breath-holding time is longer at total lung capacity than at residual volume. This lung volume effect suggests that afferent impulses arising from the lung and traveling via the vagus nerves alter the sensation during breath-holding.

During breath-holding the respiratory muscles are initially quiescent. Thereafter central respiratory activity sometimes induces diaphragmatic contractions, which may become more intense and frequent until the breath-hold breaks. The increasing tension in the isometrically contracting diaphragm and the resulting stimulation of mechanoreceptors in the muscle, presumably tendon organs, may generate the sensation during breath-holding.

Active Breathing

Dyspnea increases progressively with the level of ventilation during exercise. The intensity of the sensation, however, is not determined solely by the extent and frequency of thoracic displacement. When pulmonary function is impaired by disease the degree of dyspnea at any given level of ventilation is exaggerated; the increase in sensation is explained by the reduction in ventilatory capacity. It is now generally believed that dyspnea is based on the level and duration of the forces generated by the contracting respiratory muscles during breathing. The sensation is largely a function of the ratio of the pressures produced by the respiratory muscles to the maximum pressures that can be achieved given the contractile state, length, and velocity of shortening of the muscles.

The sensation of labored or difficult breathing may be mediated in a manner similar to that of other kinesthetic sensations, such as heaviness and force. Peripheral receptors that mediate the sensation are thought to be situated in the respiratory muscles or tendons and may include the muscle spindle. Of critical importance is the sense of effort or sensation of innervation stemming directly from the respiratory motor command that originates in the central nervous system.

During exercise and when mechanical lung function is disordered by disease, behavioral influences may affect the level and pattern of breathing. These influences tend to minimize respiratory muscle force and prevent undue respiratory muscle fatigue.

Bibliography

Journal articles

Bennett, F.M., et al.: Dynamics of ventilatory response to exercise in humans, J. Appl. Physiol. **51:**194, 1981.

Brain, J.D., et al.: Behavior of magnetic dusts in the lungs of rabbits correlates with phagocytosis, Exp. Lung Res. **6:**115, 1984.

Bureau, M.A., and Begin, R.: Postnatal maturation of the respiratory response to O_2 in awake newborn lambs, J. Appl. Physiol. **52:**428, 1982.

Cherniack, N.S.: Respiratory dysrhythmias during sleep, N. Engl. J. Med. **305:**325, 1981.

Corrales, R.J., et al.: Source of the fluid component of secretions from tracheal submucosal glands in cats, J. Appl. Physiol. **56:**1076, 1984.

Crapo, J.D., et al.: Structural and biochemical changes in rat lungs occurring during exposures to lethal and adaptive doses of oxygen, Am. Rev. Respir. Dis. **112:**123, 1980.

Davis, B., et al.: Reflex tracheal gland secretion evoked by stimulation of bronchial C-fibers in dogs, J. Appl. Physiol. **53:**985, 1982.

Deneke, S.M., et al.: Potentiation of oxygen toxicity in rats by dietary protein or amino acid deficiency, J. Appl. Physiol. **54:**147, 1983.

DeShazo, R.D., et al.: Bronchoalveolar lavage cell—lymphocyte interactions in normal nonsmokers and smokers. Analysis with a novel system, Am. Rev. Respir. Dis. **127:**545, 1983.

Eldridge, F.L., et al.: Exercise hyperpnea and locomotion: parallel activation from the hypothalamus, Science **211:**844, 1981.

Fisher, J.T., et al.: Respiration in newborns. Development of the control of breathing, Am. Rev. Respir. Dis. **125:**650, 1982.

Forman, H.J., et al.: Hyperoxia inhibits stimulated superoxide release by rat alveolar macrophages, J. Appl. Physiol. **53:**685, 1982.

Gothe, B., et al.: Effect of progressive hypoxia on breathing during sleep, Am. Rev. Respir. Dis. **126:**97, 1982.

Haddad, G.G., et al.: Determination of ventilatory pattern in REM sleep in normal infants, J. Appl. Physiol. **53:**52, 1982.

Janoff, A., et al.: Levels of elastase activity in bronchoalveolar lavage fluids of healthy smokers and nonsmokers, Am. Rev. Respir. Dis. **127:**540, 1983.

Jones, P.W., et al.: Cardiac output as a controller of ventilation through changes in right ventricular load, J. Appl. Physiol. **53:**218, 1982.

Lenfant, C., and Sullivan, K.: Adaptation to high altitude, N. Engl. J. Med. **284:**1298, 1971.

Lopes, J.M., et al.: Total airway resistance and respiratory muscle activity during sleep, J. Appl. Physiol. **54:**773, 1983.

Nathanson, I., et al.: Effect of vasoactive intestinal peptide on ion transport across the dog tracheal epithelium, J. Appl. Physiol. **55:**1844, 1983.

Oren, A., et al.: Effect of acid-base status on the kinetics of the ventilatory response to moderate exercise, J. Appl. Physiol. **52:**1013, 1982.

Shorofsky, S.R., et al.: Electrophysiology of Cl secretion in canine trachea, J. Membr. Biol. **72:**105, 1983.

Thach, B.T., et al.: Intercostal muscle reflexes and sleep breathing patterns in the human infant, J. Appl. Physiol. **48:**139, 1980.

Valberg, P.A.: Magnetometry of ingested particles in pulmonary macrophages, Science **224:**513, 1984.

Wasserman, K.: Breathing during exercise, N. Engl. J. Med. **298:**780, 1978.

White, D.P., et al.: Hypoxic ventilatory response during sleep in normal premenopausal women, Am. Rev. Respir. Dis. **126:**530, 1982.

Books and monographs

Cherniack, N.S., and Widdicombe, J.G., editors: Handbook of physiology, section 3: Respiratory system—control of breathing, vol. II, Bethesda, Md., 1986, American Physiological Society.

Fishman, A.P., and Fisher, A.B., editors: Handbook of physiology, section 3: Respiratory system—circulation and nonrespiratory functions, vol. I, Bethesda, Md., 1985, American Physiological Society.

Muir, D.C.F.: Deposition and clearance of inhaled particles. In Muir, D.C.F., editor: Clinical aspects of inhaled particles, London, 1972, William Heinemann, Ltd.

Scarpelli, E.M.: Pulmonary physiology of the fetus, newborn and child, Philadelphia, 1975, Lea & Febiger.

SECTION

VII

THE GASTROINTESTINAL SYSTEM

Howard C. Kutchai

CHAPTER

42

Gastrointestinal Motility

■ *Structure and Innervation of the Gastrointestinal Tract*

■ *Structure of the Wall of the Gastrointestinal Tract*

The structure of the gastrointestinal tract varies greatly from region to region, but there are common features in the overall organization of the tissue. Fig. 42-1 depicts the general layered structure of the wall of the gastrointestinal tract.

The *mucosa* consists of an epithelium, the lamina propria, and the muscularis mucosae. The nature of the epithelium varies greatly from one part of the digestive tract to another. The *lamina propria* consists largely of loose connective tissue containing collagen and elastin fibrils. The lamina propria is rich in several types of glands and contains lymph nodules and capillaries. The *muscularis mucosae* is the thin innermost layer of intestinal smooth muscle. In some parts of the gastrointestinal tract, the muscularis mucosae has an inner circular layer and an outer longitudinal layer. Contractions of the muscularis mucosae throw the mucosa into folds and ridges.

The *submucosa* consists largely of loose connective tissue with collagen and elastin fibrils. In some regions submucosal glands are present. The larger blood vessels of the intestinal wall travel in the submucosa.

The *muscularis externa* characteristically consists of two substantial layers of smooth muscle cells: an inner circular layer and an outer longitudinal layer. Contractions of the muscularis externa mix the contents in the lumen and propel them in a controlled fashion toward the anus.

The wall of the gastrointestinal tract contains a great many neurons that are highly interconnected. A dense network of nerve cells in the submucosa is called the *submucosal plexus* (Meissner's plexus). The prominent *myenteric plexus* (Auerbach's plexus) is located between the circular and longitudinal smooth muscle layers. The submucosal and myenteric plexuses (intramural plexuses), together with the other neurons of the gastrointestinal tract, constitute the *enteric nervous system,* which helps integrate the motor and secretory activities of the gastrointestinal system. If the sympathetic and parasympathetic nerves to the gut are cut, many of the motor and secretory activities continue to occur because of control by the enteric nervous system.

The *serosa,* or adventitia, is the outermost layer and consists mainly of connective tissue covered with a layer of squamous mesothelial cells.

■ *Innervation of the Gastrointestinal Tract*

Sympathetic innervation. Sympathetic innervation of the gastrointestinal tract is primarily via postganglionic adrenergic fibers whose cell bodies are in prevertebral and paravertebral plexuses. The celiac, superior and inferior mesenteric, and hypogastric plexuses provide postganglionic sympathetic innervation to various segments of the gastrointestinal tract. Most of the sympathetic fibers terminate in the submucosal and myenteric plexuses. Activation of the sympathetic nerves usually has an inhibitory effect on synaptic transmission in the enteric plexuses. Some sympathetic fibers innervate blood vessels of the gastrointestinal tract, causing vasoconstriction. Other sympathetic fibers innervate glandular structures in the wall of the gut. Relatively few of the sympathetic fibers terminate in the muscularis externa. Stimulation of the sympathetic input to the gastrointestinal tract inhibits motor activity of the muscularis externa but stimulates contraction of the muscularis mucosae and certain sphincters. The inhibitory effect of the sympathetic nerves on the muscularis externa is not a direct action on the smooth muscle cells, since there are few sympathetic nerve endings in the muscularis externa. Rather the sympathetic nerves act to influence neural circuits in the intrinsic plexuses

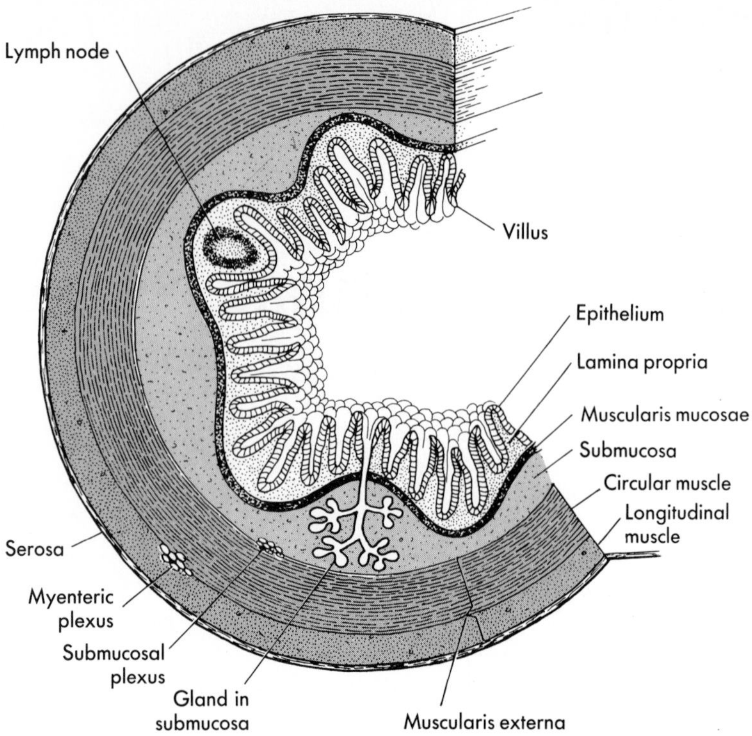

Fig. 42-1. The general organization of the layers of the gastrointestinal tract. (Modified from Ham, A.W.: Histology, ed. 3, Philadelphia, 1957, J.B. Lippincott Co.)

that provide input to the smooth muscle cells. This effect may be reinforced by the action of the sympathetic nerves in reducing blood flow to the muscularis externa. Other fibers that travel with the sympathetic nerves may be cholinergic; still others release neurotransmitters that remain to be identified. Fig. 42-2 summarizes the sympathetic innervation of the gastrointestinal tract.

Parasympathetic innervation. Parasympathetic innervation of the gastrointestinal tract down to the level of the transverse colon is provided by branches of the vagus nerve. The remainder of the colon receives parasympathetic fibers from the pelvic nerves by way of the hypogastric plexus. These parasympathetic fibers are preganglionic and predominantly cholinergic. Other fibers that travel in the vagus and its branches have other transmitters that have not been identified. Preganglionic fibers terminate predominantly on the ganglion cells in the intramural plexuses. The ganglion cells innervate the smooth muscle and secretory cells of the gastrointestinal tract. Parasympathetic input usually stimulates the motor and secretory activity of the gut. Fig. 42-3 illustrates the parasympathetic innervation of the gastrointestinal tract.

Enteric nervous system. The myenteric and submucosal plexuses are the most well-defined plexuses in the wall of the gastrointestinal tract. Other plexuses also have been identified. The plexuses are networks of nerve fibers and ganglion cell bodies. Some of the incoming axons are preganglionic parasympathetic fibers, and others are postganglionic sympathetic fibers. Interneurons in the plexuses connect afferent sensory fibers with efferent neurons to smooth muscle and secretory cells. Consequently, the myenteric and submucosal plexuses can control a good deal of coordinated activity in the absence of extrinsic innervation to the gastrointestinal tract. Axons of plexus neurons innervate gland cells in the mucosa and submucosa, smooth muscle cells in the muscularis externa and muscularis mucosae, and intramural endocrine and exocrine cells. Afferent fibers from mechanoreceptors and chemoreceptors in the mucosa or deeper in the wall of the gastrointestinal tract synapse in the plexuses, so that local reflex activity is possible (Fig. 42-4). The functions of the intramural plexuses are discussed in more detail later in this chapter.

Afferent fibers. Afferent fibers in the gut provide the afferent limbs of reflex arcs that are both local and central (Fig. 42-4). Chemoreceptor and mechanoreceptor endings are present in the mucosa and in the muscularis externa. The cell bodies of some of these sensory receptors are located in the myenteric and submucosal plexuses. The axons of some of these receptor cells synapse with other cells in the plexuses to mediate local reflex activity. Others of these receptors send their axons back to the central nervous system. The cell bodies of still other sensory neurons are located more centrally. These afferent fibers in the vagus, which greatly outnumber vagal motor fibers, have cell bodies principally in the nodose ganglion, whereas sensory fibers that travel via the sympathetic nerves have their cell bodies in dorsal root ganglia. The number of sensory afferent fibers from the gastrointestinal tract is large, and its complex

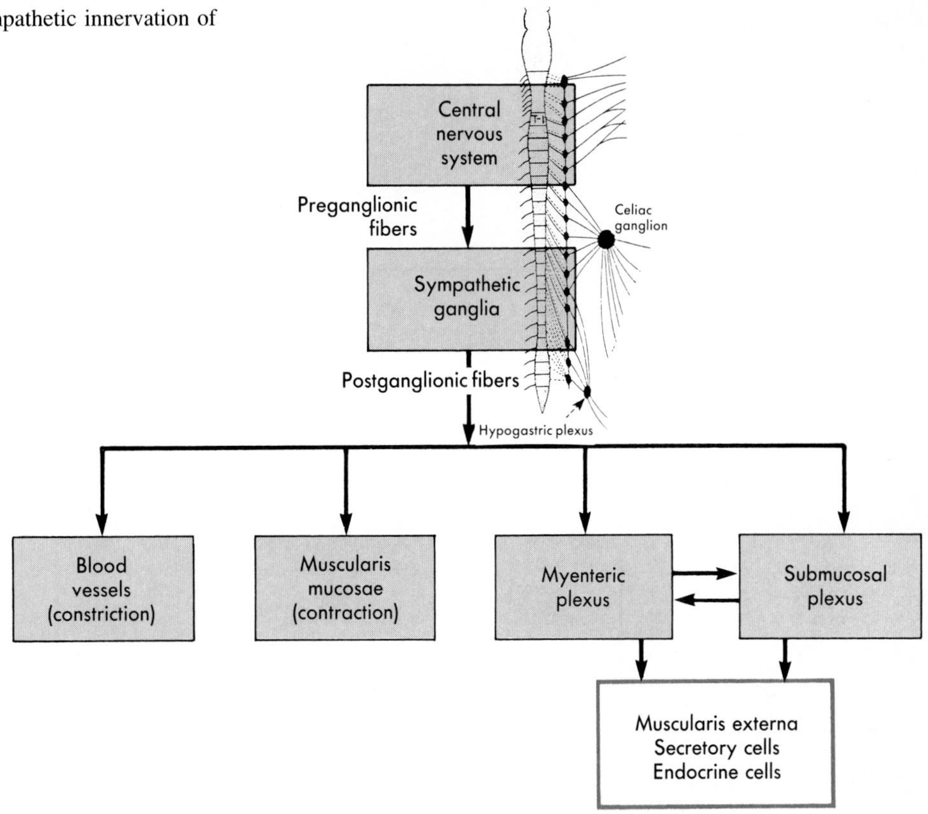

■ **Fig. 42-2.** Major aspects of the sympathetic innervation of the gastrointestinal tract.

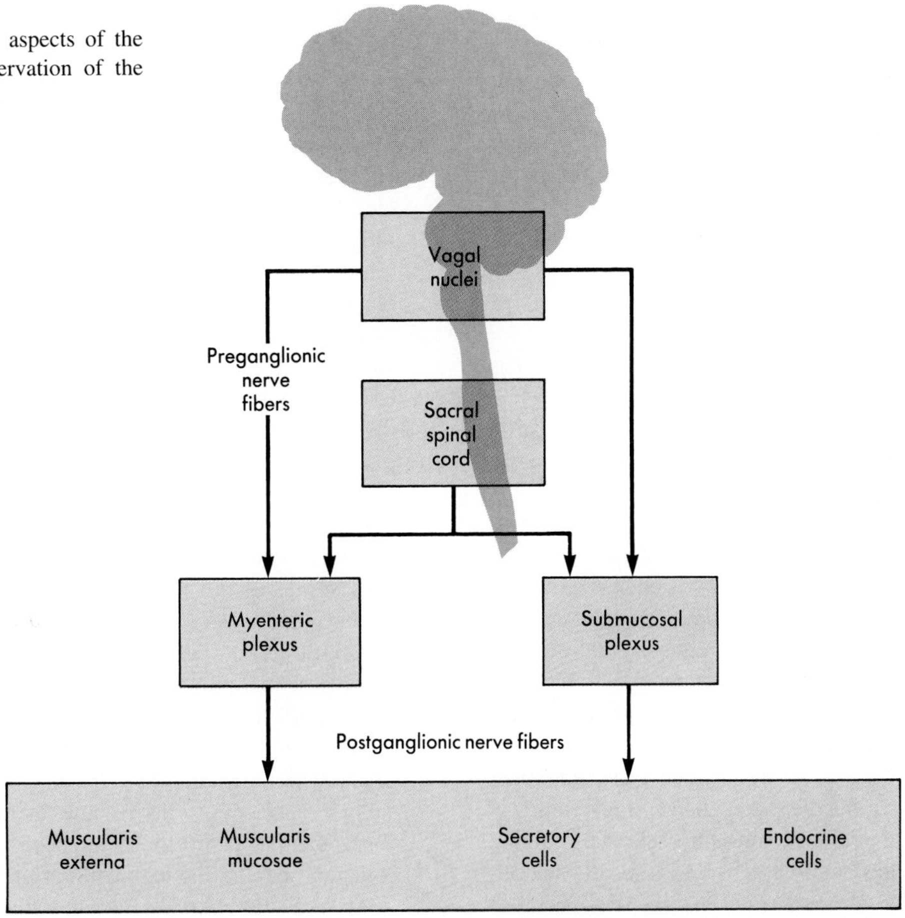

■ **Fig. 42-3.** Major aspects of the parasympathetic innervation of the gastrointestinal tract.

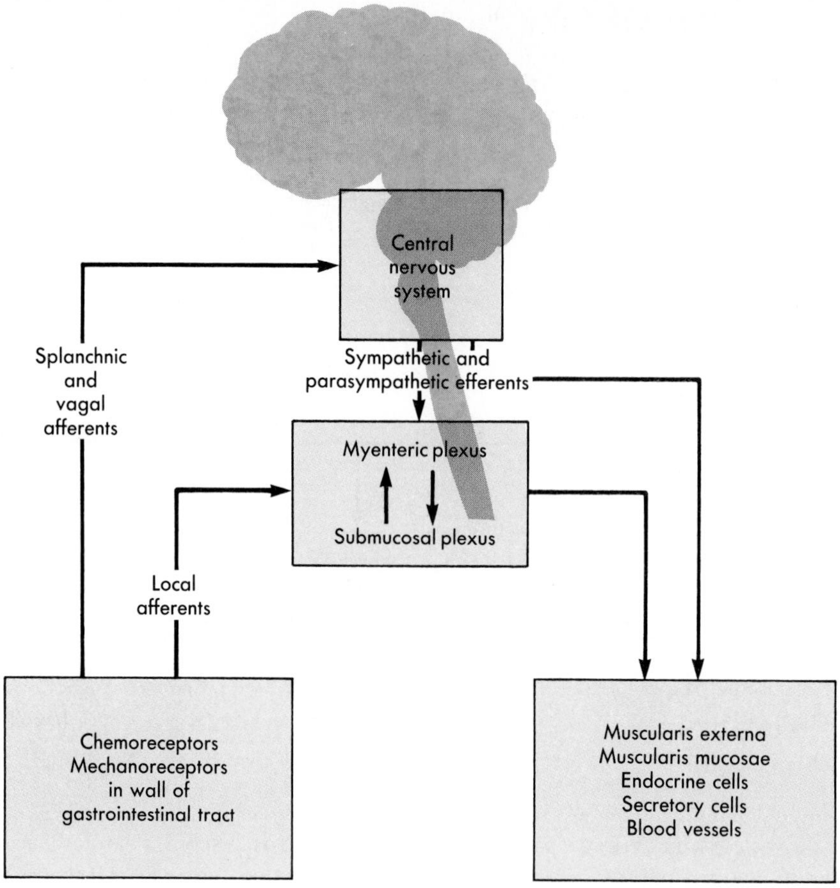

Fig. 42.4. Local and central reflex pathways in the gastrointestinal system.

afferent and efferent innervation allows for fine control of secretory and motor activities by intrinsic and central pathways.

Gastrointestinal Smooth Muscle

Smooth Muscle Cells of the Muscularis Externa

The properties of smooth muscle were discussed in Chapter 23. Some of the properties pertinent to gastrointestinal smooth muscle are reviewed here.

The smooth muscle cells of the gastrointestinal tract are long (about 500 μm in length) and slender (about 5 μm in diameter). The smooth muscle cells are arranged in bundles that are separated and defined by connective tissue. The bundles branch and frequently anastomose with other bundles. The bundles are about 200 μm thick, and a cross-section of a bundle contains several thousand cells. Because of the high degree of electrical coupling among the smooth muscle cells, the functional contractile unit is a bundle rather than a single cell.

Gastrointestinal smooth muscle cells are much smaller than skeletal muscle cells. As a result, smooth muscle cells have a much larger surface/volume ratio. The resting conductances to the major ions (Na^+, K^+, Cl^-) are considerably less than those in skeletal muscle. Therefore, in spite of the increased surface/volume ratio in smooth muscle, the energy cost of maintaining ion gradients across the plasma membrane is comparable to that of skeletal muscle. The plasma membrane contains both the Na^+, K^+-ATPase and a Ca^{++}-ATPase to accumulate K^+ and to extrude Na^+ and Ca^{++}. Cl^- is accumulated against a gradient, but the nature of the active transport responsible for Cl^- uptake remains to be elucidated.

The plasma membranes of gastrointestinal smooth muscle cells have numerous invaginations that form structures called caveolae. Caveolae may increase the surface area of the cell by 50% to 70%. The interior of the caveolae is accessible to macromolecules present in the extracellular fluid. The functional significance of caveolae is unknown.

A significant fraction of the interior surface of the plasma membrane is covered with plaques of electron-dense material called *dense bands*. Frequently the extracellular surface of the plasma membrane opposite a dense band appears to be attached via microfibrils to collagen fibers in the extracellular matrix. These attachments may function as intramuscular "microtendons" that allow the force of contraction of individual cells to be transferred to the muscle as a whole. Occasionally a dense band of one cell may be aligned with a dense band of a neighboring cell. In such cases the neighboring plasma membranes are frequently closely apposed between the dense bands, with electron-dense material bridging the gap between the cells. This structure is known as an *intermediate junction*. Intermediate junctions are believed to represent mechanical connections that allow transmission of contractile force between neighboring cells.

Neighboring smooth muscle cells in the circular muscle layer are joined by frequent *gap junctions*. The gap junctions (Chapter 1) mediate electrical coupling between the smooth muscle cells of the circular layer. The cells of the longitudinal layer are also relatively well-coupled electrically but lack demonstrable gap junctions. It is not known which structures mediate electrical coupling in the longitudinal muscle layer.

Smooth muscle cells have *thin filaments,* composed principally of actin. *Thick filaments* are also present in smooth muscle (although more difficult to preserve for microscopy) and are composed chiefly of myosin. The ratio of thin to thick filaments is much higher in smooth muscle than in striated muscle. Although smooth muscle contains less myosin than does striated muscle, smooth muscle can generate as much force per cross-sectional area as does striated muscle.

Smooth muscle cells also contain *intermediate filaments,* so named because of their 10 nm diameter. Intermediate filaments link the dense bands on the inner surface of the plasma membrane to *dense bodies* in the cytoplasm. The network of dense filaments that connects dense bodies and dense bands forms a superstructure for the smooth muscle cell. The filament network allows the transmission of contractile force to the plasma membrane. The dense bodies and dense bands contain *α-actinin,* a major protein of the Z disk in striated muscle. Numerous thin filaments are attached to dense bodies and dense bands. Thus dense bodies and dense bands may be functionally homologous to the Z disks of striated muscle. Whether smooth muscle possesses a definable structure that corresponds to the sarcomere of striated muscle is controversial.

Striated muscle develops force effectively in a relatively narrow range of muscle length. Smooth muscle in general and gastrointestinal smooth muscle in particular have a much wider range of lengths in which they can develop a large fraction of their maximal contractile force. Smooth muscle can contract to smaller fractions of its rest length than can striated muscle, and smooth muscle can still develop force when stretched to several times its rest length. The irregular arrangement of the contractile apparatus of smooth muscle may be responsible for the breadth of its length-tension curve.

Electrophysiology of Gastrointestinal Smooth Muscle Cells

The resting membrane potential. The resting membrane potential of gastrointestinal smooth muscle cells ranges from about −40 mV to around −80 mV. As discussed in Chapter 2, the relative membrane conductances of K^+, Na^+, and Cl^- are important in determining the resting membrane potential. Compared with skeletal muscle, gastrointestinal smooth muscle cells have a higher ratio of Na^+ conductance to K^+ conductance. This contributes to the somewhat lower resting membrane potential of gastrointestinal smooth muscle.

If the relative conductances and the equilibrium potentials for K^+, Na^+, and Cl^- in guinea pig teniae coli (a well-studied preparation) are used in the chord conductance equation (Chapter 2), a *predicted* resting membrane potential of about −40 mV for the teniae coli is computed. When the resting membrane potential of teniae coli is measured with microelectrodes, a value of about −60 mV is obtained. The potential difference across the plasma membrane of the resting teniae coli is about 20 mV larger than can be accounted for by diffusion of ions.

What is responsible for the extra 20 to 25 mV of polarization across the plasma membrane? The current view is that the Na^+, K^+ pump is *electrogenic,* and it contributes about 20 mV to the resting membrane potential. Since 3 Na^+ are extruded for every 2 K^+ taken up, the pump produces a net outward flow of positive charge, which contributes to the membrane potential. (A process producing net current flow is called *electrogenic*.) Experiments indicate that the electrogenic Na^+, K^+ pump is responsible for a significant part of the resting potential in gastrointestinal smooth muscle (Fig. 42-5). If the Na^+, K^+ pump of guinea pig teniae coli is inhibited with ouabain, the resting membrane potential changes from about −60 to near −40 mV. When ouabain is washed out, the resting potential returns to near −60 mV.

Variations in the membrane potential. In most other excitable tissues the resting membrane potential is rather constant in time. In gastrointestinal smooth muscle the resting membrane potential characteristically varies in time.

Slow waves. Slow waves are slow potential oscillations that are characteristic of gastrointestinal smooth muscle. The frequency of the oscillations varies from about 3 per minute in the stomach to about 12 per minute in the duodenum. Fig. 42-6 shows sinusoidal slow waves in the rabbit jejunum. *Slow waves* also are referred to as the *basic electrical rhythm* or *electrical control activity*.

In each segment of the gastrointestinal tract the basic electrical rhythm of some cells is faster than that of the

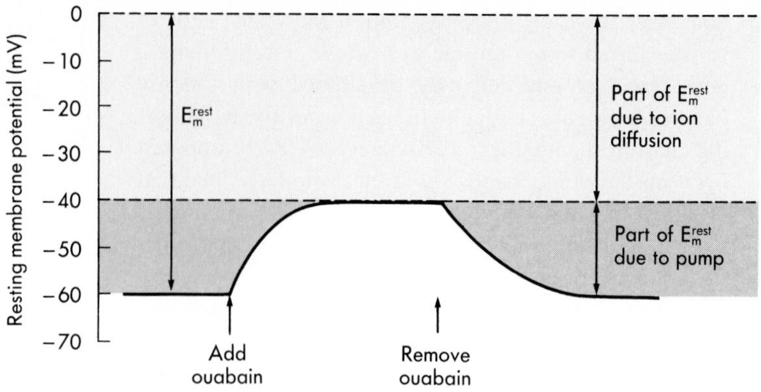

■ **Fig. 42-5.** Idealized representation of an experiment suggesting that a significant part of the resting membrane potential of gastrointestinal smooth muscle is caused by the electrogenic Na^+,K^+ pump.

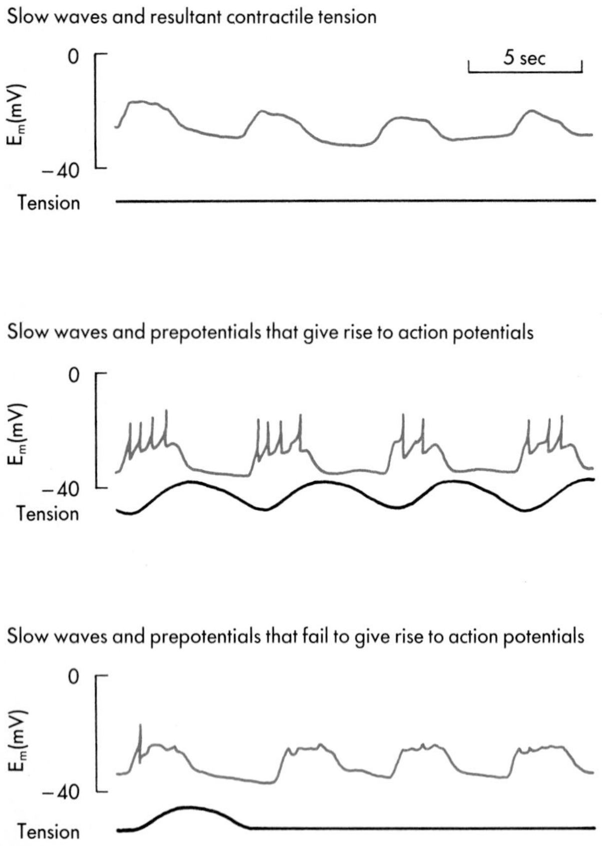

■ **Fig. 42-6.** Electrical and mechanical activity of gastrointestinal smooth muscle. The upper tracing in each panel is the transmembrane electrical potential difference (millivolts), and the lower tracing shows contractile tension. The range of different electrical behaviors in a single tissue is noteworthy. The tissue contracts only in response to action potentials. (Modified from Bortoff, A.: Am. J. Physiol. **201:**203, 1961.)

other cells. These faster cells serve as "pacemakers" for the slow waves in the region. Because the smooth muscle cells are well coupled electrically, the wave of depolarization (or repolarization) spreads throughout that segment of the gut.

The mechanism whereby the slow waves are generated is not well understood. Cyclical variation in the rate of the Na^+, K^+ pump may play a role, but cyclical changes in the membrane conductance to Na^+ may also be involved.

The amplitude and, to a lesser extent, the frequency of the slow waves can be modulated in some segments of the gastrointestinal tract by the activity of intrinsic and extrinsic nerves and by circulating hormones. In general, circulating epinephrine and norepinephrine released from sympathetic nerve terminals tend to decrease the amplitude of the slow waves or to abolish them, whereas acetylcholine tends to increase the size of the slow waves.

Depending on the exictability of the smooth muscle cells, the slow waves may induce action potentials. If the peak of the slow wave is above threshold for the cells to fire action potentials, one or more action potentials may be triggered during the peak of the slow wave (Fig. 42-6).

The site of origin of the slow waves is a matter of controversy. The generation of the slow waves does not require the participation of the intramural plexuses.

Prepotentials. Some gastrointestinal smooth muscle cells show spontaneous depolarizations that are much more rapid than the slow waves. These depolarizations resemble those in the pacemaker cells of the sinoatrial node of the heart and are known as *prepotentials* or *pacemaker potentials* (Fig. 42-6). Prepotentials often, but not always, serve to depolarize the smooth muscle cell to threshold, whereupon an action potential is fired (Fig. 42-6).

The prepotential may be caused by an increase in the

membrane conductance to Na^+, Ca^{++}, or both. The slope of the prepotential may be altered by nervous and humoral influences. Sympathetic stimulation or circulating epinephrine may decrease or abolish prepotentials. Stimulation of parasympathetic nerves to the gut may increase the slope of the pacemaker potentials and increase their frequency. The degree of stretch of the tissue also can alter the frequency of slow waves and prepotentials.

Action potentials. Action potentials in gastrointestinal smooth muscle are more prolonged (10 to 20 msec duration) than those of skeletal muscle and have little or no overshoot. The rising phase of the action potential is caused by ion flow through channels that are relatively slow to open and probably conduct both Ca^{++} and Na^+. The Ca^{++} that enters the cell during the action potential plays a significant role in initiating contraction. Repolarization is aided by a delayed increase in K^+ conductance.

When the membrane potential of gastrointestinal smooth muscle reaches a threshold value, a train of action potentials (1 to 10 per second) occurs (Fig. 42-6). The extent of depolarization of the cells and the frequency of action potentials are enhanced by acetylcholine and other excitatory compounds liberated from nerve endings. Circulating epinephrine and norepinephrine released from sympathetic nerve endings tend to hyperpolarize the smooth muscle cells and to abolish action potential spikes.

Electrical Coupling Between Smooth Muscle Cells

Neighboring cells are said to be well coupled electrically if a perturbation of the membrane potential of one cell spreads rapidly, and with little decrement, to the other cell. The smooth muscle cells of the muscularis externa are well coupled.

The smooth muscle cells of the circular layer are somewhat better coupled than those of the longitudinal layer. The cells of the circular layer are joined by frequent gap junctions that are believed to provide low-resistance pathways that allow the spread of electrical current from one cell to another.

Since the electrical resistance of membranes is much higher than the resistance of the cytoplasm, the resistance along the long axis of smooth muscle cells is less than that in the transverse direction. Thus an electrical depolarization in the circular muscle is readily conducted circumferentially and quickly depolarizes a ring of circular muscle near the original site of depolarization. The ring of depolarization then spreads more slowly along the long axis of the gut.

In the intact gastrointestinal tract the longitudinal smooth muscle of the muscularis externa also behaves as though its cells were well coupled electrically. Because of the low electrical resistance along the length of the smooth muscle cells, an electrical disturbance is rapidly propagated along the long axis of the gut in the longitudinal layer. Thus the slow waves are conducted along the longitudinal layer, and they may spread from the longitudinal layer inward to the circular layer of smooth muscle cells. Contractions of the longitudinal smooth muscle, with concomitant inhibition of the contractions of the circular muscle, shorten and expand segments of the intestine and in this way contribute to propelling intestinal contents.

Coupling between longitudinal and circular layers of smooth muscle. Typically the slow waves themselves do not elicit contractions of the smooth muscle layers. Contraction is evoked by the action potentials that are intermittently triggered near the peaks of the slow waves (Fig. 42-6). (The smooth muscle of the stomach is an exception and contracts in response to the depolarizing phases of the slow waves.) Neighboring cells of the circular and longitudinal layers may be joined by gap junctions, which may play a role in the high degree of electrical coupling between the two layers. As mentioned earlier, the site of the origin of slow waves is not certain, but regardless of which layer contains the pacemaker cells, the slow waves would be rapidly propagated to the other layer.

Excitation-Contraction Coupling

As in skeletal and cardiac muscle, the level of intracellular Ca^{++} plays a central role in regulating the contraction of smooth muscle. In smooth muscle, contraction is initiated by the Ca^{++} that crosses the plasma membrane during the action potential, as well as by Ca^{++} released from the sarcoplasmic reticulum. The less well developed the sarcoplasmic reticulum, the more the smooth muscle is dependent on extracellular Ca^{++} for contraction.

As discussed in Chapter 22, regulation of contraction by Ca^{++} in smooth muscle occurs by a different mechanism than in skeletal muscle. Smooth muscle thin filaments lack troponin, and it appears that Ca^{++} regulation in smooth muscle centers on the myosin molecule itself. At present the best evidence supports the view that increased cytoplasmic Ca^{++} binds to calmodulin, which then activates a myosin light chain kinase, which phosphorylates one of the two myosin light chains. The phosphorylation allows the head of the myosin molecule to interact cyclically with actin of the thin filaments, causing muscle shortening.

Contractility of Intestinal Smooth Muscle

The length/tension curve. The length/tension curve of gastrointestinal smooth muscle is similar in shape to that of skeletal muscle, but it has a much broader maximum. This gives gastrointestinal smooth muscle the ability to develop force effectively over a greater range of muscle length. This property may reflect the organization of muscle cells within the tissue and the organization of the contractile elements within the cells more than it reflects the intrinsic length/tension properties of the contractile elements themselves.

Relationship between membrane potential and tension. Gastrointestinal smooth muscle cells fire action potentials (Fig. 42-6), and the cells contract phasically in response to action potentials. The action potentials usually occur in bursts at the peak of the slow waves and cause phasic contractions that are superimposed on the baseline level of contraction. Because smooth muscle cells contract rather slowly (about 10 times slower than skeletal muscle), the individual contractions caused by each action potential in a burst are not visible as distinct twitches but rather sum temporally to produce a smoothly increasing level of tension. The increase in tension in response to a burst of action potentials is proportional to the number of action potentials in the burst. Gastric smooth muscle is exceptional in that it contracts in response to the depolarizing phase of the slow waves, even in the absence of action potentials.

Between bursts of action potentials the tension developed by gastrointestinal smooth muscle falls, but not to zero. This nonzero ''resting,'' or baseline, tension developed by the smooth muscle is called *tone*. The tone of gastrointestinal smooth muscle may be altered by neurotransmitters, hormones, or drugs.

Responses to stretch. Gastrointestinal smooth muscle may respond to stretch or release of stretch. In many cases rapidly stretching gastrointestinal smooth muscle will lead to an immediate increase in the frequency of action potentials and an increase in contractile tension—a phenomenon known as *stress activation*. This may be followed by a decrease in tension back toward the original level—a phenomenon known as *stress relaxation*. Both these responses to stretch play roles in the motility of the various segments of the gastrointestinal tract.

Integration and Control of Gastrointestinal Motor Activities

Control of the contractile activities of gastrointestinal smooth muscle involves the central nervous system, the intrinsic plexuses of the gut, humoral factors, and electrical coupling among the smooth muscle cells. The motor behavior of particular segments of the gastrointestinal tract is discussed in the succeeding sections of this chapter.

This section outlines some of the general structural and functional properties of gastrointestinal smooth muscle and its innervation that underlie control and integration of contractile function. A major conclusion is that the gastrointestinal tract displays a great deal of intrinsic control. The intramural plexuses can mediate control and integration of most of the contractile behavior of the gut without intervention by the central nervous system. The autonomic nervous system modulates patterns of muscular and secretory activity that are controlled by the enteric nervous system.

Neuromuscular Interactions in the Gastrointestinal Tract

Neuromuscular interactions in the gastrointestinal tract do not involve true neuromuscular junctions with specialization of the postjunctional membrane, as occurs at the motor endplate. The neurons of the intramural plexuses send axons to the smooth muscle layers, and each axon branches extensively to innervate many smooth muscle cells.

The circular smooth muscle layer of the muscularis externa is heavily innervated by nerve terminals that form close associations with the plasma membranes of the smooth muscle cells. Neuromuscular gaps of about 20 nm are typical in the circular layer. The predominant innervation is by fibers that are *inhibitory* to the electrical and contractile activity of the smooth muscle cells.

The longitudinal smooth muscle cells are much less richly innervated by the neurons of the intrinsic plexuses, and the neuromuscular contacts are not so intimate. Gaps of about 80 nm separate nerve terminals from the plasma membrane of the smooth muscle cells they innervate. In contrast to the situation in the circular layer, the longitudinal layer appears to be totally devoid of inhibitory nerve endings. The sparse excitatory nerve fibers are predominantly cholinergic.

Properties of Gastrointestinal Smooth Muscle Relevant to Control of Motility

Circular layer. In the absence of neural influences (as when the plexus neurons are inactivated by drugs) each slow wave elicits a burst of action potentials and a near-maximal contraction of the cells of the circular layer. Because of the phase relation among the slow waves in neighboring segments of the small intestine, the circumferential ring of contraction elicited by each slow wave travels along the small bowel toward the co-

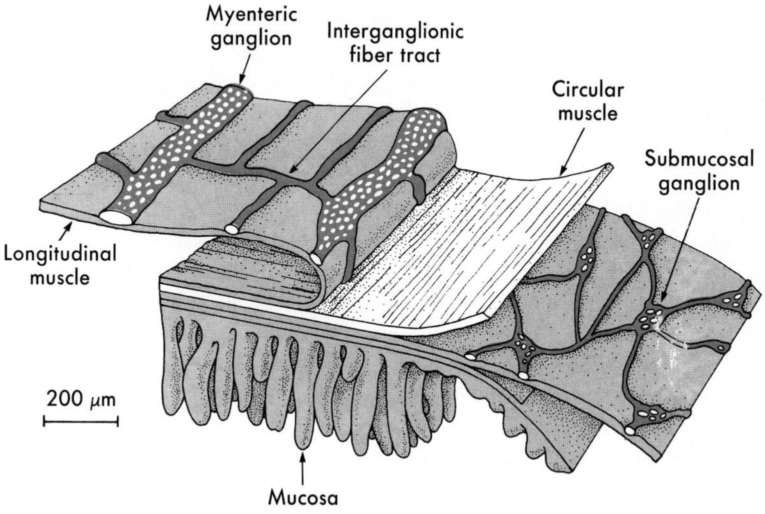

■ **Fig. 42-7.** Enteric neurons of the submucosal and myenteric plexuses in the wall of the gastrointestinal tract. The plexuses consist of ganglia interconnected by fiber tracts. Note the different shapes of ganglia in the two plexuses. (From Wood, J.D. In Johnson, R.L., editor: Physiology of the gastrointestinal tract, New York, 1981, Raven Press.)

lon. This contrasts strongly with the behavior of the intact gut with a functional intrinsic innervation. In the functionally intact gut only every third or fourth slow wave may elicit action potentials and contractions, which are usually submaximal and may be quite localized. The normal activity of the intrinsic plexuses of the gut is largely inhibitory to the circular muscle layer, preventing the burst of action potentials and the maximal propagated wave of contraction that occur in response to each slow wave in the absence of the intrinsic neurons.

Longitudinal layer. The predominant innervation of the longitudinal layer is by cholinergic excitatory fibers from the intrinsic plexuses. For this reason action potentials and resultant contractions are less frequent in the functionally denervated longitudinal layer than in a more intact preparation.

Thus neural activity of the plexus neurons tends to inhibit the circular muscle but to stimulate the longitudinal muscle. This is physiologically relevant, since the two layers are antagonistic in the sense that contraction of one layer opposes contraction of the other.

■ *The Enteric Nervous System*

In 1921 Langley proposed that the plexuses of the wall of the gastrointestinal tract function as a semiautonomous nervous system controlling the motor and secretory activities of the digestive system. The enteric nervous system of the large and small intestines alone contains about 10^8 neurons, about as many neurons as in the spinal cord. Fig. 42-7 depicts the myenteric and submucosal plexuses and their locations in the wall of the intestine. Both plexuses consist of ganglia, which are interconnected by tracts of fine, unmyelinated nerve fibers. Note that the submucosal ganglia are smaller and less elongated than the myenteric ganglia, which are elongated in a direction perpendicular to the long axis of the intestine. The neurons in the ganglia include sensory neurons, with their sensory endings in the wall of the gastrointestinal tract. Neurons sensitive to mechanical deformation, to particular chemical stimuli, and to temperature have been identified. Some of the neurons in the enteric ganglia are effector neurons that send axons to smooth muscle cells of the circular or longitudinal layers, to secretory cells of the gastrointestinal tract, or to gastrointestinal blood vessels. Perhaps most of the neurons in the enteric ganglia are interneurons that are part of a network of neurons that integrates the sensory input to the ganglia and formulates the output of the effector neurons.

Neuromodulatory substances of the enteric nervous system. The number of neuromodulatory substances present in the wall of the gastrointestinal tract is similar to the number of such compounds in the brain; in fact, most of the known neuromodulators are

■ **Table 42-1.** List of substances present in enteric neurons

Presence established	*Presence proposed*
Acetylcholine	Angiotensin II–like peptide
Adenosine triphosphate (ATP)	Dynorphin
Bombesin–gastrin-releasing peptide	γ-Aminobutyric acid
Cholecystokinin octapeptide (CCK8)	Motilin
Leu-enkephalin	Neurotensin
Met-enkephalin	Pancreatic polypeptide (PP)
5-Hydroxytryptamine (5-HT)	Physalaemin-like peptide
Noradrenaline	
Somatostatin	
Substance P	
Vasoactive intestinal polypeptide (VIP)	

From Costa, M., and Furness, J.B.: Br. Med. Bull. **38:**247, 1982.

present in both gut and brain. Table 42-1 lists some of the neuroactive substances that are present in the gastrointestinal tract. The functions of some of these neuroactive substances in the central nervous system were discussed in Chapter 4. Table 42-2 lists the densities of enteric neurons that contain particular candidate transmitters and neuromodulators. In most cases a certain neuropeptide has not been proved to be the neurotransmitter at a particular synapse in the gastrointestinal tract. However, it is likely that in many cases intestinal neuropeptides are true transmitters. Neuropeptides also probably function as neuromodulators in the gastrointestinal tract. Neuromodulators are substances that act presynaptically to alter the amount of transmitter released or that act postsynaptically to modify the response of the postsynaptic cell to transmitter.

Classification of enteric neurons. Enteric neurons have been classified according to their morphology and their electrophysiological properties. The current state of understanding of the enteric nervous system is such that correlations among enteric neurons classified in these ways cannot be made with certainty. For example, extracellular recordings reveal a population of enteric neurons, known as "steady bursters," that fire short groups of action potentials at constant intervals. However, studies with intracellular electrodes have failed to find cells that correspond to the steady bursters.

Classification based on extracellular recordings. Three different kinds of neurons have been distinguished in the plexuses on the basis of extracellular recordings: burst units, mechanosensitive units, and single-spike units.

Burst units can be further subdivided into "steady bursters" and "erratic bursters." The steady bursters (Fig. 42-8, *A*) discharge clusters of action potentials at rather regular intervals. The rhythm of the steady bursters is not dependent on their synaptic input. The steady bursters appear to play a central role in the activity patterns that are intrinsic to the gastrointestinal tract and do not depend on extrinsic innervation. The erratic bursters (Fig 42-8, *B*) depend on synaptic input, perhaps from the steady bursters as well as from other neurons.

Mechanosensitive units, which also can be subdivided into groups based on their properties, are activated by mechanical distortion of the mechanosensitive plexus neurons themselves. The mechanosensitive neurons provide afferent input to other plexus neurons and also to the central nervous system. Perhaps this afferent information modulates the activities of the erratic burst neurons, both directly and via reflex pathways.

Single-spike units fire single action potentials at infrequent and inconsistent intervals. The pattern of discharge of the single-spike units may be modulated by synaptic input, but the discharge of these units is not absolutely dependent on synaptic input.

The three types of plexus neurons presumably inter-

Table 42-2. Number and type of neurons in 1 centimeter length of guinea pig's small intestine

	Myenteric plexus	*Submucous plexus*
Total number	10,000	7,200
Substance P	350	820
VIP	240	3,060
Somatostatin	470	1,260
Enkephalin	2,450	0
5-HT	200	0
Amine handling	50	850

From Costa, M., and Furness, J.B.: Br. Med. Bull. **38:**247, 1982.

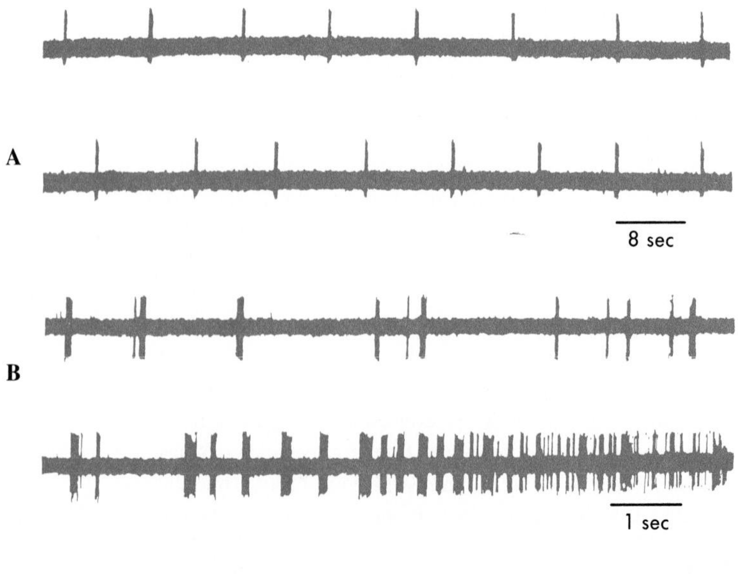

Fig. 42-8. Discharge patterns of two enteric neurons of the jejunum as recorded by extracellular electrodes. **A,** Steady-burst type of neuron from the cat. **B,** Erratic-burst type of neuron from the guinea pig. (From Wood, J.D.: Physiol. Rev. **55:**307, 1975.)

act in complex ways that allow for the high degree of local control of the motor and secretory activities of the gastrointestinal tract. Apparently the enteric brain contains pattern-generating neural circuits that are repeated along the length of the gastrointestinal tract and play a central role in the coordinated motor activities of the gut.

Classification based on intracellular recordings. At present four different types of enteric neurons have been detected with intracellular electrodes. Intracellular microelectrodes allow the investigator to probe the ionic mechanisms that underlie the electrophysiological behavior of individual enteric neurons. Only the somas of enteric neurons may be studied with intracellular microelectrodes, because their axons and dendrites are too small to be penetrated. The classification of enteric neurons based on intracellular recordings is tentative at the present time.

Type 1 or S neurons. These neurons have a low resting membrane potential and a high input resistance, so the low resting membrane potential may result from a low membrane conductance to K^+. When the type 1 neurons are depolarized they respond by firing bursts of action potentials that persist as long as the depolarization is maintained (Fig. 42-9, *B*). Action potentials are blocked by tetrodotoxin. Therefore fast Na^+ channels of the type present in nerve are probably involved in generating the action potential. Frequent excitatory postsynaptic potentials (EPSPs) of cholinergic synapses occur in type 1 cells.

Type 2 or AH neurons. These neurons have a higher resting membrane potential than type 1 neurons and a lower input resistance. These characteristics probably reflect a higher resting K^+ conductance. In response to depolarization, these neurons discharge only one or two action potentials (Fig. 42-9, *A*), which are followed by a prolonged period of hyperpolarization that inhibits further excitability (Fig. 42-10). The hyperpolarization is caused by an increased intracellular Ca^{++} concentration, which increases the K^+ conductance.

Other enteric neurons. Other neurons are less well defined than the types 1 and 2 neurons. Type 3 neurons resemble type 2 neurons in having a high resting membrane potential, apparently resulting from high resting K^+ conductance. However, unlike type 2 neurons, type 3 neurons do not fire action potentials in response to depolarization. Prominent *excitatory postsynaptic potentials* (EPSPs) are evident in type 3 neurons. Another class of enteric neurons, called type 4, shows neither action potentials nor EPSPs.

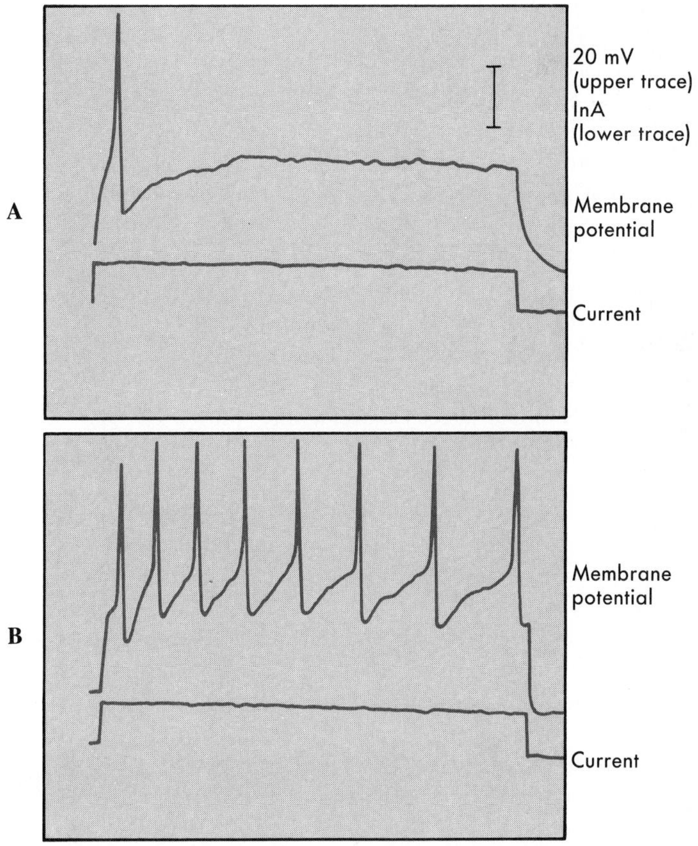

Fig. 42-9. Intracellular recordings from enteric neurons. The tracings show the response to injection of a 200 msec–duration pulse of depolarizing current into neurons of the myenteric plexus of a guinea pig's small intestine. **A,** Type 2 (AH type) neuron. **B,** Type 1 (S type) neuron. Voltage and current scale in **A** also applies to **B.** (From Wood, J.D.: In Johnson R.L.: Physiology of the gastrointestinal tract, New York, 1981, Raven Press.)

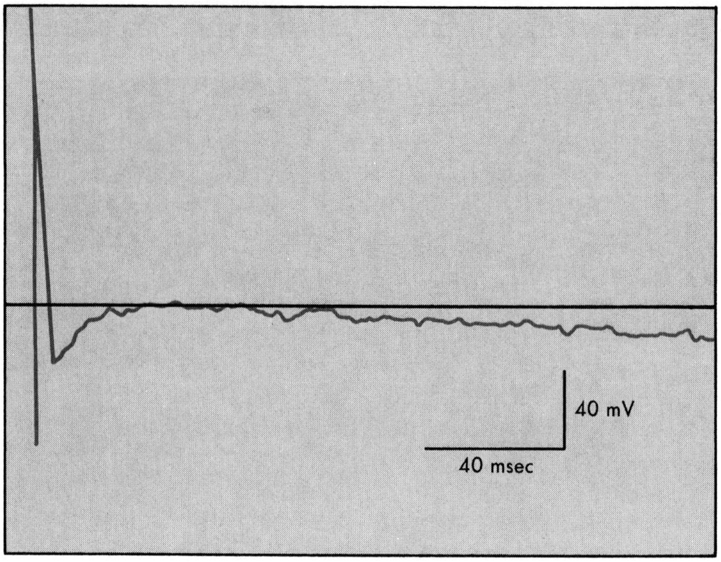

Fig. 42-10. Intracellular record from a type 2 (AH type) neuron in the myenteric plexus of a guinea pig's small intestine. Note the prolonged hyperpolarization that follows the action potential spike after a delay of about 100 msec. (From Wood, J.D.: In Johnson, R.L., editor: Physiology of the gastrointestinal tract, New York, 1981, Raven Press.)

Synaptic transmission in enteric ganglia. Axons of enteric neurons synapse with somas, dendrites, or axons of other enteric neurons. The kinds of synaptic responses seen in enteric ganglia are similar to those observed in the brain and spinal cord. Excitatory and inhibitory synaptic potentials have been observed in enteric neurons. Some synaptic events are fast (taking 50 msec or less) and other synaptic potential changes are slow (lasting several seconds). Presynaptic inhibition also occurs in enteric ganglia.

Fast EPSPs have been observed in types 1, 2, and 3 enteric neurons. Acetylcholine is probably the transmitter that evokes fast EPSPs.

Fast inhibitory postsynaptic potentials (IPSPs) are observed and are associated with a decreased membrane resistance. The fast IPSPs have a reversal potential around -90 mV. The ionic conductance responsible for the IPSP has not been identified, but increased K^+ conductance is the leading candidate.

Slow EPSPs are depolarizations that persist for seconds, or even minutes, in enteric neurons. Slow EPSPs are associated with decreased ionic conductance (probably to K^+) and increased excitability. Slow EPSPs in the submucosal plexus occur primarily in type 1 neurons and in the myenteric plexus predominantly in type 2 neurons. Slow EPSPs reduce the hyperpolarizing afterpotentials of type 2 neurons and convert them transiently to neurons that can fire trains of action potentials.

Slow IPSPs can be evoked in neurons of both myenteric and submucosal plexuses. Slow IPSPs are hyperpolarizations that persist for 1 to 5 seconds and are associated with increased ionic conductance of the membrane, perhaps to K^+. Slow IPSPs induce persistent inhibition of enteric neurons.

Presynaptic inhibition is caused by suppression of transmitter release from axons of enteric neurons. The inhibition results from depolarization of the axon terminals by inhibitory synapses on them (Chapter 4). Presynaptic inhibition has been observed at fast and slow excitatory synapses in enteric ganglia.

Neurotransmitters in the enteric nervous system. In most cases the synaptic transmitters responsible for particular synaptic events have not been unequivocally identified. The task of identifying transmitters at synapses in the enteric nervous system is complicated by the presence in enteric neurons of numerous neuroactive substances, which may function as neuromodulators or as neurotransmitters (see Tables 42-1 and 42-2).

Fast EPSPs appear to be mediated by acetylcholine acting at nicotinic receptors. Acetylcholine applied from micropipettes mimics the fast EPSP, and the fast EPSP is attenuated by nicotinic receptor blockers and potentiated by cholinesterase inhibitors.

Fast IPSPs are mediated by a synaptic transmitter that is presently unknown.

Slow EPSPs may be mediated by a transmitter whose receptor, when occupied by transmitter, activates adenylate cyclase. Both serotonin and substance P are transmitters that can bring about slow EPSPs. Serotonin has met all of the criteria for a synaptic transmitter in the myenteric plexus, and substance P has met most of the criteria. Slow EPSPs can be evoked in some myenteric ganglia that contain no serotonergic endings, suggesting that serotonin is not the only mediator of slow EPSPs.

Other substances, when applied to enteric neurons, may elicit responses that resemble slow EPSPs. Among these substances are acetylcholine, histamine, cholecystokinin octapeptide (CCK8), bombesin, caerulein, and vasoactive intestinal peptide (VIP). For various reasons

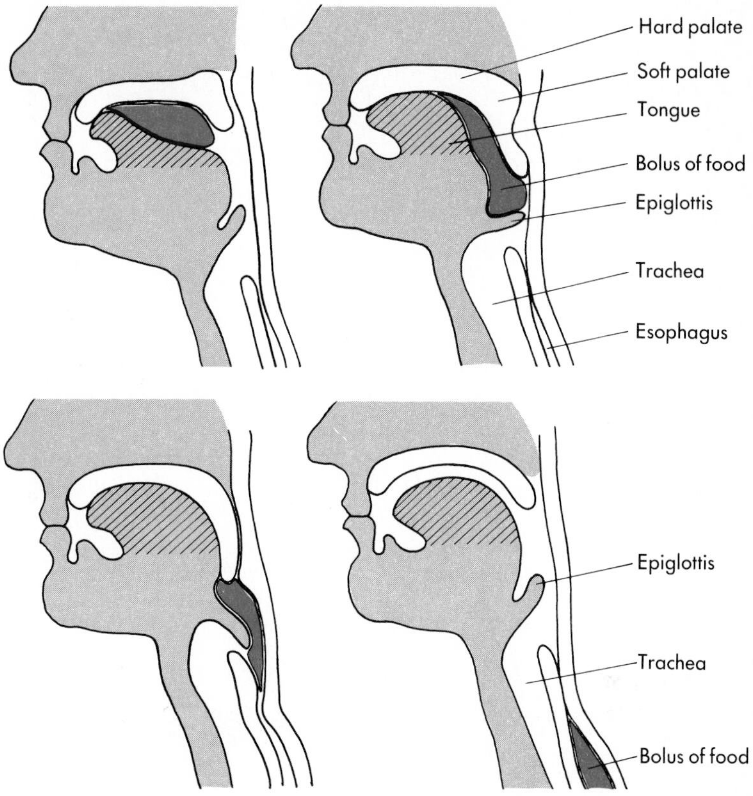

■ **Fig. 42-11.** Major events involved in the swallowing reflex. (Modified from Johnson, L.R.: Gastrointestinal physiology, ed. 2, St. Louis, 1981, The C.V. Mosby Co.)

these compounds probably do not function as neurotransmitters that mediate slow EPSPs, but some of them may modulate slow synaptic excitation in the enteric plexuses.

Responses that resemble slow IPSPs may be produced by application of norepinephrine or dopamine to certain submucosal plexus neurons. The submucosal neurons, however, do not contain the enzymes required for synthesis of dopamine and norepinephrine, so these amines are probably not the physiological transmitters. The transmitter that mediates the slow IPSP is not known, but the most likely candidates are CCK8 and somatostatin.

Presynaptic inhibition may be the major mechanism whereby the sympathetic nervous system influences the function of the enteric nervous system. Stimulation of postganglionic sympathetic fibers inhibits fast EPSPs in the myenteric plexus but does not diminish the response to application of acetylcholine. Slow EPSPs evoked by nerve stimulation are blocked by norepinephrine, but norepinephrine does not block the slow EPSPs elicited by microapplication of substance P or serotonin. Serotonin may act presynaptically to suppress the release of acetylcholine that mediates fast EPSPs. Acetylcholine itself may feed back via presynaptic muscarinic receptors to suppress its own release.

■ *Chewing (Mastication)*

Chewing can be carried out voluntarily, but it is more frequently a reflex behavior. Chewing serves to lubricate the food by mixing it with salivary mucus, to mix starch-containing food with the salivary α-amylase, and to subdivide the food so that it can be mixed more readily with the digestive secretions of the stomach and duodenum.

■ *Swallowing*

Swallowing can be initiated voluntarily, but thereafter it is almost entirely under reflex control. The swallowing reflex is a rigidly ordered sequence of events that results in the propulsion of food from the mouth to the stomach, at the same time inhibiting respiration and preventing the entrance of food into the trachea (Fig. 42-11). The afferent limb of the swallowing reflex begins with tactile receptors, most notably those near the opening of the pharynx. Sensory impulses from these receptors are transmitted to certain areas in the medulla. The central integrating areas for swallowing lie in the medulla and lower pons; they are collectively called the "swallowing center." Motor impulses travel from the

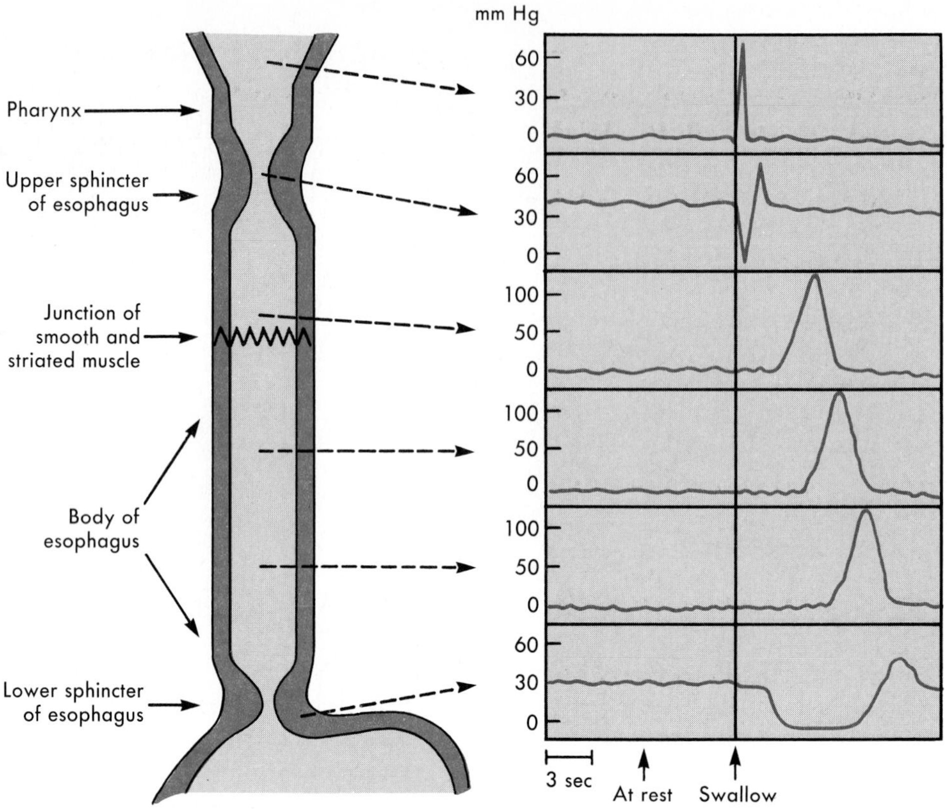

Fig. 42-12. Pressures in the pharynx, esophagus, and esophageal sphincters during swallowing. Note the reflex relaxation of the upper and lower esophageal sphincters and the timing of the relaxation. (From Christensen, J.L.: In Christensen, J., and Wingate, D.L., editors: A guide to gastrointestinal motility, Bristol, U.K., 1983, John Wright and Sons.)

swallowing center to the musculature of the pharynx and upper esophagus via various cranial nerves.

The oral or voluntary phase of swallowing is initiated by separating a bolus of food from the mass in the mouth with the tip of the tongue. The bolus to be swallowed is moved upward and backward in the mouth by pressing first the tip of the tongue and later also the more posterior portions of the tongue as well against the hard palate. This forces the bolus into the pharynx, where it stimulates the tactile receptors that initiate the swallowing reflex.

The pharyngeal stage of swallowing involves the following sequence of events, which occur in less than 1 second:

1. The soft palate is pulled upward, and the palatopharyngeal folds move inward toward one another. This prevents reflux of food into the nasopharynx and provides a narrow passage through which the food moves into the pharynx.
2. The vocal cords are pulled together, and the epiglottis covers the opening to the larynx. The larynx is moved upwards against the epiglottis. These actions prevent food from entering the trachea.
3. The upper esophageal sphincter relaxes to receive the bolus of food. Then the superior constrictor muscles of the pharynx contract strongly to force the bolus deeply into the pharynx.
4. A peristaltic wave is initiated with the contraction of the superior constrictor muscles of the pharynx, and it moves toward the esophagus. This forces the bolus of food through the relaxed upper esophageal sphincter.

During the pharyngeal stage of swallowing, respiration is reflexly inhibited.

The esophageal phase of swallowing also is partially controlled by the swallowing center. After the bolus of food passes the upper esophageal sphincter, the sphincter reflexly constricts. A peristaltic wave then begins just below the upper esophageal sphincter and traverses the entire esophagus in about 10 seconds (Fig. 42-12). This initial wave of peristalsis, called primary peristalsis, is controlled by the swallowing center. The peristaltic wave travels down the esophagus at 3 to 5 cm/second and takes about 5 seconds to reach the stomach. Should the primary peristalsis be insufficient to clear the esophagus of food, the distension of the esophagus would initiate another peristaltic wave that begins at the site of distension and moves downward. This latter type of peristalsis, termed secondary peristalsis, is partially

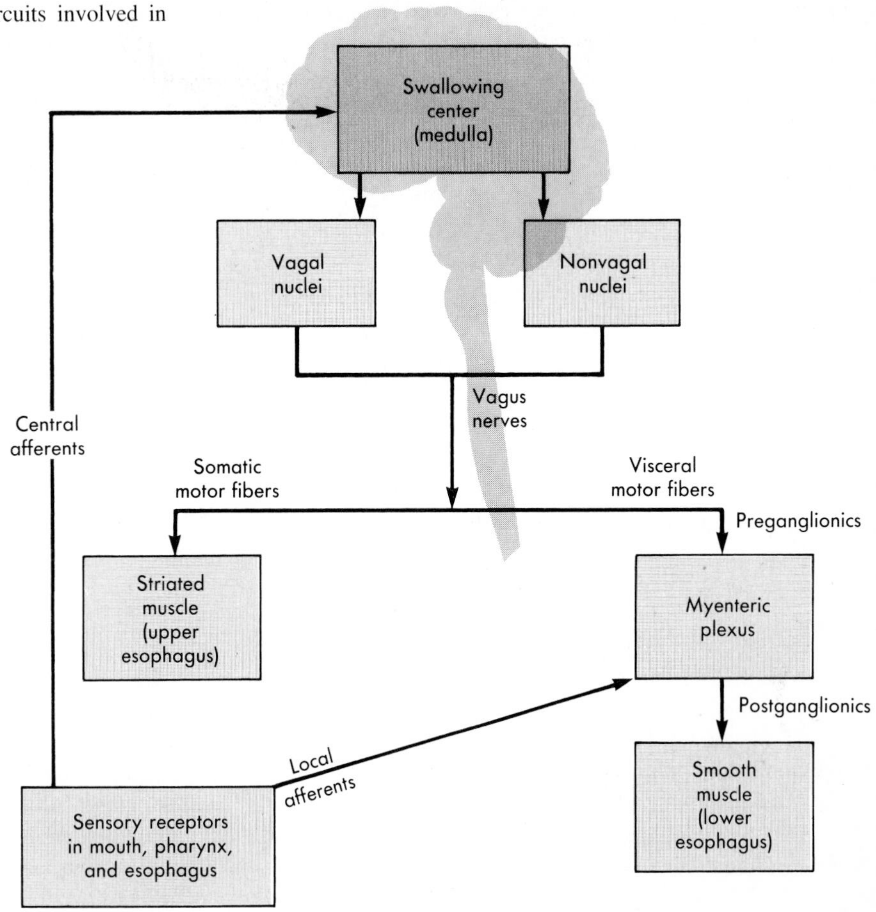

■ **Fig. 42-13.** Local and central neural circuits involved in the control of esophageal motility.

mediated by the enteric nervous system, since it occurs (albeit more weakly) in the extrinsically denervated esophagus. Input from esophageal sensory fibers to the central and enteric nervous systems is involved in modulating esophageal peristalsis.

■ *Esophageal Function*

After food is swallowed, the esophagus functions as a conduit to move the food from the pharynx to the stomach. It is important to prevent air from entering at the upper end of the esophagus and to keep corrosive gastric contents from refluxing back into the esophagus at its lower end. Reflux is particularly problematic because the pressure in the body of the resting thoracic esophagus closely approximates inthrathoracic pressure. Thus it is less than atmospheric pressure and less than intraabdominal pressure. The pressure in the short abdominal section of the esophagus mirrors the intraabdominal pressure.

The structure of the esophagus follows the general scheme described earlier in this chapter, except that in the upper third of the esophagus both the inner circular and outer longitudinal muscle layers are striated. In the lower third of the esophagus the muscle layers are composed entirely of smooth muscle cells. In the middle third, skeletal and smooth muscle coexist, there being a gradient from all skeletal above to all smooth below.

The esophageal musculature, both striated and smooth, is primarily innervated by branches of the vagus nerve. Somatic motor fibers of the vagus nerve that originate in the nucleus ambiguus form motor endplates on striated muscle fibers. Visceral motor nerves, primarily from the dorsal motor nucleus of the vagus nerve, are preganglionic parasympathetic fibers. They synapse primarily on the nerve cells of the myenteric plexus. Neurons of the myenteric plexus innervate the smooth muscle cells of the esophagus and communicate with one another. The neural circuits that control the esophagus are schematized in Fig. 42-13.

The upper and lower ends of the esophagus function as sphincters to prevent the entry of air and gastric contents, respectively, into the esophagus. The sphincters are known as the *upper esophageal* (or pharyngeoesophageal) *sphincter* and the *lower esophageal* (or gastroesophageal) *sphincter*. The upper esophageal sphincter is formed by a thickening of the circular layer of striated muscle at the upper end of the esophagus; the sphincter consists of the cricopharyngeus muscle and lower fibers of the inferior pharyngeal constrictor. The pressure in the upper esophageal sphincter is about 40

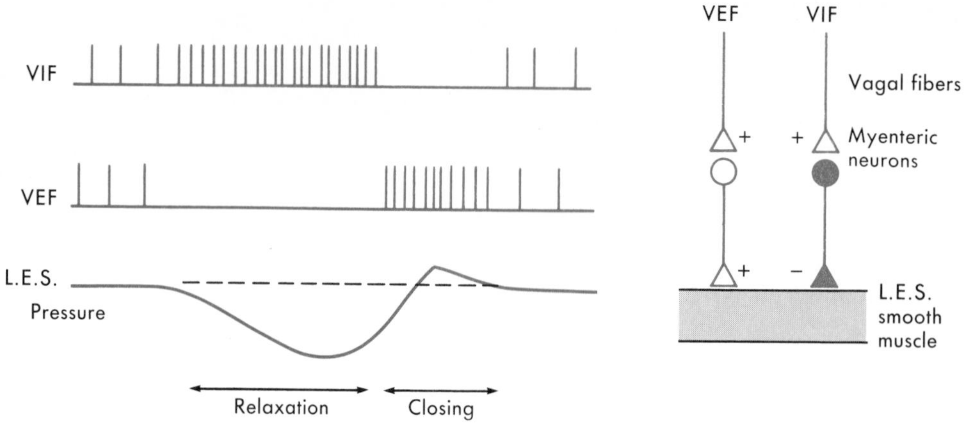

■ **Fig. 42-14.** Schematic representation of vagal control of the lower esophageal sphincter *(LES)*. Note that relaxation of the *LES* is associated with an increase in the firing rate in vagal inhibitory fibers *(VIF)* and a decreased frequency of action potentials in vagal excitatory fibers *(VEF)*. Reciprocal changes occur when the sphincter regains its resting tone. (From Miolan, J.P., and Roman, C.: J. Physiol. [Paris] **74:**709, 1978.)

mm Hg at rest. The lower esophageal sphincter is difficult to identify anatomically, but the lower 1 to 2 cm of the esophagus functions as a sphincter. In normal individuals the pressure at the lower esophageal sphincter is always greater than that in the stomach.

The lower esophageal sphincter opens when a wave of esophageal peristalsis reaches it. The opening is vagally mediated, and the major transmitter involved is probably vasoactive intestinal polypeptide. In the absence of esophageal peristalsis the sphincter must remain tightly closed to prevent reflux of gastric contents, which would cause esophagitis and the sensation of heartburn.

■ *Control of the Lower Esophageal Sphincter*

Control of LES tone. The resting pressure in the lower esophageal sphincter (LES) is about 30 mm Hg. The tonic contraction of the circular musculature of the LES is regulated by nerves, both intrinsic and extrinsic, and perhaps by hormones and neuromodulators. A significant fraction of basal tone in the LES is mediated by vagal cholinergic nerves. Stimulating sympathetic nerves to the LES also causes contraction. This effect is mediated by norepinephrine acting on α-receptors. Norepinephrine may also act presynaptically to inhibit release of acetylcholine from vagal cholinergic fibers. Significant LES tone and regulation of that tone persist when the extrinsic nerves to the LES are sectioned. Hence, the enteric nervous system plays a major role in regulating the function of the LES. A number of the gut-brain peptides can influence the tone of the LES, but the physiological roles of these substances remain to be determined.

Relaxation of the LES. The innervation, both intrinsic and extrinsic, of the LES is both excitatory and inhibitory (Fig. 42-14). A major component of the relaxation of the LES that occurs in response to primary peristalsis in the esophagus is mediated by vagal fibers that are inhibitory to the circular muscle and that probably use vasoactive intestinal polypeptide as their transmitter. A decrease in the impulse traffic in vagal excitatory nerves, which are probably mostly cholinergic, promotes relaxation of the sphincter. The enteric nervous system probably helps to control relaxation of the LES, but the mechanisms involved remain to be elucidated.

During swallowing in some individuals the LES fails to relax sufficiently to allow food to enter the stomach, a condition known as *achalasia*. Therapy for achalasia may involve surgically weakening the lower esophageal sphincter. Individuals with *diffuse esophageal spasm* have prolonged and painful contraction of the lower part of the esophagus after swallowing, instead of the normal esophageal peristaltic wave. Achalasia, incompetence of the lower sphincter, and diffuse esophageal spasm are common disorders of esophageal function.

■ *Gastric Motility*

The motility of the stomach serves the following major functions: (1) to allow the stomach to serve as a reservoir for the large volume of food that may be ingested at a single meal, (2) to fragment food into smaller particles and to mix chyme with gastric secretions so that digestion can begin, and (3) to empty gastric contents into the duodenum at a controlled rate. Fig. 42-15

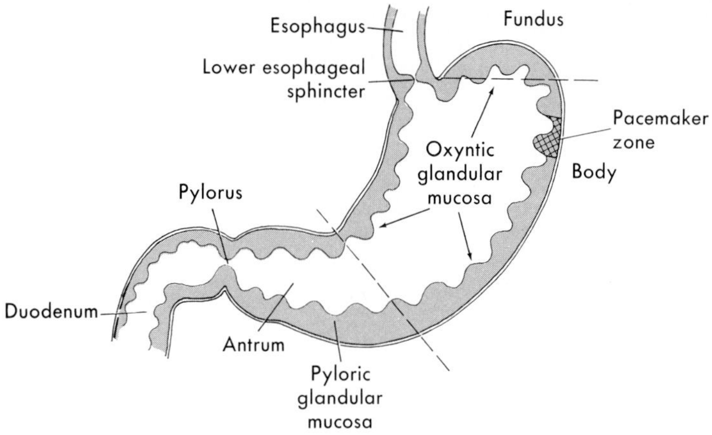

Fig. 42-15. The major anatomical divisions of the stomach.

shows the major anatomical divisions of the stomach.

The stomach has features that allow it to carry out each of these functions. The fundus and the body of the stomach can accommodate volume increases as large as 1.5 L without a marked increase in intragastric pressure. Contractions of the fundus and body are normally weak, so that much of the gastric contents remains relatively unmixed for long periods. Thus the fundus and the body serve the reservoir functions of the stomach. In the antrum, however, contractions are vigorous and thoroughly mix antral chyme with gastric juice and subdivide food into smaller particles. The antral contractions serve to empty the gastric contents in small squirts into the duodenal bulb. The rate of gastric emptying is adjusted by a number of mechanisms so that chyme is not delivered to the duodenum too rapidly. The physiological mechanisms that underlie this behavior are discussed later.

Structure and Innervation of the Stomach

The basic structure of the gastric wall follows the scheme presented in Fig. 42-1. The circular muscle layer of muscularis externa is more prominent than the longitudinal layer. An incomplete inner layer of obliquely oriented muscle cells is present only on the anterior and posterior sides of the stomach. The muscularis externa of the fundus and the body is relatively thin, but that of the antrum is considerably thicker, and it increases in thickness toward the pylorus.

The stomach is richly innervated by extrinsic nerves and by the neurons of the submucosal and myenteric plexuses. Axons from the cells of the intramural plexuses innervate smooth muscle and secretory cells.

Extrinsic innervation comes via the vagus nerves and from the celiac plexus. In general, cholinergic nerves stimulate gastric smooth muscle motility and gastric secretions, whereas adrenergic fibers inhibit these functions. A number of other gastric transmitter and modulator substances have been identified.

Numerous sensory afferent fibers leave the stomach in the vagus nerves, and some travel with sympathetic nerves. Other fibers are the afferent links of intrinsic reflex arcs via the intramural plexuses of the stomach. Some of these afferent fibers relay information to the central nervous system about intragastric pressure, gastric distension, intragastric pH, or pain. The afferent fibers signaling gastric distension and intragastric pressure have been implicated in satiety.

Responses to Gastric Filling

When a wave of esophageal peristalsis reaches the lower esophageal sphincter, the sphincter reflexly relaxes. This is followed by a relaxation, known as *receptive relaxation*, of the fundus and body of the stomach. The stomach also will relax if it is directly filled with gas or liquid. Both of these responses are greatly diminished if the vagus nerves are sectioned, so it is believed that the vagi are a major efferent pathway for reflex relaxation of the stomach. The vagal fibers that mediate this response are neither adrenergic nor cholinergic. When the stomach relaxes in response to filling, the sensory afferent fibers that report gastric distension constitute the afferent limb of the response.

The smooth muscle of the fundus and body facilitates receptive relaxation and the reservoir function of the stomach. The smooth muscle cells in the fundus have a particularly low resting membrane potential, about -50 mV. At this resting membrane potential the fundic smooth muscle cells are partly contracted. In response to hyperpolarization induced by impulses in inhibitory nerve fibers, the smooth muscle cells relax. The smooth muscle of the body of the stomach has a lower resistance to stretch than the smooth muscle of the antrum. Both of these properties enhance the ability of the fundus and body to accommodate a large increase in volume with little increase in intragastric pressure.

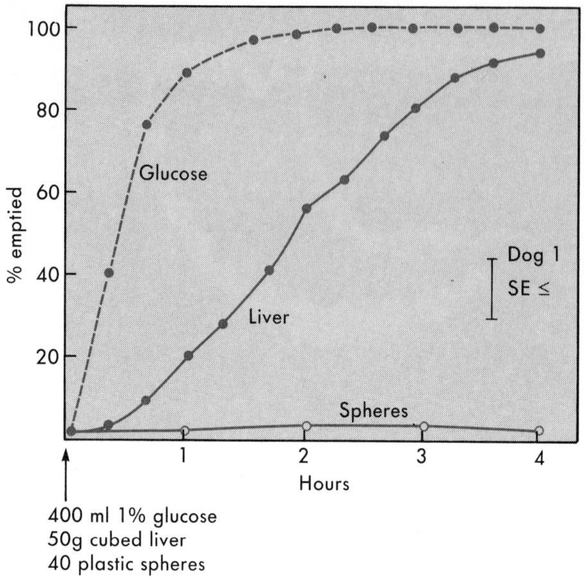

Fig. 42-16. Rates of emptying of different materials from dog stomach. A solution (1% glucose) is emptied faster than a digestible solid (cubed liver). An indigestible solid (plastic spheres, 7 mm in diameter) remains in the stomach under these conditions. (From Hinder, R.A., and Kelly, K.A.: Am. J. Physiol. **233:**E335, 1977.)

Mixing and Emptying of Gastric Contents

The muscle layers in the fundus and body are thin, and therefore weak contractions are characteristic of these parts of the stomach. As a result, the contents of the fundus and the body tend to form layers based on their density. The gastric contents may remain unmixed for as long as 1 hour after eating. Fats tend to form an oily layer on top of the other gastric contents. Fats thus are emptied later than other gastric contents. Liquids can flow around the mass of food contained in the body of the stomach and are emptied more rapidly into the duodenum. Large or indigestible particles are retained in the stomach for a longer period of time (Fig. 42-16).

Gastric contractions usually begin in the middle of the body of the stomach and travel toward the pylorus. The contractions increase in force and velocity as they approach the gastroduodenal junction. As a result, the major mixing activity occurs in the antrum, the contents of which are mixed rapidly and thoroughly with gastric secretions. Immediately after eating, the antral contractions are relatively weak, but as digestion proceeds, they become more forceful. Typically the frequency of gastric contraction is about three per minute.

Because of the acceleration of the peristaltic wave, the terminal end of the antrum and the pylorus contract almost simultaneously. This is known as *systolic contraction of the antrum.* The peristaltic wave pushes the antral contents ahead of the ring of contraction. Most waves are strong enough to squirt a small fraction of antral contents into the duodenal bulb. The squirt is terminated by the abrupt closure of the pyloric sphincter. Because of its small diameter, the sphincter closes early in the systolic contraction of the antrum. The forceful contraction of the terminal end of the antrum rapidly forces antral contents back into the more proximal part of the antrum. The vigorous backward motion of antral content, called *retropulsion,* effectively mixes the antral contents with gastric secretions.

After an animal feeds, regular contractions of the antrum occur at the frequency of the gastric slow waves. As discussed later, the rate of gastric emptying is regulated by mechanisms that feed back to diminish the force of antral contractions. Contractions of the antrum after ingestion of food vary from moderately forceful to weak. In a fasted animal a different pattern of antral contractions occurs. The antrum is quiescent for 1 to 2 hours; then a short period of intense electrical and motor activity occurs that lasts for 10 to 20 minutes. This activity is characterized by strong contractions of the antrum with a relaxed pylorus. During this period even large chunks of material that remain from the previous meal are emptied from the stomach (Fig. 42-17). The period of intense contractions is followed by another 1- to 2-hour period of quiescence. This cyclical pattern of contractions in the stomach is part of a pattern of contractile activity that periodically sweeps from the stomach to the terminal ileum. This pattern of cyclical contractile activity is known as the *migrating myoelectric complex* and is discussed below.

Electrical Activity that Underlies Gastric Contractions

The gastric peristaltic waves occur at the frequency of the gastric slow waves that are generated by a "pacemaker zone" (Fig. 42-15) on the greater curvature near the middle of the body of the stomach. These waves are conducted toward the pylorus. It is not known whether the cells that generate the slow waves reside in the circular or longitudinal layer of smooth muscle. Isolated strips of gastric smooth muscle have intrinsic rhythmicity and generate slow waves at a characteristic frequency. The pacemaker zone has the highest intrinsic frequency, and for that reason acts as the pacemaker. In the human stomach the frequency of slow waves is about three per minute. In an individual at rest the frequency of slow waves is relatively constant. Injections of gastrin can increase the frequency of slow waves to about four per minute. Secretin can slow the frequency of the slow waves or, in high doses, can even abolish them. The physiological roles of these responses to gastrin and secretin have not been established.

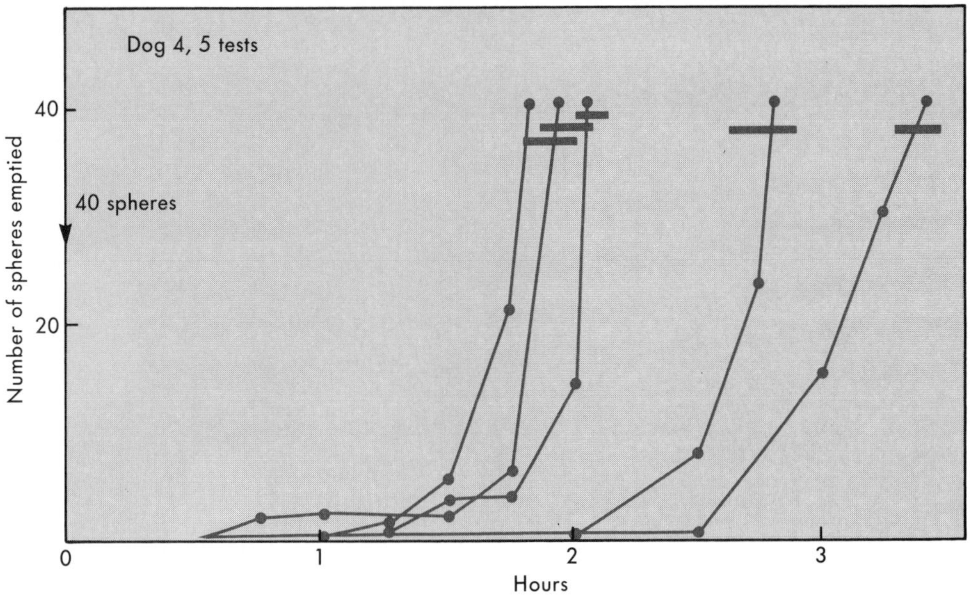

Fig. 42-17. Emptying of plastic spheres (7 mm in diameter) from dog stomach during fasting. Vigorous gastric contractions that are associated with phase III of the migrating myoelectric complex *(solid bars)* empty the plastic spheres into the duodenum. (From Mroz, C.T., and Kelly, K.A.: Surgery **145:**369, 1977.)

The gastric slow wave is triphasic (Fig. 42-18). Its shape is reminiscent of action potentials in cardiac muscle. However, the gastric slow wave lasts about 10 times longer than the cardiac action potential and does not overshoot. The initial, rapid depolarizing phase of the slow wave appears to be caused by entry of Ca^{++} via voltage-gated Ca^{++} channels. The plateau of the slow wave is believed to be caused by the entry of Ca^{++} and Na^{+} through slower voltage-gated channels. Repolarization of the membrane is associated with a delayed increase in K^{+} conductance.

Gastric smooth muscle contracts when the depolarization during the slow wave exceeds the threshold for contraction (Fig. 42-18). The greater the extent of depolarization and the longer the cell remains depolarized above the threshold, the greater the force of contraction. In this respect, gastric smooth muscle differs from intestinal smooth muscle, which contracts only in response to action potential spikes that are generated during the plateau of the slow wave (Fig. 42-6). Acetylcholine and gastrin stimulate gastric contractility by increasing the amplitude and duration of the plateau phase of the gastric slow wave. Norepinephrine and neurotensin, which reduce the force of gastric contractions, decrease the extent of depolarization and the duration of the plateau phase.

In the terminal antrum and pylorus, wavelike depolarizations and action potential spikes are generated during the plateau phase of the slow wave (Fig. 42-19). The action potential spikes are associated with increased force of contraction of the antral muscle. The force of antral contraction varies greatly from moment to moment, with periods of moderate peristaltic mixing movements being interspersed among periods of intense propulsive contractions. The more vigorous contractions occur in response to trains of action potential spikes during the plateaus of the slow waves.

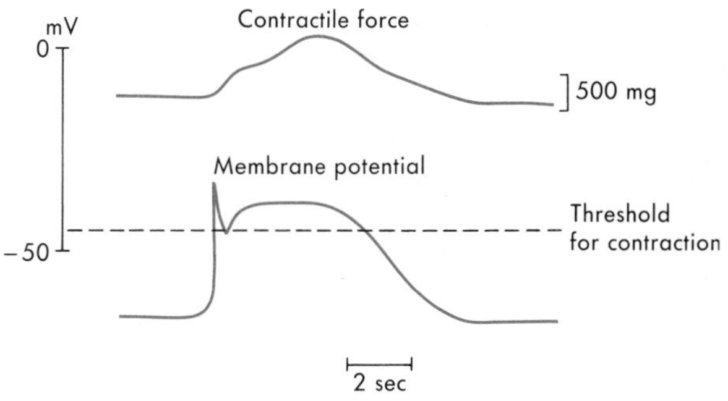

Fig. 42-18. Relationship between contraction of smooth muscle of dog stomach *(upper tracing)* and intracellularly recorded slow wave *(lower tracing)*. Note the triphasic shape of the slow wave in gastric smooth muscle. Contraction occurs when the depolarizing phase of the slow wave exceeds the threshold for contraction, even though there are no action potential spikes on the plateau of the slow wave. (From Szurszewski, J.: In Johnson, L.R., editor: Physiology of the gastrointestinal tract, New York, 1981, Raven Press.)

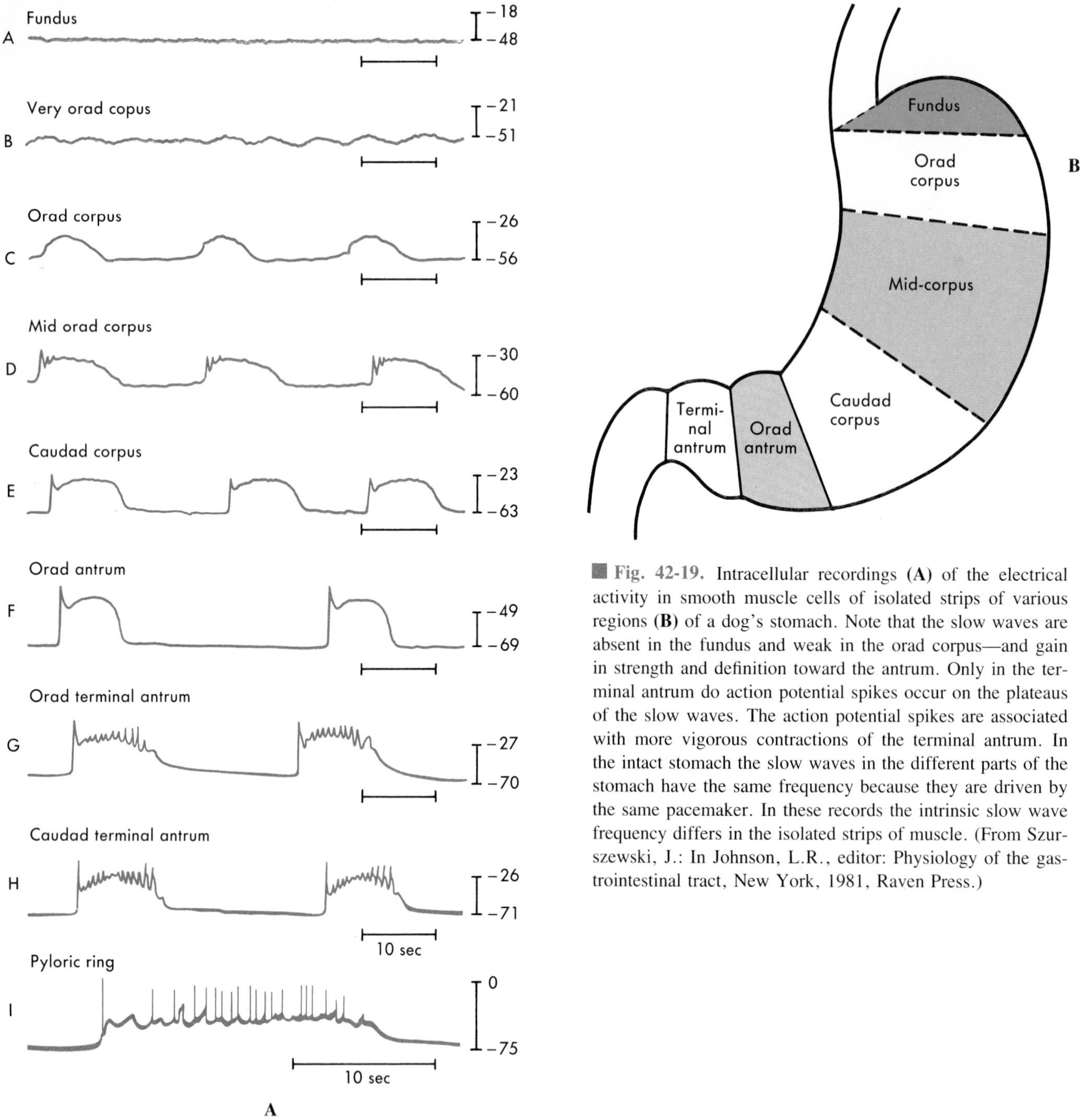

■ **Fig. 42-19.** Intracellular recordings **(A)** of the electrical activity in smooth muscle cells of isolated strips of various regions **(B)** of a dog's stomach. Note that the slow waves are absent in the fundus and weak in the orad corpus—and gain in strength and definition toward the antrum. Only in the terminal antrum do action potential spikes occur on the plateaus of the slow waves. The action potential spikes are associated with more vigorous contractions of the terminal antrum. In the intact stomach the slow waves in the different parts of the stomach have the same frequency because they are driven by the same pacemaker. In these records the intrinsic slow wave frequency differs in the isolated strips of muscle. (From Szurszewski, J.: In Johnson, L.R., editor: Physiology of the gastrointestinal tract, New York, 1981, Raven Press.)

■ *The Gastroduodenal Junction*

The pylorus separates the gastric antrum from the first part of the duodenum, the duodenal bulb (or cap). It is debatable whether the pylorus constitutes a true anatomical sphincter, but it functions physiologically as a sphincter in many respects. The circular smooth muscle of the pylorus forms two ring-like thickenings that are followed by a connective tissue ring that separates pylorus from duodenum (Fig. 42-20). The mucosa, submucosa, and muscle layers of the pylorus and duodenal bulb are separate, with the exception of a few longitudinal muscle fibers that cross over the junction. However the myenteric plexuses of the pylorus and duodenal bulb are continuous.

The duodenum has a basic electrical rhythm of 10 to 12 per minute, far faster than the 3 per minute of the stomach. The duodenal bulb is influenced by the basic electrical rhythms of both the stomach and the postbulbar duodenum. It thus contracts somewhat irregularly. The antrum and duodenum are coordinated; when the antrum contracts, the duodenal bulb is relaxed.

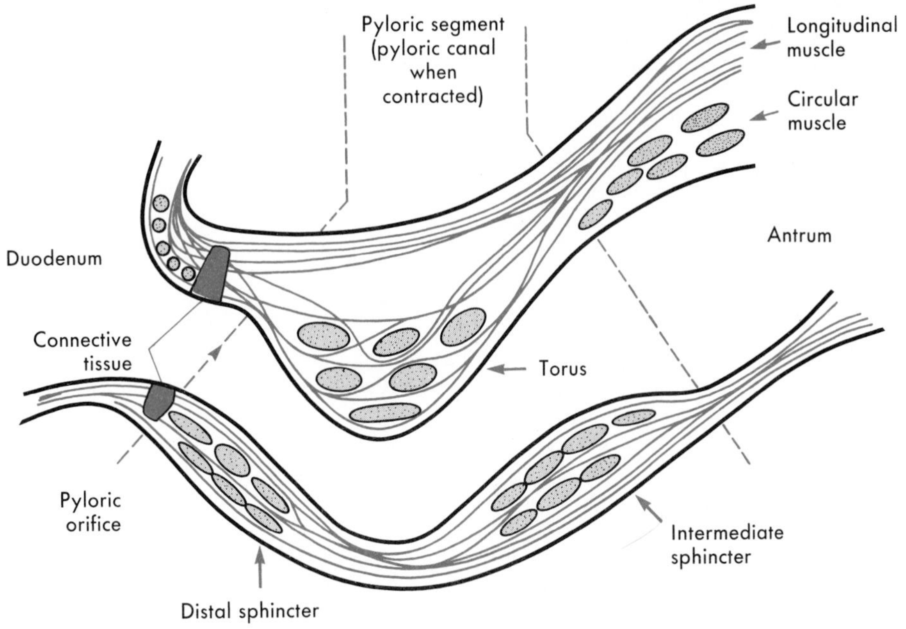

■ **Fig. 42-20.** Diagrammatic representation of a longitudinal section through the pyloric sphincter. Note the connective tissue ring that separates the pyloric sphincter from the duodenum. (Modified from Schulze-Delrier, K., et al.: In Roman, C., editor: Gastrointestinal motility, 1984, Lancaster, U.K., MTP Press.)

The essential functions of the gastroduodenal junction are (1) to allow the carefully regulated emptying of gastric contents at a rate commensurate with the ability of the duodenum to process the chyme, and (2) to prevent regurgitation of duodenal contents back into the stomach. The gastric mucosa is highly resistant to acid, but it may be damaged by bile. The duodenal mucosa has the opposite properties. Thus too rapid gastric emptying may lead to duodenal ulcers, whereas regurgitation of duodenal contents often contributes to gastric ulcers.

The pylorus is densely innervated by both vagal and sympathetic nerve fibers. Sympathetic postganglionic fibers release norepinephrine, which acts on α-adrenergic receptors to increase constriction of the sphincter. Vagal fibers are both excitatory and inhibitory to pyloric smooth muscle. Cholinergic vagal fibers stimulate constriction of the sphincter. Inhibitory vagal fibers release another transmitter, possibly vasoactive intestinal polypeptide, that mediates relaxation of the sphincter.

Cholecystokinin, gastrin, gastric inhibitory peptide, and secretin all cause constriction of the pyloric sphincter. Whether these hormones physiologically modulate the function of the pyloric sphincter is not known. Cholecystokinin is a good candidate for a physiological regulator, because it constricts the pyloric sphincter at concentrations that also activate gallbladder contraction.

■ *Regulation of the Rate of Gastric Emptying*

The emptying of gastric contents is regulated by both neural and humoral mechanisms. The duodenal and jejunal mucosa have receptors that sense acidity, osmotic pressure and fat content.

The presence of fatty acids or monoglycerides in the duodenum dramatically decreases the rate of gastric emptying. The rate of the gastric slow waves remains essentially unchanged, but the contractility of the pyloric sphincter is greatly increased so that the rate of gastric emptying is much less.

The chyme that leaves the stomach is usually hypertonic, and it becomes more hypertonic due to the action of the digestive enzymes in the duodenum. Hypertonic solutions in the duodenum slow gastric emptying.

Duodenal contents with a pH less than about 3.5 retard gastric emptying. The presence of amino acids and peptides in the duodenum also may slow gastric emptying.

As a result of these mechanisms, (1) fat is not emptied into the duodenum at a rate greater than that at which it can be emulsified by the bile acids and lecithin of the bile; (2) acid is not dumped into the duodenum more rapidly than it can be neutralized by pancreatic and duodenal secretions and by other mechanisms; and (3) in general, the rates at which the other components of chyme are presented to the small intestine correspond to the rates at which the small intestine can process those components.

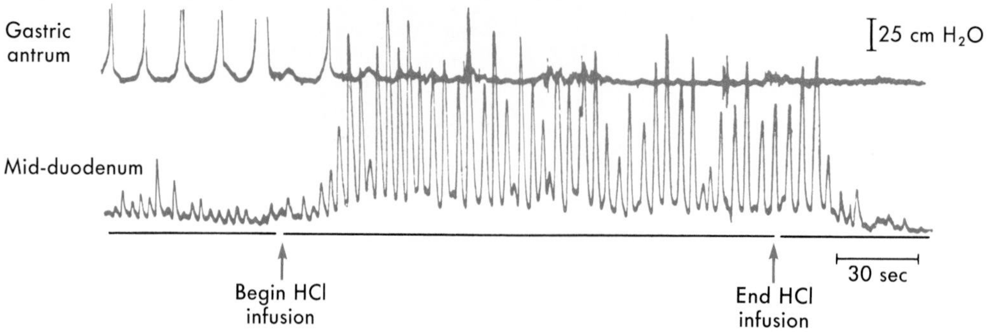

■ **Fig. 42-21.** Effect of instilling 100 mM HCl at 6 ml/min into the duodenum of a dog. Contractile activities of gastric antrum *(upper tracing)* and mid duodenum *(lower tracing)* are shown. (From Brink, B.M., et al.: Gut **6**:163, 1965.)

Mechanisms that regulate gastric emptying. The slowing of gastric emptying in response to fatty acids, low pH, or hypertonicity of duodenal contents is mediated by neural and hormonal mechanisms. The interplay between the neural and hormonal mechanisms and their relative importance is not well understood. Considerable knowledge of the mechanisms that influence the rate of gastric emptying is available, but how these mechanisms function in vivo is not well understood. As an example, consider the response of gastric emptying to the presence of acid in the duodenum.

Acid in the duodenum. Fig. 42-21 shows that when acid is instilled into the duodenum of a dog, the force of contractions of the gastric antrum promptly decreases and duodenal motility promptly increases. This response has a neural component, because cutting the branches of the vagus to the stomach and duodenum markedly decreases the response. The presence of acid in the duodenum also releases secretin, which may diminish the rate of gastric emptying by inhibiting antral contractions and stimulating contraction of the pyloric sphincter. Secretin also stimulates the output of the bicarbonate-rich secretions of the pancreas and liver. These secretions buffer the duodenal pH. Partly for this reason, the duodenal pH that elicits the response shown in Fig. 42-21 is much lower than the pH that exists in the duodenum under physiological conditions. The physiological importance of the decrease in gastric emptying rate in response to increased duodenal acidity, and the relative importance of neural and hormonal mechanisms in this response both remain to be elucidated. Neural and hormonal mechanisms that appear to control the rate of gastric emptying are schematically depicted in Fig. 42-22.

Fat-digestion products. The presence of fatty acids—as well as mono- and diglycerides—in the duodenum and jejunum activates neural and hormonal mechanisms that decrease the rate of gastric emptying. Unsaturated fatty acids elicit this response more effectively than saturated fatty acids. Fatty acids of 14 carbons or more are more effective than short- and medium-chain fatty acids. Although the decreased gastric emptying may have a neural component, this response is probably caused mainly by release of cholecystokinin from the duodenum and jejunum. Cholecystokinin stimulates contractions of the gastric antrum and constriction of the pyloric sphincter. The net effect of cholecystokinin is to decrease the rate of gastric emptying. Physiological levels of cholecystokinin have been shown to diminish the rate of gastric emptying. The presence of fatty acids in the duodenum and jejunum also elicits the release of another hormone, gastric inhibitory peptide, that decreases the rate of gastric emptying. The relative physiological importance of cholecystokinin and gastric inhibitory peptide in the response to fatty acids in the upper small intestine remains to be determined. A similar decrease in the rate of gastric emptying is elicited by the presence of fatty acids in the ileum.

Osmolarity of duodenal contents. When hyperosmolar solutions are present in the duodenum and jejunum, the rate of gastric emptying is slowed. Osmoreceptors in the mucosa or submucosa of the duodenum and jejunum may be responsible for this reaction. The decreased gastric emptying in response to hyperosmolarity may have a neural component. Hyperosmolarity in the duodenum releases an unidentified hormone that diminishes the rate of gastric emptying.

Peptides and amino acids in the duodenum. Peptides and amino acids release gastrin from G cells located in the antrum of the stomach and the duodenum. Gastrin acts, probably at physiological levels, to increase the strength of antral contractions and to increase constriction of the pyloric sphincter. Under most circumstances the net effect of gastrin is to diminish the rate of gastric emptying. The presence of tryptophan in the duodenum strongly inhibits the rate of gastric emp-

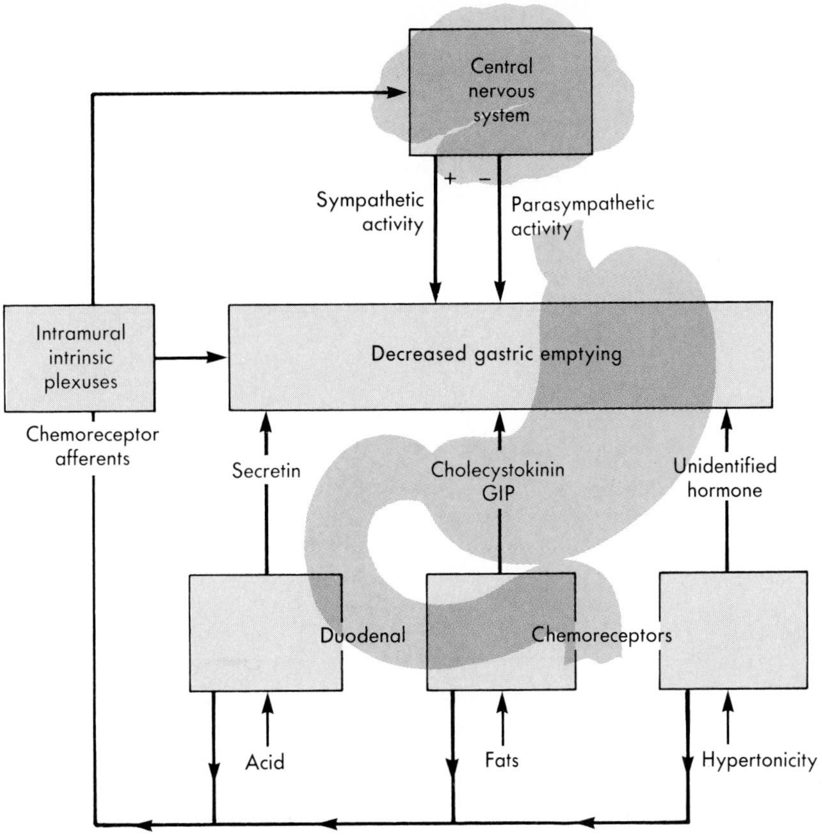

Fig. 42-22. Neural and hormonal inhibition of gastric emptying.

tying. Gastrin appears to be involved in the response to tryptophan.

Vomiting

Vomiting is the expulsion of gastric (and sometimes duodenal) contents from the gastrointestinal tract via the mouth. Vomiting often is preceded by a feeling of nausea, a rapid or irregular heart beat, dizziness, sweating, pallor, and pupillary dilation. Vomiting usually is also preceded by *retching,* in which gastric contents are forced up into the esophagus but do not enter the pharynx. A series of retches of increasing strength often precedes vomiting.

Vomiting is a reflex behavior controlled and coordinated by certain medullary centers (Fig. 42-23). Electrical stimulation of an area of the medulla known as the *vomiting center* elicits prompt vomiting without preliminary retching. Stimulation of another medullary locus leads to retching without subsequent vomiting. It appears that important interactions between these two areas in the medulla occur in typical vomiting. A large number of different areas in the body have receptors that provide afferent input to the vomiting center. Distension of the stomach and duodenum is a strong stimulus that elicits vomiting. Tickling the back of the throat, painful injury to the genitourinary system, dizziness, and certain other stimuli are among the diverse events that can bring about vomiting.

Certain chemicals, called *emetics,* can elicit vomiting. Some emetics do this by stimulating receptors in the stomach or more commonly in the duodenum. The widely used emetic ipecac works via duodenal receptors. Certain other emetics act at the level of the central nervous system on receptors in the floor of the fourth ventricle in an area known as the *chemoreceptor trigger zone.* The chemoreceptor trigger zone is in or near the area postrema, which lies on the blood side of the blood-brain barrier, and thus can be reached by most blood-borne substances.

When the vomiting reflex is initiated, the sequence of events is the same regardless of the stimulus that initiates the reflex. Early events in the vomiting reflex include a wave of reverse peristalsis that sweeps from the middle of the small intestine to the duodenum. The pyloric sphincter and the stomach relax to receive intestinal contents. The gastric relaxation is mediated by vagal inhibitory fibers. Then a forced inspiration occurs against a closed glottis. This results in a decreased in-

■ **Fig. 42-23.** Some aspects of control of vomiting.

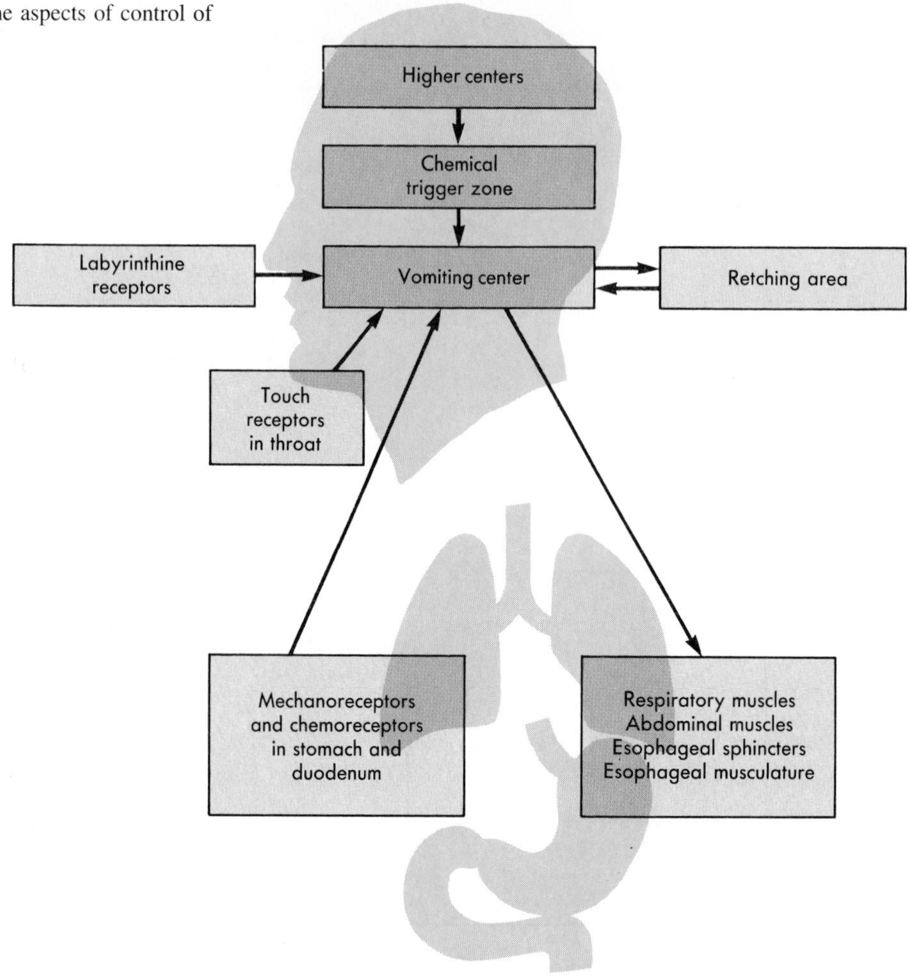

trathoracic pressure, whereas the lowering of the diaphragm causes an increase in intraabdominal pressure. Then a forceful contraction of abdominal muscles sharply elevates intraabdominal pressure, driving gastric contents into the esophagus. The lower esophageal sphincter relaxes reflexly to receive gastric contents, and the pylorus and antrum contract reflexly, which prevents orthograde flow of gastric contents.

When a person retches, the upper esophageal sphincter remains closed, preventing vomiting. When the respiratory and abdominal muscles relax, the esophagus is emptied by secondary peristalsis into the stomach. Often a series of stronger and stronger retches precedes vomiting.

When a person vomits, the rapid propulsion of gastric contents into the esophagus is accompanied by a reflex relaxation of the upper esophageal sphincter. Relaxation is accomplished by a forward movement of the hyoid bone and the larynx. This allows vomitus to be projected into the pharynx and mouth. Entry of vomitus into the trachea is prevented by approximation of the vocal chords, closure of the glottis, and inhibition of inspiration.

■ *Motility of the Small Intestine*

The small intestine, the largest segment of the gastrointestinal system, accounts for about three fourths of the length of the human gastrointestinal tract. The small intestine is about 5 m in length, and chyme typically takes 2 to 4 hours to traverse the small intestine. The first 25 cm or so of the small intestine is the duodenum, which has no mesentery and can be distinguished from the rest of the small intestine histologically. The remaining small intestine is divided into the jejunum and the ileum. The jejunum is more proximal and occupies about 40% of the length of the small bowel. The ileum is the distal part of the small intestine, and it accounts for approximately 60% of the length of the small intestine.

The small intestine, particularly the duodenum and jejunum, is the site of most digestion and absorption. The movements of the small intestine mix chyme with digestive secretions, bring fresh chyme into contact with the absorptive surface of the microvilli, and propel chyme toward the colon.

The most frequent type of movement of the small

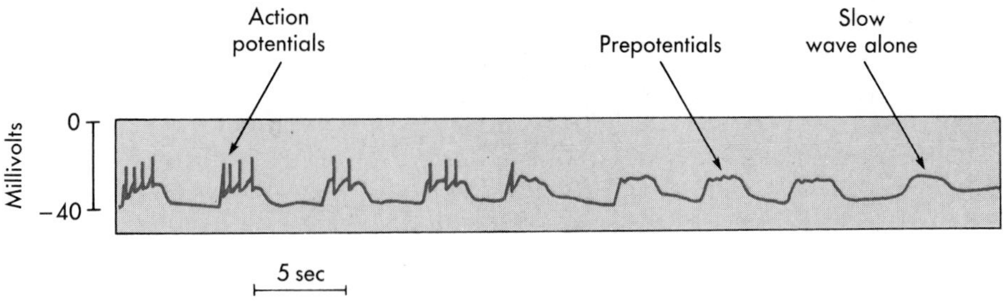

■ **Fig. 42-24.** Intracellular microelectrode recordings of electrical activity in isolated longitudinal muscle of rabbit jejunum. Note that the depolarizing phase of the slow wave may elicit action potentials, prepotentials, or neither. (Redrawn from Bortoff, A.: Am. J. Physiol. **201:**203, 1961.)

intestine is termed *segmentation*. Segmentation is characterized by closely spaced contractions of the circular muscle layer. The contractions divide the small intestine into small neighboring segments. In rhythmic segmentation the sites of the circular contractions alternate, so that a given segment of gut contracts and then relaxes. Segmentation is effective at mixing chyme with digestive secretions and bringing fresh chyme into contact with the mucosal surface. *Peristalsis* is the progressive contraction of successive sections of circular smooth muscle. The contractions move along the gastrointestinal tract in an aboral direction. *Short peristaltic waves* do occur in the small intestine, but they usually involve only a small length of intestine. Peristaltic rushes that travel along much of the length of the small bowel are rare.

As in other parts of the digestive tract, the slow waves of the smooth muscle cells determine the timing of intestinal contractions. The intramural plexuses can control segmentation and short peristaltic movements in the absence of extrinsic innervation. However, extrinsic nerves mediate certain long-range reflexes and modulate the excitability of small intestinal smooth muscle.

■ *Electrical Activity of Small Intestinal Smooth Muscle*

Regular slow waves occur all along the small intestine (Fig. 42-24). The frequency is highest (11 to 13 per minute in humans) in the duodenum, and it declines along the length of the small bowel (to a minimum 8 or 9 per minute in humans in the terminal aspect of the ileum). As in the stomach, the slow waves may or may not be accompanied by bursts of action potential spikes during the depolarizing part of the slow waves. In contrast to the behavior of gastric smooth muscle the action potentials, but not the slow waves, elicit the contractions of the smooth muscle that cause the major mixing and propulsive movements of the small intestine. Since the action potential bursts occur near the peaks of the slow waves, the slow wave frequency determines the maximum possible frequency of intestinal movements. The time interval between intestinal contractions tends to be some multiple of the time interval between slow waves. Action potential bursts are localized to short segments of the intestine. Thus they elicit the highly localized contractions of the circular smooth muscle that cause segmentation.

From the duodenum toward the ileum, there tends to be a phase lag in the small intestinal slow waves. Thus it appears that the slow waves are propagated in the orthograde direction. When peristaltic contractions do occur, they are propagated along the intestine at the apparent velocity of the slow waves.

The basic electrical rhythm of the small intestine is independent of extrinsic innervation; the extrinsically denervated intestine can carry out segmentation and short peristaltic movements. The frequency of the action potential spike bursts that elicit contractile behavior depends on the excitability of the smooth muscle cells of the small intestine. The excitability of the smooth muscle (and thus the frequency of spikes) is influenced by circulating hormones, by the autonomic nervous system, and by enteric neurons. Excitability is enhanced by preganglionic parasympathetic nerves and is inhibited by sympathetic nerves, both acting primarily in the intramural plexuses. Even though much of the direct control of intestinal motility resides in the intramural plexuses, the extrinsic innervation of the small intestine (parasympathetic via the vagus nerves and sympathetic nerves from the celiac and superior mesenteric plexuses) are important in modulating contractile activity. The extrinsic neural circuits are essential for certain long-range intestinal reflexes discussed later.

■ *Contractile Behavior of the Small Intestine*

Contractions of the duodenal bulb mix chyme with pancreatic and biliary secretions, and they propel the chyme along the duodenum. Contractions of the duodenal bulb follow antral systolic contractions. This helps prevent regurgitation of duodenal contents back into the stomach. The basic electrical rhythm of the stomach is about 3 per minute, and that in the duodenal

A

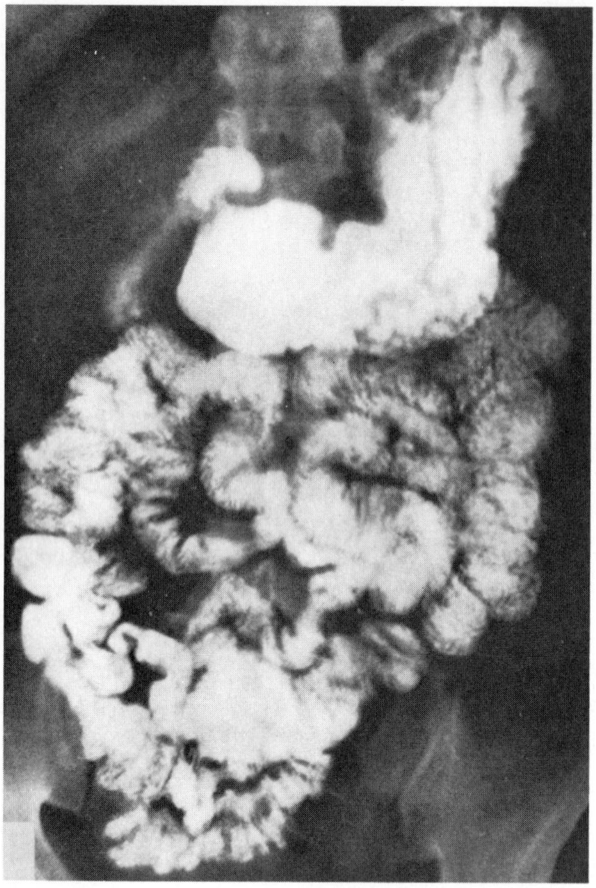

B

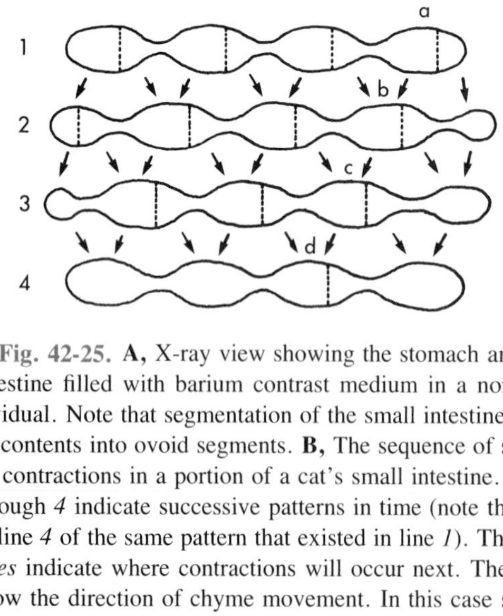

■ **Fig. 42-25.** **A,** X-ray view showing the stomach and small intestine filled with barium contrast medium in a normal individual. Note that segmentation of the small intestine divides its contents into ovoid segments. **B,** The sequence of segmental contractions in a portion of a cat's small intestine. Lines *1* through *4* indicate successive patterns in time (note the return in line *4* of the same pattern that existed in line *1*). The *dotted lines* indicate where contractions will occur next. The arrows show the direction of chyme movement. In this case segmentation occurred from 18 to 21 times per minute. (**A** from Gardner, E.M., et al.: Anatomy: a regional study of human structure, ed. 4, Philadelphia, 1975, W.B. Saunders Co. **B** redrawn from Cannon, W.B.: Am. J. Physiol. **6:**251, 1902.)

bulb is about 11 per minute. Via the longitudinal muscle fibers that cross from stomach to duodenum, and through enteric neurons, the gastric slow waves can influence the duodenal smooth muscle. Thus about every fifth duodenal slow wave is often somewhat increased in amplitude because of the propagated gastric slow wave.

Segmentation is the most frequent type of movement by the small intestine. Segmentation is characterized by localized contractions of rings of the circular smooth muscle. These contractions divide the small intestinal contents into oval segments (Fig. 42-25). When the contracted segments relax, neighboring segments contract. Segmentation occurs at a frequency similar to that of the small intestinal slow waves: about 11 or 12 contractions per minute in the duodenum and 8 or 9 contractions per minute in the ileum. The frequency of segmental contractions is not as constant as that of the slow waves. A group of sequential contractions tends to be followed by a period of rest. The decrease in segmental contraction frequency along the small intestine is illustrated in Fig. 42-26.

In the jejunum, contractile activity usually occurs in bursts, separated by an interval in which contractions are weak or absent. Because the bursts are approximately 1 minute apart, the pattern has been termed the *minute rhythm* of the jejunum (Fig. 42-27).

Segmentation causes a good deal of back-and-forth movement of intestinal contents. It mixes intestinal contents with digestive secretions and circulates the chyme across the surface of the mucosal epithelium. Because of the phase relationship among small intestinal slow waves, contraction in one segment is followed by contraction of a neighboring segment farther down the intestine. Such a sequence of contractions propels chyme in an aboral direction. The decreasing frequency of segmentation as the chyme moves down the small intestine also contributes to aboral propulsion.

Short-range peristalsis also occurs in the small intestine, although much less frequently than segmentation. A peristaltic wave rarely traverses a large part of the small intestine. Typically a peristaltic contraction will die out after traveling about 10 cm. The relatively low rate of net propulsion of chyme in the small intestine allows time for digestion and absorption. Whereas intestinal motility is increased after eating, the net rate of chyme movement is actually decreased. This response depends on the extrinsic nerves to the gut.

■ *Intestinal Reflexes*

Intestinal reflexes can occur along a considerable length of the gastrointestinal tract. These depend to some extent on the function of both intrinsic and extrinsic nerves.

When a bolus of material is placed in the small intestine, the intestine may contract oral to the bolus and

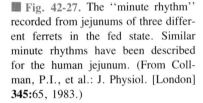

■ **Fig. 42-26.** The rate of segmentation decreases along the length of the small intestine of the rabbit. The data were collected from nearly 30 rabbits. (Redrawn from Alvarez, W.C.: Am. J. Physiol. **37:**267, 1915.)

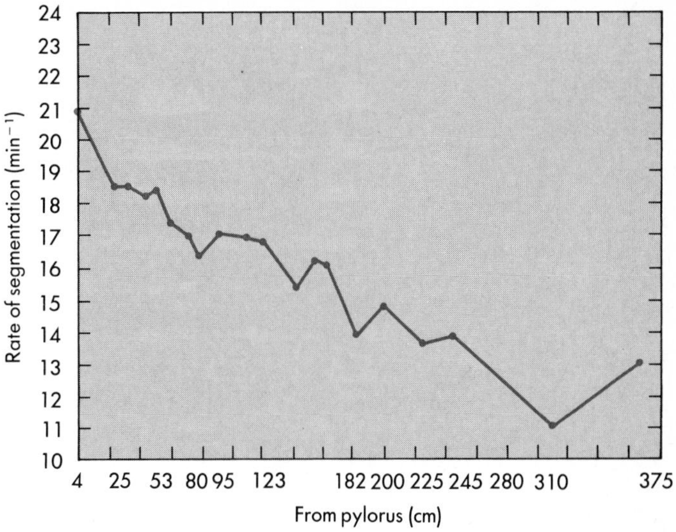

■ **Fig. 42-27.** The "minute rhythm" recorded from jejunums of three different ferrets in the fed state. Similar minute rhythms have been described for the human jejunum. (From Collman, P.I., et al.: J. Physiol. [London] **345:**65, 1983.)

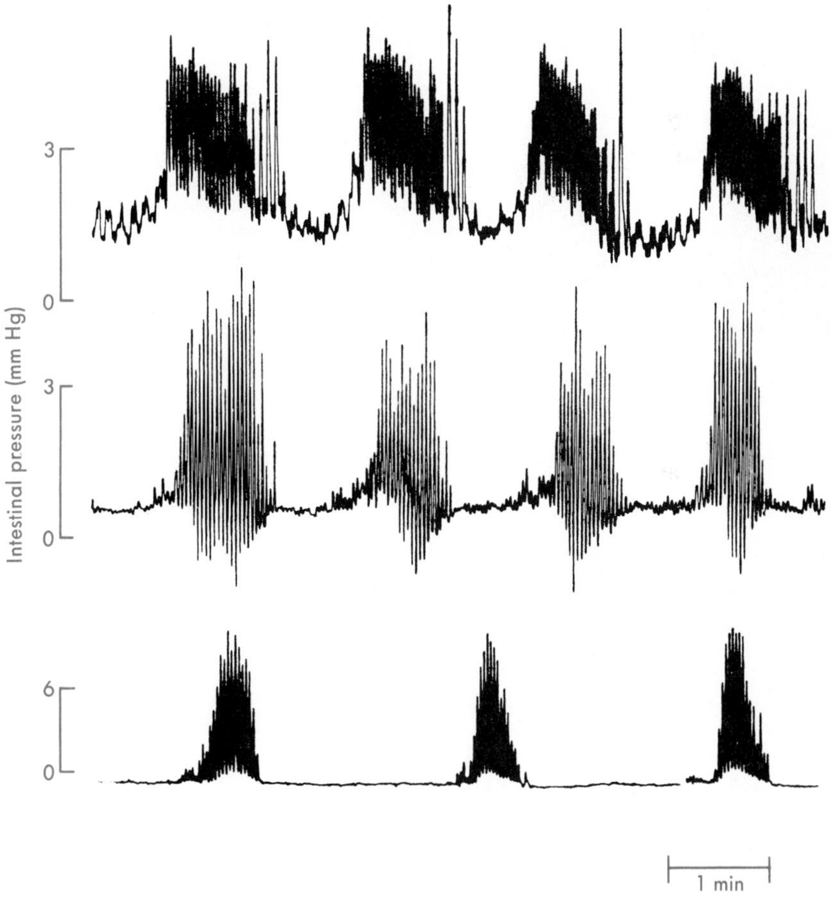

relax aboral to the bolus. This may propel the bolus in an aboral direction, like a peristaltic wave.

Overdistension of one segment of the intestine relaxes the smooth muscle in the rest of the intestine. This response is known as the *intestinointestinal reflex,* and it requires intact extrinsic innervation. The stomach and the terminal aspect of the ileum interact reflexly. Distension of the ileum decreases gastric motility, a response called the *ileogastric reflex.*

Elevated secretory and motor functions of the stomach increase the motility of the terminal part of the ileum and accelerate the movement of material through the ileocecal sphincter. This response is called the *gastroileal reflex.* The hormone gastrin may play a role in this response, since gastrin, at physiological levels, increases ileal motility and relaxes the ileocecal sphincter.

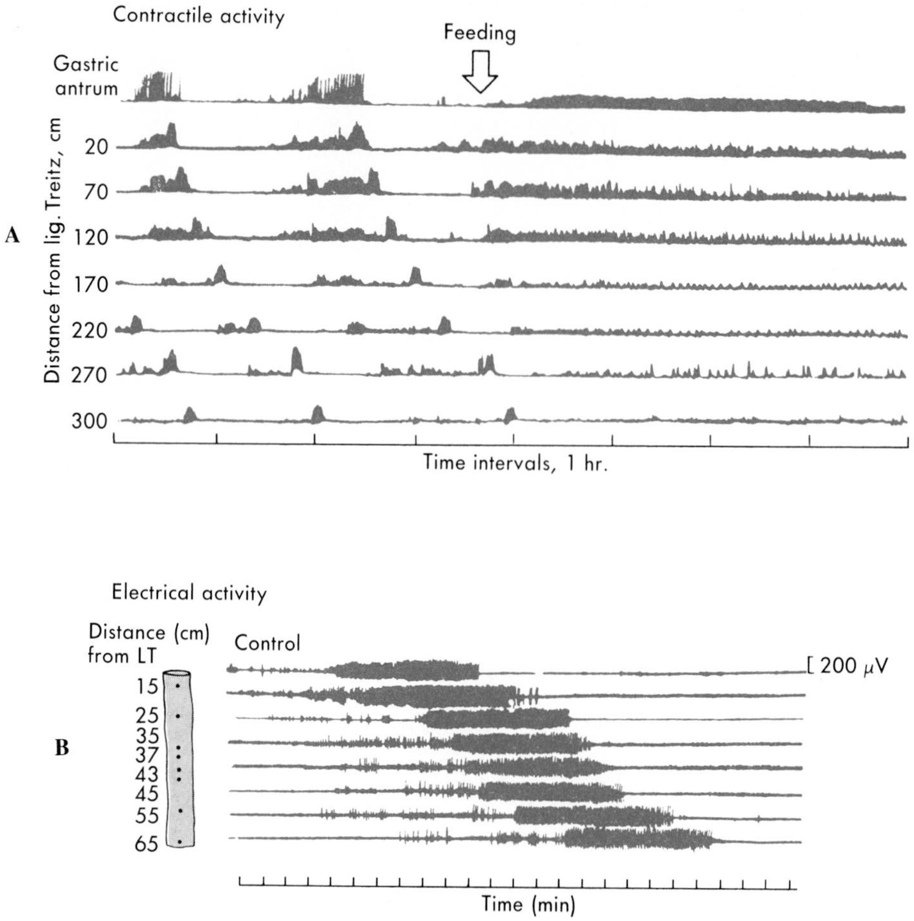

■ **Fig. 42-28.** Mechanical and electrical activity of the small intestine of the conscious dog. The recordings are taken at different distances from the ligament of Treitz *(LT),* which marks the beginning of the jejunum. **A,** Contractile activity in the small intestine of a fasted animal (initially) in which periodic migrating myoelectric complexes are present. Note that feeding converts the small intestine from intermittent to continuous contractile activity. (From Itoh, Z., and Sekiguchi, T., Scand. J. Gastroenterol. Suppl. **82:**121, 1983.) **B,** Electrical activity recorded by extracellular electrodes in the small bowel of a fasted dog. The intense bursts of action potential spikes are associated with phase III of the migrating myoelectric complex. (From Bueno, L., et al.: J. Physiol. [London] **292:**15, 1979.)

■ *The Migrating Myoelectric Complex*

The contractile behavior of the small intestine (already discussed) is characteristic of the period after ingestion of a meal. In a fasted individual or after the processing of the previous meal, small intestinal motility follows a different pattern that is characterized by bursts of intense electrical and contractile activity separated by longer quiescent periods. This pattern appears to be propagated from the stomach to the terminal ileum (Fig. 42-28) and is known as the *migrating myoelectric complex,* the migrating motor complex, or *MMC*. The MMC in the stomach was discussed previously.

The MMC repeats every 75 to 90 minutes in humans, and it occurs in four phases. Phase I, the quiescent phase, is characterized by slow waves with very few action potentials and very few contractions. In phase II irregular action potentials and contractions occur and gradually increase in intensity and frequency. Phase III is a period of intense electrical and contractile activity lasting from 3 to 6 minutes. During phase IV the electrical and contractile activity rapidly decline. Phase IV varies in duration and blends almost imperceptibly into phase I. About the time that one MMC reaches the distal ileum, the next MMC begins in the stomach.

The contractions of phase III of the MMC, both in the stomach and in the small intestine, are more vigorous and more propulsive than the contractions that occur in the fed individual. Phase III sweeps the small bowel clean and empties its contents into the cecum. Thus the MMC has been termed the "housekeeper" of the small intestine. By periodically sweeping the contents of the small intestine into the cecum, the MMC inhibits the migration of colonic bacteria into the distal ileum.

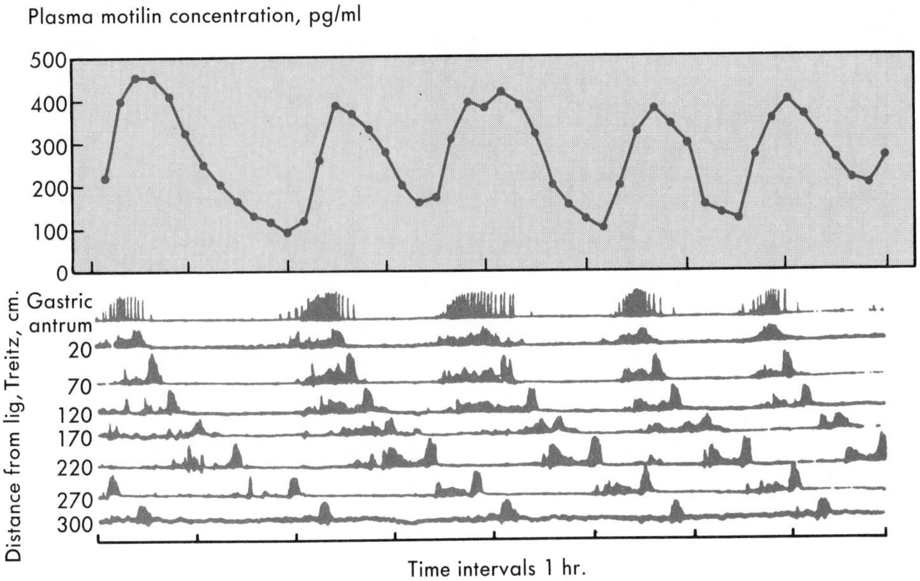

■ **Fig. 42-29.** The relationship between plasma levels of motilin *(upper tracing)* and the occurrence of migrating myoelectric complexes in the dog stomach and small intestine *(lower records)*. (From Itoh, Z., and Sekiguchi, T.: Scand. J. Gastroenterol. Suppl. **82:**121, 1983.)

■ *Propagation of the MMC*

The mechanisms by which the MMC propagates are incompletely understood. The MMC is not a rigid reflex sequence that can only originate in the stomach. Occasionally an MMC originates in the upper small bowel, and then it propagates normally. The enteric nervous system plays a major role in propagation of the MMC. If the enteric neurons are blocked at a particular level of the small intestine, the MMC is not propagated past that point. The small bowel can be cut into several segments and then reanastomosed. This procedure interrupts the enteric nervous system but disturbs the extrinsic nerves only slightly. The small intestine then behaves like a series of independent segments. Each segment generates MMCs that propagate along that segment but not to the next segment.

■ *Generation of the MMC*

The MMC usually originates in the stomach, sometimes as high as the lower esophageal sphincter. However, MMCs occasionally begin in the duodenum or jejunum and sometimes even in the ileum. The initiation of the MMC is not fully understood. Neural and hormonal mechanisms are probably both involved in initiating the MMC.

Some investigators contend that the MMC is normally initiated by vagal impulses to the stomach. A corollary of this hypothesis is that the timing of the MMC is determined by pattern-generating circuitry in the central nervous system. This viewpoint is supported by experiments in which the cervical vagi of dogs were exteriorized in loops of skin so that they could be rapidly and reversibly inactivated by cooling. Cooling the cervical vagi to 4° C caused all MMC activity in the stomach to cease. However, the MMC in the duodenum, jejunum, and ileum continued at about the same frequency, but phase II activity did not precede phase III.

Other investigators favor a hormonal mechanism for initiation of the MMC. They note that chronic extrinsic denervation of the stomach and upper small bowel does not abolish MMC activity. Motilin is currently the favored candidate as the hormone responsible for initiation of the MMC. Intravenous administration of motilin initiates MMCs that strongly resemble the naturally occuring MMC. In addition, plasma motilin levels oscillate in phase with the MMCs (Fig. 42-29). Other hormones that may influence the MMC include substance P, somatostatin, and neurotensin.

■ *Contractile Activity of the Muscularis Mucosae*

Sections of the muscularis mucosae contract irregularly at a rate of about three contractions per minute. These contractions alter the pattern of ridges and folds of the mucosa, mix the luminal contents, and bring different parts of the mucosal surface into contact with freshly mixed chyme. Especially in the proximal part of the small intestine, the villi themselves contract irregularly.

■ **Fig. 42-30.** Major anatomical subdivisions of the colon.

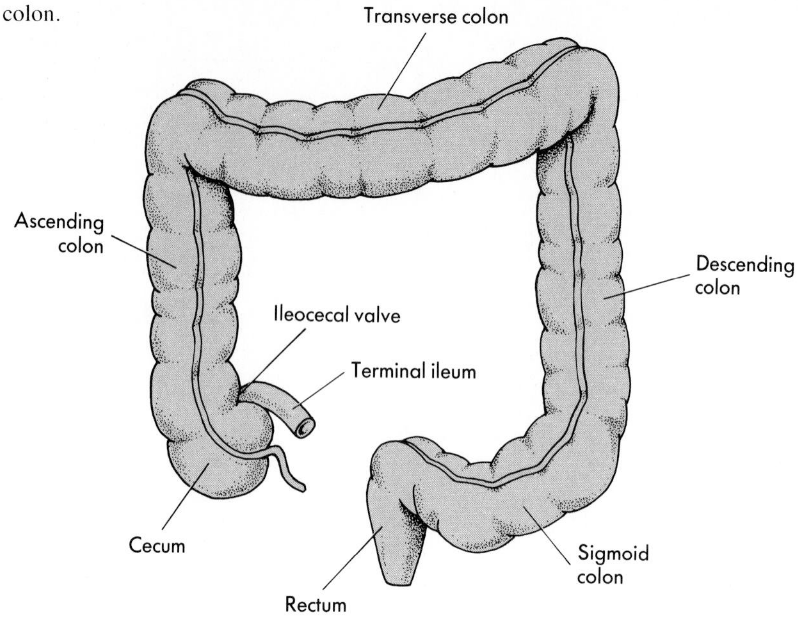

This may aid in emptying the central lacteals of the villi and thus enhance intestinal lymph flow.

■ *Emptying the Ileum*

The *ileocecal sphincter* separates the terminal end of the ileum from the cecum, the first part of the colon. Normally the sphincter is closed, but short-range peristalsis in the terminal aspect of the ileum causes relaxation of the sphincter and allows a small amount of chyme to squirt into the cecum. Distension of the terminal ileum relaxes the sphincter reflexly. Distension of the cecum causes the sphincter to contract and prevents additional emptying of the ileum. The gastroileal reflex enhances ileal emptying after eating. Under normal conditions the ileocecal sphincter allows ileal chyme to enter the colon at a slow enough rate so that the colon can absorb most of the salts and water of the chyme. The ileocecal sphincter is coordinated primarily by the neurons of the intramural plexuses.

■ *Colonic Motility*

The colon receives about 1500 ml of chyme per day from the terminal part of the ileum. Most of the salts and water that enter the colon are absorbed; the feces normally contain only about 50 to 100 ml of water each day. Colonic contractions mix the chyme and circulate it across the mucosal surface of the colon. As the chyme becomes semisolid, this mixing resembles a kneading process. The progress of colonic contents is slow, about 5 to 10 cm/hour at most. One to three times daily a wave of contraction, called a *mass movement,* occurs. A mass movement resembles a peristaltic wave in which the contracted segments remain contracted for some time. Mass movements serve to push the contents of a significant length of colon in an aboral direction.

■ *Structure and Innervation of the Large Intestine*

As shown in Fig. 42-30, the major subdivisions of the large intestine are the cecum, the ascending colon, the transverse colon, the descending colon, the sigmoid colon, the rectum, and the anal canal. The structure of the wall of the large bowel follows the general plan presented earlier in this chapter, but the longitudinal muscle layer of the muscularis externa is concentrated into three bands called the *teniae coli*. In between the teniae coli the longitudinal layer is quite thin. The myenteric plexus is more dense under the teniae coli. The longitudinal muscle of the rectum and anal canal is substantial and continuous.

The extrinsic innervation of the large intestine is predominantly autonomic. Preganglionic parasympathetic innervation of the cecum and the ascending and transverse colon is via branches of the vagus nerve; that of the descending and sigmoid colon, the rectum, and the anal canal is via the pelvic nerves from the sacral spinal cord. The preganglionic parasympathetic nerves end primarily on neurons of the intramural plexuses. Sympathetic innervation to the large intestine is postganglionic and comes to the proximal part of the bowel via the superior mesenteric plexus, to the distal part of the large intestine from the inferior mesenteric and superior hypogastric plexuses, and to the rectum and anal canal from the inferior hypogastric plexus.

Stimulation of sympathetic nerves stops colonic

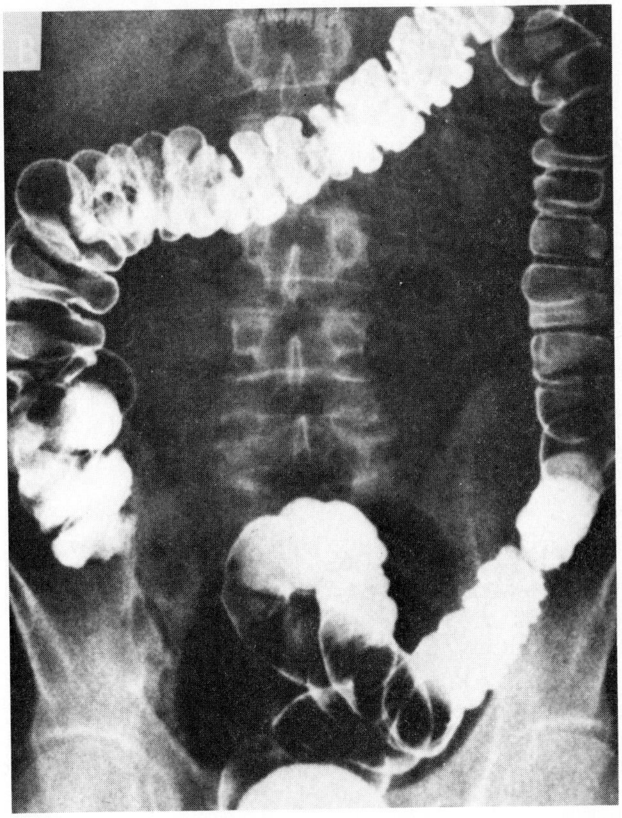

■ **Fig. 42-31.** X-ray image showing a prominent haustral pattern in the colon in a normal individual. (Reproduced with permission from Keats, T.E.: An atlas of normal roentgen variants, ed. 2. Copyright © 1979 by Year Book Medical Publishers, Inc., Chicago.)

movements. Vagal stimulation causes segmental contractions of the proximal part of the colon. Stimulation of the pelvic nerves brings about expulsive movements of the colon and sustained contraction of some segments.

The anal canal usually is kept closed by the internal and external anal sphincters. The internal anal sphincter is a thickening of the circular smooth muscle of the anal canal. The external anal sphincter is more distal, and it consists entirely of striated muscle. The external anal sphincter is innervated by somatic motor fibers via the pudendal nerves, which allow it to be controlled both reflexly and voluntarily.

At most times the tonic contraction of the puborectal muscle pulls the upper anal canal forward to cause a sharp angle between the rectum and the anal canal and prevent feces from entering the anal canal. The innervation of the puborectal muscle is the same as that of the external anal sphincter.

■ *Motility of the Cecum and Proximal Part of the Colon*

Most contractions of the cecum and proximal part of the large bowel are segmental, and they are more effective at mixing and circulating the contents than at propelling them. The mixing action facilitates absorption of salts and water by the mucosal epithelium.

Localized segmental contractions divide the colon into neighboring ovoid segments called *haustra* (Fig. 42-31). Hence segmentation in the colon is known as *haustration*. The most dramatic difference between haustration and the segmentation that occurs in the small intestine is the regularity of the segments (haustra) produced by haustration and the large length of bowel involved in haustration (Fig. 42-31). Segmental thickenings of the circular smooth muscle are probably an anatomical component of haustration. The pattern of haustra is not fixed, however. It fluctuates as segments of the large bowel relax and contract. This results in back-and-forth mixing of luminal contents, known as haustral shuttling. When a few neighboring haustra are emptied in a proximal to distal direction, net propulsion results; this is called *segmental propulsion*. Less frequently a few neighboring haustra will empty because of a concerted contraction of their smooth muscle; this is termed *systolic multihaustral propulsion*.

In the proximal colon, ''antipropulsive'' patterns are not uncommon. Reverse peristalsis and segmental propulsion toward the cecum both occur. These antipropulsive movements tend to retain chyme in the proximal colon and may thus facilitate the absorption of salts and water from the proximal colon.

The net rate of chyme flow in the proximal part of

the colon is only about 5 cm/hour in a fasting individual. Colonic propulsive motility increases after eating, and the net rate of flow increases to about 10 cm/hour. About one to three times daily a mass movement occurs that empties a large portion of the proximal part of the colon in an aboral direction. A mass movement resembles a peristaltic wave in which the contracted segments of the gut remain contracted for a longer period than in peristalsis. Just before a mass movement takes place in a segment of the colon, haustral contractions relax. Then a ring of contraction of the circular muscle begins at the proximal end of the segment and rapidly elongates until one entire segment is contracted. Then that segment of bowel relaxes, haustra reappear, and the dominant motility pattern resumes.

Motility of the Distal Part of the Colon

Normally the distal part of the colon is filled with semisolid feces by a mass movement. Segmental contractions knead the feces, facilitating absorption of remaining salts and water. About one to three times daily, mass movements occur and sweep the feces toward the rectum.

Control of Colonic Movements

As in other segments of the gastrointestinal tract, the intramural plexuses control the contractile behavior of the colon, and the extrinsic autonomic nerves to the colon serve a modulatory function. The defecation reflex, discussed later, is an exception to this rule; it requires the function of the spinal cord via the pelvic nerves. Enteric neurons are both excitatory and inhibitory to colonic smooth muscle. The net effect is inhibitory, because blocking enteric neurons with drugs leads to sustained contractions of the circular muscle of the colon. In *Hirschsprung's disease*, in which enteric neurons are congenitally absent, tonic contractions of the colonic circular muscle lead to obstruction of the colon.

Control by extrinsic nerves. Branches of the vagus nerves supply the proximal colon. Experiments that involve cooling the vagi or injecting cholinergic antagonists support the interpretation that the vagus nerves tonically stimulate the motility of the proximal colon. Some of the tonic vagal stimulation is mediated by cholinergic fibers and some by noncholinergic fibers.

Cutting the sympathetic nerves to the colon enhances contractile activity. Sympathetic nerves tonically inhibit colonic motility. This effect is mediated by norepinephrine acting on neurons of the intramural plexuses.

Electrophysiology of the colon. The colons of some mammals, including dogs and cats, show continuous slow wave activity at frequencies of 4 to 6 per minute. The colons of other mammals, humans included, show intermittent slow waves. The slow waves are generated by pacemaker regions of the circular muscle. As in the small intestine, contractions are associated with bursts of action potentials that occur near the crests of the slow waves. The action potentials usually occur in *short spike bursts,* and they do not propagate very far from the site of origin. They are associated with weak, highly localized contractions of the circular muscle (*type I* contractions). Longer bursts of actions potentials (*long spike bursts*) occur as well. They are propagated for greater distances up and down the colon and are associated with more vigorous contractions (*type II* contractions) of longer segments of the colon. A pattern of intense spiking activity that spreads in the orthograde direction has been observed. This pattern is probably the electrophysiological correlate of a mass movement.

Reflex control. Distension of one part of the colon elicits reflex relaxation of other parts of the colon. This *colonocolonic reflex* is mediated partly by the sympathetic fibers that supply the colon.

The motility of proximal and distal colon and the frequency of mass movements increase reflexly after a meal enters the stomach. This so-called *gastrocolic reflex* is ill-defined; its afferent and efferent limbs are uncertain. The gastrocolic reflex has a rapid neural component, which is elicited by stretching the stomach. The efferent limb is probably carried by parasympathetic nerve fibers to the proximal (via vagus nerves) and distal (via pelvic nerves) segments of the colon. The gastrocolic reflex may also have a slower, hormonal component, for which cholecystokinin and gastrin are the leading mediator candidates. The blood levels of both cholecystokinin and gastrin increase after a meal, and the concentrations that these hormones attain in the blood are high enough to stimulate colonic motility in vitro.

The Rectum and Anal Canal

The rectum is usually empty, or nearly so. The rectum is more active in segmental contractions than is the sigmoid colon. Thus rectal contents tend to move retrograde into the sigmoid colon. The anal canal is tightly closed by the anal sphincters. Prior to defecation the rectum is filled as a result of a mass movement in the sigmoid colon. Filling the rectum brings about reflex relaxation of the internal anal sphincter and reflex constriction of the external anal sphincter (Fig. 42-32) and causes the urge to defecate. Persons who lack functional motor nerves to the external anal sphincter defecate involuntarily when the rectum is filled. The reflex reactions of the sphincters to rectal distension are transient. If defecation is postponed, the sphincters regain their normal tone, and the urge to defecate temporarily subsides.

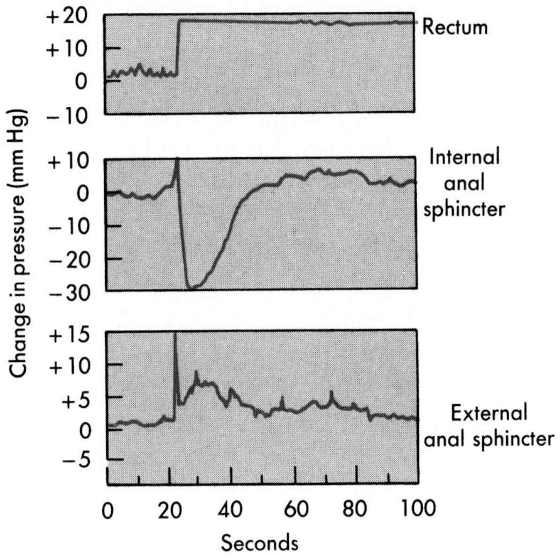

■ Fig. 42-32. Responses of internal and external anal sphincters to a prolonged distension of the rectum. Note that the responses of the sphincters are transient. (Redrawn from Schuster, M.M., et al.: Bull. Johns Hopkins Hosp. **116:**79, 1965.)

■ *Defecation*

When an individual feels the circumstances are appropriate, he or she voluntarily relaxes the external anal sphincter to allow defecation to proceed. Defecation is a complex behavior involving both reflex and voluntary actions. The integrating center for the reflex actions is in the sacral spinal cord but is modulated by higher centers. The efferent pathways are cholinergic parasympathetic fibers in the pelvic nerves. The sympathetic nervous system does not play a significant role in normal defecation.

Before defecation the smooth muscle layers of the descending colon and the sigmoid colon contract in a mass movement to force feces toward the anus. The distension of the rectum signals the urge to defecate. The internal and external sphincters relax, the internal sphincter reflexly and the external sphincter voluntarily. The puborectal muscle reflexly relaxes and allows alignment of the rectum and the anal canal.

Voluntary actions are also important in defecation. The external anal sphincter is voluntarily held in the relaxed state. Intraabdominal pressure is elevated to aid in expulsion of feces. Evacuation is normally preceded by a deep breath, which moves the diaphragm downward. The glottis is then closed, and contractions of the respiratory muscles on full lungs elevates both the intrathoracic and the intraabdominal pressure. Contractions of the muscles of the abdominal wall further increase intraabdominal pressure, which may be as large as 200 cm H_2O and helps to force feces through the relaxed sphincters. The muscles of the pelvic floor are relaxed to allow the floor to drop. This helps to straighten out the rectum and prevent rectal prolapse.

■ *Bibliography*

Journal articles

Wood, J.D.: Intrinsic control of intestinal motility, Annu. Rev. Physiol. **43:**33, 1981.

Wood, J.D.: Enteric neurophysiology, Am. J. Physiol. **247:**G585, 1984.

Books and monographs

Bertaccini: G., editor: Mediators and drugs in gastrointestinal motility, vols. 1 and 2, Berlin, 1982, Springer-Verlag. (vol. 59 of Handbook of Experimental Pharmacology)

Christensen, J., and Wingate, D.L., editors: A guide to gastrointestinal motility, Bristol, U.K., 1983, John Wright and Sons.

Davenport, H.W.: Physiology of the digestive tract, ed. 5, Chicago, 1985, Year Book Medical Publishers.

Gregory, R.A., editor: Regulatory peptides of gut and brain, Br. Med. Bull. **38(3):**219, 1982.

Grundy, D.: Gastrointestinal motility: the integration of physiological mechanisms, Lancaster, U.K., 1985, MTP Press.

Hughes, J., editor: Opioid peptides, Br. Med. Bull. **39(1):**1, 1983.

Johnson, L.R., editor: Physiology of the gastrointestinal tract, vols. 1 and 2, New York, 1981, Raven Press.

Sleisenger, M.H., and Fordtran, J., editors: Gastrointestinal disease: pathophysiology, diagnosis, management, ed. 3, Philadelphia, 1983, W.B. Saunders Co.

Weinbeck, M., editor: Motility of the digestive tract, New York, 1981, Raven Press.

CHAPTER

43

Gastrointestinal Secretions

Regulation of Secretion—General Aspects

This chapter deals with glandular secretion of fluids and compounds that have important functions in the digestive tract. In particular, the secretions of salivary glands, gastric glands, the exocrine pancreas, and the liver are considered. In each case the nature of the secretions and their functions in digestion are discussed, and the regulation of the secretory processes is emphasized.

The secretions mentioned above are elicited by the action of specific effector substances on the secretory cells. These substances may be classified as neurocrine, endocrine, or paracrine. *Neurocrine* substances are released from the endings of neurons that innervate the secretory cells. *Endocrine* modulators are produced by specific cells located some distance from their target cell, and they reach the target cell via the circulation. *Paracrine* regulatory substances are released in the neighborhood of the target secretory cell and reach the target cell by diffusion.

Most of the substances that elicit secretion (and other cellular responses as well) do so by a few basic intracellular mechanisms. For example, the secretion of ptyalin by salivary glands can be elicited by β-adrenergic agents. This process has much in common with the cellular events by which cholecystokinin elicits secretion of pancreatic enzymes. General aspects of the cellular mechanisms that evoke secretion will be discussed first. Then various digestive secretions will be considered individually.

A substance that stimulates a particular cell to secrete is called a *secretagogue*. The number of secretagogues is large, but the cellular mechanisms of action of secretagogues are few. Most secretagogues increase the intracellular level of cyclic adenosine monophosphate (cAMP) or increase the turnover of certain inositol-containing phospholipids in the plasma membrane of the secretory cell. The phospholipid turnover may result in an increase in the concentration of calcium ions in the cytosol. Ca^{++} and cAMP, sometimes termed *second messengers,* initiate chains of events that culminate in increased secretion (or in other responses in nonsecretory cells).

The β-adrenergic effects of norepinephrine and epinephrine are among the cellular responses brought about by increased cytosolic cAMP. β-Adrenergic agonists exert their effects in the following way (Fig. 43-1). The β-agonist first binds to β-adrenergic receptors. The occupied receptors interact with guanosine triphosphate (GTP)–dependent regulatory proteins in a way that causes the regulatory protein to activate adenylate cyclase, the enzyme that makes cAMP from ATP. The β-receptor, GTP-dependent regulatory protein, and adenylate cyclase are all integral proteins of the plasma membrane of the target cell. Increased activity of adenylate cyclase raises the level of cAMP in the cytosol of the cell. The increased cAMP activates certain cAMP-dependent protein kinases. The stimulated protein kinases then phosphorylate particular proteins, which are thereby activated or inactivated. In certain secretory cells the cAMP-stimulated phosphorylation of particular proteins enhances secretion. In many respects, these processes resemble the processes by which catecholamines regulate the contraction of cardiac muscle (Fig. 22-21).

Certain other agonists (which may be secretagogues) act in another way that involves the turnover of inositol-containing phospholipids (Fig. 43-2). The agonist binds to its specific receptor in the plasma membrane of the target cell. The receptor-agonist complex activates a particular phospholipase C that is specific for phosphatidylinositol-(4,5)-diphosphate (PIP_2). (PIP_2 is synthesized in the plasma membrane by two phosphorylation reactions that are catalyzed by specific kinases and that put two phosphate groups on the inositol ring of phosphatidylinositol, a normal component of plasma membranes.) The products of the cleavage catalyzed by the specific phospholipase C are a diglyceride (or diacylglycerol), which remains in the plasma membrane, and

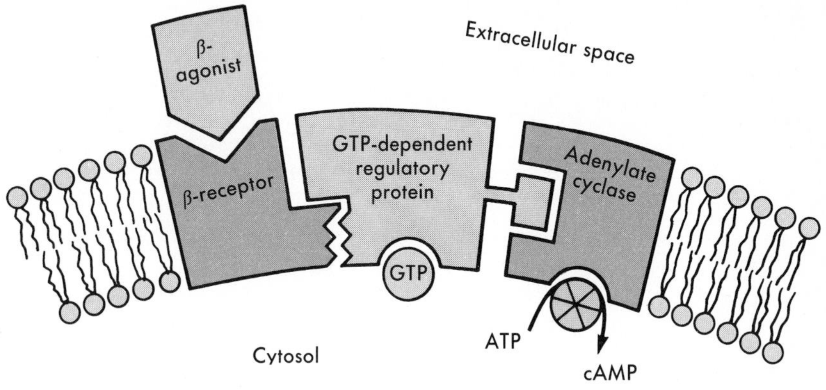

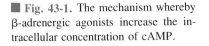

■ **Fig. 43-1.** The mechanism whereby β-adrenergic agonists increase the intracellular concentration of cAMP.

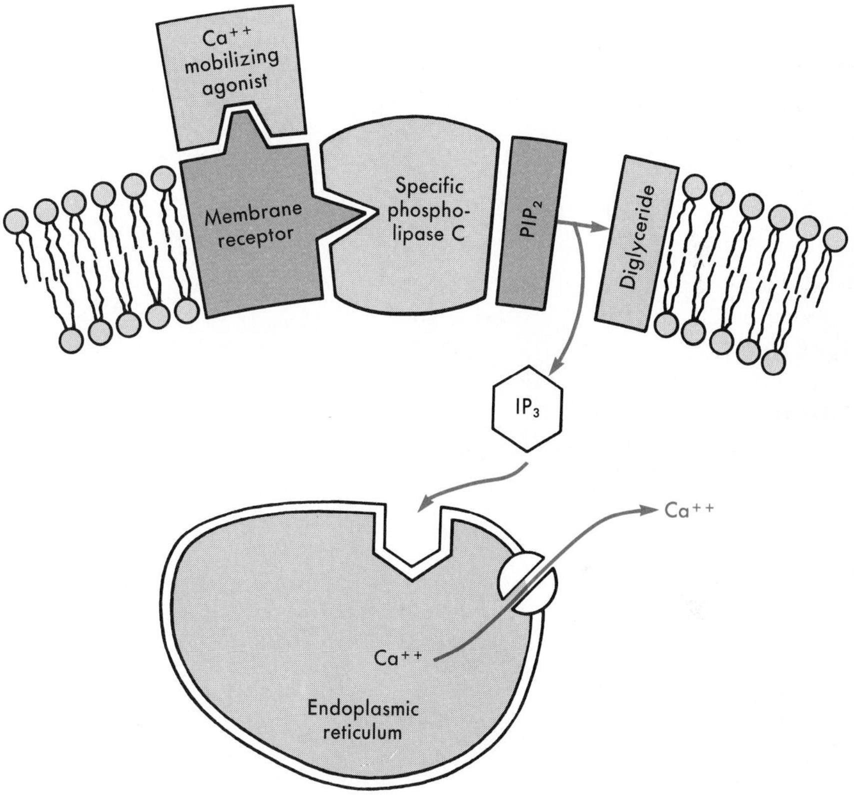

■ **Fig. 43-2.** The mechanism whereby agonists that stimulate hydrolysis of inositol phosphatides lead to an increase in the intracellular level of free Ca^{++}.

inositol-(1, 3, 5)-triphosphate (IP_3), a water-soluble compound that is released into the cytosol. IP_3 is a second messenger that may elicit a number of cellular responses. The most common effect of IP_3 is to increase cytosolic levels of Ca^{++} by releasing Ca^{++} from an intracellular store, usually the endoplasmic reticulum. Ca^{++} controls a host of cellular responses, including secretion, muscle contraction, phototransduction, growth, and development. Ca^{++} may modulate cellular processes directly, or it may act by binding to calmodulin. Calmodulin is a ubiquitous protein that modulates, in a calcium-dependent fashion, many cellular processes.

The increase in cytosolic Ca^{++} mediated by IP_3 is probably the principal mechanism by which an agonist-induced turnover of inositol-containing phospholipids elicits secretion from particular cells. The turnover of phosphoinositides, however, also has other effects on cellular regulation. In the presence of Ca^{++}, the diglyceride produced from cleavage of PIP_2 can activate a particular protein kinase, known as C kinase (Fig. 43-3). Once activated, C kinase may phosphorylate particular target proteins to bring about a cellular response.

Turnover of membrane phosphoinositides can affect the regulation of cellular processes in still another way. The diglyceride produced by cleavage of PIP_2 may be cleaved by a diglyceride lipase to produce a free fatty acid and a monoglyceride. Phosphatidylinositol frequently has arachidonic acid esterified at the number 2 carbon of the glycerol backbone. Hydrolysis of the diglyceride that results from cleavage of PIP_2 releases arachidonic acid. Arachidonic acid is the major precursor

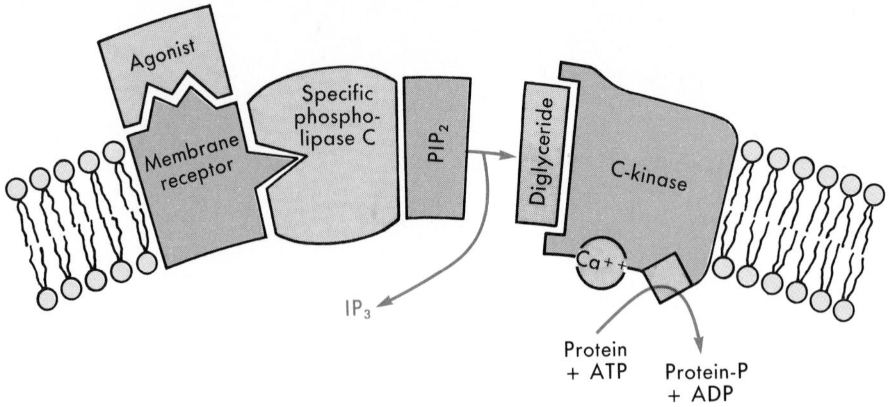

■ **Fig. 43-3.** The mechanism whereby agonists that stimulate hydrolysis of inositol phosphatides lead to the activation of protein kinase C.

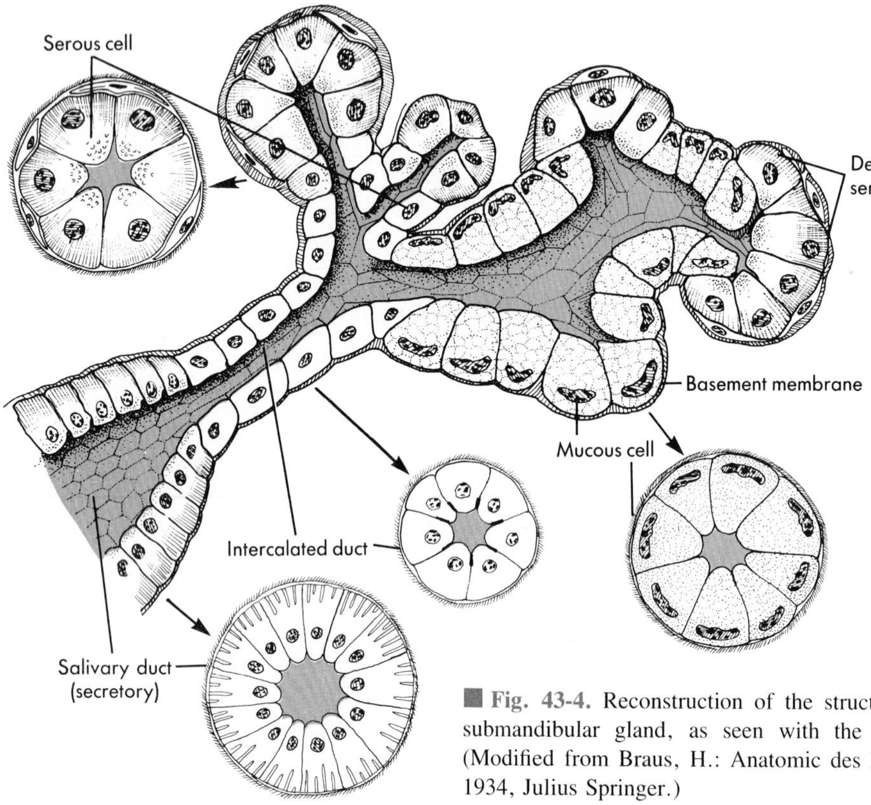

■ **Fig. 43-4.** Reconstruction of the structure of the human submandibular gland, as seen with the light microscope. (Modified from Braus, H.: Anatomic des Menschen, Berlin, 1934, Julius Springer.)

of compounds such as prostaglandins and thromboxanes, which modulate a host of biological processes.

■ *Secretion of Saliva*

In humans the salivary glands produce about 1 L of saliva each day. Saliva lubricates food for greater ease of swallowing. It also facilitates speaking. Saliva contains ptyalin, an α-amylase, which begins the digestion of starch. In people lacking functional salivary glands a condition called xerostomia (dry mouth), dental caries, and infections of the buccal mucosa are prevalent. Secretion of saliva is an active process. The ionic composition of the fluid produced by the acinar cells is similar to that of plasma. The epithelial cells that line the ducts of the salivary glands modify the ionic composition as the saliva flows by. The functions of the salivary glands are primarily controlled by the autonomic nervous system; both sympathetic and parasympathetic stimulation enhance the overall rate of salivary secretion. Physiologically the parasympathetic system is more important.

■ *Functions of Saliva*

Mucins (glycoproteins) produced by the submaxillary and sublingual glands lubricate food so that it may be more readily swallowed. The major digestive function

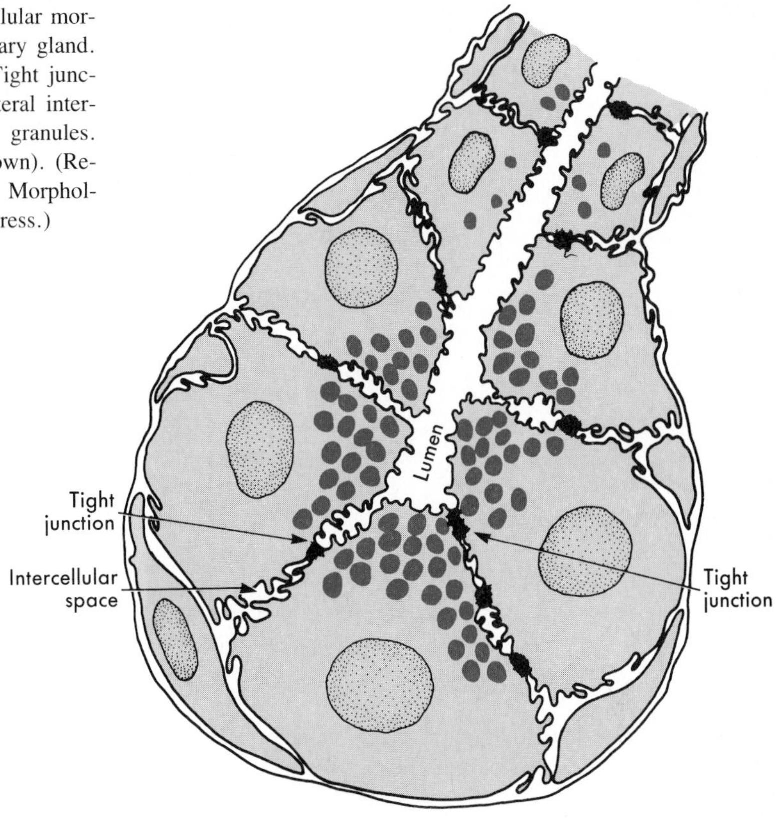

■ Fig. 43-5. A schematic representation of the cellular morphology of a secretory end-piece of a serous salivary gland. Secretory canaliculi drain into the acinar lumen. Tight junctions separate the secretory canaliculi from the lateral intercellular spaces. Orange circles represent zymogen granules. Acinar cells are coupled by gap junctions (not shown). (Redrawn from Young, J.A., and Van Lennep, E.W.: Morphology of salivary glands, London, 1978, Academic Press.)

of saliva results from the action of ptyalin on starch. Ptyalin is an enzyme that has the same specificity as the α-amylase of pancreatic juice. Ptyalin cleaves the internal α-1,4 glycosidic linkages in starch, but it cannot hydrolyze the terminal α-1,4 linkages or the α-1,6 linkages at the branch points. The major products of ptyalin action are maltose, maltotriose, and oligosaccharides containing an α-1,6 branch point (called α-limit dextrins). The pH optimum of ptyalin is about 7, but it has activity between pH levels of 4 and 11. Ptyalin action continues in the mass of food in the stomach, and it is terminated only when the contents of the antrum are mixed with enough gastric acid to lower the pH of the antral contents to below 4. More than half the starch in a well-chewed meal may be reduced to small oligosaccharides by ptyalin action. However, because of the large capacity of the pancreatic α-amylase to digest starch in the small intestine, the absence of ptyalin causes no malabsorption of starch. Other components of saliva, present in smaller amounts, include RNAase, DNAase, lysozyme, peroxidase, and IgA.

The salivary glands have an extremely high blood flow. The capillaries that supply the ducts typically coalesce into venules, which in turn break up into another set of capillaries that supply the acini. The acini are thus partly supplied by a portal circulation.

■ *The Major Salivary Glands and Their Structure*

In humans, the parotid glands, the largest glands, are entirely serous glands. Their watery secretion lacks mucins. The submaxillary and sublingual glands are mixed mucous and serous glands, and they secrete a more viscous saliva containing mucins. Many smaller salivary glands are present in the oral cavity. The microscopic structure of mixed salivary glands is depicted in Fig. 43-4. The salivary glands structurally resemble the exocrine pancreas in many respects. The serous acinar cells, located in the end-pieces, have apical zymogen granules that contain ptyalin and perhaps certain other salivary proteins as well (Fig. 43-5). Mucous acinar cells secrete glycoprotein mucins into the saliva. The intercalated ducts that drain the acini are lined with columnar epithelial cells, which process the saliva to alter its ionic composition. The intercalated ducts drain the acini into somewhat larger ducts, the striated ducts, which empty into still larger excretory ducts, and so forth. A single large duct brings the secretions of each major gland into the mouth. Numerous smaller salivary glands are also present.

The current concept is that a primary secretion is elaborated in the secretory end-pieces. The cells that

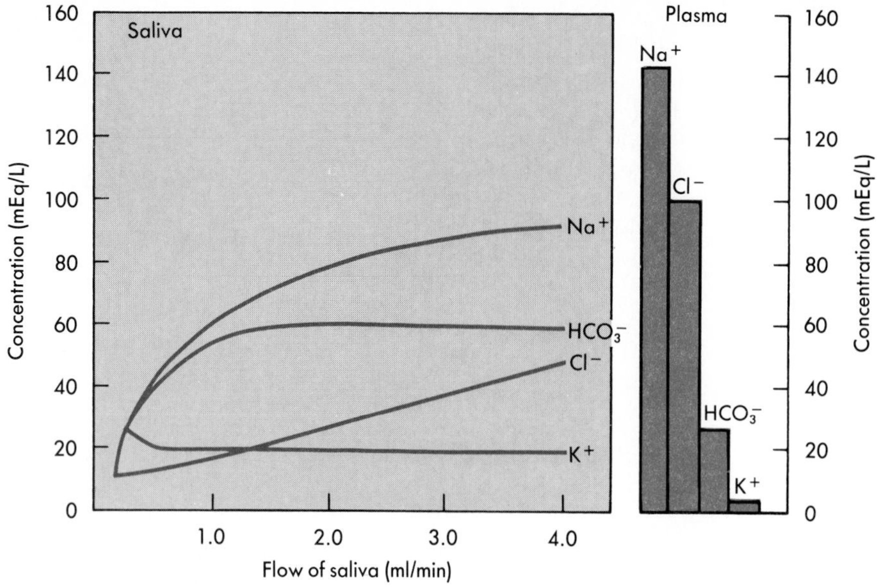

Fig. 43-6. Average composition of the parotid saliva as a function of the rate of salivary flow. (From Thaysen, J.H., et al.: Am. J. Physiol. **178:**155, 1954.)

line the intercalated ducts may modify the secretions of the end-piece cells. The basal membranes of serous acinar cells often have many thin, fingerlike projections that greatly increase the surface area of the basal membrane. Hence the basal membrane may participate in salt and water secretion. The cells that line the duct system modify the primary secretion. The striated ducts that drain the intercalated ducts are so named because of striations in the basal cytoplasm of their epithelial cells. The striations are due to numerous mitochondria aligned with regular infoldings of the basal plasma membrane. This structure suggests that the striated ducts may modify the primary salivary secretion. Such a role for the larger excretory ducts has been well established.

Metabolism and Blood Flow of Salivary Glands

For their size the salivary glands produce a prodigious volume flow of saliva: the maximal rate in humans is about 1 ml per minute per gram of gland. Salivary glands have a high rate of metabolism and a high blood flow, both being proportional to the rate of saliva formation. The blood flow to maximally secreting salivary glands is approximately 10 times that of an equal mass of actively contracting skeletal muscle. Stimulation of the parasympathetic nerves to salivary glands increases blood flow by dilation of the vasculature of the glands. The peptide hormone *vasoactive intestinal polypeptide* coexists with acetylcholine in parasympathetic nerve terminals in the salivary glands. Vasoactive intestinal polypeptide is released along with acetylcholine when the parasympathetic nerves are stimulated and may contribute to vasodilation during secretory activity.

The Ionic Composition of Saliva and Secretion of Water and Electrolytes

In humans, saliva is always hypotonic to plasma. As shown in Fig. 43-6 the Na^+ and Cl^- concentrations are less than those of plasma, but K^+ and HCO_3^- concentrations are greater than those of plasma. The tonicity of saliva and its ionic composition vary from species to species and from one salivary gland to another in the same species. The greater the rate of secretory flow, the higher is the tonicity of the saliva; at maximal flow rates the tonicity of saliva in humans is about 70% of that of plasma. The pH of saliva from resting glands is slightly acidic. During active secretion, however, the saliva becomes basic, with pH approaching 8. The increase in pH with secretory flow rate is partly caused by the increase in salivary HCO_3^- concentration. The concentration of K^+ in saliva is almost independent of salivary flow rate over a wide range of flows. However, at very low salivary flow rates the concentration of K^+ in saliva increases steeply with decreasing flow rate. K^+ concentration can reach levels of 100 mmol/L or more.

Studies in which small samples of fluid are taken from the intercalated ducts of rat salivary glands show that the intercalated ducts contain a fluid that resembles plasma in its concentration of Na^+, K^+, Cl^-, and probably also HCO_3^-. Fluid in the intercalated ducts is either isotonic or slightly hypertonic to plasma.

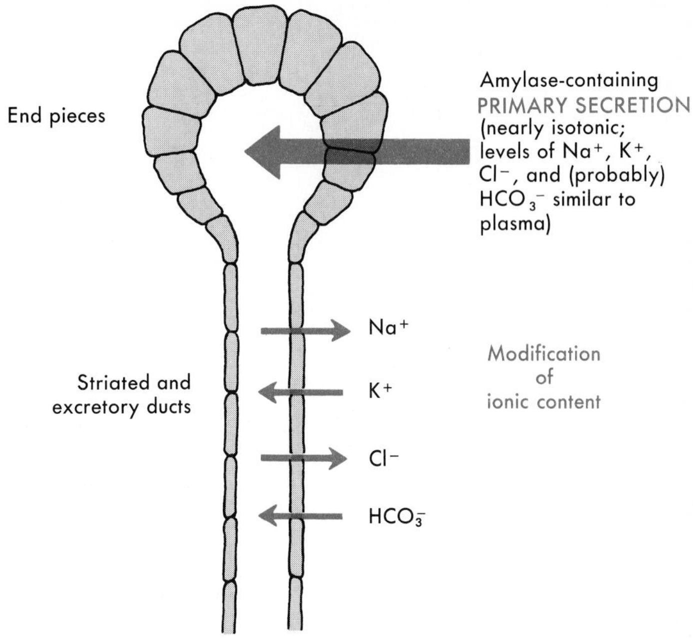

■ **Fig. 43-7.** Schematic representation of the two-stage model of salivary secretion.

Available evidence is consistent with a *two-stage model* of salivary secretion (Fig. 43-7). The two-stage model postulates the following:

1. The end-pieces, perhaps with the participation of intercalated ducts, produce an isotonic primary secretion. The amylase concentration and the rate of fluid secretion vary with the level and type of stimulation. However, the electrolyte composition of the secretion is fairly constant, and the levels of Na^+, K^+, and Cl^- are close to plasma levels.

2. The execretory ducts, and probably the striated ducts too, modify the primary secretion by extracting Na^+ and Cl^- from and adding K^+ and HCO_3^- to the saliva. The amylase concentration of saliva is at least as great as that in the primary secretion. Thus the ducts may reabsorb water, and they do not add to the volume of saliva.

As saliva flows down the ducts it becomes progressively more hypotonic. Thus the ducts remove more ions from saliva than they contribute to it.

The faster the flow rate of the saliva down the striated and excretory ducts, the closer to isotonicity is the saliva. In spite of the gradient for osmotic water flow out of the saliva, relatively little water is absorbed in the ducts, primarily because of the low permeability of the ductular epithelium to water.

■ *Secretion of Ptyalin*

In their apical cytoplasm, serous acinar cells have zymogen granules (Fig. 43-5) that contain ptyalin. Formation of zymogen granules proceeds by the following pathway. Ptyalin is synthesized on the ribosomes of rough endoplasmic reticulum and enters the cisternae of the endoplasmic reticulum. Smooth vesicles containing the newly synthesized enzyme molecules move toward the Golgi apparatus, where the enzymes become encapsulated in membrane-bound vacuoles. The contents of the vacuoles are condensed, and the resulting vesicles take up residence in the apical cytoplasm of the cell, where they are recognizable as zymogen granules. When the gland is stimulated to secrete, the zymogen granules fuse with the plasma membrane. Their contents are released into the lumen of an acinus by exocytosis. Ptyalin also may be secreted by a pathway that is independent of the zymogen granules.

■ *Nervous Control of Salivary Gland Function*

The primary physiological control of the salivary glands is by the autonomic nervous system. Stimulation of either sympathetic or parasympathetic nerves to the salivary glands stimulates salivary secretion, but the effects of the parasympathetic nerves are stronger and more long lasting. Interruption of the sympathetic nerves causes no major defect in the function of the salivary glands. This effect suggests that the essential physiological control is by way of the parasympathetic nervous system. If the parasympathetic supply is interrupted the salivary glands atrophy.

Postganglionic sympathetic fibers to the salivary glands come from the superior cervical ganglion. Preganglionic parasympathetic fibers come via branches of the facial and glossopharyngeal nerves (cranial nerves VII and IX, respectively), and they synapse with postganglionic neurons in or near the salivary glands. The

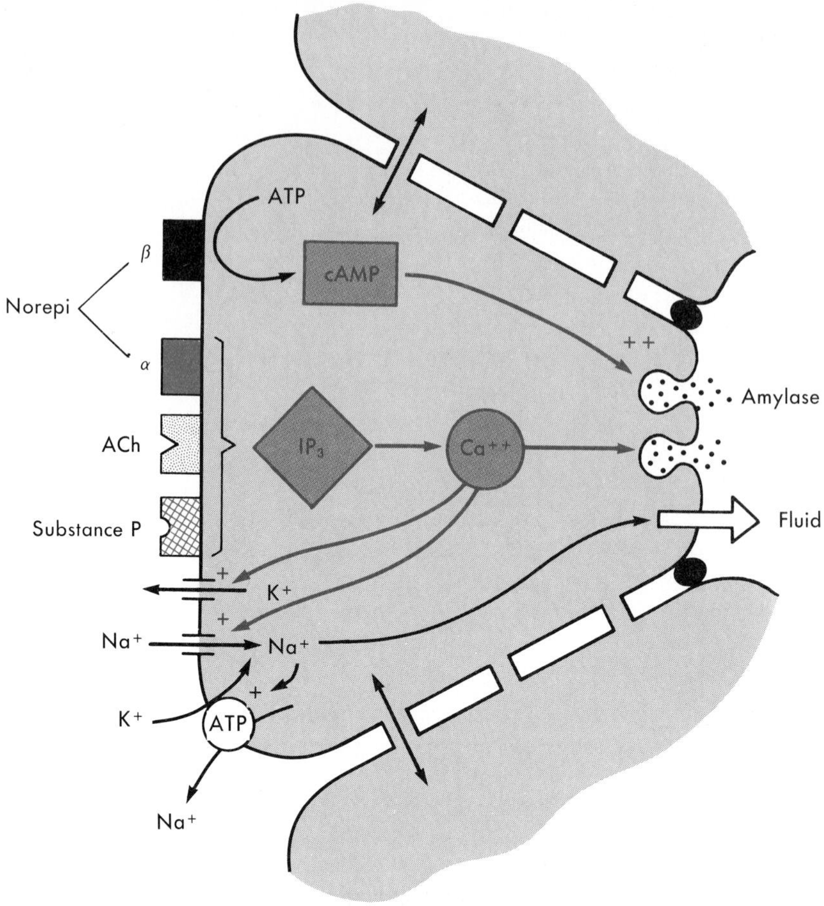

Fig. 43-8. The cellular mechanisms whereby norepinephrine, acetylcholine, and substance P evoke salivary secretion. (Modified from Petersen, O.H.: In Johnson, R.L., editor: Physiology of the gastrointestinal tract, New York, 1981, Raven Press.)

acinar cells and ducts are supplied with parasympathetic nerve endings.

Parasympathetic stimulation increases the synthesis and secretion of ptyalin and mucins, enhances the transport activities of the ductular epithelium, greatly increases blood flow to the glands, and stimulates glandular metabolism and growth.

Sympathetic stimulation and circulating catecholamines stimulate secretion of saliva rich in ptyalin, K^+, and HCO_3^- primarily via β-adrenergic receptors. Sympathetic stimulation causes contraction of myoepithelial cells around the acini and ducts and constriction of blood vessels, with consequent reductions in salivary gland blood flow. The increased salivary secretion that results from stimulation of sympathetic nerves is transient. Stimulation of parasympathetic or sympathetic nerves enhances the ion transport activities of the ducts and yields a saliva rich in K^+ and HCO_3^-.

Cellular Mechanisms of Secretion of Saliva

Duct system. Current knowledge of cellular mechanisms that control ion transport activities in the striated and excretory ducts of salivary glands is sparse. The ducts respond to both cholinergic and adrenergic agonists by increasing their rates of secretion of K^+ and HCO_3^-, but the cellular mechanisms by which these responses are elicited are unknown.

Acinar cells. Understanding of secretory mechanisms of acinar cells, particularly serous cells, has increased dramatically in recent years. The neuroeffector substances that stimulate acinar cell secretions act primarily by one of the two general mechanisms outlined at the beginning of this chapter; that is, by elevating intracellular cAMP or by increasing the level of Ca^{++} in the cytosol (Fig. 43-8).

Acetylcholine, norepinephrine, substance P, and vasoactive intestinal polypeptide are released in salivary glands by specific nerve terminals. Each of these neuroeffectors may increase the secretion of ptyalin and the flow of saliva.

Acinar cells have both α- and β-adrenergic receptors. Norepinephrine binds to both classes of receptors. By using agonists that are specific for one receptor class it has been found that activation of β-receptors (the β_1-subtype) in salivary glands elevates intracellular cAMP. Vasoactive intestinal polypeptides also increase cAMP. By contrast, acetylcholine, substance P, and norepinephrine acting on α-receptors (α_1-subtype) act by mobilizing cellular Ca^{++} via hydrolysis of PIP_2 in the plasma membrane.

In general, effectors that increase cellular cAMP elicit a primary secretion that is richer in ptyalin than the secretion evoked by agents that increase intracellular Ca^{++}. Substances that increase intracellular Ca^{++} increase the volume of acinar cell secretion to a greater extent than compounds that increase cellular cAMP. The ionic composition of acinar cell secretions is relatively independent of the mode of stimulation.

Effectors that act by elevating cAMP. β-Adrenergic agonists such as isoproterenol elevate cAMP. The elevation of cAMP correlates with the increase of amylase secretion and the volume of the primary secretion. The mechanisms whereby elevated cAMP enhances secretion are not well understood. In rat parotid gland, β-adrenergic agonists may stimulate phosphorylation of three specific proteins of the plasma membrane or endoplasmic reticulum. A relationship between phosphorylation of these proteins and enhanced secretion remains to be demonstrated.

Effectors that mobilize cellular Ca^{++}. Cholinergic agents, β-adrenergic agonists, and substance P each bind to distinct classes of receptors. However, they all act by mobilizing a common pool of intracellular Ca^{++}, probably located in the endoplasmic reticulum of the acinar cell. When the Ca^{++} pool has been discharged by one of the classes of effector in the absence of extracellular Ca^{++}, the cell cannot then respond to the other types of effectors. Apparently, refilling of the intracellular Ca^{++} pool depends on extracellur Ca^{++}.

All agonists that mobilize intracellular Ca^{++} cause electrophysiological changes in acinar cells by altering the fluxes of particular ions across the plasma membrane. Elevation of intracellular Ca^{++} opens membrane channels that conduct both Na^+ and K^+. The conductance of these Ca^{++}-gated channels for K^+ exceeds that for Na^+; the reversal potential for the channel is about −50 mV. The resting membrane potential of the acinar cell is usually more negative than −50 mV, so that opening these channels results in depolarization. However, when the resting potential is more positive than −50 mV, Ca^{++}-mobilizing agonists hyperpolarize the acinar cell. The fluxes of Na^+ and K^+ through these chan nels are large and they change the intracellular concentrations of Na^+ and K^+ (Fig. 43-9). The changes in Na^+ and K^+ concentrations stimulate the Na^+, K^+-ATPase to take up K^+ and extrude Na^+ (Fig. 43-10). Because the Na^+, K^+-pump is electrogenic its increased activity hyperpolarizes the acinar cell. Thus stimulation of acinar cells by Ca^{++}-mobilizing agonists causes a characteristic two-phase change in membrane potential, which is called the secretory potential (Fig. 43-11). This first phase is due to the opening of Ca^{++}-gated cation channels, and the second phase is due to electrogenic ion pumping. The second phase can be blocked by ouabain and other inhibitors of the Na^+, K^+-ATPase.

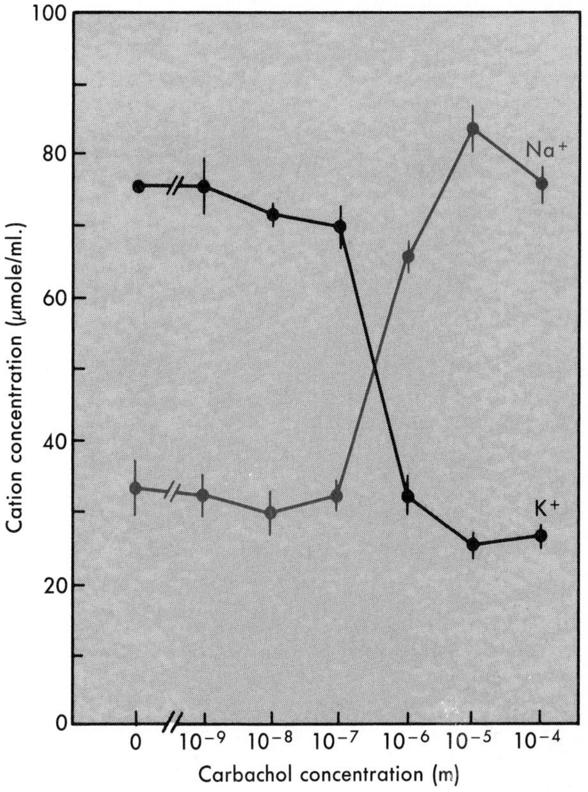

Fig. 43-9. When isolated parotid acinar cells are exposed to a cholinergic agonist (carbachol), they gain Na^+ and lose K^+. Intracellular concentrations of the cations after a 2-minute treatment with carbachol are shown as functions of the carbachol concentration. (Redrawn from Landis, C.A., and Putney, J.W.: J. Physiol. [Lond] **297:**369, 1979.)

The secretory potential appears to be a consequence rather than a cause of the events that elicit acinar cell secretion. The cellular mechanisms whereby elevated intracellular Ca^{++} enhances secretion by acinar cells are poorly understood. Exocytotic release of amylase from zymogen granules is a major mechanism of amylase secretion. Increased intracellular Ca^{++} is the trigger for exocytosis in acinar cells, as it is in nerve terminals.

How Ca^{++} evokes secretion of salts and water by acinar cells is more mysterious. Salivary secretion is an active process; it can proceed against a back pressure that exceeds the arterial blood pressure. It is presumed that the movement of water into the acinar lumen is secondary to ion fluxes. The details of the ion fluxes in acinar cells are not well understood.

Potential role of cyclic guanosine monophosphate. Stimulation of acinar cells by Ca^{++}-mobilizing effec-

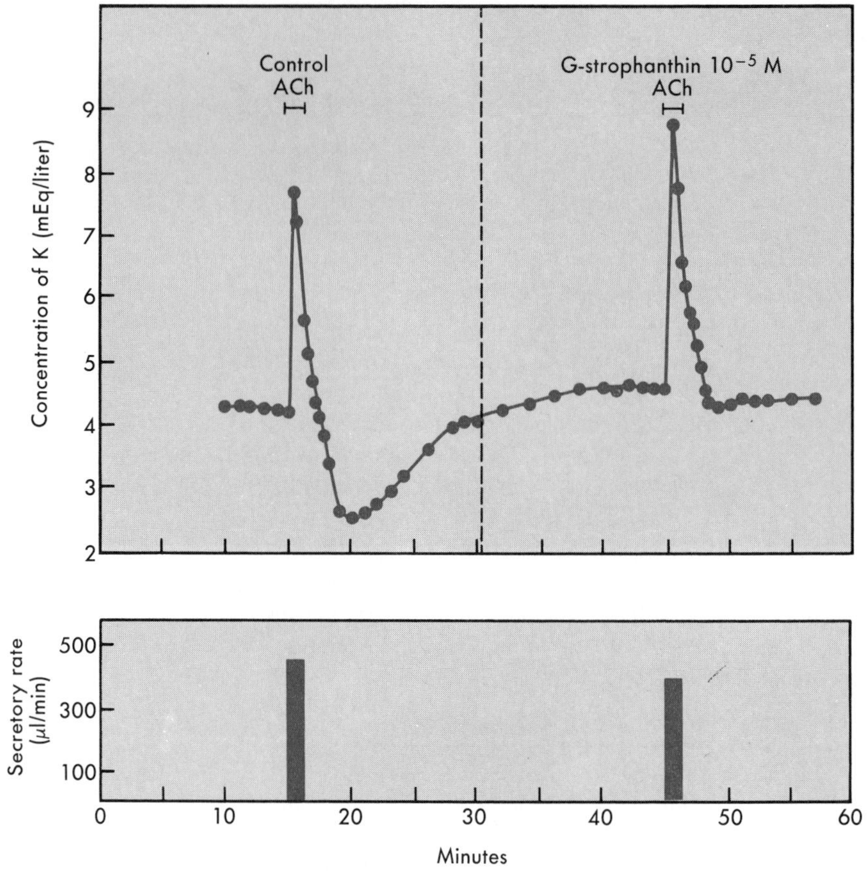

■ **Fig. 43-10.** When a salivary gland is stimulated with acetylcholine *(ACh),* an increase in the flow of saliva occurs *(bottom panel)* and the level of potassium changes in the venous effluent *(top panel).* The initial response to acetylcholine includes an increase in the concentration of K^+ in venous blood from the gland owing to K^+ efflux from the acinar cells. After the stimulus ceases K^+ in the venous effluent transiently diminishes owing to active K^+ uptake by the acinar cells. This second phase of active K^+ uptake is blocked by the cardiac glycoside strophanthidin. (Redrawn from Petersen, O.H.: In Giebisch, G., editor: Electrophysiology of epithelial cells, Stuttgart, 1971, Schattauer Verlag.)

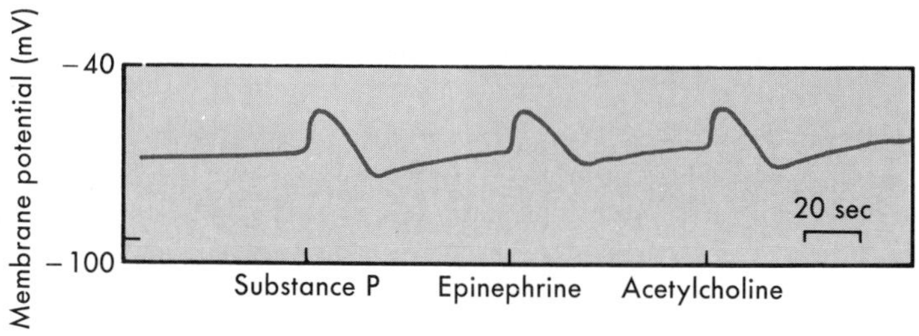

■ **Fig. 43-11.** Membrane potential responses of isolated acinar cells of rat parotid gland to application of secretagogues: substance P, epinephrine, and acetylcholine. (Redrawn from Gallacher, D.V., and Petersen, O.H.: Nature **83:**393, 1980.)

tors may also increase the intracellular levels of cyclic guanosine monophosphate (cGMP), which may correlate with the activation of secretion. By contrast, substance P, a powerful secretagogue, has no effect on cGMP levels. Thus cGMP is probably not directly involved in evoking acinar cell secretions. cGMP may play other roles, perhaps in regulation of growth or development of acinar cells.

■ *Gastric Secretion*

The stomach performs a number of functions. It serves as a reservoir, allowing the ingestion of large meals. It empties its contents into the duodenum at a rate consistent with the ability of the duodenum and small intestine to deal with them. The major secretions of the stomach are HCl, pepsinogens, intrinsic factor, and mucus. HCl keeps gastric contents relatively free of microorganisms. In the absence of HCl secretion bacterial overgrowth occurs in the stomach and upper small intestine. HCl catalyzes the cleavage of pepsinogens to active pepsins and provides a low pH, at which *pepsins* can begin digestion of proteins. Pepsin is not required for normal digestion. In its absence, ingested protein is completely digested by pancreatic proteases and small intestinal brush border peptidases. *Intrinsic factor* is a glycoprotein that binds vitamin B_{12} and allows it to be absorbed by the ileal epithelium. Intrinsic factor is the only gastric secretion required for life. The hormone *gastrin* is secreted by G cells in the antrum and helps regulate gastric acid secretion. *Mucous secretions* protect the gastric mucosa from mechanical and chemical destruction.

Fig. 43-12. Structure of the gastric mucosa. **A,** Reconstruction of part of the gastric wall. (Redrawn from Braus, H.: Anatomie des Menschen, Berlin, 1934, Julius Springer.) **B,** Two gastric glands from a human stomach. (Redrawn from Weiss, L., editor: Histology: cell and tissue biology, ed. 5, New York, 1981, Elsevier.)

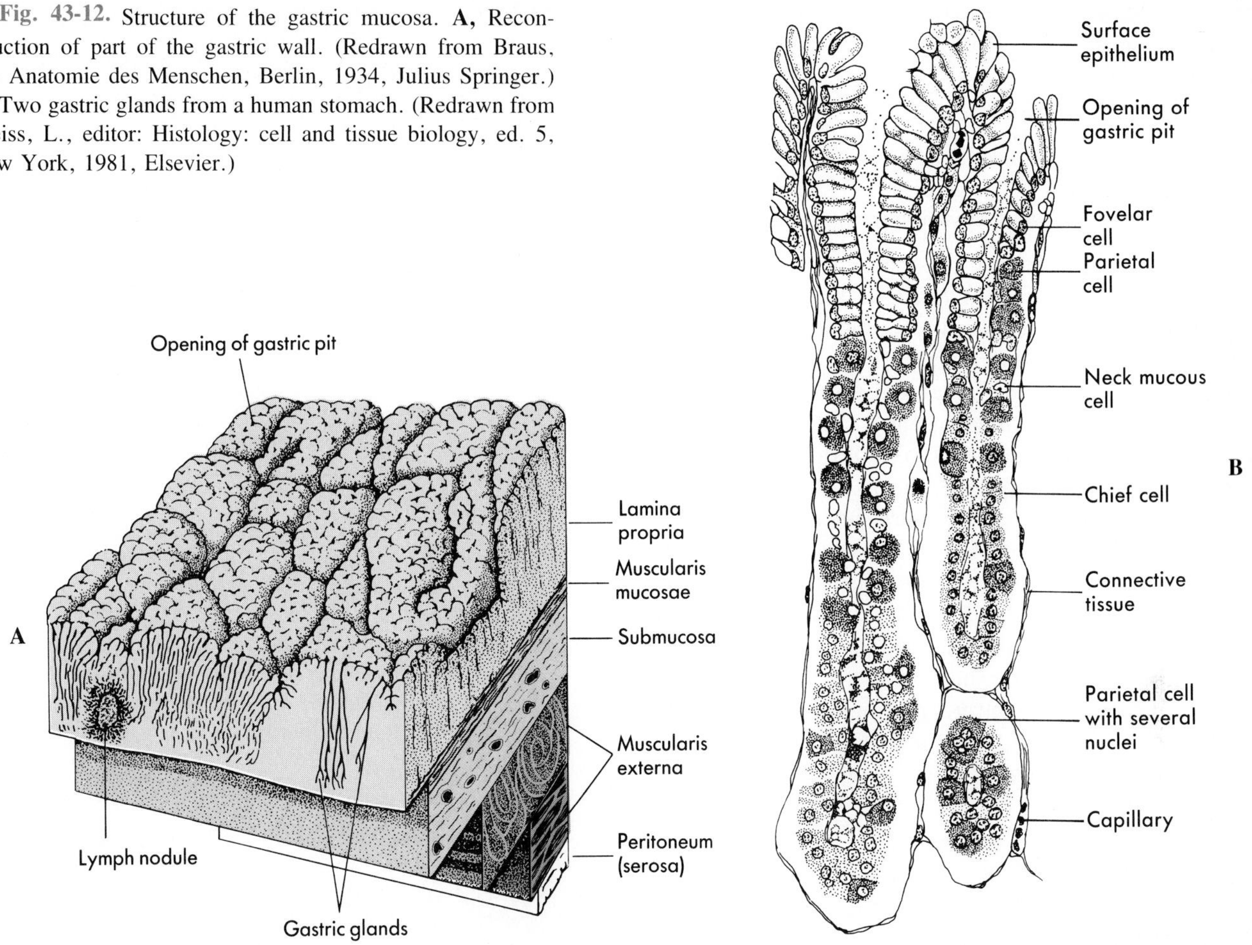

Structure of the Gastric Mucosa

The surface of the gastric mucosa (Fig. 43-12, *A*) is covered by columnar epithelial cells that secrete mucus and an alkaline fluid that protects the epithelium from mechanical injury and from gastric acid. The surface is studded with gastric pits; each pit is the opening of a duct into which one or more gastric glands empty. The gastric pits are so numerous that they account for a significant fraction of the total surface area.

The gastric mucosa can be divided into three regions, based on the structures of the glands present. The region just below the lower esophageal sphincter is called the *cardiac glandular region*. In humans the cardiac glandular region is, at most, a few centimeters wide. The glands of this region are tortuous and contain primarily mucus-secreting cells. The remainder of the gastric mucosa is divided into the oxyntic *(acid-secreting) glandular region,* above the notch, and the *pyloric glandular region,* below the notch.

The structure of a gastric gland from the oxyntic glandular region is illustrated in Fig. 43-12, *B*. The surface epithelial cells extend a bit into the duct opening. In the narrow neck of the gland are the *mucous neck cells,* which secrete mucus. Deeper in the gland are parietal or *oxyntic* cells, which secrete HCl and intrinsic factor, and *chief* or *peptic* cells, which secrete pepsinogen. Oxyntic cells are particularly numerous in glands in the fundus.

The glands of the pyloric glandular region contain few oxyntic and peptic cells; mucus-secreting cells predominate there. The pyloric glands also contain *G cells* which secrete gastrin.

Surface epithelial cells are exfoliated into the lumen at a considerable rate during normal gastric function. They are replaced by mucous neck cells, which differentiate into columnar surface epithelial cells and migrate up out of the necks of the glands. The capacity of the stomach to repair damage to its epithelial surface in this way is impressive.

Gastric Acid Secretion

The fluid secreted into the stomach is called gastric juice. Gastric juice is a mixture of the secretions of the surface epithelial cells and the secretions of gastric glands. Among the important components of gastric juice are salts and water, HCl, pepsins, intrinsic factor, and mucus. Secretion of all of these components increases after a meal.

Ionic Composition of Gastric Juice

The ionic composition of gastric juice depends on the rate of secretion. Fig. 43-13 shows that the higher the secretory rate, the higher the concentration of hydrogen ion. At lower secretory rates [H^+] diminishes, and [Na^+] increases. [K^+] in gastric juice is always higher than in plasma, and consequently prolonged vomiting may lead to hypokalemia. At all rates of secretion Cl^- is the major anion of gastric juice. At high rates of secretion the composition of gastric juice resembles that of an isotonic solution of HCl. At low rates of secretion gastric juice is hypotonic to plasma. When secretion is stimulated the gastric juice approaches isotonicity. The cellular mechanisms by which the ionic composition of gastric juice varies are not well understood. Gastric HCl converts pepsinogen to pepsin, provides an acid pH at which pepsin is active, and kills most ingested bacteria.

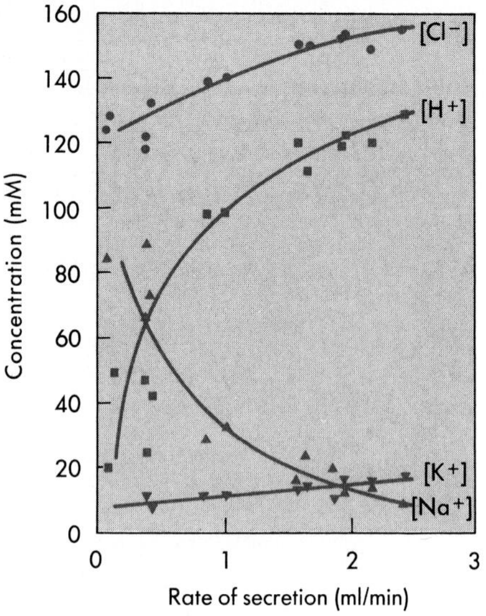

Fig. 43-13. Concentrations of major ions in the gastric juice as a function of the rate of secretion in a normal young person. (Redrawn and reproduced with permission from Davenport, H.W.: Physiology of the digestive tract, ed. 5. Copyright © 1982 by Year Book Medical Publishers, Inc., Chicago; adapted from Nordgren, B.: Acta Physiol. Scand. **58**[suppl. 202]:1, 1963.

Rate of Secretion of Gastric Acid

The rate of gastric acid secretion varies considerably among individuals, partly because of variations in the number of parietal cells. Basal (unstimulated) rates of gastric acid production typically range from about 1 to 5 mEq/hour in humans. On maximal stimulation with histamine or pentagastrin, production rises to 6 to 40 mEq/hour. Patients with gastric ulcers secrete less HCl on the average and patients with duodenal ulcers secrete more HCl on the average than do normal individuals (Fig. 43-14).

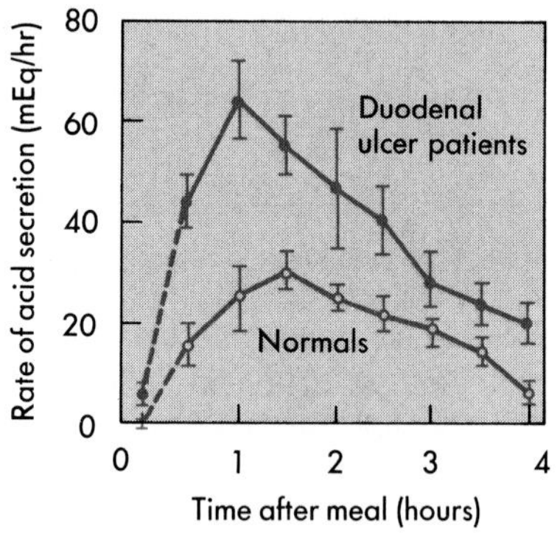

Fig. 43-14. Rate of gastric acid secretion after a meal in six normal subjects and seven patients with duodenal ulcers. (Redrawn from Fordtran, J.S., and Walsh, J.H.: J. Clin. Invest. **52**:645, 1973 by copyright permission of The American Society for Clinical Investigation.)

Morphological Changes That Accompany Gastric Acid Secretion

Parietal cells have a distinctive ultrastructure (Fig. 43-15). They have an elaborate system of branching *secretory canaliculi*, which course through the cytoplasm and are connected by a common outlet to the luminal surface of the cell. Microvilli line the surfaces of the canaliculi. In addition the cytoplasm of the parietal cells contains extensive tubules and vesicles—the *tubulovesicular system*.

When parietal cells are stimulated to secrete, a pronounced morphological change occurs (Fig. 43-15). Tubules and vesicles of the tubulovesicular system fuse with the plasma membrane of the secretory canaliculi, greatly diminishing the content of tubulovesicles and greatly increasing the surface area of the secretory canaliculi. If, as some evidence suggests, the tubulovesicles contain the HCl secretory apparatus, then the extensive membrane fusion that occurs on stimulation would greatly increase the number of HCl pumping sites available at the surface of the secretory canaliculi.

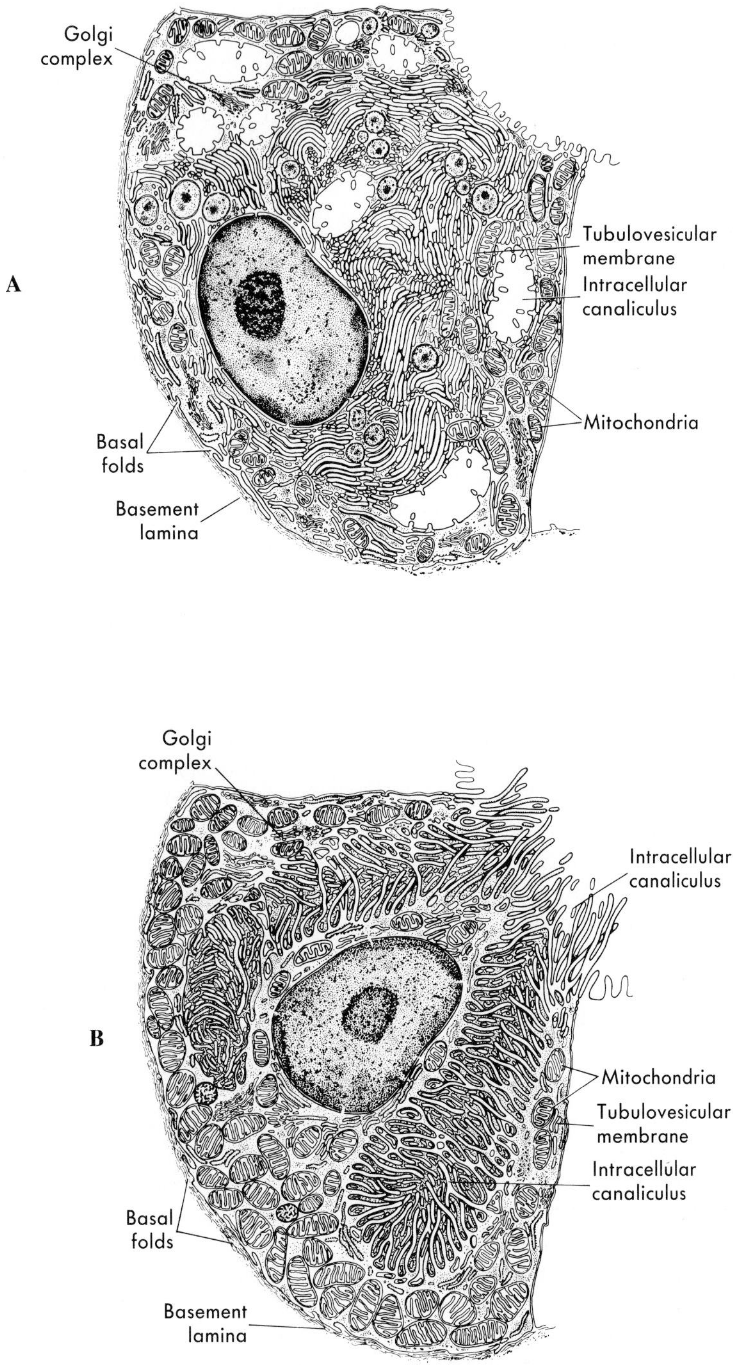

■ **Fig. 43-15. A,** Drawing of a resting parietal cell with cytoplasm full of tubulovesicles and an internalized intracellular canaliculus. **B,** An acid-secreting parietal cell. Tubulovesicles have fused with the membrane of the intracellular canaliculus, which is now open to the lumen of the gland and lined with abundant long microvilli. (Redrawn from Ito, S.: In Johnson, L.R., editor: Physiology of the gastrointestinal tract, New York, 1981, Raven Press.)

The Cellular Mechanism of Gastric Acid Secretion

The mucosal surface of the stomach is always electrically negative with respect to the serosal surface. In the resting stomach the mucosa is −60 to −80 mV (negative with respect to the serosa). When acid secretion is stimulated the potential difference falls to −30 to −50 mV. Thus Cl^-, the ultimate source of which is the plasma, is transported from the extracellular fluid into the lumen of the stomach against both electrical and concentration gradients. Hydrogen ion moves down an electrical gradient into the lumen of the stomach but against a much larger chemical concentration gradient. At maximal rates of secretion H^+ is pumped against a concentration gradient that is more than one million to one. Thus energy is required for transport of both H^+ and Cl^-.

A current model of the mechanism of gastric acid transport is shown in Fig. 43-16. The apical membrane of the parietal cell (the membrane facing the secretory canaliculus) contains an H^+, K^+-activated ATPase, which can exchange H^+ for K^+. This ATPase appears to be the primary H^+ pump. Both H^+ and K^+ are pumped against their electrochemical potential gradients.

The H^+, K^+-ATPase of the parietal cell resembles the Na^+, K^+-ATPase of the plasma membrane and the Ca^{++}-ATPase of the sarcoplasmic reticulum in important respects. Each of these ion-transporting ATPases goes through a pump cycle driven by the phosphorylation and dephosphorylation of an aspartyl residue of the protein. The three ion-transporting ATPases have similar amino acid sequences in their nucleotide-binding domains and in the neighborhood of the phosphorylated aspartate residue. The ion transport cycle of each of the transport proteins involves oscillation between two conformational states: one in which the ion-binding sites face the cytosol and one in which they face the opposite side of the membrane.

When H^+ is pumped out of the parietal cell, an excess of HCO_3^- is left behind. Bicarbonate flows down its electrochemical potential difference across the basolateral plasma membrane. The protein that mediates HCO_3^- efflux transports another anion in the opposite direction. Because Cl^- is the major anion of the extracellular fluid, Cl^- exchanges for HCO_3^-. Chloride moves against its electrochemical potential gradient into the cell, the energy for the active transport of Cl^- coming from the downhill movement of HCO_3^-. The HCO_3^- that leaves the parietal cell is carried away in the blood. During active secretion the pH of venous blood leaving the stomach is elevated. This phenomenon is called the *alkaline tide*.

Some of the entry of Cl^- into the parietal cell appears to depend on the presence of Na^+ in the serosal fluid. Some investigators feel there is an Na^+/Cl^- cotransport system in the basolateral plasma membrane. It is proposed that Na^+ enters the cell down its electrochemical potential gradient and provides the energy to bring Cl^- into the cell against its electrochemical potential gradient. The existence of the Na^+ gradient depends on the action of the Na^+, K^+-ATPase in the basolateral plasma membrane. The nature of the coupling between Na^+ and Cl^- influx across the basolateral membrane remains to be determined.

As a result of the action of the Cl^-/HCO_3^- exchange pump and the Na^+/Cl^- cotransport protein, Cl^- is concentrated in the cytoplasm of the parietal cell. It leaves the parietal cell at the apical membrane by facilitated transport. Recent experiments suggest that when parietal cells are stimulated to secrete acid, the conductances to K^+ and Cl^- of the membrane that bounds the secretory canaliculus increase dramatically. These conductance pathways may be important sites for the regulation of HCl secretion. Whether the movements of K^+ and Cl^- are coupled, as suggested in Fig. 43-16, or occur by independent pathways remains to be determined.

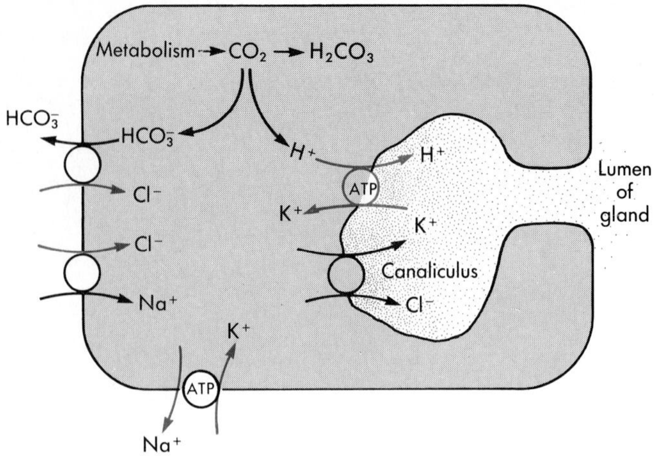

Fig. 43-16. Hypothetical scheme for HCl secretion by the parietal cell. Ion fluxes against an electrochemical potential gradient are shown in color.

Pepsins

The pepsins are a group of proteases secreted primarily by the chief cells of the gastric glands and often collectively referred to as pepsin. The pepsins fall into two electrophoretic classes: group I, secreted mostly in the oxyntic glandular mucosa, and group II, secreted throughout the stomach and by Brunner's glands of the duodenum.

Pepsins are secreted as inactive proenzymes known as *pepsinogens*. Because of gastric acidity, cleavage of acid-labile linkages converts pepsinogens to pepsins;

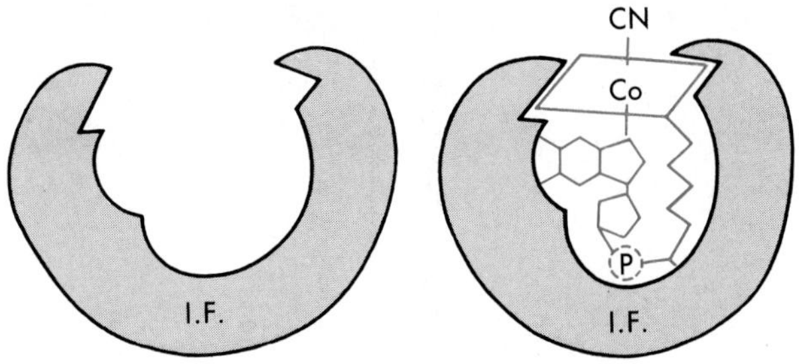

■ Fig. 43-17. The binding of vitamin B_{12} by intrinsic factor (*I.F.*). (Redrawn from Callendar, S.T.: In Badenoch, J., and Brooke, B.N., editors: Recent advances in gastroenterology, Edinburgh, 1972, Churchill.)

the lower the pH, the more rapid the conversion. Pepsins also act proteolytically on pepsinogens to form more pepsins.

The pepsins have unusually low pH optima; they have their highest proteolytic activity at pH 3 and below. Pepsins may digest as much as 20% of the protein in a typical meal. When the duodenal contents are neutralized, pepsins are inactivated irreversibly.

Pepsinogens are contained in membrane-bound zymogen granules in the chief cells. The contents of the zymogen granules are released by exocytosis when the chief cells are stimulated to secrete. Chief cells without zymogen granules also can secrete pepsinogens.

■ *Intrinsic Factor*

Intrinsic factor is a glycoprotein that has a molecular weight of about 55,000 and is secreted by the parietal cells of the stomach. Intrinsic factor is required for normal intestinal absorption of vitamin B_{12}. Vitamin B_{12} (in all its physiological forms) binds to intrinsic factor (Fig. 43-17). The intrinsic factor–B_{12} complex is highly resistant to digestion. Receptors in the mucosa of the ileum bind the complex, and the B_{12} is taken up by the ileal mucosal epithelial cells. Intrinsic factor is released in response to the same stimuli that evoke secretion of gastric acid from the parietal cells.

■ *Secretion of Mucus*

Secretions that contain glycoprotein mucins are viscous and sticky and are collectively termed *mucus*. Mucus adheres to the gastric mucosa and protects the mucosal surface from abrasion by lumps of food and from chemical damage caused by the acid and digestive enzymes.

Mucous neck cells in the necks of gastric glands secrete a clear mucus sometimes called *soluble mucus*. Soluble mucus is not present in the resting stomach. Secretion of soluble mucus is stimulated by sham feeding and by some of the same stimuli that enhance acid and pepsinogen secretion, especially by acetylcholine released from parasympathetic nerve endings near the gastric glands.

The surface epithelial cells secrete a mucus containing different mucins. This mucus is cloudy in appearance and thus is termed *visible mucus*. The surface epithelial cells also secrete watery fluid with Na^+ and Cl^- concentrations similar to plasma but with higher K^+ (four times) and HCO_3^- (two times) concentrations than in plasma. The high $[HCO_3^-]$ makes the visible mucus alkaline. Visible mucus is secreted by the resting mucosa and lines the stomach with a sticky, viscous, alkaline coat. When food is eaten the rates of secretion of visible mucus and of HCO_3^- by the surface epithelial cells increase. Among the stimuli that enhance secretion are mechanical stimulation of the mucosa and stimulation of either sympathetic or parasympathetic nerves to the stomach.

The mucus forms a gel on the luminal surface of the mucosa. The gel protects the mucosa from mechanical damage from chunks of food. The alkaline fluid it entraps protects the mucosa against damage by HCl and pepsin. The mucus and alkaline secretions are part of the *gastric mucosal barrier* that prevents damage to the mucosa by gastric contents.

The mucus layer prevents the bicarbonate-rich secretions of the surface epithelial cells from rapidly mixing with the contents of the gastric lumen. Consequently, the surfaces of gastric epithelial cells are bathed in their own bicarbonate-rich secretions. HCO_3^- buffers H^+ ions that diffuse from the lumen to the epithelial surface (Fig. 43-18). Experiments with pH microelectrodes showed that the surface of the epithelial cells can be maintained at a slightly alkaline pH, despite a luminal pH of about 2. The protection depends on both mucus and HCO_3^- secretion; either mucus alone or HCO_3^- alone could not hold the pH at the epithelial cell surface near neutral. The unstirred layer provided by the mucus retards convective mixing of epithelial cell secretions with luminal contents and slows diffusion of H^+ to the surface and HCO_3^- into the lumen. When the unstirred layer is 1 mm thick the diffusion times for H^+ and HCO_3^- are about 10 minutes, but this time delay would not prevent a rise in the concentration of H^+ at the epithelial surface without a continuous secretion of HCO_3^- to neutralize the H^+ as it arrives.

Mucus is stored in large granules in the apical cytoplasm of mucous neck cells and surface epithelial cells. It is released by exocytosis, by dissolution of the apical membrane of the cell, or by exfoliation of an entire epithelial cell into the mucous coat. The normal epithelium can replace exfoliated surface cells effectively even when extensive cell loss occurs.

The mechanism of HCO_3^- secretion by surface epithelial cells is not well understood. Carbonic anhydrase

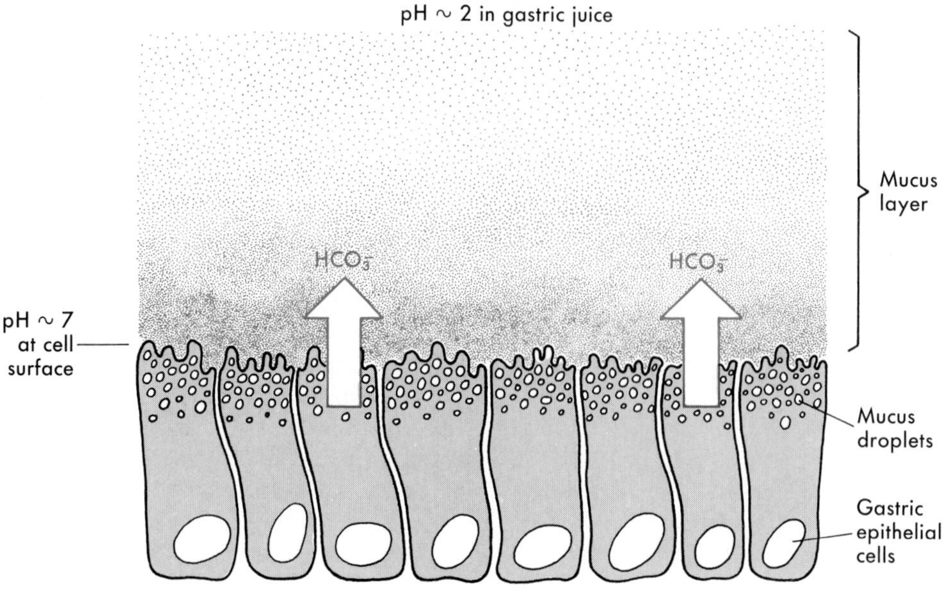

Fig. 43-18. The protection provided the mucosal surface of the stomach by the mucus layer. Buffering by the bicarbonate-rich secretions of the surface epithelial cells and the restraint to convective mixing owing to the high viscosity of the mucus layer allow the pH at the cell surface to remain near 7, whereas the pH in the gastric juice is 1 to 2.

in the epithelial cells supports HCO_3^- secretion, which suggests that the source of HCO_3^- is CO_2 produced by metabolism. Secretion of HCO_3^- may involve exchange of HCO_3^- for Cl^- across the luminal plasma membrane of the epithelial cell. The maximal rate of HCO_3^- secretion is about 10% of the highest rate of gastric acid secretion. Elevated Ca^{++} in the serosal fluid and cholinergic agonists stimulate secretion of HCO_3^-, but histamine and gastrin have little effect. α-Adrenergic agonists decrease bicarbonate secretion. This effect may play a role in pathogenesis of stress ulcers: a chronically elevated level of circulating epinephrine may suppress secretion of HCO_3^- sufficiently to diminish protection of the epithelial cell surface. Aspirin and other nonsteroidal anti-inflammatory agents inhibit secretion of both mucus and bicarbonate; prolonged use of these drugs may damage the mucosal surface. Certain prostaglandins enhance mucus and HCO_3^- secretion and may protect an individual who is susceptible to gastric ulcers.

The structure of mucus. Mucins are the major components of mucus. The mucins produced by surface epithelial cells of the stomach are glycoproteins that are about 80% carbohydrate. About 65% of the sugars present in the carbohydrate side chains of gastric mucins are galactose and *N*-acetylglucosamine. The protein backbone is rich in threonine, serine, and proline, and the carbohydrate side chains are linked to the hydroxyl groups of these amino acid residues. Intact mucins have molecular weights near 2 million. They consist of four similar monomers with molecular weights of 500,000, linked together by disulfide cross-links (Fig. 43-19). Each monomer is largely covered by carbohydrate side chains that protect it from proteolytic degradation. The portion of the monomer that participates in the disulfide cross-links is rich in cystine and free of carboydrate; thus it is vulnerable to proteolysis. Pepsins cleave bonds in the central region of the tetramer and release fragments roughly as large as monomers. The tetrameric mucins form a gel when their concentration exceeds about 50 mg/ml. The mucin monomers do not form a gel. Proteolysis of mucins by pepsin thus dissolves the gel. Maintenance of the protective mucus layer requires that new tetrameric mucins be secreted to replace mucins that are cleaved by pepsin.

Control of Acid Secretion

Control of HCl Secretion at the Level of the Parietal Cell

Acetylcholine, histamine, and gastrin each binds to a distinct class of receptors on the plasma membrane of the parietal cell and directly stimulates the parietal cell to secrete HCl (Fig. 43-20). Acetylcholine is released near parietal cells by cholinergic nerve terminals. Gastrin, a hormone, is produced by G cells in the mucosa of the gastric antrum and the duodenum and reaches parietal cells via the bloodstream. Histamine is

Fig. 43-19. Schematic representation of the structure of gastric mucus glycoprotein before and after hydrolysis by pepsin. (Redrawn from Allen, A.: Br. Med. Bull. **34:**28, 1978.)

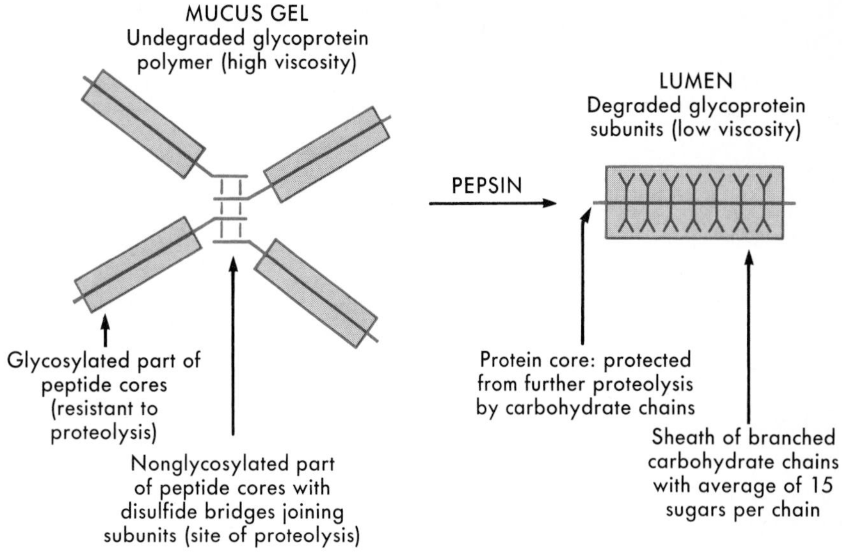

Fig. 43-20. Mechanisms whereby secretagogues elicit acid secretion from parietal cells. The action of blocking agents is also shown: atropine *(A)*, cimetidine *(CM)*, lanthanum ion *(La^{+++})*, and prostaglandin E_2 *(PGE_2)*. (Redrawn from Soll, A.H.: In Johnson, R.L., editor: Physiology of the gastrointestinal tract, New York, 1981, Raven Press.)

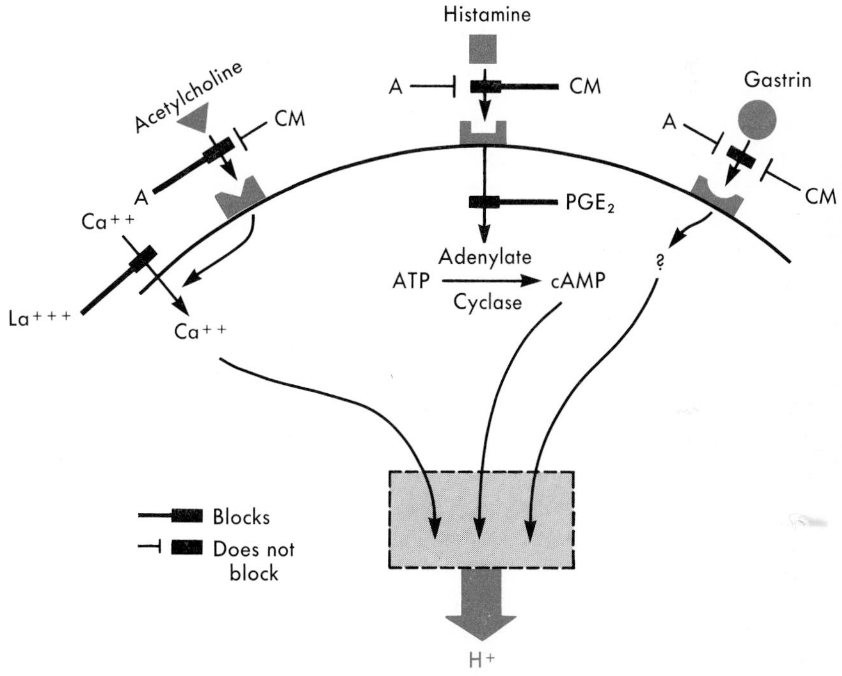

released from cells in the gastric mucosa and diffuses to the parietal cells (hence histamine is a paracrine regulator).

Acetylcholine, histamine, and gastrin are important in the regulation of HCl secretion. Specific antagonists for acetylcholine (e.g., atropine) and histamine (e.g., cimetidine) not only block the effects of acetylcholine and histamine, respectively, but also inhibit acid secretion in response to any effective stimulus. Although specific blockers of gastrin action have not been discovered, a physiological role for gastrin has been established by correlating the rate of gastric acid secretion with the level of gastrin in the blood. Under many circumstances each of the primary agonists (acetylcholine, histamine, and gastrin) is able to potentiate the acid secretion elicited by the other two mediators. The mutual potentiation of the primary mediators appears to explain the lack of specificity in vivo of atropine and cimetidine, substances that are believed to be specific receptor antagonists.

Cellular mechanisms of parietal cell agonists. Acetylcholine, histamine, and gastrin are the three natural secretagogues that act directly on parietal cells to enhance the rate of acid secretion.

Acetylcholine stimulates acid secretion by binding to muscarinic receptors on the basal membrane of the parietal cell (Fig. 43-20). Binding of acetylcholine to its receptors opens Ca^{++} channels and allows Ca^{++} to enter the cell, increasing the cytosolic level of free Ca^{++}. The increased intracellular Ca^{++} concentration somehow increases the secretion of HCl.

Histamine binds to H_2 receptors on the parietal cell membrane and activates adenylate cyclase in the plasma membrane, thereby increasing cytosolic level of cAMP. The increased cAMP probably stimulates a cAMP-dependent protein kinase that phosphorylates a protein that regulates acid secretion.

Gastrin is not as potent a direct stimulant of parietal cells as acetylcholine and histamine. The direct actions of gastrin are not blocked by muscarinic antagonists (e.g., atropine) or by H_2 receptor blockers (e.g., cimetidine), and the *second messenger* for gastrin action is not known. The physiological response to elevated levels of gastrin in the blood is markedly attenuated by cimetidine. Thus a major component of the physiological response to gastrin may be due to gastrin-stimulated release of histamine.

Histamine is certainly a major physiological mediator of HCl secretion. Cimetidine blocks a large portion of the acid secretion elicited by any known secretagogue. How histamine levels in the vicinity of parietal cells are regulated is not known. Cells that resemble mast cells are present in the gastric mucosa, and these mast-like cells may synthesize and store histamine. Upon stimulation by acetylcholine or gastrin the mast-like cells may release histamine, which would diffuse to nearby parietal cells to stimulate HCl secretion. Parietal cells can actively take up histamine and inactivate it by methylation.

In vivo control of the rate of acid secretion. When the stomach has been empty for several hours, HCl is secreted at a significant basal rate, which is about 10% of the maximal rate. The regulation of the basal secretory rate is not well understood. Background levels of acetylcholine and histamine may tonically stimulate the parietal cells. The basal rate of HCl secretion varies diurnally, being highest in the evening and lowest in the morning. The mechanisms responsible for the diurnal variation are not known. However, the level of gastrin in the blood does not vary diurnally.

After a meal the rate of acid secretion by the stomach increases promptly. There are three phases of increased acid secretion in response to food: the *cephalic phase* (elicited before food reaches the stomach), the *gastric phase* (secretion elicited by the presence of food in the stomach), and the *intestinal phase* (elicited by mechanisms originating in the duodenum and upper jejunum).

The Cephalic Phase of Gastric Secretion

The cephalic phase of gastric secretion is normally elicited by the sight, smell, and taste of food. The cephalic phase may be studied in isolation by means of *sham feeding*. In sham feeding, food is chewed but not swallowed or is swallowed and then diverted to the outside of the body by an esophageal fistula. The acid secretion rate during the cephalic phase can be as much as 40% of the maximum rate.

The cephalic phase of gastric acid secretion is mediated entirely by impulses in the vagus nerves; bilateral truncal vagotomy completely abolishes the cephalic phase. Cholinergic vagal fibers and cholinergic neurons of the intramural plexuses are the principal mediators of the cephalic phase. Acetylcholine released from these neurons directly stimulates parietal cells to secrete HCl. The acetylcholine also stimulates acid secretion indirectly by releasing gastrin from G cells in the antrum and duodenum and histamine from mast-like cells in the gastric mucosa.

Other stimuli sensed in the brain, besides those related to the presence of food, may evoke acid secretion via vagal impulses. For example, a decreased concentration of glucose in the cerebral arterial blood elicits acid secretion. Inhibition of brain glucose metabolism by 2-deoxyglucose (a nonmetabolizable analogue of glucose) augments gastric acid secretion.

Atropine blocks most but not all of the acid secretion in response to cephalic-phase stimuli. However, truncal vagotomy blocks the response completely, which suggests that noncholinergic vagal fibers also participate in the cephalic phase. The responses to cephalic stimuli that increase blood gastrin levels are completely blocked by bilateral truncal vagotomy but are not completely blocked by atropine, which suggests that a noncholinergic vagal neuroeffector evokes some gastrin release.

The role of gastrin in the cephalic phase is controversial. Plasma gastrin levels do not increase consistently in response to sham feeding in man. In certain experimental preparations large increases in blood gastrin are associated with relatively low levels of acid secretion. Vagal inhibitory mechanisms may account for the discrepancy between gastrin levels and acid secretion in these experiments.

The presence of a low pH in the antrum diminishes the amount of HCl secreted during the cephalic phase. In the absence of food in the stomach to buffer the acid secreted, the pH of the antral contents falls rapidly during the cephalic phase. The low pH limits the amount of acid secreted by a direct effect on parietal cells.

Regulation of gastric acid secretion by the central nervous system. At the present time knowledge of cerebral control of gastric acid secretion is fragmentary. The hypothalamus participates in the regulation of gastric acid secretion. Electrical stimulation of the lateral hypothalamus increases gastric acid secretion, whereas stimulation of the ventromedial hypothalamus inhibits acid secretion. Injection of 2-deoxyglucose into the lateral hypothalamus increases gastric acid secretion, suggesting that this area enhances gastric acid secretion when cerebral glucose metabolism is impeded. The amygdala and the limbic system may also regulate acid secretion.

A complex interplay between norepinephrine, γ-aminobutyric acid (GABA), and certain neuropeptides may occur in the brain in the control of gastric acid secretion. Norepinephrine appears to be the neurotransmitter in some central neural pathways that inhibit gastric acid secretion, whereas GABA is the transmitter in certain pathways that stimulate acid secretion.

A number of brain peptides may regulate gastric acid secretion. Thyrotropin-releasing hormone and gastrin may affect pathways that enhance acid secretion. Neurotensin, calcitonin, somatostatin, and certain opioid peptides (β-endorphin, Met-enkephalin, and dynorphin) may act centrally to inhibit acid secretion.

Chronic exposure to nicotine increases gastric acid secretion. Nicotine appears to exert this effect on the central nervous system.

The Gastric Phase of Gastric Secretion

The gastric phase of gastric secretion is brought about by the presence of food in the stomach. The principal stimuli are distension of the stomach and the presence of amino acids and peptides resulting from the actions of pepsins. Most of the acid secreted in response to a meal is secreted during the gastric phase.

When either the body or the antrum of the stomach is distended, mechanoreceptors are stimulated. These mechanoreceptors serve as the afferent arms of local and central reflexes that bring about secretion. Both local and central responses are largely cholinergic. Afferent and efferent pathways of the central reflexes are in the vagus nerve.

When the oxyntic glandular area is distended, local and central reflexes bring about release of acetylcholine near parietal cells and stimulate HCl secretion directly. Distension of the body (oxyntic gland area) of the stomach brings about gastrin release from the antral mucosa via a vagal reflex.

Distension of the pyloric glandular area (antrum) enhances gastrin release and, via a vagal reflex, increases acid secretion by the oxyntic glandular mucosa.

All the responses elicited by gastric distension can be blocked effectively by bathing the mucosal surface with an acid solution with pH 2 or less. Once the buffering capacity of the gastric contents is saturated, gastric pH falls rapidly and greatly inhibits further acid release. In this way the acidity of gastric contents regulates itself. In patients with duodenal ulcers, acid secretion is less inhibited by the presence of acid in the antrum.

The presence of amino acids or peptides in the stomach elicits secretion of gastric acid. Intact proteins do not have this effect. The various amino acids differ greatly in their abilities to stimulate acid secretion; tryptophan and phenylalanine are particularly potent stimuli of acid secretion.

Amino acids and peptides may directly stimulate parietal cells, but this action is controversial. Amino acids and peptides do act directly on G cells in the gastric antrum to release gastrin. This appears to be the principal way in which amino acids and peptides stimulate acid secretion during the gastric phase.

Other frequently ingested substances that elicit gastric acid secretion include calcium ions, caffeine, and alcohol. Pure caffeine stimulates acid secretion, perhaps by inhibiting the phosphodiesterase that breaks down cAMP. Acid secretion is enhanced by coffee, but caffeine is not the mediator of this effect, because decaffeinated coffee is equally effective. Ethanol in high concentrations stimulates gastric acid secretion; dilute ethanol is not an effective stimulus.

Gastric distension potentiates the stimulatory effects of chemical stimuli on acid secretion during the gastric phase.

The Intestinal Phase of Gastric Secretion

The presence of chyme in the duodenum brings about neural and endocrine responses that first stimulate and later inhibit secretion of acid by the stomach. Early in gastric emptying, when the pH of gastric chyme is above 3, the stimulatory influences predominate. Later, when the buffer capacity of gastric chyme is exhausted and the pH of chyme emptied into the duodenum falls below pH 2, inhibitory influences prevail. Tables 43-1 and 43-2 summarize the major mechanisms that control gastric acid secretion.

Stimulation of gastric secretion. Gastric secretion is brought about by distension of the duodenum and by the presence of products of protein digestion (peptides and amino acids) in the duodenum, primarily via endocrine mechanisms. The duodenum and proximal jejunum contain G cells that release gastrin when stimulated by peptides and amino acids. The gastrin is carried in the blood to the parietal cells and stimulates them to secrete acid. A second hormone is also released from the duodenum in response to the presence of chyme, and it also acts directly on the parietal cells to stimulate acid secretion and to potentiate the effect of gastrin. This hormone has been named *enterooxyntin,* and its chemical identity and its physiological role in humans remain to be defined. Amino acids in the blood can stimulate gastric acid secretion; thus absorbed amino acids may help to stimulate acid secretion during the intestinal phase.

Inhibition of gastric acid secretion. Several different mechanisms that operate during the intestinal phase inhibit gastric secretion (Table 43-2). The stimuli for these mechanisms are the presence of acid, fat digestion

Table 43-1. Major mechanisms for stimulation of gastric acid secretion

Phase	*Stimulus*	*Pathway*	*Stimulus to parietal cell*
Cephalic	Chewing, swallowing, etc.	Vagus nerve to: 1. Parietal cells 2. G cells	 Acetylcholine Gastrin
Gastric	Gastric distension	Local and vagovagal reflexes to: 1. Parietal cells 2. G cells	 Acetylcholine Gastrin
Intestinal	Protein digestion products in duodenum	1. Intestinal G cells 2. Intestinal endocrine cells	Gastrin Enterooxyntin

Modified from Johnson, L.R., editor: Gastrointestinal physiology, ed. 2, St. Louis, 1981, The C.V. Mosby Co.; adapted from M.I. Grossman.

Table 43-2. Major mechanisms for inhibition of gastric acid secretion

Region	*Stimulus*	*Mediator*	*Inhibit gastrin release*	*Inhibit acid secretion*
Antrum	Acid (pH < 3.0)	None, direct	+	
Duodenum	Acid	Secretin	+	+
		Bulbogastrone	+	+
		Nervous reflex		+
Duodenum and jejunum	Hyperosmotic solutions	Unidentified enterogastrone		+
	Fatty acids, monoglycerides	Gastric inhibitory peptide	+	+
		Cholecystokinin		+
		Unidentified enterogastrone		+

Modified from Johnson, L.R., editor: Gastrointestinal physiology, ed. 2, St. Louis, 1981, The C.V. Mosby Co.; adapted from M.I. Grossman.

products, and hypertonicity in the duodenum and proximal part of the jejunum.

Acid solutions in the duodenum cause the release of secretin into the bloodstream. Secretin inhibits gastric acid secretion in two ways: it inhibits gastrin release by G cells, and it inhibits the response of parietal cells to gastrin. Acid in the duodenum also inhibits gastric acid secretion via a local nervous reflex.

Acid in the duodenal bulb releases the hormone bulbogastrone, which has not been chemically characterized. Bulbogastrone, like secretin, inhibits gastrin-stimulated acid secretion by the parietal cells.

Fatty acids with 10 or more carbons and monoglycerides, the major products of triglyceride digestion, in the duodenum and proximal part of the jejunum release two hormones: *gastric inhibitory peptide* and cholecystokinin. Gastric inhibitory peptide inhibits acid secretion by suppressing gastrin release and by directly inhibiting secretion of acid from the parietal cells. Cholecystokinin also inhibits acid secretion by parietal cells, but this effect may not be physiologically significant.

Another hormone, as yet unidentified, may also inhibit gastric acid secretion in response to fats in the duodenum. *Hyperosmotic* solutions in the duodenum release another unidentified hormone that inhibits gastric acid secretion.

Pepsinogen Secretion

Most of the agents that stimulate parietal cells to secrete acid also elicit release of pepsinogens from chief cells. Hence the rates of release of acid and pepsinogens from the gastric glands are highly correlated. Acetylcholine is a potent stimulus for the chief cells to release pepsinogens. When cholinergic fibers that enhance acid secretion are active, so are cholinergic fibers that release acetylcholine near chief cells. Gastrin also stimulates chief cells to secrete pepsinogens. Acid in contact with the gastric mucosa enhances the output of pepsinogens by a local reflex. Secretin and cholecystokinin, hormones released by the duodenal mucosa in response to acid and fat digestion products, respectively, stimulate chief cells to secrete pepsinogens. The relative importance of these various secretagogues remains to be determined.

Studies with isolated gastric glands have provided information about the cellular mechanisms of action of the compounds that mediate release of pepsinogens. β-Adrenergic agonists release pepsinogens by increasing the level of cAMP in chief cells. Secretin shares this mechanism of eliciting secretion. Acetylcholine, cholecystokinin, and gastrin stimulate release of pepsinogens by increasing the intracellular level of Ca^{++}. How elevated cAMP and Ca^{++} levels increase the release of pepsinogens is not understood.

Pancreatic Secretion

The human pancreas weighs less than 100 g, and each day it elaborates 1 L, 10 times its mass, of pancreatic juice. The pancreas is unusual in having both endocrine and exocrine secretory functions. Its principal endocrine secretions are insulin and glucagon, whose functions are discussed in Chapter 50. The exocrine secretions of the pancreas are important in digestion. Pancreatic juice is composed of an aqueous component, rich in bicarbonate, that helps to neutralize duodenal contents, and an enzyme component that contains enzymes for digesting carbohydrates, proteins, and fats. Pancreatic exocrine secretion is controlled by both neural and hormonal signals, elicited primarily by the presence of acid and digestion products in the duodenum. *Secretin* plays the major role in eliciting secretion of the aqueous component, and *cholecystokinin* stimulates the secretion of pancreatic enzymes.

Structure and Innervation of the Pancreas

The structure of the exocrine pancreas resembles that of the salivary glands. Microscopic blind-ended tubules are surrounded by polygonal acinar cells whose primary function is to secrete the enzyme component of pancreatic juice. The acini are organized into lobules. The tiny ducts that drain the acini are called intercalated ducts (Fig. 43-21). The intercalated ducts empty into somewhat larger intralobular ducts. The intralobular ducts of a particular lobule drain into a single extralobular duct that empties that lobule into still larger ducts. The larger ducts converge into a still larger main collecting duct that drains the pancreas and enters the duodenum along with the common bile duct.

The cells of both acini and ducts are joined by junctional complexes that consist of tight junctions, zonulae adherens, and desmosomes. The tight junctions constitute a permeability barrier between the luminal fluid and the extracellular fluid that bathes the basolateral surfaces of the cells. The tight junctions are impermeable to macromolecules but relatively leaky to water and ions. Acinar and duct cells make gap junctions with their neighbors. These junctions allow rapid communication of changes in membrane potential and permit exchange of molecules with molecular weights of 1400 or less.

The pancreas is supplied by branches of the celiac and superior mesenteric arteries. The portal ve.n is the exit pathway for pancreatic blood flow. The acini and islets are supplied by separate capillary nets. Some of those capillaries that supply the islets converge into venules, which then break up into second capillary networks around the acini.

The endocrine cells of the pancreas reside in the *islets of Langerhans*. Although islet cells account for less than 2% of the volume of the pancreas (Table 43-3), their hormones are essential in regulating metabolism. Insulin, glucagon, somatostatin, and pancreatic polypeptide are hormones released from cells of the islets of Langerhans. Each of these hormones, when administered intravenously, influences the exocrine secretion of the pancreas. However, the physiological roles of these hormones in the regulation of acinar and duct cells remain to be elucidated.

The pancreas is innervated by preganglionic parasympathetic branches of the vagus. Vagal fibers synapse with cholinergic neurons that are within the pancreas and that innervate both acinar and islet cells. Postganglionic sympathetic nerves from the celiac and

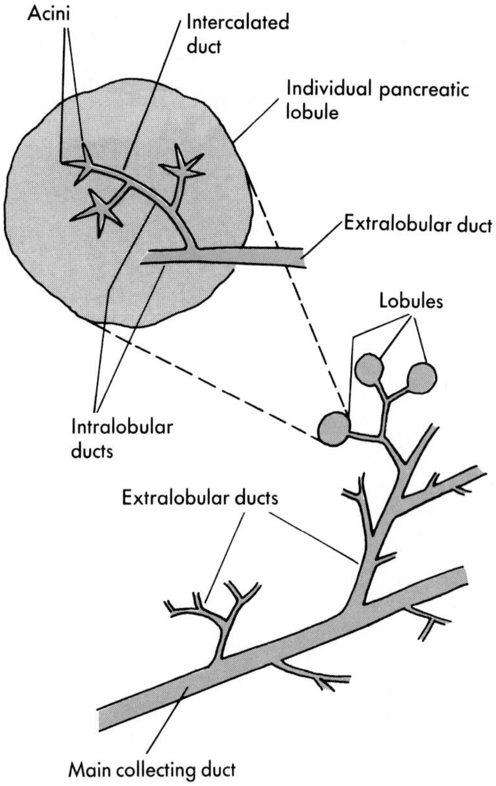

Fig. 43-21. The duct system of the pancreas. (Redrawn from Swanson, C.H., and Solomon, A.K.: J. Gen. Physiol. **62**:407, 1973 by copyright permission of The Rockefeller University Press.)

Table 43-3. Volumes occupied by various cellular elements in guinea pig pancreas

Cell type	*Volume occupied (%)*
Acinar	82.0
Duct	3.9
Endocrine	1.8
Blood vessels	3.7
Extracellular space	9.4

Data of Bolender, R.P.: J. Cell Biol. **61**:269, 1974.

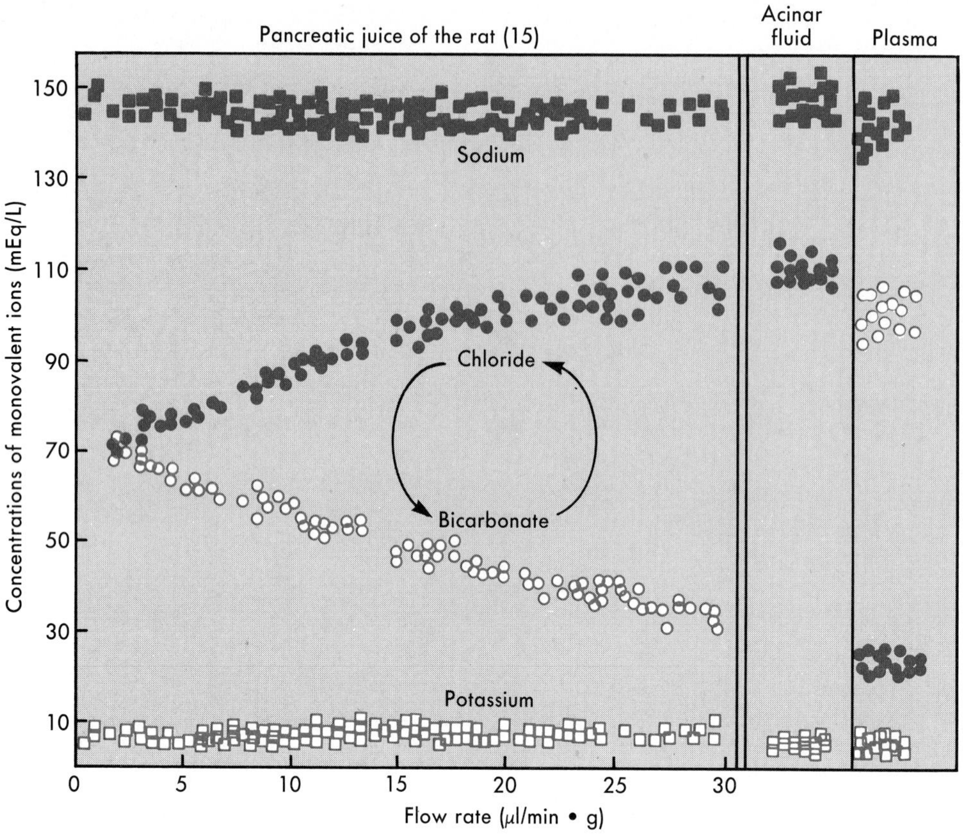

Fig. 43-22. Concentrations of the major ions in pancreatic juice as functions of the secretory flow rate. The concentrations of the ions in acinar fluid and plasma are shown for reference. Secretion was stimulated by intravenous injection of secretin. (Redrawn from Mangos, J.A., and McSherry, N.R.: Am. J. Physiol. **221:**496, 1971.)

superior mesenteric plexuses innervate pancreatic blood vessels. In general, secretion of pancreatic juice is stimulated by parasympathetic activity and inhibited by sympathetic activity.

The Aqueous Component of Pancreatic Juice

The aqueous component of pancreatic juice is elaborated principally by the columnar epithelial cells that line the ducts. The Na^+ and K^+ concentrations of pancreatic juice are similar to those in plasma. HCO_3^- and Cl^- are its major anions. The HCO_3^- concentration varies from about 30 mEq/L at low rates of secretion to over 100 mEq/L at high secretory rates (Fig. 43-22). HCO_3^- and Cl^- concentrations vary reciprocally. As secreted by the duct cells the aqueous component is slightly hypertonic and has a high HCO_3^- concentration. As it flows down the ducts, water equilibrates across the epithelium to make the pancreatic juice isotonic, and some HCO_3^- exchanges for Cl^- (Fig. 43-23). The faster the flow rate, the less is the time available for HCO_3^--Cl^- exchange, and the higher is the HCO_3^- concentration in the juice.

Under resting conditions the aqueous component is produced primarily by the intercalated and other intralobular ducts. When secretion is stimulated by secretin, however, the additional flow comes mostly from the extralobular ducts (Fig. 43-23). Secretin is the major physiological stimulus for secretion of the aqueous component. The secretin-stimulated juice secreted by the extralobular ducts resembles the resting secretion produced by the intralobular ducts, but the extralobular secretion has a slightly higher HCO_3^- concentration.

When acinar cells are stimulated by cholecystokinin to secrete proteins, a small volume of fluid is also secreted in the lumen of the acinus. This fluid is isotonic and resembles plasma in its ionic composition. It is not clear whether the acinar cells themselves are the source of the acinar fluid. The centroacinar cells that reside near the junction between the acini and the intercalated ducts, or the epithelial cells of the intercalated ducts cannot be excluded as the source of the fluid secreted into the acini.

When secretion of the aqueous component of pancreatic juice is stimulated by secretin, the composition of the acinar fluid and the fluid secreted by the intralobular ducts does not change. However, the extralobular

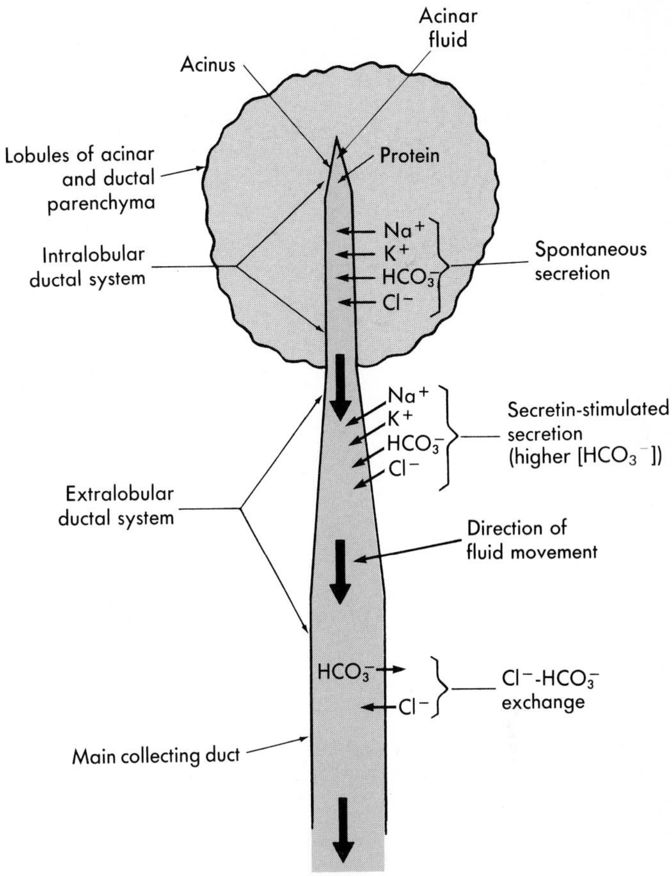

■ **Fig. 43-23.** The locations of some of the transport processes involved in the elaboration of the aqueous component of pancreatic juice. Acinar fluid: isotonic; resembles plasma in concentrations of Na^+, K^+, Cl^-, and HCO_3^-; high level of pancreatic enzymes; secretion stimulated by cholecystokinin and acetylcholine. (Redrawn from Swanson, C.H., and Solomon, A.K.: J. Gen. Physiol. **62**:407, 1973 by copyright permission of The Rockefeller University Press.)

ducts respond to secretin stimulation by elaborating a secretion of greater volume and with higher bicarbonate levels than are found in the resting secretion.

The cellular events that result in secretion of bicarbonate-rich fluid by the cells of the intra- and extralobular ducts are not well understood. Bicarbonate secretion is dependent on the presence of bicarbonate in the plasma: both the concentration of bicarbonate in pancreatic juice and its rate of production are proportional to the plasma level of bicarbonate. Thus the source of bicarbonate in pancreatic juice appears to be plasma bicarbonate, rather than CO_2 produced by duct cell metabolism.

A hypothetical view of the cellular processes that result in secretion of the aqueous component of pancreatic juice is depicted in Fig. 43-24. The Na^+,K^+-ATPase is the only process powered directly by metabolic energy in the form of adenosine triphosphate. In the basolateral membrane of the duct cell an Na^+/H^+ exchanger is postulated to actively extrude H^+ from the cell in exchange for Na^+. The Na^+,K^+-ATPase creates the electrochemical potential gradient of Na^+, and it provides the energy for the transport of H^+ out of the cell against its electrochemical potential gradient. Acidification of the extracellular fluid increases its concentration of CO_2. Carbon dioxide diffuses readily into the duct cell, where carbonic anhydrase catalyzes the hydration of CO_2 to H_2CO_3, which dissociates into H^+ and HCO_3^-. The continued extrusion of H^+ from the cell raises the intracellular concentration of HCO_3^-. Bicarbonate then leaves the apical membrane of the cell in exchange for Cl^- by means of an anion exchange protein that is present in many different cell types. Levels of Na^+ present in pancreatic juice are similar to those in plasma. Much of the Na^+ diffuses through the intercellular spaces, because the tight junctions that join the duct cells are rather leaky to water and small univalent ions.

■ *The Enzyme Component of Pancreatic Juice*

Table 43-4 lists proteins secreted by pancreatic acinar cells. The secretions of the acinar cells comprise the *enzyme component* of pancreatic juice. This component contains enzymes important for the digestion of all the major classes of foodstuffs. In the total absence of pancreatic enzymes, malabsorption of lipids, proteins, and carbohydrates occurs.

The proteases of pancreatic juice are secreted in inactive zymogen form. The major pancreatic proteases are *trypsin, chymotrypsin,* and *carboxypeptidase*. They

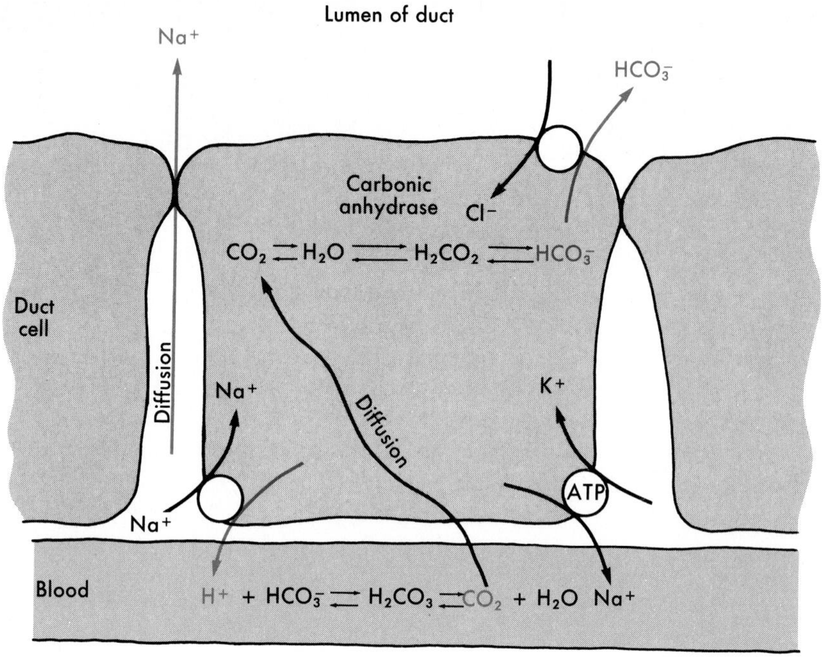

Fig. 43-24. A hypothetical mechanism for the secretion of HCO_3^- and Na^+ by salivary duct epithelial cells. Acidification of the blood (driven by the Na^+ gradient) is the primary mechanism for driving bicarbonate transport. Na^+ diffuses into the saliva through the intercellular junctions.

Table 43-4. Data on proteins of guinea pig pancreatic juice separated by two-dimensional gel electrophoresis, one dimension based on molecular weight and the other on isoelectric point

Spot no.	*Protein*	*Isoelectric point*	*Molecular weight*	*Mass proportion (%)*
1		6.4	13,000	0.2
2		7.8	13,000	0.3
3	Ribonuclease I	8.7	13,000	0.7
4		8.8	13,600	0.4
5	Trypsin(ogen)	8.7	24,400	33.0
6		>9.3	24,900	2.0
7	Chymotrypsin(ogen) II	8.7	25,850	16.4
8		7.8		0.7
9		7.6		
10	(Pro)elastase II	7.5	28,000	8.0 (10–11)
11	(Pro)elastase I	6.9	28,000	
12	Chymotrypsin(ogen) I	4.8	25,850	1.7
13	(Pro)carboxypeptidase A_1	4.6	45,000	3.5
14	(Pro)carboxypeptidase A_2	4.8	45,000	8.8
15	(Pro)carboxypeptidase B	6.6	47,700	8.8
16	Triacylglycerol lipase II	8.1	49,700	3.4
17	α-Amylase	8.4	52,000	3.6 (17–18)
18	α-Amylase	8.4	53,000	
20	Triacylglycerol lipase I	5.0	66,000	8.5

From Scheele, G.A.: Am. J. Physiol. **238:**G467, 1980.

are secreted as trypsinogen, chymotrypsinogen, and procarboxypeptidase, respectively. Trypsinogen is specifically activated by *enterokinase* (not a kinase but a protease), which is secreted by the duodenal mucosa. Trypsin then activates trypsinogen, chymotrypsinogen, and procarboxypeptidase. Trypsin and chymotrypsin cleave certain peptide bonds to reduce polypeptide chains to smaller peptides. Carboxypeptidase specifically removes amino acids from the C-terminal ends of peptide chains.

Pancreatic juice contains an α-amylase that is secreted in active form. Pancreatic amylase has the same substrate specificity as salivary amylase, and it cleaves only interior α-1,4 links in starch to yield maltose, maltotriose, and various α-limit dextrins.

Pancreatice juice also contains a number of lipid-digesting enzymes, or *lipases*. Among the major pancreatic lipases are triacylglycerol hydrolase, cholesterol ester hydrolase, and phospholipase A_2. Triacylglycerol hydrolase (sometimes called pancreatic lipase) acts on

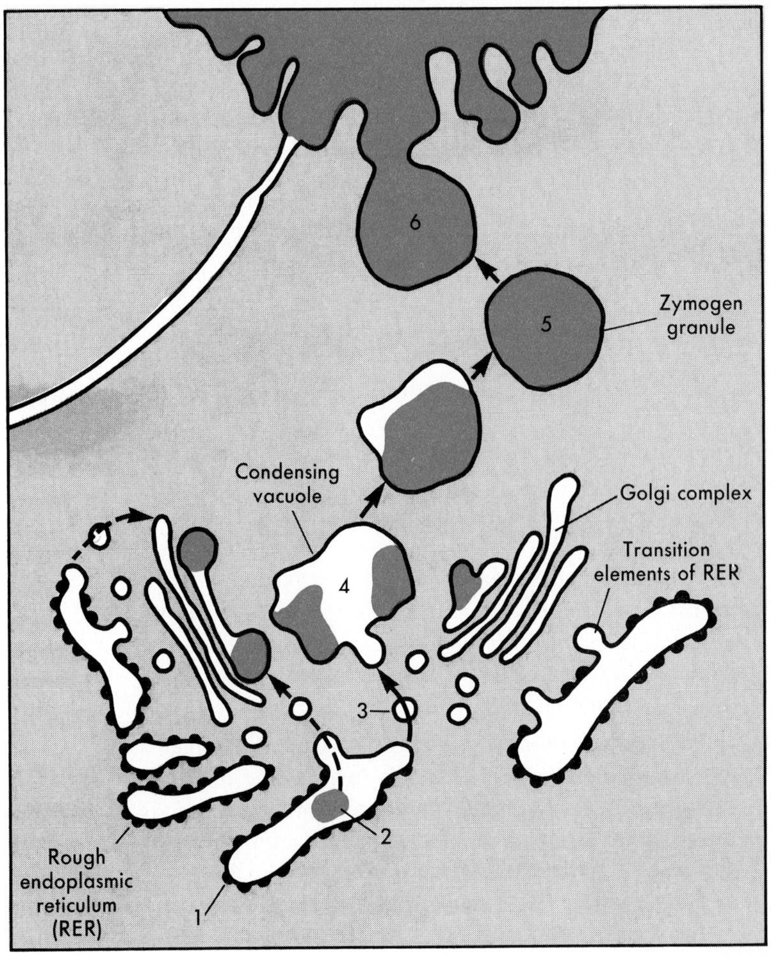

Fig. 43-25. Steps involved in processing of secretory proteins by pancreatic acinar cells: *1,* synthesis in rough endoplasmic reticulum; *2,* posttranslational modification in lumen of rough endoplasmic reticulum; *3,* transfer of proteins to the Golgi complex; *4,* modification in the Golgi complex and concentration of the protein in condensing vacuoles; *5,* storage in zymogen granules; *6,* exocytosis. (Redrawn from Gorelick, F.S., and Jamieson, J.D.: In Johnson, R.L., editor: Physiology of the gastrointestinal tract, New York, 1981, Raven Press.)

triglycerides to cleave specifically the ester bonds at the 1 and 1′ positions to release two free fatty acids and leave a 2-monoglyceride. Cholesterol ester hydrolase acts on cholesterol esters to produce cholesterol and free fatty acids. Phospholipase A_2 specifically cleaves the fatty acyl ester bond at the 2 carbon of a phosphoglyceride to produce a free fatty acid and a 1-lysophosphatide.

Among the other enzymes contained in pancreatic juice are ribonuclease and deoxyribonuclease, which reduce RNA and DNA, respectively, to their constituent nucleotides. The nucleotides are absorbed to some extent and are also reduced to nucleosides and free purine and pyrimidine bases by brush border enzymes. A trypsin inhibitor present in pancreatic juice prevents premature activation of hydrolytic enzymes within the pancreatic ducts.

Secretion of the Enzyme Component

Pancreatic acinar cells contain numerous membrane-bound zymogen granules in their apical cytoplasm. These granules contain most of the enzymes of pancreatic juice. Typically the number of zymogen granules increases between meals. After a meal the zymogen granules release their contents by exocytosis into the duct lumen, and the density of zymogen granules in the acinar cells decreases. Pancreatic enzymes and zymogens are synthesized on membrane-bound ribosomes, cross the membrane of the rough endoplasmic reticulum to enter the cisternae, and are concentrated somewhat in the cisternae of the rough endoplasmic reticulum. Smooth vesicles bud off from the rough endoplasmic reticulum. These vesicles carry the enzymes and zymogens to the Golgi apparatus, where they are packaged into zymogen granules (secretory vesicles). In response to stimulation of the acinar cell to secrete, the zymogen granules fuse with the plasma membrane to dump their contents into the lumen by exocytosis. The processes of synthesis, storage, and release of proteins by acinar cells are summarized in Fig. 43-25.

The zymogen granules are not a necessary route of secretion of pancreatic enzymes and zymogens. Acinar cells that have been completely depleted of zymogen granules by prolonged stimulation nonetheless can secrete pancreatic enzymes at high rates. We do not yet understand the way in which the zymogen granule secretory pathway and the granule-free pathway interact under physiological conditions.

Regulation of Pancreatic Exocrine Secretion

The secretory activities of duct and acinar cells of the pancreas are controlled by hormones and by substances released from nerve terminals. Stimulation of the vagal branches to the pancreas enhances the rate of secretion of the enzyme and aqueous components of pancreatic juice. Activation of sympathetic fibers inhibits pancreatic secretion, perhaps principally by decreasing blood flow to the pancreas. Secretin and cholecystokinin, hormones released from the duodenal mucosa in response to particular constituents of duodenal contents, stimulate secretion of the aqueous and enzyme components, respectively. Because the aqueous and enzyme components of pancreatic juice are separately controlled, the composition of the juice varies from less than 1% to as much as 10% protein. Substances other than secretin and cholecystokinin also modulate pancreatic exocrine function.

The cephalic phase of pancreatic secretion. Sham feeding induces the secretion of a low volume of pancreatic juice with a high protein content. Gastrin released from the mucosa of the gastric antrum in response to vagal impulses is a major mediator of pancreatic secretion during the cephalic phase. Gastrin is a member of the same class of peptides as cholecystokinin but is markedly less potent as a pancreatic secretagogue than is cholecystokinin. However, levels of gastrin that elicit acid secretion by the parietal cells also evoke enzyme secretion by pancreatic acinar cells. Acidification of antral contents, which blocks release of gastrin during the cephalic phase, also blocks the cephalic phase of pancreatic secretion.

The gastric phase of pancreatic secretion. During the gastric phase of secretion gastrin is released in response to gastric distension and to the presence of amino acids and peptides in the antrum of the stomach. The gastrin released during the gastric phase continues to stimulate secretion by the pancreas. In addition, vagovagal reflexes elicited by stretching either the fundus or the antrum of the stomach evoke secretion of small volumes of pancreatic juice with high enzyme content.

Intestinal phase of pancreatic secretion. In the intestinal phase of secretion certain components of the chyme present in the duodenum and upper jejunum evoke pancreatic secretion. Acid in the duodenum and upper jejunum elicits the secretion of a large volume of pancreatic juice rich in bicarbonate but poor in pancreatic enzymes. The hormone secretin is a major mediator of this response to acid. Secretin is released by certain cells in the mucosa of the duodenum and upper jejunum in response to acid in the lumen. Secretin is released when the pH of duodenal contents is 4.5 or below; when the pH of duodenal chyme is below about 3, the quantity of titratable acid present, rather than the pH, is the major determinant of the amount of secretin released. Secretin directly stimulates the cells of the pancreatic ductular epithelium to secrete the bicarbonate-rich aqueous component of the pancreatic juice.

The presence of peptides and certain amino acids, especially tryptophan and phenylalanine, in the duodenum brings about the secretion of pancreatic juice that is rich in protein components. In the duodenum fatty acids of chain length longer than eight carbon atoms and monoglycerides of these fatty acids also elicit secretion of protein-rich pancreatic juice. Cholecystokinin is the most important physiological mediator of this response to the digestion products of proteins and lipids. Cholecystokinin is a hormone released by particular cells in the duodenum and upper jejunum in response to these digestion products. This hormone directly stimulates the acinar cells to release the contents of their zymogen granules.

Cholecystokinin has little direct effect on the ductular epithelium of the pancreas, but it potentiates the stimulatory effect of secretin on the ducts. Secretin is a weak agonist in acinar cells, but it potentiates the effect of cholecystokinin on acinar cells.

The neural component of the intestinal phase of pancreatic secretion is not well-characterized. Vagotomy decreases the pancreatic response to the presence of chyme in the duodenum by about 50%. The pancreas responds more rapidly to the presence of acids, peptides, or fats in the duodenum than can be accounted for by the responses to secretin and cholecystokinin. The rapid component of pancreatic secretion in response to chyme in the duodenum is greatly diminished by vagotomy. The initial rapid increase in pancreatic secretion during the intestinal phase is probably mediated mainly by enteropancreatic vagovagal reflexes.

Regulatory Molecules That Influence Pancreatic Exocrine Secretion

Cholecystokinin and secretin are not the only endogenous substances that modulate pancreatic secretion. Table 43-5 lists some of the compounds that can stimulate or inhibit pancreatic secretion. The physiological significance of the actions of these substances is not yet understood. The second messenger of some of these regulatory molecules is cAMP or Ca^{++}. The second messengers of some of the regulators are not known. Specific receptors for some of the regulatory molecules in Table 43-5 are present in the pancreas.

Cellular Mechanisms of the Mediators of Pancreatic Exocrine Secretion

Fig. 43-26 shows six classes of receptors that mediate the responses of pancreatic acinar cells to secreta-

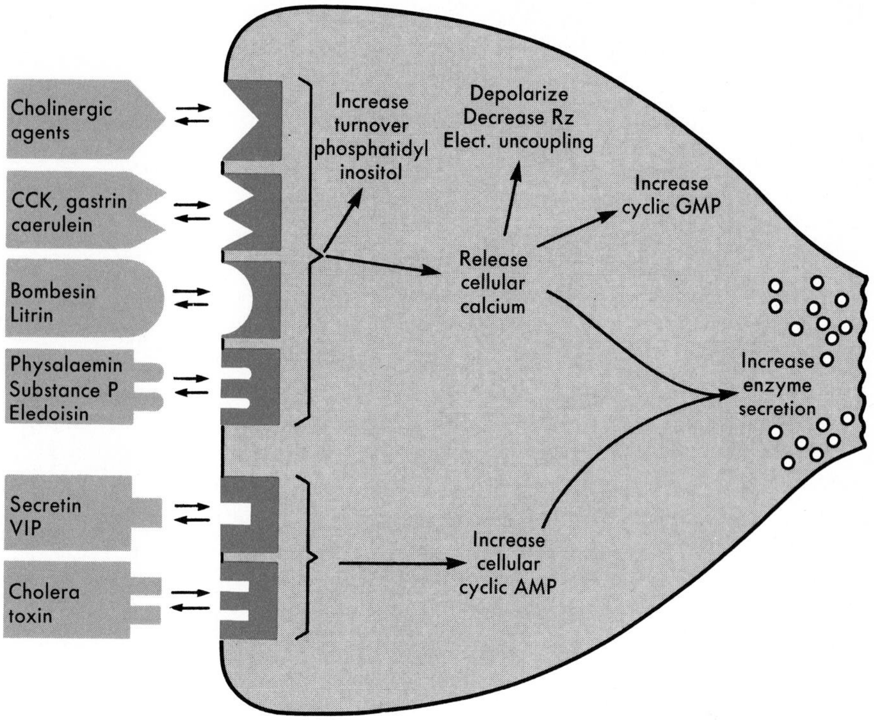

■ **Fig. 43-26.** Representation of the cellular mechanisms of action of secretagogues on pancreatic acinar cells. Six discrete classes of receptors are proposed for secretagogues. Two receptor types are linked to adenylate cyclase and to increase intracellular cAMP. The other four receptor classes are coupled to turnover of inositol phosphatides and to increased intracellular free Ca^{++}. (Redrawn from Jensen, R.T., and Gardner, J.D.: Adv. Cyclic Nucleotide Res. **17**:375, 1984.)

■ **Table 43-5.** Substances that may influence pancreatic exocrine function

Regulator	*Actions*	*Second messenger*
Acetylcholine, cholecystokinin, gastrin	Digestive enzyme secretion, Cl^--rich pancreatic juice from acini, digestive enzyme synthesis, trophic effects, increased cellular metabolism	Ca^{++}
Bombesin, substance P	Digestive enzyme secretion	Ca^{++}
Vasoactive intestinal polypeptide, secretin	HCO_3^--rich pancreatic juice from ducts, potentiation of enzyme secretion from acini, potentiation of trophic effect	cAMP
Insulin	Potentiation of digestive enzyme secretion, trophic effect on amylase, increased cellular metabolism	Unknown
IGF1, IGF2	Insulin-like effects on acinar cell metabolism	Unknown
EGF	Maintenance of differentiated function	Unknown
Somatostatin	Inhibition of pancreatic secretion in vivo (action site unknown)	Unknown

From Williams, J.A.: Annu. Rev. Physiol. **46**:361, 1984.
EGF, Epidermal growth factor, *IGF*, Insulin-derived growth factor.

gogues. Secretin and vasoactive intestinal polypeptide are related peptides that compete for receptor binding. At least two kinds of receptors for secretin and vasoactive intestinal polypeptide exist: one kind prefers secretin, and the other type prefers vasoactive intestinal polypeptide. Binding of secretin and vasoactive intestinal polypeptide to these receptors elevates intracellular cAMP, the second messenger for these secretagogues.

Four classes of secretagogues exert their effects by increasing the level of intracellular calcium by the inositol phosphate pathway (Fig. 43-26). Among the *endogenous* secretagogues that act in this way are acetylcholine, cholecystokinin, gastrin, and substance P. Acetylcholine and cholecystokinin compete for the same receptor.

Whereas the second messengers of the secretagogues shown in Fig. 43-26 have been identified, further details of their mechanisms of action are unclear. Phosphorylation of specific proteins by protein kinases activated by cAMP, Ca^{++}, or diglyceride may play a role in the stimulation of secretion brought about by these secretagogues. Acetylcholine and cholecystokinin compete for the same receptor.

Many of the secretagogues that elevate intracellular Ca^{++} also increase cytosolic levels of cGMP in the acinar cell. The increased cGMP does not directly elicit pancreatic exocrine secretion, but cGMP may play a regulatory role in the acinar cells.

Secretagogues that act via cAMP potentiate the effect of secretagogues that use Ca^{++} as second messenger, and vice versa. The maximal enzyme secretion in response to the two secretagogues acting together is greater than the maximal response to either agonist alone. Secretagogues that act via a common second messenger do not potentiate one another's effects.

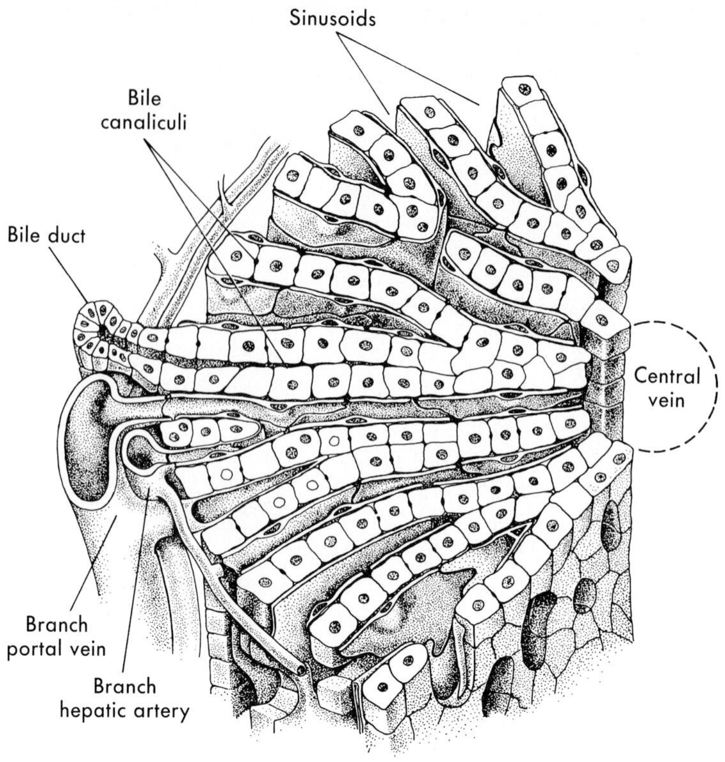

Fig. 43-27. Diagrammatic representation of a hepatic lobule. A central vein is located in the center of the lobule with plates of hepatic cells disposed radially. Branches of the portal vein and hepatic artery are located on the periphery of the lobule, and blood from both perfuses the sinusoids. Peripherally located bile ducts drain the bile canaliculi that run between the hepatocytes. (Redrawn from Bloom, W., and Fawcell, D.W.: A textbook of histology, ed. 10, Philadelphia, 1975, W.B. Saunders.)

Functions of Liver and Gallbladder

Structure of the Liver

One view of hepatic histology is shown in Fig. 43-27. Each liver lobule is organized around a central vein. At the periphery of the lobule blood enters the sinusoids from branches of the portal vein and the hepatic artery. In the sinusoids blood flows toward the center of the lobule between plates of hepatic cells that are one or two hepatocytes thick. Each hepatocyte is thus in direct contact with sinusoidal blood because of the large fenestrations between the endothelial cells that line the sinusoids. The intimate contact of a large fraction of the hepatocyte surface with blood contributes to the ability of the liver to clear the blood effectively of certain classes of compounds. Biliary canaliculi lie between adjacent hepatocytes, and the canaliculi drain into bile ducts at the periphery of the lobule.

Functions of the Liver

Bile secretion is the principal digestive function of the liver. A discussion of the composition of bile, the mechanisms of bile formation, and regulation of bile synthesis and secretion follows. The liver performs a large number of other functions that are vital to the health of the organism. This book does not have a separate chapter on the liver, but each function of the liver is discussed in the context of the individual organ system most affected. Nevertheless, it may be useful at this point to mention some of the most important activities of the liver. The liver is essential in regulating metabolism, in synthesizing certain proteins, in serving as a storage site for certain vitamins and iron, in degrading certain hormones, and in inactivating and excreting certain drugs and toxins.

The liver regulates the metabolism of carbohydrates, lipids, and proteins. Liver and skeletal muscle are the two major sites of glycogen storage in the body. When the level of glucose in the blood is high, glycogen is deposited in the liver. When blood glucose is low, liver glycogen is broken down to glucose (the process is called *glycogenolysis*), and the glucose is then released into the blood. In this way the liver helps to maintain a relatively constant blood glucose level. The liver is also the major site of *gluconeogenesis,* the conversion of amino acids, lipids, or simple carbohydrate substances (lactate, for example) into glucose. Carbohydrate metabolism by the liver is regulated by several hormones (Chapter 50).

The liver is also involved in lipid metabolism. As described in Chapter 44, lipids absorbed from the intestine leave the intestine in chylomicrons in the lymph. Lipoprotein lipase on the endothelial cell surface of blood vessels hydrolyzes some of the triglyceride in the *chylomicrons,* thereby allowing glycerol and fatty acids to be taken up by adipocytes. This results in formation of *chylomicron remnants* that are rich in cholesterol. Chylomicron remnants are taken up by hepatocytes and degraded. Hepatocytes synthesize and secrete very low-density lipoproteins. Very low density lipoproteins are then converted to the other types of serum lipoproteins: low-, intermediate-, and high-density lipoproteins. These lipoproteins are the major sources of cholesterol and triglycerides for most other tissues of the body. Cholesterol present in bile represents the only route of excretion of cholesterol. Hepatocytes are thus a principal source of cholesterol in the body and are the major site of excretion of cholesterol. These cells play an important role in regulation of serum cholesterol levels.

In certain physiological and pathological conditions, β-oxidation of fatty acids provides the major source of energy for the body. In the liver, acetyl coenzyme A liberated from fatty acids condenses to form acetoacetate. Acetoacetate is converted to β-hydroxybutyrate and acetone. These three compounds are called *ketone bodies*. Ketone bodies are released from hepatocytes and carried in the circulation to other tissues, where they are metabolized. The hepatic functions that regulate lipid metabolism are also subject to endocrine control.

The liver is centrally involved in protein metabolism. When proteins are catabolized, amino acids are deaminated to form ammonia (NH_3). Ammonia cannot be further metabolized by most tissues and becomes toxic at levels achievable by metabolism. Ammonia is dissipated by conversion to urea, mainly in the liver. The liver also synthesizes all the nonessential amino acids.

The liver synthesizes certain proteins, among them the plasma lipoproteins discussed above. Plasma contains numerous other proteins; among them are albumins, globulins, fibrinogens, and other proteins involved in blood clotting. With the exception of the γ-globulins produced by plasma cells, the liver synthesizes all the plasma proteins.

The liver stores certain substances important in metabolism. Next to hemoglobin in red blood cells the liver is the most important storage site for iron. Certain vitamins, most notably vitamins A, D, and B_{12}, are stored in the liver. Hepatic storage protects the body from limited dietary deficiencies of these vitamins.

The liver is a major site for the degradation and excretion of hormones. Epinephrine and norepinephrine are inactivated by oxidation (catalyzed by monoamine oxidase) and methylation (catalyzed by catechol-*O*-methyltransferase). Both of these enzymes are abundant in hepatocytes. Certain polypeptide hormones are degraded by liver cells. The liver inactivates and helps excrete steroid hormones. For example, cortisol, a principal glucocorticoid, is reduced in the liver to inactive tetrahydrocortisol and then conjugated to glucuronic acid. The conjugated derivative is freely soluble and enters the circulation to be excreted in the urine.

The liver transforms and excretes a large number of drugs and toxins. Drugs and toxins are frequently converted to inactive forms by reactions that occur in hepatocytes. However, some drugs and toxins are activated by hepatic transformation, and some drugs are converted to toxic products. The smooth endoplasmic reticulum of hepatocytes contains systems of enzymes and cofactors, known as mixed function oxidases, that are responsible for oxidative transformations of many drugs. Certain other enzymes in the endoplasmic reticulum catalyze the conjugation of many compounds with glucuronic acid, glycine, or glutathione. Other drug transformations that occur in hepatocytes include acetylation, methylation, reduction, hydrolysis, and oxidation. The transformations that occur in the liver render many drugs more water soluble, and thus they are more readily excreted by the kidneys. Some drug metabolites are secreted into bile. Many organic anions and cations are actively secreted into bile by an Na^+-dependent mechanism. Steroids and related molecules are secreted into bile by a carrier-mediated mechanism, perhaps by active transport.

Bile Secretion

The foregoing discussion briefly described various vital functions of the liver. The hepatic function most important to the digestive tract is the secretion of bile. The mechanisms of bile secretion and the control of bile secretion are the major topics of this section.

Bile, elaborated by hepatocytes, contains bile acids, cholesterol, lecithin, and bile pigments. These constituents are all synthesized and secreted by hepatocytes into the bile canaliculi along with an isotonic fluid that resembles plasma in its concentrations of Na^+, K^+, Cl^-, and HCO_3^-. The bile canaliculi merge into ever larger ducts and finally into a single large bile duct. The epithelial cells that line the bile ducts secrete a watery fluid that is rich in bicarbonate and contributes to the volume of bile that leaves the liver.

The secretory function of the liver resembles that of the exocrine pancreas. In both organs the major parenchymal cell type elaborates a primary secretion containing the substances responsible for the major digestive function of the organ. In both pancreas and liver the primary secretion is isotonic and contains Na^+, K^+, and Cl^- at concentrations near plasma levels, and the primary secretion is stimulated by cholecystokinin. In both pancreas and liver the epithelial cells that line the duct system modify the primary secretion. When stimulated by secretin these epithelial cells contribute an aqueous secretion that is characterized by high bicarbonate concentrations (Fig. 43-28).

In the periods between meals, bile is diverted into the gallbladder. The gallbladder epithelium extracts salts and water from the stored bile, and the bile acids are thereby concentrated five- to twentyfold. After an individual has eaten, the gallbladder contracts and empties

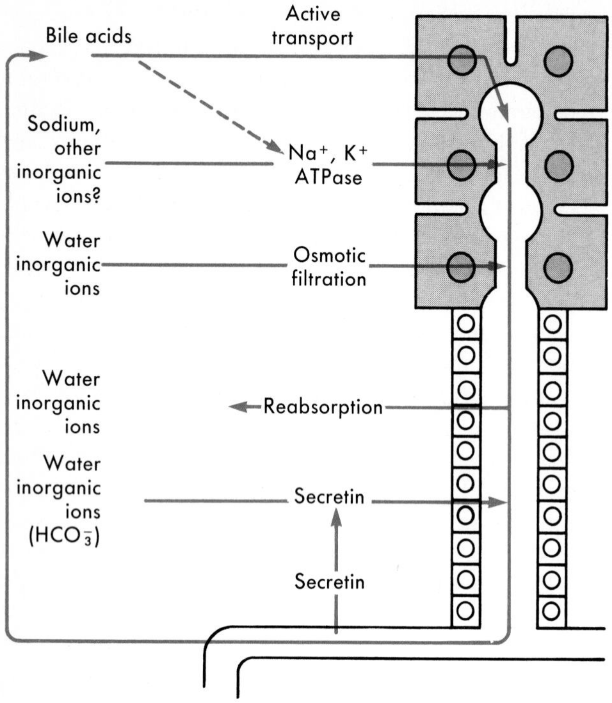

■ **Fig. 43-28.** Schematic view of bile secretion by hepatocytes and bile duct epithelium. The hepatocytes secrete bile acids, bilirubin, and protein into the bile canaliculi in an isotonic fluid with Na^+, Cl, and HCO_3^- at concentrations that resemble those in plasma. The bile duct epithelial cells contribute an aqueous secretion with a bicarbonate concentration higher and a chloride concentration lower than those of plasma. (Modified from Erlinger, S.: In Schiff, L., and Schiff, L.R., editors: Diseases of the liver, ed. 5, Philadelphia, 1982, J.B. Lippincott.)

ties its concentrated bile into the duodenum. The most potent stimulus for emptying of the gallbladder is the hormone *cholecystokinin,* which is released by the duodenal mucosa primarily in response to the presence of fats and their digestion products. From 250 to 1500 ml of bile enters the duodenum each day.

Bile acids emulsify lipids, thereby increasing the surface area available to lipolytic enzymes. Bile acids then form mixed micelles with the products of lipid digestion. This process increases the transport of lipid digestion products to the brush border surface and in this way enhances absorption of lipids by the epithelial cells. Bile acids are actively absorbed, chiefly in the terminal part of the ileum. A small fraction of bile acids escapes absorption and is excreted. The returning bile acids are avidly taken up by the liver and are rapidly resecreted during the course of digestion. The entire bile acid pool (approximately 2.5 g) is recirculated twice in response to a typical meal. The recirculation of the bile is known as the *enterohepatic circulation.* About 20% of the bile acid pool is excreted in the feces each day and is replenished by hepatic synthesis of new bile acids. Fig. 43-29 summarizes some major aspects of the enterohepatic circulation.

■ **Fig. 43-29.** Overview of the enterohepatic circulation of bile.

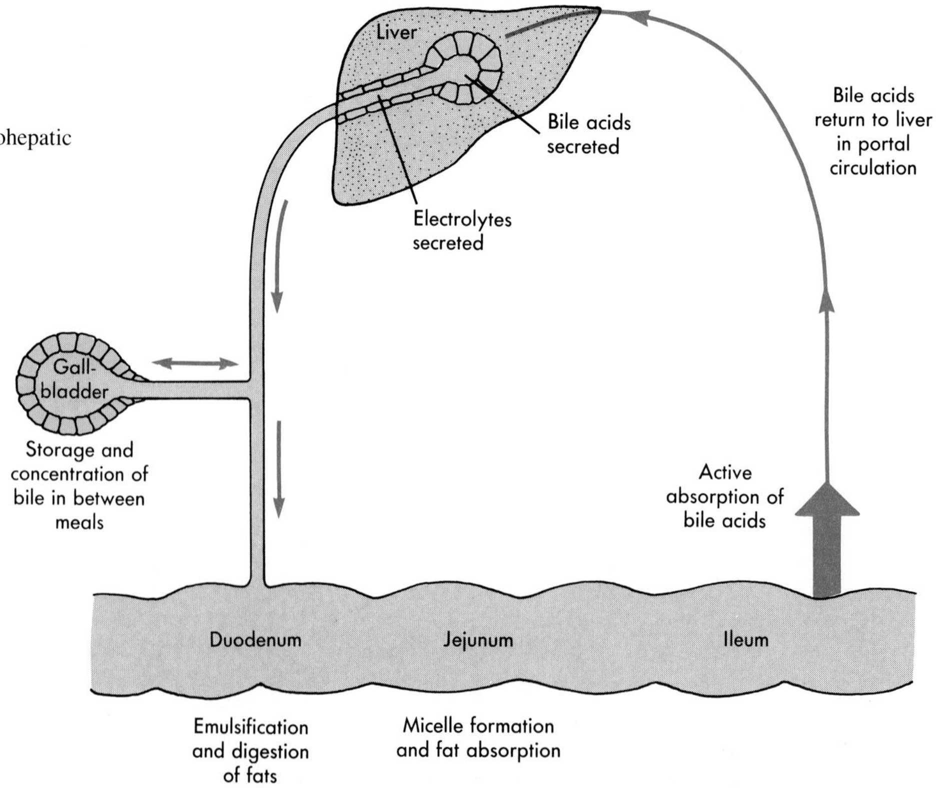

The Fraction of Bile Secreted by Hepatocytes

The bile acids. Bile acids comprise about 50% of the dry weight of bile. Other important compounds secreted by the hepatocytes into the bile include lecithin, cholesterol, bile pigments, and proteins.

Bile acids have a steroid nucleus and are synthesized by the hepatocytes from cholesterol. The major bile acids synthesized by the liver are called *primary bile acids* (Fig. 43-30). These are cholic acid (3-hydroxyl groups) and chenodeoxycholic acid (2 hydroxyl groups). The presence of the carboxyl and hydroxyl groups makes the bile acids much more water soluble than the cholesterol from which they are synthesized.

Bacteria in the digestive tract dehydroxylate bile acids to form *secondary bile acids*. The major secondary bile acids (Fig. 43-30) are deoxycholic acid (from dehydroxylation of cholic acid) and lithocholic acid (from dehydroxylation of chenodeoxycholic acid). Bile contains both primary and secondary bile acids.

Bile acids normally are secreted conjugated with glycine or taurine (Fig. 43-30). *Conjugated bile acids* contain glycine or taurine linked by a peptide bond between the carboxyl group of an unconjugated bile acid and the amino group of glycine or taurine. The pK_a's of the carboxyl groups of unconjugated bile acids are near neutral pH, but the pK_a's of conjugated bile acids are considerably lower. Thus at the near-neutral pH of the gastrointestinal tract the conjugated bile acids are more completely ionized, and thus more water soluble, than the unconjugated bile acids. Conjugated bile acids therefore exist almost entirely as salts of various cations (mostly Na^+) and hence often are called *bile salts*.

The steroid nucleus of bile acids is roughly planar. In solution bile acids have their polar (hydrophilic) groups—the hydroxyl groups, the carboxyl moiety of glycine or taurine, and the peptide bond—all on one surface of the molecule. The other surface is quite hydrophobic (Fig. 43-31, *A*). This makes the bile acid molecule amphipathic, that is, having both hydrophilic and hydrophobic domains. Conjugated bile acids are more amphipathic than unconjugated ones. Because they are amphipathic, bile acids tend to form molecular aggregates, called micelles, by turning their hydrophobic faces inside and away from water and their hydrophilic surfaces toward the water (Fig. 43-31, *B*). Whenever bile acids are present above a certain concentration, called the *critical micelle concentration,* bile acid micelles will form. Above this concentration any additional bile acid will go into the micelles exclusively and not into molecular solution. In bile, the bile acids are normally present at a concentration well above the critical micelle concentration.

Phospholipids in bile. Hepatocytes also secrete phospholipids into the bile, the most prominent class being lecithins. Cholesterol also is secreted into the bile, and this is the major route for cholesterol excretion. Being essentially insoluble in water, lecithin and cholesterol partition into (dissolve in) the bile acid micelles. Cholesterol, being apolar, partitions into the center of the micelle. Lecithin, because it is amphipathic, buries its fatty acyl chains in the micelle interior and leaves its polar head group near the micelle surface (Fig. 43-31, *B*). The lecithin increases the amount of cholesterol that can be solubilized in the micelles. If more cholesterol is present in the bile than can be solubilized in the micelles, crystals of cholesterol will form in the bile. These crystals are important in formation of cholesterol gallstones (the most common kind of gallstones) in the duct system of the liver or more commonly in the gallbladder.

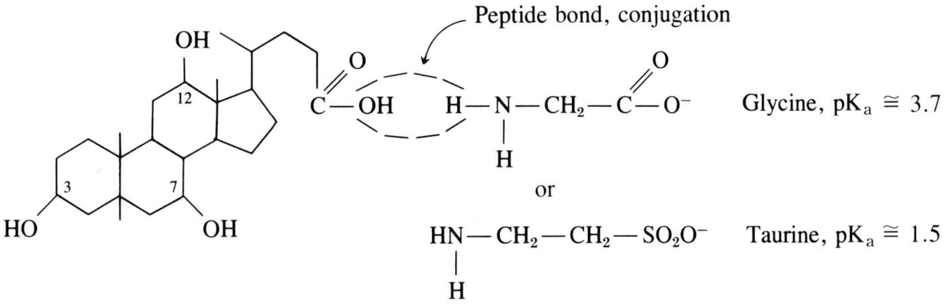

Cholic acid (primary bile acid)

No-OH on 12 = Chenodeoxycholic acid (primary)
No-OH on 7 = Deoxycholic acid (secondary)
No-OH on 7 and 12 = Lithocholic acid (secondary)

Fig. 43-30. The structure of the most common primary and secondary bile acids. Most bile acids are secreted conjugated to glycine or taurine by the formation of a peptide bond, as shown.

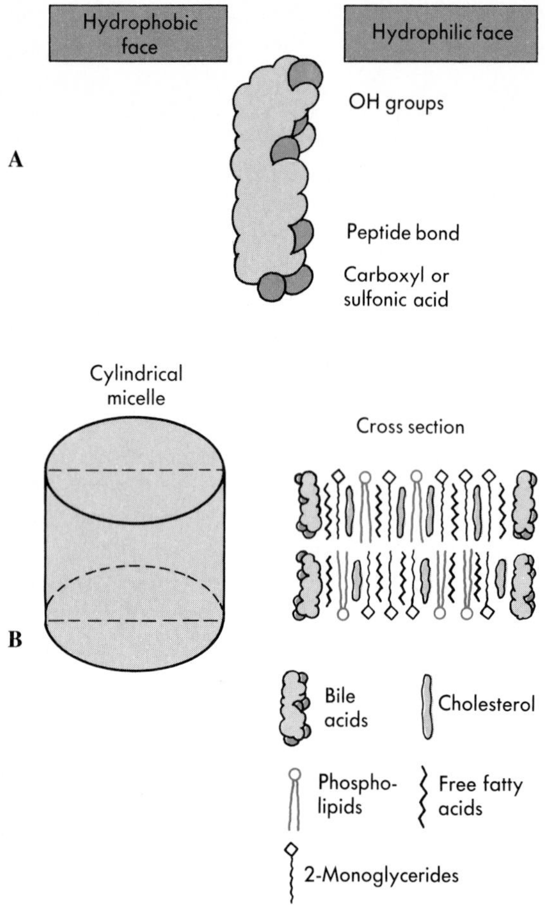

■ **Fig. 43-31.** Structure of bile acids and micelles. **A,** A bile acid molecule in solution. The molecule is amphipathic in that it has a hydrophilic face and a hydrophobic face. The amphipathic structure is key in the ability of the bile acids to emulsify lipids and to form micelles. **B,** A model of the structure of a bile acid–lipid mixed micelle.

Bile pigments. When senescent red blood cells are degraded in reticuloendothelial cells, the porphyrin moiety of hemoglobin is converted to bilirubin. Bilirubin is released into the plasma, where it is bound to albumin. Hepatocytes efficiently remove bilirubin from blood in the sinusoids via a protein-mediated transport mechanism in the hepatocyte plasma membrane that faces the sinusoids. In the hepatocytes bilirubin is conjugated with one or two glucuronic acid molecules, and the bilirubin glucuronides are secreted into the bile, probably by an active transport mechanism. Unconjugated bilirubin is not secreted into the bile. Bilirubin is yellow and contributes to the yellow color of bile. Colonic bacteria convert bilirubin to mesobilirubinogen and then to urobilinogen. Some of the urobilinogen is absorbed in the colon. A fraction of the absorbed urobilinogen is excreted in the urine, and the remainder is resecreted into bile.

Proteins in bile. After being concentrated in the gallbladder, bile contains protein at 5 to 50 mg/ml. Proteins in the bile include most of the proteins present in plasma, including immunoglobulins and the apoproteins of serum lipoproteins. A number of peptide hormones are present in bile. The hepatocytes themselves are the source of other biliary proteins, including lysosomal hydrolases and enzymes associated with other hepatocyte organelles. The mechanisms of protein secretion into bile are not well characterized, and the possible physiological functions of proteins in bile are not clear.

■ *Secretion by Bile Duct Epithelium*

The epithelial cells that line the bile ducts contribute an aqueous secretion that can account for about 50% of the total volume of bile. The secretion of the bile duct epithelium is isotonic and contains Na^+ and K^+ at levels similar to plasma. However, the concentration of HCO_3^- is greater and the concentration of Cl^- is less than in plasma. The secretory activity of the bile duct epithelium is specifically stimulated by the hormone secretin. When bile flow is stimulated by secretin alone the volume flow of bile and its bicarbonate concentration increase, but content of bile acids does not increase.

■ *Cellular Mechanism of Bile Formation*

Secretion of bile acids. Fig. 43-32 illustrates the current understanding of the cellular mechanisms responsible for secretion of bile by hepatocytes into bile canaliculi. As is the case for many epithelial cells, the plasma membrane of the hepatocyte is polarized such that the membrane facing the bile canaliculus is different from the basolateral membrane (the membrane facing the sinusoid plus the lateral cell membrane that faces adjacent hepatocytes). The basolateral membrane contains a bile acid transport protein that uses the energy of the Na^+ gradient to accumulate bile acids actively from the sinusoidal blood into the cytosol of the hepatocyte. The Na^+ gradient is in turn created by the Na^+, K^+-ATPase that resides in the basolateral plasma membrane. In the hepatocyte most of the bile acids present are bound by specific cytosolic proteins.

How the bile acids cross the plasma membrane that bounds the bile canaliculus is not known. One view is that bile acids enter the canaliculus via facilitated transport. The facilitated transport of anionic bile acids from cell to canalicular lumen is favored by the electrical potential difference between canaliculus and cytosol, about −35 mV, cytosol negative. Certain evidence favors the interpretation that bile acids are stored in vesicles in the hepatocyte and released into the canaliculi by exocytosis.

The tight junctions joining hepatocytes that surround the canaliculi are leaky to water-soluble molecules of less than macromolecular dimensions. The leakage of

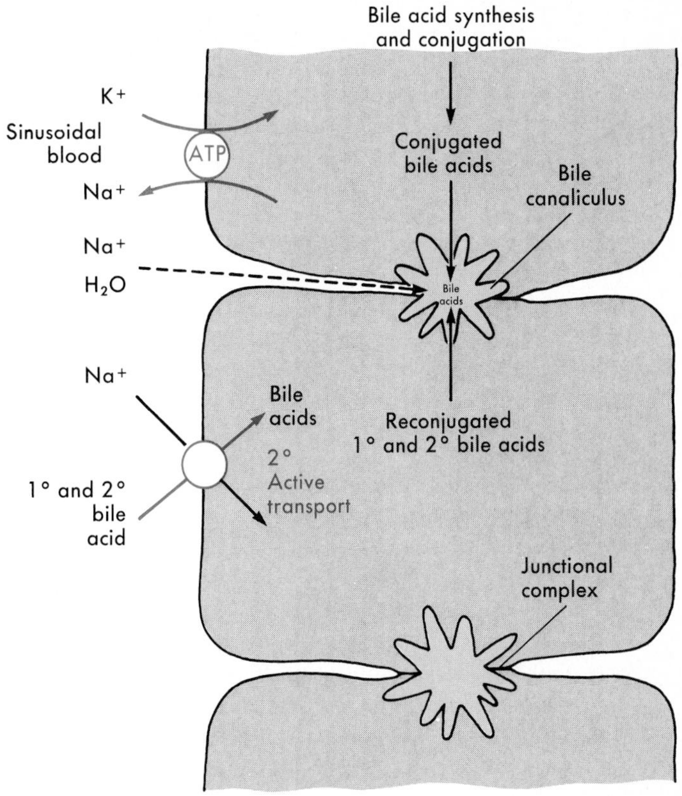

Fig. 43-32. Hypothetical view of cellular mechanisms responsible for secretion of bile acids by hepatocytes into the bile canaliculi.

bile acids from the canalicular lumen back into the sinusoidal blood is limited by micelle formation in the lumen (micelles are too large to back diffuse) and by the relative impermeability of the tight junctions to anions.

Secretion of electrolytes and water. Water and electrolytes enter the primary secretion in the bile canaliculi and also enter bile by the secretory activities of the bile duct epithelium. The mechanisms of secretion of water and electrolytes into bile are not well understood. One view of movement of water and electrolytes into the bile canaliculi is that the osmotic pressure of bile acids in the canaliculi causes the flow of water and ions through the leaky tight junctions. It is not known whether specific ion transport processes are responsible for the secretion of particular ions into the bile canaliculi. The high concentration of bicarbonate in the fluid secreted by the epithelial cells of the bile ducts suggests the involvement of protein-mediated transport sytems, but this remains to be demonstrated.

Bile Concentration and Storage in the Gallbladder

Between meals the tone of the sphincter of Oddi, which guards the entrance of the common bile duct into the duodenum, is high. Thus most bile flow is diverted into the gallbladder. The gallbladder is a small organ, having a capacity of 15 to 60 ml (average about 35 ml) in humans. Many times this volume of bile may be secreted by the liver between meals. The gallbladder concentrates the bile by absorbing Na^+, Cl^-, HCO_3^-, and water from the bile so that the bile acids can be concentrated from 5 to 20 times in the gallbladder. K^+ is concentrated in the bile when water is absorbed, and then K^+ is absorbed by simple diffusion.

The active transport of Na^+ is the primary active process in the concentrating action of the gallbladder; Cl^- and HCO_3^- are absorbed to preserve electroneutrality.

Because of its high rate of water absorption the gallbladder serves as a model for water and electrolyte transport by tight-junctioned epithelia. The *standing gradient mechanism* for fluid absorption was first proposed for the gallbladder. A key initial observation was that, when fluid was being reabsorbed by the gallbladder, the lateral intercellular spaces between the epithelial cells were large and swollen. When fluid transport was blocked, for example, by poisoning the Na^+ pumps with ouabain, the intercellular spaces almost disappeared. These observations strongly suggested that the intercellular spaces are a major route of fluid flow during absorption. A current view is that Na^+ is actively pumped into the lateral intercellular spaces. The Na^+ pumps are believed to be especially dense near the mucosal (apical) end of the channel (Fig. 43-33). Cl^-

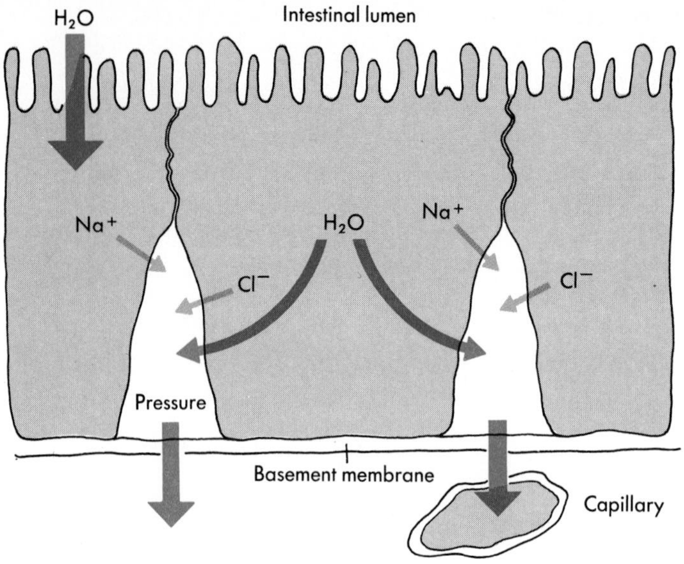

■ **Fig. 43-33.** Water absorption from the gallbladder by the mechanism of the standing osmotic gradient. Na^+ is actively pumped into the lateral intercellular spaces; Cl^- follows. Water is drawn by osmosis to enter the intercellular spaces, elevating the hydrostatic pressure. Water, Na^+, and Cl^- are filtered across the porous basement membrane and enter the capillaries.

and HCO_3^- also are transported into the intercellular space, probably because of an electrical potential created by Na^+ transport. The high ion concentration near the apical end of the intercellular space causes the fluid there to be hypertonic. This produces an osmotic flow of water from the lumen via adjacent cells into the intercellular space. Water distends the intercellular channel because of increased hydrostatic pressure. Because of water flow from adjacent cells the fluid becomes less hypertonic as it flows down the intercellular channel, and it is essentially isotonic when it reaches the serosal (basal) end of the channel. Ions and water flow across the basement membrane of the epithelium and are carried away by the capillaries.

■ *Emptying of the Gallbladder*

Emptying of the gallbladder contents into the duodenum begins several minutes after the start of a meal. Intermittent contractions of the gallbladder force bile through the partially relaxed sphincter of Oddi. During the cephalic and gastric phases of digestion gallbladder contraction and relaxation of the sphincter are mediated by cholinergic fibers in branches of the vagus nerve (Table 43-6). Stimulation of sympathetic nerves to the gallbladder and duodenum inhibits emptying of the gallbladder. Vasoactive intestinal polypeptide (VIP) inhibits gallbladder contractions; VIP-containing nerve terminals are present in the wall of the gallbladder.

The highest rate of gallbladder emptying occurs during the intestinal phase of digestion. The strongest stimulus for the emptying is the hormone *cholecystokinin*. Cholecystokinin is released by the duodenal mucosa most strongly in response to the presence of fat digestion products and essential amino acids in the duodenum. Cholecystokinin reaches the gallbladder via the circulation, and it causes strong contractions of the gallbladder and relaxation of the sphincter of Oddi. Substances that mimic the actions of cholecystokinin in promoting gallbladder emptying are called *cholecystagogues*. Gastrin and cholecystokinin share a common sequence of five amino acids at their C-terminals. Gastrin is not nearly as potent a cholecystagogue as cholecystokinin, but gastrin may play a role in eliciting gallbladder contractions during the cephalic and gastric phases.

Under normal circumstances the rate of gallbladder emptying is sufficient to keep the concentration of bile acids in the duodenum above the critical micelle concentration.

■ *Intestinal Absorption of Bile Acids and Their Enterohepatic Circulation*

The functions of bile acids in emulsifying dietary lipid and in forming mixed micelles with the products of lipid digestion are discussed in Chapter 44.

Normally, by the time chyme reaches the terminal part of the ileum, dietary fat is almost completely absorbed. Bile acids are then absorbed. The epithelial cells of the distal part of the ileum actively take up bile acids against a large concentration gradient. The active transport system has a higher affinity for conjugated bile acids.

Bile acids also have a fair degree of lipid solubility. Thus they also can be taken up by simple diffusion. Bacteria in the terminal part of the ileum and colon deconjugate bile acids and also dehydroxylate them to produce secondary bile acids. Both deconjugation and dehydroxylation lessen the polarity of bile acids, enhancing their lipid solubility and their absorption by simple diffusion.

Table 43-6. The major factors affecting gallbladder emptying and bile synthesis and secretion

Phase of digestion	Stimulus	Mediating factor	Response
Cephalic	Taste and smell of food; food in mouth and pharynx	Impulses in branches of vagus nerve Gastrin?	Increased rate of gallbaldder emptying
Gastric	Gastric distension	Impulses in branches of vagus nerve Gastrin?	Increased rate of gallbladder emptying
Intestinal	Fat digestion products in duodenum	Cholecystokinin	Increased rate of gallbladder emptying; increased rate of bile acid secretion
	Acid in duodenum	Secretin	Increased rate of secretion of bicarbonate-rich fluid by the bile duct epithelium (this effect strongly potentiated by cholecystokinin)
	Absorption of bile acids in the distal part of ileum	High concentration of bile acids in portal blood	Stimulation of bile acid secretion; inhibition of bile acid synthesis
Interdigestive period	Low rate of release of bile to duodenum	Low concentration of bile acids in portal blood	Stimulation of bile acid synthesis; inhibition of bile acid secretion

Typically about 0.5 g of bile acids escapes absorption and is excreted in the feces each day. This quantity is 15% to 35% of the total bile acid pool, and normally it is replenished by synthesis of new bile acids by the liver.

Bile acids, whether absorbed by active transport or simple diffusion, are transported away from the intestine in the portal blood, mostly bound to plasma proteins. In the liver, hepatocytes avidly extract the bile acids from the portal blood. In a single pass through the liver the portal blood is essentially cleared of bile acids. Bile acids in all forms, primary and secondary, both conjugated and deconjugated, are taken up by the hepatocytes. The hepatocytes reconjugate almost all the deconjugated bile acids and rehydroxylate some of the secondary bile acids. These bile acids are secreted into the bile along with newly synthesized bile acids.

Control of Bile Acid Synthesis and Secretion

The rate of return of bile acids to the liver is a major influence on the rate of synthesis and secretion of bile acids. Bile acids in the portal blood *stimulate the secretion* of bile acids by the hepatocytes but *inhibit the synthesis* of bile acids. This is called the *choleretic* effect of bile acids. Substances that act to enhance bile acid secretion are known as *choleretics*.

Long after a meal, when bile acids already have been returned to the liver, the level of bile acids in the portal blood is low. Hence, synthesis of new bile acids is not inhibited, and synthesis proceeds at near maximal rates. The rate of secretion of bile acids is low because secretion is not being stimulated by bile acids in the portal blood, and the gallbladder fills rather slowly with bile.

After a meal the presence of fat digestion products in the duodenum causes release of cholecystokinin. This brings about emptying of gallbladder contents into the duodenum and strongly stimulates secretion of bile acids by the hepatocytes. Secretin, released by acidic chyme in the duodenum, increases the secretion of bicarbonate-rich fluid by the bile duct epithelium. At this time the liver continues to synthesize bile acids at a high rate and to secrete them in response to cholecystokinin. This continues until bile acids are absorbed from the terminal aspect of the ileum and return to the liver in the portal blood.

Bile acids returning to the liver inhibit the synthesis of new bile acids but stimulate high rates of secretion. Bile acids that are taken up are rapidly reconjugated (some secondary bile acids are rehydroxylated) and resecreted almost immediately. So powerful is the stimulus to resecrete the returning bile acids that the entire pool of bile acids (1.5 to 31.5 g) recirculates twice in response to a typical meal. In response to a single meal with a very high fat content, the bile acid pool may recirculate five or more times. Table 43-6 summarizes major aspects of control of gallbladder emptying and bile synthesis and secretion.

Gallstones

The most common type of gallstones contains cholesterol as their major component. Cholesterol is essentially insoluble in water. When bile contains more cholesterol than can be solubilized in the bile acid–lecithin micelles, crystals of cholesterol form in the bile. Such bile is said to be supersaturated with cholesterol. The greater the concentration of bile acids and lecithin in bile, the greater is the amount of cholesterol that can be contained in the mixed micelles. Lecithin is important in this regard because lecithin-cholesterol mixed micelles can solubilize more cholesterol than can micelles of bile acids alone.

At night most normal individuals secrete bile that is supersaturated with cholesterol. This occurs because the rate of bile acid secretion is particularly low as a result of the absence of stimulation by bile acids returning in the portal blood. During the day, however, the average rate of bile acid secretion is higher, and bile is unsaturated with cholesterol in normal persons. Individuals with cholesterol gallstones tend to secrete bile that is supersaturated with cholesterol both night and day. Most cholesterol gallstones contain crystallized bilirubin at their cores. This suggests that bilirubin crystals serve as nucleation sites for the crystallization of cholesterol.

Bile pigment stones are the other major class of gallstones. Their major constituent is the calcium salt of unconjugated bilirubin. Conjugated bilirubin is quite soluble and does not form insoluble calcium salts in bile. Bile may contain elevated levels of unconjugated bilirubin because hepatocytes are deficient in forming the glucuronide or because of excessive deconjugation by glucuronidase.

Intestinal Secretions

The mucosa of the intestine, from the duodenum through the rectum, elaborates secretions that contain mucus, electrolytes, and water. The total volume of intestinal secretions is about 1500 ml/day. The mucus in the secretions protects the mucosa from mechanical damage. The nature of the secretions and the mechanisms that control secretion vary from one segment of the intestine to another.

Duodenal Secretions

The duodenal submucosa contains branching glands that elaborate a secretion rich in mucus. The glands have ducts that empty into the crypts of Lieberkühn. The duodenal epithelial cells probably also contribute to duodenal secretions, but most of the secretions are produced by the glands. The duodenal secretion contains mucus and an aqueous component that does not differ greatly from plasma in its concentrations of the major ions. Gastrin, secretin, and cholecystokinin stimulate duodenal secretion, but a physiological role for any of these hormones has not yet been established.

Secretions of the Small Intestine

Goblet cells, which lie among the columnar epithelial cells of the small intestine, secrete mucus that lubricates the mucosal surface and protects it against mechanical damage. During the course of normal digestion an aqueous secretion is elaborated by the epithelial cells at a rate only slightly less than the rate of fluid absorption by these cells. So the net absorption of fluid that normally occurs is the result of much larger unidirectional absorptive and secretory flows. As discussed on p.730, cholera toxin greatly increases the rate of aqueous secretion by the small intestine, particularly by the jejunum.This process leads to secretory diarrhea.

Secretions of the Colon

The secretions of the colon are smaller in volume but richer in mucus than small intestinal secretions. The mucus is produced by the numerous goblet cells of the colonic mucosa. The aqueous component of colonic secretions is alkaline because of secretion of HCO_3^- in exchange for Cl^-, and the secretion is rich in K^+. The production of colonic secretions is stimulated by mechanical irritation of the mucosa and by activation of cholinergic pathways to the colon. Stimulation of sympathetic nerves to the colon decreases the rate of colonic secretion.

Bibliography

Journal articles

Allen, A., and Garner, A.: Mucus and bicarbonate secretion in the stomach and their possible role in mucosal protection, Gut **21**:249, 1980.

Berglindh, T.: The mammalian gastric parietal cell in vitro, Annu. Rev. Physiol. **46**:377, 1984.

Berridge, M.J.: Cellular control through interactions between cyclic nucleotides and calcium, Adv. Cyclic Nucleotide Protein Phosphorylation Res. **17**:329, 1984.

Berridge, M.J., and Irvine, R.F.: Inositol trisphosphate, a novel second messenger in cellular signal transduction, Nature **312**:315, 1984.

Blitzer, B.L., and Boyer, J.L.: Cellular mechanisms of bile formation, Gastroenterology **82**:346, 1982.

Butcher, F.R., and Putney, Jr, J.W.: Regulation of parotid gland function by cyclic nucleotides and calcium, Adv. Cyclic Nucleotide Res. **13**:215, 1980.

Hersey, S.J., et al.: Cellular control of pepsinogen secretion, Annu. Rev. Physiol. **46**:393, 1984.

Jensen, R.T., and Gardner, J.D.: The cellular basis of action of gastrointestinal peptides, Adv. Cyclic Nucleotide Protein Phosphorylation Res. **17**:375, 1984.

Klaasen, C.D., and Watkins III, J.B.: Mechanisms of bile formation, hepatic uptake, and biliary excretion, Pharmacol. Rev. **36**:1, 1984.

LaRusso, N.F.: Proteins in bile: how they get there and what they do, Am. J. Physiol. **247:**G199, 1984.

Morley, J.E., et al.: Central regulation of gastric acid secretion: the role of neuropeptides, Life Sci. **31:**399, 1982.

Putney, Jr., J.W.: Identification of cellular activation mechanisms associated with salivary secretion, Annu. Rev. Physiol. **48:**75,1986.

Soll, A.H., and Grossman, M.I.: Cellular mechanisms in acid secretion, Annu. Rev. Med. **29:**495, 1978.

Williams, J.A.: Regulatory mechanisms in pancreas and salivary acini, Annu. Rev. Physiol. **46:**361, 1984.

Books and monographs

Davenport, H.W.: Physiology of the digestive tract, ed. 5, Chicago, 1982, Year Book Medical Publishers.

Gallacher, D.V., and Petersen, O.H.: Stimulus-secretion coupling in mammalian salivary glands. In Young, J.A., editor: International review of physiology, vol. 28, Gastrointestinal physiology IV, Baltimore, 1983, University Park Press.

Jamieson, J.D.: The exocrine pancreas and salivary glands. In Weiss, L., editor: Histology: cell and tissue biology, ed. 5, New York, 1981, Elsevier.

Johnson, R.L., editor: Physiology of the gastrointestinal tract, vols. 1 and 2, New York, 1981, Raven Press.

Petersen, O.H.: New perspectives on the molecular mechanism of action of peptide secretagogues. In Paton, W., editor: Proceedings of the Ninth International Congress of Pharmacology, London, 1984, Macmillan.

Young, J.A.: Salivary secretion of inorganic electrolytes. In Crane, R.K., editor: International review of physiology, vol. 19, Gastrointestinal physiology III, Baltimore, 1979, University Park Press.

CHAPTER

44

Digestion and Absorption

Digestion and Absorption of Carbohydrates

Carbohydrates in the Diet

Plant starch, amylopectin, is the major source of carbohydrate in most human diets. There is no nutritional requirement for carbohydrate per se, but it is usually the principal source of calories. Amylopectin is a high molecular weight ($>10^6$), branched molecule of glucose monomers. A smaller proportion of dietary starch is amylose, a smaller molecular weight ($>10^5$), linear α-1,4 linked polymer of glucose. Cellulose is a β-1,4 linked glucose polymer. Intestinal enzymes cannot hydrolyze β-glycosidic linkages; thus cellulose and other molecules with β-glycosidic linkages remain undigested. Cellulose is a major component of dietary "fiber." The amount of the animal starch, glycogen, typically ingested varies widely among cultures and among individuals within a given culture. Sucrose and lactose are the principal dietary disaccharides, and glucose and fructose are the major monosaccharides. The capacity of a healthy digestive system to digest and absorb carbohydrates greatly exceeds the amount of carbohydrate presented to it under normal circumstances.

Digestion of Carbohydrates

The structure of a branched starch molecule is depicted in Fig. 44-1. Starch is a polymer of glucose and it consists of chains of glucose units linked by α-1,4 glycosidic bonds. The α-1,4 chains have branch points formed by α-1,6 linkages, and the starch molecule is highly branched. The digestion of starch begins in the mouth with the action of the α-amylase ptyalin contained in salivary secretions. This enzyme catalyzes the hydrolysis of the internal α-1,4 links of starch but cannot hydrolyze the α-1,6 branching links. (The α-amylase secreted by the pancreas has the same specificity.) As shown in Fig. 44-1, the principal products of α-amylase digestion of starch are maltose, maltotriose, and branched oligosaccharides known as α-limit dextrins. The action of the salivary α-amylase continues until the food in the stomach is mixed with gastric acid. Considerable digestion of starch by the salivary α-amylase may occur normally, but this enzyme is not required for the complete digestion and absorption of the starch ingested. After the salivary α-amylase is inactivated by gastric acid, no further processing of carbohydrate occurs in the stomach.

The pancreatic secretions contain a highly active α-amylase. The products of starch digestion by this enzyme are the same as for the salivary α-amylase (Fig. 44-1), but the total activity of the pancreatic enzyme is considerably greater than the salivary amylase. The pancreatic α-amylase is most concentrated in the duodenum. Within 10 minutes after entering the duodenum, starch is entirely converted to the following small oligosaccharides: maltose, maltotriose, α-1,4 linked maltooligosaccharides (from four to nine glucose units long), and α-limit dextrins containing from five to nine glucose monomers.

The further digestion of these oligosaccharides is accomplished by enzymes that reside in the brush border membrane of the epithelium of the duodenum and jejunum (Fig. 44-2). The major brush border oligosaccharidases are *lactase*, which splits lactose into glucose and galactose, *sucrase*, which splits sucrose into fructose and glucose, *α-dextrinase* (which is also called isomaltase and which "debranches" the α-limit dextrins by cleaving the α-1,6 linkages at the branch points), and *glucoamylase*, which breaks maltooligosaccharides down to glucose units.

Both α-dextrinase and sucrase also cleave α-1,4 glucose–glucose linkages. The digestion of α-limit dextrins proceeds by the sequential removal of glucose monomers from the nonreducing ends. When branch points are encountered, they are cleaved by α-dextrinase (Fig. 44-3). Sucrase and isomaltase are noncovalently associated subunits of a single protein. Each en-

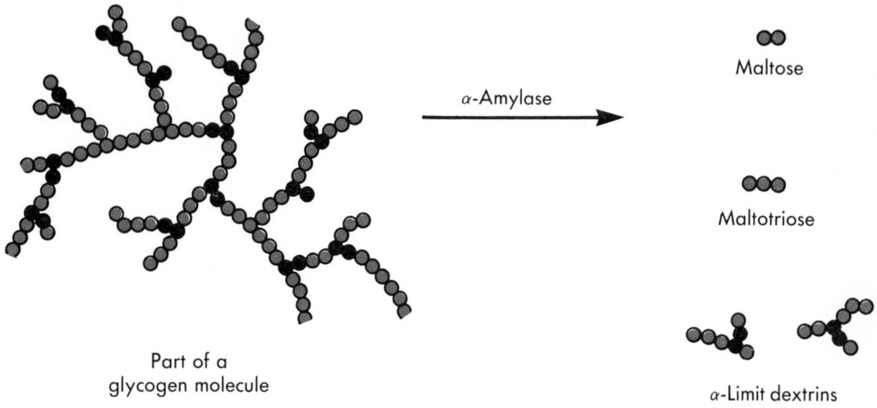

Fig. 44-1. Structure of a branched starch molecule and the action of α-amylase. The circles represent glucose monomers. The black circles show glucose units linked by α-1,6 linkages at the branch points. The α-1,6 linkages and terminal α-1,4 bonds cannot be cleaved by α-amylase. Other sugars are linked by α-1,4 linkages.

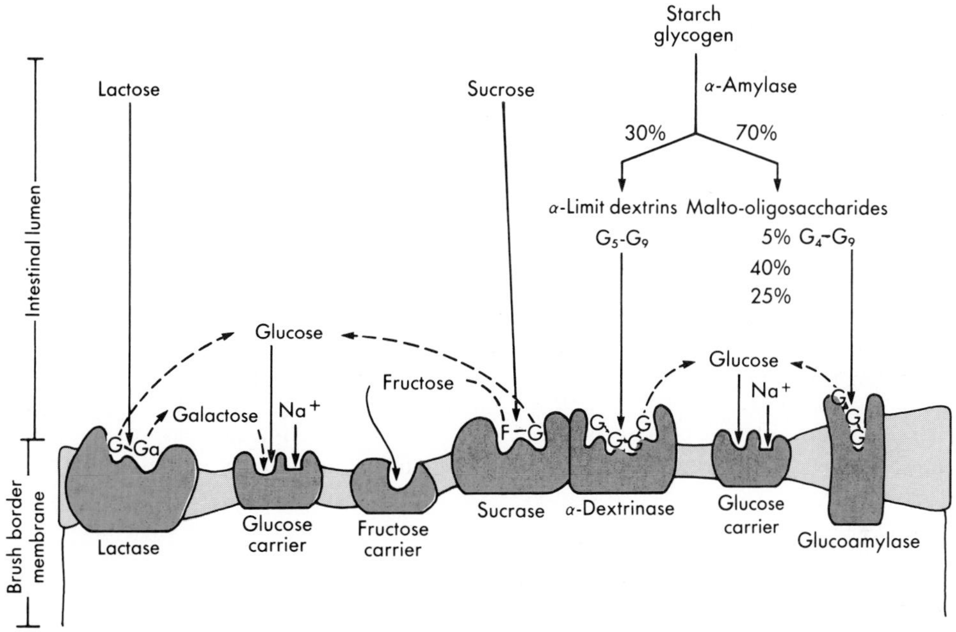

Fig. 44-2. Functions of the major brush border oligosaccharidases. The glucose, galactose, and fructose molecules released by enzymatic hydrolysis are then transported into the epithelial cell by specific transport carrier proteins. *G,* Glucose; *Ga,* galactose; *F,* fructose. (From Gray, G.M.: N. Engl. J. Med. **292:**1225, 1975. Reprinted by permission of *The New England Journal of Medicine.*)

zyme is a single polypeptide chain. The two enzyme activities are not affected by their molecular interactions, so the function for this association remains obscure. The activities of these four enzymes is highest in the brush border of the upper jejunum, and they gradually decline through the rest of the small intestine.

Absorption of Carbohydrates

The duodenum and upper jejunum have the highest capacity to absorb sugars. The capacities of the lower jejunum and ileum are progressively less. The only dietary monosaccharides that are well absorbed are glucose, galactose, and fructose. Glucose and galactose are actively taken up by the brush border epithelial cells through a fairly well-characterized transport system. Glucose and galactose compete for entry; other sugars are less effective competitors. Transport is inhibited by phlorhizin and by metabolic inhibitors.

The active entry of glucose and galactose into the intestinal epithelial cells is stimulated by the presence of Na^+ in the lumen. Similarly, the entry of Na^+ into the epithelial cell across the brush border membrane is stimulated by glucose or galactose in the lumen. The current interpretation is that Na^+ and glucose or galactose are transported into the cell by the same membrane protein, which has two Na^+-binding sites and one sugar-binding site. Na^+ enters the cell down a large electrochemical potential gradient; both concentration

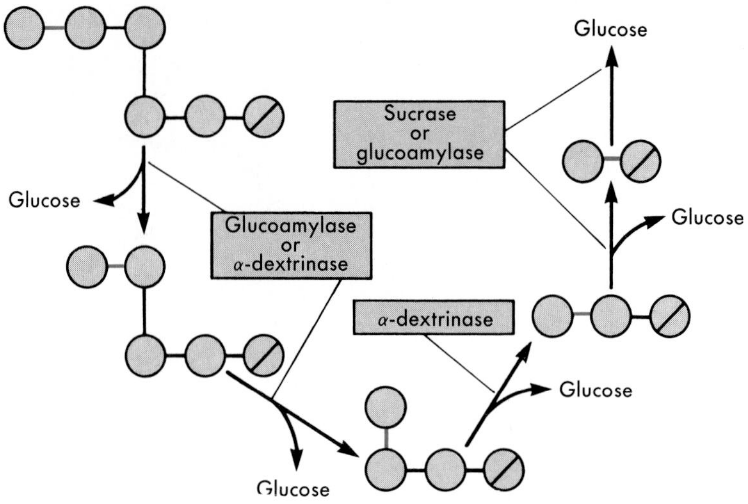

Fig. 44-3. Cleavage of an α-limit dextrin by the oligosaccharidases of the brush border plasma membrane. Glucose monomers are removed in seqeunce, beginning at the nonreducing end of the molecule. Note the overlapping specificities of glucoamylase and α-dextrinase and of sucrase and glucoamylase. α-Dextrinase (isomaltase) is the only enzyme that cleaves the α-1,6 linkages at the branch points of the α-limit dextrins. (From Gray, G.M.: Carbohydrate absorption and malabsorption. In Johnson, R.L., editor: Physiology of the gastrointestinal tract, New York, 1981, Raven Press.)

and electrical forces drive it into the cell. The energy released by the flow of Na^+ down its electrochemical potential is harnessed to force glucose or galactose into the cell *against* a concentration gradient for the sugar. The electrochemical potential gradient for Na^+ is created by Na^+, K^+-ATPase molecules in the basal and lateral plasma membranes of the intestinal epithelial cells. Na^+ is transported by *primary active transport* by the sodium-potassium pump. The sugars are transported by *secondary active transport,* because their active transport depends on the electrochemical potential gradient of another species (Na^+). Glucose and galactose leave the intestinal epithelial cell at the basal and lateral plasma membranes via facilitated transport and by simple diffusion to some extent, and they diffuse into the mucosal capillaries. Fig. 44-4 summarizes some major features of glucose and galactose absorption.

The absorption of fructose by the intestine is less well understood. Fructose does not compete well for the glucose-galactose system, and fructose transport is not linked to Na^+ absorption. Yet it is transported almost as rapidly as glucose and galactose and much more rapidly than other monosaccharides. Hence it is presumed that fructose uptake is mediated by a membrane protein. Active transport of fructose has been observed in rat intestine, but has not been demonstrated in the human intestine.

Carbohydrate Malabsorption Syndromes

Malabsorption of carbohydrates is usually caused by a deficiency in one of the digestive enzymes of the brush border.

Lactose malabsorption syndrome. This common disorder is caused by a deficiency of lactase in the brush border of the duodenum and jejunum. Undigested lactose cannot be absorbed. Thus it is passed on to the colonic bacteria, which avidly metabolize the lactose, producing gas and metabolic products that enhance colonic motility.

Individuals with this disorder are said to be *lactose intolerant*. The symptoms of this disorder, and the other carbohydrate malabsorption syndromes as well, are intestinal distension, borborygmi, and diarrhea. Borborygmi are the gurgling noises made by the intestine as it mixes gas and liquid.

More than 50% of the adults in the world are lactose intolerant. This condition seems to be genetically determined. In Oriental societies lactose intolerance among adults is almost universal. The majority of northern European adults, on the other hand, are lactose tolerant. Many black American adults are lactose intolerant. Many lactose-intolerant adults simply do not drink milk or eat certain milk products and thereby avoid the symptoms without being aware that they have the disorder. The presence of lactose in the diet may induce a higher level of intestinal lactase activity than would be present in the absence of dietary lactose.

Congenital lactose intolerance. This is rare. Infants who lack lactase have diarrhea when they are fed breast milk or formula containing lactose. The resulting dehydration and electrolyte imbalance is life threatening. Such infants do well when fed a formula containing sucrose or fructose instead of lactose.

Sucrase-isomaltase deficiency. A deficiency of sucrase and isomaltase activities in the small intestinal mucosa is an autosomal recessive, inherited disorder that results in intolerance to ingested sucrose. About 10% of Greenland's Eskimos and as many as 0.2% of North Americans have sucrase-isomaltase deficiency. In this disorder either the synthesis of sucrase and isomaltase is suppressed or the enzymes are destroyed by antibodies. Individuals with sucrase-isomaltase deficiency do well on diets low in sucrose.

Glucose-galactose malabsorption syndrome. This is a rare hereditary disorder due to a defect in the brush

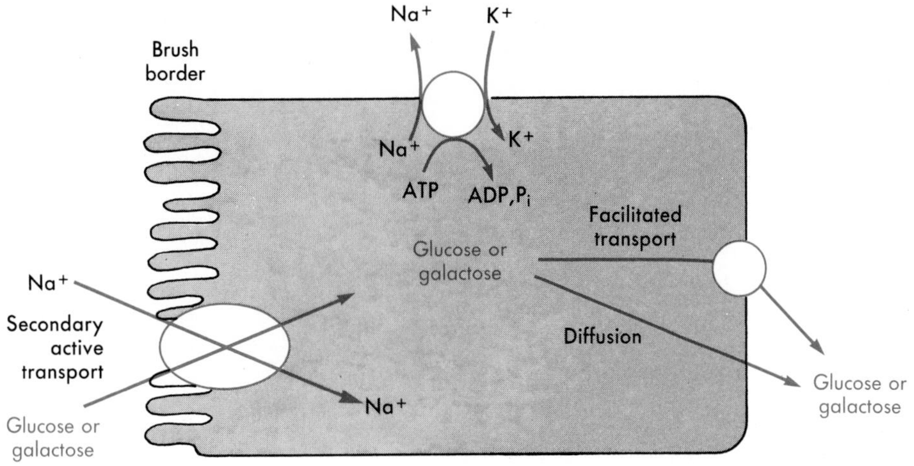

■ Fig. 44-4. Major features of glucose and galactose absorption in the small intestine. Glucose and galactose enter the epithelial cell against a concentration gradient (Na^+ entry provides the energy). Glucose and galactose leave the cell at the basolateral membrane by facilitated transport and by simple diffusion.

border active transport system for glucose and galactose. Ingestion of glucose, galactose, or starch leads to flatulence and severe diarrhea. Fructose is well tolerated and can be fed to infants with this disorder. The brush border saccharidases are normal in this disease.

Some diagnostic procedures for carbohydrate intolerance. The most common diagnostic test is the *oral sugar tolerance test*. The patient is given an oral dose of the sugar in question, and the levels of that sugar in the patient's blood and feces are monitored. If the patient is intolerant of the administered sugar, diarrhea will ensue and that sugar will fail to appear in the blood, but it will appear in the feces. In suspected saccharidase deficiency the most definitive test is to take a biopsy sample of the jejunal mucosa and assay it for the deficient enzyme.

■ Digestion and Absorption of Proteins

The amount of dietary protein varies greatly among cultures and among individuals within a culture. In poor societies it is difficult for an adult to obtain the amount of protein (0.5 to 0.7 g/day/kg of body weight) required to balance normal catabolism of proteins, and it is still more difficult for children to get the relatively greater amounts of protein required to sustain normal growth. In wealthier societies, chiefly in industrially developed countries, a typical individual may ingest protein far in excess of the nutritional requirement.

The gastrointestinal tract must deal with the 10 to 30 g of protein per day contained in digestive secretions and a similar amount of protein in desquamated epithelial cells, in addition to the protein ingested.

In normal humans essentially all ingested protein is digested and absorbed. Most of the protein in digestive secretions and desquamated cells is also digested and absorbed. The small amount of protein in the feces is derived principally from colonic bacteria, desquamated cells, and proteins in mucous secretions of the colon. In humans, ingested protein is almost completely absorbed by the time the meal has traversed the jejunum.

■ Digestion of Proteins

Digestion in the stomach. Pepsinogen is secreted by the chief cells of the stomach and is converted by hydrogen ions to the active enzyme *pepsin*. The extent to which pepsin hydrolyzes dietary protein is significant but highly variable. At most, about 15% of dietary protein may be reduced to amino acids and small peptides by pepsin. The duodenum and small intestine have such a high capacity to process protein that the total absence of pepsin does not impair the digestion and absorption of dietary protein.

Digestion in the duodenum and small intestine. Proteases secreted by the pancreas play a major role in protein digestion. The most important of these proteases are *trypsin, chymotrypsin,* and *carboxypeptidase*. The pancreatic juice contains these enzymes in inactive, proenzyme forms. The enzyme *enterokinase,* secreted by the mucosa of the duodenum and jejunum, converts trypsinogen to trypsin. Trypsin acts autocatalytically to activate trypsinogen and also converts chymotrypsinogen and procarboxypeptidase to the active enzymes (Fig. 44-5). The pancreatic proteases are present at high activities in the duodenum and rapidly convert dietary protein to small peptides. About 50% of the ingested protein is digested and absorbed in the duodenum.

The brush border of the duodenum and the small intestine contains a number of peptidases. These pepti-

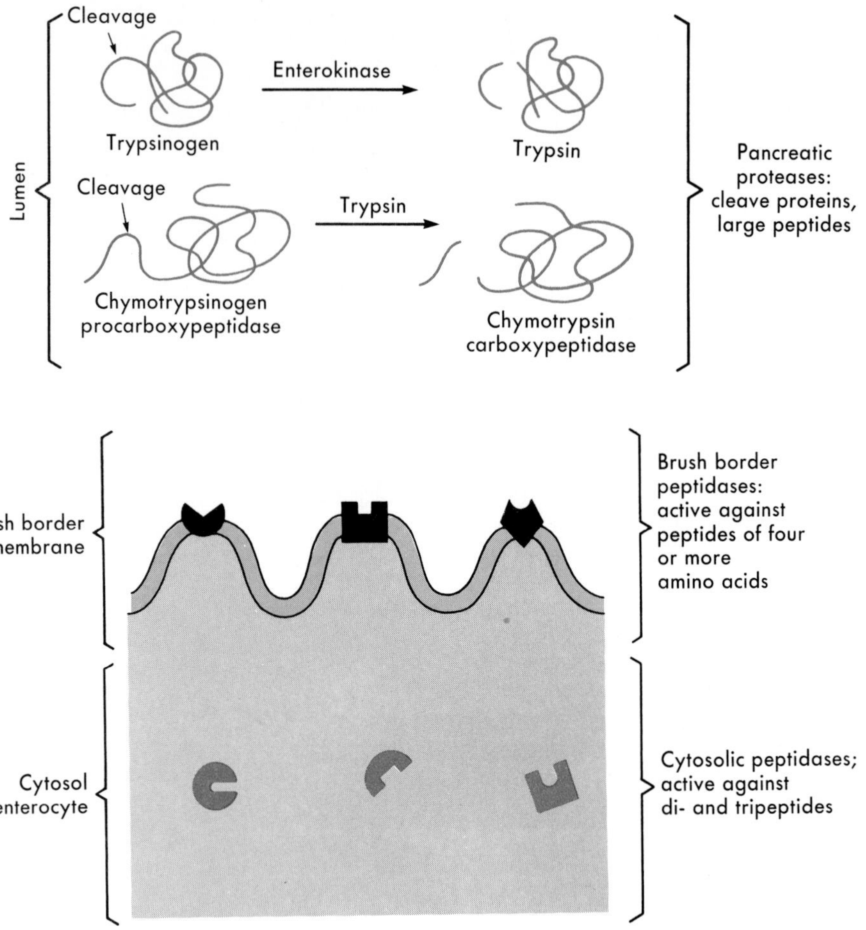

Fig. 44-5. The major proteases and peptidases present in the lumen of the small intestine, on the brush border plasma membrane, and in the cytosol of enterocytes of the small intestine.

dases are integral membrane proteins whose active sites face the intestinal lumen. The brush border enzymes are richest in the proximal jejunum. They reduce the peptides produced by pancreatic proteases to oligopeptides and amino acids. The brush border peptidases include aminopeptidases (that cleave single amino acids from the N-terminals of peptides), dipeptidases (that cleave dipeptides to amino acids), and dipeptidyl aminopeptidases (that cleave a dipeptide from the N-terminal end of a peptide). Fig. 44-5 illustrates some major proteases in the small intestine.

The principal products of protein digestion by pancreatic proteases and brush border peptidases are small peptides and amino acids. The small peptides (primarily dipeptides, tripeptides, and tetrapeptides) are about three or four times more concentrated than the single amino acids. As discussed below, small peptides and amino acids are transported across the brush border plasma membrane into intestinal epithelial cells. Small peptides are then hydrolyzed by peptidases in the cytosol of the epithelial cells; consequently, single amino acids and a few dipeptides appear in the portal blood. The cytosolic peptidases are more abundant than the brush border peptidases, and the cytosolic peptidases are particularly active against dipeptides and tripeptides (which are transported with high efficiency across the brush border membrane). The brush border peptidases, on the other hand, are mainly active against peptides of four or more amino acids.

Absorption of the Products of Protein Digestion

Intact proteins and large peptides. These are not absorbed by humans to an extent that is nutritionally significant, but amounts sufficient to trigger an immunological response can be absorbed. In ruminants and rodents, but not in humans, the neonatal intestine has a high capacity for the specific absorption of immune globulins present in colostrum. This is vital in the development of normal immune competence in ruminants and rodents. Absorption takes place by receptor-mediated endocytosis.

Absorption of small peptides. Dipeptides and tripeptides are transported across the brush border membrane. The rate of transport of dipeptides or tripeptides usually exceeds the rate of transport of individual amino acids. In the experiments shown in Fig. 44-6, glycine was absorbed by the human jejunum less rapidly as the amino acid than it was from glycylglycine or from gly-

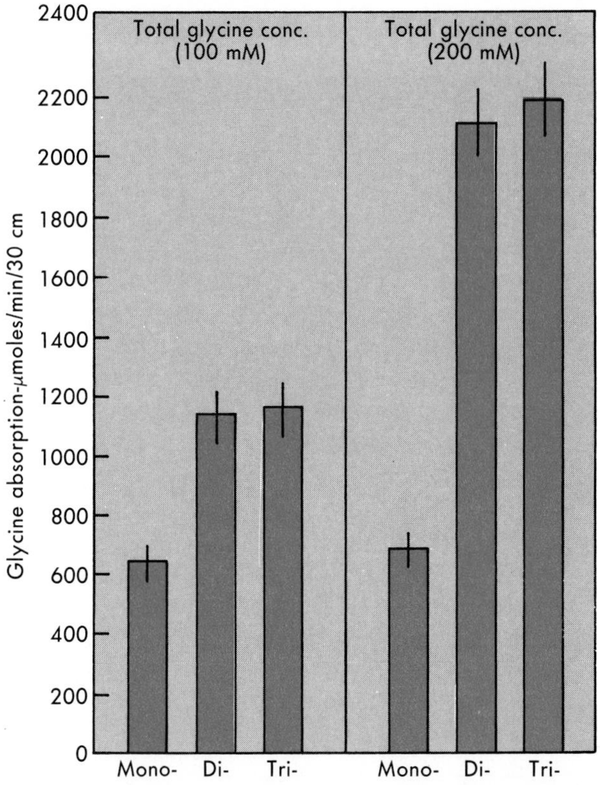

Fig. 44-6. Glycine is absorbed more rapidly by the human small intestine in the form of diglycine or triglycine than as free glycine. (From Adibi, S.A., and Morse, E.L.: J. Clin. Invest. **60:**1008, 1977.)

cylglycylglycine. In this case the transport capacity of the jejunal mucosa was greater for the dipeptide and the tripeptide than for the amino acid; the apparent affinities for glycine and glycylglycine were similar.

A single membrane transport system with broad specificity is probably responsible for absorption of small peptides. The transport system has high affinity for dipeptides and tripeptides, but very low affinities for peptides of four or more amino acid residues. The transport system is stereospecific and prefers peptides of the physiological L-amino acids. The affinity is higher for peptides of amino acids with bulky side chains. Transport of dipeptides and tripeptides across the brush border plasma membrane is a secondary active transport process powered by the electrochemical potential difference of Na^+ across the membrane. The total amount of each amino acid that enters intestinal epithelial cells in the form of dipeptides or tripeptides is considerably greater than the amount that enters as the single amino acid.

Absorption of amino acids. With respect to amino acid transport properties, the brush border plasma membrane of small intestinal epithelial cells differs considerably from the basolateral plasma membrane. Normally amino acids are transported across the brush border plasma membrane into the enterocyte by way of certain specific amino acid transport systems. Transport of amino acids out of the epithelial cell across the basolateral membrane occurs by other transporters. Some of the transporters depend on the Na^+ gradient (as previously described for glucose and galactose absorption), whereas other transport systems are independent of Na^+. The Na^+-dependent brush border transport systems are found only in epithelial cells. Brush border membranes of the small intestine and proximal renal tubule are similar with respect to the nature of Na^+-dependent amino acid transport processes. The amino acid transporters in the brush border plasma membrane that do not depend on the Na^+ gradient are similar to amino acid transport systems that have been found in nonepithelial cells.

Table 44-1. Pathways for mediated transport of amino acids in jejunal epithelial cells

Na^+-dependent system	*Na^+-independent system*
Brush border membrane	
NBB: transports most neutral amino acids	y^+: transports basic amino acids
IMINO: transports proline and hydroxyproline	L: transports most neutral amino acids, prefers hydrophobic side chains
PHE: transports phenylalanine and methionine	
Basolateral membrane	
A: transports small polar amino acids	L: transports most neutral amino acids, prefers hydrophobic amino acids
ASC: transports neutral amino acids	

Modified from Stevens, B.R., et al.: Annu. Rev. Physiol. **46:**417, 1984.

Other transport systems are responsible for transporting amino acids across the basolateral plasma membrane of the intestinal epithelial cell. As is true for the brush border membrane, some of the amino acid transporters present in the basolateral membrane depend on Na^+ and others do not. The basolateral membrane, however, is less highly differentiated than the brush border membrane, in that all of the amino acid transporters present in the basolateral membrane occur in certain nonepithelial cells. For most amino acids simple diffusion is a significant pathway across both brush border and basolateral membranes. The more hydrophobic the amino acid and the larger its concentration gradient across the membrane, the greater the importance of diffusion. Table 44-1 summarizes the processes responsible for amino acid transport in epithelial cells of the small intestine.

Brush border membranes. Three Na^+-dependent active transport systems for amino acids occur in brush border membranes. All three systems are present only in epithelial cell types, such as the epithelial cells of the proximal renal tubule. The system called *neutral brush border* (NBB) transports most of the neutral amino acids, both hydrophobic and hydrophilic. The system called *PHE* transports primarily phenylalanine and methionine. The *IMINO* acid transport system handles proline and hydroxyproline.

The *L* and y^+ systems present in the brush border do not depend on the Na^+. These systems have been described in nonepithelial cells. The L system is present in all eukaryotic cell types and transports neutral amino acids, with preference for those with hydrophobic side chains. The y^+ system transports basic amino acids such as lysine and arginine.

The basolateral membrane. Two Na^+-dependent systems transport neutral amino acids across the basolateral plasma membrane. The *A* system prefers small, hydrophilic amino acids, and the *ASC* system also prefers small neutral amino acids. The Na^+-independent *L* system (also present in brush border membranes) transports neutral amino acids across the basolateral membrane and has high affinity for hydrophobic amino acids. All of the transport systems present in the basolateral membrane also occur in nonepithelial cells. The basolateral membrane is more permeable to amino acids than is the brush border membrane. Therefore diffusion is a more important pathway for transport across the basolateral plasma membrane, especially for amino acids with hydrophobic side chains.

Defects of Amino Acid Absorption

Hartnup's disease. This rare hereditary disease involves defective renal transport of neutral amino acids. Intestinal absorption of neutral amino acids is also decreased. Hartnup's disease may involve defects in the NBB system of the epithelial cells of the jejunum and the proximal renal tubule.

Prolinuria. This condition is rare and involves defective renal and intestinal reabsorption of proline. Prolinuria appears to be due to a defect in the IMINO system.

Intestinal Absorption of Salts and Water

Under normal circumstances humans absorb almost 99% of the water and ions presented to them in ingested food and in gastrointestinal secretions. Thus net fluxes of water and ions are normally from the lumen to the blood. In most cases the net fluxes of water and ions are the differences between much larger unidirectional fluxes from lumen to blood and from blood to lumen.

Absorption of Water

Typically, about 2 L of water is ingested each day and approximately 7 L/day is contained in gastrointestinal secretions. Only about 50 to 100 ml of water per day is lost in the feces. Thus the gastrointestinal tract typically absorbs more than 8 L/day.

Very little net absorption occurs in the duodenum, but the chyme is brought to isotonicity here. The chyme that is delivered from the stomach is often hypertonic. The action of digestive enzymes creates still more osmotic activity. The duodenum is highly water permeable, and very large fluxes of water occur from lumen to blood and from blood to lumen. Usually the net flux is from blood to lumen because of the hypertonicity of the chyme. Large net water absorption occurs in the *small intestine;* the *jejunum* is more active than the *ileum* in absorbing water. The net absorption that occurs in the *colon* is relatively small, about 400 ml/day. However, the colon can absorb water against a larger osmotic pressure difference than can the rest of the gastrointestinal tract. Fig. 44-7 summarizes the handling of water by the gastrointestinal tract.

Absorption of Na^+

Na^+ is absorbed along the entire length of the intestine. tine. As is the case with water, net absorption is the result of large, unidirectional fluxes from blood to lumen and from lumen to blood. The unidirectional fluxes are greater in the proximal gut than in the distal intestine. The fluxes correlate with the greater brush border surface area per unit length in the jejunum, and this ratio diminishes toward the ileum and is still smaller in the colon. Na^+ crosses the brush border membrane down an electrochemical gradient, and it is actively extruded from the epithelial cells by the Na^+, K^+-ATPase in the basal and lateral plasma membrane. Normally the contents of the small bowel are isotonic to plasma. Luminal contents have about the same Na^+ concentration as plasma, so that Na^+ absorption normally takes place in the absence of a significant concentration gradient. Na^+ absorption is active, however, and can occur against a small electrochemical potential difference for Na^+.

In the jejunum the net rate of absorption of Na^+ is highest. Here Na^+ absorption is enhanced by the presence in the lumen of glucose, galactose, and neutral amino acids. These substances and Na^+ may cross the brush border membrane on the same transport proteins. Na^+ moves down its electrochemical potential gradient and provides the energy for moving the sugars (glucose and galactose) and neutral amino acids into the epithelial cells against a concentration gradient. Thus Na^+ enhances the absorption of sugars and amino acids and vice versa.

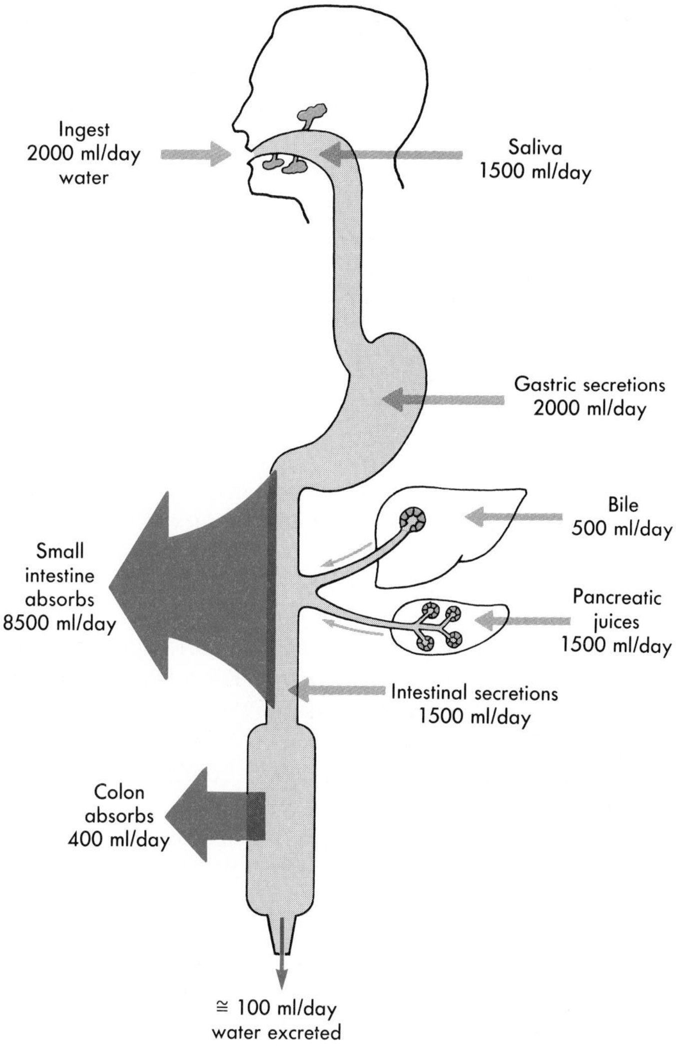

Fig. 44-7. Overall fluid balance in the human gastrointestinal tract. About 2 L of water is ingested each day, and 7 L of various secretions enters the gastrointestinal tract. Of this total of 9 L, 8.5 is absorbed in the small intestine. About 500 ml is passed on to the colon, which normally absorbs 80% to 90% of the water presented to it.

In the ileum the net rate of Na^+ absorption is smaller. Na^+ absorption is only slightly stimulated by sugars and amino acids because the sugar and amino acid transport proteins are less concentrated in the ileum. The ileum can absorb Na^+ against a larger electrochemical potential than can the jejunum.

In the colon Na^+ is normally absorbed against a large electrochemical potential difference. Sodium concentrations in the luminal contents can be as low as 25 mM, compared with about 120 mM in the plasma.

Absorption of Cl^- and HCO_3^-

In the jejunum both Cl^- and HCO_3^- are absorbed in large amounts. By the end of the jejunum most of the HCO_3^- of the hepatic and pancreatic secretions has been absorbed. *In the ileum* Cl^- is absorbed, but HCO_3^- is normally secreted. If the HCO_3^- concentration in the lumen of the ileum exceeds about 45 mM, then the flux from lumen to blood exceeds that from blood to lumen, and net absorption occurs. *In the colon* the transport of these ions is qualitatively similar to that in the ileum, in that Cl^- is absorbed and bicarbonate is usually secreted.

Absorption of K^+

As with the other ions, the net movement of potassium across the intestinal epithelium is the difference between large unidirectional fluxes from lumen to blood and from blood to lumen. *In the jejunum and in the ileum* the net flux is from lumen to blood. As the volume of intestinal contents is reduced due to the absorption of water, K^+ is concentrated, providing a driving force for the movement of K^+ across the intestinal mucosa and into the blood. Evidence for active transport of K^+ in the small intestine is lacking. *In the colon* K^+ may be secreted or absorbed. Net secretion occurs when the luminal concentration is less than about 25 mM; above 25 mM, net absorption occurs. Under most circumstances there is net secretion of K^+ in the colon; the secretory process may be active.

Because most absorption of K^+ is a consequence of its enhanced concentration in the lumen due to the absorption of water, significant K^+ loss may occur in diarrhea. If diarrhea is prolonged, the K^+ level in the extracellular fluid compartment of the body falls. Because of the importance of maintaining the normal K^+ level in the extracellular fluid compartment, especially for the heart and other muscles, life-threatening consequences such as cardiac arrhythmias may ensue. Table 44-2 summarizes the transport of Na^+, K^+, Cl^-, and HCO_3^- in the small and large intestines.

Table 44-2. Transport of Na^+, K^+, Cl^-, and HCO_3^- in the large and small intestines

Segment of intestine	Na^+	K^+	Cl^-	HCO_3^-
Jejunum	Actively absorbed; absorption enhanced by sugars, neutral amino acids	Passively absorbed when concentration rises due to absorption of water	Absorbed	Absorbed
Ileum	Actively absorbed	Passively absorbed	Absorbed, some in exchange for HCO_3^-	Secreted, partly in exchange for Cl^-
Colon	Actively absorbed	Net secretion occurs when (K^+) concentration in lumen< 25 mM	Absorbed, some in exchange for HCO_3^-	Secreted, partly in exchange for Cl^-

Mechanisms of Salt and Water Absorption by the Intestine

Structural considerations

The tight junctions. The epithelial cells that line the intestine are connected to their neighbors by tight junctions near their luminal surfaces. The tight junctions are leakiest in the duodenum, a bit tighter in the jejunum, still tighter in the ileum, and tightest in the colon.

Transcellular versus paracellular transport. Because the tight junctions are leaky, some fraction of the water and ions that traverse the intestinal epithelium passes between the epithelial cells, rather than passing through them. Transmucosal movement by passing through the tight junctions and the lateral intercellular spaces is called *paracellular transport*. Passage through the epithelial cells is termed *transcellular transport*.

Because the tight junctions in the duodenum are very leaky, significant proportions of the large unidirectional fluxes of water and ions that take place in the duodenum occur via the paracellular pathway. The proportions of water or a particular ion that pass through the transcellular and paracellular routes are determined by the relative permeabilities of the two pathways for the substance in question. Even in the ileum, where the junctions are much tighter than in the duodenum, the paracellular pathway contributes more to the total ionic conductance of the mucosa than does the transcellular pathway.

Villous versus crypt cells. Recent evidence suggests that the highly differentiated epithelial cells near the tips of the villi are specialized for absorption of water and ions, whereas the less differentiated cells in the crypts produce net secretion of water and ions.

Ion Transport by Intestinal Epithelial Cells.

The movement of water across the intestinal epithelium is secondary to the movement of ions. Significant progress in characterizing the ionic transport processes that occur in intestinal epithelial cells has recently been made. Nevertheless, important issues remain unresolved.

Ion transport in the jejunum. The basolateral plasma membrane contains the Na^+, K^+ -ATPase (Fig. 44-8). As a result of the active extrusion of Na^+ ions from the cytoplasm by the Na^+, K^+-ATPase, the electrochemical potential of Na^+ in the cytoplasm is much less than in the luminal fluid. Na^+ enters the epithelial cell by flowing across the brush border plasma membrane, moving down its large electrochemical potential gradient. Some of the Na^+ enters on the same transport proteins with sugars or amino acids, as described previously.

The luminal plasma membrane also contains a Na^+/H^+ exchange protein. Some of the energy released by Na^+ moving down its electrochemical potential gradient is harnessed to actively extrude H^+ into the lumen. Some of the H^+ in the lumen then reacts with HCO_3^- from bile and pancreatic juice to produce H_2CO_3, some of which then forms CO_2 and H_2O. The CO_2 diffuses readily across the epithelial cells and is carried away in the blood. Acidification of the luminal fluid appears to be the major mechanism for absorption of HCO_3^- in the jejunum.

The extrusion of H^+ from the cytoplasm creates a high cytoplasmic concentration of HCO_3^- in the epithelial cells. HCO_3^- leaves the cell by crossing the basolateral plasma membrane by facilitated transport.

The electrogenic effect of the Na^+, K^+-ATPase in the basolateral membrane and the electrogenic entry of Na^+ with sugars and amino acids at the luminal surface both tend to produce an electrical potential difference (lumen negative) across the epithelium. The electrical potential difference causes Cl^- to flow through the tight junctions into the lateral intercellular spaces. The tight junctions in the jejunum are leaky, so large fluxes of Cl^- occur by this route. The large paracellular flux of Cl^- almost short-circuits the electrogenic effects of Na^+ transport, and as a result the potential difference across the jejunum is only a few millivolts (lumen negative). The flow of Cl^- via the tight junctions appears

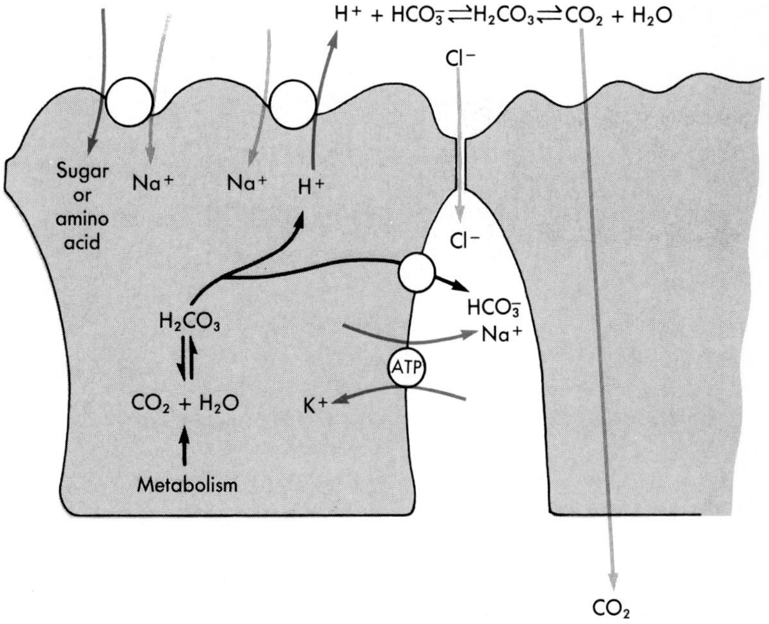

■ **Fig. 44-8.** A summary of major ion transport processes that occur in the *jejunum*.

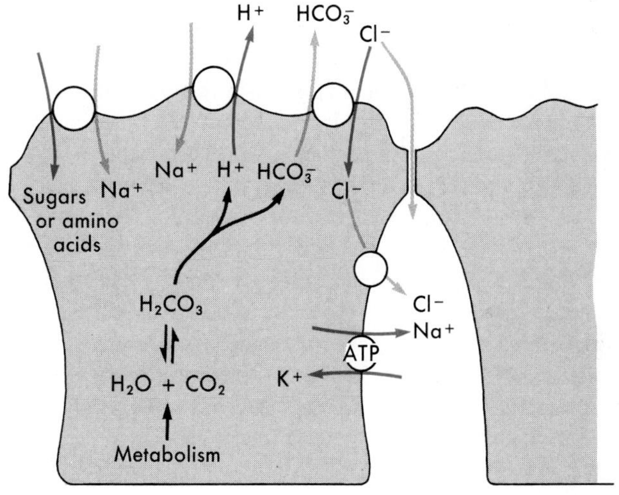

■ **Fig. 44-9.** A summary of major ion transport processes that occur in the *ileum*.

to be a major route of Cl^- absorption in the jejunum. Some Cl^- flows through the cells, but the transport processes involved are not well characterized.

Note that the only primary active transport is by the Na^+, K^+-ATPase. The resulting Na^+ gradient is responsible for the active extrusion of H^+. The extrusion of H^+ brings about absorption of HCO_3^-. The electrogenic effects of Na^+ transport create an electrical potential difference that powers Cl^- absorption in the jejunum.

Ion transport in the ileum. The ionic transport processes that occur in the ileum (Fig. 44-9) resemble those in the jejunum. Basolateral Na^+, K^+-ATPase, electrogenic Na^+ entry with sugars and amino acids (to a lesser extent than in the jejunum), and luminal Na^+/H^+ exchange are all present in the ileum.

The brush border plasma membrane of the ileum contains a Cl^-/HCO_3^- exchange protein that is not present in the jejunum. As in the jejunum, the active extrusion of H^+ into the lumen is driven by Na^+ entry. The extrusion of H^+ elevates the intracellular concentration of HCO_3^-. The HCO_3^- then flows down its electrochemical potential gradient into the lumen, and part of the energy liberated is used to bring Cl^- into the epithelial cell against its electrochemical potential gradient. Cl^- then leaves the cell at the basolateral membrane by facilitated transport.

The net effect of Na^+/H^+ exchange and the Cl^-/HCO_3^- exchange occurring together is the electroneutral entry of Na^+ and Cl^- into the epithelial cell. Because the extrusion of H^+ is driven by the downhill entry of Na^+, and the resulting downhill efflux of HCO_3^- drives

Fig. 44-10. A summary of major ion transport processes that occur in the *colon*.

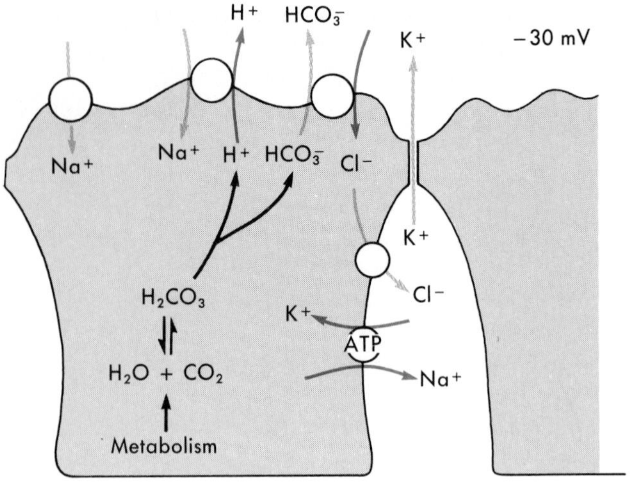

the uphill entry of Cl^-, the energy for Cl^- entry comes, albeit indirectly, from the electrochemical potential gradient of Na^+. Thus the actual linkage of the ionic transport processes to metabolism in the ileum occurs via the Na^+, K^+-ATPase.

As in the jejunum, the electrogenic Na^+, K^+-ATPase and the electrogenic entry of Na^+ at the brush border cause a small electrical potential difference across the ileal mucosa (lumen negative). The electrical potential difference causes Cl^- to flow through the tight junctions into the intercellular spaces. The electrically driven flux of Cl^- in the ileum is smaller than it is in the jejunum because the ileal junctions are tighter than those in the jejunum.

Ion transport in the colon. Ion transport in the colon (Fig. 44-10) is characterized by luminal Na^+/H^+ and Cl^-/HCO_3^- exchange pumps and by basolateral Na^+, K^+-ATPase and facilitated Cl^- transport, as is the case in the ileum.

The Na^+/sugar and Na^+/amino acid cotransport systems are not present in the colon, but another process that mediates the electrogenic entry of Na^+ into the epithelial cells is present. The latter process is inhibited by amiloride. The electrogenic Na^+, K^+-ATPase and the luminal electrogenic entry of Na^+ cause an electrical potential difference (lumen negative) across the colonic mucosa. The transmucosal potential difference in the colon is about −30 mV, much larger than in the jejunum or the ileum. The greater electrical potential difference across the colonic mucosa is due partly to the decreased leakiness of tight junctions in the colon.

The tight junctions in all regions of the intestine are more permeable to cations than to anions. The large electronegativity in the lumen of the colon causes K^+ to flow from the intercellular spaces to the lumen via the tight junctions. This may be the major mechanism for the net secretion of K^+ that usually occurs in the colon. Facilitated transport of K^+ from cytoplasm to intestinal lumen across the luminal plasma membrane may also occur in the colon.

In the colon, as in the other segments of the intestine, the Na^+, K^+-ATPase is the only transport process that is directly linked to metabolic energy.

The Mechanism of Water Absorption

The absorption of water depends on the absorption of ions, principally Na^+ and Cl^-. Under normal circumstances, water absorption in the small intestine occurs in the absence of an osmotic pressure difference between the luminal contents and the blood in the intestinal capillaries. Water absorption by the colon typically proceeds against an osmotic pressure gradient. For a long time students of the gastrointestinal tract were puzzled by the absorption of water in the apparent absence of an osmotic pressure gradient to power the absorption.

Our current understanding is that water absorption occurs by a mechanism known as *standing gradient osmosis* (Fig. 43-33). The major features of the standing gradient osmotic mechanism follow:

1. Active pumping of Na^+ into the lateral intercellular spaces by the Na^+, K^+-ATPase
2. Entry of Cl^- into the lateral intercellular spaces by flow from the lumen via the tight junctions or from the adjacent epithelial cells by facilitated transport
3. Presence of slightly hypertonic fluid near the luminal ends of the lateral intercellular spaces
4. Entry of water by osmosis into the lateral intercellular spaces
5. Hydrostatic flow of water and ions down the lateral intercellular spaces and across the epithelial basement membrane

Na^+, K^+-ATPase molecules are particularly concentrated in the basolateral plasma membrane that surrounds the luminal ends of the intercellular spaces. Because of the high rate of Na^+ pumping and the narrowness of the luminal ends of the intercellular spaces, the Na^+ concentration near the luminal ends of the intercellular spaces tends to reach levels above the luminal concentration. Chloride enters the intercellular spaces from the lumen via the tight junctions (driven by the transmucosal electrical potential difference) and/or from the adjacent epithelial cells by facilitated transport across the basolateral membrane.

The NaCl concentration in the luminal ends of the lateral intercellular spaces is high enough that the fluid there is hypertonic to the luminal contents and to the cytoplasm of the adjacent epithelial cells. Because of the hypertonicity in the intercellular space, water flows by osmosis into the intercellular space from the adjacent epithelial cells and from the lumen via the tight junctions. The inflow of water raises the hydrostatic pressure and dilates the intercellular channels. Fluid flows down the intercellular space because of the hydrostatic pressure gradient, and water and ions flow across the basement membrane of the epithelium, which offers little resistance to their passage. As the hypertonic fluid flows down the intercellular channels, water continues to enter the channels by osmosis from the adjacent epithelial cells. By the time the fluid reaches the basement membrane, it is essentially isotonic to the cytoplasm of the epithelial cells. In this way isotonic fluid is taken up at the brush border and is discharged through the basement membrane, to be carried away by the intestinal capillaries. The slight gradient of tonicity in the lateral intercellular space, hypertonic at the luminal end of the intercellular space and isotonic at the serosal end, gives the standing gradient mechanism of fluid absorption its name.

Because most water absorption takes place in the absence of a transmucosal osmotic pressure difference, the absorption of the end products of digestion, particularly sugars and amino acids, plays an important role in water absorption. The absorption of sugars and amino acids allows more water to be absorbed.

Control of Intestinal Electrolyte Absorption

Electrolyte transport in the intestine is regulated by certain hormones, neurotransmitters, and paracrine substances. The cellular mechanisms by which most of the regulatory substances act are not completely characterized.

The autonomic nervous system. Stimulation of sympathetic nerves to the intestine or an elevated level of epinephrine increases the absorption of Na^+, Cl^-, and water. Stimulation of parasympathetic nerves to the gut decreases the net rate of ion and water absorption.

Adrenal hormones. Aldosterone strongly stimulates the secretion of K^+ and the absorption of Na^+ and water by the colon and to a much lesser extent by the ileum. Aldosterone acts by increasing the number of electrogenic Na^+ channels in the luminal membrane of the colonic epithelial cells (Fig. 44-10) and the number of active Na^+, K^+-ATPase molecules in the basolateral membrane. Aldosterone has similar effects on the epithelial cells of the distal tubule of the kidney. The enhanced absorption of NaCl and water induced by aldosterone in the colon and the kidney is an important mechanism in the body's compensatory response to dehydration. Glucocorticoids also increase the content of Na^+, K^+-ATPase in the basolateral membrane and thereby enhance Na^+ and water absorption and K^+ secretion in the colon.

Opioid peptides. Enkephalins and other opioid peptides increase the absorption of Na^+, Cl^-, and water by intestinal epithelial cells. The cellular mechanism of action of opioids remains to be determined.

Somatostatin. Somatostatin, which is present in certain enteric neurons, enhances the absorption of Na^+, Cl^-, and water in the ileum and the colon.

Secretion of electrolytes and water. The normal net absorption of Na^+, Cl^-, and water is the result of large unidirectional fluxes from lumen to blood and from blood to lumen. Mature intestinal epithelial cells near the tips of the villi are active in net absorption, whereas the more immature epithelial cells in Lieberkühn's crypts function as net secretors. The secretory activities of the crypt cells are also subject to physiological and pharmacological regulation. Secretion by the crypt cells is a normal physiological function. However, because secretion is excessive in certain diarrheal diseases, this topic is discussed in the following section.

Pathophysiological Alterations of Salt and Water Absorption

The general causes of abnormalities in the absorption of salts and water include (1) deficiency of a normal ion transport system, (2) failure to absorb a nonelectrolyte normally with resultant osmotic diarrhea, (3) hypermotility of the intestine leading to abnormally rapid flow of intestinal contents past the absorptive epithelium; and (4) an enhanced rate of net secretion of water and electrolytes by the intestinal mucosa. Examples of each of these classes of abnormalities follow.

Failure to absorb an electrolyte normally. In congenital chloride diarrhea, ion transport and water absorption occur normally in the duodenum and jejunum, which normally lack the Cl^-/HCO_3^- exchanger. However, the $Cl^-HCO_3^-$ exchange transport system in the brush border plasma membrane of the ileum and colon is missing or grossly deficient. As a result, chloride absorption is severely impaired. This leads to diarrhea in which the stools contain an unusually high chloride concentration. In this disease the concentration of Cl^- in the stool exceeds the sum of the concentrations of Na^+ and K^+. The Na^+/H^+ exchange pump continues to operate, so that H^+ is eliminated in the feces without HCO_3^- to neutralize it. The net loss of H^+, with retention of HCO_3^-, contributes to a metabolic alkalosis.

Fig. 44-11. The ion transport pathways involved in secretion of Cl^-, Na^+, and water by the epithelial cells in Lieberkühn's crypts in the small intestine.

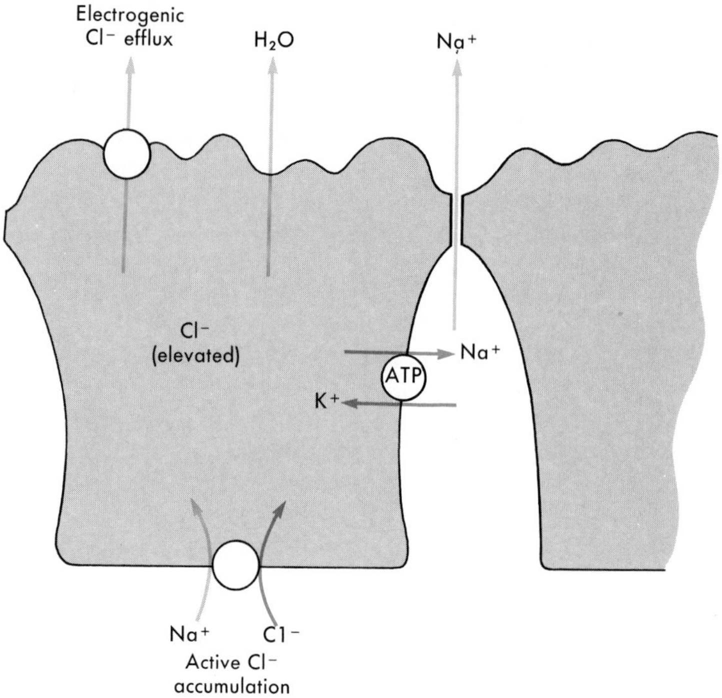

Failure to absorb a nutrient normally. In any of the carbohydrate malabsorption syndromes the sugar that is retained in the lumen of the small intestine increases the osmotic pressure of the luminal contents. Water is retained as a result, and an increased volume of chyme is passed on to the colon. The increased volume flow may overwhelm the ability of the colon to absorb electrolytes and water, resulting in pronounced diarrhea. In addition, the high level of carbohydrates provides a medium that supports increased growth and metabolism of colonic bacteria. The increased production of CO_2 by colonic bacteria contributes to gassiness and borborygmi, and certain products of bacterial metabolism may inhibit absorption of electrolytes by the colonic epithelium.

Hypermotility of the intestine. The causes of hypermotility of the intestine are not well understood. Hypermotility of the small intestine may deliver electrolytes and water to the colon at faster rates than they can be absorbed by the colonic epithelial cells. Consequently, hypermotility of the colon may result in the elimination of feces before the maximum amount of salts and water can be extracted from them. Hypermotility may add to other factors that cause diarrhea. In cases of fat malabsorption, colonic bacteria metabolize lipids and produce certain waste products, such as hydroxylated fatty acids, that enhance the motility of the colon and inhibit salt and water absorption by the colonic epithelium.

Enhanced secretion of water and electrolytes. An increased secretion of water and electrolytes is probably the most important mechanism in serious diarrheal diseases. As mentioned previously, immature epithelial cells in Lieberkühn's crypts normally function to secrete Na^+, Cl^-, and H_2O. When the secretory activities of the crypt cells are elevated, the unidirectional secretory flux may exceed the unidirectional absorptive flux, so that net secretion prevails.

The secretory mechanisms of cells in Lieberkühn's crypts are incompletely characterized. One current view is illustrated in Fig. 44-11. Chloride is believed to be actively accumulated in the crypt epithelial cells, perhaps by an Na^+ gradient–powered transport system in the basolateral membrane. Chloride passively flows through channels in the luminal membrane. The electrogenic transport of Cl^- results in Na^+ also flowing into the intestinal lumen. The osmotic activity of the secreted Na^+ and Cl^- causes water to be secreted into the lumen as well.

The chloride channel in the luminal membrane of crypt epithelial cells is controlled by the cytosolic level of cyclic AMP (cAMP) and Ca^{++}. An increase in either cAMP or Ca^{++} opens the Cl^- channel, which increases the secretion of Cl^- and thereby stimulates Na^+ and water secretion.

Cholera is perhaps the best understood type of secretory diarrhea. The toxin released by *Vibrio cholerae* in the intestine binds to receptors on the luminal surface of the crypt cells of the small intestine. The binding of cholera toxin to its receptors irreversibly activates adenylate cyclase in the plasma membranes of these cells. The resulting increase in cAMP greatly increases the secretion of Na^+, Cl^-, and water by the cells, perhaps by increasing the conductance of the luminal membrane to Cl^-, as discussed previously.

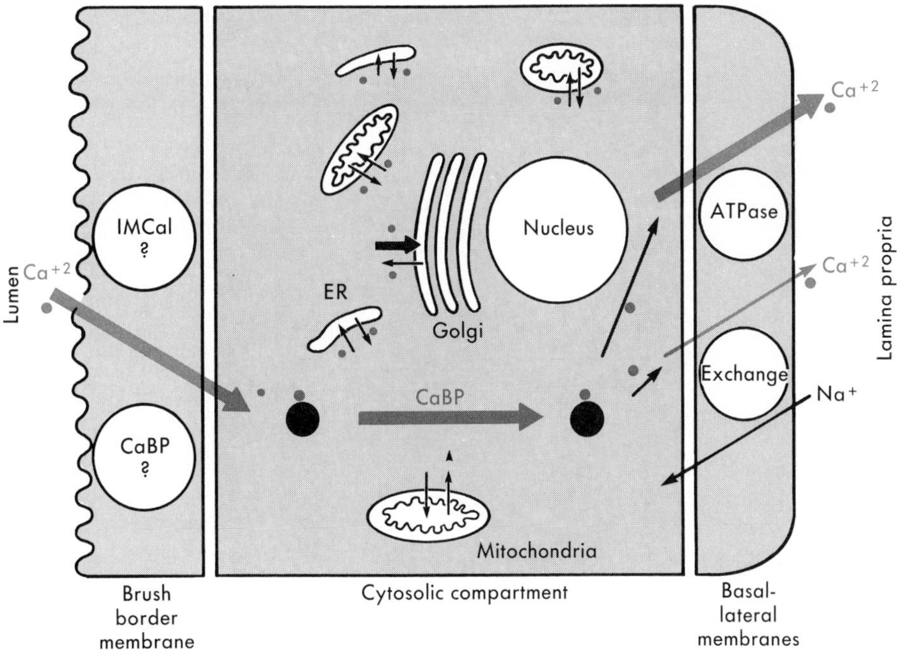

■ Fig. 44-12. Cellular mechanisms of Ca^{++} absorption in the small intestine. Ca^{++} crosses the brush border plasma membrane by pathways that are poorly characterized, but may involve intestinal membrane calcium-binding protein (IMCal) or calcium-binding protein (CaBP). In the cytosol of the enterocyte Ca^{++} is bound to a soluble CaBP. Ca^{++} is extruded across the basolateral membrane by a Ca^{++}-ATPase and a Na^{+}/Ca^{++} exchange mechanism. Orange dots represent Ca^{++}; large black dots represent cytosolic Ca^{++}-binding protein. (From Wasserman, R.H., and Fullmer, C.S.: Annu. Rev. Physiol. **45**:375, 1983.)

Other agents that elevate cAMP in intestinal epithelial cells also lead to secretion of water and electrolytes. Vasoactive intestinal polypeptide, which is present in certain enteric neurons and also circulates as a hormone, can act in this way. Certain individuals with islet cell tumors of the pancreas suffer from a watery diarrhea known as *pancreatic cholera*. In this disorder plasma levels of vasoactive intestinal polypeptide are elevated, and this may be the cause of the secretory diarrhea. Certain prostaglandins augment secretion by elevating cAMP in crypt cells.

The epithelial cells of the crypts of the small intestine are also stimulated to secrete electrolytes and water by elevated intracellular Ca^{++} levels. Acetylcholine, serotonin, substance P, and neurotensin may elicit intestinal secretion of water and electrolytes by increasing intracellular Ca^{++}. All of these agents are present in intrinsic neurons of the intestinal wall, and thus they may produce diarrhea in certain pathological situations and may modulate intestinal secretion under normal circumstances.

■ *Absorption of Calcium*

Calcium ions are actively absorbed by all segments of the intestine. The duodenum and jejunum are especially active and can concentrate Ca^{++} against a greater than tenfold concentration gradient. The rate of absorption of Ca^{++} is much greater than that of any other divalent ion, but still 50 times slower than Na^{+} absorption.

The ability of the intestine to absorb Ca^{++} is regulated. Animals receiving a calcium-deficient diet increase their ability to absorb Ca^{++}. Animals receiving high-calcium diets are less able to absorb Ca^{++}. Intestinal absorption of Ca^{++} is stimulated by vitamin D and slightly stimulated by parathyroid hormone by a mechanism that is not yet understood.

Cellular mechanism of calcium absorption

The brush border membrane. A current view of the cellular mechanism of Ca^{++} absorption by the epithelial cells of the small intestine is shown in Fig. 44-12. Ca^{++} moves down its electrochemical potential gradient across the brush border membrane into the cytosol. An integral protein of the brush border plasma membrane is called the intestinal membrane calcium-binding protein (IMCal), and it may function as a membrane transporter for Ca^{++}. Some of the cytosolic calcium-binding protein (CaBP) is associated with the brush border membrane; its function in the brush border membrane has not been established.

Epithelial cell cytosol. The cytosol of the intestinal epithelial cells contains CaBP. In mammals the CaBP has a molecular weight of about 10,000 and binds two calcium ions with high affinity. As discussed below, the

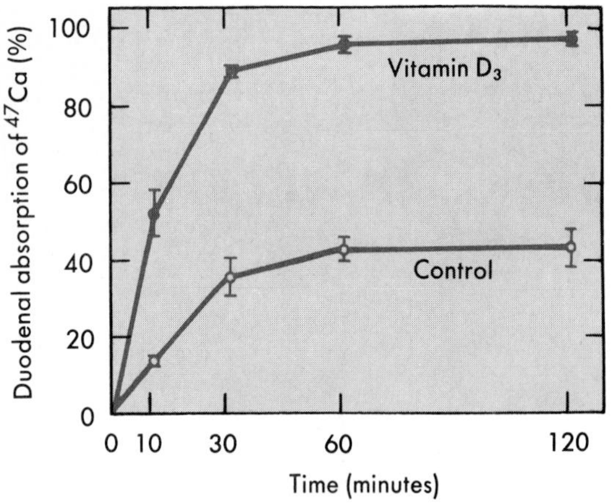

■ **Fig. 44-13.** Effects of vitamin D_3 on absorption of Ca^{++} by chick duodenum. Control animals were fed a diet deficient in vitamin D *(lower curve). Upper curve,* Ca^{++} absorption by animals fed the same diet, but administered vitamin D_3 24 hours before the experiment. (Redrawn from Wasserman, R.H.: J. Nutrition **77:**69, 1962. © J. Nutr. American Institute of Nutrition.)

level of CaBP in the epithelial cells correlates well with the capacity to absorb Ca^{++}. The function of the cytosol CaBP is not known. It has been proposed that CaBP allows large amounts of Ca^{++} to traverse the cytosol, while avoiding concentrations of free Ca^{++} high enough to form insoluble salts with intracellular anions. CaBP appears to be an essential component of Ca^{++} absorption. Free Ca^{++} and Ca^{++} bound to CaBP are in dynamic exchange with Ca^{++} in mitochondria and endoplasmic reticulum, but the role of these organelles in Ca^{++} absorption is not clear.

The basolateral membrane. The basolateral plasma membrane contains two transport proteins capable of ejecting Ca^{++} from the cell against its electrochemical potential gradient. A Ca^{++}-ATPase in the basolateral membrane is a primary active transport protein that splits ATP and uses the energy to transport Ca^{++}. The Na^+/Ca^{++} exchanger present in the basolateral membrane uses the energy of the Na^+ gradient to extrude Ca^{++} by secondary active transport. The Na^+/Ca^{++} exchanger is more effective at high levels of free intracellular Ca^{++}, whereas at low levels of free intracellular Ca^{++} the Ca^{++}-ATPase is the major mechanism for Ca^{++} extrusion. Ca^{++} itself stimulates the activity of the Ca^{++}-ATPase by binding to calmodulin. The calcium-calmodulin complex then stimulates a protein kinase, which phosphorylates the Ca^{++}-ATPase, thereby enhancing its enzymatic and transport activities.

Actions of vitamin D. Vitamin D is essential for normal levels of calcium absorption by the intestine. In *rickets*, a disease due to vitamin D deficiency, the rate of absorption of Ca^{++} is very low. Fig. 44-13 illustrates the effects of administration of vitamin D to chicks with rickets.

The actions of vitamin D are discussed in Chapter 51. Vitamin D may stimulate each phase of absorption of Ca^{++} by the epithelium of the small intestine: passage across the brush border membrane, traversal of the cytosol, and active extrusion across the basolateral membrane. Vitamin D, like other steroid hormones, exerts its major effects by binding to nuclear receptors and stimulating the synthesis of messenger RNA that codes for particular proteins. Vitamin D_3 induces the synthesis of the cytosolic CaBP (Fig. 44-14). The CaBP level correlates well with the capacity of the small intestine to absorb Ca^{++}. Vitamin D also increases the level of the intestinal membrane CaBP that may be involved in transporting Ca^{++} across the brush border plasma membrane. In addition, vitamin D increases the level of the basolateral Ca^{++}-ATPase that actively pumps calcium out of the enterocyte.

■ *Absorption of Iron*

A typical adult in Western societies ingests about 15 to 20 mg of iron daily. Of the 15 to 20 mg ingested, only 0.5 to 1 mg is absorbed by normal adult men and 1 to 1.5 mg is absorbed by premenopausal adult women. Iron depletion, due to hemorrhage for example, results in increased iron absorption. Growing children and pregnant women also absorb increased amounts of iron.

Iron absorption is limited because iron tends to form insoluble salts, such as hydroxide, phosphate, and bicarbonate, with anions that are present in intestinal secretions. Iron also tends to form insoluble complexes with other substances commonly present in food, such as phytate, tannins, and the fiber of cereal grains. These iron complexes are more soluble at low pH. Therefore HCl secreted by the stomach enhances iron absorption, whereas iron absorption is commonly low in individuals deficient in acid secretion. Ascorbate effectively promotes iron absorption. Ascorbate forms a soluble complex with iron, preventing it from forming insoluble complexes, and ascorbate reduces Fe^{3+} to Fe^{2+}. Fe^{2+} has much less tendency to form insoluble complexes than Fe^{3+} and, for this reason, Fe^{2+} is absorbed much better.

Heme iron is relatively well absorbed; about 20% of the heme ingested is absorbed. Proteolytic enzymes release heme groups from proteins in the intestinal lumen. Heme is probably taken up by facilitated transport by the epithelial cells that line the upper small intestine. In the epithelial cell, iron is split from the heme by reactions involving xanthine oxidase. No intact heme is transported into the portal blood.

Cellular mechanism of iron absorption. One current view of iron absorption is depicted in Fig. 44-15. The epithelial cells of the duodenum and jejunum release an iron-binding protein into the lumen. The protein is called transferrin, and it is very similar, but not

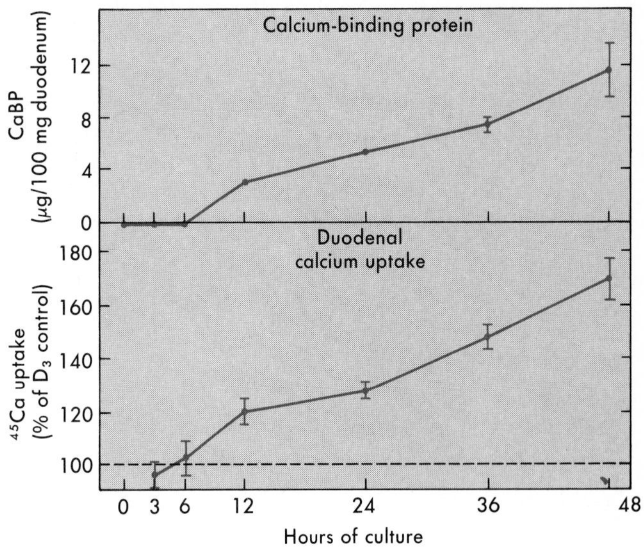

■ **Fig. 44-14.** The correlation between the levels of CaBP and the rate of calcium uptake by chick duodenum in organ culture. At time zero, vitamin D_3 was added to the culture medium. (Redrawn from Corradino, R.A.: Endocrinology **94:**1607, 1974. © 1974, The Endocrine Society.)

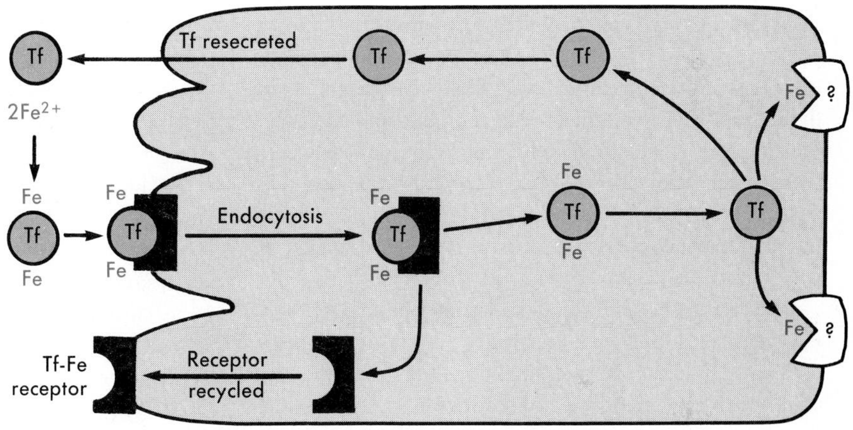

■ **Fig. 44-15.** A current view of the mechanism of iron absorption by the epithelial cells of the small intestine. A form of transferrin *(Tf)* is secreted into the lumen, where it binds Fe^{++}. The Fe_2-Tf complex is taken up by receptor-mediated endocytosis and the receptor and the Tf are recycled. Iron is transported across the basolateral membrane by a poorly understood process and appears in portal blood bound to the plasma form of transferrin.

identical, to the transferrin that is the principal iron-binding protein of plasma. In the lumen of the duodenum and jejunum, transferrin binds iron; two iron ions can be bound by each transferrin molecule. Receptors on the brush border surface of the duodenum and jejunum bind the transferrin-iron complex, and the complex is taken up into the epithelial cell, probably by receptor-mediated endocytosis. Endocytic uptake of iron-transferrin by hepatocytes and reticulocytes also has been demonstrated. In the cytosol of the intestinal epithelial cell, transferrin apparently acts as a soluble iron carrier. Much of the transferrin, after it releases its bound iron, is resecreted into the lumen. Much of the brush border receptor for the iron-transferrin complex is recycled back into the brush border membrane.

Iron ultimately appears in plasma bound to plasma transferrin. Postulated steps that occur between uptake by the enterocyte of the transferrin-iron complex and the appearance of iron in the plasma include release of iron from transferrin in the enterocyte, transport of iron across the basolateral membrane, and binding of iron to plasma transferrin. These processes are poorly understood. Lysosomes may play a role in release of iron from the secreted form of transferrin. The transport of iron across the basolateral membrane requires metabolic energy, but is otherwise undefined.

Regulation of iron absorption. Iron absorption is regulated in accordance with the body's need for iron. In chronic iron deficiency or after hemorrhage, the capacity of the duodenum and jejunum to absorb iron is elevated. The intestine also protects the body from the consequences of absorbing too much iron. The excretion of iron is limited, and thus the absorption of more iron than is needed may lead to iron overload. Iron overload can result from chronic ingestion of large amounts of absorbable iron. This condition is common in certain African tribes that regularly consume a home-brewed beer high in iron content. In the genetic disease called *idiopathic hemochromatosis,* an excessive amount of iron is absorbed from a diet that is normal in iron content.

An important mechanism for preventing excess absorption of iron is the almost irreversible binding of iron to ferritin in the intestinal epithelial cell. Iron

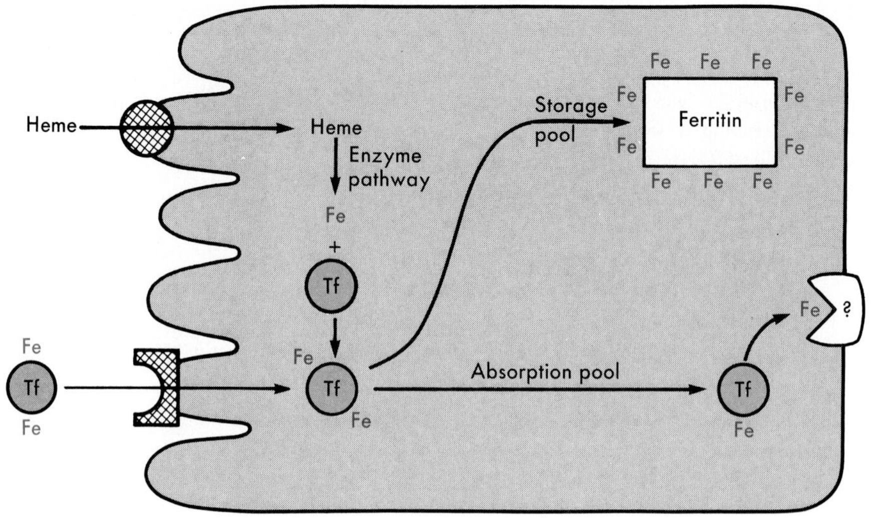

Fig. 44-16. In the enterocytes of the small intestine iron bound to transferrin *(Tf)* is available for transport across the basolateral membrane. Iron bound to ferritin is, however, unavailable for absorption and is lost into the lumen when the cell desquamates.

bound to ferritin is not available for transport into the plasma (Fig. 44-16), but it is lost into the intestinal lumen and excreted in the feces when the intestinal epithelial cell desquamates. The amount of apoferritin present in the intestinal epithelial cells may determine how much iron can be trapped in this nonabsorbable pool. The synthesis of apoferritin is stimulated at the translational level by iron, and this protects against absorption of excessive amounts of iron.

The capacity of the duodenum and jejunum to absorb iron increases after a hemorrhage, with a time lag of 3 to 4 days. The intestinal epithelial cells require this time to migrate from their sites of formation in Lieberkühn's crypts to the tips of the villi, where they are most involved in absorptive activities. The iron-absorbing capacity of the epithelial cells appears to be programmed when the cells are in Lieberkühn's crypts. The mechanisms that require the capacity of the epithelial cells to absorb iron are poorly characterized. Iron absorption by duodenal and jejunal epithelial cells is subject to multiple controls. The brush border membranes of the duodenum and jejunum of an iron-deficient animal have an increased number of receptors for the complex of iron with transferrin and thus absorb the iron-transferrin complex from the lumen more rapidly. Other regulatory factors are probably also involved.

Absorption of Other Ions

Magnesium. Magnesium is absorbed along the entire length of the small intestine, with about half the normal dietary intake being absorbed. The regulation of the rate of Mg^{++} absorption is poorly understood.

Phosphate. Phosphate is also absorbed all along the small intestine. Some phosphate may be absorbed by active transport. Little is known about the regulation of phosphate absorption.

Copper. Copper is absorbed in the jejunum, with approximately 50% of the ingested load being absorbed. Copper is secreted in the bile bound to certain bile acids, and this copper is lost in the feces. In individuals that fail to secrete sufficient amounts of copper in the bile, the body's copper pool grows and copper accumulates in certain tissues. The mechanisms by which absorption and excretion of copper are regulated are poorly characterized.

Absorption of Water-Soluble Vitamins

Most water-soluble vitamins can be absorbed by simple diffusion if they are taken in sufficiently high doses. Nevertheless, specific transport mechanisms play important roles in the normal absorption of most water-soluble vitamins. Table 44-3 summarizes current knowledge of these transport mechanisms.

Ascorbic acid (vitamin C). Vitamin C is absorbed in the proximal ileum by active transport. The active transport of ascorbate depends on the presence of Na^+ in the lumen. Na^+ and ascorbate are probably cotransported into the cell and the energy for ascorbate transport comes from the energy in the electrochemical potential gradient of Na^+.

Biotin. Biotin is actively taken up by the epithelial cells of the upper small intestine by a mechanism that depends on Na^+ in the lumen.

Folic acid (pteroylglutamic acid). Folic acid is absorbed by carrier-mediated transport (perhaps by active transport) in the jejunum. Pteroylpolyglutamates apparently share this transport pathway. The polyglutamates are cleaved intracellularly to the monoglutamate deriv-

Table 44-3. Intestinal absorption of vitamins

Vitamin	*Species*	*Site of absorption*	*Transport mechanism*	*Maximum absorption capacity in humans (per day)*	*Dietary requirements in humans (per day)*
Ascorbic acid (C)	Humans, guinea pig	Ileum	Active	> 5000 mg	< 50 mg
Biotin	Hamster	Upper small intestine	Active	?	?
Choline	Guinea pig, hamster	Small intestine	Facilitated	?	?
Folic acid					
Pteroylglutamate	Rat	Jejunum	Facilitated	> 1000 μg/dose	100-200 μg
5-Methyltetra-hydrofolate	Rat	Jejunum	Diffusion		
Nicotinic acid	Rat	Jejunum	Facilitated	?	10-20 mg
Pantothenic acid		Small intestine	?	?	(?) 10 mg
Pyridoxine (B_6)	Rat, hamster	Small intestine	Diffusion	> 50 mg/dose	1-2 mg
Riboflavin (B_2)	Humans, rat	Jejunum	Facilitated	10-12 mg/dose	1-2 mg
Thiamin (B_1)	Rat	Jejunum	Active	8-14 mg	≈ 1 mg
Vitamin B_{12}	Humans, rat, hamster	Distal ileum	Active	6-9 μg	3-7 μg

Data from Matthews, D.M., In Smyth, D.H., editor: Intestinal absorption, vol. 4B (Biomembranes), London, 1974, Plenum; and Rose, R.C.: Annu. Rev. Physiol. **42:**157, 1980.

ative, which is transported into the blood. Another dietary source of folic acid, 5-methyltetrahydrofolate, is absorbed by simple diffusion.

Nicotinic acid. Nicotinic acid is absorbed by the jejunum partly by an Na^+-dependent, saturable mechanism. The details of nicotinic acid absorption remain to be elucidated.

Pyridoxine (vitamin B_6). Evidence favors simple diffusion as the mechanism of pyridoxine absorption.

Riboflavin (vitamin B_2). Riboflavin is absorbed in the proximal small intestine, probably by facilitated transport. The presence of bile acids enhances the absorption of riboflavin by an unknown mechanism.

Thiamin (vitamin B_1). Thiamin is absorbed by an Na^+-dependent active transport mechanism in the jejunum. Some thiamin is phosphorylated in the jejunal epithelial cells, but the thiamin that appears in the blood is primarily free thiamin.

Vitamin B_{12}. A specific active transport process also has been implicated in the absorption of vitamin B_{12}. The four physiologically important forms of B_{12} are cyanocobalamin, hydroxycobalamin, deoxyadenosylcobalamin, and methylcobalamin. In the absence of vitamin B_{12} the maturation of red blood cells is retarded, and *pernicious anemia* ensues. Because of its medical importance, a good deal of attention has been paid to the absorption of vitamin B_{12}. The dietary requirement for B_{12} is fairly close to the maximal absorption capacity for the vitamin (Table 44-3). Enteric bacteria synthesize vitamin B_{12} and other B vitamins, but the colonic epithelium lacks specific mechanisms for their absorption.

Storage in liver. The liver contains a large store of vitamin B_{12} (2 to 5 mg). Vitamin B_{12} is normally present in the bile (0.5 to 5 μg daily), but about 70% of this is normally reabsorbed. Since only about 0.1% of the store is lost daily, even if absorption totally ceases the store will last for 3 to 6 years.

Gastric phase. Most of the cobalamins present in food are bound to proteins. The low pH in the stomach and the digestion of proteins by pepsin release free cobalamins. The free cobalamins are rapidly bound to a number of cobalamin-binding glycoproteins known as R proteins. R proteins are present in saliva and in gastric juice and bind cobalamins tightly over a wide pH range. The R proteins of saliva and gastric juice have molecular weights near 60,000 and are probably closely related to transcobalamins I, II, and III.

Intrinsic factor (IF) is a cobalamin-binding protein that is secreted by the gastric parietal cells. IF is a glycoprotein with 15% carbohydrate and a molecular weight of about 45,000. The rate of IF secretion usually parallels the rate of HCl secretion. IF binds cobalamins with less affinity than the R proteins, so in the stomach most of the cobalamin present in food is bound to R proteins.

Intestinal phase. Pancreatic proteases begin degradation of the complexes between R proteins and cobalamins. This degradation greatly lowers the affinities of the R proteins for cobalamins, so that cobalamins are transferred to IF. IF and IF-cobalamin complexes are very resistant to digestion by pancreatic proteases. As described below, the normal mechanism for absorption

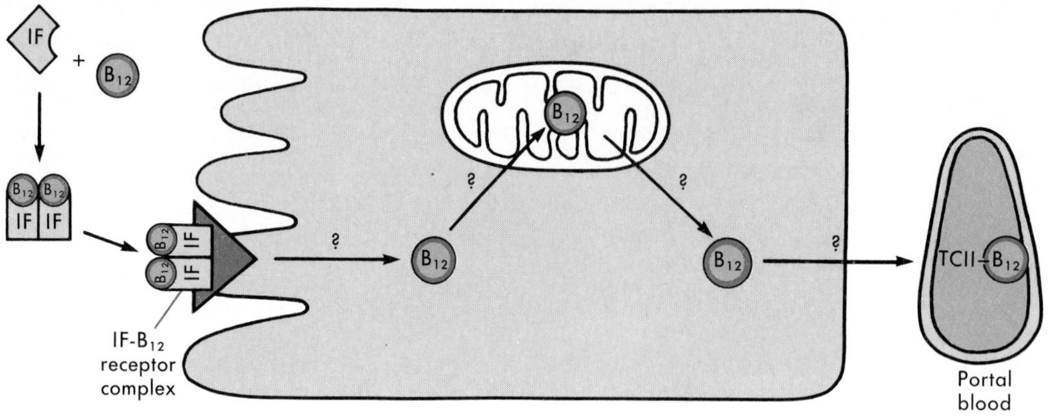

Fig. 44-17. Mechanism of vitamin B_{12} absorption by epithelial cells of the ileum.

involves brush border receptors for IF-cobalamin complexes. These receptors do not recognize the R protein–cobalamin complexes. Thus in pancreatic insufficiency, when R proteins are not degraded, cobalamins remain bound to R proteins. These complexes are not available for absorption, and therefore vitamin B_{12} deficiency may ensue.

Absorption of vitamin B_{12}. Fig. 44-17 summarizes a current view of the mechanism of vitamin B_{12} absorption. The normal absorption of cobalamins depends on the presence of IF. When IF binds vitamin B_{12} the IF undergoes a conformational change that favors the formation of dimers; each dimer binds two vitamin B_{12} molecules. The brush border plasma membranes of the epithelial cells of the ileum contain a receptor protein that recognizes and binds the IF-B_{12} dimer. Free IF does not compete for binding and the receptor does not recognize free cobalamins.

Binding to the receptor is required for uptake of vitamin B_{12} into the cell. It is not known whether the IF-B_{12} complex or B_{12} alone enters the ileal epithelial cell. The most convincing evidence favors the hypothesis that B_{12} is split from IF extracellularly and that free B_{12} enters the cell by an active transport mechanism.

After the binding of the IF-B_{12} complex to the ileal receptors, B_{12} is slowly transported through the epithelial cell and into the blood. Vitamin B_{12} does not appear in the blood until 4 hours after it is fed, and the peak B_{12} level in plasma occurs 6 to 8 hours after feeding. The reason for this delay is not well understood, but some evidence suggests that for much of the lag period B_{12} is located predominantly in the mitochondria of the epithelial cells and that some conversion of cobalamin to deoxyadenosylcobalamin occurs there.

The exit of vitamin B_{12} from the cells of the ileal epithelium is even less well understood than its entry. Facilitated or active transport is presumably involved. Most of the B_{12} absorbed appears in the portal blood bound to transcobalamin II, a globulin. Transcobalamin II may be synthesized in the liver, but the ileal epithelium may also make this protein. The transcobalamin II–B_{12} complex is rapidly cleared from the portal blood by the liver by receptor-mediated endocytosis. Other tissues that take up the transcobalamin-B_{12} complex include the kidney, spleen, heart, and placenta. Reticulocytes and fibroblasts also take up the complex.

Absorption in the absence of IF. In the complete absence of IF, about 1% to 2% of an ingested load of B_{12} will be absorbed. If massive doses of B_{12} are taken (about 1 mg/day), enough can be absorbed to treat pernicious anemia. The IF-independent mechanism shows no maximal absorptive capacity, does not appear to be limited to the ileum, and shows a much shorter lag time (about 1 hour) than IF-dependent absorption.

Pernicious anemia and other diseases involving the malabsorption of vitamin B_{12}. In the absence of sufficient levels of B_{12} the maturation of red cells is retarded, and anemia results. Pernicious anemia is due to atrophy of the gastric mucosa, with almost complete inability to secrete HCl, pepsin, and IF. Most patients with pernicious anemia have serum antibodies against parietal cells. However, it is not clear whether the antibodies cause the disease or are the response to gastric damage from some other cause.

Pernicious anemia in childhood is rare and has three forms: (1) an autoimmune type of pernicious anemia that has the characteristics just described; (2) congenital IF deficiency, in which pepsin and acid secretion are normal but IF secretion is deficient; and (3) congenital B_{12} malabsorption syndrome, in which there is normal gastric function and normal levels of IF are secreted, but B_{12} absorption is deficient because of a defect in the ileal IF-B_{12} receptors.

Digestion and Absorption of Lipids

The primary lipids of a normal diet are triglycerides. The diet contains smaller amounts of sterols, sterol esters, and phospholipids (Fig. 44-18). Because lipids are

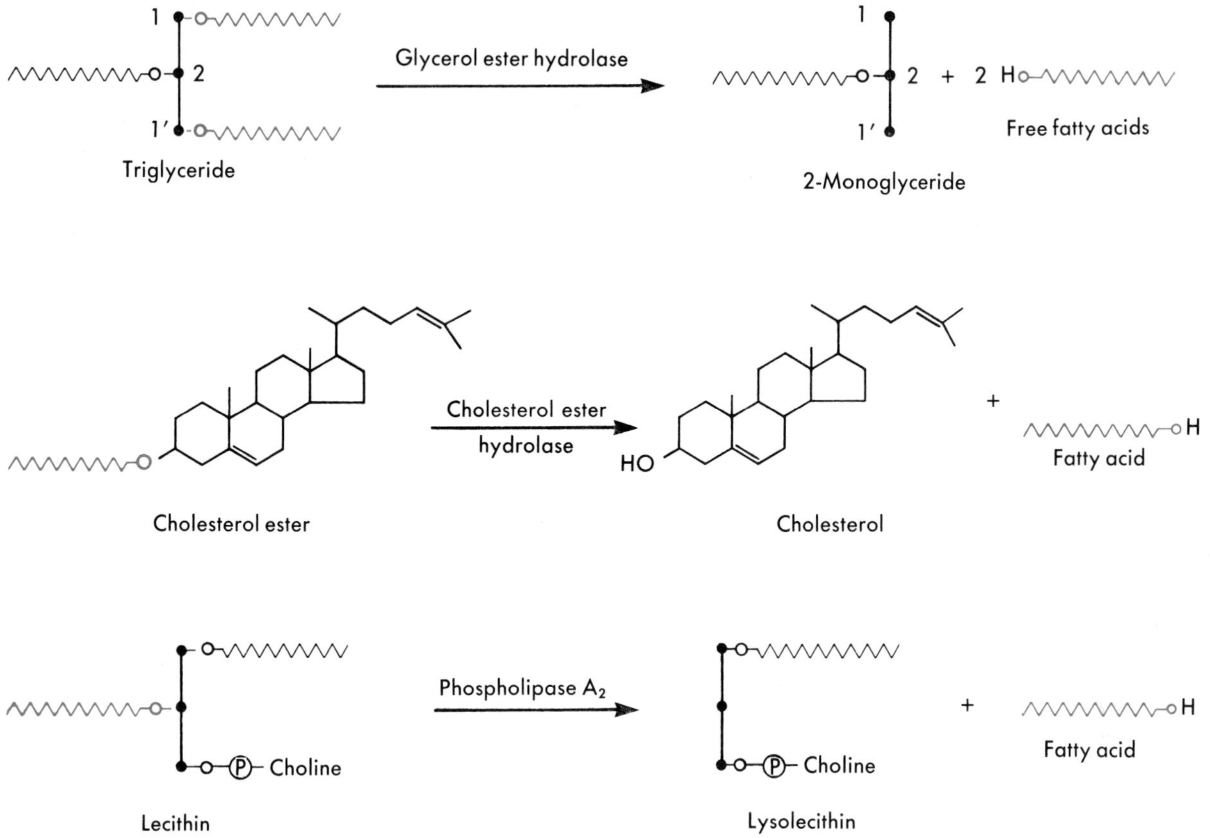

■ **Fig. 44-18.** Action of major pancreatic lipases. The cleavage of lipids by glycerol ester hydrolase (pancreatic lipase), cholesterol ester hydrolase, and phospholipase A_2 is illustrated.

only slightly soluble in water, they pose special problems to the gastrointestinal tract at every stage of their processing. In the stomach, lipids tend to separate out into an oily phase. In the duodenum and small intestine, lipids are emulsified with the aid of bile acids. The large surface area of the emulsion droplets allows access of the water-soluble lipolytic enzymes to their substrates. The digestion products of lipids form small molecular aggregates, known as micelles, with the bile acids. The micelles are small enough to diffuse among the microvilli and allow absorption of the lipids from molecular solution at the intestinal brush border. The digestion and absorption of lipids are more complex than for any other class of nutrients and are thus more frequently subject to malfunction.

In the Stomach

Because fats tend to separate out into an oily phase, they tend to be emptied from the stomach later than the other gastric contents. Despite the presence of gastric lipase, little digestion of lipids occurs in the stomach. Any tendency to form emulsions with phospholipids or other natural emulsifying agents is inhibited by the high acidity. Fat in the duodenum strongly inhibits gastric emptying. This ensures that the fat is not emptied from the stomach more rapidly than it can be accommodated by the duodenal mechanisms that provide for emulsification and digestion.

Digestion of Lipids and Micelle Formation

The lipolytic enzymes of the pancreatic juice are water-soluble molecules and thus have access to the lipids only at the surfaces of the fat droplets. The surface available for digestion is increased many thousand times by emulsification of the lipids. Bile acids themselves are rather poor emulsifying agents. However, with the aid of lecithin, which is present in high concentration in the bile, the bile acids emulsify dietary fats. The *emulsion droplets* are around 1 μm in diameter, and they have a large surface area on which the digestive enzymes can work.

The pancreatic secretions. Pancreatic juice contains the major lipolytic enzymes responsible for digestion of lipids (Fig. 44-18). The most important digestive enzymes are as follows:

1. *Glycerol ester hydrolase* (also called simply pancreatic lipase), which cleaves the 1 and 1′ fatty

acids preferentially off a triglyceride to produce two free fatty acids and one 2-monoglyceride

2. *Cholesterol esterase,* which cleaves the ester bond in a cholesterol ester to give one fatty acid and free cholesterol
3. *Phospholipase A_2,* which cleaves the ester bond at the 2 position of a glycerophosphatide to yield, in the case of lecithin, one fatty acid and one lysolecithin

The formation of micelles. Bile acids form micelles with the products of fat digestion, especially 2-monoglycerides. The micelles are multimolecular aggregates (about 5 nm in diameter) containing about 20 to 30 molecules. 2-Monoglycerides and lysophosphatides tend to have their hydrophobic acyl chains in the interior of the micelle and their more polar portions facing the surrounding water. Bile acids are flat molecules that have a polar face and a nonpolar face (Fig. 43-31). Much of the surface of the micelles is covered with bile acids, with the nonpolar face toward the lipid interior of the micelle and the polar face toward the outside. Extremely hydrophobic molecules, such as long-chain fatty acids, cholesterol, and certain fat-soluble vitamins, tend to partition into the interior of the micelle. Phospholipids and monoglycerides tend to have their more polar ends facing the outside aqueous phase. Micelles contain almost no intact triglyceride.

Bile acids must be present at a certain minimum concentration, called the *critical micelle concentration,* before micelles will form. Conjugated bile acids have a much lower critical micelle concentration than the unconjugated forms. In the normal state bile acids are always present in the duodenum at greater than the critical micelle concentration.

Lipids and lipid digestion products in the micelles are in rapid exchange with lipid digestion products in the aqueous solution surrounding the micelle. In this way the micelles keep the aqueous solution surrounding them saturated with 2-monoglycerides, various fatty acids, cholesterol, and lysophosphatides. These lipids are present in the aqueous solution at a low concentration because of their limited water solubility.

Primary Enzymes of Lipid Digestion

Glycerol ester hydrolase (pancreatic lipase). The glycerol ester hydrolase of pancreatic juice has a molecular weight of about 50,000, and it is rather specific for triglycerides. Pancreatic lipase has very low activity against triglycerides in molecular solution. It is active on droplets or emulsions of triglycerides. The enzyme operates at the interface between the aqueous phase and the triglyceride-containing oil phase. Its activity is proportional to the surface area of the oil phase. There is a huge excess of pancreatic lipase. Based on samples of duodenal contents, sufficient pancreatic lipase is present during lipid digestion to hydrolyze in 1 minute an amount of trigyceride almost as large as the average daily intake of triglyceride.

Colipase. Pancreatic lipase is essentially completely inactivated by bile salts at physiological concentrations. A protein of 10,000 molecular weight, known as *colipase,* is present in pancreatic juice. Colipase is able to relieve the inactivation of lipase by bile salts. Bile salts inhibit the activity of pancreatic lipase by binding to the surface of triglyceride-containing oil droplets, thereby preventing the pancreatic lipase from binding. Colipase displaces bile salts from the surface of oil droplets. One pancreatic lipase molecule then binds to each colipase molecule. Lipase binds to colipase more tightly in the presence of droplets of triglyceride. The lipase-colipase complex is very active in cleaving the 1 and 1′ fatty acids from triglycerides at the surface of the droplet. Colipase also has a site that binds to bile salt micelles. This suggests that a lipase-colipase complex may simultaneously bind to a fat droplet and to a bile salt micelle. This binding would allow a direct transfer of the products of lipase action, and other contents of the fat droplet, to the bile salt micelle.

Cholesterol esterase. Cholesterol esterase is probably identical to a nonspecific enzyme, known as *nonspecific lipase,* that cleaves fatty acid ester linkages in a variety of lipid substrates. In humans nonspecific lipase has a molecular weight of about 100,000. The enzyme forms dimers in the presence of bile salts, and in this form it is protected from proteolytic digestion. The dimeric enzyme is active in hydrolyzing fatty acids from cholesterol esters, lysophospholipids, triglycerides, 2-monoglycerides, and fatty acyl esters of vitamins A, D, and E. The total activity of nonspecific lipase in pancreatic juice is small compared to the activity of pancreatic lipase.

Phospholipase A_2. Phospholipase A_2 is secreted as a proenzyme by the pancreas. Tryptic cleavage activates phospholipase A_2 against phospholipids emulsified by bile salts. Phospholipase A_2 is a highly stable protein with a molecular weight of 14,000. The enzyme requires calcium ions for activity. Phospholipase A_2 hydrolyzes the fatty acid at the number 2 position of phosphatidylcholine, phosphatidylethanolamine, phosphatidylserine, phosphatidylglycerol, and cardiolipin to yield free fatty acids and lysophosphatides. Sphingolipids and glycosphingolipids are not good substrates for phospholipase A_2. The fatty acyl ester linkage at the number 1 position is not cleaved by this enzyme and is hydrolyzed poorly by pancreatic lipase; therefore lysophosphatides are the principal form in which phospholipids are absorbed. Phospholipase A_2 is much more active against phospholipids in bile acid–mixed micelles than against phospholipid bilayers.

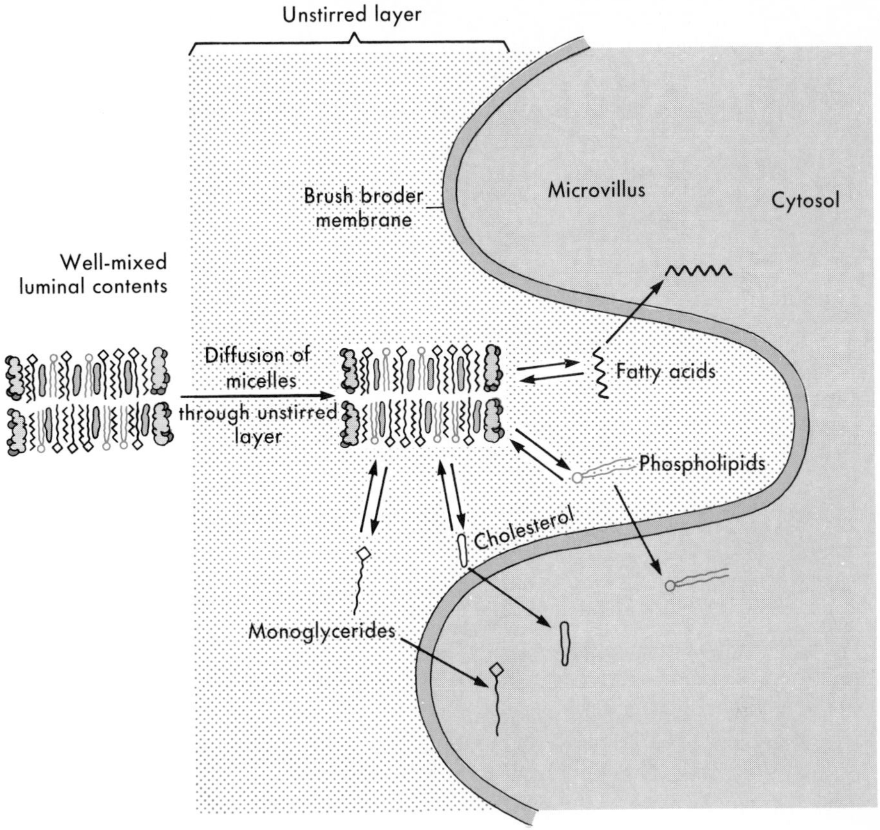

■ Fig. 44-19. Lipid absorption in the small intestine. Mixed micelles of bile acids and lipid digestion products diffuse through the unstirred layer and among the microvilli. As digestion products are absorbed from free solution by the enterocytes, more digestion products partition out of the micelles.

■ *Absorption of the Products of Lipid Digestion*

Transport into the intestinal epithelial cell. The micelles are important in the absorption of the products of lipid digestion and in the absorption of most other fat-soluble molecules (such as the fat-soluble vitamins). The micelles diffuse among the microvilli that form the brush border. The presence of micelles tends to keep the aqueous solution in contact with the brush border saturated with fatty acids, 2-monoglycerides, cholesterol, and other micellar contents. In this way the huge surface area of the brush border is made available for the absorption of the micellar contents (Fig. 44-19).

Because of their high lipid solubility, the fatty acids, 2-monoglycerides, cholesterol, and lysolecithin can readily diffuse across the brush border membrane. There are no known protein-mediated mechanisms for transporting the products of lipid digestion across the plasma membrane of the intestinal epithelial cells. Free fatty acids, 2-monoglycerides, and the other products of lipid digestion can diffuse across the brush border plasma membrane so rapidly that this step does not limit the rate of their uptake. The main limitation to the rate of lipid uptake by the epithelial cells of the upper small intestine is the diffusion of the mixed micelles through an *unstirred layer* (or diffusion boundary layer) on the luminal surface of the brush border plasma membrane. Partly because of the convoluted surface of the intestinal mucosa, the fluid in immediate contact with the epithelial cell surface is not readily mixed with the bulk of the luminal contents. The effective thickness of the unstirred layer ranges from 200 to 500 μm. Nutrients present in the well-mixed contents of the intestinal lumen must diffuse through the unstirred layer to reach the brush border plasma membrane (Fig. 44-19). A concentration gradient exists across the unstirred layer, with micelles and lipid digestion products in lower concentration at the brush border surface than in the well-mixed contents of the lumen. A pH gradient also exists across the unstirred layer, such that the fluid in immediate contact with the brush border plasma membrane is about 1 pH unit more acidic than the bulk luminal contents. The lower pH at the brush border surface may enhance absorption of fatty acids, because protonated fatty acids are more lipid soluble than ionized ones.

Cholesterol is absorbed more slowly than most of the other constituents of the micelles. Therefore, as the micelles progress down the small intestine, they become more concentrated in cholesterol. The duodenum and jejunum are most active in fat absorption, and most ingested fat is absorbed by the midjejunum. The fat present in normal stools is not ingested fat (which is com-

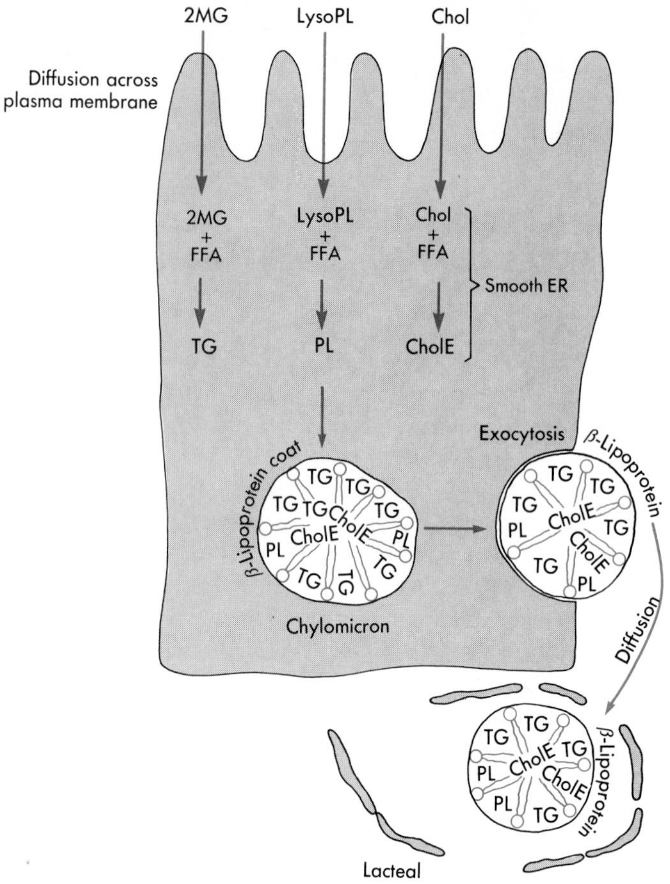

■ **Fig. 44-20.** Lipid resynthesis in the epithelial cells of the small intestine, chylomicron formation, and subsequent transport of chylomicrons. *FFA,* Free fatty acid; *2MG,* 2-monoglyceride; *TG,* triglyceride; *lysoPL,* lysophospholipid; *PL,* phospholipid; *Chol,* cholesterol; *CholE,* cholesterol ester.

pletely absorbed), but fat from colonic bacteria and from desquamated intestinal epithelial cells.

Inside the intestinal epithelial cell. The products of lipid digestion make their way to the smooth endoplasmic reticulum. A recently discovered cytoplasmic fatty acid–binding protein may play a role in transporting fatty acids to the smooth endoplasmic reticulum, or it may simply function to prevent fatty acids from forming fat droplets prematurely. The fatty acid–binding protein has a molecular weight of 12,000. It has higher affinity for unsaturated fatty acids than for saturated ones and binds long-chain fatty acids more tightly than short- or medium-chain fatty acids.

In the smooth endoplasmic reticulum, which is engorged with lipid after a meal, considerable chemical reprocessing goes on (Fig. 44-20). The 2-monoglycerides are reesterified with fatty acids at the 1 and 1′ carbons to re-form triglycerides. Lysophospholipids are reconverted to phospholipids. Cholesterol is reesterified to a considerable extent, although some free cholesterol remains. The processing of 2-monoglycerides and lysophospholipids is essentially complete, since the blood levels of these components are negligible. The intestinal epithelial cells are also capable of de novo synthesis of lipids to some extent.

Chylomicron formation and transport. The reprocessed lipids, along with those that are synthesized de novo, accumulate in the vesicles of the smooth endoplasmic reticulum. Phospholipids tend to cover the external surfaces of these lipid droplets, with their hydrophobic acyl chains in the fatty interior and their polar head groups toward the aqueous exterior. The lipid droplets, of the order of 1 nm in diameter at this point, are known as *chylomicrons*. About 10% of their surface is covered by β-lipoprotein, which is synthesized in the intestinal epithelial cells.

Chylomicrons are ejected from the cell by exocytosis (Fig. 44-20). The β-lipoprotein is important in this process; when it is absent, the intestinal epithelial cells become engorged with lipid, and lipid absorption is severely impaired. The chylomicrons leave the cells at the level of the nuclei and enter the lateral intercellular spaces. Chylomicrons are too large to pass through the basement membrane that invests the mucosal capillaries. However, they do enter the lacteals, which have sufficiently large fenestrations for the chylomicrons to pass through. The chylomicrons leave the intestine with the lymph, primarily via the thoracic duct, and are dumped into the venous circulation.

Chylomicrons are approximately spherical and vary greatly in size (60 to 750 nm). When large amounts of lipid are being absorbed, large chylomicrons are formed; when little lipid is being absorbed the chylomicrons formed tend to be small. Triglyceride is the predominant component, comprising 85% to 92% of the chylomicrons. Phospholipids cover about 80% of the surface of the chylomicrons and account for 6% to 8% of their mass. Biliary phospholipids are used preferentially for this purpose. Apolipoproteins cover the remaining 20% of the chylomicron surface. Cholesterol and cholesterol esters are present in the triglyceride-rich core of the particles, each comprising about 1% of the chylomicron mass. Apolipoproteins of the C class constitute about 50% of the protein present in chylomicrons. There are smaller proportions of apo A-I (15% to 35%), apo A-IV (10%), apo B (10%), and apo E (5%).

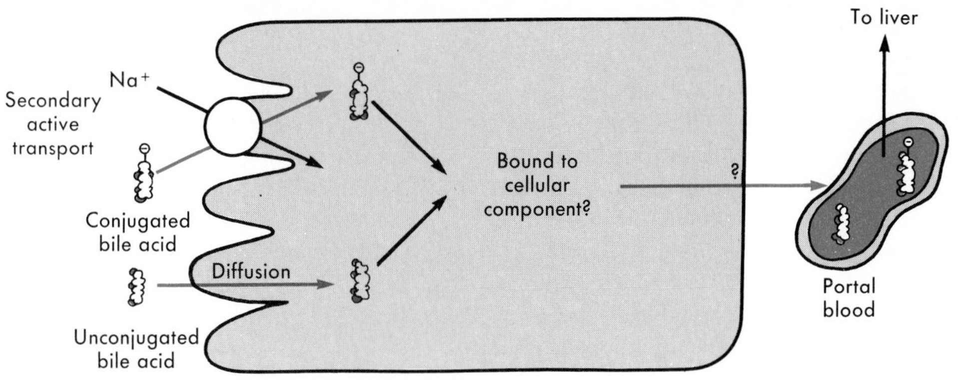

■ **Fig. 44-21.** Absorption of bile acids by epithelial cells of the terminal ileum. Bile acids are absorbed both by simple diffusion and by Na^+-powered secondary active transport. Conjugated bile acids are absorbed avidly by active transport. Unconjugated bile acids are absorbed chiefly by simple diffusion.

Synthesis by enterocytes of apolipoproteins is required for normal absorption of chylomicrons. A graphic demonstration of the role of apolipoproteins occurs in the disease known as *abetalipoproteinemia,* in which enterocytes are deficient in their ability to synthesize apo B. In this disease the transport of chylomicrons from enterocytes to intestinal lymph is greatly diminished, with the result that the enterocytes become engorged with nascent chylomicrons.

Many, but not all, of the apoproteins associated with chylomicrons present in intestinal lymph are synthesized by intestinal epithelial cells. Hepatocytes are the other major source of apolipoproteins. The major source of certain of these apoproteins is controversial. It is becoming clear that the intestine is an important source of certain apoproteins of plasma lipoproteins. Recent evidence suggests that even when fat is absent from the intestine, significant amounts of apo A-I and apo B may be secreted into intestinal lymph. The view is emerging that the intestine plays a major role in the body's synthesis and metabolism of serum lipoproteins and that these intestinal processes are subject to multiple levels of regulation.

■ *Absorption of Bile Acids*

The absorption of dietary lipids is typically complete by the midjejunum. Bile acids, by contrast, are absorbed largely in the terminal part of the ileum. As for other fat-soluble substances, the unstirred layer is an important barrier to bile acid absorption. Bile acids cross the brush border plasma membrane by two routes: by an active transport process and by simple diffusion (Fig. 44-21). The active process is secondary active transport, powered by the Na^+ gradient across the brush border membrane. In this process one Na^+ ion is cotransported across the brush border plasma membrane with one bile acid molecule. Conjugated bile acids are the principal substrates for active absorption; deconjugated bile acids have poor affinity for the transporter. Deconjugated bile acids are less polar than conjugated bile acids, and are thus better absorbed by simple diffusion. The fewer hydroxyl groups on a bile acid, the better substrate it is for active absorption. For this reason, dehydroxylation of bile acids by enteric bacteria to form secondary bile acids enhances active absorption of bile acids by diffusion.

Other aspects of the absorption of bile acids are less well understood. It is suspected that bile acids are bound to cytosolic structures or macromolecules, which remain to be identified, in intestinal epithelial cells. The process by which bile acids traverse the basolateral plasma membrane of the enterocyte has not been characterized.

Absorbed bile acids are carried away from the intestine in the portal blood. Hepatocytes avidly extract bile acids, essentially clearing them from the blood in a single pass through the liver. In the hepatocytes most deconjugated bile acids are reconjugated, and some secondary bile acids are rehydroxylated. The reprocessed bile acids, together with newly synthesized bile acids, are secreted into bile.

■ *Malabsorption of Lipids*

Malabsorption of lipids occurs more frequently than malabsorption of proteins or carbohydrates. Among the general causes of lipid malabsorption are bile deficiency, pancreatic insufficiency, and the intestinal mucosal atrophy that occurs in some disease states.

In cases of *bile deficiency* and *pancreatic insufficiency* the levels of bile acids and lipolytic enzymes, respectively, must be severely reduced before serious malabsorption occurs. In both cases the quantity of fe-cal fat is roughly proportional to the quantity ingested.

Even in the *complete absence of bile acids,* significant hydrolysis of triglyceride occurs. The rate of absorption of fatty acids from triglycerides may be 50% of normal. Cholesterol, cholesterol esters, and fat-soluble vitamins are much less water soluble than fatty acids, and their absorption is grossly deficient in the absence of bile acids.

In the *complete absence of pancreatic lipases* all lipid classes are poorly absorbed. This is probably because of the necessity of 2-monoglycerides and lysophosphatides (products of the action of pancreatic lipases) for the formation of mixed micelles with bile acids.

In *tropical sprue* and *gluten enteropathy* the intestinal epithelium is flattened and the density of microvilli is decreased. Lipid malabsorption in these diseases is probably a consequence of the marked decrease in the surface area available for lipid absorption.

Absorption of Fat-Soluble Vitamins

Because of their solubility in nonpolar solvents, the fat-soluble vitamins (A, D, E, and K) partition into the mixed micelles formed by the bile acids and lipid digestion products. As is the case for other lipids, the fat-soluble vitamins enter the intestinal epithelial cell by diffusing across the brush border plasma membrane. There is no evidence that the passage of fat-soluble vitamins across the brush border plasma membrane is mediated by membrane proteins. In general, the presence of bile acids and lipid digestion products enhances the absorption of fat-soluble vitamins. In the intestinal epithelial cell the fat-soluble vitamins enter the chylomicrons and leave the intestine in the lymph. In the absence of bile acids a significant fraction of the ingested load of a fat-soluble vitamin may be absorbed and leave the intestine in the portal blood.

Vitamin A. Vitamin A (retinol) is better absorbed than is provitamin A (β-carotene) and vitamin A aldehyde (retinal). In the epithelial cells of the intestine, β-carotene and retinal are converted to retinol. Vitamin A is present in thoracic duct lymph primarily as fatty acid esters of retinol. Vitamin A acid (retinoic acid) is absorbed independently of the bile acid–mixed micelles and leaves the intestine in the portal blood.

Vitamin D. Vitamin D is absorbed principally in the jejunum, primarily as the free vitamin. Most fatty acid esters of vitamin D are hydrolyzed in the intestinal lumen prior to absorption. With normal oral loads, 55% to 99% of the vitamin D ingested is absorbed.

Vitamin E. The absorption of vitamin E (α-tocopherol) requires the presence of bile acid–mixed micelles. Small oral doses of vitamin E are almost completely absorbed. With larger doses, a significant fraction escapes absorption and appears in the feces. In the intestinal epithelial cell, vitamin E enters the chylomicrons and leaves the intestine in the lymph.

Vitamin K. Vitamins K_1 and K_2, which have hydrophobic side chains, partition into bile acid–mixed micelles, are absorbed from free solution, enter the chylomicrons, and leave the intestine in the lymph. Vitamin K_3 (2-methylnaphthoquinone) lacks a side chain, is absorbed independently of the mixed micelles, and leaves the intestine primarily in the portal blood.

Bibliography

Journal articles

Adibi, S.A., and Morse, E.L.: The number of glycine residues which limit intact absorption of glycine oligopeptides in human jejunum, J. Clin. Invest. **60**:1008, 1977.

Binder, H.J.: The pathophysiology of diarrhea, Hosp. Pract. **19** (Oct.):107, 1984.

Bisgaier, C.L., and Glickman, R.M.: Intestinal synthesis, secretion, and transport of lipoproteins, Annu. Rev. Physiol. **45**:625, 1983.

Charlton, R.W., and Bothwell, T.H.: Iron absorption, Annu. Rev. Med. **34**:55, 1983.

Gray, G.M., et al.: Sucrase-isomaltase deficiency: absence of an inactive enzyme variant, N. Engl. J. Med. **294**:750, 1976.

Rose, R.C.: Water-soluble vitamin absorption in intestine, Annu. Rev. Physiol. **42**:157, 1980.

Schultz, S.G.: A cellular model for active sodium absorption by mammalian colon, Annu. Rev. Physiol. **46**:435, 1984.

Seetharam, B., and Alpers, D.H.: Absorption and transport of cobalamin (vitamin B_{12}), Annu. Rev. Nutr. **2**:343, 1982.

Shiau, Y.-F.: Mechanisms of intestinal fat absorption, Am. J. Physiol. **240**:G1, 1981.

Stevens, B.R., et al.: Intestinal transport of amino acids and sugars: advances using membrane vesicles, Annu. Rev. Physiol. **46**:417, 1984.

Wasserman, R.H., and Fullmer, C.S.: Calcium transport proteins, calcium absorption, and vitamin D, Annu. Rev. Physiol. **45**:375, 1983.

Wilson, F.A.: Intestinal transport of bile acids, Am. J. Physiol. **241**:G83, 1981.

Young, S., and Bomford, A.: Transferrin and cellular iron exchange, Clin. Sci. **67**:273, 1984.

Books and monographs

Crane, R.K., editor: Gastrointestinal physiology II, Baltimore, 1977, University Park Press.

Crane, R.K., editor: Gastrointestinal physiology III, Baltimore, 1979, University Park Press.

Davenport, H.W.: Physiology of the digestive tract, ed. 5, Chicago, 1982, Year Book Medical Publishers, Inc.

Greenberger, N.J.: Gastrointestinal disorders: a pathophysiologic approach, ed. 3, Chicago, 1986, Year Book Medical Publishers, Inc.

Johnson, L.R., editor: Gastrointestinal physiology, ed. 3, St. Louis, 1985, The C.V. Mosby Co.

McColl, I., and Sladen, G.E., editors: Intestinal absorption in man, New York, 1975, Academic Press, Inc.

Sleisenger, M.H., and Fordtran, J.S., editors: Gastrointestinal disease, ed. 3, Philadelphia, 1983, W.B. Saunders Co.

SECTION

VIII

THE KIDNEY

Brian R. Duling

CHAPTER

45

Components of Renal Function

Very simple organisms that live freely in aquatic surroundings have a cellular environment that is stabilized by the relatively large fluid volumes surrounding the cells. The excretory and regulatory needs of these organisms can be met by direct exchange of waste products with the environment. More complex organisms must ingest food, transform it into energy forms that can be used by various tissues, and then excrete the generated wastes. Intake of food and the unavoidable loss of salts and water through processes such as respiration, perspiration, and defecation require that the body be able to adapt to varying exchanges between the organism and its environment. This adaptation is the primary role played by the kidney in homeostasis. Stated most broadly, the function of the kidney is to regulate the composition of the extracellular fluid and thereby to provide a relatively constant environment within which the normal functions of the cells can progress. The functions of the kidney are integrated with the requirements for body homeostasis by both neural and humoral mechanisms.

The kidney serves a variety of other needs, however, especially through its role in the production of hormones such as *angiotensin II, prostaglandins,* and the *kinins,* all of which are involved in the regulation of blood pressure. The kidney also helps to regulate red cell synthesis through the formation of the hormone *erythropoietin*.

The kidney regulates the composition of the extracellular fluid by selectively adjusting the composition of the plasma that flows through the renal vasculature. Materials to be conserved are retained in the plasma, and waste products are extracted and excreted. The formation of urine involves three processes: *filtration*, *reabsorption*, and *secretion*. The first process is *filtration* of a portion of the blood through a specialized ultrafilter, the *glomerulus*. The filtrate is free of blood cells and large molecular–weight materials in the plasma. The fluid formed by filtration is modified as it flows through a series of very narrow tubes, the *nephrons,* formed of epithelial cells. Some materials are reabsorbed from the filtrate and returned to the plasma. Other materials are removed from the plasma circulating through the kidney and added to the fluid that is ultimately excreted as urine; that is, they are secreted. As a rule, the volume of the filtrate is reduced by about 99% before elimination from the body. The major fraction of filtered ions, such as sodium, also is reabsorbed. Materials such as ammonia may be secreted to produce urine that is hundreds of times more concentrated than the plasma concentration.

■ *Body Fluid Compartments*

As already stated, the primary function of the kidney is to stabilize the composition of the extracellular fluid and in so doing to regulate indirectly the composition of intracellular fluid. The kidney regulates the composition of the extracellular fluid by adjusting the composition of plasma as it passes through the renal vasculature. In turn, the circulating plasma regulates the composition of the interstitial fluid that surrounds the cells. It is therefore essential to know how the various fluid compartments within the body are interrelated to understand how a particular renal action may influence first plasma, then extracellular fluid, and finally intracellular fluid.

Fig. 45-1 illustrates the major body fluid compartments and the connections among them. The figure clearly defines the relative volumes of the compartments and the fact that they are separated by membranes with different characteristics (cell membrane, capillary endothelium, and blood-brain barrier). The volumes of the various compartments are expressed as a percentage of body weight. Total body water comprises about 60% of the body mass in a normal subject. Tissues such as fat and bone contain little water, so an increase in the quantity of either component can reduce the fractional water content of the body.

Body water is partitioned between *intracellular water,* which is 33% of body weight (colored regions in

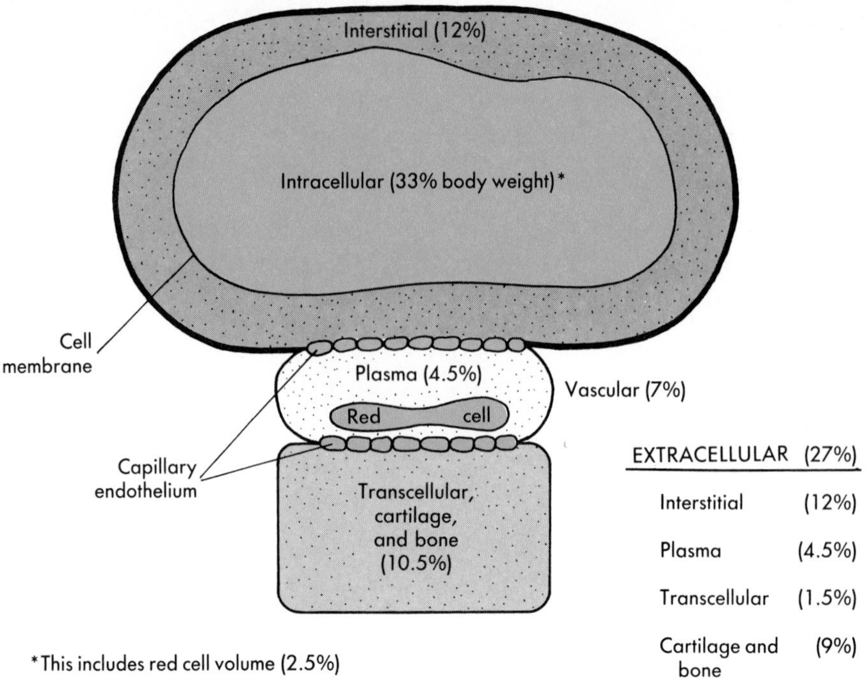

Fig. 45-1. Compartmentalization of body water. Water volumes are expressed as a percentage of body mass. Basic volumes are intracellular *(color)* and extracellular water *(gray)*. Extracellular volume is the sum of interstitial and plasma volumes (i.e., 12% + 4.5% = 16.5%) and a slowly equilibrating component consisting of transcellular elements, cartilage, and bone (10.5%). Note the special position occupied by red cells in this scheme, i.e., as part of the intracellular compartment that circulates in the blood. The membranes separating the various compartments (cell membrane and capillary endothelium) determine the type of tracer that can be used in the volume measurement (Table 45-1).

Table 45-1. Measurement of the volumes of the various fluid compartments in the body

Compartment	*Dilution indicator*	*Volume (percent of body weight)*	*Volume (percent of body water)*
Vascular (VB)	[P(100-hematocrit)] ÷ 100	7	11.7
Plasma (P)	Serum albumin	4.5	7.5
Interstitial (Int)	[ECF,F—plasma]	12	20
Extracellular, fast (ECF,F)	Inulin, thiosulfate	16.5	27.5
Transcellular (ECF,T)	ECF,S—ECF,F	10.5	17.5
Extracellular, slow (ECF,S)	Thiocyanate	27	45
Intracellular (ICF)	[BW—ECF,S]	33	55
Total body H_2O (BW)	DHO, antipyrine	60	100

Fig. 45-1), and extracellular water, which is 27% of body weight (gray regions in Fig. 45-1). The cell membranes separate intracellular water from extracellular water, and the transport characteristics of the membranes influence the concentration differences between the compartments.

Extracellular space is a combination of several volumes that are functionally different (Fig. 45-1). *Interstitial fluid,* which represents about 12% of body weight, is that fluid in intimate contact with the cells. Together with the *plasma* it comprises a portion of the extracellular fluid that rapidly equilibrates in various parts of the body. For this reason the plasma and interstitial fluid together are called the *fast extracellular fluid* (ECF,F). A significant fraction of the total extracellular fluid is relatively inaccessible, since it is contained in closed cavities such as the joints or cerebral ventricles (*transcellular water*, or ECF,T) or bound up in matrices of cartilage or bone. This inaccessible fraction amounts to about 10.5% of the body weight. The total extracellular volume is then the sum of interstitial volume, plasma volume, transcellular water, and bound water (27% of body weight). The compartment volumes are listed in Table 45-1 as fractions of body weight and as fractions of total body water.

The size of the compartments may change during adaptations to environmental stress or with disease. Therefore a variety of methods have been developed for measuring body fluid volumes. The simplest and most frequently used method for measuring short-term changes in total body water is the measurement of the body weight. Over a few days, changes in body weight can be attributed almost entirely to gains or losses in body water.

Direct measurement of other components of body water is more difficult and can be accomplished only by applying the *indicator dilution* concept, as illustrated in Fig. 45-2, *A*. To measure a compartment volume, a known quantity *(Q)* of a reference material is injected and allowed to distribute throughout the compartment

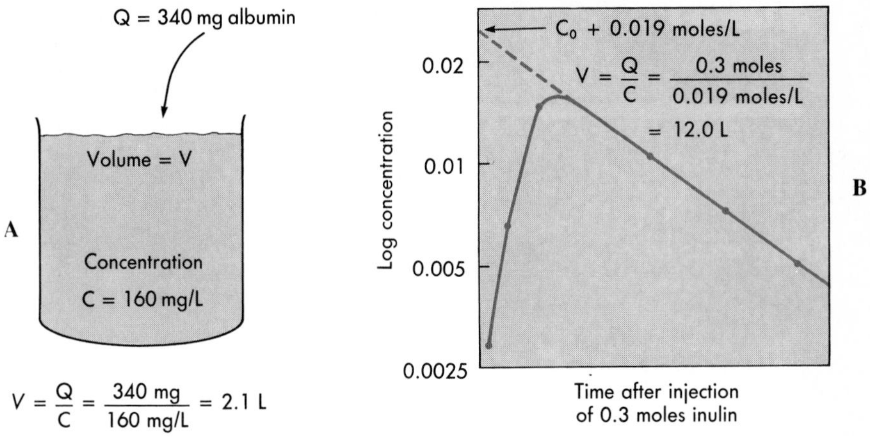

■ **Fig. 45-2.** Use of the dilution principle in the determination of compartment volumes. **A,** Simple estimation of plasma volume uncorrected for loss. **B,** Correction for loss of indicator during the measurement of extracellular volume. Inulin is assumed to be lost in the urine during the time required for equilibration. Losses are corrected for by *extrapolation back* to zero time. The extrapolated concentration at zero time (C_o) is equal to the concentration that would have been expected if there were no loss of indicator.

whose volume *(V)* is to be measured. Once the material is uniformly distributed, its concentration *(C)* must be equal to Q/V. It follows that, if Q is known and C is measured, V can be calculated from the quotient of Q and V (Fig. 45-2, *A*).

Two problems arise in the use of the indicator dilution method as shown. First, the body has several fluid compartments, and only an indicator that is confined to the appropriate compartment is entirely suitable. Second, loss of indicator by excretory processes during measurement leads to an overestimation of the compartment volume. The solution to the first problem has been achieved largely through the selection of indicators with characteristics appropriately related to the nature of the membranes that surround the compartments, as illustrated in Fig. 45-1.

Table 45-1 lists some of the indicators commonly used to measure the fluid compartments. For example, plasma volumes are often measured by using a labeled serum albumin molecule that equilibrates quickly within the plasma compartment but crosses the capillary endothelium very slowly.

The measurement of extracellular volume requires a molecule smaller than albumin so it can readily enter the interstitial space. However, the molecule must not be so small or so permeable that it can cross cell membranes and enter the intracellular fluid. Because of the complex composition of the extracellular fluid, no single molecule completely defines the extracellular fluid volume. Certain solutes, such as inulin and thiosulfate, exchange rapidly and freely between the plasma and interstitial fluid and such solutes can readily trace the volumes, which comprise the ECF,F. Smaller, more permeable tracer molecules, such as bromide and thiocyanate, not only equilibrate with the interstitial space and plasma but also enter cartilage and the transcellular compartments, such as the synovial fluid and the cerebrospinal fluid. The aggregate volume of all of these elements is referred to as the *slow extracellular volume* (ECF,S) because of the rate of equilibration.

Measurement of total body water requires a compound that readily crosses cell membranes and is distributed uniformly in body water. Deuterated or tritiated water and antipyrine are commonly used for this purpose.

The other body compartments must be computed by difference calculations (Table 45-1). Note that the total blood volume (V_B) is computed by dividing the measured plasma volume by 100 minus the measured red cell percentage (hematocrit).

The second problem with the indicator dilution method arises as a result of loss of tracer during the mixing period. The use of small molecules, such as inulin, is particularly difficult because they are excreted rapidly by the kidney; tracer water leaves the body in urine, sweat, and expired air. To compensate for these losses, serial concentrations of the indicator are measured in the plasma after injection, and the measured concentrations are extrapolated back to the time of injection (zero time). Fig 45-2, *B*, illustrates this process for estimation of extracellular volume. During an initial mixing period after injection of the tracer, concentration of the tracer rises to a peak value. Renal losses of the tracer then cause a progressive decrease in concentration; typically this decrease occurs as an exponential function of time. As shown in Fig. 45-2, *B*, when the concentrations are plotted on a log scale against time on a linear scale, a portion of the decay curve is straight, and this portion can be extrapolated to zero time. The resultant concentration at this point is used to compute the compartment volume.

■ *Osmotic Interactions Between Compartments*

Through intake or loss, total body water can be changed and water can shift between compartments from day to day. The osmotic activities of the fluids lost in the urine and sweat are variable, ranging from 0 (pure water) to 1,200 mOsm/L in the case of highly concentrated urine.

Table 45-2. Distribution of ions in body compartments

	Intracellular (mEq/L)	*Interstitial (mEq/L)*	*Plasma (mEq/L)*	*Plasma water (mEq/L)*
Cations				
Na^+	10	147	142	153
K^+	140	4.0	5.0	5.4
Ca^{++}	5	2.5	5.0	5.4
Mg^{++}	27	2.0	3.0	3.2
Anions				
HCO_3^-	10	30	24	29
Cl^-	25	114	103	111
$PO_4^=$	80	2.0	2	2.2
$SO_4^=$	20	1.0	1	1.1
Organic acids	—	7.5	6	6.5
Proteins	47	1.0	16	17.2

From Koushanpour, E.: Renal physiology: principles and functions, Philadelphia, 1976, W.B. Saunders Co.

How water is gained or lost from a fluid compartment is important to understand. This can be very complex, but, in the steady state, two basic relations govern water exchanges between compartments. First, the osmotic activities in all body compartments are almost in equilibrium. That is, the osmotic activity of intracellular fluid, interstitial fluid, and plasma is always almost equal. Thus removal of water from one compartment, with a subsequent increase in osmolality, automatically moves water osmotically from the other compartments into the one with the increased osmotic activity, and quickly yields the same osmolality throughout.

The second relation that permits one to understand fluid shifts among the body compartments is that body water moves in proportion to the volumes of the compartments. For example, if a patient loses 1 L of pure water by respiratory losses and *insensible perspiration* (evaporation from skin without the appearance of fluid on the skin surface), osmotic activity can remain the same in all compartments only if this liter of water comes proportionally from the intracellular and the extracellular spaces. Examination of Table 45-1 shows that 55% of the fluid must come from intracellular space and 45% from extracellular space if all the body fluid compartments are involved. However, in acute shifts, the water of the body cavities as well as water of bones and cartilage is not readily shifted. Therefore only interstitial water (20% of total water), plasma water (7.5% of total water), and intracellular water (55% of total body water) are involved. Thus only about 82.5% (20% + 7.5% + 55%) of the total water exchanges acutely. From this it can be calculated that about one third of the liter of lost water mentioned previously would come from extracellular fluid, and two thirds would come from intracellular fluid.

Ion Distributions and Body Compartments

Ions are not distributed uniformly among the various body compartments (see Chapter 2). Table 45-2 summarizes the distributions of cations and anions between the intracellular and extracellular spaces. Note that concentrations are expressed both as mEq/L plasma and as mEq/L plasma water. This is necessary because plasma has a high protein content (6% by weight), and thus only about 94% of the volume of plasma is actually water.

Sodium is the major extracellular cation, and potassium is the major intracellular cation. Similarly, the major intracellular anions are proteins and phosphate, whereas the major extracellular anions are chloride and bicarbonate. Chapter 2 explains the causes for these nonequilibrium distributions of the various anions and cations and relates these to the *Donnan effect* and the sodium-potassium pump.

The nonuniform distributions of ions determine the ways in which changes in body fluids may alter the volumes of the compartments. For example, if an isotonic sodium chloride solution is infused into a subject, the compartmentalization of most of the sodium in the extracellular space will cause most of the volume expansion to occur in the extracellular space.

Another important consequence of the relative sizes of the compartments and the distributions of ions is that the plasma concentration of a particular ion is often a misleading index of the status of electrolyte balance within the body. A striking example of this occurs when potassium is depleted from the organism. Since potassium is largely an intracellular cation, and since intracellular fluid represents a larger fraction of the body water, the vast bulk of body potassium is found in the cells. If potassium is lost from extracellular fluid—for example, by vomiting—the loss is partially compensated by the subsequent movement of potassium from intracellular stores to extracellular fluid. As a result, extracellular fluid potassium levels may fall little in spite of substantial losses in body potassium. Only by direct measurement of total body potassium or by a clear understanding of the source of the pathologic condition is it possible to assess the changes in potassium levels in tissue cells.

Functional Renal Anatomy

Form and function of the kidney are intricately related, and understanding of renal physiology is closely tied to understanding of renal anatomy and histology. The kidneys in the human are bean-shaped organs located in the abdomen on either side of the aorta. Each kidney is divided into lobes, or *pyramids,* which are the largest functional units of the organ (Fig. 45-3, *A*). Each of these lobes is divided into a *cortical,* or outer, portion (Fig. 45-3, *A*) and a *medullary,* or inner, portion. The cortical portion of the kidney has a granular appearance, whereas the medulla has a striated appearance. The striations are finer in the inner than in the outer portion of the medulla.

The initial steps in urine formation occur in the cortex, and the final modification of urine composition occurs in the medulla. The urine exits the kidney through a number of small pores called the *ducts of Bellini* at the end of the medullary papillae. The urine exiting through the ducts of Bellini flows into minor calyces of the renal pelvis and passes through the major calyx and the ureter into the bladder.

Tubules

The functional unit of the kidney is a long thin tubular structure, the *nephron*. Approximately 1,000,000 nephrons comprise the human kidney. The nephron begins with a filtering unit, Bowman's capsule *(BC),* which contains the vascular elements, the *glomerular capillaries* (*GC* in Fig. 45-3, *B*). Together, Bowman's capsule and the glomerular capillaries form the glomerulus (Fig. 45-4). Fluid filtered by the glomerulus initially flows through a highly convoluted section, the *proximal convoluted tubule (PT*, simplified in Fig. 45-3, *B*). Fluid then enters a straight section of the proximal tubule and flows into the medulla.

The proximal tubule can be divided into three functionally distinct sections, designated *S1, S2,* and *S3*. The S1 segment includes the major portion of the proximal convoluted tubule. The S2 segment is a bridge between the end of the proximal tubule and the beginning of the straight segment. The S3 segment includes most of the straight segment of the proximal tubule.

After penetrating a variable distance toward or into the medulla, the tubule turns back and reenters the cortex. The hairpin loop formed by this segment of the nephron is called the *loop of Henle (LH),* which is divided into a *descending limb* and an *ascending limb.*

There are two populations of nephrons in the kidney, *cortical nephrons* originating near the surface of the kidney and *juxtamedullary nephrons* originating close to the border between cortex and medulla. The length of Henle's loop varies with the site of origin of the nephron. Henle's loops of *cortical nephrons* are short and thick over much of their length. In contrast, *juxtamedullary nephrons* originate deeper within the cortex and are much longer. Segments at the middle of these nephrons are very thin (LH), and these portions of the Henle's loops of the juxtamedullary nephrons penetrate the entire length of the medulla.

The segment of the nephron distal to Henle's loop is the *distal convoluted tubule* (DT). It originates in the region of close contact between the glomerulus and the thick ascending limb of Henle's loop. The convoluted portion of the distal tubule is shorter than the convoluted portion of the proximal tubule.

The initial segment of the distal tubule lies close to the glomerulus of the same nephron. The association between the distal tubule and the glomerulus forms a complex anatomical structure, the *juxtaglomerular apparatus,* which is described in detail later in relation to the control of renal blood flow.

The distal tubule is subdivided into a proximal convoluted portion and a distal short segment, the *connecting tubule* (CT), which connects the convoluted distal tubule to the cortical collecting duct. The cell composition and length of the collecting tubule vary with the location of the nephron within the kidney and with the species. The collecting duct is divided into three segments, the cortical collecting duct (CCT), the medullary collecting duct (CD), and the papillary collecting ducts.

The nephron, including Bowman's capsule, proximal and distal tubules, loop of Henle, and collecting duct, is formed entirely of epithelial cells. However, the cell types along the tubules are considerably different. The typical ultrastructure of the different tubular epithelial cells is shown in Fig. 45-3, *C*. Cells from the proximal convoluted tubule have large concentrations of mitochondria at the basal surface, marked basal membrane infoldings, and a well-developed brush border. The cells of the more distal, straight portions of the proximal tubule and thick limbs of Henle's loop have fewer mitochondria and a sparser brush border. In the thin segment of Henle's loop the cells are flatter and have few mitochondria and only a very sparse brush border. Mitochondria and brush border reappear in the distal tubule, although not as abundantly as in the proximal

■ **Fig. 45-3.** Anatomy of the nephrons and vasculature of the mammalian kidney. **A,** Sagittal section through the kidney showing gross anatomical relations. The outer, stippled portion of the kidney is known as the *cortex*. The inner, striated portion is referred to as the *medulla*. The medulla can be subdivided into outer and inner zones. **B,** The segment from the renal pyramid shown by the colored lines in **A** is enlarged to illustrate the relations among nephrons from cortical and juxtamedullary regions. The vascular and tubular anatomies are drawn to the same scale, and, if desired, a transparent copy of the tubular anatomy can be made to overlay the vascular anatomy. There are two populations of nephrons, the cortical and the juxtamedullary. Note that only juxtamedullary nephrons possess long, thin loops, which penetrate into the interstitium of the medulla (TL). It should also be noted that the ascending thick limb of the loop of Henle returns to the afferent arteriole of the glomerulus from which the nephron originated. This site is the origin of the distal convoluted tubule, and the tubular structures in combination with the arterioles from the *juxtaglomerular apparatus (JGA)*. The nephron continues from the distal convoluted tubule via a short *connecting tubule (CT)* to empty into the cortical collecting duct *(CCT)*. The cortical collecting duct merges with a medullary collecting duct *(CD)* and finally forms a papillary collecting duct (not shown), which empties into the ducts of Bellini. *PT,* Proximal tubule; *BC,* Bowman's capsule; *LH,* loop of Henle; *TL,* thin limb; *IA,* interlobular artery; *Aa,* afferent arteriole; *Ea,* efferent arteriole; *PTC,* peritubular capillary; *GC,* glomerular capillary; *VR,* vasa recta.

A

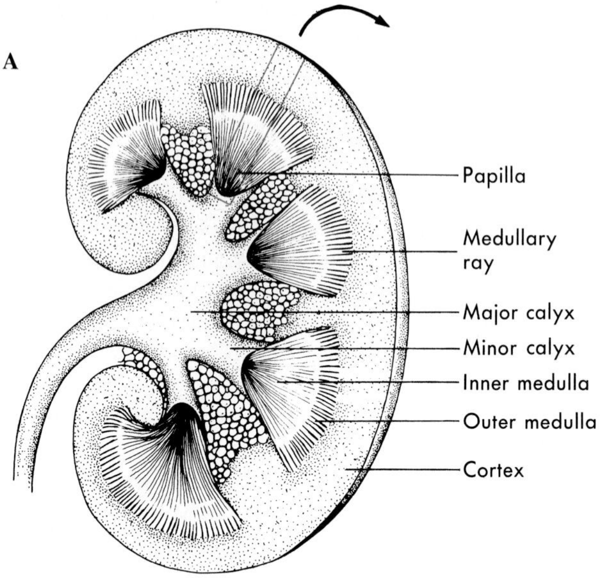

B

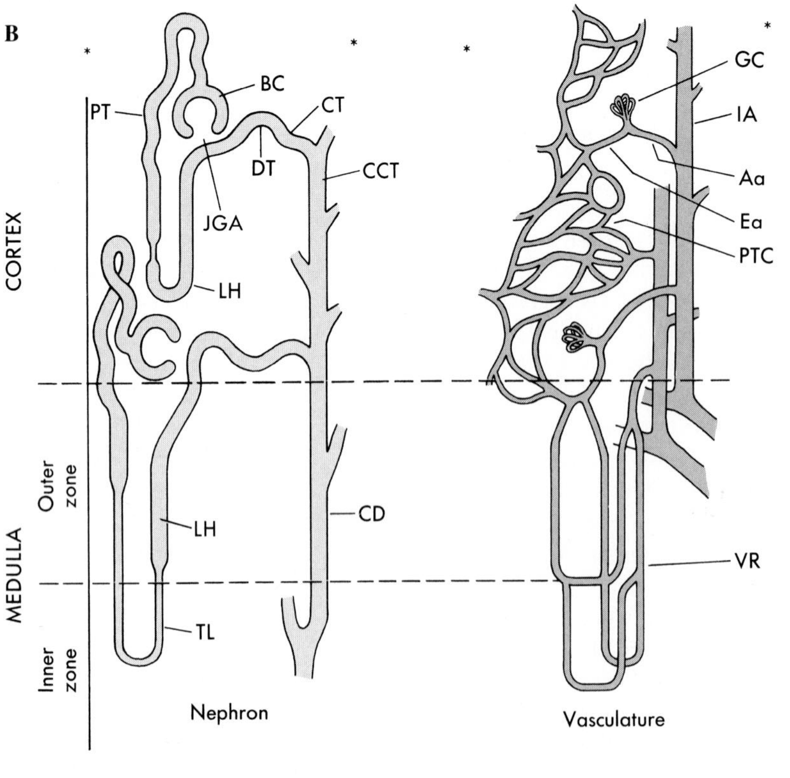

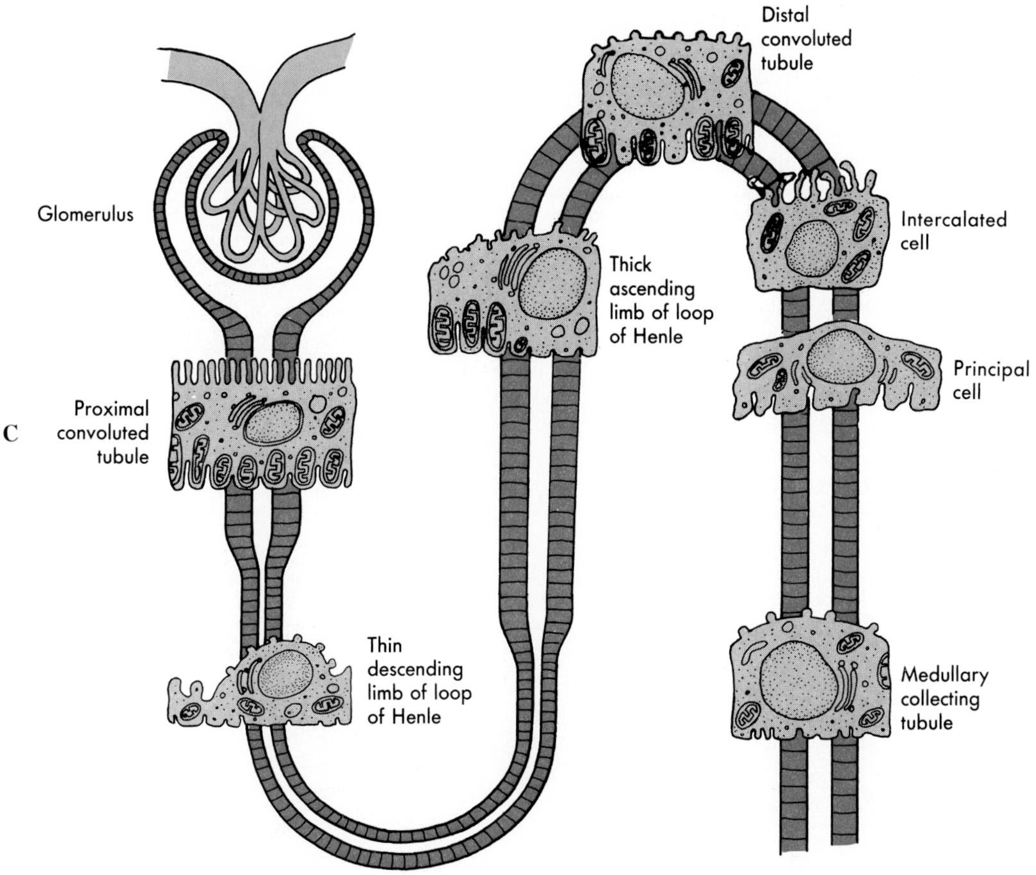

Fig. 45-3, cont'd. C, The histology of the cells of the various segments of the tubule.

convoluted tubule. The cellular makeup of the connecting tubule is similar to that of the cortical collecting duct. These nephron segments are made up of *principal cells,* with a general appearance intermediate between the cells of the distal convoluted tubule and those of the descending limb of Henle's loop. As explained later, the abundance of mitochondria correlates with the transport of large quantities of salt and water, as does the presence of a thick brush border on the epithelium. The connecting segment and distal tubule also possess a cell known as either an *intercalated cell* or a *dark cell.* These cells do not form a continuous layer but are interspersed among the principal cells of the tubules. The intercalated cells have prominent microvilli and are rounder than their neighbors. They play a major role in the secretion of acid.

Adjacent epithelial cells are joined by "tight" junctions. The cell coupling is more complete in some segments of the nephron than in others, and consequently, the permeability of the various tubule segments varies substantially. In general, the junctions of the proximal tubule are more open than those of the distal tubule, and therefore the epithelium is more permeable to solutes and water. The junctions of the ascending limb of Henle's loop are very tight; hence, permeability to salt and water is very low.

Vasculature

The pattern of the renal vasculature closely parallels the nephron structure, and the association between the vessels and nephrons has important functional implications (Fig. 45-3). Blood enters via the renal artery, which divides and ramifies throughout the kidney with three successive right-angle branches. These branches finally form a set of distributing vessels, the *interlobular arteries,* which pass radially from the corticomedullary border into the cortex and supply the glomerular circulation. The arterial blood supply to the glomerulus enters via the *afferent arteriole (Aa),* passes through the *glomerular capillaries (GC,)* and exits through the *efferent arteriole* (*Ea,* Fig. 45-3, *B*). In most of the cortex, blood leaving the glomerulus flows into a second capillary network, the *pertitubular capillaries (PTC),*

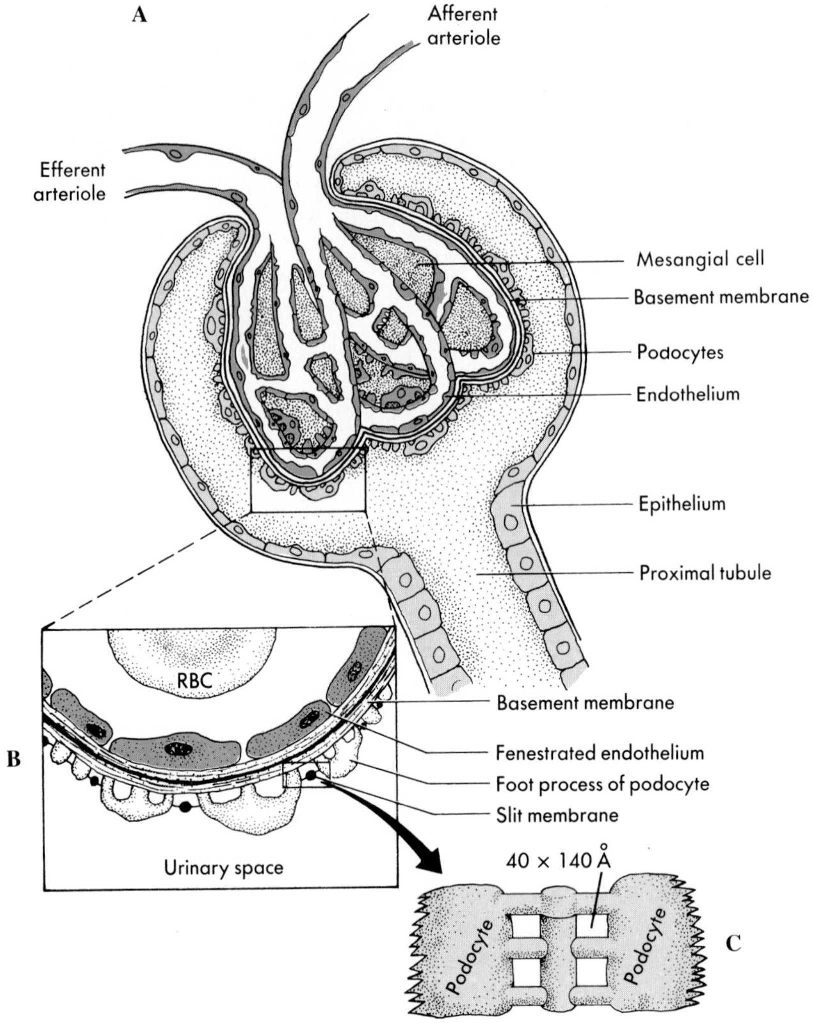

■ **Fig. 45-4.** Anatomy of the mammalian glomerulus. The dark color represents endothelium, and the light color represents epithelium. **A,** Cross section through a single glomerulus. Note that the glomerular capillaries form an invagination into a capsule of epithelial cells (Bowman's capsule). The surface of Bowman's capsule covers the glomerular capillaries in the form of specialized epithelial cells (podocytes). **B,** Expanded view of the filtration membrane of the glomerulus. **C,** View of one of the slit membranes rotated 90 degrees and viewed en face.

which are intertwined among the convolutions of the proximal and distal tubules.

The vascular pattern of the juxtamedullary nephrons is very similar to that of the superficial cortical nephrons, except that the juxtamedullary nephrons also possess a parallel capillary network that extends into the medulla. This parallel network originates distal to the efferent arterioles of the glomeruli and forms the *vasa recta*. The vasa recta are long, thin capillary loops that match the pattern of the thin loops of the medullary Henle's loop.

Blood from juxtamedullary and superficial pertitubular capillaries and from the vasa recta capillaries combines and returns to the vena cava via venous drainage that parallels the arterial supply.

Fig. 45-3, *B* and *C*, shows the reason for the difference, mentioned earlier, in the gross anatomy of the three portions of the kidney: cortex, inner medulla, and outer medulla. The cortex is mainly glomeruli, convoluted tubules (both distal and proximal), and peritubular capillaries. The outer medulla is formed primarily of vasa recta and thick segments of Henle's loop. The inner medulla is formed of vasa recta, collecting ducts, and thin limbs of Henle's loops.

■ *Glomerular Filtration*

It is possible to insert small glass pipettes into the beginning of the proximal tubule and withdraw fluid samples for chemical analysis. When this is done it becomes apparent that the material entering the proximal tubule is an ultrafiltrate of plasma, i.e., a fluid whose composition is essentially identical to that of plasma, with the exception that the large molecules, especially proteins, have been sieved out.

■ *Starling Hypothesis*

Knowing the protein concentration in the proximal tubular fluid allows one to use the Starling hypothesis for capillary exchange to yield quantitative predictions about glomerular filtration. The Starling hypothesis as-

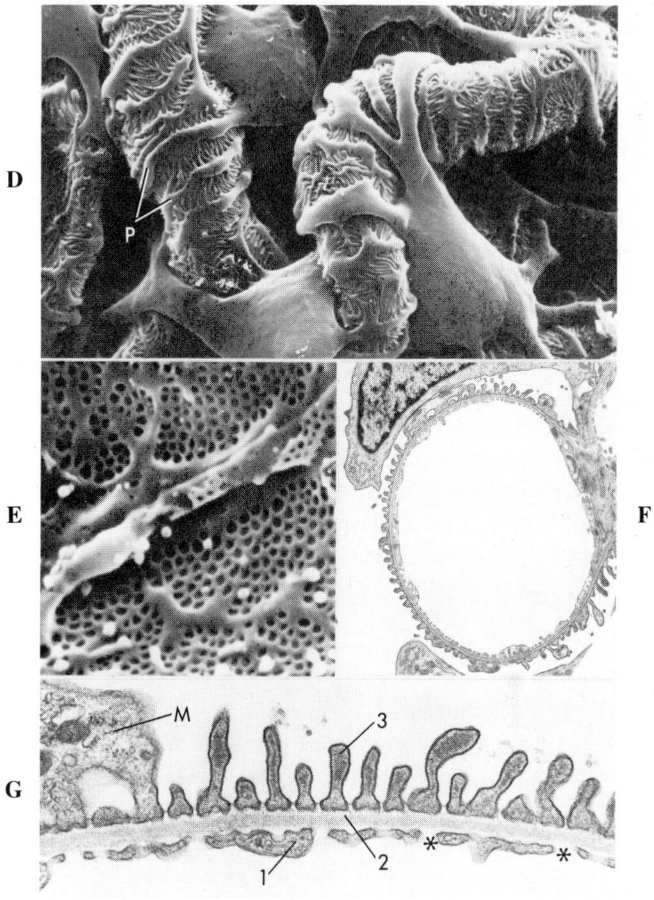

■ **Fig. 45-4, cont'd. D,** Scanning electron micrographs of a rat glomerulus corresponding to drawing **A.** *P,* Epithelial cell podocytes. **E,** Scanning electron micrograph of the endothelial surface of a glomerular capillary (i.e., the blood side). **F,** Cross-sectional transmission electron micrograph corresponding to **B. G,** Transmission electron micrograph of the glomerular filtration barrier. *1, 2,* and *3* represent endothelial cell, basement membrane, and epithelial cell, respectively. *M,* Mesangial cell;*, **D,** fenestrations of endothelium. (From Kriz, W., and Kaissling, B.: In Seldin, D.W., and Giebisch, G., editors: The kidney: physiology and pathophysiology, vol. 1, New York, 1985, Raven Press.)

serts that capillary filtration equals the product of a constant and the sum of all hydrostatic and oncotic pressures acting across the capillary wall. The algebraic statement of the Starling hypothesis (see Chapter 31) is:

$$\text{Glomerular filtration rate} = k[(P_c + \pi_t) - (P_t + \pi_c)]$$
$$= k \times \text{net filtration pressure}$$

where

P_c = glomerular capillary hydrostatic pressure
P_t = proximal tubular hydrostatic pressure
π_c = glomerular capillary oncotic pressure
π_t = proximal tubular fluid oncotic pressure

The algebraic sum of the terms in brackets is called the *net filtration pressure.* k is a coefficient that describes how fast fluid moves across the glomerulus under a given net filtration pressure. k can be estimated either for a single glomerulus or for the kidney as a whole. The combined effective k for both kidneys in the human is about 15 ml/minute/mm Hg. In other words, if the net driving pressure (the sum of the difference in hydrostatic and oncotic pressures across the glomerulus) is 1 mm Hg, a filtration rate of 15 ml/minute will be induced. This filtration rate reflects, of course, the sum of the individual filtration rates of all the approximately 1 million nephrons of each kidney. The magnitude of this filtration is remarkable, since it represents 180 L per day, or about 60 times the total plasma volume. Obviously a large fraction of the total volume is reabsorbed in the process of urine formation.

■ *Anatomical Correlates of the Filtration Coefficient*

For capillaries in other vascular beds, such as those in striated muscle, it appears that the anatomical correlate of the filtration coefficient is closely related to the space between endothelial cells. Fig. 45-4 illustrates the anatomical basis of the filtration coefficient in the kidney. The glomerular membrane consists of a *fenestrated endothelium* overlying a *basement membrane,* which is a complex structure composed of collagen and glycoprotein (Fig. 45-4, *B, G2*). Beneath the basement membrane lie the *foot processes* of the *podocytes* (Fig. 45-4, *G*). The podocytes are specialized epithelial cells that are continuations of the epithelium of the proximal tubule, and the foot processes are extensions of these cells. The spaces between the epithelial cell foot processes are spanned by a diaphragm called the *filtration slit membrane*. In transmission electron micrographs this diaphragm appears as a line connecting two podocytes with a dot midway across it. When viewed en face, that is, when rotated 90 degrees from the cross-sectional view shown in Fig. 45-4, *B*, the slit diaphragm is in fact a ladderlike structure with spaces between the rungs (Fig. 45-4, *C*).

Permeability characteristics of the glomerulus can be determined by infusing molecules of different sizes into the blood and analyzing plasma and proximal tubular fluid concentrations. Such an experiment yields data like those shown in Fig. 45-5. The figure depicts the concentrations of several molecules normally present in plasma. Note that molecules smaller than about 15 Å radius are found in the filtrate in essentially the same concentration as in plasma. The concentration of molecules larger than about 15 Å declines rapidly, and at approximately 35 to 40 Å the concentration in the filtrate approaches zero. This progressive reduction in passage of larger molecules through the glomerulus is called *molecular sieving*.

The determinants of permeability, however, reflect a more complex anatomy than the size of the podocyte slits alone. The structure of the basement membrane also contributes to the overall molecular selectivity of the glomerulus. The basement membrane is a glycosaminoglycan matrix and functions in series with the foot processes of the podocytes to filter large molecules. In addition, the basement membrane conveys a charge preference to the glomerulus. Anionic macromolecules (*dashed line* in Fig. 45-5) enter the filtrate in lower concentrations than either uncharged molecules or cations. This charge selectivity probably reflects the presence of anionic sites in the basement membrane. Such sites would repel anions, decrease their entry into the membrane, and thus reduce the rate of passage of anions into the filtrate.

All of the separate elements in the filtration system may be altered in relatively selective ways in various clinical syndromes. For example, antigen-antibody reactions or certain kinds of systemic infections may thicken the basement membrane and disrupt the foot processes, with resultant loss of molecular selectivity. Also, one of the most common histological findings in diabetes is a thickening of the glomerular basement membrane.

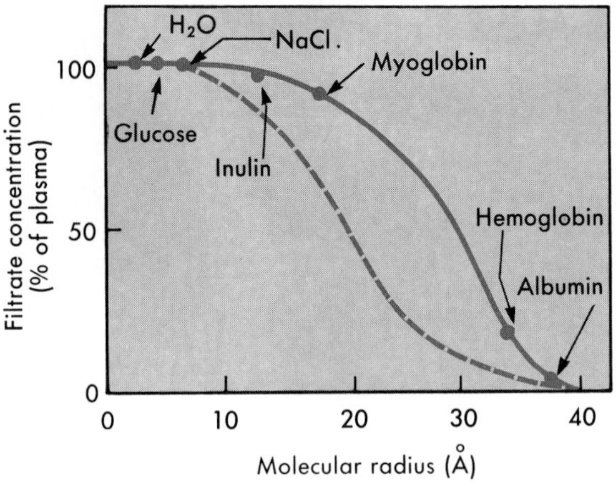

Fig. 45-5. The relation between molecular size and passage of molecules through the glomerulus. Note that molecules much larger than albumin cannot cross the glomerular membrane and do not appear in the filtrate. The dashed line represents data obtained when anionic dextran molecules of different sizes were infused and their concentration in the filtrate determined.

Filtration Pressure

Control of the glomerular filtration rate (GFR) is a multifaceted process involving each of the components of filtration. From Fig. 45-5 it should be apparent that tubular fluid oncotic pressure must be close to zero (π_t = 0) because few proteins are present in the filtrate. The proximal tubular hydrostatic pressure has been measured directly with micropipettes (inserted into the tubular lumen); it is approximately 10 mm Hg (P_t = 10). The other two factors that determine filtration pressure are the hydrostatic pressure within the glomerulus and the glomerular plasma oncotic pressure. Normally the glomerulus lies deep within the cortex of the kidney and thus a glomerular capillary is inaccessible for puncture and direct pressure measurement. However, in one strain of rats that possesses superficial glomeruli, puncture of the capillaries has been achieved, and the capillary hydrostatic pressure is about 45 mm Hg (P_c = 45). In contrast to the pressure in other capillaries, hydrostatic pressure in glomerular capillaries decreases little from afferent to efferent arterioles. This is the result of the short, wide geometrical configuration of the glomerular capillaries. Thus one can assume that the hydrostatic pressure is close to 45 mm Hg along the entire length of the glomerular capillaries.

Assigning a value for glomerular plasma oncotic pressure is difficult because substantial quantities of fluid are filtered from the capillaries, thereby elevating protein concentration and oncotic pressure in the remaining plasma. However, a representative value for the oncotic pressure in glomerular plasma is about 27 mm Hg, i.e., slightly greater than circulating plasma oncotic pressure.

In computing the magnitude of glomerular filtration, the net pressure that causes filtration in the example just cited would be 45 mm Hg (45 − 0), and the net pressure opposing filtration would be 37 mm Hg (10 + 27). Putting these numbers into the Starling equation, GFR = k (45 − 37) = 8 k. Recall that k was 15 ml/minute/mm Hg. Hence, GFR = 8 × 15 = 120 ml/minute.

As with other vascular beds, control of filtration is accomplished largely by changing hydrostatic pressure. For example, suppose that it is desirable to conserve fluid in a dehydrated subject. This could be accomplished by constricting the afferent arteriole, thereby lowering the glomerular hydrostatic pressure and reducing net filtration. Assume that after dehydration, constriction of the renal arteriole might reduce hydrostatic pressure in the glomerulus from 45 to 40 mm Hg; in that case the following holds true:

$$\text{GFR} = k(40 + 0) - (10 + 27) = 3\ k = 45\ \text{ml/minute}$$

It is striking that an almost threefold reduction in the GFR, from 120 to 45 ml/minute, was produced by a 5 mm Hg reduction in glomerular hydrostatic pressure.

A number of vasoactive materials may influence the GFR, although their physiological roles have yet to be established. These substances can have complex effects on GFR because they act primarily on either afferent or efferent arterioles. For example, angiotensin II causes vasoconstriction and raises P_c because of its predominant effect on the efferent arteriole. Similar effects are seen with norepinephrine. Dopamine is a vasodilator with approximately equal effects on afferent and efferent arterioles and therefore it has little effect on capillary hydrostatic pressure. In contrast, infusion of bradykinin into the renal circulation causes a greater vasodilation of the efferent than of the afferent arteriole and hence a fall in capillary pressure.

Regulation of GFR-Capillary Flow

Traditionally renal physiologists ascribed the control of the GFR exclusively to changes in glomerular hydrostatic pressure evoked by constriction of afferent and/or efferent arterioles. However, flow itself may influence glomerular filtration. The finding that plasma oncotic pressure rises substantially along the length of the glomerular capillary has generated significant uncertainty regarding the quantitative relations among renal blood flow, glomerular hydrostatic pressure, and oncotic pressure and regarding their influence on the GFR.

Fig. 45-6 shows the various components of the net filtration pressure at different flow rates. The line $P_c - P_t$ represents the net hydrostatic pressure along the capillary. The oncotic pressure at three levels of flow is shown, and it is seen that the oncotic pressure rises as the water is filtered from the plasma. Net filtration pressure is the difference between the hydrostatic pressures and the oncotic pressures, or the area bounded by the lines $P_c - P_t$ and π_p. The three curves—*A, B,* and *C*—are predicted curves at three progressively higher rates of flow. When the flow is increased, the capillary oncotic pressure rises less for a given distance moved along the capillary because the plasma spends less time at any particular location. Thus at higher flows, the plasma oncotic pressure is lower along the capillary, and the net filtration pressure is elevated. In other words, increasing flow tends to increase the GFR, even if the mean capillary hydrostatic pressure is unaltered.

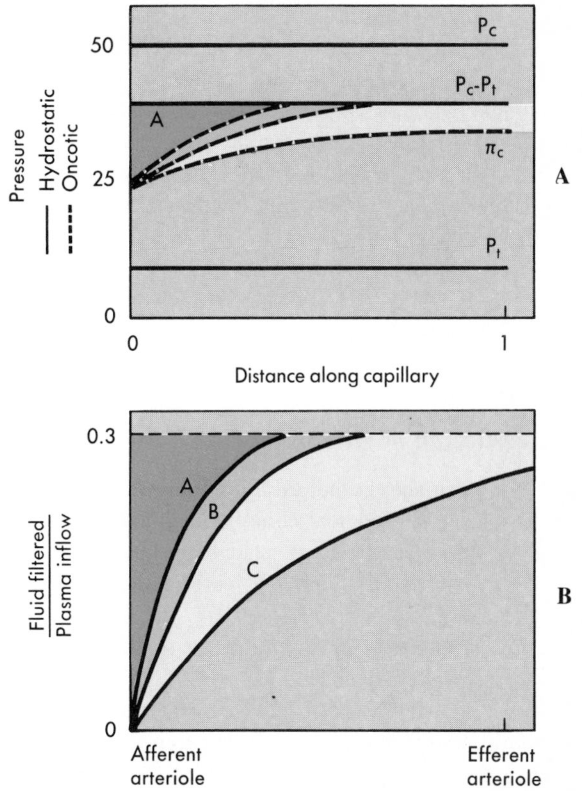

Fig. 45-6. **A,** Graphical analysis of the application of the Starling hypothesis to the glomerulus. *P,* Hydrostatic pressure; π, oncotic pressure; P_c the value in glomerular capillary; P_t, the value in the tubule. $P_c - P_t$ is the net hydrostatic pressure across the glomerulus. The three curves, *A, B,* and *C,* represent observations made at progressively higher renal plasma flows. The area bounded by the net hydrostatic pressure line ($P_c - P_t$) and the oncotic pressure line (π_p) represents the net filtration pressure that drives glomerular filtration. **B,** Effect of elevated flow rate on filtration. Note that, as flow increases from *A* to *B,* the fraction filtered does not change. However, filtration of a constant fraction of the inflowing plasma represents an increase in the absolute rate of glomerular filtration when the flow rate is higher. **C,** When flow is high relative to the filtration rate, π_c is less than $P_c - P_t$.

If flow is low enough relative to the filtration rate, then sufficient water may be filtered from the plasma to cause the oncotic pressure to rise and equal the net hydrostatic pressure. At this point, no further filtration can occur, and therefore this is referred to as a condition of *filtration equilibrium* (curves *A* and *B,* Fig. 45-6, *A*).

This process can be viewed in another way. In Fig. 45-6, *B,* the fraction of the inflowing plasma that is filtered is plotted on the ordinate. For the two cases in which equilibrium occurs, the fraction filtered is the same, since equilibrium would be attained at the same oncotic pressure and after the same amount of fluid is filtered from each unit of plasma. However, since flow is higher in case B, filtration of the same fraction must mean filtration of a larger absolute quantity of fluid. Note that at high flows an equilibrium will not be reached (Fig. 45-6, curve *C*), and π_c will be less than the hydrostatic pressure ($P_c - P_t$). It is known that filtration equilibrium occurs in some, but not all, species. Data are inadequate to permit conclusions as to whether this happens in humans.

Regulation of GFR-k

It has been shown that the filtration coefficient, k, may be under physiological control. Angiotensin II, norepinephrine, prostaglandins, and bradykinin all decrease the filtration coefficient. It is curious that vasoconstrictors (such as norepinephrine) and vasodilators (such as acetylcholine) both cause a decrease in k, the reason remains to be established.

The mechanism whereby these agents may act on k is still under study. It does not appear to depend on stimulation of the smooth muscle of the arterioles. Several lines of evidence suggest that the mesangial cells may be the effectors. These cells possess contractile filaments and can easily alter capillary perfusion by changing their dimensions and altering the number of open capillaries in the glomerulus.

In summary, control of the GFR is a complex process involving the following three modes of regulation:

1. Capillary hydrostatic pressure can be regulated by changing the relation between precapillary and postcapillary resistance.
2. Capillary oncotic pressure can be influenced by changing flow.
3. The filtration coefficient can be varied by changing the number of perfused capillaries.

Bibliography

Journal articles

Aukland, K.: Methods for measuring renal blood flow: total flow and regional distribution, Annu. Rev Physiol. **42:**543, 1983.

Bulger, R.E., and Dobyan, D.C.: Recent structure-function relationships in normal and injured mammalian kidneys, Anat. Rec. **205:**1, 1983.

Chang, R.L.S., et al.: Permeability of the glomerular capillary wall. III. Restricted transport of polyanions, Kidney Int. **8:**212, 1975.

Edelman, I.S., et al.: Body composition: studies in the human being by the dilution principle, Science **115:**447, 1952.

Farquhar, M.G.: The primary glomerular filtration barrier: basement membrane or epithelial slits? Kidney Int. **8:**197, 1975.

Karnovsky, M.J.: The ultrastructure of glomerular filtration, Annu. Rev. Med. **30:**213, 1979.

Kuntz, I.D., and Zipp, A.: Water in biological systems, N. Engl. J. Med. **297**(5):262, 1977.

Rodewald, R., and Karnovsky, M.J.: Porous substructure of the glomerular slit diaphragm in the rat and mouse, J. Cell Biol. **60:**423, 1974.

Tischer, C.C., et al.: Human renal ultrastructure. III. The distal tubule in healthy individuals, Lab. Invest. **18:**655, 1968.

Wright, F.S., and Briggs, J.P.: Feedback control of glomerular blood flow, pressure and filtration rate, Physiol. Rev. **59**(4):958, 1979.

Books and monographs

Brenner, B.M., et al.: Glomerular ultra-filtration. In Brenner, B.M., and Rector, F.C., editors: The kidney, vol. 1, ed. 3, Philadelphia, 1986, W.B. Saunders Co.

Dworkin, L.D., and Brenner, B.M.: Biophysical basis of glomerular filtration. In Seldin, D.W., and Giebisch, G., editors: The kidney: physiology and pathophysiology, vol. 1, New York, 1985, Raven Press.

Koushanpour, E.: Renal physiology: principles and functions, Philadelphia, 1976, W.B. Saunders Co.

Kriz, W., and Kaissling, B.: Structural organization of the mammalian kidney. In Seldin, D.W., and Giebisch, G., editors: The kidney: physiology and pathophysiology, vol. 1, New York, 1985, Raven Press.

Maffly, R.H.: The body fluids: volume, composition, and physical chemistry. In Brenner, B.M., and Rector, F.C., editors: The kidney, vol. 1, ed. 2, Philadelphia, 1981, W.B. Saunders Co.

Navar, L.G.: The regulation of glomerular filtration rate in mammalian kidneys. In Andreoli, T.E., et al., editors: Physiology of membrane disorders, New York, 1978, Plenum Press.

Tischer, C.C., and Madsen, K.M.: Anatomy of the kidney. In Brenner, B.M., and Rector, F.C., editors: The kidney, vol. 1, ed. 3, Philadelphia, 1986, W.B. Saunders Co.

Windhager, E.E.: Micropuncture techniques and nephron function, New York, 1968, Appleton-Century-Crofts.

Woodbury, D.M.: Physiology of body fluids. In Ruch, T.C., and Patton, H.D., editors: Physiology, and biophysics, vol. 2, ed. 20, Philadelphia, 1974, W.B. Saunders Co.

CHAPTER

46

Tubular Mechanisms

■ *Methods in Renal Physiology*

To a large degree, the history of the advancement of renal physiology has been determined by the development of the necessary methods and tools.These methods have provided the means for describing the anatomy, physiology, and biochemistry of many cell types, for explaining complex reactions among nephrons, and for characterizing the integrated functions of the various elements of the intact kidney.

One of the most powerful tools in understanding renal function in the intact animal is the *clearance concept*. As described in Fig. 47-1, the clearance concept depends on an analysis of the excretory patterns of specialized reference molecules. It allows one to measure quantitatively the filtration rate, as well as the net secretion or net reabsorption that has taken place in the kidney.

More detailed analyses of renal function are permitted by the *stop-flow technique,* which provides some insights into the location of various transport processes along the nephron in the intact kidney. To implement the stop-flow technique, a very high urine flow is established, so that the composition of the fluid along the entire nephron approaches that of the glomerular filtrate. The ureter is then occluded for several minutes, during which time the tubules act on the stationary urine column, changing its composition in accord with local tubular transport activities. The ureter is then released. As the urine exits rapidly from the kidney, it is collected in a number of small samples. The first sample represents fluid resident in the collecting duct during the period of occlusion, and the last sample represents new filtrate. Chemical analysis of the samples permits one to infer the nature of the tubular activity that took place during the occlusion period. Although subject to many difficulties, the stop-flow method represents one of the few means for assessing specific tubular function in the intact kidney.

Perhaps the most important single technical development in methodology applied to the study of renal function is the *micropuncture technique*. Small, sharp micropipettes are inserted into the tubules and samples of fluid are withdrawn and analyzed for a wide variety of compounds. Alternatively, fluid of known composition can be injected into the tubules and the resultant changes observed after subsequent withdrawal and microanalysis of the fluid. Simultaneous measurements of electrical potentials within the tubular lumen permit a complete analysis of transport phenomena. The major limitation of the technique is that only surface portions of the nephrons can be punctured, thus restricting study to the middle two thirds of the proximal and distal tubule and to the tip of Henle's loop.

Biochemists and pharmacologists have made great use of the *slice technique,* in which the accumulation of materials or metabolic modification of substrates within slices of the kidney is monitored over time. The technique lacks precision, however, in that many cell types are contained in a single slice (Fig. 45-3).

Significant new insights into epithelial cell physiology also have been gained recently with the preparation of *membrane vesicles*. After microdissection of the kidney to obtain specific portions of the nephron and subsequent sonication of cells, membrane vesicles are prepared both from the different portions of the nephron and from specific cell types or from portions of cells (e.g., brush borders of epithelial cells). The vesicles can be sealed while in fluid of a desired composition, thus fixing the internal environment. The vesicles are then placed in a defined medium, and the membrane permeabilities and transporter characteristics, as well as ATPase activities, can be assessed by noting the time-dependent alterations in vesicle contents.

■ *Characteristics of Transport Processes*

The glomerular ultrafiltrate is modified by a variety of secretory and reabsorptive processes as fluid passes

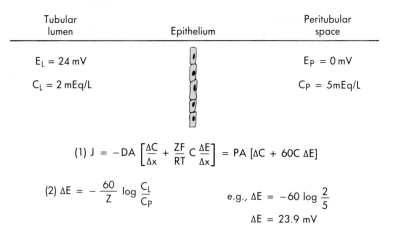

Tubular lumen | Epithelium | Peritubular space

$E_L = 24\ mV$ $\quad$ $E_P = 0\ mV$

$C_L = 2\ mEq/L$ $\quad$ $C_P = 5 mEq/L$

$$(1)\ J = -DA\left[\frac{\Delta C}{\Delta x} + \frac{ZF}{RT}C\frac{\Delta E}{\Delta x}\right] = PA\ [\Delta C + 60C\ \Delta E]$$

$$(2)\ \Delta E = -\frac{60}{Z}\log\frac{C_L}{C_P} \qquad \text{e.g., } \Delta E = -60\log\frac{2}{5}$$

$$\Delta E = 23.9\ mV$$

■ **Fig. 46-1.** Quantitative relations among the variables determining passive movement of materials across the epithelial cells. *P*, Permeability; *A*, area of tubular membrane; *E*, potential difference; *C*, concentration; *Z*, *F*, *R*, and *T*, the valence, the Faraday constant, the universal gas constant, and the absolute temperature, respectively; *J*, the rate of movement of solute under the conditions described in equation 1; $_L$ and $_P$, lumen and peritubular space, respectively. Data are typical for the potassium ion. Other variables are as described in the text.

through the tubules. This tubular processing is so extensive that about 99% of the filtered sodium and water are removed, leaving little correspondence in volume or composition between the fluid that is filtered and the urine that is excreted. The task of understanding the movement of diverse materials such as sodium chloride, glucose, and protein across the tubular epithelium can be facilitated by viewing these processes within a framework of the basic concepts of transport and according to the general principles of epithelia behavior. These principles have been discussed previously in relation to intestinal function (Chapter 44). Also, general principles of cell transport, as described in Chapter 1, apply to transport by epithelial cell layers, as well as by individual cells.

Fig. 46-1 shows an epithelial boundary separating two compartments and delineates some of the important factors determining transepithelial exchange. Transport may occur across the epithelial boundary either from the tubular lumen through the peritubular space and into capillary blood or vice versa.

The processes and forces that determine how material will traverse the epithelium are fundamentally the same as those that determine cellular transport in other locations (Chapter 1). Lipid solubility and molecular size, charge, and concentration differences are the primary determinants of how easily a molecule may move across the tubular epithelium and of the equilibrium condition. Very small, uncharged molecules, such as water, move more easily than do larger molecules, such as glucose, or charged molecules, such as sodium ions. A molecule the size of glucose, which has a low lipid solubility, can scarcely cross the tubular epithelium without some sort of specialized transport system.

The rate (J) at which an uncharged molecule diffuses across the epithelium is equal to the product of the diffusion coefficient (D), the total area available for exchange (A), and the diffusion gradient, that is, the change in concentration over distance ($\Delta C/\Delta x$). In biological systems the quotient of the diffusion coefficient and the diffusion distance are often considered together as the permeability (*P*, equation 1 in Fig. 46-1). In the case of charged molecules an additional term must be included to take into account the influence of electrical potential difference (ΔE) on the molecular motion.

If one wishes to understand the patterns of motion of materials whose transepithelial movement is determined mainly by interaction with specific transport molecules, simply knowing that the molecules are transported by active or facilitated transport systems existing within the tubular epithelium suggests a number of generalizations. Following is a partial list of such generalizations:

Characteristics of active transport systems

1. More rapid molecular transport than would be predicted for the size and lipid solubility of a molecule
2. Saturation at high concentrations
3. Molecular specificity
4. Compete for molecules of similar type
5. Sensitive to inhibitors

Typical classes of molecules transported by the renal epithelia

1. Monovalent ions: Na^+, Cl^-
2. Polyvalent ions: $PO_4^{\equiv}$, Ca^{++}
3. Aromatic acids: penicillin, aminohippuric acid (PAH)
4. Organic bases: quinine, choline, thiamine
5. Aliphatic acids: amino acids, pyruvate
6. Sugars: glucose, fructose

Understanding these characteristics of active transport systems allows one to generalize about various kinds of tubular transport processes.

Other generalizations can be made if one knows that tubular transport processes follow behavior patterns and show kinetics that are characteristic of a variety of biochemical interactions and enzyme systems. Fig. 46-2 shows an idealized version of the relation between the rate of a biochemical process (in this case the transport system) and the substrate concentration (in this case, the concentration of the molecule to be transported). As

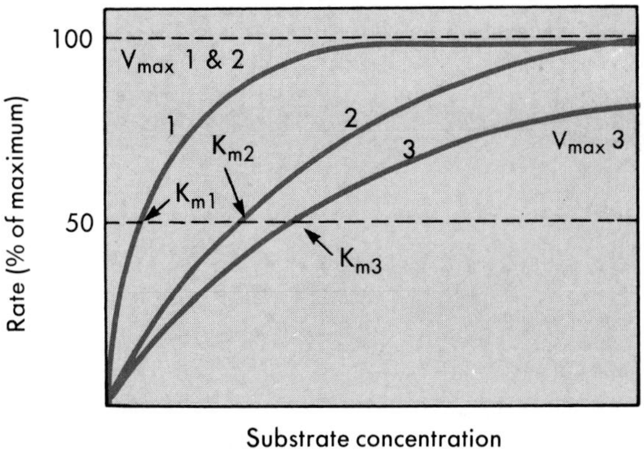

■ Fig. 46-2. Influence of substrate concentration on the rate of a biochemical reaction. In this case the rate is assumed to be the rate at which a transport carrier molecule might combine with and move a molecule across the epithelial membrane. V_{max} is the maximal rate of the reaction, and K_m is the substrate concentration that could drive the system at half the maximal rate.

substrate concentration is increased from zero, the rate of the transport reaction rises and approaches a maximal rate (V_{max}), at which point transport is said to be saturated. The substrate concentration that will raise the rate of the reaction to half the maximal rate is called the K_m. Fig. 46-2 shows two processes with identical V_{max} (*1* and *2*) but with different K_m values. Curve *1* in Fig. 46-2 represents a system in which the K_m is low, that is, the affinity of the carrier molecule for the substrate is high. Note that the reaction rate is accelerated to the V_{max} by a small increase in substrate concentration. Contrast this with line *2*, which will ultimately reach the same rate, but only if the substrate concentration is raised to much higher levels. Line *3* shows a case in which the V_{max} is lower than that for *1* and *2*. Thus a transport system depending on this enzyme pattern could not translocate molecules across the tubular epithelium as quickly as for case two, even at high substrate levels.

In part the kinetics of transport systems probably explain the fact that two patterns of transport are commonly recognized by renal physiologists. These are referred to as the *gradient-time systems* and the *transport maximum* (or T_m) *systems*. These designations have little significance insofar as specific tubular mechanisms are concerned but are commonly used terms to describe renal excretory patterns. When the plasma concentrations of various substances are changed and the resulting alterations in excretion are analyzed, the two broad categories of behavior shown in Fig. 46-3 are often apparent. Fig. 46-3 shows how the excretory rate and the filtration rate for a solute change as the plasma concentration is varied. The dashed lines show the amount of solute that is filtered per unit time, as predicted from the product of the plasma concentration and the glomerular filtration rate (GFR). This, of course, assumes that one is dealing with a small molecule that is filtered in the same concentration as is found in plasma. If there were no tubular modification of the filtrate, the excretory rate would be equal to the filtration rate. However, the solid lines in the illustration show that at all plasma concentrations excretion for both sodium and glucose is much less than filtration. The colored area between the two lines illustrates the rate of reabsorption of the substrate.

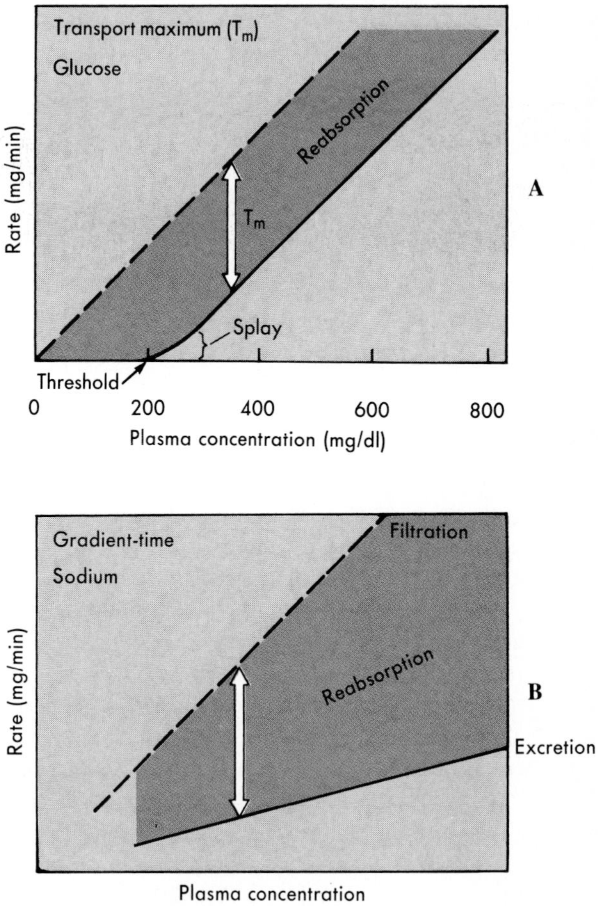

■ Fig. 46-3. Comparison of a T_m system, **A,** and a gradient-time system, **B.** The dashed line represents the rate at which a test molecule would be excreted if all filtered material passed through the tubule and was voided. The solid line represents excretion rates actually observed. The shaded areas and heavy arrows represent the rate of absorption of the solute (that is, the difference between filtration rate and excretion rate).

Note that a change in plasma concentration has a much different effect on glucose than on sodium. Glucose transport is said to be of the T_m type. For the class of molecules that display T_m characteristics, progressive elevation of plasma concentration from zero produces no change in excretion initially. In the case of glucose, excretion is zero until the concentration reaches a plasma level of about 200 mg/dl of plasma, at which

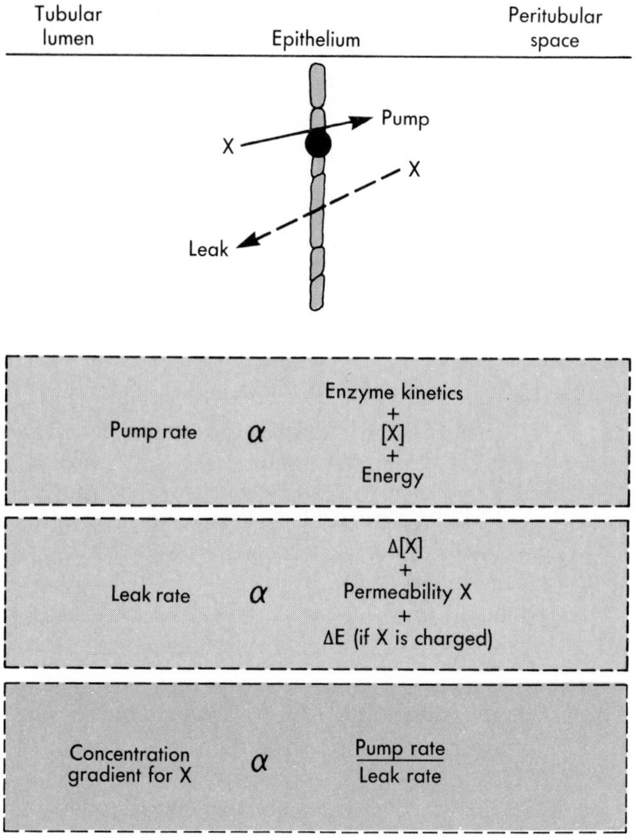

Fig. 46-4. Pump/leak relations as determinants of net tubular transport. It is assumed that the molecule in question has a finite permeativity and that pump activity will cause a concentration gradient that in turn will cause back leak into the lumen. The pump rate is assumed to be proportional to the concentration of the substrate [X]. ΔE is the potential difference across the epithelium.

point the excretion begins to rise in a characteristic curvilinear fashion. The minimal plasma concentration at which excretion begins is the *plasma threshold*.

Over a small range of plasma concentrations, the excretion varies curvilinearly with the plasma concentration, and above this range the excretion line parallels the filtration line. The curvilinear portion of the relation results from saturation of the enzyme system and also from the fact that different tubules have different transport characteristics. The breadth of the curvilinear portion is referred to as the *splay* of the transport system. Above the range of the splay, the difference between filtration and reabsorption is constant, implying that reabsorption is constant and that increases in filtration result directly in increases in excretion. The system is said to be "saturated," and the maximum reabsorptive rate is called the T_m.

Contrast the behavior of glucose with the behavior of a system displaying gradient-time characteristics, such as the sodium transport system. Excretion varies linearly with plasma concentration over a wide range, and there is no evidence of a threshold (Fig. 46-3, *B*), because in the gradient-time system (sodium) the reabsorption rate varies continuously with the plasma concentration. In the proximal tubule about 67% of filtered sodium is reabsorbed over a wide range of filtration rates.

The factors that contribute to the characteristics of the curves shown in Fig. 46-3 are complex and depend on many elements of renal function. However, a partial understanding of the processes can be developed by simply considering the enzyme kinetics likely to be associated with the transport process. The T_m system behaves as though the carrier molecules display kinetic behavior typified by a relatively low V_{max} and low K_m; that is, the carrier exhibits a high affinity for the substrate but is quickly saturated. In this case the carrier combines with the molecule to be transported at a low concentration and transports the molecule out of the tubule. High affinity is suggested by the fact that essentially all of the filtered molecule can be reabsorbed. A limiting transport rate is encountered when the V_{max} for the carrier is reached, and this corresponds in some degree to the threshold.

In contrast, we can assume that the gradient-time system has a lower affinity relative to ambient concentrations and therefore does not become saturated. The V_{max} also must be relatively greater for the gradient-time system, because the curve is approximately linear. Thus the curve probably represents only a small portion of the total range of transporter activity. Active intrarenal regulatory mechanisms also are involved in balancing the tubular reabsorption processes to the filtration rate; these are discussed later.

A second, basic tubular process that confers different behaviors on molecules of different types is *back leak* (Fig. 46-4). To a first approximation one can assume that the concentration of a solute in peritubular interstitial fluid is held approximately equal to the concentration in plasma by diffusional exchange between the interstitium and the peritubular capillary blood. This contrasts sharply with the situation in the tubule, where a relatively small volume of solutes is present. Active transport of any significant quantity of material across the tubular epithelium will alter the luminal concentration and result in the generation of a concentration gradient between tubular fluid and interstitial space. Once this occurs, material will tend to leak back across the epithelium at a rate determined by the tubular permeability and the chemical and electrical gradients. In the steady state the concentration gradient that is achieved across the tubular epithelium will be proportional to the ratio of the rate at which material is actively transported to the rate at which material leaks back (i.e., the *pump/leak ratio*).

This pump/leak concept enables one to further understand the factors that contribute to the distinction between a T_m system and a gradient-time system. In a T_m system, such as that for glucose, the leak must be small relative to the pump rate, because virtually all the material can be reabsorbed out of the tubule over a significant range of plasma concentrations (below threshold; Fig. 46-3, *B*). However, with a gradient-time system, such as that for sodium, the permeability of the tubule is high compared with the pump. Concentration gradients as great as those in T_m systems cannot be achieved because of back leak down the concentration gradient.

A second result of having a relatively low pump/leak ratio in a system in which the tubular volume is small, as in the kidney, is that the net transport becomes flow dependent. When flow is relatively low, transport of material out of the tubule will quickly establish a limiting gradient (back diffusion equals pump rate). When flow is high, however, the tubular fluid spends a shorter time at each point along the tubule. Therefore the concentration gradient rises less rapidly, back leak is minimized, and net transport is favored.

These facts suggest the basis for the behavior of the gradient-time system. When the affinity of the carrier is relatively low, the V_{max} is relatively high, and the tubular permeability is high compared with the transport rate; a limiting gradient probably will be established before the end of the tubule is reached. The conditions under which this gradient is established will be a function of flow; that is, it will be a function of the time that the fluid remains in the tubule. Thus both *concentration gradient* and *time* become important determinants of reabsorptive processes.

Special Transport Systems

Analysis of Transport

Renal physiology, as a rule, is concerned most intensely with the transport of small molecules, such as sodium, chloride, and water. However, many molecules are transported by specialized carrier systems located at various points along the nephron. This section presents general principles of the transport of these molecules and gives a few examples that have particular physiological relevance. The importance of these transport systems can be appreciated by recalling that the glomerular filtration rate is approximately 180 L/day for an average man with 45 L of total body water. All small molecules, such as glucose and sodium, are filtered in a concentration equal to that in plasma (Fig. 45-5). The tubular epithelium is not especially permeable to these molecules, and thus they would be excreted if no further processing by the kidney occurred. Continued excretion of such essential materials would be catastrophic. To prevent this, a primary activity of the renal epithelium is to *reabsorb* and return to the body those filtered materials that need to be conserved.

A second major function of the transport systems is to *secrete* materials that are not filtered in adequate quantities to permit elimination from the body at the required rates. End products of hepatic metabolism frequently appear in the form of organic anions (e.g., hippuric acid). As might be expected, appropriate transport systems in the tubular epithelium permit the kidney to excrete these anions in the final urine, rather than allowing their accumulation in the body. (These transport sites for organic anions are frequently the means whereby drugs and other foreign substances, such as penicillin, are eliminated from the body.)

The amount of material with which the epithelial transport systems of the kidney must deal is usually quantified in terms of the *solute load*. The load is the quantity of solute per unit time that is presented to the tubule. Obviously the magnitude of this load will ultimately determine whether the tubule can meet the demand for transport. For a reabsorptive system the load is simply the quantity of material that is filtered per unit time. In other words, the glomerular filtration rate determines the volume of fluid per unit time that passes the transport system, and the concentration of the solute in the glomerular filtrate determines the solute quantity in each volume. Reabsorptive load is then simply the product of glomerular filtration rate and plasma concentration for small solutes that are freely filtered. Although not usually referred to by renal physiologists, a secretory load also might be determined. The secretory load is less easy to quantitate but is conceptually the peritubular capillary blood flow times the concentration of the solute in peritubular capillary blood.

Tests of renal function often rely on the infusion of a secreted dye at progressively faster rates until the secretory system is saturated. Analysis of the urine under these conditions yields an estimate of the capacity of the tubules to transport such a molecule. This capacity is the combined result of the tubular transport capacity and the renal blood flow. As a result, the measurement yields data on both cardiovascular and tubular function.

It is important to think of transport as net transport, because a given molecule can be both secreted and reabsorbed in different portions of the tubule. A general equation for the net tubular transport rate can be written as follows:

$$\text{Net transport rate} = \text{filtration rate} - \text{excretion rate}$$

$$TR = (GFR \times P_x) - (U_x \times V) \quad (1)$$

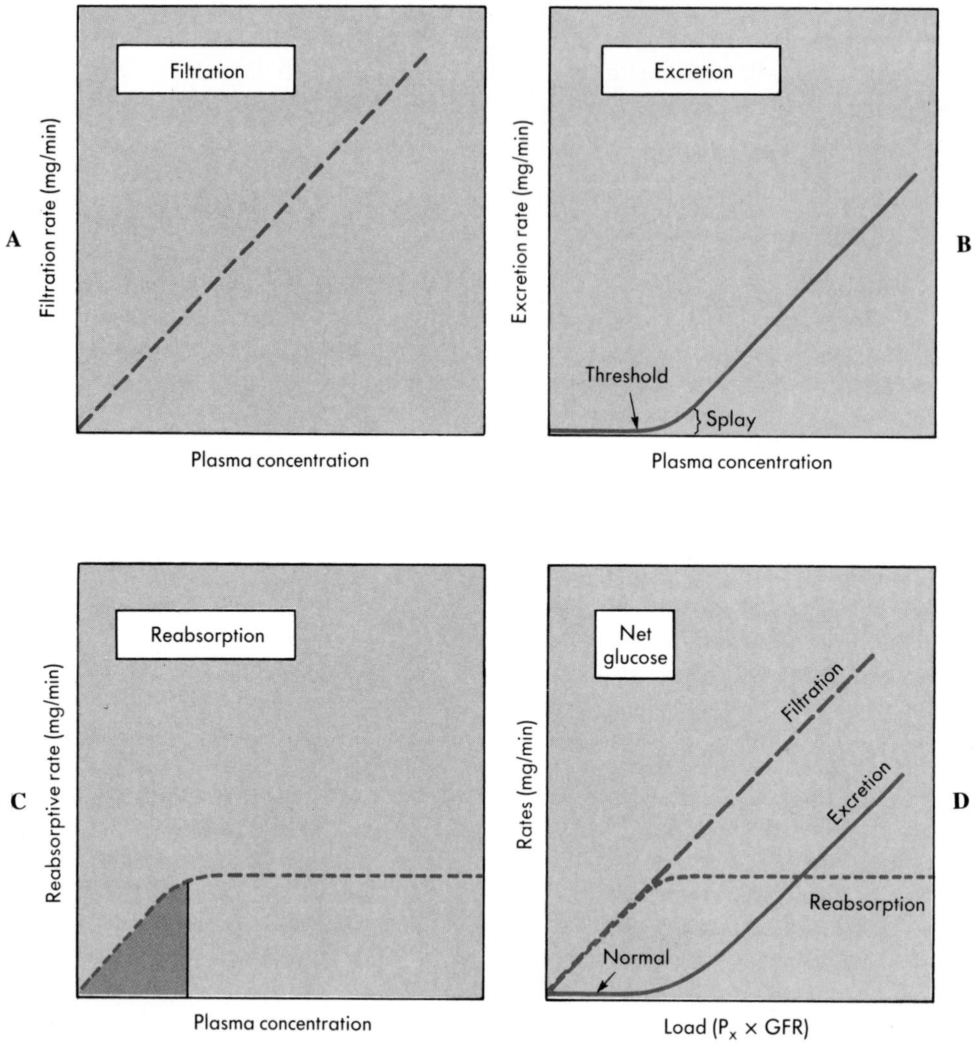

■ **Fig. 46-5.** Graphical analysis of tubular transport of a solute (*glucose*). **A,** The rate at which filtered solute enters the tubule as a function of plasma concentration. **B,** Excretion rate as a function of plasma concentration. **C,** Reabsorption, given filtration and excretion rates in **A** and **B.** The colored area represents the region in which filtration is exactly matched by reabsorption. When reabsorption cannot match filtration, excretion occurs, and the threshold is said to have been reached. **D,** The three curves plotted together.

where

P_x = the plasma concentration of a solute
U_x = the corresponding urine concentration
V = the *rate* of urine flow (Note that in renal physiology V rather than $\dot{V}$ is used to denote flow rate.)

If the transport rate has a negative value, then the excretion rate ($U_x \times V$) is in excess of the filtration rate (GFR $\times$ P_x). Hence solute must have been added to the glomerular filtrate, that is, secretion took place. Conversely, if the transport rate has a positive value, then reabsorption must have taken place, because less material appears in the final urine than had originally been filtered.

A graphical analysis is helpful in understanding the interrelations among filtration, reabsorption, and excretion of a solute. Fig. 46-5, *A,* shows the rate at which the filtered solute (glucose) enters the tubule as a function of the plasma concentration (filtration). If glomerular filtration rate is constant, the plasma concentration will determine how much solute is filtered per unit time. Excretion rate as a function of plasma concentration is shown in Fig. 46-5, *B*. Fig. 46-5, *C,* shows what *must* happen to reabsorption, given the filtration and excretion rates shown in Fig. 46-5, *A* and *B*. Since the rate of filtration of glucose must increase linearly with increases in plasma glucose (Fig. 46-5, *A*) and no glucose is excreted for plasma levels up to the threshold

Table 46-1. Special transport systems

Reabsorbed		*Secreted*	
Monosaccharides	***Amino acids****	***Organic anions†***	***Organic cations‡***
D-Glucose	Neutral (e.g., L-glycine)	Benzoates	Acetylcholine
D-Galactose	Acidic (e.g., L-glutamic acid)	Bile acids	Creatinine
D-Fructose	Basic (e.g., L-arginine)	Hippurates	Dopamine
		Prostaglandins	Morphine
		Sulfonamides	Tyramine

*Note stereospecificity for L form of amino acids versus D form for monosaccharides.
†Interaction between these compounds varies with species and site in nephron.
‡Secretory processes for these molecules differ from those for the organic anions, but their biochemical basis has yet to be determined. Transport sites for these molecules are concentrated in the S3 segment of the proximal tubule.

(Fig. 46-5, *B*), reabsorption must exactly match filtration over this range of plasma concentration (*lightly colored area,* Fig. 46-5, *C*). Once the threshold is reached and the range over which splay is observed is exceeded, a maximum rate of reabsorption (T_m) will be achieved at some predictable plasma concentration. In Fig. 46-5, *D,* the three curves are plotted together on the same graph. The load (GFR $\times$ P_x) replaces P_x as the abscissa.

Nature of Organic Solute Transport

Transporters for a variety of organic solutes reside primarily in the proximal tubule, especially in the S1 segment. There are two pathways for reabsorption of solutes. Materials can cross the tubular epithelium passively via movement through the intercellular spaces (the *paracellular pathway*). Of greater importance are a number of *secondary active transport systems.* The secondary systems are similar to those found in intestinal epithelium in which glucose shares a common carrier or transporter with sodium. Such active transport systems translocate a wide variety of molecules, including hexoses and amino acids, as well as organic acids, bases, and polyvalent cations.

Energy for the transport process comes from sodium gradients established by the Na^+, K^+-ATPase, which lowers intracellular sodium concentration. Sodium and the organic molecule to be transported combine with a carrier at the apical surface of the epithelial cell. The solute pair is translocated across the membrane as a result of changes in the affinity of the carrier for glucose and sodium at the interior membrane surface. The sodium requirement of the renal glucose carrier saturates at a sodium concentration of about 40 mEq/L. Because luminal sodium rarely falls to values as low as 100 mEq/L, sodium is usually not limiting for glucose transport in the kidney.

A similar carrier is probably present on the basolateral side of the cell. It can facilitate the extrusion of glucose into the interstitial space surrounding the tubule. Glucose can then be picked up by the blood passing through the peritubular capillaries and returned to the circulation.

Because the sodium-glucose-carrier complex formed at the apical border is charged (+1), transport of glucose will be electrogenic. Thus raising glucose concentration in the luminal fluid can both accelerate sodium reabsorption and produce a more negative intraluminal potential.

Transport systems of the nature just described exist in the proximal tubule for a variety of sugars, amino acids, and organic anions and cations. Such systems share many common properties, including stereospecificity (e.g., D-glucose versus L-glucose), molecular specificity (e.g., basic amino acids versus acidic amino acids), saturation, competition (e.g., glutamate versus aspartate), and inhibition (e.g., phloritin inhibits glucose transport). Table 46-1 summarizes the groups of molecules that share common or related carriers. Note that all of these systems show secondary active transport, and the transport occurs in many, but not all, cases in the proximal tubule. With the exception of organic cation transport, energy is derived from sodium gradients. The energy source for the transport of organic cations remains to be established but may depend on hydrogen transport. In the following sections, typical examples of specialized transport systems are present. Examples of both reabsorptive and secretory systems are given, and the possibility for physiological and pathological variation of the T_m is explored.

Calcium and Phosphate Reabsorption: Physiological Alteration in T_m

The behavior of the phosphate transport system typifies a system under humoral control. Fig. 46-6, *A,* shows the behavior of the phosphate transport system before and after the administration of *parathyroid hormone* (parathormone, or PTH). Parathormone regulates calcium and phosphate balance in the body, as described in Chapter 51. In general, these two ions are regulated

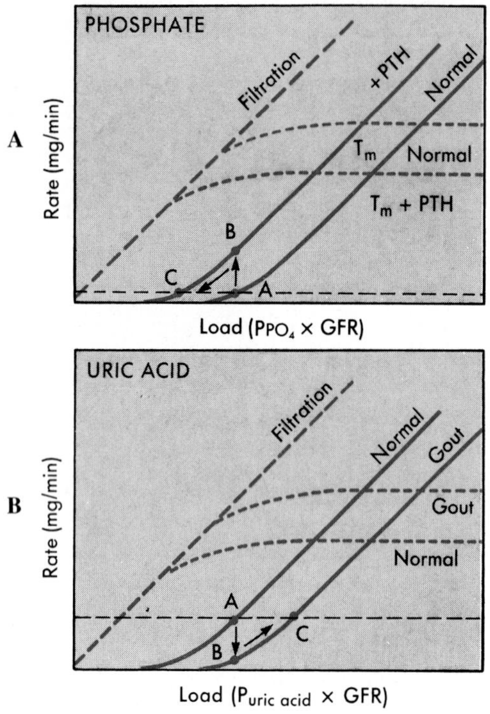

■ **Fig. 46-6. A,** Phosphate reabsorption. *PTH,* Parathormone. *A* is the normal operating point for phosphate. Solid lines represent excretion, dashed lines represent filtration, and dotted lines represent tubular transport. Curves represent responses before *(normal)* and after administration of parathormone. **B,** Uric acid reabsorption, *A* is the normal operating point for the system. Gout elevates the T_m. Dashed black lines in **A** and **B** represent excretion required by normal dietary intake.

in a reciprocal fashion. Parathormone increases the excretion of phosphate and decreases the excretion of calcium, allowing calcium levels in plasma to rise without danger of reaching a level that would precipitate salts of calcium and phosphate.

In contrast to the condition for glucose, the normal plasma concentration and filtered load for phosphate are above the threshold for renal excretion (for example, *A* in Fig. 46-6). Furthermore, the splay in the phosphate curve is somewhat wider. Thus phosphate excretion depends very much on filtered load, and changes in both glomerular filtration rate and plasma phosphate concentration can alter phosphate excretion.

Assume that the normal relation between excretion and load is at point *A* on Fig. 46-6, *A*. After administration of parathormone, the T_m for phosphate would be reduced. A reduction in the T_m increases phosphate excretion at a given phosphate load (as shown by the transition *A-B* in Fig. 46-6, *A*). As excretion rises, the plasma phosphate concentration will fall *(B-C)*. The system will stabilize at some new, lower plasma phosphate concentration, because reduced plasma phosphate levels will reduce the filtration of phosphates. The new stable point will occur when plasma levels fall low enough to cause the excretion of phosphate to equal the daily intake of this ion *(dashed black line)*. Note that in the steady state the phosphate excretion is the same before *(A)* and after *(C)* parathormone administration; the difference is that similar phosphate excretion rates are attained at lower plasma levels after the hormone has been given.

The effects of parathormone on renal function are complex. They may reflect altered glomerular filtration rate, as well as altered tubular reabsorption of phosphate, sodium, and calcium. In general, the net result of parathormone addition is to decrease the reabsorption of phosphate and to increase the reabsorption of calcium.

Phosphate and calcium transport occur largely in the proximal tubule (about 60% to 70% of the filtered ion is reabsorbed) and to a lesser degree in the more distal portions of the nephron. Parathormone modulates transport all along the nephron by binding to the basal cell border and inducing intracellular formation of cyclic AMP. Cyclic AMP formation somehow inhibits the apical phosphate transporter.

■ *Uric Acid Reabsorption: Pathological Alteration in T_m*

Special transport systems also are subject to pathological changes. One of the more notable of these is the altered uric acid transport that occurs in gout. Fig. 46-6, *B*, shows the variations in uric acid reabsorption with load in a normal individual and in an individual with gout. The basic relations between filtration, excretion, and reabsorption are similar to those shown previously for glucose and phosphate. The normal operating point *(A)* for a human is such that uric acid is continuously excreted.

Gout is a disease in which uric acid accumulates in the plasma and interstitial fluid. The uric acid subsequently precipitates in a variety of locations, especially within the joints, thereby causing inflammation *(arthritis)* and pain. The disease has a complex origin that includes changes in both the metabolism and the excretion of uric acid. Fig. 46-6, *B*, illustrates the change in tubular reabsorptive rate. In the gouty individual the reabsorptive T_m for uric acid is increased, and, as a result, an elevated plasma concentration is required to raise the filtered load to a value such that the excretion of uric acid will equal its production. A typical pathway for pathophysiological changes in renal uric acid handling is shown in the path *A-B-C* in Fig. 46-6, *B*. Beginning with a stable situation at *A*, the T_m is presumed to increase pathologically, and excretion thus falls to *B*. Assuming a constant dietary intake of purines, reduced excretion will be followed by increasing plasma concentration and increasing excretion *(B-C)*. A new, stable equilibrium will be achieved when excretion reaches a level mandated by purine intake *(C)*.

Uric acid crystals precipitate not only in the joints but also in the basement membrane of the glomerulus, reducing the filtration rate. As a result, accumulation of uric acid is exacerbated by reduced glomerular filtration superimposed on the elevated T_m.

Treatment for gout consists of the administration of antiinflammatory agents, dietary modifications, and the use of drugs that diminish uric acid formation, as well as drugs that enhance excretion of uric acid by the kidney. The latter drugs are the *uricosuric agents,* such as sulfinpyrazone. This agent is a competitive inhibitor of both uric acid transport and the transport of other organic anions.

One of the peculiar characteristics of the uric acid transport system is that, although the net activity of tubular function is reabsorption of uric acid, the molecule is *both* secreted and reabsorbed during its passage through the nephron. The secretory and reabsorptive mechanisms vary in importance along the proximal tubule, with reabsorption dominating in the S1 and S3 segments and secretion dominating in the S2 segment. As a consequence of this bidirectional transport, drugs that inhibit uric acid transport may decrease rather than increase the excretion of uric acid. This is the case with low doses of sulfinpyrazone, which are observed to inhibit rather than augment urate excretion. Obviously, such an effect compromises their therapeutic usefulness.

Penicillin Excretion—A Secretory System

There are many secretory transport systems along the length of the nephron. These transporters move material from the peritubular capillary blood into the tubular lumen. One of the most studied is the organic anion transport system, typified by the system that transports penicillin. A graphical analysis of the transport system for penicillin is shown in Fig. 46-7. As the load is increased, secretion rises and approaches a limit (T_m). In contrast to reabsorptive systems, the secreted molecules are added to the quantity of material filtered, and excretion exceeds filtration.

The organic anion secretory systems are very active. For some molecules, such as *p*-aminohippuric acid (PAH), virtually all the molecules that enter the peritubular capillary plasma are secreted and ultimately are excreted in the urine, along with the filtered components. As shown later, PAH excretion becomes an important tool in the measurement of renal blood flow because of this property.

The avidity of the penicillin transport system is such that a large fraction of an administered dose of penicillin will be rapidly secreted and then excreted. Hence, the blood concentrations tend to drop rapidly. The dose

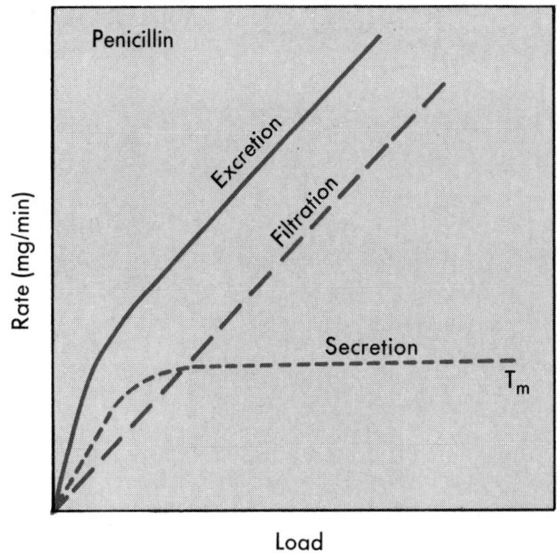

Fig. 46-7. Renal handling of penicillin. Note that excretion is the sum of filtration and secretion.

therefore must be high, and administration must be frequent if therapeutic levels are to be maintained. When penicillin was originally introduced and was expensive, therapeutic levels were achieved more economically by concomitantly administering probenecid, a competitive inhibitor of organic anion transport. Probenecid blocks the secretory system for penicillin and diminishes its excretion.

Proximal Tubular Salt and Water Transport

Transport Characteristics

The force for both secretion and reabsorption of many molecules is derived from the hydrolysis of adenosine triphosphate (ATP) by a small number of ATPases. By far the most important of these is the Na^+, K^+-ATPase on the basolateral cell borders of the epithelial cells (Fig. 46-8, *A*). This ATPase provides the energy for reabsorption of sodium, chloride, bicarbonate, potassium, small organic solutes, and water. In addition, it powers the secretion of hydrogen and a number of organic solutes. In the proximal tubule only the calcium-sensitive, magnesium-dependent ATPase has any other significant primary active transport function. The distal tubule also has an active H^+ carrier.

Fig. 46-8, *A*, summarizes the elementary salt and water transport in proximal tubular cells. This illustration shows the tubular epithelial cell, with one side facing the tubular lumen and exposed to the filtrate and the other side facing the peritubular space and close to the peritubular capillaries. Only a small distance separates epithelial cells from the peritubular capillaries, and flow through these capillaries is relatively high. As a result,

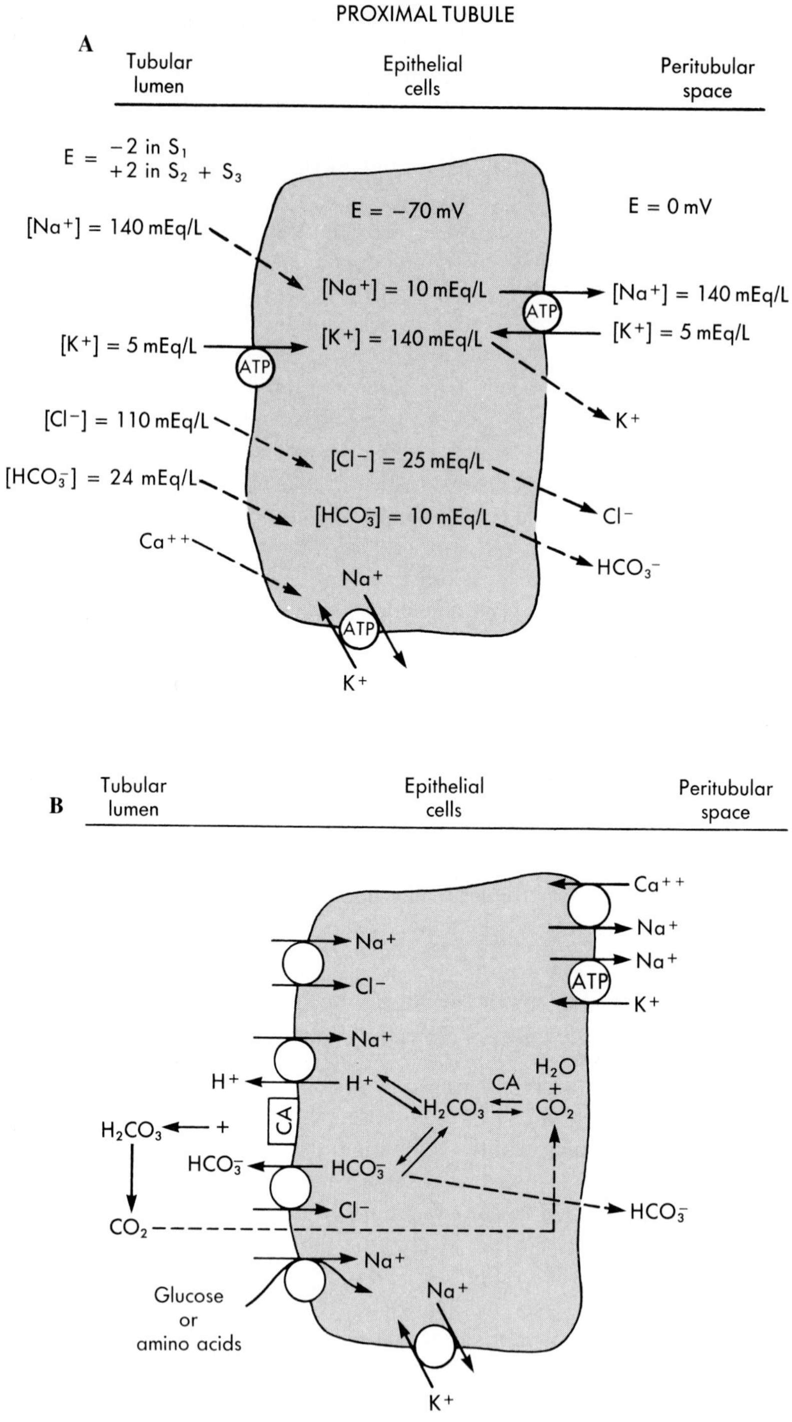

■ **Fig. 46-8. A,** Ion transport in proximal tubular epithelial cells. Dashed arrows represent passive diffusion. Solid arrows represent either primary active transport, secondary active transport, cotransport, or countertransport. Primary active transport is indicated by an ATP within the circle. Electrical potential is indicated by *E*. All potentials are referenced to interstitial fluid or peritubular space. The potential difference between peritubular space and tubular lumen varies from −2 V in the early proximal tubule to +2 in the late proximal tubule. The intracellular potential is assumed to be −70 mV. Concentrations in the early part of the proximal tubule are shown as an indication of the chemical gradients driving ion movement. **B,** Summary of cotransport and countertransport systems driven by the Na^+, K^+-ATPase. See text for additional details. *CA,* Carbonic anhydrase.

concentrations of various small molecules, such as glucose and sodium, within the cortical interstitial space will approximate plasma levels, regardless of the reabsorptive processes.

Ion transport depends on the electrochemical potential difference that exists across the membrane to be traversed (Chapter 2). Equation *1* in Fig. 46-1 describes the relation between the important variables and the movement of material in a purely passive system. For a monovalent cation this equation reduces to the Nernst equation:

$$J = PA\ (\Delta C + 60C\ \Delta E) \qquad (2)$$

The equation indicates that net passive transport depends on both the electrical potential difference (ΔE) and the concentration difference (ΔC), as well as on the permeability (P) and the area (A) of the membrane. At equilibrium, movement of material is zero, and the relation between potential difference and the concentration gradient across the epithelium reduces to:

$$\Delta E = -60 \log C_L/C_P \qquad (3)$$

C_L and C_P are concentrations in the lumen and plasma, respectively. The equation shows that a tenfold difference in concentration for a monovalent ion can be produced by a 60 mV potential difference across the membrane. From these two equations it is apparent that understanding tubular transport of small ions depends heavily on clearly defining concentration and potential differences across the tubular epithelium and across the various membranes of the tubular epithelium.

Sodium Transport

With the preceding concepts in mind, let us examine the reabsorption of filtered sodium (Fig. 46-8, *A*). The description is based on a transport model that assumes that the sodium ion traverses the apical cell membrane passively and is actively extruded into the peritubular space at the basal cell surface. Sodium in the lumen of the first portion of the proximal tubule has the same concentration as in plasma, since this fluid represents the glomerular filtrate. The luminal concentration, approximately 140 mEq/L, is much greater than the 10 mEq/L concentration existing inside the cell. This gradient drives sodium from the lumen into the cell. A potential difference also favors cell entry for the sodium ion, since the potential in the tubular lumen is always more positive than the cell interior. The luminal potential over the length of the proximal tubule ranges from −2 mV in the first segment to +2 mV in the middle and end segments. The intracellular potentials are approximately −70 mV, yielding a net electrical gradient of 68 to 72 mV, which favors passive movement of sodium into the cell. It follows, given these large electrical and chemical driving forces for sodium, that sodium entry into the epithelial cell is passive. Sodium does not diffuse into the cell equally from both cell surfaces, however, because the apical surface possesses a much larger surface area as a result of the extensive brush border. In addition, carriers that facilitate sodium entry are present at both surfaces, but are more active on the luminal surface.

The forces that drive the exit of sodium at the abluminal side of the epithelial cell are quite different from those that exist at the apical side. The basal concentration gradient is from an intracellular level of 10 to a peritubular space value of 149 mEq/L. The potential difference is between −70 mV in the cell and 0 mV in the peritubular space. Thus both chemical and electrical gradients oppose sodium exit at the basolateral surface, and active sodium extrusion from the cell is required if sodium is to be reabsorbed.

Sodium is actively extruded from the cells at the basal cell borders, as well as into the space between adjacent epithelial cells. As mentioned, the bulk of the transport is accomplished by the coupled Na^+, K^+-ATPase on the basal cell border. This enzyme transports two potassium ions into the cell for every three sodium ions removed.

Less is known of the transport system located on the lateral cell surfaces. This system requires energy, but coupling ratios and kinetics have yet to be measured. However, it is clear that transport results in net transfer of solute into the clefts between the cells. This transport system is important in determining water reabsorption across the proximal tubular epithelium and is discussed later.

Attention has been focused on the Na^+, K^+-ATPase as the pivotal element in sodium transport by the epithelium. However, the rate-limiting process for transport is probably the rate of sodium leak into the cell at the apical membrane. This leak is a major determinant of intracellular sodium concentration and, in turn, of the availability of sodium to the ATPase. The sodium leak across the apical cell membrane can be inhibited by the diuretic amiloride, which acts on the membrane sodium channels. Because sodium enters the cell less readily in the presence of amiloride, less sodium is reabsorbed and a larger portion is excreted.

Both intracellular and extracellular messengers can modulate the sodium permeability of the apical cell membrane and thus net sodium transport. Also, certain hormones, such as aldosterone, modulate the sodium permeability of the apical cell membrane.

A point of some concern in understanding epithelial transport is the coordination of the activities of the Na^+, K^+-ATPase and the ion leaks across the cell membranes. These processes must be balanced if the epithelial cells are to transport large and variable quantities of solute, while maintaining their own internal environment reasonably constant.

It is hypothesized that the intracellular Ca^{++} concentration coordinates ion fluxes at the apical and basal cell membranes. This process involves modification of the Na^+ leak rate at the apical surface and the Na^+, K^+-ATPase activity on the basal cell membrane. The ion movements are coordinated by the sodium-calcium exchange mechanisms (Fig. 46-8, *B*). Intracellular calcium appears to reduce the sodium permeability of the apical cell membrane and the potassium permeability of the basal cell membrane. Thus an increase in the Na^+, K^+-ATPase activity may decrease the intracellular Na^+ concentration by extruding sodium more rapidly. This steepens the diffusion gradient for sodium entry and thereby increases sodium influx. The increased sodium influx increases calcium efflux via the exchanger. Consequently increased Na^+, K^+-ATPase activity causes a secondary decrease in the intracellular calcium concentration and the inhibitory effect of calcium on the sodium permeability of the apical cell membrane is abolished. Sodium influx would then increase to match the higher transport rate at the basal border. Also, increased cellular calcium can simultaneously accelerate potassium flux at the basal cell border to accelerate its entry and stabilize the concentration of this ion.

Net reabsorption. The net reabsorption of sodium or any other ion will be determined by the total electrochemical gradient across the tubular epithelium. If intraluminal concentration and potential are known, as well as the concentration and potential in the interstitial space, then the Nernst equation can be applied to predict if energy will be required to drive the transport process. Whether molecular movement is active or passive at a particular cell surface needs to be considered only to determine the location of transport processes. Much of the discussion in this chapter is concerned with this net transport between lumen and renal interstitial space, rather than with a detailed examination of fluxes at particular cell borders. A more detailed treatment of epithelial transport processes is presented in Chapters 43 and 44.

Paracellular flux. One other feature that is important for proximal tubular sodium and water transport is the permeability to water and small ions of the tight junctions that connect adjacent epithelial cells. Because of this permeability, only small concentration gradients can be generated for these substances across the proximal tubular epithelium. The high permeability of the junctions means that back leaks will be large and the pump/leak ratio will be small. This low pump/leak ratio signifies that concentrations of sodium will be very close to those in plasma along the entire length of the proximal tubule. Only in the presence of osmotic diuretics are significant concentration gradients observed (as explained later).

Glomerulotubular balance. Although only small sodium concentration gradients are generated, large quantities of sodium are reabsorbed by the renal tubules. As shown in Fig. 46-3, *B*, about 67% of the filtered sodium is reabsorbed, and the fraction is constant over a wide range of sodium filtration rates. This close coupling between sodium filtration and sodium reabsorption is due to many factors. Earlier it was pointed out that the kinetics of the transport system and the relatively high tubular permeability to sodium contribute to the parallelism. The coupling is also the result of a process called *glomerulotubular feedback*, or *glomerulotubular balance*, in which the filtration rate modifies tubular reabsorption and vice versa.

Glomerulotubular balance helps to maintain body sodium levels constant in the face of substantial changes in the glomerular filtration rate. Such charges may occur during strenuous physical exertion or after a large meal. A link between sodium reabsorption and glomerular filtration tends to stabilize sodium excretion and minimize swings in plasma sodium levels.

Glomerulotubular balance also serves to match filtration and reabsorption from nephrons of different lengths. There is no *a priori* reason why the glomerulus of a given nephron should filter at a rate exactly commensurate with the length of its proximal tubule. However, glomerulotubular balance ensures that, even in the face of anatomical disparities, filtration and reabsorption will be balanced. The splay in the glucose transport system (Fig. 46-5) is caused, at least in part, by uncorrected imbalances between the filtration rate and the transport rate for glucose.

Potassium Transport

Continuing with an examination of proximal tubular function, we next consider the reabsorption of filtered potassium. In the S1 and S2 segments of the proximal tubule, the potassium concentration remains relatively constant at about 5 mEq/L. (Recall that the intracellular potassium concentration is about 140 mEq/L.) Potassium is normally reabsorbed passively into the cell against a concentration gradient, but down an electrical gradient of -70 mV. The reabsorption of potassium is, of course, driven by sodium and water reabsorption, which causes movement of potassium. Much of the potassium appears to move, along with the water, through the paracellular pathways.

Some form of active transport for potassium must exist at the luminal cell surface because during potassium deprivation, potassium reabsorption can move against an electrochemical gradient. The details of this mechanism remain to be determined. Exit from the cell on the abluminal side is the result of passive diffusion down an electrochemical gradient. In the distal portion of the proximal tubule (S3 segment), potassium is secreted rather than reabsorbed. Because of the inaccessibility of this segment for direct sampling of tubular fluid, little is known about the nature of the secretory process. The net result of reabsorption in the S1 and S2 segments of the tubule, and of secretion in the S3 segment, is the reabsorption of about 70% of the filtered potassium by the time the tubular fluid reaches the end of the proximal tubule.

Chloride and Bicarbonate Reabsorption

The net reabsorption of both chloride and bicarbonate in the proximal tubule is down electrochemical gradients and is closely related to sodium movement and, to a lesser degree, to potassium reabsorption. (See discussion on coupled ion fluxes below.)

Cation reabsorption cannot occur without accompanying charge movement to maintain electroneutrality. Chloride and bicarbonate are mainly responsible for maintaining electroneutrality in the proximal tubule. In the first part of the proximal tubule bicarbonate is the major anion accompanying sodium reabsorption. In contrast, in the middle and terminal portions of the proximal tubule, chloride movement contributes more significantly to the maintenance of charge neutrality as sodium and potassium are transported.

Coupled Solute Fluxes

The elements of epithelial function shown in Fig. 46-8, *A*, are adequate for an overview of the determinants of the net flux of salt and water. However, important factors related to coupled solute fluxes and the membrane carriers that facilitate diffusion across the epithelial cell membrane are neglected. A number of carriers are present that do not use energy directly; rather, they are driven indirectly by energy derived from sodium concentration gradients. These gradients, of course, are generated at the expense of hydrolysis of ATP by the Na^+, K^+-ATPase.

The most important carriers are shown in Fig. 46-8, *B*. As mentioned previously, sodium ion extrusion from the cell at the basolateral borders lowers intracellular sodium ion concentration and sets up a diffusion gradient for this ion. In turn, sodium entry into the cell can drive the exit of the hydrogen ion.

Chloride ion reabsorption. Sodium-hydrogen exchange and chloride reabsorption are also coupled by a complicated process. The Na^+, K^+-ATPase generates a sodium gradient that causes sodium influx. A portion of the sodium influx will occur via the sodium-hydrogen carrier and is therefore associated with transport of the cellular hydrogen ion into the lumen. Some of the hydrogen ions are derived from the intracellular carbonic acid that is in equilibrium with cellular carbon dioxide. Removal of hydrogen ions leaves an excess of the bicarbonate ions in the cell, thereby generating a concentration gradient for the bicarbonate ion between the cell and the lumen. The bicarbonate ion can then move down its concentration gradient in exchange for the chloride ion, which moves into the cell. The abluminal surface of the epithelial cell is permeable to chloride, which diffuses passively into the renal interstitial space and is reabsorbed.

The equilibrium between carbonic acid and carbon dioxide is attained slowly in saline solutions. In the kidneys the enzyme *carbonic anhydrase (CA,* Fig. 46-8, *B)* is present in the renal tubules, both on the brush border and in the cytoplasm of the epithelial cells. This enzyme accelerates the formation of carbon dioxide in the tubular lumen from the hydrogen ion and bicarbonate ion. The high permeativity of the carbon dioxide and water formed means that these molecules can traverse the epithelium rapidly.

Bicarbonate ion reabsorption. Whereas HCO_3^- can engage in countertransport with Cl^-, the dominant movement of HCO_3^- is from lumen to peritubular space (i.e., reabsorption). HCO_3^- is, of course, present in the tubular lumen as part of the original filtrate. As will be discussed later, this bicarbonate must be reabsorbed if the acid-base balance is to be maintained. The sodium-hydrogen exchanger is also important in indirectly facilitating the reabsorption of bicarbonate, an ion that has low inherent permeativity because of charge and size.

Bicarbonate is present in the glomerular filtrate at a concentration of about 25 mEq/L. Therefore a hydrogen ion that enters the lumen in exchange for a sodium ion is likely to combine with a filtered bicarbonate ion and form H_2CO_3. When this happens, H_2CO_3 dissociates to H_2O and CO_2, and the resultant carbon dioxide can easily enter the cell and dissociate to form hydrogen and bicarbonate ions. The hydrogen ion formed in the process essentially replaces the hydrogen ion originally exchanged for the sodium. The bicarbonate ion left in the epithelial cell cytoplasm is free to diffuse across the abluminal membrane in association with the sodium. The net result of the process is the reabsorption of $NaHCO_3$ by a very efficient mechanism, rather than the slower absorption of sodium chloride. Because more bicarbonate than water is reabsorbed, the luminal pH falls in the proximal tubule. The pH reaches a value of about 7.2 by the end of the proximal tubule.

Water is reabsorbed with solute in the proximal tubule (as described later). Consequently, the concentration of remaining solutes rises, thereby generating diffusion gradients for these materials. Because the chloride ion is more permeative than the bicarbonate ion, its diffusion down its concentration gradient generates a slightly positive luminal potential in the S1 and S2 portions of the proximal tubule.

Bicarbonate represents a significant portion of the filtered solute in the tubular lumen. Therefore inhibition of bicarbonate reabsorption inhibits sodium and water reabsorption and produces diuresis. A class of diuretics, the *carbonic anhydrase inhibitors,* acts via this mechanism. These drugs reduce the rates of bicarbonate reabsorption, as well as those of sodium and water.

Water Reabsorption

The tubular fluid remains isotonic while sodium, potassium, chloride, bicarbonate, and glucose are being reabsorbed in the proximal tubule. The volume of this isotonic fluid is reduced by up to 70%; this must mean that solute reabsorption and water reabsorption are in some way perfectly matched. In trying to understand this process of *isotonic fluid reabsorption,* early workers attempted to induce water flux across the tubular epithelium. They changed the osmotic gradient between tubular fluid and blood by adding hypotonic or hypertonic fluids to the tubular lumen. They found that large osmotic gradients were required to produce fluid movement equal to that observed during tubular water reabsorption. However, no such osmotic gradients could be measured in micropuncture samples of tubular fluid.

These facts led to the hypothesis that water movement is somehow coupled to sodium transport via a complex interplay between the anatomy of the epithelial cells and the localization of the transport processes. The process by which this coupling is believed to occur is called the *standing gradient hypothesis* for water transport.

The operation of the standing gradient mechanism is shown in Fig. 46-9. The key element in the process is the location of a sodium pump on the lateral cell membrane of the epithelial cell. In the tubular epithelium, especially in the proximal tubule, the spaces between epithelial cells are very long, narrow, and tortuous. The luminal end of these intercellular clefts is bounded by a *tight junction* of the epithelial cells.

Recall that tight junction is a misnomer, in that these junctions are permeable to solutes and water, thus providing a pathway for fluid movement (the paracellular pathway). In fact, a substantial fraction of the net solute and water reabsorption probably occurs via the paracellular pathway.

If sodium is transported into the intercellular spaces, the concentration of sodium can rise substantially because of the restricted nature of the channels. Therefore large osmotic gradients will be generated. These osmotic gradients can pull water from the lumen through the large surface area of the brush border of the epithelial cells or through the tight junctions into the intercellular spaces. This movement of water into the intercellular clefts raises the hydrostatic pressure and causes water and the dissolved solutes to flow out of the intercellular spaces, through the interstitial space, and into the peritubular capillary blood.

The water and sodium that enter the intercellular spaces must be removed by the peritubular capillaries. This flow of water is driven by two forces (Fig. 46-9). First, the oncotic pressure in the peritubular capillary plasma is high as a result of the prior removal of a protein-free fluid in the process of glomerular filtration—a process that concentrates the remaining proteins in the plasma (Fig. 45-6, *A*). This concentrated plasma then circulates through the peritubular capillaries with a high oncotic pressure, encouraging fluid absorption. The second factor that facilitates water and salt reabsorption from the renal interstitium is the low capillary hydrostatic pressure, created by the pressure drop across the efferent arterioles.

Because the epithelial tight junctions are permeable to water and salts, some sodium and water can leak back into the lumen rather than flow into the peritubular capillary blood. This fact, coupled with the dynamics occurring at the capillary, may contribute to the glomerulotubular balance, the process that couples glomerular capillary function and tubular reabsorption. Recall from Fig. 45-6 that when the glomerular capillary flow is high relative to the filtration rate, equilibrium between filtration forces and plasma oncotic pressure is not attained. Under this condition the oncotic pressure of the plasma exiting the glomerulus will vary inversely with the flow. An increase in glomerular filtration will cause an increased protein oncotic pressure in the peritubular capillary blood at the end of the glomerular capillary. This increase in plasma oncotic pressure will enhance reabsorption of fluid from the interstitial space surrounding the tubular epithelium and thereby will reduce the hydrostatic pressure in this region. This will in turn increase the movement of fluid and ions out of the spaces between epithelial cells. This increases net ion and water reabsorption and thus balances the elevated filtration.

Osmotic diuretics. The high permeability of the proximal tubular epithelium to water is, in part, responsible for the isotonic reabsorption of the filtrate. However, this high permeability to water also means that

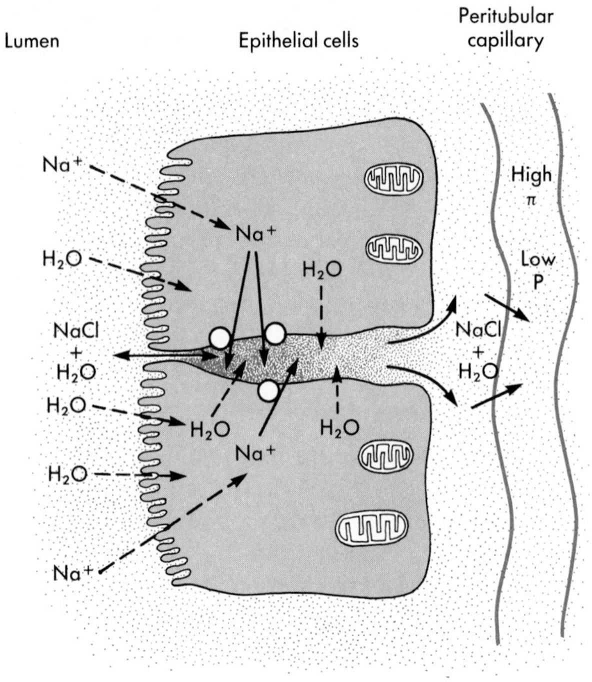

Fig. 46-9. Elements of the standing gradient hypothesis. Note that sodium is transported into the restricted intercellular spaces; thus the osmolality is raised (indicated by the density of shading). The double-headed arrow indicates the paracellular pathway for water and salt movement.

any nonreabsorbable solute present in the filtrate will severely restrict the amount of water that can be reabsorbed. As water is reabsorbed, the nonreabsorbable solute will be concentrated and will exert an osmotic pressure that will hold fluid in the tubule. In the normal process of sodium reabsorption water moves pari passu with the sodium. The isotonic fluid that is left behind contains sodium and the other normal constituents of the filtrate plus the nonreabsorbable solute. The water remaining behind balances the solute osmotically. In the process, however, it dilutes the sodium and generates a concentration gradient for this ion. As a result, the proximal tubular sodium concentration may fall to 100 mEq/L when a nonreabsorbable solute is present in the lumen. With about 100 mEq/L of sodium in the lumen and a peritubular fluid sodium concentration of 140 mEq/L, the back leak of sodium equals the transport rate. Hence further net transport can no longer occur. The fluid and solutes remaining in the tubules will then proceed through the proximal tubules and enter the distal portion of the nephron, and a relatively large fraction of the filtered water will be excreted.

Any substance that increases urine volume is called a *diuretic*. A nonreabsorbable solute that acts as a diuretic in the manner just described is called an *osmotic diuretic*. *Mannitol* is a low–molecular weight, nonmetabolizable sugar-alcohol that is used as an osmotic diuretic.

Henle's Loop Ion and Water Transport

Sodium Transport

It is very difficult to study transport in Henle's loop because much of it is not accessible for micropuncture in vivo. Thus detailed knowledge of the transport mechanisms in this structure is still scanty. It is known, however, that salt and water transport in this part of the nephron play a vital role in determining the ability of the kidney to concentrate the urine; this process is discussed later.

In general, little net active sodium transport occurs in the descending limb of Henle's loop, although large changes in composition are induced by passive exchange of water and solute with the fluid in the medullary interstitium. This portion of the nephron is highly permeable to both salts and water.

The thin ascending limb of Henle's loop has a lower permeability for salts and water, and whether significant active transport of material occurs is debatable. The epithelial cells in this portion of the nephron possess few mitochondria, a fact consistent with the idea that the thin ascending limb is simply a conduit for tubular fluid.

The major site of active sodium transport in Henle's loop is the thick ascending limb. Measurements of the luminal potential in isolated thick segments yield values ranging from approximately +10 mV in the medullary segments to +25 mV in the cortical segments. The origin of this positive potential remains controversial. The sodium concentration in the medullary segments of the loop cannot be measured because the loop is buried in the parenchyma of the medulla. However, sodium concentration in the first part of the distal tubular fluids is about 70 mEq/L. Because sodium leaves the proximal tubule at 140 mEq/L, substantial sodium reabsorption must take place in the thick ascending limb.

A class of diuretics known as *loop diuretics*, or *high-ceiling diuretics*, such as furosemide and ethacrynic acid, acts selectively on the sodium transport process in the ascending limb of Henle's loop. Furosemide acts on the apical surface of the epithelial cell, apparently by blocking chloride cotransport, and thus affects sodium reabsorption indirectly. In contrast, ethacrynic acid acts at some undetermined intracellular site and only after a prolonged delay.

Potassium Transport

About 30% of the filtered potassium leaves the proximal tubule and flows into Henle's loop. Some potassium enters the descending limb of Henle's loop, but the sources of this potassium are poorly understood. Subsequently, in the ascending limb of Henle's loop, potassium is reabsorbed to an extent that reduces potassium in samples from the beginning of the distal tubule to about 10% of the filtered potassium. Thus substantial net reabsorption of potassium occurs in the ascending limb of Henle's loop. A large part of this reabsorption is secondary active transport, dependent on sodium reabsorption. It appears that one potassium and two chloride ions share a carrier that mediates transport of the three ions in exchange for a sodium ion.

Chloride Transport

The chloride concentration in the early distal tubular fluid is approximately as low as sodium. This, combined with the positive luminal potential, led some investigators to propose that it is the chloride ion that is actively transported. Current evidence, however, indicates that here, as well as in the proximal tubule, the chloride ion movement is secondary to activity of the Na^+, K^+-ATPase and primary sodium transport.

Water Transport

The descending portion of Henle's loop appears to be permeable to water, whereas the ascending limb has a very low permeability to water. As shown later, the interstitial fluid of the medulla has a very high osmotic activity (Fig. 46-10). These facts generate a peculiar pattern for water handling in Henle's loop.

During passage of fluid from the proximal tubule down the descending limb, water is drawn osmotically out of the nephron into the medullary interstitium. This contrasts with the situation in the ascending limb of the loop, where solute transport occurs and where permeability to water is low. The reabsorption of sodium chloride from the segment with low water permeability dilutes the remaining fluid. This causes the fluids entering the distal tubule to have osmolalities less than 100 mOsm/L, as compared with the 300 mOsm/L in plasma and in the latter part of the proximal tubule. This characteristic of absorbing solute in excess of water, which takes place in the ascending limb of Henle's loop has led to its designation as the *diluting segment* of the nephron.

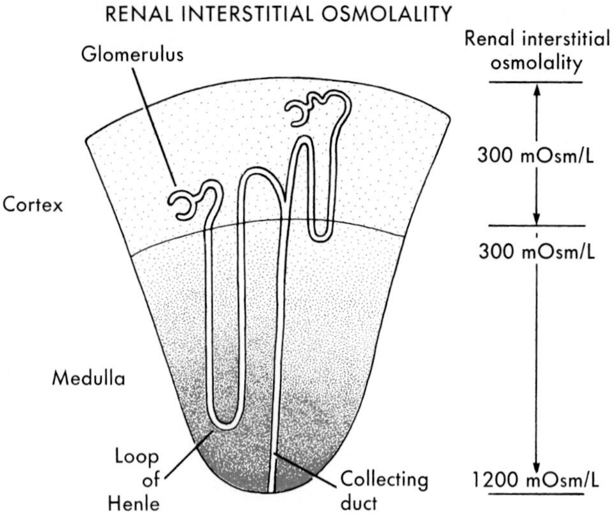

Fig. 46-10. Osmolality of the renal interstitial fluid. A cross section through a renal pyramid shows cortical and juxtamedullary nephrons. The intensity of shading is proportional to solute concentration in the interstitium. Note the accumulation of solutes in the tip of the medulla.

Distal Tubular Ion and Water Transport

Sodium Transport

Fig. 46-11, *A*, shows the general characteristics of the transport in a distal tubular cell. Much of the regulation of salt and water occurs in this portion of the nephron. Most elements of distal tubular function are qualitatively similar to those seen in the proximal tubule, and the carriers shown in Fig. 46-8, *B*, exist in the distal tubule as well. Perhaps the most significant element that distinguishes distal tubular function from proximal tubular function is a tighter epithelium with a more sparse brush border. These differences result in a lower permeability to water and to ions, and thereby in a higher pump/leak ratio. As a result, larger intraluminal potentials can be achieved and greater transepithelial concentration gradients are found. Beginning with slightly positive potentials in the earliest portion, the distal tubular potential difference becomes more negative along the course of the tubule. By the end of the distal tubule, a large negative potential of approximately -50 mV is achieved. This potential is sodium dependent and is much larger than that found in the proximal tubule (± 2 mV). Also, the sodium gradients can be much larger in the distal tubule, with tubular sodium concentrations falling to 40 mEq/L.

Fig. 46-11, *B*, shows how sodium reabsorption varies as a function of sodium load in the distal tubule. Note that here sodium transport is more of the T_m type than the gradient-time type found in the proximal tubule. This has substantial functional implications. In the distal tubule the sodium load is, in fact, the amount of sodium entering from the proximal tubule and Henle's loop, not the sodium that was originally filtered. The fact that the distal sodium reabsorptive process is a T_m system means that, over a substantial range of sodium loads entering the distal tubule from the proximal tubule, distal tubular sodium reabsorption can match sodium entry. This minimizes the escape of sodium from the tubule and loss from the body. However, when the T_m is exceeded, further elevations in sodium load allow sodium to escape into the collecting duct.

The sodium gradient depends on the state of hydration of the individual and the associated changes in the antidiuretic hormone (ADH) level. During antidiuresis, that is, when water conservation is important, the ADH level is high, and there are modest increases in the sodium pump rate and increases in the permeability of the distal tubule to water. The effects of ADH appear to be confined to the most distal segments of the distal tubule. As is explained later, the major effect of ADH with regard to water reabsorption has to do with its effect on tubular permeability to water, not on sodium transport.

Potassium Secretion

Since most of the filtered potassium is reabsorbed in the proximal tubule and Henle's loop, the urinary potassium concentration largely reflects the quantity of potassium that is secreted during the passage of tubular fluid through the distal tubule and collecting duct. Potassium secretion in the distal tubule occurs mainly in the late distal tubule or connecting tubule and is largely passive. The potential difference across the tubular epithelium is relatively large (−50 mV), and, since most of the potassium is reabsorbed in the proximal tubule and ascending limb of Henle's loop, fluid entering the distal tubule has a low potassium concentration. It follows that a substantial concentration difference drives potassium secretion. Thus the electrochemical potential that drives potassium secretion can be relatively high.

A number of factors influence potassium secretion, especially the amount of potassium in the diet. Individuals maintaining a high potassium intake show much larger tubular secretion of potassium than normal. Conversely, when potassium is restricted in the diet, potassium actually may be reabsorbed rather than secreted over the course of the distal tubule.

Sodium and potassium transport processes are closely related because the distal tubular potential depends heavily on sodium reabsorption. Under conditions where sodium reabsorption is stimulated (e.g., salt deprivation), the luminal potential becomes more negative as sodium is more avidly reabsorbed. This increase in potential enhances potassium secretion as a result of greater driving forces for transepithelial movement.

Chloride Reabsorption

A large fraction of the bicarbonate has been reabsorbed by the time the distal tubule is reached, and accordingly chloride becomes the dominant anion accompanying sodium.

Hydrogen Ion Secretion

In contrast to its handling in the proximal tubule, primary active transport of hydrogen may occur in the distal tubule. The hydrogen secretion is attributable largely to the activity of the "dark" or "intercalated" cells of the connecting segment of the distal tubule and the cortical collecting duct.

The hydrogen transporter is electrogenic; increased activity tends to reduce the luminal potential. This confers an important reciprocal relation on the pattern of hydrogen and potassium secretion. It was once believed that potassium and hydrogen competed for an exchange pump with sodium. This remains a useful device for remembering the interactions among these three ions, but it is now clear that concept is not correct. The interaction between the hydrogen ion and the potassium ion is explained more accurately by the fact that the hydrogen pump in the distal tubule is electrogenic. An increase in hydrogen secretion diminishes the magnitude of the transepithelial potential difference in the distal tubule. This reduction in potential difference reduces the electrochemical driving force for passive potassium secretion and thus reduces net potassium secretion.

Water Reabsorption

An active solute pump and low permeability to sodium, chloride, and water in the ascending limb of Henle's loop yield a hypotonic fluid in the first part of the distal tubule. This fluid may enter the distal tubule with an osmolality of about 100 mOsm/L. The connecting segment of the distal tubule is sensitive to ADH, and the amount of water that is reabsorbed over the course of the distal tubule varies greatly depending on tubular permeability to water. When ADH is absent, permeability to water is low, and hypotonic fluid entering the distal tubule from the ascending limb remains hypotonic throughout the length of the distal tubule. However, when ADH is present, the osmotic gradient between the hypotonic fluid in the lumen of the distal tubule and the isotonic fluid in the peritubular capillary blood induces osmotic reabsorption of water. The osmolality of distal tubular fluid rises progressively toward isotonicity.

Collecting Duct Ion and Water Transport

Sodium Reabsorption

Sodium reabsorption continues throughout the collecting duct. Because of the low permeability of the duct, the luminal sodium can be reduced to very low concentrations prior to excretion. The sodium reabsorptive process varies along the length of the collecting duct, with maximal sodium transport in the cortical segment, low sodium transport in the medullary segment, and, again, very active transport in the papillary segment. In concert with the late distal tubule, the cortical segment of the collecting duct is the primary site of aldosterone action.

Potassium Transport

In the cortical collecting duct, potassium secretion continues and is passive. In the final section of the nephron a small amount of potassium reabsorption may be detectable when dietary potassium is restricted.

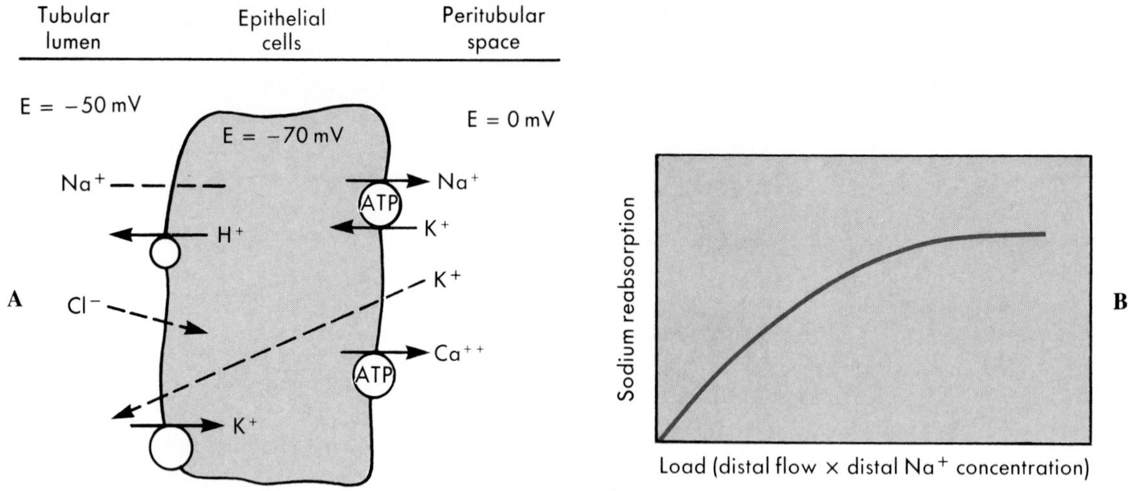

■ **Fig. 46-11. A,** Distal tubular salt and water transport. Note the more negative luminal potential than in the proximal tubule and the fact that potassium is passively *secreted*. The hydrogen pump on the luminal side of the epithelial cell is electrogenic, and its activity can diminish the luminal potential. **B,** The relation between the sodium load entering the distal tubule from the proximal tubule and distal sodium reabsorption. Compare this with Fig. 46-8.

■ *Hydrogen Ion Secretion*

The final pH and bicarbonate concentrations are established in the distal tubule and collecting ducts. Each of these segments absorbs about 5% of the total filtered bicarbonate. As will be seen later, the ability of the distal nephron to absorb bicarbonate changes with the body's acid-base status. This may limit the capacity of the kidney to regulate hydrogen ions in the body.

■ *Water Reabsorption*

In the collecting duct, water reabsorption is much the same as in the distal tubule, with one notable exception (Fig. 46-10). By a process described in detail below, the interstitial fluid of the medulla is made extremely hypertonic; osmolalities as high as 1200 mOsm/L exist at the tip of the papilla. ADH increases the permeability of the latter part of the distal tubule (connecting segment) and cortical collecting duct to water, and thereby permits greater osmotic withdrawal of fluid from these portions of the nephron. Under these conditions, osmotic extraction of water can raise the final urine concentration as high as the osmolality of the interstitium. Thus the final urine concentration can range from something less than 100 mOsm/L (that is, the osmolality established by solute reabsorption without accompanying water in the ascending limb of Henle's loop) to 1200 mOsm/L, a value limited by the concentration of fluid in the medullary interstitium.

■ *Neural and Humoral Control of the Nephron*

■ *Neural Control*

The primary influence of the renal sympathetic innervation on function is mediated via the afferent and efferent arterioles (Chapter 47). It has been shown, however, that stimulation of the renal nerves can induce sodium and water retention, and denervation can induce natriuresis and diuresis. Furthermore, the effects of nerve stimulation can be blocked by α-adrenergic blocking agents. Nerve stimulation has a direct action on the tubules and an indirect one via changes in renal blood flow or release of local hormones. Whereas the potential for neural control of tubular transport has been well established, its role in moment-to-moment regulation of sodium and water transport remains unsettled.

■ *Humoral Control*

A variety of hormones influence the way in which salt and water are handled along the course of the nephron, and a large part of this control occurs in the connecting segment and collecting duct. Aldosterone enhances sodium and water reabsorption and potassium excretion (Chapter 54). This hormone is a steroid released from the adrenal cortex under the influence of *angiotensin II* and *angiotensin III*. Release of aldosterone is also stimulated by low sodium or high potassium concentrations in the plasma. In general, conditions in which conservation of sodium is required or in which blood pressure falls stimulate the release of aldosterone.

The mechanism of action of aldosterone has been studied extensively, and several steps in the process are known. Aldosterone binds to a cytoplasmic receptor, and the combination enters the cell nucleus to stimulate DNA synthesis and production of proteins (Chapter 54). These proteins act on the apical membrane of the distal tubular cell to increase permeability to sodium and on the mitochondria to stimulate oxidative phosphorylation. The increased cell membrane permeability allows sodium to enter the epithelial cell more readily and from there to be pumped into peritubular interstitial space.

Angiotensin II plays an important role in the control of renal function in addition to its effects on aldosterone secretion. It has a modest direct stimulatory action on sodium reabsorption by the tubular epithelium, but most importantly, angiotensin II causes contraction of the smooth muscle of the renal arterioles and thereby affects renal blood flow and glomerular filtration.

Another hormone that may influence salt transport in the kidney is the *natriuretic factor*. The natriuretic factor is believed to be a small polypeptide that is released in response to expansion of extracellular fluid volume and to sodium loading. One site of natriuretic factor formation site is the atrium of the heart, where the *atrial natriuretic factor* (ANF) is produced. This factor dilates the renal vasculature, thereby inducing natriuresis. It also directly inhibits tubular sodium reabsorption. The relative importance of these effects in physiological control remains to be established. Other natriuretic factors are probably involved, but these have not yet been isolated.

The existence of the natriuretic factor is deduced from the fact that sodium loading elevates sodium excretion in animal preparations in which all other known influences on the sodium reabsorptive mechanism have been excluded. Furthermore, such a natriuretic factor has been recovered from the urine of laboratory animals and humans. It is believed that expansion of extracellular fluid volume causes release of this hormone, with a consequent inhibition of sodium reabsorption and resultant reduction in body sodium content. The reduction in body sodium content is followed by a reduction of the extracellular fluid volume.

ADH has been studied more intensively than almost any other renal hormone. Its site of action is primarily in the collecting duct and secondarily in the connecting segment of the distal tubule. ADH increases permeability to water, thereby allowing tubular fluid to equilibrate osmotically with interstitial fluid. The change in permeability is highly selective for water; other small molecules may be affected differently. For example, ADH does not increase permeability to urea in the end of the distal tubule and early collecting duct.

The mechanism of action of ADH depends on binding of the hormone to the basolateral border of the renal tubular cell (Chapter 52). Hormone binding elevates cAMP levels, which causes a rearrangement of *cytoskeletal structures,* including actin filaments and microtubules. This increases the aggregation of certain vesicular structures and insertion of these into the membrane surface. These vesicles appear to act as water-permeable pores and, as a result, they increase apical cell membrane permeability.

Intrarenal hormones. A number of hormones are synthesized within the kidney and act locally on the nephrons and renal vasculature. *Prostaglandins* are formed within the kidney and induce natriuresis when infused. Furthermore, antinatriuresis is induced by blocking the synthesis of prostaglandins with indomethacin. Thus prostaglandins appear to exert a tonic natriuretic effect. Prostaglandins may also act in concert with the *renin-angiotensin system*. Several of the prostaglandins stimulate renin release and also sensitize the juxtaglomerular cells to normal stimuli for renin release, such as sympathetic nerve activity. A number of *kinins,* notably bradykinin, are found within the kidney, and these can reduce tubular sodium reabsorption. The mechanisms by which release is modulated and the physiological role of these hormones remain to be established.

Passive Transport Mechanisms

This section illustrates how various aspects of nephron function may indirectly influence solute transport. Passive transport processes dissipate gradients that depend on the expenditure of energy used in actively transporting other materials. In the kidney the most common forces that drive passive transport are gradients generated secondary to sodium reabsorption and gradients generated by the active secretion of hydrogen ions.

Passive Reabsorption Secondary to Water Reabsorption: Urea Reabsorption

Because previous emphasis was directed so intently at active transport processes, it is easy to lose sight of the fact (1) that the permeability for small ions and small organic solutes in the nephron is finite and (2) that water reabsorption concentrates all these solutes, thereby generating a concentration gradient that favors reabsorption. Essentially all small solutes in the filtrate are influenced to some degree by the reabsorption of water and by the subsequent concentration gradients that are generated.

Urea is taken as an example of such transport because, it is uncharged and its transport is relatively easy to understand. The rate of urea reabsorption will be mainly a function of the *permeability* of the tubule, the *tubular area* available for reabsorption, and the *urea concentration gradient*. The concentration gradient, of course, is induced by the reabsorption of water.

Two general statements may be made about the excretion of any molecule that is passively reabsorbed secondary to sodium and water reabsorption: (1) excretion will vary linearly with changes in plasma concentration, and (2) excretion will vary directly, but not linearly, with urine flow. Parenthetically, these two characteristics are not exclusively the behavior of a molecule transported passively. For example, sodium excretion will vary linearly with plasma concentration under some circumstances, even though transport is active. However, this reflects glomerulotubular balance, not passive transport.

The effects of flow and plasma concentration on net molecular transport must be clearly understood, because urine flow and plasma concentrations vary independently as an individual's water intake, diet, and metabolism change.

Fig. 46-12, *A,* shows the relation between urea excretion and plasma urea concentration. The relation is shown at two different glomerular filtration rates. As plasma urea concentration is varied, the excretion changes linearly with the concentration; in other words, a constant fraction of the filtered urea is reabsorbed. This relation between plasma concentration and urea excretion might be predicted when it is recalled that absence of a specific transport protein would preclude the display of saturation kinetics. Therefore variations in plasma concentration simply produce proportional variations in the quantity of urea filtered and ultimately in the quantity of urea excreted.

Note that the slope of the line relating excretion and plasma concentration is constant. The slope of this line is equal to the ratio of the urea excretion $(U \times V)$ to the plasma concentration (P), and this ratio $(U \times V/P)$ is known as the *clearance*. The urea clearance can be used as a quantitative assessment of renal function, as described in Chapter 47.

Fig. 46-12, *A,* also shows the effect of a change in glomerular filtration rate on urea excretion. A low glomerular filtration rate decreases the slope of the line relating urea excretion to plasma urea concentration, but the linear correspondence between urea excretion and plasma urea concentration is maintained. This relationship simply reflects the fact that, for any given urine flow rate, a constant fraction of the filtered urea is excreted. Therefore doubling the glomerular filtration rate doubles the amount of urea filtered per unit time, and this in turn doubles the amount excreted.

Fig. 46-12, *B,* shows the effect of variations in urine flow on urea excretion; the form of this relationship is quite different from the relationship between urea excretion and plasma concentration, as shown in Fig. 46-12, *A*. As urine flow is reduced to levels below about 2 ml/minute, there is a disproportionate reduction in urea excretion, even when the glomerular filtration rate is constant. This implies that the fraction of the filtered water that is reabsorbed increases at low flows.

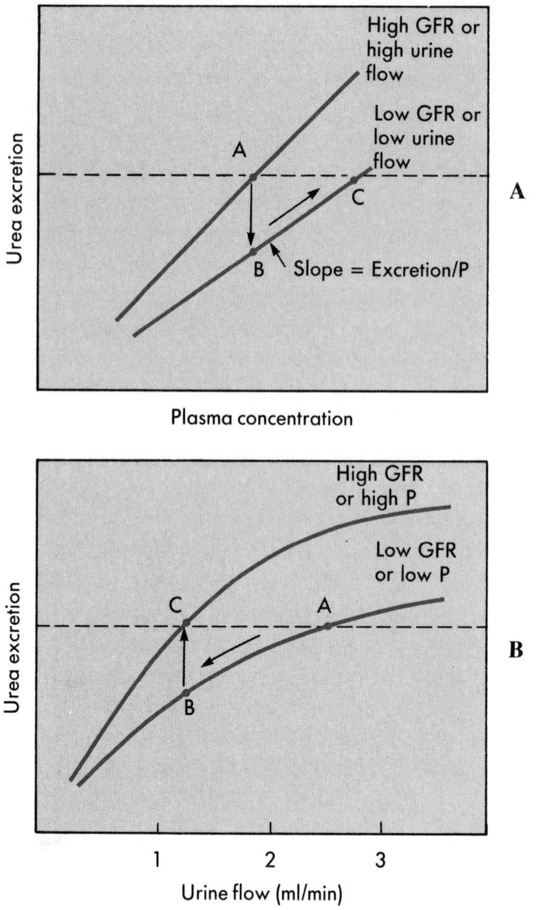

Fig. 46-12. Effects of changes in plasma concentration **(A)** and urine flow rate **(B)** on urea excretion. Excretion is filtration minus reabsorption, and thus increasing excretion means decreasing reabsorption. The dashed line represents the urea excretion rate demanded by urea production. For a steady state to occur, urea excretion must equal the sum of the urea production and urea intake *(A* and *C)*. (See text for details.)

If urea reabsorption is passive, the reabsorption rate must depend on concentration and permeability. It follows that the nonlinear excretion must reflect some nonlinear change in tubular urea concentration as the urine flow changes. An explanation for this nonlinear behavior can be derived by considering that urea reabsorption is a two-phase process: first, water is absorbed, thereby generating a urea concentration gradient; and second, a fixed time period is allowed for urea to diffuse out of the tubule down the concentration gradient.

Fig. 46-13 shows what would happen if water were

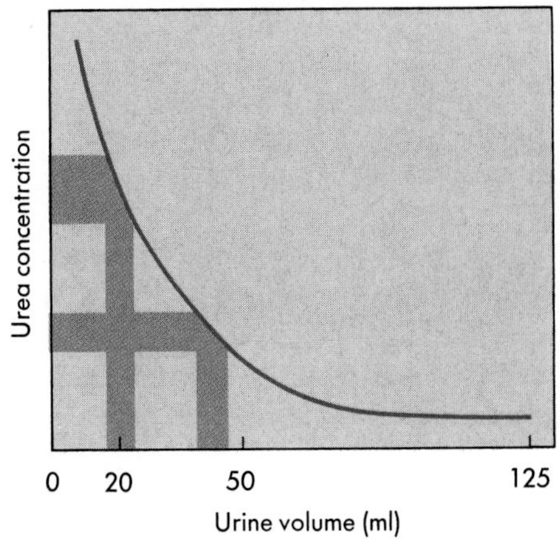

■ Fig. 46-13. Hypothetical relation between urea concentration, which drives urea reabsorption, and the urine volume. Note the sharp rise in concentration at small urine volumes. At very high urine flows, the urea concentration falls to a constant, minimal value equal to the plasma urea concentrations. This follows from the fact that no water reabsorption has occurred, implying no urea reabsorption.

selectively extracted from 125 ml of solution (i.e., the volume of water filtered in 1 minute) and the solute were left behind. As water is extracted, the concentration rises nonlinearly and increases sharply (approaching infinity) as volume approaches zero. If this process is analogous to the pattern in the tubule, then the reason for the nonlinear relationship between urine flow and urea excretion can be appreciated (Fig. 46-12, *B*). The colored areas of Fig. 46-13 show the effect on concentration of the removal of 5 ml of filtrate from a volume of 48 ml and from a volume of 22 ml. In other words, it represents the response that might be expected when urine flow is reduced by 5 ml/minute, starting at an initial flow of 48 or at an initial flow of 22 ml/minute. The concentration increase for the solute is twice as large when starting from the lower volume. Therefore equivalent reductions in urine flow would induce larger concentration gradients across the tubule when the initial urine flow is relatively small. As a result, the reduction in urine flow would induce a larger fractional reabsorption of urea (smaller excretion).

The relationships among plasma urea concentration, urine flow, and urea excretion are important to an understanding of renal function. In a steady state, *urea excretion must equal urea production*. Urea production is, of course, inextricably linked to intake and metabolism of proteins. Fig. 46-12, *A* and *B*, shows that glomerular filtration rate and urine flow both affect the urea excretory rate. Furthermore, glomerular filtration rate and urine flow vary with the state of body hydration. It follows that urea excretion must be related in some way to the state of hydration of the individual.

The net effect of an alteration in glomerular filtration rate can be easily ascertained from Fig. 46-12, *A*. Assume that a subject is operating normally, with a urea excretion rate indicated by the dashed line at the high glomerular filtration rate (point *A*). Assume also that, because of dehydration secondary to restricted fluid intake, the glomerular filtration rate is diminished. This would cause the urea excretion to diminish from *A* to *B* and thus the excretion rate becomes less than the rate of urea formation. If the imbalance between excretion and production continues, it would result in a slow increase in plasma urea concentration and a return of urea excretion toward normal; it would follow the path shown by the line *B-C*. The system then would be stabilized at a lower glomerular filtration rate, a higher plasma urea concentration, and a urea excretory rate equal to the urea production rate.

Careful examination of Fig. 46-12, *A* and *B*, permits one to use similar, although more involved, logic to predict the effect of a reduction in urine flow on urea excretion and plasma urea concentrations. Suppose urine flow were reduced from 2.5 to 1.2 ml/minute (*A-B*, Fig. 46-12, *B*). This would lead to a reduction in urea excretion. In the face of continued constant protein intake, urea would begin to accumulate. Accumulation of urea would elevate plasma urea concentration, and the elevated concentration would in turn increase urea excretion (Fig. 46-12, *A*, point *B* to *C*). Stabilization would occur at a point where urea excretion and urea production were equal, with a reduced urine flow and an elevated plasma concentration of urea (point *C*).

The importance of understanding the changes in urea excretion associated with changes in urine flow and plasma concentration resides in the fact that urea concentration frequently is used as an index of renal function. Obviously plasma urea concentration alone is an inadequate index of renal function in view of its dependence on physiological regulation of urine flow and glomerular filtration rate. Therefore measurement of plasma urea levels cannot permit a distinction between renal pathology, changes in the state of hydration of the individual, or a variety of other physiological adaptations. As will be shown below, the various possibilities can be distinguished by understanding the concepts presented here and by using the clearance concept.

■ *Passive Transport Secondary to Hydrogen Ion Secretion*

Charge commonly influences permeability of biological membranes. Because the charge of many molecules is related to pH, the permeability of the tubule to a molecule can change as a result of hydrogen ion secretion. For example, consider the behavior of a weak acid (RCOOH) when urine is made acidic with respect to plasma (Fig. 46-14, *A*). Attention is focused on the concentrations of the nonionized forms, because we assume that this species is permeative, whereas the ion-

Fig. 46-14. Effects of urinary pH on nonionic diffusion. *Colored bar,* Concentration of a weak acid in tubular fluid; *Gray bar,* concentration of weak acid in peritubular capillary blood. For simplicity it is assumed that capillary blood concentration of a weak acid is the same when the urine is acidified (**A**) and when the urine is alkalinized (**B**). The direction of acid movement is determined by the nonionized acid concentrations in the tubule and capillary blood. The pH-related changes in tubular concentration of the nonionized form result in reabsorption in the first case and secretion in the second *(gray arrows)*. See text for details.

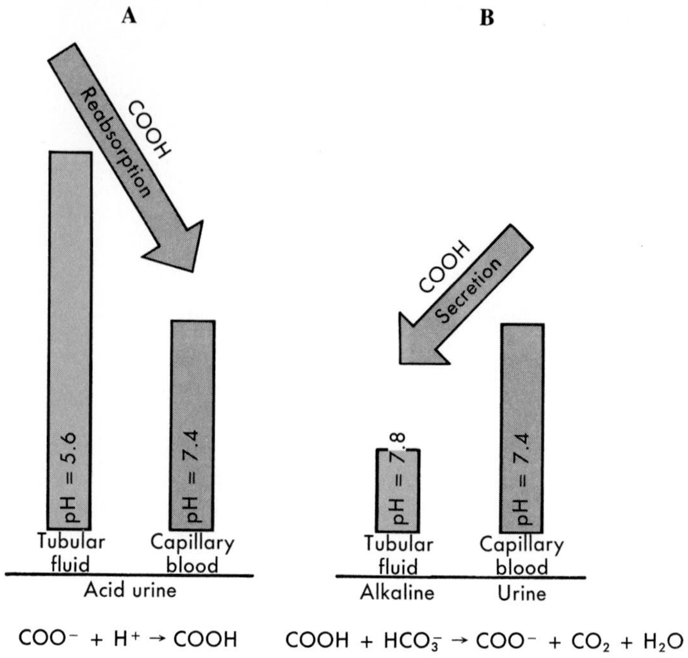

ized form is not. Furthermore, the capillary blood concentration of the nonionized form is taken as a measure of its concentration in the filtrate. As filtrate flows along the nephron, hydrogen is added. This lowers the tubular fluid pH and protonates a portion of the filtered salt, thereby making it uncharged and raising the concentration of the nonionized form in the tubular fluid (Fig. 46-14, *A*, colored bar, pH = 5.6). The resultant concentration gradient causes the nonionized form to diffuse down its concentration gradient from lumen to plasma and be *reabsorbed* (Fig. 46-14, *A, gray arrow).*

Conversely, suppose that an alkaline urine is formed (Fig. 46-14, *B*). In this case, the reduction in the hydrogen ion would cause the undissociated form of the acid to ionize, resulting in a reduction in the quantity of the nonionized form in the tubule (*colored bar*, pH = 7.8). A gradient from blood to lumen would be produced for the undissociated form, causing the acid to diffuse into the lumen. With continued alkalinization, a number of the molecules that diffuse into the lumen would in turn dissociate in the high pH environment. These ions would then be unable to leave because of their negative charge. As a result, they would be *secreted,* to be subsequently excreted. The process whereby a change in the form of a molecule generates a one-way diffusional flux is often referred to as *diffusion trapping*.

The reabsorption or secretion of an acid or base in response to a change in tubular pH can be understood by recognizing that the key factor is the establishment of a *concentration gradient for the nonionized form* of the molecule across the tubular epithelium. For this reason the process is often called *nonionic diffusion*. The concentration gradient of the nonionized species is established by changing pH, and the effect of the pH change is predictable from the Henderson-Hasselbalch equation:

$$\log \frac{\text{Base}}{\text{Acid}} = \frac{\text{Ionized form}}{\text{Nonionized form}} = \text{pH} - \text{pK} \quad (4)$$

If we assume for the simplest case that the nonionized form is protonated (an assumption not always true), then equation 4 shows that a pH change of one unit will alter the ratio of ionized to nonionized forms by a factor of 10 (i.e., 1 log unit).

Fig. 46-15 illustrates the net result of the mechanism as it might actually operate. The solid-colored reference line shows the ratio of basic (ionized) to acidic (unionized) forms for a weak acid with a pK of 6.5, as predicted by equation 4. The solid and dashed black lines represent the excretion of a weak base, quinine (pK = 8.4), and a weak acid, probenecid (pK = 3.4), as a function of urinary pH. Excretion is expressed on the lefthand ordinate as the percentage of the material that was originally filtered. Thus the colored upper half of the graph, in which excretion is greater than 100% of the filtered material, represents a region of net secretion, and the lower half of the graph represents a region of net reabsorption.

The two compounds shown are also secreted weakly by the proximal tubule via a secondary active transport mechanism. Net transport is therefore the sum of active secretion and passive secretion or reabsorption. The amount of the material that is ultimately excreted depends strongly on the urinary pH. Starting from a high pH, substantial quantities of probenecid are secreted. As the urine pH is lowered and approaches the pK for the probenecid, net secretion falls due to sharply rising passive reabsorption, and ultimately secretion is converted to net reabsorption. Conversely, at a high pH, diffusion of the nonionized form of the weak base, qui-

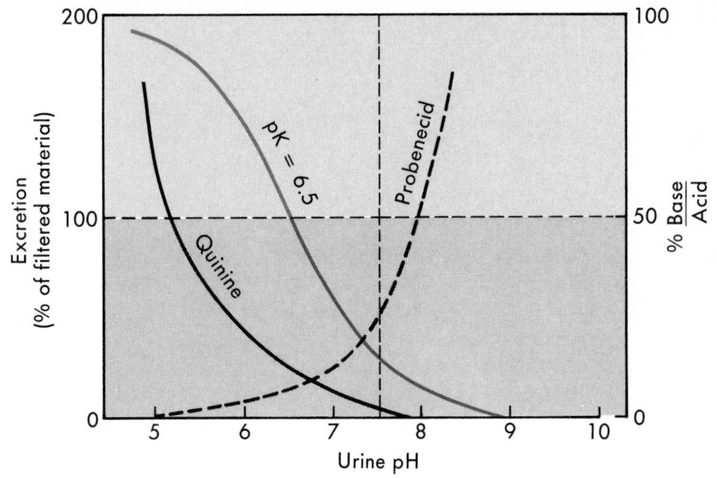

■ **Fig. 46-15.** Effect of urine pH on excretion of a weak acid (probenecid) and a weak base (quinine). The black lines show the excretory rate, expressed as the percentage of the rate at which the substance was filtered. The colored region represents a set of conditions under which there is net secretion of the molecule. The solid-colored line, based on the predictions of the Henderson-Hasselbalch equation, shows the fraction of weak acid in the base form as a function of urine pH. It is known that both quinine and probenecid are secreted to a small degree in the proximal tubule by a secondary active process. Net transport is thus a combination of reabsorption or secretion by nonionic diffusion, superimposed on active secretion. Net secretion occurs only when urine pH is well displaced from the pK of the material. (Redrawn from Mudge, G.H., et al.: Med. Clin. North Am. **59:**681, 1975).

nine, from lumen to blood converts net secretion to net reabsorption.

Knowledge of the relationships just described allows one to manipulate excretion of a weak acid or base. For example, in salicylate intoxication, it would be desirable to enhance renal elimination of the salicylate (a weak acid). Fig. 46-15 makes it obvious that alkalinization of the urine, which may happen during hyperventilation or by ingesting a diet high in fruit juices, will result in a smaller concentration of the uncharged form of salicylic acid within the tubule. Therefore secretion will occur passively, and the excretion of salicylic acid will be increased.

■ Bibliography

Journal articles

Brodsky, W.A., editor: Anion and proton transport, Ann. N.Y. Acad. Sci. **341:**608, 1980.

Crane, R.K.: The gradient hypothesis and other models of carrier-mediated active transport, Rev. Physiol. Biochem. Pharmacol. **78:**101, 1977.

DiBona, G.F.: The functions of the renal nerves, Rev. Physiol. Biochem. Pharmacol. **94:**75, 1982.

Imai, M., and Nakamura, R.: Function of distal convoluted and connecting tubules studied by isolated nephron fragments, Kidney Int. **22:**465, 1982.

Kinne, R.: Properties of the glucose transport system in the renal brush border membrane, Curr. Top. Membr. Transp. **8:**209, 1976.

Knepper, M., and Burg, M.: Organization of nephron function, Am. J. Physiol. **244:**F579, 1983.

Mudge, G.H., Silva, P., and Stibitz, G.R.: Renal excretion by nonionic diffusion: the nature of the disequilibrium, Med. Clin. North Am. **59:**681, 1975.

Sacktor, B.: Transport in membrane vesicles isolated from the mammalian kidney and intestine, Curr. Top. Bioeng. **6:**39, 1977.

Schafer, J.A.: Salt and water absorption in the proximal tubule, Physiologist **25:**95, 1982.

Schultz, S.G., and Curran, P.F.: Coupled transport of sodium and organic solutes, Physiol. Rev. **50:**637, 1970.

Wade, J.B.: Membrane structural studies of the action of vasopressin, Fed. Proc. **44:**2687, 1985.

Wright, F.S.: Potassium transport in successive segments of the mammalian nephron, Fed. Proc. **40:**2398, 1981.

Books and monographs

Bijvoet, O.L.M.: Kidney function and calcium and phosphate metabolism. In Anioli, L.V., and Krane, S.M., editors: Metabolic bone disease, New York, 1977, Academic Press, Inc.

Brenner, B.M., et al.: The renal circulation. In Brenner, B.M., and Rector, F.C., editors: The kidney, ed. 3, vol. 1, Philadelphia, 1986, W.B. Saunders Co.

Burg, M.B.: Renal handling of sodium, chloride, water, amino acids and glucose. In Brenner, B.M., and Rector, F.C., editors: The kidney, ed. 3, vol. 1, Philadelphia, 1986, W.B. Saunders Co.

Giebisch, G., et al.: Renal transport and control of potassium excretion. In Brenner, B.M., and Rector, F.C., editors: The kidney, ed. 3, vol. 1, Philadelphia, 1986, W.B. Saunders Co.

Koushanpour, E.: Renal physiology: principles and functions, Philadelphia, 1976, W.B. Saunders Co.

Schmidt-Nielsen, B., editor: Urea and the kidney, Proceedings of an International Colloquy at Sarasota, Fla., Amsterdam, 1970, Exerpta Medica.

Schultz, S.G.: Ion-coupled transport across biological membranes. In Andreoli, T.E., et al., editors: Physiology of membrane disorders, New York, 1978, Plenum Press.

Smith, H.W.: The kidney: structure and function in health and disease, New York, 1951, Oxford University Press, Inc.

Valtin, H.: Renal function: mechanisms preserving fluid and solute balance in health, ed. 2, Boston, 1983, Little, Brown & Co.

Windhager, E.E.: Micropuncture techniques and nephron function, New York, 1968, Appleton-Century-Crofts.

CHAPTER

47

Integrated Nephron Function

Clearance Concept

Clearance and Dilution Indicators

To this point the discussion of renal physiology has focused largely on specific tubular processes and the behavior of the epithelial cells in separate segments of the nephron. The essence of renal physiology, however, depends on a synthesis of the behavior of the individual nephron segments and the relations of these segments to the renal vasculature. A most important aspect of this synthesis is the development of a means of acquiring quantitative data on renal function. Especially important in this regard has been the clearance technique, because it offers a means of assessing the overall effects of tubular function on excretory patterns and of comparing the excretion of different molecules under various conditions of urine flow and plasma concentration. As an example of the usefulness of the clearance concept, consider the case of an individual with a low urea excretory rate. How does one decide whether the rate is low because the plasma urea concentration is low, the urine flow is low, or renal function is impaired? The clearance calculation offers an important tool to answer this question.

In the analysis of the renal processing of urea (Fig. 46-12) it was noted that urea excretion and plasma urea concentration are linearly related. The slope of the line relating the two provides a useful way to interpret urea excretion values at different plasma urea concentrations. Fig. 46-12, *A*, also depicts this relationship at normal urine flow and at reduced urine flow. When urine flow is reduced, urea excretion is reduced disproportionately (Fig. 46-12, *B*). Note, that at the lower urine flow, the urea excretion varies linearly with plasma concentration, but the slope (U × V/P) of the line is less (Fig. 46-12, *A*). In each case, the slopes of the two lines in Fig. 46-12, *A*, represent the ratio of the excretion of a material (U × V) to its plasma concentration (P); this ratio is called the *clearance*. Determination of the clearance of certain molecules is a powerful tool to elucidate renal function.

The basic idea of the clearance calculation is illustrated in Fig. 47-1, *A*. The excretion rate for any material is the quantity (in milligrams or milliequivalents) of material that appears in the urine in a given unit of time; it is equal to the product of urine flow rate (V) and urine concentration (U). The plasma concentration is the *quantity* of material in a unit of plasma. The ratio of the excretory rate to the plasma concentration is then equal to the *minimum* volume of plasma at the prevailing plasma concentration that *could* have supplied the material appearing in the urine in the given time interval. In other words, the clearance is the volume of plasma per unit time that would have to be *cleared* entirely of solute to provide the molecules appearing in the urine.

Clearance appears to be a peculiar concept at first because the calculation need not refer to any real volume of plasma that is actually cleared of a substance in a unit of time. The number calculated is a *virtual volume* and gives only a lower limit on the behavior of the kidney. In other words, no less plasma than the computed volume could have supplied the excreted solute, but partial removal of the excreted substance from plasma would mean that more plasma was actually involved than indicated by the clearance computation. Obviously, if 20% of the solute were removed from each unit of plasma that passed through the kidney, the calculated clearance would be a fivefold underestimate of the total volume of plasma that was acted upon.

Although the calculation of clearance does not yield a value equal to the true volume that has been stripped of a particular solute per unit of time, the calculation does provide an estimate of how rapidly concentration might change in plasma or interstitial space. For example, suppose sodium appears in the urine at a concentration of 150 mEq/L, and the urine flow rate is 1 ml/minute; sodium excretion would equal 150 × 0.001 =

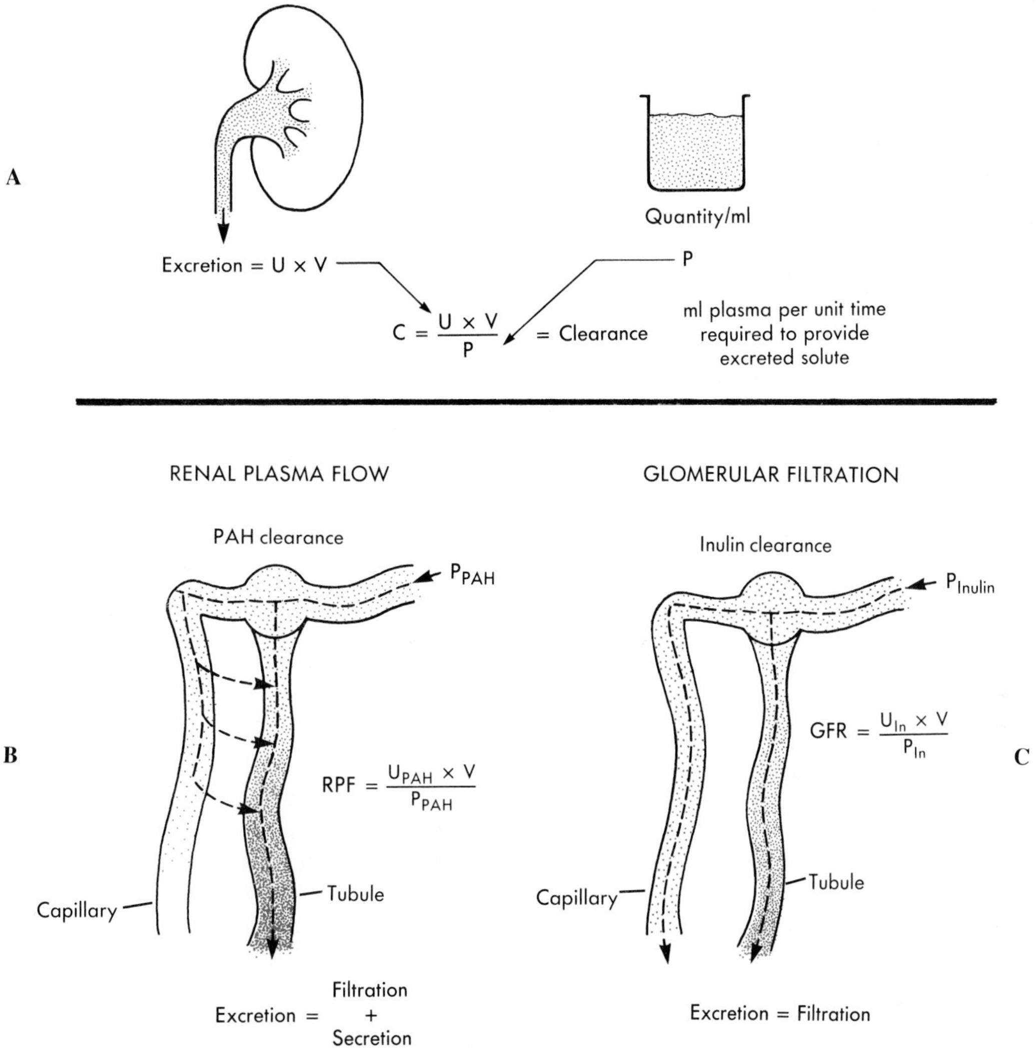

Fig. 47-1. Clearance calculation and estimation of flow. **A,** Basic concepts of the clearance calculation. The calculation shows that C milliliters of plasma per minute would be required to provide the solute at the observed rate of excretion (U × V) if the plasma concentration were P and the urine concentration were U. **B,** Measurement of renal plasma flow. The colored stipple represents *p*-aminohippuric acid (PAH). Note that all of the PAH is removed from the plasma by a combination of filtration of the solute through the glomerulus and secretion from the peritubular capillaries into the tubular lumen. Clearance of PAH approximates renal plasma flow. **C,** Measurement of glomerular filtration rate. The colored stipple represents inulin. Because inulin is a relatively small uncharged molecule, inulin concentrations in plasma and in the initial filtrate are the same. Once in the tubule there is no further exit or entry of inulin, but it is concentrated by water reabsorption. Clearance of inulin equals the glomerular filtration rate.

0.150 mEq/minute. The sodium clearance would then be equal to 0.15 mEq/minute divided by 150 mEq/L, or 0.001 L/minute or 1 ml/minute of plasma. This implies that sodium is excreted at a rate that could be matched if all of the sodium were being removed from 1 ml of plasma per minute. If the total volume of the extracellular space is about 20 L and the excreted sodium is coming from the extracellular space, then the clearance computation suggests that sodium is being excreted at a rate approximately equal to 1/20,000 or 0.005% of the total body sodium per minute. Thus we have some sense of how rapidly continued excretion of sodium by the kidney (without further replacement) might alter body composition.

Renal Blood Flow and Glomerular Filtration Rate Measurements

Despite the fact that clearance, as a rule, represents a virtual volume per unit time, not a real volume, it is possible under two circumstances to use clearance to estimate a true volume rate of flow and in so doing to obtain estimates of *renal blood flow* or *glomerular filtration rate*. The measurement of renal blood flow is based on the idea that if a specific material actually *is*, by some process, completely removed from the plasma passing through the kidney, then the rate at which plasma enters the kidney, that is, the renal plasma flow, would be equal to the clearance. In fact, a number of organic dyes are not only filtered but are also actively secreted by the kidney, and the combination of filtration and secretion results in almost total removal of the dyes from the plasma in a single pass through the kidney. As illustrated in Fig. 47-1, *B*, *p*-aminohippuric acid (PAH) is such a material, and over 90% of the dye that enters the kidney via the renal artery will normally be excreted in the final urine as the combined result of filtration and secretion. Thus simply by measuring PAH clearance, one can estimate the renal plasma flow (RPF) to within 10% by calculating the PAH clearance as follows:

$$\text{RPF} = \frac{U_{PAH} \times V}{P_{PAH}} \qquad (1)$$

Extraction of PAH from the renal arterial blood is not 100% because of (1) the lack of an ideal transport system with no back leak in the tubule, and (2) a significant fraction of the blood flowing into the kidney does not come into sufficiently close contact with the proximal tubules, where PAH secretion occurs. After passage through the efferent arterioles of the juxtamedullary nephrons, a fraction of the blood flows into the medulla via the vasa recta capillaries; no proximal tubules are in the medulla where the vasa recta capillaries are located. Therefore blood flowing through the vasa recta capillaries is never exposed to proximal tubules for secretion.

The second case in which clearance can be a true measure of a real volume rate of flow is that of glomerular filtration (Fig. 47-1, *C*). Renal physiologists found a marker that did not alter renal blood flow or filtration rate, that was filtered in a concentration exactly equal to its concentration in plasma, and that was not metabolized, reabsorbed, or secreted by the tubules. Such a material can enter the tubule only by glomerular filtration, and whatever enters the tubule must be totally excreted in the urine. Thus given the fact that the filtrate concentration of the marker is equal to its plasma concentration, the minimum volume of fluid that could have supplied the material in the final urine would be equal to the volume of fluid that was filtered with its contained solute. The substance meeting all of these criteria is *inulin*, a nonmetabolizable polysaccharide. Inulin clearance is equal to the glomerular filtration rate (GFR).

$$\text{GFR} = \frac{U_{In} \times V}{P_{In}} \qquad (2)$$

Thus the clearance computation can be used to calculate two important values, glomerular filtration rate and renal plasma flow. In each of those two special cases a molecule was chosen that has very special characteristics permitting the clearance calculation to yield a real volume rate of flow, not a virtual volume.

In practice, a clearance measurement is made by infusing the appropriate substance at a constant rate until the plasma concentration is stable. Samples of urine and plasma are then taken for chemical analysis and the computation can be made.

Corrections for Tubular Water Reabsorption—Micropuncture Analyses

Micropuncture techniques allow one to insert a micropipette into the lumen of a single tubule and withdraw fluid for analysis. Along the course of the nephron solutes move into and out of the tubules, and water is reabsorbed. To understand the meaning of a micropuncture sample, fluid composition changes that result from water reabsorption must be distinguished from changes induced by solute transport. This is usually accomplished by use of an indicator that is quantitatively retained within the tubular fluid and is filtered with a concentration equal to that in plasma. Obviously, inulin is the appropriate indicator.

The fraction of the filtrate that remains at any point in the nephron is simply equal to the tubular fluid flow rate at that point divided by the glomerular filtration rate. In the case of the excreted urine, the fraction of filtered water excreted is simply equal to V/GFR.

$$\begin{aligned} f &= \text{fraction of filtered water excreted} \\ &= \frac{V}{\text{GFR}} = \frac{V}{\dfrac{(U_{In}V)}{P_{In}}} = \frac{P_{In}}{U_{In}} \end{aligned} \qquad (3)$$

$$\text{Fraction reabsorbed} = 1 - f = 1 - \left(\frac{P_{In}}{U_{In}}\right) \qquad (4)$$

Thus simply by computing the plasma/urine ratio of inulin concentrations in a steady state, one can compute either the fraction of the filtered water that is excreted or, alternatively, the fraction that has been reabsorbed.

The observed behavior of inulin can be used to correct the various solute concentrations for water reabsorption. The quantity of substance x filtered per unit time should be P_x times GFR, or P_x times C_{In}. The rate of excretion of solute in the urine is V times U_x. It follows that the fraction (F) of the filtered material that is excreted must be

$$F = \frac{U_x \times V}{P_x \times C_{In}} = \frac{C_x}{C_{In}} \qquad (5)$$

■ Fig. 47-2. Network analysis of the renal circulation. Variable resistors represent arterioles. Black lines represent capillary networks.

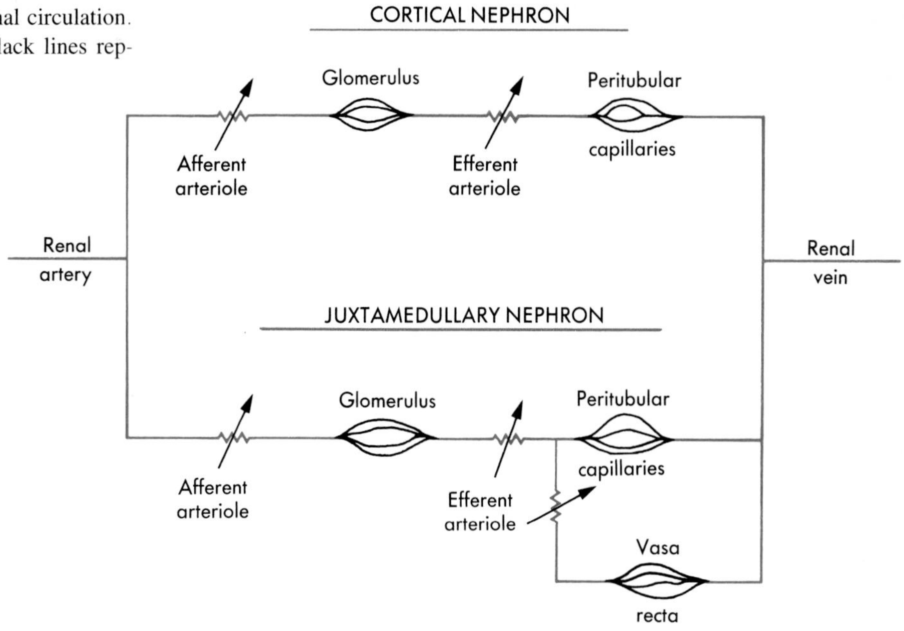

Because the urine flow term (V) is the same for substance x and inulin, equation 5 can be further simplified to:

$$\text{Fraction of filtered solute that is excreted} = \frac{U_x/P_x}{U_{In}/P_{In}} \qquad (6)$$

Thus the ratio of the clearance of an unknown substance (sodium, for example) to that of inulin gives the fraction of the filtered material that appears in the urine. If the clearance of a substance is greater than the clearance of inulin, this denotes net secretion. If the clearance is less than the clearance of inulin, this implies net reabsorption.

■ *Renal Blood Flow*

The renal circulation enables the kidney to sample and regulate the composition of extracellular fluid and electrolytes in the body. In general, the renal circulation represents the point of integration between renal function and whole body homeostasis. Fig. 47-2 is a schematic illustration of the anatomy of the renal circulation; it emphasizes the variety of pathways that blood can take through the vasculature of the kidney. There are basically three parallel networks: one supplying the cortical nephrons, one supplying the juxtamedullary nephrons, and one supplying the medulla. The networks within the cortical structures consist of two capillary beds in series, the glomerular capillaries and the peritubular capillaries. The glomerular capillaries are preceded and followed by two sets of arterioles, the afferent and efferent vessels, respectively. Blood that perfuses the vasa recta represents blood that has passed through a juxtamedullary glomerulus, into its efferent arteriole, and, finally, into a vasa recta bundle.

The series arrangement of arterioles on both sides of the glomerulus stabilizes the glomerular filtration pressure. This is easily understood in terms of the simple resistive network shown in Fig. 47-2. The flow through the circuit equals the pressure difference between the input to the afferent arteriole and the output from the efferent arteriole, divided by the resistance. The total resistance is approximately equal to the sum of the resistances of the afferent and efferent arteriolar networks because glomerular capillary resistance is low.

Increasing the resistance of either arteriole will increase the total resistance and therefore decrease the flow, if arterial pressure is constant. However, the two arterioles have directionally different effects on glomerular pressure. Constriction of the afferent arteriole will increase the pressure drop across this vessel and thus reduce pressure within the glomerulus. In contrast, constriction of the efferent arteriole will increase glomerular filtration pressure because reduced flow through the afferent arteriole will result in a smaller pressure drop before reaching the glomerulus. Constriction of either arteriole will reduce flow, but because the afferent and efferent arterioles have opposite effects on filtration pressure, renal blood flow and glomerular filtration can be controlled independently to some extent. The effects of afferent and efferent arteriolar constriction offset each other; therefore simultaneous constriction of the afferent and efferent arterioles will produce minimal changes in capillary hydrostatic pressure and glomerular filtration rate but will reduce renal blood flow. Thus changes in renal blood flow can be integrated into overall circulatory homeostasis without compromising glomerular filtration.

The presence of a high resistance (the efferent arterioles) downstream of the glomerulus but upstream of the peritubular capillaries also ensures that hydrostatic pressure within the peritubular capillaries is reduced to relatively low levels (10 to 12 mm Hg). This low pressure within the peritubular capillaries facilitates fluid reabsorption from proximal and distal tubules.

The juxtamedullary nephrons have a similar vascular pattern in parallel with the superficial cortical glomeruli, although some differences exist in the distribution of the peritubular capillary networks. Vasa recta blood flow originates distal to the efferent arteriole of the juxtamedullary nephrons. This point of origin becomes important in understanding the ability of the kidney to secrete various materials. Any blood that exits from the pathway through the juxtamedullary cortical circulation and flows into the vasa recta is not exposed to the environment surrounding the proximal and distal tubules of these nephrons. Thus blood flowing through this pathway cannot contribute its solute to the secretory activities of the juxtamedullary nephrons. This is the major reason PAH is not fully cleared from the plasma circulating through the kidney.

Renal plasma flow can be approximated by measuring the extraction of PAH by the kidney. Renal blood flow equals renal plasma flow divided by 1 minus the red cell fraction in the blood (hematocrit). A parameter that is frequently used to characterize renal function is the *filtration fraction* (FF), which is the fraction of the total renal plasma flow that is filtered:

$$FF = GFR/RPF = C_{In}/C_{PAH} \quad (7)$$

This fraction is proportional to the amount of protein-free water removed from plasma that flows through the glomeruli. Therefore the greater this fraction, the greater will be the concentration of the plasma proteins in blood entering peritubular capillaries, and the larger will be the oncotic pressure available to ensure reabsorption of tubular fluid by peritubular capillaries.

The interplay mentioned previously between afferent and efferent arterioles allows a relatively independent control of glomerular filtration rate and renal plasma flow. This independent control implies that the filtration fraction may vary significantly with different rates of renal blood flow.

The total amount of blood flowing into the kidney is not distributed uniformly. Approximately 90% of the renal blood flow passes through the cortical vessels and 10% through the medulla. The blood flow to different portions of the kidney may be selectively manipulated, either by humoral or neural mechanisms, but this possibility remains controversial.

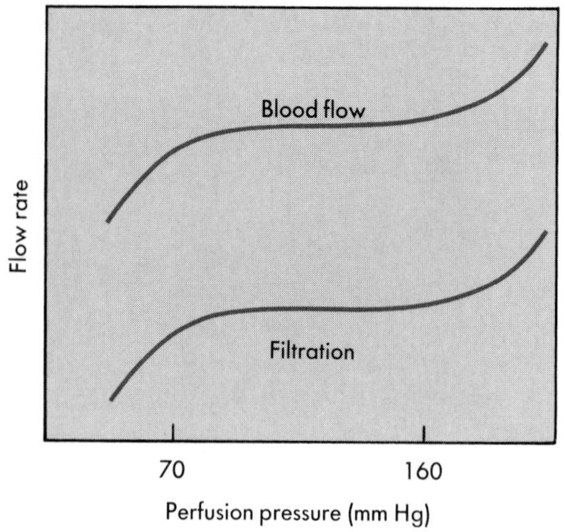

■ **Fig. 47-3.** Renal autoregulation. Blood flow and glomerular filtration rate are stabilized in the face of changes in perfusion pressure.

■ *Local Control of Blood Flow*

As with other circulations, the blood flow through the kidney is *autoregulated*. This is illustrated in Fig. 47-3, which shows that as perfusion pressure varies over a certain range, there is only a small change in the renal blood flow rate. In the kidney this autoregulatory process is particularly effective; both renal blood flow and glomerular filtration rate are regulated. The fact that both are regulated implies that the afferent arteriole is involved in the regulatory process. Renal plasma flow might be regulated by the efferent arteriole alone, but simultaneous regulation of glomerular filtration rate would not then be possible because constriction of the efferent arteriole has opposite effects on renal blood flow and on glomerular filtration rate, decreasing the former and increasing the latter. The afferent arteriole, in contrast, has similar effects on both flow and glomerular filtration rate. In fact, both the afferent and the efferent arterioles are probably involved in autoregulation.

The mechanism responsible for autoregulation is controversial but an important component of the regulation appears to be the myogenic mechanism, as described in Chapter 33. As perfusion pressure is increased, the smooth muscle of the arterioles is stretched; this stretch induces a constriction of the smooth muscle and an increase in vascular resistance that offsets the increase in perfusion pressure. Metabolic autoregulation is not a likely mechanism in the kidney because of the very high blood flow relative to the metabolic rate. Consequently, the oxygen tension in renal venous blood and renal tissue is very high and thus the stimulus for metabolic regulation appears to be negligible. Another possible mechanism to explain autoregulation involves release of renin from the juxtaglomerular cells and formation of angiotensin, as described later.

Neural Control

For purposes of simplification, the neural component in the regulation of renal blood flow can be viewed as subserving primarily the needs of the body as a whole, whereas local control mechanisms contribute to the stability of renal function. Because the renal circulation receives about 25% of the total cardiac output, renal vascular resistance can, in principle, be modulated as a means of controlling arterial blood pressure.

Neural control of renal blood flow is accomplished almost exclusively by release of norepinephrine from sympathetic nerve terminals. Thus the sympathetic nervous system makes little use of the control of the renal circulation in cardiovascular responses such as the baroreceptor reflex. However, both anesthesia and pain result in a large increase in sympathetic drive and an exaggerated neural tone in the renal circulation.

Such conditions as changes in posture, mild emotional responses, and light physical activity all increase sympathetic nerve discharge to the kidney. Moderate activity in the renal sympathetic nerves evokes a balanced constriction of the afferent and efferent arterioles. Consequently, glomerular filtration is reduced proportionately less than renal blood flow, and the filtration fraction is increased. This is important because the reduction in renal blood flow contributes to the regulation of systemic blood pressure, and renal function is not impaired, because the glomerular filtration rate is only slightly reduced.

The stability of the filtration rate is not maintained at high rates of sympathetic discharge. In response to hemorrhage or other stimuli that induce intense renal vasoconstriction, the capacity of the kidney to dissociate renal blood flow and glomerular filtration rate is exceeded, and both glomerular filtration rate and blood flow are reduced substantially.

Renin-Angiotensin System

The *renin-angiotensin system* is a unique control system that influences blood pressure, blood volume, and the intake and excretion of salt and water. It is highly specialized, both with regard to its influence on renal blood flow and as a process that can alter systemic arterial blood pressure. Renin is an enzyme that is formed and stored in granules in specialized vascular smooth muscle cells, the *juxtaglomerular cells,* of the afferent and efferent arterioles (Fig. 47-4, *A*). These cells are in contact with specialized epithelial cells of the distal tubule of the nephron, the *macula densa*. The space delineated by the macula densa and the two arterioles, as well as the space between the glomerular capillaries, is designated the *mesangial* region. The mesangial cells that occupy the space between the two arterioles and the macula densa are often referred to as the *Goormaghtigh cells*. In response to stimuli (described later), renin is released from the juxtaglomerular cells. Renin acts on an α_2-globulin, carried by the plasma, to cleave off the decapeptide angiotensin I (Fig. 47-4, *B*). In the presence of *converting enzyme* and *chloride ion, angiotensin I* is split to form the octapeptide *angiotensin II,* which is a very potent vasoconstrictor. The *converting enzyme* is selectively localized on the surface of endothelial cells. In the adrenal cortex, angiotensin II is converted by a peptidase to the heptapeptide angiotensin III, a very potent stimulus to the secretion of aldosterone.

Control of Renin Release

Regulation of renin release and thereby angiotensin II formation is a complex process that involves interactions among a number of hormones. Three key elements in the process can be clearly defined. First, low plasma sodium stimulates renin release. Second, reduced stretch of the juxtaglomerular cells, associated with reduced blood pressure in the afferent arterioles, induces renin release. Third, stimulation of the renal sympathetic nerves induces renin release via activation of β-adrenergic receptors on the juxtaglomerular cells.

Renin release and subsequent angiotensin II formation are involved in many aspects of cardiovascular control (Fig. 47-4, *B*). This system affects vascular tone, sympathetic nerve activity, adrenal steroid release, and central nervous system control of blood pressure. In general, angiotensin II formation is regulated in a manner that stabilizes blood pressure and extracellular fluid volume.

Angiotensin II has wide-ranging and complex effects in a variety of other organs in the body. Five effects, however, dominate the overall response (Fig. 47-4, *B*). First, angiotensin II and angiotensin III stimulate the release of *aldosterone* from the *adrenal cortex,* thus enhancing distal tubular and collecting duct sodium reabsorption. Second, angiotensin causes systemic vasoconstriction. Third, angiotensin acts on sympathetic nerve terminals to enhance neurotransmitter release. Fourth, angiotensin stimulates the release of antidiuretic hormone (ADH) from the posterior pituitary. Fifth, angiotensin acts within the brain to stimulate drinking and thereby increase water intake. Note that each effect represents part of a coordinated response that tends to expand extracellular fluid volume and elevate blood pressure.

The effect of a pathological constriction of a renal artery illustrates one aspect of the behavior of the renin-angiotensin system in the regulation of the circulation.

Fig. 47-4. Renin-angiotensin system. **A,** The juxtaglomerular apparatus is composed of the macula densa (specialized cells of the distal tubule) and modified smooth muscle (the juxtaglomerular cells located primarily on the afferent arteriole). (See also Fig. 45-3, *B*.) The mesangial cells (Goormaghtigh cells) are outside the glomerulus. **B,** Formation of and physiological responses to angiotensin II.

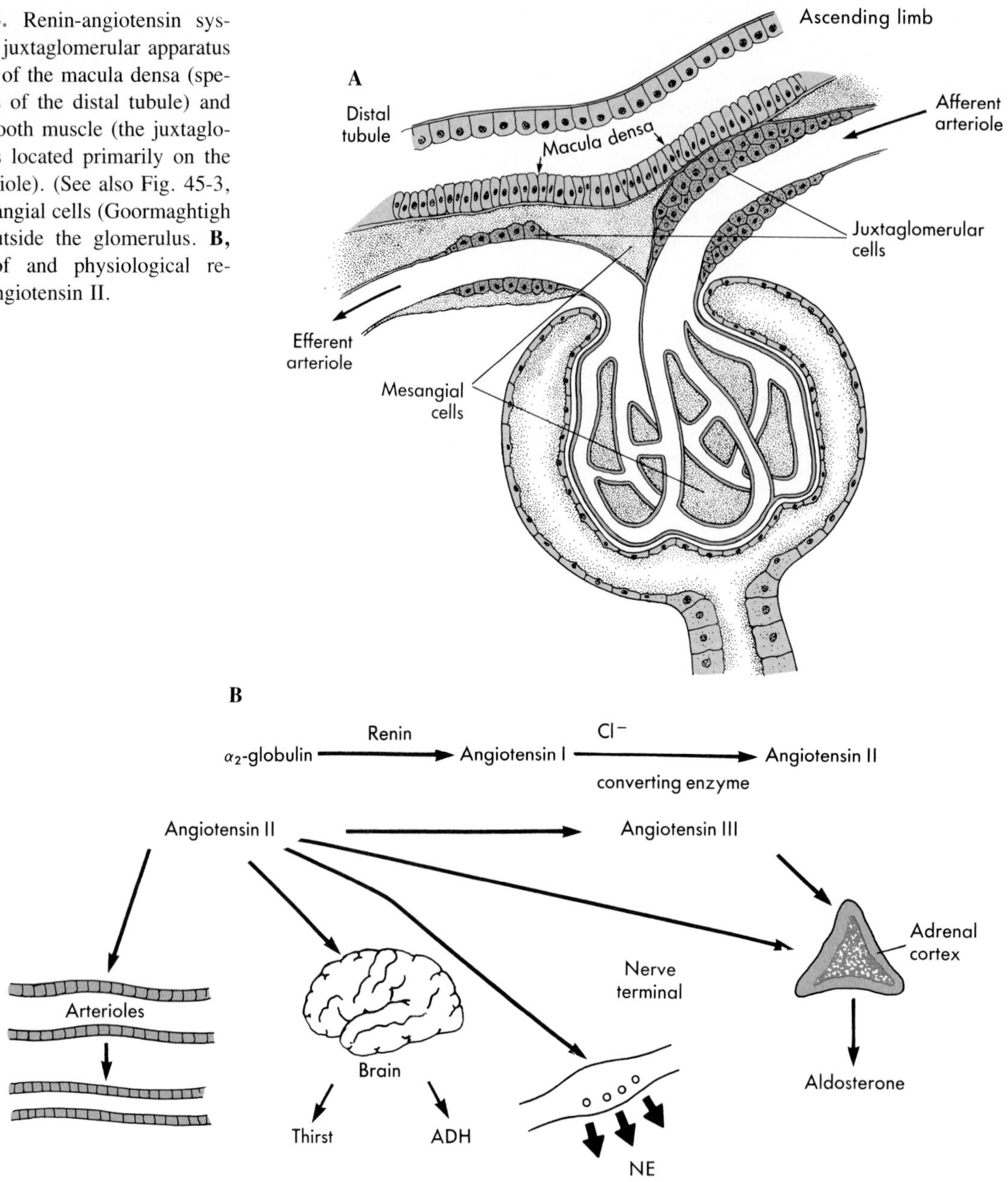

Partial occlusion of the renal artery, as may occur with an atherosclerotic lesion, will lower intravascular pressure in the renal circulation and diminish the stretch of the juxtaglomerular cells. This causes renin release, angiotensin II formation, and ultimately stimulation of aldosterone secretion, systemic vasoconstriction, and enhanced sympathetic nerve discharge. These responses increase sodium reabsorption, peripheral resistance, and extracellular volume, all of which raise the blood pressure. In other words, the kidney responds to renal artery stenosis in such a way as to raise arterial blood pressure high enough to achieve normal renal perfusion pressure distal to the stenosis.

This process is probably a factor in the etiology of some types of hypertension in humans, especially in *renovascular hypertension*. However, the process is complicated. Alleviation of a chronic renal artery constriction, in humans and in experimental animals, may fail to alleviate the hypertension. Apparently, other factors can stabilize the hypertensive state after it has been established by the renin-angiotensin system.

The renin-angiotensin system may play a very subtle role in the regulation of renal blood flow as part of the mechanism responsible for maintaining *glomerulotubular balance*. Sodium chloride concentration of the early distal tubular fluid at the *macula densa* is influenced by tubular flow rate. It has been observed that increased distal tubular flow is associated with a decrease in the filtration rate of the corresponding nephron. Based on these findings it has been proposed that if the glomerular filtration rate of the nephron is raised, the macula densa sodium chloride concentration changes and angiotensin II formation is stimulated. This results in arteriolar constriction, reduced filtration, and stabilization of the relation between filtration and tubular reabsorption.

The coupling between the changes in macula densa NaCl and angiotensin II release is not known. However, the close apposition of the Goormaghtigh cells to the macula densa epithelial cells and to the juxtaglomerular cells suggests that the Goormaghtigh cells may act as transducer cells. Intrarenal hormones (e.g., prostaglandins) also may be involved in the distal tubular feedback, either directly or by modulating the activity of the renin-angiotensin system.

Although the details of the mechanism are uncertain, the observations are clear. Elevation of distal tubular flow, as might be expected if proximal tubular or Henle's loop reabsorption were reduced, decreases the glomerular filtration rate of that nephron. Hence, filtration and tubular reabsorption are maintained in balance.

A point of clarification must be made about the implication of a high salt diet in hypertension. This is not due to effects of salt on the renin-angiotensin system, but rather to effects on volume of the extracellular fluid. In general, hypertension is not induced in a healthy person simply by the ingestion of large quantities of salt. However, if renal function is compromised, large salt intakes cause an accumulation of salt. Secondarily, water retention within the extracellular space expands the vascular volume. The expanded vascular volume tends to increase cardiac output (Chapter 34) and thus to increase systemic arterial blood pressure.

Intrarenal Hormones

The prostaglandins and kinins are hormones that are formed within and act on the kidney. Both glomerular filtration rate and sodium reabsorption may be under local hormonal controls. Also, the prostaglandins modify afferent neural sensitivity of the renal microvessels, modulate sensitivity of the vascular smooth muscle to angiotensin II, and inhibit the actions of ADH on the permeability of the collecting duct to water.

Concentration and Dilution of the Urine

Humans experience wide variations in requirements for excretion of water. High dietary intake of fluids implies a requirement for excretion of urine with low solute content, that is, low osmolality. In contrast, persons in an arid environment with little water to drink must reduce urinary water losses as far as possible while continuing to excrete necessary waste solutes. The combined need to excrete solutes and retain water means that efficient mechanisms must be developed to concentrate the urine. The mammalian kidney that has evolved in the face of these needs can produce urine osmolalities ranging from approximately 100 to 1200 mOsm/L. The renal mechanisms responsible for this ability are so efficient that some animals, such as the desert rat, can excrete urine concentrated to such a degree that no dietary water intake is necessary. They survive by acquiring necessary fluids from food and metabolic water. The process that has evolved to produce urine with such a wide range of concentrations is one of the more remarkable examples of form and function that has evolved in a biological system.

An underlying fact in the evolution of this complex process is that it is more efficient to move water by a passive osmotic process than by an active transport mechanism. Suppose that the osmotic difference between plasma and urine is 300 mOsm/L and that the difference is established by solute transport. The 300 mOsm/L difference represents 1.8×10^{23} solute molecules per liter of fluid transported (0.3 mole/L $\times$ 6 $\times$ 10^{23} molecules/mole). In contrast, primary water transport would require the movement of a larger number of water molecules because a 300 mOsm/L solution is approximately 55 M with respect to water (1000 g/L $\div$ 18 g/mole = 55 moles/L), and 330×10^{23} molecules of water would have to be moved to produce a 300 mOsm osmotic gradient. In other words, approximately 180 times as many water as solute molecules would have to be moved to produce the observed osmotic gradient. The mechanism for concentration of the urine that has evolved is one acting via salt transport rather than water transport.

The urinary concentrating mechanism is based on the existence of a concentrated fluid within the medullary interstitium, as shown in Fig. 46-10. Interstitial fluid osmolality in the medulla increases from 300 mOsm/L at the corticomedullary border to 1200 mOsm/L at the renal papilla. This concentrated fluid can be used to concentrate collecting duct fluid by osmotic withdrawal of water.

Fig. 47-5 shows the ratio of tubular fluid to plasma osmolality through the length of the nephron under conditions in which water is being conserved *(solid colored line)* and in which water excretion is required *(dashed colored line),* that is, in the presence and in the absence of ADH, respectively. It is striking that little difference is observed in the two cases until tubular fluid is well into the distal tubule and especially within the connecting segment and the collecting duct. Proximal tubular fluid is isotonic, whether ADH is present or absent. Osmolality rises sharply through Henle's loop (the origin of this will be considered later). The osmolality then falls and reaches *hypotonic* levels as the fluid passes through the ascending limb of Henle's loop.

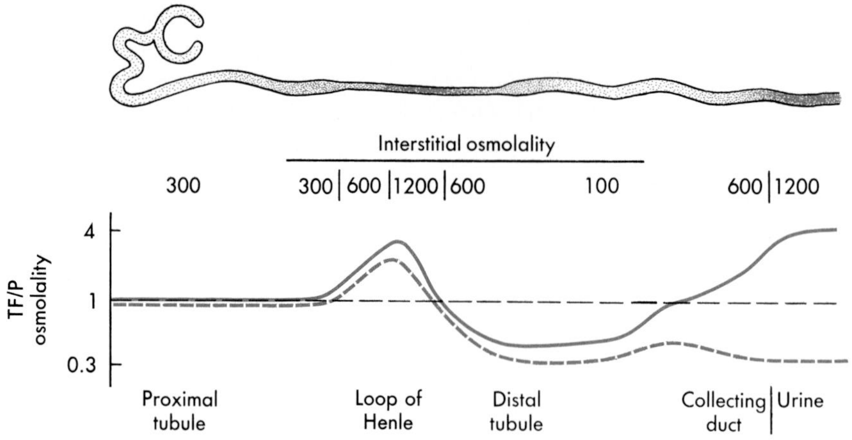

■ Fig. 47-5. Tubular fluid osmolality. TF/P represents the ratio of the osmolality in the tubular fluid at any given location to the osmolality of plasma. The colored lines show osmolality in the presence *(solid)* or the absence *(dashed)* of ADH. The osmolality at various locations relative to the nephron is shown above.

In the absence of ADH (Fig. 47-5, *dashed line*), a dilute urine is excreted. The dilution is accomplished largely by transporting solute out of the tubules in a region where the epithelium is relatively impermeable to water. This occurs mainly in the ascending limb of Henle's loop, but it continues to some extent throughout the distal tubule and collecting duct if ADH is absent.

Quite a different situation exists during the formation of concentrated urine (Fig. 47-5, *solid line*). The osmolality of the tubular fluid is similar to that observed during the formation of dilute urine until Henle's loop is reached, where a slightly higher osmolality can be observed in the ascending limb. The composition of the fluid entering the distal tubule is only slightly different from that observed during diuresis. Once fluid enters the distal tubule, however, the osmolality begins to rise, and it ultimately approaches the isotonic level observed in the proximal tubule. The rise in osmolality largely reflects increases in connecting tubule permeability to water induced by ADH. The distal tubule is surrounded by capillaries that contain plasma at an osmolality of 300 mOsm/L, that is, higher than that found in the fluid exiting Henle's loop. Therefore a high tubular water permeability will cause fluid to equilibrate osmotically between the lumen of the nephron and the lumen of the capillary.

In contrast to the environment surrounding the distal tubules, the fluid surrounding the collecting ducts in the medullary interstitium is strikingly hypertonic (up to 1200 mOsm/L). The formation of concentrated urine depends on osmotic equilibration between the fluid in the collecting ducts and the hypertonic medullary interstitium. Such osmotic equilibration occurs if ADH is present. The urine can be concentrated to a level equal to that in the surrounding interstitium.

■ *Formation of a Concentrated Fluid in the Medullary Interstitium—Countercurrent Multiplication*

Much of the difficulty in understanding the urinary concentrating mechanism is involved with the question of how the high osmolalities in the medullary interstitium are generated. These concentrations are generated in Henle's loop by a process called *countercurrent multiplication*. This mechanism has evolved to meet the demands for the formation of a urine that is as much as 900 mOsm/L more concentrated than plasma. This is accomplished by a tubule that can sustain a gradient of only about 200 mOsm/L across its epithelium. Countercurrent multiplication uses a highly selective localization of transport systems in the ascending limb of Henle's loop, combined with the unique hairpin loop geometry and the great length of Henle's loop. Four facts should be stated before the mechanism is described:

1. Henle's loop reabsorbs sodium by an active process in the ascending limb; chloride accompanies sodium as the necessary anion.
2. Permeability of the ascending limb to water is low, thus permitting the formation of a dilute fluid as chloride and sodium are removed.
3. Salt that is transported out of the ascending limb enters the medullary interstitium and raises its osmolality.
4. Descending limb permeability to water is high; water extraction combined with solute entry raises the osmotic activity of the fluid in the descending limb.

Fig. 47-6 illustrates how the geometry and transport characteristics of Henle's loop can produce a highly concentrated fluid in the medullary interstitium. The

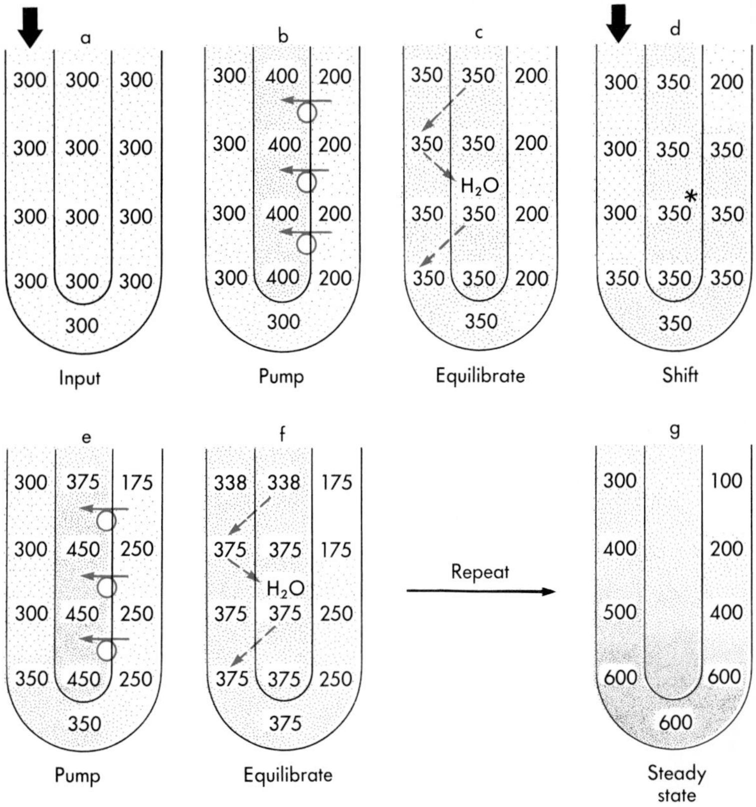

■ **Fig. 47-6.** Countercurrent multiplication. Each figure represents a schematic of Henle's loop. Note that lateral active transport combined with flow through the parallel elements of the loop results in a longitudinal concentration gradient in the system. As the process is reiterated, the peak salt concentration in the loop reaches 600 mOsm/L.

process is broken down into three steps: (1) a transport step by the tubular epithelial carriers, (2) a fluid equilibration step between the medullary interstitium and the descending limb of Henle's loop, and (3) a step in which new fluid is brought into Henle's loop from the proximal tubule.

Consider that initially the ascending limb, descending limb, and interstitium are at equilibrium, and all fluids have the same osmolality as that in the proximal tubule (Fig. 47-6, *a*). Assume that sodium chloride is transported out of the ascending limb into the interstitium, and a gradient of 200 mOsm/L is established across the epithelium *(b)*. This is followed by equilibration between the fluids in the interstitium and descending limb *(c)*. Water and solute diffusion into the descending limb of Henle's loop raises the solute concentration within the limb to 350 mOsm and simultaneously lowers the medullary interstitial concentration to the same value. The process is continued by cycling new fluid in from the proximal tubule *(d)*. This forces fluid from the descending limb around the tip of the loop and reduces the gradient across the epithelium of the ascending limb *(asterisk)*. Pump activity can then establish a new gradient of 200 mOsm/L between the ascending limb fluid and the interstitial fluid *(e)*. This is followed by reequilibration of the fluid within the interstitium and the descending limb of Henle's loop *(f)*.

With each cycle the concentration of fluid at the tip of Henle's loop rises. In the human kidney this process is reiterated until the salt concentration at the tip reaches approximately 600 mOsm/L (Fig. 47-6, *g*). This represents a steady-state condition in which the gradient across the epithelium at any level is no greater than 200 mOsm/L, but the longitudinal gradient from origin to tip is substantially larger. The importance of the length of Henle's loop can be appreciated from the fact that, for a variety of mammals, the maximal achievable urinary osmolality correlates well with the length of Henle's loop. In some species the medullary interstitial osmolality can rise to 15 times the plasma osmolality.

The osmolality generated by the sodium chloride acting in the countercurrent process is only about half the final osmolality of 1200 mOsm/L that is usually observed. The remainder of the osmotic pressure is generated by urea that is selectively deposited in the medulla by a process to be described later.

■ *Stabilization of the Medullary Interstitial Osmotic Gradient—The Vasa Recta*

The countercurrent multiplication system will generate a concentrated medullary interstitium and produce the dilute fluid that enters the distal tubule. The collecting duct permeability is then the ultimate determinant of the osmolality of the urine. However, as described to this point, the system is not complete, and it would function only briefly. Water is osmotically withdrawn from the collecting duct and deposited in the medullary intersti-

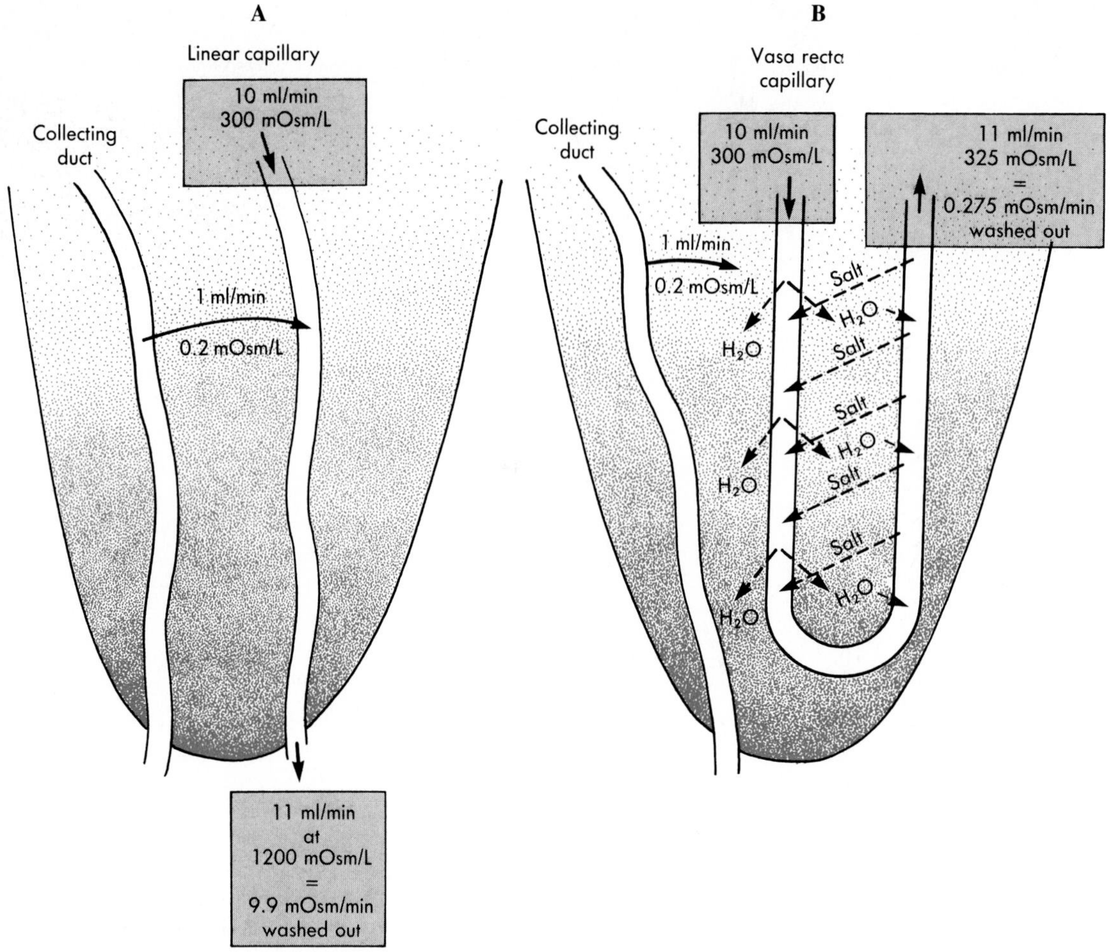

■ Fig. 47-7. Countercurrent exchange. **A** and **B** represent two possible capillary organizations passing through the renal medulla. In both cases, 1 ml/minute of dilute fluid is reabsorbed from the collecting duct and carried out in the effluent blood. Note the difference, however, in effluent solute concentration and rate of washout of solute with the flow-through linear capillary as compared to the countercurrent capillary of the vasa recta.

tium. Obviously this will result in a cumulative dilution of the concentrated salt solutions formed in the interstitial fluid. In order to prevent this, some means must be provided for removal of the water that is reabsorbed from the collecting duct fluid. This must be accomplished, however, without washing out the solute that has accumulated in the medulla through the action of Henle's loop. These tasks are accomplished by the *vasa recta,* a highly specialized capillary network that complements Henle's loop.

Fig. 47-7 shows two hypothetical capillary networks passing through the medulla. One of these *(A)* is modeled after a typical capillary, such as might be found in skeletal muscle, that is, a linear capillary that enters and courses through the tissue and exits into a venule. The other capillary (Fig. 47-7, *B*) has an anatomy similar to that of the vasa recta; that is, it has a hairpin configuration. The diagram illustrates how the hairpin configuration of the vasa recta confers the ability to remove water and salts from the interstitium with minimal disruption of the osmotic concentration in the medullary interstitium.

It is assumed that the plasma in both types of capillaries equilibrates with interstitial fluid osmolalities by the capillary midpoint and that both types of capillaries have a flow of 10 ml/minute. In Fig. 47-7, *A,* equilibration at the midpoint would mean that the fluid within the capillary would exit at 1200 mOsm/L. In addition to solute, 1 ml of water, which was reabsorbed from the collecting duct, must be taken up so that 11 ml/minute leaves in the capillary blood. An exit concentration of 1200 mOsm/L (900 mOsm/L greater than entry), would yield a washout of 9.9 mOsm/minute of solute.

Fig. 47-7, *B* shows the configuration of the vasa recta capillary and a behavior that is quite different. In this capillary network, the concentration rises to 1200 mOsm/L at the midpoint of the capillary. However, because of the close association between the ascending

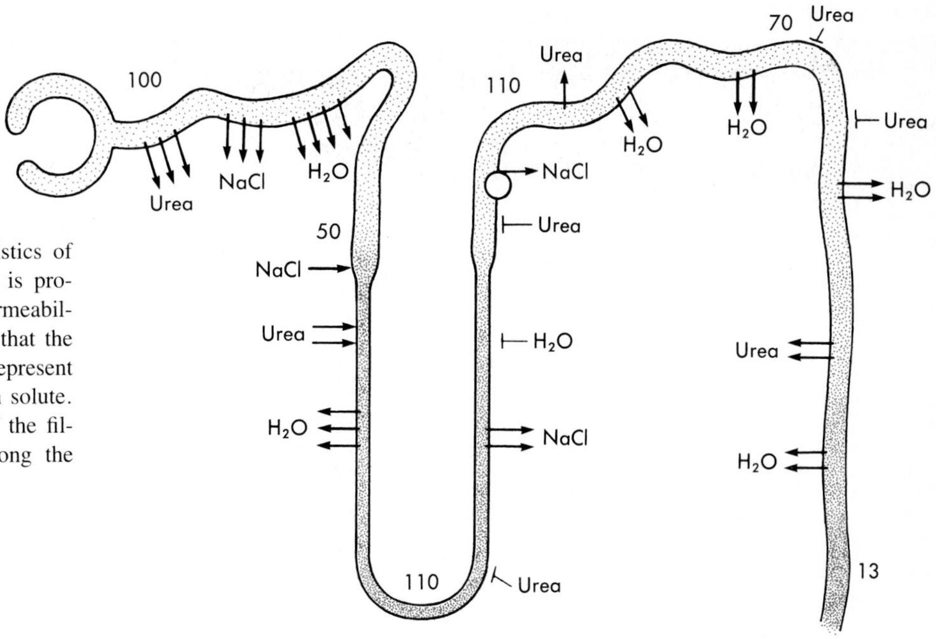

■ **Fig. 47-8.** Permeability characteristics of the nephron. The number of arrows is proportional to the magnitude of the permeability. A line with a crossbar indicates that the tubule is impermeable. The arrows represent the direction of movement of a given solute. The numbers indicate the fraction of the filtered urea present at each point along the nephron.

and descending limbs, blood comes into proximity with more dilute fluid entering from the input side of the vasa recta. Thus a concentration gradient exists across the intervening interstitium between the ascending limb and the descending limb. Solute and water movement progressively reduce the osmolality of the blood moving through the ascending limb of the vasa recta and out of the medulla. In this case, rather than exiting the organ at 1200 mOsm/L as shown in Fig. 47-7, *A*, the fluid exits at 325 mOsm/L. In the latter case, the result is a washout of 0.275 mOsm/minute, rather than the 9.9 mOsm/minute that would be expected if a linear capillary were involved (Fig. 47-7, *A*). Thus the hairpin loop configuration allows the solute and the water to be reabsorbed from the collecting duct and returned to the systemic circulation with much less dilution of the medullary interstitium than a linear capillary system would induce.

■ *Role of Urea in Concentration of the Urine*

Urea has a multifaceted effect on urine osmolality. As the end product of protein metabolism, urea must be excreted each day in a quantity determined by the rate of protein intake. Thus increased dietary protein intake requires an increased urea excretion, and this solute must be excreted in whatever volume of urine is formed. A large fraction of urinary solute is urea, and it contributes to urine osmolality.

The kidney can continue to excrete large quantities of urea despite relatively low urine flow rates. This ability derives from the differential urea permeabilities in various segments of the nephron; such characteristics cause the accumulation of urea within the medullary interstitium. In fact, with a normal protein intake, approximately half of the 1200 mOsm/L found in the medullary interstitium is contributed by the urea molecule.

The elements of tubular function that cause urea to accumulate in the interstitium of the medulla are shown in Fig. 47-8. Relative permeabilities for urea and water in the various tubular segments are shown. Permeability to urea is high in the proximal tubule, and urea follows sodium chloride and water movement rather closely. In the descending limb of Henle's loop there is a modest permeability to urea and a relatively high permeability to water. Sodium chloride and urea diffuse into the descending limb of Henle's loop from the surrounding interstitium.

A key element in the sequence of events that leads to the accumulation of urea within the medullary interstitium is the fact that permeability to water and permeability to urea are quite different over the length of the distal tubule and collecting duct. In the distal tubule and early part of the collecting duct, urea permeability is low and is relatively unaffected by ADH. Thus over this segment of the nephron, water from the dilute solution created in the ascending limb of Henle's loop will be reabsorbed into peritubular capillaries. However, because the nephron is relatively impermeable to urea in this segment, urea will fail to leave the tubule as readily as water, and therefore it becomes concentrated in tubular fluid. In the medulla, the urea permeability of the collecting duct increases and the tubular fluid urea concentration is relatively high as a result of the prior removal of water in the distal tubule. Urea can therefore diffuse from the collecting duct into the medullary interstitium. Once urea enters the medullary interstitium, it tends to accumulate because the vasa recta system removes permeable solutes inefficiently. Accordingly, the interstitial concentration of urea will rise to a level that is high enough to cause the rate of urea removal from the medullary interstitium to equal the rate of entry from the collecting ducts.

Urea may play another, more central role in the formation of concentrated urine. Fig. 47-8 illustrates the key elements of this mechanism in highly simplified form. The mechanism was proposed because the epithelial cells in the thin portions of Henle's loop have relatively few mitochondria and thus are probably not involved in highly active transport processes (Fig. 45-3, *C*). The hypothesis is based on the observation that urea tends to recirculate in the medulla of the kidney. The numbers in Fig. 47-8 show the fraction of the filtered urea present in the tubule at the indicated points. A portion of the urea exiting the collecting duct enters the descending limb of Henle's loop and then circulates back to the collecting duct.

Given the foregoing description of urea recirculation, it has been proposed that dilution of the tubular fluid in the ascending limb is accomplished by using the urea accumulated in the medullary interstitium. A proviso of the hypothesis is that the increasing concentration of tubular fluid that occurs in the descending limb of Henle's loop is largely the result of water removal in response to the osmotic gradient generated by the high concentrations of sodium, chloride, and urea in the interstitium, not by solute entry. Thus the fluid that reaches the tip of Henle's loop has high sodium, chloride, and urea concentrations. In the ascending thin limb, the tubular permeability to sodium is substantial, whereas permeabilities to both water and urea are low. The sodium, which is concentrated by osmotic withdrawal of water from the descending limb of Henle's loop, can diffuse down a concentration gradient into the medullary interstitium. This dilutes the fluid in the ascending limb of Henle's loop without requiring energy expenditure at this site. Once the fluid enters the thick segment of the ascending limb of Henle's loop, active sodium transport and passive chloride movement out of the tubule further dilute the fluid within the tubular lumen.

In this model the active transport of salt that occurs in the thick portion of the ascending limb of Henle's loop is the source of energy for the system. This transport process ultimately raises the urea concentration within the medullary interstitium. The urea acts with sodium chloride on the descending limb of Henle's loop to extract water, thereby generating the necessary concentration gradient to drive sodium reabsorption in the sodium-permeable segment of the thin ascending limb of Henle's loop. It is not known how important this process may be in urinary concentrating mechanisms relative to more classical views of the countercurrent system, in which all solute transport is assumed to be active.

Bibliography

Journal articles

Aukland, K.: Renal blood flow, Int. Rev. Physiol. **11:**23, 1976.

Baer, P.G., and McGiff, J.C.: Hormonal systems and renal hemodynamics, Annu. Rev. Physiol. **42:**589, 1980.

Blantz, R.C.: Segmental renal vascular resistance: single nephron, Annu. Rev. Physiol. **42:**573, 1980.

Dousa, T.P., and Valtin, H.: Cellular actions of vasopressin in the mammalian kidney, Kidney Int. **10:**46, 1976.

Editorial: Prostaglandins in the kidney, Lancet **2:**343, 1981.

Fitzsimons, J.T.: Thirst, Physiol. Rev. **52:**468, 1972.

Grantham, J.J.: Action of antidiuretic hormone in the mammalian kidney, Int. Rev. Physiol. **6:**247, 1974.

Kokko, J.P., and Rector, F.C., Jr.: Countercurrent multiplication system without active transport in inner medulla, Kidney Int. **2:**214, 1972.

Navar, L.G.: Renal autoregulation: perspectives from whole kidney and single nephron studies, Am. J. Physiol. **234:**F357, 1978.

Robertson, G.L.: The regulation of vasopressin function in health and disease, Recent Prog. Horm. Res. **33:**333, 1977.

Schafer, J.A., and Andreoli, T.E.: Cellular constraints to diffusion: the effect of antidiuretic hormone on water flows in isolated mammalian collecting tubules, J. Clin. Invest. **51:**1264, 1972.

Schnermann, J., et al.: Tubuloglomerular feedback: nonlinear relation between glomerular hydrostatic pressure and loop of Henle perfusion rate, J. Clin. Invest. **52:**862, 1973.

Schrier, R.W., editor: Symposium on water metabolism, Kidney Int. **10**(1):1, 1976.

Books and monographs

Beeuwkes, R., III, et al.: The renal circulation. In Brenner, B.M., and Rector, F.C., Jr., editors: The kidney, vol. 1, ed. 2, Philadelphia, 1981, W.B. Saunders Co.

Burg, M.B.: Renal handling of sodium chloride, water, amino acids and glucose. In Brenner, B.M., and Rector, F.C., Jr., editors: The kidney, vol. 1, ed. 2, Philadelphia, 1981, W.B. Saunders Co.

Jacobson, H.R., and Kokko, J.P.: Intrarenal heterogeneity: vascular and tubular. In Seldin, D.W., and Giebisch, G., editors: The kidney: physiology and pathophysiology, vol. 1, New York, 1985, Raven Press.

Jamison, R.L.: Urine concentration and dilution: the roles of antidiuretic hormone and urea. In Brenner, B.M., and Rector, F.C., Jr., editors: The kidney, vol. 1, ed. 2, Philadelphia, 1981, W.B. Saunders Co.

Jamison, R.L., and Kriz, W.: Urinary concentrating mechanisms: structure and function, New York, 1982, Oxford University Press.

Kokko, J.P., and Jacobson, H.R.: Renal chloride transport. In Seldin, D.W., and Giebisch, G., editors: The kidney: physiology and pathophysiology, vol. 2, New York, 1985, Raven Press.

Roy, D.R., and Jamison, R.L.: Countercurrent system and its regulation. In Seldin, D.W., and Giebisch, G., editors: The kidney: physiology and pathophysiology, vol. 2, New York, 1985, Raven Press.

Schnermann, J., and Briggs, J.: Function of the juxtaglomerular apparatus: local control of glomerular hemodynamics. In Seldin, D.W., and Giebisch, G., editors: The kidney:

physiology and pathophysiology, vol. 1, New York, 1985, Raven Press.

Schuster, V.L., and Seldin, D.W.: Renal clearance. In Seldin, D.W., and Giebisch, G., editors: The kidney: physiology and pathophysiology, vol. 1, New York, 1985, Raven Press.

Wilcox, C.S., and Baylis, C.: Glomerular-tubular balance and proximal regulation. In Seldin, D.W., and Giebisch, G., editors: The kidney: physiology and pathophysiology, vol. 2, New York, 1985, Raven Press.

CHAPTER

48

Regulation of the Composition of Extracellular Fluid

Regulation of Volume and Osmolality

The kidney must maintain the total volume and the osmolality of the extracellular fluid within very narrow limits. A host of regulatory processes are involved, and few control systems within the body produce such a complex network of neural, humoral, and local factors that act on a single common effector, in this case, the nephron.

The *regulation of osmolality* is accomplished almost entirely by the actions of *antidiuretic hormone* (ADH), or *vasopressin,* on the connecting tubule and collecting duct (Fig. 47-8). The rate of formation and release of ADH is directly related to plasma osmolality. In contrast, the *regulation of extracellular fluid volume* is a more complex process involving interactions among ADH, aldosterone, natriuretic hormone, and neural and local mechanisms within the kidney. In each case a feedback system is involved that consists of (1) *receptors,* which sense the appropriate parameter, (2) *control elements,* hormones, nerves, or local events within the kidney, and (3) *effectors,* the tubules or renal arterioles. The following section describes the receptors for osmolality and for extracellular volume as well as the various control loops. Four forms of control are analyzed: (1) control of osmolality, (2) volume control via primary modification of water excretion (ADH), (3) volume control secondary to changes in solute excretion, and (4) volume control as a result of vascular adjustments.

Receptors

Osmolality. Fig. 48-1 shows, in a highly schematic fashion, a sagittal section running through the optic chiasm, the hypothalamus, and the anterior and posterior pituitary gland at the base of the brain. The receptors that are sensitive to osmolality are found in association with the *supraoptic* and *paraventricular nuclei* in this region. Fibers that pass from this area to the posterior pituitary gland integrate sensing areas in the forebrain and secretory areas in the pituitary gland for the control of ADH secretion (Chapter 52).

When hypertonic solutions are injected into the carotid artery or into the cerebrospinal fluid, plasma ADH levels increase promptly. The primary stimulus appears to be osmotically induced water shifts in the brain. The hypertonic solution in the plasma pulls water across the cerebral capillaries from the region of the paraventricular and supraoptic nuclei. This reduces interstitial and extracellular fluid volume in the vicinity of these nuclei. Simultaneously, the sodium concentration in the extracellular space surrounding the nuclei is changed. One or the other of these effects is the stimulus to activate the osmoreceptors and induce propagated action potentials along the fibers of the supraoptic-hypophyseal tract. The neural activity passes along this fiber tract, through the median eminence, and into the neural lobe of the posterior pituitary gland, releasing stored ADH.

A variety of stimuli activate the system, leading to the release of ADH. Hypotension, decreased blood volume, fear, smoking, barbiturates, acetylcholine, and epinephrine all induce release of this material, whereas alcohol inhibits its release. Of these stimuli, hypotension and decreased blood volume are the most important. In fact, the responses to many of the other agents may be secondary to either shifts in volume distribution or altered blood pressure.

Extracellular volume. The receptors that regulate extracellular volume in fact are not true receptors of extracellular volume but rather are detectors of the fullness of the vascular space. The receptors are basically

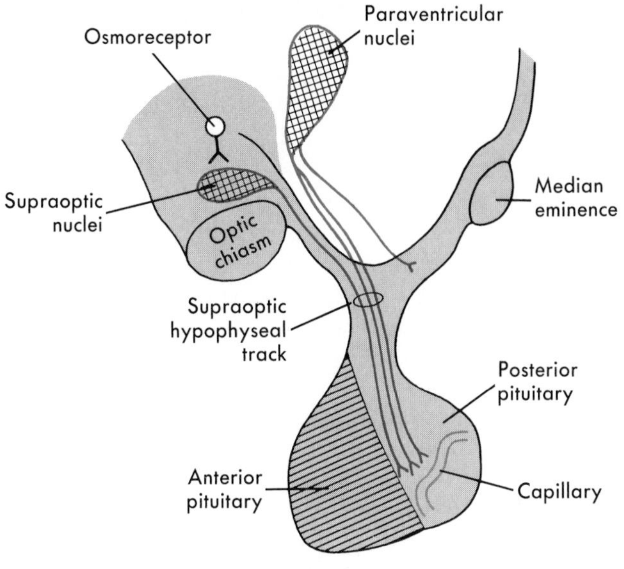

■ **Fig. 48-1.** Sagittal section through the pituitary and anterior hypothalamus. ADH is formed in the hypothalamic nuclei, migrates down the supraoptic hypophyseal tract, and is released into the capillary blood in the posterior pituitary.

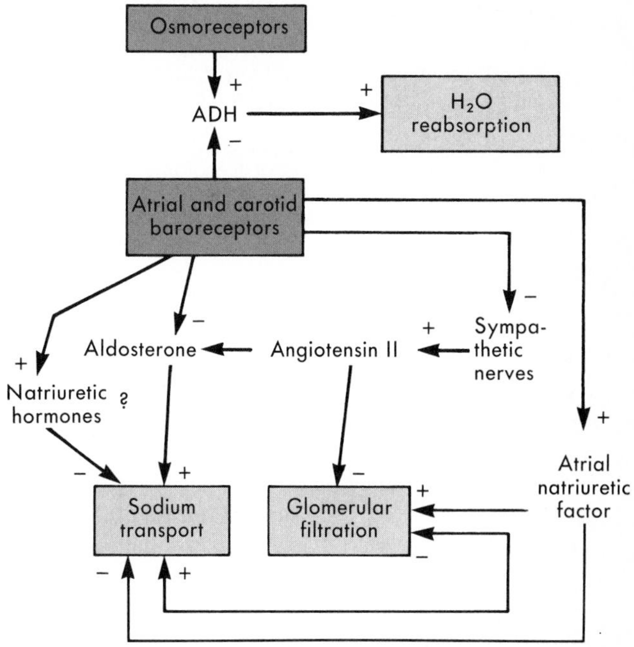

■ **Fig. 48-2.** Regulation of volume and osmolality. Compare the single factor relating ADH and osmolality with the complex web of interactions concerned with volume control. Atrial natriuretic factor is probably only one of several natriuretic factors. The dark color shows receptors; the light color shows altered function.

stretch receptors, sensitive to distension of the atria, the great veins in the pulmonary circulation, and the aorta and carotid arteries. Distension of the atria and distortion of these receptors increase neural traffic over afferent neurons that, via the *nucleus tractus solitarii* and the *area postrema,* modulate the function of the neurons of the supraoptic nucleus. The neurons of the supraoptic nucleus in turn regulate the release of various hormones from the central nervous system.

Control of extracellular volume originates from within the vascular compartment. This implies that the control of extracellular volume depends on the distribution and partition of extracellular fluid between the plasma and the interstitial space. This partition is a reflection of interactions among the factors in Starling's hypothesis for capillary exchange: the net hydrostatic and oncotic pressure difference across the capillaries (Chapter 32). The consequences of this are simple. An elevation in hydrostatic pressure within the capillaries, all other things remaining equal, will result in filtration of fluid from the vascular space into the interstitial space and fluid accumulation within this compartment. Thus a larger fraction of the total extracellular fluid will reside in the interstitial space. Because the sensors for extracellular volume reside within the vascular space, this will be interpreted as a reduction in total extracellular volume, and appropriate compensatory mechanisms will be initiated. In other words, any force that shifts fluid from the vasculature into the tissues, even without altering the total extracellular fluid volume, will be interpreted as a decrease in extracellular fluid volume by the receptors.

Because most of the receptors within the vascular compartment are in the cephalad portions of the body (thorax and neck), shifts of blood into and out of the thoracic circulation will have the same effects on volume receptors as would a true volume change. For example, immersion to the waist in water will compress cutaneous veins in the legs and abdomen and shift fluid into the thoracic vasculature. This will be interpreted as an increase in extracellular fluid volume. Conversely, shifting from the recumbent to the upright position will shift blood out of the thorax and into dependent portions of the body. This will be interpreted as a reduction in extracellular volume.

In the following section, control loops for fluid volume and osmolality are outlined. Little of this material will be new, because the control loops operate on mechanisms that have already been described. However, synthesis of the action of the control loops requires one to focus on the coordinated activity of receptor, hormone, neural activity, and the nephron response (Fig. 48-2).

■ *Control of Osmolality*

After osmolality has been increased by an intravenous injection of a hypertonic fluid, the supraoptic nuclei are stimulated and ADH is released at a greater rate from the pituitary gland. ADH acts on the distal tubules and collecting duct to increase water reabsorption. ADH also may reduce vasa recta blood flow slightly, thereby decreasing the washout of solutes from the medulla. These effects would tend to reduce water excretion and restore osmolality toward normal. The opposite series of events follows the infusion of hypotonic fluid.

Control of Volume

Control of water excretion. Expansion of extracellular fluid volume with isotonic saline increases both the plasma and interstitial volumes. The former will increase the mean arterial pressure and the pulse pressure as a result of the increased cardiac preload and an increase in stroke volume (Chapter 31). The increased arterial blood pressure will reduce the release of ADH via stimulation of the arterial baroreceptors, with a subsequent increase in water excretion.

Volume expansion acting on the low-pressure side of the circulation will increase left atrial stretch. Left atrial distension inhibits ADH release and thereby diminishes water reabsorption in the collecting ducts, which in turn will reduce the extracellular fluid volume toward normal.

Control of sodium excretion. As mentioned previously, volume control can be accomplished by affecting both sodium and water excretion. Effects on solute excretion are slower and less precise than effects on water excretion, because the hormones that influence solute excretion usually have a longer lag time before release, and the effects are less prompt than those of ADH.

An increase in extracellular fluid volume stimulates atrial receptors and pulmonary receptors on the low-pressure side of the circulation. These receptors activate central nervous system pathways that impinge on a variety of regulatory mechanisms.

Control of aldosterone release by the renin-angiotensin system is an important element in volume regulation. Stimulation of renal nerves causes direct release of renin and subsequent formation of angiotensin II and angiotensin III. These peptides stimulate aldosterone secretion and thereby influence salt reabsorption and extracellular fluid volume.

If the adrenal cortex is removed, or if animals are given large doses of aldosterone, expansion of extracellular volume still increases sodium excretion. These increases persist even when the expansion of volume is accomplished without a measurable change in glomerular filtration rate or in other aspects of renal function. This suggests that a *natriuretic* factor or hormone is involved in volume control. An increase in natriuretic hormone reduces proximal tubular sodium reabsorption and therefore increases sodium excretion.

Distension of the atria releases atrial natriuretic factor, with subsequent renal vasodilation, increased filtration, and inhibition of sodium reabsorption. In addition, atrial distension may elevate sodium excretion by reducing sympathetic neural discharge and thereby reducing tubular sodium reabsorption.

Renal vascular adjustments. Increased intravascular volume, acting via right atrial stretch receptors, can decrease sympathetic tone, increase glomerular filtration rate, and thus increase sodium and water excretion. The efficacy of a change in glomerular filtration rate probably depends on the operating level of the kidney. At low filtration rates a large fraction of the sodium load delivered to Henle's loop, the distal tubule, and the collecting duct would be reabsorbed. Because the distal sodium transport mechanism is characterized by T_m kinetics (Fig. 46-10, *B*), changes in glomerular filtration rate would have little effect on sodium excretion. However, at high filtration rates a large sodium load is delivered to the distal tubule. Saturating quantities of sodium would be delivered to the tubular mechanisms, and thus the effects of a change in filtration rate would be substantial.

Changes in renal sympathetic activity also tend to redistribute flow within the kidney. A decrease in vascular volume elicits reflexes that shift blood flow toward juxtamedullary nephrons and away from cortical nephrons. This brings a larger fraction of the glomerular filtrate through nephrons with long loops and higher concentrating ability.

Regulation of Fluid and Salt Intake

All of the mechanisms outlined so far stabilize extracellular volume and osmolality via an effect on the kidneys and subsequent alterations in the excretion of salt and water. Intake of salt and water are under equally precise control by humoral mechanisms, presumably triggered by the receptors just described. Reduction in extracellular volume or increase in osmolality via hemorrhage or hypertonic saline are associated with thirst and with the intake of appropriate quantities of water. This response appears to be triggered by specific receptors in the region of the supraoptic nuclei. Thirst is enhanced by a dryness of the oral mucosa induced by inhibition of secretion of salivary and other mucosal glands.

Angiotensin II plays a major role in triggering thirst in response to altered blood volume, hypotension, or plasma osmolality. When this peptide is injected at appropriate sites within the brain of experimental animals, a prompt increase in water intake is triggered. The sites of action of angiotensin II are in the *subfornical organ* and the *vascular organ of the lamina terminalis* regions of the brain where the blood-brain barrier is absent.

Behavior is also important in the regulation of salt intake. When animals deprived of water or salt are given access to a variety of salt solutions in various concentrations, they will select those that contain a salt concentration appropriate to their dietary or fluid balance needs. Little is known about the feedback mechanisms that regulate this behavior.

Quantitation of the Effects of Diuresis or Antidiuresis on Animals

Water excretion can be controlled by independent modification of renal tubular water reabsorption or by combined changes in solute and water reabsorption. It is important to distinguish between these two mechanisms if one wishes to determine whether the kidneys are, at a given moment, concentrating or diluting the plasma and how fast the concentration or dilution may be occurring. Important tools in understanding these processes are the computations of the *osmolar clearance* (C_{osm}) and the *free water clearance* (C_{H_2O}).

The osmolar clearance is determined simply by applying the usual clearance equation to the total solute efflux in the urine; that is:

$$C_{osm} = \frac{U_{osm} \times V}{P_{osm}} \quad (1)$$

Thus osmolar clearance reflects the rate at which the plasma is being cleared of solute particles, regardless of molecular species. For example, dehydration might lead to excretion of a hypertonic urine of 1100 mOsm/L and a urine flow of 0.5 ml/minute. Plasma concentration might be slightly elevated, to about 308 mOsm/L. The osmolar clearance then would be equal to 1100 × 0.5/308, or 1.79 ml/minute. On the other hand, a very well-hydrated subject might excrete urine with an osmolality of 27 mOsm/L and have a urine flow of 20 ml/minute. This subject probably would have a slightly dilute plasma, 298 mOsm/L. The osmolar clearance in this case would be 27 × 20/298, or 1.79 ml/minute, which is the same rate as in the previous example.

The definition of clearance is *the minimum volume of plasma that could have provided all the solute excreted in a given time*. Thus in this case both the dehydrated subject and the overhydrated subject were removing solute from plasma at the same rate (i.e., 1.79 ml/minute was being cleared). This, of course, would be expected if their dietary intake and production rates for solutes were the same.

The clearance value of 1.79 ml/minute means that, if the total solute excreted in the urine in 1 minute had been dissolved in 1.79 ml of water, the resulting urine would have been isotonic. The water and solute would have been excreted at exactly the same rate. If water is excreted in excess of solute, the excess water is called the *free water clearance,* or C_{H_2O}. The free water clearance can be computed simply as the difference between urine flow and osmolar clearance.

$$C_{H_2O} = V - C_{osm} \quad (2)$$

If the free water clearance is positive, then the urine flow is greater than that required to match the rate of solute excretion in the urine, and a dilute urine is being excreted. Excretion of a dilute urine implies that the plasma is being progressively concentrated. It follows that the magnitude of the free water clearance suggests how rapidly the plasma concentration might be changing.

If the free water clearance is less than zero, that is, if there is a *negative free water clearance*, a concentrated urine is being produced; plasma is being diluted by the return of fluid to the circulation. In the case of the dehydrated subject, the free water clearance equals 0.5 ml/minute − 1.79 ml/minute = −1.29 ml/minute. However, the overhydrated subject would have a free water clearance equal to 20 − 1.79 = 18.2 ml/minute. Water is removed from the plasma of the overhydrated subject but is restored to the plasma of the dehydrated subject by the action of the urinary concentrating process.

The extent to which the urine can be concentrated is limited by the excretory requirements for the solutes that are generated by metabolism and dietary intake. These two factors determine the required excretory rate for solute, and the solute concentration in the medullary interstitium determines the minimum quantity of fluid that must accompany the excreted solute.

Regulation of Acid-Base Balance

Metabolic Production of Hydrogen Ions and Bicarbonate Ions

The concentration of hydrogen and bicarbonate ions in plasma must be precisely regulated in the face of enormous variations in dietary intake, metabolic production, and normal excretory loss of these ions. Hydrogen ions are continuously produced as substrates are oxidized in the production of ATP. The largest contribution of metabolic acids arises from the oxidative metabolism of glucose and other carbohydrates, which leads to an average total production of approximately 13,000 mmol/day of carbon dioxide. Those oxidations place little excretory demand on the kidney, however, because carbon dioxide is excreted in the lungs not the kidneys. Because of its elimination as a respiratory gas, CO_2 is often referred to as a *volatile acid*.

For the average individual in modern Western society, the large amounts of protein and fats consumed result in a net dietary production of H^+, often referred to as the production of *nonvolatile acid*. Catabolism of protein forms hydrogen ions and urea as the end products of amino acid metabolism; approximately 20 mEq/day of hydrogen ions are formed by this process. Oxidation of sulfhydryl groups of substances such as cysteine and methionine to form sulfuric acid contributes about 35 mEq/day of H^+. Another 35 mEq/day of H^+ are produced by incomplete oxidation of glucose to lactate, and oxidation of fats forms about 5 mEq/day of H^+. Some metabolic processes use H^+. Oxidation of the salts of weak organic acids (e.g., citrate), which are common components of fruits, use approximately 35 mEq/day of H^+ in a typical diet. Thus the net production of hydrogen by an individual consuming a mixed diet is about 60 mEq/day (20 + 35 + 35 + 5 − 35), which must be continuously excreted if the acid-base balance is to be maintained. In contrast, because the removal of H^+ is equivalent to adding HCO_3^-, an individual consuming a vegetarian diet is obliged to excrete substantial quantities of HCO_3^- each day. Not only must the kidney control the excretion of dietary H^+ and HCO_3^-, it must also cope with pathological losses of these ions from the body. For example, the hydrogen ions can be lost by vomiting, and HCO_3^- can be lost by diarrhea. All of the inputs and outputs of H^+ and HCO_3^- contribute to the acid-base balance of the body, and the net result determines the pH of body fluids. Excretion is complemented by respiratory regulation of Pa_{CO_2}.

The regulation of pH is one of the more complex aspects of human physiology and involves many different processes. The following sections approach the study of acid-base regulation in three steps: the renal tubular *mechanisms of H^+ regulation,* the *mechanisms of buffering* the free hydrogen ion concentration, and the *integration of renal and respiratory function.*

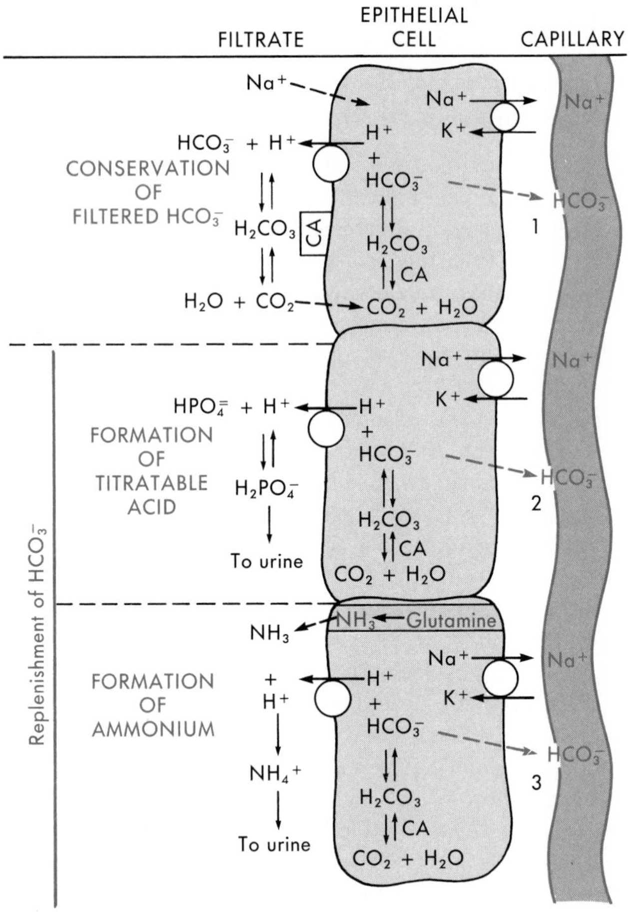

■ **Fig. 48-3.** Hydrogen ion–secretion roles in hydrogen excretion and bicarbonate reabsorption. The top epithelial cell shows the mechanism by which filtered bicarbonate is reabsorbed and thereby conserved *(colored arrow 1).* The lower two cells show the coupling between hydrogen ion secretion and replenishment of bicarbonate stores that have been used elsewhere in the body for buffering. The middle cell shows the formation of titratable acid. The secreted hydrogen ion lowers the pH of tubular fluid, thus shifting the equilibrium toward the protonated form of the monobasic phosphate molecule. It is presumed that the secreted hydrogen was derived from the dissociation of carbonic acid within the epithelial cell. Thus secretion of a hydrogen ion frees a bicarbonate to be reabsorbed passively *(colored arrow 2)* into the peritubular capillary blood along with sodium. The contribution of titratable acid formation to bicarbonate replenishment is limited by the amount of filtered buffer in the urine. The remainder of hydrogen excretion is accompanied by diffusional trapping of H^+ by the NH_3 that is produced largely by glutamine metabolism. Hydrogen, once associated with the NH_3 to form NH_4^+, cannot return to the epithelial cell, and a bicarbonate is freed for reabsorption as sodium bicarbonate *(colored arrow 3).*

■ *Hydrogen Ion Secretion and Bicarbonate Ion Reabsorption*

Tubular H^+ secretion and HCO_3^- reabsorption are closely intertwined. The central process in acid-base balance is the secretion of H^+, either by a carrier-mediated, sodium-hydrogen exchange in the proximal tubule or by primary active transport of H^+ in the distal tubule. As will be shown, tubular H^+ secretion not only eliminates H^+ directly, but it can also act indirectly to add HCO_3^- to the plasma. Secretion of hydrogen ions can thus accomplish three results. First, and most obvious, the metabolically produced H^+ is delivered to the urine. Second, the HCO_3^- that is continuously filtered in the glomerulus is returned to the circulation. Third, H^+ secretion continuously adds bicarbonate ions to the extracellular buffer pool. These ions replenish the buffer stores depleted in the process of buffering strong acids produced by metabolism or ingestion. Each of these three processes is discussed below and shown in Fig. 48-3.

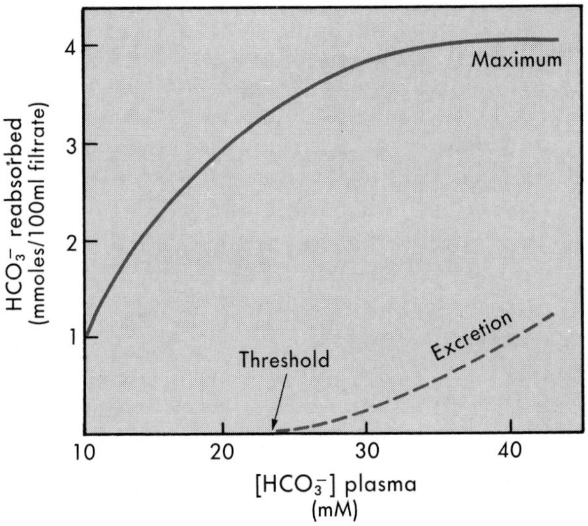

Fig. 48-4. Relation between bicarbonate reabsorption and plasma bicarbonate. The bicarbonate concentration is normalized to the quantity of bicarbonate reabsorbed per 100 ml of filtrate produced. Note the threshold of about 24 mM HCO_3^-, below which essentially no HCO_3^- is excreted. Elevating the bicarbonate concentration to levels in excess of 36 mM leads to no further increase in reabsorption.

Conservation of Filtered Bicarbonate

The upper portion of Fig. 48-3 recapitulates the basic HCO_3^- reabsorptive process described in Fig. 46-8. It shows how the normally filtered HCO_3^- is reabsorbed and returned to the plasma and thereby is *conserved.* The process involves the initial secretion of H^+, which combines with filtered HCO_3^- to form H_2CO_3 and then CO_2 and water. The diffusible CO_2 returns to the epithelial cell, where it becomes hydrated to form H_2CO_3, which then dissociates. The H^+ remains in the intracellular pool in the epithelial cell, and the HCO_3^- diffuses passively into the peritubular space and is reabsorbed (Fig. 48-3, *colored arrow 1*).

The absorption rate of filtered bicarbonate can vary widely, but usually over 99% is reabsorbed. Fig. 48-4 shows the relations among bicarbonate reabsorption, excretion of filtered bicarbonate, and plasma bicarbonate concentration. The curves look much like those that describe maximum transport behavior (T_m) of other solutes (Fig. 46-6). In reality, however, they are quite different, because the HCO_3^- reabsorption is normalized on the basis of the filtration rate.

The nonlinear relation between reabsorption of filtered HCO_3^- and the plasma HCO_3^- concentration has several important consequences. First, it means that at or below normal plasma HCO_3^- levels, essentially no HCO_3^- appears in the urine (all filtered HCO_3^- is reabsorbed). Urinary H^+ secretion and HCO_3^- reabsorption are adequate to assure recovery of essentially all filtered HCO_3^-.

The second important consequence of the relation between HCO_3^- reabsorption and the plasma HCO_3^- level is that HCO_3^- is readily excreted if present in excess. Therefore, if the diet has a sufficiently high alkali content or can produce enough alkali, the acid-base balance can be regulated simply by varying the quantity of HCO_3^- excreted. In fact, under some circumstances, HCO_3^- is secreted rather than absorbed when plasma levels of HCO_3^- are high. This behavior occurs in herbivores and in some human vegetarians.

Fig. 48-4 also shows that the HCO_3^- reabsorptive rate saturates at a plasma concentration of about 34 mM. This signifies that the ability of the kidney to augment HCO_3^- reabsorption and to compensate for respiration-induced disturbances in acid-base balance is limited, a fact that is discussed in more detail below.

Restoration of Depleted Bicarbonate Reserves

Depletion of bicarbonate by extracellular buffering. Bicarbonate may be lost from the body's extracellular fluid in a way other than via renal excretion. The H^+ appearing in the body fluids, either through ingestion or metabolism, must be buffered until it can be excreted by the kidneys. The H^+ may be buffered by various intracellular and extracellular buffers (see the isohydric principle below), but one of the major buffering schemes involves the association of the free H^+ with HCO_3^- to form CO_2 and water (Fig. 48-5, extracellular fluid, *gray*). H^+ produced by cells can be viewed as exiting the cell with its corresponding anion (HPO_4^- in Fig. 48-5). Once in the extracellular fluid (Fig. 48-5, ①), the H^+ can combine with a bicarbonate ion to form carbonic acid and sodium phosphate; carbonic acid can then dissociate into CO_2 and water.

The CO_2 so formed circulates to the lungs and is exhaled along with metabolically produced CO_2 (Fig. 48-5, *light-colored* box ③). The critical point in understanding much of the overall integration of acid-base balance is to recognize that the HCO_3^- moieties used in buffering H^+ (Fig. 48-5, box ②) are then exhaled as CO_2 (Fig. 48-5, ③) and represent a *net loss of HCO_3^- from the body.* These lost HCO_3^- molecules thus represent a *depletion of bicarbonate stores,* and a reduction in the total body buffering capacity.

The kidneys regulate acid-base balance by producing HCO_3^- in the tubular epithelial cells. This HCO_3^- diffuses to the peritubular capillary blood (Fig. 48-5, *colored arrow,* ④) and thereby raises the HCO_3^- levels in renal venous blood.

Although metabolically produced H^+ can be buffered

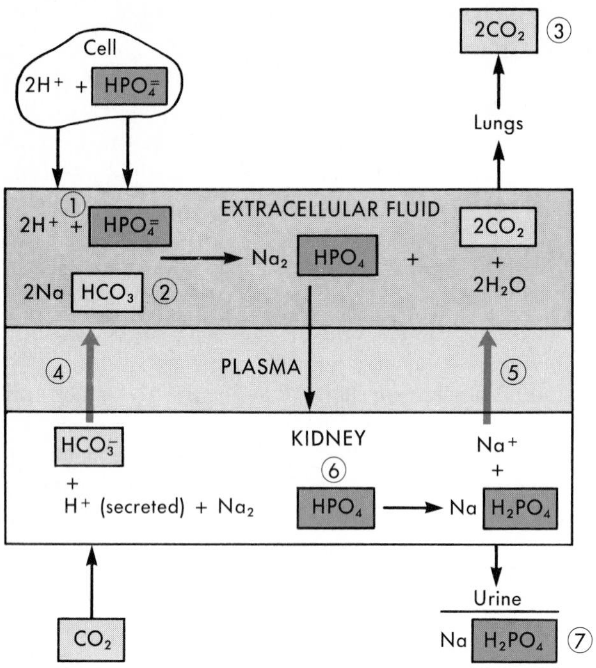

Fig. 48-5. Buffering and excretion of strong acids. Extracellular events are shown in the upper half of the diagram *(large gray rectangle),* and renal events are shown in the lower half of the diagram *(large white rectangle).* Bicarbonate is used initially to buffer metabolically produced acids (H_3PO_4, ① and ②). The bicarbonate ion forms CO_2, which is then exhaled with the loss of two bicarbonate ions in buffering each phosphoric acid ③. Hydrogen ion secretion into the tubular lumen *(lower rectangle)* and subsequent combination of the hydrogen ion with $HPO_4^=$ ⑥ in the tubular fluid frees a bicarbonate derived from CO_2 to return to the extracellular fluid ④. The bicarbonate ion is reabsorbed with a sodium ion ⑤.

temporarily, ultimately the H^+ must be excreted by secreting H^+ into the tubular lumen. However, this cannot be accomplished simply by secreting free H^+ into the urine. The 60 mEq of H^+ produced each day must be excreted in an average urine volume of 1.5 L. If H^+ were excreted as the free ion, a urinary concentration of about 40 mEq/L would result, with a urinary pH of 1.4. This is not compatible with the comfort of the individual or with life of mammalian cells. Therefore a complex series of processes has evolved that permits the mammalian kidney to excrete the necessary quantity of H^+ without exceeding a free tubular H^+ concentration of about 1.5×10^{-5}M, that is, a pH lower than about 4.8. The secretion of acid and the replenishment of HCO_3^- stores are accomplished largely by forming complexes in the tubule between H^+ and other urinary constituents. The complexes of major importance are those made with *filtered buffers* and with *ammonia.*

Titratable acid formation. The simplest method by which the secreted H^+ might be buffered in the tubules is to attach it to a weak acid. This is how a significant fraction of the H^+ is excreted and returned to the extracellular fluid. The process has a pleasing symmetry because some of the metabolically produced salts of acids are filtered and can therefore act as urinary buffers, as illustrated in Figs. 48-3 and 48-5.

The large gray rectangle in Fig. 48-5 shows the processes that occur in extracellular fluid, and the large white rectangle shows those processes that occur in the kidney. Note that the circulating plasma *(large light-colored rectangle)* acts as the mode of communication between the two. Within extracellular fluid HCO_3^- buffers the acid phosphates formed in the cell during metabolism (Fig. 48-5, boxes ① and ②). CO_2 is formed in the process and is excreted in the lungs, with a loss of two bicarbonate ions for each mole of H_3PO_4 neutralized (Fig. 48-5, ③).

The $HPO_4^=$ that results from the extracellular buffering is carried to the kidney in the circulating plasma *(large light-colored rectangle)* and filtered as Na_2HPO_4 (Fig. 48-5, ⑥). Because Na_2HPO_4 is the salt of a weak acid, it constitutes a urinary buffer. Thus it can participate in the formation of titratable acids, as shown in the middle cell of Fig. 48-3. It is assumed that either circulating or metabolically produced CO_2 within the epithelial cells acts as a source of H^+ and HCO_3^- ions. The H^+ is actively secreted into the lumen, where it is buffered by $HPO_4^=$ or other filtered buffers. The buffering process leaves an excess of bicarbonate within the epithelial cell. This HCO_3^- can be returned to the extracellular fluid (Fig. 48-3, *colored arrow 2*), and the secreted hydrogen ion can be excreted in association with the filtered $HPO_4^=$. A sodium ion is also salvaged by this process. Sodium ions are filtered with the $HPO_4^=$ and once the secreted H^+ is added to the $HPO_4^=$, the associated Na^+ is no longer required for charge neutrality. The Na^+ therefore can be returned to the plasma along with the HCO_3^-. The net result of the entire process is shown in Fig. 48-5 as the filtration of Na_2HPO_4 and subsequent recovery of $NaHCO_3$ (④ and ⑤).

To recapitulate: (1) extracellular buffering of metabolically produced strong acids leads to respiratory loss of bicarbonate as CO_2, and (2) tubular buffering of a secreted hydrogen ion by the filtered acidic salts frees a bicarbonate ion and a sodium ion for reabsorption to restore bicarbonate ion reserves.

Formation of titratable acids by the process shown in Figs. 48-3 and 48-5 is limited by the minimum urine pH that can be established, because the fractions of the phosphate ion that exist in the singly, doubly, and triply protonated form are determined by the ionization constant and the pH of the solution. (See Fig. 48-11 and discussion below for details.) The relations among the various forms of phosphate and the corresponding dissociation constants are shown below:

$$H_3PO_4 \underset{}{\overset{K_1}{\rightleftharpoons}} H^+ + H_2PO_4^- \quad pK_1 = 2.2$$

$$H_2PO_4^- \underset{}{\overset{K_2}{\rightleftharpoons}} H^+ + HPO_4^= \quad pK_2 = 6.8$$

$$HPO_4^= \underset{}{\overset{K_3}{\rightleftharpoons}} H^+ + PO_4^{\equiv} \quad pK_3 = 9.7$$

The Henderson-Hasselbalch equation predicts that, at a plasma pH of 7.4, the log of the ratio of $HPO_4^=$ to $H_2PO_4^-$ will equal the difference between the pH and the pK. For example, at a pH of 7.4 $HPO_4^=$ is about 20% of $H_2PO_4^-$ total (i.e., pH − pK = 7.4 − 6.8 = 0.6; antilog of 0.6 = 0.2). Because pK_1 and pK_3 are so far from the plasma pH, essentially none of the phosphate exists either as H_3PO_4 or as $PO_4^{\equiv}$.

For the filtered phosphate to function effectively as a urinary buffer, the urine pH must be lower than the plasma pH. This sets a limit on how much bicarbonate can be reclaimed by combining a secreted hydrogen ion with a filtered phosphate. Urine pH cannot fall much lower than about 4.8 because at this point, a maximum transepithelial gradient for H^+ is established. At a pH of 4.8, about 99% of the phosphate will be in the form of $H_2PO_4^-$, and little will be in the form of H_3PO_4. Thus one hydrogen ion can be buffered by each $HPO_4^=$ filtered and, consequently, one bicarbonate ion can be recovered. However, two bicarbonate ions are lost in the form of exhaled CO_2 used to buffer the original metabolically formed H_3PO_4 (Fig. 48-5, ② and ③). Therefore titrating the filtered phosphate with secreted hydrogen ions can recover only half the bicarbonate ions lost to the atmosphere as CO_2. Further replenishment of bicarbonate stores requires the presence of other buffers in the urine or some other molecule to combine with the hydrogen ion. Experimentally and clinically, supplementary buffers with appropriate pKs can be infused intravenously. Physiologically, however, the process of titratable acid formation is supplemented by the formation of urinary ammonium.

Ammonia secretion. Production of ammonia and diffusion trapping of ammonium within the tubular lumen provides another major means of adding bicarbonate to the plasma. Ammonia and ammonium are produced by the metabolism of glutamine and other amino acids by the renal epithelial cells. As shown in the lower cell of Fig. 48-3, ammonium, which has a high lipid solubility, diffuses into the lumen of the tubule, where it combines with a secreted H^+. Protonation of the NH_3 renders it much less permeative. As a result, both the NH_3 and the H^+ are trapped in the lumen and excreted. Excretion of the NH_3^+ leaves behind one unpaired HCO_3^-, which is then free to return to the plasma (Fig. 48-3, *colored arrow 3)* along with a sodium or a potassium ion.

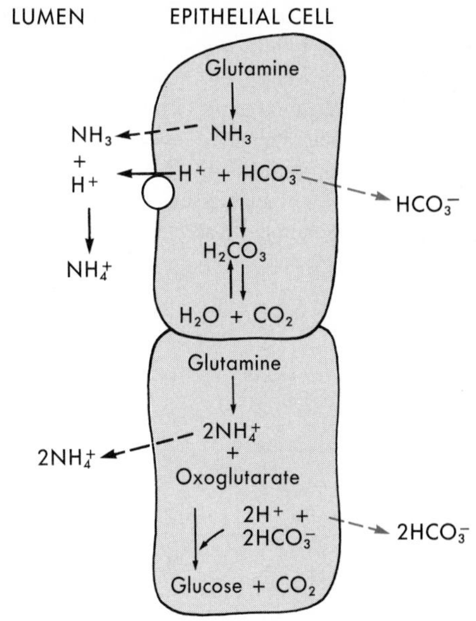

Fig. 48-6. Ammonia secretion and the conservation of bicarbonate. The upper cell shows the classical diffusion trapping mechanism for ammonia secretion. NH_3, derived from glutamine, diffuses into the tubular lumen by virtue of its high membrane permeability. Protonation, forming NH_4^+, traps the ammonium within the lumen, ensuring subsequent excretion. The lower epithelial cell shows a different biochemical process with different specific events, but nevertheless leading to hydrogen excretion and bicarbonate reabsorption. Glutamine metabolism forms NH_4^+ rather than NH_3. In addition, oxoglutarate is formed, which must subsequently be metabolized. Oxoglutarate metabolism uses two hydrogen ions, thus freeing two bicarbonate ions for subsequent reabsorption. The net result of both processes is similar, but the specific biochemical events differ.

An alternate and biochemically more accurate view of urinary ammonium excretion is based on the fact that the metabolism of glutamine produces mainly NH_4^+, not NH_3. The critical factor that links the metabolism of an amino acid and the production of ammonium to the restoration of HCO_3^- is the production of a carboxylate anion simultaneously with the NH_4^+ (Fig. 48-6, *lower cell*). The metabolism of the carboxylate anion consumes two H^+, and *the utilization of these ions frees bicarbonate ions for reabsorption and replenishment of buffer stores*. The ammonium ion, which is formed metabolically, will be distributed between plasma and tubular fluid in proportion to urinary pH, and bicarbonate can still be reabsorbed as shown in Fig. 48-6.

Note that this biochemical pathway in a sense spares the hydrogen secretory mechanism, because it obviates the need for hepatic metabolism of nitrogen compounds. If not for the renal formation of NH_4^+, the glutamine would have to be metabolized in the liver to

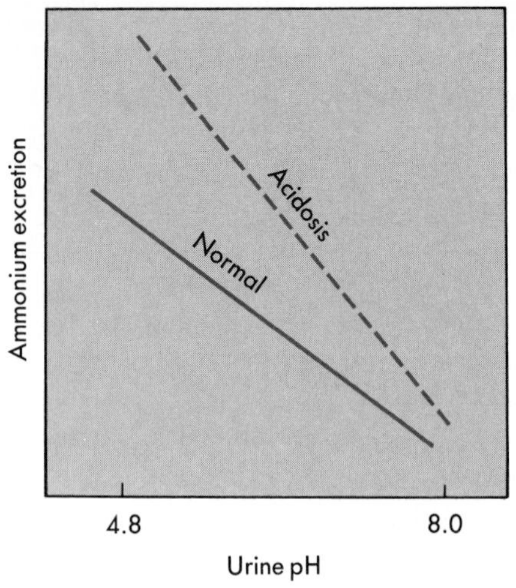

Fig. 48-7. Relations between ammonium excretion and urine pH. Increased excretion of ammonium associated with decreasing urine pH is caused by diffusion trapping of ammonium. The elevated ammonium excretion during acidosis *(dashed line)* represents increased hydrogen secretion and therefore a reduction in luminal pH. In addition, chronic acidosis increases glutamine metabolism to NH_4^+ and increases renal availability of glutamine by accelerating hepatic production of glutamine.

CO_2, urea, and H^+. The H^+ would then become part of the total renal H^+ load to be excreted.

Regardless of the means of formation, a second major facet of ammonium excretion is the fact that ammonium production provides a cation that can replace either sodium or potassium. Thus these two ions are free to be reabsorbed with bicarbonate or with chloride. The net result is the excretion of ammonium phosphate, rather than sodium or potassium phosphate, and therefore *cations are conserved.*

Urinary ammonium excretion is affected in two ways by the acid-base status of the subject, as shown in Fig. 48-7. Normally, as the pH falls, the ammonium excretion increases simply because lowering the urine pH causes more of the relatively impermeative NH_4^+ to be formed in the tubular lumen. In addition, during a period of prolonged acidosis, tubular epithelial cells can adapt their NH_4^+ synthesis, either by enzymatic adaptations or perhaps through changes in mitochondrial transport processes. Also, altered hepatic metabolism maintains higher levels of plasma glutamine. The net result of these adaptations is to increase the formation of ammonia by the epithelial cell and therefore increase the rate of ammonium excretion at any given urine pH (*dashed line,* Fig. 48-7).

Quantitative Relationships

Tubular mechanisms conserve bicarbonate by the three separate processes just described; all depend on the activity of the hydrogen ion pump. The quantitative importance of each can be evaluated by three measurements of the urine composition. First, the quantity of filtered bicarbonate reabsorbed can be calculated from the difference between the filtered and excreted bicarbonate. That is,

$$\text{Conservation of filtered bicarbonate} = (GFR \times [HCO_3^-]_p) - (V \times [HCO_3^-]_u) \quad (3)$$

Table 48-1. Values of ammonium excretion and the formation of titratable acidity

	Normal subject	*Diabetic subject*
Titratable acidity (mEq, H^+)	5	100
Ammonium (mEq, H^+)	25	300
Anion gap* (mEq, H^+)	15	25

*See text for definition.

Second, the amount of buffering of secreted hydrogen ion by filtered buffers can be evaluated quantitatively by a measure called the *titratable acidity.* The titratable acidity is defined as the milliequivalents of strong base that must be added to the final urine to restore the urine pH to that of plasma, typically 7.4. It is assumed that the hydrogen ion, being a strong acid, was added by tubular secretion to the urine in a quantity sufficient to lower the pH from 7.4 to the observed urine pH. The quantity of strong alkali required to return the pH of the urine to that of plasma must therefore equal the number of hydrogen ions added by the tubular transport process.

The third measure of urinary hydrogen ion excretion is the ammonium excretion, because this is a direct measure of the number of moles of hydrogen ion associated with the ammonium ion.

Typical values of ammonium excretion and titratable acid production for a normal subject and a diabetic subject are shown in Table 48-1. Note that ammonium excretion normally represents a larger fraction of excreted acid than does excretion of titratable acidity.

In diabetic persons both increased formation of titratable acidity and ammonium production combat the acidosis induced by excessive fatty acid metabolism. As mentioned, the ammonium excretion is augmented because of the increased production of ammonium by renal epithelial cells and by a reduced urine pH. The increased excretion of titratable acidity in the diabetic is related to the presence of ketoacids in the urine. Normally, the low pK of the ketoacids limits their usefulness as buffers (see Fig. 48-11 for details). However, because of the large metabolism of lipid in the diabetic, the ketoacids may be excreted in such high quantities that they contribute significantly to urinary buffering.

Anion gap. With evaluation of the status of acid-base balance and fluid and electrolyte balance, the *anion gap* is often computed. This computation permits inference of the presence of ketoacids or other anions in circulating plasma from data such as those shown in Table 48-1. In routine clinical assays, Na^+, K^+, Cl^-, and HCO_3^- concentrations are measured. The bulk of all plasma cations is included in the measurement of only the sodium and potassium ions. However, a number of anions other than chloride and bicarbonate normally circulate in blood. These other anions include ketoacids, lactate, and protein. Thus the sum of the major cations (Na^+, K^+) exceeds the sum of the major anions (Cl^-, HCO_3^-); the deficiency of anions is called the anion gap.

$$\text{Anion gap} = [Na^+] + [K^+] - [Cl^-] - [HCO_3^-] \quad (4)$$

Normally, the anion gap is about 15 mEq/L of plasma. This can change with different circumstances. For example, a chronically hypoxic person produces abnormally large quantities of lactate. This increases the anion gap because lactate is not accounted for in the calculation. Major losses of plasma proteins, which normally have a net negative charge, lower the anion gap. The anion gaps for the normal individual and the diabetic individual are shown in Table 48-1. As would be expected, the diabetic has a large anion gap, because of the presence of ketoacids in the plasma.

Interactions Between the Hydrogen Ion and Other Ions

Several molecules exert important effects on the processes described in the preceding section.

Carbon dioxide. An increase in plasma *carbon dioxide* tension, as might occur in respiratory acidosis, increases the fraction of the filtered bicarbonate that is reabsorbed (Fig. 48-8), at least up to the maximum reabsorptive rate for HCO_3^- (Fig. 48-4). This is caused by the elevation in carbon dioxide concentration inside the epithelial cell. This elevated carbon dioxide raises the hydrogen ion concentration and stimulates hydrogen ion secretion, which in turn enhances bicarbonate ion reabsorption. This is obviously an important compensatory mechanism, because elevation of the fraction of bicarbonate reabsorbed will increase plasma bicarbonate levels in parallel with changes in carbon dioxide levels, thereby tending to stabilize plasma pH.

Potassium ion. The potassium ion interacts in important ways with the hydrogen ion in all cells of the body. An important general relation is the reciprocity

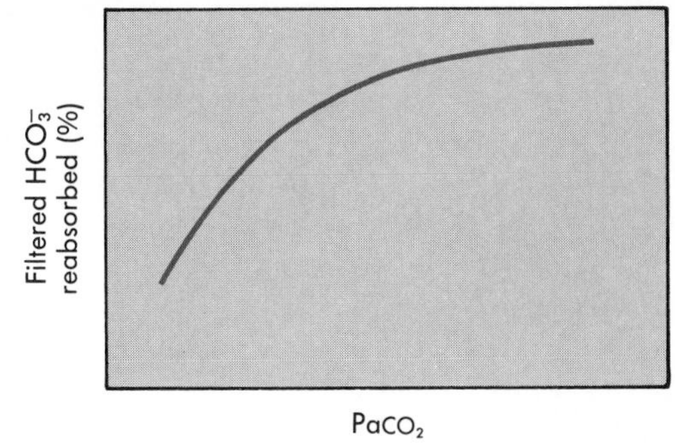

Fig. 48-8. Rate of bicarbonate reabsorption by the renal tubules as a function of the plasma carbon dioxide tension.

commonly observed between these two ions. Elevation of hydrogen ions in the extracellular fluid, as occurs, for example, in metabolic acidosis, will result in movement of hydrogen into a cell, with a reciprocal exit of potassium from the cell and a diminution of cellular potassium concentration. The significance of this exchange is that the majority of buffers within the body are located in the intracellular compartment. Thus the hydrogen-potassium exchange allows the hydrogen to achieve access to protein buffer stores and minimizes the pH change in the body. However, in the process, hyperkalemia (i.e. elevated plasma potassium) is induced by the potassium flux from cell to plasma and is often associated with metabolic acidosis. Only a small fraction of the large quantity of intracellular potassium need enter extracellular space to markedly alter the extracellular concentration. (See Tables 45-1 and 45-2.) This reciprocity between hydrogen and potassium fluxes is reflected in the cellular responses to a variety of interventions and physiological stimuli.

Elevation of the plasma potassium ion level in the tubular epithelial cells commonly induces the exchange of hydrogen ion for potassium ion and reduces the amount of H^+ available for transport by the epithelial secretory pump. The reduction in intracellular hydrogen diminishes H^+ secretion by the tubules and decreases H^+ excretion. The reduced hydrogen secretion causes systemic acidosis. On the other hand, acidosis causes reduced potassium secretion. Fig. 48-9 shows the inverse relation between hydrogen ion concentration and potassium secretion. This relation depends on the fact that the hydrogen ion secretory pump is electrogenic, and enhanced hydrogen ion secretion reduces the luminal potential *(dashed line)* and thus reduces the driving force for potassium secretion.

Sodium ion. Sodium reabsorption is also importantly related to hydrogen ion balance of the body. The interrelations between sodium and bicarbonate excretion are largely tied to the fact that sodium reabsorption carries with it the requirement for reabsorption of anions; chloride and bicarbonate represent the major extracellular

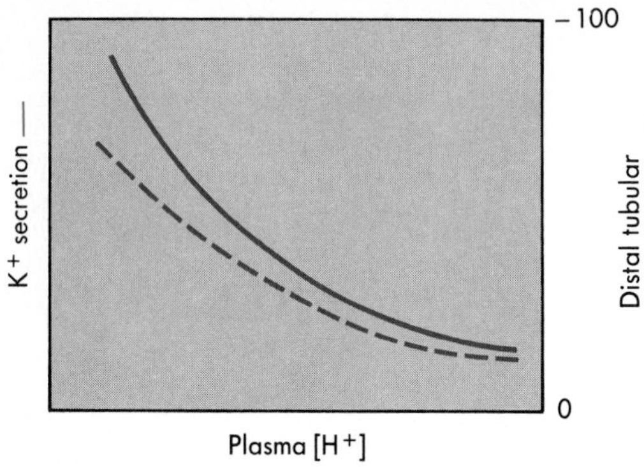

■ **Fig. 48-9.** Relation between the plasma hydrogen ion concentration and the rate of potassium secretion. The hydrogen secretory pump is electrogenic; therefore increased hydrogen pump activity causes the distal tubular potential.

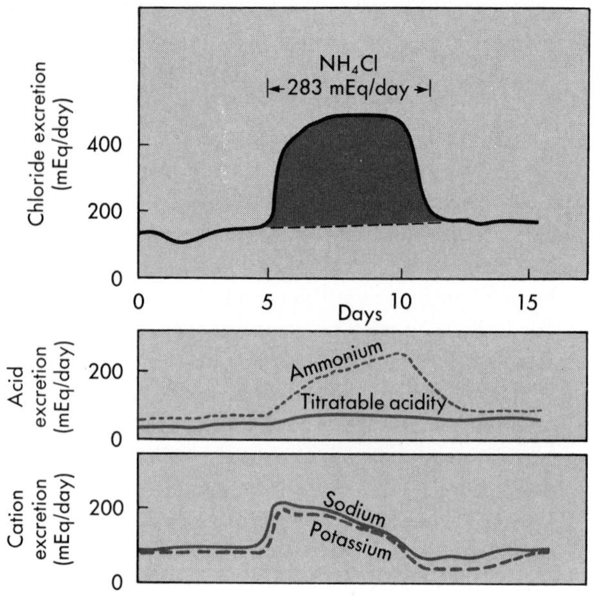

■ **Fig. 48-10.** Effect of dietary acid on hydrogen ion balance. Ammonium is converted to urea and hydrogen ion, producing an acidosis. Early in the test period, the ingested chloride must be excreted (*top panel*) either with a potassium or sodium ion *(bottom panel)*. The acidosis increases tubular hydrogen ion excretion, which in turn elevates titratable acidity slightly *(middle panel)*. Prolonged acidosis stimulates ammonium secretion, and ammonium replaces potassium and sodium ions in the urine, thus conserving these cations *(bottom panel)*. See text for details.

fluid anions. Thus an increase in sodium reabsorption tends to increase bicarbonate reabsorption. As a rule, this process does not affect plasma pH. However, if tubular sodium reabsorption is stimulated markedly, as in prolonged sodium deprivation, increased bicarbonate reabsorption may induce an alkalosis in conjunction with the hyponatremia.

Chloride ion. The chloride ion can affect acid-base status, and chloride and bicarbonate movements are linked via an exchange carrier. An increase in chloride levels in plasma means that less bicarbonate need be reabsorbed with the sodium ion. The reduced bicarbonate reabsorption leads to diminished plasma bicarbonate levels and acidosis associated with hyperchloremia.

■ *Adaptation to a Chronic Acid Load*

Fig. 48-10 shows the integration of several elements of renal acid-base function that can be observed if an experimental subject is exposed to an acid load. In this experiment the subject ingested ammonium chloride, which was converted to urea and hydrogen ion in the liver. This process added an acid load to the body in the same way that the administration of hydrochloric acid would have. The required responses to restore homeostasis are renal elimination of the excess chloride and elimination of the hydrogen ion and urea. Elimination of urea, a neutral molecule, causes no particular problem as long as urine flow is adequate, which in this case was true. Chloride elimination poses no particular problem, because the anion can simply be excreted along with whatever urinary cations are present.

In this experiment chloride excretion exceeded chloride ingestion. This loss of chloride reflects the fact that the acidosis associated with NH_4^+ ingestion required an enhanced reabsorption of bicarbonate. Because the acid-base balance required an increased bicarbonate reabsorption, the maintenance of charge neutrality required less chloride reabsorption.

The most interesting responses to the administration of ammonium chloride are those displayed by the cations sodium and potassium. Immediately after the administration of ammonium chloride, potassium and sodium excretion increased. This increase reflected the fact that cations had to be excreted in combination with the chloride ingested in the ammonium chloride. The cation portion of the original ammonium chloride was changed to a neutral compound, urea, and to a hydrogen ion. Presumably the hydrogen ion would have been largely buffered and excreted as carbon dioxide via the lungs, thus leaving a requirement for a cation to balance the chloride excreted. The cations would have been either sodium or potassium ion.

Both potassium and sodium excretion reached an early peak (Fig. 48-10, *bottom panel*) and then diminished with time. This diminution was approximately the mirror image of a rise in ammonium excretion. The rise in ammonium excretion did not reflect excretion of the ammonium ion ingested, because most of this was metabolized to hydrogen ion and urea. Rather the ammonium excretion rose because of the induced acidosis and the enhanced ammonium secretion by the tubular epithelial cells (Fig. 48-7). This ammonium traps a hydrogen ion, frees a bicarbonate to restore the bicarbonate reserves that were depleted in buffering the hydrogen ions, and conserves cations (Fig. 48-3, *arrow 2*). The latter occurs because the ammonium ion formed in the tubular lumen can replace the sodium and the potassium ions previously lost in the urine as cations that accompany the chloride. After recovery from the effects of the ingestion of ammonium chloride all variables promptly return to normal, with a slight undershoot in the excretion of potassium and sodium. This undershoot reflects the fact that some time is required for restoration of the ammonium production rates of the epithelial cells.

Titratable acidity also increases after ingestion of ammonium chloride. The titratable acidity depends on the quantity of buffers present in the urine and on the urinary pH. Ingestion of ammonium chloride does not appreciably change the quantity of filtered buffers. Thus the increase in titratable acidity reflects largely the fact that urine pH falls as a result of the acidosis.

Acid-Base Balance in the Extracellular Fluid

Disturbances in the acid-base status may originate either from altered respiratory function (respiratory acidosis or alkalosis) or altered metabolism or intake of food (metabolic acidosis or alkalosis). Compensation for a disturbance originating from altered respiratory function involves appropriate shifts in bicarbonate reabsorption by the kidney to stabilize the ratio of HCO_3^-/P_{CO_2}. Similarly, a metabolic disturbance can be compensated for by altering the rate and depth of respiration to adjust plasma P_{CO_2}. The net result of interactions between respiratory function and renal function that control blood pH is shown diagrammatically in Fig. 48-15. Before these interactions can be understood, however, a number of relations in body buffer systems must be explored.

Body Buffering

Analysis of buffers and buffer capacity. As pointed out earlier, the initial event in a disturbance of H^+ balance in the body is buffering in the intracellular or extracellular fluid. Buffering in biological fluids is complex, and only a simplified view can be presented here. As with most problems concerning acids and bases, we begin with the Henderson-Hasselbalch equation, which defines the relations among pH, proton donor (acid), and proton acceptor (base) as follows:

$$pH = pK + \log \frac{Base}{Acid} \qquad (5)$$

The chemical system stabilizes, or *buffers,* the hydro-

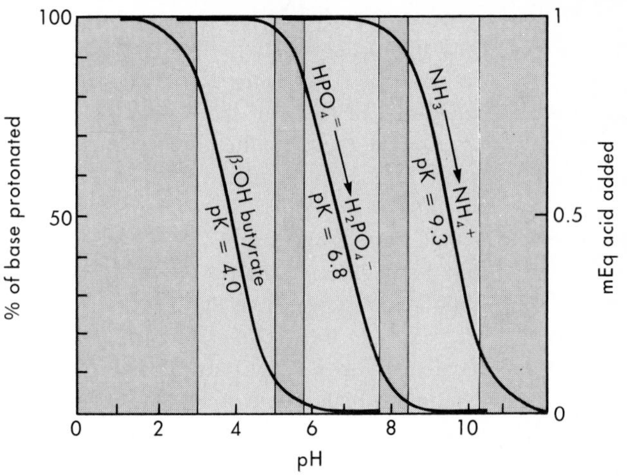

■ **Fig. 48-11.** Titration of buffers by addition of acid. Three buffers were prepared by adding 1 milliequivalent of the base form of the buffer molecule to 1 L of fluid. Acid was then added and the pH measured. The resulting fraction of the base in the protonated form (i.e., associated with an added hydrogen ion) is shown. The colored regions represent the buffer region for each of the molecules; these regions include a range of one pH unit on either side of the pK of the three species.

gen ion concentration by virtue of the fact that when protons are added, a substantial fraction of the added H^+ will combine with the base and therefore will fail to appear as free H^+ (i.e., the change in pH will be minimized). Two factors determine the capacity of a buffer system to stabilize the pH: the pK of the buffer pair in relation to the ambient pH and the quantity of buffer present.

Fig. 48-11 illustrates the significance of the pK values of possible buffers. The titration curves show what would occur with the addition of a strong acid to 1 L of a solution composed of 1 mEq/L of the base form for each of three substances with different pK values. Note that all curves have the same shape. At the pH values that correspond to the top and the bottom portions of each curve, the pH changes substantially as acid is added. However, at pH values over the central regions, the pH changes only slightly with the addition of acid. It is within the central regions, about one pH unit on either side of the pK, that the system is buffered effectively and the pH is stabilized. Thus at a plasma pH of 7.4, only the phosphate has significant buffering capacity, and even this ion is at the limits of its buffering range. In the urine, however, the pH is usually less than 7.4, and thus both phosphate and β-hydroxybutyrate can act as effective buffers.

The factor limiting the efficacy of a potential urinary buffer, such as β-hydroxybutyrate, is not only its pK but also the quantity filtered. Normally, too little β-hydroxybutyrate is filtered for it to be an effective urinary buffer. However, in diabetic acidosis, the amount filtered may rise sufficiently to allow it and other ketoacids to become significant urinary buffers.

The net effectiveness, or *buffer capacity,* of a buffer system is measured in terms of the quantity of strong acid or base required to change the pH by one unit. The buffer capacity is equal to the reciprocal of the slope of the titration curve at any specified pH (Fig. 48-11). It is apparent that, for a given concentration of base, all of the buffers have the same buffer capacity when the pH is a comparable distance from the pK.

Multicomponent buffer systems. Buffering in the body is complex because many buffers are distributed among the different body compartments. In the extracellular space, the major buffers are albumin, phosphate, and bicarbonate, although the latter functions in a rather special way, as described below. The major intracellular buffers are phosphates, organic ions, and proteins. In general, intracellular buffering is a slower process, which is limited by the rate of entry of H^+ into the cells. Hemoglobin is unique as a buffer, because it is intracellular but positioned behind the red cell membrane that is quite permeable to H^+. Therefore the hydrogen ion readily penetrates red blood cells so that short-term buffering usually occurs in extracellular fluid and in the intracellular fluid of red blood cells.

With so many buffers involved, how can the process be analyzed in a manageable way? An empirical analytical procedure has been developed based on the fact that all buffer systems are predictably linked by the *isohydric principle*. This principle states that if the ratio of the components of one buffer pair in a mixture of many pairs is known, then the ratio of all the other buffer pairs is known as well. In general, one can define a dissociation constant, K, for the relation:

$$H^+ + A^- \rightleftharpoons HA \qquad (6)$$

$$K = \frac{[H^+][A^-]}{[HA]} \qquad (7)$$

For a number of potential buffers in extracellular fluid:

$$H^+ = \frac{K_1[HA_1]}{[A_1^-]} = \frac{K_2[HA_2]}{[A_2^-]} = \frac{K_3[HA_3]}{[A_3^-]} = \frac{K_n[HA_n]}{[A_n^-]} \qquad (8)$$

Equation 8 shows that if the ratio of one of the buffer pairs is known, then given the dissociation constants of the other buffers, their ratios are predictable. From the isohydric principle, one can focus attention on any one of the important buffers (in physiological systems usually the HCO_3^- – CO_2 pair) and infer the behavior of the others.

The isohydric principle does not imply that the presence of additional buffer pairs does not influence the way the hydrogen ion behaves in the fluid. The total buffer capacity of a fluid is the sum of the buffer capacities of all the buffer pairs present. Furthermore, the total buffer capacity determines the stability of the pH during the addition of acids or bases. A large buffer capacity means that larger additions of acids or bases can be tolerated, because the pH changes will be smaller.

In practice, physiological considerations of acid-base status focus on the ratio of HCO_3^- to H_2CO_3 or, more commonly, on the ratio of HCO_3^- to CO_2. If this ratio is known, the pH and the ratios of all other buffer pairs in the system are defined. As with other buffer systems, the relations among the components—the bicarbonate ion, carbon dioxide, and hydrogen ion—are described by a modification of the Henderson-Hasselbalch equation.

$$pH = 6.1 + \log \frac{[HCO_3^-]}{0.03\ [Paco_2]} \qquad (9)$$

From equation 9 it is apparent that the coordinated activity of lungs and kidneys determines the blood pH. In general, the kidneys regulate the concentration of bicarbonate, whereas the lungs regulate the concentration of carbon dioxide. It is worth reiterating that the bicarbonate buffer system is an effective buffer only because respiratory function acts to stabilize $Paco_2$. Without this stabilization the low pK (6.1) of the bicarbonate system would make it an ineffective buffer.

Classification of Acid-Base Status

From the foregoing discussion, the significance of the four terms commonly used as descriptors of acid-base status should be apparent. First, we note that the reference point for pH is not the chemically neutral value of 7.0; rather, it is the physiologically normal value of 7.4. A blood pH lower than the normal value of 7.4 is called *acidosis,* and a blood pH higher than 7.4 is called *alkalosis*. Two additional broad subdivisions are made regarding the source of any pH disturbance—*respiratory* or *metabolic*. A change in acid-base status induced by altered respiration will produce a primary change in CO_2. Such alterations include respiratory acidosis or alkalosis. A change in acid-base status that is induced by removal or addition of acids or bases to the blood results in metabolic acidosis or alkalosis.

Empirical Analysis of Buffers

The number of buffers within the body is so large and the distribution of buffers is so complex that an analytical determination of the body buffer capacity is impossible. Therefore the analysis of acid-base status commonly begins with an empirical determination of the relations among CO_2, HCO_3^-, H^+, and the body buffer capacity. As a rule this analysis is based on the relations predicted by the Henderson-Hasselbalch equation. For convenience the analysis is usually done graphi-

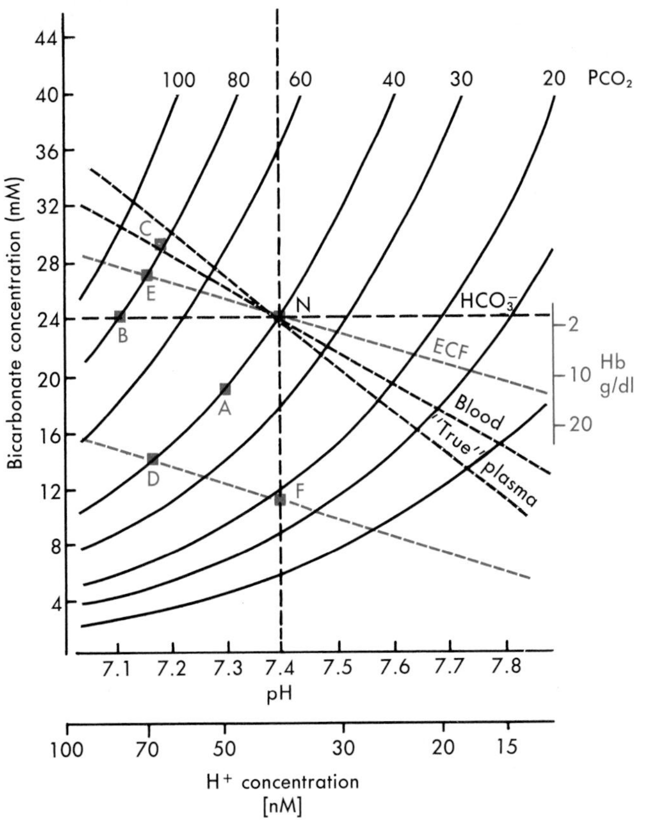

■ **Fig. 48-12.** Graphical solution to the Henderson-Hasselbalch equation and blood buffer lines. The solid black lines represent graphical displays of the Henderson-Hasselbalch equation for different P_{CO_2} values. Thus for any indicated P_{CO_2}, the corresponding bicarbonate concentration and pH are shown by the curved black lines. Any point on this graph predicts the required P_{CO_2}, bicarbonate, and pH for an equilibrium condition among these species, given a pK′ of 6.1. The dashed lines are reference lines describing the buffer capacities of a variety of fluids. The horizontal dashed line at a bicarbonate concentration equal to 24 mM shows the behavior of a pure bicarbonate buffer. The dashed lines passing through *N* show the empirically determined buffer curves for extracellular fluid *(ECF)*, blood, and true plasma. The *true plasma* line represents data that would be obtained if blood were equilibrated with different CO_2 values, and the resultant plasma composition is then analyzed after removal of the red blood cells by centrifugation. The slope of the *ECF* buffer line can be adjusted for different hemoglobin concentrations by placing a ruler through *N* and across the ordinate at the appropriate hemoglobin concentration. See text for details of points *A* through *F*.

cally, as shown in Fig. 48-12. The solid curves in the illustration represent solutions of the Henderson-Hasselbalch equation at the specified P_{CO_2} values and thus the curves represent P_{CO_2} isobars.

The relations between pH and [HCO_3^-] for three different fluids are also shown in Fig. 48-12. These fluids are plasma, blood, extracellular fluid, and a simple bicarbonate solution (HCO_3^-). For each fluid, two of the three variables in the Henderson-Hasselbalch equation were manipulated and the third was measured to determine the buffer capacity of the fluid. The resulting data are plotted as buffer lines (*dashed lines,* Fig. 48-12). A graphical solution to an acid-base problem is achieved when one of the Henderson-Hasselbalch equation lines intersects a buffer line for a particular fluid.

The simplest example to analyze is that for a pure bicarbonate solution, in which only the three components of the Henderson-Hasselbalch equation are involved (HCO_3^-, Fig. 48-12). Adding a strong acid or base at a constant P_{CO_2} will cause the composition of the solution to change in a predictable way, as indicated by the P_{CO_2} isobar. Adding 5 mEq of strong acid to 1 L of the solution will titrate 5 mEq of HCO_3^-, converting it to CO_2. The CO_2 thus generated will then be eliminated, because the CO_2 concentration in this example is held constant by bubbling the solution with a gas of 40 mm Hg CO_2 partial pressure. Starting from a normal condition (N) with a pH of 7.4, a bicarbonate concentration of 24 mEq/L, and a P_{CO_2} of 40 mm Hg, addition of 5 mEq/L of acid will shift the operating point down by 5 mEq/L of bicarbonate to a point *(A)* on the 40 mm Hg P_{CO_2} isobar. The pH at point *A* will be 7.3. The system also can be manipulated by changing the P_{CO_2} at constant [HCO_3^-]. If CO_2 is increased from 40 to 80 mm Hg, the pH moves along the line of constant [HCO_3^-] at a value of 24 mM/L to a new pH of 7.1 (point *B*).

Biological fluids contain more complex buffer systems than the bicarbonate solution just analyzed. Hence one would imagine that, for the same acid load, the pH change would be less than for a simple bicarbonate buffer (i.e., the buffer capacity would be greater). The greater buffering capacity can be illustrated by equilibrating a sample of blood or plasma with different levels of P_{CO_2}. The changes in [HCO_3^-] and pH are determined, and a buffer line for the biological fluid is thereby generated. Lines for blood, plasma, and extracellular fluid are shown in Fig. 48-12. The difference between the bicarbonate solution and blood is apparent. In blood the effect of raising the P_{CO_2} from 40 to 80 mm Hg is shown by the transition from *N* to *C* in Fig. 48-12. Because of the greater buffering capacity, the same elevation of P_{CO_2} causes a smaller reduction in the pH of blood *(C)* than was previously found in the HCO_3^- solution *(B)*.

Note that the bicarbonate concentration increases substantially when CO_2 is added to blood, but not when it is added to the pure bicarbonate solution. This disparate effect is due to the fact that other buffers in addition to HCO_3^- are present in the blood. When CO_2 is added to the blood, a portion of the H_2CO_3 that is formed dissociates to produce a free H^+ and a free HCO_3^-. If the H^+ is buffered, HCO_3^- is left in the solution. Thus the rise in HCO_3^- that occurs between *N* and *C* is a direct estimate of the quantity of H^+ buffered by the blood during addition of the volatile acid CO_2.

One might imagine that increasing CO_2 would directly raise the bicarbonate concentration by hydrating the CO_2 and dissociating the resultant H_2CO_3 into H^+ and HCO_3^-. This is true, but the quantity of HCO_3^- added in this way is insignificant compared to the millimolar concentrations of HCO_3^- present in biological fluids.

Buffer lines also can be determined by titrating plasma, but they must be interpreted with care. If a pure plasma sample were titrated, the buffer line would be intermediate between extracellular fluid and blood and would reflect the buffer capacity of the plasma proteins alone. Quite a different curve is obtained if a blood sample is equilibrated with CO_2 and the resultant pH and $[HCO_3^-]$ determined in plasma after the red blood cells are removed (Fig. 48-12, *true plasma*); the buffer line is steeper for plasma than for blood. However, this does not indicate that plasma has a greater buffer capacity. Rather, the slope of the buffer line represents the combined effects of intracellular buffering in the red cell, and unequal distribution of bicarbonate between plasma and erythrocyte. Because of the negative intracellular potential of the erythrocytes, there is a nonuniform distribution of Cl^- and HCO_3^- across the red blood membrane with a relatively low intracellular concentration of these ions and a correspondingly higher plasma $[HCO_3^-]$. The higher $[HCO_3^-]$ in plasma gives the appearance that plasma has a greater buffer capacity than does blood, but this appearance results from the complex buffering in blood. The true plasma buffer line must be used to predict the changes in HCO_3^- and pH to be expected in plasma after a change in P_{CO_2}. However, to estimate the total acid added to the system, the buffer value for blood must be used.

Buffering in extracellular fluid. In the intact organism, added acid or alkali is rapidly distributed throughout the entire extracellular volume (plasma and interstitial fluid). Thus the buffering capacity of all the extracellular fluid, not just plasma or blood, determines the shifts in pH. Buffer lines for extracellular fluid can be determined, essentially as described earlier, by manipulating the plasma P_{CO_2}. The arterial P_{CO_2} (Pa_{CO_2}) can be an experimentally elevated by adding a known quantity of CO_2 to the inspired gas; similarly, systematic reductions in Pa_{CO_2} can be produced by hyperventilation. After a period of stabilization, when CO_2 has equilibrated throughout the extracellular space, a CO_2 titration curve can be generated from the pH, Pa_{CO_2}, and $[HCO_3^-]$ values in the blood. Such a titration curve is illustrated by the line for extracellular fluid shown in Fig. 48-12. The extracellular fluid buffering line lies between the line for constant bicarbonate and the line for blood. This indicates that the interstitial fluid has much less buffering capacity (fewer proteins) than does blood; the interstitial fluid essentially dilutes the buffering capacity of red blood cells and plasma.

Because of the rapid exchange of H^+ across the red blood cell membrane, the red blood cells contribute to what is usually called extracellular buffer capacity. Account must be taken of the blood hemoglobin concentration to estimate accurately the blood or extracellular buffer line, because hemoglobin concentration and the hematocrit level can change in response to physiological and pathological processes. An empirically determined scale on the right side of Fig. 48-12 permits adjustment of the extracellular buffer value for changes in the blood hemoglobin concentration. This is accomplished by drawing a line through point *N* and the measured blood hemoglobin concentration. Note that if hemoglobin in the blood is increased, the buffer line becomes steeper, as would be anticipated with the addition of new buffer constituents, largely the histidine residues of hemoglobin.

The buffer lines in Fig. 48-12 facilitate the analysis of complex acid-base relations. Suppose a respiratory acidosis is produced by increasing the dead space, which causes the Pa_{CO_2} to increase from its normal value of 40 mm Hg *(N)* to 80 mm Hg *(E)*. The CO_2 distributes through extracellular space. The hydrogen ion concentration (pH decreases from 7.4 to 7.15) rises as a result of the increase in Pa_{CO_2} and subsequent formation of H_2CO_3 (increases from 24 to 27 mM). Part of the H^+ is buffered by extracellular fluid proteins, and the remainder is present as free H^+, which decreases the pH. The buffering of H^+ by extracellular fluid and red blood cell proteins frees HCO_3^- and raises the HCO_3^- concentration in the extracellular fluid.

Contrast the preceding respiratory acidosis with the effect of adding a metabolic acid to the body (metabolic acidosis) in sufficient quantity to lower the blood pH to the same level as in the previous example (7.15). If ventilation does not change at first, the Pa_{CO_2} would remain initially at 40 mm Hg. Thus the addition of acid would shift the operating point in Fig. 48-12 from *N* to *D*. Since point *D* lies on the isobar $PCO_2 = 40$, at a pH of 7.15, one can predict that the extracellular fluid bicarbonate concentration of the subject should fall from 24 to 13.5 mM. The bicarbonate is lost as exhaled CO_2 after being titrated by the added H^+.

An interesting contrast can be drawn between the above examples of respiratory and metabolic acidosis (points *E* and *D*). Although the pH values may be the same in both cases, the changes in bicarbonate concentration induced by respiratory and metabolic acidosis are in opposite directions. In metabolic acidosis the Pa_{CO_2} is initially held constant by ventilation. As the added hydrogen ion titrates the bicarbonate in extracellular fluid, CO_2 is formed and exhaled, thereby reducing HCO_3^- stores. In contrast, in respiratory acidosis each hydrogen ion added by the retained CO_2 is accompanied by a bicarbonate ion; when the H^+ is buffered, the HCO_3^- is left in excess. Note that the rise in

HCO_3^- is *not* the direct result of formation of H_2CO_3 and its dissociation into H^+ and HCO_3^-. It is the buffering of added H^+ that frees HCO_3^- and raises its concentration.

The total quantity of acid added in metabolic acidosis also can be estimated from Fig. 48-12. This graphical method assumes that the change in bicarbonate comes from two sources. First, the H^+ can titrate the protein buffers present in the extracellular fluid. Second, the H^+ can combine with HCO_3^- to produce CO_2. The CO_2 is then exhaled, thus reducing the body HCO_3^- stores (Fig. 48-5). If respiratory acidosis is produced by raising $Pa{CO_2}$ to 80 mm Hg, the added acid is buffered, and the rise in HCO_3^- directly reflects the quantity of acid buffered. The change of 3 mM (at *E*), however, is a change in concentration, and the actual quantity of hydrogen ions added to the extracellular fluid is the product of the extracellular fluid volume and the change in H^+ concentration in the extracellular fluid. The extracellular fluid volume is approximately 11.6 L in a 70 kg man. (See Table 45-1.) Thus the increase in $Pa{CO_2}$ from 40 to 80 mm Hg added 35 mmoles of bicarbonate to the extracellular fluid.

In metabolic acidosis (*D* in Fig. 48-12), the protein buffers are titrated and, in addition, bicarbonate is removed as exhaled CO_2 after the added H^+ has been buffered. To assess the relative contribution of the two processes (protein buffering and CO_2 exhalation), the buffer line for extracellular fluid can be shifted to various positions on the HCO_3^--pH graph while holding the slope of the line constant. For example, provided some form of compensation has not occurred, the dashed line through *D*, drawn parallel to the extracellular fluid line, represents the buffer value of the extracellular fluid at a pH of 7.15. To determine how much H^+ has been buffered by proteins, one assumes that each H^+ buffered will leave an HCO_3^- in excess. The contribution of the buffering by proteins to the change in $[HCO_3^-]$ can then be assessed by determining the change in $[HCO_3^-]$ that would have occurred had the pH not changed. This should be the $[HCO_3^-]$ observed if the extracellular fluid were back titrated to a pH equal to 7.4, that is, along the extracellular fluid buffer line to *F*. The fall in HCO_3^- must represent the release of a quantity of H^+ from the buffer equal to that originally buffered during the acidosis. Thus the difference between the $[HCO_3^-]$ at 7.15 (Fig. 48-12, *D*, 14 mM) and that predicted at a pH of 7.4 (11 mM, *F*) after the $Pa{CO_2}$ had been reduced, represents the bicarbonate buffered by extracellular fluid and red blood cell proteins (i.e., 3 mM).

The total fall in bicarbonate concentration induced by metabolic acidosis is the difference (13 mM) between the ordinate values of *N* and *F*. Therefore, in addition to the 3 mmoles/L of added acid buffered by the proteins, 10 mmoles/L were buffered by exchanging the CO_2 after the bicarbonate had buffered the added H^+. These exhaled CO_2 molecules represent the so-called depleted bicarbonate reserves that were referred to previously (Fig. 48-5); these reserves must be replenished by processes associated with tubular H^+ secretion (Fig. 48-3).

The apparent loss of bicarbonate associated with the acidosis (difference between the ordinate values of *N* and *F*) defines the *base deficit,* which is the reduction in $[HCO_3^-]$ from the control state. It estimates the amount of base that would have to be added to extracellular fluid in an acidotic patient to return the pH to normal. For the case of a bicarbonate concentration of 13 mM in a 70 kg man with 11.6 L of extracellular fluid, approximately 151 mmoles of bicarbonate must be added to return the extracellular fluid pH to normal (13 mM $\times$ 11.6 L of extracellular fluid). This assumes no other compensation and little change in intracellular fluid bicarbonate concentration, an assumption that is not true except at very short times after alteration of the acid-base status (Fig. 48-14). A similar evaluation could be made of an increase in bicarbonate concentration associated with a metabolic alkalosis condition or respiratory acidosis; the increased bicarbonate in this case is called the *base excess.*

Intracellular buffering. Up to this point, the discussion has focused entirely on the buffering capacity in extracellular fluid and erythrocytes. However, in the long term, the intracellular compartment and bone actually provide much of the buffer stores in the organism, because both the buffer capacity and the volume of intracellular fluid are greater than those of extracellular fluid. In vitro studies on muscle suggest that the intracellular buffer capacity is about five times that of a comparable volume of extracellular fluid. The intracellular buffering is mainly by proteins (the imidazole groups), various organic phosphates, anserine, and carnosine. Thus effective buffering requires movement of hydrogen across cell membranes.

To achieve access to the intracellular buffer stores, H^+ and HCO_3 must move across cell membranes. The passive permeabilities to hydrogen and bicarbonate are relatively limited, but an energy source in the form of sodium-potassium ATPase coupled to a sodium-hydrogen carrier move hydrogen into or out of the cell as the acid-base status is modified (Fig. 48-13). The hydrogen carrier maintains a pH differential across the cell membrane and yields an intracellular pH averaging 7.1. A transmembrane HCO_3^- flux is coupled to the movement of Cl^- (Fig. 48-13). Thus both H^+ and HCO_3^- can flow into and out of the cell as required for buffering.

Intracellular buffering in acid-base disturbances can be associated with transmembrane fluxes of various ions, as shown in Fig. 48-13. The arrow sizes at the tops of Fig. 48-13, *A* and *B,* indicate the relative importance of intracellular *(colored areas)* and extracellular *(gray boxes)* buffer stores. In a metabolic disturbance (acidosis or alkalosis), the buffering is divided

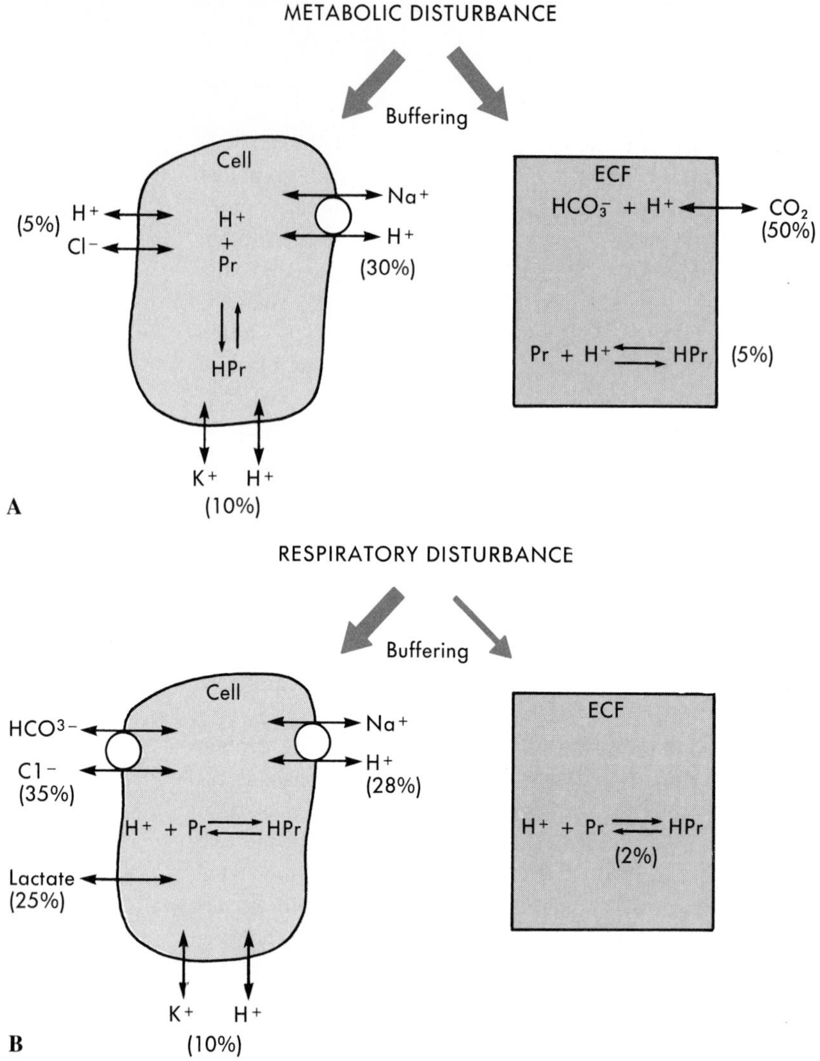

■ **Fig. 48-13.** Sources of buffering in acid-base disturbances. Sizes of arrows at the top of each panel indicate the relative importance of extracellular and intracellular buffering. The percentages are rough approximations of the buffering fraction attributable to the indicated processes. Bidirectional arrows on cells and in extracellular fluid indicate that flux may occur in either direction, depending on gradients and the type of acid-base disturbance. **A,** Metabolic disturbances. Note the approximately equal buffering in intracellular and extracellular fluid. **B,** Respiratory disturbance. The majority of buffering takes place within the cell, since the disturbance in acid-base balance originates with an alteration in Pa_{CO_2} and therefore limits the contribution of extracellular buffers (largely HCO_3^-) to protein buffers rather than bicarbonate.

approximately equally between the intracellular and extracellular space (Fig. 48-13, *A*). In a respiratory disturbance, however, buffering is accomplished almost entirely in the intracellular space (Fig. 48-13, *B*). The extracellular HCO_3^- cannot buffer a respiratory acidosis, because each carbon dioxide molecule that is retained produces as much HCO_3^- as H^+.

The relative importance of the various types of transmembrane ion fluxes associated with intracellular buffering also varies, depending on the type of acid-base disturbance. In both respiratory and metabolic imbalances, the Na^+-H^+ exchange and the K^+-H^+ exchange are significant. In respiratory derangements, the HCO_3^--Cl^- shift assumes greater importance as CO_2 enters the cell and becomes hydrated in the presence of carbonic anhydrase. The H_2CO_3 then dissociates, and the free hydrogen is buffered (Fig. 39-12). The HCO_3^- freed by the buffering process then exits the cell in exchange for Cl^-.

An important consequence of the large amount of intracellular buffering is that full buffering is delayed for some time (Fig. 48-14, ''cell buffering''). In less than 1 hour, extracellular fluid equilibrates with added acid or base. However, 4 to 6 hours are required for H^+ to equilibrate with the intracellular pool. One of the difficulties in assessing the importance of the intracellular component of buffering (Fig. 48-14) is that the time course of exchange with intracellular space overlaps the time course of renal and respiratory compensation (see below).

Release of carbonate by bone provides an additional important buffer source to extracellular fluid. In chronic acidosis, carbonate is released into extracellular fluid in significant quantities for several hours and continues for many days. In fact, in chronic acidosis release of carbonate from bone may be extensive enough to decalcify and weaken the bones.

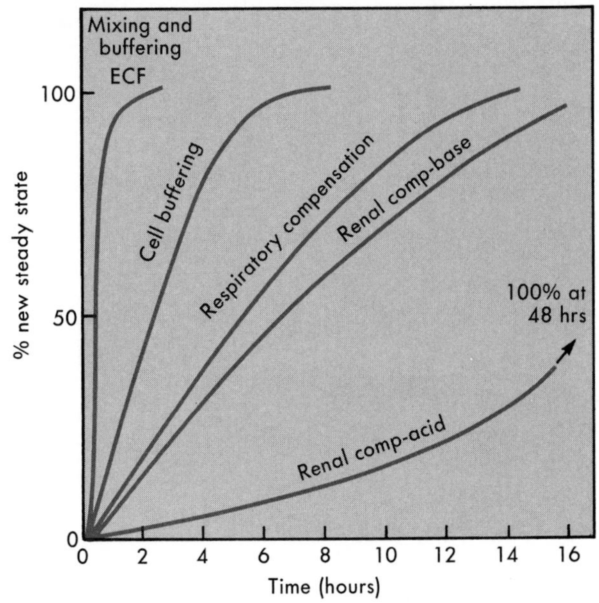

■ **Fig. 48-14.** Time course of compensation for acid-base disturbances. Mixing and buffering in the extracellular space is rapid, approaching completion within 1 hour. Cell buffering takes longer as a result of the transmembrane flux required, as shown in Fig. 48-13. Respiratory and renal compensation requires several hours to days. *Comp,* Compensation.

■ *Physiological Compensation for Acid-Base Disturbances*

Buffering, whether extracellular or intracellular, is limited by available stores of buffers. During long-term disturbances in acid-base balance, the stability of pH can be achieved only by more active compensatory responses. Physiological compensation is accomplished by modification of renal HCO_3^- reabsorption and respiratory CO_2 excretion. If a respiratory disturbance of H^+ balance occurs, a compensatory modification of renal bicarbonate reabsorption (H^+ excretion) will tend to maintain the $[HCO_3^-]/Pa_{CO_2}$ ratio constant and thereby stabilize pH. Conversely, if ingested or metabolically produced acid lowers the pH, hyperventilation can lower Pa_{CO_2} and stabilize the pH. The necessary compensatory mechanisms inherent in the respiratory and renal regulatory processes are discussed in Chapters 39 and 47.

Fig. 48-15 coalesces the material presented in Fig. 48-12 with the earlier discussions of renal and respiratory compensatory mechanisms. As in Fig. 48-12, the curved solid black lines represent P_{CO_2} isobars for the Henderson-Hasselbalch equation. In this case, the P_{CO_2} of interest is arterial P_{CO_2} (Pa_{CO_2}). The extracellular buffer line is shown as the straight solid black line that passes through *D* and *F* and is labeled *ECF*.

A predictive curve for the patterns of respiratory compensation during acid-base disturbances can be obtained by varying the blood pH and examining the ensuing respiratory responses. Such a curve for normal human subjects is depicted in Fig. 48-15 by the solid colored line. Renal compensatory patterns are also determined empirically as shown by the solid black lines labeled *renal-comp*. During respiratory acidosis, bicarbonate reabsorption is augmented in the renal tubules and plasma $[HCO_3^-]$ increases. When the blood pH is below 7.4 the compensation line follows a pattern that depends on the combined effects of $PaCO_2$ and plasma $[HCO_3^-]$ on the reabsorption of bicarbonate (Figs. 48-4 and 48-8). At high pH values, renal compensation is accomplished by the excretion of excess HCO_3^-, and the compensation line follows the extracellular fluid buffer line.

Metabolic disturbances. Metabolic disturbances occur when $[H^+]$ or $[HCO_3^-]$ is changed by altered intake or by loss or production of these ions. The resultant pH change must be compensated by altered respiratory function. The initial responses and subsequent compensations are graphically displayed in Fig. 48-15.

Suppose that metabolic alkalosis is initiated by vomiting, which raises the blood pH to *A*. Assume that the Pa_{CO_2} remains initially at 40 mm Hg. The alkalosis is associated with a fall in blood $[H^+]$ and a rise in $[HCO_3^-]$ to 35 mM.

The alkalosis depresses respiration, which in turn elevates the Pa_{CO_2}. Increasing levels of extracellular fluid CO_2 titrate the extracellular fluid along the extracellular fluid buffer line from *A* to *B*. The rising Pa_{CO_2} stimulates respiration, which offsets the initial depression of respiration induced by the alkalosis. At point *B*, a steady state is reached between the initial respiratory depression and the compensatory respiratory stimulation. The initial alkalinization of the extracellular fluid tended to raise the pH to 7.55, but the hypoventilation returned the pH to 7.5.

Point *A* in Fig. 48-15 represents an *uncompensated metabolic alkalosis,* and point *B* represents a *compensated metabolic alkalosis*. Note that the distinction between the compensated and uncompensated states is based on higher values of $[HCO_3^-]$ and Pa_{CO_2} than are predicted by the normal CO_2 isobar at Pa_{CO_2} = 40 mm Hg.

During prolonged acid-base derangements with elevated Pa_{CO_2}, respiratory compensation is assisted by an increase in the cerebrospinal fluid $[HCO_3^-]$. In this condition, the respiratory drive induced by the elevated arterial Pa_{CO_2} is diminished. This diminished drive allows the Pa_{CO_2} to rise even higher. Such compensation of chronic alkalosis is reflected by a shift in the coordinates in Fig. 48-15 from point *B* to point *C,* again parallel to the extracellular fluid buffer line.

Similar reciprocal changes are observed with a metabolic acidosis. The colored region between the isobar at Pa_{CO_2} = 40 mm Hg and the vertical line at pH =

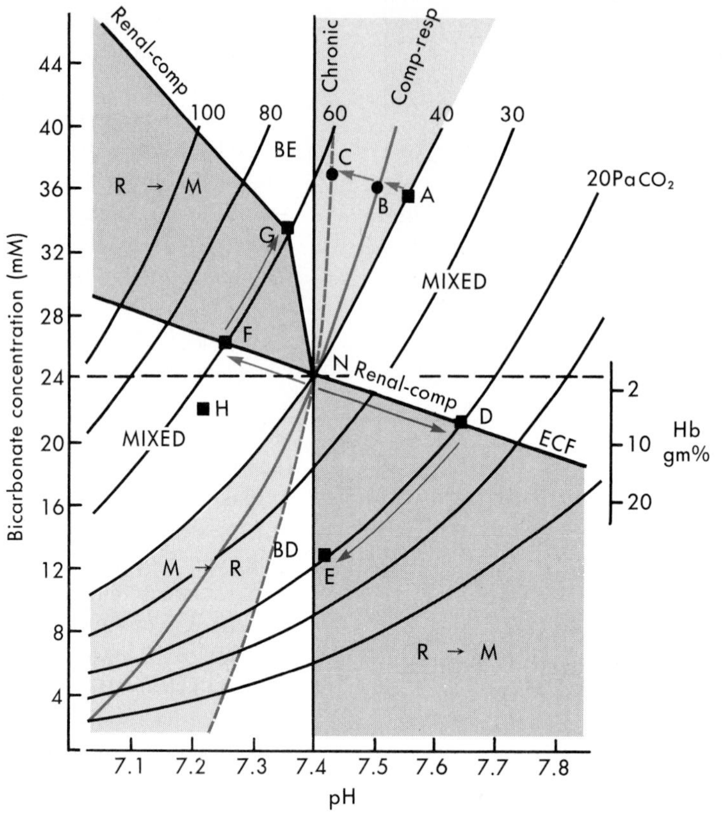

Fig. 48-15. Composite, showing disturbances in acid-base balance followed by renal and respiratory compensatory processes. The basic diagram is as described in Fig. 48-12. Several derangements of acid-base balance are shown. Metabolic alkalosis *(A)* is indicated by a Pa_{CO_2} in the normal range and a pH higher than 7.4. Respiratory compensation (*comp-resp*) for the metabolic disturbance is accomplished by hypoventilation, which raises the Pa_{CO_2} and lowers the pH along a line parallel to the extracellular fluid buffer line *(A-B)*. Chronic metabolic disturbance of acid-base balance leads to adaptation of respiratory control, so that a higher than normal Pa_{CO_2} is maintained, further compensating for the disturbance *(B-C, heavy colored dashed line)*. Respiratory alkalosis *(D)* is indicated by hypocapnia and an elevated pH. Renal compensation is accomplished by increased renal excretion of bicarbonate, thus lowering plasma bicarbonate levels to approximately 11 mM and pH to 7.42 *(E)*. Compensation is associated with a base deficit of 13 mEq/L (i.e., 24 − 11). Respiratory acidosis *(F)* is indicated by the high Pa_{CO_2} and the low pH. Compensation is accomplished by increased reabsorption of bicarbonate at a constant Pa_{CO_2}, which returns the pH toward normal *(F-G)*. Renal compensation for a respiratory acidosis is limited by the ability of the kidney to increase bicarbonate reabsorption. Above 36 mM of bicarbonate, further augmentation of HCO_3^- reabsorption by the kidney is impossible. (See Fig. 48-4.) The gray regions indicate the parameters of interest in various stages of acid-base disturbance and subsequent compensation. The colored regions represent conditions encountered when metabolic disturbances, either alkalosis or acidosis, are compensated by respiratory adjustments. The gray regions represent areas in which an initial respiratory disturbance is followed by metabolic compensation. The white regions represent mixed acid-base disturbances. For example, a respiratory acidosis with a superimposed metabolic acidosis might be represented by point *H*. *R,* Respiratory; *M,* metabolic; *ISF,* interstitial fluid; *BE,* base excess; *BD,* base deficit.

7.4 represents the possible combinations of pH and [HCO_3^-] in the plasma of a person experiencing a *compensated, chronic metabolic acid-base disturbance*. The longer the disturbance persists, the more the subject's acid-base status is reflected by a shift from the initial respiratory compensation line *(solid colored line)* toward the chronic state (*colored dashed line*).

Respiratory disturbances. Respiratory disturbances are compensated by altered renal reabsorption of HCO_3^-. An initial stage of respiratory alkalosis is represented by point *D* in Fig. 48-15. The respiratory origin is indicated by its placement on the extracellular fluid buffer line. Note that this is the change expected if nonvolatile acids have not been added to the extracellular fluid. Such a condition could be caused by hyperventilation, which would reduce the Pa_{CO_2}, as reflected by a shift of coordinates along the buffer line from *N* to *D*. Compensation is accomplished by reducing renal HCO_3^- reabsorption and increasing HCO_3^- excretion, by virtue of the effect of Pa_{CO_2} on the renal reabsorption of HCO_3^- (Fig. 48-8). Since large quantities of HCO_3^- are continuously filtered at the glomeruli, large quantities of HCO_3^- can be excreted, and compensation can effectively return the pH nearly to normal *(E)*.

The response is more complicated when respiratory acidosis *(F)* is induced by increased pulmonary dead space or by hypoventilation. Up to a Pa_{CO_2} of about 60 mm Hg, the elevations of arterial [H^+] and of Pa_{CO_2} increase the tubular reabsorption of HCO_3^- and thereby compensate for the respiratory acidosis (Fig. 48-8). However, tubular secretion of H^+ ultimately limits tubular reabsorption of HCO_3^- when the plasma [HCO_3^-] reaches about 34 mM (Fig. 48-4 and Fig. 48-15, *point G*). Above this point, compensation is limited to the maximum bicarbonate reabsorptive capacity *(heavy black line, renal-comp)*. Thus the composition of the blood of subjects with compensated respiratory acidosis will be defined by the upper segment of the heavy black line, which reflects the limits of tubular bicarbonate reabsorption.

The directions of the arrows show the sequences of buffering and subsequent compensation. Primary respiratory disturbances move the coordinates away from the normal point *(N)* along the extracellular fluid buffer line. Metabolic disturbances shift the coordinates from the normal point along the Pa_{CO_2} isobars. Respiratory compensation for metabolic disturbances is associated with shifts along lines that are parallel to the extracellular fluid buffer line. Renal compensation for respiratory disturbances tracks along the Pa_{CO_2} isobars.

For purposes of analysis, Fig. 48-15 can be divided into regions that define the acid-base status of a subject. To the left of the vertical line at pH = 7.4 lies acidosis; to the right lies alkalosis. Any point above the extracellular fluid buffer line reflects the addition of base to the body, *regardless of whether the base was added as a primary or secondary compensation*. The area below the extracellular fluid line defines those points at which base has been removed from the body. The isobar at a Pa_{CO_2} of 40 defines the normal Pa_{CO_2}, and all uncompensated metabolic disturbances will lie along this line. Similarly, all uncompensated respiratory disturbances will lie along the extracellular fluid buffer line.

The two colored wedge-shaped regions (bounded by the Pa_{CO_2} = 40 curve, the vertical pH = 7.4 line, and the chronic compensation curve) define regions of metabolic disturbances that can be compensated by respiratory adjustments *(M → R)*. The gray shaded regions (bounded by the pH = 7.4 line, the extracellular fluid buffer line, and the upper left segment of the renal compensation curve) represent regions in which a primary respiratory disturbance is compensated by altered renal function *(R → M)*. In the case of a respiratory disturbance, points lying between the ECF buffer line and either the pH = 7.4 line or the comp-renal line are said to be "partially compensated." Similarly, metabolic disturbances in the regions between the line depicting Pa_{CO_2} = 40 and the "chronic" line are "partially compensated."

Base excess or deficit. Fig. 48-15 also permits an analysis of the base excess or deficit induced by the disturbance; this refers to the change in [HCO_3^-] that is independent of the immediate effect of the pH change. Thus if a line parallel to the extracellular fluid buffer line is drawn through a point of interest, its intersection on the ordinate at pH = 7.4 will define the base deficit or base excess. In the uncompensated respiratory alkalosis *(A)*, the base excess (*BE*) is 14 mM *(38−24)*. In the compensated respiratory alkalosis, renal HCO_3^- excretion increases to compensate for the respiratory alkalosis. This depletes the stores of HCO_3^- in the extracellular fluids and creates a base deficit (*BD*) of 11 mM (24−13).

■ *The Urine-Collecting System*

■ *Anatomy*

Urine passes out of the body through a series of structures known collectively as the *urine-collecting system*. This consists of the *calyces,* the *renal pelvis,* the *ureters,* the *bladder,* and the *posterior urethra*. The ureters are tubes 4 or 5 mm in diameter and 30 cm long and are connected to the renal pelvis. They enter the bladder bilaterally at a location near its base and close to the origin of the urethra. The region comprising the two ureters and the urethra is known as the *trigone*. The bladder consists of two parts—the *fundus,* or body, and the *neck*. The neck is also known as the posterior urethra. In females the posterior urethra is the end of the urine-collecting system and the point of exit of urine from the body. In males a continuation of the urethra, the *anterior urethra,* extends through the penis. The anterior urethra, however, is embryologically part of the reproductive system rather than the urinary system.

The entire urine-collecting system is lined with transitional epithelium and is surrounded by several layers of smooth muscle. The smooth muscle cells in much of the system are electrically coupled, fire action potentials, and are capable of developing spontaneous tone and of responding to neurotransmitters. The orientation of the smooth muscle varies with location along the collecting system. In the renal calyces and pelvis the orientation is largely circular; in the ureter the orientation is more spiral. The ureters run obliquely for 2 or 3 cm within the wall of the bladder before opening into the bladder lumen. Where the ureters enter the bladder the smooth muscles of the ureters and the bladder are continuous. The smooth muscle of the bladder, known as the *detrusor muscle,* forms a basketlike meshwork that converges toward the urethra into fibers arranged in a more longitudinal pattern.

The smooth muscle and the epithelium, as well as the elastic lamina between the two, are organized to permit substantial distension of the ureters and bladder. The epithelium is arranged in multicellular layers overlying the folded elastic lamina. The smooth muscle possesses a substantial capacity to adapt its tension to altered wall stress. The inner elements of the urethra and bladder are thrown up into folds, or *rugae.* During expansion the rugae can be flattened, and the epithelial cells spread out into a single layer without disrupting the integrity of the epithelial lining. This pattern is exaggerated in the bladder. As a result the volume of this structure can vary from 10 to 400 ml with a pressure change of only 5 to 10 cm H_2O.

Innervation

The innervation of the collecting system is a major element in determining function. The dominant neural input to the collecting system is the parasympathetic innervation of the bladder and the posterior urethra. The fibers arise in the second through the fourth sacral segments of the spinal cord and travel via the *pelvic nerve* to the bladder wall. Synapses located within the bladder give rise to postganglionic fibers that innervate the smooth muscle. Control of bladder function and of the process of urination (*micturition*) also greatly depends on voluntary activity, which in the collecting system is regulated by somatic motor nerves innervating the external sphincter of the urethra. These somatic motor nerves originate in the anterior horn cells of the spinal cord and travel via the *pudendal nerves* to the sphincter.

Sensory fibers that detect distension originate in the bladder and travel via sympathetic fiber tracts to the spinal cord. Parasympathetic fibers carry information about the presence of urine in the posterior urethra to the cord. Sympathetic and parasympathetic afferent fibers synapse in the cord and participate in spinal reflexes. They also continue to higher centers, including the pons and the sensory motors trip of the cortex.

Passage of Urine

Urine flows passively out of Bellini's ducts into the calyces. Contraction of the smooth muscle of the calyces and renal pelvis drives urine into the ureter. With distension the circularly arranged smooth muscle surrounding these structures contracts and generates pressures of 6 to 10 mm Hg, which force urine into the ureter. Once urine enters the ureter, distension again triggers contraction of the ureteral smooth muscle. This contraction is sufficiently vigorous to close the *ureteropelvic junction.* At the same time, a peristaltic wave is initiated, which moves down the ureter and thus moves the urine toward the bladder. This process is autonomous and is related to the myogenic response of the smooth muscle. In other words, stretch per se initiates muscular contraction, which then propagates along the length of the ureter. Peristaltic waves are generated at frequencies of one to five per minute, with four to six segmentations along the length of the ureter. Pressures vary with the presence of peristalsis and with the level of hydration of the subject, but they range from 20 to 100 mm Hg.

After a person voids, the residual volume of the bladder is usually less than 10 ml. As urine is formed and transported through the ureters into the bladder, it accumulates and bladder volume expands. When the volume reaches 200 to 300 ml, stretch receptors in the bladder wall are activated, and a sense of fullness and an urge to urinate are felt. The maximum bladder capacity is generally about twice the volume at which the urge to urinate is perceived.

Neural Control of Micturition

Ureteral function is autonomous, that is, inherent within the smooth muscle surrounding the ureter. Conversely, control of the bladder, posterior urethra, and external sphincter depends on complex neural mechanisms. Distension of the bladder discharges over afferent fibers of the pelvic nerve and the pudendal nerve to the spinal cord and thence to the pons and cerebral cortex. Afferent discharge activates a spinal reflex involving the pelvic nerve. Pelvic nerve discharge induces contraction of the bladder and expulsion of urine. However, this reflex is easily inhibited by cortical efferent nerves. Alternatively, when so desired, the spinal reflex mechanism that leads to bladder discharge can be facilitated by descending cortical impulses.

When the volume within the bladder initiates afferent nerve discharge and cortical inhibition is released, parasympathetic discharge to the detrusor muscle is initiated; the muscle contracts and bladder pressure is raised. Because the ureters course through the wall of the bladder, an increase in bladder pressure closes the ureters and prevents reflux of urine into these structures. Contraction of the detrusor muscle also shortens and widens the posterior urethra and urine enters this structure. Entry of urine into the posterior urethra triggers parasympathetic afferents; these in turn inhibit neural traffic over the pudendal nerve, which normally maintains closure of the external sphincter. This relaxes the external sphincter, and urine can then flow through the posterior urethra and exit the body. The smooth muscle of the bladder continues to contract, thereby maintaining high intrabladder pressure and forcing urine out of the body.

Once initiated, micturition normally continues until the bladder is emptied. However, cortical influences can inhibit the process by triggering nerve activity over the pudendal nerve, which causes contraction of the external sphincter. This is then followed by relaxation of the detrusor muscle.

The process of micturition is assisted by contraction of a variety of voluntary muscles that are coordinated by the cerebral cortex. Resistance to flow through the posterior urethra is reduced at the onset of micturition by contraction of the *levator ani* and *perineal* muscles. Coordinated with this process is the contraction of the abdominal muscles and a lowering of the diaphragm, both of which elevate intraabdominal pressure, which forces urine from the bladder.

Incontinence

Voluntary control of urination develops after birth, usually sometime between the second and third year of life. Until that time, voiding is entirely reflex and depends on bladder stretch. Interference with the reflex processes just described may severely handicap the ability of an individual to control the release of urine—a common clinical problem. If the pelvic nerves are damaged, urine is retained and the bladder becomes grossly distended. If damage is greater and the pelvic as well as the pudendal nerves are involved, the afferent input to the reflex arc is eliminated and the external sphincter no longer maintains tone. In this condition urine drips continuously from the urethra as a result of the combined high pressure in the distended bladder and the lack of control of the external sphincter. This condition is referred to as *urinary incontinence*.

Bibliography

Journal articles

Bie, P.: Osmoreceptors, vasopressin, and control of renal water excretion, Physiol. Rev. **60:**961, 1980.

Bundy, H.F.: Review: carbonic anhydrase, Comp. Biochem. Physiol. **57B:**1, 1977.

Cheema-Dhadli, S., and Halperin, M.L.: Role of mitochondrial anion transporters in the regulation of ammoniagenesis in renal cortex mitochondria of the rabbit and rat, Eur. J. Biochem. **99:**483, 1979.

Davis, J.O., and Freeman, R.H.: Mechanisms regulating renin release, Physiol. Rev. **56**(1):1, 1976.

DeBold, A.J., et al.: A rapid and patent natriuretic response to intravenous injection of atrial myocardial extract in rats, Life Sci. **28:**89, 1981.

De Wardener, H.E., and Clarkson, E.M.: The natriuretic hormone: recent development, Clin. Sci. **63:**415, 1982.

Edelman, I.S., and Marver, D.: Mediating events in the action of aldosterone, J. Steroid Biochem. **12:**219, 1980.

Harmanci, M.C., et al.: Vasopressin and collecting duct intramembranous particle clusters: a dose-response relationship, Am. J. Physiol. **239:**F560, 1981.

Jamison, R.L.: Micropuncture study of segments of thin loop of Henle in the rat, Am. J. Physiol. **215:**236, 1968.

Jewell, P.A., and Verney, E.B.: An experimental attempt to determine the site of neurohypophysial osmoreceptors in the dog, Philos. Trans. R. Soc. Lond. [Biol.] **240**(672):197, 1957.

Koepke, J.P., and DiBona, G.F.: Functions of the renal nerves, Physiologist **28:**47, 1985.

Levens, N.R., et al.: Role of the intrarenal renin-angiotensin system in the control of renal function, Circ. Res. **48**(2):157, 1981.

Nasjletti, A., and Malik, K.U.: Renal kinin-prostaglandin relationship: implications for renal function, Kidney Int. **19:**860, 1981.

Schrier, R.W., editor: Symposium on water metabolism, Kidney Int. **10:**1, 1976.

Books and monographs

Alpern, R.J., et al.: Renal acidification mechanisms. In Brenner, B.M., and Rector, F.C., Jr., editors: The kidney, vol. 1, ed. 3, Philadelphia, 1986, W.B. Saunders Co.

Balagura-Baruch, S.: Renal metabolism and transfer of ammonia. In Rouiller, C., and Muller, A.F., editors: The kidney, vol. 3, New York, 1971, Academic Press, Inc.

Ballermann, B.J., et al.: Renin, angiotensin, kinins, prostaglandins and leukotrienes. In Brenner, B.M., and Rector, F.C., Jr., editors: The kidney, vol. 1, ed. 3, Philadelphia, 1986, W.B. Saunders Co.

Cogan, M.G., and Rector, F.C., Jr.: Acid-base disorders. In Brenner, B.M., and Rector, F.C., Jr., editors: The kidney, vol. 1, ed. 3, Philadelphia, 1986, W.B. Saunders Co.

Davenport, H.W.: The ABC of acid-base chemistry, ed. 5, Chicago, 1958, University of Chicago Press.

Halperin, M.L., et al.: Biochemistry and physiology of ammonium excretion. In Seldin, D.W., and Giebisch, G., editors: The kidney: physiology and pathophysiology, vol. 2, New York, 1985, Raven Press.

Hogg, R.J., and Kokko, J.P.: Urine concentrating and diluting mechanisms in mammalian kidneys. In Brenner, B.M., and Rector, F.C., Jr., editors: The kidney, vol. 1, ed. 3, Philadelphia, 1986, W.B. Saunders Co.

Robinson, J.R.: Fundamentals of acid-base regulation, ed. 5, London, 1975, Blackwell Scientific Publishers.

Valtin, H.: Renal functions: mechanisms preserving fluid and solute balance in health, Boston, 1973, Little, Brown & Co.

Woodbury, J.W.: Body acid-base state and its regulation. In Ruch, T.C., and Patton, H.D., editors: Physiology and biophysics, vol. 2, ed. 20, Philadelphia, 1986, W.B. Saunders Co.

SECTION

IX

THE ENDOCRINE SYSTEM

Saul M. Genuth

CHAPTER

49

General Principles of Endocrine Physiology

Endocrinology was classically defined as a discipline concerned with the "internal secretions of the body." The original concept—that a chemical substance called a hormone, liberated by one kind of cell, is carried by the bloodstream to act on a distant target cell—represented a major advance in physiological understanding. It suggested a basic mechanism for maintaining the stability of the internal milieu in the face of irregular nutrient, mineral, water, and thermal fluxes. A hormone was defined as a substance whose secretion was evoked by a specific alteration in the milieu; as a result of the hormone's action on its target cell(s), the alteration was counteracted, and the status quo was restored. However, this homeostatic notion has grown increasingly complex as the intricate operation of the endocrine system continues to be revealed.

The overall mission of the endocrine system currently is seen to include regulation of growth, maturation, body mass, reproduction, and behavior, as well as regulation of substrate and mineral flow for maintenance of chemical homeostasis. A diversity of endocrine cell types and of their organization and locations has been found; moreover the number of known hormones and their molecular variety has multiplied. Target cells of hormone action now include other hormone-producing cells, and some hormones now are known to require chemical modification at intermediate sites between the gland of origin and the target cells before their mission can be accomplished. A complementary group of target cell substances—hormone receptors—has been found to play an essential role in mediating hormone action. Furthermore hormone action now is known to involve several different intracellular mechanisms. Much of the control of hormone secretion has been found to depend on the feedback principle described on p. 824. In addition, an important relationship between neural function and hormonal secretion has been established. Because of this complexity and diversity, students are in danger of being overwhelmed unless they learn to relate the functional characteristics of each component of the endocrine system to a set of unifying principles. The major purpose of this chapter is to provide such a framework for the subsequent chapters.

Relationship Between Endocrine and Neural Physiology

In a conceptual sense the nervous system and the endocrine system have important functional similarities. Each is basically a system for signaling. Each operates in a stimulus-response manner. Each transmits signals that in some cases are highly localized, narrow, and unitary in purpose, and in other cases are widespread, broad, and diverse in purpose. Each system is crucial to the cooperative physiological functioning of the multiplicity of highly differentiated cells, tissues, and organs that make up the human organism. Therefore the nervous system and the endocrine system together must *integrate incoming stimuli* so as to *integrate the organism's response* to changes in its external and internal environment. It should have come as no surprise when recent advances in immunohistochemistry and molecular biology revealed an intimate relationship between neural function and hormone secretion. The very distinction between a hormone and a neurotransmitter has grown increasingly blurred. It may not be fanciful either to think of a circulating hormone as a signal molecule freed from the confines of a single axon to reach all responsive cells, or to think of a neurotransmitter as a signal molecule captured from the circulation to provide a chemical connection between two specific cells separated in space.

This ever closer relationship is illustrated by several key observations. Peptides originally discovered as hormonal products of classic endocrine cells have been found to exist in various and often widespread distributions in the brain and other neural tissue. In some instances the same cell gives rise to both a biogenic amine neurotransmitter and a peptide hormone. Neuropeptides and peptide hormones may be synthesized at the direction of common or very similar genes, with the final cell product being specified by alternative pathways of processing the gene transcript. Furthermore, peptide sequences with neurotransmitter or neuromodulatory function can be found within longer hormonal peptide molecules and vice versa.

For all of these reasons, it is now necessary to distinguish between *endocrine*, *neurocrine*, *paracrine*, and even *autocrine* effects, to describe the known spectrum of hormonal signaling (Fig. 49-1). The original concept of a hormone is that it is a substance secreted by one cell and transmitted via the blood stream to act on a target cell: endocrine function. The original concept of a neurotransmitter is that it is a substance liberated from a nerve terminal that travels a short distance across a synapse to act on another effector cell or membrane: neural function. The discovery of peptides that are synthesized in the cell body of neurons in the hypothalamus and stored in axons (like neurotransmitters) but secreted into the blood stream to act on relatively distant target cells (like hormones) led to the concept of a "neurohormone" and of neurocrine function (Fig. 49-1). It is also now appreciated that hormone molecules from one cell may be conveyed through local channels to act on a neighboring cell: paracrine function. Finally, a hormone or neurohormone may act back on its cell of origin or on similar cells after secretion into the intercellular fluid (or after transmission along an axon): autocrine function.

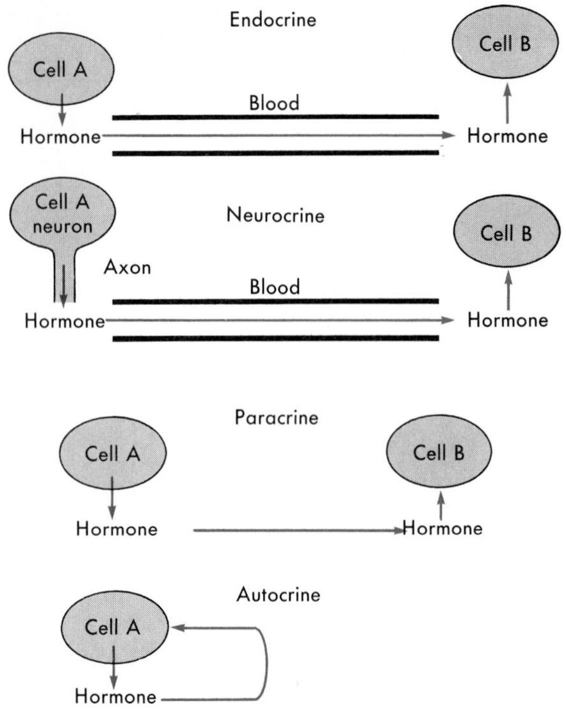

■ **Fig. 49-1.** Schematic representation of mechanisms for cell-to-cell signaling via hormone molecules. In *endocrine* function the signal is carried to a distant target via the blood stream. In *neurocrine* function the hormone signal originates in a neuron, and after axonal transport to the blood stream it is carried to a distant target cell. In *paracrine* function the hormone signal is carried to an adjacent target cell over short distances via the interstitial fluid. In *autocrine* function the hormone signal acts back on the cell of origin or adjacent identical cells.

Thus the concept of a neuroendocrine system of cells is emerging. According to the route by which it is transmitted, the same messenger molecule may function as an endocrine hormone (blood stream conveyance), as a neurotransmitter (axonal conveyance), as a neurohormone (combined axonal and blood stream conveyance), or as a paracrine or autocrine hormone (local conveyance). The effect produced by the signaling molecule then depends on the target cell and the intracellular mechanisms that are activated. For example, a hypothalamic cell may secrete the neurohormone somatostatin (see Chapter 52) into blood, which then bathes the pituitary gland and inhibits the release of a hormone that stimulates growth; a brain cell may transmit somatostatin to a second brain cell so as to alter behavior; and one cell in the pancreas may release somatostatin into the fluid bathing adjacent cells and inhibit their release of insulin (see Chapter 50).

Not only does knowledge of the relationship between neural and endocrine function enhance our understanding of physiology, but it also helps to illuminate the pathogenesis of some endocrine diseases. For example, in certain families, separate neoplasms tend to arise within seemingly disparate neural and endocrine tissue in the same individual; the connection between these coincidental events is becoming clearer. In other instances, a single neoplasm may be associated with clinical manifestations that can be traced to simultaneous production of a peptide hormone and a neurotransmitter molecule. Other diseases, originally thought to represent primary aberration of an endocrine cell, are known or suspected to be the result of abnormal neural regulation of that cell. Finally and perhaps most imporant, analogues are being developed for replacement, substitution, and selective inhibition of neurohormones involved in neuroendocrine disorders.

■ *Types of Hormones*

Hormone molecules fall into three general chemical classes. The first to be discovered—the amines—includes thyroid hormones and catecholamines. Both originate from the amino acid tyrosine and retain the

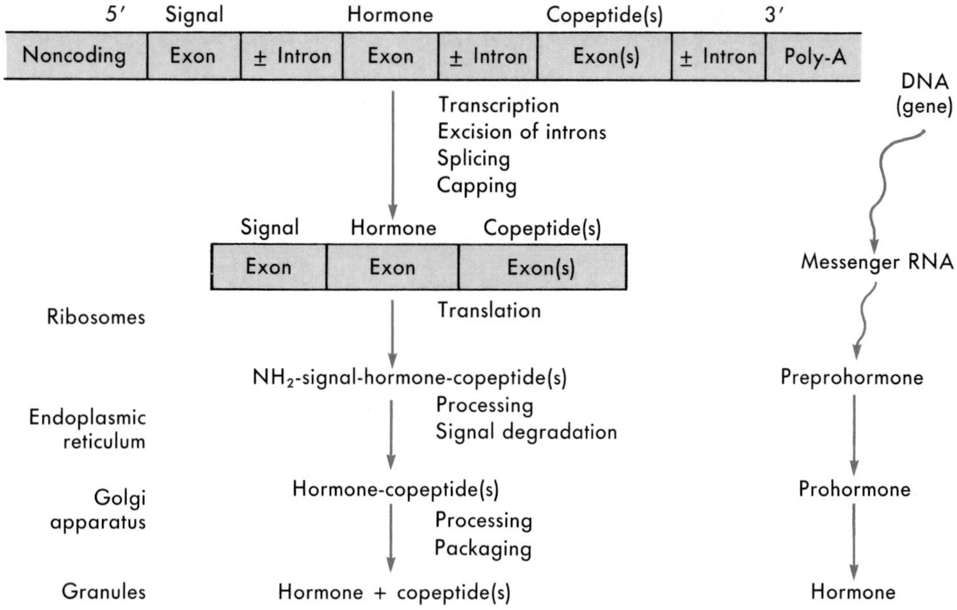

■ Fig. 49-2. Schematic representation of peptide hormone synthesis. In the nucleus the primary gene transcript undergoes excision of introns, splicing of the exons, capping of the 5′ end, and in addition of poly-A at the 3′ end. The resultant mature messenger RNA enters the cytoplasm, where it directs the synthesis of a preprohormone peptide sequence on ribosomes. In this process the N-terminal signal is removed, and the resultant prohormone is transferred vectorially into the endoplasmic reticulum. The prohormone undergoes further processing and packaging in the Golgi apparatus. After final cleavage of the prohormone within the granules, they contain the hormone and copeptides ready for secretion by exocytosis.

aliphatic α-amino group. Introduction of a second hydroxyl group in the ortho- position on the benzene ring is characteristic of the catecholamines, whereas iodination of the benzene ring distinguishes the thyroid hormones. The second group is composed of proteins and peptides. In some instances protein hormones with dissimilar missions nonetheless share identical structural sequences away from the biologically active core, suggesting a common branch point in evolution. In other instances a single progenitor protein gives rise to several hormone offspring of different sizes, with some sharing amino acid sequences and some having overlapping action. The third group to be identified chemically is the steroids, which include adrenal cortical and reproductive gland hormones and the active metabolites of vitamin D. Cholesterol is the common precursor in this class. Modification of side chains, hydroxylation at various sites, and ring aromatization confer individual biological activities on the various steroid hormones.

■ *Hormone Synthesis*

The protein and peptide hormones are synthesized on the rough endoplasmic reticulum in the same manner as other proteins. The appropriate amino acid sequence is dictated by specific messenger RNAs that originate in the nucleus. These have been isolated and characterized for a number of hormones, and the structures of the corresponding DNA molecules have been elucidated. The ability to isolate and clone a single gene from a large "library" and determine its structure has grown very rapidly. Consequently, the structure of messenger RNAs and peptides resulting from expression of the gene has often been deduced just from the DNA nucleotide sequences. As a result, "hormones" have been discovered chemically before any function can be ascribed to them. Furthermore, by use of recombinant DNA technology, bacteria are being directed to synthesize a number of authentic mammalian protein hormones for therapeutic use. In general, a single gene determines the structure and synthesis of a single peptide hormone. However, multiple genes containing the same sequence or only slightly varied sequences directing a single peptide hormone have also been described. These may have physiological and clinical implications. Alternatively, a unique gene may give rise to more than one primary RNA message by inclusion or exclusion of particular exons in the processes of excision and splicing. Thus one gene may direct the synthesis of different products in different cells.

The process of peptide hormone synthesis is illustrated in Fig. 49-2. Translation of the mature RNA message begins with an N-terminal signal peptide sequence. When this is complete, translation temporarily ceases, while the signal peptide attaches the message to the endoplasmic reticulum via "docking proteins." The endoplasmic reticulum apparently contains receptors for these docking proteins. Translation then resumes until the entire encoded peptide sequence, known at this stage as *preprohormone,* is formed. The signal peptide is then cleaved as it directs this product into the cisternal space for transport to the Golgi apparatus. This product, known as a *prohormone,* contains the hormone along with other peptide sequences. Some may function to ensure proper folding of the hormone peptide chain so as to permit formation of intramolecular linkages. Other peptide sequences within the prohormone may be coproducts of gene expression with related or independent functions of their own; they are cosecreted with the hormone. Within the Golgi apparatus, the prohormone is packaged for storage in secretory granules. The latter may also contain the proteolytic enzymes necessary for subsequent conversion of prohormone to hormone or for elimination of copeptides. In addition, many secretory granules of peptide hormones contain a soluble acidic protein of unknown function—*chromogranin.*

The possibility of genetic errors in protein and peptide hormone synthesis is appreciable. Single amino acid substitutions or deletions in active sites can alter the degree or specificity of biological action. If such errors occur in accessory peptides of preprohormones or prohormones, they may prevent normal processing to the active hormone. The difficulty in obtaining sufficient human hormone from any one individual for structural analysis has previously prevented an appreciation of the true frequency of such errors in humans, as compared, for example, with that of abnormal hemoglobins. However, a few such instances have been demonstrated, and the number undoubtedly will grow.

The amine and steroid hormones are synthesized from tyrosine and cholesterol, respectively, through a sequential series of discrete enzymatic reactions. The intermediate products of the pathway may have hormonal properties of their own. This multiplicity of distinctive steps leads to the prediction that a variety of defects in hormone synthesis (and their clinical consequences) could arise from single gene–single enzyme mutations and deletions or from selective drug-induced enzyme inhibition. In fact, numerous examples of congenital enzyme deficiencies in adrenal and gonadal steroid hormone and in thyroid hormone biosynthesis do occur. The resultant clinical states may reflect both the deficiency of product hormone and the accumulation of hormone precursors.

■ *Hormone Release*

Catecholamine and protein hormones share the property of being stored in secretory granules. Hormone release then is accomplished by the process of *exocytosis.* After migration through the cytoplasm and suitable positioning, the membrane of the secretory granule fuses with the immediately adjacent plasma cell membrane. The combined membrane material is lysed, and the contents of the granule are discharged into the extracellular space, leaving an empty core. The membrane material itself may be reprocessed. The movement of secretory

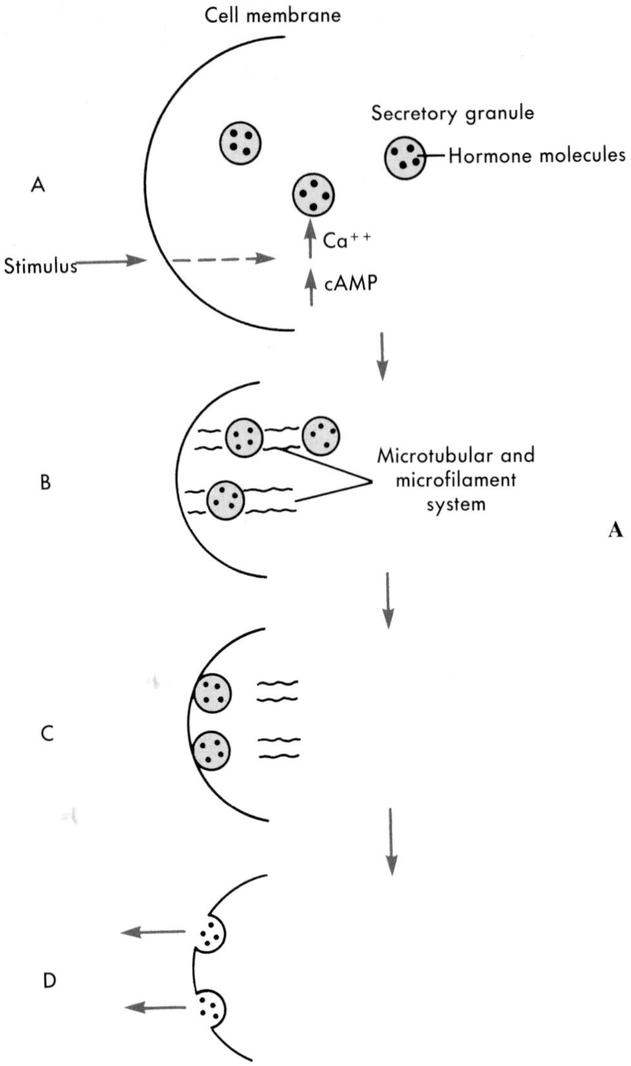

A

■ **Fig. 49-3. A,** Secretion of peptide hormones via exocytosis is initiated *(A)* by application of a stimulus that raises intracellular cAMP levels and also usually raises cytosolic Ca^{++}. The secretory granules are lined up and translocated to the plasma membrane *(B)* via activation of a microtubular and microfilament system. The membrane of the secretory granule fuses with that of the cell *(C).* The common membrane is lysed *(D),* releasing the hormone into the interstitial space. **B,** Insulin secretory granules in a β-cell being stimulated by glucose. The arrow indicates a granule undergoing exocytosis. (From Lacy, P.E.: Beta cell secretion—from the standpoint of a pathobiologist, Diabetes **19:**895, 1970. Reproduced with permission from the American Diabetes Association, Inc.)

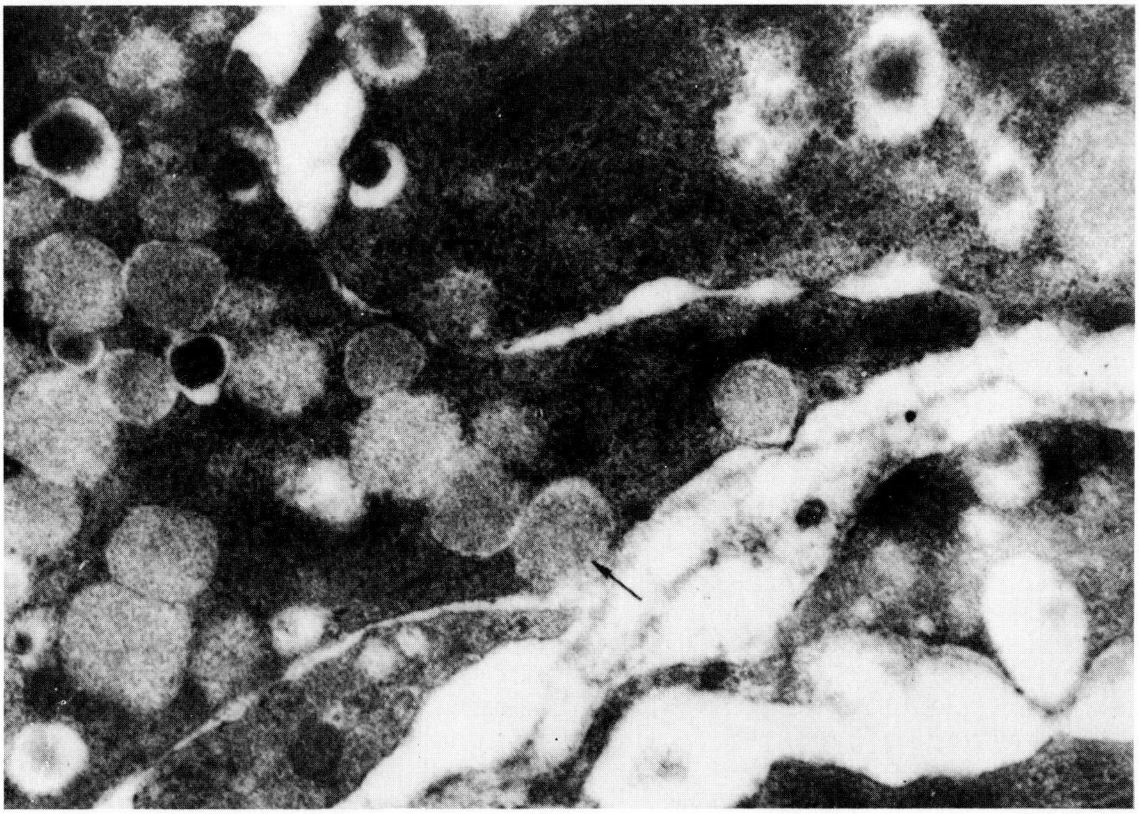

B

■ **Fig. 49-3, cont'd.** For legend see opposite page.

granules to the cell membrane probably involves the participation of microtubular elements in the cytoplasm. Their activation often is mediated by prior generation of cyclic AMP (cAMP) and/or by uptake of calcium ions from extracellular fluid (Fig. 49-3). When secretory granules contain accessory peptides from the prohormone together with the enzymes that catalyze its cleavage, these materials are released simultaneously with the hormone. In the case of thyroid and steroid hormones storage does not take place in discrete granules, although the hormones may be compartmentalized in the cell. Once the hormones have appeared in free form within the cytoplasm, they apparently leave the cell by simple transfer through the plasma membrane.

These modes of hormone synthesis and release are essentially unicellular. However, more complicated patterns of hormone production also are encountered. Two adjacent cell types in a single gland may interact so that hormone A from cell A is modified in cell B to produce hormone B with an entirely different spectrum of biological effects. In this manner, for example, estrogens are produced from androgens in the gonads. In a further extension of this principle, peripheral tissues that are ordinarily considered to be nonendocrine may carry out similar conversions. A second mode of hormone production involves modification of a precursor molecule of low activity to one of higher activity by successive steps in several tissues. For example, a sterol synthesized in the skin requires actions by the liver and kidney to produce the most potent vitamin D hormone. Lastly, production of peptide hormones can even occur in the circulation itself from a protein substrate provided by one organ and then acted on by enzymes originating in other organs. A prototype for this pattern is the synthesis of angiotensin—a peptide hormone—from a globulin precursor released by the liver and acted on sequentially by enzymes from the kidney and the lung.

■ *Regulation of Hormone Secretion*

A number of general mechanisms that govern the secretion of hormones are the following:

Feedback control
- Hormone-hormone
- Substrate-hormone
- Mineral-hormone

Neural control
- Adrenergic
- Cholinergic
- Dopaminergic
- Serotoninergic
- Endorphinergic-enkephalinergic
- Gabergic

Chronotropic control
- Diurnal rhythm
- Sleep-wake cycle
- Menstrual rhythm
- Seasonal rhythm
- Developmental rhythm

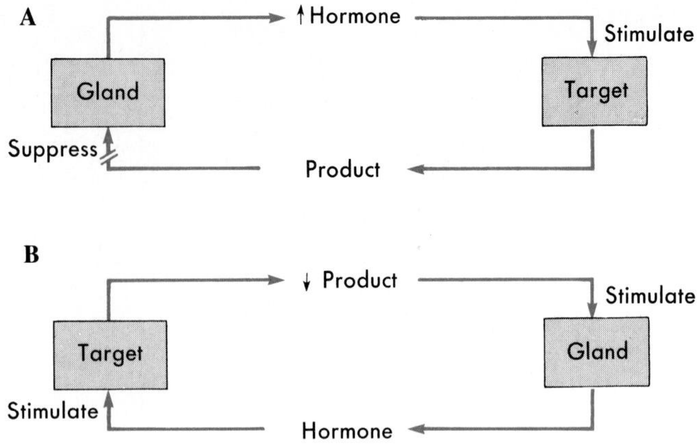

■ **Fig. 49-4.** Negative feedback principle. **A,** A primary increase in hormone secretion stimulates a greater output of product from the target cell. The product then feeds back on the gland to suppress further hormone secretion. In this fashion hormone excess is limited or prevented. **B,** A primary decrease in output of product from the target cell stimulates the gland to secrete hormone. The hormone then stimulates a greater output of product from the target cell. In this fashion the product deficiency is limited or corrected.

The feedback principle is universally operative. *Negative feedback* is most common and acts to *limit* the excursions in output of each partner in the pair (Fig. 49-4). In the simplest instance hormone A, which stimulates secretion of hormone B, then in turn will be inhibited by an excess of hormone B. This straightforward mechanism dominates the relationship between hormones of the pituitary gland and its target glands. Secretion of a hormone that either accelerates the production or retards the utilization of a particular substrate, thus increasing its concentration in plasma, then will be inhibited by the resultant circulating excess of that substrate; conversely, secretion of the hormone will be stimulated by a circulating deficit of that substrate. On the other hand, the secretion of a hormone that impedes the production or accelerates the utilization of a particular substrate, thus decreasing its plasma concentration, will be stimulated by a circulating excess but inhibited by a circulating deficit of the substrate.

Positive feedback, which is less common, acts to *amplify* the initial biological effect of the hormone. Thus hormone A, which stimulates secretion of hormone B, in turn may be initially stimulated to greater secretion rates by hormone B, but only through a limited dose response range. Once sufficient biological momentum for secretion of hormone B has been obtained, other influences, including negative feedback, will reduce the response of hormone A to fit the final biological purpose.

Neural control acts to evoke or suppress hormone secretion in response to both external and internal stimuli. These may arise from visual, auditory, olfactory, gustatory, tactile, or pressure sensations and may be perceived consciously or unconsciously. Pain, emotion, sexual excitement, fright, injury, and stress all can modulate hormone secretion through neural mechanisms. Simple examples include the release of oxytocin, which fills the milk ducts in response to the stimulus of suckling, or the release of aldosterone, which augments the volume of the circulatory system in response to upright posture. Certain patterns of hormone secretion appear to be dictated by rhythms that may be genetically encoded or acquired. Many hormones are secreted in distinct, independent diurnal rhythms. These may be tied to sleep-wake cycles or to light-dark cycles. Seasonal variation in hormone secretion also occurs, which may reflect the influence of temperature, sunlight, or tides. Some of these rhythms appear atavistic in humans, having probably served to fulfill biological needs of evolutionary ancestors. Perhaps the most intriguing of all are those patterns of hormone secretion which coincide with and are unique to developmental stages, such as the onset of puberty. A diverse and still growing number of neurotransmitter molecules carries these signals to the endocrine cells.

■ *Mechanisms of Hormone Action*

Although it is impossible to reconcile the great multiplicity of hormone effects with a single principle of action, several general mechanisms have been identified. Because these relatively few mechanisms serve a large number of hormones, the specificity of any particular hormone–target cell interaction must rest on two factors: the uniqueness with which the hormone is recognized by the target cell and the uniqueness of the intracellular pathways with which a target cell can respond to hormonal stimulation (Fig. 49-5).

■ *Receptors*

Recognition of a hormone can take place in several sites. Receptors exist in the plasma cell membrane for catecholamine, peptide, and protein hormones, in the cytoplasm and nucleus for steroid hormones, and largely in the nucleus for thyroid hormones. Receptors are large protein molecules; plasma membrane receptors, in addition, contain carbohydrate and occasionally phospholipid moieties. Recent evidence suggests that within a plasma membrane receptor the amino acid se-

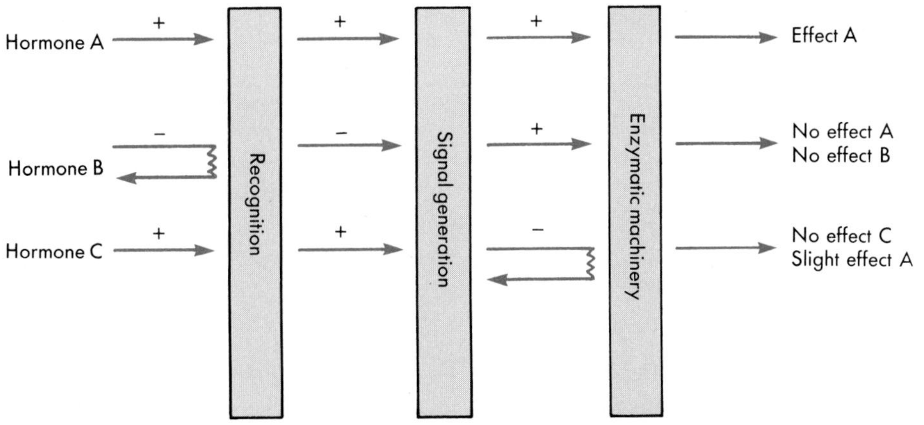

■ **Fig. 49-5.** Hormone-cell interaction. Hormone A is recognized by this cell through binding with its specific receptor. An intracellular signal is generated that stimulates the appropriate enzymatic machinery, and effect A is produced. Hormone B is not recognized because the cell lacks a receptor for it. Thus hormone B cannot produce effect A in this cell even though it might operate through an identical enzymatic machinery in its own target cell. Hormone C may be slightly recognized by this cell through individual overlap with the receptor to hormone A. Although a weak intracellular signal may be generated, no effect of C results because the cell lacks the appropriate enzymatic machinery. To a minor extent, however, hormone C may produce effect A.

quence which actually binds the peptide hormone to that receptor may be an "internal image" of a complementary amino acid sequence within the hormone molecule. It has been reported in several cases that construction of a messenger RNA complementary to the messenger RNA of a peptide hormone yields a new codon which directs the synthesis of a new peptide that can bind the hormone. This artificial binding peptide is similar chemically to binding regions on the native receptor for that hormone. Thus, via the genetic code, expression of complementary DNA molecules may be the basis for a general system of molecular recognition.

A given hormone may act through different receptor molecules in different target cells, however. It is also not yet certain whether multiple actions of a given hormone in the same target cell are always initiated by binding to a sole receptor. There are generally 2000 to 100,000 receptor molecules per cell. In many instances this number guarantees that receptor availability will not be rate limiting for hormone action.

The association of a hormone with its receptor is a reversible reaction that appears to obey the following molecular chemical kinetics:

$$[H] + [R] \rightleftharpoons [HR] \tag{1}$$

$$K_{assoc} = \frac{[HR]}{[H]\,[R]} \tag{2}$$

$$\frac{[HR]}{[H]} = K_{assoc} \times [R] \tag{3}$$

where

$[H]$ = free hormone in solution
$[R]$ = unoccupied receptor
$[HR]$ = bound hormone = occupied receptor
R_o = initial receptor capacity = $[R] + [HR]$
K_{assoc} = affinity constant

If a fixed number of cells (equal to a constant amount of receptor) is incubated in vitro with increasing concentrations of hormone, the amount of bound hormone increases until receptor occupancy reaches 100%. At this point the number of bound hormone molecules equals the total number of originally available receptor molecules, i.e., the receptor capacity (R_o). At the same time the ratio of bound hormone to free hormone has progressively decreased and approaches zero as the free hormone concentration approaches infinity and the receptor occupancy approaches 100%. That is, as

$$[H] \rightarrow \text{Infinity}$$
$$[HR] \rightarrow R_o \tag{4}$$
$$\frac{[HR]}{[H]} \rightarrow 0$$

The data obtained by incubating cell receptors with increasing amounts of hormone can be plotted in a meaningful way by simple substitution in equation 3. Since

$$[R] = R_o - [HR]$$

$$\frac{[HR]}{[H]} = K_{assoc} \times (R_o - [HR]) \quad (5)$$

$$\frac{[HR]}{[H]} = -K_{assoc}[HR] + K_{assoc}R_o \quad (6)$$

$$\frac{\text{Bound hormone}}{\text{Free hormone}} = -K_{assoc} \times \text{Bound hormone} + K_{assoc} \times \text{Receptor capacity}$$

Plotting the ratio of bound hormone to free hormone as a function of bound hormone is called a *Scatchard plot*. This theoretically yields a straight line (Fig. 49-6, *A*). The slope of the line equals the negative of the association constant (K_{assoc}), and the x intercept equals R_o (the receptor capacity).

In practice many Scatchard plots of hormone binding to receptor yield exponential curves (Fig. 49-6, *B*). One interpretation of this phenomenon is that the cell possesses two classes of receptors, R_1 and R_2. R_1 has a higher affinity than R_2 (K_{assoc1} is greater than K_{assoc2}) but a lower capacity (R_{o1} is less than R_{o2}). Often the biological action of a hormone is quantitatively accounted for by interaction with the apparent high affinity, low-capacity receptor, requiring occupancy of as little as 5% to 10% of the total receptor capacity of the cell. The function of the remaining "spare receptors" is not known for certain. However, they can act to increase the concentration of receptor-bound hormone (HR) when either hormone concentration or receptor affinity is low. A second interpretation of nonlinear Scatchard plots is that the affinity of unoccupied receptor molecules for hormone is decreased by the presence of adjacent occupied receptor molecules. This phenomenon is termed *negative cooperativity*. It has the effect of moderating the increase in receptor-bound hormone and thus moderating the increase in hormone action should hormone concentration rise too precipitously. Although Scatchard plots provide a conceptual framework for visualizing the kinetics of hormone-receptor interaction, the accuracy of the derived receptor capacities and association constants can be questioned. Therefore they are best used to assess the effects of physiological manipulations in a defined system rather than to compare values obtained in one system with values obtained in another.

Regulation of the hormone receptor provides another mechanism for regulating hormone action. Rearrangement of equation 5 to provide an expression for HR yields

$$[HR] = R_o \times \frac{K_{assoc}[H]}{K_{assoc}[H] + 1} \quad (7)$$

Thus it is seen that HR is directly proportional to R_o, the initial receptor number. An increase in receptor number (R_o) raises the maximal [HR] obtainable at saturating concentrations of hormone. This would raise the *maximal responsiveness* of the cell for those hormone effects in which receptor binding, rather than a later intracellular step in hormone action, is rate limiting. This may be seen where there are no spare receptors, as is generally the case for steroid and thyroid hormones. At submaximal hormone concentrations the in-

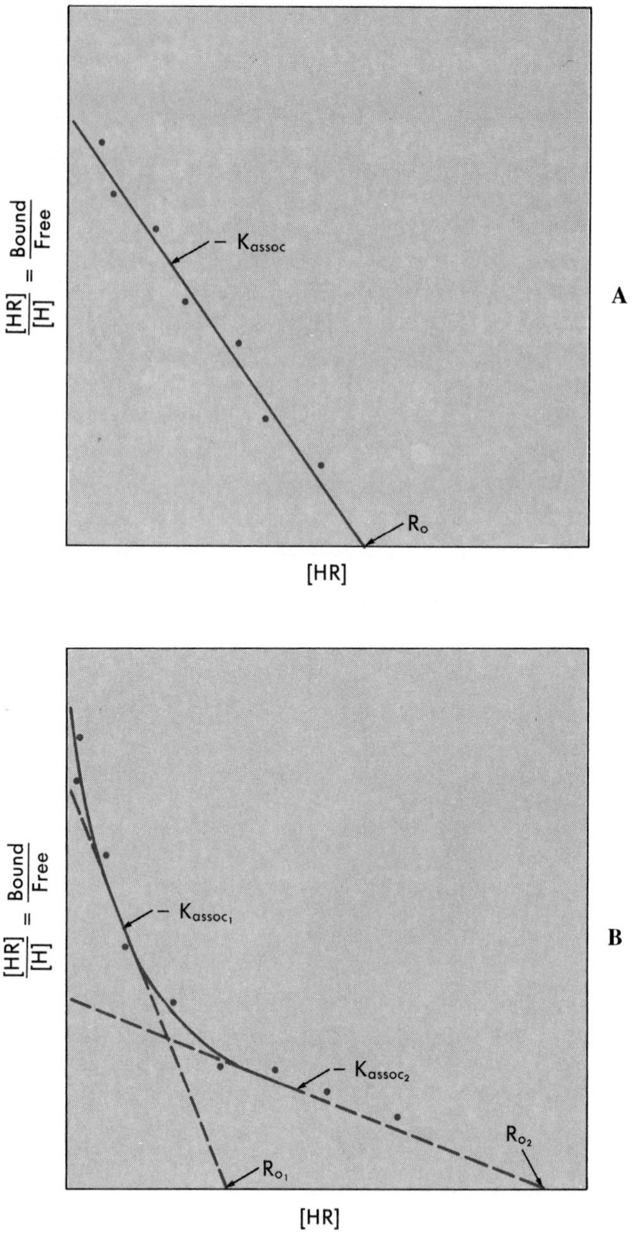

Fig. 49-6. Scatchard plot. **A,** Linear plot results when the hormone reacts with a single receptor class and no cooperativity is present. **B,** An exponential plot results when the hormone reacts with multiple classes of receptors in the same cell or when the reaction with a single class of receptor is influenced by cooperativity. The negative of the association constant, K_{assoc}, equals the slope of the line. The receptor number, R_o, equals the intercept with the x axis.

crease in [HR] produced by an increase in R_o would enhance the *sensitivity* of the cell. This is commonly seen with peptide hormones.

It is interesting that receptor capacity often is regulated by its own hormone. In the usual instance the regulation is inverse, that is, a sustained excess of hormone decreases the number of its receptors per cell, which is *down regulation*. This acts to moderate the effect of chronic exposure to excess hormone. However, in some instances, especially at low concentrations of hormone, the *relationship* is direct, i.e., the hormone appears to recruit its own receptors—*up regulation*—after a period of exposure. This amplifies the cell's response to the hormone. An increase in receptor affinity (K_{assoc}) also will increase [HR] and the sensitivity of the cell to hormone stimulation. Receptor affinity can be altered by factors such as pH, osmolality, ion concentrations, and substrate levels. In a growing number of diseases abnormalities of the hormone receptor are involved in pathogenesis. Therefore interest in receptor structure, biosynthesis, degradation, function, and regulation is currently intense.

Intracellular Pathways of Hormone Action

The binding of a hormone to its specific receptor is followed by the generation of intracellular signals or messengers, which couple hormone recognition with the specific enzymatic machinery necessary for hormone action. For a large number of diverse hormones, two universal mechanisms for coupling operate. One involves second messengers that change enzyme activities rapidly, and the other involves transcriptional or translational effects that change enzyme concentrations more slowly. For other hormones, however, the link between receptor recognition and unique intracellular action remains speculative or completely unknown.

Second Messengers

Peptide and catecholamine hormones bind to plasma membrane receptors. These receptors are large complexes, often composed of subunits, and of molecular weight 50,000 to 200,000. They all appear to be glycoproteins and in several instances, their structure resembles that of immunoglobulins (see Fig. 50-16, for example). They tend to be concentrated in cellular microvilli. Hormone binding to these receptors may change their conformation as well as their distribution within the plasma membrane. This leads to clustering of hormone-receptor complexes. After receptor activation is completed, internalization of the complexes occurs at these sites by endocytosis, sometimes via coated pits or vesicles. Within the cell, lysozomal degradation of the complex occurs, with either destruction of the individual hormone and receptor molecules or recycling of the receptor molecules back into the plasma membrane. It remains unclear whether the initially internalized hormone-receptor complex can also mediate certain intracellular actions of the hormone before its disruption.

After hormone binding to a plasma membrane receptor, at least three different systems can couple hormone recognition with specific hormone action. These coupling systems are the adenylate cyclase–cAMP system, the calcium–calmodulin system, and the membrane phospholipid system. The first to be described was the adenylate cyclase–cAMP system. Many peptide hormones as well as catecholamines stimulate an immediate rise in intracellular cAMP. The latter acts as a second messenger that activates numerous enzymes critical to the actions of these hormones. cAMP is generated from ATP by the action of adenylate cyclase, a membrane-bound enzyme. This enzyme consists of a catalytic and a regulatory subunit.

As shown in Fig. 49-7, formation of the hormone-receptor complex on the outer surface of the plasma membrane changes the regulatory subunit of adenylate cyclase by causing the latter to complex with guanosine triphosphate (GTP). This complex, in turn, activates the catalytic subunit of adenylate cyclase on the inner surface of the plasma membrane, where it can act on Mg-ATP to form cAMP. In this process, GTP may be simultaneously hydrolyzed to guanosine diphosphate (GDP); the GDP regulatory subunit complex then becomes less able to activate the catalytic subunit. In addition, the GDP regulatory subunit complex may decrease the affinity of the receptor for the hormone, providing still another mechanism for limiting the effect of excess hormone. The regulatory subunit is itself composed of two parts; the smaller binds magnesium and interacts with the hormone receptor, whereas the larger binds GTP and interacts with the catalytic subunit. A separate and distinct regulatory subunit, which inhibits rather than stimulates adenylate cyclase activity, has also been found. This accounts for the inhibitory influence of opiate peptides and some bioamine neurotransmitters on this key enzyme. The exact structural and geographical relationships among the receptor and the subunits of adenylate cyclase is not yet clear, but it is evident that all three have mobility within the plasma membrane.

An increase in cAMP stimulates the activation of *protein kinase A* (Fig. 49-7). This, in turn, activates a number of enzymes in numerous metabolic pathways by phosphorylating their kinases. Alternatively, cAMP-stimulated phosphorylation may deactivate other enzymes. Thus after hormone binding to receptor, this system generates a cascade of effects that ultimately changes the flux of metabolites in the cell. In the end either the storage or the release of an important metabolite may be facilitated. Activation or deactivation of

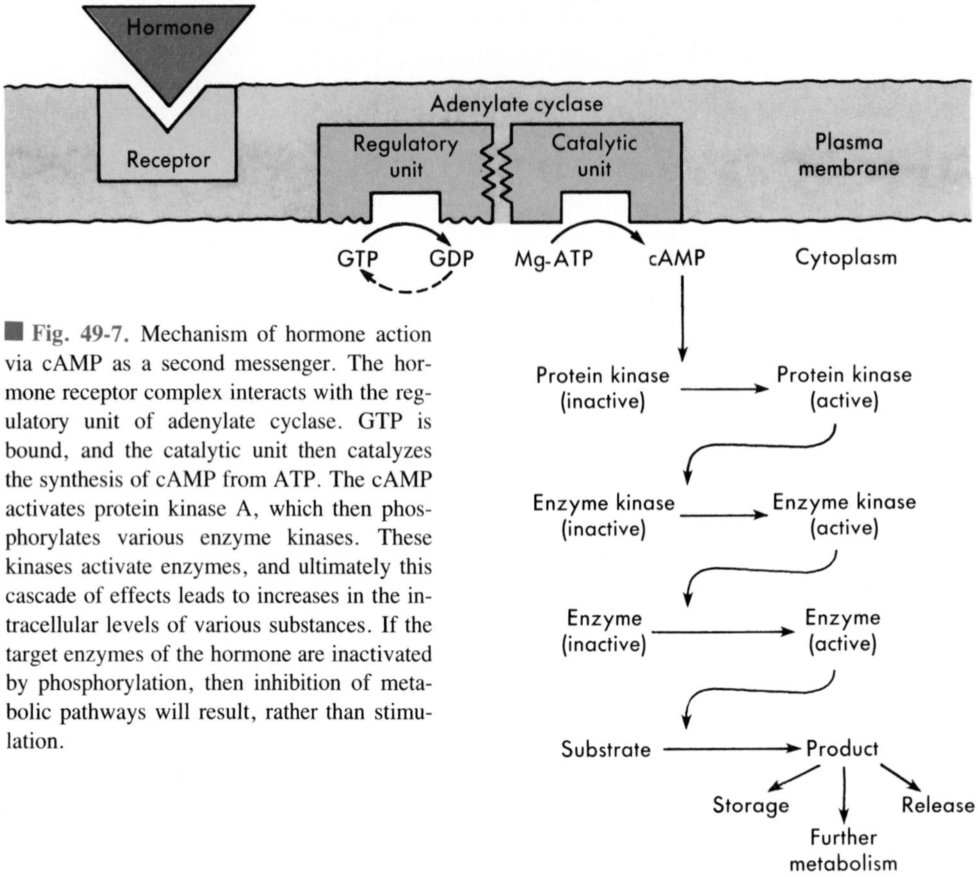

■ **Fig. 49-7.** Mechanism of hormone action via cAMP as a second messenger. The hormone receptor complex interacts with the regulatory unit of adenylate cyclase. GTP is bound, and the catalytic unit then catalyzes the synthesis of cAMP from ATP. The cAMP activates protein kinase A, which then phosphorylates various enzyme kinases. These kinases activate enzymes, and ultimately this cascade of effects leads to increases in the intracellular levels of various substances. If the target enzymes of the hormone are inactivated by phosphorylation, then inhibition of metabolic pathways will result, rather than stimulation.

reciprocal pathways by the same hormone can augment the result, for example, by simultaneously inhibiting the release pathway while stimulating the storage pathway. Adenylate cyclase and cAMP are ubiquitously distributed. Therefore the specificity of enzyme responses to a hormone also may depend on the compartmentalization of cAMP increases within the cell or on the proximity of the activated adenylate cyclase to target enzyme(s).

A second system for transduction of the hormone signal is the *calcium-calmodulin system* (Fig. 49-8). As a result of hormone occupancy of its receptor, channels are opened in the plasma membrane through which extracellular calcium can enter the cytoplasm. A specific plasma membrane protein has been identified as essential to creating the calcium channel, but its exact structure and mechanism of action are still unknown. Calcium may also be mobilized from intracellular reservoirs in the mitochondria and endoplasmic reticulum. The increased cytosolic calcium combines with its ubiquitous and specific binding protein, calmodulin, in various proportions. The different calcium-calmodulin complexes amplify the activities of a variety of calcium-dependent enzymes. In other instances, enzyme activities may be diminished by this signal. In this fashion, metabolite levels within the cell are ultimately altered.

A third system for transduction of the initial hormone-receptor signal is through intermediates generated from *plasma membrane phospholipids* (Fig. 49-9). A key phospholipid is *phosphatidylinositol*, which undergoes phosphorylation to phosphatidylinositol 4,5 biphosphate. The latter compound is split by the action of membrane-bound phospholipase A_2 and C, initiated by hormone occupancy of its receptor. One product of this reaction is diacylglycerol, of which arachidonic acid forms a major proportion of the fatty acid component. The other product is *inositol triphosphate*. The diacylglycerols are potent activators of a newly discovered *protein kinase C*. The latter is active only when incorporated within the plasma membrane and in contact with other phospholipids such as phosphatidylserine. Protein kinase C is a calcium-dependent enzyme, and the diacylglycerol markedly increases its affinity for calcium. The inositol triphosphate, released simultaneously with diacylglycerol from the plasma membrane, is believed to trigger mobilization of calcium from intracellular reservoirs. Therefore, the combined products of phosphatidylinositol biphosphate breakdown greatly amplify protein kinase C activity. In turn, protein kinase C, in further synergism with calcium, activates a number of enzymes involved in hormonal responses. Finally, the arachidonic acid derived from hydrolysis of the diacylglycerol serves as a substrate for

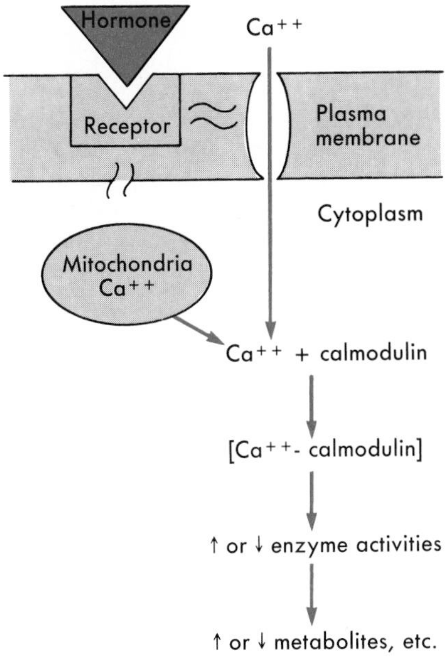

■ **Fig. 49-8.** Mechanism of hormone action via calcium as a second messenger. The binding of hormone to receptor leads to an opening of calcium channels in the cell membrane. Influx of calcium is followed by its binding to calmodulin, which produces a complex that can activate or deactivate various enzymes and metabolic pathways. The intracellular pool of calcium can also be increased after hormonal stimulation by release of calcium from mitochondria and from the endoplasmic reticulum.

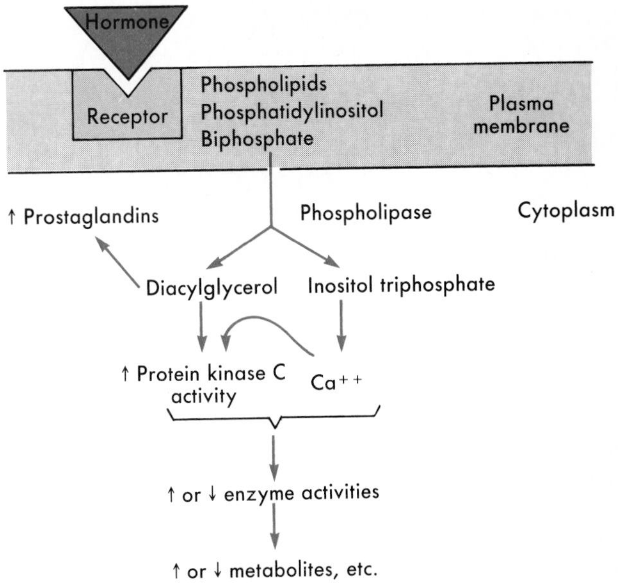

■ **Fig. 49-9.** Mechanism of hormone action via membrane phospholipids. The binding of hormone to receptor causes release of phosphatidylinositol biphosphate from the membrane. This is converted through the action of phospholipases to its diacylglycerol and inositol triphosphate components. Diacylglycerol activates protein kinase C by enhancing its binding to calcium. The latter is mobilized from intracellular stores by inositol triphosphate. Protein kinase C then activates or deactivates target enzymes via phosphorylation. As a subsidiary effect, the diacylglycerol yields arachidonic acid for synthesis of prostaglandins, which also modulate intracellular processes.

rapid synthesis of prostaglandins, which are themselves capable of mediating or modulating some hormonal responses.

As complex as each of these three systems seems individually, it must be appreciated that a single hormone may operate through one, two, or all three of the second messenger systems simultaneously or at different dose levels in sequence. Each messenger may subserve a different function of the hormone. Moreover, the three systems can interact with each other in important ways. For example, calcium-calmodulin can stimulate not only adenylate cyclase activity itself, but also the activity of phosphodiesterase, the enzyme that hydrolyzes cAMP. Thus cAMP levels, initially increased by a rapid hormone activation of adenylate cyclase, may be subsequently dampened by a later hormone response involving phosphodiesterase activity. Likewise, in some situations, cAMP and protein kinase A can inhibit the activity of the phospholipid messenger system by decreasing the generation of diacylglycerols. In this situation, an early stimulation of cell processes by hormone action through protein kinase C may be later dampened by hormone action through protein kinase A. Thus, dual regulation can result from receptor-receptor interactions involving the adenylate cyclase and phospholipid pathways.

■ *Transcriptional or Translational Effects*

Steroid and thyroid hormones generate intracellular signals primarily by modulating transcription in specific areas of chromatin (Fig. 49-10). In this second major mechanism the steroid hormones combine with their receptors in the cytoplasm, and the complexes then enter the nucleus, whereas thyroid hormones enter the nucleus in the free state and then combine with their nuclear receptors. In both cases the hormone receptor complex interacts with target DNA molecules in the chromatin.

Although details are lacking for thyroid hormones, some understanding of the mechanism for steroid hormones is emerging. Steroid hormone receptors are large proteins of 50,000 to 200,000 molecular weight. They appear to have three domains. The C-terminus domain binds the steroid hormone. The middle domain binds the receptor to target DNA molecules. The remaining N-terminus domain has no presently assignable function. The receptors exist in inactive form in the cytoplasm. Following binding of the steroid hormone, the complex undergoes "activation," a step that involves a reduction in molecular size, a change in conformation, and probably phosphorylation of the receptor protein.

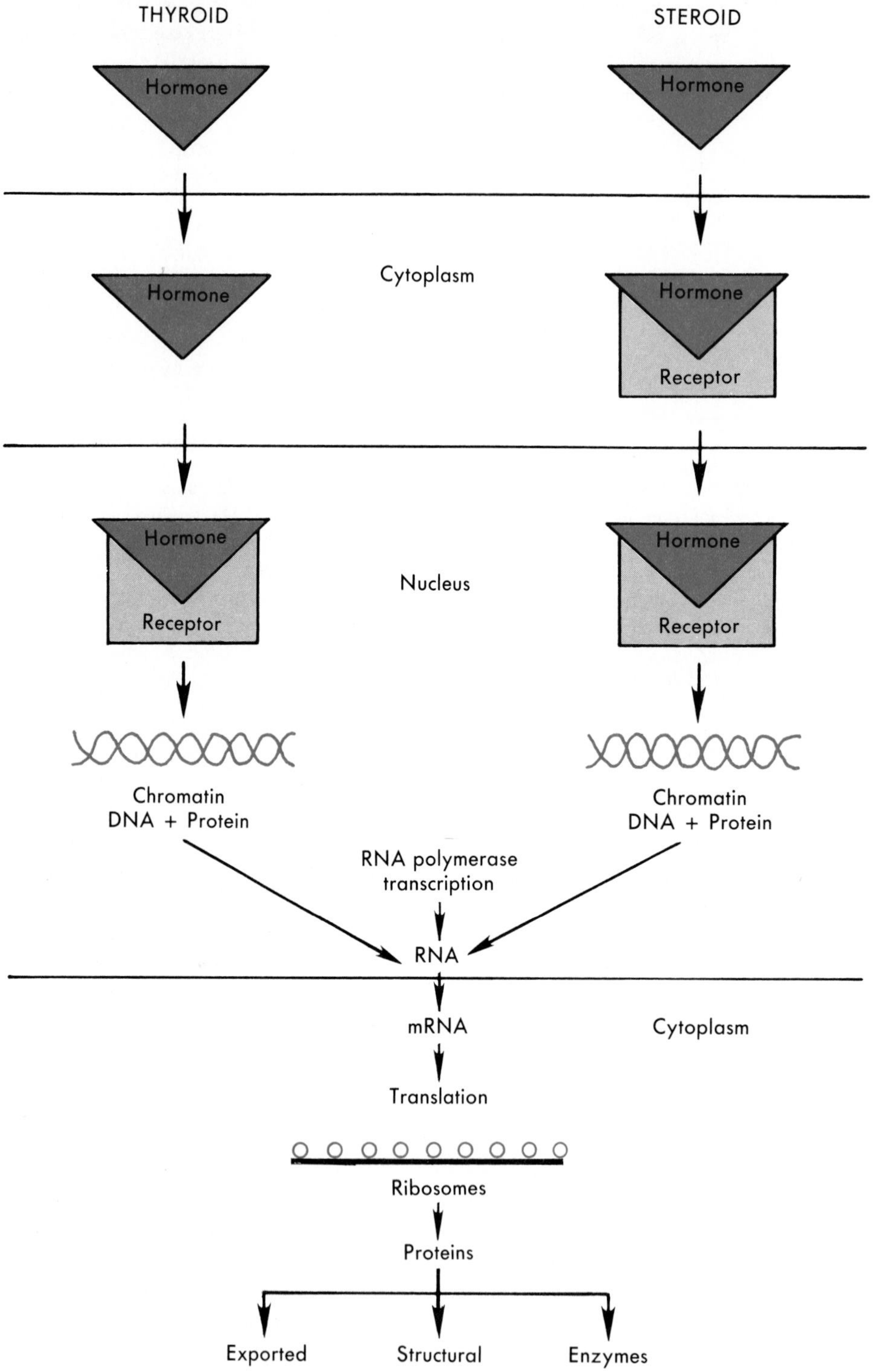

■ **Fig. 49-10.** The schema for the mechanism of thyroid hormone *(left)* and steroid hormone *(right)* actions. The hormone receptor complexes undergo some form of activation or transformation prior to effectuating changes in DNA conformation and generation of mRNAs. Either induction or suppression of the synthesis of specific proteins can result.

Binding of the hormone to its domain in the receptor in some manner frees the middle domain for interaction with DNA. The activated hormone-receptor complex is translocated into the nucleus. Within the nucleus, *acceptor* proteins, bound to the nuclear matrix, combine with the hormone receptor protein. At these sites, chromatin containing target DNA molecules are also fixed. The combination of hormone-receptor and protein-acceptor interacts with various promoter elements of the target DNA molecule upstream from its capsite at the 5′ end. Through this interaction, nucleotide sequences in promoter regions of the gene initiate transcription at the start site of the specific gene message by RNA polymerase II. The resultant "immature" RNA then undergoes maturation to messenger RNA by capping, excision of untranslated nucleotide sequences, and splicing. Translation of the RNA message in the cytoplasm results in synthesis of specific target proteins of the hormone. This mechanism of steroid hormone induction probably includes enhancement of the activity of positive regulatory elements in the gene as well as relief of repression exerted by negative regulatory elements in the gene. Since the gene and acceptor protein are present in all cells, the limitation of steroid hormone action to specific cells depends primarily on the presence of its receptor. However, within the nucleus of non–target cells, other proteins may prevent steroid hormone action by masking the acceptor protein. Finally, very early in development, steroid hormone–responsive genes may be "closed" in non–target cells by processes such as methylation.

Proteins whose synthesis is regulated by thyroid and steroid hormones may be enzymes, structural proteins, receptor proteins, or proteins that are exported by the cell. By this mechanism, enzymes are either induced or repressed, rather than activated or inactivated, as they generally are by the cAMP, calcium-calmodulin, and phospholipid messenger systems. A metabolic pathway utilizing a hormone-induced or repressed enzyme can either be accelerated or retarded. Similarly, other protein products may be either increased or decreased as a result of hormonal regulation of transcription or translation. The rate and magnitude of this type of hormone action may be limited by specific factors, such as the number of hormone receptor molecules or the number of acceptor protein molecules per cell; or by nonspecific factors, such as the concentration of RNA polymerase II, of the enzymes of protein synthesis or processing, of transfer RNA, and of amino acid substrates.

The plasma membrane mechanisms readily account for the frequently rapid (within minutes) effects produced by catecholamine, peptide, and protein hormones. The transcription mechanism logically explains the usually slower (hours to days) effects generally elicited by steroid and thyroid hormones. However, definite overlapping exists. Peptide hormones also cause effects that must invoke actions on protein synthesis at either a transcriptional or a translational level because they are blocked by selective inhibitors such as actinomycin and puromycin. Furthermore, definite increases in enzyme mass, rather than simply in enzyme activity, have been documented after exposure to protein hormones such as insulin. More recently, increases in messenger RNA levels due to peptide hormone action have been observed. Whether these effects are caused by internalization of the hormone receptor complex or by other actions of the second messengers remains to be elucidated. Conversely, steroid and thyroid hormones can produce rapid effects on membranes that cannot be readily accounted for by stimulation of transcriptional or translational events. Finally, some hormone actions cannot be explained by any of the mechanisms already detailed. These actions may involve activation of guanylate cyclase to alter cyclic GMP levels, phosphorylation of specific membrane proteins, or alterations in cellular and organelle permeability to ions other than calcium.

Responsivity to Hormones

The final outcome of the interaction of a hormone with its target cell depends on a number of factors. These include hormone concentration, receptor number, duration of exposure, intervals between consecutive exposures, intracellular conditions such as concentrations of rate-limiting enzymes, cofactors or substrates, and the concurrent effects of antagonistic or synergistic hormones. Hormonal effects are not "all or none" phenomena. The dose-response curve for the action of a hormone is generally complex and often exhibits a sigmoidal shape (Fig. 49-11, *A*). An intrinsic basal level of activity may be observed independent of added hormone and long after any previous exposure. A certain minimal *threshold* concentration of hormone then is required to elicit a measurable response. The effect that is obtained at saturating doses of hormone defines the *maximal responsiveness* of the target cell. The concentration of hormone required to elicit a half-maximal response is an index of the *sensitivity* of the target cell. Alterations in the dose-response curve in vivo can take two general forms (Fig. 49-11, *B*). (1) A decrease in maximal responsiveness could be caused by a decrease in the number of functional target cells, in the total number of receptors per cell, in the concentration of an enzyme being activated by the hormone, or in the concentration of a precursor essential to the final product of hormone action; it also could be caused by an increase in the concentration of a noncompetitive inhibitor. (2) A decrease in hormone sensitivity could be caused by a decrease in the number or affinity of hormone receptors, alterations in the concentration of modulating cofactors, an increase in the rate of hormone degradation, or increases in antagonistic hormones.

Fig. 49-11. A, The general shape of a hormone dose-response curve. Sensitivity is most often expressed as the concentration of the hormone that produces a half-maximal response. **B,** Alterations in the dose-response curve can take the form of a change in maximal responsiveness *(left panel),* a change in sensitivity *(right panel),* or both.

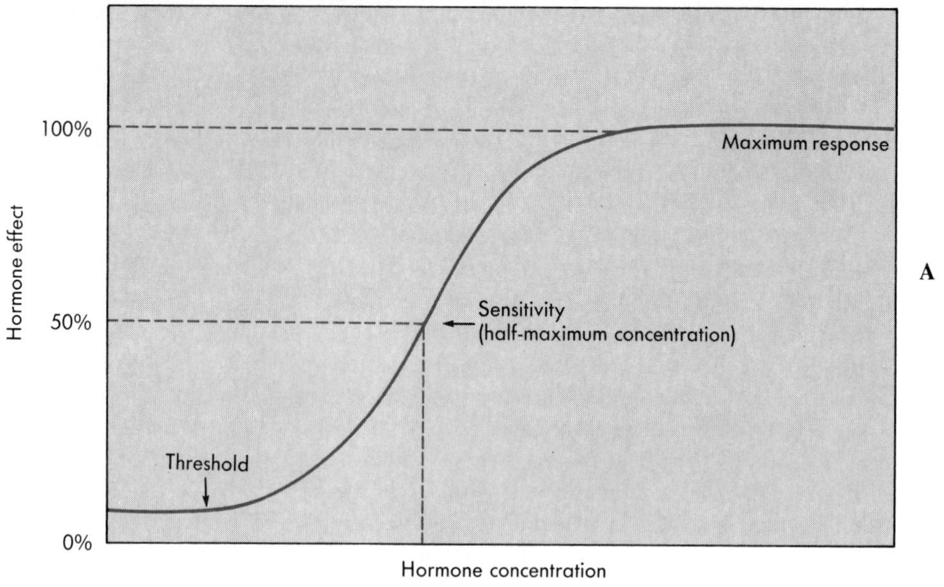

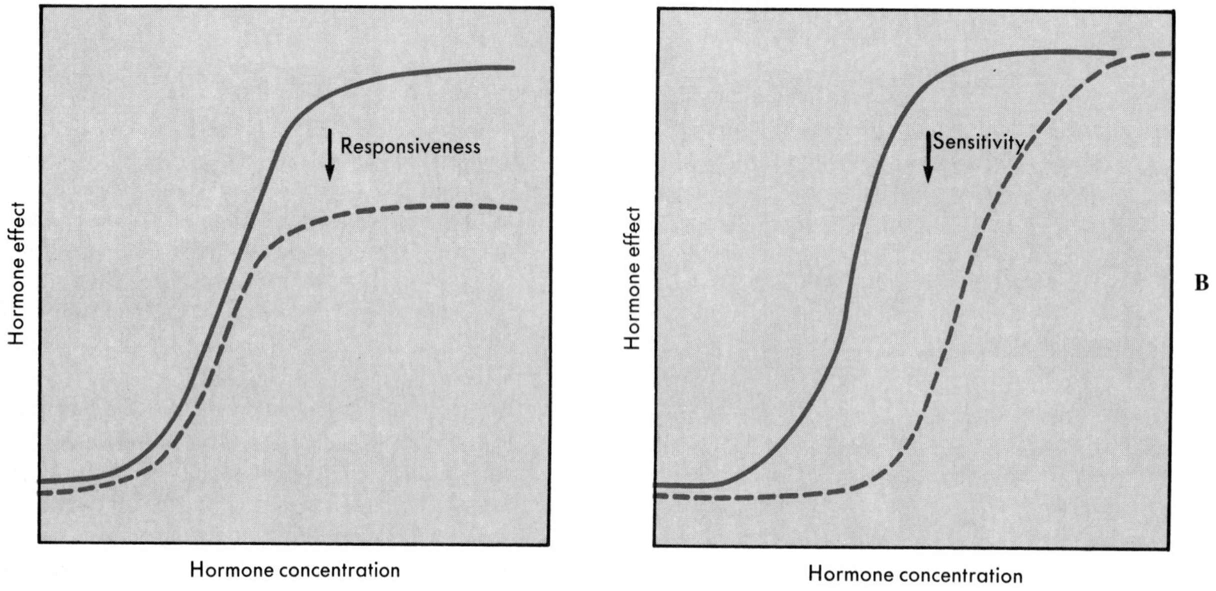

The normal range of responsiveness to a hormone is usually rather broad and is in part a result of the physiological variability created by the aforementioned factors within and among normal individuals. However, the very exquisiteness with which hormonal effects can be modulated is an important component in achieving one major objective of hormonal regulation: metabolic stability.

Hormone Transport

After secretion, hormones enter pools characterized by widely varying size, volume of distribution, degree of compartmentalization, and fractional turnover rates. Initially all hormones enter the plasma pool, where they may circulate either as free molecules or bound to specific carrier proteins. Ordinarily, catecholamine, peptide, and protein hormones circulate unbound, although exceptions are noted. In contrast, steroid and thyroid hormones largely circulate bound to specific globulins that are synthesized in the liver. The extent of protein binding markedly influences the exit rates of hormones from plasma into interstitial fluid and thence into the intracellular space. As noted in examples presented in Table 49-1, the plasma half-life of a hormone is directly correlated with the percentage of protein binding. For example, thyroxine, a thyroid hormone, is 99.95% protein bound and has a plasma half-life of 6 days, whereas aldosterone, a steroid hormone, is only 15% bound and has a plasma half-life of 25 minutes. Larger and more complex protein hormones tend to have longer half-lives than do smaller proteins and peptides. Hormone exit from plasma does not have to be entirely

Table 49-1. Correlation of plasma half-life and metabolic clearance of hormones with their structure and degree of protein binding

Hormone	*Protein binding (percent)*	*Plasma half-life (days)*	*Metabolic clearance (ml/minute)*
Thyroid			
Thyroxine	99.95	6	0.7
Triiodothyronine	99.65	1	18
Steroids			
Cortisol	94	0.07	140
Testosterone	89	0.04	860
Aldosterone	15	0.016	1100
Proteins			
Thyrotropin (molecular weight 28,000)	Nil	0.034	50
Growth hormone (molecular weight 22,000)	Nil	0.017	220
Insulin (molecular weight 6,000)	Nil	0.006	800

irreversible. In some instances hormone molecules may return to plasma from other compartments, possibly after dissociation from cell membrane receptors. The return to plasma can occur by way of the lymphatic channels.

Hormone Disposal

Irreversible removal of hormone is a result of target cell uptake, metabolic degradation, and urinary or biliary excretion. The sum of all removal processes is expressed in the term *metabolic clearance rate* (MCR). In a steady state this is defined as volume of plasma cleared/unit time, which equals mass removed/unit time divided by circulating mass/unit volume, that is,

$$\text{MCR} = \frac{\text{mg/minute removed}}{\text{mg/ml of plasma}} = \frac{\text{ml cleared}}{\text{minute}} \qquad (8)$$

MCR is one expression of the efficiency with which a hormone is removed from plasma. It is most conveniently determined by tracer techniques. A minute quantity of a valid radioactive tracer of the hormone is infused at a constant rate $\left(\frac{\text{CPM}}{\text{minute}}\right)$ until equilibrium is reached, i.e., until the plasma concentration of tracer $\left(\frac{\text{CPM}}{\text{ml}}\right)$ becomes constant. At this point the outflow of tracer from the plasma must equal the inflow of tracer. Thus

$$\text{MCR} = \frac{\text{CPM/minute infused}}{\text{CPM/ml of plasma}} \qquad (9)$$

The ratio of MCR to the volume of distribution of a hormone is a measure of its fractional turnover rate (K). The plasma half-life, which is inversely related to K, is a cruder but more conveniently determined index of hormone disappearance. As shown in Table 49-1, MCR is inversely correlated with plasma half-life and usually also inversely correlated with the percentage of protein binding.

The kidney and liver are usually the major sites of hormone extraction and degradation. Renal clearance of hormone is reduced markedly by protein binding to globulins in the plasma. For example, less than 1% of secreted cortisol appears unchanged in the urine because only the small, free fraction of plasma cortisol is filtered by the glomerulus. On the other hand, about 30% of cortisol metabolites are excreted in the urine, since they are generally unbound or only loosely bound to protein. Many peptide and smaller protein hormones are filtered to some degree by the glomerulus. Usually, however, they subsequently undergo tubular reabsorption and degradation within the kidney, so that only a minute amount appears in the final urine.

Metabolic degradation occurs by enzymatic processes that include proteolysis, oxidation, reduction, hydroxylation, decarboxylation, and methylation. Virtually all hormones are extracted from the plasma and degraded to some extent by the liver, which often is quantitatively the most important site. In addition, glucuronidation and sulfation of hormones or their metabolites may be carried out, with the conjugates subsequently excreted in the bile or the urine. Some hormonal degradation appears to take place during interaction with target tissues. As noted previously, internalization of a portion of the hormone–plasma membrane receptor complex does occur, and hormone plus receptor may be degraded together by lysosomes within the cell. It is difficult to tell at present to what extent hormone action and hormone degradation are quantitatively linked in these situations.

Hormone Measurement

In large part the history of endocrinology is the history of methodology for hormone measurement. Initially hormones all were measured by biological assay, i.e., by measuring an effect that they produced. When whole animals served as test subjects, sensitivity was necessarily low, and precision was greatly influenced by many variables that were difficult to control. As more knowledge of hormone action developed, biological assays could be carried out on organs in vivo and later on tissues in vitro. The sensitivity improved, less replication was needed as precision was enhanced, and the amount of sample required decreased. Of equal importance, specificity was increased. For example, a biological effect such as elevation of plasma glucose concentration could be produced by two very dissimilar hormones via different mechanisms. A whole animal assay with plasma glucose as an end point might not distinguish between the two. However, if one hormone worked on the liver and the other on muscle, selective tissue assays could discriminate between the two. Nevertheless tissue assays still put practical limitations on sensitivity and on the number of samples that could be handled and thus on the amount of information that could be obtained. Physicochemical methods, such as spectrophotometry and fluorometry, were developed that offered greater ease. Their sensitivity permitted their use only for catecholamine, thyroid and steroid hormones, or their metabolites. Even then often only the relatively higher concentrations in tissue, urine, or effluent blood from the gland of origin could be accurately determined. Furthermore, laborious chromatographic procedures sometimes were required to separate hormones with similar chemical groupings.

Radioimmunoassay

A dramatic advance in the acquisition of information and understanding in endocrinology followed the development of radioimmunoassay and related methods in 1957. In one stroke, sensitivity was brought to the level of normal plasma concentrations for virtually all hormones, precision to the level where changes of 25% could be reliably determined, and ease, cost, and sample volume to the point where as many as 50 values could be obtained from an individual in a single day. Specificity also was vastly enhanced initially; for example, two hormones, such as glucagon and epinephrine, which both raise blood glucose (in vivo bioassay) and tissue cAMP levels (in vitro bioassay), have absolutely no overlapping in a radioimmunoassay. It subsequently has become possible to design radioimmunoassays that can reliably discriminate between a hormone and its precursor (for example, insulin and proinsulin) or a hormone and its slightly modified metabolites (for example, thyroxine, triiodothyronine, and reverse triiodothyronine).

The principle of radioimmunoassay is shown in Fig. 49-12. The sample, which may be plasma, urine, cerebrospinal fluid, or a tissue extract, is incubated with a fixed amount of radioactively labeled hormone (the tracer) and a fixed amount of a specific hormone-binding protein, most often an antibody. The nonradioactive hormone molecules in the sample *compete* with the tracer hormone molecules for the binding sites on the protein. The concentration of binding sites is fixed and limiting. Therefore progressive increases in the number of nonradioactive hormone molecules in the sample will displace more and more of the tracer hormone molecules from the binding sites. At the end of the incubation the bound tracer hormone molecules are separated from those that are free. The radioactivity in the individual bound and free tracer fractions is counted. If the hormone concentration in the sample is relatively high, the percentage of radioactivity remaining in the bound fraction will be relatively low, and vice versa (Fig. 49-12, *A*). The absolute amount of hormone in the sample is calculated by comparison with the standard curve generated by incubating varying amounts of authentic hormone with identical amounts of tracer and binding proteins (Fig. 49-12, *B*).

The specificity of radioimmunoassay rests on the reaction of a hormone with a binding site that uniquely recognizes the hormone molecule. Most commonly the binding protein is an antibody produced by immunizing an animal. Protein hormones frequently possess sufficient antigenic power by themselves. Peptides, catecholamines, steroids, and thyroid hormones must be coupled, as haptens, to a larger protein, such as albumin, to render them sufficiently antigenic. In addition to antibodies, naturally occurring binding proteins also may serve in radioassays. These include proteins obtained from plasma and receptor proteins obtained from cells. The latter offers the special advantage that radioassay specificity almost always will parallel the specificity of biological action. Alterations in structure that render a hormone devoid of biological activity may still permit its easy recognition by an induced radioimmunoassay antibody whose determinants happen to be removed from the biologically active sites. Therefore elevated levels of the hormone may be measured by a radioimmunoassay employing such an antibody even though clinical signs of hormone deficiency exist. This is less likely to occur if a normal tissue receptor replaces the induced antibody as the binding protein in the radioassay.

Radioimmunoassays may not always distinguish completely or sufficiently between similar hormones secreted by the same gland (for example, two adrenal steroid hormones), between a peptide hormone and its prohormone, or between a hormone and its metabolic products. This may give rise to experimental or clinical

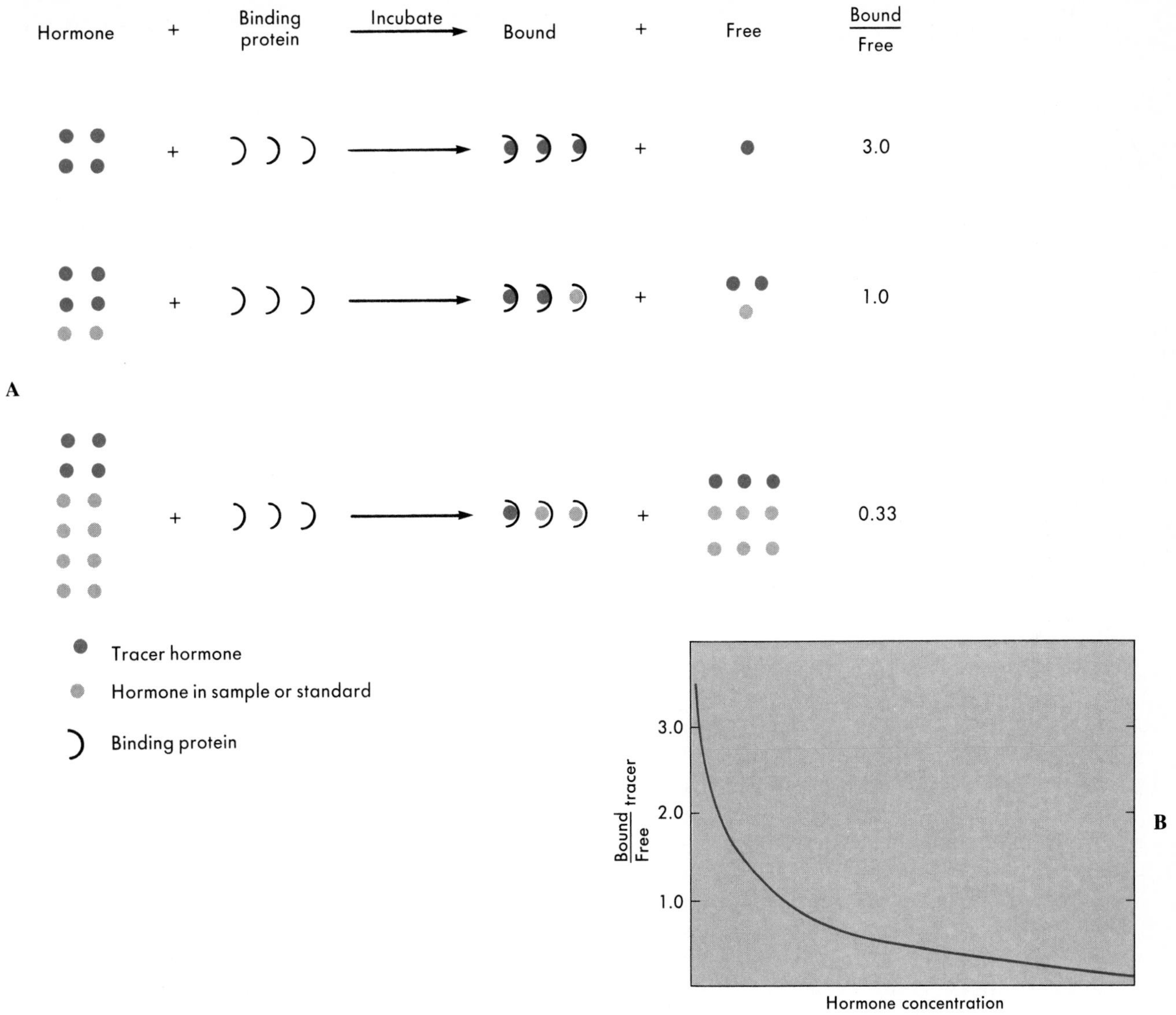

■ **Fig. 49-12. A,** The principle of radioimmunoassay. Note that concentrations of tracer hormone and binding protein are fixed. The addition of increasing amounts of unlabeled hormone in sample or standard in effect dilutes the tracer hormone molecules and displaces them from the binding protein. This results in a decreasing ratio of bound/free tracer hormone. **B,** Typical radioimmunoassay standard curve is exponential. The curve, however, often can be made linear by reciprocal or logarithmic transformation of the data.

misinterpretations unless the specificity of each assay is carefully documented. Preliminary separation procedures sometimes can obviate such problems.

The great sensitivity of radioimmunoassay rests on the high association constants of antigen-antibody reactions, which can be around 10^{10} to 10^{12} L/mole. This permits use of very low concentrations of reactants. High sensitivity also derives from the ability to measure very small amounts of radioactivity. In clinical assays, however, it is still sometimes difficult to differentiate pathologically low levels from normal unstimulated basal levels of hormone in plasma. In such instances the conclusion that glandular hypofunction exists may have to be made from the inference that the hormone level is not elevated, despite the presence of a strong negative feedback stimulus, or that it does not increase to the normal extent with pharmacological stimulation.

The precision of radioimmunoassay stems from the simplicity of handling and its suitability for automation. When samples are analyzed in replicate in a single assay, the coefficient of variation, an index of precision, is 5% to 10%. When the same sample is measured repeatedly in separate assays, a slightly higher coefficient of variation is observed. Samples with high concentration of hormone are diluted to bring them into the most precise portion of the standard curve.

Estimates of Hormone Secretion

Direct Measurement

Absolute quantitation of the output of a single hormone by an individual gland can only be accomplished in vivo by catheterization of the blood supply to the gland. The arterial (A) and venous (V) concentrations of hormone, in mass per unit volume, and the blood flow (BF) across the gland, in volume per unit time, must be measured. The rate of secretion in mass per unit time is then (V − A) × BF. This method for assessing secretion rate is suitable for animal studies but rarely can be performed in clinical circumstances and certainly not often enough to obtain daily secretion rates. However, sampling of veins that drain glands is a useful clinical procedure for localizing the source of excess hormone production. This is true for multiple glands, such as the parathyroid glands, or for hormones, such as steroids, that may be secreted from different glands, such as the adrenal glands and the ovaries.

Blood Production Rate

A less direct but frequently satisfactory method to estimate hormone secretion is the measurement of the blood production rate (BPR) in mass per unit time. This is the total amount of the hormone entering the peripheral circulation each day; in a steady state it will equal the total amount of hormone leaving the circulation. Therefore blood production rate is determined by measuring plasma concentration (P) and the MCR (p. 833):

$$BPR = P \times MCR \qquad (10)$$

Although this measurement primarily is carried out under research conditions in animals and humans, it can be used in select circumstances for clinical diagnosis.

Plasma Levels

A simple plasma concentration provides a valid index of blood production rate when the metabolic clearance of the hormone is within normal limits *and can be taken as a constant;* since BPR = P × MCR, BPR is proportional to P if MCR is a constant. This is the theoretical basis for employing plasma hormone measurements alone as an index of activity of the gland of origin. However, the release of many hormones is characterized by diurnal variation as well as by episodic spurts. It therefore may be hazardous to conclude too much from a single plasma value. Multiple daily measurements or measurements taken at different times of the day may be needed. To reduce the number of laboratory analyses, multiple samples of equal volume can be pooled, and a single careful measurement of this pool then yields the average plasma concentration over the interval in which the samples were collected.

Urinary Excretion

Measurement of urinary hormone excretion is cumbersome because it requires accurately timed collections. However, it offers the advantage of, in effect, averaging plasma fluctuations over the collection period. Furthermore, in some instances the quantity of a hormone metabolite in the urine far exceeds the plasma or urinary level of the hormone itself and is therefore more easily measurable. Urinary excretion (UV) is equal to the urinary concentration (U) times the urine flow (V). Urinary excretion of the hormone is related to blood production rate and the renal clearance (C) of the hormone (Chapter 47), as follows:

$$C = \frac{UV}{P}; P = \frac{UV}{C} \qquad (11)$$

By substitution into equation 10:

$$BPR = UV \times \frac{MCR}{C} \qquad (12)$$

Hence BPR is proportional to UV when MCR/C can be taken as a constant. Thus urinary excretion of a hormone is a valid index of blood production rate and hence secretion rate when the following two conditions are fulfilled: (1) the kidney must be contributing its usual fraction of the total metabolic clearance; and (2) within the kidney itself the usual proportioning of hormone between intrarenal degradation and urinary excretion must be maintained. The chief sources of error in employing urinary excretion as a reflection of hormone secretion are general impairment in renal function, a change in the pattern of degradation to a metabolic product excreted differently by the kidney, and incomplete collections of urine. When there is little diurnal variation in hormone secretion and degradation, incomplete collections can be partially compensated for by indexing hormone excretion to simultaneously determined creatinine excretion.

It is important that these principles of hormone measurement and of estimating hormone secretion rates be kept in mind when interpreting the results of physiological experiments or of clinical diagnostic studies.

Bibliography

Journal articles

Alford, F.P., et al.: Temporal patterns of circulating hormones as assessed by continuous blood sampling, J. Clin. Endocrinol. Metab. **36:**108, 1973.

Baxter, J.D., et al.: Hormone receptors, N. Engl. J. Med. **301:**1149, 1979.

Bost, K.L., et al.: Similarity between the corticotrophin ACTH receptor and a peptide encoded by an RNA that is complementary to ACTH mRNA, Proc. Natl. Acad. Sci. **82:**1372, 1985.

Catt, K., and Dufau, M.: Introduction: the clinical significance of peptide hormone receptors, J. Clin. Endocrinol. Metab. **12:**xi, 1983.

Chambon, P., et al.: Promoter elements of genes coding for proteins and modulation of transcription by estrogens and progesterone, Recent Prog. Horm. Res. **40:**1, 1984.

Chambon, P., et al.: Steroid hormones relieve repression of the ovalbumin gene promoter in chick oviduct tubular gland cells. In Labrie, F., and Proulx, L., editors: Endocrinology: Proceedings of the seventh International Congress of Endocrinology, Quebec City, 1984, Amsterdam, 1984, Elsevier Science Publishers, B.V. pp. 3-10.

Gordon, P., et al.: Internalization of polypeptide hormones; mechanism, intracellular localization and significance, Diabetologia **18:**263, 1980.

Lacy, P.E.: Beta cell secretion—from the standpoint of a pathobiologist, Diabetes **19:**895, 1970.

Lefkowitz, R., et al.: Mechanisms of membrane-receptor regulation: biochemical, physiological, and clinical insights derived from studies of the adrenergic receptors, N. Engl. J. Med. **310:**1570, 1984.

Nishizuka, Y.: Membrane phospholipids and the mechanism of action of hormones. In LaBrie, F., and Proulx, L., editors: Endocrinology: Proceedings of the seventh International Congress of Endocrinology, Quebec City, 1984, Amsterdam. 1984, Elsevier Science Publishers, B.V., pp. 33-40.

Sherwin, R.S., et al.: A model of the kinetics of insulin in man, J. Clin. Invest. **53:**1481, 1974.

Tait, J.F.: The use of isotopic steroids for the measurement of production rates in vivo, J. Clin. Endocrinol. **23:**1285, 1963.

Zor, U.: Role of cytoskeletal organization in the regulation of adenylate cyclase–cyclic adenosine monophosphate by hormones, Endocrine Rev. **4:**1, 1984.

Books and monographs

Yalow, R.S., et al.: Introduction and general considerations. In Odell, W.D., and Daughaday, W.H., editors: Principles of competitive protein-binding assays, Philadelphia, 1971, J.B. Lippincott Co.

CHAPTER 50

Whole Body Metabolism and the Hormones of the Pancreatic Islets

The hormones of the pancreatic islets, especially *insulin* and *glucagon,* are among the most important regulators of fuel metabolism. Therefore before a detailed consideration of the endocrine function of the islets, some fundamental aspects of whole body metabolism and substrate flux are reviewed.

A healthy adult human being ingests and expends an average of 2500 kcal/day* in the form of carbohydrate, protein, and fat and maintains a rather constant body weight. This energy turnover can vary from 1500 to 5000 kcal, depending on the individual's size, habitus, and physical activity. For adults, total daily energy expenditure averages 39 kcal/kg in men and 34 kcal/kg in women. Of this, approximately 20 kcal/kg are expended in *basal metabolism,* that is, the energy consumed in all the chemical reactions and mechanical work (such as breathing and heart action) when the individual is at complete rest.

Basal metabolism may be increased up to 15% by eating. This increase in metabolism was known in the past as the specific dynamic action of food, but it is currently referred to as *diet-induced thermogenesis.* In part it may be attributable to chemical reactions such as deamination and ureagenesis, which are initiated by absorption of amino acids. In part it is due to the reactions involved in the storage of carbohydrate as glycogen and in the conversion of carbohydrate to fat for storage in that form. The remainder of daily energy expenditure results from voluntary or involuntary activity. If daily activity is light, the values for total energy expenditure are 10% to 15% less than just noted. If activity is moderately heavy, these values are 15% to 20% greater; and if very heavy, 25% to 35% greater. Energy expenditures for a number of common activities are listed in Table 50-1.

*1 kilocalorie (kcal) = 1 calorie ("C") = 1000 calories ("c"); 1 kcal = 4184 joules.

Table 50-1. Estimates of energy expenditure in adults

Activity	*Calorie expenditure (kcal/minute)*
Basal	1.1
Sitting	1.8
Walking, 2.5 miles/hour	4.3
Walking, 4.0 miles/hour	8.2
Climbing stairs	9.0
Swimming	10.9
Bicycling, 13 miles/hour	11.1
Household domestic work	2 to 4.5
Factory work	2 to 6
Farming	4 to 6
Building trades	4 to 9

Data from Kottke, F.J.: Animal energy exchange. In Altman, P.L., editor: Metabolism, Bethesda, Md., 1968, Federation of American Society for Experimental Biology.

The approximate composition of a normal 70 kg adult human, viewed from the standpoint of energy content, is shown in Fig. 50-1. It is clear that fat forms the major energy storage depot, both because of its greater mass and because of its high caloric value (Table 50-2). Certain features of the relationship between the individual components of energy stores and their turnover are noteworthy. The average daily turnover of carbohydrate is approximately 250 g/day (1000 kcal/day). This is of the same magnitude as the total carbohydrate stores of 450 g (1800 kcal) and exceeds the rapidly mobilizable stores of 100 g contained in the extracellular fluid and liver glycogen. Thus a daily intake of at least 150 g of carbohydrate is desirable; if this ceases, a prompt mechanism for synthesis of glucose from other sources is required. The major source of endogenous glucose production is protein, with a much smaller fraction available from the glycerol contained in fat (triglycerides).

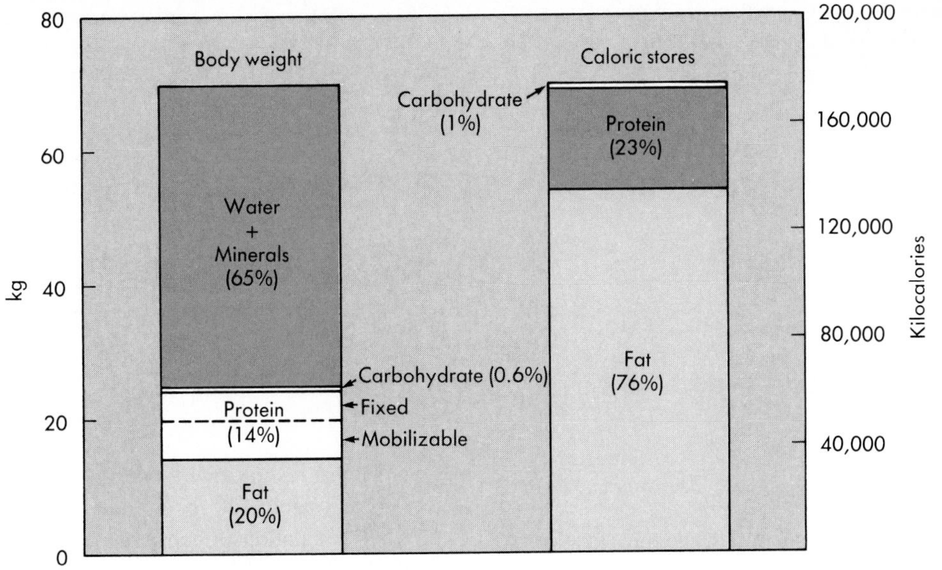

Fig. 50-1. The composition of an average 70 kg human is shown in terms of weight *(left)* and caloric stores *(right)*. Note the trivial proportion of carbohydrate stores in relative to fat stores.

Table 50-2. Energy equivalents of foodstuffs

	Kilocalories produced per gram	*O_2 used (L/g)*	*Kilocalories produced per liter of O_2*	*Respiratory quotient*
Carbohydrate	4.2	0.84	5.0	1.00
Fat	9.4	2.00	4.7	0.70
Protein	4.3*	0.96*	4.5	0.80
Typical fuel mix	—	—	4.8	0.85

*Each gram of protein oxidized yields 0.16 g of urinary nitrogen. Thus these values should be multiplied by 6.25 to express them per gram of urinary nitrogen.

In adult humans the average daily turnover of body protein is 200 to 300 g/day (800 to 1200 calories). In contrast to carbohydrate, this represents only about 2.5% of total body protein (10 kg) and about 4% of the metabolically mobilizable portion (6 kg). The daily release of amino acids from the main endogenous source, the muscle mass, is about 50 g/day. Therefore dietary protein intake of this order is sufficient for healthy adults. The minimum daily requirement for adults is less, 0.5 g/kg, because of efficient reutilization for protein synthesis of those amino acids derived from proteolysis but retained within the cells of muscle, liver, and other organs. The minimum protein requirement per unit of body weight is much higher in infants and children because of growth needs; it is estimated at 1.5 to 2 g/kg. Although the carbon skeletons of amino acids can be used as a source of oxidative energy, it is obviously advantageous to conserve the amino acids for the renewal of functional and structural protein molecules. This is particularly so in the case of the essential amino acids, which cannot be replaced by endogenous resynthesis.

The daily turnover of fat is about 100 g/day (900 calories), and it represents less than 1% of body stores. While carbohydrate and protein stores are narrowly fixed by limitations of hepatic size and function and by the adult body habitus, fat stores are variable. The percentage of body weight due to fat can vary from as little as 5% in very thin individuals to as much as 80% in pathologically obese humans. Because of the usually ample stores, there is no true minimum daily requirement of dietary fat per se. However, when endogenous lipid is depleted or its mobilization is suppressed, the essential unsaturated fatty acids may be deficient if exogenous fat is not provided.

In the resting postabsorptive state the average 70 kg human being expends approximately 1 to 1.2 kcal/minute in the process of basal metabolism. This requires the use of approximately 200 to 250 ml of oxygen per minute (Table 50-2). This basal metabolic rate is linearly related to lean body mass or fat-free mass and to body surface area. In the process of oxidizing substrates to meet basal energy needs, the proportion of carbon dioxide produced ($\dot{V}CO_2$) to oxygen used ($\dot{V}O_2$) varies according to the fuel mix. The proportion of $\dot{V}CO_2$ to $\dot{V}O_2$ is known as the *respiratory quotient* (RQ) (Chapter

39). As indicated by the following equations, RQ equals 1 for oxidation of carbohydrate (e.g., glucose), whereas RQ equals 0.70 for oxidation of fat (e.g., palmitic acid).

For carbohydrates:

$$C_6H_{12}O_6 + 6\ O_2 \rightarrow 6\ CO_2 + 6\ H_2O$$

Glucose

$$RQ = \frac{6\ CO_2}{6\ O_2} = 1$$

For fats:

$$C_{15}H_{31}COOH + 23\ O_2 \rightarrow 16\ CO_2 + 16\ H_2O$$

Palmitic acid

$$RQ = \frac{16\ CO_2}{23\ O_2} = 0.70^*$$

The RQ for protein reflects that of the individual RQs of the amino acids and averages 0.80. Under ordinary circumstances protein is a minor energy source. Its small contribution to the overall RQ can be corrected for by measuring the urinary excretion of the nitrogen that results from the metabolism of amino acids. The amount of protein oxidized equals 6.25 × grams of nitrogen. The amount of O_2 consumed by oxidation of 1 g of protein equals 0.96 L (Table 50-2); the amount of CO_2 produced equals 0.80 × the amount of O_2 consumed. Thus for each gram of urinary nitrogen excreted, 5.9 L of O_2 are consumed, and 4.8 L of CO_2 are produced.

By determining the nonprotein RQ, one can then calculate the proportion of carbohydrate and fat in the fuel mix being oxidized.

For example, when the O_2 consumed = 210 ml/minute and the CO_2 produced = 174 ml/minute:

$$RQ = \frac{174}{210} = 0.83$$

If C is the proportion of energy due to carbohydrate oxidation and F is the proportion of energy due to fat oxidation, then:

$$1 \times C + 0.7 \times F = 0.83$$

Since C + F = 1:

$$1 \times C + 0.7 \times (1 - C) = 0.83$$

$$C = 0.43$$

$$F = 0.57$$

That is, 43% of the calories are being provided by carbohydrate oxidation, and 57% of the calories are being provided by fat oxidation.

With a knowledge of the actual rate of energy expenditure in kilocalories per minute and the standard values of 4 kcal/g of carbohydrate and 9 kcal/g of fat, one can calculate the actual rates of oxidation of carbohydrate and fat, respectively, in grams per minute:

$$210 \text{ ml of } O_2 \text{ consumed per minute} = 1.05 \text{ kcal/minute}$$

$$4 \times \text{gram of carbohydrate} = 0.43 \times 1.05$$

Therefore $\text{gram of carbohydrate} = 0.113$

$$9 \times \text{gram of fat} = 0.57 \times 1.05$$

Therefore $\text{gram of fat} = 0.067$

In the resting adult the rate of glucose oxidation is approximately 120 mg/minute (170 g/day), whereas that of fat oxidation is approximately 60 mg/minute (90 g/day). Thus in caloric equivalents glucose supplies about 45% of the energy required for basal metabolism. The central nervous system is an obligate glucose consumer and uses the major portion of this fuel. Estimates of cerebral glucose utilization are about 125 g/day. In contrast, the large muscle mass oxidizes primarily fatty acids in resting individuals. However, like all tissues, the muscle mass uses at least some glucose, thereby maintaining sufficient concentrations of Krebs cycle intermediates for efficient disposal of acetyl CoA and completion of fatty acid oxidation.

Metabolic Interrelationships Among Carbohydrate, Protein, and Fat

The interrelationships among the three main classes of substrates—carbohydrates, fats, and proteins— will be discussed before they are considered individually in detail. Carbohydrate and protein can be converted to fat, and protein can be converted to carbohydrate. In humans conversion of fat to carbohydrate (via Ω-oxidation of odd-chain fatty acids) is almost nil. In a metabolic sense humans normally cycle between the fasting state, lasting 8 to 12 hours overnight, and several fed states, lasting 3 to 5 hours after each mixed meal. The general relationships among carbohydrate, fat, and protein metabolism are best considered by examining them in these two basic states.

In the fasting state the individual totally depends on endogenous substrates for energy and for resynthesis of critical molecules that have rapid turnover rates. Survival therefore depends on rapid mobilization of glucose as fuel for the central nervous system and of fat to provide as much of the oxidative needs of other tissues as possible. An increase in protein degradation to amino acids is also a fundemental feature of the response to fasting. The individual is said to be in a state of *catabolism* because carbohydrate, fat, and protein stores are

*When fat is rapidly being synthesized from glucose, CO_2, is utilized and O_2 is disproportionately released. The RQ can be greater than 1 under these conditions.

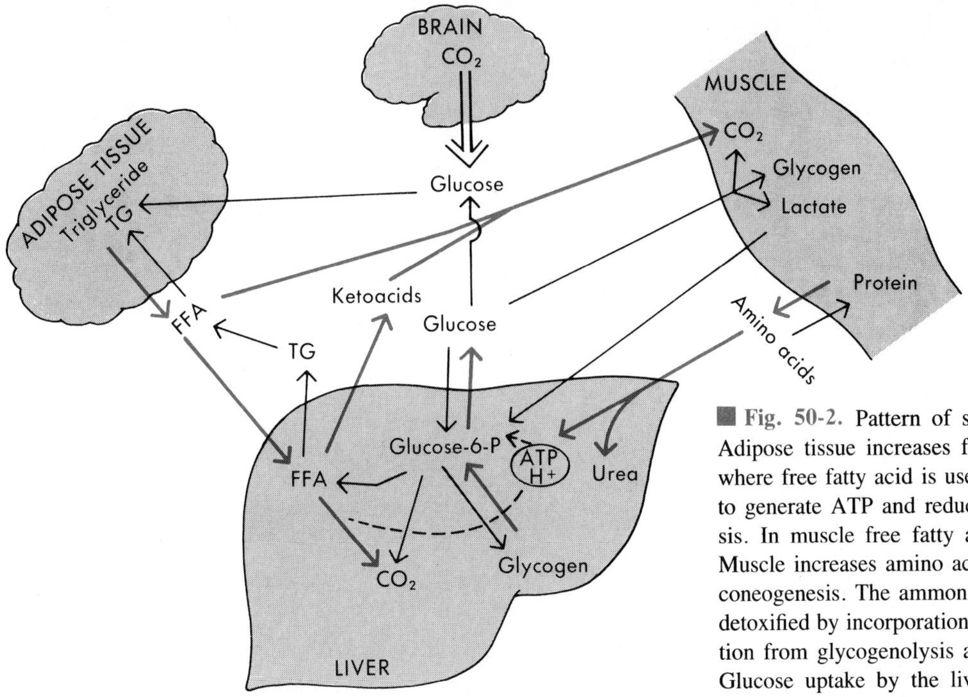

■ **Fig. 50-2.** Pattern of substrate flow in the fasting state. Adipose tissue increases free fatty acid supply to the liver, where free fatty acid is used for energy, for ketogenesis, and to generate ATP and reducing equivalents for gluconeogenesis. In muscle free fatty acid oxidation spares glucose use. Muscle increases amino acid flow to the liver for use in gluconeogenesis. The ammonia released from the amino acids is detoxified by incorporation into urea. Hepatic glucose production from glycogenolysis and gluconeogenesis is augmented. Glucose uptake by the liver, adipose tissue, and muscle is diminished, but uptake by the brain is sustained. Ketoacid production by the liver and eventually ketoacid utilization by peripheral tissues is increased. *TG,* Triglycerides; *FFA,* free fatty acids; *colored arrows,* increased flow of fuel; *black arrows,* decreased flow of fuel.

all being diminished (Fig. 50-2). In the adipose tissue, triglyceride hydrolysis to fatty acids is augmented. The fatty acids are then taken up by muscle for oxidation, during which they impede and thus spare glucose use. Fatty acids taken up by the liver are also preferentially oxidized, yielding increased amounts of the water-soluble ketoacids, β-hydroxybutyrate, and acetoacetate. The ketoacids also provide oxidative substrate to muscle and ultimately help reduce central nervous system use of glucose. In addition, the reducing equivalents and adenosine triphosphate (ATP) generated by fat oxidation are used for gluconeogenesis, and the acetyl CoA produced activates the key gluconeogenetic enzyme, pyruvate carboxylase. In muscle tissue protein degradation supplies amino acids, which are transferred to the liver for conversion to glucose. The liver supplies glucose to the circulation by augmenting first glycogenolysis and later gluconeogenesis. During fasting, glucose uptake is diminished in all organs but the brain.

In the fed state all of these processes are essentially reversed (Fig. 50-3). Calories ingested as fat, beyond that needed for ongoing energy production, are stored as triglycerides in adipose tissue. Calories ingested as carbohydrate are stored as glycogen in liver and muscle. In addition, carbohydrate is converted to triglycerides in the liver and transferred for storage in adipose tissue. Calories ingested as amino acids are used for new protein synthesis. During feeding the individual is said to be in a state of *anabolism*. Fatty acid oxidation and ketoacid production are greatly diminished in the liver, glucose uptake is enhanced in all tissues, and lactate formed from glucose in the liver and muscle dominates as a gluconeogenetic precursor.

■ *Quantitative Aspects of Carbohydrate Metabolism*

In addition to being a major energy source, carbohydrates are components of glycoproteins, glycopeptides, and glycolipids with a myriad of functional and structural roles. Examples include collagen in basement membranes, mucopolysaccharides as ground substance, myelin, hormones, and triglycerides.

Glucose is the central molecule in carbohydrate metabolism, because other ingested sugars, such as fructose and galactose, are metabolized via glucose or its three carbon products. Postabsorptive plasma glucose concentration averages 80 mg/dl (4.5 mmole/L), with a range of 60 to 115 mg/dl. Its major glycolytic products, lactate and pyruvate, circulate at average concentrations of 0.7 and 0.07 mmole/L, respectively. Under normal circumstances, the ratio of lactate to pyruvate remains about 10, even when the glycolysis rate changes. However, when tissues are deprived of oxygen, the equilibrium between the two shifts toward lactate, which is the

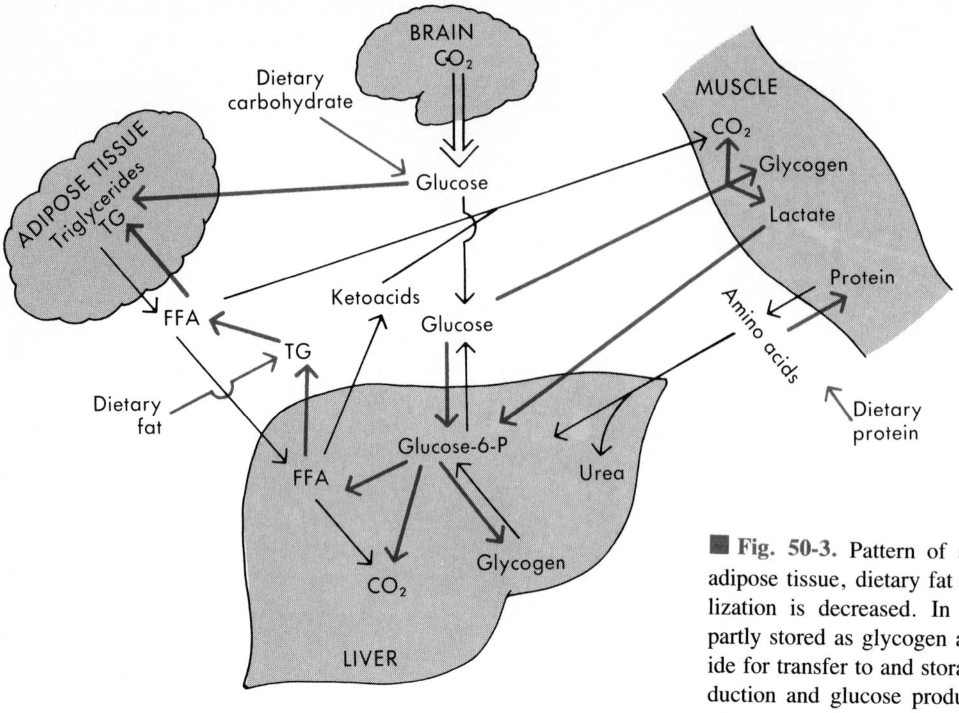

■ **Fig. 50-3.** Pattern of substrate flow in the fed state. In adipose tissue, dietary fat is stored, and free fatty acid mobilization is decreased. In the liver, dietary carbohydrate is partly stored as glycogen and is partly converted to triglyceride for transfer to and storage in adipose tissue. Ketoacid production and glucose production by the liver are suppressed. Glucose uptake in the liver, muscle, and adipose tissue is increased. Amino acids from dietary protein are taken up in muscle and liver and used for protein synthesis. *TG,* Triglycerides; *FFA,* free fatty acids; *colored arrows,* increased flow of fuel; *black arrows,* decreased flow of fuel.

reduced molecule; ratios as high as 30 may be observed in their plasma concentrations. Brain uptake of glucose and the utilization of oxygen by the brain decrease in parallel when plasma glucose falls below 60 mg/dl. Central nervous system function is progressively impaired, and death may ensue.

In the postabsorptive (overnight fasted) state, glucose turnover averages 2 mg/kg/minute (11 μmole/kg/minute) or about 200 g/day in adults. As depicted in Fig. 50-4, about 55% of glucose utilization is due to terminal oxidation, of which the brain accounts for most. Another 20% is due to glycolysis; the resulting lactate then returns to the liver for resynthesis into glucose (Cori cycle). Reuptake by the splanchnic tissues, accounts for the remainder of glucose utilization. Most of the glucose utilization in the postabsorptive state (about 70%) is independent of insulin, the major regulating hormone.

In the postabsorptive state, the only significant source of glucose is hepatic output. Therefore the normal basal rate of hepatic glucose production equals that of glucose utilization (2 mg/kg/minute). About 75% of this postabsorptive glucose production results from glycogenolysis and only 25% from gluconeogenesis (Fig. 50-4). Hepatic uptake and utilization of circulating lactate accounts for over half of the glucose supplied by gluconeogenesis. Most of the remainder is attributable to uptake and utilization of amino acids, especially alanine; circulating glycerol and pyruvate make only minimal contributions. The supply of lactate largely comes

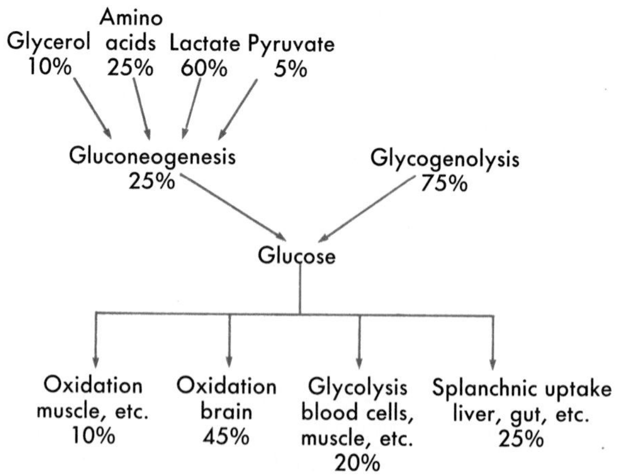

■ **Fig. 50-4.** Quantitative overview of glucose turnover in overnight fasted humans. The disposal of circulating glucose *(lower portion)* is largely by oxidation and glycolysis, with the brain acting as the single largest consumer. The production of glucose in the circulation *(upper portion)* is largely from glycogenolysis. The proportionate contribution of glucose precursors to gluconeogenesis, with lactate predominating, is also shown. The average overall rate of glucose turnover is 2.0 mg/kg/minute or 12 μmoles/kg/minute.

Table 50-3. Amino acids in humans

	Plasma concentration (μmole/L)	*Glycogenic/ketogenic*	*Glycogenic intermediate*
Essential			
Threonine	137	G	Propionyl CoA
Methionine	26	G	Propionyl CoA
Valine	197	G	Methylmalonyl CoA
Histidine*	78	G	α-Ketoglutarate
Isoleucine	65	G/K	Propionyl CoA
Tryptophan	—	G/K	Propionyl CoA
Phenylalanine	50	G/K	Fumarate
Tyrosine*	52	G/K	Fumarate
Lysine	187	G/K	α-Ketoglutarate
Leucine	116	K	—
Nonessential			
Glutamine	468	G	α-Ketoglutarate
Glutamic acid	55	G	α-Ketoglutarate
Alanine	289	G	Pyruvate
Serine	118	G	Pyruvate
Cysteine	87	G	Pyruvate
Glycine	215	G	Pyruvate
Asparagine	43	G	Oxaloacetate
Asparatic acid	19	G	Oxaloacetate
Arginine	84	G	α-Ketoglutarate
Proline	169	G	α-Ketoglutarate

Data from Simmons, P.S., et al.: J. Clin. Invest. **73**:412, 1984; Sherwin, R.S., et al.: J. Clin. Invest. **55**:1382, 1975; and Marliss, E.B., et al.: J. Clin. Invest. **50**:814, 1971.
*Histidine is essential in infants only. Tyrosine is synthesized from phenylalanine.

from glycolysis in red blood cells, white blood cells, and muscle. The amino acid precursors come from muscle proteolysis.

The distribution of an ingested carbohydrate load, given as 100 g of glucose in the morning, is different from that described for the basal glucose level. Hormonal responses, to be described later, regulate the disposition of the load. Only 20% to 30% of the glucose load is oxidized, whereas 70% is stored in tissues. Hepatic production of glucose is suppressed by 50% to 70% during the 3- to 5-hour period required for complete absorption of the exogenous glucose. Peripheral tissues assimilate 70% of the extra glucose, whereas the liver and other splanchnic tissues accept only 30%. Glucose utilization by the brain presumably remains unchanged from the basal level, and this need is met partly by some of the load and partly by the residual hepatic glucose output. The large muscle mass is the major consumer of a glucose load. Whereas the fate of this extra glucose initially taken up by muscle remains uncertain, undoubtedly some is initially converted to muscle glycogen. Glucose oxidation increases slightly to offset a corresponding decrease in lipid oxidation, consequent to suppression of lipolysis. The most important function of the ingested glucose is ultimately to rebuild the store of liver glycogen that overnight fasting has depleted (Fig. 50-3). Much evidence now suggests that conversion of exogenous glucose to glycogen via direct phosphorylation of the glucose to glucose-6-phosphate in the liver is inefficient. Instead, most glucose first undergoes glycolysis to lactate. The latter is then recycled via the gluconeogenic pathway into glucose-6-phosphate and thence to glycogen. Therefore the muscle mass may act as a temporary reservoir for exogenous glucose, which may ultimately be transferred as lactate to the liver for resynthesis into glycogen.

Quantitative Aspects of Protein Metabolism

All proteins are composed of the same 20 amino acids (Table 50-3). Half of these—the nonessential amino acids—can be synthesized endogenously by amination of the appropriate carbon skeletons built from glucose metabolites such as pyruvate, oxaloacetate, and α-ketoglutarate. The other half, composed largely of branch-chain and aromatic amino acids, are called essential because their carbon skeletons (the corresponding α-ketoacids) cannot be synthesized by humans, although once present, they can be aminated. Hence the essential amino acids must be supplied in the diet, with minimum requirements ranging from 0.5 to 1.1 g/day. All 20 amino acids are required for normal protein synthesis; therefore a deficiency of even one essential amino acid disrupts anabolism. Unlike the case for carbohy-

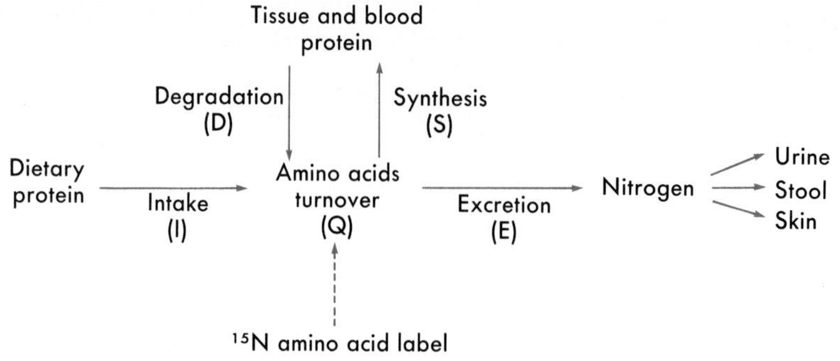

1. Label body pool
2. At isotopic steady state
3. Q = D+I = S+E
4. D = Q−I
5. S = Q−E

$$Q = \frac{\text{Tracer inflow}}{\text{Pool specific activity}} = \frac{\text{CPM}}{\text{Day}} \div \frac{\text{CPM}}{\text{GM}} = \frac{\text{GM}}{\text{Day}}$$

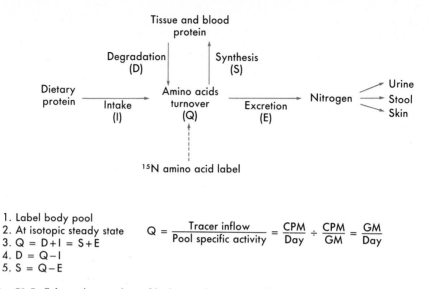

■ **Fig. 50-5.** Schematic overview of body protein turnover. Body protein turnover is assessed by labeling the amino acid pool with a stable isotope tracer, ^{15}N, and simultaneously determining nitrogen balance. Input to the amino acid pool is from dietary protein *(I)* and from tissue degradation *(D)*; output from the amino acid pool is via tissue protein synthesis *(S)* and via oxidation and excretion as nitrogen *(E)*. At an isotopic steady state, the turnover or flux *(Q)* equals input *(I + D)*, which equals output *(S + E)*. When *Q* and *I* are measured, *D* can be calculated. Similarly, when *Q* and *E* are measured, *S* can be calculated. When *I = E*, the individual is said to be in nitrogen balance. In this state, *D = S*.

drates, and more so than for fats, protein sources can vary greatly in their biological effectiveness, depending in part on the ratio of essential to nonessential amino acids. Milk and egg proteins are of the highest quality in this regard. During infancy and childhood 35% to 40% of the protein intake should yield essential amino acids to support growth. In adults this requirement falls to 20%. In addition to their incorporation into proteins, many of the amino acids are precursors for other important molecules, such as purines, pyrimidines, polyamines, phospholipids, creatine, carnitine, methyl donors, thyroid and catecholamine hormones, and neurotransmitters.

All 20 amino acids can be completely oxidized to CO_2 and water after removal of the amino group. Each traverses a specific degradative pathway, the details of which are beyond the scope of this discussion. However, these pathways converge into three general metabolic processes of physiological importance. With a single exception, all of the amino acids are gluconeogenic, that is, they can contribute carbon atoms for the synthesis of glucose. The entry point for each amino acid into the gluconeogenic pathway is shown in Table 50-3. Five amino acids give rise simultaneously to gluconeogenic precursors and to either acetoacetate or its CoA precursors; that is, these amino acids are ketogenic as well as glycogenic. For these amino acids, dietary excesses can be disposed of directly by β-oxidation or stored as fat. Leucine is the only amino acid that is solely ketogenic. Finally, the degradation of all amino acids gives rise to ammonia by oxidative deamination or by transamination. The ammonia is transported from peripheral degradation sites to the liver largely as glutamine and to a lesser extent as alanine and as NH_4^+. In the liver the ammonia is detoxified by incorporation into urea, which is synthesized via the Krebs-Henseleit cycle. The urea is then excreted by the kidney (Chapter 47). Except for minor losses of nitrogen in the feces (0.4 g/day) and skin (0.3 g/day), urinary nitrogen constitutes the major excretory route for the products of protein metabolism.

In the healthy adult under steady-state conditions, the total nitrogen excreted in the urine daily as urea plus ammonia is essentially equal to the nitrogen released during terminal metabolism of exogenous and endogenous protein. Such an individual is said to be in *nitrogen balance*. When protein intake is nil, the urea plus ammonia nitrogen in the urine reflects quantitatively the rate of endogenous protein catabolism; when protein intake is deficient or when protein catabolism is greatly accelerated by tissue trauma or disease, urinary urea plus ammonia nitrogen exceeds intake. In these three cases the individual is said to be in *negative nitrogen balance*. In a growing child or in a previously malnourished individual undergoing protein repletion with accretion of body mass, urinary nitrogen excretion is less than the intake of nitrogen as protein and the individual is said to be in *positive nitrogen balance*.

Measurements of external nitrogen balance do not themselves provide insight into the dynamic internal equilibrium between protein synthesis and protein degradation. The latter must be estimated by labeling

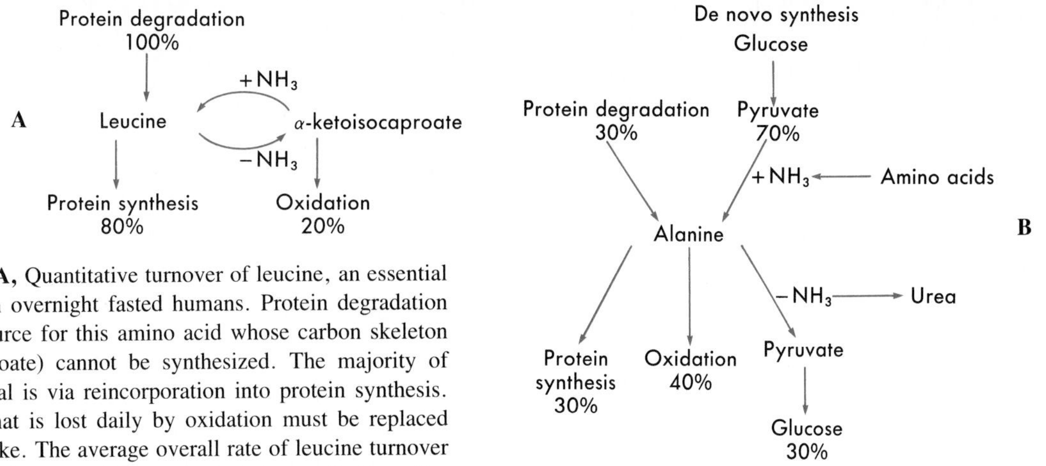

■ **Fig. 50-6. A,** Quantitative turnover of leucine, an essential amino acid, in overnight fasted humans. Protein degradation is the only source for this amino acid whose carbon skeleton (α-ketoisocaproate) cannot be synthesized. The majority of leucine disposal is via reincorporation into protein synthesis. The leucine that is lost daily by oxidation must be replaced by dietary intake. The average overall rate of leucine turnover is 0.2 mg/kg/minute or 1.5 μmoles/kg/minute. **B,** Quantitative turnover of alanine, a nonessential amino acid, in overnight fasted humans. Alanine production is mostly by de novo synthesis from pyruvate, with a lesser contribution from protein degradation. Alanine disposal takes place via reincorporation into protein, oxidation, and gluconeogenesis. Alanine molecules, which arise from pyruvate and are reconverted to pyruvate, act as carriers for ammonia transferred from other amino acids. (Data in **A** from Matthews, D., et al.: Am. J. Physiol. **238:**473, 1980; data from Chochinov, R., et al.: Diabetes **27:**287, 1978.)

body protein with an isotopic amino acid tracer, such as ^{15}N-glycine, and determining the flux of protein from measurements of ^{15}N-specific activity (Fig. 50-5). Such studies show that healthy adults, receiving isocaloric diets containing adequate protein, degrade and synthesize protein at a rate of 3 to 4 g/kg/day. Individual synthesis rates for most proteins are not known, but the hepatic synthesis of albumin accounts for about 5% of the above total. The rate of total body protein synthesis is diminished when the diet is severely deficient in energy, in total protein, or in one of the essential amino acids. In such situations, the rate of protein degradation usually also diminishes, but not to the same extent as synthesis, so that net loss of body protein results.

In addition to the flux of total body protein, each individual amino acid undergoes its own unique turnover. It is instructive to examine and compare the flux through plasma of two that have been well studied, namely, leucine, an essential amino acid, and alanine, a nonessential amino acid (Fig. 50-6, *A*). In the postabsorptive state the only source of leucine is endogenous protein degradation. (Whereas leucine can be regenerated from its α-keto analogue by transamination, α-ketoisocaproate cannot be synthesized endogenously.) Oxidation accounts for 20% of leucine disappearance from plasma (approximately 4 g/day), whereas 80% is reincorporated into new protein molecules. From a knowledge of the rate of leucine production (0.2 mg/kg/minute) and the average leucine content of protein (8%) the calculated rate of proteolysis is about 2.5 mg/kg/minute. In this way measurement of leucine production can be used as a marker for hormonal regulation of proteolysis. After ingestion of food, leucine oxidation diminishes and more is utilized for protein synthesis. However, even in the fed state, there is a minimum irreversible rate of leucine (α-ketoisocaproate) oxidation. Hence a dietary intake of approximately 1 g of leucine is required each day. Under conditions of prolonged fasting, oxidation of α-ketoisocaproate (and other branch-chain ketoacids) contributes slightly to energy needs. More important, leucine produced by catabolism of some proteins helps maintain the synthesis of other more critical proteins. Thus leucine flux through plasma allows interprotein and interorgan transfers to accommodate specific metabolic needs.

In contrast to leucine, the turnover of alanine in the postabsorptive state is threefold higher and more complex (Fig. 50-6, *B*). About 30% of alanine is produced from protein degradation. The other 70% of alanine arises from de novo synthesis: the carbon skeleton from pyruvate and ultimately from glucose and the nitrogen by transamination from other amino acids, particularly leucine and similar branch-chain amino acids. The fate of plasma alanine reflects three major routes and processes; about 40% is oxidized after reconversion to pyruvate, about 30% reappears as glucose via gluconeogenesis, and the remaining 30% is reutilized for protein synthesis. Fig. 50-6, *B,* shows that a glucose-alanine cycle, similar to the glucose-lactate (Cori cycle), can be envisioned. In this cycle alanine functions as a carrier of amino groups. Except for glutamine, alanine is the predominant amino acid released by muscle and accounts for half of the remaining output. Much of this probably represents pyruvate formed by glycolysis within the muscle and transaminated to alanine by other amino acids released during muscle proteolysis. After transfer through the plasma, the alanine is extracted by the liver. Here the amino group is removed for urea synthesis, thereby regenerating pyruvate for gluconeogenesis. It should be clear from Fig. 50-6, *B*, that although alanine is a gluconeogenic amino acid, no *net* synthesis of glucose from protein results from this particular pathway. This is so even though fully 10% of the glucose molecules can be traced back to alanine molecules. On the other hand, with somewhat more prolonged fasting, the production of alanine from glucose via pyruvate diminishes, whereas alanine arising from protein degradation increases. Gluconeogenesis from the latter molecules would then constitute net synthesis of glucose. As in the case of leucine, the supply of alanine for synthesis of critical proteins can be maintained by interprotein or interorgan transfers.

Quantitative Aspects of Lipid Metabolism

Lipid molecules serve numerous functions in the body. As already indicated, lipids, mainly in the form of long-chain saturated fatty acids, are the major substrates for oxidation in muscle, heart, and visceral organs via the β-oxidation pathway. They also give rise in the liver to the water-soluble ketoacids, which in turn can be used for fuel by these same organs and, when necessary, by the central nervous system. Triglycerides (glycerol esterified with three fatty acid molecules) constitute the storage form of lipid and the main energy reserve of the body. In addition to the large adipose tissue depot, much smaller amounts are also maintained in the muscle mass and heart. Total body triglycerides derive from (1) exogenous sources, with chylomicrons serving as the vehicle for transport from the gut to the other tissues and (2) endogenous synthesis from glucose and amino acid carbon. Endogenous synthesis takes place largely in the liver, and very low density lipoproteins (VLDL) serve as the transport vehicle to adipose tissue and other sites. Essential fatty acids are incorporated into phospholipids and along with cholesterol are key components of plasma, mitochondrial, nuclear, and neural membranes. Certain unsaturated fatty acids are also precursors for important compounds, such as the prostaglandins. Cholesterol is also important as the precursor of bile salts and steroid hormones.

The production, transport, and utilization of lipids are complex. They will be reviewed here briefly, primarily as a background for understanding their relationship to the endocrine system. Table 50-4 presents the average plasma concentrations of the most important lipids. It is apparent that, except for the water-soluble ketoacids, nonesterified or free fatty acids circulate in the lowest concentration, although their rate of utilization is by far the greatest. Therefore they turn over very rapidly. The plasma half-life is about 2 minutes, and the daily rate is about 200 g/day. This turnover represents molecules released from adipose tissue by lipolysis and taken up by muscle, liver, and other parenchyma. Terminal oxidation of the fatty acids accounts for about half the turnover, whereas the other half represents their reesterification back to triglycerides in muscle and liver.

Triglycerides and cholesterol, which form the major components of plasma lipids (Table 50-4), circulate as complex lipoprotein particles ranging in size from 75 to 1500 μm. The nonpolar core of these particles contains triglyceride and cholesterol esters, whereas their polar surfaces consist of phospholipid, cholesterol, and numerous apoproteins. The plasma lipoproteins are classified basically according to their physical densities. The major classes and their compositions are shown in Table 50-5. As would be expected, the lowest density particles contain primarily triglycerides. As the particles

Table 50-4. Average lipid concentrations in postabsorptive plasma

	mg/dl	*mmole/L*
Ketoacids	10	0.1
Free fatty acids	10	0.4
Triglycerides	100	1.2
Cholesterol (total)	185	4.8
Low density	120	
High density	50	
Very low density	15	

Table 50-5. Major lipoprotein classes

			Cholesterol			
	Density	*Triglyceride (%)*	*Free (%)*	*Esters (%)*	*Phospholipid (%)*	*Protein (%)*
Chylomicrons	<0.94	85	2	4	8	2
VLDL	0.94-1.006	60	6	16	18	10
IDL	1.006-1.019	30	8	22	22	18
LDL	1.019-1.063	7	10	40	20	25
HDL	1.063-1.21	5	4	15	30	50

increase in density, the triglyceride proportion decreases while that of phospholipid and protein increases. Thirteen types of apoproteins varying in size and charge are found among the lipoprotein classes. Their varied functions include facilitation of triglyceride transport out of the intestine and liver, activation of enzymes, such as lecithin-cholesterol acyltransferase and lipoprotein lipase, and attachment of the lipoprotein particles to specific cell surface receptors. Apoproteins also can be exchanged by lipoprotein particles, a process that facilitates the normal traffic of lipids through the blood.

Chylomicrons circulate after a meal. They are formed from dietary fat and transported across the intestinal wall into the blood, as detailed in Chapter 44. They are rapidly cleared from plasma, with a half-life of 5 minutes. On the capillary endothelial surfaces of adipose tissue, muscle, and heart, chylomicron triglyceride is partly hydrolyzed by the enzyme lipoprotein lipase (Fig. 50-7). The latter is activated by apoprotein CII. The liberated free fatty acids are transported across the endothelial cells and taken up for resynthesis and storage as intracellular triglycerides. The remaining particles, known as chylomicron remnants, now contain less triglyceride and are enriched with cholesterol esters. By interaction with high-density lipoprotein (HDL) particles, these remnants have also acquired apoprotein E, which with apoprotein B48 directs the uptake of the remnants by the liver. There the remnants are degraded back to free fatty acids, glycerol, free cholesterol, and the protein portion to amino acids. Thus the net result of chylomicron metabolism is to transfer most of the dietary fat to adipose tissue and the dietary cholesterol to the liver.

VLDL particles are the major source of plasma triglycerides in the postabsorptive state. VLDL particles are synthesized and secreted largely by the liver, with a much smaller contribution from the intestine. The liver uses free fatty acids, which are produced by de novo synthesis, derived from intrahepatic hydrolysis of triglycerides, or taken up directly from the plasma free fatty acid fraction. Under normal dietary circumstances about 15 g of VLDL are produced per day. However, this can increase threefold to sixfold to accommodate high-calorie or high-carbohydrate diets. VLDL particles have a plasma half-life of about 2 hours, turning over

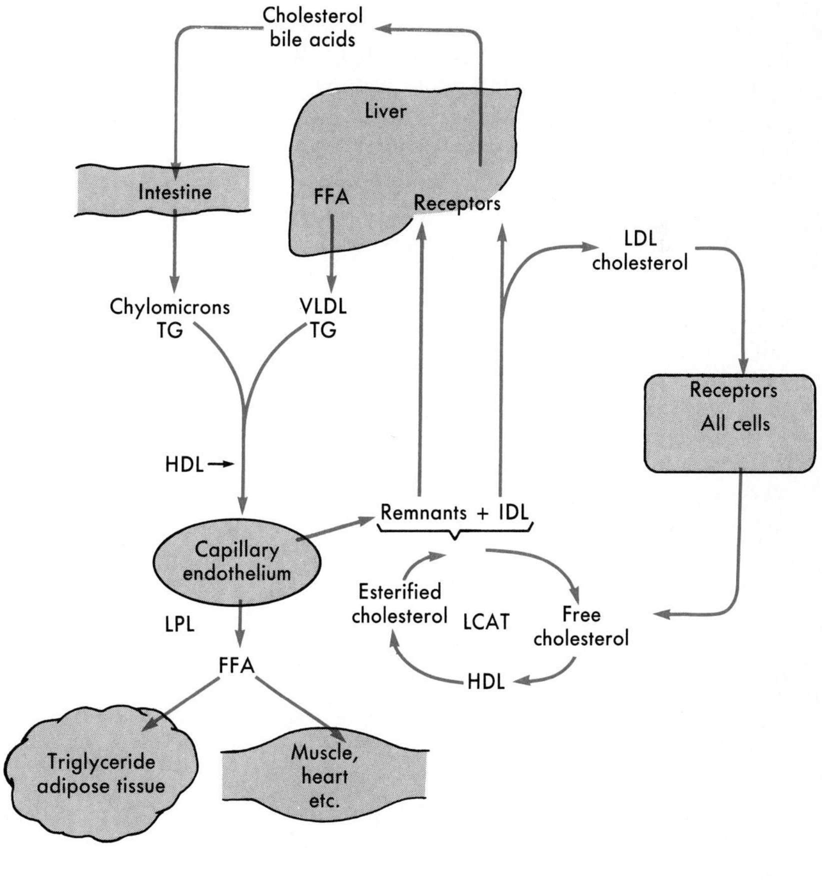

■ **Fig. 50-7.** Schematic overview of major aspects of lipid turnover in humans. Exogenous triglycerides (chylomicrons absorbed from the intestine) and endogenous triglycerides (very low density lipoproteins [VLDL] produced in the liver) both give rise to free fatty acids for storage in adipose tissue and oxidation in muscle. High-density lipoprotein particles (HDL) facilitate, and the enzyme lipoprotein lipase (LPL) directly catalyzes, liberation of the free fatty acids from triglycerides. The resultant particles, called chylomicron remnants and intermediate-density lipoproteins (IDL) from VLDL, undergo further change in the circulation, which also is facilitated by HDL particles. The ratio of esterified cholesterol to free cholesterol is increased in the remnant and IDL particles by the enzyme lecithin-cholesterol acyltransferase (LCAT). The remnant particles are then taken up by the liver for further metabolism. The IDL particles are partly taken up by the liver and partly converted to cholesterol-rich low-density lipoprotein particles (LDL). The latter are then taken up by virtually all cells after interaction with specific LDL receptors. Cholesterol, either synthesized in the liver or extracted from remnant and IDL particles, is also excreted into the intestine, in part as bile acids.

more slowly than chylomicrons. The initial phase of VLDL metabolism follows essentially the same route as chylomicrons (Fig. 50-7). The triglyceride again is hydrolyzed by endothelial lipoprotein lipase, and the fatty acids are taken up by adipose tissue and muscle. HDL particles participate in this process, as they do for chylomicrons. The resultant smaller particle, partially triglyceride depleted and cholesterol enriched, is called intermediate density lipoprotein (IDL) (Table 50-5). About half of the IDL particles are taken up by the liver, directed by apoprotein E and apoprotein B100. However, the other 50% is converted to low-density lipoprotein (LDL) particles in the blood and liver by an unknown mechanism.

LDL is the major cholesterol compound in plasma (Table 50-4), and it turns over with a half-life of 24 hours. This product of VLDL is also of major physiological importance. The cholesterol esters in LDL particles are taken up by numerous tissues after interaction of the lipoprotein particle with specific receptors that recognize its apoprotein E and apoprotein B100. The internalized cholesterol esters are hydrolyzed in lysosomes to free cholesterol, which is then utilized by different cells for various purposes. It also may be reesterified (a reaction catalyzed by acyl-CoA–cholesterol acyltransferase) and then stored. The free cholesterol level within the cell modulates itself. Cholesterol diminishes its own further uptake into the cell by down regulating the LDL receptor, and it reduces its own intracellular de novo synthesis by suppressing the rate-limiting enzyme, hydroxymethylglutaryl-CoA (HMG-CoA) reductase. The net result of VLDL metabolism is to transfer exogenously and endogenously produced cholesterol, as well as endogenously produced triglyceride, from the liver to other tissues.

HDL particles are synthesized by the liver and to a lesser extent by the intestine, and then they are secreted into the blood in a discoidal form. They circulate with a half-life of 5 to 6 days. As already indicated, they assist in the metabolism of the other lipoprotein particles by exchange of apoproteins. One subclass of nascent HDL can also pick up free cholesterol from cell membranes, as well as from chylomicron remnants and IDL and LDL particles. The free cholesterol is then esterified by transfer of a fatty acid from lecithin, a reaction catalyzed by the circulating enzyme lecithin-cholesterol acyltransferase and activated by apoprotein A1. The newly esterified cholesterol can then be transferred from the now spherical HDL particle to other remnants and IDL or LDL particles. Apoprotein D facilitates these transfers. The net result of HDL particle function is to accelerate clearance of triglycerides from the plasma and to regulate the amounts and ratios of cholesterol and cholesterol esters in plasma.

An understanding of the metabolism of triglyceride and especially cholesterol is very important to medicine. Strong epidemiological data link these lipids to the development of atherosclerosis. The atheromatous plaques that obstruct the aorta and major arteries in the heart, brain, kidneys, and extremities are rich in cholesterol and other lipids. The risk of cardiovascular disease and death is correlated positively with plasma LDL cholesterol levels and negatively with plasma HDL cholesterol levels. A high ratio of HDL to LDL cholesterol decreases this risk. This ratio is higher in association with the female gender, exercise, and estrogen or nicotinic acid administration. The net balance of cholesterol in the body is a result of the amount absorbed from the diet (approximately 300 mg) plus the amount synthesized in the liver (approximately 600 mg) versus the daily excretion from the liver as bile salts and neutral steroids. Hepatic synthesis of cholesterol varies inversely with the dietary intake. Synthesis increases to meet tissue cholesterol needs when dietary intake is low. Anything influencing cholesterol production, distribution, or disposal may be very important to an individual's health.

Major abnormalities in lipoprotein levels are caused by specific defects in the pathways of lipid metabolism and they may be very serious. A deficiency in endothelial lipoprotein lipase activity leads to excessive and prolonged chylomicron and triglyceride levels after fat-containing meals. A deficiency of cell-surface LDL receptors leads to very high plasma LDL cholesterol levels and premature coronary artery disease. Synthesis of an abnormal apoprotein E, which cannot efficiently direct lipoprotein particles to interact with cell receptors, leads to accumulation of chylomicron remnants and IDL in plasma and the production of an abnormal amount of VLDL. In this situation, both triglyceride and cholesterol levels are elevated in plasma. Conversely, a deficiency of the transfer apoprotein B leads to very low levels of chylomicrons, VLDL, and LDL in plasma, with accumulation of triglyceride in the intestines and the liver.

■ *Caloric Balance*

Changes in caloric expenditure, largely because of muscle activity, are compensated for by increases or decreases in caloric intake so as to prevent undue weight loss or gain. If normal humans are fed liquid diets whose caloric density (kilocalories per milliliter) is systematically raised or lowered without their knowledge, they will quickly adjust the volume they ingest to maintain an isocaloric intake. The mechanisms that regulate appetite and stabilize body weight are multifactorial and still are not fully understood. The crucial role of specific areas of the hypothalamus in the brain appears well established. Current concepts include a *hunger center,* which stimulates food seeking and probably is located in the ventrolateral hypothalamus, and a *satiety center,* which tonically inhibits the hunger center and probably is located in the ventromedial area of the hypothalamus.

Various pathways regulate caloric intake: (1) A decrease in plasma glucose concentration, in body glucose stores, or in intracellular glucose metabolism in the hypothalamus reduces the firing rate of the satiety center. This in turn releases the hunger center from its tonic inhibition and initiates food seeking. This would key total energy and nutrient input to the need for one critical substrate: glucose. (2) A decrease in total adipose tissue mass, sensed in the hypothalamus via neural or chemical signals, decreases satiety center activity. This would key caloric input to the level of the major energy stores. (3) Increased thermogenesis following the oxidation of fuels is sensed by the hypothalamus, which then suppresses eating. This would key caloric intake to energy expenditure and to the need to maintain body temperature. It is likely that a glucostat, a lipostat, and a thermostat all contribute to keeping caloric intake equal to caloric expenditure, when averaged over extended time periods.

Other factors also may influence appetite significantly. Increased adrenergic input to the hypothalamus may stimulate eating, relating it to "stress." Increases in endorphins—opiate polypeptides associated with pain relief—may increase eating, possibly relating it to "pleasure." Increases in brain concentrations of gastrointestinal hormones, such as cholecystokinin, may decrease eating, possibly relating it to feedback from the process of digestion and absorption of food.

Inappropriate increases or decreases in caloric intake could be compensated for by responsive changes in caloric expenditure. Long-term feeding of a caloric excess to volunteers causes weight gains that are almost entirely accounted for by fat storage. However, the rate of weight gain is less than that expected for the caloric excess that is administered. Furthermore, maintenance of the new increased body weight requires 50% more calories per day than did maintenance of the original weight. These data cannot be explained by any observable increase in physical activity or in cost of work. Therefore an adaptive increase in thermogenesis, in either the basal or postprandial state, has been suggested as a normal physiological response to overexpansion of adipose tissue mass. The dissipation of extra ingested energy as heat could come about in part from hypertrophy of "futile metabolic cycles," which waste high-energy ATP. Examples include glucose → glucose-6-phosphate → glucose, and fatty acids → triglyceride → fatty acids. The ability of islet hormones to regulate the key enzymes in such cycles (as discussed later) could be relevant to mechanisms that relate fuel turnover to fuel storage.

Anatomy of Pancreatic Islets

Insulin, glucagon, somatostatin, and *pancreatic polypeptide* are the four hormones currently known to be released by the *islets of Langerhans* in the pancreas. Each is synthesized, stored, and secreted by a different cell type. The islets are discrete bodies that are scattered throughout the pancreas and comprise 1% to 2% of its weight. They are more numerous in the tail than in the body or head of the organ. The adult human pancreas contains approximately 900,000 islets.

The β-cells, which are the source of insulin, make up the bulk of the islet mass (60%) and in humans are located in the center of the islet. α-Cells, the source of glucagon, make up 25%, and δ-cells, the source of somatostatin, comprise 10% of the islet mass. These latter two cell types are located in the periphery of the islet. However, penetration of α- and δ-cells into the central core appears to occur in lobules. This arrangement allows for more zones in which β-, α-, and δ-cells may form a functional syncytium (Fig. 50-8) with paracrine effects among them. The cells responsible for pancreatic polypeptide release are concentrated in islets located in the head of the pancreas. The islets are heavily vascularized, and the capillary endothelial cells are fenestrated, which permits rapid exchange across them. Innervation to the islets is by both adrenergic and cholinergic fibers of the autonomic nervous system.

The endocrine cells of the islets appear to develop from pancreatic ducts, which are endodermal in origin. Although they show some characteristics similar to those of neuroectodermal cells of the neural crest, there are no conclusive data that islet cells first migrated from the neural crest to the foregut. In the course of evolution of the intestinal mucosa, the β-cells were the first to agglomerate into discrete masses resembling islets in cyclostomes (e.g., hagfish). Shortly thereafter the δ-cells began to shift into the islets. Not until the phylogenetic stage of cartilaginous fish did the α-cells migrate from their diffuse intestinal loci into the islets. Thus the close anatomical association of these cells serves a relatively late evolutionary function. In the human fetus the islets are identified by 4 weeks of gestation. They are capable of synthesis and secretion of islet hormones by 10 weeks. However, responsiveness to stimulation is markedly blunted compared with that of adult islets, probably as a result of the stable supply of substrates from the mother.

The islet cells are arranged in cords along capillary channels. Their plasma membranes contain various specialized areas. Tight junctions are linear lines of fusion between plasma membranes of adjacent cells. These may serve to form intercellular compartments protected from the general interstitial fluid of the islets but permitting the secreted hormone of one cell to act on the membrane of its neighbor (Fig. 50-9). Gap junctions

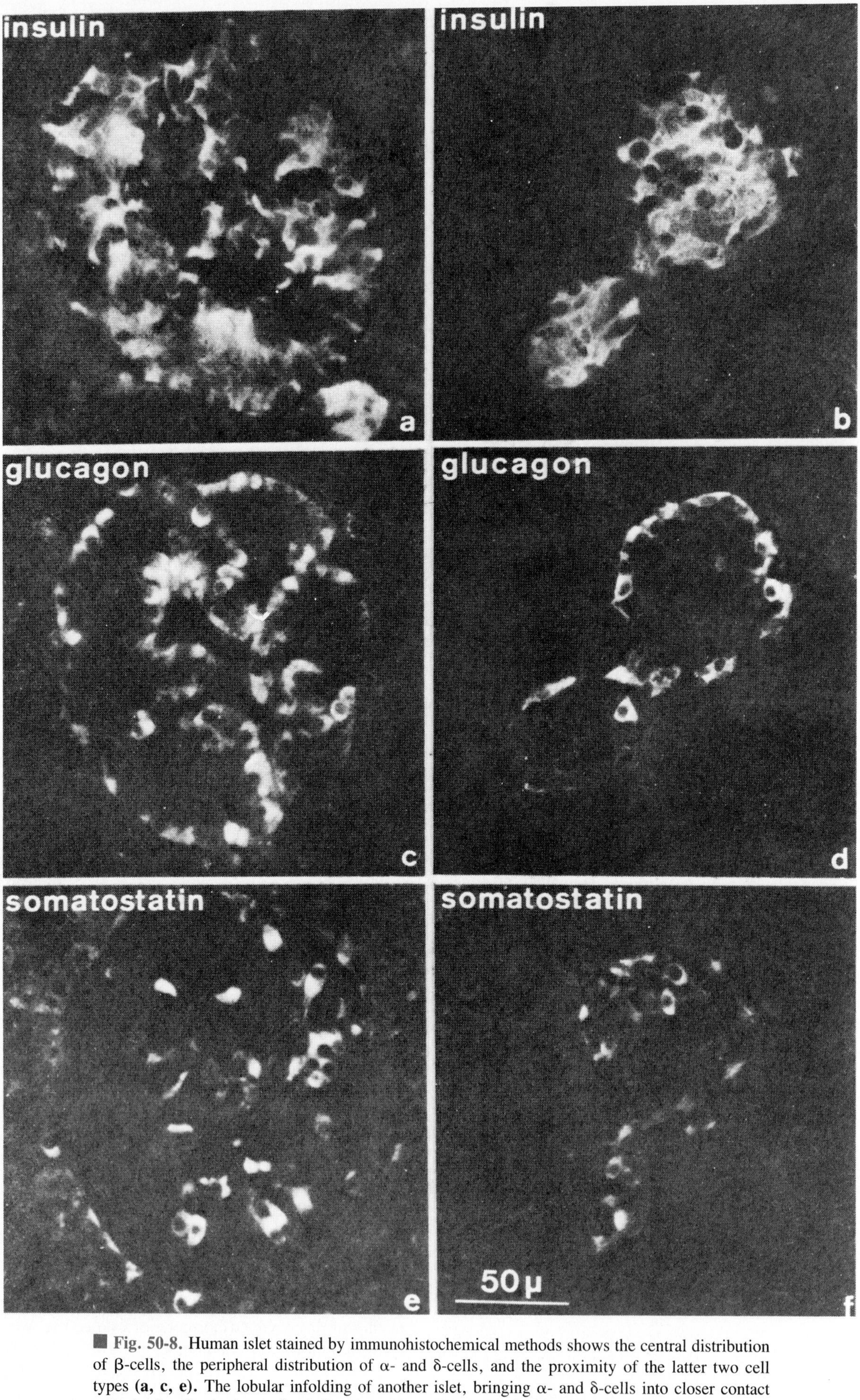

■ **Fig. 50-8.** Human islet stained by immunohistochemical methods shows the central distribution of β-cells, the peripheral distribution of α- and δ-cells, and the proximity of the latter two cell types **(a, c, e)**. The lobular infolding of another islet, bringing α- and δ-cells into closer contact with the β-cells, is shown **(b, d, f)**. (From Unger, R.H., et al.: Reproduced with permission from the *Annual Review of Physiology,* volume 40. Copyright © 1978 by Annual Reviews, Inc.)

are bridging channels through adjacent plasma membranes that may permit direct passage of low–molecular weight substances from the cytosol of one islet cell to the cytosol of its neighbor.

By electron microscopy, islet cells are seen to contain secretory granules with smooth membranes within which hormone is stored. A system of microtubules is present, often lying in parallel bundles that separate linear rows of secretory granules. In addition, microfilaments containing myosin and actin form a web adjacent to the plasma membrane and in association with the microtubules. Agents that destroy microtubules or prevent their function (such as colchicine and deuterium oxide) inhibit hormone release, whereas agents that cause hypercontraction of microfilaments (such as cytochalasin B) enhance hormone release. Therefore it has been proposed that, when an islet cell is stimulated, secretory granules are actively transported to the plasma membrane, guided by the microtubules and pulled by the microfilaments. This has been most extensively studied

A

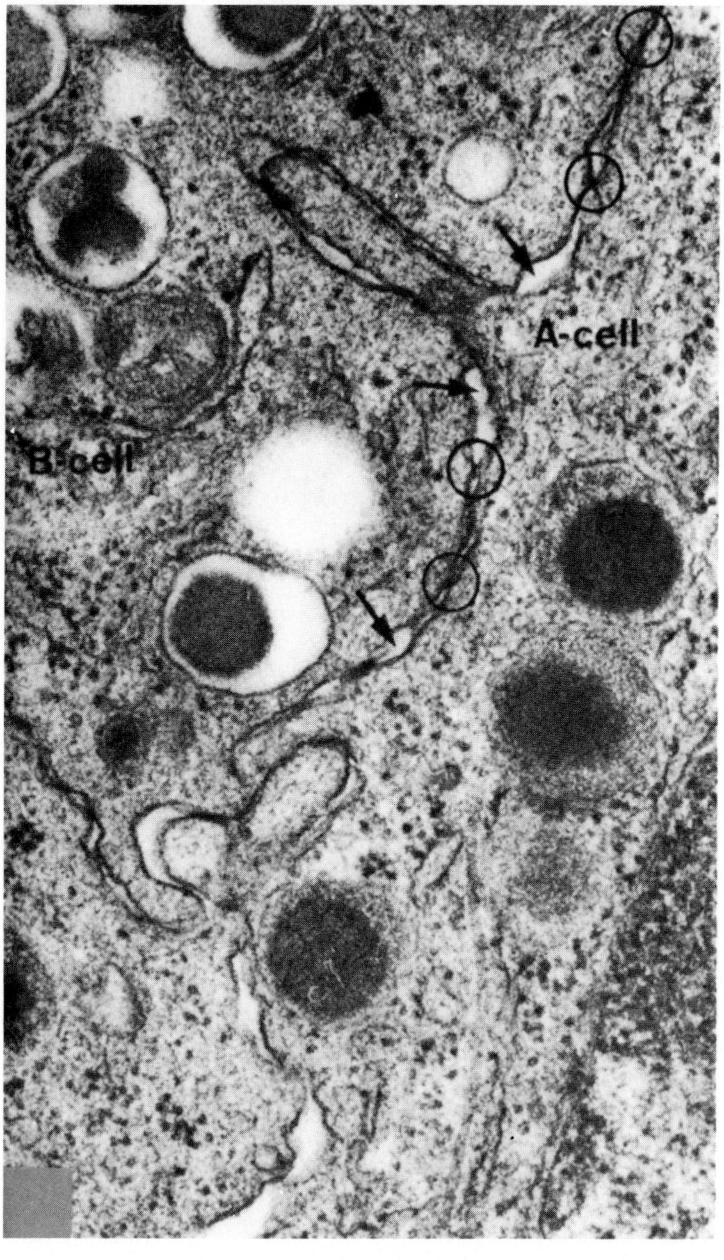

■ **Fig. 50-9.** **A,** Electron microscopic photograph of adjacent β- and α-cells, showing tight junctions *(arrows)* and gap junctions *(circled areas)*. (From Orci, L., et al.: Cell contacts in human islets of Langerhans, J. Clin. Endocrinol. Metab. **41:**841, 1975. Copyright 1975. Reproduced by permission.)

Continued.

B

■ **Fig. 50-9, cont'd. B,** Further magnification of **A** showing the discrete structural characteristics of the tight junctions *(TJ)* and gap junctions *(GJ)*.

by cinemicroscopy, which has shown movement of insulin-containing granules through the cytoplasm of β-cells at a rate of 1.5 μm/second.

■ *Insulin*

■ *Structure and Synthesis*

Among the peptide hormones, insulin has been of preeminent historical, physiological, and clinical importance. It was the first hormone to be isolated from animal sources in pure enough form to be administered therapeutically—and with dramatic life-saving effects. It was also the first to have its amino acid sequence and tertiary structure elucidated. Insulin was the first peptide hormone for which a mechanism of action on the cell membrane was demonstrated. It was also the first to be measured by radioimmunoassay; indeed, the entire concept of this landmark analytical method developed from studies of the metabolism of radiolabeled insulin in humans. The biosynthesis of peptide hormones from larger precursor molecules was observed initially in insulin. Finally, it was the first mammalian peptide hormone whose biosynthesis in bacteria by recombinant DNA technology was shown to result in a fully active product in humans. When one adds to this record the central role insulin plays in the rapid modulation of all major fuel fluxes, its importance to endocrine physiology can hardly be overstated.

Insulin is a peptide consisting of two straight chains. Its molecular weight is 6000. The A chain, containing 21 amino acids, and the B chain, containing 30 amino acids, are linked by two disulfide bridges. In addition, the A chain contains an intrachain disulfide ring (Fig. 50-10). The multiple amino acid differences among insulins from various species change the biological activity little; mammalian and most fish insulins are virtually equipotent in humans. The features most conserved in vertebrate evolution are the positions of the three disulfide bonds, the N-terminal and C-terminal amino acids

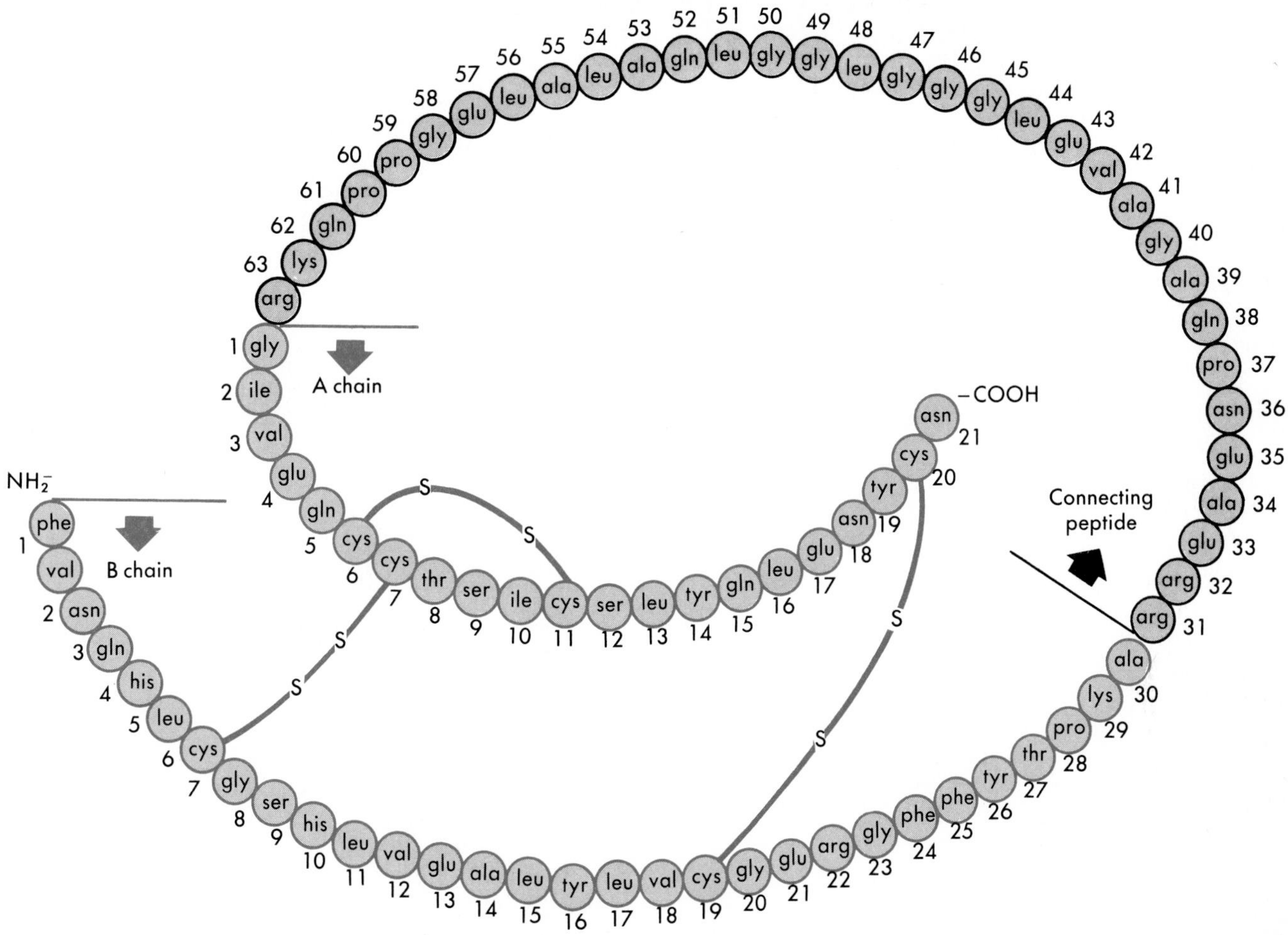

■ **Fig. 50-10.** The structure of porcine proinsulin. The solid area is the insulin molecule released by cleavage of the connecting peptide. (Redrawn from Shaw, W.N., and Chance, R.E.: Effect of porcine proinsulin in vitro on adipose tissue and diaphragm of the normal rat, Diabetes **17**:737, 1968. Reproduced with permission from the American Diabetes Association, Inc.)

of the A chain, and the hydrophobic character of the amino acids at the C-terminal of the B chain. These areas determine secondary and tertiary structure, which are critical to biological activity. Insulin monomers readily form dimers (molecular weight 12,000). In the presence of zinc three of the latter in turn form a crystalline hexameric unit with a threefold axis passing through two zinc atoms. Crystalline zinc insulin is the basic pharmaceutical preparation of greatest importance in therapy.

Synthesis of insulin occurs as shown in Fig. 50-11. The insulin gene is located on human chromosome 11, and it is composed of four exons and two introns. The gene is transcribed to a messenger RNA that directs the synthesis of preproinsulin, a precursor of molecular weight 11,500. Preproinsulin is synthesized as four sequential peptides: an N-terminal signal peptide, the B chain of insulin, a connecting peptide, and the A chain of insulin (Fig. 50-11). The N-terminal 23 amino acid signal is rapidly cleaved at the site of synthesis while the proinsulin chain is being completed. The function of the evanescent signal may be to bring the ribosomes into association with the membrane of the endoplasmic reticulum and permit guided passage of proinsulin to the Golgi apparatus. Establishment of the disulfide linkages yields the "folded" proinsulin molecule with a molecular weight of 9000. Proinsulin contains the disulfide bonded A and B chains of insulin linked to a connecting peptide (C-peptide) through two basic residues each at the C-terminal of the B chain and the N-terminal of the A chain (Fig. 50-10). During its transport from the endoplasmic reticulum and its packaging into granules by the Golgi apparatus, proinsulin is slowly cleaved. Separate and specific trypsinlike and carboxypeptidase-like enzyme activities are required to split off the arg-arg and lys-arg residues (Figs. 50-10 and 50-11). The generated insulin and C-peptide molecules are each retained in the granules and released during secretion in equimolar amounts. The association of insulin with zinc takes place as the secretory granules

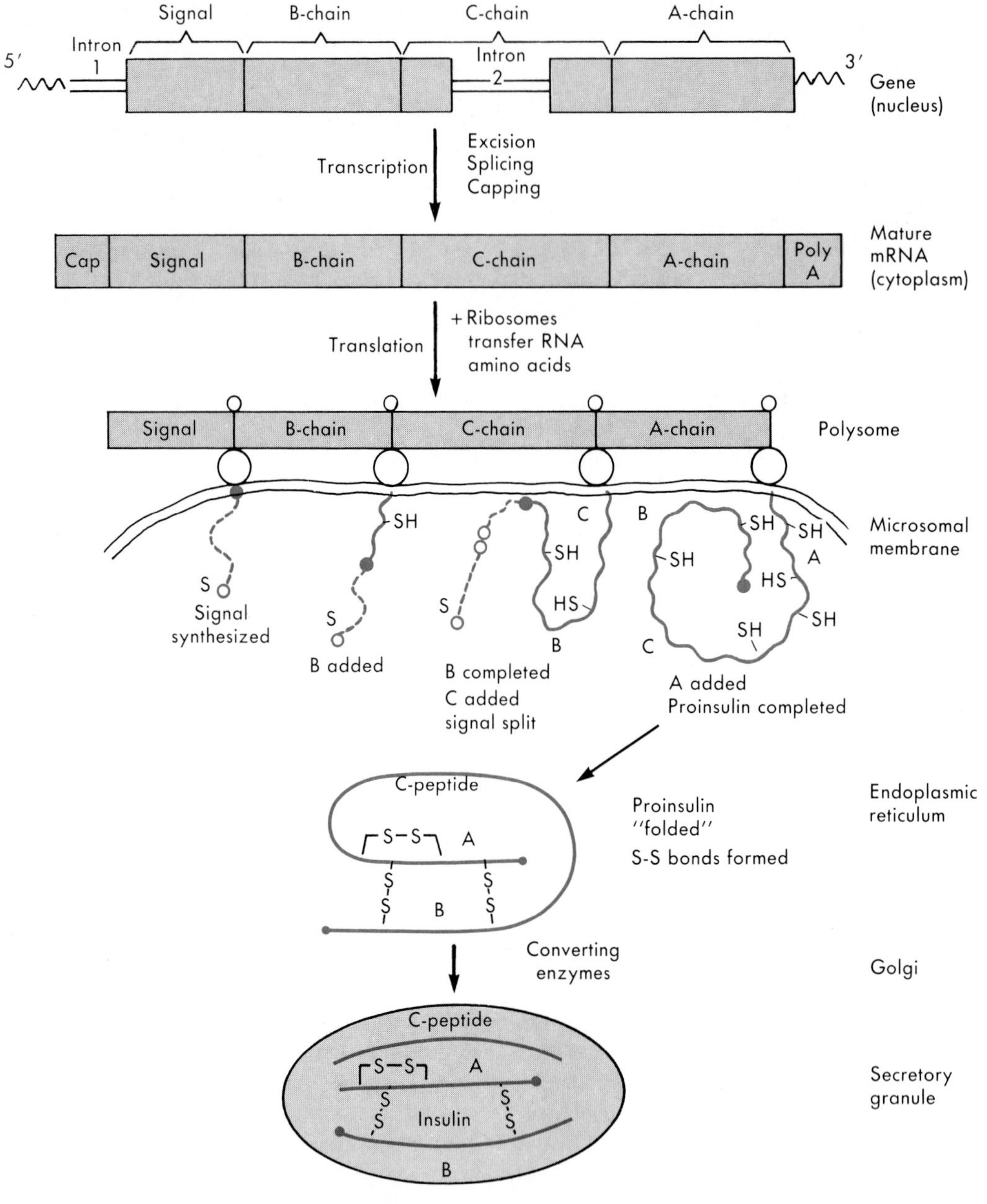

■ **Fig. 50-11.** Insulin synthesis; the insulin gene codes for preproinsulin. The mature messenger RNA initiates synthesis of the N-terminal signal peptide *(S)* in the ribosomes, followed by the B, C, and A chains. The signal is degraded during the course of completion of the proinsulin molecule. The latter is folded into a conformation that permits the disulfide linkages between the A and B chains to form. Within the Golgi and secretory granule, converting enzymes cleave off the C chain, known as the C-peptide, completing the synthesis of insulin. The insulin molecules are concentrated in the electron-dense core of the granule, whereas the C-peptide molecules are in the peripheral halo areas of the granule. (From Permutt, M., et al.: Insulin gene structure and function: a review of studies using recombinant DNA methodology, Diabetes Care **7**:386, 1984. Reproduced with permission from the American Diabetes Association, Inc.)

Table 50-6. Insulin secretion

Stimulators		*Inhibitors*
D-Glucose	Secretin	Fasting
Galactose	Cholecystokinin	Exercise
Mannose	Glucagon	Somatostatin
Glyceraldehyde	Enteroglucagon	α-Adrenergic stimuli
Protein	Calcium	Prostaglandins
Arginine	Potassium	Diazoxide
Lysine	Vagal activity	Phenytoin
Leucine	β-Adrenergic stimuli	
Alanine	Acetylcholine	
Free fatty acids	Sulfonylurea drugs	
Ketoacids		
Gastric inhibitory polypeptide		
Gastrin		

mature. The zinc insulin crystals then form the dense central core of the granule, whereas C-peptide is present in the clear space between the membrane and the core. The overall process of insulin synthesis is stimulated by glucose and by feeding, which increase both transcription and translation of the gene. Release of insulin occurs by exocytosis (Chapter 49).

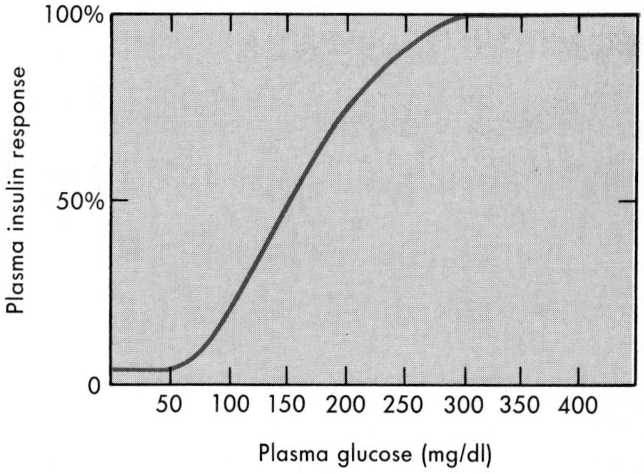

Fig. 50-12. Approximate in vivo relationship between plasma glucose and insulin secretion, the latter being assessed by the plasma insulin response to stepwise infusion of glucose in humans. No insulin is secreted below a plasma glucose level of 50 mg/dl. Half maximum secretion occurs at 125 to 150 mg/dl. (Redrawn from Karam, J.H., et al.: "Staircase" glucose stimulation of insulin secretion in obesity; measurement of beta cell sensitivity and capacity, Diabetes **23:**763, 1974. Reproduced with permission from the American Diabetes Association, Inc.)

Regulation of Secretion

A large number of factors modulate insulin secretion (Table 50-6). The major principle, however, is that *insulin secretion is stimulated under circumstances of fuel excess and is inhibited under circumstances of fuel deficiency*. Glucose is the stimulant of greatest importance in the human being. Since insulin in turn stimulates the utilization of glucose, this substrate-hormone pair forms a feedback system for close regulation of plasma glucose levels. The relationship between plasma insulin and plasma glucose is sigmoidal (Fig. 50-12). Virtually no insulin is secreted below a plasma glucose threshold of about 50 mg/dl. A half-maximum insulin secretory response occurs at a plasma glucose level of about 150 mg/dl and a maximum insulin response at levels of 300 to 500 mg/dl.

Both in vitro and in vivo insulin secretions exhibit a biphasic response to a continuous glucose stimulus (Fig. 50-13). Within seconds of exposure of the β-cells to an increased glucose concentration, there is an immediate pulse of insulin release, peaking at 1 minute and then returning toward baseline. After 10 minutes of continuous stimulation, a second phase of secretion begins with a slower rise of insulin to a second plateau, which can be maintained for many hours in normal individuals. The genesis of this biphasic response remains controversial. Although initially described for insulin, it also is seen with stimulation of other peptide hormones by appropriate agents. A number of explanations have been suggested: (1) two storage compartments exist—labile and stable—that contain granules with different sensitivities to glucose; (2) the second phase of insulin secretion represents newly synthesized insulin resulting from glucose stimulation of the synthetic process; (3) a feedback inhibitor of insulin release is rapidly generated after glucose stimulation and then slowly removed. Whatever the mechanism, it is important to note that loss of the initial rapid phase of insulin secretion is the most characteristic feature of insulin-deficient diabetes mellitus.

When glucose is given orally, a greater insulin response is elicited than when plasma glucose is comparably elevated by intravenous administration. This is accounted for by release of a number of gastrointestinal hormones that normally accompany the digestive process and that are capable of potentiating glucose-stimulated insulin secretion. Gastric inhibitory polypeptide is probably the most important of these insulinogogues, but in high concentrations gastrin, secretin, cholecystokinin, pancreatic glucagon, and enteroglucagon all share this property. This prompt gastrointestinal mechanism of insulinogenesis moderates the early rise in plasma glucose that follows the ingestion and absorption of a carbohydrate meal.

Insulin secretion also is stimulated by oral protein; this is mediated by the amino acids resulting from digestion of the protein. The basic amino acids, arginine and lysine, are the most potent stimulants; leucine, alanine, and others contribute modestly to this effect. Glucose and amino acids are synergistic in their actions, so that the insulin rise that follows a meal repre-

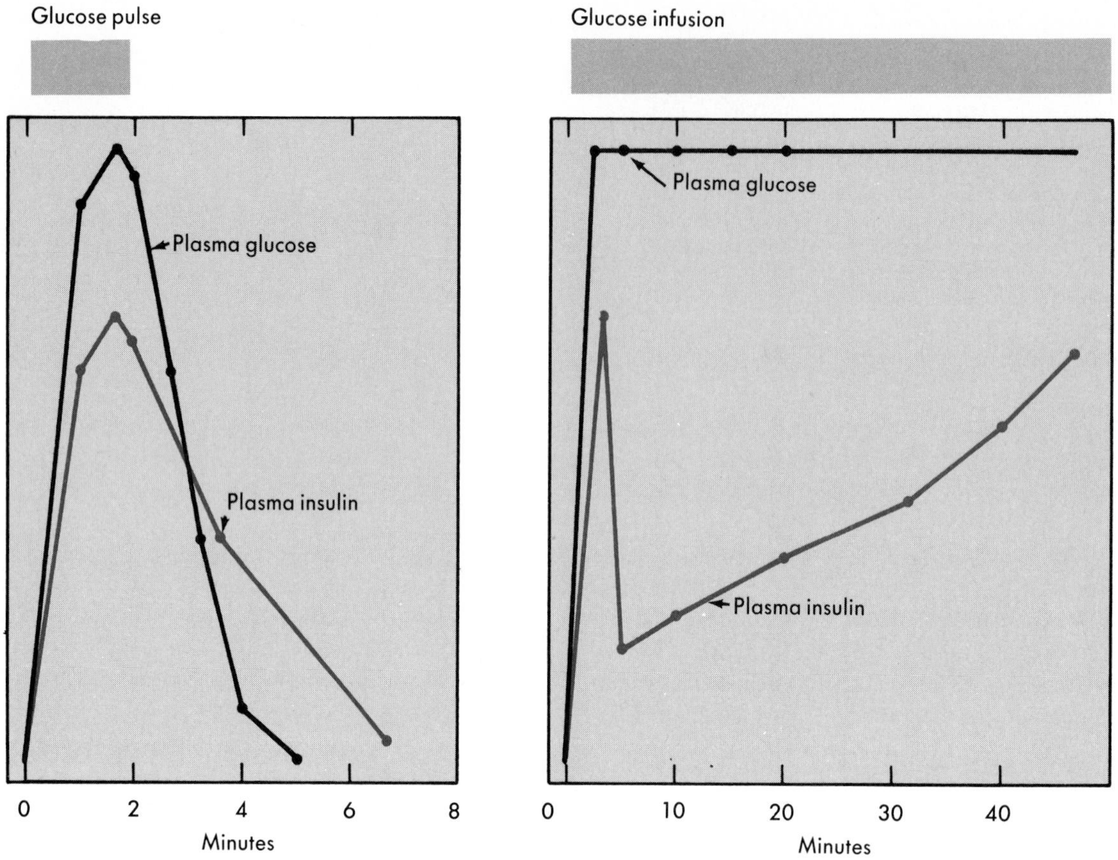

Fig. 50-13. In vitro insulin response to a glucose pulse *(left panel)* shows no β-cell memory. If glucose infusion is continued *(right panel),* the initial burst of insulin is not sustained, but a second wave of secretion develops soon after.

sents more than the additive effect of its carbohydrate and protein content. Triglycerides and fatty acids have little, if any, stimulatory effect in humans but may contribute more to sustaining insulin secretion in other species. Ketoacids at concentrations that prevail during prolonged fasting stimulate insulin secretion; although modest, this effect may help sustain a critical low level of insulin when β-cell stimulation by glucose and amino acids is reduced.

Both potassium and calcium are essential for normal insulin responses to glucose. Thus relative insulin deficiency occurs in subjects depleted of potassium or calcium, and insulin excess is seen in hypercalcemic subjects. Sympathetic nervous system activity and the adrenal medullary hormone epinephrine stimulate insulin secretion via their β-adrenergic receptors but inhibit insulin secretion via their α-adrenergic receptors. Parasympathetic activity also increases insulin release, as shown by stimulating the vagus nerve in vivo or by exposing β-cells to acetylcholine in vitro. The neural routes of regulation may mediate a cephalic phase of insulin secretion, which in animals has been demonstrated to precede the entrance of food into the gastrointestinal tract. In humans insulin is secreted cyclically with a period of 13 minutes. The generation of this frequency appears to be intrinsic to the islets and may be attributable to local neurohormone activity. A class of drugs known as sulfonylureas acutely stimulates insulin release; these drugs are useful in the treatment of some patients with diabetes mellitus.

A large number of other hormones directly or indirectly produce hyperplasia of the β-cells and lead to chronic increases in insulin secretion. The list includes cortisol, growth hormone, estrogen-progesterone, human placental lactogen, and thyroid hormones. These hormones increase insulin secretion largely by antagonizing its action or increasing its need by peripheral tissues. Of great interest is the presence of somatostatin within the islets; somatostatin is a powerful peptide inhibitor of insulin release. Insulin secretion also is inhibited by prostaglandins. A feedback effect of insulin on its own secretion has been demonstrated that is independent of its hypoglycemic action. During an insulin infusion plasma C-peptide levels (representing β-cell activity) will decrease even though plasma glucose is held constant by concomitant glucose administration.

The net result of these many physiological influences is to maintain an average basal peripheral plasma insu-

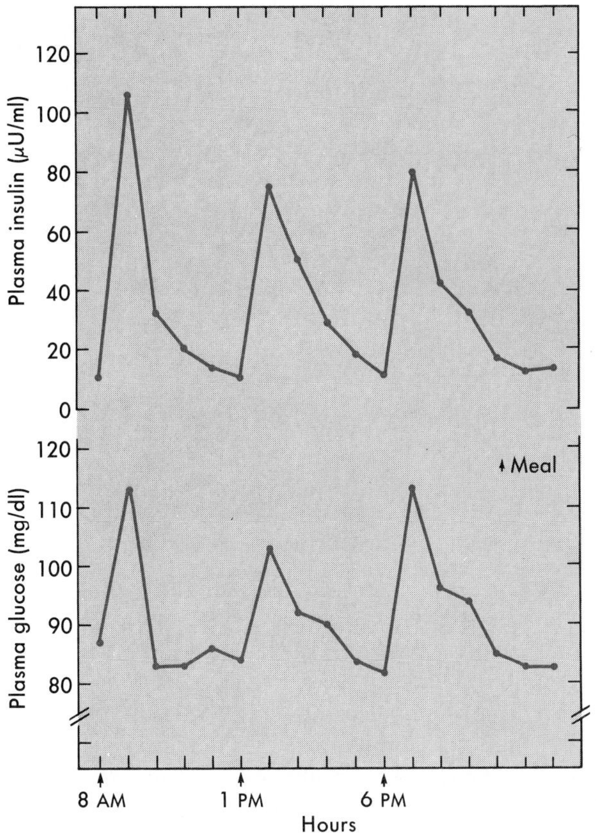

Fig. 50-14. Plasma insulin response to isocaloric meals in humans. Note slightly higher morning insulin secretion and the small excursions of plasma glucose throughout the day.

lin level of 10 μU/ml (0.2 ng/ml or 7×10^{-11} M) in normal humans. Under conditions of total fasting this value declines 50%, largely because of a decrease in plasma glucose. A similar decline also occurs during prolonged exercise. Plasma insulin increases threefold to tenfold after a typical meal, usually peaking 30 to 60 minutes after eating is initiated (Fig. 50-14). A larger prandial pulse of insulin secretion is seen with breakfast than with subsequent meals of identical carbohydrate and calorie content. Because the pancreatic vein drains into the portal vein, the liver is regularly exposed to insulin concentrations two or three times higher than those of other organs when the β-cells are in the basal state and five to ten times higher after acute β-cell stimulation. This strategic location of the liver is important to insulin's effects on metabolism.

The intimate mechanism of β-cell response to stimulants continues to be studied. Thus far the evidence for specific D-glucose or L-amino acid receptors on the β-cell membrane is indirect. Considerable evidence suggests that glucose and other nutrients must be oxidized within the β-cell for full stimulation of insulin release to occur. For example, D-glyceraldehyde, a three-carbon compound metabolized like glucose, is an equally potent stimulant. Glucose-stimulated insulin release is inhibited by mannoheptulose, a sugar that inhibits glucose phosphorylation, and by 2-deoxyglucose, an analogue that is phosphorylated but then blocks glucose utilization. Considerable evidence supports the key role of the enzyme glucokinase in regulating glucose metabolism within the β-cell and hence insulin release.

Suggested coupling molecules between glucose oxidation and insulin release include ATP or even protons. Intracellular cAMP levels increase during glucose-stimulated insulin secretion, and this nucleotide also potentiates the response to glucose. Finally, the electrical activity of the β-cell is altered by glucose that, in stimulating concentrations, depolarizes the cell membrane. More important, calcium is rapidly translocated into the cytosol of the β-cell during stimulation; this translocation is enhanced by cAMP and is essential for β-cell response to occur. Furthermore, activators of the phospholipid and calcium-dependent protein kinase C also stimulate hormone release. It has been postulated that the influx of calcium causes contraction of the microtubular-microfilament system, through activation of a calcium-calmodulin–sensitive, myosin light-chain kinase present in the islets. This moves the secretion granules to the plasma cell membrane where they fuse with it and extrude insulin and C-peptide by exocytosis. The fused membrane material returns to the cytoplasm as microvesicles, and it can either be recycled or degraded.

Once secreted, insulin normally circulates unbound to any carrier protein. Its half-life in plasma is 5 to 8 minutes, its volume of distribution is about 20% of body weight, and its metabolic clearance rate is 800 ml/minute (1200 L/day). In humans estimates of basal insulin delivery rate to the peripheral circulation are about 0.5 to 1 units/hour (20 to 40 μg/hour).* During meals the delivery rate increases up to tenfold, and the total daily peripheral delivery of insulin is about 30 units. Since the liver removes 50% of portal vein insulin on the first pass, the actual β-cell secretory rate is approximately 60 units/day. The initial hepatic extraction is decreased by glucose administration, permitting a greater escape of insulin to the periphery for stimulation of glucose uptake.

Once in the peripheral circulation, insulin is metabolized largely in the kidney and liver by a specific protease and by a glutathione-dependent transhydrogenase, which splits the disulfide bonds, producing separate A and B chains. Very little insulin is excreted unchanged in the urine. Degradation of insulin also occurs in association with its plasma membrane receptor after internalization by target cells. In patients who are treated chronically with animal insulins, circulating antibodies to insulin are produced. A majority of the plasma insulin is bound to these antibodies. Such binding influences the rate of availability of the hormone to target cells and degradative processes.

Although C-peptide is secreted in amounts that are

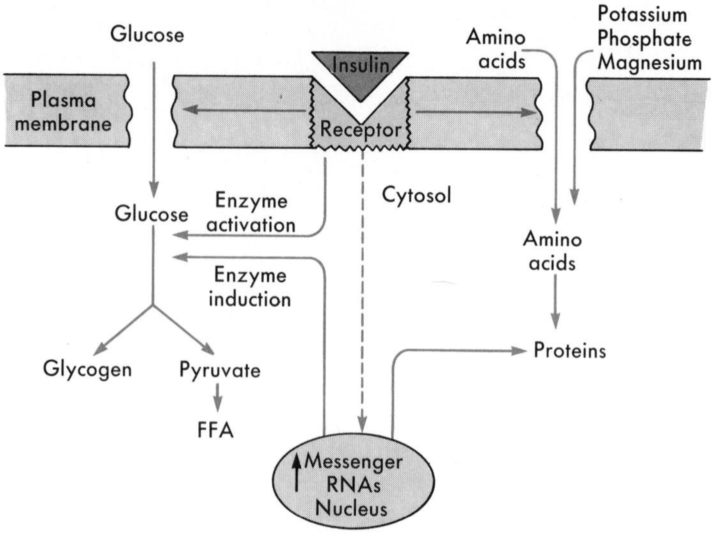

Fig. 50-15. Overall insulin actions on the cell. Consequent to insulin combination with its plasma membrane receptor, glucose enters the cell via stimulation of a carrier system operating by facilitated diffusion. Entry of amino acids and major intracellular electrolytes is also stimulated via an unknown mechanism. Activation of some enzymes and deactivation of others shift glucose metabolism toward glycogen and pyruvate and shift further metabolism of the latter toward free fatty acids. Induction of some enzymes and repression of others via effects on transcription reinforce stimulation of these metabolic routes. Protein synthesis is also generally increased.

equimolar to insulin, its basal peripheral plasma levels are approximately fourfold higher, averaging 1 ng/ml (3 $\times 10^{-10}$ M). This difference is caused by a lower rate of metabolic clearance for C-peptide. In contrast to insulin, C-peptide undergoes no significant hepatic extraction. Despite its lack of biological activity, plasma C-peptide measurements provide useful information about β-cell function. This is particularly so when a small increase in insulin secretion may not be detected by a change in peripheral plasma levels, because of its hepatic extraction, or when measurement of plasma insulin is rendered unsatisfactory by prior insulin treatment. Measurement of 24-hour urine C-peptide excretion also may be used as an index of daily insulin secretion. Proinsulin is released by the β-cell and forms up to 20% of total insulin immunoreactivity in the basal state. During normal β-cell stimulation absolute proinsulin levels increase more slowly and to a lesser degree than does insulin. Although proinsulin can exert insulin effects in vitro and in vivo, it has only 5% to 10% the activity of insulin. At present, no biological function can be ascribed to the prohormone. However, neoplastic β-cells release a greater proportion of proinsulin, a phenomenon that is of occasional diagnostic value.

Hormone Actions: Intracellular Mechanisms

Insulin exerts a multitude of actions on cell metabolism and growth. A single mechanism probably cannot account for all of them. Our current understanding is based partly on observations made either after in vivo administration of the hormone or after its selective withdrawal by destruction or antagonism of β-cell function. Other evidence consists of the direct demonstration of insulin effects on isolated or cultured cells. Therefore, some of the well-defined consequences of insulin action may prove in the final analysis to be only indirectly attributable to the hormone. For example, changes in substrate levels in one pathway that is directly regulated by insulin may, in turn, affect critically the activities of enzymes and the rate of metabolic flux in another pathway.

All actions of insulin are at least initiated by reversible association with its specific plasma cell membrane receptor (Fig. 50-15). Binding to a receptor occurs at insulin concentrations well within the physiological range of 10 to 150 μU/ml; half-maximum binding is observed at the upper end of that range. For certain actions of the hormone, full biological activity is expressed with only 5% of the receptor sites occupied; therefore spare receptors exist. Insulin receptors appear to be clustered in coated pits rather than randomly distributed on the cell surface. The receptors also exhibit mobility within the plane of the plasma membrane. The binding and affinity of insulin receptors can be modulated. Insulin itself down regulates the number of receptors. For example, the numbers of receptors are reduced in obese humans and obese animals who have elevated plasma insulin levels. When their plasma insulin levels are lowered by experimental manipulation, the number of insulin receptors increases. This effect on its own receptors is a direct action of insulin. The hormone increases the rate at which its receptors are degraded after they are internalized by the cell.

After receptor binding of insulin, a number of processes are rapidly stimulated by the hormone (Fig. 50-15). Within 1 minute, glucose transport into muscle and adipose cells is increased fivefold to twentyfold by activation of a glucose carrier system in the plasma membrane. A specific glucose transport protein of 55,000 molecular weight has been isolated from the plasma membrane, its structure determined, and its gene

*1 mg crystalline zinc insulin equals 25 units of biological activity.

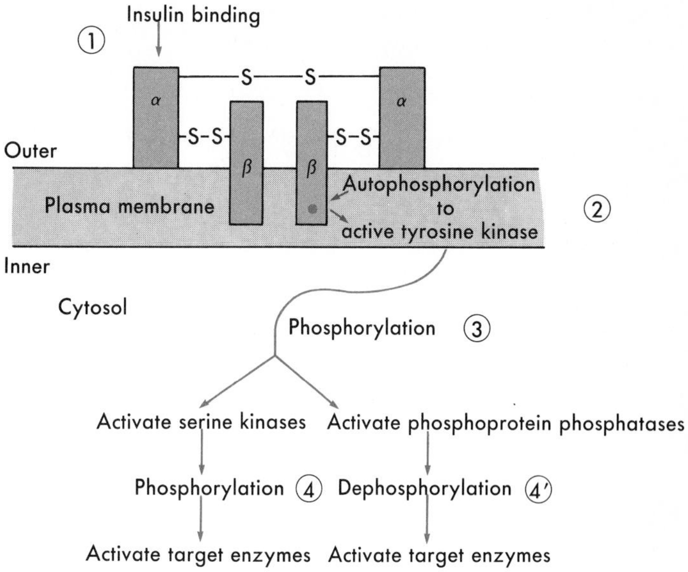

■ **Fig. 50-16.** Proposed mechanism for insulin action. Insulin binding through the α-subunit of its receptor results in autophosphorylation of a tyrosine site on the β-subunit. The resultant tyrosine kinase activity of the receptor may then itself catalyze phosphorylation of serine kinases or of phosphoprotein phosphatases. These in turn phosphorylate or dephosphorylate target enzymes, resulting in the alteration of metabolic pathways typical of insulin action.

cloned. This carrier facilitates glucose diffusion from extracellular fluid into the cytosol along an already existing, large concentration gradient. This gradient constitutes a gating mechanism for glucose that is opened by the hormone rather than one that operates as an active transport system. Therefore ATP is not required for this action of insulin to be manifest. Insulin may increase the number of glucose transport units in the outer plasma membrane by recruiting them either from the inner plasma membrane or from intracytoplasmic storage vesicles. The connecting link between the insulin receptor and the glucose transport system is not known. The importance of glucose transport to the cell is underscored by the fact that this process is normally rate limiting for glucose utilization. Interestingly, insulin also facilitates the cellular uptake of amino acids, potassium, magnesium, and phosphate (Fig. 50-15), but the mechanism has not been studied adequately.

In addition to making glucose and amino acids available, insulin directs their disposition once these compounds are within various cells. Conversion of glucose to glycogen, to pyruvate and lactate, and thence to fatty acids are all stimulated to varying degrees. From the amino acids, synthesis of specific proteins such as albumin, casein, and certain enzymes is selectively enhanced. Some of these effects are mediated rapidly by altering target enzyme activities through covalent modification. Other effects are slowly mediated by altering the number of target enzyme molecules through transcriptional or translational effects.

The key links between insulin receptor binding and the hormone's subsequent intracellular actions (i.e., the mechanism of transmembrane signaling) is under intensive investigation. The structure of the insulin receptor is now reasonably well worked out (Fig. 50-16), and the gene that codes for it has been cloned. The insulin receptor is a glycoprotein composed of two symmetrical units connected by a disulfide bond. Each unit, in turn, is composed of an α-subunit of molecular weight 135,000 and a β-subunit of molecular weight 90,000. These subunits are also connected by disulfide bonds. The entire structure resembles an immunoglobulin. The α-subunits protrude outward from the cell membrane, whereas the β-subunits extend inward through the membrane toward the cytoplasm. The insulin molecule binds to the α-subunit. Consequent to binding, a single tyrosine residue on one receptor β-subunit is immediately phosphorylated by ATP and Mn^{++}. Thus the receptor possesses catalytic activity toward itself and is considered to be an *autotyrosine kinase*. However, the tyrosine kinase activity of the hormone-occupied receptor also can be exerted on other protein substrates. Because insulin activates some enzymes by causing them to be phosphorylated and activates other enzymes by causing them to be dephosphorylated, an attractive hypothesis has developed. As outlined in Fig. 50-16, the tyrosine kinase activity of the bound receptor could itself be a second messenger for insulin activity, and it could generate parallel cascades of effects. On the one hand, the tyrosine kinase could activate other more common serine kinases that would then phosphorylate and thereby activate insulin target enzymes such as acetyl-CoA carboxylase. On the other hand, the tyrosine kinase could activate phosphoprotein phosphatases, which would in turn dephosphorylate and thereby activate insulin target enzymes, such as pyruvate dehydrogenase.

Other mechanisms may also couple receptor binding of insulin to its intracellular effects. Insulin causes the rapid generation of low–molecular weight plasma membrane products, which in turn activate insulin target enzymes, such as glycogen synthase. These may be products of receptor autoproteolysis or of membrane phospholipase C activity. One example of such a puta-

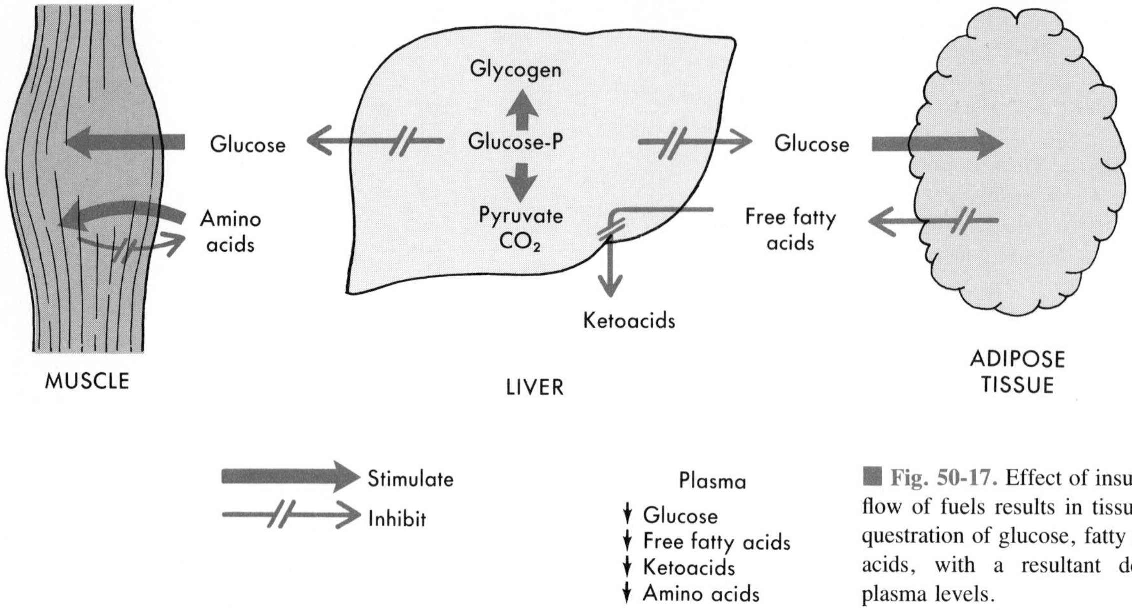

■ **Fig. 50-17.** Effect of insulin on the overall flow of fuels results in tissue uptake and sequestration of glucose, fatty acids, and amino acids, with a resultant decrease in their plasma levels.

tive mediator is a phosphatidylinositol glycan generated from a membrane glycoprotein by the action of phospholipase C. The role of cAMP and of calcium in insulin action also has been extensively studied. Insulin inhibits cAMP-dependent protein kinase A and enzymes activated by this kinase. Insulin can also decrease cAMP levels by activating phosphodiesterase. In some systems translocation of calcium into the cell also follows insulin exposure. At present, most evidence suggests that changes in intracellular cAMP and calcium levels are modulators of insulin action rather than critical second messengers.

Other effects of insulin are associated with changes in tissue concentrations of enzymes and their messenger RNAs. Some occur rapidly and some slowly. The slow effects of insulin on metabolism wax and wane gradually in adjustment to slowly developing changes in the milieu, such as occurs with prolonged fasting.

A well-studied rapid effect of insulin is suppression of the synthesis of phosphoenolpyruvate carboxykinase—a key gluconeogenic enzyme. Within 30 minutes of exposure of liver cells to insulin, the rate of transcription of the PEP-CK gene to its messenger RNA decreases markedly. Within a similar time frame, messenger RNA levels for glucokinase (the first enzyme in glucose metabolism to either glycogen or pyruvate) are greatly increased. The hormone increases levels of messenger RNA for a number of other enzymes such as fatty acid synthetase and pyruvate kinase. All of these nuclear actions of the hormone could be secondarily mediated by mechanisms described previously. However, some of the insulin receptor complex undergoes endocytosis, making the free hormone potentially available within the cell. Furthermore, insulin may bind to the Golgi apparatus, to the endoplasmic reticulum, and to nuclei. These observations suggest alternate routes whereby insulin may influence transcription and translation.

■ *Actions on Flow of Fuels*

Insulin is the hormone of abundance. When the influx of nutrients exceeds concurrent energy needs and rates of anabolism, the secreted insulin induces efficient storage of the excess nutrients while suppressing mobilization of endogenous substrates. The stored nutrients then can be made available during subsequent fasting periods to maintain glucose delivery to the central nervous system and free fatty acid delivery to the muscle mass and viscera. The major targets for insulin action are the liver, the adipose tissue, and the muscle mass. Fig. 50-17 displays the general flow of substrates produced by insulin, and Fig. 50-18 shows some of the important metabolic control points where insulin acts directly or indirectly in the liver.

Carbohydrate metabolism. Insulin stimulates glucose utilization and storage and simultaneously inhibits glucose production. Therefore insulin either lowers the basal circulating glucose concentration or limits the rise in plasma glucose that results from a dietary carbohydrate load. This is accomplished by a number of insulin effects.

1. In the *liver,* insulin increases glucose uptake and its storage in the form of glycogen. Since extracellular glucose levels equilibrate rapidly with intracellular levels in the liver, insulin facilitates uptake not by stimulating a glucose carrier system, but by inducing hepatic glucokinase and stimulating phosphorylation of glucose to glucose-6-phosphate. Insulin then promotes storage

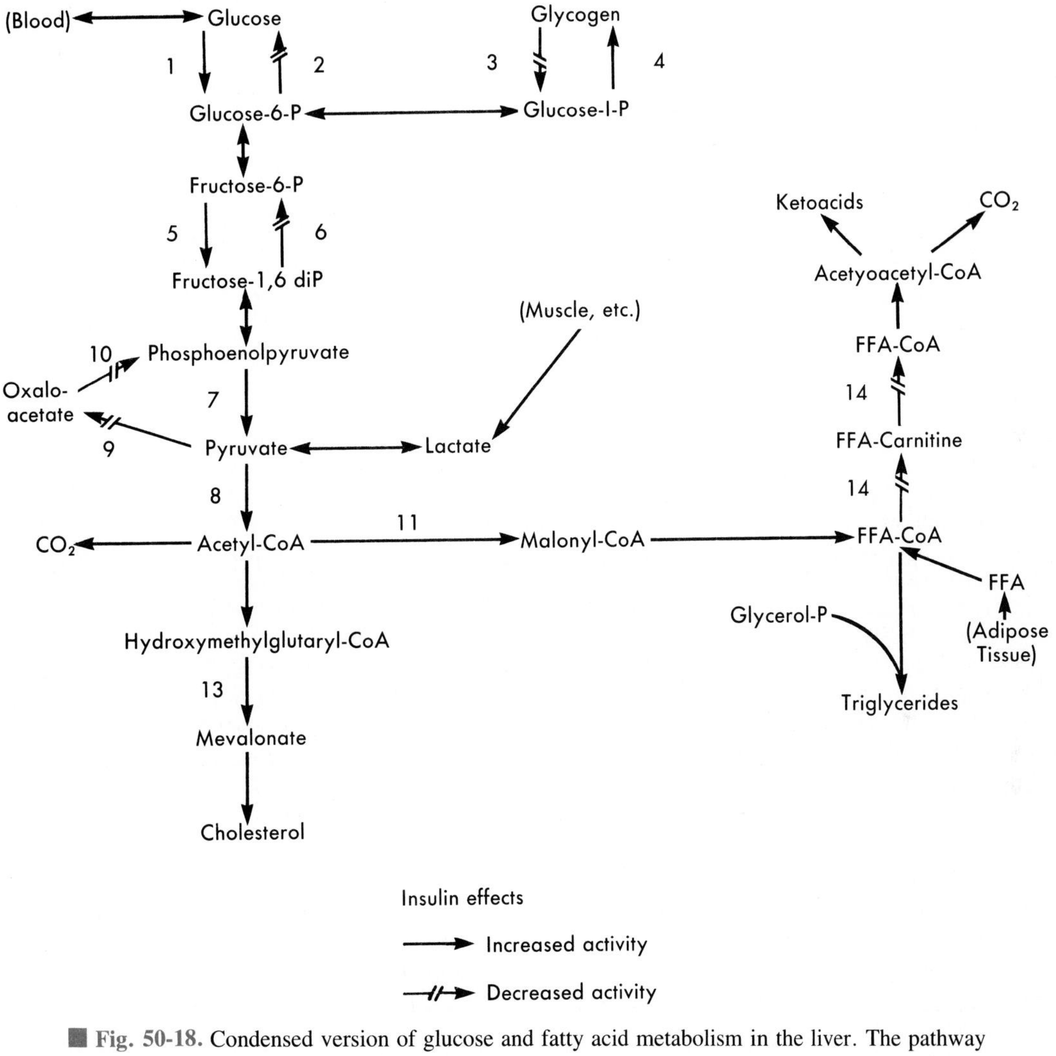

■ **Fig. 50-18.** Condensed version of glucose and fatty acid metabolism in the liver. The pathway from glucose to pyruvate and lactate constitutes glycolysis. The pathway from pyruvate to glucose via oxaloacetate constitutes gluconeogenesis. The key enzymes involved that are influenced by insulin are as follows: *1,* glucokinase; *2,* glucose-6-phosphatase; *3,* phosphorylase; *4,* glycogen synthase; *5,* phosphofructokinase; *6,* fructose-1,6-diphosphatase; *7,* pyruvate kinase; *8,* pyruvate dehydrogenase; *9,* pyruvate carboxylase; *10,* phosphoenolpyruvate carboxykinase; *11,* acetyl-CoA carboxylase; *12,* fatty acid synthetase; *13,* hydroxymethylglutaryl-CoA reductase; and *14,* carnitine acyltransferase.

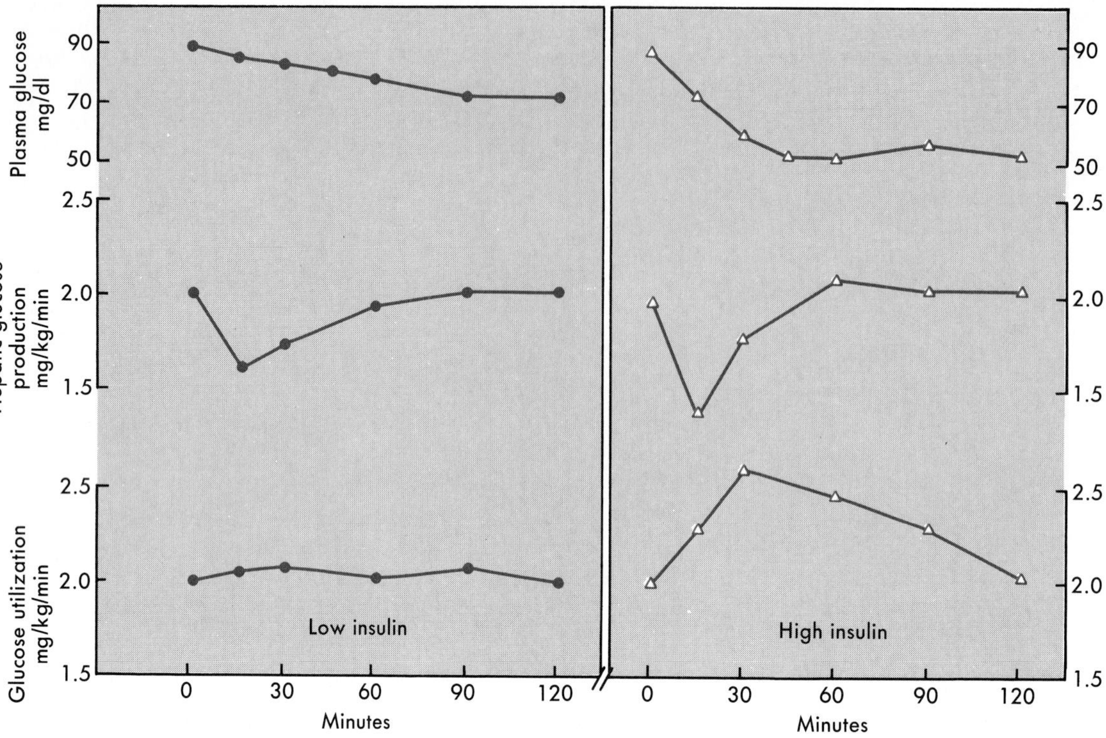

■ **Fig. 50-19.** Insulin effect on plasma glucose levels and glucose turnover in humans. Insulin was infused to a steady-state level of either 25 μU/ml *(left panel)* or 40 μU/ml *(right panel)*. The lower dose decreased plasma glucose levels by inhibiting hepatic glucose output. The larger dose caused a more marked fall in plasma glucose by stimulating glucose utilization in addition. Despite continuous hyperinsulinemia, hepatic glucose output recovers from insulin inhibition in each case, and plasma glucose stabilizes at a new lower level. Establishment of this new plasma glucose "floor" depends on secretion of hormones antagonistic to insulin, as well as on hepatic autoregulation of glucose production. (Modified from Sacca, L., et al.: Influence of continuous physiologic hyperinsulinemia on glucose kinetics and counterregulatory hormones in normal and diabetic humans, J. Clin. Invest. **63:**849, 1979. By copyright permission of The American Society for Clinical Investigation.)

of the glucose by activating the glycogen synthase enzyme complex. At the same time, insulin stimulates the flow of glucose to pyruvate and lactate (glycolysis) by increasing the activities of the committed enzymes phosphofructokinase and pyruvate kinase. Insulin also rapidly reduces hepatic glucose output by inhibiting glycogenolysis by decreasing glycogen phosphorylase activity and, gradually, by decreasing glucose-6-phosphatase levels. In addition, insulin inhibits gluconeogenesis. This is accomplished by decreasing the availability of precursor amino acids from muscle and decreasing their hepatic uptake. Insulin also decreases the levels or activities of the committed gluconeogenic enzymes pyruvate carboxylase, phosphoenolypyruvate carboxykinase, and fructose-1, 6-diphosphatase. Finally, insulin diminishes the supply of phosphoenolpyruvate for gluconeogenesis by increasing the activities of pyruvate kinase and pyruvate dehydrogenase.

Many of these hepatic effects of insulin require concurrent administration of glucose and may be reinforced by generation of products of glucose metabolism. Some also may be mediated or at least augmented by other consequences of insulin action. For example, since insulin decreases free fatty acid delivery to the liver, β-oxidation and generation of intramitochondrial acetyl CoA are decreased; by increasing the conversion of extramitochondrial acetyl CoA to malonyl CoA through activation of acetyl-CoA carboxylase, insulin further decreases hepatic levels of acetyl CoA. Since acetyl CoA is an important allosteric activator of pyruvate carboxylase, the activity of this gluconeogenic enzyme is thus indirectly lowered by insulin. Finally, as noted later, virtually all these hepatic actions of insulin are directly countered by glucagon. Since insulin also inhibits the secretion of this antagonist, the effect of increasing insulin levels on glucose handling by the liver are amplified by the concurrent restraint of glucagon release.

If insulin is infused into a subject in the basal state, plasma glucose declines; however, hepatic glucose production recovers from its insulin-induced nadir despite continuation of insulin administration (Fig. 50-19). This recovery begins after a short time and reflects three counterregulatory processes: (a) an autoregulatory response of the liver to a lowered plasma glucose concentration, probably mediated by the direct effects of glucose on phosphorylase and other enzymes; (b) a sensing of hypoglycemia by the hypothalamus, leading to stimulation of glucose output via activation of the sympathetic nervous system; and (c) a sensing of hypoglycemia by the α-cells of the pancreatic islets, which overcomes the inhibition by insulin and causes secretion of its antagonist, glucagon. Thus the actions of insulin on the liver—so important for assimilation of an excess of glucose—do not go unchecked for very long should insulin increase inappropriately in the absence of exogenous glucose. These mechanisms are essential to protection of the central nervous system from glucose deprivation.

2. In *muscle* insulin stimulates the transport of glucose into the cell by the membrane carrier system just described. Secondary to insulin action a small portion of the translocated glucose subsequently undergoes glycolysis and oxidation, while a major fraction is specifically directed into storage as muscle glycogen by activating glycogen synthase. The increased stores of muscle glycogen then are available for oxidative support of short-term exercise.

3. In *adipose tissue* insulin also stimulates the transport of glucose into the adipose cells in a manner analogous to its effect on muscle. This glucose is then metabolized to α-glycerophosphate, which is used in the esterification of fatty acids, permitting their storage as triglycerides. On subsequent lipolysis the glycerol that is released can then serve as a gluconeogenic precursor in the liver.

The overall results of insulin deprivation on carbohydrate metabolism therefore are depleted stores of glycogen in liver and muscle, high basal levels of glucose in the circulation, retarded assimilation of an exogenous glucose load, and increased mobilization of glucose precursors from peripheral tissues.

Protein metabolism. Insulin actions on protein and amino acid metabolism also are directed toward sequestration, that is, anabolism. During the assimilation of a protein meal the increase in insulin secretion limits the rise in plasma amino acid levels, especially of the branched chain amino acids leucine, valine, and isoleucine. If administered to a subject in the basal state, insulin preferentially lowers the levels of these particular amino acids. In *muscle* insulin stimulates the uptake of certain amino acids across the cell membrane, an action independent of its effect on glucose uptake. This effect is most prominently on the essential amino acids, especially leucine, valine, isoleucine, tyrosine, and phenylalanine. Insulin also stimulates the rate of protein synthesis in general. The mechanism involves effects on transcription as exemplified by insulin-stimulated increases in messenger RNA for albumin. In addition, insulin modifies the function of ribosomes that have already been charged, so as to permit more efficient translation of the RNA message. It is not clear whether insulin increases the number of ribosomes as well. The anabolic effects of insulin are reinforced by the hormone's anticatabolic effects. Insulin inhibits proteolysis and suppresses the release of the essential branched-chain and aromatic amino acids from muscle. Insulin also inhibits the oxidation of the branched-chain amino acids. In contrast, insulin in physiological concentrations has little effect on the flux of alanine. All these effects on protein metabolism, although quantitatively most important for muscle, are also operative in liver, adipose tissue, and other organs. Of great importance are similar effects on stimulation of macromolecule formation in cartilage and osseous tissue. Taken together, this spectrum firmly establishes insulin as an essential hormone for growth. The insulin-deprived young animal or human has a reduced lean body mass and is retarded in height and maturation.

Certain other peptides and proteins, variously called insulin-like growth factors (IGF-1 and IGF-2), somatomedins, multiplication-stimulating activity, nerve growth factor, and epidermal growth factor, are derived from diverse animal and human tissues and circulate in plasma. These display some of the aforementioned anabolic actions of insulin, have certain similar amino acid sequences, and often cross-react modestly with insulin receptors. However, they have distinct plasma membrane receptors and physiologic roles of their own. In general, these peptides have less insulin-like effect on carbohydrate and fat metabolism than on protein metabolism.

Fat metabolism. The metabolism of both endogenous and exogenous fat is profoundly influenced by insulin. The net overall effect is to enhance storage and to block mobilization and oxidation of fatty acids. Thus insulin lowers the circulating levels of free fatty acids and ketoacids within minutes of administration and may eventually reduce the level of triglycerides.

In *adipose tissue* the deposition of fat is stimulated by insulin in several ways. As previously noted, exogenous triglycerides from a meal and endogenous triglycerides produced by the liver circulate as lipoprotein complexes, which cannot directly enter fat cells. They first must be split within the capillary endothelium by the enzyme lipoprotein lipase, and the resultant free fatty acids associated with albumin then can enter the adipose tissue cells. Insulin induces this key lipoprotein lipase. Once in the adipose cell the free fatty acids are converted to triglycerides by esterification with α-glycerophosphate. Since this cell lacks the glycerol kinase necessary to convert free glycerol to its phosphate ester, the latter must be generated from glucose transported into the adipose cell under insulin stimulation. In addition, glucose itself is converted to fatty acids by insulin activation of glycolysis, pyruvate dehydrogenase, acetyl-CoA carboxylase, and fatty acid synthetase. Although this is a very minor quantitative pathway in the adipose tissue of humans, it is of considerable significance for storage of carbohydrate as fat in rodents.

Most important, insulin profoundly inhibits hormone-sensitive adipose tissue lipase activity. It may do this by decreasing the levels of cAMP and thereby inhibiting cAMP-dependent protein kinase. By suppressing lipolysis of stored triglycerides and the release of free fatty acids, insulin lowers their rate of delivery to the liver and to peripheral tissues. A major consequence of decreasing free fatty acid supply to the liver is a marked reduction in the generation of ketoacids. In addition, insulin stimulates the utilization of ketoacids by the peripheral tissues. Thus insulin is the major and perhaps the sole *antiketogenic* hormone.

Within the *liver,* insulin is also antiketogenic and lipogenic. Free fatty acids entering from the circulation are shunted away from β-oxidation and from ketogenesis. They are instead reesterified with glycerophosphate, derived either from insulin-stimulated glycolysis or from glycerol via the enzyme glycerophosphate kinase. Fatty acids are also synthesized from glucose under the influence of insulin, because the hormone activates pyruvate dehydrogenase, thereby generating acetyl CoA from pyruvate. The rate-limiting step in the subsequent synthesis of free fatty acids from acetyl CoA appears to be its conversion to malonyl CoA by acetyl-CoA carboxylase. As indicated previously, insulin increases the activity of this cytosolic enzyme, as well as inducing fatty acid synthetase, which catalyzes the remaining steps. When glucose is available and glycolysis is stimulated by insulin, acetyl CoA generated in the mitochondria from pyruvate is transferred to the cytoplasm via citrate. In the cytoplasm the acetyl CoA is regenerated and built up into fatty acids under the further influence of insulin. Moreover, insulin increases the activity of the hexose monophosphate shunt by inducing the enzyme glucose-6-phosphate dehydrogenase. This generates the supply of reduced triphosphopyridine nucleotide, which is also needed for fatty acid synthesis. The intrahepatic antiketogenic action of insulin may be mediated by the formation of malonyl CoA from acetyl CoA, because malonyl CoA inhibits the enzyme carnitine acyltransferase (Fig. 50-18). The latter enzyme is responsible for transferring free fatty acids from the cytoplasm into the mitochondria for oxidation and conversion to ketoacids. Finally, insulin also favors hepatic sequestration of cholesterol by activating hydroxymethylglutaryl CoA reductase, the rate-limiting step in cholesterol synthesis. Thus the net effect of insulin on lipid metabolism in the liver is to decrease free fatty acid oxidation and ketogenesis, to increase the synthesis of free fatty acids and cholesterol, and to increase the storage and decrease the release of triglycerides and cholesterol.

In summary, the insulin-deprived animal or human has high circulating levels of free fatty acids, ketoacids, cholesterol, and triglyceride and has diminished adipose tissue lipid stores. Whether hepatic lipid content is increased or decreased depends on the relative rates of augmentation of free fatty acid delivery versus free fatty acid oxidation. In a well-nourished subject the former may overwhelm the latter and lead to excess hepatic accumulation of triglyceride.

Other actions. Both protein anabolism and the storage of glucose as glycogen require concomitant cellular uptake of potassium, phosphate, and magnesium. Insulin stimulates translocation of all three minerals into muscle cells and of potassium and phosphate into the liver. The facilitation of muscle potassium uptake is mediated by an insulin action on membrane polarization that is independent of the hormone's effect on glucose transport. Insulin secreted in response to a carbohydrate load lowers serum potassium, phosphate, and magnesium levels, whereas inhibition of basal insulin secretion (by somatostatin) causes a rise in serum potassium levels. Thus insulin is considered to be one of the normal regulators of potassium balance. Another effect of insulin on electrolyte balance is to increase reabsorption of potassium, phosphate, and sodium by the tubules of the kidney. These renal effects contribute to anabolism by conserving vital intracellular electrolytes, as well as sodium, which is necessary for formation of additional extracellular fluid required when lean body mass is being expanded.

The overall consumption of glucose by the central nervous system is independent of insulin. However, selected areas of the brain, particularly the hypothalamus, are insulin responsive. Radioactively labeled insulin binds to hypothalamic cells or to the adjacent capillary endothelium, and insulin receptors can be demonstrated in hypothalamic plasma membranes. The glucose analogue, gold thioglucose, which selectively destroys the ventromedial nucleus (the satiety center), requires insulin to act. And, interestingly, insulin administration initially increases the firing rate of single neurons from the same area of the hypothalamus before the suppressive effect of insulin-induced hypoglycemia appears. All these data suggest that insulin, secreted in response to carbohydrate, also may facilitate a hyperglycemic feedback signal to the hypothalamus, which then shuts off hunger by stimulating the activity of the satiety center.

Correlation of Insulin Secretion and Action

The insulin sensitivity of tissues relates well to prevailing plasma insulin levels in various physiological states. Fuel mobilization, that is, lipolysis, ketogenesis, and proteolysis, are significantly but only partially inhibited at postabsorptive plasma insulin concentrations of 10 μU/ml. This permits a finely regulated increase in substrate flow and plasma levels of free fatty acids and ketoacids during the daily nocturnal fasting period. In addition, sufficient glycogenolysis is permitted so as to sustain the plasma glucose level. Shortly after a subject has eaten, insulin concentrations rise to 20 to 30 μU/ml due to enhanced β-cell secretion, and hepatic glucose output becomes suppressed (Fig. 50-19). With a further increase to postprandial levels of 40 to 100 μU/ml, glucose uptake (and amino acid uptake) by peripheral tissues is strongly stimulated, facilitating utilization of substrate at a time of abundance. Under maximum insulin stimulation at approximately 200 μU/ml, glucose utilization increases from 2 mg/kg/minute to 12 mg/kg/minute. Of this total increase, only one third is accounted for by oxidation, the rest being stored as glycogen. Thus the β-cell responds to the physiological need of the moment with insulin delivery rates that provide appropriate hormone concentrations for regulating substrate fluxes. In addition, the intrinsic oscillatory pattern of insulin release cited previously enhances the action of the hormone on glucose metabolism.

Glucagon

Structure and Synthesis

Glucagon is a single straight-chain peptide hormone of 29 amino acids and molecular weight 3500. The amino acid composition and sequence are identical in mammalian glucagons thus far studied, including the human, porcine, and bovine hormones. Even in lower species, amino acid substitutions are infrequent, and both the structure and function of glucagon appear highly conserved in evolution. The N-terminal residues 1 to 6 are essential for receptor binding and biological activity.

Glucagon is synthesized by α-cells under the direction of a gene located on human chromosome 2. A preprohormone of molecular weight 18,000 is processed to a prohormone, known as *glycentin,* with a molecular weight of 12,000. The latter contains both N-terminal and C-terminal extensions from the glucagon sequence. Glycentin is localized to the peripheral halo of the secretory granule, whereas glucagon is localized to its dense core; release of the hormone from its secretory granule is accomplished by exocytosis. The biosynthesis of glucagon is stimulated when glucose concentrations are low and is progressively diminished as glucose concentrations are raised. In humans glucagon is found exclusively in the pancreatic islets. However, in other species, such as the dog, glucagon also is produced by α-like cells scattered in the walls of the stomach. In addition, intestinal cells in humans and other species secrete glycentin as well as a variety of glucagon-like peptides that are produced by alternative processing of preproglucagon.

Regulation of Secretion

The most important principle governing glucagon secretion appears to be the maintenance of normoglycemia. Exactly opposite to insulin, glucagon is secreted in response to glucose deficiency and acts to increase circulating glucose levels. Basal plasma glucagon concentration in humans averages 100 pg/ml (3×10^{-11} M). Hypoglycemia causes a twofold to fourfold increase in these levels, whereas hyperglycemia lowers them approximately 50%. That glucose directly regulates α-cell secretion is shown by in vitro studies. The effect of glucose appears to be modulated by insulin. Glucagon secretion is stimulated much more by low glucose levels if insulin is absent. Conversely, the presence of insulin greatly potentiates the suppressive effect of high glucose levels on the α-cell. Neither the exact mechanism through which glucose is directly sensed by the α-cell nor how it is modulated by insulin is known. In general, the specificity of α-cell responses to glucose analogues, metabolites, and blockers of intracellular glucose metabolism is similar to that exhibited by β-cells.

Glucagon secretion also is stimulated by a protein meal and, most powerfully, by amino acids such as arginine and alanine. However, the α-cell response to protein is greatly dampened if glucose is administered concurrently. This interaction is partly via insulin; glucagon responses to amino acids are restrained by insulin excess and augmented by insulin deficiency. In species other than humans a triglyceride meal also can be shown to stimulate glucagon release modestly. However, in humans and animals free fatty acids—like glucose—exert a suppressive effect, whereas glucagon secretion increases following a rapid decline in plasma free fatty acids. Glucagon responses to orally ingested nutrients (as opposed to their intravenous delivery) may be reinforced by the release of the same gastrointestinal glucagon secretagogues that augment insulin secretion, with the exception of secretin. The sum of all these individual influences is that the ingestion of ordinary meals in humans produces much less variation in plasma glucagon than in plasma insulin levels. This is at least partly explained by the offsetting effects of the carbohydrate and protein portions of the meal on the α-cell, as compared with the synergistic effects of these nutrients on the β-cell. It is also in keeping with the secondary role of glucagon versus the primary role of insulin in the disposal of exogenous nutrients.

Fasting for 3 days increases plasma glucagon twofold; however, the mechanism may involve a reduction in the metabolic clearance of glucagon, as well as a

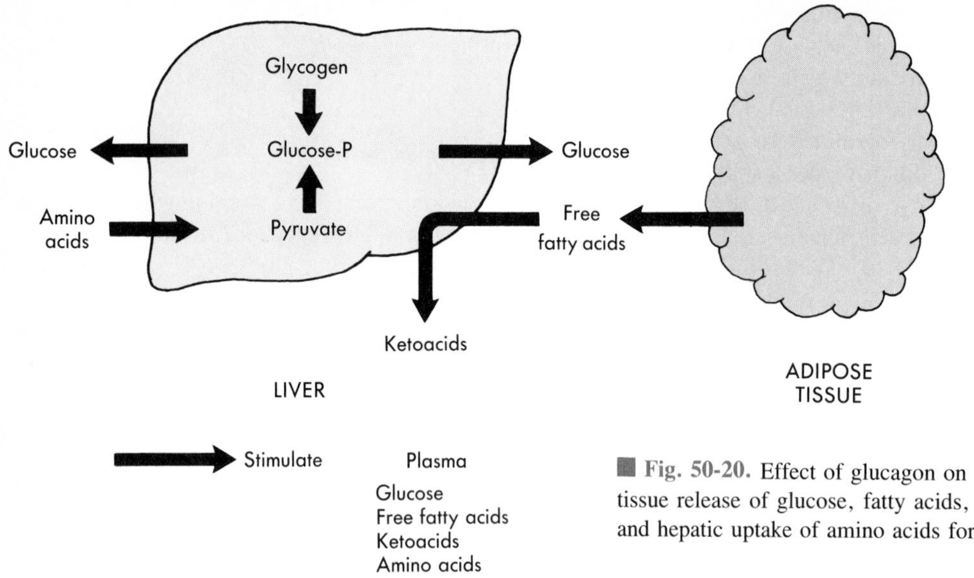

■ **Fig. 50-20.** Effect of glucagon on the overall flow of fuels results in tissue release of glucose, fatty acids, and ketoacids into the circulation and hepatic uptake of amino acids for gluconeogenesis.

possible increase in secretion. Exercise of sufficient intensity and duration also increases plasma glucagon. Neural mechanisms may mediate some of these α-cell responses. In particular, vagal stimulation and acetylcholine acutely increase glucagon secretion. Of importance to the physician is that a variety of stresses, including infection, toxemia, burns, tissue infarction, and major surgery, all increase glucagon secretion promptly. This phenomenon is probably mediated by the adrenergic nervous system via sympathetic outflow from the ventromedial hypothalamus to α-adrenergic receptors in the β-cells. The excess of glucagon often leads to clinically significant hyperglycemia. The neurohormone somatostatin inhibits the secretion of glucagon, as it does that of insulin. Finally, various other protein hormones (growth hormone, human placental lactogen) and steroid hormones (cortisol, estrogen-progesterone) affect glucagon secretion indirectly.

Monomeric glucagon circulates essentially unbound in plasma. It has a half-life of 6 minutes and a metabolic clearance rate of about 600 ml/minute. The daily secretion rate of glucagon is estimated to be 100 to 150 μg. A considerable portion of glucagon immunoreactivity is caused by higher molecular weight substances, at least one of which is an intermediate in the processing of glycentin to glucagon. The physiological significance of these fractions is unclear. The ratio of portal vein to peripheral vein glucagon concentrations is about 1.5 in the basal state. Like insulin, about 50% of glucagon is extracted by the liver on a single passage. The kidney and liver are the major loci of glucagon degradation. Less than 1% of the glucagon filtered by the glomerulus is excreted in the urine.

■ *Hormone Actions*

In almost all respects the actions of glucagon are exactly opposite to those of insulin. Glucagon promotes mobilization rather than storage of fuels, especially glucose (Fig. 50-20). Both hormones act at numerous similar control points in the liver. Indeed, some investigators believe glucagon to be the primary hormone regulating hepatic glucose production and ketogenesis, with insulin's role being that of glucagon antagonist.

Glucagon binds to plasma membrane receptors in the liver. The glucagon-receptor complex causes a rapid increase in intracellular cAMP (Chapter 49). A specific enzymatic cascade ensues. Protein-phosphokinase catalytic activity increases, converting inactive phosphorylase kinase to active phosphorylase kinase. The latter then converts inactive phosphorylase to active phosphorylase. A number of other enzymes that are important to glucose and fat metabolism and whose functional state depends on interconversion of phosphorylated and dephosphorylated forms also are regulated by glucagon. These include phosphofructokinase, pyruvate kinase, acetyl-CoA carboxylase, and hydroxymethlyglutaryl-CoA reductase. In these instances, however, the catalytic activity of the enzyme is decreased by the phosphorylation that results from glucagon action. Hence glucagon inhibits their action.

Unequivocally the most important mission of glucagon is to promote and sustain hepatic glucose output. Thus the dominant effect of glucagon is in the liver. Its actions on adipose tissue and muscle are minor unless insulin is virtually absent. Glucagon exerts an immediate and profound glycogenolytic effect through activation of glycogen phosphorylase. The glucose-1-phosphate released is prevented from undergoing resynthesis to glycogen by a simultaneous inhibition of glycogen synthase. Glucagon also stimulates gluconeogenesis by at least several mechanisms. The hepatic extraction of amino acids, especially alanine, is increased. The activities of the gluconeogenic enzymes, pyruvate carboxylase, phosphoenolpyruvate carboxykinase, and fructose-1,6-diphosphatase are increased, whereas that of the glycolytic enzymes phosphofructokinase, and pyruvate kinase are decreased. The enzyme pair phosphofructokinase/fructose-1,6-diphosphatase determines the flow between fructose-6-phosphate and fructose-1,6-diphosphate. Hence this enzyme pair determines the relative rates of glycolysis and gluconeogenesis (Fig. 50-18). The activities of these two enzymes, in turn, are reciprocally related by the level of fructose-2,6-biphosphate. This important metabolite is very sensitive hormonally. Glucagon decreases its level in the liver, thereby favoring the flow from fructose-1,6-diphosphate to fructose-6-phosphate and hence stimulating gluconeogenesis. Insulin has the opposite effect, probably by inhibiting the action of glucagon.

The crucial importance of glucagon to the maintenance of basal hepatic glucose output is shown by the marked decline of 75% that follows the selective inhibition of glucagon secretion by somatostatin. The powerful glycogenolytic and hyperglycemic action of glucagon is exhibited at plasma hormone concentrations of 150 to 500 pg/ml and occurs even in the presence of insulin levels somewhat above basal levels (20 to 30 μU/ml). However, this action is transient; the acute initial stimulation of hepatic glucose output, which occurs during continuous glucagon administration, wanes after about 30 minutes due to autoregulation by hyperglycemia and stimulation of insulin release. However, if glucagon is given in a more physiological, fluctuating pattern, each increment in the hormone causes a response. Glucagon has little or no influence on glucose utilization by peripheral tissues. Thus hyperglucagonemia has no effect on the plasma glucose levels that are generated by an exogenous glucose load, as long as the insulin response is normal.

Another intrahepatic action of glucagon is to direct incoming free fatty acids away from triglyceride synthesis and toward β-oxidation. Thus glucagon is a ketogenic as well as a hyperglycemic hormone. Recent evidence suggests that glucagon inactivates acetyl-CoA carboxylase, the rate-limiting step in free fatty acid synthesis from cytoplasmic acetyl CoA. This results in lower levels of malonyl CoA, an allosteric inhibitor of carnitine acyltransferase. In turn, this allows a faster rate of influx of fatty acyl CoA into the mitochondrion for conversion to ketoacids. Simultaneously the greater rate of β-oxidation of free fatty acids increases the level of intramitochondrial acetyl CoA, which activates pyruvate carboxylase, helping to increase gluconeogenesis. Glucagon also is capable of activating adipose tissue lipase, thereby increasing lipolysis, the delivery of free fatty acids from adipose tissue to the liver, and ketogenesis. Although glucagon's ketogenic actions are physiologically relevant, they are easily nullified by rather small amounts of insulin, particularly at the adipose tissue locus. Finally, hepatic hydroxymethylglutaryl-CoA reductase activity is inhibited by glucagon, thereby decreasing hepatic cholesterol synthesis.

Other actions of glucagon include inhibition of renal tubular sodium resorption, causing natriuresis, and activation of myocardial adenylate cyclase, causing a moderate increase in cardiac output. The latter effect has been of rare therapeutic value in refractory heart failure. The possibility that glucagon—like insulin—may act locally as a central nervous system hormone in the regulation of appetite also has been suggested.

The Insulin/Glucagon Ratio and Metabolism

From the foregoing it is apparent that fuel fluxes are finely regulated by coordinated secretion of the β-cells and α-cells in a manner appropriate to the physiological situation. Indeed, as discussed previously, many common enzyme and substrate control points for glycogenolysis, gluconeogenesis, and ketogenesis, as well as for glycogen and fat synthesis, are affected by both hormones but in exactly the opposite fashion. Therefore it may be the ratio of insulin to glucagon (I/G ratio), rather than the absolute level of each, that is of critical importance in metabolic regulation.

Fasting

When fasting is prolonged beyond the usual overnight period, there is a rapid fall in the molar ratio of I/G from 2 to 0.5, or less. This stimulates glycogenolysis, which is the initial source of support for the plasma glucose level. Since glycogen stores in the liver are only about 75 g and are quickly depleted, a rapid enhancement of gluconeogenesis also is necessary to maintain delivery of glucose for the metabolism of the central nervous system and blood cells. About 75 to 100 g of muscle protein must then be catabolized daily to provide the bulk of glucose precursors. Enhanced hepatic extraction of amino acids, especially of alanine, facilitates the increasing gluconeogenesis from protein. Another 15 to 20 g of glucose precursors are provided by glycerol arising from lipolysis. The supply of free fatty acids is dramatically increased by the lowered I/G ratio. The increased rate of hepatic fatty acid oxidation

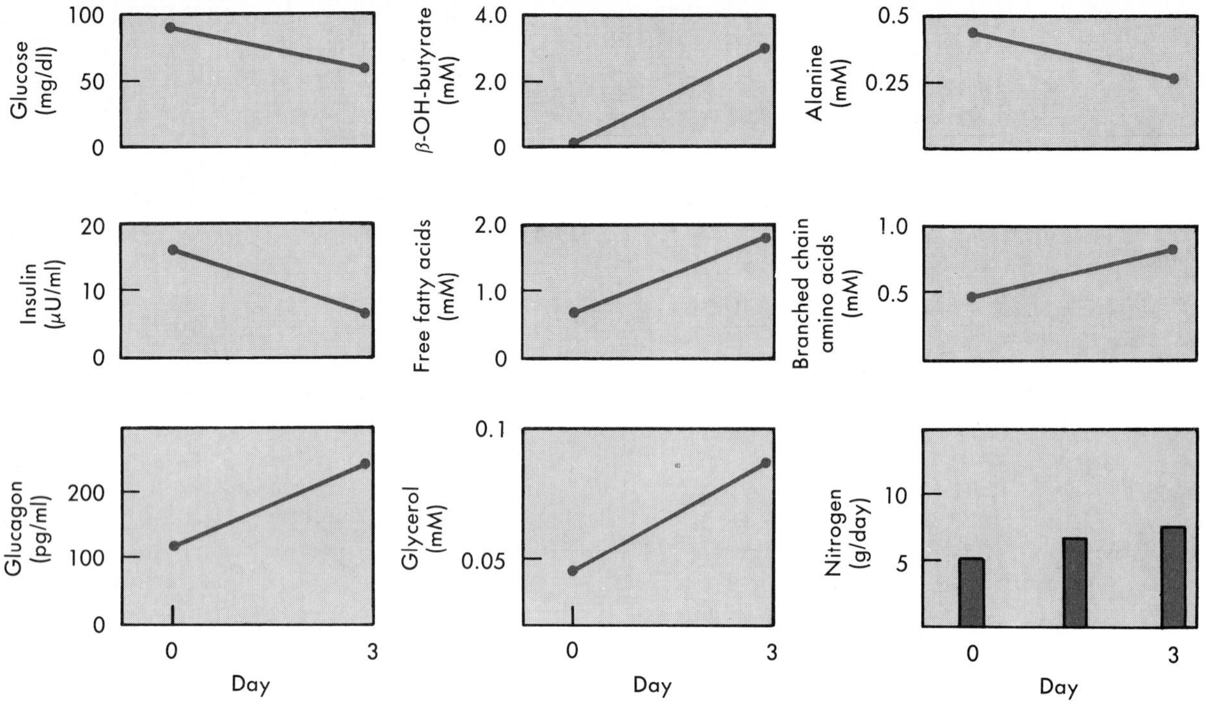

■ **Fig. 50-21.** Changes in plasma hormone and substrate levels and urine nitrogen during 3 days of fasting in humans. Note that as the ratio of insulin to glucagon declines parallel to plasma glucose, there is a rise in lipid-derived fuels and an increase in protein catabolism. (Redrawn in part from Felig, P., et al.: J. Clin. Invest. **48:**584, 1969. Reproduced by copyright permission of the American Society for Clinical Investigation.)

boosts the production of the ketoacids acetoacetate and β-hydroxybutyrate from basal levels of 10 g/day, equivalent to 60 kcal, to as much as 80 g/day, yielding 500 kcal. These ketoacids are then used by peripheral tissues, partly replacing glucose. Conservation of glucose for central nervous system use is further aided by a reduced rate of glucose utilization by the muscle mass and viscera; therefore RQ decreases. The net result of this rapid lowering of I/G is reflected in changing levels of plasma substrates and in rates of urinary nitrogen excretion (Fig. 50-21). Plasma glucose declines 15 to 30 mg/dl but then stabilizes. Plasma free fatty acids, glycerol, ketoacids, and branched-chain amino acids all increase. Plasma alanine decreases as hepatic extraction of this amino acid exceeds its release from muscle. Urinary nitrogen excretion rises consequent to the increased catabolism of endogenous protein.

If fasting continues beyond a few days, further important adaptive changes occur, although the I/G ratio remains at the same low level. The basal rate of metabolism diminishes 15% to 20%, conserving body stores of energy. This adaptation may be due to diminished sympathetic nervous system activity and to a reduction in effective thyroid hormone activity. Glucose utilization by the central nervous system declines by two thirds, to 45 g/day, and ketoacids replace glucose as the major oxidative fuel of the brain. This key change permits a large reduction in gluconeogenesis, which declines to match the lowered rate of glucose utilization. In turn, this greatly spares muscle protein, the catabolism of which now only needs to supply 20 to 25 g of glucose precursors per day. This is reflected in a much lower rate of urinary nitrogen excretion (3 to 4 g/day). Plasma substrate concentrations of glucose, free fatty acids, and glycerol remain essentially the same as at 3 days of fasting. Ketoacid levels, however, increase still further. Alanine and other amino acid levels are sharply reduced. With these additional accommodations and an ample supply of fluid, a normal-weight individual can survive up to 2 months of total fasting, whereas very obese individuals have been kept without food for up to 1 year for therapeutic purposes.

■ *Exercise*

Exercise represents a special example of rapid fuel mobilization, geared to supplying an excess of substrate for muscle oxidation while preserving a steady delivery of glucose to the central nervous system. Although glucose is generally used by resting muscle at a low rate, the situation changes dramatically with exercise. The initial response is an increase in glucose oxidation. Breakdown of muscle glycogen (present at concentrations of 9 to 16 g/kg wet weight) provides an immediate source of glucose-6-phosphate for glycolysis and oxi-

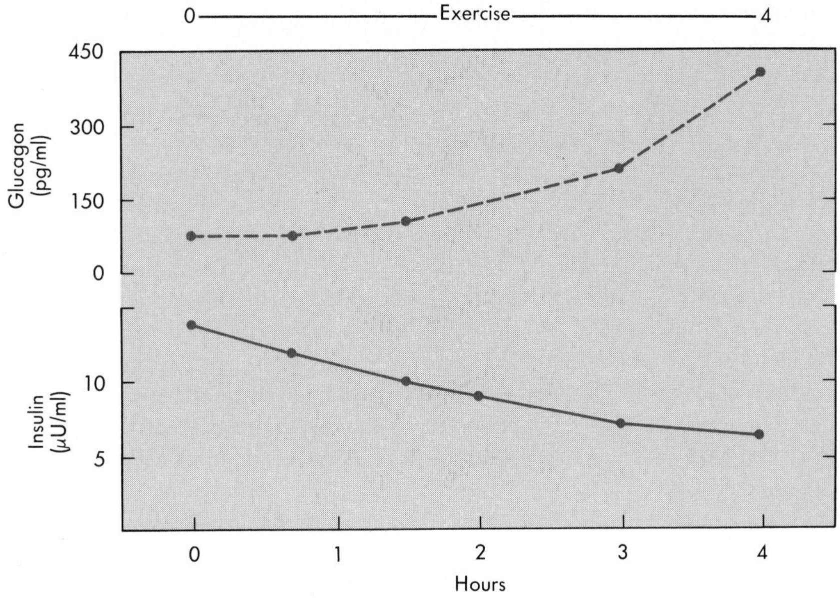

■ **Fig. 50-22.** Effect of prolonged exercise on plasma insulin and glucagon in humans. (Redrawn from Ahlborg, G., et al.: J. Clin. Invest. **53:**1080, 1974. Reproduced by copyright permission of the American Society for Clinical Investigation.)

dation. However, this must soon be augmented. Within 10 minutes glucose uptake from blood may increase sevenfold to fifteenfold, and by 60 minutes, twentyfold to thirtyfold, depending on the muscle group and the intensity of exercise. It is important to emphasize that this increased uptake of glucose by exercising muscle does not require insulin. Oxidation of fatty acids also increases with time, sustained by increased uptake of free fatty acids from plasma. During steady, moderate leg exercise about one third of the energy is supplied by oxidation of glucose and two thirds by oxidation of free fatty acids. Total energy needs may be five to ten times the basal needs.

It is to supply these needed fuels that glucagon secretion increases, insulin secretion decreases, and the I/G ratio falls during exercise (Fig. 50-22). This promotes an increase in hepatic glucose output of up to 500%, depending on the intensity and duration of the exercise. In mild exercise of short duration glycogenolysis is the dominant source of the extra glucose, with gluconeogenesis contributing only 20% to 30%. If exercise is more strenuous and glucose demand greater, absolute rates of glycogenolysis and gluconeogenesis both increase initially, but the fraction contributed by gluconeogenesis declines to 10%. If exercise is more moderate but is prolonged for hours, liver glycogen stores become depleted, and gluconeogenesis then becomes the major source of hepatic glucose production. Thus muscle proteolysis increases, and the amino acids so generated are taken up by the liver at an accelerated rate. Alanine is again the dominant amino acid that turns over in this process, although it may largely be an ammonia transporter.

The free fatty acids for muscle oxidation come from the increased rate of lipolysis generated by the low I/G ratio. Ketoacid production also increases slightly but contributes little to the energy needs of exercising muscle. Plasma substrate levels during exercise depend on the relative rates of production and utilization and on the coordinated changes in the I/G ratio. Most important, plasma glucose is maintained constant for a long time, although it may eventually decline a modest 10 to 15 mg/dl. Plasma alanine initially increases but eventually also declines below basal level. The same is true for branched-chain amino acids if exercise is strenuous. Plasma free fatty acids increase strikingly to levels up to 400% over the basal level. Plasma lactate, pyruvate, glycerol, and ketoacids all increase moderately.

■ *The Newborn Period*

The I/G ratio declines within hours of birth in the human, and it remains low for several days. This period marks a transition from dependence on maternal fuels to dependence on the endogenous stores of the neonate. A rapidly falling plasma glucose level provides the signal for the appropriate set of islet hormones. Initially glycogenolysis is stimulated, but once this limited store is gone, the low I/G ratio promotes development of the immature gluconeogenic capacity of the newborn. Until this occurs, the enhanced release of free fatty acids and their ketoacid products temporarily takes up the fuel slack, the ketoacids being of special importance to the central nervous system, as in prolonged fasting. Once regular feeding becomes established, the I/G ratio slowly returns to that characterizing the postabsorptive adult.

Carbohydrate Intake

When a pure load of carbohydrate is ingested, the I/G ratio rises strikingly from 2 to 30. This ensures efficient retention of the glucose, 30% of which is initially taken up by the liver and rapidly phosphorylated. Part of the glucose is converted to glycogen, which will sustain plasma glucose many hours later. Part is metabolized to acetyl CoA, which in turn serves as the substrate for de novo free fatty acid synthesis. In this way, extra carbohydrate calories can be more efficiently stored as triglyceride. The high I/G ratio immediately suppresses endogenous glucose output from the liver. Muscle proteolysis is also inhibited as the need to provide amino acid precursors for glucose is diminished. Likewise, lipolysis is suppressed as less free fatty acids are needed for oxidation. As previously noted, about 70% of the exogenous carbohydrate escapes immediate hepatic capture. Instead the carbohydrate is taken up by muscle for storage as glycogen and for later return to the liver as lactate and alanine. The RQ rises, reflecting the increased oxidation of glucose and the decreased oxidation of fat. Were it not for the high I/G ratio, retention of the carbohydrate load would be less efficient, plasma glucose would rise unduly, and some glucose might even be lost by excretion in the urine.

Protein Intake

When a protein load is ingested, plasma insulin and glucagon both increase; the resultant I/G ratio increases slightly. The liver preferentially removes only a small fraction of the absorbed amino acids for use in intrahepatic protein synthesis and to sustain the continuing need for gluconeogenesis. The bulk of amino acids escapes hepatic sequestration, particularly the essential branched-chain amino acids. They subsequently enter muscle, where they are synthesized into structural and contractile proteins, as well as into a labile protein pool from which glucose precursors can come during a later fasting period. Other organs also extract amino acids required for specific functions or for protein renewal. Muscle release of gluconeogenic precursors—now not needed by the liver— is greatly reduced after a protein load. The dual islet hormone response to protein smoothly coordinates these processes. The increase in insulin promotes the anabolic disposal of the amino acids, largely into muscle. The increase in glucagon prevents a fall in hepatic glucose output and the hypoglycemia that would ensue if insulin action were unopposed.

Fat Intake

A load of pure fat is seldom ingested by a human, so that the necessary hormonal influences on fat disposition are generated by the accompanying carbohydrate and protein in the meal. Mixed meals increase the I/G ratio because of the sharp rise in insulin release (Fig. 50-14). Clearance of the triglycerides in the chylomicrons from plasma is facilitated by insulin activation of the enzyme lipoprotein lipase. The fate of the released free fatty acids depends on the site of hydrolysis. In adipose tissue they are promptly reesterified with glycerophosphate and stored as triglyceride. In muscle they are largely oxidized. In the liver they may be either oxidized or reesterified into triglyceride and stored for later release as very low density lipoprotein. The increase in insulin secretion facilitates these processes, both by increasing the key enzyme lipoprotein lipase and by stimulating uptake of glucose to supply the glycerophosphate needed for intracellular triglyceride synthesis.

Somatostatin Secretion and Action

The neurohormone somatostatin was originally discovered in hypothalamic extracts as an inhibitor of growth hormone secretion (Chatper 52). However, subsequent studies showed that somatostatin is present in various sites in the gastrointestinal tract and is synthesized and secreted by the δ-cells of the pancreatic islets. Two forms of the hormone exist: a single-chain peptide containing 14 amino acids and one with an N-terminal extension to 28 amino acids. Somatostatin 14 and 28 both originate from a larger preprohormone, and somatostatin 14 arises from further processing of somatostatin 28 in the δ-cells. Many, but not all, of their biological effects are similar. Secretion of somatostatin by exocytosis is mediated through increases in intracellular cAMP. Somatostatin circulates in average concentrations of 60 pg/ml (4×10^{-11} M) in peripheral plasma with a half-life of approximately 2 minutes. The secretion of somatostatin from δ-cells is stimulated by glucose, by amino acids, by free fatty acids, by gastrointestinal hormones, such as vasoactive intestinal peptide, secretin, and cholecystokinin; by β-adrenergic and cholinergic stimuli; and by glucagon. It is inhibited by α-adrenergic stimuli, by dopamine, and probably by insulin. In humans, plasma somatostatin levels increase modestly after ingestion of a mixed meal, probably as a result of δ-cell secretion.

Somatostatin is a profound inhibitor of both insulin and glucagon secretion. The administration of somatostatin to experimental animals or humans also decreases the assimilation rate of all nutrients from the gastrointestinal tract. This is accomplished by inhibitory actions

■ **Fig. 50-23.** Schema interrelating the effects of somatostatin, insulin, and glucagon on each other's secretion with their effects on glucose and amino acid metabolism. (Modified from Unger, R.H., et al. Reproduced with permission from the *Annual Review of Physiology,* volume 40. Copyright © 1978 by Annual Reviews, Inc.)

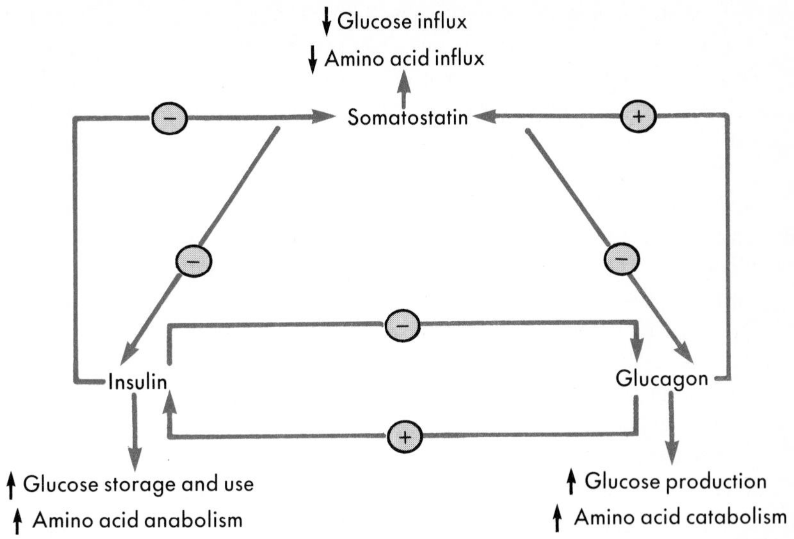

at all levels of gastrointestinal function. Somatostatin inhibits gastric, duodenal, and gallbladder motility; it reduces the secretion of hydrochloric acid, pepsin, gastrin, secretin, intestinal juices and pancreatic exocrine function. Somatostatin also inhibits the absorption of glucose, xylose, and triglycerides across the mucosal membrane. It seems reasonable to believe that somatostatin of δ-cell origin participates in a feedback arrangement, whereby entrance of food into the gut stimulates the release of the hormone so as to prevent rapid nutrient overload. The anatomical relationships between α-, β-, and δ-cells and the existence of tight junctions and gap junctions between them, as described previously, suggest that all three islet hormones may be influencing each other's secretion by local or *paracrine* effects (Fig. 50-23). This could improve the coordination between the bulk movement, digestion, and absorption of nutrients with the insulin and glucagon responses necessary for their proper disposition in the liver and other organs. In addition, by coordinating its inhibiting effects on exogenous nutrient influx with an inhibitory paracrine effect on insulin secretion, somatostatin could reduce fat deposition and act as an antiobesity hormone.

■ Pancreatic Polypeptide

Pancreatic polypeptide, an islet hormone,was isolated from pancreatic digests, its structure was analyzed, a radioimmunoassay was developed for it, and it was localized to a specific islet cell type—all without any clear idea of its function or purpose. Nonetheless it is present in plasma, and its concentration increases markedly with ingestion of food. Moreover, it is suppressed by glucose infusion and stimulated by hypoglycemia. Insulin-deficient animals and humans show hyperplasia of its cell of origin and generally exhibit augmented secretion of the hormone in response to stimuli. Although the metabolic function of pancreatic polypeptide in normal humans currently is not known, patients with pancreatic islet cell tumors often show elevated plasma levels of the hormone. It therefore may serve as a useful means of diagnosing and monitoring the therapy of such neoplasms.

■ Clinical Syndromes of Islet Cell Dysfunction

■ Insulin

Tumors of the β-cells produce excess insulin. The cardinal manifestation is a low plasma glucose level (< 50 mg/dl) in the fasting state. Because the sympathetic nervous system is actuated quickly by abrupt hypoglycemia, bursts of insulin hypersecretion produce episodes of rapid heart rate, nervousness, sweating, and hunger. With chronic insulin excess and persistent hypoglycemia, central nervous system function is disturbed, leading to bizarre behavior, defects in cerebration, loss of consciousness, convulsions, and permanent brain damage if not treated soon enough. The need to ingest large amounts of carbohydrate to offset the autonomous secretion of insulin combined with the lipogenic effect of the hormone produces weight gain. The restraining effect of insulin on glucagon secretion prevents an adequate compensatory response of its physiological antagonist, aggravating the tendency to hypoglycemia. The diagnosis is established by demonstrating plasma insulin levels inappropriately high for the prevailing glucose levels during a short-term fast. Removal of the tumor cures the condition. Failing that, drugs that inhibit insulin secretion palliate the hypoglycemia.

Primary deficiency of insulin is the consequence of β-cell destruction. This disorder, known as *diabetes mellitus,* results from a genetically conferred vulnerability to an insult by a currently unspecified environmental agent. The pathophysiological picture in many respects resembles starvation (fasting) in an accelerated and aggravated form. The exceedingly low I/G ratio is responsible for the markedly distorted metabolic state. Glucose production is augmented, and peripheral glucose utilization is reduced, until a new equilibrium between these processes is reached at a very high plasma glucose level (300 to 2000 mg/dl). An increased rate of gluconeogenesis is supported by increased proteolysis. Muscle anabolism is correspondingly reduced; thus nitrogen balance becomes negative, and lean body mass decreases. Excess mobilization of adipose tissue triglycerides elevates plasma free fatty acids and body fat decreases. Ketogenesis is markedly stimulated, whereas peripheral ketoacid use is inhibited, thereby greatly elevating plasma ketoacid levels (10 to 20 mM). These relatively strong carboxylic acids raise the hydrogen ion concentration of the blood. As they are neutralized by sodium bicarbonate, the primary plasma buffer, carbonic acid is formed, which dissociates to carbon dioxide and water. Pulmonary hyperventilation is stimulated, which removes some of the excess carbon dioxide. Despite this compensatory reduction in carbon dioxide content, the blood pH may finally fall to lethal levels (less than 6.8), and death from diabetic ketoacidosis and coma ensues.

Before this terminal point a classic constellation of symptoms is observed. Because of the high plasma glucose levels, the filtered load of glucose exceeds the renal tubular capacity for reabsorption. Glucose therefore is excreted in the urine in large quantities (up to 400 g/day), causing, by its osmotic effect, increased excretion of water and salts. The patient notes an increased frequency of urination. Thirst is stimulated by the hyperosmolality of the plasma and by the hypovolemia. The loss of glucose is also a caloric drain, which the patient attempts to recoup by eating more. This causes further exaggerated rises in plasma glucose, as the insulin-deficient individual is unable to store the influx of carbohydrate from the gut efficiently. Despite the increase in appetite and caloric intake, loss of lean body mass, adipose tissue, and body fluids ensues. Deficits of nitrogen, potassium, phosphate, magnesium, and other intracellular components develop as these are excreted in the urine in amounts that exceed their intake. Fatigue is a common symptom, and muscle performance may be impaired during exercise because of the reduction in both muscle and liver glycogen stores.

Osmotic fluid shifts secondary to the high plasma glucose, and its conversion to other hexoses, such as sorbitol, adversely affect other tissues. For example, the lens of the eye swells, causing blurred vision, and even opacification. Other cellular malfunctions whose pathogenesis is less clear include initially increased filtration through the renal glomeruli, slowed transmission of nerve impulses, decreased resistance to infectious agents, and long-term complications such as retinal disease, kidney failure, and accelerated atherosclerosis. Numerous proteins, including hemoglobin and albumin, are nonenzymatically glucosylated via formation of an adduct between the aldehyde group of glucose and free amino groups. This results in functional changes in proteins such as collagen and may contribute to the long-term tissue damage seen in individuals with diabetes.

The biochemical abnormalities and symptoms of this form of diabetes mellitus are rapidly reversed by insulin replacement. The typical pattern of biochemical response is shown in Fig. 50-24. The I/G ratio increases not only because plasma insulin rises with treatment but also because plasma glucagon decreases as the α-cell hyperactivity is inhibited by insulin. Sequentially, plasma glucose and urine glucose excretion declines, lipolysis and ketosis diminish, and protein catabolism is replaced by anabolism, as indicated by the decline in urinary nitrogen excretion. Hydration is normalized, body weight regained, and vigor restored. It should be pointed out that a modest improvement in the I/G ratio in decompensated diabetic subjects also can be effected solely by decreasing glucagon secretion, for example, by administration of somatostatin. This reduces plasma and urine glucose significantly in undertreated patients and can moderate the otherwise rapid development of diabetic ketoacidosis if insulin is withdrawn completely.

Another even more common form of diabetes mellitus often is associated with obesity. In this situation the major target tissues are resistant to the action of insulin, in part because of a decrease in the number of insulin receptors and in part because of postreceptor block in insulin action. In addition, there is a derangement in β-cell recognition of glucose as a stimulus, so that insulin secretion is delayed and inadequate. The effective I/G ratio is less markedly reduced in these patients, so that the major biochemical manifestations are those of elevated plasma and urine glucose concentrations, particularly following exogenous carbohydrate intake. Accelerated mobilization of endogenous lipids with ketogenesis is rarely seen in the absence of intercurrent illness, and nitrogen balance is generally maintained. Treatment of this form of diabetes does not ordinarily require insulin administration. Instead caloric regulation, weight reduction if obesity is present, and occasionally use of sulfonylurea drugs simultaneously improve tissue responsiveness to endogenous insulin and β-cell responsiveness to glucose. Hyperglycemia and glycosuria then abate.

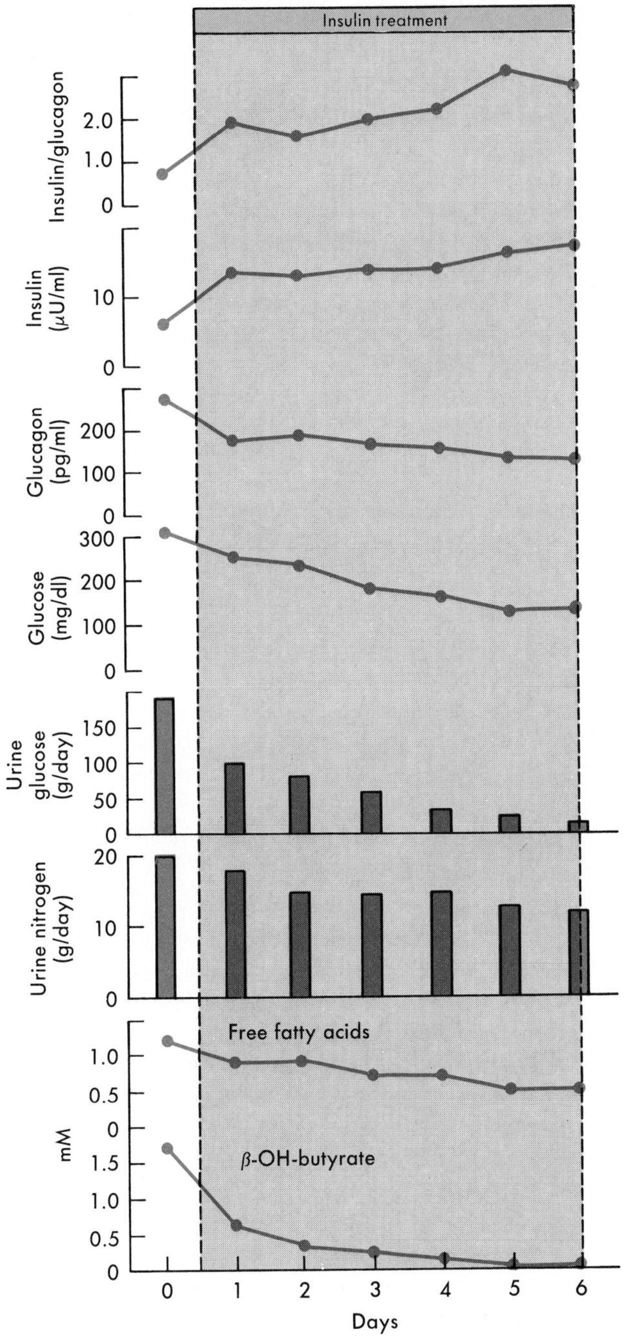

■ **Fig. 50-24.** The effects of insulin replacement on plasma hormone and substrate levels and urine losses of glucose and nitrogen in insulin-deficient diabetic humans.

■ *Obesity*

In a very small number of cases human obesity clearly results from destruction or infiltration of the hypothalamic areas that regulate appetite. In another small fraction obesity results from primary disturbances in cortisol, thyroid hormone, or growth hormone secretion. However, in the vast majority of cases no primary cause for morbid obesity can be identified. Nonetheless a considerable body of physiological and pathophysiological data does suggest a role for insulin either in the genesis or the maintenance of obesity. First, obese animals and humans are generally resistant to the actions of insulin on carbohydrate metabolism, particularly in muscle and liver. Second, they almost invariably secrete excess amounts of insulin in the basal state and in response to nutrients. Their β-cells are more numerous, and their insulin stores are enlarged. In a model strain of mice with genetic obesity, hyperinsulinism can be detected at virtually the same time as body lipid stores begin to enlarge compared with those of normal littermates. Even though caloric restriction can prevent the excess gain in weight, the percentage of body weight accounted for by fat remains abnormally high; however, chemical destruction of the β-cells of such obese mice does diminish their adiposity. Since the action of insulin is clearly to facilitate storage of calories as triglycerides, it seems reasonable to postulate that their hyperinsulinism is involved in the genesis of their obesity. Interestingly, in all models of obesity studied, as well as in human obesity, the excessive secretion of insulin is decreased by caloric deprivation and weight reduction as insulin resistance also diminishes. Thus hyperinsulinism must be deemed a phenomenon accompanying the ingestion of excess calories, and one that clearly acts to prevent dissipation of the calories and to promote their storage.

■ *Glucagon*

Primary glucagon excess is produced by α-cell tumors. The catabolic action of the hormone is exhibited clinically by loss of weight and by a peculiar destructive skin lesion. Plasma levels of glucose and of ketoacids are elevated. The marked increase in gluconeogenesis causes a generalized reduction of plasma amino acids and an increase in urinary nitrogen. Severe diabetes mellitus is uncommon because of the compensatory increase in insulin secretion. The predictable consequence of primary glucagon deficiency would be fasting hypoglycemia. In fact, however, such cases only rarely have been documented. It is likely that other counterregulatory hormones, such as epinephrine, take up the slack when glucagon is absent.

■ *Somatostatin*

A primary excess of somatostatin occurs rarely with certain islet cell tumors. The patients exhibit weight loss caused by inhibition of nutrient absorption. Plasma glucagon and insulin levels are low, as anticipated. Because of the latter deficiency, hyperglycemia is induced by carbohydrate loading.

An isolated deficiency of δ-cell somatostatin is as yet unknown. In a few cases of hyperinsulinism in infancy, hyperplasia of β-cells has been accompanied by an absence of δ-cells. It has been suggested that loss of a local inhibitory effect of somatostatin might be responsible for the unrestrained secretion of insulin.

Bibliography

Journal articles

Cahill, G.F.: Starvation in man, N. Engl. J. Med. **282**:668, 1970.

Cheng, K., and Larner, J.: Intracellular mediators of insulin action, Annu. Rev. Physiol. **47**:405, 1985.

Czech, M.P.: Insulin action, Am. J. Med. **70**:142, 1981.

Czech, M.P.: The nature and regulation of the insulin receptor: structure and function, Annu. Rev. Physiol. **47**:357, 1985.

Exton, J.H.: Gluconeogenesis, Metabolism **21**:945, 1972.

Felig, P., et al.: Amino acid metabolism during prolonged starvation, J. Clin. Invest. **48**:584, 1969.

Ferrannini, E., et al.: Effect of fatty acids on glucose production and utilization in man, J. Clin. Invest. **72**:1737, 1983.

Ferrannini E., et al.: The disposal of an oral glucose load in health subjects: a quantitative study, Diabetes **34**:580, 1985.

Foster, D.: From glycogen to ketones—and back, Banting Lecture 1984, Diabetes **33**:1188, 1984.

Frank, H.J.L., et al.: A direct in vitro demonstration of insulin binding to isolated brain microvessels, Diabetes **30**:757, 1981.

Genuth, S.M.: Plasma insulin and glucose profiles in normal, obese, and diabetic persons, Ann. Intern. Med. **79**:812, 1973.

Gottesman, I., et al.: Insulin increases the maximum velocity for glucose uptake without altering the Michaelis constant in man: evidence that insulin increases glucose uptake merely by providing additional transport sites. J. Clin. Invest. **70**:1310, 1982.

Granner, D.K., and Andreone, T.L.: Insulin modulation of gene expression, Diabetes/Metab. Rev. **1**:139, 1985.

Haymond, M., and Miles, J.: Branched chain amino acids as a major source of alanine nitrogen in man, Diabetes **31**:86, 1982.

Hornnes, P.J., et al.: Simultaneous recording of the gastroenteropancreatic hormonal peptide response to food in man, Metabolism **29**:777, 1980.

Itoh, M., et al.: Antisomatostatin gamma globulin augments secretion of both insulin and glucagon in vitro: evidence for a physiologic role for endogenous somatostatin in the regulation of pancreatic α- and β-cell function, Diabetes **29**:693, 1980.

Jefferson, L.S.: Role of insulin in the regulation of protein synthesis, Diabetes **29**:487, 1980.

Katz, L., et al.: Splanchnic and peripheral disposal of oral glucose in man, Diabetes **32**:675, 1983.

Liljenquist, J.E., et al.: Evidence for an important role of glucagon in the regulation of hepatic glucose production in normal man. J. Clin. Invest. **59**:369, 1977.

Olefsky, J.M.: Insulin resistance and insulin action: an in vitro and in vivo perspective, Diabetes **30**:148, 1981.

Reeds, P., and James, W.: Nutrition: the changing scene, protein turnover, Lancet **1**:571, 1983.

Reichlin, S.: Somatostatin, N. Engl. J. Med. **309**:1495, 1556, 1983.

Robbins, D., et al.: Biologic and clinical importance of proinsulin, N. Engl. J. Med. **310**:1165, 1984.

Saltiel, A.R., et al.: Insulin stimulated hydrolysis of a novel glycolipid generates modulators of cAMP phosphodiesterase, Science **233**:967, 1986.

Schaefer, E., et al.: Pathogenesis and management of lipoprotein disorders, N. Engl. J. Med. **312**:1300, 1985.

Sherwin, R., et al.: Effect of ketone infusions in amino acid and nitrogen metabolism in man, J. Clin. Invest. **55**:1382, 1975.

Stricker, E.M.: Hyperphagia, N. Engl. J. Med. **298**:1010, 1978.

Thiebaud, D., et al.: The effect of graded doses of insulin on total glucose uptake, glucose oxidation, and glucose storage in man, Diabetes **31**:957, 1982.

Unger, R.H., et al.: Insulin, glucagon, and somatostatin secretion in the regulation of metabolism, Annu. Rev. Physiol. **40**:307, 1978.

Unger, R.H., et al.: Glucagon and the A cell. I. Physiology and pathophysiology, N. Engl. J. Med. **304**:1518, 1981.

Verspohl, E.J., et al.: Evidence for presence of insulin receptors in rat islets of Langerhans, J. Clin. Invest. **65**:1230, 1980.

Weigle, D.S.: Pulsatile secretion of fuel regulatory hormones, Diabetes **36**:764, 1987.

CHAPTER

51

Endocrine Regulation of Calcium and Phosphate Metabolism

Calcium and Phosphate Pools and Turnover

The maintenance of calcium and phosphate homeostasis is dependent on major contributions from three organ systems: the intestinal tract, the skeleton, and the kidneys, with minor but essential contributions from the skin and the liver. Before detailing the hormonal mechanisms of regulation it is necessary to present an overall view of calcium and phosphate metabolism with special emphasis on the role played by the bone mass. Without such an initial overview it is difficult to tie together the individual endocrine components of the system.

The calcium ion is of fundamental importance to all biological systems. Calcium participates in numerous enzymatic reactions. It is a vital component in the mechanism of hormone secretion and a mediator of hormonal effects. Calcium is intimately involved in neurotransmission, in muscle contraction, and in blood clotting. It is the major cation in the crystalline structure of bone and teeth. For these and other reasons it is vital that cells be bathed with fluid in which the calcium concentration is kept within the narrow limits of physiological tolerance.

The normal range of calcium in the plasma is 8.6 to 10.6 mg/dl (4.3 to 5.3 mEq/L). For any individual, however, the variation from day to day is generally well under 10%. Approximately 50% of plasma calcium is in ionized, biologically active form; 10% is complexed in nonionic but ultrafilterable forms, such as calcium bicarbonate; and 40% is bound to proteins, mainly albumin. The equilibrium between ionized and protein-bound calcium is shifted toward increased binding as the blood pH increases.

Fig. 51-1 details in simplified fashion the normal turnover of calcium in the body. Daily dietary calcium intake can range from as little as 200 to as much as 2000 mg. The percentage of dietary calcium that is absorbed from the gut is inversely related to intake but the relation is curvilinear. An adaptive increase in fractional absorption is one important mechanism for maintaining normal body calcium stores in the face of dietary deprivation, whereas an adaptive decrease prevents overload in the face of dietary surfeit. At an average daily intake of 1000 mg about 35% is absorbed. The same amount of calcium, 350 mg, ultimately must be excreted to maintain balance. About 150 mg of calcium is secreted back into the intestine and excreted in the stools, along with the unabsorbed fraction from the diet. The remaining 200 mg is excreted in the urine. The kidney filters about 10,000 mg of calcium per day (non-protein-bound calcium concentration × glomerular filtration rate = 60 mg/L × 170 L/day). However, approximately 98% is reabsorbed in the tubules. Alterations in the small fraction of filtered calcium that is finally excreted provide another sensitive means of maintaining overall calcium balance.

Calcium enters and exits from an extracellular pool of 1000 mg. This in turn is in equilibrium with a rapidly exchanging pool of several times that size. This pool probably represents the surface of recently or partially mineralized bone. Although it has been estimated that 30,000 to 40,000 mg of calcium may move back and forth in these internal pools daily, only 500 mg is "irreversibly" removed from the extracellular space by bone formation, and 500 mg is returned to it by bone resorption in the process that constitutes normal bone remodeling. The tremendous store of skeletal calcium (approximately 1 kg) makes it quantitatively the most important depot in maintaining the stability of plasma calcium.

The phosphate ion is also of critical importance to all biological systems. Phosphate is an integral component of all glycolytic compounds from glucose-1-phosphate

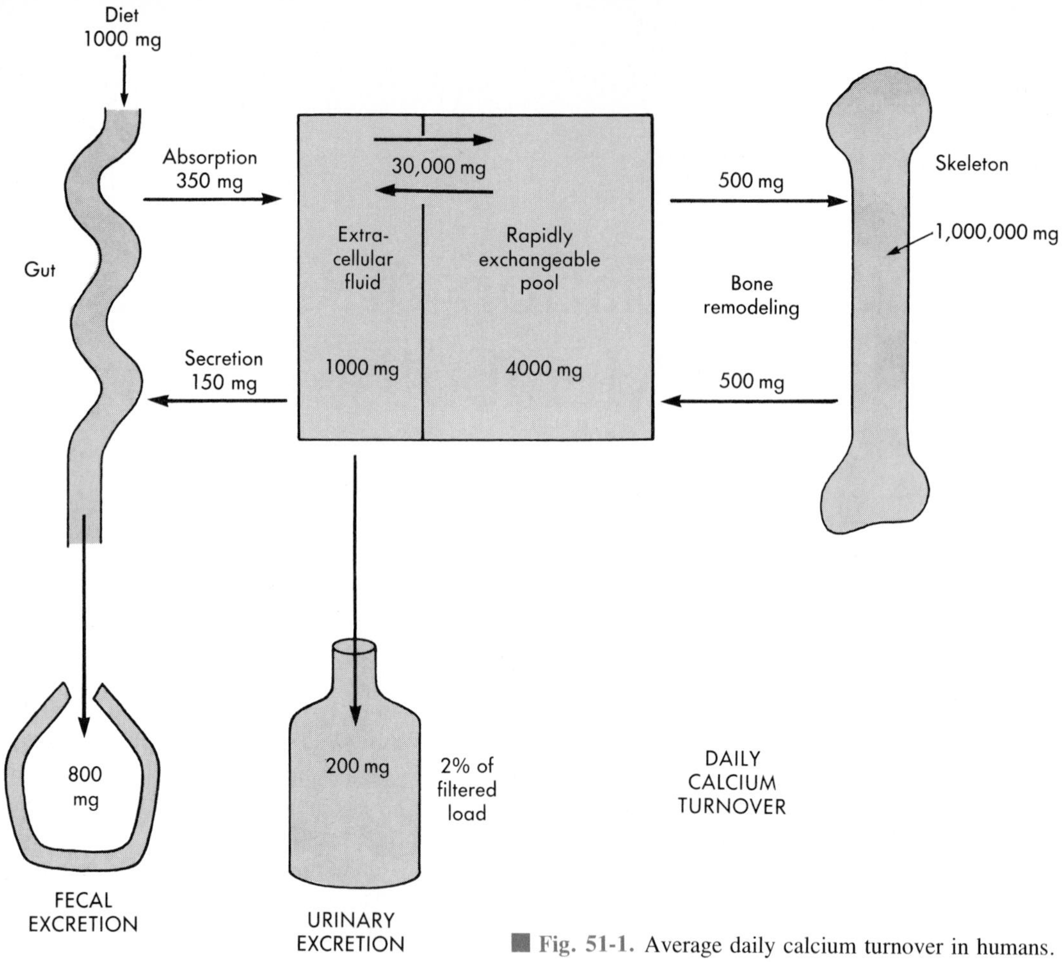

Fig. 51-1. Average daily calcium turnover in humans.

to phosphoenolpyruvate. It is part of the structure of high-energy transfer compounds, such as ATP and creatine phosphate, of cofactors such as NAD, NADP, and thiamin pyrophosphate, and of lipids such as phosphatidylcholine. Phosphate functions as a covalent modifier of numerous enzymes. And of course phosphate is a major anion in the crystalline structure of bone.

The normal concentration of phosphate in the plasma is 2.5 to 4.5 mg/dl (0.81 to 1.45 mmole/L.) Because the valence of phosphate changes with pH, it is less meaningful to express phosphate concentration in milliequivalents per liter. The turnover of phosphate is shown in Fig. 51-2. In contrast to calcium, the percentage of phosphate absorbed from the diet is relatively constant. Thus net absorption of phosphate from the gut is linearly related to intake over a wide range, and adaptive regulation at this site is of minor importance. Therefore urinary excretion provides the major mechanism for preserving phosphate balance. Of the daily filtered load of approximately 6000 mg (plasma concentration × glomerular filtration rate = 36 mg/L × 170 L/day), renal tubular reabsorption can vary from 70% to 100%, with an average of 90%. This provides the needed flexibility to compensate for large swings in dietary intake. Soft tissue stores of phosphate, such as in the muscle mass, are large. Transfer between these stores and the extracellular fluid is an important factor in minute-to-minute regulation of plasma phosphate concentration. About 250 mg, or half the total extracellular fluid pool of 500 mg, enters and leaves the bone mass daily in the process of remodeling.

The divalent cation magnesium (Mg^{++}) is related in some metabolic respects to calcium and phosphate. Magnesium is essential in neuromuscular transmission and serves as a cofactor in numerous enzyme reactions, most notably those involving energy transfers via ATP and those concerned with ribosomal protein synthesis. The normal range of magnesium in plasma is 1.8 to 2.4 mg/dl (1.5 to 2.0 mEq/L). One third of plasma magnesium is bound to protein. The body content is about 25,000 mg, of which 50% is present in the skeleton. Virtually all the rest is present in the intracellular fluid, where magnesium functions with potassium as a major cation. The average daily intake is about 300 mg. Forty percent of this is absorbed, and in a steady state the same amount, 120 mg, is excreted in the urine.

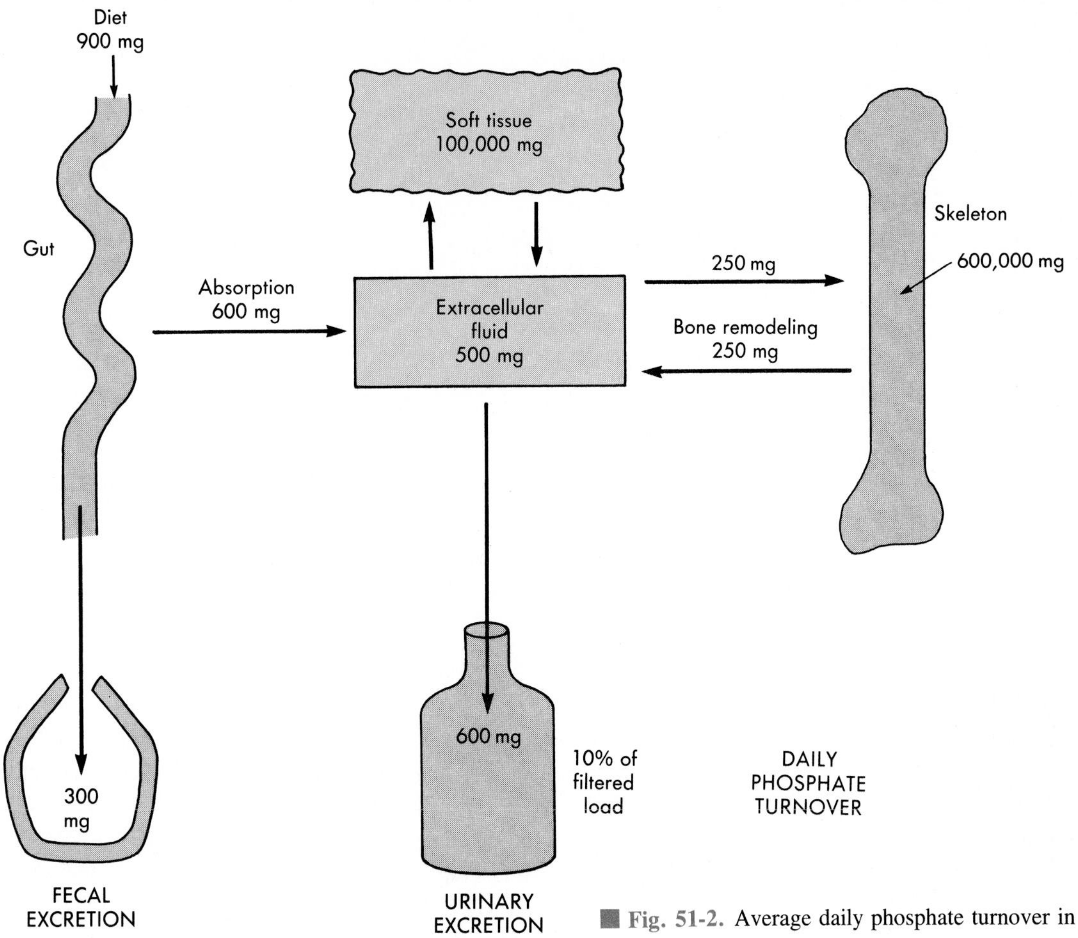

■ **Fig. 51-2.** Average daily phosphate turnover in humans.

■ *Bone Dynamics*

A detailed account of the development of the skeleton, its structural properties, and its mechanical function is beyond the purview of endocrine physiology. However, a brief review of the organization of bone in relation to its function as a mineral reservoir is essential to understanding hormonal regulation of calcium and phosphate metabolism.

About 80% of the skeletal mass consists of cortical bone and 20% of trabecular bone, as illustrated in Fig. 51-3. Most of the trabecular bone is actually contained in the axial skeleton (vertebrae, skull, and ribs). Though of lesser mass, trabecular bone has five times as much total surface area as cortical bone, an important consideration in calcium dynamics and bone turnover. Throughout life the bone mass is continuously turning over by a well-regulated coupling of the processes of formation and resorption. This coupling is achieved in all types of bone within individual microscopic units, called osteons. The chemical or mechanical signals that coordinate local rates of formation and resorption remain mysterious. During the growth years formation exceeds resorption, and skeletal mass increases. Linear growth occurs at the ends of long bones by replacement of cartilage in specialized areas known as epiphyseal plates. These close off at the end of puberty when adult height is reached. Increase in bone width occurs by apposition to the outer surfaces under the connective tissue covering, the periosteum. A peak in bone mass is reached between ages 20 and 30 years. Thereafter equal rates of formation and resorption stabilize the bone mass until age 40 to 50 years, at which time resorption begins to exceed formation, and the total mass slowly decreases. This process of bone turnover in the adult is known as *remodeling;* it is one of the major mechanisms for maintaining calcium homeostasis. As much as 15% of the total bone mass normally turns over each year in the remodeling process. Endocrine diseases that disrupt the coupling of formation and resorption are more florid in their manifestations when they are superimposed on the natural disequilibrium that exists during the early phase of growth and the late phase of senescence. Women are severely affected by such diseases more often than men because women have a smaller peak bone mass and a more rapid physiological rate of senescent loss.

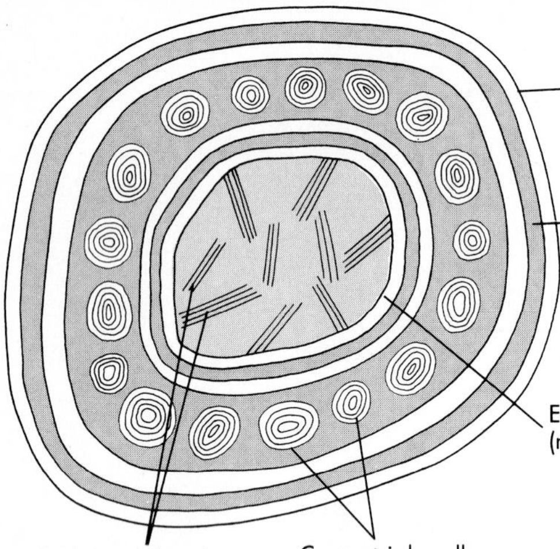

Fig. 51-3. Cross section of a long bone. The outer cortex of bone consists of circumferential lamellae within which other concentric lamellae (haversian canals) are present. These contain nutrient vessels. The inner portion consists of bridging trabeculae or spicules of bone in lamellar arrays, between which bone marrow elements and connective tissue cells are found.

Three major cell types are recognized in histological sections of bone: *osteoblasts*, *osteocytes*, and *osteoclasts* (Fig. 51-4). The first two arise from primitive cells, called osteoprogenitor cells, within the connective tissue of the mesenchyme. The primitive cells are similar to fibroblast precursors. A variety of proteins in the bone attract osteoprogenitor cells, direct their differentiation into osteoblasts, and stimulate their growth. The osteoclasts arise from the same precursors that give rise to circulating monocytes and tissue macrophages (i.e., promonocytes or monoblasts).

Bone formation is carried out by active osteoblasts, which synthesize and extrude collagen into the adjacent extracellular space. The collagen fibrils line up in regular arrays and produce an organic matrix known as *osteoid* within which calcium is then deposited as amorphous masses of calcium phosphate. About ten days elapse between osteoid formation and mineralization. Once initiated, however, most of the calcium phosphate is deposited within a period of 6 to 12 hours. Thereafter hydroxide and bicarbonate ions are added to the mineral phase and mature hydroxyapatite crystals are formed. These have a calcium/phosphate ratio of 2.2 by weight and 1.7 by moles. As this completely mineralized bone accumulates and surrounds the osteoblast, that cell decreases its synthetic activity and becomes an interior osteocyte (Fig. 51-4). Osteoblastic activity, therefore, is observed only along the surfaces of bone, whether they are the concentric lamaellae of cortical bone or the linear lamellae of the interior bridging trabeculae (Fig. 51-3). Along these surfaces are resting osteocytes that appear to be potential osteoblasts.

The mineralization process requires normal plasma concentrations of calcium and phosphate and is dependent on vitamin D. The enzyme alkaline phosphatase and possibly other macromolecules from the osteoblast also participate in this process. Of note are *osteonectin*, a protein of 32,000 molecular weight, and *osteocalcin*, a protein of 6,000 molecular weight. These two proteins form 1% to 2% of all bone protein. Osteonectin binds to collagen, and the complex, in turn, binds hydroxyapatite crystals. Osteocalcin is distinguished by the presence of γ-carboxyglutamate residues, which have an affinity for calcium and a strong avidity for uncrystalized hydroxyapatite. Both alkaline phosphatase and osteocalcin circulate in plasma and their levels are markers of osteoblastic activity.

Within each bone unit minute fluid-containing channels, called canaliculi, traverse the mineralized bone; within these channels the interior osteocytes remain connected with surface osteocytes and osteoblasts via syncytial cell processes (Fig. 51-4). This arrangement provides an enormous surface area for transfer of calcium from the interior to the exterior of the bone units and thence into the extracellular fluid. This transfer process, which is carried out by the osteocytes, is known as *osteocytic osteolysis*. It probably does not actually decrease bone mass but simply removes calcium from the most recently formed crystals.

In contrast, the process of resorption of bone does not merely extract calcium; it also destroys the entire matrix, thereby diminishing the bone mass. The cell responsible for this is the osteoclast, which is a giant multinucleated cell formed by fusion of several precursor cells (Fig. 51-4). The osteoclast contains large numbers of mitochondria and lysosomes. It attaches to the surface of the bone unit. At the point of attachment a ruffled border is created by infolding of the plasma membrane. In this zone the process of bone dissolution is carried out by collagenase, lysosomal enzymes, and phosphatase. Calcium, phosphate, and amino acids, such as hydroxyproline and hydroxylysine, that are unique to collagen, are released into the extracellular fluid. In this way the osteoclast literally tunnels its way

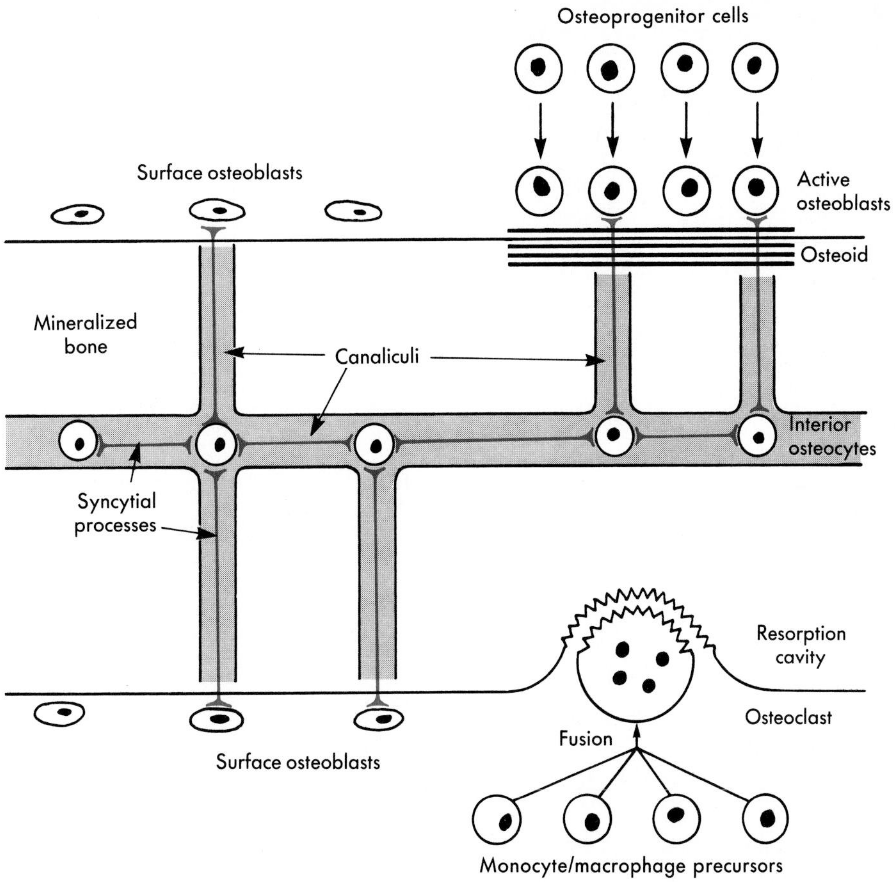

■ **Fig. 51-4.** The relationships between bone cells and bone surfaces. The canaliculi provide a huge interface between the interior surfaces of mineralized bone and intercellular fluid. This permits efficient osteolysis with transfer of calcium and phosphate to the exterior via syncytial processes connecting interior and surface osteocytes. (Redrawn from Avioli, L.V., et al.: Bone metabolism and disease. In Bondy, P.K., and Rosenberg, L.E.: Metabolic control and disease, Philadelphia, 1980, W.B. Saunders Co.)

■ **Table 51-1.** Major effects of various hormones on bone

Bone formation	*Bone resorption*
Stimulated by	*Stimulated by*
Growth hormone (somatomedins)	Parathyroid hormone
Insulin	Vitamin D
Androgens	Thyroid hormone
Vitamin D (?)	Cortisol
Inhibited by	Prostaglandins
Cortisol	Lymphokines
Parathyroid hormone	*Inhibited by*
	Calcitonin
	Estrogens

into the mineralized bone. As noted previously, resorption and formation are closely coordinated locally. Therefore the resorption cavity created by the osteoclast is often the site of subsequent osteoblastic activity, which fills in the cavity with new bone.

The recruitment of osteoblasts and osteoclasts, as well as the activity of each cell type, is greatly influenced by an array of hormones (Table 51-1). As a general principle, whether the primary effect of a hormone is on formation or resorption of bone, the phenomenon of coupling will secondarily alter the other process in the same direction. For example, whereas the primary effect of an excess of parathyroid hormone (PTH) is to increase bone resorption, a compensatory increase in bone formation follows. The net effect of any endocrine abnormality will depend on the degree to which the total bone mass is defended by the process of coupling.

■ *The Parathyroid Glands*

The parathyroid glands are major regulators of plasma calcium concentration and calcium flux in the body. The paramount effect of PTH is to increase or sustain the plasma calcium level. This is accomplished by causing entry of calcium into the plasma from the bone mass, the tubular urine, and the intestinal tract. The four parathyroid glands develop at 5 to 14 weeks of gestation from the third and fourth branchial pouches. They descend to lie posterior to the thyroid gland. The

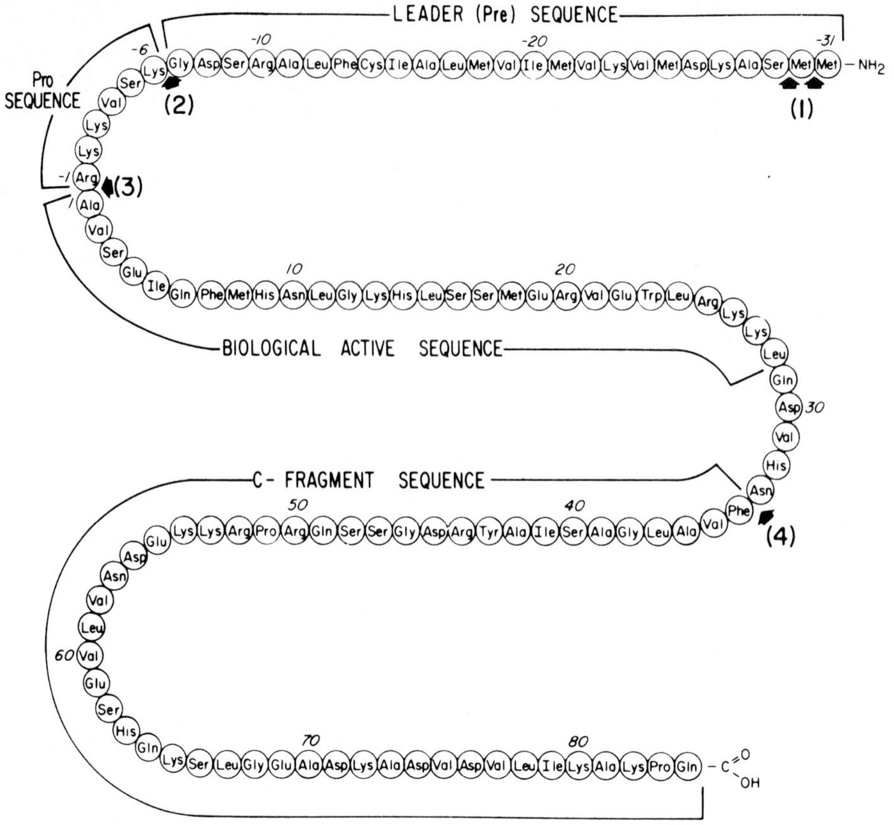

Fig. 51-5. Structure and processing of preproparathyroid hormone. The N-terminal leader (signal) sequence is cleaved in two steps (*steps 1* and *2*) leaving proparathyroid hormone. Removal of 6 more amino acids (*step 3*) yields the 1-84 parathyroid hormone structure. Further cleavage within the gland as well as in peripheral tissues (*step 4*) yields as the major metabolic product of parathyroid hormone, a C-terminal fragment that is biologically inactive. (Redrawn from Habener, J., et al.: Parathyroid hormone: biochemical aspects of biosynthesis, secretion, action, and metabolism, Physiol. Rev. **64:**985, 1984.)

lower pair actually is derived from the third branchial pouch and traverses a longer embryological descent. Ectopic locations in the neck and mediastinum are not rare, a fact that surgeons must appreciate in searching for abnormal parathyroid glands. The total weight of adult parathyroid tissue is about 130 mg, and that of any one gland is 30 to 50 mg. Their blood supply is from the thyroid arteries, and venous drainage is into the jugular and inferior thyroid veins. Samples drawn from veins on the left and the right can localize the site of unilateral parathyroid gland hyperfunction.

The histological appearance of the parathyroid glands changes with age. The predominant cell, known as the *chief cell,* is present throughout life and is the normal source of PTH. The appearance of individual chief cells on electron microscopy varies with their state of activity. Resting cells have abundant glycogen, an involuted Golgi apparatus, and a few clusters of secretory granules. Active cells have little glycogen, a large convoluted Golgi apparatus with vacuoles and vesicles, and a granular endoplasmic reticulum. During active hormone secretion numerous granules may be seen undergoing exocytosis. A second cell type—the *oxyphil cell*—is distinguished by an eosinophilic cytoplasm. This cell first appears at puberty. Its normal function is not known, but its ultrastructural appearance suggests energy production and storage rather than hormone synthesis. Nonetheless under some circumstances it does appear to be capable of hormone release. Either of these cell types may undergo hyperplasia or form neoplasms, giving rise to hypersecretion of PTH.

Synthesis and Release of PTH

PTH is a single-chain protein of 9000 molecular weight, containing 84 amino acids (Fig. 51-5). Although there is species variation in structure, PTH from bovine, porcine, and human glands shows a great deal of immunological and biological cross-reactivity. The biological activity of PTH resides in the N-terminal portion of the molecule within amino acids 1 to 34. This fragment has 77% of the activity of native hormone. Alteration, elimination, or backward extension of the N-terminal amino acid markedly decreases biological activity. The function of the remainder of the molecule with its carboxy terminus is not known.

The synthesis of PTH begins with prepro-PTH, a 115–amino acid protein that contains all the structural information encoded in the gene for PTH (Fig. 51-5). This conclusion has been strongly supported by reversed synthesis of the DNA transcript from messenger RNA for PTH. As the peptide chain of prepro-PTH grows to its complete length on the ribosomes, first 2 and then 23 amino acids are enzymatically removed from the N-terminal, leaving pro-PTH. This 90–amino acid molecule is then transported to the Golgi apparatus, within which another 6 amino acids are removed by a tryptic enzyme. Processing of pro-PTH to PTH is very efficient, as only 7% of the glandular content of PTH immunoreactivity is accounted for by the precursor. Some of the resulting PTH is packaged for storage in mature secretory granules and is released later by the process of exocytosis. There is also evidence that on acute stimulation some newly synthesized PTH may be transported, still in Golgi vesicles, directly through the cell for immediate release. Cleavage of PTH between amino acids 33 and 40 (Fig. 51-5) also occurs within the gland, and therefore not all of the synthesized molecules reach the circulation.

The dominant regulator of parathyroid gland activity is the plasma calcium level. The hormone and the ion form a negative feedback pair. Secretion of PTH is inversely related to the plasma calcium concentration in a sigmoidal fashion (Fig. 51-6). Maximum secretory rates are achieved below a calcium concentration of 7 mg/dl (3.5 mEq/L). As calcium concentration increases to 11 mg/dl (5.5 mEq/L), PTH secretion is progressively diminished. However, above 11 mg/dl there is a persistent basal rate of PTH secretion that is not suppressible by further elevation of plasma calcium. It is actually the ionized fraction of plasma calcium that regulates PTH secretion. Alterations in PTH secretion in response to changes in plasma calcium occur within minutes. In vitro, after exposure to hypocalcemia, exocytosis of secretory granules is mediated by activation of adenylate cyclase and a resultant rise in intracellular cyclic AMP levels. Exocytosis also is promoted by inhibition of phosphodiesterase, the enzyme that degrades cyclic AMP. Suppression of PTH secretion by calcium influx into the parathyroid chief cells represents a notable exception to the general rule that calcium influx into endocrine cells stimulates hormonal secretion. The mechanism underlying this paradoxical effect in the parathyroid glands is unknown.

In addition to rapidly regulating secretion, calcium also modulates PTH synthesis and degradation within the glands. Prolonged exposure to a high ambient calcium concentration eventually depresses the rate of PTH synthesis; the exact control point is not known. However, careful studies have thus far failed to show significant effects of calcium on the transcriptional, translational, or processing steps in PTH synthesis. Calcium does regulate parathyroid cell size and number, and this may prove to be the mechanism whereby total hormone synthesis is regulated by the ion. Opposite to its inhibitory effect on PTH synthesis (and cell growth), calcium stimulates the intraglandular degradation of PTH. The net effect, then, of increasing ambient calcium is to decrease both the glandular stores and the release rates of PTH. Conversely, hypocalcemia increases PTH stores and secretory rates, and ultimately stimulates hyperplasia of the glands as well.

The divalent cation Mg^{++} modulates PTH secretion in a manner analogous to that of Ca^{++}, although

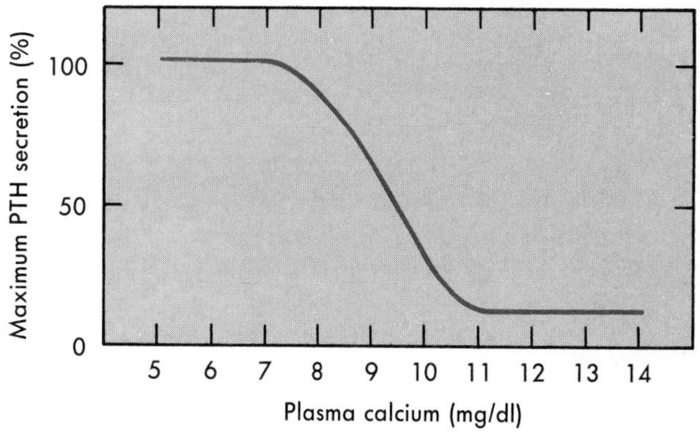

■ Fig. 51-6. The relationship between PTH secretion and plasma calcium level. Note maximum stimulation occurs at 7 mg/dl and that a low level of secretion persists above 11 mg/dl. (Redrawn from Mayer, G.P.: Effect of calcium and magnesium on parathyroid hormone secretion rate in calves. In Talmadge, R.V., et al.: Calcium regulating hormones, Amsterdam, 1975, Excerpta Medica.)

Mg^{++} is only half as effective on a molar basis. Therefore Mg^{++} is of much less importance in its normal physiological range (1.5 to 2.5 mEq/L). However, chronic hypomagnesemia strongly inhibits PTH synthesis. In severely magnesium-depleted individuals this phenomenon often predominates, leading to a reduced rate of PTH release. In addition, hypomagnesemia impairs the response of target tissues to PTH.

Because of the close physiological relation of phosphate to calcium, much work has been done to determine the influence of phosphate on PTH secretion. No direct effects of phosphate on the parathyroid glands have been demonstrated. However, a rise in plasma phosphate causes an immediate fall in ionized calcium levels, which in turn does stimulate PTH secretion. As is elaborated later, this resultant increase in PTH helps moderate hyperphosphatemia by increasing renal phosphate excretion.

Direct regulation of PTH secretion by vitamin D also occurs. $1,25\text{-}(OH)_2\text{-}D$, the most active form of the vitamin (see below) sharply decreases the level of prepro-PTH messenger RNA and eventually reduces the secretion of PTH. Epinephrine—a catecholamine hormone—stimulates PTH secretion through its β-adrenergic receptor, probably by activating adenylate cyclase. Histamine also stimulates PTH release via specific histamine (H_2) receptors. Finally, some circadian variation in PTH secretion occurs, probably independent of plasma calcium concentration, with the highest levels of PTH occurring in the evening.

Under normal circumstances neither prepro-PTH nor pro-PTH is secreted from the glands. Furthermore neither of these molecules appears to have intrinsic biological activity. One or more products of the intraglandular degradation of PTH, however, are released. In addition, PTH undergoes rapid metabolism in the peripheral tissues, with the kidney and liver being the major sites. The native hormone is predominantly split in the liver, producing as the major species a 6,000 molecular weight carboxy-terminus fragment that is biologically inactive (see Fig. 51-5). The latter fragment is probably acted upon further in the kidney. The circulating fragments often cross-react in radioimmunoassays that employ antibodies raised against intact hormone. Thus studies of plasma concentration and turnover of PTH have been hampered by the fact that often less than half the total measured immunoreactivity is accounted for by intact hormone. The plasma concentration of total immunoreactive PTH in humans is around 0.2 ng/ml. By a very sensitive cytochemical assay the concentration of biologically active hormone may be as low as 0.01 ng/ml (approximately 10^{-12} M). Whether amino terminal fragments generated during cleavage of PTH also circulate is controversial, as is any putative biological activity they might have in various target tissues.

There is no evidence for carrier protein binding of circulating PTH or its metabolites. The plasma half-life of the intact hormone is 20 to 30 minutes. The carboxy terminal fragment has a much longer plasma half-life (6 to 12 hours). For this reason radioimmunosassays that are specifically directed toward either the carboxy terminus or another portion of the 6000 molecular weight species have been very useful for determining chronic states of hyperparathyroidism and hypoparathyroidism. Assays directed at the N-terminus of the molecule have been especially useful for dynamic studies of parathyroid gland function, since they measure only the biologically active hormone.

PTH Actions

PTH action is initiated by binding to plasma cell membrane receptors; the determinants of binding lie in a restricted zone of the biologically active 1 to 34 sequence. In all target cells activation of adenylate cyclase raises intracellular cAMP levels. Most of the effects of PTH in vivo can be mimicked by infusions of cAMP or a suitable analogue that can enter cells. The subsequent intracellular events mediated by cAMP are not known in detail. Presumably cAMP triggers a protein kinase cascade (Fig. 49-6) that ultimately leads to phosphorylation of proteins necessary for enhanced transport of calcium and other ions.

Independent of its action on cAMP levels, PTH also stimulates the uptake of calcium by the cytosol from the bathing bone fluid and from mitochondrial stores. Whether this calcium is itself another intracellular second messenger or whether it only acts to modulate the adenylate cyclase response by counterregulatory inhibition remains moot. The initial uptake of calcium is reflected in a slight transient hypocalcemia, which follows PTH administration and precedes the classic hypercalcemic response. Of great significance is that the presence of an active vitamin D metabolite is required for the exhibition of the full spectrum of PTH actions. It remains to be determined whether vitamin D is an intracellular cofactor, whether it stimulates synthesis of an essential intracellular protein mediator, or whether it only augments the pool of extracellular calcium from which the rapid early uptake occurs. A sufficient intracellular concentration of magnesium is also necessary for maximum PTH responsiveness.

The overall effect of PTH is to increase plasma calcium and decrease plasma phosphate by acting on three major target organs: bone, kidney, and gastrointestinal tract. All three actions ultimately increase calcium influx into the plasma, raising the concentration. In contrast, the actions on bone and gut, which increase phosphate influx, are overwhelmed by the action on the kidney, which increases phosphate efflux, so that plasma phosphate concentration falls (Fig. 51-7).

Bone. PTH accelerates removal of calcium from bone by at least two processes. Its initial effect is to stimulate osteolysis by surface osteocytes. This process causes transfer of calcium from the bone canalicular fluid into the osteocyte and thence out the opposite side of the cell into the extracellular fluid. Replenishment of calcium in the canalicular fluid probably then occurs from the surface of partially mineralized bone. Phosphate does not appear to be mobilized with calcium in this process. A second, more slowly developing effect of PTH is to stimulate the osteoclasts to resorb completely mineralized bone. In this process both calcium and phosphate are released for transfer into the extracellular fluid and the organic bone matrix is hydrolyzed by increased activity of collagenase and of lysosomal enzymes. PTH initially increases the ruffled border of the osteoclast and the clear resorptive zone that develops between it and the mineralized bone. This is followed by PTH stimulation of osteoclast size, RNA synthesis, and number of nuclei. Ultimately proliferation of osteoclasts is increased by PTH. The giant osteoclasts create large resorption cavities in both cortical and trabecular bone. PTH also induces increases in acid phosphatase and carbonic anhydrase and accumulation of lactic acid and citric acid. The resultant lowering of ambient pH contributes to the resorptive process. One consequence of bone destruction is the release of hydroxyproline into the plasma. From the plasma it is then excreted into the urine.

PTH receptors are also found on osteoblasts. When exposed to PTH osteoblasts change their shape. This has been attributed to changes in phosphorylation of myosin light chains with subsequent aggregation of actin-myosin fibrils. Later, PTH inhibits the synthesis of collagen by the osteoblasts, probably at the level of transcription, since collagen messenger RNA levels are

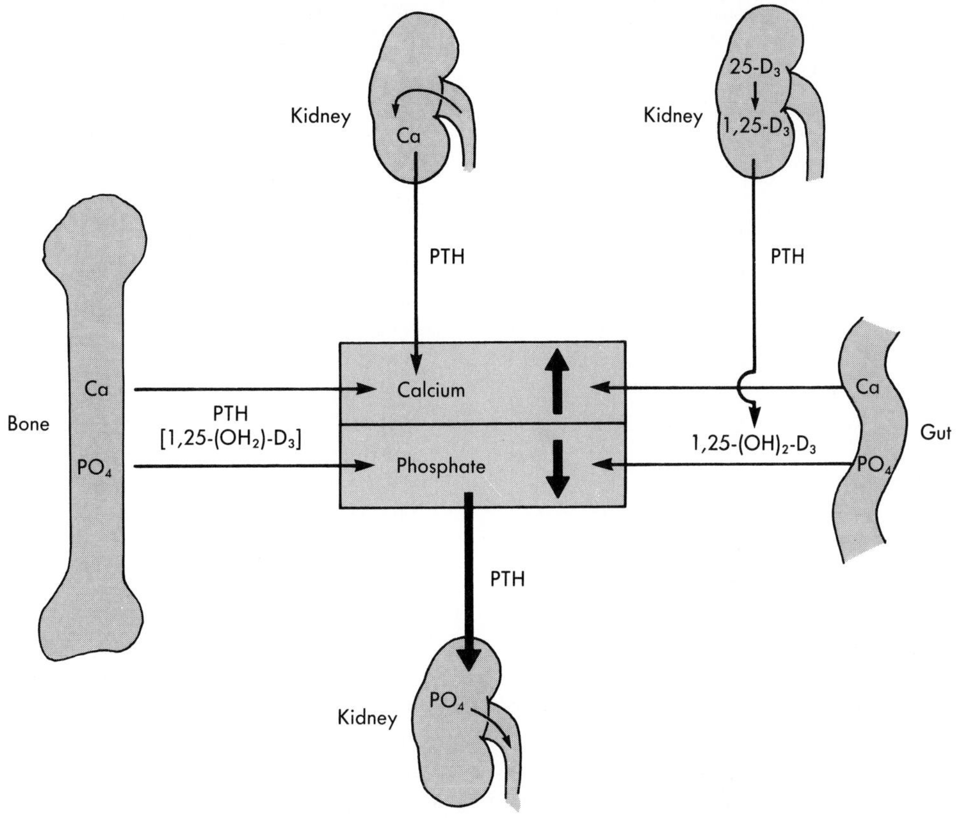

Fig. 51-7. The sites of PTH actions that lead to a net increase in serum calcium level and a net decrease in serum phosphate level. Note that PTH stimulates renal formation of 1,25-$(OH)\text{-}_2D_3$, and in turn the latter potentiates PTH action on bone and increases absorption of both minerals from the gut.

reduced by the hormone. The catabolic effects of PTH on bone are achieved by the elevated concentrations of hormone that result from stimulation of the parathyroid glands by hypocalcemia; thus they rapidly restore the plasma calcium level to normal. These actions generally require the concomitant presence of a specific metabolite of vitamin D, 1,25-$(OH)_2$-D_3.

At lower concentrations PTH also has an anabolic action on bone. This occurs under circumstances in which plasma calcium is normal or even increased, but compensatory hypersecretion of PTH has existed for a long time. In bone cultures low doses of PTH increase the number of osteoblasts and collagen synthesis. In bone biopsies there may be seen an increase in woven (as opposed to lamellar) bone and a proliferation of fibroblasts adjacent to the areas of increased resorptive activity. The plasma level of alkaline phosphatase, an osteoblastic enzyme whose activity parallels bone formation, is often increased by PTH. Thus the net effect of sustained increases in PTH may be either a decrease or an increase in total skeletal mass. This undoubtedly depends on concomitant factors that affect bone remodeling, such as the availability of calcium, phosphate, and vitamin D.

Kidney. PTH increases the reabsorption of calcium from the distal tubule of the kidney. By this mechanism the PTH that is secreted in response to hypocalcemia helps raise a depressed plasma calcium concentration. This effect is mediated by PTH stimulation of cAMP production at the capillary surface of the renal tubular cell. The cAMP is transported to the luminal surface where it activates unidentified protein kinases that are located in the brush border and are involved in calcium resorption. The relationship between urinary calcium excretion and plasma calcium concentration is shifted to the right by PTH (Fig. 51-8). Therefore suppression of PTH secretion by a sudden calcium load will help prevent hypercalcemia by permitting a greater fraction of the filtered calcium to escape into the urine. The net effect of prolonged alteration in PTH secretion on urinary calcium excretion, however, is dominated by the influence of PTH on bone and gut. Deficiency of PTH eventually will lower plasma calcium and, with it, the filtered load of calcium. Therefore the absolute amount of calcium excreted in the urine decreases. The converse sequence of events occurs with a prolonged excess of PTH.

The most dramatic effect of PTH on the kidney is to inhibit the reabsorption of phosphate in the proximal tubule. Within minutes of its administration PTH in-

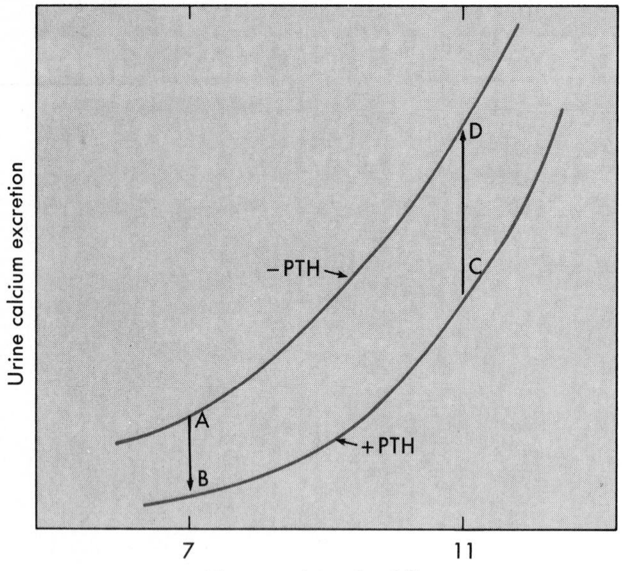

Fig. 51-8. Effect of PTH on the relationship between urine calcium excretion and plasma calcium level. At a low plasma calcium level, PTH secretion is stimulated, and the hormone shifts urine calcium excretion from point *A* to point *B*, thus conserving calcium. At a high plasma calcium level, PTH secretion is suppressed, and urine calcium excretion is shifted from point *C* to point *D*, thus disposing of excess calcium. (Redrawn from Nordin, B.E.C., et al.: Lancet **2:**1280, 1969).

creases urinary phosphate excretion. As befits its second messenger role in this action, cAMP excretion into the urine also increases just before that of phosphate. The phosphaturic effect of PTH allows disposition of the extra phosphate released by PTH-stimulated bone resorption. Otherwise simultaneous plasma elevation of calcium and phosphate would occur, with the potential danger of precipitating calcium-phosphate complexes in critical tissues. In contrast, under circumstances of primary phosphate deprivation, serum calcium tends to rise, thereby suppressing PTH secretion. This in turn increases tubular phosphate reabsorption, conserving this essential mineral.

PTH also inhibits the reabsorption of sodium and bicarbonate in the proximal tubule in a manner parallel to that of phosphate reabsorption inhibition. This action may prevent the occurrence of metabolic alkalosis, which could result from the release of bicarbonate during the dissolution of hydroxyapatite crystals in bone.

PTH also stimulates the synthesis in the renal tubules of vitamin D metabolites with bone-reabsorbing action. The decrease in plasma and renal phosphate content caused by PTH fortifies its direct action on renal vitamin D metabolism (as explained later); this contributes significantly to the major function of PTH: to increase plasma calcium.

Intestine. The absorption of calcium as well as phosphate from the gastrointestinal tract is enhanced by PTH. This action, however, is indirect. It comes about because PTH stimulates the synthesis of 1,25-$(OH)_2$-D_3, a potent vitamin D metabolite that directly increases active transport of calcium across the intestinal mucosa.

Miscellaneous. Alterations in the function of the central nervous system, of peripheral nerves, of muscles, and of other endocrine glands are seen in states of PTH excess or deficiency. Most of these effects can be attributed to the concomitant changes in plasma calcium levels. However, the possibility exists that PTH or one of its metabolites may have direct action on these tissues, either by stimulating calcium transfer across their cell membranes or by some calcium-independent mechanism still to be discovered.

Overall action of PTH. The summated biochemical effects of PTH are illustrated in Fig. 51-9, which demonstrates the results of administering the hormone to a hypoparathyroid patient. There is a prompt increase in plasma calcium and a decrease in plasma phosphate. The renal tubular reabsorption of phosphate falls. Urinary excretion of calcium initially declines as its tubular reabsorption increases. However, as plasma calcium continues to increase, the filtered load goes up, and urinary calcium excretion also rises. Urinary excretion of hydroxyproline increases as a result of PTH-stimulated bone resorption.

Calcitonin

Synthesis and Release

The parafollicular, or C, cells of the thyroid gland secrete another protein hormone, *calcitonin*, which influences calcium metabolism. Whereas PTH acts to increase plasma calcium, calcitonin acts to lower it. Although uncertainty exists about the significance of its physiological role in humans, calcitonin is likely to be an important regulator of plasma calcium in lower animals that live in an aquatic environment high in calcium. The parafollicular cells are of neural crest origin. In lower animals they form a discrete ultimobranchial gland from the sixth branchial pouch. In humans they are concentrated in the central portion of the lateral lobes of the thyroid gland, forming 0.1% of the epithelial cells. They are relatively large cells with a pale cytoplasm that contains small secretory granules enclosed in membranes. Parafollicular cells give rise to thyroid gland neoplasms, which also contain large amounts of a nondescript protein called amyloid. Pathological secretion of calcitonin also occurs with other tumors that are of neural crest origin and that secrete amine or peptide hormones.

Calcitonin is a straight-chain peptide composed of 32 amino acids. It has a molecular weight of 3,400. The

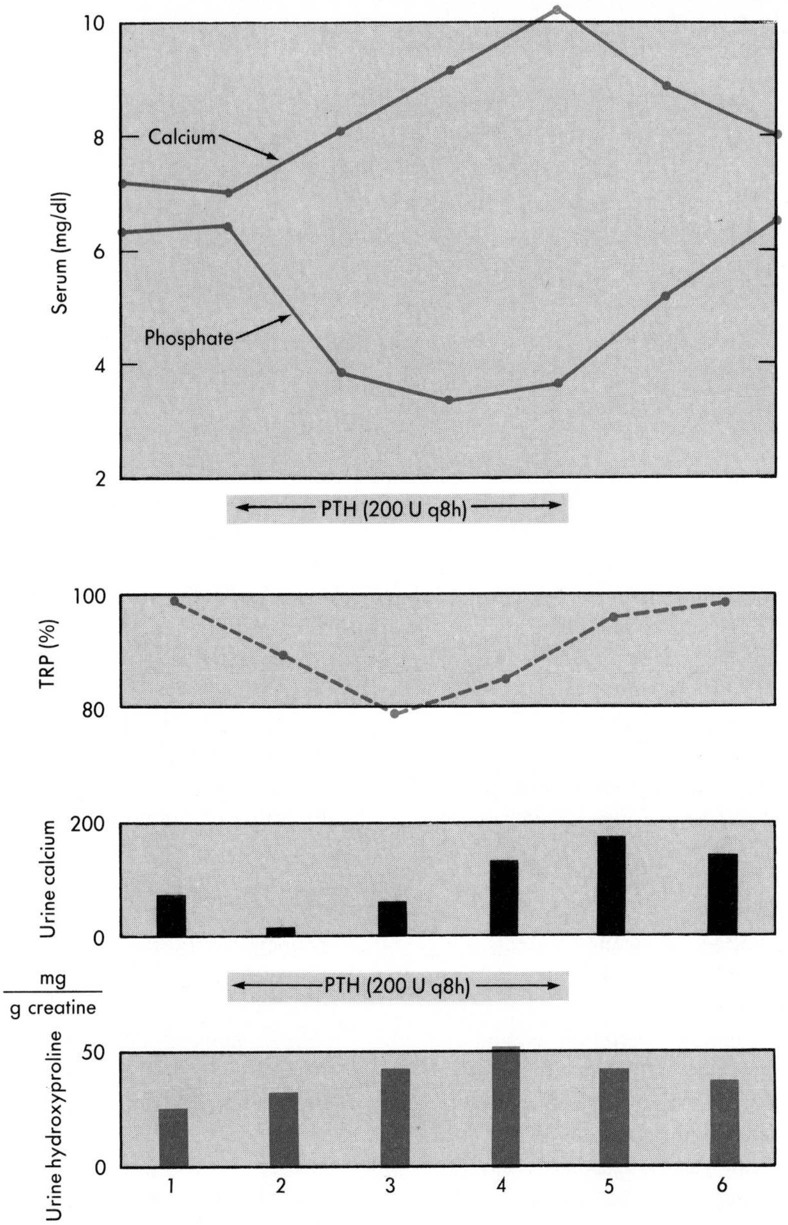

Fig. 51-9. Effect of PTH administration to a PTH-deficient human. Serum calcium level increases and serum phosphate level decreases. Urine calcium initially declines because of PTH action on the renal tubule; urine calcium excretion subsequently increases as serum calcium level rises and the filtered load of calcium rises in parallel. Urine hydroxyproline increases because of enhanced bone resorption. Tubular resorption of phosphate (TRP) falls so that urine phosphate (not shown) excretion increases. Numbers at bottom refer to days.

hormone contains a seven-membered disulfide ring at the N-terminus and prolineamide at the C-terminus. There is considerable species variation in structure, but both fish and animal calcitonins are active in humans. The biologically active core of the molecule probably resides in its central region, although some evidence suggests that the entire peptide sequence is required. As in the case of PTH, the synthesis of calcitonin proceeds from a preprohormone of 17,000 molecular weight through a prohormone to the hormone, which is packaged in granules along with N-terminal and C-terminal copeptides. The latter, which is named *katacalcin,* also has a calcium-lowering action and is present in the plasma of normal humans. Its role in physiology is now being explored.

The gene for calcitonin illustrates the significant relationship between the endocrine and nervous systems. In some cells the primary RNA transcript of the gene is processed to a messenger RNA that encodes for preprocalcitonin and ultimately directs synthesis of calcitonin. However, in other cells (both in thyroid and nervous tissue) the same primary RNA transcript is processed to a different messenger RNA that encodes the precursor and directs the synthesis of an entirely different peptide. This molecule, known as *calcitonin gene related peptide* (CGRP), also circulates in human plasma, and it probably arises from perivascular nerves. CGRP is a potent vasodilator and cardiac inotropic agent. The evolutionary path by which a hormone and a neuropeptide arose from *alternate* expression of the same gene and the functional significance of this relationship remain to be determined.

The major stimulus to secretion of calcitonin is a rise in plasma calcium. However, the degree of response seen in various species is related to their need to prevent hypercalcemia. Vertebrates that originated in fresh water (of low calcium concentration) but migrated into the sea (with a calcium concentration of 40 mg/dl) were the first to require and to develop a calcium-lowering hormone. When vertebrates moved to land, the emphasis in calcium economy shifted toward defense against hypocalcemia rather than against hypercalcemia. PTH was developed, and the importance of calcitonin in plasma calcium regulation probably declined. Calcitonin circulates in humans at concentrations of 10 to 100 pg/ml (10^{-11}M) and increases two- to tenfold after an acute increase in serum calcium level of as little as 1 mg/dl. Much larger responses of the hormone to calcium infusion are elicited in patients with calcitonin-secreting tumors. Likewise, in such patients a sharp reduction in ionized calcium lowers plasma calcitonin levels. The stimulating effect of calcium on calcitonin secretion has been verified in vitro. It appears to involve an increase in intracellular cAMP levels. Ingestion of food stimulates calcitonin secretion without elevating plasma calcium. This is mediated by several gastrointestinal hormones. of which gastrin is the most potent. The physiological significance of this phenomenon is not yet clear, but gastrin responsiveness provides a useful diagnostic test for states of calcitonin hypersecretion.

Circulating calcitonin is heterogeneous; immunoreactive molecules both larger and smaller than the native hormone have been found in the plasma of patients with tumors. Whether precursors are secreted normally is not known. Calcitonin appears to be largely degraded and cleared by the kidney. Its plasma half-life is less than 1 hour.

Actions

Binding of calcitonin to plasma membrane receptors is followed by an elevation of the intracellular cAMP level. This second messenger initiates at least a portion of calcitonin action in all target cells. The subsequent intracellular events are obscure. Sequestration of calcium in the mitochondria, thus lowering cytosol calcium concentration, has been advanced as a mechanism for reducing the efflux of calcium from bone cells into the extracellular fluid. The major effect of calcitonin administration is a rapid fall in plasma calcium levels. The magnitude of this decrease is directly proportional to the baseline rate of bone turnover. Thus young growing animals are most affected, whereas in adults, who have more stable skeletons, only a minimal response is seen. Inhibition of osteolysis by osteocytes and inhibition of bone resorption by osteoclasts, particularly when these are stimulated by PTH, are the immediate mechanisms of the hypocalcemic action. An escape phenomenon often is noted after 12 hours, possibly caused by down regulation of calcitonin receptors. However, continued provision of calcitonin eventually decreases the number of osteoclasts and alters their morphology, as well as their activity. More dense bone with fewer resorption cavities eventually results. Calcitonin has minor effects, the reverse of those of PTH, on calcium handling by the kidney and gut.

Calcitonin is clearly a physiological antagonist to PTH with respect to calcium. However, with respect to phosphate, it has the same net effect as PTH; i.e. it causes a decrease in plasma phosphate level. The fate of the phosphate leaving the extracellular fluid under the action of calcitonin is still not clear, but most likely it enters bone. There is also a small, somewhat inconsistent increase in urinary phosphate excretion after calcitonin administration. The hypophosphatemic effect is independent of the hypocalcemic effect. Nonetheless the enhanced entrance of phosphate into bone may mediate the hypocalcemic action of calcitonin, at least in part, by complexing calcium with phosphate within the bone cells.

The importance of calcitonin to normal human calcium economy is unclear. Under ordinary circumstances the absorption of dietary calcium loads produces little, if any, elevation of the plasma calcium level. Whether the increase in calcitonin provoked by eating helps prevent the development of postprandial hypercalcemia is not clear. Calcitonin deficiency resulting from complete removal of the thyroid gland does not lead to hypercalcemia, although disposition of an acute calcium load may be retarded. A chronic excess of calcitonin, generated either by tumor secretion or by exogenous administration, does not produce hypocalcemia. At present it may be most reason able to conclude that any effects of calcitonin deficiency or excess are easily offset by appropriate adjustment of PTH and vitamin D levels.

On the other hand, the possibility that calcitonin participates significantly in the regulation of bone remodeling cannot be easily excluded. The fact that plasma calcitonin is lower in women than in men and also declines with aging may imply a functional role for the hormone in the common development of accelerated bone loss after the menopause. In lower vertebrates and mammals a role for calcitonin appears likely in diverse calcium-related processes, such as lactation, production of egg shells, and protection of the skeleton from the calcium drain of pregnancy. Calcitonin has found clinical uses in the acute treatment of hypercalcemia and in certain bone diseases where a sustained reduction in osteoclastic resorption is therapeutically beneficial. Finally, the discovery of calcitonin in a number of locations throughout the body—in the hypothalamus and within cells of neural crest origin that contain amine neurotransmitters—has raised the possibility that calcitonin also may have a neurotransmitter function. In this regard, calcitonin has analgesic properties independent of the opioid system.

Vitamin D and Its Metabolism

There are two sources of vitamin D in humans: that produced in the skin by ultraviolet irradiation (D_3) and that ingested in the diet (D_2). In this sense vitamin D is not a classic hormone of endocrine gland origin. But its intimate and hormonelike relationship with calcium and phosphate metabolism, its pathway of molecular modification to yield active metabolites, and its mechanism of action similar to that of bona fide steroid hormones all justify classifying it as a hormone. Certainly no discussion of the endocrine regulation of calcium and phosphate metabolism would be possible without including the role of vitamin D.

The structures of vitamin D_3, its precursor, and its metabolites are shown in Fig. 51-10. Vitamin D_2, which differs only in having an additional double bond at the 21 to 22 position, is derived from the plant sterol, ergosterol, by ultraviolet radiation and is a major dietary source in many countries. It has identical biological actions to vitamin D_3, and henceforth the term *vitamin D* is used to indicate both forms. Under the influence of sunlight 7-dehydrocholesterol is photoconverted to a previtamin D_3 of unknown structure, which spontaneously transforms to vitamin D_3. The process occurs in the epidermis, but a specific cell of origin has not been identified. In summer, endogenous vitamin D is plentiful, and exogenous sources are unimportant. In winter or in sunlight-poor climates dietary vitamin D is essential for health. The most important sources are fish, liver, and irradiated milk. The minimum daily dietary requirement is approximately 2.5 μg (100 units), and the recommended daily intake is 10 μg. Because of its fat solubility, absorption of vitamin D from the gut is mediated by bile salts and occurs via the lymphatic glands.

Once vitamin D has entered the circulation, it is concentrated in the liver. There it is hydroxylated by a microsomal and mitochondrial enzyme to 25-OH-D_3 in a reaction dependent on NADPH and O_2. The rate of 25-hydroxylation appears to be modulated by vitamin D availability. When vitamin D is deficient, 25-hydroxylation is enhanced, thereby increasing the efficiency of conversion to active metabolites. Although 25-OH-D_3 is two to five times more potent than vitamin D in vivo and can be shown to have intrinsic activity in vitro, its physiological function in calcium regulation appears to be largely that of a precursor to still more active metabolites. From the liver 25-OH-D_3 is transported to the kidney (and possibly to other sites), where it undergoes alternative fates (Fig. 51-10). Hydroxylation in the 1 position to 1,25-$(OH)_2$-D_3 occurs in the mitochondria, requiring the participation of NADPH, flavoprotein, cytochrome P-450, and O_2. 1,25-$(OH)_2$-D_3 is at least 10 times as potent as vitamin D in vivo. It is unquestionably the metabolite that expresses much, if not all, the stimulatory activity of vitamin D on absorption of calcium from the intestine and on resorption of bone. Alternatively, 25-OH-D_3 may be hydroxylated in the 24 position to 24,25-$(OH)_2$-D_3 in a mitochondrial reaction that again requires NADPH and O_2. The biological activity of 24,25-$(OH)_2$-D_3 is an important subject of controversy. It is only a twentieth as potent as 1,25-$(OH)_2$-D_3 in most assay systems and may only represent a means of inactivating vitamin D. However, some evidence suggests that 24,25-$(OH)_2$-D_3 could have a separate specific role in expressing vitamin D actions.

Endocrine control of vitamin D actions occurs through regulation of renal 1-hydroxylase and 24-hydroxylase activities. Numerous animal studies have contributed to the development of the concepts illustrated in Fig. 51-11. 25-OH-D_3 is preferentially directed toward the highly active metabolite 1,25-$(OH)_2$-D_3 whenever there is a lack of vitamin D, a lack of calcium, or a lack of phosphate. In vitamin D–deficient states it may be partly the lack of 1,25-$(OH)_2$-D_3 itself that enhances 1-hydroxylation, since 1,25-$(OH)_2$-D_3 is a suppressor of 1-hydroxylase activity. When vitamin D is sufficient, the 1-hydroxylase enzyme also is subject to regulation by calcium and phosphate. Calcium deprivation leads to hypocalcemia, which in turn stimulates PTH hypersecretion. The lowered plasma calcium and the elevated PTH concentration each independently stimulates 1-hydroxylase activity. In addition, the PTH excess leads to a phosphate diuresis; the resultant hypophosphatemia and lowered renal cortical phosphate content also stimulate 1-hydroxylase activity. Finally, in experiments in which calcium and PTH are held constant, the imposition of phosphate deprivation leads to hypophosphatemia and directly to an increase in 1-hydroxylase activity. The mechanisms underlying regulation of 1-hydroxylation by these various factors are still unknown. In the case of PTH, stimulation of the enzyme NADPH diaphorase may be important in generating a needed cofactor.

The regulation of 24-hydroxylase activity (present in numerous tissues) is essentially the mirror image (Fig.

Fig. 51-10. Structures of vitamin D_3, its precursor, and its metabolites.

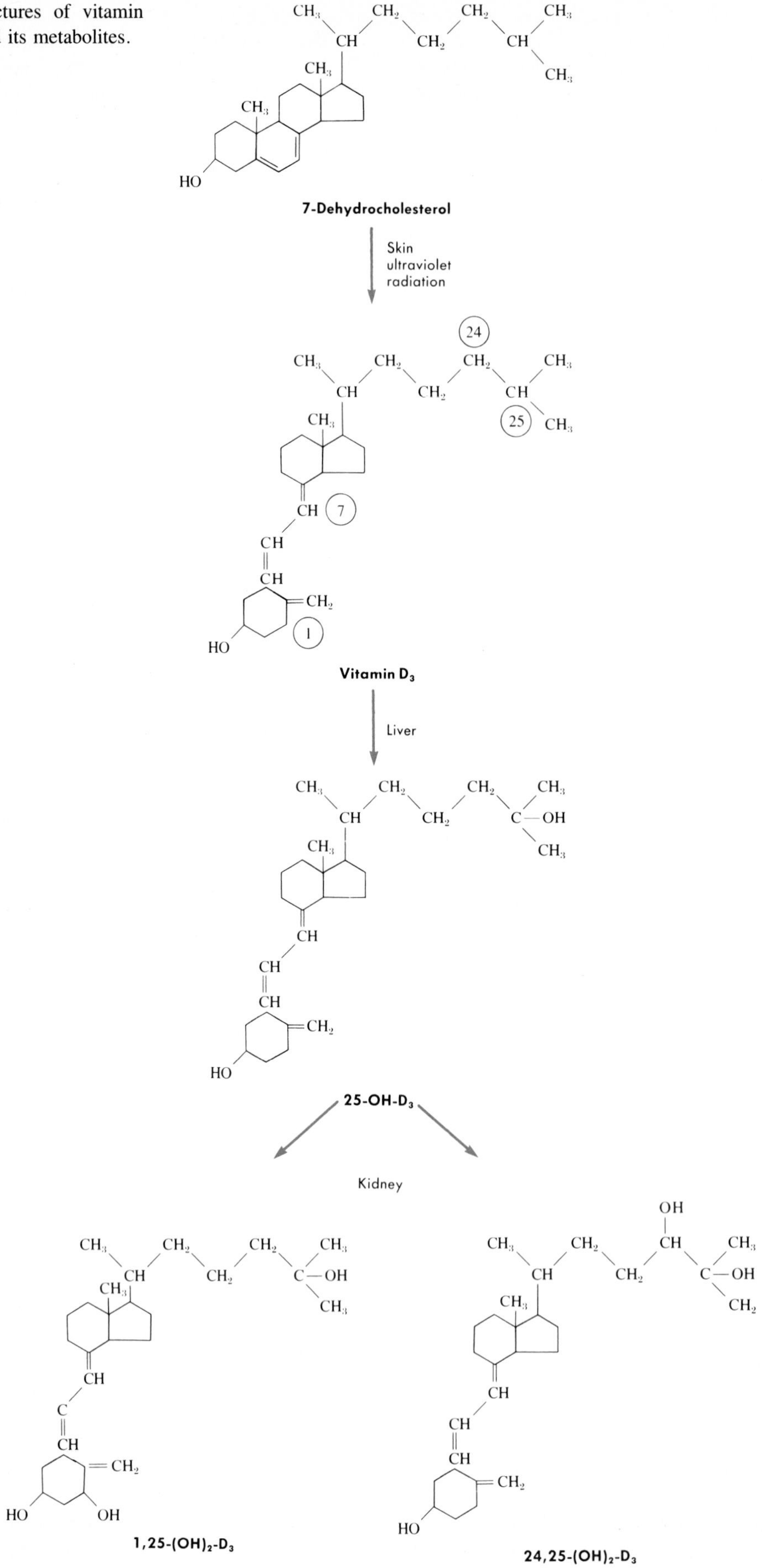

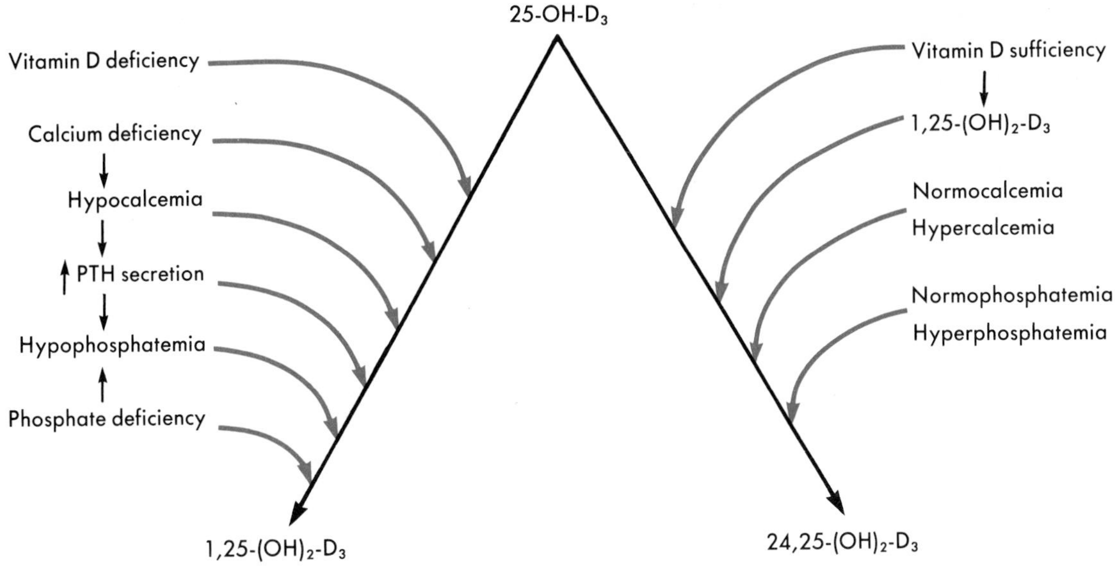

Fig. 51-11. Factors that regulate conversion of 25-OH-D_3 to either 1,25-$(OH)_2$-D_3 or 24,25-$(OH)_2$-D_3. The former is increased by either calcium or phosphate lack.

TABLE 51-2. Vitamin D metabolism in humans

	Plasma concentration (μg/L)	*Plasma half-life (days)*	*Estimated production rate (μg/day)*
1.25-$(OH)_2$-D_3	0.03	1 to 3	1
24,25-$(OH)_2$-D_3	2	15 to 40	1
25-OH-D_3	20	5 to 20	10

51-11). A sufficiency of vitamin D, via 1,25-$(OH)_2$-D_3, and normal or supernormal calcium and phosphate concentrations all stimulate the synthesis of 24,25-$(OH)_2$-D_3. Thus the supply of active 1,25-$(OH)_2$-D_3 is augmented whenever the mobilization of calcium or phosphate from intestine and bone into the extracellular fluid is needed. Conversely, augmentation of the supply of relatively inactive 24,25-$(OH)_2$-D_3 is favored when calcium and phosphate are plentiful and bone accretion can be sustained. The intracellular mechanism whereby 24-hydroxylation is regulated has not been defined. However, many hours are required for changes to be effected, implying that enzyme synthesis probably is involved.

Vitamin D, 25-OH-D_3, 1,25-$(OH)_2$-D_3, and 24,25-$(OH)_2$-D_3 all circulate bound to an α-globulin of about 50,000 molecular weight. The concentrations, approximate half-lives, and estimated daily production rates for the key metabolites in humans are shown in Table 51-2. It is evident that 1,25-$(OH)_2$-D_3 has by far the lowest concentration and the shortest half-life of the three. The production rate of 1,25-$(OH)_2$-D_3 is approximately equal to the amount that must be administered to anephric individuals in order to increase calcium absorption. Certain relationships among the plasma concentrations are of physiological significance. 1,25-$(OH)_2$-D_3 concentration is ordinarily independent of 25-OH-D_3 concentration. Except in severe vitamin D–deficiency states, the regulatory factors previously outlined appear to maintain the appropriate concentration of this active metabolite irrespective of the supply of precursor. In contrast, 24,25-$(OH)_2$-D_3 concentration is ordinarily directly proportional to 25-OH-D_3 concentration. This suggests that 24-hydroxylation is a key sluice for disposing of the excess precursor. Both 1,25-$(OH)_2$-D_3 and 24,25-$(OH)_2$-D_3 can be cross-hydroxylated to 1,24,25-$(OH)_3$-D_3, a compound of little activity. Thus if an excess of 1,25-$(OH)_2$-D_3 should develop (for example, in response to hypocalcemia) it can still be inactivated by 24-hydroxylation. Although a separate 24-hydroxylase is involved, its activity is regulated in similar fashion to the 24-hydroxylase that acts on 25-OH-D_3 (see above). Further inactivation of vitamin D metabolites occurs in the liver by 23- and 26-hydroxylation and lactone formation. Both active and inactive vitamin D metabolites undergo biliary excretion and enterohepatic recycling. A loss of recycling because of intestinal disease may contribute to the development of vitamin D deficiency.

$1,25\text{-}(OH)_2\text{-}D_3$ acts through the general mechanism outlined for steroid hormones (see Fig. 49-10). After combination with a cytosolic receptor the hormone receptor complex enters the nucleus where it stimulates transcription of mRNA for at least one identified product. This is a calcium-binding protein that can be isolated from the intestinal mucosa and from bone and kidney cells. This 10,000 molecular weight protein has significant homology with calmodulin and with the light chain of myosin. It has a high affinity for calcium in the cytosol of target cells. It is probably not a direct mediator of vitamin D effects on calcium transport. Rather, it may act as a buffering protein that protects the cells from the large transcellular Ca^{++} fluxes brought about by vitamin D.

Absorption of calcium from the intestinal lumen against a concentration gradient is the major action of $1,25\text{-}(OH)_2\text{-}D_3$. This metabolite localizes to the nuclei of intestinal villus and crypt cells but not to goblet or submucosa cells. It acts on the brush order, possibly by changing the lipid concentration of the membrane. Calcium entry from the intestinal lumen is stimulated, and the ion is subsequently extruded from the opposite side of the cell and enters the bathing capillary blood. This action of $1,25\text{-}(OH)_2\text{-}D_3$ is responsible for the adaptation, described previously, whereby intestinal calcium absorption adjusts to alteration in dietary calcium. By an independent, unknown mechanism, active absorption of phosphate across the intestinal cell membrane is also stimulated by vitamin D.

$1,25\text{-}(OH)_2\text{-}D_3$, but not $25\text{-}OH\text{-}D_3$, stimulates bone resorption in vitro and in vivo. A cytosolic receptor for $1,25\text{-}(OH)_2\text{-}D_3$ has been found in bone cells, and the resorptive effect can be blocked by actinomycin, a DNA transcription inhibitor. In this action $1,25\text{-}(OH)_2\text{-}D_3$ synergizes critically with PTH by an unknown mechanism. Exceedingly low concentrations of $1,25\text{-}(OH)_2\text{-}D_3$ also can stimulate osteolysis and mobilization of bone calcium, independent of PTH.

The normal mineralization of newly formed osteoid along a calcification front is also critically dependent on vitamin D. In the absence of the vitamin, excess osteoid accumulates from osteoblastic activity, and the bone so formed is weakened. It remains enigmatic whether this action of vitamin D is entirely accounted for by augmentation of the supply of calcium and phosphate in the fluid bathing the osteoblast, or whether $1,25\text{-}(OH)_2\text{-}D_3$ or another metabolite, acts in a direct and specific manner on the bone cells themselves to hasten mineralization. The presence of high affinity receptors for $1,25\text{-}(OH)_2\text{-}D_3$ in osteoblasts supports a direct action, as does the observation that, in vitro, $1,25\text{-}(OH)_2\text{-}D_3$ inhibits collagen synthesis by these cells. In any case the first observable effect of vitamin D replacement on the bone from vitamin D–deficient animals and humans is the reappearance of a normal mineralized front.

In vitro, $24,25\text{-}(OH)_2\text{-}D_3$ stimulates cartilage development and bone mineralization. On the other hand, if the 24 position of $25\text{-}(OH)\text{-}D_3$ is blocked by fluorination, this compound still completely restores calcium and phosphate balance when administered in vivo to vitamin D–deficient animals. This casts doubt on the importance of 24-hydroxylation (as opposed to the importance of 1-hydroxylation) in the production of active vitamin D metabolites.

Despite much contradictory evidence, it is likely that $1,25\text{-}(OH)_2\text{-}D_3$ has direct renal effects similar to those of PTH. However, these probably are far outweighed by the action of PTH and the effects of ambient calcium and phosphate concentrations themselves on renal tubular function. $1,25\text{-}(OH)_2\text{-}D_3$ also accumulates in the parathyroid glands, where a vitamin D–dependent calcium-binding protein can be found. Feedback inhibition of PTH synthesis and secretion by $1,25\text{-}(OH)_2\text{-}D_3$ occurs.

Another storage site for vitamin D is in muscle tissue. Profound muscle weakness is a prominent result of vitamin D deficiency. Although no direct effect of vitamin D on muscle cells has yet been discovered, there is some evidence that $25\text{-}OH\text{-}D_3$ may play a role in the maintenance of muscle metabolism and function.

Recent studies have broadened the spectrum of possible physiological roles for vitamin D. Binding to pancreatic islet cells and to pituitary cells is mirrored by stimulatory effects of $1,25\text{-}(OH)_2\text{-}D_3$ on the secretion of insulin and on the secretion of prolactin, a lactogenic pituitary hormone. Of special interest is the relationship of vitamin D to immunoregulation. $1,25\text{-}(OH)_2\text{-}D_3$ induces differentiation of monocytic leukemia cells. Receptors for this vitamin D metabolite have been found in macrophages and in transformed lymphocytes. As previously noted, macrophage precursors are the source of osteoclasts. Furthermore, in certain pathological states, macrophages and monocytes possess 1-hydroxylase activity, producing $1,25\text{-}(OH)_2\text{-}D_3$ from $25\text{-}(OH)\text{-}D_3$. Thus $1,25\text{-}(OH)_2\text{-}D_3$ could mediate an autocatalytic system of recruitment of tissue macrophages either when they are needed for tissue response to injury or for enhanced bone resorption via osteoclasts.

Integrated Hormonal Regulation of Calcium and Phosphate

From all the foregoing it should be clear that a complex interplay of several hormones acting on a number of tissues is responsible for maintenance of normal concentrations of calcium and phosphate in body fluids. This integrated system is best visualized by tracing the compensatory responses to deprivation of calcium and of phosphate (Figs. 51-12 and 51-13).

Calcium deprivation, with hypocalcemia as the signal, primarily stimulates PTH secretion. PTH increases urinary phosphate excretion, thereby decreasing serum

Fig. 51-12. The compensatory response to calcium deprivation. See text for explication.

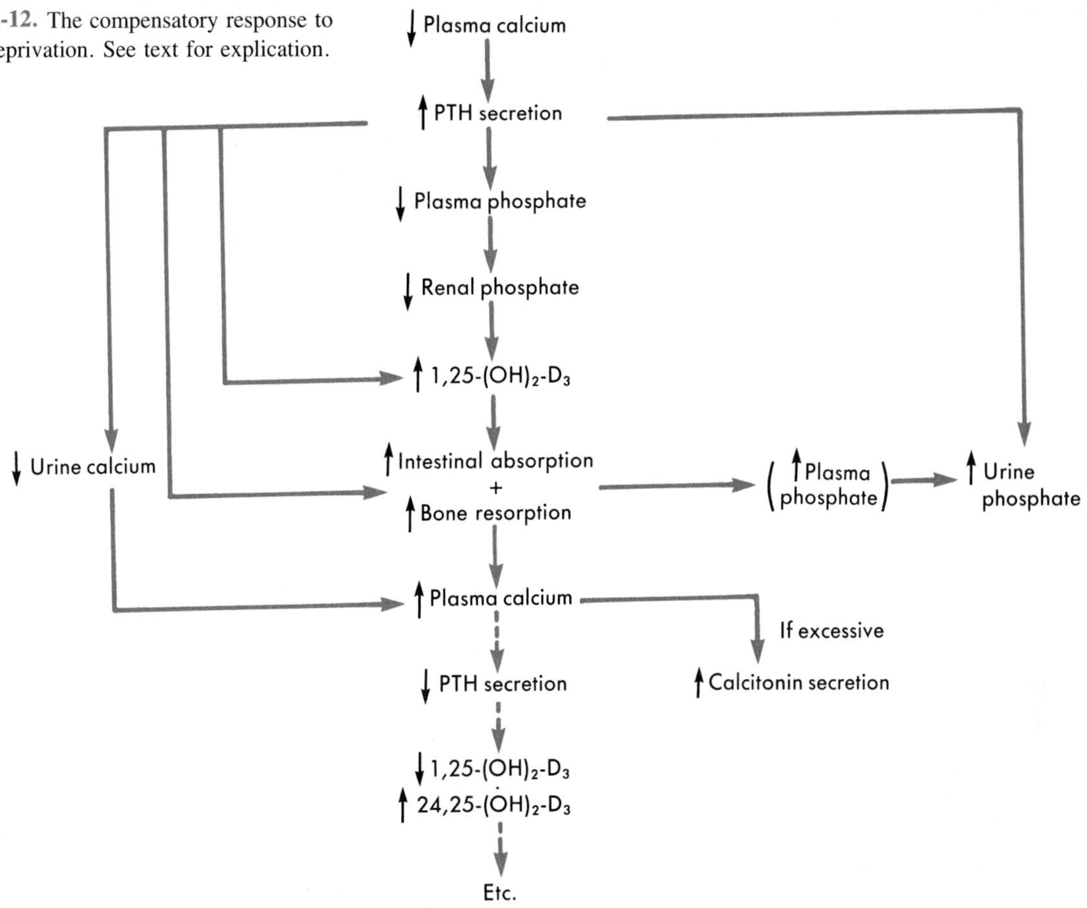

Fig. 51-13. The compensatory response to phosphate deprivation. See text for explication.

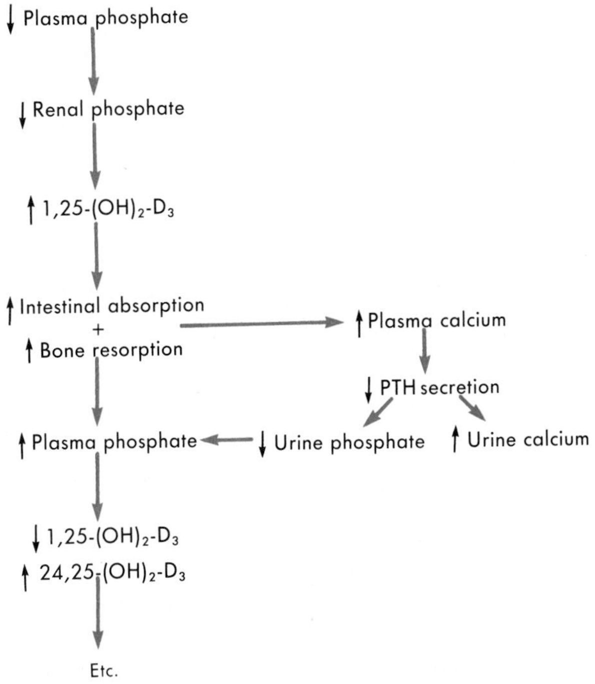

phosphate level and renal cortical phosphate content. All three factors—hypocalcemia, excess PTH, and hypophosphatemia—act to stimulate the production of 1,25-$(OH)_2$-D_3. The latter raises plasma calcium levels toward normal by increasing absorption of calcium from the gastrointestinal tract and, in concert with PTH, by increasing osteocytic and osteoclastic bone resorption. PTH further acts to return calcium to the plasma by increasing its reabsorption from the renal tubular urine. Thus this beautifully integrated response to calcium deprivation increases the flux of calcium into the extracellular fluid. Simultaneously the extra phosphate that enters with the calcium from the bone and gut is disposed of by excretion in the urine. The recovery of plasma calcium concentration to normal will shut off PTH hypersecretion by negative feedback. 1,25-$(OH)_2$-D_3 synthesis will then decline. 24,25-$(OH)_2$-D_3 synthesis will increase, and the whole sequence will diminish. As a further safety valve, should the compensatory rise in plasma calcium concentration exceed normal levels, stimulation of calcitonin secretion would act to moderate it as well.

In a contrasting sequence phosphate deprivation via hypophosphatemia will directly stimulate 1,25-$(OH)_2$-

D_3 production. The latter increases the flux of phosphate into the extracellular fluid by stimulating its absorption from the gut and by stimulating bone resorption. The extra calcium that simultaneously enters the extracellular fluid raises plasma calcium. Hypophosphatemia, by retarding bone formation, also directly helps raise plasma calcium. The increase in plasma calcium suppresses PTH secretion. The absence of PTH causes tubular reabsorption of phosphate to increase, conserving urinary phosphate and aiding in the restoration of plasma phosphate levels to normal. At the same time the lack of PTH permits easier disposal of the extra calcium that was mobilized by diminishing its renal tubular reabsorption and increasing its excretion in the urine. As plasma phosphate returns to normal, 24,25-$(OH)_2$-D_3 production will be favored, 1,25-$(OH)_2$-D_3 levels will decline, and the whole process will be reversed.

Thus this combined arrangement of dual hormone regulation and dual hormone action permits selective defense of either plasma calcium or plasma phosphate level, without creating a circulatory excess of the other. Obviously the same principles apply in reverse to imposition of excess calcium or excess phosphate loads of either endogenous or exogenous sources.

Certain characteristics of the homeostatic systems for adjustment of body calcium and phosphate stores deserve emphasis. The renal responses of PTH provide the most rapid (within minutes) defense against perturbations of both calcium and phosphate stores. As PTH secretion ranges from very high to very low levels, the rate of urinary calcium excretion can rise 25-fold from approximately 0.05 to 1.2 mg/minute, and that of phosphate can fall from 2 to 0 mg/minute. A sudden 2 to 3 mg/dl increase in either calcium or phosphate concentration in the extracellular fluid can be corrected within 24 hours by the kidney, acting under the appropriate alteration in PTH levels. In the face of complete phosphate and calcium deprivation, renal conservation of phosphate is complete, whereas that of calcium is not quite so. The gastrointestinal component of this homeostatic system is both slower and narrower in range. As a result of variations in 1,25-$(OH)_2$-D_3, the absorption of dietary calcium increases from 20% to 70% as calcium intake decreases from 2,000 to 200 mg/day. Thus absorbed calcium can effectively range from 140 to 400 mg/day. Hormonal effects on phosphate absorption are even less striking, since the latter is virtually a linear function of dietary intake. Bone responses to regulatory fluctuations in both PTH and 1,25-$(OH)_2$-D_3 are rapid when produced by osteocytic osteolysis and relatively slow when caused by osteoclastic resorption. However, the capacity for compensatory calcium and phosphate uptake and release is enormous. For example, in humans tenfold variations in calcium turnover have been observed. Finally, an important major difference between the renal and gastrointestinal mechanisms on the one hand and the bone mechanisms on the other hand must be borne in mind. The compensatory responses of the kidney and the gut have the virtue of defending total body *and bone* stores of calcium and phosphate against erosion or inundation. In contrast, the skeletal mechanisms of defense against perturbations of plasma calcium and phosphate have the disadvantage that their long-term employment eventually sacrifices the chemical and structural integrity of the bone mass.

Clinical Syndromes of Hormone Dysfunction

PTH

Hyperparathyroidism most commonly results from a single parathyroid adenoma and less often from hyperplasia of all four glands. The clinical picture is largely dominated by the effects of hypercalcemia, which depresses neuromuscular excitability. This produces symptoms of dulled mentation, lethargy, anorexia, constipation, and muscle weakness. If urinary excretion of calcium is excessive, renal stones may form. Peptic ulcers can result from stimulation of gastric acid secretion by calcium. In long-standing cases with marked hypersecretion of PTH, massive irregular bone resorption, demonstrable on x-ray films, causes weakening of the bones, pain, fractures, and deformities. Persistent hypercalcemia also can lead to reduced renal function and to irreversible renal failure, secondary to deposition of calcium in the kidney cells. Hyperparathyroidism must be distinguished from numerous other causes of hypercalcemia. The diagnosis is established by demonstrating an elevated plasma total calcium or ionized calcium level, a decreased plasma phosphate level, and an elevated PTH level. A mild hyperchloremic acidosis is also frequently present. Treatment is surgical removal of the hyperfunctioning tissue.

Hypoparathyroidism most often is caused by autoimmune atrophy or inadvertent surgical removal of the glands. In rare cases the problem is end organ resistance to PTH action, in at least one instance attributable to a deficiency in the regulatory component of adenylate cyclase. Calcium absorption from the gut is diminished, because of defective synthesis of 1,25-$(OH)_2$-D_3, and there is also decreased bone reabsorption. The resultant hypocalcemia increases neuromuscular excitability. This produces hyperactive reflexes, spontaneous muscle contractions, convulsions, and laryngeal spasm with airway obstruction. Formation of ocular cataracts is common but not well explained. Low plasma total calcium or ionized calcium levels and elevated plasma phosphate levels are essential to the diagnosis. Plasma PTH levels are low when hypoparathyroidism is caused by destruction of the glands. Plasma PTH level is elevated when target tissue unresponsiveness is the problem. In the former case administration of exogenous PTH for diagnostic purposes causes prompt increases in urinary excretion of cAMP and phosphate; in the latter case it does not. Although treatment with PTH would be logical, this is impractical because of the development of serum antibodies to the bovine hormone preparations currently available. Instead treatment consists of replacement with 1,25-$(OH)_2$-D_3 and calcium supplements.

Vitamin D

Toxicity occurs from excessive administration of vitamin D or, in a few diseases, from abnormal sensitivity to ordinary amounts of the vitamin. Absorption of calcium from the gut and resorption of bone are enhanced. This elevates plasma and urinary calcium levels, with the same clinical consequences noted previously for hyperparathyroidism. In contradistinction to the latter, however, vitamin D excess raises, rather than lowers, serum phosphate because the hypercalcemia shuts off PTH secretion. Tubular reabsorption of phosphate therefore is increased, and phosphate is retained. The duration of toxicity is greatest from exogenous vitamin D itself because of the body's large storage capacity for it and its low turnover rate; toxicity is least for 1,25-$(OH)_2$-D_3, which is rapidly removed from the body. Treatment consists of blocking the effects of vitamin D by administration of calcitonin or cortisol while waiting for the excess vitamin D to be cleared.

Deficiency of vitamin D action can result from inadequate sunlight, lack of dietary intake, diminished absorption, or defective hydroxylation in the liver or kidney. This leads to decreased gastrointestinal absorption of calcium and phosphate. The resultant fall in plasma calcium level is buffered by stimulation of PTH secretion. This in turn strongly accentuates the fall in plasma phosphate level by decreasing its tubular resorption in the kidney and increasing its excretion in the urine. Urinary calcium excretion, on the other hand, is reduced by the PTH excess. The strongly negative phosphate balance and more modestly negative calcium balance together decrease the rate of bone mineralization. This is further aggravated by the lack of the putative direct action of a vitamin D metabolite on calcification of osteoid. An excess of osteoid accumulates on bone surfaces, and no regular line of calcification can be seen.

Two clinical pictures result from vitamin D deficiency. In children the centers of endochondral ossification at the epiphyseal plates are most critically affected, producing growth failure and the characteristic deformities and x-ray pictures of *rickets*. In adults softening and bending of long bones, with fracture lines along nutrient arteries, creates the condition known as *osteomalacia*. If lack of either sunlight or dietary vitamin D is the cause, plasma 25-(OH)-D_3 levels (and in severe cases, 1,25-$(OH)_2$-D_3 levels) will be low. If resistance to vitamin D action is the cause, 25-(OH)-D_3 and 1,25-$(OH)_2$-D_3 levels will be elevated. If kidney failure is the cause, plasma 25-(OH)-D_3 will be high, whereas plasma 1,25-$(OH)_2$-D_3 levels will be low. In general, plasma calcium will be reduced, as will plasma phosphate levels (except in cases of kidney failure), and plasma PTH levels will be elevated. The level of alkaline phosphatase is also high in the plasma because of overactivity of the osteoblasts. In adults bone biopsy may be needed to clarify subtle cases. Treatment of simple deficiency consists of replacement doses of vitamin D itself, with supplemental calcium. In cases caused by biosynthetic defects (for example, hepatic or renal disease) or target tissue unresponsiveness large doses of vitamin D, 25-OH-D_3, or 1,25-$(OH)_2$-D_3 may be needed.

Bibliography

Journal articles

Austin, L.A., et al.: Calcitonin: physiology and pathophysiology, N. Engl. J. Med. **304:**269, 1981.

Avioli, L., and Haddad, J.: The vitamin D family revisited, N. Engl. J. Med. **311:**47, 1984.

Bordier, P., et al.: Vitamin D metabolites and bone mineralization in man, J. Clin. Endocrinol. Metab. **46:**284, 1978.

Burger, E., et al.: In vitro formation of osteoclasts from long-term cultures of bone marrow mononuclear phagocytes, J. Exp. Med. **156:**1604, 1982.

Canalis, E.: The hormonal and local regulation of bone formation, Endocr. Rev. **4:**62, 1983.

Cheung, W.Y.: Calmodulin plays a pivotal role in cellular regulation, Science **207:**19, 1979.

DeLuca, H.F.: Vitamin D: revisited 1980, Clin. Endocrinol. Metab. **1:**3, 1980.

DeLuca, H., and Schnoes, H.: Vitamin D: recent advances, Annu. Rev. Biochem. **52:**411, 1983.

Hillyard, C., et al.: Katacalcin: a new plasma calcium-lowering hormone, Lancet **1:**846, 1983.

Holtrop, M., and Raisz, L.: Comparison of the effects of 1,25dihydroxycholecalciferol, prostaglandin E_2, and osteoclast-activating factor with parathyroid hormone on the ultrastructure of osteoclasts in cultured long bones of fetal rats, Calcif. Tissue Int. **29:**201, 1979.

Kumar, R.: Metabolism of 1,25-dihydroxyvitamin D_3, Physiol. Rev. **64:**478, 1984.

Mahoney, C., and Nissenson, R.: Canine renal receptors for parathyroid hormone: down-regulation in vivo by exogenous parathyroid hormone, J. Clin. Invest. **72:**411, 1983.

Mawer, E.B.: Clinical implications of measurements of circulating vitamin D metabolites, Clin. Endocrinol. Metab. **9:**63, 1980.

Norman, A., et al.: The vitamin D endocrine system: steroid metabolism, hormone receptors, and biological response (calcium binding proteins), Endocr. Rev. **3:**331, 1982.

Parfitt, A.M.: The actions of parathyroid hormone on bone: relation to bone remodeling and turnover, calcium homeostasis, and metabolic bone disease. I. Mechanisms of calcium transfer between blood and bone and their cellular basis: morphologic and kinetic approaches to bone turnover, Metabolism **25:**809, 1976.

Parfitt, A.M.: The actions of parathyroid hormone on bone: relation to bone remodeling and turnover, calcium homeostasis, and metabolic bone disease. II. PTH and bone cells: bone turnover and plasma calcium regulation, Metabolism **25:**909, 1976.

Parthemore, J.G., et al.: Calcitonin secretion in normal human subjects, J. Clin. Endocrinol. Metab. **47:**184, 1978.

Raisz, L.G.: Direct effects of vitamin D and its metabolites on skeletal tissue, Clin. Endocrinol. Metab. **9:**27, 1980.

Russell, R.G.G., et al.: Physiological and pharmacological aspects of 24,25-dihydroxycholecalciferol in man, Adv. Exp. Med. Biol. **103:**487, 1978.

Silver, J., et al.: Regulation by vitamin D metabolites of parathyroid hormone gene transcription in vivo in the rat, J. Clin. Invest. **78:**1296. 1986.

Books and monographs

Avioli, L.V., et al.: Bone metabolism and disease. In Bondy, P.K., and Rosenberg, L.E., editors: Metabolic control and disease, Philadelphia, 1980, W.B. Saunders Co.

Mayer, G.P.: Effect of calcium and magnesium on parathyroid hormone secretion rate in calves. In Talmadge, R.V., et al., editors: Calcium regulating hormones, Amsterdam, 1975, Excerpta Medica.

Neer, R.M.: Calcium and inorganic phosphate homeostasis. In Degroot, L.J., editor: Endocrinology, New York, 1979, Grune & Stratton Inc.

Rosenfeld, M., et al.: Tissue specific patterns of RNA processing regulation and developmental specificity of neuroendocrine gene expression. In Labrie, F., and Proulx, L., editors: Endocrinology, Proceedings of the 7th International Congress of Endocrinology, Quebec City, 1984, Amsterdam, 1984. Elsevier Science Publishers V.B., 923–929.

CHAPTER

52

The Hypothalamus and the Pituitary Gland

The hypothalamus-pituitary consortium forms the most complex and dominant portion of the entire endocrine system. Its internal anatomical and functional relationships are elaborate and subtle. The output of the hypothalamus-pituitary unit regulates the function of the thyroid, adrenal, and reproductive glands and also shares in the control of somatic growth, lactation, milk secretion, and water metabolism. Two hormones, *antidiuretic hormone (ADH* or *vasopressin)* and *oxytocin,* are synthesized in the hypothalamus but are stored and secreted by the *posterior pituitary gland,* or *neurohypophysis.* A group of *tropic hormones,* adrenocorticotropic hormone (ACTH), thyroid-stimulating hormone (TSH), luteinizing hormone (LH), follicle-stimulating hormone (FSH), growth hormone, and prolactin, are synthesized, stored, and secreted by the *anterior pituitary gland,* or *adenohypophysis.* However, a set of releasing and inhibiting hormones that are synthesized in the hypothalamus and travel to the adenohypophysis regulates the synthesis and secretion of these tropic hormones. When one considers that all of this emanates from a mass of only 500 mg of pituitary in association with 10 g of adjacent hypothalamus, the performance is truly awesome.

■ *Anatomy*

A knowledge of the embryological development of the pituitary gland is crucial to understanding its anatomy and function. The fully developed gland is really an amalgam of hormone-producing glandular cells (the adenohypophysis) and neural cells with secretory function (the neurohypophysis). The anterior endocrine portion of the pituitary develops from an upward outpouching of ectodermal cells from the roof of the oral cavity (Rathke's pouch). This pouch eventually pinches off and becomes separated from the oral cavity by the sphenoid bone of the skull. The lumen of the pouch is reduced to a small cleft. The posterior neural portion of the pituitary develops from a downward outpouching of ectoderm from the brain in the floor of the third ventricle. The lumen of this pouch is obliterated inferiorly as the sides fuse into the infundibular process. Superiorly, the lumen remains continguous with and forms a recess in the adult third ventricle. The upper portion of this neural stalk expands to invest the lowest portion of the hypothalamus and is called the *median eminence.* The cleftlike remnant of Rathke's pouch demarcates the interwoven anterior and posterior portions of the pituitary. In some animals, but not in humans, cells in the area of Rathke's pouch and adjacent to the neurohypophysis form a distinct intermediate lobe of variable size. The entire pituitary gland sits in a socket of sphenoid bone called the *sella turcica.* A reflection of the dura mater, called the *diaphragm,* extends across the top of the sella turcica and separates the bulk of the pituitary gland from the brain. However, the neural stalk penetrates the diaphragm, maintaining its continuity with the brain. These anatomical relationships are shown in Fig. 52-1, *B*. The recent ability to visualize the human pituitary gland by computerized axial tomography (CAT scanning) and nuclear magnetic resonance imaging has greatly facilitated detection of disease in this area (Fig. 52-1, *A*).

The blood supply to this complex of neural and endocrine tissue is equally important. In the posterior pituitary, the neural tissue of the infundibular process derives its blood mostly from the inferior hypophyseal artery, and the capillary plexus thereof drains into the dural sinus. The neural tissue of the upper stalk and of the median eminence is supplied largely by the superior hypophyseal artery. After investing the axons in these areas, the capillary plexus emanating from this artery forms a set of long portal veins that carry the blood downward into the anterior pituitary. There these portal

A

B

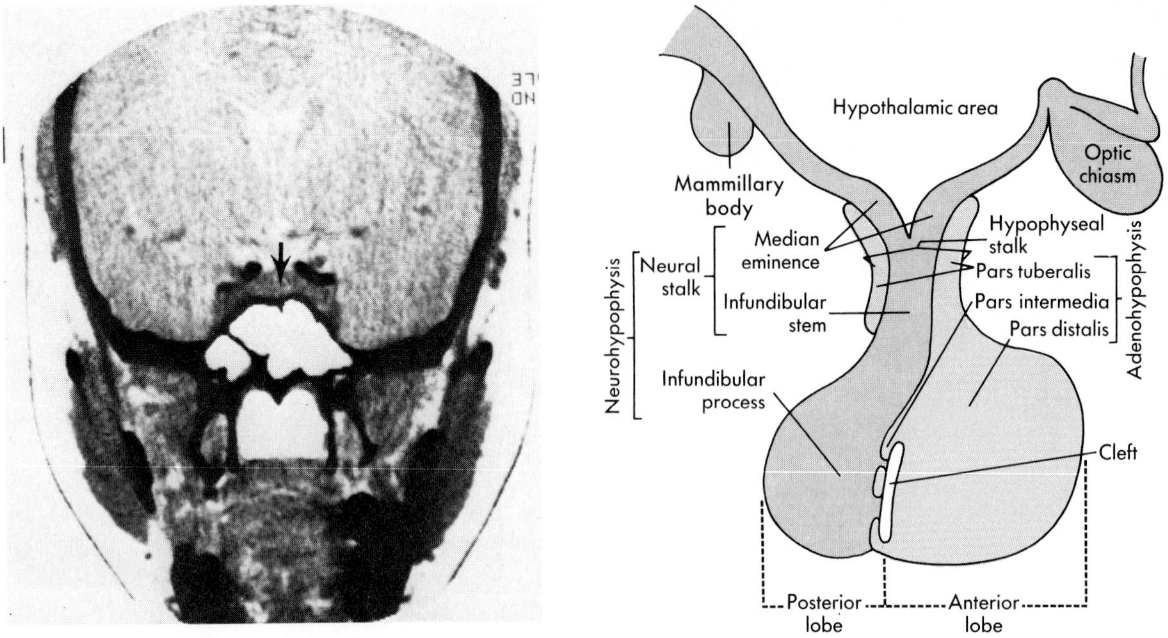

■ **Fig. 52-1.** **A,** CAT scan of the skull showing the location of the pituitary gland (arrow) in the sella turcica above the sphenoid sinus. **B,** Diagram of the pituitary gland showing its division into the adenohypophysis and neurohypophysis. (**B** adapted from an original painting by Frank H. Netter, M.D., from The CIBA Collection of Medical Illustrations copyright by CIBA Pharmaceutical Co., Division of CIBA-Geigy Corporation.)

veins give rise to a second capillary plexus that supplies the endocrine cells with the majority of their blood, which is then drained off into the dural sinus. The anterior pituitary receives its remaining blood via a set of short portal veins originating in the capillary plexus of the inferior hypophyseal artery within the neural stalk. Thus there is little or no direct arterial blood supply to the adenohypophyseal cells. Furthermore, it should be noted that the pituitary gland lies outside the blood-brain barrier.

The functional implications of this involved anatomical arrangement become apparent when the functional relationships are examined. These are illustrated in block diagram form in Fig. 52-2. The entire neurohypophysis represents a collection of axons whose cell bodies lie in the hypothalamus. Peptide hormones synthesized in the cell bodies of these hypothalamic neurons travel down their axons in neurosecretory granules to be stored in the nerve terminals lying in the posterior pituitary gland. After neural stimulation of the cell bodies, the granules are released from the terminals by exocytosis; the peptide hormones then enter the peripheral circulation via the capillary plexuses of the inferior hypophyseal artery. Thus a single cell performs the entire process of hormone synthesis, storage, and release in a classic example of neurocrine function.

In contrast, the adenohypophysis is a collection of glandular cells that are regulated by bloodborne stimuli originating in neural tissue. Cell bodies of hypothalamic neurons synthesize *releasing hormones* and *inhibiting hormones,* which travel in packets down their axons only as far as the median eminence. Here they are stored as neurosecretory granules in the nerve terminals (synaptosomes). After stimulation of these hypothalamic cells, the releasing or inhibiting hormones are discharged into the median eminence and enter the capillary plexus of the superior hypophyseal artery. Having now left neural tissue, they are transported down the long portal veins and exit from a secondary capillary plexus to reach their specific glandular target cells in the anterior pituitary. These cells then react to the releasing hormones or inhibiting hormones by increasing or decreasing their output of tropic hormones. The latter enter the same second capillary plexus through which they ultimately reach the peripheral circulation. Thus two cells, one neural and one glandular, participate in the processes leading to synthesis and release of the anterior pituitary tropic hormones, a combination of neurocrine and endocrine function. Key evidence in support of this functional arrangement includes the following observations:

1. Neural tracts containing hypothalamic peptides can be traced by immunohistochemical techniques down to the median eminence where they end in proximity to capillaries (Fig. 52-3).
2. Direct measurement of hypothalamic peptides in

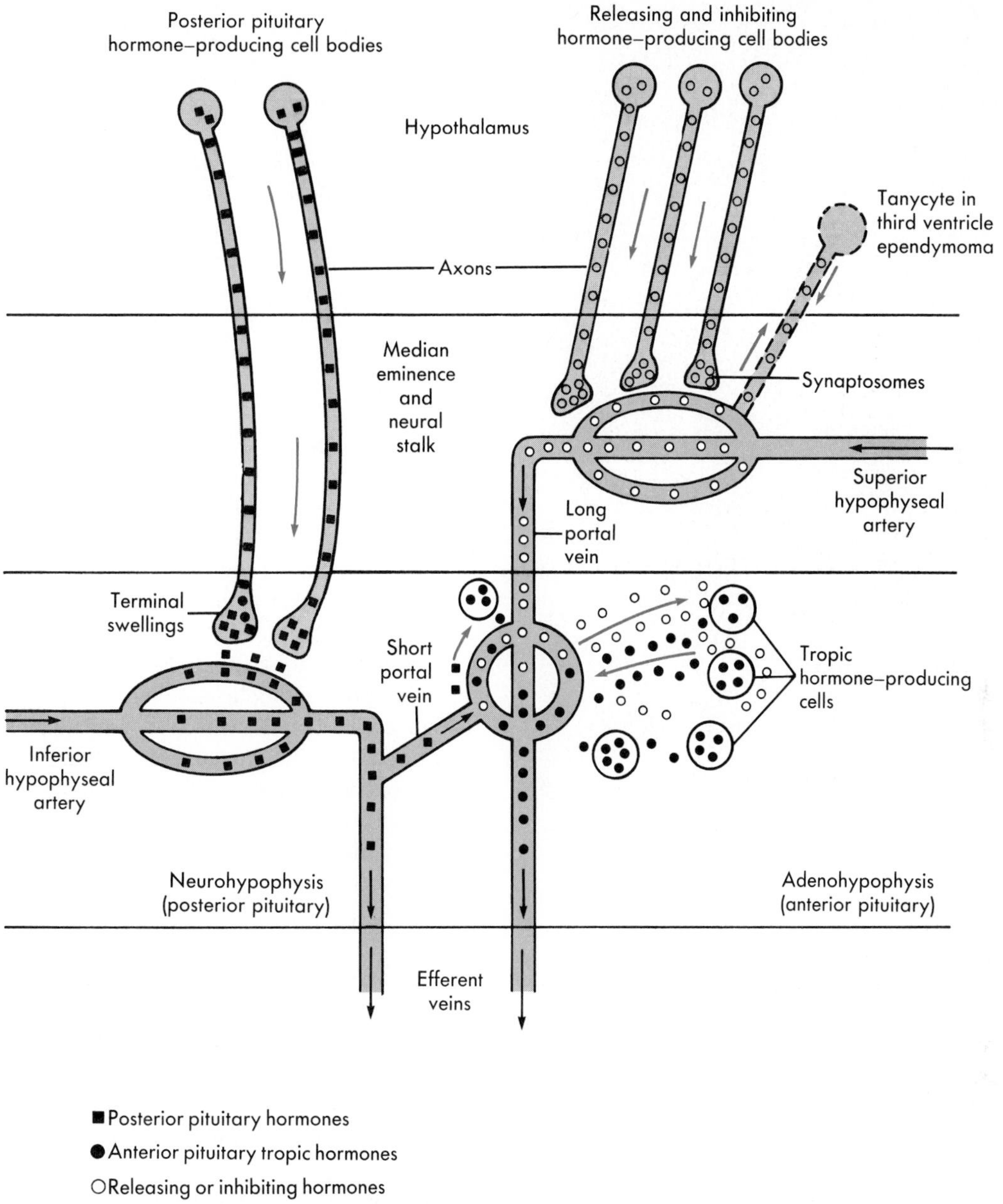

Fig. 52-2. Anatomical and functional relationships between the hypothalamus, the pituitary gland, and its blood supply. Arrows indicate direction of movement of hormone molecules. Note that the adenohypophysis has no direct arterial supply but receives blood from the median eminence, which contains hypothalamic releasing and inhibiting hormones.

pituitary portal venous blood shows concentrations ten-fold to twenty-fold higher than in peripheral blood.

3. Exposure of anterior pituitary tissue in perfusion systems or in tissue culture to individual hypothalamic peptides causes specific patterns of stimulation or inhibition of the synthesis and release of the corresponding tropic hormones.

The preceding description implies an entirely unidirectional arrangement. However, recent studies have suggested that this may not be entirely the case. It has been shown that not all the venous drainage from the anterior pituitary necessarily empties directly into the dural sinus and systemic circulation. The short portal veins may act as conduits for reverse flow of blood from the anterior pituitary cells through the neurohypophyseal capillary plexus back up to the neurons in either the median eminence or the hypothalamus itself. This direction of flow would permit anterior pituitary tropic hormones to bathe the neurons in high concentration

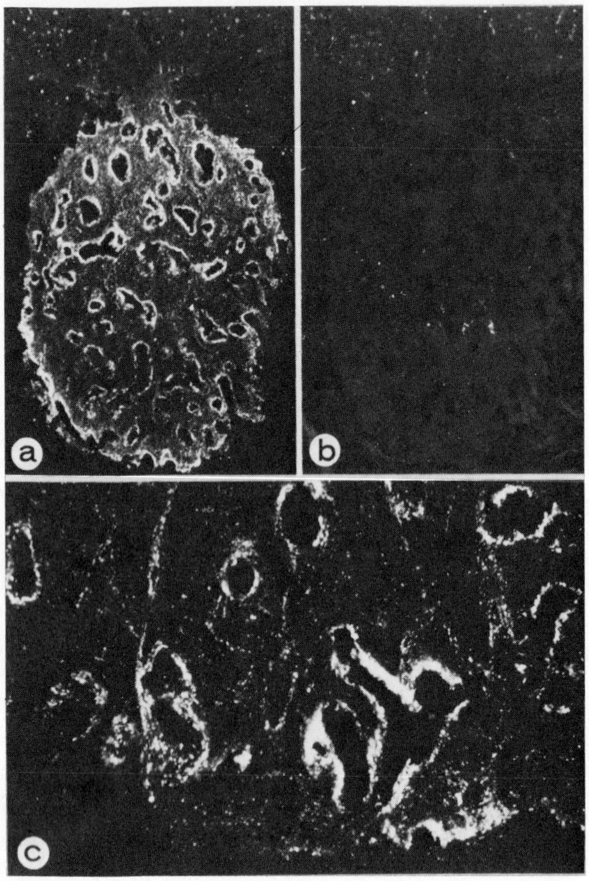

■ **Fig. 52-3.** Immunohistochemical localization of growth hormone–releasing hormone (GHRH) in the median eminence of the squirrel monkey. **A,** Section stained with fluorescent-labeled antibody to GHRH shows localization of GHRH around capillaries of the median eminence. **B,** Section stained with fluorescent-labeled control serum shows little reaction, demonstrating specificity of the antibody to GHRH. **C,** Higher power of **A** showing GHRH in axonal tracts ending in the vicinity of the capillaries. (Reprinted by permission from Bloch, B., et al.: Nature **301:**607, 1983. Copyright 1983 Macmillan Journals Limited.)

without impedance from the blood-brain barrier. It is also possible that two-way traffic between the cerebrospinal fluid and both the neurohypophysis and adenohypophysis may exist. Specialized ependymal cells in the interior recess of the third ventricle send long processes down into the portal veins of the median eminence and pituitary gland (Fig. 52-2). These cells, known as *pituitary tanycytes,* could facilitate transfer of regulatory substances from the cerebrospinal fluid to the pituitary. They could also allow posterior pituitary peptide hormones, hypothalamic releasing or inhibiting hormones, or even anterior pituitary tropic hormones to have access to the brain via the cerebrospinal fluid. Further definition of the physiological significance of these possible relationships is awaited.

■ *Hypothalamic Function*

The hypothalamus clearly plays a key role in regulating pituitary function. It can be considered a central relay station for collecting and integrating signals from diverse sources and funneling them to the pituitary. The hypothalamus receives afferent nerve tracts from the thalamus, the reticular activating substance, the limbic system (amygdala, olfactory bulb, hippocampus, and habenula), the eyes, and remotely from the neocortex. Some of the connections are multisynaptic. Through this input, pituitary function can be influenced by pain, sleep or wakefulness, emotion, fright, rage, olfactory sensations, light, and possibly even thought. It can be coordinated with patterned behavior and mating responses. The proximity of other hypothalamic nuclei that govern autonomic nervous system function also allows coordination between the output of pituitary hormones and either augmentation or reduction of sympathetic or parasympathetic activities.

Hypothalamohypothalamic tracts exist that may help to integrate multiple simultaneous pituitary responses with each other, as well as to regulate pituitary function in accordance with change in temperature, caloric status, or fluid balance. The neurotransmitters involved in afferent impulses to the hypothalamus are largely norepinephrine and serotonin. Dopamine (along with acetylcholine or gamma aminobutyric acid) and the opioid peptide, β-endorphin, act as the chief neurotransmitters for efferent nerve tracts to the median eminence. These impulses regulate the discharge of releasing hormones or inhibiting hormones from the nerve terminals in the median eminence (Fig. 52-2). In addition, neurotransmitters from the hypothalamus may reach the portal vein blood and directly influence the output of anterior pituitary tropic hormone. The presence of neurotransmitter receptors in adenohypophyseal cells supports such an additional regulatory route. Finally, dopamine and β-endorphin also modulate efferent outflow by transmitting signals between neurons within different areas of the hypothalamus.

The hypothalamus-pituitary axis is under the influence of blood-borne substances and neural input. Virtually all of the tropic hormones from the adenohypophysis cause changes either in the concentrations of peripheral gland hormones (thyroid, adrenal, gonadal) or of substrates, such as glucose or free fatty acids. Conditions exist for at least three levels of *humoral feedback,* as illustrated in Fig. 52-4. Peripheral gland hormones or substrates arising from tissue metabolism can exert feedback control on both the hypothalamus and the anterior pituitary gland. This is known as *long-loop feedback* and is usually negative, although it can occasionally be positive. Negative feedback can also be exerted by the tropic hormones themselves through ef-

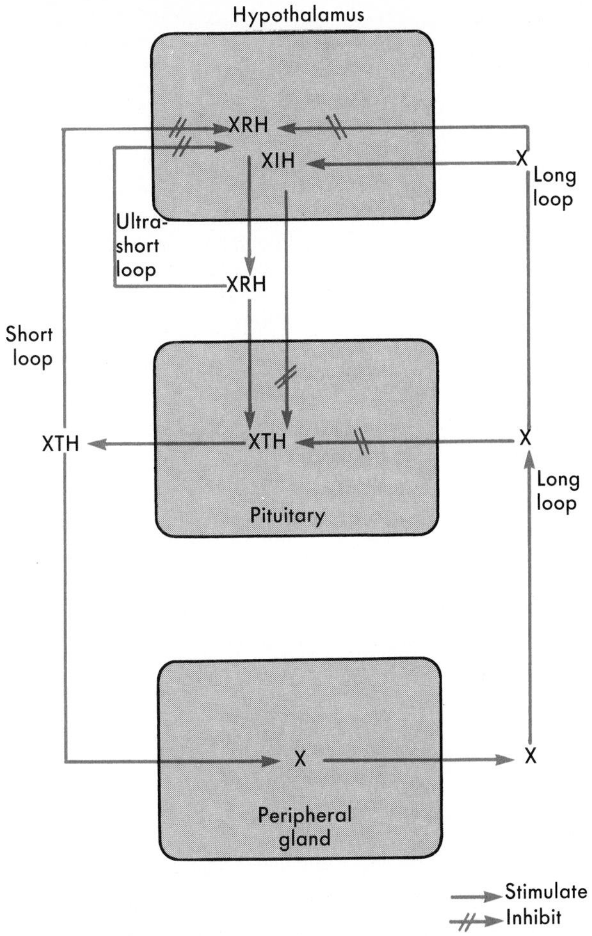

Fig. 52-4. Negative feedback loops regulating hormone secretion in a typical hypothalamus–pituitary–peripheral gland axis. *X,* Peripheral gland hormone; *XTH,* pituitary tropic hormone; *XRH,* hypothalamic releasing hormone; *XIH,* hypothalamic inhibiting hormone.

fects on the synthesis or discharge of the related hypothalamic releasing or inhibiting hormones. This is known as *short-loop feedback*. Since tropic hormones do not ordinarily cross the blood-brain barrier, short-loop feedback may occur either by specialized transport across fenestrated endothelial cells of the capillaries that bathe hypothalamic neurons or by retrograde flow through the short portal veins, as just described. Finally, hypothalamic releasing hormones may even inhibit their own synthesis and discharge or stimulate the discharge of a paired hypothalamic inhibiting hormone. This is called *ultrashort-loop feedback*. It could occur under unusual circumstances in which hypothalamic releasing hormones might gain access to the peripheral circulation in high concentrations, or it could be mediated by transport of the releasing hormone to the cerebrospinal fluid via the pituitary tanycytes. Examples of each of these feedback loops will be noted as the individual pituitary hormones are discussed.

Table 52-1 lists the currently known or suspected hypothalamic releasing or inhibiting hormones.

To some extent, the hypothalamus can be subdivided endocrinologically into functionally distinct areas. The lateral hypothalamus is largely concerned with accepting afferent impulses and relaying them to the neurosecretory nuclei of the anterior and medial basal portions. The anterior segment of the hypothalamus contains two well-defined collections of large neurons, the supraoptic and paraventricular nuclei, which are responsible for the synthesis of the two posterior pituitary peptide hormones. Their axons project primarily to the posterior pituitary, though some fibers project to the median eminence and to other neurons in the floor of the third ventricle and the brain stem. In the arcuate nucleus and the periventricular nucleus of the medial basal hypothalamus immediately beneath the third ventricle, small, neurons, responsible for synthesis of the various hypothalamic releasing and inhibiting hormones, tend to be clustered. Some are also located in the paraventriccular nucleus. The axons of these neurons project to the median eminence. However, a scattering of cells, also containing the known hypothalamic releasing hormones, is present in numerous other areas of the hypothalamus. In these neurons, the peptides may have particular neurotransmitter roles related to or distinct from their known endocrine functions. By immunohistochemical mapping, the anatomical separation between dense collections of the hypothalamic peptides is sufficient to indicate that, in general, only one cell type produces each neurohormone. However, in at least one instance, two peptides (corticotropin-releasing hormone and antidiuretic hormone) are co-localized within certain hypothalamic neurons.

All hypothalamic peptides have been named on the basis of the anterior pituitary hormone whose secretion they were originally discovered to influence. Although it was initially presumed that each tropic hormone might be under the control of a unique hypothalamic releasing or inhibiting hormone and that each hypothalamic hormone would have only one target anterior pituitary cell, the actual physiology has proven to be much more complex. The tripeptide conventionally known as *thyrotropin-releasing hormone (TRH)* can also stimulate secretion of prolactin from normal cells and secretion of growth hormone from neoplastic pituitary cells. Somatostatin, discovered as a growth hormone–inhibiting factor, can also inhibit the secretion of thyrotropin and prolactin. Perhaps other examples of overlap will be discovered. It should also be remembered that virtually all the hypothalamic releasing and inhibiting hormones have also been found outside the hypothalamus, in such diverse areas as the cerebral cortex, limbic area, spinal cord, autonomic ganglia, sensory neurons, pancreatic islets, and throughout the gastrointestinal tract. In these areas, they may serve

Table 52-1. Hypothalamic hormones and factors

	Identity	*Target tropic hormones*
Thyrotropin-releasing hormone (TRH)	pGLU-HIS-PRO-NH_2	Thyrotropin Prolactin Growth hormone (pathological)
Luteinizing hormone–releasing hormone (LHRH)	pGLU-HIS-TRP-SER-TYR-GLY-LEU-ARG-PRO-GLY-NH_2	Luteinizing hormone Follicle-stimulating hormone Growth hormone (pathological)
Corticotropin-releasing hormone (CRH)	SER-GLN-GLU-PRO-PRO-ILE-SER-LEU-ASP-LEU-THR-PHE-HIS-LEU-LEU-ARG-GLU-VAL-LEU-GLU-MET-THR-LYS-ALA-ASP-GLN-LEU-ALA-GLN-GLN-ALA-HIS-SER-ASN-ARG-LYS-LEU-LEU-ASP-ILE-ALA-NH_2	Adrenocorticotropin β-Lipotropin Endorphins
Growth hormone–releasing hormone (GHRH)	TYR-ALA-ASP-ALA-ILE-PHE-THR-ASN-SER-TYR-ARG-LYS-VAL-LEU-GLY-GLN-LEU-SER-ALA-ARG-LYS-LEU-LEU-GLN-ASP-ILE-MET-SER-ARG-GLN-GLN-GLY-GLU-SER-ASN-GLN-GLU-ARG-GLY-ALA-ARG-ALA-ARG-LEU-NH_2	Growth hormone
Growth hormone–inhibiting hormone (somatostatin)	ALA-GLY-CYS-LYS-ASN-PHE-PHE-TRP-LYS-THR-PHE-THR-SER-CYS	Growth hormone Prolactin Thyrotropin Adrenocorticotropin (pathological)
Prolactin-inhibiting factor (PIF)	Dopamine	Prolactin
Prolactin-releasing factor (PRF)	Not established	Prolactin

neurotransmitter roles related to or independent of their endocrine function.

The hypothalamic peptides are synthesized via preprohormones in classic fashion (Chapter 49). In several instances the gene responsible has already been cloned and sequenced. Further studies are still needed to elucidate the mechanisms for processing the primary gene product, for packaging the hormones in granules, for transporting them to the median eminence, and for coordinating their synthesis with that of other neurotransmitters within the same cell. Many common features characterize the functional behavior of hypothalamic releasing peptides, as follows:

Characteristics of hypothalamic releasing hormones

1. Are secreted in pulses
2. Induce effects through calcium, cyclic AMP, and membrane phospholipid mediators
3. Stimulate synthesis of target pituitary hormones at transcriptional level
4. Modify biological activity of target pituitary hormones by posttranslational effects
5. Stimulate release of target pituitary hormones
6. Modulate effects by regulation of own receptors

The secretion of hypothalamic releasing hormones into the pituitary portal veins is pulsatile, a pattern apparently dependent on an intrinsic neural oscillator within the cells of origin. This pulsatility appears to be critical for maintaining the appropriate level of secretion of their target anterior pituitary hormones. The releasing hormones react with plasma membrane receptors in the anterior pituitary cells, following which cytosolic calcium and then cAMP levels increase. In addition, diacylglycerols, inositol phosphates, and arachidonic acid from membrane phospholipids help mediate the intracellular effects that follow. The specific proteins that are phosphorylated by activated protein kinase A or C are not known. Subsequent to binding of the releasing hormone by the target pituitary cell, granule exocytosis is rapidly stimulated, with release of their tropic hormone content. In addition, tropic hormone synthesis is stimulated by increasing the levels of its specific messenger RNA. In some instances, the biological activity of the target pituitary hormones may also be increased after translation by modifying their content of sugars or sialic acid.

The anterior pituitary contains at least five endocrine cell types (Table 52-2). These cannot be completely distinguished by conventional histological staining or by localization to specific areas. However, the development of techniques employing hormone-specific antisera has permitted each type to be identified specifically in normal pituitary tissue and in functioning pituitary tumors. Fig. 52-5 demonstrates the results of immunohistochemical staining of serial sections of a normal anterior pituitary gland and gives some idea of cell distri-

Table 52-2. Anterior pituitary cells and hormones

Cell	*% Pituitary population*	*Products/molecular weight*	*Targets*
Corticotroph	15-20	Adrenocorticotropin (ACTH), 4,500 β-Lipotropin, 11,000	Adrenal gland Adipose tissue Melanocytes
Thyrotroph	3-5	Thyrotropin (TSH), 28,000	Thyroid gland
Gonadotroph	10-15	Luteinizing hormone (LH), 28,000 Follicle-stimulating hormone (FSH), 33,000	Gonads
Somatotroph	40-50	Somatotropin, growth hormone (HGH), 22,000	All tissues
Mammotroph	10-25	Prolactin, 23,000	Breasts Gonads

A

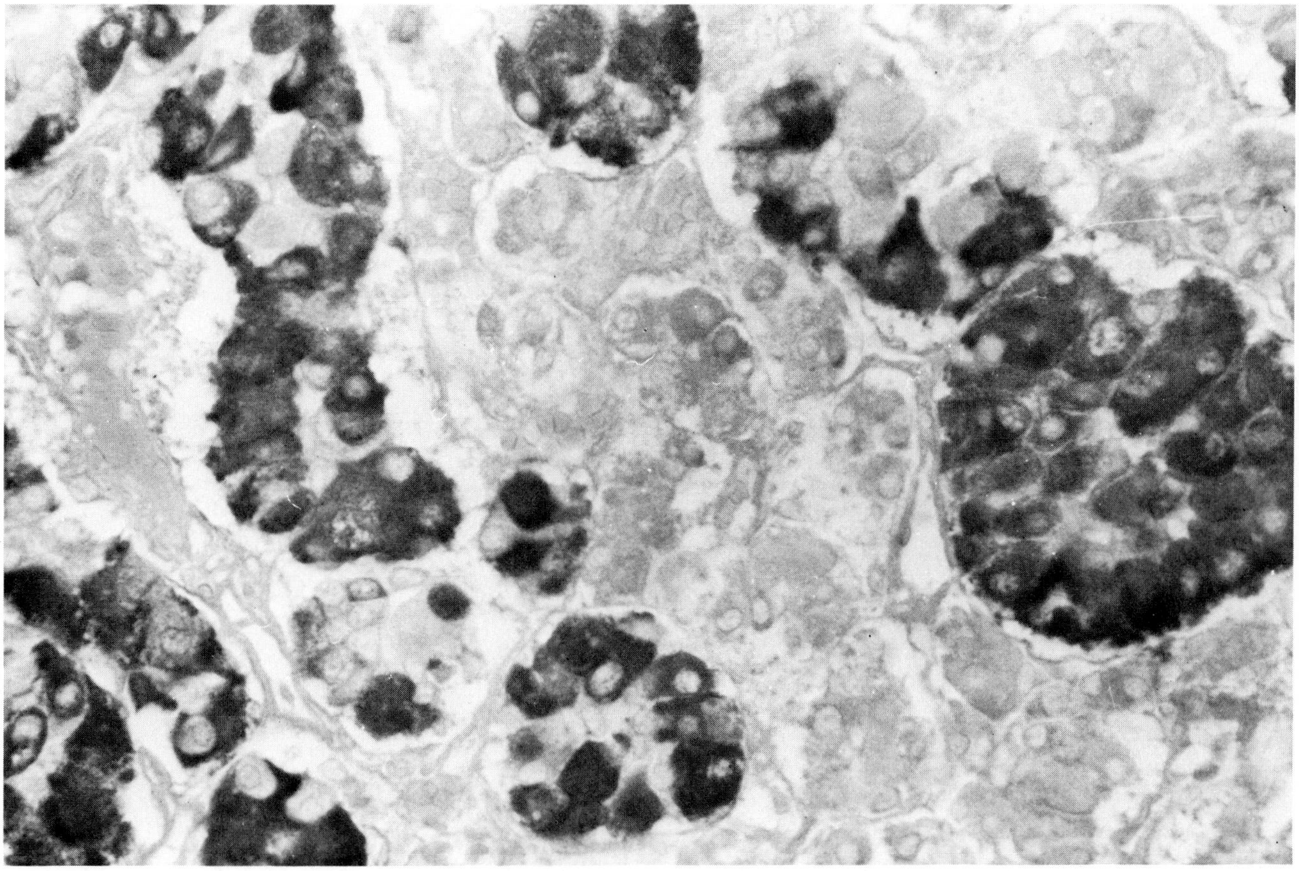

Fig. 52-5. Adjacent sections of a human anterior pituitary gland stained by immunohistochemical techniques for ACTH **(A)**, HGH **(B)**, and prolactin **(C)**. Note the different patterns of distribution of the three hormones. (Courtesy Dr. Manuel Velasco, Case Western Reserve University, School of Medicine, Cleveland, Ohio.) *Continued.*

■ **Fig. 52-5, cont'd** For legend see p. 901.

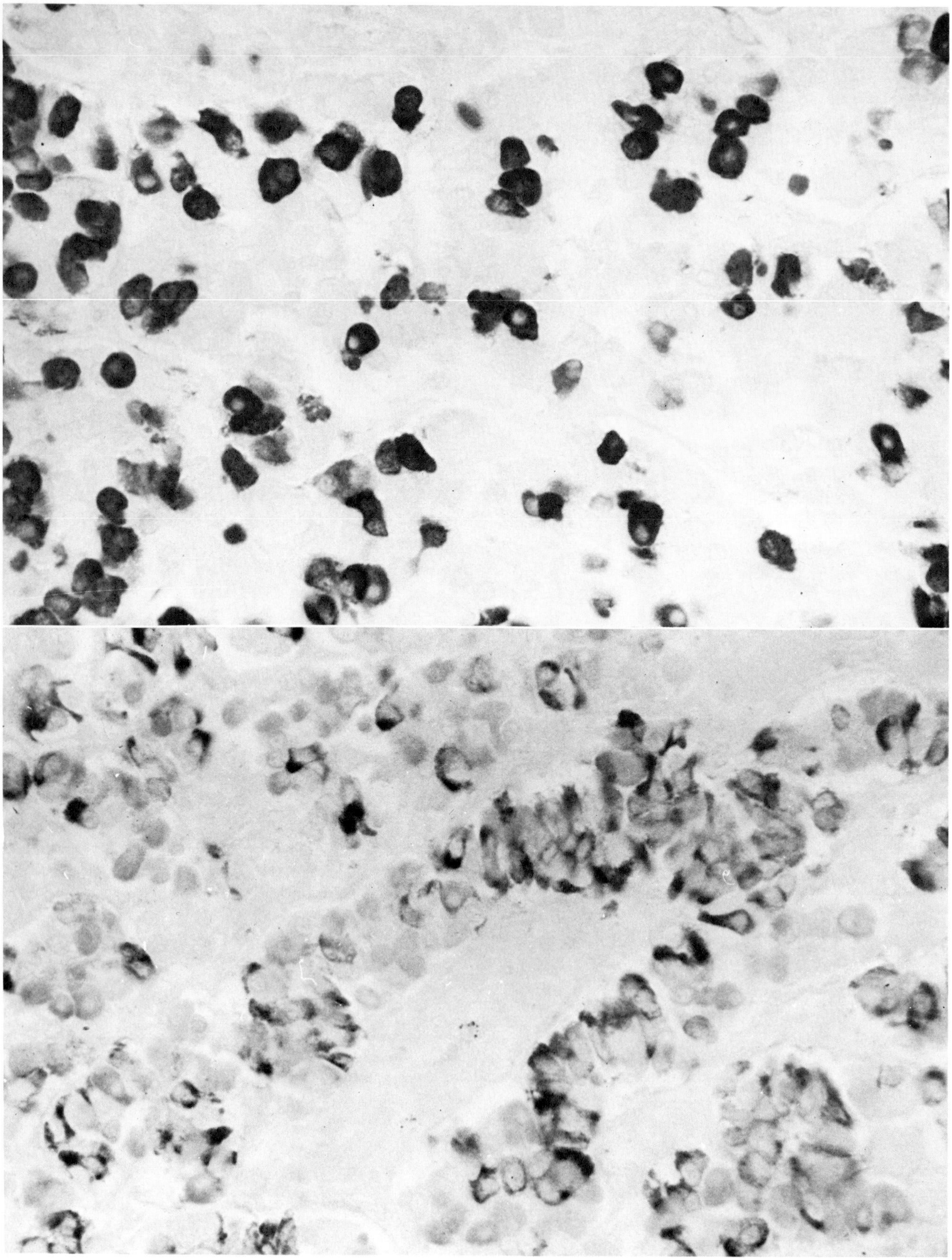

B

C

butions. The possibility of paracrine interactions among these closely juxtaposed cells is now being explored, as methods of sorting and separating them from pituitary tissue are being developed.

Anterior Pituitary Hormones

Thyrotropic Hormone

TSH is a glycoprotein hormone whose function is to regulate the growth and metabolism of the thyroid gland and the secretion of its hormones, thyroxine (T_4) and triiodothyronine (T_3). The TSH-producing cells normally form 3% to 5% of the adult human anterior pituitary population, and they are found predominantly in the anteromedial area of the gland. These cells develop at about 13 weeks of gestation at the same time that the fetal thyroid gland is beginning to secrete thyroid hormone in response to TSH. In the adult pituitary, TSH is stored in small secretory granules of 125 to 200 nm in diameter.

TSH has a molecular weight of 28,000 and contains 15% carbohydrate bound covalently to the peptide chains. The hormone is made of two subunits associated by noncovalent forces. The α-subunit of 96 amino acids is nonspecific, being a component also of two other anterior pituitary hormones (FSH and LH), as well as of a placental hormone (human chorionic gonadotropin). The β-subunit of 110 amino acids confers the specific biological activity on the TSH molecule; however, by itself it is essentially inactive. Species variation among β-subunits exists, but biological activity overlaps considerably. Bovine TSH, for example, is very active in humans. Separate genes, located on different chromosomes, code for the individual messenger RNAs of the α and β subunits. In each instance, during translation, a signal N-terminal peptide is eliminated from the primary translation product (termed a prehormone). Subsequently, the N-glycosidic-linked sugar moieties are added. During transport from the rough endoplasmic reticulum and packaging in the Golgi apparatus, the carbohydrate units are further modified and intramolecular disulfide bonds are formed. These changes assure the proper conformation that permits the two individual subunits to combine in the mature TSH molecule. Expression of the α and β subunit genes is separately regulated, though there must be coordination between them. Ordinarily an excess of the nonspecific α subunit is produced, but selective addition of an extra oligosaccharide renders the unneeded α subunits incapable of combining with β subunits. TSH synthesis is stimulated by the hypothalamic thyrotropin-releasing hormone (TRH) and is suppressed by thyroid hormone, in each instance by altering subunit messenger RNA levels. In addition, TRH and thyroid hormone also modulate the glycosylation process.

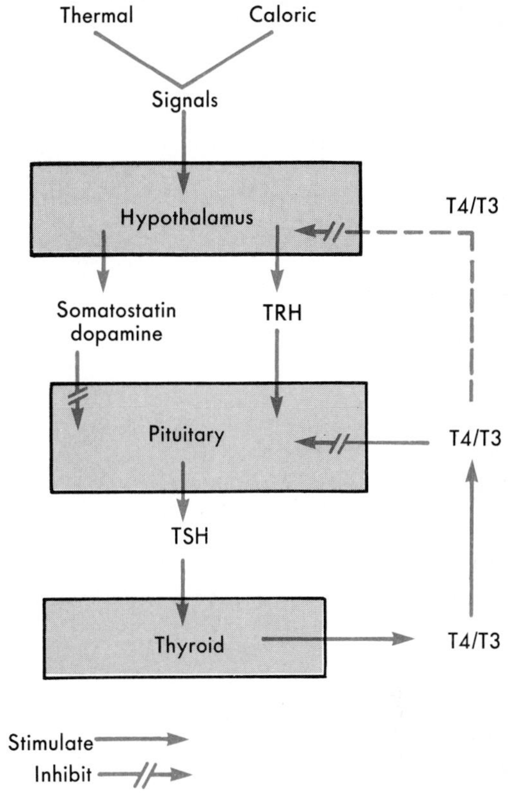

■ Fig. 52-6. Regulation of TSH secretion. Thyroxine (T_4) and triiodothyronine (T_3) exert negative feedback on the pituitary by blocking the action of TRH. Negative feedback of T_4 and T_3 at the level of the hypothalamus is less well established. Somatostatin and dopamine each inhibit TSH secretion tonically.

The secretion of TSH is regulated by two factors. TRH increases the rate of secretion, whereas thyroid hormone decreases it in negative feedback fashion (Fig. 52-6). As a result of the balance between TRH stimulation and thyroid hormone inhibition, TSH is secreted in a steady fashion. This is in contrast to certain other hormones like growth hormone or ACTH, whose secretion fluctuates greatly. This befits the role of TSH, which is to stimulate a target gland whose own output is meant to be steady because the actions of its hormones (thyroxine and triiodothyronine) wax and wane slowly. In other words, although the hypothalamic-pituitary-thyroid gland axis is set with great precision, it moves to higher or lower levels in hours or days rather than in minutes. In humans, within minutes of intravenous administration of TRH, plasma TSH levels rise as much as ten-fold and return toward baseline levels by 60 minutes (Fig. 52-7). With repeated TRH injections, the TSH response diminishes over time primarily because the secondarily stimulated thyroid gland increases its output of T_4 and T_3 (Fig. 52-7). This demonstrates vividly the negative feedback regulation of TSH secretion. These small increments of thyroid hormone con-

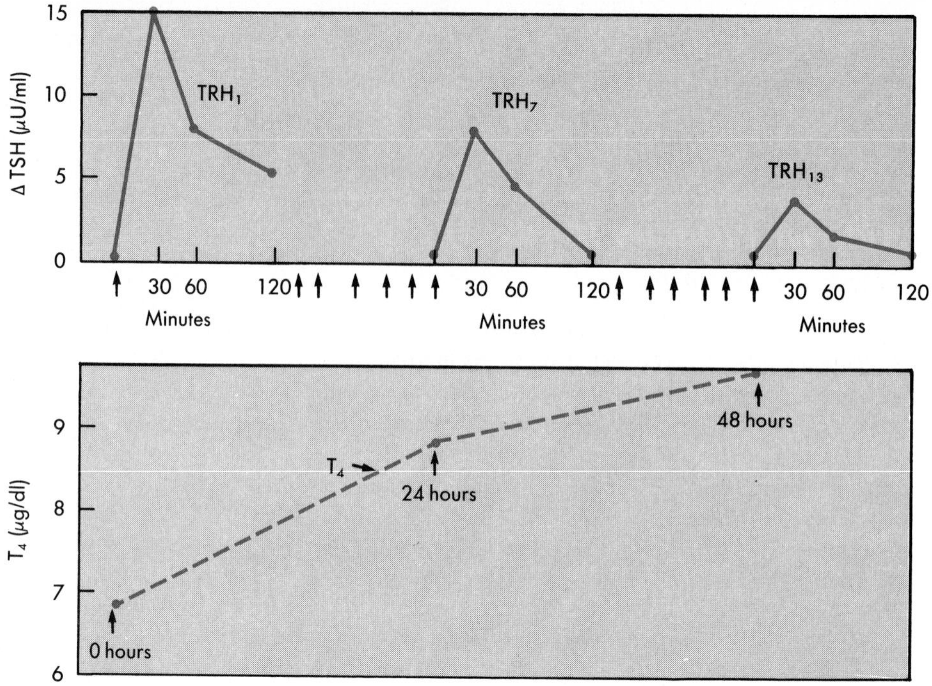

Fig. 52-7. Pituitary and thyroid gland responses to repetitive injections of TRH every 4 hours for 48 hours in humans. Note that as plasma thyroxine (T_4) increases as a result of stimulation by TSH, the TSH responses to TRH are progressively blunted. (Redrawn from Snyder, P.J.: J. Clin. Invest. **52:**2305, 1973. Reproduced by copyright permission of The American Society for Clinical Investigation.)

centration suppress TSH secretion by blocking the stimulatory action of TRH; conversely, small decrements of thyroid hormone augment TSH responsivity to TRH. Significant modulation of TSH secretion is associated with variations in plasma thyroid hormone concentrations of only 10% to 30% above or below the individual's baseline level. The thyrotroph's response to continuous TRH stimulation may also be limited by down regulation of the TRH receptor.

The intracellular mediator of thyroid hormone's effect on TSH is likely T_3. Furthermore, it appears that T_3 generated within the pituitary cell from T_4 is more effective and important in this regard than is the T_3 that enters from the circulation. The acute suppressive effect of thyroid hormone on TSH release has a half-life of days, and the suppressive effect is prevented if protein synthesis by the pituitary thyrotroph is inhibited. This suggests that T_3 induces the synthesis of a protein with TSH-suppressing properties. In addition, T_3 decreases the number of TRH receptors. Although thyroid hormone clearly inhibits TSH secretion and synthesis at the pituitary level, evidence from microinjection directly into the hypothalamus indicates that T_3 may reduce the synthesis or release of TRH as well. Because of negative feedback, individuals with thyroid diseases that result in chronic deficiency of thyroid hormone *(hypothyroidism)* have very high plasma TSH levels, whereas the opposite is true of patients with *hyperthyroidism*. Hyperplasia of the thyrotrophs sufficient to produce enlargement of the adenohypophysis can also result from hypothyroidism.

Physiological modulation of TSH secretion (and consequently of thyroid hormone output) occurs in at least two circumstances, namely, fasting and exposure to cold. TSH responsiveness to TRH and possibly TRH release itself are diminished during total fasting. This coincides with a decrease in metabolic rate and appears teleologically useful. In animals, TSH secretion is augmented by exposure to cold, but this has been demonstrated only infrequently in humans. Since TSH will increase thermogenesis via stimulation of the thyroid gland, this is a logical response. Other hormonal and neural influences have been noted. A slight diurnal variation in TSH secretion has been observed, with the highest levels occurring at night. A tonic inhibitory effect on TSH secretion is exerted by the hypothalamic peptide somatostatin and the neurotransmitter dopamine. Furthermore, thyroid hormone may suppress TSH secretion partly by stimulating the release of these hypothalamic inhibitors. Cortisol (a hormone from the adrenal cortex) decreases both TRH and TSH secretion; growth hormone also reduces TSH secretion. The importance of these latter effects remains to be elucidated.

TSH normally circulates in plasma at a concentration of 0.5 to 5 μU/ml. Because of discrepancies and variations in bioassays, an exact equivalent molar concentration cannot be given, but it is about 10^{-11}M, similar to other protein hormones. Daily TSH production is about 165,000 μU, which is equivalent to the entire content of one normal pituitary gland. The metabolic clearance rate of TSH is 50 L/day. In normal individuals, the α-subunit is also individually secreted and circulates at low levels. When TSH secretion is chronically hyperstimulated in response to deficient function of the thyroid gland, both β- and α-subunits circulate individually in elevated amounts.

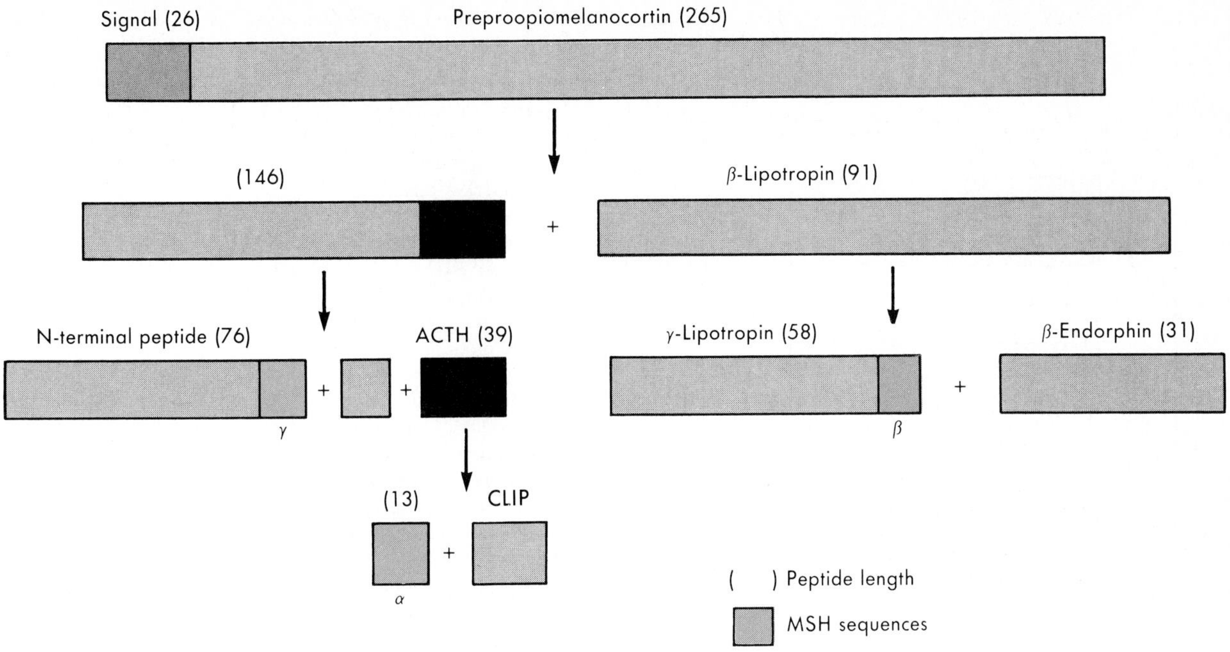

Fig. 52-8. The processing of preproopiomelanocortin. In the anterior lobe of the human pituitary, ACTH, β-lipotropin, γ-lipotropin, β-endorphin, and a 76-amino acid N-terminal fragment are end products that are released. In other species, ACTH is further cleaved to α-MSH and corticotropin-like intermediate peptide (CLIP) in the neural intermediate lobe.

The only TSH actions of importance are those exerted on the thyroid gland. TSH, as its name implies, is tropic; that is, it promotes growth of the gland and stimulates all aspects of its function. The glandular uptake of iodide, its organification, the completion of thyroid hormone synthesis, and the subsequent release of thyroid gland products are all stimulated by TSH. These effects are described in detail in Chapter 53. TSH binds to a plasma membrane receptor, and cAMP is the intracellular mediator of many of the hormone's effects by activating a protein kinase in the thyroid gland. The substrates for this enzyme and their subsequent connection with all the steps in thyroid hormone synthesis and release are not yet known. In tissue culture preparations, cAMP may also mediate the TSH-induced differentiation of thyroid cells into follicles.

The sole pathological effects of an excess or deficiency of TSH are those of increased or decreased thyroid gland function, as described in Chapter 53.

Adrenocorticotropic Hormone

ACTH is an anterior pituitary polypeptide hormone whose function is to regulate the growth and secretion of the adrenal cortex. Its most important target gland hormone is cortisol. The ACTH-producing cells form 20% of the anterior pituitary population. Although these are largely localized to the pars distalis of the anterior lobe (Fig 52-1), there is some evidence in animals for existence of ACTH-producing cells in the pars intermedia as well. The hormone is stored in secretory granules that are about 375 to 550 nm in diameter and have a clear space or halo between the contents and the membrane. The ACTH-producing cells are distinguished ultrastructurally by the presence of large numbers of microfilaments. In the human fetus, ACTH synthesis and secretion begin at about 10 to 12 weeks of gestation, just before the subsequent rapid enlargement and development of the fetal adrenal cortex.

ACTH is a straight-chain peptide with 39 amino acids and a molecular weight of 4500. The N-terminal 1 to 24 sequence contains full biological activity; the remaining C-terminal portion probably only prolongs the hormone's action by protecting it against enzymatic degradation. Sequence 5 to 10 of ACTH is critical for stimulating the adrenal cortex.

Synthesis of ACTH. Our developing knowledge of the biosynthesis of ACTH has formed a fascinating chapter of modern endocrinology, which has given clues to the basic meaning of the close connection between the nervous system and the endocrine system. As shown in Fig. 52-8, a single gene controls the transcription of a mature messenger RNA that directs the synthesis of a 31,000 molecular weight protein known as preproopiomelanocortin. Sequential processing of the latter in the human gives rise to several anterior pituitary cell products that are cosecreted into the plasma. These products include ACTH, β-lipotropin, γ-lipotropin, β-endorphin, and the N-terminal peptide. Melano-

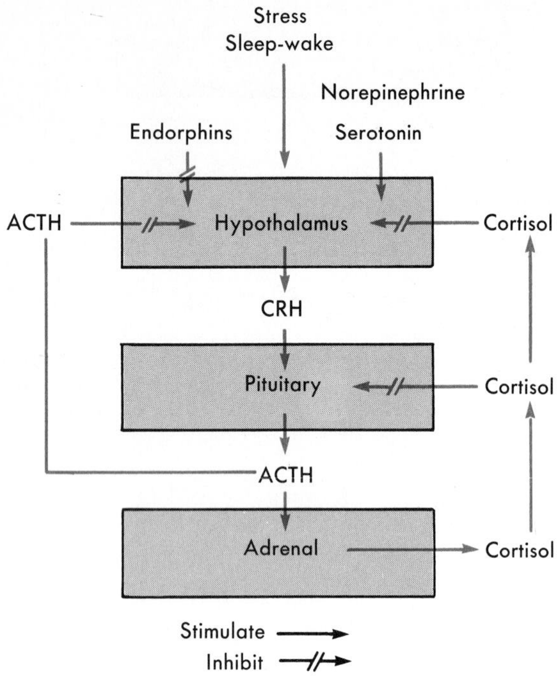

Fig. 52-9. Regulation of ACTH secretion. Cortisol exerts negative feedback (1) at the pituitary level by blocking CRH action, and (2) at the hypothalamus level by inhibiting CRH release. Norepinephrine and serotonin are positive modulators, whereas endorphins and ACTH itself are negative modulators of CRH release.

Table 52-3. Regulation of ACTH secretion

Stimulation	*Inhibition*
Cortisol decrease	Cortisol increase
Adrenalectomy	Enkephalins
Metyrapone	Opioids
Sleep-wake transition	ACTH
Stress	
Hypoglycemia	
Anesthesia	
Surgery	
Trauma	
Infection	
Pyrogens	
Psychiatric disturbance	
Anxiety	
Depression	
Antidiuretic hormone	
α-adrenergic agonists	
β-adrenergic antagonists	
Serotonin	
γ-aminobutyric acid	

cyte-stimulating hormone (MSH) activity is also contained within several of these peptides. Strong immunohistological evidence indicates that these peptides are present in the same cell, and they are frequently, if not always released together. In certain animal species but not in humans, the cells of the intermediate neural lobe further cleave ACTH to α-MSH and a second product known as corticotropin-like intermediate peptide (CLIP). In addition, these cells may cleave γ-MSH from the N-terminal peptide and β-MSH from γ-lipotropin (Fig. 52-8). A similar pattern of processing from the same precursor molecule does take place in human extrapituitary tissue (brain, hypothalamus, gastrointestinal tract, pancreatic islets, adrenal medulla) where ACTH and these smaller peptides may subserve completely different functions. Since there is little or no evidence for the existence of α-, β-, or γ-MSH in human plasma, the expression of biological melanocyte-stimulating activity may depend on the presence of these amino acid sequences within the parent molecules.

Finally, the N-terminal pentapeptide of β-endorphin is identical to metenkephalin, with which it shares the analgesic and mood-modifying effects of opioids. The enkephalins, however, are synthesized from a different gene-directed precursor and do not arise by cleavage of β-endorphin. Both preproopiomelanocortin products and enkephalin can, however, be synthesized in the same extrapituitary cells, again emphasizing the intimate relationship between endocrine and neural function.

Secretion of ACTH. The regulation of ACTH secretion is among the most complex of all the pituitary hormones (Fig. 52-9). The hormone exhibits circadian rhythms, cyclic bursts, feedback control, and responses to a wide variety of stimuli (Table 52-3). Although the mechanisms for each form of control are not completely clear, the hypothalamic corticotropin-releasing hormone (CRH) is the important mediator. CRH is a peptide with 41 amino acids (Table 52-1), and it stimulates the release of ACTH and its proopiomelanocortin coproducts (Fig. 52-8). Antidiuretic hormone (ADH) also exhibits corticotropin-releasing activity in vivo and in vitro. Under particular physiological circumstances (e.g., stress), ADH may augment the primary effect of CRH. The gene that directs the synthesis of prepro-CRH has considerable homology with the genes for prepro-ADH and preproopiomelanocortin itself, suggesting a common evolutionary starting point for these molecules.

One dominant rhythm of ACTH secretion is clearly diurnal. As shown in Fig. 52-10 a peak occurs 2 to 4 hours before awakening. Thereafter the average level decreases to a nadir just before or after falling asleep. A rise and fall in the major adrenocortical hormone, cortisol, is entrained in this ACTH pattern. The clock time of the diurnal pattern can be shifted by systematically altering the sleep-wake cycle for a number of days; however, the ACTH peak is not entrained with a specific stage of sleep. The circadian rhythm is diminished or abolished by loss of consciousness, blindness, or by constant exposure to either dark or light. It is clear that the nocturnal ACTH peak is primarily gener-

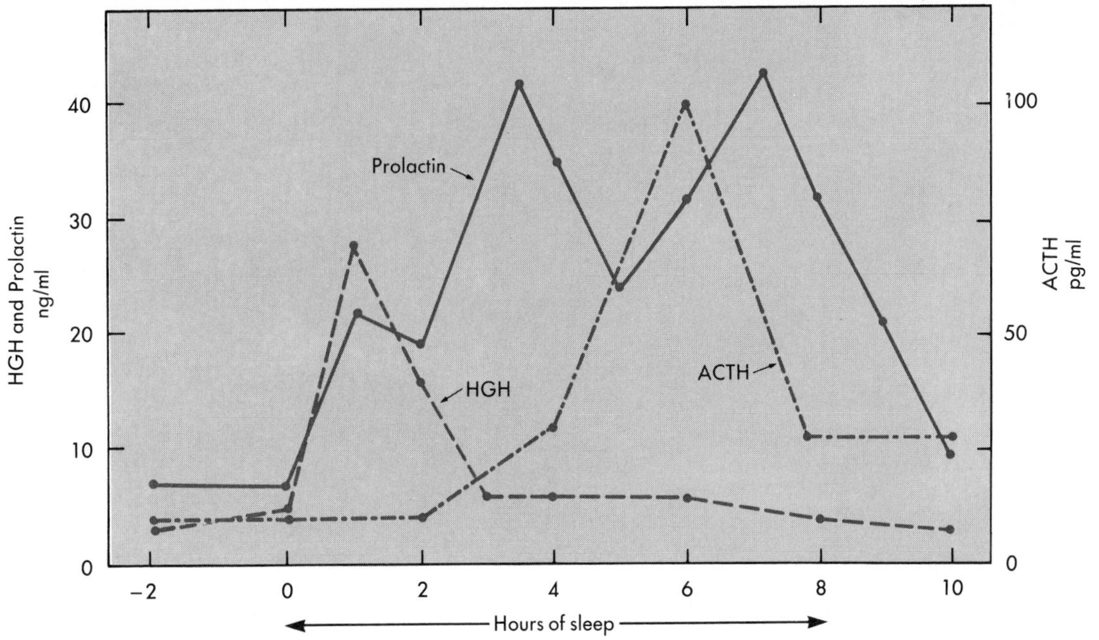

■ **Fig. 52-10.** Nocturnal release of ACTH, HGH, and prolactin. Note the distinctive pattern for each hormone. (Redrawn from Takahashi, Y., et al.: J. Clin. Invest. **47:**2079, 1968; Berson, S.A., et al.: J. Clin. Invest. **47:**2725, 1968; reproduced by copyright permission of The American Society for Clinical Investigation; and Sassin, J.F., et al.: Science **177:**1205, 1972. Copyright 1972 by the American Association for the Advancement of Science.)

ated by the hypothalamus and does not depend on negative feedback from its target, the adrenal gland. Nevertheless, the nocturnal peak is augmented by prior cortisol deficiency and can be completely suppressed by either exogenous or endogenous excess cortisol. The overall diurnal pattern is composed of short cyclic bursts of ACTH release lasting 10 to 20 minutes. It is an increase in the frequency of these bursts that accounts for the rise of mean plasma ACTH levels in the early hours of the morning.

Feedback inhibition of ACTH secretion is effected by its peripheral target hormone, cortisol, or by any cortisol-like steroid with a potency proportional to its glucocorticoid activity (Fig. 52-9). The suppressive action often outlives the duration of plasma glucocorticoid elevation. Conversely, when cortisol action is blocked by an antagonist, or when cortisol secretion is reduced by disease, or when cortisol synthesis and release are acutely inhibited by the drug metyrapone (Fig. 52-11), ACTH secretion is stimulated. Cortisol modulates ACTH secretion at the pituitary level by blocking the stimulatory action of CRH (Fig. 52-12). Cortisol also decreases the synthesis of ACTH and its coproducts by inhibiting transcription of preproopiomelanocortin. In addition, cortisol blocks hypothalamic release of CRH, although apparently not its synthesis. ACTH may also inhibit its own secretion by decreasing CRH release, an example of short-loop feed back. Chronic deficiency of cortisol leads to persistent elevation of plasma ACTH, but the diurnal and pulsatile patterns are preserved, indicating their basic nonfeedback origin. Chronic autonomous hypersecretion of cortisol or long-term therapeutic administration of cortisol analogues for various diseases leads to functional atrophy of the CRH-ACTH axis. Several months may be required for recovery of this axis after the suppressive influence has been removed.

ACTH secretion responds most strikingly to stressful stimuli, a response that is critical to survival. Numerous factors that elicit the stress reaction are noted in Table 52-3. All have been demonstrated in humans. The response to insulin-induced hypoglycemia is illustrated in Fig. 52-13. In some instances, such as major abdominal surgery or severe psychiatric disturbance, the stress-induced hypersecretion of ACTH completely overrides negative feedback, and it cannot be suppressed by even the maximum level of cortisol secretion of which the adrenal cortex is capable. Stress also often obliterates the regular diurnal variation of ACTH, although secretory spurts may still be observed. The pathways by which each stress is signaled, sensed, and then stimulates CRH secretion vary. Among the monoamine neurotransmitters, both norepinephrine (via α-adrenergic receptors) and serotonin have been implicated in modulating the ACTH response to insulin-induced hypoglycemia. Other stress responses may be augmented by ADH; some may be modulated by acetylcholine, since they can be blocked by atropine.

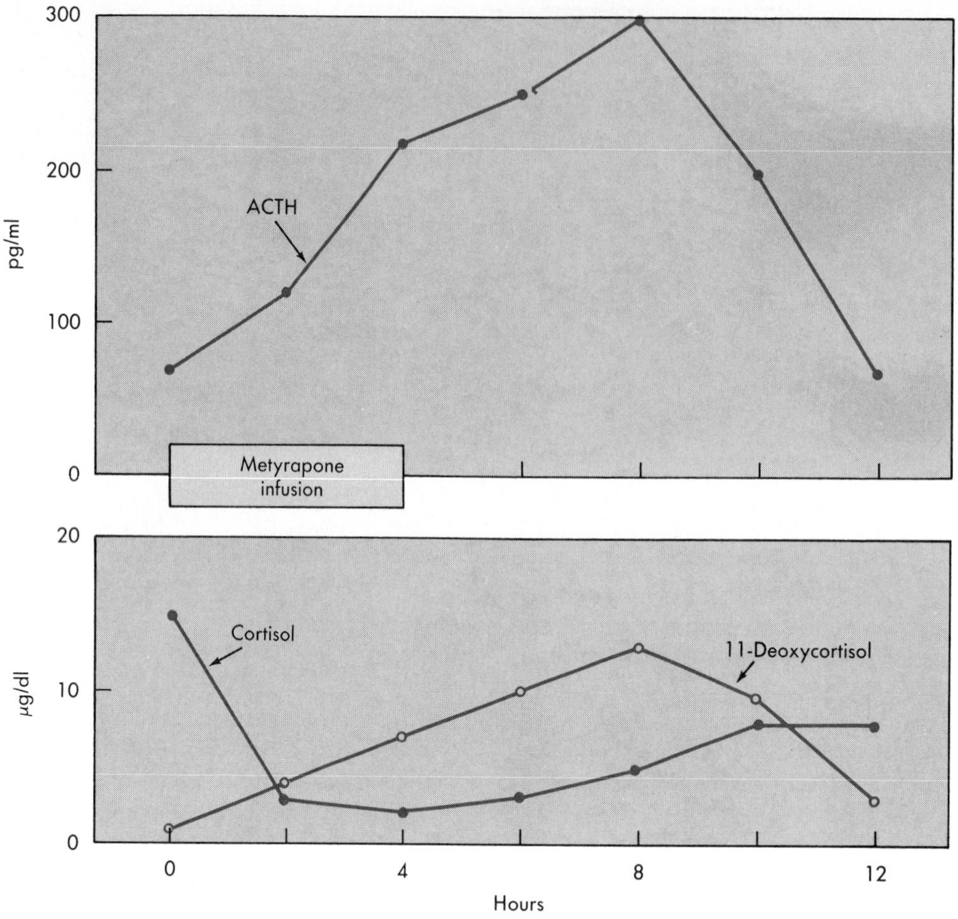

Fig. 52-11. Negative feedback stimulation of ACTH release by metyrapone, a drug that blocks the conversion of 11-deoxycortisol to cortisol. (Redrawn from Jubiz, W., et al.: Arch. Intern. Med. **125:**468, 1970. Copyright 1970, American Medical Association.)

Once secreted, ACTH circulates in plasma unbound to protein. Basal concentrations at 6 AM range from 20 to 100 pg/ml (average 50 pg/ml or 10^{-11}M). Accurate measurements of daily production rate are not available. A rate of 100 to 300 µg/day has been estimated. This estimate compares to an average adult human pituitary content of 250 to 500 µg. However, the bulk of the ACTH is secreted in a limited period of each day, which allows time for recovery of stores.

Action of ACTH. ACTH stimulates the growth of those specific zones of the adrenal cortex concerned with secretion of cortisol and androgenic steroids. In this respect, the effect is more to increase the size rather than the number of adrenal cells. In the absence of ACTH, profound atrophy of the relevant adrenal zones occurs. ACTH action follows binding to a specific plasma membrane receptor and activation of adenylate cyclase. A number of steps in the synthesis of adrenal steroids are thereby stimulated by ACTH, and these are detailed in Chapter 54. Because of the rapidity of synthesis, ACTH promptly causes secretion of adrenocortical hormones. Adrenal responsiveness to ACTH is attenuated and delayed by prior chronic underexposure to the tropic hormone; conversely, responsiveness is accentuated by prior chronic overexposure.

In animals, a number of extraadrenal actions of intact ACTH or of fully active fragments such as the 1 to 24 peptide, have been described. These require much larger doses than those needed for adrenal stimulation and are difficult to demonstrate in humans. Such effects include stimulation of lipolysis and stimulation of insulin secretion with consequent hypoglycemia. These may reflect the regulatory influences of extrapituitary ACTH molecules, which are generated within these respective sites of action.

ACTH, because of its MSH sequences, increases skin pigmentation. In amphibians, MSH acts on melanocytes, causing the dispersal of melanin pigment granules within these cells and their dendrites. This action, which is probably mediated by cAMP, results in darkening of the skin. In humans, it is more likely that peptides with MSH activity cause hyperpigmentation by stimulating melanin synthesis and the transfer of melanin from the melanocytes to epidermal cells.

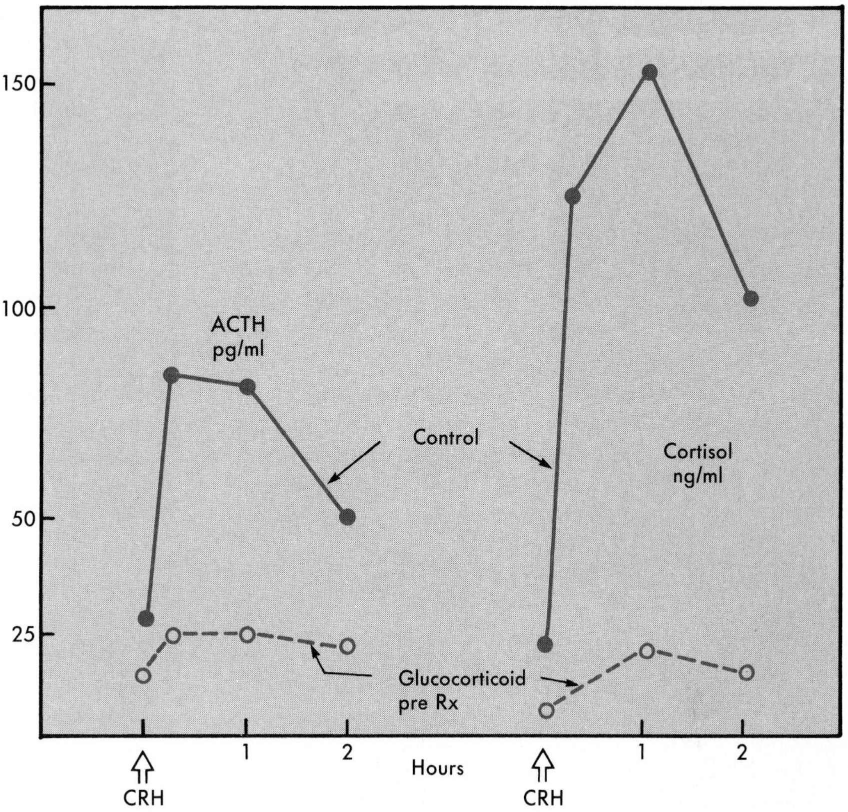

Fig. 52-12. Plasma ACTH and cortisol responses to administration of CRH. Pretreatment with a synthetic glucocorticoid (analog of cortisol) suppresses the action of CRH on the pituitary. The diminished ACTH response leads secondarily to a diminished secretion of cortisol by the adrenal glands. (Redrawn from Copinschi, G., et al.: J. Clin. Endocrinol. Metab. **57;**1287, 1983.)

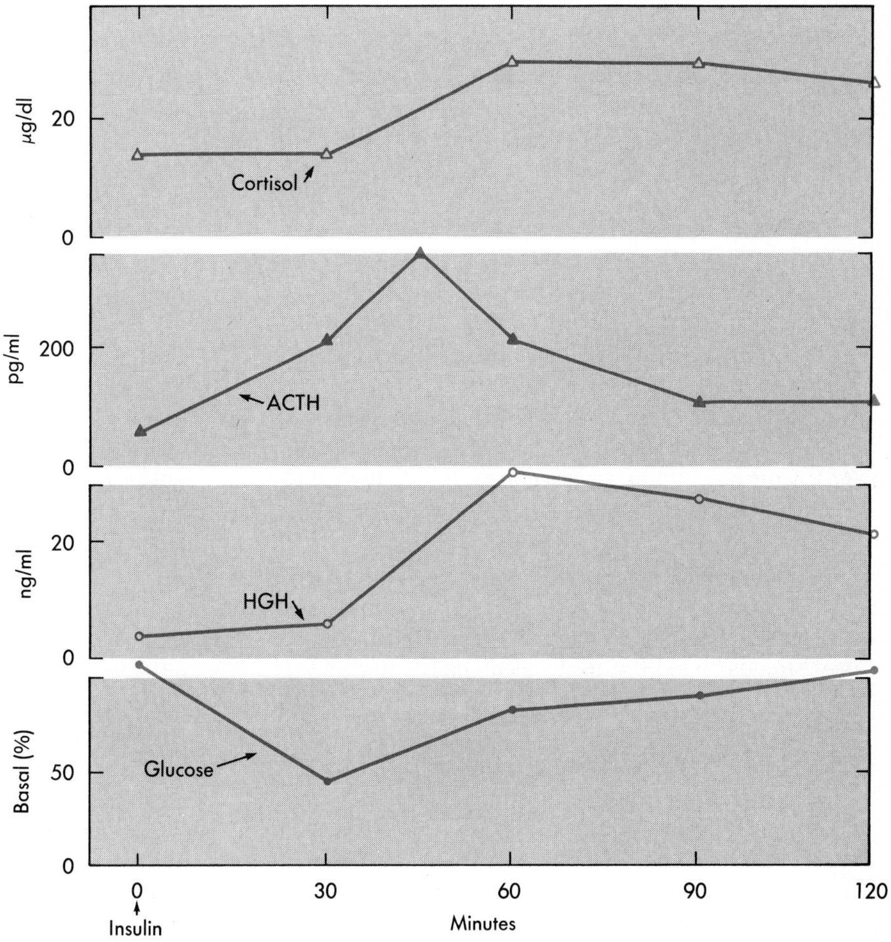

Fig. 52-13. Stimulation of ACTH, cortisol, and *HGH* secretion by insulin-induced hypoglycemia in humans. (Redrawn from Ichikawa, Y., et al.: Plasma corticotropin, cortisol and growth hormone responses to hypoglycemia in the morning and evening, J. Clin. Endocrinol. Metab. **34:**895, 1972. Reproduced by permission.)

Secretion and Actions of Other Proopiomelanocortin Peptides

As already noted, the remaining peptides in proopiomelanocortin are under the same transcriptional and translational control as ACTH. The functional significance of this fact is still not well understood, though it could reflect a coordinated physiological response to stress. A four-fold increase in the plasma level of β-lipotropin follows stimulation of the corticotroph by insulin hypoglycemia, paralleling the concurrent rise in plasma ACTH. β-Lipotropin is also suppressed by exogenous cortisol and is stimulated by acute or chronic cortisol deficiency, indicating a pattern of negative feedback control identical to that of ACTH. Although the synthetic sequence outlined previously predicts equimolar secretion of these two peptides, the molar ratio of β-lipotropin to ACTH in plasma is only 0.3. This suggests that β-lipotropin has a higher metabolic clearance rate. β-Lipotropin was originally discovered as a distinct lipolytic factor in the pituitary gland. Its physiological role in fatty acid mobilization from human adipose tissue requires further study. γ-Lipotropin also circulates in plasma, but its actions are unknown.

β-Endorphin is found in very low concentrations in human plasma. However, its level clearly increases in parallel with ACTH when the corticotroph is stimulated by either stress or cortisol lack. The function, or even the source, of circulating β-endorphin is unclear. Since it penetrates the blood-brain barrier poorly, it is unlikely that β-endorphin secreted by the pituitary contributes to the peptide present in the brain and responsible for analgesic activity. When β-endorphin or enkephalins are administered systemically in pharmacological amounts, they inhibit ACTH secretion and inhibit gonadotropin secretion. Conversely, they stimulate prolactin, insulin, and glucagon secretion and raise plasma glucose levels. These observations may reflect neurocrine actions of endogenous opioids that are generated within the hypothalamus or paracrine actions of opioids generated within the pancreatic islets. They may also explain some previously described effects of morphine.

The N-terminal peptide of proopiomelanocortin origin is also present in human plasma. Its function is uncertain, but preliminary studies suggest that this peptide may promote growth of the adrenal cortex.

Clinical syndromes of ACTH dysfunction. The pathological effects of a primary excess or deficiency of ACTH are essentially those due to increased or decreased secretion of adrenocortical hormones. These are described in Chapter 54. Hyperpigmentation of the skin characterizes those diseases in which large increases in ACTH secretion occur.

Gonadotropic Hormones

LH and FSH are glycoproteins whose function is to regulate the growth, pubertal maturation, reproductive processes, and sex steroid hormone secretion of the gonads of either sex. Both hormones are secreted by a single cell type, the gonadotroph, which forms about 10% to 15% of the anterior pituitary population and which is scattered throughout the gland. By immunohistology, the same cell generally stains for both FSH and LH, but occasional cells contain only one or the other gonadotropin. Granule diameters are 275 to 375 nm. Both hormones are present by 10 to 12 weeks of fetal life; however, neither is required for initial intrauterine gonadal development or for sexual differentiation.

LH, with a molecular weight of 28,000, and FSH, with a molecular weight of 33,000, have similar structures. Each is composed of the common pituitary hormone α-subunit (molecular weight, 14,000; 96 amino acids) and a unique β-subunit, which differentiates the two hormones from each other, as well as from TSH and human chorionic gonadotropin (HCG) (Fig. 52-14). The α- and β-subunits are held together by noncovalent forces. The carbohydrate moieties are about 15% by weight and contain oligosaccharides composed of mannose, galactose, fucose, galactosamine, and sialic acid. The carbohydrate groups seem to function in receptor attachment, whereas the sialic acid residues protect the hormone from rapid degradation in the circulation. Neither the β-subunit of LH nor that of FSH is biologically active by itself.

The details of LH and FSH biosynthesis are similar to those already described for TSH. Ribosomal assembly of the two peptide chains is followed by addition of the carbohydrate moieties in the endoplasmic reticulum and Golgi apparatus. In women, the pituitary stores of both LH and FSH fluctuate throughout the menstrual cycle, being highest just before ovulation (Chapter 55). Only a single releasable pool of FSH appears to exist; in contrast, the responses of LH to its releasing hormone strongly suggest the existence of a rapidly releasable pool and a slowly releasable pool. Whether these reflect two types of granules differing in quality, size, or proximity to the cell membrane is not apparent. Hormone secretion occurs by exocytosis.

Secretion of LH and FSH. The regulation of LH and FSH secretion is highly complex, embodying pulsatile, periodic, diurnal, and cyclic elements. It is also different in women and men. The main factors controlling gonadotropin secretion will be discussed in this chapter, and their reproductive function in both sexes and their relationship to the menstrual cycle will be reiterated and amplified in Chapter 55. Both the secretion of LH and that of FSH are stimulated primarily by a single hypothalamic hormone, *luteinizing hormone–releasing hormone (LHRH)*. As its name implies, it causes

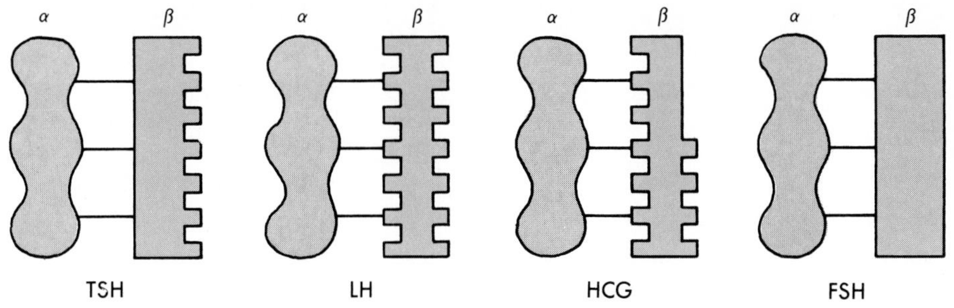

Fig. 52-14. Structural similarities among TSH, LH, HCG, and FSH are depicted schematically. Note all share the same α-subunit.

a much greater increase in LH than in FSH secretion, but because of the dual response, the term *gonadotropin-releasing hormone (GnRH)* is also sometimes used. Whether a separate hypothalamic releasing hormone exists that has a greater specificity for FSH remains uncertain. Human LHRH is a decapeptide (Table 52-1) that is synthesized from a large prohormone of which the gene has been cloned and other products identified. The cells of origin of LHRH are found predominantly in the arcuate nucleus and the preoptic area of the hypothalamus. Whether these two clusters have different functional roles is not yet known. After transport to the median eminence, LHRH is stored in small granules.

Detailed chemical and anatomical studies have shown that LHRH neurons are under the influence of dopaminergic, serotonergic, and noradrenergic input to the hypothalamus, as well as under endorphinergic influence. In particular, LHRH neurons are closely associated with dopamine neurons within the arcuate nucleus of the hypothalamus. Dopamine clearly inhibits LH secretion, both indirectly by decreasing LHRH release and directly by an action on the gonadotroph. Since administration of opioid antagonists increases LH secretion and in vitro endorphins decrease LHRH release, endorphins also inhibit LH secretion. Norepinephrine input to LHRH neurons is probably stimulatory in humans. Much clinical experience indicates that neurotransmitters regulate human gonadotropin secretion. Neural input from the retina to the hypothalamus probably accounts for the influence of light-dark cycles on gonadotropin secretion. Connections from the olfactory bulb probably transfer reproductive signals from another individual. These signals are transmitted via *pheromones,* which are airborne or waterborne chemical exciters or inhibitors. Menstrual function in women and sperm production in men is commonly lost during prolonged physical or psychic stress. This may be mediated by CRH, which inhibits LH release by augmenting endorphinergic tone.

LHRH binds to specific plasma membrane receptors in the gonadotroph and operates, through calcium, phospholipids, and cyclic AMP as second messengers. In humans, an infusion of LHRH causes a biphasic response in plasma LH; the initial peak is reached at 30 minutes, followed by a secondary rise beginning at 90 minutes and continuing for hours thereafter (Fig. 52-15). Since LHRH also stimulates synthesis of LH, this action may account for the second phase of secretion. In contrast, LHRH infusion causes only a uniphasic progressive rise in FSH (Fig. 52-15). In normal women, LH is spontaneously secreted in pulses. These are characterized by a 15-minute upsurge and a falloff with a half-life of 60 minutes. Thus peaks of the plasma LH level are produced with periodicity, varying from 1 to 7 hours, depending on the phase of the menstrual cycle. The amplitude of the pulses can be equivalent to 100% changes in the plasma LH level, except at the time of ovulation, when it may be significantly greater. Normal men also exhibit 8 to 10 secretory bursts of LH per day (Fig. 52-16).

Much evidence indicates that the pulsatile secretion of LH is due primarily to pulsatile secretion of LHRH into the portal veins of the anterior pituitary gland rather than to a rapidly fluctuating sensitivity of the gonadotrophs to the releasing hormone (Fig. 52-17). Moreover, pulsatility does not depend on the presence of sex steroid hormones from target glands, since agonadal individuals and postmenopausal women exhibit, if anything, even sharper spikes of the plasma LH level. Pulsatile secretion of LH is not found in young children but it makes its appearance just before puberty, at first occurring only at night. During the initial stages of puberty, this produces a nocturnal peak of LH. Although this diurnal pattern lasts only 1 or 2 years, disappearing as puberty is completed, the pulsatility of LH secretion becomes fixed. The most striking feature of LH secretion in women, as opposed to men, is its monthly cyclicity. This results from a complex interaction between the LHRH neuron–gonadotroph unit and sequential changes in ovarian steroid secretion.

FSH secretion also exhibits a pulsatile pattern, usually synchronized with LH, but of lesser magnitude (Fig. 52-16). The possibility of a separate hypothalamic releasing hormone for FSH has been suggested by the

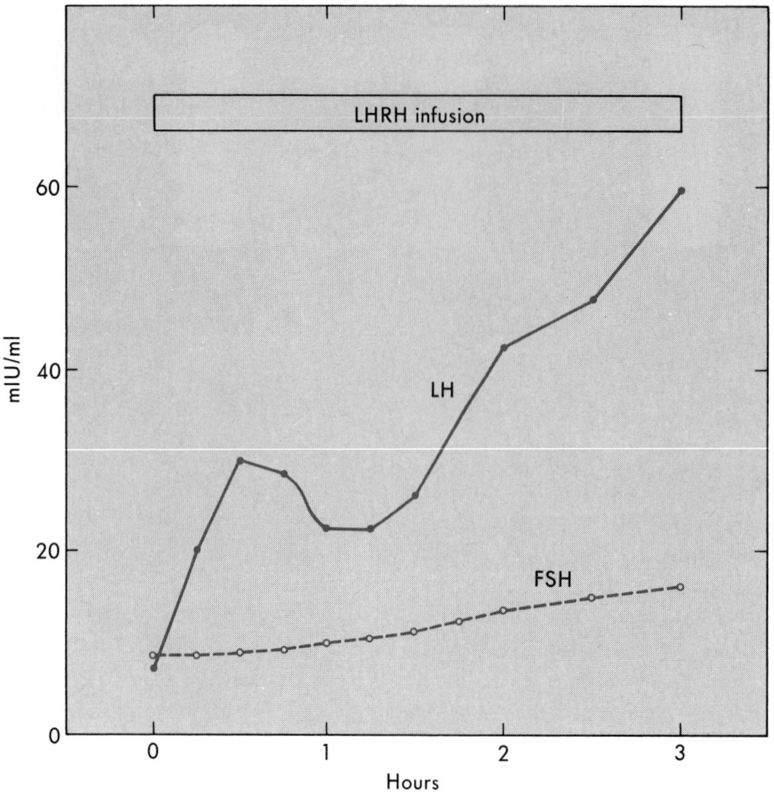

■ **Fig. 52-15.** Stimulation of gonadotropin release by LHRH. Note biphasic response of LH and uniphasic response of FSH. (Redrawn from Wang, C.F., et al.: The functional changes of the pituitary gonadotrophs during the menstrual cycle, J. Clin. Endocrinol. Metab. **42:**718, 1976. Copyright 1976. Reproduced by permission.)

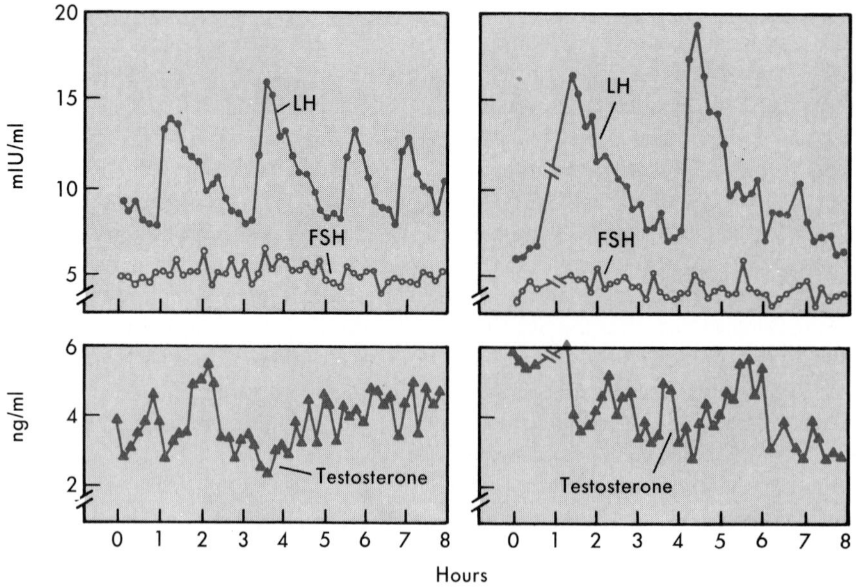

■ **Fig. 52-16.** Pulsatile fluctuations in plasma LH levels and its target hormone, testosterone, in men. (Redrawn from Naftolin, F., et al.: Pulsatile patterns of gonadotropins and testosterone in man: the effects of clomiphene with and without testosterone, J. Clin. Endocrinol. Metab. **36:**285, 1973. Copyright 1973. Reproduced by permission.)

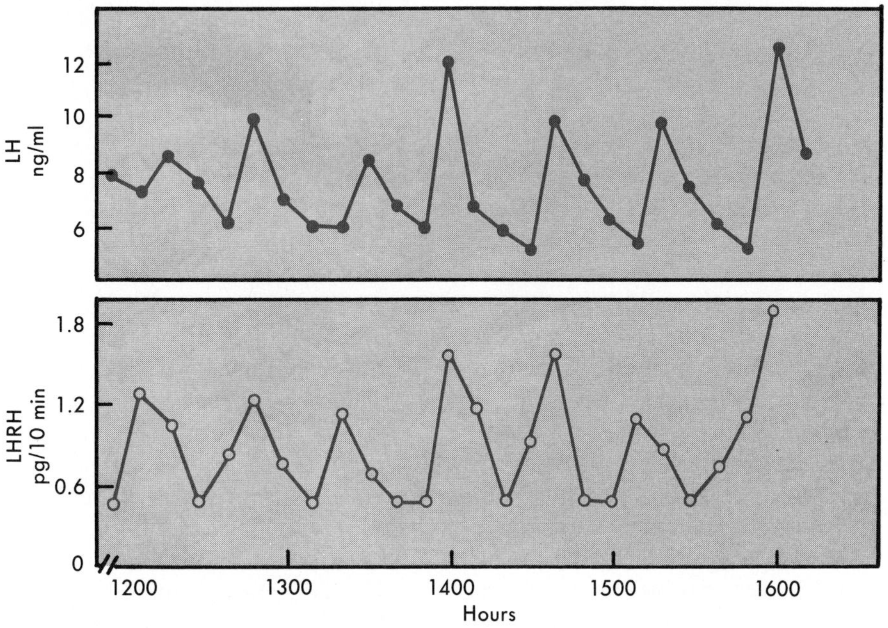

■ **Fig. 52-17.** Fluctuation of peripheral vein plasma LH levels and portal vein plasma LHRH levels in unanesthetized ovariectomized female sheep. Each pulse of LH is coordinated with a pulse of LHRH. This supports the view that pulsatility of LH release is dependent on pulsatile stimulation of the pituitary by LHRH. (From Levine, J., et al.: Endocrinology **111:**1449, 1982.)

fact that the ratio of FSH to LH levels in plasma fluctuates. It is much less than one during the ovulatory gonadotropin spike, but it is greater than one in other phases of the menstrual cycle and in functionally agonadal individuals of either sex. These changes have not been consistently or completely reproduced simply by sex steroid modulation of pituitary responsiveness to LHRH. However, the frequency of LHRH pulses may be a factor, since a decreased frequency is associated with an increased FSH/LH ratio.

Feedback regulation of gonadotropins. The secretion of both LH and FSH are regulated by gonadal products. However, the patterns and mechanisms are more complex than those which have thus far been described for TSH and ACTH. It is best to consider first the basic framework, which is of the classic negative feedback type. The existence of negative feedback is proven by the observation that plasma levels of FSH and LH are both elevated by surgical or functional removal of the gonads in either humans or experimental animals. FSH, however, is usually increased proportionally more than LH. A number of gonadal products from at least two gonadal cell types normally act to restrain the secretion of each gonadotropin. The basic schema is depicted in Fig. 52-18.

The major androgen, testosterone, from the Leydig cells of the testis and the interstitial cells of the ovary, inhibits the release of LH. It does so more by decreasing the frequency of the LH pulses than their amplitude. The major estrogen, estradiol, which arises from the

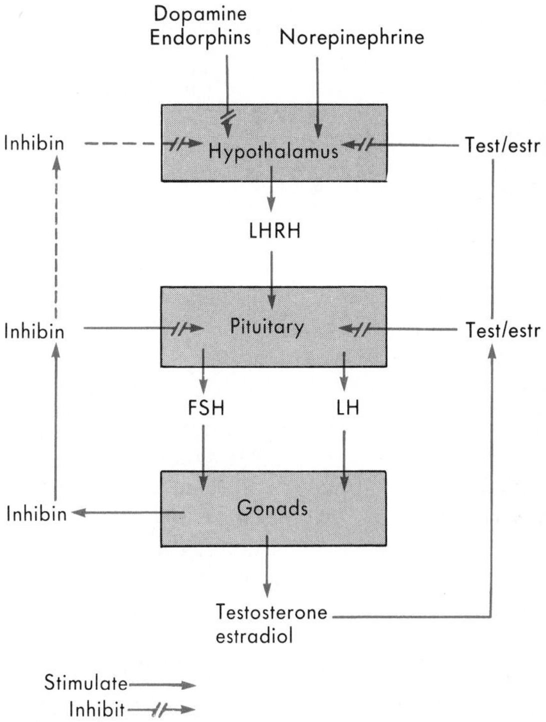

■ **Fig. 52-18.** Regulation of LH and FSH secretion. The gonadal steroids estradiol in women and testosterone in men exert negative feedback (1) at the pituitary level by blocking LHRH stimulation of LH and FSH secretion and (2) at the hypothalamus level by inhibiting LHRH release. A separate gonadal product, inhibin, feeds back to selectively suppress FSH release. Negative modulation by endorphins and dopamine and positive modulation by norepinephrine are also important in regulating LH and FSH secretion.

granulosa cells of the ovary and the Sertoli cells of the testis as well as by conversion from testosterone in peripheral tissues, also inhibits the release of LH. Estradiol, however, decreases the amplitude of the LH pulses more than their frequency. Both estradiol and testosterone administration blunt the response of the gonadotroph to acute administration of LHRH. Conversely, in estradiol-deficient women and in testosterone-deficient men, LH responses to LHRH are exaggerated. In addition to inhibiting release of LH (and FSH), estradiol decreases their synthesis by lowering the levels of messenger RNA for the common α subunit and for the specific β subunits of the gonadotropins. Some of these actions of estradiol may be mediated by altering the number of LHRH receptors on the gonadotroph. In addition to these pituitary effects, estradiol and testosterone also act on the hypothalamus to decrease LHRH secretion. The sex steroids probably do not interact directly with LHRH neurons, but rather with endorphin neurons in the hypothalamus. The latter then complete this pathway of negative feedback by suppressing discharge of LHRH from the median eminence into the portal blood.

FSH secretion is also inhibited by estradiol and testosterone, which act to block the pituitary response to LHRH. However, feedback inhibition of FSH secretion is specifically carried out by another gonadal product, a glycoprotein called *inhibin*. This substance was originally discovered in fluid from ovarian follicles and testicular seminiferous tubules and has been detected in cultures of testicular Sertoli cells. Purified preparations of inhibin reduce the synthesis and basal secretion of FSH in cultured pituitary cells and diminish their response to LHRH stimulation. In contrast, inhibin has little effect on the same features of LH secretion. Preliminary reports also indicate that inhibin reduces LHRH secretion by hypothalamic cell cultures.

Engrafted on this negative feedback framework are striking *positive feedback* effects of sex steroids. This is most clearly seen with estradiol in women. When estradiol is administered in an appropriate dose range and for a sufficient number of days, LH response to LHRH is *augmented* rather than reduced. Furthermore, if LHRH is administered repetitively to properly estradiol-primed women, both the response to the first LHRH pulse and the cumulative increments following the subsequent LHRH pulses are amplified. This has been interpreted to mean that both the sensitivity of the gonadotroph and its reserve capacity (LH stores) have been enhanced by estradiol treatment. Moreover, in women so treated, a rapid further increase in estradiol itself causes a significant rise in plasma levels of LH. Aspects ...sitive feedback and negative feedback can be ob- ...ultaneously. This occurs when estradiol-de- ...l women are given initial estradiol re- ...erapy. After 7 days of treatment, the originally elevated basal levels of LH (and FSH) decline (negative feedback) yet the capacity to respond to subsequent repetitive doses of LHRH actually increases (positive feedback).

Progesterone, another major steroid product of the ovary, also modulates LH release. Progesterone, administered acutely, can increase plasma LH levels 24 to 48 hours later. Progesterone can also either enhance or blunt the positive feedback effects of estradiol on LHRH responsiveness, depending on the timing with which the two hormones are administered. However, continuous administration of progesterone (or analogs) inhibits gonadotropin secretion. This is one basic mechanism of action of the oral contraceptives. All of the complex interactions between gonadal steroids and gonadotropin secretion help to shape the typical patterns observed throughout the menstrual cycle.

Prolactin, a mammotropic hormone from the anterior pituitary, also inhibits LHRH release and lowers basal secretion of LH and FSH. Although the physiological significance of the effect is not well understood, it helps explain many cases of infertility and pathological loss of menses. Finally, animal experiments suggest that LH can inhibit secretion of its own releasing hormone, LHRH, via short-loop negative feedback. The route of access of LH to the hypothalamic LHRH neuron may be retrograde flow in the pituitary portal veins or via the tanycytes.

LH and FSH both circulate unbound to plasma proteins. The concentrations of both are in the range of 5 to 25 mIU/ml in men and in reproductive-age women. In the latter, the levels of both hormones are higher in the first half of the menstrual cycle than in the second half; in addition, both hormones show sharp, single-day peaks at the time of ovulation. Basal plasma concentrations of each hormone are of the order of 10^{-11}M. The metabolic clearance rates of LH and FSH, respectively, are 36 and 20 L/day, and their half-lives in plasma are approximately 1 and 3 hours. Estimates of daily secretory rates are 1100 IU for LH and 200 IU for FSH except on the day of ovulation. Metabolism of gonadotropins largely occurs in the liver and kidney. In contrast to the trivial excretion of other peptide hormones, 10% of the daily production of LH appears in the urine. This permits employment of urinary LH measurements as a reflection of integrated plasma concentrations. Such measurements are particularly useful when plasma levels are low, as in children. The α-subunit common to LH and FSH is secreted separately and circulates in most normal individuals. In contrast, the specific β-subunits of LH and FSH are secreted only by hyperstimulated gonadotrophs or by neoplastic cells.

Actions of gonadotropins. Both LH and FSH bind to plasma membrane receptors. At least part of the gonadotropins' actions are mediated by activation of adenylate cyclase with cAMP as the second messenger. The latter activates a protein kinase that is important in one or more steps of steroid hormone synthesis. Each gonadotropin has specific primary target cells in the gonads of women and men. Granulosa cells in the ovary and Sertoli cells in the testis are stimulated by FSH. Interstitial cells in the ovary and Leydig cells in the testis are stimulated by LH. In addition, in the ovary, LH also acts on FSH-primed granulosa cells and on luteal cells in the ovary. The major effect of FSH on the testis is to stimulate spermatogenesis. Its major effect on the ovary is to stimulate follicle development and estradiol synthesis and to induce responsiveness to LH. The major effect of LH in the testis is to stimulate testosterone (and estradiol) synthesis. In the ovary, LH stimulates estradiol, progesterone, and testosterone synthesis and ovulation. The relationship between the specificity and timing of each gonadotropin's actions on ovarian and testicular function is complex and will be discussed in the sections on the gonads.

Abnormalities in secretion. Disorders (usually neoplasms) of the hypothalamus or pituitary gland may produce deficiency of one or both gonadotropins. With rare exceptions, this leads to a loss of reproductive capacity in adults and can cause some regression of already established secondary sexual characteristics. If gonadotropin deficiency occurs before the onset of puberty, the development of secondary sexual characteristics, the expected rapid growth spurt, and skeletal maturation are all prevented. Slow growth may continue for a long time, producing the eunuchoidal habitus, that is, a tall, juvenile-appearing adult. A primary excess of FSH or LH secretion is exceedingly rare, and a unique clinical picture cannot be described. In some women a persistently high ratio of LH to FSH secretion exists. This is associated with noncycling and elevated estradiol levels, increased androgen secretion, loss of ovulation, infrequent or absent menses, and multiple cysts of the ovary. This condition, known as the *polycystic ovary syndrome,* is a common cause of infertility.

Growth Hormone (Somatotropin)

Growth hormone stimulates postnatal somatic growth and development. In addition, it has numerous actions on protein, carbohydrate, and fat metabolism. The hormone originates in anterior pituitary cells that make up 40% to 50% of the cell population of the gland in adult humans. It is stored in large dense granules, 350 to 500 nm in diameter. Typically, somatotrophs stain with acidophilic dyes, such as eosin. In humans, these cells may form tumors that hypersecrete growth hormone and produce a highly distinctive disease called *acromegaly.*

Human growth hormone (HGH) is a single-chain polypeptide with a molecular weight of 22,000. It contains 191 amino acids and 2 disulfide bridges (Fig. 52-19). Neither the three-dimensional structure nor the exact sites or sequences essential for biological activity are known, but considerable activity is retained by products of tryptic digestion. HGH is active in many animals, but only primate growth hormones are active in humans. Until recently, this required that HGH–deficient individuals be treated with hormone extracted from human pituitaries, thereby limiting the available therapeutic supply. The development of recombinant DNA technology has now led to the production of HGH in bacteria.

Synthesis and Release of HGH

The synthesis of growth hormone is characteristic of secretory proteins. The human genome contains multiple genes coding for a family of closely related HGH molecules. Only one of these genes is expressed as normal HGH, though variant molecules could be expressed in selected circumstances or in disease. The gene transcribes a messenger RNA that directs synthesis of a prehormone. Subsequently, a signal peptide is removed and the hormone in final form is stored in granules. The synthesis of HGH is increased by its specific hypothalamic releasing hormone, GHRH (Table 52-1). The latter causes rapid stimulation of gene transcription. Thyroid hormone and cortisol individually and synergistically also induce HGH synthesis by increasing levels of the messenger RNA.

The release of HGH requires activation of a microtubular system that brings the granules to the plasma membrane prior to exocytosis. Release is stimulated by GHRH, after the latter binds to a plasma membrane receptor. GHRH causes activation of adenylate cyclase with increase in cAMP levels, influx of calcium, and increased membrane phosphatidylinositol turnover. The exact participation of each of these mediators in th
release mechanisms remains to be worked out, but t
role of calcium is essential. Prostaglandins are also
tent stimulators of HGH release in vitro.

Another hypothalamic peptide, somatostatin (
52-1), is a powerful inhibitor of HGH release.
tostatin blocks GHRH stimulation in a nonco
manner. In addition to somatostatin-14, the
extended form, known as somatostatin-28,
iologically active, but it exhibits different
characteristics. Somatostatin acts thr
plasma membrane receptor, in part by
cellular cAMP and calcium influx.
creted in pulses, which are due to
of GHRH into the portal blood.
suppress this pulsatility, but di
which the somatotroph is able
pulses.

Secretion of HGH. As

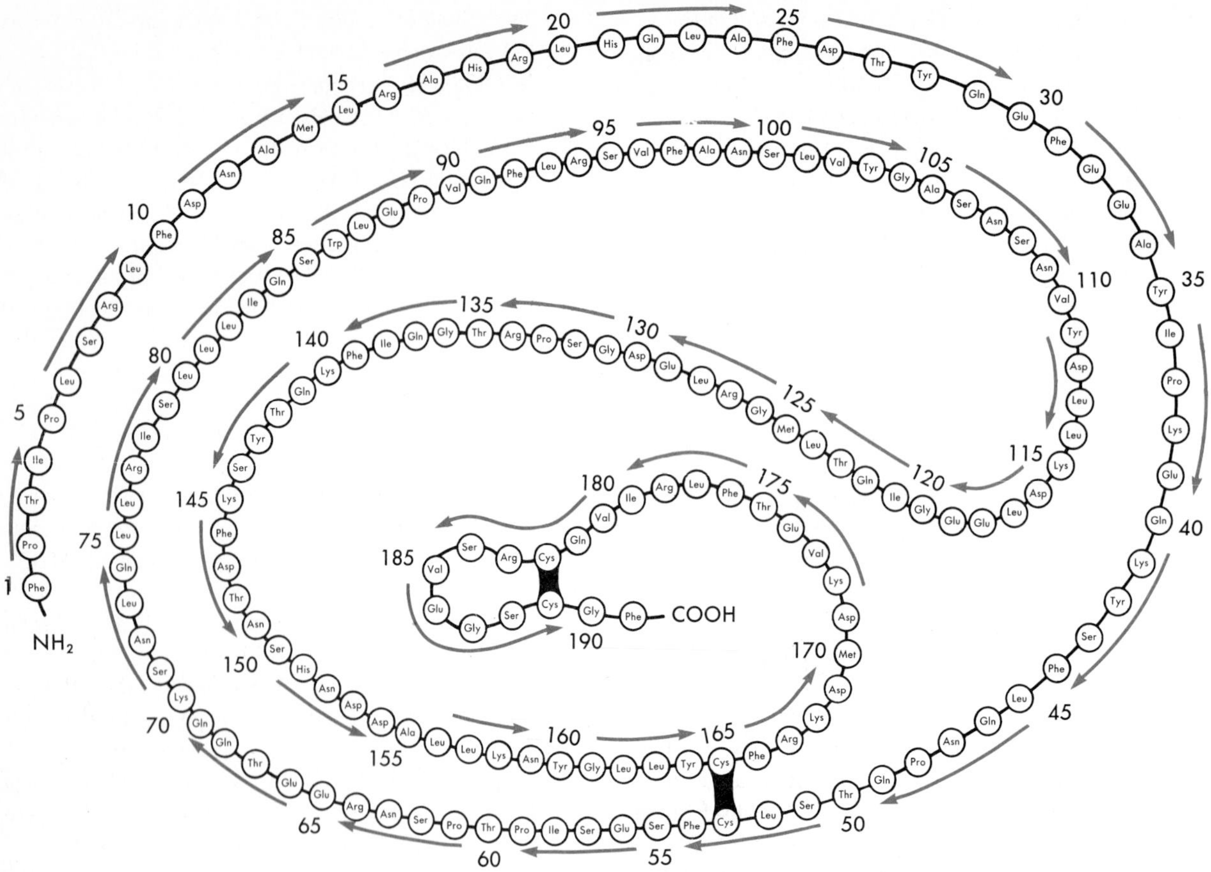

■ **Fig. 52-19.** Structure of human growth hormone. Though consisting of nearly 200 amino acids and containing two disulfide bridges, growth hormone is a single polypeptide chain (From Li, C., et al.: Proc. Natl. Acad. Sci. **74:**1016, 1977.)

cretion is under many different influences. An acute fall in plasma levels of either of the major energy-yielding substrates, glucose or free fatty acids, produces an increase in the plasma level of HGH. For example, when ...sulin is administered intravenously, the plasma level ...GH rises two-fold to ten-fold 30 to 60 minutes after ...ma glucose level has declined to below 50 mg/ ...2-11). Conversely, a carbohydrate-rich meal ...ucose load causes a prompt decrease in the H...evel of at least 50%. Responses to alter- so...ty acid levels are generally slower and ...cose and free fatty acids suppress amin... by reducing the response of the the m... prolong... the infusion of a mixture of also stim...ma HGH level; arginine is mechanis...id stimulator. However, and variou... thesia, feve...vation or total fasting HGH secreti...as yet undetermined ...ning and cycling, ...drawing, anes- ..., all increase ...spikes in the

■ **Table 52-4.** Regulation of growth hormone secretion

Stimulation	*Inhibition*
Glucose decrease	Glucose increase
Free fatty acid decrease	Free fatty acid increase
Amino acid increase (arginine)	Cortisol
Fasting	Obesity
Prolonged caloric deprivation	Pregnancy
Stage IV sleep	Somatostatin
Exercise	Growth hormone
Stress (Table 3)	
Estrogens	
Dopamine	
Serotonin	
Alpha adrenergic agonists	
Gamma amino butyric acid	
Enkephalins	

plasma levels of HGH produced by all of these factors, a regular nocturnal peak occurs 1 or 2 hours after the onset of deep sleep (Fig. 52-10). This correlates with stage 3 or stage 4 sleep, as indicated by the electroencephalogram.

The neurotransmitters dopamine, norepinephrine, and

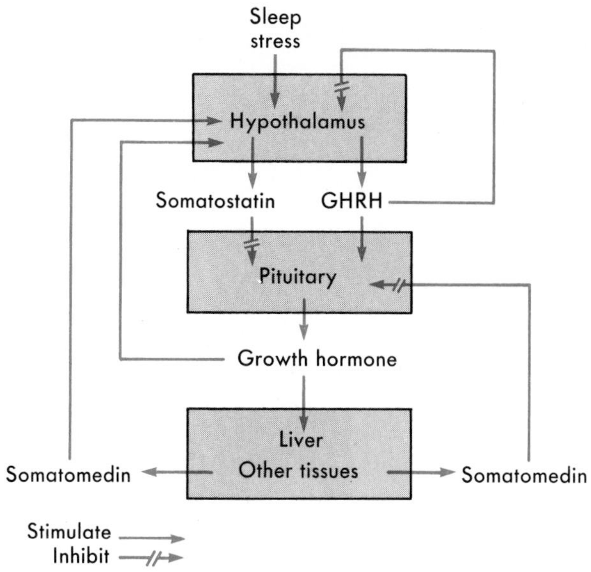

■ **Fig. 52-20.** Regulation of human growth hormone (HGH) secretion. The hypothalamic peptide (GRH) stimulates growth hormone release, whereas the hypothalamic peptide somatostatin inhibits it. Negative feedback is by the peripheral mediator of HGH action: somatomedin. Negative feedback occurs both via somatomedin *inhibition* of GRH action and by somatomedin *stimulation* of somatostatin release. HGH inhibits its own secretion by short-loop feedback. In addition, GRH inhibits its own release via ultra short-loop feedback. In both of these cases, the negative feedback is probably via increasing somatostatin release.

serotonin all increase HGH secretion. They may act at the hypothalamic or median eminence level by stimulating GHRH. The HGH responses to exercise, stress, hypoglycemia, and arginine are reduced by α-adrenergic blockade and are augmented by β-adrenergic blockade. This suggests that these responses are facilitated by central α-adrenergic receptors and inhibited by central β-adrenergic receptors in hypothalamic neurons. In contrast, the sleep-induced rise in HGH is unaffected by the adrenergic system and is more likely stimulated by serotoninergic pathways from the brainstem. The latter may also modulate the response to hypoglycemia. The physiological role of dopamine pathways is still unclear. HGH responsiveness is greater in women than in men and is greatest just before ovulation. This is explained by an augmenting effect of estrogens on HGH secretion. Daily HGH secretion is somewhat greater in childhood than in adult life.

Feedback regulation of HGH is also complex (Fig. 52-20). For example, HGH inhibits its own secretion. Administration of exogenous HGH dampens subsequent endogenous HGH responsiveness to a number of stimuli, such as hypoglycemia and stage IV sleep. The mechanisms may involve a short-loop feedback, inasmuch as HGH stimulates the synthesis and release of somatostatin in vitro (Fig. 52-20). However, somatostatin synthesis and release may also be increased by somatomedins, peptide mediators of HGH action that are generated in the periphery, as described later. Somatomedins also can act at the pituitary level to decrease responsiveness to GHRH (Fig. 52-20). Finally, recent evidence suggests that GHRH itself, in doses too small to stimulate the somatotrophs but administered in a manner that allows access to the hypothalamus, will decrease rather than increase HGH secretion (Fig. 52-21). This paradoxical effect likely results from a stimulation of somatostatin release and may reflect the existence of axonal connections between GHRH neurons and somatostatin neurons.

Other negative regulatory influences are also noted. Cortisol decreases HGH secretion, an action that may contribute modestly to the steroid's adverse effect on growth. An unexplained decline in HGH secretion occurs during the latter part of pregnancy, despite the presence of high estrogen levels. Finally, obese animals and humans exhibit dampened HGH responses to all stimuli, including GHRH itself. These responses are increased by a return to normal weight.

The normal basal plasma HGH concentration is 1 to 5 ng/ml (about 10^{-10}M). This may increase as much as ten-fold to fifty-fold under various stimuli. The plasma half-life of HGH is 20 minutes and the metabolic clearance rate is 350 L/day. Daily secretion in normal adults is approximately 500 μg, or 5% of the average pituitary content. A minor portion of the circulating immunoreactive HGH consists of larger forms whose biological activity is unknown. A specific HGH binding protein has also been described.

HGH actions. HGH is a hormone with profound anabolic action. In its absence, animals and humans show stunted growth. When it is administered to hypophysectomized animals, it causes nitrogen retention, hypoaminoacidemia, and decreased urea production because the amino acids are diverted from oxidation to protein synthesis as growth ensues.

The multiplicity of HGH targets and effects is indicated in Fig. 52-22. The most striking and specific effect is the stimulation of linear growth that results from HGH action on the epiphyseal cartilage plates of long bones. All aspects of the metabolism of the cartilage-forming cells, the chondrocytes, are stimulated. This includes the incorporation of proline into collagen and its conversion to hydroxyproline and the incorporation of sulfate into the proteoglycan chondroitin, which, together with collagen, forms the resilient extracellular matrix of cartilage. In addition, HGH stimulates the general synthesis of proteins, RNA and DNA in chondrocytes, as well as their proliferation. In support of the accelerated rate of protein synthesis, HGH also stimulates cellular uptake of amino acids. Other tissues share in the anabolic response to HGH. The width of bones increases as a result of enhanced growth at the periosteal surface. Visceral organs (liver, kidney, pancreas, intestines), endocrine glands (adrenals, parathyroids,

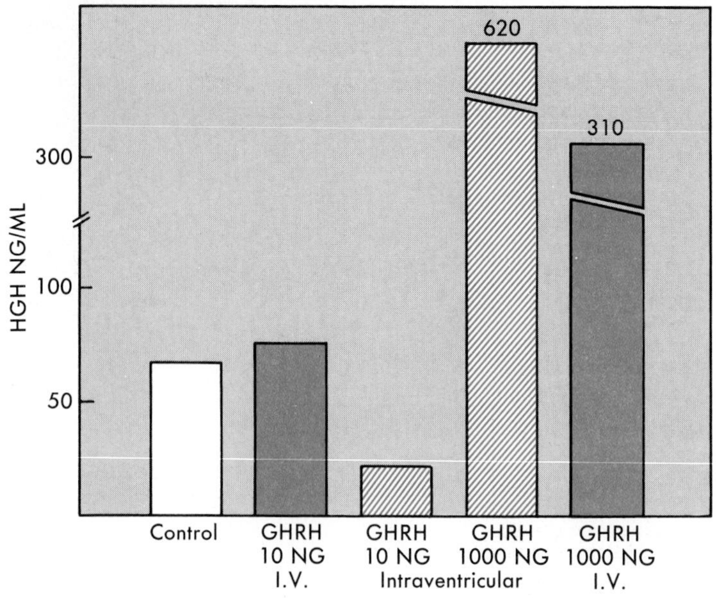

■ Fig. 52-21. Effect of GHRH administration on plasma HGH in the rat. If a very small 10 ng dose of GHRH is administered intravenously, there is no stimulation of HGH release *(second bar from left).* If the same minute dose is administered directly into the ventricle of the rat with subsequent access to the hypothalamus, this dose paradoxically inhibits HGH release *(third bar from left).* When a large 1000 ng dose of GHRH is administered intravenously, it stimulates HGH release as expected *(fifth bar from left).* The same large dose administered into the ventricle also stimulates HGH release because a sufficient quantitity of GHRH reaches the pituitary. The paradoxical inhibition of HGH release by GHRH can be considered a form of ultra short–loop feedback, which probably reflects interaction between GHRH and somatostatin neurons. (Redrawn from Katakami, A., and Frohman, L.: Dual action of growth-hormone releasing factor on spontaneous growth hormone secretion in conscious male rats. Abstracts of the 7th International Congress of Endocrinology, Excerpta Medica, International Congress Series **652:**876, 1984.)

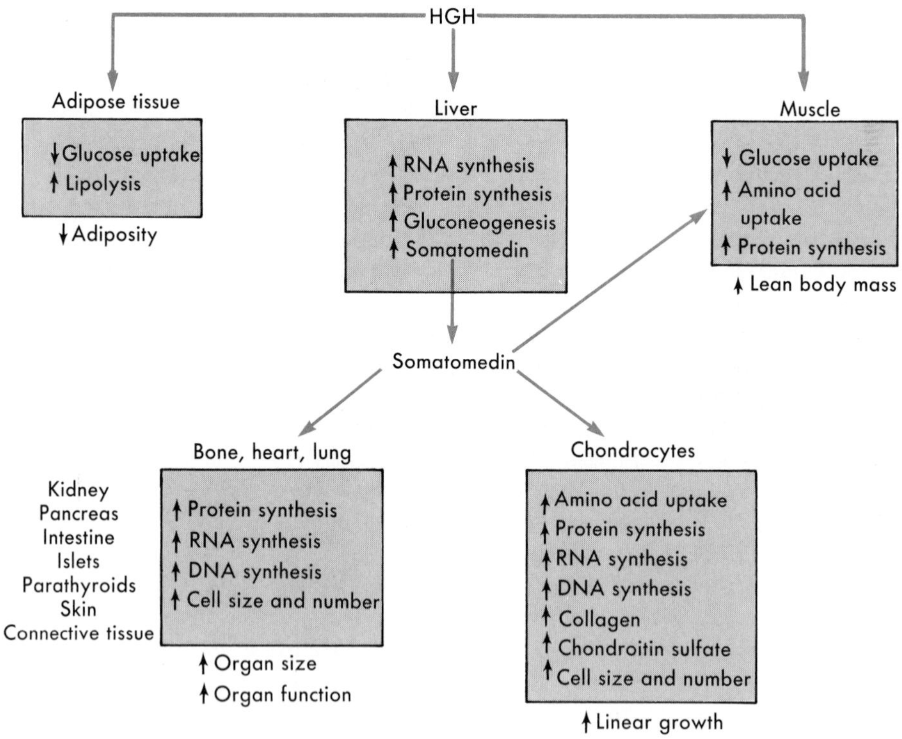

■ Fig. 52-22. Biological actions of HGH. The effects on linear growth, organ size, and lean body mass are mediated by somatomedin produced in the liver.

pancreatic islets), skeletal muscle, heart, skin, and connective tissue all undergo hypertrophy and hyperplasia in response to HGH. In most instances, this is reflected in an enhanced functional capacity of the enlarged organ. For example, glomerular filtration, cardiac output, and hepatic clearance of test substances are all increased by HGH.

The actions of HGH on carbohydrate and lipid metabolism are bimodal. As noted previously, HGH has a tropic effect on the pancreatic islets; in its absence, insulin secretion in response to β-cell stimulation declines. However, the predominant effect of a prolonged growth hormone *excess* is to increase plasma glucose levels, despite a compensatory increase in insulin secretion. This occurs because HGH induces resistance to the action of insulin at a postreceptor site, inhibiting glucose uptake by muscle and adipose cells and stimulating hepatic gluconeogenesis. HGH also enhances lipolysis (in the presence of cortisol) and antagonizes insulin-stimulated lipogenesis. These actions lead to increased plasma free fatty acid levels and to a generalized decrease in adipose tissue. If insulin secretion is deficient, HGH can even cause ketosis. Thus on balance, HGH is a classic diabetogenic hormone in animals and humans.

HGH increases intestinal calcium absorption, urinary calcium excretion, and urinary hydroxyproline excretion. Sodium retention by the kidney is increased, an action that leads to an expansion of fluid volume in states of growth hormone excess. HGH has sufficient structural overlap with prolactin to give it definite mammotropic activity and may be involved in spermatogenesis.

Mechanisms of HGH action. The mechanism of HGH action on cells is incompletely understood. A variety of specific plasma membrane HGH receptors has been demonstrated in different target tissues, but thus far no intracellular second messenger has been clearly identified. Earlier observations suggested that the anabolic, growth-promoting effects of HGH required an intermediary. For example, HGH seemed to have little or no direct effect on cartilage in vitro. When the hormone was administered in vivo, a time lapse of 12 hours was necessary before its action could be demonstrated. However, at that time, the plasma harvested from the HGH-treated animal did stimulate cartilage metabolism in vitro.

Many of the activities of HGH require the prior generation of one or more of a family of peptides known as *somatomedins*. These compounds have a molecular weight of about 7000 and bear considerable structural resemblance to proinsulin. In fact, the term *insulin growth factors* (IGF) has been applied to them. IGF-1 is a 70–amino acid straight chain peptide with 50% homology to the A-chain and B-chain domains of proinsulin (see Fig. 50-10). IGF-2 has 70% homology with IGF-1 in these same domains. Both somatomedins, however, have distinctive C-chain domains. Circulating somatomedins originate primarily, if not solely, in the liver. Several hours after HGH administration to the whole animal, somatomedin release by the perfused liver increases. In contrast to the fluctuation of plasma levels of HGH, somatomedin concentrations are relatively stable. For example, they do not change significantly when the plasma glucose level is varied, whereas HGH does. In general, plasma concentrations of IGF-2 are three-fold to four-fold higher than those of IGF-1. The plasma half-life of somatomedins is longer than that of HGH. This is due in part to the fact that the somatomedins circulate bound to a large carrier protein. This protein is also synthesized in the liver and appears to be under HGH regulation.

At present, the growth-promoting effects of HGH can be largely accounted for by the somatomedins, which have been shown to stimulate typical HGH responses in cartilage, muscle, adipose tissue, fibroblasts, and tumor cells in vitro. Some evidence also suggests that somatomedin molecules generated locally within clones of cells recruited by HGH in its target tissues may act in autocrine and paracrine fashion, in addition to or in preference to somatomedin molecules derived from the plasma. Somatomedins bind to specific plasma membrane receptors in responsive tissues. The receptor for IGF-1 is structurally similar to the insulin receptor (see Fig. 50-16). It binds both insulin and IGF-2, though with lower affinities. The receptor for IGF-2 is dissimilar to both those of IGF-1 and those of insulin; it binds IGF-1 with lower affinity and insulin not at all. These cross-reactivities may assume biological importance when very high concentrations of either somatomedins or of insulin exist. For example, some patients with tumors that secrete somatomedins develop spontaneous hypoglycemia, presumably because of insulin receptor activation by the somatomedins. Other patients with insulin receptor deficiency who have (in compensation) extremely high plasma insulin levels develop excessive soft tissue growth, presumably because of somatomedin receptor activation by insulin. As is the case with insulin, the roles of cAMP and cGMP as intracellular mediators of somatomedin action are still obscure, and no other second messengers have been definitely identified.

Plasma somatomedins are increased by administration of HGH, with a time lag of 12 to 18 hours, and they disappear from HGH-deficient animals or humans. IGF-1 is clearly HGH dependent, and its plasma levels are very sensitive to changes in HGH availability. IGF-2 levels do not reflect HGH status nearly as well. During adolescence, levels of GHRH and of HGH increase, in turn producing an increase in plasma levels of IGF-1. The levels of IGF-1 correlate well with the progression of pubertal growth. Although HGH itself is not necessary for fetal growth, one or more of the somatomedins may be. Presumably, their production in utero is stimulated by non-HGH factors. Somatomedin production is reduced by factors that can override HGH. Fasting, protein deprivation, and insulin deficiency all lead to diminished liver production of somatomedins and to a decrease in their plasma levels, despite increases in growth hormone secretion. Indeed, in these pathophysiological states, the lack of somatomedin may be the cause of the elevated HGH levels through negative feedback. Estrogens also decrease somatomedin production; this may account for their antagonism to HGH action despite their stimulation of growth hormone secretion. Cortisol also diminishes somatomedin levels.

Overall role of HGH in substrate flow. It is useful to review the interactions between HGH and insulin in common physiological circumstances, as presented in Fig. 52-23. When protein and energy intake are ample, the absorbed amino acids are used for protein synthesis and to stimulate growth. Hence, both HGH and insulin

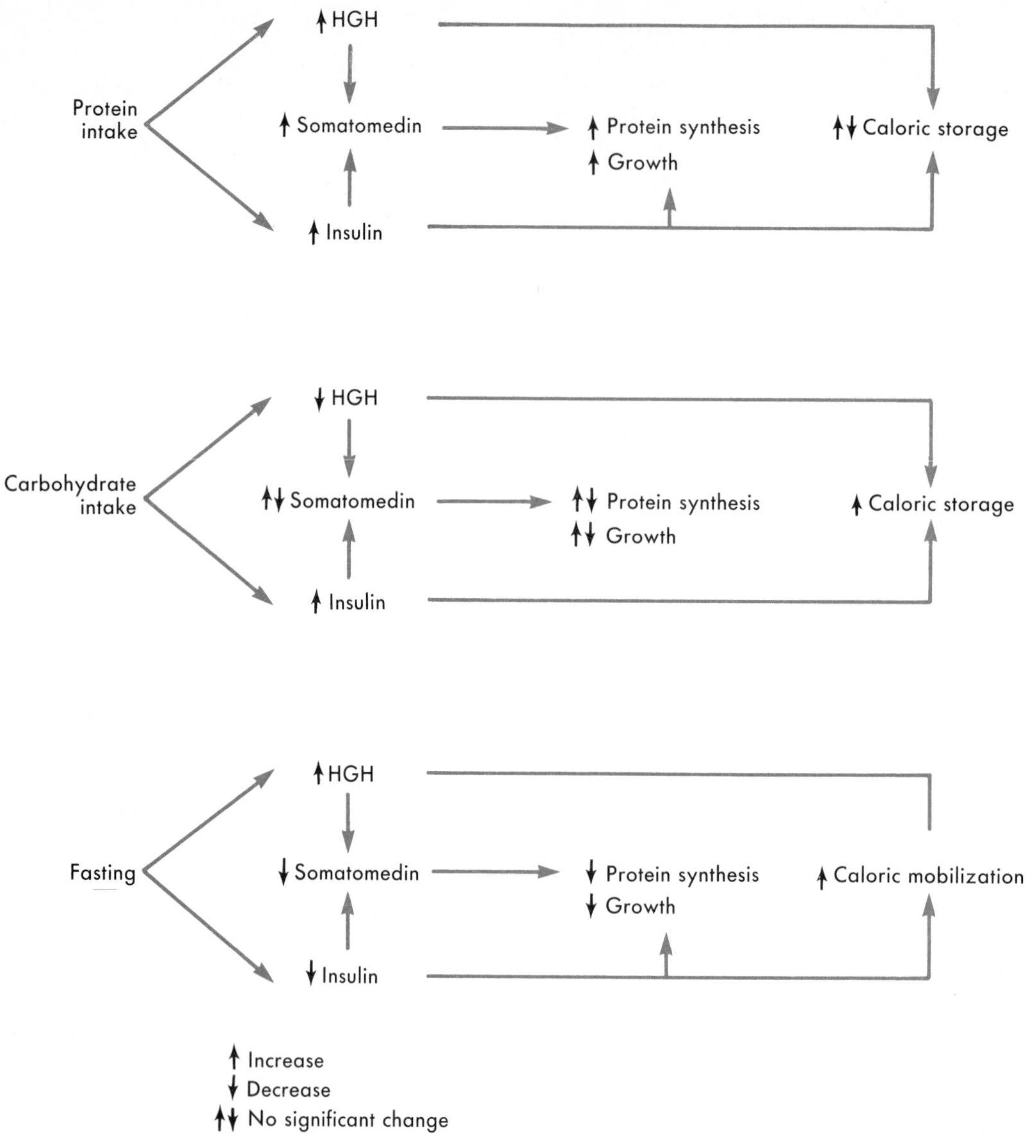

■ **Fig. 52-23.** Complementary regulation of HGH and insulin secretion coordinate nutrient availability with anabolism and caloric flux. Note that both hormones are increased by protein, and both stimulate protein synthesis.

secretion are stimulated by amino acids, and together they augment the production of somatomedins. The latter, in turn, stimulate accretion of cartilage and bone. (These actions are probably directly enhanced by insulin as well.) The insulin antagonistic effect of the HGH molecule itself on carbohydrate metabolism is also useful at this time; it helps to prevent hypoglycemia, which might result from insulin stimulation in the absence of carbohydrate. On the other hand, when a carbohydrate load is ingested and insulin secretion is correspondingly increased, HGH secretion is suppressed. In this circumstance, accelerated generation of somatomedins is not needed because protein anabolism is not advantageous in the absence of amino acid inflow. Neither is insulin antagonism necessary; on the contrary, unrestrained expression of insulin action permits efficient storage of the excess carbohydrate calories in the liver, muscle, and adipose tissue. Finally, when an individual is fasting, insulin secretion falls, partly because of a fall in plasma glucose levels. Although this increases HGH secretion, the calorie deficit and significant deficiency of insulin lead to a decrease in somatomedin production. Again, this is appropriate in a situation where an increase in protein anabolism is disadvantageous and protein catabolism is essential. However, the increase in HGH may still be beneficial during fasting, since it enhances lipolysis, decreases peripheral glucose utilization, and increases gluconeogenesis. Thus, like glucagon, HGH may help to provide glucose for central nervous system needs.

Clinical syndromes of HGH dysfunction. Deficiency of HGH in children can result from hypothalamic dysfunction, pituitary tumors, a biologically incompetent HGH molecule, a failure to generate somatomedins normally, or receptor deficiency. Short stature and correspondingly delayed bone maturation are the consequences. Mild obesity is common and puberty is usually delayed. In the adult, no physical signs are evident. In both children and adults, the lack of insulin antagonism from HGH may lead to rare episodes of hypoglycemia. The diagnosis of absolute HGH deficiency is established by demonstrating low plasma HGH levels, which fail to rise after stimulation with insulin hypoglycemia, arginine, exercise, or at night. This is confirmed by low levels of somatomedins. Replacement treatment with HGH causes nitrogen retention and an increase in growth velocity. Pubescence occurs and fertility is established.

Sustained hypersecretion of HGH results from pituitary tumors and produces a unique syndrome called *acromegaly*. If hypersecretion of HGH begins before puberty is completed and before the bony epiphyses are closed, the individual grows very tall and has long arms and legs. This condition has been referred to as *gigantism*. In adults, only periosteal bone growth can be increased by HGH, leading to enlarged fingers, toes, hands, and feet, prominent bony ridges above the eyes, and a prominent lower jaw. Facial features are coarsened by accumulation of excess soft tissue. The nose is bulbous, the tongue enlarged, and the skin is thick, whereas subcutaneous fat is sparse. The overall appearance is usually so characteristic as to allow diagnosis at a glance. Nerves may be entrapped and compressed by the excess soft tissue and accumulation of interstitial fluid. Virtually all organ sizes are increased; in some, such as the kidney, this is accompanied by increased function. Enlargement of the heart and accelerated atherosclerosis often lead to a shortened life span. Finally, the insulin antagonistic effect of HGH produces an abnormal tolerance to carbohydrate or even frank diabetes mellitus requiring treatment with insulin. The diagnosis is confirmed by demonstrating elevated plasma HGH levels, which are not suppressed when glucose is administered. Plasma levels of somatomedins (IGF-1) are also high. The pituitary tumor can often be visualized. Definitive treatment requires surgical removal of the tumor or ablation by irradiation. Somatostatin analogues that inhibit neoplastic somatrophs and diminish HGH hypersecretion are also therapeutically useful.

Prolactin

Prolactin is a protein hormone principally concerned with stimulating breast development and milk production. In addition, it exerts an influence on reproductive function; in humans this is most notable when prolactin is present in excess. Prolactin originates in specific anterior pituitary cells with secretory granules of 275 to 350 nm in diameter. These cells comprise 10% to 25% of the normal pituitary population. They increase in number during pregnancy, lactation, and with estrogen treatment. They also give rise to the commonest tumor of the human pituitary gland.

Table 52-5. Regulation of prolactin secretion

Stimulation	*Inhibition*
Pregnancy	Dopamine
Estrogen	Dopaminergic agonists
Nursing—breast manipulation	Bromergocriptine
Sleep	Apomorphine
Stress (Table 52-3)	*L*-Dopa
TRH	Prolactin
Dopaminergic antagonists	γ aminobutyric acid
Phenothiazines	
Metoclopramide	
Opioids	
Serotonin	
Histamine antagonists (H_2)	
Cimetidine	
Adrenergic antagonists	
Reserpine	
α-methyl DOPA	

Prolactin is a single-chain protein of molecular weight 23,000 with 198 amino acids and three disulfide bridges. It is structurally similar to HGH (Fig. 52-19), but does not cross-react with it in specific radioimmunoassays. The biological and immunological potency of various mammalian prolactins is similar. Synthesis of prolactin proceeds in the manner described for HGH, via a prehormone. The N-terminal signal peptide is cleaved before secretion. Transcription of the prolactin gene is regulated by factors that also regulate secretion of the hormone. Thus TRH increases prolactin messenger RNA, whereas dopamine decreases it.

Secretion of prolactin. Table 52-5 lists the most important influences on prolactin secretion. Consistent with its essential role in lactation, prolactin secretion increases steadily during pregnancy and causes the plasma levels to rise twenty-fold at term. This is probably mediated by the large increase in estrogen, which stimulates hyperplasia of prolactin-producing cells and synthesis of the hormone by inducing transcription of the gene. In addition, although estrogen does not itself stimulate the release of prolactin, it enhances responsiveness to other stimuli. If a new mother fails to nurse her child, the plasma level of prolactin declines 3 to 6 weeks after delivery to the normal range of nonpregnant women. Nursing, however, maintains elevated levels of prolactin secretion, especially for the first 8 to 12 weeks (Fig. 52-24). This effect may be mimicked by the use of a breast pump or by other nipple manipulation.

Like other tropic hormones, prolactin secretion rises at night, possibly because of entrainment with sleep (Fig. 52-10). The first peak appears 60 to 90 minutes

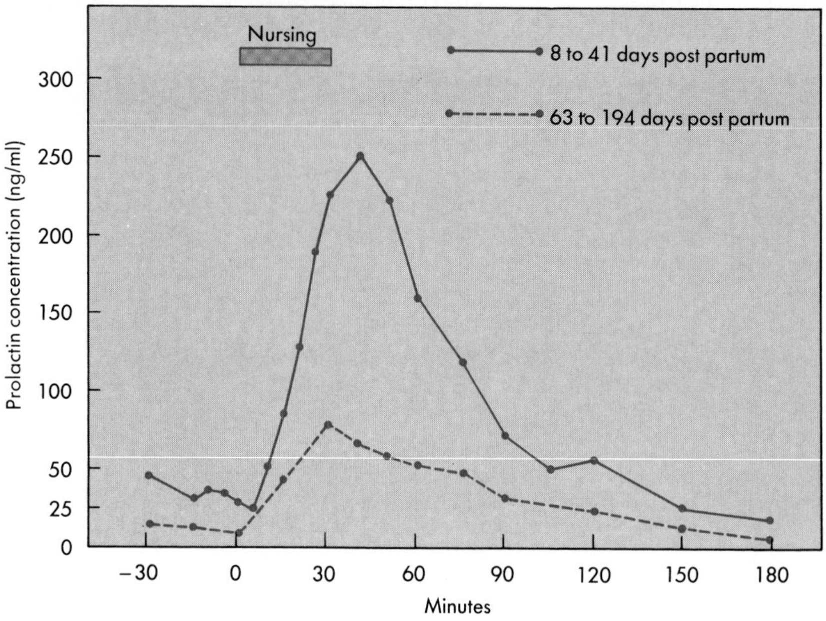

■ **Fig. 52-24.** Stimulation of prolactin secretion by nursing. Note decreased responses with increasing interval of time from delivery. (Redrawn from Noel, G.L., et al.: Prolactin release during nursing and breast stimulation in postpartum and nonpostpartum subjects, J. Clin. Endocrinol. Metab. **38:**413, 1974. Copyright 1974. Reproduced by permission.)

after the onset of slow-wave sleep and subsequent peaks occur later, after cycles of REM (rapid eye movement) sleep. The function of this association with sleep is unknown. Stress of various sorts, including anesthesia, surgery, insulin-induced hypoglycemia, fear, and mental tension, all cause prolactin release. Whether this represents a ''spillover'' phenomenon or whether the prolactin participates in accommodating to stresses is unclear.

The pathways for regulating prolactin release in each physiological circumstance remain to be worked out. However, uniquely among the pituitary hormones, prolactin secretion appears to be tonically *inhibited* by the hypothalamus (Fig. 52-25). Disruption of the hypothalamic-pituitary connection induces prompt and enduring increases in plasma prolactin, often with biological consequences. Dopamine has many characteristics that qualify it for the role of primary *prolactin-inhibiting factor* (PIF), although it is not a hypothalamic peptide. This catecholamine strongly inhibits prolactin release, either when generated within the brain in vivo, or when applied to pituitary tissue in vitro. Its action is mediated, at least in part, by lowering cAMP levels in the mammotroph. A dopaminergic tract runs from the hypothalamus to the median eminence, and dopamine concentrations in the pituitary portal veins are elevated to levels capable of inhibiting prolactin release in vitro. However, a separate hypothalamic peptide with prolactin-inhibiting properties (PIF) has been found among the peptides produced from processing the gene transcript of LHRH. Prolactin inhibits its own secretion via a short-loop feedback. It does so by directly increasing the synthesis and release of dopamine, its hypothalamic inhibitor. The profound prolactin-inhibiting effect of dopamine is exploited therapeutically by

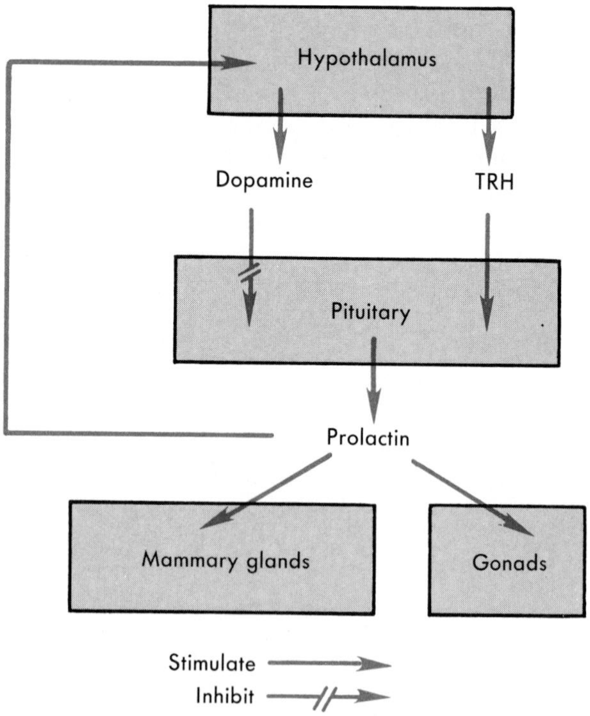

■ **Fig. 52-25.** Regulation of prolactin secretion. The predominant mode of hypothalamic regulation is tonic inhibition via dopamine. Although TRH stimulates prolactin release, its physiological role is uncertain, and evidence suggests another hypothalamic peptide may be more physiologically important. Prolactin exerts short-loop feedback on its own secretion by stimulating production of the hypothalamic inhibitor, dopamine.

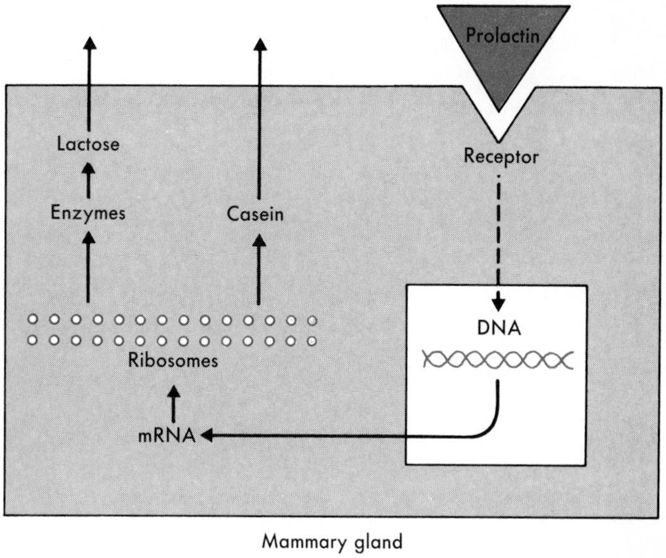

Fig. 52-26. Schema of prolactin effect on lactogenesis. A second messenger coupling the prolactin-plasma membrane receptor interaction with the nuclear effect has been proposed. Its structure has not been identified.

the use of agonists, such as bromergocriptine, in pathological states of prolactin excess. Conversely, hyperprolactinemia and sometimes milk secretion may result from the use of dopaminergic antagonists, such as phenothiazine compounds, in the treatment of psychiatric disorders.

The hypothalamus clearly has positive as well as negative effects on prolactin. TRH strongly stimulates prolactin synthesis and release, acting through specific membrane receptors in the mammotroph and probably via cAMP. Within minutes of intravenous administration of TRH, plasma prolactin increases two-fold to five-fold. When TSH secretion is chronically increased because of negative feedback from the thyroid gland, prolactin also tends to increase modestly, probably because of increased endogenous TRH. It is doubtful, however, that TRH is the sole or perhaps even the most important positive regulator of prolactin. In nursing women, for example, an acute rise in plasma levels of TSH does not accompany that of prolactin, as would be expected if TRH were the mediator. Another candidate for the role of prolactin-releasing hormone is vasoactive intestinal peptide (VIP). This peptide, originally discovered in the gastrointestinal tract, is present in high concentrations in the hypothalamus and portal veins and is a potent stimulator of prolactin secretion. The physiological significance of the positive effects of opioid peptides and serotonin and of the negative effects of norepinephrine on prolactin release remains to be established.

Normal basal plasma concentrations of prolactin are about 10 ng/ml (5×10^{-10}M). In addition to the natural hormone, which accounts for 70% to 90% of immunoreactivity in plasma, larger molecular weight species, termed "big prolactin," also circulate. Whether this fraction represents the incompletely processed prehormone remains unsettled. There is no evidence for protein binding of plasma prolactin. The half-life of the hormone is 20 minutes. Its metabolic clearance rate is 110 L/day, and the daily production is estimated at 350 μg. Its metabolic fate is not worked out, but the kidney is one likely organ of degradation, since patients with renal failure often have high plasma prolactin levels. It is interesting that relative to plasma levels, concentrations of prolactin are high in amniotic fluid and milk and low in cerebrospinal fluid.

Biological effects of prolactin. Prolactin participates in stimulating the original development of breast tissue and its further hyperplasia during pregnancy. It is the principal hormone responsible for lactogenesis. During prepubertal and postpubertal life, prolactin, together with estrogens, progesterone, cortisol, and growth hormone, stimulates the proliferation and branching of ducts in the female breast. During pregnancy, prolactin, along with estrogen and progesterone, causes development of lobules of alveoli within which milk production will occur. Finally, after parturition, prolactin with cortisol stimulates milk synthesis and secretion.

The action of prolactin begins by combination with a plasma cell membrane receptor (Fig. 52-26). Present evidence does not support an essential role for cAMP, cGMP, calcium, or membrane phospholipid products as mediators of the hormone's action. A peptide second messenger, generated from the membrane, has been described but not confirmed. Prolactin rapidly induces transcription of messenger RNA for the milk protein, casein. Increased synthesis of casein itself quickly follows. Galactosyl transferase and N-acetyllactosamine synthetase, enzymes necessary for synthesis of lactose, the major sugar in milk, are concurrently induced. Prolactin augments the number of its own receptors, this being an example of up-regulation. Estrogen also increases the number of prolactin receptors; nontheless, it directly antagonizes the stimulatory effect of prolactin on milk synthesis.

The second major effect of prolactin is on the reproductive axis. In women, prolactin blocks the synthesis and release of LHRH, causing a loss of normal spurts of LH and preventing ovulation. In female rodents and in women, low concentrations of prolactin help to sustain ovarian progesterone secretion; but at higher concentrations, inhibiting effects on gonadal steroidogenesis appear to predominate in women and men. Certain behavioral effects of prolactin have been described, such as inhibition of libido in humans and stimulation of parental protective behavior toward the newborn in animals.

Clinical syndromes of prolactin dysfunction. In women prolactin deficiency produces the inability to lactate. No other clinical consequences are known for certain. Prolactin excess results from hypothalamic dysfunction or from pituitary tumors. In women, this causes infertility due to anovulation and even complete loss of menses. Less often, lactation unassociated with pregnancy (galactorrhea) occurs. In men, decreased testosterone secretion and sperm production result from prolactin excess. Stimulation of breast development is uncommon and lactation is rare. In both sexes, decreased libido is noted. The diagnosis is established by demonstrating a high plasma prolactin level that often fails to increase further after stimulation with TRH or a dopamine antagonist. Therapy may consist of surgical ablation of tumor tissue. However, in many instances, treatment with dopaminergic drugs reduces prolactin secretion to normal, reversing all adverse effects on reproduction and galactorrhea.

Posterior Pituitary Hormones

Two nonapeptides of homologous structure (Fig. 52-27), *antidiuretic hormone* (ADH), also known as arginine vasopressin (AVP), and *oxytocin,* are secreted from the posterior pituitary gland. The primary role of ADH is to conserve body water and regulate the tonicity of body fluids. The primary role of oxytocin is to eject milk from the lactating mammary gland. Although their functions are different, the synthesis, storage, and mode of secretion of the two hormones are similar and will be discussed together.

Both hormones are synthesized in the cell bodies of hypothalamic neurons. ADH largely originates in the supraoptic nucleus and oxytocin largely in the paraventricular nucleus of the hypothalamus, although small amounts of each hormone are synthesized in the alternate site. Each nonapeptide is synthesized as part of a much larger precursor. The genes directing synthesis of the respective preprohormones are remarkably similar

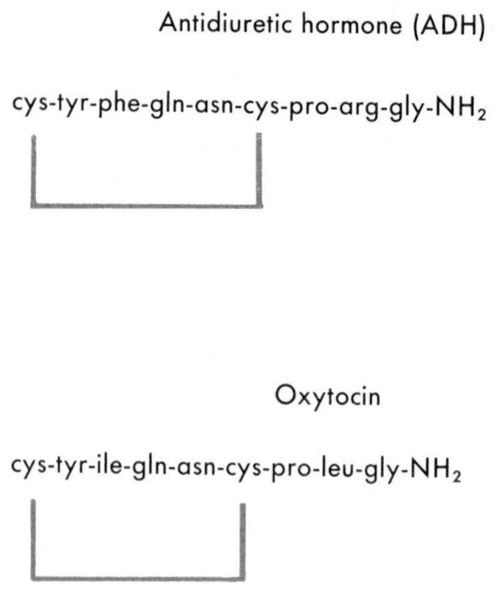

Fig. 52-27. Structures of posterior pituitary peptides. The alternate term for ADH is arginine vasopressin (AVP).

in each case (Fig. 52-28). In addition to ADH or oxytocin, the gene products include distinctive proteins, known as *neurophysins,* of molecular weight 10,000. Neurophysin-1 for oxytocin and neurophysin-2 for ADH are virtually identical in their large central cores (corresponding to a single exon in each gene). The neurophysins differ in their N-terminal portions, which are coded for by exons that also code for the hormones themselves (Fig. 52-28) and in their C-terminal portions. In the case of ADH, an additional glycopeptide is coded for by the same exon that codes for the C-terminal portion of neurophysin-2. After processing of the preprohormones, ADH and oxytocin are packaged together with their respective neurophysins in neurosecretory granules. The neurophysins appear to serve as low-affinity carrier proteins in the process of transport of the neurohormones down the axons. The latter end in the posterior pituitary as terminal swellings, known as *Herring bodies*.

Release of ADH or oxytocin occurs when a nerve impulse is transmitted from the cell body in the hypothalamus down the axon, where it depolarizes the neurosecretory vesicles within the terminal Herring body. An influx of calcium into the neurosecretory vesicle then results in hormone secretion by exocytosis. During this process, the nonapeptide hormone dissociates from its neurophysin, and separately, each enters the closely adjacent capillary. Subsequent passage of the hormone into the bloodstream is by endocytosis into the endothelial cell and by diffusion through pores in the fenestrated capillary endothelium.

Secretion of ADH

Consonant with its role in water metabolism, secretion of ADH is primarily regulated by osmotic and vol-

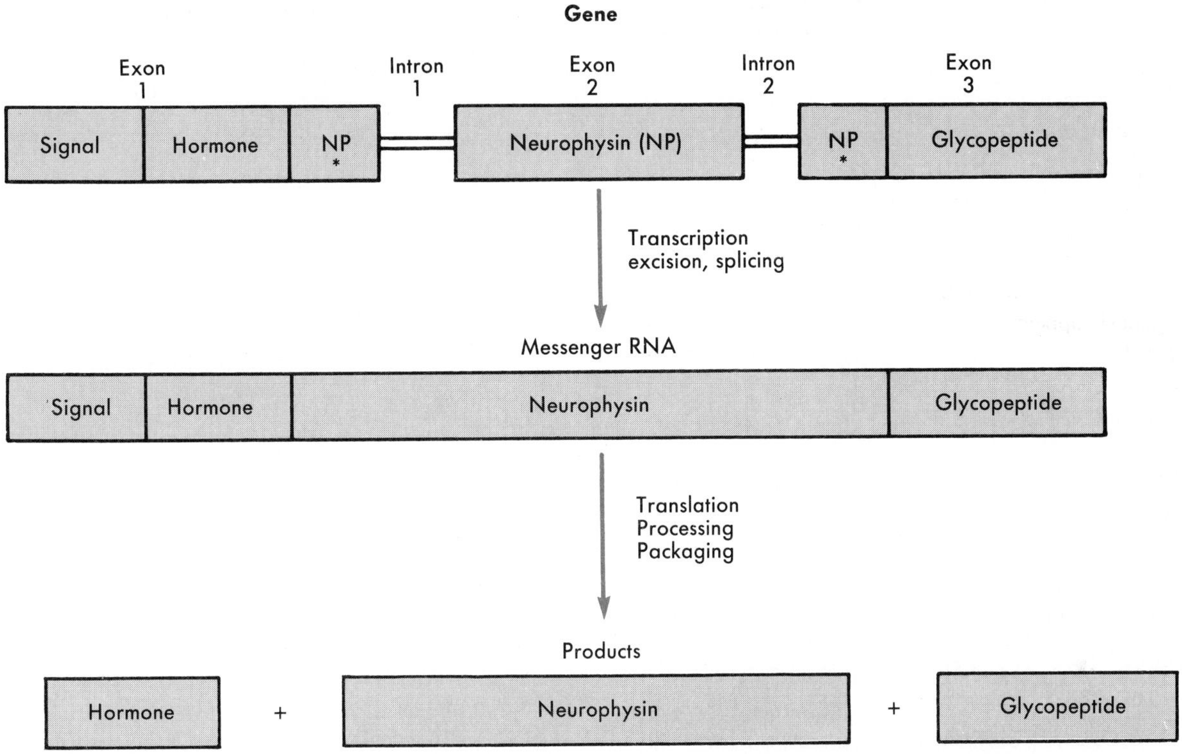

■ **Fig. 52-28.** Schematic representation of the synthesis of the two posterior pituitary peptides ADH and oxytocin. The two gene structures are similar. In each case, exon-1 codes for the signal peptide, the hormone, and a variable portion of its corresponding neurophysin. Exon-2 is virtually identical in the two genes and codes for the homologous large central core of each neurophysin. Exon-3 codes for the remaining variable portion of each neurophysin. In the case of ADH only, exon-3 contains a base sequence extension that codes for a C-terminal glycopeptide, which is coreleased with ADH. (Redrawn from Richter, D., and Ivell, R.: Gene organization, biosynthesis, and chemistry of neurohypophyseal hormones. In Imura, H.: The pituitary gland, New York, 1985, Raven Press.)

ume stimuli (Table 52-6). Water deprivation produces an increase in the osmolality of plasma and hence of the fluids bathing the brain. This causes the loss of intracellular water from osmoreceptor neurons in the hypothalamus. Although these could be identical with the magnocellular neurons that secrete ADH, current evidence favors the presence of a distinct population of osmoreceptor neurons with connections to the ADH neurons. In either case, the shrinkage of cell volume causes ADH to be released immediately in the manner just described. Conversely, water ingestion causes a decrease in plasma osmolality, which suppresses osmoreceptor firing and consequently shuts off ADH release. If plasma osmolality is directly increased by administration of solutes, only those which do not freely or rapidly penetrate cell membranes, such as sodium, cause ADH release. Substances, such as urea, that enter cells

■ **Table 52-6.** Regulation of ADH secretion

Stimulation	*Inhibition*
Extracellular fluid osmolality increase	Temperature decrease
Volume decrease	α-adrenergic agonists
Pressure decrease	Ethanol
Cerebrospinal fluid sodium increase	Cortisol
Pain	Thyroid hormone(?)
Nausea and vomiting	
Stress (Table 52-3)	
Temperature increase	
Drugs	
Nicotine	
Opiates	
Barbiturates	
Sulfonylureas	
Antineoplastic agents	

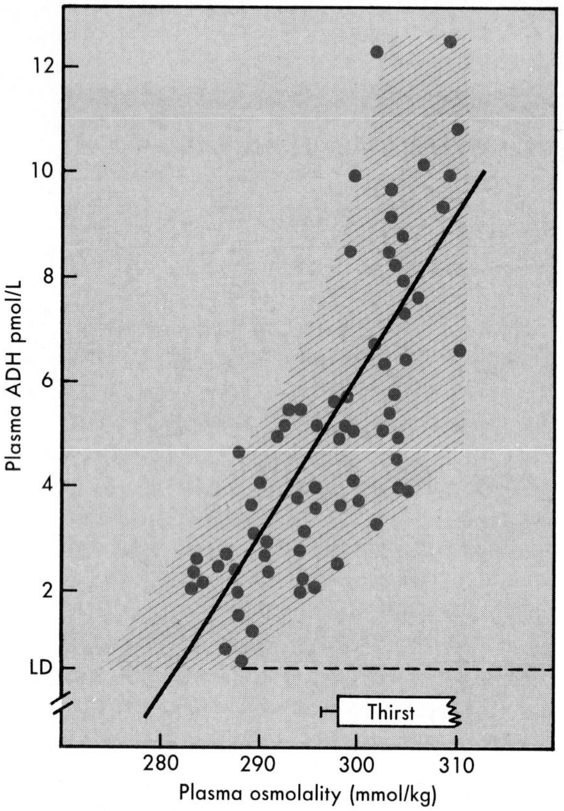

■ **Fig. 52-29.** Correlation between plasma ADH and plasma osmolality in humans. As plasma osmolality is increased by infusing hypertonic sodium chloride, ADH secretion is stimulated and plasma ADH rises over a linear concentration range. This response to hyperosmolality precedes the response of thirst. Both ADH release and thirst lead to increases in body water that limit further increments in the osmolality of body fluids. (From Baylis, P.: Clin. Endocrinol. Metab. **12:**747, 1983.)

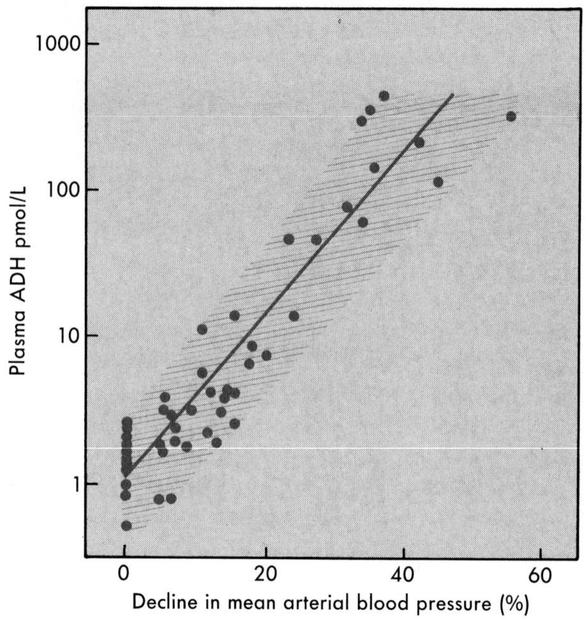

■ **Fig. 52-30.** Correlation between plasma ADH and declining blood pressure in humans. As blood pressure is *decreased* by infusing an agent that blocks sympathetic ganglion function, ADH secretion is stimulated. In this response, plasma ADH rises over an exponential concentration range. (From Bayliss, P.: Clin. Endocrinol. Metab. **12:**747, 1983.)

rapidly do not stimulate ADH secretion, because they do not produce osmotic disequilibrium between extracellular and intracellular fluids. An increase in sodium concentration of the cerebrospinal fluid also increases ADH secretion, possibly through a specific sodium-sensing membrane. The hypothalamic osmoreceptors are extraordinarily sensitive, being responsive to changes in osmolality of only 1% to 2% (Fig. 52-29). If water deprivation is prolonged, ADH synthesis and secretion are increased. In response to plasma hyperosmolarity, osmoreceptor neurons also stimulate thirst. However, in humans, the threshold for this action is 290 to 295 mOsm/kg, which is distinctly higher than the threshold for ADH release of around 280 mOsm/kg (Fig. 52-29). This fact emphasizes the preeminence of ADH in maintaining normal body water content.

ADH release is also stimulated by a decrease of 5% to 10% in total circulating blood volume, central blood volume, cardiac output, or blood pressure (Fig. 52-30). Hemorrhage is a potent stimulus to ADH release. Quiet standing and positive pressure breathing, both of which reduce cardiac output and central blood volume, also increase ADH secretion. Conversely, administration of blood or isotonic saline solution, which increases total circulating blood volume, or immersion to the neck in water, which increases central blood volume, all suppress ADH release. Hypovolemia is perceived by a number of pressure (rather than volume) sensors (see Chapters 29 and 33). These include carotid and aortic baroreceptors, stretch receptors in the walls of the left atrium and pulmonary veins, and possibly the juxtaglomerular apparatus of the kidney. The afferent impulses of this neurohumoral arc are carried by the ninth and tenth cranial nerves to their respective nuclei in the medulla and then by way of the midbrain via adrenergic neurotransmitters to the supraoptic nuclei of the hypothalamus. Normally the pressure receptors *tonically inhibit* ADH release by modulating the inhibitory flow of adrenergic impulses from the medulla to the hypothalamus. A decrease in circulating volume reduces the flow of impulses from the baroreceptors to the brainstem; this relieves the inhibition on the hypothalamus, resulting in increased ADH secretion. Hypovolemia also stimulates the generation of renin and angiotensin directly within the brain; this local angiotensin, in addition to stimulating thirst, also enhances the release of ADH. Peripherally generated circulating angiotensin plays only a minor role, if any, in the response to hypovolemia. Plasma ADH rises to much higher levels in

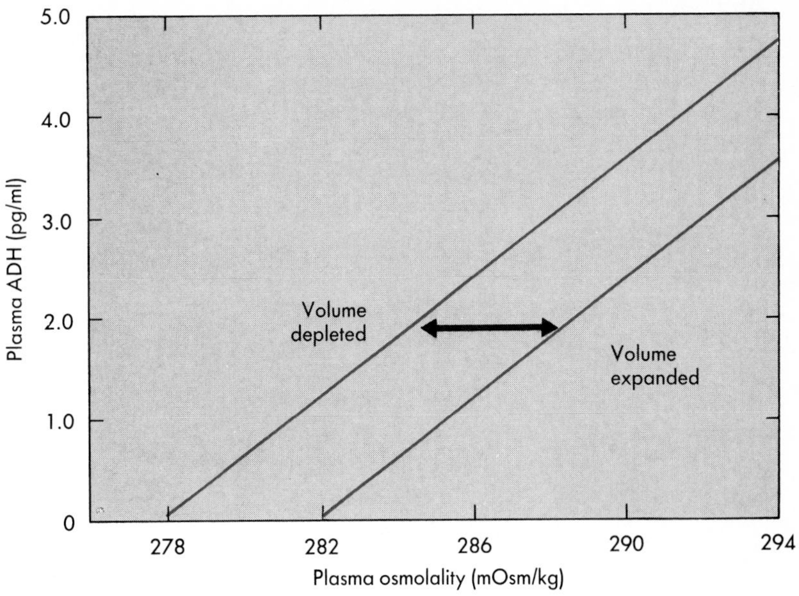

Fig. 52-31. Regulation of ADH secretion by the interaction between plasma osmolality and plasma volume in humans. Increases or decreases in plasma volume, respectively, increase or decrease the threshold for ADH release in response to osmolality. (Redrawn from Robertson, G.L., et al.: J. Clin. Endocrinol. Metab. **42:**613, 1976. Copyright 1976, The Endocrine Society.)

response to hypotension than in response to hyperosmolarity (compare Fig. 52-30 with Fig. 52-29). This is consistent with the fact that the vascular system is less sensitive to the hormone than is the kidney.

The interaction between the two major stimuli of ADH release is shown in Fig. 52-31. Increases or decreases in volume reinforce the osmolar responses by raising or lowering, respectively, the threshold for osmotic release of ADH. Thus hypovolemia sensitizes the system to hyperosmolarity. An increase in plasma osmolality of 3 mOsm/kg increases plasma levels of ADH by about 1 pg/ml. Under circumstances of marked hypovolemia, baroregulation overrides osmotic regulation and ADH secretion is stimulated, even though plasma osmolality may be well below 270 mOsm/kg.

Secretion of ADH is also influenced by a number of other conditions (Table 52-6). Pain, emotional stress, heat, and a variety of drugs are stimulators; nausea and vomiting are especially potent. Ethanol is a commonly encountered inhibitor; as little as 30 to 90 ml of whiskey is sufficient to suppress secretion completely. Cortisol and thyroid hormones appear to restrain ADH release in a permissive manner; in their absence, ADH may be secreted even though plasma osmolality is low.

ADH circulates at concentrations of about 1 pg/ml (10^{-12}M). The plasma half-life, determined by radioimmunoassay, is 6 to 10 minutes, although the half-life of biological effect may be up to 20 minutes. Degradation occurs in the kidney and liver. Estimates of daily secretion under ordinary circumstances are of the order of 1 μg, or about 4% of the posterior pituitary content. During water deprivation for 48 hours, secretion increases three-fold to five-fold. Transient increases of fifty-fold can occur with hemorrhage (Chapter 36), severe pain, or nausea. Neurophysin-2 also circulates in plasma and its levels rise and fall parallel with ADH. Thus measurement of neurophysin-2, as well as ADH, can reflect hypothalamic posterior pituitary function. The C-terminal glycopeptide of prepro-ADH origin is also present in plasma. No functional role for these peptides in peripheral tissues has been identified, despite extensive search. Their physiological purpose may lie within extrapituitary areas of the brain into which some ADH neurons project.

Actions of ADH

The major action of ADH is on renal cells that are responsible for reabsorbing free (i.e., osmotically unencumbered) water from the glomerular filtrate (Chapters 41 and 47). These ADH-responsive cells line the distal convoluted tubules and collecting ducts of the renal medulla. ADH binds to a specific plasma membrane receptor (known as the V_2 receptor) on the capillary side of the cell where it activates adenylate cyclase. The increase in intracellular cAMP activates a protein kinase on the luminal side of the cell. This phosphorylates a currently unidentified membrane protein, following which the permeability of the cell membrane to water is enhanced. The participation of microtubular and microfilamentous elements of the cell in this process has also been suggested. The increase in membrane permeability permits back diffusion of water along an osmotic gradient from the hypotonic tubular urine that emerges from the loop of Henle to the hypertonic interstitial fluid of the renal medulla. The mechanisms for establishing this gradient are discussed in Chapter 46. Recent evidence suggests that ADH may also act on the loop of Henle and participate in creation of the osmotic

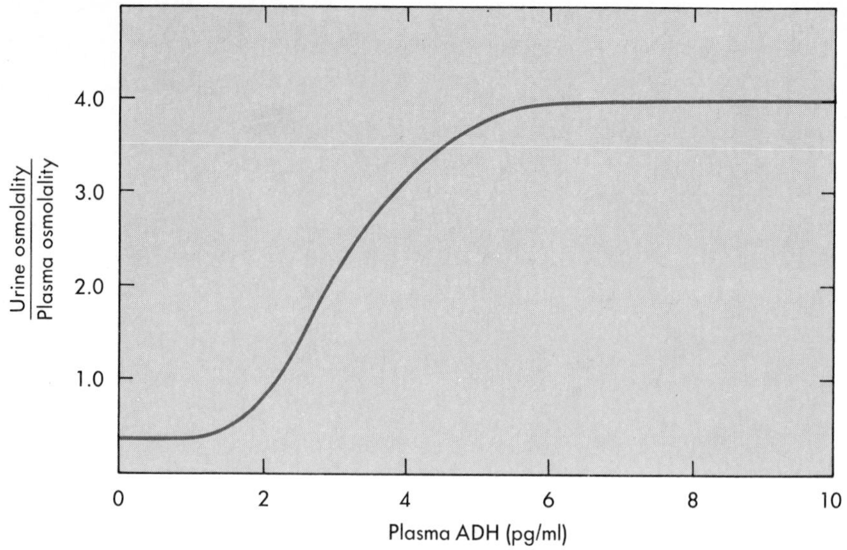

Fig. 52-32. Dose-response curve for the effect of ADH to increase renal tubular reabsorption of free water, expressed as the ratio of urine to plasma osmolality. (Data from Moore, W.W.: Fed. Proc. **30:**1387, 1971.)

gradient. The net result of ADH action is to increase the osmolality of urine to a maximum that is four-fold greater than that of the glomerular filtrate or plasma. In other words, ADH significantly reduces free water clearance by the kidney.

Water deprivation stimulates ADH secretion, resulting in decreased free water clearance and enhanced conservation of water. Thus ADH and water form a negative feedback loop. A water load decreases ADH secretion, resulting in increased free water clearance and more efficient excretion of the load. The dose-response relation between plasma levels of ADH and urine osmolality is shown in Fig. 52-32. The relationship is sigmoidal, the majority of biological effect being observed between 2 and 5 pg/ml of ADH. In this range, urine osmolality correlates directly with plasma ADH concentrations. The generation of much higher plasma levels of ADH in humans may be related to other functions.

A number of factors blunt the action of ADH on the tubular cell: solute diuresis, chronic water loading (which reduces medullary hyperosmolarity), prostaglandin E (which interferes with ADH activation of adenylate cyclase), cortisol, potassium deficiency, calcium excess, and lithium. Certain sulfonylureas, used in the treatment of diabetes mellitus, are prominent among a number of agents that potentiate ADH action.

ADH may subserve other functions in addition to its primary role in water metabolism. As noted, it contributes to increasing vascular tone in response to hemorrhage by binding to arteriolar smooth muscles via the ADH V_1 receptor and causing them to constrict. This action is not mediated by the adenylate cyclase system. When ADH is administered systemically in appropriate doses, it elevates the blood pressure and constricts the coronary and splanchnic beds. The last effect has been exploited therapeutically in controlling serious gastrointestinal bleeding. ADH also functions as a corticotropin releasing factor via axons that transmit the peptide to the median eminence. In addition, ADH may serve as a neurotransmitter within the brain to facilitate long-term memory.

Oxytocin Secretion and Actions

Suckling is the major stimulus for oxytocin release. Afferent impulses are carried from sensory receptors in the nipple via nerves to the spinal cord, where they ascend in the spinothalamic tract. From relays in the brain stem and midbrain, they reach the paraventricular nuclei of the hypothalamus, and, via a cholinergic synapse, they trigger oxytocin release within seconds from the neurosecretory vesicles in the posterior pituitary. As suckling is continued, oxytocin synthesis and transfer down the hypothalamic axon are also stimulated. As shown in Fig. 52-33, the stimulus of suckling is specific for oxytocin, since no release of ADH is noted. Correspondingly, the various stimuli for ADH secretion do not stimulate appreciable oxytocin release in humans. Oxytocin circulates unbound and exhibits a plasma half-life of 3 to 5 minutes. It is degraded by the kidneys and liver.

The unique effect of oxytocin is to cause contraction of the myoepithelial cells of the alveoli of the mammary glands. This forces milk from the alveoli into the ducts, from where it is evacuated by the infant. No other hormone has this action. Oxytocin combines with plasma membrane receptors in the breast, and receptor binding is increased by estrogen. Catecholamines, on the other hand, can block the action of oxytocin. At present, no analogous action of oxytocin has been identified in

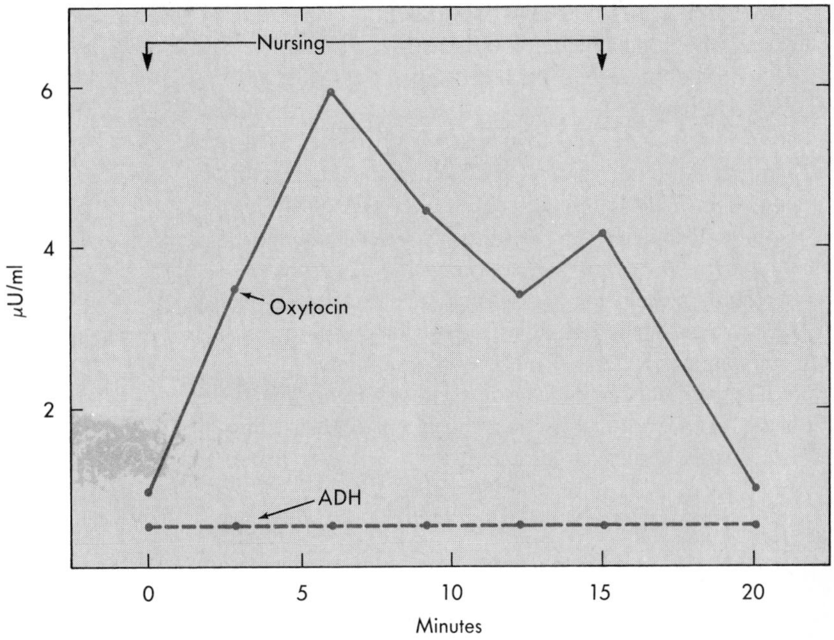

■ **Fig. 52-33.** Stimulation of oxytocin secretion by nursing in humans. Note specificity of response as no release of ADH is observed. (Redrawn from Weitzman, R.E., et al.: The effect of nursing on neurohypophyseal hormone and prolactin secretion in human subjects, J. Clin. Endocrinol. Metab. **41:**836, 1980. Copyright 1980. Reproduced by permission.)

men, although their plasma levels of the hormone are similar to those in women.

Oxytocin also has a powerful contracting action on the uterus. Rhythmic contractions of the myometrium are stimulated by very small doses, which act by slightly lowering the threshold for membrane depolarization in the uterine smooth muscle. Large doses lower the threshold still further, prevent repolarization and spiking discharges, and induce a sustained tetanic contraction. Despite the ability to stimulate rhythmic uterine contractions, there is little evidence that oxytocin is an essential hormone of labor, although it probably contributes to the process. It may, however, play an important role in the sustained postpartum contractions that help to maintain hemostasis after evacuation of the placenta. Oxytocin and its receptor have recently been found in the human ovary. A role for the hormone in terminating the corpus luteum at the end of the menstrual cycle (Chapter 55) has been suggested. Since oxytocin has only 0.5% to 1% the antidiuretic activity of ADH, it has no role in the therapy of ADH deficiency. Oxytocin is used clinically to induce labor in women who are physiologically ready, and it is also used therapeutically to decrease immediate postpartum bleeding.

■ *Clinical Syndromes of Posterior Pituitary Hormone Dysfunction*

ADH deficiency is caused by destruction or dysfunction of the supraoptic and paraventricular nuclei of the hypothalamus. The posterior pituitary gland alone can be removed without seriously affecting the availability of the hormone because the sectioned axons acquire the ability to secrete ADH. The inability to produce a concentrated urine is the hallmark of ADH deficiency, a condition called *diabetes insipidus*. In normal humans, water deprivation can be compensated for by an increase in urine osmolality to 1000 to 1400 mOsm/kg. Individuals who totally lack ADH cannot achieve osmolalities higher than that of plasma (290 mOsm/kg) and often not higher than 50 to 200 mOsm/kg. Since a typical diet generates 600 to 900 mOsm of solute per day for obligatory excretion by the kidney, urine volumes as high as 10 to 12 L may be required in diabetes insipidus, in contrast to the usual 1 to 3 L. The patient therefore urinates frequently both day and night and must also drink fluids constantly to replace the loss of water. Laboratory studies generally show a chronic elevation of serum osmolality (greater than 290 mOsm/kg) and of serum sodium (greater than 145 mEq/L). If, for any reason, the patient loses access to fluids, hyperosmolarity, severe dehydration, circulatory collapse, and death may ensue. Replacement of ADH (available as arginine vasopressin or analogues) prevents this sequence and relieves the symptoms of thirst and frequent urination.

ADH excess, inappropriate to either the osmolarity or volume of the body fluids, can result from a variety of conditions: (1) increased secretion because of central nervous system disease, trauma, or psychosis; (2) ectopic production of the hormone by tumors of the lung and pancreas; or (3) potentiation of hormone action by drugs. The reduction in free water clearance caused by ADH, combined with voluntary or involuntary fluid intake, leads to water retention. Plasma sodium concentration and osmolality are significantly lowered, whereas urine osmolality is increased. Characteristically, sodium excretion in the urine is also increased, despite the hyponatremia. Both intracellular and extracellular body fluids are expanded. The swelling of brain cells and hypoosmolality cause headache, nausea, lethargy, somnolence, convulsions, and coma. These symptoms are not usually seen until plasma osmolality has declined to below 250 mOsm/kg, and plasma sodium to below 120 to 125 mEq/L. Water restriction is a logical and effective acute treatment. It may have to be supplemented in very sick patients with agents that induce a water diuresis or with hypertonic sodium chloride solutions to raise osmolality more rapidly.

Oxytocin deficiency leads to difficulty in nursing because of poor milk ejection. Oxytocin excess is unknown as a clinical syndrome.

Bibliography

Journal articles

Argente, J., et al.: Relationship of plasma growth hormone–releasing hormone levels to pubertal changes, J. Clin. Endocrinol. Metab. **63:**680, 1986.

Berson, S.A., et al.: Radioimmunoassay of ACTH in plasma, J. Clin. Invest. **47:**2725, 1968.

Clemmons, D., and VanWyk, J.: Factors controlling blood concentration of somatomedin C, Clin. Endocrinol. Metab. **13:**113, 1984.

Estivariz, F., et al.: Stimulation of adrenal mitogenesis by N-terminal proopiocortin peptides, Nature **297:**419, 1982.

Frohman, L.: Growth hormone–releasing factor: a neuroendocrine perspective, J. Lab. Clin. Med. **103:**819, 1984.

Gelato, M., et al.: Effects of a growth hormone releasing factor in man, J. Clin. Endocrinol. Metab. **57:**674, 1983.

Grossman, A., et al.: New hypothalamic hormone, corticotropin-releasing factor, specifically stimulates the release of adrenocorticotropic hormone and cortisol in man, Lancet **1:**921, 1982.

Grossman: Neuroendocrinology of opioid peptides, Br. Med. Bull. **39:**83, 1983.

Hall, K., and Sara, V.: Somatomedin levels in childhood, adolescence and adult life, Clin. Endocrinol. Metab. **13:**91, 1984.

Hökfelt, T., et al.: Peptidergic neurons, Nature **284:**515, 1980.

Isaksson, O., et al.: Growth hormone stimulates longitudinal bone growth directly, Science **216:**1237, 1982.

Isley, W., et al.: Dietary components that regulate serum somatomedin-C concentrations in humans, J. Clin. Invest. **71:**175, 1983.

Jackson, R., et al.: Synthetic ovine corticotropin-releasing hormone: simultaneous release of propiolipomelanocortin peptides in man, J. Clin. Endocrinol. Metab. **58:**740, 1984.

Krieger, D.T., et al.: Human plasma immunoreactive lipotropin and adrenocorticotropin in normal subjects and in patients with pituitary-adrenal disease, J. Clin. Endocrinol. Metab. **48:**566, 1979.

Krieger, D.T., et al.: Brain peptides. I., N. Engl. J. Med. **304:**876, 1981.

Krieger, D.T., et al.: Brain peptides. II., N., Engl J. Med. **304:**944, 1981.

Matsumoto, A., and Bremner, W.: Modulation of pulsatile gonadotropin secretion by testosterone in man, J. Clin. Endocrinol. Metab. **58:**609, 1984.

Mendelson, W., et al.: Negative feedback suppression of sleep-related growth hormone secretion, J. Clin. Endocrinol. Metab. **56:**486, 1983.

Pelletier, G., et al.: Identification of human anterior pituitary cells by immunoelectron microscopy, J. Clin. Endocrinol. Metab. **46:**534, 1978.

Phillips, L.S., et al.: Somatomedins. I., N. Engl. J. Med. **302:**371, 1980.

Philips, L.S., et al.: Somatomedins. II., N. Engl. J. Med. **302:**438, 1980.

Scott, R., and Burger, H.: An inverse relationship exists between seminal plasma inhibin and serum follicle-stimulating hormone in man, J. Clin. Endocrinol. Metab. **52:**796, 1981.

Shibahara, S., et al.: Isolation and sequence analysis of the human corticotropin-releasing factor precursor gene, EMBO J. **2:**775, 1983.

Shurmeyer, T., et al.: Human corticotropin-releasing factor in man: pharmacokinetic properties and dose-response of plasma adrenocorticotropin and cortisol secretion, J. Clin. Endocrinol. Metab. **59:**1103, 1984.

Sklar, A., and Schrier, R.: Central nervous system mediators of vasopressin release, Physiol. Rev. **63:**1243, 1983.

Snyder, S.H.: Brain peptides as neurotransmitters, Science **209:**976, 1980.

Taylor, T., and Weintraub, B.: Thyrotropin (TSH)-releasing hormone regulation of TSH subunit biosynthesis and glycosylation in normal and hypothyroid rat pituitaries, Endocrinology **116:**1968, 1985.

Taylor, T., et al.: β-Endorphin suppresses adrenocorticotropin and cortisol levels in normal human subjects, J. Clin. Endocrinol. Metab. **57:**592, 1983.

Vale, W., et al.: Characterization of a 41-residue ovine hypothalamic peptide that stimulates secretion of a corticotropin and β-endorphin. Science **213:**1394, 1981.

Veldhuis, J., et al.: Endogenous opiates modulate the pulsatile secretion of biologically active luteinizing hormone in man, J. Clin. Invest. **72:**2031, 1983.

Weitzman, R.E., et al.: The effect of nursing on neurohypophyseal hormone and prolactin secretion in human subjects, J. Clin. Endocrinol. Metab. **41:**836, 1980.

Books and monographs

Chin, W.: Organization and expression of glycoprotein hormone genes. In Imura, H.: The pituitary gland, New York, 1985, Raven Press.

deKretzer, D., et al.: The pituitary and testis, clinical and experimental studies, Berlin, 1983, Springer-Verlag.

Frohman, L.A., et al.: The physiological and pharmacological control of anterior pituitary hormone secretion. In Dunn, A., and Nemeroff, C., editors: Behavioral neuroendocrinology, New York, 1983, Spectrum Publications, Inc.

Guillemin, R.: Neuroendocrine interrelations. In Bondy, P., and Rosenberg, L.E., editors: Metabolic control and disease, Philadelphia, 1980, W.B. Saunders Co.

Kato, Y., et al.: Regulation of prolactin secretion. In Imura, H.: The pituitary gland, New York, 1985, Raven Press.

Keith, L.D., and Kendall, J.W.: Regulation of ACTH secretion. In Imura, H.: The pituitary gland, New York, 1985, Raven Press.

Numa, S., and Imura, H.: ACTH and related peptides: gene structure and biosynthesis. In Imura, H.: The pituitary gland, New York, 1985, Raven Press.

Savoy-Moore, R.T., et al.: Differential control of FSH and LH secretion. In Greep, R.O., editor: Reproductive physiology, III, Baltimore, 1980, University Park Press.

Seo, H.: Growth hormone and prolactin: chemistry, gene organization, biosynthesis, and regulation of gene expression. In Imura, H.: The pituitary gland, New York, 1985, Raven Press.

Verbalis, J., and Robinson, A.: Neurophysin and vasopressin: newer concepts of secretion and regulation. In Imura, H.: The pituitary gland, New York, 1985, Raven Press.

CHAPTER

53

The Thyroid Gland

The symptom complex associated with hyperfunction of the thyroid gland was recognized as early as 1825, in part because the location of the gland provided visible and palpable appreciation of its enlargement. Shortly thereafter the appearance of a distinctive symptom complex of opposite character, which developed with the gland's surgical absence or spontaneous atrophy, was noted. The early speculation that this was caused by the deficiency of an internal secretion was subsequently borne out when crude extracts of the thyroid gland became the first example of hormonal replacement therapy in 1891. Thyroid hormone secretion is unusually steady, and its actions are very general. Its most important mission is to regulate the overall rate of body metabolism; in addition, it is critical for normal growth and development.

The thyroid gland develops from endoderm associated with the pharyngeal gut. It descends to the anterior part of the neck, where it lies on either side of the trachea. Abnormalities in its developmental descent may lead to final locations anywhere from the base of the tongue to the anterior mediastinum. By 11 to 12 weeks of gestational age the gland is capable of synthesizing and secreting its own thyroid hormones under the stimulus of fetal thyroid-stimulating hormone (TSH). Both are absolutely required for subsequent normal intrauterine development of the central nervous system and skeleton (although not for body growth), because neither maternal TSH nor maternal thyroid hormone can reach the fetus to any significant extent.

The two lobes of the adult thyroid gland together weigh approximately 20 g. They receive a rich blood supply from the thyrocervical arteries and innervation from the autonomic nervous system. The histological structure is shown in Fig. 53-1, *A*. The hormone-producing, cuboidal epithelial cells form circular follicles 200 to 300 μm in diameter. Within their lumens newly synthesized hormone is stored in the form of a colloid material. The base of each cell is covered by a basement membrane, and tight junctions connect adjacent cells at both their basal and apical (luminal) portions. When the gland is under intensive stimulation, the endocrine cells enlarge and assume a more columnar shape with their nuclei at the base (Fig. 53-1, *B*). The lumens of the follicles then appear scalloped because of active proteolysis of the hormone-containing colloid (Fig. 53-1, *B*). Scattered within the gland in close association with the epithelial cells is a separate line of parafollicular cells, called *C-cells*. These are the source of the polypeptide hormone, calcitonin, which is discussed in the section dealing with calcium metabolism in Chapter 51.

Synthesis and Release of Thyroid Hormones

The secretory products of the thyroid gland are *iodothyronines*, a series of compounds resulting from the coupling of two iodinated tyrosine molecules. From 80% to 90% of the output is 3,5,3′,5′-tetraiodothyronine (*thyroxine*, or T_4), 10% is 3,5,3′-triiodothyronine (T_3), and less than 1% is 3,3′,5′-triiodothyronine (reverse T_3, or rT_3). Normally these three compounds are secreted in the same proportions as they are stored in the gland.

Because of the unique role of *iodide* in thyroid physiology, a description of thyroid hormone synthesis properly begins with a consideration of iodide (Fig. 53-2). An average of 400 μg of iodide per person is ingested daily in the United States. In a steady state virtually the same amount is excreted in the urine. Iodide is actively concentrated in the thyroid gland, the salivary glands, and gastric glands. About 70 to 80 μg of iodide is taken up daily by the thyroid gland. This represents 8% to 35% of the circulating pool of iodide, which ranges from 250 to 750 μg. If this extrathyroidal iodide pool is labeled with a small dose of radioactive iodine (^{123}I or ^{131}I), the percentage of uptake of this tracer in 24 hours (8% to 35%) gives a clinically useful dynamic index of thyroid gland activity. The total io-

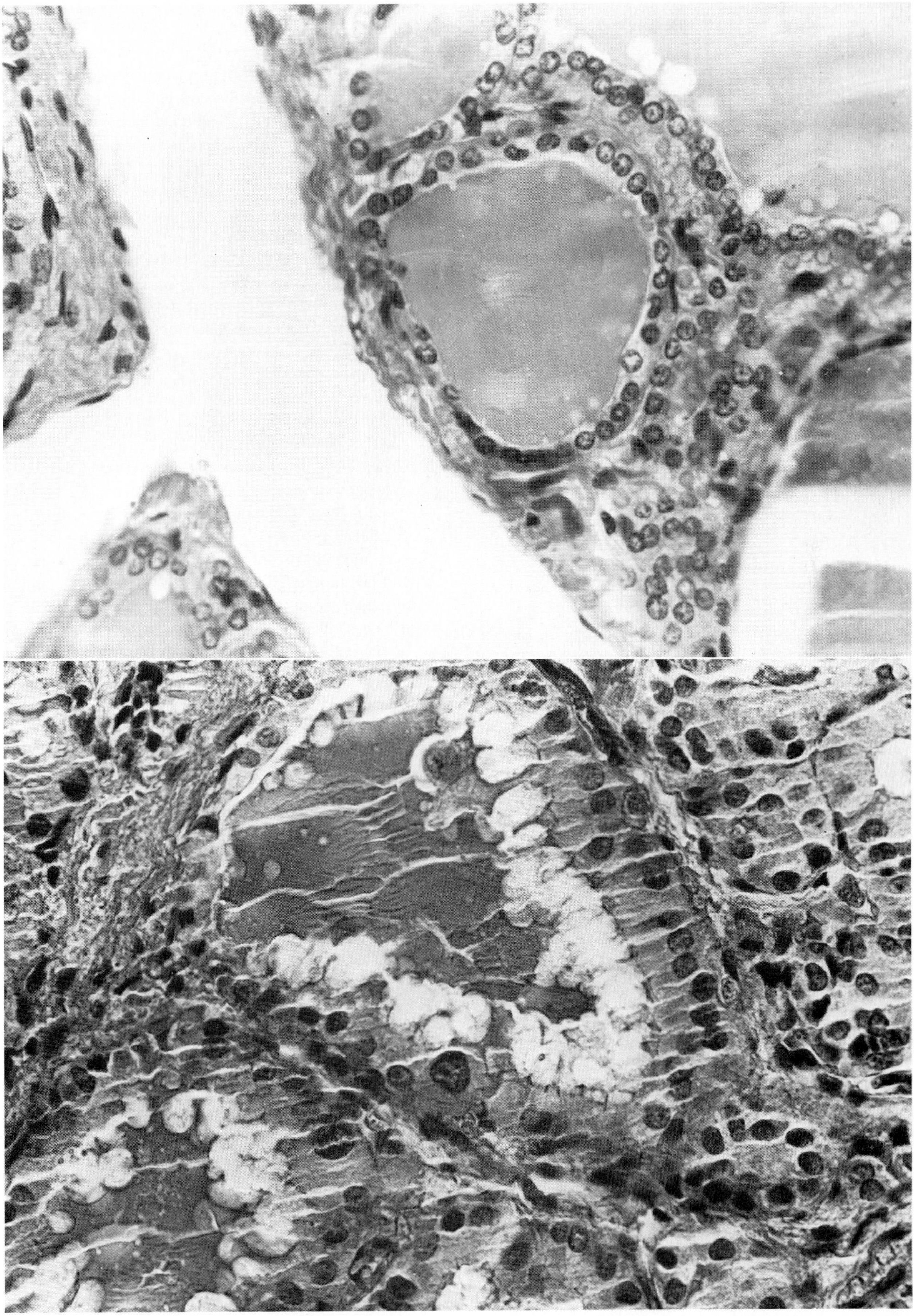

A

B

■ **Fig. 53-1.** Photomicrographs of human thyroid gland follicles. **A,** Normal follicle lined by cuboidal cells and filled with colloid. **B,** TSH-stimulated follicle lined by hypertrophied columnar cells with scalloping of the colloid caused by active proteolysis. (Courtesy Dr. William Hawk).

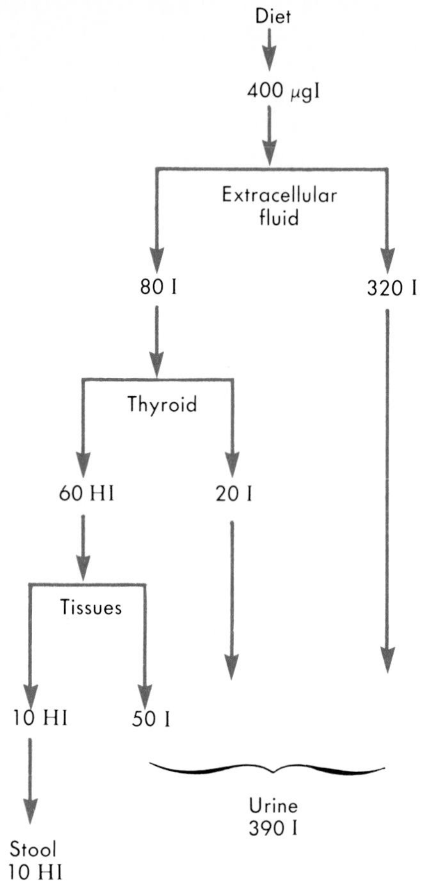

■ **Fig. 53-2.** Average daily iodide turnover in humans (United States). Note that 20% of the intake is taken up by the thyroid gland, and 15% turns over in hormone synthesis and disposal. The unneeded excess is excreted in the urine. *I*, Iodide; *HI*, hormonal iodide.

dide content of the thyroid gland averages 7500 μg, virtually all of which is in the form of iodothyronines. In a steady-state condition 70 to 80 μg of iodide, or about 1% of the total, also is released from the gland daily. Of this amount, 75% is secreted as thyroid hormone, and the remainder is free iodide. The very large ratio (100:1) of iodide stored in the form of hormone to the amount turned over daily, initially protects the individual from the effects of iodide deficiency for at least 2 months. Further conservation of iodide then is effected by a marked reduction in its renal excretion as the circulating concentration and filtered load fall.

Iodide is actively transported into the gland against a normal thyroid/plasma free iodide ratio of 30. This so-called iodide trap requires energy generation via oxidative phosphorylation, but the chemical nature of the transport system remains obscure. Some evidence links it to a Na,K-ATPase. The iodide trap is markedly stimulated by TSH. There is evidence that this is mediated through cAMP and also that it involves stimulation of the synthesis of a specific protein, possibly the iodide carrier itself. The trap displays saturation kinetics. However, the K_m of this transport system is above the estimated normal plasma concentration of iodide (2×10^{-8}M). Thus small to moderate increases in iodide intake lead to increases in thyroidal iodide, and the gland/plasma ratio of 30 is maintained. Initially this acts to increase the rate of thyroid hormone synthesis to above normal. However, as the dose of iodide exceeds 2 mg/day, an intraglandular concentration of iodide or of some organic iodide product is reached that is inhibitory to the iodide trap and to the biosynthetic mechanism, so that hormone production declines back to normal. This autoregulatory phenomenon is known as the *Wolff-Chaikoff effect*. In unusual instances the inhibition of hormone synthesis by iodide can be great enough to induce thyroid hormone deficiency. A primary reduction in dietary iodide intake depletes of the iodide pool and greatly enhances the activity of the thyroid iodide trap. Under these circumstances the percentage uptake of a radioactive iodide tracer can increase to 80% to 90%. If the lack of iodide is severe enough, thyroid hormone deficiency results.

A number of anions, such as thiocyanate (CNS^-) and perchlorate $HClO_4^-$), act as competitive inhibitors of active iodide transport. If for any reason iodide cannot be rapidly incorporated into tyrosine after its uptake by the cell, then administration of one of these competitive anions will, by blocking further uptake, cause a rapid discharge of the iodide from the gland in accordance with the high thyroid/plasma concentration gradient. This discharge can be demonstrated by monitoring the thyroid gland in vivo after labeling the iodide pool with radioactive tracer, a maneuver that may assist in the diagnosis of biosynthetic defects. Another competitive inhibitor, the pertechnetate ion, also is taken up like iodide, but it cannot be significantly incorporated into organic molecules. In its radioactive form as $^{99m}TcO_4$, it is a useful substitute for radioactive iodide in the measurement of the trapping function and in the visualization of thyroid gland anatomy by external isotope scanning with a photon detector.

Once within the gland, iodide rapidly moves to the apical surface of the cell and into the lumen. Iodide (I^-) is immediately oxidized to iodine (I^0) as a postulated intermediate and is immediately incorporated into tyrosine molecules (Fig. 53-3). The latter are not free in solution, but they exist in peptide linkages as components of *thyroglobulin*. This is a 19S glycoprotein with a molecular weight of 670,000. Thyroglobulin is synthesized on the rough endoplasmic reticulum as peptide units of molecular weight 330,000 (the primary translation product of its messenger RNA). These units combine and then carbohydrate moieties are added in transit to the Golgi apparatus. The completed protein, incorporated in small vesicles, moves to the apical plasma membrane and then into the adjacent lumen of the follicle.

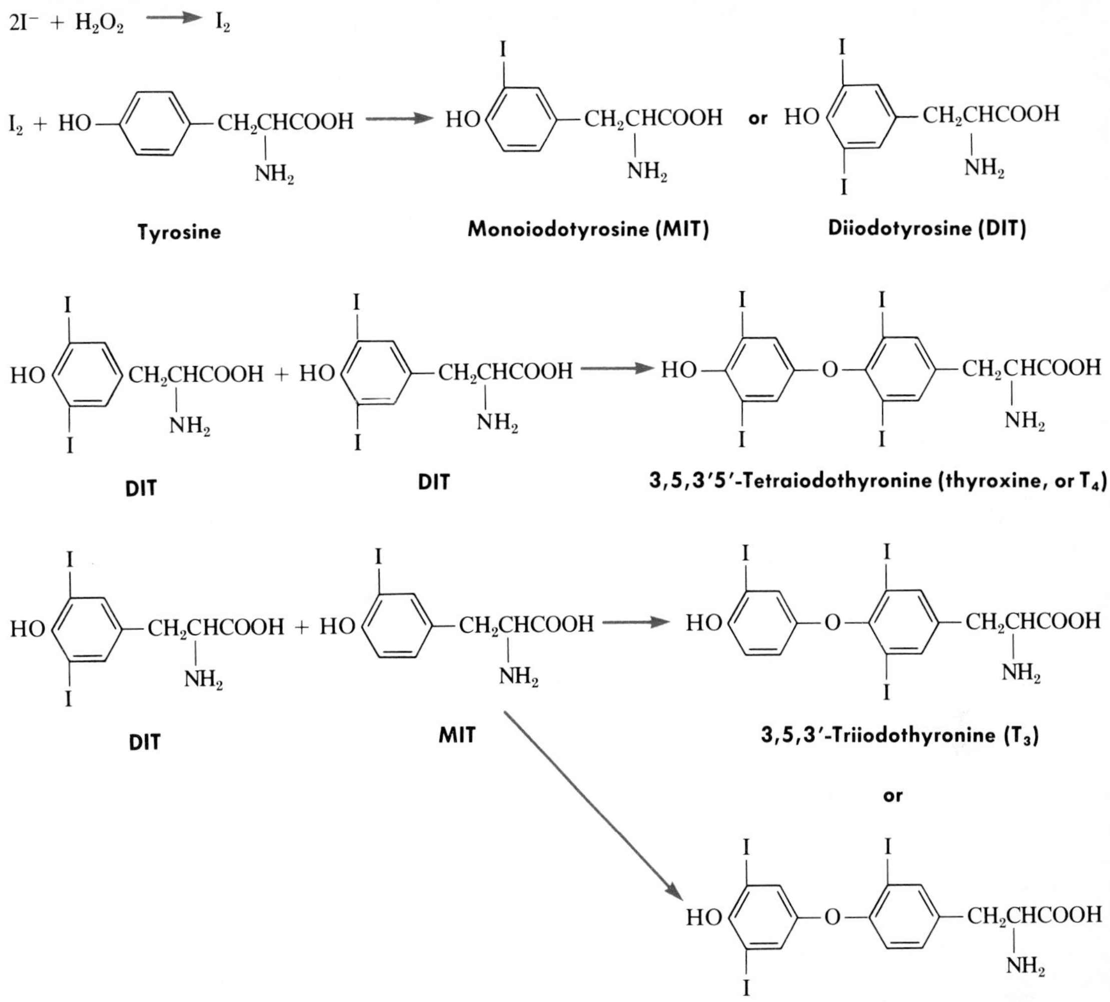

■ **Fig. 53-3.** Overall pathway of thyroid hormone synthesis.

3,3′5′-Triiodothyronine (reverse T_3)

Just within the follicle, iodide is incorporated into the thyroglobulin. As a result of iodination, both monoiodotyrosine (MIT) and diiodotyrosine (DIT) are formed (Fig. 53-3). Thereafter two DIT molecules are coupled to form T_4, or one MIT and one DIT molecule are coupled to form T_3. Very little rT_3 is synthesized. This entire sequence of reactions is catalyzed by *thyroid peroxidase,* an enzyme complex adjacent to or within the cell membrane bordering the follicular lumen. The immediate oxidant (electron acceptor) for the reaction iodide → iodine is hydrogen peroxide. The mechanism whereby hydrogen peroxide is itself generated in the thyroid gland remains debatable, but evidence suggests reduction of oxygen by NADPH or NADH via cytochrome reductases. The intimate chemistry of tyrosine iodination and iodotyrosine coupling likewise is unsettled.

A single tyrosine, located at the fifth position from the N-terminus of each of the two constituent peptides of thyroglobulin (site A), may be a preferential but not an exclusive site of synthesis of T_4 or T_3. Both iodide and this tyrosine may be initially oxidized to free radicals that are complexed to the peroxidase enzyme and then combined to form MIT or DIT. Coupling may subsequently be facilitated by cleaving a smaller peptide containing MIT or DIT at site A and orienting it into juxtaposition with another MIT or DIT molecule buried deeper within thyroglobulin. The latter MIT or DIT donates its iodinated phenolic ring to the MIT or DIT at site A. This produces T_4 or T_3 at site A and leaves dehydroalanine in peptide linkage within the deeper portion of thyroglobulin.

The entire sequence of thyroglobulin iodination is carried out very rapidly; labeled iodide appears in hormone molecules 1 minute after in vivo administration, and by 1 hour 90% to 95% of iodide is organically bound. The usual distribution of iodoaminoacids, as residues per molecule of thyroglobulin, is MIT, 7; DIT, 6; T_4, 2; and T_3, 0.2. Approximately one third of the iodine in thyroglobulin is in the form of calorigenic hormone (T_4 and T_3). Certain factors regulate the proportion of T_3 to T_4 that is synthesized. When iodide avail-

A

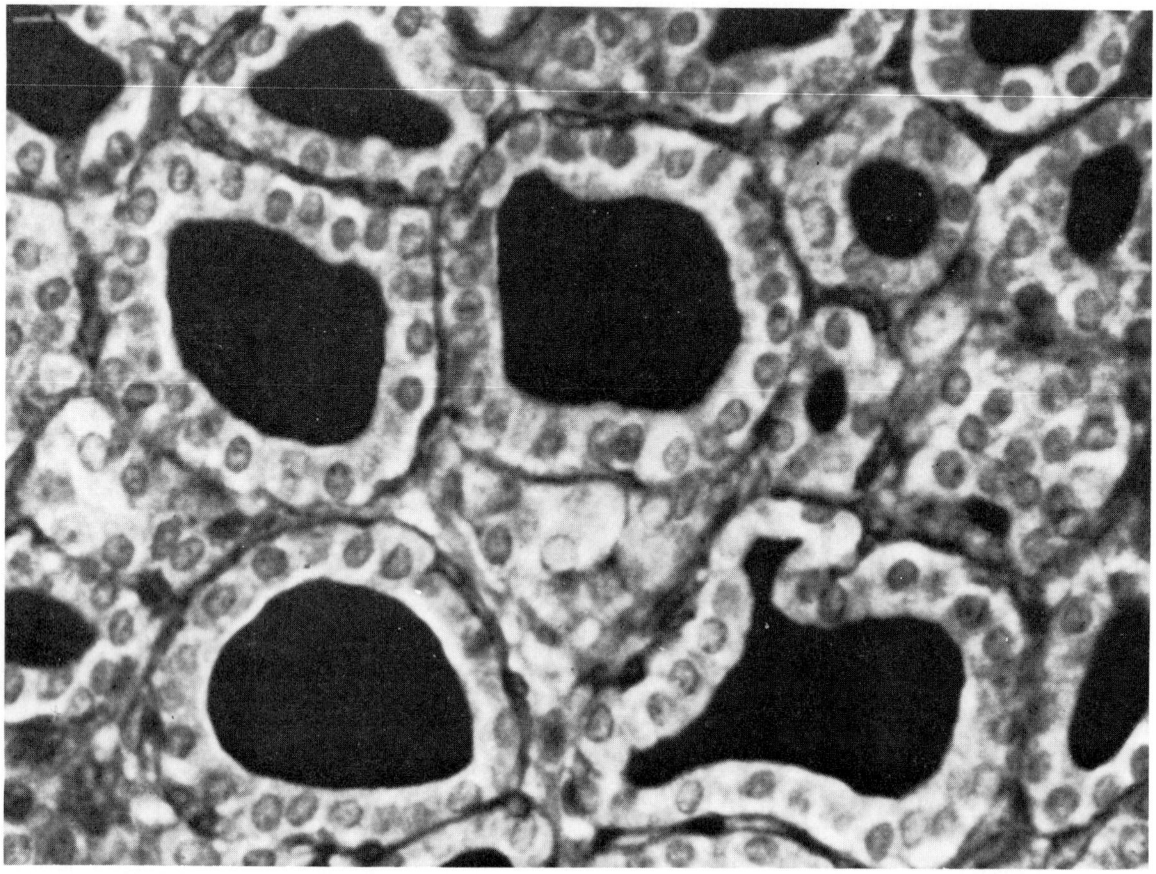

■ **Fig. 53-4.** Histological demonstration of the process of resorption of colloid. **A,** Unstimulated follicles, **B,** Within minutes of TSH administration colloid droplets are seen inside the follicular cells. (From Wollman, S.H., et al.: J. Cell Biol. **21:**191, 1964. Reproduced by copyright permission of The Rockefeller University Press.)

B

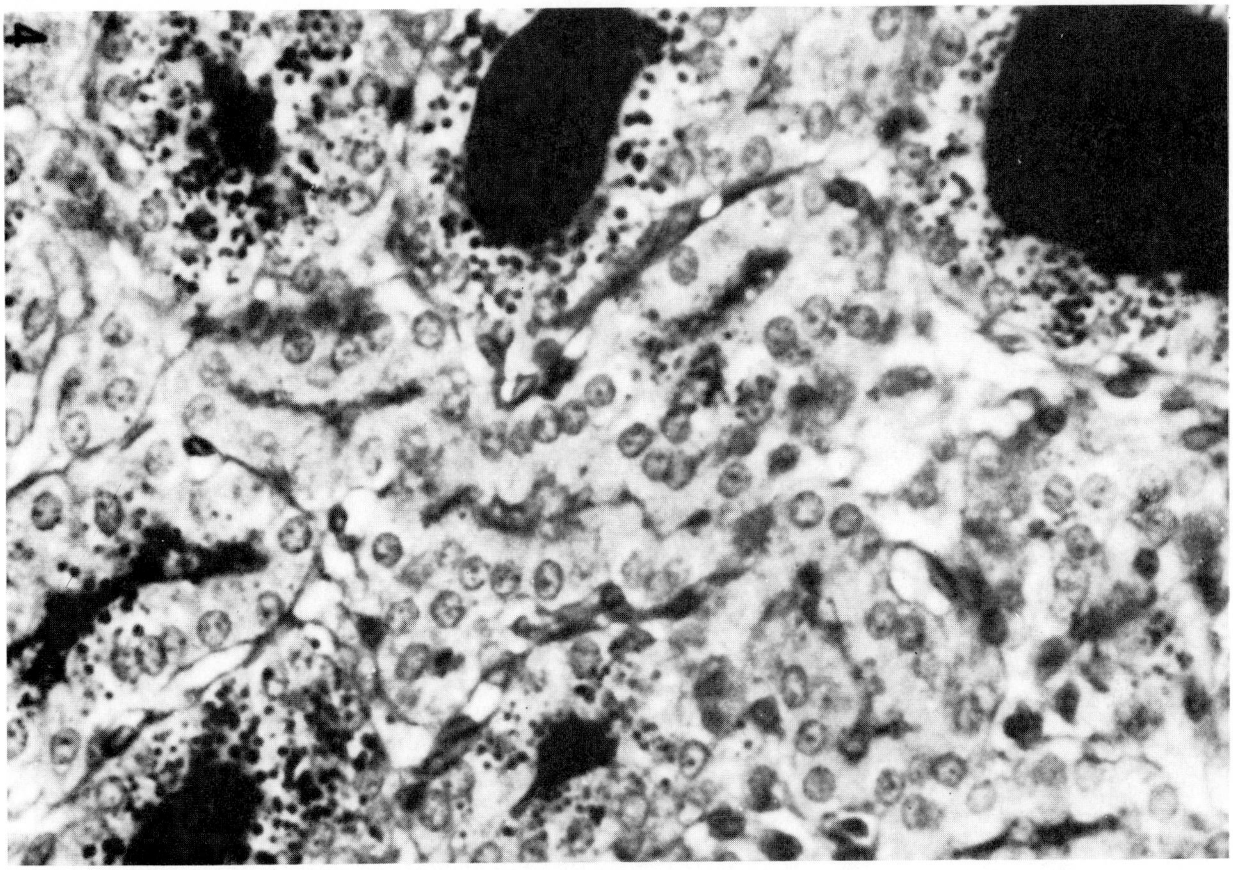

■ **Fig. 53-4.** For legend see opposite page.

ability is restricted, the formation of T_3 is favored. Because T_3 is three times as potent as T_4, this response provides more active hormone per molecule of organified iodide. The proportion of T_3 also is increased when the gland is hyperstimulated by TSH or other activators. Finally, certain autonomous thyroid adenomas preferentially secrete T_3 for unknown reasons.

Once thyroglobulin has been iodinated, it is translocated into the lumen of the follicle, where it is stored as colloid. Release of the peptide-linked T_4 and T_3 into the bloodstream requires proteolysis of the thyroglobulin. Histochemical and radiographic studies have demonstrated that the colloid is retrieved from the lumen of the follicle by the epithelial cell through the process of endocytosis. The plasma cell membrane forms pseudopods that engulf a pocket of colloid. After this portion of the luminal content has been pinched off by the plasma cell membrane, it appears as a colloid droplet within the cytoplasm (Fig. 53-4). The droplet moves through the cytoplasm in a basal direction, probably as a result of microtubule and microfilament function. At the same time lysosomes move from the base toward the apex of the cell and fuse with the colloid droplets that are moving to meet them. The action of the lysosomal proteases then releases free T_4 and T_3, which leave the cell through the plasma membrane at the basal end and enter the bloodstream via the adjacent rich capillary plexus.

The MIT and DIT molecules, which also are released on proteolysis of thyroglobulin, are rapidly deiodinated within the follicular cell by the enzyme *deiodinase*. Since these compounds are metabolically useless and would be lost in the urine if secreted, their deiodination retrieves the iodide for recycling into T_4 and T_3 synthesis. Only minor amounts of intact thyroglobulin leave the follicular cell under normal circumstances. However, in certain pathological processes and in hyperthyroid states thyroglobulin escapes at a greater rate and circulating concentrations of the storage protein increase. Each of the preceding steps in the sequence of thyroid hormone synthesis and release is discrete, so that specific congenital biosynthetic defects occur that result in thyroid hormone deficiency. The entire process of secretion is summarized in Fig. 53-5.

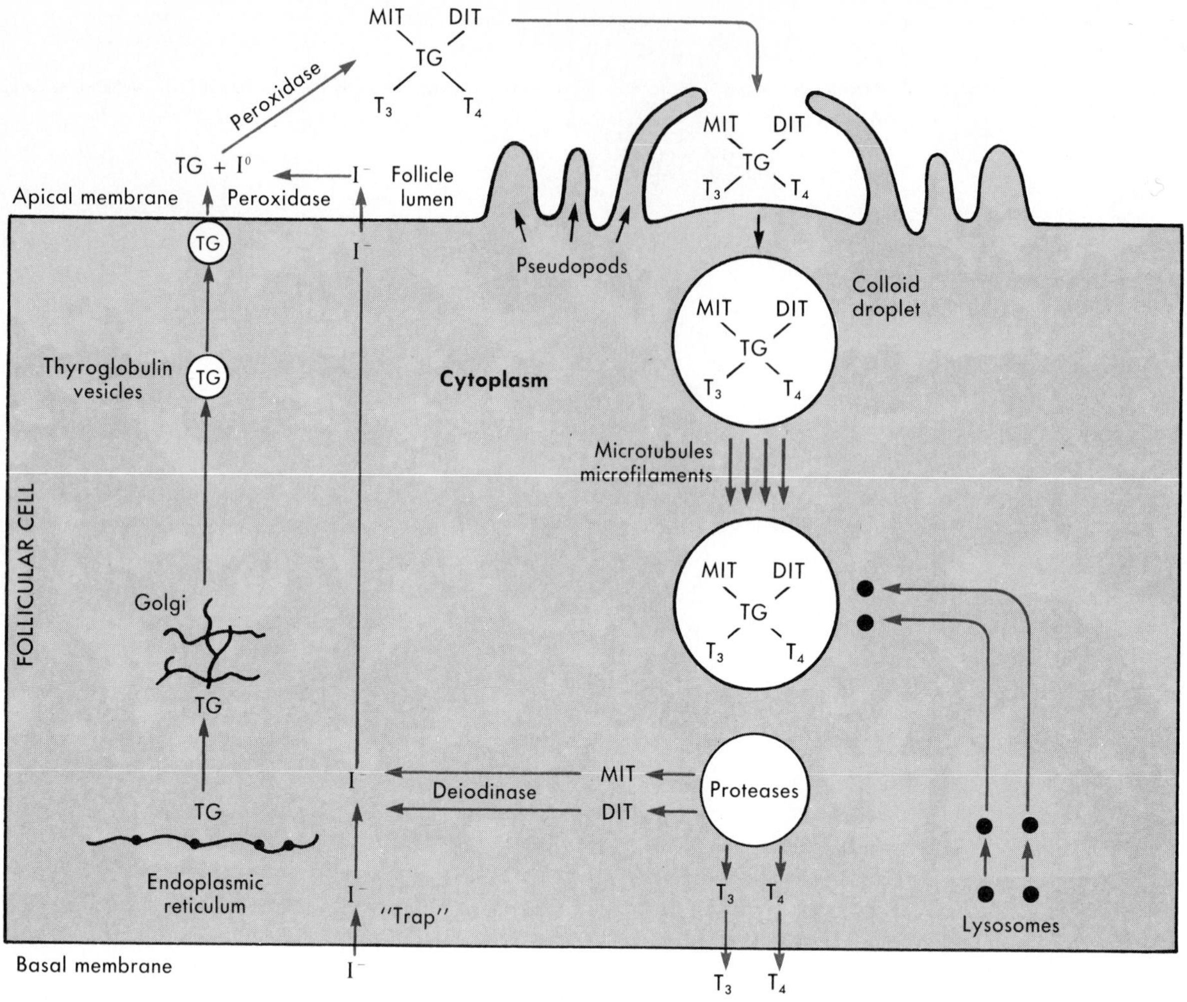

■ **Fig. 53-5.** Overall schema of thyroid hormone synthesis and release. Note that iodide is incorporated into tyrosine molecules, which are in peptide linkage in thyroglobulin *(TG)*. The stored T_4 and T_3 are retrieved by endocytosis of colloid from the follicular lumen followed by proteolysis with lysosomal enzymes. Iodide is recovered from MIT and DIT by the action of deiodinase.

■ *Regulation of Thyroid Gland Activity*

The most important regulator of thyroid gland function and growth is TSH. Because TSH secretion shows only minimal diurnal or day-to-day variation, thyroid hormone secretion and plasma concentrations are also relatively constant. Small nocturnal increases in secretion of TSH and release of T_4 have been described; they are of unknown physiological significance. TSH stimulates the process of iodide trapping and of each step in T_4 and T_3 synthesis. It also stimulates the endocytosis of colloid, the proteolysis of thyroglobulin, and the release of T_4 and T_3 from the gland. Sustained TSH stimulation leads to hypertrophy and hyperplasia of the follicular cells (Fig. 53-1, *B*). The enlarged cells show an increased volume of endoplasmic reticulum, increased numbers of ribosomes, and a larger and more complex Golgi apparatus. An increase in DNA synthesis is shown by increased incorporation of tritiated thymidine. Proliferation of capillaries also is observed, and thyroid blood flow increases. In the absence of TSH, marked atrophy of the gland occurs. However, a low basal level of thyroid hormone production and release usually continues in humans, seemingly independent of TSH.

The effects of TSH are exerted through a multiplicity of actions. The initial step in TSH action is binding to a plasma membrane receptor. This transmembrane molecule consists of two TSH binding components, a glycoprotein and a ganglioside. The latter is functionally linked to adenylate cyclase; binding at this site increases cAMP, which then mediates TSH stimulation of iodide uptake by the cell. The glycoprotein site acting through the phosphatidylinositol system as second messenger mediates TSH stimulation of the subsequent steps in thyroid hormone synthesis. Within minutes of thyroid cell exposure to TSH, thyroglobulin, stored within the follicular lumen, undergoes endocytosis, and colloid droplets appear in the cytoplasm. Shortly thereafter an increase in iodide uptake and in peroxidase activity can be demonstrated. Coincident with these actions on hormone synthesis and release, TSH also stimulates glucose oxidation, especially via the hexose monophosphate shunt. This may be the means for generating the NADPH needed for the peroxidase reaction. Further effects of TSH on the thyroid gland are seen after a delay of hours to days. Nucleic acid and protein synthesis is generally increased, and effects on both transcription and translation have been described. Phospholipid synthesis also is stimulated. The mechanism of these actions remains to be further elucidated. They probably relate to the effects of TSH on promoting growth of the gland.

The regulation of thyroid hormone secretion by TSH is under exquisite feedback control (Figs. 52-6 and 52-7). Circulating T_4 and T_3 each produce feedback to the pituitary to decrease TSH secretion; as their levels fall, TSH secretion increases. It is free T_4 and T_3, not the protein-bound portions, that regulate pituitary TSH output. Since the pituitary gland is capable of deiodinating T_4 to T_3, the latter may be the final effector molecule in turning off TSH. In situations in which the diseased thyroid gland produces thyroid hormones autonomously or in which exogenous thyroid hormone is administered chronically, plasma TSH will be low.

Another important regulator of thyroid gland function is iodide itself, which has a biphasic action. At relatively low levels of iodide, thyroid hormone synthesis is positively correlated with iodide availability. However, beyond the levels needed for maximum synthesis, an important negative effect of iodide on hormone secretion occurs. A large excess of iodide strongly inhibits the synthesis of T_4 and T_3. In addition, probably by an independent mechanism, an excess of iodide eventually also inhibits its own binding by the follicular cells. Additional modes of autoregulation may play a role in preventing excessive responses to TSH stimulation. Thyroglobulin inhibits binding of TSH to its receptors, as well as the response of adenylate cyclase to the tropic hormone. In addition, T_4 and T_3 also directly inhibit the thyroid gland in vitro. The close proximity of sympathetic nerve fibers to the capillary plexuses in the gland and to the basal membrane of the epithelial cells suggests a regulatory influence. Thyroid hormone secretion is somewhat stimulated by epinephrine, via β-adrenergic receptors. However, the physiological significance of this, as well as an inhibitory influence of acetylcholine, is unknown. Similar clarification is awaited for the role of prostaglandins, which are present in the thyroid and can mimic some effects of TSH.

Thyroid hormones increase energy expenditure and heat production. Therefore it would be logical to expect that thyroid hormone availability would respond to the body's changing caloric and thermal status. In fact ingestion of an excess of calories, particularly in the form of carbohydrate, increases T_3 production and plasma concentration and the metabolic rate, whereas prolonged fasting leads to corresponding decreases. However, similar fluctuations in T_4 do not occur. Therefore, since most T_3 arises from circulating T_4 (Table 53-1), peripheral mechanisms are more important in mediating these changes than are alterations in thyroid gland secretion. In animals exposure to cold increases thyroid gland activity. In humans this is clearly seen only in the neonatal period when the infant suddenly becomes responsible for maintaining his or her own body temperature. At birth an acute rise in TSH secretion is followed by a rise in plasma T_4 to levels well above those of adults. Over the ensuing weeks or months plasma T_4 then subsides to a range that remains stable throughout life.

Table 53-1 Thyroid hormone turnover

	T_4	T_3	rT_3
Daily production (μg)	90	35	35
From thyroid (%)	100	25	5
From T_4 (%)	—	75	95
Extracellular pool (μg)	850	40	40
Plasma concentration			
Total (μg/dl)	8.0	0.12	0.04
Free (ng/dl)	2.0	0.28	0.20
Half-life (days)	7	1	0.8
Metabolic clearance (L/day)	1	26	77
Fractional turnover per day (%)	10	75	90

Pharmacological inhibition of thyroid gland activity is of considerable therapeutic importance. A class of drugs known as *thiouracils* (for example, propylthiouracil and methimazole) blocks the synthesis of T_4 and T_3 by acting on the peroxidase complex. After continuous administration for weeks the stores of thyroid hormone (and of iodide) become depleted. Because of the inhibition of organification, iodide taken up by activity of the trap is rapidly discharged again, as shown by studies with radioactive iodine. The thiouracils are very effective in the treatment of *hyperthyroidism*. Lithium salts, in the same doses used to treat manic depressive illness, inhibit the release of thyroid hormones and probably, secondarily, their synthesis. Lithium may act by blocking adenylate cyclase and cAMP accumulation. Finally, a large excess of iodide, in addition to the effects previously noted, can also promptly inhibit thyroid hormone release. Although this is a transient action, the administration of iodide may benefit severely thyrotoxic individuals may produce an important substantially.

Metabolism of Thyroid Hormones

Table 53-1 shows the daily production rates, pool sizes, plasma concentrations, half-lives, metabolic clearances, and fractional turnovers of T_4, T_3, and rT_3. T_4 is clearly the dominant secreted and circulating form of thyroid hormone. It is important to note that the major portion of T_3 and virtually all of rT_3 come secondarily from circulating T_4, rather than primarily from thyroid gland secretion. Thus T_4 serves as a prohormone for T_3, in addition to providing some intrinsic intracellular action of its own. This "storage" function of plasma T_4 also is reflected in its much lower metabolic clearance and fractional turnover rates, compared with those of T_3 or rT_3. The concentration of intact thyroglobulin in plasma, averaging about 5 ng/ml, is much lower than that of the thyroid hormones.

Secreted T_4 and T_3 circulate in the bloodstream almost entirely bound to proteins. Normally only 0.03% of total plasma T_4 and 0.3% of total plasma T_3 are in the free state (Table 53-1). However, these are the critical fractions that are *biologically active,* not only in exerting thyroid hormone effects on peripheral tissues but in pituitary feedback as well. The major binding protein is *thyroxine-binding globulin* (TBG). This is a glycoprotein α-globulin with a molecular weight of 63,000 which is synthesized in the liver. Each TBG molecule binds one molecule of T_4; at the normal TBG concentration of 1.5 ng/dl, 20 μg of T_4 can be bound per deciliter.

About 70% to 80% of circulating T_4 and T_3 is bound to TBG; the remainder is bound to albumin and to a thyroid-binding prealbumin (TBPA). Compared with TBG, these additional binding proteins have much lower affinities but much higher capacities for T_4 and T_3. Ordinarily, however, only alterations in TBG concentration significantly alter total plasma T_4 and T_3 levels.

Two biological functions have been ascribed to TBG. First, it maintains a large circulating reservoir of T_4, which buffers against acute changes in thyroid gland function. Even the addition to the plasma of the calorigenic hormone needed for an entire day would cause only a 10% increase in the total T_4 concentration (Fig. 53-6). Conversely, after removal of the thyroid gland it would take 1 week for the plasma T_4 concentration to fall as much as 50%. Second, the binding of plasma T_4 and T_3 to large proteins prevents the loss of these relatively small hormone molecules into the urine and thereby helps conserve iodide.

The reservoir function of TBG is best understood by examining the chemical equilibrium between T_4 and TBG. This equilibrium governs the distribution of the hormone between the free (T_4) and bound (T_4·TBG) forms.

$$T_4 + TBG \leftrightharpoons T_4 \cdot TBG \qquad (1)$$

$$Keq = \frac{[T_4 \cdot TBG]}{[T_4]\,[TBG]} \qquad (2)$$

$$\frac{[T_4]}{[T_4 \cdot TBG]} = \frac{\text{Free } T_4}{\text{Bound } T_4} = \frac{1}{Keq\,[TBG]} \qquad (3)$$

$$[T_4] = [T_4 \cdot TBG] \times \frac{1}{Keq\,[TBG]} \qquad (4)$$

A temporary decrease in free T_4, caused by a decrease in thyroid gland output or accelerated uptake by target cells, can be rapidly compensated for by a dissociation of bound T_4 (T_4·TBG), until the new ratio of T_4/(T_4·TBG) returns to that required by Keq (equation 3). A temporary increase in free T_4, caused by endogenous secretion or exogenous administration, can be rapidly compensated for by association of the excess with TBG, because normally only 30% of the available T_4 binding sites on TBG are occupied. Of course sustained decreases or increases in T_4 supply, resulting from thyroid disease, must eventually lead to sustained decreases or increases in free T_4, because the latter is directly proportional to T_4·TBG (equation 4).

It is important to note that a primary change in TBG concentration will also disturb the ratio of free to bound T_4 (equation 3). In this circumstance the normal thyroid gland must increase or decrease its rate of hormone secretion appropriately, until the new equilibrium state restores the free T_4 level to normal. TBG concentration can decrease because of reduced hepatic synthesis (liver disease) or excessive loss in the urine (kidney disease). Free T_4 then will increase temporarily. In compensation, pituitary TSH secretion will be suppressed by negative feedback. T_4 output by the thyroid gland then will decrease until the new, lower steady-state level of T_4·TBG yields a normal level of free T_4 (equation 4). Hepatic synthesis of TBG also can be stimulated (for example, by estrogen administration or pregnancy), leading to an increase in TBG concentration. Free T_4 will decrease temporarily; this will stimulate pituitary secretion of TSH, and consequently T_4 output by the thyroid gland will increase. This will continue until the elevated level of T_4· TBG is sufficient to restore the free T_4 level to normal in a new steady-state condition.

Identical qualitative considerations govern the circulating levels of free and bound T_3. However, the buffering action of TBG is less effective, because the Keq for T_3 is an order of magnitude lower than that for T_4 (2×10^9 versus 2×10^{10}, respectively), and because the total extrathyroidal pool of T_3 is only a twentieth that of T_4 (Table 53-1). Thus rapid addition of T_3 equivalent to the calorigenic hormone needed for an entire day produces greater swings in the concentrations of total and free T_3 (Fig. 53-6). Although alterations in TBG are not usually caused by thyroid gland disease, they must be taken into account when plasma thyroid

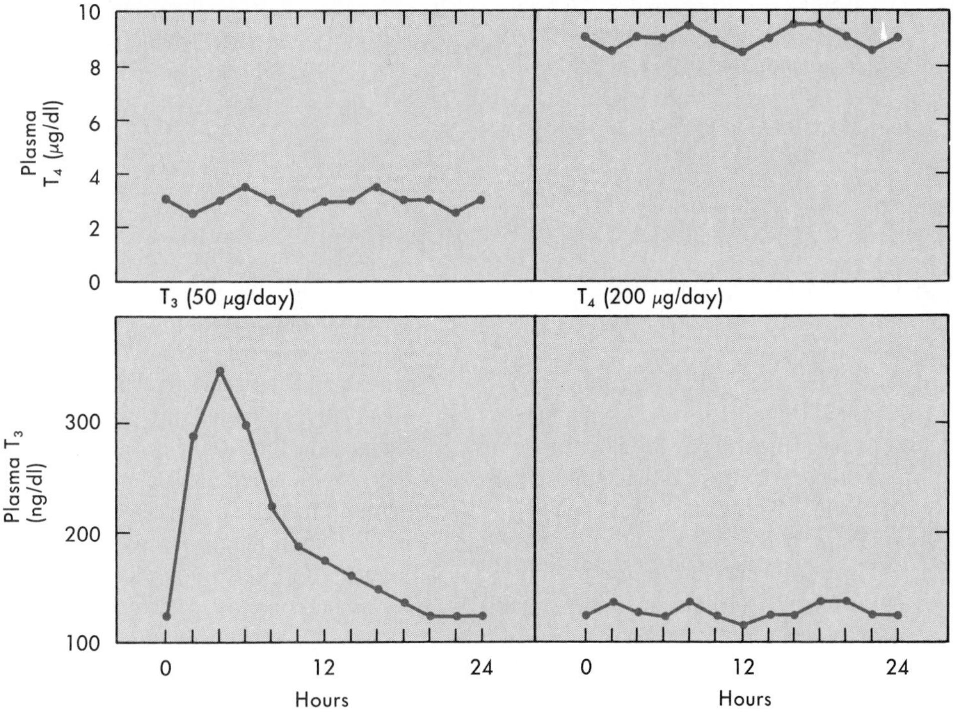

■ **Fig. 53-6.** Effect of administering a single day's supply of calorigenic hormone by mouth to hypothyroid individuals. Note that 50 μg of T_3 elevates plasma T_3 levels for many hours, whereas 200 μg of T_4 causes no significant change in either plasma T_4 or T_3. This occurs because of the much larger pool size and tighter protein binding of T_4 than of T_3. (Redrawn from Saberi, M., et al.: Serum thyroid hormone and thyrotropin concentrations during thyroxine and triiodothyronine therapy, J. Clin. Endocrinol. Metab. **39:**923, 1974. Copyright 1974. Reproduced by permission.)

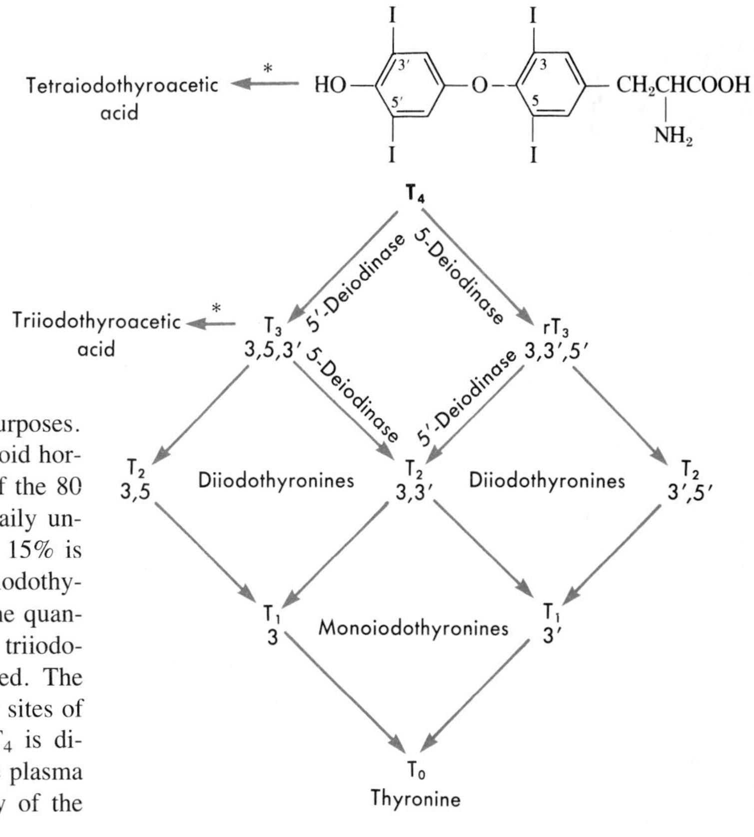

■ **Fig. 53-7.** Peripheral metabolism of thyroxine (T_4) by successive deiodinations. A key regulatory step is the proportion of T_4 undergoing the initial deiodination to metabolically active T_3 versus metabolically inactive rT_3. Asterisks signify oxidative deamination and decarboxylation.

hormone levels are measured for diagnostic purposes.

The peripheral metabolism of circulating thyroid hormones is outlined in Fig. 53-7. The majority of the 80 to 90 μg of T_4 that is released by the gland daily undergoes deiodination. However, approximately 15% is irreversibly excreted in the bile as the various iodothyronines in glucuronide or sulfate conjugates. The quantitative importance of tetraiodoacetic acid and triiodoacetic acid as metabolites remains to be defined. The liver, kidney, and skeletal muscle are the major sites of degradation. The overall rate of disposal of T_4 is directly related to the free T_4 concentration in the plasma and to the intracellular T_4 content, particularly of the liver. Thus T_4 increases its own degradative metabolism. The entire cascade of products—T_3, T_2, and T_1—of the sequential deiodination steps is regularly increased in the plasma in *hyperthyroidism* (T_4 excess) and is usually decreased in *hypothyroidism* (T_4 deficiency).

Relationship between Hormone Metabolism and Hormone Action

The initial step in T_4 metabolism, the intracellular conversion of T_4 to either T_3 or rT_3, is of critical importance to thyroid hormone action. T_3 is the hormone of greatest biological activity, whereas rT_3 has almost no apparent calorigenic action. Therefore factors that regulate the relative rates of inner ring versus outer ring monodeiodination (Fig. 53-7) also determine the final biological effect of secreted T_4. In humans the normal distribution of T_4 products is 45% T_3 and 55% rT_3. An increase in T_4 concentration leads to a decrease in its conversion to T_3. Thus the biological effects of T_4 excess or deficiency are automatically mitigated to a slight extent by accelerated or retarded metabolic inactivation, respectively. Other states and factors are associated with reduced conversion of T_4 to T_3 and often with a reciprocally enhanced conversion of T_4 to rT_3. These include the gestational period, fasting, stressful states, catabolic diseases, hepatic disease, renal failure, glucocorticoid excess, thiouracil drugs, and β-adrenergic blockade. A single common denominator responsible for switching T_4 away from T_3 and toward rT_3 has not been identified with certainty in all these conditions. However, in many instances the inhibition of the enzyme *5′-monodeiodinase* appears to explain this switch. As seen in Fig. 53-7, such inhibition would decrease production of T_3 from T_4 (reducing plasma T_3) and simultaneously decrease degradation of rT_3 to $3,3'T_2$ (increasing plasma rT_3).

The biological effects of T_4 are largely a result of its intracellular conversion to T_3. When administered exogenously, T_3 is three to four times more potent than T_4 in humans. However it is still debatable how much intrinsic biological activity of its own is possessed by T_4. Evidence favoring activity of T_4 is found in situations in which a low plasma T_4 level is accompanied by a state of biological thyroid deficiency despite a normal plasma T_3, and in situations in which a clinically normal state exists with a normal plasma T_4 concentration despite a low plasma T_3. Furthermore, in the absence of endogenous thyroid gland function the maintenance of a euthyroid state requires exogenous doses of T_3 that sustain supranormal plasma T_3 levels, whereas only doses of T_4 that sustain normal plasma T_4 levels are required. The T_3 that has been generated from T_4 intracellularly might be more efficient than the T_3 reaching its intracellular sites of action from the circulation; rT_3 is almost without calorigenic activity.

Intracellular Actions of Thyroid Hormone

T_4 and T_3 appear to enter cells freely by passive diffusion. Within the cell most if not all the T_4 undergoes 5′-monodeiodination to T_3 (or rT_3). Unlike steroid hormones, T_3 and T_4 require no cytoplasmic receptor for subsequent transport into the nucleus (Fig. 49-10). Instead they are bound directly to a specific nuclear receptor protein that has a molecular weight of 50,000 and that is associated with chromatin, usually template-inactive. Recent evidence suggests that the structure of the receptor may vary somewhat in a tissue-specific manner. T_3 binds to this receptor with much greater affinity than does T_4. Histones may enhance this affinity, though the receptor is not itself a histone protein. The T_3 receptor complex interacts with DNA to stimulate or inhibit transcription of messenger RNAs. The latter then direct increased or decreased synthesis of specific proteins. This sequence of events has been well-documented in several in vitro systems, such as the synthesis of growth hormone by cultured pituitary cells. T_3 stimulates parallel increases in the nuclear content of the primary RNA transcript of the growth hormone gene and the cytoplasmic content of messenger RNA for growth hormone, followed by an increase in synthesis of the protein hormone itself. In any one tissue, e.g., the liver, a large number of messenger RNA levels are altered by thyroid hormone. Furthermore, the effect of the hormone on a particular gene product, e.g., myosin ATPase, can vary in direction from tissue to tissue and even at different stages of development. Thus complex factors must interact in determining the overall regulatory action of thyroid hormone on gene expression. In addition to nuclear receptors, thyroid hormone–binding sites have been identified in ribosomes, mitochondria, and the plasma membrane. These may mediate posttranscriptional and pretranslational events, such as association of messenger RNA with ribosomes, or posttranslational processes, such as membrane transport.

The responsiveness of tissues to T_3 correlates well with their nuclear receptor capacity and with the degree of receptor saturation. In euthyroid subjects about half the available T_3 receptor sites are occupied. In humans several indices of T_3 action have been shown to correlate linearly with the percentage occupancy of nuclear receptors; occupancy was calculated from serum T_3 concentration and the known affinities of human liver and kidney T_3 receptors. Such observations strongly support the physiological importance of nuclear receptor binding for T_3 action. Furthermore, in some tissues T_3 may downregulate its own receptor by inhibiting its synthesis, providing still another means for receptor modulation of T_3 action. Since thyroid hormone appears to act largely through influencing transcription, many of its effects can be blocked by inhibitors of protein synthesis, such as puromycin. This mechanism accounts for the 12- to 48-hour delay before most of the hormone's effects become evident in vivo. Indeed several weeks of T_4 replacement are required before all the consequences of the hypothyroid state are eliminated.

A voluminous amount of experimental data has been obtained in the attempt to explain the multitude of thyroid hormone actions on an intracellular basis. A large catalogue of several hundred thyroid-induced enzyme and substrate changes can be listed. To date, however, no single biochemical mechanism or final common pathway can be offered incontrovertibly as a unifying theory of hormone action. More likely, actions exist at multiple loci, which may vary in different tissues. Effects in the nucleus include stimulation of RNA polymerase and phosphoprotein kinases and the synthesis of other nuclear proteins. These nuclear effects are followed or paralleled by an increase in the size and number of mitochondria with a corresponding increase in their rate of respiration. Direct effects of T_4 and T_3 on mitochondrial RNA and protein synthesis also have been noted. Key respiratory enzyme activities, such as NADPH cytochrome C reductase and cytochrome oxidase, are increased. α-Glycerophosphate dehydrogenase and pyridine nucleotide transhydrogenases, important in regulating the levels of pyridine nucleotide cofactors, are likewise increased. The level of malic enzyme, involved in providing NADPH for fatty acid synthesis, and its messenger RNA, is also greatly augmented by T_3. In this interesting example, T_3 acts synergistically with high carbohydrate feeding to provide an enzyme important in converting an excess of glucose to triglycerides. Here T_3 may be acting as a multiplier of a primary carbohydrate-generated signal. The activities of a host of other enzymes concerned with glucose oxidation and gluconeogenesis also are augmented by thyroid hormone.

The clinical observation that thyroid excess appears to increase the rate of oxygen use without increasing useful work output suggests that T_4 and T_3 decreases the efficiency with which high-energy phosphate bonds are formed during aerobic respiration. Early in vitro studies support the hypothesis that thyroid hormone acts like dinitrophenol to uncouple oxidative phosphorylation. However, supraphysiological doses of thyroid hormone often were required for such an effect. Also this uncoupling effect could be obtained with thyroxine analogues that had no thyroid hormone activity in vivo. Furthermore normal P/O ratios of approximately 3 subsequently were observed in muscle from hyperthyroid humans—an observation weakening the support for this theory.

A more recent and tenable explanation for increased oxygen consumption is the observation that thyroid hormone increases the activity and amount of plasma membrane Na^+, K^+-ATPase, an enzyme essential for membrane cation transport (see also Chapter 1). Ouabain, an inhibitor of this enzyme, also blocks the action of thyroid hormone on respiration. Since the sodium pump is responsible for 15% to 80% of energy turnover in various tissues, large amounts of ADP would be generated by augmenting its activity. The extra ADP resulting from the increased Na^+, K^+-ATPase activity would then stimulate oxygen utilization in the mitochondria.

In those tissues, such as the brain, in which oxygen consumption is not stimulated, the action of thyroid hormones to increase the synthesis of specific structural or functional proteins via transcriptional or translational effects must be invoked. In brain and other tissues thyroid hormone stimulates the transport of amino acids across the cell membrane, thereby facilitating protein synthesis. On the other hand, proteolytic and lysosomal enzyme activities are also increased by thyroid hormone, especially in muscle.

■ *Whole Body Actions of Thyroid Hormone on Metabolism*

The most obvious in vivo effect of thyroid hormone is to increase the rate of oxygen consumption and heat production (Fig. 53-8). This action is demonstrable in all tissues except the brain, gonads, and spleen. Oxygen use at rest in humans ranges from about 150 ml/min in the hypothyroid state to about 400 ml/min in the hyperthyroid state. The average level in the euthyroid subject is 250 ml/min. When converted to its caloric equivalent and standardized to body surface area, the basal metabolic rate ranges from −40% to +80% of normal at the clinical extremes of thyroid function. Thyroid hormone also augments the increase in oxygen consumption that accompanies exercise. Of necessity, thermogenesis increases concomitantly with oxygen use. Thus changes in body temperature parallel fluctuations in thyroid hormone availability. The potential increase in body temperature, however, is moderated by a compensatory increase in heat loss through appropriate thyroid hormone–mediated increases in blood flow, sweating, and ventilation.

Thyroid hormone could not augment oxygen utilization for long without also influencing oxygen supply to the tissues. Thus T_4 and T_3 increase the resting rate of ventilation sufficiently to maintain a normal arterial Po_2 in the face of increased oxygen utilization and a normal Pco_2 in the face of increased carbon dioxide production. Additionally, a small increase in red blood cell mass is effected—enhancing the oxygen-carrying capacity—probably through stimulation of erythropoietin production.

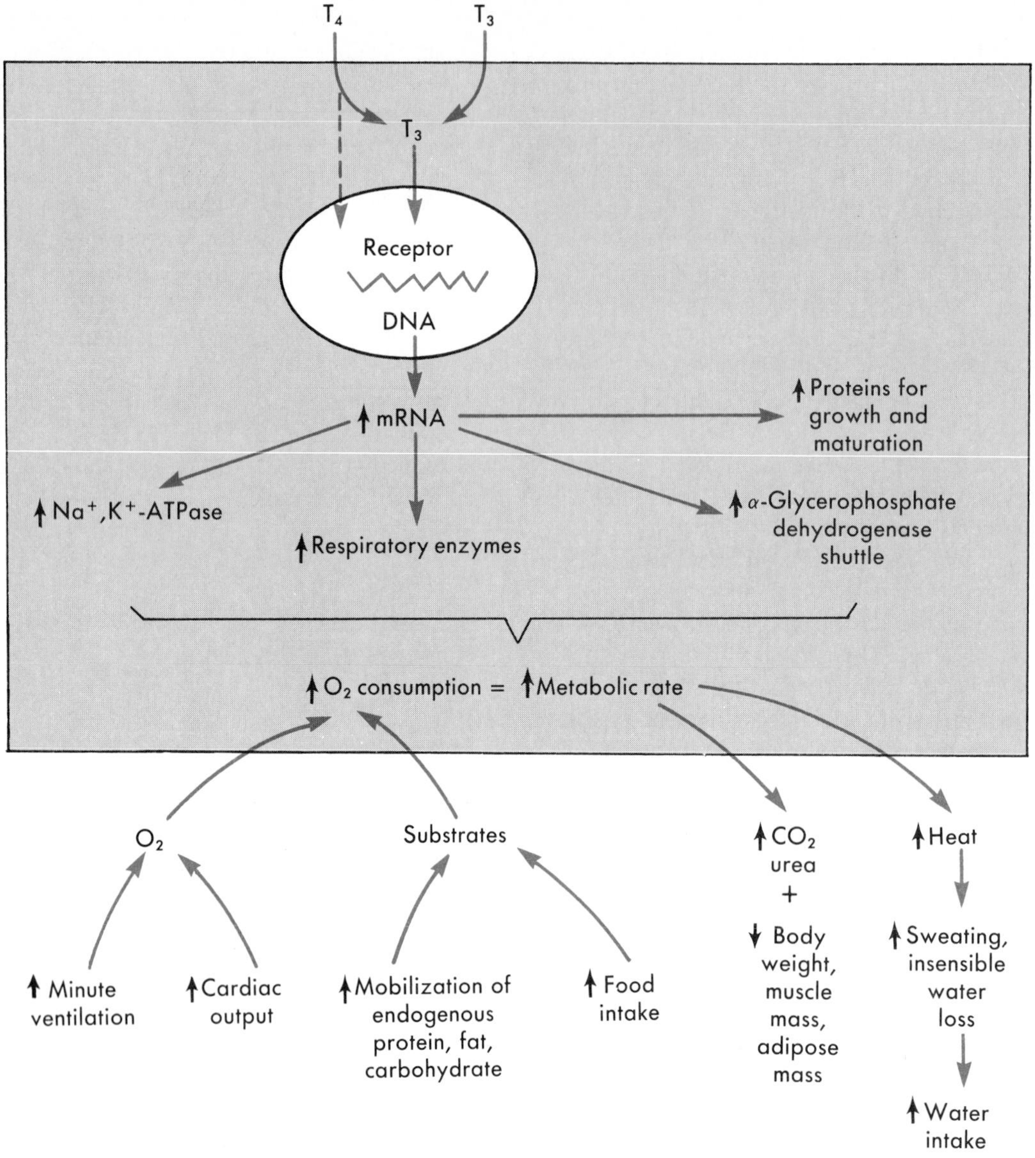

■ **Fig. 53-8.** Overall schema of thyroid hormone effects. The upper portion represents intracellular actions, the lower portion whole body actions.

Most important, thyroid hormone increases cardiac output, ensuring sufficient oxygen delivery to the tissues (Chapter 29). The resting heart rate and the stroke volume are both increased. The speed and force of myocardial contractions are enhanced. These effects are indirect, via adrenergic stimulation, and direct, via thyroid hormone-induced increases in myocardial Ca^{++} uptake and adenylate cyclase activity.

Recent evidence also shows that thyroid hormone increases cardiac concentrations of calcium-stimulated ATPase, which regulates the maximum force developed by myocardial tissue, as well as the active form of myosin-stimulated ATPase, which regulates the velocity of muscle fiber shortening. Systolic blood pressure is augmented and diastolic blood pressure is decreased, so that the net result is a widened pulse pressure. This reflects the combined effects of the increased stroke volume and a substantial reduction in peripheral vascular resistance resulting from blood vessel dilation. The latter in turn is secondary to the increase in tissue metabolism that thyroid hormone induces (Chapters 31 and 33).

Stimulation of oxygen utilization must ultimately depend on the provision of necessary substrates for oxidation. T_4 and T_3 potentiate the stimulatory effects of other hormones on glucose absorption from the gastrointestinal tract, on gluconeogenesis, on lipolysis, on ketogenesis, and on proteolysis of the labile protein pool. The respective actions of epinephrine, norepinephrine, glucagon, cortisol, and growth hormone on these processes are all enhanced. The overall metabolic effect of thyroid hormone therefore may be aptly described as accelerating the response to starvation. In addition, thyroid hormone stimulates both the biosynthesis of cholesterol and its oxidation, its conversion to bile acids, and its biliary secretion. The net effect is to decrease the body pool and plasma level of cholesterol.

The metabolic disposal of steroid hormones, of B vitamins, and of many administered drugs is increased by thyroid hormone. Therefore the endogenous secretion rates of hormones, such as cortisol, the dietary requirements of vitamins, such as thiamin, and the administered doses of drugs, such as digoxin, that are necessary to maintain normal or effective plasma levels of these substances are all increased by thyroid hormone.

Thyroid Hormone and Sympathetic Nervous System Activity

One of the prominent but incompletely understood features of thyroid hormone is its interaction with the sympathetic nervous system. Certain effects, such as the increases in metabolic rate, heat production, heart rate, motor activity, and central nervous system excitation, also are produced by the adrenergic catecholamines epinephrine and norepinephrine. An indisputable explanation for this striking similarity remains to be found. Thyroid hormone does not increase the levels of catecholamine hormones or their metabolites in blood, urine, or tissues. But increased levels of cAMP, are found in plasma, urine, and muscle from hyperthyroid persons. Furthermore the cAMP response to epinephrine in cultured myocardial cells is augmented by T_3. T_3 increases the number of β-adrenergic receptors in heart muscle. Synergism between catecholamines and thyroid hormones also may be required for maximum thermogenesis, lipolysis, glycogenolysis, and gluconeogenesis to occur. The importance of this physiological issue is underscored by the fact that adrenergic blockade, particularly of the β-receptors, attenuates some of the cardiovascular and central nervous system manifestations of hyperthyroidism.

Thyroid Hormone Effects on Growth and Development

Another major effect of thyroid hormone is on growth and maturation. Perhaps the most spectacular example is seen in the process of metamorphosis. Endogenous thyroid hormone levels are very low in amphibians until just before the major stage of metamorphosis. At this point, the hormone levels increase sharply, paralleling the rapid change from the larval to the adult form, after which they again decline. Addition of exogenous thyroid hormone to the fluid bathing tadpoles accelerates all aspects of their metamorphosis, including limb growth, tail resorption, shortening of the gastrointestinal tract, and induction of hepatic ureagenesis. Biochemically the thyroid hormone increases protein and nucleic acid synthesis in the limb buds, increases proteolytic and hydrolytic enzyme activities in the tail, and increases the hepatic content of carbamyl phosphate synthase, the rate-limiting enzyme in the urea cycle.

In humans and other mammals, thyroid hormone stimulates endochondral ossification, linear growth of bone, and maturation of the epiphyseal bone centers. A direct effect of T_3 on the maturation and activity of chondrocytes in the cartilage growth plate (Fig. 53-9) may initiate this process. In addition, T_3 may accelerate growth by facilitating the synthesis and secretion of growth hormone, as previously noted. Interestingly, thyroid hormone is not required for linear growth until after birth, whereas it is already essential for normal maturation of growth centers in the bones of the developing fetus. The regular progression of tooth development and eruption is dependent on thyroid hormone, as is the normal cycle of growth and maturation of the epidermis and its hair follicles. Because the normal degradative processes in these structural and integumentary tissues also are stimulated by thyroid hormone, excess exposure to T_4 and T_3 also can cause reabsorption of bone, rapid desquamation of skin, and hair loss. The synthesis of mucopolysaccharides in the intercellular ground substance is inhibited by thyroid hormone.

A critical set of actions of thyroid hormone is on the development of the central nervous system. If thyroid hormone is deficient in utero, growth of the cerebral and cerebellar cortex, proliferation of axons and branching of dendrites, and the process of myelinization are all decreased. Irreversible brain damage can result when the deficiency of thyroid hormone is not recognized and treated promptly after birth. The anatomical defects listed are paralleled by a number of biochemical abnormalities. In various areas of the brain of hypothyroid embryos the cell size, the RNA and protein content, the rates of protein synthesis, the enzymes necessary for nucleic acid synthesis, the protein and lipid content of myelin, and the enzymes essential for energy generation, such as succinic dehydrogenase, are all decreased. Some of these effects of thyroid hormone may be mediated by increasing the levels of nerve growth factor. The crucial role of thyroid hormone in central nervous system development is underscored by a number of adaptive phenomena seen specifically in the neonatal brain. T_3 receptors in the cortex and other areas are increased. The activity of brain 5′-monodeiodinase is augmented, enhancing local conversion of T_4 to T_3, whereas the activity of brain 5-monodeiodinase is diminished, reducing the local degradation of T_3 to $3,3'T_2$ (Fig. 53-7). All of these alterations increase the biological effectiveness of thyroid hormone at this critical time. In children and adults, thyroid hormone also enhances wakefulness, alertness, responsiveness to various stimuli, awareness of hunger, memory, and learn-

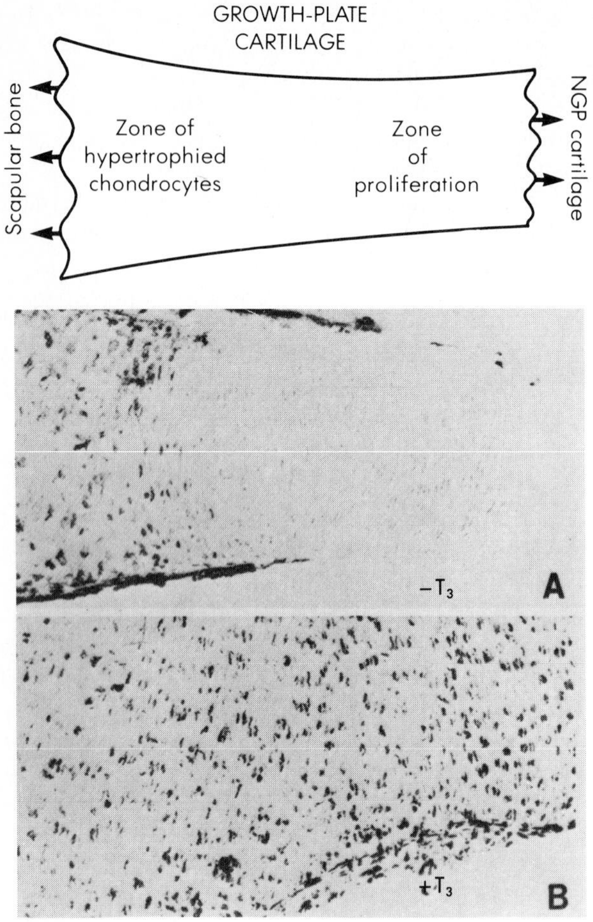

■ **Fig. 53-9.** Effect of thyroid hormone on cartilage growth. Cartilage from the growth plates of pigs was incubated in vitro with or without T_3 for 3 days. Chondrocytes were then identified by staining sections for the enzyme alkaline phosphatase. As seen in **B,** addition of T_3 caused a marked increase in the rate of proliferation of new chondrocytes as compared to **A.** (From Burch, W.M., and Lebovitz, H.E.: Triiodothyronine stimulates maturation of porcine growth-plate cartilage in vitro, J. Clin. Invest. **70:**496, 1982.)

ing capacity. Normal emotional tone also depends on proper thyroid hormone availability. Furthermore, the speed and amplitude of peripheral nerve reflexes are increased by thyroid hormone, as is motility of the gastrointestinal tract.

In both women and men thyroid hormone plays an important permissive role in the regulation of reproductive function. The normal ovarian cycle of follicular development, maturation, and ovulation, the homologous testicular process of spermatogenesis, and the maintenance of the healthy pregnant state all are disrupted by significant deviations of thyroid hormone from the normal range. In part these may be caused by alterations in the metabolism or interconversion of steroid hormones.

Normal function of skeletal muscles also requires normal amounts of thyroid hormone. This may well be related to the regulation of energy production and storage in this tissue. Concentrations of creatine phosphate are reduced by an excess of T_4 and T_3; the inability of muscle to take up and phosphorylate creatine leads to its increased urinary excretion.

Kidney size, renal plasma flow, glomerular filtration rate, and tubular transport maximums for a number of substances also are increased by thyroid hormone.

■ *Clinical Syndromes of Thyroid Dysfunction*

Hyperthyroidism results from a number of causes. Most commonly the entire gland undergoes hyperplasia and increases its secretion of T_4 and T_3 *(Graves' disease).* This occurs because of development of autoantibodies (thyroid-stimulating immunoglobulins), which bind to the normal TSH receptor and initiate all the biological activities of the natural tropic hormone. The production of this ''antibody to self'' is thought to reflect dysfunction of thymus-derived lymphocytes, which normally suppress the development of such autoantibodies. Next most commonly, one or more areas of the thyroid form benign neoplasms or autonomous nodules, which escape from the normal hypothalamic-pituitary regulation. Least common causes are inflammation of the thyroid, excessive pituitary secretion of TSH, or ingestion of exogenous T_4 or T_3.

The patient suffering from an excess of thyroid hormone presents one of the most striking pictures in clinical medicine. The large increase in metabolic rate causes the highly characteristic combination of weight loss concomitant with an increased intake of food. The increased heat that is generated in the hypermetabolic state causes discomfort in warm environments, fever when the condition is severe, excessive sweating, and a greater intake of water. The increase in adrenergic activity is manifested by a rapid heart rate, atrial arrhythmias, hyperkinesis, tremor, nervousness, and a wide-eyed stare. Weakness is due to loss of muscle mass as well as a specific thyrotoxic myopathy. Other symptoms include diarrhea, emotional lability, and breathlessness during exercise. The patient also may have difficulty swallowing or breathing because of compression of the esophagus or trachea by the enlarged thyroid gland *(goiter).*

The diagnosis is established by demonstrating an elevated serum T_4 or T_3 level (appropriately corrected for any abnormalities in TBG concentrations). In most instances the thyroid uptake of iodine (labeled with ^{131}I and ^{123}I) is excessive. Serum TSH levels are low and cannot be increased by administration of thyroid-releasing hormone (TRH), since the pituitary is inhibited by the high levels of T_4 and T_3. Various modalities of

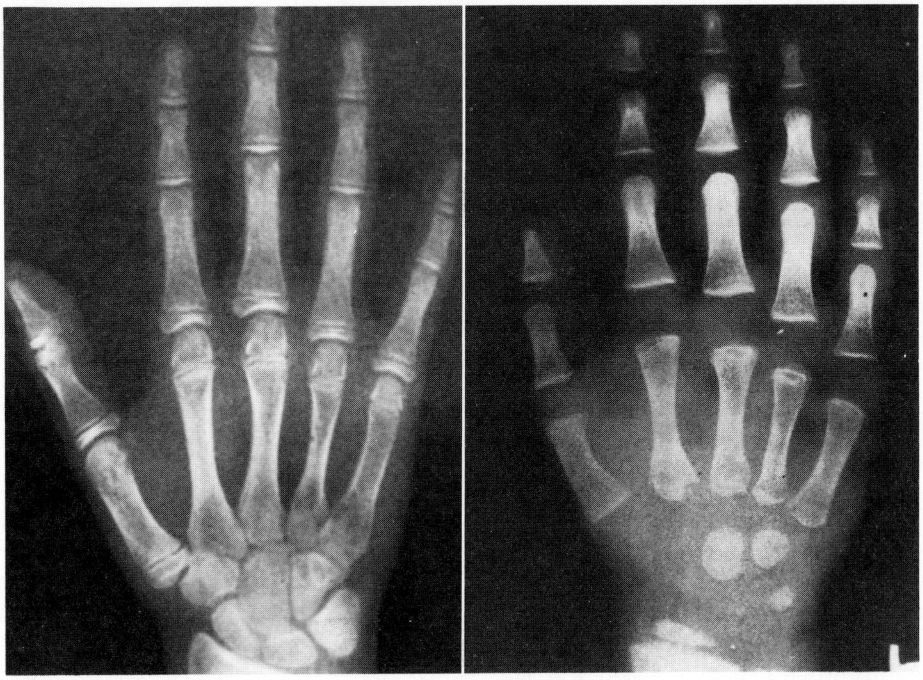

■ **Fig. 53-10.** Hand x-ray films of a 13-year-old normal child **(A)**, and a 13-year-old hypothyroid child **(B)**. Note that the hypothyroid child has a marked delay in development of the small bones of the hands, in growth centers at either end of the fingers, and in the growth center of the distal end of the radius. (**A,** from Tanner, J.M., et al.: Assessment of skeletal maturity and prediction of adult height (TW2 method), New York, 1975, Academic Press, Inc; **B,** from Andersen, H.J.: Nongoitrous hypothyroidism. In Gardner, L.I., editor: Endocrine and genetic diseases of childhood and adolescence, Philadelphia, 1975, W.B. Saunders Co.)

treatment are available. β-Adrenergic antagonists, such as propranolol, provide relief from tachycardia, tremor, and nervousness. Iodide inhibits release of thyroid hormones, but it is only useful temporarily. Thiouracil drugs block hormone synthesis and generally are given for at least 1 year. The most definitive treatment is ablation of thyroid tissue, either by radiation effects of ^{131}I or by surgery.

Hypothyroidism most often results from idiopathic atrophy of the gland—often thought to be preceded by a chronic inflammatory reaction. This form of *lymphocytic thyroiditis* is also an autoimmune condition. In this case the antibodies are produced to various thyroid cell antigens, including thyroglobulin, and some may have cytotoxic properties. Other causes of hypothyroidism include radiation damage, surgical removal, nodular goiters, hypothalamic or pituitary destruction, and, in certain areas of the world, iodide deficiency. Rarest of all is resistance to the action of thyroid hormones, possibly because of deficiency of receptors.

The clinical picture is, in many respects, the exact opposite of that seen in hyperthyroidism. The lower than normal metabolic rate leads to weight gain without appreciable increase in caloric intake. The decreased thermogenesis lowers body temperature and causes intolerance to cold, decreased sweating, and dry skin. There is decreased adrenergic activity, with bradycardia; a generalized slowing of movement, speech, and thought; lethargy; sleepiness; and a lowering of the upper eyelids (ptosis). An accumulation of mucopolysaccharides—ground substance—in the tissues causes an accumulation of fluid, as well. This *myxedema* produces puffy features, an enlarged tongue, hoarseness, joint stiffness, effusions in the pleural, pericardial, and peritoneal spaces, and entrapment of peripheral and cranial nerves with consequent dysfunction. Constipation, loss of hair, menstrual dysfunction, and anemia are other signs. Notably hypothyroidism in infancy or childhood causes marked retardation of growth and even greater slowing in the maturation of the epiphyseal growth centers of the bone (Fig. 53-10). If hypothyroidism is present at birth and remains untreated for even 2 to 4 weeks, the central nervous system will not undergo its normal maturation process in the first year of life. Developmental milestones, such as sitting, standing, and walking, will be late, and severe irreversible mental retardation can result. Such individuals are known as *cretins*.

The diagnosis of hypathyroidism is made by finding a low serum T_4 level. Serum TSH will be elevated because of negative feedback unless the hypothyroidism is caused by hypothalamic or pituitary disease. If the pi-

Table 53-2. Some congenital defects in thyroid hormone synthesis

Defect	*Diagnostic pattern*
Iodide trap	Decreased uptake of radioactive iodine; decreased salivary/blood ratio of radioactive iodine
Peroxidase	Increased early uptake of radioactive iodine*; rapid discharge by perchlorate
Deiodinase	Increased uptake of radioactive iodine*; increased MIT and DIT in urine
Coupling	Increased uptake of radioactive iodine*; increased MIT and DIT and decreased T_4 and T_3 in thyroid tissue

*Radioactive iodine uptake is increased because of increased TSH secretion, which stimulates the iodide trap.

tuitary is at fault, TSH levels will be low and will not respond to administration of TRH. Other common laboratory abnormalities include elevation of serum cholesterol, because of slowed degradation, and elevation of muscle enzymes such as creatine phosphokinase in the blood. Replacement therapy with T_4 is curative. T_3 is not needed, since it will be generated intracellularly from the administered T_4.

The stepwise nature of thyroid hormone synthesis offers multiple possibilities for congenital hypothyroidism caused by specific enzyme deficiencies. Clinically these syndromes are characterized by the symptoms of childhood hypothyroidism noted previously, plus thyroid gland enlargement (congenital goiter) resulting from persistent hypersecretion of TSH. Table 53-2 lists the best understood of these rare syndromes, with the biochemical findings that point to the lesion. Inability to carry out active transport of iodide into the gland is deduced from the finding of a very low uptake of radioactive iodine. Adequate synthesis of T_4 can be achieved by administration of large amounts of iodine, which is effective treatment. In peroxidase deficiency, iodide is concentrated by the gland but not incorporated into tyrosine. Therefore, after radioactive iodine is rapidly taken up, it is gradually discharged again as radioactive iodide levels in the blood fall. This discharge of iodine is markedly accelerated by administration of perchlorate ion, which competes for iodide in the trap, thus preventing replenishment of the thyroid free iodide level after it begins to fall. Treatment of peroxidase deficiency requires T_4 replacement.

A deficiency of thyroid deiodinase acts to produce a state of intraglandular iodine deficiency. The MIT and DIT molecules released during proteolysis of thyroglobulin leak out of the gland, carrying with them iodine; they then are irretrievably lost in the urine. Thus this defect also is corrected by administration of excess iodine. Radioactive iodine uptake is high, as it would be in iodide-deficient states. The diagnosis is established by finding large amounts of MIT and DIT in the urine.

The diagnosis of a coupling defects is suspected when radioactive iodine uptake by the gland is high but there is no evidence for peroxidase or deiodinase deficiency. The diagnosis can be established only by demonstrating the presence of large amounts of MIT and DIT but very little T_4 and T_3 in a specimen of thyroid tissue. Treatment with T_4 is required.

Bibliography

Journal articles

Bantle, J.P., et al.: Common clinical indices of thyroid hormone action: relationships to serum free 3,5,3'-triiodothyronine concentration and estimated nuclear occupancy, J. Clin Endocrinol. Metab. **50:**286, 1980.

Crowley, W.F., Jr., et al.: Noninvasive evaluation of cardiac function in hypothyroidism; response to gradual thyroxine replacement, N. Engl. J. Med. **296:**1, 1977.

Danforth, E., Jr.: The role of thyroid hormone and insulin in the regulation of energy metabolism, Am. J. Clin. Nutr. **38:** 1006, 1983.

Everett, A.W., et al.: Change in synthesis rates of α and β-myosin heavy chains in rabbit heart after treatment with thyroid hormone, J. Biol. Chem. **258:**2421, 1983.

Gavaret, J.M., et al.: Formation of dehydroalanine residues during thyroid hormone synthesis in thyroglobulin, J. Biol. Chem. **255:**5281, 1980.

Izumo, S., et al.: All members of the MHC multigene family respond to thyroid hormone in a highly tissue-specific manner, Science **231:**597, 1986.

Klein, I., and Levey, G.S.: New perspective on thyroid hormone, catecholamines, and the heart, Am. J. Med. **76:**167, 1984.

Larsen P.R., et al.: Inhibition of intrapituitary thyroxine to 3,5,3'-triiodothyronine conversion prevents the acute suppression of thyrotropin release by throxine in hypothyroid rats, J. Clin. Invest. **64:**117, 1979.

Larsen, P.R., et al: Relationships between circulating and intracellular thyroid hormones, physiological and clinical implications, Endocrine Rev. **2:**87, 1981.

Martial, J.A., et al.: Regulation of growth hormone gene expression; synergistic effects of thyroid and glucocorticoid hormones, Proc. Natl. Acad. Sci. USA **74:**4293, 1977.

Oppenheimer, J.H., et al.: Stimulation of hepatic mitochondrial α-glycerophosphate dehydrogenase and malic enzyme by L-triiodothyronine; characteristics of the response with specific nuclear thyroid hormone binding sites fully saturated, J. Clin. Invest. **59:**517, 1977.

Oppenheimer, J.H.: Thyroid hormone action at the nuclear level, Ann. Intern. Med. **102:**374, 1985.

Schimmel, M., et al.: Thyroidal and peripheral production of thyroid hormones, Ann. Intern. Med. **87:**760, 1977.

Sestoft, L.: Metabolic aspects of the calorigenic effect of thyroid hormone in mammals, Clin. Endocrinol. **13:**489, 1980.

Shapiro, L.E., et al.: Thyroid and glucocorticoid hormones synergistically control growth hormone mRNA in cultured GH_1 cells, Proc. Natl. Acad. Sci. USA **75:**45, 1978.

Smith, T.J., et al.: Regulation of glycosaminoglycan synthesis by thyroid hormone in vitro, J. Clin. Invest. **70:**1066, 1982.

Stocker, W.W., et al.: Coupled oxidative phosphorylation in muscle of thyrotoxic patients, Am. J. Med. **44:**900, 1968.

Vagenakis, A.G., et al.: Effect of starvation on the production and metabolism of thyroxine and triiodothyronine in euthyroid obese patients, J. Clin Endocrinol. Metab. **45:**1305, 1977.

Books and monographs

Degroot, L.J., et al.: Post-receptor mechanisms in thyroid hormone action. In Labrie, F., and Prouix, L.: Endocrinology: proceedings of the 7th International Congress of Endocrinology, Quebec City, 1984, Amsterdam, 1984, Elsevier Science Publishers, B.V.

Galton, V.A.: Thyroid hormone action in amphibian metamorphosis. In Oppenheimer, J.H., and Samuels, H.H.: Molecular basis of thyroid hormone action, New York, 1983, Academic Press, Inc.

Greer, M.A., et al.: Thyroid secretion. In Handbook of physiology; Section 7, Enddocrinology, vol. III, Thyroid, Baltimore, 1974, American Physiological Society.

Schwartz, H.L.: Effect of thyroid hormone on growth and development. In Oppenheimer, J.H., and Samuels, H.H.: Molecular basis of thyroid hormone action, New York, 1983, Academic Press, Inc.

CHAPTER

54

The Adrenal Glands

The adrenal glands are complex, multifunctional endocrine organs, which are essential for life. Severe illness results from their atrophy, and death follows their complete removal.

The adrenal glands are located in the retroperitoneum just above each kidney and within that organ's investing fat. Their total weight is 6 to 10 g. Each adrenal gland is really a combination of two separate functional entities (Fig. 54-1). The outer zone, or *cortex,* which comprises 80% to 90% of the gland, is derived from mesodermal tissue and is the source of steroid hormones. The inner zone, or *medulla,* comprising the other 10% to 20%, is derived from neuroectodermal cells of the sympathetic ganglia and is the source of catecholamine hormones.

The adrenal glands are exceedingly well vascularized, receiving arterial blood from branches of the aorta, the renal arteries, and the phrenic arteries; they have one of the body's highest rates of blood flow per gram of tissue. Arterial blood enters sinusoidal capillaries in the cortex.The blood then drains into medullary venules, an arrangement that exposes the medulla to relatively high concentrations of corticosteroids from the cortex. The right adrenal vein drains directly into the inferior vena cava, whereas the left drains into the renal vein on that side. Catheterization of each adrenal vein for individual sampling or radiographic visualization is possible, but catheterization is more difficult on the right. The human adrenal glands can be well visualized by computed tomography (CT scanning).

■ The Adrenal Cortex

The major hormones of the cortex are (1) the *glucocorticoids, cortisol* and *corticosterone,* which are critical to life by virtue of their effects on carbohydrate and protein metabolism, (2) a *mineralocorticoid, aldosterone,* which is vital to maintaining sodium and potassium balance, and (3) the *sex steroids, androgens* and *estrogens* and their precursors, which contribute to establishing and maintaining secondary sexual characteristics. Particular interest in cortisol and in glucocorticoids in general was heightened greatly by the discovery of their potent antiinflammatory effects. In supraphysiological doses, the glucocorticoids have been used to treat a wider variety of disease than has any other single group of drugs. This has produced a far greater incidence of iatrogenic disorders of glucocorticoid excess and deficiency than that resulting from spontaneous adrenal diseases.

The cortical portion of the adrenal gland differentiates by 8 weeks of intrauterine life and is initially much larger than the adjacent kidney. At this time the cortex contains two zones. The *peripheral neocortex,* comprising 15% of the cortex, is undifferentiated and relatively inactive. The inner 85%, known as the *fetal cortex,* is highly active and is responsible for fetal adrenal steroid production throughout almost all intrauterine life. Shortly after birth, the fetal cortex begins to involute; hemorrhage and necrosis occur, and this zone disappears completely in 3 to 12 months. Concurrently, the thin outer zone enlarges, differentiates, and becomes the permanent three-layered adrenal cortex of the normal human (Fig. 54-1).

The histological appearance of the three adult cortical zones differs. The outermost *zona glomerulosa* is the narrowest zone and consists of small cells with numerous elongated mitochondria that possess lamellar cristae. The middle *zona fasciculata* is the widest zone and consists of columnar cells that form long cords. The cytoplasm is highly vacuolated and contains lipid droplets. The mitochondria are distinguished by their large size and numerous vesicular cristae within their membranes or matrices. The innermost *zona reticularis* contains networks of interconnecting cells with fewer lipid droplets. Their mitochondria are similar to those of fasciculata cells. Under stimulation by adrenocorticotropic hormone (ACTH), the size and the number of cells in the fasciculata and reticularis increase. The mitochondria of these cells become more numerous and larger and develop central ribosomes, vesicular cristae, and polylamellar membranes that extend to nearby choles-

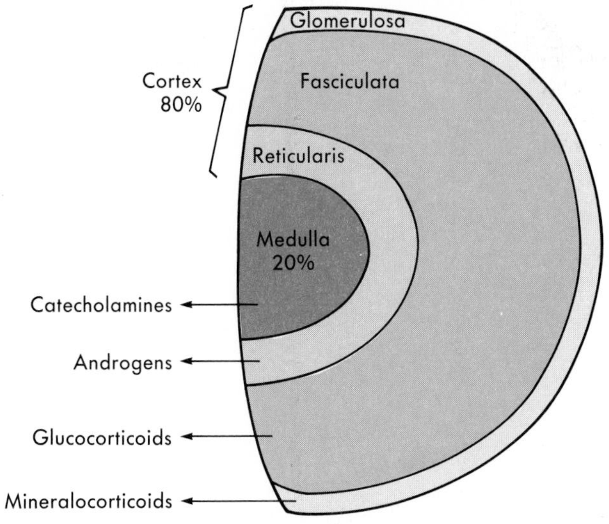

Fig. 54-1. Schematic representation of the zones of the adrenal gland and their main secretory products.

terol-containing vacuoles. In addition, the endoplasmic reticulum increases. These changes relate to ACTH effects on steroid hormone synthesis as detailed later.

Synthesis of Adrenocortical Hormones

All hormones of the adrenal cortex represent chemical modifications of the steroid nucleus shown in Fig. 54-2. Potent glucocorticoids require the presence of a ketone at the 3 position and hydroxyl groups at the 11 and 21 positions. Potent mineralocorticoids require an oxygenated carbon at the 18 position. Potent androgens are characterized by the elimination of the $C_{20\text{-}21}$ side chain and the presence of an oxygenated carbon at the 17 position. Estrogens are characterized by aromatization of the A ring.

The precursor for all adrenocortical hormones is *cholesterol*. This 27-carbon steroid molecule is actively taken up from the plasma by adrenal cells. Specific plasma membrane receptors bind circulating low-density lipoproteins rich in cholesterol. After transfer into the cell by endocytosis, the cholesterol is largely esterified and stored in cytoplasmic vacuoles. Cholesterol is also synthesized in adrenal cells from acetylcoenzyme A (acetyl CoA) by the usual pathway. Under basal circumstances, free cholesterol from plasma is the most direct and major source used for hormone synthesis. When production of corticosteroids is stimulated, however, stored cholesterol becomes an increasingly important precursor. Cholesterol synthesized de novo is always a minor component of the pool.

Most of the synthetic reactions from cholesterol to active hormones involve cytochrome P-450 enzymes, which are mixed oxygenases that catalyze steroid hydroxylations. These hydrophobic hemoproteins are lo-

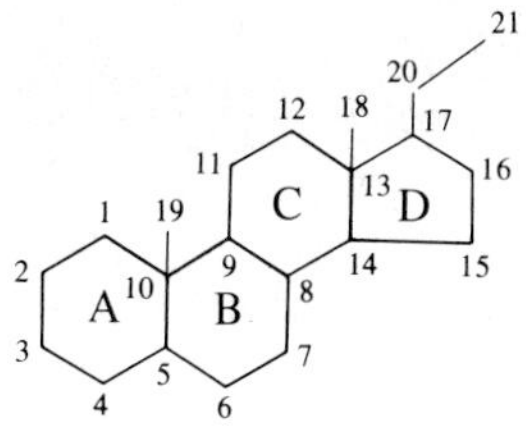

Fig. 54-2. The adrenocorticosteroid nucleus.

calized in the lipophilic membranes of the endoplasmic reticulum and mitochondrial cristae. Molecular oxygen is split so that one oxygen atom interposes between the carbon and hydrogen of the steroid site; the other oxygen atom is reduced by a hydrogen to H_2O. NADPH and, to some extent, NADH, which are generated by oxidation of a variety of substrates, are the ultimate donors of the hydrogen. A flavoprotein, adrenoxin reductase, and an iron-containing protein, adrenoxin, are intermediates in the transfer of hydrogen from NADPH to the P-450 enzymes.

Glucocorticoids. The synthesis of glucocorticoids occurs largely in the zona fasciculata with a smaller contribution from the zona reticularis. The sequence of reactions for synthesis of cortisol and corticosterone is shown in Fig. 54-3. The intracellular localization of the various steps is illustrated in Fig. 54-4. Cortisol is the glucocorticoid for humans and most mammals, whereas corticosterone serves this function in some rodents. However, if cortisol synthesis is blocked but the pathway to corticosterone is open, increased synthesis of the latter can provide the necessary glucocorticoid activity in humans. The initial reaction converting cholesterol to Δ^5-pregnenolone is catalyzed by a P-450 enzyme complex known as *desmolase*. This intramitochondrial complex, which carries out successive hydroxylations followed by cleavage of the cholesterol side chain, is rate limiting for adrenal steroidogenesis in general. The Δ^5-pregnenolone is then converted to 11-deoxycortisol by successive steps within the endoplasmic reticulum. The latter steroid is hydroxylated in the 11 position after transfer back to the mitochondria. The end product, cortisol, rapidly diffuses out of the organelle and then out of the cell. The last and critical step for glucocorticoid synthesis, 11-hydroxylation, is very efficient in humans; 95% of the 11-deoxycortisol formed is converted to cortisol.

The order of hydroxylations from Δ^5-pregnenolone to 11-deoxycortisol is not invariant. Even the 3β-ol-dehydrogenase and $\Delta^{4,5}$-isomerase reactions can occur after all the hydroxylations, rather than before. However, the order of hydroxylation presented in Fig. 54-3 *(17-21-11)* is compatible with the usual pattern of precursor accumulation that occurs when the various hydroxylases are either chemically blocked or congenitally deficient. It is important to note that the critical hormone, corti-

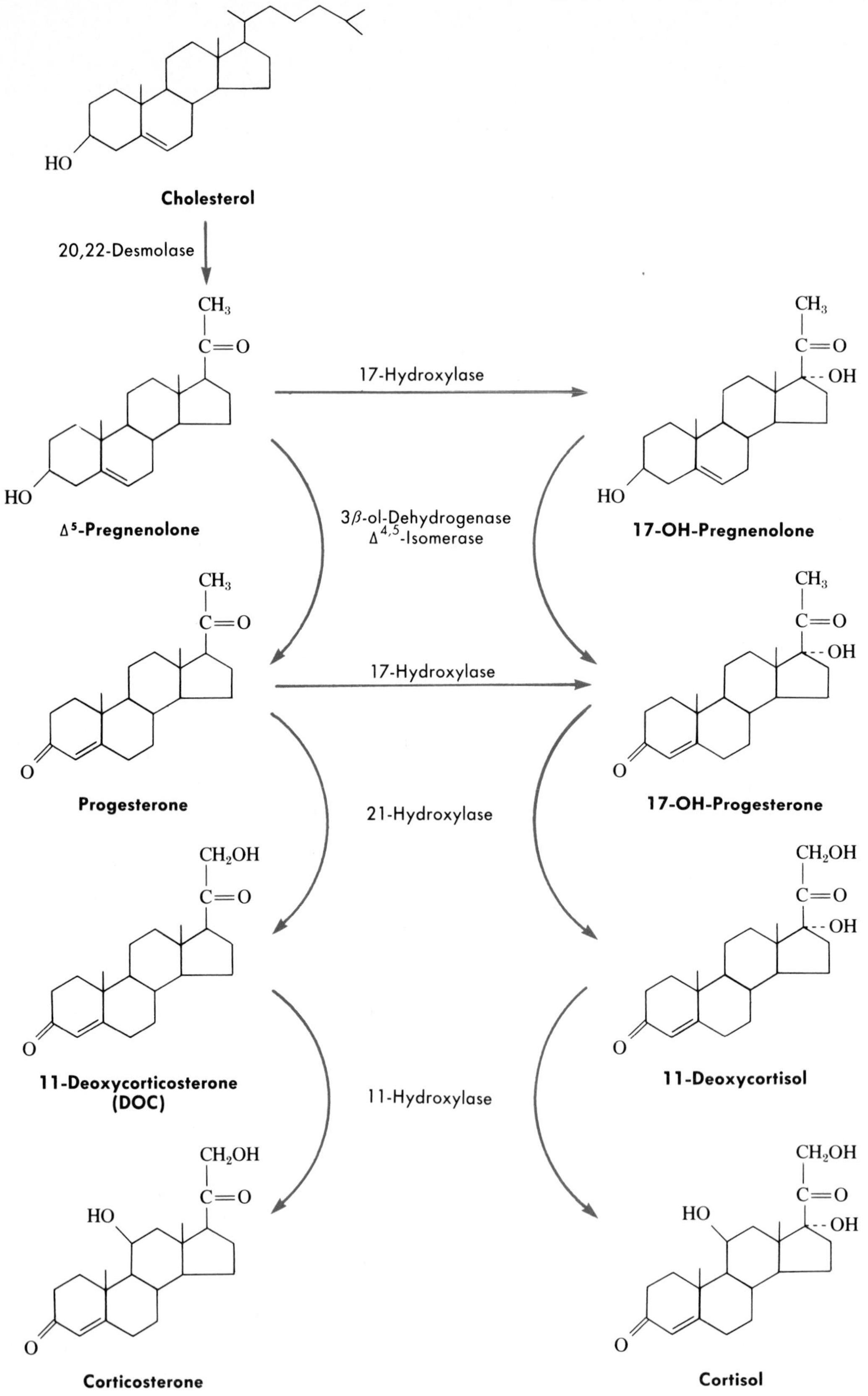

■ **Fig. 54-3.** Synthesis of glucocorticoids in the zona fasciculata. Cortisol is the major product in humans.

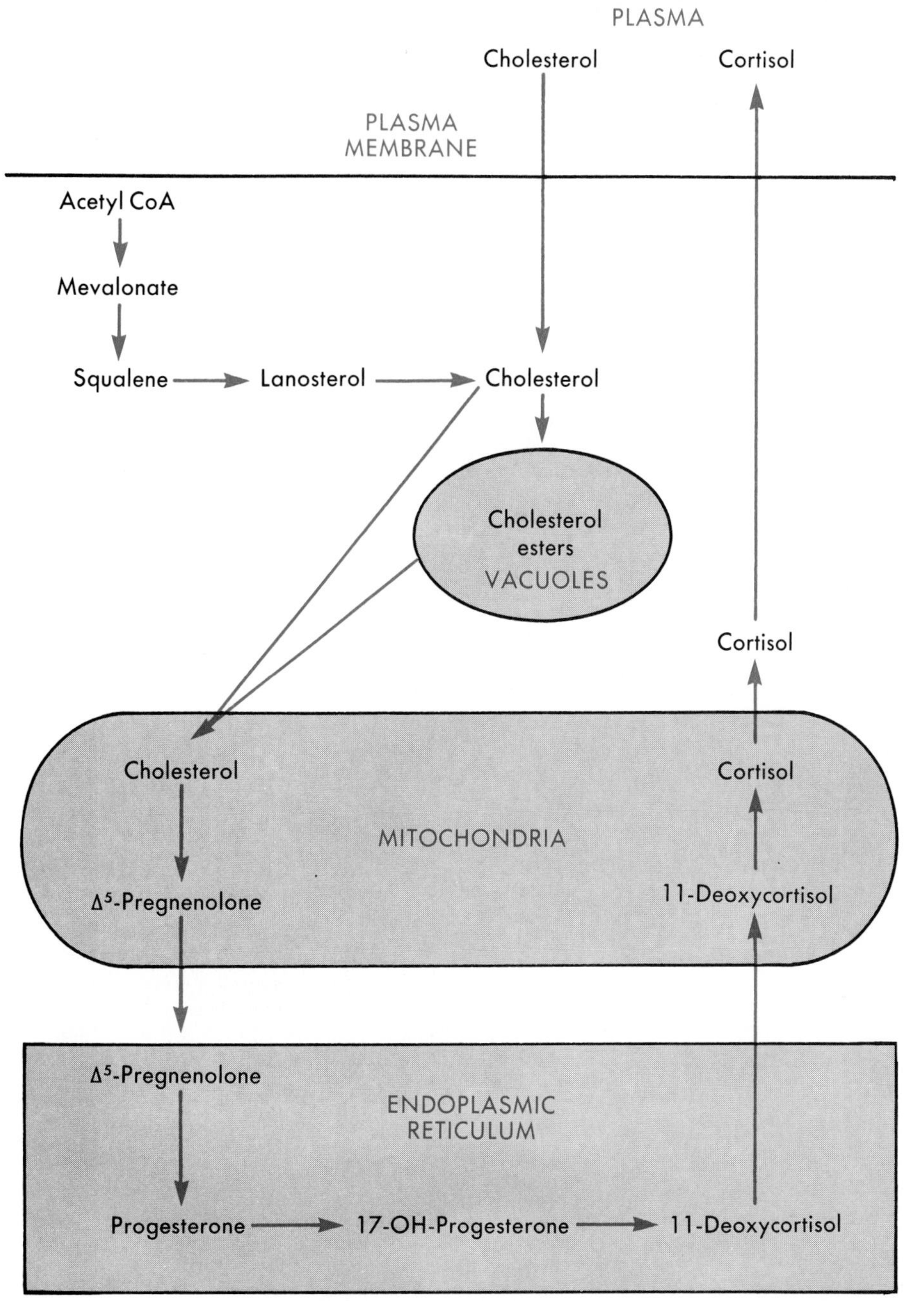

■ **Fig. 54-4.** Intracellular localization of the reactions involved in cortisol biosynthesis.

sol, and its precursors are not stored in the adrenocortical cell to any significant extent. Hence, an acute need for increased amounts of circulating cortisol requires rapid activation of the entire synthetic sequence and most particularly that of the rate-limiting desmolase reaction.

Androgens and estrogens. The synthesis of sex steroids occurs largely in the zona reticularis, although the zona fasciculata possesses the same capabilities. The 17-hydroxylated derivatives of Δ^5-pregnenolone and progesterone are the starting points for androgen and estrogen synthesis. Removal of the C_{20-21} side chain by a microsomal desmolase-like reaction is the key step yielding dehydroepiandrosterone (DHEA) and androstenedione, respectively, as shown in Fig. 54-5. This reaction is catalyzed by the same enzyme that catalyzes the previous 17-hydroxylation step. DHEA is sulfated by a specific enzyme; the sulfate donor is 3'-phosphoadenosine 5'-phosphosulfate. Dehydroepiandrosterone sulfate (DHEA-S) and DHEA are the major androgenic products of the adrenal glands. Although rather weak androgens themselves, they are converted to the potent androgen *testosterone* in peripheral tissues. Smaller amounts of androstenedione, an androgen of limited potency, and tiny amounts of testosterone itself are synthesized and secreted by the zona reticularis.

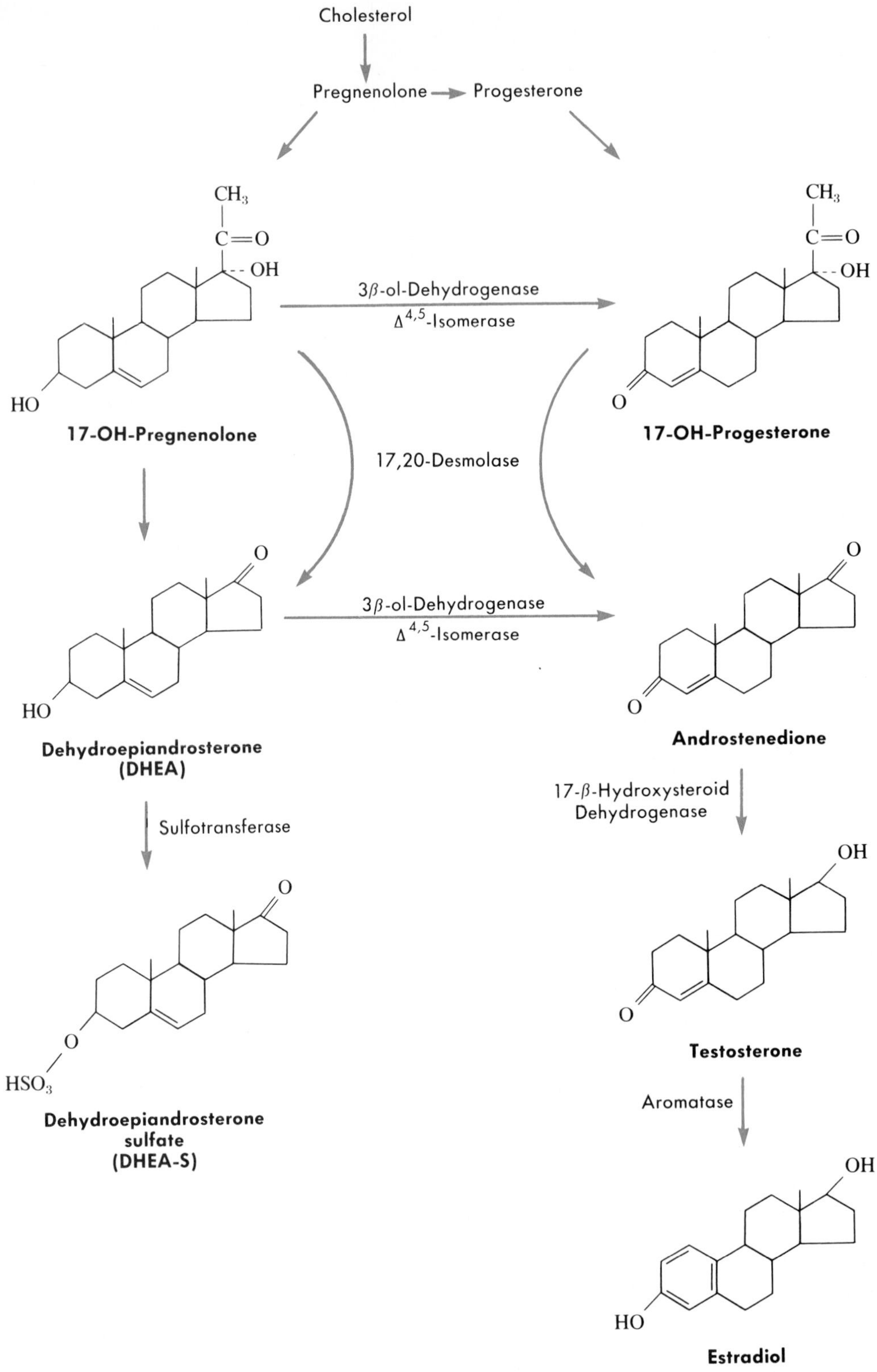

Fig. 54-5. Synthesis of androgens and estrogens in the zona reticularis. DHEA-S is the major product. Normally only trivial amounts of testosterone and estradiol are produced.

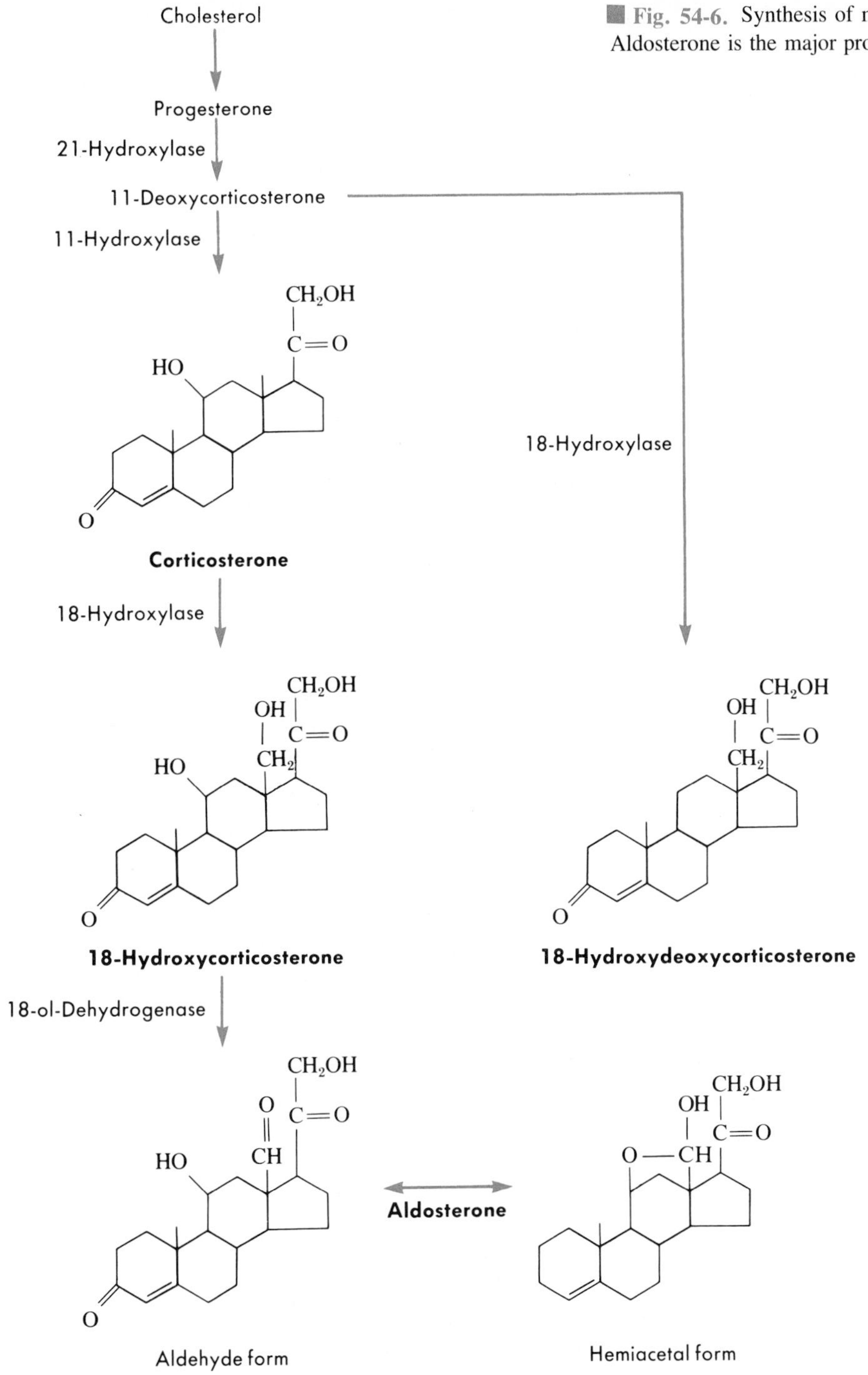

■ Fig. 54-6. Synthesis of mineralocorticoids in the zona glomerulosa. Aldosterone is the major product.

In women the adrenals supply 50% to 60% of the androgenic hormone requirements. Adrenal androgens are of little biological importance to men because the testes produce a large quantity of testosterone. The further conversion within the adrenal cortex of testosterone and androstenedione to estradiol and estrone, respectively, is of little quantitative significance in women until the ovaries are removed or cease to function after menopause. Then estrogens secreted directly from the adrenal glands or arising in the periphery from adrenal precursors become the only source for this biological activity. It is very important to note that 17-hydroxylation is the last reaction common to the synthesis of cortisol and the adrenal androgens (compare Fig. 54-3 with Fig. 54-5). For this reason, when cortisol synthesis is impaired at any point beyond this step, the accumulation of 17-hydroxypregnenolone and 17-hydroxyprogesterone leads to greatly increased androgen synthesis.

Mineralocorticoids. The synthesis and secretion of aldosterone, the major mineralocorticoid, are carried out exclusively by the zona glomerulosa (Fig. 54-6). The sequence from cholesterol to corticosterone is identical to that in the zona fasciculata. The C_{18} methyl group of corticosterone is then oxidized by mitochondrial P-450 enzymes to yield aldosterone, which is rapidly released. The conversion of the 18 methyl group to an aldehyde is carried out by the same enzyme that is responsible for 11-hydroxylation in the zona fasciculata. Corticosterone, deoxycorticosterone, and 18-hydroxydeoxycorticosterone also have some mineralocorticoid activity and can be synthesized in the zona fasciculata under ACTH stimulation. However, only in rare circumstances are they secreted in sufficient quantity to provide either physiologically adequate action or a pathological excess.

Inhibitors of adrenocortical hormone synthesis. A number of drugs have been developed that are capable of blocking steroid synthesis at various steps. These have both diagnostic and therapeutic usefulness. Of particular note is metyrapone, which inhibits 11-hydroxylation, the last step in cortisol synthesis. Administration of this drug creates acute cortisol deficiency, thereby stimulating ACTH secretion via negative feedback. As a result of the increase in ACTH, adrenal production of the immediate precursor to cortisol, 11-deoxycortisol (Fig. 54-3), increases markedly, demonstrating the reserve capacity of the normal hypothalamic–pituitary ACTH axis (Chapter 52). Failure to respond indicates a hypothalamic-pituitary disease that has diminished or ablated this function. Aldosterone secretion is also decreased by metyrapone by virtue of its additional inhibitory effect on 18-oxygenation.

Aminoglutethimide is a potent inhibitor of the desmolase reaction, thereby decreasing all adrenal steroid synthesis and causing accumulation of cholesterol in the gland. The aromatase reaction is also blocked by aminoglutethimide. This drug has been useful in treating patients with excessive cortisol and androgen output, as well as in diminishing estrogen production in women with breast cancer.

Another drug, 2,2-*bis*(2-chlorophenyl, 4-chlorophenyl)-1-1-dichloroethane, known as *o,p'*-DDD, predominantly destroys the zona fasciculata and reticularis, resulting in cortisol deficiency. This drug is effective in therapy of adrenocortical cancer. Ketoconazole, an antifungal agent in current use, also inhibits adrenocortical function.

Metabolism of Adrenocorticosteroids

Cortisol circulates in plasma 75% to 80% bound to a specific corticosteroid-binding α_2-globulin called *transcortin*. This glycoprotein has a molecular weight of 52,000 and binds a single molecule of cortisol with a K_{assoc} of 3×10^7 moles/L. The normal concentration of transcortin is 3 mg/dl, with a binding capacity of 20 μg cortisol/dl. An additional 15% of plasma cortisol is bound to albumin, leaving only 5% to 10% free. The concentration of transcortin is increased during pregnancy and by estrogen administration; thus total plasma cortisol is increased in these conditions. However, because bound cortisol is biologically inactive, the physiological effects of an increase in transcortin are determined by principles similar to those discussed with regard to thyroxine binding (Chapter 53). The plasma half-life of cortisol is about 70 minutes, and the metabolic clearance rate averages 200 L/day. The free fraction of cortisol in plasma is filtered by the kidney, but only about 0.3% of the total daily secretion, or approximately 50 μg, is excreted in the urine in this manner.

Cortisol is in equilibrium with its 11-keto analogue cortisone. This interconversion is catalyzed by 11β-ol-dehydrogenase, which is present in many tissues. The glucocorticoid activity of cortisone is dependent on its conversion to cortisol. The vast majority of cortisol and cortisone is metabolized in the liver; the metabolites then are conjugated and excreted in the urine as glucuronides. Each undergoes a parallel series of reactions that reduces ring A to tetrahydro derivatives and the 20-keto group to a hydroxyl group. About half of these excretory products are normally derived from cortisol and half from cortisone. The sequence is illustrated in Fig. 54-7.

The measurement of urinary metabolites provides a reliable index of cortisol secretion as long as hepatic and renal functions are intact. Of particular use is the 17,21-dihydroxy-20-ketone configuration of tetrahydrocortisol and tetrahydrocortisone, which forms the basis for a long-employed procedure (Fig. 54-7). The Porter-Silber method measures the urinary "17-hydroxycorticoids," which represent up to 50% of the total daily cortisol secretion. Normal values of 17-hydroxycorticoids range from 2 to 12 mg/day. The values are slightly higher in men than in women, as are cortisol secretion rates. Daily urinary 17-hydroxycorticoid excretion may be corrected for differences in lean body mass by expressing the result as a ratio to the urinary creatinine excretion. Measurement of urinary 17-hydroxycorticoids (or of urinary free cortisol) is used to assess adrenocortical responsiveness to ACTH stimulation or its suppressibility by exogenous synthetic glucocorticoid. These are valuable tests for adrenocortical insufficiency or for autonomous adrenocortical hyperfunction, respectively. In addition, because 11-deoxycortisol also possesses the Porter-Silber reaction grouping and is excreted as a tetrahydrometabolite, the 17-hydroxycorticoid fraction will normally increase in the urine following administration of the 11-hydroxylase inhibitor metyrapone, as previously described. Hence, the integrity of the negative feedback response of the corticotropin-releasing hormone (CRH)–ACTH axis may also be assessed in this manner.

The cortisol precursors, progesterone and 17-hydroxy-

Fig. 54-7. Major pathways of cortisol metabolism. The analogous compounds are formed from cortisone. The ratio of cortisol to cortisone metabolites is normally about 1:1.

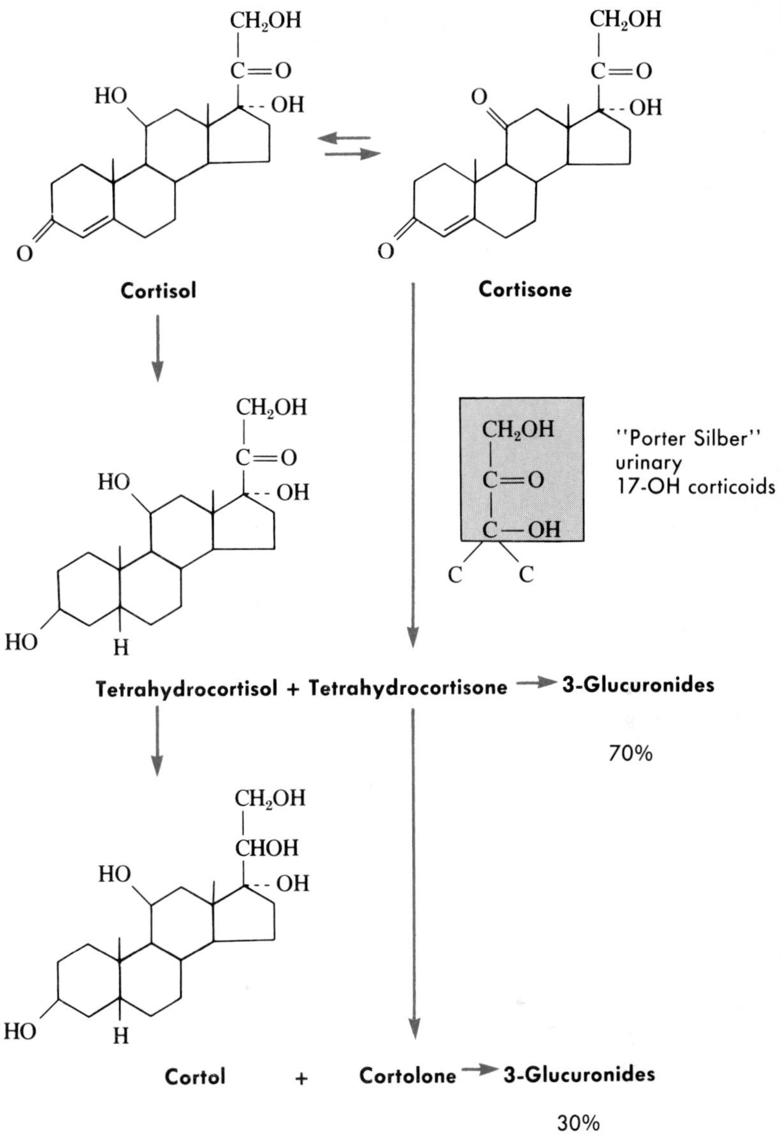

progesterone, are metabolized to cortols analogous to those of cortisol. The excretory products are known as pregnanediol and pregnanetriol, respectively. In adult females, measurement of these urinary metabolites reflects both adrenal and ovarian secretion. In prepubertal children, however, elevation of urinary pregnanetriol specifically indicates increased secretion of adrenal 17-hydroxyprogesterone and is, therefore, a valuable marker for congenital blocks in cortisol secretion.

Aldosterone circulates in plasma bound to a specific aldosterone-binding globulin, to transcortin, and to albumin. Overall binding is weaker than for cortisol. Hence, the plasma half-life is only 20 minutes, and the metabolic clearance rate is 1,600 L/day. Ninety percent of aldosterone is cleared by the liver in a single passage. In the liver aldosterone is reduced to tetrahydroaldosterone, the major metabolite that is excreted in the urine as the 3-glucuronide (Fig. 54-8). A smaller portion of aldosterone is excreted simply as the 18-glucuronide. The latter, however, is the urinary form that is commonly measured. The values in subjects with a normal sodium diet range from 5 to 20 μg/day.

The metabolism of androgens, in general, involves reduction of the 3-ketone group and the A ring in the liver. The two isomers formed, androsterone and etiocholanolone (Fig. 54-9), are then excreted in the urine. These metabolites, however, are not specific for the adrenal gland because they arise from gonadal androgens as well. DHEA-S is entirely excreted directly in the urine.

Androsterone, etiocholanolone, and DHEA-S together make up the major part of a urinary fraction called 17-ketosteroids. They are measured by a single colorimetric procedure known as the Zimmerman reaction. Normal values range from 5 to 14 mg/day in women and 8 to 20 mg/day in men. Two thirds of this fraction is normally derived from adrenal and one third from gonadal androgen secretions. A large increase in urinary 17-ketosteroid excretion almost always indicates

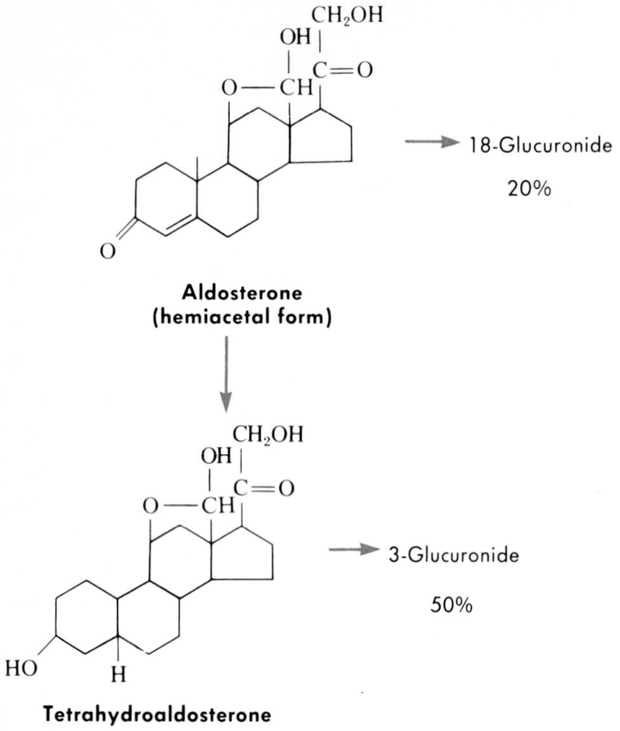

■ **Fig. 54-8.** Major pathway of aldosterone metabolism.

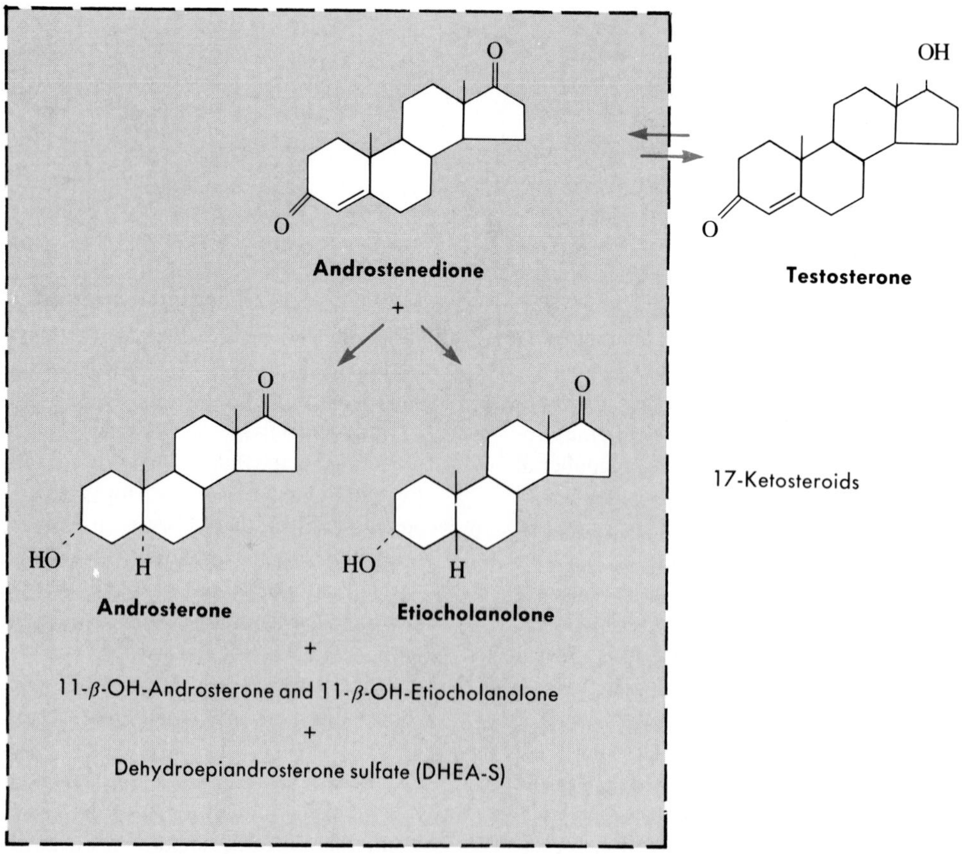

■ **Fig. 54-9.** Major pathways of androgen metabolism. The urinary metabolites of testosterone comprise part of the 17-ketosteroid fraction. Note, however, that the potent androgen testosterone is not itself a 17-ketosteroid.

an adrenal abnormality. Measurement of urinary 17-ketosteroids is especially useful diagnostically in women or children who show evidence of virilization. Similar information can be obtained by measurement of plasma or urinary DHEA-S.

Regulation of Zona Fasciculata and Zona Reticularis Functions

The secretion of cortisol by the zona fasciculata is, for all practical purposes, under the exclusive physiological control of pituitary ACTH. The secretion of adrenal androgens is likewise influenced by ACTH, but some evidence suggests the possible existence of a separate tropic hormone for the zona reticularis. In the complete absence of ACTH, adrenocortical secretion in humans virtually ceases.

ACTH initiates its action by binding to an adrenal plasma membrane receptor. The composition of the receptor includes protein, carbohydrate, sialic acid, and phospholipid. Smaller analogues of ACTH have been synthesized that react well with the receptor but have virtually no biological activity. This suggests that sites on the ACTH molecule other than those involved in receptor binding may be required to couple the occupied receptor to subsequent steps in hormone action. ACTH activates adenylate cyclase in a Ca^{++}-dependent manner, and cAMP levels rise. It is not clear, however, that all the actions of ACTH are mediated by cAMP. There is evidence suggesting that one subsequent event is the phosphorylation of ribosomal protein or proteins; the latter, in turn, may control the translation from messenger RNA of another labile regulatory protein capable of activating a key cytochrome P-450 enzyme. The nature of the regulatory protein is unknown. ACTH also increases adrenal phosphatidylinositol turnover, and this second messenger system may participate in mediating the stimulation of steroid synthesis.

The most important action of ACTH is to stimulate the rate-limiting desmolase reaction (cholesterol $\rightarrow$ Δ^5-pregnenolone). In addition, 11-hydroxylase activity is increased by ACTH. The tropic hormone also increases availability of cholesterol by stimulating its active uptake from plasma, its de novo synthesis from acetyl CoA, and cholesterol ester hydrolase, the enzyme that liberates cholesterol from its stored esterified form. The activity of this enzyme is enhanced by phosphorylation by a cAMP-dependent protein kinase. Finally, it has been suggested that ACTH also directs mitochondrial uptake of cholesterol and its binding to the cytochrome P-450 desmolase system. In this connection, ACTH has been shown to induce changes in adrenal cell shape through actions on the microtubules and microfilaments of the cytoskeleton. These actions may bring the cholesterol-containing vacuoles into association with the mitochondria.

As seen in Fig. 54-10, plasma levels of cortisol, adrenal androgens, and their precursors rise within minutes of intravenous ACTH administration to humans. A sustained increase in cortisol secretion of two- to fivefold is obtained when 250 μg of the fully active 1 to 24 sequence of ACTH is infused over 8 hours. Plasma ACTH concentrations of about 300 pg/ml,

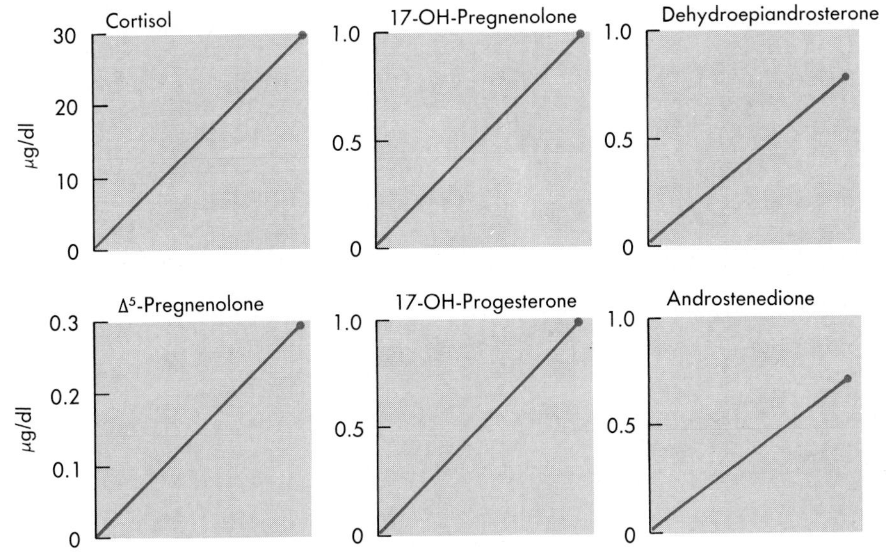

Fig. 54-10. Plasma adrenocortical hormone responses to a 1-hour infusion of ACTH in humans. Note increase of precursors as well as hormonally active products. These changes indicate that ACTH activates the entire biosynthetic sequence. (Redrawn from Lachelin, G.C.L., et al.: Adrenal function in normal women and women with the polycystic ovary syndrome, J. Clin. Endocrinol. Metab. **49**:892, 1979. Copyright 1979. Reproduced by permission.)

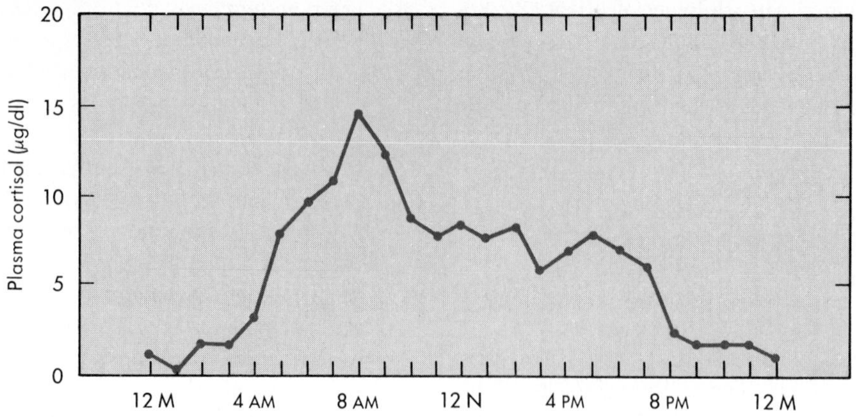

■ **Fig. 54-11.** Diurnal variation in plasma cortisol as assessed from mean values of hourly sampling. Note the early morning peak and compare with the plasma ACTH profile in Fig. 52-10. (Redrawn from Weitzman, E.D., et al.: Twenty-four hour pattern of the episodic secretion of cortisol in normal subjects, J. Clin. Endocrinol. Metab. **33:**14, 1971, Copyright 1971. Reproduced by permission.)

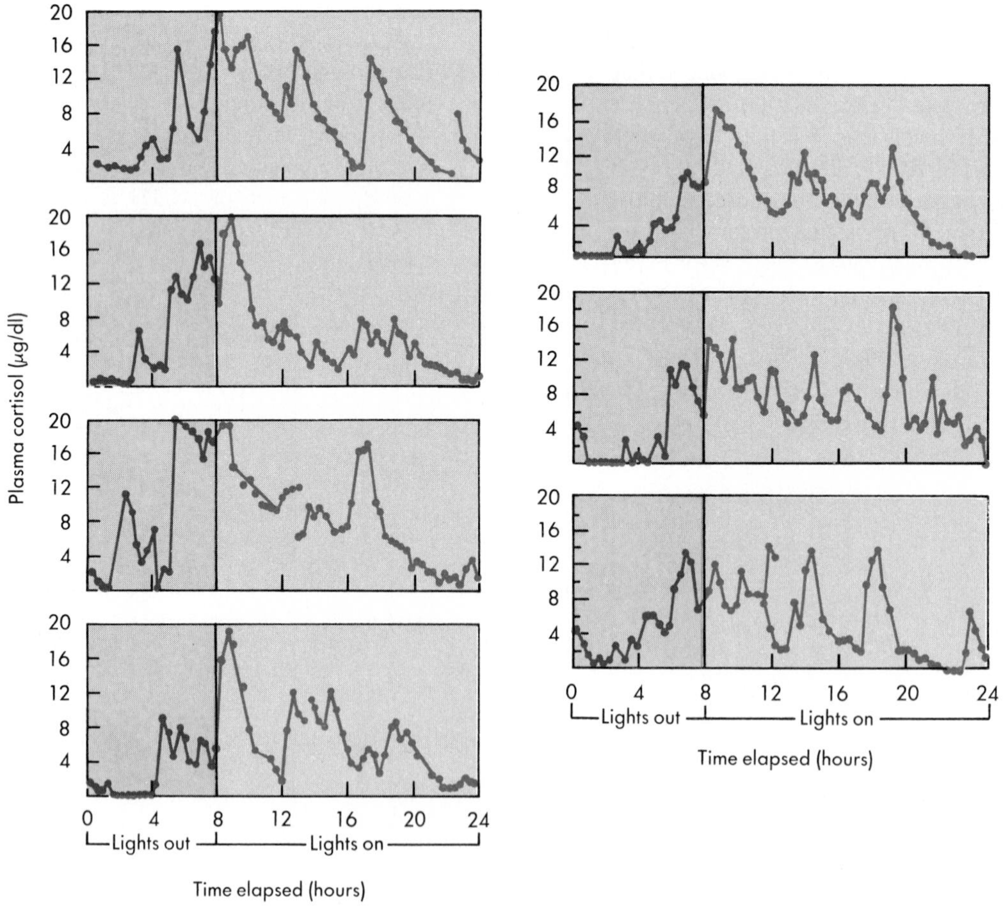

■ **Fig. 54-12.** Pulsatile nature of cortisol secretion is shown by individual plasma cortisol profiles obtained by sampling every 20 minutes. (Redrawn from Weitzman, E.D., et al.: Twenty-four hour pattern of the episodic secretion of cortisol in normal subjects, J. Clin. Endocrinol. Metab. **33;**14, 1971. Copyright 1971. Reproduced by permission.)

or 6 times the basal level, are maximally effective in the short term. However, when the adrenal gland is chronically hyperstimulated by endogenous or exogenous ACTH, it undergoes hyperplasia. With the increase in adrenal mass, the capacity for cortisol secretion rises up to twentyfold.

As detailed in Chapter 52, all of the factors that influence ACTH secretion likewise affect cortisol secretion; plasma levels of the latter generally follow those of the former by 15 to 30 minutes. Thus cortisol secretion, like that of ACTH, exhibits distinct diurnal variation with a peak just before the subject awakes in the morning and a nadir at or near zero just after the subject falls asleep (Fig. 54-11). However, when plasma cortisol concentrations are measured every 20 minutes instead of every hour (Fig. 54-12), the deceptively smooth diurnal curve is seen to consist of 7 to 13 pulses or episodes of secretion per day. Within the major nocturnal pulse before the subject awakes, 50% of the day's total cortisol is secreted.

The reason for the daytime bursts of cortisol is unknown. Animal studies and at least one human observation suggest cortisol secretion may be entrained with feeding patterns, although not necessarily in response to the resultant plasma substrate changes. Calculations based on cortisol half-life and volume of distribution have suggested that plasma cortisol peaks are determined by the frequency and duration of secretory bursts rather than by changes in the absolute rate of cortisol secretion in milligrams per minute. This interpretation implies that the basal unstimulated rate of secretion must be near zero and that ACTH produces an all-or-none adrenal response. Other adrenal steroids, such as DHEA, show profiles parallel to that of cortisol when their plasma levels are measured frequently; any disparities reflect differences in metabolic clearances.

Plasma cortisol levels are increased by the stress of surgery, burns, infection, fever, psychosis, electroconvulsive therapy, acute anxiety, prolonged and strenuous exercise, and hypoglycemia. If attendant pain is prevented by disruption of the sensory input to the hypothalamus or by opioid analgesia, the cortisol response is blocked. Plasma cortisol levels are decreased promptly by the administration of synthetic glucocorticoids, such as dexamethasone (Table 54-1), which suppresses ACTH secretion by negative feedback. A single dose of dexamethasone that is biologically equivalent to twice the daily secretion rate of cortisol is sufficient to block completely the nocturnal ACTH peak and the morning rise in plasma cortisol level. Under circumstances of stress, the normal diurnal pattern of cortisol secretion may be lost, and feedback suppressibility may be impaired.

The average normal 8 AM plasma levels of cortisol and other adrenal steroids in humans, as well as their estimated secretion rates, are given in Table 54-2. The dominance of cortisol over corticosterone as a glucocorticoid and of DHEA-S as a source of androgenic activity is evident. Under severe stress, the maximal rate of cortisol secretion is 300 to 400 mg/day. Therefore this amount is usually provided to patients who lack adrenal

Table 54-2. Average 8 AM plasma concentration and secretion rates of adrenocortical steroids in adult humans

	Plasma concentration (μg/dl)	*Secretion rate (mg/day)*
Cortisol	13	15
Corticosterone	1	3
11-Deoxycortisol	0.16	0.40
Deoxycorticosterone	0.07	0.20
Aldosterone	0.009	0.15
18-OH Corticosterone	0.009	0.10
Dehydroepiandrosterone sulfate	115	15
Dehydroepiandrosterone	0.5	15

Table 54-1. Relative glucocorticoid and mineralocorticoid potency of natural corticosteroids and some synthetic analogues in clinical use*

	Glucocorticoid	*Mineralocorticoid*
Cortisol	1.0	1.0
Cortisone (11-keto)	0.8	0.8
Corticosterone	0.5	1.5
Prednisone (1,2 double bond)	4	<0.1
6α-Methylprednisone (Medrol)	5	<0.1
9α-Fluoro-16α-hydroxyprednisolone (triamcinolone)	5	<0.1
9α-Fluoro-16α-methylprednisolone (dexamethasone)	30	<0.1
Aldosterone	0.25	500
Deoxycorticosterone	0.01	30
9α-Fluorocortisol	10	500

*All values are relative to the glucocorticoid and mineralocorticoid potencies of cortisol, which have each been set at 1.0 arbitrarily. Cortisol actually has only 1/500 the potency of the natural mineralocorticoid aldosterone.

function and are either acutely ill or must undergo surgical stress; without the steroid they may not survive. There is little variation of plasma cortisol concentration with age, but secretion rates are correlated with lean body mass. The age-dependent changes in adrenal androgens are detailed later.

Actions of Glucocorticoids

Cortisol is one of the few hormones essential for life. Although provision of carbohydrate and of a pure mineralocorticoid or sodium chloride can postpone death, human beings cannot survive total adrenalectomy for long without glucocorticoid replacement. The exact reasons for this—the most critical site of life-preserving action—are still difficult to explain, despite knowledge of many important effects of cortisol. The hormone is involved in maintaining glucose production from protein, facilitating fat metabolism, supporting vascular responsiveness, and modulating central nervous system function. In addition, the hormone affects skeletal turnover, hematopoiesis, muscle function, immune responses, and renal function. The net effect of its metabolic actions is catabolic or antianabolic. The term *permissive* has been used to describe many of cortisol's actions, implying that the hormone may not directly *initiate* so much as *allow* certain processes to occur. This concept may help to explain the need for the hormone's critical yet unfocused presence. Several illustrations may serve to better define this permissive role:

1. Cortisol may amplify the effect of another hormone on a process that it does not affect directly. For example, cortisol does not itself stimulate glycogenolysis. However, cortisol augments the stimulation of glycogenolysis by glucagon.
2. Cortisol and glucagon individually increase the activity of the enzyme phosphoenolpyruvate carboxykinase. However, their combined effect is more than additive, i.e., cortisol synergizes with glucagon in this important regulatory step of gluconeogenesis.
3. The enzyme tyrosine transaminase is inducible experimentally by cortisol but not normally by its substrate tyrosine. However, in the presence of submaximal doses of cortisol, tyrosine administration will now induce the enzyme.

In vitro and in vivo, the effects of cortisol may be evident within minutes (inhibition of ACTH release) or hours (increase in plasma glucose) or may require days to be expressed (induction of glucose 6-phosphase). Cortisol enters target cells freely and is then bound to a specific receptor which can be found in both the cytoplasm and the nucleus. Its point of origin is uncertain. This receptor, which consists of four identical subunits with molecular weights of 85,000, has been demonstrated in virtually every tissue, consistent with the hormone's widespread actions. The gene for the receptor has also been cloned. After a transformation process, the hormone-bound receptor is translocated from the cytoplasm to the nucleus, where it interacts with DNA, as illustrated in Fig. 49-10. In a model system, the glucocorticoid receptor binds to specific sequences on the target DNA molecules (Fig. 54-13); these binding sites are upstream from the gene sequences themselves.

The mechanisms whereby gene expression is effected remains to be elucidated. Phosphorylation and acetylation of chromatin protein and increases in messenger RNA, transfer RNA, and ribosomal RNA have all been observed after exposure to cortisol in various systems. However, none of these effects has been proven critical to the hormone's actions. Increased synthesis of a number of enzymes follows, and these partly explain the effects of cortisol on amino acid turnover and glycogen accumulation. Induction of the phosphoprotein *lipocortin*, an inhibitor of phospholipase A_2, is an important component in the antiinflammatory and immunoregulatory effects of cortisol. The magnitude of hormone action is directly correlated with the amount of cytoplasmic receptor bound by the hormone and with the nuclear binding of the hormone receptor complex. In addition, cortisol induction of some enzymes, e.g., tyrosine aminotransferase, can only be demonstrated in the G_1 and S phases of the target cell cycle. Cortisol downregulates its own receptor, adding another dimension to the modulation of hormone action.

The specificity of cortisol responses is not so much dependent on interaction with the cytosolic receptor as on the subsequent nuclear events and the particular messenger RNAs that have been augmented or diminished. For example, progesterone binds to the cortisol receptor, but this steroid does not initiate most glucocorticoid effects; indeed, progesterone can inhibit certain effects by tying up the cortisol receptor. Furthermore, the *progesterone receptor* can bind to some DNA sites for the glucocorticoid receptor, but this also does not initiate the glucocorticoid action. Thus the unique combination of cortisol, its cytosolic receptor, and its DNA binding site are required to produce the specific glucocorticoid actions.

Other intracellular mechanisms of cortisol action also probably exist. Cortisol does not generally alter intracellular cAMP levels. It does, however, synergize with the nucleotide in a number of situations, and some actions of cortisol can be mimicked by cAMP. Cortisol may also act by altering the phospholipid component of various intracellular membranes, particularly those to which enzymes are bound. Even in a single cell type, various enzyme changes produced by cortisol are not necessarily linked. This observation bespeaks multiple mechanisms of action.

Effects on metabolism. The most important overall action of cortisol is to facilitate the conversion of pro-

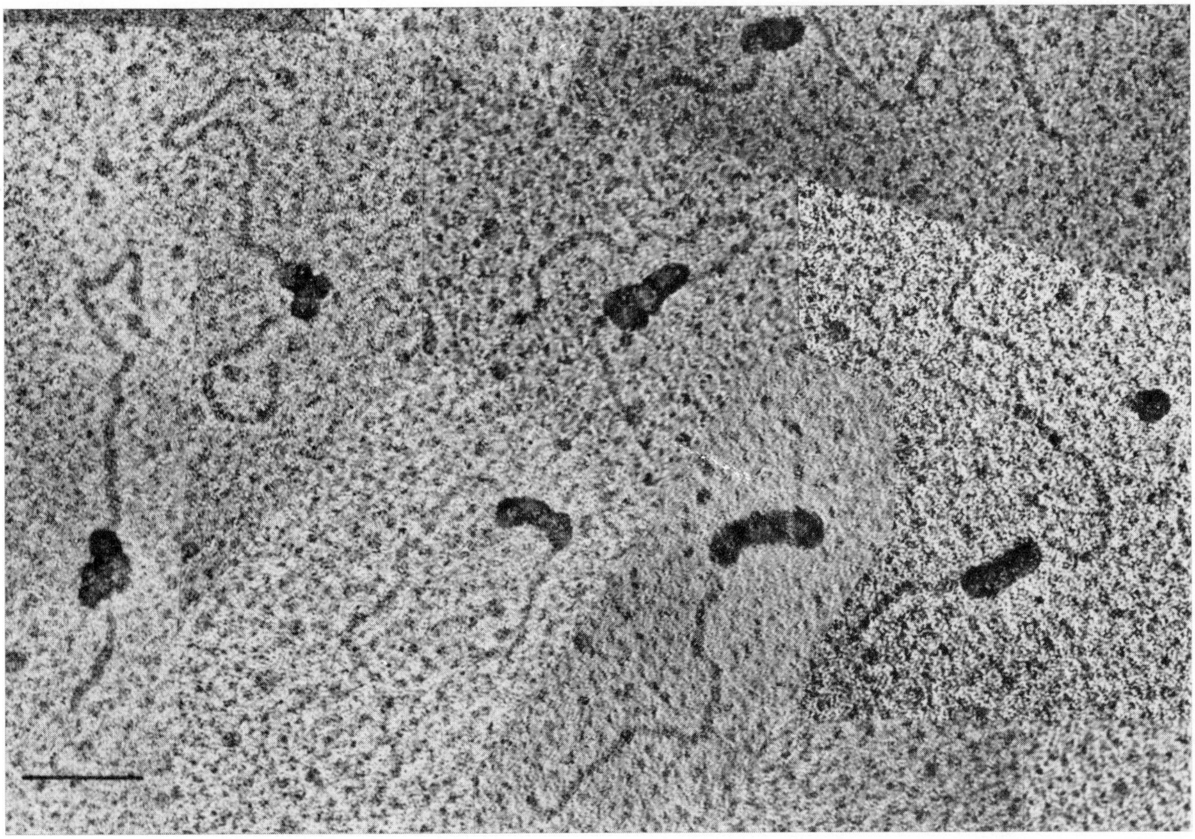

■ Fig. 54-13. Electron micrographs of DNA molecules from murine mammary tumor virus. The DNA molecules were incubated with purified glucocorticoid receptor complexes that appear as dark areas along the DNA molecules. Note that the glucocorticoid receptor complexes bind to the same site near one end of each of the DNA molecules. Bar at lower left represents 100 nm. (From Payvar, F., et al.: Cell **35**:381, 1983.)

tein to glycogen (Fig. 54-14). In animal experiments, cortisol enhances the mobilization of muscle protein for gluconeogenesis by accelerating protein degradation and inhibiting protein synthesis (Fig. 54-15). In humans the blood production rates of both the essential amino acid, leucine, and the nonessential amino acid, alanine, are increased by cortisol, indicating protein breakdown (Fig. 54-16). The increased production rates raise the plasma concentrations of the branch chain amino acids; however, alanine levels do not rise because the conversion of alanine to glucose is also markedly increased by cortisol (Fig. 54-16).

Although the combined catabolic and antianabolic action of cortisol in normal amounts is physiologically beneficial, an excess of endogenous cortisol or any exogenous glucocorticoid produces a continuous drain on body protein stores, most notably in muscle, bone, connective tissue, and skin. This drain cannot be compensated for by dietary protein because of the inhibition of protein synthesis. Cortisol further stimulates the transformation of the proteolytically derived amino acids into glucose precusors (Fig. 54-14). Table 54-3 lists a number of liver enzymes that are involved in the processes of gluconeogenesis and that are induced by cortisol.

Glucocorticoids are critical for the survival of a fasting animal or human. Without them there is no increase in proteolysis during fasting, as evidenced by lack of increase in urinary nitrogen excretion. Therefore when liver glycogen stores are depleted, gluconeogenesis from protein is deficient, and death from hypoglycemia may ensue. The secretion of cortisol is not increased to any great degree, if at all, by fasting. It is the *previous* exposure to normal levels of cortisol and the maintenance of those levels during fasting that "permit" amino acid mobilization to be augmented.

Cortisol plays a role in the defense against hypoglycemia similar to that evoked by insulin. Although the rapid release of the glycogenolytic hormones, glucagon and epinephrine, is primarily responsible for the rapid recovery of plasma glucose levels, the *prior* action of cortisol leads to the buildup of sufficient glycogen stores on which the other hormones can act. In addition, cortisol amplifies both the glycogenolytic and glu-

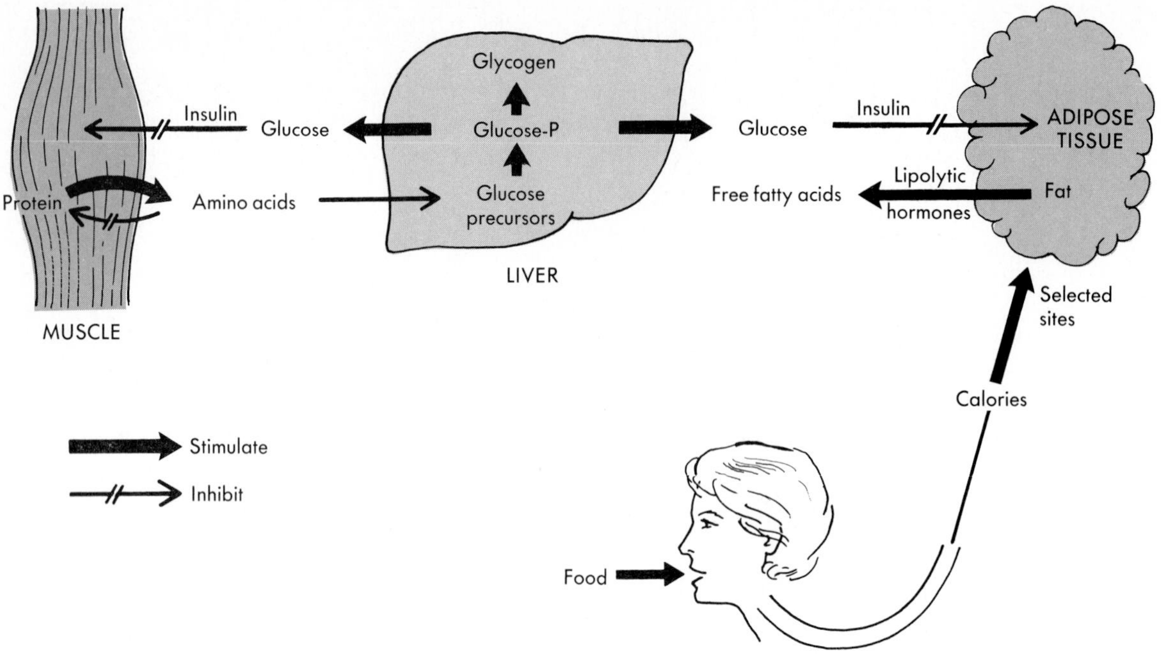

■ **Fig. 54-14.** Effect of cortisol on the overall flow of fuels facilitates release of amino acids as well as both storage and release of glucose and fatty acids.

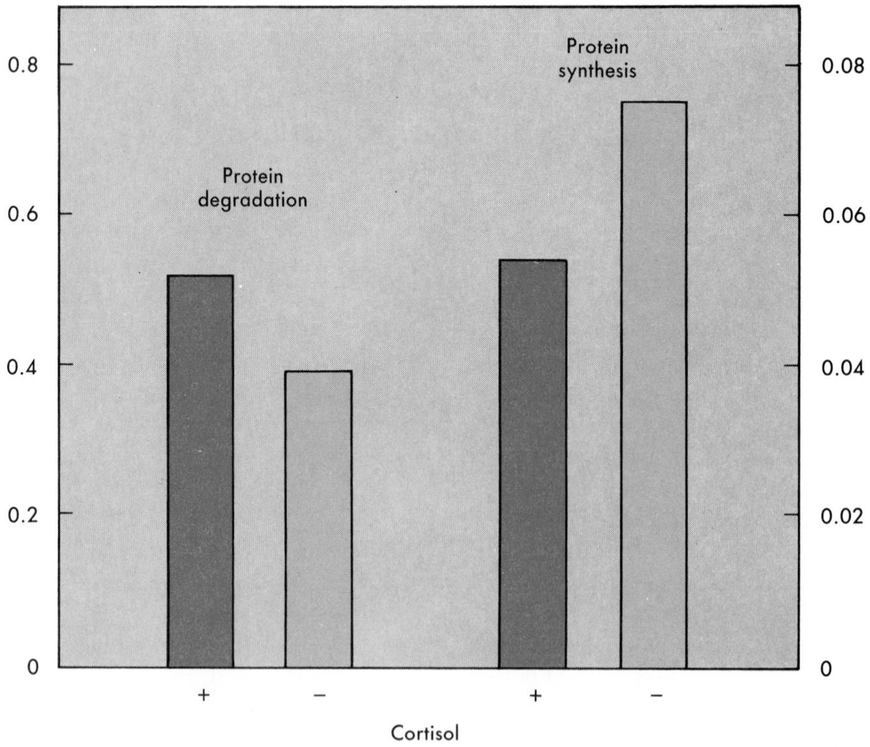

■ **Fig. 54-15.** The effect of cortisol replacement on muscle protein turnover in adrenalectomized fasted rats. Cortisol increases protein degradation and decreases protein synthesis. +, With cortisol; −, without cortisol. (Data from Goldberg, A.L., et al.: Fed. Proc. **39**:31, 1980.)

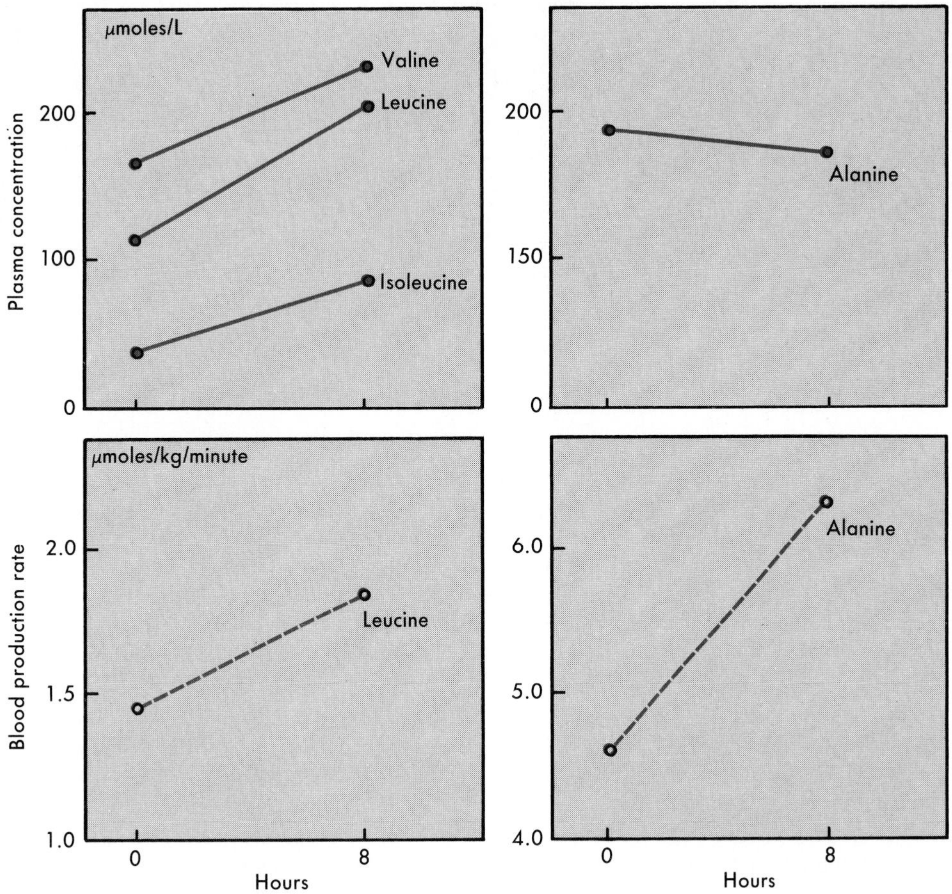

■ **Fig. 54-16.** Effect of cortisol infusion for 8 hours in normal humans on the plasma concentrations and blood production rates of amino acids. Production rates were determined isotopically as outlined in Chapter 49. The plasma concentrations of leucine and two other branch chain amino acids increase. The increase of leucine is accomplished by an increase in its production rate and suggests that cortisol stimulates proteolysis. Although the blood production rate of alanine also increases, its plasma concentration does not, because cortisol simultaneously stimulates alanine use by augmenting its conversion to glucose. (Redrawn from Simmons, P.S., et al.: J. Clin. Invest. **73**:412, 1984.)

■ **Table 54-3.** Enzymes whose activities are increased by cortisol*

Provide carbon precursors	*Convert pyruvate to glycogen*	*Release glucose*	*Dispose of ammonia liberated from amino acids in urea cycle*
Alanine transaminase	Pyruvate carboxylase	Glucose 6-phosphatase	Arginine synthetase
Tyrosine transaminase	Phosphoenolpyruvate carboxykinase		Arginosuccinase
Tryptophan pyrrolase	Phosphoglyceraldehyde dehydrogenase		Arginase
Threonine dehydrase	Aldolase		
Serine dehydrase	Fructose 1,6-diphosphatase		
	Phosphohexoisomerase		
	Glycogen synthetase		

*Increase varies from 130% for glycogen synthetase to 1,000% for alanine transaminase. Induction time varies from 3 hours for tryptophan pyrrolase to 4 days for aldolase.

coneogenic actions of glucagon and epinephrine. Finally, cortisol enhances glucagon release by the alpha cells of the pancreatic islets. Although the major impact of cortisol is on liver glycogen, an excess of the hormone eventually increases plasma glucose levels. This occurs because cortisol also antagonizes the actions of insulin on glucose metabolism by inhibiting insulin-stimulated glucose uptake in adipose tissue and muscle and by reversing insulin suppression of hepatic glucose production. As shown in Fig. 54-17, cortisol decreases

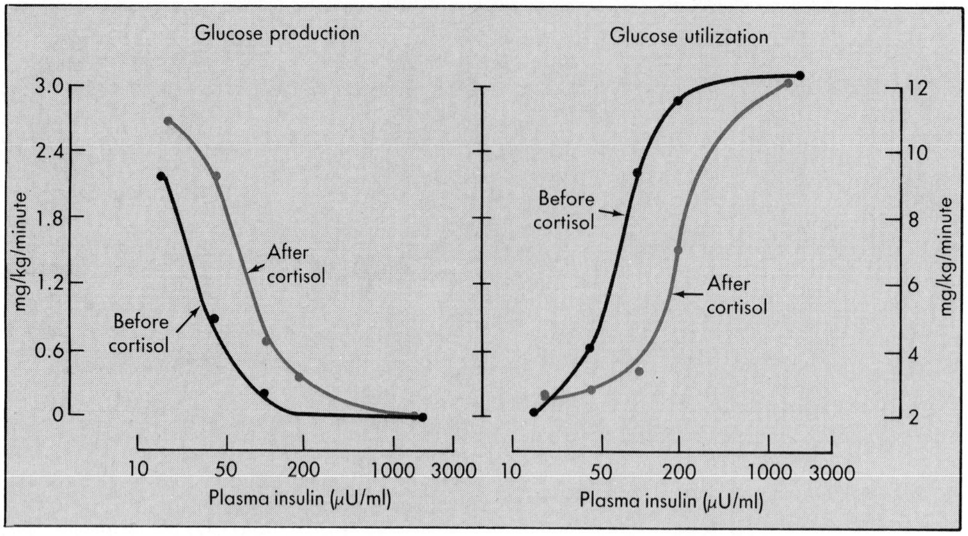

Fig. 54-17. The effect of cortisol on glucose turnover in response to increasing levels of insulin in a 24-year-old man. Cortisol decreases the sensitivity to insulin (the dose response curve is shifted to the right) both with regard to insulin inhibition of glucose production and insulin stimulation of glucose use. (Redrawn from Rizza, R.A., et al.: Am. J. Med. **70:**169, 1981).

tissue sensitivity but not maximal responsiveness to insulin. This antagonism takes place at both the receptor and postreceptor steps in insulin action.

In part, cortisol plays an analogous role in fat metabolism (Fig. 54-14). Although only weakly lipolytic itself, the presence of cortisol is necessary for maximal stimulation of fat mobilization by epinephrine, growth hormone, and other lipolytic peptides. Thus during fasting, cortisol permits accelerated release of energy stores as well as protein stores. Along with the free fatty acids, the glycerol released from adipose tissue provides yet another gluconeogenic substrate. Total fat stores are increased in cortisol-deficient adrenalectomized animals, and fasting ketogenesis is reduced. However, cortisol actions on body fat are quite complex. The hormone increases appetite and caloric intake and stimulates lipogenesis, especially in certain areas of the adipose tissue mass. Therefore an excess of cortisol finally results in obesity with a peculiar redistribution of fat that favors the trunk and face but spares the extremities. The uptake and degradation of low density lipoproteins are inhibited by cortisol in vitro, an action that may influence cholesterol metabolism.

Overall, cortisol is an important diabetogenic, antiinsulin hormone. Its hyperglycemic, lipolytic, and ketogenic actions are usually exhibited only when its secretion is stimulated by stress. With stress cortisol potentiates and extends the duration of the hyperglycemia evoked by glucagon, epinephrine, and growth hormone, while further accentuating loss of body protein. These diabetogenic and catabolic actions are markedly amplified when insulin secretion is deficient and are commonly encountered in that clinical setting.

Other effects. Cortisol has actions on the structure and function of numerous organs. Many of these have been deduced from observations in patients with disorders of cortisol secretion or in experimental animals that have been rendered either cortisol-deficient or have been treated with excess hormone.

Cortisol exerts a dual action on muscle function. In the absence of the hormone, the contractility and work performance of skeletal and cardiac muscle decline. The inotropic action of cortisol on skeletal muscle may be exerted at the myoneural junction via an increase in acetylcholine synthesis. In addition, cortisol may improve cardiac function by increasing myocardial β-adrenergic receptors. Yet an excess of cortisol causes decreased muscle protein synthesis, increased muscle catabolism, protein wastage, a consequent reduction of muscle mass, and muscle weakness. The myocardium, however, is spared these deleterious effects.

Bone. Cortisol inhibits bone formation by several mechanisms. First, the synthesis of type 1 collagen, the fundamental component of bone matrix, is reduced by cortisol. Second, the rate of differentiation of osteoprogenitor cells to active osteoblasts is decreased (Chapter 51). Third, cortisol decreases the absorption of calcium from the intestinal tract by antagonizing the action of 1,25-$(OH)_2$-D_3 and possibly also by diminishing the synthesis of this active vitamin D metabolite. In addition, the excretion of calcium by the kidney is enhanced by cortisol. The net result of the latter effects is to reduce the availability of calcium for bone mineralization. Finally, cortisol increases the rate of bone resorption. Thus one major consequence of a cortisol excess is an overall reduction in bone mass.

Connective tissue. Inhibition of collagen synthesis by cortisol produces thinning of the skin and of the walls of capillaries. The consequent fragility of the capillaries leads to their easy rupture and intracutaneous hemorrhage.

Vascular system. Cortisol is required for the maintenance of blood pressure. Besides sustaining myocardial performance, the hormone permits normal responsiveness of arterioles to the constrictive action of catecholamines. Cortisol also helps to maintain blood volume by decreasing the permeability of the vascular endothelium and by a weak mineralocorticoid action that enhances sodium conservation. The mass of red blood cells is also increased by cortisol.

Kidney. The rate of glomerular filtration is increased by cortisol. The hormone is also essential for the rapid excretion of a water load. In the absence of cortisol, the secretion of antidiuretic hormone and its action on the renal tubule are enhanced. Therefore free water clearance is diminished, and maximal dilution of the urine cannot occur (Fig. 54-18).

Central nervous system. Receptors for cortisol are present in the brain, especially in the limbic system. In an unknown manner cortisol modulates perceptual and emotional functioning. Auditory, olfactory, and gustatory acuity are accentuated by cortisol deficiency; this suggests that the hormone normally has a damping effect. That may be important in preventing sensory overload and disorganized responses. However, an excess of cortisol interferes with normal sleep and can either elevate the mood strikingly or depress it. In addition, the threshold for seizure activity may be lowered.

Fetus. Cortisol has important permissive effects that facilitate in utero maturation of the central nervous system, retina, skin, gastrointestinal tract, and lungs. The latter two have been studied best. The digestive enzyme capacity of the intestinal mucosa changes from a fetal pattern to a mature adult pattern under the influence of cortisol. This permits the newborn to use disaccharides present in milk. Timely preparation of the fetal lung to permit satisfactory breathing immediately after birth is facilitated by cortisol. The rate of development of the alveoli, flattening of the lining cells, and thinning of the lung septa are increased by the hormone. Most importantly, during the last weeks of gestation, the synthesis of surfactant, a phospholipid vital for maintaining alveolar surface tension, is increased. The effect is mediated by increasing the activity of key enzymes in the surfactant biosynthetic pathway, including phosphatidyl acid phosphatase and choline phosphotransferase.

Inflammatory and immune responses. Cortisol has a very special and important influence on the set of reactions evoked by tissue trauma, chemical irritants, foreign proteins, and infection. The immediate local reaction to injury consists of dilation of capillaries and changes in the endothelial cell membranes that enhance the trapping of circulating leukocytes at the site of injury. These reactions, mediated by prostaglandins, thromboxanes, and leukotrienes, are profoundly inhibited by all glucocorticoids through global depression of these mediators. This results because cortisol decreases the activity of the enzyme phospholipase A_2, which re-

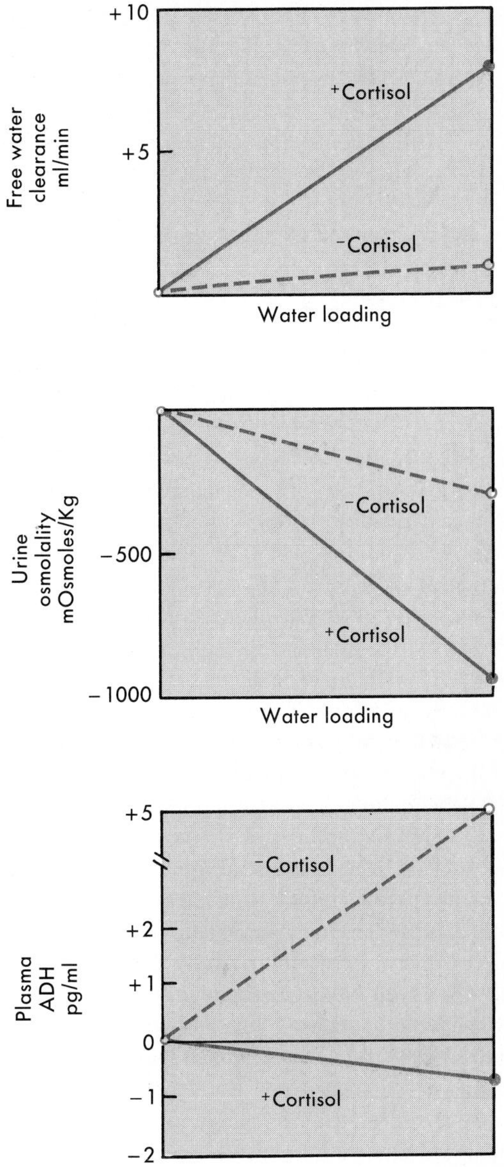

Fig. 54-18. Effect of cortisol replacement on the response of adrenalectomized dogs to water loading. In the absence of cortisol, water loading leads to little increase in free water clearance or decrease in urine osmolality, in part caused by lack of suppression of ADH release. In the presence of cortisol, the water load is promptly excreted as free water clearance rises sharply, urine osmolality falls markedly, and ADH levels decline. The differences are specifically caused by cortisol, since extracellular fluid volume was well maintained in these experiments by administration of sodium chloride and a mineralocorticoid. (Redrawn from Boykin, J., et al.: J. Clin. Invest. 62:738, 1978.)

leases arachidonic acid from its linkage to phosphatidyl choline. Arachidonic acid is the immediate precursor to prostaglandins, thromboxanes, and leukotrienes, and its production is the rate-limiting step in the synthesis of these mediators of inflammation. In addition, glucocorticoids stabilize lysosomes, thereby reducing the local release of proteolytic enzymes and hyaluronidase.

The recruitment of circulating neutrophils to the site of trauma or infection is inhibited by cortisol. The hormone decreases margination of leukocytes from blood vessels and their adherence to capillary endothelium. This process requires interaction between chemotactic peptides that attract the leukocytes and specific cell surface receptors. Cortisol inhibits the binding of these peptides to their receptors. Cortisol probably also decreases the phagocytic and bactericidal activity of leukocytes. Because cortisol increases the release of neutrophils from bone marrow, their circulating number actually increases, although their effectiveness decreases. Cortisol also decreases the proliferation of fibroblasts, the synthesis of collagen, and the deposition of fibrils, which form the basis for the more chronic response to injury. The net result of these actions is to impede the ability to deal locally in an effective manner with irritants or organisms and to prevent the walling off of infection.

Cortisol also influences immune responses to foreign substances. The hormone decreases the number of circulating thymus-derived lymphocytes, their transport to the site of antigenic stimulation, and their function. This cell-mediated immunity, as manifested by the classical delayed hypersensitivity response to the products of bacteria that cause tuberculosis or by the rejection of transplanted tissue, is markedly inhibited by the hormone.

The mechanism of inhibition is multifactorial (Fig. 54-19). When a foreign protein, or *antigen,* intrudes into the body, it is picked up by a macrophage. This cell presents the antigen to thymus-derived lymphocytes (T-cells) and simultaneously elaborates interleukin-1, a peptide lymphokine that activates a subset of T-cells with helper or inducer function. In turn, the helper T-cell secretes interleukin-2, a peptide that stimulates the proliferation of still more T-cells. These can then either activate or suppress bursa-derived lymphocytes (B-cells). These B-cells then produce antibodies directed against the original antigen. Cortisol inhibits the production of both interleukin-1 by macrophages and interleukin-2 by helper T-cells, thus limiting T-cell responses.

To the extent that suppressor T-cells are diminished, antibody production by B-cells could actually increase. More commonly, however, helper T-cells are diminished, leading to a secondary decrease in B-cells and antibody production. Once antibody molecules are present, neither their degradation nor their specific reaction with antigen molecules is affected by cortisol. The an-

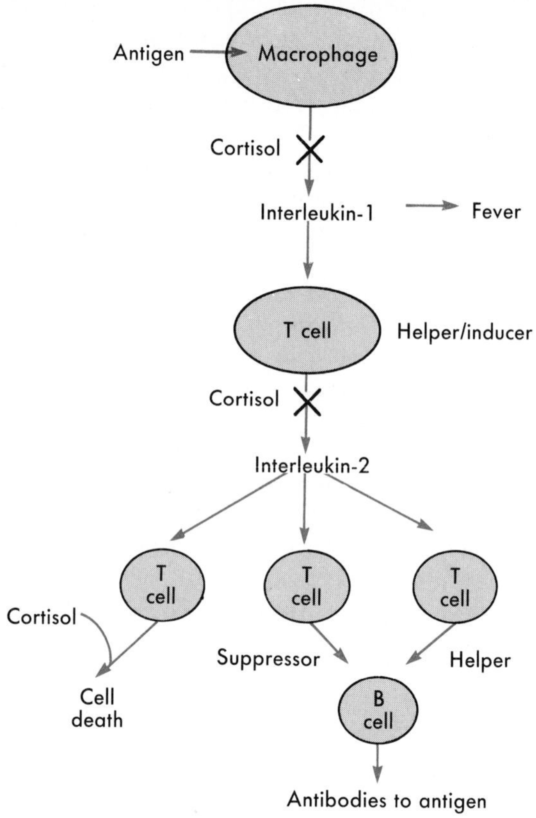

■ **Fig. 54-19.** Cortisol effects on cell-mediated immunity. Cortisol inhibits antigen-stimulated production of the peptide interleukin-1 by macrophages. This decreases the initial recruitment of T-lymphocytes. Cortisol also inhibits production of interleukin-2 and this reduces secondary proliferation of helper and suppressor T-lymphocytes. Depending in part on the ratio of these two cell populations, production of antibody to the original antigen may either be facilitated or retarded. In some species, large doses of cortisol are also lymphocytotoxic, causing cell death.

tiinflammatory action of glucocorticoids also includes suppression of the usual febrile response to infections or tissue injury. This likely comes about through decreased production of interleukin-1, which acts as an endogenous pyrogen.

At this point, a paradox is evident. On the one hand, the metabolic actions of cortisol are essential to the survival of the severely stressed, traumatized or infected individual. On the other hand, many of the defense mechanisms incorporated in the response to injury are inhibited by elevated levels of glucocorticoids. Although a final explanation cannot be given, the following has been suggested. Permissive lower levels of cortisol may be required for the initial metabolic (and possibly immunological) responses to stress. The evoked higher levels of cortisol may later serve to limit cellular and tissue reactions so that they do not themselves seriously damage the organism, for example by autoimmune reactions.

The therapeutic actions of cortisol in clinical medicine generally require administration of high doses for some time. Therefore, they represent a two-edged sword. When the symptoms of tissue injury resulting from disease are severe, functionally disabling, or life threatening, or when the rejection of transplanted organs or tissues must be prevented, glucocorticoids are dramatically beneficial. This is exemplified by their employment in severe asthma, in virulent forms of dermatitis, and in the preservation of kidney transplants. However, if glucocorticoids are administered therapeutically for very long, they may increase the susceptibility to infections or allow their dissemination, and they may prevent normal wound healing after injury. These serious adverse effects, along with the tendency for glucocorticoids to induce diabetes, osteoporosis, and psychiatric disorders, enjoin physicians to prescribe glucocorticoids cautiously and only when no safer form of treatment can succeed. Naturally this injunction does not apply to the use of cortisol as replacement therapy for patients lacking endogenous adrenocortical function.

Action of Adrenal Androgens

The adrenal steroids DHEA-S, DHEA, and androstenedione are relatively weak androgens. Their physiological function is largely expressed by their peripheral conversion to the potent androgen testosterone. In females, testosterone of ultimate adrenal origin sustains normal pubic and axillary hair. It may possibly also contribute to red blood cell production. When present in excess, adrenal androgens are of considerable importance in females (p. 974). In males, testosterone of testicular origin far exceeds that of adrenal origin, rendering the latter physiologically unimportant.

Regulation of Zona Glomerulosa Function

The principal function of *aldosterone,* the major product of the zona glomerulosa, is to sustain extracellular fluid volume by conserving body sodium. Hence, aldosterone is largely secreted in response to signals arising from the kidney when a reduction in circulating fluid volume is sensed. As shown in Fig. 54-20, when sodium depletion is produced, for example, by dietary restriction, the fall in extracellular fluid and plasma volume causes a decrease in renal arterial blood flow and pressure. The juxtaglomerular cells of the kidney respond to this change by secreting the enzyme *renin* into the peripheral circulation. As detailed in Chapter 47, renin acts on its substrate, *angiotensinogen* (an α_2-globulin of hepatic origin), to form the decapeptide *angiotensin I.* The latter is then further cleaved by a converting enzyme of pulmonary origin to the octapeptide *angiotensin II.* This extremely potent vasoconstrictor binds to specific receptors in the zona glomerulosa and directly stimulates the synthesis and release of aldosterone (Fig. 54-21). Calcium and the phosphatidylinositol messenger system is likely involved. Both the desmolase and 18-hydroxylation steps in the synthesis of aldosterone are stimulated by angiotensin.

After 5 days of only 10 mEq sodium intake, aldosterone secretion rates increase four- to eightfold. Conversely, when excess sodium is ingested and extracellular fluid volume expands, renin release, angiotensin II generation, and aldosterone secretion are all suppressed. The renin and aldosterone response to hypovolemia can also be evoked rapidly by hemorrhage, by assumption of the upright posture for several hours, or by an acute diuresis. Such maneuvers increase plasma aldosterone two- to fourfold. Thus the juxtaglomerular cells and the zona glomerulosa form a physiological feedback system. Sodium deprivation induces aldosterone hypersecretion via renin and angiotensin. When the additional aldosterone has caused sufficient sodium retention to restore extracellular fluid and plasma volume to normal, renin release is dampened, and aldosterone hypersecretion ceases. In this manner, daily aldosterone secretion ranges from 50 μg with a dietary sodium intake of 150 mEq to 250 μg with a dietary sodium intake of 10 mEq.

The release of renin from the juxtaglomerular cells is additionally enhanced by increased sympathetic neural activity, which is induced by hypovolemia. Norepinephrine, released at the nerve endings, reacts with β-adrenergic receptors in the kidney to stimulate renin release. Hence, β-adrenergic antagonists depress renin and aldosterone responses to sodium depletion. The release of renin also appears to be mediated by local prostaglandins; therefore prostaglandin synthesis inhibitors, such as indomethacin, also reduce aldosterone responses. Short-loop feedback inhibition of renin release is exerted by angiotensin II, but there is no direct feedback on the juxtaglomerular cells by aldosterone. The newly described natriuretic atrial peptides, or *atriopeptins,* reinforce the volume-related effects of the renin-angiotensin system on aldosterone secretion. These natriuretic hormones are synthesized and released by atrial myocytes in response to changes in distension of the atria. An increase in vascular volume causes secretion of atriopeptins into the circulation. They are bound by specific receptors in the zona glomerulosa and inhibit the synthesis and release of aldosterone. In addition, the peptides decrease renin release and have direct natriuretic effects in the kidney.

Aldosterone also participates in a vital physiological feedback relationship with potassium (Fig. 54-20). Another major function of aldosterone is to facilitate the clearance of potassium from the extracellular fluid, and concordantly potassium acts as an important stimulator of aldosterone secretion. In humans an acute infusion of potassium that raises plasma levels only 0.5 mEq/L

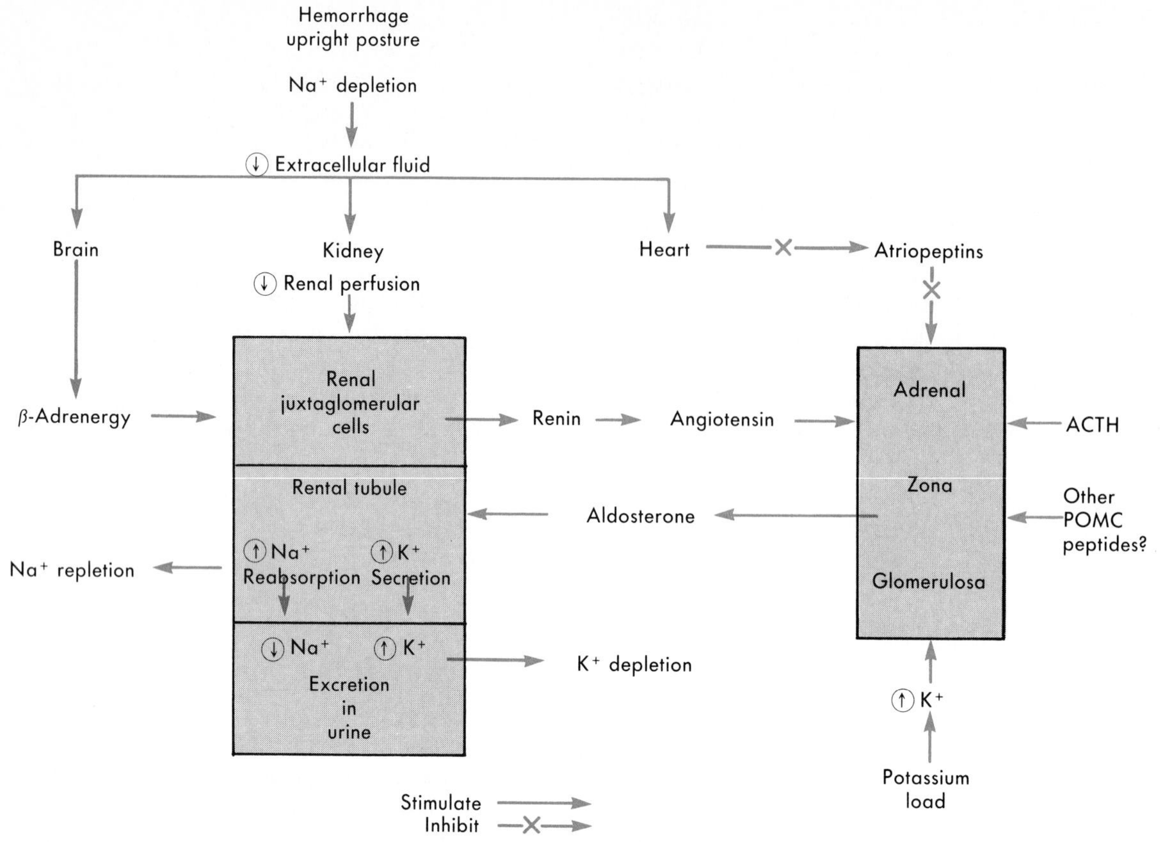

■ **Fig. 54-20.** Regulation of aldosterone secretion. Sodium deprivation and hypovolemia stimulate aldosterone release via several mechanisms. Decreased renal perfusion, reinforced by β-adrenergic signals, stimulates the renin-angiotensin system, thereby augmenting aldosterone secretion. In addition, cardiac myocytes respond to hypovolemia by decreasing atriopeptin release, thereby relieving the inhibitory effect of these peptides on aldosterone secretion. The increase in aldosterone causes renal sodium retention and restoration of extracellular volume. This closes the feedback loop between sodium and aldosterone. An increase in plasma potassium directly stimulates aldosterone release and the steroid hormone causes potassium excretion. When sufficient potassium has been lost, feedback on the zona glomerulosa shuts off aldosterone secretion. ACTH and possibly other proopiomelanocortin peptides facilitate aldosterone responses to mineral and fluid regulation.

immediately increases plasma aldosterone threefold (Fig. 54-21). An increase in dietary potassium from 40 to 200 mEq/Day increases plasma aldosterone sixfold in supine resting subjects. Conversely, potassium depletion lowers aldosterone secretion (Fig. 54-20). The effect of potassium on the zona glomerulosa is a direct one, and is demonstrable in vitros. Several steps in aldosterone biosynthesis are enhanced by potassium. However, potassium also directly inhibits the release of renin. Thus, to some extent, the stimulation of aldosterone secretion by potassium loading will be limited by a simultaneous reduction in renin release (provided sodium intake and volume are held constant). Conversely, the diminution in aldosterone secretion caused by potassium depletion will be moderated by an increase in renin release.

Zona glomerulosa function is also stimulated by ACTH. In doses similar to those that increase cortisol release from the zona fasciculata, ACTH raises plasma aldosterone levels (Fig. 54-21). ACTH stimulates aldosterone synthesis at the desmolase step, probably acting through an increase in cAMP levels. However, in vivo ACTH stimulation of aldosterone secretion wanes after several days. The mechanism of this escape is not completely understood; it appears that as sodium is retained and extracellular fluid volume rises (as a result of the action of the mineralocorticoid) renin release is suppressed, and its action on the zona glomerulosa is lost.

ACTH also tends to inhibit the conversion of deoxycorticosterone (DOC) to aldosterone. The physiological role of ACTH in regulating aldosterone ouput appears limited to a tonic one; that is, when ACTH is deficient, the zona glomerulosa is less able to respond to the primary stimulus of sodium depletion. This debility is seldom critical in patients with hypopituitarism. ACTH

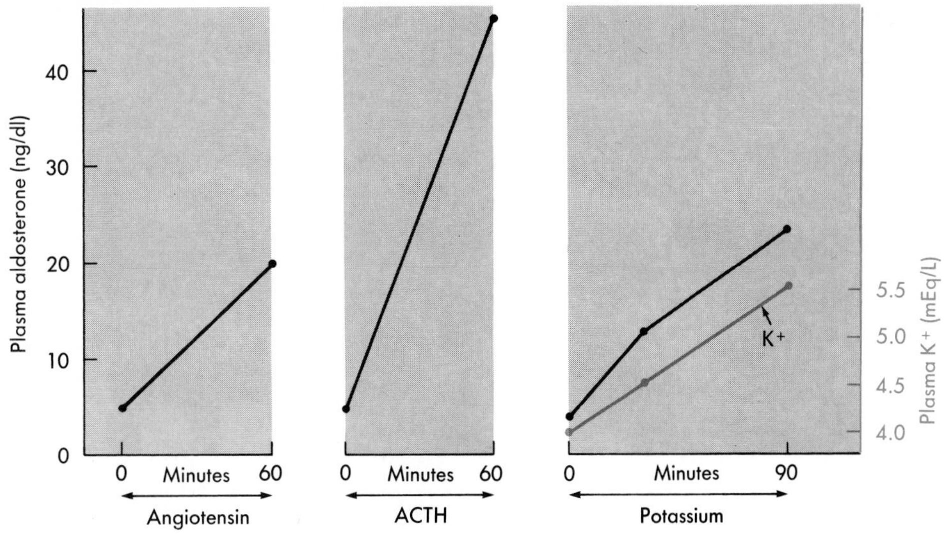

Fig. 54-21. Plasma aldosterone responses to stimulators of zona glomerulosa function in humans. Note that an increase in plasma potassium in the physiological range stimulates aldosterone secretion. (Redrawn from Dluhy, R.G., et al.: J. Clin. Invest. **51:**1950, 1972; Horton, R.: J. Clin. Invest **48:**1930, 1969. Reproduced by copyright permission of The American Society for Clinical Investigation.)

also stimulates the secretion of DOC from the zona fasciculata, as well as from the zona glomerulosa, and the secretion of 18-hydroxy DOC (18-OH-DOC) from the latter (Fig. 54-6). Under rare pathological circumstances, these steroids can generate clinical syndromes of mineralocorticoid excess.

Stimulation of aldosterone secretion by the three major factors noted earlier is interrelated. A low sodium intake potentiates aldosterone responsiveness to angiotensin, potassium, and ACTH. The increased sensitivity to angiotensin that results from sodium depletion is explained partly by increased binding of the peptide hormone to its receptors in the zona glomerulosa and partly by enhanced activity of the biosynthetic pathway. Conversely, if the potassium content of the adrenal cell is depleted, the responses to angiotensin, ACTH, and cAMP are diminished.

There is evidence for still other regulatory factors of physiological significance. The existence of a pituitary tropic hormone other than ACTH that is specific for the zona glomerulosa has long been suspected. A glycopeptide of anterior pituitary origin and likely a product of proopiomelanocortin other than ACTH (Chapter 52) may stimulate aldosterone secretion, but this requires clarification. Among neural transmitters, serotonin is stimulatory, whereas dopamine tonically inhibits aldosterone secretion by a direct action on the zona glomerulosa. Because of this effect of dopamine, aldosterone responses to other stimuli are enhanced by dopamine antagonists.

In humans the plasma aldosterone level shows a definite diurnal fluctuation, with the highest concentration occurring at 8 AM and the lowest at 11 PM. Although this profile correlates with similar directional changes in plasma renin and plasma cortisol levels, the diurnal pattern of aldosterone arises independently because it is not affected by variation in sodium intake, posture, ACTH suppression by exogenous glucocorticoids, or serum potassium.

Actions of Aldosterone and Other Mineralocorticoids

The kidney is the major site of mineralocorticoid activity. Aldosterone binds to a cytosolic receptor (different from the cortisol receptor) in renal tubular cells and is transferred with the receptor to the nucleus, where the steroid effects transcriptional changes. Messenger RNAs and proteins of still undetermined structure are induced, which apparently mediate the hormone's effects. A lag of 1 to 2 hours is therefore required between exposure to aldosterone and its onset of action.

Aldosterone stimulates the active reabsorption of sodium from the distal tubular urine; the sodium is transported through the tubular cell and back into the capillary blood. Thus net urinary sodium excretion is diminished, and the vital extracellular cation is conserved (Fig 54-22). Because water is passively reabsorbed with the sodium, there is little increase in serum sodium concentration, and extracellular fluid volume expands in an isotonic fashion. Although only 3% of total sodium reabsorption is regulated by aldosterone, deficiency of this hormone produces a significant negative sodium balance.

Aldosterone acts at the following loci in distal renal

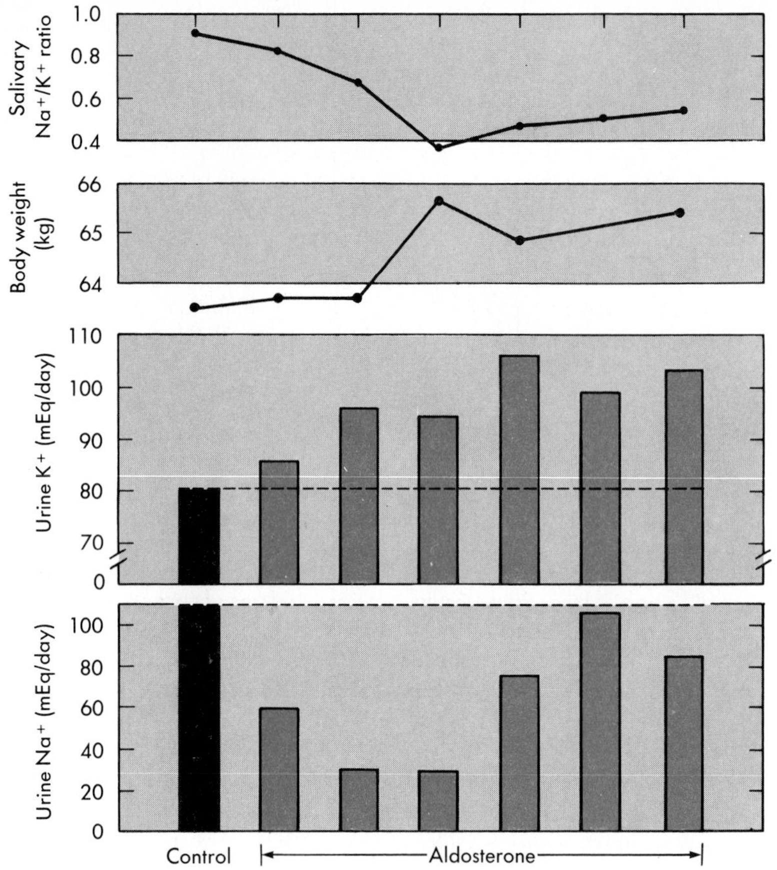

Fig. 54-22. The effects of aldosterone administration in a normal human. Note eventual escape from sodium retention with stabilization of body weight yet continuing loss of potassium in the urine. Dashed lines represent levels of sodium and potassium intake. (Redrawn from August, J.T., et al.: J. Clin. Invest. **37:**1549, 1958. Reproduced by copyright permission of The American Society for Clinical Investigation.)

tubule and collecting duct cells: (1) at the apical (luminal) surface to increase the number of membrane channels through which sodium enters the cell along an electrochemical gradient; (2) at the basal (capillary) surface of the cell to activate Na^+,K^+-ATPase, which pumps the sodium out; (3) in the mitochondria, stimulating Krebs cycle reactions, such as citrate synthase, that help generate the needed energy for extrusion of sodium into the interstitial fluid and capillary blood; and (4) in the cytosol to increase phospholipase activity and fatty acid synthesis, possibly for membrane regulation.

Aldosterone stimulates the active secretion of potassium out of the tubular cell and into the urine (Fig. 54-22) concurrently with sodium reabsorption. This does not constitute a stoichiometric *exchange* of potassium for sodium. Nonetheless, the active reabsorption of sodium is thought to create an electronegative condition in the tubular lumen, which facilitates the passive transfer of potassium into the tubular urine. Therefore the extent of kaluresis is greatly dependent on the delivery of sodium to the distal tubule. Aldosterone does not increase potassium excretion in a sodium-depleted subject because virtually all the sodium will be reabsorbed proximally. Conversely, a high sodium intake will greatly exacerbate urinary potassium losses caused by aldosterone. Ordinarily most of the potassium that is excreted daily results from distal tubular secretion. Hence, the presence of aldosterone or some other mineralocorticoid is critical for disposal of the daily dietary potassium load. Potassium flux, unlike that of sodium, does not entrain the movement of water; therefore, in the absence of aldosterone, potassium retention can cause a dangerous rise in serum potassium levels. An excess of the hormone causes a significant decrease in serum potassium levels. Continued administration of aldosterone in the face of a normal sodium intake produces sodium retention, weight gain (Fig. 54-22), and an increase in blood pressure as a result of the expanded extracellular fluid volume. However, after several days and an accumulation of 200 to 300 mEq of sodium, retention ceases, balance is achieved, and body weight stabilizes. This escape is thought to be caused by a depression in proximal tubule sodium reabsorption as a result of expansion of the extracellular fluid and probably mediated by atriopeptins. Nonetheless, the mineralocorticoid-induced potassium loss continues because sodium delivery to the distal tubule is maintained.

In addition to its effects on potassium secretion, aldosterone enhances tubular secretion of hydrogen ion as sodium is reabsorbed. Therefore aldosterone excess leads to the development of a mild systemic metabolic alkalosis, which can be further aggravated by the depletion of potassium. Ammonium excretion is also increased; however, the final urine pH is usually alkaline because the expansion of extracellular fluid volume inhibits bicarbonate reabsorption. In contrast, a deficiency of aldosterone produces a metabolic acidosis. Finally, aldosterone also stimulates the excretion of magnesium.

Aldosterone additionally affects mineral transport in other organs. The hormone stimulates sodium reabsorption from the gastrointestinal tract while enhancing potassium excretion in the feces. Similarly, the hormone decreases the ratio of sodium to potassium in perspiration and in saliva (Fig. 54-22). These actions, however, have little importance in overall cation balance. Aldosterone significantly affects sodium and potassium exchange between the extracellular fluid and the intracellular fluid. The net result is to increase potassium content of the intracellular space.

An effect of clinical importance is the increased blood pressure that results from an excess of aldosterone. This is an indirect consequence of the retention of sodium, expansion of the extracellular fluid volume, and a slight increase in cardiac output. In addition, the sodium and water content of the arteriolar cells may increase; the resultant swelling narrows the arteriolar lumen and increases peripheral resistance. In contrast, blood pressure falls below normal in aldosterone-deficiency states as a result of hypovolemia, although there is usually a compensatory renin-stimulated increase in the vasoconstrictor angiotensin II.

Aldosterone actions in the kidney are blocked by progesterone and 17-hydroxyprogesterone, which therefore have natriuretic activity. An important inhibitor in clinical usage as a diuretic and antihypertensive agent is spironolactone. The mechanism of inhibition by these steroids appears to involve competition with aldosterone for binding to its renal tubular receptor.

Clinical Syndromes of Adrenocortical Dysfunction

Hypofunction. Destruction of the adrenal cortex, *Addison's disease,* is ultimately incompatible with life. Except in cases of surgical removal or sudden infarction of the gland, Addison's disease usually progresses slowly and leads to the gradual development of glucocorticoid, adrenal androgen, and mineralocorticoid deficiencies. A lack of cortisol leads to loss of appetite with weight loss, malaise, fatigue, lethargy, muscle weakness, nausea, vomiting and abdominal pain, fever, poor tolerance of minor medical or surgical stress, fasting hypoglycemia, and an increase in circulating lymphocytes and eosinophils with a reduction in neutrophils. A loss of adrenal androgens may contribute to anemia and, in females, a reduction in pubic and axillary hair. Because of negative feedback, the secretion of ACTH and all proopiomelanocortin products increases as the cortisol levels decline, and their melanocyte-stimulating activity produces striking hyperpigmentation of the skin. The diagnosis of destruction of the zona fasciculata and zona reticularis is established by demonstrating low plasma cortisol levels and decreased urinary excretion of 17-hydroxycorticoids and 17-ketosteroids. Plasma ACTH levels are elevated, and if exogenous ACTH is administered, the levels of cortisol and its metabolites fail to increase. Deficiency of aldosterone is marked by polyuria (which is caused by natriuresis), dehydration, hypotension, hyperkalemia, hyponatremia, and metabolic acidosis. Plasma and urinary aldosterone levels are low. Plasma renin and angiotensin levels are elevated consequent to the stimulus of sodium depletion. In most cases of adrenal disease, the entire adrenal cortex is involved. However, primary selective loss of either glucocorticoid or mineralocorticoid function can occur.

Adrenal insufficiency may be caused by ACTH deficiency, resulting from disease of the hypothalamus or pituitary. The clinical picture, then, is that described earlier for loss of cortisol and adrenal androgen. Hyperpigmentation, however, does not occur because ACTH and copeptides are not present. When ACTH is replaced exogenously, plasma cortisol and urinary 17-hydroxycorticoid levels gradually increase.

Treatment of acute life-threatening adrenal insufficiency (adrenal crisis) consists of large doses of intravenous cortisol, glucose, and sufficient isotonic sodium chloride infusion to restore normal extracellular fluid volume and lower serum potassium levels. For lifetime maintenance, patients require oral cortisol, a generous salt intake, and usually a synthetic mineralocorticoid such as 9α-fluorocortisol (Table 54-1).

Hyperfunction. The most common cause of endogenous adrenocortical hormone excess is bilateral hyperplasia of the adrenal cortex. The hyperplasia results from hypersecretion of ACTH. Tumors that give rise to autonomous hormone secretion also occur in the various adrenocortical zones. The major manifestations of increased glucocorticoids include (1) obesity with a peculiar distribution of fat involving the cheeks (moon facies), the supraclavicular areas, the posterior cervicothoracic junction (buffalo hump), the trunk, and the abdomen—the extremities being spared; (2) a loss of bone mass (osteoporosis), associated back pain, vertebral fractures, and necrosis of the hips; (3) a loss of connective tissue integrity, associated with fragile capillaries, easy bruisability, and thin skin through which the underlying blood vessels may be seen (purple striae); (4) increased protein catabolism, resulting in atrophy and weakness of the muscles of the trunk and extremities, poor wound healing, and stunted growth in children; (5) abnormal carbohydrate metabolism, sometimes sufficient to produce frank diabetes; (6) impaired response to infections, particularly those produced by staphylococci, *Mycobacterium tuberculosis,* and fungi; and (7) headache, insomnia, and psychosis. All these pathological consequences can be produced iatrogenically, as well, by the therapeutic administration of synthetic glucocorticoids.

Adrenal androgen hypersecretion in adult females is manifested by different degrees of masculinization. This includes loss of regular menses, regression of breast tissue, increased body hair, acne, deepening of the voice, enlargement of the clitoris, increased muscularity, and heightened libido. There are virtually no clinically detectable changes in men.

The diagnosis of endogenous glucocorticoid excess depends on demonstrating elevated plasma levels of cortisol and loss of its normal diurnal variation. In parallel, urinary excretion of free cortisol and its 17-hydroxycorticoid metabolites increases. The concurrent or independent existence of adrenal androgen excess is marked by elevated urinary excretion of 17-ketosteroids and elevated plasma levels of DHEA, DHEA-S, androstenedione, and, in women, testosterone. Negative feedback suppression of plasma gonadotropins may be observed.

Once it has been determined that endogenous glucocorticoid or adrenal androgen excess exists, the next step is to determine the cause. If plasma ACTH is modestly elevated, then a pituitary (or hypothalamic) origin is indicated, usually a small ACTH-producing neoplasm. The dependence of the adrenocortical excess on pituitary ACTH is shown by demonstrating that a large dose of exogenous glucocorticoids (8 mg dexamethasone) suppresses endogenous cortisol secretion, whereas a smaller dose (2 mg dexamethasone) does not; however, the latter dose is sufficient to suppress cortisol secretion in normal persons. If plasma ACTH levels are low and adrenocortical secretion cannot be suppressed by dexamethasone, then an autonomous adrenal neoplasm is present. If plasma ACTH levels are very high but the adrenal gland cannot be suppressed, then an extraadrenal, extrapituitary neoplasm secreting the ACTH is implicated.

A primary excess of aldosterone produces a clinical syndrome characterized by hypertension, an expanded extracellular fluid volume with minimal edema, headache, hypokalemia with metabolic alkalosis, and slight hypernatremia. The diagnosis is established by demonstrating that plasma and urinary aldosterone (or rarely, DOC or 18-OH-DOC) are elevated when the patient has a high sodium intake (200 to 300 mEq). In the majority of cases, the lesion is a neoplasm arising in the zona glomerulosa; a lesser number have bilateral hyperplasia. Hence, the secretion of aldosterone is essentially autonomous, and plasma renin levels are low, being suppressed by the expanded extracellular fluid. It is important to distinguish primary from secondary aldosteronism. In secondary aldosteronism the hypersecretion of the mineralocorticoid is a result of an excess of renin. This may be generated by obstructive lesions of the renal arteries, which reduce perfusion pressure, or by secretion from a tumor of the juxtaglomerular cells.

Treatment of excess adrenocortical hormone secretion by adrenal tumors is best accomplished by removal of the neoplasm. If the basic lesion is within the pituitary gland, modern microsurgical techniques permit removal of the small adenomas, usually without disturbing the remaining function of the pituitary gland. Ectopic production of ACTH by malignancies requires treatment of the primary lesion, but palliative treatment may be afforded by chemical blockade of adrenocortical secretion with drugs such as aminoglutethimide. Definitive treatment of hyperaldosteronism resulting from renal artery stenosis requires restoration of renal perfusion by vascular grafting or removal of the ischemic kidney. Symptomatic relief from any form of hyperaldosteronism is obtained by treatment with the aldosterone antagonist spironolactone.

Biosynthetic defects. A number of congenital enzyme deficiencies in the pathways of adrenocortical hormone synthesis occur. These include deficiencies of desmolase, 3β-ol-dehydrogenase, 21-hydroxylase, 11-hydroxylase, 17-hydroxylase, and 18-hydroxylase. *Adrenogenital syndrome* is the term given to this group of disorders. The consequences of each biosynthetic defect can be predicted (Fig. 54-3); the product of the reaction will be deficient, and the precursor or precursors will be increased enormously as a result of negative feedback to the pituitary gland or, occasionally, to the juxtaglomerular cells as well. Detailed descriptions of each of these syndromes is beyond the scope of this discussion. However, a single example will be presented to emphasize the importance of understanding the detailed pathways of adrenal hormone biosynthesis.

The most common biochemical lesion encountered clinically is that of 21-hydroxylase deficiency. Plasma levels of cortisol and 11-deoxycortisol tend to be low, whereas plasma levels of the immediate precursor, 17-hydroxyprogesterone, are greatly elevated (Fig. 54-3). The corresponding urinary metabolites show the same pattern; i.e., 17-hydroxycorticoid levels are low whereas pregnanetriol is increased. The block in cortisol synthesis leads to increased ACTH secretion, which stimulates all open pathways and causes immense hyperplasia of the zona fasciculata and zona reticularis. As seen in Fig 54-5, 17-hydroxyprogesterone serves as the major precursor for adrenal androgens. Therefore in 21-hydroxylase–deficient patients, androgens will be secreted in great excess as a result of hyperstimulation by ACTH. Plasma androgen levels (DHEA, DHEA-S, androstenedione, testosterone) will be high, as will the excretion of their urinary metabolites (17-ketosteroids).

The clinical consequences of this block and its overflow are dramatic. The high adrenal androgen levels in female fetuses cause a masculinized pattern of development of the external genitalia. Thus they have penile-like clitorides and scrotumlike labia. The ambiguous genitalia can lead to incorrect gender assignment at birth. If not treated promptly, the androgen excess will cause early acceleration of linear growth and early appearance of pubic and axillary hair but suppression of gonadal function and normal puberty in females and males. Eventually the patient will be short as a result of premature closure of the growth centers in the bones. If the 21-hydroxylase block is severe, cortisol and aldosterone secretion may be so impaired as to cause episodes of adrenal crisis (p. 973). This entire sequence can be reversed by supplying appropriate amounts of both cortisol and an aldosterone substitute.

The Adrenal Medulla

The adrenal medulla is the source of the circulating catecholamine hormone *epinephrine*. It also secretes small amounts of *norepinephrine*–nominally a neurotransmitter—which in select circumstances may also function as a hormone. These compounds have diverse effects on metabolism as well as on virtually all organ systems in the body. The adrenal medulla represents essentially an enlarged and specialized sympathetic ganglion. However, the neuronal cell bodies of the medulla do not have axons; instead they discharge their catecholamine hormones directly into the bloodstream, thus functioning as endocrine rather than nerve cells. The adrenal medulla is formed in parallel with the peripheral sympathetic nervous system. At about 7 weeks of gestation, neuroectodermal cells from the neural crest invade the anlage of the primitive adrenal cortex. There they develop into the medulla, which begins to secrete during gestation and by birth is completely functional. The development of sympathetic nervous tissue and induction of neural hormone synthesis is stimulated by *nerve growth factor,* one of the insulin-like growth factors (Chapter 50).

Adrenal medullary tissue in the adult weighs about 1 g and consists of chromaffin cells (so named for their affinity for chromium stains). These are organized in cords and clumps in intimate relationship with venules that drain the adrenal cortex and with nerve endings from cholinergic preganglionic fibers of the sympathetic nervous system. Within the chromaffin cells are numerous granules of 100 to 300 nm diameter, similar to those found in postganglionic sympathetic nerve terminals. These granules consist of the catecholamine hormones epinephrine and norepinephrine (20% by weight), adenosine triphosphate and other nucleotides (15%), protein (35%), and lipid (20%). They also contain enkephalins, β-endorphin, and other proopiomelanocortin peptides. About 85% of the chromaffin granules store epinephrine and 15% norepinephrine. Small clumps of similar chromaffin cells are also found outside the adrenal medulla along the aorta and the chain of sympathetic ganglia. Although such cells are physiologically inconsequential, they can give rise to functioning chromaffin tumors.

The adrenal medulla is usually activated in association with the rest of the sympathetic nervous system and acts in concert with it. Many of the actions of the neurotransmitter norepinephrine (which is released locally at the effector site of the postganglionic sympathetic nerve endings) are duplicated and amplified by the hormone epinephrine, which reaches similar sites via the circulation. However, epinephrine has unique effects of its own, some of which modulate those of norepinephrine. Furthermore, under certain circumstances, e.g., during hypoglycemia, the adrenal medulla is probably activated selectively. A complete description of the sympathetic nervous system and its function is presented in Chapter 20.

Synthesis and Storage

The catecholamine hormones are synthesized within the chromaffin cell by a series of reactions shown in Fig. 54-23. The first, catalyzed by the enzyme *tyrosine hydroxylase,* is the rate-limiting step in the sequence and occurs in the cytoplasm. The conversion of tyrosine to dihydroxyphenylalanine (dopa) requires molecular oxygen, a tetrahydropteridine, and NADPH. Norepinephrine inhibits this reaction by negative feedback. The conversion of dopa to dopamine is catalyzed by a nonspecific aromatic L-amino acid decarboxylase that employs pyridoxal phosphate as a cofactor. The dopamine thus formed in the cytoplasm must be taken up by the chromaffin granule before it can be acted on further.

The next enzyme in the sequence, *dopamine β-hydroxylase,* is present exclusively within the granule, both in membrane-bound and soluble form. In the presence of molecular oxygen and a hydrogen donor, it catalyzes the formation of norepinephrine from dopamine. In approximately 15% of the granules, the sequence ends here, and the norepinephrine is stored. In 85% of the granules, norepinephrine diffuses back into the cytoplasm. There it is N-methylated by phenylethanolamine N-methyltransferase using S-adenosylmethionine as the methyl donor. The resultant epinephrine is then taken back up into the chromaffin granule where it is stored as the predominant adrenal medullary hormone. The uptake of dopamine, norepinephrine, and epinephrine by the secretory granules is an active process that requires adenosine triphosphate (ATP) and magnesium. The storage of the catecholamine hormones at such high intragranular concentrations also requires energy in the form of ATP. One mole of the nucleotide is present in a complex with 4 moles of catecholamine and a specific protein known as *chromogranin.*

The synthesis of epinephrine and norepinephrine is regulated by several factors. Acute stimulation of the sympathetic innervation to the medulla activates tyrosine hydroxylase, possibly by decreasing cytoplasmic catecholamine levels and relieving the feedback inhibition. Chronic stimulation of the preganglionic fibers induces increased concentrations of both tyrosine hydroxylase and dopamine β-hydroxylase, thus helping to

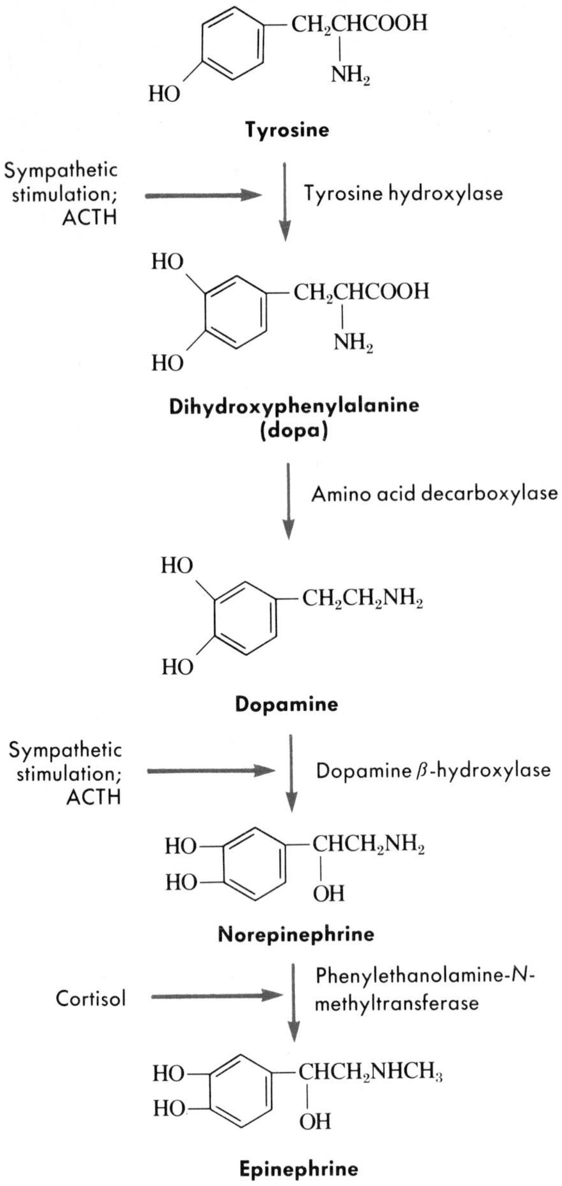

Fig. 54-23. Pathway of catecholamine hormone synthesis in the adrenal medulla. The dopamine β-hydroxylase reaction occurs within the secretory granule. Note stimulatory effects of ACTH, cortisol, and sympathetic nerve impulses at various points.

ensure maintenance of catecholamine output in the face of continuous demand. The mechanism of induction may involve a cAMP–dependent protein kinase. ACTH, acting directly, helps to sustain the levels of the same two enzymes under stressful conditions. In contrast, cortisol specifically induces the N-methyltransferase and therefore selectively stimulates epinephrine synthesis. The anatomical relationship between the medulla and the cortex subserves this action, since blood from the cortex with a high concentration of cortisol directly perfuses the chromaffin cells.

Metabolism of Cathecholamines

Essentially all the circulating epinephrine is derived from adrenal medullary secretion. In contrast, most of the circulating norepinephrine is derived from sympathetic nerve terminals and from the brain, having escaped immediate local re-uptake from synaptic clefts. However, the metabolic fate of epinephrine and norepinephrine merges into one or two major excretory products.

Epinephrine and norepinephrine have extremely short life spans in the circulation, allowing rapid turnoff of their dramatic effects. Half-lives are in the range of 1 to 3 minutes. The metabolic clearance rate of epinephrine is reported to be 3.5 to 6.0 L/minute, and that of norepinephrine is 2.0 to 4.0 L/minute. Both hormones increase their clearance rates still further by activating β-adrenergic receptors, another mechanism that helps to limit their actions. Only 2% to 3% of catecholamines are disposed of unchanged in the urine. The normal total daily excretion is about 50 μg, of which 20% is epinephrine and 80% is norepinephrine. Another 100 μg is excreted as sulfate or glucuronide conjugates. The vast majority of epinephrine is metabolized within the adrenal medullary chromaffin cell when synthesis exceeds the capacity for storage. Circulating epinephrine and norepinephrine are metabolized in many tissues, but predominantly in the liver and kidney.

The catecholamine hormones are metabolized by the reaction sequences shown in Fig. 54-24. The key enzymes are catecholamine *O-methyltransferase* and the combination of *monamine oxidase* and *aldehyde oxidase*. O-Methylation and oxidative deamination can be carried out in either order, giving rise to several products that are then excreted in the urine. O-Methylation alone yields an average daily excretion of metanephrine (from epinephrine) plus normetanephrine (from norepinephrine) of 300 μg. In contrast, the excretion of the common deaminated products, vanillylmandelic acid (VMA) and methoxyhydroxyphenylglycol (MOPG), averages 4000 μg and 2000 μg, respectively. It is important to note that under normal circumstances epinephrine accounts for only a very minor proportion of urinary VMA and MOPG. Because the majority is derived from norepinephrine, urinary VMA and MOPG levels largely reflect activity of the sympathetic nervous system rather than that of the adrenal medulla. Activity of the adrenal medulla can only be assessed specifically by measurement of urinary free epinephrine or plasma

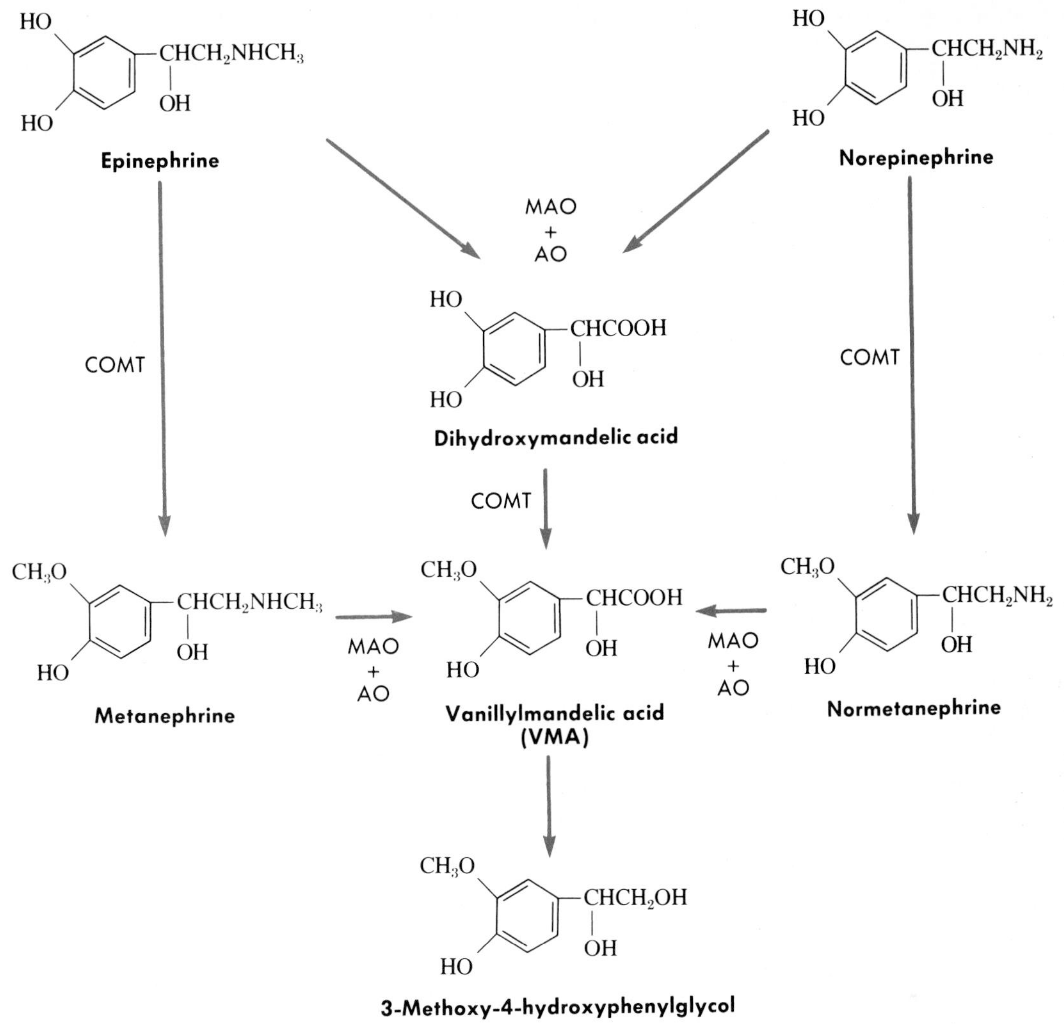

Fig. 54-24. Metabolism of catecholamine hormones. VMA is quantitatively the main product. *MAO,* Monamine oxidase; *AO,* aldehyde oxidase; *COMT,* catecholamine O-methyltransferase.

epinephrine levels. However, measurement of urinary VMA and total metanephrines is important in the detection of hyperfunctioning adrenal medullary neoplasms that secrete either epinephrine or norepinephrine.

Regulation of Adrenal Medullary Secretion

Secretion from the adrenal medulla forms an integral part of the "fight-or-flight" reaction evoked by stimulation of the sympathetic nervous system. Thus perception or even anticipation of danger or harm (anxiety), trauma, pain, hypovolemia from hemorrhage or fluid loss, hypotension, anoxia, extremes of temperature, hypoglycemia, and severe exercise cause rapid secretion of epinephrine (and probably norepinephrine) from the adrenal medulla. These stimuli are sensed at various higher levels in the sympathetic nervous system and responses are initiated in the hypothalamus and brain stem (Chapter 20). Usually activation of the adrenal medulla follows activation of the sympathetic nervous system and is evoked when the stimuli are more intense. In general, the hormonal response of the adrenal medulla reinforces that of the sympathetic nervous system. In certain circumstances, however, epinephrine secretion increases, whereas sympathetic nervous system activity decreases (see below).

The final common effector pathway activating the adrenal medulla consists of cholinergic preganglionic fibers in the greater splanchnic nerve. On stimulation, acetylcholine is released from the nerve terminals. This neurotransmitter depolarizes the chromaffin cell membrane by increasing its permeability to sodium. This, in turn, induces an influx of calcium ions, which probably produce microfilament contraction and draw the chromaffin granules to the cell membrane. The granules fuse with the membrane and discharge their contents by exocytosis. Epinephrine, norepinephrine, ATP, the en-

Table 54-4. Comparison of circulating concentrations of catecholamine hormones with biologically effective concentrations

Physiological state	Relevant biological action	Plasma epinephrine (pg/ml) Observed	Plasma epinephrine (pg/ml) Effective range for relevant biological action	Plasma norepinephrine (pg/ml) Observed	Plasma norepinephrine (pg/ml) Effective range for relevant biological action
Basal	—	34	—	228	—
Upright position	↑ Heart rate and blood pressure	73	50-125	526	1800
↓ Plasma glucose	↑ Plasma glucose	230	150-200	262	1800
Severe hypoglycemia	—	1500	—	770	—
Diabetic ketoacidosis	↑ Lipolysis and ketosis ↓ Insulin	510	100-400	1270	1800

Data from Clutter, W.E., et al: J. Clin Invest **66:**94, 1980; Silverberg, A.B., et al.: Am. J. Physiol. **234:**E252, 1978; and Christensen, N.J.: Diabetes **23:**1, Jan. 1974.

exocytosis. Epinephrine, norepinephrine, ATP, the enzyme dopamine β-hydroxylase, opioid peptides, chromogranin and soluble granule proteins are all simultaneously released into the circulation. The membranous material of the granule is retained and probably recycled. Exocytosis can also be induced by histamine and glucagon in chromaffin cell tumors.

Basal plasma epinephrine levels are 25 to 50 pg/ml (6×10^{-10}M). A daily basal delivery rate of 150 μg can be estimated, and this rate of epinephrine release can increase greatly with physiological stimuli (Table 54-4). For example, under the stimulation of a modest fall in plasma glucose from 90 to 60 mg/dl, adrenal medullary secretion causes a rise of endogenous epinephrine concentrations to 230 pg/ml. If epinephrine is infused exogenously at a rate sufficient to maintain its plasma level at 150 to 200 pg/ml, a rise in plasma glucose results. Hence, the adrenal medulla is quite capable of secreting physiologically meaningful amounts of epinephrine to contribute to glucose homeostasis.

The same is true of cardiovascular responses to the hormone. An increase in heart rate and systolic blood pressure can be produced by the concentrations of epinephrine that are generated endogenously by assumption of the upright position (Table 54-4). In addition, the high concentrations of epinephrine that occur in such illnesses as diabetic ketoacidosis are quite capable of contributing to the pathological state by stimulation of lipolysis and ketosis. Thus epinephrine functions as a true hormone in all these situations.

In contrast, circulating norepinephrine levels do not generally increase with stimulation to levels sufficient to produce relevant biological actions (Table 54-4). Therefore, under normal circumstances, norepinephrine does not appear to function in an endocrine fashion, although it may do so in severe, stressful illnesses such as myocardial infarction. Instead, norepinephrine's effects on similar metabolic processes, such as glucose production or lipolysis, result from its role as a neurotransmitter, wherein the necessary high concentrations are generated locally at the effector site.

Actions of Catecholamines

Epinephrine and norepinephrine exert their effects on a group of plasma membrane receptors designated β-1, β-2, α-1, and α-2. The existence of multiple receptors was originally inferred from the differences in biological activity exhibited by epinephrine, norephinephrine, and other agonists, as well as from differential effects of various antagonist molecules. Epinephrine reacts well with β-1 and β-2 receptors, but its potency is less with α-receptors. Norepinephrine reacts predominantly with α receptors, less strongly with β-1 receptors, and only weakly with β-2 receptors. β-1, β-2, and α-2 receptors are structurally similar. They are single-unit glycoproteins with molecular weights around 64,000. α-1 receptors differ from the others, and have molecular weights of 80,000. β-1 and β-2 receptors are coupled to and stimulate adenylate cyclase, so that cAMP is the second messenger for these biological effects (Fig. 49-7). Protein kinase A is activated and a cascade of changes in enzyme activities follows.

The hormone binding site and the regulatory site, though distinct from each other, reside within a single peptide in the β-receptors. The α-2 receptor, in contrast, contains an adenylate cyclase *inhibitory* regulatory unit, so that hormone binding decreases cAMP levels and protein kinase A activity.* The α-1 receptor is coupled to the phosphatidylinositol membrane system (Fig. 49-9). Protein kinase C, along with calcium, mediates hormone effects as a result of the α-1 receptor.

Continuous stimulation of catecholamine release or exposure to synthetic catecholamine agonists such as isoproterenol downregulates adrenergic receptor numbers and induces partial refractoriness to hormone action. Conversely, sympathectomy increases receptor number and enhances sensitivity to catecholamines. Acute exposure to catecholamine hormones produces rapid *desensitization* to subsequent doses. This effect is associ-

*Thus the α-2 receptor and the β-1 plus β-2 receptors form an antagonist pair.

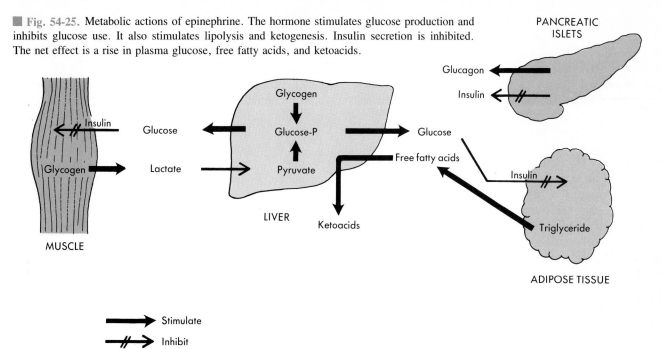

Fig. 54-25. Metabolic actions of epinephrine. The hormone stimulates glucose production and inhibits glucose use. It also stimulates lipolysis and ketogenesis. Insulin secretion is inhibited. The net effect is a rise in plasma glucose, free fatty acids, and ketoacids.

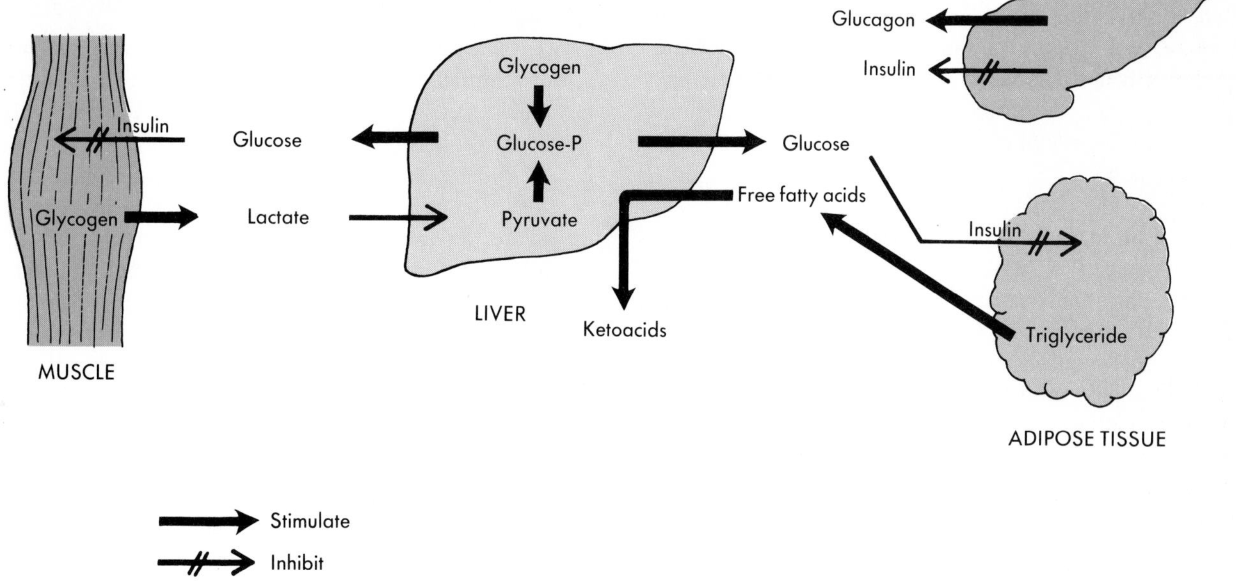

ated with phosphorylation of the various receptors, rendering them inaccessible to further hormone binding. If, as suspected, receptor phosphorylation is catalyzed by protein kinase A or C, then this desensitization process may be viewed as a form of rapid intracellular negative feedback, which limits hormone actions.

A listing of important relevant actions of epinephrine and norepinephrine is presented in Table 54-5. Both catecholamine hormones increase glucose production. They stimulate glycogenolysis in the liver via β-receptors by activating phosphorylase through the same cAMP–initiated cascade as is produced by glucagon. Glycogen synthase activity is concurrently restrained. The adrenal medullary epinephrine response to hypoglycemia is not needed as long as glucagon secretion is intact. However, if the secretion of glucagon is blocked, e.g., by somatostatin, the catecholamine action becomes essential for recovery from hypoglycemia.

Epinephrine and norepinephrine also stimulate gluconeogenesis, by activation of α-receptors on the liver cells. In addition, they stimulate muscle glycogenolysis, which increases plasma lactate levels and provides additional substrate to the liver. Simultaneously, epinephrine inhibits insulin-mediated glucose uptake by muscle and adipose tissue. The catecholamines also stimulate glucagon secretion while inhibiting insulin secretion. All these actions help restore plasma glucose and its delivery to the central nervous system. At the same time, epinephrine activates adipose tissue lipase, thereby increasing plasma free fatty acids, their β-oxidation in muscle and liver, and ketogenesis. These actions in humans are demonstrated in Fig. 54-25.

Table 54-5. Some actions of catecholamine hormones

β *epinephrine > norepinephrine*	α *norepinephrine > epinephrine*
↑ Glycogenolysis	↑ Gluconeogenesis (α1)
↑ Lipolysis and ketosis (β1)	
↑ Calorigenesis (β1)	
↓ Glucose utilization	
↑ Insulin secretion (β2)	↓ Insulin secretion (α2)
↑ Glucagon secretion (β2)	
↑ Muscle K^+ uptake (β2)	
↑ Cardiac contractility (β1)	
↑ Heart rate (β1)	
↑ Conduction velocity (β1)	
↑ Arteriolar dilation; ↓ BP (β2) (muscle)	↑ Arteriolar vasoconstriction; ↑ BP (α1) (Splanchnic, renal, cutaneous, genital)
↑ Muscle relaxation (β2) Gastrointestinal Urinary Bronchial	↑ Sphincter contraction (α1) Gastrointestinal Urinary Platelet aggregation (α2) Sweating (''adrenergic'') Dilation of pupils
Antagonists	Antagonists
Propranolol Timolol Atenolol (β1) Butoxamine (β2)	Phentolamine Phenoxybenzamine Prazosin (α1) Yohimbine (α2)

When the catecholamine hormones are secreted during exercise, they promote (1) use of muscle glycogen stores by stimulating phosphorylase, (2) efficient hepatic reutilization of lactate released by the exercising muscle, and (3) provision of free fatty acids as alternate fuels. When epinephrine secretion is stimulated by "stress," such as during illness or surgery, its actions on fuel turnover contribute significantly to induction of hyperglycemia and ketosis; i.e., it is a diabetogenic hormone. These diabetogenic effects are generally dependent on the presence of cortisol through its permissive actions.

Epinephrine also increases the basal metabolic rate. This action increases nonshivering thermogenesis as well as diet-induced thermogenesis. It is, therefore, an important part of the response to cold exposure and helps to regulate overall energy balance and stores. The mechanism may involve enhancement of Na^+, K^+-ATPase activity and increased mobilization of fuels. In neonates of many species, brown adipose tissue is an important site where catecholamines increase heat production. Here these amines stimulate proton conductance into the mitochondria, thereby uncoupling ATP synthesis from oxygen utilization.* In most metabolic effects, epinephrine is much more potent than norepinephrine. The latter, nonetheless, contributes importantly to metabolism via the activity of the sympathetic nervous system. For example, sympathetic nervous system activity, with norepinephrine as the mediator, decreases with fasting and increases after feeding. In this way, norepinephrine helps to adapt total energy utilization to its availability by modulating thermogenesis. In contrast, epinephrine secretion increases modestly during prolonged fasting and also 4 to 5 hours after a meal, in both cases in response to a declining plasma glucose level. This serves the purpose of helping to sustain glucose production for use by the central nervous system.

The cardiovascular effects of secreted epinephrine reinforce its metabolic actions. Cardiac output rises as a result of an increase in heart rate and contractile force; on the other hand, arteriolar constriction is selectively produced in the renal, splanchnic, and cutaneous beds, whereas the muscle arterioles are dilated. Systolic blood pressure increases, whereas diastolic blood pressure remains unchanged or decreases slightly. The net effect of these changes is to shunt blood toward exercising muscles while maintaining coronary and cerebral blood flow (Chapters 33 and 36). This guarantees delivery of substrate for energy production to the critical organs in the fight-or-flight situation.

During exposure to cold, constriction of cutaneous vessels helps to conserve heat, reinforcing epinephrine's thermogenic action. The inhibition of gastrointestinal and genitourinary motor activity, the relaxation of bronchioles to prevent expiratory airway obstruction and improve gas exchange, and the dilation of pupils to permit better distant vision are of obvious benefit to the endangered person.

The cardiovascular consequences of catecholamines also initially benefit the individual who has suffered major trauma, circulatory failure, or hypoxia. Without them, death may rapidly ensue. However, prolonged secretion of catecholamines eventually becomes deleterious. Reduced blood flow to the kidneys leads to renal failure; reduced blood flow to the splanchnic bed leads to hepatic failure as well as intestinal paralysis and necrosis; reduced blood flow to many tissues leads to decreased oxygenation with increased lactate production, which, in the face of decreased lactate utilization in the liver, produces metabolic (lactate) acidosis (Chapter 36).

Catecholamines exhibit a number of other diverse actions. By interacting with β-receptors in the vascular tree and lowering blood pressure, they may stimulate ADH release and water retention; conversely, by interacting with α receptors and raising blood pressure, they may inhibit ADH release and cause water loss (Chapter 52). Catecholamines increase renin release by stimulation of β-receptors in the kidney. This increases aldosterone secretion, which in turn enhances sodium retention. This action is augmented by local catecholamine effects in the kidney on the distribution of blood flow and on tubular function. Epinephrine stimulates influx of potassium into muscle cells by stimulation of β-2 receptors. This effect, which is reinforced by the increase in aldosterone release, helps to prevent hyperkalemia. Epinephrine increases parathyroid hormone, calcitonin, growth hormone, and gastrin secretion. Whether these effects are produced in physiological circumstances remains to be determined.

The interaction between catecholamine and thyroid hormone function remains an important area of investigation. Thyroid hormone secretion is enhanced by catecholamines under some circumstances. The peripheral conversion of T_4 to T_3 is stimulated by catecholamines, probably by stimulation of β-2 receptors. The effects of thyroidal status on adrenal medullary function and sympathetic nervous activity have been previously discussed (Chapter 53). In general, thyroid hormone sensitizes animals to catecholamine actions, but the mechanisms are complex and translation of these results to humans remains controversial.

Therapeutic Usage and Endocrine Implications

Catecholamine agonists and antagonists are in widespread use in medicine. A group of agonists called *amphetamines* are used as nasal decongestants, appetite suppressants, and general stimulants. All of these may

*The role of this specialized adipose tissue in human physiology is controversial.

be prescribed or sold over the counter, but their illicit availability has become a public health problem as well. They may cause hypertension; exacerbate tachycardia, palpitations and nervousness in hyperthyroid patients; or, rarely, increase plasma glucose in diabetic patients. In large doses, they can produce life-threatening "highs." Both β-adrenergic antagonists and α-adrenergic antagonists are used in the treatment of hypertension, and the former are also of value in the treatment of coronary artery disease. When β-antagonists are administered to diabetic patients taking insulin, they can mask the epinephrine-generated symptoms of hypoglycemia and impede epinephrine's contribution to the recovery from hypoglycemia. β-antagonists are also used effectively to counteract the hyperactive adrenergic state in hyperthyroid patients.

Pathological Secretion of Catecholamines

Spontaneous deficiency of epinephrine is unknown as an adult disease, and adrenalectomized patients do not require epinephrine replacement. However, in some young children idiopathic hypoglycemia may result from epinephrine deficiency, and it may respond to treatment with longer-acting preparations of the hormone.

Hypersecretion of epinephrine and norepinephrine from tumors of the chromaffin cells *(pheochromocytomas)* results in well-defined syndromes. In two thirds of the patients, sustained hypertension is produced; in one third blood pressure elevation is sporadic and accompanies dramatic clinical episodes produced by sudden spurts of excess catecholamines. These bursts of secretion can result from any stress, such as anesthesia, or from a rapid change in posture.

Typically, the patient notes sudden severe headache, palpitations, chest pain, extreme anxiety with a sense of impending death, cold perspiration, pallor of the skin caused by vasoconstriction, and blurred vision. When the patient is examined during an attack, blood pressure may be extremely high, for example, 250/150. If primarily epinephrine is being secreted, the heart rate will be increased; if norepinephrine is the predominant hormone, the heart rate will be decreased reflexly in response to the marked hypertension. Attacks may end spontaneously in minutes or may last hours. In addition to these episodes, chronic catecholamine excess may produce weight loss as a result of an increased metabolic rate and hyperglycemia caused by inhibition of insulin secretion.

The diagnosis is established by detecting high levels of plasma epinephrine or norepinephrine when the patient is recumbent and at rest. In addition, urinary excretion of free catecholamines, metanephrines, and VMA will usually be increased.

Definitive treatment requires removal of the adrenal medullary tumor. Symptomatic treatment during attacks is provided by α-adrenergic antagonists such as phentolamine or prazosin, which rapidly lower the elevated blood pressure. A β-adrenergic antagonist, such as propranolol, is used to reduce tachycardia, but it may exacerbate hypertension unless simultaneous α-adrenergic blockade is provided.

Bibliography

Journal articles

Axelrod, J., et al: Catecholamines, N. Engl. J. Med. **287:**237, 1972.

Cantin, M., and Genest, J.: The heart and the atrial natriuretic factor, Endocrine Rev. **6:**107, 1985.

Clutter, W., et al: Epinephrine plasma metabolic clearance rates and physiologic thresholds for metabolic and hemodynamic actions in man, J. Clin. Invest. **66:**94, 1980.

Cryer, P.E.: Physiology and pathophysiology of the human sympathoadrenal neuroendocrine system, N. Engl. J. Med. **303:**436, 1980.

Eigler, N., et al.: Synergistic interactions of physiologic increments of glucagon, epinephrine, and cortisol in the dog: a model for stress-induced hyperglycemia, J. Clin. Invest. **63:**114, 1979.

Fauci, A.S., et al: Glucocorticosteroid therapy: mechanisms of action and clinical considerations, Ann. Intern. Med. **84:**304, 1976.

Goldberg, A.L., et al: Hormonal regulation of protein degradation and synthesis in skeletal muscle, Fed. Proc. **39:**31, 1980.

Gustafsson, J., et al.: Biochemistry, molecular biology and physiology of the glucocorticoid receptor, Endocrine Rev. **8:**185, 1987.

Hamburg, S., et al.: Influence of small increments of epinephrine on glucose tolerance in normal humans, Ann. Intern. Med. **93:**566, 1980.

Landsberg, L., and Young, J.B.: The role of the sympathetic nervous system and catecholamines in the regulation of energy metabolism, Am. J. Clin. Nutr. **38:**1018, 1983.

Lefkowitz, R.J., and Caron, M.G.: Adrenergic receptors: molecular mechanisms of clinically relevant recognition, Clin. Res. **33:**395, 1985.

Munck, A., et al.: Physiological functions of glucocorticoids in stress and their relation to pharmacological actions, Endocrine Rev. **5:**25, 1984.

Parker, L.N., et al: Evidence for existence of cortical androgen-stimulating hormone, Am. J. Physiol. **236:**E616, 1979.

Rizza, R.A., et al: Cortisol-induced insulin resistance in man: impaired suppression of glucose production and stimulation of glucose utilization due to a postreceptor defect of insulin action, J. Clin. Endocrinol. Metab. **54:**131, 1982.

Santiago, J.V., et al: Epinephrine, norepinephrine, glucagon, and growth hormone release in association with physiological decrements in the plasma glucose concentration in normal and diabetic man, J. Clin. Endocrinol. Metab. **51:**877, 1980.

Silverberg, A., et al.: Norepinephrine: hormone and neurotransmitter in man, Am. J. Physiol. **234:**E252, 1978.

Thompson, E.B., et al.: Unlinked control of multiple glucocorticoid-induced processes in HTC cells, Mol. Cell Endocrinol. **15:**135, 1979.

Wortsman, J., et al.: Adrenomedullary response to maximal stress in humans, Am. J. Med. **77:**779, 1984.

Books and monographs

Baxter, J.D., et al: Glucocorticoid hormone action, New York, 1979, Springer-Verlag New York, Inc.

Nelson, D.H.: The secretion of the adrenal cortex and steroid biosynthesis. In Smith, L.H., Jr., editor: The adrenal cortex: physiological function and disease, vol. 18, Major problems in internal medicine, Philadelphia, 1980, W.B. Saunders Co.

CHAPTER

55

The Reproductive Glands

The endocrine glands that have been discussed thus far are essential to the maintenance of the life and well-being of the individual. In contrast, the endocrine function of the gonads is primarily concerned with maintaining the life and well-being of the species. The evolution of sexual reproduction has required the development of highly complex patterns of gonadal function. These are concerned with the nurturance and maturation of the individual male and female germ cells, their successful union, and the subsequent growth and development of the newly created individual within the body of the mother. There are many obvious differences between male and female gonadal function, but there are also important basic conceptual similarities and operational homologies. Therefore the subject of human gonadal endocrinology is approached as a single unit in the following sequence: (1) sexual differentiation, (2) common aspects of gonadal structure and function, (3) testicular function, (4) ovarian function, and (5) endocrine aspects of pregnancy.

Sexual Differentiation

Any discussion of reproductive endocrine physiology should begin with a consideration of the process of sexual differentiation, that is, the pattern of development of the gonads, genital ducts, and external genitalia. It is convenient and logical to divide sexual differentiation into three components: genetic sex, gonadal sex, and genital, or phenotypic, sex.

Genetic Sex

The normal female chromosome complement is 44 autosomes and two sex chromosomes, XX. Both of the X chromosomes are active in germ cells and are essential for the genesis of a normal ovary (Fig. 55-1). In addition, evidence suggests that there is participation of autosomes in ovarian development, since rare individuals with a normal complement of XX sex chromosomes inherit defective gonads as an autosomal recessive trait. In contrast, differentiation of the genital ducts and external genitalia along normal female lines requires that only a single X chromosome be active in directing transcription within the cell. The second X chromosome of a normal XX female is genetically inactive in all extragonadal tissues. Therefore individuals with an XO sex chromosome complement have very abnormal gonads but still develop as normal phenotypic females.

The presence of a Y chromosome is the single most consistent determinant of maleness. The normal male has a chromosome complement of 44 autosomes and two sex chromosomes, X and Y (Fig. 55-2). The presence of additional X chromosomes does not alter the fundamental maleness dictated by the Y chromosome, even though the gonads usually are rendered dysfunctional. With rare exceptions, the absence of the Y chromosome precludes normal testicular development, as well as masculinization of the genital ducts and external genitalia. However, even though the Y chromosome is virtually essential, it is not in itself sufficient for maleness. Genetic material located on the X chromosome may participate in directing the organization of the gonad into a testis and in sensitizing the genital ducts and external genitalia to the masculinizing effects of androgenic hormones (Fig. 55-2).

Chromatin material on the short Y_p arm and possibly in the pericentric regions of the Y chromosome directs the development of the testes. Chromatin on the long Y_q arm directs the development of spermatogenesis. In addition, the X chromosome must be *inactive* for normal spermatogenesis to occur, implying that this chromosome contains material inhibitory to spermatogenesis.

Gonadal Sex

During the initial 5 weeks of fetal life the gonads develop along sexually indistinguishable lines. The primordial germ cells migrate to the genital ridge and associate themselves with mesonephric tissue, an

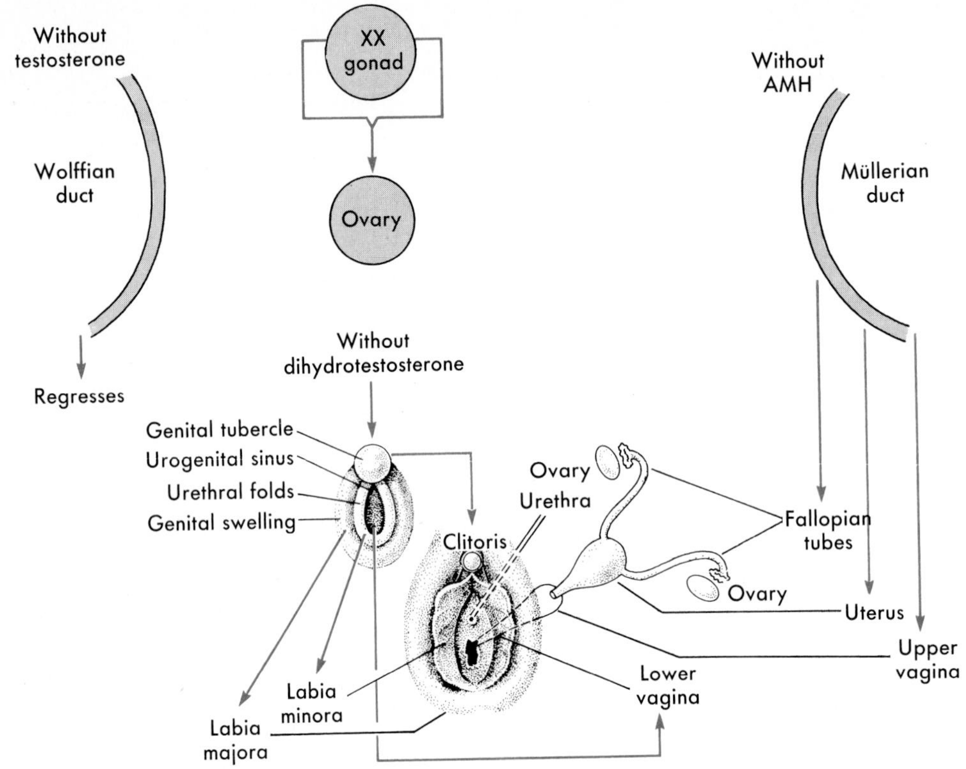

■ **Fig. 55-1.** Development of the human female reproductive organs and tract. Note the independence from hormonal products of the gonad. In the absence of any gonads the female format results. *AMH,* Anti-Müllerian hormone.

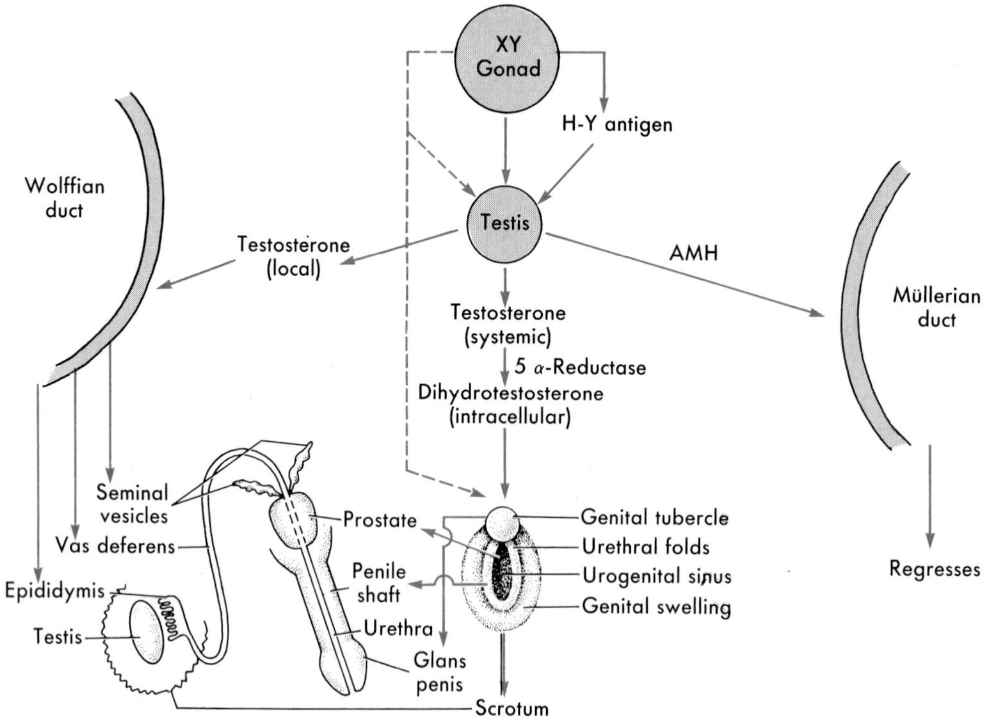

■ **Fig. 55-2.** Development of the human male reproductive organs and tract. Note the dependence on hormone products of the gonad. *AMH,* Anti-Müllerian hormone.

Table 55-1. Time of onset of the development of male reproductive system

Time (weeks)	*Testicular cells*	*Hormone produced*	*Ducts*	*External genitalia Masculinizes*
6	Seminiferous tubules			
6 to 7	Sertoli cells	Antimüllerian hormone (AMH)	Müllerian regresses	
8 to 9	Leydig cells	Testosterone		
9 to 10			Wolffian grows	Urogenital sinus*
10 to 11				Urogenital tubercule,* swelling, and folds*;
12 to 39				genitalia enlarge; testes descend

*Dihydrotestosterone synthetic capacity is present in these tissues at 6 to 10 weeks.

assembly that consists of an outer cortex and inner medulla. In a normal boy at 6 weeks of age the seminiferous tubules begin to form, followed by the Sertoli cells at 7 weeks and the Leydig, or interstitial, cells at 8 to 9 weeks (Table 55-1). At 9 weeks a recognizable testis is present, and testosterone secretion is established. The medulla of the early testis encloses the germ cells while the cortex regresses. The organization of the gonadal anlage into the characteristic seminiferous or spermatogenic tubules of the male is directed by a specific substance, known as the H-Y antigen (Fig. 55-2). This surface glycoprotein is found in all male cells, and its presence correlates almost perfectly with the presence of a testis in all species. Synthesis of the H-Y antigen is normally under the control of the Y chromosome, although there may also be regulatory influences by the X chromosome or autosomes. There are rare instances of XX individuals with testes. In these cases there is either sex chromosome mosaicism (XX/XY), or the H-Y antigen may have been translocated to an X chromosome or even to an autosome.

In the normal female, differentiation of the indifferent gonad into an ovary does not start until 9 weeks of age. At this time *both X chromosomes* within the germ cells become activated (Fig. 55-1). The germ cells begin to undergo mitosis, giving rise to oogonia, which continue to proliferate. Shortly thereafter meiosis is initiated in some oogonia, and they become surrounded by granulosa cells and stroma, from which interstitial cells subsequently appear. The germ cells are now known as primary oocytes and remain in the first stage of meiosis (prophase) until ovulation many years later. In contrast to the male gonad, the cortex predominates in the developed ovary while the medulla regresses. The capacity of the primitive ovary to synthesize estrogenic hormones develops at about the same time that testosterone synthesis begins in the testis.

Genital (Phenotypic) Sex

During the initial stages of fetal development sexual differentiation does not require any known hormonal products. However, differentiation of the genital ducts and of the external genitalia does require hormones. The guiding principle is that positive hormonal influences from the gonad are required to produce the masculine format. In the absence of any gonadal hormonal input the feminine format will result.

During the sexually indifferent stage, from 3 to 7 weeks, two genital ducts develop on each side. In the male at about 9 to 10 weeks, the wolffian, or mesonephric, ducts begin to grow and eventually give rise to the epididymis, the vas deferens, the seminal vesicles, and the ejaculatory duct (Fig. 55-2). This constitutes the system for delivering sperm from the testis to the penis. The differentiation of the wolffian ducts is preceded by the appearance of Leydig cells in the testis and the production of testosterone by those cells (Table 55-1). It is the steroid hormone *testosterone* that stimulates the growth and differentiation of the wolffian ducts in the male. Furthermore the testosterone produced by each testis acts unilaterally on its own wolffian duct (Fig. 55-2), as shown by gonadal transplantation experiments or by testosterone implantations. Testosterone does not have to be converted to its hormonally active product, dihydrotestosterone, to act within the wolffian duct cells, as it does in some other tissues (described later). Indeed these cells do not develop the 5α-reductase activity necessary for this conversion until after they have fully differentiated. In the female the wolffian ducts, lacking testosterone stimulation, begin to regress at 10 to 11 weeks.

The müllerian ducts arise parallel to and in part from the wolffian ducts on each side. In the male these ducts begin to regress at 7 to 8 weeks, about the same time that the Sertoli cells of the testis appear. These cells produce a glycoprotein hormone called *müllerian-inhibiting factor* (MIF), or *antimüllerian hormone* (AMH), which causes atrophy of the müllerian ducts. It may do so by blocking the stimulatory effect of epidermal growth factor on the müllerian system. In addition, AMH may initiate descent of the testes into the inguinal area. Although AMH is found in the postnatal testis, it has no certain role after birth. The secretion of AMH may be inhibited by the appearance of fetal follicle-stimulating hormone (FSH).

In the female who lacks AMH, the müllerian ducts grow and differentiate into fallopian tubes at the upper ends, whereas at the lower ends they join to form the uterus, cervix, and upper vagina (Fig. 55-1). This process is not completed until 18 to 20 weeks of age. It does not require any ovarian hormone.

The external genitalia of both sexes begin to differentiate at 9 to 10 weeks. They are derived from the same anlage: the genital tubercle, the genital swelling, the urethral, or genital, folds, and the urogenital sinus. In the normal female or in the absence of any gonads these develop without apparent hormonal influences into the clitoris, labia majora, labia minora, and lower vagina, respectively (Fig. 55-1). The possibility that steroid hormones derived from the placenta might be involved in imprinting the normal female phenotype cannot be excluded at present. If the normal female fetus is exposed to an excess of testosterone or other potent androgens during differentiation of the external genitalia, a male pattern can result. However, once the female pattern of differentiation has been achieved, exposure to testosterone cannot change the external genitalia to the male pattern, although it can cause enlargement of the clitoris.

In the male, testosterone must be secreted into the circulation of the fetus and subsequently must be converted to *dihydrotestosterone* within the cells of the anlage tissues for normal differentiation of the external genitalia to occur. As a result of dihydrotestosterone stimulation, the genital tubercle grows into the glans penis, the genital swellings fold and fuse into the scrotum, the urethral folds enlarge and enclose the penile urethra and corpora spongiosa, and the urogenital sinus gives rise to the prostate gland. Although the intracellular mechanism of action of dihydrotestosterone is discussed later in more detail, it is pertinent to note here that this steroid hormone requires a cytoplasmic receptor for its action.

The hormone production necessary for sexual differentiation does not depend on fetal pituitary gonadotropins. An LH-like hormone, chorionic gonadotropin from the placenta, undoubtedly stimulates testosterone production by the Leydig cells. Placental steroid hormone precursors, such as pregnenolone, might serve as a source of fetal gonadal androgens or estrogens. This obviates the necessity for gonadotropin stimulation of the reactions from cholesterol to pregnenolone (see later discussions).

Other aspects of phenotypic sexual differentiation are not evident until long after birth. These include the constant pattern of gonadotropin secretion in the male versus the cyclic pattern in the female, the different degree of breast development, and the psychological identification with a unique gender. It is difficult at present to be certain what factors imprint or regulate these traits in humans. Evidence from rodent studies suggests that circulating androgens induce the fetal hypothalamus to set a constant pattern of gonadotropin secretion in the postpubertal male. To do so, they may require metabolism to estrogens within the target neurons. In the absence of androgens the cyclic pattern of the female results. This would constitute another instance in which the female pattern was the "neutral pattern," whereas the male pattern required an action ultimately derived from the Y chromosome.

Mammary gland development in the rodent embryo also is clearly under androgen regulation. In its absence a normal female breast develops; in its presence the ductal system is suppressed. However, in the human, male-female differences in breast development are not apparent before puberty. At that time the hormonal milieu in the female induces growth and differentiation of breast tissue, whereas that in the male suppresses it. A large body of clinical evidence suggests that psychological gender identification is independent of hormonal regulation or even of the phenotype of the genitalia; instead it appears to depend on rearing cues. However, exceptions to this are noted in certain cases of male pseudohermaphrodites raised as girls. In such individuals significant growth of the penis under pubertal testosterone stimulation seems to cause a reversal of psychosocial gender from female to male.

Abnormalities of Sexual Differentiation

Anatomical aberrations that result from genetic errors are listed in Table 55-2. Sexual differentiation can be distorted by abnormalities in either sex chromosomes or autosomes.

The XO chromosomal karyotype produces individuals with only a vestigial gonadal streak, because they do not have either the ovarian organizational input of two active X chromosomes or the testicular organizational input of the Y chromosome. The absence of testicular function in turn leads to müllerian duct development, female external genitalia, and wolffian duct regression.

XY individuals who completely lack the capacity to respond to androgenic hormones because of receptor deficiency (the X-linked testicular feminization syndrome) will develop testes because of the presence of the Y chromosome. They will demonstrate müllerian duct regression, caused by the presence of AMH. However, they will show no growth or development of the

Table 55-2. Examples of abnormal development of the reproductive system

Genetic state	*Gonad*	*Müllerian duct*	*Wolffian duct*	*External genitalia*
XY, normal ♂	Testis	Regressed	Developed	♂
XX, normal ♀	Ovary	Developed	Regressed	♀
XO, Turner's syndrome	Streak*	Developed	Regressed	♀
XY, loss of X-linked gene for androgen receptor	Testis	Regressed	Regressed	♀
XY, deficient testosterone synthesis	Testis	Regressed	Regressed to variably developed	♀/♂
XY, deficient 5α-reductase	Testis	Regressed	Developed	♀/♂
XXY, Klinefelter's syndrome	Dysgenetic testis	Regressed	Developed	♂
XX, adrenal 21- or 11-hydroxylase deficiency	Ovary	Developed	Regressed	♀/♂

*A fibrous streak essentially devoid of germ cells.

wolffian ducts nor masculinization of their external genitalia, since without receptors there is no effective testosterone or dihydrotestosterone action.

XY individuals who have any one of five known defects in testosterone biosynthesis will develop testes because of the presence of the Y chromosome, and will show müllerian duct regression because of the presence of AMH. However, depending on the severity of the testosterone lack, the wolffian duct structures will show variable degrees of underdevelopment, and the external genitalia will vary from a completely female pattern to simple failure of the urethral folds to fuse completely.

XY individuals who are deficient only in the conversion of testosterone to dihydrotestosterone will have a normal testis because of the presence of the Y chromosome. They will demonstrate müllerian duct regression because of the presence of AMH. The development of the epididymis, vas deferens, and seminal vesicles will be normal because of the presence of testosterone. However, they will have external genitalia that vary from a partial to a complete female pattern because of the lack of dihydrotestosterone.

XX individuals with a deficiency of adrenal 21- or 11-hydroxylase enzymes will overproduce androgens in utero (Chapter 54). They will have ovaries because of the presence of two X chromosomes and normal müllerian duct development because of the absence of AMH. Their wolffian structures regress because of the absence of local gonadal testosterone and the relatively late exposure to adrenal androgen excess. However, depending on the severity of the enzymatic block and the degree of resultant androgen hypersecretion, the external genitalia of such XX individuals show variable degrees of the male pattern. This ranges from mild enlargement of the clitoris to complete scrotal fusion of the labia and a persistent urogenital sinus. Similar although less severe masculinization of otherwise normal XX fetuses has resulted from transplacental passage of excess androgenic hormones from the mother.

Individuals with supernumerary X chromosomes develop testes if a Y chromosome is also present and ovaries if it is not. Their genital ducts and external genitalia develop normally. However, spermatogenesis and seminiferous tubule development are markedly deficient in individuals (males) with XXY chromosomes *(Klinefelter's syndrome)*. Individuals (females) with XXX chromosomes may have shortened reproductive lives. The mechanisms whereby extra X chromosomes damage germ cell function are unknown.

Common Aspects of Gonadal Structure and Function

It is helpful to review certain homologous aspects of gonadal structure and function in the two sexes before discussing the numerous differences that exist. As seen in Fig. 55-3, the primordial germ cells generate the oogonia and spermatogonia that undergo eventual reductional division and maturation into large numbers of ova and sperm, respectively. Only a few of each will eventually unite with each other to reproduce the species, in a manner guaranteeing an almost infinite variety of individual characteristics. One cell line of the indifferent gonad becomes the *granulosa cells* of the ovarian follicle and the *Sertoli cells* of the seminiferous tubules. The function of these cells is homologous: to sustain or "nurse" the germ cells, foster their maturation, and guide their movement into the genital duct system. This cell line is probably the main source of estrogenic hormones in both sexes. Another cell line, the *interstitial cells,* gives rise to *theca cells* in the ovary and *Leydig cells* in the testis. The primary function of this cell line is to secrete androgenic hormones. These are essential for sperm production in the testis and as precursors for estrogen synthesis in the ovary.

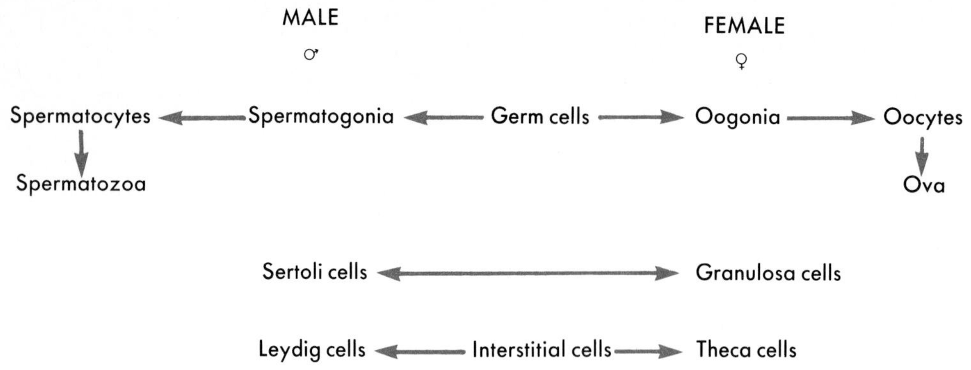

■ Fig. 55-3. Homologous cell lines of the male and female gonads.

Pathway of Gonadal Steroid Synthesis

Both sexes use a common pathway of steroid hormone biosynthesis in gonadal tissue. It is essentially identical to that of the adrenal cortex. The enzyme characteristics and cofactor requirements are also those previously described for the adrenal glands in Chapter 54. As shown in Fig. 55-4, cholesterol, either generated by de novo synthesis from acetyl CoA or taken up from the circulating plasma pool, is the starting compound. In the gonads, in situ cholesterol synthesis may be quantitatively more important than in the adrenal glands. Cholesterol side chain cleavage (the 20, 22-desmolase step) is localized to the mitochondria, and it appears to be rate limiting for synthesis of progesterone, of the androgens testosterone, dihydrotestosterone, and androstanediol, and of the estrogens estradiol and estrone. The 17-hydroxylase, 17, 20-desmolase, 3-β-ol-dehydrogenase, and $\Delta^{4,5}$-isomerase activities are all located in the microsomes of the smooth endoplasmic reticulum. The first two and the second two form closely associated pairs within the microsomal membranes.

Two parallel pathways to testosterone are evident (Fig. 55-4). Oxidation of the A ring by the 3-β-ol-dehydrogenase-isomerase complex can take place at any level from pregnenolone to androstenediol. The factors that determine which route will be taken are not known. A small quantity of testosterone undergoes 5α-reduction to dihydrotestosterone and a further α-reduction of the 3-ketone position to 5α-androstanediol (Fig. 55-4).

The conversion of androgens to estrogens occurs by aromatization of the A ring. The aromatase enzyme complex is a cytochrome P450 that sequentially hydroxylates the 19-methyl group, oxidizes it to the aldehyde, hydroxylates the 2 position, and then creates a 1-2 double bond by reduction. Following these steps, the 19-carbon is removed by decarboxylation, and the characteristic benzene ring is formed.

This series of steps occurs in the endoplasmic reticulum. Estradiol and estrone result from testosterone and androstenedione, respectively. The two estrogens also may be interconverted via 17-hydroxysteroid dehydrogenase.

Mechanisms of Gonadotropin Action

Luteinizing hormone (LH) stimulates the interstitial cell line of male and female gonads to secrete androgens and possibly estrogens. The mechanism of action of LH begins with binding to its membrane receptor, a glycoprotein of 200,000 molecular weight and consisting of 2 subunits. The complex then activates adenylate cyclase and raises intracellular cAMP levels.

The interaction of LH with its receptors is exquisitely sensitive. As little as 1% receptor occupancy may be sufficient for stimulation, and 5% to 10% occupancy may produce maximal cellular responsiveness to the hormone. cAMP formation and LH receptor occupancy usually correlate well. Continued stimulation of gonadal cells by LH leads to downregulation of its receptors and reduced responsivity to the hormone. Prostaglandins have been implicated as additional intermediaries in LH action, possibly potentiating cAMP effects.

As a result of protein kinase activation by cAMP, one or more specific regulatory proteins concerned with steroidogenesis are activated. The major locus of acute LH action is on the mitochondrial conversion of cholesterol to pregnenolone. This may involve facilitating the movement of cholesterol from its esterified storage form into the mitochondrial matrix and the activation of the 20,22-desmolase reaction. Longer-term stimulation with LH may increase the microsomal steroidogenic enzymes, as well.

FSH acts on ovarian granulosa cells and testicular Sertoli cells also via plasma membrane receptors. The subsequent increase in cAMP is followed by an increase in steroidogenesis and in the rate of the aromatase reaction. The mechanism of these FSH actions is less well understood. Another important effect of FSH is to

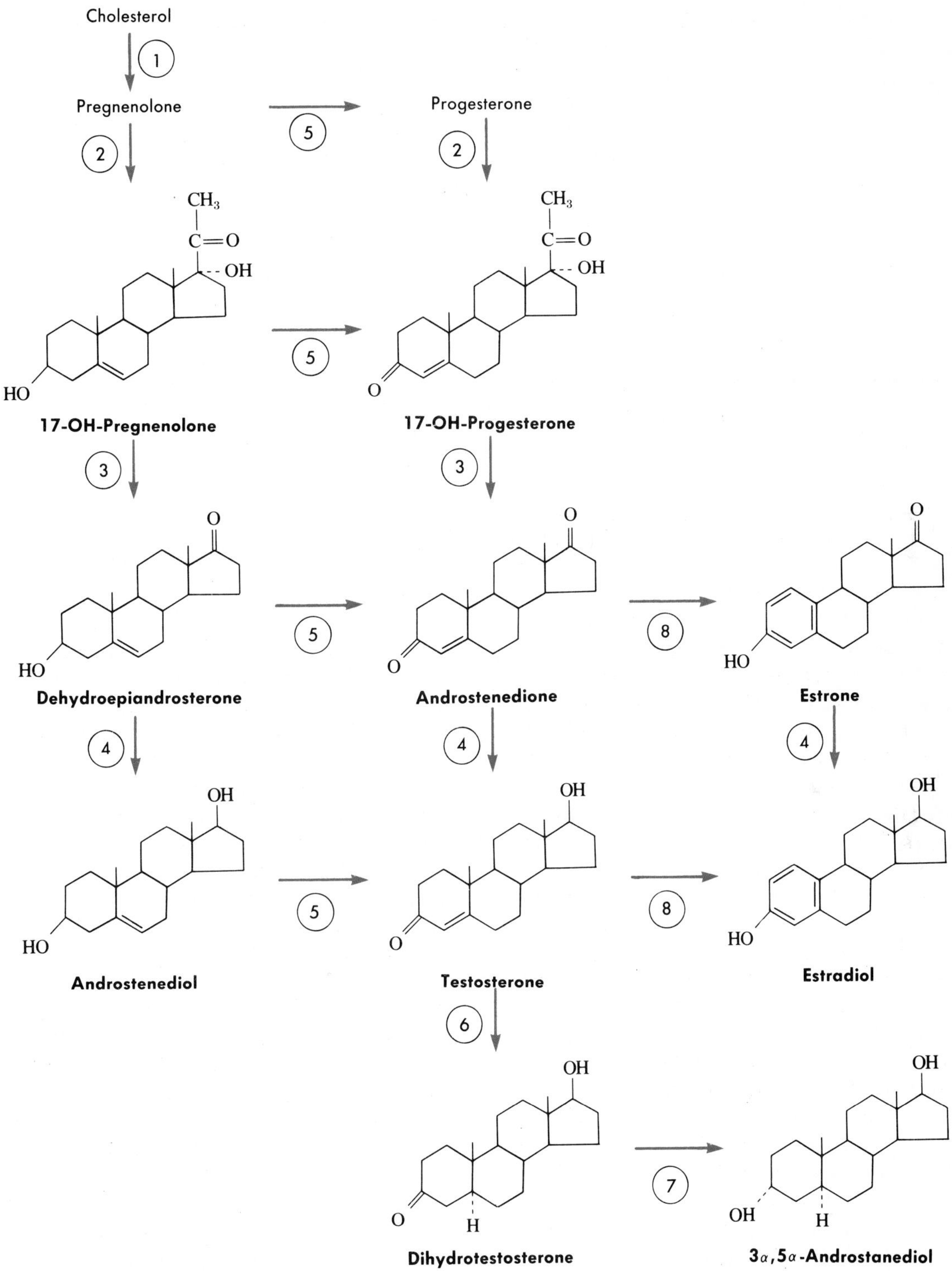

■ **Fig. 55-4.** Pathways of synthesis of gonadal steroid hormones. Testosterone is the major secretory product of the testis. Estradiol and progesterone are the major secretory products of the ovary. Enzymes are 20,22-desmolase *(1)*, 17-hydroxylase *(2)* 17,20-desmolase (3), 17β-OH-steroid dehydrogenase *(4)*, 3β-ol-dehydrogenase and $^{\Delta 4,5}$-isomerase (5), 5α-reductase *(6)*, 3α-reductase *(7)*, and aromatase (8).

increase the number of LH receptors in target cells, thereby amplifying their sensitivity to LH.

In addition to their actions on steroidogenesis, LH and FSH produce diverse metabolic effects on their target gonadal cells. Glucose oxidation and lactic acid production are increased, and this may lead to local vasodilation. The long-term tropic effects of the two hormones depend on stimulation of amino acid transport, RNA synthesis, and general protein synthesis.

Both testosterone and estradiol have negative feedback effects on LH and FSH secretion in both sexes (Chapter 52). In addition, the natural cessation of ovarian function in menopausal women, as well as the slower decline of testicular function in men in the seventh and eighth decades of life, is associated with increases in both plasma gonadotropins. In all these instances loss of inhibin secretion plays a specific role in the elevation of plasma FSH (Chapter 52). Therefore, in individuals whose gonads are surgically removed or functionless because of disease, plasma FSH and LH levels will be elevated.

The Testes

Anatomy

The human testes are normally situated in the scrotum, where they are maintained at a temperature somewhat below that of the body core temperature. Each testis weighs about 40 g and has a long diameter of 4.5 cm. The testes receive blood from the spermatic arteries, which arise directly from the aorta. The right spermatic vein drains into the inferior vena cava, whereas the left drains into the ipsilateral renal vein. Eighty percent of the adult testis is made up of the seminiferous tubules; the remaining 20% is composed of supportive connective tissue, throughout which the Leydig cells are scattered (Fig. 55-5). The seminiferous tubules are a coiled mass of loops; each loop begins and ends in a single duct, the *tubulus rectus*. The tubuli recti, in turn, anastomose in the *rete testis* and eventually drain via the *ductuli efferentes* into the *epididymis*. The latter constitutes a storage and maturation depot for spermatozoa. From the epidiymis the spermatozoa are carried via the *vas deferens* and *ejaculatory duct* into the penis, to be emitted during copulation.

The structure of the adult seminiferous tubule is shown in Fig. 55-6. Each seminiferous tubule is bounded by a basement membrane, separating it from the Leydig cells and the surrounding connective tissue. Immediately next to the basement membrane are spermatogonia and Sertoli cells. As the spermatogonia divide and develop into spermatocytes and spermatids, a column of cells is formed that reaches from the basement membrane to the lumen of the tubule and culminates in the spermatozoa. In contrast, the cytoplasm of each Sertoli cells extends all the way from the basement membrane to the lumen. This cytoplasm invests the spermatogonia and its germ cell line successors (Fig. 55-6). Special processes of the Sertoli cell cytoplasm fuse into *tight junctions*, which create two compartments of intercellular space between the basement membrane and the lumen of the tubule. The spermatogonia lie within the *basal compartment*, whereas the spermatocytes and subsequent stages in spermatozoon development lie in the *adluminal compartment*. This compartmentalization accomplishes two things. In effect the Sertoli cell cytoplasm forms a blood-testis barrier that excludes a variety of circulating substances from the fluid bathing the maturing germ cells and from the seminiferous tubular fluid. Conversely, products from the later stages of spermatogenesis are prevented from diffusing back into the bloodstream and, if recognized as foreign, are prevented from producing antibodies.

In short, the testis may be considered to consist of two functional elements. The first element is the Leydig cells, which are pure steroid-secreting cells whose major product, testosterone, has both vital local effects on germ cell replication and actions on distant target cells. The second element is the seminiferous tubules, which carry out the process of spermatogenesis while bathed in locally generated testosterone and Sertoli cell products.

The Biology of Spermatogenesis

The production of sperm is an ongoing process throughout the reproductive life of the male. Approximately 100 to 200 million sperm are produced daily. In generating this large number of sperm, the spermatogonia must renew themselves by cell division. This situation differs fundamentally from that in the female, who at birth has a fixed number of oocytes, which decrease throughout her life.

The extraordinary metamorphosis from spermatogonium to spermatozoon is depicted in Fig. 55-7. The first two mitotic divisions of a spermatogonium give rise to four cells: a single resting cell (*Ad*) that will eventually serve as the ancestor of a later generation of sperm and three active cells (*Ap*). The latter divide by further mitoses to yield type B spermatogonia, which then give rise to a number of primary spermatocytes. These cells enter the prophase of meiosis, the first reduction division, in which they remain for about 20 days.

The complex process of chromosomal reduplication, synapsis, crossover, division, and separation is reflected histologically in the changing appearance of the primary spermatocytes (Fig. 55-7). After completion of meiosis (late pachytene spermatocyte, Fig. 55-7), their daughter cells, the secondary spermatocytes, immediately divide again. The products, called spermatids, each contain 22

A

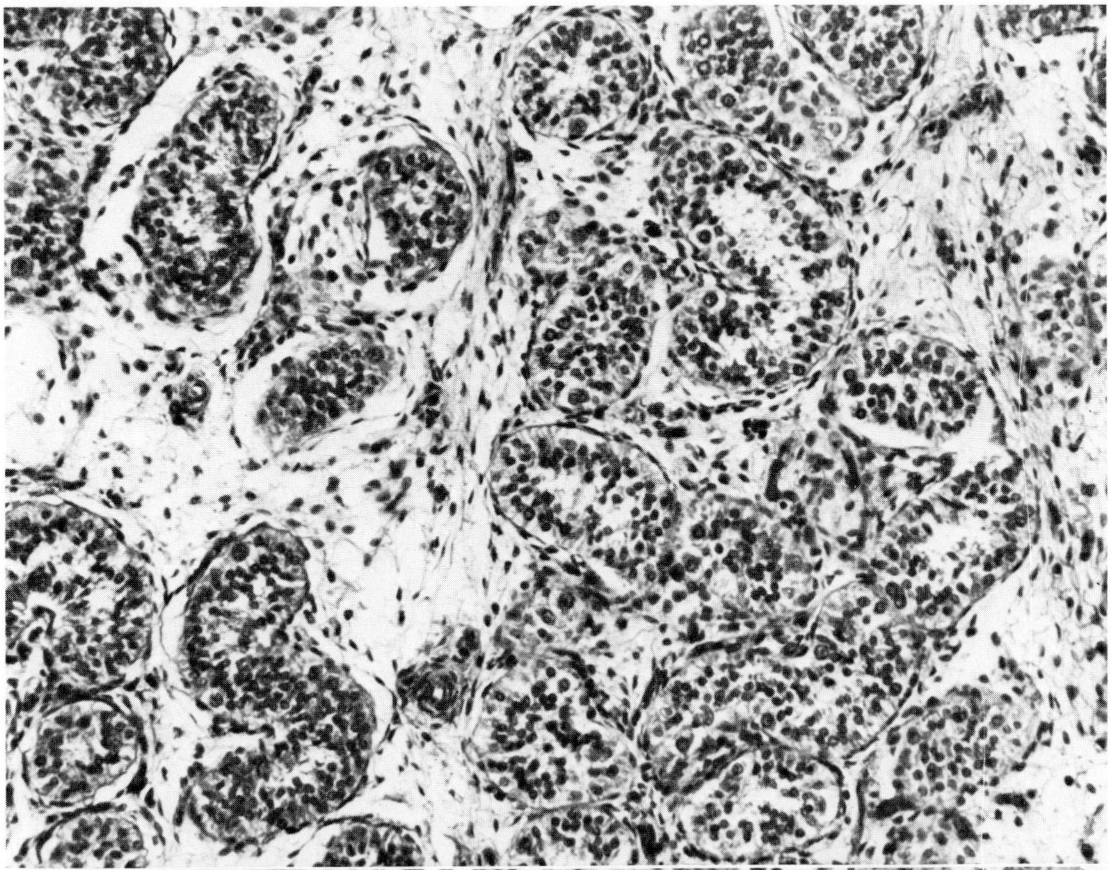

B

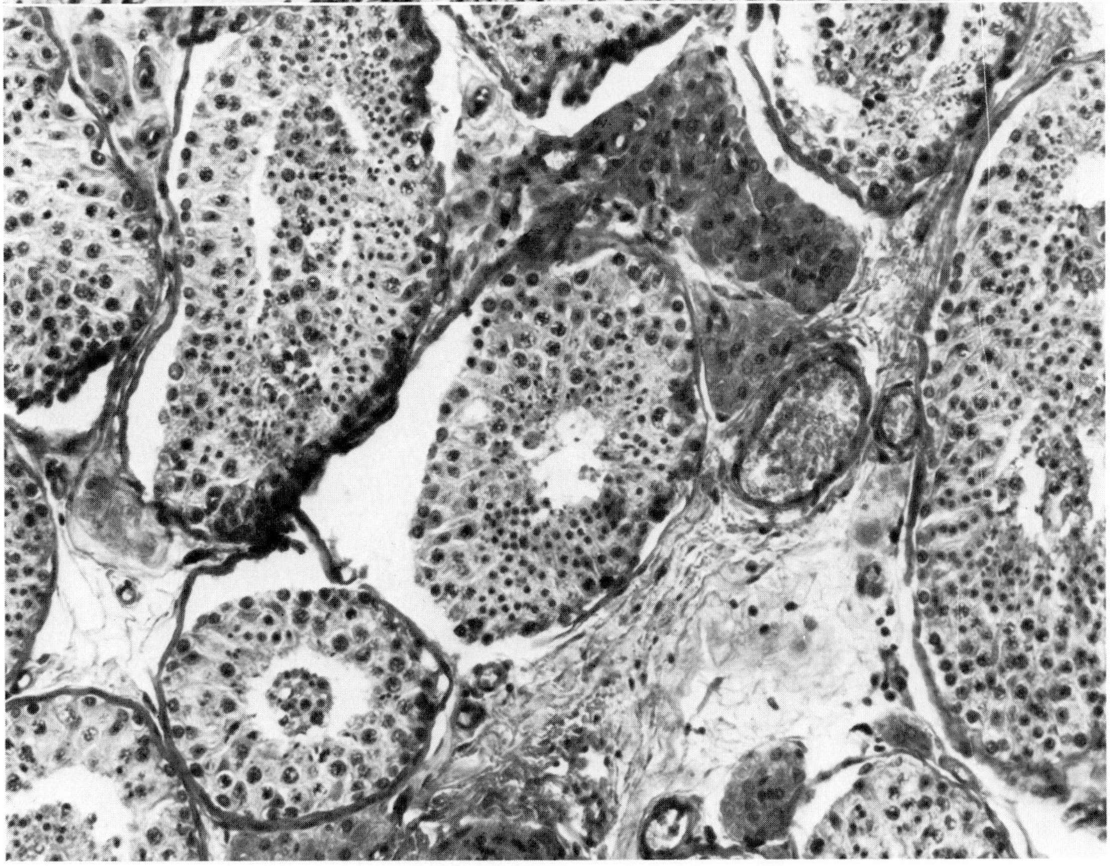

■ Fig. 55-5. Histological sections of the testis at low (**A** and **B**) and high (**C** and **D**) magnification. **A** and **C** are from a prepubertal testis. Note absence of Leydig cells and absence of active spermatogenesis, with only spermatogonia being evident in the tubules. **B** and **D** are from a postpubertal testis. Note clumps of Leydig cells adjacent to tubules and the presence of active spermatogenesis, as evidenced by the progression of cell types between the basement membrane and lumen of the tubules and the many mitotic figures. (Courtesy Dr. Howard Levin).

Continued.

C

D

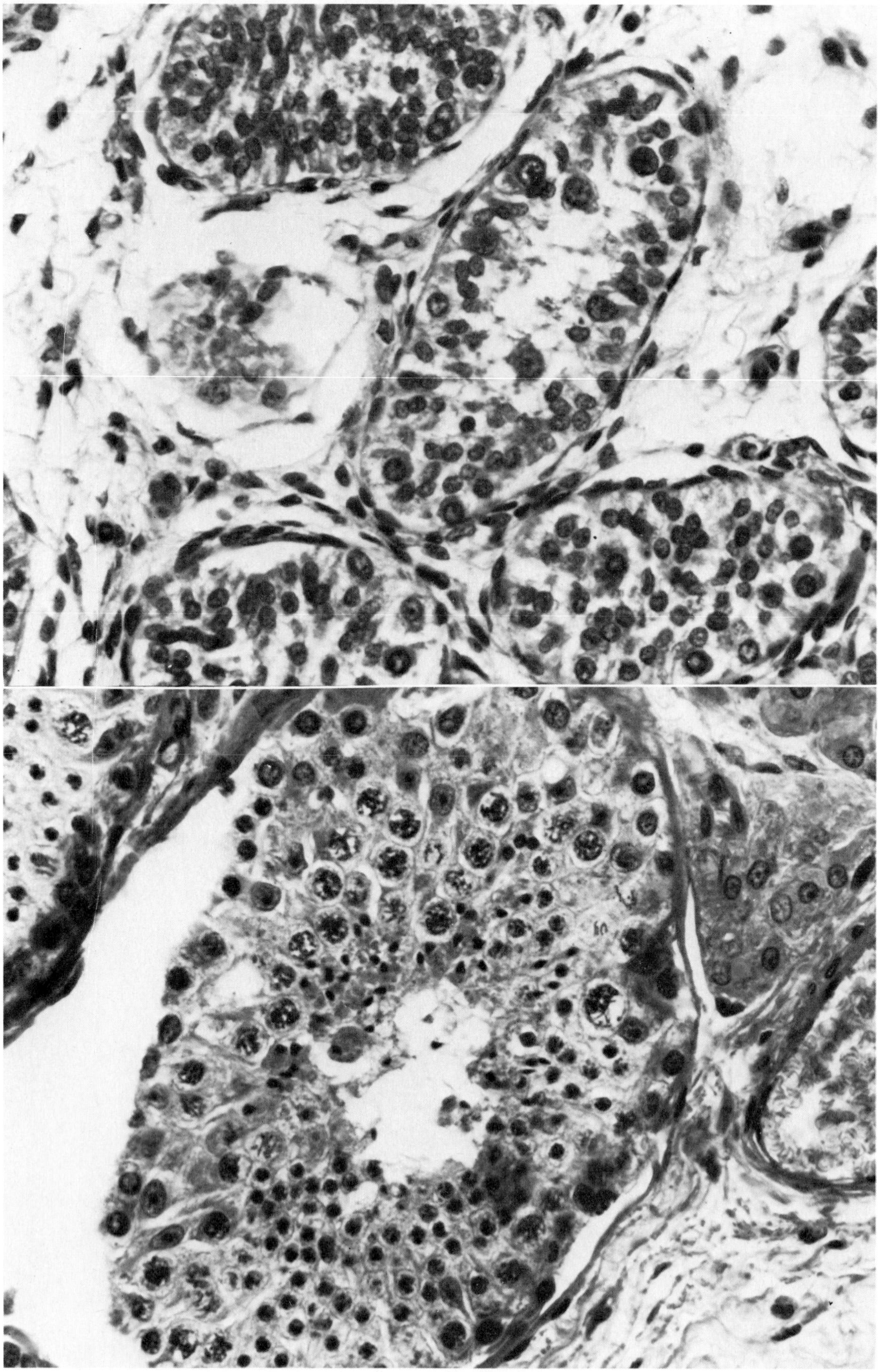

■ **Fig. 55-5, cont'd.** For legend see page 991.

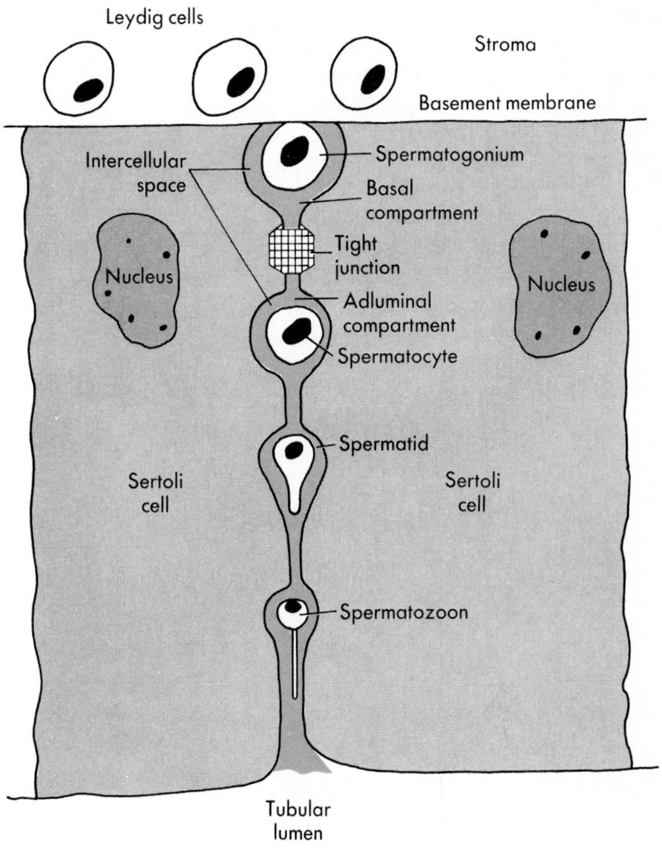

■ **Fig. 55-6.** Geographical relationships between Leydig cells, Sertoli cells, and germ cells in the testis. Note that the tight junction between the two Sertoli cells seals off the intercellular space surrounding the spermatocytes and spermatids from contact with the general intercellular space and thus with plasma fluid contents. (Redrawn from Fawcett, D.W.: Ultrastructure and function of the Sertoli cell. In Hamilton, D.W., and Greep, R.O., editors: Handbook of physiology; Section 7, vol. 5, Bethesda, Md., 1975, The American Physiological Society.)

autosomes and either an X or a Y sex chromosome. The spermatids lie near the lumen of the seminiferous tubule. They are attached to the abutting Sertoli cells and are connected with each other through intercellular bridges. In a process termed *spermiogenesis*, they undergo nuclear condensation, shrinkage of cytoplasm, formation of an acrosome, and development of a tail to emerge as flagellated spermatozoa. The spermatozoa are then extruded into the lumen of the tubule by a process called *spermiation*, during which most of the cytoplasm of the spermatozoa remains embedded in the cytoplasm of a Sertoli cell.

Once the spermatazoa have entered the seminiferous tubules, they consist of linear structures with several components. The head contains the nucleus and an acrosomal cap in which are concentrated hydrolytic and proteolytic enzymes, which facilitate penetration of the ovum and possibly also the mucous plug of the female cervix. The middle piece, or body, contains mitochondria, which generate the motile energy of the spermatozoon. The chief piece of the tail contains stored ATP and pairs of contractile microtubules down its entire length, one pair in the center and nine pairs around the circumference. An ATPase releases the stored energy in a way that allows the microtubules to impart flagellar motion to the spermatozoa.

In man approximately 70 days are required for the entire sequence of development from spermatogonia to spermatozoa. However, individual resting spermatogonia do not start into the process of spermatogenesis randomly. Cycles of spermatogenesis exist with distinct cycle times. Groups of adjacent resting spermatogonia initiate a new cycle about every 16 days, thus constituting one "generation." At about the same time that the primary spermatocytes of one cycle enter prophase, a second cycle of spermatogonia is activated. A third cycle begins approximately synchronously with the appearance of the spermatids from the first cycle. By the time these spermatids have completed their transformation into spermatozoa, a fourth cycle of spermatogonia has been started.

Around the circumference of any individual seminiferous tubule several spermatogenic cycles may be in process simultaneously. Within each cycle approximately six stages of cellular development can be identified histologically. This gives rise to several specific cellular constellations that exist side by side (Fig. 55-5, *B*). In some mammals, but not in man, spermatogenic cycles are repeated in a defined topographical relationship to each other along the length of each seminiferous tubule. This has been termed the *wave of spermatogenesis*.

There may not be total separation of the individual germ cells that constitute the successive descendents of type B spermatogonia and which lie within the adluminal compartment of the tubule. Continuity of cytoplasm and possibly cell-to-cell intercommunication may exist. Because of these possibilities and because of the regular topographical association of particular stages of spermatogenesis in neighboring cycles, products of germ cells in one stage of spermatogenesis may initiate or regulate events in other stages.

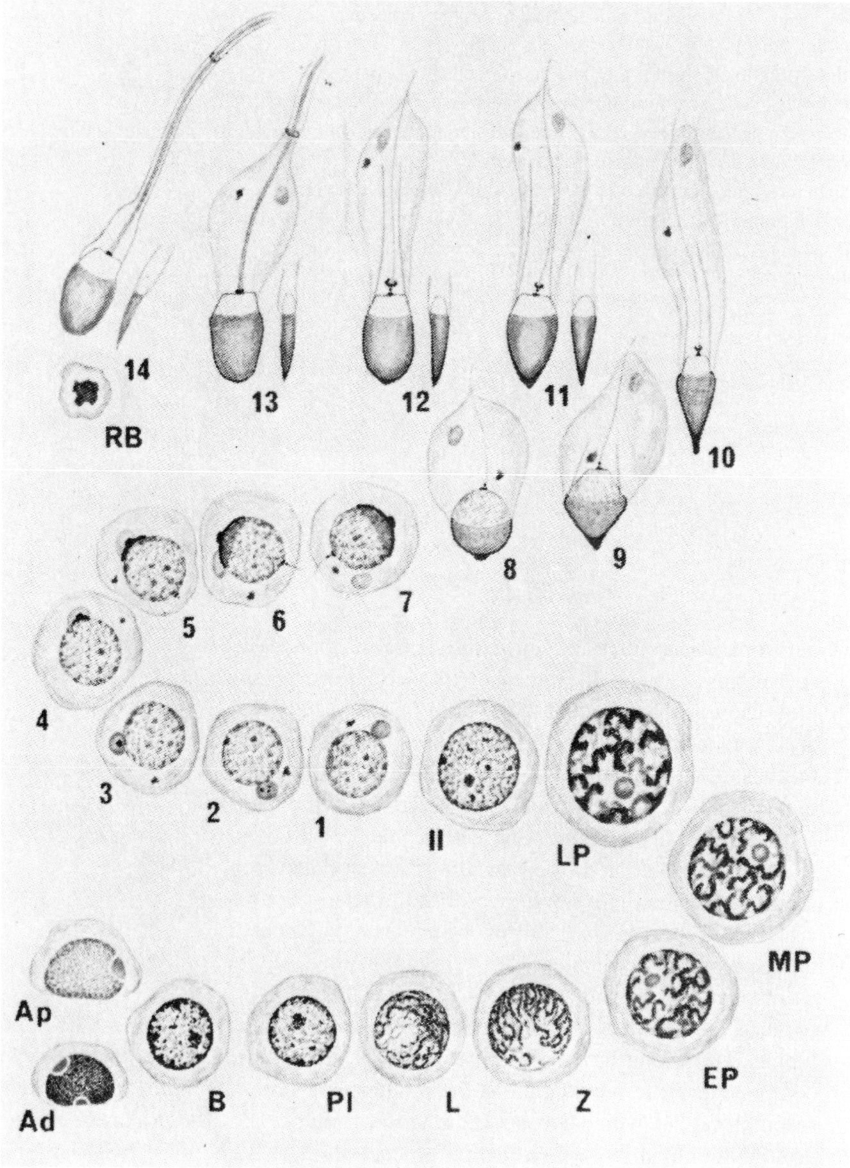

Fig. 55-7. Development of spermatozoa from spermatogonia in the subhuman primate. *Ad,* Dark spermatogonium; *Ap,* pale spermatogonium; *B,* type B spermatogonium; *PI,* preleptotene primary spermatocyte; *L,* Leptotene spermatocyte; *Z,* Zygotene spermatocyte; *EP, MP, LP,* early, middle, and late pachytene spermatocytes; *II,* secondary spermatocyte; *1-7,* early spermatids; *8-13,* late spermatids with progressive formation of flagellum; *14,* spermatozoon; *RB,* residual body. (Redrawn from Clermont, Y.: Physiol. Rev. **52:**198, 1972).

Sexual Functioning

After spermiation, the spermatozoa reach the epididymis, which they traverse at a variable rate (1 to 21 days). During this time, they undergo further maturation. It is still unclear to what extent this is preprogrammed within the spermatozoa or to what extent it depends on specific secretions of the epididymal cells. By the time they reach the vas deferens, the spermatozoa have lost the last droplet of cytoplasm and have acquired motility. They may then be stored viably in the vas deferens for several months.

Delivery of spermatozoa into the female genital tract occurs by ejaculation from the vas deferens. To the contents of the vas deferens are added successive fluids. The initial secretions from the prostate gland contain citrate, calcium, zinc, and acid phosphatase. The alkalinity of prostatic fluid helps neutralize the acid pH of the vaginal and cervical secretions. The terminal portion of the ejaculate is composed largely of secretions from the seminal vesicles. These contain fructose, an important oxidative substrate for the spermatozoa. Seminal vesicle secretions also contain prostaglandins, which may stimulate contractions of the uterus and fallopian tubes—helping propel the spermatozoa toward the ovum.

Seminal fluid also contains LH, FSH, prolactin, testosterone, estradiol, inhibin, and endorphins. Often the concentrations of these substances in the seminal fluid are higher than those in plasma. Their exact source and role in fertilization remain to be determined. Relaxin, an insulin-like peptide, is also present in seminal fluid and may be important in promoting motility of spermatozoa.

Fig. 55-8. The individual effects of FSH and LH on human sperm production. Normal men were given sufficient testosterone to suppress endogenous FSH and LH secretion by negative feedback. As a result, sperm density declined to very low levels. Selective restoration of either FSH or LH individually raised sperm levels. However, neither gonadotropin alone could return sperm production to normal. (**A** Redrawn from Matsumoto, A.M., et al.: J. Clin. Invest. **72:**1005, 1983. and from Matsumoto, A.M., et al.: J. Clin. Endocrinol. Metab. **59:**882, 1984.)

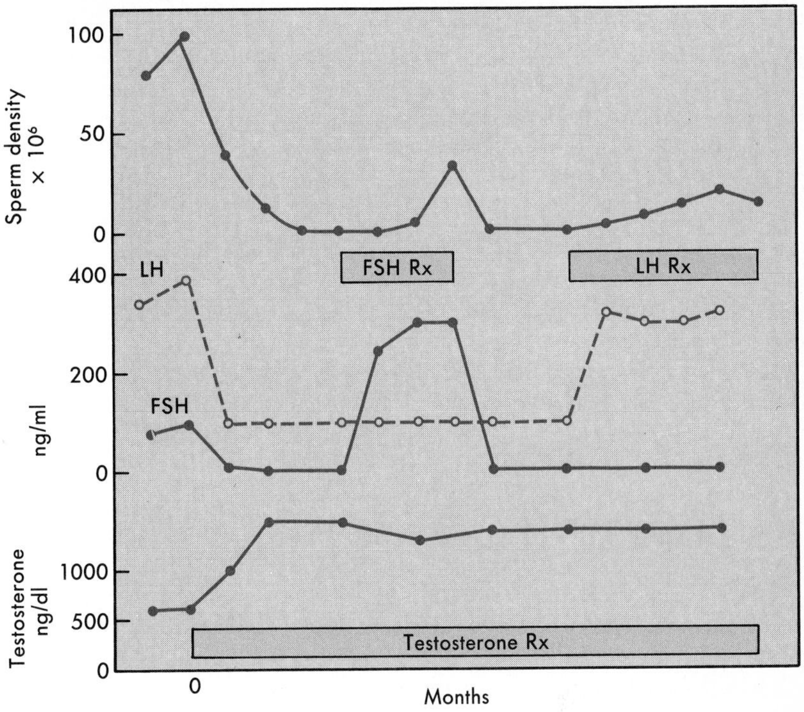

Ejaculated sperm cannot immediately fertilize an ovum. In vivo fertilization can take place only after the sperm has been in the milieu of the female reproductive tract for several hours, a process termed *capacitation*. In vitro human fertilization can take place after the spermatozoa have been washed free of seminal fluid.

Although the process of capacitation is poorly understood, it results in unique patterns of sperm motility that may enhance the penetration of the ovum. In addition, and most important, capacitation results in the acrosomal reaction, which consists of fusion of the acrosomal membrane with that of the outer sperm membrane. The membrane fusion creates pores through which the acrosomal hydrolytic and proteolytic enzymes can escape. These then create a path for penetration of the sperm through the protective investments of the ovum.

Delivery of the sperm into the vagina requires erection of the penis, which is caused by filling of the venous sinuses with blood. This converts the penis into a firm organ for penetration. The venous sinuses are filled by the coupling of arteriolar dilation with venous constriction; the erection is under parasympathetic control. Ejaculation is effected by sympathetic nervous system impulses. A typical emission contains 200 million to 400 million spermatozoa in a volume of 3 to 4 ml. Once within the vagina, the spermatozoa's rate of flagellated movement is up to 44 mm/minute. The life span of sperm in the female genital tract is approximately 2 days.

Hormonal Control of Spermatogenesis

FSH, LH, testosterone, and probably estradiol, all coordinate in the regulation of spermatogenesis. Fig. 55-8 shows that in men with normal spermatogenesis, withdrawal of both gonadotropins by suppression of the pituitary gland with testosterone results in rapid and nearly complete loss of sperm production. Selective replacement of either FSH or LH in such men (Fig. 55-8) reinitiates sperm production but does not restore it to normal levels. Prolactin and possibly growth hormone may also be involved in human spermatogenesis. Because our present knowledge is still insufficient to pinpoint the function of each hormone in the human spermatogenic cycle, only a tentative description can be offered.

Testosterone may stimulate the transformation of the primordial germ cells to the primitive type A spermatogonia during ontogenesis of the fetal testes at 8 to 20 weeks. From then until puberty, the spermatogonia normally remain in a resting stage, presumably because gonadotropin secretion is low and nonpulsatile throughout childhood. Whether the spermatogonia have already been conditioned by the initially higher FSH and LH levels of midgestation and early neonatal life is not clear.

Spermatogenesis begins with the initial division into type B spermatogonia. This step and possibly the subsequent divisions into primary spermatocytes are presently thought to be independent of gonadotropins or gonadal steroid hormones but may be facilitated by growth hormone. The completion of the long prophase of the primary spermatocytes does appear to depend on the high intratesticular concentration of testosterone (a hundredfold higher than in plasma). The high concentration is produced locally by LH action on the Leydig cells.

The importance of testosterone locally is illustrated in prepubertal boys with Leydig cell tumors, which produce the hormone. Spermatogenic activity occurs only in the testis the harbors the tumor. In men lacking LH or Leydig cells, administration of testosterone in amounts sufficient to raise its plasma levels to normal and to restore its other biological actions fails to promote spermatogenesis.

The terminal steps of the differentiation of spermatids into spermatozoa in man may require the continuous presence of FSH. In other species, such as rodents, only an initial period of exposure to FSH early in life is required. The effects of FSH are probably mediated solely through stimulation of Sertoli cell secretions, which then condition the spermatids. Spermiation appears to be precipitated by LH, acting either directly or indirectly. The testis may possess prolactin receptors, and prolactin appears to synergize with LH in some of its action.

Finally, the role of inhibin, which is produced by the Sertoli cells, remains to be clarified. Selective destruction of the spermatogenic cells increases FSH secretion, and the high level of FSH in plasma is associated with reciprocally low levels of inhibin in seminal fluid. It is therefore presumed that some product from the germ cell line normally directs the Sertoli cells to secrete inhibin. FSH also increases the synthesis of inhibin, completing a feedback loop with the pituitary gonadotroph that may help coordinate the availability of FSH with the needs of a particular phase of the spermatogenic cycle.

Despite the cyclic nature of spermatogenesis locally, the testis as a whole is continuously releasing spermatozoa. Furthermore, even though gonadotropin release is pulsatile, the mean daily plasma levels of FSH and LH are essentially constant in adult men. It is conjectural whether local differences in gonadotropin sensitivity or regional differences in gonadotropin distribution contribute to the cyclic and topographical nature of the spermatogenic process. The situation is clearly different from that of the ovary, in which a phasic pattern of gonadotropin secretion, highlighted by a single distinct burst, produces the single monthly release of an ovum.

The Role of the Sertoli Cells

In early fetal life the Sertoli cells secrete AMH. Their subsequent function until puberty is not known for certain, although presumably they are secreting inhibin, thereby helping to keep pituitary FSH secretion low. After puberty the Sertoli cells do not undergo any further cell divisions. In association with the cycle of spermatogenesis they do undergo regular changes in the activity and shape of the nucleus, in the size, shape, and branching of the cytoplasmic processes, in the concentrations of lipid and glycogen, in mitochondrial function, and in enzyme content. In some way these changes relate to the processing of the germ cells, but whether this is in a controlling, a facilitative, or a reactive manner still remains to be determined.

The cytoplasmic processes of the Sertoli cells, extending from the basement membrane to the lumen of the seminiferous tubule, act as conduits, between which the various stages of germ cells move in their passage to the lumen (Fig. 55-6). In some regular fashion the tight junctions must open to permit the maturing primary spermatocytes to pass, and then the junctions must close again behind them. The Sertoli cell cytoplasm also acts as a filter, permitting only certain substances, presumably those advantageous to spermatogenesis, to reach the spermatocytes.

Sertoli cells probably synthesize estradiol and estrone in response to FSH stimulation. Testosterone and androstenedione, which have diffused in from the Leydig cells, serve as precursors (Fig. 55-4). A unique, FSH-dependent function of the Sertoli cells is the secretion of androgen-binding protein. This is a protein with a molecular weight of 90,000 that has very similar properties to those of plasma sex steroid-binding globulin (SSBG) (as discussed later). The androgen-binding protein complexes testosterone, dihydrotestosterone, and estradiol with high affinity. It effectively concentrates testosterone in the Sertoli cell, creating a storage form of the hormone for controlled release by this cell during appropriate stages of spermatogenesis.

Androgen-binding protein also is secreted into the fluid of the seminiferous tubules, where it may serve to prevent reabsorption of testosterone from this fluid. Such an action would ensure the availability of testosterone (possibly of estradiol, as well) to the spermatozoa during their maturational sojourn in the epididymis. Androgen-binding protein, by complexing estradiol, may prevent this steroid from diffusing back into the Leydig cell and inhibiting testosterone synthesis.

In addition to androgen-binding protein and inhibin, several other proteins have been identified as Sertoli cell secretory products. These include transferrin, an iron-binding globulin, and plasminogen activator, which is important in fibrinolysis. An intriguing observation is the discovery of a family of Sertoli cell products termed *gonadocrinins*. These peptides are structurally similar to LHRH and inhibit Leydig cell function. Finally, both α-MSH and β-endorphin are also secreted by Sertoli cells and are thought to have paracrine and autocrine functions.

Either by mechanical or chemical means, the Sertoli cell also is clearly responsible for the ejection of the spermatozoa into the lumen. In this process the nucleus of the spermatozoon is oriented toward the base of the tubule. The bulk of the cytoplasm then is squeezed out past the nucleus and shed as the residual body, while the spermatozoon is cast free. The residual body and other fragments are then phagocytosed by the Sertoli cells and subsequently degraded. Finally, the Sertoli cells generate the fluid drive that sweeps the spermatozoa through the rete testis into the epididymis.

Androgenic Hormones

Secretion. Testosterone, the major androgenic hormone, is synthesized as described previously (Fig. 55-4). Its synthesis and release by the Leydig cells are regulated by LH. In adult men LH is secreted in a pulsatile pattern produced by luteinizing hormone–releasing hormone (LHRH); correspondingly, plasma testosterone levels also show small pulses throughout the day (Fig. 52-16). There is, in addition, a superimposed diurnal trend, such that plasma testosterone is about 25% lower at 8:00 PM than at 8:00 AM. The response to prolonged stimulation with exogenous LH is a rapid rise in plasma testosterone, followed by a brief decline and then a second plateau (Fig. 55-9). The temporary decrease may be caused by downregulation of LH receptors by the peak concentrations of gonadotropin.

Testosterone also gives rise to two other potent androgens: dihydrotestosterone and 5α-androstanediol (Fig. 55-4). Although both of these may be secreted by the testis in small amounts, the major fraction of circulating dihydrotestosterone and androstanediol is derived from the reduction of testosterone in peripheral tissues. The plasma levels, blood production rates, and metabolic clearances of these androgens are shown in Table 55-3. The testosterone precursor, androstenedione, also is secreted in major amounts by the Leydig cells (Fig. 55-9), but it contributes little per se to androgen action. The two estrogens—estradiol and estrone— are produced in significant amounts in men. However, only 10% to 20% of the daily production is by direct testicular secretion in response to LH (Fig. 55-8). The majority is derived from circulating testosterone and androstenedione by aromatization in various peripheral sites, most notably in the adipose tissue and liver.

Leydig cell function varies distinctively during the life span of the individual. As shown in Fig. 55-10, plasma testosterone rises to levels of 400 ng/dl in the fetus at the time that the external genitalia are undergoing differentiation to the masculine pattern. By birth, however, these levels have declined to less than 50 ng/dl. Very soon thereafter plasma testosterone begins to rise, reaching a peak of 150 to 200 ng/dl at 4 to 8 weeks of age. This peak is stimulated by a corresponding rise in plasma LH. Its physiological significance is not known. After age 2 months plasma testosterone and LH again fall to low levels, and thereafter throughout childhood Leydig cells cannot be identified in the testes (Fig. 55-5, *C* and *D*). At about age 11 plasma testoster-

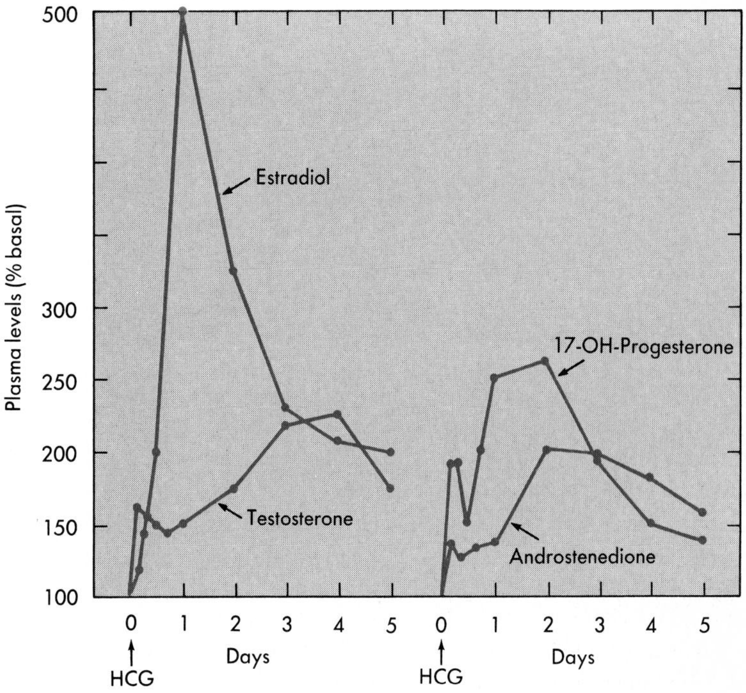

Fig. 55-9. Plasma hormone responses to stimulation of testes in normal men by human chorionic gonadotropin (HCG), with LH-like activity. Note biphasic response of testosterone level. Note also the increase in precursor levels on right, indicating activation of an early step in the synthetic pathway. (Redrawn from Forest, M.G., et al.: Kinetics of human chorionic gonadotropin–induced steroidogenic response of the human testis. J. Clin. Endocrinol. Metab. **49:**284, 1979. Copyright 1979. Reproduced by permission.)

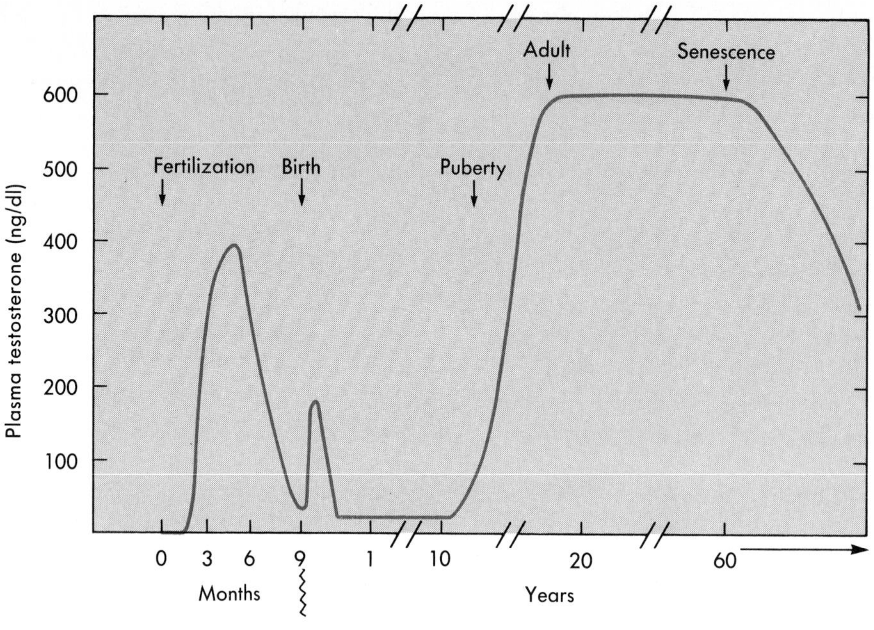

Fig. 55-10. Plasma testosterone profile during the life span of a normal male. (Redrawn from Griffin, J.E., et al.: The testis. In Bondy, P.K., and Rosenberg, L.E.: Metabolic control and disease, Philadelphia, 1980, W.B. Saunders Co.; and Winter, J.S.D., et al.: Pituitary-gonadal relations in infancy, J. Clin. Endocrinol. Metab. **42:**679, 1976. Copyright 1976. Reproduced by permission.)

Table 55-3. Turnover of gonadal steroids in adult men

Steroid	*Plasma concentration (ng/dl)*	*Blood production rate (μg/day)*	*Metabolic clearance rate (L/day)*
Testosterone	650	7000	1100
Dihydrotestosterone	45	300*	600
5α-Androstanediol	12	200*	1800
Androstenedione	120	2400	2000
Estradiol	3.0	50†	1700
Estrone	2.5	60†	2500

*About 60% to 80% produced peripherally from testosterone.
†About 80% to 90% produced peripherally from testosterone and androstenedione, respectively.

one begins a steep rise, reaching an adult plateau at about age 17 (Fig. 55-10). This is sustained for about 50 years. The pubertal increase and adult plateau of testosterone secretion correspond with the function of the hormone in helping to initiate and then to maintain spermatogenesis. During the seventh and eighth decades of life plasma testosterone gradually declines about 50% because of loss of Leydig cell responsiveness to stimulation. Because of negative feedback, plasma LH levels slowly rise at this time. Although decreasing testosterone levels may be associated with a decline in libido and a decrease in bone and muscle mass, spermatogenesis itself is remarkably well preserved in most octogenarians.

Metabolism. Testosterone circulates largely bound to a sex steroid–binding globulin (SSBG), also known as testosterone-estradiol-binding globulin. This is a β-globulin glycoprotein with a molecular weight of 94,000. Fifty to sixty percent of circulating testosterone is complexed to SSBG, and most of the remainder is bound to albumin and other proteins. SSBG also binds dihydrotestosterone and 5α-androstanediol. Only the free and the loosely bound albumin fractions of testosterone and the other androgens are biologically active. The SSBG-bound fractions serve as circulating reservoirs, similar to those of thyroid hormone and cortisol.

The concentration of SSBG is increased by estrogens and is decreased by androgens. Reciprocally, then, estrogen reduces the percentage of free testosterone, whereas androgen increases it. About 1% of the daily production of testosterone (70 μg) is excreted daily in the urine as glucuronide. Most of the remainder is metabolized to products that are excreted in the urine as part of the 17-ketosteroid fraction (Chapter 54). In men only 30% of the total urinary 17-ketosteroids derives from testosterone; most arises from adrenal sources. Therefore measurement of plasma testosterone (and occasionally urine testosterone) is the mainstay for assessing Leydig cell function.

Actions. The extratesticular effects of testosterone and related androgens can be divided into two major categories: those pertaining specifically to reproductive function and secondary sexual characteristics and those pertaining more generally to stimulation of tissue growth and maturation. At present the evidence suggests that similar intracellular mechanisms are involved in both categories. In general, the model for steroid hormone effects is applicable (Fig. 49-10).

Testosterone diffuses freely into cells. In many but not all target cells it rapidly undergoes reduction to dihydrotestosterone and, in some, to 5α-androstanediol (Fig. 55-4). The relevant hydroxysteroid dehydrogenases are microsomal in location and employ NADPH as the reductant. A single cytoplasmic protein receptor binds all three steroids. The absolute requirement for the androgen receptor is best illustrated by the syndrome of *testicular feminization* in humans and mice. Lacking the gene for androgen receptor synthesis, XY individuals with testes show complete failure to masculinize their genital ducts or external genitalia.

The steroid receptor complex moves into the nucleus, where it interacts with chromosomal DNA and nuclear proteins. The result is a significant stimulation of RNA polymerase, of various messenger RNAs, and of the synthesis of proteins. In addition, the activities of enzymes concerned with DNA synthesis, such as thymidine kinase and DNA polymerase, are increased. Virtually all actions of androgens are blocked by inhibitors of RNA or protein synthesis. Therefore they require induction of new enzyme molecules, as opposed to allosteric or covalent activation of existing enzyme molecules.

In target tissues, such as the prostate gland and seminal vesicles, polyamine (for example, spermine and putrescine) synthesis is stimulated by androgens, and these compounds in turn enhance RNA synthesis. Androgens also stimulate remarkable growth of these accessory organs of reproduction, characterized by hypertrophy and hyperplasia of the epithelial cells, stromal components, and blood vessels.

By far the major circulating androgen is testosterone (Table 55-3). Testosterone in part can be considered a prohormone for dihydrotestosterone and 5α-androstanediol, much as thyroxine is a prohormone for triiodothyronine. However, unlike thyroxine, testosterone definitely has intrinsic hormonal activity of its own in tissues that lack the enzyme 5α-reductase. An interesting and major source of such information comes from studies of individuals who are genetic and gonadal males but who have congenital 5α-reductase deficiency. These individuals have feminized external genitalia at birth, but during puberty they undergo selective masculinization in response to the rising testosterone secretion.

Table 55-4. Major actions of androgenic hormones

Testosterone	*Dihydrotestosterone*
Fetal development of	Fetal development of
Epididymis	Penis
Vas deferens	Penile urethra
Seminal vesicles	Scrotum
Pubertal growth of	Prostate
Penis	Pubertal growth of
Seminal vesicles	Scrotum
Musculature	Prostate
Skeleton	Sexual hair
Larynx	Sebaceous glands
Spermatogenesis	Prostatic secretion

A presumptive classification of androgen effects, according to the probable actual effector hormone, is shown in Table 55-4. Dihydrotestosterone is specifically required in the fetus for the differentiation of the genital tubercle, genital swellings, genital folds, and urogenital sinus into the penis, scrotum, penile urethra, and prostate, respectively. It is required again during puberty for growth of the scrotum and prostate and in adult life for the stimulation of prostatic secretions. Dihydrotestosterone or 5α-androstanediol stimulates the hair follicles and produces the typical male pattern. This consists of beard growth, a diamond-shaped pubic escutcheon, relatively large amounts of body hair, and the recession of the temporal hairline, which in some men culminates in baldness. Increased production of sebum by the sebaceous glands also is brought about by dihydrotestosterone or 5α-androstanediol.

Testosterone, on the other hand, specifically stimulates the differentiation of the wolffian ducts into the epididymis, vas deferens, and seminal vesicles. During puberty testosterone causes enlargement of the penis and of the seminal vesicles to adult size. It also causes enlargement of the larynx and thickening of the vocal cords, resulting in a deeper voice. Most important, testosterone is the local hormone required for initiation and maintenance of spermatogenesis. This requirement is demonstrated by the fact that spermatozoa production is normal in postpubertal individuals who lack dihydrostestosterone because of 5α-reductase deficiency.

Testosterone itself also first stimulates the pubertal growth spurt and then causes cessation of linear growth by closure of the epiphyseal growth centers. Testosterone causes enlargement of the muscle mass in boys during puberty. In subsequent adult life administration of testosterone causes nitrogen retention in both sexes, reflecting protein anabolism. It is noteworthy that the hypothalamus lacks significant 5α-reductase activity. Thus suppression of gonadotropin secretion by negative feedback is largely a direct function of testosterone, with a

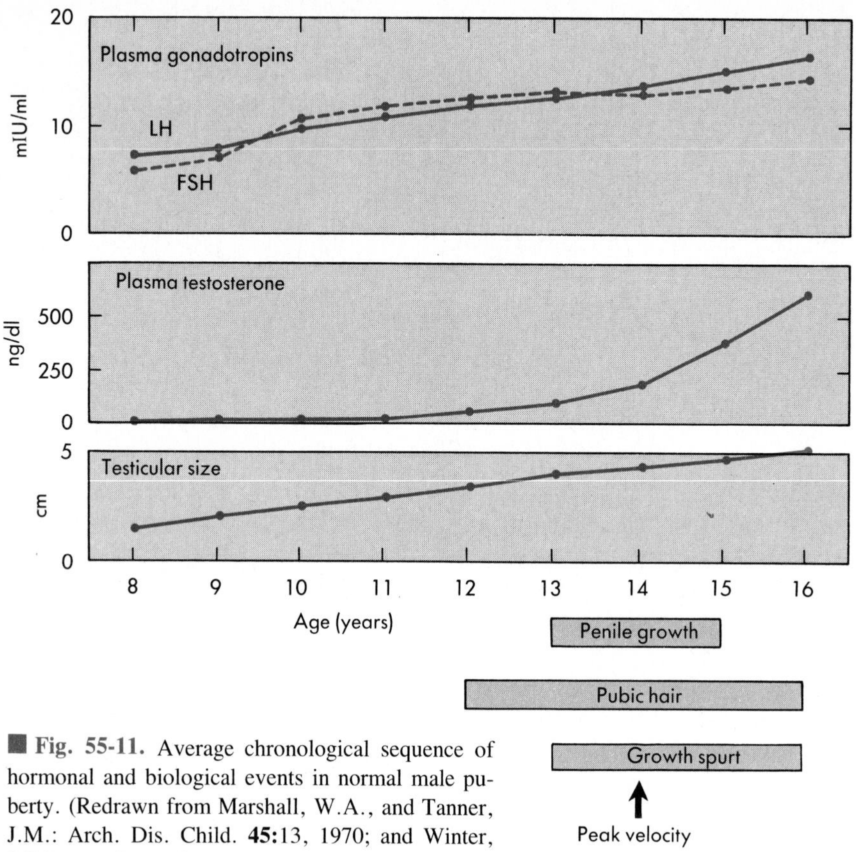

■ **Fig. 55-11.** Average chronological sequence of hormonal and biological events in normal male puberty. (Redrawn from Marshall, W.A., and Tanner, J.M.: Arch. Dis. Child. **45:**13, 1970; and Winter, J.S.D., et al.: Pediatr. Res. **6:**126, 1972.)

possible small additional effect from circulating dihydrotestosterone.

Certain other diverse androgenic actions can be ascribed to testosterone. These include (1) initiation of sexual drive (libido) and the ability to achieve a physiologically complete erection (potency), (2) suppression of mammary gland growth, (3) stimulation of hematopoiesis and maintenance of a normal red blood cell mass, (4) stimulation of renal sodium reabsorption, (5) stimulation of aggressive behavior, and (6) suppression of hepatic synthesis of SSBG, cortisol-binding globulin, and thyroxine-binding globulin.

■ *Male Puberty*

From the early newborn period to the onset of puberty the testis is endocrinologically dormant (Fig. 55-5, *C* and *D*). Beginning at an average age of 10 to 11 and ending at an average age of 15 to 17, males develop full reproductive function, Leydig cell proliferation, and adult levels of androgenic hormones (Fig. 55-5, *A* and *B*). Secondary to activation of the testis, males acquire adult size and function of the accessory organs of reproduction, complete secondary sexual characteristics, and adult musculature. They undergo a linear growth spurt, and the epiphyses close when they attain adult height. A composite picture of the measurable and visible portions of this sequence is shown in Fig. 55-11. It must be stressed that this process can start as early as age 8 and as late as age 20, without any evidence of disease.

The events of puberty begin with the secretion of increasing amounts of gonadotropin. Before age 10, plasma gonadotropin levels are low, despite the very low plasma concentration of testicular steroids. Therefore either the negative feedback system is inoperative or the hypothalamus and pituitary gland are exquisitely sensitive to testosterone or estradiol. Thus one explanation offered for the onset of puberty is the gradual maturing of a hypothalamic gonadostat, leading to increased synthesis and release of LHRH. The rate of this maturational process may well be genetically preprogrammed. However, a simple resetting of the negative feedback axis does not explain all the observed phenomena.

For example, as puberty approaches, a pulsatile pattern of LH and FSH secretion emerges. As the pulse frequency gradually increases, the ratio of LH to FSH rises. Furthermore, during early and midpuberty but at no other time of life, a nocturnal peak in LH secretion is observed (Fig. 55-12). These events suggest some other central nervous sytem influences on the initiation of puberty. Various animal and clinical studies implicate input from the amygdala, the pineal gland, or the olfactory bulb as triggers of pubertal hypothalamic functioning.

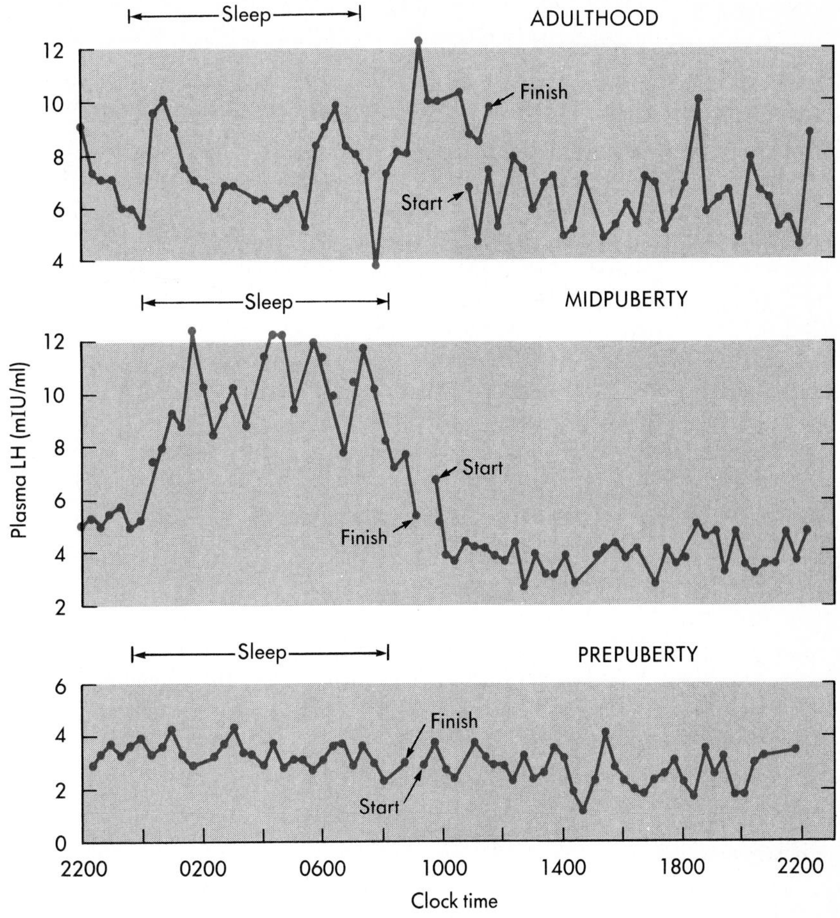

■ **Fig. 55-12.** The development of pulsatile secretion of LH and the transient appearance of a nocturnal peak during puberty. The same sequence occurs in males and females. (Redrawn from Boyar, R.M., et al.: Reprinted, by permission of the New England Journal of Medicine **287:**582, 1972.)

The gonad itself is not necessary for activation of gonadotropin secretion and the appearance of pulsation at the usual age of puberty. The fact that dehydroepiandrosterone sulfate (DHEA-S) levels increase, coincident with the onset of puberty, has suggested that this adrenal steroid may play an important role in the maturing of the hypothalamus. However, this seems unlikely, because boys with adrenal cortical insufficiency do undergo normal pubescence (with proper cortisol replacement), whereas boys with high adrenal androgen secretion as a result of congenital adrenal hyperplasia or adrenal tumors generally do not.

The responsiveness of the pituitary gland also changes. In childhood the response of LH secretion to acute LHRH stimulation is less than that of FSH secretion, but it gradually increases during puberty. This may be partly explained by an increasing synthesis and storage of LH in response to activation of LHRH secretion. Responsiveness of the Leydig cells to LH stimulation, although present in childhood, also is augmented during puberty. Thus the pubescent period can be viewed as a cascade of increasing responsiveness from the hypothalamic level to the pituitary gland to the testis.

Enlargement of the testis is the first and most important clinical sign of puberty. This represents principally an increase in the volume of the seminiferous tubules, and it is preceded by small increases in plasma FSH. Leydig cells appear, and testosterone secretion is stimulated, as plasma LH increases. Plasma testosterone then climbs rapidly over a 2-year period, during which time pubic hair appears, the penis enlarges, and peak velocity in linear growth is achieved (Fig. 55-11). Sometime during this interval—at a median age of 13 years—sperm production begins. In about one third of boys, breast growth and tenderness appear transiently. This probably reflects increased production of estradiol secondary to LH stimulation. As testosterone levels continue to climb, the breast tissue regresses. One to two years after adult testosterone levels are reached, closure of the epiphyseal growth centers ends puberty.

■ *The Ovaries*

The ovaries, fallopian tubes, and uterus comprise the internal reproductive organs of the female and are situated in the pelvis. Each adult ovary weighs approximately 15 g and is attached to the lateral pelvic wall and to the uterus by ligaments, through which run the

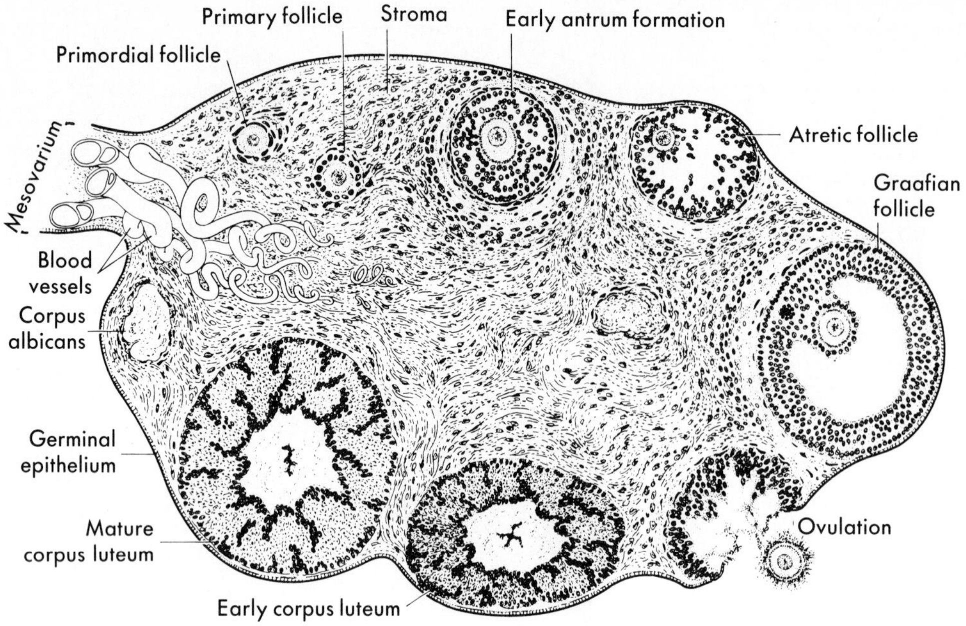

Fig. 55-13. The microscopic anatomy of the human ovary. (From Ham, A.W., and Leeson, T.S.: Histology, ed. 4, Philadelphia, 1961, J.B. Lippincott Co.)

ipsilateral ovarian artery, vein, lymphatic vessels, and nerve supply.

The ovary consists of three distinct zones (Fig. 55-13). The dominant zone is the cortex, which is lined by germinal epithelium and contains all the oocytes, each enclosed within a follicle. Follicles in various stages of development and regression can be seen through the cortex during the reproductive years (Fig. 55-13). Interposed between the follicles is the stroma, composed of supporting connective tissue elements and interstitial cells. The other two zones of the ovary are the medulla, consisting of a heterogeneous group of cells, and the hilum, at which the blood vessels enter. These zones contain scattered steroid-producing cells, whose normal function is unknown.

Neoplasms of the ovary can arise from any one of its three zones and their individual cell lines. During physical examination the ovaries can be palpated manually through the abdominal wall and also can be well visualized by ultrasonography and computerized axial tomography (CAT scanning).

As a hormone-secreting organ, the ovary functions in two ways. First, the ovarian sex steroids function locally to modulate the complex events in the development and extrusion of the ova. Second, these steroids are secreted into the circulation and act on diverse target organs, including the uterus, vagina, breasts, hypothalamus, pituitary gland, adipose tissue, bones, and liver. Many but not all of the distant effects are closely related to the reproductive sequence.

Oogenesis

The primordial germ cells migrate from the yolk sac of the embryo to the genital ridge at 5 to 6 weeks of gestation. There, in the developing ovary, they produce oogonia by mitotic division until 20 to 24 weeks, when the total number of oogonia has reached a maximum of 7 million. Beginning at 8 to 9 weeks some oogonia start into the prophase of meiosis, becoming primary oocytes. This process continues until 6 months after birth, when all oogonia have been converted to oocytes. Almost from the start, however, there is also a process of attrition, so that by birth only 2 million primary oocytes remain, and by the onset of puberty the number falls to 400,000. Thus, in contrast to the man, who is continuously producing spermatogonia and primary spermatocytes, the woman cannot manufacture new oogonia and must function with a continuously declining number of primary oocytes from which ova can mature. At or soon after menopause there are few, if any, oocytes left, and reproductive capacity ends.

At the time meiosis begins, oocytes are 10 to 25 μm in diameter. They grow to 50 to 120 μm at maturity, the size of the nucleus and cytoplasm having increased proportionately. The first meiotic division is not completed until the time of ovulation; thus primary oocytes have life spans up to 50 years. The lengthy suspension of the oocyte in prophase apparently depends on the hormonal milieu provided by its surrounding sustaining cells.

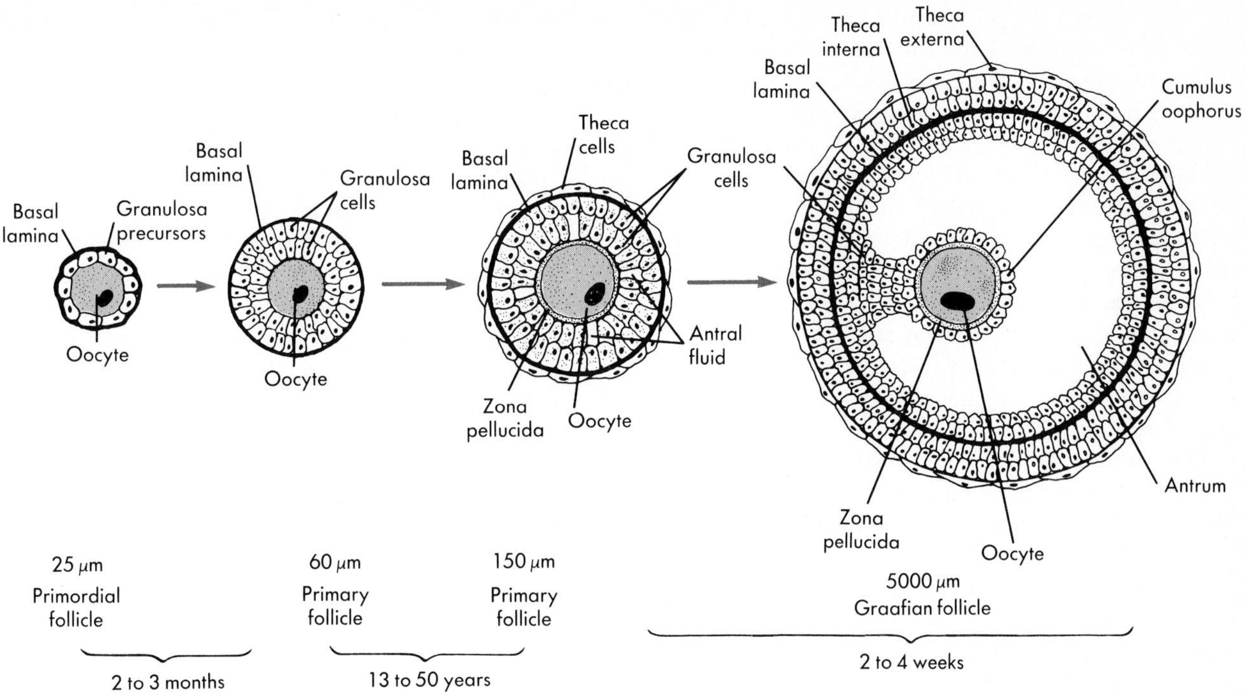

■ **Fig. 55-14.** The development of an ovarian follicle (not to scale). The graafian follicle has more layers of granulosa and theca cells than shown here.

First stage. The first stage of follicular development occurs very slowly, over a period that is usually not less than 13 years but that may be as long as 50 years. As an oocyte enters meiosis, it induces a single layer of spindle cells from the stroma to surround it completely. These cells are the precursors of the female follicular cells, the *granulosa cells*. Cytoplasmic processes from these cells attach to the plasma membrane of the oocyte. In addition, a membrane called the *basal lamina* forms outside the spindle cells, delimiting the complex from the surrounding stroma. This constitutes the *primordial follicle*, which is about 25 μm in diameter (Fig.55-14).

Beginning at 5 to 6 months of gestation some of these follicles enter the next phase of development. The spindle-shaped cells become cuboidal and begin to divide, creating several layers of granulosa cells around the oocyte. This complex is called the *primary follicle*. The granulosa cells secrete mucopolysaccharides, which form a protective halo, the *zona pellucida*, around the oocyte (Fig. 55-14). The cytoplasmic processes of the granulosa cells, however, continue to penetrate the zona pellucida, evidently providing nutrients and chemical signals to the maturing primary oocyte within. Thus cytoplasm of the granulosa cells, like that of the male Sertoli cells, forms a filter through which plasma substances must pass before reaching the germ cell.

The primary follicle continues to grow, reaching a diameter of 150 μm. At this point the oocyte has reached its maximal size, averaging 80 μm in diameter. There are two other concurrent developments. Another layer of spindle interstitial cells is recruited outside the basal lamina and forms the *theca interna*, while the granulosa cells begin to extrude small collections of fluid between themselves. This completes the first stage of follicular development. It is also the maximal degree of development ordinarily found in the prepubertal ovary.

Second stage. In contrast to the first stage, the second stage of follicular development is much more rapid, requiring only 2 to 4 weeks for completion. This stage takes place after menarche, that is, after the onset of menses. During each menstrual cycle approximately six to twelve primary follicles enter the next sequence. The small collections of follicular fluid coalesce into a single central area called the *antrum* (Fig. 55-14). The fluid in the antrum contains mucopolysaccharides, plasma proteins, electrolytes, gonadal steroid hormones, FSH, inhibin, and other factors. The steroid hormones reach the antrum by direct secretion from granulosa cells and by diffusion from the theca cells outside the basal lamina. A nonsteroidal substance that is capable of inhibiting oocyte meiosis probably also is secreted into the antral fluid and is believed to be of granulosa cell origin. AMH has recently been reported to have such an action and could be this factor.

The granulosa cells continue to proliferate and displace the oocyte into an eccentric position on a stalk, where it is surrounded by a distinctive layer, two to three cells thick, called the *cumulus oophorus*. The theca cells also proliferate, and those nearest the basal lamina are transformed into cuboidal steroid-secreting cells—the *theca interna*. Additional peripheral layers of spindle cells from the stroma form around the theca interna and together with vascular spaces comprise the *theca externa*. By the end of this second stage the entire complex, called a *graafian follicle* (Fig. 55-14), has reached an average diameter of 5000 μm (5 mm). Although graafian follicles may rarely be found in premenarchal ovaries, they never achieve this size or degree of development.

Third stage. The third and final stage of follicular development is the most rapid, being completed within 48 hours. It occurs only in the postmenarchal reproductive ovary. Several days before ovulation a single graafian follicle from that cycle achieves dominance. In this particular graafian follicle the granulosa cells significantly increase the production of antral fluid. The colloid osmotic pressure of the fluid also increases because of depolymerization of the mucopolysaccharides. The granulosa cells spread apart, and the cumulus oophorus loosens. At the same time the vascularity of the theca increases greatly. The total size of this follicle reaches 10 to 20 mm within 48 hours. The portion of the basal lamina adjacent to the surface of the ovary then is subjected to proteolysis. The follicle gently ruptures, releasing the oocyte with its adherent cumulus oophorus into the peritoneal cavity. At this time the initial meiotic division is completed. The resultant secondary oocyte is drawn into the closely approximated fallopian tube, and the first polar body is discarded. In the fallopian tube fertilization causes completion of the second meiotic division, resulting in the haploid (23 chromosome) ovum and the second polar body.

Corpus luteum formation. The residual elements of the ruptured follicle next form a new endocrine structure, the *corpus luteum* (Fig. 55-13). This unit will provide the necessary balance of gonadal steroids that optimizes conditions for implantation of the ovum, should fertilization occur, and for subsequent maintenance of the zygote until the placenta can assume this function. The corpus luteum is made up of granulosa cells, theca cells, thecal capillaries, and fibroblasts. The granulosa cells comprise 80% of the corpus luteum. They hypertrophy to a diameter of 30 μm, become arranged in rows, and undergo striking changes. The mitochondria develop dense matrices with tubular cristae, numerous lipid droplets form within the cytoplasm, and the smooth endoplasmic reticulum proliferates. This process, called *luteinization*, is precipitated by the exit of the oocyte from the follicle.

The remaining 20% of the corpus luteum consists of theca cells arranged in folds along its outer surface. The theca cells exhibit similar although less dramatic changes of luteinization. The basal lamina between the theca and granulosa cells disappears, allowing direct vascularization of the latter.

The antrum may become temporarily engorged with blood from hemorrhaging thecal vessels, but a clot quickly forms and is subsequently lysed. If fertilization and pregnancy do not ensue, the corpus luteum begins to regress after a 14-day life span. In this process, known as *luteolysis*, the endocrine cells undergo necrosis, and the structure is invaded by leukocytes, macrophages, and fibroblasts. Gradually the former corpus luteum is replaced by an avascular scar known as the *corpus albicans* (Fig. 55-13).

Atresia of follicles. During the reproductive life span of the average woman only 400 to 500 oocytes (one per month) will undergo the complete sequence of events culminating in ovulation. The remaining millions disappear in a process called *atresia*, which begins almost as soon as the first primordial follicles appear in the fetal ovary.

In first-stage follicles atresia is a relatively simple process. The oocyte becomes necrotic, its nucleus becomes pyknotic, and the granulosa cells degenerate. This accounts for the vast majority of oocytes. In more advanced follicles atresia is a more complex process. In some follicles the granulosa cells furthest from the oocyte first undergo necrotic changes. Loss of their function may actually precipitate a resumption of meiosis in the oocyte to the point of extrusion of the first polar body. Eventually the granulosa cells in the cumulus oophorus also die, the protective zona pellucida disappears, and the oocyte degenerates. In other follicles degeneration of the oocyte may initiate the process. Eventually fibroblasts invade the follicle, and everything inside the basal lamina collapses into an avascular scar, called a *corpus albicans*. Outside the basal lamina the theca cells dedifferentiate and return to the pool of interstitial cells from which they came.

Hormonal Patterns during the Menstrual Cycle

The menstrual cycle is divided physiologically into three sequential phases. The *follicular phase* begins with the onset of menstrual bleeding and is of variable length. The *ovulatory phase* is of 1 to 3 days' duration and culminates in ovulation. The *luteal phase* has a rather constant length of 13 to 14 days and ends with the onset of menstrual bleeding. The overall duration of a normal menstrual cycle can vary from 21 to 35 days, depending mostly on the length of the follicular phase.

A series of cyclic changes in gonadal steroid hormone production characterizes adult ovarian function (Fig. 55-15, Table 55-5). This monthly steroid hormone profile results from cyclic changes in pituitary gonadotropins (Fig. 55-15). However, the pattern of gonado-

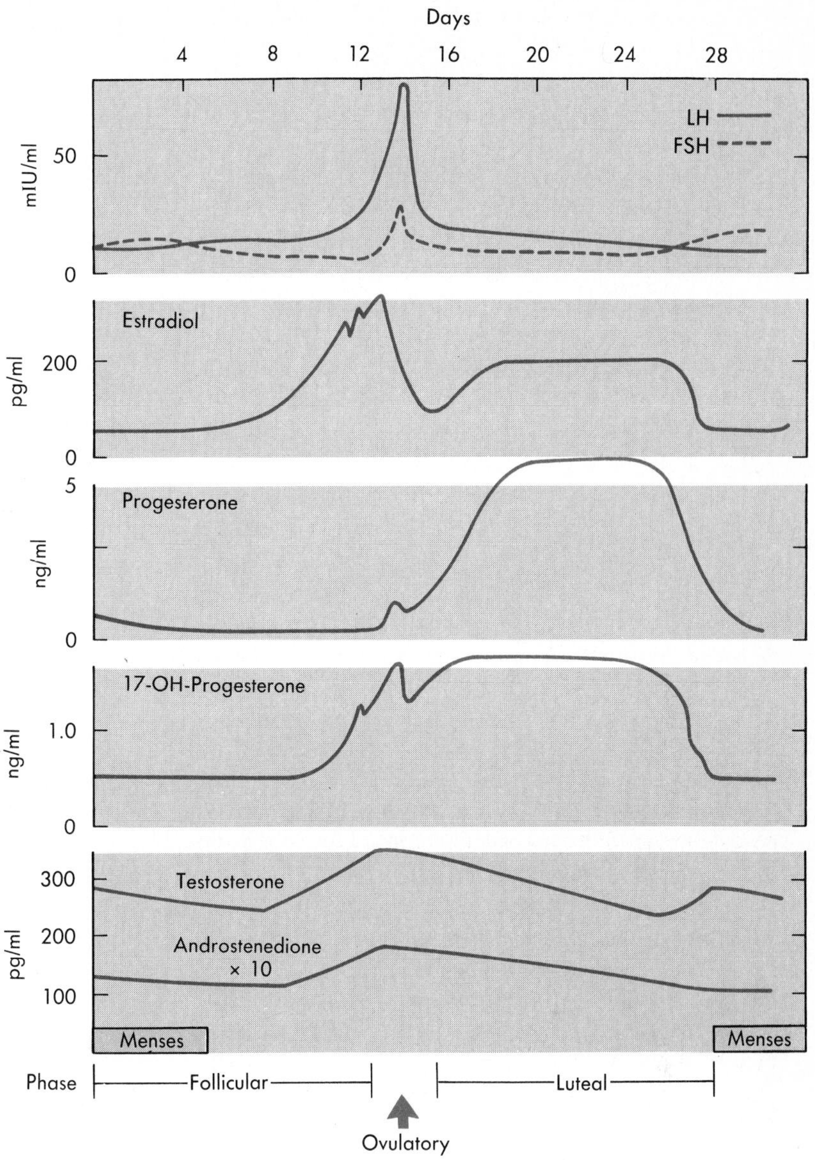

■ **Fig. 55-15.** Plasma hormone profile of the human menstrual cycle. Note that increase in estradiol precedes ovulatory LH peak. (Redrawn from Yen, S.S.C., and Jaffe, R.B.: Reproductive endocrinology, Philadelphia, 1978, W.B. Saunders Co.)

tropin secretion in turn is critically regulated by both negative and positive feedback from gonadal steroids (Chapter 52). In a sense the menstrual cycle might be viewed as a stone rolling up and down a series of hills; the momentum derived from each phase of the cycle powers the next phase, and so on into the next cycle.

Toward the end of the luteal phase and beginning of the follicular phase, plasma FSH and LH are at their lowest levels (Fig. 55-15). The LH/FSH ratio is slightly greater than 1. During the first half of the follicular phase FSH levels rise, followed somewhat later by LH levels. The estrogens (estradiol and estrone) increase almost imperceptibly, whereas progesterone and 17-hydroxyprogesterone levels remain constant and low during this critical 6 to 8 days. The same is true for the androgens androstenedione and testosterone.

During the second half of the follicular phase FSH levels fall modestly, whereas LH levels continue to rise very slowly. The LH/FSH ratio therefore increases to about 2. Concurrently, estradiol and estrone production and plasma levels rise sharply, reaching peaks fivefold to ninefold higher just before the ovulatory phase. The estradiol in the plasma is secreted directly by the ovaries, whereas the estrone largely arises from peripheral conversion of estradiol and androstenedione. Plasma progesterone and 17-hydroxyprogesterone remain low until just before the ovulatory phase, when progesterone begins to increase due to ovarian secretion. Androstenedione and testosterone also rise modestly in parallel with 17-hydroxyprogesterone. About half the plasma androgens are derived from ovarian androstenedione secretion and half from adrenal androstenedione secretion.

The large increment in estradiol secretion during the latter half of the follicular phase arises from the granulosa cells of the dominant follicle, which previously were stimulated by FSH and then by LH. The rising plasma estradiol levels, augmented by follicular inhi-

Table 55-5. Turnover of gonadal steroids in adult women

Steroids	*Plasma concentration (ng/dl)*	*Production rate (μg/day)*	*Metabolic clearance rate (L/day)*
Estradiol			
Early follicular	6	80	1400
Late follicular	50	700	
Middle luteal	20	300	
Estrone			
Early follicular	5	100	2200
Late follicular	20	500	
Middle luteal	10	250	
17-Hydroxyprogesterone			
Early follicular	30	600	2000
Late follicular	200	4000	
Middle luteal	200	4000	
Progesterone			
Follicular	100	2000	2200
Luteal	1000	25000	
Testosterone	40	250	700
Dihydrotestosterone	20	50	400
Androstenedione	150	3000	2000
DHEA	500	8000	1600

Modified from Lipsett, M.B.: In Yen, S.S.C., and Jaffe, R.B., editors: Reproductive endocrinology, Philadelphia, 1978, W.B. Saunders Co.

bin, suppress FSH secretion by negative feedback. The rise in ovarian androgen secretion toward the end of the follicular phase results from LH stimulation of the theca cells.

The succeeding ovulatory phase is characterized by a very sharp, transient spike in plasma gonadotropin levels. LH increases much more than FSH (Fig. 55-15), so that the LH/FSH ratio rises to about 5. Plasma estradiol levels plummet from their peak at the same time that LH and FSH are on their ovulatory upswing. Estrone, 17-hydroxyprogesterone, androstenedione, and testosterone also now decrease but much more gradually than estradiol. In contrast, a small rise in progesterone begins during the ovulatory phase.

After ovulation LH and FSH both continue to decline during the luteal phase, reaching their lowest points in the cycle toward its end, before the onset of menses. Inhibin is also present in plasma throughout the cycle, reaching a maximum in the luteal phase. Plasma inhibin tends to be inversely related to plasma FSH, in congruence with the negative feedback effect of inhibin on FSH secretion. The most distinctive and important feature of the luteal phase is a tenfold increase in progesterone, which emanates from the corpus luteum. Estradiol, estrone, and 17-hydroxyprogesterone, partly of corpus luteum origin, also increase, providing broad second peaks of these steroids through the middle of the luteal phase. Androstenedione and testosterone, however, continue to decline during the luteal phase. If pregnancy does not occur, the menstrual cycle ends as the gonadal steroid levels decrease dramatically to their lowest values and bleeding starts.

Hormonal Regulation of Oogenesis

The entire first stage, from the primordial follicle to the primary follicle, *can* proceed in the absence of the pituitary gland. However, studies of the ovaries of individuals with congenital gonadotropin deficiencies suggest that the midgestation surge of fetal FSH and LH secretion, as well as the low levels of gonadotropins secreted during childhood, may increase the rate of growth or the number of follicles undergoing development. Nonetheless first-stage growth appears to be largely a local phenomenon in which one or more factors from the oocyte stimulate granulosa cell development. In turn the granulosa cells probably initiate thecal development and arrest the size and maturation of the oocyte once it has reached 80 μm in diameter. Estradiol and an oocyte maturation–inhibiting factor are the probable granulosa cell hormones involved in these functions.

Second-stage follicular development is initiated by FSH. Two crops of follicles begin antral formation during each menstrual cycle. The first group follows the FSH surge of the ovulatory phase, reaches its peak size of 3 to 5 mm about 8 to 9 days later, possibly contributes to ovarian estrogen production in the luteal phase, and then undergoes atresia by the end of that phase. The second group of follicles is initiated by the rise in FSH that begins just before the start of the follicular phase. This group, plus a few survivors of the first group, reaches a peak size of 5 mm at 10 to 14 days after the onset of menstrual bleeding. It is from this group that the single dominant follicle, from only one of the two ovaries, ordinarily emerges to undergo ovulation.

The initial action of FSH on the primary follicles is to stimulate growth of the granulosa cells (Fig. 55-16). Additionally, aromatase activity is increased, so that estrogen synthesis from androgen precursors is enhanced. The increasing *local estradiol* causes proliferation of its own receptors and reinforces FSH actions. FSH action is reinforced by increasing FSH receptors and by synergizing with the gonadotropin to stimulate further granulosa cell hyperplasia and hypertrophy (Fig. 55-16). This in turn further boosts estradiol production. Thus the initiation of second-stage follicular development may be viewed as a selfpropelling mechanism that involves fine coordination between the pituitary gland and ovary and that yields exponential rates of follicular growth and estradiol production.

Two other important actions, which develop somewhat later, contribute to this autocatalytic process. (1) FSH, along with estradiol, induces LH receptors on the granulosa cells. (2) The slowly rising plasma estradiol levels condition the hypothalamic gonadotropin axis so as to maintain or slightly increase plasma LH while plasma FSH is decreasing. Furthermore, pituitary LH stores are enhanced by estradiol. This is reflected in the

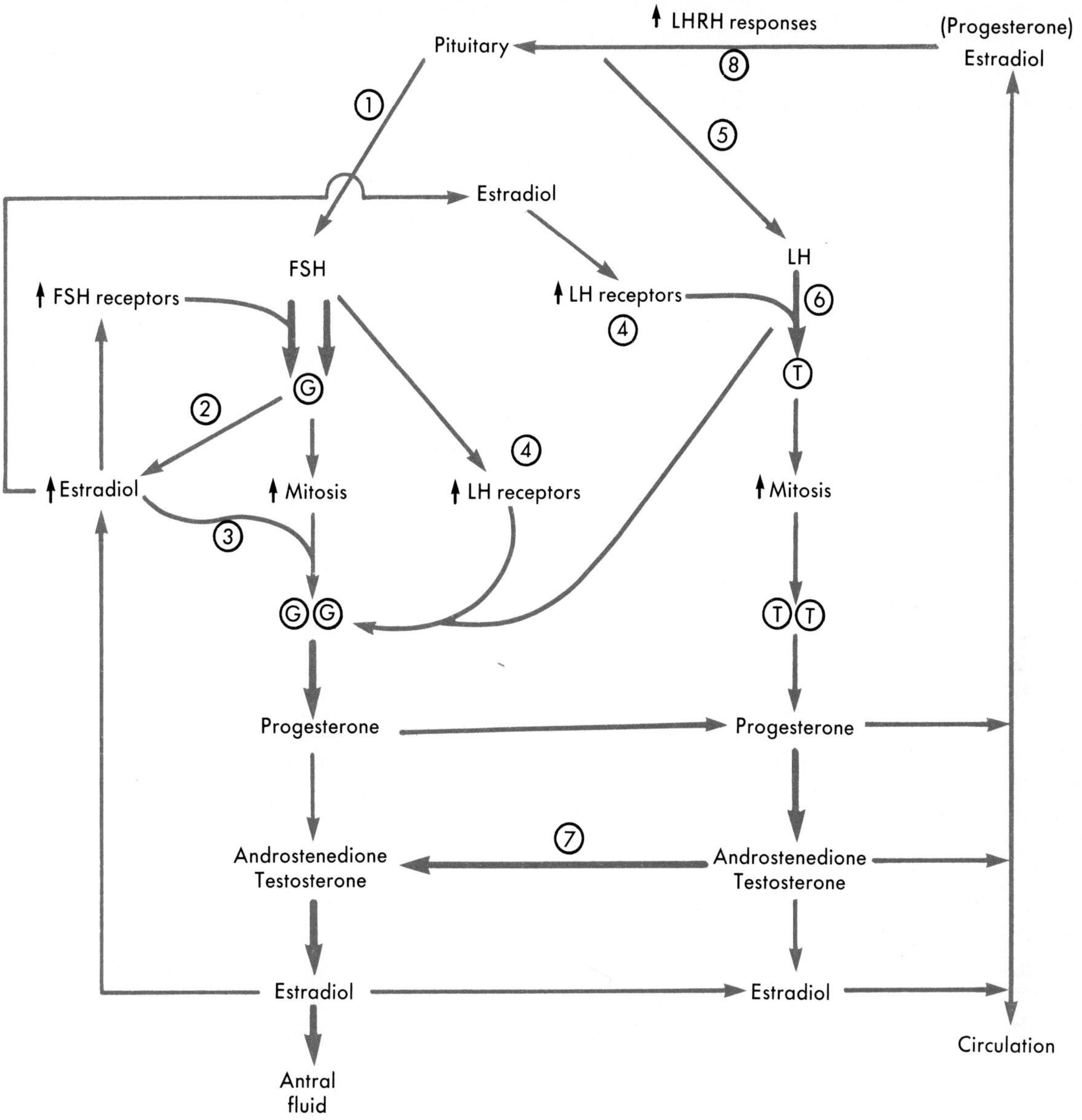

■ **Fig. 55-16.** Hormonal regulation of follicular development in the human. Note self-reinforcing effects of FSH and estradiol on the granulosa cells ② and ③, two-way steroid hormone traffic between the granulosa and theca cells, facilitating estradiol production by the former ⑦, and eventual positive feedback effect of estradiol on the pituitary gland, leading to augmented LHRH responsiveness prior to ovulation ⑧. *G,* Granulosa cell; *T,* theca cell.

fact that administration of exogenous pulses of LHRH produces greater LH responses in the second half of the follicular phase than in the first half (Fig. 55-17).

LH stimulates the theca cells to produce increasing amounts of androstenedione and testosterone. These steroids diffuse across the basal lamina, where they serve as substrates for granulosa cell aromatase and sustain the augmented estradiol production (Fig. 55-16). This is of considerable importance because isolated granulosa cells are relatively inefficient producers of estradiol. In addition, LH stimulates the granulosa cells to produce progesterone, some of which diffuses back into the theca cells to serve as a substrate for androgen synthesis (Fig. 55-16). Thus, although individually granulosa cells and theca cells can synthesize both androgens and estrogens to some extent, their proximity and the two-way traffic of steroids between them increase the overall efficiency of the follicle greatly.

During the latter half of the follicular phase a single follicle outstrips the others. Exactly how this follicle is selected remains unknown. One important characteristic may be a greater density of FSH receptors at the outset, giving a particular follicle the ability to stay ahead of the others in its production of estrogen. The key may be in the balance of factors and hormones within the follicular fluid. The remaining follicles of the group un-

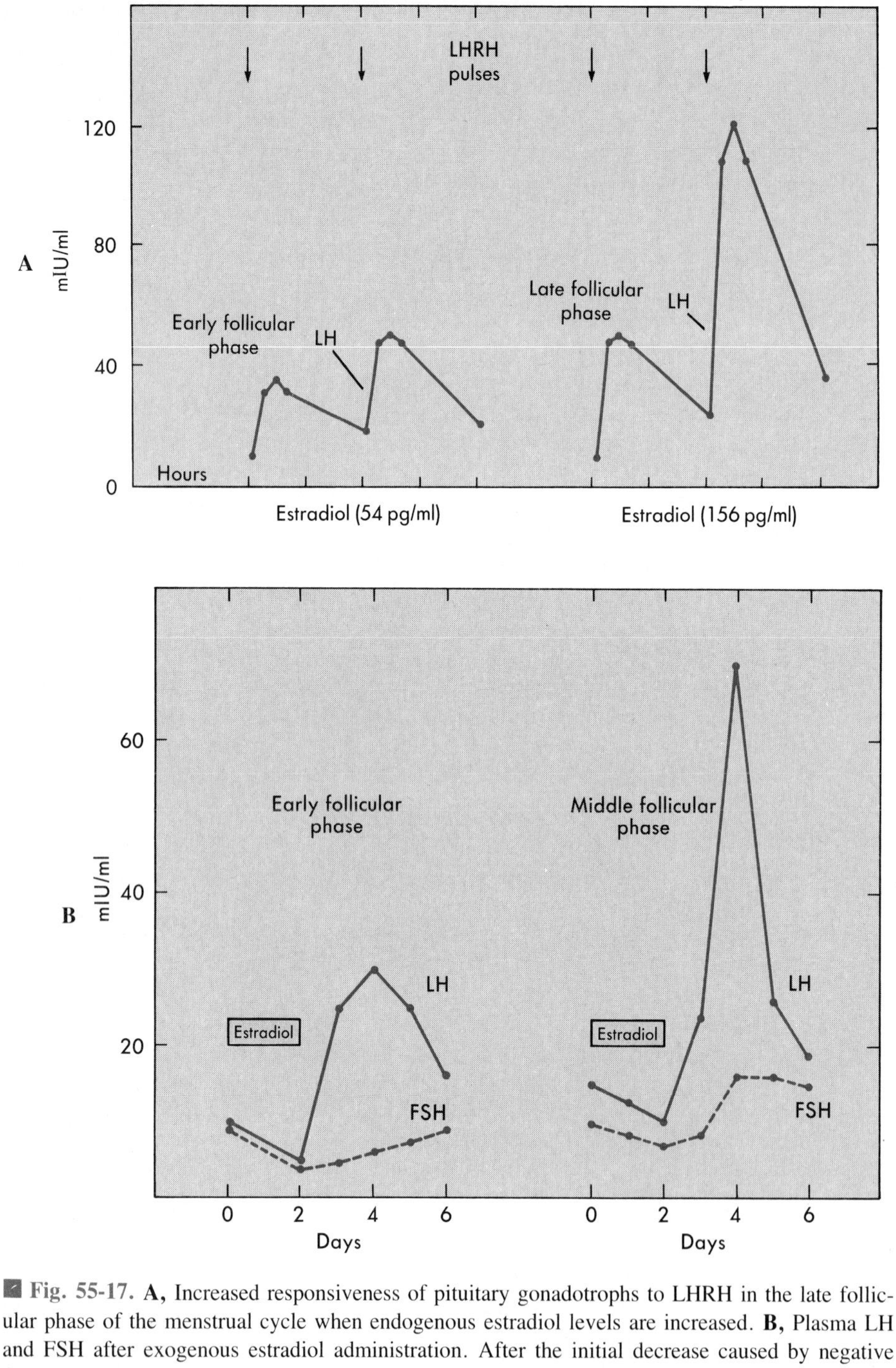

Fig. 55-17. A, Increased responsiveness of pituitary gonadotrophs to LHRH in the late follicular phase of the menstrual cycle when endogenous estradiol levels are increased. **B,** Plasma LH and FSH after exogenous estradiol administration. After the initial decrease caused by negative feedback, plasma LH rebounds well above the baseline when estradiol is discontinued. This positive effect of estradiol also is accentuated as the follicular phase of the menstrual cycle progresses. (**A** redrawn from Wang, C.F., et al.: The functional changes of the pituitary gonadotrophs during the menstrual cycle, J. Clin. Endocrinol. Metab. **42:**718, 1976. Copyright 1976. Reproduced by permission. **B** redrawn from Yen, S.S.C., et al.: Causal relationship between the hormonal variables. In Ferin, M., et al., editors: Biorhythms and human reproduction, New York, 1974, John Wiley & Sons, Inc.)

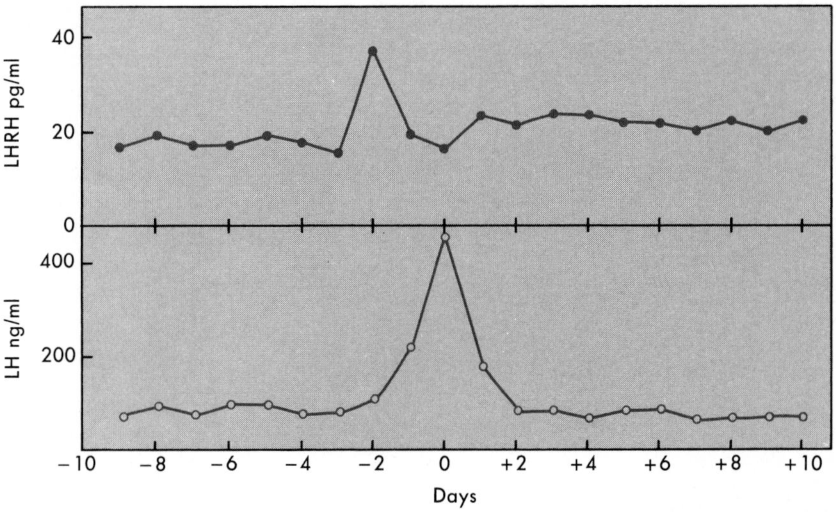

■ **Fig. 55-18.** Daily variation in plasma LHRH and LH levels in normal women. A distinct brief increase in peripheral plasma LHRH is seen to precede the ovulatory surge of LH. (Redrawn from Sarda, A.K., et al.: J. Clin. Endocrinol. **15:**265, 1981.)

dergo atresia. The most consistent finding in fluid sampled from smaller or atretic follicles is a relatively high concentration of androgens and relatively low concentrations of estradiol and FSH.

An atretic fate may be determined by insufficient FSH receptors, coupled with decreased FSH availability, as estrogen from the dominant follicle feeds back to inhibit pituitary FSH secretion. A lack of FSH action would lead to a relative inability of the granulosa cells in the atretic follicle to synthesize estrogens by aromatizing the supply of androgens from its own theca cell partners as well as by the diffusion in of androgens from the more rapidly growing dominant follicle. Thus the estradiol/testosterone ratio in the atretic follicle would become low. In any event, estrogen secretion by the dominant follicle is rapidly augmented after 8 to 12 days.

Plasma estradiol rises abruptly (Fig. 55-15). This surge triggers a marked release of LH accompanied by a lesser release of FSH. A critical plasma estradiol level of at least 200 pg/ml sustained for at least 2 days is required to elicit this positive feedback effect on LH (*pg*-picograms). The much smaller preovulatory increase in progesterone, although not absolutely required, synergizes with estradiol by amplifying and prolonging the gonadotropin surge. The loci of this positive feedback on gonadotropin secretion are in the pituitary and the hypothalamus. The pituitary gonadotrophs, appropriately primed by the preceding pattern of gonadal steroid exposure, respond at this time to LHRH with heightened sensitivity (Fig. 55-17, *A*). The LH molecules released are more active, probably because of posttranslational modulation of their sialic acid content by estradiol. In addition, a peak of peripheral plasma LHRH levels precedes the LH/FSH peak (Fig. 55-18). This suggests an augmented flow of LHRH from the hypothalamus to the pituitary. The increase in LHRH is attributable to gonadal steroids. In vivo, estrogen administration raises plasma LHRH levels (Fig. 55-19), and in vitro, progesterone stimulates LHRH production by estrogen-primed hypothalamic tissue.

Thus the hypothalamic pituitary unit is conditioned by the output of gonadal steroids to provide a sudden increase in LH stimulation of the dominant follicle. This surge of LH with FSH then triggers ovulation by a multicomponent mechanism, as already described. LH stimulation of the granulosa cell neutralizes the action of an oocyte maturation inhibitor, allowing completion of meiosis. Stimulation of progesterone levels enhances proteolytic enzyme activity and increases distensibility of the follicle. This permits a rapid increase in follicular fluid volume. Local prostaglandin synthesis also markedly increases, and these compounds are required for follicular rupture. FSH stimulates production of the plasminogen activator needed for generation of the proteolytic enzyme plasmin. The latter catalyzes breakdown of the follicular wall. FSH also participates in the process whereby the oocyte-cumulus complex becomes detached and free-floating just before extrusion. Finally, immediately after the LH surge, LH receptors are lost by downregulation. This desensitizes the granulosa and thecal cells to LH. The resultant rapid fall in androgen and estradiol production contributes to loss of integrity of the follicle. The LH surge also neutralizes the activity of the luteinization-inhibiting factor found in preovulatory fluid, thereby stimulating luteinization of the granulosa cells.

The dispatched oocyte is not under any other immediate hormonal influence. However, the organization and growth of the corpus luteum and its secretory pattern probably are under hormonal control. Current evidence is conflicting, and a great deal of species variation exists.

In the human, LH is essential for normal corpus lutein function and a high rate of progesterone production. Exposure to proper amounts of FSH in the preced-

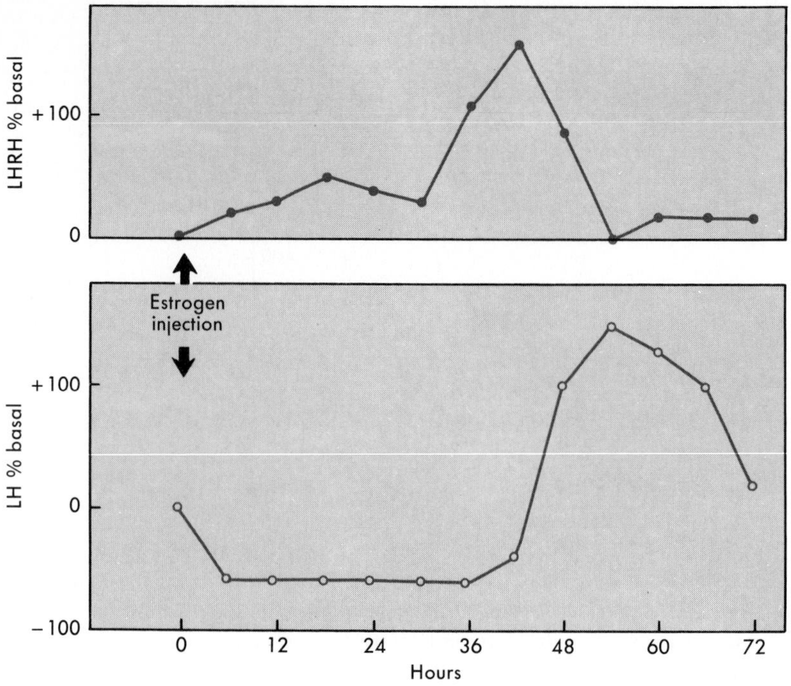

Fig. 55-19. The effect of exogenous administration of estrogen on plasma LHRH and LH levels. After an initial phase of suppression, the positive feedback action of estrogen on LH secretion is seen at 48 hours. This delayed rise in LH levels is preceded by an increase in LHRH levels, suggesting hypothalamic locus of estrogen action. (Redrawn from Miyake, A., et al.: J. Clin. Endocrinol. Metab. **56:**1100, 1983.)

ing follicular phase is also important to ensure the presence of sufficient receptors for the LH. If the declining LH levels of the late luteal phase are not replaced by the equivalent placental hormone, HCG, the corpus luteum regresses, and its secretion of progesterone and estradiol ceases completely by 14 days. Prolactin may contribute to sustaining progesterone output by increasing the LH receptors. However, there is little systematic variation in plasma prolactin throughout the menstrual cycle, and corpus luteum function has been observed in prolactin-deficient women. Estradiol also may play a role in maintaining the corpus luteum.

The steadily increasing progesterone and estradiol output of the corpus luteum exerts negative feedback on the pituitary gland. The resultant gradual decline in both LH and FSH to their low luteal phase levels withdraws support from the postovulatory group of follicles that had entered second-stage development. This, along with the high local progesterone concentration that antagonizes estradiol action, probably causes their atresia.

After the eighth postovulatory day the corpus luteum in the nonpregnant female begins to regress. This process is marked by increasing corpus luteum concentrations of cholesterol as progesterone synthesis declines. Luteolysis is mediated by prostaglandins made within the corpus luteum. By the twelfth postovulatory day progesterone and estradiol levels have fallen low enough to release the pituitary gland from negative feedback inhibition, and the FSH rise of the next cycle begins.

The hormonal regulation of the female reproductive cycle, as just described, has left open an important question: what determines the monthly cyclicity of the LH/FSH surge and the resultant ovulation? Although the concept of a primary central nervous sytem clock is inherently attractive, considerable evidence suggests that in the human it is the ovary that determines the basic rhythm.

Five observations support this view. (1) No cyclic release of LH/FSH is observed in women whose ovaries never functioned, in women whose ovaries were removed in the midst of their reproductive years, or in postmenopausal women after follicular development had ceased. (2) The follicular phase of the cycle has a variable length. The determinant of this interval seems to be the rate of maturation of the follicle through the second stage of development; that is, the ovulatory gonadotropin pulse does not occur until the dominant follicle has reached the appropriate stage of development. (3) In some otherwise normal women who are consistently not ovulating, brief administration of clomiphene citrate—an antiestrogen drug—stimulates a small rise in FSH and LH by negative feedback. About 10 to 14 days later, after sufficient time has elapsed for follicular development, a spontaneous LH/FSH surge and ovulation ensue. This suggests that the pattern of gonadal steroid secretion that characterizes a normal follicular phase conditions the hypothalamus and the pituitary gland each month to respond at the appropriate time with an acute LHRH and gonadotropin discharge. The ovarian steroids may recruit the function of a cycle center in the hypothalamus, in contradistinction to a center for tonic gonadotropin control. (4) Administration of estradiol with or without progesterone in the proper format induces a rise in LH and FSH levels that resembles the preovulatory surge (Fig. 55-19). (5) In a monkey whose pituitary gland has been severed from its hypothalamus, any central nervous system regulation of gonadotropin secretion is prevented. However, if pulsatile LHRH infusions are used to replace and sustain basal FSH and LH secretion, cyclic ovarian function with augmented gonadotropin release is observed at the proper time, even though the ongoing LHRH infusion profile is not changed. This indicates an inherent ability of the pituitary gland to produce a preovulatory surge of LH in response to proper signals from the ovary.

However, ovarian signals can be either overridden or reinforced by other influences on and from the hypothalamus. Loss of cyclic gonadotropin secretion can occur in a number of situations, which suggests that the hypothalamus is responding to a caloric, thermal, photic, olfactory, or emotional signal. Such a secretory loss occurs in women who are calorically deprived and who thus lose considerable amounts of adipose tissue and lean body mass. They would be incapable of sustaining a fetus if they became pregnant. It is also seen in women who undergo physical translocation, climatic change, or emotional deprivation or who suffer from chronic diseases.

Such inhibitory influences may be mediated by hypothalamic endorphins or dopamine or by changes in the levels of adrenal androgen, cortisol, or thyroid hormone. Conversely, it has been observed that women living in physical proximity can adopt a common timing of their menstrual cycles, possibly because of pheromones. These are chemical signals emitted by one individual that affect another. In the human there is no evidence that ovulation is stimulated by sexual behavior. However, some, but not all, reports suggest that sexual activity may increase in women around the time of ovulation, possibly due to increased levels of ovarian androgens.

Extraovarian Actions of Gonadal Steroids

The cyclic changes in estradiol and progesterone secretion produce effects on the uterus, fallopian tubes, vagina, and breasts. These coordinate precisely with the expectation of conception and the institution of a pregnancy.

Uterus. The function of the uterus is to house and nurture the developing fetus and then evacuate him or her safely at the appropriate time. It is a muscular organ, enclosing a cavity that is lined with a special mucous membrane called the *endometrium*. At the beginning of the follicular phase in each menstrual cycle the uterus is in the process of shedding its lining and is therefore incapable of receiving a conceptus. The endometrium is thin, and its glands are sparse, straight, and narrow lumened and exhibit few mitoses (Fig. 55-20). After the menstrual slough has ceased, the increase in estradiol secretion during the follicular phase produces a threefold to fivefold increase in endometrial thickness. Mitoses appear in the glands and stroma, the glands become tortuous and the spiral arteries that supply the endometrium elongate. This is termed the *proliferative phase* of the endometrium. The mucus elaborated by the cervix also changes dramatically during this phase from a scant, thick, viscous material to a copious, more watery, but more elastic substance that can be stretched into a long, fine thread. It also produces a characteristic fernlike pattern when dried on a glass slide. In this estrogen-stimulated condition the cervical mucus creates a myriad of channels in the opening of the cervix that facilitate the entrance of the sperm and direct their motion forward into the uterine cavity.

Shortly after ovulation the rise in the plasma progesterone level produces marked alterations (Fig. 55-20). Rapid proliferation of the endometrium is slowed, and mitotic activity is reduced. The uterine glands become much more tortuous, and they begin to accumulate glycogen in large vacuoles at the base of each cell. As the luteal phase of the cycle progresses, the vacuoles move toward the lumen, and the glands greatly increase their secretions. The stroma of the endometrium becomes edematous, the originally straight spiral arteries elongate further and become coiled. This is termed the *secretory phase* of the endometrium. These changes all enhance its ability to support an implanted conceptus. At the same time progesterone decreases the quantity of the cervical mucus, causes it to return to its original thick, nonelastic state, and inhibits the ferning pattern.

If pregnancy does not occur and the corpus luteum regresses, the abrupt loss of estradiol and progesterone causes spasmic contractions of the spiral arteries, probably mediated locally by prostaglandins. The resultant ischemia produces necrosis, the stroma condenses and degenerates, and the superficial endometrial cells are sloughed along with sludged blood. This comprises the *menstrual period*.

Fallopian tubes. The fallopian tubes are the normal site of fertilization These bilateral structures are 10 cm long and emerge from the uterus. Each tube ends in fingerlike projections called *fimbriae*, which lie close to the ipsilateral ovary. The fallopian tube consists of a muscular layer surrounding a mucosa lined by an epithelium that contains both ciliated and secretory cells. The cilia beat toward the uterus. During the follicular phase estradiol causes an increase in the number of cilia and in their rate of beating, as well as in the number of actively secreting epithelial cells. These effects are maximal at the time of ovulation. In addition, the fimbria become more vascularized. As ovulation approaches, tubal contractions increase, and the fimbria undulate so as to draw the shed ovum into the tube. During the luteal phase, progesterone probably maximizes the ciliary beat, enhancing movement of any fertilized ovum toward the uterus. Progesterone also increases the secretion of materials nutritious to the

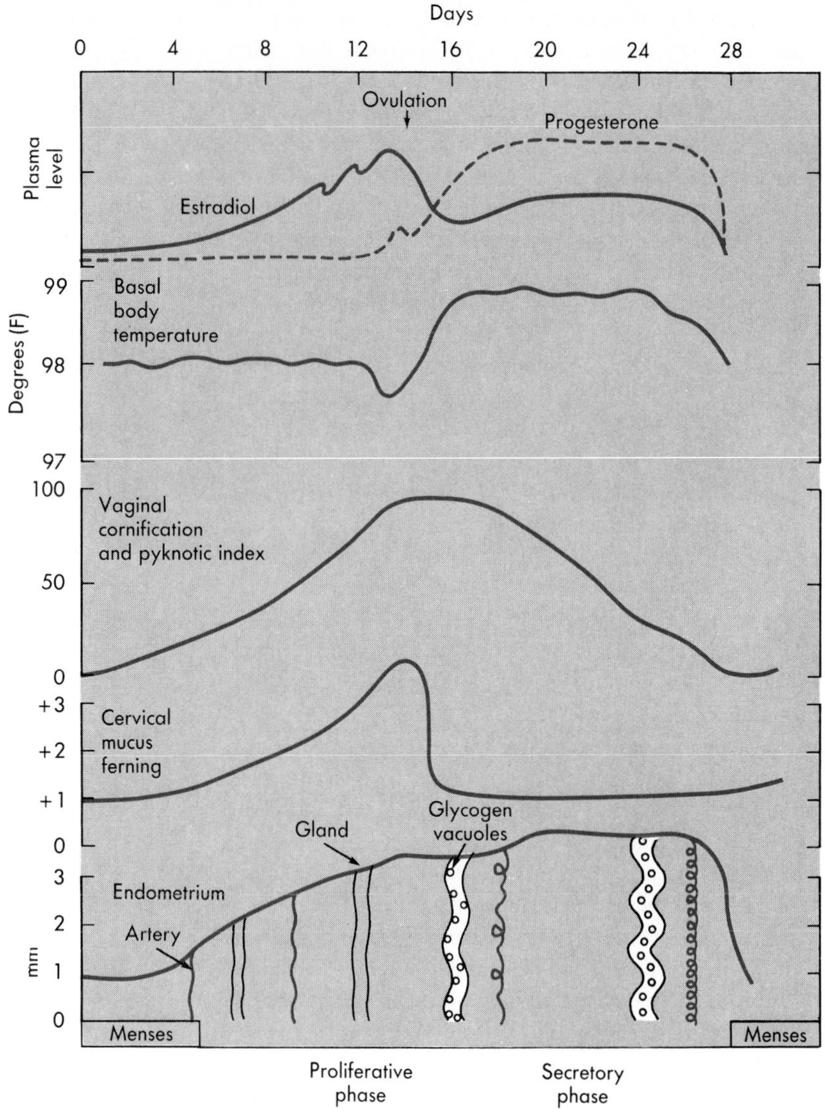

■ **Fig. 55-20.** Correlation of biological changes throughout the menstrual cycle with the profiles of plasma estradiol and progesterone levels. (Redrawn by permission from Odell, W.D.: The reproductive system in women. In Degroot, L.J., et al., editors: Endocrinology, vol. 3, New York, 1979, Grune & Stratton, Inc.)

ovum, to any incoming sperm, and to the zygote, should fertilization occur. A well-studied example of such a progesterone effect is the secretion of ovalbumin by the oviduct of the hen.

Vagina. The vaginal canal is lined with a stratified squamous epithelium that is highly sensitive to estradiol. In the absence of the hormone there is only a thin layer of basal and parabasal cells. For the first few days of the follicular phase of the menstrual cycle the epithelium is relatively thin, and smears taken from the surface show cells with vesicular nuclei from the intermediate layer.

As the cycle progresses to the ovulatory phase, more layers of epithelium are added, and the maturing cells accumulate glycogen. Vaginal smears at this point show many large, eosinophilic staining, cornified cells with small pyknotic or absent nuclei. The percentage of these cells on a vaginal smear is a sensitive index of estrogenic activity (Fig. 55-20). Progesterone, on the other hand, reduces the percentage of cornified cells. Vaginal secretions are increased by estradiol, and they too form an important element in the events leading to fertilization.

Breasts. The mammary glands consist of a large series of lobular ducts lined by an epithelium that is capable of secreting milk. These ducts empty into larger milk-conveying ducts that converge at the nipple. These glandular structures are embedded in supporting adipose tissue, and the breasts are separated into lobules by connective tissue.

The development of adult-sized mammary glands is absolutely dependent on estrogens. Before puberty the breasts grow only in proportion to the rest of the body. After the onset of increased estrogen secretion that occurs with pubescence, the growth of the lobular ducts is accelerated, and the area around the nipple (the areola) enlarges. Estrogens also selectively increase the adipose tissue of the breast, giving it its distinctive female shape. The lobular ducts are capable of outpouching to form numerous secretory alveoli. This process is stimulated by progesterone. During the menstrual cycle proliferation of the lobules occurs primarily in parallel with estradiol levels but possibly is further fortified by progesterone. This causes swelling of the breasts; however, by the end of the luteal phase breast size and tenderness diminish.

Other tissues. During puberty estradiol is to the female what testosterone is to the male. Estradiol causes almost all the changes that result in the normal adult female phenotype. In addition to stimulating growth of the internal reproductive organs and breasts, estrogens cause pubertal enlargement of the labia majora and labia minora. Linear growth is accelerated by estradiol. However, because the epiphyseal growth centers are more sensitive to estradiol than to testosterone, they close sooner. For this reason the average height of women is less than that of men. The hips enlarge, and the pelvic inlet widens, facilitating future pregnancy. The predominance of estradiol over testosterone as a gonadal steroid in women is responsible for the fact that their total body adipose mass is twice as large as that of men, whereas their muscle and bone mass is only two thirds that of men. The specific deposition of fat about the hips is another effect of estradiol.

The adult skeleton, the kidney, and the liver are also target tissues of estrogens. Estrogen inhibits bone resorption; the loss of this hormone action after menopause contributes to a declining bone mass (osteoporosis) and a resultant increased frequency of fractures. Reabsorption of sodium from the renal tubules is stimulated by estradiol, and this may contribute to the cyclic fluid retention noted by some women. The hepatic synthesis of a number of circulating proteins is increased, including thyroxine-binding globulin, cortisol-binding globulin, sex steroid–binding globulin, the renin substrate, angiotensinogen, and very–low density lipoproteins. Estrogens also elevate plasma glucose in susceptible patients.

Only a few systemic actions of progesterone are known. Body temperature is increased by this steroid, accounting for the 0.5° C rise that occurs shortly after ovulation (Fig. 55-20). Central nervous system actions include an increase in appetite, a tendency to somnolence, and a heightened sensitivity of the respiratory center to stimulation by carbon dioxide. Because progesterone is an aldosterone antagonist, it can induce natriuresis.

Mechanism of Action of Gonadal Steroids

Estradiol, estrone, other estrogens, and progesterone all enter cells freely and bind to cytoplasmic receptors. The estrogen receptor is a single unit of 65,000 molecular weight, whereas the progesterone receptor is composed of two subunits totaling 200,000 in molecular weight. No further intracellular metabolism of estrogens or progesterone is required for their actions. However, the hormone receptor complex must undergo a conformational change, possibly involving phosphorylation, which is essential for activity. After this, the complex enters the nucleus, where it initiates transcriptional changes that increase or decrease synthesis of specific proteins. The mechanism for inducing or suppressing specific messenger RNA molecules is not yet completely elucidated. Current data suggest that subunit B of the progesterone receptor complex combines with an acceptor protein associated with a target DNA molecule. Subunit A of the progesterone receptor then associates with the target DNA molecule in a manner that may permit DNA-dependent RNA polymerase to transcribe a particular sequence of nucleotides into a precursor RNA molecule. Processing of the latter by cleavage of its introns and splicing of the exons yields the specific message. Either absence of the acceptor protein or its masking by still another protein may account for lack of transcription from nontarget DNA molecules.

Spare receptors generally are not present, so the sensitivity and the maximum responsiveness of various tissues to gonadal steroids are proportional to receptor concentrations and binding affinities. An important clinical exception is found in a group of synthetic steroids that have a high affinity for estrogen receptors but which act as antiestrogen agents by blocking the access of estradiol and estrone to their receptors. Although the antiestrogen-receptor complex is translocated to the nucleus, its conformation must be abnormal because it maintains prolonged association with the nuclear acceptor site without initating any biological action. Two important examples are tamoxifen, a compound widely used in the treatment of estrogen-sensitive breast cancer, and clomiphene, an agent used to stimulate LH and FSH release. The latter action results from negative feedback because clomiphene blockade of the estrogen receptor in hypothalamic cells mimics deficiency of the hormone.

Receptors are important sites for estrogen and progesterone interactions. Estrogen increases the number of progesterone receptors, thus priming target tissues (for example, the uterus) for sequential estrogen and progesterone effects. In contrast, progesterone can decrease the number of estrogen receptors, accounting in part for its antiestrogen actions.

Metabolism of Gonadal Steroids

Estradiol and estrone bind to SSBG, but their affinities are much lower than that of testosterone. Therefore the estrogens circulate bound loosely to albumin, and they have relatively high metabolic clearance rates (Table 55-5). In menstruating women most of the circulating estradiol is derived from ovarian secretion; a minor fraction is formed from testosterone in adipose tissue,

Fig. 55-21. Metabolism of estrogens.

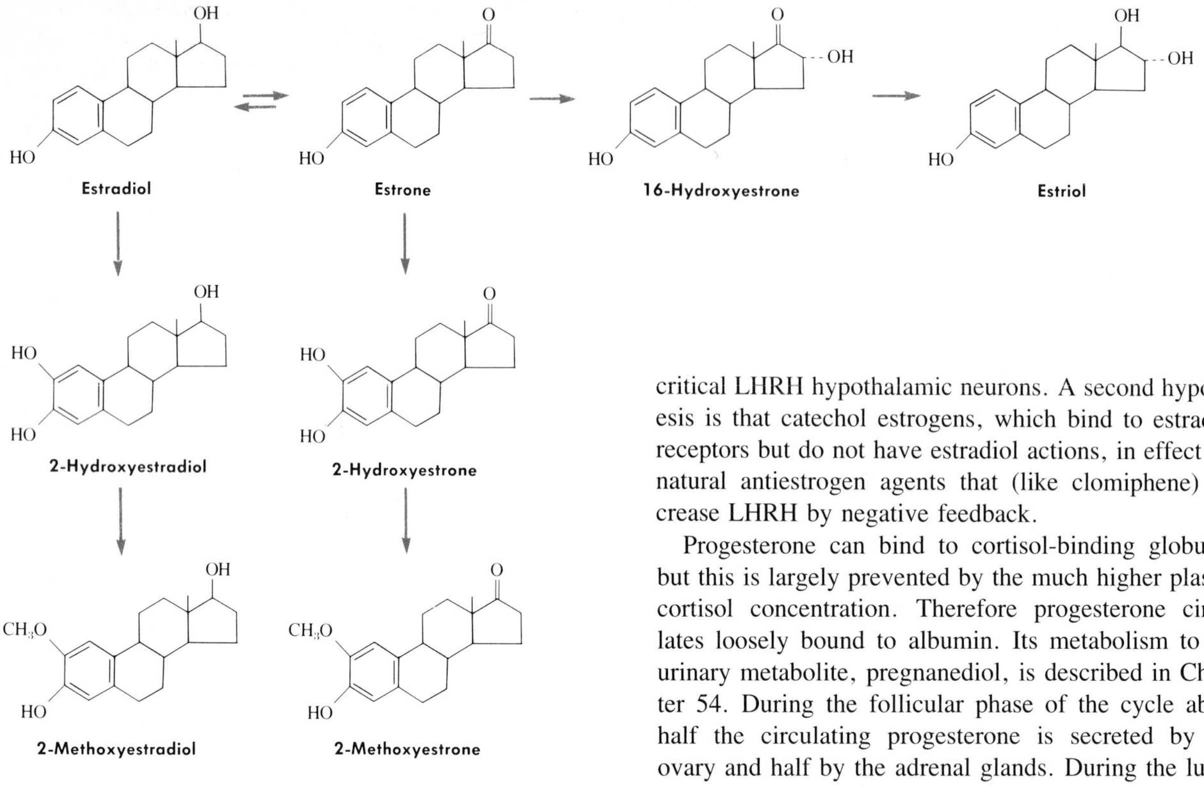

liver, and other sites. Most of the circulating estrone is derived from estradiol by peripheral 17-hydroxysteroid dehydrogenases. In postmenopausal women estrone is the dominant circulating estrogen, and it is formed from adrenal and theca cell androgens. Estrone is further metabolized to estriol, a compound with some estrogenic activity (Fig. 55-21).

Sulfated and glucuronidated derivatives of all three estrogens are excreted in the urine. Total daily urinary estrogen excretion reflects ovarian function. Values range from 20μg during the early follicular phase to 65μg at the preovulatory peak. An additional pathway of estrogen metabolism involves 2-hydroxylation and produces the so-called catechol estrogens (Fig. 55-21). These compounds resemble the catecholamine neurotransmitters norepinephrine and dopamine in their hydroxylated benzene rings. Because 2-hydroxylase activity is present in the hypothalamus, it has been suggested that catechol estrogens generated within the brain might modulate estradiol effects on LHRH release. Two separate hypotheses have been proposed. The first is that catechol estrogens competitively inhibit catecholamine methyltransferase, an enzyme that inactivates norepinephrine by methylation (see Chapter 54). Such inhibition could raise norepinephrine to stimulating levels in critical LHRH hypothalamic neurons. A second hypothesis is that catechol estrogens, which bind to estradiol receptors but do not have estradiol actions, in effect are natural antiestrogen agents that (like clomiphene) increase LHRH by negative feedback.

Progesterone can bind to cortisol-binding globulin, but this is largely prevented by the much higher plasma cortisol concentration. Therefore progesterone circulates loosely bound to albumin. Its metabolism to the urinary metabolite, pregnanediol, is described in Chapter 54. During the follicular phase of the cycle about half the circulating progesterone is secreted by the ovary and half by the adrenal glands. During the luteal phase, however, the vast majority originates in the ovary.

In women 70% to 80% of circulating testosterone is derived from peripheral conversion of DHEA and androstenedione. About half the daily production stems from adrenal activity and half from ovarian precursors. Therefore relatively little testosterone or dihydrotestosterone is normally secreted by the ovary. In pathological situations, however, ovarian cells can secrete amounts of these potent androgens that approach those of normal men and cause virilization.

Female Puberty

The possible initiating factors of puberty, the development of pulsatile gonadotropin secretion, and the transient phenomenon of nocturnal gonadotropin peaks are all similar in females to those already discussed in males. Reproductive function begins after an increase in gonadotropin secretion from the low levels of childhood (Fig. 55-22). Females differ from males in more clearly demonstrating an earlier rise in FSH than in LH (compare Figs. 55-11 and 55-22). Budding of the breasts is the first observable physical sign of puberty and coincides with the first detectable increase in plasma estradiol, as ovarian secretion commences. The onset of menses occurs approximately 2 years later, after LH levels have risen more sharply.

Because the positive feedback effect of estradiol on

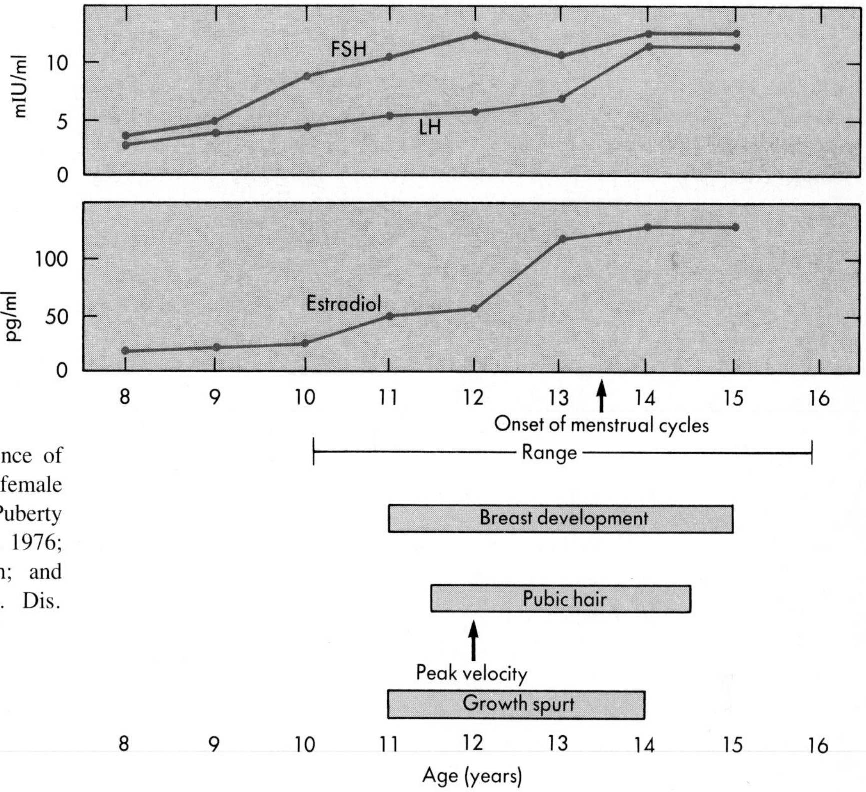

■ Fig. 55-22. Average chronological sequence of hormonal and biological events in normal female puberty. (Redrawn from Lee, P.A., et al.: Puberty in girls, J. Clin. Endocrinol. Metab. **43:**775, 1976; copyright 1976; reproduced by permission; and Marshall, W.A., and Tanner J.M.: Arch. Dis. Child. **45:**13, 1970.)

gonadotropin secretion is the last step in the maturation of the hypothalamic pituitary ovarian unit, ovulation usually does not occur in the first few cycles. Initial irregularity of menstrual cycles is therefore common, as the menstrual bleeding is induced by withdrawal of estrogen from graafian follicles undergoing atresia.

The growth spurt and the peak velocity of growth are characteristically earlier in girls than in boys. Further increase in height usually ceases 1 to 2 years after the onset of menses. The development of pubic hair precedes menses, and it correlates best with rising levels of adrenal androgens, especially DHEA-S. All stages of female puberty have wide ranges that may be influenced by factors of race, climate, individual heredity, and degree of adiposity.

■ *Sexual Functioning*

The desire for sexual activity is increased by androgens, but there is little evidence that the normal variation in plasma testosterone or dihydrotestosterone is of physiological significance in this regard. During sexual intercourse vascular erectile tissue beneath the clitoris is activated by parasympathetic impulses. This causes the introitus to be tightened around the penis. Simultaneously these impulses stimulate copious secretion of mucus by glands located beneath the labia minora and in the vagina. The secretions lubricate the vagina and help it produce a massaging effect on the penis.

Female orgasm results from spinal cord reflexes similar to those involved in male ejaculation. Involuntary contractions of the skeletal muscle of the perineum, of the musculature of the vagina, uterus, and tubes, and of the rectal sphincter occur. The clitoris retracts against the symphysis pubis. After orgasm the cervix remains widely patent for 20 to 30 minutes, permitting entrance of sperm into the uterus. The first wave of sperm may reach an ovum in the fallopian tube within 10 minutes. However, orgasm is not required for this rapid internal propulsion.

Many spermatozoa are trapped and eventually destroyed in the vagina within a few hours. The remainder reach the cervix, where they dwell in storage sites formed by the convoluted mucosa (cervical crypts) and its mucus. From this reservoir, spermatozoa migrate into the uterine cavity and fallopian tubes over 24 to 48 hours, undergoing tremendous losses along the way. Fewer than one in every 100,000 eventually reach an ovum.

■ *Menopause*

The reproductive capacity of women begins to wane in the fifth decade of life, and menses terminate at an average age of 50. For several years before menopause, the frequency of ovulation decreases. The menses occur at variable intervals and with decreased flow, caused by irregular peaks of estradiol without adequate secretion of progesterone in the luteal phase. With the disappearance of virtually all follicles, ovarian secretion of estrogens declines to low levels, eventually to cease completely. From then on, maintenance of the lowered plasma estradiol concentrations (characteristic of menopause) depends on peripheral conversion of androgen precursors secreted by ovarian stromal cells and by the adrenal glands. Furthermore, the dominant estrogen becomes estrone rather than estradiol (estrone 3 to 7 ng/dl versus estradiol 1 to 2 ng/dl; compare with Table 55-5).

During the last few years of reproductive life follicular sensitivity to gonadotropin stimulation diminishes, and plasma FSH and LH gradually increase in compensation. Once menopause occurs in response to the loss of negative feedback from estradiol and inhibin, gonadotropin levels average 4 to 10 times those of the normal follicular phase, and the LH/FSH ratio falls to less than 1. Although the cyclicity of gonadotropin secretion is lost, pulsatile secretion persists.

The programmed decline in available estrogenic biological activity causes thinning of the vaginal epithelium and loss of its secretions, a decrease in breast mass, and an accelerated loss of bone. Phenomena such as vascular flushing, emotional lability, and an increase in the incidence of coronary vascular disease also are related to estrogen deficiency. Because adipose tissue contains aromatase activity, it is an important site of production of estrogen from stromal and adrenal androgens. Obese women may therefore suffer somewhat less from estrogen deprivation. Hormone production does not stop completely in the postmenopausal ovary. The stromal interstitial cells continue to secrete androstenedione and testosterone. In some women proliferation of these cells leads to enough increase in plasma androgens that mild stimulation of hair growth in a masculine pattern can occur.

Pregnancy

Fertilization and Implantation

After ovulation and capture by the widened proximal end of the fallopian tube (ampulla), the ovum usually must encounter sperm in 12 to 24 hours for fertilization to occur. The sperm, in turn, probably must reach the ovum within 48 hours of deposit in the vagina. Contact between the sperm and ovum is thought to be on a random basis. The granulosa cells of the cumulus surrounding the ovum are dispersed by the action of hyaluronidase and a corona-dispersing enzyme, which are contained in the acrosomal cap of the sperm.

The zona pellucida of the ovum (Fig. 55-14) contains species-specific receptors for sperm. The fertilizing sperm penetrates this acellular barrier by releasing *acrosin,* a proteolytic enzyme. Penetration of the zona by the first sperm blocks other sperm by releasing materials contained in granules within the ovum. This important step prevents polyploidy, the production of an organism with more than two sets of homologous chromosomes. The polar body resulting from the second reduction-division of meiosis is then released, leaving the ovum in a haploid state, i.e., with 23 chromosomes. After fusion of their respective membranes, the chromatin material of the sperm head is engulfed by the ovum and forms the haploid male pronucleus. The two pronuclei generate a spindle on which the chromosomes are arranged, and a new diploid individual with 46 chromosomes is created.

The zygote, now in the blastocyst stage, traverses the tube in about 3 days. After another 2 to 3 days in the uterus the zygote initiates implantation. The requisite dissolution of the zona pellucida is brought about by alternate contraction and expansion of the blastocyst as well as by lytic substances in the uterine secretions. Although their nature is unknown, these and other factors necessary for implantation depend on adequate luteal phase progesterone levels.

From the initial solid mass of cells, a layer of *trophoblasts* separates. Microvilli of these cells interdigitate with those of endometrial cells, and junctional complexes form between the respective cell membranes. Once firmly attached, trophoblast cells intrude between and burrow beneath endometrial cells, lysing the intercellular matrix with a variety of enzymes. In addition, the trophoblasts phagocytize and digest dead endometrial cells. Prostaglandins, probably of endometrial cell origin, and histamine may also participate in the implantation process.

Penetration by the trophoblasts is limited by concurrent changes in the stroma of the uterus. Late in the normal luteal phase, fibroblast-type stromal cells near uterine blood vessels enlarge and accumulate glycogen and lipid. These *decidual* cells disappear unless pregnancy supervenes and the corpus luteum is maintained. In the latter case, however, continuing estrogen and especially progesterone stimulate widespread decidualization, rapidly changing the entire stroma into a sheet of compact decidual cells. At the same time, the endometrial glands progressively atrophy. The decidua functions initially as a source of essential nutrients for the embryo until the central circulation has been established. Thereafter, the decidua may provide a mechanical and an immunological barrier to further invasion of the uterine wall. Finally, the decidua also functions as an endocrine organ, secreting prolactin, relaxin, and prostaglandins.

Pregnancy is marked by the development of a unique organ with a limited life span, the *placenta.* In addition to its vital nutritive role, the placenta functions as an extraordinarily versatile endocrine gland, capable of synthesizing and secreting a wide variety of protein and

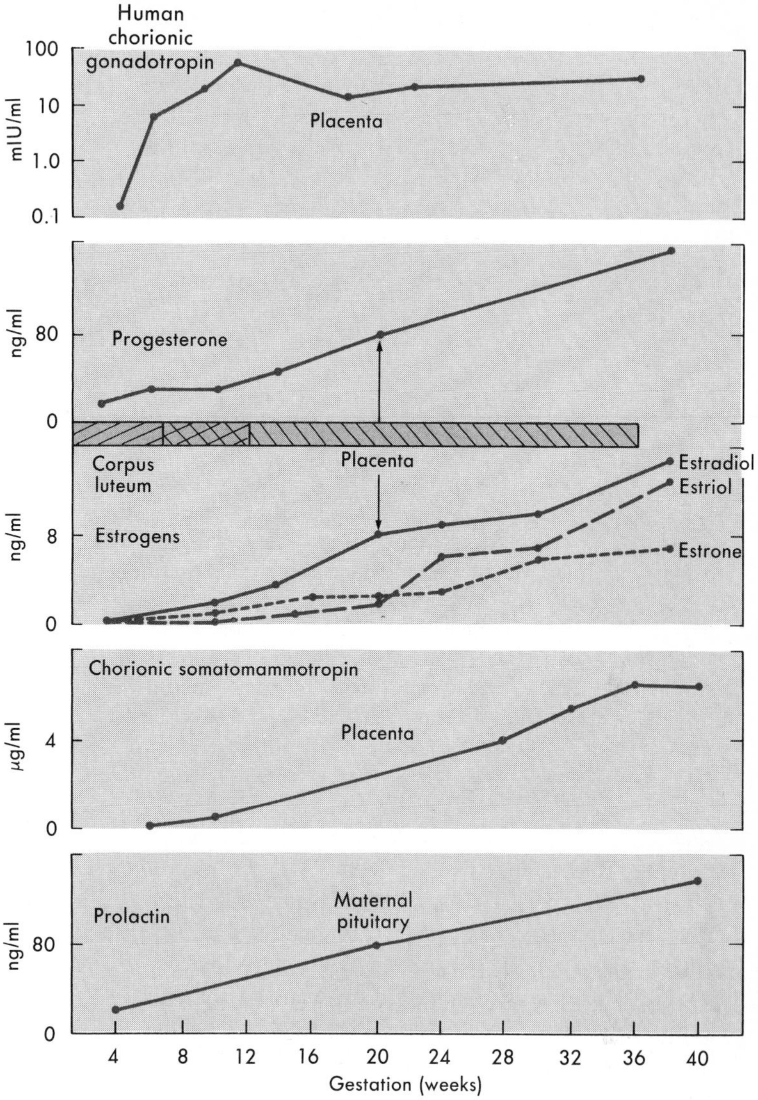

■ **Fig. 55-23.** Profile of plasma hormone changes during normal human pregnancy. Note logarithmic scale for HCG. Also note shift from corpus luteum to placenta as the source of estrogens and progesterone between 6 and 12 weeks of gestation. (Redrawn from Goldstein, D.P., et al.: Am. J. Obstet. Gynecol. **102:**110, 1968; Rigg, L.A., et al.: Am. J. Obstet. Gynecol. **129:**454, 1977; Selenkow, H.A., et al.: Measurement and pathophysiologic significance of human placental lactogen. In Pecile, A., and Finzi, C.: The foetoplacental unit, Amsterdam, 1969, Excerpta Medica; and Tulchinsky, D., et al.: Am. J. Obstet. Gynecol. **112:**1095, 1972.)

steroid hormones. These hormones affect both maternal and fetal metabolism; they exhibit regular concentration profiles in maternal plasma and also can be found in fetal plasma and amniotic fluid. Fig. 55-23 summarizes the temporal pattern of gestational hormone concentrations in maternal plasma.

■ *Human Chorionic Gonadotropin*

Human chorionic gonadotropin (HCG) is the first key hormone of pregnancy. Secreted by the syntiocytotrophoblast cells, it can be detected in maternal plasma and urine within 9 days of conception. The secretion of HCG may be stimulated by LHRH produced in adjacent cytotophoblasts. HCG is a glycoprotein of 39,000 molecular weight with two subunits. The α subunuit is identical to that of thyroid-stimulating hormone (TSH), LH, and FSH, whereas the β subunit has 80% homology with that of LH. However, specific radioimmunoassays, developed to measure the β subunit of HCG permit reliable discrimination of it from LH. Detection of HCG is the most commonly employed and most specific test for pregnancy. Maternal plasma levels of HCG increase at an exponential rate, reach a peak at 9 to 12 weeks, and then decline to a stable plateau for the remainder of pregnancy (Fig. 55-23). After delivery, HCG disappears from maternal plasma with a half life of 12 to 24 hours.

HCG acts to maintain the function of the corpus luteum beyond its usual life span of 14 days. The placental gonadotropin stimulates ovarian secretion of progesterone and estrogens by mechanisms essentially identical to those previously described for LH. When the placenta assumes the synthesis of these steroids, thereby relieving the fetus of its dependence on the corpus luteum, HCG secretion declines. The role of the gonadotropin during the remainder of pregnancy is unclear, but other actions are attributed to it. HCG has an inhibitory effect on maternal pituitary LH secretion. Be-

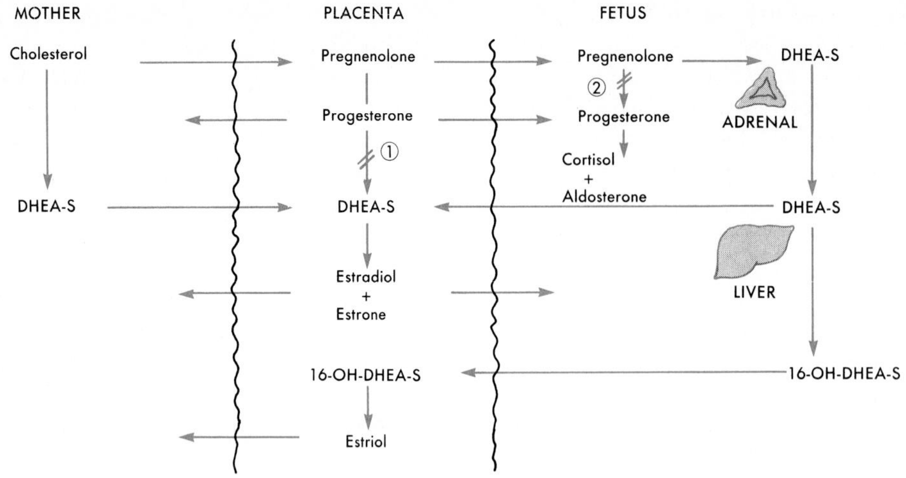

Fig. 55-24. The maternal-fetal-placental unit in steroid hormone synthesis. Progesterone in the maternal and fetal circulations is synthesized in the placenta from maternal cholesterol. Cortisol and aldosterone in the fetus are synthesized from progesterone derived from the placenta. Maternal and fetal estradiol and estrone are synthesized in the placenta from DHEA-S derived from both the mother and the fetus. Estriol in the maternal circulation is synthesized in the placenta from 16-OH-DHEA-S. This precursor must be provided by the combined action of the fetal adrenal gland and liver on pregnenolone supplied by the placenta. *DHEA-S,* Dehydroepiandrosterone sulfate; *16-OH-DHEA-S,* 16 α-hydroxydehydroepiandrosterone sulfate; *1,* 17-hydroxylase; 17,20 desmolase; *2,* 3 β-OH-dehydrogenase; $\Delta^{4,5}$ isomerase.

cause of its structural overlap with TSH, the plasma concentrations of HCG in normal pregnancy may stimulate an increase in maternal thyroid gland activity. In pathological excess HCG can even induce hyperthyroidism. HCG also may stimulate the production of relaxin (as discussed later). Finally, HCG that reaches the fetus may stimulate DHEA-S production by the fetal zone of the adrenal gland and testosterone production by the Leydig cells of the testis.

Progesterone

Progesterone is the hormone most directly responsible for the establishment and sustenance of the fetus in the uterine cavity. As just noted, during the first 2 weeks of pregnancy, progesterone stimulates the tubal and endometrial glands to secrete nutrients on which the free-floating zygote depends. Thereafter, it maintains the decidual lining of the uterus. Progesterone produced by the placenta is the principal substrate for synthesis of cortisol and aldosterone by the fetal adrenal gland (Fig. 55-24). Lacking the 3β-OL dehydrogenase-$\delta^{4,5}$ isomerase enzyme complex (Fig. 54-3), the fetal adrenal gland cannot itself make progesterone. Finally, progesterone may participate in inhibiting maternal immune responses to antigens from the fetus, preventing its rejection.

Two other actions of progesterone are important during pregnancy. The steroid inhibits uterine contractions, and hence prevents premature expulsion of the fetus. It also stimulates the development of the alveolar pouches of the mammary glands, greatly magnifying their eventual capacity to secrete milk.

The placenta begins to synthesize progesterone at about 6 weeks, and by 12 weeks it is producing enough to replace the corpus luteum for this purpose. During this transition period the otherwise progressive rise in plasma progesterone reaches a temporary plateau (Fig. 55-23). Cholesterol, present in low-density lipoproteins, is extracted from maternal plasma and serves as the major precursor for placental progesterone. The synthetic pathway is like that of the adrenal gland and the ovary. By term, progesterone production reaches a level of 250 mg/day, which is tenfold greater than the peak rates seen during the luteal phase of the menstrual cycle. Urinary pregnanediol excretion also rises markedly, reflecting this huge increase in production.

Estrogens

Augmented production of estrogens (estradiol, estrone, and estriol) also occurs throughout pregnancy, resulting in several important actions. Estrogens stimulate (1) the continuous growth of the uterine myometrium, preparing it for its role in labor, (2) the further growth of the ductal system of the breast, out of which the alveoli will develop, and (3) the enlargement of the external genitalia. In addition, estrogen, along with relaxin, causes relaxation and softening of the pelvic ligaments and the symphysis pubis of the pelvic bones, allowing better accommodation of the expanding uterus.

Like progesterone, estrogens are initially produced by the corpus luteum under stimulation by HCG. The placenta then assumes this role, but requires steroid hormone precursors from both the maternal and fetal compartments to complete the synthesis of estrogens. This unique example of coordinated maternal-placental-fetal function is depicted in Fig. 55-24. The placenta lacks significant 17-hydroxylase and 17-20 desmolase activity and therefore cannot generate the androgens that serve as substrates for aromatization (Fig. 55-4). Instead, the placenta extracts DHEA-S, derived from the maternal and fetal adrenal glands, removes the sulfate and synthesizes estradiol and estrone. The fetal source of DHEA-S predominates as pregnancy progresses. Placental synthesis of estriol, a 16-hydroxylated estrogen, almost entirely depends on precursors from the fetus (Fig. 55-24). The fetal adrenal gland synthesizes DHEA-S from placental pregnenolone, the fetal liver hydroxylates it in the 16 position, and the placenta then desulfates it and aromatizes it to estriol.

One third of the unconjugated estrogens circulating in maternal plasma at term is accounted for by estriol (Fig. 55-23). In the form of sulfate and glucuronic acid conjugates, estriol represents 90% of the total estrogen excreted in maternal urine. Daily urinary estriol levels correlate quite well with placental weight. Because estriol is derived almost entirely from the fetal placental unit, measurement of this estrogen in maternal plasma or urine provides a valuable clinical index to the state of the fetus. A rapid decrease in urinary estriol suggests fetal distress or placental insufficiency. In conjunction with the results of other monitoring procedures, low urinary estriol levels may dictate active intervention and prompt delivery of a fetus at risk.

Human Chorionic Somatomammotropin

Another protein hormone, unique to pregnancy, is *human chorionic somatomammotropin* (HCS), also called human placental lactogen (HPL). The synthesis of HCS by the placental trophoblasts can be detected at about 4 weeks of gestation. The maternal plasma concentration rises steadily to a peak of 6 μg/ml at term (Fig. 55-23). The HCS production rate of 1 to 2 g/day far exceeds that of any other human protein hormone. HCS concentrations correlate well with placental weight. Therefore measurement of maternal plasma HCS provides another indicator of placental function, and it has been used to monitor threatened pregnancies. After delivery the hormone rapidly disappears from maternal plasma with a half-life of 20 minutes.

HCS has a very close structural similarity to human growth hormone (HGH) and prolactin. In several in vitro and in vivo test systems HCS exhibits anabolic nitrogen-retaining activity like that of HGH, although at only 1% of the potency. HCS also has lactogenic activity similar to that of prolactin, and it has some luteotrophic activity as well. Most important, HCS stimulates lipolysis and, like HGH, exerts antagonism to insulin actions on carbohydrate metabolism; this tends to raise plasma glucose. Plasma HCS is slightly increased by fasting or by insulin-induced hypoglycemia, suggesting that placental secretion may be under feedback regulation by glucose. As is detailed later, HCS appears to direct maternal metabolism to maintain a continuous flow of substrates, especially glucose, to the fetus. HCS levels in fetal plasma are far below those in maternal plasma, and there is no known direct role for HCS in the fetus. However, somatomedins (IGF-2), produced in the placenta as the result of HCS action, may help to stimulate fetal growth.

Prolactin

Another hormone secreted in excess during normal pregnancy is maternal pituitary prolactin. Plasma levels rise linearly, by term reaching values eightfold to tenfold higher than those of nonpregnant women (Fig. 55-23). Prolactin is essential for expression of the mammotropic effects of estrogen and progesterone. More specifically, prolactin stimulates the lactogenic apparatus (Chapter 52). Limited formation of milk begins at about 5 months of gestation, but significant lactation is inhibited by the great excess of estrogen and progesterone. Lactation is initiated after delivery by the precipitous drop in steroid hormones, and it is maintained thereafter in the nursing mother by prolactin through the stimulus of suckling. Although basal prolactin concentrations gradually decline over the next 4 to 8 weeks, they are acutely elevated during each period of suckling (Chapter 52), helping to sustain milk secretion.

Prolactin also suppresses reproductive function in the nursing mother. During the first 7 to 10 days postpartum, plasma FSH and LH levels remain low. FSH then rises to above-normal follicular phase levels, but LH still remains low. The responsiveness of the ovaries to FSH probably is reduced by prolactin, and LH secretion by the pituitary probably is inhibited by prolactin.

A decrease in circulating prolactin, because of either cessation of nursing or administration of a dopaminergic agonist, will trigger LH release and initiate cycling. It should also be noted that prolactin is synthesized by the decidual cells of the pregnant uterus under stimulation by progesterone. This is the source of prolactin in the amniotic fluid; the function of this prolactin is currently unknown.

Relaxin

In addition to gonadal steroids, the corpus luteum of pregnancy secretes a polypeptide hormone called *relaxin*. Its structure is insulin-like. Plasma levels of relaxin rise early, peak in the first trimester, and then decline somewhat. The production of relaxin by the corpus luteum appears to be stimulated by HCG. Relaxin may also be produced by uterine tissue and by the placenta. In addition to its relaxing effect on pelvic bones and ligaments, relaxin inhibits myometrial contractions and softens the cervix. Thus it may function to ensure uterine quiescence and prevent early abortion of the pregnancy.

Other Hormones

Other hormones, including ACTH, CRH, TRH, LHRH, inhibin, somatostatin, β-endorphins, and a thyrotropin-like substance, have been detected in placental extracts. These hormones may actually also be synthesized by trophoblastic cells. It also remains to be established whether they are secreted into maternal or fetal plasma and whether they function as local hormones within the placenta itself or act on distant maternal or fetal target cells.

Other Maternal Hormonal Changes

The pregnant state induces a characteristic series of changes in a number of other maternal hormones. One of the most significant alterations is in pancreatic islet β-cell function. Insulin secretion, in response to glucose challenge or to meals, increases after the third month of pregnancy. This hypersecretion reaches its peak during the last trimester, coinciding with the peaks of placental weight and of plasma HCS. During this same period maternal sensitivity to insulin is greatly diminished. Therefore, to some extent, insulin hypersecretion may be considered compensatory. In contrast, basal glucagon levels and responses to stimulation do not change significantly, although suppressibility of the α-cells by glucose is somewhat enhanced.

Aldosterone secretion increases significantly throughout pregnancy, reaching a sixfold to eightfold elevation by term. This is because of increased function of the renin angiotensin system; plasma renin and the renin substrate (angiotensinogen) are both augmented by the high estrogen levels of pregnancy. Various reasons have been advanced for the hyperaldosteronism of pregnancy. It may be stimulated by a reduction in *effective* circulating blood volume that results from the large placental blood pool. It also could represent a response to the antialdosterone action of progesterone on the renal tubules. Hyperaldosteronism contributes to the positive sodium balance that is needed to maintain a high total maternal plasma volume and to build the extracellular fluid of the fetus. In this connection, another mineralocorticoid, desoxycorticosterone (Table 54-1), is also present in thousandfold excess in plasma during pregnancy. It is synthesized in the maternal kidneys by 21-hydroxylation of progesterone originating from the placenta.

The plasma total cortisol level is elevated because of the estrogen-induced increase in cortisol-binding globulin. However, plasma free cortisol also rises modestly, possibly due to stimulation of the maternal adrenal gland by ACTH of placental origin and/or stimulation of the maternal pituitary by CRH of placental origin. Maternal plasma levels of ACTH and CRH both increase during pregnancy. The enhanced glucocorticoid activity may contribute to maternal adipose tissue gain and to mammary gland development. It also may be responsible for the plethoric face, thin skin, and susceptibility to bruising of the pregnant woman.

Parathyroid hormone secretion is increased in response to the continuous drain on maternal calcium created by the growing fetal skeleton. The hyperparathyroidism augments plasma levels of the active vitamin D metabolite, 1,25-$(OH)_2$-D_3, which in turn increases dietary calcium absorption. The enhanced supply of calcium is especially important during the third trimester.

Total plasma thyroid hormones, thyroxine (T_4) and triiodothyronine (T_3), are elevated because of estrogen-induced increases in thyroid-binding globulin. Plasma free T_4 concentration remains within the normal range. Nonetheless there are increases during pregnancy in maternal thyroid gland size, radioactive iodine uptake, basal metabolic rate, and resting pulse rate, all of which are compatible with some augmentation of thyroid gland activity. This may be caused by the thyrotropic activity of HCG or secretion of another placental thyrotropin.

Growth hormone secretion decreases in response to various stimuli during pregnancy, possibly because the anabolic functions of maternal pituitary HGH are carried out by placental HCS. Maternal LH and FSH secretion also is suppressed by the high levels of estrogen and progesterone.

Maternal-Fetal Metabolism

During normal pregnancy the average gain in maternal weight is 11 kg. About half of this is attributable to changes in maternal tissues and half to the conceptus. The typical distribution of the excess weight is shown in Fig. 55-25. Approximately 75,000 extra calories (250 to 300 kcal/day) must be ingested to support this weight gain; 65,000 kcal support fetal metabolism and growth, and 10,000 kcal are stored in maternal fat. An extra protein intake of 30 g/day ensures adequate sup-

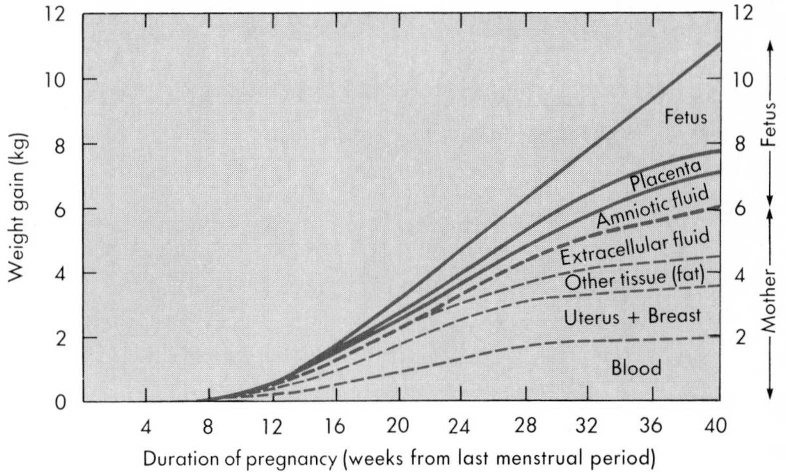

■ **Fig. 55-25.** Pattern and components of maternal weight gain during normal pregnancy. (Redrawn from Pitkin, R.M.: Obstetrics and gynecology. In Schneider, H.A., Anderson, C.E., and Coursin, D.B.: Nutritional support of medical practice, New York, 1977, Harper & Row Publishers, Inc.)

plies for maternal needs and for the accumulation of fetal protoplasm. At birth the protein content of the fetus has reached 400 to 500 g.

From the metabolic standpoint, pregnancy can be divided into two phases. For approximately the first half the mother herself is in an anabolic phase, whereas the conceptus represents an insignificant nutritional drain. During the second half, and especially the final third, fetal and placental weight increases at an accelerated rate. These demands cause the mother to shift into a state aptly described as "accelerated starvation."

The maternal anabolic phase is characterized by normal or even increased sensitivity to insulin. Progesterone and estrogen may facilitate insulin actions at this stage. Maternal plasma levels of glucose, amino acids, free fatty acids, and glycerol are normal or slightly reduced. Carbohydrate and amino acid loads are readily assimilated. Lipogenesis is favored, and lipolysis is braked in maternal adipose tissue, glycogen stores are increased in liver and muscle, and protein synthesis is enhanced. The net effect is to stimulate growth of the breasts, uterus, and essential musculature in the mother, while preparing her to withstand the metabolic demands of later fetal growth.

During the second phase of pregnancy the metabolism of the mother shifts into a mode that effectively accommodates the accelerating needs of the fetus. Insulin sensitivity is replaced by insulin resistance. The assimilaton of dietary carbohydrate, protein, and fat by maternal tissues is slowed, so that postprandial plasma levels of glucose and amino acids are elevated. This increases the rate of glucose diffusion and of facilitated amino acid transport across the placenta into the fetus. Glucose is the major fuel of the fetus, and the amino acids are required for fetal protein synthesis. By term, the fetus, who is using glucose at a rate of 5 mg/kg/minute compared with the maternal rate of 2.5 mg/kg/minute, must be supplied with up to 25 g/day. During fasting intervals maternal plasma glucose and amino acid levels fall more rapidly than in nonpregnant women, because of continued fetal siphoning of these substances. Conversely, lipolysis is accelerated, and maternal plasma free fatty acid, glycerol, and ketoacid levels rise more rapidly than in nonpregnant women. This ensures alternate oxidative fuels for the mother; in addition, ketoacids and, to a lesser extent, free fatty acids cross to the fetus, where they may be used instead of some glucose. Placental HCS is probably the key hormone responsible for insulin resistance and for facilitating lipid mobilization during fasting in the latter stage of pregnancy. The increase in plasma free cortisol also may contribute to these actions.

Along with the other changes in maternal metabolism, plasma cholesterol and triglyceride levels rise throughout pregnancy. The cholesterol is partly used for estrogen and progesterone synthesis. The increased circulating triglycerides are largely the result of an increased hepatic synthesis of very-low density lipoprotein, which is stimulated by estrogens. Some of the triglycerides are stored in the breasts in preparation for milk production. They are shifted away from less specific storage elsewhere by a marked reduction in adipose tissue lipoprotein lipase levels.

■ *Parturition*

Just as the maintenance of the pregnant state depends on a unique hormonal milieu, its termination probably also depends on specific hormonal changes. Roles for glucocorticoids, estrogen, progesterone, relaxin, oxytocin, prostaglandins, and catecholamines in the initiation and final achievement of uterine evacuation have been suggested by various lines of evidence. Because much species variation exists, it is difficult to extrapolate the results of the numerous studies in subprimates to humans. Fig. 55-26 shows current notions of the endocrine regulation of parturition.

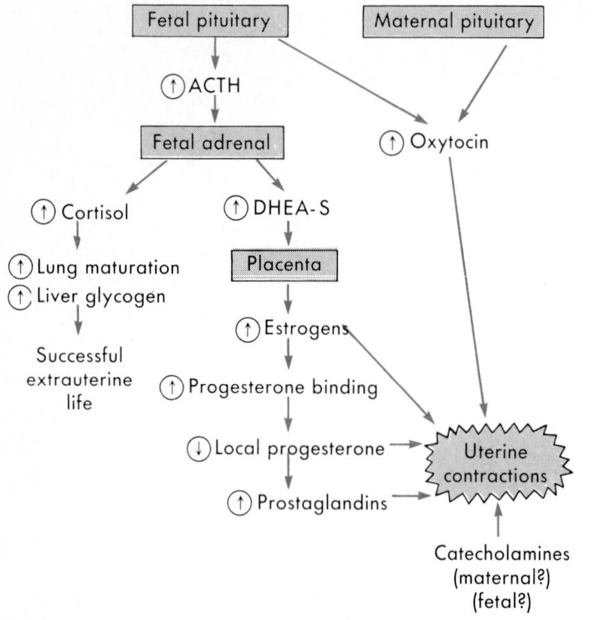

■ **Fig. 55-26.** Endocrine regulation of parturition. Near term, the fetal-pituitary-adrenal axis initiates signals that lead to an increased ratio of estrogens to progesterone in the myometrium. This, along with increased prostaglandins, causes uterine contractions. Augmenting effects of oxytocin from maternal and fetal sources as well as catecholamines probably contribute to the normal course of labor but are not essential. Oxytocin is important in sustaining uterine contractions after delivery of the conceptus so as to minimize maternal blood loss. Cortisol is needed to prepare the fetus to maintain its own supply of oxygen and glucose after birth.

Once the conceptus has reached a critical size, distension of the uterus itself and stretching of the muscle fibers increase their contractility. In the human, uncoordinated uterine contractions begin at least 1 month before the end of gestation. The inherent contractility of the uterus probably in itself would cause eventual evacuation of the conceptus. Some signal from the fetus probably initiates active labor. In sheep, a fetal adrenal product—probably cortisol—has been strongly implicated. This agrees with the fact that fetal cortisol production rises sharply during the last few weeks of gestation. In the sheep, cortisol directs placental output more toward estrogens and less toward progesterone, by increasing the activities of 17-hydroxylase, 17-20 desmolase, and aromatase (Fig. 55-4). Although gestation is prolonged in humans when the fetus lacks an intact hypothalamic pituitary adrenal unit, the evidence for a surge of fetal cortisol immediately preceding human parturition is contradictory. Irrespective of its role in initiating parturition, the late gestational increase in cortisol secretion is vital for preparing the fetus for the abrupt transition to extrauterine life by stimulating lung maturation and by increasing stores of liver glycogen (Fig. 55-26).

The relative availability of estrogens and progesterone is thought to be important in maintaining uterine quiescence throughout pregnancy. In some species a sharp drop in maternal plasma progesterone precedes parturition. Since progesterone has been shown to inhibit transmission of impulses through the myometrium, a decrease in progesterone secretion could facilitate the onset of labor. In humans plasma progesterone does not fall significantly before delivery, although estrogen levels continue to rise near the end of gestation. Some evidence suggests the existence of a placental progesterone-binding protein whose concentration may be increased by estrogen near term, causing an effective removal of local progesterone from the myometrium and thereby decreasing uterine quiescence. Progesterone may act to inhibit uterine contractions by preventing the release from lysosomes of phospholipase A_2, the rate-limiting enzyme in prostaglandin synthesis. With a decrease in progesterone the concentrations of prostaglandins would rise.

The increased levels of estrogen near term augment uterine prostaglandin synthesis. In addition, prostaglandins synthesized in the fetal membranes may reach the uterine musculature. Prostaglandins increase free intracellular calcium concentrations in the myometrium and trigger uterine contractions. A key role for prostaglandins in parturition is supported by the high levels found in maternal plasma and amniotic fluid during labor. Also, these compounds are effective for inducing abortion.

Relaxin probably synergizes with progesterone in suppressing uterine contractility throughout gestation. However, relaxin also may assist in parturition in two ways. First, it acts to soften the cervix, which is critical for permitting eventual passage of the fetus. Second, relaxin causes an increase in oxytocin receptors, which may contribute to the greatly increased sensitivity of the uterus to stimulation by oxytocin at term.

Once labor is initiated, both maternal and fetal oxytocin may help sustain uterine activity and increase the strength of uterine contractions. However, labor can proceed in the absence of this peptide. The most important role of oxytocin may be to cause maximal contraction of the uterus after it has been emptied. Such strong contractions minimize blood loss during parturition.

Both α- and β-adrenergic receptors are present in the myometrium; α-stimuli cause contraction, and β-stimuli cause relaxation. Therefore circulating catecholamines also may contribute to the final hormonal cascade of parturition. Finally, a sharp increase in maternal plasma CRH occurs early in labor and peaks at the time of delivery, possibly reflecting the mother's adaptation to the stress of parturition.

Once labor has begun, it proceeds in three clinically recognized stages. In the first stage, lasting a variable number of hours, the uterine contractions, which originate at the fundus and sweep downward, force the head of the fetus against the cervix. This progressively widens and thins the opening to the vaginal canal. In the second stage, lasting less than 1 hour, the fetus is forced out of the uterine cavity and through the cervix and is delivered from the vagina. In the third stage, lasting 10 minutes or less, the placenta is separated from the decidual tissue of the uterus and forcefully evacuated. Myometrial contractions at this point act to constrict the uterine vessels and prevent excessive bleeding. Once the placenta has been removed, all its hormonal products disappear from the maternal plasma according to their characteristic half-lives. In general, by 48 to 72 hours the steroid and protein hormone concentrations have reached nonpregnant levels.

Endocrine State of the Fetus

No maternal protein or peptide hormone effectively traverses the placenta, with the exception of TRH and possibly antidiuretic hormone. The same is essentially true for maternal thyroid hormones. Steroid hormones, in contrast, can move readily from the mother to the fetus, and vice versa. Catecholamines also cross to the fetus.

The ontogeny of each of the endocrine glands is individually described in previous sections. Fetuses depend on their own insulin for anabolism and for deposition of adipose tissue. Their own thyroid hormones are required for normal central nervous system and skeletal maturation. Linear growth of fetuses does not depend on their own growth hormone, even though fetal plasma HGH levels are very high. Somatomedins and other insulin-like growth factors also are present in fetal plasma. They may originate in the placenta, and IGF-2 in particular may be a major stimulator of fetal growth.

Fetal ACTH probably is not essential for the first 12 to 20 weeks, although later it definitely stimulates production of steroids by the fetal zone of the adrenal cortex. The newborn mounts an immediate stress response, as shown by high levels of cortisol in umbilical cord plasma. If endogenous ACTH and cortisol cannot be secreted at this time, death will ensue unless replacement therapy is provided. The roles of fetal FSH and LH, the plasma levels of which peak in midgestation, remain to be worked out. In their absence, however, development of primordial follicles into primary follicles is reduced in the fetal ovary.

Prolactin concentrations are high in fetal plasma and in amniotic fluid. Prolactin may contribute to fetal growth, and the hormone also may stimulate adrenocortical secretion of DHEA-S and cortisol. The free diffusion of calcium across the placenta negates the need for fetal parathyroid hormone secretion, but fetal calcitonin levels are relatively elevated. A role for parathormone in bone formation has been postulated. Finally, current studies hint at the existence of other placental protein hormones with specific functions in fetal development.

Bibliography

Journal articles

Belchetz, P.E., et al.: Hypophysial responses to continuous and intermittent delivery of hypothalamic gonadrotropin-releasing hormone, Science **202:**631, 1978.

Bryant-Greenwood, G.D.: Relaxin as a new hormone, Endocr. Rev. **3:**62, 1982.

Forest, M.G., et al.: Kinetics of human chorionic gonadotropin–induced steroidogenic response of the human testis. II. Plasma 17 α-hydroxyprogesterone, δ^4-androstenedione, estrone, and 17 β-estradiol: evidence for the action of human chorionic gonadotropin on intermediate enzymes implicated in steroid biosynthesis, J. Clin. Endocrinol. Metab. **49:**284, 1979.

Goebelsmann, U.: Protein and steroid hormones in pregnancy, J. Reprod. Med. **23:**166, 1979.

Harman, S.M., et al.: Reproductive hormones in aging men. I. Measurement of sex steroids, basal luteinizing hormone, and Leydig cell response to human chorionic gonadotropin, J. Clin. Endocrinol. Metab. **51:**35, 1980.

Hoff, J.D., et al.: Hormonal dynamics at midcycle: a reevaluation, J. Clin. Endocrinol. Metab. **57:**792, 1983.

Lipsett, M.B.: Physiology and pathology of the Leydig cell, N. Engl. J. Med. **303:**682, 1980.

Liu, J.H., and Yen, S.S.: Induction of midcycle gonadotropin surge by ovarian steroids in women: a critical evaluation, J. Clin. Endocrinol. Metab. **57:**797, 1983.

McCarty, K.S., Jr., et al.: Oestrogen and progesterone receptors: physiological and pathological considerations, J. Clin. Endocrinol. Metab. **12:**133, 1983.

McNatty, K.P., et al.: The production of progesterone, androgens, and estrogens by granulosa cells, thecal tissue, and stromal tissue from human ovaries in vitro, J. Clin. Endocrinol. Metab. **49:**687, 1979.

McNatty, K.P., et al.: The microenvironment of the human antral follicle: interrelationships among the steroid levels in antral fluid, the population of the granulosa cells, and the status of the oocyte in vivo and in vitro, J. Clin. Endocrinol. Metab. **49:**851, 1979.

Naftolin, F., and Butz, E.: Sexual dimorphism, Science **211:**1263, 1981.

Nathanielsz, P.W.: Endocrine mechanisms of parturition, Annu. Rev. Physiol. **40:**411, 1978.

Ohno, S.: The role of H-Y antigen in primary sex determination, J.A.M.A. **239:**217, 1978.

Overstreet, J.W., and Blazak, W.F.: The biology of human male reproduction: an overview, Am. J. Indust. Med. **4:**5, 1983.

Pardridge, W.M., et al.: Androgens and sexual behavior, Ann. Intern. Med. **96:**488, 1982.

Pohl, C.R., and Knobil, E.: The role of the central nervous system in the control of ovarian function in higher primates, Annu. Rev. Physiol. **44:**583, 1982.

Quagliarello, J., et al.: Induction of relaxin secretion in nonpregnant women by human chorionic gonadotropin, J. Clin. Endocrinol. Metab. **51:**74, 1980.

Rasmussen, D.D., and Yen, S.S.: Progesterone and 20-α-hydroxyprogesterone stimulate the in vitro release of GnRH by the isolated mediobasal hypothalamus, Life Sci. **32:**1523, 1983.

Rich, B.H., et al: Adrenarche: changing adrenal response to adrenocortiocotropin, J. Clin. Endocrinol. Metab. **52:**1129, 1981.

Richelson, L.S., et al.: Relative contributions of aging and estrogen deficiency to postmenopausal bone loss, N. Engl. J. Med. **311:**1273, 1984.

Simpson, E.R., and McDonald, P.C.: Endocrine physiology of the placenta, Annu. Rev. Physiol. **43:**163, 1981.

Thoburn, G.D., et al: Endocrine control of parturition, Physiol. Rev. **59:**863, 1979.

Tsang, B.K., et al.: Androgen biosynthesis in human ovarian follicles: cellular source, gonadotropic control, and adenosine 3′,5′-monophosphate mediation, J. Clin. Endocrinol. Metab. **48:**153, 1979.

Wilson, J.D.: Sexual differentiation, Annu. Rev. Physiol. **40:**279, 1978.

Books and monographs

Bardin, C.W.: Pituitary-testicular axis. In Yen, S.S.C., and Jaffe, F.B.: Reproductive endocrinology, Philadelphia. 1978, W.B. Saunders Co.

Fawcett, D.W.: Ultrastructure and function of the Sertoli cell. In Hamilton, D.W., and Greep, R.O., editors: Handbook of physiology; Section 7, vol. 5, Bethesda, Md., 1975, The American Physiological Society.

Fisher, D.A.: Fetal endocrinology: endocrine disease and pregnancy. In Degroot, L.J., et al., editors: Endocrinology, vol. 3, New York, 1979, Grune & Stratton, Inc.

Griffin, J.E., et al.: The testis. In Bondy, P.K., and Rosenberg, L.E.: Metabolic control and disease, Philadelphia, 1980, W.B. Saunders Co.

Odell, W.D.: The reproductive system in women. In Degroot, L.J., et al., editors: Endocrinology, vol. 3, New York, 1979, Grune & Stratton, Inc.

Ross, G.T., et al.: The ovary. In Yen, S.S.C., and Jaffe, R.B.; Reproductive endocrinology, Philadelphia, 1978, W.B. Saunders Co.

Yen, S.S.C.: The human menstrual cycle (integrative function of the hypothalamic-pituitary-ovarian-endometrial axis). In Yen, S.S.C., and Jaffe, R.B.: Reproductive endocrinology, Philadelphia, 1978, W.B. Saunders Co.

INDEX

U